THE GENETIC CODE (mRNA)

Second position

First position	U	C	A	G	Third position
U	UUU UUC } Phe UUA UUG } Leu	UCU UCC UCA UCG } Ser	UAU UAC } Tyr UAA Stop UAG Stop	UGU UGC } Cys UGA Stop UGG Trp	U C A G
C	CUU CUC CUA CUG } Leu	CCU CCC CCA CCG } Pro	CAU CAC } His CAA CAG } Gln	CGU CGC CGA CGG } Arg	U C A G
A	AUU AUC } Ile AUA AUG Met/start	ACU ACC ACA ACG } Thr	AAU AAC } Asn AAA AAG } Lys	AGU AGC } Ser AGA AGG } Arg	U C A G
G	GUU GUC GUA GUG } Val	GCU GCC GCA GCG } Ala	GAU GAC } Asp GAA GAG } Glu	GGU GGC GGA GGG } Gly	U C A G

SOME PREFIXES USED IN THE INTERNATIONAL SYSTEM OF UNITS

10^9	giga	
10^6	mega	
10^3	kilo	
10^{-1}	deci	
10^{-2}	centi	
10^{-3}	milli	
10^{-6}	micro	μ
10^{-9}	nano	n
10^{-12}	pico	p
10^{-15}	femto	f

CONVERSION FACTORS

Energy: 1 joule = 10^7 ergs = 0.239 cal
$\quad\quad\quad$ 1 cal = 4.184 joule

Length: 1 nm = 10 Å = 1×10^{-7} cm = 1×10^{-9} m

Mass: 1 kg = 1000 g = 2.2 lb
$\quad\quad$ 1 lb = 453.6 g

Pressure: 1 atm = 760 torr = 1.013 bar = 14.696 psi
$\quad\quad\quad\quad$ 1 bar = 100 kPa = 0.987 atm = 750.1 torr
$\quad\quad\quad\quad$ 1 torr = 1 mm Hg

Temperature: K = °C + 273
$\quad\quad\quad\quad\quad$ °C = (5/9)(°F − 32)

Volume: 1 L = 1×10^{-3} m^3 = 1000 cm^3

PHYSICAL CONSTANTS

Name	Symbol	SI Units	cgs Units
Avogadro's number	N	6.022137×10^{23}/mol	6.022137×10^{23}/mol
Boltzmann constant	k	1.38066×10^{-23} J/K	1.38066×10^{-16} erg/K
Curie	Ci	3.7×10^{10} d/s	3.7×10^{10} d/s
Electron charge	e	1.602177×10^{-19} coulomb[b]	4.80321×10^{-10} esu
Faraday constant	F	96485 J/V·mol	9.6485×10^{11} erg/V·m
Gas constant[a]	R	8.31451 J/K·mol	8.31451×10^7 erg/K·m
Light speed (vacuum)	c	2.99792×10^8 m/s	2.99792×10^{10} cm/s
Planck's constant	h	6.626075×10^{-34} J·s	6.626075×10^{-27} erg·s

[a]Other values of R: 1.9872 cal/K·mol = 0.082 L·atm/K·mol
[b]1 coulomb = 1 J/V

USEFUL EQUATIONS

Henderson–Hasselbalch equation	$pH = pK_a + \log([A^-]/[HA])$
Michaelis–Menten equation	$v = V_{max}[S]/(K_M + [S])$
Free energy change under non-standard-state conditions	$\Delta G = \Delta G° + RT \ln([C][D]/[A][B])$
Free energy change and standard reduction potential	$\Delta G°' = -nF\Delta E_0'$
Reduction potentials in a redox reaction	$\Delta E_0' = E_0'(\text{acceptor}) - E_0'(\text{donor})$
Proton motive force	$\Delta p = \Delta\Psi - 2.3RT\,\Delta pH/F$
Passive diffusion of a charged species	$\Delta G = G_2 - G_1 = RT \ln(C_2/C_1) + ZF\Delta\Psi$

FOURTH EDITION

BIOCHEMISTRY

CHRISTOPHER K. MATHEWS
Oregon State University

K. E. VAN HOLDE
Oregon State University

DEAN R. APPLING
The University of Texas at Austin

SPENCER J. ANTHONY-CAHILL
Western Washington University

Toronto

Vice-President, Editorial Director: Gary Bennett
Executive Acquisitions Editor: Cathleen Sullivan
Marketing Manager: Julia Jevmenova
Senior Developmental Editor: John Polanszky
Project Managers: Marissa Lok and Tracy Duff (PreMediaGlobal)
Manufacturing Manager: Susan Johnson
Production Editor: Tracy Duff (PreMediaGlobal)
Copy Editor: Kelly Birch
Proofreader: Stephany Craig
Compositor: PreMediaGlobal
Photo and Permissions Researcher: Heather Jackson
Art Director: Julia Hall
Cover and Interior Designer: Miriam Blier
Cover Image: C. Spiegel and S. Anthony-Cahill

Credits and acknowledgments borrowed from other sources and reproduced, with permission, in this textbook appear on the appropriate page within text.

Earlier editions published by Pearson Education, Inc., Upper Saddle River, New Jersey, USA. Copyright © 2000, 1996, 1990 Pearson Education, Inc.

10 9 8 7 6 5 4 3 CKV

Library and Archives Canada Cataloguing in Publication
    Biochemistry / [edited by] Christopher K. Mathews ... [et al.]. — 4th U.S. ed.

Includes bibliographical references and index.
ISBN 978-0-13-800464-4

    1. Biochemistry. I. Mathews, Christopher K., 1937-
QH345.B43 2012   572'.3   C2011-902175-7

ISBN 978-0-13-800464-4

ABOUT THE AUTHORS

Christopher K. Mathews is Distinguished Professor Emeritus of Biochemistry at Oregon State University. He earned his B.A. in chemistry from Reed College (1958) and Ph.D. in biochemistry from the University of Washington (1962). He served on the faculties of Yale University and the University of Arizona from 1963 to 1978, when he moved to Oregon State University as chair of the Department of Biochemistry and Biophysics, a position he held until 2002. His major research interest is the enzymology and regulation of DNA precursor metabolism and the intracellular coordination between deoxyribonucleotide synthesis and DNA replication. From 1984 to 1985, Dr. Mathews was an Eleanor Roosevelt International Cancer Fellow at the Karolinska Institute in Stockholm, and in 1994–1995 he held the Tage Erlander Guest Professorship at Stockholm University. Dr. Mathews has published over 175 scientific papers dealing with molecular virology, metabolic regulation, nucleotide enzymology, and biochemical genetics. He is the author of *Bacteriophage Biochemistry* (1971) and coeditor of *Bacteriophage T4* (1983) and *Structural and Organizational Aspects of Metabolic Regulation* (1990). He was a coauthor of the three previous editions of *Biochemistry*. His teaching experience includes undergraduate, graduate, and medical school biochemistry courses.

He has backpacked and floated the mountains and rivers, respectively, of Oregon and the Northwest. As an enthusiastic birder he has served as President of the Audubon Society of Corvallis and is President of the Great Basin Society, which operates the Malheur Field Station.

K. E. van Holde is Distinguished Professor Emeritus of Biophysics and Biochemistry at Oregon State University. He earned his B.A. (1949) and Ph.D. (1952) from the University of Wisconsin. Over many years, Dr. van Holde's major research interest has been the structure of chromatin; his work resulted in the award of an American Cancer Society research professorship in 1977. He has been at Oregon State University since 1967, and was named Distinguished Professor in 1988. He is a member of the National Academy of Sciences and the American Academy of Arts and Sciences, and has received Guggenheim, NSF, and EMBO fellowships. He is the author of over 200 scientific papers and four books in addition to this volume: *Physical Biochemistry* (1971, 1985), *Chromatin* (1988), *Principles of Physical Biochemistry* (1998), and *Oxygen and the Evolution of Life* (2011). He was also coeditor of *The Origins of Life and Evolution* (1981). His teaching experience includes undergraduate and graduate chemistry, biochemistry and biophysics, and the physiology and molecular biology course at the Marine Biological Laboratory at Woods Hole.

Dean R. Appling is the Lester J. Reed Professor of Biochemistry and Associate Dean for Research and Facilities for the College of Natural Sciences at the University of Texas at Austin, where he has taught and done research for the past 26 years. Dean earned his B.S. in biology from Texas A&M University, and his Ph.D. in biochemistry from Vanderbilt University. The Appling laboratory studies the organization and regulation of metabolic pathways in eukaryotes, focusing on folate-mediated one-carbon metabolism. The lab is particularly interested in understanding how one-carbon metabolism is organized in mitochondria, as these organelles are central players in many human diseases. In addition to coauthoring this book, Dean has published over 60 scientific papers and book chapters.

As much fun as writing a textbook might be, Dean would rather be outdoors. He is an avid fisherman and hiker. Recently, Dean and his wife, Maureen, have become entranced by the birds on the Texas coast. They were introduced to bird-watching by coauthor Chris Mathews and his wife, Kate—an unintended consequence of working on this book!

Spencer J. Anthony-Cahill is a Professor in the Department of Chemistry at Western Washington University (WWU), Bellingham, WA. Spencer earned his B.A. in chemistry from Whitman College, and his Ph.D. in bioorganic chemistry from the University of California, Berkeley. His graduate work, in the lab of Peter Schultz, focused on the biosynthetic incorporation of unnatural amino acids into proteins. Spencer was an NIH postdoctoral fellow in the laboratory of Bill DeGrado (then at DuPont Central Research), where he worked on *de novo* peptide design and the prediction of the tertiary structure of the HLH DNA-binding motif. He then worked for five years as a research scientist in the biotechnology industry, developing recombinant hemoglobin as a treatment for acute blood loss. In 1997, Spencer decided to pursue his long-standing interest in teaching and moved to WWU, where he is today.

Research in the Anthony-Cahill laboratory is directed at the protein engineering of heme proteins. The primary focus is on circular permutation of human β-globin as a means to develop a single-chain hemoglobin with desirable therapeutic properties. The lab is also pursuing the design of self-assembling protein nanowires.

Outside the classroom and laboratory, Spencer is a great fan of the outdoors—especially the North Cascades and southeastern Utah, where he has often back-packed, camped, climbed, and mountain biked. Spencer also holds the rank of 3rd Dan in Aikido, and instructs children and adults at the Kulshan Aikikai Dojo in Bellingham, WA.

PREFACE

A NEW EDITION

What factors might explain the re-emergence of a well-received biochemistry textbook (*Biochemistry*, Third Edition, 2000, by C. K. Mathews, K. E. van Holde, and K. G. Ahern), some 12 years after publication of the previous edition? In a rapidly evolving field like biochemistry, textbooks are typically revised every four or five years to retain their educational value.

Still, biochemistry instructors and students continued to ask when and whether a fourth edition might appear. While Chris Mathews was interested in revising and updating the book, his previous coauthors were unable to commit to a project of this magnitude, and so the search began for a new author team. After a long and careful selection process, two new coauthors joined Chris Mathews: Dr. Dean R. Appling, Lester J. Reed Professor of Biochemistry and Associate Dean for Research and Facilities for the College of Natural Sciences at the University of Texas at Austin, and Dr. Spencer J. Anthony-Cahill, Professor of Chemistry at Western Washington University, Bellingham.

Dean Appling is an enzymologist with interests in regulation and organization of metabolic pathways, with particular emphasis upon folate cofactors and the metabolism of single-carbon units. Much of his work uses NMR and molecular genetics to probe metabolic compartmentation and control. Spencer Anthony-Cahill's chief interest is protein folding and design, with current emphasis upon folding patterns in protein variants that have circularly permuted sequences. Before assuming his present faculty position, Spencer worked for five years in the biotechnology industry, an experience that gives him a valuable perspective in teaching biochemistry. Both Dean and Spencer have used previous editions of *Biochemistry* in their own teaching, so they were well aware of the strengths of this book and areas where fresh attention was needed.

The research interests of the new author team created a natural division of writing responsibilities. Spencer's writing was focused upon biomolecular structure and mechanisms, Dean dealt with metabolism and its control, and Chris put his major effort into genetic biochemistry. However, the project was truly a team effort. Each chapter draft was scrutinized by all three authors, with revisions made by each principal draft author before submission to our editors and outside reviewers. We found our fellow authors to be our strongest critics. And, although Ken van Holde was not actively involved with this edition, he did review some drafts and much of his graceful writing remains in this new edition. We are proud to include him as a coauthor of this new edition.

EVOLUTION OF THE TEXT
Major Changes

In addition to dealing with the vast amount of new information appearing since the publication of the third edition in 2000, this new edition introduces three significant changes. First is more emphasis upon biochemical reaction mechanisms in the enzymes and metabolism chapters. Second is a significant reorganization in the chapters dealing with intermediary metabolism. The coverage of carbohydrate metabolism has been unified, so that we now present glycolysis, gluconeogenesis, glycogen metabolism, and the pentose phosphate pathway in one chapter (Chapter 13). To accomplish this without excessive expansion of the chapter, we moved the section on complex carbohydrate metabolism to Chapter 9; instructors can present this material as part of the metabolism section of the course, if they prefer. Redox thermodynamics has been moved from Chapter 15 (Biological Oxidations) to Chapter 3 (Bioenergetics), where it more properly belongs. The material on interorgan coordination in mammalian metabolism has been split into two chapters—Chapter 18 (Interorgan and Intracellular Coordination of Energy Metabolism in Vertebrates) and Chapter 23 (Signal Transduction).

The third major change is the reorganization of genetic biochemistry in the last major section of the book. As in previous editions, we introduce processes in biological information transfer early, in Chapter 4, with details presented later. In addition, we have integrated prokaryotic and eukaryotic informational metabolism, rather than presenting them in separate chapters, as in previous editions. The four genetic biochemistry chapters in previous editions are now six— Chapters 24 (Genome Organization), 25 (DNA Replication), 26 (Information Restructuring), 27 (Transcription and Its Control), 28 (Protein Synthesis and Processing), and 29 (Control of Gene Expression).

New Topics

A special challenge in writing a new edition after an interval of so many years was incorporating the most important of the many spectacular new developments in molecular life sciences. A partial list of new or significantly revised topics includes:

- Phosphorothioate bonds in DNA (Chapter 4)
- Gene sequence analysis, phylogenetic analysis, proteomic analysis, and amino acid sequencing by mass spectroscopy (Chapter 5)
- New approaches to classifying protein secondary structure, protein structure prediction, and protein folding energy landscapes (Chapter 6)
- Dynamics of myoglobin, roles of heme proteins in nitric oxide physiology, and antibody–drug conjugates as anticancer agents (Chapter 7)
- Biological imaging of complex glycoproteins (Chapter 9)
- Lipid rafts (Chapter 10)
- Organic chemical mechanisms of the common biochemical reaction types (Chapter 12)
- Coordination of energy homeostasis, including mTOR, AMPK, and sirtuins and protein acetylation (Chapter 18)
- Evolution of metabolic pathways (several chapters); regulation of cholesterol metabolism (Chapter 19)
- Ubiquitin and regulated protein turnover (Chapter 20)
- Methyl group metabolism (Chapter 21)
- Pharmacogenetics (Chapter 22)

- A kinase anchoring proteins (Chapter 23)
- Restriction fragment length polymorphisms, single-nucleotide polymorphisms and genome mapping, chromatin structure, and the centromere (Chapter 24)
- Double-strand DNA break repair (Chapter 26)
- Structure and function of RNA polymerases (Chapter 27) and of ribosomes (Chapter 28)
- Apoptosis (Chapter 28)
- The role of Mediator in transcription complexes, DNA methylation and epigenetics, functional significance of histone modifications, RNA interference, and riboswitches (Chapter 29)

Biochemistry Applications

One feature requested by students and instructors alike is practical applications of biochemical knowledge—particularly, applications to the health sciences. Unlike some other textbooks, we prefer to integrate applications with the main text, instead of setting them apart in boxes. We believe that this makes the text flow more smoothly.

New applications discussed in this edition include:

- Influenza virus neuraminidase and the action of Tamiflu (Chapter 9)
- Biofuels (Chapter 13)
- Mitochondrial diseases (Chapter 15)
- Artificial photosynthesis (Chapter 16)
- Diabetes, obesity (Chapter 18)
- Calorie restriction and lifespan extension (Chapter 18)
- Methylenetetrahydrofolate reductase variants and disease susceptibility (Chapter 21)
- Chromosomal translocations and targeted cancer drugs (Chapter 23); mapping disease genes (Chapter 24)
- Patterns of oncogene mutations in cancer (Chapter 23)

Keeping What Works Best

Not everything is new in this edition. We have worked hard to retain and improve the best-loved features of previous editions, such as an emphasis upon the physico-chemical concepts upon which biochemical processes and mechanisms are based, and an emphasis upon the experimental nature of biochemistry. This latter emphasis is realized with our continued use of the popular Tools of Biochemistry feature.

TOOLS OF BIOCHEMISTRY

As in past editions, we emphasize the importance of incisive experimental techniques as the engine that drives our increasing understanding of the molecular nature of life processes. This is accomplished through end-of-chapter essays on the most important techniques in biochemistry and molecular biology research. Most of the Tools sections in this edition have been updated or introduced for the first time. New or significantly modified Tools sections include:

- Introduction to Proteomics; Tandem Mass Spectrometry (Chapter 5)
- Nuclear Magnetic Resonance Spectroscopy (Chapter 6)

- *In Vitro* Evolution of Protein Function (Chapter 11)
- Metabolomics (Chapter 12)
- Gene Targeting by Homologous Recombination; Single-Molecule Biochemistry (Chapter 26)
- Microarrays; Chromatin Immunoprecipitation (Chapter 27)

Several Tools sections on manipulating DNA have been combined and moved earlier in the book, to Chapter 4. The Tools section on radioisotopes in Chapter 12 has been considerably shortened. The material on kinetic isotope methods in analysis of enzyme mechanisms has been strengthened in the Tools section in Chapter 11.

END-OF-CHAPTER PROBLEMS

Wherever possible, we have removed problems that emphasize rote learning and retained or added problems that require analytical or quantitative thought to be solved. Several new problems have been added to each chapter. Importantly, we now include complete solutions to each problem, as well as the answers, at the back of the book.

ABOUT THE COVER

The cover illustration depicts the structure of the yeast 80S ribosome at 4.15 Ångstrom resolution, based upon X-ray crystallography. This complex RNA- and protein-containing particle is an enormous molecular machine, which binds the components of protein synthesis—messenger RNA, transfer RNAs containing activated amino acids, and soluble protein factors that aid in all phases of translation—initiation, polypeptide chain elongation, and termination.

Tremendous insight into mechanisms of protein synthesis was gained beginning in 2000, when crystal structures for prokaryotic and archaeal ribosomes were reported. This work was recognized in 2009, with the Nobel Prize in Chemistry to V. Ramakrishnan, T. A. Steitz, and A. Yonath. Although basic processes in translation are similar in all cells, protein synthesis in eukaryotic cells is much more complex than in bacteria, particularly with regard to steps in initiation, where many more soluble protein factors must participate. The eukaryotic ribosome is correspondingly larger and more complex—about 40% larger than the bacterial ribosome, with correspondingly more different proteins and larger RNA components. These factors make solving the eukaryotic ribosome structure an even more formidable problem. This feat was accomplished in several laboratories, beginning in late 2010.

The structure of the yeast ribosome shown here was described by A. Ben-Shem, L. Jenner, G. Yusupova, and M. Yusupov, in *Science* 330:1203–1209 (2010). The image on the cover was created by C. Spiegel and S. Anthony-Cahill, working from atomic coordinates deposited by Ben-Shem and coauthors in the Brookhaven Protein Database (PBD). Color scheme: 40S particle (PDB ID: 3O30): RNA is in orange; proteins are in slate blue; 60S particle (PDB ID: 3O5H): RNA is in raspberry red; proteins are in forest green.

SUPPLEMENTS

For Instructors

Instructor resources are password protected and available for download via the Pearson online catalog at www.pearsonhighered.com. For your convenience, many of these resources are also available on the **Instructor's Resource CD-ROM (IRCD)** (ISBN 978-0-13-279159-5).

- **Test Item File.** The fourth edition features a brand new testbank created by Scott Lefler, Senior Lecturer, Arizona State University, with more than 700 thoughtful questions in editable Word format. The Test Item File can be found on the IRCD or downloaded from the online catalog.

- **PowerPoint® Presentations.** Two sets of PowerPoint® slides are available for the text. The first consists of all the figures and photos in the textbook in PowerPoint® format. The second set, created by Bruce Burnham, Associate Professor of Chemistry, Rider University, consists of PowerPoint® lecture slides that provide an outline to use in a lecture setting, presenting definitions, key concepts, and figures from the textbook. Both sets of PowerPoint® slides can be found on the IRCD or downloaded from the online catalog.

- **Complete Solutions Manual.** As in previous editions, we have created a solutions manual to complement chapter problems in the current edition of our text. The complete solutions manual, prepared by Sara Codding and Tim Rhoads of Oregon State University, contains fully worked solutions for those questions that may benefit from explanations beyond those provided at the back of the book. Instructors can arrange with the publisher to make this material available to students (ISBN 978-0-13-292628-7).

- **CourseSmart for Instructors.** CourseSmart goes beyond traditional expectations, providing instant, online access to the textbooks and course materials you need at a lower cost for students. And even as students save money, you can save time and hassle with a digital eTextbook that allows you to search for the most relevant content at the very moment you need it. Whether it's evaluating textbooks or creating lecture notes to help students with difficult concepts, CourseSmart can make life a little easier. See how when you visit www.coursesmart.com/instructors.

- **Technology Specialists.** Pearson's Technology Specialists work with faculty and campus course designers to ensure that Pearson technology products, assessment tools, and online course materials are tailored to meet your specific needs. This highly qualified team is dedicated to helping schools take full advantage of a wide range of educational resources by assisting in the integration of a variety of instructional materials and media formats. Your local Pearson Education sales representative can provide you with more details on this service program.

For Students

- **The Chemistry Place for Biochemistry, Fourth Edition.** The Chemistry Place is an online tool that provides students with tutorial aids to help them succeed in biochemistry. This Website includes animations of key concepts and processes and self-quizzing created by Scott Napper, Associate Professor, University of Saskatchewan, to allow students to check their understanding of subject matter. TheChemistryPlace also contains our Pearson eText. Please visit the site at www.chemplace.com.

- **Pearson eText** gives students access to the text whenever and wherever they have access to the Internet. eText pages look exactly like the printed text, offering powerful new functionality for students and instructors. Users can create notes, highlight text in different colors, create bookmarks, zoom, click hyperlinked words and phrases to view definitions, and view in single-page or two-page view. Pearson eText allows for quick navigation to key parts of the eText using a table of contents and provides full-text search. The eText may also offer links to associated media files, enabling users to access videos, animations, or other activities as they read the text.

- **CourseSmart for Students.** CourseSmart goes beyond traditional expectations, providing instant, online access to the textbooks and course materials you need at an average savings of 60%. With instant access from any computer and the ability to search your text, you'll find the content you need quickly, no matter where you are. And with online tools like highlighting and note-taking, you can save time and study efficiently. See all the benefits at www.coursesmart.com/students.

ACKNOWLEDGMENTS

Although our names appear as authors on the cover of this book, and we expect to receive most of the credit or criticism resulting therefrom, the book in fact was created by a large team, with many participants whose contributions rivaled ours. To begin, this book would never have come into existence but for the enterprise of Michelle Sartor, now Acting Editor-in-Chief for Humanities and Social Sciences for Pearson Canada. In her former assignment, she became aware of a continuing interest in our book, particularly in Canada. Even though a previous attempt at a fourth edition had aborted, Michelle exercised quiet but effective persistence until the present author team had been assembled. After Michelle's reassignment, Cathleen Sullivan, Executive Editor for Engineering, Science, and Mathematics, took over, and she has held a steady hand on the tiller through calm seas that occasionally, but only briefly, became rough.

Our day-to-day contact through the writing and development phase was John Polanszky, Senior Developmental Editor. We appreciated that he gave us much independence and in general was a calming influence and a source of useful advice. And we credit him with securing some extremely helpful reviewers, whose names are listed separately.

When the writing, reviewing, and revisions were completed, we maintained our valuable contacts with Cathleen and John, but began interacting with a much larger team of dedicated and skilled professionals, particularly Marissa Lok, in-house project manager, and Tracy Duff, project manager at PreMediaGlobal. We exchanged e-mails and phone calls with these ladies nearly every day, and consider them friends, even though we have yet to meet them in person. Signal contributions were made during this phase by Kelly Birch, copy editor; Stephany Craig, proofreader; Heather Jackson, permissions researcher; and Greg Miller, Scott Napper, Mark Jonklaas, Masoud Jelokhani-Niaraki, technical reviewers. Katy Mehrtens, Publishing Services Director, oversaw the efforts at PreMediaGlobal. Julia Jevmenova, Marketing Manager at Pearson Canada, impressed us with her quiet insistence at learning about the substance and content of our book, so that she could become a truly effective advocate.

We owe a great deal to friends and colleagues in science for advice, updated information, and graphics. At the risk of neglecting to recognize all who helped us, the following deserve mention.

Gary Carlton (Arcsine Graphics) produced many of the new figures in Chapters 6–11. Gary's attention to detail and quality yielded spectacular results. Thanks to those researchers who provided new figures for this edition of the text: Shing Ho (Colorado State University), Figures 4A.6 and 26.24; Jack Benner (New England Biolabs), Figure 5D.1; Andy Karplus (Oregon State University), Figures 6.12 and 6.13; Scott Delbecq and Rachel Klevit (University of Washington), Figure 6A.10b; Serge Smirnov (Western Washington University), Figure 6A.12; Vlado Gelev (FBReagents Inc.), Figure 6A.13b; Stephan Grzesiek (University of Basel), Figure 6A.13a; Richard Harris, Figure 6A.14; John Olson (Rice University), Figure 7.6; Marjorie Longo (University of California, Davis), Figure 10.27; Vamsi Mootha (MIT), Figure 12B.3; Adrian Keatinge-Clay (University of Texas at Austin), Figure

17.37B; Rowena Matthews (University of Michigan), Figure 21.12; John Tesmer (University of Michigan), Figure 23.10; Lawrence Loeb (University of Washington), Figure 23.22; Mike O'Donnell (Rockefeller University), Figures 25.25 and 25.34; Whitney Yin (University of Texas at Austin), Figure 25.35; David Josephy (University of Guelph), Figure 26.7; D. G. Vassylyev (University of Alabama at Birmingham), Figure 27.11; Robin Gutell (University of Texas at Austin), Figure 28.19.

P. Clint Spiegel (Western Washington University) rendered the structure of the eukaryotic ribosome that appears on the cover.

Special thanks to several colleagues who generously provided feedback on material in the draft stage, and thereby improved the quality of the final text. Rachel Klevit (University of Washington), Andrew Baldwin (University of Toronto), and Serge Smirnov (Western Washington University) reviewed the material describing optical and NMR spectroscopy in Chapter 6. Tom Brittain (University of Auckland), John Olson (Rice University), and Antony Mathews (Pfizer, Inc.) reviewed the material in Chapter 7 describing the structure and function of globin proteins. Heather Van Epps (Seattle Genetics) provided critical feedback on the sections of Chapter 10 describing excitable membranes. Jack Benner (New England Biolabs) reviewed material in Chapter 5 describing proteomics. Andrew Hanson (University of Florida) provided helpful feedback on the section in Chapter 16 describing photorespiration. John Denu (University of Wisconsin) shared unpublished data and provided helpful feedback on the section in Chapter 18 describing protein acetylation and sirtuins. Jon Huibregtse (University of Texas at Austin) provided helpful feedback on the section in Chapter 20 describing ubiquitin function. Ralph Green (University of California, Davis) provided helpful feedback on the section in Chapter 20 describing vitamin B12 and pernicious anemia. JoAnne Stubbe (MIT) provided helpful feedback on the discussion of ribonucleotide reductase in Chapter 22. John Tesmer (University of Michigan) provided helpful feedback on the section in Chapter 23 describing G protein structure and function. Michael Freitag (Oregon State University) offered valuable information about centromeres, kinetochores, and epigenetics (Chapters 24 and 29).

As always, our most effective critics were our wives—Kate Mathews, Maureen Appling, and Yvonne Anthony-Cahill. Barbara van Holde is greatly missed. But Kate, Maureen, and Yvonne were constant sources of love and support, with occasional pungent advice and criticism. Their patience and enduring support were the most important elements in seeing this project to a timely and satisfying conclusion.

Christopher K. Mathews   Dean R. Appling
Spencer J. Anthony-Cahill   K. E. van Holde

REVIEWERS

The following reviewers provided valuable feedback on the manuscript at various stages throughout the writing process.

Nahel Awadallah, Sampson Community College
Stephen L. Bearne, Dalhousie University
Roberto Botelho, Ryerson University
John Brewer, University of Georgia
Robert Brown, Memorial University of Newfoundland
Bruce Burnham, Rider University
Danielle Carrier, University of Ottawa
Lisa Carter, Athabasca University
Amanda Cockshutt, Mount Allison University
Betsey Daub, University of Waterloo
Richard Epand, McMaster University
Eric Gauthier, Laurentian University
Dara Gilbert, University of Waterloo
Masoud Jelokhani, Wilfrid Laurier University
Mark Jonklaas, Baylor University
David Josephy, University of Guelph
Lana Lee, University of Windsor
Elke Lohmeier-Vogel, University of Calgary
Derek McLachlin, University of Western Ontario
Vas Mezl, University of Ottawa
Scott Napper, University of Saskatchewan
Arnim Pause, McGill University
Dorothy Pocock-Goldman, Concordia University
Shauna Reckseidler-Zenteno, Athabasca University
Jim Sandercock, Northern Alberta Institute of Technology
Anthony Siame, Trinity Western University
Anthony Serianni, University of Notre Dame
Ron Smith, Thompson Rivers University
Lakshmaiah Sreerama, St. Cloud State University
David Villeneuve, Canadore College
William Willmore, Carleton University
Boris Zhorov, McMaster University

BRIEF CONTENTS

DETAILED CONTENTS

CHAPTER 5

Introduction to Proteins: The Primary Level of Protein Structure   136

PART 2
Molecular Architecture of Living Matter   89

CHAPTER 4

Nucleic Acids   90

CHAPTER 6

The Three-Dimensional Structure of Proteins   177

CHAPTER 7

Protein Function and Evolution   234

PART 4
Dynamics of Life: Energy, Biosynthesis, and Utilization of Precursors   517

CHAPTER 13

Carbohydrate Metabolism: Glycolysis, Gluconeogenesis, Glycogen Metabolism, and the Pentose Phosphate Pathway   518

CHAPTER 14

Citric Acid Cycle and Glyoxylate Cycle   591

CHAPTER 26

DNA Restructuring: Repair, Recombination, Rearrangement, Amplification   1079

CHAPTER 27

Information Readout: Transcription and Post-transcriptional Processing   1125

TOOLS OF BIOCHEMISTRY

PART 1

The Realm of Biochemistry

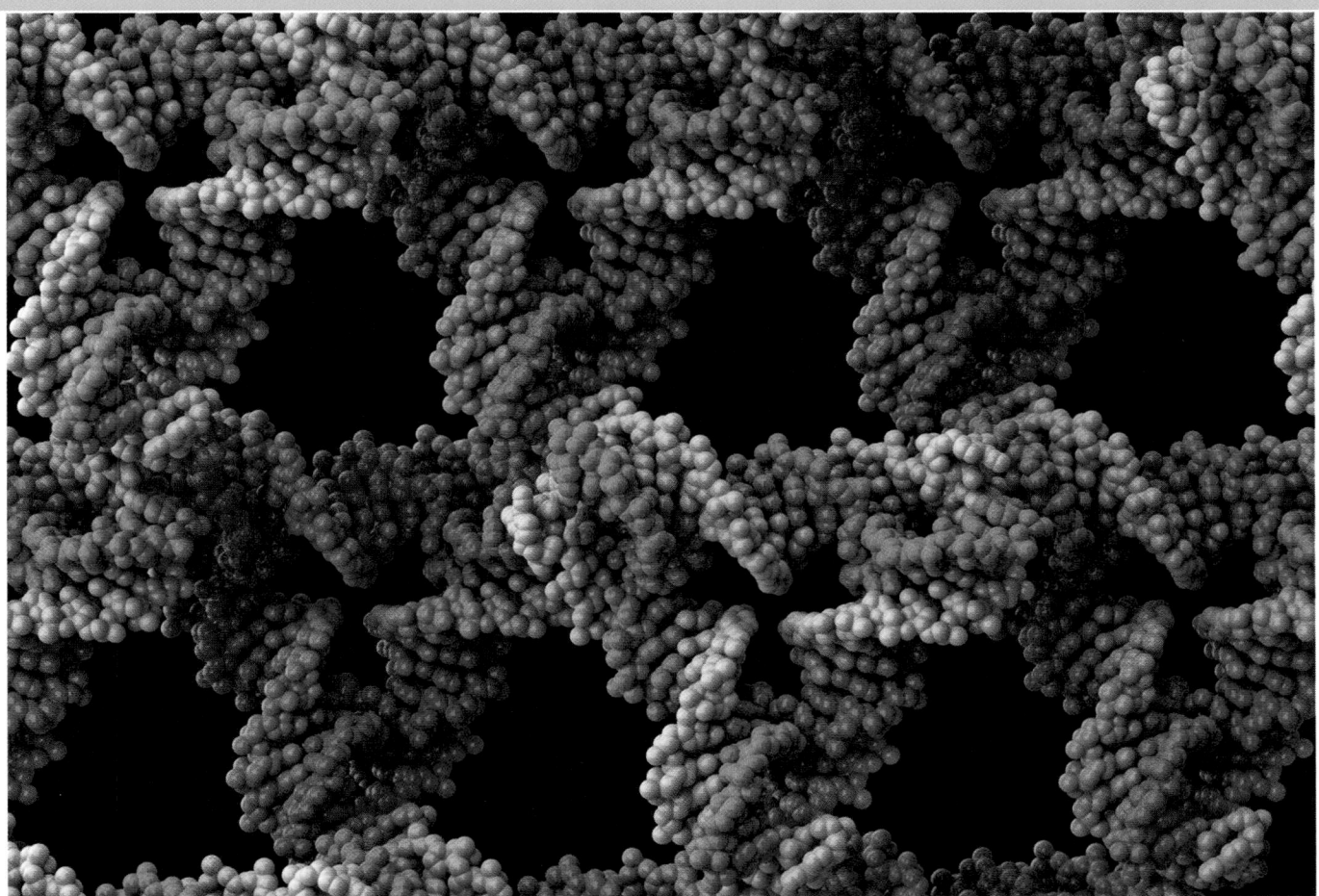

A regular three-dimensional DNA crystal, formed from chemically synthesized DNA fragments. Source: Visual Science/Science Photo Library.

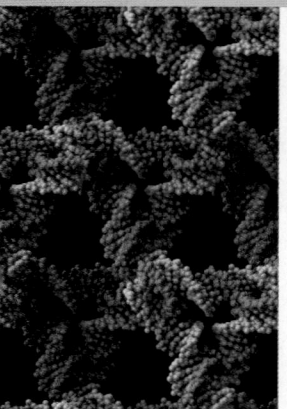

CHAPTER 1

The Scope of Biochemistry

Welcome to the study of biochemistry, a branch of science that seeks to describe the structure, organization, and functions of living matter in molecular terms. What are the chemical structures of the components of living matter? How do interactions among these components give rise to organized supramolecular structures, cells, multicellular tissues, and organisms? How does living matter extract energy from its surroundings in order to remain alive? How does an organism store and transmit the information it needs to grow and reproduce? What chemical changes accompany the reproduction, aging, and death of cells and organisms? These are the kinds of questions asked by biochemists, the answers to which provide insight into the chemical nature of life.

Biochemistry can be divided into three principal areas, and that is how this book is arranged: (1) the **structural chemistry** of the components of living matter and relationships of biological function to chemical structure (Chapters 1–11); (2) **metabolism**, the totality of chemical reactions that occur in living matter (Chapters 12–23); and (3) **genetic biochemistry**, the chemistry of processes and substances that store and transmit biological information (Chapters 24–29). This third area is also the province of **molecular genetics**, a field that seeks to understand heredity and the expression of genetic information in molecular terms.

Biochemistry and the Biological Revolution

Biology was transformed in 1953 when Watson and Crick proposed the double-helical model for DNA structure.

The biological sciences are undergoing a revolution, and biochemistry is at the heart of that revolution. The revolution began in the mid-twentieth century and received great impetus in 1953, when Watson and Crick proposed the now-familiar double-helical structure for DNA. The revolution continues today, with no letup in the rate at which fundamental knowledge is being revealed about the physics and chemistry of life processes.

The years since the Watson–Crick revelation have seen milestones at intervals of approximately a decade, milestones that help us appreciate how the pace of discovery has been maintained and accelerated. Ten years after the seminal Watson–Crick paper, we saw in the early 1960s the elucidation of the genetic code and the discovery of messenger RNA, keys to understanding how genetic information is expressed. Ten years later, the first recombinant DNA molecules were created, opening the door as no other discovery had done, to practical applications of biological information in health, agriculture, forensics, and environmental science. A decade later scientists had learned how to chemically synthesize any nucleic acid fragment and had discovered how to amplify minute amounts of DNA by the polymerase chain reaction, so that any gene could be cloned and subjected to mutagenesis, allowing any desired change to be made in the structure of that gene. After another decade, by the early 1990s, scientists had learned not only how to introduce new genes into the germ line of plants and animals but also how to disrupt or delete any existing gene, allowing analysis of the metabolic function of any gene product. A decade later the nearly complete nucleotide sequence of the human genome was announced—2.9×10^9 base pairs of DNA, representing more than 20,000 different genes. Shortly afterward the scientific community was rocked by the discovery of genetic regulation by small interfering RNAs. The wealth of information from genomic sequence analysis and gene regulation by RNA continues to transform the biochemical landscape well into the twenty-first century.

During the half century between discovery of the double helix and completion of the human genome sequence, powerful new research techniques in biochemistry and biophysics—X-ray crystallography, mass spectrometry, nuclear magnetic resonance, high-resolution separation techniques, automated protein and nucleic acid sequence determination, techniques for visualizing individual molecules—have fueled the biological revolution. Figure 1.1 gives a timeline for the introduction of these technologies, most of which you will encounter throughout this book.

Our timeline begins with the introduction of radioisotopes as research reagents, at the end of World War II. In the process, the biological sciences have been transformed. It is now clear that biochemistry underlies all of the life sciences, and that understanding any biological process—sensory perception, evolutionary selection, immune mechanisms, energy transformations—demands comprehension of the underlying chemistry and physics. It is also clear that the biology of the twenty-first century involves the gathering, integration, and analysis of enormous sets of data, beginning with the billions of DNA base pairs in the genomes of most animals and plants. A goal of this book is to show how we reached this point—to understand the concepts and principles of biochemistry in terms of the experimental approaches and research technologies upon which those concepts are based and to consider how huge data sets come into existence and are analyzed in the new science called bioinformatics, the quantitative and computational analysis of biological data.

The Roots of Biochemistry

Biochemistry is a young science, although it began well before Watson and Crick's work in 1953. However, much or most of what you will learn by reading this book has been discovered only in the past half-century—since the first shots, so to speak, were fired in the biological revolution. But let us briefly explore the earlier origins of the field.

Biochemistry had its roots as a distinct field of study in the early nineteenth century, with the pioneering work of Friedrich Wöhler. Prior to Wöhler's time it was believed that the substances in living matter were somehow qualitatively different from those in nonliving matter and did not behave according to the known laws of physics and chemistry. In 1828 Wöhler showed that urea, a substance of biological origin, could be synthesized in the laboratory from the inorganic compound ammonium cyanate. As Wöhler phrased it in a letter to a colleague, "I must

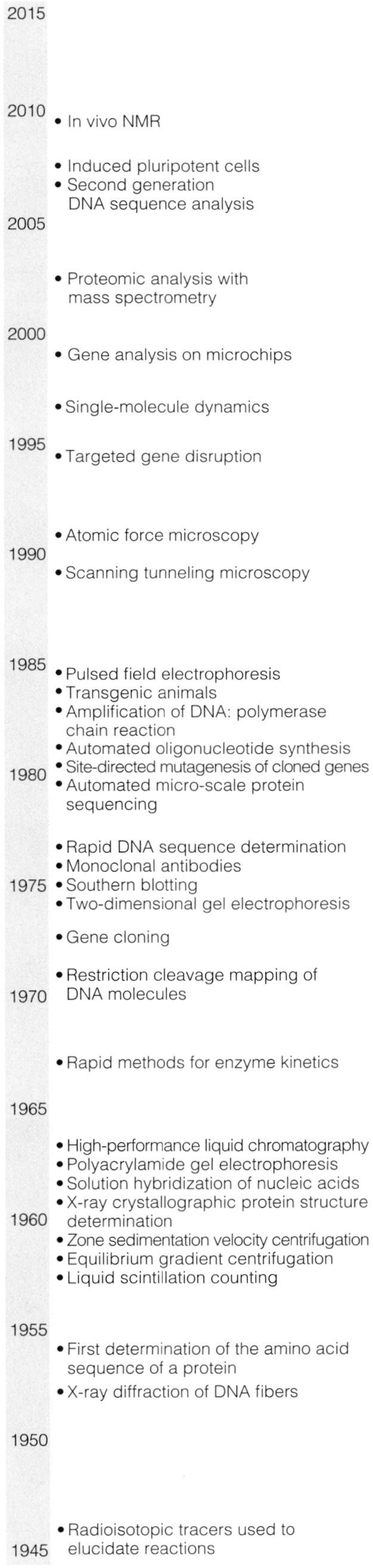

FIGURE 1.1

The recent history of biochemistry as shown by the introduction of new research techniques. The timeline begins with the introduction of radioisotopes as biochemical reagents, immediately following World War II.

Timeline (Figure 1.1):

- 2015
- 2010 — In vivo NMR
  - Induced pluripotent cells
  - Second generation DNA sequence analysis
- 2005
  - Proteomic analysis with mass spectrometry
- 2000
  - Gene analysis on microchips
  - Single-molecule dynamics
- 1995 — Targeted gene disruption
  - Atomic force microscopy
- 1990 — Scanning tunneling microscopy
- 1985 — Pulsed field electrophoresis
  - Transgenic animals
  - Amplification of DNA: polymerase chain reaction
  - Automated oligonucleotide synthesis
- 1980 — Site-directed mutagenesis of cloned genes
  - Automated micro-scale protein sequencing
  - Rapid DNA sequence determination
  - Monoclonal antibodies
- 1975 — Southern blotting
  - Two-dimensional gel electrophoresis
  - Gene cloning
  - Restriction cleavage mapping of DNA molecules
- 1970
  - Rapid methods for enzyme kinetics
- 1965
  - High-performance liquid chromatography
  - Polyacrylamide gel electrophoresis
  - Solution hybridization of nucleic acids
  - X-ray crystallographic protein structure determination
- 1960 — Zone sedimentation velocity centrifugation
  - Equilibrium gradient centrifugation
  - Liquid scintillation counting
- 1955 — First determination of the amino acid sequence of a protein
  - X-ray diffraction of DNA fibers
- 1950
  - Radioisotopic tracers used to elucidate reactions
- 1945

Exceedingly powerful new chemical and physical techniques have accelerated the pace at which biological processes have become understood in molecular terms.

$$\overset{+}{N}H_4NCO^- \longrightarrow H_2N-\overset{\overset{\displaystyle O}{\|}}{C}-NH_2$$

Ammonium   **Urea**
cyanate

Early biochemists had to overcome the view that living matter and nonliving matter were fundamentally different.

tell you that I can prepare urea without requiring a kidney or an animal, either man or dog." This was a shocking statement in its time, for it breached the presumed barrier between the living and the nonliving.

Even after Wöhler's demonstration, a pervasive viewpoint called vitalism held that, if not the compounds, at least the reactions of living matter could occur only in living cells. According to this view, biological reactions took place through the action of a mysterious "life force" rather than physical and chemical processes. The vitalist dogma was shattered in 1897, when two German brothers, Eduard and Hans Buchner, found that extracts from broken—and thoroughly dead—yeast cells could carry out the entire process of fermentation of sugar into ethanol. This discovery opened the door to analysis of biochemical reactions and processes **in vitro** (Latin, "in glass"), meaning in a test tube rather than **in vivo**—in living matter. In succeeding decades many other metabolic reactions and reaction pathways were reproduced in vitro, allowing identification of reactants and products and of the enzymes, or biological catalysts, that promoted each biochemical reaction.

The nature of biological catalysis remained the last refuge of the vitalists, who held that the structures of enzymes (or "ferments") were far too complex to be described in chemical terms. But in 1926 James B. Sumner showed that the protein urease, an enzyme from jack beans, could be crystallized like any organic compound. Although proteins have large and complex structures, they are just organic compounds, and their structures can be determined by the methods of chemistry and physics. This discovery marked the final fall of vitalism.

In parallel with developments in biochemistry, cell biologists had been continually refining knowledge of cellular structure. Beginning with Robert Hooke's first observation of cells in the seventeenth century, steady improvements in microscopic techniques led to the understanding that the cell was a complex, multicompartmented structure (see Figure 1.2). *Chromosomes* were discovered in 1875 by Walter Flemming and identified as genetic elements by 1902. The development of the electron microscope, between about 1930 and 1950, provided a new level of insight into cellular structure. At this point subcellular organelles like *mitochondria* and *chloroplasts* could be studied, and it was realized that specific biochemical processes were localized in these subcellular particles. Figure 1.3 illustrates the parallel developments leading to modern biochemistry and cell biology, and their interweaving with the newer science of genetics.

Although developments in the first half of the twentieth century revealed in broad outline the chemical structures of biological materials, identified the reactions in many metabolic pathways, and localized these reactions within the cell, biochemistry remained an incomplete science. We knew that the uniqueness of an organism is determined by the totality of its chemical reactions. However, we had little understanding of how those reactions are controlled in living tissue or of how the information that regulates those reactions is stored, transmitted when cells divide, and processed when cells differentiate.

The idea of the **gene**, a unit of hereditary information, was first proposed in the mid-nineteenth century by Gregor Mendel, from his studies on the genetics of pea plants. By about 1900, cell biologists realized that genes must be found in chromosomes, which are composed of proteins and nucleic acids. Subsequently, the new science of genetics provided increasingly detailed knowledge of the patterns of inheritance and development. However, until the mid-twentieth century no one had isolated a gene or determined its chemical composition. Nucleic acids had been isolated in 1869 by Friedrich Miescher, but their chemical structures were poorly understood, and in the early 1900s they were thought to be simple substances, fit only for structural roles in the cell. Most biochemists believed that only the proteins were sufficiently complex to carry genetic information.

That belief was incorrect. Experiments in the 1940s and early 1950s proved conclusively that **deoxyribonucleic acid (DNA)** is the bearer of genetic information (see Chapter 4). As noted earlier, 1953 was a landmark year, when Watson

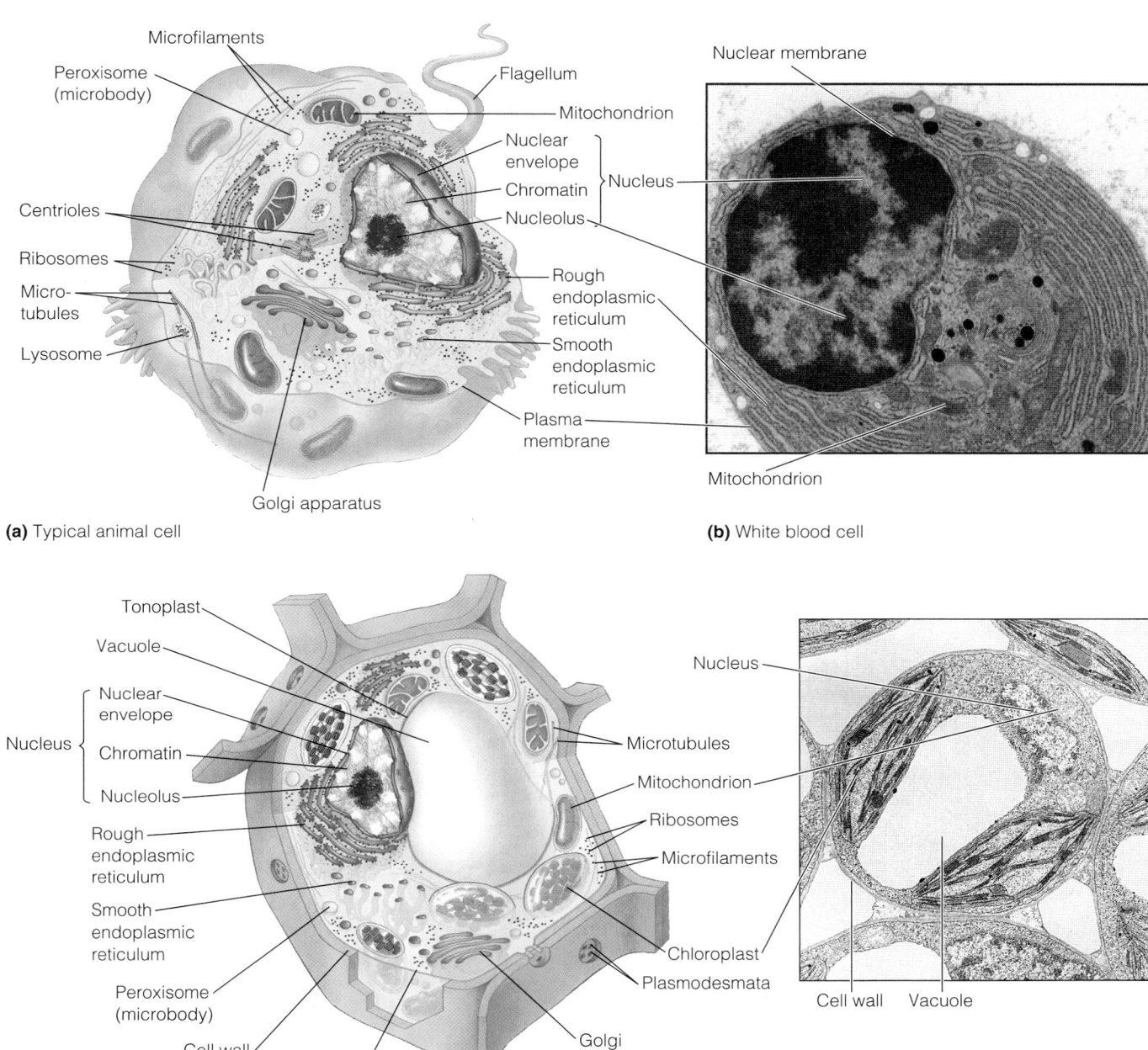

(a) Typical animal cell

(b) White blood cell

(c) Typical plant cell

(d) Cell from *Coleus* leaf

FIGURE 1.2

The complexity of cells. Shown in **(a)** and **(c)** are schematic diagrams of a typical animal and plant cell, respectively. **(b)** An electron micrograph of a representative animal cell, a white blood cell. **(d)** An electron micrograph of a representative plant cell, a thin section of a *Coleus* leaf cell.

(a, c) *Biology*, 5th ed., Neil A. Campbell, Jane B. Reece, and Lawrence A. Mitchell. © 1999. Reprinted by permission of Pearson Education Inc., Upper Saddle River, NJ; (b) Steve Gschmeissner/Science Photo Library

and Crick described the double-helical structure of DNA. This concept immediately suggested ways in which information could be encoded in the structure of molecules and transmitted intact from one generation to the next.

At about this point the strands of scientific development shown in Figure 1.3—biochemistry, cell biology, and genetics—became inextricably interwoven, and the new science of molecular biology emerged. The distinction between molecular biology and biochemistry is not always clear because both

Molecular biology is a fusion of biochemistry, cell biology, and genetics.

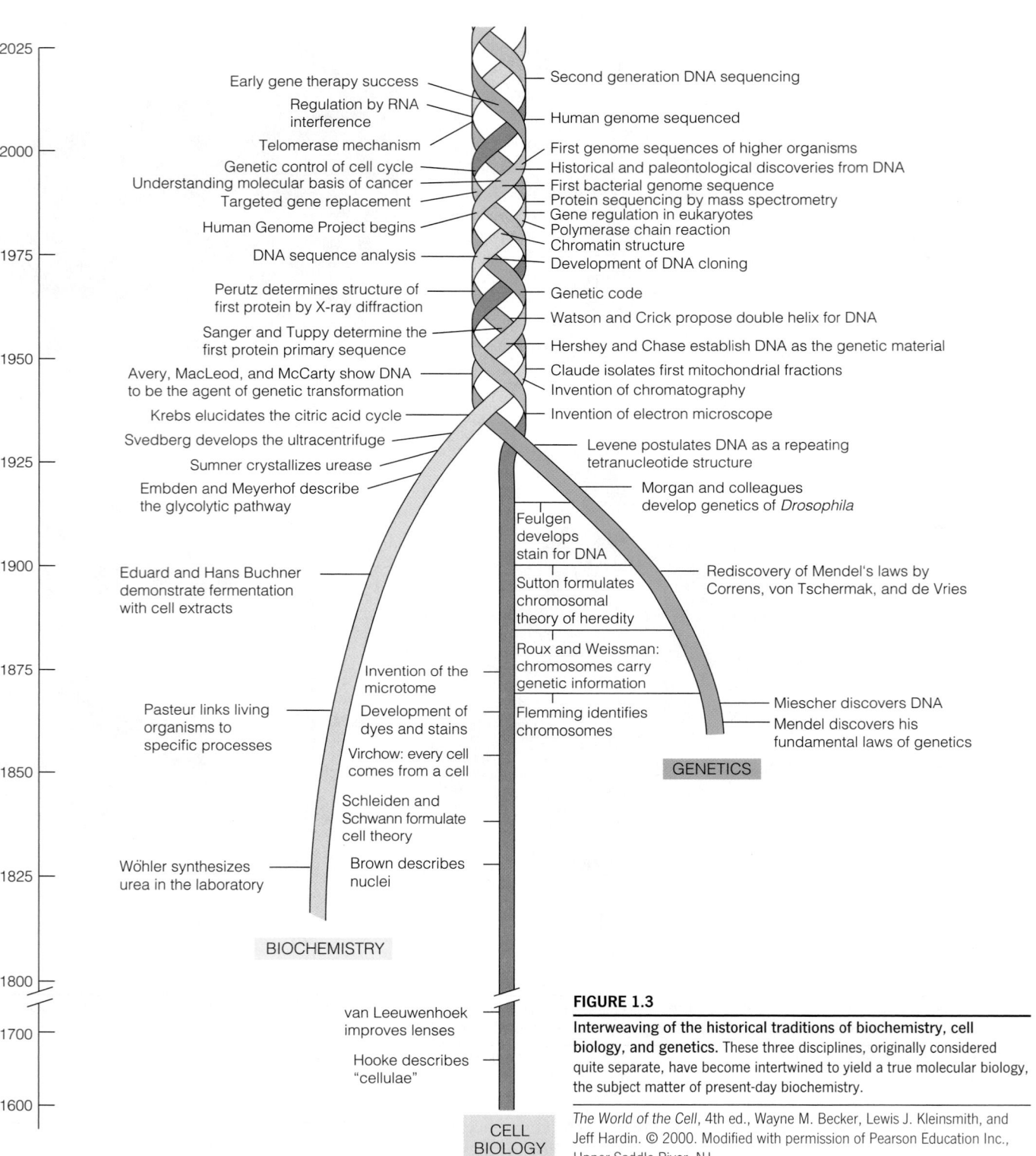

MOLECULAR BIOLOGY

Early gene therapy success

Regulation by RNA interference

Telomerase mechanism

Genetic control of cell cycle

Understanding molecular basis of cancer

Targeted gene replacement

Human Genome Project begins

DNA sequence analysis

Perutz determines structure of first protein by X-ray diffraction

Sanger and Tuppy determine the first protein primary sequence

Avery, MacLeod, and McCarty show DNA to be the agent of genetic transformation

Krebs elucidates the citric acid cycle

Svedberg develops the ultracentrifuge

Sumner crystallizes urease

Embden and Meyerhof describe the glycolytic pathway

Eduard and Hans Buchner demonstrate fermentation with cell extracts

Invention of the microtome

Pasteur links living organisms to specific processes

Development of dyes and stains

Virchow: every cell comes from a cell

Schleiden and Schwann formulate cell theory

Brown describes nuclei

Wöhler synthesizes urea in the laboratory

BIOCHEMISTRY

Second generation DNA sequencing

Human genome sequenced

First genome sequences of higher organisms

Historical and paleontological discoveries from DNA

First bacterial genome sequence

Protein sequencing by mass spectrometry

Gene regulation in eukaryotes

Polymerase chain reaction

Chromatin structure

Development of DNA cloning

Genetic code

Watson and Crick propose double helix for DNA

Hershey and Chase establish DNA as the genetic material

Claude isolates first mitochondrial fractions

Invention of chromatography

Invention of electron microscope

Levene postulates DNA as a repeating tetranucleotide structure

Morgan and colleagues develop genetics of *Drosophila*

Feulgen develops stain for DNA

Sutton formulates chromosomal theory of heredity

Rediscovery of Mendel's laws by Correns, von Tschermak, and de Vries

Roux and Weissman: chromosomes carry genetic information

Flemming identifies chromosomes

Miescher discovers DNA

Mendel discovers his fundamental laws of genetics

GENETICS

van Leeuwenhoek improves lenses

Hooke describes "cellulae"

CELL BIOLOGY

FIGURE 1.3

Interweaving of the historical traditions of biochemistry, cell biology, and genetics. These three disciplines, originally considered quite separate, have become intertwined to yield a true molecular biology, the subject matter of present-day biochemistry.

The World of the Cell, 4th ed., Wayne M. Becker, Lewis J. Kleinsmith, and Jeff Hardin. © 2000. Modified with permission of Pearson Education Inc., Upper Saddle River, NJ.

disciplines take as their ultimate aim the complete definition of life in molecular terms. The term *molecular biology* is often used in a narrower sense, to denote the study of nucleic acid structure and function and the genetic aspects of biochemistry—an area we might more properly call *molecular genetics*. In the early twenty-first century, however, the distinction between biochemistry and molecular biology has become somewhat artificial because successful scientists in either field must use the approaches of all relevant disciplines, including chemistry, biology, and physics. In fact, three of the most powerful research techniques used by biochemists were developed by physicists: **electron microscopy**, which has revealed remarkable details of cellular structure (see Tools of Biochemistry 1A), and **X-ray diffraction** and **nuclear magnetic resonance**, which have revealed the precise three-dimensional structures of huge biological molecules (see Tools of Biochemistry 4A and 6A).

Biochemistry as a Discipline and an Interdisciplinary Science

Biochemistry draws its major themes from many disciplines—from organic chemistry, which describes the properties of biomolecules; from physical chemistry, which describes thermodynamics, properties of water, and electrical parameters of oxidation-reduction reactions; from biophysics, which applies the techniques of physics to study the structures and functions of biomolecules; from medical science, which increasingly seeks to understand disease states in molecular terms; from nutrition, which has illuminated metabolism by describing the dietary requirements for maintenance of health; from microbiology, which has shown that single-celled organisms and viruses are ideally suited for the elucidation of many metabolic pathways and regulatory mechanisms; from physiology, which investigates life processes at the tissue and organism levels; from cell biology, which describes the metabolic and mechanical division of labor within a cell; and from genetics, which describes mechanisms that give a particular cell or organism its biochemical identity. Biochemistry draws strength from all of these disciplines, and it nourishes them in return; it is truly an interdisciplinary science.

Biochemistry is also a distinct discipline, with its own identity. It is distinctive in its emphasis on the structures and reactions of biomolecules, particularly on enzymes and biological catalysis; on the elucidation of metabolic pathways and their control; and on the principle that life processes can be understood through the laws of physics and chemistry. As you read this book, keep in mind both the uniqueness of biochemistry as a separate discipline and the absolute interdependence between biochemistry and other physical and life sciences.

Biochemistry as a Chemical Science

Though we often describe biochemistry as a life science and relate its developments to the history of biology, it remains first and foremost a chemical science. In order to understand the impact of biochemistry on biology, you must understand the chemical elements of living matter and the complete structures of many biological compounds—amino acids, sugars, lipids, nucleotides, vitamins, and hormones—and their behavior during metabolic reactions. You will need to know the stoichiometry and mechanisms of a large number of reactions. In addition, an understanding of basic thermodynamic principles is essential for learning how plants derive energy from sunlight, how animals derive energy from food, and how biomolecules self-assemble to form complex structures.

To understand biochemistry, one must first understand basic chemistry.

All forms of life, from the smallest bacterial cell to a human being, are constructed from the same chemical elements, which in turn make up the same types of molecules. The chemistry of living matter is similar throughout the biological world. Undoubtedly, this continuity in biochemical processes reflects the common evolutionary ancestry of all cells and organisms. Let us begin a preliminary examination of the composition of living matter, starting with the chemical elements.

The Chemical Elements of Living Matter

Life is a phenomenon of the second generation of stars. This rather strange-sounding statement is based on the fact that life, as we conceive it, can come into being only when certain elements—carbon, hydrogen, oxygen, nitrogen, phosphorus, and sulfur (C, H, O, N, P, and S)—are abundant. The very early universe was made almost entirely of hydrogen and helium, for only these simplest elements were produced in the condensation of matter following the primeval explosion, or "big bang." The first generation of stars contained no heavier elements from which to form planets. As these early stars matured over the next seven to eight billion years, they burned their hydrogen and helium in thermonuclear reactions. These reactions produced heavier elements—first carbon, nitrogen, and oxygen and later all the other members of the periodic table. As large stars matured, they became unstable and exploded as novas and supernovas, spreading the heavier elements through the cosmic surroundings. This matter condensed again to form second-generation stars, complete with planetary systems rich in the heavier elements. Our universe, which is now rich in second-generation stars, has an elemental composition compatible with life as we know it.

Why are elements heavier than H and He essential for life? The answer is that life, as we can imagine it, requires large and complex molecular structures. These can be formed only from certain elements and can be stable only under restricted environmental conditions. A universe of hydrogen and helium has no chemistry. Nor is chemistry possible in the heat of stars, where no chemical compounds exist, only elements. In environments as cold as the moon or space, a slow, simple chemistry may occur, but we cannot envision the formation of molecules as complex as proteins or nucleic acids. Only in the temperate environment of an appropriate planet, enriched with elements capable of forming complicated compounds, can life arise.

Few elements are involved in the creation of living matter. Living creatures on the earth are composed primarily of just four—carbon, hydrogen, oxygen, and nitrogen. These elements are also, with the addition of helium and neon, the most abundant elements in the universe. Helium and neon, inert gases, are not equipped for a role in life processes; they do not form stable compounds, and they are readily lost from planetary atmospheres.

The abundance of oxygen and hydrogen in organisms is explained partly by the major role of water in life on earth. We live in a highly aqueous world, and, as we shall see in Chapter 2, the solvent properties of water are indispensable in biochemical processes. The human body, in fact, is about 70% water. The elements C, H, O, and N are important to life because of their strong tendencies to form covalent bonds. In particular, the stability of carbon–carbon bonds and the possibility of forming single, double, or triple bonds give carbon the versatility to be part of an enormous diversity of chemical compounds.

But life is not built on these four elements alone. Many other elements are necessary for terrestrial organisms, as you can see in Table 1.1. A "second tier" of essential elements includes sulfur and phosphorus, which form covalent bonds, and the ions Na^+, K^+, Mg^{2+}, Ca^{2+}, and Cl^-. Sulfur is a constituent of nearly all proteins, and phosphorus plays essential roles in energy metabolism and the structure of nucleic acids. Beyond the first two tiers of elements (which correspond roughly to the most abundant elements of the first two rows of the periodic table), we come to those that play quantitatively minor, but often indispensable, roles. As Table 1.1 shows, most of these third- and fourth-tier elements are metals, and some serve as aids to catalysis of biochemical reactions. In succeeding chapters we shall encounter many examples of the importance of these trace elements to life.

Biological Molecules

The complexity of life processes requires that many of the molecules participating in these processes be enormous. Consider, for instance, the DNA molecules released from one human chromosome, as shown in Figure 1.4. The long, looped

---

Life depends primarily on a few elements (C, H, O, N, S, and P), although many others play essential functions as well.

TABLE 1.1   Elements found in organisms

Element	Comment
First Tier	
Carbon (C)	Most abundant
Hydrogen (H)	in *all*
Nitrogen (N)	*organisms*
Oxygen (O)	
Second Tier	
Calcium (Ca)	Much less
Chlorine (Cl)	abundant but
Magnesium (Mg)	found in *all*
Phosphorus (P)	*organisms*
Potassium (K)	
Sodium (Na)	
Sulfur (S)	
Third Tier	
Cobalt (Co)	Metals present
Copper (Cu)	in small
Iron (Fe)	amounts in
Manganese (Mn)	*all organisms*
Zinc (Zn)	and essential
	to life
Fourth Tier	
Aluminum (Al)	Found in or
Arsenic (As)	required
Boron (B)	by *some*
Bromine (Br)	*organisms*
Chromium (Cr)	in trace
Fluorine (F)	amounts
Gallium (Ga)	
Iodine (I)	
Molybdenum (Mo)	
Nickel (Ni)	
Selenium (Se)	
Silicon (Si)	
Tungsten (W)	
Vanadium (V)	

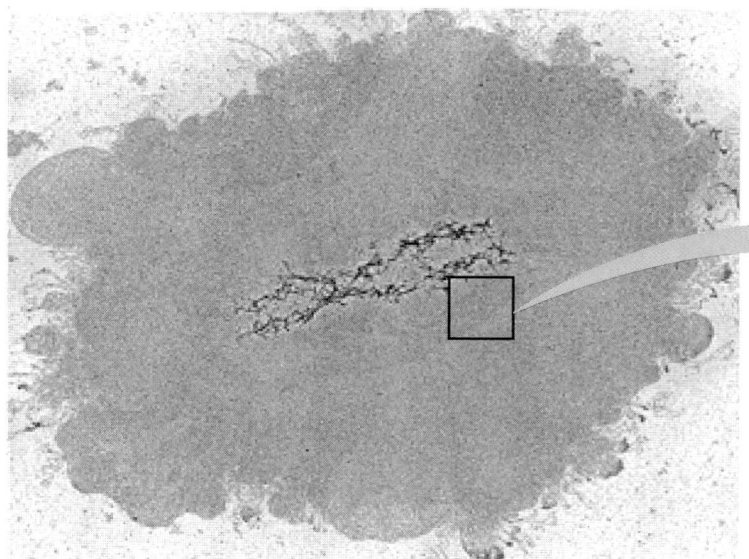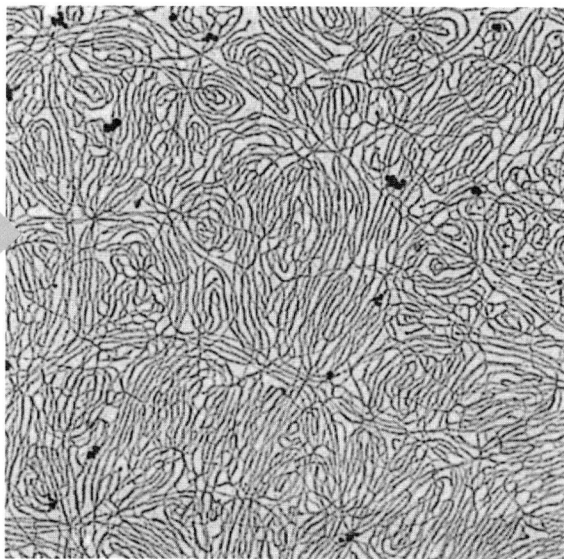

FIGURE 1.4

The DNA from a single human chromosome. Most of the chromosomal proteins have been removed, leaving only a protein "skeleton" from which enormous loops of DNA emerge. There are just two DNA molecules in this chromosome. The enlargement to the right shows the long DNA fiber in more detail. The fiber is about 2 nm thick.

Reprinted from *Cell* 12:817–828, J. Paulson and U. K. Laemmli, The structure of histone-depleted metaphase chromosomes. © 1977, with permission from Elsevier.

thread you see corresponds to just two huge molecules, each with a molecular mass of about 20 billion daltons (a dalton [Da] is 1/12 the mass of a ^{12}C atom, 1.66×10^{-24} g). Even a simple organism such as the single-celled bacterium *Escherichia coli* contains a DNA molecule with a molecular mass of about 2 billion Da; this molecule is more than one millimeter long. Protein molecules are generally much smaller, but they are still large, with molecular masses ranging from about 10,000 to one million Da. The complexity of these molecules can be seen from the three-dimensional structure of even a fairly small protein. Figure 1.5 illustrates the structure of myoglobin, an oxygen storage protein of muscle, which has a molecular weight of about 16,000 Da.

These giant molecules, or **macromolecules**, constitute a large fraction of the mass of any cell. As we shall see in detail in later chapters, there are good reasons for some biological materials to be so large. DNA molecules, for example, can be thought of as tapes or computer files, from which genetic information is read out in a linear fashion. Because the amount of information needed to specify the structure of a multicellular organism is enormous, these tapes must be extremely long. In fact, the DNA molecules in a single human cell, if stretched end to end, would reach a length of about 2 m. As revealed in the early twenty-first century through the Human Genome Project, the information contained in human DNA could encode about 100,000 proteins, although the actual number of functional genes is far smaller.

The synthesis of such large molecules poses an interesting challenge to the cell. If the cell functioned like an organic chemist carrying out a complex laboratory synthesis bit by bit, millions of different types of reactions would be involved, and thousands of intermediates would accumulate. Instead, cells use a modular approach for constructing large molecules. All such structures are **polymers**, made by joining prefabricated units, or **monomers**. The monomers of a given type of macromolecule are of limited diversity and are linked together, or **polymerized**, by identical mechanisms, each involving **condensation**, or removal of a molecule of water in the joining reaction. A simple example is the carbohydrate **cellulose** (Figure 1.6a), a major constituent of the cell walls of plants. Cellulose is a polymer made by joining thousands of molecules of glucose, a simple sugar; in this polymer all of the chemical linkages between the monomers are identical. Covalent links between glucose units are formed by removing the elements of a water molecule between two adjoining glucose molecules; the portion of a glucose molecule remaining in the chain is called a glucose **residue**. Whereas

Many of the important molecules in cells are enormous. The major biopolymers include the nucleic acids, the proteins, and the polysaccharides. All are polymers, made up of one or more kinds of monomer units.

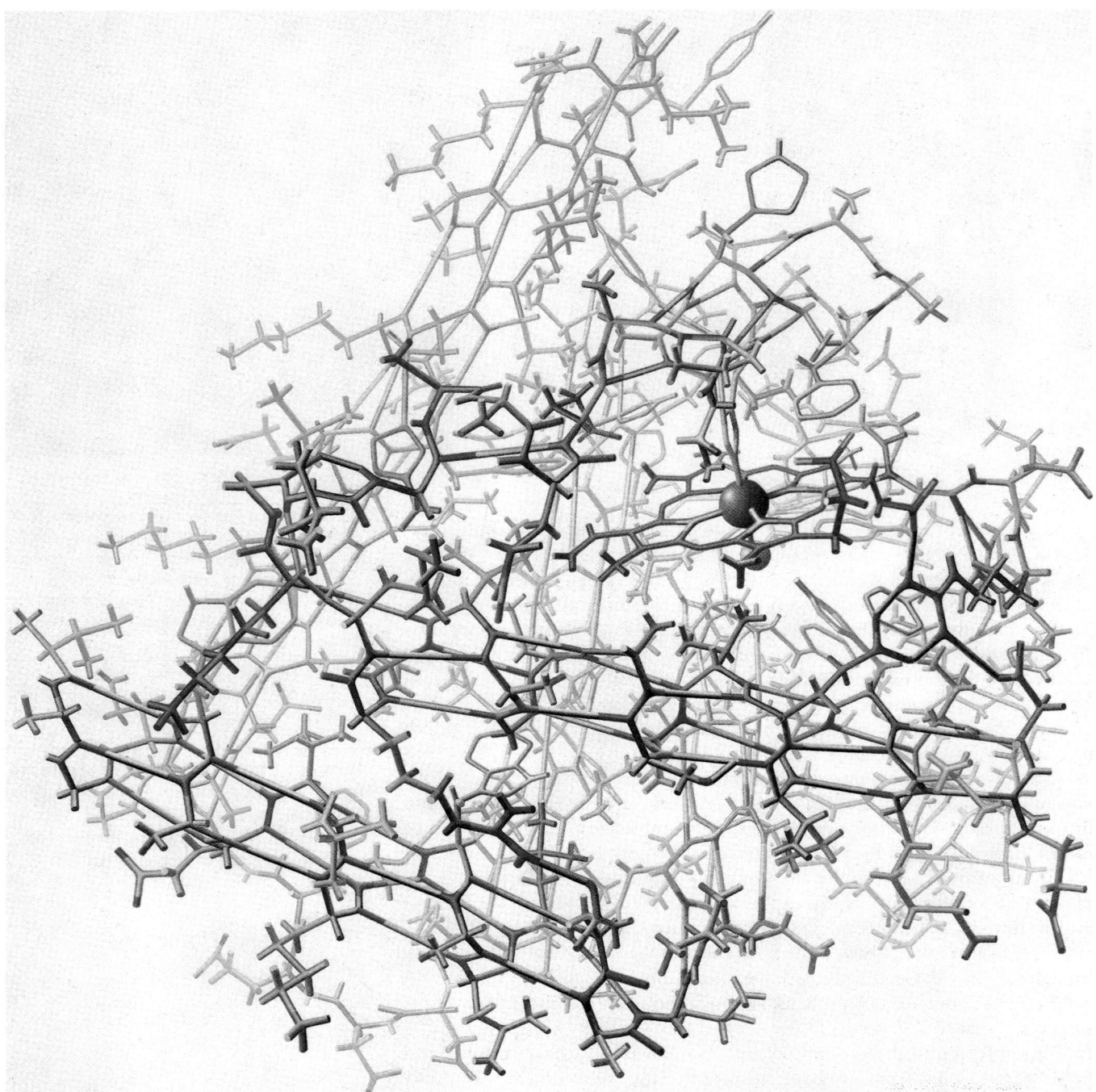

FIGURE 1.5

The three-dimensional structure of myoglobin. This historic painting, by Irving Geis in collaboration with John Kendrew, portrays as a stick model the first protein whose structure was deduced by X-ray diffraction— sperm whale myoglobin. It depicts, therefore, our first indication of the complexity and specificity of the three-dimensional structure of proteins. Such structures are now routinely displayed by computer graphics.

Cells use a modular approach for constructing large molecules.

all residues in a cellulose chain are identical, the sequences of nucleotides, in nucleic acids, and of amino acids, in proteins, are far more complex.

Because cellulose is a polymer of a simple sugar, or **saccharide**, it is called a **polysaccharide**. This particular polymer is constructed from identical monomeric units, so it is called a **homopolymer**. In contrast, many polysaccharides and all nucleic acids and proteins are **heteropolymers**, polymers constructed from a number of different kinds of monomer units. Nucleic acids (Figure 1.6b) are polymers of four **nucleotides**, so nucleic acids are also called **polynucleotides**. Similarly, proteins (Figure 1.6c) are assembled from combinations of 20 different **amino acids**. Protein chains are called **polypeptides**, a term derived from the **peptide bond** that joins two amino acids together.

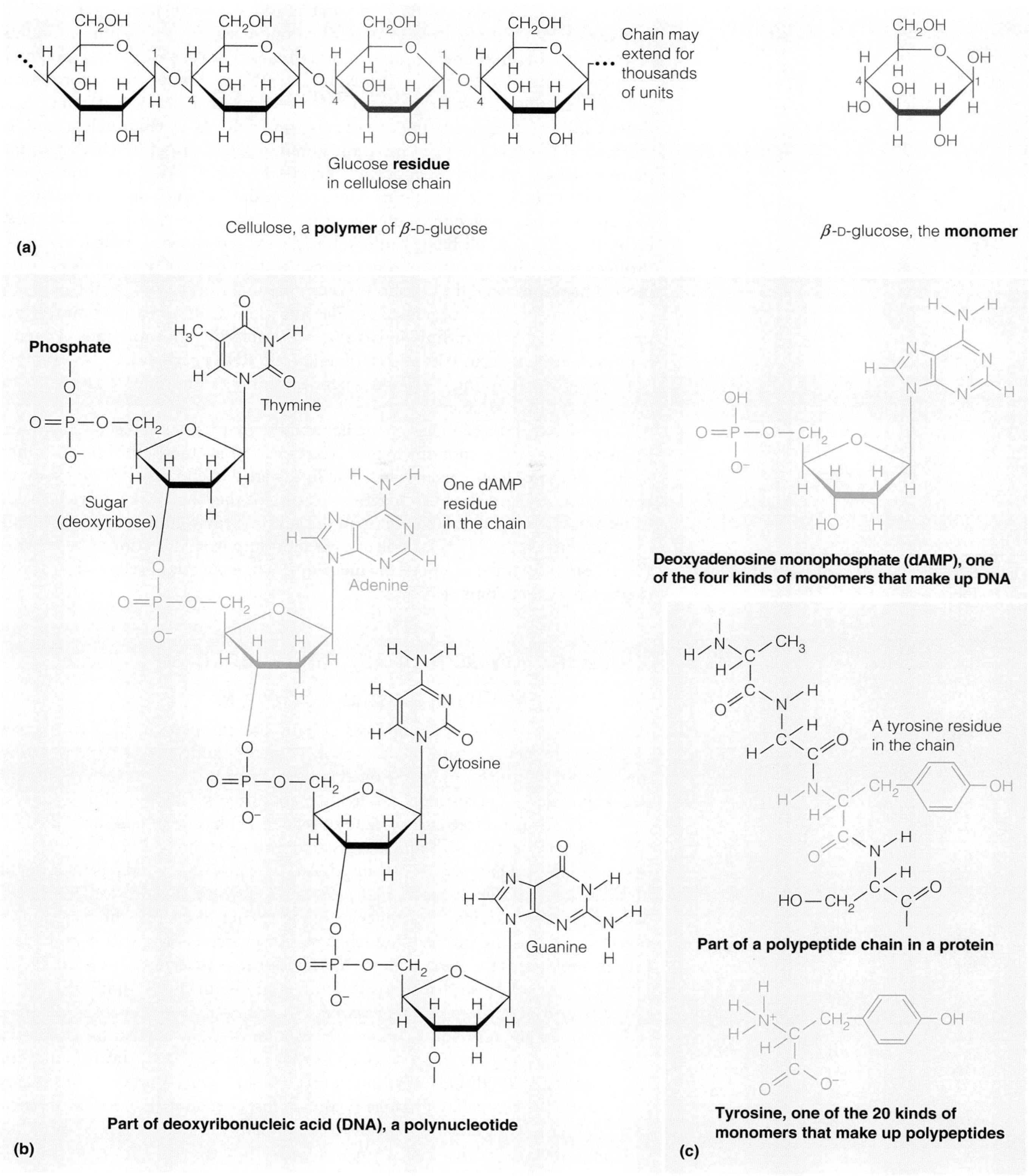

(a)

Glucose **residue**
in cellulose chain

Cellulose, a **polymer** of β-ᴅ-glucose

β-ᴅ-glucose, the **monomer**

Chain may extend for thousands of units

Phosphate

Thymine

Sugar (deoxyribose)

One dAMP residue in the chain

Adenine

Cytosine

Guanine

Part of deoxyribonucleic acid (DNA), a polynucleotide

(b)

Deoxyadenosine monophosphate (dAMP), one of the four kinds of monomers that make up DNA

A tyrosine residue in the chain

Part of a polypeptide chain in a protein

Tyrosine, one of the 20 kinds of monomers that make up polypeptides

(c)

FIGURE 1.6

Examples of biological polymers, or biopolymers. **(a)** A carbohydrate. The carbohydrate cellulose is a polymer of β-ᴅ-glucose monomers, with a molecule of water split out in each joining reaction. **(b)** A nucleic acid. The nucleic acids, DNA and RNA, are polymers of nucleotides. Part of a DNA molecule is shown, along with one of its monomers, deoxyadenosine monophosphate. **(c)** A polypeptide. Protein chains, or polypeptides, are polymers, assembled from 20 different amino acids. Part of a polypeptide is shown, along with one of its monomers, tyrosine.

Polymers form much of the structural and functional machinery of the cell. Polysaccharides serve both as structural components, such as cellulose, and as reserves of biological energy, such as **starch**, another type of glucose polymer found in plants. The nucleic acids, DNA and RNA, participate in information storage, transmission, and expression. DNA serves principally as a storehouse of genetic information, while the chemically similar **RNA**, or **ribonucleic acid**, is involved in the readout of information stored in DNA. Proteins, which have far more structural diversity than polysaccharides or nucleic acids, perform a more diverse set of biological functions. Some play structural roles, such as keratin in hair and skin or collagen in connective tissue. Others act as transport substances, an outstanding example being hemoglobin, the oxygen-carrying protein of blood. Proteins may transmit information between distant parts of an organism, as do protein **hormones** and the cell surface **receptors** that receive the hormone signals, or they may defend an organism against infection, as do the **antibodies**. Most important of all, proteins function as **enzymes**, catalyzing the thousands of chemical reactions that occur within an individual cell. **RNA polymerase** is an example of an enzyme, catalyzing the DNA-dependent joining of nucleotide molecules to synthesize RNA (see Chapter 27).

In addition to these macromolecules and the many small molecules involved in metabolism and as monomers in macromolecular synthesis, there is one other major class of cellular constituents. The **lipids** are a chemically diverse group of compounds that are classified together because of their hydrocarbon-rich structures, which give them low solubility in the aqueous environment of the cell. This low solubility equips lipids for one of their most important functions—to serve as the major structural element of the **membranes** that surround cells and partition them into various compartments.

Lipids form the major constituents of biological membranes.

Biochemistry as a Biological Science

Distinguishing Characteristics of Living Matter

The chemistry of *life* is what concerns us here. The complex chemical substances and reactions that we have mentioned have their significance as parts of living matter and life processes. To see biochemistry from this perspective, we should begin by asking, what distinguishes living matter from nonliving matter?

Daniel Koshland (see end-of-chapter references) has described seven distinctive attributes, or "pillars of life," the essential principles on which a living system operates. First is a *program*, or organized plan for constitution and regeneration of an organism. For life on Earth, that program is the information stored in DNA. Second is *improvisation*, the ability of living matter to change the program to assure survival as the surroundings change. The processes of mutation and selection ensure that, as the environment changes (aqueous to terrestrial, for example), organisms can adjust so that survival continues under the new conditions.

The third pillar is *compartmentation*, the ability of an organism to separate itself from the environment (with membranes, for example) so that the chemistry needed to carry out the program, such as enzyme-catalyzed reactions, can occur under favorable conditions of temperature, pH, and concentrations of reactants and products. The smallest organisms contain just one compartment, the interior of a single cell, while larger organisms contain many cells, specialized to carry out different functions, such as sensory perception or movement. In addition, the cells of multicellular organisms contain subcompartments, the organelles that allow for division of labor within the basic functional unit, the cell itself. The fourth pillar is *energy*. Thermodynamics tells us that spontaneous processes occur in the direction of simplicity and randomness. Yet living matter must create complexity in order to sustain the program and the other pillars of life. To do this, cells and organisms carry out reactions that yield energy, such as the oxidation of nutrients, and they couple some of that energy to energy-requiring reactions that create complexity, such as the synthesis of nucleic acids and proteins or the transmission of nerve impulses. The ultimate source of that energy is the sun, which is used either directly, by photosynthetic plants and bacteria to drive the synthesis of biomolecules, or indirectly, by organisms that consume other organisms and derive their energy from the breakdown of dietary nutrients.

Fifth is *regeneration*, the ability to compensate for the inevitable wear involved in maintaining a physical state far from equilibrium. For example, all of the proteins in a cell are subjected to continuous degradation, either because they suffer environmental damage or because, like digestive enzymes, they undergo degradation as part of their normal function. The ability to continuously replace damaged molecules of this type is a distinguishing characteristic of life. The sixth pillar is *adaptability*, the capacity of an organism to respond to environmental changes. For example, when nutrient stores within an animal are depleted, the animal becomes hungry and seeks food. Adaptability, a property of individual organisms, is distinguished from improvisation, the capacity of populations of organisms to respond to environmental change over a time scale of many generations.

The final pillar is *seclusion*, which means that metabolic processes and pathways must operate in isolation from one another, even though they may take place within the same compartment of a cell. When we digest carbohydrate, a consequence is a rise in intracellular glucose concentration. In liver or muscle that glucose can either be consumed to provide energy or polymerized into glycogen, a glucose polymer, comparable to starch, which is stored for later release when there is an energy demand. Glucose polymerization and the initial steps in breakdown occur within the same cell compartment. Yet intracellular control processes and the specificities of the enzymatic catalysts involved ensure that one of the pathways is favored and the other inhibited, in response to the cell's needs.

Interconnected with all seven pillars of life is the function of semipermeable membranes, which surround cells and intracellular **organelles**, such as mitochondria, maintaining **homeostasis**, a condition in which the chemical composition of a biological system is held constant. Throughout this book we will be concerned with the difference between equilibrium and homeostasis, a non-equilibrium steady state. Cells and organisms maintain constancy of their chemical environment, which may suggest that those systems are at equilibrium. However, the steady-state characteristic of living systems can be maintained only by continuous infusion of energy. A protein molecule at equilibrium will have broken down to amino acids. Proteins in cells, however, are synthesized and maintained in their polymeric state only through the energy-dependent maintenance of a nonequilibrium steady state. As another example, membranes are responsible for maintaining ion concentration gradients. Most cells actively transport potassium inward against a concentration gradient. At equilibrium, the intra- and extracellular potassium concentrations would be equivalent.

How did this remarkable process we call life arise? We do not know, but we do know that it is truly ancient, almost as old as the earth itself. The earth condensed from cosmic dust about 4.5 billion years ago, but recognizable traces of living microbes have been dated to 3.8 billion years, only 700 million years later. It is possible that some of these earliest organisms (or proto-organisms) utilized preformed chemical building blocks. For example, traces of amino acids have been found in meteorites, strong evidence that such substances can be generated **abiotically**, without the involvement of a living system.

Whatever their origin, we know that the earliest organisms must have lived an anaerobic existence, for the earth was devoid of free oxygen. Indeed, it is believed that all of the oxygen now present in the earth's atmosphere is the product of photosynthesis by algae and plants. It probably required 3–4 billion years for the present oxygen level to accumulate. Life has not only occupied this planet, it has rebuilt it.

Life depends on creating and duplicating order in a chaotic environment. This ordering requires energy.

All living creatures are composed of cells. Most cells are similar in size.

The Unit of Biological Organization: The Cell

A seminal early discovery in biology was Robert Hooke's observation (1665) that plant tissues (in this case, cork) were divided into tiny compartments, which he called *cellulae*, or **cells**. By 1840, improved observations on many tissues led Theodor Schwann to propose that all organisms exist as either single cells or aggregates of cells. That hypothesis is now firmly established.

Furthermore, cells, from whatever organism, are of rather similar size. Most bacterial cells are about 1–2 μm in diameter, and most cells of higher organisms are only about 10- to 20-fold larger, plant cells being somewhat larger than animal cells. There are, to be sure, exceptions: There are very small bacteria (0.2 μm), and there are unusual cells like those in the nervous systems of vertebrates, some of which may be over 1 m long. But compared with the overall range of sizes of natural organisms, most cell sizes are much alike.

In both plants and animals, the size of the cells bears no relationship to the size of the organism. An elephant and a flea have cells of about the same size; the elephant just has more of them. Why is such uniformity in cell size maintained? A clue can be found in the fact that the surface/volume ratio for an object of a given shape depends on its size. For example, a cube 20 μm on a side has a surface-to-volume ratio of 0.3, while 1000 cubes, each 2 μm on a side, have the same total volume, but a surface-to-volume ratio of 3.0, 10 times larger. The ability to exchange substances with its surroundings is crucial for the chemical processes that occur within a cell, meaning that a critical surface-to-volume ratio must be maintained. Evidently, cellular function is sufficiently similar between elephants and fleas, for example, that evolution has selected cells exhibiting similar surface-to-volume ratios.

FIGURE 1.7

Molecular tree of life, based upon ribosomal RNA sequence comparisons. Not all lines of descent are shown. The branch points represent common evolutionary origins, and the distance between two branches represents genetic relatedness. Some lineages represent "environmental sequences," *i.e.*, sequences detected in the environment from which the relevant organisms have not yet been isolated in culture.

Reprinted with permission from *Microbe* 3:15–20, N. R. Pace, The molecular tree of life changes how we see, teach microbial diversity. © 2008 American Society for Microbiology.

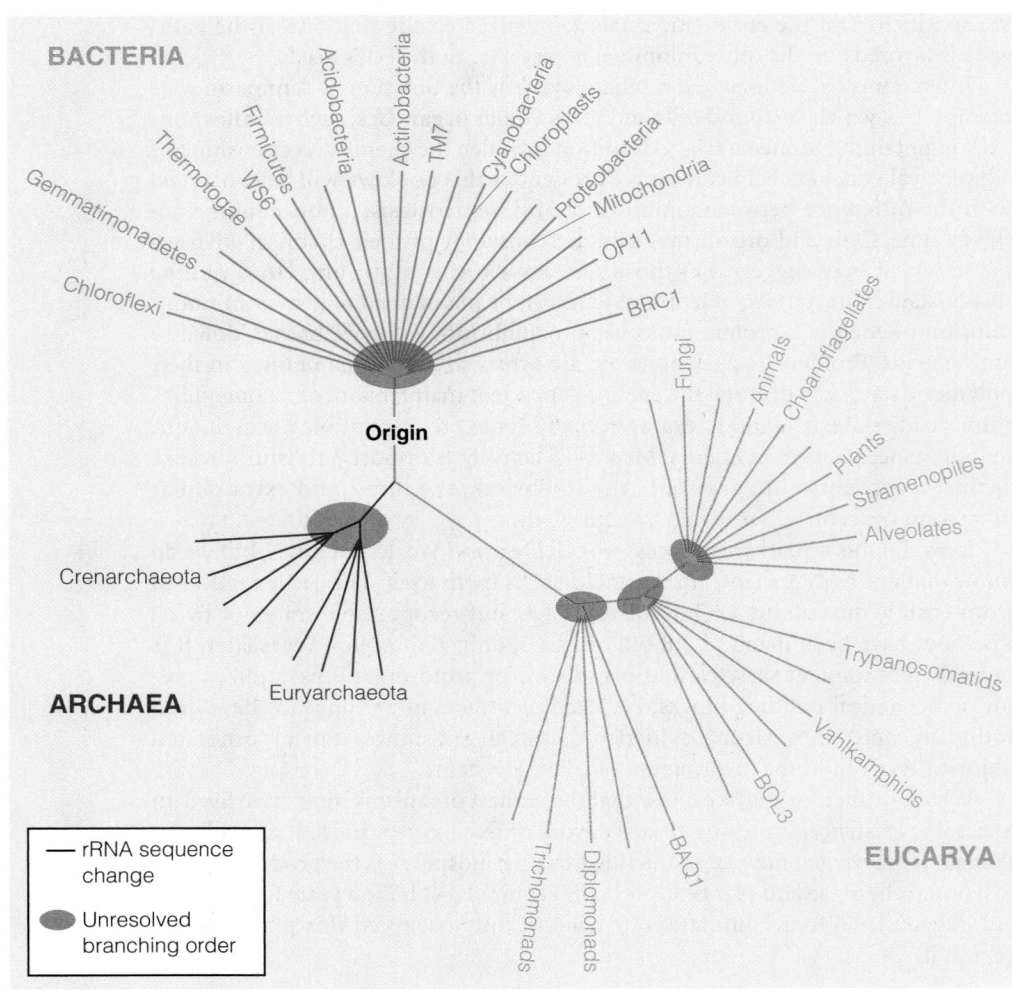

Because cells are the universal units of life, let us examine them more closely. Major differences between cell structures define the two great classes of organisms— prokaryotic and eukaryotic. The **prokaryotes**, which are always unicellular, include the true **bacteria** (eubacteria) and an ancient class called **archaea**. Although organisms were originally classified in terms of morphological or structural criteria, current classification is based upon biochemical analysis, primarily DNA nucleotide sequence determination. Figure 1.7 shows the three great branches of life as an evolutionary "tree," with relationships based upon similarities in ribosomal RNA sequences. The closer together are two branches, the more closely related are organisms in those two branches.

A typical prokaryotic organism is shown schematically in Figure 1.8. Prokaryotic cells are surrounded by a plasma membrane and usually by a rigid cell wall as well. Within the membrane is the **cytoplasm**, which contains the **cytosol**—a semiliquid concentrated solution or gel—and the structures suspended within it. In prokaryotes the cytoplasm is not divided into compartments, and the genetic information is in the form of one or more DNA molecules that exist free in the cytosol. Also suspended in the cytosol are the **ribosomes**, which constitute the molecular machinery for protein synthesis. The surface of a prokaryotic cell may carry **pili**, which aid in attaching the organism to other cells or surfaces, and **flagellae**, which enable it to swim.

All other organisms are called **eukaryotes**. These include the multicellular plants and animals, as well as the unicellular and simple multicellular organisms called protozoans, fungi, and algae. A few of the many differences between eukaryotes and prokaryotes are listed in Table 1.2. Most eukaryotic cells are larger (by 10- to 20- fold) than prokaryotic cells, but they compensate for their large size by being *compartmentalized*. Their specialized functions are carried out in **organelles**— membrane-surrounded structures lying within the surrounding cytoplasm.

> The two great classes of organisms have different cell types. Prokaryotic cells are uncompartmented; eukaryotes have membrane-bound organelles.

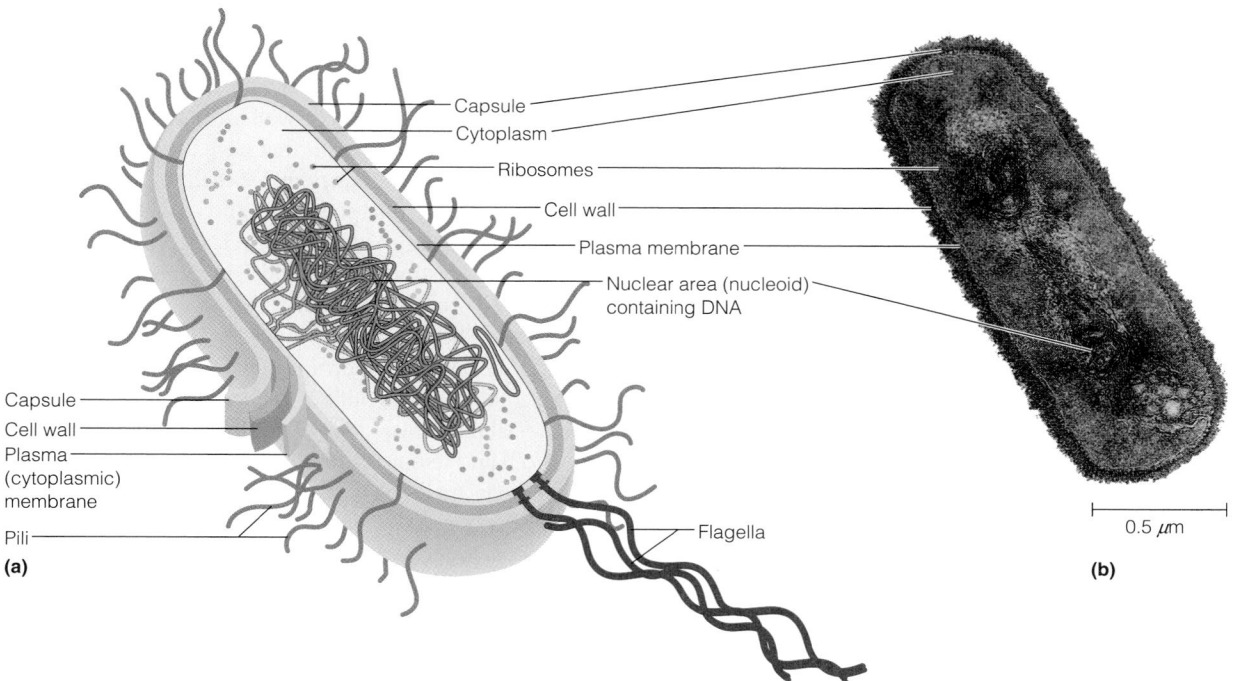

FIGURE 1.8

Prokaryotic cells. **(a)** Schematic view of a representative bacterial cell. The DNA molecule that constitutes most of the genetic material is coiled up in a region called the nucleoid, which shares the fluid interior of the cell (the cytoplasm) with ribosomes (which synthesize proteins), other particles, and a large variety of dissolved molecules. The cell is bounded by a plasma membrane, outside of which is usually a fairly rigid cell wall. Many bacteria also have a gelatinous outer capsule. Projecting from the surface may be pili, which attach the cell to other cells or surfaces, and one or more flagellae, which enable the cell to swim through a liquid environment. **(b)** Electron micrograph of a bacterial cell, a thin section of a dividing cell of the bacterium *Bacillus coagulans*. The light areas represent the two nucleoids, and the dark granules are ribosomes.

TABLE 1.2 Comparison of some properties of prokaryotic and eukaryotic cells

	Prokaryotic Cells	Eukaryotic Cells
Size	0.2–5 μm in diameter	Most are 10–50 μm in diameter
Internal compartmentation	No	Yes, with several different kinds of organelles
Containment of DNA	Free in cytoplasm as nucleoid	In nucleus, condensed with proteins into multiple chromosomes
Ploidy[a]	Usually haploid	Almost always diploid or polyploid
Mechanism of cell replication	Simple division following DNA replication	Mitosis in somatic cells, meiosis in gametes[b]

[a]The term *ploidy* refers to the number of copies of the genetic information carried by each cell. Haploid cells have one copy, diploid cells two, and polyploid cells more than two.

[b]In mitosis the diploid state is retained by chromosome duplication. This occurs in most *somatic*, or "body," cells of organisms. In the cells that produce gametes (sperm or ova) the process of meiosis leads to a haploid state.

The cell can be thought of as a factory, with organelles and compartments specialized to perform different functions.

Benzene   Toluene

vs.

An androgen

vs.

An estrogen

Schematic views of idealized animal and plant cells were shown earlier in Figure 1.2. Major organelles common to most eukaryotic cells are the **mitochondria**, which specialize in oxidative metabolism; the **endoplasmic reticulum**, a folded membrane structure rich in ribosomes, where much protein synthesis occurs; the **Golgi complex**, membrane-bound chambers that function in secretion and the transport of newly synthesized proteins to their destinations; and the **nucleus**. The nucleus of a eukaryotic cell contains the cell's genetic information, encoded in DNA that is packaged into **chromosomes**. A portion of this DNA is subpackaged into a dense region within the nucleus called the **nucleolus**. Surrounding the nucleus is a **nuclear envelope**, pierced by pores through which the nucleus and cytoplasm communicate.

There are also organelles specific to plant or animal cells. For example, animal cells contain digestive bodies called **lysosomes**, which are lacking in plants. Plant cells have **chloroplasts**, the sites of photosynthesis; and usually a large, water-filled **vacuole**. Furthermore, whereas most animal cells are surrounded only by a **plasma membrane**, plant cells often have a tough cellulosic **cell wall** outside the membrane. **Basal bodies** act as anchors for cilia or flagellae in cells that have those appendages.

It is useful to think of the cell as a factory, an analogy we shall frequently use in later chapters. Membranes enclose the whole structure and separate different organelles, which can be thought of as departments with specialized functions. The nucleus, for example, is the central administration. It contains in its DNA a library of information for cellular structures and processes, and it issues instructions for proper regulation of the business of the cell. The chloroplasts and mitochondria are power generators (the former being solar, the latter fuel-burning). The cytoplasm is the general work area, where protein machinery carries out the formation of new molecules from imported raw materials. Special molecular channels in the membranes between compartments and between the cell and its surroundings monitor the flow of molecules in appropriate directions. Like factories, cells tend to specialize in function; for example, many of the cells of higher organisms are largely devoted to the production and export of one or a few molecular products. Examples are pancreatic beta cells, which secrete digestive enzymes, and white blood cells, each of which is specialized to synthesize just one of the several million different antibody molecules that can be synthesized by humans as part of the immune response.

Biochemistry as a Biological Science: Form and Function

Function and evolution lie at the heart of biology. Evolution occurs from selection of function. Because biochemistry deals in part with the evolution of proteins and genes, biochemistry is, as we have noted, a biological science. The question of function, however, does not arise in the physical sciences. Function and purpose are irrelevant concepts in the nonliving world. A chemist, for example, explores the relationship between structure and reactivity of molecules by examining the effects of substitutions on reaction rates and electronic structure. He or she may inquire whether toluene is more reactive than benzene toward nitronium ions (NO_2^+). However, the chemist cannot ask, "What is the function of the methyl group in toluene?" Nor can a geographer ask, "What is the function of the Florida peninsula?" But a biologist may ask why there are species differences in the coloration and pattern of butterfly wings. Similarly, the biochemist can, and does, attempt to explain functional roles of methyl groups; for example, he or she may investigate why methyl groups are introduced into DNA following its replication, or why there are differences in DNA methylation patterns between normal cells and cancer cells.

Or, the biochemist may note that in the biosynthetic pathways to sex hormones, a methyl group is removed from the steroid ring system as an androgen (male sex hormone) is converted to an estrogen (female sex hormone). Because the sex hormones convey different signals in males and females, the methyl group is clearly fulfilling a function.

Therefore, because the biochemist studies living things and he or she deals with functional molecules, in contrast to the chemist, who studies nonliving things and is not concerned with function, biochemists have had to shape their own approach. They use the research techniques of chemistry, but the questions that they seek to answer are unique.

The functioning of each molecule in an organism cannot be properly understood in isolation. Consider an enzyme. Enzymes are proteins, and protein molecules are functionally designed. "Design" here implies the intimate relation of structure to function. Proteins are not random sequences of amino acids. The amino acid sequence dictates the folding of the polypeptide into a particular functional conformation. However, during evolution, the three-dimensional structure is conserved, but not necessarily the linear (one-dimensional) sequence of amino acids.

X-ray crystallography enables us to "see" the exquisite three-dimensional structure of an enzyme. However, X-ray analysis alone cannot show all the events that occur during catalysis. In fact, without prior knowledge that the molecule under study is in fact a catalyst, the crystallographer can only appreciate the beauty of the folded structure. He or she could not identify which crevice holds the substrate or how substrate binding triggers a conformational change, or what adaptations of the enzyme reflect its place in a metabolic pathway in different cell types with varying metabolic needs. In short, the structure of the enzyme assumes its full significance only in relation to its activity, not simply as a catalyst but also as a control device within the functioning of an organism.

Windows on Cellular Function: The Viruses

In analyzing metabolism and the processing of genetic information, biochemists have been aided immeasurably by the **viruses**. Viruses are not cellular and are therefore described as "biological entities" rather than organisms; they are intracellular parasites that can grow only by invading cells. Viruses usually consist of one molecule of nucleic acid (either DNA or RNA) wrapped in an envelope made largely or completely of protein. The envelope is specialized to allow the virus particle to enter particular plant, animal, or bacterial cells. Figure 1.9 shows electron micrographs of several representative viruses.

Because viruses contain no metabolic machinery of their own, they must use that of the host cell to reproduce. By studying the conditions necessary for virus replication, we can learn how this cellular machinery operates. Viruses thus provide useful windows onto the cellular functions that they co-opt during infection. For example, the smallest DNA-containing viruses replicate their DNA using only host-cell enzymes. Because of their small size, these viral DNA molecules can be isolated and characterized much more easily than the giant DNA molecules in cellular chromosomes. Larger viruses stimulate the formation of new enzymes after infection, with viral genes specifying the structures of these enzymes. Studies of these enzymes have illuminated genetic regulatory mechanisms, and they have identified targets for treating viral diseases, including AIDS, influenza, and herpes virus infections.

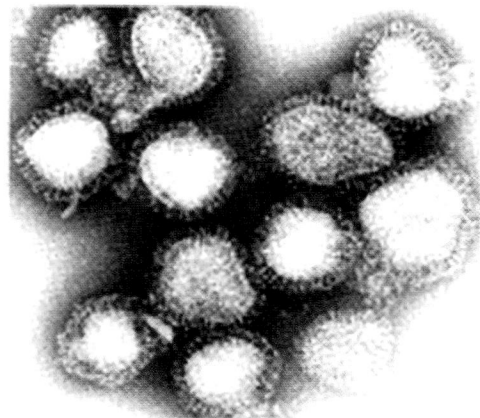

(a) Influenza virus

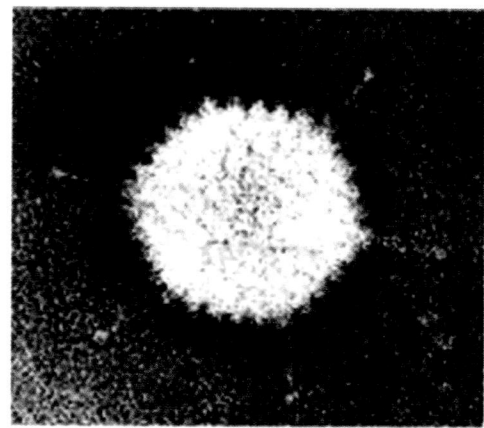

(b) Adenovirus

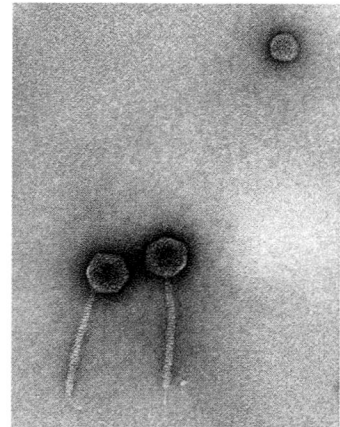

(c) Bacteriophage lambda

FIGURE 1.9

Examples of viruses.

(a, b) Courtesy of Frederick A. Murphy, Centers for Disease Control and Prevention; (c) Dr. M. Wurtz/Biozentrum, University of Basel/Science Photo Library.

Viruses are intracellular parasites that use cellular machinery and energy sources to replicate.

Much of today's biochemistry looks at the cell globally, attempting to understand its functions in terms of the expression of all of the genes in the genome.

Biochemistry and the Information Explosion

As you read this book, you will see that our understanding of biochemistry has taken place by a process involving single molecules and individual reactions. To understand a metabolic pathway, such as the citric acid cycle, which is responsible for oxidation of all nutrients, we must identify all of the intermediate molecules in the pathway, isolate each of the enzymes that catalyzes a reaction in the pathway (there are eight), and characterize each reaction in terms of identifying substrates, products, stoichiometry, and regulation. Ultimately, we would want to learn how each enzyme works, by determining the atomic structure of the enzyme and learning the molecular mechanism of catalysis.

At the same time, we realize that all of the metabolic pathways described in this book represent but a small fraction of the total potential of a cell or organism to carry out chemical reactions in support of its existence and function. The amount of DNA in the 23 pairs of human chromosomes is sufficient to encode 100,000 different proteins, although sequence analysis indicates that far fewer proteins are actually encoded. The ability of scientists to carry out ultra-large-scale DNA sequence analysis has generated biological data on a scale that far exceeds our capacity to integrate it through analysis of individual reactions and pathways. This has given rise to new fields of science that can be considered areas within biochemistry or molecular biology—**bioinformatics**, **genomics**, **proteomics**, and **metabolomics**.

Bioinformatics can be considered as information science applied to biology. Bioinformatics can include mathematical analysis of DNA sequence data, computer simulation of metabolic pathways, or analysis of potential drug targets (enzymes or receptors) for structure-based design of new drugs. Whereas genetics concerns itself with the location, expression, and function of individual genes or small groups of genes, genomics concerns itself with the entire genome, or the totality of genetic information in an organism. This includes not only determining the nucleotide sequence of the whole genome but also assessing the expression and function of each gene, as well as evolutionary relationships among genes in the same genome and with genomes of different organisms. The use of "gene chips," or microarrays, allows the collection of enormous amounts of data concerning the genes expressed in a given cell or tissue. For this analysis, DNA fragments representing many or most of the genes of an organism are immobilized on a glass slide. Many thousands of DNA fragments, each in a minute amount, are arrayed individually on such a slide. The slide is then incubated with a preparation of total RNA from a cell under conditions where a particular RNA molecule can bind to DNA containing sequences from the gene specifying that RNA (for details see Tools of Biochemistry 27C). The use of fluorescent reagents to monitor binding (see Figure 1.10) allows one to generate a snapshot of the global, or genome-wide, pattern of gene expression within a cell, so that one can ask, for example, how the pattern of gene expression changes as the result of hormonal stimulation of that cell or in the transformation of a normal cell to a cancer cell.

The products of most genes are proteins, which carry out nearly all of the chemistry of a cell, including catalysis of reactions, signaling between cells, and movement of material both inside and outside of cells. A goal of proteomics is to identify all of the proteins present in a given cell, and the amount and function of each protein. The **proteome**, or entire complement of proteins in a cell, can be displayed by two-dimensional gel electrophoresis, as described in Tools of Biochemistry 5D. In this technique, the proteins in a cell are resolved by techniques that separate them on the basis of molecular weight and electric charge. In Figure 1.11 each spot represents one protein, with the size and intensity of each spot determined by the amount of that protein in the cell. The challenge of proteomics is to analyze data like this electrophoretic pattern, such that each protein can be identified and quantitated. Then, as in microarray analysis, one can determine how the proteome changes, for example, in the transformation of a normal cell to a cancer cell, so as to help understand the chemistry of cancer.

FIGURE 1.10

A DNA microchip array. In this experiment, using the MICROMAX detection system, 2 μg of human messenger RNA was annealed with immobilized cDNAs representing 2400 human genes. The cDNAs (complementary DNAs) are DNA copies of individual messenger RNA molecules. Spot intensities and colors indicate the abundance of particular gene-specific mRNAs (for further information see page 1166, Tools of Biochemistry 27A).

Courtesy of InCyte Pharmaceuticals, Inc., Palo Alto, CA.

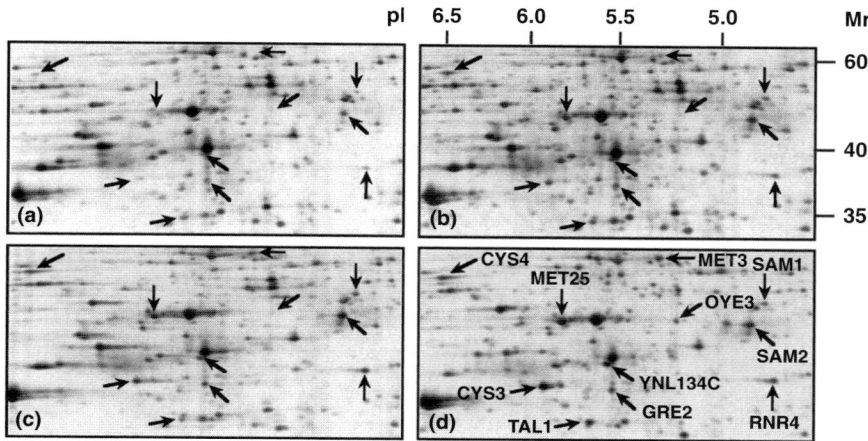

FIGURE 1.11

A display of the proteome by two-dimensional gel electrophoresis. An extract containing all of the proteins of a cell or tissue is applied to one corner of a slab of gel-like material, and the proteins are resolved along one edge of the gel by isoelectric focusing (Tools of Biochemistry 2A), which separates proteins on the basis of their relative acidity (horizontally in the figure). The gel is rotated by 90 degrees, and resolution is carried out by denaturing polyacrylamide gel electrophoresis (Tools of Biochemistry 5D), which separates proteins on the basis of molecular weight. The example shown here is part of a study of cadmium toxicity in yeast. Proteins were labeled by growth of yeast in a medium containing ^{35}S, a radioisotope of sulfur. After electrophoresis the gel was placed in contact with photosensitive film. Each protein darkens the film where it was deposited on the gel. The arrows denote proteins whose concentration, as determined by spot intensity, was changed as a result of the cadmium treatment. **(a)** untreated; **(b)**, **(c)**, and **(d)**, varying cadmium treatment protocols.

The Journal of Biological Chemistry, 276:8469–8474, K. Vido, D. Spector, G. Lagniel, S. Lopez, M. B. Toledano and J. Labarre, A proteome analysis of the cadmium response in *Saccharomyces cerevisiae*. Reprinted with permission. © 2001 The American Society for Biochemistry and Molecular Biology. All rights reserved.

Finally, many of the proteins in a cell are enzymes, and the intracellular rates of enzyme-catalyzed reactions can be estimated by analysis of the intracellular concentrations of the substrates and products of each reaction. A goal of metabolomics is to determine the intracellular concentrations of all small molecules that serve as intermediates in metabolic pathways (see Tools of Biochemistry 12B). Metabolite pool analysis, of course, is at the heart of much of clinical diagnosis. A prime example is the use of cholesterol levels to assess risk of cardiovascular disease. At present there is no single technique that would allow analysis of the metabolome, or complete complement of low-molecular-weight metabolites, but that does represent an important goal of much of contemporary biochemistry. Mass spectrometry is an approach that has the potential to describe the complete metabolome, but formidable challenges remain.

Bioinformatics and the allied fields have the potential to yield enormous amounts of information about the function of cells and organisms. However, effective use of that information requires understanding of the thousands of individual chemical reactions that occur in metabolism, the biological functions of each reaction, and the processes that regulate the expression of genetic information, both as DNA sequences that are read out as RNA and as RNA sequences that are translated into proteins. These latter topics are the focus of this book, but you should be aware that much of the biochemistry of today and tomorrow will integrate this information into a much broader context.

SUMMARY

The aim of the science of biochemistry is to explain life in molecular terms because we now realize that living matter and nonliving matter obey the same fundamental laws of physics and chemistry. Modern biochemistry draws on knowledge from chemistry, cell biology, and genetics and uses techniques adapted from physics. Discoveries in all of these sciences have contributed to the development of a true molecular biology.

Although biochemistry deals with organisms, cells, and cellular components, it is fundamentally a chemical science. The basic chemistry involved is that of carbon, hydrogen, oxygen, nitrogen, phosphorus, and sulfur, but organisms use many other elements in smaller quantities. Many of the important biological substances are giant molecules that are polymers of simpler monomer units. Such biopolymers include polysaccharides, proteins, and nucleic acids. Lipids form the fourth major group of biologically important substances.

The distinguishing feature of living matter is the use of energy to create and duplicate orderly structure. All living organisms are composed of one or more cells, and these cells are quite similar in size. However, the two great classes of organisms, prokaryotes and eukaryotes, have fundamentally different cellular structures: Prokaryotic cells are uncompartmented, lacking the membrane-bound organelles characteristic of eukaryotic cells. Viruses are intracellular parasites, essentially carriers of nucleic acid; to duplicate themselves, they use the reproductive machinery and energy sources of the host cell.

Biochemistry is an experimental science, and the remarkable recent advances in biochemistry are due in large part to the development of powerful new laboratory techniques. Some of these techniques are generating significant information far more rapidly than it can be integrated and understood without recourse to new approaches in information technology.

REFERENCES

Aebersold, R., and B. F. Cravatt (eds.) (2002) A TRENDS Guide to Proteomics. One member of a series of annual supplements to *Trends in Biotechnology*.

Jasry, B. R., and D. Kennedy (eds.) (2001) The human genome. *Science* 291:1148–1432. A special issue of *Science*, reporting and analyzing the near-completion of the sequence determination of human DNA.

Jasry, B. R., and L. Roberts (eds.) (2003) Building on the DNA revolution. *Science* 300:277–296. A series of articles in a special issue of *Science* commemorating the 50th anniversary of the Watson–Crick discovery.

Koshland, D. E. (2002) The seven pillars of life. *Science* 295:2215–2216. A two-page essay outlining seven distinctive attributes of living matter.

Lander, E. S. (2011) Initial impact of the sequencing of the human genome. *Nature* 470:187–197. A review of the various ways in which knowledge of the human genome impacted human biomedicine and the prospective impact of genomics upon medicine.

TOOLS OF BIOCHEMISTRY 1A

MICROSCOPY AT MANY LEVELS

The Light Microscope and Its Limitations

All students of science are familiar with the light microscope (Figure 1A.1), the instrument that made cell biology possible, beginning with Hooke's pioneering microscopic studies. Generations of biologists followed Hooke with steadily improved instruments. But as biology has looked deeper into the details of life, the light microscope has reached its limits.

To understand why these limits exist, consider the **resolution** of a microscope. The resolution (r) is quantitatively defined as the minimum distance between two objects that can just be distinguished as separate. It is given by the equation

$$r = \frac{0.61\lambda}{n \sin \alpha} \qquad (1A.1)$$

Here λ is the wavelength of the radiation used, n is the refractive index of the medium between the sample and the objective lens, and α is the **angular aperture** of the objective lens. The quantity $\sin \alpha$ is basically a measure of the radiation-gathering power of the lens system. Resolution depends primarily on wavelength because the objects must be comparable in size to the wavelength in order to perturb the waves sufficiently to convey information.

The angular apertures of the best light microscopes are about 70°, so even if deep blue light of wavelength 450 nm is used and the medium between the sample and the objective lens is air ($n = 1$), we get

$$r = \frac{0.61 \times 450}{1.0 \sin 70°} \backsim 300\,\text{nm} = 0.3\,\mu\text{m} \qquad (1A.2)$$

This value represents the practical limit of resolution for light microscopy. A bit more resolution can be gained by going into the near ultraviolet, but absorption of this light by cellular materials limits its usefulness. Photographic images can be enlarged, but there is no sense in magnifying an image beyond the point where its resolution is just what the eye can resolve. Because our eyes can resolve images about 0.3–0.6 mm apart, the best light microscopes have a useful maximum magnifying power of about 1000–2000 (magnifying 0.3 μm by 2000 gives 0.6 mm). Further magnification of the image does not help—the fuzziness just gets bigger. To make a major advance, it was necessary to use radiation of much shorter wavelength, radiation that we cannot see but that can produce a photographic image. Thus the **electron microscope** was born in the 1930s.

Transmission Electron Microscopy

There are several types of electron microscopes. The first type to be used was the **transmission electron microscope (TEM)**, so called because it detects electrons that have been transmitted through a sample. The transmission electron microscope is compared with the light microscope in Figure 1A.1. An electron beam is emitted from a tungsten filament and accelerated by an electric field. Magnetic lenses focus the beam, as glass lenses focus a beam of light in the conventional microscope. The key to the higher resolution is that electrons, like the photons of light, have both a particle-like and a wavelike nature. A photon or an electron moving with an energy E is characterized by a wavelength where h is Planck's constant (6.626×10^{34} J · s) and

$$\lambda = \frac{hc}{E} \qquad (1A.3)$$

c is the speed of light (3×10^8 m/s). When electrons are accelerated by 50,000–100,000 volts between the cathode and the anode, their wavelengths are much shorter than that of visible light—in

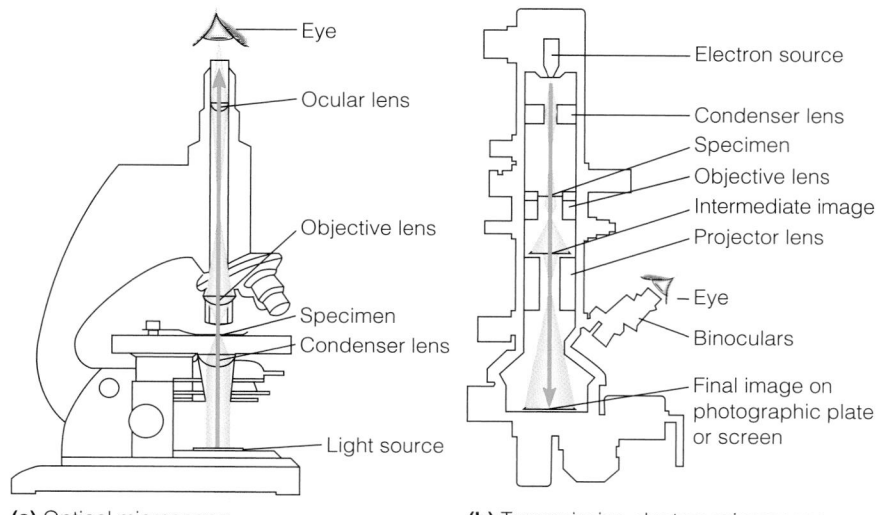

(a) Optical microscope   (b) Transmission electron microscope

FIGURE 1A.1

Structure of the optical microscope and the transmission electron microscope. The two images are not to the same scale; the electron microscope is much larger than a conventional light microscope.

fact, less than 1 nm. This wavelength would predict a resolution of better than 1 nm for a transmission electron microscope. Practical considerations give an operational limit of about 2 nm for most instruments. Still, this resolution is about 100 times finer than even the best optical microscope can accomplish; a good transmission electron microscope can usefully magnify to over 100,000 times.

Clear as this advantage may be, transmission electron microscopy has some disadvantages. The electron beam requires that a high vacuum be maintained throughout the instrument, including the sample chamber. This, in turn, means that only completely dried samples can be examined. Although methods for sample fixation and drying have been devised, there is always the possibility of inducing changes in samples as a result of their dehydration. Living structures, of course, cannot be examined.

Some of the methods used to prepare samples for transmission electron microscopy are shown in Figure 1A.2. The electron energies in most transmission microscopes do not allow penetration of thick samples (>100 nm). Thus, cell samples must be fixed, stained, and sliced very thin, by using an **ultramicrotome** (Figure 1A.2a). Particles like viruses and large molecules can be deposited directly on a thin film supported by a copper grid. But the contrast between such a particle and the background is not sufficient, so the sample is usually **negatively stained** (Figure 1A.2b) or **shadowed** (Figure 1A.2c). Other techniques such as *freeze fracturing* and *freeze etching* are discussed in later chapters.

Scanning Electron Microscopy

A quite different technique is called **scanning electron microscopy (SEM)**. A schematic diagram of a scanning electron microscope is shown in Figure 1A.3. Here, the electron beam is scanned back and forth across the sample, in a pattern generated by the scan generator and beam deflector, and secondary electrons emitted from the point at which the beam impinges on the

sample surface are picked up by a detector. The image is then displayed on a video screen, whose surface is scanned in register with the scanning of the sample. SEM does not have the resolution of TEM, but it is excellent for obtaining extremely clear views of the surfaces of minute objects, as you can see in Figure 1A.4. Preparation for SEM studies does not require sectioning, but the specimen must be fixed and dried to be stable at high vacuum and is usually coated with a thin layer of gold to aid in emission of secondary electrons.

Another technique should be mentioned: **scanning transmission electron microscopy (STEM)**. In this method the electron beam is scanned over the specimen, as in SEM, but it is detected in transmission. The method has the advantage that unstained, unfixed specimens can sometimes be used. Furthermore, the absorption of electrons of different energies provides information about the composition of different portions of the sample.

Laser Scanning Confocal Microscopy

Aside from the problems of resolution mentioned on page 20, there exists another fundamental limitation to conventional light microscopy for studying internal structure in cells and other biological samples. At high resolution ($\sim 0.3 \mu$m) the depth of focus of a light microscope is about 3μm. Superposition of images of material in this thick slice will obscure detail. To get around this problem, the confocal microscope was developed. As shown in Figure 1A.5, a light beam (preferably from a laser) is focused into a very small volume at the desired level within the sample. Reflected or fluorescent light from this spot is brought back to a detector, through a pinhole that excludes light scattered from other regions. The position of the illuminated spot is scanned back and forth through the sample, always at the same level. The image that is electronically built up in this way represents a very thin, highly resolved "slice" through the sample. It can also be repeated at different levels to build up a three-dimensional image.

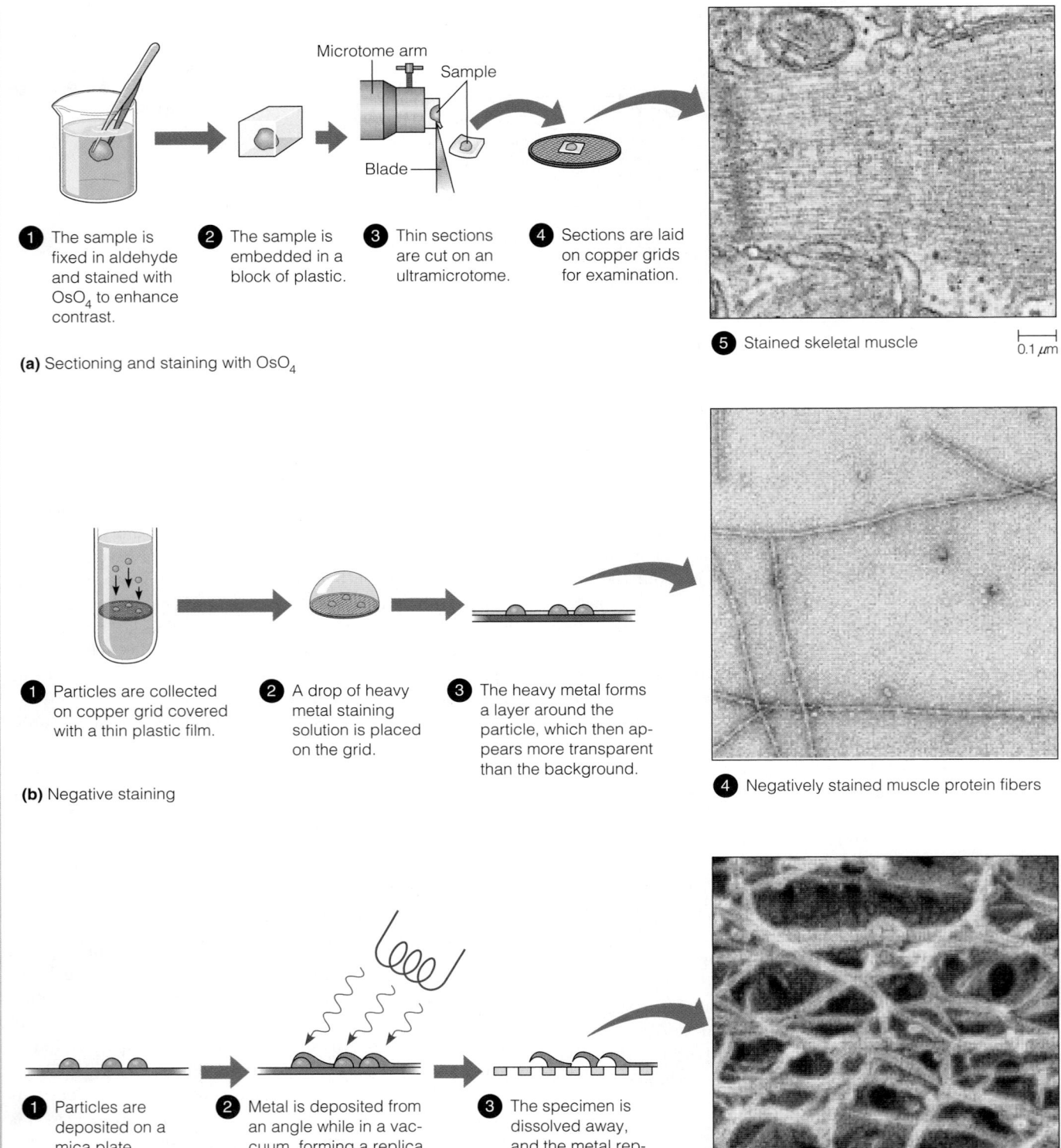

(a) Sectioning and staining with OsO_4

① The sample is fixed in aldehyde and stained with OsO_4 to enhance contrast.

② The sample is embedded in a block of plastic.

③ Thin sections are cut on an ultramicrotome.

④ Sections are laid on copper grids for examination.

⑤ Stained skeletal muscle   0.1 μm

(b) Negative staining

① Particles are collected on copper grid covered with a thin plastic film.

② A drop of heavy metal staining solution is placed on the grid.

③ The heavy metal forms a layer around the particle, which then appears more transparent than the background.

④ Negatively stained muscle protein fibers

(c) Shadowing

① Particles are deposited on a mica plate.

② Metal is deposited from an angle while in a vacuum, forming a replica of the specimen.

③ The specimen is dissolved away, and the metal replica is placed on a grid for examination.

④ Shadowed muscle protein fibers   0.1 μm

FIGURE 1A.2

Three methods of preparing samples for transmission electron microscopy.

Electron Microscopy in Biology, Vol. II, T. Pollard and P. Maupin; J. D. Griffith, ed. © 1982 John Wiley & Sons, Inc. Reproduced with permission from John Wiley & Sons, Inc.

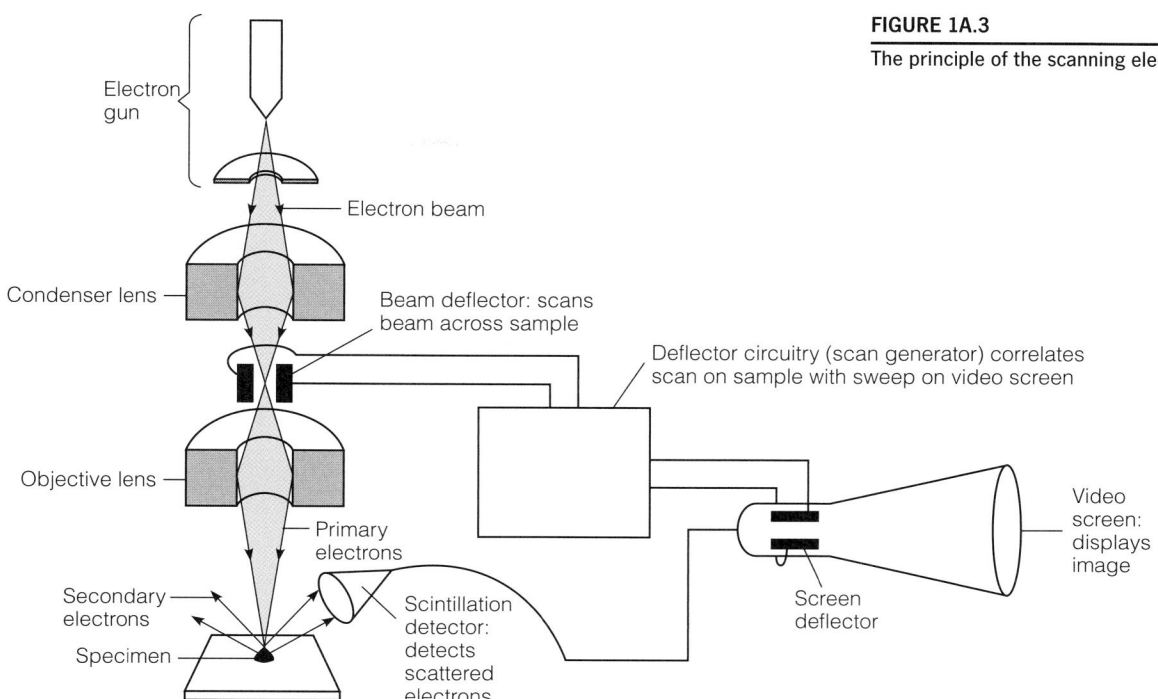

The method is most powerful when fluorescence detection is used, for then specifically labeled structures or substances can be precisely located within a cell. Because the method is relatively nondestructive, it can be used to follow dynamic processes in living cells. It has been employed, for example, to pinpoint the places within a cell nucleus where active replication of DNA is taking place. The rapid development of more versatile and discriminating fluorescent probes is making confocal microscopy a major technique in cellular biochemistry (See Figure 1A.6).

Scanning Tunneling and Atomic Force Microscopy

Recently, a remarkable new kind of microscope has been developed. **Scanning tunneling microscopy** uses a very fine, electrically charged metal tip, which is scanned across the sample. As electrons leak (tunnel) between the tip and the surface supporting the sample, the resistance they encounter varies according to the height of microscopic objects lying on the surface. The resulting fluctuations in current produce a video display of the surface with a resolution comparable to that of an electron microscope. In **atomic force microscopy** (Figure 1A.7), an extremely sharp tip is either dragged or tapped back and forth across the sample, and its up-and-down motion is detected by the deflection of a laser beam reflected off the cantilever which holds the tip. This

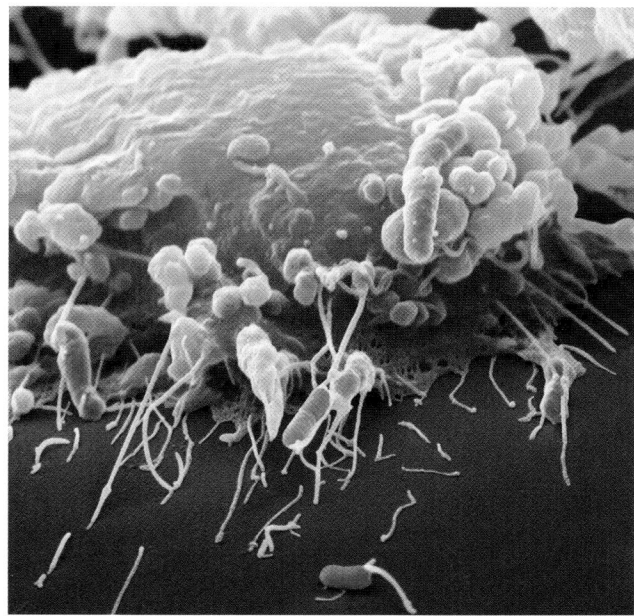

FIGURE 1A.4

A scanning electron micrograph showing phagocytosis. A macrophage is engulfing several sausage-shaped *E. coli* cells in this image, which is magnified 4300X.

Eye of Science/Science Photo Library.

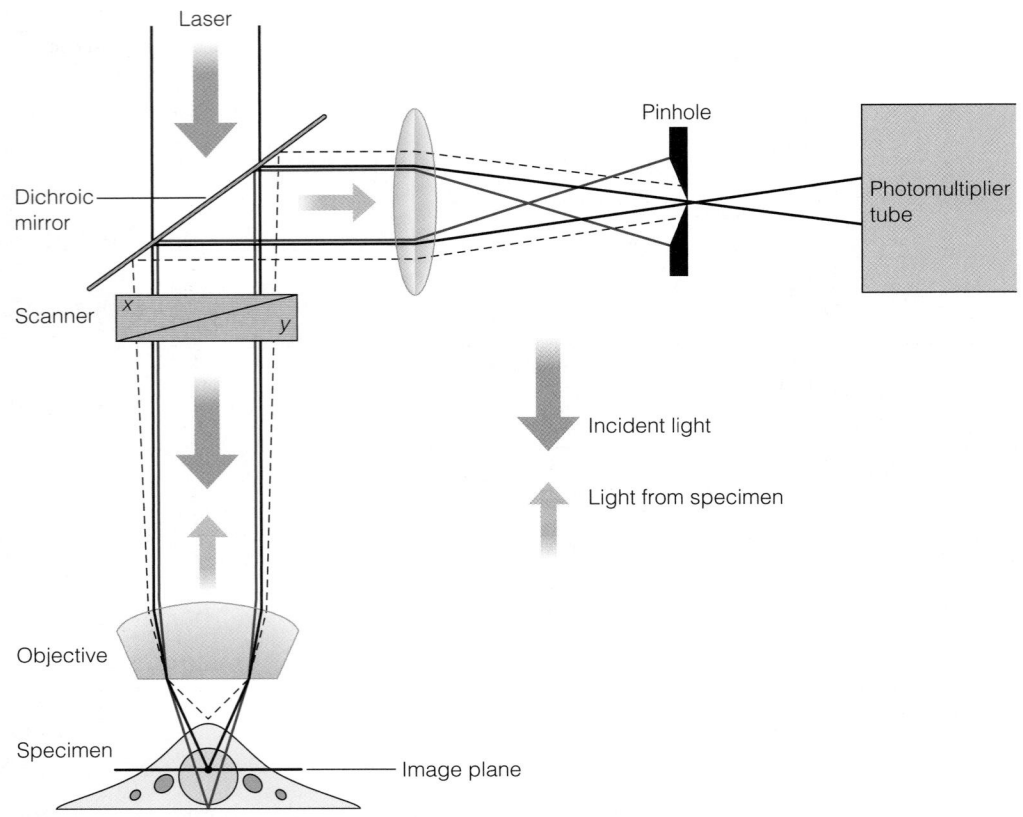

FIGURE 1A.5

Diagram illustrating the principle of laser scanning confocal microscopy. A laser beam is passed through an x–y scanner, collimated to a small spot by the objective lens, and scanned across the specimen. Fluoresced light is collected by the objective and directed by a dichroic mirror (a mirror that reflects the fluorescent light but not the shorter-wavelength laser light) to a pinhole aperture placed in the conjugate image plane. Light originating from the specimen plane of focus passes through the pinhole to a photomultiplier detector. Light from above or below the specimen focal plane strikes the walls of the aperture and is not transmitted.

Reprinted from *Optical Microscopy: Emerging Methods and Applications*, Brian Herman and John J. Lemasters, eds., pp. 339–354. © 1993, with permission from Elsevier.

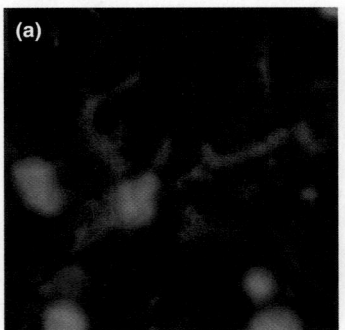
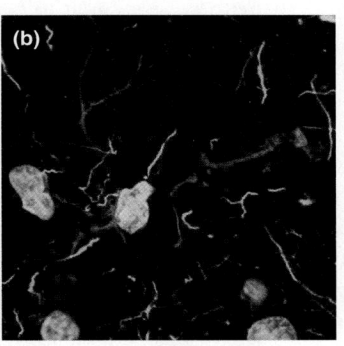

FIGURE 1A.6

Image of a thick slice of hippocampus from a mouse brain visualized either by conventional optical microscopy (a) or by laser scanning confocal microscopy (b). The preparation was treated with fluorescent-tagged antibodies to glial fibrillary acidic protein (red) and neurofilaments H (green), and with a fluorescent DNA-binding dye, Hoechst 33342, to label nuclei (blue).

Michael Davidson, The Florida State University/Molecular Expressions™.

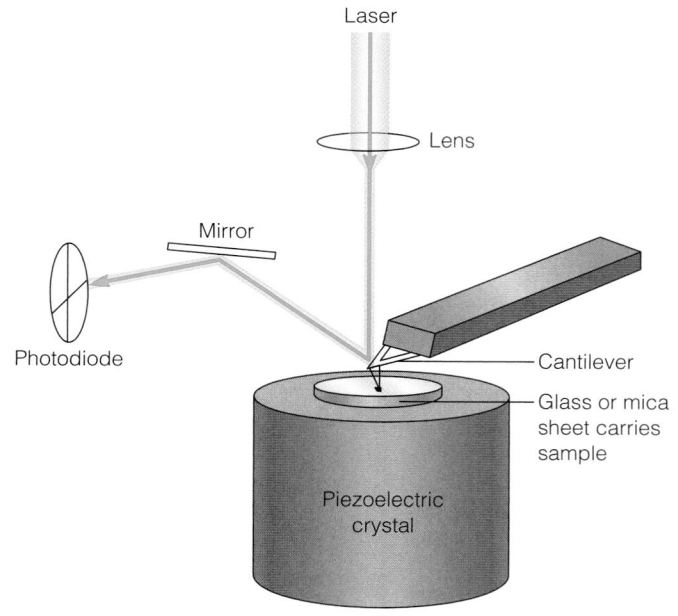

FIGURE 1A.7

The principle of the atomic force microscope. Power to the piezoelectric crystal is regulated to move the sample up and down and keep the tip at constant height as the sample is scanned.

motion is greatly amplified to give a contour map of the object. Both of these techniques have the enormous advantage over electron microscopy that wet, even immersed, samples can be studied, And the resolution allows visualization of single macromolecules.

References

Claxton, N. S., T. J. Fellers, and M. W. Davidson (2006) Laser Scanning Confocal Microscopy. http://www.olympusfluoview.com/theory/LSCMIntro.pdf. A 37-page web archive that describes the theory and applications of this technique.

Corle, T. R., and G. S. Kino (eds.) (1998) *Confocal Scanning Optical Microscopy and Related Imaging Systems.* Academic Press, San Diego.

Egerton, R. F. (2005) *Physical principles of electron microscopy: An introduction to TEM, SEM, and AEM.* Springer, New York.

Engel, A. (1991) Biological application of scanning probe microscopy. *Annu. Rev. Biophys. & Biophys. Chem.* 20:79–108.

Herman, B., and J. J. Lemasters (eds.) (1996) *Optical Microscopy: Emerging Methods and Applications.* Academic Press, San Diego. A collection of short papers on a wide variety of new microscopic methods.

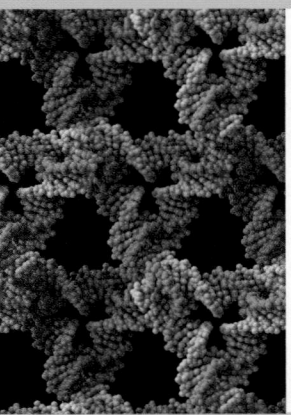

CHAPTER 2

The Matrix of Life: Weak Interactions in an Aqueous Environment

Noncovalent interactions are critically important determinants of biomolecular structure, stability, and function.

The macromolecules that participate in the structural and functional matrix of life are immense structures held together by strong, covalent bonds. Yet covalent bonding alone cannot begin to describe the complexity of molecular structure in biology. Much weaker interactions are responsible for most of the elegant cellular architecture visible in the electron micrographs of Chapter 1. These are the **noncovalent interactions**, also called *noncovalent forces* or *noncovalent bonds,* between ions, molecules, and parts of molecules.

Consider the macromolecules we discussed in Chapter 1. The linear sequence of the nucleotide residues in a strand of DNA is maintained by covalent bonds. But DNA also has a highly specific three-dimensional structure, which is stabilized by noncovalent interactions between different parts of the molecule. Similarly, every kind of protein is made up of amino acids linked by covalent **peptide bonds**; but each protein is also folded into a specific molecular conformation that is stabilized by noncovalent interactions. Proteins interact with macromolecules, such as other proteins or nucleic acids or lipids, to form still higher levels of organization, ultimately leading to cells, tissues, and whole organisms. All of this complexity is accounted for by a myriad of noncovalent interactions within and between macromolecules.

What makes noncovalent interactions so important in biology and biochemistry? The key is seen in Figure 2.1, which compares noncovalent and covalent bond energies. The *covalent* bonds most important in biology (such as C — C and C — H) have bond energies in the range of 300–400 kJ/mol. Biologically important *noncovalent* bonds are 10 to 100 times weaker. It is their very weakness that makes noncovalent bonds so essential, for it allows them to be continually broken and re-formed in the dynamic molecular interplay that is life. This interplay depends on rapid exchanges of molecular partners, which could not occur if intermolecular forces were so strong as to lock the molecules in conformation and in place.

If we are to understand life at the molecular level, we must know something about noncovalent interactions. Furthermore, we must know how such interactions behave in an aqueous environment, for every cell in every organism on earth is bathed in and permeated by water. This is as true for creatures living in the most arid deserts as for those in the depths of the sea. Water is the major constituent of organisms—70% or more of the total weight, in most cases.

This chapter first describes noncovalent interactions, and then shows that the properties of water have a profound effect on those interactions.

The Nature of Noncovalent Interactions

Molecules and ions can interact noncovalently in a number of different ways, as described in this section and summarized in Figure 2.2. All of these noncovalent interactions are fundamentally electrostatic in nature; thus, they all depend on the forces that electrical charges exert on one another. Table 2.1 lists ranges for the energies of some of the noncovalent interactions prevalent in biomolecules.

Charge–Charge Interactions

The simplest noncovalent interaction is the electrostatic interaction between a pair of charged particles. Such charge–charge interactions are also referred to as ionic bonds or salt bridges. Many of the molecules in cells, including macromolecules like DNA and proteins, carry a net electrical charge. In addition to these molecules, the cell contains an abundance of small ions, both cations like Na^+, K^+, and Mg^{2+} and anions like Cl^- and HPO_4^{2-}. All of these charged entities exert forces on one another (see Figure 2.2a). The force between a pair of charges q_1 and q_2, separated *in a vacuum* by a distance r, is given by **Coulomb's law**:

$$F = k \frac{q_1 q_2}{r^2} \qquad (2.1)$$

where k is a constant whose value depends on the units used.[*] If q_1 and q_2 have the same sign, F is positive, so a positive value corresponds to repulsion. If one charge is $+$ and the other $-$, F is negative, signifying attraction. It is such charge–charge interactions that stabilize a crystal of a salt, like that shown in Figure 2.3.

The biological environment, of course, is not a vacuum. In a cell, charges are always separated by water or by other molecules or parts of molecules. The existence of this **dielectric medium** between charges has the effect of screening them from one another, so that the actual force is always less than that given by Equation 2.1. This screening effect is expressed by inserting a dimensionless number, the **dielectric constant** (ε), in Equation 2.1:

$$F = k \frac{q_1 q_2}{\varepsilon r^2} \qquad (2.2)$$

Every substance that acts as a dielectric medium has a characteristic value of ε; the higher this value, the weaker the force between the separated charges. The dielectric constant of water is high, approximately 80, whereas organic liquids usually have much lower values, in the range of 1 to 10. We shall see presently the reason for the high value of ε in water, but its major consequence is that charged

---

[*]In the c.g.s. (centimeter–gram–second) system, with charges in electrostatic units, k is unity. In this book we use the SI, or international, system of units. Here q_1 and q_2 are in coulombs (C), r is in meters (m), and $k = 1/(4\pi\varepsilon_0)$. The quantity ε_0 is the *permittivity of a vacuum* and has the value 8.85×10^{-12} $J^{-1}C^2m^{-1}$, where J is the energy unit joules. F is in newtons (N).

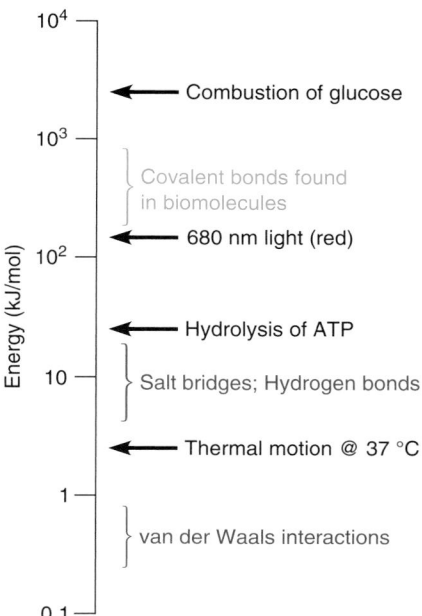

FIGURE 2.1

Covalent and noncovalent bond energies. Energies typical of noncovalent bonds (0.5–20 kJ/mol; red text) are about one to two orders of magnitude weaker than energies of the covalent bonds commonly found in biochemical compounds (150–600 kJ/mol; blue text). The energies available from thermal motion, ATP hydrolysis, red light, and aerobic glucose metabolism are also shown as reference points (discussed in detail in later chapters). Note that values are plotted on a log scale.

Noncovalent interactions always involve electrical charges.

TABLE 2.1	Energies of noncovalent interactions in biomolecules

Type of interaction	Approximate energy (kJ/mol)
Charge–charge	−13 to −17
Charge–dipole (H–bond)	−13 to −21
Dipole–dipole (H–bond)	−2 to −8
van der Waals	−0.4 to −0.8

Values reprinted from *Advances in Protein Chemistry* 39:125–189, S. K. Burley and G. A. Petsko, Weakly polar interactions in proteins. © 1988, with permission from Elsevier.

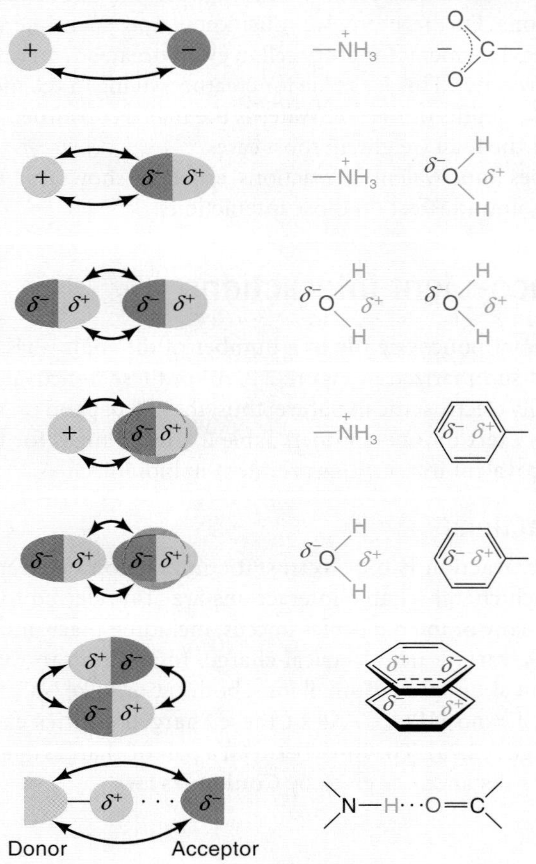

Type of Interaction	Model	Example	Dependence of Energy on Distance
(a) Charge–charge Longest-range force; nondirectional			$1/r$
(b) Charge–dipole Depends on orientation of dipole			$1/r^2$
(c) Dipole–dipole Depends on mutual orientation of dipoles			$1/r^3$
(d) Charge–induced dipole Depends on polarizability of molecule in which dipole is induced			$1/r^4$
(e) Dipole–induced dipole Depends on polarizability of molecule in which dipole is induced			$1/r^5$
(f) Dispersion (van der Waals) Involves mutual synchronization of fluctuating charges			$1/r^6$
(g) Hydrogen bond Charge attraction + partial covalent bond	Donor Acceptor		Length of bond fixed

FIGURE 2.2

Types of noncovalent interactions. The induced dipole **(d, e)** and the dispersion forces **(f)** depend on a distortion of the electron distribution in a nonpolar atom or molecule. The symbols δ^- and δ^+ denote a fraction of an electron or proton charge.

particles interact rather weakly with one another in an aqueous environment unless they are very close together (i.e., within 0.4 to 1.0 nm).

Coulomb's law is an expression of *force*; that is, it is a quantitative description of an interaction. However, every interaction involves a change in *energy,* and because we are concerned with the energy changes in biological processes, we are particularly interested in the **energy of interaction** (**E**). This is the energy required to separate two charged particles from a distance r to an infinite distance—in other words, to pull them apart working against the electrostatic force. The energy of interaction is given by Equation 2.3, which is similar to Equation 2.2:

$$E = k\frac{q_1 q_2}{\varepsilon r} \tag{2.3}$$

As with force, the energy of an oppositely charged pair q_1 and q_2 is always negative, signifying attraction, but E approaches zero as r becomes very large. For charge–charge interactions, the energy of interaction is inversely proportional to the first power of r; thus, these interactions are relatively strong over greater distances compared to the other noncovalent interactions listed in Figure 2.2. Charge–charge interactions often occur within or between biomolecules—for example, in the attraction between amino and carboxylate groups, as shown in Figure 2.2a. As will be discussed in Chapter 5, charge–charge interactions can play an important role in the purification of a protein from a complex mixture of cellular components.

Permanent and Induced Dipole Interactions

Molecules that carry no *net* charge may nevertheless have an asymmetric internal distribution of charge. For example, the electron distribution of the uncharged carbon monoxide molecule is such that the oxygen end is slightly more negative than the carbon end (Figure 2.4a). Such a molecule is called *polar*, or a **permanent dipole**, and is said to have a **permanent dipole moment** (μ). The dipole moment expresses the magnitude of a molecule's polarity. If a linear molecule like CO has fractional charges δ^+ and δ^-, separated by a distance x, the dipole moment is a vector directed toward δ^+, whose magnitude is

$$\mu = qx \qquad (2.4)$$

where q is the magnitude of the charge (or fractional charge). In molecules with a more complex shape, like water, the dipole moment for the entire molecule is a vector sum of the dipole moments along each polar bond (Figure 2.4b). Water has a significant μ, because electrons are drawn from the hydrogen atoms toward the oxygen atom, due to the much greater electronegativity of the oxygen atom.

Some dipole moment values are given in Table 2.2. Notice the large values for glycine and glycylglycine. At neutral pH, the amino acid glycine exists as the ion $^+NH_3CH_2COO^-$, which has both a positive ammonium group and a negative carboxylate group. Thus, whole electron charges are separated by the length of the molecule, accounting for the large μ. In glycylglycine, which is made by covalently linking two glycine molecules, the dipole moment is nearly twice as large because the charge separation is almost doubled. Molecules with large dipole moments are said to be highly polar.

FIGURE 2.3

Charge–charge interactions in an ionic crystal. Ionic crystals are held together by charge–charge interactions between positive and negative ions. In a sodium chloride crystal, each sodium ion is surrounded by six chloride ions, and each chloride ion is surrounded by six sodium ions.

Marcel Clemens/Shutterstock.

Some molecules interact because they possess dipole moments.

TABLE 2.2	Dipole moments of some molecules	
Molecule	Formula	Dipole Moment (D)[a]
Carbon monoxide	C≡O	0.12
Carbon dioxide	O=C=O	0
Water	(H₂O structure)	1.83
para-Dichlorobenzene	Cl—⬡—Cl	0
ortho-Dichlorobenzene	(ortho-dichlorobenzene structure)	2.59
Glycine	H₃N⁺—CH₂—COO⁻	16.7
Glycylglycine	H₃N⁺—CH₂—C(=O)—N(H)—CH₂COO⁻	28.6

[a]The common units of dipole moment are *debyes*; 1 debye (D) equals 3.34×10^{-30} C·m.

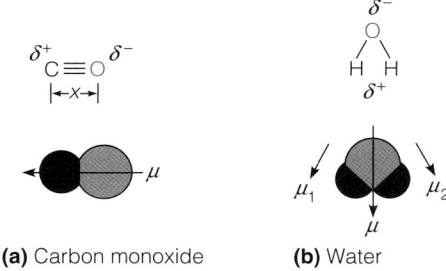

(a) Carbon monoxide   **(b)** Water

FIGURE 2.4

Dipolar molecules. (a) Carbon monoxide: the partial negative charge (δ^-) on the oxygen together with corresponding partial positive charge (δ^+) on the carbon produces a dipole moment directed along the O—C axis. **(b)** Water: the partial negative charge on O together with the partial positive charge on each H produces two moments, μ_1 and μ_2, directed along the O—H bonds. Their vector sum (μ) represents the net dipole moment of the molecule.

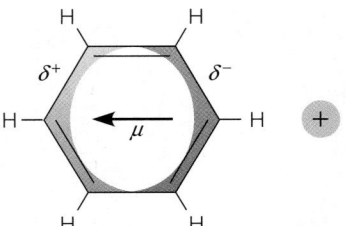

(a) Induction of a dipole in benzene by a positively charged ion

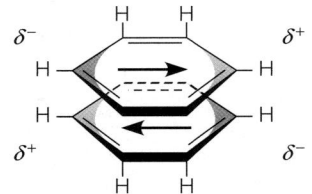

(b) Dispersion forces between two benzene molecules

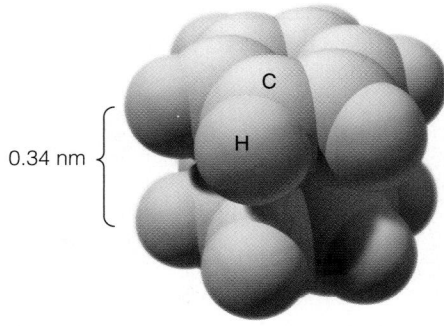

0.34 nm

(c) Space-filling model of molecules in (b)

FIGURE 2.5

Induced dipoles and dispersion forces. **(a)** Benzene has neither a net charge nor a permanent dipole moment, but a nearby charge can induce a redistribution of electrons within the benzene ring, producing an induced dipole moment (μ). **(b)** Planar molecules like benzene have a strong tendency to stack because fluctuations in the electron clouds of the stacked rings interact with one another, producing a dispersion force. **(c)** Although the molecules approach closely, they do not interpenetrate.

Molecules may attract one another by noncovalent forces but cannot interpenetrate: van der Waals radii determine molecular surfaces.

Note also in Table 2.2 that molecules must possess an appropriate geometry to have dipole moments: compare C≡O with O=C=O, or o-dichlorobenzene with p-dichlorobenzene. In carbon dioxide and p-dichlorobenzene, the dipole vectors are equal in magnitude but oppositely directed so that their effects cancel each other, leaving no net dipole moment.

In the aqueous environment of a cell, a permanent dipole can be attracted by a nearby ion (a *charge–dipole interaction*) or by another permanent dipole (a *dipole–dipole interaction*). These **permanent dipole interactions** are described in Figure 2.2b and c. Unlike the simple charge–charge interactions described earlier, dipole interactions depend on the orientation of the dipoles. Furthermore, they are shorter-range interactions: The energy of a charge–dipole interaction is proportional to $1/r^2$ and that of a dipole–dipole interaction to $1/r^3$. Thus, a pair of permanent dipoles must be quite close together before the interaction becomes strong.

Molecules that do not have permanent dipole moments can become dipolar in the presence of an electric field. The field may be externally imposed, as in a laboratory instrument, or it may be produced by a neighboring charged or dipolar particle. A molecule in which a dipole can be so induced is said to be **polarizable**. Aromatic rings, for example, are very polarizable because the electrons can easily be displaced in the plane of the ring, as shown in Figure 2.5a. Interactions of polarizable molecules are called **induced dipole interactions**. An anion or a cation may induce a dipole in a polarizable molecule and thereby be attracted to it (a *charge–induced dipole interaction*, Figure 2.2d), or a permanent dipole may do the same (a *dipole–induced dipole interaction*, Figure 2.2e). These induced dipole interactions are even shorter range than permanent dipole interactions, with energies of interaction proportional to $1/r^4$ and $1/r^5$, respectively.

Even two molecules that have neither a net charge nor a permanent dipole moment can attract one another if they are close enough (Figure 2.2f). The distribution of electronic charge in a molecule is never static, but fluctuates. When two molecules approach very closely, they synchronize their charge fluctuations so as to give a net attractive force. Such intermolecular forces, which can be thought of as mutual dipole induction, are called **van der Waals**, or **dispersion**, **forces**. Their attractive energy varies as the inverse sixth power of the distance, so van der Waals forces are significant only at very short range. They can become particularly strong when two planar molecules can stack on one another, as shown in Figure 2.5b and c. We shall encounter many examples of such interactions in the internal packing of molecules like proteins and nucleic acids. As discussed in Chapter 6, van der Waals forces are individually weak; but, collectively they make significant contributions to the stability of biomolecules.

Molecular Repulsion at Extremely Close Approach: The van der Waals Radius

When molecules or atoms that do not have covalent bonds between them come so close together that their outer electron orbitals begin to overlap, there is a mutual repulsion. This repulsion increases very steeply as the distance between their centers (r) decreases; it can be approximated as proportional to r^{-12}. If we combine this repulsive energy with one or more of the kinds of attractive energy described previously, we see that the total energy of noncovalent interaction (E) for a pair of atoms, molecules, or ions will vary with distance of their separation (r) in the manner depicted in Figure 2.6. Two points on the graph should be noted. First, there is a minimum in the energy curve, at position r_0. This minimum corresponds to the most stable distance between the centers of the two particles. If we allow them to approach each other, this is how close they will come. Second, the repulsive potential rises so steeply at shorter distances that it acts as a "wall," effectively barring approach closer than the distance r_v. This distance defines the

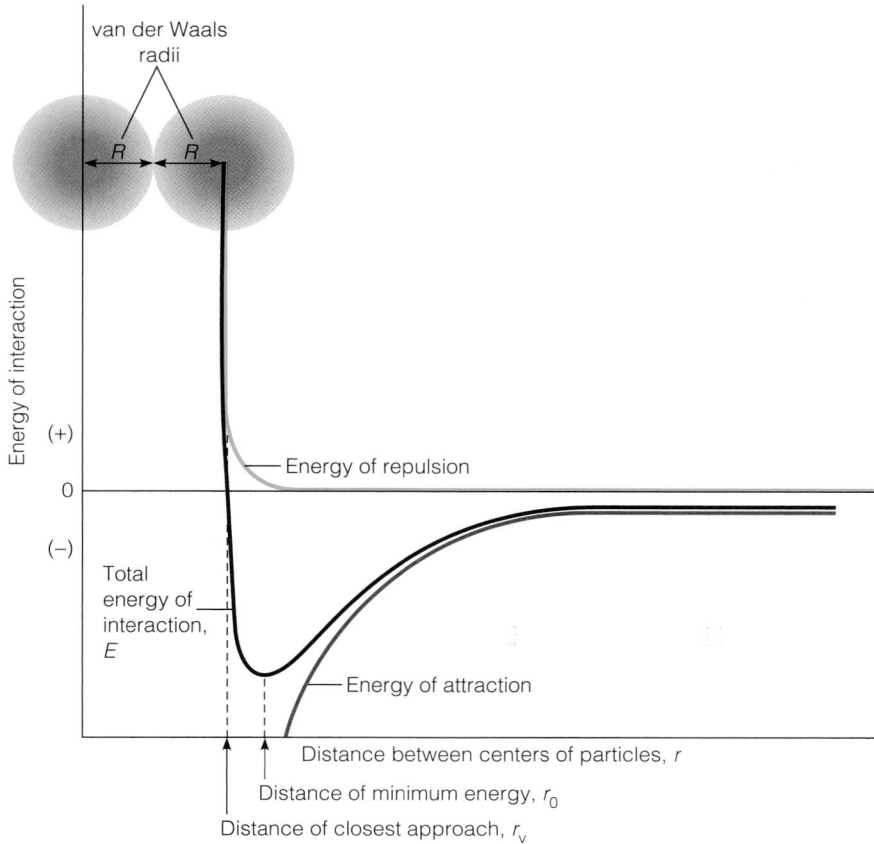

FIGURE 2.6

Noncovalent interaction energy of two approaching particles. The interaction energy of two atoms, molecules, or ions is graphed versus the distance between their centers, r. The total interaction energy (E) at any distance is the sum of the energy of attraction and the energy of repulsion. As the distance between the particles decreases (reading *right to left* along the x-axis), both the attractive energy (<0) and the repulsive energy (>0) increase, but at different rates. At first the longer-range attraction dominates, but then the repulsive energy increases so rapidly that it acts as a barrier, defining the distance of closest approach (r_v) and the van der Waals radii (R, described by the orange spheres). The position of minimum energy (r_0) is usually very close to r_v.

so-called **van der Waals radius**, **R**, the effective radius for closest molecular packing. For a pair of identical spherical molecules, $r_v = 2R$; for molecules with van der Waals radii R_1 and R_2, $r_v = R_1 + R_2$.

Real molecules, of course, are not spherical objects like those depicted in Figure 2.6. Because large biological molecules all have complicated shapes, it is useful to extend the concept of the van der Waals radius to atoms or groups of atoms within a molecule. The values for van der Waals radii given in Table 2.3 represent the distances of closest approach for another atom or group. When we depict complex molecules in a so-called "space filling" manner, we represent each atom by a sphere with its appropriate van der Waals radius (see Figure 2.5c). In this case, the van der Waals radii of the carbon atoms (0.17 nm) mean that the planes of the two stacked rings cannot be closer than 0.34 nm.

Hydrogen Bonds

One specific kind of noncovalent interaction, the **hydrogen bond**, is of the greatest importance in biochemistry. The structure and properties of many biological molecules and of water, the universal biological solvent, are determined largely by this type of bond. A hydrogen bond is an interaction between a covalently bonded hydrogen atom on a *donor* atom (e.g., —O—H or $>$N—H) and a pair of nonbonded electrons on an *acceptor* atom (e.g., O=C— or N$\leqslant$), as shown in Figure 2.2g and Figure 2.7. The atom to which hydrogen is covalently bonded is called the *hydrogen-bond donor*, and the atom with the nonbonded electron pair is called the *hydrogen-bond acceptor*. The interaction between the donor and acceptor is typically represented by a dotted line between the acceptor atom and the shared H.

The ability of an atom to function as a hydrogen-bond donor depends greatly on its electronegativity. The more electronegative the donor atom, the more

TABLE 2.3	van der Waals radii of some atoms and groups of atoms	
		R (nm)
Atoms		
	H	0.12
	O	0.14
	N	0.15
	C	0.17
	S	0.18
	P	0.19
Groups		
	—OH	0.14
	—NH$_2$	0.15
	—CH$_2$—	0.20
	—CH$_3$	0.20
	Half-thickness of aromatic ring	0.17

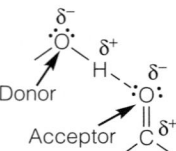

FIGURE 2.7

The hydrogen bond. The figure shows an idealized H bond that might exist, for example, between an alcohol (the donor) and a keto compound (the acceptor). The H bond is represented by a dotted line between the H and the acceptor atom, and a solid line between the H and the donor atom.

Hydrogen bonds are among the strongest, most specific, noncovalent interactions.

negative charge it withdraws from the hydrogen to which it is bonded; thus, the hydrogen becomes more positive and is more strongly attracted to the electron pair of the acceptor. Among the atoms encountered in biological compounds, only O and N have appropriate electronegativities to serve as strong donors. Thus, $\equiv$C—H groups do not form strong hydrogen bonds, but —O—H groups do.

The hydrogen bond has features in common with both covalent and noncovalent interactions. In part, it is like a charge–charge interaction between the partial positive charge on H and the negative charge of the electron pair. But it is also true that there is electron sharing (as in a covalent bond) between H and the acceptor. This double character is reflected in the bond length of the hydrogen bond. The distance between the hydrogen atom and the acceptor atom in a hydrogen bond is considerably less than would be expected from their van der Waals radii. For example, the distance between H and O in the bond $>$N—H $\cdots$ O$=$C$<$ is only about 0.19 nm, whereas we would predict about 0.26 nm from the sum of the van der Waals radii given in Table 2.2. On the other hand, a *covalent* H—O bond has a length of only 0.10 nm. The distance between the hydrogen-bond donor and acceptor is about 0.29 nm. The donor–acceptor distances of some particularly strong hydrogen bonds are listed in Table 2.4. Note that these distances are fixed, as they are for covalent (but not for other noncovalent) bonds.

The energy of hydrogen bonds is considerably higher than that of most other noncovalent bonds, in keeping with their partially covalent character (see Figure 2.1). Hydrogen bonds are also like covalent bonds in being highly directional. Computational studies predict that hydrogen bonds are strongest when the angle defined by the donor atom, the shared H atom, and the acceptor atom is 180° (i.e., the three atoms are colinear). The majority of hydrogen bond angles in proteins are within 30° of 180°. The importance of this directionality is seen in the role that hydrogen bonds play in organizing a regular biochemical structure such as the α-helix in proteins (Figure 2.8, discussed in greater detail in Chapter 6). This is but one example of many we shall encounter in which hydrogen bonds stabilize ordered structure in large molecules.

We finish this section by reiterating the important point that the various noncovalent interactions we have described are individually weak, but when many are present within a given macromolecule, or between macromolecules, their energies can sum to an impressive total—often several hundreds of

TABLE 2.4	Major types of hydrogen bonds found in biomolecular interactions	
Donor$\cdots$ Acceptor	Distance between Donor and Acceptor (nm)	Comment
—O—H $\cdots$ O$\big\backslash^{/}_{H}$	0.28 ± 0.01	H bond formed in water
—O—H $\cdots$ O$=$C$\big\backslash^{/}$	0.28 ± 0.01	Bonding of water to other molecules often involves these
$>$N—H $\cdots$ O$\big\backslash^{/}_{H}$	0.29 ± 0.01	Bonding of water to other molecules often involves these
$>$N—H $\cdots$ O$=$C$\big\backslash^{/}$	0.29 ± 0.01	Very important in protein and nucleic acid structures
$>$N—H $\cdots$ N$\big\backslash^{/}$	0.31 ± 0.02	Very important in protein and nucleic acid structures
$>$N—H $\cdots$ S$\big\backslash^{/}$	0.37	Relatively rare; weaker than above

kilojoules. Thus, such interactions can account for the stability of macromolecular structures. At the same time, the ease with which individual noncovalent bonds can be broken and re-formed gives these structures a dynamic flexibility necessary to their function.

The Role of Water in Biological Processes

The chemical and physical processes of life require that molecules be able to move about, encounter one another, and change partners frequently in the complicated processes of metabolism and synthesis. A fluid environment allows molecular mobility, and water not only is the most abundant fluid on earth, it is also admirably suited to this purpose. To see why, we must examine the properties of water in some detail.

The Structure and Properties of Water

Although we tend to take its properties for granted, water is really a most curious substance. Table 2.5, which contrasts H_2O with other hydrogen-rich compounds of comparable molecular weight, reveals a remarkable fact. These compounds are gases at room temperature and have much lower boiling points than water does. Why is water unique? The answer lies mainly in the strong tendency of water molecules to form hydrogen bonds with other water molecules.

The electron arrangement of a single water molecule is shown in Figure 2.9. Of the six electrons in the outer orbitals of the oxygen atom, two are involved in covalent bonds to the hydrogens. The other four electrons exist in nonbonded pairs, which are excellent hydrogen-bond acceptors. The —OH groups in water are strong hydrogen-bond donors. Thus each water molecule is both a hydrogen-bond donor and a hydrogen-bond acceptor, capable of forming up to *four* hydrogen bonds simultaneously. A sample of water is a dynamic network of H-bonded molecules (Figure 2.10c). The strength of this extensive H-bond network gives water an unusually large heat capacity, and the vaporization of water requires a large amount of energy for a molecule of its size. Both the heat of vaporization and the boiling point of water are therefore unusually high (Table 2.5), and water remains in the liquid state at temperatures characteristic of much of the earth's surface.

The hydrogen bonding between water molecules becomes most regular and clearly defined when water freezes to ice, creating a rigid tetrahedral molecular lattice in which each molecule is H-bonded to four others (Figure 2.10a,b). The lattice structure is only partially dismantled when ice melts, and some long-range order persists even at higher temperatures. On average there are 15% fewer hydrogen bonds in liquid water than in ice (i.e., 3.4 H-bonds per water molecule vs. 4 per ice molecule). The structure of liquid water has been described as "flickering clusters" of hydrogen bonds, with remnants of the ice lattice continually breaking and re-forming as the molecules move about (Figure 2.10c). The rather open

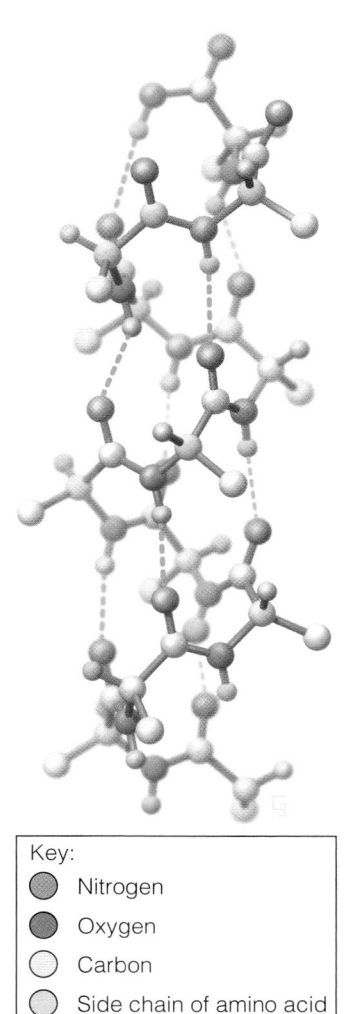

Key:
- ● Nitrogen
- ● Oxygen
- ○ Carbon
- ○ Side chain of amino acid
- ○ Hydrogen
- ⋯ Hydrogen bond

FIGURE 2.8

Hydrogen bonding in biological structure. This example is a portion of a protein in an α-helical conformation. The α helix, a common structural element in proteins, is maintained by N—H ⋯ O=C hydrogen bonds between groups in the protein chain. See Chapter 6 for more details on the structure of the α helix.

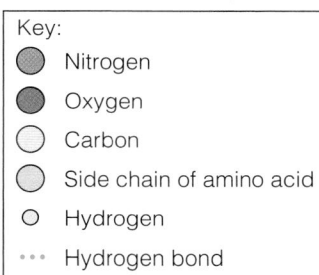

bond angle = 104.5°

FIGURE 2.9

Hydrogen-bond donors and acceptors in water. The two nonbonded electron pairs on O act as H-bond acceptors and the two O—H bonds act as H-bond donors. The angle between the two O—H bonds is 104.5°; thus, water has a net dipole moment.

TABLE 2.5	Properties of water compared to some other low-molecular-weight compounds			
Compound	Molecular Weight	Melting Point (°C)	Boiling Point (°C)	Heat of Vaporization (kJ/mol)
CH_4	16.04	−182	−164	8.16
NH_3	17.03	−78	−33	23.26
H_2O	18.02	0	+100	40.71
H_2S	34.08	−86	−61	18.66

FIGURE 2.10

Water as a molecular lattice. **(a)** A space-filling model of the structure of ice, where O atoms are red and H atoms are blue. Ice is a molecular lattice formed by indefinite repetition of a tetrahedral hydrogen-bonding pattern. Each molecule acts as a hydrogen-bond donor to two others and as an acceptor from two others. Because of the length of the hydrogen bonds, the structure is a relatively open one, which accounts for the low density of ice. **(b)** A skeletal model of the ice lattice. **(c)** The structure of liquid water. When ice melts, the regular tetrahedral lattice is broken, but substantial portions of it remain, especially at low temperatures. In liquid water, flickering clusters of molecules are held together by hydrogen bonds that continually break and re-form. In this schematic "motion picture," successive frames represent changes occurring in picoseconds (10^{-12} s).

(a, b) Courtesy of Gary Carlton.

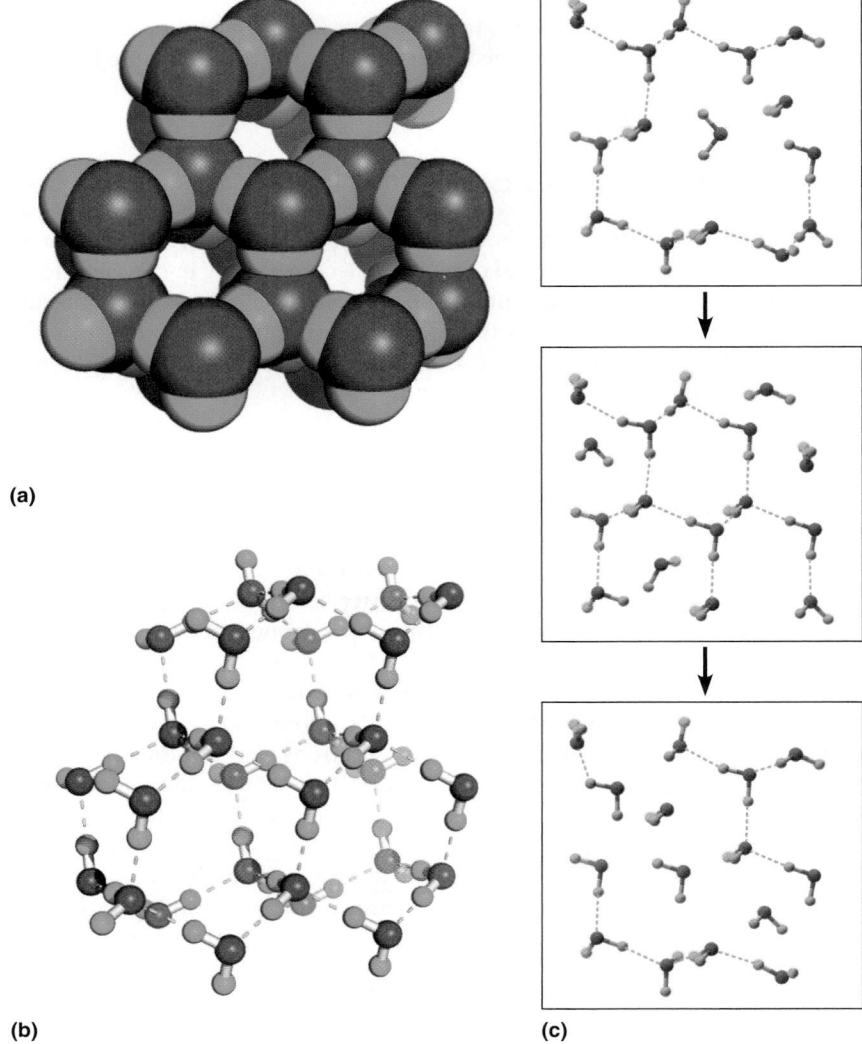

(a)

(b)

(c)

Most of the unique properties of water are due to its hydrogen-bonding potential and its polar nature.

structure of the ice lattice accounts for another of water's unusual properties—liquid water is more dense than its solid form because when the lattice breaks down, molecules can move closer together. This seemingly trivial fact is of the utmost importance for life on earth. If water behaved like most substances and became more dense when solid, the ice formed on lake and ocean surfaces each winter would sink to the bottom. There, insulated by the overlying layers, it would have accumulated over the ages, and most of the water on earth would by now have become locked up in ice, leaving little liquid water to support life. It seems doubtful that life could have evolved if liquid water at 0 °C were 9% less dense than it is, for then the ice would sink. Another consequence of this remarkable property of water is that the more dense liquid state is favored over the less dense solid state at high pressures. Thus, even at the high pressures in the depths of the ocean, water remains liquid.

Other unusual properties of water, listed in Table 2.6, are also readily explained in terms of its molecular structure. Relative to most organic liquids, water has a high *viscosity*—a consequence of the interlocked, hydrogen-bonded structure. This cohesiveness also accounts for the high *surface tension* of water. The high dielectric constant of water, mentioned previously, results from its dipolar character. An electric field generated between two ions causes extensive orientation

Property	Water	n-Pentane
TABLE 2.6 Important properties of liquid water compared with those of n-pentane, a nonpolar, non-hydrogen-bonding liquid[a]		
Molecular weight (g/mol)	18.02	72.15
Density (g/cm^3)	0.997	0.626
Boiling point (°C)	100	36.1
Dielectric constant	78.54	1.84
Viscosity (g/cm·s)	0.890×10^{-2}	0.228×10^{-2}
Surface tension (dyne/cm)	71.97	17

[a]All data are for 25 °C.

of intervening water dipoles and a significant amount of induced polarization. These oriented dipoles contribute to a counterfield, reducing the effective electrostatic force between the two ions.

Water as a Solvent

The processes of life require a wide variety of ions and molecules to move about in proximity, that is, to be soluble in a common medium. Water serves as the universal intracellular and extracellular medium, thanks primarily to the two properties of water we have discussed: its tendency to form hydrogen bonds and its dipolar character. Substances that can take advantage of these properties so as to readily dissolve in water are called **hydrophilic**, or "water loving."

Water is an excellent solvent because of its hydrogen-bonding potential and its polar nature.

Hydrophilic Molecules in Aqueous Solution

Molecules with groups capable of forming H bonds tend to hydrogen-bond with water. Thus, water tends to dissolve molecules, such as proteins and nucleic acids, which display on their solvent-accessible surfaces groups that can form H bonds. These include, for example, hydroxyl, carbonyl, and ester groups that are uncharged but polar, as well as charged groups such as amines, carboxylic acids, and phosphate esters. In addition, when molecules that contain internal hydrogen bonds (such as the α helix shown in Figure 2.8) dissolve in water, some or all of their internal H bonds may be in dynamic exchange for H bonds to H$_2$O (Figure 2.11).

In contrast to most organic liquids, water is an excellent solvent for ionic compounds. Substances like sodium chloride, which exist in the solid state as very stable lattices of ions, dissolve readily in water. The explanation lies in the dipolar nature of the water molecule. The interactions of the negative ends of the water dipoles with cations (e.g., Figure 2.12) and the positive ends with anions in aqueous solution cause the ions to become **hydrated**, that is, surrounded by shells of water molecules called **hydration shells** (Figure 2.12). The propensity of many ionic compounds like NaCl to dissolve in water can be accounted for largely by two factors. First, the formation of hydration shells is energetically favorable. Second, the high dielectric constant of water screens and decreases the electrostatic force between oppositely charged ions that would otherwise pull them back together.

The dipolar nature of the water molecule also contributes to water's ability to dissolve such nonionic, but polar, organic molecules as phenols, esters, and amides. These molecules often have large dipole moments, and interaction with the water dipole promotes their solubility in water.

Hydrophobic Molecules in Aqueous Solution

The solubility of hydrophilic substances depends on their energetically favorable interaction with water molecules. It is therefore not surprising that

FIGURE 2.11

Exchange of internal hydrogen bonds for water hydrogen bonds. A section of a protein molecule like that in Figure 2.8 is shown here replacing some of its intramolecular hydrogen bonds with hydrogen bonds to solvent water. This dynamic (transient) exchange of H bonds is observed most often at the ends, rather than in the middle, of the helix.

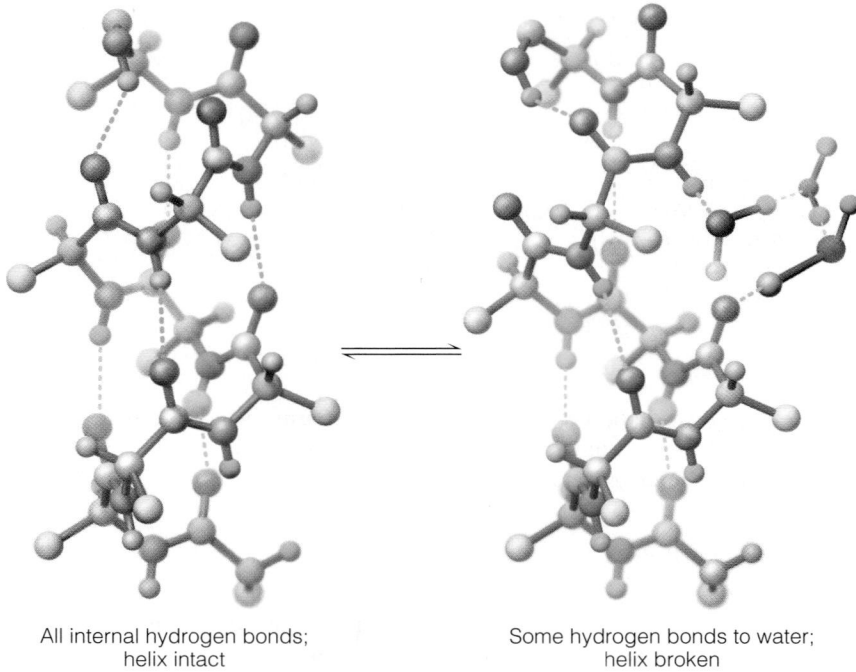

All internal hydrogen bonds;
helix intact

Some hydrogen bonds to water;
helix broken

FIGURE 2.12

Hydration of ions in solution. A salt crystal is shown dissolving in water. As sodium and chloride ions leave the crystal, the noncovalent interaction between these ions and the dipolar water molecules produces a hydration shell around each ion. The energy released in this interaction helps overcome the charge–charge interactions stabilizing the crystal.

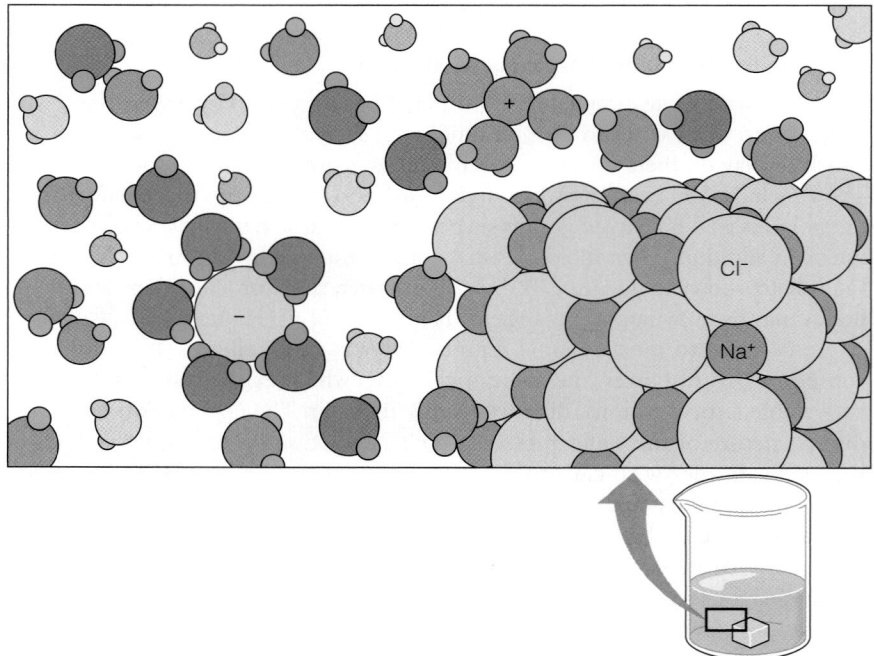

substances like hydrocarbons, which are nonpolar and nonionic and cannot form hydrogen bonds, show only limited solubility in water. Molecules that behave in this way are called **hydrophobic**, or "water fearing." However, bond energy is not the only factor limiting their solubility. When hydrophobic molecules do dissolve, they are not surrounded by hydration shells such as those formed around hydrophilic substances. Instead, the regular water lattice forms ice-like **clathrate** structures, or "cages," about nonpolar molecules (Figure 2.13). This ordering of water molecules, which extends well beyond the cage, corresponds to a decrease in the *entropy*, or randomness, of the mixture (see Chapter 3). The decrease in entropy contributes to the low solubility of hydrophobic substances in water. As is discussed in greater detail in Chapter 3, decreasing entropy is unfavorable thermodynamically; thus, the dissolving of a hydrophobic substance in water is entropically unfavorable. This accounts for the well-known tendency of hydrophobic substances to self-associate, rather than dissolve, in water—we have all seen oil form droplets when we shake it with vinegar. Surrounding two hydrophobic molecules with two *separate* cages requires more ordering of water in clathrates than surrounding both hydrophobes within a *single* cage. Thus, the hydrophobic molecules tend to aggregate because doing so releases some water molecules from the clathrates, thereby increasing the entropy of the system. This phenomenon, called the **hydrophobic effect**, plays an important role in the folding of protein molecules (see Chapter 6) as well as in the self-assembly of lipid bilayers (see Figure 2.15a and Chapter 10).

Amphipathic Molecules in Aqueous Solution

A most interesting and important class of molecules exhibits both hydrophilic and hydrophobic properties simultaneously. Such **amphipathic** substances include fatty acids, lipids, and detergents (Figure 2.14). This class of amphipathic molecules has a "head" group that is strongly hydrophilic, coupled to a hydrophobic "tail"—usually a hydrocarbon. When we attempt to dissolve them in water, these substances form one or more of the structures shown in Figure 2.15a. For example, they may form a **monolayer** on the water surface, with only the head groups immersed. Alternatively, if the mixture is vigorously stirred, **micelles** (spherical structures formed by a single layer of molecules) or **bilayer vesicles** may form. In such cases the hydrocarbon tails of the molecules tend to lie in roughly parallel arrays, which allows them to interact via van der Waals forces.

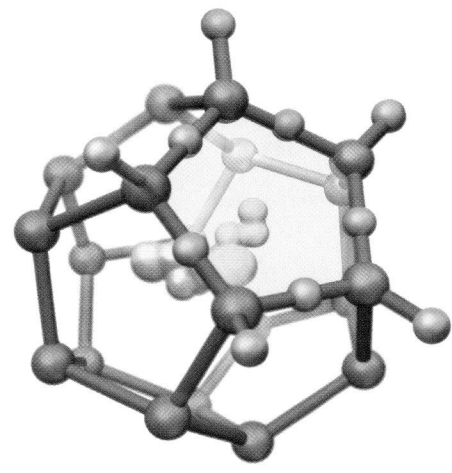

FIGURE 2.13

One unit of clathrate structure surrounding a hydrophobic molecule (yellow). Oxygen atoms are shown in red. Hydrogens are shown for one pentagon of oxygens. The ordered structure may extend considerably further into the surrounding water.

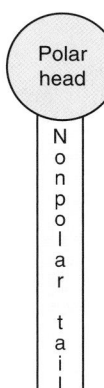

A simplified representation of an amphipathic lipid molecule

A molecule is "amphipathic" if significant parts of the molecular surface are hydrophilic and other parts of the surface are significantly hydrophobic.

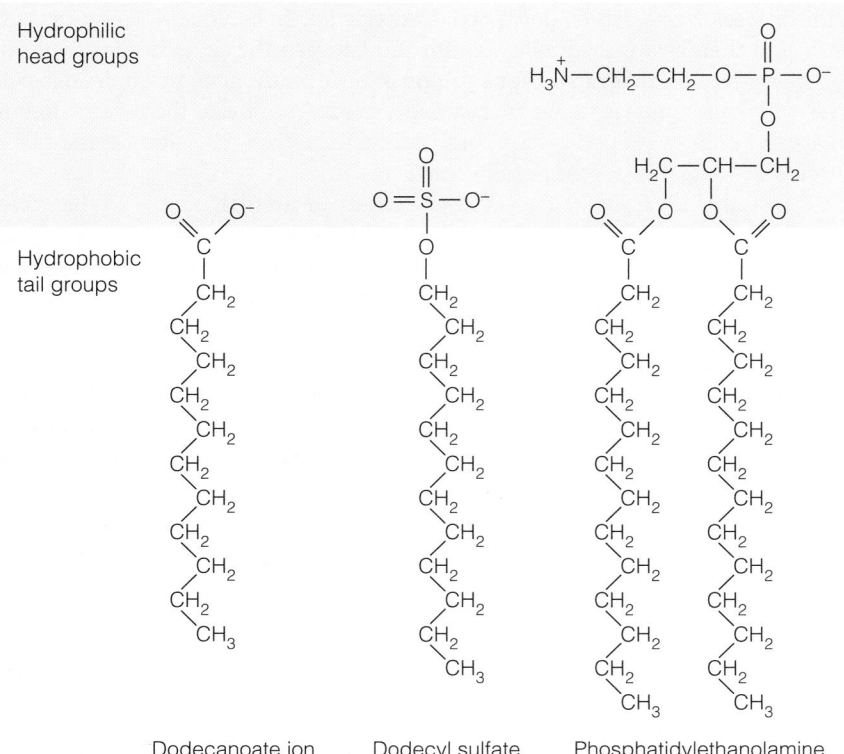

FIGURE 2.14

Amphipathic molecules. These three examples illustrate the dual nature of amphipathic molecules, which have a hydrophilic head group attached to a hydrophobic tail.

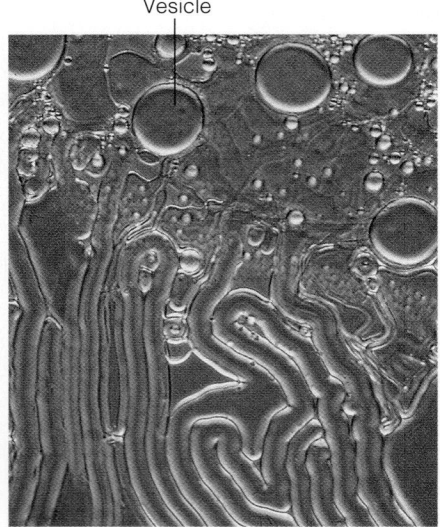

(a) Structures formed in water

Vesicle

(b) Vesicle formation

FIGURE 2.15

Interactions of amphipathic molecules with water.
(a) Structures that can occur when amphipathic substances are mixed with water include a monolayer on the water surface, a micelle, and a bilayer vesicle, a hollow sphere with water both inside and out. In each case, the hydrophilic head groups are in contact with the aqueous phase, whereas the hydrophobic tails associate with one another. **(b)** When phospholipids are mixed with water, the amphipathic molecules aggregate to form films similar to biological membranes. Agitation causes the film to break up into vesicles.

(b) Courtesy of D. W. Deamer and P. B. Armstrong, University of California, Davis.

Many biological molecules are weak acids or weak bases.

The polar or ionic head groups are strongly hydrated by the water around them. Most important to biochemistry is the fact that amphipathic molecules are the building blocks of the biological **membrane bilayers** that surround cells and form the partitions between cellular compartments. These bilayers are made primarily from phospholipids, such as that shown in Figure 2.14. Figure 2.15b shows the formation of synthetic membranes from phospholipids. We have much more to say about phospholipids and membranes in Chapter 10.

Ionic Equilibria

Most biochemical reactions occur in an aqueous environment; the exceptions are those that occur within the hydrophobic interiors of membranes. The many substances dissolved in the aqueous cytosol and extracellular body fluids include free ions like K^+, Cl^-, and Mg^{2+}, as well as molecules and macromolecules carrying ionizable groups. The behavior of all these molecules in biochemical processes depends strongly on their state of ionization. Thus, it is important that we review briefly some aspects of ionic equilibrium, particularly acid–base equilibria and the ionization of water.

Acids and Bases: Proton Donors and Acceptors

A useful definition of acids and bases in aqueous systems is the Brønsted-Lowry definition: that acids are proton donors and bases are proton acceptors. A **strong acid** dissociates almost completely into a proton and a weak conjugate base. For example, HCl dissociates almost completely in water to yield H^+ (actually, hydronium ion, H_3O^+; see below) and Cl^-; thus, the H^+ concentration in the solution is almost exactly equal to the molar concentration of HCl added. Similarly, NaOH is considered a **strong base** because it ionizes entirely, releasing OH^- ion, which is a powerful proton acceptor.

Most of the acidic and basic substances encountered in biochemistry are **weak acids** or **weak bases**, which only partially dissociate. In an aqueous solution of a weak acid there is a measurable equilibrium between the acid and its *conjugate base*, the substance that can accept a proton to re-form the acid. Examples of weak acids and their conjugate bases are given in Table 2.7. Note that these bases do not necessarily contain —OH groups, but they increase the OH^- concentration of a solution by extracting a proton from water.

The weak acids listed in Table 2.7 vary greatly in strength, that is, in their tendency to donate protons. This variation in acid strength is indicated by the range of values of K_a and pK_a, which we will discuss shortly. The stronger the acid, the weaker its conjugate base; in other words, the more strongly an acid tends to donate a proton, the more weakly its conjugate base tends to accept a proton and re-form the acid.

Ionization of Water and the Ion Product

Although water is essentially a neutral molecule, it does have a slight tendency to ionize; in fact, it can act as both a very weak acid and a very weak base. The most correct way to understand this ionization reaction is to note that one water molecule can transfer a proton to another to yield a hydronium ion (H_3O^+) and a hydroxide ion (OH^-) so that water is both the proton donor and the proton acceptor:

$$H_2O + H_2O \rightleftharpoons H_3O^+ + OH^-$$

This is really an oversimplification because the transferred proton may be associated with different clusters of water molecules to yield species like $H_5O_2^+$ and $H_7O_3^+$. Protons in aqueous solution are very mobile, with the proton hopping from one water molecule to another with a period of about 10^{-15} second.

TABLE 2.7 Some weak acids and their conjugate bases

Acid (Proton Donor)		Conjugate Base (Proton Acceptor)		pK_a	K_a (M)
HCOOH Formic acid	$\rightleftharpoons$	$HCOO^-$ Formate ion	$+ H^+$	3.75	1.78×10^{-4}
CH_3COOH Acetic acid	$\rightleftharpoons$	CH_3COO^- Acetate ion	$+ H^+$	4.76	1.74×10^{-5}
OH | $CH_3CH-COOH$ Lactic acid	$\rightleftharpoons$	OH | $CH_3CH-COO^-$ Lactate ion	$+ H^+$	3.86	1.38×10^{-4}
H_3PO_4 Phosphoric acid	$\rightleftharpoons$	$H_2PO_4^-$ Dihydrogen phosphate ion	$+ H^+$	2.14	7.24×10^{-3}
$H_2PO_4^-$ Dihydrogen phosphate ion	$\rightleftharpoons$	HPO_4^{2-} Monohydrogen phosphate ion	$+ H^+$	6.86	1.38×10^{-7}
HPO_4^{2-} Monohydrogen phosphate ion	$\rightleftharpoons$	PO_4^{3-} Phosphate ion	$+ H^+$	12.4	3.98×10^{-13}
H_2CO_3 Carbonic acid	$\rightleftharpoons$	HCO_3^- Bicarbonate ion	$+ H^+$	6.3*	5.1×10^{-7}*
HCO_3^- Bicarbonate ion	$\rightleftharpoons$	CO_3^{2-} Carbonate ion	$+ H^+$	10.25	5.62×10^{-11}
C_6H_5OH Phenol	$\rightleftharpoons$	$C_6H_5O^-$ Phenolate ion	$+ H^+$	9.89	1.29×10^{-10}
$\overset{+}{N}H_4$ Ammonium ion	$\rightleftharpoons$	NH_3 Ammonia	$+ H^+$	9.25	5.62×10^{-10}

▓ Phosphoric acid series   ░ Carbonic acid series

*Apparent pK_a and K_a values (see text for explanation).

For the purposes of the following discussion, it suffices to describe the ionization process in a much simpler way,

$$H_2O \rightleftharpoons H^+ + OH^- \tag{2.5}$$

as long as we remember that a proton *never* exists in aqueous solution as a free ion—it is always associated with one or more water molecules. Whenever we write a reaction involving aqueous H^+, we are really referring to a *hydrated* proton.

The equilibrium described by the previous equation can be expressed in terms of K_w, the **ion product** of water, which is 10^{-14} at 25 °C:

$$K_W = \frac{(a_{H^+})(a_{OH^-})}{(a_{H_2O})} = 10^{-14} \tag{2.6a}$$

As discussed in greater detail in Chapter 3, the proper terms to use in a mass action expression such as K_w are unitless numbers, called "*activities*" (designated by *a*), equal to the effective concentrations of the chemical species. In practice, the distinction between *molar concentration* (i.e., moles of solute per liter of solution, or M) and *activity* is almost always neglected in biochemistry. This approach is appropriate in most biochemical experiments, which are usually conducted in dilute solutions, where activity and molar concentration become nearly equal. The activities of pure liquids and solids have a value of one, and in biochemical reactions the activity of the solvent water is often assumed to have an activity of one. Thus, an alternative expression for the equilibrium shown in Equation 2.5 is

$$K_W \cong \frac{[H^+][OH^-]}{1} = 10^{-14} \tag{2.6b}$$

To correctly evaluate Equation 2.6b we must use unitless values for $[H^+]$ and $[OH^-]$ that are equivalent to their concentrations expressed in units of *molarity* (moles per liter). Throughout this text we will adopt the convention of indicating molar concentration for a solute with square brackets around the symbol for that solute.

Because K_w is a constant, $[H^+]$ and $[OH^-]$ cannot vary independently. If we change either $[H^+]$ or $[OH^-]$ by adding acidic or basic substances to water, the other concentration must change accordingly. A solution with a high $[H^+]$ has a low $[OH^-]$ and vice versa. In pure water to which no acidic or basic substances have been added, all the H^+ and OH^- ions must come from the dissociation of the water itself. Under these circumstances the concentrations of H^+ and OH^- must be equal; thus, at 25 °C

$$[H^+] = [OH^-] = 1 \times 10^{-7}\ \text{M} \tag{2.7}$$

and the solution is said to be *neutral*, that is, neither acidic nor basic. However, the ion product depends on temperature, so a neutral solution does not always have $[H^+]$ and $[OH^-]$ of exactly 10^{-7} M. For example, at human body temperature (37 °C) the concentrations of H^+ and OH^- ions in a neutral solution are each 1.6×10^{-7} M.

The pH Scale and the Physiological pH Range

To avoid working with negative powers of 10, we almost always express hydrogen ion concentration in terms of pH, defined as

$$pH = -\log(a_{H^+}) \cong -\log[H^+] \tag{2.8}$$

The higher the $[H^+]$ of a solution, the lower the pH, so a low pH describes an acid solution. On the other hand, a low $[H^+]$ must be accompanied by a high $[OH^-]$, as indicated by Equation 2.6b, so a high pH describes a basic solution.

A diagrammatic scale of pH values is shown in Figure 2.16, with the values for some well-known solutions indicated. Note that most body fluids have pH values in the range 6.5–8.0, which is often referred to as the **physiological pH range**. Most biochemistry occurs in this region of the scale.

Because of the sensitivity of biochemical processes to even small pH changes, controlling and monitoring pH are essential in most biochemical experiments. The control of pH is achieved by using solutions that are "buffered." The composition of buffered solutions is described in detail later in this chapter. The pH of a solution is most conveniently measured with glass electrode pH meters. The electrode generates an electrical potential, which depends on the H^+ concentration; this is converted by the instrument into a pH reading.

Weak Acid and Base Equilibria

Many biologically important compounds contain weakly acidic and basic groups. Very large protein molecules, for example, carry on their surfaces both acidic (e.g., carboxylate) and basic (e.g., amino) groups. The response of such groups to changes in pH is often of considerable importance to their function. For example, the catalytic efficiency of many enzymes depends critically on the ionization state of certain groups, so these catalysts are effective only in defined pH ranges (see Chapter 11). We will see that the overall charge on a protein changes as a function of pH; thus, intermolecular recognition events based on complementary charge–charge interactions (e.g., substrate binding to a receptor) are also sensitive to pH. The dissociation equilibria of weak acids and bases can be used to describe the molecular basis for such effects. We begin this discussion by considering a few examples given in Table 2.7.

K_a and pK_a

Each of the reactions shown in Table 2.7 can be written as the dissociation of an acid. This dissociation may take several forms, depending on the substance involved:

$$HA^+ \rightleftharpoons H^+ + A$$
$$HA \rightleftharpoons H^+ + A^-$$
$$HA^- \rightleftharpoons H^+ + A^{2-}$$

Most biological reactions occur between pH 6.5 and pH 8.0.

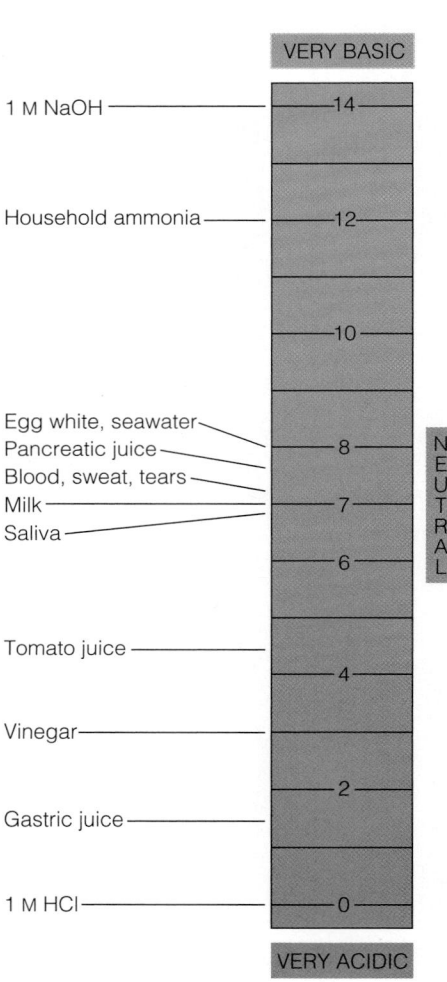

FIGURE 2.16

The pH scale and the physiological pH range. The pH values of some common substances and body fluids are listed, with NaOH at the basic end of the range shown here and HCl at the acidic end. The pH of most body fluids is in the neutral range, between 6.5 and 8.0, where most physiological processes occur. There are a few exceptions, like gastric juice, which has a pH between 1 and 2.

Note that in some cases the conjugate base has a negative charge and in other cases it does not, but in *all* cases it has one proton fewer than the acid. For convenience, we will always write such reactions as $HA \rightleftharpoons H^+ + A^-$. The equilibrium constant for the dissociation of a weak acid (often called the **dissociation constant**) is defined as

$$K_a = \frac{[H^+][A^-]}{[HA]} \tag{2.9}$$

The larger K_a is, the greater is the tendency for the acid to dissociate. Thus, the larger K_a is, the stronger the acid.

The strength of acids is usually expressed in terms of the pK_a value:

$$pK_a = -\log K_a \tag{2.10}$$

Because pK_a is the negative logarithm of K_a, a numerically small value of pK_a corresponds to a strong acid, and a large value corresponds to a weak acid. Both K_a and pK_a values are given for the acids listed in Table 2.7.

Some acids, such as the phosphoric and carbonic acids in Table 2.7, are capable of losing more than one proton. These acids are called **polyprotic acids**. The successive dissociations involve separate steps, with separate pK_a values; thus polyprotic acids exist in several different ionization states.

A Closer Look At pK_a Values: Factors Affecting Acid Dissociation

The tendency of a particular acid to dissociate results from a specific balance of the factors favoring and opposing dissociation. We can understand some of the factors affecting pK_a in light of our earlier discussion of the solvent properties of water.

The dissociation of an acid results in hydration of the proton and, in most cases, of the conjugate base as well. Because hydration is energetically favorable and decreases the attraction between the ions, it favors the dissociation of most acids. Exceptions are positively charged acids like NH_4^+, which dissociate to produce an uncharged conjugate base. In these cases it is the acid that is hydrated and thus stabilized; this is one reason NH_4^+ is such a weak acid.

Opposing the dissociation of an acid is the electrostatic attraction between the proton and a negatively charged conjugate base. This effect can be seen by comparing the successive pK_a values for phosphoric acid dissociation in Table 2.7. As the charge on the conjugate base rises in going from $H_2PO_4^-$ to PO_4^{3-}, the pK_a rises as well; HPO_4^{2-} is a very weak acid.

These examples are important, for they show how environmental effects can influence pK_a values. When we investigate proteins, we will find that the pK_a values for supposedly identical groups can vary widely, depending on the local molecular environment.

Titration of Weak Acids: The Henderson–Hasselbalch Equation

As will be illustrated throughout this text, the structures and functions of many biomolecules show a strong dependence on pH. As pH changes, so does the level of protonation of acidic and basic groups displayed on the surface of a biomolecule. The important consequence of changing pH is that the *overall charge on the molecule changes*. How does changing the pH of a solution alter the charge on a molecule? We can answer this question by considering the *Henderson–Hasselbalch equation* (Equation 2.12), which is derived by taking the negative logarithm of both sides of Equation 2.9 and rearranging

$$-\log[H^+] = -\log K_a + \log \frac{[A^-]}{[HA]} \tag{2.11}$$

Substituting pH for $-\log[H^+]$ and pK_a for $-\log K_a$ gives

$$pH = pK_a + \log \frac{[A^-]}{[HA]} \tag{2.12}$$

A convenient way to express the strength of an acid is by its pK_a; the lower the pK_a, the stronger the acid.

The Henderson–Hasselbalch equation shows the direct relation between the pH of a solution and the ratio of concentrations of the *deprotonated* form [A⁻] to the *protonated* form [HA] of some ionizable group. In the case of a carboxylic acid $[A^-] = [RCOO^-]$ and $[HA] = [RCOOH]$, and in the case of a primary amine base $[A^-] = [RNH_2]$ and $[HA] = [RNH_3^+]$. For example, if we have a buffered solution of formic acid (where the R group = H), Equation 2.12 becomes

$$pH = pK_a + \log \frac{[HCOO^-]}{[HCOOH]} \qquad (2.13)$$

The Henderson–Hasselbalch equation provides a theoretical basis for understanding (1) how the charge on a molecule at a given pH is determined by the ratio of [A⁻]/[HA] (see Problem 16b) and (2) how the ratio of [A⁻]/[HA] can be used to calculate the pH of a buffered solution made up of some acid HA and its conjugate base A⁻ (see Problems 9 and 10).

The Henderson–Hasselbalch equation also shows how pH changes during a titration. Suppose we want to titrate a 1 M solution of formic acid with sodium hydroxide. First, we must ask: What is the pH of the solution made by dissolving 1 mol formic acid in sufficient water to make 1 L of solution? This can be calculated from Equation 2.9, if we note that virtually all of the H⁺ in such a solution comes from the formic acid, rather than from water, and that dissociation of one formic acid molecule gives one H⁺ and one HCOO⁻ ion. If we denote their molar concentrations by x, then Equation 2.9 becomes

$$K_a = 1.78 \times 10^{-4} = \frac{[H^+][HCOO^-]}{[HCOOH]} = \frac{x^2}{1 - x} \qquad (2.14)$$

To get an exact answer, we would have to solve this quadratic equation (see Problem 4). For weak acids, however, the quantity x is often much less than the concentration of total acid added (in this case 1 M). As a result, in such cases we can neglect x in the denominator, giving as a good approximation

$$K_a \approx x^2 \qquad (2.15)$$

In this example, this yields

$$x = [H^+] = [HCOO^-] = 1.33 \times 10^{-2} M \qquad (2.16)$$

Note that only about 1% of the acid has dissociated, so our approximation is quite reasonable. It would not be if the acid were more dilute, for then a larger fraction would dissociate.

The preceding calculation tells us that the initial pH is about 1.9. Now, what happens as a solution of NaOH is added to the formic acid solution? As NaOH is added, it dissociates completely into Na⁺ and OH⁻; however, the hydroxide ions are in equilibrium with protons according to the relation $K_w = [H^+][OH^-]$, so addition of OH⁻ reduces the concentration of protons in the solution. According to Le Chatelier's principle, as [H⁺] decreases, more formic acid must dissociate to satisfy the equilibrium relationship given by Equation 2.14. This means that the ratio [HCOO⁻]/[HCOOH] increases as NaOH is added. Applying the Henderson–Hasselbalch equation (2.13), we see that the pH must also increase continuously as the titration proceeds. At the midpoint of the titration, half of the original formic acid has been neutralized. That means that half is still present in the acid form and half is present as the conjugate base, so [A⁻]/[HA] = 1. The Henderson–Hasselbalch equation then becomes

$$pH = pK_a + \log 1 = pK_a \qquad (2.17)$$

Thus, the pH of a weak acid at the midpoint of its titration curve has the same value as its pK_a. This can be confirmed experimentally, as shown by the titration curves of two acids, formic acid and ammonium ion, in Figure 2.17. The titration curves in Figure 2.17 plot the measured pH against *moles of base added per mole of*

The Henderson–Hasselbalch equation describes the change in pH during titration of a weak acid or a weak base.

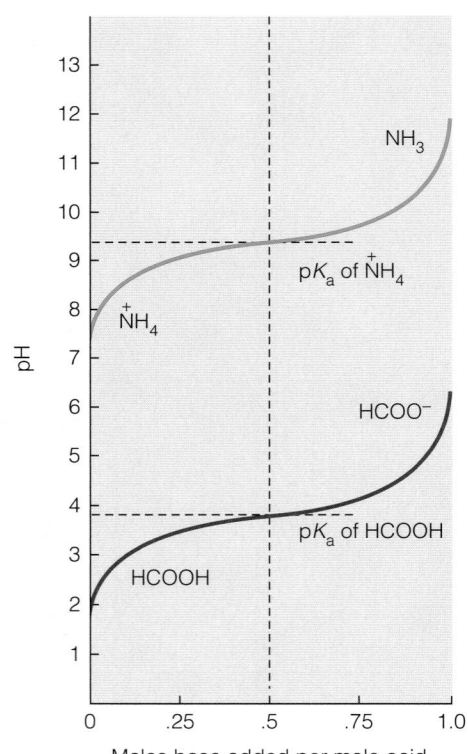

FIGURE 2.17

Titration curves of weak acids. The curves for titration of formic acid (HCOOH) and ammonium ion (ṆH₄⁺) show the change in pH as moles of base are added. Note that pH = pK_a at the half-titration point of each substance. Most of the titration occurs in a range of about one pH unit on either side of this value. In this range, the change in pH with moles of base added is minimal; thus, this is the best buffering range.

acid originally present. Note that over much of the titration curve the pH lies within one pH unit below or above the pK_a.

It should be emphasized that titration curves are reversible. If we were to take the final solution, at high pH, and begin adding a strong acid like HCl, the same curve would be retraced in the opposite direction.

Buffer Solutions

If we look at Figure 2.17 in a different way, another important point emerges. In the pH range near the pK_a, the pH changes only a little with each increment of base or acid added. In fact, the pH is least changed per increment of acid or base just at the pK_a. This is the principle behind **buffering** of solutions by the use of weak acid–base mixtures, a technique used in virtually every biochemical experiment.

Buffered solutions are able to minimize the change in pH following addition of H^+ or OH^- because the conjugate acid (HA) and conjugate base (A^-) of the buffering compound (commonly called the **buffer** or **buffer salt**) are present in sufficient concentration to combine with the added H^+ or OH^- and neutralize them:

Neutralization of OH^- by conjugate acid: $HA + OH^- \rightleftharpoons A^- + H_2O$

Neutralization of H^+ by conjugate base: $A^- + H^+ \rightleftharpoons HA$

Suppose a biochemist wishes to study a reaction at pH 4.00. The reaction may be one that generates or consumes protons. To prevent the pH from drifting too much during the reaction, the experimenter should use a buffer solution consisting of a nearly equal mixture of a weak acid and its conjugate base. Recall that the Henderson–Hasselbalch equation predicts that an equimolar ratio of A^- to HA is achieved when the pH of the solution equals the pK_a of the buffer. In this example a formic acid–formate buffer would be a good choice because the pK_a of formic acid (3.75) is close to the pH value required for the experiment. An acetic acid–acetate mixture would not be as satisfactory because the pK_a of acetic acid (4.76) is nearly 1 pH unit away. The ratio of formate ion to formic acid required to make a buffered solution with a pH of 4.00 can be calculated from the Henderson–Hasselbalch equation:

$$4.00 = 3.75 + \log \frac{[HCOO^-]}{[HCOOH]} \tag{2.18}$$

To calculate the base/acid ratio, we simply subtract 3.75 from 4.00 and take the antilogarithm of both sides of Equation 2.18:

$$\frac{[HCOO^-]}{[HCOOH]} = 10^{(4.00-3.75)} = 10^{0.25} = 1.78 \tag{2.19}$$

This result shows that at pH = 4.00 there will be a ratio of 1.78 mol of $HCOO^-$ in solution to every 1.00 mol of HCOOH in solution. Such a buffer could be made, for example, by mixing equal volumes of 0.1 M formic acid and 0.178 M sodium formate. Alternatively, a buffer solution could be prepared by titrating a solution of formic acid to pH 4.00 with sodium hydroxide.

Because it is desirable to study many biochemical reactions near physiological pH, there is a particular need for mixtures that buffer the pH in the pH range 6.5–8.0. Of the acid–base pairs listed in Table 2.7, only mixtures of dihydrogen phosphate ion ($H_2PO_4^-$) and hydrogen phosphate ion (HPO_4^{2-}), or possibly carbonic acid (H_2CO_3) and bicarbonate ion (HCO_3^-), would be satisfactory. Phosphate buffers are often used in experiments, but they cannot serve under all circumstances because phosphate is consumed or produced in some biochemical reactions. Furthermore, both phosphate- and carbonate-containing solutions precipitate some ions (e.g., Ca^{2+}) that may be needed in the reaction. Therefore, a number of other naturally occurring and synthetic compounds are employed as buffers in this range. Examples are given in Table 2.8.

Organisms must maintain the pH inside cells and in most bodily fluids within the narrow pH range of about 6.5 to 8.0. For example, the normal pH of human

Buffer solutions function because the pH of a weak acid–base solution is least sensitive to added acid or base near the pK_a, where the conjugate acid and conjugate base of the buffer are both present in nearly equimolar concentrations.

TABLE 2.8 Some buffers commonly employed for biochemical studies

Buffer Substance (Acid Form)	Common Name	pK_a
Cacodylic acid	—	6.2
2,2-Bis(hydroxymethyl)2,2′,2″ nitrilotriethanol	BISTRIS	6.5
Piperazine-N,N'-bis(2-ethanesulfonic acid)	PIPES	6.8
Imidazole	—	7.0
N'-2-Hydroxyethylpiperazine-N',2-ethanesulfonic acid	HEPES	7.6
Tris(hydroxymethyl)aminomethane	Tris	8.3

Organisms use buffer systems to maintain the pH of cells and body fluids in the appropriate range.

blood is 7.4, which is also the pH inside most human cells. We have already mentioned two buffer systems of great importance for biological pH control. The dihydrogen phosphate–hydrogen phosphate system, with a pK_a of 6.86, plays a major role in controlling intracellular pH because phosphate is abundant in cells. In blood, which contains dissolved CO_2 as a waste product of metabolism, the carbonic acid–bicarbonate system provides considerable buffering capacity. Carbonic acid has a pK_a of 3.8; however, because it readily decomposes into water and dissolved CO_2, the concentration of H_2CO_3 in solution is very low. As a consequence, the *apparent* pK_a (designated pK_a') of carbonic acid is 6.3 (this is the value listed in Table 2.7). The relationship between dissolved CO_2 and proton concentration (i.e., pH) can be appreciated by considering the following chemical equations:

$$CO_2 + H_2O \rightleftharpoons H_2CO_3 \qquad \text{(reaction to form carbonic acid from } CO_2\text{)}$$
$$H_2CO_3 \rightleftharpoons HCO_3^- + H^+ \qquad \text{(first proton dissociation of carbonic acid)}$$
$$CO_2 + H_2O \rightleftharpoons HCO_3^- + H^+ \qquad \text{(sum of the two equations)} \qquad (2.20)$$

Le Chatelier's principle and Equation 2.20 predict that as the concentration of dissolved CO_2 increases, the pH will decrease as a result of increasing H^+ concentration. We will see in Chapter 7 that a drop in pH in actively respiring tissues results in increased oxygen delivery to these tissues by the oxygen-carrying protein hemoglobin.

From Equation 2.20 we can write an expression that describes the apparent dissociation of carbonic acid to bicarbonate ion and a proton:

$$K_a' = \frac{[H^+][HCO_3^-]}{[CO_2][H_2O]} \qquad (2.21)$$

The value of the pK_a' for carbonic acid can be derived from the combination of the two relevant equilibrium expressions for the chemical equations that were used to derive Equation 2.20. The first of these is for the equilibrium between dissolved CO_2, water, and H_2CO_3:

$$CO_2 + H_2O \rightleftharpoons H_2CO_3 \qquad K_{eq} \cong 3 \times 10^{-3} = \frac{[H_2CO_3]}{[CO_2][H_2O]} \qquad (2.22)$$

The second is for the dissociation of H_2CO_3:

$$H_2CO_3 \rightleftharpoons HCO_3^- + H^+ \qquad K_a \cong 1.7 \times 10^{-4} = \frac{[H^+][HCO_3^-]}{[H_2CO_3]} \qquad (2.23)$$

When these two mass action expressions are multiplied they yield Equation 2.21:

$$\frac{[H_2CO_3]}{[CO_2][H_2O]} \times \frac{[H^+][HCO_3^-]}{[H_2CO_3]} = \frac{[H^+][HCO_3^-]}{[CO_2][H_2O]} = K_a'$$

$$K_a' = K_{eq} \times K_a = 5.1 \times 10^{-7} \text{ and } pK_a' = -\log(5.1 \times 10^{-7}) = 6.3$$

In addition to phosphate and bicarbonate buffers, proteins play a major role in the control of pH in organisms. As shown in Chapter 5, proteins contain many weakly acidic or basic groups, and some of these have pK_a values near 7.0. The abundance of proteins in cells and body fluids such as blood and lymph contributes to the buffering capacity of these biological solutions.

Molecules with Multiple Ionizing Groups: Ampholytes, Polyampholytes, and Polyelectrolytes

So far, we have considered molecules containing only one or a few weakly acidic or basic groups. But many molecules contain multiple ionizing groups and display more complex behavior during titration.

A molecule that contains groups with both acidic and basic pK_a values is called an **ampholyte**. Consider, for example, glycine: H_2N — CH_2 — $COOH$. Glycine is an α-amino acid, one of a group of important amino acids encountered in Chapter 5 as constituents of proteins. The pK_a values of the α-carboxylate and α-amino groups on glycine are 2.3 and 9.6, respectively. If we dissolved glycine in a very acidic solution (e.g., pH 1.0), both the α-amino group and the α-carboxylate group would be protonated, and the molecule would have a net charge of +1. If the pH was increased (e.g., by adding NaOH), proton dissociation would occur in the following sequence:

Increasing pH ⟶

| pH | 1.0 | 6 | 14 |

$$H-\overset{\overset{\displaystyle H}{|}}{\underset{\underset{\displaystyle H}{|}}{\overset{+}{N}}}-CH_2-COOH \rightleftharpoons H-\overset{\overset{\displaystyle H}{|}}{\underset{\underset{\displaystyle H}{|}}{\overset{+}{N}}}-CH_2-COO^- \rightleftharpoons H-\overset{\overset{\displaystyle H}{|}}{\underset{\underset{\displaystyle H}{|}}{N}}-CH_2-COO^-$$

Net charge: +1   0   −1

Thus, the titration of glycine occurs in two steps as the more acidic α-carboxylate and the less acidic α-amino groups successively lose their protons. Glycine can therefore serve as a good buffer in two quite different pH ranges, as shown in Figure 2.18. In each range we can describe the titration curve by applying the Henderson–Hasselbalch equation to the appropriate ionizing group. At low pH the predominant form of glycine has a net charge of +1; at high pH the predominant form has a net charge of −1. The relative concentrations of the three forms are shown in Figure 2.19, which illustrates the important relationship between pH and overall

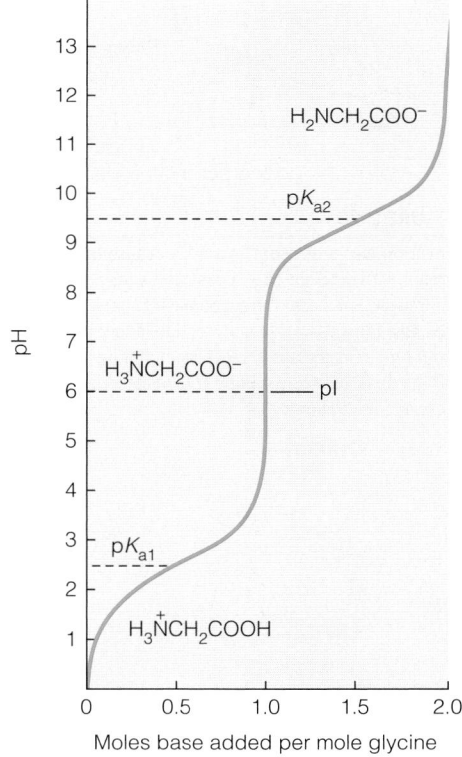

FIGURE 2.18

Titration of the ampholyte glycine. Because two groups with quite different pK_a values can be titrated, this is a two-step titration curve. The calculated isoelectric point, pI, is shown.

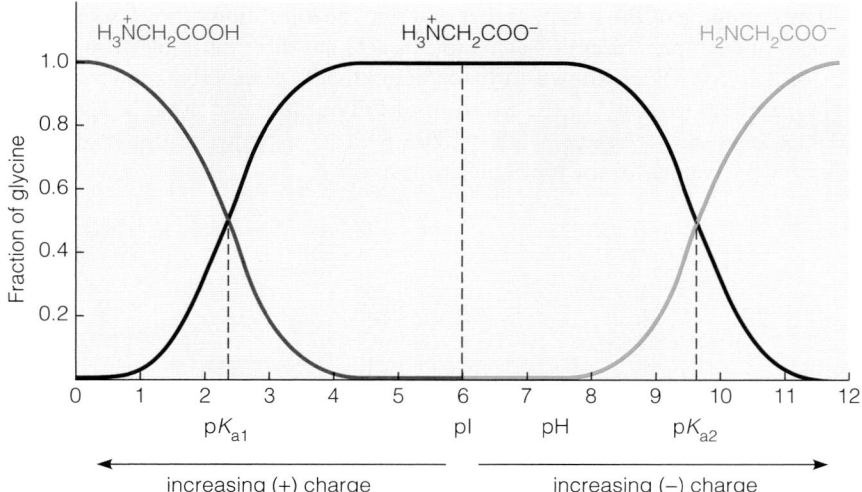

FIGURE 2.19

The relative concentrations of the three forms of glycine as a function of pH. The three forms are $H_3\overset{+}{N}CH_2COOH$, shown in red; $H_3\overset{+}{N}CH_2COO^-$, black; and $H_2NCH_2COO^-$, blue. The two pK_a values and the isoelectric point (pI) are indicated. As pH increases, the molecule becomes more negatively charged, and as pH decreases, the molecule becomes more positively charged.

charge on a molecule: *as pH drops, a molecule will become more positively charged, and as pH increases, a molecule will become more negatively charged* (see Problem 15).

The situation near neutral pH is an interesting one. In this region, most glycine is in the form $H_3\overset{+}{N}—CH_2—COO^-$, which has a net charge of zero. An ampholyte in this state, with equal numbers of positive and negative charges, is called a *zwitterion*. However, there is only one point within this pH region where the *average* charge on glycine is zero. At this pH, called the **isoelectric point (pI)**, most of the glycine molecules are in the zwitterion form, with very small but exactly equal amounts of $H_3\overset{+}{N}—CH_2—COOH$ and $H_2N—CH_2—COO^-$ molecules. We can calculate the isoelectric point by applying the Henderson–Hasselbalch equation to both of the ionizing groups. If we call the pH at the isoelectric point pI, we have

and

$$pI = pK_{COOH} + \log \frac{[H_3\overset{+}{N}CH_2COO^-]}{[H_3\overset{+}{N}CH_2COOH]} \quad (2.24)$$

$$pI = pK^+_{NH_3} + \log \frac{[H_2NCH_2COO^-]}{[H_3\overset{+}{N}CH_2COO^-]} \quad (2.25)$$

Adding these equations (remembering that the sum of the logarithms of two quantities is the logarithm of their product) gives

$$2\,pI = pK_{COOH} + pK^+_{NH_3} + \log \frac{[H_2NCH_2COO^-]}{[H_3\overset{+}{N}CH_2COOH]} \quad (2.26)$$

But because the definition of pI requires that $[H_2NCH_2COO^-] = [H_3\overset{+}{N}CH_2COOH]$, the term to the far right is equal to log 1, or zero, so

$$pI = \frac{pK_{COOH} + pK^+_{NH_3}}{2} \quad (2.27)$$

The result is simple in this case: For a molecule with only two ionizable groups, the pI is simply the average of the two pK_as. If we insert the pK_a values given previously, we obtain pI = 5.95 for glycine. Actually, as Figure 2.19 shows, glycine will be almost entirely in the zwitterion form from about pH 4 to about pH 8; thus, the average charge on glycine is very close to zero throughout this pH range.

For molecules that have three or more ionizable groups, the pI can be calculated by averaging of the two pK_as that describe the ionization of the "isoelectric" species. For example, the amino acid aspartic acid has three ionizable groups with pK_as of 2.1 (α-COOH, shown in black below), 3.9 (β-COOH, shown in blue below), and 9.8 (α-NH$_3^+$). The titration of fully protonated aspartic acid with NaOH occurs in three steps (Figure 2.20), which we can write in the order of *increasing* pK_a values for the ionizable groups:

An ampholyte's isoelectric point, pI, is the pH at which average charge, for all forms of the molecule, is zero. At pH < pI, a molecule will be positively charged, and at pH > pI, the molecule will be negatively charged.

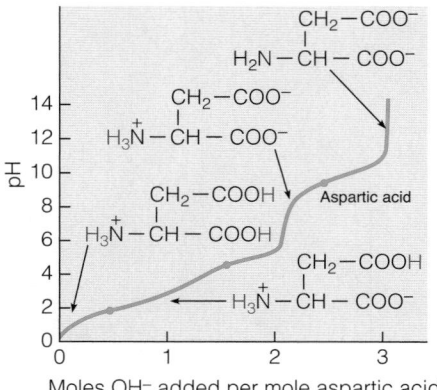

FIGURE 2.20

Titration curve of aspartic acid. Because three groups with different pK_a values can be titrated, this is a three-step process. The predominant species in solution as a function of added hydroxide ion are shown (with titratable protons in red). The buffering regions for the three titratable groups are indicated by the blue dots on the titration curve.

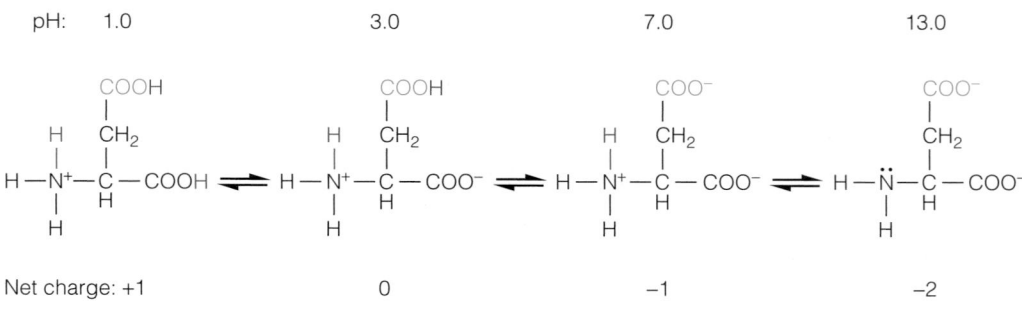

If we sum the charges on each of the four possible ionization states for aspartic acid, we see that one of these, the so-called "isoelectric" species, carries no net charge. As stated above, the two pK_as that describe the ionization of the isoelectric species, $pK_{\alpha COOH}$ and $pK_{\beta COOH}$, have values of 2.1 and 3.9, respectively. Thus, the pI for aspartic acid is

$$pI = \frac{pK_{\alpha COOH} + pK_{\beta COOH}}{2} = \frac{(2.1 + 3.9)}{2} = 3.0 \qquad (2.28)$$

Careful analysis of the ionization of each functional group at pH = 3.0 would show that the α-amino group is fully protonated and therefore carries a full (+1) charge, whereas the two carboxylic acids are *partially* deprotonated; thus, each carries only a partial ($-$) charge. The sum of these partial ($-$) charges $= -1$, which balances the +1 charge on the α–amino group; thus, the net charge on the molecule at pH = 3 is zero. For an example of such an analysis, see Problem 16 at the end of this chapter.

Large molecules such as proteins can have *many* acidic and basic groups. Such molecules are called **polyampholytes**. With anything from tens to hundreds of charged groups present, the calculation of pI becomes more complicated. However, as long as the molecule has both positively and negatively charged groups, it always has an isoelectric point, at which the average charge is zero. For example, human hemoglobin has 148 ionizable groups on its surface, and its pI is 6.85. If acidic groups predominate, the pI will be low; if basic groups predominate, the pI will be high. In Chapter 5 we find that this is an important consideration in working with solutions of proteins.

Three important conclusions can be drawn from this discussion of pI. First, when the pH of a protein solution equals the pI for the protein there is no net charge on the protein because the sum of all the negative charges, $\Sigma(-)$, is exactly offset by the sum of all the positive charges, $\Sigma(+)$, or

$$at\ pH = pI \qquad \left| \sum(-) \right| = \left| \sum(+) \right| \qquad (2.29)$$

thus, the net charge of the molecule = 0. Second, when the pH is greater than the pI, the molecule carries a net negative charge because the sum of the negative charges dominates the sum of the positive charges:

$$at\ pH > pI \qquad \left| \sum(-) \right| > \left| \sum(+) \right| \qquad (2.30)$$

thus, the net charge of the molecule is ($-$). Third, when the pH is below the pI, the molecule carries a net positive charge because the sum of the negative charges is dominated by the sum of the positive charges:

$$at\ pH < pI \qquad \left| \sum(-) \right| < \left| \sum(+) \right| \qquad (2.31)$$

thus, the net charge of the molecule is (+).

This behavior can be understood by considering the chemistry involved. Ionization by H^+ transfer is a *dynamic* process, in which ionizable groups are constantly gaining or losing a proton. Using the Henderson–Hasselbalch equation we can predict, on average, what fraction of these ionizable groups is protonated. As the pH changes, this fraction will also change. For example, protein molecules in aqueous solution become increasingly protonated as the pH decreases. As a consequence, proteins become more positively charged because carboxylic acids, the groups largely responsible for the negative charge density on the surfaces of proteins, become *less negatively charged* in response to a drop in pH, whereas nitrogenous bases (e.g., amines), which are largely responsible for the positive charge density on the surfaces of proteins, become *more positively charged*. A similar logic explains the observation that proteins become more negatively charged as pH increases. Under conditions of increasing pH, the ionizable groups become more deprotonated. Thus, acidic groups become more negatively charged while the basic groups become less positively charged. The effects of pH on the charges of carboxylic acids and amine bases are summarized in Table 2.9.

Significant fluctuations in the charges on proteins will result in loss of function; thus, the concepts summarized in Table 2.9 and Equations 2.29–2.31 not only explain the basis for molecular recognition via charge complementarity but also the critical need to maintain intracellular pH within a narrow range such that charges on molecules remain relatively constant.

TABLE 2.9   Effects of pH on the ionization and charges of amines and carboxylic acids

pH Relative to pK_a	Ratio of $[A^-]$ to $[HA]$	Effect on H^+ Transfer (Ionization)	Net Charge on Group as a Consequence of Ionization
$pH = pK_a$	$[A^-] = [HA]$	A^- is protonated 50% of the time	
Acidic group	$[-COO^-] = [-COOH]$	$-COO^-$ is protonated 50% of the time	net charge $= -0.5$
Basic group	$[-NH_2] = [-NH_3]^+$	$-NH_2$ is protonated 50% of the time	net charge $= +0.5$
$pH < pK_a$	$[A^-] < [HA]$	A^- is protonated $>$ 50% of the time	
Acidic group	$[-COO^-] < [-COOH]$	$-COO^-$ is protonated $>$ 50% of the time	$0 \geq$ net charge > -0.5
Basic group	$[-NH_2] < [-NH_3]^+$	$-NH_2$ is protonated $>$ 50% of the time	$+1 \geq$ net charge $> +0.5$
$pH > pK_a$	$[A^-] > [HA]$	A^- is protonated $<$ 50% of the time	
Acidic group	$[-COO^-] > [-COOH]$	$-COO^-$ is protonated $<$ 50% of the time	$-1 <$ net charge < -0.5
Basic group	$[-NH_2] > [-NH_3]^+$	$-NH_2$ is protonated $<$ 50% of the time	$0 \leq$ net charge $< +0.5$

Although the theory described above can be used to predict a pI value from the amino acid composition of a protein, there are many caveats associated with such predictions. In practice, pI is determined experimentally using the simple method of *electrophoresis*. When an electric field is applied to a solution of charged ampholytes, positively charged molecules migrate toward the cathode and negatively charged ones toward the anode. At its isoelectric point, an ampholyte moves in *neither* direction because it has zero net charge. In the method called **isoelectric focusing**, ampholytes move through a pH gradient, each coming to rest at its own isoelectric point. In this way, the ampholytes are separated and their isoelectric points determined (see Tools of Biochemistry 2A).

Some macromolecules, called **polyelectrolytes**, carry multiples of only positive or only negative charge. Strong polyelectrolytes, like the negatively charged nucleic acids (see Chapter 4), are ionized over a wide pH range. In addition, there are weak polyelectrolytes, like polylysine, a polymer of the amino acid lysine.

When a number of weakly ionizing groups are carried on the same molecule, the pK_a of each group is influenced by the state of ionization of the others. In a molecule like polylysine, the first protons are more easily removed than the last because the strong positive charge on the fully protonated molecule makes deprotonation, and the associated reduction in repulsive charge–charge interactions, more favorable. Conversely, a molecule that develops a strong negative charge as protons are removed gives up the last ones less readily, as reflected in the increasing values of pK_a for successive deprotonations (see entries for H_3PO_4 and H_2CO_3 in Table 2.7).

Interactions Between Macroions in Solution

Large polyelectrolytes such as nucleic acids and polyampholytes such as proteins are classed together as **macroions**. Depending on the solution pH, they may carry a substantial net charge. The electrostatic forces of attraction or repulsion between such charged particles play a major role in determining their behavior in solution.

Solubility of Macroions and pH

Because macroions of like net charge repel one another, nucleic acid molecules tend to remain separated in solution (Figure 2.21a). For the same reason, proteins tend to be soluble when they have a net charge, that is, at pH values above or below their isoelectric points. On the other hand, if positively and negatively charged macromolecules are mixed, electrostatic attraction makes them tend to associate with one another (Figure 2.21b). Many proteins interact strongly with DNA; most of these turn out to be positively charged. A striking example is found in the chromosomes of higher organisms, in which the negatively charged DNA is

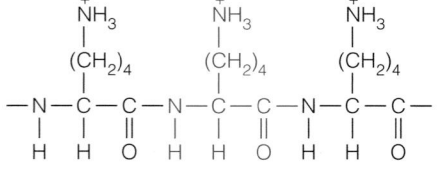

Polylysine

strongly associated with positively charged proteins called *histones* to form the complex called chromatin (discussed in Chapter 24).

A more subtle type of electrostatic interaction may cause the molecules of a particular protein to self-associate at the isoelectric pH (Figure 2.22). For example, the common milk protein β-lactoglobulin is a polyampholyte with an isoelectric point of about 5.3. Above or below this pH, the molecules all have either negative or positive charges and repel one another. This protein is therefore very soluble at either acidic or basic pH. At the isoelectric point the net charge is zero, but each molecule still carries surface patches of both positive and negative charge. The charge–charge interactions, together with other kinds of intermolecular interactions such as van der Waals forces, make the molecules tend to clump together and precipitate. Therefore, β-lactoglobulin, like many other proteins, has minimum solubility at its isoelectric point (Figure 2.22d).

The Influence of Small Ions: Ionic Strength

The interactions of macroions are strongly modified by the presence of small ions, such as those from salts dissolved in the same solution. Each macroion collects about it a **counterion atmosphere** enriched in oppositely charged small ions, and this cloud of ions tends to screen the molecules from one another (Figure 2.23a). Obviously, the larger the concentration of small ions present, the more effective this electrostatic screening will be; however, the precise relationship of screening to concentration is rather complex. A quantitative expression of the screening

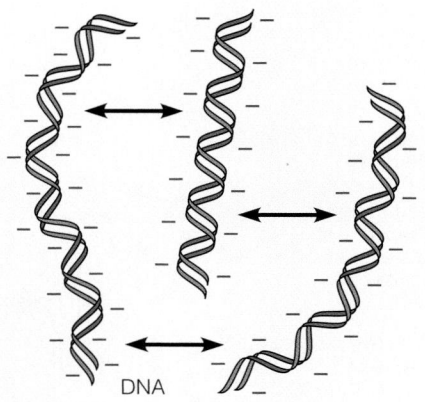

(a) Repulsion

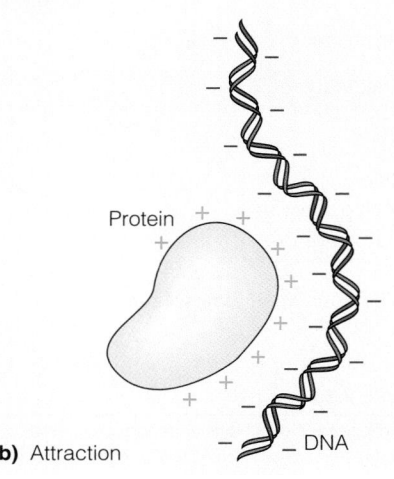

(b) Attraction

FIGURE 2.21

Electrostatic interactions between macroions.
(a) Repulsion. DNA molecules, with many negative charges, strongly repel one another in solution.
(b) Attraction. If DNA is mixed with a positively charged protein, these molecules have a strong tendency to associate.

Interactions between macroions are greatly influenced by pH and the small ions in the solution.

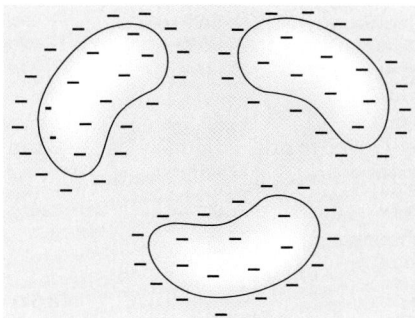

(a) High pH: protein soluble (deprotonated)

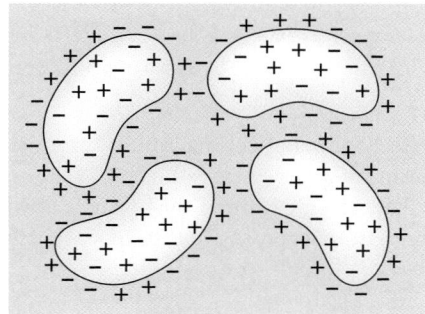

(b) Isoelectric point: protein aggregates

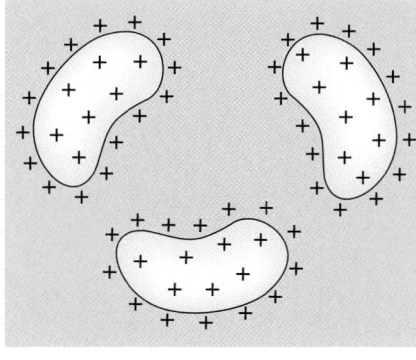

(c) Low pH: protein soluble (protonated)

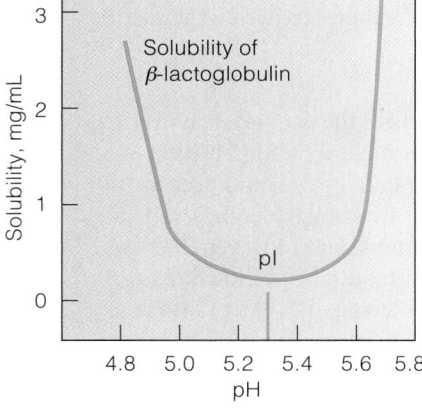

(d) Solubility of β-lactoglobulin

FIGURE 2.22

Dependence of protein solubility on pH. **(a)** Most proteins are very soluble at high pH, where all of their molecules are negatively charged. **(b)** At the isoelectric point, where a protein has no *net* charge, its molecules retain regions of positive and negative charge on their surfaces, resulting in aggregation and precipitation. **(c)** At low pH the proteins are soluble because of their positive charge. **(d)** The solubility of β-lactoglobulin with varying pH; the lowest solubility occurs at the isoelectric point.

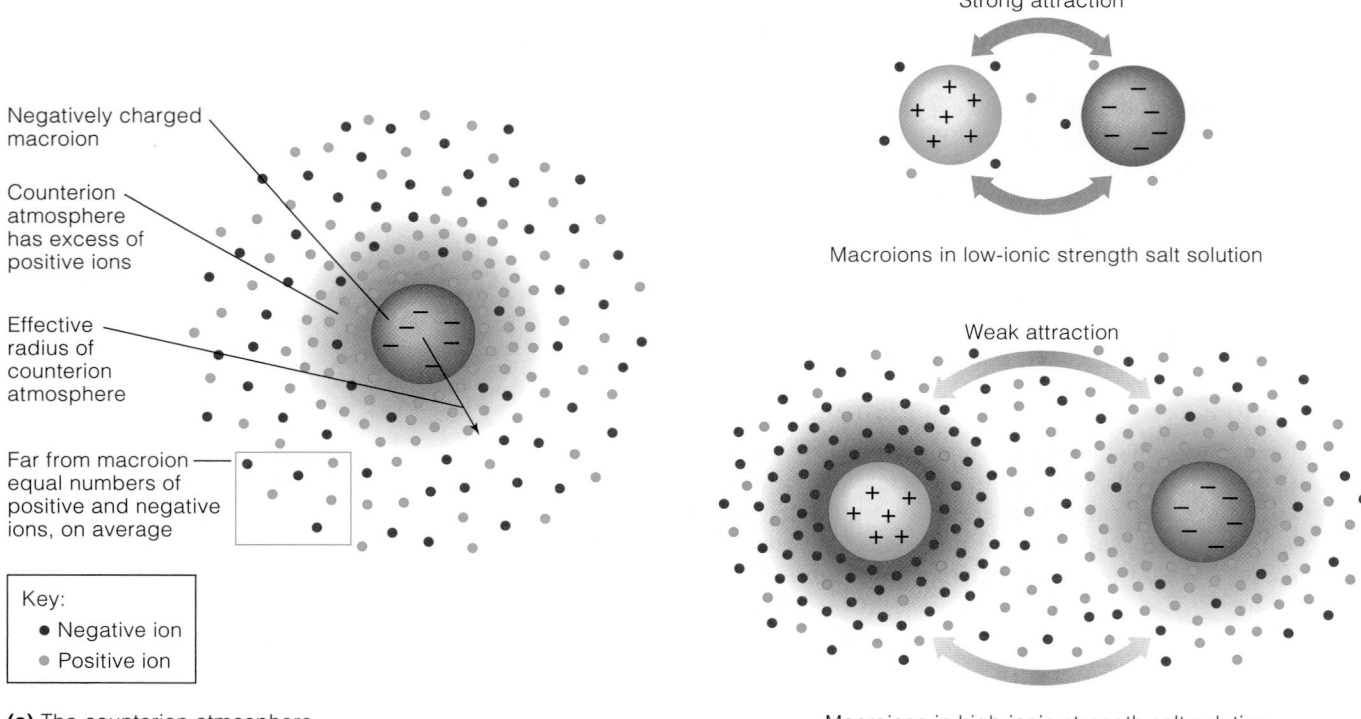

Negatively charged macroion

Counterion atmosphere has excess of positive ions

Effective radius of counterion atmosphere

Far from macroion — equal numbers of positive and negative ions, on average

Key:
● Negative ion
● Positive ion

(a) The counterion atmosphere

Strong attraction

Macroions in low-ionic strength salt solution

Weak attraction

Macroions in high-ionic strength salt solution

(b) The influence of ionic strength

FIGURE 2.23

The influence of small ions on macroion interactions. (a) When a macroion (in this example, negatively charged) is placed in an aqueous salt solution, small ions of the opposite sign tend to cluster about it, forming a counterion atmosphere. There are more cations than anions near the macroanion shown here; far away from the macroion, the average concentrations of cations and anions are equal. **(b)** At low ionic strength, the counterion atmosphere is diffuse and interferes little with the interactions of the macroions. At high ionic strength, the counterion atmosphere is concentrated about the macroions and greatly reduces their interactions.

effect for spherical macroions, proposed by P. Debye and E. Hückel, is stated in terms of an effective radius (r) of the counterion atmosphere. This radius may be taken as a measure of the distance at which two macroions "sense" one another's presence. According to the Debye–Hückel theory,

$$r = \frac{K}{I^{1/2}} \tag{2.32}$$

where K is a constant that depends on the dielectric constant of the medium and the temperature, and I is a function of the concentration called the **ionic strength**. The ionic strength is defined as follows:

$$I = \tfrac{1}{2}\sum_i M_i Z_i^2 \tag{2.33}$$

where the sum is taken over all small ions in the solution. For each ion type, M_i is its molarity and Z_i is its charge. For a 1:1 electrolyte like NaCl we have $Z_{Na^+} = +1$ and $Z_{Cl^-} = -1$, and because $M_{Na^+} = M_{Cl^-} = M_{NaCl}$, we find that $I_{NaCl} = M_{NaCl}$. Thus, ionic strength equals salt molarity for 1:1 electrolytes, but this is not true if multivalent ions (e.g., Mg^{2+} or SO_4^{2-}) are involved. Multivalent ions make greater individual contributions to the ion atmosphere than do monovalent ions, as reflected in the fact that the *square* of the ion charge is included in calculating the ionic strength. For these electrolytes, $I > M$.

The effects of the ionic strength of the medium on the interaction between charged macroions can be summarized as shown in Figure 2.23b. At very low ionic strength, the counterion atmosphere is highly expanded and diffuse, and screening is ineffective. In such a solution, macroions attract or repel one another strongly. If the ionic strength is increased, the counterion atmosphere shrinks and becomes concentrated about the macroion, and the attractive interactions between positive and negative groups are effectively screened.

The screening effect of the counterion atmosphere helps explain a general observation concerning protein solubility: increasing ionic strength (up to a point) increases solubility even at the isoelectric point. This effect of putting proteins into solution by increasing the salt concentration is called **salting in**.

Raising the salt concentration to very high levels (several molar, for example) introduces another, opposite effect. In very concentrated salt solutions, much of the water that would normally solvate and help solubilize the protein molecule is bound up in the hydration shells of the numerous salt ions, preventing sufficient hydration of the protein. Thus, at extremely high salt concentration, the solubility of a protein again decreases, an effect called **salting out**. Because different proteins respond differently to these two effects, salting in and salting out can be used to separate proteins in a complex mixture.

The effects of ionic interactions on the behavior of biological macromolecules mean that the biochemist must pay close attention to both ionic strength and pH. In addition to using a buffer to control pH, experimenters will usually add a neutral salt (like NaCl or KCl) to control the ionic strength of a solution. In determining the amount of salt to add, they often try to mimic the ionic strengths of cell and body fluids. Although ionic strengths vary from one cell type or fluid to another, a value of 0.1 to 0.2 M is often appropriate in biochemical experiments.

SUMMARY

The major concepts discussed in this chapter are summarized in two equations: Coulomb's law (Equation 2.1), and the Henderson–Hasselbalch equation (Equation 2.12).

Coulomb's law provides the theoretical foundation for describing the variety of weak, noncovalent interactions that occur among ions, molecules, and parts of molecules in the cell. These interactions, which are 10 to 100 times weaker than most covalent interactions, include charge–charge interactions and the interactions of permanent and induced dipoles. Molecules never interpenetrate because attraction between them is countered by repulsion when their electron orbitals begin to overlap; a molecule's van der Waals radius is half its distance of closest approach to another molecule. The hydrogen bond is a strong noncovalent interaction, and it shares some features (directionality, specificity) with covalent bonds.

Water is the essential milieu of life. Most of the unique properties of water as a substance are accounted for by its polarity and hydrogen bonding, properties that also make it an excellent solvent. Polar, hydrogen-bonding, and ionic substances dissolve easily in water and are called hydrophilic, whereas other compounds dissolve in water to only a limited extent and are called hydrophobic. Amphipathic molecules, which have both polar and nonpolar parts, form distinctive structures such as monolayers, vesicles, and micelles when in contact with water. Such molecules form the membrane bilayers that surround cells and cellular compartments.

The ionization of weak acids and bases is of major importance in biochemistry because it establishes the charges on biomolecules. Most of the important processes occur in the pH range between 6.5 and 8.0, called the physiological pH range. The behavior of weak acids and their conjugate bases is described by the Henderson–Hasselbalch equation, which relates the conjugate base/undissociated acid ratio to pH and pK_a. Titration curves show that the pH change with added acid or base is most gradual in the range near the pK_a of the acid; this is the basis for preparation of buffer solutions.

An ampholyte has both acidic and basic ionizing groups; the molecules can have a net positive, zero, or net negative charge, depending on solution pH. A polyampholyte has many acidic and basic groups. The isoelectric point of an ampholyte or a polyampholyte is the pH at which the average net charge of the molecules is zero. Polyelectrolytes have multiple ionizing groups with a single kind of charge. The behavior of macroions (polyampholytes and polyelectrolytes) in solution depends on pH and on the presence of small ions, which screen the macroions from each other's charges. The magnitude of screening depends on the ionic strength of the solution and is described quantitatively by the Debye–Hückel theory.

REFERENCES

Noncovalent Interactions

Burley, S. K., and G. A. Petsko (1988) Weakly polar interactions in proteins. *Adv. Protein Chem.* 39:125–189. Contains an excellent treatment of weak interactions in general.

Creighton, T. E. (2010) *The Physical and Chemical Basis of Molecular Biology.* Helvetian Press, United Kingdom. See Chapters 2 and 3.

Eisenberg, D., and D. Crothers (1979) *Physical Chemistry with Applications to the Life Sciences.* Benjamin/Cummings, Redwood City, Calif. In addition to a thorough description of covalent bonding, Chapter 11 contains an excellent discussion of dipole moments, polarizability, and noncovalent interactions. Chapter 8 contains some useful material on electrolyte solutions, at a more advanced level than this book.

Leckband, D., and J. Israelachvili (2001) Intermolecular forces in biology. *Quart. Rev. Biophys.* 34:105–267. Extensive review of the theory for predicting forces and the practice of measuring forces in biological systems.

van Holde, K. E., W. C. Johnson, and P. S. Ho (2006) *Principles of Physical Biochemistry* (2nd ed.). Prentice Hall, Upper Saddle River, N.J. Covers most of the topics in this chapter in considerably more depth.

Water

Eigen, M., and L. DeMaeyer (1959) Hydrogen bond structure, proton hydration, and proton transfer in aqueous solutions. In: *The Structure of Electrolyte Solutions,* edited by W. J. Hamer, pp. 64–85. Wiley, New York. Although not recent, this remains an excellent, interesting review.

Hagler, A. T., and J. Moult (1978) Computer simulation of solvent structure around biological macromolecules. *Nature* 272:222.

Kamb, B. (1968) Ice polymorphism and the structure of water. In: *Structural Chemistry and Molecular Biology,* edited by A. Rich and N. Davidson, pp. 507–542. Freeman, San Francisco. A review of theories of water structure.

Moore, F. G., and G. L. Richmond (2008) Integration or segregation: How do molecules behave at oil/water interfaces? *Accts. of Chem. Res.* 41:739–748. A detailed study of interfacial regions between water and nonpolar fluids.

Tanford, C. (1980) *The Hydrophobic Effect. Formation of Micelles and Biological Membranes.* Wiley, New York. A classic study of hydrophobicity.

Ionic Equilibria

Edsall, J. T., and J. Wyman (1958) *Biophysical Chemistry,* Vol. 1. Academic Press, New York. An excellent in-depth treatment. Here you can find extensive discussions of polyprotic dissociation, isoelectric points for polyampholytes, and so forth.

Phillips, R., J. Kondev, and J. Theriot (2009) *Physical Biology of the Cell.* Garland Science, New York. Chapter 9 has a good discussion of water and electrostatics in ionic solutions.

Tossell, J. A. (2006) H_2CO_3 and its oligomers: Structures, stabilities, vibrational and NMR spectra, and acidities. *Inorganic Chemistry* 45:5061–5970. Computational analysis of stability and pK_a of "carbonic acid."

PROBLEMS

In this and following chapters, more difficult problems are indicated by an asterisk.

1. A chloride ion and a sodium ion are separated by a center–center distance of 0.5 nm. Calculate the interaction energy (the energy required to pull them infinitely far apart) if the medium between them is (a) water and (b) *n*-pentane (see Table 2.6). Express the energy in joules per mole of ion pairs. [Note: The charge on an electron is 1.602×10^{-19} C, Avogadro's number is 6.02×10^{23} molecules/mol.]

2. Rank the following in terms of expected dipole moment, and explain your choice:

$H_2S, CCl_4, H_3N^+ \!\!-\!\! CH_2 \!\!-\!\! COO^-, H_3N^+ \!\!-\!\! CH_2 \!\!-\!\! CH_2 \!\!-\!\! CH_2 \!\!-\!\! COO^-$

3. The accompanying graph depicts the interaction energy between two water molecules situated so that their dipole moments are parallel and pointing in the same direction. Sketch an approximate curve for the interaction between two water molecules oriented with *antiparallel* dipole moments.

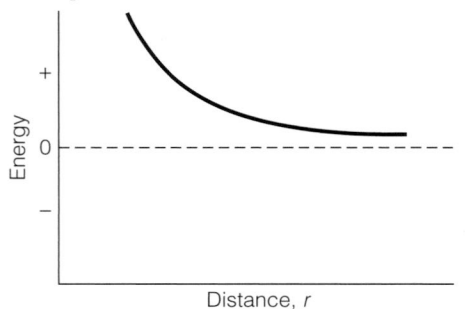

*4. (a) Solve the quadratic equation (referred to in Equation 2.14) for the more general case where the total concentration of acid ($A_0 = [HA] + [A^-]$) has any value A_0.
(b) One can avoid solution of the quadratic equation by using a method of successive approximations, starting with $x^2 \approx A_0 K_a$. Explain how this would be done.

5. What is the pH of each of the following solutions?
(a) 0.35 M hydrochloric acid
(b) 0.35 M acetic acid
(c) 0.035 M acetic acid
[Note that the approximate method used in Equation 2.15 will not give a good answer at low A_0.]

6. The weak acid HA is 2% ionized (dissociated) in a 0.20 m solution.
(a) What is K_a for this acid?
(b) What is the pH of this solution?

*7. Calculate the pH values and draw the titration curve for the titration of 500 mL of 0.010 m acetic acid (pK_a 4.76) with 0.010 M KOH.

8. What is the pH of the following buffer mixtures?
(a) 100 mL 1 M acetic acid plus 100 mL 0.5 M sodium acetate
(b) 250 mL 0.3 M phosphoric acid plus 250 mL 0.8 M KH_2PO_4

*9. (a) Suppose you wanted to make a buffer of exactly pH 7.00 using KH_2PO_4 and Na_2HPO_4. If the final solution was 0.1 M in KH_2PO_4, what concentration of Na_2HPO_4 would you need?
(b) Now assume you wish to make a buffer at the same pH, using the same substances, but want the total phosphate molarity ($[HPO_4^{-2}]$ + $[H_2PO_4^-]$) to equal 0.3. What concentrations of the KH_2PO_4 and Na_2HPO_4 would be required?

10. A 500-mL sample of a 0.100 M formate buffer, pH 3.75, is treated with 5 mL of 1.00 m KOH. What is the pH following this addition?

11. You need to make a buffer whose pH is 7.0, and you can choose from the weak acids shown in Table 2.7. Briefly explain your choice.

12. Describe the preparation of 2.00 L of 0.100 m glycine buffer, pH 9.0, from glycine and 1.00 M NaOH. What mass of glycine is required and what volume of 1.00 NaOH is required? The appropriate pK_a of glycine is 9.6.

13. Carbon dioxide is dissolved in blood (pH 7.4) to form a mixture of carbonic acid and bicarbonate. Neglecting free CO_2, what fraction will be present as carbonic acid? Would you expect a significant amount of carbonate (CO_3^{2-})?

*14. The efficiency of a buffer in resisting changes in pH upon addition of base or acid is referred to as the *buffer capacity*. One way to define this quantity, which we shall call B, is $B = dx/d\text{pH}$, where x is the number of moles of base added to a weak acid. Note that the Henderson–Hasselbalch equation may be written as

$$\text{pH} = pK_a + \log \frac{x}{A_0 - x}$$

where A_0 is the total concentration of acid.
(a) Obtain an expression for B, in terms of A_0 and the fraction of acid titrated f, where $f = x/A_0$. [Hint: It is easiest to determine $d\text{pH}/dx$ and take the reciprocal.]
(b) At what value of f is buffer capacity maximal? (The easiest way to determine this is to make a graph of B versus f.)
(c) What is the maximal value of B?
(d) What is the effect of A_0 on B?

15. What is the molecular basis for the observation that the overall charge on a protein becomes increasingly positive as pH drops and more negative as pH increases?

*16. The amino acid *arginine* ionizes according to the scheme below:
(a) Calculate the isoelectric point of arginine. You can neglect contributions from form I. Why?
(b) Calculate the average charge on arginine when pH = 9.20 (Hint: find the average charge for each ionizable group and sum these together; see Table 2.8 for some guidance).
(c) Is the value of average charge you calculated in part (b) reasonable, given the pI you calculated in part (a)? Explain your answer.

17. It is possible to make a buffer that functions well near pH 7, using citric acid, which contains only carboxylate groups. Explain.

$$
\begin{array}{c}
CH_2 - CO_2H \\
| \\
HO - C - CO_2H \\
| \\
CH_2 - CO_2H
\end{array}
$$

Citric acid

18. A student is carrying out a biological preparation that requires 1 M NaCl to maintain an ionic strength of 1.0. The student chooses to use 1.0 M ammonium sulfate instead. Why is this a serious error?

Problems 19 and 20 refer to material presented in Tools of Biochemistry 2A.

*19. What is the optimum pH to separate a mixture of lysine, arginine, and cysteine using electrophoresis? Draw the structures of the three amino acids *in the protonation state that would predominate at the pH you have chosen*. See Table 5.1 on page 138. For each amino acid, indicate the net charge at the chosen pH as well as the direction of migration and relative mobility in the electric field.

Apply the sample mixture here

20. Suppose you have two genetic variants of a large protein that differ only in that one contains a histidine (side chain $pK_a = 6.4$) when the other has a valine (uncharged side chain).
(a) Which would be better for separation: gel electrophoresis or isoelectric focusing? Why?
(b) What pH would you choose for the separation?

$$
I \xrightarrow[+H^+]{\substack{pK_a = 1.82 \\ -H^+}} II \xrightarrow[+H^+]{\substack{pK_a = 8.99 \\ -H^+}} III \xrightarrow[+H^+]{\substack{pK_a = 12.5 \\ -H^+}} IV
$$

I

II

III

IV

TOOLS OF BIOCHEMISTRY 2A

ELECTROPHORESIS AND ISOELECTRIC FOCUSING

General Principles

When an electric field is applied to a solution, solute molecules with a net positive charge migrate toward the cathode, and molecules with a net negative charge move toward the anode. This migration is called **electrophoresis**. The velocity of the molecules depends on two factors. Driving the motion is the force $q\mathscr{E}$ exerted by the electric field on the particle, where q is the molecule's charge (in coulombs) and $\mathscr{E}$ is the electrical field strength (in volts per meter). Resisting the motion is the frictional force fv exerted on the particle by the medium, where v is the velocity of the particle and f is the **frictional coefficient**, which depends on the size and shape of the molecules. Large or asymmetric molecules encounter more frictional resistance than small or compact ones and consequently have larger frictional coefficients.

When the electric field is turned on, the molecule quickly accelerates to a velocity at which these forces balance and then moves steadily at this rate. The steady velocity is determined by the balance of forces:

$$fv = q\mathscr{E} \tag{2A.1}$$

We can rewrite this equation as $v/\mathscr{E} = q/f$ in order to express the rate of motion per unit of field strength, $v/\mathscr{E}$. This ratio is called the **electrophoretic mobility** (μ) of the molecule:

$$\mu = \frac{v}{\mathscr{E}} = \frac{q}{f} = \frac{Ze}{f} \tag{2A.2}$$

On the right-hand side of this equation, we have expressed the charge on the molecule as the product of the unit of electron (or proton) charge (e) times the number of unit charges, Z (a positive or negative integer). Because f depends on molecular size and shape, Equation 2A.1 tells us that the mobility of a molecule depends on its charge and on the molecular dimensions.*

Because ions and macroions differ in both respects, their behavior in an electric field provides a powerful way of separating them. Electrophoretic separation is one of the most widely used methods in biochemistry.

Paper Electrophoresis and Gel Electrophoresis

Although electrophoresis can be carried out free in solution, it is more convenient to use some kind of *supporting medium*. The two most commonly used supporting media, paper and gel, are shown in Figures 2A.1 and 2A.2. **Paper electrophoresis** (Figure 2A.1) can be used for separating mixtures of small charged molecules. A piece of filter paper, moistened with a buffer solution to control the pH, is stretched between two electrode vessels. A drop of the mixture to be analyzed is placed on the paper, and the electric current is turned on. After the molecules have migrated for a sufficient time—usually several hours—the paper is removed, dried, and stained with a dye that colors the substances to be examined. Each kind of charged molecule in the mixture will have migrated a certain distance toward either the anode or the cathode, depending on its charge and dimensions, and will show up as a stained spot on the paper at the new position. Usually the spots can be identified by comparison with a set of standards run on the same paper. If the unknown substances are radioactive, the spots can be cut out and their radioactivity measured by scintillation counting (see Tools of Biochemistry 12A).

Gel electrophoresis (Figure 2A.2) is a technique commonly used with proteins and nucleic acids. A gel containing the appropriate buffer solution is cast as a thin slab between glass plates. Common gel-forming materials are polyacrylamide, a water-soluble, cross-linked polymer, and agarose, a polysaccharide. The

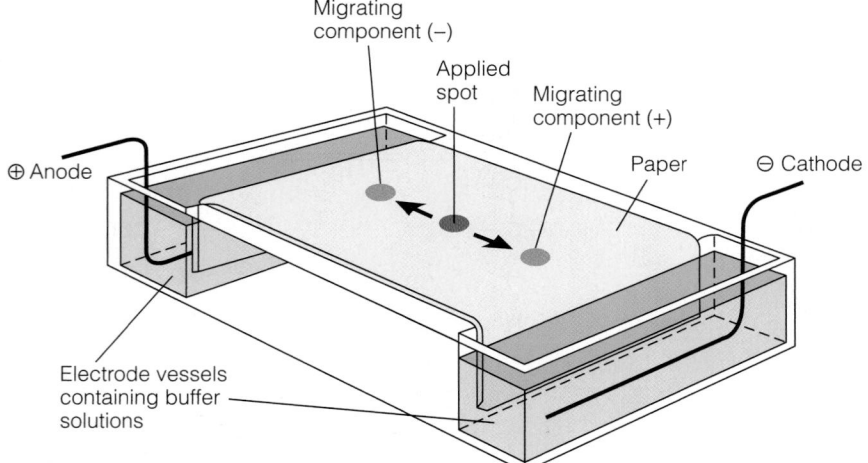

Migrating component (−)
Applied spot
Migrating component (+)
Paper
⊕ Anode
⊖ Cathode
Electrode vessels containing buffer solutions

FIGURE 2A.1

Paper electrophoresis.

---

*Equation 2A.2 is actually an approximation, for it neglects effects of the ion atmosphere. See van Holde, Johnson, and Ho (References) for more detail.

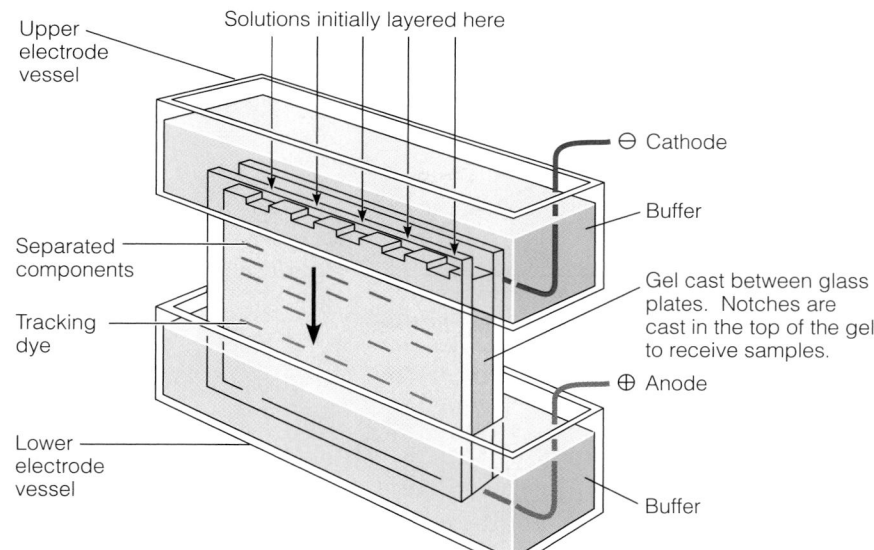

Upper electrode vessel

Solutions initially layered here

⊖ Cathode

Buffer

Separated components

Tracking dye

Gel cast between glass plates. Notches are cast in the top of the gel to receive samples.

⊕ Anode

Lower electrode vessel

Buffer

FIGURE 2A.2

Gel electrophoresis.

slab is placed between electrode compartments, with the bottom selected as anode or cathode, depending on whether anions or cations are being separated. A small amount of a solution of each sample is carefully pipetted into one of several precast notches, called "wells", on top of the gel. Usually glycerol and a water-soluble cationic or anionic "tracking" dye (e.g., bromophenol blue) are added to the sample. The glycerol makes the sample solution dense so that it sinks into the well and does not mix into the buffer solution in the upper electrode chamber. The dye migrates faster than most macroions, so the experimenter is able to follow the progress of the experiment. The current is turned on until the tracking dye band is near the bottom of the slab. The gel is then removed from between the glass plates and is usually stained with a dye that binds to proteins or nucleic acids. At this point, a photograph is taken of the gel for a permanent record. Because the protein or nucleic acid mixture was applied as a narrow band at the top of the gel, components migrating with different mobilities appear as narrow bands on the gel, although the bands may be broadened somewhat by diffusion. Certain techniques (see Tools of Biochemistry References) make it possible to sharpen the bands even further so that individual types of macroions appear as narrow lines on the gel; Figure 2A.3 shows an example of separation of DNA fragments by this method. The *relative mobility* of each component is calculated from the distance it has moved relative to the tracking dye.

Principles of Separation in Gel Electrophoresis

When electrophoresis is carried out in a gel or other supporting medium, the mobility is lower than would be expected from Equation 2A.1 because the gel or other matrix exhibits a molecular sieving effect. This can be seen by graphing mobility as a function of the concentration of the gel-forming material (Figure 2A.4a). A graph of log μ versus weight-percent gel composition is usually linear; this is called a **Ferguson plot**. The limiting mobility approached as percent gel approaches zero is called the **free mobility**; it should be given (approximately) by

Equation 2A.1. The steepness of the Ferguson plot depends on the size and shape of the macroion, for it reflects the difficulty the macroion experiences in passing through the molecular mesh of the gel.

As a result of these several factors, different kinds of molecules can exhibit widely different behaviors in gel electrophoresis (Figure 2A.4b). However, certain simple cases are of great importance. Polyelectrolytes like DNA or polylysine have one unit charge on each residue, so each molecule has a charge (Ze) proportional to its molecular length. But the frictional coefficient (f) also increases with molecular length, so to a first approximation, a macroion whose charge is proportional to its length has a *free* mobility almost independent of its size. In a mixture of such molecules, the molecular sieving effect determines the relative mobilities at any given gel concentration (Figure 2A.4c), and the

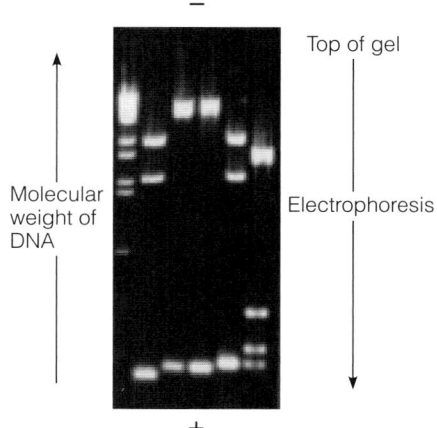

Molecular weight of DNA

Top of gel

Electrophoresis

FIGURE 2A.3

Gel showing separation of DNA fragments. Following electrophoretic separation of the different length DNA molecules, the gel is mixed with ethidium bromide, a fluorescent dye that binds DNA (page 1078). The unbound dye is then washed off and the stained DNA molecules are visualized under ultraviolet light.

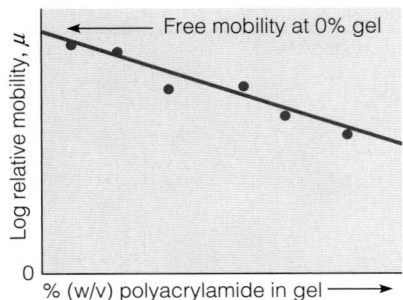

(a) Ferguson plot: mobility of a single type of molecule

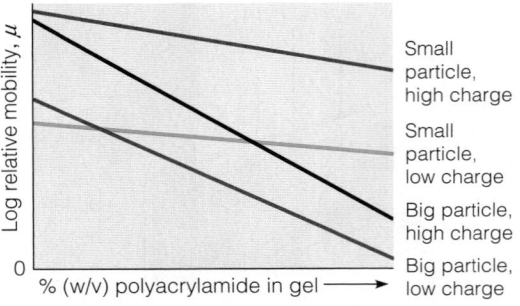

(b) Representative Ferguson plot for different kinds of molecules

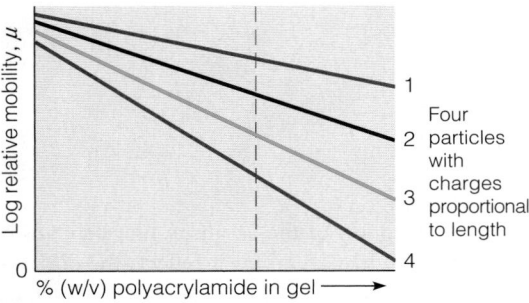

(c) Ferguson plot observed when charge is proportional to length

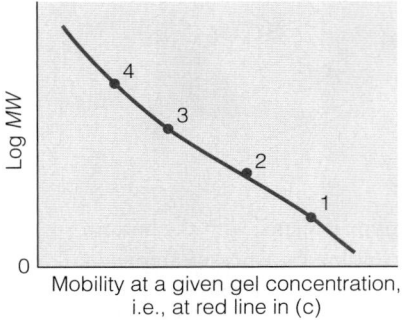

(d) Relationship between molecular weight (*MW*) and mobility, at a given gel concentration for molecules like those shown in (c)

FIGURE 2A.4

Mobility of the particles in gel electrophoresis. The mobility of the particles varies with the concentration of the gel. A Ferguson plot graphs the log of the relative mobility (μ) against the percent gel in the matrix. **(a)** Ferguson plot for a single type of molecule. Extending the plot to 0% gel gives the theoretical free mobility of the molecule. **(b)** Ferguson plot for four molecules of different size and charge. Note that free mobilities depend more on charge than on size but the slope depends mainly on size. **(c)** Ferguson plot for molecules with charges proportional to their lengths. The molecules are numbered in order of increasing length and charge. The free mobilities of such molecules are almost the same, but the longer the molecule, the more it is slowed by increasing gel concentration. **(d)** Plot of the relationship between molecular weight and mobility. The log of the molecular weight (*MW*) of the four molecules shown in (c) is plotted against their mobilities at a single gel concentration. When this type of graph is prepared from standards, it can be used to determine the molecular weights of separated molecules.

sieving effect is proportional to molecular length or molecular weight. This means that we can neatly separate molecules of this kind on the basis of *size alone* by gel electrophoresis, as shown in Figure 2A.3. For extended molecules like nucleic acids, the relative mobility is often approximately a linear function of the logarithm of the molecular weight (Figure 2A.4d). Usually, standards of known molecular weight are electrophoresed in one or more lanes on the gel. The molecular weight is then read from a graph like that in Figure 2A.4d prepared from the standards. For proteins, a similar sieving effect is achieved by coating the denatured molecule with the anionic detergent sodium dodecylsulfate (SDS) before electrophoresis. This important technique is discussed further in Tools of Biochemistry 6B.

Isoelectric Focusing

Yet another gel electrophoresis technique allows separation of molecules purely on the basis of their charge characteristics. A polyampholyte will migrate in an electric field like other ions if it has a net positive or negative charge. At its isoelectric point, however, its net charge is zero, and it is attracted to neither the anode nor the cathode. If we use a gel with a stable pH gradient covering a wide pH range, each polyampholyte molecule migrates to the position of its isoelectric point and accumulates there. We can

establish such a gradient by using a mixture of low-molecular-weight ampholytes as the gel buffer. This method of separation, called isoelectric focusing, produces distinct bands of accumulated polyampholytes and can resolve molecules with very small differences in isoelectric point (Figure 2A.5). Because the pH of each portion of the gel is known, isoelectric focusing can also be used to determine the isoelectric point of a particular polyampholyte.

What we have presented here is only a brief overview of a widely applied technique. For further information on gel electrophoresis, consult the references.

Capillary Electrophoresis

The term "capillary electrophoresis" (CE) broadly defines several related techniques that achieve high efficiency separations of ions, including macroions. The techniques all employ narrow-bore capillaries (20-200 μm inner diameter) in which the ions are separated as they migrate from one buffer chamber, past a detector, to a destination buffer chamber (see Figure 2A.6). The equipment required to perform CE is more expensive than, and therefore not as accessible as, that required for gel electrophoresis. In spite of this, for analysis of biomolecules, CE offers many advantages over traditional electrophoresis. The requirement to use low potentials (e.g., 25 V/cm) to avoid heat damage to the sample is a

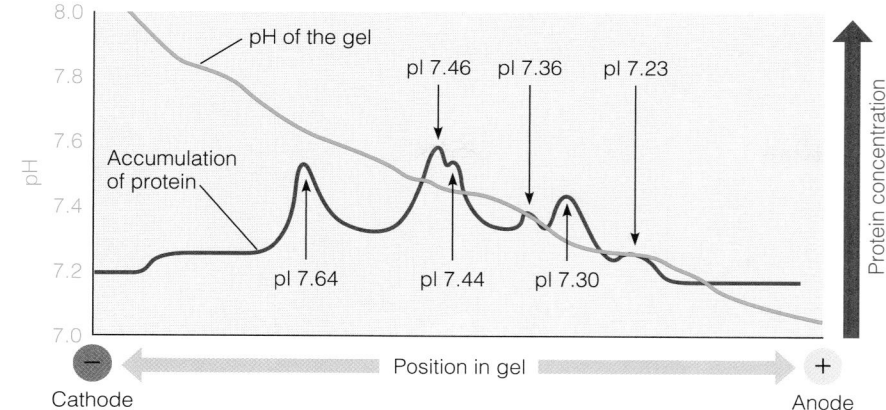

FIGURE 2A.5

Isoelectric focusing of polyampholytes. A mixture of variants of the polyampholyte hemoglobin is placed on a gel with a pH gradient. When an electric field is applied, each variant protein migrates to its own isoelectric point.

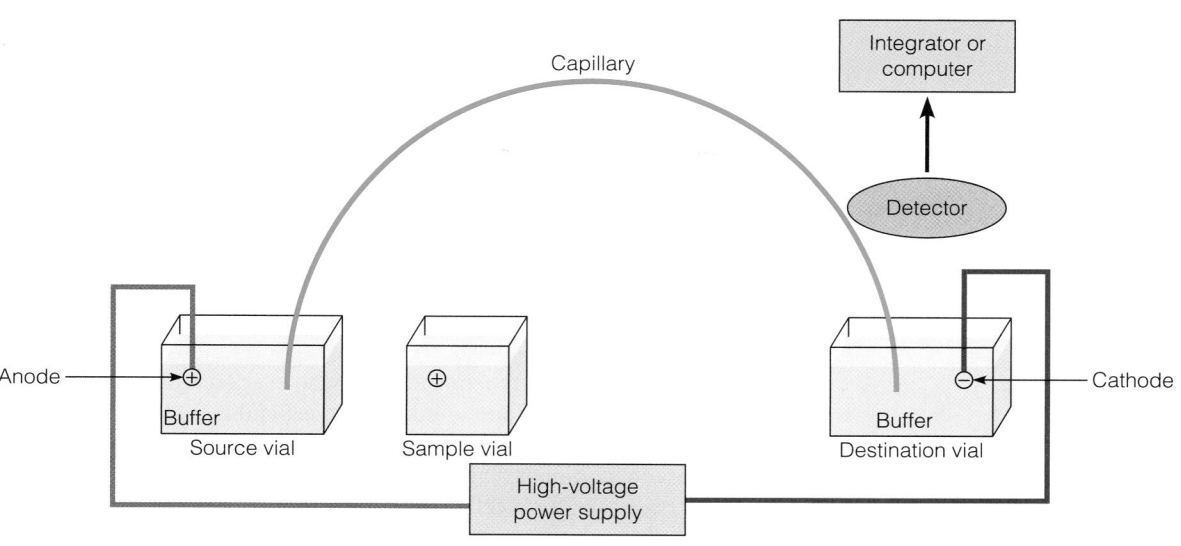

FIGURE 2A.6

A simple capillary electrophoresis experiment. Sample is introduced into the capillary from the sample vial. The capillary end with the sample is then placed in the source vial and high voltage is applied. The sample components will separate as they migrate toward the destination vial.

major limitation to the efficiency of separations achieved by the gel electrophoresis methods described above. Because CE capillaries have a large surface area to volume ratio, heat is efficiently dissipated and much higher potentials can be used (e.g., 500 V/cm). The efficiency of separations is dramatically increased. Analysis times (typically 5–30 minutes) are short, only 10–100 nL of sample is required, and detection is immediate and easily automated. For these reasons, CE has been used as an analytical technique in genomics studies to achieve rapid sequence analysis of large DNA molecules (discussed further in Chapter 4).

The simplest CE experiments use fused silica capillaries filled with an appropriate buffer and the cathode as the destination electrode (see Figure 2A.6). Capillaries with different coatings, as well as gel-filled capillaries, allow for more sophisticated experiments, including isoelectric focusing and SDS polyacrylamide gel CE.

References

Hames, B. D., and D. Rickwood, eds. (1981) *Gel Electrophoresis of Proteins.* IRL Press, Oxford, Washington, D.C.; and Rickwood, D., and B. D. Hames, eds. (1982) *Gel Electrophoresis of Nucleic Acids.* IRL Press, Oxford, Washington, D.C. These two volumes are extremely useful laboratory manuals for gel electrophoresis techniques.

Osterman, L. A. (1984) *Methods of Protein and Nucleic Acids Research,* Vol. 1, Parts 1 and 2. Springer-Verlag, New York. A comprehensive summary of electrophoresis and isoelectric focusing.

Schmitt-Kopplin, P., ed. (2008) *Capillary Electrophoresis: Methods and Protocols.* Humana Press, Totowa, N.J. An introduction to many methods of CE analysis, geared toward newcomers to the field.

van Holde, K. E., W. C. Johnson, and P. S. Ho (2006) *Principles of Physical Biochemistry* (2nd ed.). Prentice Hall, Upper Saddle River, N.J. Chapter 5 contains a more detailed discussion than given here.

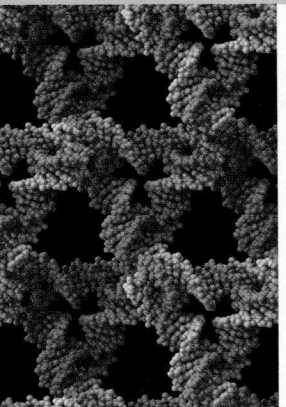

CHAPTER 3

The Energetics of Life

A living cell is a dynamic structure. It grows, it moves, it synthesizes complex macromolecules, and it selectively shuttles substances in and out and between compartments. All of this activity requires energy, so every cell and every organism must obtain energy from its surroundings and expend it as efficiently as possible. Plants gather radiant energy from sunlight; animals use the chemical energy stored in plants or other animals that they consume. The processing of this energy to do the things necessary for a cell or organism to maintain the living state is what much of biochemistry is about. Much of the elegant molecular machinery that exists in every cell is dedicated to this task.

Bioenergetics describes how organisms capture, transform, store, and utilize energy.

Because of the central role of energy in life, it is appropriate that we begin a study of biochemistry with an introduction to **bioenergetics**—the quantitative analysis of how organisms capture, transform, store, and utilize energy. Bioenergetics may be regarded as a special part of the general science of energy transformations, which is called **thermodynamics**. In this chapter we shall review just a bit of that field, focusing attention on fundamental concepts, such as enthalpy, entropy, and free energy, that are important to the biochemist or biologist.

We will introduce in this chapter the basic approaches for determining changes in free energy in biochemical systems. In subsequent chapters these basic approaches will be discussed further in the context of processes such as protein folding, transport of ions across membranes, and extraction of chemical energy from nutrients to make ATP.

Energy, Heat, and Work

A word we shall often use in our discussion is **system**. In this context, a system is any part of the universe that we choose for study. It can be a single bacterial cell, a Petri dish containing nutrient and millions of cells, the whole laboratory in which this dish rests, the earth, or the entire universe. A system must have defined boundaries, but otherwise there are few restrictions. Anything not defined as part of the system is considered to be the **surroundings**. The system may be *isolated,* and thus unable to exchange energy and matter with its surroundings; it may be *closed,* able to exchange energy but not matter; or it may be *open,* so that both energy and matter can pass between the system and surroundings. For example, our planet displays the essential features of a closed system: Earth can exchange energy (e.g., in the form of electromagnetic radiation) with its surroundings, but except for a few bits of metal (spacecraft and satellites) and some geological samples (meteorites and lunar rocks), material is not exchanged between the planet and its surroundings.

Internal Energy and the State of a System

Any system contains a certain amount of **internal energy**, which we denote by **U**. It is important for our understanding to be specific as to what this internal energy includes. The system's atoms and molecules have energy of vibration and rotation, and kinetic energy of motion. In addition, we include all of the energy stored in the chemical bonds between atoms and the energy of noncovalent interactions between molecules. We also include any kind of energy that might be changed by chemical or nonnuclear physical processes. We need not include energy stored in the atomic nucleus, for this is unchanged in any chemical or biochemical reaction. The internal energy is a function of the **state** of a system. The thermodynamic state is defined by describing the amounts of all substances present and any two of the following three variables: the temperature (T), the pressure on the system (P), or the volume of the system (V). It is essentially a recipe for producing the system in a defined way. For example, a system composed of 1 mole of O_2 gas in 1 liter at 273 K has a defined state and therefore a definite internal energy value. This value is independent of any past history of the system.

> The internal energy of a system includes all forms of energy that can be exchanged via simple (nonnuclear) physical processes or chemical reactions.

Unless a system is isolated, it can exchange energy with its surroundings and thereby change its internal energy; we define this change as ΔU. For a closed system, this exchange can happen in only two ways. First, **heat** may be transferred to or from the system. Second, the system may do **work** on its surroundings or have work done on it. Work can take many forms. It may include expansion of the system against an external pressure such as expansion of the lungs, electrical work such as that done by a battery or required for the pumping of ions across a membrane, expansion of a surface against surface tension, flexing of a flagellum to propel a protozoan, or lifting of a weight by contraction of a muscle. In all of these examples, a force is exerted against a resistance to produce a displacement, so work is done.

Note that heat and work are not *properties* of the system. They may be thought of as "energy in transit" between the system and its surroundings. Certain conventions have been adopted to describe these ways of exchanging energy:

1. We denote heat by the symbol q. A positive value of q indicates that heat is absorbed by the system from its surroundings. A negative value means that heat flows from the system to its surroundings.

2. We denote work by the symbol w. A positive value of w indicates that work is done by the system on its surroundings. A negative value means that the surroundings do work on the system.

All of this may seem excessively abstract, yet it bears the most direct relationship to the everyday functioning of our bodies. When we ingest a nutrient like glucose, we metabolize it, ultimately oxidizing it to CO_2 and water. A defined energy change (ΔU) is associated with oxidizing a gram of glucose, and some of the released energy is available for our use. We expend a significant portion of this energy as heat (a by-product of mitochondrial metabolism that allows insulated birds and mammals to maintain their normal body temperatures) and the remainder performing various kinds of work. These latter kinds of work include not only the obvious ones, like walking and breathing, but other more subtle kinds—sending impulses along nerves, pumping ions across membranes, and so forth.

Energy is conserved. According to the first law of thermodynamics, in a closed system, the internal energy (U) can change only by the exchange of heat or work with the surroundings; however, energy can be converted from one form to another.

FIGURE 3.1

Exchange of heat and work in constant-volume and constant-pressure reactions. A single reaction, oxidation of 1 mole of a fatty acid, is carried out under two sets of conditions. **(a)** The reaction occurs in a sealed vessel, or "bomb." Heat (q) is transferred to the surrounding water bath and is measured by the small increase in temperature of the water. No work is done because the system is at constant volume. **(b)** The reaction vessel is fitted with a piston held at 1 atm pressure. During the reaction, heating of the gas in the vessel causes the piston to be pushed up. However, the reaction results in a decrease in the number of moles of gas, so after the vessel and gas have cooled to the water temperature, the volume of the gas is smaller than the initial volume. Thus, net work is done on the system, and the total amount of heat delivered to the bath is slightly more than in (a).

The First Law of Thermodynamics

Because the internal energy of a closed system can change only by heat or work exchanges with the surroundings, the *change in internal energy* must be given by

$$\Delta U = q - w \tag{3.1}$$

This equation, which holds true for all processes, expresses the **first law of thermodynamics**. This first law is simply a bookkeeping rule, a statement of the conservation of energy. When a physical process or a chemical reaction has occurred, we can total up the incomes and expenditures of energy, and the books must balance. Energy can be gained and released in different ways, but at least in chemical processes, it can neither be created nor destroyed. Consider, for example, some process in which a certain amount of heat is absorbed by a system, while the system does an exactly equivalent amount of work on its surroundings. In this case, both q and w are positive, and $q = w$, so $\Delta U = 0$. This agrees with common sense: If some amount of energy went in as heat and an equal quantity came out as work, then the energy within the system must be unchanged.

Changes in internal energy, as for any function of state, depend only on the initial and final states of a system and are independent of the path; however, the amounts of heat and work exchanged in any process depend very much on the pathway taken between the initial and final states. To make this idea concrete, let us consider a specific chemical reaction—the complete oxidation of 1 mole of a fatty acid, palmitic acid:

$$CH_3(CH_2)_{14}COOH \text{ (solid)} + 23O_2 \text{ (gas)} \rightarrow 16CO_2 \text{ (gas)} + 16H_2O \text{ (liquid)}$$

The oxidation of palmitic acid is an important biochemical reaction that takes place, in a much more indirect way, in our bodies when we metabolize fats. We shall consider running this reaction in two different ways, as shown in Figure 3.1.

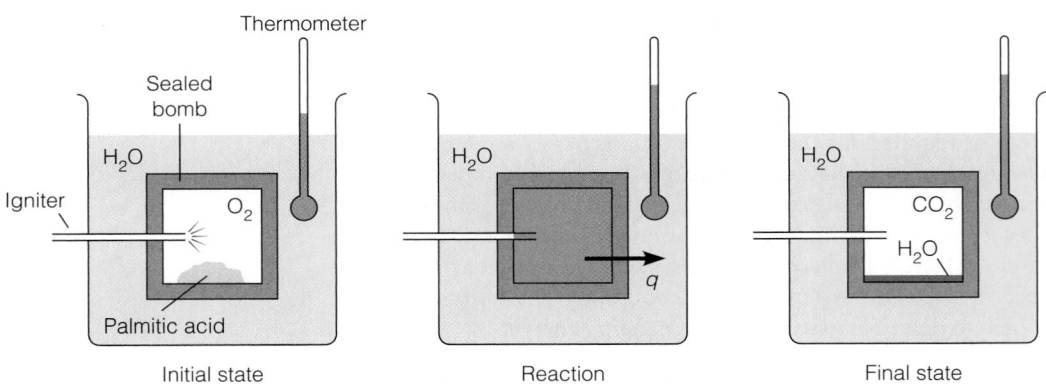

(a) Reaction at constant volume

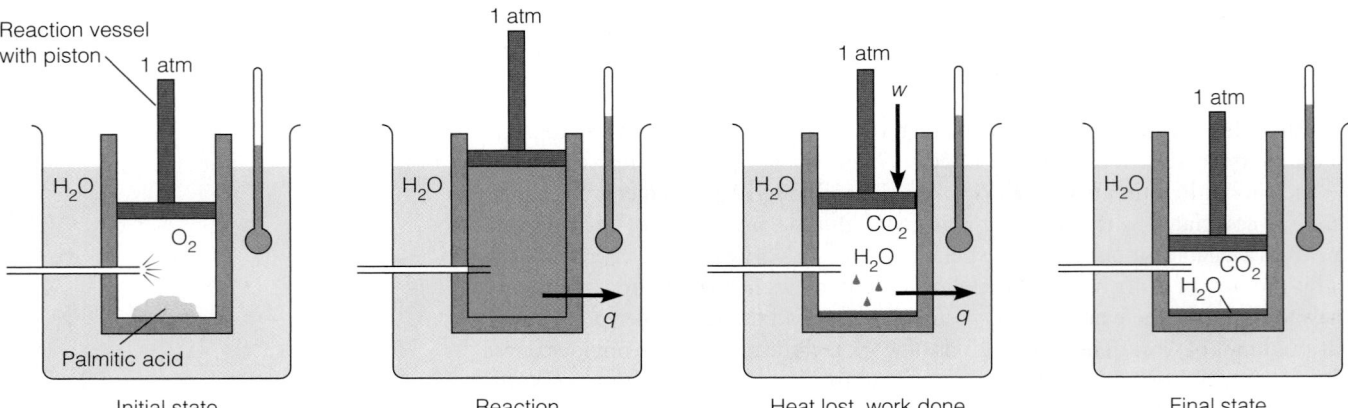

(b) Reaction at constant pressure

In Figure 3.1a the reaction is carried out by igniting the mixture in a sealed vessel (a "bomb" calorimeter) immersed in a water bath. The reaction, under these conditions, is being carried out at constant volume. We can measure the heat passed from the reaction vessel (the system) to the water bath (the surroundings) by the temperature change in the bath, knowing the mass of water and the heat capacity (per gram) of water. Because the reaction vessel has a fixed volume, no work has been done against the surroundings or by the surroundings; therefore, $w = 0$, and from equation (3.1),

$$\Delta U = q \qquad (3.2)$$

The total heat that is transferred from the reaction vessel to the surroundings equals the change in internal energy, and that energy change results mainly from the changes in chemical bonding that occurred during the reaction. The presently accepted unit for heat, work, and energy is the **joule (J)**.* For the above reaction, the value observed for ΔU is -9941.4 kJ/mol. The negative sign indicates that the reaction *releases* energy stored in chemical bonds. The energy within the system *decreased* as this bond energy was transferred as heat to the surroundings.

> The heat evolved in a reaction at constant volume is equal to ΔU.

Now suppose the same reaction is carried out at a constant pressure of 1 atmosphere, as shown in Figure 3.1b. In this case, the system is free to either expand or contract, and it finally contracts by an amount proportional to the decrease in the number of moles of gas, which went from 23 to 16 moles during the reaction. (We neglect the relatively tiny volume of solids and liquids.) The decrease in gas volume means that a certain amount of work has been done by the surroundings on the system. This can be calculated in the following way.

When volume (V) is changed against a constant pressure (P),

$$w = P \, \Delta V \qquad (3.3)$$

To calculate w, we may make an approximation. We assume that the initial and final temperatures of the system are essentially the same (say 25 °C, or 298 K) and that the gases are "ideal." We may then use the ideal gas law, $PV = nRT$. This gives

$$\Delta V = \Delta n \frac{RT}{P} \qquad (3.4)$$

where R is the gas constant, T the absolute temperature in kelvin, and Δn the change in number of moles of gas per mole of palmitic acid oxidized. Then, inserting (3.4) in (3.3), we obtain

$$w = \Delta n \, RT \qquad (3.5)$$

Because we would like w in joules per mole, we use $R = 8.315$ J/K·mol in Equation 3.5, resulting in $w = -17,300$ J/mol (or -17.3 kJ/mol) palmitate.

The heat evolved in this constant-pressure combustion will then be

$$q = \Delta U + w = \Delta U + P \, \Delta V = \Delta U + \Delta n \, RT$$
$$= (-9941.4 \text{ kJ/mol}) + (-17.3 \text{ kJ/mol}) = -9958.7 \text{ kJ/mol} \qquad (3.6)$$

A slightly greater amount of heat is released to the surroundings under these *constant-pressure* conditions than under the *constant-volume* conditions of

---

*In the past, biochemists tended to express energy, heat, and work in *calories* or *kilocalories*. However, the International System of Units (SI units) *joules* and *kilojoules* are now replacing these. To convert these units: 1 cal = 4.184 J. Similarly, 1 kcal (kilocalorie, or 10^3 calories) = 4.184 kJ (kilojoules). A complication arises from the fact that the "calorie" ("C") referred to in dietetics is really a *kilocalorie*.

Figure 3.1a. Under constant-pressure conditions the surroundings can do work on the system and this work (called *PV* work) reappears as extra heat released from the system to the surroundings (which is required to keep the temperature of the system at a constant value of 298 K).

Although the heat and work exchanged with the surroundings depend on path, it is important to remember that ΔU does not—it depends only on the initial and final states.

Enthalpy

Most chemical reactions in the laboratory and virtually all biochemical processes occur under conditions more nearly approximating constant pressure than constant volume. If we are interested in the heat obtainable by oxidizing palmitic acid in an animal, then the heat evolved at constant pressure is what we want to know. As we showed in Equation 3.6, this heat is not exactly equal to U because of the PV work done; thus, to express the heat change in a constant-pressure reaction, we need another function of state. We define a new quantity, the **enthalpy**, which we give the symbol *H*:

$$H = U + PV \qquad (3.7)$$

Because U and PV are functions of state, H is also a function of state. The change ΔH depends only on the initial and final states of the process for which it is calculated. For reactions at constant pressure, ΔH is defined as follows:

$$\Delta H = \Delta U + P\Delta V \qquad (3.8)$$

The value of ΔH is the same as the amount of heat (q) calculated in Equation 3.6. In other words, *when the heat of a reaction is measured at constant pressure, it is ΔH that is determined.*

The energy changes you will find tabulated throughout this book and other books on biochemistry will almost always be given as ΔH values. That is most appropriate, for in vivo these reactions occur under nearly constant-pressure conditions. If a nutritionist wishes to know the energy available from the oxidation of palmitic acid in the body, ΔH is the appropriate quantity.

Measuring changes in energy such as ΔU and ΔH in a calorimeter is of practical use to biochemists and dieticians even though the oxidation of a substance like palmitic acid occurs very differently in the human body than it does in a reaction vessel like that shown in Figure 3.1. The values of ΔU and ΔH for the oxidation of palmitic acid are *exactly* the same in both pathways because a quantity like ΔU or ΔH depends only on the final and initial states. Thus, the calorimeter provides an exact measurement of the energy available to a human from each gram of palmitic acid oxidized completely to CO_2 and H_2O.

The average human requires the expenditure of about 6000 kJ per day (roughly 1500 kcal or 1500 of the "calories" used in dietetics) just to sustain basal metabolic rates. With moderate exercise, this need for metabolic energy may easily double.

Although we have pointed out the distinction between ΔU and ΔH, we should emphasize that for most biochemical reactions the quantitative difference between them is of little consequence. Most of these reactions occur in solution and do not involve the consumption or formation of gases. The volume changes are thus exceedingly small, and $P\Delta V$ is a tiny quantity relative to ΔU or ΔH. Even for the example given, the oxidation of palmitic acid, the difference between ΔH and ΔU is only 0.2%. Thus, we are justified in most cases in thinking of ΔH as a direct measure of the energy change in a process, and we commonly refer to ΔH as the energy change.

The heat evolved in a reaction at constant pressure is equal to the change in enthalpy, ΔH.

The enthalpy change in a reaction is the energy change of most interest to biochemists.

Entropy and the Second Law of Thermodynamics

The Direction of Processes

However useful the first law may be for keeping track of energy changes in processes, it cannot give us one very important piece of information: What is the *favored direction* for a process? The first law cannot answer questions like these:

> We place an ice cube in a glass of water at room temperature. It melts. Why doesn't the rest of the water freeze instead?
>
> We place an ice cube in a jar of supercooled water. All the water freezes. Why?
>
> We touch a lit match to a piece of paper. The paper burns to carbon dioxide and water. Why can't we mix carbon dioxide and water to form paper?

One characteristic of such processes is their *irreversibility* under the conditions described above. An ice cube in a glass of room-temperature water at 1 atm will continue to melt—there is no way to turn that process around without major changes in the conditions. But there is a *reversible* way to melt ice—to have it in contact with water at 0 °C and 1 atm. Under these conditions, adding a bit of heat to the glass will result in a small amount of ice melting, whereas removing a little heat will cause a small amount of the water to freeze. A **reversible** process like melting ice at 0 °C is always near a state of **equilibrium**. The defining feature of the equilibrium state is that it is the lowest energy state for a system. As discussed below, lower energy states are favored over those of higher energy; thus, systems tend to adopt states of lower energy. The **irreversible** processes we just described happen when systems are set up far from an equilibrium state. They then drive *toward* a state of equilibrium.

In the jargon of thermodynamics, an irreversible process is often called a "spontaneous" process, but we prefer the word *favorable*. The word *spontaneous* tends to imply, perhaps falsely, that the process is rapid. Thermodynamics has nothing to say about how fast processes will be (this is described by "kinetics"—see Chapter 11), but it does indicate which direction for a process is favored. The *melting* of ice, rather than *freezing*, is favored at 25 °C and 1 atm. Here, the result is intuitive; you would not expect the ice cube to grow, or even remain unmelted, when placed in 25 °C water.

Knowing whether a process is reversible, favorable, or unfavorable is vital to bioenergetics. This information can be expressed most succinctly by the *second law of thermodynamics*, which tells us which processes are thermodynamically favorable. To present the second law, we must consider a new concept—entropy.

Entropy

Why do chemical and physical processes have thermodynamically favored directions? A first guess at an explanation might be that systems simply go toward a lowest-energy state. Water runs downhill, losing energy as it spontaneously falls in the earth's gravitational field; the oxidation of palmitic acid, like the burning of paper, releases energy as heat. Certainly, energy minimization is the key to the favored direction for *some* processes. But such an explanation cannot account for the melting of ice at 25 °C; in fact, energy is *absorbed* in that process. Another, very different factor must be at work, and a simple experiment gives a clear indication of what this factor may be. If we carefully layer pure water on top of a sucrose solution, we will observe as time passes that the solution becomes more and more uniform (Figure 3.2). Eventually the sucrose molecules will be evenly distributed throughout the solution. Though there is practically no energy change, in terms of heat and work, the process is clearly a

Reversible processes occur always near a state of equilibrium; irreversible processes drive toward equilibrium.

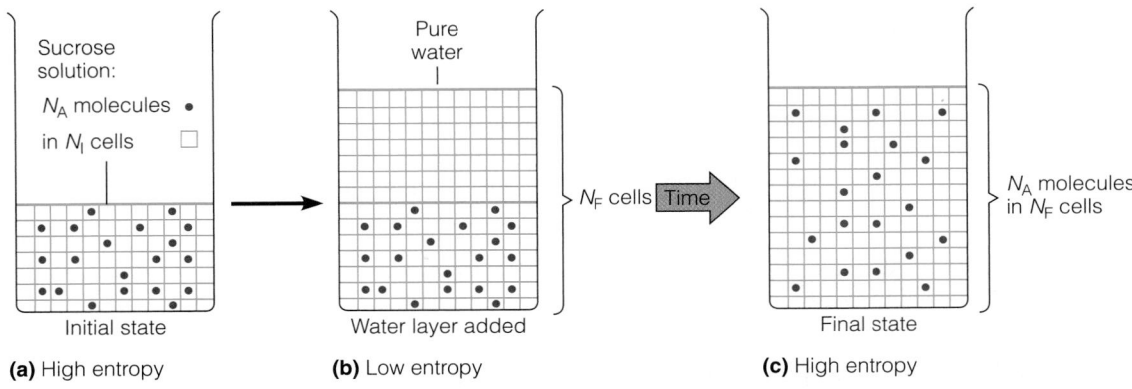

(a) High entropy   **(b)** Low entropy   **(c)** High entropy

FIGURE 3.2

Diffusion as an entropy-driven process. The gradual mixing of a dilute sucrose solution and pure water is the result of random movement of their molecules. We can visualize the increase in entropy if we imagine the volume of the two liquids to be made up of cells, each big enough to hold one sucrose molecule. **(a)** Initially, the sucrose solution is at equilibrium because its N_A molecules are distributed randomly into its N_I cells. **(b)** When a layer of pure water is added without mixing, the system is no longer at equilibrium. It has become more ordered, with all the occupied cells located in one-half of the solution. **(c)** As sucrose and water molecules continue to move randomly, their arrangement becomes less ordered because every cell has an equal chance of being occupied. Eventually, the solution reaches a new equilibrium, with sucrose molecules randomly distributed throughout. The drive toward equilibrium is a consequence of the tendency for entropy to increase. A system would never go spontaneously from state (c) to state (b).

favorable one. We know from experience that the opposite process (self-segregation of the sucrose molecules into a portion of the solution volume) never occurs. What is clearly important here is that *systems of molecules have a natural tendency to randomization*.

The degree of randomness or disorder of a system is measured by a function of state called the **entropy (S)**. There are several ways of defining entropy, but the most useful for our applications depends on the fact that a given thermodynamic state may have many *substates* of equal energy. Those substates correspond, for example, to different ways in which molecules can be arranged or distributed within the system (see Figure 3.2). If the thermodynamic state has a number (W) of substates of equal energy, the entropy (S) is defined as

$$S = k_B \ln W \qquad (3.9)$$

where k_B is the **Boltzmann constant**, the gas constant R divided by Avogadro's number. A consequence of this definition is that entropy is seen to be a measure of disorder. There will always be many more ways of putting a large number of molecules into a disorderly arrangement than into an orderly one; therefore, the entropy of an ordered state is lower than that of a disordered state of the same system. In fact, the minimal value of entropy (zero) is predicted *only* for a perfect crystal at the absolute zero of temperature (0 K or −273.15 °C). The process of diffusion evens out the concentrations in our sucrose solution simply because there are more ways to distribute molecules over a large volume than over a small one. In Figure 3.2 the state shown in panel (c) has more substates of equal energy (i.e., different random arrangements of the sucrose molecules in the cells) than does the state shown in panel (b). We could also say that state (c) has more **degrees of freedom** than state (b); thus, the value of W is greater for state (c). Rotations around bonds increase the degrees of freedom available to a molecule. We will consider the entropy associated with bond rotations when we examine the folding of proteins (Chapter 6). To make the concept of entropy a bit more familiar, consider the examples given in Table 3.1.

Entropy is a measure of the randomness or disorder in a system.

TABLE 3.1	Examples of low-entropy and high-entropy states
Low Entropy	**High Entropy**
Ice, at 0 °C	Water, at 0 °C
A diamond, at 0 K	Carbon vapor, at 1,000,000 K
A protein molecule in its regular, native structure	The same protein molecule in an unfolded, random coil state
A Shakespearean sonnet	A random string of letters
A bank manager's desk	A professor's desk

The Second Law of Thermodynamics

The preceding example shows that the driving force toward equilibrium for the sucrose solution in Figure 3.2 is just the increase in entropy. This observation can be generalized as the **second law of thermodynamics**. *The entropy of an isolated* *system will tend to increase to a maximum value.* The entropy of such a system will not decrease—sucrose will never "de-diffuse" into a corner of the solution. This simply reflects our commonsense understanding that things, if left alone, will not get more orderly.

Free Energy: The Second Law in Open Systems

The form of the second law as stated in the previous section is not very useful to biologists or biochemists because we never deal with isolated systems. Every biological system (cell, organism, or population, for example) can exchange energy and matter with its environment. Because living systems can exchange energy with their surroundings, both energy *and* entropy changes will take place in many reactions, and both must be of importance in determining the direction of thermodynamically favorable processes. For such systems we need *a function of state that includes both energy and entropy.* There are several such functions, but the one of importance in biochemistry is the **Gibbs free energy (G)** or, as we shall call it, the **free energy**. This function of state combines an enthalpy term, which measures the energy change at constant pressure, and an entropy term, which takes into account the importance of randomization. The Gibbs free energy is defined as

$$G = H - TS \qquad (3.10)$$

where T is the absolute temperature measured in kelvin. For a free energy change ΔG in a system at constant temperature and pressure, we can write

$$\Delta G = \Delta H - T\Delta S \qquad (3.11)$$

Why is the quantity ΔG called *free* energy? The reason is that ΔG represents the portion of an energy change (ΔH) that is *available*, or *free*, to do **useful work** (that is, work beyond the $P\Delta V$ work of expansion). We can gain an insight into the meaning of free energy by considering the factors that make a process favorable. We said that a decrease in energy (ΔH is negative) and/or an increase in entropy (ΔS is positive) are typical of favorable processes. Either of these conditions will tend to make ΔG negative. In fact, another way to state the second law of thermodynamics—the most important way for our purposes—is this: *The criterion for a favorable process in a nonisolated system, at constant temperature and pressure, is that ΔG be negative.* Conversely, a positive ΔG means that a process is *not favorable*; rather, the *reverse* of that process is favorable.

The importance of quantitating free energy lies in this *predictive* power of ΔG. Given a set of conditions for a process (e.g., temperature, concentrations of reactants and products, pH, etc.), we can calculate the value of ΔG and thereby determine whether or not the process is favorable. Processes accompanied by negative free energy changes are said to be **exergonic**; those for which ΔG is positive are **endergonic**.

Suppose that the ΔH and $T\Delta S$ terms in the free energy equation just balance one another. In this case $\Delta G = 0$, and the process is not favored to go either forward or backward. In fact, the system is at equilibrium. Under these conditions, the process is reversible; that is, it can be displaced in either direction by an infinitesimal push one way or the other. These simple but important rules about free energy changes are summarized in Table 3.2.

An Example of the Interplay of Enthalpy and Entropy: The Transition Between Liquid Water and Ice

To make these ideas more concrete, let us consider in detail a process we mentioned before, the transition between liquid water and ice. This familiar example demonstrates the interplay of enthalpy and entropy in determining the state of a system. In an ice crystal there is a maximum number of hydrogen bonds between the water

The second law of thermodynamics states that the entropy of an isolated system will tend to increase to a maximum value.

The free energy change for a process at constant temperature and pressure is $\Delta G = \Delta H - T\Delta S$.

A thermodynamically favored process tends in the direction that minimizes free energy (results in a negative ΔG); this is one way of stating the second law of thermodynamics.

TABLE 3.2 Free energy rules

If ΔG is …	The process is …
Negative	Thermodynamically favorable
Zero	Reversible; at equilibrium
Positive	Thermodynamically unfavorable; the reverse process is favorable

molecules (see Figure 2.10b in Chapter 2). When ice melts, some of these bonds must be broken. The enthalpy difference between ice and water corresponds almost entirely to the energy that must be put into the system to break these hydrogen bonds. As Figure 3.3 shows, the enthalpy change for the transition ice ⟶ water is positive, as would be expected from the preceding argument.

The entropy change in melting arises primarily because liquid water is a more disordered structure than ice. In an ice crystal, each water molecule has a fixed place in the lattice and binds to its neighbor in the same way as every other water molecule. On the other hand, molecules in liquid water are continually moving, exchanging hydrogen-bond partners as they go (compare Figures 2.10b and 2.10c). Figure 3.3 shows that the entropy change for melting ice to liquid water is a positive quantity, due to the increase in randomness. If we calculate the free energy change ($\Delta G = \Delta H - T\Delta S$) for the ice ⟶ water transition, we find the following: At low temperatures, ΔH dominates and ΔG is positive. For example, at 263 K (−10 °C) we find $\Delta G = +213$ J/mol. This means that the transition ice ⟶ water is *not* favorable under these conditions. The opposite transition (water ⟶ ice) *is* favorable and therefore irreversible, as we can see from the behavior of supercooled water. If we disturb supercooled water, or add a minute ice crystal to start the process of freezing, the entire sample will freeze. No infinitesimal change we can make will reverse this process.

At a temperature above the melting point of ice, say 283 K (+10 °C), an ice cube will irreversibly melt. Again, we would predict this outcome from the data, for if we calculate ΔG at 283 K, we obtain $\Delta G = -225$ J/mol. The sign of ΔG is now negative because the $T\Delta S$ term dominates when T becomes large enough. At 283 K the process (ice ⟶ water) is favorable and irreversible.

At the melting temperature, 273 K, the ΔH and $T\Delta S$ terms exactly balance and $\Delta G = 0$. A zero value for ΔG is the condition for equilibrium, and we know that ice and liquid water are in equilibrium at 273 K (0 °C). The change is now reversible; when ice and liquid water are together at 273 K, we can melt a bit more ice by adding an infinitesimal amount of heat. Alternatively, we can take a minute amount of heat away from the system and freeze a bit more water. The melting point of a substance is simply the temperature at which the curves for ΔH and $T\Delta S$

At the melting point of any substance, enthalpy and entropy contributions to ΔG balance, and $\Delta G = 0$.

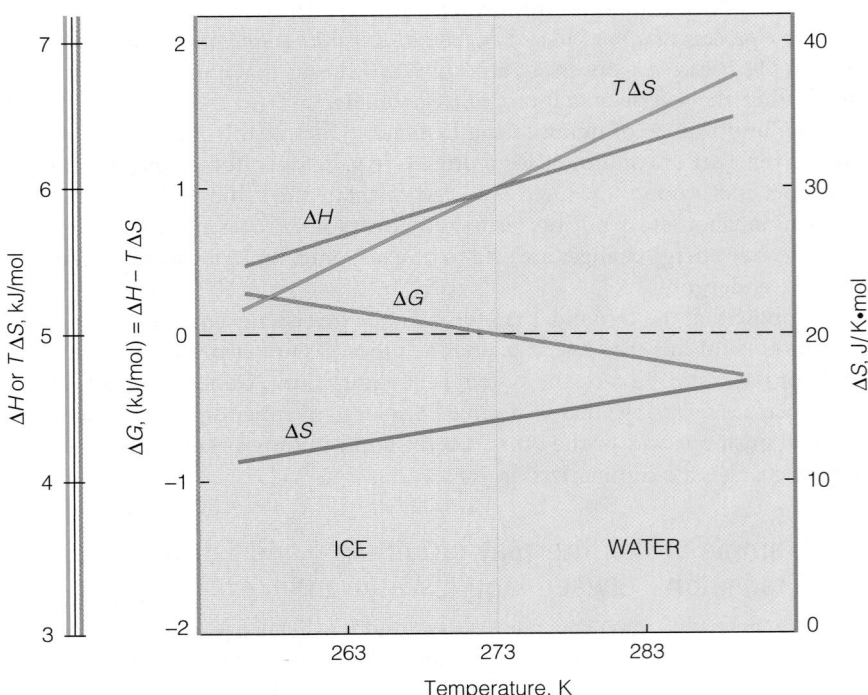

FIGURE 3.3

Interplay of enthalpy and entropy in the ice-to-water transition. For the ice-to-water transition, ΔH and ΔS are both positive and approximately constant over a large temperature range. The increase in $T\Delta S$ with increasing temperature means that ΔG decreases from a positive to a negative value. At the melting temperature of ice, 273 K (0 °C), the ΔH and $T\Delta S$ curves intersect and ΔG is zero.

intersect; at this temperature, the energetically favored process of freezing is in balance with the entropically favored process of melting. Neither ΔH nor ΔS alone can tell us what will happen, but their combination, expressed as $\Delta H - T\Delta S$, defines exactly which form of water is stable at any temperature.

The Interplay of Enthalpy and Entropy: A Summary

For all chemical and physical processes it is the competition between enthalpy and entropy terms that determines the favorable direction. As Figure 3.4 shows, in some processes the enthalpy change dominates; in others, the entropy change is more important. Furthermore, because ΔS is multiplied by T in Equation 3.11, the favorable direction will depend on the temperature. We have seen one example, the melting of ice, but quite different scenarios are possible depending on the signs of ΔH and ΔS. Table 3.3 lists the possibilities. Note that when ΔH is negative and ΔS is positive, ΔG must always be negative, so the reaction is favorable at all temperatures. The reverse is true when ΔH is positive and ΔS is negative; ΔG is always positive, and the reaction is not favorable at any temperature.

Two matters that frequently cause confusion should be cleared up at this point. First, we must emphasize a point we have mentioned already: The favorability of a process has nothing to do with its rate. Students frequently assume that favorable processes are rapid, but this is not necessarily so. A reaction may have a large negative free energy change but still proceed at a slow rate (for reasons discussed in Chapter 11). A surprising example of this situation is the simple reaction C (diamond) $\longrightarrow$ C (graphite). The free energy change for this transformation, at room temperature, is -2.88 kJ/mol; thus, relative to graphite, diamond is unstable. Yet the reaction is imperceptibly slow because it is very difficult for the rigid crystal lattice to change its form. A **catalyst** may increase the rate for some reactions, but the favored direction is always dictated by ΔG and is independent of whether or not the reaction is catalyzed. We will see in Chapter 11 that the protein catalysts called **enzymes** selectively increase the rates for specific *thermodynamically favorable* reactions.

Second, the entropy of an open system can *decrease*. We have just seen that this happens whenever water freezes. More important to biochemists is that decreases in entropy happen all the time in living organisms. An organism takes in nutrients, often in the form of disorganized small molecules, and from them it builds enormous, complex, highly ordered macromolecules like proteins and nucleic acids. From these macromolecules, it constructs elegantly structured cells, tissues, and organs. All of this activity involves a tremendous entropy decrease. The implication of Equation 3.11 is

Favorable processes are not necessarily rapid.

FIGURE 3.4

Contribution of enthalpy and entropy to several favorable processes. Each of these processes has a negative free energy change, but the change is accomplished in different ways. (Note that the arrows in the diagrams are not to scale.) **(a)** ΔH negative, $-T\Delta S$ negative (because ΔS is positive). When glucose is fermented to ethanol, enthalpy decreases and entropy increases, so both the enthalpy and the entropy changes favor this reaction. **(b)** ΔH negative, $-T\Delta S$ positive (because ΔS is negative). When ethanol burns, enthalpy and entropy both decrease. The negative ΔH favors this reaction, but the negative ΔS opposes it. **(c)** ΔH and ΔS both positive. When nitrogen pentoxide decomposes, enthalpy and entropy both increase. The positive ΔH opposes this reaction, but the positive ΔS favors it.

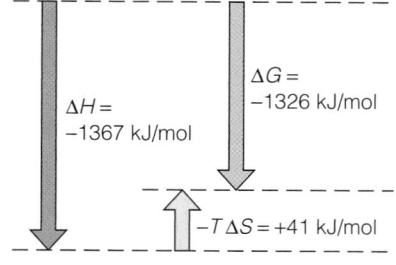

(a) Fermentation of glucose to ethanol

$$C_6H_{12}O_6(s) \rightarrow 2C_2H_5OH(l) + 2CO_2(g)$$

Both enthalpy and entropy changes favor the reaction.

(b) Combustion of ethanol

$$C_2H_5OH(l) + 3O_2(g) \rightarrow 2CO_2(g) + 3H_2O(l)$$

Enthalpy favors this reaction, but entropy opposes it. We could call this an "enthalpy-driven" reaction. If water *vapor* were the product, an entropy increase would favor the reaction as well.

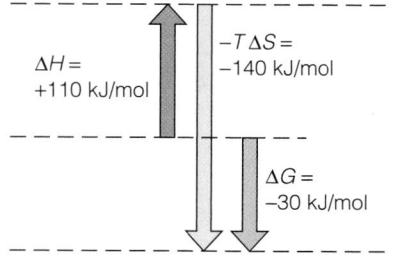

(c) Decomposition of nitrogen pentoxide

$$N_2O_5(s) \longrightarrow 2NO_2(g) + 1/2O_2(g)$$

This is a somewhat unusual chemical reaction in that it is "entropy-driven." The reaction actually absorbs heat but is favored by the large entropy increase resulting from the formation of gaseous products.

TABLE 3.3 The effect of temperature on ΔG for a process or reaction depends on the signs of ΔH and ΔS.

ΔH	ΔS	Low T	High T
+	+	ΔG positive; not favored	ΔG negative; favored
+	−	ΔG positive; not favored	ΔG positive; not favored
−	+	ΔG negative; favored	ΔG negative; favored
−	−	ΔG negative; favored	ΔG positive; not favored

that entropy can decrease in a favored process, but only if this change is accompanied by a large enthalpy decrease. *Energy must be expended to pay the price of organization.* This exchange really is what life is all about. Living organisms spend energy to overcome entropy. For these life processes to proceed, the *overall* free energy changes in the organism must be negative; thus, life is an irreversible process. An organism that comes to equilibrium with its surroundings is dead.

There is an even deeper philosophical implication of bioenergetics. The universe as a whole is an isolated system. *The entropy of the whole universe must be increasing.* It follows that each of us, as a living organism that locally and temporarily decreases entropy, must produce somewhere in the world around us an increase in entropy. As we metabolize food, for example, we give off heat and increase random molecular motion around us. In a sense, we sustain our lives through the entropic death of the universe.

We will revisit the interplay between enthalpy and entropy in Chapter 6, where protein folding and stability are discussed. The active forms of biomolecules, particularly proteins and RNAs, are generally associated with a well-defined and highly ordered structure. The process whereby a protein goes from an inactive, unstructured ("unfolded") state to its active, structured ("folded") state is of central importance in biochemistry and the biotechnology industry. The folding and stability of proteins are related and both are well-described by Equation 3.11, where ΔH is a measure of the change in bonding interactions and ΔS is a measure of the change in order of the system going from one state to the other.

Free Energy and Useful Work

Knowing that ΔG measures the maximum amount of *useful* work that can be obtained from a chemical process is of great importance to biochemistry because useful work includes, for example, the work involved in muscle contraction and cell motility, in transport of ions and molecules, in signal transmission, and in tissue growth.

Recall that ΔH is the total energy change in a reaction, including the $P\Delta V$ work. The equation $\Delta G = \Delta H - T\Delta S$ indicates that a part of ΔH is always dissipated as heat, as expressed by the $T\Delta S$ term, and is therefore not available to do other things. Whatever the process may be, at least the amount of energy represented by $T\Delta S$ must be unavailable. The remainder, ΔG, is potentially available for other needs, although how much of it is actually used for work depends on the path of the process. The **efficiency** of a biochemical process is defined as the ratio of the work actually accomplished to the theoretical maximum work expected from the change in free energy.

Free Energy and Concentration

The sign of ΔG for a process tells us whether that process, or its reverse, is thermodynamically favorable. The magnitude of ΔG is an indication of how far the process is from equilibrium, and how much useful work may be obtained from it. Clearly, ΔG is a quantity of fundamental importance in determining which processes will or will not occur in a cell, and for what they may be used.

Life involves a temporary decrease in entropy, paid for by the expenditure of energy.

The folding of proteins is well-described by $\Delta G = \Delta H - T\Delta S$, where ΔH is a measure of the change in bonding interactions and ΔS is a measure of the change in order of the system going from the unfolded state to the folded state.

The free energy change, ΔG, is a measure of the maximum useful work obtainable from any reaction.

To express these ideas quantitatively we need to answer this question: How does the free energy of a system depend on the concentrations of various components in a mixture? In the next two sections of this chapter we derive an alternative expression for ΔG that allows us to answer that question:

$$\Delta G = \Delta G^\circ + RT \ln Q \tag{3.12}$$

where ΔG° is the **standard free energy change** for a reaction (defined in the next section), R is the gas constant, T is the absolute temperature in kelvin, and Q is the **mass action expression** for the reaction of interest. In essence, Q describes the relative concentrations of the reactants and products. Equation 3.12 allows us to quantify the energy available from a biochemical reaction to do work, and we will use it frequently in the chapters describing metabolic chemistry.

Chemical Potential

The relationship between free energy and the concentrations of components in a mixture can be expressed very simply: If we have a mixture containing a moles of component A, b moles of component B, and so on, we may write

$$G = a\overline{G}_A + b\overline{G}_B + c\overline{G}_C + \cdots \tag{3.13}$$

The quantities $\overline{G}_A$, $\overline{G}_B$, and so forth are called the **partial molar free energies** or **chemical potentials** of the various components. Each represents the contribution, per mole, of a particular component to the total free energy of the system (in some texts, chemical potential is given the symbol μ). We shall make the approximation, which is usually valid for dilute solutions, that each of the chemical potentials depends only on the concentration of the substance in question. For dilute solutions, $\overline{G}_A$, $\overline{G}_B$, etc. turn out to be simple logarithmic functions of the *activities* of the corresponding substance. As described in Chapter 2 (see Equation 2.6a), the **activity (a)** is a dimensionless quantity that measures the effective concentration of a substance, describing its contribution to the free energy of the system. We find

> The chemical potential of a substance measures the contribution of that substance to the free energy of the system.

$$\overline{G}_A = G_A^\circ + RT \ln a_A$$
$$\overline{G}_B = G_B^\circ + RT \ln a_B \tag{3.14a}$$
$$\dots \textit{etc.}$$

For dilute solutions the activity of each solute component can be taken as a number approximately equal to the *molar concentration* of that component. At very low concentration, *activity* and *molar concentration* are numerically the same. This simplifying approximation is satisfactory for almost all biochemical applications. We can then rewrite Equation 3.14a in terms of concentrations:

$$\overline{G}_A = G_A^\circ + RT \ln [A]$$
$$\overline{G}_B = G_B^\circ + RT \ln [B] \tag{3.14b}$$
$$\dots \textit{etc.}$$

where $[A]$, $[B]$, etc. have (dimensionless) values equal to the molar concentrations of the components. Note what happens when the concentration equals 1 M. The logarithmic term then vanishes (because $\ln 1 = 0$) and, for example, $\overline{G}_A = G_A^\circ$. This shows that G_A°, G_B°, etc. are reference, or standard state, values of the chemical potential. We always express the chemical potentials with respect to a **standard state**. In solutions, the standard state for each *solute* component is 1 M concentration; the standard state for the solvent is pure solvent. In each case, $a = 1$ in the standard state. In the following examples of thermodynamic calculations we will convert concentration terms to activities using the **standard state concentrations** for each component of the system.

The importance of Equations 3.14a-b is that they allow us to apply general thermodynamic principles to practical problems. In particular, they allow us to predict the favored directions for real processes under defined conditions. The calculated value of ΔG, which predicts the favored direction for a biochemical process, is a measure of the difference in free energy between the initial and final states for that process. To calculate the change in any state function (e.g., ΔT, ΔH, ΔS, ΔG, etc.) we subtract the initial value from the final value for that state function; thus, we define ΔG as follows:

$$\Delta G = G_{Final} - G_{Initial} \tag{3.15}$$

One biochemical process to which these principles are particularly relevant is diffusion through membranes, a process we shall use as an example here and discuss in much more detail in Chapters 10 and 15.

An Example of How Chemical Potential Is Used: A Close Look at Diffusion Through a Membrane

We know from experience that if a substance can diffuse across a membrane, it will do so in such a direction as to make the concentrations on the two sides equal. Now we will see whether our thermodynamic arguments justify this.

Suppose we have two solutions of substance A separated by a membrane through which A can pass (Figure 3.5). Assume that in region 1 the concentration is initially $[A]_1$ and in region 2 the concentration is $[A]_2$. We must define the direction of transfer for the process to evaluate ΔG. Let us consider transferring some quantity of A from region 1 (the "initial" state of A) to region 2 (the "final" state of A). By the definition of ΔG given in Equation 3.15, the overall free energy change is determined by the difference in molar free energies (or chemical potentials) of A in regions 1 and 2, or

$$\Delta G = G_{A_2} - G_{A_1} \tag{3.16}$$

We expect these molar free energies to be different because the chemical potential for a solute depends on the concentration of that solute (Equation 3.14b). The free energy change for moving A from region 1 to region 2 can be calculated by applying Equations 3.14b and 3.16:

$$G_{A_1} = G_A^\circ + RT \ln[A]_1 \tag{3.17a}$$

$$G_{A_2} = G_A^\circ + RT \ln[A]_2 \tag{3.17b}$$

$$\Delta G = G_{A_2} - G_{A_1} = (G_A^\circ + RT \ln[A]_2) - (G_A^\circ + RT \ln[A]_1) \tag{3.18}$$

FIGURE 3.5

Equilibration across a membrane. Two solutions of A, with concentrations $[A_1]$ and $[A_2]$, are separated by a membrane through which A can pass in either direction. If the initial concentration is higher in region 1, the chemical potential of A will also be higher in that region, and net transport will occur from region 1 to region 2 until equal concentrations (and chemical potentials) are obtained. Blue arrows indicate direction of travel, which is random.

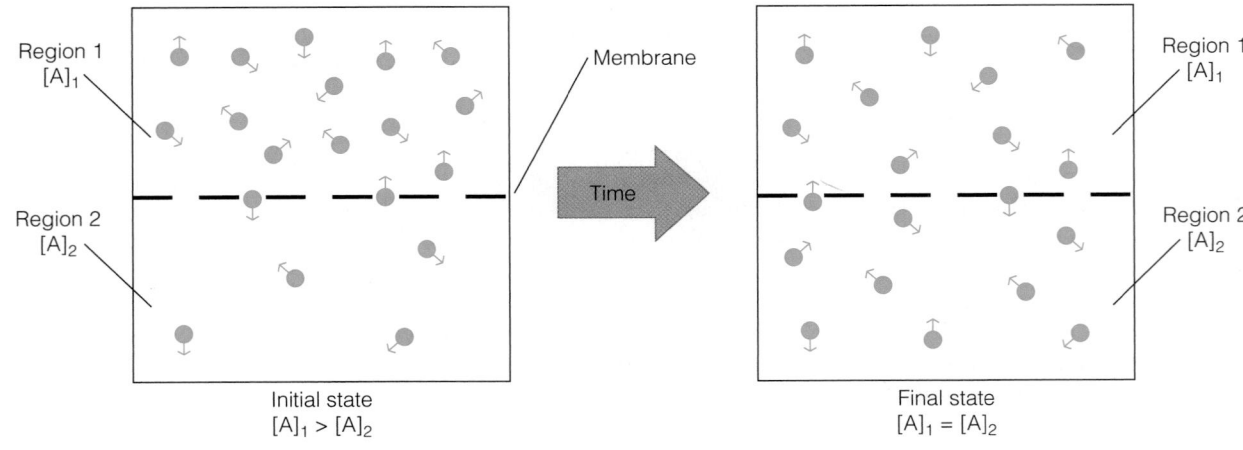

Region 1 $[A]_1$ Membrane   Region 1 $[A]_1$

Region 2 $[A]_2$ Time   Region 2 $[A]_2$

Initial state
$[A]_1 > [A]_2$

Final state
$[A]_1 = [A]_2$

After canceling G_A° terms, Equation 3.18 simplifies to

$$\Delta G = RT \ln \frac{[A]_2}{[A]_1} \tag{3.19}$$

Equation 3.19 predicts the following:

1. If $[A]_2 < [A]_1$, ΔG is negative; thus, transfer from region 1 to region 2 is favorable (this is the situation described in the "initial state" of Figure 3.5).

2. Conversely, for a system where $[A]_2 > [A]_1$, ΔG is positive; thus, transfer from region 1 to region 2 would not be favorable (but transfer in the opposite direction would be favored).

3. If $[A]_2 = [A]_1$, ΔG is zero; thus, there is no net driving force in either direction for the transfer of A. The system is at equilibrium (this is the situation described in the "final state" of Figure 3.5).

From this analysis, we can conclude that if a substance is able to pass through a membrane, the direction of favorable transfer will always be from the region of higher concentration to the region of lower concentration. More generally, we can say that a substance will diffuse from a region where its chemical potential is higher to one where it is lower. Chemical potential plays much the same role for chemical substances that electrical potential does for electrons—it is the driving force. This driving force moves the system from whatever distribution of $[A]$ exists initially, toward the equilibrium state of equal concentrations across the membrane where there is no more driving force because $\Delta G = 0$.

There are cases in which substances pass readily from regions of lower concentration to regions of higher concentration; but, in such circumstances the necessary free energy price is paid by *coupling* the transport process to one or more thermodynamically favorable chemical reactions. Two (or more) reactions are "coupled" when one reaction cannot occur without the other reaction(s) also occurring. We will see throughout this text that the coupling of biochemical reactions is mediated predominantly by enzymes. However, we must show how chemical potential affects ΔG for biochemical reactions before we can explore further the concept of reaction coupling.

Free Energy and Chemical Reactions: Chemical Equilibrium

The Free Energy Change and the Equilibrium Constant

Perhaps the most important use of the chemical potential is to describe quantitatively the free energy changes accompanying chemical reactions under various conditions. We are thereby able to predict the favorable direction for a reaction. Suppose we have a reaction like

$$aA + bB \rightleftharpoons cC + dD$$

Even though the reaction can proceed in either direction, we have written it with C and D on the right, so we call them *products* and A and B *reactants*. We wish to calculate the change in free energy that occurs when *a* moles of A and *b* moles of B form *c* moles of C and *d* moles of D, each at some given concentration.*

---

*How do we carry out a reaction while keeping concentrations of both reactants and products constant? Two ways can be imagined. First, the total amounts of reactants and products could be so enormous that reaction of some finite amount of A and B to produce C and D would not appreciably change the concentrations of these species. Alternatively, we could imagine hypothetical processes that would remove products and add reactants so as to keep concentrations unchanged. This often happens in living cells, where the concentrations of metabolites remain within a narrow, nearly constant, concentration range *far from equilibrium values* (this characteristic of the living state is called "homeostasis" and must not be confused with thermodynamic equilibrium).

For any chemical reaction we can equate the reactant state as the "initial state" and the product state as the "final state." According to Equation 3.15, the free energy change for the reaction must be the free energy of the products minus that of the reactants.

$$\Delta G = G(\text{products}) - G(\text{reactants}) \tag{3.20}$$

By Equation 3.13, we can write these free energies in terms of the chemical potentials of the substances, each multiplied by the number of moles involved. For our example,

$$\Delta G = c\overline{G}_C + d\overline{G}_D - a\overline{G}_A - b\overline{G}_B \tag{3.21}$$

Here we are simply stating that the driving force for the reaction is the total free energy of the products minus that of the reactants.

Now, using Equation 3.14b, we insert the appropriate expressions for $\overline{G}_C$ and so forth in terms of standard concentrations and obtain

$$\Delta G = cG_C^\circ + cRT\ln[C] + dG_D^\circ + dRT\ln[D] - aG_A^\circ - aRT\ln[A] - bG_B^\circ - bRT\ln[B] \tag{3.22a}$$

or

$$\Delta G = cG_C^\circ + dG_D^\circ - aG_A^\circ - bG_B^\circ + RT\ln[C]^c + RT\ln[D]^d - RT\ln[A]^a - RT\ln[B]^b \tag{3.22b}$$

or

$$\Delta G = \Delta G^\circ + RT\ln\left(\frac{[C]^c[D]^d}{[A]^a[B]^b}\right) \tag{3.23}$$

> The free energy change in a chemical reaction depends on the standard state free energy change (ΔG°) and on the activities of reactants and products (described by $RT\ln Q$).

In going from Equation 3.22a to 3.23, we have done two things: grouped the G° terms into ΔG° and made use of rearrangements like: $aRT\ln[A] = RT\ln[A]^a$. The group of G° terms (ΔG°) has a simple meaning; because G° is the free energy per mole of a substance in the standard state (1 M), ΔG° represents the **standard state free energy change** in the reaction. It is the free energy change that would be observed if a moles of A and b moles of B, each at 1 M concentration, formed c moles of C and d moles of D, each at 1 M.

The ratio of products to reactants given inside the parentheses in Equation 3.23 is a *mass action expression*, which can also be represented by the symbol Q:

$$\frac{[C]^c[D]^d}{[A]^a[B]^b} = Q \tag{3.24}$$

If we combine equations 3.23 and 3.24 we get Equation 3.12: $\Delta G = \Delta G^\circ + RT\ln Q$.

The quantity ΔG represents the free energy change when a moles of A (at concentration [A]) and b moles of B (at [B]) react to produce c moles of C (at [C]) and d moles of D (at [D]). These concentrations may be anything we want them to be. When all are 1 M, Equation 3.23 reduces to $\Delta G = \Delta G^\circ$. The importance of Equation 3.23 is that it allows us to calculate ΔG under any conditions we wish.

Suppose that the reaction has come to equilibrium. In that case, two things must be true. First, the concentrations in the mass action expression (Equation 3.24) must be equilibrium concentrations. At equilibrium, the factor Q has a value that is identical to the *equilibrium constant K* for the reaction:

$$K = \left(\frac{[C]^c[D]^d}{[A]^a[B]^b}\right)_{eq} \tag{3.25}$$

Second, if the system is at equilibrium, ΔG must equal zero. In this case, Equation 3.23 reduces to

$$0 = \Delta G^\circ + RT\ln\left(\frac{[C]^c[D]^d}{[A]^a[B]^b}\right)_{eq} = \Delta G^\circ + RT\ln K \tag{3.26}$$

> The equilibrium constant K can be calculated from the standard state free energy change ΔG°, and vice versa.

or

$$-\Delta G^\circ = RT\ln K \tag{3.27}$$

This can be rearranged as

$$K = e^{-\Delta G^\circ / RT} \tag{3.28}$$

Equations 3.27 and 3.28 express an important relationship between the standard free energy change, ΔG°, and the equilibrium constant, K, that we shall use many times. These equations make it possible, for example, to use data from tables of standard state free energy changes to predict equilibrium constants for reactions.

Equation 3.23 may best be thought of in the following way. ΔG° represents a reference value for the free energy change, whereby the intrinsic free energy changes for different reactions can be compared under equivalent circumstances (i.e., 1 M concentrations). The magnitude of this term tells us the equilibrium constant. The (concentration-dependent) $RT \ln Q$ term in Equation 3.23 represents the extra free energy change ($+$ or $-$) involved if we were to carry out the reaction at some other, nonequilibrium, set of concentrations. Tabulations of ΔG° values for different reactions are common, but in applying such data to biochemical problems, we must always keep in mind that it is ΔG, as determined by the concentrations of reactants and products found in the cell, rather than ΔG° that determines whether or not a reaction is favorable in vivo.

> It is ΔG, as determined by the actual concentrations of reactants and products in the cell, rather than ΔG° that determines whether or not a reaction is favorable in vivo.

Free Energy Calculations: A Biochemical Example

To illustrate the application of these somewhat abstract ideas, let us consider an example—a very simple but important biochemical reaction, the isomerization of glucose-6-phosphate (G6P) to fructose-6-phosphate (F6P) shown in Figure 3.6:

$$\text{Glucose-6-phosphate} \rightleftharpoons \text{Fructose-6-phosphate} \quad \Delta G^\circ = +1.7 \text{ kJ/mol}$$

which may be written more compactly as

$$\text{G6P} \rightleftharpoons \text{F6P} \quad \Delta G^\circ = +1.7 \text{ kJ/mol}$$

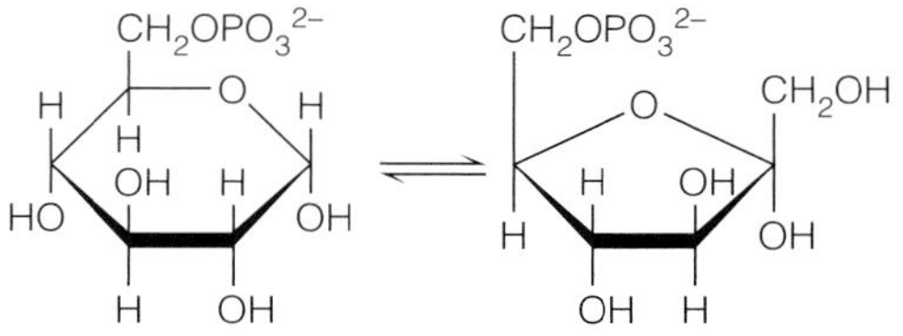

This is the second step in the **glycolytic pathway**, which is discussed in Chapter 13. The reaction is clearly endergonic under standard conditions. In other words, the system is not at equilibrium when G6P and F6P are both at 1 M because ΔG° is positive ($+1.7$ kJ/mol) and the reverse reaction is favored under standard conditions. Therefore, the equilibrium must lie to the left, with a higher concentration of G6P than F6P. We can express this quantitatively by calculating the equilibrium constant using Equation 3.28. Using the given value for ΔG° and assuming a temperature of 25 °C (298 K), we obtain:

Glucose-6-phosphate **Fructose-6-phosphate**

FIGURE 3.6

Isomerization of glucose-6-phosphate (G6P) to fructose-6-phosphate (F6P).

$$K = e^{\left(\frac{-\Delta G^\circ}{RT}\right)} = e^{\left(\frac{-\left(1700\,\frac{J}{mol}\right)}{\left(8.315\,\frac{J}{mol \cdot K}\right)(298\,K)}\right)} = 0.504 = \left(\frac{[F6P]}{[G6P]}\right)_{eq} \tag{3.29}$$

where $([F6P]/[G6P])_{eq}$ is the equilibrium ratio of the concentration of fructose-6-phosphate to that of glucose-6-phosphate. The fact that $K < 1$ shows that the equilibrium lies to the left, or favors the reactants (in this case, G6P).

Living Cells Are Not at Equilibrium

The concentrations of metabolites for any process in *living* cells are not typically the equilibrium concentrations defined by ΔG° for that process. The state of chemical equilibrium is characteristic of dead cells, not living ones. As stated at the beginning of this chapter, characteristics of the living state include the capture, transformation, storage, and utilization of energy (all of which are discussed in great detail in the following chapters of this textbook). Such processes can only occur when $\Delta G < 0$. In other words, a living organism is not at equilibrium.

Life occurs within relatively narrow ranges of temperature, pH, and concentrations for ions and metabolites. This set of conditions is referred to as **homeostasis** or the **homeostatic condition**. Because the concentrations of certain solutes inside cells remain relatively constant, homeostasis is frequently confused

The homeostatic condition, which is characteristic of living cells, must not be confused with true thermodynamic equilibrium.

with true thermodynamic equilibrium; however, *homeostasis must not be confused with equilibrium*! The critical distinction between the two is that under homeostatic conditions, ΔG for numerous vital processes is < 0, whereas, at equilibrium, ΔG for any process $= 0$. In addition, energy is required to maintain homeostasis; hence, the need to capture, transform, and store energy.

We know from Le Chatelier's principle that any system that is not at equilibrium has a driving force ($\Delta G < 0$) to move in the direction that re-establishes the equilibrium state. We can understand the magnitude of these driving forces by comparing equilibrium and nonequilibrium systems. For the reaction

$$a\text{A} + b\text{B} \rightleftharpoons c\text{C} + d\text{D}$$

at equilibrium, Le Chatelier's principle predicts that a perturbation of the equilibrium by increasing the concentration of *either* reactant, [A] or [B], would result in an increase in the concentrations of *both* products, [C] and [D], as the system moves back to the equilibrium state. Likewise, if the equilibrium was perturbed by an increase in either [C] or [D], the system would respond by increasing [A] and [B]. We can understand the basis of Le Chatelier's principle using the thermodynamic arguments developed above. Let us reconsider Equation 3.12:

$$\Delta G = \Delta G° + RT \ln Q$$

and substitute $-RT \ln K$ for $\Delta G°$ (Equation 3.27) to give:

$$\Delta G = -RT \ln K + RT \ln Q \tag{3.30}$$

For a system at equilibrium $Q = K$, thus

$$\Delta G = -RT \ln K + RT \ln K = 0 \tag{3.31}$$

Because $\Delta G = 0$, there is no driving force in either direction. On the other hand, for a system that is not at equilibrium $Q \neq K$, thus

$$\Delta G = -RT \ln K + RT \ln Q \neq 0 \tag{3.32}$$

In this case, $\Delta G \neq 0$, or, there is a driving force for the reaction, which Le Chatelier predicts will favor the direction that re-establishes equilibrium. We can predict which direction, toward reactants or toward products, will be favored by considering the relative values of K and Q. This is most easily seen if we rearrange Equation 3.30:

$$\Delta G = RT(\ln Q - \ln K) = RT \ln\left(\frac{Q}{K}\right) \tag{3.33}$$

Equation 3.33 predicts that a reaction will proceed as written whenever the ratio $Q/K < 1$ (see Table 3.4). We can imagine two ways that a living cell could maintain $Q/K < 1$. The first is by consuming products as fast as they are formed, such that the homeostatic concentrations of the products are relatively low compared to the reactant concentrations. This is a common strategy employed in multistep metabolic pathways. For example, for the hypothetical pathway

$$\text{A} \longrightarrow \text{B} \longrightarrow \text{C} \longrightarrow \text{D} \longrightarrow \text{E}$$

B is the "product" of the first step; but it is also the "reactant" for the second step. The concentration of B in the cell will be low if B is rapidly converted to C. By Le Chatelier's principle, the reaction A → B will be driven to the right as a result of removing B from the system. The second way to maintain $Q/K < 1$ is by keeping the concentrations of the "reactant" species relatively high. Again, Le Chatelier's principle predicts that as [A] increases there is a greater tendency to drive the reaction A → B to the right. The magnitude and sign of that driving force is succinctly expressed as ΔG and can be readily calculated using Equation 3.23.

In summary, we have described how thermodynamically favored (irreversible) processes are related to equilibrium. Whenever a system is displaced from equilibrium, it will proceed in the direction that moves toward the equilibrium state because the direction leading toward equilibrium will have a $\Delta G < 0$. Because living

TABLE 3.4	Relationships between K, Q, and ΔG for a reaction	
Value of Q	Value of ΔG	Favored direction
$< K$	< 0	Forward reaction (formation of products)
$= K$	$= 0$	Neither (system at equilibrium)
$> K$	> 0	Reverse reaction (formation of reactants)

cells are not at equilibrium, most cellular reactions are driven to proceed in either the forward or reverse direction.

We are now prepared to answer a question of fundamental importance in bioenergetics: "How are unfavorable reactions driven forward in living cells?" Thermodynamically unfavorable reactions can be driven by either, or both, of the following strategies:

1. Maintaining $Q < K$
2. Coupling an unfavorable reaction to a highly favorable reaction

The theoretical basis for strategy #1 has been described above. In the following sections we will describe the basis for strategy #2. In later chapters we will illustrate the implementation of these strategies in numerous biochemical reactions. We will see that highly favorable processes such as ion transport across membranes and hydrolysis of **adenosine triphosphate (ATP)** are commonly used to drive unfavorable reactions. The processes of ATP consumption and ATP production feature prominently in biochemistry; thus, we will introduce here the features of ATP (and similar compounds) that make it suitable as the central energy-transduction molecule in cells.

> Thermodynamically unfavorable reactions become favorable when $Q < K$, and/or when coupled to strongly favorable (i.e., highly exergonic) reactions.

High-Energy Phosphate Compounds: Free Energy Sources in Biological Systems

Understanding the central role of free energy changes in determining the favorable directions for chemical reactions is of great importance in the study of biochemistry because every biochemical process (e.g., protein folding, metabolic reactions, DNA replication, muscle contraction) must, overall, be a thermodynamically favorable process. Very often, a particular reaction or process that is necessary for life is in itself endergonic. As stated above, such intrinsically unfavorable processes can be made thermodynamically favorable by *coupling* them to strongly favorable reactions. Suppose, for example, we have a reaction that is part of an essential pathway but is endergonic:

$$A \rightleftharpoons B \qquad \Delta G° = +10 \text{ kJ/mol}$$

At the same time, suppose another process is highly exergonic:

$$C \rightleftharpoons D \qquad \Delta G° = -30 \text{ kJ/mol}$$

If the cell can manage to *couple* these two reactions, the $\Delta G°$ for the overall process will be the algebraic sum of the values of $\Delta G°$ for the individual reactions:

$$
\begin{array}{ll}
A \rightleftharpoons B & \Delta G° = +10 \text{ kJ/mol} \\
\underline{C \rightleftharpoons D} & \underline{\Delta G° = -30 \text{ kJ/mol}} \\
\text{Overall: } A + C \rightleftharpoons B + D & \Delta G° = -20 \text{ kJ/mol}
\end{array}
$$

Equilibrium for the overall process now lies far to the right; the consequence is that B is now more favorably produced from A.

Coupling of endergonic reactions or processes to exergonic reactions is used not only to drive countless reactions but also to transport materials across membranes, transmit nerve impulses, contract muscles, and carry out other physical changes.

High-Energy Phosphate Compounds as Energy Transducers

Driving an unfavorable process by coupling it to a favorable one requires the availability in cells of compounds (like the hypothetical C in our previous example) that can undergo reactions with large negative free energy changes. Such substances can be thought of as energy transducers in the cell. The most important of these high-energy compounds are certain phosphates, which can undergo hydrolytic release of their phosphate groups in aqueous solution. A number of such compounds and their hydrolysis reactions are shown in Figure 3.7 and Table 3.5. You will encounter

> "High-energy" phosphate compounds have very large negative free energies of hydrolysis.

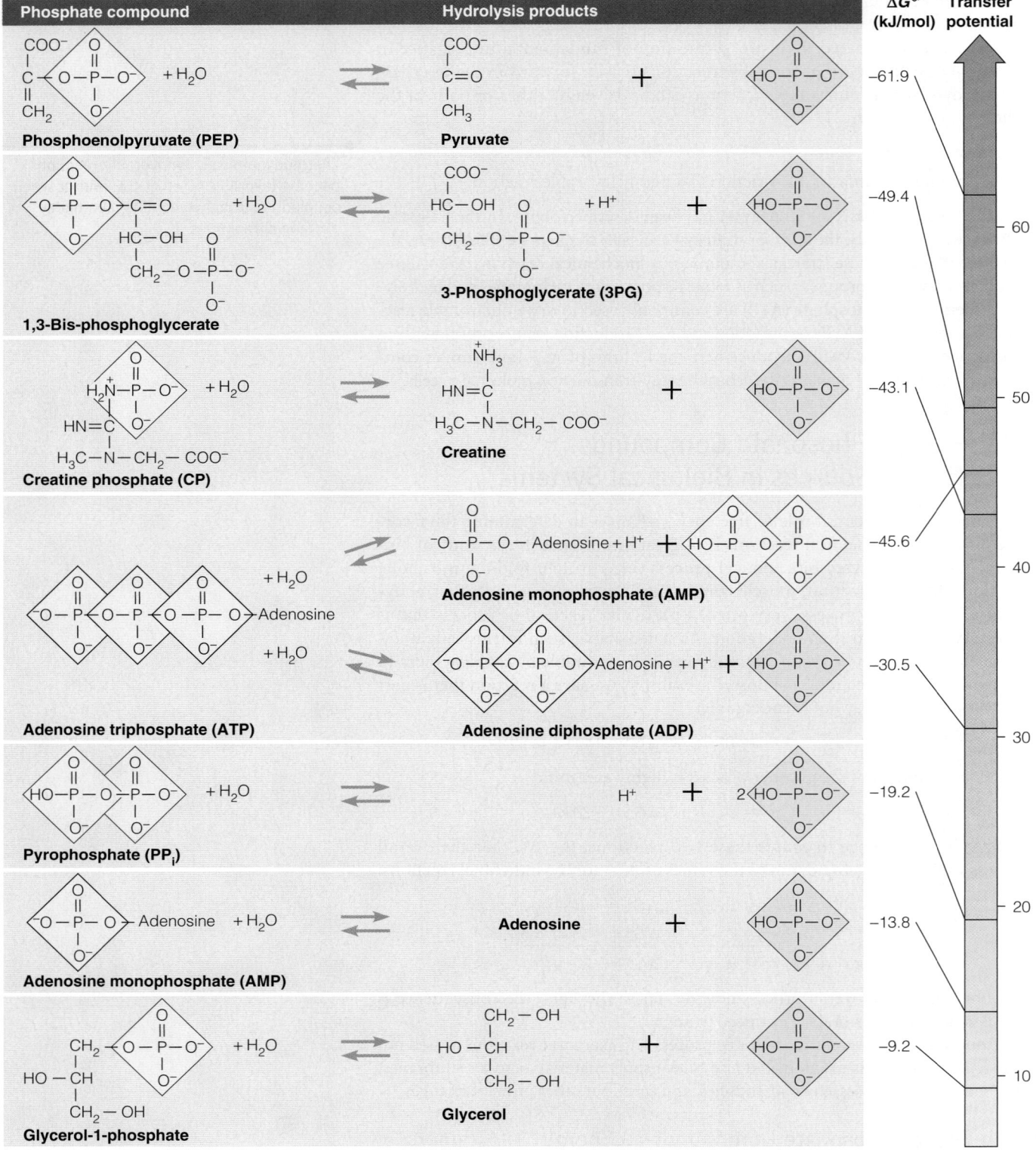

FIGURE 3.7

Hydrolysis reactions for some biochemically important phosphate compounds. The labile phosphate group of each compound is shown in yellow. The more stable reaction product P_i is in gray. A scale of phosphate transfer potential (see text) is shown to the right.

TABLE 3.5 $\Delta G^{\circ\prime}$ for hydrolysis of some phosphate compounds

Hydrolysis reaction	$\Delta G^{\circ\prime}$ (kJ/mol)	pH	$[Mg^{2+}]$
Phosphoenolpyruvate + $H_2O \longrightarrow$ pyruvate + P_i	−61.9	7	NS
1,3-Bisphosphoglycerate + $H_2O \longrightarrow$ 3-phosphoglycerate + P_i + H^+	−49.4	7	NS
Acetyl phosphate + $H_2O \longrightarrow$ acetate + P_i + H^+	−43.1	7	NS
Creatine phosphate + $H_2O \longrightarrow$ creatine + P_i	−43.1	7	NS
ATP + $H_2O \longrightarrow$ AMP + PP_i + H^+	−45.6	7	NS
ATP + $H_2O \longrightarrow$ ADP + P_i + H^+	−30.5	7	excess
PP_i + $H_2O \longrightarrow$ $2P_i$ + H^+	−33.5	NS	NS
PP_i + $H_2O \longrightarrow$ $2P_i$ + H^+	−18.8	NS	0.005
PP_i + $H_2O \longrightarrow$ $2P_i$ + H^+	−19.2	7	0.001
Glucose-1-phosphate + $H_2O \longrightarrow$ glucose + P_i	−20.9	NS	NS
Glucose-6-phosphate + $H_2O \longrightarrow$ glucose + P_i	−13.8	NS	NS

NS = not specified.

W. P. Jencks (1976) Free energies of hydrolysis and decarboxylation in *Handbook of Biochemistry and Molecular Biology*, 3rd ed., G. Fasman ed., CRC Press, Boca Raton, FL.

P. Frey and A. Arabshahi (1995) Standard free energy change for the hydrolysis of the α, β–phosphoanhydride bridge in ATP. *Biochemistry* 34:11307–11310.

all of these important substances in later chapters on metabolism. Some of these substances, like ATP, **phosphoenolpyruvate (PEP)**, **creatine phosphate (CP)**, and **1,3-bisphosphoglycerate (1,3-BPG)**, have large negative standard free energies of hydrolysis. ATP is perhaps the most important of these compounds, and the one you will encounter most often in this book. The structure and hydrolysis reactions of ATP are shown in Figure 3.8. Hydrolysis of ATP to ADP is highly exergonic, with $\Delta G^{\circ\prime} = -30.5$ kJ/mol (see page 79 for the definition of $\Delta G^{\circ\prime}$). This value corresponds to an equilibrium constant greater than 10^5. Such an equilibrium lies so far to the right that ATP hydrolysis can be considered essentially irreversible.

Figure 3.7 also shows that whereas some of these phosphate hydrolysis reactions are truly highly exergonic processes, others are not. For example, hydrolysis of phosphoanhydride (as in ATP, ADP, and pyrophosphate) and mixed anhydride (1,3-BPG) bonds is much more exergonic than is hydrolysis of phosphate esters (AMP, glycerol-1-phosphate). A variety of factors can explain these differences. Those that seem to be most important are described next.

Resonance Stabilization of the Phosphate Products

The **orthophosphate** ion (HPO_4^{2-}), which is often abbreviated as P_i (inorganic phosphate), is capable of a wide variety of resonance forms. Both the bound proton and the oxygen bonding should be thought of as delocalized, so a more appropriate way to write the structure is as shown in Figure 3.9. The multiple forms, which are of equal energy, contribute to the high entropy of such a resonance structure (see Equation 3.9). Not all of these forms are possible when the phosphate is bound in an ester. Consequently, release of the orthophosphate results in an entropy increase in the system and is therefore favored. Resonance stabilization applies in *all* of the phosphate hydrolysis reactions described in Figure 3.7.

Additional Hydration of the Hydrolysis Products

Release of the phosphate residue from its bonded state allows greater opportunities for hydration, especially when both products are charged. Recall from

78 CHAPTER 3 THE ENERGETICS OF LIFE

FIGURE 3.8

The ATP molecule and its hydrolysis reactions.
Throughout this book ⟨P⟩ represents the tetrahedral orthophosphate group. Hydrolysis of ATP or ADP cleaves a phosphoanhydride bond, whereas hydrolysis of AMP cleaves a phosphate ester bond.

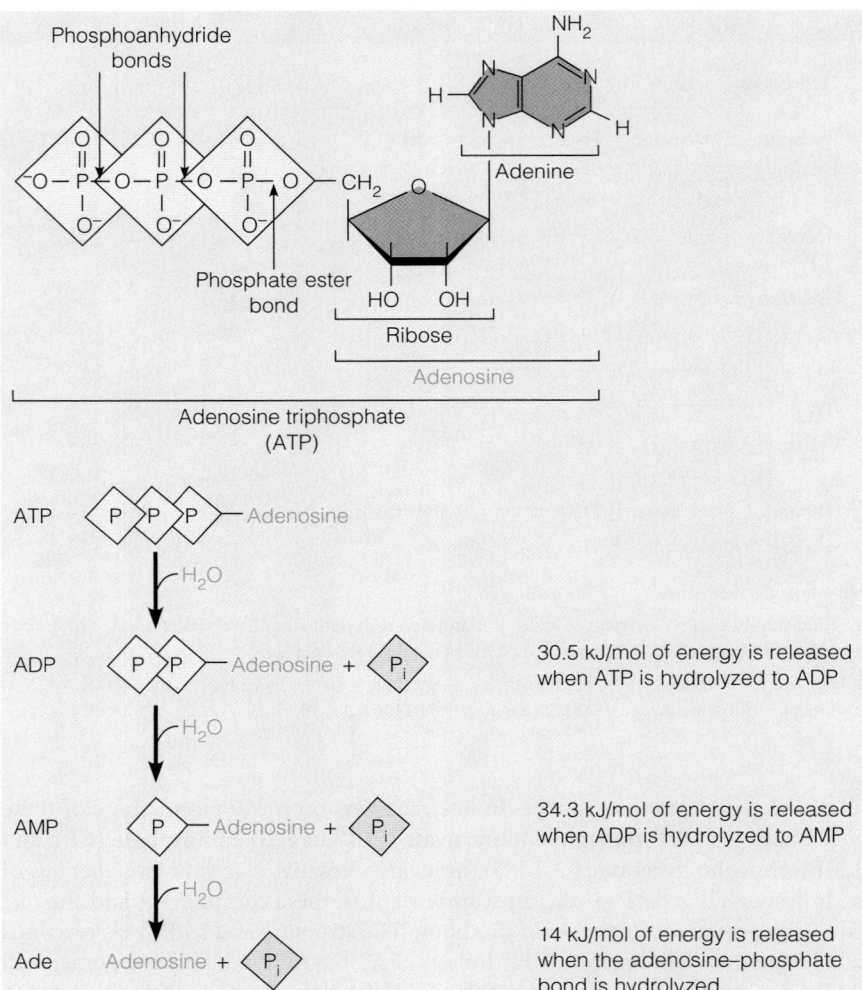

FIGURE 3.9

Resonance stabilization of orthophosphate, HPO_4^{2-} (P_i). The delocalization of the charge of the orthophosphate ion is depicted here in two ways. **(a)** In this depiction, the four resonance forms of the phosphate ion are shown with the H^+ not assigned permanently to any one of the four oxygen atoms. **(b)** This depiction shows the physical significance of the delocalized P=O double bond: the phosphate ion is a tetrahedral structure with four equivalent phosphorous–oxygen bonds.

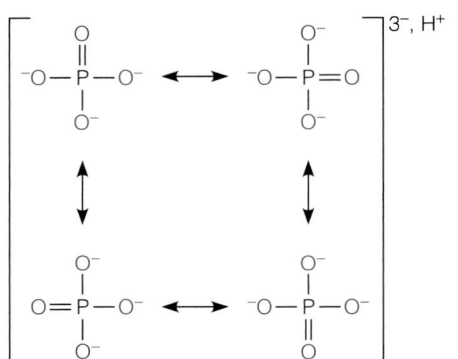

(a) Structures of phosphate ion contributing to resonance stabilization

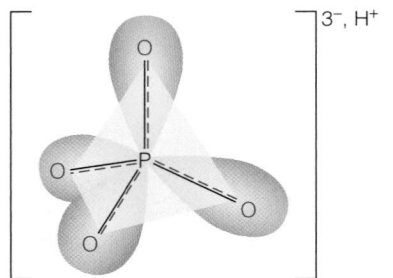

(b) Hybrid atomic orbitals of tetrahedral phosphate ion

Chapter 2 that ions are highly hydrated in aqueous solution and that such hydration is an energetically favored state.

Electrostatic Repulsion Between Charged Products

In the hydrolysis of phosphoenolpyruvate, 1,3-bisphosphoglycerate, adenosine triphosphate, and pyrophosphate, both products of hydrolysis carry a negative charge. The repulsion between these ionic products strongly favors the hydrolysis reaction.

Tautomerization of Product Molecules

$\Delta G^{\circ\prime}$ for the hydrolysis of phosphoenol pyruvate, -61.9 kJ/mol, is significantly more favorable than one would predict for the hydrolysis of a simple phosphate ester (e.g., compare to $\Delta G^{\circ\prime}_{\text{hydrolysis}} = -13.8$ kJ/mol for AMP). The explanation for this unexpected reactivity of PEP lies in the structural isomerization, in this case called "tautomerization," of the pyruvate product that occurs following the release of P_i from PEP. The direct product of phosphate hydrolysis from PEP is the *enol* form of pyruvate, which rapidly tautomerizes to the thermodynamically favored *keto* form:

$\Delta G^{\circ\prime}$ (kJ/mol)

Phosphoenolpyruvate + H_2O → (Hydrolysis) → enol form + HPO_4^{2-} −16

enol form → (Tautomerization) → keto form   −46

Phosphoenolpyruvate + H_2O → (Net reaction) → Pyruvate + HPO_4^{2-} −61.9

The equilibrium for this tautomerization lies far toward the keto form ($K_{eq} \approx 6 \times 10^7$); thus, it is this essentially irreversible isomerization reaction that provides most of the driving force for the overall hydrolysis of PEP to pyruvate and P_i.

Water, Protons in Buffered Solutions, and the "Biochemical Standard State"

In some of the reactions listed in Figure 3.7, a proton is released. Therefore, the hydrogen ion concentration (i.e., the pH) will affect ΔG for the reaction. For the hydrolysis of ATP to ADP:

$$ATP + H_2O \rightleftharpoons ADP + P_i + H^+$$

we can calculate ΔG for the reaction using Equation 3.23:

$$\Delta G = \Delta G^\circ + RT \ln \left(\frac{[ADP][P_i][H^+]}{[ATP][H_2O]} \right) \qquad (3.34)$$

Because biochemical reactions typically occur in a relatively dilute aqueous solution, buffered near pH 7, it is appropriate to treat the chemical potentials of H_2O and H^+ differently than is done for thermodynamic calculations based on the standard state conditions described previously. In a dilute solution, the activity of the solvent (water) is not significantly changed by reactions that consume or produce H_2O; thus, for the "biochemical standard state" the *activity* of water is defined as "unity." For chemical reactions the standard state for solutes is defined as 1 M; however, in living cells the concentration of H^+ is roughly 10^{-7} M, much lower than the standard value of 1 M. It is therefore appropriate to define the chemical potential of H^+ in biochemical reactions relative to the H^+ concentration found in the living state (i.e., 10^{-7} M), rather than the value 1 M defined by the chemical standard state. Recall that when a solute in a dilute solution has a concentration of 1 M, the activity of that solute is unity. For the *biochemical standard state* we define the activity of H^+ to be unity when $[H^+] = 10^{-7}$ M. We distinguish values of ΔG° that are referenced to the chemical standard state from those referenced to the biochemical standard state by a *superscript prime*: $\Delta G^{\circ\prime}$. If we assume the activity of $H_2O = 1$ and we calculate the activities of the other solutes[†] relative to their

Standard free energy changes for biochemical reactions are defined by $\Delta G^{\circ\prime}$, where the water concentration is assumed to be constant (thus, the activity of water is defined as unity), and the activity of H^+ is defined as unity when pH = 7.0.

[†]Recall that in the derivation of Equation 3.14a we made the simplifying assumption that the activities of solutes in dilute solutions are not significantly different from the values of molar concentration for each solute. This is not rigorously true; but, it is a common assumption in biochemistry, where the concentrations of solutes are typically well below 1 M. For many reactions described in this text the assumption that activity $\cong$ molar concentration will hold; but, students should be aware that a rigorous calculation of chemical potential requires the use of a more sophisticated definition of activity.

biochemical standard state concentrations, Equation 3.34 can be written as follows:

$$\Delta G = \Delta G^{\circ\prime} + RT \ln \left(\frac{\frac{[ADP]}{(1\,\text{M})} \frac{[P_i]}{(1\,\text{M})} \frac{[H^+]}{(10^{-7}\,\text{M})}}{\frac{[ATP]}{(1\,\text{M})}(1)} \right) \tag{3.35}$$

In this example Equation 3.35 does not include other factors, such as Mg^{2+} concentration and ionic strength, that can affect significantly the value of ΔG for ATP hydrolysis (as well as many other biochemical reactions; see Table 3.5). In principle, temperature, Mg^{2+} concentration, and ionic strength ("I") must also be specified for a given value of $\Delta G^{\circ\prime}$. For our purposes it will be sufficient to define the biochemical standard state more simply (i.e., pH = 7.0 and the activity of H_2O = 1); however, the International Union of Pure and Applied Chemistry has recommended the following set of conditions as a standard for the study of biochemical reactions under approximately "physiological conditions": Temperature = 37 °C (310.15 K), pH = 7.0, $\left[Mg^{2+} \right]$ = 0.001 M, and I = 0.25 M.

Equation 3.35 illustrates two key points regarding the calculation of ΔG for the reactions you will encounter throughout this text:

1. $\Delta G^{\circ\prime}$ is used, signifying the biochemical standard state.

2. The mass action expression Q is unitless. We can strip the units for each concentration term in Q by dividing each term by its proper standard concentration (e.g., 1 M for all solutes except H^+; 10^{-7} M for H^+; 1 bar for gases, etc.)

The significance of these two points is illustrated in the following example. Let us calculate ΔG for the hydrolysis of ATP at pH 7.4, 25 °C, where the concentrations of ATP, ADP, and P_i are, respectively, 5 mM, 0.1 mM, and 35 mM. Under these conditions Equation 3.34 becomes

$$\Delta G = -30.5\,\frac{\text{kJ}}{\text{mol}} + \left(0.008315\,\frac{\text{kJ}}{\text{mol} \cdot \text{K}} \right)(298\,\text{K}) \ln \left(\frac{\frac{(0.0001\,\text{M})}{(1\,\text{M})} \frac{(0.035\,\text{M})}{(1\,\text{M})} \frac{(10^{-7.4}\,\text{M})}{(10^{-7}\,\text{M})}}{\frac{(0.005\,\text{M})}{(1\,\text{M})}(1)} \right) \tag{3.36a}$$

or

$$\Delta G = -30.5\,\frac{\text{kJ}}{\text{mol}} + \left(2.478\,\frac{\text{kJ}}{\text{mol}} \right) \ln \left(\frac{(0.0001)(0.035)(0.398)}{(0.005)} \right) \tag{3.36b}$$

$$\Delta G = -30.5\,\frac{\text{kJ}}{\text{mol}} + -20.3\,\frac{\text{kJ}}{\text{mol}} = -50.8\,\frac{\text{kJ}}{\text{mol}} \tag{3.36c}$$

Note that in Equation 3.36b, the Q term has no units. In addition, the value calculated for ΔG is much more negative (i.e., more favorable) than the standard free energy change $\Delta G^{\circ\prime}$. This last point underscores the fact that $\Delta G^{\circ\prime}$ alone is not always a reliable predictor of reaction favorability in vivo.

Ultimately it is ΔG and not $\Delta G^{\circ\prime}$ that determines the driving force for a reaction; however, to evaluate ΔG using Equation 3.23 we must be given, or be able to calculate, $\Delta G^{\circ\prime}$ for the reaction of interest. Recall that $\Delta G^{\circ\prime}$ is easily calculated from K_{eq} using Equation 3.27. In the remaining pages of this chapter we will use examples relevant to biochemistry to illustrate two alternative methods for calculating $\Delta G^{\circ\prime}$.

Phosphate Transfer Potential

The phosphate transfer potential shows which compounds can phosphorylate others under standard conditions.

There is another useful way in which we can think about the $\Delta G^{\circ\prime}$ values for hydrolysis of phosphate compounds. As Figure 3.7 shows, these values form a scale of **phosphate transfer potentials**. The potential is defined simply as $-\Delta G^{\circ\prime}$

of hydrolysis; thus, of the compounds listed in Table 3.5, PEP has the highest phosphate transfer potential (61.9 kJ/mol) and glycerol-1-phosphate has the lowest (9.2 kJ/mol). Each compound is capable of driving the phosphorylation of compounds lower on the scale, provided that a suitable coupling mechanism is available. Consider, for example, the following reactions:

(1) Hydrolysis of phosphoenolpyruvate	$PEP + H_2O \rightleftharpoons$ pyruvate $+ P_i$	$\Delta G^{\circ\prime} = -61.9$ kJ/mol
(2) Phosphorylation of adenosine diphosphate	$ADP + P_i + H^+ \rightleftharpoons ATP + H_2O$	$\Delta G^{\circ\prime} = +30.5$ kJ/mol
(1) + (2): Coupled phosphorylation of ADP by PEP	$PEP + ADP + H^+ \rightleftharpoons ATP +$ pyruvate	$\Delta G^{\circ\prime} = -31.4$ kJ/mol

Thus, PEP, which has a greater phosphate transfer potential than ATP, is capable of adding a phosphate group to ADP in a thermodynamically favored process. ATP, in turn, can phosphorylate glucose because the phosphate transfer potential of glucose-6-phosphate (G-6-P) lies still farther down the scale:

(1) Hydrolysis of ATP	$ATP + H_2O \rightleftharpoons ADP + P_i + H^+$	$\Delta G^{\circ\prime} = -30.5$ kJ/mol
(2) Phosphorylation of glucose	glucose $+ P_i \rightleftharpoons$ G-6-P $+ H_2O$	$\Delta G^{\circ\prime} = +13.8$ kJ/mol
(1) + (2): Coupled phosphorylation of glucose by ATP	glucose $+ ATP \rightleftharpoons ADP +$ G-6-P $+ H^+$	$\Delta G^{\circ\prime} = -16.7$ kJ/mol

These examples emphasize how ATP can act as a versatile phosphate transfer agent through coupled reactions. In each case, the coupling is accomplished by having the reactions take place on the surface of a large protein molecule (i.e., an *enzyme*). We shall discuss in Chapter 11 how enzymes both facilitate such coupling and accelerate reactions.

The examples above also show how $\Delta G^{\circ\prime}$ can be calculated for a reaction of interest by summing the $\Delta G^{\circ\prime}$ values for two (or more) reactions that, when added together, yield an overall chemical equation that is *identical* to the reaction of interest. Note that when the chemical equation for a reaction is reversed, the sign of $\Delta G^{\circ\prime}$ is also reversed. For more practice with the calculation of $\Delta G^{\circ\prime}$ by this method, see Problem 17b at the end of this chapter. Finally, we consider the calculation of $\Delta G^{\circ\prime}$ for the important class of oxidation/reduction reactions.

$\Delta G^{\circ\prime}$ for Oxidation/Reduction Reactions

The oxidation of nutrients such as carbohydrates or fats provides cells with substantial free energy for the synthesis of ATP. As we shall see in Chapter 15, the transfer of electrons from nutrients to O_2 releases energy that is stored in the form of a proton gradient across a membrane that, in turn, provides the driving force for the synthesis of ATP in mitochondria. The electron transfer occurs via a series of linked oxidations and reductions, or "redox" reactions. To understand the processes by which metabolic energy is extracted from nutrients, one must understand how to calculate the free energy available from a redox reaction.

Quantitation of Reducing Power: Standard Reduction Potential

Redox chemistry is comparable in many ways to acid–base chemistry, which we discussed in Chapter 2. In an acid–base equilibrium we have an acid (HA) and its conjugate base (A^-), which represent a proton donor and a proton acceptor, respectively.

$$HA \rightleftharpoons H^+ + A^-$$

e.g., $\qquad CH_3COOH_{(aq)} \rightleftharpoons H^+_{(aq)} + CH_3COO^-_{(aq)}$

Similarly, in a redox reaction we have a donor and acceptor of *electrons*.

reduced compound (e^- donor) $\rightleftharpoons$ oxidized compound (e^- acceptor) $+ e^-$

e.g., $\qquad Fe^{2+} \rightleftharpoons Fe^{3+} + e^-$

Free protons and free electrons exist at negligible concentrations in aqueous media, so these equilibrium expressions are merely half-reactions in an overall acid–base or redox reaction scheme. A complete redox reaction must show one reactant as an electron acceptor, which becomes reduced by gaining electrons, and another reactant as an electron donor, which becomes oxidized by losing electrons. One well-known mnemonic for the definitions of **reduction** and **oxidation** is "OIL-RIG": Oxidation Is Loss (of electrons); Reduction Is Gain (of electrons). Of the two reactants in a redox reaction, the electron donor is the **reductant** (or **reducing agent**), which becomes oxidized while transferring electrons to the other reactant, the **oxidant** (or **oxidizing agent**). The general form of a redox reaction is then:

$$\text{reductant} + \text{oxidant} \rightleftharpoons \text{oxidized reductant} + \text{reduced oxidant}$$

$$\text{or: } A_{(red)} + B_{(ox)} \rightleftharpoons A_{(ox)} + B_{(red)}$$

$$\text{e.g., } Cu^{1+} + Fe^{3+} \rightleftharpoons Cu^{2+} + Fe^{2+}$$

Cu^{1+} is the reductant in this reaction because it is the electron donor. The reductant is thus analogous to the acid in an acid–base equilibrium.

Critical to our understanding of acid–base chemistry is the concept of pK_a, which represents a quantitative measure of the tendency of an acid to lose a proton. In the same sense, our understanding of biological oxidations demands a comparable measure of the tendency of a reductant to lose electrons (or of an oxidant to gain electrons). Such an index is provided by the **standard reduction potential**, or E^0. In acid–base equilibria we arbitrarily define water, with a pK_a of 7.0, as neutral. Redox chemistry also employs a reference standard, namely, the *standard hydrogen electrode* in an electrochemical cell.

An electrochemical cell consists of two **half-cells**, each containing an electron donor and its conjugate acceptor (see margin). In the diagram the left-hand beaker constitutes the reference half-cell, a standard hydrogen electrode, with H^+ at 1 M and H_2 at 1 bar (the standard unit of pressure, which is equal to 100 kPa, or ~750 torr). The right-hand beaker contains the test half-cell, with the solution containing the test electron donor and its conjugate acceptor, each at 1 M concentration. In this example, the solution contains Fe^{2+} and Fe^{3+}, each at 1 M. The half-cells are connected with an agar salt bridge, which maintains charge neutrality by allowing ions to flow between the cells as electrons move through the completed circuit. A galvanometer placed between the two half-cells measures the **electromotive force**, or **emf**, in volts. Electromotive force is a measure of the potential, or "pressure," for electrons to flow from one half-cell to the other. The electrons may flow either toward or away from the reference half-cell, depending on whether H_2 or the test electron donor has the greater tendency to lose electrons. Because H_2 loses electrons more readily than Fe^{2+}, the electrons in our example will flow from the reference half-cell to the test half-cell, oxidizing H_2, reducing Fe^{3+}, and causing the galvanometer to record a positive emf. If the test electron donor loses electrons more readily than H_2, then electrons flow in the reverse direction, reducing H^+ to H_2 in the reference half-cell and causing a negative emf to be recorded. The stronger oxidant, whether H^+ or the test electron acceptor, will draw electrons away from the other half-cell and become reduced.

By convention, E^0 for the standard hydrogen electrode is set at 0.00 volts. Any redox couple that tends to donate electrons to the standard hydrogen electrode has a negative value of E^0. A positive E^0 means that electrons from H_2 are flowing toward the test cell and reducing the electron acceptor, or, that the test cell acceptor is oxidizing H_2. *The higher the value of E^0 for a redox couple, the stronger an oxidant is the electron acceptor of that couple.*

As discussed earlier, standard conditions for biochemists include a pH value of 7.0, a condition far from that seen in the standard hydrogen electrode, which contains H^+ at 1.0 M. Therefore, biochemists use a modified term, $E^{0\prime}$, which is the standard reduction potential for a half-reaction measured at 10^{-7} M H^+. These

E^0 is the tendency of a reductant to lose an electron, in the same sense that pK_a is the tendency of an acid to lose a proton.

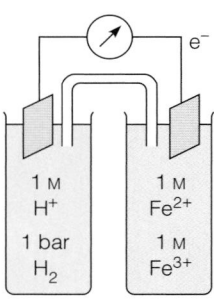

The greater the standard reduction potential, the greater the tendency of the oxidized form of a redox couple to attract electrons.

TABLE 3.6 A few standard reduction potentials ($E^{0'}$) of interest in biochemistry

Oxidant	Reductant	n	$E^{0'}$ (V)
$H^+ + e^-$	½ H_2	1	−0.421
$NAD^+ + H^+ + 2e^-$	NADH	2	−0.315
1,3-Bisphosphoglycerate $+ 2H^+ + 2e^-$	Glyceraldehyde-3-phosphate $+ P_i$	2	−0.290
$FAD + 2H^+ + 2e^-$	$FADH_2$	2	−0.219
Acetaldehyde $+ 2H^+ + 2e^-$	Ethanol	2	−0.197
Pyruvate $+ 2H^+ + 2e^-$	Lactate	2	−0.185
$Fe^{3+} + e^-$	Fe^{2+}	1	+0.769
½$O_2 + 2H^+ + 2e^-$	H_2O	2	+0.815

Note: $E^{0'}$ is the standard reduction potential at pH 7 and 25 °C, n is the number of electrons transferred, and each potential is for the partial reaction written as follows:

$$\text{Oxidant} + ne^- \rightarrow \text{reductant}$$

The entry for the H^+/H_2 couple ($E^{0'} = -0.421$ V) is not zero because it is measured with $[H^+] = 1\,\text{M}$ in the reference cell (i.e., the standard hydrogen electrode) and $[H^+] = 10^{-7}\,\text{M}$ in the test cell.

are the values used in this book and most other biochemical references. $E^{0'}$ values for a few biochemically important redox pairs are recorded in Table 3.6 (a more complete table of $E^{0'}$ values is given in Chapter 15, page 629). This table is organized with the strongest oxidizing agents listed at the bottom of the "Oxidant" column, and the strongest reducing agents listed at the top of the "Reductant" column. Thus, O_2 in the O_2/H_2O couple is the strongest oxidant, and H_2 in the H^+/H_2 couple is the strongest reducing agent. Another way to express this relationship is to say that H_2O in the O_2/H_2O couple is the weakest reducing agent (just as the strongest acid has the weakest conjugate base). There is very little tendency for water to give up electrons and become oxidized to O_2, because none of the common biological oxidants has a higher $E^{0'}$ than O_2/H_2O. Photosynthesis, which does oxidize H_2O to O_2, requires considerable energy in the form of sunlight to accomplish this feat (more about this in Chapter 16). The information in Table 3.6 is useful because the favorable direction of electron flow in a redox reaction is from the reductant in one redox couple to the oxidant in another couple listed lower in the table.

In the next section we illustrate how we use the information in Table 3.6 to calculate $\Delta G^{\circ\prime}$ and ΔG using examples of reactions involving the important biological electron carrier **nicotinamide adenine dinucleotide**, which is stable in both the oxidized (NAD^+) and reduced (**NADH**) form. We will discuss the structure and function of $NAD^+/NADH$ in greater detail in later chapters.

Free Energy Changes from Oxidation–Reduction Reactions

To recapitulate, the greater the value of $E^{0'}$ for a redox couple, the greater is the tendency for that couple to participate in oxidation of another substrate. We can describe this tendency in quantitative terms because free energy changes are directly related to differences between the reduction potentials listed in Table 3.6:

$$\Delta G^{\circ\prime} = -nF\Delta E^{0'} = -nF\left[E^{0'}_{\text{(acceptor)}} - E^{0'}_{\text{(donor)}}\right] \qquad (3.37)$$

where n is the number of electrons transferred in the half-reactions, F is Faraday's constant (96.5 kJ mol^{-1}V^{-1}), and $\Delta E^{0'}$ is the difference in standard reduction potentials between the two redox couples. Note that $\Delta E^{0'}$ is to ΔE^0 as $\Delta G^{\circ\prime}$ is to ΔG°; the "prime" has the same significance in both cases: each reactant and product (except H^+) is at 1M and the reaction is carried out at a pH of 7.0.

Free Energy Changes Under Standard Conditions

As an example, consider the oxidation of ethanol by NAD^+. This reaction is catalyzed by the enzyme alcohol dehydrogenase:

$$\text{ethanol} + NAD^+ \rightleftharpoons \text{acetaldehyde} + NADH + H^+$$

The two half-reactions given in Table 3.6 are written in the direction of reduction (i.e., as half-reactions for electron acceptors):

(a) $NAD^+ + H^+ + 2e^- \rightleftharpoons NADH$ $E^{0\prime} = -0.315$ V

(b) $\text{acetaldehyde} + 2H^+ + 2e^- \rightleftharpoons \text{ethanol}$ $E^{0\prime} = -0.197$ V

Because ethanol becomes oxidized in the reaction, we reverse the second half-reaction to give the desired half-reaction for the electron donor:

(c) $\text{ethanol} \rightleftharpoons \text{acetaldehyde} + 2H^+ + 2e^-$

Now the overall redox reaction is the sum of half-reactions (a) and (c), in which NAD^+ is the electron acceptor, ethanol is the electron donor, and $\Delta E^{0\prime}$ is calculated as shown in Equation 3.37:

$$\Delta E^{0\prime} = E^{0\prime}{}_{(acceptor)} - E^{0\prime}{}_{(donor)} = (-0.315\text{V}) - (-0.197\text{V}) = -0.118\text{V}$$

The standard free energy change is thus:

$$\Delta G^{\circ\prime} = -nF\Delta E^{0\prime} = -(2)\left(96485\frac{\text{J}}{\text{mol}\cdot\text{V}}\right)(-0.118\text{V}) = +22.8\frac{\text{kJ}}{\text{mol}}$$

Note from this example that a negative $\Delta E^{0\prime}$ value gives a positive $\Delta G^{\circ\prime}$ and hence corresponds to a reaction that, under standard conditions, is *not* favorable in the direction written. Note also that if we were to calculate $\Delta G^{\circ\prime}$ for the reverse reaction (reduction of acetaldehyde by NADH), $\Delta E^{0\prime}$ would be +0.118 V, and $\Delta G^{\circ\prime}$ would be −22.8 kJ/mol.

Calculating Free Energy Changes for Biological Oxidations Under Nonequilibrium Conditions

The values given in Table 3.6 (and also in Table 15.1, page 629) allow calculation of the biochemical standard state free energy changes. To calculate ΔG for a redox reaction under nonstandard conditions, we must use Equation 3.23. Let us consider the free energy change associated with the transfer of electrons from NADH to O_2, which is one of the significant energy producing reactions in mitochondria. The overall electron transport process, which actually occurs in several steps, is given by the following equation:

$$O_2 + 2NADH + 2H^+ \rightleftharpoons 2H_2O + 2NAD^+$$

According to Equation 3.23, ΔG for the reaction is given by

$$\Delta G = \Delta G^{\circ\prime} + RT\ln\left(\frac{[H_2O]^2[NAD^+]^2}{[O_2][NADH]^2[H^+]^2}\right) \tag{3.38a}$$

Let us assume that inside the mitochondrion the temperature is 37 °C, pH = 8.4, the partial pressure of oxygen is 2 torr, and the concentrations of NAD^+ and NADH are 10 mM and 100 μM, respectively. The value of the $RT\ln Q$ term can be calculated as described previously; however, in this example we will enter the concentration term for oxygen relative to the standard concentration for a gas, which is 1 bar (or 750 torr), rather than as a molar concentration:

$$\Delta G = \Delta G^{\circ\prime} + \left(8.315\frac{\text{J}}{\text{mol}\cdot\text{K}}\right)(310\text{K})\ln\left(\frac{(1)^2\left(\frac{0.010\,\text{M}}{1\,\text{M}}\right)^2}{\left(\frac{2\,\text{torr}}{750\,\text{torr}}\right)\left(\frac{0.00010\,\text{M}}{1\,\text{M}}\right)^2\left(\frac{10^{-8.4}\,\text{M}}{10^{-7.0}\,\text{M}}\right)^2}\right)$$

$$\tag{3.38b}$$

or

$$\Delta G = \Delta G^{\circ\prime} + \left(2.58\,\frac{kJ}{mol}\right)\ln\left(\frac{(1)(10^{-4})}{(2.66 \times 10^{-3})(10^{-8})(1.58 \times 10^{-3})}\right) \qquad (3.38c)$$

$$\Delta G = \Delta G^{\circ\prime} + \left(2.58\,\frac{kJ}{mol}\right)\ln\left(2.37 \times 10^{9}\right) = \Delta G^{\circ\prime} + \left(55.6\,\frac{kJ}{mol}\right) \qquad (3.38.d)$$

Now we turn our attention to the calculation of $\Delta G^{\circ\prime}$. Because this is a biochemical redox reaction we can use the information in Table 3.6 and Equation 3.37. Which values of $E^{0\prime}$ should be used to calculate $\Delta E^{0\prime}$ (and thereby $\Delta G^{\circ\prime}$)? To answer that question we must first identify the relevant "half-reactions." In this case, the acceptor is O_2, so we select the half-reaction for the reduction of oxygen to water

$$\frac{1}{2}O_2 + 2H^+ + 2e^- \rightleftharpoons H_2O \qquad\qquad E^{0\prime} = +0.815\text{ V}$$

The donor is NADH, so we select the half-reaction for the oxidation of NADH to NAD^+

$$NAD^+ + H^+ + 2e^- \rightleftharpoons NADH \qquad\qquad E^{0\prime} = -0.315\text{ V}$$

then reverse this half-reaction to show that NADH loses electrons to O_2:

$$NADH \rightleftharpoons NAD^+ + H^+ + 2e^-$$

The half-reactions are adjusted for stoichiometry (such that the e^- terms cancel) and then added together to recreate the overall reaction of interest:

$$O_2 + 4H^+ + 4e^- \rightleftharpoons 2H_2O$$
$$\underline{2NADH \rightleftharpoons 2NAD^+ + 2H^+ + 4e^-}$$
$$\text{net: } O_2 + 2H^+ + 2NADH \rightleftharpoons 2NAD^+ + 2H_2O$$

Note that when we adjust stoichiometry (i.e., by multiplying each half-reaction by a factor of 2), we do not make any adjustments to the value of $E^{0\prime}$. The adjustment for changes in stoichiometry is accounted for in the value of n used in Equation 3.37. When we multiply each half reaction by 2 we also double the number of electrons transferred; thus, the value of n is also doubled, and

$$\Delta G^{\circ\prime} = -nF\Delta E^{0\prime} = -(4)\left(96485\,\frac{J}{mol \cdot V}\right)(+1.130V) = -436.2\,\frac{kJ}{mol} \qquad (3.39)$$

We can now use this value of $\Delta G^{\circ\prime}$ in Equation 3.38d to calculate ΔG for the reaction

$$\Delta G = \left(-436.2\,\frac{kJ}{mol}\right) + \left(+55.6\,\frac{kJ}{mol}\right) = -380.6\,\frac{kJ}{mol} \qquad (3.40)$$

This calculation shows that under the specified conditions, the oxidation of NADH by O_2 is highly favorable. As will be described in greater detail in Chapter 15, a significant portion of this energy is stored in the form of a proton concentration gradient across the inner mitochondrial membrane. The free energy available in the proton gradient (see Equation 3.19) is sufficient to drive the synthesis of ATP from ADP in mitochondria. We will revisit these concepts in Chapter 10 when we discuss membrane transport.

SUMMARY

Bioenergetics is that branch of thermodynamics that deals with energy acquisition, exchange, and utilization in organisms. The internal energy (U) of a system includes all energy that can be exchanged by nonnuclear processes: the energy of motion of atoms and molecules and the energy of chemical bonds and noncovalent interactions. U is determined by the state of the system and can be changed only by exchange of heat or work with the surroundings ($\Delta U = q - w$). This is the first law of thermodynamics. Under constant-volume conditions, $q = \Delta U$. Under constant-pressure conditions, $q = \Delta U + P\Delta V = \Delta H$, where H denotes the enthalpy ($H = U + PV$). In biochemistry, ΔH is more relevant than ΔU.

Processes may be reversible (near equilibrium) or irreversible (far from equilibrium). The thermodynamically favored direction of a reaction (the direction that leads toward equilibrium) is determined by changes in both the enthalpy (H) and the entropy (S, a measure of randomness). The free energy, $G = H - TS$, takes both into account. The criterion for a favorable process is that the free energy change, $\Delta G = \Delta H - T\Delta S$, be negative ("exergonic"), rather than positive ("endergonic"); this is one statement of the second law of thermodynamics. The ice-to-water transition demonstrates the importance of temperature (T) in determining reaction direction. At the melting point, solid and liquid are in equilibrium ($\Delta G = 0$). The entropy of an open system can decrease, as in freezing of water, but only if the enthalpy decreases. Thus, organisms must constantly expend energy to maintain organization. In every energy transfer, some part of the energy (ΔH) is lost as heat ($T\Delta S$), so ΔG is a measure of the energy that is potentially available for useful work.

To apply thermodynamic relationships in biochemical problems, we use chemical potential, which relates the concentration of each substance to its contribution to the total free energy of the system. From the chemical potential we obtain the equation governing free energy changes in chemical reactions:

$$\Delta G = \Delta G^{\circ\prime} + RT \ln Q$$

To evaluate the mass action term Q, biochemists commonly use molar concentrations for dilute aqueous solutes, which approximate the *activities* of the solutes in the reaction and are expressed relative to the "biochemical" standard concentrations listed in the text.

There are three common methods for the evaluation of $\Delta G^{\circ\prime}$:

1. $\Delta G^{\circ\prime}$ can be calculated from the equilibrium constant, K, using $\Delta G^{\circ\prime} = -RT \ln K$.

2. $\Delta G^{\circ\prime}$ can be calculated from tables of standard reduction potentials, $E^{0\prime}$, using $\Delta G^{\circ\prime} = -nF\Delta E^{0\prime}$.

3. $\Delta G^{\circ\prime}$ for a reaction of interest can be calculated from the values of $\Delta G^{\circ\prime}$ for two or more reactions that sum to give the chemical equation of interest.

Reactions that are not thermodynamically favored may nevertheless be driven forward if coupled to reactions that have large negative ΔG values. In living systems, the hydrolysis of certain phosphate compounds is frequently used for this purpose. The phosphate transfer potential ranks these compounds according to their ability to phosphorylate other compounds under standard conditions. ATP, the most important of these compounds, is generated in the energy-producing metabolic pathways and is used to drive many reactions.

REFERENCES

This chapter has presented an abbreviated treatment of thermodynamics. For the student who wishes a more rigorous background in this field and more information about its applications to biochemistry, we recommend the following books:

Dill, K. A., and S. Bromberg (2010) *Molecular Driving Forces* (2nd ed.). Garland Science, New York. An excellent resource for those desiring a clear and comprehensive presentation of thermodynamic principles.

Eisenberg, D., and D. Crothers (1979) *Physical Chemistry with Applications to the Life Sciences*. Benjamin/Cummings, Redwood City, Calif. A very fine physical chemistry text, written by two physical biochemists.

Strongly recommended, as it contains many biochemical applications of physical–chemical principles not found in most physical chemistry texts.

Klotz, I. (1986) *Introduction to Biomolecular Energetics*. Academic Press, New York. A brief introduction to thermodynamics for biochemists. Some excellent examples and explanations.

Phillips, R., J. Kondev, and J. Theriot (2009) *Physical Biology of the Cell*. Garland Science, N.Y. Chapters 5 and 6 provide greater detail on many of the topics covered here.

Tinoco, I., K. Sauer, J. C. Wang, and J. D. Puglisi (2002) *Physical Chemistry: Principles and Applications in Biological Sciences* (4th ed.). Prentice

Hall, Upper Saddle River, N.J. Many explicit applications of physical chemistry to the study of biological systems, with many excellent practice problems.

van Holde, K. E., W. C. Johnson, and P. S. Ho (2006) *Principles of Physical Biochemistry* (2nd ed.). Prentice Hall, Upper Saddle River, N. J. Chapters 2–4 extend the applications of thermodynamics to biochemistry.

For a sophisticated discussion of the effect of ionic conditions on the free energy changes in phosphate ester hydrolysis, see the following article:

Alberty, R. A. (1992) Equilibrium calculations on systems of biochemical reactions at specified pH and pMg. *Biophys. Chem.* 42:117–131.

For more on the biochemical standard state and standard free energies see these sources:

Alberty, R. A. (1994) Recommendations for nomenclature and tables in biochemical thermodynamics. *Pure Appl. Chem.* 66:1641–1666.

Frey, P., and A. Arabshahi (1995) Standard free energy change for the hydrolysis of the α,β–phosphoanhydride bridge in ATP. *Biochemistry* 34:11307–11310.

Lundblad, R. L., and F. M. MacDonald (eds.) (2010) *Handbook of Biochemistry and Molecular Biology* (4th ed.). CRC Press, Boca Raton, FL.

Méndez, E. (2008) Biochemical thermodynamics under near physiological conditions. *Biochem. Mol. Biol. Educ.* 36:116–119.

PROBLEMS

1. The enthalpy change (heat of fusion, ΔH_f) for the transition ice $\rightarrow$ water at 0 °C and 1 atm pressure is $+6.01$ kJ/mol. The change in volume when 1 mole of ice is melted is -1.625 cm^3/mol $= -1.625 \times 10^{-6}$ m^3/mol. Calculate the difference between ΔH_f and ΔU_f for this process, and express it as a percentage of ΔH_f. [Note: 1 atm $= 1.013 \times 10^5$ N/m^2 in SI units, where N = newton.]

*2. Given the following reactions and their enthalpies:

	ΔH (kJ/mol)
$H_2(g) \rightarrow 2H(g)$	$+436$
$O_2(g) \rightarrow 2O(g)$	$+495$
$H_2(g) + \frac{1}{2}O_2(g) \rightarrow H_2O(g)$	-242

(a) Devise a way to calculate ΔH for the reaction

$$H_2O(g) \rightarrow 2H(g) + O(g)$$

(b) From this, estimate the H—O bond energy.

3. The decomposition of crystalline N_2O_5

$$N_2O_5(s) \rightarrow 2NO_2(g) + \frac{1}{2}O_2(g)$$

is an example of a reaction that is thermodynamically favored even though it absorbs heat. At 25 °C we have the following values for the standard state enthalpy and free energy changes of the reaction:

$$\Delta H° = +109.6 \text{ kJ/mol}$$

$$\Delta G° = -30.5 \text{ kJ/mol}$$

(a) Calculate $\Delta S°$ at 25 °C.
(b) Why is the entropy change so favorable for this reaction?
(c) Calculate $\Delta U°$ for this reaction at 25 °C.
(d) Why is $\Delta H°$ greater than $\Delta U°$?

4. The oxidation of glucose to CO_2 and water is a major source of energy in aerobic organisms. It is a reaction favored mainly by a large negative enthalpy change.

$$C_6H_{12}O_6(s) + 6O_2(g) \rightarrow 6CO_2(g) + 6H_2O(l)$$

$$\Delta H° = -2816 \text{ kJ/mol} \qquad \Delta S° = +181 \text{ J/mol·K}$$

(a) At 37 °C, what is the value for $\Delta G°$?

(b) In the overall reaction of aerobic metabolism of glucose, 32 moles of ATP are produced from ADP for every mole of glucose oxidized. Calculate the standard state free energy change for the *overall* reaction when glucose oxidation is coupled to the formation of ATP at 37 °C.
(c) What is the *efficiency* of the process in terms of the percentage of the available free energy change captured in ATP?

*5. The first reaction in glycolysis is the phosphorylation of glucose:

$$P_i + \text{glucose} \rightarrow \text{glucose-6-phosphate} + H_2O$$

This is a thermodynamically unfavorable process, with $\Delta G° = +13.8$ kJ/mol.
(a) In a liver cell at 37 °C the concentrations of both phosphate and glucose are normally maintained at about 5 mM each. What would be the *equilibrium* concentration of glucose-6-phosphate, according to the above data?
(b) This very low concentration of the desired product would be unfavorable for glycolysis. In fact, the reaction is coupled to ATP hydrolysis to give the overall reaction

$$\text{ATP} + \text{glucose} \rightarrow \text{glucose-6-phosphate} + \text{ADP} + H^+$$

What is $\Delta G°$ for the coupled reaction?
(c) If, in addition to the constraints on glucose concentration listed previously, we have in the liver cell ATP concentration = 3 mM and ADP concentration = 1 mM, what is the theoretical concentration of glucose-6-phosphate at equilibrium at pH = 7.4 and 37 °C? The answer you will obtain is an absurdly high value for the cell and in fact is never approached in reality. Explain why.

6. In another key reaction in glycolysis, dihydroxyacetone phosphate (DHAP) is isomerized into glyceraldehyde-3-phosphate (GAP):

Because $\Delta G°'$ is positive, the equilibrium lies to the left.
(a) Calculate the equilibrium constant, and the equilibrium fraction of GAP from the above, at 37 °C.
(b) In the cell, depletion of GAP makes the reaction proceed. What will ΔG be if the concentration of GAP is always kept at 1/100 of the concentration of DHAP?

7. Assume that some protein molecule, in its folded native state, has *one* favored conformation. But, when it is denatured, it becomes a "random coil," with many possible conformations.
(a) What must be the sign of ΔS for the change: native $\rightarrow$ denatured?
(b) How will the contribution of ΔS for native $\rightarrow$ denatured affect the favorability of the process? What *apparent* requirement does this impose on ΔH if proteins are to be stable structures?

8. When a hydrophobic substance like a hydrocarbon is dissolved in water, a clathrate cage of ordered water molecules is formed about it (see Figure 2.13 in Chapter 2). What do you expect the sign of ΔS to be for this process? Explain your answer.

*9. It is observed that as temperature is increased, most protein molecules go from their defined, folded state into a random-coil, denatured state which exposes more hydrophobic surface area than is exposed in the folded state.
(a) Given what you have learned so far about ΔH and ΔS, explain why this is reasonable. [Hint: Consider Problem 7.]
(b) Sometimes, however, proteins denature as temperature is *decreased*. How might this be explained? [Hint: Consider Problem 8.]

*10. Suppose a reaction has $\Delta H°$ and $\Delta S°$ values independent of temperature. Show from this, and equations given in this chapter, that

$$\ln K = \frac{-\Delta H°}{RT} + \frac{\Delta S°}{R}$$

where K is the equilibrium constant. How could you use values of K determined at different temperatures to determine $\Delta H°$ for the reaction?

11. The following data give the ion product, K_w (see Equation 2.6b on page 39), for water at various temperatures:

T (°C)	K_w (M^2)
0	1.14×10^{-15}
25	1.00×10^{-14}
30	1.47×10^{-14}
37	2.56×10^{-14}

(a) Using the results from Problem 10, calculate $\Delta H°$ for the ionization of water.
(b) Use these data, and the ion product at 25 °C, to calculate $\Delta S°$ for water ionization (hint: use the chemical standard state for the activity of water rather than that for the biochemical standard state activity of water).

12. The phosphate transfer potentials for glucose-1-phosphate and glucose-6-phosphate are 20.9 kJ/mol and 13.8 kJ/mol, respectively.
(a) What is the equilibrium constant for this reaction at 25 °C?

Glucose-1-phosphate   Glucose-6-phosphate

(b) If a mixture was prepared containing 1 M glucose-6-phosphate and 1×10^{-3} M glucose-1-phosphate, what would be the thermodynamically favored direction for the reaction?

13. As you take each breath, you expel about 0.5 L against a pressure of 1 atm.
(a) If you breathe about 30 times per minute, how much work do you do in this way each day? (Neglect any work involved in inhaling.)
(b) Using the $\Delta G°'$ value for ATP hydrolysis, what is the minimum number of moles of ATP you will utilize per day just in breathing?

14. Undergoing moderate activity, an average person will generate about 350 kJ of heat per hour. Using the heat of combustion of palmitic acid (Equation 3.6) as an approximate value for fatty substances, estimate how many grams of fat would be required per day to sustain this level, if all were burned for heat.

*15. The major difference between a protein molecule in its native state and in its denatured state lies in the number of conformations available. To a first approximation, the native, folded state can be thought to have one conformation. The unfolded state can be estimated to have three possible orientations about each bond between residues.
(a) For a protein of 100 residues, estimate the entropy change per mole upon denaturation.
(b) What must be the enthalpy change accompanying denaturation to allow the protein to be half-denatured at 50 °C?
(c) Will the fraction denatured increase or decrease with increasing temperature?

16. Suppose the concentration of glucose inside a cell is 0.1 mM and the cell is suspended in a glucose solution of 0.01 mM.
(a) What would be the free energy change involved in transporting 10^{-6} mole of glucose from the medium into the cell? Assume $T = 37$ °C.
(b) What would be the free energy change involved in transporting 10^{-6} mole of glucose from the medium into the cell if the intracellular and extracellular concentrations were 1 mM and 10 mM, respectively?
(c) If the processes described in parts (a) and (b) were coupled to ATP hydrolysis, how many moles of ATP would have to be hydrolyzed in order to make *each* process favorable? (Use the standard free energy change for ATP hydrolysis.)

17. For parts (a) and (b) of this problem use the following standard reduction potentials, free energies, and nonequilibrium concentrations of reactants and products:

ATP = 3.10 mM	P_i = 5.90 mM	ADP = 220 μM
glucose = 5.10 mM	pyruvate = 62.0 μM	
NAD^+ = 350 μM	NADH = 15.0 μM	CO_2 = 15.0 torr

half reaction	$E^{0'}$ (V)
$NAD^+ + H^+ + 2e^- \rightarrow NADH$	−0.315
$2Pyruvate + 6H^+ + 4e^- \rightarrow glucose$	−0.590

pyruvate + NADH + $2H^+ \rightarrow$ ethanol + NAD^+ + CO_2
$\Delta G°' = -64.4$ kJ/mol

ATP + $H_2O \rightarrow$ ADP + P_i + H^+ $\Delta G°' = -30.5$ kJ/mol

(a) Consider the last two steps in the alcoholic fermentation of glucose by brewer's yeast:

pyruvate + NADH + $2H^+ \rightarrow$ ethanol + NAD^+ + CO_2

Calculate the nonequilibrium concentration of ethanol in yeast cells, if $\Delta G = -38.3$ kJ/mol for this reaction at pH = 7.4 and 37 °C when the reactants and products are at the concentrations given above.
(b) Consider the degradation of glucose to pyruvate by the glycolytic pathway:

glucose + 2ADP + $2P_i$ + $2NAD^+ \rightarrow$ 2 pyruvate + 2ATP + $2H_2O$ + 2NADH + $2H^+$

Calculate ΔG for this reaction at pH = 7.4 and 37 °C.

PART 2

Molecular Architecture of Living Matter

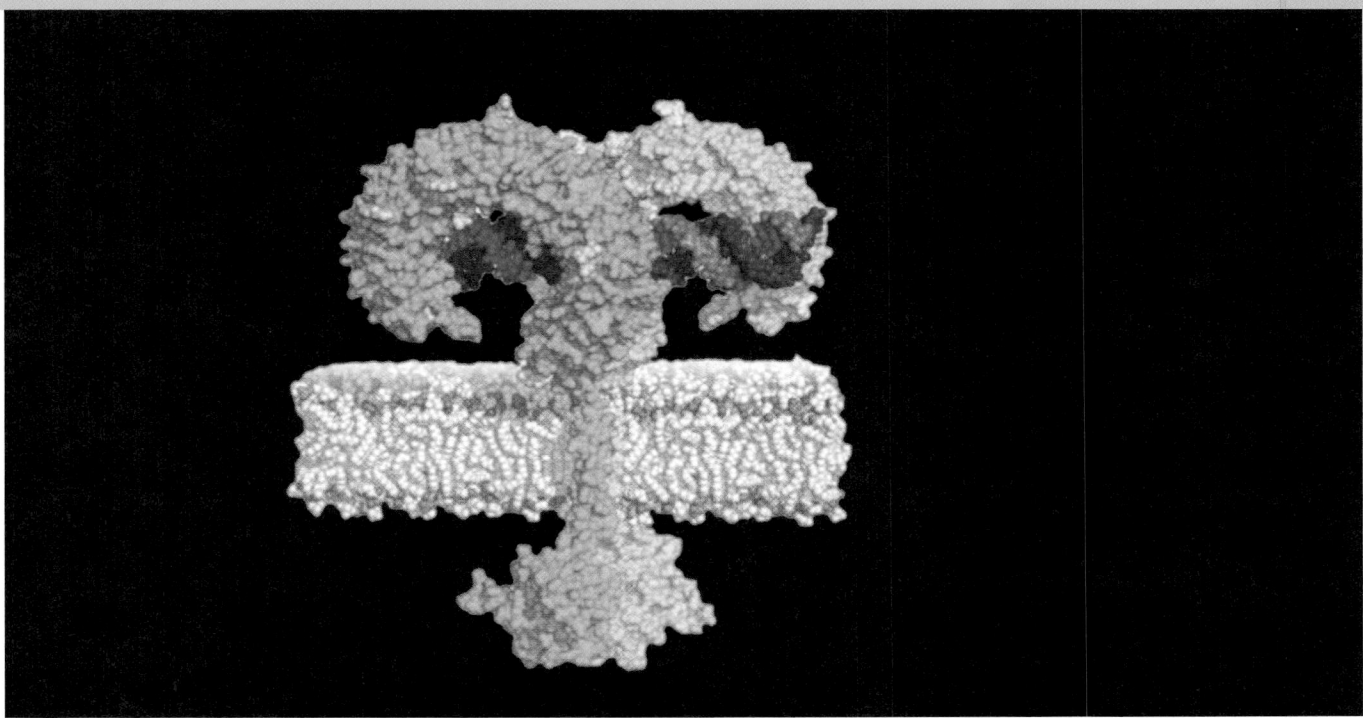

A toll-like receptor protein (green and light blue) binding to double-stranded RNA (red and blue), a signature of many viruses, triggering an inflammatory response that helps to limit the spread of virus infection. Source: From *Science* 320:379–381 L. Liu, I. Botos, Y. Wang, J. N. Leonard, J. Shiloach, D. M. Segal, and D. R. Davies, Structural basis of toll-like receptor 3 signaling with double-stranded RNA. © 2008. Reprinted with permission from AAAS.

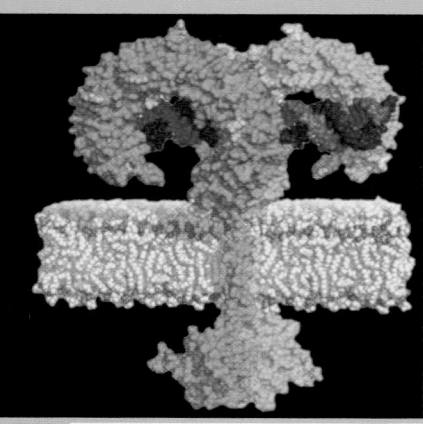

CHAPTER 4

Nucleic Acids

Recall from Chapter 1 the names of the principal organic constituents of cells and organisms—proteins, nucleic acids, carbohydrates, and lipids. Together, they make up a large portion of all living matter. Although they differ in structure, certain design features are held in common. Proteins, nucleic acids, and polysaccharides are all polymers of monomeric constituents, held in linear order by hydrolyzable chemical bonds. Although individual lipid molecules differ in not being polymeric, lipids usually exist as components of giant complexes—membranes and lipoproteins.

The Nature of Nucleic Acids

We begin our treatment of biomolecular architecture with nucleic acids because, with their roles in storage and transmission of biological information, these can be considered the most fundamental of biological molecules. It seems probable that life began its evolution with nucleic acids, for only they, of all biological substances, carry the remarkable potential for self-duplication. The blueprint for an organism is encoded in its nucleic acid, in gigantic molecules like that shown in Figure 1.4 (page 9). Much of an organism's physical development throughout life is programmed in these molecules. The proteins that its cells will make and the functions that they will perform are all recorded on these molecular tapes.

In this chapter, and in the several that follow, we first describe the nucleic acids structurally and then provide a brief introduction to the ways in which they preserve and transmit genetic information. We review events in DNA replication, transcription, and translation, processes that you might have studied previously and which are covered in more detail later in this book. Also, because of the tremendous impact of recombinant DNA techniques upon biomolecular science and technology, some of the most fundamental and informative techniques are briefly presented at the end of this chapter, in Tools of Biochemistry 4B.

The Two Types of Nucleic Acid: DNA and RNA

DNA was discovered in 1869 by Friedrich Miescher, a military surgeon during the Franco-Prussian War. Miescher found quantities of an acidic substance in the pus

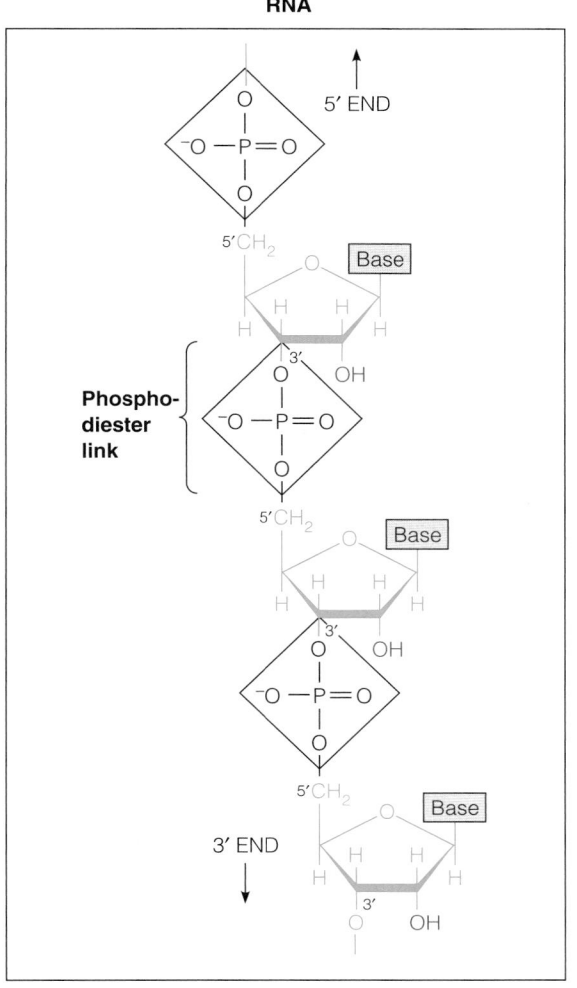
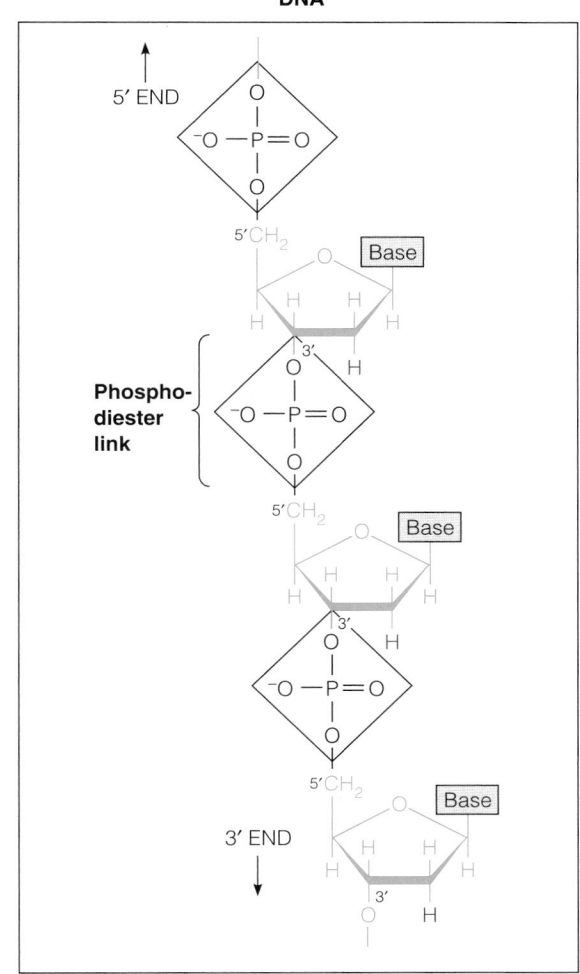

RNA

DNA

FIGURE 4.1

Chemical structures of ribonucleic acid (RNA) and deoxyribonucleic acid (DNA). The ribose–phosphate or deoxyribose–phosphate backbone of each chain is shown in detail. The bases shown schematically here are detailed in Figure 4.2.

from discarded surgical dressings. Because this material was predominantly in the nuclei of the white blood cells that constituted much of this material, he and his student named it (in German) *nukleinsäure*, or nucleic acid. Chemically, nucleic acid was found to consist of organic nitrogenous bases, a pentose sugar, and phosphate. Later, it was recognized that there are two chemical species of nucleic acid, differing in the nature of the sugar component. Miescher discovered **DNA, deoxyribonucleic acid**, and the other major form is **RNA, ribonucleic acid**. As Figure 4.1 shows, each is a polymeric chain in which the monomer units are connected by covalent bonds. Structures of the monomer units of RNA and DNA are shown here:

**Repeating unit of
ribonucleic acid (RNA)**

**Repeating unit of
deoxyribonucleic acid (DNA)**

Both DNA and RNA are polynucleotides. RNA contains the sugar ribose; DNA has deoxyribose.

The nucleic acid bases are of two kinds: the purines, adenine and guanine, and the pyrimidines, cytosine, thymine, and uracil. RNA and DNA use three of the same bases, with uracil being used by RNA, where DNA uses thymine.

Purine

Pyrimidine

FIGURE 4.2

Purine and pyrimidine bases found in DNA and RNA. DNA always contains the bases A, G, C, T, whereas RNA always contains A, G, C, U. Thymine is simply 5-methyluracil.

In each case the monomer unit contains a five-carbon sugar, **ribose** in RNA and **2-deoxyribose** in DNA (shown in blue in the structures). The carbon atoms are designated by primes (1′, 2′, etc.) to distinguish them from atoms in the bases. The difference between the two sugars lies solely in the 2′ hydroxyl group on ribose in RNA, which is replaced by hydrogen in DNA. Each successive monomer unit in both nucleic acids is connected by a phosphate group attached to the hydroxyl on carbon 5′ of one unit and the hydroxyl on carbon 3′ of the next one. This forms a **phosphodiester link** between two sugar residues (Figure 4.1). The name refers to the fact that hydrolysis of this phosphate diester link yields one acid (phosphoric acid) and two alcoholic sugar hydroxyls. In this way, long nucleic acid chains, containing up to hundreds of millions of units, are built up. The phosphate group is a strong acid, with a pK_a of about 1; this is why DNA and RNA are called nucleic *acids*. Every **residue**, or polymerized monomer unit, in a DNA or an RNA molecule carries a negative charge at physiological pH.

The phosphodiester-linked sugar residues form the backbone of the nucleic acid molecule. By itself, the backbone is a repetitious structure, incapable of encoding information. The importance of the nucleic acids in information storage and transmission derives from their being **heteropolymers**. Each monomer in the chain carries a heterocyclic base, always linked to the 1′ carbon of the sugar (see Figure 4.1). The structures of the major bases found in nucleic acids are shown in Figure 4.2. There are two types of heterocyclic bases, which are derivatives of **purine** and of **pyrimidine**.

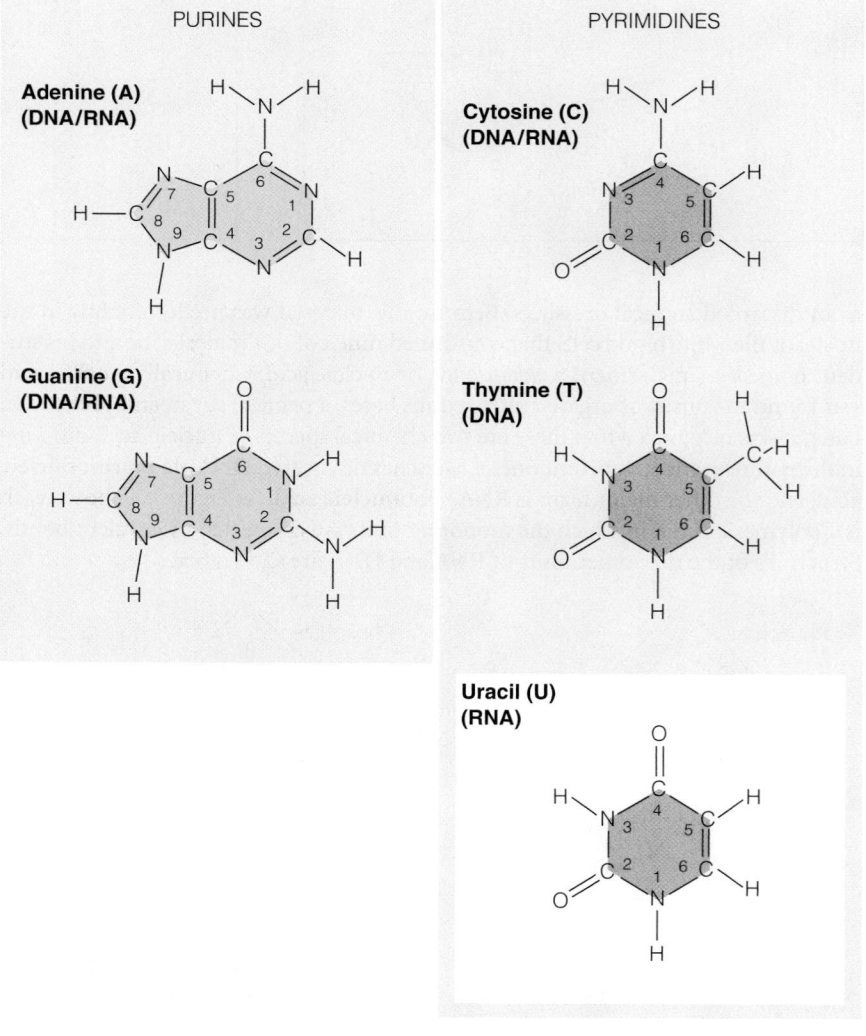

DNA has two purine bases, **adenine (A)** and **guanine (G)**, and two pyrimidines, **cytosine (C)** and **thymine (T)**. RNA has three of the same bases, and the fourth, **uracil (U)**, replaces thymine. RNA, particularly the species called **transfer RNA** (page 108), contains several chemically modified bases, which serve in part to stabilize the molecule, as discussed in Chapter 28. Most eukaryotic DNAs contain a small proportion of cytosines methylated at carbon-5. The biological significance of DNA methylation is discussed in Chapter 26. 5-Hydroxymethylcytosine, with —CH$_2$OH replacing —CH$_3$ (Chapter 22, page 950), has recently been found in some eukaryotic DNAs.

DNA and RNA can each be regarded as a polymer made from four kinds of monomers. The monomers are phosphorylated ribose or deoxyribose molecules with purine or pyrimidine bases attached to their 1′ carbons. In purines the attachment is at nitrogen 9, in pyrimidines at nitrogen 1. The bond between the carbon 1′ of the sugar and the base nitrogen is referred to as a **glycosidic bond**. These monomers are called **nucleotides**. Each nucleotide can be considered the 5′-monophosphorylated derivative of a sugar-base adduct called a **nucleoside** (Figure 4.3). Thus, these nucleotides are also called *nucleoside 5′-monophosphates*. You have already encountered one of these molecules in Chapter 3: adenosine 5′-monophosphate, or AMP.

FIGURE 4.3

Nucleosides and nucleotides. The ribonucleosides and ribonucleotides are shown here; the deoxyribonucleosides and deoxyribonucleotides are identical, except that they lack the 2′OH, and except that T substitutes for the U found in RNA. Each nucleoside is formed by coupling ribose or deoxyribose to a base. The nucleotides, which can be considered the monomer units of nucleic acids, are the 5′-monophosphates of the nucleosides. Nucleoside phosphates with phosphorylation on other hydroxyl groups exist, but they are not found in nucleic acids. The glycosidic bond linking adenine to ribose is shown for adenosine.

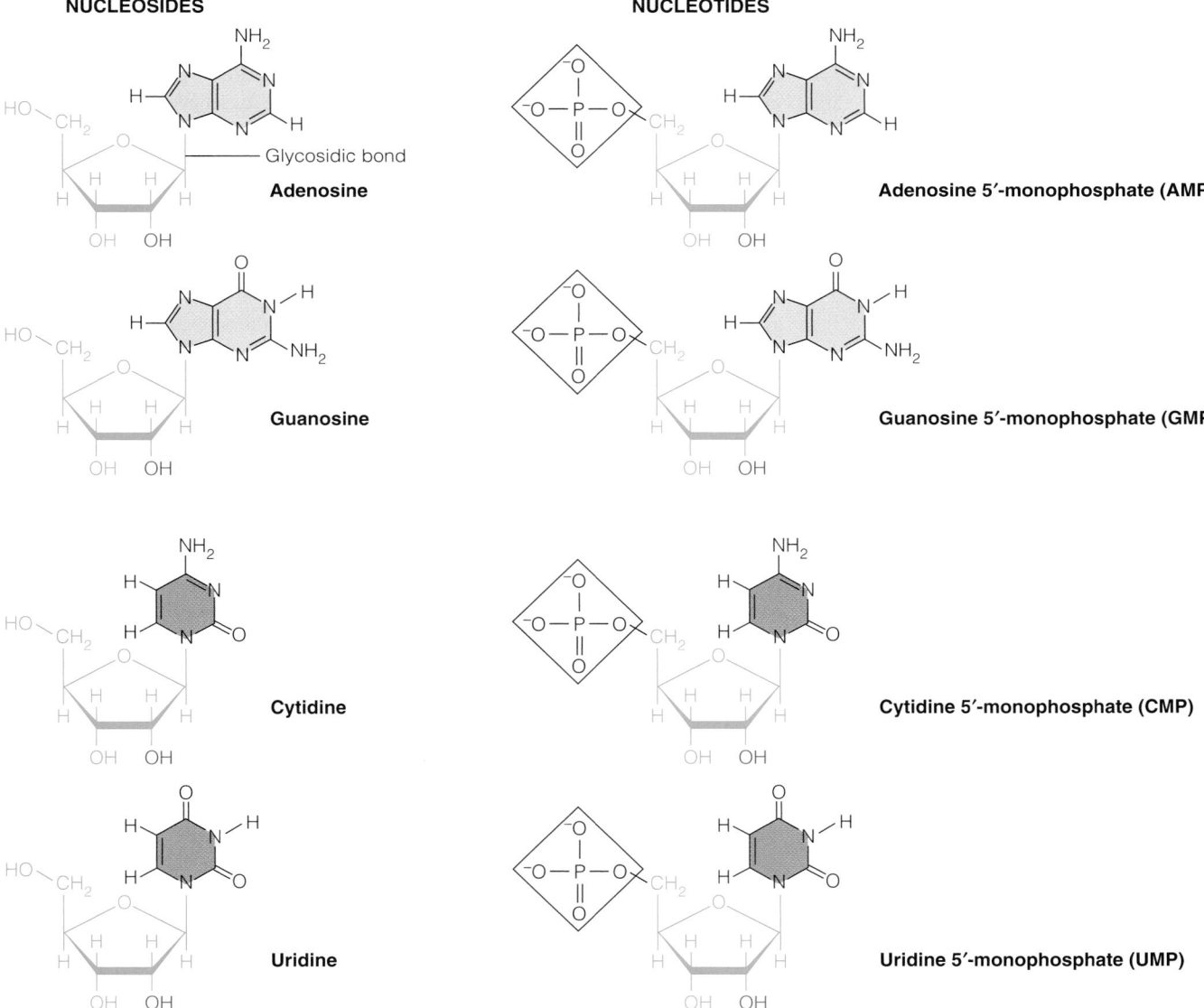

NUCLEOSIDES   NUCLEOTIDES

Adenosine — Glycosidic bond

Adenosine 5′-monophosphate (AMP)

Guanosine

Guanosine 5′-monophosphate (GMP)

Cytidine

Cytidine 5′-monophosphate (CMP)

Uridine

Uridine 5′-monophosphate (UMP)

TABLE 4.1   Ionization constants of ribonucleotides expressed as pK_a values

	Phosphate				Base	
	Primary Ionization		Secondary Ionization			
	pK_{a1} + H$^+$		pK_{a2} + H$^+$		pK_a	Reaction (as Loss of Proton from)
5′ AMP	0.9		6.1		3.8	N-1
5′ GMP	0.7		6.1		2.4	N-7
					9.4	N-1
5′ UMP	1.0		6.4		9.5	N-3
5′ CMP	0.8		6.3		4.5	N-3

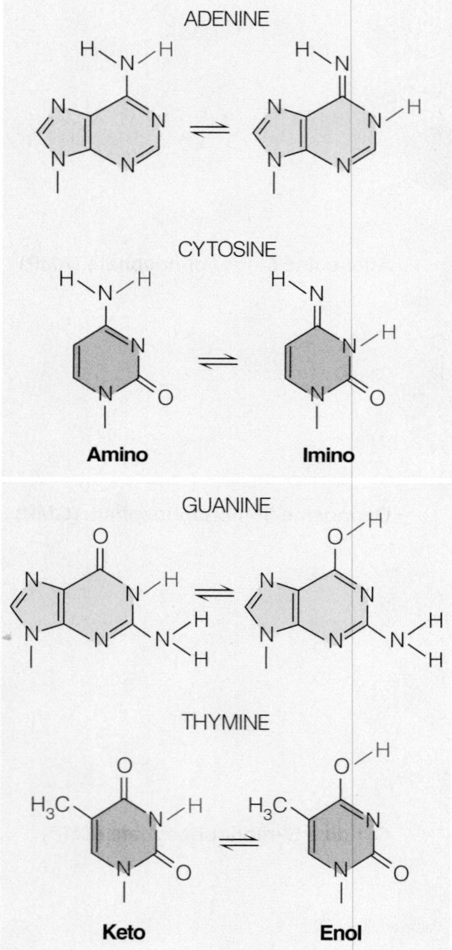

ADENINE

Amino   Imino

CYTOSINE

Amino   Imino

GUANINE

THYMINE

Keto   Enol

FIGURE 4.4

Tautomerization of the bases. The most stable (and therefore common) forms are shown at the left. The less common imino and enol forms, shown on the right, are found in some special base interactions. Still other tautomers (not shown here) are possible.

Because all of the nucleic acids may be regarded as polymers of nucleotides, they are often referred to by the generic name **polynucleotides**. Small polymers, containing only a few residues, are called **oligonucleotides**.

The presence of the 2′ hydroxyl groups is of far more than academic interest—it gives RNA a functionality lacking in DNA. As discussed in Chapter 11, Thomas Cech and Sidney Altman, working independently, discovered **ribozymes**, or RNA molecules capable of catalyzing chemical reactions. The 2′ hydroxyl groups are critically involved in the catalytic mechanisms. Hence, RNA molecules have the capacity both for information storage and for catalysis. Whether or not this is the reason, many biochemists believe that RNA came into existence earlier than DNA in the primordial environment in which the first living organisms are thought to have evolved. Studies on prebiotic chemistry suggest that ribose was among the earliest organic compounds to be formed. DNA is chemically more stable than RNA, which permits the generation and maintenance of longer genomes. For these reasons, biochemists postulate an initial **RNA world**, which gave way to DNA-based organisms once a mechanism had appeared, allowing conversion of ribose-containing compounds to their deoxyribose-containing counterparts. You should be aware that DNA-based catalysts also exist, but to date these are all laboratory constructs. No "DNAzymes" have yet been found in nature. See Chapter 11 for further discussion of this topic.

Properties of the Nucleotides

Nucleotides are strong acids; the primary ionization of the phosphate occurs with a pK_a of approximately 1.0. Both secondary ionization of the phosphate and protonation or deprotonation of the amino groups on the bases within the nucleotides can be observed at pH values close to neutrality (Table 4.1). The bases are also capable of conversion between **tautomeric** forms. Tautomeric forms, or **tautomers**, are structural isomers differing only in the location of their hydrogen atoms and double bonds. The major forms are those shown in Figure 4.2, but G, T, and U can partially isomerize to enol forms, and A and C to imino forms, as shown in Figure 4.4.

As a consequence of the conjugated double-bond systems in the purine and pyrimidine rings, the bases and all of their derivatives (nucleosides, nucleotides, and nucleic acids) absorb light strongly in the ultraviolet region of the spectrum. This absorption depends somewhat on pH because of the ionization reactions in the bases; representative spectra at neutral pH for ribonucleotides are depicted in Figure 4.5. This strong absorbance is often used for quantitative determination of nucleic acids because it allows measurement of nucleic acid concentrations at the microgram/mL level by measurement of light absorption at 260 nm in a spectrophotometer (see Tools of Biochemistry 6A). Ultraviolet light can also have chemically damaging effects on DNA, leading, for example, to skin cancer.

Stability and Formation of the Phosphodiester Linkage

If we compare the structures of the nucleotides shown in Figure 4.3 with the polynucleotide chains depicted in Figure 4.1, we see that, in principle, a polynucleotide could be generated from its nucleotide monomers by elimination of a water molecule between each pair of monomers. That is, we might imagine adding another nucleotide residue to a polynucleotide chain by the dehydration reaction shown in Figure 4.6. However, the free energy change in this hypothetical reaction is quite positive, about +25 kJ/mol; therefore equilibrium lies far to the side of hydrolysis of the phosphodiester bond in the aqueous environment of the cell. Hydrolysis of polynucleotides to nucleotides is the thermodynamically favored process.

We encounter here the first of many examples of the **metastability** of biologically important polymers. Metastable compounds are thermodynamically favored to break down, but do so only very slowly unless the reaction is catalyzed. According to the free energy change involved, polynucleotides should hydrolyze under conditions existing in living cells, but their hydrolysis is exceedingly slow unless catalyzed. This characteristic is of greatest importance, for it ensures that the DNA in cells is sufficiently stable to serve as a useful repository of genetic information. In dehydrated conditions, DNA is so stable that fragments of DNA molecules have been recovered from some ancient fossils. When catalysts *are* present, however, hydrolysis can be exceedingly rapid in aqueous solution. Acid catalysis leads to hydrolysis of the phosphodiester bonds in RNA, yielding a mixture of nucleotides. In both RNA and DNA, the glycosidic bond between the base and the sugar is also hydrolyzed; a mixture of bases, phosphoric acid, and ribose (or deoxyribose) is produced. RNA, but not DNA, is also labile in alkaline solution; treatment with 0.1 M alkali yields a mixture of 2′- and 3′-nucleoside monophosphates. Finally, and most important biologically, enzymes called **nucleases** catalyze the hydrolysis of phosphodiester bonds in both RNA and DNA. Your body is able to break down and utilize polynucleotides in the food you eat because your digestive system contains nucleases. Examples of such enzymes are described in Chapter 22.

The unfavorable thermodynamics of the hypothetical dehydration reaction shown in Figure 4.6 leads us to ask: If polynucleotides cannot be synthesized in vivo by the direct elimination of water, how are they actually made? The answer is that their synthesis involves the energy-rich nucleoside or deoxynucleoside *triphosphates*. Although the process as it occurs in cells is quite complex, the basic reaction is simple. Instead of the dehydration reaction of Figure 4.6, what happens in living cells is the reaction shown in Figure 4.7. The nucleoside monophosphate being added to the growing chain is presented as a *nucleoside triphosphate*, like ATP or deoxy ATP (dATP), and pyrophosphate is released in the reaction. We can calculate the free energy change for this reaction by noting that it can be considered the sum of two reactions—hydrolysis of a nucleoside triphosphate and formation of a phosphodiester link by elimination of water:

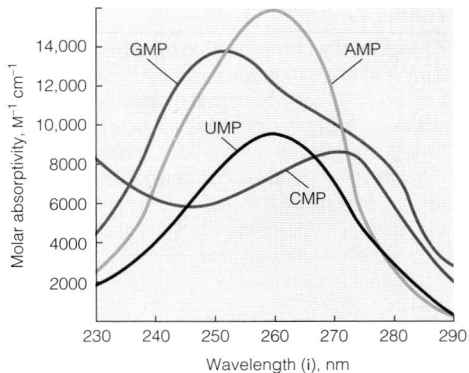

FIGURE 4.5

Ultraviolet absorption spectra of ribonucleotides. The dimensions of the absorption coefficients are $M^{-1}cm^{-1}$. Thus a 10^{-4} solution of UMP would have an absorbance of 0.95 at 260 mm in a 1-cm-thick cuvette. (Absorbance = molar absorptivity × light path in cm × molar concentration; see Tools of Biochemistry 6A).

Data from *Principles of Biochemistry*, 2nd ed., A. L. Lehninger, D. L. Nelson, and M. M. Cox. © 1993, 1982, Worth Publishers, Inc., New York.

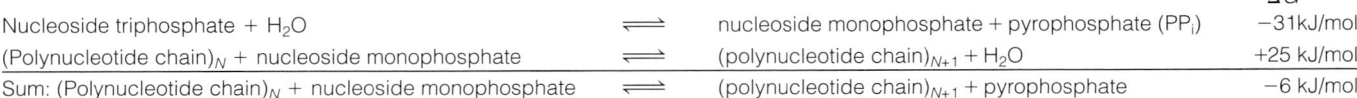

			$\Delta G^{\circ\prime}$
Nucleoside triphosphate + H_2O	$\rightleftharpoons$	nucleoside monophosphate + pyrophosphate (PP_i)	−31 kJ/mol
(Polynucleotide chain)$_N$ + nucleoside monophosphate	$\rightleftharpoons$	(polynucleotide chain)$_{N+1}$ + H_2O	+25 kJ/mol
Sum: (Polynucleotide chain)$_N$ + nucleoside monophosphate	$\rightleftharpoons$	(polynucleotide chain)$_{N+1}$ + pyrophosphate	−6 kJ/mol

The coupled reaction is favorable because the *net* $\Delta G^{\circ\prime}$ is negative. The reaction is further favored because the hydrolysis of the pyrophosphate product (PP_i) to orthophosphate, or inorganic phosphate (P_i), has a $\Delta G^{\circ\prime} = -19$ kJ/mol. Thus, the pyrophosphate is readily removed, driving the synthesis reaction even further to the right and yielding an overall $\Delta G^{\circ\prime}$ of −25 kJ/mol. Polynucleotide synthesis is an example of a principle we emphasized in Chapter 3—the use of favorable reactions to drive thermodynamically unfavorable ones.

It is important to appreciate how the energetics of such processes fit into the overall scheme of life. An organism obtains energy—from photosynthesis, if it is a plant, or from inorganic compounds if it is a chemolithotroph, or in all cases from

FIGURE 4.6 (*left*)

Formation of a polynucleotide by a hypothetical dehydration reaction. We might imagine that a polynucleotide could be formed directly from nucleoside monophosphates by removal of water, as shown here, but the dehydration reaction is thermodynamically unfavorable. The reverse reaction, hydrolysis, is favored. Note that in this and subsequent figures we adopt a somewhat more compact way of representing the sugar-phosphate backbone.

FIGURE 4.7 (*right*)

How polynucleotides are actually formed. In this reaction, each monomer is presented as a nucleoside triphosphate to be added to the chain. Cleavage of the nucleoside triphosphate provides the free energy that makes the reaction thermodynamically favorable. The enzymes catalyzing such reactions are called **polymerases**.

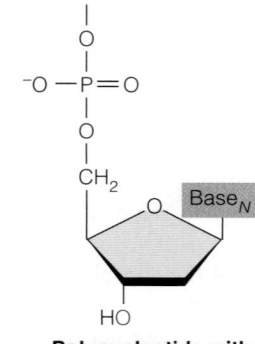

Polynucleotide with *N* residues

Polynucleotide with *N* residues

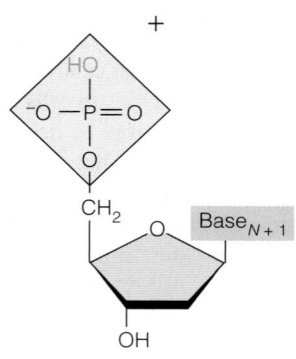

Deoxynucleoside monophosphate

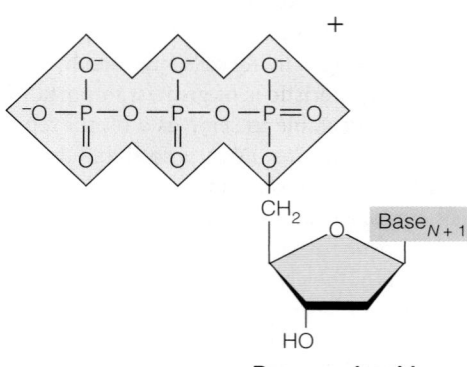

Deoxynucleoside triphosphate

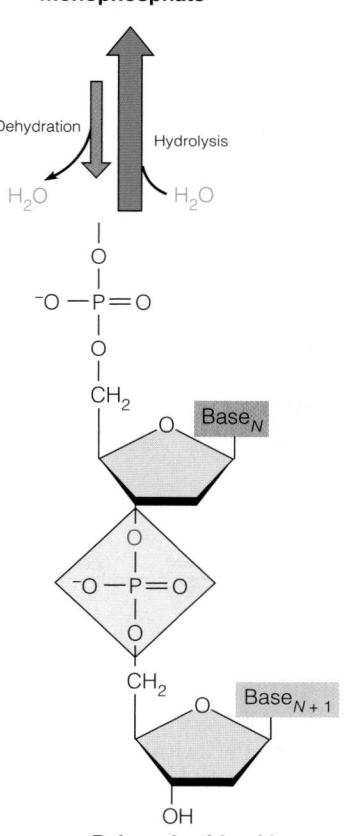

Dehydration

Hydrolysis

H_2O

H_2O

Polynucleotide with *N* + 1 residues

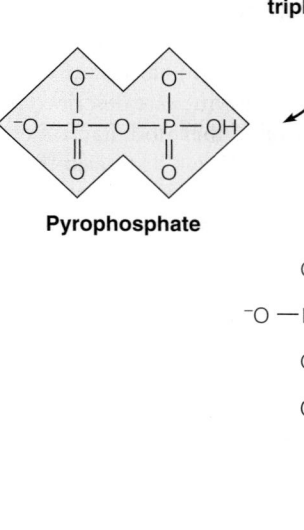

Pyrophosphate

Polynucleotide with *N* + 1 residues

metabolism of nutrients—and stores part of this energy by generating ATP, GTP, dATP, dGTP, and similar energy-rich compounds. It uses these compounds in turn as energy sources to drive the synthesis of macromolecules like DNA, RNA, and proteins. This use of triphosphates as the energy currency of the cell is a theme that you will see repeated throughout this book.

Primary Structure of Nucleic Acids

The Nature and Significance of Primary Structure

A closer examination of Figure 4.1 reveals two important features of all polynucleotides:

1. A polynucleotide chain has a *sense* or *directionality.* The phosphodiester linkage between monomer units is between the 3′ carbon of one monomer and the 5′ carbon of the next. Thus, the two ends of a linear polynucleotide chain are distinguishable. One end normally carries an unreacted phosphate, the other end an unreacted 3′ hydroxyl group.

2. A polynucleotide chain has *individuality,* determined by the sequence of its bases—that is, the *nucleotide sequence.* This sequence is called the **primary structure** of that particular nucleic acid.

If we want to describe a particular polynucleotide sequence (either DNA or RNA), it is awkward and unnecessary to draw the molecule in its entirety as in Figure 4.1. Accordingly, some compact conventions have been devised. If we state that we are describing a DNA molecule or an RNA molecule, then most of the structure is understood. We can then abbreviate a small DNA molecule as follows:

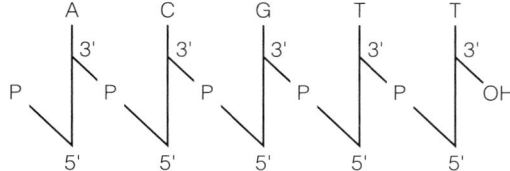

This notation shows (1) the sequence of nucleotides, by their letter abbreviations (A, C, G, T); (2) that all phosphodiester links are between 3′ hydroxyls and 5′ phosphates; and (3) that this particular molecule has a phosphate group at its 5′ end and an unreacted hydroxyl at its 3′ end. It also tells us it is a DNA sequence, not RNA because it has T, not U.

If all of the phosphodiester bonds can be assumed to link a 3′ hydroxyl to a 5′ phosphate (as is usually the case), a more compact notation is possible for the same molecule:

pApCpGpTpT

The 3′ — OH group is understood to be present and unreacted. Were there a phosphate on the 3′ end and an unreacted hydroxyl on the 5′ end, we would write

ApCpGpTpTp

Finally, if we are concerned *only* with the sequence of bases in the molecule, as will often be the case, we can write it still more compactly as

ACGTT

Note that the sequence of a polynucleotide chain is usually written, by convention, with the 5′ end to the left and the 3′ end to the right.

The main importance of primary structure, or sequence, is that *genetic information is stored in the primary structure of DNA.* A *gene* is nothing more than a particular DNA sequence, encoding information in a four-letter language in which each "letter" is one of the bases.

Every naturally occurring polynucleotide has a defined sequence, its primary structure.

The primary structure of DNA encodes genetic information.

DNA as the Genetic Substance: Early Evidence

The search for the substance of which genes are made has a long history. In the late 1800s, shortly after Friedrich Miescher had first isolated DNA from white blood cells, some scientists suspected that DNA might be the genetic material. But subsequent studies showing that DNA contained only four kinds of monomers seemed to deny that it could have such a complicated role. Early researchers thought it more likely that genes were made of proteins, for proteins were beginning to be recognized as extremely complex molecules. For most of the first half of the twentieth century, nucleic acids were considered to be merely some kind of structural material in the cell nucleus.

Between 1944 and 1952 a series of crucial experiments clearly pointed to DNA as the genetic material. In 1944 Oswald Avery, Colin MacLeod, and Maclyn McCarty found that the DNA from pathogenic strains of the bacterium *Pneumococcus* could be transferred into nonpathogenic strains, making them pathogenic (Figure 4.8a). This **transformation** was genetically stable; succeeding generations of bacteria retained the new characteristics. However, it was an elegant experiment by Alfred Hershey and Martha Chase that finally convinced most scientists. Hershey and Chase studied infection of the bacterium *Escherichia coli* by a bacterial virus, the bacteriophage, or phage, T2. Making use of the fact that phage proteins contain sulfur but little phosphorus and that phage DNA contains phosphorus but no sulfur, they labeled T2 bacteriophage with the radioisotopes ^{35}S and ^{32}P (Figure 4.8b). They then showed that when the phage attached to *E. coli,* it was mainly the ^{32}P (and hence the phage DNA) that was transferred into the bacteria. Even if the residual protein part of the bacteriophage was shaken off the bacteria, the inserted DNA alone was sufficient to direct the formation of new bacteriophage.

Through these and similar experiments, it was generally recognized by 1952 that DNA must be the genetic substance. But how could it carry the enormous amount of information that a cell needed, how could it transmit that information to the cell, and, above all, how could it be accurately replicated in cell division? The answers to these questions came only after one of the most momentous discoveries in the history of science. In 1953 James Watson and Francis Crick proposed a structure for DNA that opened a whole new world of molecular biology.

Secondary and Tertiary Structure of Nucleic Acids

The Double Helix

Watson and Crick sought answers to the questions posed above in the three-dimensional structure of DNA. For some time, investigators had been studying fibers drawn from concentrated DNA solutions, using the technique of **X-ray diffraction** (see Tools of Biochemistry 4A). Watson and Crick, working at Cambridge University in England, had access to DNA diffraction patterns photographed by Rosalind Franklin, a researcher in the laboratory of Maurice Wilkins at King's College, London. The critical photographs clearly showed that DNA in the fibers must have some kind of regular, repetitive three-dimensional structure. We refer to such regular folding in polymers as *secondary structure,* as distinguished from the primary structure, which, as noted earlier, is simply the sequence of individual nucleotide residues. Watson and Crick were also aware of Erwin Chargaff's data, which showed regularities in DNA base composition (see page 101).

Watson and Crick quickly recognized that the DNA fiber diffraction exhibited a cross pattern typical of a helical secondary structure (Figure 4.9). They noted that because the layer line spacing was one-tenth of the pattern repeat, there must be 10 residues per turn (see Tools of Biochemistry 4A). Data on the density of the fibers suggested that there must be *two* DNA strands in each helical molecule. At that point, only direct scientific deductions had been made from the data. The great leap of intuition made by Watson and Crick was their realization that a two-strand helix

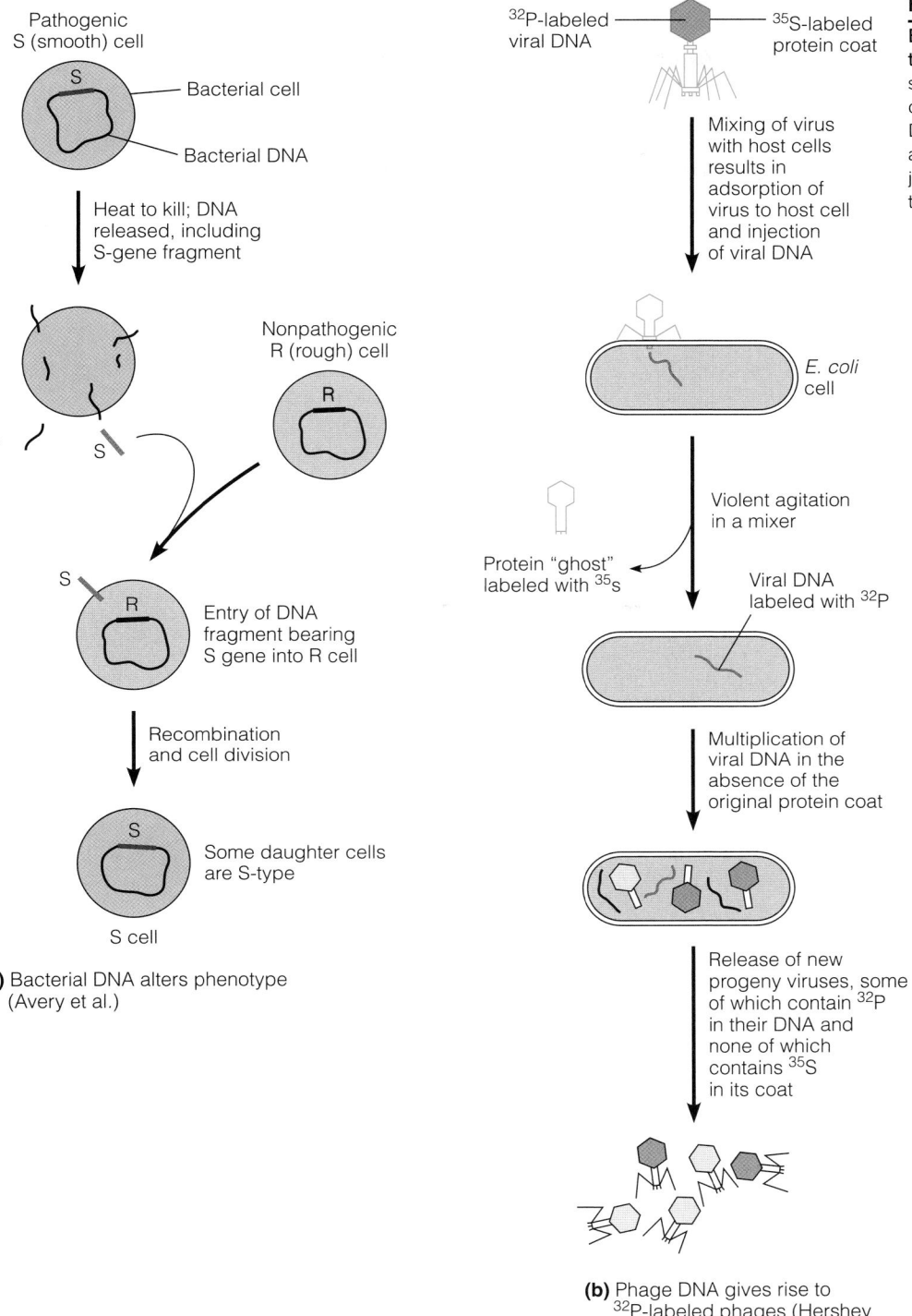

Pathogenic
S (smooth) cell

Bacterial cell

Bacterial DNA

Heat to kill; DNA
released, including
S-gene fragment

Nonpathogenic
R (rough) cell

Entry of DNA
fragment bearing
S gene into R cell

Recombination
and cell division

Some daughter cells
are S-type

S cell

(a) Bacterial DNA alters phenotype
(Avery et al.)

^{32}P-labeled
viral DNA

^{35}S-labeled
protein coat

Mixing of virus
with host cells
results in
adsorption of
virus to host cell
and injection
of viral DNA

E. coli
cell

Violent agitation
in a mixer

Protein "ghost"
labeled with ^{35}s

Viral DNA
labeled with ^{32}P

Multiplication of
viral DNA in the
absence of the
original protein coat

Release of new
progeny viruses, some
of which contain ^{32}P
in their DNA and
none of which
contains ^{35}S
in its coat

(b) Phage DNA gives rise to
^{32}P-labeled phages (Hershey
and Chase)

FIGURE 4.8

Experiments that showed DNA to be the genetic substance. (a) Avery et al. showed that nonpathogenic pneumococci could be made pathogenic by transfer of DNA from a pathogenic strain. **(b)** Hershey and Chase showed that it is the transfer of just the viral DNA from a virus to a bacterium that gives rise to new viruses.

could be stabilized by hydrogen bonding between bases on opposite strands if the bases were paired in *one* particular way—the A-T and G-C pairs shown in Figure 4.10. With this pairing, strong hydrogen bonds are formed between the bases. Furthermore, distances between the 1′ carbons of the deoxyribose moieties of A-T and of G-C are the same—about 1.1 nm in each case (Figure 4.10a). This pairing arrangement meant that the double helix could be regular in diameter, an impossibility if purines paired with purines or pyrimidines with pyrimidines.

The Watson–Crick model for DNA is a two-strand, antiparallel double helix with 10 base pairs per turn. Pairing is A-T and G-C.

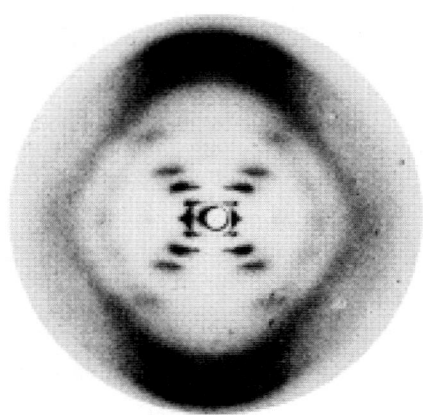

FIGURE 4.9

Evidence for the structure of DNA. This photograph, taken by Rosalind Franklin, shows the X-ray diffraction pattern produced by wet DNA fibers. It played a key role in the elucidation of DNA structure. The cross pattern indicates a helical structure, and the strong spots at top and bottom correspond to a helical rise of 0.34 nm. The layer line spacing is one-tenth of the distance from the center to either of these spots, showing that there are 10 base pairs per repeat.

Reprinted by permission from Macmillan Publishers Ltd. *Nature* 171:740–741, R. E. Franklin and R. Gosling, Molecular configuration in sodium thymonucleate. © 1953.

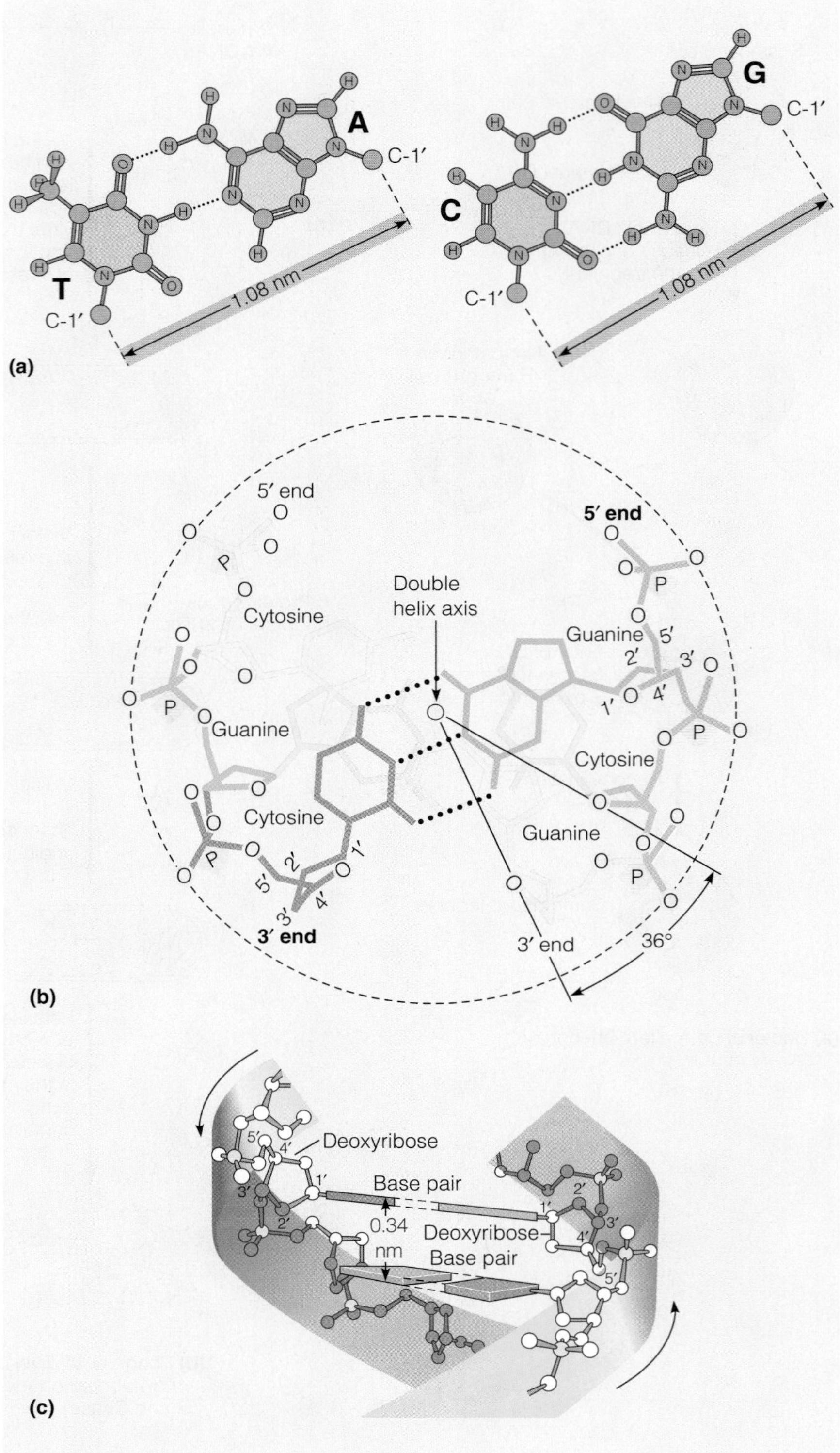

FIGURE 4.10

Fundamental elements of structure in the DNA double helix. **(a)** Base pairing. A-T and G-C are the base pairs in the Watson–Crick model of DNA. This pairing allows the C-1′ carbons on the two strands to be exactly the same distance apart for both pairs. **(b)** Stacking of the base pairs. This view down the helix axis shows how the base pairs stack on one another, with each pair rotated 36° with respect to the next. **(c)** Distance between the base pairs. A side view of the base pairs shows the 0.34-nm distance between them. This distance is called the rise of the helix.

In the Watson–Crick model, the hydrophilic phosphate–deoxyribose backbones of the helix were on the outside, in contact with the aqueous environment, and the base pairs were stacked on one another with their planes perpendicular to the helix axis. Two views of such a structure are shown in Figure 4.10b and c. (The figure shows a recent, refined model, based on better data than Watson and Crick had available: The bases are not exactly perpendicular to the helix axis, and the sugar conformation is slightly different from that proposed by Watson and Crick.) Stacking of the bases, as shown in Figure 4.10b, allows strong interactions between them, probably of van der Waals type; this is usually referred to as "stacking interaction" (see also Figure 2.2, page 28). Each base pair is rotated by 36°, that is, 1/10 of a 360° rotation with respect to the next, to accommodate 10 base pairs in each turn of the helix (later structural studies showed the number to be closer to 10.5; see page 106). The diffraction pattern showed the repeat distance to be about 3.4 nm, so the helix *rise,* that is, the distance between base pairs, had to be about 0.34 nm (Figure 4.10c). This distance is just twice the van der Waals thickness of a planar ring (see Table 2.2 on page 29), so the bases are closely packed within the helix, as shown in a space-filling model (Figure 4.11).

The model also shows that although the bases are inside, they can be approached through two deep spiral grooves called the *major* and *minor* grooves. The major groove gives more direct access to the bases; the minor groove faces the sugar backbone. Building molecular models of two-strand DNA structures soon convinced Watson and Crick that the DNA strands must run in opposite directions. In other words, the two strands are antiparallel, as indicated in Figure 4.10c. The model Watson and Crick presented was for a right-handed helix, although at that time evidence for the sense (direction of the turn) of the helix was weak. Their guess proved to be correct.

As is often the case with a good theory or model, the Watson–Crick structure also explained other data that had not been understood until then. Erwin Chargaff, who had measured the relative amounts of A, T, G, and C in DNAs from many organisms, had noted the perplexing fact that A and T were almost always present in nearly equal quantities, as were G and C (Table 4.2). If most DNA in cells was double-stranded, with the Watson–Crick base pairing, then *Chargaff's rule* followed as a natural consequence.

The Watson–Crick model not only explained the structure of DNA and Chargaff's rule but also carried implications that went to the very heart of biology. Because A always pairs with T, and G always pairs with C, the two strands are *complementary*. If the strands could be separated and new DNA synthesized along each, following the same base-pairing rule, two double-stranded DNA molecules would be obtained, each an *exact* copy of the original (Figure 4.12). This **self-replication** is precisely the property that the genetic material must have: When a cell divides, two complete copies of the genetic information carried in the original cell must be produced. In their 1953 paper announcing the model (see References), Watson and Crick expressed this idea in what may be the most understated scientific prediction ever made: "It has not escaped our notice that the specific pairing we have postulated immediately suggests a possible copying mechanism for the genetic material."

Semiconservative Nature of DNA Replication

The DNA copying mechanism we have mentioned involves unwinding the two strands of a parental DNA duplex, with each strand serving as template for the synthesis of a new strand, complementary to and wound about that parental strand. Complete replication of a DNA molecule would yield two "daughter" duplexes, each consisting of one-half parental DNA (one strand of the original duplex) and one-half new material. This mode of replication is called **semiconservative** because half of the original material is conserved in each of the two copies (Figure 4.13). It is distinguished from two other possible modes: **conservative**, in which one of the two daughter duplexes is the conserved parental duplex while the other is synthesized *de novo*, and **dispersive**, in which parental material is scattered through the structures of the daughter duplexes.

The complementary, two-strand structure of DNA explains how the genetic material can be replicated.

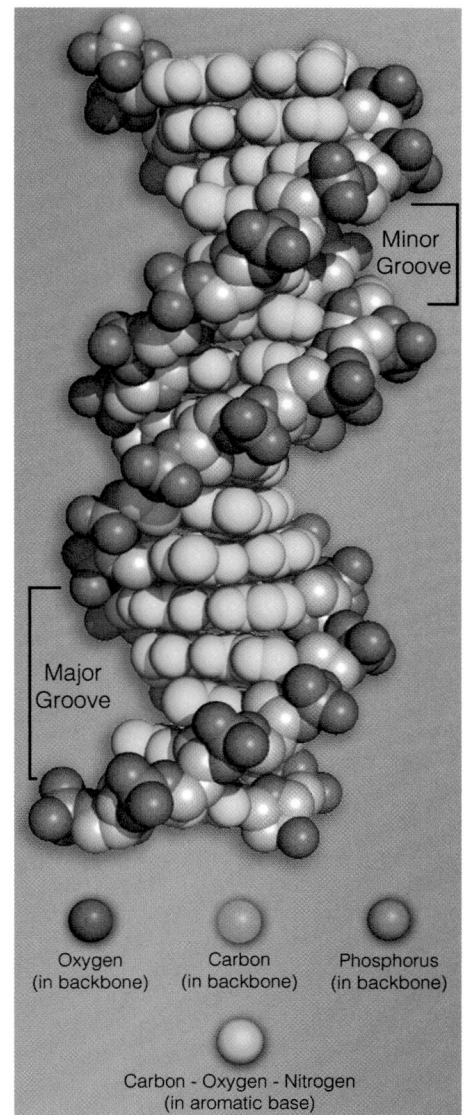

FIGURE 4.11

A space-filling model of DNA. The DNA molecule as modeled by Watson and Crick is shown here with each atom given its van der Waals radius. This model shows more clearly than Figure 4.10 how closely the bases are packed within the helix. The major and minor grooves are indicated.

Courtesy of Gary Carlton.

TABLE 4.2   Base compositions of DNAs from various organisms

| | Mol % of Bases | | | | | Ratios | |
Source	Adenine (A)	Guanine (G)	Cytosine[a] (C)	Thymine (T)	(G + C)	A/T	G/C
Bacteriophage φX174	24.0	23.3	21.5	31.2	44.8	0.77[b]	1.08[b]
Bacteriophage T7	26.0	23.8	23.6	26.6	47.4	0.98	1.01
Escherichia coli B	23.8	26.8	26.3	23.1	53.2	1.03	1.02
Neurospora	23.0	27.1	26.6	23.3	53.8	0.99	1.02
Corn (maize)	26.8	22.8	23.2	27.2	46.1	0.99	0.98
Tetrahymena	35.4	14.5	14.7	35.4	29.2	1.00	0.99
Octopus	33.2	17.6	17.6	31.6	35.2	1.05	1.00
Drosophila	30.7	19.6	20.2	29.5	39.8	1.03	0.97
Starfish	29.8	20.7	20.7	28.8	41.3	1.03	1.00
Salmon	28.0	22.0	21.8	27.8	44.1	1.01	1.01
Frog	26.3	23.5	23.8	26.8	47.4	1.00	0.99
Chicken	28.0	22.0	21.6	28.4	43.7	0.99	1.02
Rat	28.6	21.4	21.6	28.4	42.9	1.01	1.00
Calf	27.3	22.5	22.5	27.7	45.0	0.99	1.00
Human	29.3	20.7	20.0	30.0	40.7	0.98	1.04

Data taken from H. E. Sober, ed. (1970) *Handbook of Biochemistry,* 2nd ed. Chemical Rubber Publishing Co. Values for higher organisms vary slightly from one tissue to another, probably as a result of experimental error.

[a]Amount includes, for some organisms, a few percent of a modified base, 5-methylcytosine.

[b]This bacteriophage has a single-strand DNA, which need not follow Chargaff's rule.

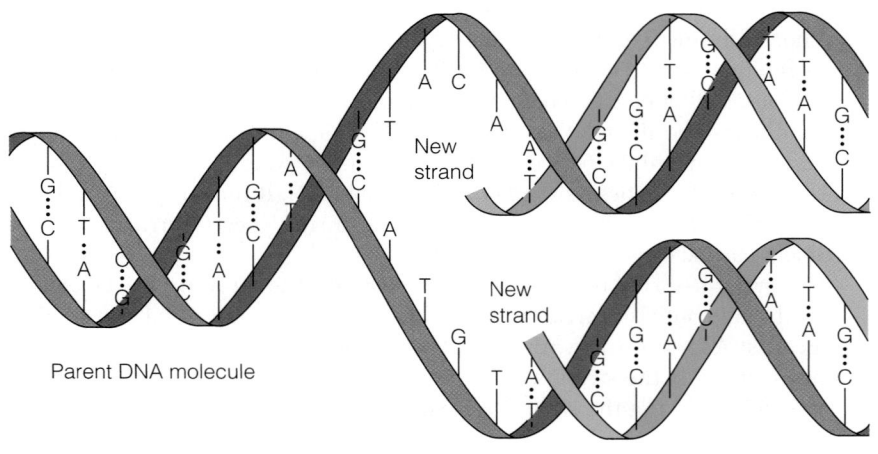

Parent DNA molecule

New strand

New strand

Daughter DNA molecules

FIGURE 4.12

A model for DNA replication. Each strand acts as a template for a new, complementary strand. When copying is complete, there will be two double-stranded daughter DNA molecules, each identical in sequence to the parent molecule. The actual process is rather complicated (see Chapter 25), but the basic principle is illustrated here.

Meselson and Stahl proved that DNA replicates semiconservatively.

The first experimental test of this model came in 1958, when Matthew Meselson and Franklin Stahl realized that molecules differing in density by very small amounts could be separated from each other by centrifugation in density gradients. In this method, a density gradient is created by centrifuging to equilibrium a concentrated solution of a heavy metal salt like cesium chloride (CsCl). Nucleic acid molecules of different densities suspended in such a gradient will each migrate to the point where the solution density equals its own. This technique allowed Meselson and Stahl to follow the fate of density-labeled DNA through several rounds of replication, with the results shown in Figure 4.14. The density label was applied by growing the bacterium *E. coli* in a medium containing the heavy isotope of nitrogen, ^{15}N, for many generations, so that DNA achieved a higher density through extensive substitution of ^{15}N for ^{14}N in its purine and pyrimidine bases. When isolated and centrifuged to equilibrium at pH 7.0, this DNA formed a single band in a region of the gradient corresponding to a density of 1.724 g/mL (Figure 4.14a, first profile). By contrast, when DNA from bacteria grown in a light medium (containing ^{14}N) was similarly analyzed, it banded at a density of 1.710 g/mL (Figure 4.14a, second profile). When density-labeled bacteria grown in a heavy medium were transferred to a light medium, the DNA isolated after one generation of growth banded exclusively at an intermediate density, 1.717 g/mL (Figure 4.14a, third profile). This result is expected if the newly replicated DNA is a *hybrid* molecular species, consisting of one-half parental material and one-half new DNA (synthesized in light medium). If these bacteria were cultured for an additional generation in light medium, two equal-sized bands were seen, one light and one of hybrid density (Figure 4.14a, fourth graph), as expected if the hybrid-density DNA underwent a second round of semiconservative replication.

These results were consistent with the idea that each replicated chromosome contains one parental strand and one daughter strand, but the data did not preclude alternative forms of semiconservative replication, involving the breaking of DNA strands. These models were ruled out by centrifugal analysis of the density-labeled DNAs at pH 12, where DNA strands separate (Figure 4.14b). When centrifuged to equilibrium after bacterial growth in heavy medium followed by one generation in light medium, the DNA formed two bands: one light (equal to that seen on analysis of ^{14}N-grown bacteria) and one heavy (Figure 4.14b, third graph). DNA analyzed after a second round of replication in light medium showed three-fourths light and one-fourth heavy material (Figure 4.14b, fourth graph). The inescapable conclusion was that the replicative hybrid contains one complete strand of parental DNA and one complete strand of newly synthesized DNA.

Alternative Nucleic Acid Structures: B and A Helices

At the time Watson and Crick proposed their model, two quite different X-ray diffraction patterns had already been obtained for DNA, indicating that the molecule can exist in more than one form. The **B form**, which is seen in DNA fibers prepared under conditions of high humidity, is shown in Figures 4.10, 4.11, 4.15a, and 4.15b. Watson and Crick chose to study the B form because they correctly expected it to be the predominant form in the aqueous milieu of the cell. DNA fibers prepared under conditions of low humidity have a different structure, the so-called **A form** (Figure 4.15c and d). Although a B helix is indeed the form of DNA found in cells, the A helix is also important biologically. Double-stranded RNA molecules always form the A structure, and so do **DNA–RNA hybrid molecules**, which are formed by the pairing of one DNA strand with one RNA strand. Thus, two major kinds of secondary structures exist in polynucleotides. As we shall see later in this chapter, a number of other kinds of secondary structures are possible under special circumstances, including the left-handed Z form mentioned in Table 4.3.

As Figure 4.15 and Table 4.3 show, the A and B forms are very different, although both are right-handed helices. In the B helix the bases lie close to the helix axis, which passes between the hydrogen bonds (note the end-on views of the helices in Figure 4.15a and c). In the A helix the bases lie farther to the outside and are strongly tilted with respect to the helix axis. The surfaces of the helices are also different. In the B helix the major and minor grooves are quite distinguishable, whereas in the A helix the two grooves are more nearly equal in width.

All fiber diffraction studies, including those that provided the information just described, suffer from a major limitation. In analyzing fiber patterns, researchers do not directly determine the details of a nucleic acid secondary structure. Instead, they propose models that best account for the positions and intensities of the spots on the diffraction pattern (see Tools of Biochemistry 4A). This approach is necessary because fibers are never perfect crystals, and there is always some ambiguity in the interpretation of their diffraction patterns. Therefore, a major advance was made when R. E. Dickerson and his colleagues succeeded in *crystallizing* a small double-stranded DNA fragment, which had the sequence

$^{5'}$CGCGAATTCGCG$^{3'}$
$^{3'}$GCGCTTAAGCGC$^{5'}$

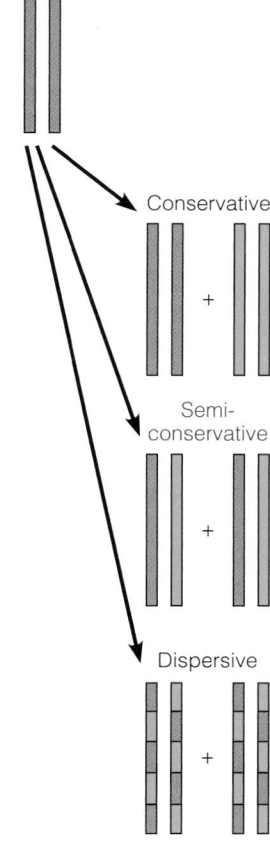

FIGURE 4.13

Three models of DNA replication. Experimental evidence supports the semiconservative model. Brown, parental DNA; blue, new DNA.

The two major forms of polynucleotide secondary structure are called A and B. Most DNA is in the B form; double-stranded RNA and DNA-RNA hybrids are in the A form.

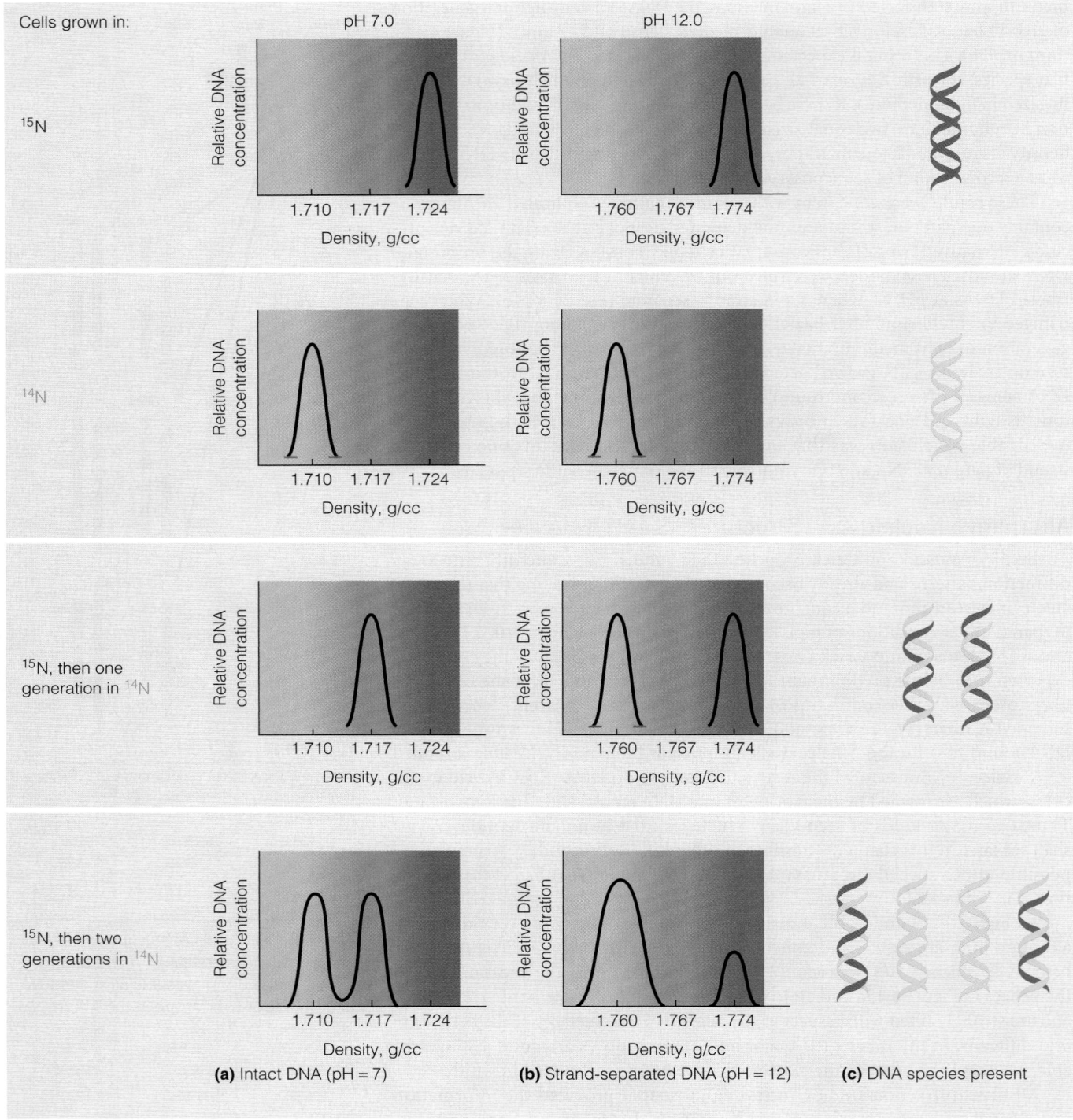

FIGURE 4.14

The Meselson–Stahl experiment proves DNA replicates semiconservatively. The banding patterns shown were obtained in density gradients under two pH conditions. At pH 7, the DNA is double-stranded; at pH 12 the strands are separated, and DNA is in the random-coil configuration.

Molecular crystallography of this fragment and other small DNA fragments has given us much more detailed information concerning the secondary structure of polynucleotides. The results of such a study on B-DNA are shown in Figure 4.16. Because the smaller size of this oligonucleotide pair, relative to polymeric DNA, gives it more structural uniformity, we can show DNA molecules with the position of every atom clearly specified.

A first major point emerging from the molecular crystal studies is that the models drawn from the fiber patterns represent oversimplifications of the structures.

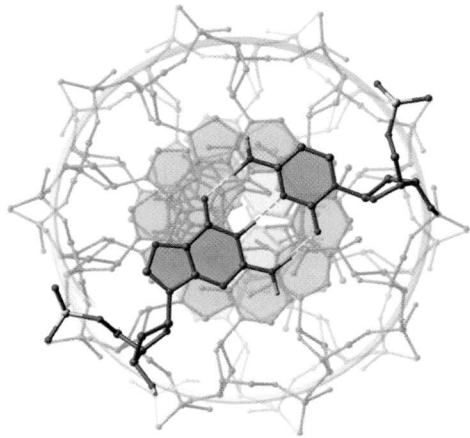

(a) B-DNA, end-on view

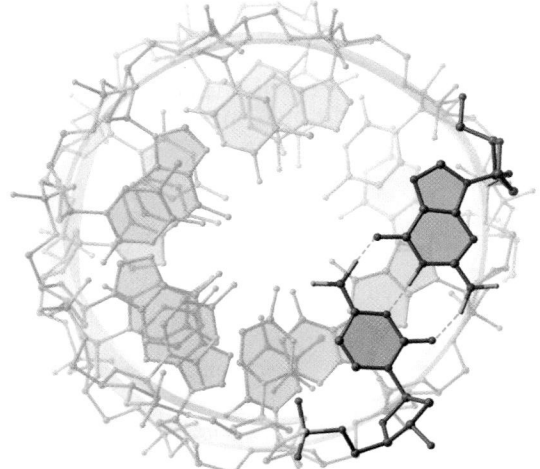

(c) A-DNA, end-on view

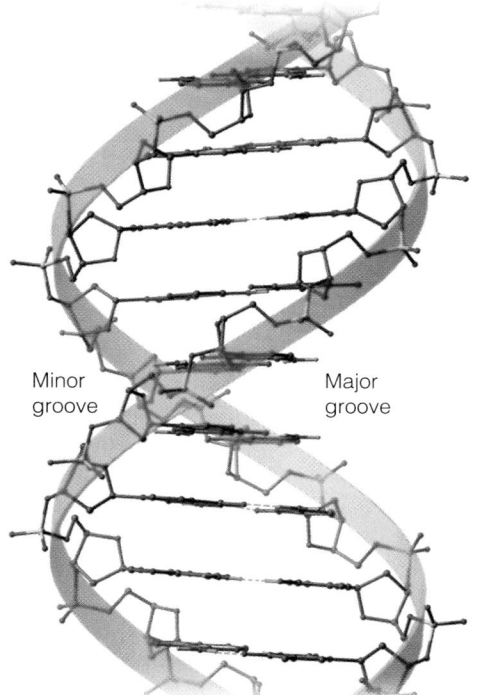

Minor groove   Major groove

(b) B-DNA, side view

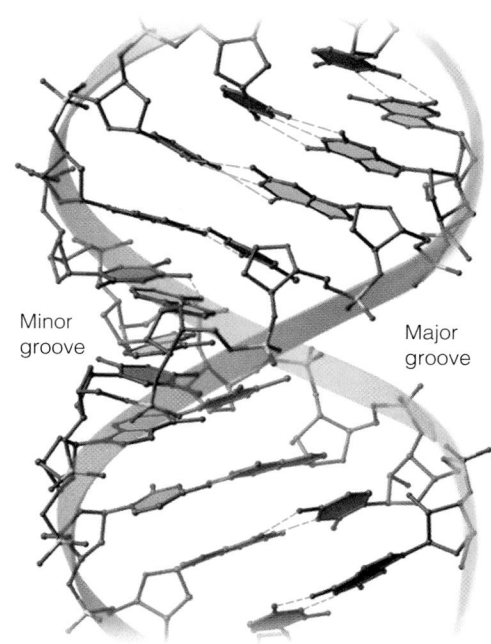

Minor groove   Major groove

(d) A-DNA, side view

FIGURE 4.15

Comparison of the two major forms of DNA.
The structures of B-DNA and A-DNA, as deduced from recent fiber diffraction studies, are shown here in both end-on and side views.

TABLE 4.3   Parameters of polynucleotide helices

	A Form	B Form	Z Form
Direction of helix rotation	Right	Right	Left
Number of residues per turn (n)	11	10	12 (6 dimers)
Rotation per residue ($= 360°/n$)	33°	36°	$-60°$ per dimer; $\sim-30°$ per residue
Rise[a] in helix per residue (h)	0.255 nm	0.34 nm	0.37 nm
Pitch[a] of helix ($= nh$)	2.8 nm	3.4 nm	4.5 nm

[a]For definitions of *rise* and *pitch* of a helix, see Tools of Biochemistry 4A.

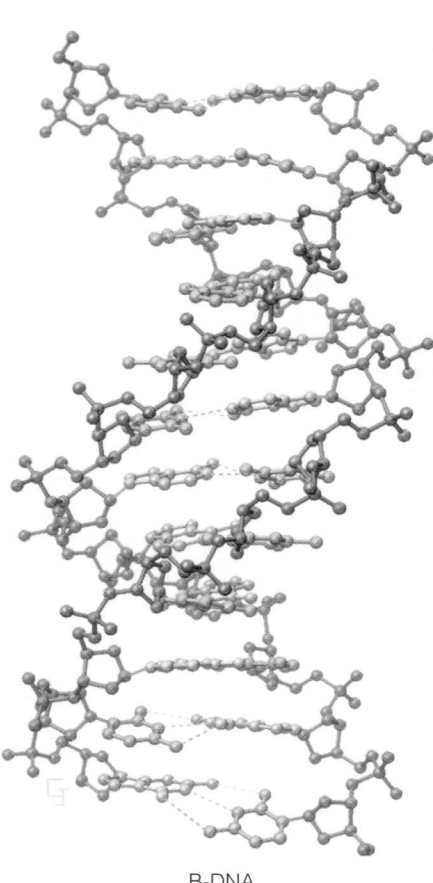

B-DNA

FIGURE 4.16

The structure of B-DNA from studies of molecular crystals. Note the local distortions of the idealized structure shown in Figures 4.11 and 4.15a and b.

Ilustration, Irving Geis. Adapted by Richard E. Dickerson. Image from Irving Geis Collection/Howard Hughes Medical Institute. Rights owned by HHMI. Not to be reproduced without permission.

Most DNA molecules in vivo are double-stranded; many are closed circles. Most double-stranded circular DNA molecules are supercoiled.

The real structure of B-DNA involves local variations in the angle of rotation between base pairs, the sugar conformation, the tilt of the bases, and even the rise distance. If you examine Figure 4.16 carefully, you can see many distortions from the idealized structures. Nucleic acid secondary structure is not homogeneous. It varies in response to the local sequence and can be changed by interaction with other molecules. The parameters given for various forms of DNA in Table 4.3 should therefore be thought of as *average* values, from which considerable local deviation is possible.

If one closely examines structures like that shown in Figure 4.16, another departure from the original Watson–Crick model becomes apparent. Many DNA molecules are slightly *bent;* that is, the helix axis does not follow a straight line. The degree and directions of bending depend in a complicated way on DNA sequence. They can also be strongly influenced by the interaction of DNA with various protein molecules; we see examples of this in later chapters.

The molecular crystallographic studies also provide a possible explanation of why B-DNA is favored in an aqueous environment. The B form of DNA, but not A-DNA, can accommodate a spine of water molecules lying in the minor groove. The hydrogen bonding between these water molecules and the DNA may confer stability to the B form. According to this hypothesis, when this water is removed (as in fibers at low humidity), the B form becomes less stable than the A form.

Why, then, do double-stranded RNA and DNA–RNA hybrids always adopt the A form? The answer probably lies in the extra hydroxyl group on the ribose in RNA. This hydroxyl interferes sterically in the B form by lying too close to the phosphate and carbon 8 on the adjacent base. Therefore, RNA *cannot* adopt the B form, even under conditions in which hydration might favor it. In DNA, the hydroxyl is replaced by hydrogen, and such steric hindrance does not occur.

DNA and RNA Molecules in Vivo

We have described some of the major features of DNA and RNA. But in what forms do these molecules exist in the living cell? Most of the DNA in most organisms is double-stranded, with the two strands being complementary, although some DNA viruses carry single-stranded DNA molecules (Table 4.4). The proportions of B and A forms of polynucleotides in vivo are as you might expect from the conditions under which these conformations are stable. Because cells contain much water, we would expect most of the double-stranded DNA to be in the B form or something much like it. There is evidence that B-DNA dissolved in solution is only a little different in conformation from the B-form seen in fiber preparations, having about 10.5 base pairs per turn instead of the expected 10.0. Double-stranded RNA, as mentioned earlier, always exists in the A form.

The DNA molecules found in organisms vary over an enormous range of sizes. The double-stranded DNA of human mitochondria is only 16,569 base pairs long, while some bacterial *plasmids*, or small extrachromosomal DNA molecules, are much smaller. On the other hand, some DNAs, like those in the chromosomes of eukaryotes, are immense molecules. The DNA from one *Drosophila* (fruit fly) chromosome has a molecular weight of about 4×10^{10} g/mol and would be 2 cm long if fully extended.

Circular DNA and Supercoiling

Another important feature of naturally occurring DNA molecules is illustrated in Figure 4.17a and b: Many of them are *circular*. This means that they do not have free 5′ or 3′ ends. The circles may be small, as in bacteriophage ϕX174 DNA (Figure 4.17a), or immense, as in *E. coli* DNA (Figure 4.17b), and they may involve either a single strand or two strands intertwined in a B-form double helix. Not all DNA molecules are circular, however. Figure 4.17c shows the linear DNA of a virus, bacteriophage T2. Human chromosomes also contain giant linear DNA molecules.

TABLE 4.4   Properties of some naturally occurring DNA molecules

Source	Single Strand (SS) or Double Strand (DS)	Circular or Linear	Number of Base Pairs (bp) or Bases (b)	Molecular Mass (Da)	Length[b]	% (G + C)
Simian virus 40 (genome)[a]	DS	Circular	5243 bp	3.293×10^6	1.78 μm	40.80
Bacteriophage ϕX174 (genome)	SS	Circular	5386 b	1.664×10^6	—[d]	44.76
Bacteriophage M13 (genome)	SS	Circular	6407 b	1.977×10^6	—[d]	40.75
Cauliflower mosaic virus (genome)	DS	Circular	8031 bp	4.962×10^6	2.73 μm	40.19
Adenovirus AD-2 (genome)	DS	Linear	35,937 bp	2.221×10^7	12.2 μm	55.20
Epstein–Barr virus (genome)	DS	Circular	172,282 bp	1.065×10^8	58.6 μm	59.94
Bacteriophage T4 (genome)	DS	Linear	168,899 bp	1.062×10^8	57.4 μm	35.30
Bacterium *E. coli* (genome)	DS	Circular	4,639,221 bp	2.869×10^9	1.57 mm	50.80
Fruit fly (*Drosophila melanogaster*) (one chromosome)[c]	DS	Linear	$\sim 6.5 \times 10^7$ bp	$\sim 4.3 \times 10^{10}$	~2 cm	~40

[a]The term *genome* designates the total DNA to specify the genetic information for an organism.
[b]Calculated for double-strand DNA of known sequence: 0.34 nm $\times$ the number of base pairs (assumes B form).
[c]This molecule has not been completely sequenced, so numbers of base pairs, molecular weights, and % (G + C) cannot be given exactly.
[d]The lengths of single-strand DNAs are not well defined; they depend very much on solvent conditions.

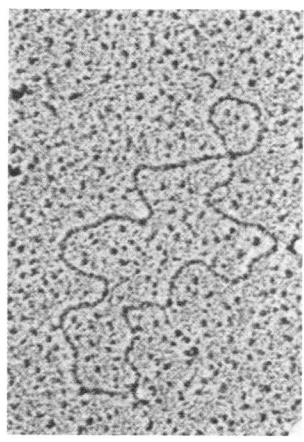

(a) Viral single-strand DNA (circular)

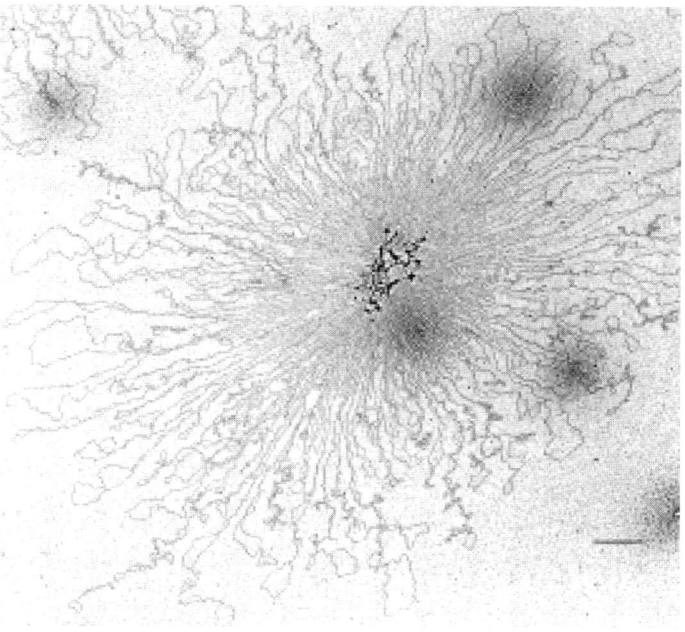

(b) Bacterial double-strand DNA (circular)

FIGURE 4.17

Circular and linear nucleic acid molecules as seen by the electron microscope. **(a)** The circular single-stranded DNA of the small bacteriophage ϕX174. **(b)** The large circular double-stranded DNA of *E. coli*. This molecule exists as a number of supercoiled loops bound to a protein matrix. **(c)** The single linear double-stranded DNA molecule of bacteriophage T2. The bacteriophage has been lysed, and its DNA has spilled out. One of the two ends is off screen to the lower right. The other is presumably buried in the phage head.

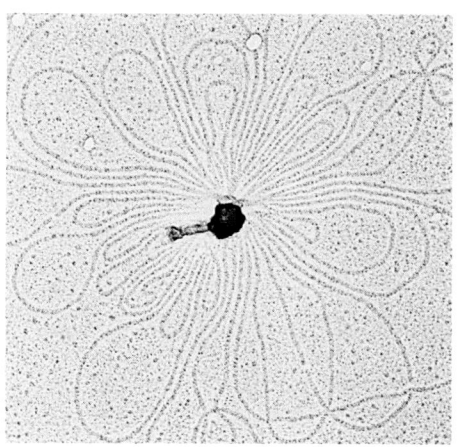

(c) Bacteriophage double-strand DNA (linear)

(a) *Journal of Virology* 24:673-684, D. P. Allison, A. T. Ganesan, A. C. Olson, C. M. Snyder, and S. Mitra, Electron microscopic studies of bacteriophage M13 DNA replication. © 1977 American Society for Microbiology. (c) © Biology Media/Photo Researchers.

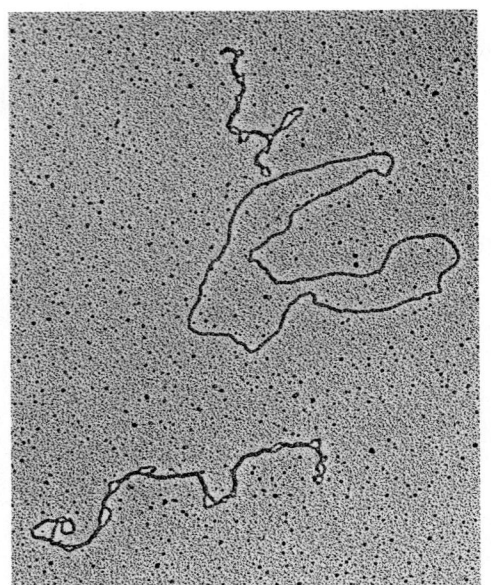

FIGURE 4.18

Relaxed and supercoiled DNA molecules. Electron micrograph showing three human mitochondrial DNA molecules. All three are of identical sequence and contain 16,569 bp each. However, the molecule in the center is relaxed, whereas those at top and bottom are tightly supercoiled.

Courtesy of Dr. David A. Clayton.

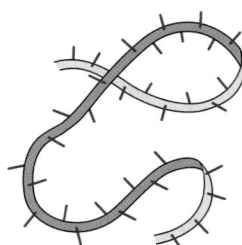

(a) Random coil

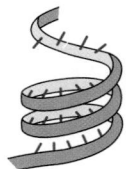

(b) Stacked-base structure (single-strand helix)

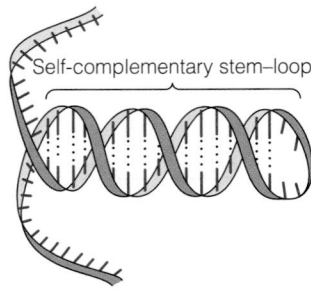

Self-complementary stem–loop

(c) Hairpin formation in self-complementary region (double helix)

There is a special aspect to circular DNA molecules that one might not readily expect: Most are **supercoiled**. What this means can be seen by considering the group of molecules shown in Figure 4.18. The figure shows both relaxed and supercoiled forms of mitochondrial DNA molecules. A relaxed circle can lie flat on a plane surface. A supercoiled molecule cannot. In addition to the twist of the DNA strands about one another, a supercoiled molecule has extra twists in the helix axis itself—the helix axis crosses over itself one or more times. Three-dimensional structure, such as supercoiling, that involves a higher-order folding of elements of regular secondary structure is called the **tertiary** structure of a polymer.

Supercoiling is by no means rare. Rather, it is the usual state of closed circular DNA molecules. Most naturally occurring circular DNA molecules have left-handed superhelical twists, but it is possible to form DNA molecules with right-handed superhelicity. By convention, we call right supercoiling **positive** and left supercoiling **negative**.

The DNA molecules shown in Figure 4.18 differ only in their topology, meaning that they can be superimposed only by breaking bonds; therefore, they are called **topoisomers**. Topoisomers can be interconverted *only* by cutting and resealing the DNA. Cells have enzymes capable of doing this. These enzymes are called **topoisomerases**, and they regulate the superhelicity of natural DNA molecules.

When superhelical turns (either left-handed or right-handed) are introduced into a previously relaxed DNA molecule, the molecule is put under strain. Therefore, energy must be expended to make a DNA molecule supercoiled. Prokaryotic cells like *E. coli* have a special topoisomerase called **DNA gyrase**. This enzyme introduces left-handed superhelical turns in a reaction driven by ATP hydrolysis. Some topoisomerases, which can only relax supercoiled DNA, do not require ATP. We discuss topoisomerases in more detail in Chapter 25. The energy stored in circular DNAs by twisting them into supercoils may have major effects on DNA conformation; we describe such effects later in this chapter after we present a quantitative theory of supercoiling.

Structure of Single-Stranded Polynucleotides

Single-stranded polynucleotides can exhibit a variety of structures, depending on their sequences and the solution conditions. At high temperature or in the presence of denaturing substances, most will be largely in a random coil form, as depicted in Figure 4.19a. Such a structure is characterized by flexibility and freedom of rotation about backbone bonds, which leads to a floppy, constantly changing form. However, under conditions closer to those found in vivo, the stacking interactions will tend to form regions of single-chain, stacked-base helix (Figure 4.19b). Furthermore, most naturally occurring nucleic acid sequences contain regions of self-complementarity between which base pairing is possible. Here the molecule can loop back upon itself to form a double-stranded structure, as diagrammed in Figure 4.19c. A more complex example is given in Figure 4.20, which portrays the structure of a **transfer RNA (tRNA)**, an RNA involved in protein synthesis (see next section). Here we find not only A-type secondary structure from the folding of the chain back on itself but also more complex folding of such helices together. Thus, the tRNA molecule possesses a defined *tertiary* structure, a higher-order folding that gives it a defined shape and internal arrangement necessary for its function. The much larger RNA molecules from ribosomes have similar secondary structure features, but much more complex tertiary structures, as we shall see in Chapter 28.

FIGURE 4.19

Conformations of single-stranded nucleic acids. (a) The random coil structure of denatured single strands. There is flexibility of rotation of residues and no specific structure. **(b)** Stacked-base structure adopted by non–self-complementary single strands under "native" conditions. Bases stack to pull the chain into a helix, but there is no H-bonding. **(c)** Hairpin structures formed by self-complementary sequences; the chain folds back on itself to make a stem–loop structure.

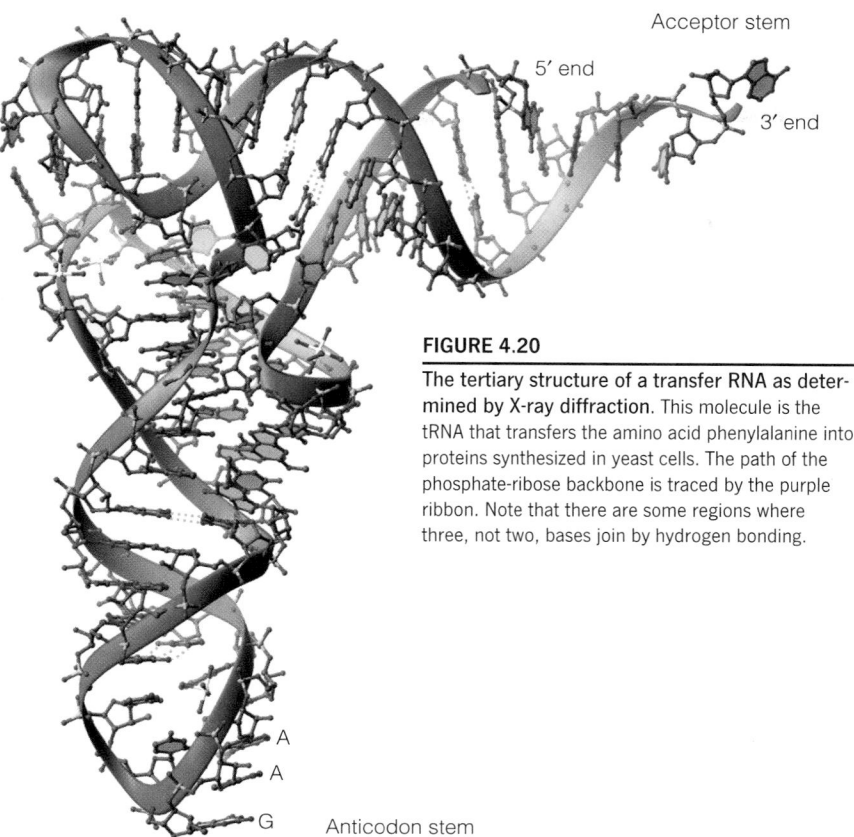

Acceptor stem

5′ end

3′ end

FIGURE 4.20

The tertiary structure of a transfer RNA as determined by X-ray diffraction. This molecule is the tRNA that transfers the amino acid phenylalanine into proteins synthesized in yeast cells. The path of the phosphate-ribose backbone is traced by the purple ribbon. Note that there are some regions where three, not two, bases join by hydrogen bonding.

A
A
G Anticodon stem

The higher-order folding of a biopolymer's secondary structure is called its tertiary structure.

Most DNA molecules found in vivo are left-handed supercoils.

RNA molecules are usually single-stranded, but most have self-complementary regions that form hairpin structures, and some have well-defined tertiary structures.

The Biological Functions of Nucleic Acids: A Preview of Molecular Biology

We have emphasized that the fundamental role of the nucleic acids is the storage and transmission of genetic information, and we will continue to develop this theme throughout the book. In Part V we describe in detail how nucleic acids are passed from parent cell to daughter cell (or from an organism to its descendents) and how they direct biochemical processes, specifically, the synthesis of proteins. In this chapter and the next, we present a preliminary overview of these nucleic acid functions. Even though this may not be your first exposure to these topics, you should gain some further appreciation of the relationships between nucleic acid and protein structures, of evolution at the molecular level, and of our ability to modify microbes, plants, and animals through genetic engineering.

Genetic Information Storage: The Genome

Every organism carries in each of its cells at least one copy of the total genetic information possessed by that organism. This is referred to as the *genome*. Usually, the genomic information is coded in the sequence of double-stranded DNA, but some viruses use single-stranded DNA or even RNA (see Table 4.4). Genomes vary enormously in size; the smallest viruses need only a few thousand bases (b) or base pairs (bp), whereas the human genome consists of about 3×10^9 bp of DNA, distributed in 23 chromosomes. Other organisms have far more (Chapter 24).

Recent years have seen remarkable advances in our ability to determine DNA or RNA sequences. In 1976, Maxam and Gilbert (see References) devised a chemical sequencing method involving selective cleavage at A, T, G, or C residues, followed by separation of the fragments on the basis of fragment length, using gel electrophoresis. This ingenious technique allowed researchers to begin the exploration of genomic information. The Maxam–Gilbert method has been largely supplanted by a technique that uses enzymes to generate oligonucleotide fragments started and terminated at specific base positions. This method, developed by Fred Sanger, is described in detail in Tools of Biochemistry 4B. Automation of this technique, and development of much faster "high-throughput" methods, has led to the complete genome sequence analysis of hundreds of species, including near-completion of the human genome sequence in 2001, and complete genome sequences for many individual humans within the subsequent few years.

In every organism a significant fraction of the genomic DNA is capable of being transcribed, or "read," to allow expression of its information in directing synthesis of RNA and protein molecules. Each DNA segment that encodes one protein or one RNA molecule is a gene. The DNA in each cell of every organism contains at least one copy (and sometimes several) of the gene carrying the information to make each protein that the organism requires. In addition, there are genes (often reiterated manyfold) for the many specific functional RNA molecules, such as the transfer RNA (tRNA) shown in Figure 4.20. Like proteins, these RNAs play specific roles as part of the cell's machinery. (See Table 4.5.) We briefly mention the functions of messenger RNA, ribosomal RNA, and transfer RNA in this chapter, and the more recently discovered micro RNAs and their role in *RNA interference* in Chapter 29.

TABLE 4.5 Properties of some naturally occurring RNA molecules

Source (Organism)	Designation	Function	Size (b or bp)
tRNA (transfer RNA)			
E. coli	tRNA^{Leu}	Transfers leucine in protein synthesis	87 b
Yeast	tRNA^{Phe}	Transfers phenylalanine in protein synthesis	76 b
Rat	tRNA^{Ser}	Transfers serine in protein synthesis	85 b
rRNA (ribosomal RNA)			
E. coli	5S RNA	Part of ribosome structure	120 b
	16S RNA	Part of ribosome structure	1542 b
	23S RNA	Part of ribosome structure	2904 b
mRNA (messenger RNA)			
Chicken	mRNA$_{LYS}$	Messenger RNA for protein lysozyme	584 b
Rat	mRNA$_{SA}$	Messenger RNA for protein serum albumin	~2030 b
vRNA (viral RNA)			
Polio virus	Polio RNA	Genome of the virus	7440 b
Cytoplasmic polyhedrosis virus of tussock moth	CPV RNA	Genome of the virus	Ten double-stranded molecules, ~890 to ~5150 bp
miRNA (microRNA), siRNA (small interfering RNA)			
All or most eukaryotes	miRNA, siRNA	Control of gene expression	21–24 b

Replication: DNA to DNA

Replication passes on the genetic information from cell to cell and from generation to generation. The essence of the process is depicted in Figure 4.12—a complementary copy of each strand of duplex DNA is made, usually resulting in two identical copies of the original duplex. The process is highly accurate in copying—making less than 1 error in 10^8 bases—but occasionally mistakes are made. These contribute to the mutations that have allowed the evolution of life to ever more complex forms.

The replication of DNA is accomplished by a complex of enzymes, acting in concert like a finely tuned machine. These enzymes are described in detail in Chapter 25. Each enzyme complex, or **replisome**, centered on a protein called *DNA polymerase,* has multiple functions. As parental DNA strands unwind, forming a *replication fork* (see Figure 4.12), DNA polymerase guides the pairing of incoming deoxyribonucleoside triphosphates, each with its complementary partner on the strand being copied. It then catalyzes the formation of the phosphodiester bond to link this residue to the new growing chain. Thus, each of the parental DNA strands serves as a **template**, specifying the sequence of a daughter strand. DNA polymerase adds nucleotides, one at a time, to the growing daughter strand, which can be considered a **primer** to which nucleotides are added as the daughter DNA strand grows from its 5′ end toward its 3′ end, as shown in Figure 4.21. In most cases, the enzyme complex also checks, or "proofreads," the addition before proceeding to add the next residue; proofreading contributes to the high overall accuracy of replication. Because the two DNA strands run in opposite directions, one daughter strand is elongated in the same direction as that of the replication fork while the other is formed in the reverse direction. This and other aspects of DNA replication are discussed in more detail in Chapter 25.

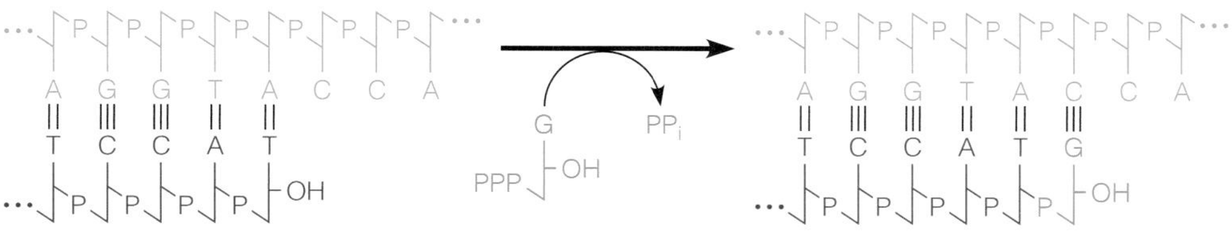

> Replication is the copying of both strands of a duplex DNA to produce two identical DNA duplexes.

FIGURE 4.21

The DNA polymerase reaction.
DNA polymerase fits a deoxyribonucleoside triphosphate molecule to its complementary nucleotide in the template strand (blue) and catalyzes formation of a phosphodiester link between the incoming nucleotide and the 3′-hydroxyl of the 3′-terminal nucleotide in the growing daughter strand (red), with pyrophosphate being split out.

Transcription: DNA to RNA

Expression of genetic information always involves as a first step the **transcription** of genes into RNA molecules of complementary nucleotide sequence. This production of specific RNA molecules is easy to visualize. Just as a DNA strand can direct replication, it can equally well direct transcription, the formation of a complementary RNA strand (Figure 4.22). Of course, the monomers required in transcription are different from those required in replication. Instead of the

> Transcription is the copying of a DNA strand into a complementary RNA molecule.

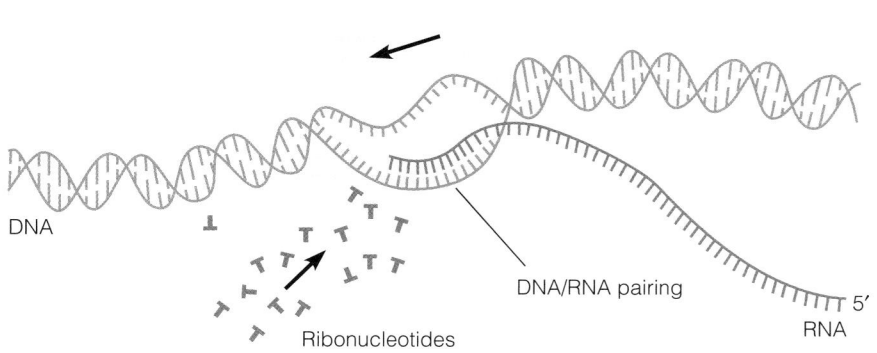

FIGURE 4.22

The basic principle of transcription. An enzyme (RNA polymerase) travels along a DNA molecule, opening the double strand and making an RNA transcript by adding one ribonucleotide at a time. It copies the oligonucleotide sequence from only one of the two DNA strands. After the enzyme passes, the DNA rewinds.

deoxyribonucleoside triphosphates (dNTPs), the ribonucleoside triphosphates ATP, GTP, CTP, and UTP are needed to make RNA. (Note that U in the new RNA pairs with A in the DNA template.) Another distinction from DNA replication is that only one of the two DNA strands, the **template strand**, is copied. The other strand is called the sense strand, or coding strand. DNA transcription, like DNA replication, requires a special set of enzyme catalysts known as *RNA polymerases*. Transcription is presented in detail in Chapter 27.

Translation: RNA to Protein

Transcription is the central process in producing the many functional RNA molecules of the cell, such as the transfer RNAs or ribosomal RNAs listed in Table 4.5. The synthesis of specific proteins, under the direction of specific genes, is a more complex matter. The problem, as we see in Chapter 5, is that proteins are polymers made from 20 different kinds of amino acid monomers. Because there are only four different nucleotide monomer types in DNA, there cannot be a one-to-one relationship between the sequence of nucleotides in a DNA molecule and the sequence of amino acids in a protein. Rather, the linear sequence of bases that constitutes the protein-coding information is "read" by the cell in blocks of three nucleotide residues, or **codons**, each of which specifies a different amino acid. The set of rules that specifies which nucleic acid codons correspond to which amino acids is known as the **genetic code**. We describe this code in Chapter 5 after we have described the amino acids and the structure of proteins.

Although the information for all protein sequences is coded in DNA, the production of proteins does not proceed directly from DNA. For the information to be converted from the DNA sequences of the genes into amino acid sequences of proteins, special RNA molecules are needed as intermediates. Complementary copies of the genes to be expressed are transcribed from DNA in the form of **messenger RNA (mRNA)** molecules (see Table 4.5), so called because they carry information from DNA to the protein-synthesizing machinery of the cell. The machinery includes tRNA molecules, special enzymes, and **ribosomes**, which are RNA–protein complexes upon which the assembly of new proteins takes place. This **translation** of RNA information is outlined in Figure 4.23. (We present the main features of translation in Chapter 5 and describe it in detail in Chapter 28.) The flow of genetic information in the cell can be summarized by the simple schematic diagram shown in Figure 4.24.

In translation, an RNA nucleotide sequence dictates a protein amino acid sequence.

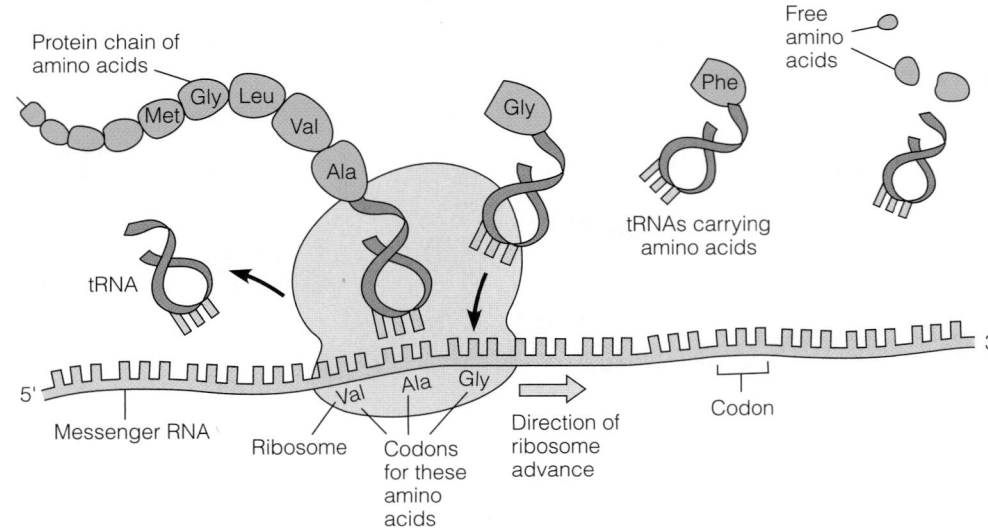

FIGURE 4.23

The basic principle of translation. A messenger RNA molecule is bound to a ribosome, and transfer RNA molecules bring amino acids to the ribosome one at a time. Each tRNA identifies the appropriate codon on the mRNA and adds this amino acid to the growing protein chain. The ribosome travels along the mRNA so that the genetic message can be read and translated into a protein.

As we show in the following chapters, proteins are the major structural and functional molecules in most cells. What a cell is like and what it can do depend largely on the proteins it contains. These, in turn, are dictated by the information stored in the cell's DNA, transcribed and processed into mRNA, and expressed by the protein-synthesizing machinery. If we use the analogy of a cell as a factory, the proteins are the working machinery. The master plans for this machinery are stored in a central repository (the DNA of the cell nucleus). Copies of certain plans (mRNAs) are sent out from time to time as new protein machinery or replacements are needed. Those plans find their way to ribosomes where these RNA- and protein-containing particles, in conjunction with transfer RNA molecules, catalyze the template-dependent assembly of proteins. Overlaid on these processes are recently described small RNA molecules, barely 20 nucleotides long, which help regulate the process. These **microRNAs** are discussed further in Chapter 29.

Plasticity of Secondary and Tertiary DNA Structure

In preceding sections we described the B and A forms of DNA, and we briefly mentioned supercoiling of circular DNA. We now recognize that a number of special secondary and tertiary structures exist. Some of these, notably supercoiling, are involved in packing giant DNA molecules into cells. Some depend on the presence of special primary structures, and in some cases they can be stabilized by the presence of supercoiling. To begin, then, we must look more closely into the nature of supercoiled DNA.

Changes in Tertiary Structure: A Closer Look at Supercoiling

In order to get a quantitative feeling for what supercoiling means, examine the "thought experiment" described in Figure 4.25. Consider the linear DNA molecule shown in Figure 4.25a. It contains enough base pairs to make exactly 10 helical turns; we say it has a **twist** $(T) = 10.0$. Now suppose we took this DNA molecule, laid it flat on a surface, and brought the strand ends together $5'$ to $3'$

FIGURE 4.24

The flow of genetic information in a typical cell. DNA can both replicate and be transcribed into RNA. Messenger RNAs are translated into protein amino acid sequences.

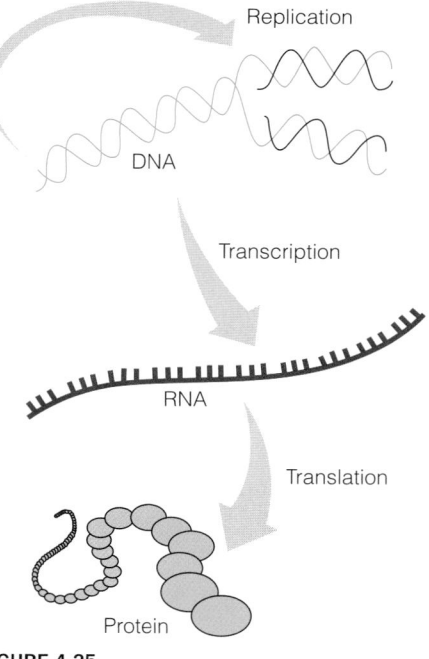

FIGURE 4.25

Forming a DNA supercoil. We can imagine making a linear DNA circular in several ways: **(a)** A linear DNA is laid flat. It has 105 base pairs and 10.5 bp/turn, so there are 10 turns; the twist $T = 10$. **(b)** Because there is an integral number of turns, we can link the $5'$ and $3'$ ends together without twisting, to form a relaxed circle that lies flat. The linking number $L = 10$ and the writhe $W = 0$. **(c)** If we reduce the number of turns before joining the ends, by unwinding the DNA one turn, the closed circle will have $L = 9$. If we still require the DNA molecule to lie flat, the twist of the helix must change. We now have $L = 9$, $T = 9$, and $W = 0$. With $T = 9$ in 105 bp, the DNA is forced into a conformation with $105/9 = 11.67$ bp/turn. **(d)** Rather than change its twist, the strained DNA molecule may writhe, or supercoil, so that $L = 9$, $T = 10$, and $W = -1$. Strained molecules usually supercoil rather than change their twist.

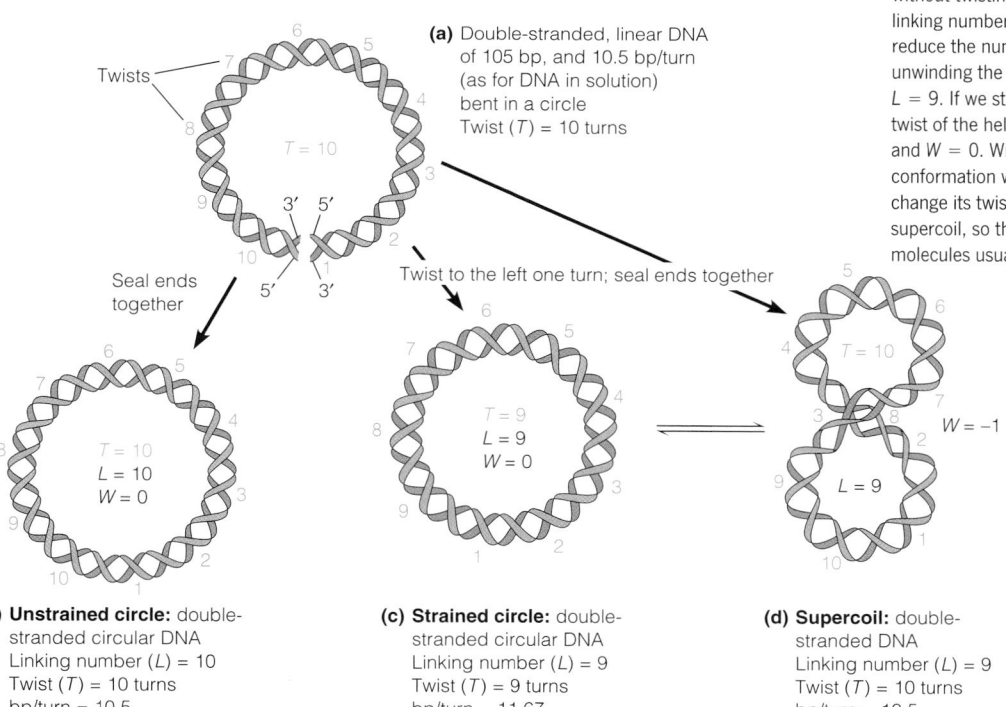

(b) Unstrained circle: double-stranded circular DNA
Linking number $(L) = 10$
Twist $(T) = 10$ turns
bp/turn = 10.5
Writhe $(W) = 0$

(c) Strained circle: double-stranded circular DNA
Linking number $(L) = 9$
Twist $(T) = 9$ turns
bp/turn = 11.67
Writhe $(W) = 0$

(d) Supercoil: double-stranded DNA
Linking number $(L) = 9$
Twist $(T) = 10$ turns
bp/turn = 10.5
Writhe $(W) = -1$

(Figure 4.25a). Because the DNA molecule contains just enough base pairs to make 10 complete turns of the helix, the 5′ and 3′ ends of each strand would be in position to meet one another, and we could covalently join them. (Enzymes called *ligases* carry out this joining.) This connection would make a relaxed circle that could still lie flat on the surface (Figure 4.25b). The covalently closed molecule now has another property—the strands are interlinked 10 times. We say that it has a **linking number** (L) of 10.

Now suppose instead that before joining the ends of this right-handed double helix, we made one turn to the left, unwinding by one turn and thereby reducing the linking number to 9. This change would place a strain on the circular DNA, which could respond in two different ways:

1. The molecule could remain flat on the surface with its twist reduced to 9 ($T = 9$, as shown in Figure 4.25c). The normal 10.5 bp per turn of the helix would then be increased to 11.67. The strain on this DNA molecule is expressed in its "unnatural" twist; it is said to be *underwound*. (If we had added an extra turn to the right instead of left, the circle would also be strained, but it would be *overwound*, with 11 twists and 9.54 bp/turn.)

2. The molecule could restore its original twist ($T = 10$) by wrapping about itself in one negative (left-handed) superhelical turn, as shown in Figure 4.25d. The number of such superhelical turns is called the **writhe** (W); in this case $W = -1$. (If we had added one twist to the right, the molecule could form a positive, or right-hand, writhe with $W = +1$; in the first response, with no supercoiling, $W = 0$.)

When we untwisted one turn and then joined, we produced a molecule in which the two strands were interlinked only 9 times; $L = 9$. The quantity L will remain unchanged no matter how we might distribute the strain of underwinding between twist and writhe. Thus, the two forms found in Figure 4.25c and 4.25d both have $L = 9$. Note that in every case, the linking number is the algebraic sum of T and W:

$$L = T + W \tag{4.1}$$

> The linking number is always the algebraic sum of twist and writhe.

The only way in which L can be changed is by cutting, twisting, and rejoining the circular molecule—by means of topoisomerases, for example. When we added the extra twist to the left at the beginning of our experiment, we changed the value of L by -1 ($\Delta L = -1$). The strain imposed by putting L extra turns in the DNA helix distributes itself between a change in twist and a change in writhe: where L is positive or negative, depending on whether the extra twists (T) and superhelical turns (W) are right-handed (positive) or left-handed (negative).

> The superhelix density, σ, is a quantitative measure of the intensity of supercoiling.

$$\Delta L = \Delta T + \Delta W \tag{4.2}$$

The superhelicity of DNA molecules is often expressed in terms of the **superhelix density**, $\sigma = \Delta L / L_0$, where L_0 is the linking number for the DNA in the relaxed state. Many naturally occurring DNA molecules have superhelix densities of about 0.06. To get an idea of what this means, consider a hypothetical DNA molecule of 10,500 bp, which is in the solution B form, with 10.5 bp/turn. Then L_0 is 10,500 bp/(10.5 bp/turn), or 1000 turns. Each DNA strand crosses the other 1000 times in the relaxed circle. If the topoisomerase DNA gyrase (see page 1055) twisted the molecule to a superhelical density of -0.06, then $L = -0.06 \, L_0$, or $L = -60$. This change could be accommodated, for example, by the helix axis writhing about itself 60 times in a left-handed sense, which would correspond to $\Delta W = -60$, $T = 0$; the molecule would have 60 left-handed superhelical turns. Alternatively, the twist of the molecule could change so that it had 940 turns in 10,000 bp ($T = 940$) or

10,500/940 = 11.2 bp/turn. This would correspond to $\Delta W = 0$, $\Delta T = -60$. Although any combination of ΔT and ΔW that sums to -60 could occur, real molecules release strain mainly by writhing into superhelical turns because it is easier to bend long DNA than to untwist it. This explains the contorted structures of the double-stranded circular DNA molecules in Figures 4.17b and 4.18.

Differences in supercoiling can be detected by gel electrophoresis. As described in Tools of Biochemistry 2A, the rate at which a molecule migrates through a gel matrix in an electric field depends on its dimensions; hence the more compact superhelical forms will move faster than the relaxed form. Figure 4.26 shows electrophoretic patterns for supercoiled DNA molecules that are being progressively relaxed by the action of the enzyme topoisomerase. Thus, gel electrophoresis allows us to separate topoisomers of a given DNA.

Unconventional Secondary Structures of DNA

Most of the DNA and RNA in cells can be described as having one of the three secondary structures—random coil (which is really a lack of secondary structure), B form, or A form. But these three do not nearly exhaust the conformational possibilities of these remarkable molecules. In the remainder of this chapter we consider some of the alternative structures that have been recognized in recent years, and we will examine the conditions under which conformational transitions between them take place. We will see that supercoiling often plays a dominant role in these transitions.

Left-Handed DNA (Z-DNA)

Because both the A and B forms of polynucleotide helices are right-handed (see Figure 4.15), the discovery of a left-handed form in 1979 caused considerable surprise. Alexander Rich and his colleagues carried out X-ray diffraction studies of crystals of the small oligodeoxynucleotide

$$^{5'}CGCGCG^{3'}$$
$$^{3'}GCGCGC^{5'}$$

and determined that it is a double-stranded helix with G-C base pairing, as they had expected. However, the data were consistent only with a peculiar *left-handed* structure they called Z-DNA. A model for a long DNA molecule in the Z conformation is shown in Figure 4.27.

In addition to the reverse sense of the helix, Z-DNA exhibits other structural peculiarities. There are, in polynucleotides, two most stable orientations of the bases with respect to their deoxyribose rings. They are called *syn* and *anti*:

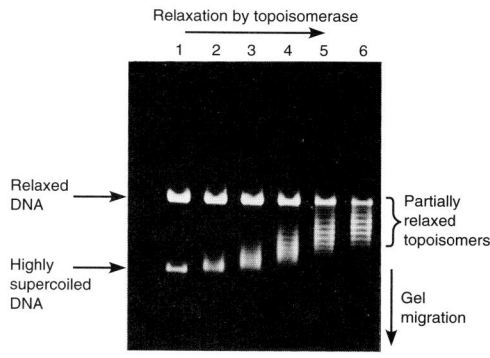

FIGURE 4.26

Gel electrophoresis demonstrating DNA supercoiling. Lane 1: A mixture of relaxed and highly supercoiled DNA. Lanes 2 to 6: The progress of relaxation catalyzed by the enzyme topoisomerase. Samples have been taken at successive times after adding the enzyme. Individual topoisomers are resolved as individual bands on the gel. The highly supercoiled material, which forms a densely packed series of overlapping bands at the bottom, gradually disappears. The DNA species are resolved by electrophoresis in an agarose gel and then visualized by adding the dye ethidium bromide (page 1078) to the gel, a treatment that makes each DNA species fluorescent.

Courtesy of J. C. Wang.

Z-DNA is a left-handed helix with alternate purine/pyrimidine bases in alternate *syn/anti* conformation.

Syn   *Anti*

Deoxyadenosine

Syn   *Anti*

Deoxycytidine

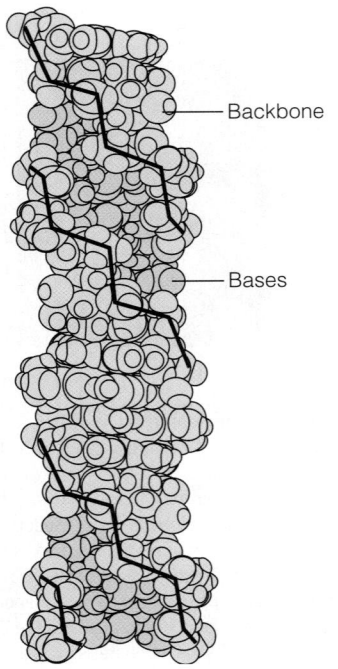

FIGURE 4.27

Z-DNA. The structure of Z-DNA as determined by single-crystal X-ray diffraction studies. Compare this left-hand form of DNA with the similar, space-filling model of B-DNA in Figure 4.11. The single groove of Z-DNA is shown in green. The black line follows the zigzag phosphate backbone.

Cold Spring Harbor Symposia on Quantitative Biology 47:41, Dr. A.H.-J. Wang, Right-handed and left-handed double-helical DNA: Structural studies. © 1982 Cold Spring Harbor Laboratory Press.

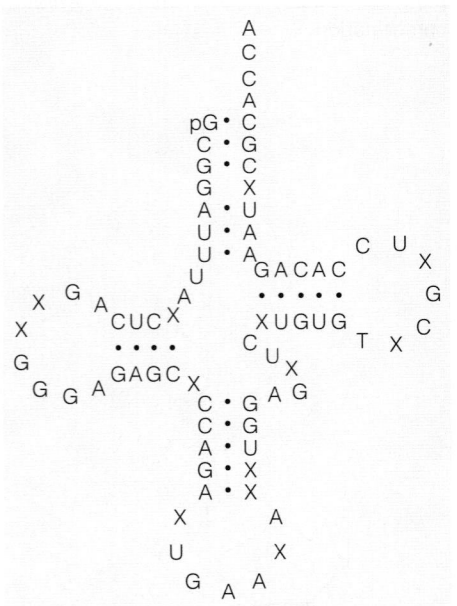

FIGURE 4.28

How self-complementarity dictates the tertiary structure of tRNA. A schematic drawing of the tRNA shown in Figure 4.20. X's represent unusual or modified bases (see Chapter 28). The molecule has three hairpin arms resulting from self-complementarity. These arms fold into the tertiary structure of Figure 4.20.

In both A- and B-form polynucleotides, all bases are in the *anti* orientation. In Z-DNA, however, the *pyrimidines* are always *anti*, but the *purines* are always *syn*. Because Z-DNA is most often found in polynucleotides with alternating purines and pyrimidines in each strand (such as that shown earlier), the base orientations will alternate. Parameters for Z-DNA (see Table 4.3, page 105) reflect this characteristic, in that the repeating unit is not one base pair but two base pairs. Furthermore, this alternation gives a zigzag pattern to the phosphates; hence the name Z-DNA (see Figure 4.27).

There is now abundant evidence that Z-DNA exists in living cells. However, the exact role played by Z-DNA in vivo is still an open question, partly because of the difficulty of quantitating Z-DNA in living cells. It is perhaps significant that methylation of cytosines on carbon 5, a common modification in vivo, favors Z-DNA.

Hairpins and Cruciforms

We have already encountered examples of "hairpin" structures—first in Figure 4.19c and then in the transfer RNA shown in Figure 4.20. In each of these single-stranded molecules, self-complementarity in the base sequence allows the chain to fold back on itself and form a base-paired, antiparallel helix. The schematic structure of a tRNA shown in Figure 4.28 depicts how a particular base sequence accomplishes this folding.

Double hairpins, often called **cruciform** (cross-like) structures, can be formed in some DNA sequences. To form this structure, the sequence must be **palindromic**. As mentioned also on page 129, the word *palindrome* is of literary origin and usually refers to a statement that reads the same backward and forward, such as "Able was I ere I saw Elba." As used in descriptions of DNA, the word refers to segments of complementary strands that are the exact (or almost exact) reverse of one another. Such a DNA sequence is shown in Figure 4.29, along with the two conformations available to it. In most instances, formation of the cruciform structure leaves a few bases unpaired at the ends of the hairpins. This means that under normal circumstances the cruciform will be less stable than the extended structure. As we shall see later in this chapter, one effect of superhelical strain is to stabilize cruciforms.

Triple Helices and H-DNA

It has long been recognized that certain homopolymers can form **triple** helices. The first to be discovered was the synthetic RNA structure

$$\text{poly(U)} \cdot \text{poly(A)} \cdot \text{poly(U)}$$

Later, it was observed that deoxy triplets like T · A · T and C$^+$ · G · C (where C$^+$ is a protonated cytosine) can also be formed. Such structures involve, in addition to the normal Watson–Crick base pairing, the *Hoogsteen-type* base pairing shown in Figure 4.30. It is now recognized that many polynucleotide strands, including some of nonrepeating sequence, can enter into such triple helices. An unusual structure that can incorporate a triple helix is shown in Figure 4.31. It is called H-DNA and requires a molecule that has one all-pyrimidine strand (Pyr) and one all-purine strand (Pur). In the example shown, the structure

$$\cdots \text{TCTCTCTCTCTC} \cdots$$

$$\cdots \text{AGAGAGAGAGAG} \cdots$$

can exist either in the normal, double-stranded form, or with one strand looped back to form a triple helix (with C$^+$ · G · C and G · A · G triplets), leaving the other (Pur) strand unpaired. Again, the loss of base pairing should make this structure unstable, except in situations where strain in the DNA molecule favors it.

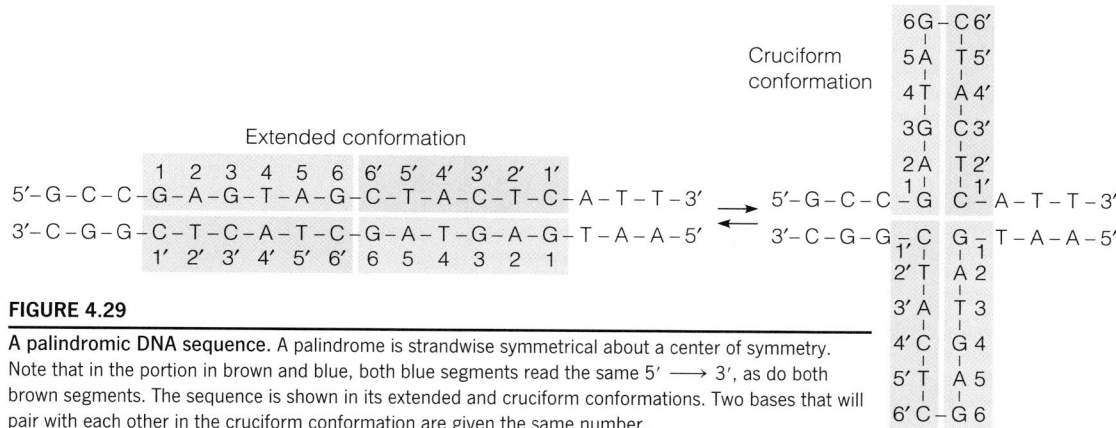

FIGURE 4.29

A palindromic DNA sequence. A palindrome is strandwise symmetrical about a center of symmetry. Note that in the portion in brown and blue, both blue segments read the same 5' ⟶ 3', as do both brown segments. The sequence is shown in its extended and cruciform conformations. Two bases that will pair with each other in the cruciform conformation are given the same number.

G-Quadruplexes

As long ago as 1962 David R. Davies realized that four guanine molecules could fit together in a planar hydrogen-bonded structure, and he proposed that such structures might form naturally in guanine-rich sections of DNA. Figure 4.32 shows schematically how four guanines can link to form a **G-quartet** (panel A), and how a single strand of guanine-rich DNA can fold to form a **G-quadruplex** (panel B), in the example shown consisting of three G-quartets. A G-quadruplex structure can form from one DNA strand, as shown, or from as many as four strands. G-quadruplexes actually form in living cells and exist in **telomeres**, special sequences at the ends of linear eukaryotic chromosomes (see Chapter 25). Figure 4.32c shows the likely folding pattern of a human telomere. More recent studies show that G-quadruplexes exist in the transcriptional control sites, or **promoters**, of several biologically important genes, including the **oncogene** cMYC (Chapter 23). Current efforts are aimed at targeting such structures with anti-cancer drugs.

An Unexpected Primary Structure Modification: DNA Phosphorothioation

All of the unconventional DNA structures described thus far affect secondary and/or tertiary structure, with the phosphodiester link being invariant. Accordingly, there was great surprise in 2007 when Peter Dedon's group at MIT described an enzyme system in bacteria that converts a phosphate group in DNA to a

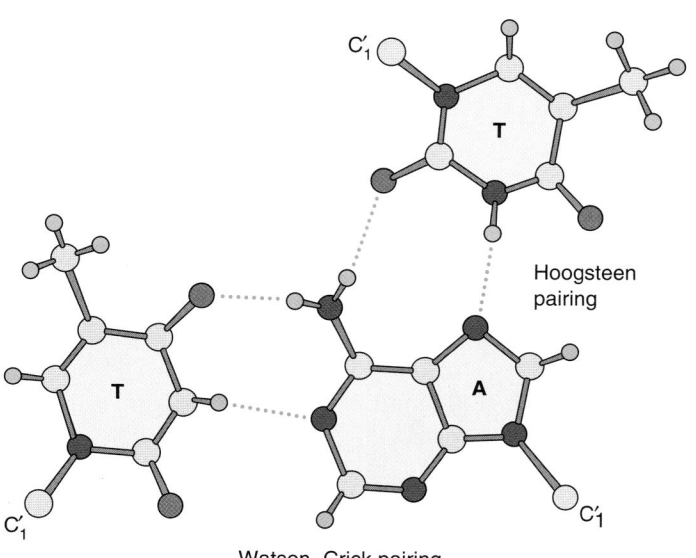

Watson–Crick pairing

FIGURE 4.30

Base-pairing in one type of DNA triple helix. Both normal Watson–Crick pairing and the unusual Hoogsteen pairing occur on the same A residue.

FIGURE 4.31

H-DNA. An H-DNA region has one all-purine (blue) and one all-pyrimidine (brown) strand, allowing it to form a triple-stranded helix by doubling back. Some segments contain both purines and pyrimidines (green). **(a)** The nucleotide sequence shown here is one that could give rise to H-DNA. A segment of the purine strand is shown bonded to two different segments of the pyrimidine strand. **(b)** The bonding shown in (a) gives rise to a triple-stranded helix, shown schematically here.

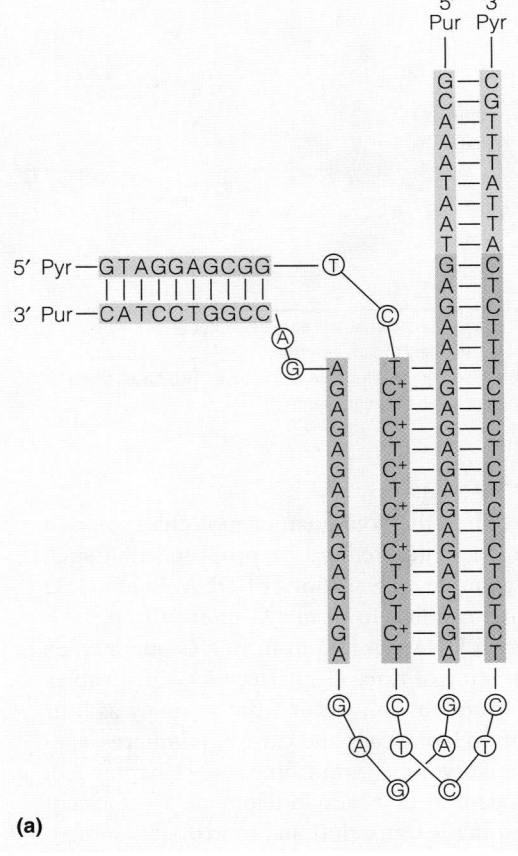

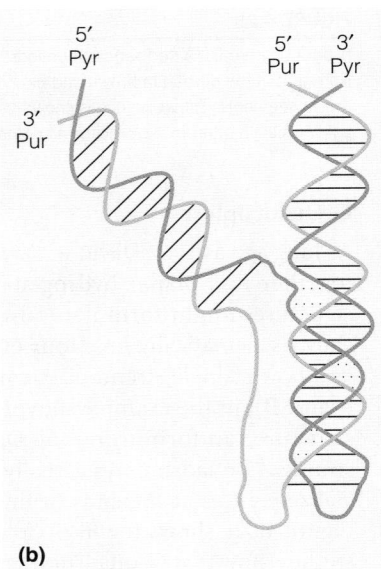

(a)

(b)

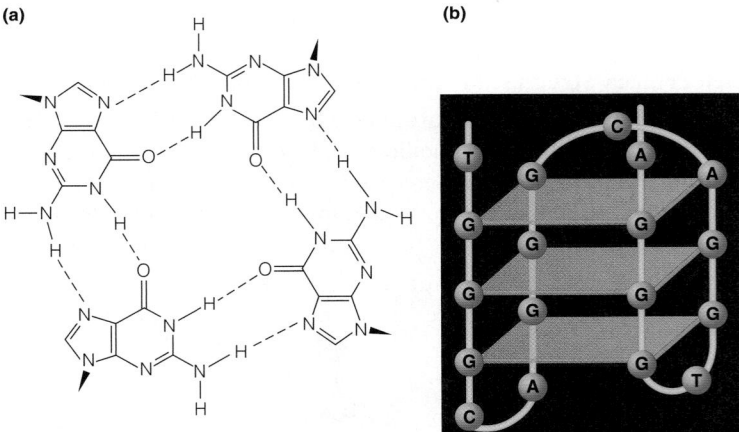

(a)

(b)

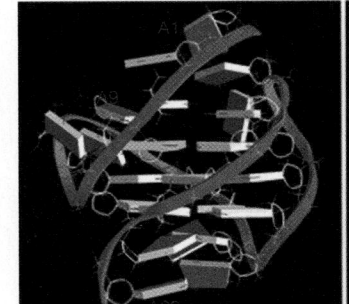

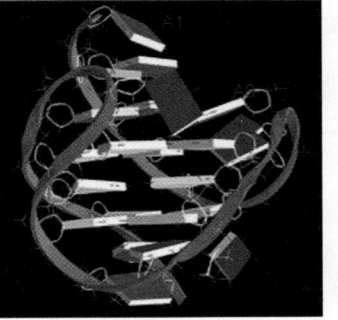

(c)

FIGURE 4.32

G-quartets and quadruplexes. (a) Arrangement of bases in a G-quartet, with four Hoogsteen-bonded guanines surrounding a central metal ion (not shown); **(b)** Folding of a single DNA strand to give a G-quadruplex, consisting in this example of three planar G-quartets. **(c)** Two views of the G-quadruplex formed by the DNA sequence in human telomeres. Yellow, guanine; red, adenine; blue, thymine.

(c) J. Dai, C. Punchihewa, A. Ambrus, D. Chen, R. A. Jones, and D. Yang, Structure of the intramolecular human telomeric G-quadruplex in potassium solution: A novel adenine triple formation, *Nucleic Acids Research* 35:2440–2450, © 2007, by permission of Oxford University Press.

phosphorothioate, as shown in Figure 4.33. The reaction is stereospecific (note that introduction of a sulfur creates an asymmetric center about the phosphorus atom). So far the biological function of this modification is unknown, but it might confer resistance to some foreign invaders, such as bacteriophages, which could degrade unmodified DNA. Recent sequence analysis shows that phosphorothioate nucleotides are clustered in many bacteria, consistent with the idea that phosphorothioation could be part of a restriction-modification system (Chapter 26).

Stability of Secondary and Tertiary Structure

The Helix-to-Random-Coil Transition: Nucleic Acid Denaturation

The major polynucleotide secondary structures (the A and B forms) are relatively stable for RNA and DNA, respectively, under physiological conditions. Yet they must not be *too* stable because important biochemical processes—DNA replication and transcription, for example—require that the double-helix structure be opened up. When it extends over large regions, this loss of secondary structure is called **denaturation** (Figure 4.34). Competing factors create a balance between structured and unstructured forms of nucleic acids.

Two major factors favor dissociation of double helices into randomly coiled single chains. First is the electrostatic repulsion between the chains. At physiological pH, every residue on a DNA or an RNA molecule carries a negative charge on the phosphate group. Even though this charge is partially neutralized by small

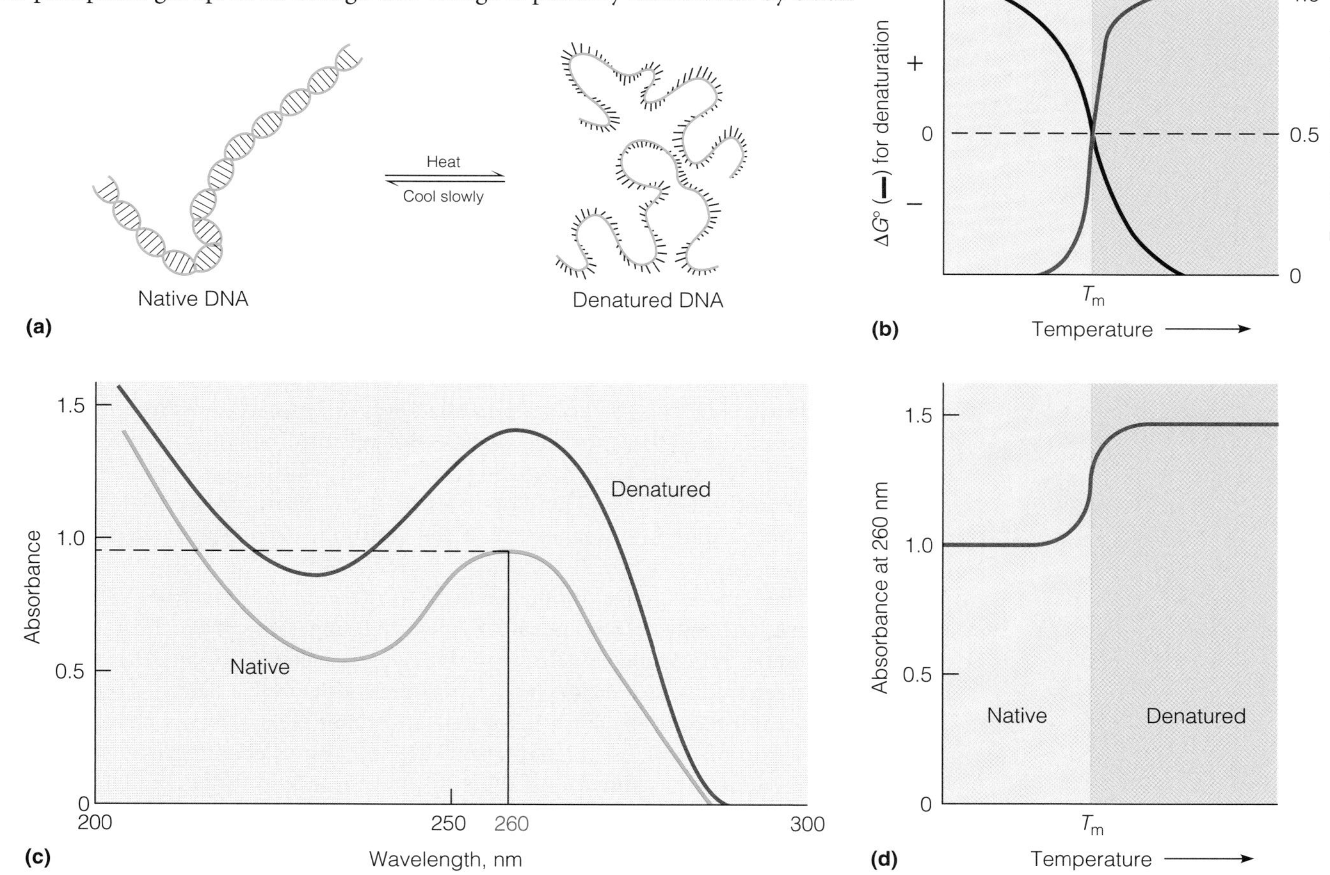

FIGURE 4.33

A phosphorothioate link in DNA, showing the correct stereoisomeric form. The source of sulfur is the amino acid cysteine. Nucleotide modification occurs after nucleotide polymerization.

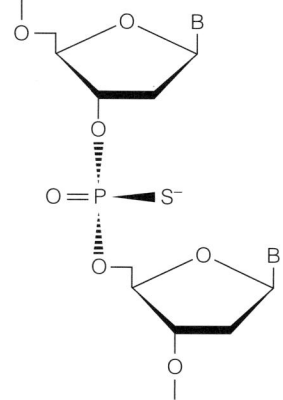

FIGURE 4.34

Denaturation of DNA. **(a)** When native (double-stranded) DNA is heated above its "melting" temperature, it is denatured (separates into single strands). The two random-coil strands have a higher entropy than the double helix. **(b)** At low T, ΔG is positive and denaturation of DNA is not favored. As T increases, $-T\Delta S$ overcomes ΔH, making ΔG negative and denaturation favorable. The midpoint of the curve marks the "melting" temperature, T_m, of DNA. **(c)** Absorption spectra of native and denatured DNA show that native DNA absorbs less light than denatured DNA, with the maximum difference occurring at a wavelength of 260 nm. This **hypochromicity** of double-stranded DNA can be used to distinguish between native and denatured forms. **(d)** The change in absorbance can be used to follow the denaturation of DNA as temperature increases. An abrupt increase in absorbance, corresponding to the sudden "melting" of DNA, is seen at T_m.

counterions (like K^+, Na^+, and Mg^{2+}) present in the medium, a substantial net negative charge remains on each chain in the helix and tends to drive the two chains apart. Therefore, high ionic strength tends to stabilize the double helix.

A more subtle factor favoring denaturation is that the random-coil structure has a higher entropy, resulting from the greater randomness of the denatured form, with its many possible configurations. Consider Equation 3.9 on page 64 ($S = k \ln W$): If a rigid double helix separates into two flexible random coils, the number of configurations accessible to the molecule greatly increases (Figure 4.34a); therefore, the entropy increases. The free energy change in going from a regular two-stranded polynucleotide secondary structure (such as B-form DNA) to individual random-coil strands is given by the usual formula:

$$\Delta G = \Delta H - T\Delta S \quad \text{(helix} \rightleftharpoons \text{random coil)} \tag{4.3}$$

Because ΔS is positive, the term ($T\Delta S$) makes a negative contribution to the free energy change, favoring denaturation.

Thus, two factors favor the helix $\longrightarrow$ coil transition: the higher randomness of the random coil ($\Delta S > 0$) and the electrostatic repulsion between chains ($\Delta H_{el} < 0$). If the double-helical structure is to be stable under any conditions, ΔG for the unfolding reaction must be positive. Therefore, we must look for a large contribution from ΔH to compensate for the factors just mentioned. The sources of such a positive ΔH are the hydrogen bonds between the base pairs and van der Waals interactions between stacked bases. In fact, the planar bases stack upon one another in van der Waals contact. Much energy must be expended to break these bonds and interactions, and hence the total ΔH is positive.

Because ΔH and ΔS in Equation 4.3 are both positive, the sign of ΔG will change as T is increased. At low temperature, the term $T\Delta S$ will be less than ΔH; ΔG will be > 0, and the helix will be stable. But as the temperature is increased, $T\Delta S$ will become greater than ΔH, and ΔG will become negative. Thus, at higher temperatures the double-stranded structure becomes unstable and will fall apart (Figure 4.34a).

One can follow this denaturation process by observing the absorbance of ultraviolet light of wavelength about 260 nm in a DNA solution. As mentioned on page 94, all nucleotides and nucleic acids absorb light strongly in this wavelength region. When the nucleotides are polymerized into a polynucleotide, and the bases packed into a helical structure, the absorption of light is reduced (Figure 4.34c). This phenomenon, called *hypochromism*, results from close interaction of the light-absorbing purine and pyrimidine rings. If the secondary structure is lost, the absorbance increases and becomes closer to that of a mixture of the free nucleotides. Therefore, raising the temperature of a DNA solution, with accompanying breakdown of the secondary structure, will result in an absorbance change like that shown in Figure 4.34d.

The remarkable feature of this helix-to-random-coil transition is that it is so abrupt. It occurs over a very small temperature range, almost like the melting of ice into water, as described in Chapter 3. Therefore, nucleic acid denaturation is sometimes referred to as a *melting* of the polynucleotide double helix, even though the term is not technically correct. We shall encounter similar abrupt changes in configuration of proteins in Chapter 6. They are always characteristic of what are called **cooperative transitions**. What this term means in the case of DNA or RNA is that a double helix cannot melt bit by bit. If you examine the kinds of structure shown in Figures 4.11 and 4.20, you will see that it would be very difficult for a single base to pop out of the stacked, hydrogen-bonded structure. Rather, the whole structure holds together until it is at the verge of instability and then denatures over a very narrow temperature range.

The "melting temperature" (T_m) of a polynucleotide depends on its (G + C)/(A + T) ratio. Because each G-C base pair forms three hydrogen bonds and each A-T pair only two, ΔH is greater for the melting of GC-rich

AT-rich regions melt more easily than GC-rich regions.

polynucleotides. The greater stacking energy of G-C pairs also contributes to the difference. The value of T_m corresponds to the temperature at which $\Delta G = 0$ (see Figure 4.34b and d). Thus,

$$0 = \Delta H - T_m \Delta S \qquad (4.4)$$

or

$$T_m = \frac{\Delta H}{\Delta S} \qquad (4.5)$$

On a per-base-pair basis, ΔS is about the same for all polynucleotides, but ΔH depends on base composition, as just described. This is why T_m increases with increasing (G + C) content. Figure 4.35 shows a graph of T_m versus percent (G + C) for a number of naturally occurring DNAs.

DNA denaturation is reversible. For example, when heat-denatured DNA is cooled, DNA duplexes can reform. The rate of cooling must be slow, allowing time for complementary strands to find one another and pair up, or **renature** (also called **annealing**). Similarly, an RNA molecule can form a duplex with a DNA of complementary base sequence, creating a DNA–RNA hybrid, consisting of one strand each of DNA and RNA. DNA–DNA renaturation and DNA–RNA hybridization are at the root of several important research techniques, as we shall see later in this book.

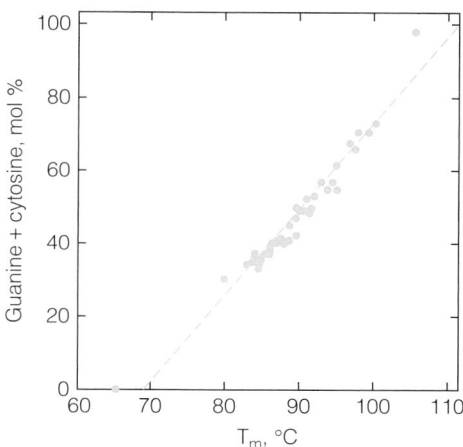

FIGURE 4.35

Effect of base-pair composition on the denaturation temperature of DNA. The graph shows the rise in "melting" temperature of DNA as its percent (G + C) increases.

Data from *Journal of Molecular Biology* (1962) 5:120, J. Marmur and P. Doty.

Superhelical Energy and Changes of DNA Conformation

The storage of energy in DNA supercoiling is analogous to the energy required to wind up (supercoil) rubber bands. That is, the first few turns are easy, but the energy required, per turn, increases as the winding gets tighter. In fact, for DNA the amount of free energy stored in supercoiling (ΔG_{sc}) is proportional to the *square* of the superhelical density σ:

$$\Delta G_{sc} = K\sigma^2 \qquad (4.6)$$

where K is a constant. Note that ΔG_{sc} is zero when the DNA is relaxed ($\sigma = 0$) and increases with either positive or negative supercoiling. Furthermore, because the square of σ is involved, the energy required to add an extra turn increases as each turn is added, and it becomes about equal to the energy that one ATP can provide when $\sigma = \pm 0.06$. This is at least one reason why superhelix density is limited to about this value in vivo. Highly supercoiled DNA has a lot of stored energy, which can be reduced by *any* process that decreases the superhelix density. For example, suppose we have a circular DNA molecule with negative superhelical turns. If one repeat of the DNA helix (10 base pairs) were to unwind and locally melt, this change would be the equivalent of a ΔT of -1. Then, to compensate, one negative superhelical turn could be removed ($\Delta W = +1$); that is, some of the writhing has been compensated by local unwinding. Converting a segment of B-DNA to Z-DNA would be even more efficient, requiring less energy. Each 10 bp of DNA that goes from B to Z goes from a twist of $+1$ (one right-handed twist) to approximately -1 (one left-handed twist). This single change amounts to $T = -2$, which would allow relaxation of two negative superhelical turns ($\Delta W = +2$).

Formation of cruciform structures also relaxes superhelical DNA because every base pair put into a cruciform hairpin is essentially removed from superhelical strain. Likewise, H-DNA formation, which leaves part of one strand unpaired, has the same effect.

In other words, imposing high levels of superhelical torsion on a DNA molecule can promote any one of the following changes: local melting, Z-DNA formation, cruciform extension, formation of H-DNA regions, and quite likely

Energy stored in supercoiled DNA can be used to drive structural transformation.

formation of some other special conformations not yet discovered. Which of these will happen depends on what special sequences are present in the DNA circle under stress. For example:

- The presence of AT-rich regions, which melt more easily than GC-rich regions, may favor local melting.
- Alternate purine/pyrimidine tracts (like $\cdots$ CGCGCG $\cdots$) favor Z-DNA formation, especially if the Cs are methylated on carbon 5.
- Palindromes allow cruciform extension.
- A segment that is primarily purines in one strand and pyrimidines in the other may permit H-DNA formation.

Superhelical stress is both widespread and controlled in DNA molecules found in cells. As we shall see in later chapters, the special structures described here play diverse roles in the regulation of gene expression. The idea that genes may be turned on and off by changes in supercoiling is an intriguing one.

SUMMARY

There are two kinds of nucleic acids, DNA and RNA. Each is a polynucleotide, a polymer of four kinds of nucleoside 5'-phosphates, connected by links between 3' hydroxyls and 5' phosphates. RNA has the sugar ribose; DNA has deoxyribose. The phosphodiester linkage is inherently unstable, but it hydrolyzes only very slowly in the absence of catalysts. Each naturally occurring nucleic acid has a defined sequence, or primary structure. Early evidence indicated that DNA might be the genetic material, but it was not until Watson and Crick elucidated its two-stranded secondary structure in 1953 that it became obvious how DNA might direct its own replication. The structure they proposed involved specific pairing between A and T and between G and C. The helix is right-handed, with about 10.5 base pairs (bp) per turn in the B form. Such a structure can replicate in a semiconservative manner, as demonstrated by Meselson and Stahl in 1958. Other forms of polynucleotide structures exist, of which the most important is the A form, found in RNA–RNA and DNA–RNA double helices. In vivo, most DNA is double-stranded; some molecules are circular. Most of the circular DNA molecules found in nature are supercoiled. Most RNA is single-stranded, but it may fold back to form hairpins and other well-defined tertiary structures.

The biological functions of nucleic acids may be briefly summarized as follows: DNA contains stored genetic information, which is transcribed into RNA. Some of these RNA molecules act as messengers to direct protein synthesis. The messenger RNA is translated on a ribosome, using the genetic code, to produce proteins. Modern molecular biological techniques allow us to manipulate DNA to make new proteins and modify existing ones.

Supercoiling of DNA can be expressed in terms of twist (T) and writhe (W). These terms are related to the linking number (L) by $L = T + W$. To form superhelical coiling requires the expenditure of ATP energy, using an enzyme called DNA gyrase. Gyrase is one of a class of topoisomerases; others relax supercoiled DNA.

Polynucleotides can form a number of unconventional structures, including left-handed DNA (Z-DNA), cruciforms, in some cases triple helices, and G-quadruplexes. The secondary structures of polynucleotides can be changed in various ways. The helix can "melt," which involves strand separation. This change is easiest for regions rich in A-T pairs. Energy stored in superhelical DNA may promote local DNA melting or changes to a variety of alternative structures, such as Z-DNA, cruciforms, or a particular triple-helical structure called H-DNA.

REFERENCES

General

Bates, A. D., and A. Maxwell (1993) *DNA Topology*. Oxford University Press, New York. A clear, helpful little book.

Saenger, W. (1984) *Principles of Nucleic Acid Structure*. Springer-Verlag, New York. This reference provides much greater detail concerning nucleic acid structure than is given in this book.

van Holde, K. E., W. C. Johnson, and P. S. Ho (2006) *Principles of Physical Biochemistry* (2nd ed.). Pearson/Prentice Hall, Upper Saddle River, N.J. Has much more on nucleic acid stability and structural transitions.

Historical

Avery, O. T., C. M. MacLeod, and M. McCarty (1944) Studies on the chemical transformation of pneumococcal types. *J. Exp. Med.* 79:137–158. The pioneering study that lent credence to the idea that DNA is the genetic substance.

Hershey, A. D., and M. Chase (1952) Independent function of viral protein and nucleic acid on growth of bacteriophage. *J. Gen. Physiol.* 36:39–56. The convincing evidence that DNA is the genetic material.

Judson, H. (1979) *The Eighth Day of Creation*. Simon & Schuster, New York. A detailed, fascinating account of the development of modern ideas about nucleic acids.

Manchester, K. L. (2007) Historical opinion: Erwin Chargaff and his "rules" for the base composition of DNA: Why did he fail to see the possibility of complementarity? *Trends Biochem. Science* 33:65–70. A fresh look at historical aspects of DNA structure.

Meselson, M., and F. Stahl (1958) The replication of DNA in *Escherichia coli*. *Proc. Natl. Acad. Sci. USA* 44:671–682. An example of a beautifully designed and executed experiment.

Sayre, A. (1978) *Rosalind Franklin and DNA*. W. W. Norton, New York. An account of the contributions of the scientist who created the best early X-ray diffraction patterns of DNA fibers.

Watson, J. D. (1968) *The Double Helix*. Atheneum, New York (trade and paperback editions); New American Library, New York (paperback). An outspoken account of the elucidation of DNA structure by one of the central characters.

Watson, J. D., and F. H. C. Crick (1953) Molecular structure of nucleic acids. A structure for deoxyribose nucleic acid. *Nature* 171:737–738. Two pages that shook the world.

Specialized Papers of Importance

Bacolla, A., and R. D. Wells (2004) Non-B DNA conformations, genomic rearrangements, and human diseases. *J. Biol. Chem.* 279:47411–47414. A minireview dealing with unconventional DNA structures.

Burge, S., G. N. Parkinson, P. Hazel, A. K. Todd, and S. Neidle (2006) Quadruplex DNA: Sequence, topology, and structure. *Nucleic Acids Research* 34:5402–5415. Chemistry and biology of this unusual DNA structure.

Castro, C. E., F. Kilchherr, D-N. Kim, E. L. Shiao, J. Wauer, P. Wortmann, M. Bathe, and H. Dietz (2011) A primer to scaffolded DNA origami. *Nature Methods* 8:221–229. A recent instruction manual for creating three-dimensional DNA assemblies.

Deweese, J. E., M. A. Osheroff, and N. Osheroff (2009) DNA topology and topoisomerases. Teaching a "knotty" subject. *Biochem. Mol. Biol. Education* 37:2–10. An exceptionally clearly written short review, with discussion of topoisomerases as drug targets.

Dietz, H., S. M. Douglas, and W. H. Shih (2009) Folding DNA into twisted and curved nanoscale shapes. *Science* 325:725–730. Careful design and annealing of synthetic oligonucleotides allows DNA to be folded into precise shapes, such as a miniature gear wheel.

Han, D., S. Pai, J. Nangreave, Z. Deng, Y. Liu, and H. Yan (2011) DNA origami with complex curvatures in three-dimensional space. *Science* 332:342–346. A recent paper describing the use of synthetic DNA to make curved three-dimensional shapes, including a flask 70 nm high.

Joyce, G. F. (2002) The antiquity of RNA-based evolution. *Nature* 418:214–221. Thoughts about a primordial RNA world.

Khuu, P., M. Sandor, J. DeYoung, and P. S. Ho (2007) Phylogenomic analysis of the emergence of GC-rich transcription elements. *Proc. Natl. Acad. Sci. USA* 104:16528–16533. Comparative DNA sequence analysis indicating that Z-DNA-forming sequences arose at specific stages in evolution.

Mardis, E. R. (2008) The impact of next-generation sequencing technology on genetics. *Trends Genet.* 24:133–141. A discussion of three new high-volume DNA sequencing technologies and some of their potential applications.

Sharma, J., R. Chhabra, A. Cheng, J. Brownell, Y. Liu, and H. Yan (2009) Control of self-assembly of DNA tubules through integration of gold nanoparticles. *Science* 323:112–116. More about the use of DNA molecules in nanotechnology.

Vologodskii, A. V., and N. R. Cozzarelli (1994) Conformational and thermodynamic properties of supercoiled DNA. *Annu. Rev. Biophys. Biomol. Struct.* 23:609–643.

Wang, L., S. Chen, T. Xu, K. Taghizadeh, J. S. Wishnok, X. Zhou, D. You, Z. Deng, and P. C. Dedon (2007) Phosphorothioation of DNA in bacteria by *dnd* genes. *Nature Chem. Biol.* 3:709–710. Surprising news about a new internucleotide link in DNA.

Wing, R. M., H. R. Drew, T. Takano, C. Brodka, S. Tanaka, K. Itakura, and R. E. Dickerson (1980) Crystal structure analysis of a complete turn of B-DNA. *Nature* 287:755–758. First crystallographic study of a B-DNA structure.

Wong, L., and 14 coauthors (2011) DNA phosphorothioation is widespread and quantized in bacterial genomes. *Proc. Natl. Acad. Sci. USA* 108:2963–2968. Use of a mass spectrometric technique for sequence analysis of the phosphorothioate modification.

PROBLEMS

1. A viral DNA is analyzed and found to have the following base composition, in mole percent: A = 32, G = 16, T = 40, C = 12.
   (a) What can you conclude about the structure of this DNA?
   (b) What kind of secondary structure do you think it would have?

2. Given the following sequence for one strand of a double-stranded oligonucleotide:

   5'ACCGTAAGGCTTTAG3'

   (a) Write the sequence for the complementary DNA strand.
   (b) Suppose you knew that the strand shown above had phosphate on both ends. Using an accepted nomenclature, write the sequence so as to show this.
   (c) Write the sequence of the RNA complementary to the strand shown above.

*3. Some naturally occurring polynucleotide sequences are *palindromic;* that is, they are self-complementary about an axis of symmetry. Such a sequence is

   TCAAGTCCATGGACTTGG

   AGTTCAGGTACCTGAACC

   Show how this structure might form a double hairpin, or cruciform, conformation. Indicate the center of symmetry in the sequence and the bounds of the cruciform.

4. The *E. coli* genome has a superhelical density in vivo of about 0.06. Assuming the DNA has 10.5 bp/turn, what is the expected writhing number of the *E. coli* genome?

*5. Given the following sequence for an RNA molecule, find a secondary structure that will be maximally stable.

   GUCCAGCCAUUGCGUUCGCAAUGGC

6. The largest of the double-stranded RNA molecules of cytoplasmic polyhedrosis virus contains 5150 bp (see Table 4.5). How long do you expect this molecule to be if extended?

7. A circular, double-stranded DNA contains 2100 base pairs. The solution conditions are such that DNA has 10.5 bp/turn.
   (a) What is L_0 for this DNA?
   (b) The DNA is found to have 12 left-handed superhelical turns. What is the superhelix density σ?

*8. In a supercoiled DNA, a stretch of about 20 base pairs changes from the B form to the Z form. What is the change in (a) T, (b) L, and (c) W?

9. Of the DNA molecules listed in Table 4.4, which would you expect to have the highest and lowest T_m?

10. A scientist isolates the DNA genome from a virus. She attempts to carry out a melting analysis but finds only 10% hypochromicity.
    (a) Suggest an explanation for the low value.
    (b) Why do you think she finds *this* much?

11. A particular double-stranded DNA has, under the conditions used in Figure 4.32, a melting point of 94 °C. Estimate the base composition (in mole percent) of this DNA.

*12. A variant of B-form DNA has been reported to exist in the presence of Li⁺ ions. This form, called C-DNA, is found by X-ray diffraction to have 9⅓ base pairs per turn.
    (a) How many base pairs are contained in one repeat of this structure? How many turns in one repeat?
    (b) Is C-DNA twisted more or less tightly than B-DNA?
    (c) Would high superhelix density favor or disfavor C-DNA over B-DNA?

*13. A closed circular supercoiled DNA is relaxed by treatment with topoisomerase. No matter how much enzyme is used, or how long the experiment is run, the experimenter always finds a gel electrophoresis pattern indicating some DNA with one, two, and three superhelical turns in addition to the relaxed (nicked) circle (see figure). Suggest an explanation for this observation.

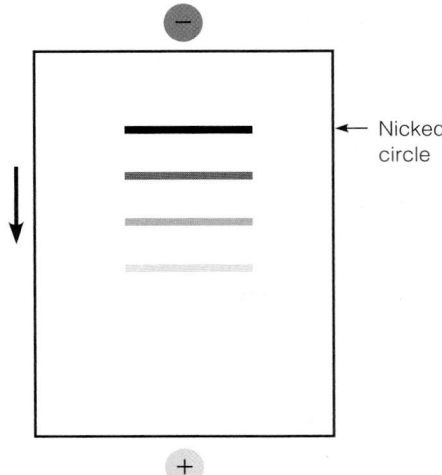

Nicked circle

*14. The dye ethidium is a planar molecule that can *intercalate* into double-stranded DNA. This means that it slips between adjacent stacked base pairs. In doing so, it unwinds the DNA helix by about 26° for every ethidium bound.
    (a) If ethidium were added to relaxed, closed circular DNA, would negative or positive writhing be expected? Explain.
    (b) What would be the effect on writhing if the DNA were nicked in one strand?
    (c) It is observed that progressive addition of ethidium to negatively supercoiled DNA has a peculiar effect: First the electrophoretic mobility decreases, but with further addition of ethidium it again increases. Explain.

*15. Explain why DNA is stable in the presence of alkali (0.3 M KOH), while RNA is quantitatively degraded to 2'- and 3'-nucleoside monophosphates under these conditions.

*16. Refer to Table 4.1 for the pK_a values for ionization of the four ribonucleoside 5'-monophosphates. Select a pH value at which each of the four nucleotides has a different net charge on the molecule, and predict the direction and relative rate of migration of each nucleotide in an electrophoretic field.

TOOLS OF BIOCHEMISTRY 4A

AN INTRODUCTION TO X-RAY DIFFRACTION

Only a few decades ago, virtually nothing was known about the three-dimensional structures of nucleic acids, proteins, and polysaccharides. Today, largely as a result of the technique of X-ray diffraction, many of these molecules are understood at the atomic level—detail that would have astounded the biochemists of 1950. The method is complicated and it is possible to give only a brief introduction here, describing what is measured and what can be obtained.

When radiation of any kind passes through a regular, repeating structure, *diffraction* is observed. This means that radiation scattered by the repeating elements in the structure shows reinforcement of the scattered waves in certain specific directions and weakening of the waves in other directions. A simple example is given in Figure 4A.1, which shows radiation being scattered from a row of equally spaced atoms. Only in certain directions will the scattered waves be in phase and therefore constructively interfere with (reinforce) one another. In all other directions they will be out of phase and destructively interfere with one another. Thus, a **diffraction pattern** is generated. For the diffraction pattern to be sharp, it is essential that the wavelength of the radiation used be somewhat shorter than the regular spacing between the elements of the structure. This is why X-rays are used in studying molecules, for X-rays typically have a wavelength of only a few tenths of a nanometer. If the regular spacing in the object being studied is large (as in a window screen), we can observe exactly the same phenomenon with visible light, which has a wavelength thousands of times longer than X-rays. We will find that a point source, seen through a window screen, gives a rectangular diffraction pattern of spots.

The rule relating the periodic spacings in object and diffraction pattern is simple: Short spacings in the periodic structure correspond to large spacings in the diffraction pattern, and vice versa. In addition, by determining the relative intensities of different spots, we can tell how matter is distributed within each repeat of the structure.

Fiber Diffraction

We consider first the diffraction from helical molecules, aligned approximately parallel to the axis of a stretched fiber. A helical molecule, like the one shown schematically in Figure 4A.2, is characterized by certain parameters:

The *repeat* (c) of the helix is the distance parallel to the axis in which the structure exactly repeats itself. The repeat contains some integral number (m) of polymer residues. In Figure 4A.2, $m = 4$.

The *pitch* (p) of the helix is the distance parallel to the helix axis in which the helix makes one turn. If there is an integral number of residues per turn (as here), the pitch and repeat are equal.

The *rise* (h) of the helix is the distance parallel to the axis from the level of one residue to the next, so $h = c/m$. If we think of a spiral staircase as an example of a helix, the rise is the height of each step and the pitch is the distance from where one is standing to the corresponding spot directly overhead.

Suppose we wish to investigate a polymer with the helical structure shown in Figure 4A.2. A fiber is pulled from a concentrated solution of the polymer. Stretching the fiber further will produce approximate alignment of the long helical molecules

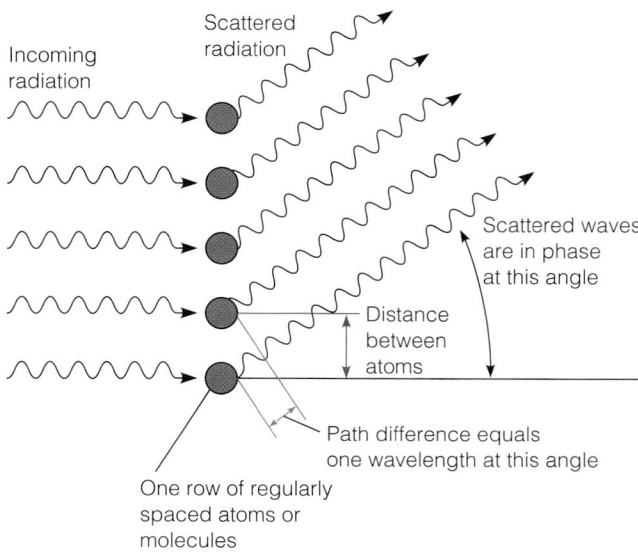

FIGURE 4A.1

Diffraction from a very simple structure—a row of atoms or molecules.

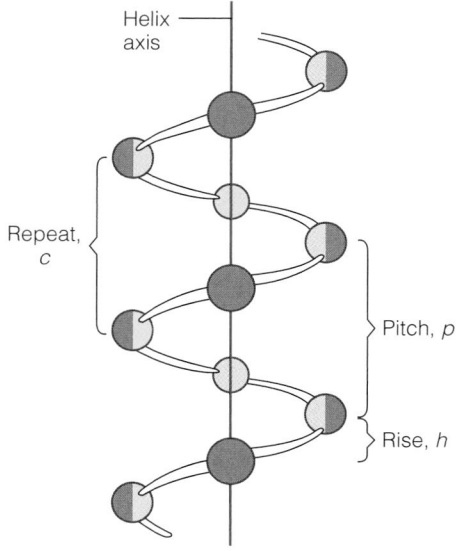

FIGURE 4A.2

A simple helical molecule.

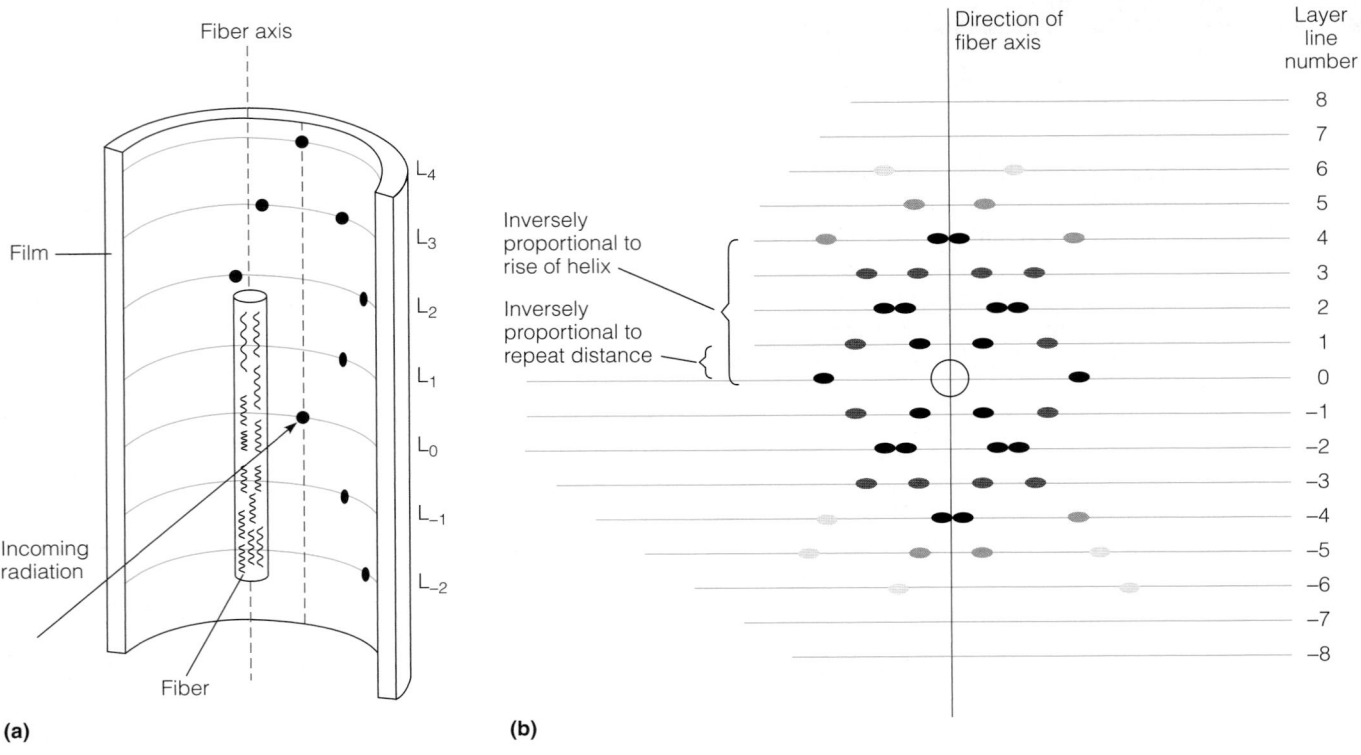

FIGURE 4A.3

Diffraction from fibers. **(a)** A fiber in an X-ray beam. **(b)** The diffraction pattern.

with the fiber axis. The fiber is then placed in an X-ray beam, and a photographic film or comparable detection medium is positioned behind it, as shown in Figure 4A.3a. The diffraction pattern, which consists of spots or short arcs, will look like that in Figure 4A.3b. It can be read as follows: According to the mathematics of diffraction theory, a helix always gives rise to this kind of cross-shaped pattern. Therefore, we know we are dealing with a helical structure. The spots all lie on lines perpendicular to the fiber axis; these are called **layer lines**. The spacing between these lines is inversely proportional to the repeat of the helix, c, which in this case equals the pitch. Note that the cross pattern repeats itself on every fourth layer line. This repetition pattern tells us that there are exactly 4.0 residues per turn in the helix. Thus, the rise in the helix is $c/4$. This is the kind of evidence telling Watson and Crick that B-DNA was a helix with 10 residues per turn.

The information above is given directly by the pattern. To find out exactly how all of the atoms in each residue are arranged in each repeat, a more detailed analysis is necessary. Usually, a model is made using the correct repeat, pitch, and rise. Model making is simplified because we know approximate bond lengths and the angles between many chemical bonds. The model must also be inspected to see that no two atoms approach closer than their van der Waals radii. From such a model, the intensities of the various spots can be predicted. These predictions are compared with the observed intensities, and the model is readjusted until a best fit is obtained. The initial determination of the structure of DNA was done in just this way. As you can see from Figure 4.9, real fiber diffraction patterns are not as neat as the idealized example, mainly because of incomplete alignment of the molecules.

Crystal Diffraction

To study molecular crystals such as those formed by small oligonucleotides, molecules like tRNA, and globular proteins, the experimenter faces a rather different problem from the study of helical fibers and proceeds in a quite different way. A schematic drawing of such a crystal is shown in Figure 4A.4.

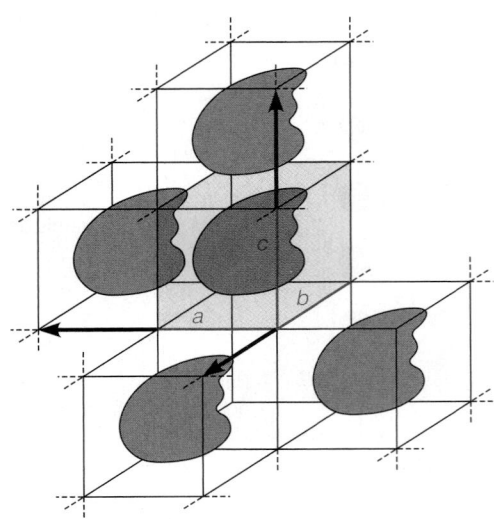

FIGURE 4A.4

Schematic drawing of a molecular crystal.

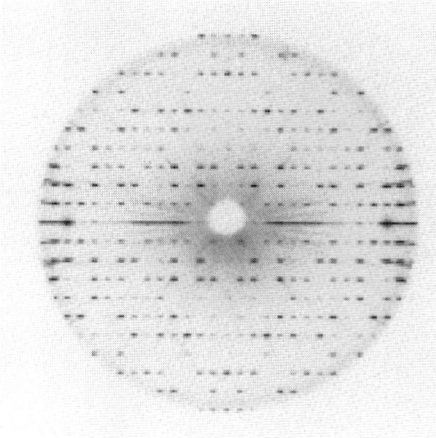

FIGURE 4A.5

Diffraction pattern produced by a molecular crystal of a small DNA.

P. S. Ho, Colorado State University.

The repeating unit is now the **unit cell**, which may contain one, two, or more molecules. The unit cell may be thought of as the basic building block of the crystal. Repetition of the unit cell in three dimensions (marked by arrows on the figure) creates the whole crystal. A simple two-dimensional analog of the crystal unit cell is the repeating pattern in wallpaper. No matter how random a wallpaper pattern may seem, if you stare at it long enough, you can always find a unit that, by repetition, fills the entire wall.

Just as in fiber diffraction, passing an X-ray beam through a molecular crystal produces a diffraction pattern. The pattern shown in Figure 4A.5 was obtained from a crystal of a small DNA. Again, the spacing of the spots allows us to determine the repeating distances in the periodic structure—in this case the x, y, and z dimensions of the unit cell labeled a, b, and c in Figure 4A.4. But the important information in crystal diffraction studies is just how the atoms are arranged *within* each unit cell, for that arrangement describes the molecule. Again, this information is contained in the relative intensities of the diffraction spots in a pattern like that shown in Figure 4A.5. But in crystal diffraction, more exact information can be extracted than from a fiber diffraction pattern because the corresponding molecules in each unit cell of the crystal are of the same shape and are oriented in the same way. In fiber diffraction, the helical molecules may all have their long axes pointed in the same direction, but they are rotated randomly about these axes. This difference in exactness of arrangement can be appreciated by comparing the sharpness of the crystal diffraction pattern shown in Figure 4A.5 with the fiber pattern depicted in Figure 4.9.

After obtaining the diffraction pattern from a molecular crystal, the experimenter measures the intensities of a large number of the spots. If the molecule being studied is a small one, it is possible to proceed in much the same manner as with fiber patterns. A structure is guessed at, and expected intensities are calculated and compared with the observed intensities. The structure is refined until the relative intensities of all spots are correctly pre-

dicted. However, such a procedure won't work with a molecule as complex as the tRNA shown in Figure 4.20—there is simply no way to guess such a structure.

Why not proceed directly from spot intensities to the structure? The difficulty is that some of the information contained in the spot intensities is hidden. To greatly simplify a complex problem, we may say that it is as if the quantities that the experimenter needed in order to deduce the structure (which are called **structure factors**) were the square roots of the intensities.* If the intensity has a value of, say, 25, the investigator knows that the number needed is +5 or −5. But which? This sort of quandary is the essence of the *phase problem,* which prevented progress in large-molecule crystallography for many years. One way of solving the problem was discovered in the early 1950s. Suppose a heavy metal atom, such as mercury, can be introduced into some point in the molecule in such a way that the molecule and crystal are otherwise unchanged. This process is called an **isomorphous replacement**. Now suppose the heavy metal contributes a value of +2 to the structure factor for the spot we were just discussing. If the original value was +5, its new value is +7 and its square is 49. If the original value was −5, the value now becomes 3 and its square is 9. The investigator takes a diffraction photograph of the crystal with the heavy metal inserted. If the new crystal has an intensity of 9 for this spot, the original structure factor must have been −5, not +5. Although an oversimplification, this example gives the essence of the method. Usually multiple isomorphous replacements are necessary to determine the phases of the structure factors.

Given structure factors for all of the spots, the investigator can calculate the positions of all atoms in the unit cell. What is actually calculated is an **electron density** distribution (Figure 4A.6), but this amounts to the same thing, for regions of high electron density are where the atoms are. In the particular view shown in Figure 4A.6, we are looking at a two-dimensional "slice" through the three-dimensional electron density distribution.

It is now appropriate to review the steps that must be taken to determine the three-dimensional structure of a macromolecule from crystal diffraction studies:

1. Obtain satisfactory crystals. This step is often the hardest part of the procedure, for the crystals must be of good quality and at least a few tenths of a millimeter in minimum dimension. Crystals that are too small will not give sharp diffraction patterns. Getting macromolecules to crystallize well is still more of an art than a science.
2. Record the diffraction pattern from the crystal, and measure the intensities of many of the spots.
3. Find some way to make isomorphous replacements in the molecule. Usually two or more replacements are required.

---

*For the more mathematically sophisticated reader, we note that the structure factors are usually complex numbers and can thus be represented as vectors in the complex plane. What are determined from the intensities are their *amplitudes*, but what is not known are their *phases*.

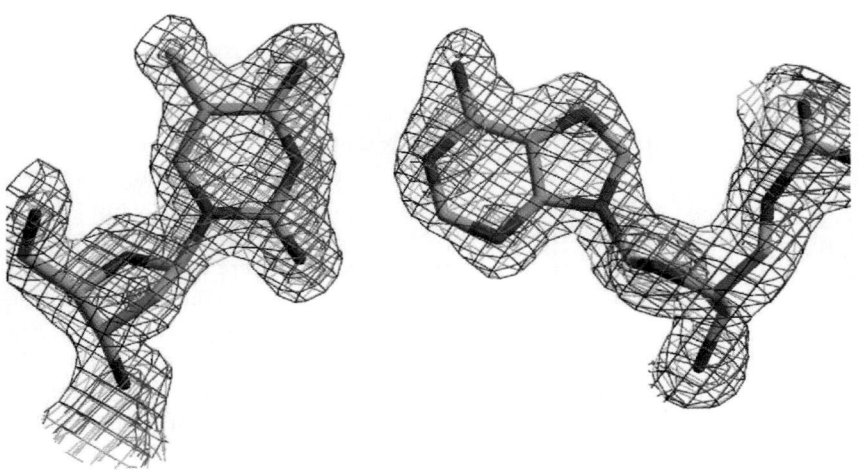

FIGURE 4A.6

Part of an electron density map derived from the DNA crystal diffraction pattern in Figure 4A.5.

P. S. Ho, Colorado State University.

4. Repeat steps 1 and 2 for each isomorphous derivative.
5. Calculate structure factors and, from them, the electron density distribution. These calculations are usually done on a large computer.

In most cases, the investigator will first carry through this analysis with a relatively small number of spots. This procedure will give a *low-resolution* structure. If all is going well, more spots will be measured and the calculations refined to give higher resolution. With the best crystals, it is now possible to obtain resolutions of about 1Å. This resolution is sufficient to identify individual groups and even some atoms and to show how they interact with one another. The detail in the phenolic ring of a protein side chain revealed at different resolutions is shown in Figure 4A.7.

Most of the detailed three-dimensional structures of biological macromolecules shown in this book have been determined by X-ray diffraction studies of crystals. At present, tens of thousands of such structures are known. This knowledge represents an enormous amount of labor in many laboratories, but the results allow us to understand macromolecular function at a level that would have been unbelievable only a short time ago.

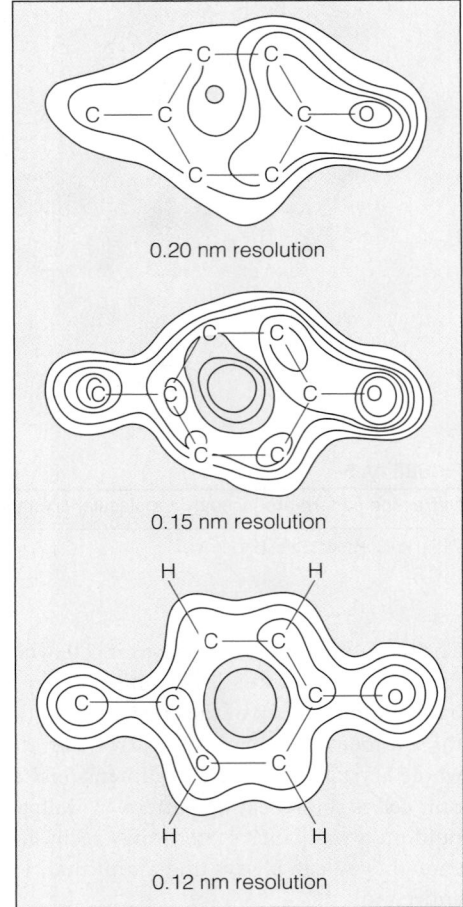

0.20 nm resolution

0.15 nm resolution

0.12 nm resolution

FIGURE 4A.7

Effect of increased resolution on molecular detail observed by X-ray diffraction. The amino acid shown in this illustration is tyrosine.

Reprinted from *Journal of Molecular Biology* 138:615–633, K. D. Watenpaugh, L. K. Sieker, and L. H. Jensen, Crystallographic refinement of rubredoxin at 1·2 Å resolution. © 1980, with permission from Elsevier.

References

van Holde, K. E., W. C. Johnson, and P. S. Ho (2006) *Principles of Physical Biochemistry* (2nd ed., Chapter 6). Pearson/Prentice Hall, Upper Saddle River, N.J. A more detailed treatment of X-ray diffraction of biopolymers.

TOOLS OF BIOCHEMISTRY 4B

MANIPULATING DNA

The biological sciences have been revolutionized by laboratory techniques that allow the researcher to isolate any desired gene, amplify the gene for determination of its nucleotide sequence, express the gene at a high level for production and analysis of its protein product, and introduce any desired mutation into the gene, either for analysis of structure–function relationships within the protein or to create some desirable property, such as increased stability. In addition, a modified gene can be introduced into the genome of a living organism for determination of its metabolic function. The key discoveries, made in the 1970s and early 1980s, spawned a large biotechnology industry, which has created hundreds of new drugs and genetically modified organisms, such as drought-resistant crop plants. Because these methods have taught us so much about the structure and function of the proteins and enzymes we will be discussing in the next several chapters, we introduce some of the benchmark techniques here, even though much of the genetic biochemistry upon which they are based is presented later in this book. Within this family of **recombinant DNA** techniques we present here outlines of (1) gene cloning, (2) chemical synthesis of oligonucleotides for use as primers, (3) DNA sequence analysis, and (4) site-directed mutagenesis. A fifth technique, **polymerase chain reaction** (PCR), which allows amplification of any DNA sequence, starting from minute amounts of material, is presented in Chapter 24 (Tools of Biochemistry 24A).

Gene Cloning

In 1973, Stanley Cohen and Herbert Boyer realized that two recent developments had set the stage for them to **clone**, or isolate, a single gene. In classical biology a clone is a population of organisms that are genetically homogeneous because they were derived from a single ancestor. For example, all of the bacterial cells in a colony represent a clone because they were derived from a single cell that was deposited at that location on a Petri plate.

The first of the developments leading to cloning of single genes was the characterization of **plasmids** as small circular DNA molecules capable of independent replication within bacterial cells. Clinical resistance to antibiotics is often caused by mutations in genes carried on plasmid DNA molecules. For example, resistance to penicillin results from a plasmid-encoded enzyme called β-lactamase (page 353), which cleaves penicillin to an inactive derivative. The second development was the discovery of a class of bacterial enzymes called **restriction endonucleases** and, more important, the fact that many of these enzymes catalyze cleavage of DNA at specific sites (Chapter 26). For example, the enzyme **EcoRI** cleaves DNA whenever it recognizes the sequence

$$5' \ldots \text{GAATTC} \ldots 3'$$
$$3' \ldots \text{CTTAAG} \ldots 5'$$

A sequence of this type is a *palindrome* (see page 116), a message that reads the same in both directions. This sequence is sym-

metrical, meaning that the two strands have identical sequences running in opposite directions. The enzyme cuts between G and A on both strands. Thus, each of the cleavage products has a short (four-base) single-stranded end, 3' . . . .TTAA 5' (see Figure 4B.1). What this means is that any two DNA molecules that have been cut by *Eco*RI can rejoin end-to-end when **annealed**, or subjected to DNA renaturation conditions, by base-pairing between the two 5'-terminated AATT sequences, also called "cohesive ends" or "sticky ends." Once this pairing has occurred, an enzyme called **DNA ligase** (Figure 4B.1 and page 1047) can form a covalent bond in the openings between the 3'-terminal G and the 5'-terminal A. If the two DNA sequences being rejoined come from different sources, the product is called a recombinant DNA molecule, after the fact that genetic recombination in vivo involves cutting and rejoining DNA from different chromosomes (Chapter 26).

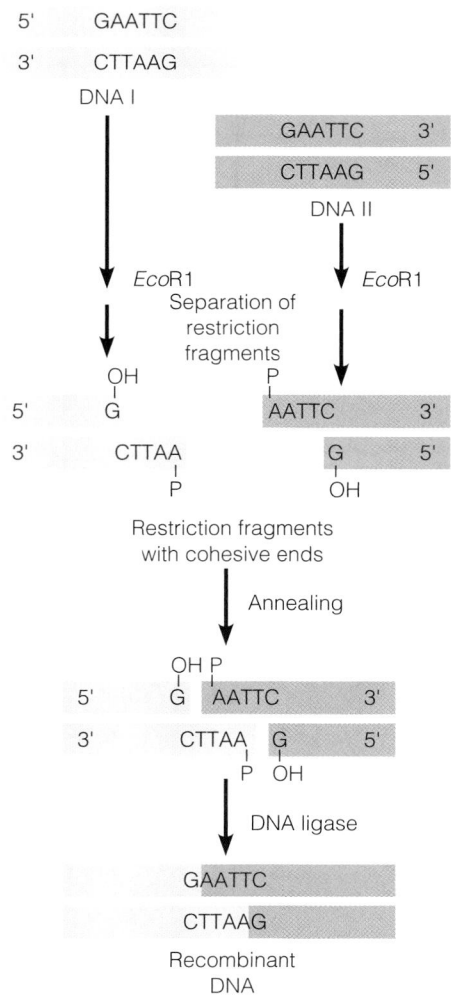

FIGURE 4B.1

Creation of recombinant DNA molecules in vitro.

Cohen and Boyer also devised a technique for high-frequency *transformation*, so that recombinant DNA molecules could be introduced into living bacterial cells. This originally involved treating bacteria with calcium chloride followed by heat shock, but this has been supplanted by more efficient methods, notably *electropo-*

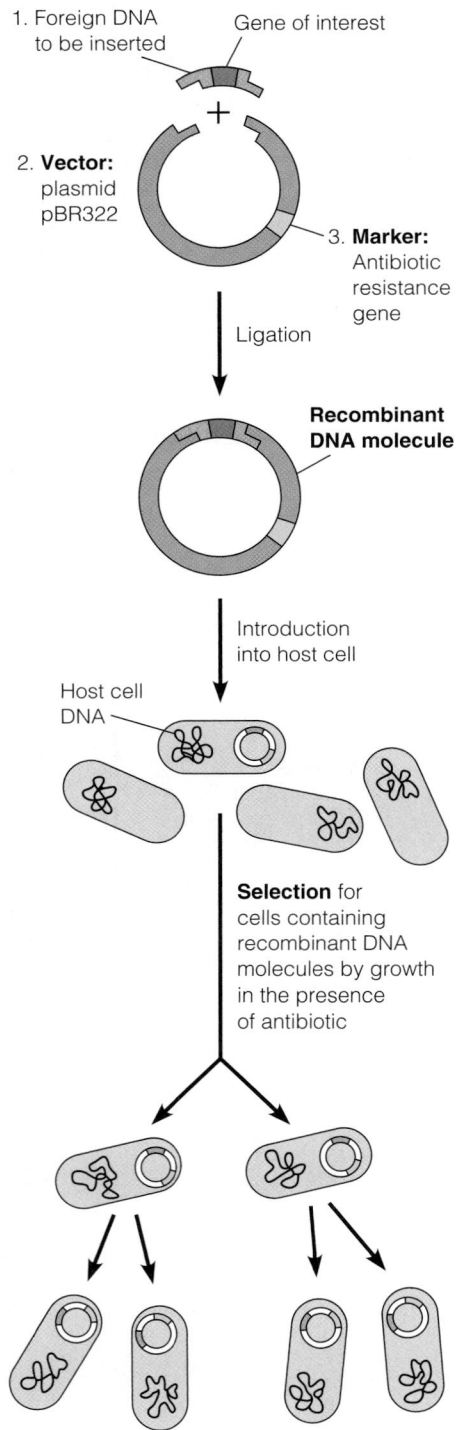

ration (application of an electric charge). The **vector** is a DNA molecule into which the gene to be cloned is inserted. As schematized in Figure 4B.2, the vector is a plasmid (small circular DNA) containing a gene that specifies resistance to a particular antibiotic, such as ampicillin. The plasmid contains just one site for cleavage by the restriction enzyme to be used. Both the plasmid and a DNA molecule or chromosome containing the gene of interest are cleaved with the same restriction enzyme, such as *Eco*RI. Annealing, or renaturation, followed by enzymatic ligation, yields recombinant DNA molecules in which the gene of interest has been spliced into the vector. Note that the vector need not be a plasmid. Any DNA capable of independent replication within a cell, such as a virus genome, can serve as a vector.

After annealing and ligation, the DNA, containing a mixture of recombinant and nonrecombinant DNAs, is introduced into recipient bacteria by transformation. Any transformed bacteria can be identified by plating in the presence of ampicillin. Only those bacteria that have been transformed will grow because they now contain the antibiotic resistance gene. The investigator must then carry out additional experiments to establish that a particular transformed cell contains the gene being cloned. A variety of methods are available, including additional antibiotic-resistance screens, DNA-DNA renaturation reactions, or analysis of expression of the cloned gene with an antibody or activity assay.

Figure 4B.3 shows the structure of pBR322, a plasmid engineered for gene cloning in 1977 and still in occasional use

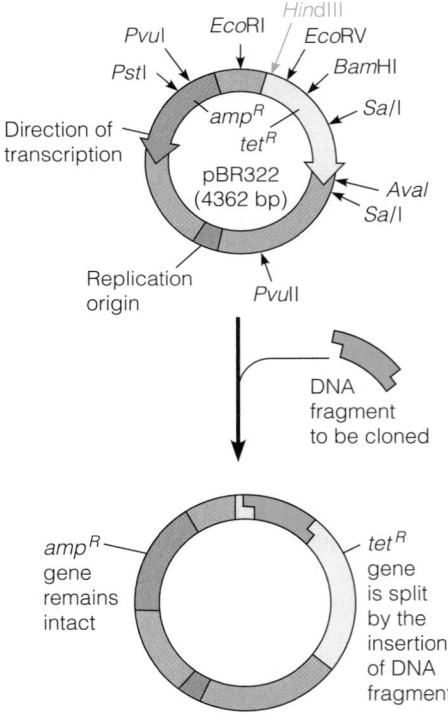

FIGURE 4B.3

pBR322, one of the earliest cloning vectors. Some of the restriction sites are shown, as well as the direction of transcription of the ampicillin and tetracycline resistance genes. The bottom diagram shows the effect of cloning a novel sequence into the *Hind*III site.

FIGURE 4B.2

Cloning a fragment of DNA into a plasmid vector and introducing the recombinant molecule into bacteria.

today. This small recombinant DNA molecule, 4632 base pairs in length, contains two antibiotic resistance genes, one for ampicillin resistance (amp^R) and one for tetracycline resistance (tet^R). Note that each of these genes contains restriction cleavage sites within its sequence. A site for the restriction enzyme *Hin*dIII lies within the tet^R sequence. Hence, if this site is used for cloning, the tet^R gene is split and, therefore, inactive. Because the amp^R gene is intact, all bacteria that have acquired a plasmid, whether recombinant or not, can be selected on the basis of their resistance to ampicillin. But now, recombinant plasmids can be identified because these bacteria are sensitive to tetracycline as well as ampicillin-resistant, whereas bacteria that acquired the original plasmid, without an insert, are resistant to both drugs.

Many variations have been devised on this original approach. Blunt-ended DNA molecules, containing no sticky ends, can now be cloned. Many cloning vectors have been devised, usually as recombinant DNA molecules themselves. One such vector, bacteriophage M13, is particularly useful because the recombinant DNA can be isolated as a single-stranded molecule, readily amenable to DNA sequence analysis (page 1039). Particularly widely used are expression vectors, in which signals to regulate and activate high-level expression of the cloned gene are built into the vector. Other modifications include sequences that aid in purification of the recombinant protein (see Figure 5A.1). In the biotechnology industry, techniques such as these led to the introduction as early as 1982 of human insulin as a recombinant gene product, used for diabetes treatment. Other recombinant proteins later approved for clinical use include blood-clotting factors, clot-dissolving enzymes used to treat heart attack victims, pituitary growth hormone, and interferons.

Automated Oligonucleotide Synthesis

In 1976 the British biochemist Fred Sanger introduced **dideoxy DNA sequencing**, the rapid DNA sequence analysis that led to completion of the 3 billion-base-pair human genome sequence in 2001. In 1978 the Canadian biochemist Michael Smith devised a method for site-directed mutagenesis, which allows an investigator to introduce any desired mutation into a cloned gene. Both researchers received the Nobel Prize for their contributions.

Both dideoxy DNA sequencing and site-directed mutagenesis depend upon the availability of oligodeoxyribonucleotide molecules of precisely known sequence to serve as primers in DNA polymerase-catalyzed reactions. The oligonucleotide synthesis process most widely used at present is the **phosphoramidite** method. This method is popular because it can be carried out in automated fashion, with the oligonucleotide linked to a solid support as nucleotides are added one at a time.

The method is illustrated in Figure 4B.4. Intermediates in this chemical synthesis are nucleoside 3′ phosphoramidites, with the phosphorus in the reactive trivalent form. The process begins with attachment of the 3′ hydroxyl of the first nucleotide to a silica matrix. The 5′-hydroxyl group is protected with a dimethyl-

trityl (DMTr) blocking group. Amino groups on the purine or pyrimidine base are also protected. The next nucleotide is introduced as the blocked derivative, with a phosphoramidite group on position 3′. Tetrazole is present, leading to protonation of the diisopropylamine moiety on the incoming nucleotide and facilitating its loss upon reaction with the unprotected 5′-hydroxyl. Oxidation with iodine converts the trivalent phosphorus to a phosphotriester. This process is repeated in stepwise fashion until up to 150 nucleotides have been added in a precisely determined sequence. Finally, the blocking groups are removed, and the finished chain is removed from the solid support, followed by chromatographic purification if necessary. Each step proceeds with about 98% efficiency, meaning that a 20-mer, containing 20 nucleotide residues, can be produced at about 80% final yield. Note that the phosphoramidite synthetic method proceeds from the 3′ end of the polymer toward the 5′ end, whereas the enzymatic synthesis of DNA by DNA polymerase proceeds in the opposite direction.

Because of the ease with which oligonucleotides of defined sequence can be synthesized, and because of the regularity of DNA secondary structure, scientists have devised numerous methods for creating defined DNA-based nanostructures. Figure 4B.5 illustrates one example. In this example synthetic oligonucleotides were designed to fold into the form of a perfect tetrahedron. By means such as these, synthetic DNAs have been used to form structures with possible applications in nanotechnology, including gears, tubes, and even mechanical devices.

Dideoxynucleotide Sequence Analysis

On page 110 we mentioned the Maxam-Gilbert method for DNA-sequence analysis, which involves treating DNA with reagents that cleave at specific nucleotides, yielding a population of molecules that can be resolved on the basis of molecular weight by gel electrophoresis. The enzymatic method introduced by Sanger similarly involves electrophoretic analysis of DNA fragments terminated at specific nucleotides, but Sanger's method allows analysis of longer stretches of DNA, and it lends itself more readily to automation. The method, as originally carried out, uses bacteriophage M13 as a cloning vector for the DNA sequence being analyzed. M13 DNA, as isolated from virus particles, is circular and single-stranded. However, the molecule replicates through a two-stranded intermediate form, called the replicative form, or RF. With introduction of useful restriction cleavage sites, M13 RF becomes a cloning vector comparable to pBR322. As shown in Figure 4B.6, the DNA fragment to be analyzed is cloned into M13 RF. After ligation the recombinant DNA is introduced into *E. coli* by transformation, and the infected cells are incubated and allowed to produce phage particles. The phage particles are purified, and DNA is isolated from each, as single-stranded DNA containing the sequence for analysis.

Isolated single-stranded circular DNA from phage genomes carrying the desired insert becomes the template for four DNA polymerase-catalyzed reactions. The primer is an oligonucleotide that is complementary to an M13 sequence lying just to the

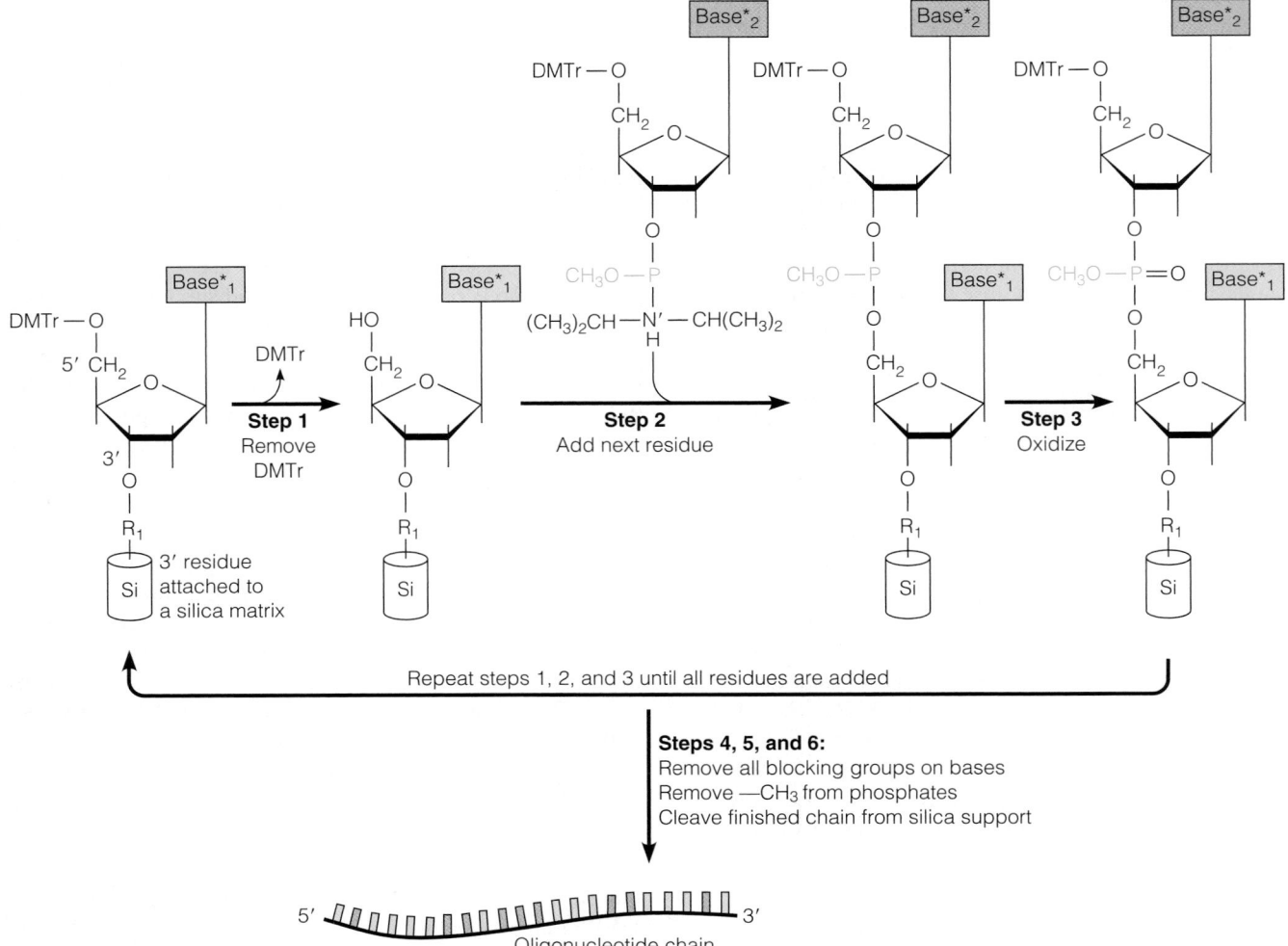

Repeat steps 1, 2, and 3 until all residues are added

Steps 4, 5, and 6:
Remove all blocking groups on bases
Remove —CH₃ from phosphates
Cleave finished chain from silica support

Oligonucleotide chain

*Reactive groups on all bases are blocked by chemical reagents

$R_1 = -C-(CH_2)_2-C-NH-(CH_2)_3-O-Si$

FIGURE 4B.4

Solid-phase synthesis of oligonucleotides by the phosphoramidite method.

3′ side of the insert. Extension of this primer by DNA polymerase copies the insert. The polymerase reactions are run in the presence of deoxyribonucleoside triphosphate analogs, the 2′,3′-**dideoxyribonucleoside triphosphates** (ddNTPs), which serve as inhibitors of chain extension because they lack 3′ hydroxyl termini. The dideoxy analog (ddATP) of deoxyadenosine triphosphate is shown here.

2′,3′-Dideoxyadenosine triphosphate

To generate a series of A-terminated fragments, the DNA polymerase reactions are run in the presence of equal concentrations of dATP, dCTP, dGTP, and dTTP, plus 1/10th that concentration of ddATP. When T is in the template strand, DNA polymerase occasionally inserts ddAMP instead of dAMP. When that happens, DNA replication stops and the fragment is released from the enzyme. Thus, a series of fragments of varying lengths accumulates, with a common 5′ end (the primer) and variable 3′ ends, and with each 3′ end identifying a T residue in the insert sequence that is being analyzed. Similarly, sites terminated by C, G, and T are identified simply by running polymerase reactions with the other three dideoxy analogs, one at a time. Inclusion of a radioactive nucleotide in the polymerization mixture and gel electrophoresis followed by radioautography yields four "sequencing ladders," as shown in Figure 4B.6. Each band in the radioautographic image of the electrophoretic gel identifies one of the four bases at that site.

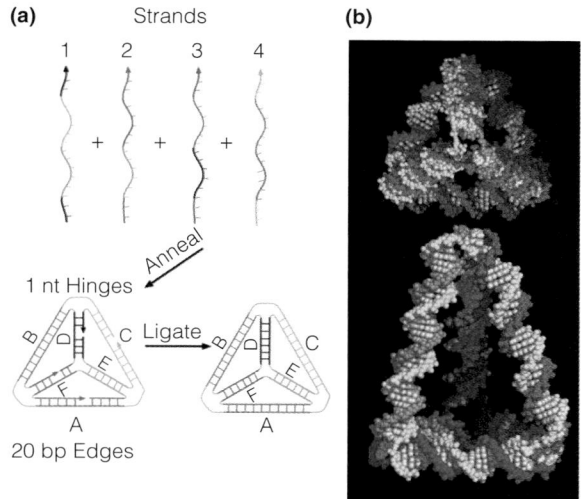

(a) Strands

1 nt Hinges

Anneal

20 bp Edges

FIGURE 4B.5

Design and synthesis of a three-dimensional DNA nanostructure, in this case a DNA tetrahedron. **(a)** Design of four synthetic oligonucleotides with complementary base sequences indicated by matching colors. The four oligonucleotides were annealed, by heating and slow-cooling, followed by ligation—treatment with an enzyme (Chapter 25) that creates covalent bonds between the DNA ends. **(b)** Two images of a space-filling representation of a tetrahedron with three 30-nucleotide sides (A, B, and C) and three 20-nucleotide sides (D, E, and F).

(a) From *Science* 310:1661–1665, R. P. Goodman, I. A. T. Schaap, C. F. Tardin, C. M. Erben, R. M. Berry, C. F. Schmidt, and A. J. Turberfield, Rapid chiral assembly of rigid DNA building blocks for molecular nanofabrication. © 2005. Reprinted with permission from AAAS.

As noted above, M13 vectors were originally used because they permitted DNA synthesis on single-stranded templates. Modifications of polymerase chain reaction (Tools of Biochemistry 24A) now permit the preparation of single-stranded DNA molecules for sequencing without the need for M13-based vectors.

Sanger sequencing is now done automatically. Each ddNTP is derivatized with a fluorescent dye, each of a different color. Thus, each fragment has a distinct color based upon the identity of the ddNTP that terminated the sequencing reaction. This allows all four reaction mixtures to be resolved in one lane of a sequencing gel, permitting analysis of far more DNA in one sequencing operation. The gel is scanned fluorometrically, and a computer reads the DNA sequence directly from the resulting pattern of differently colored peaks (Figure 4B.7).

Further refinements of this method have greatly increased its "throughput," or the amount of sequencing information derived from one operation. These methods yielded complete genomic sequences for several bacteria in the mid-1990s, and the near-completion of the human genome sequence in 2001. Since then, further modifications have yielded several approaches that greatly expand the speed and accuracy of DNA sequencing operations. Several of the most prominent of these "second-generation" sequencing technologies are described in articles cited at the end of this section. As detailed later in this book, enormous amounts of information about health and disease, biological individuality, and evolutionary relationships have come from these developments.

Site-directed Mutagenesis

Analysis of the function of a protein involves altering the structure of the protein and then determining whether and how the biological function of the protein has been altered. Two methods have been used classically. One is to alter certain residues chemically by treatment with protein-modifying reagents. The approach lacks specificity because all residues of a given amino acid may be altered, not just the one or two of special interest. Another approach is to mutagenize an organism with ultraviolet light, ionizing radiation, or a chemical mutagen and then select for surviving organisms containing the mutation of interest, followed by isolation of the mutant protein. The problem with this approach is the difficulty in targeting the action of the mutagen to a specific region of the gene of interest, typically the catalytic site of an enzyme or a region involved in regulatory interactions with DNA or with other proteins.

Once it became possible to clone the gene encoding a protein of interest, it became possible to systematically alter the gene at specific sites to generate virtually any desired mutation, a technique known as **site-directed mutagenesis**. Introduction of the cloned mutant gene into a host cell, followed by its expression, could then yield the mutant protein for study of its altered function.

The most powerful and widely used method for site-directed mutagenesis, conceived by Michael Smith, allows the introduction of practically any mutation at any site, including single-base substitutions, short deletions, or insertions. The approach, illustrated in Figure 4B.8, requires that the gene first be cloned into a single-stranded vector, such as phage M13, as discussed for DNA sequence analysis. Next an oligodeoxynucleotide, about 20 nucleotides long, is synthesized that is complementary in sequence to the cloned gene at the site of the desired mutation, *except in the center of the sequence.*

Here the oligonucleotide sequence contains one or two deliberate mistakes—either single nucleotides that do not pair with the template or insertions or gaps of a few nucleotides. Upon annealing to the cloned gene, these alterations create either non–Watson–Crick base pairs or bases that have no partners and therefore form a "looping out." The correctly matched bases on both sides of the mismatch cause it to remain annealed, despite the mismatch. DNA polymerases are then used to synthesize around the circular vector from this primer, followed by enzymatic ligation to create a closed circular duplex. After introduction of this circular molecule into bacteria by transformation, both strands replicate and yield phage. In principle, 50% of the phage should contain the desired mutation within the inserted sequence. In practice, that percentage is considerably less, but it can be increased by various techniques (see reference by Kunkel et al.). In any event, the mutant gene is then cleaved out of the modified phage genome by restriction nuclease treatment, and it can be recloned into an expression vector for large-scale preparation and subsequent isolation of the mutant protein.

Although single-stranded DNA phages, such as M13, were originally used to prepare templates for site-directed mutagenesis, such molecules are now readily prepared by polymerase

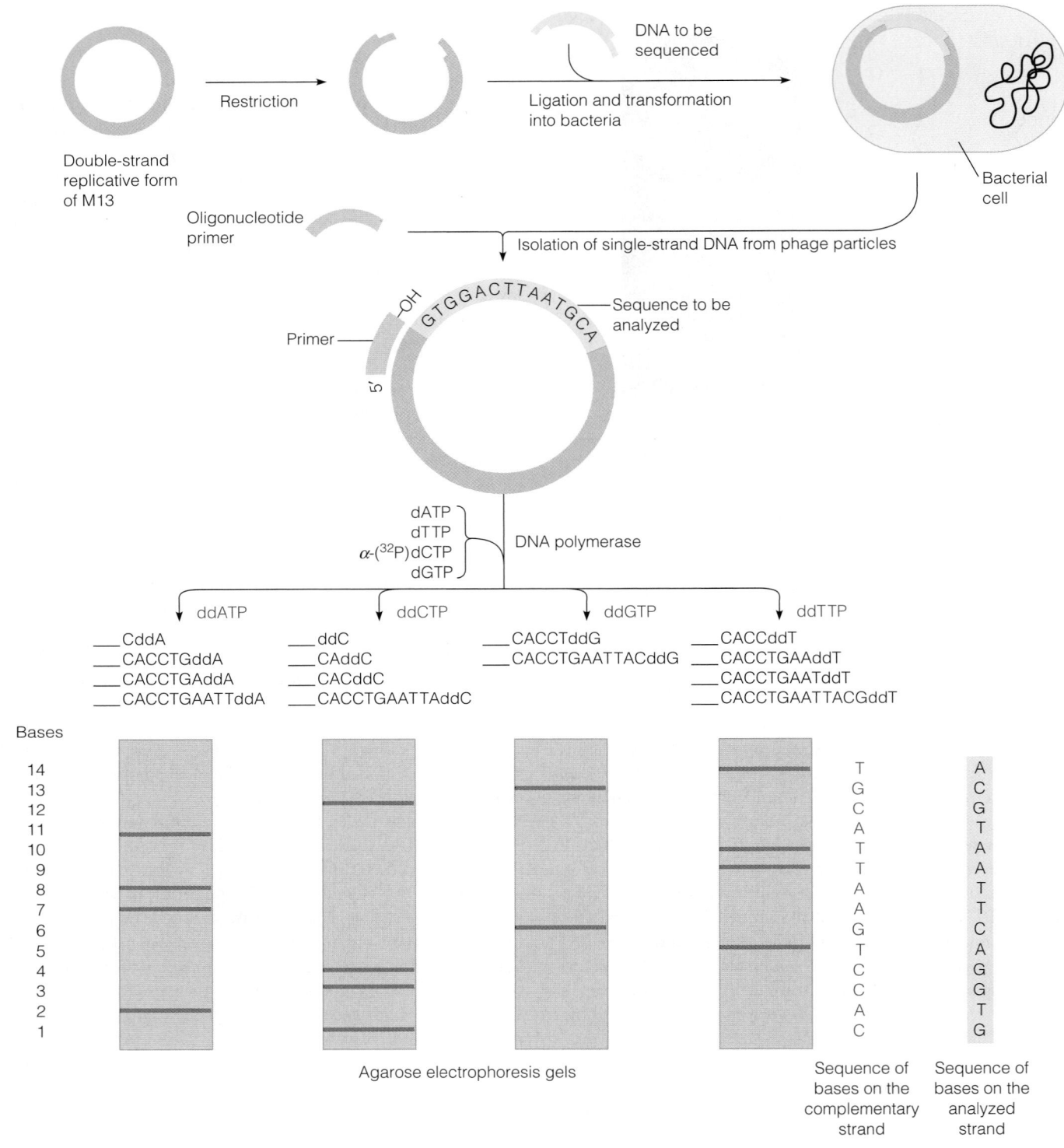

FIGURE 4B.6

Cloning into M13 and sequencing by the Sanger method.

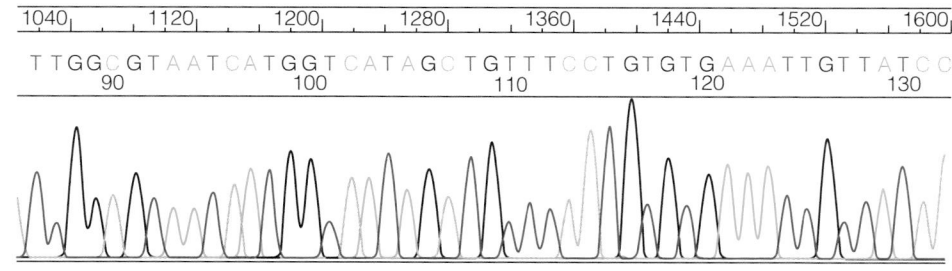

FIGURE 4B.7

Data from a DNA sequencing gel.

Courtesy of Dr. Robert H. Lyons, The University of Michigan's DNA Sequencing Core.

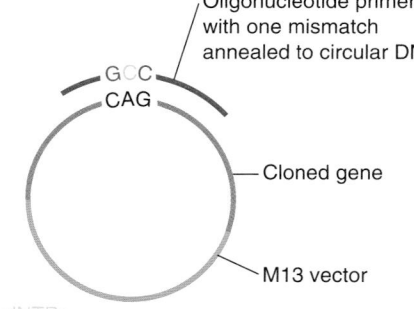

Oligonucleotide primer with one mismatch annealed to circular DNA

Cloned gene

M13 vector

dNTPs

DNA polymerase, DNA ligase

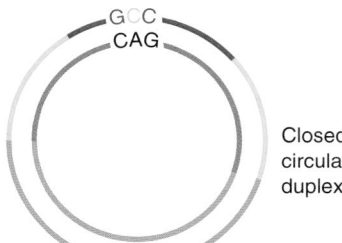

Closed circular duplex

Replication in *E. coli*

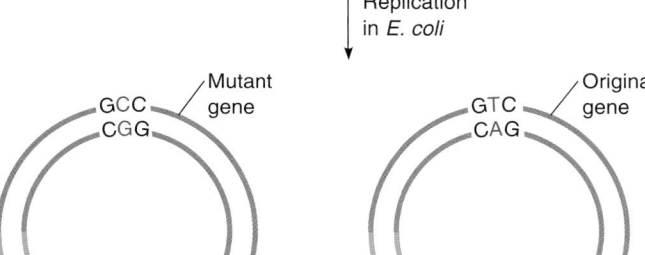

Mutant gene

Original gene

FIGURE 4B.8

Use of a mismatched synthetic oligonucleotide primer to introduce mutations in a gene cloned into a single-stranded vector.

chain reaction (PCR; Tools of Biochemistry 24A), and that has made an already relatively straightforward technique even simpler.

References

Ding, B., and N. C. Seaman (2006) Operation of a DNA robot arm inserted into a 2D DNA crystalline substrate. *Science* 314:1583–1585. An early application in DNA nanotechnology.

Douglas, S. M., H. Dietz, T. Liedl, B. Högberg, F. Graf, and W. M. Shih (2009) Self-assembly of DNA into nanoscale three-dimensional shapes. *Nature* 459:414–418. Both two- and three-dimensional shapes can be designed from DNA.

Drmanac, R., and 66 coauthors (2010) Human genome sequencing using unchained base reads on self-assembling DNA nanoarrays. *Science* 327:78–81. Using second- and third-generation technology, these workers sequenced three human genomes at a cost of $4400 per genome and accuracy of one error per 100 kb.

Endo, M., and H. Sugiyama (2009) Chemical approaches to DNA nanotechnology. *ChemBioChem* 10:2420–2443. A detailed and informative review.

Kunkel, T. A., J. D. Roberts, and R. A. Zakour (1989) Rapid and efficient site-specific mutagenesis without phenotypic selection. In: *Recombinant DNA Methodology*, edited by R. Wu, L. Grossman, and K. Moldave, pp. 587–601. Academic Press, San Diego, CA. Laboratory instructions for the most widely used method of site-directed mutagenesis.

Mardis, E. R. (2011) A decade's perspective on DNA sequencing technology. *Nature* 470:198–203. A recent review of the sequencing methods arising since the Sanger technique.

Mattencci, M. D., and M. H. Caruthers (1981) Synthesis of deoxyoligonucleotides on a polymer support. *J. Am. Chem. Soc.* 103:3185–3191. More detailed description of the chemistry involved.

Metzker, M. L. (2010) Sequencing technologies—the next generation. *Nature Reviews Genetics* 11:31–46. This review article discusses the principles and applications of six "second-generation" high-throughput DNA sequencing technologies.

Sambrook, P. J., and D. W. Russell (2001) *Molecular Cloning, A Laboratory Manual, Volumes 1–3*, 3rd ed. Cold Spring Harbor Laboratory, Cold Spring Harbor, N.Y. The definitive laboratory handbook of molecular biological methods.

Zheng, J., and eight coauthors (2009) From molecular to macroscopic via the rational design of a self-asssembled 3D DNA crystal. *Nature* 461:74–77. A triangular DNA structure formed from synthetic oligodeoxyribonucleotides forms large crystals, well beyond nanoscale.

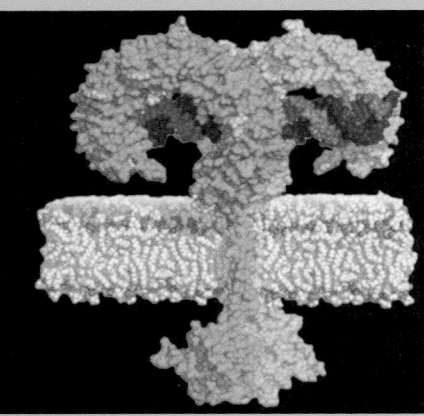

CHAPTER 5

Introduction to Proteins: The Primary Level of Protein Structure

We have seen that one class of biopolymers, the nucleic acids, stores and transmits the genetic information of the cell. Much of that information is expressed in another class of biopolymers, the **proteins**. Proteins play an enormous variety of roles: Some carry out the transport and storage of small molecules; others make up a large part of the structural framework of cells and tissues. Muscle contraction, the immune response, and blood clotting are all mediated by proteins. An important class of proteins is the **enzymes**—the catalysts that promote the tremendous variety of reactions that are required to support the living state. Each type of cell in every organism has several thousand kinds of proteins to serve these many functions.

In keeping with the multiplicity of their functions, proteins are extremely complex molecules. This complexity is illustrated in Figure 5.1, which depicts the molecular structure of myoglobin, a relatively small protein that functions primarily in oxygen binding and storage in animal tissues. In this and the following three chapters, we analyze in detail the structures and functions of a handful of proteins, including myoglobin. We will see that although there are general features of protein structure shared by most proteins, each protein has a distinct structure that is optimally suited to its function. Protein structures may appear at first glance to be hopelessly complex; however, there is an elegant and readily comprehensible logic to protein structure, which we will describe here and in Chapter 6. We begin with a description of the simple "building blocks" that are found in all proteins: the amino acids.

Amino Acids

Structure of the α-Amino Acids

Protein function is determined by protein structure, which, in turn, is determined by the structures and properties of the various amino acids which make up the protein.

All proteins are polymers, and the monomers that combine to make them are **α-amino acids**. A general representation of an α-amino acid is shown in Figure 5.2a. The amino group is attached to the α-carbon, the carbon next to the carboxylic acid group; hence the name α-amino acid. To the α-carbon of every amino acid are also

FIGURE 5.1

The three-dimensional structure of the globular protein **myoglobin.** This molecular model was generated from the X-ray crystal structure determined by H. C. Watson and J. C. Kendrew (PDB ID: 1MBN), and it shows the heavy atoms (i.e., non-hydrogen atoms) in sperm whale myoglobin. The amino acid atoms are shown as sticks, where carbon atoms are gray, oxygen atoms are red, nitrogen atoms are blue, and sulfur atoms are yellow. The heme atoms are shown in a space-filling display that represents the van der Waals surface of each atom. The orange sphere in the center of the heme is the Fe^{2+} ion that binds a molecule of oxygen (the oxygen-binding site is obscured by the heme). The orientation of the protein in this figure is the same at that shown in the painting by Irving Geis in Chapter 1 (Figure 1.5).

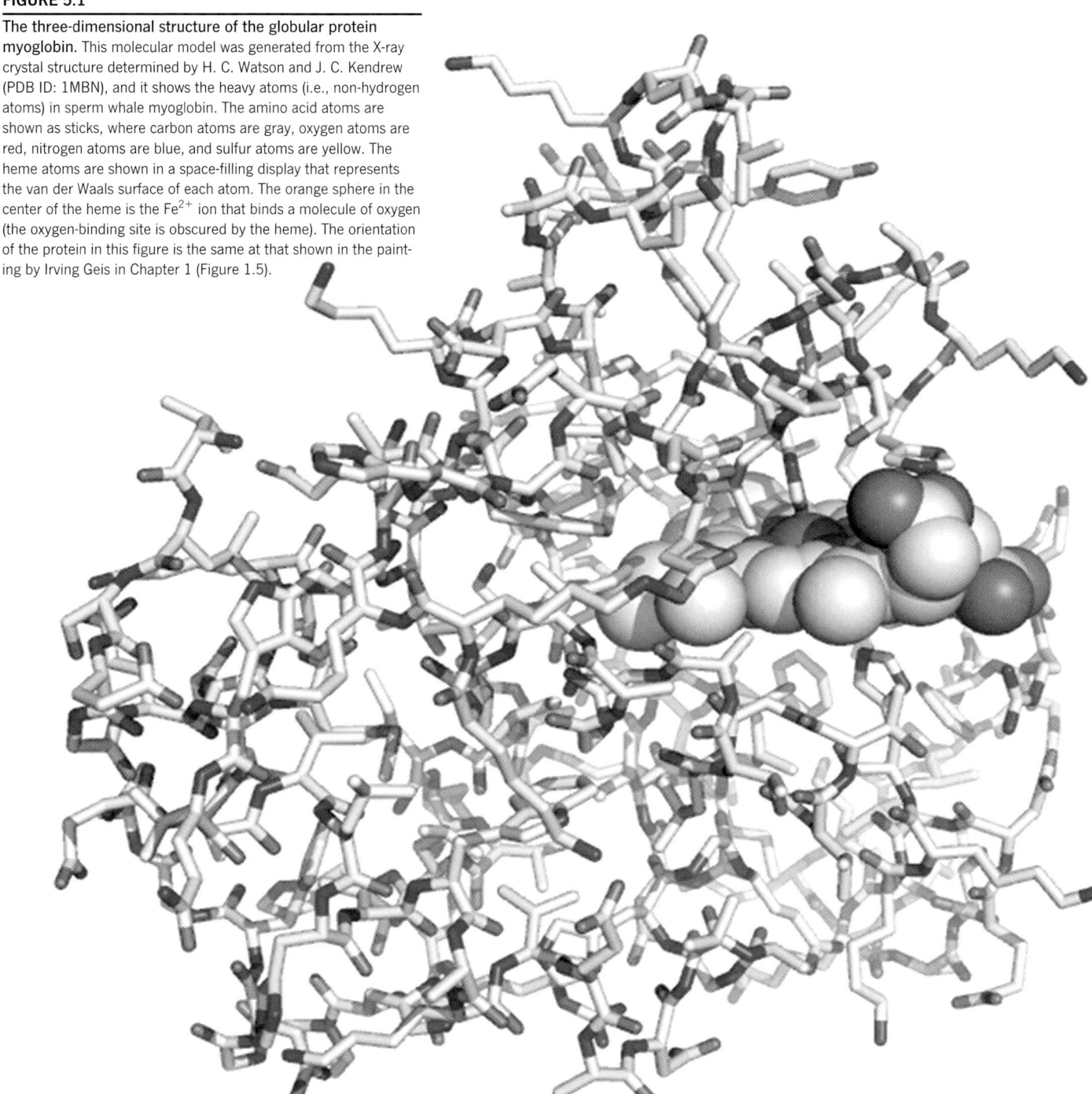

attached a *hydrogen atom* and a *side chain* ("R" group). Different α-amino acids are distinguished by their different side chains. We can write the general structure for an α-amino acid as shown in Figure 5.2a. This representation, although chemically correct, ignores the conditions in vivo. As pointed out in Chapter 2, most biochemistry occurs in the physiological pH range near neutrality. The pK_a's of the carboxylic acid and amino groups of the α-amino acids are about 2 and 10, respectively. Therefore, near neutral pH the carboxylic acid group will have lost a proton, and the amino group will have picked up a proton, to yield the **zwitterion** form shown in Figure 5.2b. This is the form in which we will customarily write amino acid structures.

Twenty different kinds of amino acids are commonly incorporated into proteins during the process of translation (see Figure 4.23 page 112). The complete structures

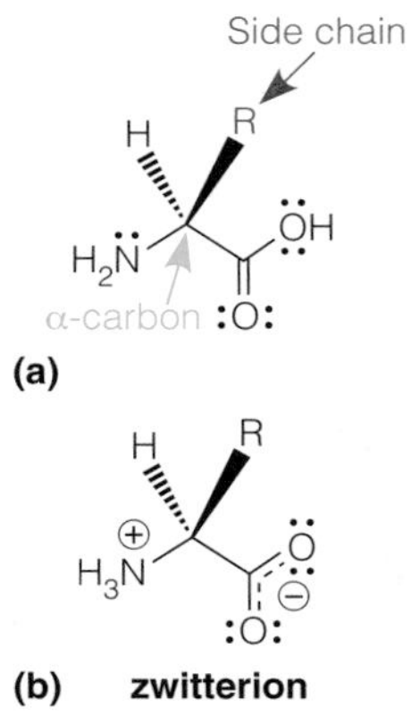

FIGURE 5.2

The structure of an α-amino acid. **(a)** A general representation of a nonionized α-amino acid showing the carboxylic acid group, the α-amino group, and a hydrogen bonded to the α-carbon, as well as the side chain (R group) that gives the amino acid its unique properties. **(b)** An amino acid shown as a zwitterion at neutral pH. Under physiological conditions, amino acids exist as zwitterions in which the α-carboxylic acid group has lost a proton and the α-amino group has gained one. Note that the negative charge on the α-carboxylate is delocalized over the two oxygen atoms. The stereochemistry shown in this figure is that for the α-amino acids found in biosynthetic proteins.

of these amino acids are shown in Figure 5.3 and other important data are given in Table 5.1. At least two additional amino acids, selenocysteine and pyrrolysine, are encoded genetically and incorporated into proteins; however, they are found in a relatively small number of proteins. For the purposes of this introductory discussion we will focus our attention on the twenty common amino acids shown in Figure 5.3.

Stereochemistry of the α-Amino Acids

The asymmetry of biomolecules plays a critical role in determining their structures and functions; thus, familiarity with the basic stereochemistry of amino acids is necessary for an understanding of the biochemistry of proteins.

The four groups shown in Figure 5.2a are bonded to the central α-carbon in a tetrahedral arrangement, as is predicted for an sp^3 hybridized carbon atom. In Figure 5.2 the projection of these groups around the α–carbon (C_α) is represented as follows: The lines represent bonds in the plane of the page, the solid wedges represent bonds projecting forward from the page, and the dashed wedges represent bonds projecting behind the page. When a carbon atom has four different substituents attached to it, it is said to be **chiral**, or a **stereocenter**, or, preferably, an **asymmetric carbon**. In Figure 5.3, the stereochemistry of the amino acids is shown by a convention known as the **Fischer projection**. In a Fischer projection the bonds are all represented as solid lines, where the horizontal bonds project forward from the page and the vertical bonds project behind the page. To help you visualize the Fischer projection convention, we have drawn in Figure 5.4 the general structure of an amino acid in a ball-and-stick rendering as well as a Fischer projection that includes solid and dashed wedges. Note that the spatial orientation of the four groups bound to the C_α is the same in Figures 5.2, 5.3, and 5.4.

TABLE 5.1   Properties of the common amino acids found in proteins

Name	Abbreviations 1- and 3-letter codes	pK_a of α-COOH Group	pK_a of α-NH$_3^+$ Group	pK_a of Ionizing Side Chain[a]	Residue[b] Mass (daltons)	Occurrence[c] in Proteins (mol %)
Alanine	A, Ala	2.3	9.7	—	71.08	8.7
Arginine	R, Arg	2.2	9.0	12.5	156.20	5.0
Asparagine	N, Asn	2.0	8.8	—	114.11	4.2
Aspartic acid	D, Asp	2.1	9.8	3.9	115.09	5.9
Cysteine	C, Cys	1.8	10.8	8.3	103.14	1.3
Glutamine	Q, Gln	2.2	9.1	—	128.14	3.7
Glutamic acid	E, Glu	2.2	9.7	4.2	129.12	6.6
Glycine	G, Gly	2.3	9.6	—	57.06	7.9
Histidine	H, His	1.8	9.2	6.0	137.15	2.4
Isoleucine	I, Ile	2.4	9.7	—	113.17	5.5
Leucine	L, Leu	2.4	9.6	—	113.17	8.9
Lysine	K, Lys	2.2	9.0	10.0	128.18	5.5
Methionine	M, Met	2.3	9.2	—	131.21	2.0
Phenylalanine	F, Phe	1.8	9.1	—	147.18	4.0
Proline	P, Pro	2.0	10.6	—	97.12	4.7
Serine	S, Ser	2.2	9.2	—	87.08	5.8
Threonine	T, Thr	2.6	10.4	—	101.11	5.6
Tryptophan	W, Trp	2.4	9.4	—	186.21	1.5
Tyrosine	Y, Tyr	2.2	9.1	10.1	163.18	3.5
Valine	V, Val	2.3	9.6	—	99.14	7.2

[a]Approximate values found for side chains on the *free* amino acids.

[b]To obtain the mass of the amino acid itself, add the mass of a molecule of water, 18.02 daltons. The values given are for neutral side chains; slightly different values will apply at pH values where protons have been gained or lost from the side chains.

[c]Average for a large number of proteins. Individual proteins can show large deviations from these values. Data from *Journal of Chemical Information and Modeling* 50:690–700, J. M. Otaki, M. Tsutsumi, T. Gotoh, and H. Yamamoto, Secondary structure characterization based on amino acid composition and availability in proteins. © 2010 American Chemical Society.

W. P. Jencks and J. Regenstein (1976) Ionization constants of acids and bases in *Handbook of Biochemistry and Molecular Biology*, 3rd ed., G. Fasman (ed.), CRC Press, Boca Raton, FL.

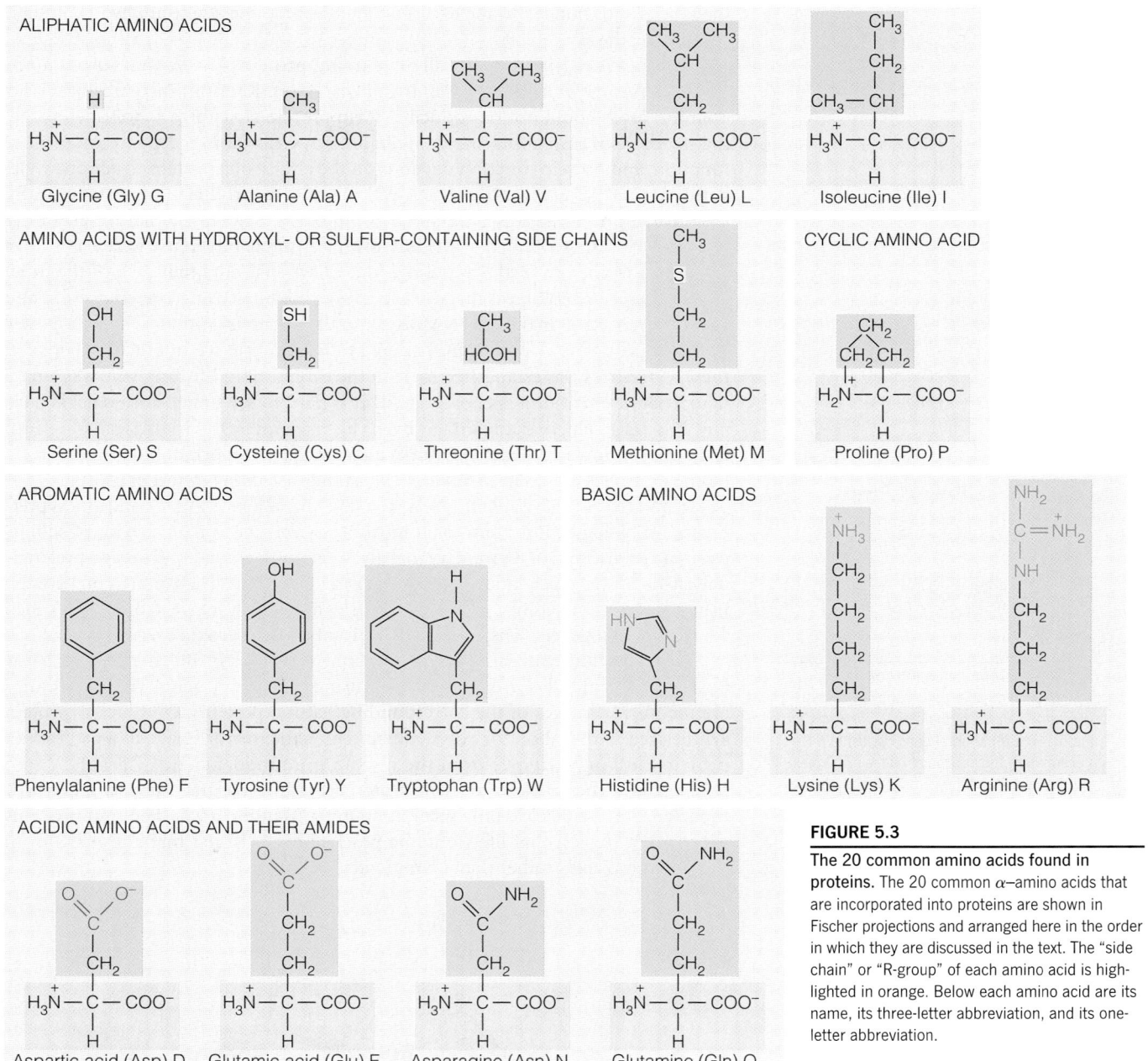

FIGURE 5.3

The 20 common amino acids found in proteins. The 20 common α–amino acids that are incorporated into proteins are shown in Fischer projections and arranged here in the order in which they are discussed in the text. The "side chain" or "R-group" of each amino acid is highlighted in orange. Below each amino acid are its name, its three-letter abbreviation, and its one-letter abbreviation.

If a molecule contains one asymmetric carbon, two distinguishable **stereoisomers** exist; these are nonsuperimposable mirror images of one another, or **enantiomers,** as shown in Figure 5.5. The stereoisomers of **alanine** shown in Figure 5.5 are called the L and D enantiomers.* The L and D enantiomers can be distinguished from one another experimentally because their solutions rotate plane polarized light in opposite directions. For this reason, enantiomers are sometimes called **optical isomers.** All amino acids except glycine can exist in D and L forms because in each case the α-carbon is asymmetric. Glycine is the sole exception because two of

———
*Those who are familiar with modern organic chemistry will know that there are two commonly used systems for distinguishing stereoisomers—the older D–L system and the newer, more comprehensive R–S system (Cahn-Ingold-Prelog system). Both are discussed in more detail in Chapter 9.

Proteins are polymers of α-amino acids. There are 20 common α-amino acids that are the major building blocks of proteins.

All α-amino acids except glycine contain an asymmetric α-carbon, thus both L and D enantiomers are possible. However, in the vast majority of proteins, only the L-enantiomers are found.

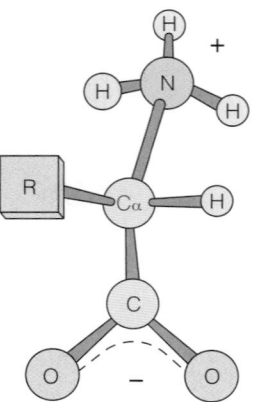

(a) α-Amino acid

(b) Representation of an amino acid in Fischer projection

FIGURE 5.4

Three-dimensional representations of α-amino acids. **(a)** This ball-and-stick model shows the three-dimensional arrangement of the atoms. The α-carbon is asymmetric, with tetrahedral bonding. **(b)** In a Fischer projection (left) the horizontal bonds are projecting toward the viewer and the vertical bonds are projecting away from the viewer. This orientation of bonds in the Fischer projection is represented on the right by solid and dashed wedges.

the four groups bonded to the α-carbon are the same (i.e., there are two H atoms), eliminating the asymmetry.

Chemical analysis of naturally occurring proteins shows that *nearly all of the amino acids found in proteins are of the L form.* Random mixtures of D-and L-amino acids could not reproducibly form protein structures as well-defined as the structure of myoglobin shown in Figure 5.1. Cell viability depends on protein function, and protein function depends on the ability of a protein to adopt a well-defined *active* structure; thus, the need for cells to produce many structurally identical copies of a given protein is absolute. For this reason, nature uses only L-amino acids in protein biosynthesis. How the absolute preference for the L-isomer over the D-isomer evolved is puzzling. Indeed, we shall find that each of the three major classes of biological macromolecules exhibits a strong preference for one stereoisomer class or the other. Most naturally occurring polysaccharides prefer D-sugars, as do DNA and RNA. It may be that productive interaction between these substances was established early in the evolution of life; but, why was a particular set of enantiomers chosen at all? It is hard to see how L-amino acids have any inherent selective advantage over the D-isomers for biological function. Indeed, D-amino acids exist in nature, and some play important biochemical roles (some examples are given in Table 5.2), but they are rarely found in proteins.

Many scientists have attempted to provide explanations for this "handedness preference" in biology. Most point to an intrinsic asymmetry in the behavior of subnuclear particles, a kind of asymmetry that gives electrons emitted in β decay a preferential left-hand spin. Such influences are very weak but might, in a competition between primitive organisms using L- or D-proteins, give a slight advantage to one or the other. After billions of generations, even a small advantage can become overwhelming.

Using peptide synthesis methods described in Tools of Biochemistry 5C, it is possible to chemically synthesize proteins using all D-amino acids. These structures are the mirror images of the corresponding natural proteins. One such D-protein synthesized in the laboratory of Stephen Kent is the mirror image of a protease (a protein-cleaving enzyme) from the human immunodeficiency virus, HIV (see References). Whereas its natural L-counterpart cleaves natural L-proteins, this synthetic enzyme will cleave only those containing D-amino acids. The results of this experiment suggest that life would be possible for cells that made proteins from only D-amino acids rather than L-amino acids.

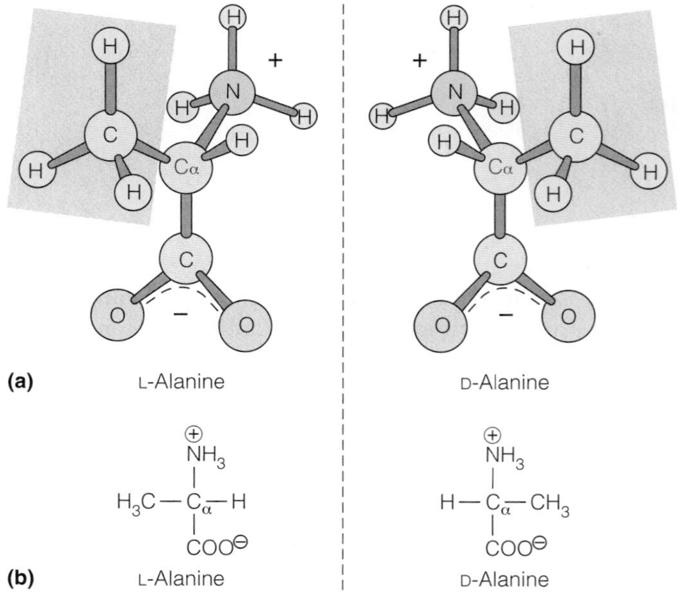

FIGURE 5.5

Stereoisomers of α-amino acids. **(a)** L-alanine and its enantiomer D-alanine are shown as ball-and-stick models. The alanine side chain is —CH₃. The two models are mirror images, which are not superimposable. The plane of mirror symmetry is represented by the vertical dashed line (red). **(b)** The same two enantiomers in a Fischer projection.

(a)   L-Alanine   D-Alanine

(b)   L-Alanine   D-Alanine

TABLE 5.2 Some biologically important amino acids not typically found in proteins

Name	Formula	Biochemical Source, Function
β-Alanine	$H_3\overset{+}{N}-CH_2-CH_2-COO^-$	Found in the vitamin pantothenic acid and in some important natural peptides
D-Alanine	$\begin{array}{c} COO^- \\ | \\ H-C-\overset{+}{N}H_3 \\ | \\ CH_3 \end{array}$	In polypeptides in some bacterial cell walls
γ-Aminobutyric acid	$H_3\overset{+}{N}-CH_2-CH_2-CH_2-COO^-$	Brain, other animal tissues; functions as neurotransmitter
D-Glutamic acid	$\begin{array}{c} COO^- \\ | \\ H-C-\overset{+}{N}H_3 \\ | \\ CH_2 \\ | \\ CH_2-COO^- \end{array}$	In polypeptides in some bacterial cell walls
L-Homocysteine	$\begin{array}{c} COO^- \\ | \\ H_3\overset{+}{N}-C-H \\ | \\ CH_2-CH_2SH \end{array}$	Many tissues; precursor for methionine biosynthesis
L-Ornithine	$\begin{array}{c} COO^- \\ | \\ H_3\overset{+}{N}-C-H \\ | \\ CH_2-CH_2-CH_2\overset{+}{N}H_3 \end{array}$	Many tissues; an intermediate in arginine synthesis
Sarcosine	$\begin{array}{c} CH_3-N-CH_2-COO^- \\ | \\ H \end{array}$	Many tissues; intermediate in amino acid synthesis
L-Thyroxine	$\begin{array}{c} COO^- \\ | \\ H_3\overset{+}{N}-C-H \\ | \\ CH_2 \end{array}$ (with iodinated diphenyl ether ring system, ending in OH)	Thyroid gland; is thyroid hormone (I = iodine)

Reza Ghadiri and coworkers have shown that a homochiral 32-residue peptide (i.e., composed of either all L- or all D-amino acids) *preferentially* catalyzes replication of homochiral products from a racemic mixture of peptide fragments. These experiments do not explain why L-amino acids are preferred over D-amino acids; but, they show that once the preference is established, there is a natural tendency to amplify homochiral sequences (in this case the sequences containing only L-amino acids).

The preference for L-amino acids in natural proteins has two important consequences, which we will discuss further in subsequent chapters:

1. The surface of any given protein, which is where the interesting biochemistry occurs, is asymmetric. This asymmetry is the basis for the highly specific molecular recognition of binding targets by proteins.

2. The stereochemistry of the amino acids plays an important role in the formation of so-called "secondary structure" (i.e., α helices and β strands) and thereby the overall structure of proteins.

The stereochemistry of the amino acids plays an important role in the formation of the structure of proteins.

Properties of Amino Acid Side Chains: Classes of α-Amino Acids

The 20 common amino acids contain, in their 20 different side chains, a remarkable collection of chemical groups. It is this diversity of the side chains that allows proteins to exhibit such a great variety of structures and properties. If we examine Figure 5.3, it becomes evident that there are several different classes of side chains, distinguished by their dominant chemical features. These features include hydrophobicity or hydrophilicity, polar or nonpolar character, and the presence or absence of ionizable groups. Many ways have been proposed to group the amino acids into classes, but none is wholly satisfactory. We shall discuss the amino acids in the order shown in Figure 5.3, which proceeds from the simplest to the more complex.

The variety of side chains on amino acids allows proteins enormous versatility in structure and function.

Amino Acids with Aliphatic Side Chains

Glycine, alanine, valine, leucine, and **isoleucine** have aliphatic side chains. As we progress from left to right along the top row of Figure 5.3, the R group becomes more extended and more hydrophobic. Isoleucine, for example, has a much greater tendency to transfer from water to a hydrocarbon solvent than does alanine. The more hydrophobic amino acids such as isoleucine are usually found *within the core* of a protein molecule, where they are shielded from water. **Proline**, which has a secondary α–amino group, is difficult to fit into any category. It is the only amino acid in this group in which the side chain forms a covalent bond with the α–amino group. The proline side chain has a primarily aliphatic character; however, it is frequently found on the *surfaces* of proteins due to its unique structural constraints. The rigid ring of proline is well-suited to those sites in a protein structure where the protein must fold back on itself (so-called "turns").

Amino Acids with Hydroxyl- or Sulfur-Containing Side Chains

In this category we can place **serine, cysteine, threonine**, **methionine,** and **tyrosine**. Although methionine and tyrosine are fairly hydrophobic, these amino acids, because of their more polar side chains, display more hydrophilic character than their aliphatic analogs. As we will see in Chapter 11, the —OH group of serine and the —SH group of cysteine are good nucleophiles and often play key roles in enzyme activity. Cysteine is noteworthy in two additional respects. First, the side chain can ionize at moderately high pH:

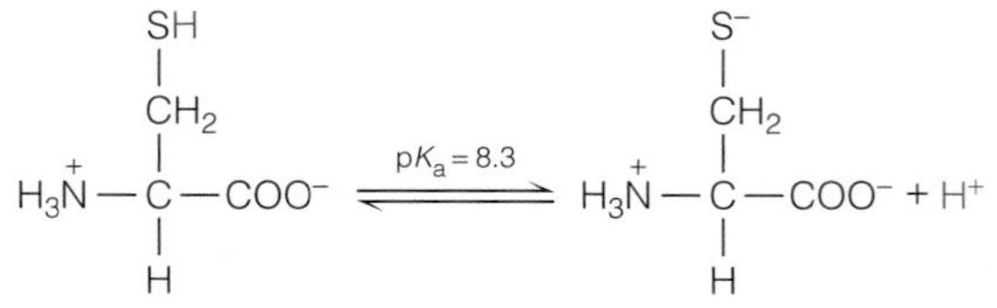

Second, oxidation of two cysteine side chains yields a **disulfide bond:**

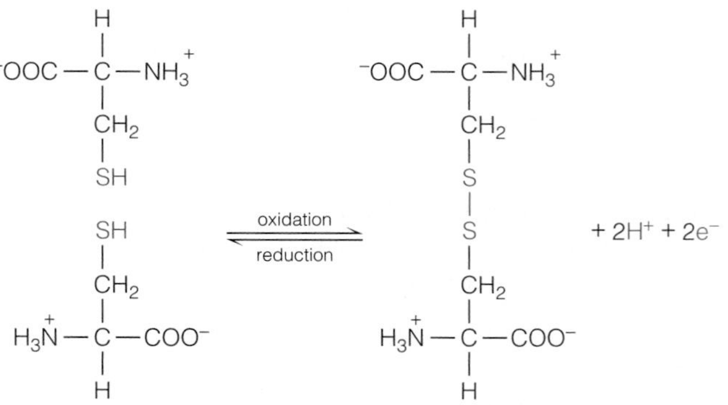

Cysteine **Cystine**

The product of this oxidation is given the name **cystine**. We do not list it among the 20 amino acids because cystine is always formed by oxidation of two cysteine side chains and is not coded for by DNA. Such disulfide bonds often play an important role in stabilizing the structure of a protein.

Aromatic Amino Acids

Three amino acids, **phenylalanine, tyrosine**, and **tryptophan**, carry aromatic side chains. Phenylalanine, together with the aliphatic amino acids valine, leucine, and isoleucine, is one of the most hydrophobic amino acids. Tyrosine and tryptophan have some hydrophobic character as well, but it is tempered by the polar groups in their side chains. In addition, tyrosine can ionize at high pH:

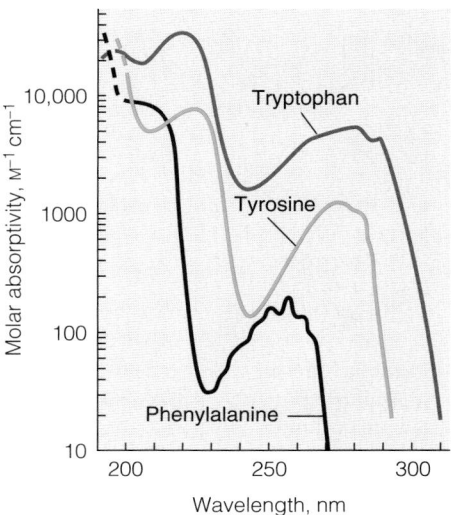

The aromatic amino acids, like most highly conjugated compounds, absorb light in the near-ultraviolet region of the spectrum (Figure 5.6). This characteristic is frequently used for the detection and/or quantitation of proteins, by measuring absorption at 280 nm.

Basic Amino Acids

Histidine, lysine, and **arginine** carry basic groups in their side chains. They are represented in Figure 5.3 in the form that predominates at pH 7. Histidine is the least basic of the three, and as its titration curve (Figure 5.7) shows, the imidazole ring in the side chain of the free amino acid loses its proton at about pH 6 (pK_a values for the side chains of free amino acids are given in Table 5.1.). When histidine is incorporated into proteins, the pK_a typically ranges from 6.5–7.4 (Table 5.3). The value of pK_a for an ionizable side chain is sensitive to the proximity of other charged groups. In the folded structures of proteins the local electrostatic environment can perturb the pK_a of an ionizable side chain by $\pm$ three pH units. Because the histidine side chain has a pK_a near physiological pH, it often plays a role in enzymatic catalysis involving proton transfer. Lysine and arginine are more basic amino acids, and as their pK_a values indicate (Tables 5.1 and 5.3), their side chains are almost always positively charged under physiological conditions. The guanidino group of arginine is a particularly strong base due to the resonance stabilization of the protonated side chain.

The basic amino acids are strongly polar, and as a consequence they are usually found on the exterior surfaces of proteins, where they can be hydrated by the surrounding aqueous environment.

Acidic Amino Acids and Their Amides

Aspartic acid and **glutamic acid** typically carry negative charges at pH 7; they are depicted in the anionic forms in Figure 5.3. The titration curve of aspartic acid is shown in Figure 2.20 (page 46). The pK_a values of the acidic amino acids are so low (see Table 5.3) that even when the amino acids are incorporated into proteins, the negative charge on the side chain is typically retained under physiological conditions. Hence, these amino acid residues are often referred to as **aspartate** and **glutamate** (i.e., the conjugate bases rather than the acids).

FIGURE 5.6

Absorption spectra of the aromatic amino acids in the near-ultraviolet region. Tryptophan (red; λ_{max} = 278 nm) and tyrosine (blue; λ_{max} = 274 nm) account for most of the UV absorbance by proteins in the region around 280 nm. Phenylalanine (black; λ_{max} = 258 nm) does not absorb at 280 nm. Note that the absorptivity scale is logarithmic. Compared with nucleic acids, amino acids absorb only weakly in the UV; see Figure 4.5 for comparison.

Reprinted from *Advances in Protein Chemistry* 17:303–390, D. B. Wetlaufer, Ultraviolet spectra of proteins and amino acids. © 1962, with permission from Elsevier.

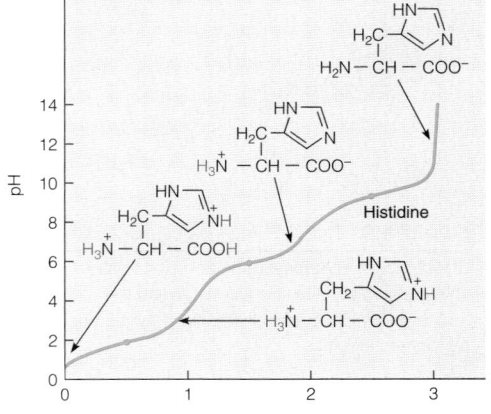

FIGURE 5.7

Titration curve of histidine. The dots correspond to pK_a values, and the forms predominating at different pH values are shown. Ionizable hydrogens are shown in red. It is presumed that the starting solution was adjusted to pH < 2 by addition of H^+ to the dissolved amino acid. See also Figure 2.20 (page 46) for the titration curve of aspartic acid.

TABLE 5.3 Typical ranges observed for pK_a values of groups in proteins	
Group Type	Typical pK_a Range[a]
α-Carboxyl	3.5–4.0
Side chain carboxyls of aspartic and glutamic acids	4.0–4.8
Imidazole (histidine)	6.5–7.4
Cysteine (—SH)	8.5–9.0
Phenolic (tyrosine)	9.5–10.5
α-Amino	8.0–9.0
Side chain amino (lysine)	9.8–10.4
Guanidinyl (arginine)	~12

[a]Values outside these ranges are observed. For example, side chain carboxyls have been reported with pK_a values as high as 7.3.

Nonpolar amino acids are typically found in the interiors of soluble proteins, whereas polar and charged amino acids are typically found on the surfaces of proteins.

Selenocysteine

Pyrrolysine

FIGURE 5.8

Structures of selenocysteine and pyrrolysine. The selenium atom in selenocysteine and the 4-methyl-pyrroline-5-carboxylic acid in pyrrolysine are highlighted in red.

Sometimes amino acid side chains are modified after incorporation into a protein—a process called "post-translational modification."

Oligopeptides and polypeptides are formed by polymerization of amino acids. All proteins are polypeptides.

Companions to aspartic and glutamic acids are their amides, **asparagine** and **glutamine**. Unlike their acidic analogs, asparagine and glutamine have *uncharged* polar side chains. Like the basic and acidic amino acids, they are hydrophilic and tend to be on the surface of a protein molecule, in contact with the surrounding water.

Rare Genetically Encoded Amino Acids

We have considered the 20 common amino acids that are coded for in DNA and are incorporated directly into proteins by ribosomal synthesis. There are two other amino acids encoded in gene sequences: selenocysteine, which is widely distributed but found in few proteins, and pyrrolysine, which is restricted to a few archaea and eubacteria. Selenocysteine ("Sec") and pyrrolysine ("Pyl") are sometimes referred to, respectively, as the 21st and 22nd amino acids (Figure 5.8). Selenocysteine is a structural analog of cysteine in which the sulfur atom is replaced by a selenium atom. In prokaryotes selenoproteins are typically involved in catabolic processes, whereas in eukaryotes the roughly 25 selenoproteins characterized to-date appear to be anabolic and/or anti-oxidant catalysts. Pyrrolysine is a derivative of lysine in which a 4-methyl-pyrroline-5-carboxylic acid forms an amide bond with the ε–amino group of the lysine side chain. Pyrrolysine is found in the active sites of several archaeal enzymes involved in the catabolism of methylamine.

Modified Amino Acids

The repertoire of side chain groups in proteins can be expanded beyond the 20 canonical structures described above by the chemical modification of certain amino acids after they are assembled into proteins. The structures of four such *post-translationally modified amino acids* are depicted below, with the modifying group shown in red.

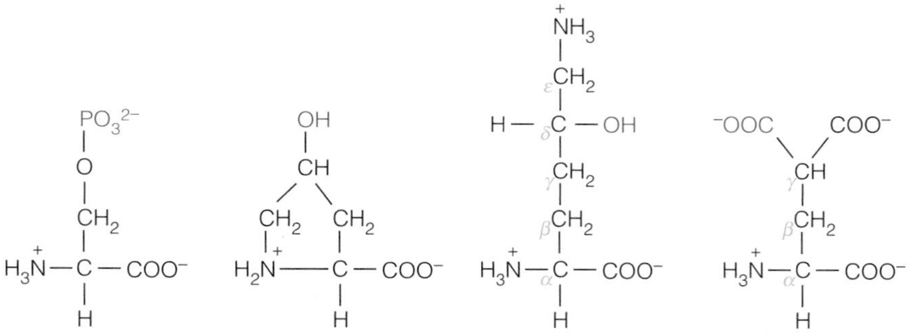

Phosphoserine   **4-Hydroxyproline**   **δ-Hydroxylysine**   **γ-Carboxyglutamic acid**

We shall consider these again when we encounter specific proteins in which such modification has occurred.

The amino acids found in proteins are by no means the only ones to occur in living organisms. Many other amino acids play important roles in metabolism. A partial list is given in Table 5.2. Note that not all of them are α-amino acids, or the L-enantiomers. We shall encounter all of the amino acids in this table in later chapters.

Peptides and the Peptide Bond

Peptides

Amino acids can be covalently linked together by formation of an **amide bond** between the α-carboxylic acid group on one amino acid and the α-amino group on another. This bond is often referred to as a **peptide bond**, and the products formed by such a linkage are called **peptides**. The formation of a peptide bond between glycine and alanine is shown in Figures 5.9 and 5.10. The product in this case is called a **dipeptide** because two amino acids have been combined. As illustrated in Figure 5.10, the reaction can be viewed as a simple elimination of a

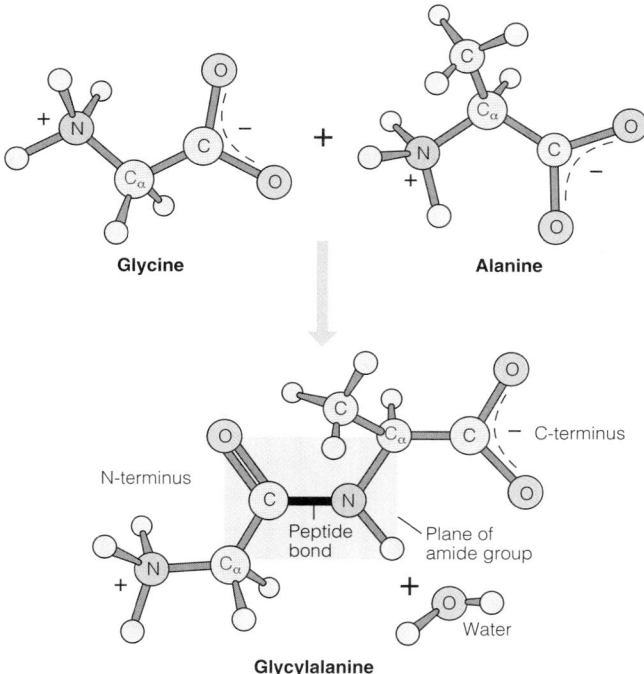

FIGURE 5.9

Formation of a dipeptide. Here the dipeptide glycylalanine (Gly–Ala) is depicted as being formed by removal of a water molecule as glycine is linked to alanine (See Figure 5.10).

water molecule between the carboxylic acid of one amino acid and the amino group of the other. Note that amide bond formation leaves an H_3N^+— group available on one end of the dipeptide and a —COO^- group on the other; thus, the reaction could in principle be continued by adding, for example, glutamic acid to one end and lysine to the other to yield the **tetrapeptide** shown in Figure 5.11. As each amino acid is added to the chain, another molecule of water must be eliminated. The portion of each amino acid remaining in the chain is called an **amino acid residue**. When specifying an amino acid residue in a peptide, the suffix **–yl** may be used to replace –ine or –ate in the name of the amino acid (e.g., glycyl for glycine, aspartyl for aspartate; tryptophanyl and cysteinyl are exceptions to this general rule). Thus, the *alanyl residue* in the tetrapeptide in Figure 5.11 is

$$-\underset{\underset{H}{|}}{N}-\underset{\underset{H}{|}}{\overset{\overset{CH_3}{|}}{C}}-\overset{\overset{O}{\|}}{C}-$$

In the structure of a peptide we distinguish the side chains (i.e., the R groups in Figure 5.3) from the **main chain** (or **peptide backbone**), which is composed of the atoms that make up the peptide bonds: the α-NH, the C_α, and the α-C=O groups of each amino acid residue in the peptide. The N-terminal amino group and C-terminal carboxylate are also part of the main chain.

Chains containing only a few amino acid residues (like a tetrapeptide) are collectively referred to as **oligopeptides**. If the chain is longer ($> \sim$15–20 residues), it is called a **polypeptide**. Polypeptides greater than $\sim$50 residues are generally referred to as **proteins** (note that most globular proteins contain 250–600 amino acid residues). As shown in Figure 5.11, most oligopeptides and polypeptides retain an unreacted amino group at one end (called the **amino terminus** or **N-terminus**) and an unreacted carboxylic acid group at the other end (the **carboxyl terminus** or **C-terminus**). Exceptions are certain small cyclic oligopeptides, in which the N- and C-termini have been linked covalently. In addition, many proteins have N-termini blocked by

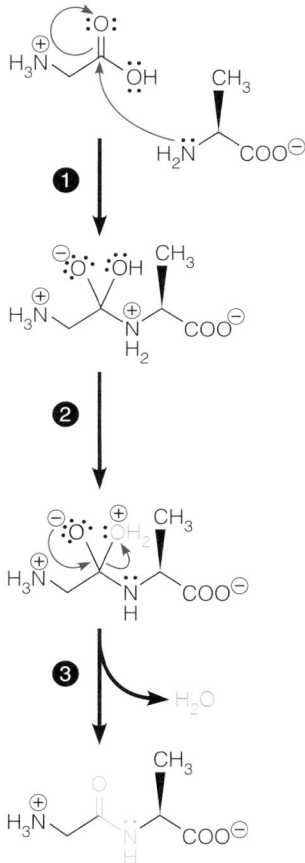

FIGURE 5.10

Elimination of water during the formation of a peptide bond. The formation of a peptide bond between glycine and alanine is shown. Step 1 involves attack of the deprotonated amino group on the carboxylic acid to form a tetrahedral intermediate. Step 2 shows proton transfer to the leaving group (H_2O, shown in green). Step 3 shows collapse of the tetrahedral intermediate and elimination of water to yield the planar amide bond. The newly formed peptide bond is highlighted in blue.

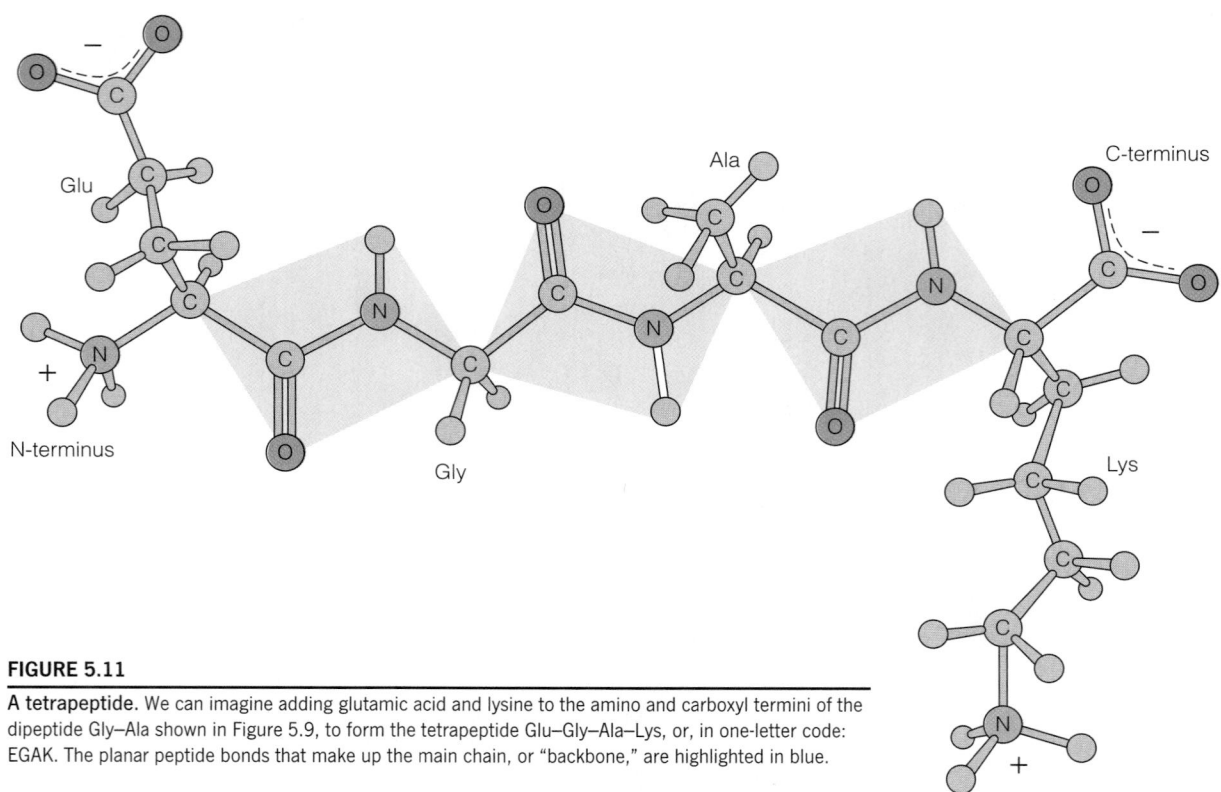

FIGURE 5.11

A tetrapeptide. We can imagine adding glutamic acid and lysine to the amino and carboxyl termini of the dipeptide Gly–Ala shown in Figure 5.9, to form the tetrapeptide Glu–Gly–Ala–Lys, or, in one-letter code: EGAK. The planar peptide bonds that make up the main chain, or "backbone," are highlighted in blue.

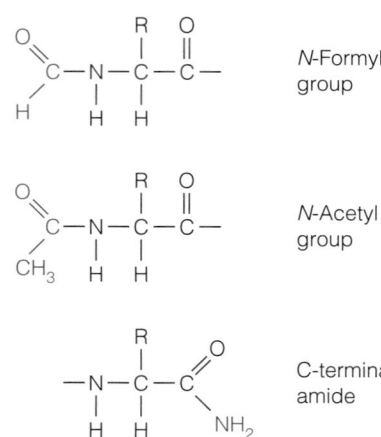

FIGURE 5.12

Groups that may block N- or C-termini in proteins. Blocking of the N-terminus by a formyl or acetyl group is more common than modification of the C-terminus to an amide.

N-formyl or *N*-acetyl groups, and a few have C-terminal carboxylates that have been modified to amides (Figure 5.12).

In writing the sequence of an oligopeptide or polypeptide, it would be awkward to spell out all the amino acid residue names. Therefore, biochemists usually write such sequences in terms of either the three-letter abbreviations or the one-letter abbreviations given in Figure 5.3. For example, the oligopeptide shown in Figure 5.11 could be written as either

Glu–Gly–Ala–Lys

or

EGAK

Note that the convention is to *always* write the N-terminal residue to the left and the C-terminal residue to the right. Because amino acids are asymmetric and there is only one free N-terminal amino group (and one free C-terminal carboxylate), the sequence EGAK ≠ KAGE.

Polypeptides as Polyampholytes

In addition to the free amino group at the N-terminus and the free carboxylate group at the C-terminus, polypeptides usually contain some amino acids that have ionizable groups on their side chains. These various groups have a wide range of pK_a values, as shown in Table 5.3, but are all weakly acidic or basic groups; thus, polypeptides are excellent examples of the polyampholytes described in Chapter 2. As stated above, amino acid side chains display a range of pK_a values in different proteins due to differences in the local electrostatic environment around a given amino acid. For example, an aspartic acid side chain near another negatively charged group (i.e., Glu or another Asp) will likely have a higher pK_a value than an aspartic acid near a positively charged group (i.e., Lys, His, or Arg). This topic will be explored in more detail in Chapters 7, 11, and 15.

The kind of behavior that is seen during the titration of an oligopeptide or a polypeptide is exemplified by the tetrapeptide (Glu-Gly-Ala-Lys) in Figure 5.11. We can imagine starting with the tetrapeptide in a very acidic solution with pH = 0. At this pH, which is below the pK_a of any of the groups present, all of the ionizable residues will be in their protonated forms:

As predicted from the information in Table 2.9 (page 48), all amino groups will be positively charged, and each carboxyl will have zero charge; thus, the tetrapeptide has an overall charge of +2 at this pH.

If we now imagine raising the pH of the solution (e.g., by titrating with NaOH), the various ionizable groups will lose protons at pH values in the vicinity of their pK_a values. The progress of this titration is shown in Figure 5.13. As the pH of the solution increases, more groups become deprotonated; thus, the net positive charge decreases and passes through zero at the isoelectric point (see Figure 2.19, page 45). As more base is added, the molecule becomes increasingly negatively charged, ultimately reaching a net charge of −2 at very high pH.

We will find the effects of changing pH on the overall charge on a protein to be of considerable importance in biochemistry. For example, even a small shift in pH can alter the constellation of charges displayed on the surface, or in the active site, of a protein and will thereby significantly affect its stability and/or functional properties. Also, the solubility of many proteins is minimal at the isoelectric point because the molecules no longer repel one another when their net charge is zero (see Figure 2.22, page 49). Finally, the fact that different proteins and oligopeptides have different net charges at a given pH is often used to advantage in their separation, either by electrophoresis (see Tools of Biochemistry 2A) or by ion-exchange chromatography (see Tools of Biochemistry 5A).

> Amino acids, peptides, and proteins are ampholytes; each has an isoelectric point.

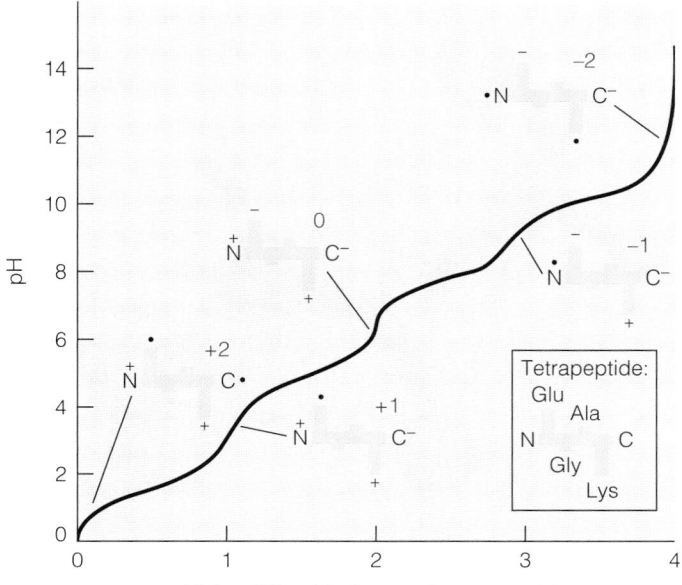

FIGURE 5.13

Polyampholytic behavior of a tetrapeptide. This titration curve for the tetrapeptide Glu-Gly-Ala-Lys (Figure 5.11) shows the major forms present at several pH values. The tetrapeptide is shown schematically here in blue (see inset, which indicates locations of N-and C-termini, as well as locations of side chains), and ionizable groups that may be charged are indicated by +, −, or · , depending on whether they carry a positive, negative, or zero charge at a given pH. Net charges for the different ionization states of the tetrapeptide are shown in red.

Moles OH⁻ added per mole tetrapeptide

The Structure of the Peptide Bond

Now let us examine the nature of the peptide bond that is formed between covalently linked amino acids. In the dipeptide of Figure 5.9 (Gly–Ala), the blue-shaded portion contains the peptide bond. This substituted amide bond, which is found between every pair of residues in a protein, has properties that are important for defining the structures of proteins. For example, almost invariably the amide carbonyl (C=O) and amide N—H bonds are nearly parallel; in fact, the six atoms shown in the blue rectangle in Figure 5.14 are usually coplanar. There is little twisting possible around the peptide bond because the C—N bond has a substantial fraction of double-bond character. The peptide bond can be considered a resonance hybrid of two forms:

A schematic depiction of the electron density about the peptide bond is shown in Figure 5.14a, and bond lengths and angles are given in Figure 5.14b.

X-ray crystallography data of proteins shows that the group of atoms about the peptide bond exist in two possible configurations, *trans* and *cis*, which are related by rotation around the C_{CO}—N bond (blue):

trans   **cis**

In fact, the *trans* form is usually favored because in the *cis* configuration the R groups on adjacent α-carbons can sterically interfere. The major exception is the bond in the sequence X—Pro, where X is any other amino acid. In this bond the *cis* configuration is sometimes allowed, although the *trans* configuration is still favored by about 4:1.

> The peptide bond is nearly planar, and the *trans* form is favored.

FIGURE 5.14

Structure of the peptide bond. (a) Delocalization of the π-electron orbitals over the three atoms O—C—N accounts for the partial double-bond character of the C—N bond. **(b)** The presently accepted values for bond angles and lengths are shown here. Bond lengths are in nanometers (nm). The six atoms shown in (a) and those in the blue rectangle in (b) are nearly coplanar.

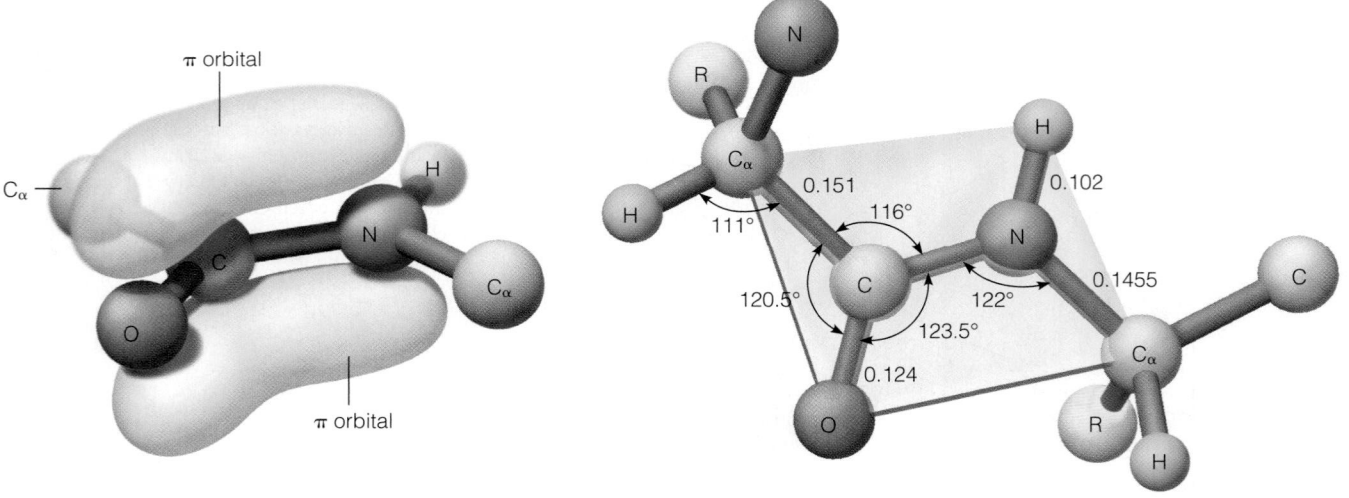

(a) Partial double-bond character of peptide bond

(b) Bond angles and lengths

Stability and Formation of the Peptide Bond

Figure 5.10 implies that a peptide bond could be formed by the elimination of a water molecule between two amino acids. In fact, in an aqueous environment this process is not thermodynamically favored. The free energy change for peptide bond formation at room temperature in aqueous solution is about $+10 \, \text{kJ/mol}$; therefore, the favored reaction under these conditions is the *hydrolysis* of the peptide bond:

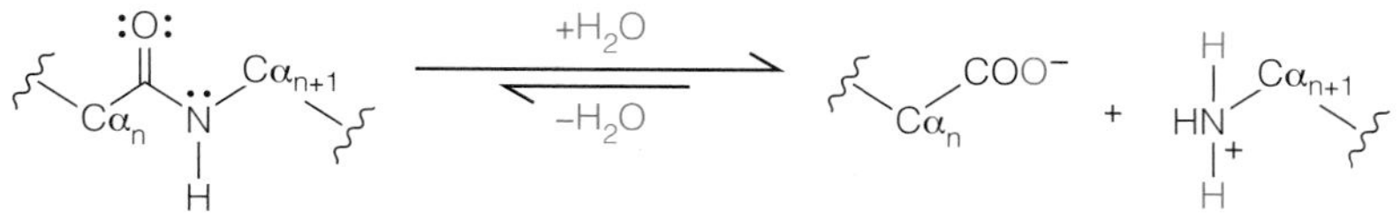

with equilibrium lying well to the right. However, the uncatalyzed reaction is exceedingly slow at physiological pH and temperature. Like polynucleotides, polypeptides are metastable, hydrolyzing rapidly only under extreme conditions or when catalysts are present.

Peptide hydrolysis can be achieved in several ways. A general method, which cleaves all peptide bonds, is boiling in strong mineral acid (usually 6 M HCl). More specific cleavage is provided by **proteolytic enzymes** or **proteases**. Many such enzymes exhibit specificity as to which peptide bonds they will cleave; a number of them are listed in Table 5.4. Some of these enzymes are secreted into the digestive tracts of animals, where they catalyze the initial breakdown of proteins to smaller peptides, which are then further digested to amino acids by other enzymes. Others, such as papain, are found in certain plant tissues. The existence of several proteolytic enzymes with different sequence specificities is of great utility to the biochemist, as they allow the cleavage of polypeptides in well-defined ways. A nonenzymatic reaction that cleaves a specific peptide bond uses the reagent cyanogen bromide ($BrC \equiv N$). Cyanogen bromide specifically cleaves the peptide bond at the carboxyl side of methionine residues (Figure 5.15). We will see later how these specific cleavage reactions can aid in determining the sequence of residues in a protein.

As with polynucleotides, the thermodynamic instability of polypeptides raises the question of how they can be synthesized in the aqueous medium of the cell. You may already have guessed the answer—that coupling of the unfavorable synthetic

The peptide bond is metastable. Peptide bonds hydrolyze in aqueous solution when a catalyst is present.

Peptides can be cleaved at specific sites by proteolytic enzymes or by treatment with chemical agents such as cyanogen bromide.

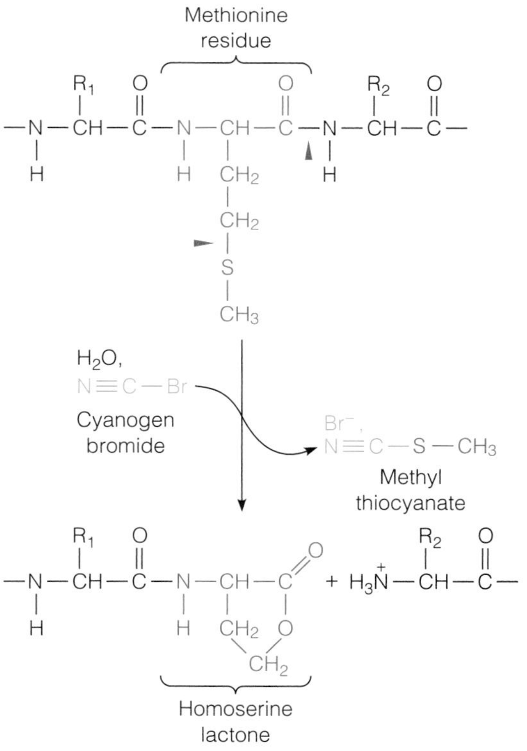

FIGURE 5.15

The cyanogen bromide reaction. This reaction specifically cleaves the peptide bond to the carboxyl side of methionine in any polypeptide and converts the Met residue to homoserine lactone. The cleavage sites are indicated by ▲.

TABLE 5.4   The sequence specificities of some proteolytic enzymes

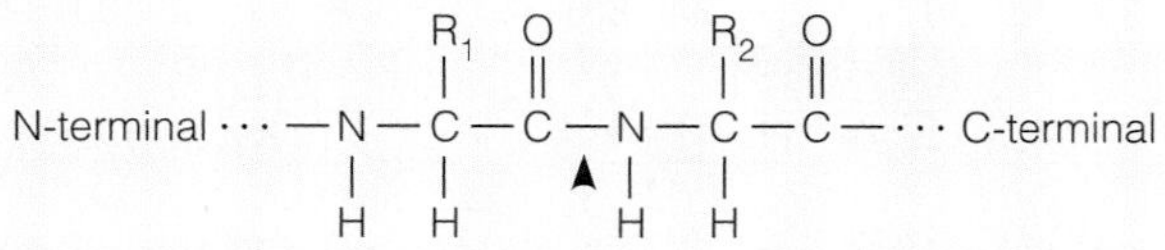

Enzyme	Preferred Site[a]	Source
Trypsin	R_1 = Lys, Arg	From digestive systems of animals, many other sources
Chymotrypsin	R_1 = Tyr, Trp, Phe, Leu	Same as trypsin
Thrombin	R_1 = Arg	From blood; involved in coagulation
V-8 protease	R_1 = Asp, Glu	From *Staphylococcus aureus*
Prolyl endopeptidase	R_1 = Pro	Lamb kidney, other tissues
Subtilisin	Very little specificity	From various bacilli
Carboxypeptidase A	R_2 = C-terminal amino acid	From digestive systems of animals
Thermolysin	R_2 = Leu, Val, Ile, Met	From *Bacillus thermoproteolyticus*

[a]The residues indicated are those next to which cleavage is most likely. Note that in some cases preference is determined by the residue on the N-terminal side of the cleaved bond (R_1) and sometimes by the residue to the C-terminal side (R_2). Generally, proteases do not cleave where proline is on the other side of the bond. Even prolyl endopeptidase will not cleave if R_2 = Pro.

reaction to the hydrolysis of high-energy phosphate compounds is required. In fact, every amino acid must be activated by an ATP-driven reaction before it can be incorporated into proteins. We shall give a brief outline of the process later in this chapter.

Proteins: Polypeptides of Defined Sequence

Proteins are polypeptides of defined sequence. Every protein has a defined number and order of amino acid residues. As with the nucleic acids, this sequence is referred to as the *primary structure* of the protein. In later chapters we shall see that it is this fundamental level of structure upon which higher levels of structural organization are based.

Figure 5.16 shows the primary structure of sperm whale *myoglobin,* the protein whose three-dimensional structure is shown in Figure 5.1. Also listed is the primary structure of human myoglobin, the protein that serves the same O_2 binding function in humans. The two sequences are aligned to show the maximum amino acid similarity between them. Two points are immediately obvious from

Every protein has a unique, defined amino acid sequence—its primary structure.

Number	1	2	3	4	5	6	7	8	9	10	11	12	13	14	15
Human	G	L	S	D	G	E	W	Q	L	V	L	N	V	W	G
Whale	V	L	S	E	G	E	W	Q	L	V	L	H	V	W	A

Number	16	17	18	19	20	21	22	23	24	25	26	27	28	29	30
Human	K	V	E	A	D	I	P	G	H	G	Q	E	V	L	I
Whale	K	V	E	A	D	V	A	G	H	G	Q	D	I	L	I

Number	31	32	33	34	35	36	37	38	39	40	41	42	43	44	45
Human	R	L	F	K	G	H	P	E	T	L	E	K	F	D	K
Whale	R	L	F	K	S	H	P	E	T	L	E	K	F	D	R

Key:
- Identical amino acids
- Conservative substitutions
- Nonconservative substitutions

Number	46	47	48	49	50	51	52	53	54	55	56	57	58	59	60
Human	F	K	H	L	K	S	E	D	E	M	K	A	S	E	D
Whale	F	K	H	L	K	T	E	A	E	M	K	A	S	E	D

Number	61	62	63	64	65	66	67	68	69	70	71	72	73	74	75
Human	L	K	K	H	G	A	T	V	L	T	A	L	G	G	I
Whale	L	K	K	H	G	V	T	V	L	T	A	L	G	A	I

Number	76	77	78	79	80	81	82	83	84	85	86	87	88	89	90
Human	L	K	K	G	H	H	E	A	E	I	K	P	L	A	
Whale	L	K	K	G	H	H	E	A	E	L	K	P	L	A	

Number	91	92	93	94	95	96	97	98	99	100	101	102	103	104	105
Human	Q	S	H	A	T	K	H	K	I	P	V	K	Y	L	E
Whale	Q	S	H	A	T	K	H	K	I	P	I	K	Y	L	E

Number	106	107	108	109	110	111	112	113	114	115	116	117	118	119	120
Human	F	I	S	E	C	I	I	Q	V	L	Q	S	K	H	P
Whale	F	I	S	E	A	I	I	H	V	L	H	S	R	H	P

Number	121	122	123	124	125	126	127	128	129	130	131	132	133	134	135
Human	G	D	F	G	A	D	A	Q	G	A	M	N	K	A	L
Whale	G	N	F	G	A	D	A	Q	G	A	M	N	K	A	L

Number	136	137	138	139	140	141	142	143	144	145	146	147	148	149	150	151	152	153
Human	E	L	F	R	K	D	M	A	S	N	Y	K	E	L	G	F	Q	G
Whale	E	L	F	R	K	D	I	A	A	K	Y	K	E	L	G	Y	Q	G

FIGURE 5.16

The amino acid sequences of sperm whale myoglobin and human myoglobin. Single-letter abbreviations are used here for the amino acids; numbering of the amino acids starts at the N-terminus. Of the 153 amino acid residues, 128 (84%) are identical in humans and whales. If we include the 13 conservative substitutions (e.g., isoleucine for leucine), the two proteins are 92% similar.

examination of these sequences. First, proteins are *long* polypeptides. Sperm whale myoglobin contains 153 amino acids, as does human myoglobin; yet, these are among the smaller proteins. Some proteins have sequences extending for many hundreds or even thousands of amino acid residues. Second, although the two myoglobin sequences are similar, they are not identical. Their similarity is sufficient for each to serve the same biochemical function; therefore, we call each a *myoglobin*. But they are not quite the same, for many millions of years have passed since sperm whales and humans had a common ancestor. Proteins evolve, and they evolve by changes in their amino acid sequences. Some of these are called **conservative** changes; they conserve the chemical properties and/or size of the side chain (Asp for Glu, for example). Other, **nonconservative** changes (Asp for Ala, for example) may have more serious consequences. It must be noted that structural context is of great importance in determining the effects of changing the identity of an amino acid at a given position in the sequence of the protein. If the amino acid side chain is in the interior of the protein, even a "conservative" substitution could have dramatic consequences for the stability and/or function of the protein. A solvent exposed site in the protein may tolerate many side chain substitutions with little effect on the structure or function of the protein.

We must also emphasize the uniqueness of a given protein in a particular species of organism. Every sample of sperm whale myoglobin, taken from any sperm whale, has the same amino acid sequence (unless, by rare chance, a sample is taken from a whale that carries a mutated myoglobin gene).

Biochemists have come to understand the complex structure of the myoglobin molecule bit by bit, by analyzing successively higher levels of complexity in the protein structure. To begin any such study of a protein, it is necessary to prepare the protein in a pure form, free of contamination by other proteins or other cellular substances. Methods for doing this are described in Tools of Biochemistry 5A. Historically, the next step after purifying a protein would be to determine its amino acid composition, that is, the relative amounts of the different amino acids in the protein, and then the sequence of the amino acids in the polypeptide chain. These steps have been superseded by the routine mass determination of purified proteins by mass spectrometry. Modern mass spectrometers can also be used to determine the amino acid sequence of a purified protein. These procedures are described in Tools of Biochemistry 5B at the end of this chapter.

A complication arises in the sequence determination if the protein being studied contains more than one polypeptide chain. These chains may be held together by noncovalent interactions, as in the protein *hemoglobin,* which is made up of four myoglobin-like chains (see Chapter 7). Alternatively, chain connection can be by covalent bonds such as disulfide bonds. An example of such a protein is the hormone **insulin** (Figure 5.17). Tools of Biochemistry 6B describes methods for detecting such multichain proteins and how to separate them.

Although direct sequencing of purified proteins provided much of the early protein sequence information, biochemists are more and more turning to gene sequencing for such information. As described in Chapter 4, the primary structure of every protein is dictated by a particular gene. Because we now know the code that relates DNA sequence to protein sequence, determination of the

Some proteins contain two or more polypeptide chains held together by noncovalent or covalent bonds.

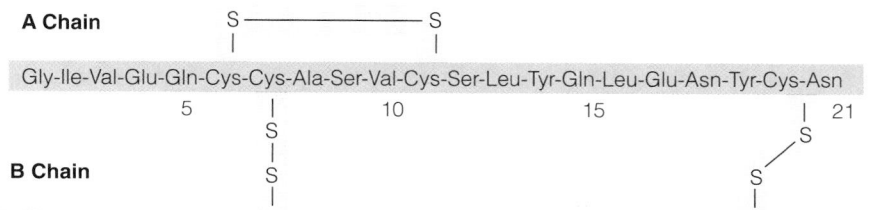

FIGURE 5.17

The primary structure of bovine insulin. This protein is composed of two polypeptide chains (A and B) joined by disulfide bonds. The A chain also contains an internal disulfide bond.

nucleotide sequence of a gene (or more often, the sequence of the messenger RNA transcribed from that gene) allows us to translate the nucleic acid sequence into the corresponding protein sequence. It should be kept in mind, however, that gene sequencing tells us only the sequence *as synthesized*. As we will see, there are often post-translational modifications of the polypeptide chain that are not revealed in this way and can be found only by analysis of the purified protein.

The techniques for identifying protein-encoding sequences in the genome and for retrieving, cloning, and sequencing the genes are discussed in Part V of this text (see also Tools of Biochemistry 4B).

From Gene to Protein

The Genetic Code

In Chapter 4 we stated that the DNA sequences of genes are transcribed into messenger RNA molecules, which are in turn translated into proteins. But there are only four kinds of nucleotides in DNA, each of which transcribes to a particular nucleotide in RNA, and there are 20 kinds of amino acids. Obviously, a 1:1 correspondence between nucleotide and amino acid is impossible. In fact, triplets of nucleotides (**codons**) are used to code for each amino acid, allowing 4^3, or 64, different combinations. This number is more than enough to code for 20 amino acids, so most amino acids have multiple codons. We can consider the information shown in Figure 5.18 to be the "standard" genetic code because nearly all terrestrial organisms use the same codons to translate their genomes into proteins. The few exceptions are scattered through the biological kingdoms. We discuss these and other details of the genetic code in Chapter 28.

Figure 5.18 depicts the genetic code in terms of the mRNA triplets that correspond to the different amino acid residues. Three triplets—UAA, UAG, and UGA—

The genetic code specifies RNA triplets that correspond to each amino acid residue.

FIGURE 5.18

The genetic code. The table is arranged so that users can quickly find any amino acid from the three bases of the mRNA codon (written in the 5′ → 3′ direction). For example, to find the amino acid corresponding to the codon 5′ AUC 3′, we look first in the A row, then in the U column, and finally in the C space, to find the amino acid to be Ile.

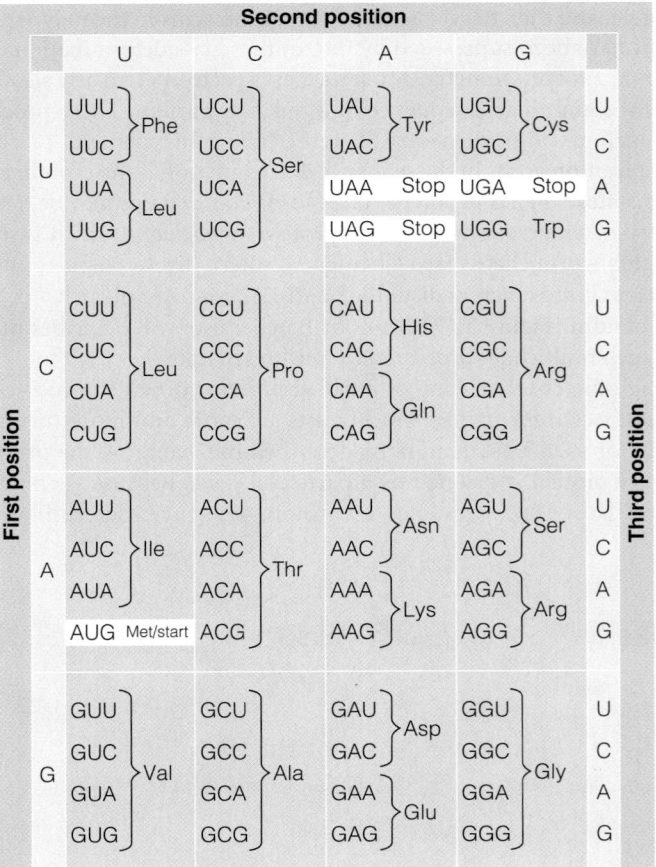

typically do not code for any amino acids but serve as "stop" signals to end translation at the C-terminus of the chain. In rare cases the codon UGA codes for seleno-cysteine and UAG codes for pyrrolysine. The codon AUG, which normally codes for methionine, also serves as a translation "start" signal that directs the placement of N-formylmethionine (in prokaryotes and eukaryotic organelles, see Figure 5.19) or methionine (in eukaryotes) at the N-terminal position. The implication is that all prokaryotic proteins should start with N-formylmethionine ("fMet"), and most eukaryotic proteins with methionine. Often this is true, though in many cases the N-terminal residue or even several residues are cleaved off in the cell by specific proteases during, or immediately after, translation. Figure 5.20 shows the relationship between DNA, mRNA, and polypeptide sequences for the N-terminal portion of human myoglobin. In this case the N-terminal methionine is removed.

Post-translational Processing of Proteins

When a polypeptide chain is released from a ribosome following translation, its synthesis is not necessarily finished. It must fold into its correct three-dimensional structure, and in some cases disulfide bonds must form. Certain amino acid residues may be acted upon by enzymes in the cell to produce, for example, the kinds of modifications shown on page 144.

Many proteins are further modified by specific proteolytic cleavage, which shortens the chain length. A remarkable example is found in the synthesis of insulin (Figure 5.21). We have encountered insulin as a two-chain protein held together by disulfide bonds (Figure 5.17). Insulin is actually synthesized as a single, much longer polypeptide chain, called **preproinsulin** (Figure 5.21, step 1). The residues at the N-terminus of the molecule (the exact number varies with the species) serve as a "signal peptide" (also called a **leader sequence**) to aid the transport of the preproinsulin molecule across the hydrophobic cell membranes. This transport is essential, for insulin is one of a class of proteins that function outside the cells in which they are synthesized. The leader sequence is then cut off by a specific protease, leaving **proinsulin** (step 2). Proinsulin folds into a specific three-dimensional structure, which helps it form the correct disulfide bonds (step 3). The connecting sequence between the A chain and the B chain is then cut out by further protease action, yielding the finished insulin molecule (step 4). This achieves an important physiological advantage. Because proinsulin is not an active hormone, it can be produced and stored in tissues at high concentrations where such high levels of active insulin would be toxic. The proinsulin can be quickly converted to insulin by proteolysis to allow for rapid secretion of active insulin when needed by the body.

The primary structure of a protein molecule is a sequence of *information*. The amino acid side chains can be thought of as words in a long sentence. These words

Translation of mRNA into protein sequence begins with an AUG codon, which encodes Met in eukaryotes and fMet in prokaryotes and eukaryotic organelles.

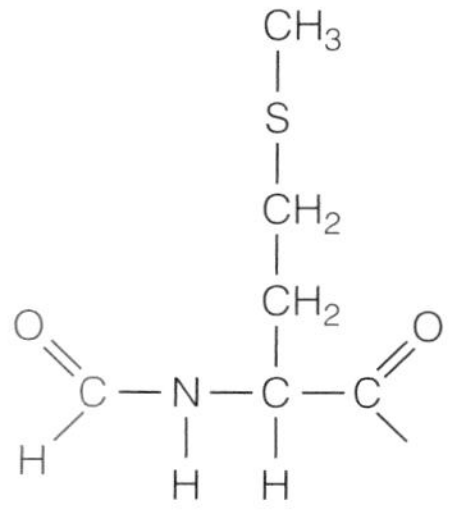

FIGURE 5.19

N-Formylmethionine. This amino acid residue initiates prokaryotic translation as well as in eukaryotic organelles. It is coded by AUG when that triplet appears at the start of a gene encoding a protein.

After being translated from mRNA, a protein may be modified in many ways—including cleavage of specific peptide bonds.

FIGURE 5.20

Relationships of DNA to mRNA to polypeptide chain. These relationships are shown for the first twelve residues of human myoglobin. Note that the DNA strand that is transcribed is the strand complementary to the final mRNA message.

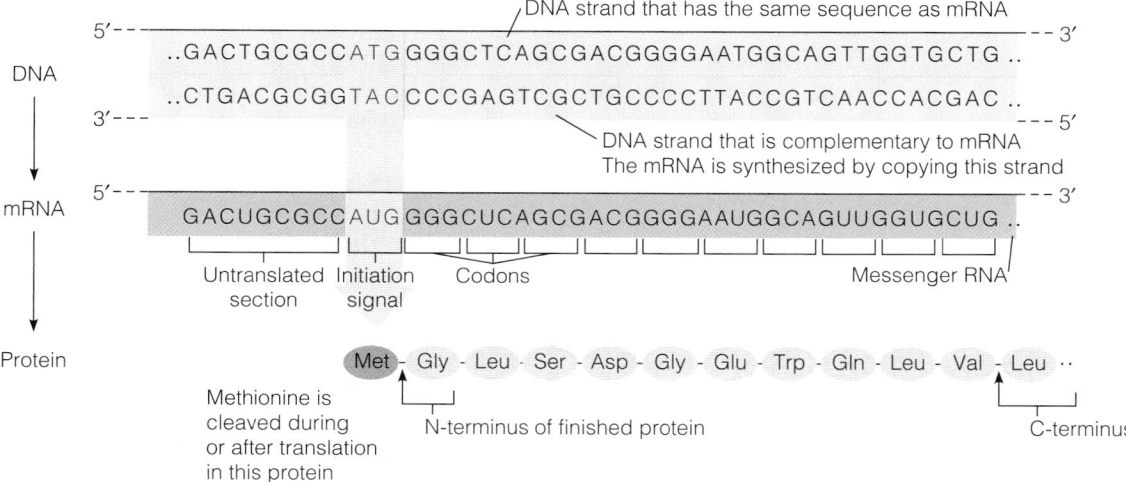

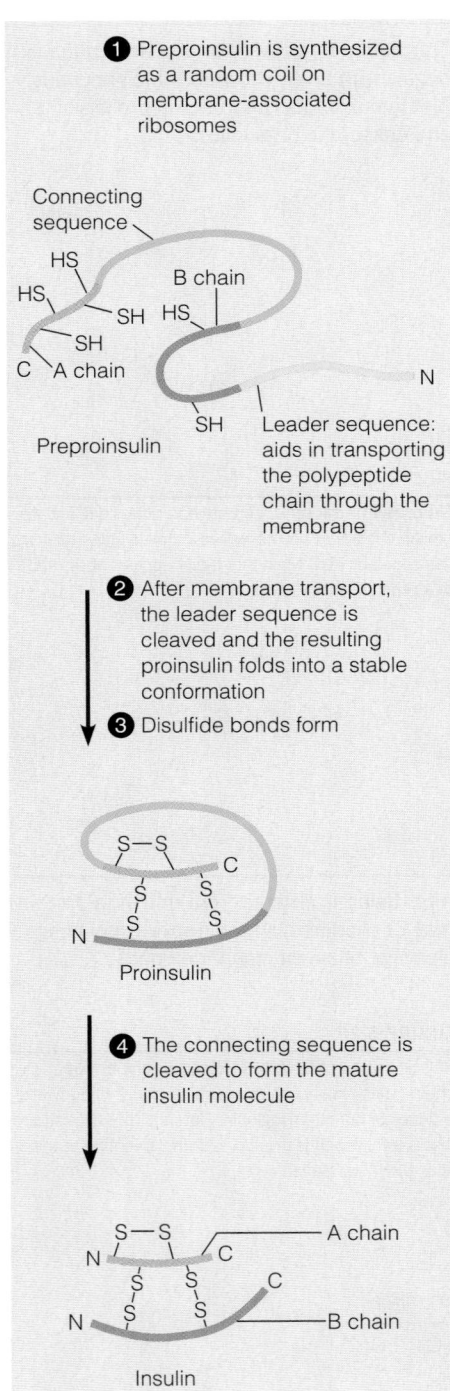

FIGURE 5.21

Structure of preproinsulin and its conversion to insulin.

have been translated from another language, the language of nucleic acid sequences stored in the genes and copied into messenger RNA. After translation, the sentence has been edited, with certain words modified and others deleted in the post-translational processing. In the next chapter we shall see that the information contained in the "sentence" of a protein sequence dictates how that protein folds in three dimensions. This folding, in turn, determines the function of the protein—how it interacts with small molecules and ions, with other proteins, and with substances like nucleic acids, carbohydrates, and lipids. The information expressed in protein sequences plays a primary role in determining how cells and organisms function.

From Gene Sequence to Protein Function

The development of **systems biology** has created a revolution in the molecular biosciences. The impact of systems biology on biochemistry will be described in much greater detail in later chapters of this textbook; however, we briefly introduce the topic here because vast amounts of protein primary structure information have been generated as a result of the genomics projects that have spurred the rapid growth of systems biology.

Historically, biochemistry has been a discipline in which the functions and/or structures of highly purified macromolecules are studied in isolation. Systems biology is directed at understanding the functions of biomolecules in the context of intact cells where, for example, several thousand different proteins, nucleic acids, and smaller molecules can potentially interact with, and thereby affect, cellular processes. Systems biology is a relatively new discipline that has grown out of the development of recent technologies. Mapping cellular interaction networks is a complex task that relies on large-scale computation resources for information management, as well as on laboratory instruments and techniques for generating large amounts of data.

To map possible interactions between biomolecules in a cell it is necessary to know what molecules are present in the cell. For this reason one of the earliest goals of systems biology was the sequencing of an entire genome for an organism. In 1995 a team led by J. Craig Venter at the Institute for Genome Research sequenced the first genome of a free-living organism, *Haemophilus influenzae* (1.8 million base pairs), and in 2001 two large teams, one led by Venter (then at a privately held company) and the other by Francis Collins (at the National Institutes of Health), jointly announced the draft sequence of the human genome (3 billion base pairs). The sequencing of the human genome is considered to be one of the most significant achievements in science. Since then, an exponential growth in genome sequencing has occurred. As of December 2011, complete sequences for 3334 microbial and 320 eukaryotic genomes had been reported.

Nearly all DNA sequences are deposited in public databases such as GenBank, and there are several web-based tools that researchers can use to search these databases (some of these are cited in the references at the end of this chapter). In addition, the Protein Data Bank (PDB) is a repository of protein and nucleic acid structures.[†] Throughout this text you will see representations of biomolecular structures that have been generated from the data in the PDB. The exponential growth of gene sequence information has had a profound effect on many areas of biochemistry. One such area is the prediction of protein function from translated gene sequences.

Because the technology to determine DNA sequence has developed to the point where entire genomes can be determined in a relatively short period of time,

---

[†]As of December 2011, there were more than 77,000 structures of biomolecules deposited in the PDB and over 132 billion nucleotide bases from more than 144 million individual sequences (obtained from more than 300,000 different organisms) deposited in GenBank (about 11 million of the sequences in GenBank are of human origin).

there is tremendous interest in identifying how much of a given genome encodes functional properties. For example, how much of the human genome encodes metabolic function, or tissue differentiation and growth, or cell signaling?

To attempt to answer these questions researchers can translate putative gene sequences into protein sequences, and then search protein sequence databases to find similar sequences. In most cases proteins with similar amino acid sequence possess similar structural and functional properties (e.g., the two myoglobins in Figure 5.16); thus, it is reasonable to use sequence similarity to propose functional properties for an uncharacterized protein.

What does it mean for two proteins to have "similar" amino acid sequence? We must first distinguish **sequence identity** from **sequence similarity**. In this context "identity" refers to those parts of the amino acid sequence that are an exact match (e.g., the amino acids highlighted in blue in Figure 5.16). The definition of "similarity" is less clear cut; but, as discussed earlier in this chapter, is based on the classification of the chemical properties of the side chains such as hydrophobicity, polarity, and charge. The classification of some amino acids is ambiguous because they possess more than one of these properties. Consider lysine, which has a charged amino group at the end of four hydrophobic methylene groups; thus, in one context lysine might be best characterized as hydrophobic and in a different context it would be considered charged.

Once a definition of "similarity" has been adopted, how is sequence similarity assessed? Sequence similarity is based on finding the best **alignment** of the sequences that are to be compared. For example, one can imagine a process in which two (or more) sequences are aligned and an amino acid similarity score for the alignment is calculated based on some scoring rubric (e.g., a "high" score for a perfect amino acid match, a "medium" score for a chemically similar amino acid, and a "low" score for a chemically dissimilar amino acid). The registry of the alignment is then changed and the similarity score is once again calculated. This iterative process continues until the alignment is found that gives the highest similarity score. For example, the alignment of the two myoglobin sequences in Figure 5.16 shows the maximum value for sequence similarity. This example is rather straightforward given the high degree of similarity between these two sequences. In many cases it is necessary to allow gaps in the alignment to get the best overall alignment score.

The alignment optimization process occurs many times during a protein database search. The output of such a search is a listing of several proteins from the database that are best matches based on a statistical analysis of sequence similarity. One widely used protocol for conducting database searches is the Basic Local Alignment Search Tool (BLAST). The suite of BLAST programs can be used to find the best alignment between the sequences of any two proteins, or more commonly, between the sequence of a "new" protein and sequences of proteins in several public databases. The results of a BLAST search can be used to infer potential function for an uncharacterized protein sequence; however, any proposed function must be validated experimentally.

To summarize, given a newly discovered gene sequence it is possible to make a best guess as to the function of the gene product based on a search of protein sequence databases using BLAST. The BLAST search generates sequence alignments and returns those alignments with the statistically significant sequence similarity scores. Similarities in sequence appear to be correlated with similarities in function and structure. As a rule of thumb, if two aligned protein sequences share at least 25% amino acid sequence identity they will have similar structure and, very likely, similar function. An example of a BLAST alignment between human myoglobin and human α globin is shown in Figure 5.22. The use of sequence information obtained from gene sequencing (Tools of Biochemistry 4B) or direct sequencing of proteins (Tools of Biochemistry 5B) to identify a protein is described further in Tools of Biochemistry 5D: A Brief Introduction to Proteomics.

The degree of similarity between protein sequences is determined from a procedure called "alignment."

Primary sequence analysis can be used to predict protein function because similarities in aligned sequences are correlated with similarities in protein function and structure.

Score = 30.8 bits (68), **Expect = 6e-06**, Method: Compositional matrix adjust.
Identities = 32/133 (25%), Positives = 48/133 (37%), Gaps = 40/133 (30%)

Human Mb   2   LSDGEWQLVLNVWGKVEADIPGHGQEVLIRLFKGHPETLEKFDKFKHLKSEDEMKASEDL   61
                 LS +   V WGKV A   +G E LR+F   PT   F   F
Human α   2   LSPADKTNVKAAWGKVGAHAGEYGAEALERMFLSFPTTKTYFPHF -------------------------   46

Human Mb   62   KKHGATVLTALGGILKKKGHHEAEIKPLAQSHATKHKI–PVKYLEFISECIIQVLQSKHP   120
                     L+AL   I   HA K ++ PV + + +S C++ L +   P
Human α   47   ----------- ALSALSDI------------------------------ HAHKLRVDPVNF–KLLSHCLLVTLAAHLP   82

Human Mb   121 GDFGADAQGAMNK   133
                 +F   +++K
Human α   83   AEFTPAVHASLDK   95

FIGURE 5.22

BLAST alignment between human myoglobin and human α globin sequences. The sequence of myoglobin is above that for the α globin, and between them are shown identical amino acids (blue) and those that are considered similar (green "+"). Gaps in the alignment are shown by red text and red dashes. In this alignment there is 25% sequence identity, suggesting a high degree of structural similarity between the proteins. The significance of the alignment is given by the "Expect" value, which is a measure of the likelihood that the alignment is due to chance. The lower this value, the more significant is the alignment.

Protein sequence analysis can be used to propose evolutionary relationships between organisms.

Protein Sequence Homology

Protein sequence similarity is also used to map evolutionary relationships between organisms. Two organisms that have a common evolutionary ancestry are likely to have gene sequences, and therefore protein sequences, that are related. Protein sequences are classified as "homologous" in cases where any sequence similarity is thought to be the result of a common evolutionary ancestry. Note that sequence similarity can also arise without common ancestry via convergent evolution (e.g., two proteins may have evolved independently to bind to the same peptide sequence found on a particular transcription factor). Two such protein sequences would be *similar* but not *homologous*; hence, "sequence similarity," which is based on sequence alignment, is distinct from "sequence homology," which is used only to indicate an evolutionary relationship between two sequences.

For organisms that have a common ancestor, a greater degree of protein sequence homology indicates a closer evolutionary relationship; whereas a lesser degree of homology indicates greater divergence. Figure 5.23a shows an alignment of sequences for cytochromes *c* from 27 different organisms. In this figure the hydrophobic (gray), positively charged (blue), negatively charged (red), and polar uncharged (green and magenta) amino acids are highlighted to illustrate the clustering of these similar amino acid properties in this alignment. Figure 5.23b shows a phylogenetic tree for the sequences, where branches show points of proposed evolutionary divergence based on differences in the amino acid sequences of the aligned proteins. The construction of such trees involves complex analyses of gene mutation rates and polymorphism within populations—topics beyond the scope of this introduction to protein sequence analysis. The important point for this discussion is that protein sequence analysis can be used to propose evolutionary relationships between organisms, such as those shown in Figure 5.23b.

Finally, when several homologous protein sequences are aligned, one can determine the so-called **consensus sequence**. A common representation of the consensus sequence lists the amino acid that is most frequently found at a given position in the sequence; however, this representation can be an oversimplification. If a certain amino acid is found at a specific position in 80% of the aligned sequences, that means that 20%, a significant percentage, of the sequences do not have that amino acid at that position. A more information-rich representation is the **sequence logo** in which the relative size of the one-letter code is correlated with the relative frequency of an amino acid at a given position. Figure 5.24 shows a sequence logo for an alignment of 412 sequences of proteins from the

(a)

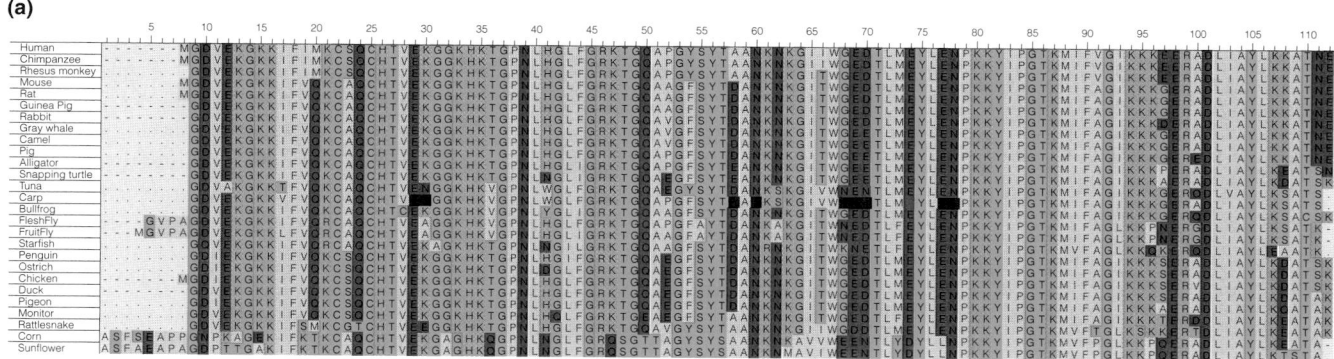

(b)

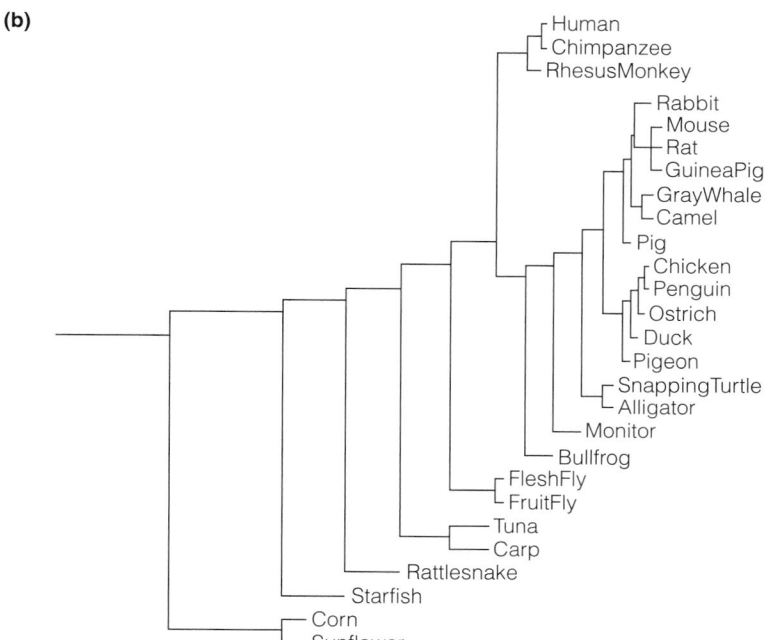

FIGURE 5.23

Sequence alignment and a phylogenetic tree for cytochromes *c* from different organisms. **(a)** Alignment of cytochrome *c* sequences from 27 organisms, where hydrophobic amino acids are highlighted in gray, basic amino acids in blue, acidic amino acids in red, and polar uncharged amino acids in green (except for Asn and Gln, which are magenta). **(b)** A phylogenetic tree for the sequences shown in (a). Branches indicate points of evolutionary divergence based on differences in the amino acid sequences of the aligned proteins. Both the alignment and the phylogenetic tree were created using CLUSTALW2 and CINEMA (both available via the ExPASY Website; see References).

cytochrome *c* family, where the amino acid found most frequently at any position is shown at the top of each column.

The sequence logo shows better the degree of **amino acid conservation** within the aligned sequences. If only one amino acid is found at a given position within an alignment of homologous proteins, that amino acid is said to be absolutely "conserved" (e.g., cysteine at position 17 and histidine at position 18 in Figure 5.24). An amino acid type can also be conserved at a given position. For example, only leucine, valine, isoleucine, and methionine are found at position 80 in the 412 aligned cytochrome *c* sequences; thus, hydrophobic amino acids are conserved at this position. As we will discuss further in Chapter 6, the most highly conserved amino acids tend to be those that serve a critical structural and/or functional role in the protein.

FIGURE 5.24

Sequence logo from the alignment of 412 sequences from the cytochrome *c* family.

The amino acid found most frequently at a given position is shown on top and the relative size of the one letter code is correlated with the relative frequency of that amino acid in that position. The sequence logo was generated using the WEBLOGO program from data available in the Prosite database (matrix alignment #51007; see the ExPASY Website).

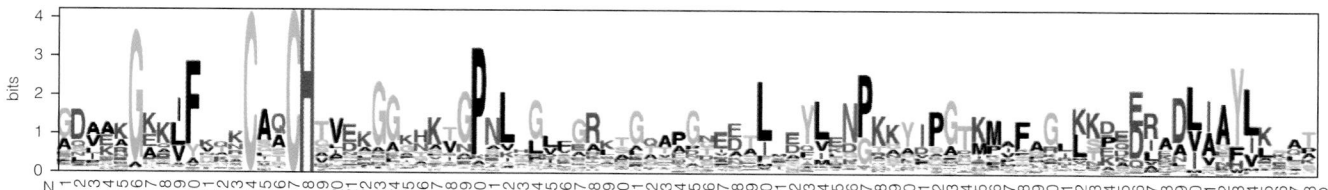

SUMMARY

Proteins are polymers of L-α-amino acids. Twenty common amino acids and two rare amino acids are coded for in genes and incorporated into proteins by the process of translation. Other, less common, amino acid structures are found in nature (e.g., in bacterial cell walls, antibiotics, and venoms). The variety of side chains—hydrophilic, hydrophobic, acidic, basic, or neutral—allows much functional complexity in proteins. Additional variation is made possible by post-translational modification of some amino acids. The presence of both positive and negative charges on side chains makes proteins polyampholytes.

Oligopeptides and polypeptides are produced by polymerization of amino acids via peptide bond formation. The peptide bond is nearly planar, and the *trans* form is favored. This bond is metastable and can be readily hydrolyzed in the presence of catalysts. The unique, defined sequence of amino acids in each protein constitutes its primary structure, dictated by its gene. Some proteins contain more than one polypeptide chain, held together by either covalent (disulfide) or noncovalent bonds. Proteins are synthesized in the cell by an ATP-dependent process called translation. The genetic code is made up of a standard set of codons of three nucleotides, each of which specifies a particular amino acid. Specific "start" and "stop" triplets in the genetic code specify the chain length for a particular protein.

Even after translation has been completed, the protein molecule can be further covalently modified, either by cleavage of particular portions or by modification of certain amino acid side chains.

Genome sequencing has generated vast amounts of gene sequence information. The functions of most of these genes are unknown; however, sequence similarity analysis can be used to predict the function, and in some cases the structure, of a gene product.

REFERENCES

General

A number of excellent books provide more detailed or supplementary information on protein structure and function. We particularly recommend the following to supplement our Chapters 5, 6, and 7.

Brändén, C., and J. Tooze (1999) *Introduction to Protein Structure* (2nd ed.). Garland, New York. Contains much information on all levels of structure. Excellent illustrations.

Creighton, T. E. (1993) *Proteins: Structure and Molecular Properties* (2nd ed.). Freeman, San Francisco. An elegant, thorough exposition of all aspects of protein chemistry. Many good references throughout the text.

Fersht, A. (1999) *Structure and Mechanism in Protein Science: A Guide to Enzyme Catalysis and Protein Folding*. Freeman, New York. An excellent and very readable introduction to the fundamental theoretical principles of protein folding and catalysis.

Kyte, J. (1995) *Structure in Protein Chemistry*. Garland, New York. An excellent treatise on protein structure.

Liljas, A., L. Liljas, J. Piskur, G. Lindblom, P. Nissen, and M. Kjeldgaard (2009) *Textbook of Structural Biology*. World Scientific Publishing, Singapore. Brief treatment on basics of protein structure; but gives a broad and reasonably detailed overview of protein structures and functions.

Petsko, G. A., and D. Ringe (2004) *Protein Structure and Function*. New Science Press, London. Concise and clearly written. Excellent illustrations complete with Protein DataBank ID codes.

Reviews and Papers on Amino Acid Properties

Greenstein, J. P., and M. Winitz (1961) *Chemistry of the Amino Acids*. Wiley, New York.

Hegstrom, R. A., and D. K. Kondepudi (1990) The handedness of the universe. *Sci. Am.* January:98–105. A clear discussion of theories of stereopreference.

Rose, G. D., A. R. Geselowitz, G. J. Lesser, R. H. Lee, and M. H. Zehfus (1985) Hydrophobicity of amino acid residues in globular proteins. *Science* 229:834–838.

Saghatelian, A., Y. Yokobayashi, K. Soltani, and M. R. Ghadiri (2001) A chiroselective peptide replicator. *Nature* 409:797–801. A paper suggesting there is a natural tendency to favor homochiral products in self-replicating processes (see also the News and views article: J. S. Siegel (2001) *Nature* 409:777–778).

Wilbur, P. J., and A. Allerhand (1977) Titration behavior and tautomeric states of individual histidine residues of myoglobin. *J. Biol. Chem.* 252:4968–4975.

Uncommon Amino Acids in Proteins

Blight, S. K., R. C. Larue, A. Mahapatra, D. G. Longstaff, E. Chang, G. Zhao, P. T. Kang, K. B. Green-Church, M. K. Chan, and J. A. Krzycki (2004) Direct charging of tRNA_{CUA} with pyrrolysine *in vitro* and *in vivo*. *Nature* 431:333–335. See also the News and Views article: Schimmel, P., and K. Beebe (2004) Genetic code seizes pyrrolysine. *Nature* 431:257–258.

Diwadkar-Navsariwala, V., and A. M. Diamond (2004) The link between selenium and chemoprevention: A case for selenoproteins. *J. Nutr.* 134:2899–2902. A review of the putative anticancer effects of seleno-proteins in humans.

Hatfield, D. L., and V. N. Gladyshev (2002) How selenium has altered our understanding of the genetic code *Mol. Cell. Biol.* 22:3565–3576. Describes the mechanism for insertion of selenocysteine into proteins.

Milton, R. C. deL., S. C. F. Milton, and S. B. H. Kent (1992) Total chemical synthesis of a D-enzyme: The enantiomers of HIV-1 protease show demonstration of reciprocal chiral substrate specificity. *Science* 256:1445–1448.

Pisarewicz, K., D. Mora, F. C. Pflueger, G. B. Fields, and F. Mari (2005) Polypeptide chains containing D-γ-hydroxyvaline. *J. Am. Chem. Soc.* 127:6207–6215.

Sandman, K. E., D. F. Tardiff, L. A. Neely, and C. J. Noren (2003) Revised *Escherichia coli* selenocysteine insertion requirements determined by *in vivo* screening of combinatorial libraries of SECIS variants. *Nucleic Acids Res.* 31:2234–2241. A more detailed description of the downstream gene sequence requirements for selenocysteine insertion into proteins.

Srinivasan, G., C. M. James, and J. A. Krzycki (2002) Pyrrolysine encoded by UAG in Archaea: Charging of a UAG-decoding specialized tRNA. *Science* 296:1459–1462. See also the Perspectives article: Atkins, J. F., and R. Gesteland (2002) The 22nd amino acid. *Science* 296:1409–1410.

Stadtman, T. C. (1987) Specific occurrence of selenium in enzymes and amino acid tRNAs. *FASEB J.* 1:375–379.

Stadtman, T. C. (2002) A gold mine of fascinating enzymes: Those remarkable, strictly anaerobic bacteria, *Methanococcus vannielii* and *Clostridium sticklandii*. *J. Biol. Chem.* 277:49091–49100. An historical reflection on selenium in proteins with many references.

Turanov, A. A., A. V. Lobanov, D. E. Fomenko, H. G. Morrison, M. L. Sogin, L. A. Klobutcher, D. L. Hatfiled, and V. M. Gladyshev (2009) Genetic code supports targeted insertion of two amino acids. *Science* 323:259–261.

Wolosker, H., E. Dumin, L. Balan, and V. N. Foltyn (2008) D-Amino acids in the brain: D-Serine in neurotransmission and neurodegeneration. *FEBS J.* 275:3514–3526.

Zhang, Y., and V. N. Gladyshev (2007) High content of proteins containing 21st and 22nd amino acids, selenocysteine and pyrrolysine, in a symbiotic deltaproteobacterium of gutless worm *Olavius algarvensis*. *Nucleic Acids Res.* 35:4952–4963. See also: Atkins, J. F., and P. V. Baranov (2007) Duality in the genetic code. *Nature* 448:1004–1005.

Zinoni, F., W. Birkmann, W. Leinfelder, and A. Böck (1987) Cotranslational insertion of selenocysteine into formate dehydrogenase from *Escherichia coli* directed by a UGA codon. *Proc. Nat'l Acad. Sci. USA* 84:3156–3160.

Sequencing of Genomes

Fleischmann, R. D., et al. (1995) Whole-genome random sequencing and assembly of *Haemophilus influenzae* Rd. *Science* 269:496–512. Describes the first entire genome sequence of a free-living organism.

Lander, E. S., et al. (2001) Initial sequencing and analysis of the human genome. *Nature* 409:860–921. One of two simultaneous reports—this one from the International Human Genome Sequencing Consortium. This issue of *Nature* contains several articles of interest regarding the significance and interpretation of the results of this project.

Roberts, L. (2001) Controversial from the start. *Science* 291:1182–1188. An interesting review of some of the controversies surrounding the human genome sequencing project.

Venter, J. C., et al. (2001) The sequence of the human genome. *Science* 291:1304–1351. One of two simultaneous reports—this one from a privately held company.

URLs for access to public sequence databases, sequence alignment tools, and other protein analysis tools (e.g., mass and/or pI calculations, DNA sequence translation, etc.):

GenBank:	www.ncbi.nlm.nih.gov/Genbank/index.html
Sequences of entire genomes:	www.ncbi.nlm.nih.gov/Genomes/index.html
Sequence alignment (BLAST):	blast.ncbi.nlm.nih.gov/Blast.cgi
Several proteomics tools:	www.expasy.ch

Database Searching, Sequence Alignment, and Similarity Scoring

Altschul, S. F., W. Gish, W. Miller, E. W. Myers, and D. J. Lipman (1990) Basic local alignment search tool. *J. Mol. Biol.*, 215:403–410.

Altschul, S. F., M. S. Boguski, W. Gish, and J. C. Wooten (1994) Issues in searching molecular sequence databases. *Nature Genet.* 6:119–129.

Altschul, S. F., T. L. Madden, A. A. Schäffer, J. Zhang, Z. Zhang, W. Miller, and D. J. Lipman (1997) Gapped BLAST and PSI-BLAST: A new generation of protein database search programs. *Nucleic Acids Res.* 25:3389–3402.

Benson, D. A., I. Karsch-Mizrachi, D. J. Lipman, J. Ostell, and E. W. Sayers (2009) GenBank. *Nucleic Acids Res.* 37:D26–D31. Every January, *Nucleic Acids Research* publishes a review of current molecular sequence databases. This article describes the sources and quality of the sequences deposited in GenBank.

Crooks, D. E., G. Hon, J. M. Chandonia, and S. E. Brenner (2004) Weblogo: A sequence logo generator. *Genome Res.* 14:1188–1190. (see: **weblogo.berkeley.edu**)

Gonnet, G. H., M. A. Cohen, and S. A. Benner (1992) Exhaustive matching of the entire protein sequence database. *Science* 256:1443–1445.

Henikoff, S., and J. G. Henikoff (1992) Amino acid substitution matrices from protein blocks. *Proc. Natl. Acad. Sci. USA* 89:10915–10919.

Sayers, E. W., et al (2009) Database resources of the National Center for Biotechnology Information. *Nucleic Acids Res.* 37:D5–D15. This article describes the various databases available at NCBI (which is the host for BLAST).

Schneider, T. D., and R. M. Stephens (1990) Sequence logos: A new way to display consensus sequences. *Nucleic Acids Res.* 18:6097–6100.

Ye, J., S. McGinnis, and T. L. Madden (2006) BLAST: Improvements for better sequence analysis. *Nucleic Acids Res.* 34:6–9.

PROBLEMS

Note that some of these problems refer to information presented in the Tools of Biochemistry 5A-D.

1. Using the data in Table 5.1, calculate the *average* amino acid residue weight in a protein of typical composition. This is a useful number to know for approximate calculations.

2. The melanocyte-stimulating peptide hormone *α-melanotropin* has the following sequence:

    Ser–Tyr–Ser–Met–Glu–His–Phe–Arg–Trp–Gly–Lys–Pro–Val

    (a) Write the sequence using the one-letter abbreviations.
    (b) Calculate the molecular weight of *α*-melanotropin, using data in Table 5.1.

*3. (a) Sketch the titration curve you would expect for *α*-melanotropin (Problem 2). Assume the pK_as of the N- and C-termini are, respectively, 7.9 and 3.8. For side chains, assume the pK_a values given in Table 5.1.
    (b) Calculate to 3 decimal places the charge on *α*-melanotropin at pH values of 11, 5, and 1.
    (c) Calculate the isoelectric point of *α*-melanotropin.

    For parts (b) and (c) of this problem refer to the logic used to solve Problem 2.16. You may find a spreadsheet useful to solve part (c).

4. What peptides are expected to be produced when *α*-melanotropin (Problem 2) is cleaved by (a) trypsin, (b) cyanogen bromide, and (c) thermolysin? (Refer to Table 5.4.)

5. There is another melanocyte-stimulating hormone called *β-melanotropin*. Cleavage of *β*-melanotropin with trypsin produces the following peptides plus free aspartic acid.

    WGSPPK   DSGPYK   MEHFR

    If you assume maximum sequence similarity between *α*-melanotropin and *β*-melanotropin, what must be the sequence of the latter?

6. Given the following peptide

    SEPIMAPVEYPK

    (a) Estimate the net charge at pH 7 and at pH 12. Assume the pK_a values given in Table 5.1.
    (b) How many peptides would result if this peptide were treated with (1) cyanogen bromide, (2) trypsin, and (3) chymotrypsin?
    (c) Suggest a method for separating the peptides produced by chymotrypsin treatment.

*7. A mutant form of polypeptide hormone angiotensin II has the amino acid *composition*

    (Asp, Arg, Ile, Met, Phe, Pro, Tyr, Val)

    The following observations are made:
    • Trypsin yields a dipeptide containing Asp and Arg, and a hexapeptide with all the rest.
    • Cyanogen bromide cleavage yields a dipeptide containing Phe and Pro, and a hexapeptide containing all the others.
    • Chymotrypsin cleaves the hormone into two tetrapeptides, of composition

    (Asp, Arg, Tyr, Val) and (Ile, Met, Phe, Pro)

    • The dipeptide of composition (Pro, Phe) cannot be cleaved by either chymotrypsin or carboxypeptidase.

    What is the sequence of angiotensin II?

8. A protein has been sequenced after cleavage of disulfide bonds. The protein is known to contain 3 Cys residues, located as shown below. Only one of the Cys has a free —SH group and the other two are involved in an —S—S— bond.

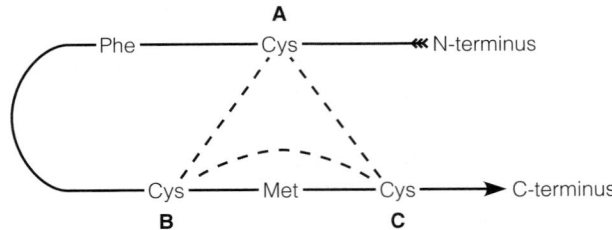

The only methionine and the only aromatic amino acid (Phe) in this protein are in the positions indicated. Cleavage of the *intact* protein (i.e., with disulfide bonds intact) by either cyanogen bromide or chymotrypsin does *not* break the protein into two peptides. Where is the —S—S— bond (AB, BC, or AC)?

9. *Apamine* is a small protein toxin present in the venom of the honeybee. It has the sequence

    CNCKAPETALCARRCQQH

    (a) It is known that apamine does not react with iodoacetate (see page 169). How many disulfide bonds are present?
    (b) Suppose trypsin cleavage gave two peptides. Where is(are) the S—S bond(s)?

10. (a) Write a possible sequence for an mRNA segment coding for apamine.
    (b) Do you think apamine is synthesized in the form shown in Problem 9, or is it more likely a product of proteolytic cleavage of a larger peptide? Explain.

11. Assume the following portion of an mRNA. Find a start signal, and write the amino acid sequence that is coded for.

    5′ GCCAUGUUUCCGAGUUAUCCCAAAGAUAAAAAAGAG 3′

12. A researcher has isolated an oligopeptide of unknown sequence and unknown amino acid composition. All attempts to sequence it by Edman degradation fail; the reaction just will not go.
    (a) Suggest two quite different possible explanations for the problem.
    (b) Suggest an experiment that will at least distinguish between these two.

*13. Suppose you had separated the A and B chains of insulin by disulfide reduction. What chromatographic method should allow the isolation of pure A and B chains? Explain your choice of separation method.

14. Sickle cell disease is caused by a so-called "point mutation" in the human *β*–globin gene. A point mutation is the result of a single base substitution in the DNA encoding a gene. The sickle cell mutation results in substitution of Val for Glu at position 6 in the *β*–globin protein.
    (a) Using the information in Figure 5.18 explain how a point mutation could change a codon for Glu to a codon for Val.
    (b) Do you expect the pI for the sickle cell *β*–globin to be higher or lower than the pI for wild-type *β*–globin? Explain.

*15. You have discovered a novel protein that has a pI = 5.5. To study the functional properties of this new protein, your research group has made a mutant that contains two amino acid changes: a surface phenylalanine residue in the normal protein has been replaced by histidine (side chain pK_a = 6.1), and a surface glutamine has been replaced by glutamic acid (side chain pK_a = 6.0). Is the pI of the mutant protein predicted to *be greater than, less than,* or *the same as* the pI of normal protein? Support your answer with the appropriate calculation.

TOOLS OF BIOCHEMISTRY 5A

PROTEIN EXPRESSION AND PURIFICATION

Much of the information presented in this text was gained through the study of highly purified protein molecules. To determine the structure and/or functional properties of a specific protein it is necessary to separate that protein from the other biomolecules in the cell (lipids, nucleic acids, saccharides) and, of course, all other proteins. In a typical cell the protein of interest is usually a minor component; thus, the isolation of that protein from such a complex mixture presents a challenge. Modern methods of gene expression and protein purification have simplified the problem by increasing the concentration of the desired protein within the cell and exploiting specific interactions between the protein and materials used in the purification process.

Recombinant Protein Expression

To begin, let's consider the problem of protein concentration. A typical enzyme may represent only 0.01% of the soluble protein in a cell. Thus, a 10,000-fold enrichment is necessary to purify that protein to homogeneity. If recombinant DNA technology (see Tools of Biochemistry 4B) can increase the intracellular abundance of that protein to 1%, then only a 100-fold enrichment is required; if to 10%, then a 10-fold enrichment will suffice. Historically, proteins were purified from natural sources, such as animal or plant tissues. The first proteins to be studied in detail were those with high abundance in particular tissues (e.g., hemoglobin in red blood cells). For proteins that are present in low concentration in their natural tissues, it is necessary to harvest large amounts of the tissue to isolate a useful amount of the desired protein. This unfortunate scenario was a fact of life for biochemists for many years until the tools of "recombinant protein expression" were developed and widely adopted in the late 1970s and early 1980s. Recombinant protein expression allows researchers to produce proteins of interest at relatively high concentrations within cells and also enables the production of so-called **site-directed mutants**, which are protein variants with designed amino acid sequence alterations (see Tools of Biochemistry 4B). Frequently, the mutant proteins exhibit changes in structure, function, and/or stability relative to the "wild-type" (i.e., naturally occurring) protein; thus, the mutants are of great scientific interest. Another important feature of recombinant technology is the ability to express a wide variety of foreign proteins in host cells. For example, many proteins of animal or plant origin have been successfully expressed in *Escherichia coli* cells. Because *E. coli* can be easily programmed to produce foreign proteins and they grow quickly compared to most plants and animals, *E. coli* cells can be viewed as convenient "factories" for protein production; however, *E. coli* cells are limited in the types of post-translational modifications they carry out. It is usually necessary to express proteins in eukaryotic systems if some post-translational modification (e.g., glycosylation) is required for activity.

Recombinant protein expression technology is based on the observation that the amino acid sequence for a protein is determined by the sequence of the DNA in the gene that encodes that protein, as described in Chapters 4 and 5 (see Figures 4.23, page 112, and 5.18, page 152). In theory, any protein sequence can be expressed in a cell that contains a copy of the gene encoding that protein. *E. coli* can be made to take up small circular DNA molecules, called "expression vectors", that are on the order of 2–10 kilobases in length. An expression vector is a modified form of a natural extrachromosomal DNA, such as a plasmid, which is capable of autonomous replication in a bacterial cell. Recombinant DNA technology allows a researcher to cut open that plasmid at a desired site and splice in a gene encoding the protein of interest. As shown in Figure 5A.1, the gene encoding the wild-type or mutant protein to be expressed is within each vector, along with a gene encoding a so-called **selection marker**. The selection marker is usually a protein that confers resistance to an antibiotic that is included in the cell growth medium; thus, only those cells that have taken up the vector, and are thereby capable of expressing the desired protein, will survive in the growth medium. With many tens to hundreds of copies of the vector in each cell, production of the desired protein is maximized. Using this approach, even those proteins that occur in low intracellular concentrations in nature can be produced in sufficient yield to allow biochemical characterization and/or commercial production.

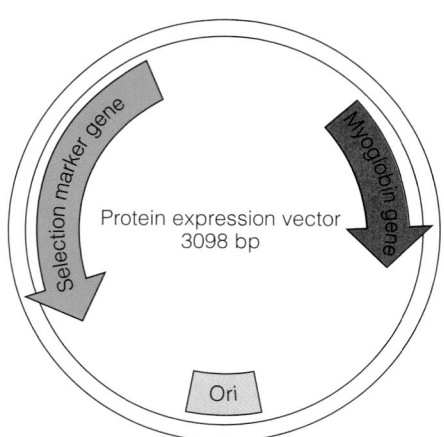

FIGURE 5A.1

Schematic representation of a generic protein expression vector. The circle represents the double-stranded DNA sequence for the entire vector. The box marked "ori" is the "origin of replication", which determines how many copies of the vector will be made in the cell. The arrows represent locations of protein genes in the vector DNA sequence. This vector will express two proteins: the recombinant protein of interest (red arrow, in this case, myoglobin) and a so-called selection marker (green arrow).

The Purification Process

Although recombinant technology can increase the concentration of a specific protein inside the cell, the problem of separating the desired protein from all the other cellular components remains. The sequence of steps taken to purify a given protein will be unique to that protein because proteins vary in chemical properties; however, many features of the purification process are common. For example, most purification steps begin with lysis, or rupture, of the cells. Cell lysis can be achieved by sonic disruption ("sonication"), mechanical rupture using a homogenizer, or by enzymatic digestion of the cell wall. This is followed by centrifugation to remove unbroken cells and insoluble cell parts (e.g., membranes) to yield an extract, called the "cell lysate," which contains the soluble proteins and other biomolecules in a cell. The desired protein is then purified from the other proteins in the lysate by one or more of the following commonly used steps: (1) affinity chromatography, (2) ion exchange chromatography, or (3) size exclusion chromatography. As illustrated in Figure 5A.2, purification of the desired protein by chromatography is the result of differential interactions between the various proteins in the mixture loaded onto the chromatography column and the matrix within the column. In general, the more strongly some protein interacts with the matrix, the later it will elute from the column. Proteins are generally detected by UV absorbance at 280 nm, or 220 nm, as they elute from the column.

Affinity Chromatography

Affinity chromatography relies upon selective adsorption of a protein to a natural or synthetic ligand, typically a substrate or inhibitor, which is immobilized by covalent attachment to an inert solid support. The support that displays the bound ligand is called the "affinity matrix." The interactions between the desired protein

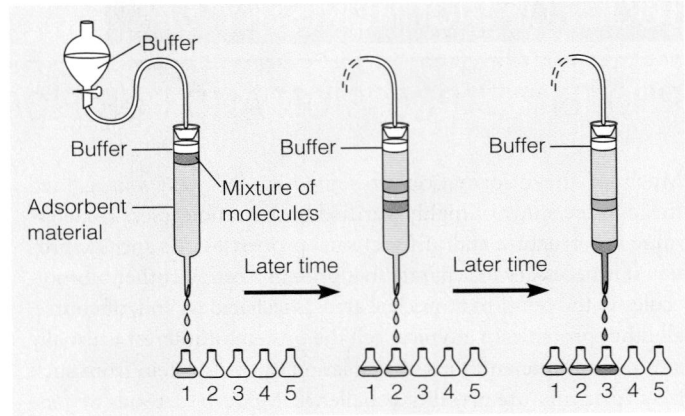

FIGURE 5A.2

The principle of column chromatography.

and the affinity matrix are expected to be highly specific; thus, most of the contaminants in the mixture will not interact with the affinity matrix, whereas the desired protein is expected to bind tightly. As shown in Figure 5A.3, when a complex mixture flows through the affinity matrix the desired protein will bind tightly and remain bound until most contaminants are washed through the column. The bound protein can then be eluted using a variety of methods that preserve the structure and activity of the protein. In some exceptional cases the elution methods require extreme conditions to disrupt the bonds between the protein and the matrix, including denaturation of the protein.

Specific, complementary, noncovalent interactions are the basis for much interesting biological chemistry, including antibody binding, enzyme-substrate recognition, enzymatic catalysis, gene regulation, cell signaling, and muscle contraction, just to name a

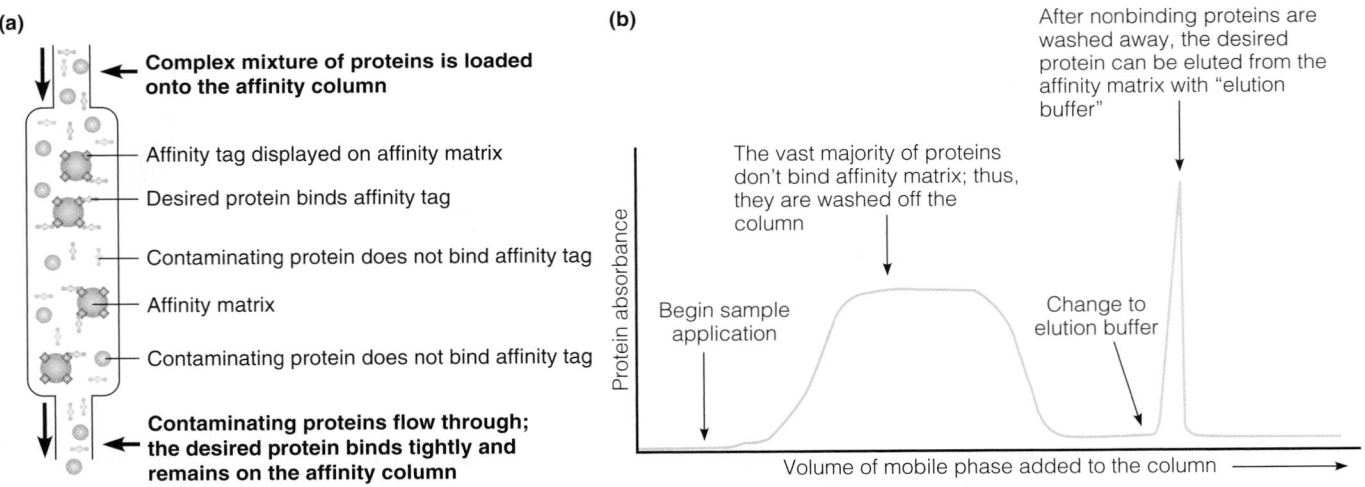

FIGURE 5A.3

(a) A simplified view of affinity chromatography. Specific binding of the desired protein (blue shapes) to the affinity matrix is shown. The contaminating proteins (green shapes, beige circles) wash through the column without binding, resulting in a significant purification of the desired protein. The bound protein can then be eluted by any one of several methods discussed in the text. **(b)** A schematic representation of an affinity chromatogram. The proteins that don't bind the affinity matrix are washed off and elute early, giving rise to a large protein absorbance. After the contaminants are washed off, the desired protein is eluted, giving a smaller protein absorbance.

few examples. Early in the development of affinity purification methods, many investigators immobilized the natural binding targets ("ligands") for a protein on a solid chromatography support to take advantage of such specific binding interactions. A related technique, called "immunoaffinity" chromatography, takes advantage of the high binding specificity of antibodies for their ligands. Antibodies raised against a given protein can be covalently attached to a solid support to make an affinity matrix that binds selectively and reversibly to that protein. Although immunoaffinity columns are efficient, the costs and time required to produce the antibodies are significant; thus, this method has largely given way to faster and less expensive techniques.

One of the most prevalent affinity methods, immobilized metal affinity chromatography (IMAC), takes advantage of the strong interactions between a Ni^{2+}, Zn^{2+}, or Co^{2+} ion and a string of six sequential histidine residues. The amino acid sequence $(His)_6$ is called a "hexahistidine-tag," (or "His-tag") and it is quite rare in nature; thus, few naturally occuring proteins will bind to the IMAC matrix. Using recombinant DNA technology, the His-tag can be appended to the gene encoding the desired protein, as shown in Figure 5A.4. When this protein is expressed it will include the His-tag sequence and therefore it will bind tightly to the IMAC matrix. The bound protein can be eluted from the IMAC column with a buffer containing imidazole (an analog of the His side chain), or a low pH buffer (which protonates the His side chains and reduces metal ion binding), or with a buffer containing the metal chelator EDTA. EDTA effectively removes the metal ion from the column and thereby disrupts the bonding interactions between the protein and the affinity matrix.

The His-tag affinity method described above is used extensively to produce purified proteins for biochemical characterization. This example illustrates the power of recombinant technol-ogy not only for increasing protein concentration in cells, but also for improving the protein-purification process. One possible limitation to this method is that the resulting protein sequence carries some modification compared to the wild-type, in this case, an extra six histidine amino acids.

Affinity chromatography is so efficient that in many cases no further purification steps are required; however, further steps may be required to achieve separation of the desired protein from those contaminants that are closely related (e.g., proteins that differ by post-translational modifications).

Ion Exchange Chromatography (IEC)

Ion-exchange chromatography is used to separate molecules on the basis of their electrical charge. The strength of interaction between a protein molecule and an ion exchange matrix depends on (1) the charge density on the protein and (2) the ionic strength of the mobile phase, which is always a buffered solution. The charge density on the protein is modulated by altering the pH of the solution (see Chapter 2). Recall that the overall charge on a protein is zero when the pH of a protein solution is equal to the isoelectric point (pI) for the protein, and that as the pH of a protein solution increases, the charge on the protein molecules becomes increasingly negative. Conversely, as the pH of the solution decreases, the charge on the protein molecules becomes increasingly positive. This behavior is a consequence of the fact that the ionizable groups on the surface of a protein are either carboxylic acids or amines (see Problem 15 in Chapter 2).

There are two main types of ion exchange matrices: (1) *anion* exchangers such as diethylaminoethyl (DEAE) cellulose and quaternary ammonium ("Q") resins, which carry a positive charge and therefore bind to negatively charged proteins, and (2) *cation* exchangers such as carboxymethyl (CM) cellulose and sulfonic acid ("S") resins, which carry a negative charge and bind to positively charged proteins. DEAE and CM exchangers are considered "weak" ion exchangers because they carry functional groups that can lose their charges at pH > ~10 (DEAE) or pH < ~4 (CM). The "Q" and "S" resins are effectively always charged in aqueous buffers; thus, they are considered "strong" ion exchangers. It is critical to match the IEC resin and buffer to the pI of the protein of interest. For example, a protein will bind to a column of DEAE-cellulose when the pH of the mobile phase is above the pI for that protein, but will not bind the column if buffer pH < pI.

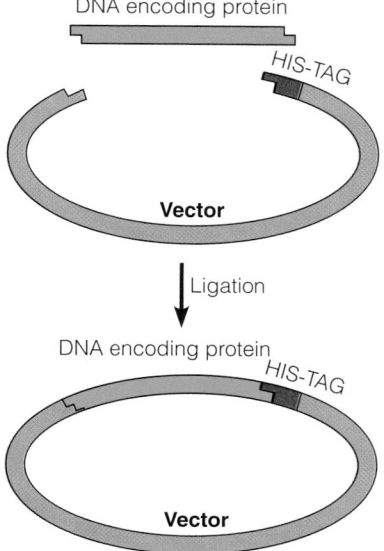

FIGURE 5A.4

Insertion of a protein gene into one of several commercially available expression vectors results in addition of the affinity "His-tag" (i.e., a sequence of six His residues) to the protein sequence.

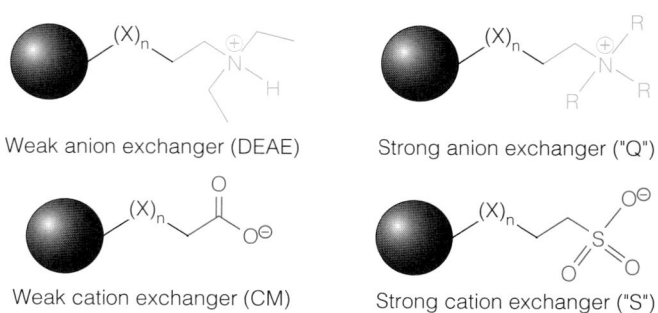

Weak anion exchanger (DEAE)   Strong anion exchanger ("Q")

Weak cation exchanger (CM)   Strong cation exchanger ("S")

Elution of a protein bound to an IEC matrix can be achieved by changing the pH of the mobile phase such that the charge on the protein is reduced, thereby weakening its binding to the support and/or increasing the ionic strength of the mobile phase (e.g., by addition of some salt to the elution buffer). Soluble ions compete for binding to the charged functional groups on the matrix. As the concentration of soluble ions increases, the ions will out-compete, and thereby displace, the protein from the matrix.

In summary, IEC allows for separation of proteins based on differences in charge density. The theoretical basis for this separation technique was presented in Chapter 2 and includes the following concepts: (1) relative strengths of electrostatic interactions, as described by Coulomb's law, (2) the isoelectric point or pI for a given protein, and (3) modulation of charge density on a protein as a function of pH, as described by the Henderson–Hasselbalch equation.

Size Exclusion Chromatography (SEC)

Size exclusion chromatography, which is also known as "**gel filtration chromatography**," differs from the two methods above in that noncovalent interactions between the protein and support are negligible. As shown in Figure 5A.5, SEC separates proteins on the basis of apparent size, or "hydrodynamic radius." The apparent size of a protein molecule is approximately correlated with the length of the protein amino acid sequence. This rule of thumb assumes, to a first approximation, that soluble folded proteins behave as spheres. The distinguishing feature of the SEC stationary phase is the porous structure of the matrix (Figure 5A.5b). These matrices are generally spherical beads with pores in the surface of the bead that are on a size scale close to that of protein molecules. Different SEC matrices have larger or smaller sized pores in the beads.

The principles at work in SEC are diffusion by Brownian motion and "excluded volume." The total volume in an SEC column includes the volume of the porous beads and the volume of the mobile phase, which is in the space between the beads and also inside the beads. The volume of buffer solution required to elute a protein from the column depends on what fraction of the column

volume that protein can occupy: the more volume the protein can occupy, the more buffer will be needed to elute the protein, and the later it will elute from the column. The size of the pores in the beads determines which proteins can occupy volume inside the bead and which proteins cannot. Because smaller proteins are more likely to diffuse into the interior of a bead, they will occupy more of the total column volume than a large protein, which can only occupy the volume between—but not inside—the beads. In other words, larger proteins are "excluded" from the volume found inside the beads, and, consequently, larger proteins elute from an SEC column earlier than smaller proteins. Note that this order of elution by size is the reverse of the order of migration in an SDS-PAGE experiment (see Tools of Biochemistry 2A).

SEC works best as a purification technique when the differences in size between the desired protein and the contaminants are a factor of two or greater; thus, for SEC to be effective the complexity of the protein mixture must be relatively low. For this reason SEC is often the last step of a purification process. In addition, SEC is a convenient way to "desalt" or change the buffer composition of a protein solution because the salts that make up a buffer are low molecular weight and elute well after the desired protein. For example, it is convenient to run affinity columns and IEC columns in nonvolatile buffers such as phosphate-buffered saline (PBS); however, nonvolatile buffers are not compatible with lyophilization (freeze-drying) and many mass spectrometry techniques. Mass spectrometry is a powerful tool for protein analysis and is often used to confirm the identity of a purified protein (see Tools of Biochemistry 5B). To prepare a protein for lyophilization or mass spectrometry, an SEC column is equilibrated with a mobile phase that uses a volatile buffer, such as ammonium acetate. As a protein, which is loaded onto the SEC column in PBS, moves through the column it becomes separated from the phosphate buffer salts and is surrounded instead by the ammonium acetate buffer solution. In this way a "buffer exchange" from PBS to ammonium acetate occurs, and the protein is now ready for lyophilization and/or mass spectrometric analysis.

FIGURE 5A.5

The principle of size-exclusion chromatography. **(a)** As a mixture of proteins flows down the column, the smaller proteins are retarded; thus, the larger proteins elute first. **(b)** A close-up of an SEC bead (gray). Proteins larger (green) than the pores in the beads flow around the beads; but, the smaller proteins (red) can enter the pores in the beads, resulting in retarded mobility through the column.

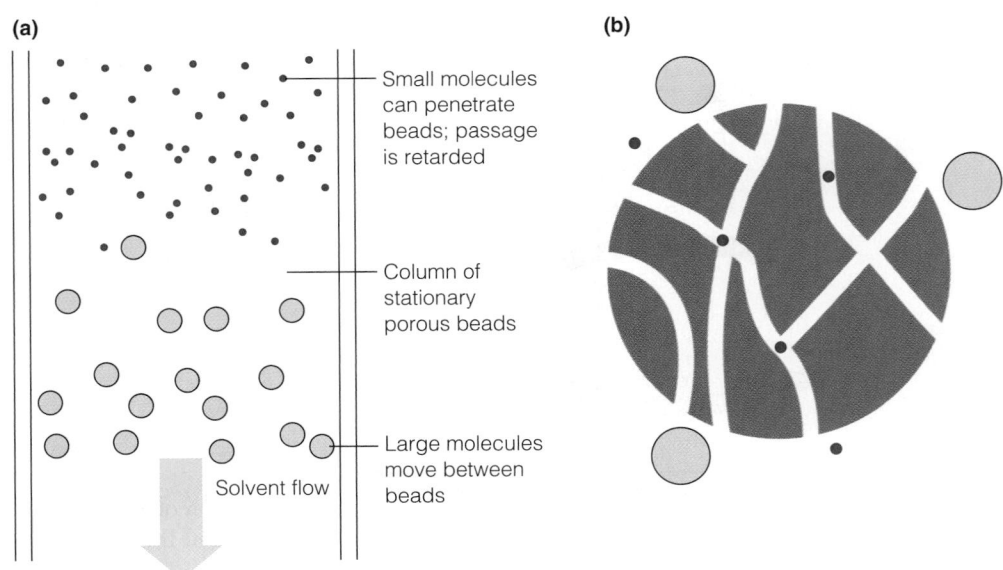

(a)

Small molecules can penetrate beads; passage is retarded

Column of stationary porous beads

Large molecules move between beads

Solvent flow

(b)

Example: Purification of a Recombinant Myoglobin Mutant

Figure 5A.6 illustrates a scheme for the purification of a mutant form of the eukaryotic muscle protein myoglobin expressed in *E. coli* bacteria. This particular mutant includes a 16 amino acid insertion into the protein sequence. Following the period of protein production in bacterial cells, the cells are lysed so that the contents of the cytoplasm are released into a buffered solution. The soluble material can then be separated from the insoluble material (e.g., membranes, precipitated protein aggregates) by centrifugation. The resulting supernatant is a complex mixture of nucleic acids and proteins, as shown in Lane 1 of Figure 5A.7. An efficient purification scheme will significantly reduce the complexity of the mixture in the early steps of the process; thus, affinity chromatography is a good choice for the first chromatography step. In this case an IMAC purification step achieves significant purification of the mutant myoglobin, as shown in Lane 2 of Figure 5A.7. After IMAC, all the remaining contaminants appear to be of higher molecular weight than the mutant myoglobin; thus, SEC can be

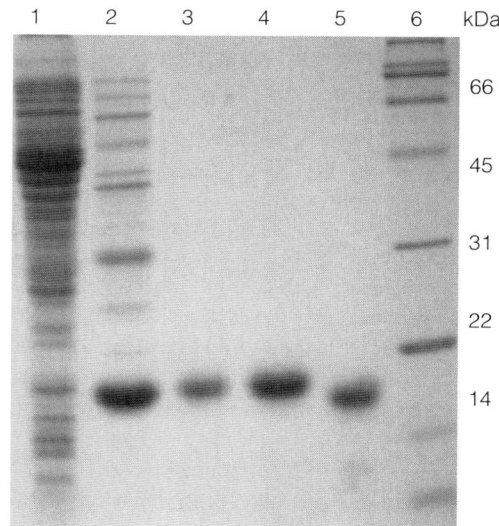

FIGURE 5A.7

The myoglobin mutant is purified to greater than 95% homogeneity by a two-column procedure. A Coomassie-stained 15% SDS-PAGE gel is shown. Lane 1: *E. coli* lysate (after centrifugation). Lane 2: IMAC purified proteins; the prominent band is the desired myoglobin mutant. Lane 3: 2 μg of the mutant myoglobin after SEC. Lane 4: 10 μg of the mutant myoglobin after SEC. Lane 5: SEC-purified wild-type myoglobin. Lane 6: Protein molecular weight markers.

Reprinted with permission from *Biochemistry* 41:13318–13327, A. L. Fishburn, J. R. Keeffe, A. V. Lissounov, D. H. Peyton, and S. J. Anthony-Cahill, A circularly permuted myoglobin possesses a folded structure and ligand binding similar to those of the wild-type protein but with a reduced thermodynamic stability. © 2002 American Chemical Society.

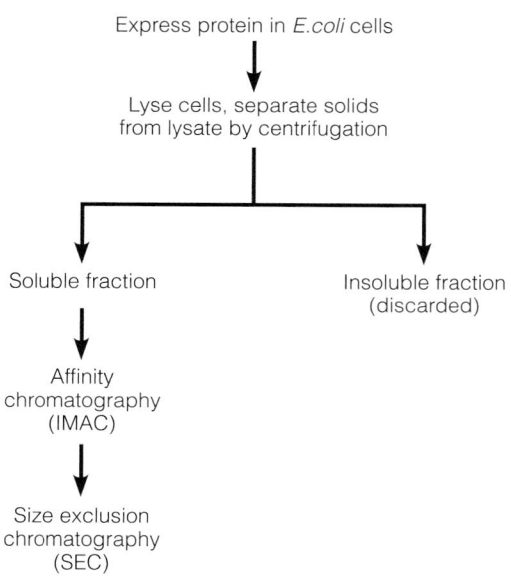

Recombinant Myoglobin Purification Scheme

Express protein in *E.coli* cells

↓

Lyse cells, separate solids from lysate by centrifugation

Soluble fraction

Insoluble fraction (discarded)

↓

Affinity chromatography (IMAC)

↓

Size exclusion chromatography (SEC)

FIGURE 5A.6

Flow chart for the purification of recombinant myoglobin.

used to separate the mutant from the contaminants, as shown in Lanes 3 and 4 of Figure 5A.7. This is an efficient purification scheme because only two chromatography steps are required to achieve greater than 95% homogeneity for the desired protein.

References

Janson, J.-C., and L. Rydén (1998) *Protein Purification: Principles, High Resolution Methods and Applications* (2nd ed.). Wiley-VCH, New York.

Roe, S. (ed.) (2001) *Protein Purification Techniques: A Practical Approach* (2nd ed.). Oxford University Press, Oxford.

Rosenberg, I. M. (2005) *Protein Analysis and Purification: Benchtop Techniques* (2nd ed.). Birkhauser, Boston.

Scopes, R. K. (1994) *Protein Purification: Principles and Practice* (3rd ed.). Springer, New York.

TOOLS OF BIOCHEMISTRY 5B

MASS, SEQUENCE, AND AMINO ACID ANALYSES OF PURIFIED PROTEINS

Mass Determination

Once a protein has been purified, how does a researcher convince her/himself that the correct target protein has been obtained in a purified form? The first indication comes typically from an SDS-PAGE gel (see Tools of Biochemistry 2A and Figure 5A.7), which shows (1) the purity of the protein and (2) an approximate molecular weight estimated by comparing the migration of the target protein in the gel to protein molecular weight standards. For example, the molecular weight of the mutant myoglobin shown in Figure 5A.7 is predicted to be 18,232 Da based on the amino acid sequence of the translated gene. Figure 5A.7 shows that the mutant protein migrates between 14 and 22 kDa as would be expected. The purified protein appears to have the correct mass by SDS-PAGE; however, this is not a high resolution technique. The actual mass could differ from the expected mass by several hundred daltons and still appear reasonably close to the expected mass by SDS-PAGE. Mass spectrometry (MS) provides the most accurate mass measurements of large biomolecules. For this reason it is desirable to obtain high resolution mass data via MS to confirm that the protein has no unexpected post-translational modifications (e.g., proteolytic cleavage and/or covalent modifications).

Protein MS has become an indispensable analytical tool since ionization techniques compatible with protein analysis were developed in the late 1980s. The application of MS to significant problems in biochemistry continues to grow as the technology improves. We will present more advanced MS techniques in later chapters; here we will focus on the application of MS to accurate protein mass determination and peptide sequencing.

Figure 5B.1 shows a simplified diagram of a mass spectrometer that contains a single mass analyzer, sufficient for the routine determination of accurate protein masses using **electrospray ionization** (ESI) or **matrix-assisted laser desorption/ionization** (MALDI) techniques. In ESI, a fine mist of protein solution is accelerated toward a mass analyzer. By the time the mist reaches the analyzer most of the solvent has evaporated, leaving protein molecules with a varying number of charges to be separated in the mass analyzer. The detector records the ratio of mass to charge (m/z, where m = mass and z = charge). The ESI-MS mass spectrum is a collection of peaks with different m/z ratios, where m is constant and z varies (Figure 5B.1 top). In the MALDI technique, the protein is imbedded in a large excess (~10,000-fold) of some matrix that absorbs UV light. When a laser pulse hits the matrix, it absorbs the energy of the laser light

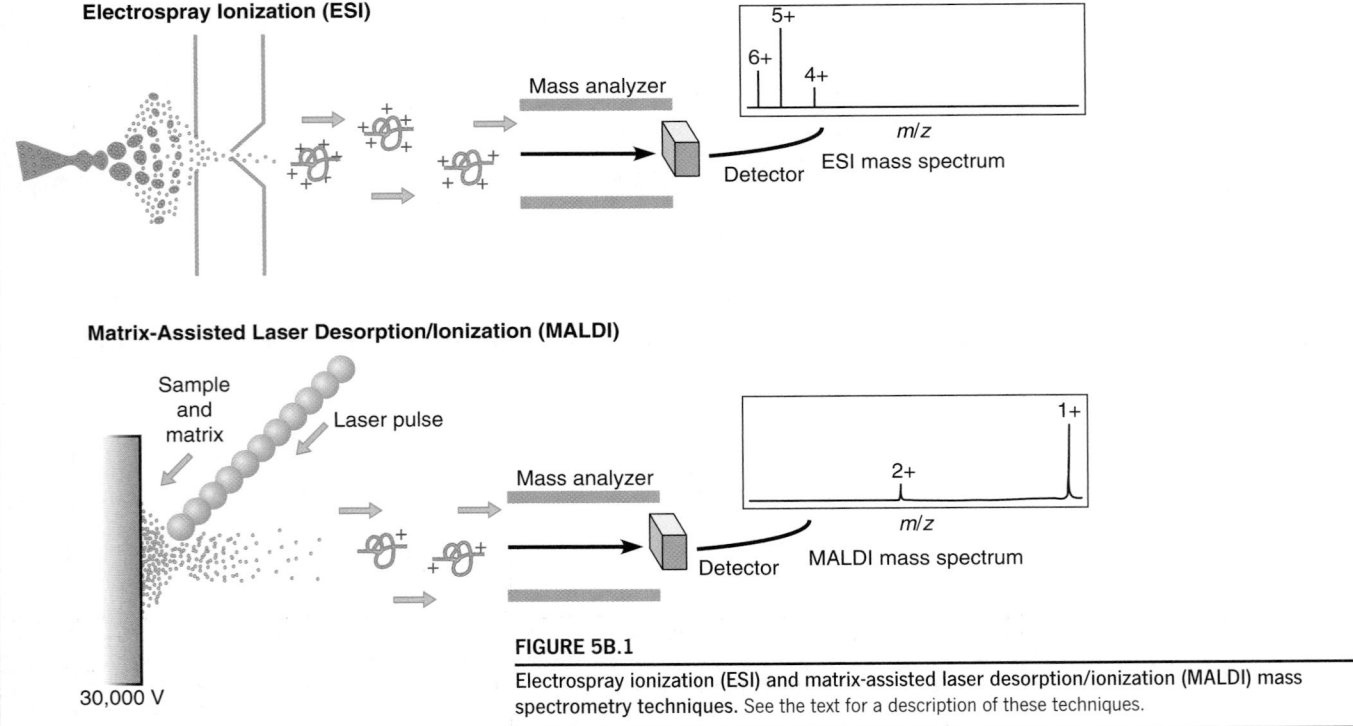

FIGURE 5B.1

Electrospray ionization (ESI) and matrix-assisted laser desorption/ionization (MALDI) mass spectrometry techniques. See the text for a description of these techniques.

Reprinted with permission from *Accounts of Chemical Research* 33:179–187, J. J. Thomas, R. Bakhtiar, and G. Suizdak, Mass spectrometry in viral proteomics. © 2000 American Chemical Society.

and is vaporized. The vaporized matrix carries intact protein molecules into the gas phase and toward the mass analyzer. The MALDI-MS spectrum shows *m/z* for predominantly the parent ion (Figure 5B.1 bottom).

Sequence Determination

An accurate protein mass is usually sufficient to confirm the identity of a known protein; however, if an unknown protein is the target of some purification scheme, the mass alone is not typically sufficient to identify the protein. In this case, sequence information is also desirable. If the function of the protein is also unknown, the sequence will allow potential identification of the function by similarity searching.

There are several ways in which the amino acid sequence can be determined. As mentioned on page 154, determination of the gene sequence is one of the easiest methods. Indeed, as the entire genomes of many organisms have been determined, we have amino acid sequence information for hundreds of thousands of proteins, many of them of still unknown function. Protein sequences translated from cloned genes do not provide us with information concerning modification of amino acids or the existence of intramolecular cross-links such as disulfide bonds. To find these, we must sequence the protein itself. Here we present two methods for obtaining peptide sequences: tandem MS, which was developed in the mid-1980s and is now the method of choice for most labs, and Edman sequencing, which was developed 20 years earlier by Pehr Edman and is still in use today.

We will present the procedure for Edman sequencing first because there is much useful protein chemistry involved, and the logic for reconstructing the overall protein sequence is the same for both the Edman and tandem MS methods.

The Edman method is based on the stepwise removal of amino acids from the N-terminus of a peptide by a series of chemical reactions called the "Edman degradation" (Figure 5B.2). The compound phenylisothiocyanate (PITC) is reacted in alkali with the N-terminal amino group to yield a phenylthiocarbamyl (PTC) derivative of the peptide (Figure 5B.2, step 1). This derivative is then treated with a strong anhydrous acid, which results in cleavage of the peptide bond between residues 1 and 2 (step 2). The derivative of the N-terminal residue then rearranges to yield a phenylthiohydantoin (PTH) derivative of the amino acid (step 3). Two important things have been accomplished: (1) the N-terminal residue has been marked with an identifiable label, and (2) the rest of the polypeptide has not been destroyed; it has simply been shortened by one residue. The whole sequence of reactions can now be repeated and the second residue determined. By continued repetition, a long polypeptide can be "read," starting from the N-terminal end. This procedure can be performed automatically with an instrument known as a sequenator, which is able to carry out the entire set of reactions shown in Figure 5B.2 over and over again. The sequenator will accumulate in a separate tube the PTH derivative of each amino acid residue in the polypeptide, starting with the N-terminal residue and proceeding for as many cycles as the operator desires or precision allows. The PTH

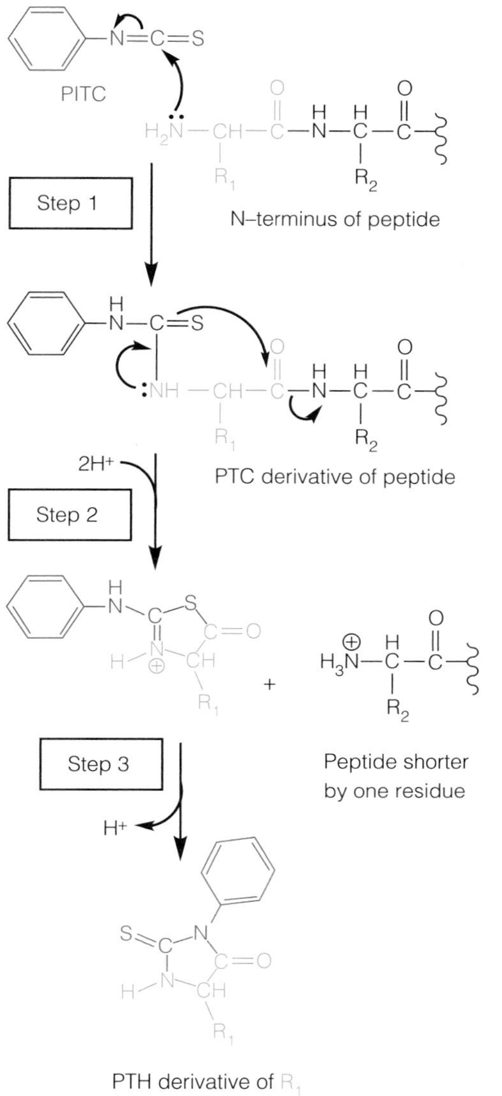

FIGURE 5B.2

The Edman degradation.

derivatives are then identified by high-performance liquid chromatography (HPLC) and/or MS.

In practice, 30–40 residues can be read reliably by Edman sequencing; thus, it is necessary to fragment longer proteins and then sequence the smaller peptides obtained from the fragmentation reaction. Fragmentation is achieved using proteases such as those listed in Table 5.4 and/or cyanogen bromide. Once this first set of fragments has been sequenced, a second fragmentation is carried out using a protease with a different specificity such that a second set of peptides is obtained that overlaps the set obtained in the first fragmentation. We will use the sequencing of bovine insulin as an example to illustrate this process. This choice is appropriate because it was the first protein ever sequenced, by Frederick Sanger and his coworkers in the early 1950s (work for which Sanger won his *first* Nobel Prize). The example is also more complicated than most because we must deal with two covalently

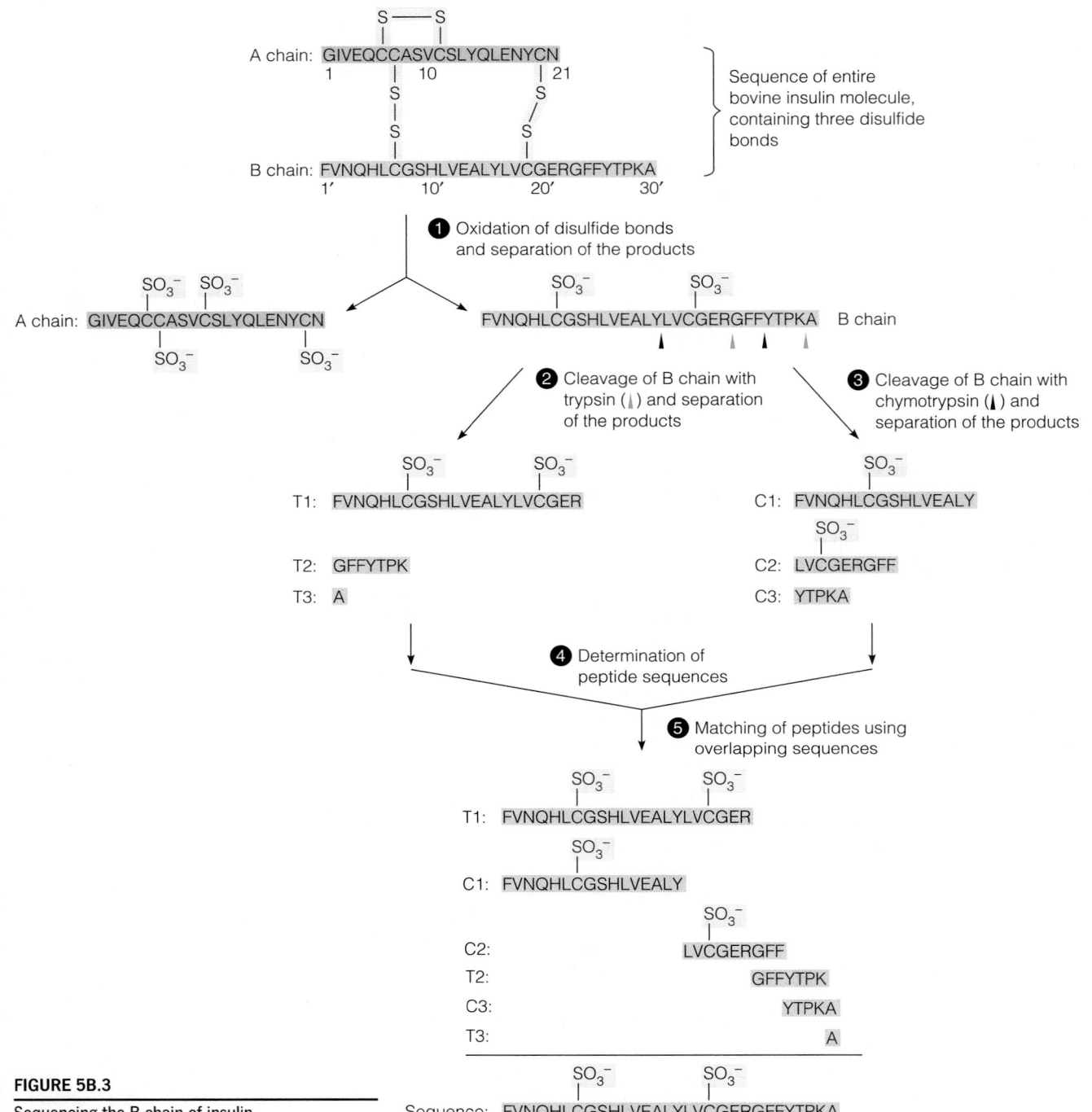

FIGURE 5B.3

Sequencing the B chain of insulin.

connected chains and locate disulfide bonds. The steps of the procedure are outlined in Figure 5B.3.

The researcher intending to sequence a protein by Edman degradation must first make sure that the material is pure. The protein can be separated from other proteins by some combination of the methods described in Tools of Biochemistry 5A and checked for purity by means of electrophoresis and/or isoelectric focusing. Next, it must be determined whether the material contains more than one polypeptide chain because, in some cases disulfide bridges

covalently bond chains together. SDS-PAGE in the presence and absence of reducing agents can answer this question (see Tools of Biochemistry 6B). In the insulin example, there are two chains, A and B, as shown in Figure 5B.3. These chains must be separated and sequenced individually because the Edman degradation would release two sets of PTH derivatives simultaneously if the peptides were not separated. To break disulfide bonds and thus separate the chains, several reactions are available. Descriptions of two common procedures follow.

Performic acid oxidation is the technique used in Figure 5B.3, step 1. The strong oxidizing agent performic acid will *irreversibly* react with cystine to yield cysteic acid residues:

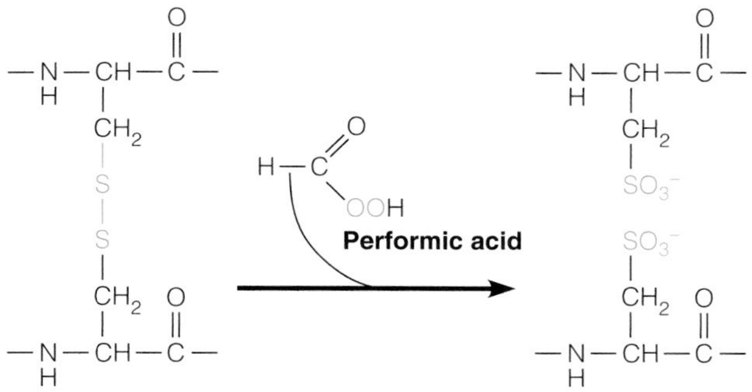

Performic acid

Reduction with β-mercaptoethanol is a milder, and *reversible*, technique.

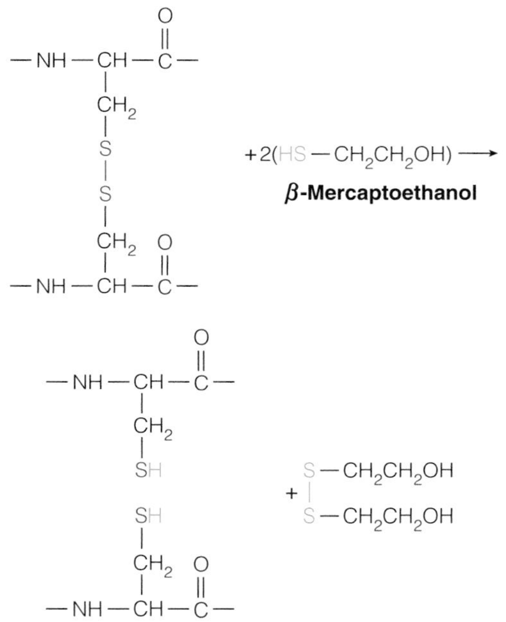

β-Mercaptoethanol

Reduction leaves free sulfhydryl groups, often positioned so that reoxidation to re-form the disulfide bond is likely. Therefore, the sulfhydryls are usually blocked to prevent this. A common blocking reagent is iodoacetate:

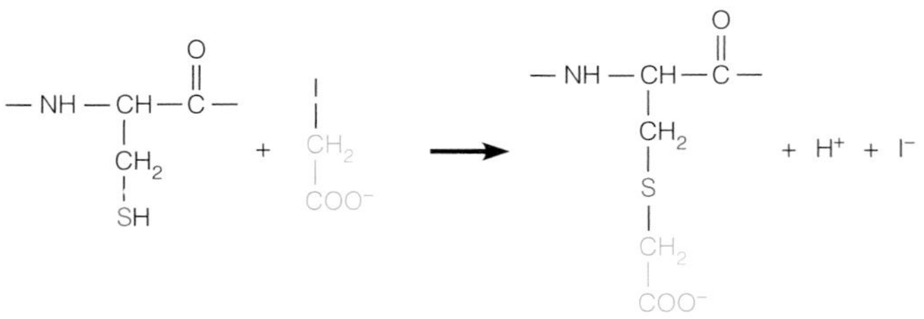

| Cysteine residue | Iodoacetate | δ-Carboxymethyl-cysteine residue |

If either of these methods is carried out with insulin, the intact protein is cleaved into A and B chains. These chains can then be separated by chromatographic methods.

Before Edman sequencing of the individual chains is started, their amino acid composition is usually determined (see below). This determination may point to unusual compositions and thus warn the operator of potential problems. Furthermore, composition data will serve as a check on the sequencing results because the sequence determined must be consistent with the amino acid composition.

In bovine insulin, the A and B chains are so short that modern instrumentation could sequence either in one sequenator run; however, to demonstrate the methods needed for larger proteins, we assume that the investigator must cleave the insulin chains into shorter polypeptides (this was indeed the case in Sanger's pioneering sequencing studies on insulin). Suppose the insulin B chain is to be sequenced. A first step would be to cleave separate aliquots of the chain with two or more of the specific cleavage reagents described in Table 5.4. Trypsin and chymotrypsin, for instance, would yield the sets of peptides shown in steps 2 and 3, respectively, of Figure 5B.3. The individual peptides would then be isolated from each of the two mixtures, using, for example, ion-exchange chromatography, and their sequences could be determined (Figure 5B.3, step 4).

Suppose each of the peptides shown in Figure 5B.3 has been sequenced (step 4). Although the tryptic peptides alone cover the whole sequence, they are not sufficient to allow us to write down the sequence of the insulin B chain because we do not know the order in which they appear in the intact chain. To overcome this problem we also have the chymotryptic peptides, which overlap the tryptic peptides; therefore, all ambiguity is removed. Only one arrangement of the whole chain is consistent with the sequences of these two sets of peptides, as can be seen by matching overlapping sequences (step 5).

Finally, a complete characterization of the covalent structure of a protein requires that the positions of any disulfide bonds be located. In preparation for sequencing, these bonds would have been destroyed, but the positions of all cysteines, some of which *might* have been involved in bonding, would have been determined. How can we determine which cysteines are linked via disulfide bonds in the native protein?

To determine the arrangement of disulfide bonds, the experimenter again starts with the native protein—insulin in the example shown in Figure 5B.4. Reaction with radioactively labeled iodoacetate marks any free cysteine residues, and fragmentation of the protein into the same peptides used in sequencing allows the positions of these nonbonded cysteines to be identified (step 1). Then samples of the intact protein are cut with various cleavage reagents but now without first cleaving disulfide bonds (steps 2 and 3). Some peptides, which are connected by these bonds, are attached to one another. These can then be isolated and their disulfide bonds cleaved to map the location of each disulfide-bonded cysteine in the protein.

Mass spectrometry can also be used to obtain peptide-sequence information. To do this the mass spectrometer must have a collision cell and two mass analyzers rather than the single mass analyzer shown in Figure 5B.1. Figure 5B.5 shows a schematic diagram of a tandem mass spectrometer (MS-MS) capable of peptide sequencing, where the two mass analyzers are labeled "quadrupole analyzer" and "time-of-flight analyzer." The role of each mass analyzer will be described below.

As in Edman sequencing, MS-MS sequencing works best on smaller peptide fragments. The fragments can be generated

FIGURE 5B.4

Locating the disulfide bonds in insulin.

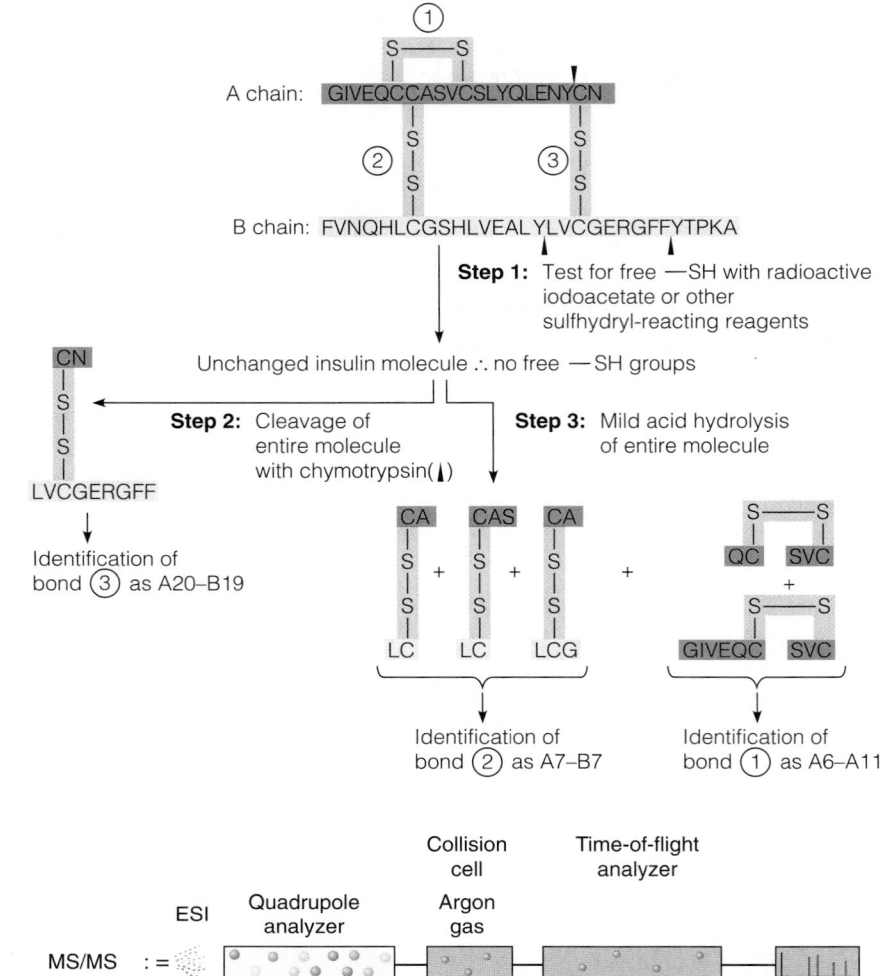

Step 1: Test for free —SH with radioactive iodoacetate or other sulfhydryl-reacting reagents

Unchanged insulin molecule ∴ no free —SH groups

Step 2: Cleavage of entire molecule with chymotrypsin(↓)

Step 3: Mild acid hydrolysis of entire molecule

Identification of bond ③ as A20–B19

Identification of bond ② as A7–B7

Identification of bond ① as A6–A11

FIGURE 5B.5

Peptide sequencing by MS-MS.

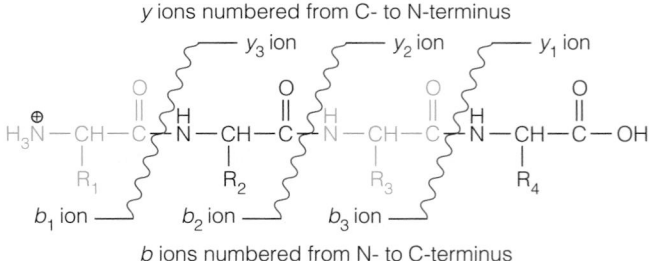

using proteases, or in the spectrometer itself. Let us consider the case in which the fragments are generated using a protease and then introduced into the MS-MS instrument using electrospray. Electrospray is advantageous for sequencing because the fragments tend to have multiple charges distributed along their lengths. Recall that the mass detector records m/z; thus, without a charge the fragment is not detectable. As depicted in Figure 5B.5 the mixture of peptide fragments is introduced by electrospray into the first mass analyzer (the quadrupole analyzer). The quadrupole analyzer can be tuned to select a specific fragment (i.e., a specific m/z range) for introduction into the collision cell. In the course of the analysis each fragment in turn will be directed from the quadrupole analyzer into the collision cell.

In the collision cell the selected fragment is fragmented further by collisions with argon atoms. To a large extent the fragmentations in the collision cell result in cleavage of the peptide backbone as shown in Figure 5B.6, where cleavage of the first peptide bond gives two subfragments: the N-terminal subfragment that includes the residue R_1, and the C-terminal subfragment that includes residues R_2–R_4. By convention, the N-terminal subfragments are called "b ions" and the C-terminal subfragments are called "y ions."

y ions numbered from C- to N-terminus

b ions numbered from N- to C-terminus

FIGURE 5B.6

Principal ions generated by low energy collision-induced dissocation. The wavy red lines indicate sites of peptide bond cleavage in the collision cell (see Figure 5B.5).

In the collision cell two series of subfragments are generated simultaneously—a series of b ions and the corresponding set of y ions. Within each series the masses of the ions differ from one another by the mass of a single amino acid residue (i.e., in Figure 5B.6 the masses of b_2 and b_3 differ by the mass of residue R_3). The m/z ratio for each ion is determined in the time-of-flight analyzer and recorded to generate a complex spectrum with peaks for each ion. Because the masses for amino acid residues are known (see Table 5.1), the amino acids present in each fragment can be reliably identified. Modern MS-MS instruments include software that can rapidly identify fragmentation patterns consistent with a specific amino acid sequence. This process is repeated until every fragment that enters the quadrupole analyzer is sequenced (a matter of a few minutes), thereby generating a set of peptide sequences for the protein. To find the order of the peptides a second MS-MS analysis is performed on a series of different fragments generated by a different protease (or CNBr, etc.).

Both MS-MS and Edman sequencing are used to determine peptide sequences and each has its set of strengths and weaknesses. Edman sequencing requires a few micrograms $(10^{-6}\,g)$ of purified protein. It gives nearly complete sequence coverage; however, if the N-terminus of the intact protein is modified, the Edman degradation is blocked. This problem is somewhat ameliorated by fragmentation methods that generate unmodified N-termini for each fragment. MS-MS methods are very sensitive; thus, only picograms $(10^{-12}\,g)$ of protein are needed. Also, separation of chains or fragments is not required because the MS-MS method achieves fragment separation in the mass spectrometer; thus, the MS-MS method is rapid. MS-MS does not usually give complete sequence coverage because some amino acid sequences are difficult to ionize; however, 70–80% sequence coverage is typical (recall that only ions can be detected in the MS instrument—if a peptide carries no charge it will not be detected).

We have described how the entire amino acid sequence, or primary structure, of a protein can be determined. Such analyses have been carried out on several thousand different proteins in the years since Sanger first determined the sequence of insulin. Today it is rare that an entire protein sequence would be determined using these methods because the sequencing of genes is much more rapid (and frequently precedes the isolation of the protein of interest); however, MS-MS is often used to determine a portion of the sequence of an unknown protein. With a sequence of only 6–10 amino acids one can often identify a protein by searching databases of protein sequences. This use of protein sequence information is the basis for the field of **proteomics**, which we discuss briefly in Tools of Biochemistry 5D.

Amino Acid Analysis

Finally, we turn our attention to the determination of the amino acid composition (or "amino acid analysis") of a purified protein. Given recent developments in gene sequencing and mass spectrometry of proteins, the determination of the amino acid composition of a protein is no longer a common analytical procedure; however, it remains a standard method for the accurate quantitation of protein in an analytical sample.

Amino acid analysis (AAA) involves three basic steps:

1. *Hydrolysis* of the protein to its constituent amino acids.
2. *Separation* of the amino acids in the mixture.
3. *Quantitation* of the individual amino acids.

A small sample of the protein is first purified, perhaps by some combination of the methods described in Tools of Biochemistry 5A. The purified protein is dissolved in 6 M HCl, and the solution is sealed in an evacuated ampoule. It is then heated at 105–110 °C for about 24 hours. Under these conditions, the metastable peptide bonds between the residues are completely hydrolyzed.

The hydrolyzed sample is then separated into the constituent amino acids on a cation-exchange column. The kinds of resin typically used are sulfonated polystyrenes:

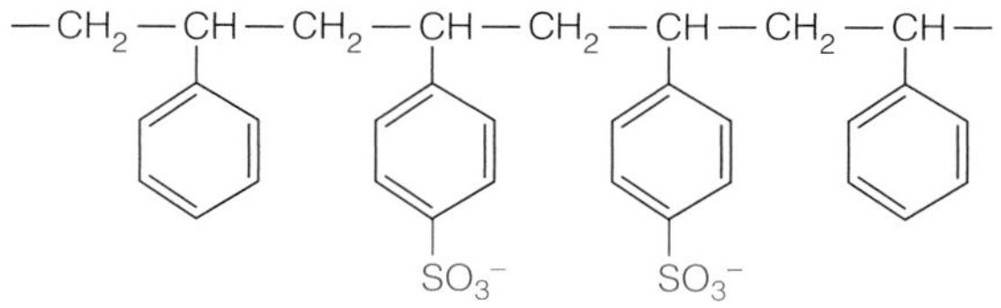

Such a resin separates amino acids in two ways. First, because it is negatively charged, it tends to pass acidic amino acids first and retain basic ones. The pH of the eluting buffer is increased during elution to facilitate this separation. Second, the hydrophobic nature of the polystyrene itself tends to hold up the more hydrophobic amino acids such as leucine and phenylalanine. An example of such an analysis is shown in Figure 5B.7. Note the order of appearance of the amino acids, proceeding from the more acidic to the more basic. Modern amino acid analyzers are completely automated and carry out both the chromatographic separation of the amino acids and their quantitation.

There are many methods for detection and quantitation of the amino acids eluting from the column; but, fluorescence is commonly used due to its sensitivity. For example, the amino acids may be reacted with o-phthalaldehyde to yield a fluorescent complex:

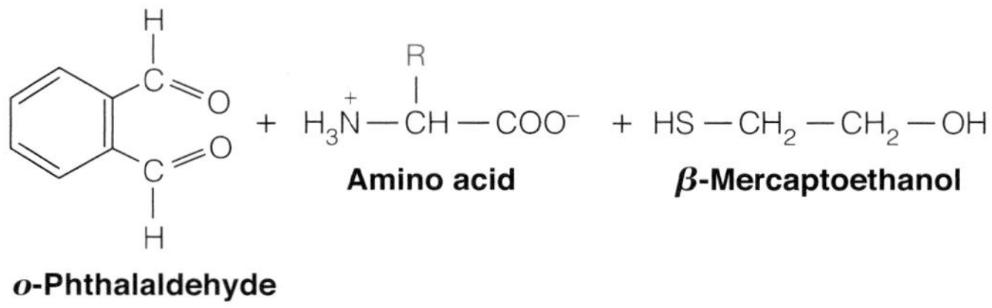

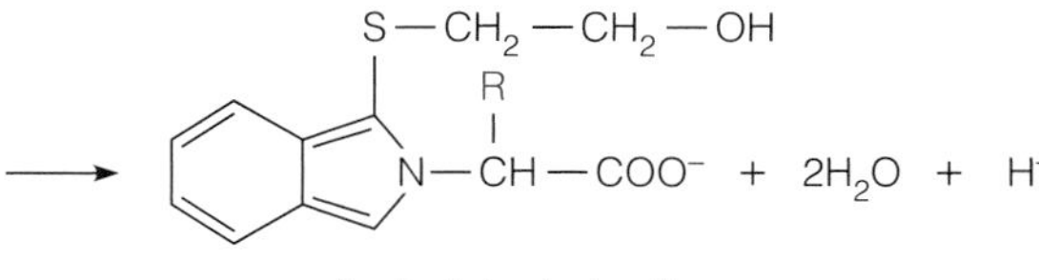

o-Phthalaldehyde　　　**Amino acid**　　　**β-Mercaptoethanol**

**Isoindole derivative
of amino acid**

Such detection techniques easily give sensitivity to the picomole (pmol, or 10^{-12} mol) range. Microelectrophoresis systems and fluorescence detection have extended this sensitivity to the *attomole* (amol, or 10^{-18} mol) range. This amount corresponds to only a few thousand molecules. Indeed, amino acid analysis techniques have proceeded to the point that the amount of protein contained in one spot in two-dimensional gel electrophoresis (see Figure 1.11, page 19) can be analyzed easily.

Of course, these procedures are not as simple and trouble free as the foregoing discussion might imply. Some amino acids give problems in reaction with the compounds used for detection; proline in particular because it is a secondary amino acid, often reacts

FIGURE 5B.7

Analysis of a protein hydrolysate on a single-column amino acid analyzer. The chromatogram shows the order of elution of hydrolyzed amino acids on a polystyrene column. Free amino acids are detected by absorbance at 220 nm.

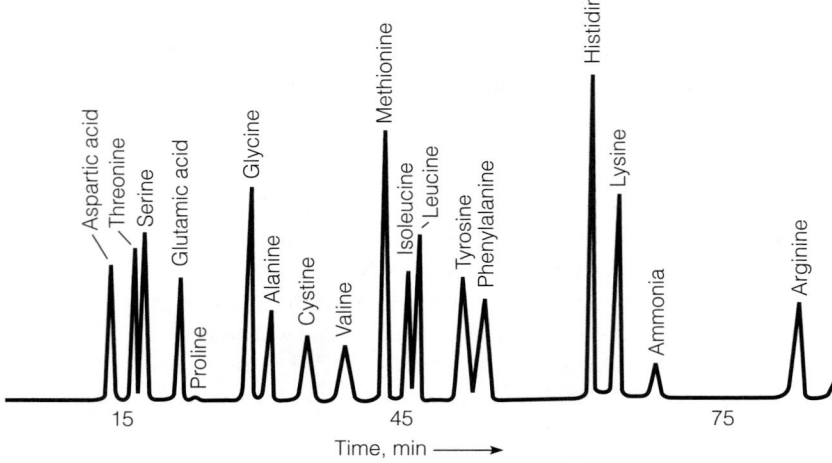

slowly or not at all. Furthermore, some amino acids tend to be partially destroyed during the severe hydrolysis. Tryptophan is troublesome in this respect and must be determined by hydrolysis with base and detection by ultraviolet absorbance (see Figure 5.6). Serine, threonine, and tyrosine also tend to be degraded during long hydrolysis. To a considerable extent, these difficulties can be circumvented either by carrying out protective reactions first or by measuring the apparent content of the amino acid at different hydrolysis times and extrapolating to zero hydrolysis time. Asparagine and glutamine are invariably hydrolyzed to aspartic and glutamic acids, so that the total content of these acids observed includes the amides. This reaction, as well as the other degradation reactions mentioned above, can be avoided by using an enzymatic hydrolysis, with a mixture of proteolytic enzymes, in place of the acid hydrolysis. However, this method also has its drawbacks because it is sometimes difficult to achieve complete hydrolysis and the enzymes themselves must be removed before analysis. Despite such complications, amino acid analysis, using automated analyzers, has become a routine operation in protein characterization.

References

Cañas, B., D. López-Ferrer, A. Ramos-Fernández, E. Camafeita, and E. Calvo (2006) Mass spectrometry technologies for proteomics. *Brief. Funct. Genom. Proteom.* 4:295–320.

Cheng, Y.-F., and N. Dovichi (1988) Subattomole amino acid analysis by capillary zone electrophoresis and laser-induced fluorescence. *Science* 242:562–564.

Edman, P., and G. Begg (1967) A protein sequenator. *Eur. J. Biochem.* 1:80–91. The first automated method.

Liu, T.-Y. (1972) Determination of tryptophan. *Methods Enzymol.* 25:44–55.

Thomas, J. J., R. Bakhtiar, and G. Suizdak (2000) Mass Spectrometry in Viral Proteomics. *Acc. Chem. Res.* 33:179–187.

Walsh, K. A., Ericsson, L. H., Parmelee, D. C., and K. Titani (1981) Advances in protein sequencing. *Annu. Rev. Biochem.* 50:261–284.

See also this Website, maintained by A. E. Ashcroft, describing mass spectrometry: www.astbury.leeds.ac.uk/facil/MStut/mstutorial.htm

TOOLS OF BIOCHEMISTRY 5C

HOW TO SYNTHESIZE A POLYPEPTIDE

Chemical synthesis of peptides of defined sequence is of great importance in medicine and molecular biology. Some synthetic hormones can be made with non-natural amino acids that make them more stable in vivo, and therefore better therapeutics. Synthetic peptides can be used to elicit antibodies against portions of specific proteins; such antibodies are useful in studying the interaction of proteins with other molecules.

To synthesize a peptide of defined sequence, several criteria must be met:

1. It should be possible to add amino acids one at a time, preferably in an automated reactor.

2. Because of peptide-bond metastability, the amino acids must be activated in some way such that peptide-bond formation is both favorable and efficient ($>$98% per cycle).

3. To avoid side reactions, all reactive groups (in this case good nucleophiles) must be protected (i.e., blocked from reacting) other than the carboxyl and amino groups that are meant to form the desired peptide bond.

4. The protecting groups used for side chain groups must be stable for the entire synthesis; but, the protecting group for the α-amino group must be removed selectively for each cycle of peptide bond formation.

There are two common synthetic schemes used to make peptides. The solid-phase peptide synthesis (SPPS) chemistry developed by Bruce Merrifield (and recognized by the Nobel Prize) uses the *t*-butyloxycarbonyl (Boc) group to protect the α-amino group. A second popular scheme uses the 9-fluorenylmethoxylcarbonyl (Fmoc) group to protect the α-amino group. Both schemes are amenable to automated solid-phase synthesis and are still widely used. In some cases, one scheme will give a better yield for a given sequence than another, so most labs that synthesize peptides have machines for both Boc and Fmoc methods.

N-terminal residue. All reactions are carried out automatically, with the growing chains attached to the resin. In the final step, HF is added to remove any side-chain protecting groups and simultaneously cleave the peptide from the resin.

Using these methods, peptides of 50 residues in length can be routinely synthesized in good yield. Expert laboratories can make chains of roughly 150 amino acids. Merrifield, for example, synthesized an active enzyme (ribonuclease) of 124 residues, and Stephen Kent and coworkers have made small proteins of 140–160 amino acids. There are also methods for condensing a synthetic peptide with a sequence derived from an intact biosynthetic protein. In this way non-natural amino acids can be incorporated into larger proteins. A description of these methods, called "native chemical ligation" and "expressed protein ligation," is beyond the scope of this discussion. The interested reader can find more details in the citations given at the end of this section.

N-α-*t*Boc-alanine

N-α-Fmoc-alanine

A simplified scheme based on the Merrifield solid-phase chemistry is shown in Figure 5C.1. The advantage of the solid-phase method is that the growing peptide chain remains attached to a solid matrix (called the "SPPS resin") until the last step of the synthesis; thus, during each step, excess reactants and contaminants can be washed away. To begin the synthesis, the C-terminal amino acid of the desired peptide sequence is covalently attached to a bead of SPPS resin, with its α-amino group exposed. A series of three steps is then carried out to complete a cycle of peptide-bond formation. These steps include: (1) deprotonation of the α-amino group to make it a better nucleophile; (2) activation of the carboxylic acid group of the next amino acid in the desired sequence, followed by covalent addition of the activated amino acid to the growing peptide chain; and (3) deprotection of the new N-terminal α-amino group. The new peptide bond is formed in step 2. Steps 1–3 are repeated until the desired sequence has been synthesized. Finally, the N-terminal α-amino group and all the protected side chains are deprotected, and the peptide is cleaved from the resin.

A brief description of each step in a peptide-bond synthesis cycle follows. The amino acid to be added to the growing peptide bound to the resin has a free carboxylic acid group, a Boc-protected α-amino group and, if necessary, a protected side chain group. The carboxylic acid is converted in situ to a more reactive carboxylic ester by reaction with a carbodiimide reagent (step 2). The coupling reaction yields a new peptide bond and a peptide, longer by one amino acid, that has a Boc-protected α-amino group. The Boc protecting group is removed selectively using trifluoroacetic acid (step 3), and the next activated residue is then added (repeats steps 1–3). Note that the last residue added is the

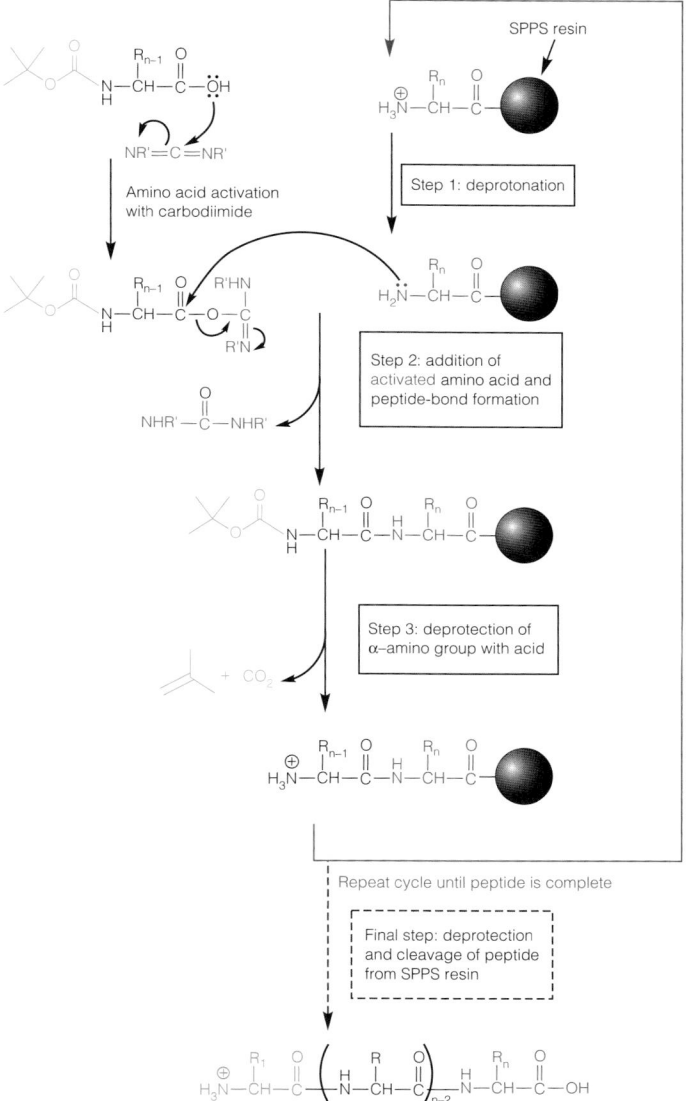

FIGURE 5C.1

Reactions in solid-phase peptide synthesis. Individual steps are described in the text.

Spatially Organized Combinatorial Peptide Arrays

Frequently, it is necessary to test simultaneously a large number of different peptides for some kind of biological activity. One might want to know, for example, which member(s) of a large family of similar oligopeptides is(are) the antigen(s) reacting with a specific antibody. Formerly, this was an extremely laborious process, involving perhaps hundreds of separate syntheses.

Using techniques borrowed from photolithography and inkjet printing, it is now possible to prepare microscopic, two-dimensional arrays containing many combinations of peptides grown on a solid surface. The photolithographic technique is illustrated in Figure 5C.2. The amino acids to be used are each blocked on the N-terminus with a photolabile protecting group and carry activated carboxyl groups. First, one class of amino acids (in this case Leu) is reacted with a surface coated with amino groups. The whole surface is then illuminated, which removes the protective groups. A second activated amino acid can then be added to each chain. In this example, after four rounds, the peptide GGFL has been grown on each site. To generate sequence diversity, a rectangular mask is placed over the surface, so that only half the squares in a checkerboard pattern are illuminated. This allows coupling Tyr residues in the illuminated portion. The other portion is then illuminated and coupled with Pro. Thus, in this example, a simple checkerboard pattern is obtained, with PGGFL and YGGFL alternating. Figure 5C.3 shows the reaction of a fluorescent antibody reactive to YGGFL on such a surface. The example shown is simple: much more complex patterns can easily be generated by use of overlapping masks, allowing thousands of different peptides to be generated, in a prescribed pattern, on one surface.

Using a different synthetic strategy, protein microarrays with greater than 10,000 different proteins immobilized on a single glass slide have been developed for the rapid detection of protein–protein interactions and the determination of intracellular protein expression levels. These "protein chips" are created by

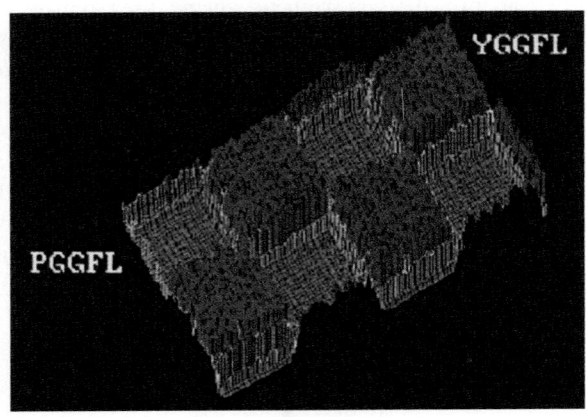

FIGURE 5C.3

Three-dimensional representation of the checkerboard array of YGGFL and PGGFL. Fluorescence intensity data were converted into spike heights that are proportional to the number of counts detected from 2.5-mm square pixels. The spikes are also color coded.

From *Science* 251:767–773, S. P. Fodor, J. L. Read, M. C. Pirrung, L. Stryer, A. T. Lu, and D. Solas, Light-directed, spatially addressable parallel chemical synthesis. © 1991. Reprinted with permission from AAAS and Stephen P. A. Fodor.

depositing a few nanoliters of a protein solution at a precise location on the slide. Each spot on such a slide can be a different protein. The proteins in these microarrays are capable of interacting with other proteins and smaller molecules such as drug candidates and enzyme substrates; thus, this technology allows for direct measurement of protein–protein and protein–ligand interactions.

References

Clark-Lewis, I., R. Aebersold, H. Ziltener, J. W. Schrader, L. E. Hood, and S. B. H. Kent (1986) Automated chemical synthesis of a protein growth factor for hemopoietic cells, interleukin-3. *Science* 231:134–139.

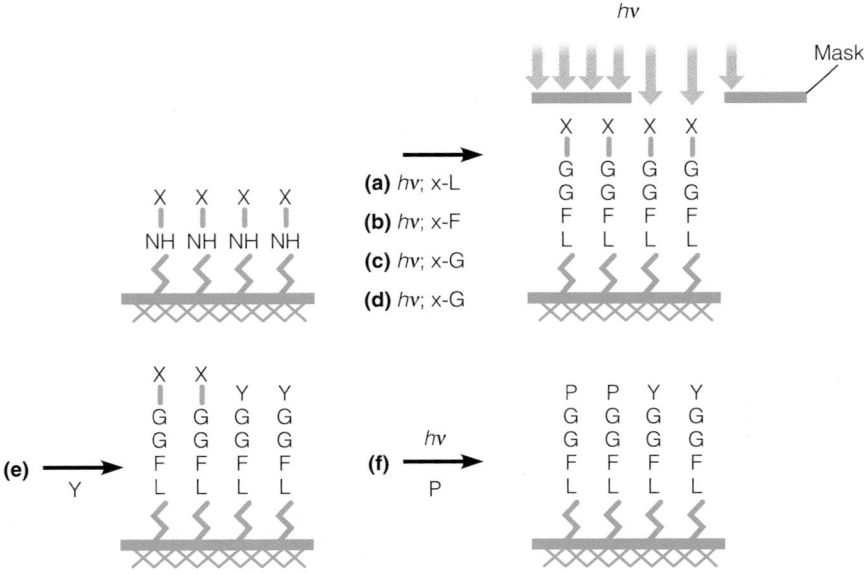

FIGURE 5C.2

An example of light-directed, spatially patterned oligopeptide synthesis. X indicates a photolabile blocking group attached to each amino acid residue added. In steps a–d, the tetrapeptide GGFL is built up on the surface. A mask is then used to illuminate and cleave blocking groups in defined areas, to allow addition of tyrosine (step e). An additional round of photo deprotection and coupling is used to add proline in the remaining areas.

From *Science* 251:767–773, S. P. Fodor, J. L. Read, M. C. Pirrung, L. Stryer, A. T. Lu, and D. Solas, Light-directed, spatially addressable parallel chemical synthesis. © 1991. Reprinted with permission from AAAS and Stephen P. A. Fodor.

Dawson, P. E., T. W. Muir, I. Clark-Lewis, and S. B. H. Kent (1994) Synthesis of proteins by native chemical ligation. *Science* 266:776–779.

Flavell, R. R., and T. W. Muir (2009) Expressed protein ligation (EPL) in the study of signal transduction, ion conduction, and chromatin biology. *Accts Chem. Res.* 42:107–116.

Fodor, S. P. A., Reed, J. L., Pirrung, M. C., Stryer, L., Lu, A. T., and D. Solas (1991) Light-directed spatially addressable parallel chemical synthesis. *Science* 251:767–773.

Kochendoerfer, G. G., et al. (2003) Design and chemical synthesis of a homogeneous polymer-modified erythropoiesis protein. *Science* 299:884–887.

MacBeath, G., and S. L. Schreiber (2000) Printing proteins as microarrays for high-throughput function determination. *Science* 289:1760–1763.

Merrifield, B. (1986) Solid phase synthesis. *Science* 232:341–347.

Schnolzer, M., and S. B. Kent (1992) Constructing proteins by dovetailing unprotected synthetic peptides: Backbone-engineered HIV protease. *Science* 256:221–225.

Vila-Perelló, M., and T. W. Muir (2010) Biological applications of protein splicing. *Cell* 143:191–200.

Zhu, H., M. Bilgin, R. Bangham, D. Hall, A. Casamayor, P. Bertone, N. Lan, R. Jansen, S. Bidlingmaier, T. Houfek, T. Mitchell, P. Miller, R. A. Dean, M. Gerstein, and M. Snyder (2001) Global analysis of protein activities using proteome chips. *Science* 293:2101–2105.

TOOLS OF BIOCHEMISTRY 5D

A BRIEF INTRODUCTION TO PROTEOMICS

The complement of proteins present in a given cell make up the so-called **proteome** of that cell. As mentioned in Chapter 1, **proteomics** is the field of study that attempts to understand the complex relationships between proteins and cell function through global analysis of the proteome rather than investigating the properties of purified protein in isolation. Proteomics includes, among other things, efforts to understand how protein expression and/or post-translational modification levels change in cells, and the consequences of such changes. For example, how does the proteome of a normal pancreatic cell differ from the proteome of a cancerous pancreatic cell? How might a researcher begin to address this question since every cell has the potential to express many thousands of different proteins at any given time? The protein chips described in Tools of Biochemistry 5C offer one method for obtaining such information; but, there are other methods which are more commonly used.

A proteomics experiment can, for example, rapidly identify differences in the levels of cellular expression, or post-translational modification (e.g., phosphorylation) between the proteomes of normal and diseased cells. The key is to correctly identify the affected proteins. The best technique for achieving this is mass spectrometry (see Tools of Biochemistry 5B). Once the affected proteins have been identified, the putative role of the protein in the disease can be researched.

A typical proteomics experiment includes the following steps: (1) separation and isolation of proteins, or protein fragments, from cells or an organism; (2) identification by MS-MS sequencing (Figures 5B.5 and 5B.6) of a particular protein within the complex mixture; and (3) database searching to identify the target protein, and its putative function. In practice, 2-D electrophoresis can be used to separate peptides (see Figure 1.11, page 19) as well as to identify potential protein targets for proteomic analysis. However, the extraction of peptides from 2-D gels is laborious, and direct analysis of complex peptide mixtures using tandem mass spectrometry is preferred for large-scale proteomic analyses.

A basic proteomics experiment is illustrated in Figure 5D.1. A complex mixture of proteins is digested—in this case with the protease trypsin. This mixture gives rise to a complex mixture of peptides that can be separated, for example, by HPLC (see Tools of Biochemistry 5B). Here, the HPLC effluent is injected directly into a mass spectrometer. The complexity of this peptide mixture is represented by the many peaks in the total ion current (TIC) mass spectrum [Panel (a)]. If a very complex sample (e.g., a cellular extract) is the starting material for this experiment, it is likely that each of the peaks in the TIC mass spectrum will contain several peptides; however, these peptides can be separated within the mass spectrometer based on differences in mass-to-charge ratio between the peptides. The separated peptides are shown in a much less complex "parent ion" mass spectrum [Panel (b)]. Each parent ion can be analyzed by MS-MS to yield amino acid sequence (see Figures 5B.5 and 5B.6). Panel (c) of Figure 5D.1 shows the MS-MS spectrum of one of the parent ions from the mass spectrum in Panel (b). The experimentally determined amino acid sequence is then used as input to search a protein sequence database. In the example here, the sequence of the peptide is a fragment of bovine serum albumin.

MS-MS is compatible with the analysis of complex peptide mixtures because only one peptide fragment from the mixture is selected for sequencing by the mass analyzer. Thus, it is still possible to make a positive identification of a protein from complex mixtures of many proteins. In theory, all the proteins in the mixture can be identified, assuming the sequences are listed in some database.

Mass spectrometry is particularly suited to the detection of post-translational modifications. A common modification is protein phosphorylation, which confers changes in both mass and charge on the phosphorylated protein. The presence of some enzymatic activity can also be detected by mass analysis by

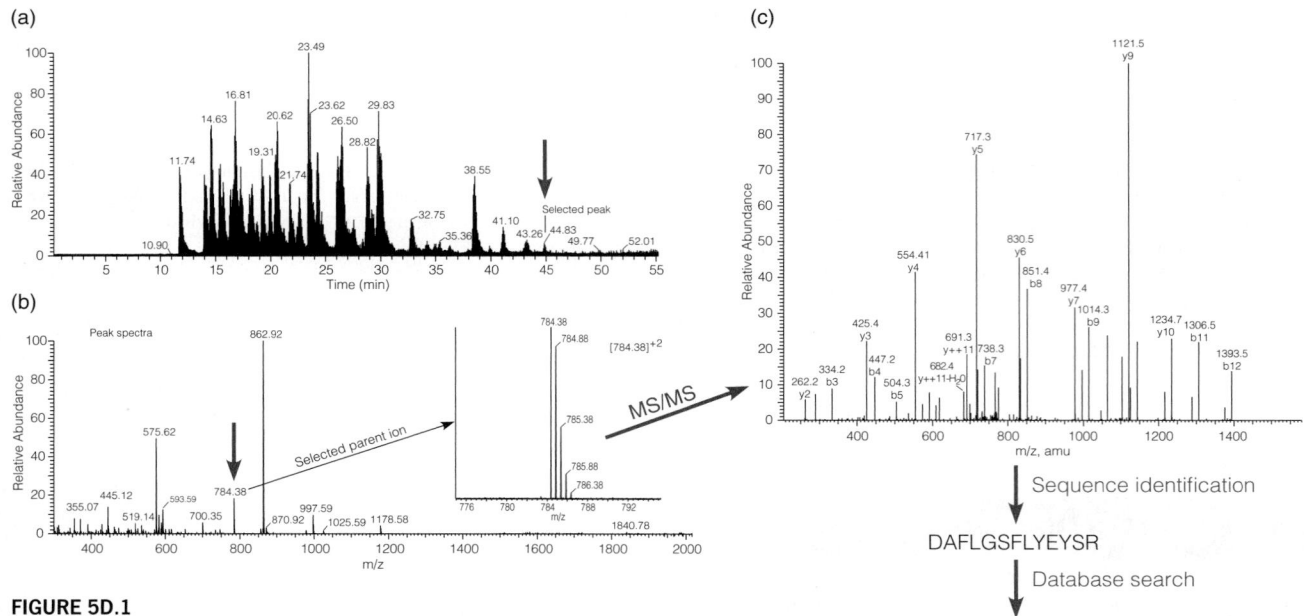

FIGURE 5D.1

Identification of a protein of interest using proteomics methods.
Panel **(a)** A total ion current (TIC) mass spectrum for a complex mixture of
peptides separated by HPLC. The peptide ions from any portion of the TIC
mass spectrum can be selected within the mass spectrometer and subjected to further analysis by tandem mass spectroscopy (e.g., the peak
under the red arrow). Panel **(b)** Fragmentation of the ions selected in
Panel (a) gives a set of ions that can be further separated by mass-to-
charge ratio in the mass spectrometer. Panel **(c)**: An MS-MS spectrum for
one of the parent ions in Panel (b) [red arrow in Panel (b)]. The 13 amino
acid sequence obtained from MS-MS analysis is used in a database search.

Panels (a–c) courtesy of Jack Benner. Panel (d) courtesy of SIB Swiss
Institute of Bioinformatics.

Sequence identification

DAFLGSFLYEYSR

Database search

P02769 Serum albumin precursor (Allergen Bos d 6) (BSA)
MKWVTFISLLLLFSSAYSRGVFRRDTHKSEIAHRFKDLGEEHFKGL
VLIAFSQYLQQCPFDEHVKLVNELTEFAKTCVADESHAGCEKSLHT
LFGDELCKVASLRETYGDMADCCEKQEPERNECFLSHKDDSPDL
PKLKPDPNTLCDEFKADEKKFWGKYLYEIARRHPYFYAPELLYYAN
KYNGVFQECCQAEDKGACLLPKIETMREKVLASSARQRLRCASIQ
KFGERALKAWSVARLSQKFPKAEFVEVTKLVTDLTKVHKECCHGD
LLECADDRADLAKYICDNQDTISSKLKECCDKPLLEKSHCIAEVEK
DAIPENLPPLTADFAEDKDVCKNYQEAK**DAFLGSFLYEYSR**RHPEY
AVSVLLRLAKEYEATLEECCAKDDPHACYSTVFDKLKHLVDEPQNL
IKQNCDQFEKLGEYGFQNALIVRYTRKVPQVSTPTLVEVSRSLGKV
GTRCCTKPESERMPCTEDYLSLILNRLCVLHEKTPVSEKVTKCCTE
SLVNRRPCFSALTPDETYVPKAFDEKLFTFHADICTLPDTEKQIKKQ
TALVELLKHKPKATEEQLKTVMENFVAFVDKCCAADDKEACFAVEG
PKLVVSTQTALA

adding a chemical labeling reagent that is either covalently
attached to some substrate by the target enzyme, or is converted
to a lower molecular weight product. These types of proteomics
experiments have been used to detect metabolic disorders in
newborns (see citations).

There are many challenges to proteomic analysis. For example, the proteins present at low levels in a cell lysate can be difficult to detect. In some eukaryotic cells, the concentration differences between the most- and least-abundant proteins can be
10^6-fold, and many proteins that are interesting targets (e.g., for
drug development) are low-abundance proteins. For these reasons a fractionation step prior to mass analysis may be included
to either remove highly-expressed proteins and/or increase the
concentrations of low-level proteins.

References

Dunn, M. J. (2000) Studying heart disease using the proteomic approach.
Drug Discov. Today 5:76–84.

Gavin, A.-C., et al. (2001) Functional organization of the yeast proteome by
systematic analysis of protein complexes. *Nature* 415:141–147.

Goh, W. W. B., Y. H. Lee, R. M. Zubaidah, J. Jin, D. Dong, Q. Lin, M. C. M.
Chung, and L. Wong (2011) Network-based pipeline for analyzing MS
data: An application toward liver cancer. *J. Proteome Res.* 10:2261–2272.

Graves, P. R., and T. A. Haystead (2002) Molecular biologist's guide to proteomics. *Microbiol. Mol. Biol. Rev.* 66: 39–63.

Nagaraj, S. H., R. B. Gasser, and S. Ranganathan (2006) A hitchhiker's guide to
expressed sequence tag (EST) analysis. *Brief. Bioinform.* 8:6–21.

Ning, Z., H. Zhou, F. Wang, M. Abu-Farha, and D. Figeys (2011) Analytical
aspects of proteomics: 2009–2010. *Anal. Chem.* 83:4407–4426.

Rain, J. C., L. Selig, H. De Reuse, V. Battaglia, C. Reverdy, S. Simon, G. Lenzen,
F. Petel, J. Wojcik, V. Schachter, Y. Chemama, A. Labigne, and P. Legrain
(2001) The protein-protein interaction map of *Helicobacter pylori*.
Nature 409:211–215.

Spacil, Z., S. Elliott, L. Reeber, M. H. Gelb, C. R. Scott, and F. Turecek (2011)
Comparative triplex tandem mass spectrometry assays of lysosomal
enzyme activities in dried blood spots using fast liquid chromatography: Application to newborn screening of Pompe, Fabry, and Hurler
diseases. *Anal. Chem.* 83:4822–4828.

Sutton, C. W., N. Rustogi, C. Gurkan, A. Scally, M. A. Loizidou, A. Hadjisavvas,
and K. Kyriacou (2010) Quantitative proteomic profiling of matched
normal and tumor breast tissues. *J. Proteome Res.* 9:3891–3902.

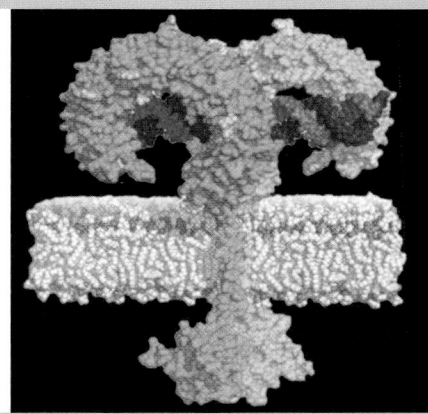

CHAPTER 6

The Three-Dimensional Structure of Proteins

In Chapter 5 we introduced the concept of protein primary structure. We emphasized that this first level of organization, the amino acid sequence, is dictated by the DNA sequence of the gene for the particular protein. However, nearly all proteins exhibit higher levels of structural organization as well. It is the three-dimensional structure of each protein that specifies its function in a particular biological process.

Figure 5.1 (page 137) shows a well-defined spatial location for every heavy atom within the protein sperm whale myoglobin. Figure 6.1 depicts another representation of the three-dimensional conformation of the myoglobin molecule, and illustrates that there exist two distinguishable levels of three-dimensional folding of the polypeptide chain. First, the chain appears to be locally coiled into regions of helical structure. Such local *regular* folding is called the **secondary structure** of the molecule. The helically coiled regions themselves are, in turn, folded into a specific compact structure for the entire polypeptide chain. We call this further level of folding the **tertiary structure** of the molecule. Later in this chapter we shall find that some proteins consist of several folded polypeptide chains, arranged in a regular manner. This arrangement we designate as the **quaternary** level of organization.

This chapter is devoted to an examination of the levels of protein structure, how folded proteins are stabilized, the mechanisms by which protein folding is thought to occur, and emerging computational methods for predicting tertiary structure from primary sequence.

Protein molecules have four levels of structural organization: primary (sequence), secondary (local folding), tertiary (overall folding), and quaternary (subunit association).

Secondary Structure: Regular Ways to Fold the Polypeptide Chain

Theoretical Descriptions of Regular Polypeptide Structures

Our understanding of the protein secondary structure had its origins in the remarkable work of Linus Pauling, one of the greatest chemists of the twentieth century. As early as the 1930s, he had begun X-ray diffraction studies of amino

FIGURE 6.1

Three-dimensional folding of the protein myoglobin. This "cartoon" rendering was generated from the X-ray crystal structure determined by H. C. Watson and J. C. Kendrew (PDB ID: 1MBN), and it shows the polypeptide main chain as helices connected by thick lines. Side chains are shown as thin lines. Individual helical regions are color-coded, with the peptide N-terminus shown in blue and the C-terminus shown in red. This protein binds a heme group (shown in space-filling display). The orientation of the protein in this figure is the same as that shown in Figure 5.1.

Of the several possible secondary structures for polypeptides, the most frequently observed are the α helix and the β sheet.

acids and small peptides, with the aim of eventually analyzing protein structure. In the early 1950s, Pauling and his collaborators used these data together with remarkable scientific intuition to begin a systematic analysis of the possible regular conformations of the polypeptide chain. They postulated several principles that any such structure must obey:

1. The bond lengths and bond angles should be distorted as little as possible from those found through X-ray diffraction studies of amino acids and peptides, as shown in Figure 5.14b (page 148).

2. No two atoms should approach one another more closely than is allowed by their van der Waals radii.

3. The amide group must remain planar and in the *trans* configuration, as shown in Figure 5.14b. (This feature had been recognized in the earlier X-ray diffraction studies of small peptides.) Consequently, rotation is possible only about the two bonds adjacent to the α-carbon in each amino acid residue, as shown in Figure 6.2.

4. Some kind of noncovalent bonding is necessary to stabilize a regular folding. The most obvious possibility is hydrogen bonding between amide protons and carbonyl oxygens:

$$\diagup N-H \cdots O=C \diagdown$$

Such a concept was familiar to Pauling, who had had much to do with the development of the idea of H-bonds. In summary, the preferred conformations must be those that allow a maximum amount of hydrogen bonding, yet satisfy criteria 1–3.

α Helices and β Sheets

Working mainly with molecular models, Pauling and his associates were able to arrive at a small number of regular conformations that satisfied all of these criteria. Some were helical structures formed by a single polypeptide chain, and some were sheet-like structures formed by adjacent chains. The two structures they proposed as most likely—the right-handed **α helix**, and the **β sheet**—are shown in Figure 6.3a and b. These two structures are, in fact, the most commonly observed secondary structures in proteins. Figure 6.3c shows the so-called **3₁₀ helix**, which is observed in some proteins but is not as common as the α helix. All of the protein secondary structures shown in Figure 6.3 satisfy the criteria listed earlier. In particular, in each structure the peptide group is planar, and every amide proton and every carbonyl oxygen (except a few near the ends of helices) is involved in hydrogen bonding. The arrangement of the main chain H-bonds in the α helix along the helical axis orients the amide N—H and C=O such that the dipole moments for each of these polar bonds align and gives rise to a **helical dipole moment** (also called a "macrodipole"). In effect, the

FIGURE 6.2

Rotation around the bonds in a polypeptide backbone. Two adjacent amide planes are shown in light green. Rotation is allowed only about the N_{amide}—C_α and C_α—$C_{carbonyl}$ bonds. The angles of rotation about these bonds are defined as ϕ (phi) and ψ (psi), respectively, with directions defined as positive rotation as shown by the arrows; positive rotation is clockwise as seen from the α-carbon. The extended conformation of the chain shown here corresponds to $\varphi = +180°$, $\psi = +180°$.

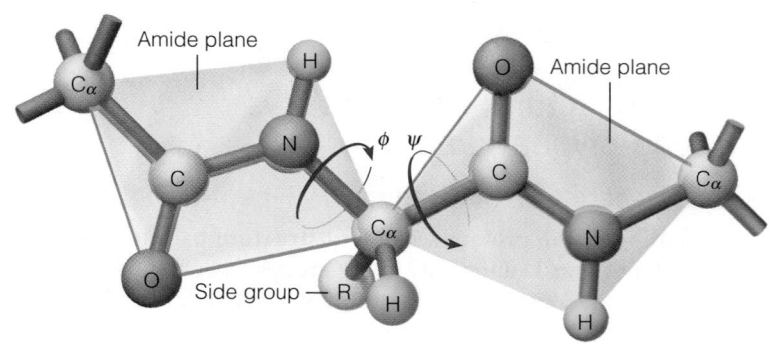

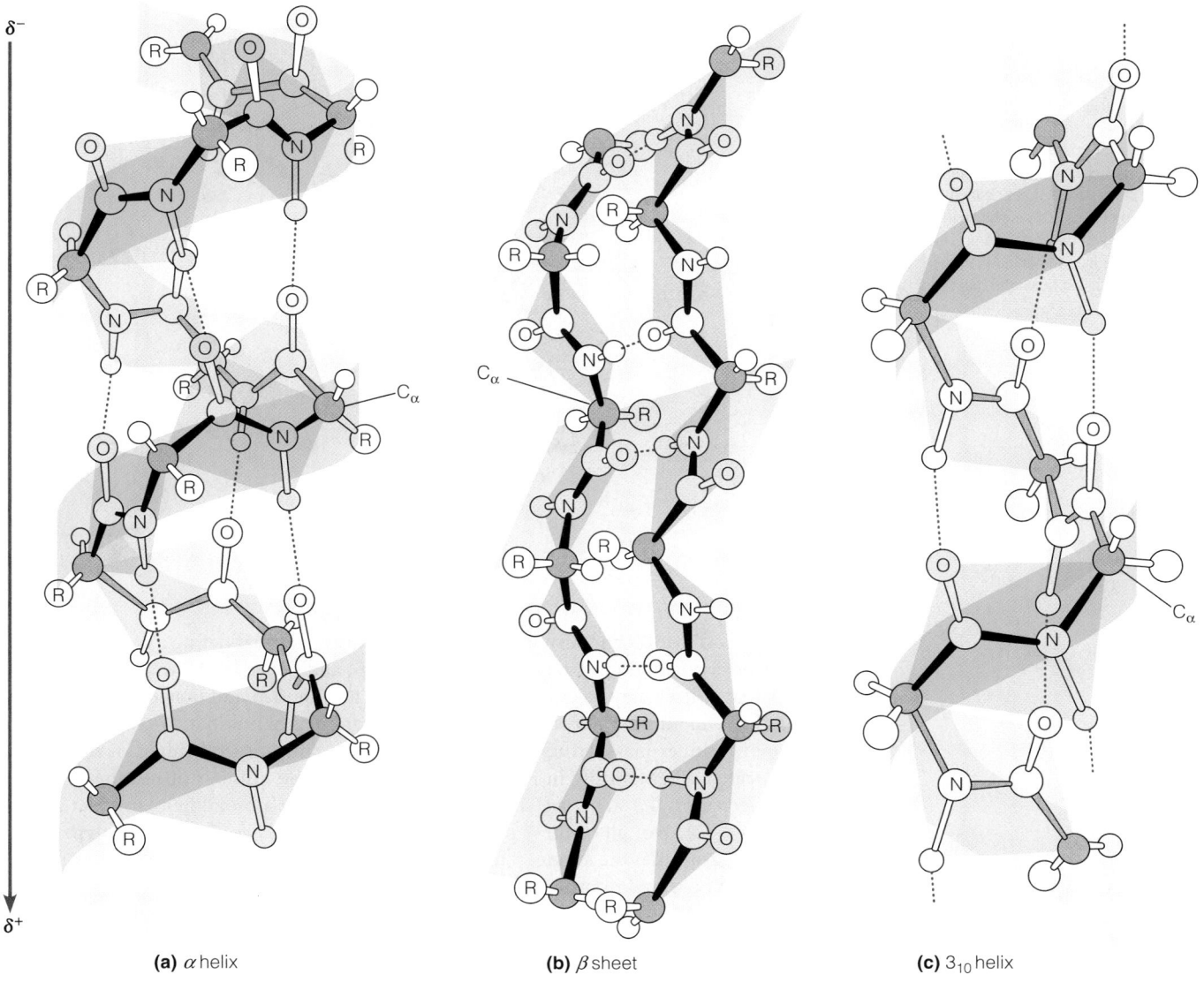

δ^-

δ^+

(a) α helix

(b) β sheet

(c) 3_{10} helix

FIGURE 6.3

The right-handed α helix, β sheet, and 3_{10} helix.
The right-handed α helix and β sheet are the two most frequently observed regular secondary structures of polypeptides. **(a)** In the α helix the hydrogen bonds (red dotted lines) are within a single polypeptide chain and are almost parallel to the helix axis. The alignment of amide bonds in the helix gives rise to a helical macrodipole moment shown by the red arrow (see Figure 2.4, page 29). The N-terminal end of the helix has partial (+) charge character and the C-terminal end has partial (−) charge character. **(b)** In the β sheet, the hydrogen bonds are between adjacent chains, of which only two are shown here. In this structure, the hydrogen bonds are nearly perpendicular to the chains. **(c)** The 3_{10} helix is found in proteins, but is less common than the α helix.

An amphiphilic α helix will have side chains of similar polarity every 3–4 residues, whereas a β strand in an amphiphilic β sheet will have alternating polar and nonpolar side chains.

N-terminus of the helix has partial (+) charge character and the C-terminus has partial (−) charge character as shown by the red arrow in Figure 6.3.

Amphiphilic Helices and Sheets

In an α helix the side chains are pointing away from the center of the helix (Figure 6.4). In a β sheet, a network of main chain H-bonds connects the β strands. If we consider the H-bonded main chains to be a "sheet," the side chains are located on opposite faces of this sheet, as shown in the right-hand panels of Figure 6.4. Side chains of similar polarity are frequently clustered together to form extended hydrophilic or hydrophobic surfaces, or "faces," on one side of a helix or sheet. Secondary structures that display a predominantly hydrophobic face opposite a predominantly hydrophilic face are said to be "**amphiphilic**" (or "**amphipathic**"). Many α helices and β sheets have this characteristic because it allows two or more secondary structures to associate via contacts between the hydrophobic surfaces, while projecting the hydrophilic surfaces toward the aqueous solvent. An amphiphilic α helix will have side chains of similar polarity every 3–4 residues, whereas a β strand in an amphiphilic β sheet will have alternating polar and nonpolar side chains. These distinct patterns of side chain polarity are the basis for many secondary structure prediction algorithms (discussed below).

FIGURE 6.4

The positions of side chains in the α helix and β sheet. An idealized right-handed α helix and a two-stranded β sheet are shown with the main chain atoms colored gray/red/blue and side chain atoms colored green, red, blue, yellow. In the α helix the side chains radiate away from the center of the helix (lower left panel). In the β sheet the side chains are located on opposite faces of the sheet defined by the H-bonded main chain amides. This is shown in the upper right panel, looking at the main chain strands from the side (one strand is hidden behind the other), and the lower right panel, looking down the main chains of both strands.

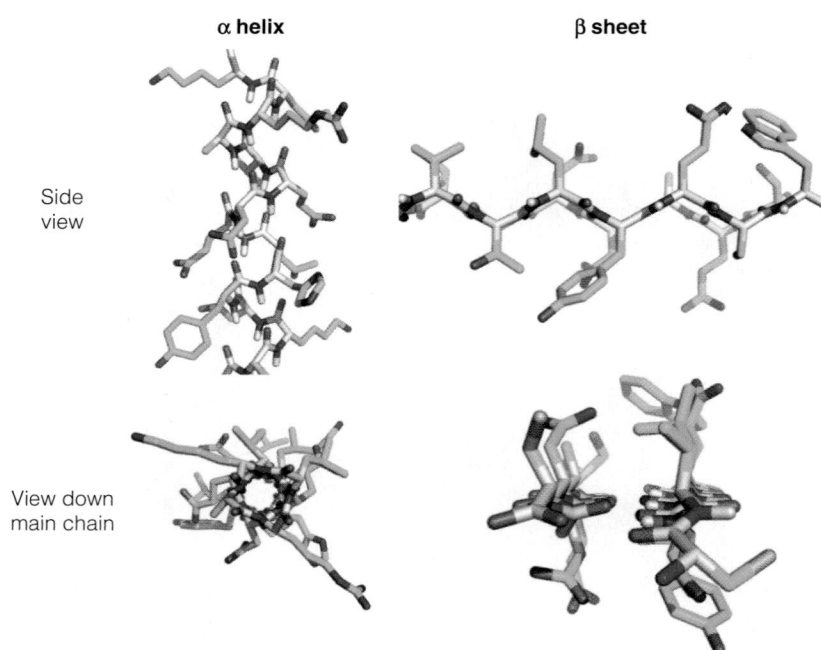

α helix   β sheet

Side view

View down main chain

Describing the Structures: Helices and Sheets

In Tools of Biochemistry 4A, we listed the distances that define a molecular helix: the crystallographic repeat (c), the pitch (p), and the rise (h). We also pointed out that helices may be either right-handed or left-handed and may contain either an integral number of residues per turn or a nonintegral number. We call the number of residues per turn n. Some idealized helices with integral values of n are illustrated schematically in Figure 6.5. Note that as the number of residues per turn decreases, the structure changes progressively from a broad helix to a flat ribbon ($n = 2$). Not all of these proposed helical structures are found in polypeptides. For example, the single-chain $n = 2$ structure shown in Figure 6.5 has not yet been observed in proteins.

One of Pauling's major insights was to recognize that polypeptide helices are not required to have an integral number of residues per turn. For example, the α helix repeats after exactly 18 residues, which amounts to 5 turns. It has, therefore, 3.6 residues per turn. Because the pitch of a helix is given by $p = nh$, we have for the α helix, with a rise of 0.15 nm/residue, $p = (3.6 \text{ res/turn}) \times (0.15 \text{ nm/res}) = 0.54$ nm/turn. Parameters for the secondary structures shown in Figures 6.3 and 6.4 are listed in Table 6.1.

FIGURE 6.5

Idealized helices. These hypothetical structures show the effect of varying the number (n) of polypeptide residues per turn of a helix. The white balls represent α-carbon atoms. In each case the pitch (p) is indicated, and for $n = 2$ the rise (h) is also shown. The $n = 4$ and $n = 3$ helices are right-handed, the $n = -3$ helix is left-handed, and $n = 2$ (a flat ribbon) has no handedness. The right-handed α helix (not shown here), with $n = 3.6$, is intermediate between the $n = 3$ and $n = 4$ structures.

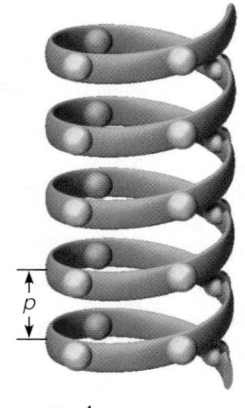

$n = 4$
Helix (right-handed)

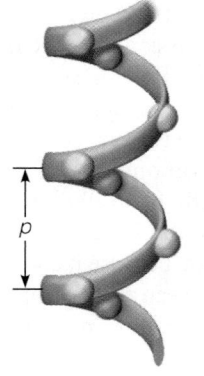

$n = 3$
Helix (right-handed)

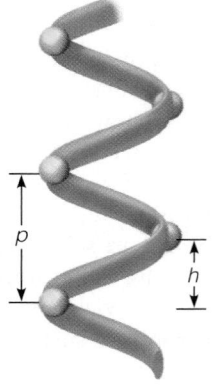

$n = 2$
Flat ribbon

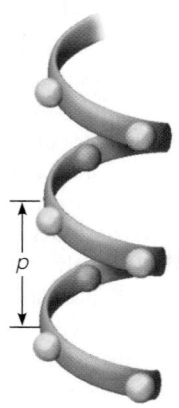

$n = -3$
Helix (left-handed)

When you examine the model for the α helix (Figure 6.3a), you will note that a given carbonyl oxygen, on residue i, is hydrogen-bonded to the amido proton that is four residues removed in the direction of the C-terminus (i.e., on residue $i + 4$). Thus, if we include the hydrogen bond, a loop of 13 atoms is formed. Figure 6.6 shows this schematically for the α and 3_{10} helices. Each helix type has a different number of atoms in such a hydrogen-bonded loop. We shall call this number N. Rather than using the parameters n, h, and p, an alternative way to describe a polypeptide helix is to combine n and N in the shorthand n_N. The 3_{10} helix fits this description; it has exactly 3.0 residues per turn and a 10-member hydrogen-bonded loop. The α helix could also be called a 3.6_{13} helix.

Because hydrogen bonds tend to be linear, the atoms $N—H\cdots O$ in polypeptide helices should lie on a straight line. Figure 6.3 shows that this requirement is approximately satisfied for the 3_{10} and α helices. On the other hand, it is very difficult to make helices with only two residues per turn *and* linear hydrogen bonds between residues in the same chain; thus, the only $n = 2$ structure that is found in proteins is *not* the flat ribbon shown in Figure 6.5 but the β sheet structure shown in Figure 6.3b.

A β sheet is composed of two or more **β strands**. Each residue in the strand is rotated by 180° with respect to the preceding one, which makes each β strand an $n = 2$ "helix." If the chains are also folded in the pleated fashion shown in Figure 6.3b, hydrogen bonds can occur *between* adjacent β strands. Forming interchain bonds allows correct bond angles with minimal strain when $n = 2$. There are two ways in which β strands can be oriented in a β sheet. The β sheet shown in Figure 6.3b shows two β strands arranged such that the N-terminus to C-terminus orientations of the two strands are in opposite directions. Such an arrangement of strands is called "antiparallel," whereas the arrangement with both strands oriented in the same direction is called "parallel" (see Figure 6.7). The hydrogen bonds between antiparallel strands are linear, whereas those between parallel strands are not.

In addition to the helices and sheets described above, there is one more regularly repeating conformation that is commonly observed in protein structures—the so-called **polyproline II helix**. This particular conformation was not predicted on theoretical grounds because it does not satisfy Pauling's requirement for H-bonding. Nevertheless, it is a common motif in protein structures. Unlike the α and 3_{10} helices, this structure does not have stabilizing H-bonds between mainchain groups, and it is left-handed. Roughly a third of the amino acid residues found in this conformation are prolines, leading to the designation "polyproline II helix"; however, glycine is often found in this conformation, as are, albeit to a much lesser extent, several other amino acids. Because this conformation is not restricted to proline residues, we will refer to this secondary structure motif by the more general term "polypeptide II helix" (Figure 6.8).

TABLE 6.1 Parameters of some polypeptide secondary structures

Structure Type	Residues/ Turn	Rise (h) per residue	Pitch (p)
β Strand (antiparallel)	2.0	0.34 nm	0.68 nm
β Strand (parallel)	2.0	0.32 nm	0.64 nm
α helix	3.6	0.15 nm	0.54 nm
3_{10} helix	3.0	0.20 nm	0.60 nm
Polypeptide II helix ("polyproline II helix")	3.0	0.47 nm	0.94 nm

FIGURE 6.6

Hydrogen bonding patterns for the α and 3_{10} helices. The structures are represented in a diagrammatic way to simplify counting the atoms in each H-bonded loop. For example, there are 13 atoms in the H-bonded loop corresponding to the α (3.6_{13}) helix. The significance of the π helix is discussed later in this chapter.

Illustration, Irving Geis. Image from Irving Geis Collection/Howard Hughes Medical Institute. Rights owned by HHMI. Not to be reproduced without permission.

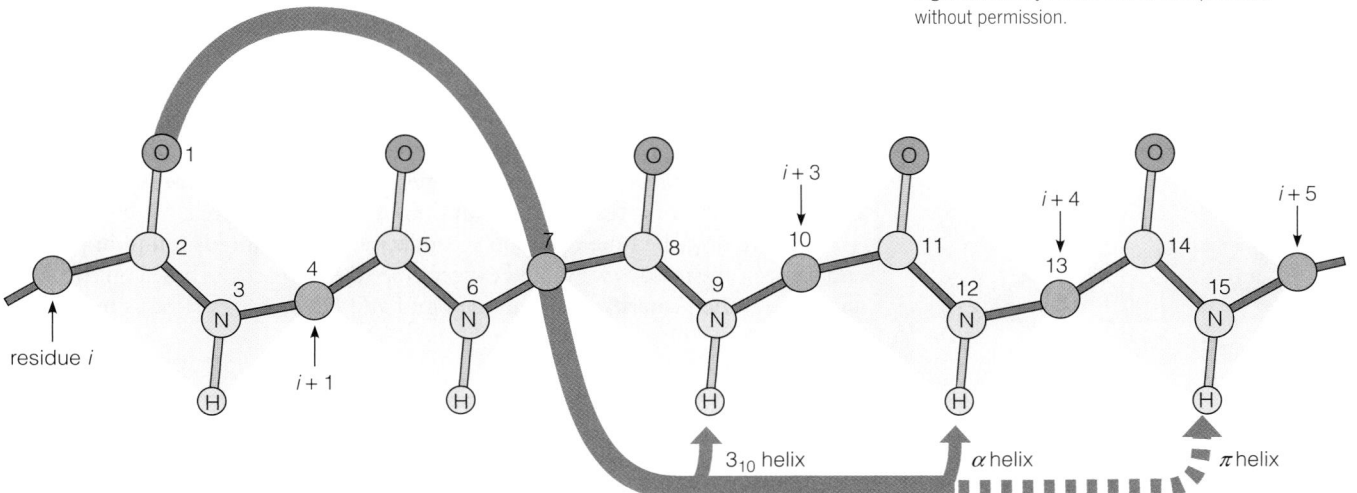

(a) Antiparallel

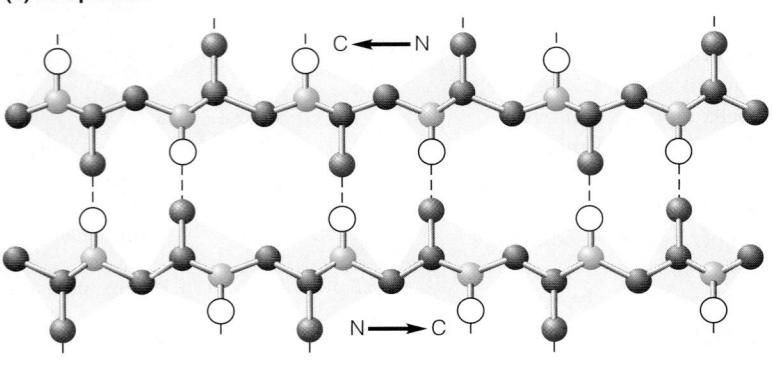

(b) Parallel

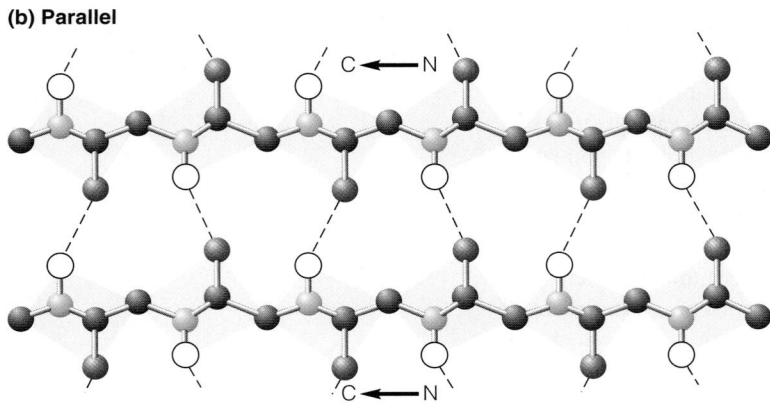

FIGURE 6.7

β Sheets. **(a)** An antiparallel arrangement of β strands. **(b)** A parallel arrangement of β strands. Only main chain atoms are shown (side chains omitted for clarity); H-bonds between strands are represented by dashed lines.

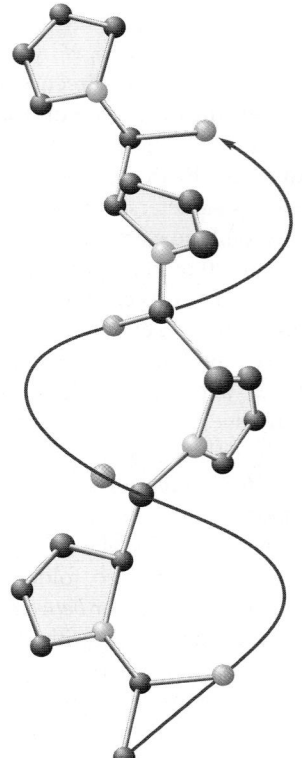

FIGURE 6.8

The polypeptide II helix. A polyproline sequence is shown; however, polyglycine also adopts this conformation. The left-handed helical twist is indicated by the curved gray arrow.

To close this section we describe a conformation called the **π helix**. The π helix is also known as an "α bulge" or "α aneurism" or "π bulge." It is widespread, as it is found in ~15% of protein sequences in the Protein Data Bank, although it is infrequent, as it typically occurs only once in any given sequence. Most (~85%) π helix conformations appear to be the result of a mutation event that results in the insertion of an amino acid into an α helix (Figure 6.9). This disrupts the normal H-bonding in the α helix, and two or more residues form H-bonds with the $i + 5$ residue, rather than the $i + 4$ residue (see Figure 6.6), creating a bulge in the helical structure. The short stretch of π helix shown in Figure 6.9 is typical of π helices found in most proteins—most are only one turn in length, and extended π helices of greater than two turns are not observed in proteins. For this reason, we will not consider the π helix as a regular repeating conformation. Nevertheless, the π helix is noteworthy because the inserted amino acid frequently confers some new functional properties on the resulting protein; thus, it is a potential marker for tracing the evolution of protein function.

We have described the common secondary structural motifs found in proteins and some of the reasons we should expect to see these structures based on consideration of (1) the planarity of the amide bond and (2) steric restrictions to rotation

FIGURE 6.9

The π helix conformation. On the left, a main-chain rendering of the C-terminal α helix from *Staphylococcus aureus* nuclease is shown (PDB ID: 1EYD). The amino acid sequence of this helix is shown below in one-letter code. The site of Gly insertion is underlined. On the right, the analogous helix from a Gly insertion mutant is shown (PDB ID: 1STY). The inserted Gly is highlighted in green and the adjacent four residues that adopt the π helix conformation are highlighted in magenta. Note that the inserted Gly carbonyl does not form an intrahelical H-bond.

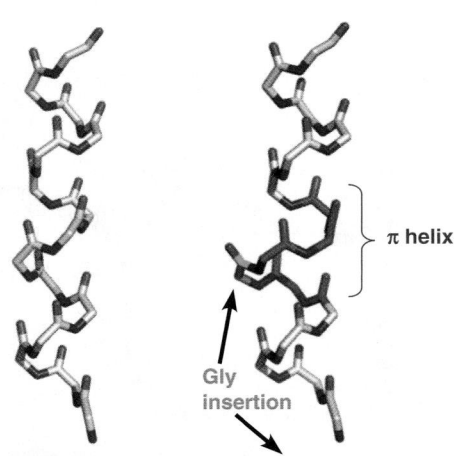

π helix

Gly insertion

HEQHLRKSEAQAKKE   **HEQHLRGKSEAQAKKE**

around the angles ϕ and ψ. We have also described two ways to specify the regular repeating unit of the motif– by n_N, or by listing the parameters in Table 6.1. Yet another way to describe the regular repeat of a secondary structural motif is to specify the values of the angles ϕ and ψ. As illustrated in Figure 6.10a, some combinations of ϕ and ψ (e.g., $\phi = 0°$ and $\psi = 0°$) are not allowed due to steric crowding. These steric constraints on peptide conformation can be appreciated by considering space-filling models for helices and sheets. For example, as shown in Figure 6.10b, the main chain atoms in the α helix are closely packed, with R groups projecting away from the helical axis. The combinations of ϕ and ψ angles that are most favorable (or "allowed")—because they relieve steric crowding—are shown in a systematic description of polypeptide backbone conformation called a **Ramachandran plot**.

Ramachandran Plots

As is shown in Figure 6.2, each residue in a polypeptide chain has two backbone bonds about which rotation is permitted. The angles of rotation about these bonds, defined as ϕ and ψ, describe the backbone conformation of any particular residue in any protein. To make the definition meaningful, we must specify what we mean by a positive direction of rotation and the zero-angle conformation of each. The conventions chosen for directions of positive rotation about ϕ and ψ are given by the arrows in Figure 6.2—that is, clockwise when looking in either direction from the α-carbon. The conformation shown in that figure corresponds to $\phi = +180°$ and $\psi = +180°$, the fully extended form of the polypeptide chain.

With these conventions, the backbone conformation of any particular residue in a protein can be described by a point on a map (Figure 6.11) with coordinates ϕ and ψ. Such maps are called Ramachandran plots, after the biochemist G. N. Ramachandran, who first made extensive use of them. For any regular

FIGURE 6.10

Steric interactions determine peptide conformation. **(a)** A sterically nonallowed conformation. The conformation $\phi = 0°$, $\psi = 0°$ is not allowed in any polypeptide chain because of the steric crowding between the carbonyl oxygen and amido proton. **(b)** The atoms in a helix are closely packed. Here, a segment of an α helix in sperm whale myoglobin (the longer green helix in Figure 6.1; PDB ID: 1MBN) is shown as a space-filling model.

(a)

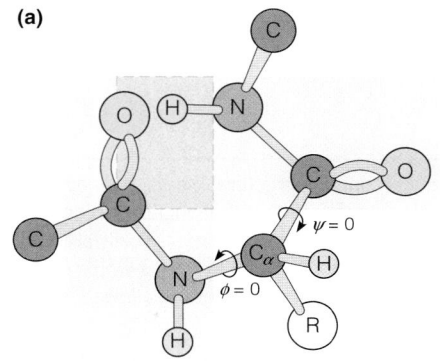

(b)

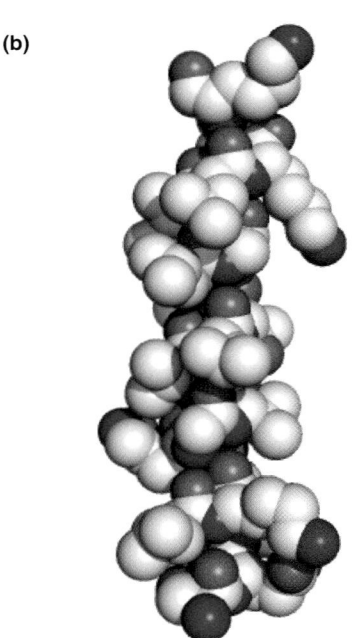

FIGURE 6.11

A Ramachandran plot for poly-L-alanine. A map of this type can be used to describe the backbone conformation of any polypeptide residue as well as the secondary structures of proteins. The coordinates are the values of the bond angles ϕ and ψ, defined as in Figure 6.2. The white areas correspond to sterically allowed conformations for poly-L-alanine (i.e., a peptide consisting of only L-Ala residues). The colored lines running across the graph correspond to various values of n (residues per turn); the line bisecting the graph corresponds to $n = 2$. The helix is right-handed when n is positive (pink and red regions), left-handed when n is negative (blue regions). The circles with the following symbols correspond to the secondary structures discussed in the text: α_R, right-handed α helix; 3, 3_{10} helix; β, β sheet; P_{II}, polypeptide II helix; α_L, left-handed α helix.

TABLE 6.2	Ranges of allowed ϕ and ψ angles for some polypeptide secondary structures	
Structure Type	Φ	Ψ
β strand	$-150°$ to $-100°$	$+120°$ to $+160°$
α helix	$-70°$ to $-60°$	$-50°$ to $-40°$
3_{10} helix	$-70°$ to $-60°$	$-30°$ to $-10°$
Polypeptide II helix ("polyproline II helix")	$-80°$ to $-60°$	$+130°$ to $+160°$

Data from *Protein Science* 18:1321–1325 (2009), S. A. Hollingsworth, D. S. Berkholz, and P. A. Karplus, On the occurrence of linear groups in proteins.

Many secondary structural motifs, defined by regular repeats of ϕ and ψ angles, can be imagined; but, only a few are sterically allowed. The Ramachandran plot illustrates which combinations of ϕ and ψ angles are allowed.

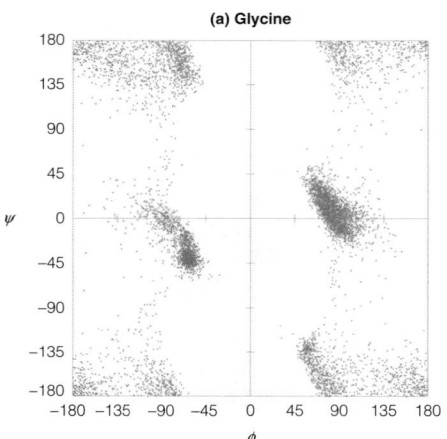

(a) Glycine

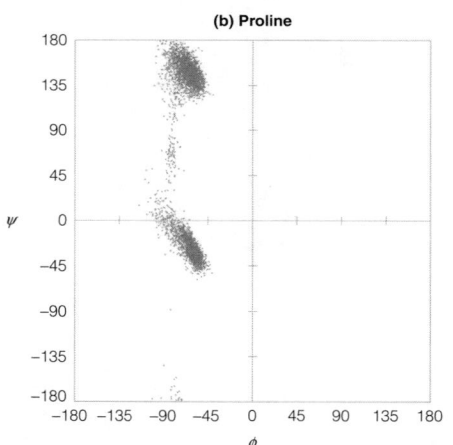

(b) Proline

FIGURE 6.12

Ramachandran plots for glycine and proline. The data shown are for glycine and proline residues found in high-resolution X-ray crystal structures of proteins. The favorable combinations of the bond angles ϕ and ψ are shown by the darker regions. Glycine has the greatest number of allowed ϕ, ψ angle combinations, whereas proline has the fewest.

Plots courtesy of S. A. Hollingsworth and P. A. Karplus, Oregon State University.

repeating secondary structure (e.g., α helix, β sheet, etc.), all the residues that are part of the structure are in nearly equivalent conformations and have nearly equivalent ϕ and ψ angles; thus, the points on a Ramachandran plot that correspond to those residues will cluster within a narrow range of sterically allowed ϕ and ψ angles for a given secondary structure. Table 6.2 lists the ranges of ϕ and ψ angles that correspond to the various helices and β sheets described above.

One of the most useful features of Ramachandran plots is that they allow us to describe very simply which structures are sterically possible and which are not. For many of the conceivable combinations of ϕ and ψ values, some atoms in the chain would approach closer than their van der Waals radii would allow (an example is shown in Figure 6.10). Such conformations are unfavorable because they are sterically crowded. Ramachandran and other researchers have examined the entire map surface, using models and computers to determine which conformations are actually allowed. The allowed ϕ, ψ combinations for poly-L-alanine correspond to the white regions in Figure 6.11. Clearly, only a small fraction of the conceivable conformations is actually favorable. All of the regular secondary structures we have discussed fall within or very close to these regions.

Although Figure 6.11 shows the left-handed α helix lying on the edge of an allowed region, it is, in fact, not nearly as favored as the right-handed form. This difference is a consequence of the fact that all amino acids in proteins are of the L-form. With L-amino acids, steric interference between the side chains and the backbone of the helix is less with a right-handed helix than with a left-handed helix. This principle can be understood from a careful inspection of Figure 6.3a. Note that each R group is approximately *trans* to the adjacent carbonyl oxygen. If the amino acid were D instead of L, the orientation would be *cis*, with more likelihood of steric crowding. Recall from Chapter 5 that chemists have synthesized proteins with all D-amino acids. These proteins have, as expected, left-handed α helices. The importance of such side-chain effects depends on the bulkiness of the side chain. The map shown in Figure 6.11 was drawn with the assumption that all residues are L-alanine (that is, all have CH_3 side chains). If bulkier side chains were considered, the "allowed" region would shrink. Conversely, glycine, with its —H side chain, allows more conformations, as shown in Figure 6.12a. Proline has fewer combinations of allowed ϕ and ψ angles due to restricted rotation around ϕ (Figure 6.12b).

The foregoing analysis of protein structure is based on a relatively simple consideration of probable steric interactions. How does it compare with observations of actual protein structures? In this case, the correspondence between theory and observation is remarkably good. Figure 6.13 shows a plot of ϕ, ψ pairs for 30,692 residues found in 209 different polypeptide chains for which high resolution (≤0.12 nm) X-ray crystallography data exist. As is evident in this figure, the majority of the observed ϕ, ψ pairs (gray dots in Figure 6.13) cluster in the regions of the Ramachandran plot that are predicted to give the most favorable combinations of ϕ and ψ (white spaces in Figure 6.11). Those residues that are classified by H-bonding patterns and geometry to be helix or sheet are identified by color in Figure 6.13 and illustrate the range of ϕ and ψ values that correspond to each of the secondary structure types discussed above (see Table 6.2).

Most of the points on this Ramachandran plot fall close to the right-hand α helix or the β sheet positions; but, they do not correspond exactly to these points, testifying to the existence of distortions of these structures in folded proteins, and to the existence of regions of structure different from either β sheet or α helix. Although most of the points fall in "allowed" regions, a few lie in "nonallowed" regions. These are mainly glycines, for which a much wider range of ϕ and ψ angles is allowed because the side chain is so small (Figure 6.12a).

Historically, the ideal and observed values of ϕ and ψ angles for parallel and antiparallel β sheets have been listed as distinct. That distinction is no longer apparent in the high-resolution structural data shown in Figure 6.13. Because the observed ϕ and ψ angles for these two β structure types overlap significantly (due to distortion of ideal angles in actual protein structures), we have chosen to list in Table 6.2 a single range of values that describes both parallel and antiparallel β sheet conformations.

Our discussion so far provides a background for understanding the basics of protein structure. It is now time to consider some specific cases. We begin with the observation that two major classes of proteins exist. These are called *fibrous* and *globular* proteins and are distinguished by major structural differences. Let us first consider the fibrous proteins.

Fibrous Proteins: Structural Materials of Cells and Tissues

Fibrous proteins are distinguished from globular proteins by their filamentous, or elongated, form. Most of them play structural roles in animal cells and tissues—they hold things together. Fibrous proteins include the major proteins of skin and connective tissue and of animal fibers like hair and silk. The amino acid sequence of each of these proteins favors a particular kind of secondary structure, which confers on each protein a particular set of appropriate mechanical properties. Table 6.3 lists

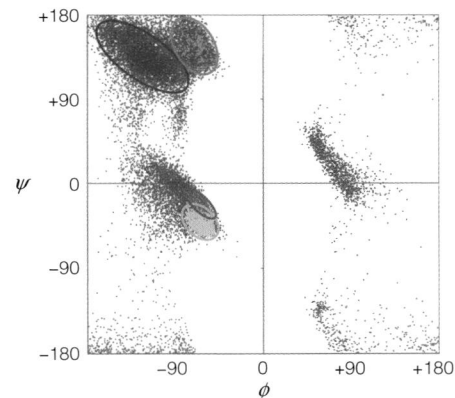

FIGURE 6.13

Observed values of ϕ and ψ from protein structural data. ϕ, ψ pairs for 30,692 residues observed in high resolution ($\leq$ 0.12 nm) crystal structures of proteins are shown (gray dots). Different colors highlight those residues defined to be right-handed α helix (cyan), right-handed 3_{10} helix (purple), β strand (blue), and left-handed polypeptide II helix (orange). The median values of ϕ and ψ for each secondary structure type are (ϕ, ψ): α helix (-63, -43), 3_{10} helix (-62, -22), β strand (-116, $+129$), and polypeptide II helix (-65, $+145$). See also Table 6.2.

Courtesy of P. A. Karplus, Oregon State University.

TABLE 6.3 Amino acid compositions of some fibrous proteins

Amino Acid	α-Keratin (wool)	Fibroin (silk)	Collagen (Bovine tendon)	Elastin (Pig aorta)	All proteins[f]
Gly	8.1	44.6	32.7	32.3	7.9
Ala	5.0	29.4	12.0	23.0	8.7
Ser	10.2	12.2	3.4	1.3	5.8
Glu + Gln	12.1	1.0	7.7	2.1	6.6 (3.7)
Cys	11.2	0	0	—[e]	1.3
Pro	7.5	0.3	22.1[a]	10.7[c]	4.7
Arg	7.2	0.5	5.0	0.6	5.0
Leu	6.9	0.5	2.1	5.1	8.9
Thr	6.5	0.9	1.6	1.6	5.6
Asp + Asn	6.0	1.3	4.5	0.9	5.9 (4.2)
Val	5.1	2.2	1.8	12.1	7.2
Tyr	4.2	5.2	0.4	1.7	3.5
Ile	2.8	0.7	0.9	1.9	5.5
Phe	2.5	0.5	1.2	3.2	4.0
Lys	2.3	0.3	3.7[b]	3.6[d]	5.5
Trp	1.2	0.2	0	—[e]	1.5
His	0.7	0.2	0.3	—[e]	2.4
Met	0.5	0	0.7	—[e]	2.0

Note: The three most abundant amino acids in each protein are indicated in red. Values given are in mole percent.

[a] About 39% of this is hydroxyproline.
[b] About 14% of this is hydroxylysine.
[c] About 13% of this is hydroxyproline.
[d] Most (about 80%) is involved in cross-links.
[e] Essentially absent.
[f] Reprinted from *Journal of Chemical Information and Modeling* 50:690–700, J. M. Otaki, M. Tsutsumi, T. Gotoh, and H. Yamamoto, Secondary structure characterization based on amino acid composition and availability in proteins. © 2010 American Chemical Society.

Fibrous proteins are elongated molecules with well-defined secondary structures. They usually play structural roles in the cell.

the amino acid composition of four examples of fibrous proteins: α-keratin, fibroin, collagen, and elastin. Compared to the typical distributions of the 20 amino acids in globular proteins (see "All proteins" column in Table 6.3) each of these fibrous proteins is significantly enriched in 3–4 particular amino acids, which stabilize the extended secondary structures typical of the fibrous proteins.

The Keratins

Two important classes of proteins that have similar amino acid sequences and biological function are called α- and β-keratins. The **α-keratins** are the major proteins of hair and fingernails and comprise a major fraction of animal skin. The α-keratins are members of a broad group of **intermediate filament proteins**, which play important structural roles in the nuclei, cytoplasm, and surfaces of many cell types. All of the intermediate filament proteins are predominantly α-helical in structure; in fact, it was the characteristic X-ray diffraction pattern of α-keratin that Pauling and his colleagues sought to explain by their α helix model.

α-Keratin is built on a coiled-coil α-helical structure.

The structure of a typical α-keratin, that of hair, is depicted in Figure 6.14. The individual molecules contain long sequences—over 300 residues in length— that are wholly α-helical. Pairs of these right-handed helices twist about one another in a left-handed **coiled-coil** structure. This pairing of α helices appears to be a consequence of a peculiarity of the amino acid sequence of α-keratin. Every third or fourth amino acid has a nonpolar, hydrophobic side chain. Because the α helix has 3.6 residues/turn, this means there is a strip of contiguous hydrophobic surface area along one face of each helical chain (the hydrophobic surface area makes a shallow spiral around the helix because 4.0 does not exactly equal 3.6). As we noted in Chapter 2, hydrophobic surfaces tend to associate in aqueous

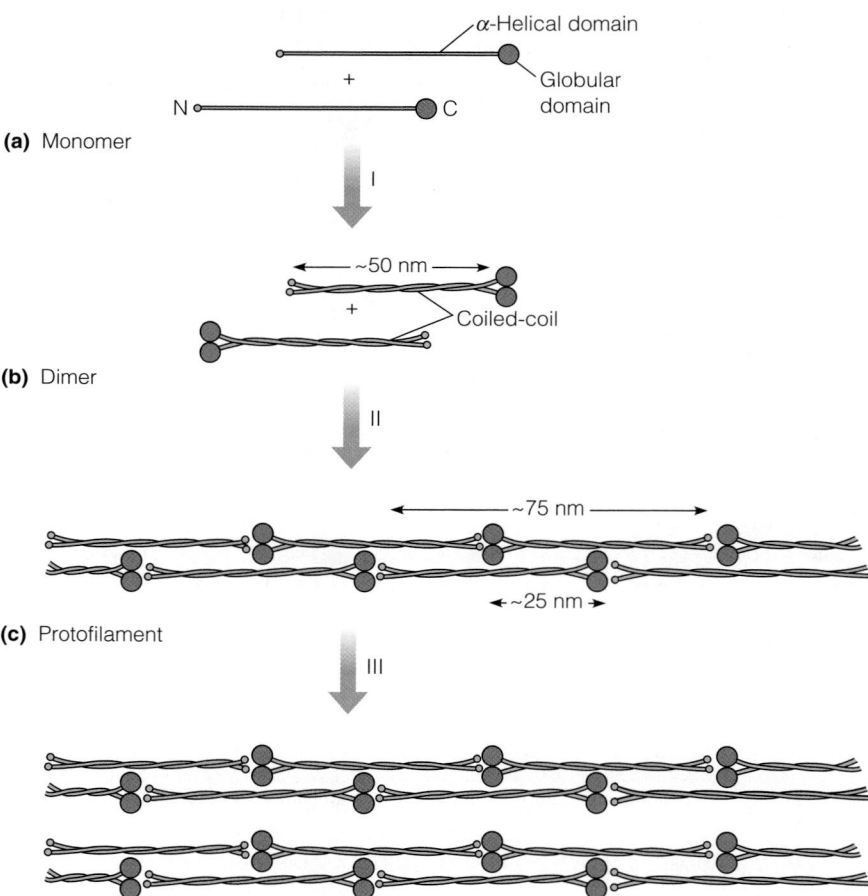

FIGURE 6.14

Proposed structure for keratin-type intermediate filaments. Two monomers **(a)** pair via a parallel coiled-coil to form the 50-nm-long dimer **(b)**. These then associate to form first a 4-strand protofilament **(c)** and then an 8-strand protofibril **(d)**. The regular spacing of 25 nm along the fibers is accounted for by the overlap.

medium; thus, two α-keratin helices are noncovalently bonded by hydrophobic interactions between the entwined helices.

In intermediate filaments, pairs of coiled coils themselves tend to associate into a four-chain protofilament (Figure 6.14c), and two of these in turn pack together to form a protofibril (Figure 6.14d). The details of these higher levels of association are still unclear. Such twisted cables can be very stretchy and flexible, but in different tissues α-keratin is hardened, to differing degrees, by the introduction of disulfide cross-links within the several levels of fiber structure (note that α-keratin has an unusually high content of cysteine—see Table 6.3). Fingernails have many cross-links in their α-keratin, whereas hair has relatively few. The process of introducing a "permanent wave" into human hair involves reduction of these disulfide bonds, rearrangement of the fibers, and reoxidation to "set" the tight curls thus introduced.

The β-keratins, as their name implies, contain much more β-sheet structure. Indeed, they represented the second major structural class described by Pauling and his coworkers. The β-keratins are found mostly in birds and reptiles, in structures like feathers and scales.

Fibroin

The β-sheet structure is most elegantly utilized in the fibers spun by silkworms and spiders. Silkworm fibroin (Figure 6.15) contains long regions of antiparallel β sheet, with the polypeptide chains running parallel to the fiber axis. The β-sheet regions comprise almost exclusively multiple repetitions of the sequence

$$[\,\text{Gly-Ala-Gly-Ala-Gly-Ser-Gly-Ala-Ala-Gly-(Ser-Gly-Ala-Gly-Ala-Gly)}_8\,]$$

In silkworm fibroin almost every other residue is Gly, which is usually followed by Ala or Ser residues. In other species, the residues following the Gly residues are different, leading to differences in physical properties. This alternating pattern of residues allows the sheets to fit together and pack on top of one another in the manner shown in Figure 6.15. This arrangement of sheets results in a fiber that is strong and relatively inextensible because the covalently bonded chains are stretched to nearly their maximum possible length. Yet the fibers are very flexible because bonding between the sheets involves only the weak van der Waals interactions between the side chains, which provide little resistance to bending.

Not all of the fibroin protein is in β sheets. As the amino acid composition in Table 6.3 shows, fibroin contains small amounts of other, bulky amino acids like valine and tyrosine, which would not fit into the structure shown. These are carried in compact folded regions that periodically interrupt the β-sheet segments,

Fibroin is a β sheet protein. Almost half of its residues are glycine.

FIGURE 6.15

The structure of silk fibroin. (a) A three-dimensional view of the stacked β sheets of fibroin, with the side chains shown in color. The region shown contains only alanine and glycine residues. **(b)** Interdigitation of alanine or serine side chains and glycine side chains in fibroin. The plane of the section is perpendicular to the folded sheets.

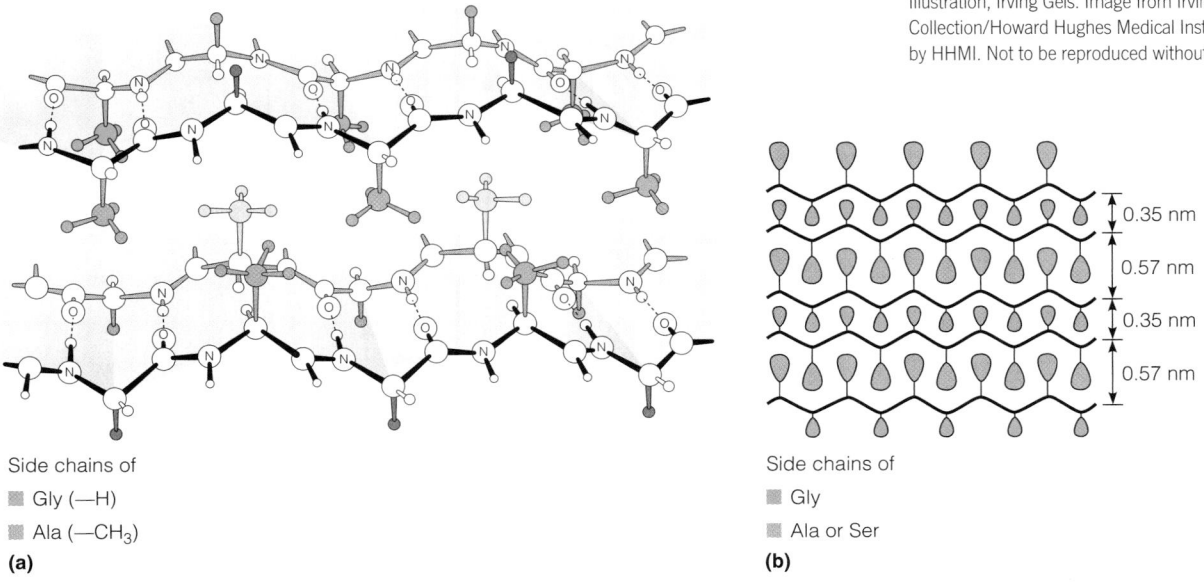

Side chains of
- Gly (—H)
- Ala (—CH$_3$)

(a)

Side chains of
- Gly
- Ala or Ser

0.35 nm
0.57 nm
0.35 nm
0.57 nm

(b)

and they probably account for the amount of stretchiness that silk fibers have. In fact, different species of silkworms produce fibroins with different extents of such non–β-sheet structure and corresponding differences in elasticity. The overall fibroin structure is a beautiful example of a protein molecule that has evolved to perform a particular function—to provide a tough, yet flexible, fiber for the silkworm's cocoon or the spider's web.

Collagen

Because it performs such a wide variety of functions, **collagen** is the most abundant single protein in most vertebrates. In large animals, it may make up a third of the total protein mass. Collagen fibers form the *matrix* material in bone, on which the mineral constituents precipitate; these fibers constitute the major portion of tendons; and a network of collagen fibers is an important constituent of skin. Basically, collagen holds most animals together.

Collagen fibers are built from triple helices of polypeptides rich in glycine and proline.

Collagen Structure

The basic unit of the collagen fiber is the **tropocollagen** molecule, a *triple helix* of three polypeptide chains, each about 1000 residues in length. This triple helical structure, shown in Figure 6.16a and b, is unique to collagen. The individual chains are *left-handed* helices, with about 3.3 residues/turn. Three of these chains wrap around one another in a right-handed sense, with hydrogen bonds extending between the chains. Examination of the model reveals that every third residue, which must lie near the center of the triple helix, can be *only* glycine (see Figure 6.16a, and Table 6.3). Any side chain other than —H would be too bulky. Formation of the individual helices of the collagen type is also favored by the presence of proline or hydroxyproline in the tropocollagen molecule. A repetitive motif in the sequence is of the form Gly–X–Y, where X is often proline and Y is

FIGURE 6.16

The structure of collagen fibers. The protein collagen is made up of tropocollagen molecules packed together to form fibers. The tropocollagen molecule is a triple helix. **(a)** and **(b)** Stick and space-filling views of the tropocollagen triple helix. **(c)** A lower magnification model emphasizes the interwoven triple-helical secondary structure. **(d)** Tropocollagen triple helices align side by side in a staggered fashion to form the collagen fiber. This regular arrangement leads to periodic pattern of bands separated by 64 nm (blue lines) **(e)** An electron micrograph of collagen shows the crisscrossing of fibers, with the 64 nm periodic pattern clearly visible in each.

(e) J. Gross, Biozentrum/Science Photo Library.

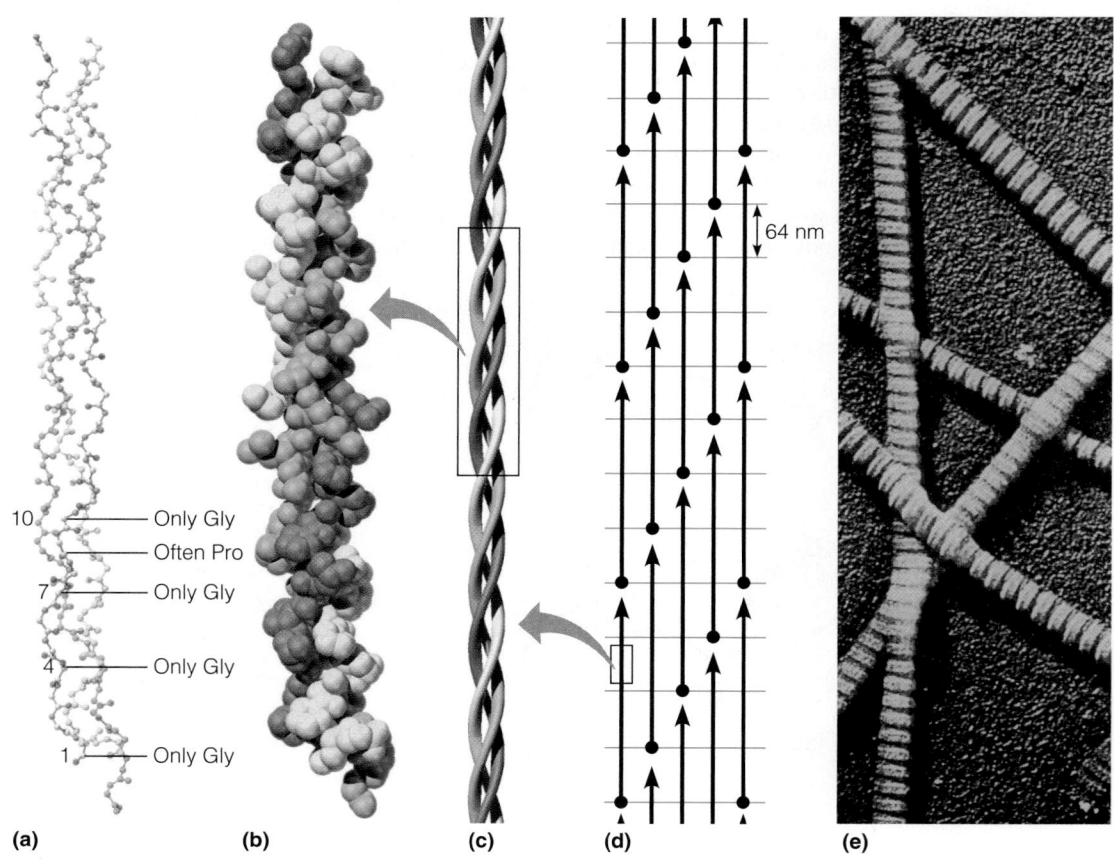

| 10 — Only Gly |
| — Often Pro |
| 7 — Only Gly |
| — Only Gly |
| 4 — Only Gly |
| 1 — Only Gly |

64 nm

(a)   (b)   (c)   (d)   (e)

proline or hydroxyproline (see Table 6.3). However, other residues are often tolerated in these positions. Like silk fibroin, collagen is a good example of how a particular kind of repetitive sequence dictates a particular structure. In order to properly serve the multiple functions it does, collagen exists in a larger number of genetic variants in higher organisms.

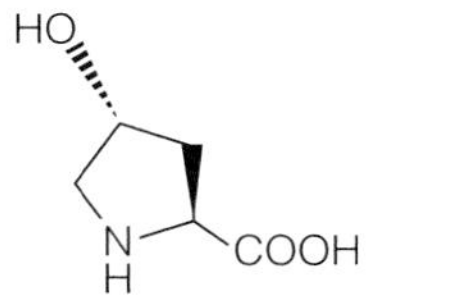

Collagen is also unusual in its widespread modification of proline to hydroxyproline. Most of the hydrogen bonds between chains in the triple helix are from amide protons to carbonyl oxygens, but the —OH groups of hydroxyproline also seem to participate in stabilizing the structure. Hydroxylation of lysine residues in collagen also occurs, but is much less frequent. It plays a different role, serving to form attachment sites for polysaccharides.

The enzymes that catalyze the hydroxylations of proline and lysine residues in collagen require **vitamin C**, L-ascorbic acid (see Figure 21.33, page 906). A symptom of extreme vitamin C deficiency, called **scurvy**, is the weakening of collagen fibers caused by the failure to hydroxylate these side chains, which results in reduced H-bonding between chains. Consequences are as might be expected: Lesions develop in skin and gums, and blood vessels weaken. The condition quickly improves with administration of vitamin C.

The individual tropocollagen molecules pack together in a collagen fiber in a specific way (Figure 6.16c). Each molecule is about 300 nm long and overlaps its neighbor by about 64 nm, producing the characteristic banded appearance of the fibers shown in Figure 6.16e. This structure contributes remarkable strength: Collagen fibers in tendons have a strength comparable to that of hard-drawn copper wire.

Part of the toughness of collagen is due to the cross-linking of tropocollagen molecules to one another via a reaction involving lysine side chains. Some of the lysine side chains are oxidized to aldehyde derivatives, which can then react with either a lysine residue, or with one another via an aldol condensation and dehydration, to produce a cross-link:

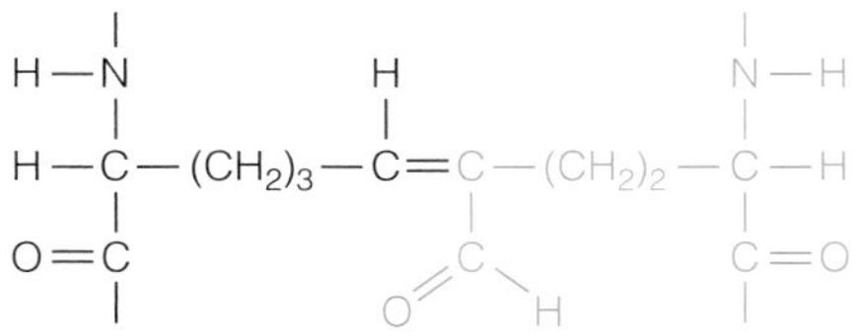

This process continues through life, and the accumulating cross-links make the collagen steadily less elastic and more brittle. As a result, bones and tendons in older individuals are more easily snapped, and the skin loses much of its elasticity. Many of the signs we associate with aging are consequences of this simple cross-linking process.

Collagen Synthesis

As you will have judged by now, collagen is a protein that undergoes extensive modification. Indeed, it can be considered an almost complete example of the post-translational modification pathways we discussed at the end of Chapter 5. The tropocollagen triple helix that ends up cross-linked into an extracellular collagen fiber is very different from the molecule that is first synthesized on a ribosome. The steps in this transformation are shown in Figure 6.17, which begins with translation (step 1). The newly translated polypeptide is hydroxylated (step 2), and then sugars are attached to some of the newly hydroxylated lysine side chains (step 3) to yield **procollagen** (step 4). Procollagen contains almost 1500 residues, of which about 500 are in N-terminal and C-terminal regions that do not have the typical collagen fiber sequence described earlier. Three molecules of procollagen wrap their central regions into a triple helix, while the N- and C-terminal regions fold into globular protein structures. The procollagen triplexes are then exported into the extracellular space (step 5), at which point the N- and C-terminal regions

(2S,4R)-4-hydroxyproline

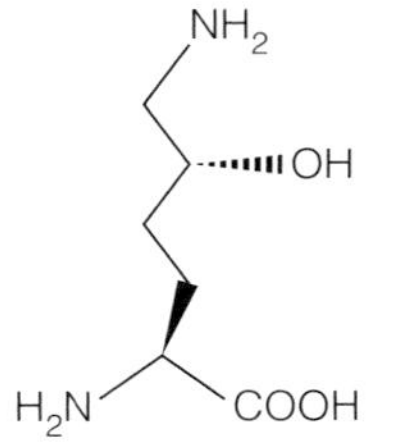

(2S,5R)-5-hydroxylysine

Scurvy is caused by failure to hydroxylate prolines and lysines in collagen.

Collagen undergoes extensive post-translational modification.

FIGURE 6.17

Biosynthesis and assembly of collagen. The process can be visualized in several steps. Steps 1–4 occur in the endoplasmic reticulum and cytosol of collagen-synthesizing cells; steps 6 and 7 occur in the extracellular region. Gal = galactose, Glc = glucose.

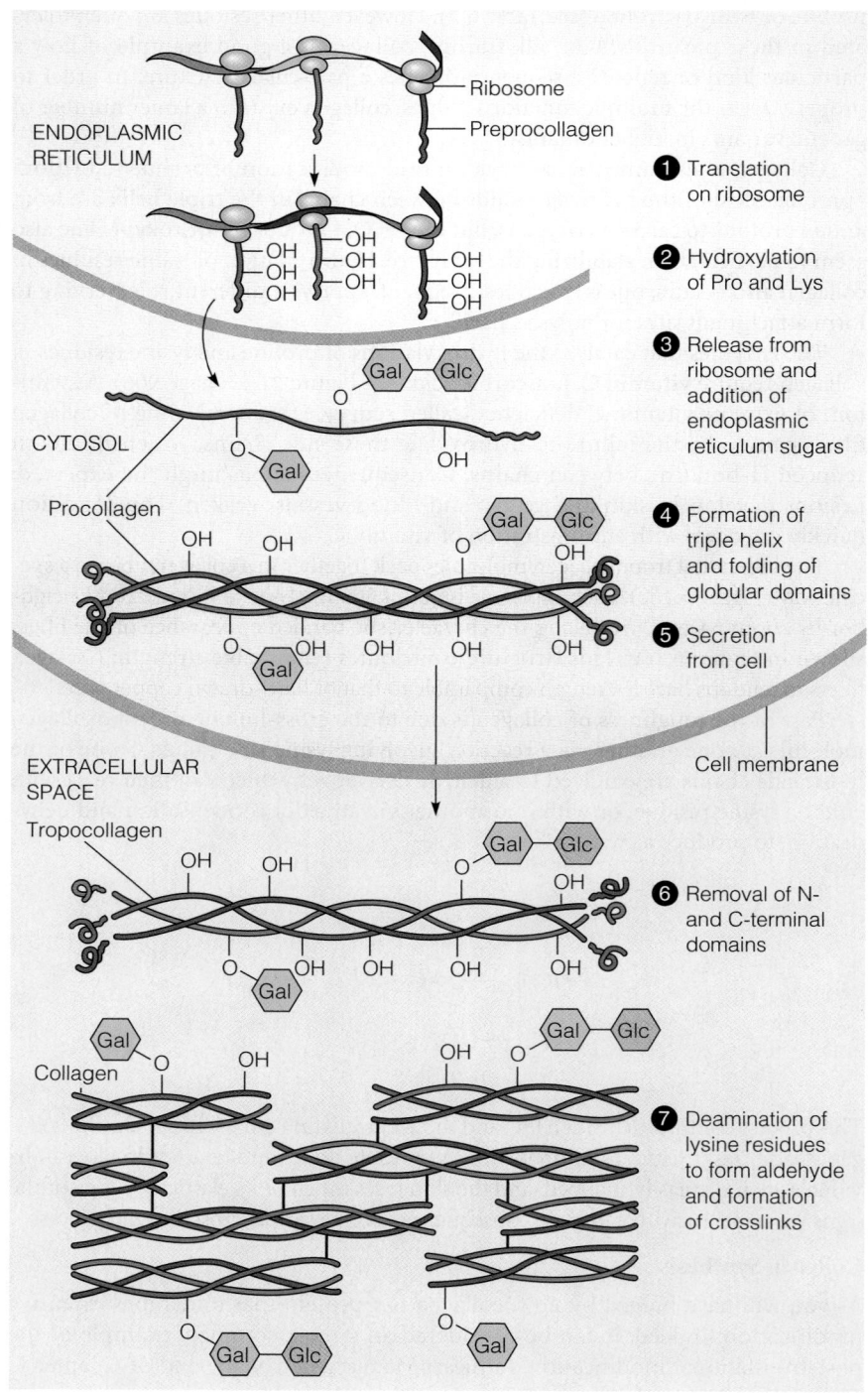

① Translation on ribosome

② Hydroxylation of Pro and Lys

③ Release from ribosome and addition of endoplasmic reticulum sugars

④ Formation of triple helix and folding of globular domains

⑤ Secretion from cell

⑥ Removal of N- and C-terminal domains

⑦ Deamination of lysine residues to form aldehyde and formation of crosslinks

are cleaved off by specific proteases, leaving only the tropocollagen triple helix, about 1000 residues long (step 6). These molecules then assemble into the staggered arrays shown in Figure 6.16d. Finally, cross-linking cements the molecules together into a tough collagen fiber.

Elastin

Collagen is found in tissues where strength or toughness is required, but some tissues, such as ligaments and arterial blood vessels, need highly elastic fibers. Such tissues contain large amounts of the fibrous protein **elastin**.

The protein elastin forms elastic fibers found in ligaments and blood vessels.

The polypeptide chain of elastin is rich in glycine, alanine, and valine and is very flexible and easily extended. In fact, its conformation probably approximates that of a **random coil**, with little secondary structure at all. However, the sequence also contains lysine side chains, which can be involved in cross-links. These cross-links prevent the elastin fibers from being extended indefinitely, causing the fibers to snap back when tension is removed. Exactly the same principle accounts for the elasticity of vulcanized rubber, the flexible chains of which are also held by cross-links. The cross-links in elastin are rather different from those in collagen, for they are designed to hold several chains together. Four lysine side chains can be combined to yield a **desmosine** cross-link as shown in the margin. Because four separate chains are connected, only a small amount of such cross-linking is needed to convert elastin fibers into a highly interconnected, rubbery network.

Desmosine

Summary

This brief overview of a few of the structural proteins brings out several points. First, proteins can evolve to serve an almost infinite diversity of functions. Second, the structural fibrous proteins do this by taking advantage of the propensities of particular repetitive sequences of amino acid residues to favor one kind of secondary structure or another. Finally, post-translational modification of proteins, including cross-linking, is an important adjunct in tailoring a protein to its function. We say more about the cellular sites for such modification in Chapter 28 when we consider the whole process of protein synthesis in detail.

These few examples do not exhaust the list of structural proteins. There are other important ones, such as actin and myosin of muscle and tubulin in microtubules. But these proteins are constructed in a quite different way and are discussed in Chapter 8.

Globular Proteins: Tertiary Structure and Functional Diversity

Different Folding for Different Functions

Abundant and essential as the structural proteins may be in any organism, they constitute only a small fraction of the *kinds* of proteins an organism possesses. Most of the chemical work of the cell—the synthesizing, transporting, and metabolizing—is carried out with the aid of an enormous class of **globular proteins**. These proteins are so named because their polypeptide chains are folded into compact structures very unlike the extended, filamentous forms of the fibrous proteins. Myoglobin (see Figure 6.1) is a typical globular protein. A glance at its three-dimensional structure, when compared with that of, say, collagen, immediately reveals this qualitative difference.

We now know a great deal about the structural details of many globular proteins, largely through the use of X-ray diffraction methods (see Tools of Biochemistry 4A) and **nuclear magnetic resonance spectrospcopy** (NMR, see Tools of Biochemistry 6A). Both of these methods provide structural information with atomic-level details.

Within the protein molecule, the polypeptide chain is often locally folded into one or another of the kinds of secondary structure (α helix, β sheet, and so forth) that we have already discussed. But to make the structure globular and compact, these regions must themselves be folded on one another. This folding is referred to as the **tertiary structure** of the protein; it is this folding that gives the molecule its overall three-dimensional shape. The distinction between secondary and tertiary structures can be clearly seen in the structure of myoglobin, as shown in Figure 6.1. About 70% of the myoglobin chain is α helical, and the eight helices in myoglobin pack against one another to form a compact molecule. A pocket within this structure holds a **prosthetic group**, the heme. Many globular proteins carry prosthetic groups—small molecules that may be noncovalently or covalently bonded to the protein and enable the protein to fulfill special functions. In this case, the noncovalently bound heme group carries the oxygen-binding site of myoglobin (discussed in Chapter 7).

Globular proteins not only possess secondary structures but also are folded into compact tertiary structures.

FIGURE 6.18

The tertiary structure of BPTI. Bovine pancreatic trypsin inhibitor, or BPTI, binds to trypsin and prevents it from catalyzing peptide hydrolysis. BPTI contains only 58 amino acid residues and is one of the most completely studied of all proteins. **(a)** A "stick" model, showing atomic positions (excluding H atoms) from X-ray diffraction (PDB ID: 4PTI). C atoms are green, N atoms are blue, O atoms are red, and S atoms are yellow. **(b)** A "cartoon" model of the backbone. There are short α helices (cyan), near the terminii of the molecule, as well as a two-stranded antiparallel β sheet (red). The positions of the three disulfides are shown by yellow sticks. **(c)** A "surface" model, showing the solvent-accessible surface of the molecule. Atom coloring is the same as in panel (a).

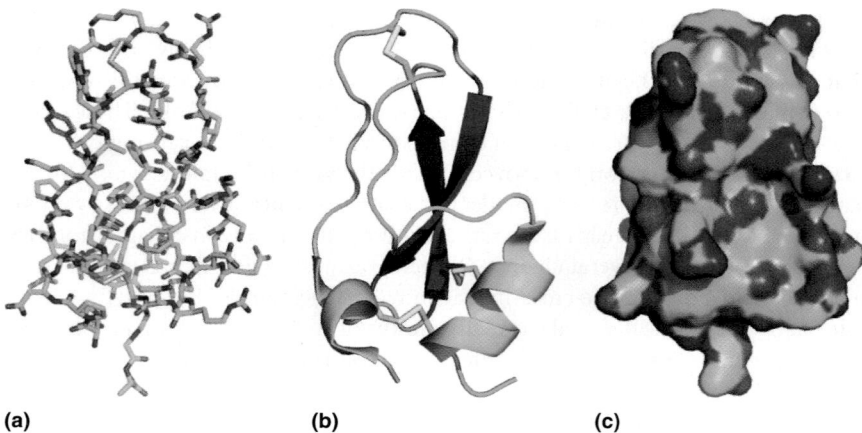

(a)   **(b)**   **(c)**

To show the great structural variation among globular proteins, let us consider another example. Figure 6.18 depicts one of the smallest and simplest globular proteins—bovine pancreatic trypsin inhibitor (BPTI). This protein, synthesized in the cow pancreas, is one of a number of proteins whose sole function is to bind to and inhibit proteolytic enzymes like trypsin. It is important to health, as it prevents damage to the pancreas if trypsin is activated prematurely. We shall use it repeatedly as an example of a globular protein because it is relatively simple and, as a consequence, it has been studied in great detail.

BPTI is depicted in three ways in Figure 6.18. A stick model showing the positions of all atoms (except hydrogens), which have been determined to high resolution by X-ray diffraction, is given in Figure 6.18a. Figure 6.18b shows a cartoon model of the main chain. In this model you can easily trace the path of the polypeptide backbone and can clearly see both α helix and β sheet structures present in this molecule; also shown are the positions of the three disulfide bonds in BPTI. Finally, Figure 6.18c depicts the surface of the molecule, illustrating the important point that globular proteins are tightly packed structures and are often, as their name implies, rather globular in shape.

Comparing Figure 6.18 with Figure 6.1, you can see that the structure of BPTI is wholly different from that of myoglobin. Whereas myoglobin is mostly α helix, BPTI has both helix and sheet regions, connected by bends in the chain. The important point is this: Every globular protein has a characteristic tertiary structure, made up of secondary structure elements (helices, β sheets, nonregular regions) folded in a specific way. As we examine proteins, we will find that each such conformation is suited to the particular functional role that the protein plays.

Varieties of Globular Protein Structure: Patterns of Folding

At first glance, it might seem that there would be an almost infinite number of ways in which globular proteins could fold. If we examine every possible detail of folding, this is true. Yet when a large number of the known structures is examined, certain common motifs and principles emerge. The first principle is that many proteins are made up of more than one domain. A **domain** is a compact, locally folded region of tertiary structure. Domains are interconnected by the polypeptide strand that runs through the whole molecule. Multiple domains are especially common in the larger globular proteins, whereas very small proteins like BPTI tend to be single folded domains. As we shall see in later sections, different domains often perform differing functions, and a given domain type can sometimes be recognized in several different proteins.

Among the domain varieties, several hundred distinct structural motifs have been classified and organized into searchable online databases (see References). These databases allow identification of potential functional and evolutionary relationships among proteins that share similar structural domains.

A protein "domain" is a compact, locally folded region of tertiary structure. Smaller proteins typically contain a single domain. Larger proteins may contain several domains.

To illustrate the immense structural variation in globular proteins, we can consider the ~1,300 distinct domain structures, which are used to classify ~130,000 protein domain entries in the CATH (**C**lass, **A**rchitecture, **T**opology, and "**H**omologous superfamily") database. Figure 6.19 shows seven distinct domain folds, and there are over 1,200 others cataloged in CATH. However, this immense variation can be distilled down to four basic folding patterns in globular proteins: those that are built about a packing of α helices, those that are constructed on a framework of β sheets, those that include both helices and sheets, and those that contain little helix or sheet structure.

In the CATH classification system, "Class" refers to one of four categories based on the prevalent 2° structure in the domain fold (top row Figure 6.19). If we look more closely at the "$\alpha + \beta$" class, we would find 15 different general shapes, or "architectures," for domain folds that include α helix and β sheet structure. Two of these architectures, "3-layer $\alpha/\beta/\alpha$ sandwich" and "α/β barrel" are shown in the second row of Figure 6.19. In the context of protein structure, **topology** refers to the order in which

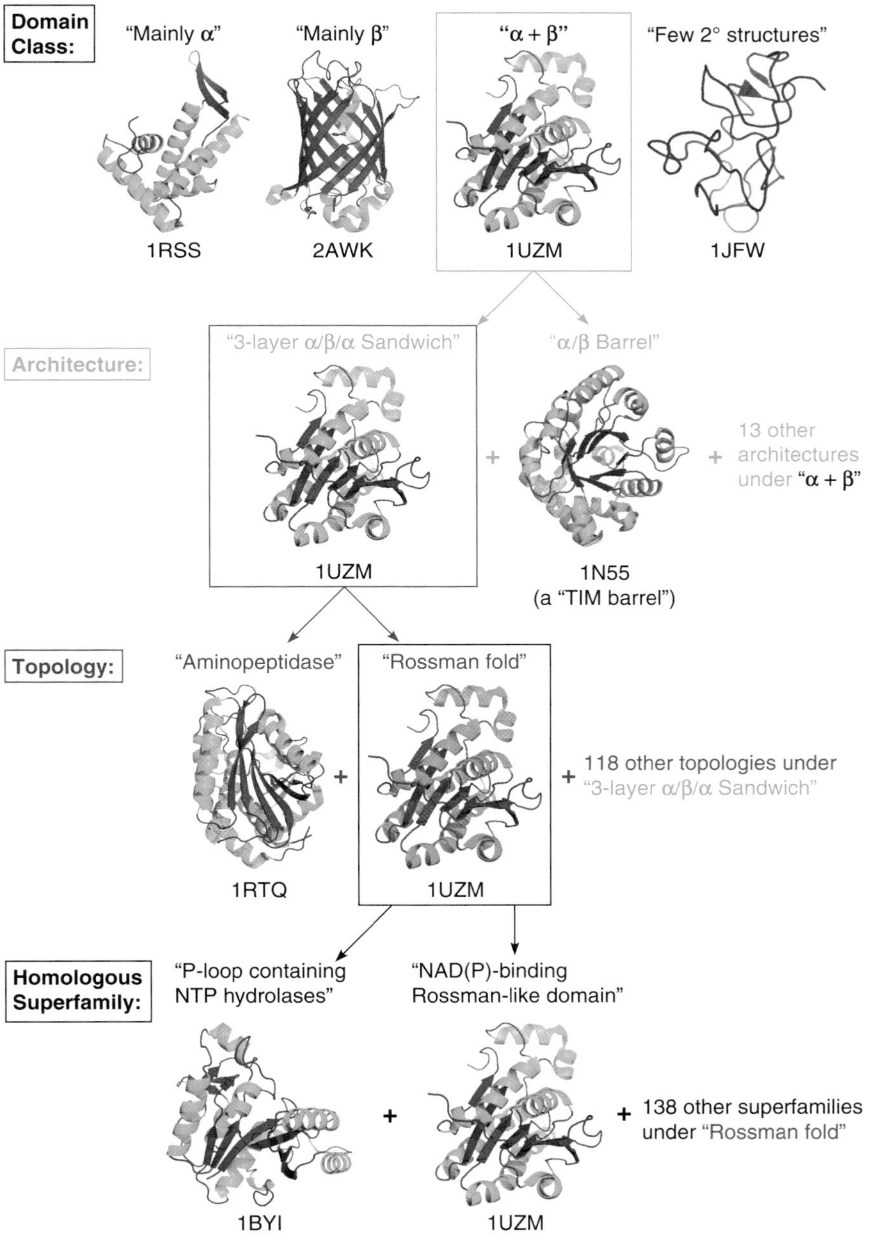

The majority of globular protein structures can be broadly classified as: "mainly α," "mainly β," and "$\alpha + \beta$." A small number of globular proteins has little α or β secondary structure.

FIGURE 6.19

Structural diversity in globular protein domains. See text for a more detailed description of this figure. Top row: Representatives of the four main classes of domain structure are shown in cartoon rendering with α helices in cyan, β strands in red, and connecting loops in magenta. Second row: two representatives of the 15 architectures within the "$\alpha + \beta$" class. Third row: two representatives of the 120 topologies found in the "3-layer $\alpha/\beta/\alpha$ sandwich" architecture. Fourth row: two representatives of the 140 homologous superfamilies found in the "Rossman fold" topology. PDB IDs: Ribosomal protein S7 (1RSS); green fluorescent protein (2AWK); β-ketoacyl ACP reductase (1UZM); HIV1 transactivator protein (1JFW); triosephosphate isomerase (1N55); leucyl aminopeptidase (1RTQ); dethiobiotin synthase (1BYI).

the 2° structural features are connected in the protein amino acid sequence. For example, in different proteins, two α helices and one β strand might be linked "helix-helix-strand" or "helix-strand-helix" or "strand-helix-helix." Each of these represents a different topology for these three secondary structural elements. Row three of Figure 6.19 shows two examples of the 120 different topologies for the "3-layer α/β/α sandwich" architecture: "amino peptidase" and the "Rossman fold." The **Rossman fold** is a common domain structure for an important class of enzymes that bind a **nicotinamide adenine dinucleotide** (NAD) cofactor (see page 446). There are 140 distinct "superfamilies" of proteins that include a Rossman fold topology. The proteins within each superfamily appear to be homologous (i.e., they have a common evolutionary ancestor). There are 3078 different domain entries within the "NAD(P)-binding Rossman-like domain" superfamily (~2% of total entries in the CATH database), and one of these is the structure shown in Figure 6.19 (1UZM). In summary, the CATH designation for this domain is: "α + β" → "3-layer α/β/α sandwich" → "Rossman fold" → "NAD(P)-binding Rossman-like domain".

Studying the details of domain structural variation can be overwhelming; however, it is in these structural details that we find a deeper understanding of protein function. Fortunately for students of biochemistry, the analysis of the structures of thousands of globular proteins has led to the formulation of a few general rules governing tertiary folding:

- *All globular proteins have a defined inside and outside.* If we examine the amino acid sequences of globular proteins, we find no particular distribution pattern of hydrophobic or hydrophilic residues (Figure 6.20a). But when we look at the positions of the amino acids in the three-dimensional structure, we invariably find that tertiary structure causes hydrophobic residues to be packed mostly on the inside (Figure 6.20b), whereas the hydrophilic residues are on the surface, in contact with water (Figure 6.20c).

- *β sheets are usually twisted, or wrapped into barrel structures.* Examples can be seen in Figure 6.19 (e.g., 1RTQ). It is probable that the structure of silk fibroin is not exactly planar, as depicted in Figure 6.15, but slightly twisted.

- *The polypeptide chain can turn corners in a number of ways,* to go from one β segment or α helix to the next. One kind of compact turn is called a β turn (Figure 6.21). There are several varieties of β turn, each able to accomplish a

FIGURE 6.20

The distribution of hydrophilic and hydrophobic residues in globular proteins. **(a)** The amino acid sequence of sperm whale myoglobin. Hydrophobic (green), hydrophilic (red), and ambivalent (black) residues appear to be scattered throughout the sequence. The two His residues in blue play a special role in heme and oxygen binding—see Chapter 7. **(b)** The three-dimensional structure of the same protein. Note how the hydrophobic side chains (shown in green) cluster about the hydrophobic heme cofactor (orange with iron ion in gray) and on the inside of the molecule. **(c)** In this view of the myoglobin structure the hydrophilic side chains are shown in red. Note how they tend to lie on the solvent-exposed surface of the protein.

(a)

VLSEGEWQLV LHVWAKVEAD VAGHGQDILI RLFKSHPETL EKFDRFKHLK
TEAEMKASED LKKHGVTVLT ALGAILKKKG HHEAELKPLA QSHATKHKIP
IKYLEFISEA IIHVLHSRHP GDFGADAQGA MNKALELFRK DIAAKYKELG
YQG

(b)   **(c)**

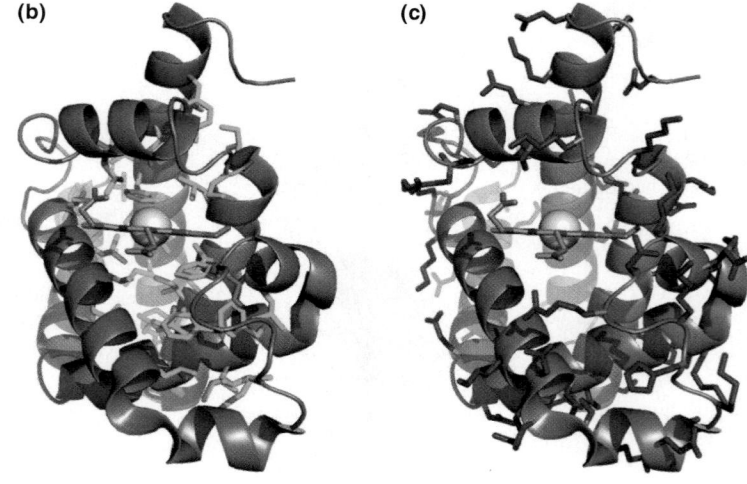

complete reversal of the polypeptide chain direction in only four residues; in each case the carbonyl of residue i hydrogen-bonds to the amide hydrogen of residue $i + 3$. In the even tighter γ turn, bonding is to residue $i + 2$ (Figure 6.22). Proline often plays a role in turns, as in Figure 6.22, and also as a breaker of α helices because this residue cannot be easily accommodated in the helix. Bends and turns most often occur at the surface of proteins.

- *Not all parts of globular proteins can be conveniently classified as helix, β sheet, or turns.* Examination of Figure 6.19, for example, reveals many strangely contorted loops and folds in the chains (the regions shown in magenta). These have sometimes been referred to as "random coil" regions, but this term is a misnomer because such sections of the chain are not flexible in the same way that a true random coil is (see page 108). Rather, X-ray diffraction and NMR data indicate that such regions do, in fact, possess a well-defined fold. We might call these *irregularly structured regions*. Several proteins also have intrinsically unstructured regions, a feature that is particularly important in a class of signaling proteins that bind to several different targets.

In some proteins the folding is dominated by the need to bind a prosthetic group. Myoglobin is an example (Figure 6.20). Although myoglobin might be roughly described as an α helix bundle, the tertiary structure of this protein has been distorted to form a hydrophobic pocket about the heme group.

Factors Determining Secondary and Tertiary Structure

The Information for Protein Folding

What ultimately determines the complex mixture of secondary and tertiary folding that characterizes each globular protein? Much evidence indicates that *most of the information for determining the three-dimensional structure of a protein is carried in the amino acid sequence of that protein.* This can be demonstrated by experiments in which the *native,* or folded, three-dimensional structure is perturbed by changing the environmental conditions. If we raise the temperature sufficiently, or make the pH extremely acidic or alkaline, or add to the solvent certain kinds of organic molecules, such as alcohols or urea, the protein structure will unfold (Figure 6.23). As with nucleic acids, this process is called **denaturation** because the natural structure of the protein has been lost, along with many of its specific functional properties. In diagrams such as Figure 6.23a, the unfolded chain is often drawn as a *random coil,* with freedom of rotation about bonds in both the polypeptide backbone and the side chains. Much recent evidence suggests that this is a gross oversimplification. In many cases the unfolded state for a protein is a dynamic ensemble of largely unstructured, extended conformations; however, even in the "unfolded" state, regions of structure may persist.

In the classic experiment by Chris Anfinsen depicted in Figure 6.23a, and recognized with the Nobel Prize in 1972, the enzyme ribonuclease A (RNase A) was denatured by addition of urea, and the four native disulfide bonds were reduced by addition of β–mercaptoethanol (BME). RNase A catalyzes the hydrolysis of ribonucleic acids. When RNase A is denatured, its tertiary and secondary structures are lost, and it can no longer catalyze RNA cleavage. The denaturation process can be tracked by various physical measurements, as shown in Figure 6.23b. Remarkably, the total disruption of RNase A structure was shown by Anfinsen to be reversible. If the urea is removed by dialysis, the reduced RNase A will spontaneously refold into its native structure. Oxidation of the refolded protein in air restores the native disulfide bonds and full enzymatic activity. Oxidation of urea-denatured RNase A, followed by removal of the urea, yields a mixture of protein molecules with randomly formed disulfide bonds. This mixture has ~1% of the original enzymatic activity, suggesting that molecules with native disulfide bonds account for ~1% mixture. Random formation of four

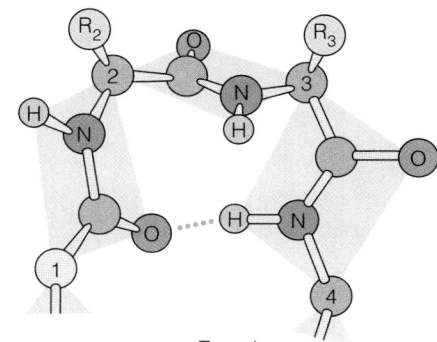

Type I

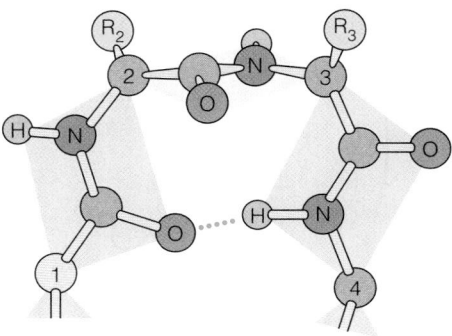

Type II

FIGURE 6.21

Examples of β turns. Each of these turn types allows an abrupt change in polypeptide chain direction. In the type II turn, residue 3 is usually glycine, presumably because a bulky R group would clash with the carbonyl oxygen of residue 2.

Amino acid sequence (primary structure) determines secondary and tertiary structure.

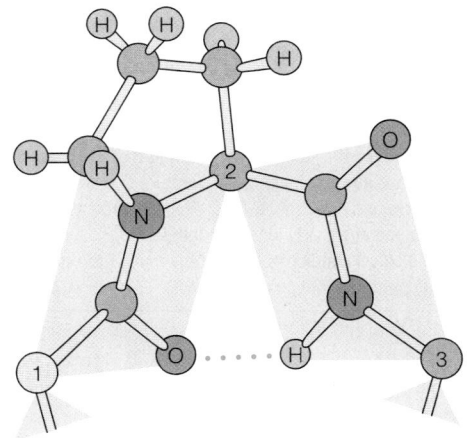

FIGURE 6.22

A γ turn. Only one residue is out of the hydrogen-bonding sequence. In this case it is a proline, which cannot act as an H-bond donor in any case. Note that no H atom is bonded to the α–amino group of the proline (the H that appears to be bonded to the α-N atom is, in fact, bonded to the δ–carbon of the proline side chain).

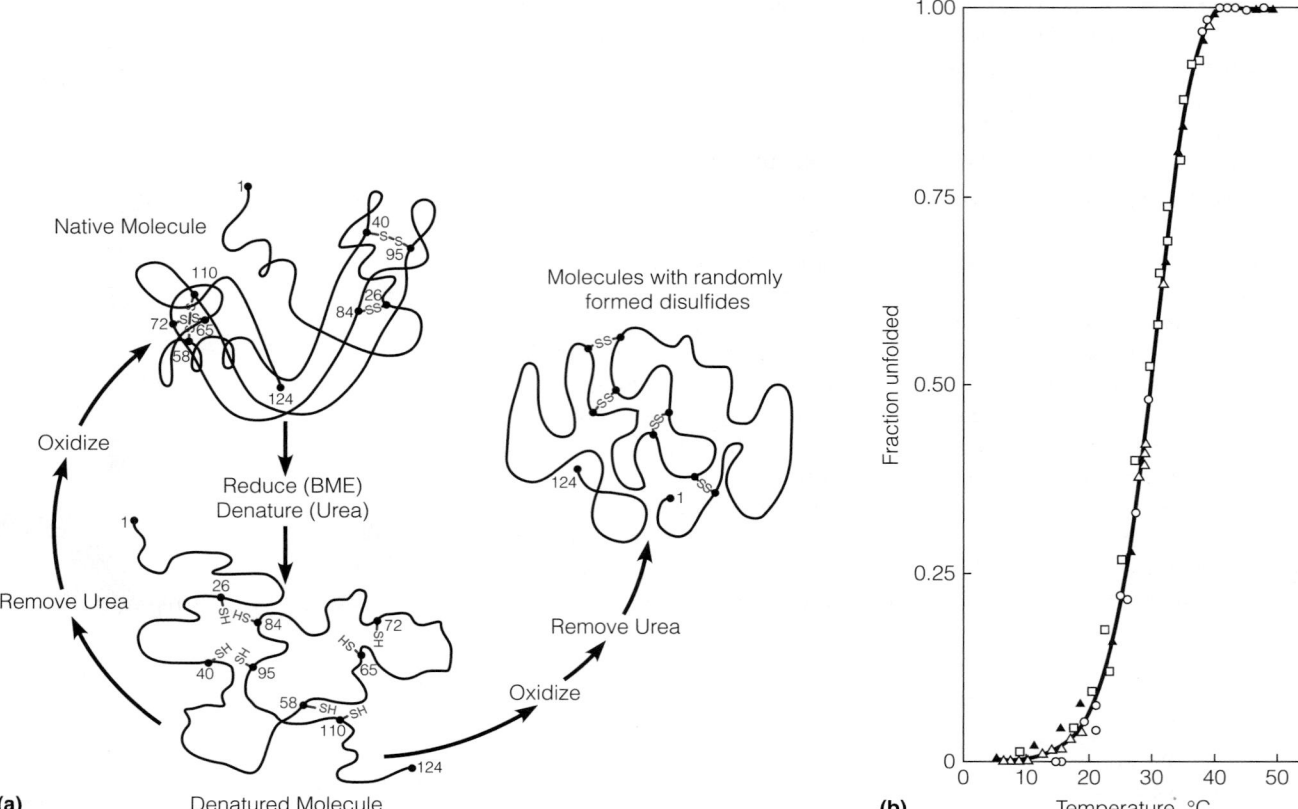

FIGURE 6.23

The denaturation and refolding of ribonuclease A.
(a) This schematic drawing depicts the classic RNase A refolding experiment of Anfinsen. See text for details. **(b)** Thermal denaturation of RNase A monitored by various physical methods. Differences between the native and denatured conformations can be detected by several of the spectroscopic methods discussed in Tools of Biochemistry 6A. This graph shows the fraction of protein that is denatured, as measured by the increase in solution viscosity (□), change in optical rotation at 365 nm (○), or change in UV absorbance at 287 nm (△). All three techniques indicate the same fraction unfolding as a function of increasing temperature. Measurements of a second denaturation after cooling (▲) produce the same curve, showing that the process is reversible with a melting temperature (T_m; see Tools of Biochemistry 5C) of 29.0 °C. The experiments were conducted at pH 2.1, ionic strength 0.019 M. Under physiological conditions, RNase A is much more stable, not denaturing until about 70–80 °C.

(a) Courtesy of Gary Carlton; (b) Data from *Biochemistry* 4:2159–2174 (1965), A. Ginsburg and W. R. Carroll, Some specific ion effects on the conformation and thermal stability of ribonuclease

disulfides from eight cysteines predicts 105 possible different combinations—one of which is the set of native disulfide bonds ($1/105 \approx 1\%$).

Anfinsen's work showed that a protein can self-assemble into its functional conformation, and it needs no information to guide it other than that contained in its sequence. The same phenomenon has been observed experimentally for many other proteins; thus, we might expect a newly synthesized polypeptide in a cell to spontaneously fold into the proper active conformation. As we shall see later in this chapter, the actual process in vivo is more complicated. To avoid misfolding or aggregation, some proteins interact with cellular machinery that guides the folding process. Nonetheless, the basic principle of self-assembly of secondary and tertiary structure seems to be the general rule.

The Thermodynamics of Folding

The folding of a globular protein is clearly a thermodynamically favorable process under physiological conditions. In other words, the overall free energy change for folding must be negative. But this negative free energy change is achieved by a balance of several thermodynamic factors, which we will now describe.

Conformational Entropy

The folding process, which involves going from a multitude of "random-coil" conformations to a *single* folded structure[†], involves a decrease in randomness and thus a decrease in entropy. This change is termed the **conformational entropy** of folding.

random coil (higher entropy) $\longrightarrow$ folded protein (lower entropy)

[†]For the sake of simplicity we will present an idealized case for which there is a single well-defined native state conformation. Because most proteins require some conformational flexibility to perform their function, it is more proper to acknowledge that the native state comprises a limited number of well-defined conformations with similar values of free energy. In any event, the entropy of the native state ensemble of conformations is much less than that of the denatured state ensemble.

The free energy equation, $\Delta G = \Delta H - T\Delta S$, shows that this negative ΔS makes a *positive* contribution to ΔG. In other words, the conformational entropy change works *against* folding. To seek the explanation for an overall negative ΔG, we must seek features of protein folding that yield either a large negative ΔH or some other *increase* in entropy on folding. Both can be found.

The major source for a negative ΔH is energetically favorable interactions between groups within the folded molecule. These include many of the noncovalent interactions described in Chapter 2.

Charge–Charge Interactions

Charge–charge interactions can occur between positively and negatively charged side chain groups. For example, a lysine side chain ε-amino group may be placed close to the γ-carboxyl group of some glutamic acid residue. At neutral pH, one group will be charged positively and the other negatively, so an electrostatic attractive force exists between them. We could say that such a pair forms a kind of salt within the protein molecule; consequently, such interactions are sometimes called **salt bridges**. These ionic bonds are broken if the protein is taken to pH values high enough or low enough that either side chain loses its charge. This loss of salt bridges is a partial explanation for acid or base denaturation of proteins. The mutual repulsion between pairs of the numerous similarly charged groups that are present in proteins in very acidic or basic solutions contributes further to the instability of the folded structure under these conditions.

Internal Hydrogen Bonds

Many of the amino acid side chains carry groups that are either good hydrogen bond donors or good acceptors. Examples are the hydroxyl groups of serine or threonine, the amino groups and carbonyl oxygens of asparagine or glutamine, and the ring nitrogens in histidine. Furthermore, if amide protons or carbonyls in the polypeptide backbone are not involved in secondary structure formation, they are potential candidates for interaction with side chain groups. A network of several types of internal hydrogen bonds is seen in the portion of the molecule of the enzyme lysozyme shown in Figure 6.24. As we have seen before, hydrogen bonds are relatively weak in aqueous solution, but their large number can add a considerable contribution to stability.

van der Waals Interactions

As discussed in Chapter 3, the weak induced dipole-induced dipole interactions between nonpolar groups can also make significant contributions to protein stability because in the folded protein these groups are densely packed and thus make a large number of van der Waals contacts. Detailed analyses of native protein structures show that the nonpolar interior of a globular protein is indeed very tightly packed, allowing maximum contact between nonpolar side chain atoms.

The contribution of any of these interactions to the negative enthalpy of folding is diminished by the fact that when a protein molecule folds from an expanded and highly solvated random coil, it has to give up some favorable interactions with water. As we pointed out in Chapter 2, an unfolded α helical peptide forms hydrogen bonds with water instead of forming internal bonds. It is not clear by how much this exchange modifies the overall value of ΔH; however, we do know that a substantial energy difference in favor of the unfolded state would result if internal H-bonds were *not* formed when H-bond acceptors or donors were buried by folding.

The change in enthalpy for folding, $\Delta H_{U \to F}$, is dominated by the differences in noncovalent bonding interactions between the unfolded and folded states:

$$\Delta H_{U \to F} = H_{\text{folded}} - H_{\text{unfolded}} \qquad (6.1)$$

The decrease in conformational entropy when a protein folds disfavors folding. This is compensated in part by energy stabilization through internal noncovalent bonding.

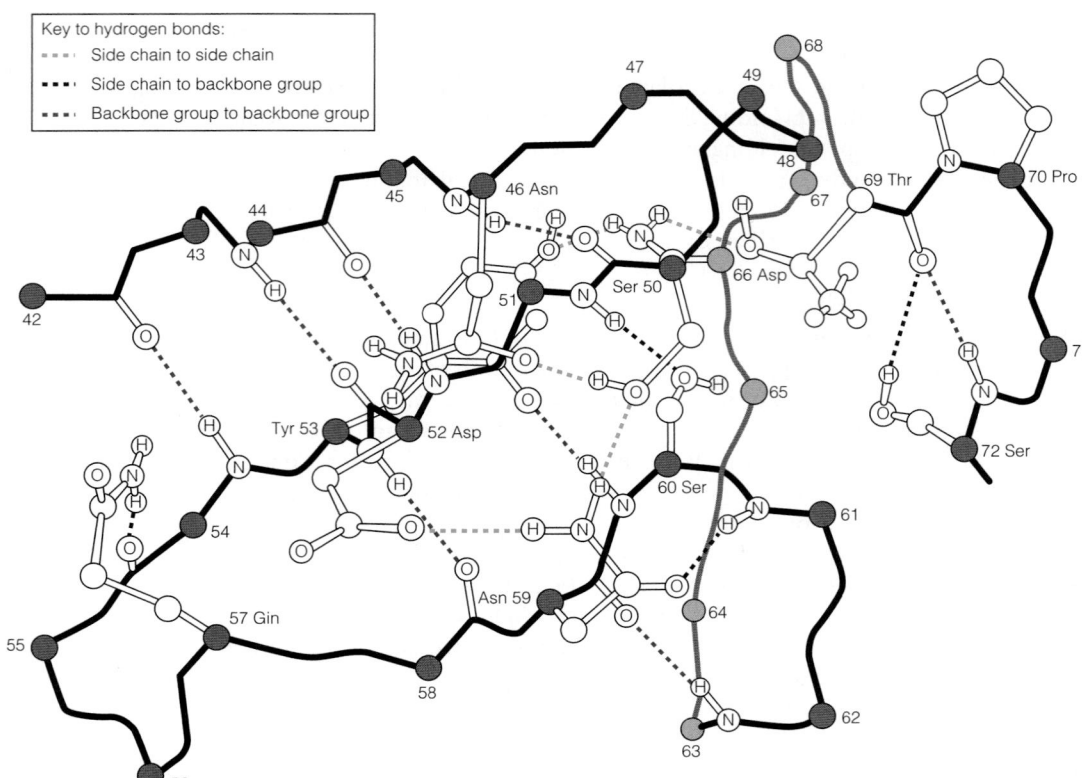

Key to hydrogen bonds:
- - - Side chain to side chain
- - - Side chain to backbone group
- - - Backbone group to backbone group

FIGURE 6.24

Details of hydrogen bonding in a typical protein. A network of hydrogen bonds within the enzyme lysozyme is illustrated. Three kinds of H bonds are distinguished: bonds between side chain groups, bonds between backbone groups, and bonds between a side chain group and a backbone amide hydrogen or carbonyl oxygen.

Illustration, Irving Geis. Image from Irving Geis Collection/Howard Hughes Medical Institute. Rights owned by HHMI. Not to be reproduced without permission.

TABLE 6.4   Thermodynamic parameters for folding of some globular proteins at 25 °C in aqueous solution

Protein	ΔG (kJ/mol)	ΔH (kJ/mol)	ΔS (kJ/K·mol)
Ribonuclease	−46	−280	−0.79
Chymotrypsin	−55	−270	−0.72
Lysozyme	−62	−220	−0.53
Cytochrome c	−44	−52	−0.027
Myoglobin	−50	0	+0.17

Note: Data adapted from *Journal of Molecular Biology* (1974) 86:665–684, P. L. Privalov and N. N. Khechinashvili, A thermodynamic approach to the problem of stabilization of globular protein structure: A calorimetric study. Each data set has been taken at the pH value where the protein is maximally stable; all are near physiological pH. Data are for the folding reaction: Denatured (unfolded) ⇌ native (folded).

where the unfolded state is characterized by noncovalent interactions between the extended protein chain and solvent water molecules and the folded state includes many fewer interactions with solvent and many more intramolecular interactions instead. Typically the only new *covalent* bonds that form upon folding are disulfide bonds (note, however, that most proteins do not contain disulfide bonds). Formation of disulfides will contribute to ΔH; however, as discussed below, the entropic effects of disulfides on protein stability are likely more significant than are their enthalpic effects.

Each individual interaction can contribute only a small amount (at most only a few kilojoules per mole) to the overall negative enthalpy of interaction. But the sum of the contributions of many interactions can yield significant stabilization to the folded structure. Examples of the total enthalpy changes for folding are given for some representative proteins in Table 6.4. In many cases, a favorable energy contribution from the sum of intramolecular interactions more than compensates for the unfavorable entropy of folding.

The Hydrophobic Effect

Yet another factor makes a major contribution to the thermodynamic stability of many globular proteins. Recall from Chapter 2 that hydrophobic substances in contact with solvent water cause the water molecules to form clathrate, or cage-like, structures around them. This ordering corresponds to a loss of randomness in the solvent; thus, the entropy of the solvent is decreased. Suppose a protein contains, in its amino acid sequence, a substantial number of residues with hydrophobic side chains (for example, leucine, isoleucine, and phenylalanine). When the polypeptide chain is in an unfolded form, these residues are in contact with water and cause ordering of the surrounding water structure into clathrates. When the chain folds into a globular structure, the hydrophobic residues become buried within the molecule (see Figure 6.20b), and the water molecules that were ordered around the solvent-exposed hydrophobic surfaces in the protein denatured state are now released, gaining freedom of motion. Thus, the randomness of the solvent is increased by internalizing hydrophobic groups via folding.

Recall that the *overall* change in entropy is the sum of the change in entropy for the system (in this case the polypeptide chain) *and* the change in entropy of the surroundings (in this case the solvent water molecules):

$$\Delta S_{universe} = \Delta S_{system} + \Delta S_{surroundings} \tag{6.2a}$$

$$\Delta S_{U \to F} = \Delta S_{protein} + \Delta S_{solvent} \tag{6.2b}$$

In equation 6.2b, $\Delta S_{protein}$ is the conformational entropy change for folding, which is *negative*; however, this unfavorable entropy change is offset by the favorable entropy increase for the solvent ($\Delta S_{solvent}$). In summary, the burial of hydrophobic surface area in the solvent-inaccessible core of the protein acts to stabilize the folded state by making the value of $\Delta S_{U \to F}$ more positive.

The term *hydrophobic bonding* has sometimes been used to describe the stabilization resulting from the burial of hydrophobic groups, but this term is a misnomer. The stabilization is not primarily the result of bonds forming between hydrophobic groups, although van der Waals interactions surely do make favorable enthalpic interactions upon folding. Rather, the folded protein is stabilized by a solvent entropy effect. The more appropriate term to describe this source of protein stabilization is **the hydrophobic effect**. Examples of the importance of the hydrophobic effect can be seen in Table 6.4. The very small negative ΔS for cytochrome *c* and the positive value for myoglobin are a consequence of the hydrophobic effect in these proteins. Indeed, the stability of myoglobin comes mainly from the hydrophobic effect.

Different amino acid residues make larger or smaller contributions to the hydrophobic effect. Studies of the transfer of amino acids from water to organic solvents, as well as theoretical considerations, have led to the construction of various **hydrophobicity scales**. Two examples are shown in Table 6.5. The two scales, based on somewhat different premises, are similar but not identical. Nevertheless, such scales can be used to predict the importance of the hydrophobic effect in stabilizing any particular protein. In Chapter 10 we shall see how they can also be used to predict which portions of a protein will insert into a lipid membrane.

In summary, the stability of the folded structure of a globular protein depends on the interplay of three factors:

1. The unfavorable conformational entropy change, which favors the unfolded state
2. The favorable enthalpy contribution arising from intramolecular interactions
3. The favorable entropy change arising from the burying of hydrophobic groups within the molecule

Thus, factor 1 works against folding, whereas factors 2 and 3 favor folding. A picture of the way these components might contribute to the free energy of folding for some protein is shown in Figure 6.25. In different proteins, the contributions to the stability of the folded protein from enthalpic interactions and the hydrophobic effect differ (see Table 6.4), but the overall consequence is the same: Some particular folded structure corresponds to a free energy minimum for the polypeptide under physiological conditions. This is why the protein folds spontaneously. Methods for determining thermodynamic parameters for protein stability are described in Tools of Biochemistry 6C.

Examination of the data in Table 6.4 reveals another important aspect of protein stabilization. Proteins are relatively unstable molecules. The free energy difference between the native and denatured states is modest—typically, 20–60 kJ/mol. The relatively small ΔG corresponding to the folding reaction is usually the difference between large ΔH and $T\Delta S$ terms. These large values arise because a protein (or protein domain) on the order of 150–200 residues tends to fold in a

The burying of hydrophobic groups within a folded protein molecule produces a stabilizing entropy increase known as the hydrophobic effect.

TABLE 6.5	Two examples of hydrophobicity scales	
Amino Acid	Scale of Engelman, Steitz, and Goldman[a]	Scale of Kyte and Doolittle[b]
Phe	3.7	2.8
Met	3.4	1.9
Ile	3.1	4.5
Leu	2.8	3.8
Val	2.6	4.2
Cys	2.0	2.5
Trp	1.9	−0.9
Ala	1.6	1.8
Thr	1.2	−0.7
Gly	1.0	−0.4
Ser	0.6	−0.8
Pro	−0.2	−1.6
Tyr	−0.7	−1.3
His	−3.0	−3.2
Gln	−4.1	−3.5
Asn	−4.8	−3.5
Glu	−8.2	−3.5
Lys	−8.8	−3.9
Asp	−9.2	−3.5
Arg	−12.3	−4.5

[a]Data from *Annual Review of Biophysics and Biophysical Chemistry* (1986) 15:321–353, D. M. Engelman, T. A. Steitz, and A. Goldman, Identifying nonpolar transbilayer helices in amino acid sequences of membrane proteins.
[b]Data from *Journal of Molecular Biology* (1982) 157:105–132, J. Kyte and R. F. Doolittle, A simple method for displaying the hydropathic character of a protein.

FIGURE 6.25

Contributions to the free energy of folding of globular proteins. The conformational entropy change works against folding, but the enthalpy of internal interactions and the entropy change from the hydrophobic effect favor folding. Summing these three quantities makes the total free energy of folding negative (favorable); thus, the folded structure is stable.

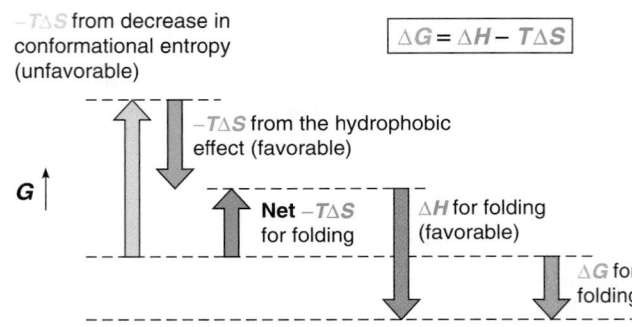

$$\Delta G = \Delta H - T\Delta S$$

$-T\Delta S$ from decrease in conformational entropy (unfavorable)

$-T\Delta S$ from the hydrophobic effect (favorable)

G

Net $-T\Delta S$ for folding

ΔH for folding (favorable)

ΔG for folding

predominantly *cooperative* manner with few intermediate structures; thus, many folding interactions within a molecule form or break in concert. As a consequence of this cooperativity, the thermal denaturation of a protein usually occurs over a narrow temperature range, as is shown in Figure 6.23b.

The Role of Disulfide Bonds

Once folding has occurred, the three-dimensional structure is in some cases further stabilized by the formation of disulfide bonds between cysteine residues. An extreme example of this bonding is found in the bovine pancreatic trypsin inhibitor (BPTI) protein, depicted in Figure 6.18. With three —S—S— bridges in 58 residues, this molecule is one of the most stable proteins known. Pioneering work on the role of disulfide bond formation in the folding of a protein was carried out on BPTI in the lab of Thomas Creighton. The BPTI is quite inert to unfolding reagents like urea and exhibits thermal denaturation below 100 °C only in very acidic solutions; the half-point for reversible denaturation is about 80 °C at pH 2.1 (Figure 6.26). However, if only *one* of the disulfide bonds (that between cysteine residues 14 and 38—the uppermost disulfide bond shown in Figure 6.18b) has been reduced and carboxymethylated, the midpoint is decreased to 59 °C. When all the disulfide bonds in BPTI are reduced, the protein is unfolded at room temperature; yet, upon reoxidation of the sulfhydryls, native protein with the three correct disulfide pairings is efficiently formed. This re-formation is not what would be expected by chance. Suppose a BPTI molecule has been reduced, yielding 6 cysteine residues, and we now randomly reoxidize the —SH groups. The first —SH group to pick a partner will have 5 choices, the second group 3, and the last only 1, so there are 5 × 3 × 1, or 15, equally probable combinations. Thus, we would expect only about 7% of reduced BPTI to refold to the correct native state *by chance.* Many studies of this and other proteins (e.g., RNAse A) containing disulfide bonds indicate that correct pairing is regained in almost 100% of the molecules if sufficient time is allowed. This finding must mean that it is the preferred folding of the protein that places the —SH groups in position for correct pairing. The corollary of this statement is that the disulfide bonds are not themselves essential for correct refolding. They do, however, contribute to the stability of the structure once it is folded.

A molecule containing disulfide bonds has a smaller number of conformations available in the unfolded state than does a comparable protein without the bonds. This dramatic reduction in conformational entropy for the unfolded state results in an increased free energy for the disulfide-bonded unfolded state, relative to the unfolded state in the absence of disulfide bonds (Figure 6.27). In contrast, the conformational entropy of the folded state is not significantly perturbed by the presence or absence of disulfide bonds because the folded state has much lower conformational entropy than does the unfolded state. These differential effects on conformational entropy result in a larger difference in the free energies of the folded and unfolded states for the disulfide-bonded protein. The net result is to make $\Delta G_{F\rightarrow U}$ more *unfavorable* for a protein containing disulfide bonds than it would be in the absence of the disulfide bonds.

Some folded proteins are stabilized by internal disulfide bonds, in addition to noncovalent forces.

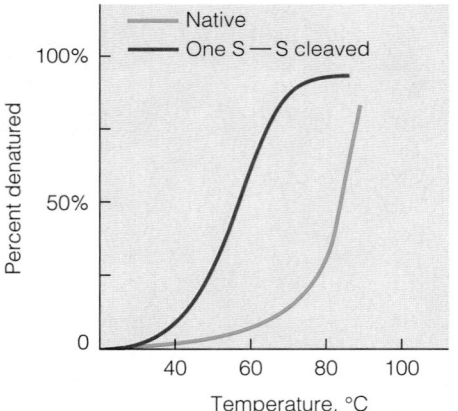

100%

50%

0

Percent denatured

— Native
— One S — S cleaved

40   60   80   100

Temperature, °C

FIGURE 6.26

Thermal denaturation of BPTI. The percent denaturation as a function of temperature at pH 2.1 is indicated for the native protein and for the protein in which the Cys 14–Cys 38 disulfide bond has been reduced and carboxymethylated.

Data from *European Journal of Biochemistry* (1971) 23:401–411 J. P. Vincent, R. Chicheportiche, and M. Lazdunski, The conformational properties of the basic pancreatic trypsin-inhibitor.

Figure 6.27 illustrates the concept that a larger free energy difference between the folded and unfolded states corresponds to a more stable protein. Thus, one can imagine two ways to increase protein stability: (1) stabilize the folded state (i.e., make G_F more negative), or (2) destabilize the unfolded state (i.e., make G_U more positive). The major effect of disulfide bond formation is to dramatically reduce the conformational entropy of the unfolded state, which destabilizes the unfolded state relative to the folded state.

The apparent advantage of disulfide-bond formation raises a question: Why don't *most* proteins have disulfide bonds? In fact, such bonds are relatively rare and are found primarily in proteins that are exported from cells, such as ribonuclease, BPTI, and insulin. One explanation is that the environment inside most cells is reducing and tends to keep sulfhydryl groups in the reduced state. External environments, for the most part, are oxidizing and stabilize —S—S— bridges.

A powerful method for studying the sources of protein structural stabilization has been developed from recombinant DNA techniques. Because the genes for proteins can be cloned and expressed in bacterial carriers (see Tools of Biochemistry 5A), it is possible to make specific changes at desired positions. This method of **site-directed mutagenesis** (see Tools of Biochemistry 4B) allows us to test the effect of changing one or more amino acid residues on protein folding and stability. The method has already been used, for example, to delete specific disulfide bonds in BPTI by replacing cysteines with serines or alanines, then measuring the effect of removing a disulfide bond on the stability of the mutant protein. Conversely, site-directed mutagenesis has been used to introduce into proteins appropriately placed cysteine residues to promote the formation of one (or more) new disulfide bonds in the folded structure. For example, in work from the lab of Brian Matthews, the X-ray crystal structure of the enzyme bacteriophage T4 **lysozyme** was used to guide the selection of three sites in that protein for the introduction of new disulfides. The introduction of one new disulfide increased the thermal stability of T4 lysozyme by 11 °C, and when all three disulfides where introduced the thermal stability increased by 23 °C compared to the enzyme lacking disulfide bonds.

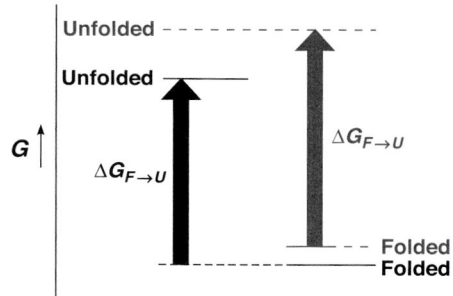

FIGURE 6.27

The effect of disulfide bonds on protein stability. The free energies of the folded and unfolded states of a protein are shown schematically for the case of a protein with (dashed red lines) or without (unbroken black lines) disulfide bonds. The presence of disulfide bonds destabilizes (i.e., raises the free energy of) the unfolded state more than the folded state due to greater restriction of conformational entropy in the unfolded state. This results in a larger difference in the free energies of the folded and unfolded states for the disulfide-bonded protein; thus, the presence of disulfide bonds makes unfolding *less favorable* ($\Delta G_{F \to U}$ becomes more positive, or more "uphill").

Dynamics of Globular Protein Structure

The structural and thermodynamic descriptions we have developed so far tend to give too static a picture of globular proteins. The models of folded proteins produced from structural studies often give the impression of a rigid structure. Likewise, the thermodynamic analysis of folding concentrates on the initial (unfolded) and final (folded) states; but, it is now recognized that globular proteins fold via complex *kinetic* pathways and that even the folded structure, once attained, is a dynamic structure. The following section explores some of these dynamic aspects of globular proteins.

Kinetics of Protein Folding

The folding of globular proteins from their denatured conformations is a remarkably rapid process, often complete in less than a second. This observation has been of profound interest to biochemists, for at first glance, the attainment of the well-defined structure typical of globular proteins would seem to be very difficult. This point of view was dramatically expressed in "Levinthal's paradox," first enunciated by Cyrus Levinthal in 1968: A rough estimate would say that about 10^{50} conformations are possible for a polypeptide chain such as RNase A (124 residues). Even if the molecule could try a new conformation every 10^{-13} second, it would still take about 10^{30} years to try a significant fraction of them! Yet RNase A is experimentally observed to fold, in vitro, in about 1 minute. Clearly, something is wrong with such a calculation.

Levinthal's paradox assumes that folding is a random process; thus, a protein must sample a vast number of *possible* conformations to find the desired native

FIGURE 6.28

A simplified representation of the folding pathway for a protein. "U" indicates the unfolded or denatured state, "F" the folded or native state, and "I" "on-pathway" intermediate states. Off-pathway states include aggregates and other non-native states that may be kinetic or thermodynamic "dead-ends"; thus, the paths to these states are generally shown as irreversible. In fact, not all pathways leading to such states are irreversible.

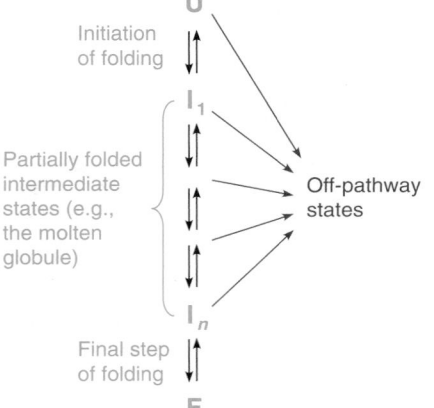

conformation. Years of experimental and theoretical studies have shown that protein folding is not a completely random search through a vast conformational space. Rapid kinetic studies, using a variety of physical techniques to monitor different aspects of protein structure, show that folding takes place through a series of intermediate states. One particularly well-studied intermediate is the so-called **molten globule**. The molten globule is a compact, partially folded intermediate state that has native-like secondary structure and backbone folding topology, but lacks the defined tertiary structure interactions of the native state. These observations led initially to the classical "pathway" model of folding, depicted in a simplified fashion in Figure 6.28.

The "Energy Landscape" Model of Protein Folding

Several proteins are known to fold via well-characterized intermediate states as shown in Figure 6.28, yet this simple model of protein folding is not sufficient to resolve Levinthal's paradox. That can only be achieved if the number of conformational states that the protein samples during folding is significantly restricted. The so-called **energy landscape**, or **folding funnel** model, described by José Onuchic and Peter Wolynes, explains how conformational restriction can be achieved during folding. Imagine that the energetics of folding is described by a funnel-shaped landscape (Figure 6.29) for which the width of the funnel corresponds to the number of conformational states and the depth of the funnel corresponds to free energy. As a protein molecule follows a "downhill" (i.e., thermodynamically favorable) trajectory

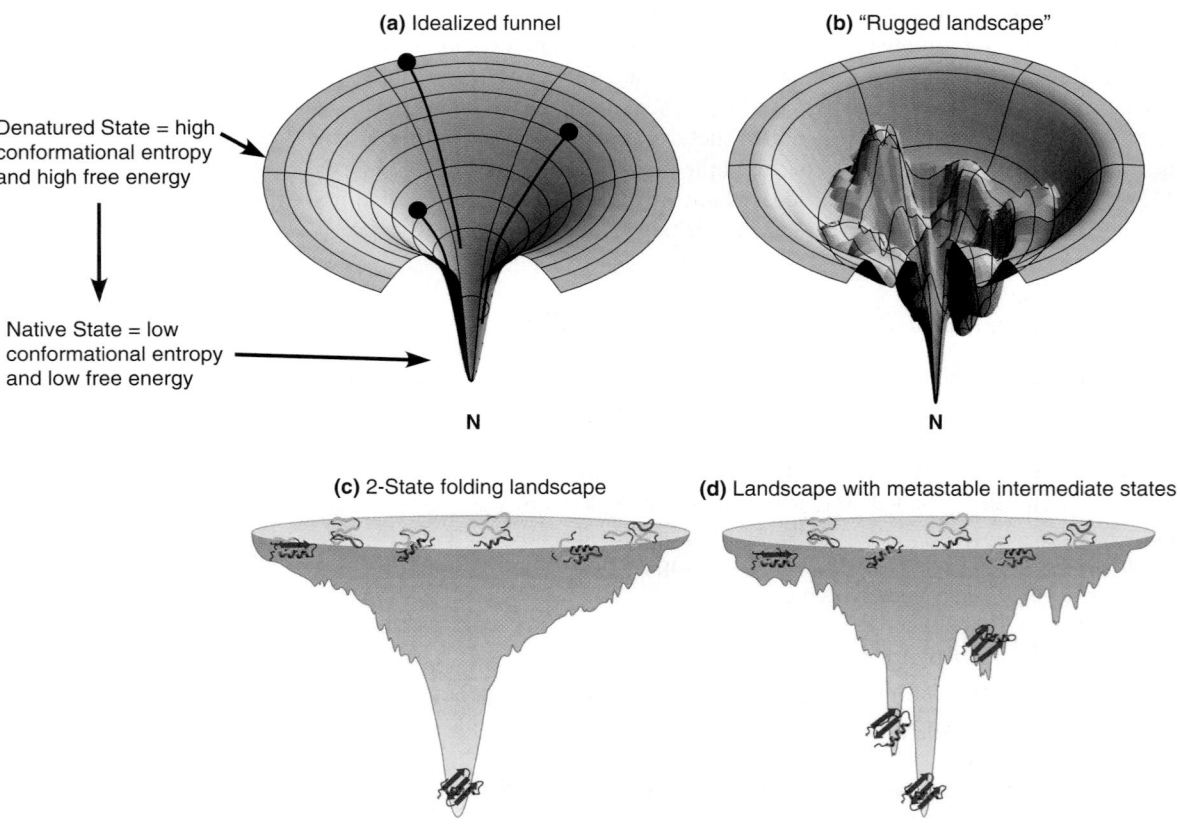

FIGURE 6.29

Protein folding energy landscapes. For all folding funnels shown, the widest part of the funnel corresponds to the denatured state and the deepest and narrowest part of the funnel corresponds to the native state. **(a)** Highly idealized energy landscape for which all trajectories lead to productive folding to the native state. **(b)** A more "rugged" energy landscape. Here, many different paths are possible, some of which lead "downhill" with no local energy minima, and give rapid folding. Others may lead to conformations corresponding to local energy minima, which may slow folding. These conformations represent stable intermediate states along the folding pathway. **(c)** A relatively smooth energy landscape with no significant energy barriers between the denatured and native states. This landscape describes so-called two-state folding because the only significantly populated states are the denatured and native states. **(d)** A more rugged energy landscape showing local minima for metastable intermediate states. This landscape describes a multistate folding process.

Adapted by permission of Macmillan Publishers Ltd. *Nature Structural and Molecular Biology* 4:10–19, K. Dill and H. S. Chan, From Levinthal to pathways to funnels. © 1997; *Nature Structural and Molecular Biology* 16:582–588, A. I. Bartlett and S. E. Radford, An expanding arsenal of experimental methods yields an explosion of insights into protein folding mechanisms. © 2009.

on the energy landscape, the funnel becomes narrower and the number of conformations accessible to the protein decreases. According to this model, any one molecule need sample only an infinitesimal fraction of the total conformations possible, and thereby, Levinthal's paradox is averted.

The classical pathway model and the folding funnel are compatible if we consider the U and I states in Figure 6.28 to be made up of ensembles of conformations, rather than unique conformations. Recent studies have shown that, in some cases, the F state is also an ensemble of closely related conformations.

The folding of several small proteins has been studied in great detail, and much experimental evidence suggests the energy landscape model is robust. Some of the earliest events in protein folding are "nucleation" of secondary structure (typically α helix; see Problem 6.18) and "hydrophobic collapse." Nucleation events are critical because it is much more difficult to initiate an α helix, for example, than to extend it (note that at least four residues must fold properly to make the first stabilizing H bond). Hydrophobic collapse is driven by the hydrophobic effect and typically involves several highly hydrophobic sidechains (see Table 6.5) rapidly associating to create a desolvated hydrophobic core. Whether secondary structure nucleation or hydrophobic collapse (or both) is the first event in protein folding, the formation of *any* partially folded structure imposes significant constraints on the conformational entropy of the peptide chain, as indicated by the landscape model. This model proposes that there is not just one but *many* possible paths from the denatured state to the folded state, and each path leads downhill in energy. It is now recognized that nucleation may start at a number of points and that all of these partially folded structures will be "funneled" toward the final state along trajectories that minimize the energy of the peptide.

Intermediate and Off-Pathway States in Protein Folding

On the folding trajectory toward the "global" free energy minimum, there may be "local" energy minima corresponding to metastable intermediates, just as in the classical pathway model. There is also evidence for "off-pathway" states—those in which some key element is incorrectly folded. Such states also correspond to local free energy minima in the funnel and may temporarily, or permanently, trap the protein. As described below, cells contain specialized proteins, and protein complexes, to assist incorrectly folded proteins to find their proper conformations.

One of the most common of folding errors results from the incorrect *cis–trans* isomerization of the amide bond adjacent to a proline residue:

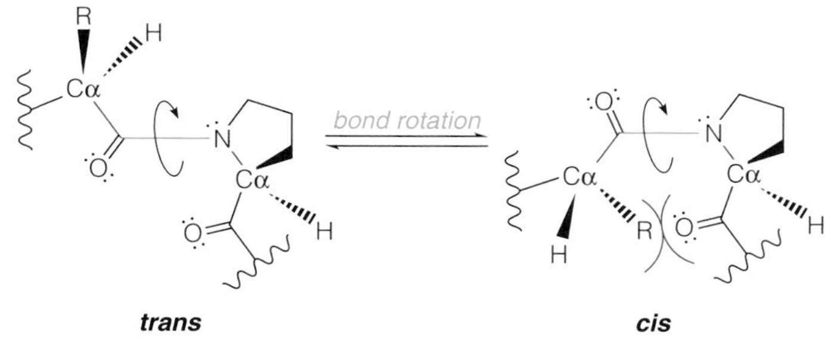

trans   **cis**

Unlike other peptide bonds in proteins, for which the *trans* isomer is highly favored (by a factor of about 1000), proline residues favor the *trans* form in the preceding peptide bond by a factor of only about 4. Therefore, there is a significant chance that the conformationally incorrect *cis* isomer will form first. Conversion to the *trans* configuration may involve a significant chain rearrangement and hence may be slow in vitro. Cells have an enzyme, called **prolyl isomerase** (or peptideprolyl isomerase, or PPIase), to catalyze this *cis–trans* isomerization, and thus speed in vivo folding.

Similarly, for proteins containing disulfide bonds, some disulfide bonds that are not found in the native structure may be formed in intermediate stages of the folding process. The protein can utilize a number of alternative pathways to fold; but, it ultimately finds both its proper tertiary structure and the correct set of

---

Protein folding can be rapid but seems to involve well-defined intermediate states.

The molten globule is a compact, partially folded intermediate state that has native-like secondary structure and backbone folding topology, but lacks the defined tertiary structure interactions of the native state.

In the "energy landscape" model, the trajectory of protein folding is "downhill"— it proceeds with a decrease in free energy.

Folding can be delayed by trapping of molecules in "off-pathway" states.

disulfide bonds. This process is aided in vivo by catalysis of —S—S— bond rearrangement by the enzyme **protein disulfide isomerase** (or PDI). If a non-native disulfide forms during folding in the endoplasmic reticulum of eukaryotes (or the periplasmic space in prokaryotes), PDI will reduce the incorrect disulfide and thereby allow the protein to continue folding toward its native structure.

Backbone Topology and Contact Order

Two factors appear to play dominant roles in determining the rate of protein folding. The first is size. In general, smaller proteins fold faster than larger ones, although the presence of multiple disulfide bonds (e.g., as in BPTI) can slow folding significantly. The second factor is backbone topology, or the arrangement of the peptide main chain in 3-D space. Figure 6.30 shows two different main chain topologies, one for two mainly β proteins (6.30a and b) and another for two $\alpha + \beta$ proteins (6.30c and d). Any given topology will define a set of local interactions (i.e., those that occur between residues that are close in sequence) and nonlocal interactions (i.e., those that occur between residues that are far apart in the primary sequence). The relative percent of nonlocal interactions is reflected in the **contact order** for a protein. Contact order is calculated as the average sequence separation between residues that physically contact in the native state, divided by the total number of residues in the sequence. This somewhat abstract idea is illustrated graphically in Figure 6.30e. A protein in which the N- and C-termini are in contact will have a relatively high contact order because the termini represent the maximum possible separation in sequence. Of interest to the current discussion is the observation of the remarkable correlation between contact order and folding rate (k_f in Figure 6.30f). As contact order, or the number of nonlocal interactions, *increases*, the folding rate *decreases*. Note that the values of folding rates in Figure 6.30f span a million-fold range. Another general trend from data like those in Figure 6.30f is that all-helical proteins tend to fold faster than all β-sheet proteins.

Proteins with lower contact order tend to fold faster than those with high contact order.

Chaperones

The fact that a protein can, by itself, find its proper folded state in vitro does not necessarily mean that the same events occur in vivo. We have already noted two of the catalytic aids to folding that are present in cells, the enzymes that accelerate the *cis–trans* isomerization at proline residues, and those that catalyze disulfide bond rearrangement; however, some proteins require the action of specialized proteins called **molecular chaperones** to achieve proper folding. As the name implies, the function of these chaperones is to keep the newly formed protein out of trouble. Trouble, in this case, means either improper folding or aggregation.

Protein folding and assembly in vivo is sometimes aided by chaperone proteins.

FIGURE 6.30

Protein topology and folding rates. Panels **(a)** and **(b)** show two proteins with very different primary sequences but similar topologies. Likewise for panels **(c)** and **(d)**. The two SH3 domains fold with similar rates, as do ADAh2 and acyl phosphatase, suggesting that protein topology is an important factor in determining the rate of protein folding. **(e)** Examples of low and high contact order (see text). **(f)** Correlation of folding rate to contact order. Proteins with higher contact orders tend to fold with slower rates. Data are shown for all-helical proteins (red circles), all-sheet proteins (green squares), and proteins containing both helix and sheet (orange diamonds and blue triangles). The blue triangles correspond to those proteins that are structurally similar to acyl phosphatase (Panel (d)).

Adapted by permission from Macmillan Publishers Ltd. *Nature* 405:39–42, D. Baker, A surprising simplicity to protein folding. © 2000.

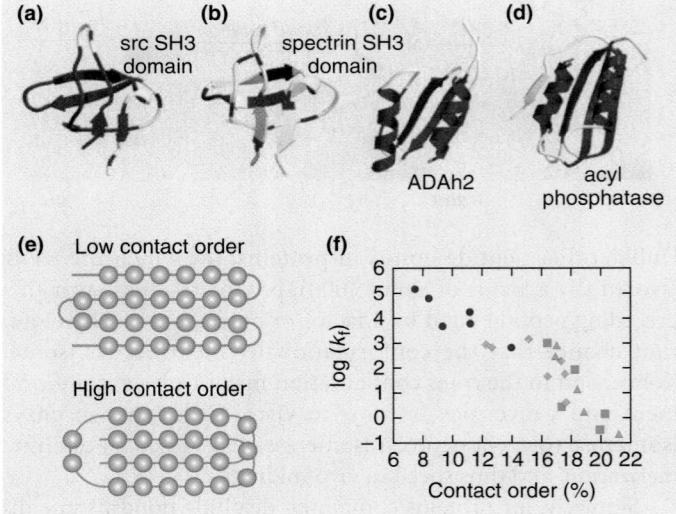

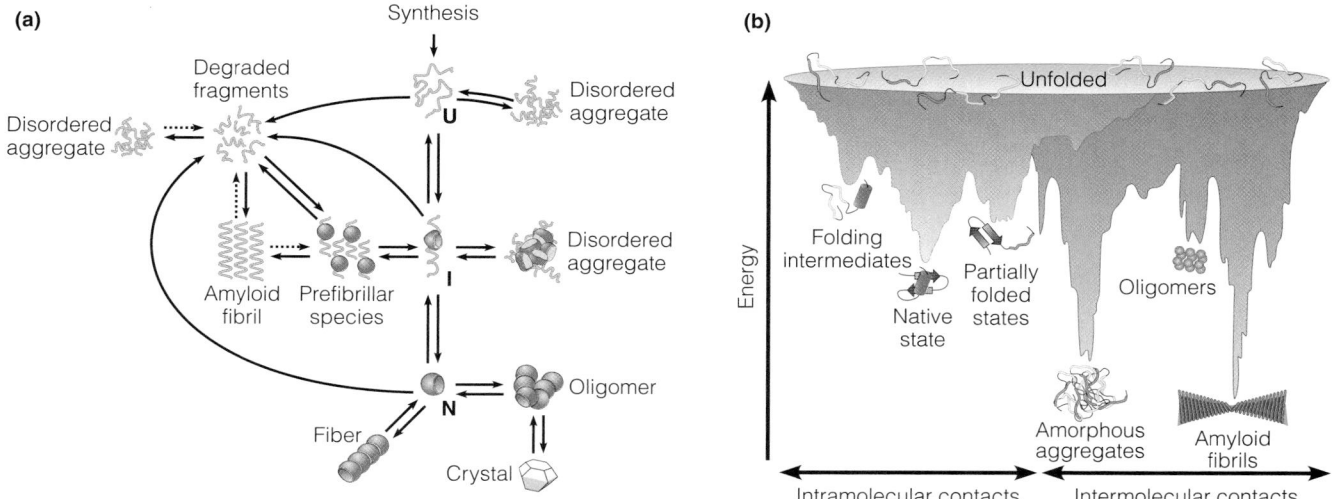

FIGURE 6.31

Protein folding and aggregation. **(a)** A pathway model showing various protein conformations and their interconversions. **(b)** The same information shown in an energy landscape model. The purple portion of the folding funnel shows trajectories leading to the native conformation. The pink portion of the funnel shows trajectories leading to amorphous and ordered aggregates.

Adapted by permission from Macmillan Publishers Ltd. *Nature Reviews Drug Discovery* 2:154–160, C. M. Dobson, Protein folding and disease: A view from the first Horizon Symposium. © 2003. *Nature Structural and Molecular Biology* 16:574–581, F. U. Hartl and M. Hayer-Hartl, Converging concepts in protein folding in vitro and in vivo. © 2009.

Improper folding may correspond to being trapped in a deep local minimum on the energy landscape (see Figure 6.31). Aggregation is often a danger because the protein, released from the ribosome in an unfolded state, will have hydrophobic groups exposed. These can form *intermolecular* hydrophobic contacts with other polypeptide strands and thereby lead to aggregation.

A molecular chaperone can be defined as an accessory protein that binds to and stabilizes a non-native protein against aggregation and/or helps it achieve folding to its native structure; but, is not part of the final functional structure of the correctly folded protein. A number of chaperone systems have been discovered, and we discuss their variety in Chapter 28, where we describe in detail the processing that proteins receive in different cell types. For now, we consider, as an example, the best studied of all chaperones: the GroEL-ES complex from *E. coli*. The structure of this enormous and remarkable complex (see Figure 6.32) consists of two basic portions—GroEL and GroES. GroEL is made of two rings each consisting of seven protein molecules, or "subunits." The center of each ring is an open cavity, accessible to the solvent at the ends. Either cavity can be "capped" with GroES, a seven-membered ring of smaller protein subunits. Such double-ringed chaperone complexes are also known as **chaperonins**.

The cavity in the GroEL-GroES complex provides a favorable environment in which non-native protein chains of up to ~60 kDa can fold properly. The chaperonin does not stipulate the folding pattern; that is determined by the protein sequence. However, insulation from the environment prevents aggregation or misfolding. The folding cycle experienced by a protein molecule is schematically shown in Figure 6.32e. The unfolded protein enters into an open form of the GroEL ring lined with hydrophobic residues. Subsequent to ATP and then GroES binding, the enclosed cavity changes conformation, presenting a hydrophilic, negatively charged surface. This releases the protein from the walls of the cavity, whereupon it folds and is then released from the chaperonin. Note that ATP is required, presumably to drive the process in one direction.

Why are chaperones needed in cells? First, the intracellular environment is very crowded. The total concentration of proteins and other macromolecules inside cells is on the order of 300–400 grams per liter, or roughly 1000 times more concentrated than is typical for in vitro studies of folding. Intermolecular interactions are more likely to occur in such a concentrated intracellular "stew"; thus, aggregation of misfolded or incompletely folded proteins is problematic in vivo. Second, chaperones play a critical role in protecting cellular proteins during times of stress. In fact, many chaperones were discovered during studies of the heat shock response. When the temperature inside cells is raised by a few degrees

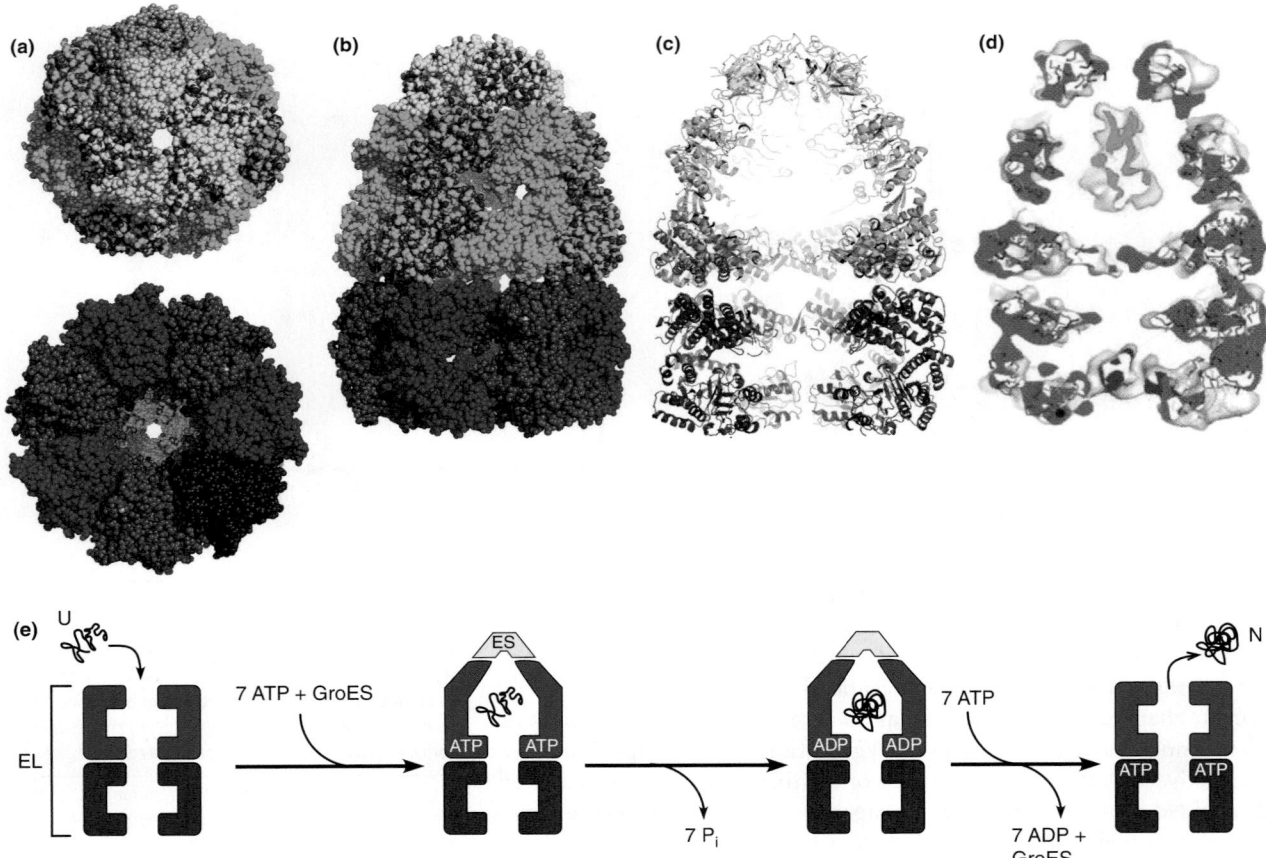

FIGURE 6.32

The GroEL-GroES chaperonin. (a) Spacefilling top and bottom views of the X-ray diffraction structure of the GroEL · ES(ADP)$_7$ complex (PDB ID: 1AON). GroES subunits are highlighted in gold or orange, the subunits in the *trans* ring of GroEL (i.e., the ring opposite the GroES cap) are highlighted in magenta and red, and those in the *cis* GroEL ring (bound by GroES) are highlighted in green. **(b)** Spacefilling side view of the chaperonin, colored as in (a). **(c)** Cartoon view showing the enclosed cavity formed by the *cis* ring of GroEL and GroES (top), as well as the compaction of the *trans* ring compared to the *cis* ring (bottom). **(d)** Electron density map obtained from cryo-electron miscroscopy of a chaperonin complexed with bacteriophage T4 coat protein gp23. The gp23 bound to the *trans* ring of GroEL is shown in red and appears to be denatured. The gp23 bound to the *cis* ring is shown in green and appears to have a native-like conformation. **(e)** A schematic of chaperonin function: U = unfolded protein, N = native protein. Coloring matches that in panel (c).

Panel (d) adapted by permission of Macmillan Publishers Ltd. *Nature* 457:107–110, D. K. Clare, P. J. Bakkes, H. van Heerikhuizen, S. M. van der Vies, H. R. Saibil, Chaperonin complex with a newly folded protein encapsulated in the folding chamber. © 2009; Panel (e) adapted from *Trends in Biochemical Science* 23:68–73, W. S. Netzer and F. U. Hartl, Protein folding in the cytosol: Chaperonin-dependent and -independent mechanisms. © 1998, with permission from Elsevier.

Celsius, so-called **heat shock proteins** (Hsp proteins) are produced by cells. For example, GroEL is also known as Hsp60. We now know that the Hsp proteins are preventing irreversible denaturation of cellular proteins. Once the temperature is restored to normal, any temperature-sensitive proteins can be refolded; thus, the viability of the cell is preserved through the action of molecular chaperones. Finally, it has been suggested that chaperones provide a critical mechanism for cells to survive mutations that might render a protein less stable and/or more susceptible to misfolding.

Understanding the mechanisms of protein folding and how folding occurs in cells is of profound importance given the number of diseases associated with protein misfolding. In addition, with the vast amount of protein sequence information available from genomics studies, the prediction of protein tertiary structure from sequence remains an outstanding challenge in molecular biology and biophysics.

Protein Misfolding and Disease

A broad range of diseases is associated with protein misfolding. A few misfolding diseases, such as cystic fibrosis, are the consequence of mutations that reduce the stability of a critical protein, thereby leading to its misfolding and subsequent clearance by cellular quality control processes (e.g., via the **proteasome**, which is discussed in Chapter 20). However, most human misfolding diseases are associated with the formation of highly ordered protein aggregates called **amyloid fibrils** or **amyloid plaques**. Proteins that misfold to form amyloid structures are said to be "amyloidogenic." Table 6.6 lists a few examples of amyloidogenic proteins associated with human disease states.

TABLE 6.6 Examples of amyloid-related human diseases

Disease	Associated Protein
Alzheimer's disease	Amyloid β or "Aβ" peptide
Parkinson's disease	α-Synuclein
Spongiform encephalopathies (e.g., Creutzfeldt-Jakob disease; kuru; etc.)	prion protein
Amyotrophic lateral sclerosis (Lou Gehrig's disease)	Superoxide dismutase I
Huntington's disease	Huntingtin with polyQ tracts
Cataract	γ-Crystallins
Type II diabetes	Islet amyloid polypeptide (IAPP)
Injection-localized amyloidosis	Insulin

Protein misfolding is the basis for several diseases, including Alzheimer's disease and Parkinson's disease.

Although some proteins, or fragments thereof, are predisposed to form amyloid fibrils under physiological conditions, many other proteins can be induced to form amyloid structures under extreme conditions (e.g., high or low pH, elevated temperature, etc.). This observation led Chris Dobson to propose the hypothesis that the amyloid structure is a generic low-energy conformation accessible to all peptides. A corollary to this hypothesis is that the folding pathways of soluble proteins have evolved to avoid amyloid formation, even though amyloid may be the most thermodynamically stable conformation for proteins (see Figure 6.31b).

Amyloid fibrils are characterized by highly organized arrays of β structure. Figure 6.33a shows two electron microscopy images of insulin amyloid fibrils. These fibrils are formed from a right-handed helix of four "protofibrils" (Figure 6.33b). Each protofibril, in turn, is formed from a highly ordered array of parallel β sheets separated by roughly 1.0 nm (Figure 6.33c). This so-called **cross-β** structure is a characteristic of amyloids. Studies on amyloidogenic peptides indicate the amyloid formation under physiological conditions is slow; however, protofibrils and fibrils form rapidly once the cross-β structure is nucleated.

The structure of amyloid and its mechanism of formation have been studied extensively in the hope of finding therapeutic interventions for misfolding

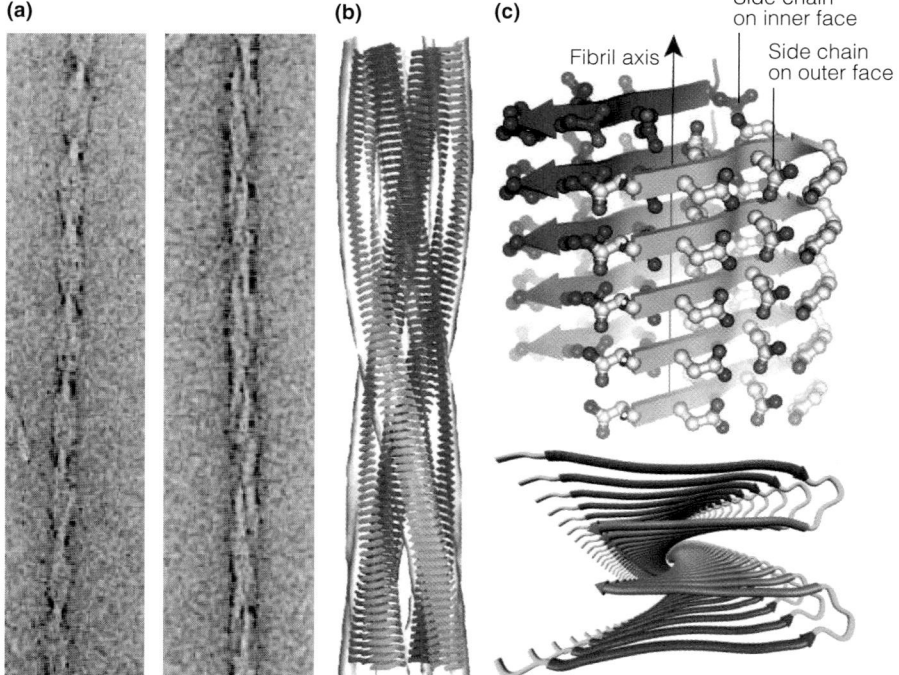

(a) **(b)** **(c)**

Fibril axis

Side chain on inner face

Side chain on outer face

FIGURE 6.33

Amyloid fibril structure. **(a)** Electron micrographs of amyloid fibrils of insulin. **(b)** A model of insulin amyloid fibrils fitted to electron density data from cryo-electron microscopy. Four protofibrils wind together to form a right-handed helix. **(c)** The cross-β structure of the protofibril. The top figure shows the X-ray crystal structure of protofibrils formed from the GNNQQY peptide from the yeast prion protein Sup35. The bottom figure shows a model of the protofibril formed from residues 1–40 of the Aβ peptide. Solid-state NMR data were used to derive this model.

(a, b) Adapted from *Proceedings of the National Academy of Sciences of the United States of America* 99:9196–9201, J. L. Jiménez, E. J. Nettleton, M. Bouchard, C. V. Robinson, C. M. Dobson, and H. R. Saibil, The protofilament structure of insulin amyloid fibrils. © 2002 National Academy of Sciences, U.S.A.; (c, top) Adapted by permission from Macmillan Publishers Ltd. *Nature* 435:773–778, R. Nelson, M. R. Sawaya, M. Balbirnie, A. Ø. Madsen, C. Riekel, R. Grothe, and D. Eisenberg, Structure of the cross-β spine of amyloid-like fibrils. © 2005; (c, bottom) Robert Tycko, National Institutes of Health, Bethesda, MD.

diseases. The formation of amyloid fibrils (Figure 6.33a) is thought to occur from non-native conformations of proteins such as folding intermediates or disordered aggregate states (Figure 6.31a). Recent experiments suggest that the amyloid fibrils *per se* are not responsible for causing disease; rather, it appears that the toxic species may be the prefibrillar or disordered aggregates that are precursors to amyloid formation. In some cases, increasing the intracellular concentration of chaperones reduces formation of these pre-fibrillar aggregates and reduces the toxicity associated with protein misfolding. Similar effects have been observed for low molecular weight compounds that can bind to, and thereby stabilize, the native structures of amyloidogenic proteins (see Tools of Biochemistry 6C).

Among the amyloidogenic proteins, **prions** are remarkable. Until very recently, virtually all researchers believed that the only ways in which diseases could be transmitted from one organism to another were via viruses or microorganisms. However, there is now evidence that a class of diseases is transmitted by a protein and nothing more. The most notorious of these diseases is *bovine spongiform encephalopathy,* or "mad cow disease," but they also include *scrapie* in sheep and certain neuropathologies in humans (see entries in Table 6.6). The infectious agent has been termed *prion,* and the protein believed to be responsible is called *prion-related protein,* or *PrP.* The PrP is normally present in many animals (including humans) in a nonpathological form called PrP^c (*prion-related protein cellular*). Under certain circumstances PrP^c can change conformation to a different structure called PrP^{sc}, or *prion-related protein scrapie.* It is this form, in which the intrinsically disordered N-terminal portion of PrP appears to at least partially fold into a β sheet, that wreaks havoc with the nervous system. Even more remarkable is the fact that ingesting PrP^{sc} can induce conversion of PrP^c in the recipient to PrP^{sc}; thus the condition is transmitted. How this conversion is catalyzed is unknown, but it strongly suggests that PrP^{sc} represents an especially stable off-path folding of the type hypothesized in the preceding section. All known prion diseases are untreatable and fatal. In 1997 Stanley Prusiner was awarded the Nobel Prize in Physiology or Medicine for his work in defining the relationship of PrP to these diseases.

Motions Within Globular Protein Molecules

Much evidence, particularly from NMR studies, indicates that various kinds of motions are continually occurring within folded protein molecules. A protein molecule undergoes continued, rapid fluctuations in its energy, as a consequence of interactions with its environment. The resulting motions can be roughly grouped into three classes, as shown in Table 6.7. Class 1 motions occur even within protein molecules in crystals and account, in part, for the limits of resolution obtainable in X-ray diffraction studies. The larger, slower motions in classes 2 and 3 are more likely to occur in solution. Some of them, like the opening and closing of clefts in molecules, are thought to be involved in the catalytic functions of enzymes. As we shall see in later chapters, how long it takes for a protein to bind or release a small molecule may depend on the time required for the protein to open or close a cleft. Similarly, the protein channels that pass molecules and ions through membranes

> Globular proteins are not static but continually undergo a wide variety of internal motions.

TABLE 6.7   Motions in protein molecules

Class	Type of Motion	Amplitude (nm)	Time (s)
		Approximate Range	
1	Vibrations and oscillations of individual atoms and groups	0.2	10^{-15}–10^{-12}
2	Concerted motions of structural elements, like α helices and groups of residues	0.2–1	10^{-12}–10^{-8}
3	Motions of whole domains; opening and closing of clefts	1–10	$\geq 10^{-8}$

rapidly change from open to closed states. It seems likely that the dynamic behavior of proteins is at least as important in their function as are the static details of their structure. In fact, an emerging area of protein research is the characterization of so-called intrinsically unstructured proteins. This class of proteins possesses significant regions of sequence that are highly dynamic and remain unstructured until one of several possible partner proteins is bound.

Prediction of Secondary and Tertiary Protein Structure

Can protein structure be predicted? In one sense, the answer to this question must surely be yes. We know that the molecular information necessary to determine the secondary and tertiary structures is carried in the amino acid sequence itself; thus, the gene "predicts" the structure. The implication of this fact is that if we fully understood the rules of folding, we could describe the entire three-dimensional conformation of any protein, starting with nothing more than a knowledge of its sequence. This kind of prediction cannot yet be done completely. We can predict secondary structure fairly reliably. Prediction of tertiary folding is a much more complicated problem; however, significant progress in this area has been made in recent years.

Protein secondary structure can now be predicted with good accuracy. The *de novo* prediction of more complex tertiary structure is not yet as accurate.

Prediction of Secondary Structure

Although a number of approaches have been applied to the problem of predicting secondary structure, the most successful method is entirely empirical. From analysis of the *known* structures of a number of proteins, tables have been compiled to show the relative frequency (P_α, P_β, P_t) with which a particular kind of amino acid residue lies in α helices, β sheets, or turns. An example of such a compilation is given in Table 6.8. From these data, certain clear distinctions can be

TABLE 6.8 Correspondence of amino acid residues to protein secondary structure

Relative probabilities of amino acid residue occurrence in different globular protein secondary structures[a]

Amino Acid	α Helix (P_α)	β Sheet (P_β)	Turn (P_t)	
Ala	1.29	0.90	0.78	
Cys	1.11	0.74	0.80	
Leu	1.30	1.02	0.59	
Met	1.47	0.97	0.39	Favor α helices
Glu	1.44	0.75	1.00	
Gln	1.27	0.80	0.97	
His	1.22	1.08	0.69	
Lys	1.23	0.77	0.96	
Val	0.91	1.49	0.47	
Ile	0.97	1.45	0.51	
Phe	1.07	1.32	0.58	Favor β sheets
Tyr	0.72	1.25	1.05	
Trp	0.99	1.14	0.75	
Thr	0.82	1.21	1.03	
Gly	0.56	0.92	1.64	
Ser	0.82	0.95	1.33	
Asp	1.04	0.72	1.41	Favor turns
Asn	0.90	0.76	1.23	
Pro	0.52	0.64	1.91	
Arg	0.96	0.99	0.88	

[a]Data adapted from *Biochemistry* 17:4277–4285, M. Leavitt, Conformational preferences of amino acids in globular proteins. © 1978 American Chemical Society.

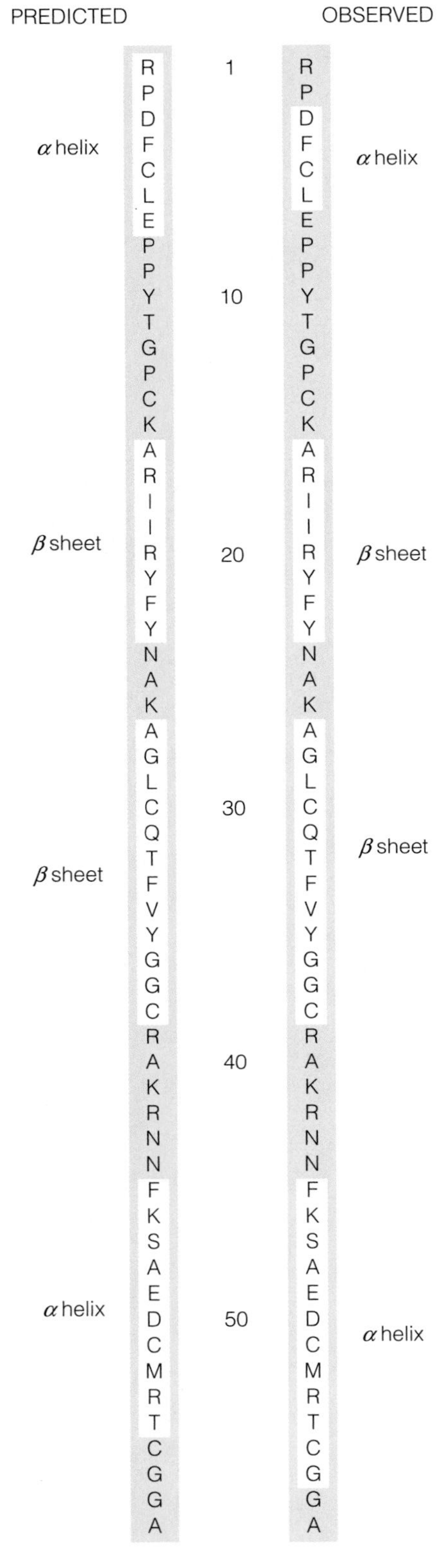

FIGURE 6.34

Prediction of the secondary structure of BPTI. The sequence on the left shows the secondary structural elements predicted by P. Y. Chou and G. D. Fasman for bovine pancreatic trypsin inhibitor. The sequence on the right shows the results of X-ray diffraction studies of the same protein. The exceptionally good agreement found between the predicted and observed structures is somewhat better for BPTI than for most proteins.

PREDICTED			OBSERVED
	R	1	R
	P		P
	D		D
α helix	F		F
	C		C
	L		L α helix
	E		E
	P		P
	P		P
	Y	10	Y
	T		T
	G		G
	P		P
	C		C
	K		K
	A		A
	R		R
	I		I
	I		I
β sheet	R	20	R β sheet
	Y		Y
	F		F
	Y		Y
	N		N
	A		A
	K		K
	A		A
	G		G
	L		L
	C	30	C
	Q		Q β sheet
	T		T
β sheet	F		F
	V		V
	Y		Y
	G		G
	G		G
	C		C
	R		R
	A	40	A
	K		K
	R		R
	N		N
	N		N
	F		F
	K		K
	S		S
	A		A
α helix	E		E
	D	50	D α helix
	C		C
	M		M
	R		R
	T		T
	C		C
	G		G
	G		G
	A		A

made. For example, Ala, Leu, Met, and Glu are all strong helix formers; Gly and Pro do not favor helices. Similarly, Ile, Val, and Phe are strong β sheet formers, whereas Pro does not fit well into β sheets. Gly is frequently found in β turns, whereas Val is not. We have already mentioned that Pro tends to lie in turns. Why other residues are generally found in one structure or another is not so clear. Various rules have been proposed to use *P* values to predict structures; those developed in the 1970s by P. Y. Chou and G. D. Fasman were used to predict the secondary structure of BPTI from its sequence (Figure 6.34).

In addition to *P* values, patterns of side chain polarity are used to predict secondary structure. A stretch of alternating polar and nonpolar residues is diagnostic for a β strand that is part of an amphiphilic β sheet. Similarly, a pattern of side chains with similar polarity every 3–4 residues indicates amphiphilic α helix.

The structures predicted by applying these guidelines often agree fairly well, although never exactly, with secondary structures determined experimentally (Figure 6.34). Differences are most frequently found in definition of the precise location of the ends of α and β regions.

As the amount of structural data stored in public databases grows, so too does the sophistication of the computational methods used to make structural predictions from sequence data. As of 2007 the accuracy of the best predictions was ~80% (up from ~70% just a few years previously). Today, many different algorithms exist online to make rapid predictions of secondary structure from amino acid or gene sequence information (see the References section at the end of this chapter).

Tertiary Structure Prediction: Computer Simulation of Folding

The prediction of tertiary structure has proved much more difficult, probably because the higher-order folding depends so critically on specific noncovalent interactions, often between residues far removed from one another in the sequence. In spite of the difficulty of the problem, recent efforts in de novo structure prediction have met with spectacular success. Figure 6.35 shows the results of two recent de novo predictions that were achieved at high resolution *before* the X-ray crystal structures for the protein sequences were released publicly. The close agreement between the predicted structures and the crystal structures is remarkable, and bodes well for the future of de novo structure prediction from amino acid sequence.

The prediction of tertiary structure from amino acid sequence is fundamentally a problem in energy minimization requiring sophisticated computation. The underlying assumption is that the native functional conformation for a polypeptide is the conformation with the lowest value of free energy.* Thus, during folding, proteins are seeking a global free energy minimum. To understand the process, let us consider the oversimplified case of a computer simulation in which a random-coil chain is allowed to undergo a very large number of small permutations in its conformation, through rotation about individual bonds. For each conformation generated by the computer, the total energy of all noncovalent interactions is calculated using a table of parameters called a "force field," or "potential function." The values of total energy for each conformation are then compared, and the conformation with the lowest energy is considered to be a better candidate for the

---

*This is a reasonable assumption for most proteins; however, there are a few notable exceptions (e.g., see: D. Baker, J. L. Sohl, and D. Agard, *Nature* (1992) 356:263–265), and more are likely to be discovered in the future.

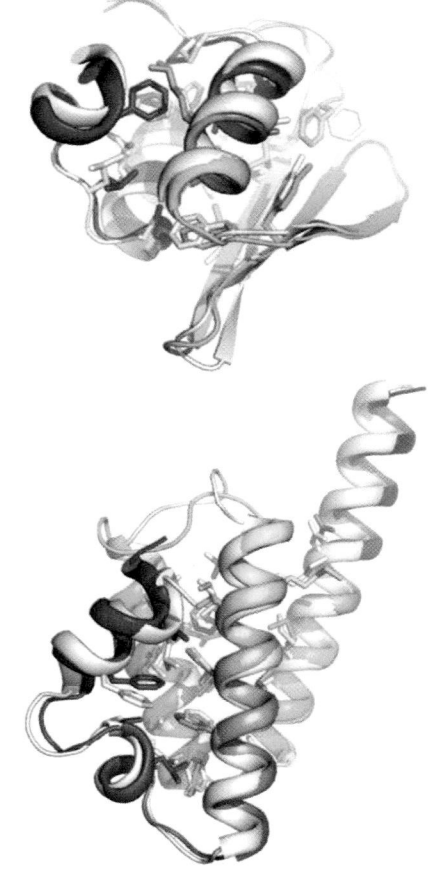

FIGURE 6.35

Comparison of de novo predictions to X-ray crystal structures. Top panel: Rosetta prediction (shown in light gray) for the Critical Assessment of Techniques for Protein Structure Prediction (CASP) 6 target, T0281, superimposed on the *subsequently* released crystal structure (PDB ID: 1whz; shown with rainbow coloring from the N-terminus [blue] to the C-terminus [red]). The agreement between the backbones of the two structures is within 0.16 nm over 70 residues. Bottom panel: The Rosetta prediction for the CASP7 target, T0283, (light gray) agrees with the subsequently released crystal structure (PDB ID: 2hh6; shown in rainbow coloring) with a backbone accuracy of 0.14 nm over 90 residues.

From *Annual Review of Biochemistry* 77:363–382, R. Das and D. Baker, Macromolecular modeling with Rosetta. © 2008 Annual Reviews.

native structure. This process may be repeated for up to a million conformations before the computation converges on a single lowest energy conformation.

These methods have two basic requirements. The first is a potential function that calculates energies with good accuracy. Most of the current potential functions are based on quantum mechanics and make certain simplifying assumptions about solvent interactions. The second requirement is a tractable method for sampling the vast number of possible conformations that ensures the true global minimum is ultimately found, rather than some local minimum on the energy landscape. A simple random search through all possible conformations is an impossible task; it is, in essence, confronting the Levinthal paradox with a computer. The massive computational power required for such a random search simply does not exist. Somehow, a search for the true energy minimum must be more directed.

Many researchers have devised elegant solutions to this problem. For the purposes of our discussion we will focus on only two of these. The first is the use of empirically derived folding "fragments," or short structural motifs, employed by the Rosetta suite of algorithms, developed in the lab of David Baker. Rosetta has not only been used to successfully predict structure de novo, it has also been successfully applied to de novo protein-design problems. Rosetta first matches local regions of amino acid sequence to a library of structural fragments derived from high-resolution crystal structures (see Figure 6.36a). In essence, Rosetta assigns the most likely local secondary structure based on sequence homology to proteins with known structures (see Tools of Biochemistry 5D). These assembled structural fragments are then packed together in ways that give low-energy conformations. For this part of the calculation Rosetta doesn't consider atomic detail for the side chains; rather, a simpler classification in terms of polarity and charge is used (Figure 6.36b). Finally, the side chains are added back to the simulation, and the conformation with the

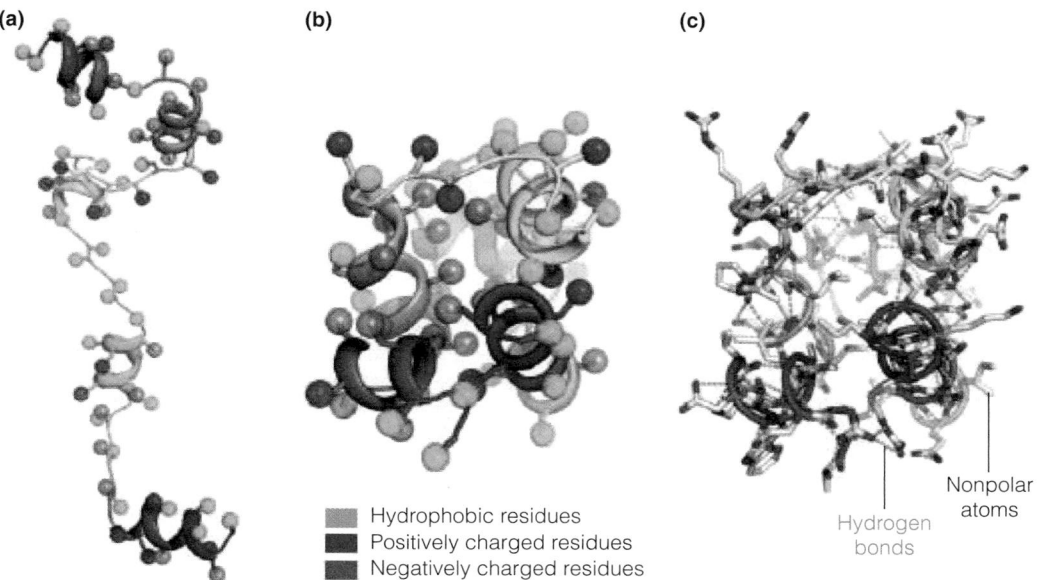

(a) **(b)** **(c)**

▨ Hydrophobic residues
▨ Positively charged residues
▨ Negatively charged residues
▨ Polar residues

Hydrogen bonds

Nonpolar atoms

FIGURE 6.36

Schematic of de novo structure prediction using Rosetta. All backbone "heavy atoms" (i.e., all atoms other than H atoms) are shown as a ribbon (rainbow coloring). Side chains are represented initially at low resolution as spheres. Conformations are evaluated with an energy function favoring burial of hydrophobic residues (gray) and the exposure of positively charged (dark blue), negatively charged (red), or other polar (green) residues. **(a)** Assembly of fragments of local secondary structure. **(b)** Final low-energy conformation produced by fragment packing. **(c)** All-atom model produced after high-resolution refinement. For clarity, hydrogen atoms are not shown.

From *Annual Review of Biochemistry* 77:363–382, R. Das and D. Baker, Macromolecular modeling with Rosetta. © 2008 Annual Reviews.

Experimental data from NMR and cryo-electron microscopy can be used to improve the successful prediction of protein tertiary structure by computational methods.

...TPPVQAAYQKVVAGVANA...

(a) Primary Structure

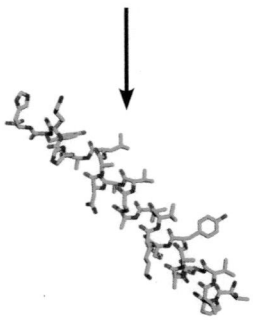

(b) Secondary Structure

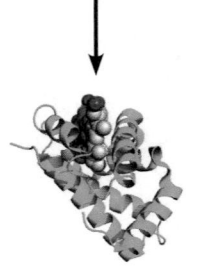

(c) Tertiary Structure

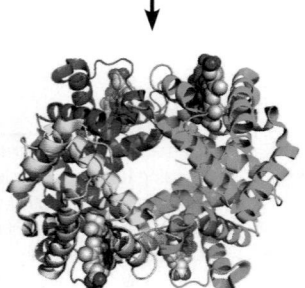

(d) Quaternary Structure

FIGURE 6.37

The four levels of protein structure. This summary of the structural levels of protein uses the hemoglobin molecule (PDB ID: 2HHB), a tetramer of myoglobin-like chains, as an example.

Association of polypeptide chains to form specific multisubunit structures is the quaternary level of protein organization.

lowest calculated energy is sought, as described in the preceding paragraphs (Figure 6.36c). This entire process is repeated, starting from many hundred assembled secondary structure fragments, until Rosetta converges on a final low-energy conformation, which is calculated using a potential function that accounts for the energies associated with the hydrophobic effect (i.e., burial of hydrophobic side chains), close packing of interior groups in the final structure (i.e., maximizing van der Waals contacts), and intramolecular H-bond formation (i.e., no unsatisfied H-bond donors or acceptors for buried groups). In summary, Rosetta solves the conformational search problem by making educated guesses for local structure, thereby dramatically restricting the number of possible conformations that must be sampled.

A second approach, which shows great promise, is the use of experimental data to restrict the conformational search. NMR chemical-shift data (which report on local amino acid conformation and the chemical environment of an atom; see Tools of Biochemistry 6A), information from chemical modification studies (which can be used to identify solvent-accessible side chains), and cryo-electron microscopy data (which provide constraints on backbone topology) have all been used to augment computational structure prediction. Such approaches are likely to further improve the success of de novo prediction. The need for reliable methods of de novo structure prediction is tremendous, given the fact that structures are known for only 1% of all known protein sequences.

Quaternary Structure of Proteins

In Chapter 5 and in this chapter, we have explored increasingly complex levels of protein structure, from primary to secondary to tertiary (Figure 6.37a–c). Functional protein organization can reach at least one more level—**quaternary structure** (Figure 6.37d). Many proteins exist in the cell (and in solution, under physiological conditions) as specific complexes of two or more folded polypeptide chains, or *subunits*. Methods for determining whether or not a protein is composed of multiple subunits are described in Tools of Biochemistry 6B. Such quaternary organization can be of two kinds—association between identical or nearly identical polypeptide chains (**homotypic**) or interactions between subunits of very different structures (**heterotypic**). In either case, *multisubunit proteins* are formed.

Multisubunit Proteins: Homotypic Protein–Protein Interactions

The interactions between the folded polypeptide chains in multisubunit proteins are of the same kinds that stabilize tertiary structure—salt bridges, hydrogen bonding, van der Waals forces, the hydrophobic effect, and sometimes disulfide bonding. These interactions provide the energy to stabilize the multisubunit structure.

Each polypeptide chain is an asymmetric unit in the complex, but the overall quaternary structure may exhibit a wide variety of symmetries, depending on the geometry of the interactions. For purposes of illustration, we shall use an asymmetric object familiar to everyone—a right shoe. Think of this shoe as a polypeptide chain folded into a compact three-dimensional form. We can stick shoes together in many ways. If the interacting surfaces (A and B) were at toe and heel, a linear complex could form:

Because the two interacting groups lie in entirely different regions of the subunit, we would call this interaction **heterologous**. Heterologous interactions must be specially oriented to give a truly linear complex. More often, the interaction is such that each unit is twisted through some angle with respect to the preceding one. This twisting gives rise to a helical structure. Figure 6.38a shows an arrangement of shoes

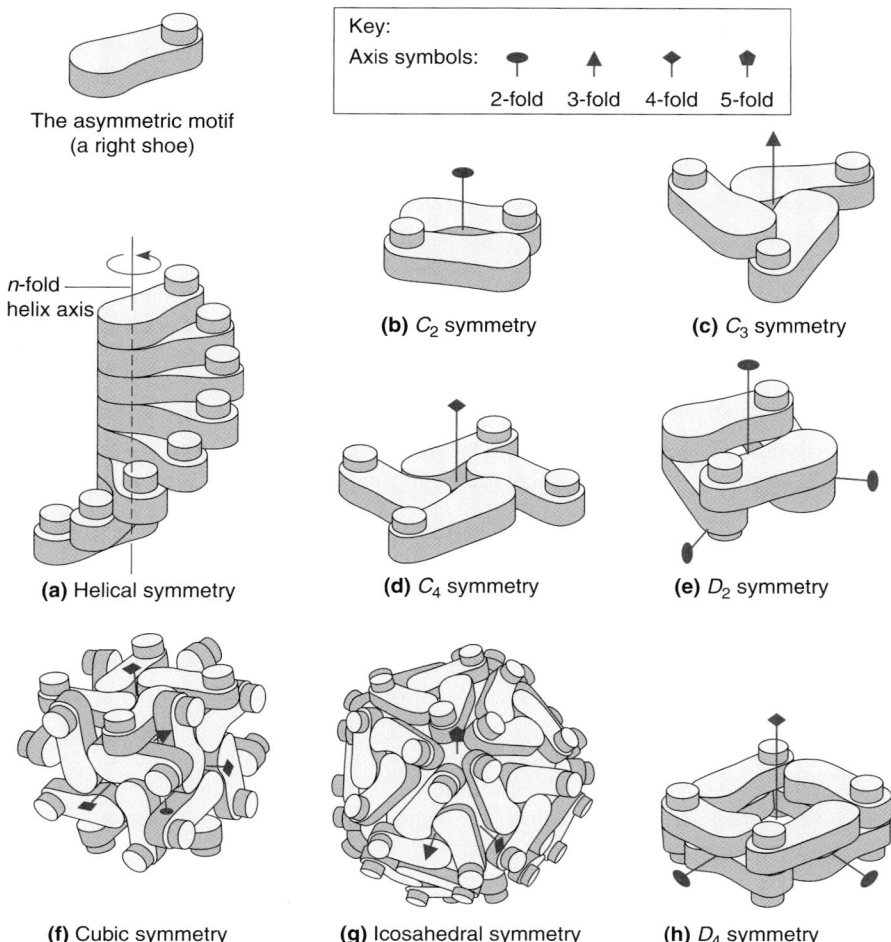

The asymmetric motif (a right shoe)

Key:
Axis symbols:

| 2-fold | 3-fold | 4-fold | 5-fold |

(a) Helical symmetry — n-fold helix axis

(b) C_2 symmetry

(c) C_3 symmetry

(d) C_4 symmetry

(e) D_2 symmetry

(f) Cubic symmetry

(g) Icosahedral symmetry

(h) D_4 symmetry

FIGURE 6.38

Symmetries of protein quaternary structures. Although composed of asymmetric polypeptides, proteins adopt many symmetrical patterns in forming quaternary structures. In this figure, a right shoe represents the asymmetric structural unit. **(a)** A helix formed by rotation of each unit by $360/n$ degrees with respect to the preceding one. Such rotation produces an n-unit-per-turn helix of indefinite length. **(b)** A dimer with C_2 symmetry: one 2-fold axis. **(c)** A trimer with C_3 symmetry: one 3-fold axis. **(d)** A tetramer with C_4 symmetry: one 4-fold axis. **(e)** A tetramer with D_2 symmetry: three 2-fold axes. **(f)** A 24-mer exhibiting cubic symmetry. This structure has 4-fold, 3-fold, and 2-fold axes. **(g)** An icosahedral 60-mer, the kind of structure found in the protein coat of a number of viruses. This structure has 5-fold, 3-fold, and 2-fold axes. **(h)** An octamer with D_4 symmetry: one 4-fold axis and two 2-fold axes.

that forms a right helix with n units per turn. The top of the toe of each shoe is attached to the sole of the toe of the next, with a rotation of $360/n$ degrees. Two biological examples, both of **helical symmetry**, are shown in Figure 6.39: the helix of the muscle protein actin and the helical coat of tobacco mosaic virus. Note that both linear and helical arrays are potentially capable of indefinite growth by addition of more subunits. Actin filaments can be thousands of units in length.

Most assemblies of protein subunits are based not on helical symmetry but on one of the classes of **point-group symmetry** (Figure 6.38b–h). The classes of point-group symmetry involve a defined number of subunits arranged about one or more **axes of symmetry**. An n-fold axis of symmetry corresponds to rotation of each subunit by $360/n$ degrees with respect to its neighbor; thus, a 2-fold axis corresponds to 180° rotation. The simplest kinds of point-group symmetry are the cyclic symmetries C_n, shown in Figure 6.38b–d. These rings of subunits involve heterologous interactions where $n = 3$ or greater.

A special situation of great importance arises when two subunits are related to one another by a 2-fold axis (also called a **dyad** axis) to give C_2 symmetry. That is, each subunit is rotated by 180° about this axis with respect to the other:

Dyad axis perpendicular to paper

(This arrangement is visualized in three dimensions in Figure 6.38b.) Imagine that there are interacting groups at A and B. For example, A could be a hydrogen-bond donor, B an acceptor. Note that in this case the 2-fold symmetry means that

Two general classes of symmetry—helical and point-group—characterize most quaternary structures.

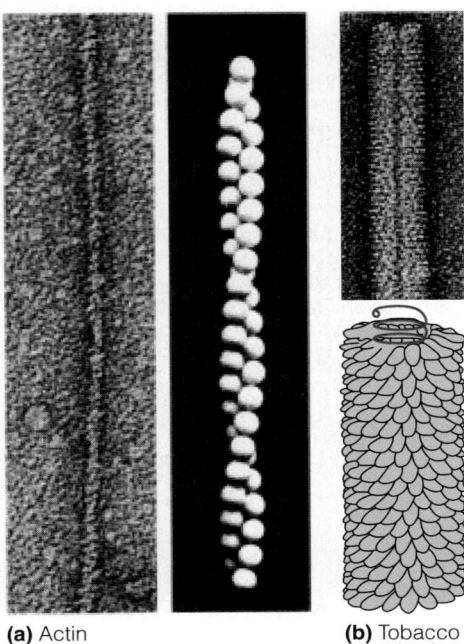

(a) Actin   **(b)** Tobacco mosaic virus

FIGURE 6.39

Two helical proteins. Beside each electron micrograph is a diagrammatic representation of the helical aggregate structure. **(a)** Actin. **(b)** Tobacco mosaic virus. In the virus, protein subunits form a helical array about a helically coiled RNA (red).

(a) © The Rockefeller University Press. *The Journal of Cell Biology*, 1981, 91:156s–165s, T. Pollard, Cytoplasmic contractile proteins; (b) Dr. Timothy Baker/Visuals Unlimited, Inc.

All higher levels of structure of a protein are dictated by its genes.

FIGURE 6.40

The prealbumin dimer. In the prealbumin dimer the two monomers combine to form a complete β sandwich, or flattened β barrel. The dimer has 2-fold symmetry about an axis perpendicular to the paper. The isologous interactions are mostly hydrogen bonds between specific β sheet strands: F to F' and H to H'. Prealbumin can also form a tetramer from two of these dimers by a second set of isologous interactions.

Reprinted from *Journal of Molecular Biology* 88:1–12, C. C. F. Blake, M. J. Geisow, I. D. A. Swan, C. Rerat, and B. Rerat, Structure of human plasma prealbumin at 2.5 A resolution: A preliminary report on the polypeptide chain conformation, quaternary structure and thyroxine binding. © 1974, with permission from Elsevier.

two identical interactions occur, symmetrically placed about the dyad axis. Such a symmetric interaction is called **isologous**, as opposed to the asymmetric heterologous interaction. Isologous interactions will be found whenever 2-fold axes are present. Dimers are the most common of all quaternary structures, and they are almost always bound together in this way. An example is shown in Figure 6.40. Further isologous interactions can easily give rise to more complex quaternary structures of higher symmetry. An example is **dihedral** symmetry (Figure 6.38e), the most common structure for tetrameric proteins. It has three mutually perpendicular 2-fold axes and, therefore, involves three pairs of isologous interactions. In fact, any D_n symmetry can be generated by isologous interactions alone.

Other, more complex point-group symmetries exist in some protein quaternary structures; examples are shown in Figure 6.38f and g. Note that the more complex structures involve both 2-fold axes and axes with $n > 2$. Such molecules exhibit both isologous and heterologous interactions. Most important is that molecules exhibiting any of the point-group symmetries are always constrained to a definite number of subunits. Most multisubunit proteins exhibit this kind of association geometry, rather than the linear or helical aggregation that can lead to indefinite growth. This is why most protein molecules, even if they contain multiple subunits, have a well-defined size, shape, and mass.

Two exceptions to this simple characterization of possible structures must be noted. First, although most dimers have 2-fold symmetry and isologous binding, it is possible to construct dimers in which the binding is heterologous but indefinite association is still sterically blocked. Such dimers do not have 2-fold symmetry (Figure 6.41). A second complication is encountered whenever more than one type of polypeptide chain is incorporated into a specific multisubunit structure. In such cases the symmetry is reduced from the level you might expect from the total number of polypeptide chains because two or more different chains will now form one asymmetric unit. We consider one such example, *hemoglobin*, in Chapter 7.

Note that each level of protein structure is built on the lower levels, as summarized in Figure 6.37. Tertiary structure can be thought of as a folding of elements of secondary structure, and quaternary structure is established by combining folded subunits. *All* of this higher-level structuring is dictated by primary structure and

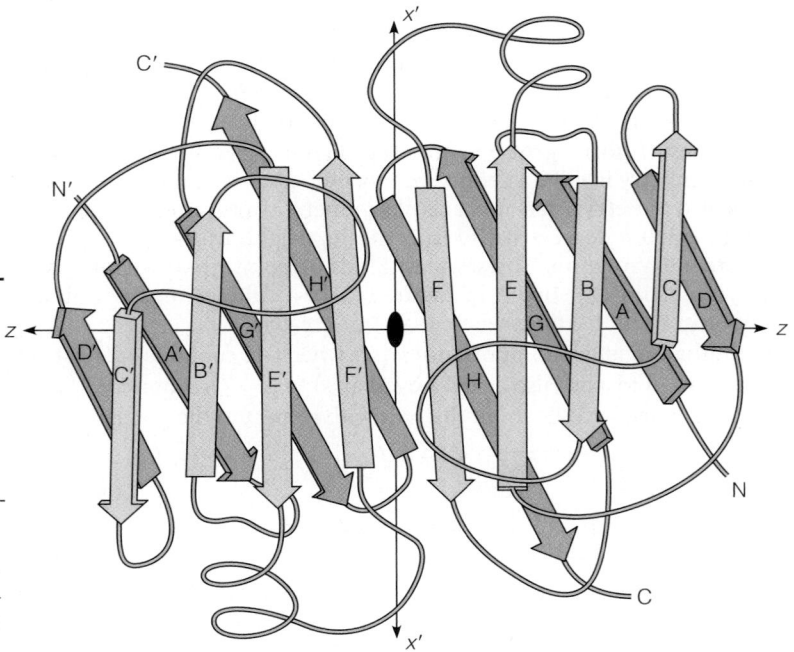

ultimately by the gene. This understanding of the relationship between protein structure and gene sequence is one of the most important ideas in molecular biology.

Heterotypic Protein–Protein Interactions

The preceding section has centered on the association of identical or near-identical protein subunits. However, the range of protein–protein interactions is much greater; specific associations between entirely different protein molecules are common. Sometimes these associations lead to organized structures containing a dozen or more different subunit types. The interactions that form these assemblies are of the same kind we described earlier—noncovalent forces at complementary protein surfaces.

Complementary interactions also determine the specific interactions between a protein, or protein complex, and its target. A simple example involves BPTI, a protein we have described in detail in this chapter. Bovine pancreatic trypsin inhibitor is so named because it forms a tight, specific complex with the enzyme trypsin, thereby inhibiting trypsin proteolytic activity in the pancreas. Figure 6.42 shows that the two protein surfaces fit one another closely.

Computational methods like those described earlier in this chapter can predict such complementary protein–protein interactions with good success. For example, the computational "docking" of BPTI with trypsin closely matches the actual structure of the complex as determined by X-ray diffraction. The accuracy of such predictions suggests that it may be easier to predict protein–protein interactions (including protein quaternary structure interactions) than it is to predict tertiary structure.

SUMMARY

Protein molecules typically have several levels of organization. The first, or primary, level is the amino acid sequence, dictated by the gene. This sequence in turn dictates local folding (secondary structure), global folding (tertiary structure), and organization into multisubunit structures (quaternary structure).

Of many conceivable secondary structures, only a limited number are sterically allowed and can be stabilized by hydrogen bonds. These include the α helix, the β sheet, and the 3_{10} helix. There are also specific structures that allow a polypeptide chain to make sharp turns. Ramachandran diagrams provide a way to visualize the possibilities, and describe various secondary structures in terms of allowed ϕ and ψ angles.

Proteins can be grouped into two broad categories—fibrous and globular. Fibrous proteins are elongated, usually of regular secondary structure, and perform structural roles in the cell and organism. Important examples include the keratins (α helix), the fibroins (β sheet), collagen (triple helix), and elastin (cross-linked random coils). Globular proteins have more complex tertiary structures and fold into compact shapes that often contain defined domains. Several classes of folding motifs have been recognized, such as α helix bundles, twisted β sheets, and β barrels.

A number of factors determine globular protein stability—conformational entropy, enthalpy from internal noncovalent bonding, the hydrophobic effect, and disulfide bonds. The folding of many globular proteins occurs spontaneously and rapidly under "native" conditions. In the cell, proteins called chaperones help prevent formation of incorrectly folded structures or undesired intermolecular interactions leading to aggregation. Even when folded, globular proteins are dynamic structures undergoing several kinds of internal motions. Protein secondary structure can be predicted with good accuracy, but prediction of tertiary structure, although improving, remains challenging.

Many (perhaps most) globular proteins exist and function as multisubunit assemblies forming a quaternary level of structure. A few of these proteins are elongated structures with helical symmetry; most have a small number of subunits (often 2, 4, or 6) and exhibit point-group symmetry. All of the levels of protein structure are determined by the gene sequence.

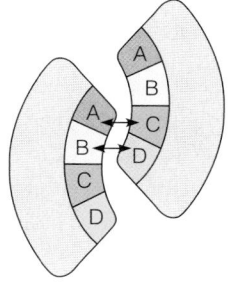

FIGURE 6.41

A dimer without symmetry. This schematic picture shows how two subunits can associate by heterologous interactions and still not yield an indefinite chain. The interaction sites (A, B, C, D) are blocked from reaction with more monomers by the close fit of the surfaces.

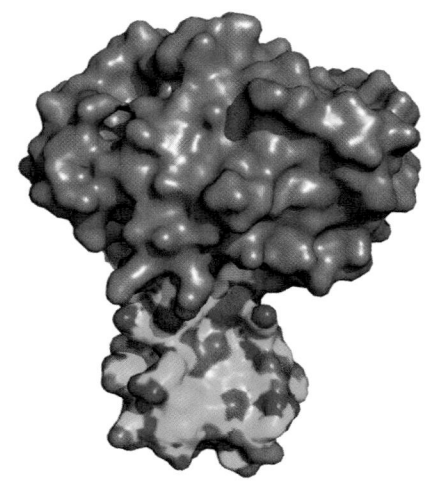

FIGURE 6.42

Interaction of BPTI with trypsin. The BPTI molecule (green, red, and blue) fits snugly onto the surface of the trypsin molecule (slate blue), blocking the active site of trypsin. The orientation of the BPTI is the same as that in Figure 6.18. PDB ID:2RA3.

REFERENCES

General

Brändén, C., and J. Tooze (1991) *Introduction to Protein Structure.* Garland, New York.

Creighton, T. E., ed. (1992) *Protein Folding.* Freeman, San Francisco.

Creighton, T. E. (2010) *The Biophysical Chemistry of Nucleic Acids and Proteins.* Helvetian Press, UK.

Liljas, A., L. Liljas, J. Piskur, G. Lindblom, P. Nissen, and M. Kjeldgaard (2009) *Textbook of Structural Biology.* World Scientific Publishing, Singapore.

Petsko, G. A., and D. Ringe (2004) *Protein Structure and Function.* New Science Press, London.

Shirley, B., ed. (1995) *Protein Stability and Folding.* Humana Press, Totowa, NJ.

Historical

Anfinsen, C. B. (1973) Principles that govern the folding of protein chains. *Science* 181:223–230.

Kauzmann, W. (1959) Some factors in the interpretation of protein denaturation. *Adv. Protein Chem.* 14:1–63. Discussion of the hydrophobic effect.

Pauling, L., R. B. Corey, and H. R. Branson (1951) The structure of proteins: Two hydrogen bonded helical conformations of the polypeptide chain. *Proc. Natl. Acad. Sci. USA* 37:205–211.

Ramachandran, G. N., and V. Sassiekharan (1968) Conformation of polypeptides and proteins. *Adv. Protein Chem.* 28:283–437. Introduction of Ramachandran plots.

Fibrous Proteins

Kaplan. D., W. W. Adams, B. Farmer, and C. Viney (1994) *Silk Polypeptides.* American Chemical Society Press, New York.

vanderRest, M., and P. Bruckner (1993) Collagens: Diversity at the molecular and supramolecular levels. *Curr. Opin. Struct. Biol.* 3:430–436.

Globular Proteins: Secondary and Tertiary Structure

Cooley, R. B., D. J. Arp, and P. A. Karplus (2010) Evolutionary origin of a secondary structure: π-Helices as cryptic but widespread insertional variations of α-helices that enhance protein functionality. *J. Mol. Biol.* 404:232–246.

Hollingsworth, S. A., D. S. Berkholz, and P. A. Karplus (2009) On the occurrence of linear groups in proteins. *Protein Sci.* 18:1321–1325.

Richardson, J. S. (1981) The anatomy and taxonomy of protein structure. *Adv. Protein Chem.* 34:167–339.

Databases of Domain Structure and Classification

Class Architecture Topology Homologous Superfamily (CATH): http://www.cathdb.info/

Families of Structurally Similar Proteins (FSSP): ftp://ftp.ebi.ac.uk/ pub/databases/fssp/

Structural Classification of Proteins (SCOP): http://scop.mrc-lmb.cam.ac.uk/scop/

Protein Folding and Stability

Baase, W. A., L. Liu, D. E. Tronrud, and B. W. Matthews (2010) Lessons from the lysozyme of phage T4. *Prot. Sci.* 19:631–641.

Baker, D. (2000) A surprising simplicity to protein folding. *Nature* 405:39–42.

Bartlett, A. I., and S. E. Radford (2009) An expanding arsenal of experimental methods yields an explosion of insights into protein folding mechanisms. *Nature Struct. Mol. Biol.* 16:582–588.

Brooks III, C. L., M. Gruebele, J. N. Onuchic, and P. G. Wolynes (1998) Chemical physics of protein folding. *Proc. Natl. Acad. Sci. USA* 95:11037–11038. A clear introduction of the energy landscape theory of protein folding.

De Sancho, D., U. Doshi, and V. Muñoz (2009) Protein folding rates and stability: How much is there beyond size? *J. Amer. Chem. Soc.* 131:2074–2075.

Dill, K. A. (1990) Dominant forces in protein folding. *Biochemistry* 29:7133–7155.

Dill, K. A., and H. S. Chan (1997) From Levinthal to pathways to funnels. *Nature Struct. Biol.* 4:10–19.

Onuchic, J. N., and P. G. Wolynes (2004) Theory of protein folding. *Curr. Opin. Struct. Biol.* 14:70–75.

Plaxco, K. W., K. T. Simons, and D. Baker (1998) Contact order, transition state placement and the refolding rates of single domain proteins. *J. Mol. Biol.* 277:985–994.

Rose, G. D., P. J. Fleming, J. R. Banavar, and A. Maritan (2006) A backbone-based theory of protein folding. *Proc. Natl. Acad. Sci. USA* 103:16623–16633. A thought-provoking review of protein folding theory that also proposes that main chain interactions, rather than side chain interactions, are the dominant factors in folding.

Chaperones

Clare, D. K., P. J. Bakkes, H. van Heerikhuizen, S. M. van der Vies, and H. R. Saibil (2009) Chaperonin complex with a newly folded protein encapsulated in the folding chamber. *Nature* 457:107–110.

Hartl, F. U., and M. Hayer-Hartl (2009) Converging concepts of protein folding in vitro and in vivo. *Nature Struct. Mol. Biol.* 16:574–581. A concise review of chaperone action in cells.

Xu, Z., A. L. Horwich, and P. B. Sigler (1997) The crystal structure of the asymmetric GroEL-GroES-(ADP)$_7$ chaperonin complex. *Nature* 388:741–750.

Prediction of Protein Structure

Secondary Structure Prediction:

Chou, P. Y., and G. D. Fasman (1978) Empirical predictions of protein structure. *Annu. Rev. Biochem.* 47:251–276.

Rost, B. (2009) Prediction of protein structure in 1D- Secondary structure, membrane regions and solvent accessibility. In J. Gu and P. E. Bourne, eds., *Structural Bioinfomatics* (2nd ed.). Wiley-Blackwell, Hoboken, NJ.

Access to several secondary structure prediction programs is available from: www.expasy.ch

Tertiary Structure Prediction:

Bradley, P., K. M. S. Misura, and D. Baker (2005) Toward high-resolution de novo structure prediction for small proteins. *Science* 309:1868–1871.

Das, R., and D. Baker (2008) Macromolecular modeling with Rosetta. *Annu. Rev. Biochem.* 77:363–382.

Kaufmann, K. W., G. H. Lemmon, S. L. Deluca, J. H. Sheehan, and J. Meiler (2010) Practically useful: What the Rosetta protein modeling suite can do for you. *Biochemistry* 49:2987–2998.

Protein Dynamics

Boehr, D. D., R. Nussinov, and P. E. Wright (2009) The role of dynamic conformational ensembles in biomolecular recognition. *Nature Chem. Biol.* 5:789–796.

Gsponer J., and M. Babu (2009) The rules of disorder or why disorder rules. *Prog. Biophys. Mol. Biol.* 99:94–103. Review of intrinsically unstructured proteins.

Painter, A. J., N. Jaya, E. Basha, E. Vierling, C. V. Robinson, and J. L. P. Benesch (2008) Real-time monitoring of protein complexes reveals their quaternary organization and dynamics. *Chem. and Biol.* 15:246–253.

Russel, D., K. Lasker, J. Phillips, D. Schneidman-Duhovny, J. A. Velázquez-Muriel, and A. Sali (2009) The structural dynamics of macromolecular processes. *Curr. Op. Cell Biol.* 21:1–12.

Prions and Misfolding Disease

Chiti, F., and C. M. Dobson (2006) Protein misfolding, functional amyloid, and human disease. *Annu. Rev. Biochem.* 75:333–366.

Cohen, F. E., and J. W. Kelly (2003) Therapeutic approaches to protein-misfolding diseases. *Nature* 426:905–909.

Prusiner, S. B. (1997) Prion diseases and the BSE crisis. *Science* 278:245–251.

Silviera, J. R., et al. (2005) The most infectious prion protein particles. *Nature* 437:257–261.

Uversky, V. N., and A. L. Fink (2006) *Protein Misfolding, Aggregation and Conformational Diseases,* Part A: Protein Aggregation and Conformational Disease. Springer, New York.

Quaternary Structure

Klotz, I. M., N. R. Langerman, and D. W. Darnall (1970) Quaternary structure of proteins. *Annu. Rev. Biochem.* 39:25–62.

Matthews, B. W., and S. A. Bernhard (1973) Structure and symmetry of oligomeric enzymes. *Annu. Rev. Biophys. Bioeng.* 2:257–317.

PROBLEMS

1. Polyglycine, a simple polypeptide, can form a helix with $\phi = -80°$, $\psi = +150°$. From the Ramachandran plot (see Figure 6.11), describe this helix with respect to (a) handedness and (b) number of residues per turn.

2. Certain polypeptide sequences that show a pronounced tendency to dimerize in an antiparallel coiled-coil structure exhibit an exact repetition of leucine every 7 residues. If you also know that the α helix is distorted somewhat from its usual 3.6 residues/turn in coiled coils, propose a mechanism for the dimerization. What does this suggest that the residue/turn value might be in a coiled coil?

3. A schematic structure of the subunit of hemerythrin (an oxygen-binding protein from invertebrate animals) is shown below.

   (a) It has been found that in some of the α-helical regions of hemerythrin, about every third or fourth amino acid residue is a hydrophobic one. Suggest a structural reason for this finding.

   (b) What would be the effect of a mutation that placed a proline residue at point A in the structure?

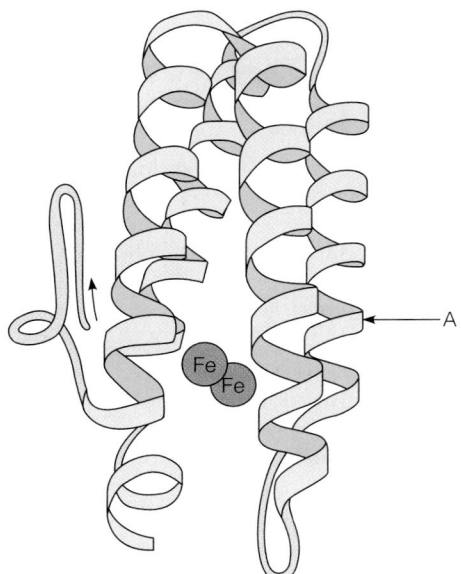

4. In the protein *adenylate kinase*, the C-terminal region is α-helical, with the sequence

   Val–Asp–Asp–**Val**–**Phe**–Ser–Gln–**Val**–Cys–Thr–His–**Leu**–Asp–Thr–**Leu**–Lys–

   The hydrophobic residues in this sequence are presented in boldface type. Suggest a possible reason for the periodicity in their spacing.

5. Although the bond energy for the hydrogen bond in a vacuum is estimated to be about 20 kJ/mol, we find that each hydrogen bond in a folded protein contributes much less—probably less than 5 kJ/mol—to the enthalpy of protein stabilization. Suggest an explanation for this difference.

6. Consider a small protein containing 101 amino acid residues. The protein will have 200 bonds about which rotation can occur. Assume that three orientations are possible about each of these bonds.

   (a) Based on these assumptions, about how many *random-coil* conformations will be possible for this protein?

   (b) The estimate obtained in (a) is surely too large. Give one reason why.

*7. (a) Based on a more conservative answer to Problem 6 (2.7×10^{92} conformations), estimate the conformational entropy change on folding a mole of this protein into a native structure with only one conformation.

   (b) If the protein folds *entirely* into α helix with H bonds as the only source of enthalpy of stabilization, and each mole of H bonds contributes -5 kJ/mol to the enthalpy, estimate $\Delta H_{folding}$. Note that you cannot form 4 H bonds at one end.

   (c) From your answers to (a) and (b), estimate $\Delta G_{folding}$ for this protein at 25 °C. Is the folded form of the protein stable at 25 °C?

*8. The following sequence is part of a globular protein. Using Table 6.8 and the Chou–Fasman rules, predict the secondary structure in this region.

   · · · RRPVVLMAACLRPVVFITYGDGGTYYHWYH · · ·

9. (a) A protein is found to be a tetramer of identical subunits. Name two symmetries possible for such a molecule. What kinds of interactions (isologous or heterologous) would stabilize each?

   (b) Suppose a tetramer, like hemoglobin, consists of two each of two types of chains, α and β. What is the highest symmetry now possible?

*10. Under physiological conditions, the protein hemerythrin exists as an octamer of eight chains of the kind shown in Problem 3.
(a) Name two symmetries possible for this molecule.
(b) Which do you think is more likely? Explain.
(c) For the more likely symmetry, what kinds of interactions (isologous, heterologous, or both) would you expect? Why?

*11. A researcher studies the thermal denaturation of hemerythrin by two methods: (1) Using circular dichroism, which measures the α helix content (see Tools of Biochemistry 6A) and (2) using differential scanning calorimetry (see Tools of Biochemistry 6C). She observes a considerably larger ΔH by calorimetry than she finds using circular dichroism. Suggest a reason for this.

12. The peptide hormone *vasopressin* is used in the regulation of saltwater balance in many vertebrates. Porcine (pig) vasopressin has the sequence

Asp–Tyr–Phe–Glu–Asn–Cys–Pro–Lys–Gly

(a) Using the data in Figure 5.6 and Table 5.1, estimate the extinction coefficient ε (in units of cm^2/mg) for vasopressin, using radiation with $\lambda = 280$ nm.
(b) A solution of vasopressin is placed in a 0.5-cm-thick cuvette. Its absorbance at 280 nm is found to be 1.3. What is the concentration of vasopressin, in mg/cm^3? (See Tools of Biochemistry 6A.)
(c) What fraction of the incident light is passed through the cuvette in (b)? (See Tools of Biochemistry 6A.)

*13. A protein gives, under conditions of buffer composition, pH, and temperature that are close to physiological conditions, a molecular weight by sedimentation equilibrium measurements of 140,000 g/mol. When the same protein is studied by SDS gel electrophoresis in the absence or presence of the reducing agent β-mercaptoethanol (BME), the patterns seen, respectively, in lanes A and B are observed. Lane C contains standards of molecular weight indicated. From these data, describe the native protein, in terms of the kinds of subunits present, the stoichiometry of subunits, and the kinds of bonding (covalent, noncovalent) existing between subunits. (See Tools of Biochemistry 6B.)

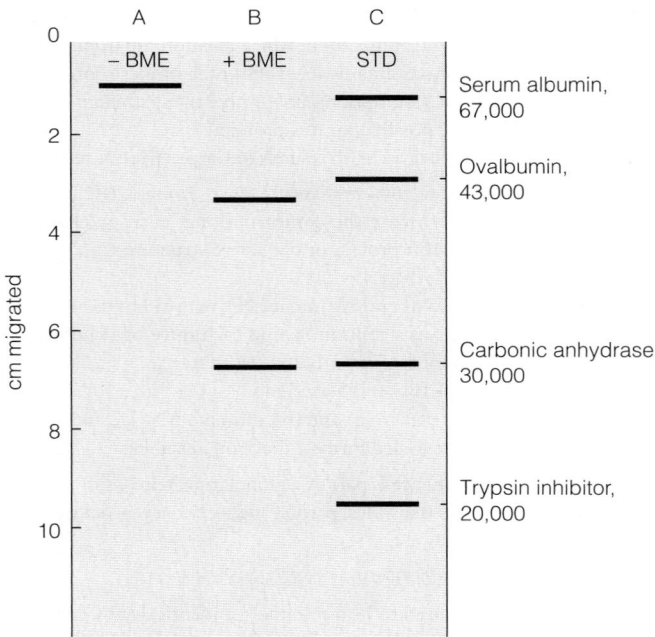

*14. A protein gives a single band on SDS gel electrophoresis, as shown in lanes 1 and 2 below. There is little if any effect from adding

β-mercaptoethanol (BME) in the gel; if anything, the protein runs a little bit slower. When treated with the proteolytic enzyme thrombin (see Chapter 5) and electrophoresis in the absence of BME, the protein migrates a bit more rapidly (lane 3). But if BME is present, two much more rapidly migrating bands are found (lane 4). Explain these results in terms of a model for the protein.

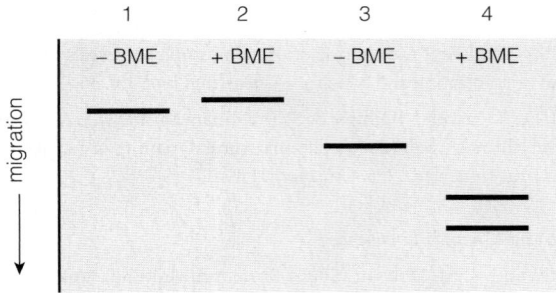

15. It has been postulated that the normal (noninfectious) form of prion differs from the infectious form only in secondary/tertiary structure.
(a) How might you show that changes in secondary structure occur?
(b) If this model is correct, what are the implications for structural prediction schemes?

16. Below are shown two views of the backbone representation of the Myc-Max complex binding to DNA (PDB ID: 1NPK). Myc and Max are members of the basic helix-loop-helix (bHLH) class of DNA-binding proteins (see Chapter 29). Myc (red) and Max (blue) associate via a coiled-coil interaction and bind DNA as a dimer.

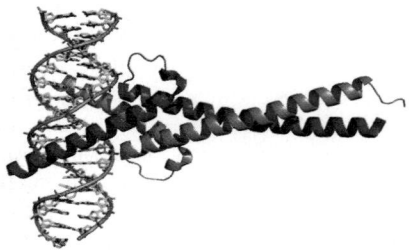

Side view of Myc-Max bound to DNA

Top view of Myc-Max bound to DNA
(looking down the DNA helical axis)

(a) Are the helices bound to the DNA likely to be amphiphilic? Explain.
(b) Where do you predict the N- and C-termini are located for Max?

17. Do you expect a Pro $\rightarrow$ Gly mutation in a surface-loop region of a globular protein to be stabilizing or destabilizing? Assume the mutant folds to a native-like conformation. Explain your answer in terms of the predicted *enthalpic* and *entropic* effects of the mutation on the ΔG for protein folding compared to ΔG of folding for the wild-type protein.

18. Rank the following in terms of predicted rates: the nucleation of an α helix; the nucleation of a *parallel* β sheet; the nucleation of an *antiparallel* β sheet. Justify your predictions.

TOOLS OF BIOCHEMISTRY 6A

SPECTROSCOPIC METHODS FOR STUDYING MACROMOLECULAR CONFORMATION IN SOLUTION

X-ray diffraction (see Tools of Biochemistry 4A) is a very powerful method for determining the details of the three-dimensional structure of globular proteins and other biopolymers; however, this technique has the fundamental limitation that it can be employed only when the molecules are crystallized, and crystallization is not always easy or even possible. For example, proteins containing significant regions of sequence that are intrinsically unstructured are notoriously difficult to crystallize. Furthermore, X-ray diffraction cannot easily be used to study conformational changes in response to changes in the molecules' environment. Other methods, however, allow us to study molecules in the solution state. A number of these methods can be grouped in the category of **spectroscopic techniques**.

Absorption Spectroscopy

Proteins, carbohydrates, and nucleic acids are complex molecules and can absorb electromagnetic radiation over a wide spectral range; however, the basic principles of such absorption phenomena can be explained in terms of the simplest kind of molecule, a diatomic molecule.

When two atoms interact to form a molecule, the potential energy curve for the lowest-energy electronic state (the **ground state**) will look like the lower curve in Figure 6A.1a. **Excited electronic states** will have similar curves for energy vs. interatomic distance, but at higher energies. For each electronic state of the molecule, there will be a series of allowed **vibrational states,** with energy levels indicated by horizontal lines in the figure. The basics of molecular absorption spectroscopy can be understood by two simple rules: (1) Transitions are possible only between allowed energy states of the molecule (energy levels are **quantized**), and (2) the energy (ΔE) that is absorbed, or emitted, in any transition between allowed states determines the wavelength (λ) of the radiation that is absorbed, or released, to accomplish that transition. The energy in a **quantum** (or **photon**) of radiation is inversely proportional to λ

$$E_{\text{final state}} - E_{\text{initial state}} = \Delta E = \frac{hc}{\lambda} \qquad (6A.1)$$

Here h is Planck's constant (6.626×10^{-34} J s), and c is the velocity of light in a vacuum (2.998×10^{8} m s^{-1}). According to equation 6A.1, transitions with smaller energy differences between

FIGURE 6A.1

The principles of absorption spectroscopy.
(a) Electronic and vibrational transitions in a diatomic molecule.
(b) The electromagnetic spectrum.

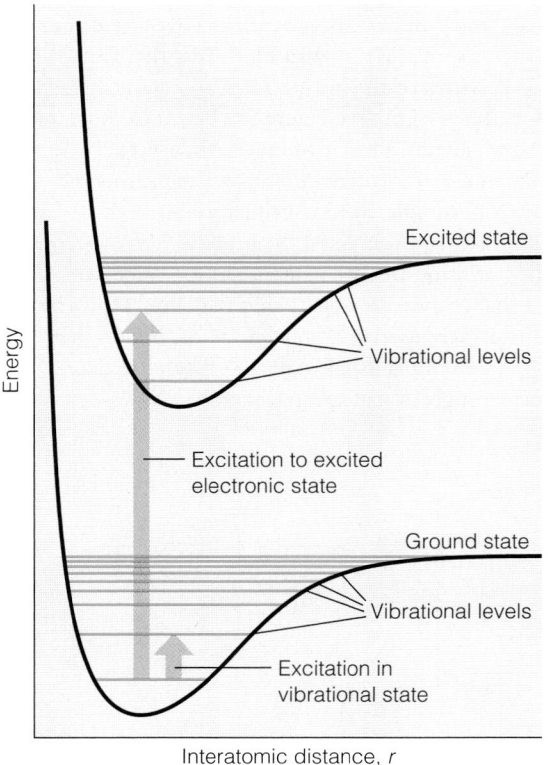

(a)

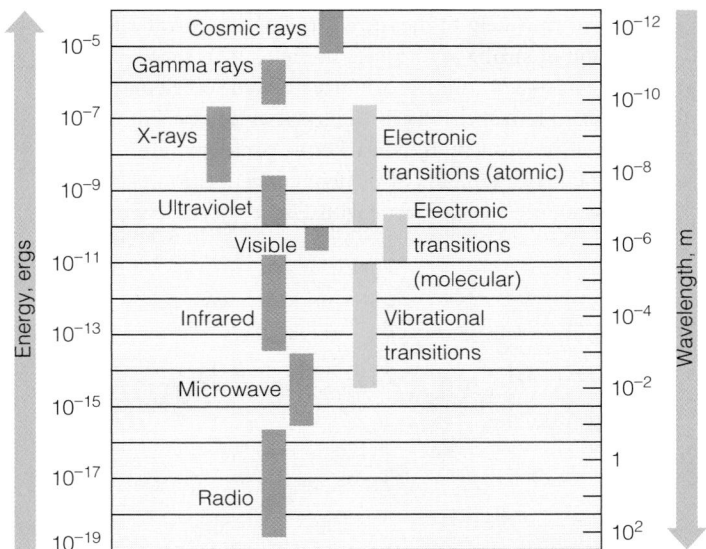

(b)

states correspond to absorption (or release) of longer wavelength radiation, and transitions with larger energy differences correspond to absorption (or release) of shorter wavelength radiation. This relationship is in accord with Figure 6A.1b, which indicates that the high-energy transitions between electronic states of a molecule lead to absorption in the visible or ultraviolet region of the spectrum, whereas the low-energy transitions between different vibrational energy levels correspond to absorption of infrared energy.

Complex biopolymers like proteins and nucleic acids can undergo many kinds of molecular vibrations and oscillations. **Infrared spectroscopy** can therefore provide direct information concerning macromolecular structure. For example, the exact positions of infrared bands corresponding to vibrations in the polypeptide backbone are sensitive to the conformational state of the protein backbone (α helix, β sheet, and so forth). Thus, studies in this region of the spectrum are often used to investigate the secondary structural features in protein molecules.

Most biopolymers do not absorb visible light to a significant extent. Some proteins are colored, but they invariably contain prosthetic groups (such as the heme in myoglobin) or metal ions (such as copper) that confer the visible absorption. Blood and red meat owe their color to the heme groups carried by hemoglobin, myoglobin, and other heme proteins. Such absorption can often be exploited to investigate changes in the molecular environment of the prosthetic group. An example is the use of absorption spectroscopy in the visible spectrum to follow the oxygenation of myoglobin or hemoglobin (see page 238 in Chapter 7). The most common uses of spectroscopic techniques in biochemistry involve **ultraviolet spectroscopy**. In the ultraviolet region both proteins and nucleic acids absorb strongly (Figure 6A.2). The strongest protein absorbances are found in two wavelength ranges within the ultraviolet region, at approximately 280 and 220 nm. In the range 270–290 nm, we see absorbance by the aromatic side chains of phenylalanine, tyrosine, and tryptophan. Because this region of the spectrum is easy to study, absorbance at 280 nm is used routinely to measure protein concentrations. The second region of strong absorbance in the protein spectrum lies in the range 180–220 nm. Absorbance at such wavelengths arises from electronic transitions in the polypeptide backbone itself and is therefore sensitive to the backbone conformation.

Spectroscopic measurements of protein concentration use a **spectrophotometer,** in which a cuvette with a light path of length l

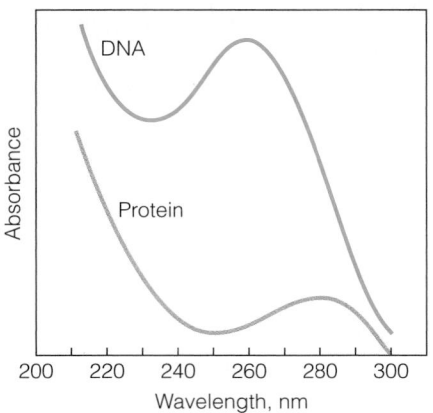

FIGURE 6A.2

The near-ultraviolet absorbance spectra of a typical protein and of DNA. Absorbance at 280 nm is commonly used to measure protein concentrations, whereas absorbance at 260 nm is more sensitive for nucleic acids.

containing a solution of the protein is placed in a beam of monochromatic radiation of intensity I_0 (Figure 6A.3). The intensity of the emerging beam will be decreased to a value I because the solution absorbs some of the radiation. The **absorbance** at wavelength λ is defined as $A_\lambda = \log(I_0/I)$ and is related to l and the concentration c by the **Beer-Lambert law:**

$$A = \varepsilon_\lambda l c \qquad (6A.2)$$

Here ε_λ is the **extinction coefficient** (or **molar absorptivity**) at wavelength λ for the particular substance being studied (see Problem 12). The dimensions of ε_λ depend on the concentration units employed. When protein concentration is measured in molarity (M) and l in cm, then ε_λ must have the dimensions $M^{-1}cm^{-1}$, because A is a dimensionless quantity. Note that the molar absorptivities of the aromatic amino acids differ in the order: Trp > Tyr $\gg$ Phe (see Figure 5.6, page 143. Note that Cys also absorbs at 280 nm, albeit 10-fold less than Tyr).

Once the molar absorptivity for a particular protein has been determined (for example, by measuring the absorbance of a solution containing a known mass of the protein), the concentration of any other solution of that protein can be calculated from a simple absorbance measurement, using Equation 6A.2. The same method is routinely used with nucleic acids, but in that case a wavelength of 260 nm is usually employed because nucleic acids absorb most strongly in that spectral region.

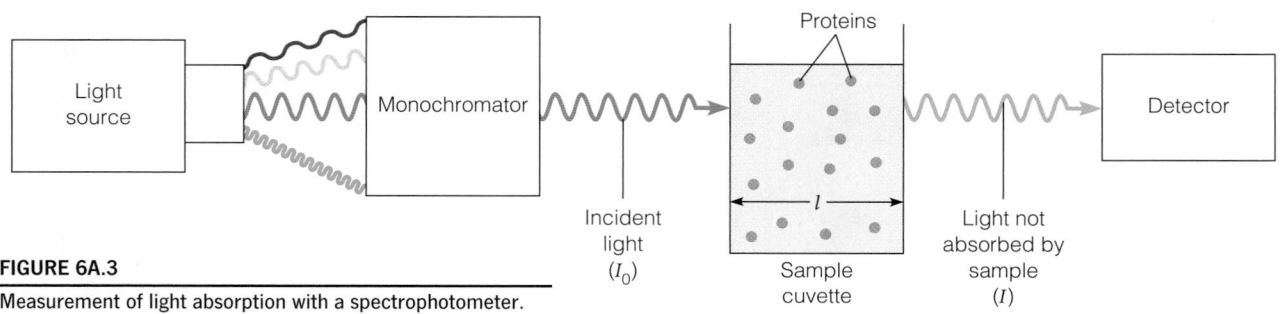

FIGURE 6A.3

Measurement of light absorption with a spectrophotometer.

Fluorescence

In most cases, molecules raised to an excited electronic state by absorption of radiant energy return, or "relax," to the ground state by **radiationless transfer** of the excitation energy to surrounding molecules. In short, the energy of relaxation most often is manifested as heat rather than an emitted photon. As shown in Figure 6A.4a, an excited state molecule can lose a small part of its energy of excitation by radiationless transfer (yellow arrow) and will lose the larger part in the form of an emitted photon (red arrow). This gives rise to the phenomenon called **fluorescence**. Because, as Figure 6A.4a shows, the quantum of energy re-emitted as fluorescence is always of lower energy than the quantum that was initially absorbed (blue arrow), the wavelength of the emitted light will be longer than the wavelength of the light used for excitation. The **fluorescence emission spectrum** of tyrosine is contrasted with its absorbance (or **excitation**) spectrum in Figure 6A.4b. In proteins, tyrosine and tryptophan are the major fluorescent groups. The local environment of these residues can greatly modify the intensity and wavelength of maximum fluorescence (the so-called $\lambda_{\max}$). For example, the fluorescence $\lambda_{\max}$ shifts to shorter wavelength (a "blue-shift") and the intensity of the fluorescence signal increases as the polarity of the solvent surrounding a tryptophan residue decreases. Tryptophan residues buried in the hydrophobic cores of proteins can have values for $\lambda_{\max}$ that are blue-shifted by 10 to 20 nm compared to tryptophans in solvent-accessible locations; thus, fluorescence spectroscopy of tryptophan can be used to monitor changes in protein conformation, such as the transition from a folded to a denatured state.

Further, excitation of fluorescence by plane polarized light (see the next section) provides a way of studying the dynamics of protein structure. If the excited residues are able to move or rotate appreciably between the excitation and emission events, the fluorescence will be depolarized to some extent. Measurement of the extent of this depolarization provides a measure of group or molecule rotational mobility.

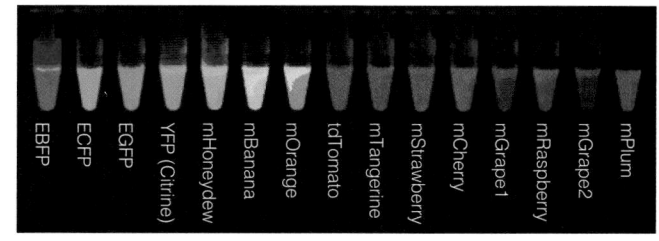

FIGURE 6A.5

Fifteen different fluorescent proteins, each with a distinct fluorescence spectrum, are shown. The emission spectra of some of these proteins are shown in Figure 6A.6.

Roger Tsien Lab/Composite by Paul Steinbach.

Because fluorescence spectroscopy is a technique that can detect a small number of fluorescent molecules, it has become widely used as a tool for precisely locating proteins in cells or subcellular organelles. Confocal microscopy (see Tools of Biochemistry 1A) allows such location, if the protein can be specifically labeled. Sometimes this can be done by covalently attaching fluorescent dyes, but this is challenging in vivo. A powerful new technique uses a highly fluorescent protein found in some jellyfish, called **green fluorescent protein** (GFP, Figure 6A.5). The intense fluorescence is due to an unusual chromophore, generated by oxidation of the amino acid sequence Ser-Tyr-Gly. GFP is used most effectively as a **fusion protein**; the gene for GFP is fused to the gene for the protein being studied, and the fused product is expressed in the organism of interest. In many cases the fusion protein functions and localizes like the native protein, and provides a brilliant marker in microscopy. Several variants of GFP have been developed that absorb and fluoresce across the visible spectrum (Figure 6A.5 and 6A.6). The impact of fluorescent-protein fusion technology on our understanding of the timing and location within cells of gene expression, protein movement, and changes in pH and/or Ca^{2+} levels was recognized by the 2008 Nobel Prize in Chemistry, shared by Osamu Shimomura, Martin Chalfie, and Roger Tsien.

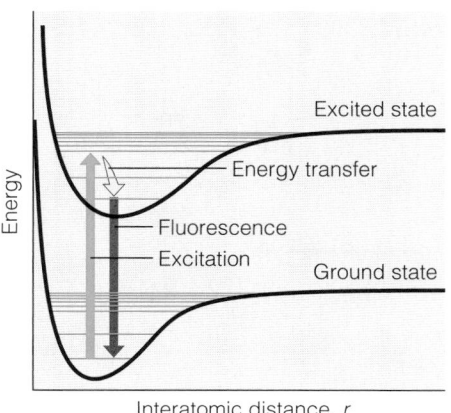

(a)

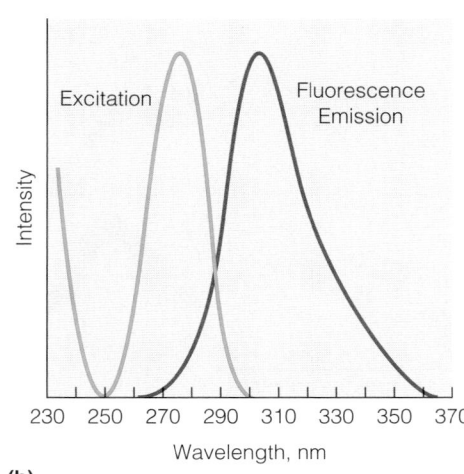

(b)

FIGURE 6A.4

Fluorescence. **(a)** The principle of fluorescence. **(b)** Excitation and fluorescence-emission spectra of tyrosine (note: the *y*-axes for excitation and emission are different).

Förster Resonance Energy Transfer (FRET)

Using fluorescent fusion proteins it is possible to detect protein–protein interactions or protein conformational change by **Förster resonance energy transfer** (**FRET**). In FRET an excited fluorescent "donor" transfers energy to a nearby ground-state "acceptor" in a nonradiative process. The donor returns to its ground state without fluorescence, but the excited acceptor can now emit the excess energy received from the donor by fluorescence. In effect, excitation of the *donor* results in fluorescence emission by the *acceptor* at a wavelength that is longer than the emission λ_{max} of the donor (Figure 6A.7). The energy transfer in FRET is achieved via long-range dipole–dipole interactions between the donor and acceptor; thus, the efficiency of FRET depends on the distance between the donor and acceptor as well as the overlap between the emission and excitation spectra, respectively, of the donor and acceptor. The efficiency of FRET depends on $1/r^6$, where r is the distance between the donor and acceptor, which must be closer than 10 nm. A commonly used donor/acceptor pair is cyan fluorescent protein (CFP; the donor) with yellow fluorescent protein (YFP; the acceptor). FRET can be used to detect interactions in whole cells between proteins fused to CFP or YFP (Figure 6A.8a).

Using a "sensor" protein labeled at its termini with an appropriate donor/acceptor pair, FRET can also monitor, in vivo, changes in Ca^{2+} or H^+ ion concentrations, or the presence of some protein ligand. In the absence of the ligand or ion there is no FRET between the donor and acceptor. Upon binding the ligand, the protein sensor undergoes a conformational change that brings the donor and acceptor in proximity and FRET is observed (Figure 6A.8b).

Circular Dichroism

Although visible absorption spectroscopy and fluorescence are useful for monitoring significant changes in protein tertiary conformation due to local or global unfolding, such measurements are difficult to interpret directly in terms of changes in secondary structure. For this purpose, infrared spectroscopy (see above) and techniques involving polarized light are more informative.

Light can be polarized in various ways. Most familiar is **plane polarization** (Figure 6A.9a, top), in which the oscillating electric

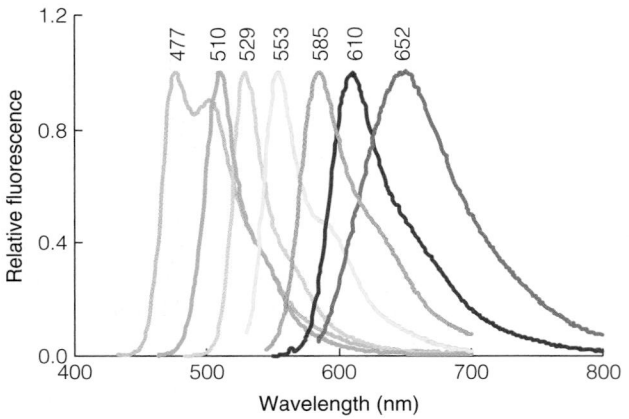

FIGURE 6A.6

Fluorescence emission spectra of several fluorescent proteins. The emission λ_{max}(nm) is listed above each peak. The spectra were generated from data available at: http://www.tsienlab.ucsd.edu/Documents/REF-FluorophoreSpectra.xls.

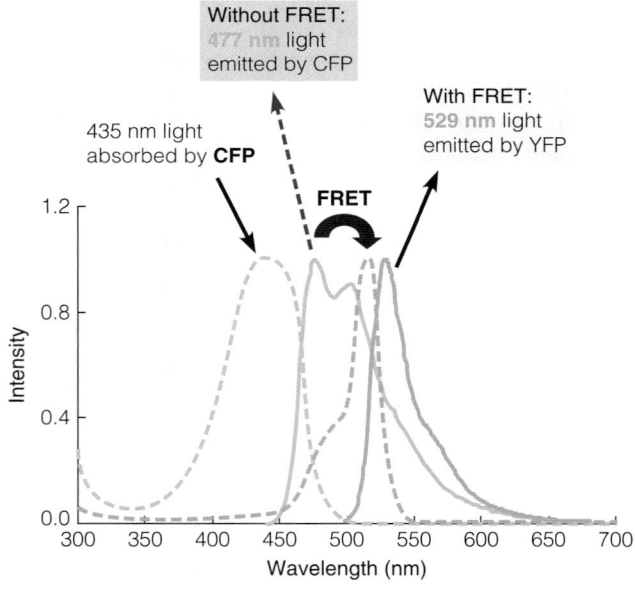

FIGURE 6A.7

Förster resonance energy transfer (FRET). FRET is a nonradiative transfer of energy from an excited donor to a nearby ground-state acceptor. Excitation (dotted lines) and emission (solid lines) spectra are shown for cyan fluorescent protein (CFP) and yellow fluorescent protein (YFP). Note the extensive overlap between the emission spectrum of CFP and the excitation spectrum of YFP.

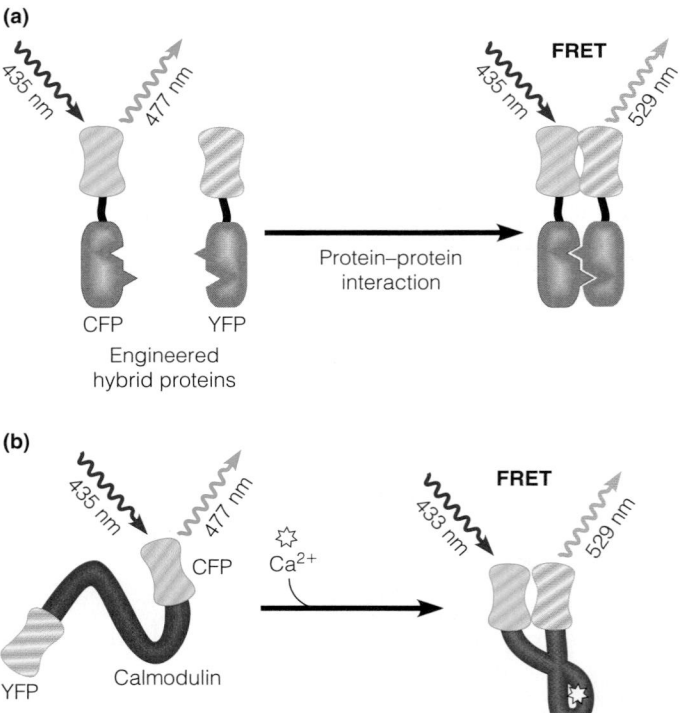

FIGURE 6A.8

Using FRET to measure protein–protein interactions, or protein conformational change.

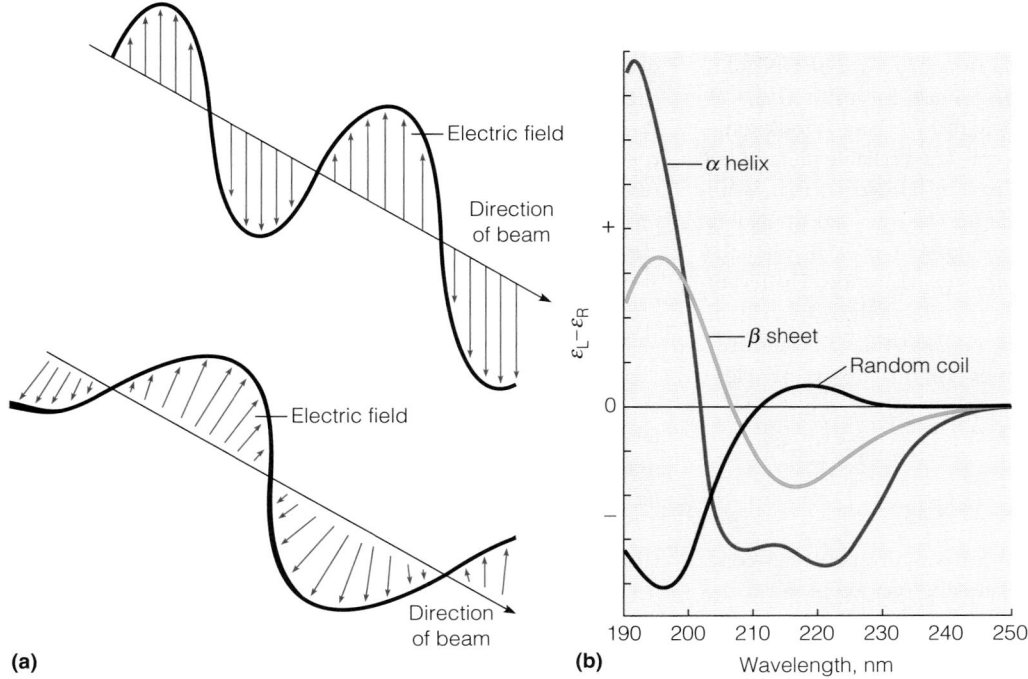

(a)

(b)

Wavelength, nm

FIGURE 6A.9

Circular dichroism. **(a)** Polarization of light. Above: Plane polarized light, in which the amplitude of the electric field oscillates in a single plane. Below: In circularly polarized light, the oscillation of the electric field follows a helical path around the axis describing the direction of the beam. **(b)** Circular dichroism spectra for polypeptides in various conformations. Here the y-axis records differences in molar absorptivity (ε) between left- and right-circularly polarized light.

field of the radiation has a fixed orientation in a single plane. In contrast, *unpolarized* light consists of waves vibrating in *all* planes perpendicular to the direction of travel. Less familiar, but equally important, is **circular polarization,** in which the direction of polarization *rotates* with the frequency of the radiation (Figure 6A.9a, bottom). If you observe a circularly polarized beam coming toward you, the electric field can be rotating in either a clockwise or counterclockwise direction; thus, as the light moves toward you, the oscillation of the electric field describes a right-handed or a left-handed helix. The former is called *right circularly polarized light,* the latter *left circularly polarized light.*

Most of the molecules studied by biochemists are asymmetric—for example, L- and D-amino acids, right- and left-handed protein helices, and right- and left-handed nucleic acid helices. Such molecules exhibit a preference for the absorption of either left or right circularly polarized light. For example, a right circularly polarized beam interacts differently with a right-handed α helix than does a left circularly polarized beam. This difference in absorption, called **circular dichroism,** is defined as

$$\Delta A = \frac{A_L - A_R}{A}$$
(6A.3)

where A_L is the absorbance for left circularly polarized light, A_R is the corresponding quantity for right circular polarization, and A is the absorbance for unpolarized light. Because ΔA can be either positive or negative, a **circular dichroism spectrum** (or **CD spectrum**) is unlike a normal absorption spectrum in that both (+) and (−) values are allowed.

Figure 6A.9b shows CD spectra for polypeptides in the α helix, β sheet, and random-coil conformations. The three spectra in the figure are very different, so circular dichroism is a sensitive technique for following conformational changes of proteins in

solution. For example, if a protein is denatured so that its native structure, containing α helix and β sheet regions, is transformed into an unfolded, random-coil structure, this transformation will be reflected in a dramatic change in its CD spectrum.

Circular dichroism can also be used to estimate the content of α helix and β sheet in native proteins. The contributions of these different secondary structures to the circular dichroism at different wavelengths are known, so we may attempt to match an observed spectrum for a protein by a linear combination of such contributions. This kind of analysis frequently turns out to agree with the secondary structure composition as determined by X-ray crystallography and NMR studies, and it has given support to the idea that the structures of globular proteins observed in crystals are preserved when the crystals are dissolved in buffer solutions at physiological pH.

Although circular dichroism is an extremely useful technique for monitoring global changes in protein or nucleic acid structures, it is not a high-resolution one. That is, it cannot provide insight into biomolecular structure at the level of atomic detail.

Two methods that can provide the details of structure at the atom level are X-ray crystallography (introduced in Tools of Biochemistry 4A) and **nuclear magnetic resonance spectroscopy** (NMR; described in the next section). The influence of these two structure-determination methods on the field of biochemistry is profound. Our detailed understanding of the functions of biomolecules is enhanced by, and in some cases based on, the knowledge of the high-resolution structures obtained by these methods. The first report of an X-ray crystal structure of a protein (myoglobin), published in 1958 by Nobel laureate John Kendrew and coworkers, established crystallography as the standard technique for the determination of high-resolution structures. As of 2011, more than 12 X-ray crystallographers had received Nobel prizes for their significant achievements in the determination of protein and/or nucleic acid structures. The first structure of a protein determined by NMR methods (proteinase inhibitor IIa) was published in 1985 by Kurt Wüthrich and coworkers. It took some time for NMR to become widely accepted as a reliable method for high-resolution structure determination. However, that is no longer in doubt, and in 2002 Wüthrich was awarded the Nobel Prize for his contributions to the development of NMR methods for the determination of protein structures in solution. The recent development of powerful methods for the analysis of protein structure, dynamics, and function by NMR justifies a brief introduction to some of the basic experiments in this important field.

Nuclear Magnetic Resonance Spectroscopy (NMR)

General Principles: One-Dimensional NMR

The nuclei of certain elemental isotopes have a property referred to as **spin**, which makes these nuclei behave like microscopic magnets. A limited number of isotopes have this property; some that are particularly useful to biochemists are listed in Table 6.A.1. The most useful nuclei for NMR have spin states of $-\frac{1}{2}$ or $+\frac{1}{2}$. If an external magnetic field is applied to a sample containing such nuclei, the different nuclear spin states will align with or against the magnetic field and therefore have different energies. As Figure 6A.10a shows, the difference in energy (ΔE) between these two spin states increases as the strength of the external magnetic field increases. For most NMR spectrometers, a pulse of radio frequency (RF) radiation can change the orientation of the nuclear spin (or "magnetization") in the external field, a phenomenon called nuclear magnetic resonance. Such reorientation of the nuclear magnetization is analogous to the "excited" electronic states described above. Whereas the techniques of absorption spectroscopy described above record the wavelengths of light absorbed during "allowed" *electronic* transitions, NMR spectroscopy records the frequencies of RF radiation that are absorbed when the spin states of different nuclei in a molecule become reoriented in the external field.

The energy of a nuclear spin in a magnetic field is very sensitive to the chemical environment surrounding the atom in question. The chemical environment of a nucleus is defined by factors such as the polarity, hydrophobicity, and charge state of the surroundings. For example, due to differences in chemical environments, different hydrogen nuclei in a compound will reach resonance at different field strengths. These differences are recorded in the **NMR spectrum** and are expressed in terms of **chemical shifts (δ)** defined with respect to a reference material:

$$\delta = \frac{B_{ref} - B}{B_{ref}} \qquad (6A.4)$$

Here B is the field strength at which the nucleus in question reaches resonance, and B_{ref} is that for a reference nucleus. The differences between B_{ref} and B are quite small, and this fact is reflected in the units for chemical shift, which are *parts per million* or ppm. A common reference compound for protein NMR is 4,4-dimethyl-4-silapentane-1-sulfonic acid (DSS), which has a strong ^1H resonance for its nine methyl protons. This resonance is assigned the value of 0 ppm. The electronic absorption spectra described earlier in this section are obtained by plotting signal intensity vs. wavelength. Similarly, an NMR spectrum is obtained by plotting signal intensity (for each resonance) vs. chemical shift. For example, a ^1H NMR spectrum will record a peak (or "line") for every ^1H nucleus in the molecule.

As illustrated in Figure 6A.10b, the ^1H NMR spectrum of a protein is extremely complex, containing many peaks because even relatively small proteins have several hundred protons with many overlapping resonances (e.g., human ubiquitin, with only 76 amino acid residues, contains >600 protons). Most of the ^1H resonances for aliphatic side chains occur between 0.5 and 5 ppm, whereas the backbone amide proton resonances generally occur between 6.5–10 ppm, and aromatic side chain resonances occur between 6–8 ppm. In cases where specific ^1H resonances in a protein can be identified, it becomes possible to monitor changes in the chemical environment of a specific amino acid residue. An example of such an experiment is given in Figure 6A.11, which uses NMR to generate titration curves of individual histidine residues

TABLE 6A.1	Nuclei most often used in biochemical NMR experiments			
Isotope	Spin	Natural Abundance[a] (%)	Relative Sensitivity[b]	Applications
^1H	½	99.98	(1.000)	Almost every kind of biochemical study
^2H	1	0.02	0.0096	Studies of selectively deuterated compounds; structure determination of proteins >20 kDa
^{13}C	½	1.11	0.0159	Multidimensional NMR; residue assignment
^{15}N	½	0.37	0.0104	Multidimensional NMR; residue assignment; protein backbone dynamics
^{19}F	½	100.00	0.834	Substituted for H (e.g., ^{19}F-Tyr) to probe local structure
^{31}P	½	100.00	0.0664	Studies of nucleic acids and phosphorylated compounds

[a]The number represents the percentage of this isotope in the naturally occurring mix of isotopes of each element. Isotopes with figures that are close to 100% can be studied directly in the naturally occurring biopolymers. Rare isotopes, such as ^2H (deuterium), ^{13}C, and ^{15}N usually must be artificially enriched in the biomolecules to be studied. This is achieved by including one or more of these isotopes in the medium used to grow the organism containing the molecule of interest (usually a recombinant protein).

[b]Indicates the sensitivity (relative to ^1H) of conventional NMR instruments to each isotope, when that isotope has been enriched to 100%. Low values mean that the experiment will be more difficult or time-consuming.

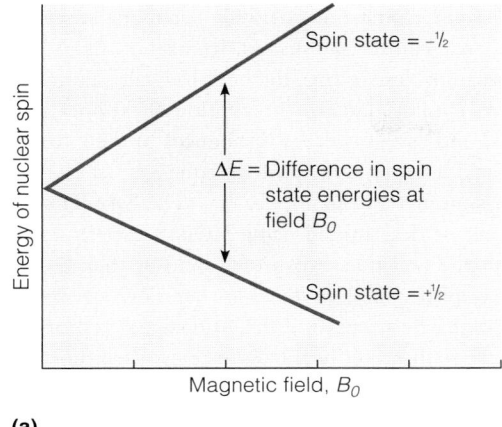

(a)

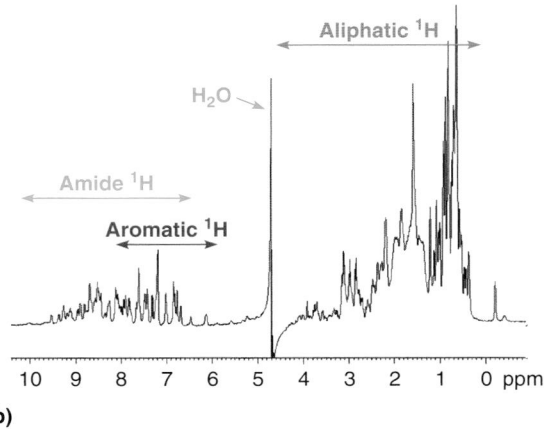

(b)

FIGURE 6A.10

Nuclear magnetic resonance spectroscopy (NMR). **(a)** The effect of magnetic field strength on the energies of nuclear spin states (e.g., ^{1}H, ^{13}C). **(b)** A 500 MHz ^{1}H NMR spectrum of human ubiquitin (1 mM ubiquitin in 25 mM sodium phosphate, 150 mM NaCl, pH 7.0). This protein has 76 residues, which give rise to ~600 peaks in the ^{1}H NMR spectrum. The x-axis is the chemical shift, δ, in parts per million (ppm).

Courtesy of S. Delbecq and R. Klevit, University of Washington.

in the protein ribonuclease A. The figure also graphically illustrates a principle described in Chapter 5: Individual side chains of a given amino acid type can show quite different pK_a values because of their different chemical environments within the protein molecule.

Changes in chemical shift are also correlated with conformational changes in protein structure; thus, NMR can be used to monitor dynamic motions of proteins, such as those that occur upon local or global unfolding of the polypeptide.

With modern NMR instruments, it is possible to resolve resonances for many of the ^{1}H nuclei in proteins as large as 30 kDa using **multidimensional techniques** (discussed below). Structure determination by solution phase NMR for proteins larger than 50 kDa is challenging. This is the major limitation on structure determination by NMR compared to X-ray crystallography; however, NMR is the more powerful technique for studying dynamic processes in solution.

Multidimensional NMR Spectroscopy

The spectrum shown in Figure 6A.10b is also known as a *one-dimensional*, or 1-D, NMR spectrum. Such 1-D data can tell us much about the behavior of individual atoms in a protein; however, the true power of NMR lies in more sophisticated *multidimensional* experiments. Discussion of the details of multidimensional NMR is beyond the scope of this basic introduction. The interested reader can find more on NMR theory in the references cited at the end of this section. Here, we present an overview of the application of multidimensional NMR to the determination of protein structure.

Multidimensional NMR techniques exploit the fact that spins on different nuclei interact or "couple" with one another either *through bonds* or *through space*. Using multiple pulses of RF energy to re-orient nuclear spins and transfer magnetization from one nucleus to its coupled partner(s), it is possible to perturb the spin of one nucleus and detect its effect on the spin states of other nuclei. Importantly, such experiments correlate the chemical shifts of the coupled nuclei, such that a detailed network of spin–spin interactions can be mapped out. It is this information that leads to the solution of the 3-D structure of the molecule under investigation. Two critical sets of data are needed to solve the 3-D structure of a protein using NMR methods: (1) A set of chemical shift

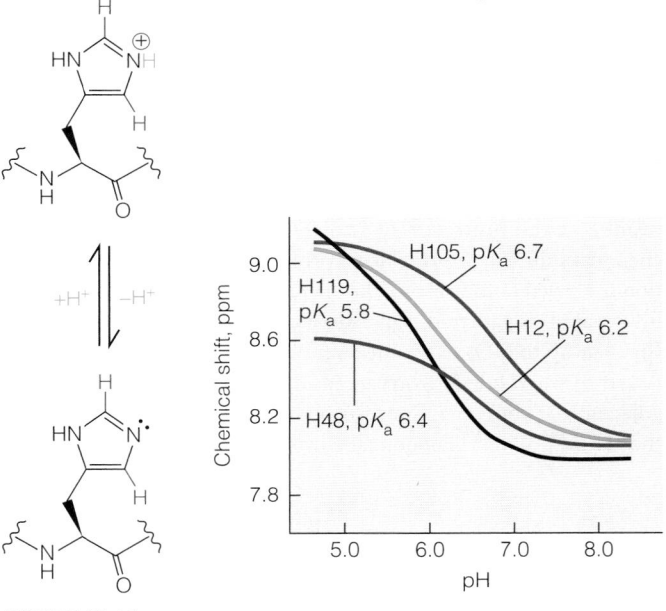

FIGURE 6A.11

Titration of the four histidine residues in ribonuclease A by NMR. The y-axis is the ^{1}H chemical shift and each curve follows the titration of an individual histidine group, as detected by the NMR chemical shift of a ^{1}H bound to either one of the two C atoms in the imidazole ring (red H atoms on ring). The labels such as H12 and H48 indicate the positions of the histidines in the primary sequence. The two histidines with lowest pK_a values (H12 and H119) are involved directly in the catalytic process.

assignments that correlate specific backbone amide ^{1}H and side chain ^{1}H resonances to at least 85% of the amino acid sequence of the protein, and (2) several hundred to thousands of distance measurements between coupled ^{1}H nuclei for which chemical shift assignments have been made. The chemical shift assignments are obtained from multidimensional **correlation spectroscopy** (COSY) experiments that determine through-bond coupling, and the distance measurements are obtained from the determination of through-space couplings by **nuclear Overhauser effect spectroscopy** (NOESY).

The methods used to make chemical shift assignments typically require proteins that have been enriched in the NMR-active nuclei ^{15}N and ^{13}C, which are present in low natural abundance (see Table 6A.1). Such isotopically labeled proteins are obtained by expressing recombinant proteins in bacteria grown in medium containing, for example, [^{15}N]-NH$_4$Cl and [^{13}C]-glucose such that all the N and C atoms in the protein are labeled. Such labeling is required to resolve resonances that overlap in one dimension by projecting them in a second (or third) dimension whereby the overlap is reduced.

With ^{15}N-labeled protein a 2-D NMR experiment called ^{1}H/^{15}N **heteronuclear single quantum coherence** (or **HSQC**) can be used to resolve resonances for every backbone amide N-H group in a protein (Figure 6A.12). In this experiment, magnetization is transferred from the ^{1}H nucleus to the ^{15}N nucleus, then back to the original ^{1}H nucleus. Each magnetization transfer step records the chemical shift of the receiving nucleus and thereby establishes a "dimension" in which these data can be plotted (in this case there are two dimensions—one corresponding to the ^{1}H chemical shift of the amide N-H, and the other corresponding to the ^{15}N chemical shift of the same N-H group). Just as 2-D electrophoresis of proteins achieves greater resolution of individual proteins in a large mixture of proteins (see Figure 1.12), this 2-D NMR spectrum achieves greater resolution of the multitude of protein amide resonances because each individual spot or 2-D peak corresponds to the amide group of an individual amino acid residue in the protein. ^{1}H/^{15}N-HSQC spectra display a useful chemical shift "fingerprint" for a protein; however, more information is needed to fully assign the ^{1}H resonances in the protein. This information is gathered using some combination of standard 3-D (or even up to 6-D) experiments in which magnetization is transferred through bonds from an amide ^{1}H to the amide ^{15}N to an adjacent backbone ^{13}C, and so on. At each step in this process the ^{1}H or ^{15}N or ^{13}C chemical shifts are recorded such that the covalent, *through-bond*, connections between backbone atoms in the primary sequence can be established. In essence, the NMR spectroscopist uses multidimensional methods to "walk" along the covalently bonded spin systems in the peptide backbone (and some side chains) and assign specific resonances to their corresponding amino acid residues in the protein sequence.

With these assignments in hand, the *through-space* ^{1}H spin couplings can then be determined using NOESY experiments, which depend on the fact that two protons closer than about 0.5 nm will have coupled spins *even if they are not closely bonded in the primary structure*. The strength of the NOE signal is proportional to the inverse sixth power of the distance between the coupled ^{1}H nuclei; thus, NOESY shows only those protons which are in close proximity in the three-dimensional structure of the protein. A typical protein NOESY spectrum is shown in Figure 6A.13a, where the ^{1}H chemical shifts are plotted along each axis and the spots on the diagonal (from upper right to lower left) correspond to the peaks in the 1-D ^{1}H NMR spectrum. The spots off this diagonal are called **crosspeaks,** and they represent the through-space interactions between two protons that have different chemical shifts. When combined with chemical shift assignments and bond constraints (i.e., "allowed" ϕ and ψ angles), such NOESY crosspeaks provide critical information that allows the calculation of accurate models for the 3-D structures of proteins in solution.

To illustrate this process, let us consider Figure 6A.13b, where each yellow sphere represents a proton in ubiquitin. The red lines in Figure 6A.13c represent the predicted NOE interactions between all the protons that are within 0.5 nm of each other. This set of red lines, which represents a set of distance constraints between the protons, can be used to locate each proton in 3-D space. The crosspeaks in the NOESY spectrum provide the experimental data that

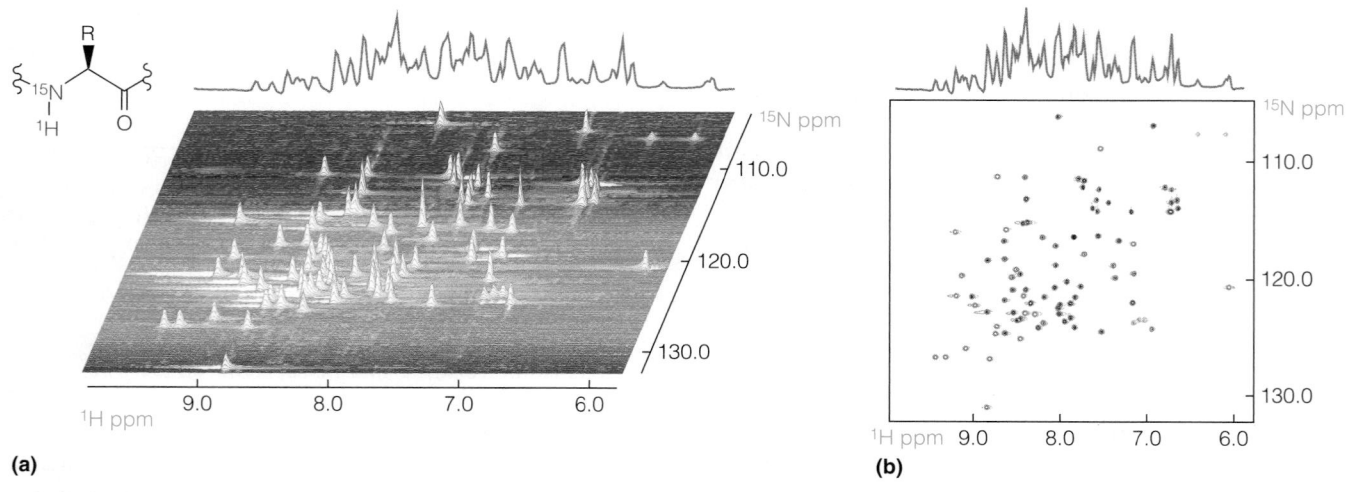

(a)   **(b)**

FIGURE 6A.12

^{1}H/^{15}N HSQC spectra of human ubiquitin. The peaks shown are for ^{1}H nuclei bonded to ^{15}N nuclei (upper left corner). **(a)** The 1-D ^{1}H spectrum of ubiquitin (brown) and chemical shifts (bottom) for the amide proton region are shown along the x-axes. The ^{15}N chemical shifts for the amide nitrogen region are shown along the rightmost y-axis. Each trace shows a ^{1}H spectrum taken at a specific ^{15}N resonance. The sum of all the traces is the 1-D ^{1}H spectrum shown in brown at the top of the panel. This image was prepared with MestRe Nova software. **(b)** ^{1}H/^{15}N HSQC spectrum of ubiquitin. This is the common way to plot such spectra. Each spot in the HSQC spectrum corresponds to a single amide N-H group with a characteristic combination of ^{1}H and ^{15}N chemical shifts (i.e., unique x and y coordinates in the spectrum). The HSQC spectrum resolves the overlapping resonances in the 1-D ^{1}H spectrum by plotting them in a second dimension (in this case the second dimension is the ^{15}N chemical shift for each amide N-H group). Plots were generated from data available at http://www.biochem.ucl.ac.uk/bsm/nmr/ubq/ (see R. Harris and P. C. Driscoll [2007] The ubiquitin NMR resource, in *Modern NMR Spectroscopy in Education*, D. Rovnyak and R. A. Stockland eds., ACS Symposium Series vol. 969, pp. 114–127).

Courtesy of Serge Smirnov, Western Washington University.

UBIQUITIN 2D NOESY

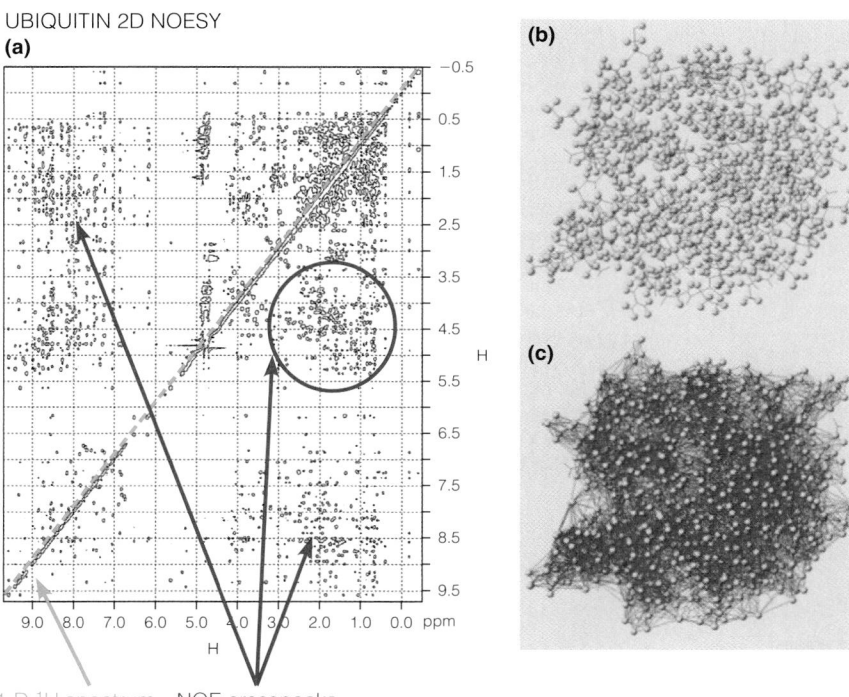

1-D ¹H spectrum along diagonal

NOE crosspeaks off diagonal

FIGURE 6A.13

Detection of through-space spin-spin coupling by NOESY. **(a)** A NOESY spectrum for human ubiquitin showing the 1-D ¹H spectrum along the diagonal and the off-diagonal crosspeaks. **(b)** A model showing the locations of ¹H nuclei (yellow spheres) in ubiquitin. **(c)** The same model shown in (b) with all the predicted NOE interactions for nuclei closer than ~0.5 nm indicated by red lines. In theory, the NOESY spectrum would show a crosspeak for each red line, and the intensity of each crosspeak would be correlated with the distance between the nuclei.

Panel (a) courtesy of Stephan Grzesiek, University of Basel; Panels (b) and (c) courtesy of Vlado Gelev, fbreagents.com, Cambridge MA.

are used to define this network of distance constraints. In other words, each NOE crosspeak in Figure 6A.13a corresponds to one of the red lines in Figure 6A.13c. In practice, not all of the red lines in Figure 6A.13c are needed to fix the locations of the protons in our model protein, but ~10 NOE crosspeaks are required for *each residue* to solve the structure at high resolution.

The NMR data are used to perform a computer-based molecular dynamics simulation (see Figures 6.35 and 6.36) that provides multiple structural models, each of which must satisfy the distance constraints obtained from the NOESY experiments, as well as other tests for "reasonableness." For example, a reasonable structural model can't have several highly unfavorable ϕ and ψ angles, or expose large hydrophobic surfaces to solvent.

Recent work has shown it is possible to accurately determine the angles ϕ and ψ from NMR chemical-shift data alone. Thus, ϕ and ψ angles can be mapped to specific residues, allowing the

solution of a low-resolution backbone structure for a protein without NOESY data. For a high-resolution structure, NOESY data are still needed, but the combination of chemical shift assignments and correlated ϕ and ψ data allows a rapid determination of protein backbone structure.

Whereas a structural model derived from X-ray crystallography shows a single best fit to the X-ray data, NMR data typically yield several very similar structures that all fit the data equally well. A set of NMR structures for human ubiquitin is shown in Figure 6A.14. The NMR data emphasize the dynamic nature of protein structures in solution because the more mobile regions of the protein are less constrained in these models (i.e., fewer NOE crosspeaks are detected for more mobile protons). To a first approximation, this means that the group of reasonable structures derived from the molecular dynamics simulation will overlap most in the regions of the protein structure that are least dynamic and show greater

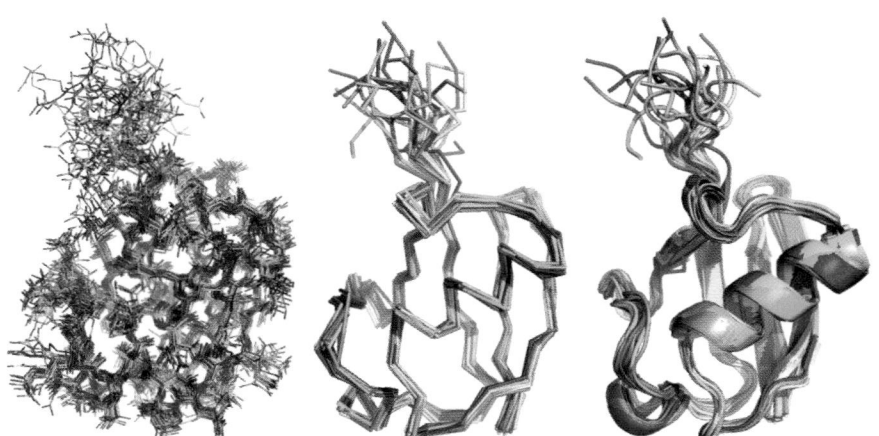

FIGURE 6A.14

NMR solution structure of human ubiquitin. In each panel, 20 structural models derived from NMR data are superimposed. Left: All atoms except H are shown. Middle: A backbone rendering showing the excellent overlap of the peptide chain in the helical and sheet regions. The C-terminus (top) is more dynamic. Right: A cartoon rendering.

Data for this figure provided courtesy of R. Harris. See R. Harris and P. C. Driscoll (2007) The ubiquitin NMR resource, in *Modern NMR Spectroscopy in Education*, D. Rovnyak and R. A. Stockland eds., ACS Symposium Series vol. 969, pp. 114–127.

divergence in those regions that are probably more dynamic. Such conclusions should be verified by other NMR experiments (not described here) that can *directly* measure protein dynamics.

This introduction to the methods of structure determination by NMR has only scratched the surface of describing applications of this versatile tool for the study of protein structure, dynamics, and ligand binding. More detail is given in the references listed below.

References

Campbell, I. D., and R. A. Dwek (1984) *Biological Spectroscopy.* Benjamin/Cummings, Menlo Park, CA.

Cavalli, A., X. Salvatella, C. M. Dobson, and M. Vendruscolo (2007) Protein structure determination from NMR chemical shifts. *Proc. Natl. Acad. Sci. USA* 104:9615–9620.

Cavanagh, J., W. J. Fairbrother, A. G. Palmer, M. Rance, and N. J. Skelton (2007) *Protein NMR Spectroscopy: Principles and Practice.* Academic Press, San Diego, CA. A comprehensive and detailed treatise.

Giepmans, B. N. G., S. R. Adams, M. H. Ellisman, and R. Y. Tsien (2006) The fluorescent toolbox for assessing protein location and function. *Science* 312:217–224.

Johnson, W. C., Jr. (1990) Protein secondary structure and circular dichroism: A practical guide. *Proteins Struct. Funct. Genet.* 7:205–214.

Neuhaus, D., and M. P. Williamson (2000) *The Nuclear Overhauser Effect in Structural and Conformational Analysis.* Wiley and Sons, New York, NY.

Sapsford, K. E., L. Berti, and I. L. Medintz (2006) Materials for fluorescence resonance energy transfer analysis: Beyond traditional donor-acceptor combinations. *Angew. Chem. Int. Ed.* 45:4562–4588.

Shen, Y., et al. (2008) Consistent blind protein structure generation from NMR chemical shift data. *Proc. Natl. Acad. Sci. USA* 105:4685–4690.

Tsien, R. Y. (2009) Constructing and exploiting the fluorescent protein paintbox (Nobel Lecture). *Angew. Chem. Int. Ed.* 48:5612–5626.

Wagner, G., W. Braun, T. F. Havel, T. Schaumann, G. Nobuhiro, and K. Wüthrich (1987) Protein structures in solution by nuclear magnetic resonance and distance geometry. The polypeptide fold of the bovine pancreatic trypsin inhibitor determined using two different algorithms, DISGEO and DISMAN. *J. Mol. Biol.* 196:611–639.

TOOLS OF BIOCHEMISTRY 6B

DETERMINING MOLECULAR MASSES AND THE NUMBER OF SUBUNITS IN A PROTEIN MOLECULE

When a new protein has been identified and purified, three questions immediately arise:

1. Does the protein exist under physiological conditions as a single polypeptide chain, or is it made up of multiple subunits?
2. If the functional protein has more than one subunit, are the subunits identical, or are there several kinds?
3. If the functional protein has more than one subunit, are the subunits covalently linked by disulfide bonds or not?

The answers to these questions can usually be obtained by first determining the **molecular mass** (sometimes called the "molecular weight"), M_W, of the protein under native conditions (i.e., in a buffer that approximates the relevant physiological pH and ionic strength) and then subjecting it to conditions under which dissociation into subunits should occur. If subunits are held together by noncovalent interactions, changing the solvent environment will often promote dissociation. For example, the pH might be raised or lowered well outside the physiological range. Alternatively, nonnative (i.e., denaturing) solvents like concentrated solutions of urea

or guanidine hydrochloride (GnHCl) might be used. These compounds, which are excellent hydrogen-bond formers, disrupt the regular water structure. For this reason they are also called **chaotropic** ("chaos-forming") agents. Disruption of the regular water structure decreases the hydrophobic effect and thereby promotes the unfolding and dissociation of protein molecules. Detergents like sodium dodecyl sulfate (SDS), which form micelle-like structures (see Figure 2.15) about individual polypeptide chains, are even more effective at unfolding proteins. By determining the molecular masses of the dissociated subunits under nonnative solvent conditions and comparing them with the native M_W, we can tell how many subunits are present in the native protein.

Determining the Molecular Mass of the Native Structure

To determine the molecular masses of proteins in their physiological states, several techniques are available. Recall from Tools of Biochemistry 5A that size exclusion chromatography (SEC) separates proteins in a mixture based on differences in hydrodynamic

Urea   **Guanidine hydrochloride (guanidinium chloride)**   **Sodium dodecyl sulfate (SDS)**

radius. The hydrodynamic radius is related to the M_W; thus, SEC can be used to estimate the native M_W of a protein. The shape of the protein (i.e., rod-like vs. spherical) has a significant impact on hydrodynamic radius and this can skew SEC data. Nevertheless, SEC provides a simple means to estimate, with reasonable accuracy, the M_W of a protein under native conditions. This is done by making a calibration curve of log M_W vs. retention time for a series of protein standards of known M_W.

A more exact, but less commonly available, way to measure M_W uses the technique of **sedimentation equilibrium**. If a protein solution is sedimented for many hours at low rotor speed in an analytical ultracentrifuge, an equilibrium will be established between the tendency of the molecules to sediment and their tendency to diffuse back into the solution. Details of sedimentation equilibrium and other physical techniques that can be used to determine the molecular masses of native proteins are given in van Holde et al. (see References).

Mass spectrometry (see Tools of Biochemistry 5B) is routinely employed to determine masses of protein subunits with great accuracy. In most cases it is possible to obtain masses for each subunit of a multisubunit complex following the isolation of the native complex by SEC. More recently, it has been reported that quaternary structural features of large protein complexes (e.g., chaperonins) can be maintained in the gas phase and are therefore amenable to mass spectrometric analysis. Work in the lab of Carol Robinson has recently demonstrated that it is possible to use these techniques to identify proteins associated in complexes of >1 MDa (1 MDa = 10^6 g/mol), as well as to monitor dynamic processes such as subunit association in real time.

Determining the Number and Approximate Masses of Subunits: SDS Gel Electrophoresis

Once the native M_W has been determined, the easiest way to estimate the molecular masses of the subunits is to use gel electrophoresis in the presence of SDS. Under these conditions, quaternary, tertiary, and secondary structures of proteins are all broken down. The chain is unfolded and coated by SDS molecules. The numerous negative charges carried by the many SDS molecules bound noncovalently to the protein make the intrinsic charge carried by the protein insignificant. The folded polypeptide chain is therefore transformed into an elongated object, the length and charge of which are each proportional to the number of amino acid residues in the chain (and hence the polypeptide M_W). As pointed out in Tools of Biochemistry 2A and 5A, such particles will migrate in gel electrophoresis with relative mobilities depending only on their lengths. This phenomenon is demonstrated by the graph shown in Figure 6B.1. If electrophoresis of an unknown protein chain is carried out on the same gel as such a set of standards, the M_W of the unknown can be measured by interpolation using a graph like Figure 6B.1.

In investigating the subunits of a protein by this technique, it is advisable to do two experiments: one in the presence of a disulfide-bond–reducing agent like β-mercaptoethanol (HSCH$_2$CH$_2$OH, see page 169) and one in its absence. This will distinguish between subunits that are bonded by —S—S— bridges and those that are held together only by noncovalent forces. If a single band is found on each of these SDS gels, corresponding to the M_W of the native protein, we

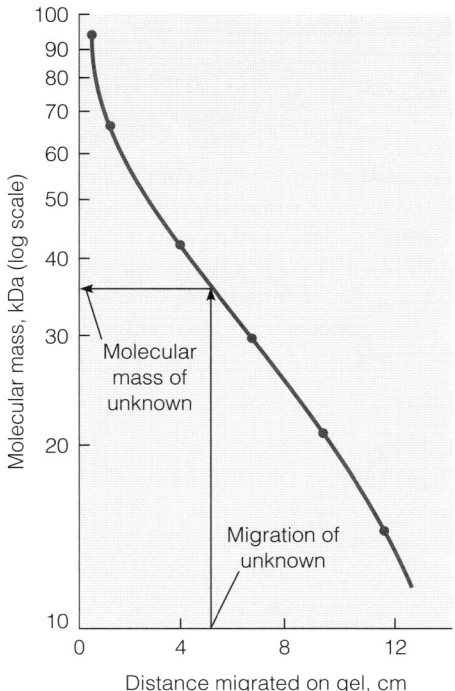

FIGURE 6B.1

SDS gel electrophoresis. The graph plots log M_W versus relative electrophoretic mobility for a series of proteins dissolved in a solution containing the detergent SDS. The resulting curve is used to interpolate data for an unknown protein.

may conclude that the protein exists under physiological conditions as a single polypeptide chain. If the band or bands observed are of much lower molecular mass, a multisubunit structure is indicated. Even though the M_W values obtained from gel electrophoresis may be approximate, they should be sufficiently accurate for a good guess as to the number of subunits (see Problem 13).

Assuming that multiple subunits are indicated, is there only one kind or are there several? More than one band on the SDS gel is a clear indication of multiple types of subunits. But finding only one band does not prove that subunits are identical. There may, in fact, be several kinds of subunits with distinct amino acid sequences but nearly identical molecular masses; these different subunits usually cannot be resolved on SDS gels. To be satisfied that only one type of chain is present, the researcher must turn to other methods. Again, mass spectrometry is the method of choice for such determinations.

References

Benesch, J. L. P., B. T. Ruotolo, D. A. Simmons, and C. V. Robinson (2007) Protein complexes in the gas phase: Technology for structural genomics and proteomics. *Chem. Rev.* 107:3544–3567.

Chait, B. T., and S. B. H. Kent (1992) Weighing naked proteins: Practical, high-accuracy mass measurement of peptides and proteins. *Science* 257: 885–1893.

Hames, B. D., and D. Rickwood, eds. (1990) *Gel Electrophoresis of Proteins*, 2nd ed. IRL Press, Oxford, Washington, D.C.

Heck, A. J. R., and R. H. H. van den Heuvel (2004) Investigation of intact protein complexes by mass spectrometry. *Mass Spectrom. Rev.* 23:368–389.

van Holde, K. E., W. C. Johnson, and P. S. Ho (2006) *Principles of Physical Biochemistry*, 2nd Ed. Prentice Hall, Upper Saddle River, NJ.

TOOLS OF BIOCHEMISTRY 6C

DETERMINING THE STABILITY OF PROTEINS

Our knowledge of the relative strengths of the noncovalent forces that stabilize protein structure has come largely from the determination of thermodynamic parameters for proteins that display apparent **two-state folding**. For such proteins the transition between the folded and unfolded states is highly cooperative (see Figure 6.23b); thus, no intermediates are detectable, and the unfolding process is described as

$$\text{Folded} \rightleftharpoons \text{Unfolded} \qquad (6C.1)$$

Because the covalent structure of the protein doesn't change upon folding or unfolding, protein unfolding is treated as a *phase transition*—such as the conversion of ice to liquid water, or liquid water to steam. For these processes the chemical identity of the molecule in question is not changed; rather, noncovalent interactions are formed, or broken, during the transition from one phase (e.g., "folded state") to the other (e.g., "unfolded state").

Here we present basic introductions to the theories behind two common methods to measure thermodynamic stabilities of proteins: differential scanning calorimetry and chemical denaturation. In both cases we will assume the two-state model applies.

Differential Scanning Calorimetry

To extract values for ΔH and ΔS, such as those reported in Table 6.4, the amount of heat required to unfold a protein must be accurately measured. Such measurements can be made in a **differential scanning calorimeter** (DSC) operating at constant pressure. Recall from page 62 of Chapter 3 that heat transferred at constant pressure corresponds to ΔH; thus, a DSC allows the direct determination of ΔH for protein unfolding.

Figure 6C.1 shows the basic features of a DSC. A buffered solution of protein is placed in the "sample" cell, and the buffer alone is placed in the "reference" cell. The initial temperature of the cells is below the denaturation temperature of the protein (also called the **transition temperature** or **melting temperature**, T_m). The temperature of both cells is then raised in a controlled fashion—generally at a rate of 0.5 to 2.0 °C per minute until the final temperature is above the T_m of the protein. As the protein unfolds, it will absorb the heat required to complete the phase transition from the native to the denatured state (Equation 6C.1). The absorption of excess heat by the protein as it denatures will require that more heat is added to the sample cell to keep it at the same temperature as the reference cell. The DSC records the difference in heat absorbed by the buffered protein solution vs. the buffer alone. This excess absorbed heat is manifested as a change in *heat capacity at constant pressure*, C_p, as a function of temperature. Figure 6C.2 shows an idealized plot of C_p vs. temperature for protein denaturation, called a "thermogram."

The thermogram provides several key pieces of information. Here we focus upon three. The first of these is the peak value of

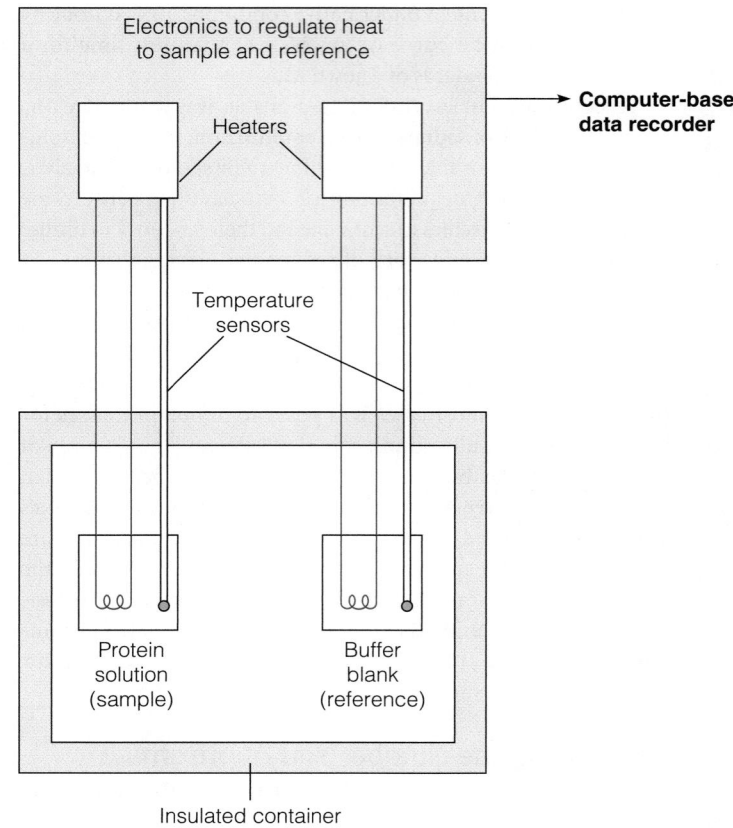

FIGURE 6C.1

Schematic diagram of a differential scanning calorimeter.

C_p, which corresponds to the T_m of the protein. For a protein that unfolds in a two-state fashion, T_m represents the temperature at which the free energies of the native and denatured states are equal; thus, at T_m

$$\Delta G_{F \to U} = G_U - G_F = 0 \qquad (6C.2)$$

and

$$\Delta G_{F \to U} = \Delta H_{F \to U} - T_m \Delta S_{F \to U} = 0 \qquad (6C.3)$$

or

$$\Delta H_{F \to U} = T_m \Delta S_{F \to U} \qquad (6C.4)$$

The second piece of key information is the value of $\Delta H_{F \to U}$, which corresponds to the area under the C_p vs. temperature curve in the transition region. To determine the enthalpy change for the unfolding transition alone, we must subtract out the contributions to C_p from the heat capacities of the native ($C_{p,f}$) and denatured ($C_{p,u}$) states in the transition region. In practice, this correction is

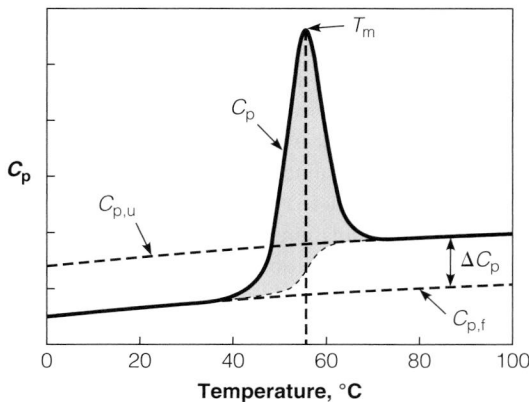

FIGURE 6C.2

An idealized DSC thermogram of protein denaturation. C_p is plotted as a solid black line. The pre-transition baseline, $C_{p,f}$ is the constant-pressure heat capacity of the native (folded) state; the post-transition baseline, $C_{p,u}$ is the constant-pressure heat capacity of the denatured (unfolded) state; $\Delta C_p = C_{p,u} - C_{p,f}$; T_m is the transition temperature; $\Delta H_{F \to U}$, shown in blue shading, is the area under the curve in the transition region corrected for differences in $C_{p,f}$ and $C_{p,u}$.

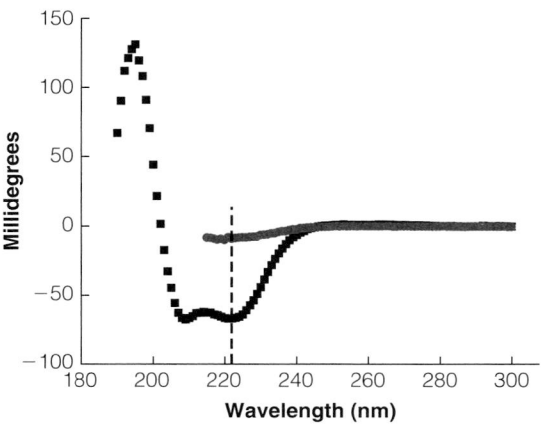

FIGURE 6C.3

Circular dichroism (CD) spectra of native and chemically denatured myoglobin. The CD spectrum of native state sperm whale myoglobin (sw Mb) in pH 7.0 buffer is shown as solid black squares. The red data show the CD spectrum of denatured sw Mb in buffered 8 M urea (note: urea absorbs strongly below 210 nm). The vertical dashed line shows the difference in the CD signal at 222 nm, which can be used to distinguish between the folded and unfolded states.

made by the data analysis software provided with the DSC; however, it requires that sufficient data are recorded for pre- and post-transition baselines, such that these baselines can be properly evaluated in the transition region. With the value of $\Delta H_{F \to U}$ in hand, we can calculate $\Delta S_{F \to U}$ using Equation 6C.4 and T_m.

You will notice that the data in Table 6.4 report values of ΔG, ΔH, and ΔS for folding at the reference temperature of 25 °C. Because most intracellular proteins are stably folded at 37 °C, how can stability parameters at 25 °C be determined? This is done by taking the temperature dependence of ΔH and ΔS into account. Recall from the thermodynamics of phase changes:

$$\Delta H_T = \Delta H_{T_m} + \Delta C_p(T - T_m) \tag{6C.5}$$

$$\Delta S_T = \Delta S_{T_m} + \Delta C_p \ln\left(\frac{T}{T_m}\right) \tag{6C.6}$$

$$\Delta G_T = \Delta H_T - T\Delta S_T \tag{6C.7}$$

where ΔH_{T_m} and ΔS_{T_m} are, respectively, the enthalpy change and entropy change at T_m, and T is any desired value of temperature, such as 298 K (25 °C). Thus, we can calculate values of ΔH, ΔS, and ΔG at any temperature if we have ΔH_{T_m} (which is the same as $\Delta H_{F \to U}$), T_m, and ΔC_p.

As shown in Figure 6C.2, ΔC_p is the third bit of critical information provided by the DSC experiment; thus, DSC provides all the information we need to derive the thermodynamic parameters that allow comparisons of protein stability to be made.

The foregoing treatment assumes two-state unfolding behavior *and* **reversibility**. In other words, can the original folded state be recovered after the thermally unfolded protein is cooled and allowed to refold? For a true two-state unfolding process one ought to get the same thermogram for a second (or third) DSC trial following cooling of the sample cell. This is an important control in DSC because a protein may undergo irreversible chemical modification (e.g., thermal deamidation of Asn or Gln side chains)

or aggregation at the higher temperatures required to get a good post-transition baseline. Evidence for either of these events precludes application of the simple two-state model described above.

It is possible to use DSC to derive thermodynamic parameters for proteins that display more complex unfolding behavior; however, the discussion of such methods is beyond the scope of this introduction (see the References cited below).

Chemical Denaturation Methods

An alternative method for comparing relative stabilities between different proteins is the method of linear extrapolation from chemical denaturation data. This is a widely used method because it uses reagents and equipment that are available in most protein chemistry labs. It is used to calculate $\Delta G^\circ_{F \to U}$ (note: this is for the process of *unfolding*); however, it gives no information about the enthalpy and entropy changes, as does DSC.

The method is based on the observation that increasing concentrations of chaotropes (i.e., chemical denaturants) such as urea and guanidine hydrochloride (GnHCl) perturb the equilibrium of Equation 6C.1 toward the unfolded state. Given any of the spectroscopic methods discussed in Tools of Biochemistry 6A that can distinguish the folded from the unfolded state (e.g., see Figure 6C.3), it is possible to determine the equilibrium constant for the protein unfolding as a function of denaturant concentration.

Assuming two-state unfolding, the equilibrium constant, K_{eq}, for protein unfolding is

$$K_{eq} = \left(\frac{[U]}{[F]}\right) \tag{6C.8}$$

and it can be shown that

$$K_{eq} = \left(\frac{A_F - A_{obs}}{A_{obs} - A_U}\right) \tag{6C.9}$$

where A_F is the spectroscopic signal of the native state; A_U is the spectroscopic signal of the denatured state; and A_{obs} is the spectroscopic signal of the protein sample (the derivation of Equation 6C.9 is given in the References cited below). These values of K_{eq} can then be used to calculate values of $\Delta G°_{F \to U}$ for each sample of protein solution containing denaturant:

$$\Delta G°_{F \to U} = -RT \ln K_{eq} \qquad (6C.10)$$

In practice, one needs 15–20 protein solutions where each sample contains the same concentration of protein but a different concentration of denaturant. The spectroscopic signal of each sample is then recorded and plotted as a function of the denaturant concentration (open triangles in Figure 6C.4). In the past, the data points in the transition region would be converted to values of $\Delta G°_{F \to U}$ and then used to *extrapolate* a value of $\Delta G°_{F \to U}$ at zero

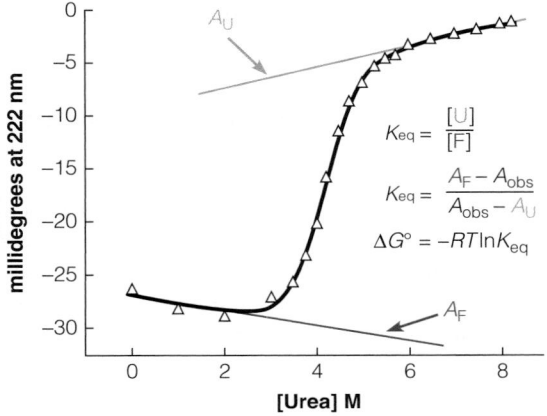

FIGURE 6C.4

Circular dichroism at 222 nm for samples of mutant sperm whale myoglobin in buffered urea solutions. The open triangles show CD signal at 222 nm. The red line shows the slope of the pre-transition baseline, corresponding to the CD signal for the mutant in the native state. The blue line shows the slope of the post-transition baseline, corresponding to the CD signal for the mutant in the denatured state. The solid black line is a nonlinear curve fit of a two-state unfolding model to the data. From the curve fit, a value of $\Delta G°_{H_2O}$ is obtained.

Reprinted with permission from *Biochemistry* 41:13318–13327, A. L. Fishburn, J. R. Keeffe, A. V. Lissounov, D. H. Peyton, and S. J. Anthony-Cahill, A circularly permuted myoglobin possesses a folded structure and ligand binding similar to those of the wild-type protein but with a reduced thermodynamic stability. © 2002 American Chemical Society.

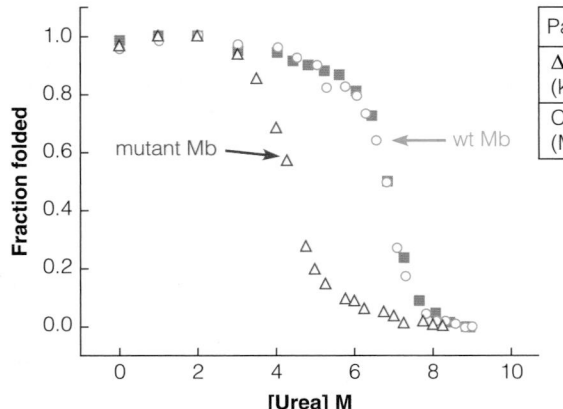

denaturant concentration. This step is the source of the method's name (linear extrapolation). The value of $\Delta G°_{F \to U}$ at zero denaturant is more commonly referred to as $\Delta G°_{H_2O}$. With the availability of software packages that perform nonlinear curve fitting, it is now possible to fit an equation for two-state unfolding directly to the spectroscopic data (solid black line in Figure 6C.4). Such curve-fitting procedures give the value of $\Delta G°_{H_2O}$ without the need for tedious calculations and extrapolation; however, it is important to have sufficient pre- and post-transition data to adequately constrain the curve-fitting with reliable values of A_F and A_U.

Using this procedure it is relatively straightforward to compare the stabilities of two or more proteins. The concentration of denaturant at the midpoint of the transition (C_m) is a crude, but quite useful, indicator of relative stabilities. The curve fitting that gives $\Delta G°_{H_2O}$ also yields C_m. More stable proteins have more positive values of $\Delta G°_{H_2O}$ and also tend to have higher values of C_m. In other words, it generally takes higher concentrations of denaturant to unfold more-stable proteins (Figure 6C.5).

As with DSC, it is important to perform controls that support use of the two-state model. A common control with chemical denaturation studies is to perform multiple trials using different spectroscopic techniques. Frequently, complementary techniques such as CD (which measures secondary structure) and fluorescence (which measures tertiary structure, i.e., packing of Trp and Tyr residues) are used. If folding is cooperative and two-state, secondary and tertiary structures should denature in concert, and the denaturation curves obtained by different spectroscopic techniques will superimpose. An example of this is shown in Figure 6C.5 for the unfolding of wild-type myoglobin monitored by CD at 222 nm (monitors α helix content) and heme absorbance at 409 nm (monitors tertiary folding around heme group). The superposition of the two data sets is consistent with, but doesn't prove, two-state behavior. If the data did not superimpose, the existence of intermediate states would be implied and application of the two-state model would not be warranted.

Chemical denaturation has also been used to screen for compounds that stabilize protein native structure. The Ala 4 to Val (A4V) mutant of superoxide dismutase (SOD1) is linked to familial amyotrophic lateral sclerosis (ALS; see Table 6.6), and predisposes SOD1 to form amyloid fibrils. Several potential drug candidates were screened for ability to stabilize the A4V mutant of SOD1 using chemical denaturation (Figure 6C.6). Fifteen compounds were found to stabilize the native state dimer and

Parameter	mutant Mb	wt Mb
$\Delta G°_{H_2O}$ (kcal/mol)	7.2 ± 0.5	12.4 ± 1.6
C_m (M urea)	4.19 ± 0.05	6.90 ± 0.06

FIGURE 6C.5

Comparison of chemical denaturation curves for wild-type myoglobin and a destabilized mutant. The data are plotted as fraction protein folded vs. denaturant concentration. The data for the mutant (red open triangles) are the same as those shown in Figure 6C.4. The open blue circles are CD data at 222 nm for wild-type myoglobin, and the solid green squares are heme absorbance data for wild-type myoglobin. The inset gives values of $\Delta G°_{H_2O}$ and C_m obtained from non-linear curve fitting of the CD data.

Reprinted with permission from *Biochemistry* 41:13318–13327, A. L. Fishburn, J. R. Keeffe, A. V. Lissounov, D. H. Peyton, and S. J. Anthony-Cahill, A circularly permuted myoglobin possesses a folded structure and ligand binding similar to those of the wild-type protein but with a reduced thermodynamic stability. © 2002 American Chemical Society.

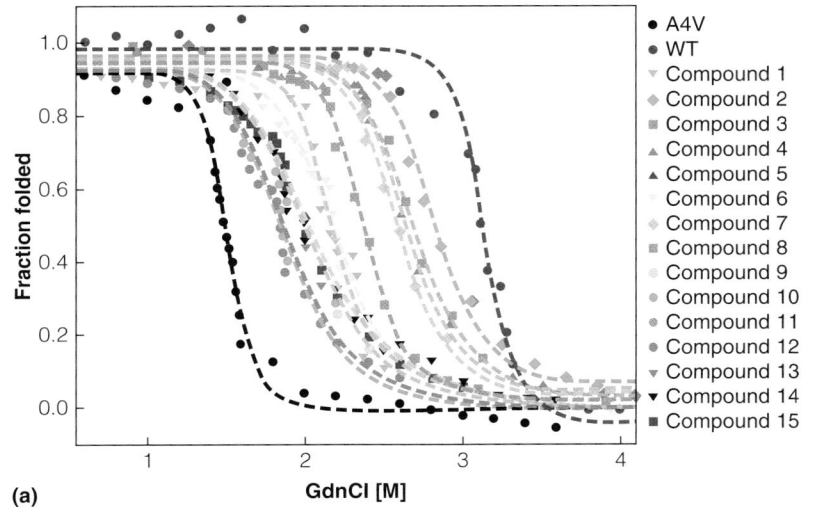

(a)

FIGURE 6C.6

Use of chemical denaturation to screen potential drugs.
(a) The A4V mutation of SOD1 (black circles) is much less stable to chemical denaturation than is wild-type SOD1 (red circles). Addition of several small molecule compounds increases the stability of the A4V mutant. **(b)** The small molecules are thought to bind to and stabilize the SOD1 dimer such that the unfolding equilibrium is perturbed in the direction of the native state. The stabilization provided by the potential drugs is shown as the green arrow in the rightmost panel.

Reprinted from *Proceedings of the National Academy of Sciences of the United States of America* 102:3639–3644, S. S. Ray, R. J. Nowak, R. H. Brown, Jr., and P. T. Lansbury, Jr., Small-molecule-mediated stabilization of familial amyotrophic lateral sclerosis-linked superoxide dismutase mutants against unfolding and aggregation. © 2005 National Academy of Sciences, U.S.A.

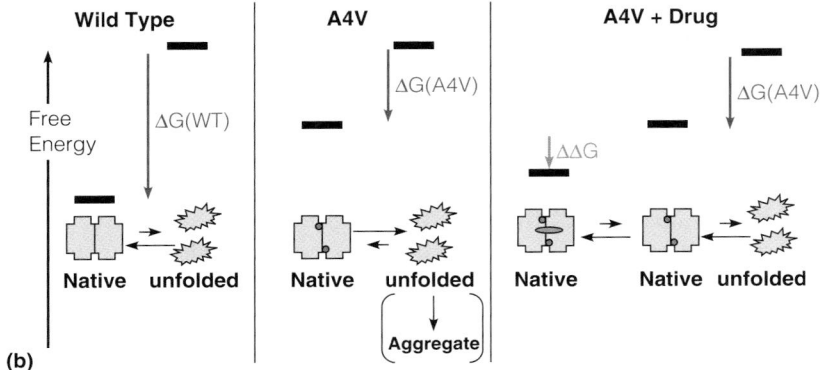

(b)

reduce aggregation in vitro. These compounds will therefore serve as lead structures in the search for drugs that could be of therapeutic value in treating familial ALS. An alternative view, based upon the finding that mutant and wild-type SODs can hybridize to form a toxic enzyme, is that the goal of therapy should be to destabilize the mutant enzyme, so that it can be degraded and removed from the cell.

References

Cooper, A. (2010) Protein heat capacity: An anomaly that maybe never was. *J. Phys. Chem. Lett.* 1:3298–3304. A very clear discussion of heat capacity in proteins.

Creighton, T. E. (2010) *The Biophysical Chemistry of Nucleic Acids and Proteins.* Helvetian Press, UK. See Chapter 11.

Friere, E. (1995) Differential Scanning Calorimetry. In *Protein Stability and Folding* (B.A. Shirley, ed.). Humana Press, Totowa, NJ.

Ibarra-Molero, B., and J. M. Sanchez-Ruiz (2006) Differential scanning calorimetry of proteins: An overview and some recent advances. In *Advanced Techniques in Biophysics* (J. L. R. Arrondo and A. Alonso, eds). Elsevier, Amsterdam.

Pace, C. N. (1986) Determination and analysis of urea and guanidine hydrochloride denaturation curves. *Methods Enzymol.* 131:266–280.

Privalov, P. L., and A. I. Dragan (2007) Microcalorimetry of biological macromolecules. *Biophysical Chemistry* 126:16–24.

Ray S. S., R. J. Nowak, R. H. Brown, Jr., and P. T. Lansbury, Jr. (2005) Small-molecule-mediated stabilization of familial amyotrophic lateral sclerosis-linked superoxide dismutase mutants against unfolding and aggregation. *Proc. Natl. Acad. Sci. USA* 102 :3639–3644.

Sahawneh, M. A., K. C. Rickart, B. R. Roberts, V. C. Bomben, M. Basso, Y. Ye, J. Sahawneh, M. C. Franco, J. S. Beckman, and A. G. Estevez (2010) Cu, Zn-Superoxide dismutase increases toxicity of mutant and zinc-deficient superoxide dismutase by enhancing protein stability. *J. Biol. Chem.* 285:33885–33897.

Santoro, M. M., and D. W. Bolen (1988) Unfolding free energy changes determined by the linear extrapolation method. 1. Unfolding of phenyl-methylsulfonyl alpha-chymotrypsin using different denaturants. *Biochemistry* 27:8063–8068.

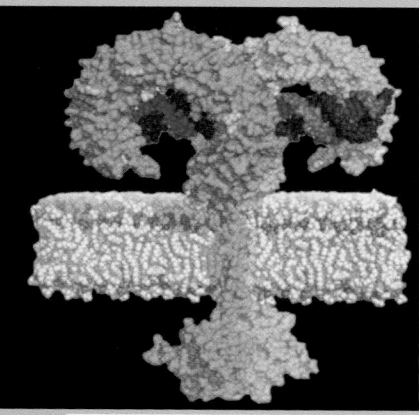

CHAPTER 7

Protein Function and Evolution

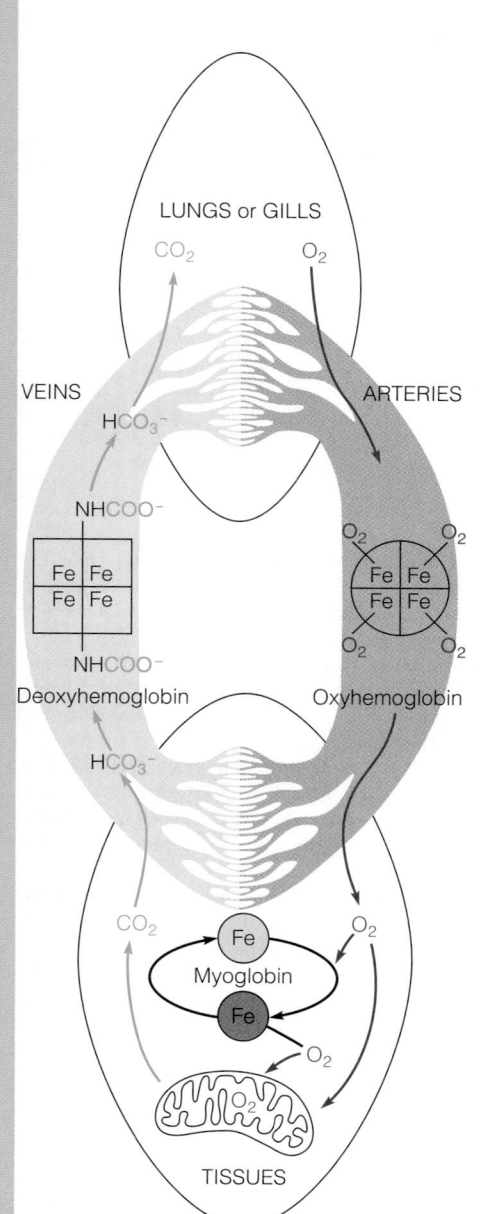

LUNGS or GILLS

CO_2 O_2

VEINS   ARTERIES

HCO_3^-

NHCOO⁻ → NHCOO$^-$

Fe Fe / Fe Fe O_2 O_2 Fe Fe / Fe Fe O_2 O_2

NHCOO$^-$

Deoxyhemoglobin   Oxyhemoglobin

HCO_3^-

CO_2 O_2

Fe
Myoglobin
Fe

O_2

TISSUES

With an understanding of the complex, folded structures of globular proteins, we can look more closely at how such structures are related to the molecules' functions and how they may have evolved to fulfill those functions. In this chapter we will use as examples two groups of proteins whose main functions are binding other molecules: the globins and immunoglobulins.

We will start with **myoglobin** (abbreviated **Mb**) and its molecular relative **hemoglobin** (**Hb**), members of a family of proteins collectively termed **globins**. We have chosen these examples for a number of reasons. First, the hemoglobins and myoglobins play vital roles in one of the most important aspects of animal metabolism—the acquisition and utilization of molecular oxygen (O_2). As described in detail in Chapter 15, the most efficient energy-generating mechanisms in animal cells require O_2 for the oxidation of nutrients; therefore, proteins that transport O_2 to respiring cells are essential for any higher organism. Myoglobin is an O_2 binding protein found primarily within the muscle tissue of animal species; hemoglobin is used for oxygen transport in all vertebrates and some invertebrates. Second, hemoglobin plays a role in removing CO_2 from tissues. CO_2 is a major product of metabolite oxidation and must be continually removed and exhaled. The role of hemoglobin in O_2 and CO_2 transport is schematically illustrated in Figure 7.1. Because O_2 binding and CO_2 removal must be carefully regulated to meet tissue needs, these examples illustrate regulation of protein function. Third, the close structural relationship between hemoglobin and myoglobin yields important insights into evolution of protein function. Finally, the structure, function, and evolution of the globin family have been studied more than for any other

FIGURE 7.1

Role of the globins in oxygen transport and storage. Vertebrate animals use hemoglobin and myoglobin to provide their tissues with a continuous O_2 supply. Hemoglobin transports O_2 from the lungs or gills to the respiring tissues, where it is used for aerobic metabolism in the mitochondria. Inside cells, dissolved O_2 diffuses freely or is bound to myoglobin, which aids transport of O_2 to the mitochondria. Myoglobin can also store O_2 for later use (as in deep-diving mammals). CO_2 produced by oxidative processes in the tissues is carried back to the lungs or gills by hemoglobin and released.

group of proteins. The study of hemoglobin has played a significant role in defining the early history of biochemistry and molecular biology. Indeed, the properties of hemoglobin have been investigated extensively because the protein was successfully crystallized in the first half of the nineteenth century.

Whereas the hemoglobins and myoglobins bind to a few specific molecules reversibly (described below), the **immunoglobulins** (or **antibody** molecules) are protein structures that can be produced in a multitude of variations, and each variant binds with exquisite specificity, and essentially nonreversibly, to a unique target. Our primary defenses against infectious disease depend on the ability of immunoglobulins to recognize and bind to "foreign" molecules (i.e., of "nonself" origin) as part of the **immune response**.

Structure–function relationships for another diverse and important group of proteins, the **enzymes**, are discussed in Chapter 11.

Oxygen Transport: The Roles of Hemoglobin and Myoglobin

Hemoglobin and myoglobin are proteins that have evolved to carry out the specialized function of O_2 transport in animals. Any animal larger than a few millimeters in diameter faces a serious problem in carrying out **aerobic** (oxygen-requiring) metabolism. It must ensure a steady supply of O_2 to cells throughout its body and remove metabolic waste products such as CO_2. These gases will diffuse through tissues; but transport by diffusion becomes very slow if appreciable distances must be crossed. Insects solve the problem by having **tracheae**, tubular networks that lead from the body surface down into the tissues. They have, in effect, increased their body surface area to the extent that diffusion is practicable. This mechanism works because insects are small. (Alternatively, we might conclude that insects are small because they rely on this mechanism for obtaining O_2.*)

Almost all other animals pick up O_2 in lungs or gills and pump it in the blood through arteries to the tissues (see Figure 7.1). Carbon dioxide is returned in the venous blood and released in the lungs or gills. In some primitive organisms, the gases are simply dissolved in the blood, but this mechanism is very inefficient because O_2 has low solubility in plasma (plasma is the liquid portion of the blood). Much plasma must be pumped, at great metabolic expense, to deliver even a little dissolved O_2. Evolution of all higher organisms has been accompanied by the development of **oxygen transport proteins**, which allow the blood to carry a 100-fold higher concentration of O_2 than would be permitted by solubility of the gas alone. Oxygen transport proteins may be either dissolved in the blood (as in some invertebrates) or concentrated in specialized cells, like the human **erythrocytes** (red blood cells) shown in Figure 7.2. In all vertebrates the O_2 transport protein is hemoglobin, a protein that can bind O_2 in lungs or gills and deliver it to other tissues.

Once the O_2 is transported to the tissues, it must be released for utilization. Some tissues, such as skeletal and heart muscle, have high energy demands and require rapid transport of O_2 from hemoglobin in the circulation to mitochondria (where ATP is produced) inside respiring cells. The relatively high concentration (~2 mg/g of human muscle tissue) of myoglobin in all mammalian skeletal muscle is thought to facilitate diffusion of O_2 in cells so it reaches mitochondria efficiently. In deep-diving mammals the concentration of myoglobin in skeletal muscle is 10–30 times higher than that of terrestrial mammals. Thus, myoglobin also acts as an O_2 storage molecule, providing a substantial reserve of O_2 to

All but the smallest animals need a protein—hemoglobin in vertebrates—to transport O_2 from gills or lungs to tissues.

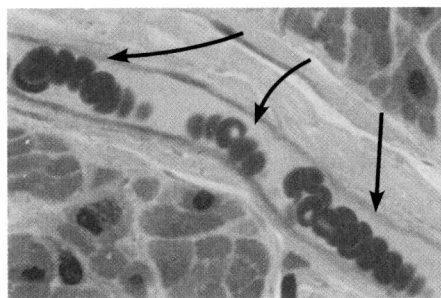

FIGURE 7.2

Human erythrocytes. Arrows point to red blood cells, or erythrocytes, shown moving in a capillary. Each erythrocyte contains about 300 million hemoglobin molecules.

© Ed Reschke/Peter Arnold/Getty Images.

---

*This hypothesis raises an intriguing problem: How do we explain the giant dragonflies of the Carboniferous period, which had wingspans up to 2 feet? Was the O_2 content of the atmosphere higher then (about 300 million years ago), or did they use some other mechanism for O_2 uptake? Recent evidence suggests that the first explanation may be correct.

FIGURE 7.3

Comparison of myoglobin and hemoglobin. These drawings illustrate the backbone structures of the two oxygen-binding molecules as revealed by X-ray crystallography. Each of the four chains in hemoglobin has a folded structure similar to that of myoglobin, and each carries a heme (shown in red). Hemoglobin contains two identical α chains and two identical β chains. The letters A–H indicate α-helical regions. The α and β chains are very similar but have distinct primary structures and folds (note that the α chain does not have a "D" helix).

Illustration, Irving Geis. Image from Irving Geis Collection/Howard Hughes Medical Institute. Rights owned by HHMI. Not to be reproduced without permission.

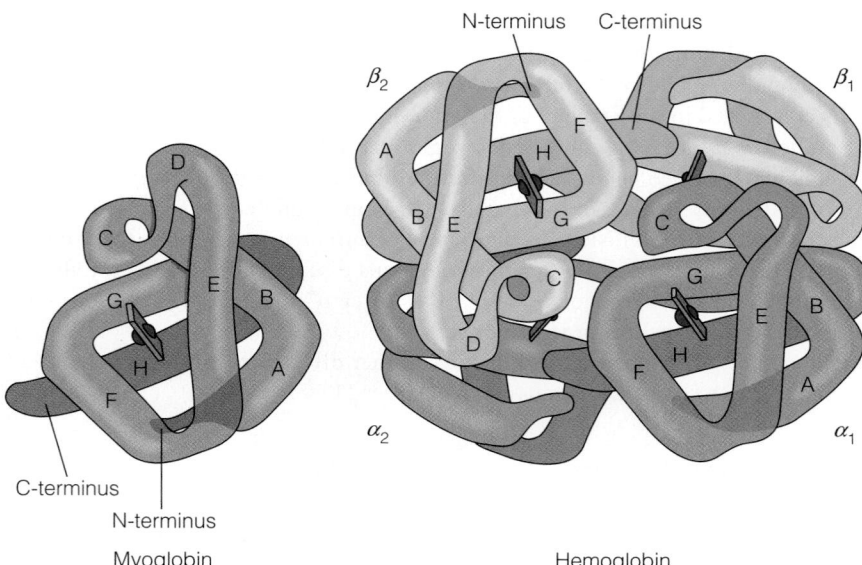

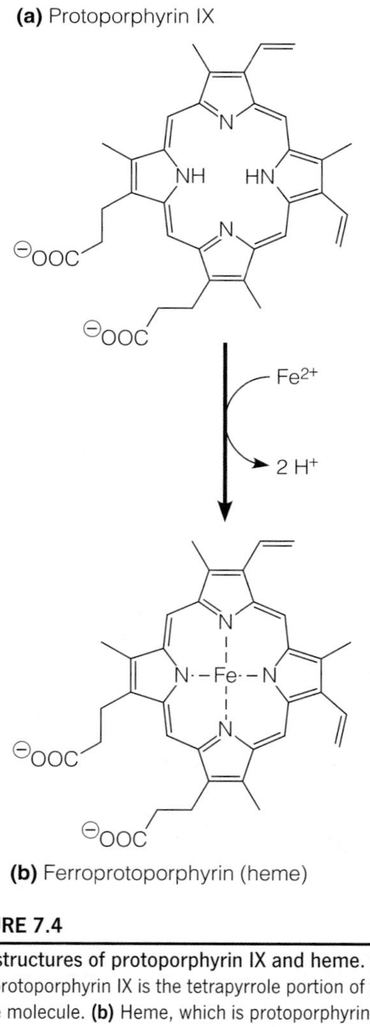

Myoglobin

Hemoglobin

(a) Protoporphyrin IX

(b) Ferroprotoporphyrin (heme)

FIGURE 7.4

The structures of protoporphyrin IX and heme. (a) The protoporphyrin IX is the tetrapyrrole portion of the heme molecule. **(b)** Heme, which is protoporphyrin IX complexed with Fe(II), is the prosthetic group of hemoglobin and myoglobin. Because of resonance delocalization of the electrons in the porphyrin ring, all N–Fe bonds within the heme are equivalent.

support the demand for ATP production while the animal is submerged. In larger whales this can be 30 minutes or longer! As is described later in this chapter, recent studies have shown that myoglobin and hemoglobin play other important physiological roles beyond O_2 transport and storage.

Myoglobin and hemoglobin are built on a common structural motif, as shown in Figure 7.3. In myoglobin, a single polypeptide chain is folded about a prosthetic group, the **heme** (Figure 7.4), which contains the O_2 binding site. Hemoglobin is a tetrameric protein, made up of four polypeptide chains each of which binds a heme group, and closely resembles myoglobin in structure. We begin the discussion of how these structures serve the function of O_2 transport from lungs to mitochondria with a description of O_2 binding in myoglobin.

The Mechanism of Oxygen Binding by Heme Proteins

An oxygen storage or transport molecule must be able to bind O_2 reversibly and protect it from reaction with any other substance that can reduce O_2, thereby rendering it useless for ATP production in mitochondria. How do the globins achieve this? To answer this question we must consider how the peptide and the prosthetic group interact. The peptide portion of any protein without its prosthetic group bound is called the **apoprotein**, whereas with its prosthetic group bound, it is called a **holoprotein**.* The apoglobins are unable to bind O_2 themselves; however, certain transition metals in their lower oxidation states—particularly Fe(II) and Cu(I)—have a strong tendency to bind O_2. The globin proteins have evolved such that Fe(II) is bound to the proteins to produce a site at which O_2 binds reversibly.

The Oxygen Binding Site

Various iron-containing proteins can hold Fe(II) in a number of possible ways. Throughout the myoglobin–hemoglobin family, the iron is chelated by a tetrapyrrole ring system called **protoporphyrin IX** (Figure 7.4a), one of a large class of **porphyrin** compounds. We will encounter other porphyrins in chlorophyll (Chapter 16), the cytochrome proteins (Chapter 15), and some natural pigments. Like most compounds with large conjugated ring systems, the porphyrins are strongly colored. The iron-porphyrin in the globin accounts for the red color of blood and meat, and the magnesium-porphyrin in chlorophyll is responsible for the green color of plants.

---

*For enzymes the corresponding terms are "apoenzyme" (without prosthetic group) and "holoenzyme" (prosthetic group present).

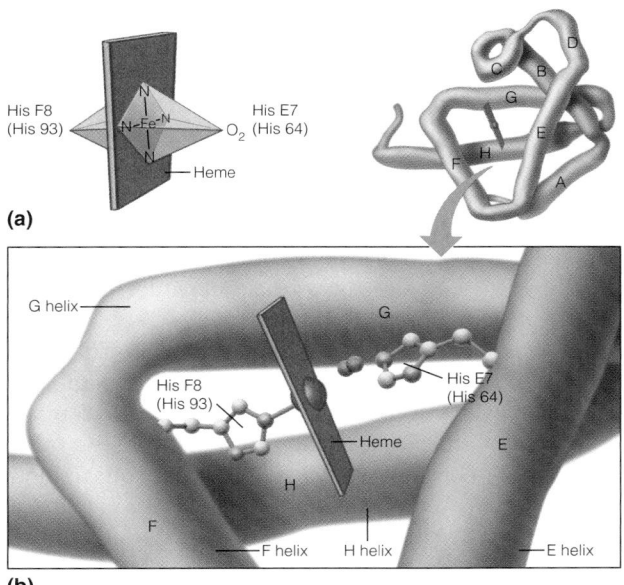

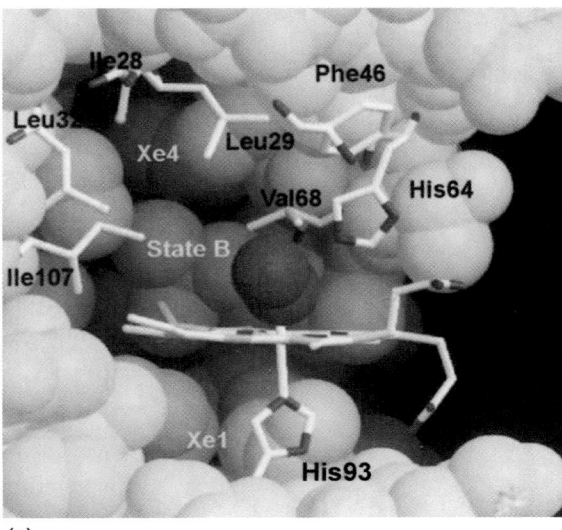

FIGURE 7.5

The geometry of iron coordination in oxymyoglobin. **(a)** The octahedral coordination of the iron ion. The iron and the four nitrogens from protoporphyrin IX lie nearly in a plane. A histidine (F8, or His 93) occupies one of the axial positions, and O_2 the other. **(b)** Schematic drawing of the heme pocket, showing the proximal (F8; His93) and distal (E7; His64) histidine side chains. **(c)** The crystal structure of the ligand-binding site in sperm whale oxymyoglobin (PDB ID: 2MGM). Amino acids in the O_2 binding pocket are shown as stick models and labeled. The heme group and the proximal histidine (His93) are also rendered as stick models, with O_2 in red (spacefilling model). The rest of the protein is drawn as a space-filling model in yellow. Here the D helix and the CD corner have been deleted to show the ligand-binding site. Residues Phe46 (or CD1) and Val68 (or E11) are highly conserved (Phe46 is invariant), and both make critical contributions to the binding affinity between the heme and the heme pocket. The sites labeled Xe1 and Xe4 indicate hydrophobic cavities that can be occupied by Xenon under pressure and apparently define channels for ligand entry and exit (see Figure 7.8).

(c) Courtesy of Dr. Jeffry Nichols.

The complex of protoporphyrin IX with Fe(II) is called heme (Figure 7.4b). This prosthetic group is bound in a hydrophobic crevice in the myoglobin or hemoglobin molecule (see Figure 7.3). The binding of oxygen to heme is illustrated in Figure 7.5, which shows the oxygenated form of myoglobin. Ferrous iron (Fe^{2+}) is normally octahedrally coordinated, which means it should have six **ligands**, or binding groups, attached to it. As shown in Figure 7.5a, the nitrogen atoms of the porphyrin ring account for only four of these ligands. Two remaining coordination sites are available, and they lie along an axis perpendicular to the plane of the heme. In both the deoxygenated and the oxygenated forms of myoglobin, one of these remaining coordination sites is occupied by the ε-nitrogen of histidine residue number 93 (see margin). The eight helical segments in the globins are designated A through H (as shown in Figure 7.3), and residue 93 is located in the F helix (Figure 7.5b). Using a nomenclature that allows for meaningful comparisons between the homologous sequences of different globins, this residue is called histidine F8 (it is the eighth residue in the F helix). Because it is in direct contact with the Fe^{2+}, it is also called the **proximal histidine**. In **deoxymyoglobin**, the remaining coordination site, on the other side of the iron, is unoccupied. When oxygen is bound, making **oxymyoglobin**, the O_2 molecule occupies this site.

The Fe—O_2 bond is highly polar due to the good orbital overlap between d orbitals on the Fe^{2+} and the π^* molecular orbital of the electronegative O_2. This backbonding increases the electron density on the O_2 and gives the formal Fe(II)—O_2 complex some Fe(III)—$O_2^{\cdot}$ character. Stabilization of this Fe(III)—$O_2^{\cdot}$-like structure is achieved by a hydrogen bond between the bound O_2 and another important His residue located in the O_2 binding pocket—the so-called **distal histidine** (His 64, or E7; see Figure 7.5c). The H-bond between His E7 and O_2 selectively increases the affinity of Mb for O_2 vs. CO, which doesn't make a similar bond to His E7. Even so, CO binds ~200 times more tightly to Mb than does O_2; however, without the E7 H-bond, that ratio would be ~6000:1 in favor of CO. Thus, the distal histidine plays a critical role in promoting O_2 binding over other ligands.

A similar mode of oxygen binding, with histidines at the homologous F8 and E7 positions, is found in each subunit of hemoglobin.

Normally, an O_2 molecule in such close contact with a ferrous ion would oxidize the latter to the ferric [Fe(III)] state. The heme alone does not protect the iron because heme dissolved free in solution is readily oxidized by O_2. However, in the hydrophobic (and anhydrous) environment provided by the heme-binding

The ε-tautomer of histidine

Coordination of Fe(II) in a porphyrin (heme) within a hydrophobic globin pocket allows reversible O_2 binding without iron oxidation.

cleft of the myoglobin or hemoglobin molecule, the iron is not easily oxidized; thus, when the O_2 is released, the iron remains in the ferrous state, able to bind another O_2. The distal histidine also plays a role by inhibiting this autooxidation reaction, which is acid-catalyzed. The globins provide environments for ferrous heme in which the first step of an oxidation reaction (the binding of oxygen) is permitted, but the final step (oxidation) is blocked.

In fact, the ferrous heme in myoglobin and hemoglobin can become oxidized to the ferric (Fe^{3+}) state, to form **metmyoglobin** or, respectively, **methemoglobin**. The met-globins do not bind O_2; a water molecule occupies the O_2 binding site instead. For this reason red blood cells contain enzymes that can reduce the Fe^{3+} in methemoglobin to Fe^{2+} and thereby restore its O_2-binding activity.

Although myoglobin and hemoglobin are ideally adapted to reversibly bind an O_2 molecule, they also bind other diatomic gases such as carbon monoxide (CO) and nitric oxide (NO). The toxicity of CO is due to its ability to block respiration by binding tightly to the Fe^{2+}-hemes in globins as well as those in other critical respiratory proteins called **cytochromes**. Nitric oxide also inhibits respiratory proteins (primarily **cytochrome-c oxidase**; discussed in Chapter 15), and is released by macrophages to destroy invading organisms as part of the immune response. At low levels, NO is also a cell-signaling molecule. The interaction of the globins and NO are discussed later in this chapter.

Analysis of Oxygen Binding by Myoglobin

The binding of O_2 by myoglobin must meet certain physiological requirements. As Figure 7.1 shows, myoglobin located inside muscle cells binds O_2, which diffuses into cells from hemoglobin circulating in capillaries, then delivers the O_2 to the mitochondria. To understand these functions on a quantitative basis, we must examine how the binding of a ligand like O_2 depends on its concentration in the surroundings.

First, a way to measure the concentration of dissolved O_2 is needed. According to Henry's law, the concentration of any gas dissolved in a fluid is proportional to the *partial pressure* of that gas above the fluid. Therefore, we can conveniently regulate (and measure) the concentration of dissolved O_2 by regulating the partial pressure of O_2 above the myoglobin solution being studied. In fact, we can express O_2 concentration as this partial pressure: P_{O_2}.

To study ligand binding, we must have a way of measuring the fraction of myoglobin molecules carrying O_2. When myoglobin or hemoglobin are oxygenated they change colors (the absorption spectrum changes due to alteration of the electronic structure of the heme iron—see Tools of Biochemistry 6A). This allows a spectrophotometric determination of the fraction of binding sites that are oxygenated (Figure 7.6). The results of such analysis, using myoglobin in solution

FIGURE 7.6

Changes in the visible spectrum of hemoglobin. Spectra for hemoglobin in the deoxygenated state (blue trace) and the O_2-bound state (red trace) are shown. Hemoglobin in the deoxygenated state is a venous purple, whereas completely oxy-Hb is bright red. As more O_2 binds to Hb, the visible spectrum shifts from the blue to the red trace (several spectra for partially bound Hb are shown). Thus, the ligand-binding behavior of the globins is easily monitored by visible spectroscopy (see Tools of Biochemistry 6A) due to the distinctive spectral differences between the various forms of the globins.

Courtesy of John S. Olson, Rice University.

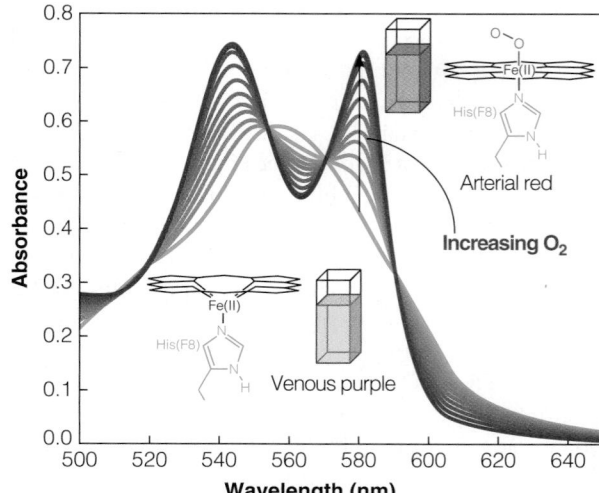

at neutral pH, are shown in Figure 7.7. Such a graph is called a *binding curve* because it describes how the fraction of the myoglobin sites that have O_2 bound to them (Y_{O_2}) depends on the concentration of free O_2 (P_{O_2}).

We can describe the binding of a ligand (in this case: O_2) to myoglobin by the following reaction

$$Mb + O_2 \xrightarrow{k_{on}} MbO_2 \qquad (7.1)$$

and ligand dissociation by

$$MbO_2 \xrightarrow{k_{off}} Mb + O_2 \qquad (7.2)$$

where k_{on} and k_{off} are the rate constants for binding and dissociation, respectively. Thus, reversible ligand binding is described by the following equilibrium

$$MbO_2 \underset{k_{on}}{\overset{k_{off}}{\rightleftharpoons}} Mb + O_2$$

and

$$K_d = \frac{[Mb][O_2]}{[MbO_2]} \qquad (7.3)$$

where the equilibrium constant K_d is called a **dissociation constant**. The quantities in brackets denote the *activities* (see Chapter 3) of oxymyoglobin [MbO_2], deoxymyoglobin [Mb], and free oxygen [O_2]. The fraction of myoglobin sites occupied is defined as follows:

$$Y_{O_2} = \frac{\text{sites occupied}}{\text{total sites available}}$$

Each myoglobin molecule has only one site, so the total number of potentially available sites is proportional to the total concentration of myoglobin species = [MbO_2] + [Mb]. Therefore,

$$Y_{O_2} = \frac{[MbO_2]}{[Mb] + [MbO_2]} = \frac{\dfrac{[Mb][O_2]}{K_d}}{[Mb] + \dfrac{[Mb][O_2]}{K_d}} \qquad (7.4)$$

where we have used [MbO_2] = [Mb][O_2]$/K_d$ from equation (7.3) to obtain the expression on the right. The concentration of deoxymyoglobin, [Mb], can be factored out of the numerator and denominator to give

$$Y_{O_2} = \frac{\dfrac{[O_2]}{K_d}}{1 + \dfrac{[O_2]}{K_d}} = \frac{[O_2]}{K_d + [O_2]} \qquad (7.5)$$

Equation 7.5 shows that K_d describes the oxygen concentration at which half the Mb molecules have O_2 bound. You can check this relationship by setting Y_{O_2} = 1/2 in Equation 7.5. Because oxygen concentration is proportional to oxygen partial pressure, Equation 7.5 can also be written as

$$Y_{O_2} = \frac{P_{O_2}}{P_{50} + P_{O_2}} \qquad (7.6)$$

where P_{50} is the oxygen partial pressure required for 50% O_2 saturation. The value of P_{50} is an indicator of the relative binding affinity for a ligand. In the case of myoglobin, when $P_{O_2} = P_{50}$, [Mb] = [MbO_2]. For a myoglobin with high O_2-binding affinity, half-saturation occurs at *low* P_{O_2}; thus, the value of P_{50} is *low*. For a myoglobin which has a low O_2-binding affinity, half-saturation would occur at *high* P_{O_2}; thus, the value of P_{50} would be *high*.

We see that Equation 7.6 describes the *hyperbolic* binding curve shown in Figure 7.7; Y_{O_2} starts at zero at P_{O_2} = 0 and approaches 1 as P_{O_2} increases. The P_{50}

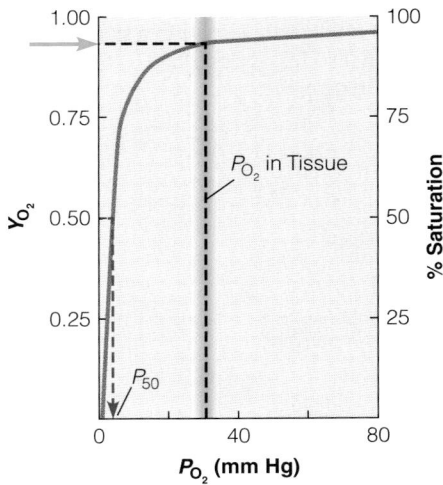

FIGURE 7.7

Oxygen-binding curve for myoglobin. The free oxygen concentration is expressed as P_{O_2}, the partial pressure of oxygen. The proportion of myoglobin binding sites that are occupied is expressed as a fraction (Y_{O_2}, on the left) or as percent saturation (on the right). As P_{O_2} becomes large, 100% saturation is approached asymptotically, as described by Equation 7.6. The value of P_{50}, the partial pressure of oxygen at 50% saturation, is indicated on the graph (magenta arrow). The dashed blue lines show that at P_{O_2} of 30 mm Hg, Mb would be >90% saturated with O_2 (blue arrow on y-axis).

P_{50} is an indicator of the relative binding affinity of a globin for a ligand: For a globin with higher O_2-binding affinity, the value of P_{50} is lower. For a globin with lower O_2-binding affinity, the value of P_{50} is higher.

Binding of a ligand (like O_2) to a single site on a protein (like Mb) is described by a hyperbolic binding curve.

for myoglobin is very low (3–4 mm Hg), signifying that myoglobin has a high affinity for O_2. This characteristic is appropriate for a protein that must extract O_2 from the blood in capillaries. At the P_{O_2} existing in the arterial capillaries (about 30 mm Hg), the myoglobin in adjacent tissues would be nearly saturated (blue arrow in Figure 7.7). When cells are metabolically active, their internal P_{O_2} falls to much lower levels (3–18 mm Hg). Under these conditions, myoglobin will release its O_2.

The myoglobin molecule must not only provide the heme a protein environment suitable for the reversible binding of O_2, it must also ensure that the O_2-binding affinity (which is indicated by the value of P_{50}) is of the right magnitude to balance the requirements for both uptake and release of O_2. We can gain some insight into how this ideal binding affinity is achieved if we recall that the O_2 dissociation constant, K_d, is an equilibrium constant and therefore must be the ratio of two rate constants, k_{on} for the *association reaction* and k_{off} for the *dissociation reaction*. Recall that at equilibrium, where the rates of ligand binding and release are equal, $k_{on}[Mb][O_2] = k_{off}[MbO_2]$ and

$$K_d = \frac{[Mb][O_2]}{[MbO_2]} = \frac{k_{off}}{k_{on}} \qquad (7.7)$$

Thus, the oxygen affinity can be controlled at the molecular level by regulating the rates of binding or release.

Rapid kinetic studies of several myoglobin mutants, performed in the laboratory of John Olson, have revealed that the binding of O_2 is mediated in large part by the distal histidine. The imidazole side chain acts both as a gate, between the solvent environment and the heme iron, as well as a hydrogen-bonding discriminator for certain ligands. As described above, the ε—NH tautomer of the distal histidine can H-bond to an O_2 bound at the heme in myoglobin (and the α and β subunits of hemoglobin). Recent kinetic and high-resolution structural studies have shown that in Mb, this H-bond has a strength of ~15 kJ/mol, whereas in the subunits of hemoglobin it is somewhat weaker, at ~8 kJ/mol. To investigate further the role of the distal His in ligand binding and release, CO-myoglobin was studied. X-ray crystallography data collected after photolytic cleavage of the Fe-CO bond in CO-myoglobin reveal that the distal histidine undergoes relatively large movements within 100 ps (1ps = 10^{-12} seconds) after ligand photolysis (Figure 7.8a). This finding is consistent with its role as the "gate." Computer simulations also suggest that the major route in ligand dissociation from the protein is through a narrow tunnel, gated by the distal histidine. However, as shown in Figure 7.8c, ligands can also exit the protein at other sites—some of them quite distant from the heme-binding pocket! In fact, the ligand may spend some time

FIGURE 7.8

Dynamics of CO release by myoglobin. **(a)** Time-resolved X-ray diffraction data comparing the positions of atoms before (magenta) and 100 ps after (green) photolysis of the L29F mutant of Mb-CO (PDB ID: 2GOV). Where there is overlap the electron densities are shown in white. 100 ps after photolysis the CO appears to be located about 0.2 nm from the heme iron (green density inside the yellow circle numbered 1). The largest displacements (yellow arrows) in Mb are for side chains in the distal pocket: His64 and Phe29. **(b)** As in (a), but 3.16 μs after photolysis (PDB ID: 2G14). At this point the CO can be detected on the proximal side of the heme (within yellow circles labeled 4 and 5), and the distal pocket side chains have adopted positions similar to those in the deoxy-state. **(c)** Computational modeling of ligand migration pathways in Mb are shown in blue. The major pathways are near the heme (labeled 1-5) and the minor pathways are on the opposite side of the protein (labeled 6-9). Significant internal motions of the protein are required for a bound ligand to dissociate via the minor pathways.

(a, b) From *Science* 300:1944–1947, F. Schotte, M. Lim, T. A. Jackson, A. V. Smirnov, J. Soman, J. S. Olson, G. N. Phillips Jr., M. Wulff, and P. A. Anfinrud, Watching a protein as it functions with 150-ps time-resolved x-ray crystallography. © 2003. Reprinted with permission from AAAS; (c) Reprinted from *Proceedings of the National Academy of Sciences of the United States of America* 105:9204–9209, J. Z. Ruscio, D. Kumar, M. Shukla, M. G. Prisant, T. M. Murali, A. V. Onufriev, Atomic level computational identification of ligand migration pathways between solvent and binding site in myoglobin. © 2008 National Academy of Sciences, U.S.A.

(c)

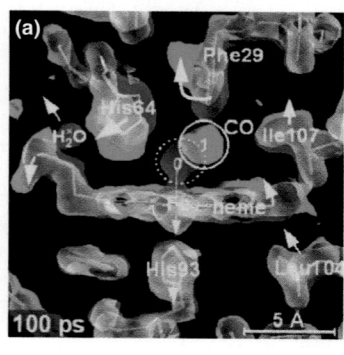

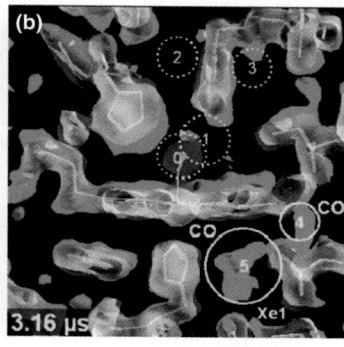

rattling in its protein cage—and perhaps rebinding at the heme iron—before it escapes to the solvent. This process is an explicit example of a principle set forth in the preceding chapter—the dynamic internal motions of globular protein molecules play important roles in the functions of proteins.

In summary, we observe in myoglobin a molecular structure that has been selected through evolution to produce an optimized environment for the binding *and* release of oxygen under physiological conditions of relatively low P_{O_2}. Recent evidence suggests that myoglobin plays another critical role in animal physiology, namely, protection from toxic levels of NO (nitric oxide) in cells. We will return to this topic later in the chapter, after we complete the discussion of O_2 transport.

> Dynamic motions of myoglobin facilitate ligand binding and release.

> Myoglobin has evolved to bind and release O_2 under conditions of relatively low oxygen concentration.

Oxygen Transport: Hemoglobin

All higher animals contain some kind of an oxygen transport protein. In vertebrates and some invertebrates this protein is hemoglobin. Nearly all hemoglobins are found to be multisubunit proteins, in contrast to the single-subunit myoglobins. Why should this be so? Investigating this question reveals new aspects of globin protein function.

> Higher animals use O_2-binding proteins to transport oxygen from lungs or gills to respiring tissues where it is needed to support metabolism.

Hemoglobin is not the only oxygen-binding protein used by animals. Most mollusks and some arthropods have a quite different protein, hemocyanin, which contains copper at the O_2-binding site. Still other invertebrates use an unrelated iron-containing protein called hemerythrin. Even the hemoglobins among the invertebrates are diverse in structure, ranging in size from dimers to complexes with 180 subunits (Figure 7.9). This diversity shows that the same function can often be arrived at by several independent evolutionary routes. In the following discussion we will focus on the tetrameric $\alpha_2\beta_2$ hemoglobin of higher vertebrates shown in Figure 7.3.

Cooperative Binding and Allostery

Consider the considerable demands placed on an O_2 transport protein. It must bind O_2 efficiently at the partial pressure found in lungs or gills (approximately 100 mm Hg) and then release a significant fraction of the O_2 to tissues. At rest,

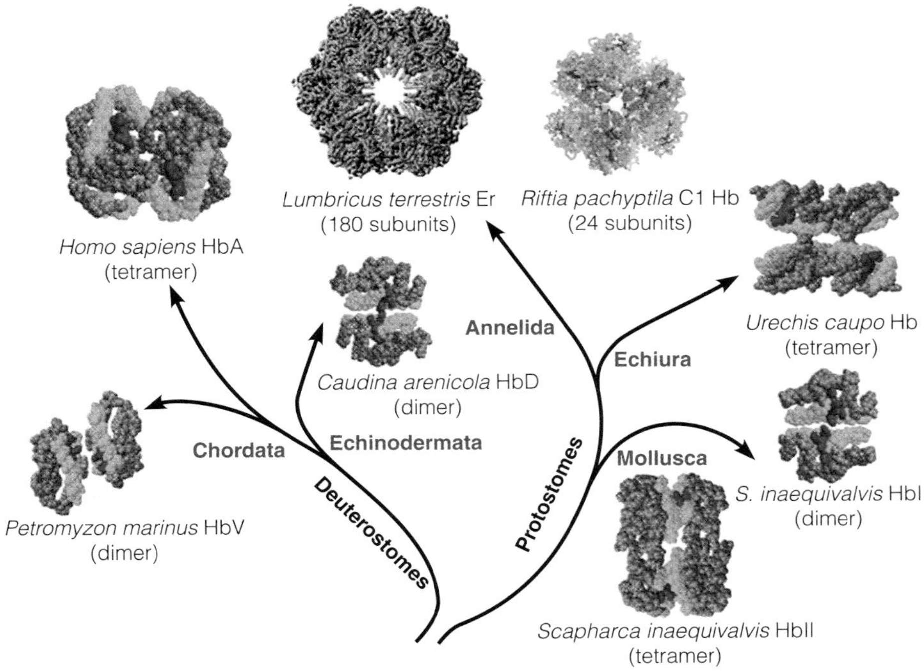

FIGURE 7.9

Diversity in hemoglobin structures. Space-filling models of Hb dimers and tetramers are shown with heme groups in *red*, E and F helices in *cyan*, and the rest of the main chain in *gray*. The main chains of the 24 subunits of *Riftia* C1 hemoglobin are shown in *green* and *blue* according to subunit type, with hemes in *red*. The Hb subunits of the 180-subunit *Lumbricus* erythrocruorin are rendered in *magenta* from a 5.5-Å electron density map with nonglobin linker chains in *blue* and *gold*. PDB IDs: human HbA (2hhb), deoxy lamprey HbV (3lhb), *Caudina* HbD (1hlm), *Riftia* C1 Hb (1yhu), *Urechis* Hb (1ith), *Scapharca* HbI (3sdh), and *Scapharca* HbII (1sct).

Journal of Biological Chemistry 280:27477–27480, We. J. R. Royer, H. Zhu, T. A. Gorr, J. F. Flores, and J. E. Knapp, Allosteric hemoglobin assembly: Diversity and similarity. Reprinted with permission.

Homo sapiens HbA (tetramer)

Lumbricus terrestris Er (180 subunits)

Riftia pachyptila C1 Hb (24 subunits)

Annelida

Urechis caupo Hb (tetramer)

Echiura

Caudina arenicola HbD (dimer)

Echinodermata

Chordata

Mollusca

S. inaequivalis HbI (dimer)

Deuterostomes

Protostomes

Petromyzon marinus HbV (dimer)

Scapharca inaequivalvis HbII (tetramer)

the P_{O_2} in capillaries is about 30 mm Hg and the difference in fractional saturation between the lungs and capillaries, ΔY_{O_2}, is roughly 0.4. Under conditions of extreme metabolic demand for O_2 (such as chasing prey or flight from predators), P_{O_2} can drop to 10 mm Hg and ΔY_{O_2} is roughly 0.85. In other words, to achieve optimum O_2 delivery to tissues, an ideal oxygen transport protein would be nearly saturated at 100 mm Hg, deliver sufficient O_2 to tissues at rest, yet maintain a significant O_2 reserve for periods of high demand. If the transport protein had a hyperbolic binding curve like that of myoglobin, such behavior would be impossible to achieve. With such a binding curve, the protein would be inefficient in supporting either basal or extreme metabolism. The advantages of enhanced aerobic capacity translate to greater survival for both predators and prey and provide the selective pressure for the evolution of optimized O_2 transporters.

Efficiency in O_2 transport is achieved by cooperative binding in multisite proteins, described by a sigmoidal binding curve.

This problem has been solved through the evolution of O_2 transport proteins, such as hemoglobin, that have the *sigmoidal* binding curve shown in Figure 7.10c. A sigmoidal curve is very efficient because it allows nearly full saturation of the protein in the lungs or gills, as well as optimal release in the capillaries. You can understand how such a curve can be possible by examining Figure 7.10d. At low oxygen pressures hemoglobin binds O_2 with low affinity;

FIGURE 7.10

Cooperative vs. noncooperative O_2-binding curves. These graphs show why an O_2 transport protein such as hemoglobin is more efficient if it switches cooperatively between lower and higher affinity states. The vertical blue and red bars represent P_{O_2} in capillaries and lungs, respectively. The difference in O_2 saturation of Hb in lungs vs. capillaries (ΔY_{O_2}) is represented by the gap between horizontal dashed lines. **(a)** If the transport protein were noncooperative and had a high O_2 affinity, ensuring saturation in the lungs, transfer of O_2 to myoglobin in the tissues would be inefficient. In this case ΔY_{O_2} is very small. **(b)** If a noncooperative transport protein had a lower O_2 affinity, it would efficiently transfer O_2 to myoglobin in tissues; but, it would not be saturated in the lungs. In this case ΔY_{O_2} is somewhat greater than it is for the high affinity noncooperative Hb. **(c)** A transport protein that binds efficiently in the lungs and unloads efficiently in the tissues requires a sigmoidal binding curve, and delivers the largest fraction of its bound O_2 (i.e., it has the largest value of ΔY_{O_2}). **(d)** The sigmoidal binding curve represents the transport protein's switch from a lower affinity state at low oxygen pressures to a higher affinity state at high oxygen pressures.

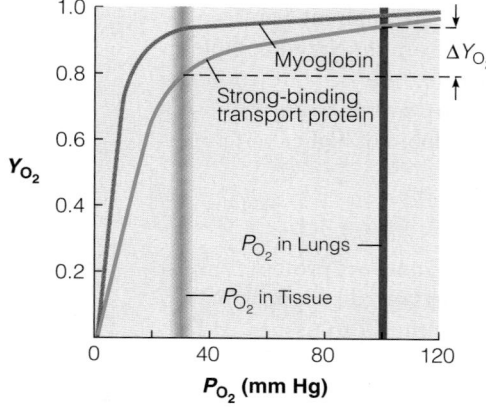

(a) Transport protein efficient in binding but inefficient in unloading (hyperbolic binding curve).

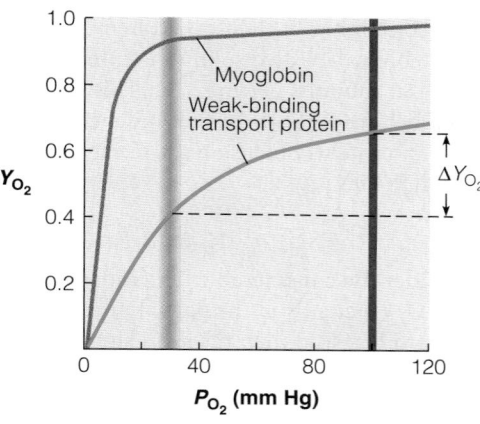

(b) Transport protein efficient in unloading but inefficient in binding (hyperbolic binding curve).

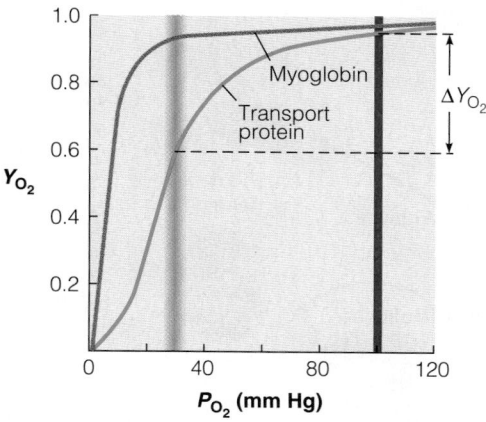

(c) Transport protein efficient in both binding and unloading because it can switch between higher and lower affinity states (sigmoidal binding curve).

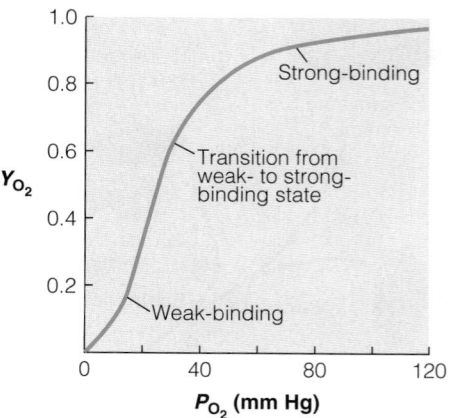

(d) Switch from lower to higher affinity states yields the sigmoidal curve.

but, as more oxygen is bound, the affinity for oxygen becomes greater. This behavior must mean that a *cooperative interaction* exists among O_2-binding sites. Ligand binding to the first empty site somehow increases the O_2-binding affinity of the remaining sites, thus promoting complete saturation of the protein with O_2. We can also express this idea the other way around, by saying that dissociation of one O_2 from the protein makes it easier for the remaining O_2 to dissociate, thus promoting the fully deoxygenated state. This behavior requires some kind of intramolecular communication between the binding sites. A single-subunit protein, such as myoglobin, cannot achieve this sort of modulation of ligand-binding affinity; however, such communication *is* possible between the subunits of a multisubunit protein, such as hemoglobin. Thus, the answer to the question "Why are hemoglobins multisubunit proteins?" lies in the evolutionary advantages conferred by cooperative switching between high- and low-affinity states.

Vertebrate hemoglobin has evolved from the *monomeric* structure of myoglobin into the *tetrameric* structure shown in Figure 7.3. Hemoglobin can bind four O_2 molecules—one in each of the four subunits. Although each of the subunits has primary, secondary, and tertiary structures similar to those of myoglobin, the amino acid side chains in hemoglobin also provide other necessary interactions—salt bridges, hydrogen bonds, and hydrophobic interactions—to stabilize a particular quaternary structure.

The functional difference between hemoglobin and myoglobin lies in the cooperativity exhibited by the ligand-binding sites in hemoglobin. This cooperativity is possible because the oxygenation state (filled or empty) of one site can be communicated to another.

As described later in this chapter, there is a structural basis for the cooperative ligand binding in hemoglobin: the lower-affinity state has a protein conformation that is distinct from that of the higher-affinity state. When binding curves are represented as in Figure 7.10c or 7.10d, the important parameters that describe the ligand binding (i.e., the difference in binding affinities of the conformational states and the degree of interaction between sites) are not obvious. A rearrangement of Equation 7.6 provides a simple method to derive some of these parameters from ligand-binding data. If we calculate the quantity $Y_{O_2}/(1 - Y_{O_2})$, we obtain

$$\frac{Y_{O_2}}{1 - Y_{O_2}} = \frac{P_{O_2}}{P_{50}} \tag{7.8}$$

Or, taking logarithms of both sides,

$$\log\left(\frac{Y_{O_2}}{1 - Y_{O_2}}\right) = \log P_{O_2} - \log P_{50} \tag{7.9}$$

Graphing $\log[Y_{O_2}/(1 - Y_{O_2})]$ versus $\log P_{O_2}$ produces what is called a **Hill plot** (Figure 7.11), after Archibald Hill, who proposed in 1910 that the binding of O_2 by hemoglobin could be empirically described by the function

$$Y_{O_2} = \frac{P_{O_2}^h}{P_{50}^h + P_{O_2}^h} \tag{7.10}$$

This gives the general form of the **Hill equation:**

$$\log\left(\frac{Y_{O_2}}{1 - Y_{O_2}}\right) = h \log P_{O_2} - h \log P_{50} \tag{7.11}$$

In the original analysis made by Hill (Equations 7.10 and 7.11), he gave no special physical meaning to the parameter **h**, which is called the **Hill coefficient.** We now know that, in fact, h is related to the number of ligand-binding sites (n) and the energy of their interaction, such that the value of h approaches the value n

Cooperativity in binding requires communication between binding sites.

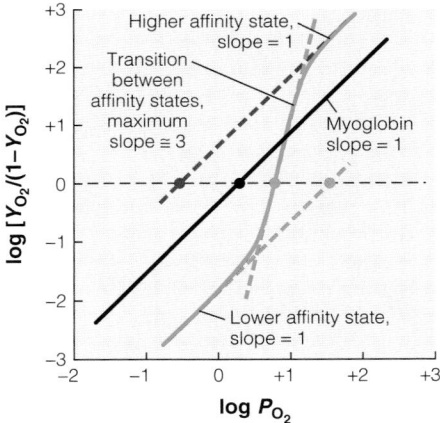

FIGURE 7.11

Hill plots of oxygen binding for myoglobin and hemoglobin under physiological conditions. The circles show the P_{50} of each protein and binding state. The plot for myoglobin, which binds oxygen noncooperatively, is a solid black line with a slope of 1. The plot for hemoglobin (in green), which binds cooperatively, shows the switch from a lower-affinity state (larger P_{50}; blue circle) to a higher-affinity state (smaller P_{50}; red circle) and has a Hill coefficient of about 3. Values of the ligand binding affinities are derived from extrapolation of the upper and lower arms of the binding curve to the x-axis at $\log[Y_{O_2}/(1 - Y_{O_2})] = 0$. The Hill coefficient h is determined by the slope at half-saturation (the green circle in the figure).

as the interaction between sites increases. Note that in the case of myoglobin, $h = 1$, and the Hill plot (Figure 7.11, black curve) yields a straight line as predicted by Equation 7.11. The Hill plot for hemoglobin does not yield a straight line as predicted by Equation 7.11—so the model doesn't fit the data. In 1925, Gilbert Adair published a more rigorous treatment of cooperative ligand binding to hemoglobin that allows for *sequential* binding of ligand with affinities for each site that are not equal and which generally increase as more sites are bound.

In spite of its limitations, the Hill equation is still widely used to analyze ligand binding to hemoglobin because it fits the data in the transition region well, and it yields values for parameters that can be used to compare the ligand binding behaviors of different hemoglobins (e.g., comparing normal to mutant hemoglobin). As shown in Figure 7.11, these parameters are derived from the Hill plot and include the value of P_{50}, estimates of P_{50} for the lower and higher affinity states, and a qualitative description of the degree of cooperativity in terms of the Hill coefficient.

On a Hill plot, the distinction between cooperative and noncooperative systems is clear. As predicted by the Hill equation, the Hill plot for noncooperative binding yields a straight line, with a slope $= 1$. Now consider the Hill plot for a cooperative ligand-binding protein such as hemoglobin. In this case, the binding curve is sigmoidal rather than linear. When deoxyhemoglobin binds the first O_2 molecule (at low P_{O_2}), its Hill plot has a slope $\cong 1$ and describes binding to the lower-affinity state (characterized by a larger P_{50} value). As progressively more O_2 binds, the curve switches over to approach another, parallel, straight line that describes binding to the higher-affinity state (which has a lower value of P_{50}). For both cooperative and noncooperative systems, the Hill plot gives the value of h as the slope at $\log[Y_{O_2}/(1 - Y_{O_2})] = 0$. Four cases may be considered for a molecule with n binding sites:

1. $h = 1$: There is no interaction between the sites; thus, the molecule binds ligands noncooperatively (e.g., as for myoglobin in Figure 7.11). This situation may also be observed for a multisite protein if the binding sites do not interact with one another.

2. $1 < h < n$: There is interaction between the sites. This situation is the usual one for a protein that binds ligands with so-called positive cooperativity, as depicted for hemoglobin in Figure 7.11. The Hill coefficient must be greater than unity for the curve to switch over from the lower-affinity line to the higher-affinity line.

3. $h = n$: The energy of interaction between sites approaches infinity. In this hypothetical situation the molecule is wholly, or *infinitely*, cooperative. In such a situation, only wholly unliganded and wholly liganded molecules would be present at any point in the binding process. This represents the upper limit on cooperativity and is not observed in real proteins. For example, the Hill coefficient measured under physiological conditions for hemoglobin ($n = 4$) is ~3.

4. $h < 1$: In this case, ligand binding at one site *reduces* binding affinity at other binding sites and is called "negative cooperativity." This situation is also predicted when a protein contains noninteracting sites with different binding affinities; thus, well-documented examples of true negative cooperativity are rare.

The cooperative binding of oxygen by hemoglobin is one example of what is referred to as an **allosteric** effect. In allosteric binding, the uptake of one ligand by a protein influences the affinities of remaining unfilled binding sites. The ligands may be of the same kind, as in the case of O_2 binding to hemoglobin, or they may be different. As discussed in Chapter 11, allostery is also an important mechanism for regulating the activity of enzymes.

A Hill plot can distinguish cooperative ligand binding from noncooperative binding. For cooperative binding, the value of the Hill coefficient is greater than one. For noncooperative binding, the Hill coefficient is equal to one.

Models for the Allosteric Change in Hemoglobin

How do allosteric transitions from lower-affinity binding states to higher-affinity binding states actually occur? A number of theories have been developed to describe the allosteric transitions in hemoglobin. These can be grouped into four classes.

1. *Sequential models:* The prototype for such models is one proposed by Gilbert Adair in 1925, then further developed in 1966 by Koshland, Nemethy, and Filmer (KNF; Figure 7.12a). The Adair-KNF model assumes that the subunits can change their tertiary conformation one at a time in response to the binding of oxygen. Positive cooperativity arises because the binding of O_2 in one subunit favors the higher-affinity conformational state in adjacent subunits whose sites are not yet filled. Thus, as oxygenation progresses, almost all the sites adopt the higher-affinity conformation. Such models are characterized by the existence of molecules containing subunits in both high-affinity and low-affinity conformational states.

2. *Concerted, or Symmetry, models:* At the opposite extreme lies the theory of Monod, Wyman, and Changeux published in 1965 (MWC; Figure 7.12b). According to the MWC model, the hemoglobin tetramer exists in an equilibrium between two distinct quaternary conformations. In the deoxy state, all subunits in each molecule are in the lower-affinity conformation (also called the T state), and in the oxy state, all are in the higher-affinity conformation (also called the R state). The symbols T and R stand for "tense" and "relaxed"; the significance of this will be seen in the next section. An

FIGURE 7.12

Two classical models for the cooperative ligand binding by hemoglobin. **(a)** The Koshland, Nemethy, and Filmer (KNF) model. As each subunit binds a ligand, it promotes a change in an adjacent subunit to the higher-affinity conformation. **(b)** The Monod, Wyman, and Changeux (MWC) model. The entire molecule has two different quaternary states—tense (T) and relaxed (R)—which are in equilibrium. Binding of ligands shifts the equilibrium toward the higher-affinity (R) state.

 Hb subunit in lower-affinity state

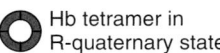 Hb subunit in higher-affinity state

Hb tetramer in T-quaternary state

Hb tetramer in R-quaternary state

O_2 bound to Hb subunit in lower-affinity state

O_2 bound to Hb subunit in higher-affinity state

● O_2 molecule

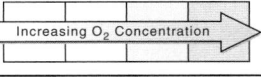 Increasing O_2 Concentration

(a) KNF (sequential) Model

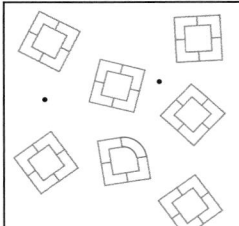 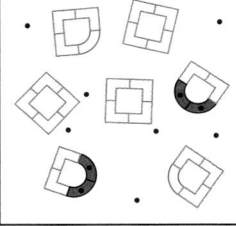 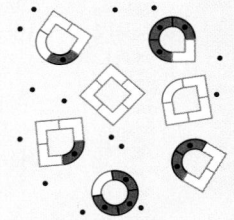 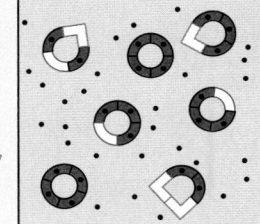

No O_2 bound. Almost all subunits are in the lower-affinity state. Few are in the higher-affinity state.

At low O_2 concentrations, most subunits are in the lower-affinity state. O_2 binding at one subunit favors the transition to the higher-affinity state in an adjacent subunit.

As O_2 concentration increases, more of the higher-affinity state subunits bind ligand and promote the transition to the higher-affinity state in adjacent subunits.

At high O_2 concentrations, the Hb approches saturation. Almost all ligand binding sites are filled and in the higher-affinity conformational state.

(b) MWC (symmetry) Model

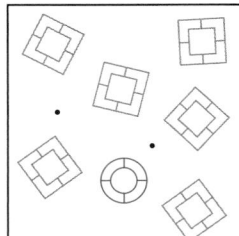 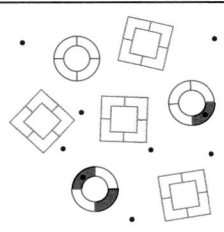 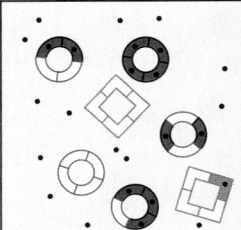 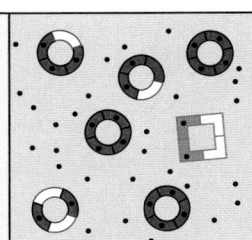

No O_2 bound. Almost all Hb tetramers are in the T-state. Only a few are in the R-state.

At low O_2 concentrations, most O_2 is bound to the higher-affinity R-state tetramers. Ligand binding shifts the T ⇌ R equilibrium toward the R-state.

As O_2 concentration increases, more of the tetramers are in the R-state. Note that the T-state tetramers can also bind O_2 but do so with lower affinity.

At high O_2 concentrations, the Hb approaches saturation. Almost all Hb tetramers are in the higher-affinity R-state.

Hemoglobin switches between conformational states with lower and higher O_2-binding affinities. In the O_2-rich environment of the lungs or gills the higher-affinity state is favored, and oxygen binds to hemoglobin. In the O_2-poor environment of respiring tissues the lower-affinity state is favored, and oxygen is released from hemoglobin.

Vertebrate hemoglobins are tetramers $(\alpha_2\beta_2)$ made up of two kinds of myoglobin-like chains.

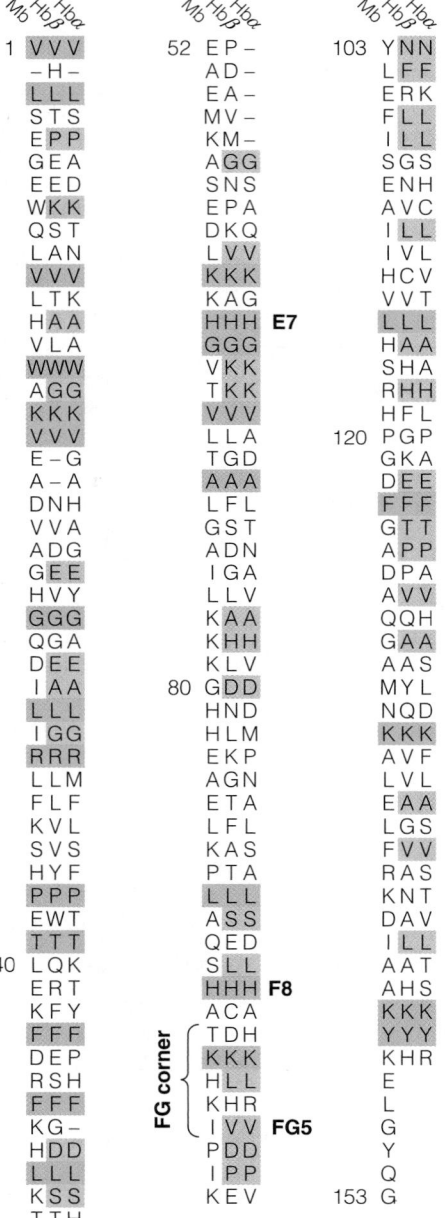

equilibrium between these states is presumed to exist, and ligand binding shifts that equilibrium toward the R state. The shift is a *concerted* one so that molecules with some subunits in the T state and some in the R state are specifically excluded. Historically, this has been the most widely accepted model for allostery in hemoglobin.

3. *Multistate models:* It has become clear since the early 1990s that neither the KNF nor the MWC model completely explains the allosteric behavior of proteins, including hemoglobin. Consequently, more complex models have been devised, though most retain some elements of the KNF and/or MWC models.

4. *Dynamics models:* A completely different proposal for allostery attributes changes in functional properties to changes in the dynamic behavior of proteins, rather than conformational changes *per se*.

Multistate and dynamic models of hemoglobin allostery are discussed further in the Appendix at the end of this chapter.

Changes in Hemoglobin Structure Accompanying Oxygen Binding

To understand the different models for allosteric behavior in hemoglobin, it is necessary to examine the protein structure in more detail. The hemoglobins of higher vertebrates are made up of two types of subunits, referred to as α and β. The primary structures of human hemoglobins are compared with that of sperm whale myoglobin in Figure 7.13. As you can see, the human α and β sequences have 44% amino acid identity to one another and 18% identity to the sequence of whale myoglobin. Essential residues, like the proximal and distal histidines (F8 and E7, respectively), are strictly conserved, and those critical to the tertiary structure appear to be conserved as well, for the hemoglobin subunits and myoglobin all have very similar tertiary structures. As a rule of thumb, sequences with greater than 30% amino acid identity are predicted to have the same tertiary structure. The hemoglobin molecule contains two copies of each type of subunit, so the whole molecule can be described as an $\alpha_2\beta_2$ tetramer. The subunits are placed in a roughly tetrahedral arrangement as shown schematically in Figure 7.3. When hemoglobin is dissolved in mildly denaturing solutions, it dissociates into $\alpha\beta$ dimers, suggesting that the strongest subunit contacts are between α and β chains, rather than $\alpha-\alpha$ or $\beta-\beta$. In other words, the molecule can be thought of as a dimer of $\alpha\beta$ dimers. Figure 7.3 also shows that the hemes, with their O_2 binding sites, are all close to the surface but *not* close to one another. Therefore, we cannot seek the source of cooperative binding in anything as simple as direct heme–heme interaction.

General features of the protein conformational changes that occur upon ligand binding/release can be seen in Figure 7.14, which illustrates two views of the crystal structural differences between the deoxygenated state and the fully oxygenated hemoglobin molecule. Early X-ray diffraction studies suggested that these differences arise primarily as a result of changes in the quaternary structure, accompanied by much smaller tertiary structure changes. One $\alpha\beta$ dimer pair rotates 15 degrees and slides with respect to the other, as seen in Figure 7.14a. Upon O_2 binding, this movement brings the β subunits closer

FIGURE 7.13

Comparison of sequences of myoglobin and the α and β subunits of hemoglobin. The aligned sequences are those of whale myoglobin and the two human hemoglobin subunits. Gaps (indicated by dashes) have been inserted where necessary to provide maximum alignment of the sequences; the residue numbers to the left of the chains are for the myoglobin sequence. A residue critical to the functioning of these proteins is indicated to the right of the sequences; F8 and E7 are the proximal and distal histidines, respectively (see Figure 7.5). Brown indicates the residues that are identical in all three sequences, and purple the residues identical in both hemoglobin sequences.

together and narrows a central cavity in the molecule, as can be seen in Figure 7.14b. To a first approximation, then, we can regard the hemoglobin molecule as having two quaternary states, one characteristic of the lower-affinity deoxy conformation (the T state) and the other favored by the higher-affinity oxy conformation (the R state).

The MWC view of hemoglobin allostery interprets the Hill plot shown in Figure 7.11 in terms of a *concerted* shift between these two protein quaternary conformations. Wholly deoxygenated hemoglobin molecules are predominantly in the T state conformation, so O_2 binding initially occurs along the line corresponding to the lower-affinity state. The key feature of the MWC model is that the switch between T and R states occurs over a narrow O_2 concentration range, where oxygenation favors transition to the higher-affinity R state. The remaining unliganded sites are in hemoglobin molecules that have predominantly the R-state conformation; thus, the binding curve switches to the line for the higher-affinity state. This leaves, however, many important questions unanswered. What triggers the change? Which models, if any, suffice to explain the results?

A Closer Look at the Allosteric Change in Hemoglobin

A stereochemical, nonquantitative mechanism to explain the cooperativity in oxygen binding was proposed in 1970 by M. F. Perutz, founder of the Laboratory of Molecular Biology at Cambridge University, a pioneer in the field of protein X-ray crystallography and Nobel laureate. Perutz and his colleagues solved crystal structures of both the deoxy and the oxy states of hemoglobin and then used these structures as the basis for a model of allostery in hemoglobin. Given the symmetries of the deoxy-hemoglobin and oxy-hemoglobin structures, it was reasonable for Perutz to describe his stereochemical model of hemoglobin allostery in terms of the symmetric MWC model. In the MWC model, the more reactive (i.e., higher affinity) conformation is called the *relaxed* (R) state and the less reactive (lower affinity) quaternary conformation is the *tense* (T) state. Before we present the Perutz model of allostery, let us consider some of the important differences between the T and R crystal structures.

The transition from the deoxy to the oxy conformation involves significant changes in the subunit–subunit interactions. Some understanding of this process can be gained from a closer study of Figure 7.14b. Note the region, to the lower left, where the β_2 subunit interacts with the α_1 chain. In the deoxy form, the C-terminus of β_2 (residue 146) lies atop the C helix of α_1 (residues 36–42) and is held in this position by a network of hydrogen bonds and salt bridges. His 97 in the FG corner of β_2 is pushed against the CD corner of α_1, between Thr 41 and Pro 44. In the oxy form, rotation and sliding of the subunits pulls the C-termini of β chains away from α contacts (Figure 7.14b). The salt bridges and hydrogen bonds holding the C-terminus in the deoxy state (see Figure 7.15) are broken, and His 97 of β_2 now lies between Thr 38 and Thr 41 of α_1. Because of the symmetry of the structure, an exactly equivalent set of changes occurs at the $\alpha_2\beta_1$ interface. The molecule "switches" into a new set of interactions in the oxy state whereby a number of strong interactions in the deoxy state must be broken.

Figure 7.15 shows more details of the key salt bridge interactions that are broken going from the T state to the R state conformation. In the T state each of the β–globin C-termini (His146) makes two salt bridges: an inter-subunit interaction between the His146 C-terminal $-COO^-$ and the $-NH_3^+$ side chain of Lys40 on a nearby α–globin, and an intra-subunit interaction between the *protonated* His146 side chain and the $-COO^-$ of Asp94. The latter interaction plays a major role in stabilizing the T state under conditions of low pH, where protonation of the His146 side chain is favored (see the *Bohr Effect* below). In the T state each α–globin C-terminus (Arg141) is involved in four critical inter-subunit interactions.

R-state hemoglobin has a higher O_2-binding affinity (lower P_{50}). T-state hemoglobin has a lower O_2-binding affinity (higher P_{50}).

Oxygenation causes hemoglobin quaternary structure to change: One $\alpha\beta$ dimer rotates and slides with respect to the other.

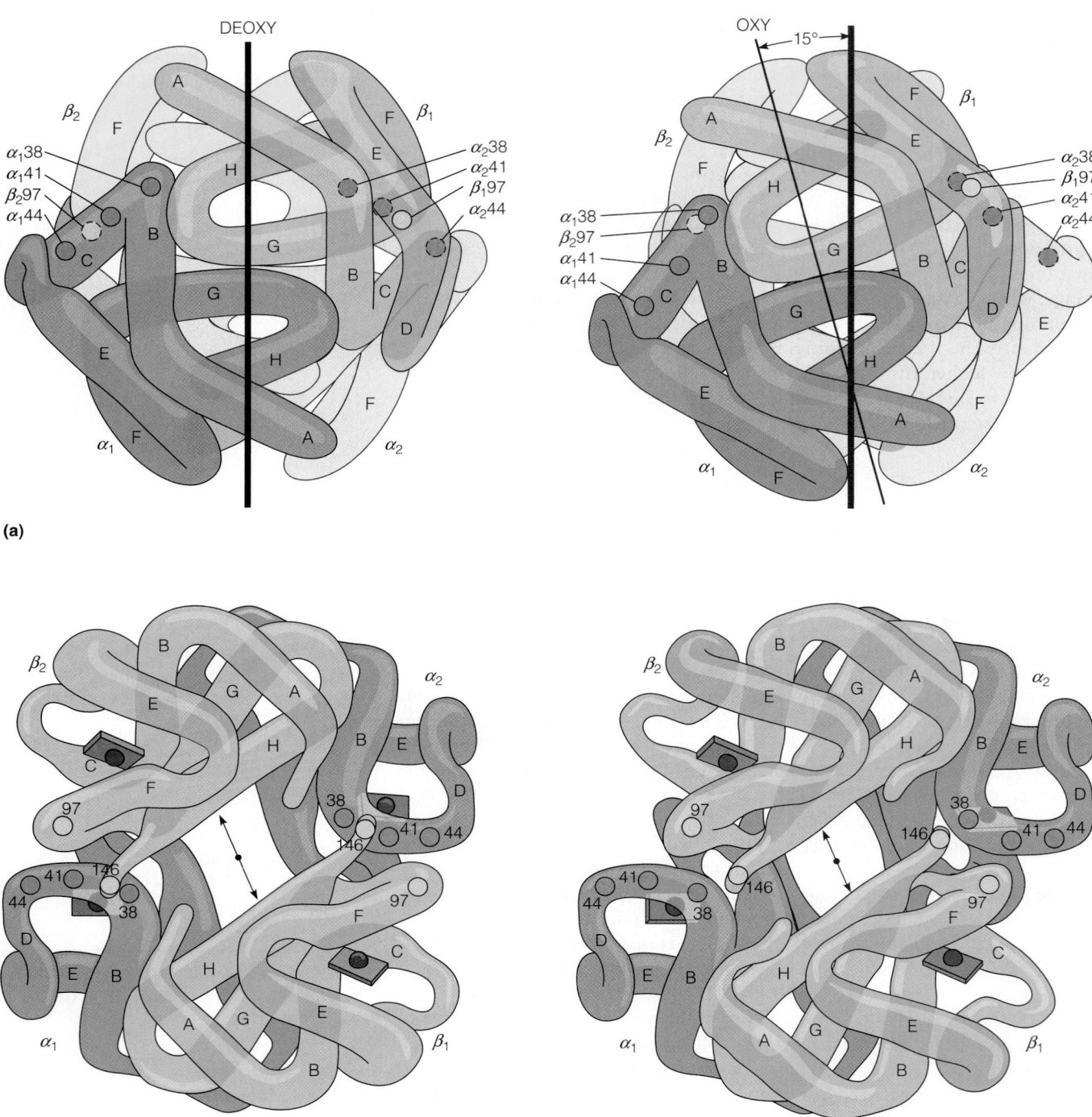

FIGURE 7.14

The change in hemoglobin quaternary structure during oxygenation. (a) The transition viewed along an axis perpendicular to the two-fold axis, with the $\alpha_1\beta_1$ dimer (darker blue and red areas) in front of the $\alpha_2\beta_2$ dimer. Deoxyhemoglobin is shown on the left, and oxyhemoglobin on the right. Note the rotation of $\alpha_1\beta_1$ with respect to $\alpha_2\beta_2$ and the shift of $\beta97$ with respect to $\alpha41$ and $\alpha44$. The rotation of about 15° is accompanied by sliding because the center of rotation is not centrally located. **(b)** Top views of hemoglobin, looking down the two-fold axis (dot in center). The two β subunits are in the foreground; α subunits are in the background. Note in deoxyhemoglobin that the central cavity is broad and that residue 97 on one β chain lies between residues 41 and 44 on the adjacent α chain. The shift from the deoxy to the oxy state is evident by the shrinkage of the central cavity and the shift in the contact of residue $\beta97$ with the α chain.

The C-terminal $—COO^-$ of Arg141 of one α globin interacts with the $—NH_4^+$ of Lys127 on the other α globin chain. The guanidinium side chain of Arg141 forms a salt bridge with the $—COO^-$ of Asp126 and a hydrogen bond with the amide carbonyl group of Val34 on a nearby β globin. Finally, the Arg141 side chain makes a bridging inter-subunit interaction with a chloride ion and the N-terminal $—NH_3^+$ of Val1. As described in the next section, increasing $[Cl^-]$ also stabilizes the T state by promoting this bridging interaction.

All of the interactions shown in the lower left of Figure 7.15 are disrupted when hemoglobin switches from the T to the R state (see lower right of Figure 7.15). These interactions are enthalpically stabilizing; thus, in the absence of bound ligand, the T state is favored thermodynamically over the R state (Figure 7.16). The thermodynamic price for switching to the R state (which requires breaking the bonding interactions in the T state) is paid by the energy provided by binding of O_2

The energetic cost of breaking stabilizing interactions in the deoxy state is paid by the formation of $Fe—O_2$ bonds in the oxy state.

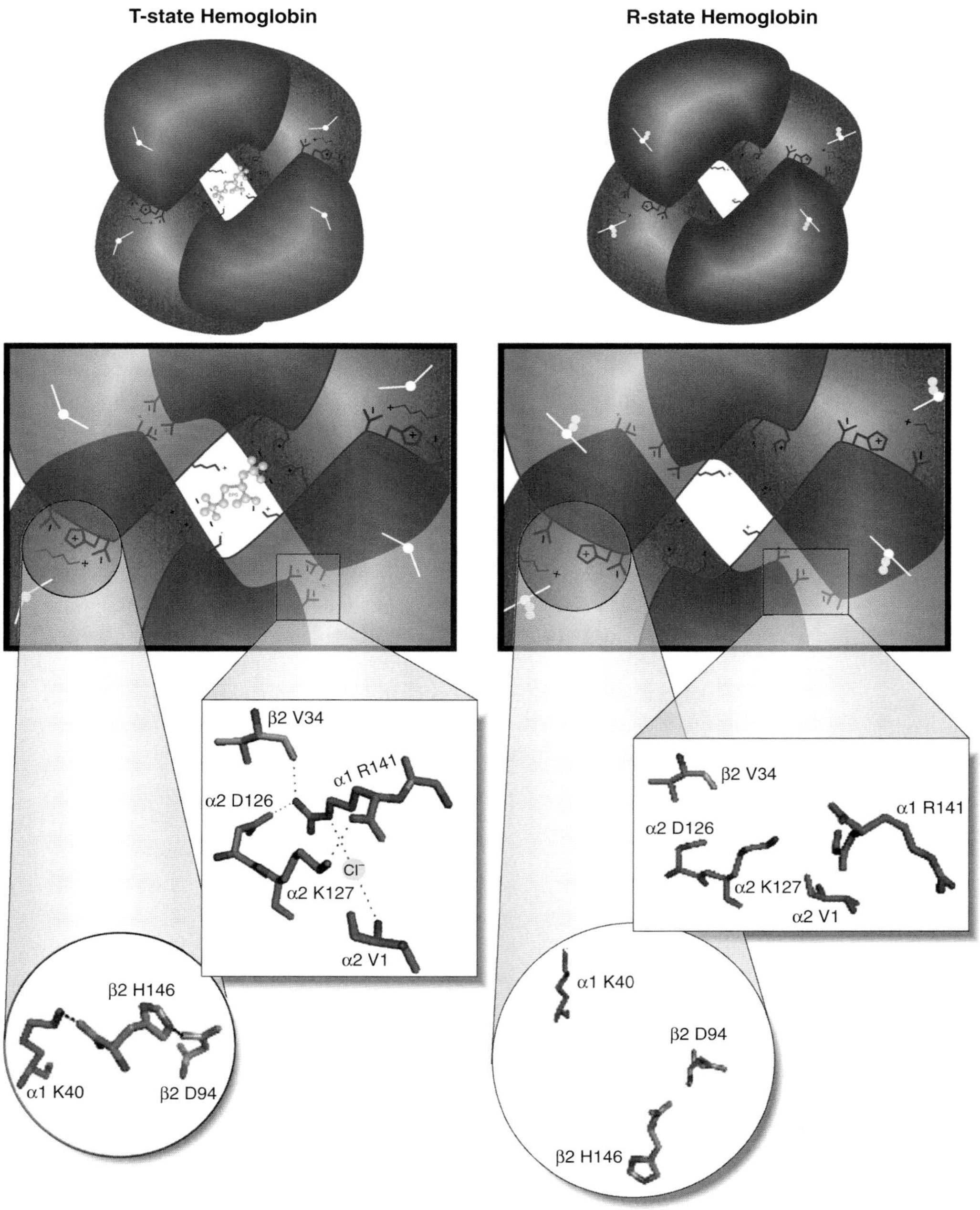

T-state Hemoglobin

R-state Hemoglobin

β2 V34
α2 D126
α1 R141
Cl⁻
α2 K127
α2 V1

β2 V34
α2 D126
α1 R141
α2 K127
α2 V1

β2 H146
α1 K40
β2 D94

α1 K40
β2 D94
β2 H146

FIGURE 7.15

Key salt bridge interactions disrupted during the switch between T and R state hemoglobin quaternary structures. Top: A schematic drawing of the four subunits in hemoglobin. β-globins are blue, α-globins are red, each heme is colored white, 2,3-BPG (see next section) is yellow and shown in a "ball and stick" rendering. Middle: Zoom showing a highly schematic drawing of the key salt bridges that are broken going between the T and R state crystal structures. Here the β-globins (blue) are transparent so interactions at the α-globin (red) N-termini can be visualized. Bottom: Zoom showing crystal structure data of the key interactions at the β-termini (circles) and α-termini (boxes). See the text for a detailed description of the residues involved. Salt bridge interactions in the T state are indicated by black dashed lines. All of these interactions are disrupted during the transition to the R state. Side chains of β-globins are highlighted in blue and those of α-globins are highlighted in magenta. The chloride ion that bridges the C-terminus of one α-globin (α_1 R141) to the N-terminus of the other α-globin (α_2 V1) is shown in orange. Note that identical sets of symmetry-related interactions occur at the other α-globin and β-globin termini. PDB IDs: T state, 1HGA; R state, 1BBB.

Courtesy of Gary Carlton.

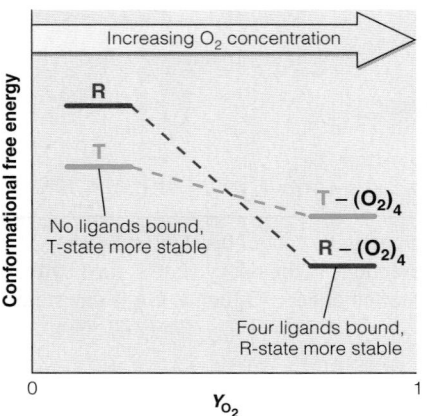

FIGURE 7.16

A simplified view of ligand binding and conformation energies in hemoglobin. The deoxy (T) conformation is favored when no ligands are bound, due to the increased number of noncovalent interactions in the T state (see Figure 7.15). As Y_{O_2} increases (i.e., more ligands are bound) the energy provided by formation of the Fe—O_2 bond stabilizes the R conformation relative to the T conformation.

FIGURE 7.17

The essential features of the "Perutz mechanism" of the T $\longrightarrow$ R transition in hemoglobin.
The binding of oxygen to deoxyhemoglobin causes conformational changes in the heme. **(a)** In the deoxy state, heme has a dome shape, exaggerated in this figure. **(b)** Binding of the O_2 ligand pulls the iron into the heme plane, flattening the heme and causing strain. **(c)** A shift in the orientation of His F8 relieves the strain, partly because Val FG5 is pushed to the right. In this way, the tertiary change in heme is communicated to the FG corner.

to the protein-bound heme iron ion. When the O_2 is released, the hemoglobin reverts to the lower-energy deoxy conformation.

Exactly how is the energy of O_2 binding communicated to induce this conformational switching? The essential features of the model proposed by Perutz are shown in Figure 7.17, which shows the relationship of His F8 and the neighboring Val (FG5) to the heme in deoxy- and oxyhemoglobins. The figure includes an important fact not mentioned previously: Not only is the iron ion in the deoxy conformation a bit outside the heme plane, but also the heme itself is not quite flat; it is distorted into a dome shape. Furthermore, in both deoxymyoglobin and deoxyhemoglobin, the axis of His F8 is not exactly perpendicular to the heme but is tilted by about 8°. When oxygen binds to the other side, it pulls the iron ion a short distance down into the heme and flattens the heme (Figure 7.17b,c). This change cannot happen without molecular rearrangement because such motion would bring both the ε-hydrogen of His F8 and the side chain of Val FG5 too close to the heme. Thus, the histidine shifts its orientation toward the perpendicular, pulling on the F helix and the FG corner as it does so. This movement in turn distorts and weakens the complex of H-bonds and salt bridges that connect FG corners of one subunit with C helices of another. Consequently, the conformational rearrangements shown in Figures 7.14 and 7.15 occur.

In the simplest terms, the binding of O_2 pulls the Fe^{2+} a fraction of a nanometer into the heme, which levers the F helix and thereby produces a much larger shift in the surrounding protein structure, particularly at the critical α–β interfaces (which are 1.7–2.3 nanometers distant from the heme). This remodeling of the α–β interfaces provides the physical link between the ligand-binding sites, which explains the observed cooperativity.

This mechanism to explain the cooperativity in oxygen binding was proposed by Perutz based on the insights he gained from identifying differences between the deoxy- and oxyhemoglobin crystal structures. As such it represents a brilliant example of the application of structural biology to explain the physiologically relevant behavior of a protein. But does it correspond to reality? Site-directed mutagenesis studies (see Tools of Biochemistry 4B) and rapid ligand binding studies (see Tools of Biochemistry 11A) have allowed many features of the Perutz mechanism to be tested, and although it is not entirely correct, much experimental data supports the general features of the model and it remains widely accepted. For example, Barrick, et al. (see References) utilized site-directed mutagenesis to replace the proximal histidine residues with glycines in α and β chains. The mutant proteins were then studied in the presence of 10 mM imidazole; the imidazole molecule can substitute for the proximal histidine side chain and bind the heme iron *but is not covalently linked to the F helix* (see Figure 7.18). As a consequence, although oxygen binding can still flatten the heme, it does not move the

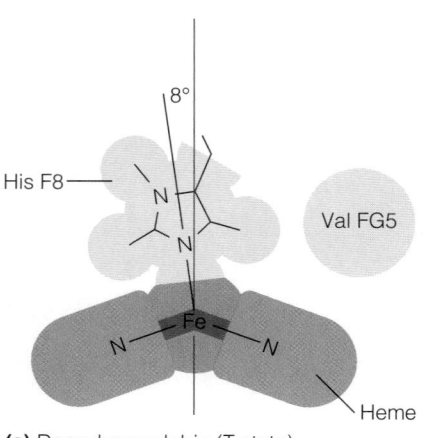

(a) Deoxyhemoglobin (T state)

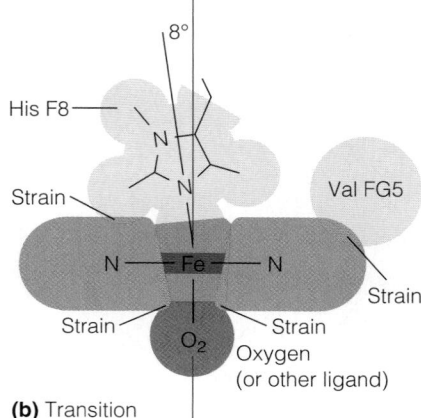

(b) Transition

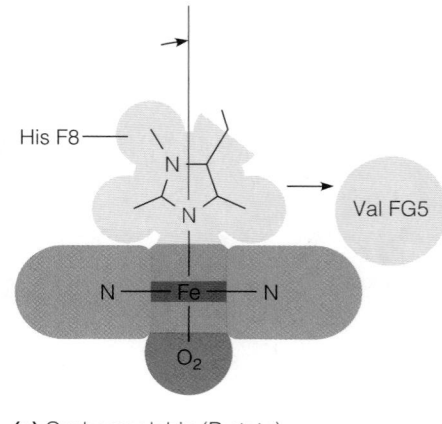

(c) Oxyhemoglobin (R state)

F helix. Barrick, et al. observed that cooperativity in binding is largely lost, and although the subunits appear to remain in the T state, the ligand-binding affinity of the mutant is *increased* compared to wild-type hemoglobin. These findings support the main features of the Perutz model and demonstrate that the proximal histidine plays a significant role in hemoglobin cooperativity. The observation that cooperativity is not completely abolished in these hemoglobin mutants suggests that other features of hemoglobin structure, perhaps residues in the distal heme pocket, also contribute to cooperative ligand binding.

The mechanism of hemoglobin allostery remains a matter of vigorous debate (see Appendix). However, to discuss the effects of specific allosteric effectors on hemoglobin function, we will continue to use the central ideas of the Perutz and MWC models presented above. We do this for two reasons: First, the general features of these models explain much of the allosteric behavior in hemoglobin. Second, these are the models to which all others are compared; thus, any attempt to follow the criticisms/refinements of these models requires familiarity with the original proposals.

Allosteric Effectors of Hemoglobin

Cooperative binding and transport of oxygen are only part of the allosteric behavior of hemoglobin. The details of animal physiology impose further demands. For example, as oxygen is utilized in tissues, carbon dioxide is produced and must be transported back to the lungs or gills. As described in Chapter 2 (page 44), accumulation of CO_2 also lowers the pH in erythrocytes through the *bicarbonate reaction,*

$$CO_2 + H_2O \rightleftharpoons HCO_3^- + H^+ \qquad (7.12)$$

This reaction in erythrocytes is catalyzed by the enzyme *carbonic anhydrase.* At the same time, the high demand for oxygen, especially in muscle involved in vigorous activity, can result in oxygen deficit, or **hypoxia**. As we shall see in Chapter 13, a consequence of this deficit is the production of lactic acid, which also lowers the pH. The falling pH in tissue and venous blood signals a demand for more oxygen delivery. Hemoglobin functions efficiently to meet these requirements. It does so through its allosteric transition between high-affinity oxy (R) and low-affinity deoxy (T) states.

As we shall see throughout this text, allostery is a general mechanism for the regulation of many important proteins; thus, we give here some general definitions of terms related to allostery. A given protein may bind specifically to several different molecules in more than one location on that protein's surface. The **active site** of a protein is where the protein must bind one or more **substrate** molecules to carry out its primary function. We can view the heme pocket as the "active site" of hemoglobin because that is where the O_2 ligand is bound. In addition to the active site, there may be other sites, called **regulatory sites**, which bind specifically to molecules that regulate the function of the protein. The molecules that regulate protein function in this way are called **effectors**, and they typically exert their effects through an allosteric mechanism.

There are many ways to characterize allosteric effectors. Those that increase protein activity are called **positive effectors**, and those that decrease activity are **negative effectors**. Effectors can also be differentiated by the site on the protein to which they bind. We have seen that O_2 binding *at the heme* has a positive cooperative effect on subsequent O_2 binding events. Because O_2 affects its own binding, by binding to the protein active site, it is called a **homotropic** effector. Effectors that bind at regulatory sites (which are typically distant from the active site) are called **heterotropic** effectors.

We now discuss the effects of four heterotropic negative effectors of hemoglobin: H^+, CO_2, Cl^-, and 2,3-BPG.

In the Perutz mechanism, a small movement of the heme iron upon O_2-binding is translated into a larger movement of the F helix by the covalent connection between the F helix and the proximal histidine.

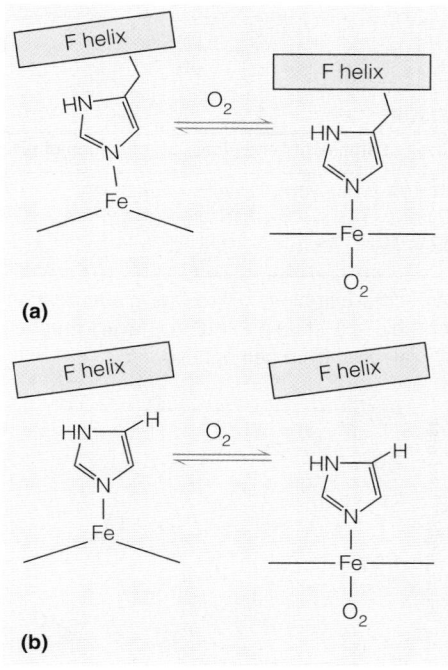

FIGURE 7.18

The effect of replacing the proximal histidine in hemoglobin with a glycine residue and adding a noncovalently bonded imidazole. **(a)** The effect of O_2 binding according to the Perutz model: the F helix is drawn toward the heme. **(b)** Now lacking a connection to the heme, the F helix is not disturbed by O_2 binding and there is significantly reduced cooperativity.

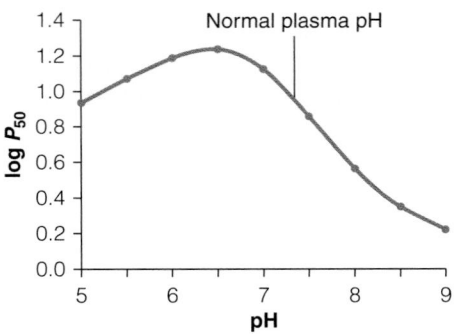

FIGURE 7.19

Oxygen affinity of hemoglobin as a function of pH.

Adapted from *Journal of Inorganic Biochemistry*
99:120–130, T. Brittain, Root effect hemoglobins.
© 2005, with permission from Elsevier.

A decrease in blood pH results in stabilization of the deoxy state and thereby favors greater O_2 released from hemoglobin.

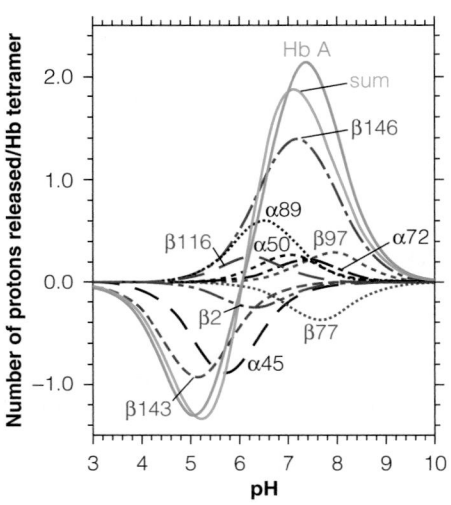

FIGURE 7.20

Contributions of various His residues to the Bohr effect in hemoglobin. The net Bohr effect for Hb is shown by the green curve. The positive portion of the curve corresponds to the alkaline Bohr effect and is of the greatest physiological relevance. Contributions of some individual His residues are shown in red (β globin residues) or black (α globin residues) and are calculated using pK_a values measured by NMR titration techniques (see Figure 6A.11). The blue curve is the summation of the contributions from all 26 surface His residues. The summation does not include contributions from the N-termini.

Reprinted with permission from *Chemical Reviews*
104:1219–1230, J. A. Lukin, and C. Ho, The structure–function relationship of hemoglobin in solution at atomic resolution. © 2004 American Chemical Society.

Hemoglobin also transports CO_2 from tissues to gills or lungs. CO_2 acts as a negative allosteric effector of O_2 binding.

Response to pH Changes: The Bohr Effect

Blood plasma has a pH that is normally 7.4. As shown in Figure 7.19, a pH drop initially has the effect of raising the P_{50} of hemoglobin (i.e, lowering O_2-binding affinity), thereby facilitating greater release of O_2. This response of hemoglobin to pH change is called the **Bohr effect**[*] after Christian Bohr (father of physicist Niels Bohr), who reported it in 1904. The overall reaction may be written

$$\text{Hb} \cdot 4O_2 + n\text{H}^+ \rightleftharpoons \text{Hb} \cdot n\text{H}^+ + 4O_2$$

where n has a value somewhat greater than 2. Physiologically, this reaction has two consequences. First, in the capillaries, H^+ ions promote the release of O_2 by driving the reaction to the right. Then, when the venous blood recirculates to the lungs or gills, the oxygenation has the effect of releasing the H^+ by shifting the equilibrium to the left. This, in turn, releases CO_2 from bicarbonate dissolved in the blood plasma by the reversal of the bicarbonate reaction (Equation 7.12). The free CO_2 can then be exhaled.

A stereochemical mechanism to explain the Bohr effect was first proposed by Perutz and coworkers in 1970. Perutz argued that certain proton binding sites in hemoglobin are of higher affinity in the deoxy form than in the oxy form, and predicted that a major contribution comes from histidine residue 146 at the C-terminus of each β chain. A change in proton affinity for some ionizable group is manifested as a change in pK_a. How can the pK_a of an amino acid side chain be altered? This can be achieved by altering the chemical environment of the ionizable side chain. Histidine β146 in the R-state tetramer has a pK_a of roughly 6.4 and is therefore predominantly deprotonated at the normal pH of blood, 7.4. As shown in Figure 7.15, when hemoglobin is in the T state, the side chain of β Asp 94 moves close enough to β His 146 to make a salt bridge *if the histidine is protonated.* Because this salt bridge stabilizes the proton against dissociation, the pK_a of β His146 is increased. This change in pK_a has been experimentally confirmed by NMR studies in the laboratory of Chien Ho. Ho and coworkers have measured the pK_a of β His146 in the deoxy state to be 7.93 but 6.42 when bound to CO (which mimics the oxy state). Thus, as the proton concentration increases, protonation of β His146 is favored, which in turn favors the *deoxy* conformation and thereby promotes the release of oxygen.

Although β His146 makes the greatest contribution to the alkaline Bohr effect (Figure 7.20), other residues are also involved, including the N-terminal amino groups of the α chains. Proton binding to these residues favors the deoxy conformation, in a manner analogous to that for β His 146. The overall effect of lowering pH on the O_2 binding affinity of hemoglobin is illustrated in Figure 7.21. Note that a decrease in pH of only 0.8 unit shifts the P_{50} from less than 20 mm Hg to over 40 mm Hg, greatly increasing the amount of O_2 released to respiring tissues.

Carbon Dioxide Transport

Release of carbon dioxide from respiring tissues lowers the oxygen affinity of hemoglobin in two ways. First, as mentioned above, most of the CO_2 is rapidly converted to bicarbonate in erythrocytes, releasing protons that contribute to the Bohr effect. Most of this bicarbonate is transported out of the erythrocytes and is carried dissolved in the blood plasma. A small portion of the CO_2 (estimated to be 5–13%) reacts directly with hemoglobin, binding to the N-terminal amino groups of the chains to form **carbamates**:

$$-\overset{+}{\text{N}}\text{H}_3 + \text{HCO}_3^- \rightleftharpoons -\overset{\overset{\text{H}}{|}}{\text{N}}-\text{COO}^- + \text{H}^+ + \text{H}_2\text{O}$$

---

[*]The effect observed above pH ~6.5 is more properly called the "alkaline Bohr effect" to distinguish it from the "acid Bohr effect," which describes an *increase* in O_2 affinity as pH drops below a value of 6. The physiological relevance of the acid Bohr effect (at least in mammals) is not clear.

This *carbamation reaction* allows hemoglobin to aid in the transport of CO_2 from tissues to lungs or gills, and the protons released on carbamate formation contribute to the Bohr effect.

We may summarize the effects of H^+ and CO_2 in terms of the respiratory cycle shown in Figure 7.1: In the lungs or gills of an animal O_2 is abundant. Oxygenation favors the oxy conformation of hemoglobin, which stimulates the release of CO_2. As the blood then travels via arteries into the tissue capillaries, the lower pH and high CO_2 content favor the deoxy form, promoting O_2 release and binding of CO_2. Carbon dioxide, both in forming bicarbonate and in reacting with hemoglobin, contributes to decreasing the pH, further stimulating O_2 release.

The role of increasing CO_2 in stimulation of O_2 release is seen in hyperventilation. If a person breathes too rapidly, plasma CO_2 concentration is significantly reduced, and consequently, release of oxygen into the tissues is impaired. This condition leads to dizziness and, in extreme cases, unconsciousness. Hyperventilation can be easily corrected by breathing into a paper bag—this brings exhaled CO_2 back into the blood, thereby increasing its concentration.

Response to Chloride Ion at the α-Globin N-Terminus

The bicarbonate ions formed inside erythrocytes are transported across the erythrocyte membrane into the surrounding plasma. To maintain charge neutrality within the red cell, Cl^- ions are exchanged for HCO_3^- ions (ion transport is discussed in detail in Chapter 10). Chloride ion binds to deoxyhemoglobin, between the terminal residues in each α globin, forming a bridge between the positively charged amino-terminus of Val1 and the side chain of Arg141. This is possible due to the proximity of these groups in the deoxy conformation (see Figure 7.15). Chloride ion binding favors protonation of the N-terminal amino group of Val1, thereby increasing its pK_a. Val1 and Arg141 don't interact in the oxy conformation; thus, both the bound Cl^- and the H^+ are released. In this way chloride ion binding augments the Bohr effect.

2,3-Bisphosphoglycerate

H^+ and CO_2 are the effectors that function rapidly to facilitate the exchange of O_2 and CO_2 in the respiratory cycle. One other major effector operates over longer periods to permit organisms like humans to adapt to gradual changes in oxygen availability. It is a common observation that people who move to high altitudes at first experience some distress but gradually acclimate to the lower oxygen pressure. In the short term (1–2 days), this acclimation results from increased concentration of red cells in plasma due to a reduction in the volume of the blood plasma; but, a more significant short-term adaptive effect is due to changes in the concentration in red cells of the allosteric effector **2,3-bisphosphoglycerate** (or **2,3-BPG**; Figure 7.22a). Within 2 days of moving to higher altitude, the concentration of 2,3-BPG in red cells nearly doubles (from 4.5 mM to ~7.6 mM), resulting in increased binding of this effector to Hb. Longer-term adaptation to higher altitude, which requires 2–3 months, is the result of increased red cell production.

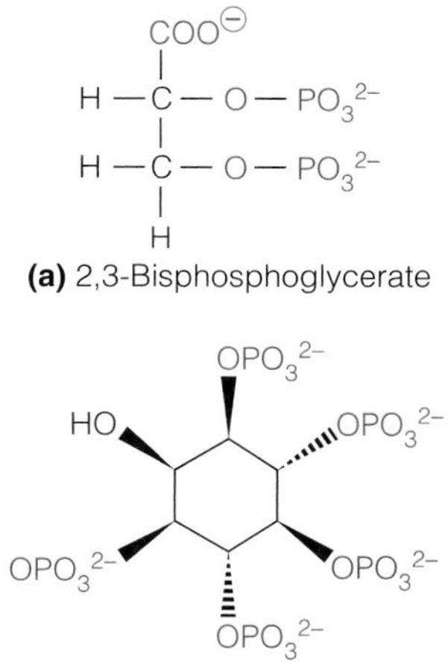

Like the effects of H^+ and CO_2, binding of 2,3-BPG acts to lower the oxygen affinity of hemoglobin. At first glance this may seem like a strange way to adapt to lower O_2 pressure, but in fact, the more efficient unloading of oxygen in the tissues more than compensates for the slight decrease in loading efficiency in the lungs (see Problem 12). The action of 2,3-BPG is illustrated in Figure 7.23. 2,3-BPG binds via ionic interactions with positively charged groups lining the cavity between the β chains in the deoxy state. Comparison of the two hemoglobin conformations shown in Figure 7.14b shows that this cavity is much narrower in oxyhemoglobin than in deoxyhemoglobin. In fact, 2,3-BPG cannot be bound in the oxy form. The higher the 2,3-BPG content in red blood cells, the more the

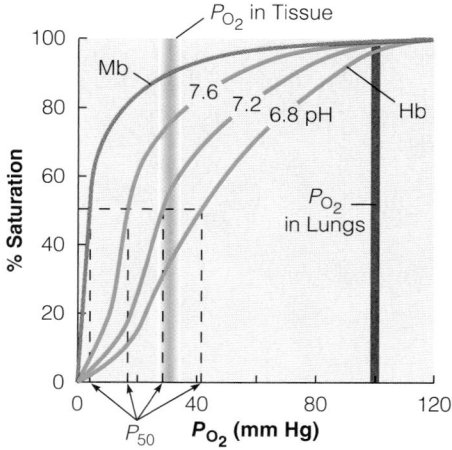

FIGURE 7.21

The Bohr effect in hemoglobin. Oxygen-binding curves for hemoglobin (green) are shown for pH 7.6, 7.2, and 6.8. Note that the efficiency of oxygen unloading, as measured by the differences in the curves at $P_{O_2} = 30$ mm Hg (vertical blue line), increases greatly as the pH drops. As the hemoglobin circulates from lungs to tissues, the lower pH favors the lower-affinity conformation (this is also reflected in the increase in P_{50} values as pH drops). Myoglobin displays little Bohr effect, so its oxygen-binding curve (orange) is approximately the same at all three pH values.

(a) 2,3-Bisphosphoglycerate

(b) myo-Inositol-1,3,4,5,6-pentaphosphate

FIGURE 7.22

Two anionic compounds that bind to deoxyhemoglobin. **(a)** 2,3-Bisphosphoglycerate (2,3-BPG), found in mammals. 2,3-BPG was formerly known as diphosphoglycerate (DPG); you will sometimes find the older term in the literature. **(b)** myo-Inositol-1,3,4,5,6-pentaphosphate (IPP), found in birds.

2,3-Bisphosphoglycerate (2,3-BPG) is found inside red blood cells and is a potent allosteric effector that lowers the O_2 affinity of hemoglobin.

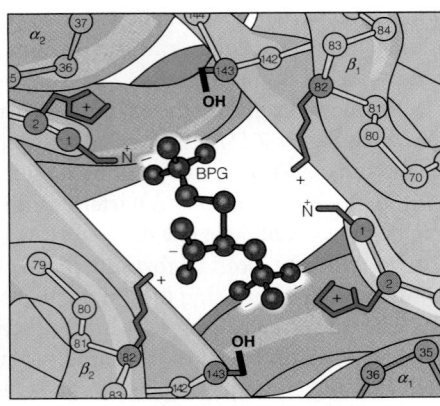

2,3-BPG binding in adult Hb

2,3-BPG binding in fetal Hb

FIGURE 7.23

Binding of 2,3-bisphosphoglycerate to deoxyhemoglobin. The binding site, in the central cavity of the adult hemoglobin (HbA) tetramer (see Figures 7.14 and 7.15), is lined with eight positively charged groups that help bind the negatively charged 2,3-BPG molecule (center panel). Note the His residues (β143) that are replaced by Ser in fetal hemoglobin (HbF; rightmost panel).

Illustration, Irving Geis. Image from Irving Geis Collection/Howard Hughes Medical Institute. Rights owned by HHMI. Not to be reproduced without permission.

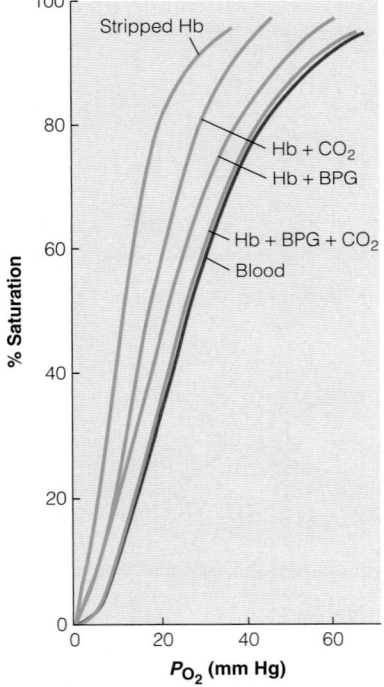

FIGURE 7.24

Combined effects of CO_2 and BPG on oxygen binding by hemoglobin. Hemoglobin that has been stripped of both CO_2 and BPG has a high oxygen affinity. When both substances are added to hemoglobin at the levels found in blood emerging from the capillaries, the hemoglobin displays almost exactly the same binding curve as observed for whole blood.

British Medical Bulletin 32:209–212, J. V. Kilmartin, Interaction of haemoglobin with protons, CO_2 and 2,3-disphosphoglycerate. © 1976, by permission of Oxford University Press.

deoxy structure is favored. Once again, a decrease in O_2 affinity is explained by preferential stabilization of the deoxy structure. Increased 2,3-BPG levels are also found in the blood of smokers, who because of the carbon monoxide in smoke, also suffer from limitation in O_2 transport.

2,3-BPG plays one other subtle, but important, role in the respiration of humans and other mammals. Consider the problem faced by a fetus, which must obtain O_2 from the mother's blood by exchange through the placenta. For this exchange to work well, fetal blood must have a higher O_2 affinity than the mother's blood. In fact, the human fetus has a hemoglobin different from the adult form. Whereas adult hemoglobin (HbA) has two α and two β chains ($\alpha_2\beta_2$), in the fetus the β chains are replaced by similar, but distinctly different, polypeptides. These are called γ chains; thus, fetal hemoglobin (HbF) has an $\alpha_2\gamma_2$ structure. The intrinsic oxygen affinity of HbF is very similar to that of HbA, but HbF has a much lower affinity for 2,3-BPG than does HbA. This difference is largely due to the replacement of His 143 in the adult β chain by a Ser in the fetal γ chain. As Figure 7.23 shows, loss of the positively charged His 143 in HbF reduces the binding affinity for 2,3-BPG. The concentration of 2,3-BPG is about the same in the circulatory systems of mother and fetus. Under these conditions, HbF will have less 2,3-BPG bound than will HbA and therefore tends to favor the R state; thus, HbF will have a higher oxygen affinity and efficiently extracts O_2 from the lower-affinity maternal HbA.

The use of effectors that facilitate oxygen release is not restricted to mammals. The blood of birds contains **inositol pentaphosphate** (see Figure 7.22b), and fish use ATP for a similar purpose. All of these molecules have a high negative charge and bind in the central cleft of deoxyhemoglobin. All of these allosteric effectors, including H^+, CO_2, Cl^-, and 2,3-BPG, act in the same general manner—by biasing the conformational equilibrium in hemoglobin toward the deoxy form. However, they interact at distinctly different sites, and therefore their effects can be additive, as illustrated for CO_2 and 2,3-BPG in Figure 7.24.

Other Functions of the Heme Globins: Reactions with Nitric Oxide

Francis Crick is reputed to have quipped "hemoglobin has a *bore* effect," reflecting a widely held view that there are no more profound discoveries to be made in the continued study of the most-studied protein on Earth. Given the current debate over hemoglobin allostery (see Appendix), it is clear that not everything about hemoglobin function is understood. Likewise, the long-held view of myoglobin as an essential O_2 transport protein was challenged by the startling result that mice lacking a gene for myoglobin appear to be normal, even under conditions of exercise-induced stress. Careful investigation showed that multiple compensatory mechanisms developed in these mice to facilitate oxygenation of their tissues (e.g., increasing the density of capillaries in heart muscle). In the past decade, both myoglobin and hemoglobin have been shown to possess essential functions beyond those associated with reversible O_2 binding, and many open questions remain regarding the mechanisms whereby these other functions are carried out. The most active areas of research are those focused on the roles of myoglobin and hemoglobin in the physiological effects of the gas nitric oxide (NO), which is both an important signaling molecule and a mediator of cytotoxicity during phagocytosis. Greater detail on the biosynthesis and action of NO is presented in Chapters 21 and 23, respectively.

At the low concentrations normally found in endothelial cells ($\sim 10^{-7}$ M), NO signals the relaxation of smooth muscle cells. Before it was identified as NO, the agent that caused this relaxation was known as EDRF, or endothelium-derived relaxation factor. Oxyhemoglobin is an efficient scavenger of NO because it reacts very rapidly ($\sim 7 \times 10^7$ M^{-1}s^{-1}) with NO and oxidizes it to nitrate in a *dioxygenation* reaction:

$$Hb(Fe^{2+}\!\!-\!\!O_2) + NO \longrightarrow Hb(Fe^{3+}) + NO_3^- \qquad (7.13)$$

Work by Paul Gardner and coworkers has shown that both oxygen atoms in O_2 are added to NO when this reaction is carried out in the presence of oxymyoglobin or oxyhemoglobin, but not when the reaction is carried out in the absence of myoglobin or hemoglobin. This led Gardner to propose a *nitric oxide dioxygenase* activity for the globin family of proteins, including bacterial **flavohemoglobins**, which are ancestors of myoglobin and hemoglobin (see below). Flavohemoglobins contain, in addition to a heme binding domain, domains that bind the redox cofactors **flavin adenine dinucleotide** (FAD/FADH$_2$) and **nicotinamide adenine dinucleotide** (NAD$^+$/NADH). The chemistry performed by these cofactors is described in greater detail in subsequent chapters covering metabolic reactions. In flavohemoglobin they act as reducing agents to convert the ferric (Fe^{3+}) heme generated in Equation 7.13 to ferrous (Fe^{2+}) heme. Why would globins have this NO dioxygenation activity, and why do microbes possess globins when they are small enough to receive sufficient O_2 by diffusion (under aerobic conditions)?

Clues to the likely answers to these questions came from the observation that in the several bacteria that have been tested, those that lack genes for flavohemoglobin are killed by exposure to NO at concentrations of $\sim 10\,\mu$M. In addition, some bacteria that express higher than normal amounts of flavohemoglobin are resistant to killing by macrophages (which produce NO as part of their cytotoxic action). Thus, it seems that in these microbes, flavohemoglobin mediates resistance to the cytotoxic effects of NO. What of higher organisms? Compared to wild-type mice, mice lacking myoglobin genes show a phenotype for increased damage to cardiac tissue following periods of NO exposure. Among its various modes of cytotoxicity, NO inhibits aerobic respiration. These observations prompted Maurizio Brunori to propose in 2001 that myoglobin, as a potent NO dioxygenase, plays a critical role in protecting cells against such NO-mediated inhibition of cellular energy production.

The well-established dioxygenase activity of myoglobin may explain another recent observation, namely, that myoglobin is expressed at low levels in many different nonmuscle tissue types. In addition, two "new" additions to the heme globin family, **cytoglobin** and **neuroglobin**, may play important roles in NO scavenging. Cytoglobin appears to be expressed in most tissue types, whereas neuroglobin is found in neuronal tissue. Although the functions of cytoglobin and neuroglobin are not known definitively, they may act as dioxygenase enzymes to prevent tissue damage following hypoxic events (e.g., stroke or heart attack). Other proposed roles for these globins include reduction in hypertension and prevention of **apoptosis** (programmed cell death, described in more detail in Chapter 28).

Cytoglobin and neuroglobin may help to prevent tissue damage following a stroke or heart attack.

In summary, myoglobin and hemoglobin represent sophisticated molecular machines, each finely tuned to carry out its functions. In the following section, we explore how these structures might have evolved. We will focus on the O_2 transport functions of vertebrate myoglobin and hemoglobin; however, it should be noted that globins are ancient proteins that are found in greater structural and functional variety in bacteria and archaea, where the NO dioxygenase activity appears to be of primary importance. Thus, O_2 transport appears to be a relatively recent functional adaptation to the ancestral globin fold, which has been present on Earth since atmospheric levels of O_2 were much lower than they are at present.

Protein Evolution: Myoglobin and Hemoglobin as Examples

Each polypeptide sequence produced by an organism is encoded by a gene. The nucleotide sequence in that gene dictates the amino acid sequence of the protein, which in turn defines the protein's secondary, tertiary, and quaternary structures. Evolution of proteins occurs through accumulated changes in the nucleotide sequences of genes. To explore this process, we will use as an example the evolutionary development of the myoglobin–hemoglobin family of proteins. First, however, we must examine in a bit more detail the structure of eukaryotic genes and the mechanisms through which mutation can occur.

The Structure of Eukaryotic Genes: Exons and Introns

In previous chapters we stated that a direct correspondence exists between the nucleotide sequence in a gene and (via mRNA) the amino acid sequence of the polypeptide chain it encodes. For most genes in prokaryotic organisms, this concept is true. But investigation of the genomes of higher organisms has produced a surprising result: Within most eukaryotic genes are DNA sequences that are never expressed in the polypeptide chain. These noncoding regions, called **introns**, alternate with regions called **exons** that include the DNA encoding the translated polypeptide sequence and untranslated regions at the 5′ and 3′ ends that are important in regulating transcription and translation. Figure 7.25 shows how the exon–intron structure of the β globin gene is related to the structure of β globin.

Eukaryotic genes are discontinuous, containing regulatory and protein-encoding sequences (exons) and intervening sequences (introns).

FIGURE 7.25

Coding and noncoding regions of the β hemoglobin gene. The gene for the human β hemoglobin chain has regulatory (blue boxes) and coding (purple, red, and green boxes) regions, or exons, alternating with noncoding regions, or introns (yellow boxes). This figure follows the transcription and translation of the gene to yield the final β hemoglobin chain. **Step 1, transcription:** A primary transcript (pre-mRNA) containing complementary copies of the exons and introns is produced from the gene. **Step 2, splicing:** The intron sequences are removed and the exons spliced together to yield the final mRNA. Note: the mechanism of mRNA splicing is given in Chapter 27. **Step 3, translation:** The coding regions of the spliced mRNA produce a β chain, which adopts its favored three-dimensional structure and incorporates a heme group. Note that the entire heme-binding region (red) is coded for by one exon.

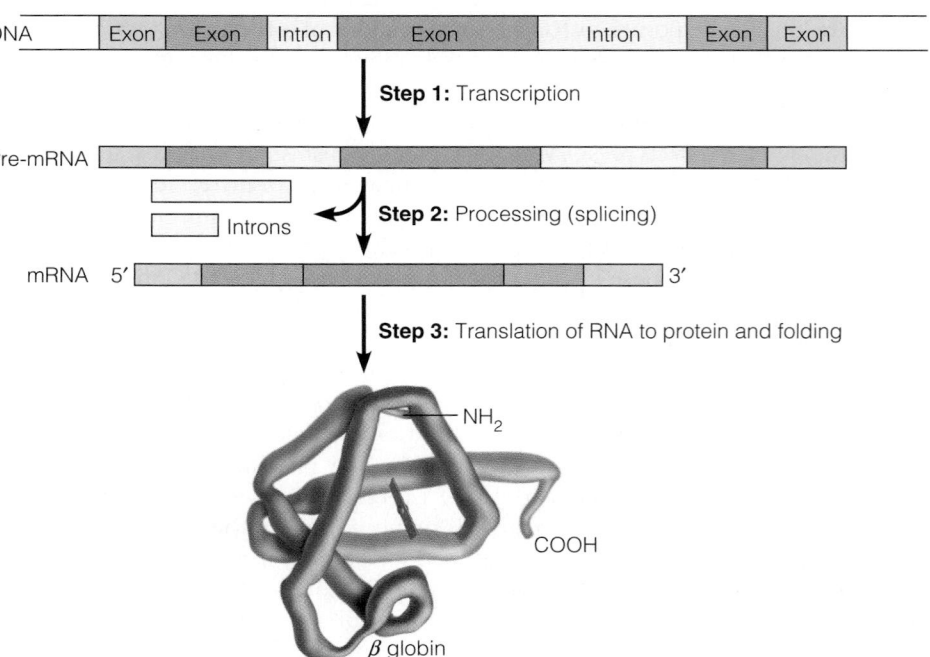

Clearly, this remarkable situation means that mRNA production in eukaryotes must be a more complex process than had been assumed at first. As Figure 7.25 shows, transcription first produces a primary transcript, or **pre-mRNA**, corresponding to the whole gene—exons, introns, and portions of flanking regions. The pre-mRNA, while still in the cell nucleus, is cut and spliced to remove the regions corresponding to introns, thereby producing an mRNA that codes correctly for expression of the polypeptide chain. We describe the details of this process in Chapter 27. For now, keep in mind that most eukaryotic genes are "patchwork" structures, containing extensive regions that do not correspond to any part of the protein sequence.

Mechanisms of Protein Mutation

When organisms reproduce, they copy their DNA, and occasionally mistakes are made. These mistakes may be random errors that occur during copying, or they may be results of damage the DNA has sustained from radiation or chemical **mutagens**, substances that produce mutations (discussed in greater detail in Chapter 26). In any event, these alterations will appear as **mutations** in the DNA of the next and subsequent generations of progeny. There are two basic kinds of changes in the DNA sequence that may give rise to mutations in proteins: (1) replacements of DNA bases by others and (2) deletion or insertion of bases in the gene.

Replacement of DNA Bases

Replacement of bases can have several possible consequences. First, the base change may not affect the protein sequence at all. The change may occur in an intron, for example. But even if it is in a protein-coding exon, this **silent**, or **synonymous mutation** (Figure 7.26a), will make no difference in the protein sequence if the new codon codes for the same amino acid as the original one. The redundancy of the genetic code (see Figure 5.18, page 152) is such that frequently a base change does not alter the protein product. Alternatively, an amino acid residue in the original protein may be replaced by a different one in the mutated protein; this type of replacement is called a **missense**, or **nonsynonymous mutation** (Figure 7.26b). Occasionally, the codon for an amino acid residue within the original protein will be changed to a *stop* codon. We call this a **nonsense mutation** because the protein will be terminated prematurely and usually be nonfunctional (Figure 7.26c). Sometimes the opposite happens—a stop codon mutates into a codon for an amino acid residue. In this case translation continues, elongating the chain.

Nucleotide Deletions or Insertions

Deletions or insertions in the gene may be large or small. Such mutations outside the coding regions will generally have no effect, unless they modify sites of

Mutations result from changes in the DNA sequence of genes, including base substitutions, deletions, or additions.

FIGURE 7.26

Mutation types. Some of the ways in which mutations can occur in the β hemoglobin chain are shown here. The first 10 residues of the normal human β chain, together with their DNA codons, are shown at top. **(a)** A silent, or synonymous, mutation has occurred in the codon for residue 2 (CAC to CAT). **(b)** A missense, or nonsynonymous, mutation has occurred in the codon for residue 6 (GAG to GTG). This is the sickle-cell mutation. **(c)** A nonsense mutation has introduced a stop signal after the codon for residue 7 (AAG to TAG), terminating the chain prematurely. **(d)** A frameshift mutation has occurred by deletion of a single T residue. The rest of the chain, with a completely altered sequence, will continue to be produced until a stop signal is encountered in the new frame. Both (c) and (d) would result in β-thalassemia (see page 266).

Residue number		1	2	3	4	5	6	7	8	9	10
Normal β gene	...A T G	G T G Val	C A **C** His	C T G Leu	A C **T** Thr	C C T Pro	G A **G** Glu	G A G Glu	A **A** G Lys	T C T Ser	G C C... Ala
(a) Silent, or synonymous mutation		G T G Val	C A T His	C T G Leu	A C T Thr	C C T Pro	G A G Glu	G A G Glu	A A G Lys	T C T Ser	G C C... Ala
(b) Missense, or nonsynonymous mutation		G T G Val	C A C His	C T G Leu	A C T Thr	C C T Pro	G T G Val	G A G Glu	A A G Lys	T C T Ser	G C C... Ala
(c) Nonsense mutation		G T G Val	C A C His	C T G Leu	A C T Thr	C C T Pro	G A G Glu	G A G Glu	T A G Stop	T C T	G C C...
(d) Frameshift mutation by deletion		G T G Val	C A C His	C T G Leu	A C □ Thr	C C T Leu	G A G Arg	G A G Arg	A A G Ser	T C T Leu	G C C...

transcriptional control (e.g., the untranslated regions of the spliced mRNA). Large insertions or deletions in coding regions almost invariably prevent the production of useful protein; sometimes even whole genes may be deleted. The effect of short deletions or insertions depends on whether they involve multiples of three bases. If one, two, or more *whole codons* are removed or added, the consequence is the deletion or addition of a corresponding number of amino acid residues. However, a deletion or insertion in a coding region of any number of bases *other* than a multiple of three has a much more profound effect: It causes a shift in the reading frame during translation. Such **frameshift mutations** result in a complete change in the amino acid sequence in the C-terminal direction from the point of mutation (Figure 7.26d).

The effects of these kinds of mutations on the functionality of the protein product, and therefore on the organism itself, can be quite varied. Base substitutions may, in some cases, be neutral in effect, either not changing the amino acid coded for or changing it to another that functions equally well at that position in the protein. Frequently, the result is deleterious. Occasionally, such mutations confer some functional advantage to the protein, and the mutated organisms may be selected for in future generations. Nonsense mutations and frameshift mutations, by contrast, almost always result in loss of protein function. If the protein function is critical to the life of the organism, such mutations are strongly selected against in the course of evolution—those who inherit them do not survive to reproduce.

Gene Duplications and Rearrangements

By accumulating many small mutational changes over millions of years, proteins gradually evolve. The diversity of functions that they can collectively perform is increased by two other phenomena: **gene duplication** and **exon recombination**.

Very occasionally, replication of the genome occurs in such a way that some DNA sequence, containing a particular gene, is copied twice. Initially, the only result of such duplication is that the descendants of the organism have two copies of the same gene. This mutation may be advantageous if the protein is needed in large amounts because the capacity for its production will be increased. In such cases, there will be selective pressures to maintain two or even more copies of the same gene. Alternatively, the two copies may evolve independently. One copy may continue to express the protein fulfilling the original function, but the other may evolve through mutations into an entirely different protein with a new function. Recall from Chapter 5 that proteins that are related by a common evolutionary origin are called **homologs**. Homologous genes within an organism that arise from gene duplication are called **paralogs**. This is in contrast to **orthologs**, which are homologous genes in different species that originated from a common ancestor.

Another way in which the diversity of proteins can increase is through the *fusion* of two or more initially independent genes. Such fusion may lead to the production of multidomain proteins exhibiting new combinations of functions.

The intervening sequences in eukaryotic genes (introns) offer a further possibility for diversification of protein structure and function. Because these regions are not used for coding, they represent positions where genes can be safely cut and recombined in the process of **genetic recombination**. The mechanisms of recombination are described in Chapter 26; at this point we are concerned only with its consequences. Suppose that an exon from one gene, which codes for a protein region with physiological function B, is inserted into an intron region in a gene for a protein carrying function A. The new hybrid protein is now capable of both functions A and B and may serve a new physiological function.

Through the combined effects of mutations, gene duplication, and genetic rearrangement, organisms can develop new abilities, adapt to new environments, and become new species. The process of organismal evolution, which we see exhibited in the fossil record and in the incredible variety of existing plants, animals, and microorganisms, is largely a consequence of this molecular evolution of proteins.

Genomes can also be modified by gene duplication, gene fusion, or exon recombination.

Evolution of Protein Function at the Level of Molecular Detail

As exemplified by bacteria that rapidly develop resistance to antibiotics, protein function evolves in response to challenges that test the "fitness" of an organism for its environment. How does this happen at the level of protein structure? Genetic mutation can result in variation in amino acid sequence that may affect protein structure and function. But, we may ask, how many mutations are required to establish a new protein function that confers a selective advantage?

By combining the tools of evolutionary and molecular biology it has been possible to predict the sequences of ancestral genes and then test potential evolutionary pathways leading from the ancestral gene sequence to the modern gene sequence for a protein. To illustrate this approach, let us consider a hypothetical example of an ancestral gene and a modern homolog that differ by 4 amino acid mutations. In this case, 16 genes encoding the possible ($2^4 = 16$) combinations of the ancestral and modern amino acid sequences can be synthesized and expressed. Detailed structural and functional characterization of the 16 individual mutants allows feasible evolutionary pathways to be identified at the level of atomic detail. As shown in Figure 7.27, what emerges from such analysis is a picture of the *order* in which mutations must occur to avoid nonfunctional (e.g., unstable) intermediate species as well as those pathways that are most likely to recapitulate the actual mutational events in the evolution of the modern protein from its ancestor.

Numerous studies of protein evolution indicate that the majority of mutations that confer new functional properties tend to be destabilizing to the parent protein. There is, therefore, a limited number of such mutations that can be tolerated along the evolutionary pathway. Not all mutations are destabilizing, and some so-called **permissive mutations** do confer greater thermodynamic stability to a protein—typically without effect on the functional properties. Another conclusion drawn from in vitro evolution studies is that there is a threshold stability that a protein must possess to be "evolvable." In other words, greater thermodynamic stability for a protein is correlated with a greater tolerance for a mutation that, in turn, leads to the evolution of new function. The requisite stability may be achieved in the course of evolution by incorporation of stabilizing (permissive) mutants.

An emerging view of protein evolution is focused on the dynamic nature of proteins and suggests that the greater conformational variability associated with a more dynamic structure would allow a protein to evolve more rapidly than if it were conformationally rigid. A protein with flexible loops making up its active site is likely to be promiscuous in terms of its substrate binding specificity—a situation that inherently predisposes the protein to evolve multiple new functions. Panels (a) and (b) of Figure 7.28 show examples of conformationally flexible proteins. What evidence is there for such dynamic structures in proteins? The number of well-characterized "natively unstructured" or "intrinsically disordered" proteins has grown significantly in the past ten years, particularly among proteins involved in signaling. How, then, would such a dynamic protein evolve a new function? One hypothesis suggests that the relative population of a minor conformer that binds some non-native ligand could be increased by a mutation that stabilizes that conformer over the major conformer that normally binds the native ligand. This process is illustrated in Figure 7.28c, where two mutations were shown to be sufficient to bias the conformational ensemble of a dynamic protein to favor one conformer over another.

Irrespective of the actual pathway that results in evolution of new function, mutation plays a key role; thus, the rate of mutation in a genome is an important parameter in determining the rate of evolution of new protein function. In mammalian genomes, the mutational frequency has been estimated to be $1.1-12.4 \times 10^{-9}$ per base pair per year. The slow rate in humans of protein evolution for improved function can be understood given (1) the size of the human genome (~3×10^9 base pairs), (2) that less than 2% of the genome

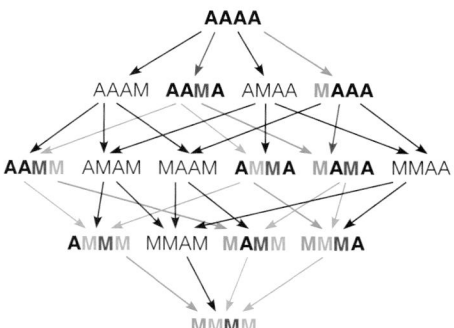

FIGURE 7.27

Mapping possible evolutionary pathways between an ancestral protein sequence and a modern homolog. In this hypothetical scenario, there are 16 possible combinations of mutants for an ancestral protein (AAAA) and a modern homolog (MMMM), which differ by 4 mutations and 24 pathways (shown by arrows) linking AAAA to MMMM. Proteins that are stable are shown in bold type. Red type indicates a permissive mutation that increases protein stability, and therefore promotes evolvability; green type indicates a mutation that is neither stabilizing nor destabilizing; and blue type indicates a destabilizing mutation. In this example the red (stabilizing) mutation must occur prior to either of the blue (destabilizing) mutations. This requirement for a specific order in which mutations occur limits the number of likely pathways to the eight possibilities shown by different combinations of the colored arrows.

FIGURE 7.28

Examples of conformational flexibility that might confer improved evolvability. **(a)** Local conformational changes allow promiscuous substrate binding to an enzyme. The open conformation of P450-CYP2B4, a detoxification protein found in liver (orange ribbon), allows binding to a large substrate (illustrated in red), and the closed one (light blue ribbon) binds the smaller substrate (darker blue). **(b)** Lymphotactin, a small protein that is part of the immune response, is a so-called metamorphic protein that exists in equilibrium between two native conformations: a β−α monomer (top) and an all-β dimer (bottom). **(c)** Two different topologies (mediated by three different disulfide bridges) are found in two naturally occurring cysteine-rich domains (CRDs) from proteins that are found in the venomous cells of jellyfish and sea anemones (NW1 and Mcol1C). These two CRDs show almost no sequence identity beyond the conserved cysteines. Conversion between these topologies is achieved by a single mutation Lys21 → Pro21 (K21P) in NW1 that yields a dynamic intermediate which exists in both conformations. A second mutation Gly11 → Val11 (G11V) completed the transition from the NW1 topology to that of Mcol1C.

From *Science* 324:203–207, N. Tokuriki and D. S. Tawfik, Protein dynamism and evolvability. © 2009. Reprinted with permission from AAAS.

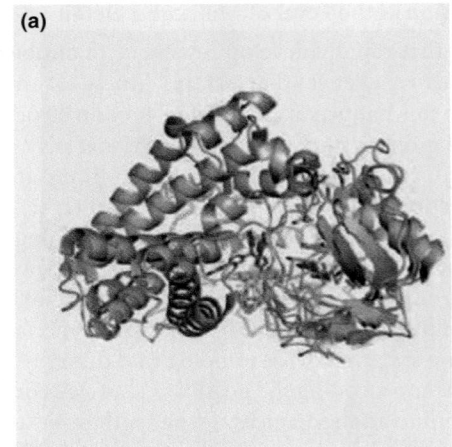

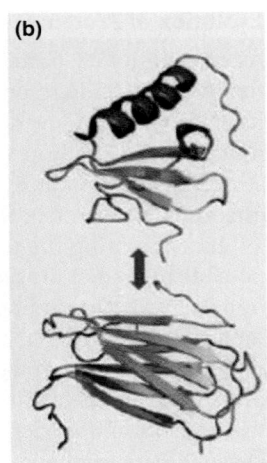

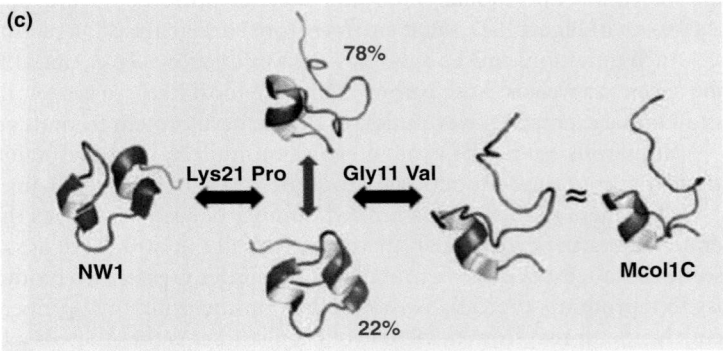

encodes protein genes, (3) that the majority of these mutations do not occur in the germ line (thus, they are not passed on to progeny), and (4) the apparent need for a defined temporal order of the multiple mutations required to confer the new function (as illustrated in Figure 7.27). Much more prevalent is the observation of deleterious, pathological mutations, which may require only a single nucleotide change. We discuss a few of these for human hemoglobin later in this chapter.

Evolution of the Myoglobin–Hemoglobin Family of Proteins

We have already seen an example of the process of protein evolution. If we compare the sequences of sperm whale and human myoglobin (see Figure 5.16), we find 25 amino acid changes. Because fossil evidence indicates that the evolutionary lines that led to sperm whales and humans diverged from a common mammalian ancestor about 100 million years ago, we can gain an idea of the rate of this process. If the rate was uniform, there has been an average of one amino acid replacement every 4 million years.

If we compare human myoglobin with that of the shark, we find about 88 differences. Because these evolutionary lines diverged about 400 million years ago, the accumulated differences are about what we would expect from the preceding example. In other words, the number of amino acid substitutions in two related proteins is roughly proportional to the evolutionary time that has elapsed since the proteins (and the species) had a common ancestor. Using this principle, we can compare the sequences of both hemoglobins and myoglobins and attempt to construct a "family tree" of globin proteins. The tree is complicated by the fact that higher eukaryotes, including humans, carry genes for both myoglobin and several *different* hemoglobin chains. These different genes are expressed at different times in human development (Figure 7.29). The α and β chains, as mentioned

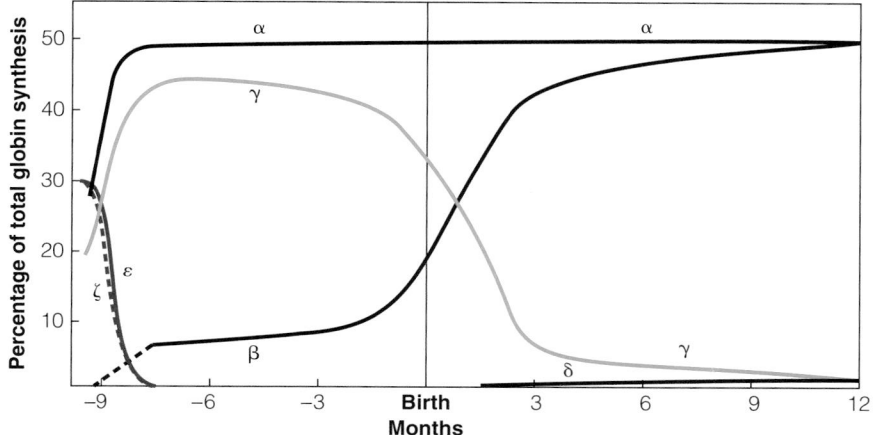

FIGURE 7.29

Expression of human globin genes at different stages in development. The human ζ and ε genes make the $\zeta_2\varepsilon_2$ hemoglobin found in the very early embryo. This is soon supplanted by the $\alpha_2\gamma_2$ hemoglobin of the fetus. At about the time of birth, transcription of the γ gene ceases, and the β gene begins to be transcribed. By six months, the infant will have almost all $\alpha_2\beta_2$ (adult) hemoglobin. The δ gene is never transcribed at high rates. There are two copies of the α gene: α_1 and α_2. Both contribute to the production of α chains.

British Medical Bulletin 32:282–287, W. G. Wood, Haemoglobin synthesis during human fetal development. © 1976, by permission of Oxford University Press.

earlier, are normally present in adults. But in the early embryo, the hemoglobin genes expressed are those for the embryonic chains, ζ (zeta) and ε. As the fetus develops, these chains are replaced by α and γ chains, ensuring efficient oxygen transfer from mother to fetus. Finally, at about the time of birth, the γ chains are replaced by β chains. In addition, after birth a small amount of a δ chain is produced. These developmental types of hemoglobin chain are slightly different, and each is coded for by a separate gene in the human genome.

Comparison of the sequences of many globins from many different species yields the evolutionary tree shown in Figure 7.30. According to these results, very primitive animals had only a myoglobin-like, single-chain ancestral globin for oxygen storage. Most of these animals, like protozoans and flatworms, were so small that they did not require a transport protein. Roughly 500 million years ago, an important event occurred: The ancestral myoglobin gene was duplicated. One of the copies became the ancestor of the myoglobin genes of all higher organisms; the other evolved into the gene for an oxygen transport protein and gave rise to the hemoglobins.

Along the evolutionary line leading to vertebrates and mammals, the most primitive animals to possess hemoglobin are the lampreys. Lamprey hemoglobin can form dimers but not tetramers and is only weakly cooperative; it represents a first step toward allosteric binding. But subsequently a *second* gene duplication occurred, giving rise to the ancestors of the present-day α and β hemoglobin chain families. Reconstruction, from sequence comparison, indicates that this

Myoglobin and hemoglobin evolved from an ancestral myoglobin-like protein.

FIGURE 7.30

Evolution of the globin genes. The arrangement of human globin genes is shown at the top. Note that they are found in five different chromosomes. Functional genes are shown in color; pseudogenes, which are nontranscribed variants of a gene, are in gray. The diagram underneath shows the probable evolution of the globin gene family, based on sequence differences among the various globin genes in humans and other animals. The times at which gene duplications occurred are inferred from a combination of sequence and fossil evidence and are only approximate. The two α genes and the two γ genes are too similar in sequence to allow us to judge the time of their divergence. We know only that it must have happened relatively recently.

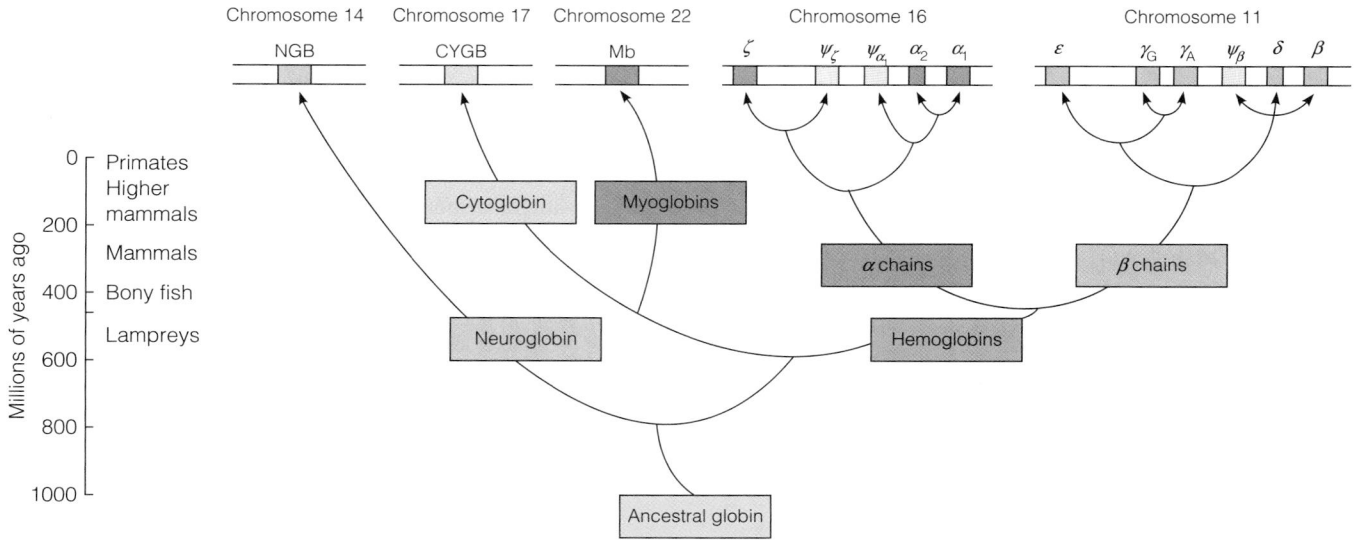

Evolution of globins has retained the common "globin fold" that holds the heme. Evidence for continuing evolution is found in the many variant proteins in existing species.

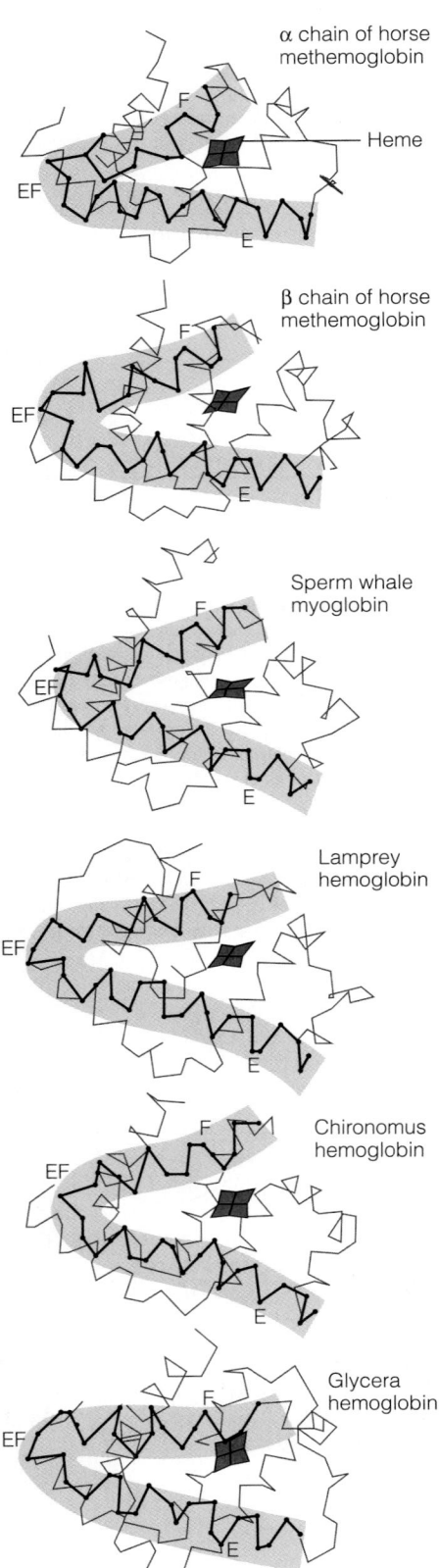

α chain of horse methemoglobin

Heme

EF

E

F

β chain of horse methemoglobin

EF

E

F

Sperm whale myoglobin

EF

E

F

Lamprey hemoglobin

EF

E

F

Chironomus hemoglobin

EF

E

F

Glycera hemoglobin

EF

E

F

must have happened about 400 million years ago, at about the time of divergence of the sharks and bony fish. The evolutionary line of the latter led to the reptiles, and eventually to the mammals, all carrying genes for both α and β globins and capable of forming tetrameric $\alpha_2\beta_2$ hemoglobins. Further gene duplications have occurred in the hemoglobin line, leading to the embryonic forms ζ and ε and the fetal γ. As Figure 7.30 shows, the duplications that led to a distinction between adult and embryonic subtypes coincide fairly well with the development of placental mammals, about 200 million years ago. This concurrence is functionally appropriate because in these mammals the later stages of embryo development occur within the mother, and a special hemoglobin, adapted to promote oxygen transfer through the placenta from mother to fetus, is essential (see page 254).

During the long evolution of the myoglobin–hemoglobin family of proteins, only a few amino acid residues have remained invariant. These *conserved residues* may mark the truly essential structural features of the molecule. As Figure 7.13 shows, they include the histidines proximal and distal to the heme iron (F8 and E7; see Figure 7.5b). Interestingly, Val FG5, which has been implicated in the hemoglobin deoxy–oxy conformation change described above, is invariant in hemoglobins, replacing the isoleucine found at this position in most myoglobins. Other regions highly conserved in hemoglobins are those near the $\alpha_1-\beta_2$ and $\alpha_2-\beta_1$ contacts. These contacts are most directly involved in the allosteric conformational change.

Despite the major changes that have occurred in the primary structure of the myoglobin–hemoglobin family over hundreds of millions of years, the secondary and tertiary structures of these proteins have remained surprisingly unchanged. Figure 7.31 shows the backbone structure of members of this family, ranging from insect to horse. All are recognizable as the same basic fold, and the similarity is particularly strong in the region that binds the heme. At first glance this similarity seems inconsistent with our earlier statements that primary structure determines secondary and tertiary structure. However, careful examination of many sequences shows that many of the replacements have been *conservative*—that is, an amino acid has been replaced by another of the same general class (e.g., polar vs. nonpolar, etc.). Obviously, evolution of these proteins has proceeded not at random but under the constraint of maintaining a physiologically functional structure. Survival of mutant proteins in the globin family has been restricted to those that maintain the basic "globin fold."

Hemoglobin Variants: Evolution in Progress

Variants and Their Inheritance

Evidence for the ongoing evolution of hemoglobin genes can be seen in the existence of hemoglobin variants or, as they are often called, abnormal hemoglobins. Today, several hundred recognized mutant hemoglobins exist within the human population. A number of mutation positions on the tetramer are shown in Figure 7.32. Most proteins in existing plants and animals probably show comparable diversity, but few of them have been as thoroughly studied as human hemoglobins. Each of the mutant forms of hemoglobin exists in only a small fraction of the total human population; some forms have been recognized in only a few

FIGURE 7.31

Evolutionary conservation of the globin folding pattern. As these drawings emphasize, the overall tertiary structure of myoglobin and hemoglobin chains has remained nearly constant despite extensive changes in the primary structure. The shaded regions delineate the E and F helices, which surround the heme. Note that they are almost invariant and that changes tend to be concentrated near the ends of the chains. The most primitive proteins shown are the single-chain hemoglobins of the marine worm *Glycera* and the fly *Chironomus*. The lamprey is the most primitive creature to have distinct myoglobin and hemoglobin. Lamprey hemoglobin forms dimers and exhibits some cooperativity in binding. The α and β chains of horse hemoglobin are almost identical to those of all other mammals.

individuals. Some of these mutant forms are deleterious and give rise to recognized pathologies; under conditions of natural selection they would eventually disappear. Most are, as far as we can tell, harmless, and are often referred to as neutral mutations. A very few may have as yet unrecognized advantages and therefore may come, in time, to dominate in the population.

We shall consider only a few of these abnormal hemoglobins. First, it is necessary to review a bit of genetics. All human cells, except for the germline cells (sperm and ova), are **diploid**; that is, they carry two copies of each chromosome. Therefore, they carry two copies of each gene, one on each of the paired chromosomes. Suppose we consider a gene such as the adult β globin gene, which can exist in two forms—the "normal" type, β, and a variant (mutant) type, β^*. An individual can have three possible combinations of these genes in his or her paired chromosomes:

A. $\beta+\beta$: **homozygous** (same genes) in the normal type

B. $\beta+\beta^*$: **heterozygous** (mixed genes)

C. $\beta^*+\beta^*$: homozygous in the variant type

Having genes for only the normal β globin, individual A will produce only normal β globin chains. Individual C, who has genes for only the variant type, will produce only variant β^* globin chains. Individual B, with genes for both types, will produce both. If the mutation is deleterious, C will be expected to manifest severe disease symptoms. B, on the other hand, may be asymptomatic, or show less severe symptoms because he or she will make normal protein chains along with the variant ones.

When two individuals produce offspring, each parent donates to a child one copy of the β globin gene, the selection of which will be random. If both parents carry only the normal gene, the child must receive two copies of the same. If both carry only the variant gene, the child must also be homozygous for that gene. If both parents are heterozygous for the gene, Figure 7.33 shows that the child has one chance in four of being homozygous normal, one in four of being homozygous for the variant gene, and two in four of being heterozygous. Because most variant hemoglobin genes are rare in the human population, only occasionally do we find an individual homozygous for the variant type.

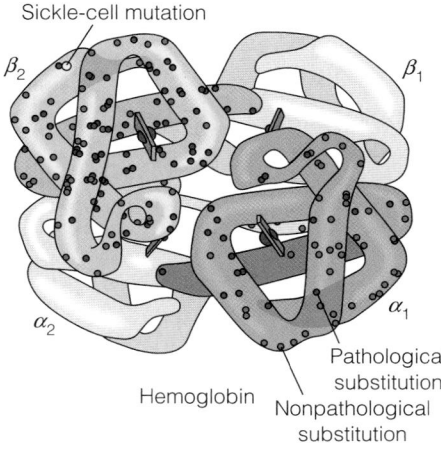

FIGURE 7.32

Distribution of mutations in human hemoglobins. The blue and red dots represent all positions at which amino acid substitutions have been found in the α and β chains (only one pair is illustrated, for clarity). Those substitutions that have known pathological effects are shown in red. At many of these positions, more than one substitution has been observed. Position 6 in the β chain, at which the sickle-cell mutation occurs, is shown in yellow.

Illustration, Irving Geis. Image from Irving Geis Collection/Howard Hughes Medical Institute. Rights owned by HHMI. Not to be reproduced without permission.

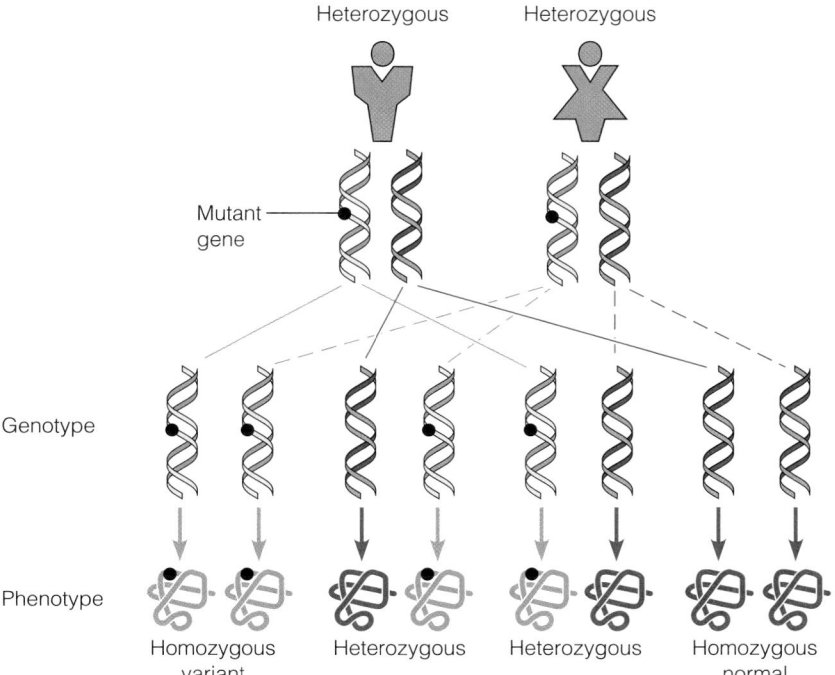

FIGURE 7.33

Inheritance of normal and variant proteins in a heterozygous cross. Diploid organisms can exist as one of three types with respect to any gene: homozygous normal, homozygous variant, or heterozygous. The offspring of a heterozygous pair may be any of the three types, as shown here: homozygous normal, heterozygous, or homozygous variant, with a probability ratio of 1:2:1. We leave it as an exercise for you to work out other possibilities—for instance, the offspring of one homozygous normal parent and one homozygous variant parent. This pattern is referred to as classical Mendelian inheritance.

Pathological Effects of Variant Hemoglobins

Of the large number of hemoglobin mutations, a significant fraction have deleterious effects. As Figure 7.32 shows, the known deleterious mutations are mostly clustered about the heme pockets and in the vicinity of the $\alpha-\beta$ contact region that is so important in the allosteric transition. A few of the well-studied pathological missense mutations are listed in Table 7.1. For example, a class of variants known as *hemoglobins M* tend to be readily oxidized to methemoglobin, which cannot bind O_2. Many of these mutations involve replacement of either the proximal or the distal histidine by other residues. Individuals carrying such mutations experience difficulty in transporting enough O_2 to tissues. Other variants that involve changes at the subunit interfaces can have several kinds of effects. Some, like *hemoglobin St. Lukes,* destabilize the hemoglobin tetramer, whereas others (e.g., *hemoglobin Suresnes*) tend to stabilize either the oxy or the deoxy conformation, inhibiting the allosteric switch. Finally, there are those like *hemoglobin Hammersmith* in which the tertiary structure of the molecule is unstable.

The most infamous of all variant hemoglobins, *sickle-cell hemoglobin* (HbS), is a source of misery and early death to many humans. The variant has gained its name because it causes red blood cells to adopt an elongated, sickle shape at low oxygen concentrations (Figure 7.34a). This "sickling" is a consequence of the tendency of the mutant hemoglobin, in its deoxygenated state, to aggregate into long, rodlike structures (Figure 7.35). The elongated cells tend to block capillaries, causing inflammation and considerable pain. Even more serious is that the sickled cells are fragile (Figure 7.34b); their breakdown leads to an anemia that leaves the victim susceptible to infections and diseases. Individuals who are homozygous for the sickle-cell mutation often do not survive into adulthood, and those who do are seriously debilitated. Heterozygous individuals, who can still produce some normal hemoglobin, usually suffer

Although most hemoglobin mutations appear to be neutral, some are deleterious.

TABLE 7.1   Selected list of missense mutations in human hemoglobins

Effect	Residue Changed	Change	Name	Consequences of Mutation	Explanation
Sickling	$\beta6$ (A3)	Glu ⟶ Val	S	Sickling	Val fits into EF pocket in chain of another hemoglobin molecule.
	$\beta6$ (A3)	Glu ⟶ Ala	G Makassar	Not significant	Ala probably does not fit the pocket as well.
	$\beta121$ (GH4)	Glu ⟶ Lys	O Arab, Egypt	Enhances sickling in S/O heterozygote	$\beta121$ lies close to residue $\beta6$; Lys increases interaction between molecules.
Change in O_2 affinity	$\alpha87$ (F8)	His ⟶ Tyr	M Iwate	Forms methemoglobin, decreased O_2 affinity	The His normally ligated to Fe has been replaced by Tyr.
	$\alpha141$ (HC3)	Arg ⟶ His	Suresnes	Increases O_2 affinity by favoring R state	Replacement eliminates bond between Arg 141 and Asn 126 in deoxy state.
	$\beta74$ (E18)	Gly ⟶ Asp	Shepherds Bush	Increases O_2 affinity by decrease in BPG binding	The negative charge at this point decreases BPG binding.
	$\beta146$ (HC3)	His ⟶ Asp	Hiroshima	Increases O_2 affinity, reduced Bohr effect	Disrupts salt bridge in deoxy state and removes a His that binds a Bohr-effect proton.
	$\beta92$ (F8)	His ⟶ Gln	St. Etienne	Loss of heme	The normal bond from F8 to Fe is lost, and the polar glutamine tends to open the heme pocket.
Heme loss	$\beta42$ (CD1)	Phe ⟶ Ser	Hammersmith	Unstable, loses heme	Replacement of hydrophobic Phe with Ser attracts water into heme pocket.
Dissociation of tetramer	$\alpha95$ (G2)	Pro ⟶ Arg	St. Lukes	Dissociation	Chain geometry is altered in subunit contact region.
	$\alpha136$ (H19)	Leu ⟶ Pro	Bibba	Dissociation	Pro interrupts helix H.

distress only under conditions of prolonged oxygen deprivation. For example, flying may be dangerous for HbS heterozygotes because of the lower oxygen level in the cabin of the aircraft.

Linus Pauling first suggested in 1949 that sickle-cell disease was a "molecular disease" resulting from a mutation in the hemoglobin molecule. Remarkably, sickling stems from what we might expect to be an innocuous mutation in a part of the molecule far from the critical regions mentioned earlier. The glutamic acid residue normally found at position 6 in β chains is replaced by a valine (see Figure 7.26b). This hydrophobic valine can fit into a pocket at the EF corner of a β chain in another hemoglobin molecule, and thus, as shown in Figure 7.35c, adjacent hemoglobin molecules can fit together into a long, rodlike helical fiber. Why sickling occurs with deoxyhemoglobin, but not with the oxygenated form, is simply explained: In the oxy form the rearrangement of subunits makes the EF pocket inaccessible to Val 6.

Sickle-cell disease is confined largely to populations originating in tropical areas of the world. At first glance, this distribution seems unexpected. Why should a *genetic* disease be climate-related? The answer tells us something about the persistence of what seem to be unfavorable traits. A high incidence of sickle-cell disease in a population generally coincides with a high incidence of malaria, a parasitic disease carried by a tropical mosquito. Individuals *heterozygous* for sickle-cell hemoglobin have a higher resistance to malaria than those who do not carry the sickle-cell mutation. The malarial parasite spends a portion of its life cycle in human red cells, and the increased fragility of the sickled cells, even in heterozygous individuals, tends to interrupt this cycle. In addition, the distortion of the cell membrane of intact sickled cells leads to a loss of potassium ions from these cells, providing a less favorable environment for the parasite. Heterozygous individuals have a higher survival rate—and therefore a better chance of passing on their genes—in malaria-infested regions. However, the high incidence of these genes in the population leads to the birth of many people who are homozygous for the mutant trait.

Many hope that sickle-cell disease will eventually be treatable by gene therapy. Introduction of functional β globin genes into an individual homozygous for the sickle-cell mutation would effectively render them heterozygous, with greatly increased chances for survival. However, it is important to keep in mind that gene therapy is not yet established for treatment of any disease, and it has been associated with considerable risk to patients. In 1998, the U.S. Food and Drug Administration approved hydroxyurea as a treatment for sickle-cell disease. Hydroxyurea appears to induce expression of the HbF $\alpha_2\gamma_2$ tetramer to levels of 10%–15% of total hemoglobin, which is sufficient to reduce some of the symptoms of the disease. Unfortunately, not all patients are responsive to hydroxyurea, and its long-term safety is not known.

Thalassemias: Effects of Misfunctioning Hemoglobin Genes

The human hemoglobin variants we have mentioned so far are all consequences of missense, or nonsynonymous, mutations. Because of a base substitution in a gene coding for one of the chains, one amino acid is substituted for another. There are, however, other genetic defects involving hemoglobin in which one or more of the chains are simply not produced or are produced in an insufficient amount. The pathological condition that arises is called **thalassemia**. The condition of thalassemia can arise in several ways:

1. One or more of the genes coding for hemoglobin chains may have been deleted.

2. All genes may be present, but one or more may have undergone a nonsense mutation that produces a shortened chain or a frameshift mutation that produces a nonfunctional chain (see Figure 7.26c and d).

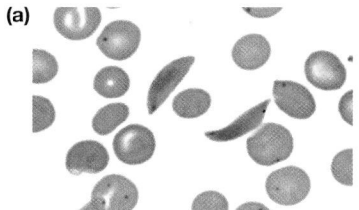

(a)

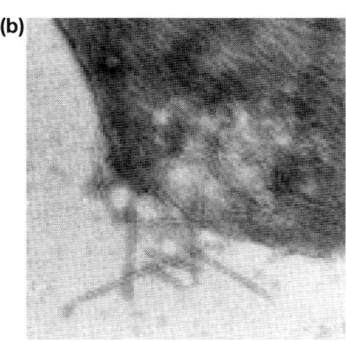

(b)

FIGURE 7.34

Erythrocytes in sickle-cell disease. **(a)** Typical sickled cells, together with some normal, rounded red blood cells. **(b)** Scanning electron micrograph of a sickled cell that has ruptured, with hemoglobin fibers spilling out.

(a) Dr. Gladden Willis/Visuals Unlimited, Inc.;
(b) Courtesy of T. Wellems and R. Josephs.

Sickle-cell disease results from a single base substitution in the β chain.

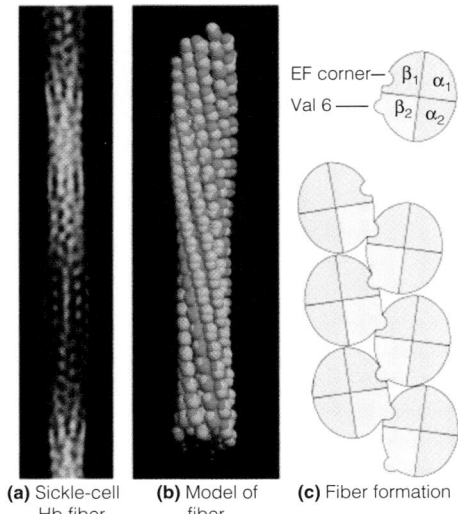

(a) Sickle-cell Hb fiber **(b)** Model of fiber **(c)** Fiber formation

FIGURE 7.35

Sickle-cell hemoglobin. Molecules of sickle-cell hemoglobin tend to aggregate, forming long fibers. **(a)** An electron micrograph of one sickle-cell fiber. **(b)** A computer-graphic depiction of one fiber. **(c)** A schematic model of fiber formation. Deoxyhemoglobin S molecules lock together to form a two-stranded cluster because Val 6 in the β chain of one hemoglobin molecule fits into a pocket in an adjacent molecule. Interaction of these two-stranded structures with one another produces the multistrand fibers shown in (a) and (b).

Courtesy of B. Carragher, D. Bluemke, M. Potell, and R. Josephs.

Thalassemias are hemoglobin mutations in which one or more genes are wholly or partially nonfunctional.

3. All genes may be present, but a mutation may have occurred outside the coding regions, leading to a block in transcription or to improper processing of the pre-mRNA so that the protein is not produced or is not functional.

In case 1 or 2, the gene produces no functional protein. In case 3, limited transcription and translation of the correct polypeptide sequence may occur.

The human genome contains a number of globin genes, corresponding to the protein chains used at different developmental stages, so there are many varieties of thalassemia. We describe here only two major classes—those involving loss or misfunction of genes for the adult β and α chains.

β-Thalassemia

If the β globin gene is lost or cannot be expressed, a most serious condition arises in individuals homozygous for this defect. They can make *no β* chains and must rely on continued production of the fetal γ chains to make a functional hemoglobin, $\alpha_2\gamma_2$ (see Figure 7.29). Such individuals may produce γ chains well into childhood, but they usually die before reaching maturity. Much less serious is the heterozygous state, in which one β gene is still functioning. There are also milder thalassemias (called β^+) in which transcription or processing of the β genes is partially inhibited; thus β globin production is limited but not entirely blocked.

α-Thalassemias

Thalassemias involving the α chain present a more complicated situation. Two copies of the gene (α_1 and α_2) are next to each other on the human chromosome 16 (Figure 7.30). The α_1 and α_2 chains differ by only one amino acid, and one can replace the other in the assembled hemoglobin tetramer. An individual can have 4, 3, 2, 1, or 0 copies of an α gene. Only if three or more genes are nonfunctional are serious effects observed. Individuals with only one α gene are anemic because their total hemoglobin production is low. The low level of α hemoglobin is partially compensated for by formation of β_4 tetramers (*hemoglobin H*) and γ_4 tetramers (*hemoglobin Bart's*). These tetramers can bind and carry oxygen, but they do not exhibit the allosteric transition (they remain always in the R state), nor do they exhibit a Bohr effect. So unloading of oxygen to tissues is inefficient. In the condition known as *hydrops fetalis,* all four α gene copies are missing. Individuals with this condition are inevitably stillborn. They can form only a γ_4 hemoglobin, and because the supply of chains falls near birth, not enough hemoglobin is available to support the near-term fetus.

Because there are two copies of the α gene but only one of the β gene, most of the deleterious mutations in mammalian hemoglobins occur in the β chains (see Figure 7.32). This phenomenon may suggest a functional role for gene duplication: If two or more copies of a gene are present, the species is somewhat protected from the harmful effects of mutations.

We have concentrated our discussion of protein mutation on hemoglobin, but it must be understood that the same principles apply to all other proteins. Although our knowledge of hemoglobin mutations is the most complete, missense mutations and deletions are found in many other proteins as well—those that are deleterious give rise to the wide class of *genetic diseases,* of which hemoglobin pathologies are only one subclass. Many other examples will be encountered in the chapters on metabolism.

Immunoglobulins: Variability in Structure Yields Versatility in Binding

In the remainder of this chapter, we turn to the **immunoglobulin** proteins, whose primary function is specific, and essentially *irreversible*, binding of substances that appear to be of nonself origin, such as bacterial or viral pathogens.

The Adaptive Immune Response

When a foreign substance—a virus, a bacterium, or even a foreign molecule—invades the tissues of a higher vertebrate (like a human), the organism defends itself by the **immune response**. The innate and adaptive immune responses are multilayered and complex, operating through several distinct mechanisms that can occur independently or in concert. Even a superficial description of the various modes of cellular self-defense would require several chapters in a textbook; thus, we restrict our attention to only a narrow range of topics in this vast and fascinating area.

We start with a brief description of the **adaptive immune response**, which includes humoral and cellular components. In the **humoral immune response**, lymphatic cells called **B lymphocytes** synthesize specific immunoglobulin molecules that are secreted from the cell and bind to the invading substance. This binding aggregates the foreign substance and marks it for destruction by cells called **macrophages**. In the **cellular immune response**, lymphatic cells called **T lymphocytes**, bearing immunoglobulin-like molecules on their surfaces, recognize and kill foreign or aberrant cells. In this section we shall be mainly concerned with the humoral immune response.

The substance that elicits an immune response is called the **antigen**, and a specific immunoglobulin that binds to this substance is called the **antibody**. If the invading particle is large, like a cell, a virus, or a protein, many different antibodies may be elicited, each type binding specifically to a given **antigenic determinant** (or **epitope**) on the surface of the particle (Figure 7.36a). Such antigenic determinants may be, for example, groups of sugar residues in a carbohydrate or groups of amino acids on a protein surface; an example is shown in Figure 7.36b.

The immune response, in all its various forms, is our major line of defense against infection and against cancer cells as well. It is the crippling of the immune system by the human immunodeficiency virus (HIV) that makes **AIDS (acquired immune deficiency syndrome)** a disease that has proved so devastating. Victims of AIDS do not die of the direct effects of the virus—they perish from infectious diseases or cancers that their immune system is no longer able to defend against.

The adaptive immune response has some remarkable features. First, it is incredibly versatile, being able to respond to an enormous number of different foreign substances. These foreign substances range from cells of another individual of the same species (the basis of tissue graft or organ transplant rejection) to synthetic molecules that could never have been encountered in nature. Second, the adaptive immune response has a so-called *memory*: After an initial exposure to a given antigen, a second exposure at a later date will result in rapid and much more massive production of the specific antibodies.

The immune response involves the defense of the body against foreign substances or pathogens and operates via many different cellular mechanisms.

In the humoral immune response, B lymphocytes secrete antibodies (immunoglobulins) that react with specific antigens.

FIGURE 7.36

Antigenic determinants. **(a)** A foreign object, or antigen (such as a virus, a bacterial cell, or a foreign protein molecule), may elicit the production of antibodies to several different antigenic determinants on its surface. When the antigen is mixed with this collection of antibodies, precipitation occurs because each antibody molecule has two binding sites for its determinant. Thus, a crosslinked network is formed. **(b)** The antigenic determinants of sperm whale myoglobin. The purple portions represent segments of the polypeptide chain that act as antigens. Some antigenic determinants involve portions of chain that are far apart in the primary sequence but close together in the tertiary structure, a so-called discontinuous epitope.

(b) Reprinted from *Immunochemistry* 12:435, M. Z. Atassi, Antigenic structure of myoglobin: The complete immunochemical anatomy of a protein and conclusions relating to antigenic structures of proteins. © 1975, with permission from Elsevier.

(a) Precipitation

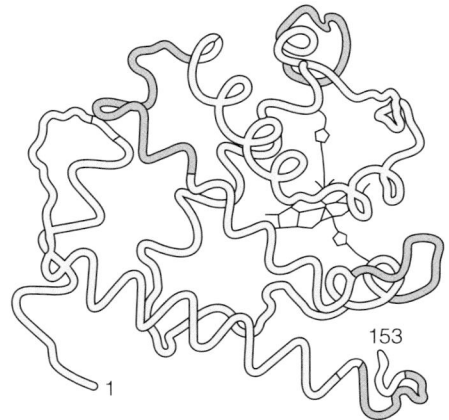

(b) Antigenic determinants, sperm whale myoglobin

Modern explanations for the adaptive immune response are based on the clonal selection theory.

As described by the so-called **clonal selection theory**, the body has an inherent ability to produce an immense diversity of antibodies with different amino acid sequences that are able to bind an enormous range of antigens. The basic postulates of the clonal selection theory, illustrated in Figure 7.37, are as follows:

1. **B stem cells** in the bone marrow differentiate to become B lymphocytes, each producing a single type of immunoglobulin molecule, each type with a binding site that will recognize a specific molecular shape. These immunoglobulins, or antibodies, are attached to the cell membrane and exposed on the outer surfaces of the B lymphocytes.

2. Binding of an antigen to one of these antibodies stimulates the cell carrying it to replicate, generating a **clone** (a collection of cells with identical genetic information). This *primary response* is aided by a special class of T cells called **helper T cells**. If a helper T cell recognizes a bound antigen, it binds to the appropriate B lymphocyte and transmits to it a signal protein (**interleukin-2**) that stimulates B-cell reproduction. Thus, only those clones of B cells that recognize antigens are stimulated to continued cell division.

3. As shown in Figures 7.37 and 7.38, two classes of cloned B cells are produced. **Effector B cells**, or **plasma cells**, now produce *soluble* antibodies, which are secreted into the circulatory system. These antibodies have the same antigen binding sites as the surface antibodies of the B lymphocyte from which the effector cells arose, but they lack the hydrophobic tail that bound the surface antibodies to the lymphocyte membrane. The other class of cells in the clone—**memory cells**—will persist for some time, even after antigen is no longer present. This persistence constitutes the immune memory: It allows a rapid *secondary response* to a second stimulation by the same antigen, as shown in Figure 7.38.

FIGURE 7.37

The clonal selection theory of the adaptive immune response. Stem cells in the bone marrow (B stem cells, top) differentiate and migrate to the lymphoid tissue. Each of the differentiated cells (B lymphocytes) synthesizes a unique kind of antibody, which it carries on its surface. When the B lymphocytes encounter antigens (shown as an orange hexagon; middle), the B cells that carry antibodies to the antigenic determinants are stimulated by helper T cells ("Th2" cell) to multiply, forming B-cell clones. In this case, the cells stimulated to form clones are those bound to either the "red" or "green" epitope on the foreign antigen. Th2 cells stimulate the B-lymphocytes by secretion of interlekin-2 (lightning bolt symbol). Some of the cloned B cells, called plasma cells, produce soluble antibodies, with each clone producing antibody against a single antigenic determinant (epitope). Other cloned B cells are called memory cells; their role is illustrated in Figure 7.38.

Courtesy of Gary Carlton.

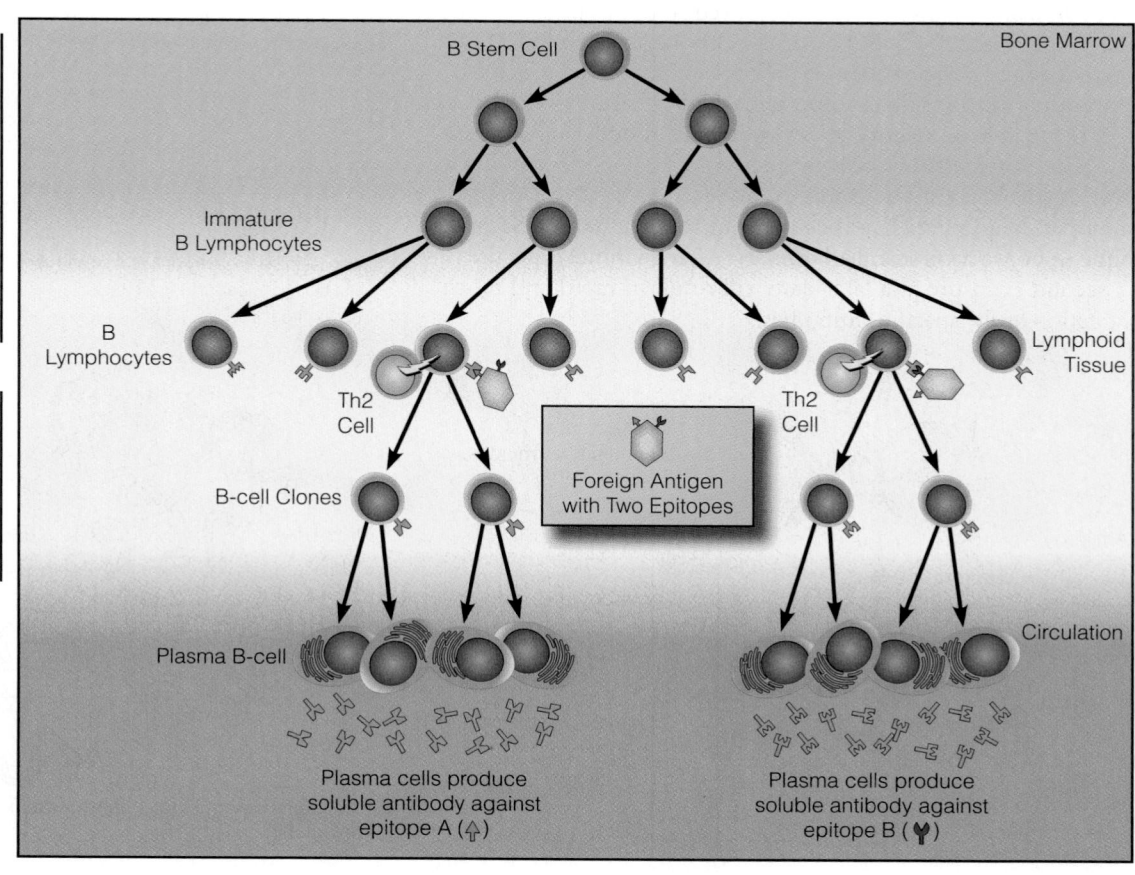

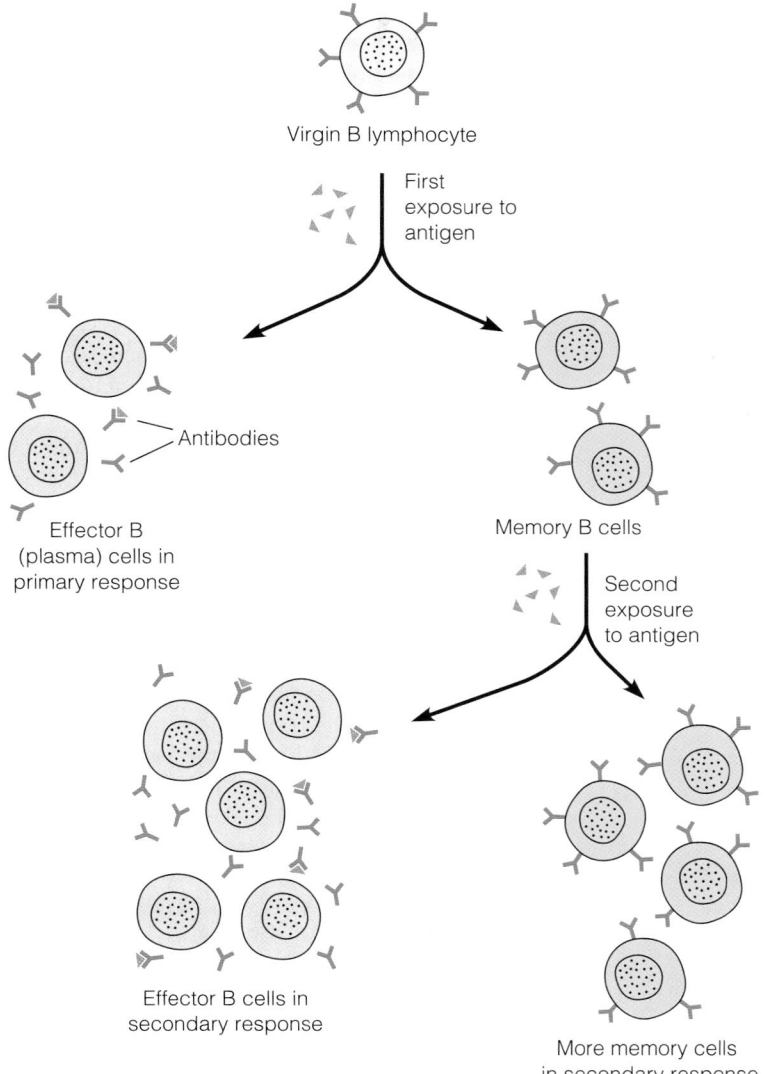

Virgin B lymphocyte

First exposure to antigen

Antibodies

Effector B (plasma) cells in primary response

Memory B cells

Second exposure to antigen

Effector B cells in secondary response

More memory cells in secondary response

FIGURE 7.38

Two developmental paths for stimulated B lymphocytes. Exposure to antigen causes two kinds of cells to develop from B lymphocytes. Cells of one type (effector B cells, or plasma cells) synthesize soluble antibody (see Figure 7.37). Cells of the second class (memory cells) carry membrane-bound antibody to allow a rapid and enhanced response to a second exposure of the same antigen.

The clonal selection theory explains many features of the adaptive immune response, but a critical question may have occurred to you: Why do we not find clones producing antibodies against *our own* proteins and tissues? The answer is a fascinating one that tells much about how biochemical "self" is established. When immature B cells in the fetus encounter substances that bind to their surface antibodies, they are *not* stimulated to replicate. Rather, these fetal B cells are destroyed. Thus, B cells producing antibodies against all of the potential "self" antigens to which we might react are eliminated before birth. The only B cells that mature are those that produce antibodies against "nonself" substances.

Occasionally, the immune system goes awry and produces antibodies against the normal tissues of an adult. The causes for such **autoimmunity** are not wholly understood, but the resulting diseases can be devastating. In *lupus erythematosus*, for example, the individual's own nucleic acids become the object of attack. Other autoimmune diseases include rheumatoid arthritis, multiple sclerosis, type 1 diabetes mellitus, and psoriasis.

In autoimmune diseases, the immune system attacks normal tissues.

The Structure of Antibodies

To see how clonal selection actually works at the molecular level, we must explore the structure of the immunoglobulin molecules that constitute the antibody arsenal. There are five classes of immunoglobulin molecules, which carry out various functions in the immune system (Table 7.2). However, all are built from the same basic immunoglobulin structure, which is shown schematically in Figure 7.39. Different kinds of antibodies may contain from one to five immunoglobulin molecules; when more than one is present, the monomers are linked by an additional polypeptide, called a J chain (see Table 7.2).

Each immunoglobulin monomer consists of four chains, two identical **heavy chains** ($M_W = 53{,}000$ Da each) and two identical **light chains** ($M_W = 23{,}000$ Da each), held together by disulfide bonds. In each chain are **constant domains** (identical in all antibodies of a given class) and a **variable domain**. It is variation in the amino acid sequence (and therefore the tertiary structure) of the variable domains of the light and heavy chains that confers the multitudinous specificities of antibodies to different antigens. Note that the four variable domains are carried at the ends of the Y-like fork of the molecule, where they form two identical binding sites for antigens.

A large protein, a virus, or a bacterial cell has on its surface many different potential antigenic determinants. Antibodies may be generated to several of these determinants, binding many antigen molecules together and thereby aggregating the antigen (see Figure 7.36a). If the antigen is so small that it has only one determinant, binding will occur but aggregation will not. Antibody-mediated aggregation, also called **immunoprecipitation**, requires the antibody to be *bivalent* (to have two binding sites). In the laboratory it is possible to cleave antibodies at the *hinge* regions (see Figure 7.39) with specific protein-cleaving enzymes, to produce an F_c **fragment** and two F_{ab} **fragments** with only one binding site each. The F_{ab} fragments will bind but not precipitate antigen.

The constant domains of the heavy chains in the base of the Y-shaped molecule serve to hold the chains together. More important, these regions also function

Immunoglobulin molecules contain both constant and variable regions. The variable regions are the antigen binding sites.

TABLE 7.2   The five classes of immunoglobulins

IgM is produced during the early response to an invading microorganism. It is the largest immunoglobulin, containing five Y-shaped units of two light and two heavy chains each. The units are held together by a component called a J chain. The relatively large size of IgM restricts it to the bloodstream. It is also effective in triggering an important mechanism for foreign cell destruction, called the complement system.

IgM
(pentamer)
— J chain

IgG molecules, also known as *γ-globulin,* are the most abundant of circulating antibodies. A variant is attached to B-cell surfaces. IgG molecules consist of a single Y-shaped unit and can traverse blood vessel walls rather readily; they also cross the placenta to carry some of the mother's immune protection to the developing fetus. Specific receptors allow such passage. IgG also triggers the complement system.

IgG
(monomer)

IgA is found in body secretions, including saliva, sweat, and tears, and along the walls of the intestines. It is the major antibody of colostrum, the initial secretion from a mother's breasts after birth, and of milk. IgA occurs as a monomer or as double-unit aggregates of the Y-shaped protein molecule. IgA molecules tend to be arranged along the surface of body cells and to combine there with antigens, such as those on a bacterium, thus preventing the foreign substance from directly attaching to the body cell. The invading substance can then be swept out of the body together with the IgA molecule.

IgA
(monomer or dimer)
— J chain

Less is known about the IgD and IgE immunoglobulins. IgD molecules are found on the surface of B cells, though little is known about their function. IgE is associated with some of the body's allergic responses, and its levels are elevated in individuals who have allergies. The constant regions of IgE molecules can bind tightly to mast cells, a type of epithelial and connective tissue cell that releases histamines as part of the allergic response. Both IgD and IgE consist of single Y-shaped units.

IgD
(monomer)   **IgE**
(monomer)

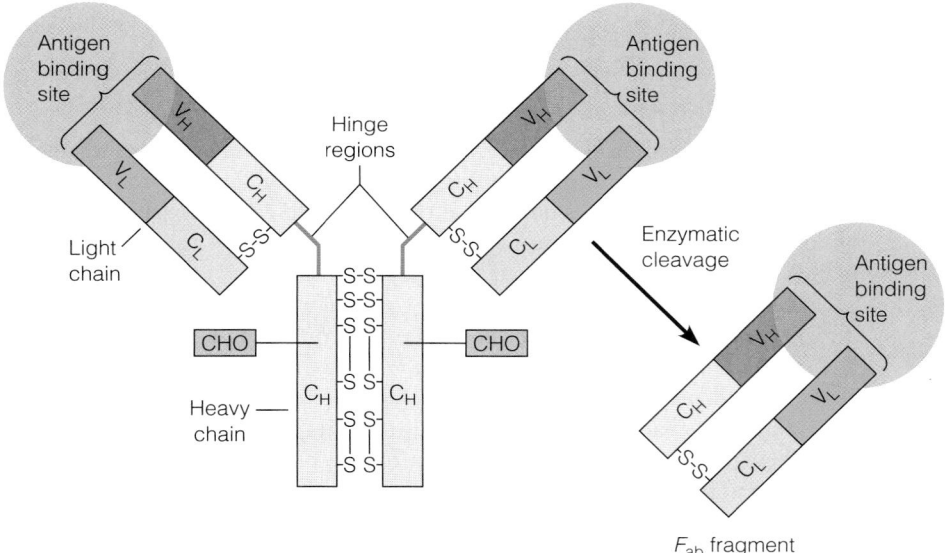

FIGURE 7.39

Schematic models of an IgG antibody molecule and an F_{ab} fragment. The IgG is made from two identical heavy chains and two identical light chains, all held together by disulfide bonds. Each chain contains both *constant* domains (C) and *variable* domains (V). Constant domains are the same in all antibody molecules of a given class (see Table 7.2), whereas variable domains confer specificity to a given antigenic determinant. Cleavage by certain proteolytic enzymes such as papain at the *hinge regions* allows production of two identical monovalent F_{ab} fragments and one F_c fragment (see Figure 7.40). The carbohydrate (CHO) attached to the heavy chains aids in determining the destinations of antibodies in the tissues and in stimulating secondary responses such as phagocytosis. The crystal structure of an immunoglobulin molecule is shown in Figure 7.40.

as effectors, to signal other cells of the immune response, such as killer T-cells or macrophages, to attack particles or cells that have been marked for destruction by antibody binding. Macrophages are large white blood cells that are specially adapted to engulf and digest foreign particles. In addition, differences in heavy chains identify immunoglobulin types for delivery to different tissues or for secretion (see Table 7.2).

The antigen binding sites lie at the extreme ends of the variable domains (Figures 7.39 and 7.40) and involve amino acid residues from the variable regions of both heavy and light chains. Different sequences in these variable regions give rise to different local secondary and tertiary structure and can thereby define binding sites to fit different antigens with exquisite specificity.

The domains in immunoglobulins and other members of the **immunoglobulin superfamily** are built on a common motif—the *immunoglobulin fold,* in which two antiparallel β sheets lie face to face (Figure 7.41). This structure probably represents the primitive structural element in the evolution of the adaptive immune response. Indeed, the immunoglobulin fold is also found in a number of other proteins that are involved in cell recognition. The immunoglobulin fold is a stable scaffold on which to display the hypervariable loops that determine the shape and charge complementarity of the antigen binding site (Figure 7.42). These hypervariable loops are known as **complementarity determining regions**, or **CDRs.** Figure 7.42 shows the results of X-ray diffraction studies of the interaction of an F_{ab} fragment with a viral protein antigen, neuraminidase, an enzyme which facilitates viral infection

The diversity as well as the exquisite specificity of antigen binding sites is determined by the hypervariable complementarity determining regions from both the light and the heavy chains.

FIGURE 7.40

The crystal structure of an IgG molecule from mouse. The identical heavy chains are colored yellow and cyan; the identical light chains are magenta and green. A cartoon model, illustrating the high degree of β secondary structure, is shown on the left. On the right is a surface rendering showing the intimate contact between the chains. The carbohydrate attached to the heavy chain is indicated in the leftmost figure. PDB ID: 1IGT.

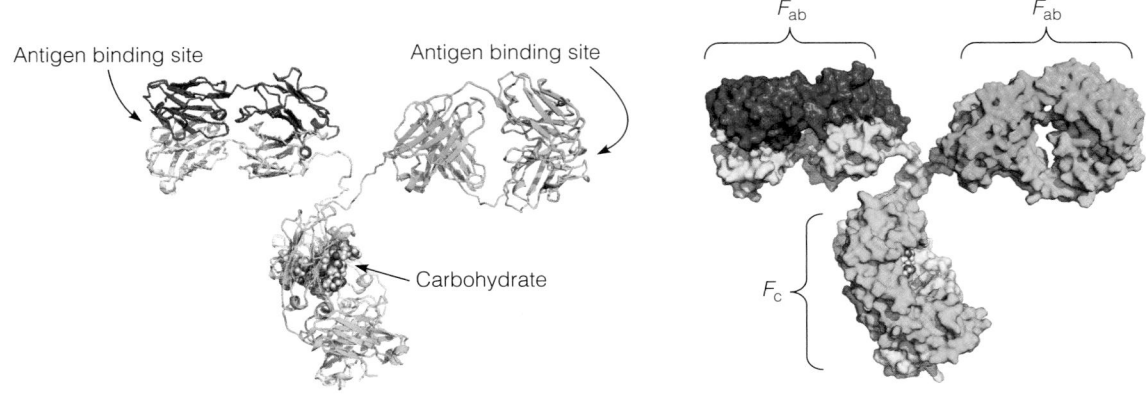

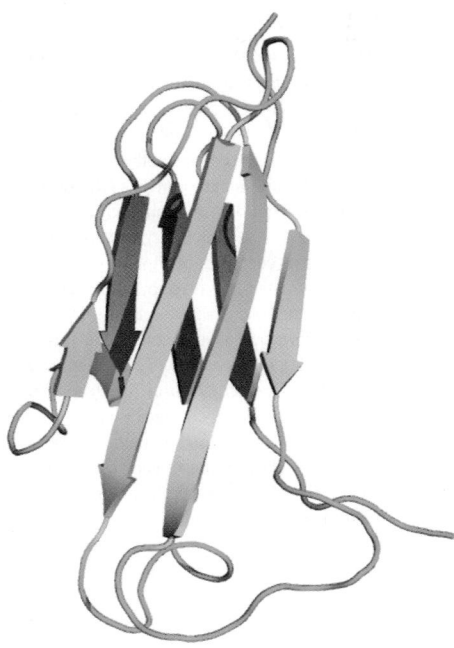

FIGURE 7.41

The immunoglobulin fold. The immunoglobulin fold is a common structure in domains of many proteins in the immunoglobulin superfamily (see text). Two antiparallel β sheets (cyan and orange) are stacked face to face and covalently bonded by a disulfide bond (not shown). This folding motif is found 12 times in the IgG molecule (see leftmost panel of Figure 7.40) and 4 times within an F_{ab}. PDB ID: 1IGT.

Through somatic recombination and rapid mutation, a human can generate over 10 billion different antibodies.

The cellular immune response uses killer T cells to destroy foreign or infected cells.

(discussed in Chapter 9). Note that the antigen and antibody surfaces fit together in a highly complementary fashion.

Generation of Antibody Diversity

How can the enormous diversity of immunoglobulin molecules be generated so that antibodies to an almost unlimited range of antigens are provided? The human genome simply does not have enough room to encode a gene for each of the millions of different immunoglobulin molecules occurring in B stem cells. Instead, two special processes occur in these cells.

The major source of antibody diversity is *recombination of exons*. The genomes of higher vertebrates contain "libraries" of exons corresponding to different portions of the immunoglobulin molecule. In antibody-producing cells, these exons are rearranged and spliced to create different sequence combinations in both the heavy and light chains. We have already mentioned that such rearrangements, when they occur in germline cells, play a role in protein evolution. The same process, when it occurs in B cells, creates new immunoglobulins in individual cells. The details and mechanism of this process are described in Chapter 26, as is an additional source of antibody diversity: *somatic* hypermutation. Such mutations are not inherited because they occur in somatic cells (cells that are not germline). In the cells that generate antibodies, the portions of the variable regions corresponding to the CDR loops in the immunoglobulin genes mutate at an unusually high rate. This process, together with recombination of gene fragments, can account for the generation of an immense diversity of immunoglobulin molecules. It has been estimated that about 10 billion combinations can be made from the library of immunoglobulin gene fragments available in the human genome.

T Cells and the Cellular Response

Whereas the humoral immune response of the adaptive immune system is based on antibody-produced aggregation, usually followed by digestion by macrophages, the cellular immune response involves a quite different mechanism for killing of foreign cells. The cellular response plays a major role in tissue rejection and in destroying virus-infected cells. It can also destroy potential cancer cells before they have a chance to propagate. Although the mechanisms of the humoral and cellular processes are quite different, structurally similar protein molecules (from the immunoglobin superfamily) are involved in both cases, pointing to a common evolutionary origin for the humoral and cellular responses.

The major participants in the cellular immune response are **cytotoxic T cells**, also referred to as **killer T cells**. These cells carry on their surfaces receptor molecules that are structurally similar to the F_{ab} fragments of antibody molecules (Figure 7.43c). Like antibodies, these fragments have a wide range of binding specificities, mostly directed toward short oligopeptide sequences. Such oligopeptides might be produced, for example, by a virus-infected cell when it partially digests virus particles within it. The T-cell receptor does not recognize free oligopeptides. Instead, the oligopeptides must be presented on the surface of the infected cell and bound to another class of immunoglobulin-like molecules, proteins of the **major histocompatibility complex**, or **MHC** (see Figure 7.43a–c). When a killer T cell identifies (via its receptor) an appropriate antigen carried on the surface of another cell by an MHC protein, it releases a protein called **perforin**. This protein forms pores in the plasma membrane of the cell being attacked, allowing critical ions to diffuse out and thereby killing the cell.

It is instructive to compare the immunoglobulin family of proteins with the myoglobin–hemoglobin family. In both cases the primary function of the proteins is binding. In the myoglobin–hemoglobin family we see evidence for the progressive evolution of increasingly sophisticated methods for regulating the binding of a particular molecule—oxygen—and for coupling oxygen binding to the removal of

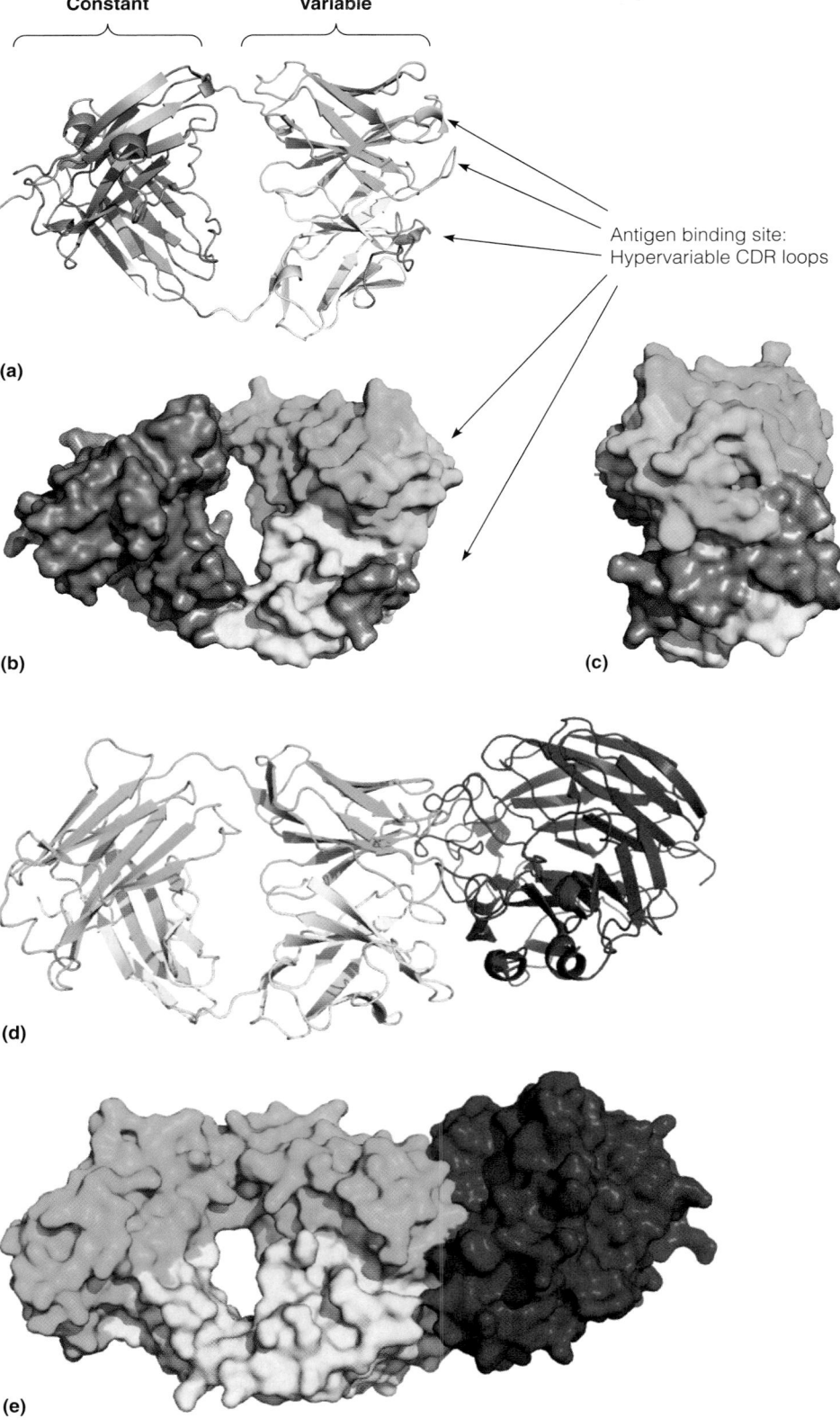

Constant **Variable**

(a)

(b)

(c)

Antigen binding site:
Hypervariable CDR loops

(d)

(e)

FIGURE 7.42

Antigen binding by an F_{ab} fragment.
Panel **(a)**: Backbone structure of an F_{ab} fragment, which contains four immunoglobulin fold domains: two on the light chain (forest and lime green), and two on the heavy chain (tan and yellow). The constant domains on each chain are to the left and the variable domains are on the right. The CDRs are shown in cyan (light chain) and orange (heavy chain). The CDRs from both the heavy and light chains are hypervariable in sequence and determine the shape and specificity of the antigen binding site. PDB ID: 1AQK. Panel **(b)**: Surface rendering of F_{ab} fragment shown in (a). Panel **(c)**: Same rendering as in (b), but rotated 90 degrees to view the surface of the antigen binding site formed by the CDR loops. Panels **(d)** and **(e)**: The close contact that occurs between antigen and antibody surfaces is shown in the backbone and surface renderings of a murine F_{ab} fragment bound to the viral protein neuraminidase (PDB ID: 1NCA). The antibody light chain is shown in green, the heavy chain fragment in yellow, and the neuraminidase molecule in purple. The surfaces of the antigen and the antibody binding site fit together with a high degree of shape and charge complementarity.

FIGURE 7.43

Structural similarity of proteins from the immunoglobulin superfamily. This figure shows crystal structures for a few members of the immunoglobulin superfamily of proteins, which includes not only the immunoglobulin family but many related cell-surface and soluble proteins involved cell recognition and binding. Notice the occurrence of several immunoglobulin domains in all these proteins. (a) Human major histocompatibility complex (MHC) class I protein (cyan and green) bound to a fragment of an HIV protein (magenta). PDB ID: 1A1M. (b) Human MHC class II protein (green and cyan) bound to an influenza virus peptide (magenta). PDB ID: 1DLH. (c) Human T-cell receptor (rose and yellow) binding to an MHC class I molecule (cyan and green) displaying a viral peptide (magenta). PDB ID: 1BD2. (d) Murine F_{ab} fragment (cyan and green) of IgG bound to neuraminidase (magenta; see Figure 7.42, PDB ID: 1NCA).

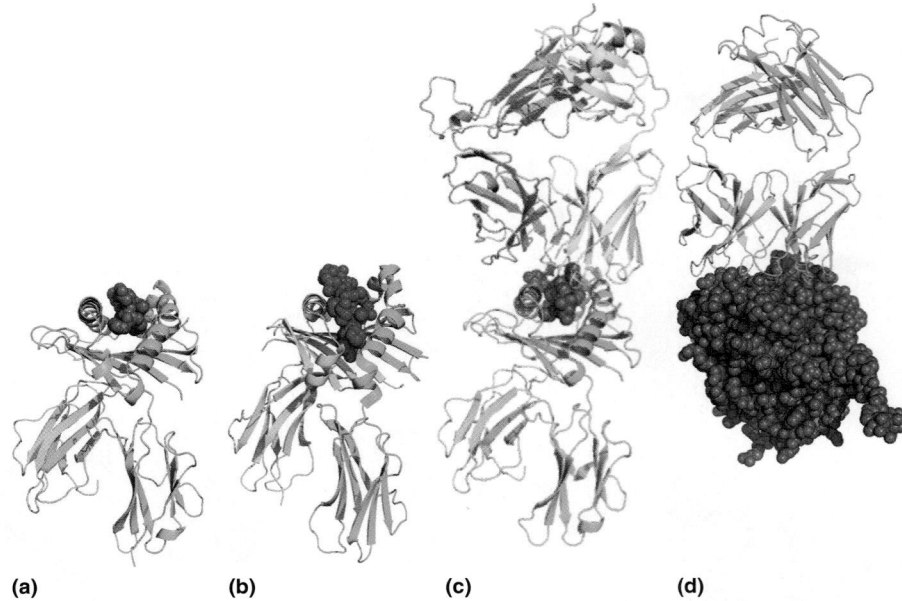

(a)   (b)   (c)   (d)

CO_2. In the immunoglobulin family, evolution from a simple motif has led to an enormous diversification in binding function. A mechanism has evolved that allows the production of an immense range of molecules with specific binding capacities.

The Innate Immune Response

In 1989 Charles A. Janeway, Jr. hypothesized that cells of the so-called **innate immune response**, distinct from the B and T cells described above, detect bacterial infection by recognizing metabolites specific to microbial pathogens. Extensive research since then has validated Janeway's hypothesis and led to the identification of a myriad of proteins and effectors that recognize bacterial, viral, fungal, and protozoan invaders and mount an immune response. Among the many important proteins in the innate immune response are the so-called **toll-like receptors** (TLRs), which recognize a diverse set of foreign molecules. These include double-stranded RNA (typical of many viruses) and organic molecules, such as **lipopolysaccharide** (LPS), which is found in the membranes of gram-negative bacteria. Figure 7.44 shows crystal structures of two different TLRs bound to their targets. Note that the TLR structures and modes of target recognition are very different from those described above for the immunoglobulin superfamily. Although the innate immune response can act independently, it can also stimulate the adaptive immune response.

FIGURE 7.44

Structure of two toll-like receptors (TLRs) bound to foreign molecules. (a) The extracellular domain of human TLR-4 (cyan and green) combines with myeloid differentiation factor-2 (purple and yellow) to bind *E. coli* lipopolysaccharide (spacefilling) PDB ID: 3FXI. (b) A model of an entire TLR. The model is a composite of the extracellular domain of murine TLR-3 (cyan and green) bound to a dsRNA (red and blue), which is connected via a trans-membrane helix to the intracellular domain of a human TLR. PDB IDs: 3CIY and 2J67.

Panel (b) from *Science* 320:379–381, L. Liu, I. Botos, Y. Wang, J. N. Leonard, J. Shiloach, D. M. Segal and D. R. Davies, Structural basis of toll-like receptor 3 signaling with double-stranded RNA. © 2008. Reprinted with permission from AAAS.

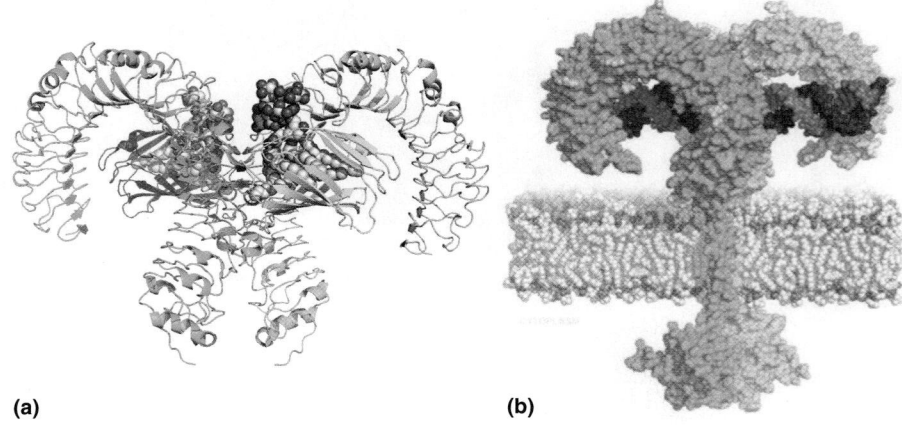

(a)   (b)

AIDS and the Immune Response

AIDS (acquired immune deficiency syndrome) is a disease of the immune system. It is caused by the **human immunodeficiency virus**, or HIV (Figure 7.45), which attacks a number of kinds of cells but is particularly virulent toward a class of helper T cells. The virus wages a long battle with rapidly replicating T cells, but eventually the rate of cell destruction exceeds the rate of replication. The consequence is a deterioration of the whole immune response, in particular the ability of B cells to proliferate in response to antigen stimulation. In addition, there is a general failure in T-cell activation. Most AIDS patients succumb either to diseases they could have easily resisted before contracting AIDS or to certain kinds of cancer. AIDS is so deadly because it attacks our most fundamental defenses against all disease.

Since 1983 more than 60 million people, predominantly in the developing world, have been infected with HIV, and nearly half of those have died. Because AIDS poses such a grave threat to world health, efforts to develop a vaccine are being intensely pursued. Such research entails unusual problems because the AIDS virus has an unparalleled capacity to mutate and thus develop strains resistant to any vaccine. Mutations occur in the HIV genome at a rate many times higher than in the human genome. The genetic variation is increased by the error-prone viral replication cycles (discussed in Chapter 25), resulting in a high level of amino acid mutation in the viral coat proteins which would be the typical immunological targets in a vaccine development strategy. The magnitude of the problem can be grasped by considering our experience with the influenza virus. We have never been able to produce a lifelong "flu" vaccine because of the great variability of the influenza virus. HIV mutates about 60 times faster than the influenza virus.

Unfortunately, no effective vaccine has yet been developed for HIV. Antiviral drugs have been developed that target virus-specific enzymes involved in replication of the viral genome or processing of the gene products. These therapies can slow the progression of AIDS; but they do not constitute a cure for the disease. One may hope that improved therapeutics, and an eventual vaccine, will halt the AIDS pandemic.

Antibodies and Immunoconjugates as Potential Cancer Treatments

Current cancer treatments frequently include radiation and/or chemotherapy with highly toxic drugs. Chemotherapy is associated with many undesirable side effects, in large part due to the broad cytotoxicity of the drugs. The efficacy of chemotherapeutics could be increased if they could be delivered specifically to cancer cells

In AIDS, the causal virus attacks helper T cells, destroying the body's immunological defense system.

FIGURE 7.45

(a) Electron micrograph of the human immunodeficiency virus (HIV) that is responsible for AIDS. **(b)** A schematic model of HIV. The surface protein gp160 is composed of two fragments, gp41 and gp120. The RNA genome is transcribed into DNA by an error-prone reverse transcriptase. This DNA integrates into the host cell genome, and is then retranscribed to produce new viral RNA. **(c)** False-color scanning electron micrograph of budding HIV-1 virus particles (green spheres) on the surface of a human lymphocyte (red).

(a) Courtesy of Hans Gelderblom; (b) From *Hospital Practice* (27)9:154, Hoth, Jr., Myers, and Stein, with permission from JTE Multimedia. Illustration © Alan D. Iselin (c) Centers for Disease Control / C. Goldsmith, P. Feorino, E. L. Palmer, and W. R. McManus.

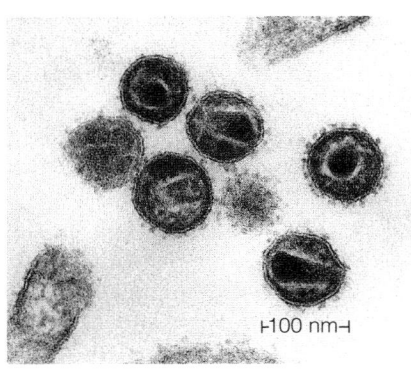

(a)

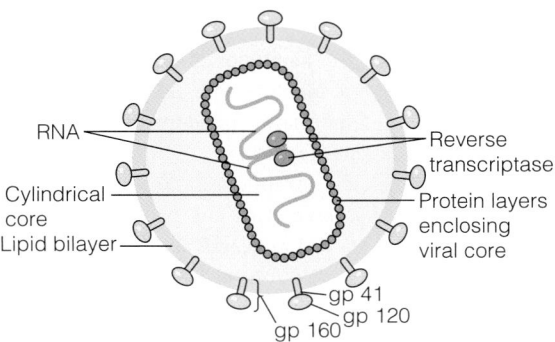

RNA
Cylindrical core
Lipid bilayer
Reverse transcriptase
Protein layers enclosing viral core
gp 41
gp 160 gp 120

(b)

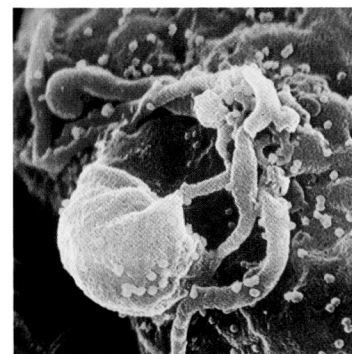

(c)

(a)

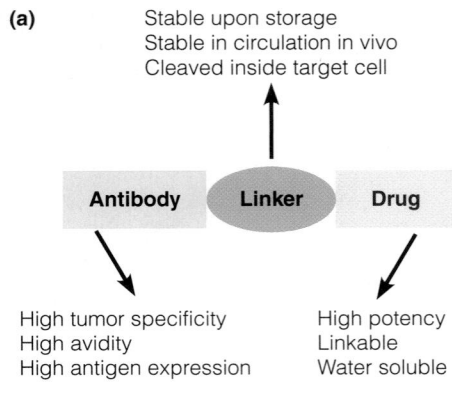

Stable upon storage
Stable in circulation in vivo
Cleaved inside target cell

| Antibody | Linker | Drug |

High tumor specificity
High avidity
High antigen expression

High potency
Linkable
Water soluble

(b)

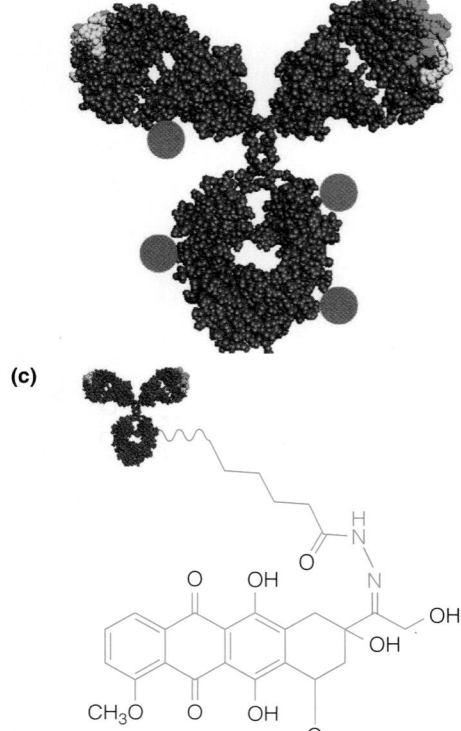

(c)

mAb-doxorubicin (hydrazone)

FIGURE 7.46

Immunoconjugate drugs for targeted chemotherapy. **(a)** The desirable features for each component of the immunoconjugate: targeting antibody, linker, and cytotoxic drug. **(b)** Common sites of attachment of drugs to the antibody constant regions are shown as orange spheres. The tumor-specific antigen binding sites are shown in green and yellow. **(c)** A schematic (not to scale) of the immunoconjugate. The acid-labile hydrazone linker is highlighted in green. The linker is stable in circulation in blood (pH 7.4), but is cleaved in the acidic environment of the endosome following endocytosis into the tumor cell.

(a, b) Reprinted with permission from *Accounts of Chemical Research* 41:98–107, R. V. Chari, Targeted cancer therapy: Conferring specificity to cytotoxic drugs. © 2008 American Chemical Society.

rather than systemically. To that end, hybrid drugs, called **immunoconjugates**, have been designed which link a cytotoxic agent to an antibody that has been raised against a tumor-specific antigen (Figure 7.46). In principle, the antibody specifically delivers the drug to the target tumor cell, where it is taken up by the cell. Once inside the cell, the covalent link between the antibody and drug is cleaved, releasing the drug to exert its cell-killing effect. In addition to cytotoxic drugs, such as taxol, radioactive isotopes have also been conjugated to antibodies. A few such immunoconjugates have been approved for clinical use, with more in clinical trials.

Because antibody binding is the first step in the recruitment of a cytotoxic response, antibodies that specifically recognize tumor antigens have been actively developed as anti-cancer drugs that can selectively target tumors for destruction. As of 2010, roughly 25 antibodies or immunoglobulin derivatives have been approved for use as human therapeutics, primarily as anti-cancer and anti-inflammatory agents.

SUMMARY

Most organisms need oxygen. Vertebrates use hemoglobin for oxygen transport between lungs/gills and respiring tissues. In the heme globins, O_2 is bound at an Fe(II)-porphyrin (heme); the heme is carried in a hydrophobic pocket, inhibiting oxidation of the iron. Myoglobin carries a single oxygen binding site and consequently exhibits a noncooperative, hyperbolic binding curve. Hemoglobin binds O_2 cooperatively, with a sigmoidal binding curve that leads to more efficient O_2 delivery. Binding O_2 to hemoglobin sites causes tertiary structure changes. When strain from these changes accumulates, a quaternary (T → R) transition occurs, shifting the molecule from the lower-affinity form to the higher-affinity form. The allosteric transition also allows allosteric effectors (H^+, CO_2, Cl^-, and 2,3-BPG) to modify oxygen binding, leading to more efficient O_2 and CO_2 transport. Globins are also potent nitric oxide (NO) dioxygenases and can protect cells from the toxic effects of elevated NO levels.

Myoglobin and hemoglobin, like other proteins, are evolving via mutations, duplications, and recombinations in their genes. Both types of globin evolved from a myoglobin-like ancestral protein, with the development of a true hemoglobin coinciding approximately with the emergence of vertebrates. Evolution of these proteins continues, as evidenced by the existence of a multitude of variant hemoglobins in the human population. Most base substitution (missense) mutations are neutral, but some, like the sickle-cell hemoglobin mutation, are deleterious. Thalassemias are hemoglobin pathologies that involve either deletion or faulty expression of whole genes or sets of genes.

The innate and adaptive immune responses are among the body's main defenses against infection. In the humoral response of the adaptive immune system, antibodies (specific immunoglobulin molecules) that will bind with specific antigens are generated and secreted. This process occurs because recognition of the antigen by a few cells leads to clonal selection of a large number of cells producing the appropriate antibody. Immense antibody diversity is achieved through multiple somatic recombinations and rapid mutation in antibody-producing cells. The cellular immune response involves receptor-bearing killer T cells. AIDS is a disease of the immune system; HIV attacks T cells essential to the growth of B-cell clones.

APPENDIX

A Brief Look at Multistate and Dynamic Models of Hemoglobin Allostery

The conformational changes described in the Perutz model of hemoglobin allostery constitute a rearrangement of the *tertiary* structure of each subunit upon oxygen binding, which is presumed to trigger the observed changes in *quaternary* structure between the fully deoxy (T) and fully oxy (R) states of the hemoglobin tetramer (see Figure 7.14). For the discussion below we will use upper-case T and R to represent quaternary structures, and lower-case *t* and *r* to represent the tertiary structures, associated with the deoxy and oxy conformations. The Perutz model is based on the two hemoglobin conformations that were crystallized at that time, but more recent data have called into question many of the assumptions of the MWC and Perutz models. For example, the central tenet of the symmetry model—that the *t* conformation is not tolerated in the R-state quaternary structure, nor is the *r* conformation allowed in the T-state quaternary structure—has been disproven in the case of hemoglobin by X-ray crystal and solution NMR structures of hemoglobin that are in the T quaternary state but have subunits in the *r* tertiary state. Ultrafast kinetic studies of ligand binding to hemoglobin encapsulated in gels, performed on timescales where tertiary conformation switching can occur but quaternary structure switching cannot, have also shown that the *r* state can exist in T tetramers. How, then, are the tertiary and quaternary structural changes correlated? Not surprisingly, there are several competing theories. We briefly summarize three, which have been chosen to illustrate the breadth of current vigorous debate over the structural basis for cooperative ligand binding in hemoglobin. The interested reader will find citations to more complete descriptions of these models in the References section.

The *symmetry rule* model proposed in 1992 by Gary Ackers and coworkers suggests that the tertiary structure changes that accompany oxygen binding can be tolerated up to a certain point before the T → R switch occurs. Specifically, whenever one site is occupied on *each* of the two *αβ* dimers, the molecule as a whole adopts the R quaternary structure. This model proposes that hemoglobin acts as a dimer of cooperative dimers, in a fashion that is in accord with the KNF model but not with the MWC model shown in Figure 7.12.

The *tertiary two-state* model proposed in 2002 by William Eaton and colleagues assumes that the quaternary T and R states of hemoglobin can be populated by both the *t* and *r* tertiary states. As shown schematically in Figure 7.47, the relative populations of *r* and *t* states within a tetramer are influenced by quaternary structure, ligand binding, and the binding of allosteric effectors. In this model, the ligand-binding affinity of a subunit is determined by its tertiary structure and is independent of the hemoglobin quaternary structure. In other words, the ligand-binding affinity of the *r* state is the same whether it resides in a T or an R tetramer. This model explains the *biphasic* kinetics of ligand binding to hemoglobin trapped in the T state within gels. The faster phase is attributed to binding to *r* state subunits in the tetramer and the slower phase is due to binding to the *t* state subunits.

The *dynamic allostery* model proposed in 2008 by Takashi Yonetani and coworkers was developed in part to explain their observations that (1) the binding of some allosteric effectors dramatically reduces the ligand-binding affinities of both R and T state hemoglobins and (2) in extreme cases the ligand-binding affinities of the T and R states have the *same* low value, even though they have *different* quaternary and tertiary conformations. How is this possible, given the foregoing discussion of the Perutz mechanism, which correlates structural differences to differences in binding affinity? Yonetani and his colleagues argue that differences in ligand-binding affinity are determined by differences in protein dynamics, where increased dynamic motions of the hemoglobin molecule are the basis for reduced ligand-binding affinity. Recall from our earlier discussion that heme-pocket dynamics appears to play a significant role in the binding and release of ligands in myoglobin (see Figure 7.8). In particular, the dynamic allostery model focuses on the changes in dynamic behavior of the E and F helices, which include the proximal and distal His residues. Molecular dynamics simulations of oxy and deoxy hemoglobins yield the counterintuitive result shown in Figure 7.48, namely, that the binding of 2,3-BPG

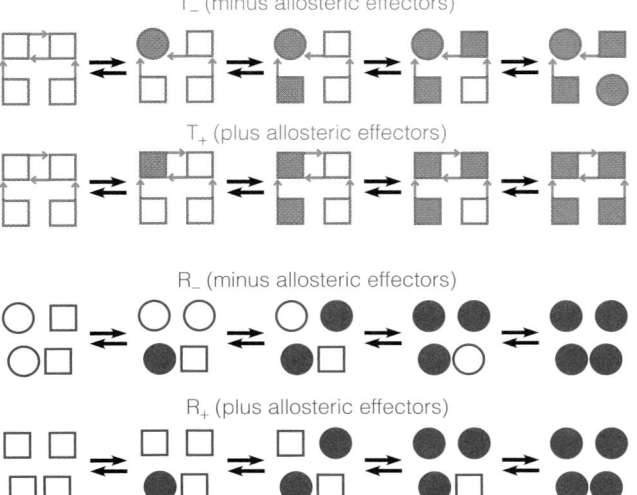

T_ (minus allosteric effectors)

T_+ (plus allosteric effectors)

R_ (minus allosteric effectors)

R_+ (plus allosteric effectors)

FIGURE 7.47

The tertiary two-state model for the cooperative transition of hemoglobin. Subunits are depicted here with $\alpha_1\beta_1$ dimers on the left, $\alpha_2\beta_2$ on the right. R state tetramers are magenta, T state tetramers are brown. The salt bridge interactions that characterize the T state (see Figure 7.15) are shown as green arrows. Subunits with ligand bound are shown as solid shapes. In this model both *r* (circles) and *t* (squares) tertiary structures are found within the T and R quaternary structures. In the presence of allosteric effectors (2,3-BPG, H^+, CO_2 ...) the $t \rightleftarrows r$ equilibrium is perturbed in the direction of the *t* state.

IUBMB Life 59:586–599, W. A. Eaton, E. R. Henry, J. Hofrichter, S. Bettat, C. Viappiani and A. Mozzarelli, Evolution of allosteric models for hemoglobin. © 2007 John Wiley & Sons, Inc. Reproduced with permission from John Wiley & Sons, Inc.

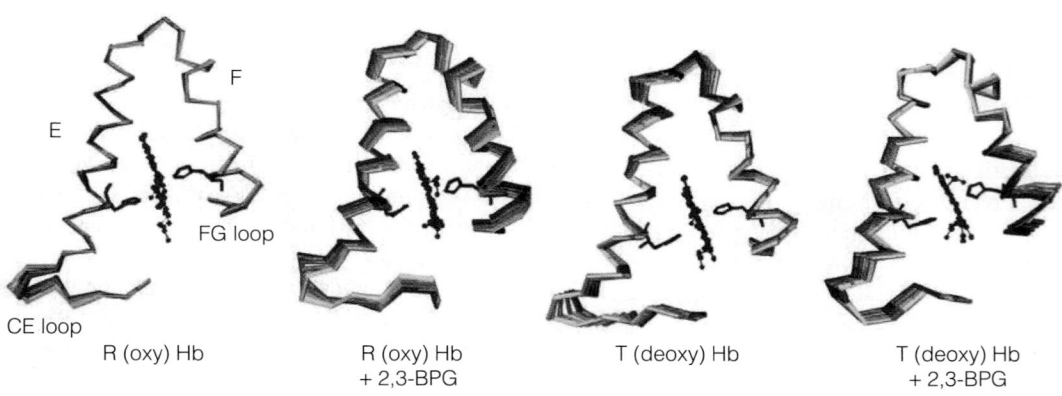

	F		
E			

FG loop

CE loop

| R (oxy) Hb | R (oxy) Hb
+ 2,3-BPG | T (deoxy) Hb | T (deoxy) Hb
+ 2,3-BPG |

FIGURE 7.48

Molecular dynamics of oxy and deoxy hemoglobins. The results of a 2-ns all-atom simulation are shown in each panel. For clarity, only the E and F helices of one α subunit are shown. The starting structure ($t = 0$) in each case is shown in blue and the last structure ($t = 2$ ns) is shown in red. Eight intermediate structures are shown in the order: blue $\rightarrow$ green $\rightarrow$ yellow $\rightarrow$ red. OxyHb (R) is less dynamic than is deoxyHb (T). The addition of 2,3-BPG to the simulations resulted in increased motions in both oxy and deoxy molecules.

Reprinted from *Biophysical Journal* 94:2737–2751, M. Laberge and T. Yonetani, Molecular dynamics simulations of hemoglobin A in different states and bound to DPG: Effector-linked perturbation of tertiary conformations and HbA concerted dynamics. © 2008, with permission from Elsevier.

increases the dynamic motions of the E and F helices in both the oxy (R) and deoxy (T) states. This is surprising, as 2,3-BPG binds to hemoglobin in a cleft *between* the β_1 and β_2 subunits (see Figure 7.23) and therefore might be expected to *reduce* dynamic motions in the hemoglobin molecule.

REFERENCES

General

Bunn, H. F., and B. G. Forget (1986) *Hemoglobin: Molecular, Genetic and Clinical Aspects.* WB Saunders, Philadelphia, PA.

Dickerson, R. E., and I. Geis (1983) *Hemoglobin: Structure, Function, Evolution, and Pathology.* Benjamin/Cummings, Redwood City, CA.

Ordway, G. A., and D. J. Garry (2004) Myoglobin: An essential hemoprotein in striated muscle. *J. Exp. Biol.* 207:3441–3446.

van Holde, K. E., W. C. Johnson, and P. S. Ho (2006) *Principles of Physical Biochemistry* (2nd ed.). Prentice Hall, Upper Saddle River, N.J. Chapter 15 contains a more detailed discussion of binding equilibrium than is presented here.

Allosteric Models

Ackers, G. K., and J. M. Holt (2006) Asymmetric cooperativity in a symmetric tetramer: Human hemoglobin. *J. Biol. Chem.* 281:11441–11443.

Adair, G. S. (1925) The hemoglobin system. VI: The oxygen dissociation curve of hemoglobin. *J. Biol. Chem.* 63:529–545.

Barrick, D., N. T. Ho, V. Simplaceanu, F. Dahlquist, and C. Ho (1997) A test of the role of the proximal histidines in the Perutz model for cooperativity in haemoglobin. *Nature Struct. Biol.* 4:78–83.

Eaton, W. A., E. R. Henry, J. Hofrichter, S. Bettati, C. Viappiani, and A. Mozzarelli (2007) Evolution of allosteric models for haemoglobin. *IUBMB Life* 59:586–599.

Koshland, D. E., G. Nemethy, and D. Filmer (1966) Comparison of experimental binding data and theoretical models in proteins containing subunits. *Biochemistry* 5:365–385.

Monod, J., J. Wyman, and J. P. Changeux (1965) On the nature of allosteric transitions: A plausible model. *J. Mol. Biol.* 12:88–118.

Perutz, M. F., A. J. Wilkinson, M. Paoli, and G. G. Dodson (1998) The steroechemical mechanism of cooperative effects in hemoglobin revisited. *Annu. Rev. Biophys. Biomol. Struct.* 27:1–34.

Tsai, C. J., A. del Sol, and R. Nussinov (2009) Protein allostery, signal transmission and dynamics: A classification scheme of allosteric mechanisms. *Mol. BioSyst.* 5:207–216.

Yonetani, T., and M. Laberge (2008) Protein dynamics explain the allosteric behaviors of haemoglobin. *Biochem. Biophys. Acta* 1784:1146–1158.

Mechanism of Oxygen Binding and Release, and Protein Structure and Dynamics

Birukou, I., R. L. Schweers, and J. S. Olson (2010) The distal histidine stabilizes bound O_2 and acts as a gate for ligand entry in both subunits of human HbA. *J. Biol. Chem.* 285:8840–8854.

Jensen, F. B. (2004) Red blood cell pH, the Bohr effect, and other oxygenation-linked phenomena in blood O_2 and CO_2 transport. *Acta Physiol. Scand.* 182:215–227.

Lukin, J. A., and C. Ho (2004) The structure-function relationship of hemoglobin in solution at atomic resolution. *Chem. Rev.* 104:1219–1230.

Ruscio, J. Z., D. Kumar, M. Shukla, M. G. Prisant, T. M. Murali, and A. V. Onufriev (2008) Atomic level computational identification of ligand migration pathways between solvent and binding site in myoglobin. *Proc. Natl. Acad. Sci. USA.* 105:9204–9209.

Schotte, F., M. Lim, T. A. Jackson, A. V. Smirnov, J. Soman, J. S. Olson, G. N. Phillips Jr., M. Wulff, and P. A. Anfinrud (2003) Watching a protein as it functions with 150-ps time-resolved x-ray crystallography. *Science* 300:1944–1947.

Song, X. J., Y. Yuan, V. Simplaceanu, S. C. Sahu, N. T. Ho, and C. Ho (2007) A comparative NMR study of the polypeptide backbone dynamics of hemoglobin in the deoxy and carbonmonoxy forms. *Biochemistry* 46:6795–6803.

NO Dioxygenation and Other Proposed Functions of Heme Globins

Brunori, M. (2001) Nitric oxide moves myoglobin centre stage. *Trends Biochem. Sci.* 26:209–210.

Burmester, T., and T. Hankeln (2009) What is the function of neuroglobin? *J. Exptl. Biol.* 212:1423–1428.

Fago, A., A. J. Mathews, and T. Brittain (2008) A role for neuroglobin: Resetting the trigger level for apoptosis in neuronal and retinal cells. *IUBMB Life* 60:398–401.

Gardner, P. R., A. M. Gardner, W. T. Brashear, T. Suzuki, A. N. Hvitved, K. D. R. Setchell, and J. S. Olson (2006) Hemoglobins dioxygenate nitric oxide with high fidelity. *J. Inorg. Biochem.* 100:542–550.

Gardner, P. R., A. M. Gardner, L. A. Martin, and A. L. Salzman (1998) Nitric oxide dioxygenase: An enzymatic function for flavohemoglobin. *Proc. Natl. Acad. Sci. USA* 95:10378–10383.

Kim-Shapiro, D. B., A. N. Schechter, and M. T. Gladwin (2006) Unraveling the reactions of nitric oxide, nitrite, and hemoglobin in physiology and therapeutics. *Arterioscler. Thromb. Vasc. Biol.* 26:697–705.

Olson, J. S., E. W. Foley, C. Rogge, A. L. Tsai, M. P. Doyle, and D. D. Lemon (2004) NO scavenging and the hypertensive effect of hemoglobin-based blood substitutes. *Free Radical Biol. Med.* 36:685–697.

Poole, R. K., and M. N. Hughes (2000) New functions for the ancient globin family: Bacterial responses to nitric oxide and nitrosative stress. *Mol. Microbiol.* 36:775–783.

Wittenberg, J. B., and B. A. Wittenberg (2003) Myoglobin function revisited. *J. Exptl. Biol.* 206:2011–2020.

Evolution of Globin Proteins and Theories of Protein Evolution

Arnheim, N., and P. Calabrese (2009) Understanding what determines the frequency and pattern of human germline mutations. *Nature Rev. Genet.* 10:478–488.

Bloom, J. D., S. T. Labthavikul, C. R. Otey, and F. H. Arnold (2006) Protein stability promotes evolvability. *Proc. Natl. Acad. Sci. USA* 103:5869–5874.

Dean, A. M., and J. W. Thornton (2007) Mechanistic approaches to the study of evolution: The functional synthesis. *Nature Rev. Genet.* 8:675–688.

Kumar, S., and S. Subramanian (2002) Mutation rates in mammalian genomes. *Proc. Natl. Acad. Sci. USA* 99:803–808.

LeComte, J. T. J., D. A. Vuletich, and A. M. Lesk (2005) Structural divergence and distant relationships in proteins: Evolution of the globins. *Curr. Op. Struct. Biol.* 15:290–301.

Meier, S., P. R. Jensen, C. N. David, J. Chapman, T. W. Holstein, S. Grzesiek, and S. Özbek (2007) Continuous molecular evolution of protein-domain structures by single amino acid changes. *Curr. Biology* 17:173–178.

Royer Jr., W. E., H. Zhu, T. A. Gorr, J. F. Flores, and J. E. Knapp (2005) Allosteric haemoglobin assembly: Diversity and similarity. *J. Biol. Chem.* 280:27477–27480.

Tokuriki, N., and D. S. Tawfik (2009) Protein dynamism and evolvability. *Science* 324:203–207.

Vinogradov, S. N., and L. Moens (2008) Diversity of globin function: Enzymatic, transport, storage, and sensing. *J. Biol. Chem.* 283:8773–8777.

Variant Hemoglobins and Hemoglobin Pathologies

Embury, S. H. (1986) The clinical pathophysiology of sickle cell disease. *Annu. Rev. Med.* 37:361–376.

Honig, G. R., and J. G. Adams (1986) *Human Hemoglobin Genetics.* Springer-Verlag, Berlin, New York.

Ingram, V. M. (1957) Gene mutation in human haemoglobin: The chemical difference between normal and sickle cell haemoglobin. *Nature* 180:326–328.

Pauling, L., H. A. Itano, S. J. Singer, and I. C. Wells (1949) Sickle cell anemia: A molecular disease. *Science* 110:543–548.

Schechter, A. N. (2008) Hemoglobin research and the origins of molecular medicine. *Blood* 112:3927–3938.

See also the HbVar database for human hemoglobin variants and thalassemias: http://globin.bx.psu.edu/hbvar/menu.html

The Immune Response

Barouch, D. H. (2008) Challenges in the development of an HIV-1 vaccine. *Nature* 455:613–619.

Chaplin, D. D. (2010) Overview of the immune response. *J. Allergy Clin. Immunol.* 125:S3–23.

Dömer, T., and A. Radbruch (2007) Antibodies and B cell memory in viral immunity. *Immunity* 27:384–392.

Medzhitov, R., and C. A. Janeway Jr. (2002) Decoding the patterns of self and nonself by the innate immune system. *Science* 296:298–300.

Pichlmair, A., and C. Reis e Sousa (2007) Innate recognition of viruses. *Immunity* 27:370–383.

Antibody-based Therapeutics

Chan A. C., and P. J. Carter (2010) Therapeutic antibodies for autoimmunity and inflammation. *Nature Rev. Immun.* 10:301–316.

Chari, R. V. J. (2008) Targeted cancer therapy: Conferring specificity to cytotoxic drugs. *Acct. Chem. Res.* 41:98–107.

Weiner, L. M., R. Surana, and S. Wang (2010) Monoclonal antibodies: versatile platforms for cancer immunotherapy. *Nature Rev. Immun.* 10:317–327.

Wu, A. M., and P. D. Senter (2005) Arming antibodies: Prospects and challenges for immunoconjugates. *Nature Biotech.* 23:1137–1146.

PROBLEMS

1. The following data describe the binding of oxygen to human myoglobin at 37 °C.

P_{O_2} (mm Hg)	Y_{O_2}	P_{O_2} (mm Hg)	Y_{O_2}
0.5	0.161	6	0.697
1	0.277	8	0.754
2	0.434	12	0.821
3	0.535	20	0.885
4	0.605		

From these data, estimate (a) P_{50} and (b) the fraction saturation of myoglobin at 30 mm Hg, the partial pressure of O_2 in venous blood.

2. What qualitative effect would you expect each of the following to have on the P_{50} of hemoglobin?
   (a) Increase in pH from 7.2 to 7.4
   (b) Increase in P_{CO_2} from 20 to 40 mm Hg
   (c) Dissociation into monomer polypeptide chains

*3. Measurements of oxygen binding by whole human blood, at 37 °C, at pH 7.4, and in the presence of 40 mm Hg of CO_2 and normal physiological levels of BPG (5 mmol/L of cells), give the following:

P_{O_2} (mm Hg)	% Saturation ($= 100 \times Y_{O_2}$)
10.6	10
19.5	30
27.4	50
37.5	70
50.4	85
77.3	96
92.3	98

(a) From these data, construct a binding curve, and estimate the percent oxygen saturation of blood at (1) 100 mm Hg, the approximate partial pressure of O_2 in the lungs, and (2) 30 mm Hg, the approximate partial pressure of O_2 in venous blood.
(b) Under these conditions, what percentage of the oxygen bound in the lungs is delivered to the tissues?
(c) Using the data in Figure 7.16 (page 253), repeat the calculation of part (b) if the pH drops to 6.8 in capillaries but goes back to 7.4 as CO_2 is unloaded in the lungs.

4. Crocodile hemoglobin does not bind 2,3-BPG. Instead, it binds bicarbonate ion, which is a strong negative allosteric effector. Why might crocodiles have a hemoglobin that is responsive to HCO_3^- instead of 2,3-BPG? Recall that crocodiles hold their prey underwater to kill them.

5. Precise data have been obtained for the oxygen binding of stripped human hemoglobin at 25 °C:

P_{O_2} (mm Hg)	% Saturation ($= 100 \times Y_{O_2}$)	P_{O_2} (mm Hg)	% Saturation ($= 100 \times Y_{O_2}$)
0.10	0.315	5.75	76.0
0.350	0.990	7.94	90.9
0.794	3.06	12.88	96.9
1.748	9.09	29.51	99.0
2.884	24.0	67.60	99.7
4.467	50.0		

Use a Hill plot to determine (a) P_{50}, (b) h (maximum slope), and (c) the P_{50} values corresponding to the T and R states.

*6. G. Ackers, M. L. Johnson, F. C. Mills, et al., in *Biochemistry* (1975) 14:5128–5134, have observed that the P_{50} of purified hemoglobin decreases as the concentration of hemoglobin in solution is decreased. Suggest an explanation.

7. Oxygen binding by the hemocyanin of the shrimp *Callianassa* has been measured. Using the following data, prepare a Hill plot and determine (a) P_{50}, (b) h (the Hill coefficient), and (c) the *minimum* number of oxygen binding sites on the protein molecule.

P_{O_2} (mm Hg)	Y_{O_2}	P_{O_2} (mm Hg)	Y_{O_2}
1.1	0.003	136.7	0.557
7.7	0.019	166.8	0.673
10.7	0.035	203.2	0.734
31.7	0.084	262.2	0.794
71.9	0.190	327.0	0.834
100.5	0.329	452.0	0.875
123.3	0.487	736.7	0.913

8. Suggest probable consequences of the following real or possible hemoglobin mutations. [Note: Consult Figures 7.15 (page 249), 7.17 (page 250), and 7.23 (page 254).]
   (a) At β146 (HC3) His $\rightarrow$ Asp
   (b) At β92 (F8) His $\rightarrow$ Leu
   (c) At β2 (NA2) His $\rightarrow$ Asp

In each case, tell whether a single base change is sufficient for the mutation.

9. Suppose each of the mutants listed in Problem 8 were electrophoresed in comparison with native hemoglobin (pI = 7.0) at pH 8.0. Which would move faster toward the anode than native protein, and which would move more slowly?

*10. In principle, an allosteric molecule could exhibit *negative* cooperativity; that is, binding of the first ligands could decrease affinity for additional ones.
   (a) What would a Hill plot look like for negative cooperativity?
   (b) It has been noted that the KNF theory allows negative cooperativity but the MWC theory doesn't. Explain.

11. In the experiments of Barrick, et al. (see Figure 7.18, page 251) it was observed that replacement of histidine by a noncovalently bonded imidazole not only reduced cooperativity but also increased the oxygen affinity of the hemoglobin. Suggest an explanation.

12. Suppose you visit the Dalai Lama in Dharamsala, India (elevation 1460 m), and you begin to ponder the "big questions," such as, "What is the fractional saturation of the Dalai Lama's hemoglobin?" **(a)** Assuming the Dalai Lama's hemoglobin has a Hill coefficient = 3.2, and a P_{50} = 31 torr, calculate the change in fractional O_2 saturation of his hemoglobin going from his lungs (where P_{O_2} = 85 torr) to his capillaries (where P_{O_2} = 25 torr). **(b)** Why do you suppose the Dalai Lama's hemoglobin has a P_{50} higher than normal (where "normal" = 27 torr)?

13. Assume that a new oxygen-transport protein has been discovered in certain invertebrate animals. X-ray diffraction of the deoxy protein reveals that it has the dimeric structure shown here in part (a) with a salt bridge between residues histidine 13 and aspartic acid 85.

The two monomers interact by salt bridges between the C- and N-termini. The O_2-binding site lies *between* the two iron atoms shown, which are rigidly linked to helices A and C (see part (b)). In the deoxy form, the space between the iron atoms is too small to hold O_2, and so the Fe atoms must be forced apart when O_2 is bound.

Answer the following questions, explaining your answer in each case in terms of the structure shown below.
(a) Is this molecule likely to show cooperative oxygen binding?
(b) Is this molecule likely to exhibit a Bohr effect?
(c) Predict the likely effect of a mutation which replaced aspartic acid 85 by a lysine residue?

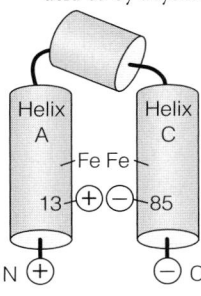

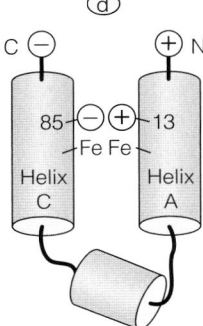

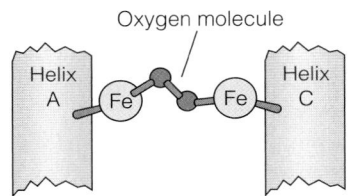

Oxygen molecule

(a) Structure of deoxy protein. N and C denote N-termini and C-termini. The dyad axid (d) is perpendicular to this page.

(b) Detail of the oxygen-binding site in the oxy form.

*14. The binding of antigens to antibodies can be thought of in ways similar to those we have used for discussing oxygen binding. Assuming we have monovalent antigens and n-valent antibodies, we can write the following equation for the number (r) of antigen molecules bound to an antibody at a concentration (c) of free antigen:

$$r = \frac{nKc}{1 + Kc}$$

where K is the equilibrium constant for the binding (the affinity constant).
(a) Show that this equation can be rearranged to give

$$\frac{r}{c} = Kn - Kr$$

This is called the *Scatchard equation*. It predicts that a graph of r/c versus r will be a straight line.
(b) Use the following data and a graph according to the Scatchard equation to obtain n and K for an antibody–antigen reaction.

c (M)	r
1.43×10^{-5}	0.50
2.57×10^{-5}	0.77
6.00×10^{-5}	1.20
1.68×10^{-4}	1.68
3.70×10^{-4}	1.85

15. What physiological effect would you predict from a mutation that replaced with serine the cysteine in the constant part of the immunoglobulin light chain that is involved in disulfide-bond formation with the heavy chain? (See Figure 7.39 page 271.)

TOOLS OF BIOCHEMISTRY 7A

IMMUNOLOGICAL METHODS

Because of the ease with which antibodies against biological materials can be prepared in the laboratory, and because of their great specificity, antibodies form the core of many important analytical and preparative biochemical procedures. We shall discuss here some of the methods most important to biochemists.

Experiments in which an antigen is injected into an animal show that one antigen can elicit formation of several different antibodies. Each of these antibodies recognizes one particular portion of the antigen molecule, called an *antigenic determinant,* or *epitope.* Figure 7.36b shows the epitopes that have been identified in sperm whale myoglobin. Each epitope is a region encompassing five or six residues in the myoglobin sequence. Thus, a myoglobin **antiserum** (that is, serum from an animal immunized against myoglobin) carries at least five different antimyoglobin antibodies, each directed against one of the five epitopes shown in the figure. Most antibodies that are useful in biochemistry are of the IgG type (see Table 7.2 and Figure 7.40). Each of these Y-shaped monomers has two antigen-combining sites. In an antigen–antibody reaction, each site usually binds to a different antigen molecule if sufficient antigen is present.

Generally speaking, immunogenic substances (substances that elicit an immune reaction, such as synthesis of antibodies) are macromolecules. However, some of the most useful immunological reagents are antibodies against low-molecular-weight substances. These low-molecular-weight compounds are not typically immunogenic per se; however, when covalently coupled to an antigenic carrier protein, such as bovine serum albumin (BSA) or keyhole limpet hemocyanin (KLH), the resultant conjugate is immunogenic. Some of the antibodies produced in response to the conjugate are directed against the low-molecular-weight, nonprotein constituent, which is called a **hapten**. The antibodies that bind specifically to the hapten can be isolated by affinity chromatography (see Tools of Biochemistry 5A) using a solid support to which the hapten has been covalently linked.

One of the earliest uses of immunoglobulins as analytical tools was in **radioimmunoassay**, in which soluble IgG is used to test for levels of steroid hormones or drugs in complex biological fluids such as blood plasma. This is a binding competition assay between radioactively labeled and unlabeled antigen for the IgG. A known amount of radioactive antigen (usually labelled with ^{125}I or ^{131}I) is incubated with the antibody in the presence of the unlabeled antigen in the sample. As the concentration of the unlabeled antigen in the sample increases, it will more effectively compete for binding to the antibody and displace the radioactive antigen from the IgG. Once the binding equilibrium has been established, the antibody is precipitated from solution. This can be achieved by adding a synthetic bead that is coated with **protein A**, or a second anti-IgG antibody (a "secondary antibody"). The secondary antibody and protein A both work in the same way—by binding specifically to the F_c region of the first IgG (or "primary" antibody) and thereby precipitating it from the solution. The ratio of radioactivity in the supernatant vs. that in the precipitate is then compared to a standard curve to determine the concentration of the unlabeled antigen in the biological sample. This technique was developed by Rosalyn Yalow and Solomon Berson in the 1950s at the Bronx Veteran's Administration Hospital, and it revolutionized the study of several hormone-mediated processes (including type I diabetes). In 1977, Yalow was awarded the Nobel Prize in Medicine in recognition of the impact of RIA in diagnostic medicine (Berson had passed away in 1972).

A technique widely used to quantify antigen–antibody reactions is the **enzyme-linked immunosorbent assay (ELISA)**. The ELISA has largely replaced RIA due to its convenience (it can be easily carried out in 96-well plates for high-throughput applications) and the fact that it does not require the use of radioactivity. We illustrate the widely used **indirect ELISA** in Figure 7A.1. In the first step, a 96-well plate is coated with a solution containing the sample mixture. Proteins stick nonspecifically to the polystyrene surfaces of the well via van der Waals forces. Next, a blocking protein is added to bind to any bare polystyrene surfaces (this prevents nonspecific binding of the antibody to the well). A primary antibody that specifically binds to the target antigen is then added (step 3). This is the "immunosorbent" part of the assay in which the detection of a specific antibody–antigen interaction occurs. The detection of this antigen–antibody interaction is achieved using a secondary antibody that recognizes the F_c region of the primary antibody (steps 4–6). If, for example, a murine antibody was used as the primary antibody, an antibody from goats or rabbits raised against murine antibodies could be used as the secondary antibody. The secondary antibody is covalently crosslinked to an enzyme (typically, horseradish peroxidase) whose activity can be easily assayed spectrophotometrically using a chromogenic substrate—this is the "enzyme-linked" part of the assay. The development of color in the wells of the plate indicates the presence of the target antigen (step 6). Because the readout is based on the acitivty of the enzyme it is *indirect* detection; however, it is also *general* because the secondary antibody will react with *any* murine primary antibody (so, the same equipment can be used to run assays for several different target antigens by altering the primary antibody in step 3, in Figure 7A.1). Although this method has many variations, the principle is to assay for bound antibody by analyzing for the activity of the conjugated enzyme. This technique forms the basis of many clinical diagnostic tests, such as the most widely used current test for HIV infection. An enzyme-linked antibody to one of the surface proteins of the human immunodeficiency virus is used to detect the presence of the viral antigen in human blood samples. The presence of the enzyme activity on antigen–antibody complexes can be detected with great sensitivity.

Another analytical technique useful to characterize proteins is **western blotting**, so called because of its superficial resemblance to a nucleic acid analytical technique called Southern

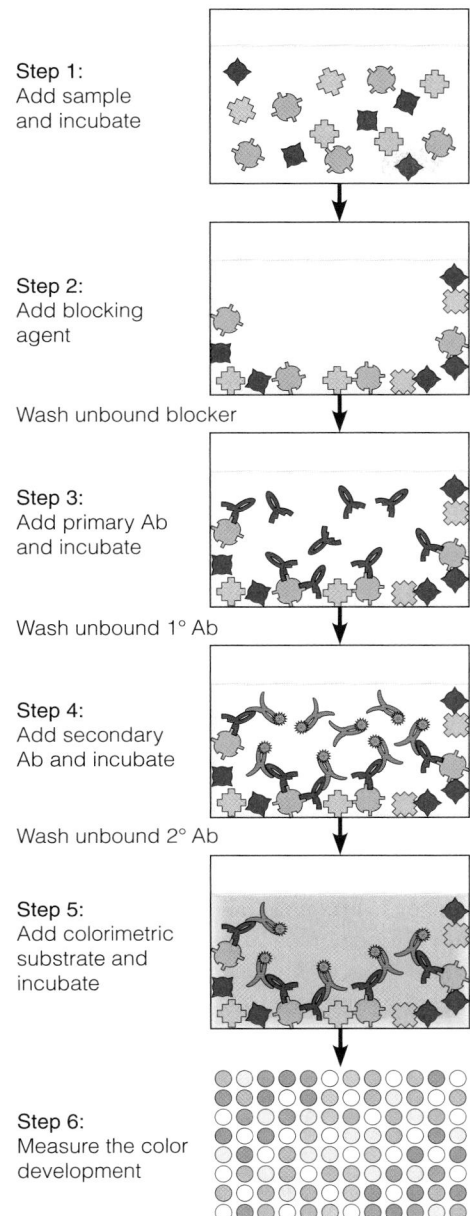

Step 1:
Add sample
and incubate

- Protein of Interest
- Other Protein
- Other Protein
- Y Primary Antibody (1° Ab)
- Y Secondary Antibody (2° Ab)
- Reporter Enzyme

Step 2:
Add blocking
agent

Wash unbound blocker

Step 3:
Add primary Ab
and incubate

Wash unbound 1° Ab

Step 4:
Add secondary
Ab and incubate

Wash unbound 2° Ab

Step 5:
Add colorimetric
substrate and
incubate

Step 6:
Measure the color
development

FIGURE 7A.1

The indirect ELISA assay. Detection of one protein in a complex mixture is shown. The assay is typically carried out in a 96-well plate. Steps 1–5 show a close-up view of a single well. In step 5 color develops in the well due to the action of the enzyme linked to the 2° antibody. The enzyme causes a chromogenic substrate to change color. In step 6, the entire 96-well plate is analyzed (each circle represents one of the 96 wells). Those wells with color are presumed to contain the target antigen, where a darker color indicates a higher concentration of the antigen in the sample well.

Biochemists, like cell biologists, must be concerned with intracellular organization and with the location of enzymes that catalyze reactions of interest. An array of techniques with the generic term **immunocytochemistry** uses antibodies to help localize particular antigens in cytological preparations. In the simplest form, an antibody is conjugated with a fluorescent dye such as fluorescein. A thin section of cell or tissue is then immersed in a solution of the fluorescent antibody. After the excess is washed off, the bound antibody can be visualized by fluorescence microscopy. It is possible to visualize different antigens within a cell simultaneously using multiple dye-linked antibodies. Figure 7A.3 shows a cell stained with three different antibodies—each linked to a different fluorescent dye. Each fluorescent antibody binds to a different mac romolecular complex, in this case

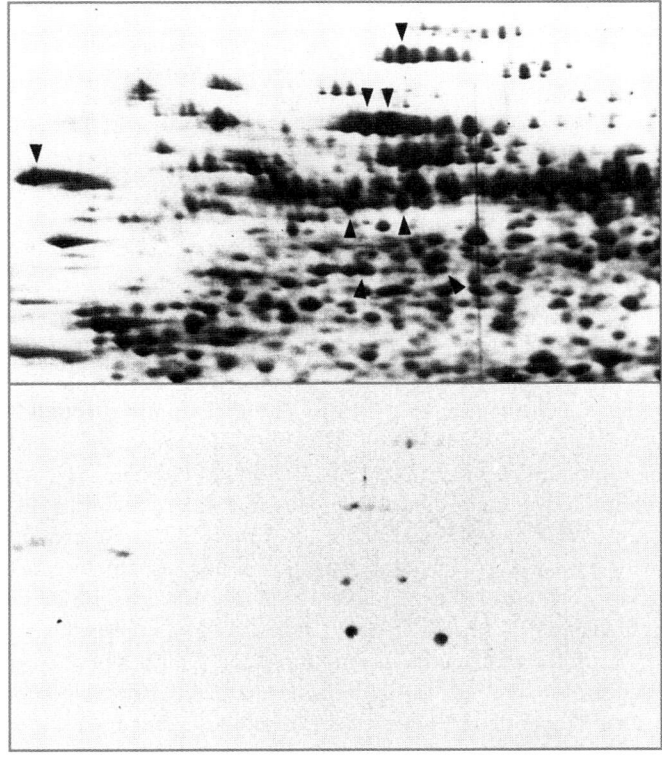

FIGURE 7A.2

Western blotting. On the top is a 2-D gel of total protein from tobacco leaf. On the bottom is the same gel, blotted with an antibody against proteins containing phosphothreonine residues.

The Plant Journal 2:723–732, J. A. Traas, A. F. Bevan, J. H. Doonan, J. Cordewener and P. J. Shaw, Cell-cycle-dependent changes in labelling of specific phosphoproteins by the monoclonal antibody MPM-2 in plant cells. © 1992 John Wiley & Sons, Inc. Reproduced with permission from John Wiley & Sons, Inc.

blotting (see Figure 24.14, page 1016). Western blotting (also called **immunoblotting**) is used to detect, in a mixture of proteins or fragments of proteins, those that react with the same antibody. It can be used, for example, in studying the post-translational cleavage of proteins that often occurs as part of protein maturation. In this technique the antibody-reactive proteins in a mixture are analyzed by first resolving the proteins in that mixture by denaturing gel electrophoresis. Often, 2-D gels are used. After electrophoresis, the gel is placed in contact with a sheet of nitrocellulose, and the proteins are transferred (or "blotted") to the nitrocellulose by an electric current. The proteins are bound irreversibly to the nitrocellulose sheet, so the antigen–antibody reactions can be visualized after treatment of the sheet with primary and secondary antibodies, as described above for indirect ELISA. Alternatively, the target can be detected by autoradiography using a primary (or secondary) antibody labeled with radioactive [125]I. An example is shown in Figure 7A.2.

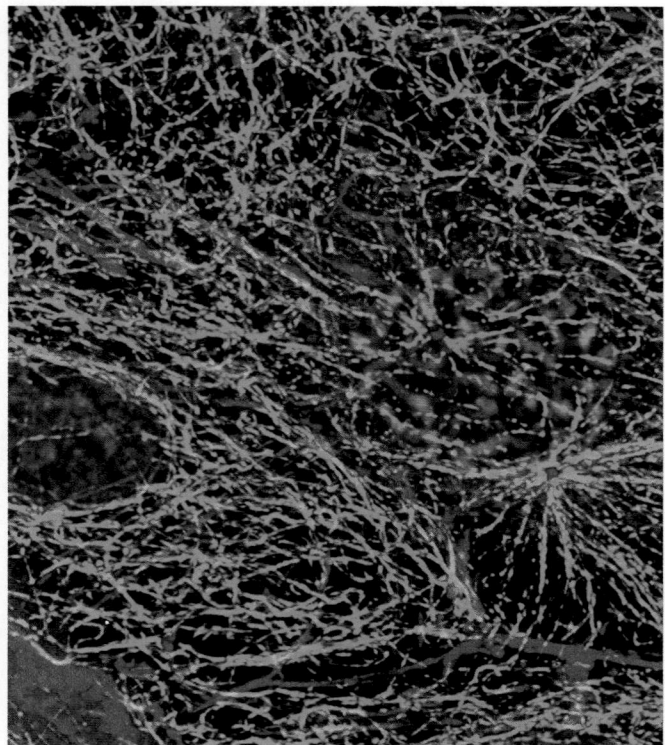

FIGURE 7A.3

Immunofluorescent light micrograph of a rat kangaroo kidney epithelial cell during mitotic cell division. The chromosomes (blue, center) are condensed, after replication. The actin microfilaments (red) and tubulin microtubules (green) of the cytoskeleton maintain the structure of the cell. Antibodies have been used to attach different fluorescent dyes to the chromatin, actin, and tubulin.

R. Alexey Khodjakov/SciencePhoto Library.

chromatin, actin filaments, and microtubules. Alternatively, the antibody can be linked to the iron-binding protein ferritin, and the bound iron can be visualized from its high electron density in the electron microscope.

Because of their high specificity in protein binding, antibodies can also be used to purify proteins. In this technique, called immunoaffinity chromatography (see Tools of Biochemistry 5A), the antibody is coupled to a chromatographic support and a column of this material is used to adsorb selectively the protein being purified. The protein is then desorbed, usually by a pH adjustment in the eluting solution (typically as low as pH 2.5 or as high as pH 11.5), and often in a state close to homogeneity.

As noted earlier, an antiserum prepared against a pure antigen, such as a protein, usually contains several different antibodies against that antigen. Moreover, the serum also contains all other antibodies that the animal carried in its bloodstream at the time that it was immunized against the protein of interest. Clearly the specificity, sensitivity, and reproducibility of all of the foregoing techniques would be increased greatly if they could be carried out

with purified antibodies that have identical sequence, and therefore identical binding specificity. Because of the chemical similarities among different IgGs, their purification by standard protein fractionation techniques is virtually impossible. Fortunately, the use of **monoclonal antibodies** (mAbs) is an alternative way to attain a highly purified antibody with a unique binding specificity in unlimited quantity.

Each antibody-forming B lymphocyte is specialized for the synthesis and secretion of one and only one antibody. If such cells could be grown in cell culture, it would be possible in principle to isolate a clone, or population derived from one cell, that synthesized one antibody directed against one site on the protein of interest. Although B lymphocytes themselves do not grow in culture, Georges Köhler and César Milstein in 1975 discovered a way to propagate antibody-forming lymphocytes with cultured cells from a mouse with multiple myeloma, a cancerous proliferation of white cells (Figure 7A.4). A mouse is immunized with the antigen of interest, and then lymphocytes from its spleen are fused with the myeloma cells. This fusion gives rise to cell lines called **hybridomas**—cells that proliferate endlessly in culture, like cancer cells, but synthesize only one antibody (see Chapter 22 for more detail on hybridoma culture). By screening a large number of hybridoma clones, several can usually be isolated that synthesize antibody to the antigen of interest, each making antibody to a different antigenic determinant. Antibodies can easily be purified from cultures of the appropriate hybridomas, and they have many uses in addition to the techniques already described. Note that this method does not require a homogeneous antigen. An investigator who has the patience to examine many hybridomas can immunize the animal at the outset with a crude preparation of the antigen of interest. Later, after the desired monoclonal antibody has been obtained, it can be used as the basis of an immunoaffinity purification of the antigen.

The ability to isolate mAbs has had a profound impact on biochemistry, molecular biology, and biopharmaceutical development. In 1984, the work of Milstein and Köhler was recognized with the Nobel Prize in Physiology or Medicine.

A particularly useful application of mAbs is in the identification of proteins in a multiprotein complex using **co-immunoprecipitation** (**co-IP**) or "**pull-down**." In this technique, an antibody specific for a protein that is thought to be part of a multiprotein complex is used to precipitate the entire complex. The precipitating antibody is typically bound to a magnetic particle, an agarose bead, or a protein A-coated bead. The various proteins "pulled down" by this procedure can be identified by SDS-PAGE followed by mass spectrometry (see Tools of Biochemistry 5A, 5B, and 5D). To verify the results of a co-IP, several different antibodies against different proteins in the complex should be used. In this way, it is possible to identify the proteins with which some target of interest interacts.

Because it can be challenging to get suitable antibodies for IP experiments, a variation of the technique is often used. In this

FIGURE 7A.4

Production of monoclonal antibodies.

Microbiology: Introduction, 5th ed., Gerard J. Tortora. © 1995. Reprinted by permission of Pearson Education Inc., Upper Saddle River, NJ.

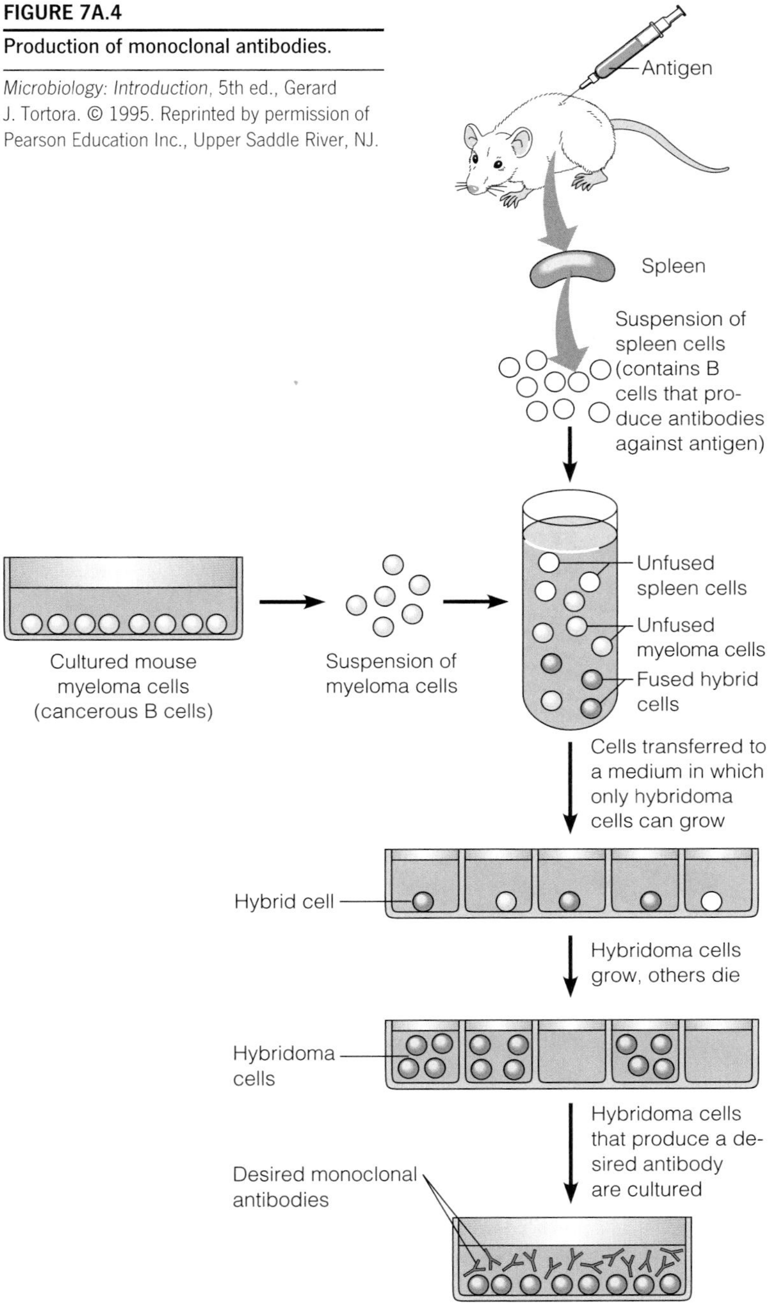

case, a protein is "tagged" with a short peptide epitope, such as $(His)_6$, (see Figure 5A.4) or the "FLAG" sequence (DYKDDDDK), and an antibody that binds the tag is used in the pull-down. It is important to show that the presence of the tag doesn't alter the properties of the tagged protein.

References

Harlow, E., and D. Lane (1988) *Antibodies: A Laboratory Manual.* Cold Spring Harbor Laboratory, Cold Spring Harbor, New York.

Weir, D. M., ed. (1986) *Handbook of Experimental Immunology.* Oxford University Press, London, New York.

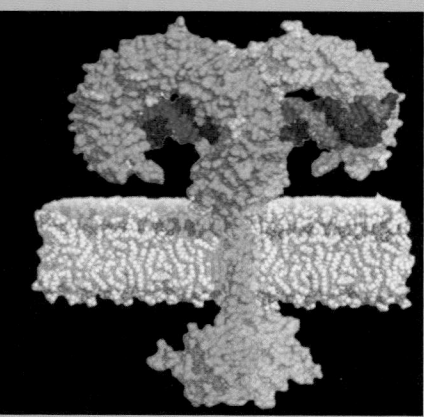

CHAPTER 8

Contractile Proteins and Molecular Motors

In the previous two chapters, we have seen how proteins, either as monomers or in defined multisubunit structures, carry out a variety of functions. To continue to explore this idea, we now turn to examples in which proteins organize into large, complex structures involving many kinds of polypeptide chains. Such supramolecular structures perform many cellular functions; the one we shall consider here is the mechanical work of motion carried out by **motor proteins**. This motion may involve the whole organism, parts thereof, cells, or subcellular constituents. We will see that protein conformational change mediated by the binding, hydrolysis, and release of ATP is a key feature of motor protein function.

Of the many kinds of motion that living systems exhibit, the one we are most aware of is the muscle contraction required for bodily movement. However, muscle contraction accomplishes a remarkable variety of other things as well. Even the emission of sound is a muscular action, as is the injection of venom by an insect or a snake. Equally important muscular motions maintain the animal's internal world, including the beating of its heart, the breathing motions of its lungs or gills, and the peristaltic motions of its digestive system. Each of these kinds of movements is produced by a specific muscular tissue.

All muscles, as well as some other contractile systems, are based on the interaction of two major proteins, **actin** and **myosin**. We often refer to these systems as **actin–myosin contractile systems**. However, certain kinds of directed motions exist—motions of individual cells and parts of cells—that do not depend on the actin–myosin system at all but use other protein mechanisms. For example, the beating of cilia and flagella and the movement of chromosomes and organelles within cells are accomplished by the interactions of a number of proteins with **microtubules**, filamentous structures made of a protein called **tubulin**. Even more remarkable are the variety of "molecular motors" that have been described in increasing detail in recent years. Some of these serve to carry molecules and vesicles along microtubules and other filaments; others produce rotation of flagella and are true microscopic motors. Also, some very well-known protein complexes are now recognized as molecular motors. These include **ATP synthase**

(Chapter 15), which uses a proton gradient to drive ATP synthesis in mitochondria, and **RNA polymerase** (Chapters 26 and 28), which uses nucleoside triphosphate hydrolysis to move along and untwist a DNA template.

The biological systems that produce movement share one common feature—they hydrolyze ATP. The energy released by the hydrolysis of ATP is converted into work by producing motions in parts of protein molecules. Thus, proteins can act as **energy transducers**. That is, some proteins can convert the chemical energy of ATP hydrolysis into mechanical work. When the motions of proteins are properly coordinated, directed macroscopic motion occurs.

Muscles and Other Actin–Myosin Contractile Systems

The major proteins in muscle tissue are actin and myosin; however, actin and myosin are also found in many other types of cells and are involved in several kinds of cellular and intracellular motions. To understand how muscles and other actin–myosin systems work, we must consider the properties of these two proteins.

Actin and Myosin

Actin

Under physiological conditions, actin exists as a long, helical polymer (filamentous actin, or **F-actin**) of a globular protein monomer (**G-actin**). The G-actin monomer, shown in Figure 8.1, is a four-domain molecule with a molecular weight (M_W) of 42 kDa. The binding of ATP by a G-actin monomer leads to polymerization; the ATP is subsequently hydrolyzed, but the ADP is held in the actin filament. In F-actin filaments, the G-actin monomers are arranged in a two-strand helix (Figure 8.2; see also Figure 6.39a, page 214). Because of the asymmetry of the subunits, the F-actin filament has a defined directionality, and the two ends have been called the "**barbed**" or "**plus**" **end**, and the "**pointed**" or "**minus**" **end**. The polymerization reaction exhibits a preferred direction, and the plus end is defined as that end that grows more rapidly under physiological conditions. The sites on the F-actin filament that bind to myosin are located on domain 1 of each actin subunit.

Myosin

Six of the 20 or so forms of myosin found in cells, and their various functions, are listed in Table 8.1. The most studied myosin molecule is myosin II from striated muscle. In the following discussion we will refer to myosin II simply as "myosin." The functional myosin molecule (Figure 8.3) is composed of six polypeptide chains: two identical heavy chains ($M_W = 230$ kDa) and two each of two kinds of light chains ($M_W = 20$ kDa). Together, they form a complex of molecular weight

Certain proteins act as energy transducers, using free energy from ATP hydrolysis or free energy stored in ion gradients to do the mechanical work of motion.

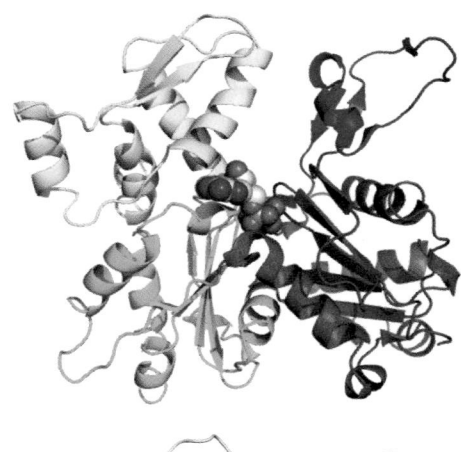

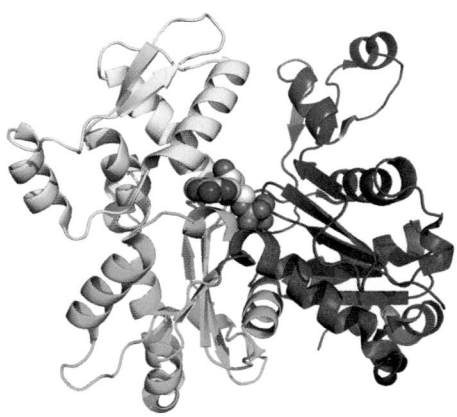

FIGURE 8.1

G-actin. Cartoon representations of the X-ray crystal structures of the G-actin monomer are shown. **Top:** with ATP bound (PDB ID: 1ATN); **bottom:** with ADP bound (PDB ID: 1J6Z). Domain 1 is red, domain 2 is blue, domain 3 is green, and domain 4 is yellow. Nucleotides are shown as spheres. The "plus" end is down in each panel.

FIGURE 8.2

A model for F-actin filaments. Left: The α-carbon backbones of five G-actin monomers are shown. Individual monomers are distinguished by colors, with bound ATP shown in red. The green residues on domain 1 of the gray actin monomer show the myosin-binding site. **Right:** a model of the myosin S1 fragment (see text) bound to F-actin. The myosin motor domain is shown in green and the essential and regulatory light chains are, respectively, magenta and cyan (PDB ID: 1ALM).

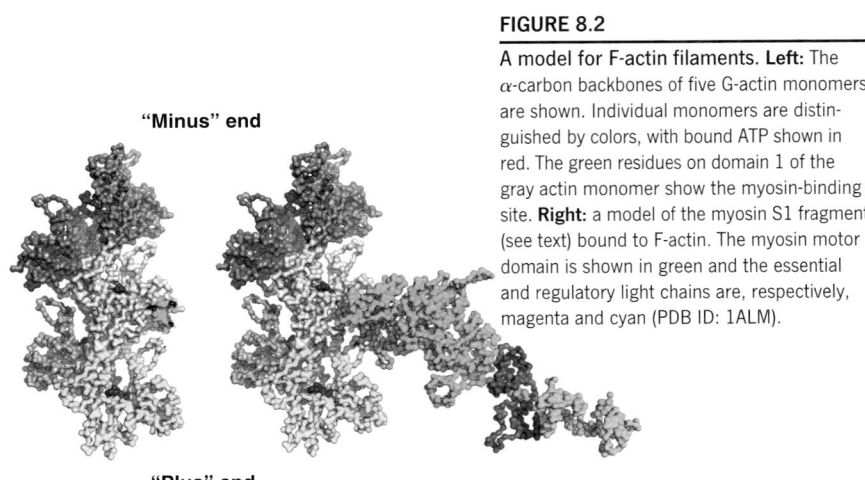

"Minus" end

"Plus" end

TABLE 8.1	Myosin types and their functions
Myosin Type	Primary Functions
I	Vesicle transport
II	Muscle contraction; cytokinesis; cell motility
V	Vesicle transport and localization at cell periphery
VI	Transport of endocytotic vesicles into cell center
VIII	Cell division in plants
XV	Part of acoustic sensor in inner ear

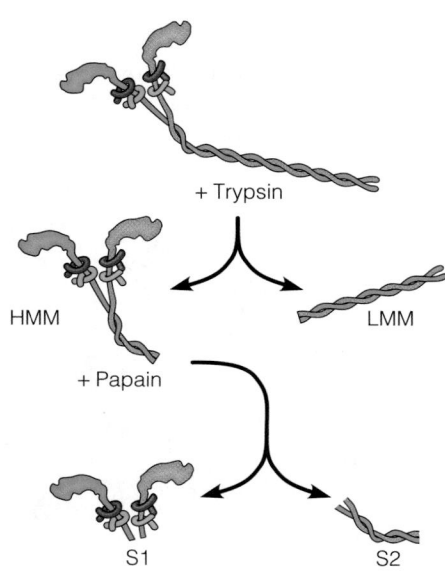

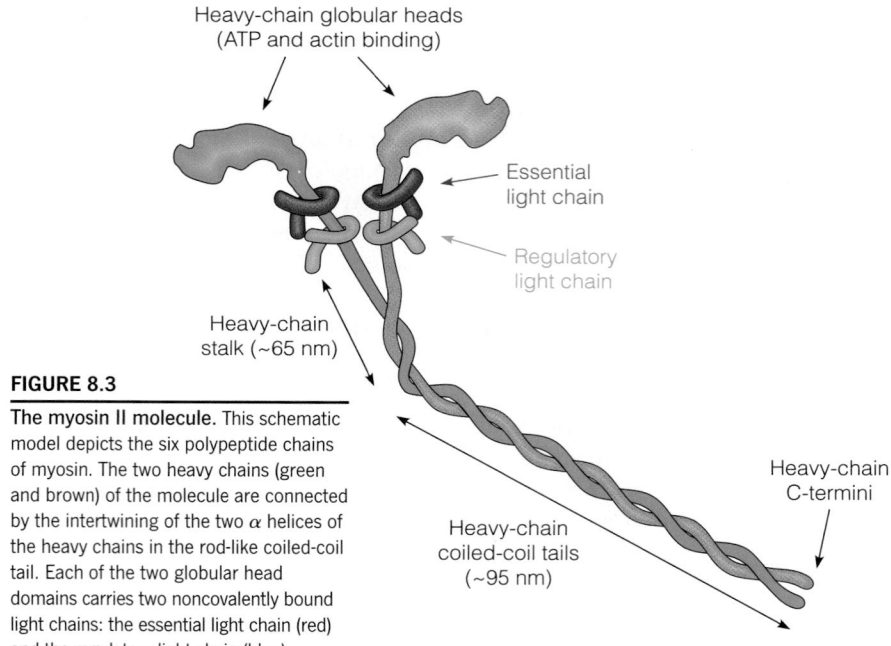

FIGURE 8.3

The myosin II molecule. This schematic model depicts the six polypeptide chains of myosin. The two heavy chains (green and brown) of the molecule are connected by the intertwining of the two α helices of the heavy chains in the rod-like coiled-coil tail. Each of the two globular head domains carries two noncovalently bound light chains: the essential light chain (red) and the regulatory light chain (blue).

FIGURE 8.4

Dissection of myosin II by proteases. Trypsin cleavage cuts through the myosin tail to produce light meromyosin (LMM) and heavy meromyosin (HMM). Treatment of heavy meromyosin with the protease papain digests part of the stalk structure, allowing separation of the two S1 fragments from the S2 fragment (see text).

The major muscle systems of animals are based on the proteins actin and myosin.

Myosins are motor proteins that move toward the (+) end of the actin filament.

540 kDa. The heavy chains have long α-helical tails, which form a two-stranded coiled coil, and globular head domains to which the light chains are bound. Between each head domain and tail domain the heavy chain acts as a flexible stalk. The coiled-coil structure of the tails is reminiscent of the structure of α-keratin (see Figure 6.14).

The myosin molecule can be cleaved by proteases, as shown in Figure 8.4. The tail domain can be cleaved at a specific point by trypsin to yield fragments called **light meromyosin** and **heavy meromyosin**. Further cleavage of heavy meromyosin by papain cuts the stalks to yield two **S1 fragments**, each consisting of a head domain carrying the light chains. The stalk removed by papain is called an S2 fragment. The ability to cleave the myosin molecule in these specific ways has helped researchers understand the functions of its several parts. Myosin exhibits aspects of both fibrous and globular proteins, and its functional domains play quite different roles. The tail domains have a pronounced tendency to aggregate, causing myosin molecules to form the kind of thick bipolar filaments shown in Figure 8.5. The S1 fragment, or **headpiece**, includes the

(a)

FIGURE 8.5

A thick filament of myosin II molecules. (a) An electron micrograph. The zone that is bare of headpieces is indicated by ℓ; some myosin headpieces are indicated by arrowheads. **(b)** A drawing of the filament structure, showing dimensions in nanometers. The projections are the pairs of headpieces on each myosin molecule.

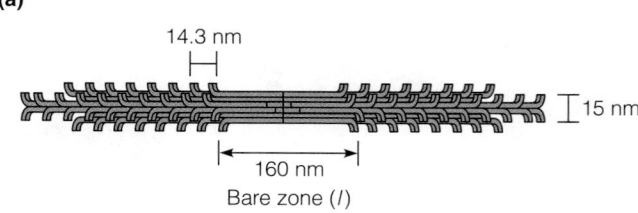

(b)

globular **motor domain**, which binds ATP and actin and two light chains—the **essential** and the **regulatory** light chains. The crystal structure of an S1 fragment is shown in Figure 8.6.

The Reaction of Myosin and Actin

If an actin filament is allowed to react with isolated S1 fragments, the filament will become decorated with these myosin headpieces, producing an asymmetric arrowhead pattern that demonstrates the polarity of the actin filament (Figures 8.7 and 8.2). Most myosin types are motor proteins that move toward the (+) end of the actin filament. As we shall see, myosin's **ATPase activity** (i.e., ATP binding and hydrolysis), as well as its ability to bind and release F-actin, are essential parts of the multistep mechanism of muscle contraction.

The Structure of Muscle

In muscle tissue, the actin and myosin filaments interact to produce the contractile structure. Vertebrates have three morphologically distinct kinds of muscle. *Striated muscle* is the kind most often associated with the term *muscle*, for it is the striated muscles in arms, legs, eyelids, and so forth that make voluntary motions possible. *Smooth muscle* surrounds internal organs such as the blood vessels, intestines, and gallbladder, which are capable of slow, sustained contractions that are not under voluntary control. *Cardiac muscle* can be considered a specialized form of striated muscle, adapted for the repetitive, involuntary beating of the heart. In the following discussion we will consider only the structure of striated muscle.

Figure 8.8 shows successive levels of organization of a typical vertebrate striated muscle. The individual muscle fibers, or **myofibers**, are very long (1–40 mm) multinucleate cells, formed by the fusion of muscle precursor cells. Each myofiber contains a bundle of protein structures called **myofibrils**. A myofibril exhibits a periodic structure when examined under the light microscope. Dark **A bands** alternate with lighter **I bands**. The I bands are divided by thin lines called **Z disks** (or sometimes **Z lines**). At the center of the A band is found a lighter region called the **H zone**. The repeating unit of muscle structure can be taken as extending from one Z disk to the next. It is called the **sarcomere** and is about 2.3 μm long in relaxed muscle.

The molecular basis for this periodic structure of the myofibril can be seen by electron microscopic studies of thin sections of muscle, as shown in Figure 8.9. **Thin filaments** of actin extend in both directions from the Z disks, interdigitating with myosin **thick filaments**. The regions in which the thick and thin filaments overlap form the dark areas of the A band. The I bands contain only thin filaments, which extend to the edges of the H zones. Within the H zones, only thick filaments are found. The dark line at the center of the H zone (sometimes called the M band) is believed to indicate positions where thick filaments associate with one another.

The composition of the thick and thin filaments has been demonstrated by extracting myofibrils with appropriate salt or detergent solutions to remove virtually all of the myosin. Because the A bands are abolished in this process, the thick A-band filaments must be composed of myosin. The myosin thick filaments are bipolar structures of the kind shown in Figure 8.5, in which the helical tails of the myosin molecules join together, with the headpieces projecting with a regular 14.3-nm spacing at either end. This spacing appears to be generated by a 14.3-nm periodicity in the amino acid sequence in the myosin tails.

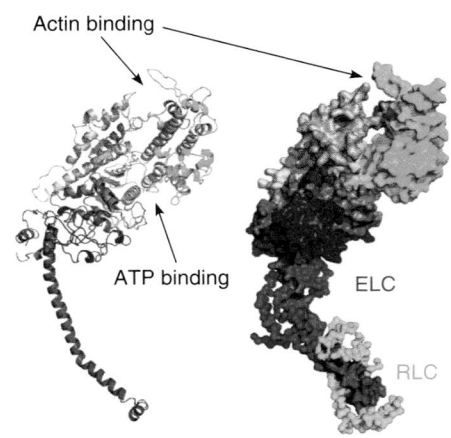

FIGURE 8.6

The X-ray crystal structure of an S1 fragment of myosin II. **Left:** A cartoon rendering of the heavy chain in rainbow coloring. The "neck" of the S1 fragment is shown by the extended red helix. **Right:** The heavy chain in a surface representation with the α-carbon backbone of the essential (ELC, magenta) and regulatory (RLC, cyan) light chains. The position of ATP binding is shown, as well as the point of contact with actin. PDB ID: 2MYS.

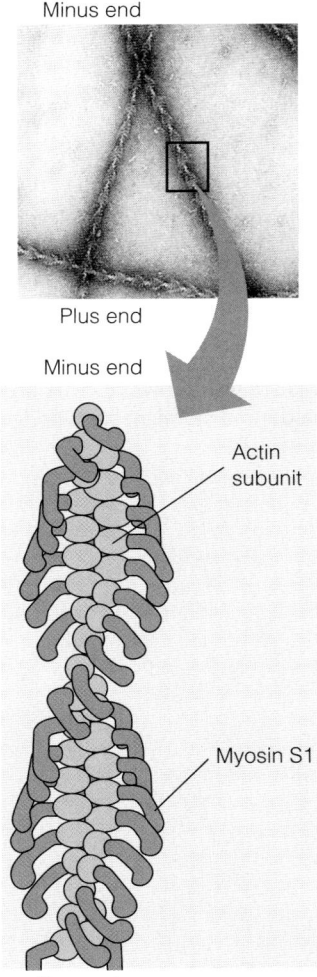

FIGURE 8.7

An actin filament decorated with myosin headpieces. **Top:** An electron micrograph. **Bottom:** A schematic at higher magnification. The S1 headpieces, indicated in the schematic, produce the arrowhead pattern visible in the micrograph.

FIGURE 8.8

Levels of organization in striated muscle.

John Bavosi/Science Photo Library.

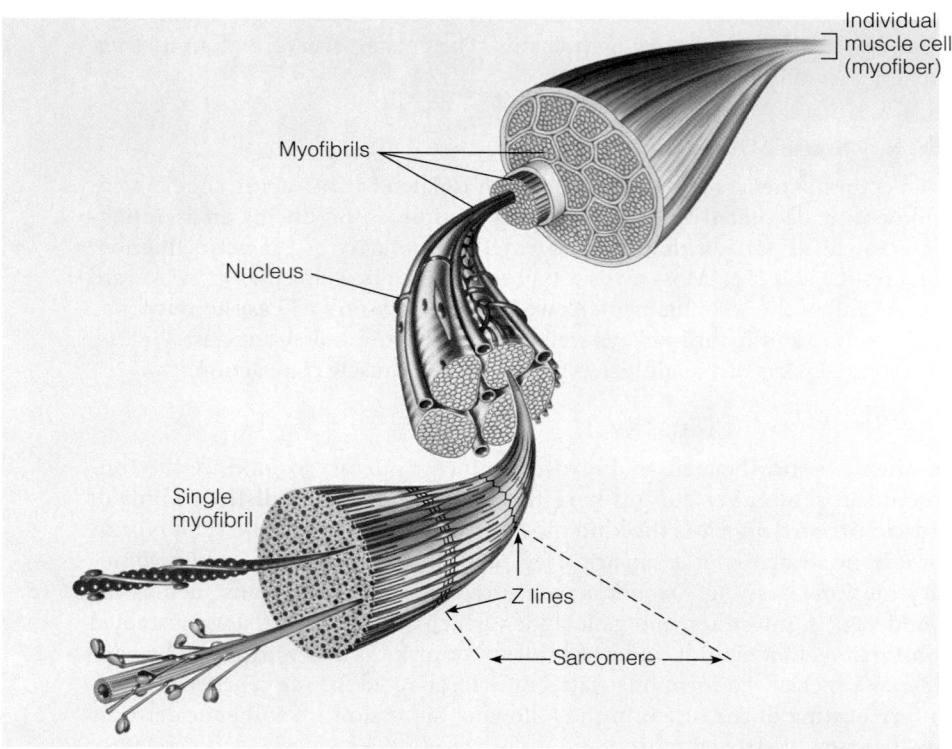

If myofibrils from which myosin has been removed are perfused with a solution of S1 fragments, the thin filaments are decorated in the arrowhead pattern (Figure 8.7); thus, the thin filaments contain actin. Furthermore, the arrowhead patterns always point outward from the Z disks, demonstrating the polarity of these thin filaments (i.e., the "plus" end of each F-actin filament is anchored to the Z disk). However, the thin filaments are not composed entirely of F-actin. As we shall see later in this chapter, they contain other important proteins as well.

If we look closely at electron micrographs of myofibrils (see Figure 8.9d), we can see small projections extending from the thick (myosin) filaments, often contacting the thin (actin) filaments. The projections correspond to the headpieces of the myosin molecules. These *cross-bridges* between myosin and actin filaments are the key to muscle contraction.

The organization of actin, myosin, and other muscle proteins into this elaborate yet specific structure found in the sarcomere is a remarkable example of how several kinds of protein can combine in a specific way to form a functional structure. We will now examine how this structure works.

The sarcomere is the basic repeating unit of a muscle myofibril.

Thin filaments are mainly actin; thick filaments are mainly myosin. They are connected by noncovalent cross-bridges.

The Mechanism of Contraction: The Sliding Filament Model

Our understanding of the mechanism of muscle contraction has come from observation both of the fine details of muscle structure and of changes in the sarcomere-banding pattern during contraction. The muscle sections shown in Figures 8.8 and 8.9, and the top of Figure 8.10, are in the relaxed, or extended, state. In a fully contracted muscle, each sarcomere shortens from a length of about 2.3 μm to 2.0 μm. During this process, the I bands and H zones disappear, and the Z disks move right up against the A bands (bottom of Figure 8.10). Such observations led two independent (and unrelated) investigators, Hugh Huxley and Andrew Huxley, to propose in the 1950s the **sliding filament model** of muscle contraction depicted in Figure 8.10. According to this model, the myosin headpieces are presumed to "walk" along the interdigitated actin filaments, pulling them past, and thereby shortening the sarcomere.

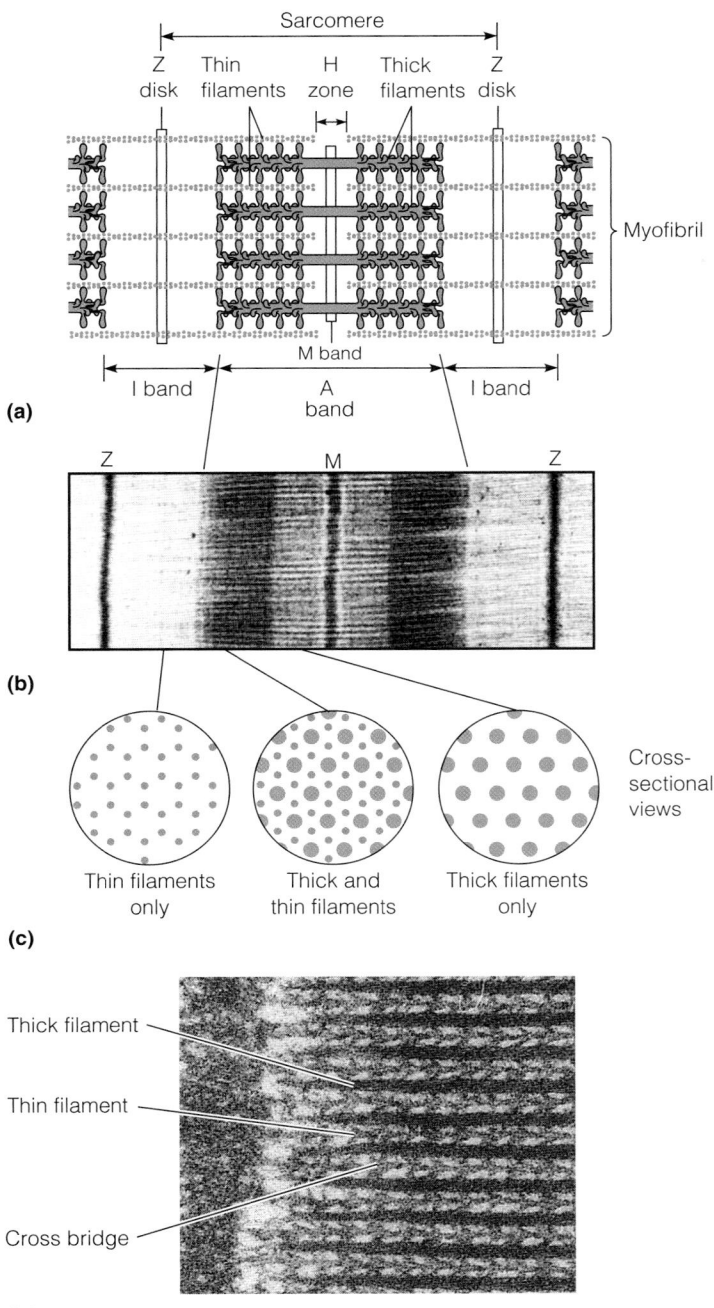

(a)

(b)

(c)

(d)

FIGURE 8.9

Muscle structure seen at the EM level. (a) A model of the sarcomere, the repeating unit in striated muscle. The I bands, A bands, and Z disks shown in Figure 8.8 are identified, and structural elements of the sarcomere are indicated. **(b)** An electron micrograph showing the same features. **(c)** A schematic drawing of cross sections of a sarcomere in the various regions shown in (a) and (b). Thick filaments are indicated by heavy brown dots, thin filaments by small purple dots. **(d)** A higher magnification within an A band showing cross-bridges between actin and myosin filaments.

(b) Courtesy of H. E. Huxley; (d) courtesy of Mary Reedy.

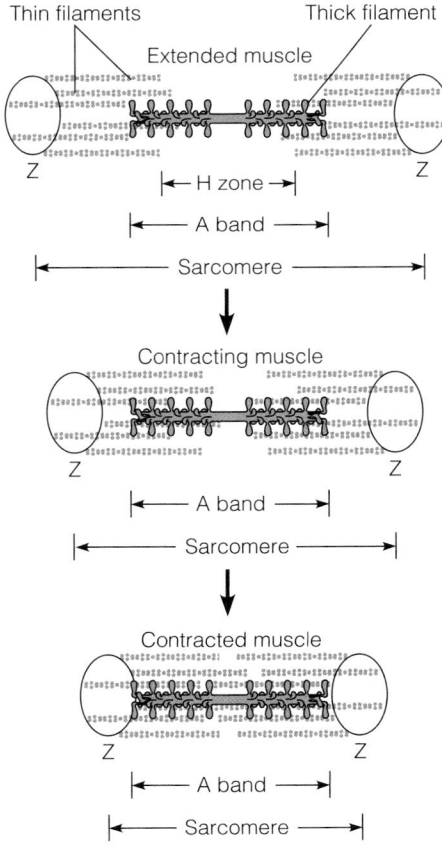

FIGURE 8.10

The sliding filament model of muscle contraction. Contraction of striated muscle occurs when the myosin headpieces pull the actin filaments toward the center of the sarcomere.

To produce such directed motion against an opposing force on the muscle, energy must be expended. You might expect that energy would be derived in some way from ATP hydrolysis, and our previous mention of the ATPase activity of the actin–myosin complex alludes to how this energy could be gained: According to a refinement of the sliding filament model called the **swinging cross-bridge model**, each myosin headpiece takes part in a repetitive cycle of making and breaking cross-bridges to an adjacent thin filament. We imagine the cycle starting with the myosin attached to actin, as shown at the top of Figure 8.11. Binding of ATP leads to release of the myosin cross-bridge (step 1). Hydrolysis of ATP then causes a conformational change, "cocking" the headpiece (step 2). Myosin, because it has now cocked, binds to the thin filament at a site that is closer to the Z disk—where the (+) end of the actin filament is anchored (step 3).

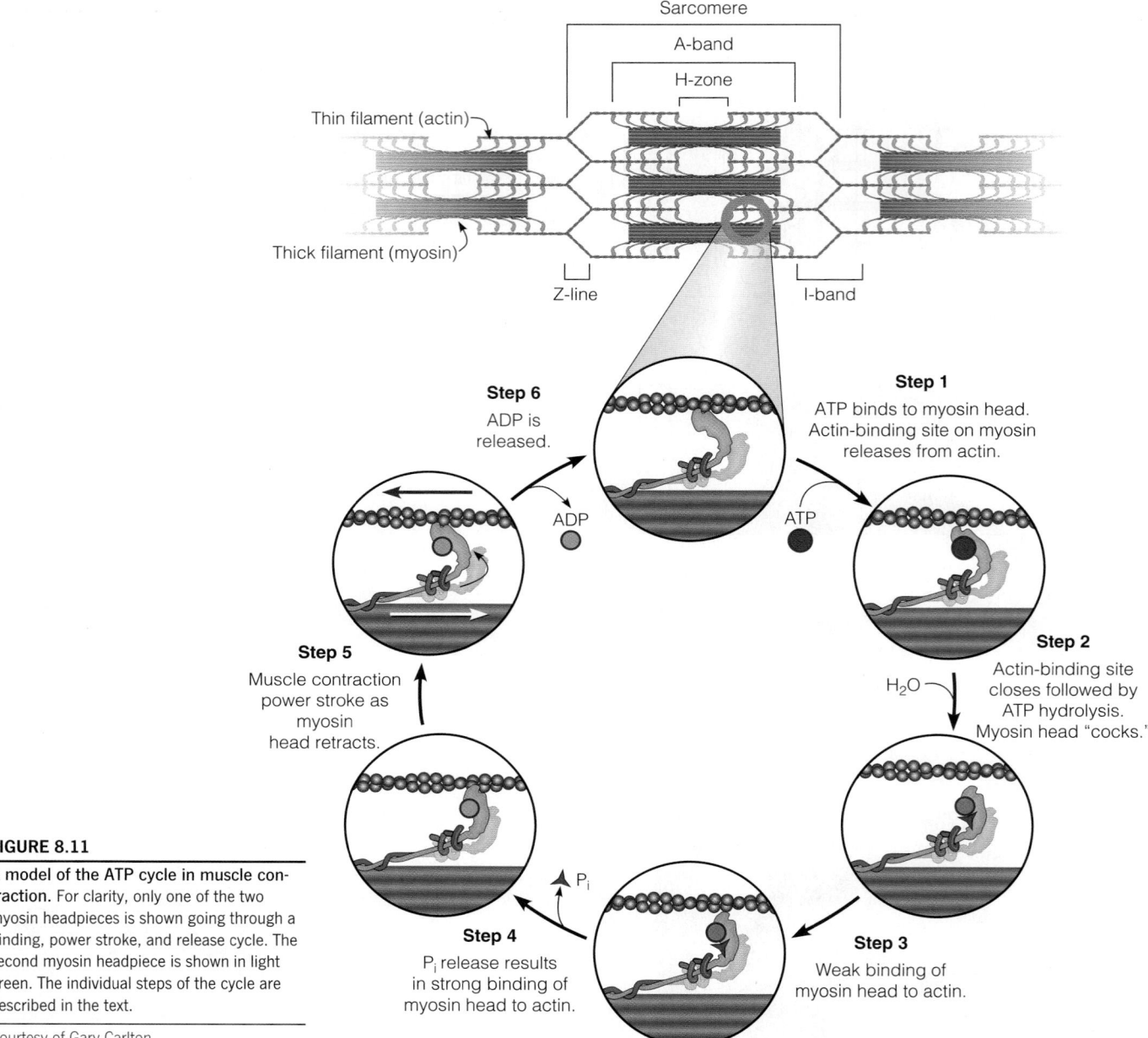

FIGURE 8.11

A model of the ATP cycle in muscle contraction. For clarity, only one of the two myosin headpieces is shown going through a binding, power stroke, and release cycle. The second myosin headpiece is shown in light green. The individual steps of the cycle are described in the text.

Courtesy of Gary Carlton.

In the swinging cross-bridge model, the periodic attachment and release of cross-bridges, with a cross-bridge conformational change, slide the thin and thick filaments past one another.

This initial binding is weak. Phosphate release (step 4) results in strong binding to the thin filament prior to the power stroke (step 5), which pulls the thin filament toward the center of the sarcomere. Release of ADP (step 6) and binding a new ATP will then restart the cycle.

It is the "neck region" of the S1 fragment (Figure 8.6), where the light chains are bound, that undergoes the greatest conformational change during the power stroke. Based on comparisons of X-ray crystal structures of myosin in the ADP-bound and nucleotide-free states, it is thought that the C-terminal part of the myosin neck moves about 10 nm during the power stroke (Figure 8.12).

At the end of each cycle, the actin filament has been moved with respect to the myosin so that each headpiece makes successive steps along the thin filament. The "walking" is rather like that of a millipede—there is always contact between some of the thick filaments "legs" (i.e., the many S1 headpieces) and the thin filament.

Thus, the thick filament doesn't slip back during muscle contraction. A thick filament displays several hundred S1 fragments and each contacts the thin filament 5 times per second during muscle contraction.

The force developed and the distance moved in each power stroke have been measured experimentally. The force is found to depend upon the load placed on the myosin–actin complex, but averages about 5 pN (piconewtons) at high load. This corresponds to an energy expenditure of about 10^{-20} J per power stroke—approximately one-fifth of the energy released when a single ATP molecule is hydrolyzed. At lower loadings, there is evidence that several power strokes can be accomplished per ATP cycle. The distance the actin filament is moved per power stroke is about 10–20 nm, and this is what can be accomplished per ATP cycle at high loading. When working against weaker resistance, the multiple power strokes can propel the actin for as much as 100 nm per ATP.

Regulation of Contraction: The Role of Calcium

The critical substance in stimulating contraction is not ATP, which is generally available in the myofibril; rather, it is a sudden increase in Ca^{2+} concentration. To understand how calcium regulates muscle contraction, we must examine the molecular structure of the thin filament in a bit more detail.

A thin filament, as found in striated muscle, is more than just an F-actin polymer. Four other proteins, shown in Figure 8.13, are essential to the thin filament's contractile function. One of these proteins is **tropomyosin**, a fibrous coiled-coil protein that overlaps the myosin-binding site in the F-actin helix. Bound to each tropomyosin molecule are three small proteins called **troponins I, C,** and **T.** Troponin C (TnC) is homologous to the Ca^{2+}-binding protein **calmodulin**, and

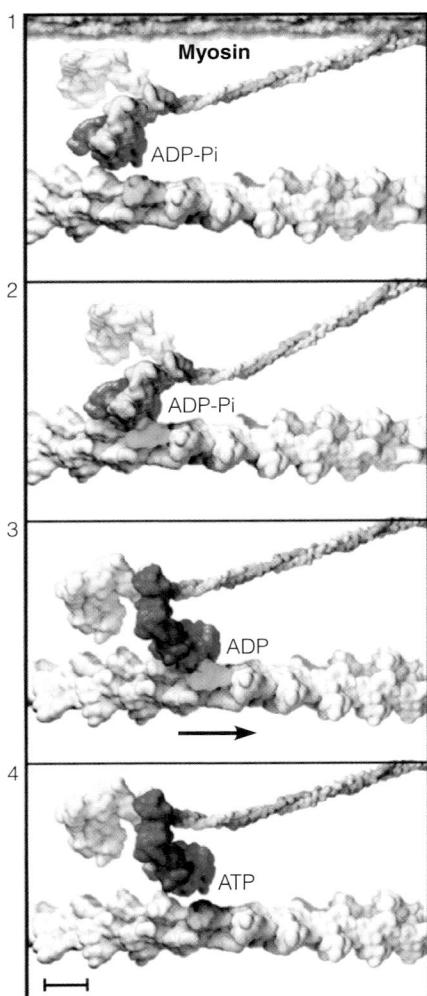

FIGURE 8.12

Model for the motility cycle of myosin II. The coiled coil (gray) extends from a myosin filament. Frame 1: The two identical S1 headpieces from a single myosin II dimer are shown, with motor domains in blue; lever arms (including the light chains) in the pre-stroke state are yellow. With ADP and P_i bound, the motor domain binds weakly to actin (see Figure 8.11, step 3). Frame 2: One motor domain forms a strong complex with an actin binding site (green). Frame 3: Upon phosphate release, the power stroke occurs and moves the actin filament by ~10 nm (see Figure 8.11, steps 4 and 5). The lever arm in the post-stroke state is colored red. Frame 4: Following the power stroke, ADP dissociates and ATP binds to the motor domain, releasing it from actin. Following hydrolysis of ATP to $ADP \cdot P_i$, the lever arm reverts back to its pre-stroke state (i.e., Frame 1). Scale bar 6 nm. The (+) end of the thin filament is to the left.

From *Science* 288:88–95, R. D. Vale and R. A. Milligan, The way things move: Looking under the hood of molecular motor proteins. © 2000. Reprinted with permission from AAAS.

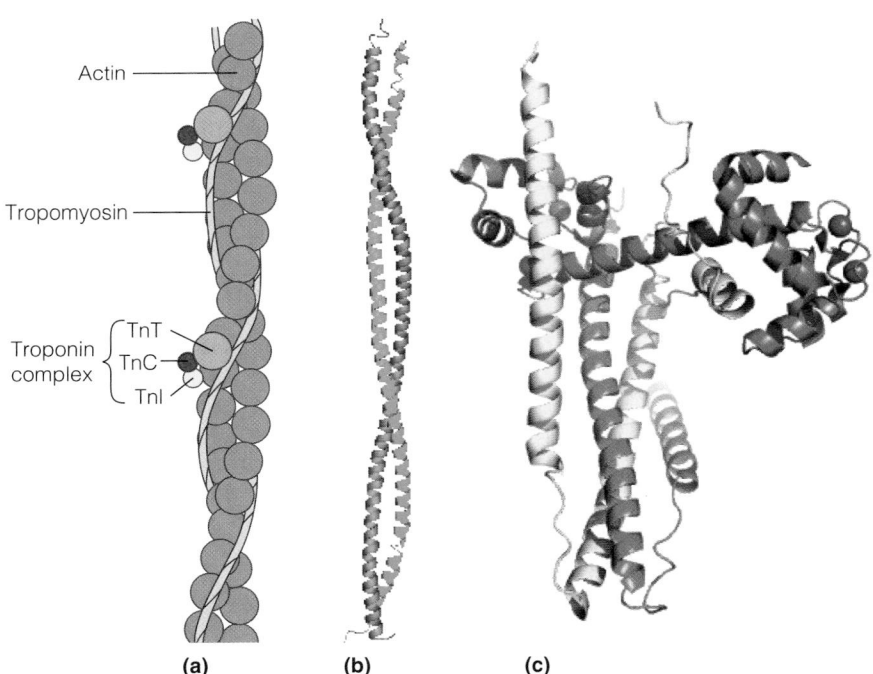

(a)   **(b)**   **(c)**

FIGURE 8.13

F-actin and its associated proteins. Panel **(a)**: This schematic drawing shows the proteins present in the thin filaments of striated muscle: F-actin, tropomyosin, and troponins (Tn) I, C, and T. Panel **(b)**: Crystal structure of a fragment of rat skeletal tropomyosin showing the coiled coil (PDB ID: 2B9C). Panel **(c)**: The troponin complex from chicken skeletal muscle (PDB ID: 1YTZ). TnI is shown in yellow, TnT in orange, and TnC in red with four bound Ca^{2+} ions shown as blue spheres. Panels (b) and (c) are not drawn to scale.

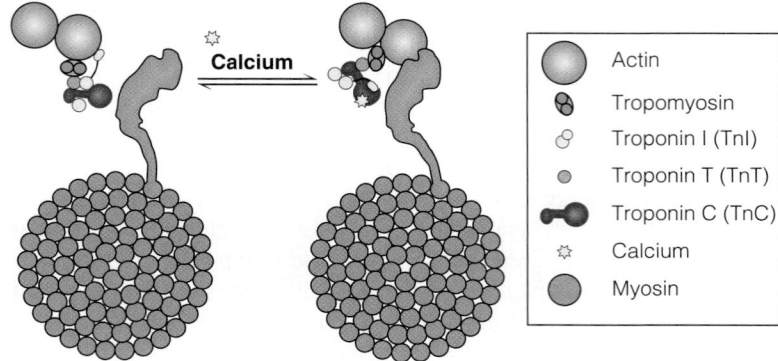

FIGURE 8.14

The regulation of muscle contraction by calcium. A single myosin headpiece is shown next to a thin filament, in cross-sectional view. **Left**: Relaxed muscle. At low Ca^{2+} levels, the configuration of actin, tropomyosin, and the troponin complex in the thin filament blocks most myosin headpieces from contacting the thin filament. **Right**: The binding of Ca^{2+} to TnC opens a binding site on TnC for the inhibitory domain of TnI, which also binds actin (see left panel). The release of TnI from actin allows the tropomyosin–troponin complex to slide closer to the central groove in F-actin, thereby exposing the myosin-binding site. Cross-bridge formation (step 3 of Figure 8.11) can then occur, and the muscle contracts.

Courtesy of Gary Carlton.

Muscle contraction is stimulated by an influx of Ca^{2+} into the sarcomere. Ca^{2+} binding by troponin C causes a rearrangement of the troponin–tropomyosin–actin complex, allowing actin–myosin cross-bridges to form.

like calmodulin, TnC undergoes a conformational change upon binding Ca^{2+}. The structure and function of calmodulin are presented in greater detail in Chapter 13 (see Figure 13.33, page 570). The presence of tropomyosin and the troponins inhibits the binding of myosin heads to actin *unless calcium is present* at a concentration of about 10^{-5} M. In resting muscle, Ca^{2+} concentrations are approximately 10^{-7} M, so new cross-bridges cannot be formed. An influx of Ca^{2+} stimulates contraction because the ion is bound by troponin C, causing a rearrangement of the troponin–tropomyosin complex, which moves the tropomyosin ~1.0 nm closer to the central groove of the F-actin helix. This shift reveals the sites on actin to which the myosin headpieces bind. The postulated mechanism, shown in Figure 8.14, permits step 3 and the subsequent steps in the cycle of Figure 8.11 to take place.

We have now traced the activation of muscle contraction to the influx of calcium into the myofibrils. But why does this influx occur? In particular, how can it be brought about by the nerve impulses that excite muscles to contract? The answer can be found from a closer examination of the myofiber, or muscle cell (Figure 8.15). Within the cell, each myofibril is surrounded by a structure called the **sarcoplasmic reticulum**, formed of membranous tubules. In resting muscles the level of Ca^{2+} in the myofibrils is maintained at about 10^{-7} M, whereas the Ca^{2+} level within the lumen of the sarcoplasmic reticulum may be 10,000-fold higher. Impulses from motor nerves depolarize the membrane of the sarcoplasmic reticulum, opening gated Ca^{2+} channels (see Chapter 10), and Ca^{2+} pours out of the lumen into the myofibrils, stimulating contraction. The signal is rapidly transmitted to the entire sarcoplasmic reticulum of a myofiber via **transverse tubules**, invaginations of the plasma membrane that connect at periodic intervals with the reticulum. Following the contraction, a Ca^{2+}-specific transport protein pumps

FIGURE 8.15

Structure of a myofiber (muscle cell). The sarcoplasmic reticulum (SR, white, see solid arrows) is a network of specialized endoplasmic reticulum tubules that surrounds the myofibrils within the myofiber. In resting muscle, the SR accumulates Ca^{2+}, which it discharges into the myofibrils when a neural signal reaches the plasma membrane. The transverse tubules (T system, see dashed arrows) are invaginations of the plasma membrane that make contact with the SR at many points, ensuring uniform response to the signal.

MedicalRF.com/Visuals Unlimited, Inc.

Ca^{2+} ions out of the sarcomere to restore the resting $[Ca^{2+}]$ to 10^{-7} M. The activities of such ion transporters are described in Chapter 10.

Although the abrupt change in Ca^{2+} level is the universal *signal* for muscle contraction, it obviously cannot in itself provide the necessary energy. What are the sources of the energy required for muscular work?

Energetics and Energy Supplies in Muscle

Basically, muscle is a mechanism for converting the chemical free energy released in ATP hydrolysis into mechanical work. The conversion can be remarkably efficient, approaching values of 80% under optimal circumstances.

How is the ATP for muscle contraction generated? Even in striated muscle the answer can vary, depending on the particular *kind* of muscle and its function. Striated muscles can be divided into two categories—*red muscle,* designed for relatively continuous use, and *white muscle,* employed for occasional, often rapid, motions. Red muscle owes its dark color to its abundant heme proteins: It is well supplied with blood vessels and therefore hemoglobin, it has many mitochondria with cytochromes, and it has large stores of myoglobin. Red muscle depends heavily on aerobic metabolism in mitochondria, so the primary energy source in red muscle is the oxidation of fat. White muscle, on the other hand, relies on glycogen as a primary energy source. Glycogen is excellent for quick energy production but cannot sustain activity for long periods (see Table 8.2 for a more detailed comparison of red and white muscle, and Chapters 13 and 18 for a more extensive discussion of glycogen metabolism).

The functional differences between the two types of striated muscle are clearly revealed in birds. In the domestic chicken, the flight muscles of the breast, used only for brief fluttering or short flights, are white, whereas the heavily used leg muscles are red. Wild flying birds, which make sustained flights but rarely walk, have just the opposite distribution of light and dark meat.

Careful measurement of ATP levels in red striated muscle has shown that the provision of energy is more complicated than it might appear at first. The amount of ATP needed for a single contraction may be greater than all the ATP immediately available to a sarcomere. Yet even after relatively long exercise, ATP levels in the sarcomeres remain essentially constant. This finding suggests that ATP is an intermediary, and not an energy *storage* compound in muscles. Indeed, aerobic muscle tissue continuously resynthesizes ATP from ADP and P_i to support prolonged muscular activity. This requires a steady import of fuel molecules, which are oxidized by O_2 in mitochondria to drive the unfavorable synthesis of ATP (see Chapter 15). Bursts of intense activity may exceed the capacity of aerobic metabolism to supply the needed ATP. To bridge the short interval between the moderate demand for ATP met by aerobic metabolism and the high demand met by anaerobic ATP production, muscles carry a small reserve of the high-energy compound *creatine phosphate.* As its high phosphate transfer potential suggests (see

TABLE 8.2 Comparison of red and white striated muscle

	Red	White
Relative fiber size	Small	Large
Mode of contraction	Slow twitch	Fast twitch (about 5 times faster)
Vascularization	Heavy	Lighter
Mitochondria	Many	Few
Myoglobin	Much	Little
Major stored fuel	–	Glycogen in muscle
Main source of ATP	Fatty acid oxidation	Glycolysis

Figure 3.7), this compound is capable of phosphorylating ADP very efficiently. The reaction is catalyzed by the enzyme *creatine kinase*.

$$^-OOC-CH_2-\underset{\underset{PO_3^{2-}}{\overset{|}{N}}}{\overset{\overset{CH_3}{|}}{N}}-C=\overset{+}{N}H_2 + ADP + H^+ \underset{\text{kinase}}{\overset{\text{Creatine}}{\rightleftharpoons}} \; ^-OOC-CH_2-\underset{\underset{NH_2}{\overset{|}{N}}}{\overset{\overset{CH_3}{|}}{N}}-C=\overset{+}{N}H_2 + ATP \qquad \Delta G^{\circ\prime} = -12.6 \text{ kJ/mol}$$

Creatine phosphate   **Creatine**

An alternate energy source in red muscle is creatine phosphate, which regenerates ATP continually as it is depleted by muscle contraction.

Because the equilibrium lies well to the right, virtually all of the muscle adenylate is maintained in the ATP form, rather than as ADP or AMP, as long as creatine phosphate is available. The consumption of creatine phosphate during exercise, while an almost constant level of ATP is maintained, is clearly demonstrated by the NMR studies of human muscle shown in Figure 12.13 (page 503).

Nonmuscle Actin and Myosin

The functions of actin and myosin are not limited to muscle contraction. Indeed, members of the actin and myosin families are found in most eukaryotic cells, even those that are in no way involved in muscular tissues (see Table 8.1). Actin and myosin play significant roles in cell motility and changes of cell shape. Actin is the major component of the **microfilaments** of the **cytoskeleton**—the fibrous array that pervades almost every kind of cell and gives it a specific shape (Figure 8.16a). Staining with fluorescent antibodies reveals that an isoform of myosin II (**nonmuscle myosin II**; Figure 8.16b) and other types of myosin (see Table 8.1) are also associated with this network. The nonmuscle myosin II (NM II) in such intracellular networks is different in sequence from muscle myosin II, and its assembly into filaments is regulated by phosphorylation of the associated regulatory light chains. The bipolar filaments formed by NM II contain roughly 14–20 myosin molecules, making them significantly smaller than the thick filaments in myofibrils. Several cellular proteins are involved in anchoring microfilaments to the cell membrane, and we will not mention them here. Rather, we point out that organized contraction and relaxation of such anchored microfilament and NM II networks (Figure 8.16c) can lead to a wide variety of cell distortions and movements, including amoeboid crawling.

A nonmuscle actin–myosin system generates the motion and shape changes of many types of cells.

FIGURE 8.16

Cytoskeletal actin and nonmuscle myosin II in fibroblasts. **(a)** Actin fibers as detected by fluorescent phalloidin, which binds specifically to actin. **(b)** Myosin in the same cell detected by an injected fluorescent antibody. **(c)** A mechanism for contractility in the cytoskeleton. An NM II filament (green and olive) will bring together two anchored actin microfilaments (red). NM II moves toward the "+" end of the microfilament. The direction of movement of the actin microfilaments is shown by the arrows.

(a, b) From *Journal of Cell Science*, Suppl. (1991) 14: 41–47, B. M. Jockusch et al.; R. A. Cross and J. Kendrick-Jones, eds. The Company of Biologists, Ltd., Cambridge. (c) Courtesy of Gary Carlton.

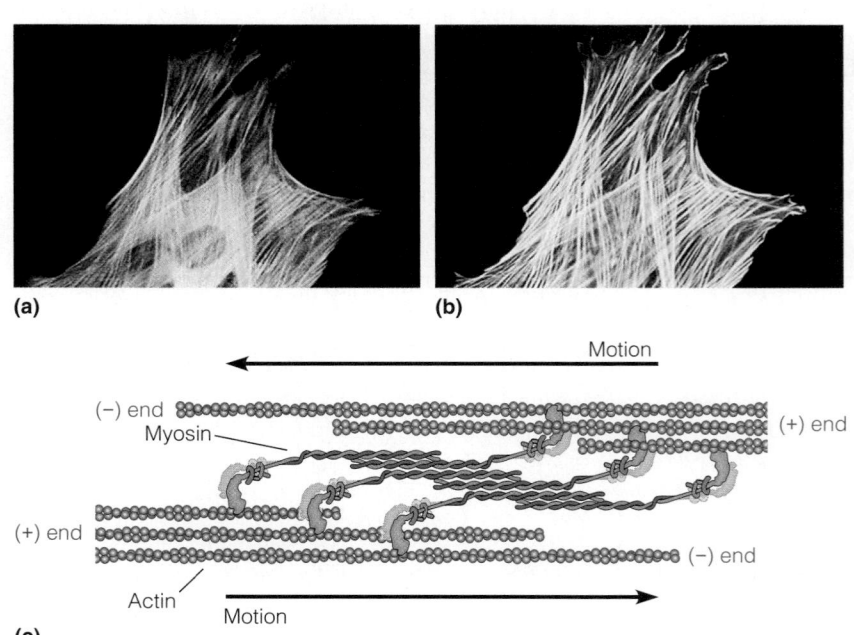

Another intracellular process involving an actin–myosin contractile complex is **cytokinesis**, the division of cells in the last stages of mitosis (Chapter 24). The process can be seen clearly in the sea urchin egg, a favorite model system for such studies (Figure 8.17). Toward the end of mitosis, when the daughter nuclei are clearly separated at the two poles of the cell, a ring of indentation is observed in the cell surface, defining a plane perpendicular to the mitotic spindle. This ring then contracts, as proposed in Figure 8.16c, forming the **cleavage furrow** and eventually cutting the cell in two. Electron microscopy shows that the ring consists of fibers, and staining with fluorescent antibodies shows that the fibers contain both actin and myosin.

The involvement of myosin in cytokinesis has been most elegantly demonstrated by S. Inoué and his colleagues in the experiment depicted in Figure 8.18. After the sea urchin egg had undergone one division, the daughter cell on the right was microinjected with antimyosin antibodies, which should make myosin nonfunctional. After 10 h, the control cell (left) had undergone many divisions to construct half of an embryo. In the treated cell, mitosis continued, as evidenced by the many nuclei present, but *cytokinesis had been completely blocked.* Therefore, myosin is essential for cytokinesis.

The experiment demonstrates one other important fact. Because mitosis continued even in the cell treated with antimyosin antibodies, the contractile process in the mitotic spindle apparently does *not* require the participation of myosin. There have been many reports of the presence of actin and myosin in the spindle, and it was long believed that they were essential for chromosome separation. But these and other experiments show that some other contractile system must be involved. It is to this second general class of motility-generating systems that we now turn.

Microtubule Systems for Motility

A class of motile systems completely different from, and unrelated to, the actin–myosin contractile systems is used in places as diverse as the mitotic spindle, protozoan and sperm flagella, and nerve axons, to name only a few. These systems are constructed from **microtubules**, very long, tubular structures built from a helical arrangement of the protein **tubulin** (Figure 8.19). There are two homologous tubulin subunits, α and β, each of molecular weight 55 kDa. They are present in equimolar quantities in the microtubule, which can be considered a helical array of $\alpha\beta$ dimers. Alternatively, we can view the microtubule as consisting of 13 rows, or **protofilaments**, of alternating α and β subunits. Because the α and β units are asymmetrical proteins, with a defined and reproducible orientation in the fiber, the microtubule has directionality as indicated by the (+) and (−) ends in Figure 8.19.

The assembly of microtubules bears resemblance to that of actin, but GTP, rather than ATP, is required. The α and β subunits each bind GTP and then associate to form $\alpha\beta$ dimers and oligomers. These oligomers form nucleation sites for the growth of microtubules (Figure 8.20). One end, called the plus end, grows more

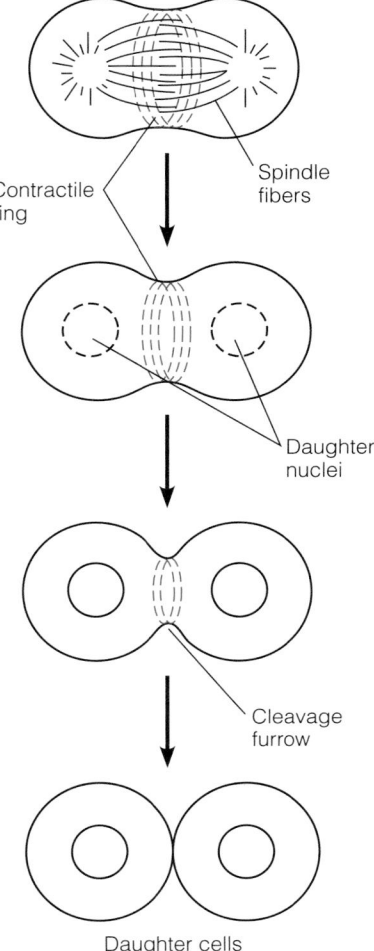

FIGURE 8.17

Actin and nonmuscle myosin II in cytokinesis. This drawing depicts cytokinesis in a fertilized sea urchin egg. The contractile ring that cleaves the cell in two is an actin–myosin complex located just beneath the plasma membrane.

Microtubules are helical, tubular polymers made from α and β tubulin heterodimers.

FIGURE 8.18

Myosin is essential to cytokinesis but not to mitosis. **(a)** A sea urchin embryo at the two-cell stage. The cell to the right has been injected with antimyosin antibodies; the oil droplet shows the site of injection. **(b)** After 10 h, the cell on the left has undergone many divisions, forming half of a normal embryo. In the cell on the right cytokinesis has been completely blocked, but the many new nuclei show that mitosis has continued.

© The Rockefeller University Press. *The Journal of Cell Biology,* 1982, 94:165, D. P. Kiehart, I. Mabuchi, and S. Inoué, Evidence that myosin does not contribute to force production in chromosome movement.

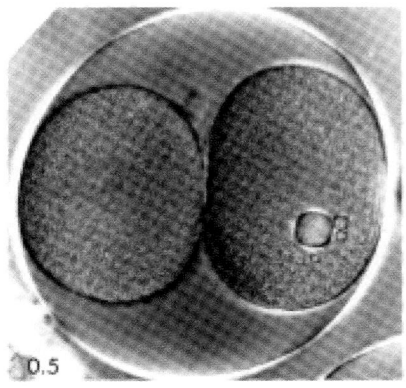
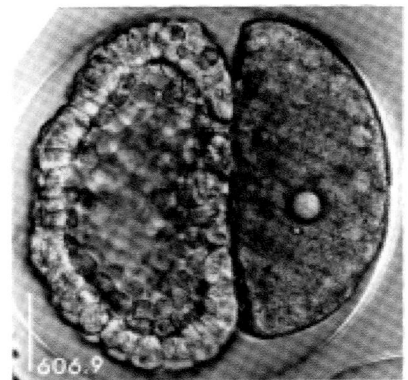

(a)　(b)

FIGURE 8.19

A helical array of α and β tubulin forms a microtubule. **Left:** The crystal structure of an $\alpha\beta$ dimer is shown. The three domains in each subunit are color-coded (blue, green, and orange). The α subunit is on top and has GTP (red spheres) bound. The β subunit has GDP (purple) and the anticancer drug paclitaxel (white, blue, and red spheres) bound. With the subunits in this orientation the (−) end is up and the (+) is down. **Right:** A schematic model of a microtubule. The α and β tubulin subunits are shown in tan and blue, respectively. The microtubule can be thought of either as a helical array of $\alpha\beta$ dimers or as 13 parallel rows of $\alpha\beta$ dimers. These rows are referred to as protofilaments.

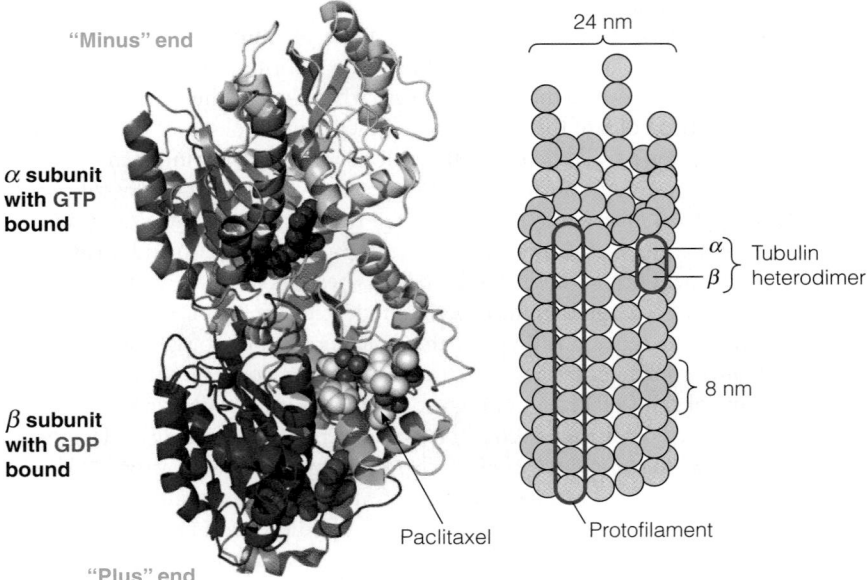

rapidly than the other, minus end. The β subunit has GTPase activity, and as in actin polymerization, the nucleotide is hydrolyzed but is held in the filament. Both microfilaments and microtubules are dynamic structures that can be disassembled and reassembled from either end. In fact, this is a critical feature of the cytoskeleton, and compounds that interfere with cytoskeleton dynamics are toxic to cells.

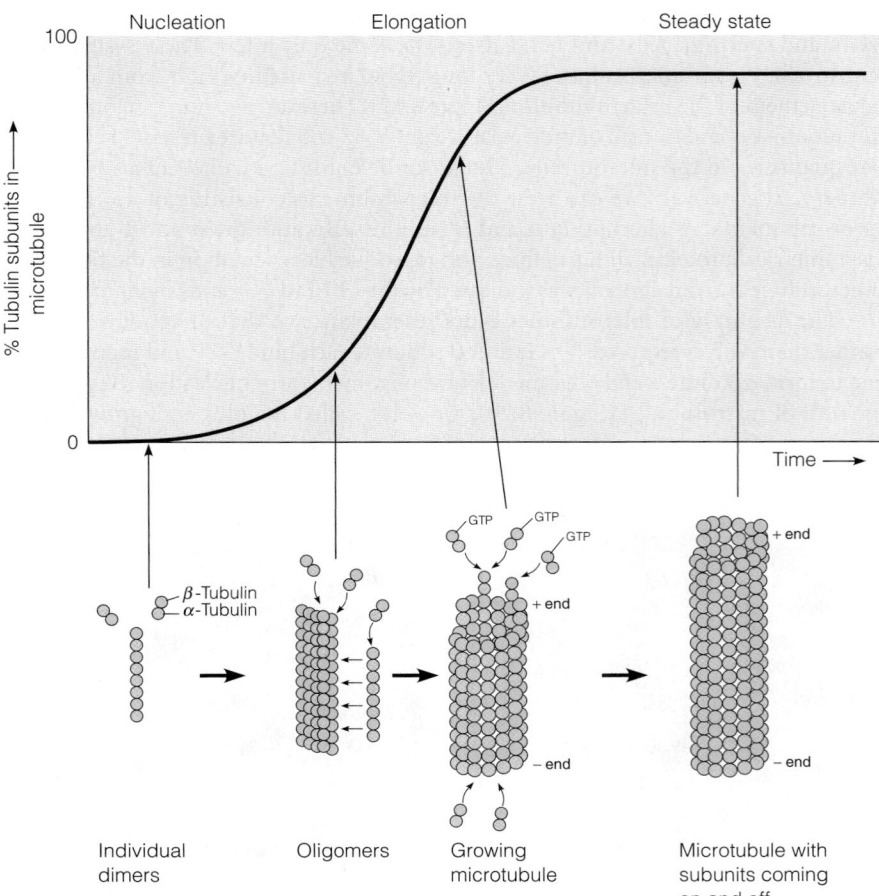

FIGURE 8.20

Assembly of microtubules. During an initial lag period, $\alpha\beta$ dimers form oligomers large enough to nucleate fiber formation. The microtubules then grow until most of the free dimers are used up and equilibrium between growth and dissociation is reached. For simplicity, we show growth only at one end because growth is more rapid at the plus end.

Because cancer cells are rapidly dividing, the disruption of microtubule assembly during mitosis is the target of several anti-cancer treatments. For example, colchicine and paclitaxel (Taxol) both affect microtubule dynamics and thereby interfere with cell division. Colchicine binds to tubulin subunits and thereby prevents microtubule polymerization. Paclitaxel binds to the β subunit of tubulin (see Figure 8.19) and stabilizes the microtubule structure to the point that the breakdown and recycling of the microtubules is inhibited, which ultimately kills the cell. Nonmalignant cells are also affected adversely by such chemotherapeutic treatment, but the rapidly dividing cancer cells are more susceptible to cell-killing by such drugs.

Microtubules are larger and appear to be more mechanically stable than actin microfilaments; thus, microtubules are better suited to high-stress functions. The final assembly of a functional microtubule usually involves the binding of other proteins to its surface. Some of these **microtubule-associated proteins (MAPs)** play functional roles, as we shall see. Other MAPs stabilize microtubule structure and/or promote the association of microtubules into bundles. A member of this family that has drawn considerable attention is the tau (τ) MAP found in neuronal tissue. Phosphorylation of tau results in its dissociation from microtubules, leading to destabilization of the microtubule structure. Hyperphosphorylation has a much more dramatic effect, resulting in the formation of tangles of τ-filaments in neural axons, one of the major cellular symptoms of Alzheimer's disease. This in turn may imply that inappropriate synthesis or activation of the enzymes (kinases) involved in tau phosphorylation may trigger the onset of the disease. We shall now consider some of the things that microtubules do.

Motions of Cilia and Flagella

Many kinds of prokaryotic and eukaryotic cells are propelled by the motions of cilia or flagella. Cilia are shorter than flagella and exert a coordinated rowing motion to move a microorganism through solution (Figure 8.21a). Eukaryotic flagella, such as sperm tails, are longer and propel the cell by an undulatory motion (Figure 8.21b). The structures of the two types of motile appendages have many elements in common (Figure 8.22). Each of them contains a highly organized bundle of microtubules

Colchicine and paclitaxel act as anti-cancer agents by disrupting microtubule dynamics.

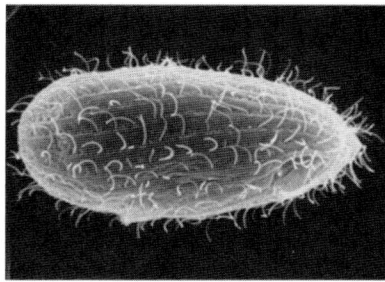

(a)

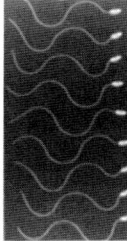

(b)

FIGURE 8.21

Cilia and flagella. **(a)** The protozoan *Tetrahymena* is covered with rows of cilia. **(b)** A sperm of the tunicate *Ciona*. A series of time-lapse photographs shows how undulations of the flagellum propel the sperm.

(a) © The Rockefeller University Press. *The Journal of Cell Biology*, 1983, 96:1610, U. W. Goodenough, Motile detergent-extracted cells of *Tetrahymena* and *Chlamydomonas*; (b) Courtesy of C. J. Brokaw, California Institute of Technology.

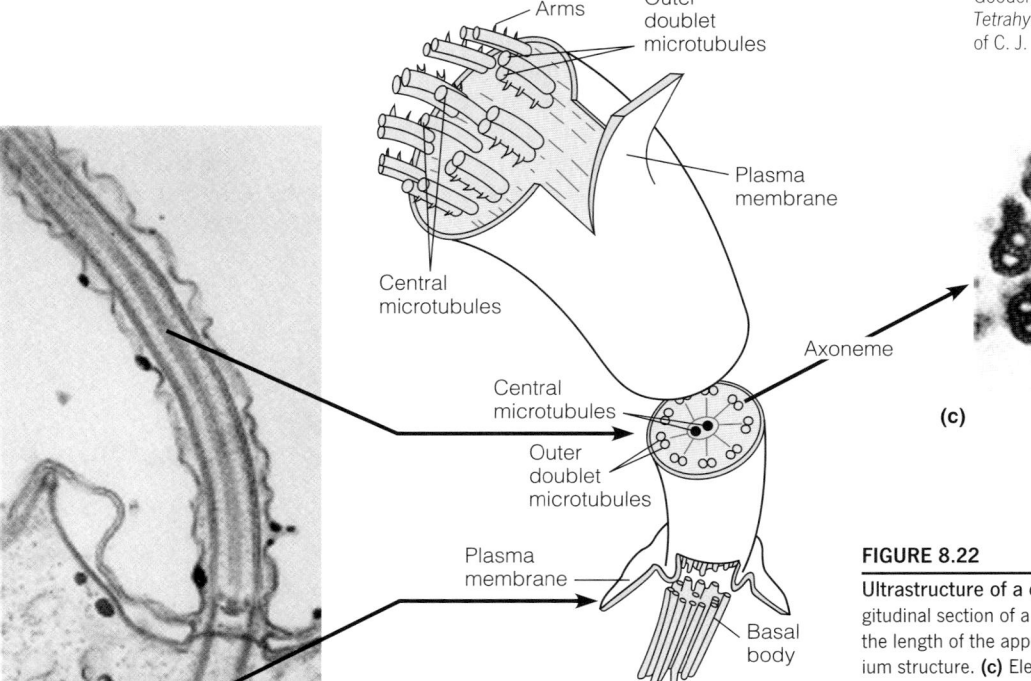

Arms
Outer doublet microtubules
Plasma membrane
Central microtubules
Central microtubules
Outer doublet microtubules
Axoneme
Plasma membrane
Basal body

(a)   **(b)**

(c)

FIGURE 8.22

Ultrastructure of a cilium. **(a)** Electron micrograph of a longitudinal section of a cilium, showing microtubules running the length of the appendage. **(b)** Schematic drawing of cilium structure. **(c)** Electron micrograph of a cross section of the axoneme, showing the 9 + 2 arrangement of outer doublets and inner tubules.

(a) © Photo Researchers; (c) Courtesy of W. L. Dentler, University of Kansas.

Many kinds of cells are propelled by cilia or flagella containing microtubules.

Bending of cilia and flagella is accomplished by the dynein-driven sliding of microtubules past one another.

called an **axoneme**, enveloped by an extension of the plasma membrane and connected to a **basal body**, an anchoring structure within the cell.

The internal structure of the axoneme is truly remarkable. As the cross-section in Figure 8.22c shows, the most obvious feature is the arrangement of microtubules known as a 9 + 2 array: two central microtubules ringed by nine microtubule doublets. The single microtubules in the center are complete, each having 13 protofilaments of $\alpha\beta$ tubulin dimers. By contrast, each of the nine surrounding doublets is composed of one complete microtubule (the **A fiber**) to which is fused an incomplete microtubule, carrying only 10 or 11 protofilaments (the **B fiber**). Closer inspection of electron micrographs reveals even greater complexity, as diagrammed in Figure 8.23. The outer doublets are periodically interconnected by a protein called *nexin* and carry at regular intervals *sidearms* composed of the motor protein **dynein**. In addition, *radial spokes*, each consisting of a head and an arm, project from the outer doublets to connect with the central pair.

The full complexity of axoneme structure is revealed by gel electrophoresis studies of isolated axonemes. About 200 polypeptides can be resolved. Analysis indicates at least 6 proteins in the spoke heads and 11 others in the arms of the spokes. Much of this apparatus seems to be directly involved in the beating motions of cilia and flagella. If ATP is added to isolated axonemes, adjacent doublets can be seen to slide past one another. The best current model holds that this sliding occurs by "walking" of the dynein sidearms along the adjacent doublet (Figure 8.24a). Doublets slide past each other first on one side of the axoneme and then the other, with the length of the slide limited by the central spokes and nexin connectors. In this way, the sliding of doublets is transformed into back-and-forth bending of the whole cilium or flagellum (Figure 8.24b). If connections within the axoneme are removed by careful proteolysis, ATP simply causes axonemes to extend and thin, as the outer doublets slide past one another with no stopping point.

It has been demonstrated that dynein has ATPase activity, with binding of ATP associated with the *breaking* of dynein cross-bridges. Thus, an obvious similarity exists between the mechanisms of beating of cilia and flagella and the ATP-driven walking of myosin heads along the actin fiber.

Intracellular Transport

At one time, all transport of materials within the cytoplasm of cells was thought to occur through simple diffusion. It is now known that some proteins and organelles are rapidly transported over long distances along microtubules, which serve as tracks that direct and facilitate the motion.

The clearest evidence comes from studies of transport in **axons**, the long projections that allow one nerve cell to contact another (see Figure 10.45, page 397). Because nerve axons can be many centimeters in length, sufficiently rapid movement

FIGURE 8.23

Diagram of the cross-section of an axoneme. The dynein arms have ATPase activity and can cause adjacent doublets to slide with respect to each other. The nexin connections between doublets and the radial spoke system give stability to the whole assembly.

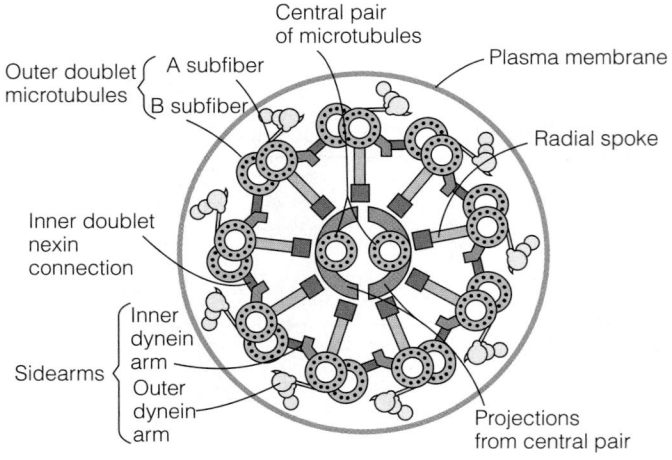

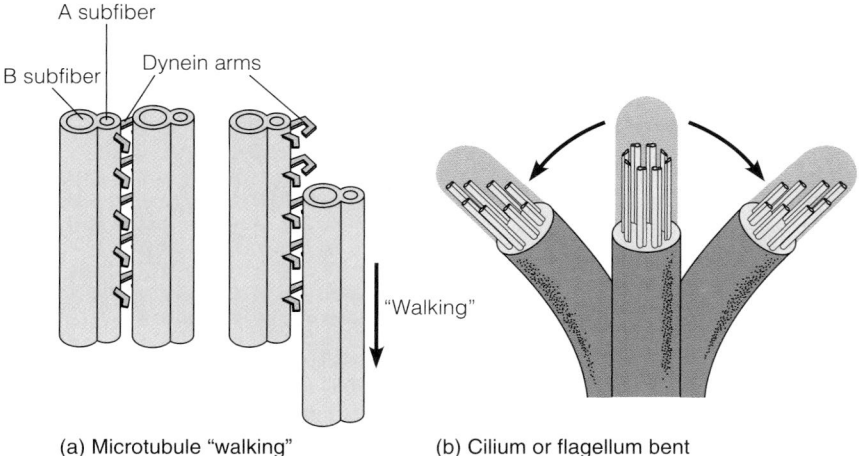

A subfiber
B subfiber
Dynein arms
"Walking"

(a) Microtubule "walking"

(b) Cilium or flagellum bent
by microtubule walking

FIGURE 8.24

Model for the bending of cilia and flagella.
(a) Isolated microtubule doublets can "walk" past one another by making and breaking dynein arm contacts, if ATP is present. **(b)** The back-and-forth bending of a cilium or flagellum is produced by synchronized short "walks" by microtubules on opposite sides of the appendage. Long walks are prevented by the cross-connections between microtubules.

Organelles and individual molecules are transported by motor proteins within cells, along molecular "tracks" of microtubules or actin microfilaments.

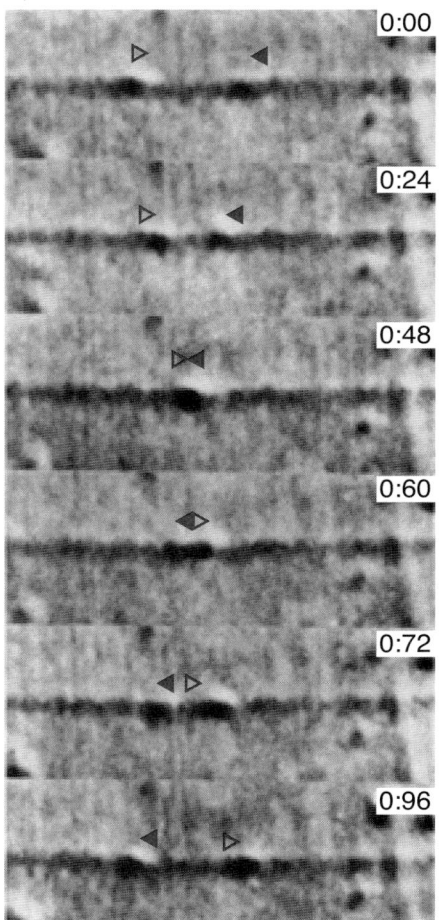

0:00
0:24
0:48
0:60
0:72
0:96

FIGURE 8.25

Movement on microtubule tracks. A series of time-lapse video micrographs shows the bidirectional motion of organelle vesicles on a single microtubule filament from squid giant axon. The two organelles (pointed to by the open and solid red triangles) move in opposite directions and pass each other. Elapsed time (s) is given in the upper right corner of each frame.

Reprinted from *Cell* 40:455–462, B. J. Schnapp, R. D. Vale, M. Sheetz, and T. S. Reese, Single microtubules from squid axoplasm support bidirectional movement of organelles. © 1985 with permission from Elsevier.

of materials between the cell body and the end of the axon cannot be accomplished by diffusion. The problem can be studied directly by using the mammalian sciatic nerve, which has very long axons extending from the cell body in the spinal cord. If radiolabeled amino acids are injected into the cell body, they are incorporated into proteins by the cell's ribosomes. After some time, the axon can be sectioned, and the location of newly synthesized, radiolabeled proteins can be determined. This method reveals that although transport rates vary greatly, some proteins, especially those associated with lipid vesicles, move as rapidly as 40 cm/day, much faster than could be accounted for by diffusion.

As shown in Figure 8.25, small vesicles or whole organelles can actually be seen moving along microtubule bundles in an axon. Transport along the microtubules occurs in both directions, in each case by the attachment of "molecular motors" to the objects to be transported. These motors are of two kinds. One, called **cytoplasmic dynein**, resembles the dynein involved in the motion of cilia and flagella (see previous section) and is responsible for transport from the plus end of the microtubule toward the minus end. The other, called **kinesin**, is used to transport objects in the opposite direction. Kinesin and cytoplasmic dynein represent different families of motor proteins, with similar but distinct transport functions in a wide variety of cell types. The two proteins have different structures (Figure 8.26); however, kinesin bears similarity to the myosin family in both structure and function. The helical tail of each molecule connects it to whatever "cargo" the dynein or kinesin is carrying, perhaps through associated proteins (shown in green in Figure 8.26).

Similarities in function between myosin and kinesin can be seen by comparing an ATP-hydrolysis cycle for each protein, as depicted in Figures 8.12 and 8.27. Both proteins have a globular motor domain that binds both its target filament and ATP. A significant difference between myosin and kinesin is seen in their respective "duty ratios," or the time each motor domain spends bound to its target. Because at any given time only a few myosin headpieces in a myosin filament will be attached to the actin microfilament, an individual myosin motor domain spends most of its time in the unbound state (see Figure 8.12, panels 1 and 4). Kinesin doesn't form oligomeric filaments; thus, it must always have one of its motor domains securely bound to the microtubule as it carries its cargo. Once bound to a microtubule, kinesin hydrolyzes roughly 100 ATP before it dissociates. Thus, the motor domains of kinesin work in concert, whereas the motor domains of myosin work independently.

The mechanism of dynein movement is not yet understood in detail; however, binding and hydrolysis of ATP in the motor domain results in large conformational changes between the motor domain and the linker to the cargo domain.

Careful studies of the motion of kinesin and dynein on microtubules indicate that they move stepwise along the microtubule track, with a step size of about

FIGURE 8.26

Structural models of kinesin I and cytoplasmic dynein. Models for kinesin and dynein were built from crystal structures of several isolated domains. Motor domains are shown in blue/purple. The microtubule-binding portions are on the left and the cargo-binding portions are on the right. The motor domain of kinesin resembles the S1 headpiece of muscle myosin. The motor domain of dynein binds ATP and contains six different subunits arranged in a circle. The microtubule-binding stalk and the linker to the cargo domain are covalently linked to the hexameric motor domain.

Adapted from *Cell* 112:467–480, R. D. Vale, The molecular motor toolbox for intracellular transport. © 2003, with permission from Elsevier.

Kinesin is a motor protein that transports its cargo toward the (+) end of a microtubule. Dynein is a motor protein that transports its cargo toward the (−) end of a microtubule.

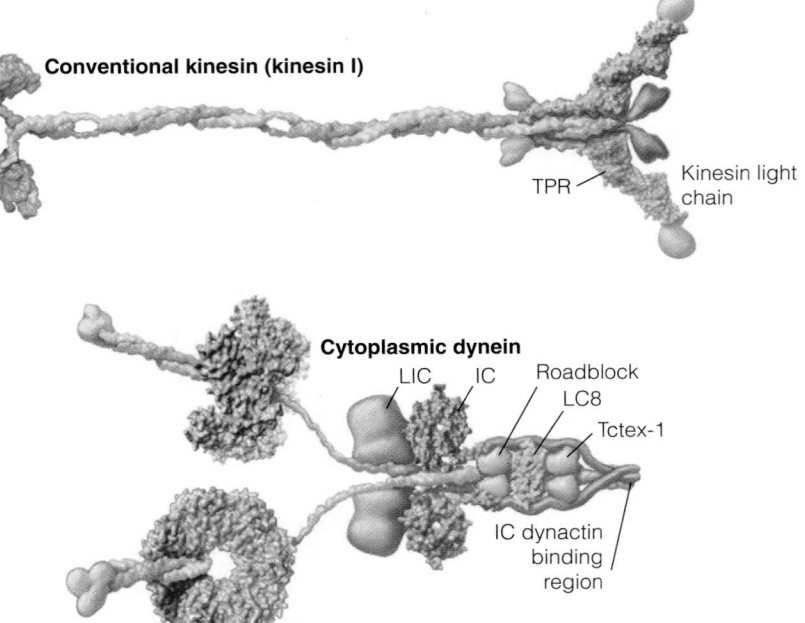

Conventional kinesin (kinesin I)

TPR

Kinesin light chain

Cytoplasmic dynein

LIC   IC   Roadblock
LC8
Tctex-1

IC dynactin binding region

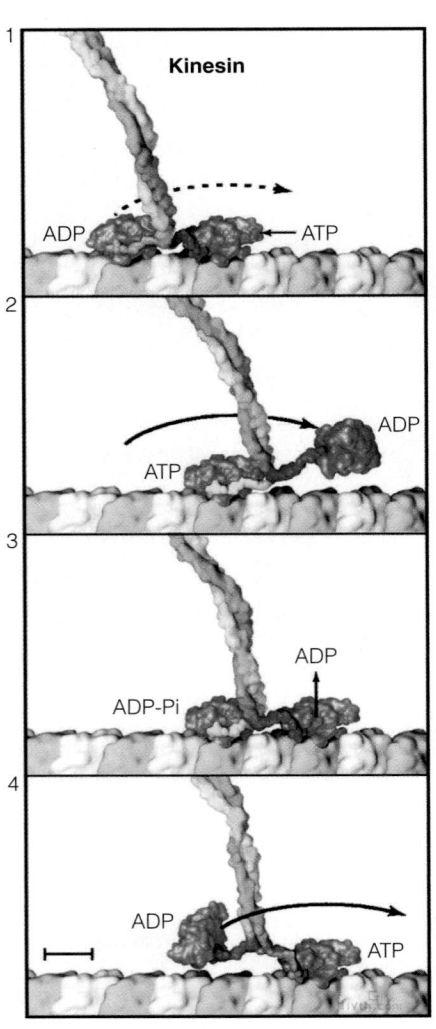

Kinesin

1
ADP ATP

2
ATP ADP

3
ADP-Pi ADP

4
ADP ATP

8 nm. This is exactly the distance from one tubulin dimer to the next. This means that kinesin can transport its cargo for ~100 steps, or about 1 micron, before it falls off the microtubule. Such behavior is defined as **processive** and is a critical feature of several proteins that travel along large linear biopolymers as part of their function (e.g., the cellular transport proteins and the polymerases that copy genomic DNA). In the current model for the motility cycle of kinesin (Figure 8.27), the molecule is believed to pivot, binding one head group and then the other to the tubulin monomers in a "hand-over-hand" fashion.

There is also clear evidence for yet another mechanism of transport in the cytoplasm. Careful observation of motions of cellular organelles revealed that some organelles periodically make short linear movements, with abrupt changes in direction, in regions where *no microtubules can be observed*. This research led to the conclusion that cellular transport also occurs on some or all of the actin microfilaments in cells. Further studies have shown that the motors in this case are myosins different from myosin II (e.g., myosins I, V, VI, etc.—see Table 8.1).

Myosin V is the best studied of these, and differs from myosin II in three important ways: (1) myosin V is a dimer with a shorter coiled-coil tail that binds cargo rather than forming filaments; (2) the lever arm of myosin V binds six calmodulin-like light chains, making it three times longer than the lever arm in myosin II; and (3) myosin V binds actin processively. Like kinesin, myosin V

FIGURE 8.27

Model for the motility cycle kinesin I. Kinesin I is modeled on a single tubulin protofilament. The coiled coil (gray) extends toward the kinesin cargo. Motor domains are attached to the coiled coil by "neck linkers." Frame 1: Each motor domain (blue) is bound to tubulin (green = β subunit; white = α subunit) along a microtubule. In this state the neck linker (orange) points *forward* (right) on the trailing head and *rearward* on the leading head (red). ATP binding to the leading head initiates conformational change in the leading neck linker, and weakens microtubule binding by the trailing head. Frame 2: The conformation change in the neck linker of the leading head (yellow) pitches the unbound trailing head forward by 16 nm (arrow) toward the next tubulin binding site. Frame 3: The new leading head docks tightly onto the microtubule, moving the attached cargo 8 nm closer to the "plus" end. During this time, the trailing head hydrolyzes ATP to ATP-P$_i$, and then ADP dissociates from the leading motor domain. Frame 4: ATP binds to the leading head and the cycle begins again. Scale bar 4 nm.

From *Science* 288:88–95, R. D. Vale and R. A. Milligan, The way things move: Looking under the hood of molecular motor proteins. © 2000. Reprinted with permission from AAAS.

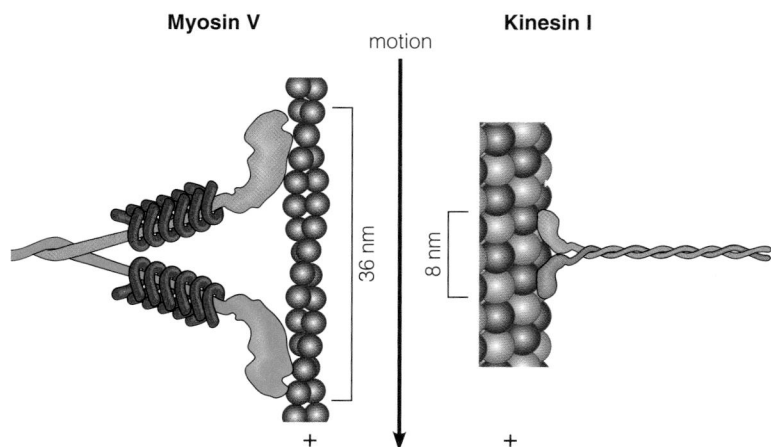

Myosin V

Kinesin I

motion

36 nm

8 nm

+ +

FIGURE 8.28

Myosin V and kinesin I. Like kinesin, myosin V is a processive motor. The motor domains of each protein are shown attached to their respective scaffold: the actin filament (brick red) or the microtubule (steel blue). The direction of movement is toward the (+) end of each filament type, and step sizes are indicated by the brackets. The six calmodulin-like light chains bound to the neck region of myosin V are shown in bright red.

Courtesy of Gary Carlton.

moves in a stepwise, hand-over-hand fashion; however, myosin V takes much larger (36 nm) steps than kinesin (see Figure 8.28).

In the cell, we can think of the microtubules as major roads for cargo transport and microfilaments as side streets. The successful delivery of cargo requires both. Experimental evidence suggests that kinesin and myosin V can both bind to the same cargo and work together to deliver it to its final destination. Although only the kinesin moves the cargo along the microtubule, the bound myosin V can interact nonspecifically with the tubulin. In the event the kinesin falls off the microtubule, the myosin V holds it in place long enough for the kinesin to rebind and continue the journey. On the actin microfilament, these roles reverse. Cargo bound to both carriers is transported farther on each filament type than when one or the other carrier is absent (Figure 8.29).

Table 8.3 summarizes the essential features of the activities of the different motor proteins we have discussed in this chapter. Each carries out contractile or transport functions by binding to a polymeric protein target and undergoing ATP-driven conformational change. We now turn to a protein motor that functions in a completely different manner.

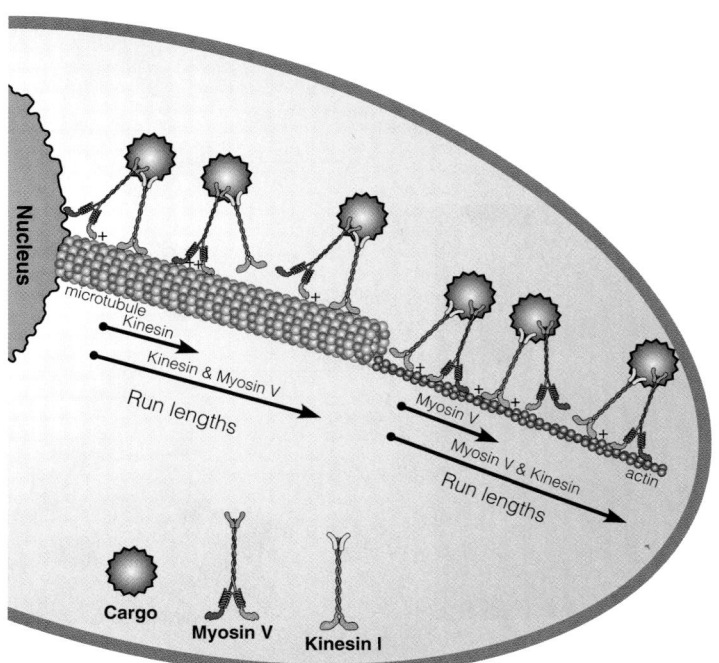

Nucleus

microtubule
Kinesin

Kinesin & Myosin V

Run lengths

Myosin V

Myosin V & Kinesin

Run lengths

actin

Cargo

Myosin V

Kinesin I

FIGURE 8.29

Myosin V and kinesin I work together to deliver cargo. The distance a cargo is transported along a microtubule or microfilament is significantly greater when both carriers are present. Nonspecific interaction between myosin V and tubulin, and kinesin and actin, keep the active carrier close to its target filament in the event it falls off too early.

Courtesy of Gary Carlton.

TABLE 8.3	Motor proteins and their functions	
Motor Protein	Primary Function	Binding Target/Direction of Movement
Muscle myosin II	Muscle contraction	Actin filament; moves toward (+) end
Nonmuscle myosin II	Cytokinesis; cell motility	Actin filament; moves toward (+) end
Myosin V	Cargo transport	Actin filament; moves toward (+) end
Kinesin	Cargo transport	Microtubule; moves toward (+) end
Cytoplasmic dynein	Cargo transport	Microtubule; moves toward (−) end
Axonemal dynein	Bending of cilia/flagella	Microtubule; moves toward (−) end

Bacterial Motility: Rotating Proteins

It is appropriate to close this chapter by examining a complex, self-assembling macromolecular machine that is almost without rival in its elegance. The bacterial flagellum is a right-handed helical fiber, composed almost entirely of one fibrous protein, **flagellin**. It does not contain microtubules, actin, myosin, or any contractile system. Yet for many years it was assumed that the bacterial flagellum underwent in-plane bending motions, like those of sperm tails. Thus, researchers were surprised to learn that, in fact, it *rotates*. This mechanism was demonstrated most simply when the flagellum of a bacterium was stuck to a glass plate by antiflagellin antibodies. Because the flagellum could no longer rotate, the bacterium did.

The remarkable structure that attaches the flagellum to the bacterium and generates the rotation is shown in Figure 8.30. The fiber of the flagellum is attached through a hook structure to a rod that passes through a "bushing" (L- and P-rings in Figure 8.30) in the outer bacterial membrane and into the

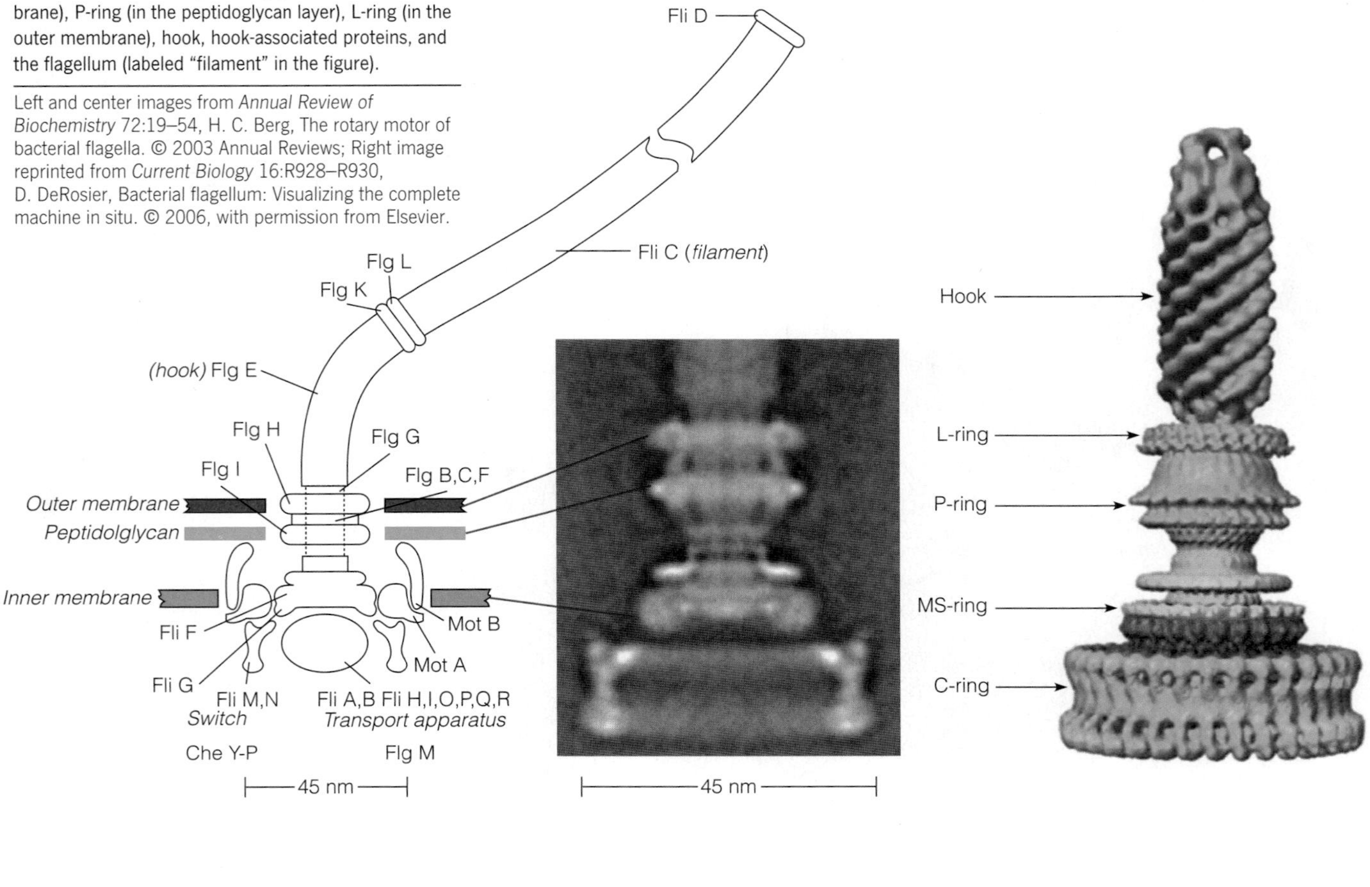

Some bacteria move by the rotation of flagella, using molecular rotating motors set in the cell membrane.

FIGURE 8.30

Structure of the bacterial flagellar motor from *Salmonella enterica*. A model of the flagellar motor, drawn to scale, compared to reconstructed images from electron microscopy. Labels indicate the different proteins (named for their genes) that make up the various components of the complex. The general morphological features, starting below the inner membrane and working out, are: C-ring (located on the cytoplasmic side of the inner membrane), MS-ring (in the inner membrane), P-ring (in the peptidoglycan layer), L-ring (in the outer membrane), hook, hook-associated proteins, and the flagellum (labeled "filament" in the figure).

Left and center images from *Annual Review of Biochemistry* 72:19–54, H. C. Berg, The rotary motor of bacterial flagella. © 2003 Annual Reviews; Right image reprinted from *Current Biology* 16:R928–R930, D. DeRosier, Bacterial flagellum: Visualizing the complete machine in situ. © 2006, with permission from Elsevier.

inner membrane. There it terminates in a multisubunit "rotor" (MS- and C-rings) that is surrounded by a "stator" ring (Mot A and B). Each of these components is made up of protein molecules, and most have been characterized. In other words, the flagellum is made to rotate by an ultramicroscopic motor consisting entirely of protein subunits. In a sense, it is an electric motor because the driving force comes from protons moving across the bacterial inner membrane. Ion gradients across membranes are a source of free energy to drive many critical processes in cells (see Chapters 10 and 15). In vivo, the motor runs at ~100−200 revolutions per second and requires the passage of about 1000 protons per revolution.

Such rotary motors are widely used in cells and exhibit a variety of functions and mechanisms. Certain marine bacteria propel themselves using motors driven by sodium ion fluxes rather than proton fluxes. On the other hand, proton flux through a rotary device is used to generate ATP from oxidative metabolism in most organisms, as we describe in Chapter 15.

The flagellar motor has still one more remarkable property—it can be reversed without changing the direction of H^+ movement through the motor. That is, it can rotate the flagellum in either a clockwise or a counterclockwise direction. This ability is important to the bacterium, for it allows for both steady, linear motion and changes in direction. If the multiple flagella are all rotating counterclockwise (Figure 8.31a), their right-handed helical sense makes them *push* together. They tend to be drawn together in a bundle and propel the bacterium in a straight line, a movement known as *running*. But if the flagella rotate clockwise (Figure 8.31b), they fly out from the surface and *pull* in all directions. The result is that the bacterium *tumbles* and thereby, changes direction.

Escherichia coli and a number of other flagellated bacteria demonstrate a response to chemicals, called **chemotaxis**. The general phenomenon of **taxis**, widespread throughout the animal and plant worlds, comprises movements in

FIGURE 8.31

Effect of the direction of flagellar rotation. These electron micrographs and accompanying diagrams show how the direction of flagellar rotation affects a bacterium with several flagella. **(a)** If the flagella rotate counterclockwise, their right-helical structure makes them be pulled together into a bundle and drive the bacterium in a straight line (running). **(b)** When rotation is clockwise, the flagella fly out in all directions, and the bacterium tumbles randomly.

Reprinted from *Journal of Molecular Biology* 112:1–30, R. M. McNab and M. K. Ornston, Normal-to-curly flagellar transitions and their role in bacterial tumbling: Stabilization of an alternative quaternary structure by mechanical force. © 1977, with permission from Elsevier.

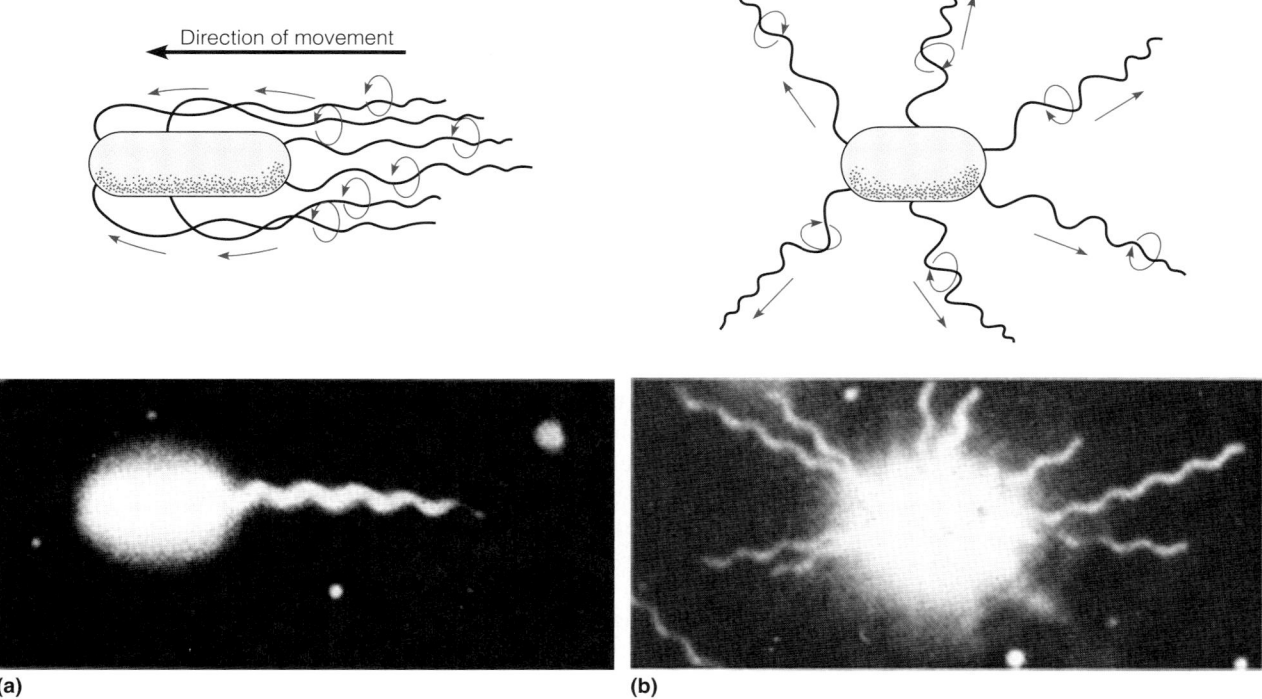

Direction of movement

(a)   (b)

FIGURE 8.32

Chemotactic motion of bacteria. **(a)** In the absence of either attractants or repellents, the bacterium stops and tumbles frequently, each time starting out in a new, random direction. **(b)** When a gradient of an attractant is present, a bacterium heading toward the attractant tends to run for longer periods without tumbling. **(c)** A gradient of a repellent has the opposite effect, favoring long runs away from the repellent source.

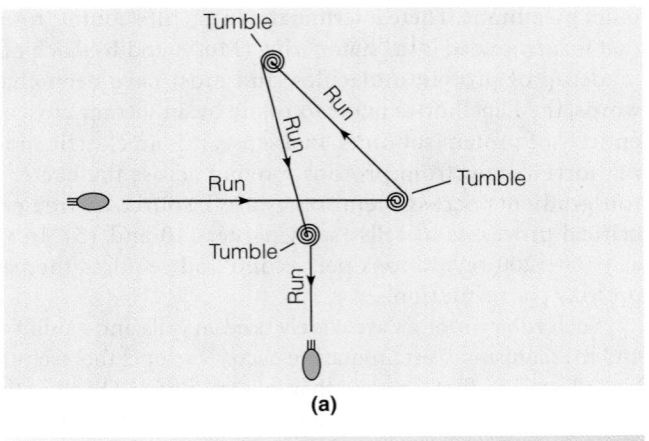

(a)

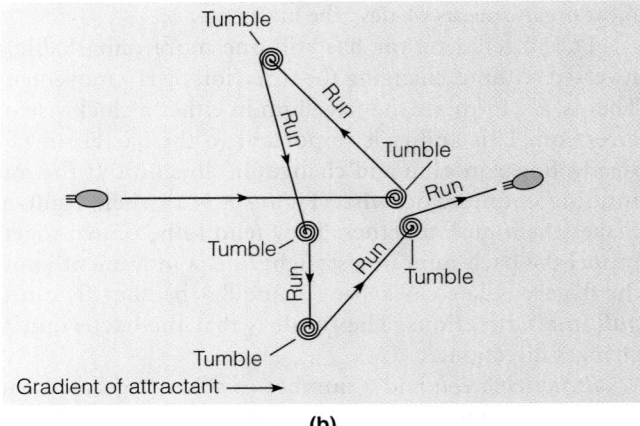

(b)

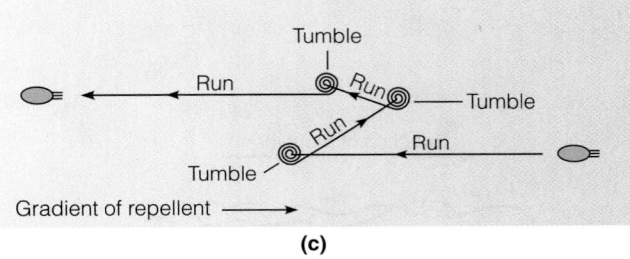

(c)

response to external stimuli. Chemotactic bacteria move preferentially toward attractants, such as nutrients, and away from repellents, such as poisons. We can now describe bacterial chemotaxis in terms of running and tumbling motions (Figure 8.32). In a neutral and uniform environment, periods of running lasting a few seconds alternate with brief periods of tumbling, and the bacterium wanders about randomly (Figure 8.32a). The presence of a gradient of either a nutrient or a noxious repellent biases this distribution of running and tumbling. If a bacterium is moving up a gradient of nutrient, tumbling is delayed, resulting in net motion toward the source of nutrition (Figure 8.32b). Conversely, a bacterium moving away from a repellent continues to do so for longer than usual before tumbling, resulting in an avoidance reaction (Figure 8.32c).

These observations imply that flagellated bacteria must have some mechanism for sensing gradients of attractants or repellents and relaying this information to the motors of their flagella. In fact, they do. We do not have space to describe this remarkable mechanism here. The interested reader should consult the references at the end of the chapter (e.g., L. D. Miller et al.; C. V. Rao, et al.; and Shimizu et al.).

Finally, the bacterial flagellum has received attention as a putative example of "irreducible complexity," a concept that has been used to argue against the possibility that this remarkable molecular machine evolved from simpler precursors by Darwinian principles. In fact, many of the components of the bacterial flagellum are homologous to other bacterial proteins—in particular, to proteins involved in the secretion of bacterial toxins; thus, the flagellum is not "irreducibly complex." Entire books have been devoted to this discussion, and we do not have the space to do justice to this wide-ranging debate. Again, we refer the interested reader to the leading references cited at the end of this chapter (e.g., the papers by K. Miller; and M. J. Pallen and N. J. Matzke).

SUMMARY

A number of macromolecular protein systems exist in cells to convert ATP energy into mechanical work. A major example is the actin–myosin system of muscle. In muscle, interdigitating filaments of actin and myosin are driven past one another by attachment, motion, and detachment of myosin cross-bridges. Muscle contraction is stimulated by the influx of calcium ions, which causes rearrangement of actin-associated proteins. The direct source of contractile energy is ATP, with creatine phosphate being a short-term energy store.

There also exist many other nonmuscle systems for producing motion and doing work. Many kinds of cell motion, including amoeboid crawling and cytokinesis, use nonmuscle actin and myosin. On the other hand, flagella and cilia are driven by the ATP-driven sliding of microtubules, filaments formed by the polymerization of tubulin. Microtubules do many other things, including acting as "roadways" for transport of organelles and proteins within cells and accomplishing the separation of chromosomes in mitosis.

A remarkable molecular motor drives the rotation of bacterial flagella. This motor, by reversing its direction, can cause the bacterium either to move in a straight line toward nutrients and away from toxic substances, or to tumble randomly and change direction.

REFERENCES

General

Vale, R. D. (2003) The molecular motor toolbox for intracellular transport. *Cell* 112:467–480.

Vale, R. D., and R. A. Milligan (2000) The way things move: Looking under the hood of molecular motor proteins. *Science* 288:88–95.

Van den Heuvel, M. G. L., and C. Dekker (2007) Motor proteins at work for nanotechnology. *Science* 317:333–336.

Muscle

Cooke, R. (2004) The sliding filament model: 1972–2004. *J. Gen. Physiol.* 123:643–656.

Himmel, D. M., S. Gourinath, L. Reshetnikova, Y. Shen, A. G. Szent-Györgyi, and C. Cohen (2002) Crystallographic findings on the internally uncoupled and near-rigor states of myosin: Further insights into the mechanics of the motor. *Proc. Natl. Acad. Sci. USA* 99:12645–12650.

Murakami, K., F. Yumoto, S. Ohki, T. Yasunaga, M. Tanokura, and T. Wakabayashi (2005) Structural basis for Ca^{2+}-regulated muscle relaxation at interaction sites of troponin with actin and tropomyosin. *J. Mol. Biol.* 352:178–201.

Rayment, I., and H. M. Holden (1994) The three-dimensional structure of a molecular motor. *Trends Biochem. Sci.* 19:129–134.

Stroud, R. M. (1996) Balancing ATP in the cell. *Nature Struct. Biol.* 3:567–569.

Vinogradova, M. V., D. B. Stone, G. G. Malanina, C. Karatzaferi, R. Cooke, R. A. Mendelson, and R. J. Fletterick (2005) Ca^{2+}-regulated structural changes in troponin. *Proc. Natl. Acad. Sci. USA* 102:5038–5043.

Yanagida, T., and A. Ishijima (1995) Forces and steps generated by single myosin molecules. *Biophys. J.* 68:312s–320s.

Microtubules, Dynein, and Kinesin

Carter, A. P., C. Cho, L. Jin, and R. D., Vale (2011) Crystal structure of the dynein motor domain. *Science* (doi:10.1126/science.1202393).

Carter, A. P., J. E. Garbarino, E. M. Wilson-Kubalek, W. E. Shipley, C. Cho, R. A. Milligan, R. D. Vale, and I. R. Gibbons (2008) Structure and functional role of dynein's microtubule-binding domain. *Science* 322:1691–1695.

Gennerich, A., and R. D. Vale (2009) Walking the walk: How kinesin and dynein coordinate their steps. *Curr. Op. Cell Biol.* 21:59–67.

Hirokawa, N., R. Nitta, and Y. Okada (2009) The mechanisms of kinesin motor motility: Lessons from the monomeric motor KIF1A. *Nature Rev. Molec. Cell. Biol.* 10:877–884.

Kodera, N., D. Yamamoto, R. Ishikawa, and T. Ando (2010) Video imaging of walking myosin V by high-speed atomic force microscopy. *Nature* 468:72–76. See also: http://www.s.kanazawa-u.ac.jp/phys/biophys/M5_movies.htm

Movassagh, T., K. H. Bui, H. Sakakibara, K. Oiwa, and T. Ishikawa (2010) Nucleotide-induced global conformational changes of flagellar dynein arms revealed by in situ analysis. *Nature Struct. Mol. Biol.* 17:761–767.

Nogales, E., S. C. Wolf, and K. H. Downing (1998) Structure of the $\alpha\beta$ tubulin dimer by electron crystallography. *Nature* 391:199–203.

Sablin, E. P., and R. J. Fletterick (2004) Coordination between motor domains in processive kinesins. *J. Biol. Chem.* 279:15707–15710.

Sindelar, C. V., and K. H. Downing (2010) An atomic-level mechanism for activation of the kinesin molecular motors. *Proc. Natl. Acad. Sci. USA* 107:4111–4116.

Vicente-Manzanares, M., X. Ma, R. S. Adelstein, and A. R. Horwitz (2009) Non-muscle myosin II takes centre stage in cell adhesion and migration. *Nature Rev. Mol. Cell Biol.* 10:778–790.

Walker, M. L., S. A. Burgess, J. R. Sellers, F. Wang, J. A. Hammer III, J. Trinick, and P. J. Knight (2000) Two-headed binding of a processive myosin to F-actin. *Nature* 405:804–807.

Yildiz, A., J. N. Forkey, S. A. McKinney, T. Ha, Y. E. Goldman, and P. R. Selvin (2003) Myosin V walks hand-over-hand: Single fluorophore imaging with 1.5 nm localization. *Science* 300:2061–2065.

Yildiz, A., M. Tomishige, R. D. Vale, and P. R. Selvin (2004) Kinesin walks hand-over-hand. *Science* 303:676–678.

See also the following animations created by G. Johnson:

http://www.scripps.edu/cb/milligan/research/movies/myosin_text.html
http://www.scripps.edu/cb/milligan/research/movies/kinesin_text.html
http://valelab.ucsf.edu/images/movies/mov-procmotconvkinrev5.mov

The Bacterial Motor and Chemotaxis

Berg, H. C. (2003) The rotary motor of bacterial flagella. *Ann. Rev. Biochem.* 72:19–54.

Chevance, F. F. V., and K. T. Hughes (2008) Coordinating assembly of a bacterial macromolecular machine. *Nature Rev. Microbiol.* 6:455–465.

DeRosier, D. (2006) Bacterial flagellum: Visualizing the complete machine in situ. *Current Biology* 16:R928–R930.

Erhardt, M., K. Namba, and K. T. Hughes (2010) Bacterial nanomachines: The flagellum and type III injectosome. *Cold Spring Harb. Perspect. Biol.* 2:a000299 (doi:10.1101/cshperspect.a000299).

Miller, K. R. (2004) The flagellum unspun: The collapse of "irreducible complexity" pp. 81–97 in *Debating Design: From Darwin to DNA*, eds. W. Dembski and M. Ruse, Cambridge University Press, New York.

Miller, L. D., M. H. Russell, and G. Alexandre (2009) Diversity in bacterial chemotactic responses and niche adaptation. *Adv. App. Microbiol.* 66:53–75.

Minamino, T., K. Imada, and K. Namba (2008) Molecular motors of the bacterial flagella. *Curr. Op. Struct. Biol.* 18:693–701. See also: http://www.fbs.osaka-u.ac.jp/labs/namba/npn/

Murphy, G. E., J. R. Ledbetter, and G. J. Jensen (2006) In situ structure of the complete *Treponema primitia* flagellar motor. *Nature* 442:1062–1064.

Pallen, M. J., and N. J. Matzke (2006) From the *Origin of the Species* to the origin of bacterial flagella. *Nature Rev. Microbiol.* 4:784–790.

Rao, C. V., and G. W. Ordal (2009) The molecular basis of excitation and adaptation during chemotactic sensory transduction in bacteria. *Contrib. Microbiol.* 16:33–64.

Shimizu, T. S., Y. Tu, and H. C. Berg (2010) A modular gradient-sensing network for chemotaxis in *Escherichia coli* revealed by responses to time-varying stimuli. *Mol. Syst. Biol.* 6:382 (doi:10.1038/msb.2010.37).

Sowa, Y., and R. M. Berry (2008) Bacterial flagellar motor. *Quart. Rev. Biophys.* 41:103–132.

PROBLEMS

1. A typical relaxed sarcomere is about 2.3 μm in length and contracts to about 2 μm in length. Within the sarcomere, the thin filaments are about 1 μm long and the thick filaments are about 1.5 μm long.
   (a) Describe the overlap of thick and thin filaments in the relaxed and contracted sarcomere.
   (b) An individual "step" by a myosin head in one cycle pulls the thin filament about 15 nm. How many steps must each actin fiber make in one contraction?

2. Each gram of mammalian skeletal muscle consumes ATP at a rate of about 1×10^{-3} mol/min during contraction. Concentrations of ATP and creatine phosphate in muscle are about 4 mM and 25 mM, respectively, and the density of muscle tissue can be taken to be about 1.2 g/cm³.
   (a) How long could contraction continue using ATP alone?
   (b) If all creatine phosphate were converted into ATP and utilized as well, how long could contraction continue?
   (c) What do these answers tell you?

3. The drug cytocholasin is known to bind to the ends of actin fibers. It does not affect striated muscle contraction, but it completely inhibits motility and changes in cell shape in eukaryotic cells. What do these findings suggest?

*4. Tubulin binds GTP, and the polymerization of a microtubule is greatly stimulated when a tubulin molecule at the growing end carries GTP. Tubulin also has a GTPase activity, with a low-turnover number.

Microtubules carrying GDP at their ends are more likely to lose tubulin monomers. Show how these facts can explain the remarkable observation that at certain GTP levels some microtubules in a mixture will grow while others simultaneously shrink.

5. A few hours after the death of an animal the corpse will stiffen as a result of continued contraction of muscle tissue (this state is called *rigor mortis*). This phenomenon is the result of the loss of ATP production in muscle tissue.
   (a) Consult Figure 8.11 (page 292) and describe, in terms of the six-step model of muscle contraction, how a lack of ATP in sarcomeres would result in *rigor mortis*.
   (b) The Ca²⁺ transporter in sarcomeres that keeps the [Ca²⁺]~10⁻⁷ M requires ATP to drive transport of Ca²⁺ ions across the membrane of the sarcoplasmic reticulum. How would a loss of this Ca²⁺ transport function result in the initiation of *rigor mortis*?
   (c) *Rigor mortis* is maximal at ~12 hrs after death, and by 72 hrs is no longer observed. Propose an explanation for the disappearance of *rigor mortis* after 12 hrs.

*6. The distance actin and myosin filaments can slide past one another as a consequence of hydrolysis of one ATP molecule is variable. At high loads, the result appears to be consistent with the stroke length (about 10 nm), but at low loads much greater distances have been claimed by some researchers. Discuss this problem in terms of the model proposed in Figure 8.11.

CHAPTER 9

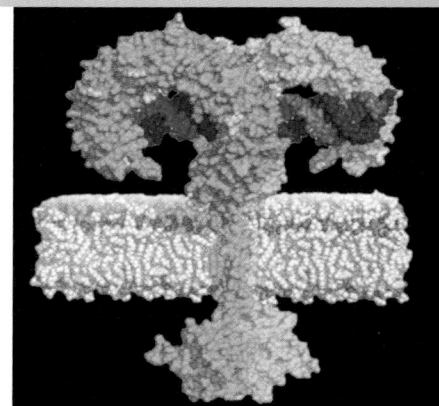

Carbohydrates: Sugars, Saccharides, Glycans

We turn now to the third major class of biological molecules, the **carbohydrates**, or **saccharides**. Like the nucleic acids and proteins, carbohydrates play biological roles both as their constituent monomeric units, such as glucose or ribose, and as polymers, such as starch or glycogen. Unlike proteins and nucleic acids, which are strictly linear polymers, macromolecular polysaccharides also exist as branched polymers. For much of this book we concern ourselves with roles of carbohydrates in generating and storing biological energy. However, carbohydrates play numerous additional functions—as diverse as molecular recognition (as in the immune system), cellular protection (as in bacterial and plant cell walls), cell signaling, cell adhesion, as biological lubricants, in controlling protein trafficking, and in maintaining biological structure (e.g., cellulose).

Many of these substances are already familiar to you. The simplest carbohydrates are small, monomeric molecules—the **monosaccharides**, typically containing from three to nine carbon atoms, which include simple sugars such as *glucose* (Figure 9.1a). Other important carbohydrates are formed by linking such monosaccharides together. If only a few monomer units are involved, we call the molecule an **oligosaccharide**. An example is *maltose* (Figure 9.1b), a **disaccharide** made by linking two glucose molecules together. Long polymers of the monosaccharides, like the starch *amylose* (Figure 9.1c), are called **polysaccharides**. Many kinds of polysaccharides exist, some of which are complex polymers made from many types of sugar monomers. Oligosaccharides and polysaccharides are also referred to as **glycans**.

Saccharides are often called by the more familiar name *carbohydrates* because many of them can be represented by the simple stoichiometric formula $(CH_2O)_n$. The name was first given when chemists knew only the stoichiometry of saccharides and thought of them as "hydrated carbon." The formula is an oversimplification, however, because many saccharides are modified, and some contain amino, sulfate, and phosphate groups. Nevertheless, all of the compounds described in this chapter either have this formula or can be derived from substances that do.

Carbohydrates are compounds with the empirical formula $(CH_2O)_n$, while saccharides include carbohydrates and all of their derivatives.

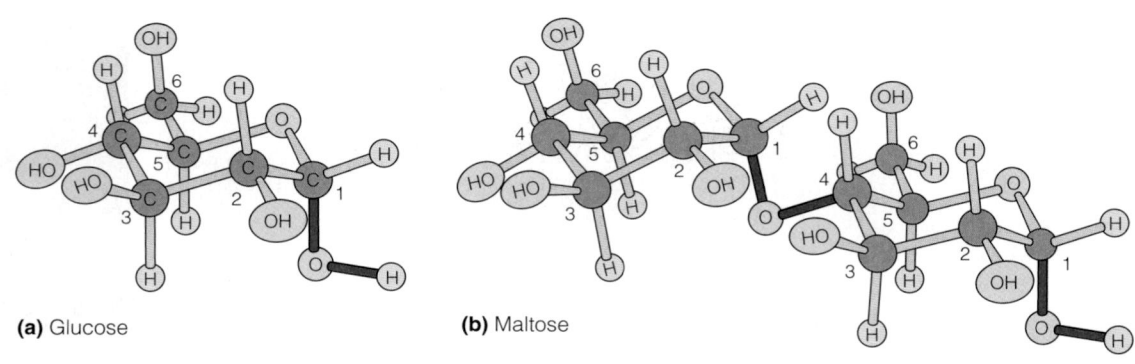

(a) Glucose

(b) Maltose

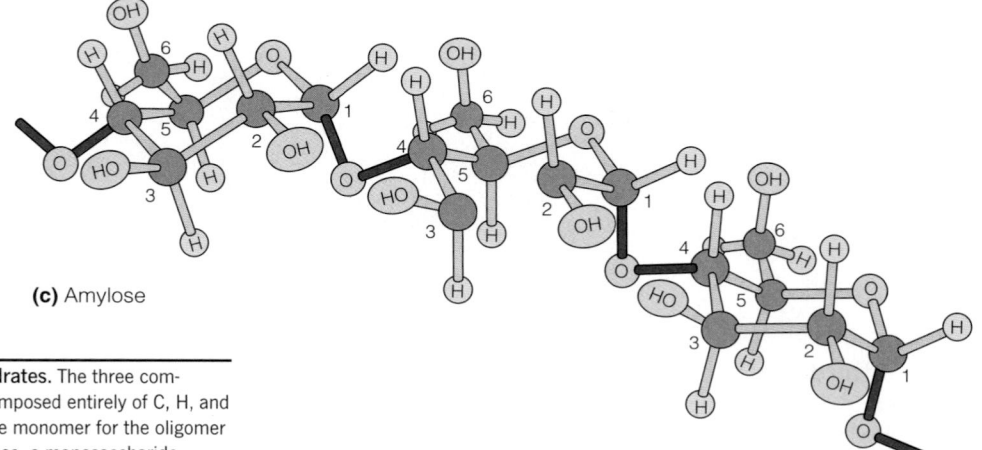

(c) Amylose

FIGURE 9.1

Representative carbohydrates. The three compounds shown here are composed entirely of C, H, and O, with glucose forming the monomer for the oligomer and the polymer. **(a)** Glucose, a monosaccharide. **(b)** Maltose, a disaccharide containing two glucose units. **(c)** A portion of a molecule of amylose, a glucose polymer found in starch.

Carbohydrate formation in photosynthesis and its oxidation in metabolism together constitute the major energy cycle of life.

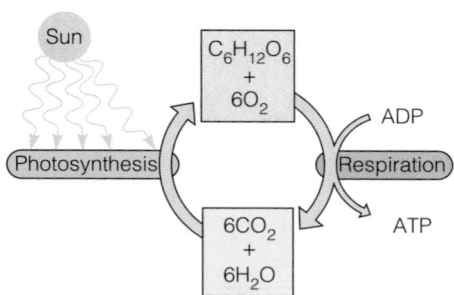

FIGURE 9.2

The energy cycle of life. In photosynthesis, plants use the energy of sunlight to combine carbon dioxide and water into carbohydrates, releasing oxygen in the process. In respiration, both plants and animals oxidize the carbohydrates made by plants, releasing energy and re-forming CO_2 and H_2O.

Strictly speaking, the term carbohydrate is reserved for compounds with the $(CH_2O)_n$ empirical formula, while the term saccharide covers both these compounds and all derivatives of carbohydrates. We will occasionally stray from strict usage and use the terms carbohydrate and saccharide interchangeably. The term *sugar* generally refers to underivatized monosaccharides and small oligosaccharides, such as *sucrose*, a disaccharide containing glucose and fructose. As noted earlier, an oligosaccharide or polysaccharide is also called a glycan.

As mentioned above, the saccharides play diverse roles in organisms. Indeed, the major energy cycle of the biosphere depends largely on carbohydrate metabolism. Before we turn to carbohydrate structure, let us look briefly at this cycle, as schematized in Figure 9.2. In *photosynthesis*, plants take up CO_2 from the atmosphere and "fix" it into carbohydrates. The basic reaction can be described in a simplified way as the light-driven reduction of CO_2 to carbohydrate, here represented as glucose. Much of this carbohydrate is stored in the plants as polymeric starch or cellulose. Animals obtain their carbohydrates by eating plants or plant-eating animals. Thus, plant-synthesized carbohydrates ultimately become the principal sources of carbon in all animal tissues. In the other half of the cycle, both plants and animals carry out, via oxidative metabolism, a reaction that is essentially the reverse of photosynthesis, to yield once again CO_2 and H_2O (Figure 9.2). This oxidation of carbohydrates is the primary energy-generating process in metabolism. The central role of carbohydrates is obvious when we consider that the staple nutrient of most humans is the starch in such plant foods as rice, wheat, or potatoes. Even the meat consumed by many of us ultimately can be traced in large part to the carbohydrates eaten by grazing animals.

As critical as energy storage and generation are, they are not the only functions of carbohydrates. As mentioned above, many biological structural materials are polysaccharide in whole or in part. Important examples are the cellulose of plant fibers, the cell walls of bacteria, and the exoskeletons of insects and other arthropods. Additionally, polysaccharides on cell surfaces or attached to proteins aid in molecular recognition. Examples include highly specific processes such as the binding of viruses or antibodies on particular cells. Thus, like proteins, carbohydrates are extremely versatile molecules, essential to all organisms.

Monosaccharides

We begin our discussion of carbohydrates with the simple, monomeric sugars—the monosaccharides. The simplest compound with the empirical formula of the class $(CH_2O)_n$ is found when $n = 1$. However, *formaldehyde,* $H_2C=O$, has little in common with our usual concept of sugars; indeed, it is a noxious, poisonous gas. The smallest molecules usually regarded as monosaccharides are the **trioses**, with $n = 3$. (The suffix *ose* is commonly used to designate compounds as saccharides.) Monosaccharides are generally characterized by the presence of one carbonyl group (aldehyde or ketone) and one or more hydroxyl groups.

Aldoses and Ketoses

There are two trioses: *glyceraldehyde* and *dihydroxyacetone* (Figure 9.3). These molecules, as simple as they are, exhibit certain features that we shall encounter again and again in discussing sugars. In fact, they represent the two major classes of monosaccharides. Glyceraldehyde is an aldehyde, one of a class of monosaccharides called **aldoses**. Dihydroxyacetone is a ketone; such monosaccharides are called **ketoses**. Note that glyceraldehyde and dihydroxyacetone each have one carbonyl carbon and both have the same atomic composition. They are *tautomers* (structural isomers differing in location of hydrogen atoms and double bonds) and can interconvert via an unstable *enediol* intermediate, as shown also in Figure 9.3. Such tautomeric interconversions occur to a certain extent between all such pairs of aldose and ketose monosaccharides, but the reactions are usually very slow unless catalyzed. Thus, glyceraldehyde and dihydroxyacetone can each exist as a stable compound.

The two major classes of monosaccharides are aldoses and ketoses.

Enantiomers

An important feature of monosaccharide structure can be seen by examining the formula for glyceraldehyde a bit more carefully. The second carbon atom carries four different substituents, so it is *chiral*, like the α-carbon in most α-amino acids. Therefore, glyceraldehyde has two stereoisomers of the type called **enantiomers**, which are nonsuperimposable mirror images of one another. Three-dimensional drawings of the two forms, designated as D- and L-glyceraldehyde, are shown in

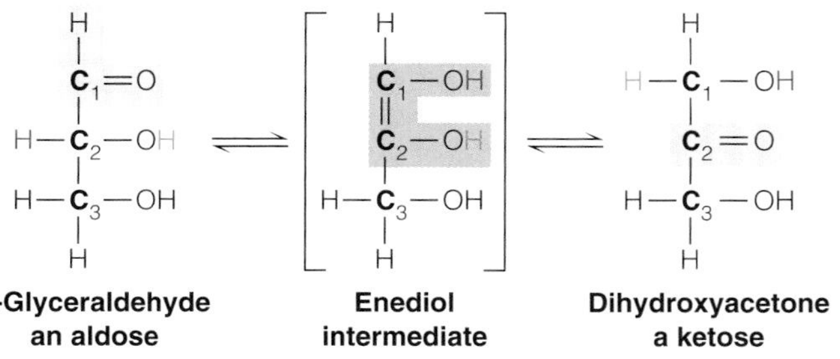

FIGURE 9.3

Trioses, the simplest monosaccharides. The two triose tautomers illustrate the difference between aldose and ketose monosaccharides, also called more descriptively aldotriose and ketotriose, respectively. Carbon numbering begins in all aldoses with the aldehyde carbon and in ketoses with the end carbon closest to the ketone group. (Because dihydroxyacetone has only three carbons, the two end carbons are equivalent and either of them could be designated number one.) The enediol intermediate through which they are interconverted is unstable and cannot be isolated.

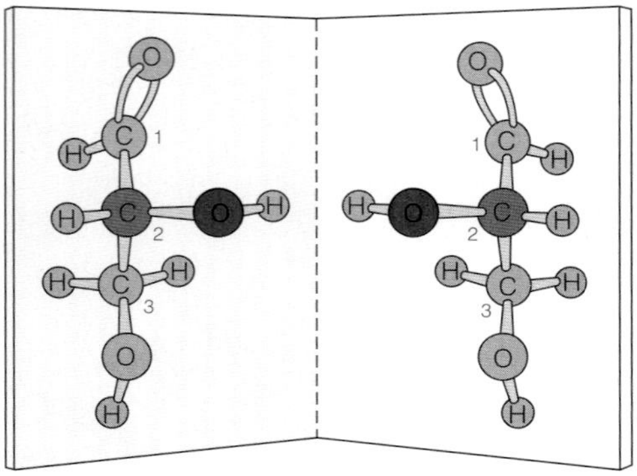

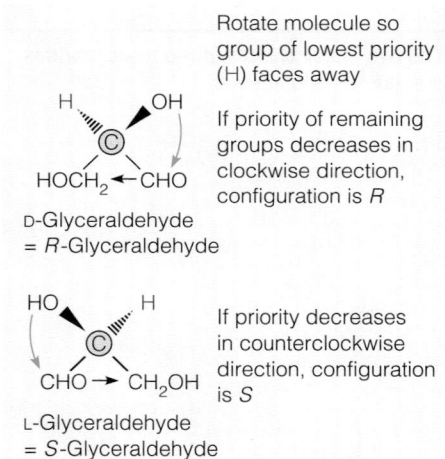

FIGURE 9.4

The enantiomers of glyceraldehyde. The configuration of groups around the chiral carbon 2 (shown in darker gray) distinguishes D-glyceraldehyde from L-glyceraldehyde. The two molecules are mirror images and cannot be superimposed on one another.

D and L forms of a monosaccharide are nonsuperimposable mirror images and are called enantiomers.

Rotate molecule so group of lowest priority (H) faces away

If priority of remaining groups decreases in clockwise direction, configuration is *R*

D-Glyceraldehyde = *R*-Glyceraldehyde

If priority decreases in counterclockwise direction, configuration is *S*

L-Glyceraldehyde = *S*-Glyceraldehyde

FIGURE 9.5

R–S **nomenclature.** The *R–S* system describes absolute stereochemical configuration, as shown in this example. Each type of group attached to a chiral carbon (gray) is given a priority, according to a set of defined rules. Priorities for groups common in carbohydrate chemistry are SH > OR > OH > NH₂ > CO₂H > CHO > CH₂OH > CH₃ > H. We view the molecule with the group of lowest priority away from us (H in our example). If the priority of the remaining three groups *decreases clockwise*, the absolute configuration is called *R* (from Latin *rectus*, "right"). If priority *decreases counterclockwise*, the configuration is *S* (from Latin *sinister*, "left"). In this notation, D-glyceraldehyde is *R*-glyceraldehyde, and L-glyceraldehyde is *S*-glyceraldehyde.

Figure 9.4. Such three-dimensional orientation about the asymmetric carbon can also be represented by the bond convention we used in Chapter 5:

$$\text{D-Glyceraldehyde} \qquad \text{L-Glyceraldehyde}$$

Note that we do not need to draw the spatial orientation of atoms about carbons 1 or 3 because these carbons are not chiral centers.

The most compact way to represent enantiomers is to use a **Fischer projection**. In a Fischer projection the bonds that are drawn horizontally are imagined as coming toward you; those drawn vertically are receding (see also Figure 5.3, page 139). Thus, for D-glyceraldehyde and L-glyceraldehyde, we have

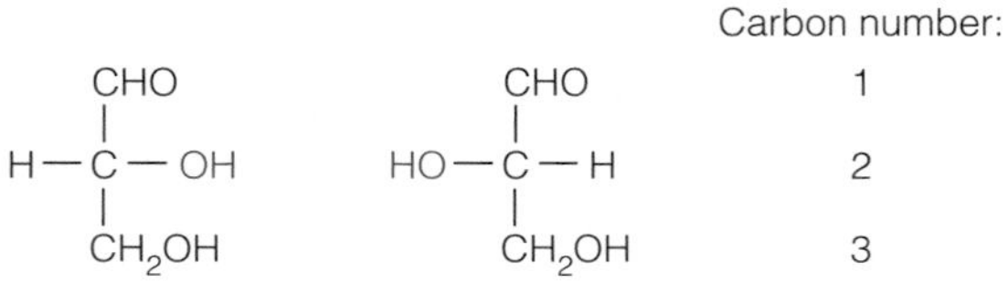

	Carbon number:
CHO	1
H—C—OH HO—C—H	2
CH₂OH CH₂OH	3

D-Glyceraldehyde   L-Glyceraldehyde

Alternative Designations for Enantiomers: *D–L* and *R–S*

Originally, the terms D and L were meant to indicate the direction of rotation of the plane of polarization of polarized light: D for right (dextro), L for left (levo). It is true that a solution of D-glyceraldehyde does rotate the plane of polarization to the right, as do many other D-monosaccharides, but this correspondence does not always hold because the magnitude and even the direction of optical rotation are a complicated function of the electronic structure surrounding the chiral center. Another disadvantage of the D–L nomenclature is that it is not absolute; the designation is always with respect to some reference compound. Accordingly, a systematic convention has been developed that allows us to assign a stereochemical designation to any compound from examination of its three-dimensional structure. This *R–S* convention, shown in Figure 9.5, is also called the Cahn–Ingold–Prelog convention, after its inventors. Although the *R–S* convention is more general and it gives the absolute configuration about a chiral center, the D-L convention is often preferred by biochemists and will be used in this chapter.

Monosaccharide Enantiomers in Nature

Just as in the case of amino acids, one enantiomeric form of monosaccharides dominates in living organisms. In proteins it is the L-amino acids; in carbohydrates it is the D-monosaccharides. Again, there is no obvious reason why this preference was established in nature. But once fixed in early evolution, it has persisted, for most of the cellular machinery has become geared to operate with D-sugars. However, just as D-amino acids are sometimes found in living organisms, so are L-monosaccharides. Like the "abnormal" D-amino acids, the L-monosaccharides play rather specialized roles. Table 9.1 includes, along with D-monosaccharides, the most commonly occurring L-monosaccharides, with examples of their occurrence and functions.

Diastereomers

When we consider monosaccharides with more than three carbons, a further structural complication appears. Such a monosaccharide may have more than one

TABLE 9.1   Examples of occurrence and biochemical roles of monosaccharides

Monosaccharides	Natural Occurrence	Physiological Role[a]
Trioses		
Glyceraldehyde	Widespread (as phosphate)	The 3-phosphate is an intermediate in glycolysis
Dihydroxyacetone	Widespread (as phosphate)	The 1-phosphate is an intermediate in glycolysis
Tetroses		
D-Erythrose	Widespread	The 4-phosphate is an intermediate in carbohydrate metabolism
Pentoses		
D-Arabinose	Some plants, tuberculosis bacilli	Plant glycosides, cell walls
L-Arabinose	Widely distributed in plants, bacterial cell walls	Constituent of cell walls, plant glycoproteins
D-Ribose	Widespread, in all organisms	Constituent of RNA and ribonucleotides
D-Xylose	Woody materials	Constituent of plant polysaccharides
Hexoses		
D-Galactose	Widespread	Milk (as part of lactose); structural polysaccharides
L-Galactose	Agar, other polysaccharides; component of lactose	Polysaccharide structures
D-Glucose	Widespread	Major energy source for animal metabolism; structural role in cellulose
D-Mannose	Plant polysaccharides, animal glycoproteins	Polysaccharide structures
D-Fructose	A major plant sugar; part of sucrose	Intermediate in glycolysis (phosphate esters)
Heptoses		
D-Sedoheptulose	Many plants	Intermediate in Calvin cycle in photosynthesis and pentose phosphate pathway

[a]Some of these monosaccharides have additional roles that are not listed.

chiral carbon, which results in its having two types of stereoisomers. These types are *enantiomers* (mirror image isomers), which we have already discussed, and *diastereomers*, which we first encounter in the tetrose monosaccharides.

> The most important naturally occurring saccharides are the D-enantiomers.

Tetrose Diastereomers

Tetroses, with the empirical formula $(CH_2O)_4$, have two chiral carbons in the aldose forms. Therefore, an aldotetrose will have four stereoisomers, as shown in Figure 9.6. In general, a molecule with n chiral centers will have 2^n stereoisomers because there are two possibilities at each chiral center. The following convention attempts to give a rational method for naming and distinguishing the

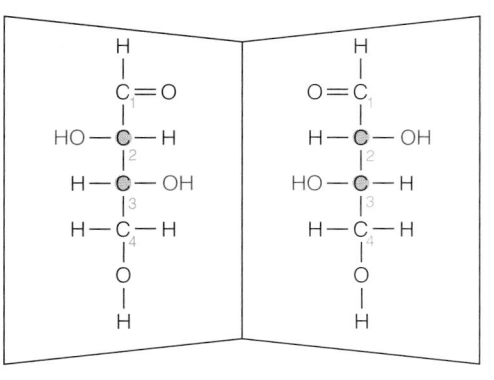

D-Threose   L-Threose

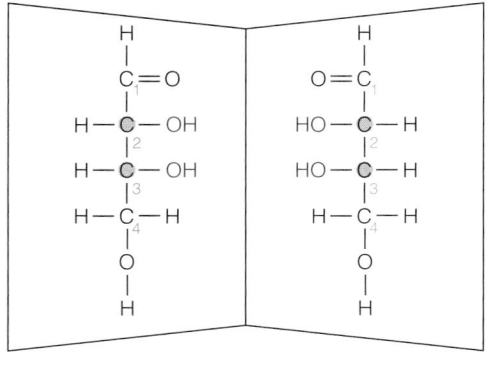

D-Erythrose   L-Erythrose

FIGURE 9.6

Stereochemistry of aldotetroses. These molecules have two chiral carbons (2 and 3) and thus have two diastereomeric forms, threose and erythrose, each with a pair of enantiomers. Note that the threose enantiomers have *opposite* configuration about carbons C2 and C3, whereas erythrose enantiomers have the *same* configuration about these two carbons.

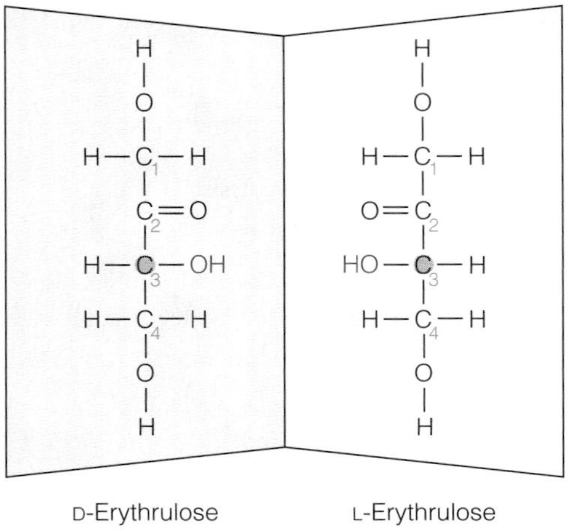

D-Erythrulose   L-Erythrulose

FIGURE 9.7

The two enantiomers of erythrulose. When converting from an aldose to a ketose, the number of chiral carbons decreases by one. Hence, unlike the four-carbon aldoses (see Figure 9.6), the four-carbon ketose has only one chiral carbon (C3) and only one pair of enantiomers.

When monosaccharides contain more than one chiral carbon, the prefix D or L designates the configuration about the carbon farthest from the carbonyl group. Isomers differing in orientation about other carbons are called diastereomers and given different names.

stereoisomers of such a molecule: The prefix D or L is used to designate the orientation about the chiral carbon *farthest* from the carbonyl group—carbon number 3 in this case. Molecules with different orientations about the carbons preceding this reference carbon are given separate names. Thus, *threose* and *erythrose* are two aldotetroses with opposite orientations about carbon 2. Stereoisomers of this kind, which are *not* mirror images, are called **diastereomers**. Threose and erythrose are diastereomers, and each has two enantiomers (D and L) that are nonsuperimposable mirror images. Unfortunately, there is no general logical rule for forming the specific names (such as threose and erythrose); they must simply be learned, like the names of the amino acids.

The four-carbon ketose, which is called *erythrulose,* has only one pair of enantiomers because this monosaccharide has only one chiral carbon (Figure 9.7). Another naming convention appears at this point: Often the ketose name is derived from the corresponding aldose name by insertion of the letters *ul.* Thus *erythrose* becomes *erythrulose.* As with glyceraldehyde (and other monosaccharides), the ketose and aldose forms are interconvertible via tautomerization in dilute alkali. The aldose–ketose conversion also provides a route for interconversion of aldose diastereomers, using the ketose as an intermediate.

Pentose Diastereomers

Adding one more carbon, we obtain the **pentoses**. The **aldopentoses** have three chiral centers; therefore we expect 2^3, or eight, stereoisomers—in four pairs of enantiomers. The D forms of the pentoses are shown in Figure 9.8a, which provides a summary of the aldoses containing from three to six carbons. Note that each of the aldopentoses shown has the D orientation about carbon 4 and that all possible combinations of orientations about carbons 2 and 3 are included. (From here on, in our illustration of carbohydrate structure we will show only the D forms; you can easily draw the L forms from the rules given above.) **Ketopentoses**, as shown in Figure 9.8b, have two chiral carbons, so four isomers (two pairs of enantiomers) must exist. The D diastereomers are called D-*ribulose* and D-*xylulose.*

Hexose Diastereomers

Monosaccharides containing six carbon atoms are called **hexoses**. As you might imagine, there is a large number of possible hexoses. To keep their structures in mind, it is useful to relate them to the simpler pentoses, tetroses, and trioses. Figure 9.8 provides a compact summary of these relationships. The hexoses we will most frequently encounter are *glucose* and *fructose*. However, *mannose* and *galactose* are also widespread in nature (see Table 9.1). In fact, almost all of the D-hexoses play some significant biological role.

Aldose Ring Structures

With the pentoses and hexoses, another feature of monosaccharide chemistry assumes critical importance. Having five or six carbons in the chain gives these compounds the potential to form very stable ring structures via internal *hemiacetal* formation. A hemiacetal results from reaction of an aldehyde with an alcohol. Hemiketals form similarly, from a ketone and an alcohol.

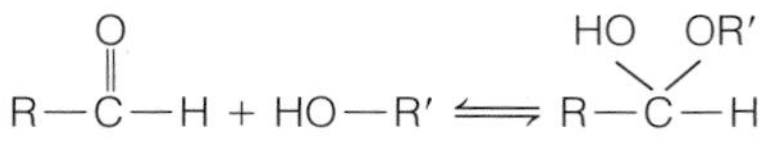

The bond angles characteristic of carbon and oxygen bonding are such that rings containing fewer than five atoms are strained to a considerable extent, whereas five- or six-membered rings are easily formed. In principle, aldotetroses can also form five-membered ring structures, but they rarely do.

FIGURE 9.8

Stereochemical relationships of the D-aldoses and D-ketoses. This figure shows relationships between pairs of diastereomers in the D-aldose series **(a)** and the D-ketose series **(b)**. Each series is generated by successive additions of one CHOH group (shaded) just below the carbonyl carbon. In each case, the two possible orientations of the added group generate a pair of diastereomers. The L forms are not shown; they are just the mirror images of the D forms.

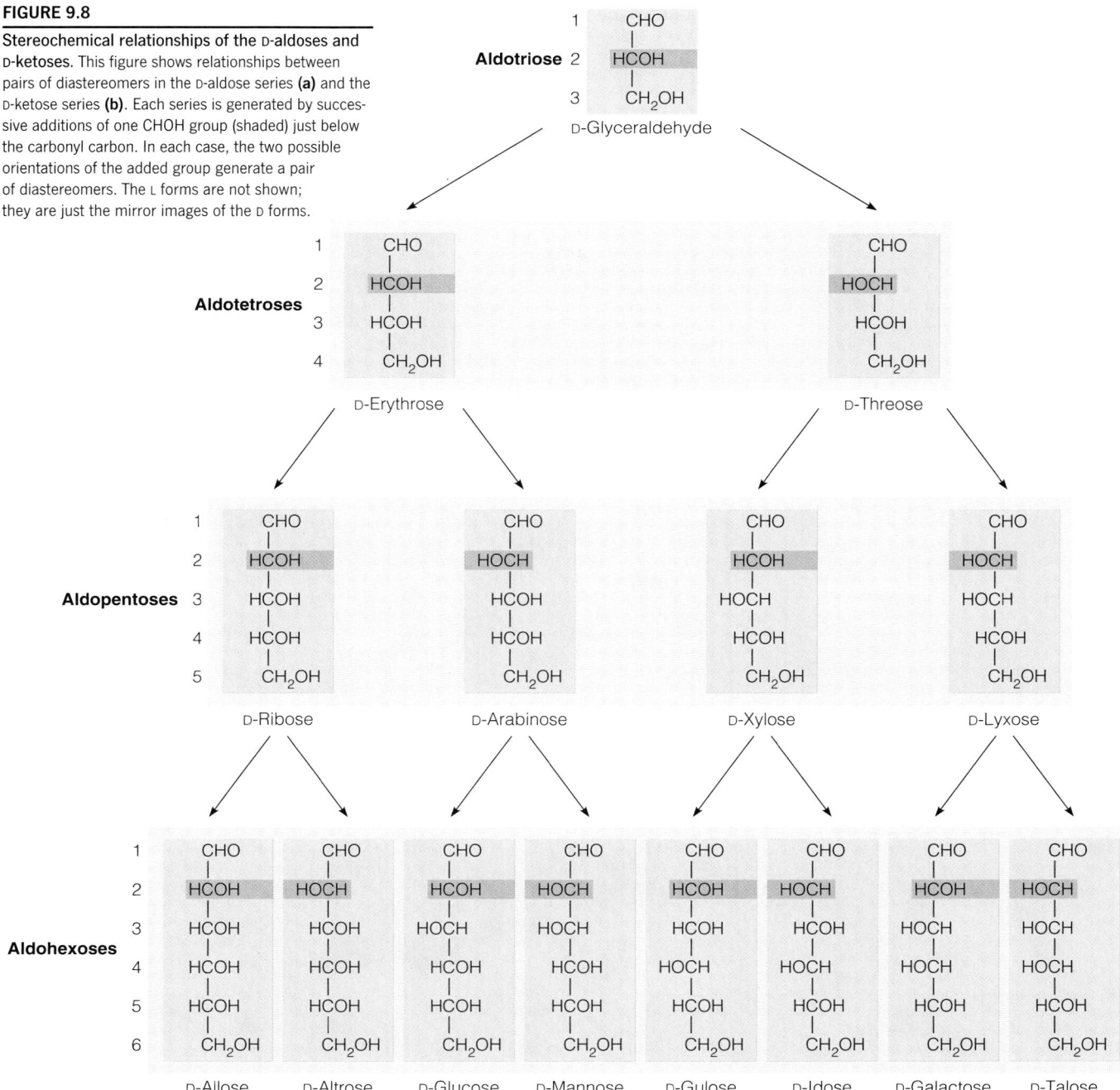

(a) D-Aldoses

Pentose Rings

Consider this hemiacetal ring formation in an aldopentose, such as D-ribose (see Figure 9.8a). Two modes of ring closure are possible, as shown in Figure 9.9. Reaction of the C-1 of D-ribose with the C-4 hydroxyl produces a five-membered ring structure called a **furanose;** the name reflects its structural similarity to the heterocyclic compound furan. Alternatively, a six-membered ring is obtained if the reaction occurs with the C-5 hydroxyl. Such a six-membered ring is called a **pyranose**, to indicate its relation to the heterocyclic compound pyran.

Monosaccharides with five or more carbons exist preferentially in five- or six-membered ring structures, resulting from internal hemiacetal formation.

FIGURE 9.8 (continued)

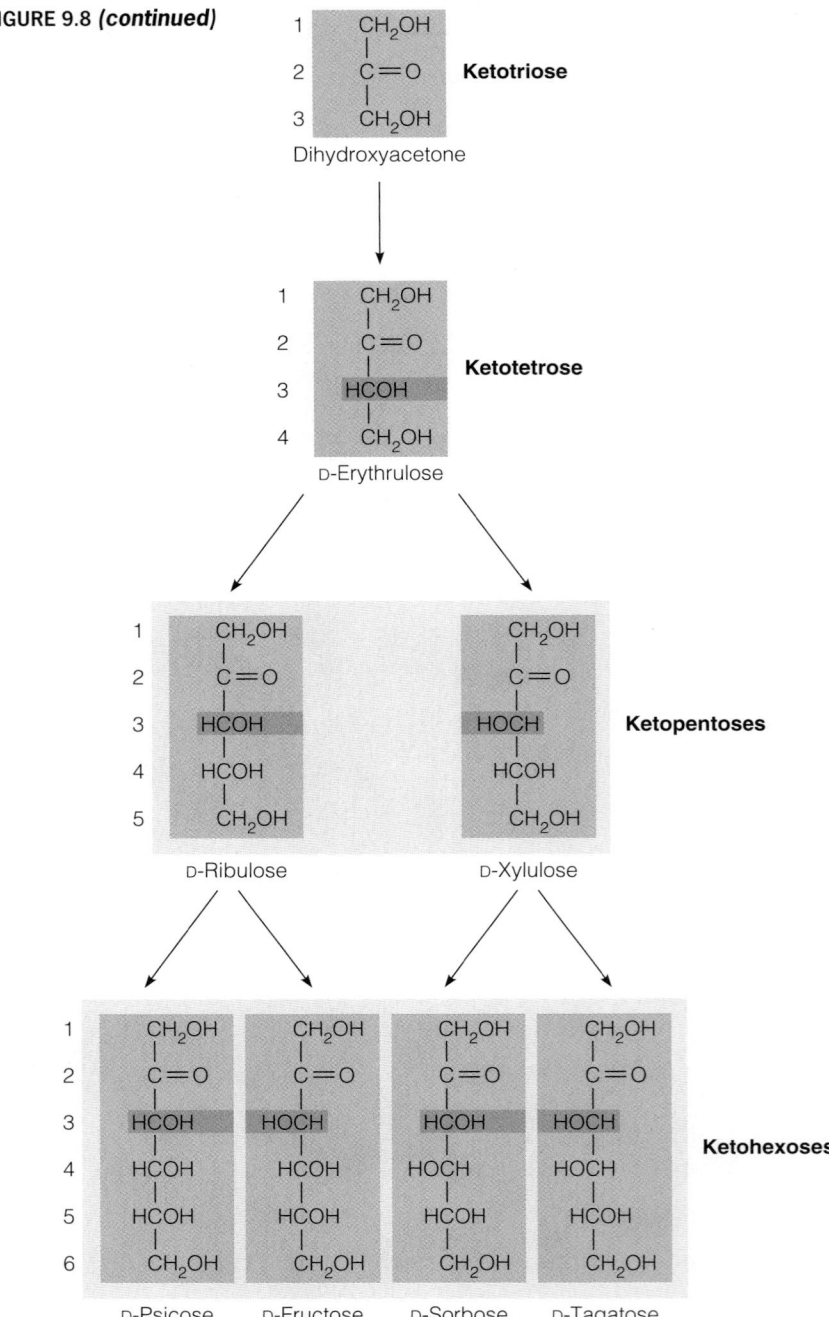

(b) D-Ketoses

Both of the reactions shown in Figure 9.9 have equilibria that highly favor the cyclic structures for pentoses or larger sugars. Under physiological conditions in solution, monosaccharides with five or more carbons exist typically more than 99% in the ring forms. The distribution between pyranose and furanose forms depends on the particular sugar structure, the pH, the solvent composition, and the temperature. Representative data obtained from nuclear magnetic resonance studies are shown in Table 9.2. When the monomers are incorporated into polysaccharides, the structure of the polymer may also influence the ring form chosen. For example, as Table 9.2 shows, D-ribose exists in solution as a mixture of the two ring forms. But in

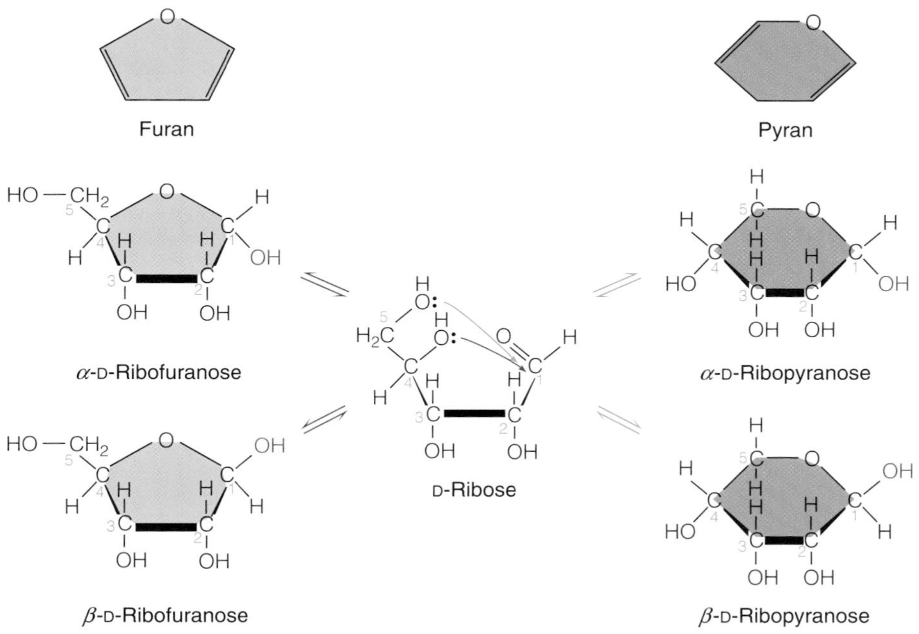

FIGURE 9.9

Formation of ring structures by pentoses. The example shown here is D-ribose, which can form either a five-membered furanose ring or a six-membered pyranose ring. The reactions involve formation of hemiacetals from the aldehyde group. In each case, two anomeric forms, α and β, are possible. (Anomers differ in conformation only at the carbonyl carbon, which in this case is carbon 1.) The sugar rings are depicted here as *Haworth projections,* with bonds closer to the viewer drawn more darkly to suggest perspective.

biological compounds, specific forms are stabilized. RNA, for example, contains ribose exclusively as ribofuranose, whereas some plant cell wall polysaccharides have pentoses entirely in the pyranose form.

Let us look more closely at the ring structures shown in Figure 9.9. Cyclization has created a new asymmetric center at carbon 1. That is why we have had to draw two stereoisomers of D-ribofuranose, referred to as α-D-ribofuranose and β-D-ribofuranose, as well as a corresponding pair of ribopyranoses. Like other kinds of stereoisomers, these α and β forms rotate the plane of polarized light differently and can be distinguished in that way. Such isomers, differing in configuration only at the carbonyl carbon (C1 in this case), are called **anomers**, and carbon 1 is often referred to as the *anomeric carbon atom*. The monosaccharides can undergo interconversion between the α and β forms, using the open-chain structure as an intermediate. This process is referred to as **mutarotation**. A purified anomer, dissolved in aqueous solution, will approach the equilibrium mixture, with an accompanying change in the optical rotation of the solution. Enzymes called **mutarotases** catalyze this process in vivo.

TABLE 9.2 Relative amounts of tautomeric forms for some monosaccharide sugars at equilibrium in water at 40 °C

Monosaccharide	Relative Amount (%)				
	α-Pyranose	β-Pyranose	α-Furanose	β-Furanose	Total Furanose
Ribose	20	56	6	18	24
Lyxose	71	29	_a	_a	<1
Altrose	27	40	20	13	33
Glucose	36	64	_a	_a	<1
Mannose	67	33	_a	_a	<1
Fructose	3	57	9	31	40

Note: In all cases, the open-chain form is much less than 1%. For data on other sugars, see S. J. Angyal, The composition and conformation of sugars in solution, *Angew. Chem.* (1969) 8:157–226.

[a]Much less than 1%.

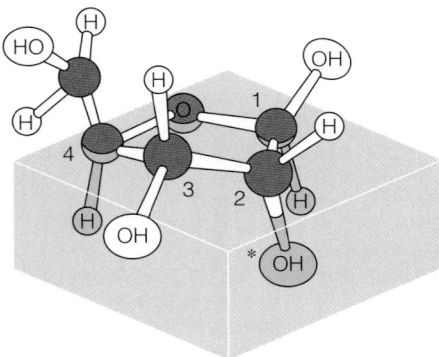

(a) β-ᴅ-Ribofuranose, C-2 endo

(b) β-ᴅ-Ribofuranose, C-3 endo

FIGURE 9.10

Conformational isomers. These models show two of
the possible ring conformations for β-ᴅ-ribofuranose. In
both of them, C-1, O, and C-4 define a plane. In the C-2
endo conformation **(a)**, C-2 is above the plane. In the C-3
endo conformation **(b)**, C-3 is above the plane. These isomers are the two most common conformations for ribose
and deoxyribose in nucleic acids. (In DNA the hydroxyl at
carbon 2, indicated here by *, is replaced by hydrogen.)
A C-3 exo conformation would look like the figure in
(b), but C-3 would be flipped below the plane.

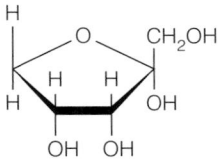

α-ᴅ-Ribulose

The representation of a cyclic sugar structure we have used in Figure 9.9 is called
a **Haworth projection**. Imagine that you are seeing the ring in perspective, and the
groups attached to the ring carbons (H, OH, CH_2OH) are pictured as being above
or below the ring. In all ᴅ-monosaccharides, the —CH_2OH is above the ring. The
relationship between hydroxyl orientations in a Fischer and a Haworth projection is
straightforward. Those represented to the right of the chain in a Fischer projection
are shown below the ring in a Haworth. For example, Fischer projections of α-ᴅ-
ribofuranose and β-ᴅ-ribofuranose would look like this:

$$\text{α-ᴅ-Ribofuranose} \qquad \text{β-ᴅ-Ribofuranose}$$

Even Haworth projections do not accurately depict the three-dimensional
structure of molecules like ribofuranose or ribopyranose. Saturated five- and six-
membered rings cannot be planar because the C—C—C bond angles are about
109° and the C—O—C angle is about 118°. Furthermore, the ring can pucker
out of plane in many different ways. The different ring conformations produced by
slightly different bond angles are called **conformational isomers**. Ball-and-stick
models of two of the several possible conformational isomers of β-ᴅ-ribofuranose
are shown in Figure 9.10.

We have already encountered β-ᴅ-ribofuranose (and its close relative β-ᴅ-2-
deoxyribofuranose) in Chapter 4. These sugars play critical roles in biochemistry,
for they are part of the backbone structures of RNA and DNA, respectively. Only
the β anomers are involved in nucleic acid structure, and the C-2 endo and C-3
endo conformations shown in Figure 9.10 are favored. However, there is some
variation in ring conformation, even locally, along DNA and RNA chains, with
resulting changes in secondary structure. This flexibility points up a fundamental
difference between *conformation* and *configuration*. Conformational isomers can
interchange by a simple deformation of the molecule. But *configurational* isomers,
such as the various kinds of stereoisomers described earlier, can interconvert only
through the breaking and re-formation of covalent bonds.

Like the aldopentoses, the ketopentoses exist almost entirely in the ring form
under physiological conditions. However, only the furanose form is possible for
ketopentoses. An example is *α-ᴅ-ribulose*, which is an intermediate in the carbon
fixation processes in photosynthesis.

Hexose Rings

The hexoses also exist primarily in ring forms under physiological conditions. As
with the aldopentoses, two kinds of rings are found: five-membered furanoses and
six-membered pyranoses. In each case, α and β anomers are possible. An example,
illustrated by Haworth projections, follows:

α-ᴅ-Glucopyranose   **β-ᴅ-Glucopyranose**

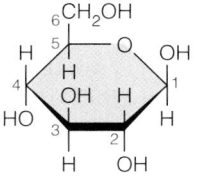 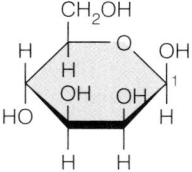 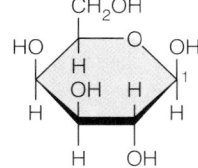

β-D-Glucopyranose   β-D-Mannopyranose   β-D-Galactopyranose   β-D-Fructofuranose

FIGURE 9.11

The four most common hexoses. These Haworth projections represent the D enantiomers; only the β anomers are shown.

Figure 9.11 shows Haworth projections of the structures of the four most common hexoses in their usual configurations. Table 9.2 gives the fractions of the furanose and pyranose forms found at equilibrium for several hexoses. Clearly, which forms are favored depends greatly on the structure of the particular sugar and its environment, although we can make the generalization that hexoses prefer the pyranose ring structure when in aqueous solution. This preference is also true of fructose, but we have depicted D-fructose in Figure 9.11 in its furanose configuration because that is how it is found in its most common biological source, the disaccharide sucrose. Elucidation of the distribution of anomeric and tautomeric forms of the sugars existing in solutions has been greatly facilitated by nuclear magnetic resonance spectroscopy (see Tools of Biochemistry 6A), which allows determination of molecular conformation in solution.

There is something else to note from Figure 9.11. Glucose and mannose differ from each other only in the configuration about C2. Sugars of this type, differing in configuration about only one carbon, are called **epimers**. Similarly, glucose and galactose are epimers, for they differ in configuration only about C4.

We have already shown that Haworth projections of the furanoses do not depict the actual three-dimensional structure correctly. The same is true for the pyranoses. Two major classes of pyranose conformations exist for the 6-carbon sugars—the more stable "chair" form and the less favored "boat" form. These two conformations are depicted as ball-and-stick models in Figure 9.12a. We will frequently depict them by skeletal diagrams in the ways shown in Figure 9.12b. For

Hexoses can exist in boat and chair conformations. Usually the chair is more stable.

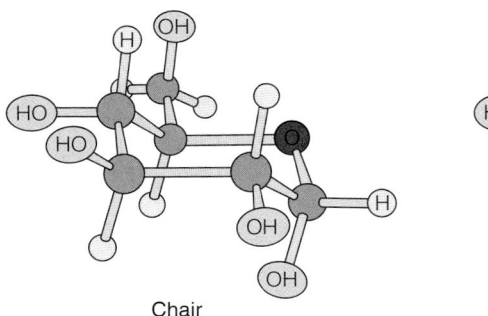 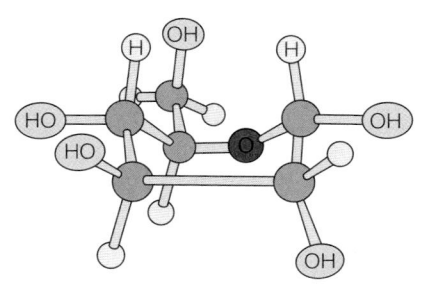

Chair   Boat

(a)

FIGURE 9.12

The pyranose ring in chair and boat conformations. Three-dimensional representations of α-D-glucopyranose in the chair form (left) and the boat form (right). **(a)** Ball-and-stick models. **(b)** Skeletal diagrams of the bonding. Axial bonds **(a)** and equatorial bonds **(e)** are indicated.

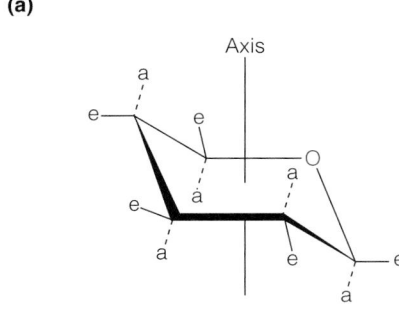 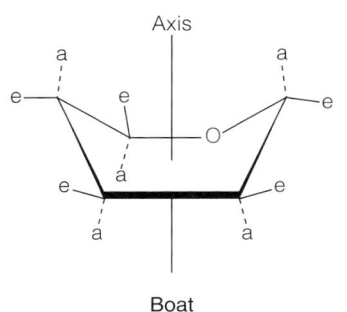

Chair   Boat

(b)

α-D-**Sedoheptulopyranose**

both the boat and chair forms of pyranose rings, a molecular axis can be defined perpendicular to the central plane of the molecule. Bonds to substituents on ring carbons can then be classed as *axial* (a) or *equatorial* (e), depending on whether they are approximately parallel or perpendicular to the axis (Figure 9.12b). For most sugars, the chair form is more stable because substituents on axial bonds tend to be more crowded in the boat form.

Sugars with More than Six Carbons

Monosaccharides with seven or even more carbons exist in nature, but most are of minor importance. However, one heptose, called *sedoheptulose,* plays a major role in the fixation of CO_2 in photosynthesis (see Table 9.1 and Chapter 16) and also in the pentose phosphate pathway (see Chapter 13).

At this point we have introduced several terms used to describe the structures of sugar molecules—enantiomers, diastereomers, anomers, epimers, and ring conformations. For review, this terminology is summarized in Figure 9.13.

FIGURE 9.13

Terminology describing the structure of sugar molecules. Conformational isomers are distinguished from configurational isomers in that the former can interconvert without breaking and re-forming bonds. Not shown are epimers, stereoisomers differing in their configuration about only one asymmetric carbon atom.

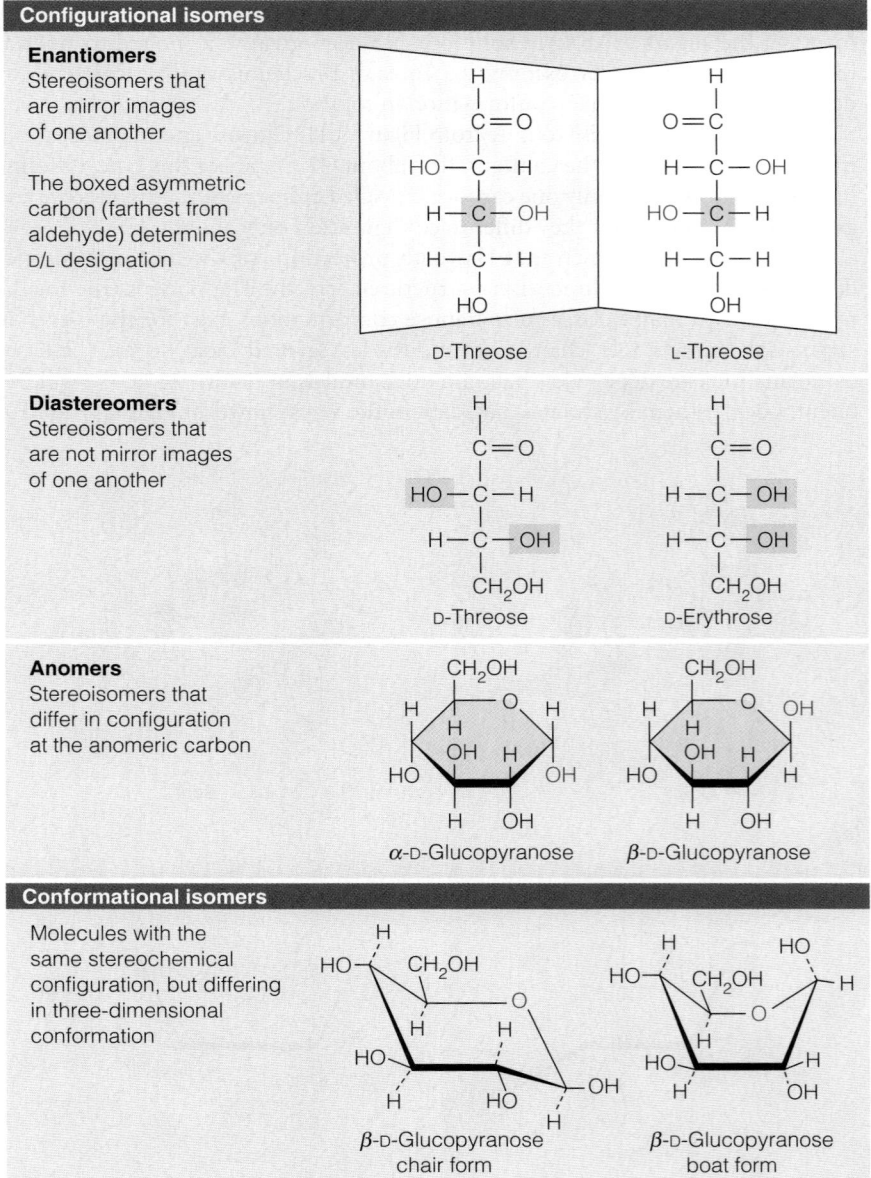

Derivatives of the Monosaccharides

The monosaccharides each carry a number of hydroxyl groups to which substituents might be attached or which could be replaced by other functional groups. An enormous number of sugars are in fact modified in this way. We shall describe here only a few of them—primarily those that play biologically important roles.

Phosphate Esters

We have already encountered sugar phosphorylation, in compounds like AMP, ATP, and the nucleic acids. As we see in later chapters, the phosphate esters of the monosaccharides themselves are major participants in many metabolic pathways. Table 9.3 illustrates some of the more important phosphate esters and includes

Sugar phosphates are important intermediates in metabolism, functioning as activated compounds in syntheses.

TABLE 9.3	Some biochemically important phosphate esters of monosaccharides				

Name	Structure	$\Delta G^{o\prime a}$ (kj/mol)	pK_{a1}	pK_{a2}
D-Glyceraldehyde-3-phosphate		~ -12	2.10	6.75
β-D-Glucose-1-phosphate		-20.9	1.10	6.13
β-D-Glucose-6-phosphate		-13.8	0.94	6.11
α-D-Fructose-6-phosphate		-13.8	0.97	6.11

[a]Free energy of hydrolysis at pH 7.0 and 37 °C.

values for the standard state free energies of hydrolysis. In all cases, these values are less negative than the free energy of hydrolysis of ATP (-31 kJ/mol); thus, ATP can act as a phosphate donor to monosaccharides. On the other hand, because hydrolysis of the phosphate esters of sugars is thermodynamically favorable, these derivatives can behave as "activated" compounds in many metabolic reactions.

Sugar phosphate esters are quite acidic, with pK_a values for the two stages of phosphate ionization of about 1–2 and 6–7, respectively (see Table 9.3).

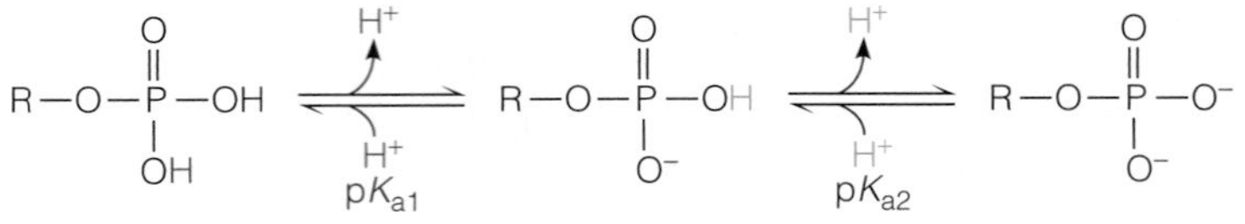

Consequently, these compounds exist under physiological conditions as a mixture of the monoanions and dianions.

In addition to the sugar phosphates, a large number of other derivatives of monosaccharides play varied and important roles in biochemistry. We consider here a few of them and the reactions by which they may be generated from monosaccharides.

Acids and Lactones

Oxidation of monosaccharides can proceed in several ways, depending upon the oxidizing agent used. For example, mild oxidation of an aldose with alkaline Cu(II) (Fehling's solution) produces the **aldonic acids**, as in the following example:

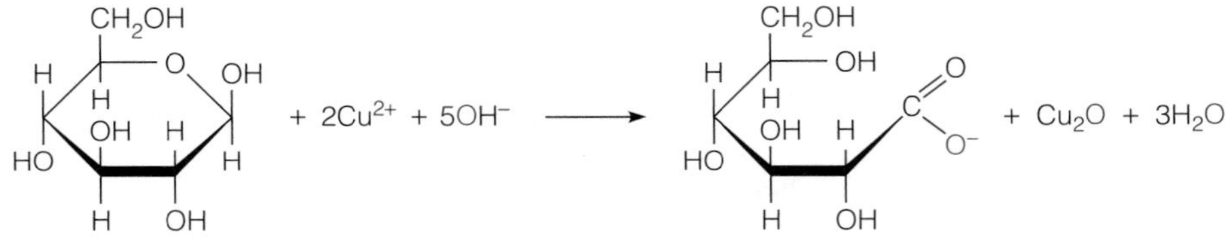

β-D-Glucopyranose   **D-Gluconic acid**

The production of a red precipitate of Cu_2O is a classic sugar test and was used formerly to test for excess sugar in the urine of persons thought to have diabetes. Another, similar reaction involves the use of Ag^+ ion as an oxidant; its reduction to metallic silver leaves a characteristic "mirror" on the glassware. These older methods have now been replaced by more specific enzyme assays. Free aldonic acids, such as gluconic acid, are in equilibrium in solution with **lactones**, which are cyclic esters, in this case involving the C1 carboxyl and the C5 hydroxyl.

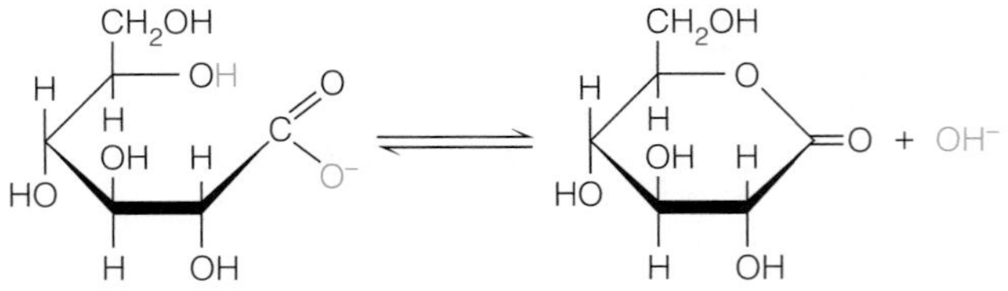

D-Gluconic acid   **D-δ-Gluconolactone**

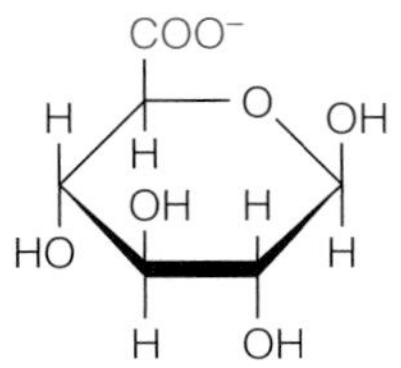

β-D-Glucuronic acid

Enzyme-catalyzed oxidation of monosaccharides gives other products, including **uronic acids** such as **glucuronic acid**, in which oxidation has occurred at carbon 6. Uronic acids are, as we see later in this chapter, important constituents of certain natural polysaccharides.

Alditols

Reduction of the carbonyl group on a sugar gives rise to the class of polyhydroxy compounds called **alditols**. Important naturally occurring ones are *erythritol*, D-*mannitol*, and D-*glucitol*, often called *sorbitol*.

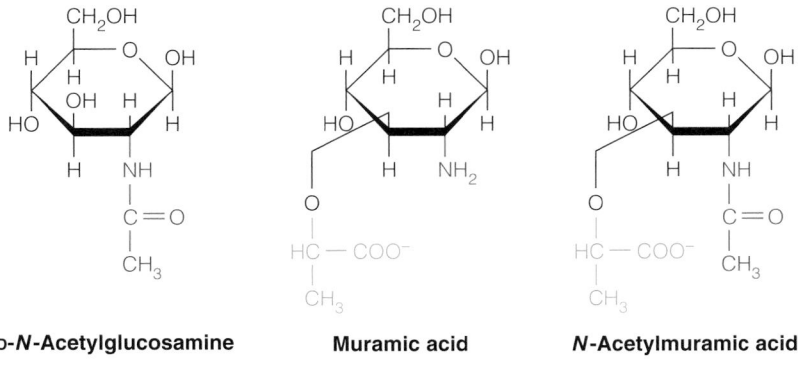

Erythritol* 　　　 **D-Mannitol** 　　　 **D-Glucitol (sorbitol)**

Each is named from the corresponding monosaccharide. When sorbitol accumulates in the lens of the eye of a person with diabetes, it can lead to the formation of cataracts.

Amino Sugars

Two amino derivatives of simple sugars are widely distributed in natural polysaccharides: *glucosamine* and *galactosamine,* derived from glucose and galactose, respectively. Further modifications of these amino sugars are common. For example, the following compounds are derived from β-D-glucosamine:

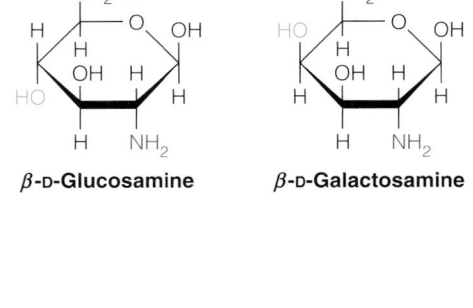

β-D-Glucosamine 　　　 **β-D-Galactosamine**

β-D-N-Acetylglucosamine 　　　 **Muramic acid** 　　　 **N-Acetylmuramic acid**

These sugar derivatives are important constituents of many natural polysaccharides. Two others we shall encounter are the following:

Amino sugars are found in many polysaccharides.

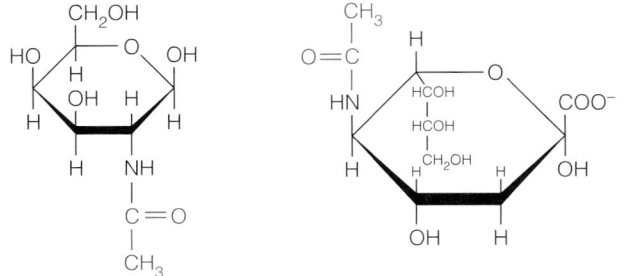

β-D-N-Acetylgalactosamine 　　　 **N-Acetylneuraminic acid (sialic acid)**

---

*Erythritol, although it contains chiral carbons, is not optically active because it has a plane of symmetry, between C-2 and C-3.

TABLE 9.4	Abbreviations for some common monosaccharide residues

Monosaccharides

Arabinose	Ara
Fructose	Fru
Fucose	Fuc
Galactose	Gal
Glucose	Glc
Lyxose	Lyx
Mannose	Man
Ribose	Rib
Xylose	Xyl

Monosaccharide Derivatives

Gluconic acid	GlcA
Glucuronic acid	GlcUA
Galactosamine	GalN
Glucosamine	GlcN
N-Acetylgalactosamine	GalNAc
N-Acetylglucosamine	GlcNAc (or NAG)
Muramic acid	Mur
N-Acetylmuramic acid	MurNAc (or NAM)
N-Acetylneuraminic acid (or sialic acid)	NeuNAc (or Sia)

O-glycosides are formed by elimination of a water molecule between a hydroxyl group on a saccharide and a hydroxyl on another compound.

The modified sugars—especially the amino sugars—are most often found as monomer residues in complex oligosaccharides and polysaccharides. To aid in writing the structures of such molecules, it is useful to have a shorthand notation, as is used in describing nucleic acid and protein structure. Therefore, a set of abbreviations has been defined for the simple sugars and their derivatives. A number of the most important ones are listed in Table 9.4.

Glycosides

Elimination of water between the anomeric hydroxyl of a cyclic monosaccharide and the hydroxyl group of another compound yields an **O-glycoside** (the O signifying attachment at a hydroxyl). The acetal bond formed is referred to as a **glycosidic bond**. A simple example is the formation of methyl-α-D-glucopyranoside:

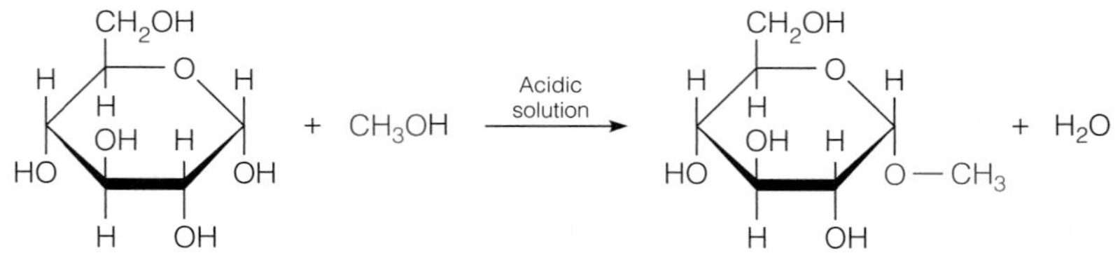

α-D-**Glucopyranose**   Methyl-α-D-**glucopyranoside**

Unlike the anomers of the sugars themselves, the anomeric glycosides (e.g., methyl-α-D-glucopyranoside in the example shown, and methyl-β-D-glucopyranoside) do not interconvert by mutarotation in the absence of an acid catalyst, a property that makes them useful in the determination of sugar configurations.

Many glycosides are found in plant and animal tissues. Some are toxic substances, in most cases because they act as inhibitors of enzymes involved in ATP utilization. Two toxic glycosides, ouabain and amygdalin, are shown in Figure 9.14. *Ouabain* inhibits the action of the enzymes that pump Na^+ and K^+ ions across cell membranes to maintain necessary electrolyte balance. It comes from an African shrub and was discovered when it was observed that Somali hunters dipped arrowheads in an extract from the plant. Ouabain now finds use in treatment of some cardiac conditions (see page 393). *Amygdalin* is toxic for a quite different reason. Found in the seeds of bitter almonds, this glycoside yields hydrogen cyanide (HCN) upon hydrolysis. It is for this reason that HCN gas is said to have the odor of bitter almonds.

Oligosaccharides

Just as monosaccharides can form glycosidic bonds with other kinds of hydroxyl-containing compounds, they can do so with one another. Such bonding gives rise to glycans—the oligosaccharides and polysaccharides.

FIGURE 9.14

Two naturally occurring glycosides. Ouabain and amygdalin are highly toxic glycosides produced by plants.

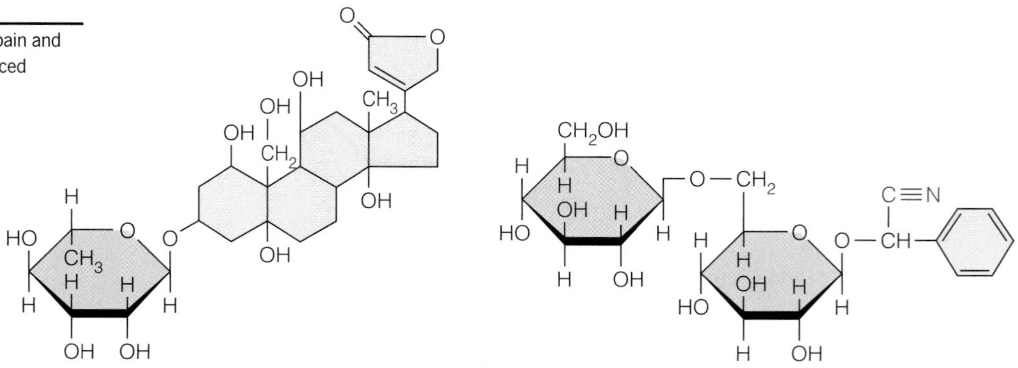

Ouabain   **Amygdalin**

Oligosaccharide Structures

The simplest and biologically most important oligosaccharides are the *disaccharides,* made up of two residues. As Table 9.5 shows, the disaccharides play many biological roles. Some, like *sucrose, lactose,* and *trehalose,* are soluble energy stores in plants and animals. Others, like *maltose* and *cellobiose,* can be regarded primarily as intermediate products in the degradation of much longer polysaccharides. Still others, like *gentiobiose,* are found principally as constituents of more complex, naturally occurring substances. The structures of these disaccharides are depicted in Figure 9.15.

Distinguishing Features of Different Disaccharides

Four major features distinguish disaccharides from one another:

1. *The two specific sugar monomers involved, and their stereoconfigurations.* The monomers may be of the same kind, as the two D-glucopyranose residues in maltose, or they may be different, as the D-glucopyranose and D-fructofuranose residues in sucrose.

2. *The carbons involved in the linkage.* Although many possibilities exist, the most common linkages are $1 \rightarrow 1$ (as in trehalose), $1 \rightarrow 2$ (as in sucrose), $1 \rightarrow 4$ (as in lactose, maltose, and cellobiose), and $1 \rightarrow 6$ (as in gentiobiose). Note that all of these disaccharides involve the anomeric hydroxyl of at least one sugar as a participant in the bond.

3. *The order of the two monomer units, if they are different kinds.* The glycosidic linkage involves the anomeric carbon on one sugar, but in most cases the other is free. Thus, the two ends of the molecule can be distinguished by their chemical reactivity. For example, the glucose residue in lactose, having a free anomeric carbon and thus a potential free aldehyde group, could be oxidized by Fehling's solution; the galactose residue could not be. Lactose is therefore a reducing sugar, and the glucose residue is at its *reducing end.* The other end is called the *nonreducing end.* In sucrose, neither residue has a potential free aldehyde group; both anomeric carbons are involved in the glycosidic bond. Therefore, sucrose is a nonreducing sugar.

4. *The configuration of the anomeric hydroxyl group of each residue.* This feature is especially important for the anomeric carbon(s) involved in the glycosidic bond. The configuration may be either α (as in the disaccharides shown in Figure 9.15a) or β (as in those in Figure 9.15b). This difference may seem small, but it has a major effect on the shape of the molecule, and the difference in shape is recognized readily by enzymes. For example, different enzymes are needed to catalyze the hydrolysis of maltose and

TABLE 9.5 Occurrence and biochemical roles of some representative disaccharides

Disaccharide	Structure	Natural Occurrence	Physiological Role
Sucrose	Glcα($1 \rightarrow 2$)Fruβ	Many fruits, seeds, roots, honey	A final product of photosynthesis; used as primary energy source in many organisms
Lactose	Galβ($1 \rightarrow 4$)Glc	Milk, some plant sources	A major animal energy source
α, α-Trehalose	Glcα($1 \rightarrow 1$)Glcα	Yeast, other fungi, insect blood	A major circulatory sugar in insects; used for energy
Maltose	Glcα($1 \rightarrow 4$)Glc	Plants (starch) and animals (glycogen)	The dimer derived from the starch and glycogen polymers
Cellobiose polymer	Glcβ($1 \rightarrow 4$)Glc	Plants (cellulose)	The dimer of the cellulose
Gentiobiose	Glcβ($1 \rightarrow 6$)Glc	Some plants (e.g., gentians)	Constituent of plant glycosides and some polysaccharides

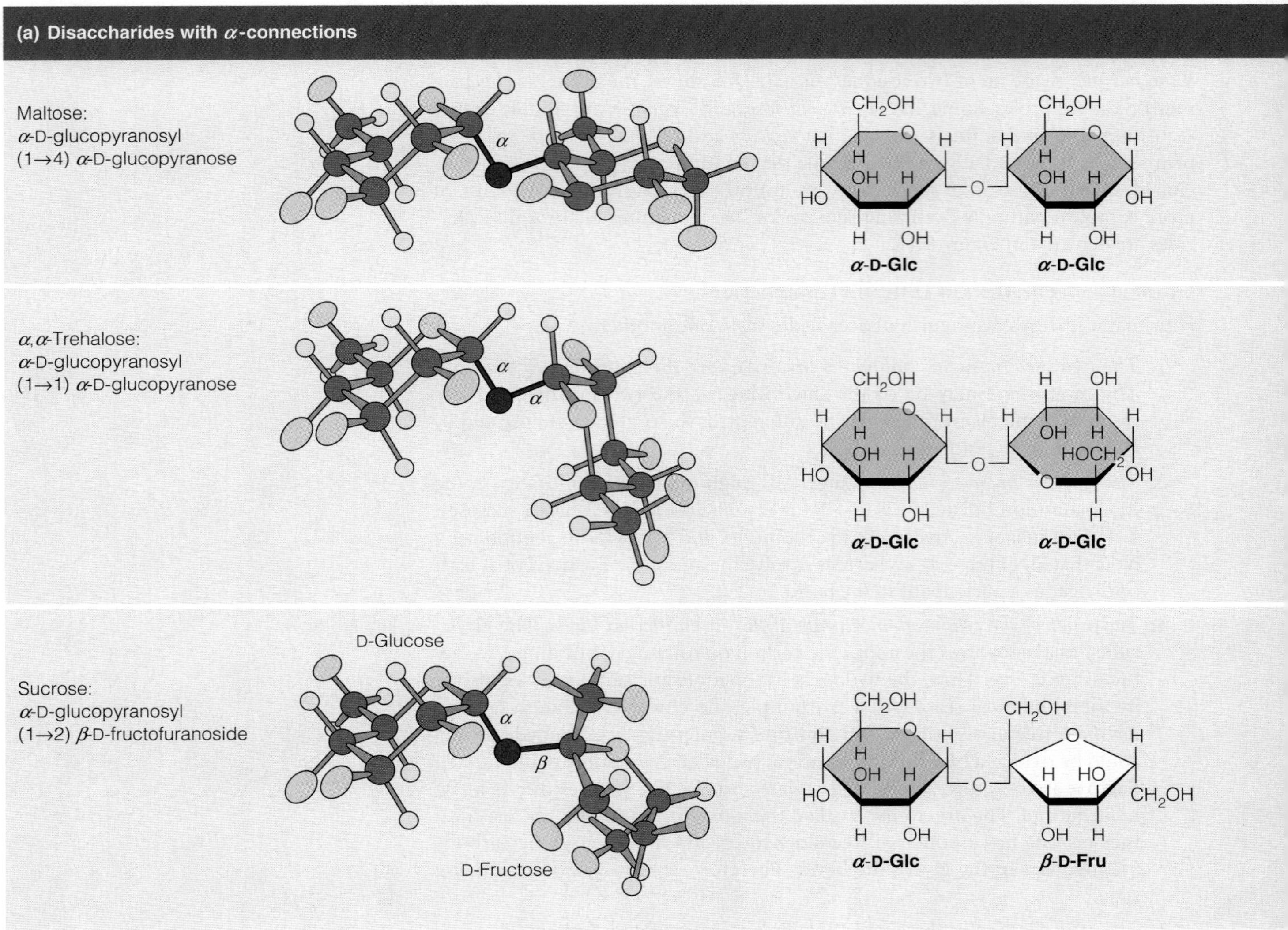

(a) Disaccharides with α-connections

Maltose:
α-D-glucopyranosyl
(1→4) α-D-glucopyranose

α-D-Glc   α-D-Glc

α,α-Trehalose:
α-D-glucopyranosyl
(1→1) α-D-glucopyranose

α-D-Glc   α-D-Glc

Sucrose:
α-D-glucopyranosyl
(1→2) β-D-fructofuranoside

D-Glucose

D-Fructose

α-D-Glc   β-D-Fru

FIGURE 9.15

Structures of some important disaccharides. Ball-and-stick models are shown on the left, with anomeric oxygens in red. On the right are Haworth projections of the same molecules, with color-coded monomers: blue = glucose, yellow = fructose, teal = galactose. **(a)** Disaccharides linked through the C-1 of the α anomer: maltose, trehalose, and sucrose. **(b)** Disaccharides with β linkage: cellobiose, lactose, and gentiobiose. Note the convention used to draw glycosidic bonds between monomers in disaccharides. The "bent bonds" allow the Haworth projections of the monomers to be drawn in parallel. The "corners" do *not* imply extra carbon atoms, as they often do in organic structure representations.

cellobiose, even though both are dimers of D-glucopyranose. Furthermore, we shall see that in polysaccharides the anomeric orientation plays a critical role in determining the secondary structures adopted by these polymers.

Writing the Structure of Disaccharides

A convenient way to describe the structures of these and more complex oligosaccharides has been devised. The rules are as follows:

1. The sequence is written starting with the nonreducing end at the left, using the abbreviations defined in Table 9.4.

2. Anomeric and enantiomeric forms are designated by prefixes (e.g., α-, D-).

3. The ring configuration is indicated by a suffix (*p* for pyranose, *f* for furanose).

4. The atoms between which glycosidic bonds are formed are indicated by numbers in parentheses between residue designations (e.g., (1 → 4) means a bond from carbon 1 of the residue on the left to carbon 4 of the residue on the right).

(b) Disaccharides with β connections

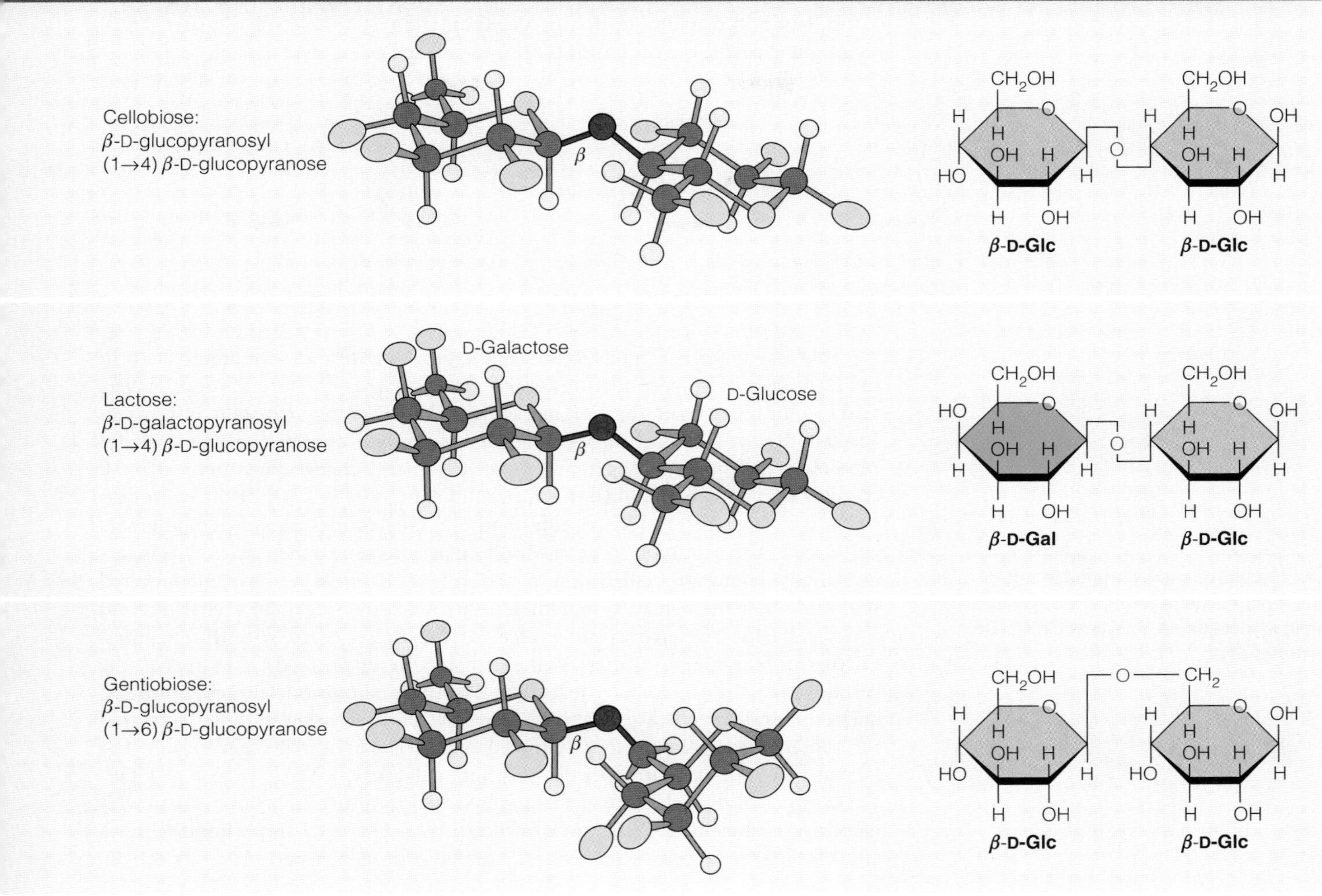

Cellobiose:
β-D-glucopyranosyl
(1→4) β-D-glucopyranose

β-D-Glc β-D-Glc

Lactose:
β-D-galactopyranosyl
(1→4) β-D-glucopyranose

D-Galactose

D-Glucose

β-D-Gal β-D-Glc

Gentiobiose:
β-D-glucopyranosyl
(1→6) β-D-glucopyranose

β-D-Glc β-D-Glc

As an example, we can write the structure of sucrose as

$$\alpha\text{-D-Glc}p\,(1 \rightarrow 2)\text{-}\beta\text{-D-Fru}f$$

In many cases, the nomenclature is further shortened by omitting the D and L designations (except in the unusual cases in which L enantiomers are encountered) and by omitting the p and f suffixes when the monomers have their usual ring forms. Thus, we would more likely write sucrose in the way shown in Table 9.5. The system can be applied to oligosaccharides of any length and can include branched structures, as we will see in the discussion of starch later in this chapter. If only one carbon involved in the linkage between two residues is anomeric, the representation can be even more condensed because the anomeric configuration at the reducing end will equilibrate in solution. For example, maltose can be represented as Glcα(1 → 4)Glc.

The list of biologically important oligosaccharides is by no means restricted to dimeric structures. Many trimers, tetramers, and even larger, yet specifically constructed, molecules are known. Examples of these compounds will be encountered later in this chapter, when we examine the oligosaccharides attached to certain proteins and to cell surfaces. Tools of Biochemistry 9A describes techniques used to sequence oligosaccharides.

Stability and Formation of the Glycosidic Bond

Formation of the glycosidic bond between two monomers in an oligosaccharide is a condensation reaction, involving the elimination of a molecule of water. Thus, we might expect the synthesis of lactose to proceed as follows:

β-D-Galactose   β-D-Glucose   Lactose

> Like the phosphodiester bond in nucleic acid and amide bond in proteins, the glycosidic bond is metastable. Enzymes control its hydrolysis.

This reaction is analogous to the elimination of water between amino acids in the formation of polypeptides or between nucleotides in the formation of nucleic acids. As in those cases, the reaction as written is thermodynamically unfavored. Instead, the hydrolysis of oligosaccharides and polysaccharides is favored under physiological conditions by a standard state free energy change of about 15 kJ/mol, corresponding to an equilibrium constant of about 800 in favor of the hydrolysis products. Nevertheless, like peptides and oligonucleotides, saccharide polymers are sufficiently metastable to persist for long periods unless their hydrolysis is catalyzed by enzymes or acid. So the situation is the same as we have encountered with the other important biopolymers: The breakdown of oligosaccharides and polysaccharides in vivo is controlled by the presence of specific enzymes. Furthermore, *synthesis* of these sugar polymers never proceeds in living cells by reactions like the one we have just shown. As in protein or nucleic acid synthesis, activated monomers are required. For glycan biosynthesis those activated monomers are usually nucleotide-linked sugars. The activated sugar molecule in lactose biosynthesis is **uridine diphosphate galactose** (UDP-galactose or UDP-Gal), a nucleotide-linked sugar formed by reaction of uridine triphosphate with galactose-1-phosphate. Here and elsewhere in this chapter a single bond with no substitutent shown represents a single-bonded hydrogen atom.

β-D-galactose-1-phosphate   uridine triphosphate   uridine diphosphate galactose   pyrophosphate

From the principles outlined in Chapter 3, you will probably recognize UDP-galactose as a high-energy compound. Hence, there is a thermodynamic tendency for one of the bonds to break, and in the process the activated galactosyl moiety can be transferred to a carbohydrate acceptor. In the reaction catalyzed by the enzyme **lactose synthase** that acceptor is glucose, as shown in Figure 9.16, and the product is lactose (Galβ(1 → 4)Glc). Note the specificity of the enzyme. Although glucose has five different hydroxyl groups to which the galactosyl moiety could be transferred, only the hydroxyl at carbon 4 serves as an acceptor for this enzyme. Also shown in Figure 9.16 is the fact that the synthesis of UDP-Gal is driven in part by the subsequent hydrolysis of the other product, inorganic pyrophosphate, which is also a high-energy compound.

Because different disaccharides (and oligosaccharides and polysaccharides) are distinguished both by the kinds of monomers involved and by the precise glycosidic linkages between them, the enzymes needed for their breakdown must also be specific. For example, hydrolysis of the common nutritional disaccharides maltose, lactose, and sucrose, which takes place in cells lining the wall of the small intestine, requires three different and specific enzymes (see page 560). None will substitute for another.

Lactose synthase is an example of a **glycosyltransferase**. Reactions of this type all involve transfer of an activated glycosyl moiety to an acceptor. The first known glycosyltransferase was the enzyme responsible for synthesis of **glycogen**, a carbohydrate storage polymer in animals (see page 330 and Chapter 13). This enzyme uses **UDP-glucose** (UDPG or UDP-Glc) as the activated glycosyl donor. Although both glycosyltransferases mentioned so far use uridine nucleotides for activation, there are numerous exceptions. For example, the biosynthesis of starch in plants uses **adenosine diphosphate glucose** (ADP-glucose or ADPG) as the activated nucleotide.

Another glycosyltransferase is responsible for the synthesis of sucrose in plants:

$$\text{UDP-glucose + fructose-6-phosphate} \rightarrow \text{sucrose-6-phosphate + UDP}$$

The product, sucrose-6-phosphate, is subsequently hydrolyzed to sucrose plus phosphate. An important feature of sucrose metabolism is that sucrose is involved in a glycan biosynthetic reaction that does not involve nucleotide-linked sugars. Some bacteria carry out the synthesis of **dextran**, an $\alpha(1 \rightarrow 6)$-linked polymer of glucose with $\alpha(1 \rightarrow 2), \alpha(1 \rightarrow 3),$ or $\alpha(1 \rightarrow 4)$ branch points. The polymerization, catalyzed by **dextran sucrase**, uses sucrose itself as the substrate:

$$n\,\text{sucrose} \rightarrow \text{glucose}_n\,(\text{dextran}) + n\,\text{fructose}$$

Several bacteria growing in the human oral cavity synthesize large quantities of dextran, which contributes toward formation of dental plaque—hence, another concern that nutritionists have about excessive sucrose consumption in the population, in addition to their concern about obesity.

As we see later in this chapter, the larger oligosaccharides and the polysaccharides can exhibit complex structures. There is one important way in which oligosaccharide and polysaccharide synthesis differs from synthesis of nucleic acids and proteins. These sugar polymers are never copied from template molecules. Instead, in the formation of glycans, a different enzyme is employed to catalyze the addition of each kind of monomer unit. Clearly, a vast array of plant and animal enzymes must be devoted to the synthesis and degradation of saccharide polymers.

Polysaccharides

Polysaccharides fulfill numerous biological functions. Some, like starch and glycogen (sometimes called animal starch), serve mainly to store sugars for energy in plants and animals, respectively. Others, like **cellulose, chitin**, and the polysaccharides of bacterial cell walls, are structural materials analogous to the structural proteins. It is simplest to consider these molecules in terms of their functional categories.

As with polypeptides and polynucleotides, the sequence of monomer residues in a polysaccharide defines its primary structure. Whereas proteins usually have complex sequences, polysaccharides often have rather simple primary structures. In some cases (e.g., cellulose), the polymer is made from only one kind of monomer residue (β-D-glucose for cellulose); these kinds of polymers are referred to as **homopolysaccharides**. If two or more different monomers are involved, the

FIGURE 9.16

Enzymatic formation of lactose. The reaction shown occurs in the formation of milk in mammary tissue. Galactose is phosphorylated by ATP, then transferred to uridine diphosphate (UDP). UDP-galactose transfers galactose to glucose, with the accompanying cleavage of a phosphate bond. The reaction is catalyzed by the enzyme lactose synthase. In keeping with the convention adopted in Chapter 4, a P within a diamond shape denotes phosphate. When shown in yellow, it is a high-energy phosphate.

Glycan biosynthesis is carried out by glycosyltransferases, enzymes that transfer an activated glycosyl moiety, such as UDP-glucose, to a specific position on a carbohydrate acceptor.

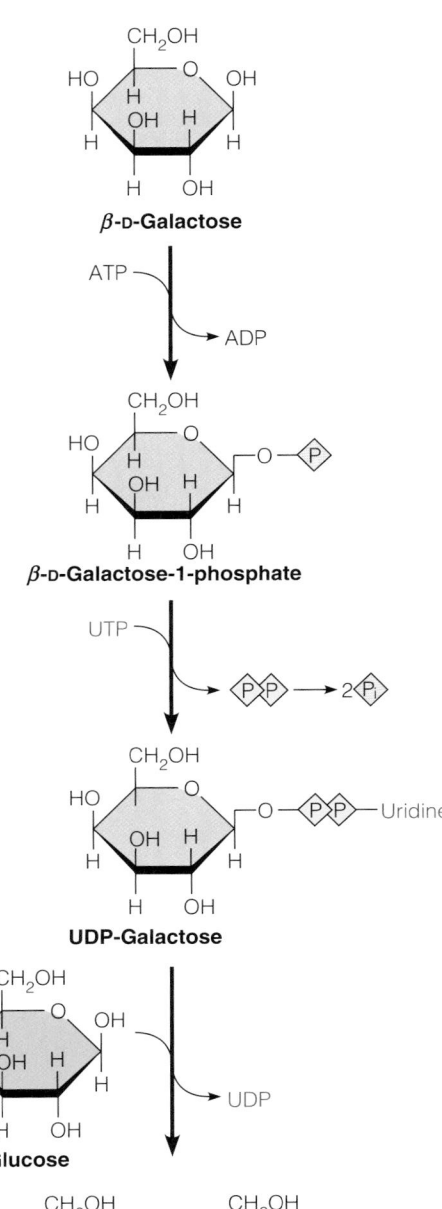

β-D-Galactose

β-D-Galactose-1-phosphate

UDP-Galactose

β-D-Glucose

Lactose

polymer is called a **heteropolysaccharide**. Even those storage and structural polysaccharides that are heteropolymers are rarely complex; usually no more than two kinds of residues are involved. In further contrast to protein and nucleic acid molecules, which are almost always of defined length, polysaccharide chains grow to random lengths. And, as mentioned earlier, glycans are distinctive in that they can form branched chains. The ability to branch, plus the numbers of functional groups on each monomer that can participate in bond formation, give polysaccharides amazing structural diversity, and this undoubtedly contributes to the large number of roles played by polysaccharides.

The functional reasons for the distinctions from proteins and nucleic acids are not hard to find. A storage material, such as starch, needs neither to convey information nor to adopt a complicated three-dimensional form. It is simply a bin in which to put away glucose molecules for future use. Many structural polysaccharides (like structural proteins) form extended, regular secondary structures, well suited to the formation of fibers or sheets. Often a regular repetition of some simple monosaccharide or disaccharide motif will serve this function. (Recall, for comparison, the simple and repetitive amino acid sequences of collagen and silk fibroin described in Chapter 6.) The only glycan polymers in which well-defined and complex sequences are found are some of the oligosaccharides attached to cell surfaces or those attached to specific glycoproteins. Because these oligomers serve to identify cells or molecules, they must convey information. This function requires precisely defined "words" in the polysaccharide language, just as nucleic acid sequences spell out information in their own language.

Storage Polysaccharides

The principal storage polysaccharides are **amylose** and **amylopectin**, which together constitute starch in plants, and **glycogen**, which is stored in animal and microbial cells. Both starch and glycogen are stored in granules within cells (Figure 9.17). Starch is found in almost every kind of plant cell, but grain seeds, tubers, and unripe fruits are especially rich in this material. Glycogen is deposited in the liver, which acts as a central energy storage organ in many animals. Glycogen is also abundant in muscle tissue, where it is more immediately available for energy release.

Glycogen and the components of starch—amylose and amylopectin—are storage polysaccharides. Amylose is linear; amylopectin and glycogen are branched.

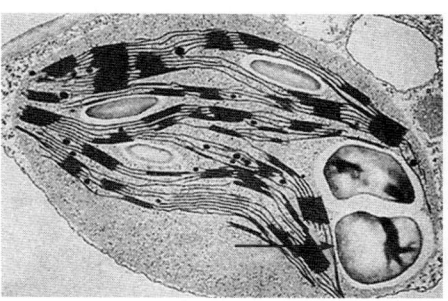

(a) Chloroplast granules

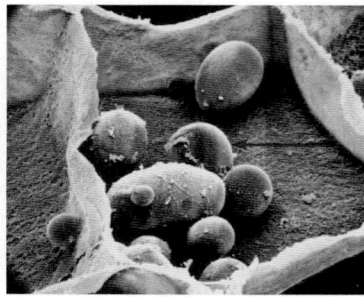

(b) Tuber cell granules

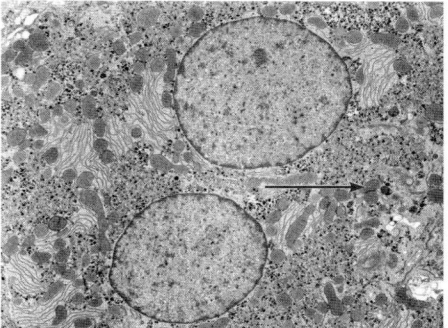

(c) Liver granules

FIGURE 9.17

Storage of starch and glycogen in granules. In each case a representative granule is indicated by an arrow. **(a)** Starch granules in a plant leaf chloroplast. **(b)** Starch granules in potato tuber cells. **(c)** Glycogen granules in liver.

FIGURE 9.18

Amylopectin, a branched glucan. **(a)** The primary structure of amylopectin. Nonreducing ends (N) and reducing ends (R) are indicated. **(b)** Detailed structure of a branch point. To simplify the figure some ring hydroxyls are not shown.

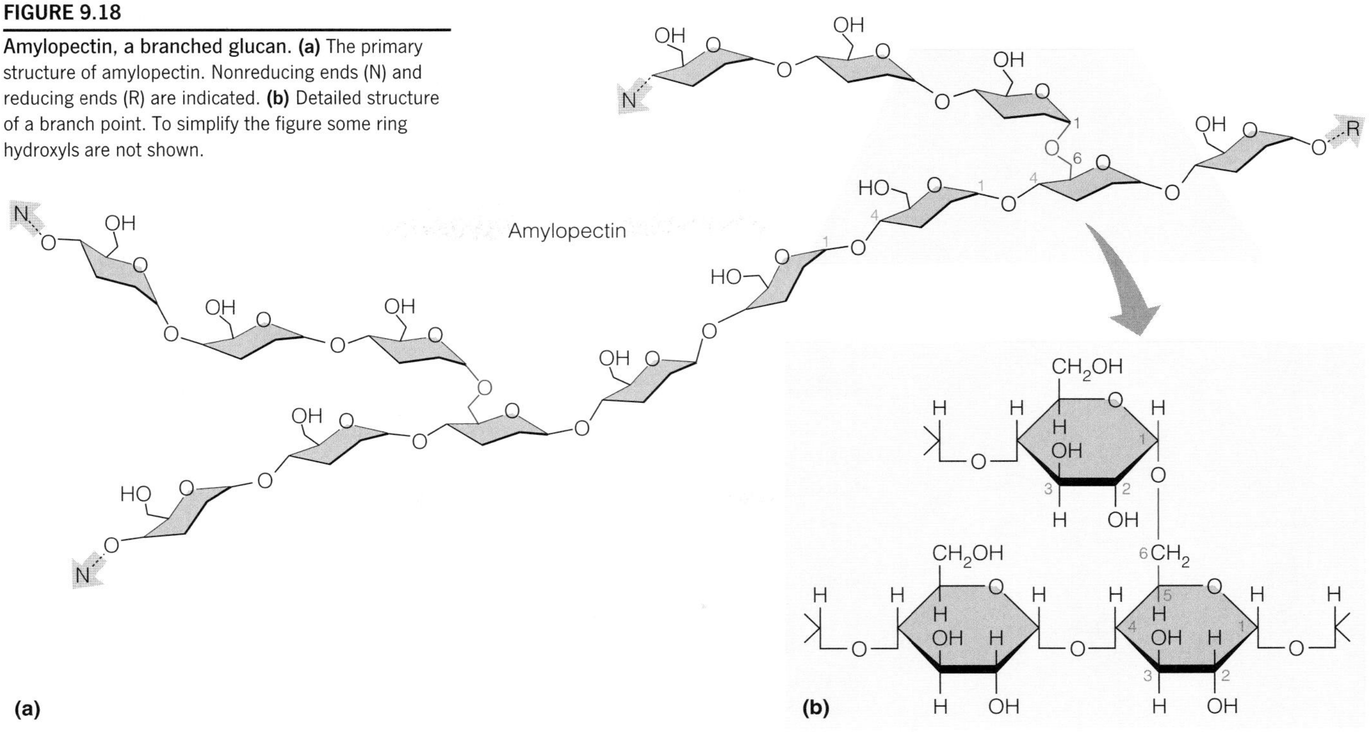

Amylose, amylopectin, and glycogen are all polymers of α-D-glucopyranose. They are homopolysaccharides of the class called **glucans**, the polymers of glucose. The polymers differ in size and the kinds of linkages between glucose residues. Amylose is a linear polymer involving exclusively $\alpha(1 \rightarrow 4)$ links between adjacent glucose residues. Amylopectin (Figure 9.18) and glycogen are both branched polymers, because they contain, in addition to the $\alpha(1 \rightarrow 4)$ links, some $\alpha(1 \rightarrow 6)$ links as well. The branches in glycogen are somewhat more frequent and shorter than those in amylopectin, and glycogen is usually of higher molecular weight, but in most respects the structures of these two polysaccharides are very similar.

The regular and simple primary structure of amylose allows a regular secondary structure for this molecule. As with polynucleotides and polypeptides, the details of this structure initially came from X-ray diffraction studies. In fact, amylose was the first biopolymer whose structure was elucidated by this method. Because of the $\alpha(1 \rightarrow 4)$ link, each residue is angled with respect to the preceding residue, favoring a regular helical conformation (Figure 9.19). The branched nature of amylopectin and glycogen inhibits the formation of helices because the helix requires 6 residues for each turn; there is a branch point about every 20–30 residues in amylopectin and about every 8–10 in glycogen.

The storage polysaccharides are admirably designed to serve their function. Glucose and even maltose are small, rapidly diffusing molecules, which are difficult to store. Were such small molecules present in large quantities in a cell, they would give rise to a very large cell osmotic pressure, which would be deleterious in most cases. Therefore, most cells build the glucose into long polymers so that large quantities can be stored in a manner that prevents its diffusion and loss. Whenever glucose is needed, it can be obtained by selective degradation of the polymers by specific enzymes. These processes are discussed in detail in Chapter 13, but one aspect should be mentioned now. Most of the enzymes employed attack the chains at their nonreducing ends, releasing one glucose residue at a time. Such "end-nibbling" (as opposed to internal cutting) prevents the continual breakup of the long polymers, which would lead to their complete solubilization. The

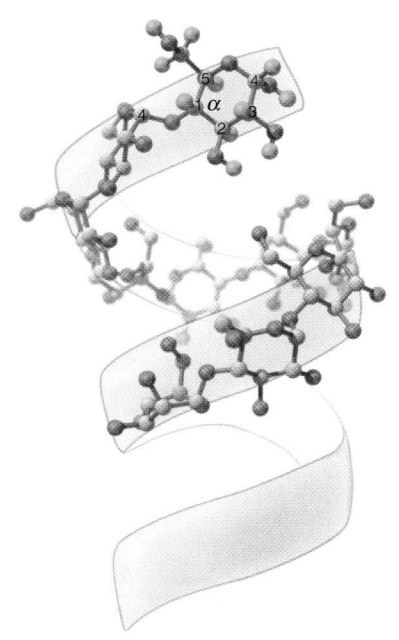

FIGURE 9.19

The secondary structure of amylose. The orientation of successive glucose residues favors helix generation. Note the large interior core. Hydrogen bonds (not shown) stabilize the helix.

branched structure of both amylopectin and glycogen is such that each molecule has *many* nonreducing ends that can be attacked simultaneously (see Figure 9.18), allowing rapid mobilization of glucose when it is needed. On the other hand, the linear chain of amylose with its single nonreducing end is used mainly for long-term storage of glucose.

Structural Polysaccharides

Plants seem not to synthesize or use fibrous structural proteins (like keratin and collagen) but instead rely entirely on special polysaccharides. Animals use both kinds of materials. Because each structural use requires different properties, a great variety of structural polysaccharides exists. We begin by considering those from plants.

Cellulose

The major polysaccharide in woody and fibrous plants (like trees and grasses), cellulose is the most abundant single polymer in the biosphere. Like amylose, cellulose is a linear polymer of D-glucose (and hence is also a glucan), but in cellulose the sugar residues are connected by $\beta(1 \rightarrow 4)$ linkages (Figure 9.20). This seemingly small distinction from starch (specifically amylose) has remarkable structural consequences. Cellulose can exist as fully extended chains, with each glucose residue flipped by 180° with respect to its neighbor in the chain. In this extended form, the chains can form ribbons that pack side by side with a network of hydrogen bonds within and between them. This arrangement is reminiscent of the β-sheet structure in silk fibroin, and as in fibroin, the fibrils of cellulose have great mechanical strength but limited extensibility.

The same small difference between cellulose and starch has another important consequence: Animal enzymes that are able to catalyze cleavage of the $\alpha(1 \rightarrow 4)$ link in starch cannot cleave cellulose. For this reason, humans, even if starving, are unable to utilize the enormous quantities of glucose all around them in the form of cellulose. Ruminants such as cows can digest cellulose only because their digestive tracts contain symbiotic bacteria that produce the necessary **cellulases**. Termites manage to eat woody substances in a somewhat more complicated fashion—their guts harbor protozoans capable of cellulose digestion but their salivary glands also produce a cellulase. Many fungi also produce such enzymes, which is why some mushrooms can live on wood as a carbon source.

Although humans don't digest cellulose, high-fiber foods containing cellulose are important nutritionally. The bulk in fiber produces a feeling of satiety, or

FIGURE 9.20

Cellulose structure. The $\beta(1 \rightarrow 4)$ linkages of cellulose generate a planar structure. The parallel cellulose chains are linked together by a network of hydrogen bonds. Hydrogens involved in such bonds are shown in blue. For clarity, all of the carbons are numbered in only one glucose residue. Not all hydrogen atoms are shown.

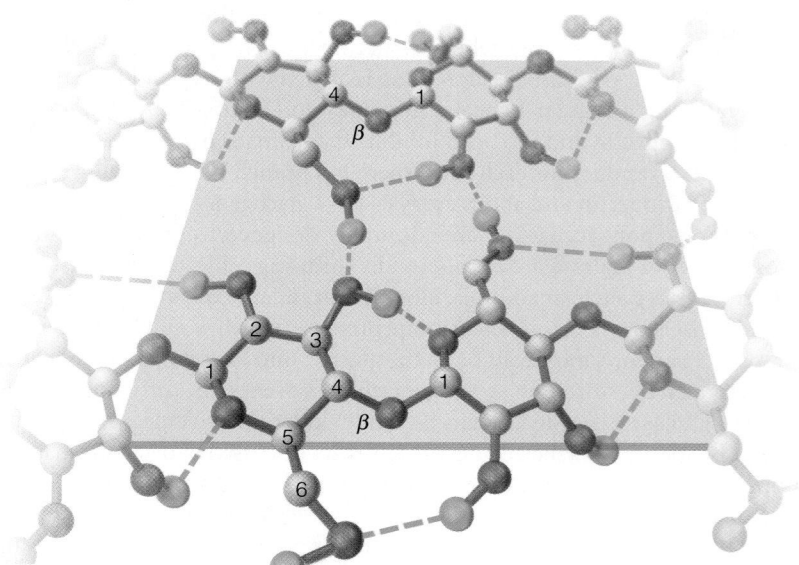

fullness, signaling when we have probably had enough to eat. Insoluble fiber increases the rate of transport of digestion products through the alimentary tract, by increasing its bulk, and is thought to reduce exposure to potential toxins or carcinogens in the diet.

A major goal of the biofuels industry is to develop efficient means for industrial-scale conversion of cellulose in plant waste materials to glucose or other substrates that can be fermented to ethanol or other potential fuels. A particular challenge is that cellulose in plant tissues is closely associated with *lignin*, a complex polymer, primarily in woody tissue, that resists breakdown much more strongly than cellulose (Chapter 21). So cellulose must be separated from lignin, or else means must be found to degrade lignin also to potential fuel molecules.

Fibrous parts of plants are not made exclusively from cellulose. A variety of other polysaccharides are present in plant cell walls. These include the **xylans**, which are polymers with $\beta(1 \rightarrow 4)$-linked D-xylopyranose, often with substituent groups attached; the **glucomannans;** and many other polymers. Often these polysaccharides are grouped together under the term **hemicellulose**.

$$\cdots \beta\text{-D-Xyl}p(1 \rightarrow 4)[\beta\text{-D-Xyl}p(1 \rightarrow 4)]_7\text{-}\beta\text{-D-Xyl}p(1 \rightarrow 4)\text{-}\beta\text{-D-Xyl}p(1 \rightarrow 4)\cdots$$
$$\quad\quad\quad\quad\quad\quad\quad | \quad\quad\quad\quad\quad\quad\quad\quad\quad\quad\quad\quad\quad | $$
$$\text{Acetyl at C-2 or C-3} \quad\quad\quad \text{4-O-Me-}\alpha\text{-D-Glc}p(1 \rightarrow 2)$$

A typical xylan structure

$$\cdots \beta\text{-D-Glc}p(1 \rightarrow 4)\text{-}\beta\text{-D-Man}p(1 \rightarrow 4)\text{-}\beta\text{-D-Man}p(1 \rightarrow 4)\text{-}\beta\text{-D-Man}p(1 \rightarrow 4)\cdots$$
$$\quad\quad\quad\quad\quad\quad\quad\quad\quad\quad\quad | \quad\quad\quad\quad\quad\quad | $$
$$\quad\quad\quad \beta\text{-D-Gal}p(1 \rightarrow 6) \quad \text{Acetyl at C-2 or C-3}$$

A typical glucomannan structure

The cell wall of a plant is a complex structure, made up of several layers. Microfibrils of cellulose are laid down in a crosshatched pattern (Figure 9.21) and impregnated with a matrix of the other polysaccharides and some proteins. The same principle is used when glass fibers are embedded in a tough resin to produce strong, durable sheets of fiberglass.

Cellulose is not confined exclusively to the plant kingdom. The marine invertebrates called *tunicates*, such as the sea squirt, contain considerable quantities of cellulose in their hard outer mantle. There are even reports of small amounts of cellulose in human connective tissue. However, as a structural material, cellulose seems to have been largely passed over in animal evolution. In the fungi, extensive use is made of other glucans, with $\beta(1 \rightarrow 3)$ or $\beta(1 \rightarrow 6)$ linkages between glucose residues, as structural polysaccharides.

Chitin

A homopolymer of *N*-acetyl-β-D-glucosamine, chitin has a structure basically similar to that of cellulose, except that the hydroxyl on carbon 2 of each residue is replaced by an acetylated amino group.

Chitin

Cellulose and chitin are examples of structural polysaccharides. Unlike starches, which have $\alpha(1 \rightarrow 4)$ links, these fibrous polymers have $\beta(1 \rightarrow 4)$ linkages.

FIGURE 9.21

Organization of plant cell walls. Microfibrils of cellulose are embedded in a matrix of hemicellulose. Note that the fibers are laid down in a crosshatched pattern to give strength in all directions.

The World of the Cell, 4th ed., Wayne M. Becker, Lewis J. Kleinsmith, and Jeff Hardin. © 2000. Reprinted by permission of Pearson Education Inc., Upper Saddle River, N.J.; (inset) Carolina Biological Supply Co./Visuals Unlimited, Inc.

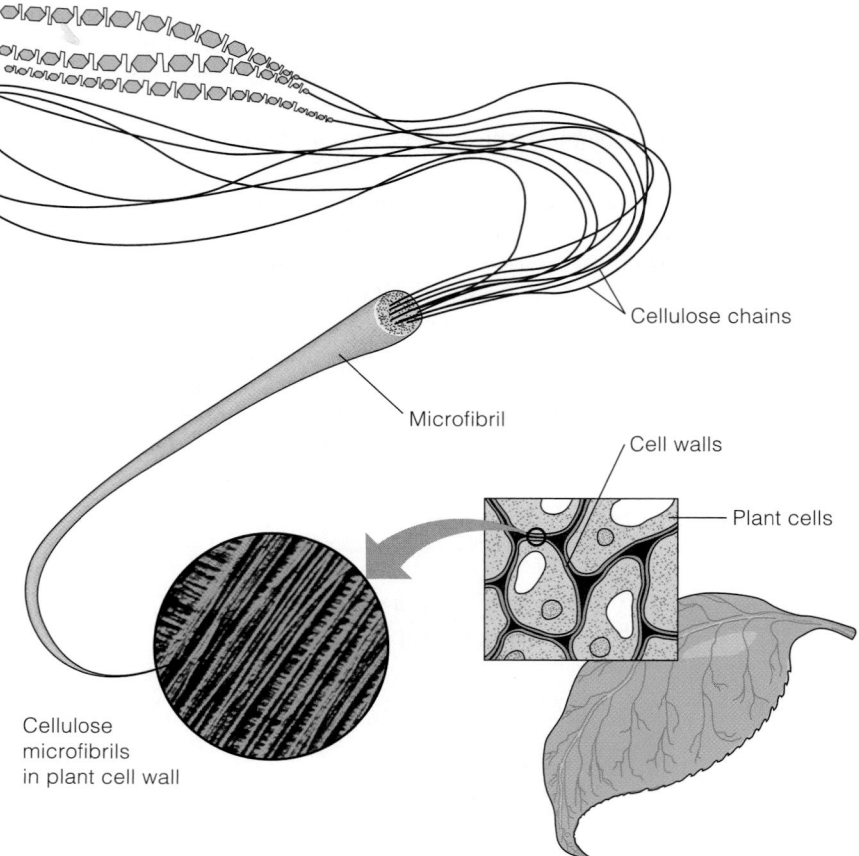

Chitin is widely distributed among all kingdoms. It is a minor constituent in most fungi and some algae, where it often substitutes for cellulose or other glucans. In dividing yeast cells, chitin is found in the septum that forms between the separating cells. The best known role of chitin, however, is in invertebrate animals; it constitutes a major structural material in the exoskeletons of many arthropods and mollusks. In many of these exoskeletons, chitin forms a matrix on which mineralization takes place, much as collagen acts as a matrix for mineral deposition in vertebrate bones. The evolutionary implications are interesting. As animals evolved to the size that made rigid body parts essential, quite different paths were taken. The ancestors of the vertebrates developed a mineral skeleton on a collagen matrix. Annelids such as earthworms also use collagen, but in a segmented exoskeleton. The arthropods and mollusks also developed exoskeletons, but theirs were built on chitin—a carbohydrate rather than a protein matrix.

Glycosaminoglycans

One group of polysaccharides is of major structural importance in vertebrate animals—the **glycosaminoglycans**, formerly called *mucopolysaccharides*. Important examples are the *chondroitin sulfates* and *keratan sulfates* of connective tissue, the *dermatan sulfates* of skin, and *hyaluronic acid*. All are polymers of repeating disaccharide units, in which one of the sugars is either *N*-acetylgalactosamine or *N*-acetylglucosamine or one of their derivatives. All are acidic, through the presence of either sulfate or carboxylate groups. Representative structures of glycosaminoglycans are shown in Figure 9.22.

The Proteoglycan Complex

A major function of the glycosaminoglycans is the formation of a matrix to hold together the protein components of skin and connective tissue. An example is given

Glycosaminoglycans are negatively charged heteropolysaccharides that serve a number of structural functions in animals.

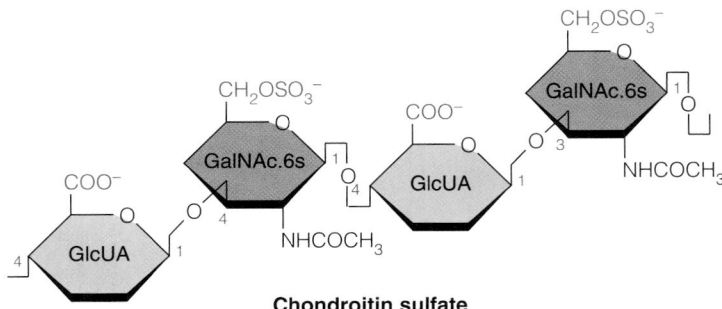

Chondroitin sulfate

FIGURE 9.22

Repeating structures of some glycosaminogly-cans. In each case, the repeating unit is a disaccha-ride, of which two are shown for each structure. Abbreviations of residues (6s means sulfonated at carbon 6) are in Table 9.4. To simplify the figure, hydrogens and nonreacted hydroxyls are not shown.

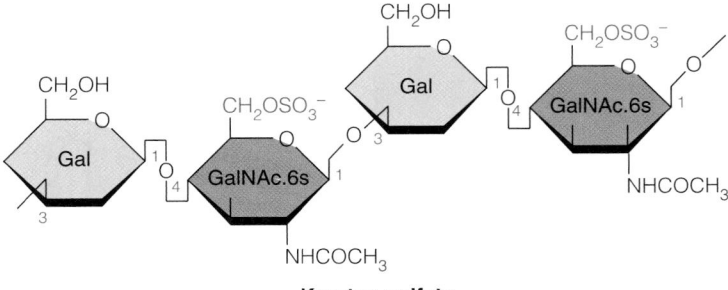

Keratan sulfate

Hyaluronic acid

in Figure 9.23, which illustrates the protein–carbohydrate, or **proteoglycan**, complex in cartilage. The filamentous structure is built on a single long hyaluronic acid mole-cule, to which extended core proteins are attached noncovalently. The core proteins, in turn, have chondroitin sulfate and keratan sulfate chains covalently bound to them through serine side chains. In cartilage, this kind of structure binds collagen (see Chapter 6) and helps hold the collagen fibers in a tight, strong network. The binding apparently involves electrostatic interactions between the sulfate and/or carboxylate groups of the proteoglycan complex and the basic side chains in collagen.

The proteoglycan complexes of connective tissue constitute one of the few examples in which the element silicon enters into biology. Some of the carbohy-drate chains are cross-linked by bridges of the type

$$R-O-\underset{\underset{OH}{|}}{\overset{\overset{OH}{|}}{Si}}-O-R'$$

where R and R′ are sugar monomers of adjacent chains. There is about one silicon atom for every 100 sugar monomers.

Nonstructural Roles of Glycosaminoglycans

Hyaluronic acid has other functions in the body besides being a structural compo-nent. The polymer is highly soluble in water and is present in the synovial fluid

FIGURE 9.23

Proteoglycan structure in bovine cartilage. **(a)** An electron micrograph of a proteoglycan aggregate. **(b)** A schematic drawing of the same structure. Keratan sulfate and chondroitin sulfate are covalently linked to extended core protein molecules. The core proteins are noncovalently attached to a long hyaluronic acid molecule with the aid of a link protein.

(a) Reprinted from *Collagen and Related Research* 3:489–504, J. A. Buckwalter and L. Rosenberg. © 1983, with permission from Elsevier.

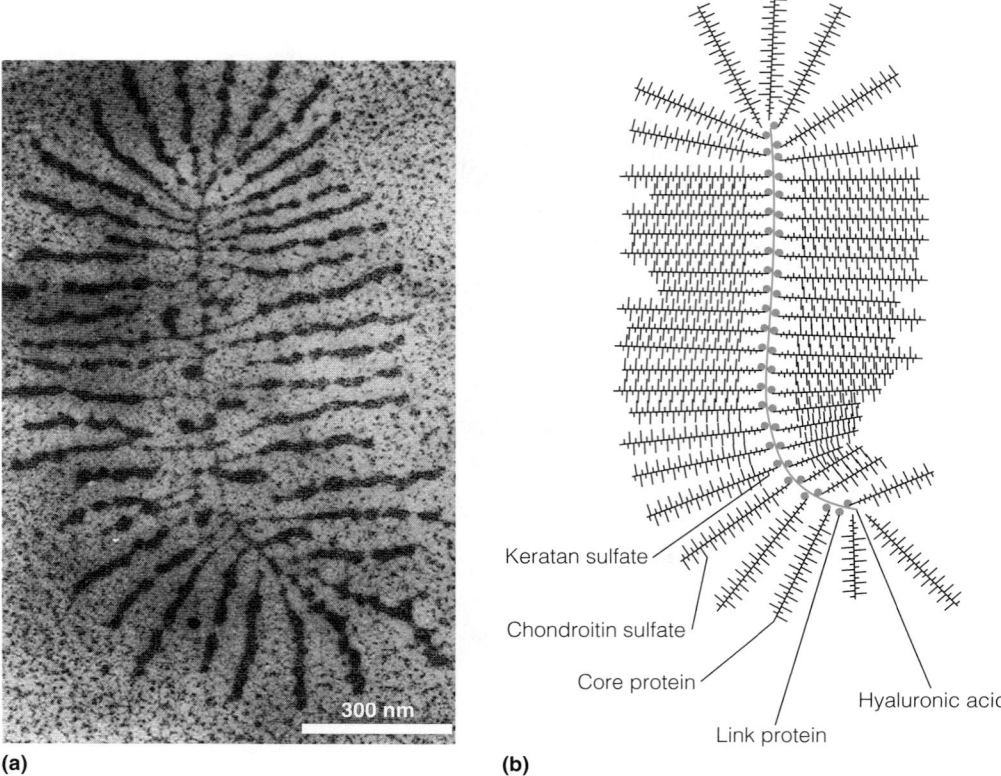

Keratan sulfate

Chondroitin sulfate

Core protein

Link protein

Hyaluronic acid

(a) **(b)**

of joints and in the vitreous humor of the eye. It appears to act as a viscosity-increasing agent or lubricating agent in these fluids, possibly related to electro-static repulsion among the many carboxylate groups in the polymer.

Another highly sulfated glycosaminoglycan is **heparin**. One repeat unit of its complex chain is shown below. Heparin appears to be a natural anticoagulant and is found in many body tissues. It binds strongly to a blood protein, antiprothrombin III, and the complex inhibits enzymes of the blood clotting process (see Chapter 11). Therefore, heparin is used medicinally to inhibit clotting in blood vessels.

Heparin

The glycosaminoglycans are interesting examples of how sugar residues can be modified to provide polymers with a wide variety of properties and functions.

Bacterial Cell Wall Polysaccharides

In Chapter 1 we noted that bacteria and most other unicellular organisms possess a *cell wall*. The nature of this cell wall is the basis for categorizing bacteria into two major classes: those that retain the Gram stain (a dye–iodine complex), which are called *Gram-positive* bacteria, and those that do not, which are termed *Gram-negative* (Figure 9.24). Gram-positive bacteria have a cell wall with a cross-linked, multilayered polysaccharide–peptide complex called **peptidoglycan** at the surface, outside the lipid cell membrane (Figure 9.24a). Gram-negative

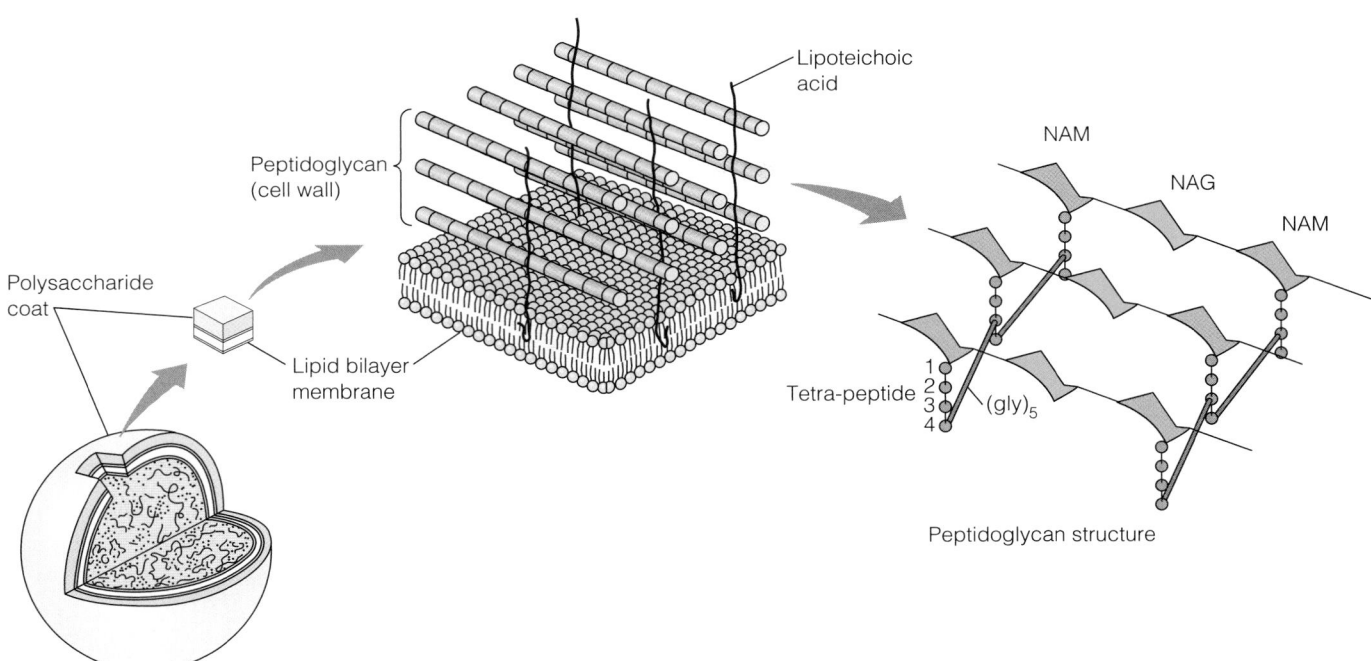

(a) Gram positive:
Staphylococcus aureus

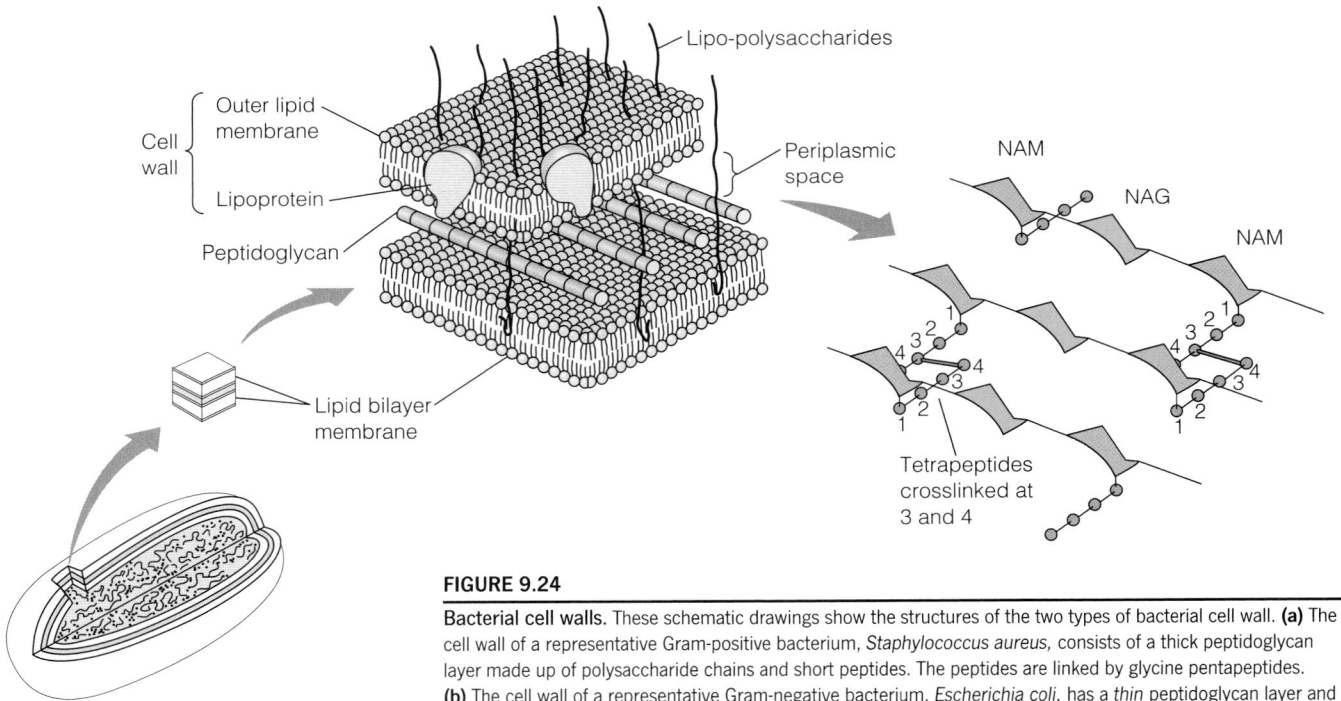

(b) Gram negative:
Escherichia coli

FIGURE 9.24

Bacterial cell walls. These schematic drawings show the structures of the two types of bacterial cell wall. **(a)** The cell wall of a representative Gram-positive bacterium, *Staphylococcus aureus,* consists of a thick peptidoglycan layer made up of polysaccharide chains and short peptides. The peptides are linked by glycine pentapeptides. **(b)** The cell wall of a representative Gram-negative bacterium, *Escherichia coli,* has a *thin* peptidoglycan layer and an outer lipid membrane. The cross-links here are between tetrapeptides attached to the *N*-acetylmuramic acid (NAM; page 323) residues in adjacent chains.

bacterial cell walls also contain peptidoglycan, but it is single-layered and covered by an outer lipid membrane layer (Figure 9.24b). This difference allows the Gram stain to be washed from Gram-negative bacteria.

The chemical structure of the peptidoglycan of a Gram-positive bacterium is shown in Figure 9.25. Long polysaccharide chains, which are strictly alternating copolymers of *N*-acetylglucosamine (NAG) and *N*-acetylmuramic acid (NAM),

> The cell walls of many bacteria are constructed of peptidoglycans, composite polymers of polysaccharides and oligopeptides.

FIGURE 9.25

The peptidoglycan layer of Gram-positive bacteria. Cross-links between the peptides are formed by pentaglycine chains between the ε-amino group of the lysine (*) on one chain and the C-terminal carboxyl group of the alanine (**) on an adjacent chain. To simplify the figure, hydrogens and nonreacted hydroxyls are not shown.

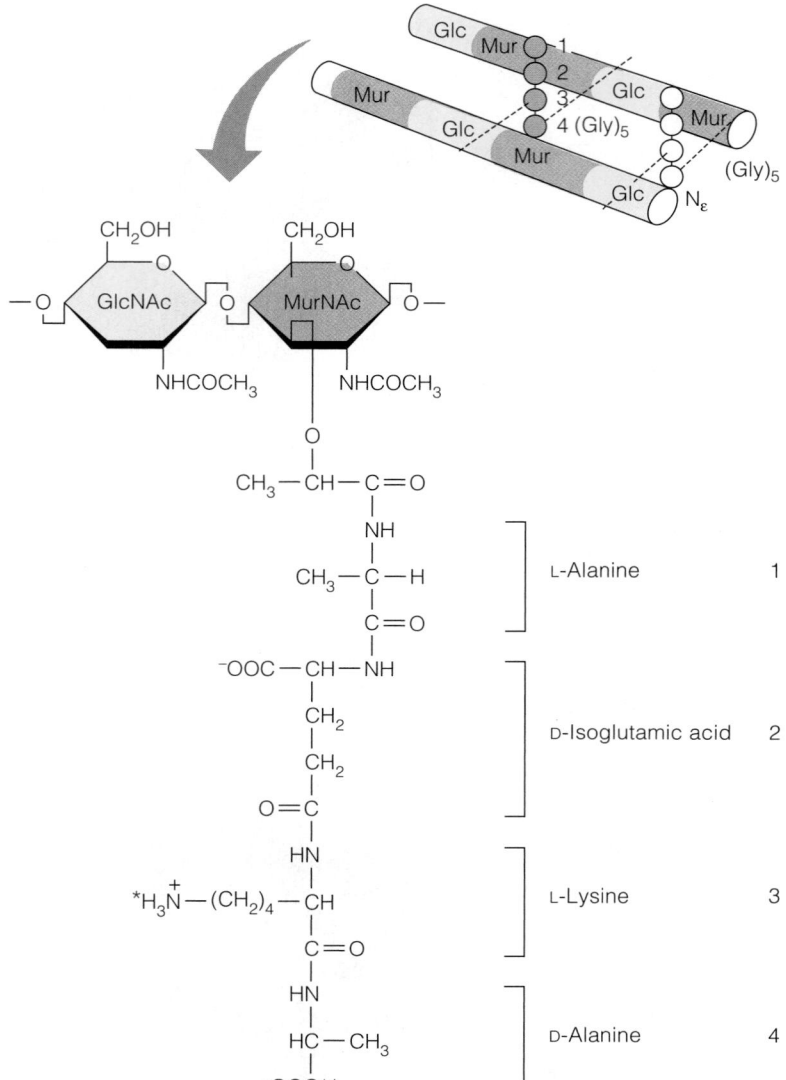

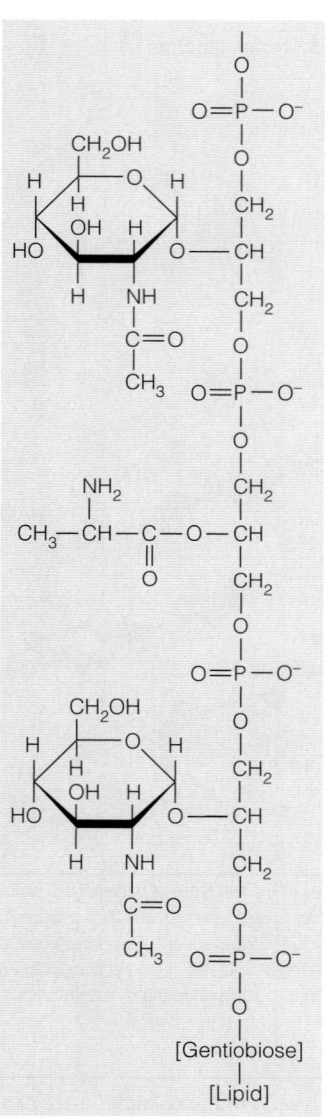

FIGURE 9.26

The structure of a lipotechoic acid. D-Alanyl and NAG groups are arranged irregularly on the chain, which is anchored in the membrane by lipid.

are cross-linked through short peptides (see Figure 9.24a). These peptides have unusual structures. Attached to the lactic acid moiety of the *N*-acetylmuramic acid is a tetrapeptide with the sequence

$$[(\text{L-Ala}) - (\text{D-Glu}) - (\text{L-Lys}) - (\text{D-Ala})]$$

This peptide is unusual in two respects: It contains some D-amino acids, and the glutamic acid residue is linked into the chain through its γ-carboxyl instead of the usual α-carboxyl linkage. To the ε-amino group of each lysine residue is attached a glycine pentapeptide, which is bonded at its other end to the terminal D-Ala residue of an adjacent chain. The result is the formation of a covalently cross-linked structure that envelops the bacterial cell. The entire cell wall can be regarded as a single enormous molecule made up of multiple layers of cross-linked peptidoglycan strands. In addition to the components mentioned above, elongated lipid–oligosaccharide complexes called **lipotechoic acids** (Figure 9.26) protrude from the membrane through the peptidoglycan wall. The cell wall protects bacteria from lysis when they are in the blood of host animals.

In the Gram-negative bacteria, the peptidoglycan layer is much thinner. Although the same basic polysaccharide structure is present, the peptide chains and their linkage are somewhat different (see Figure 9.24b).

Clearly, assembly of a structure as complex as the bacterial cell wall requires a battery of enzymes and reactions. A number of antibiotics (for example, penicillin) inhibit bacterial growth by interfering with the formation of the peptidoglycan layer. Synthesis of peptidoglycan involves chemistry quite alien to the human or animal host to a pathogenic bacterium, making cell wall biosynthesis a specific target for drug action. We return to this topic later in this chapter. However, we should note that one class of naturally occurring antibiotic substances acts not by interfering with cell wall synthesis but by attacking the peptidoglycan layer itself. These substances are the **lysozymes**, enzymes with wide distribution—they are found in bacteriophages, egg white, and human tears, for example. In egg white and tears they help maintain asepsis; in bacteriophage they help the phage to rupture an infected bacterium, releasing progeny phage at the conclusion of a lytic cycle of growth. Lysozymes (see Chapter 11 for mechanism) catalyze hydrolysis of the glycosidic links between GlcNAc and MurNAc residues in the polysaccharide. Thus, they dissolve the cell wall, resulting in osmotic rupture of the membrane and bacterial death.

Glycoproteins

More than half of all eukaryotic proteins carry covalently attached oligosaccharide or polysaccharide chains. There is an astonishing variety of these modified proteins, which are known as **glycoproteins**, and they serve multiple functions, including cell adhesion and the recognition of eggs by sperm cells.

N-Linked and O-Linked Glycoproteins

The saccharide chains, or glycans, can be linked to proteins in two major ways. *N-linked glycans* are attached, usually through *N*-acetylglucosamine, or sometimes through *N*-acetylgalactosamine, to the side chain amide group in an asparagine residue. A common sequence surrounding the asparagine is –Asn–X–Ser/Thr–, where X may be any amino acid residue. *O-linked glycans* are usually attached by an O-glycosidic bond between *N*-acetylgalactosamine and the hydroxyl group of a threonine or serine residue, although in a few cases— collagen, for example (Chapter 6)—hydroxylysine or hydroxyproline is employed.

Oligosaccharides and proteins can be linked to form glycoproteins in two ways: O-linked glycans are attached via threonine or serine hydroxyls and N-linked glycans via asparagine amino groups.

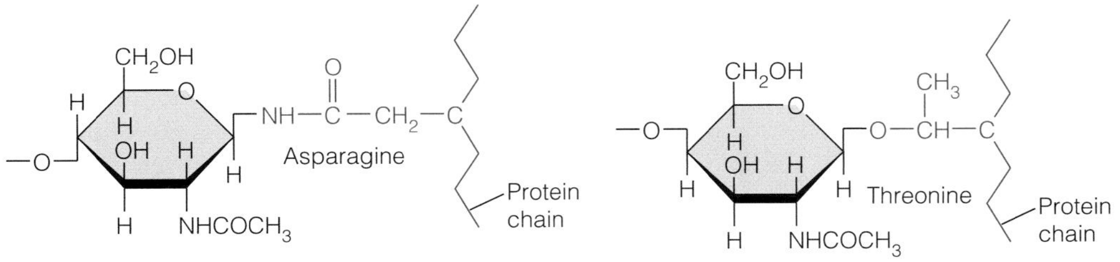

N-Acetylglucosamine   **N-Acetylgalactosamine**

N-Linked Glycans

Careful study of many glycoproteins has revealed an enormous variety of N-linked oligosaccharide side chains, often exhibiting a complex branched structure. However, a common motif is often seen. The following structure often serves as a foundation for further elaboration:

Manα(1 → 6)
          Manβ(1 → 4)GlcNAcβ(1 → 4)GlcNAcβ(1 → N)Asp
Manα(1 → 3)

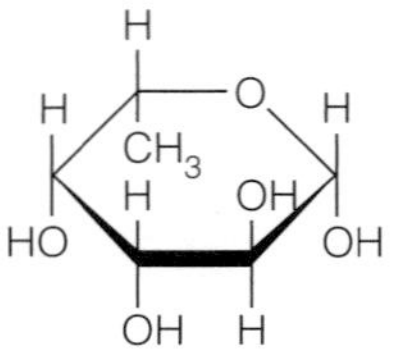

α-L-Fucose

This motif can be seen, for example, in the glycan moieties of ovalbumin in egg whites, and in the immunoglobulins. The structure of the oligosaccharide found attached to a human immunoglobulin G (IgG) is shown in simplified form below.

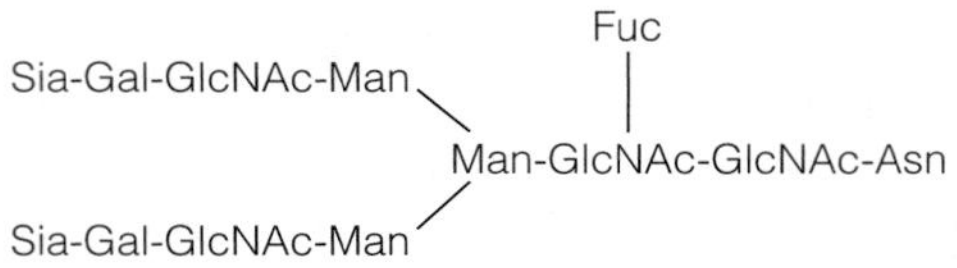

The residue denoted Fuc is α-L-fucose, a residue often attached near the protein connection of N-linked glycans. The immunoglobulins represent an important example of the informational function of the glycan chains on glycoproteins. Recall from Chapter 7 that every immunoglobulin has carbohydrate attached to the constant domain of each heavy chain. The different types of immunoglobulins must be recognized, both for proper tissue distribution and for interaction with phagocytic cells, which will destroy the antigen–immunoglobulin complex. A part, at least, of this recognition is based on differences in the oligosaccharide chains.

An important further use of N-linked oligosaccharides is in intracellular targeting in eukaryotic organisms. Later in this chapter, and also in Chapter 28, we see how proteins destined for certain organelles or for secretion from the cell are marked specifically by oligosaccharides during post-translational processing. This marking ensures that they pass appropriately through the endoplasmic reticulum, organelle membranes, the Golgi complex, and/or the plasma membrane so that each glycoprotein arrives at its proper destination.

O-Linked Glycans

Many proteins carry O-linked oligosaccharides that serve a variety of functions. Antarctic fish contain a glycoprotein that serves as an "antifreeze," preventing the freezing of body fluids even in extremely cold water. The **mucins**, glycoproteins found extensively in salivary secretions, contain many short O-linked glycans. The highly extended and highly hydrated mucins increase the viscosity of the fluids in which they are dissolved. Some O-linked glycans also function in intracellular targeting and molecular and cellular identification. An example significant to human biology is found in the *blood group antigens.*

Blood Group Antigens

An important group of oligosaccharides is the **blood group antigens**. On some cells these antigens are attached as O-linked glycans to membrane proteins. Alternatively, the oligosaccharide may be linked to a lipid molecule to form a **glycolipid** (see Chapter 10). The lipid portion of the molecule helps anchor the antigen to the outside surface of erythrocyte membranes. It is these oligosaccharides that determine the blood group types in humans. Their presence in a blood sample is detected by blood typing—determining whether antibodies to a particular antigen cause the red cells of that blood sample to clump, or agglutinate. Although the system consisting of blood types A, B, AB, and O is probably most familiar to you, it is just one of 14 genetically characterized blood group systems, with more than 100 different blood group antigens. These substances are also present in many cells and tissues other than blood, but we often focus on blood because of the widespread use of typing in establishing familial relationships and selecting blood for transfusion.

For simplicity, we will take the ABO system as an example. Figure 9.27 depicts cell surface oligosaccharides corresponding to each of these blood types. Almost all humans can produce the type O saccharide, but addition of either galactose (to make type B) or N-acetylgalactosamine (to make type A) requires an additional enzyme. Synthesis of these oligonucleotides is outlined on page 346. Some individuals possess one of these enzymes, some possess the other, and a few are heterozygous and can produce both. The heterozygous individuals have type AB blood, with both A and B oligosaccharides present on cell surfaces.

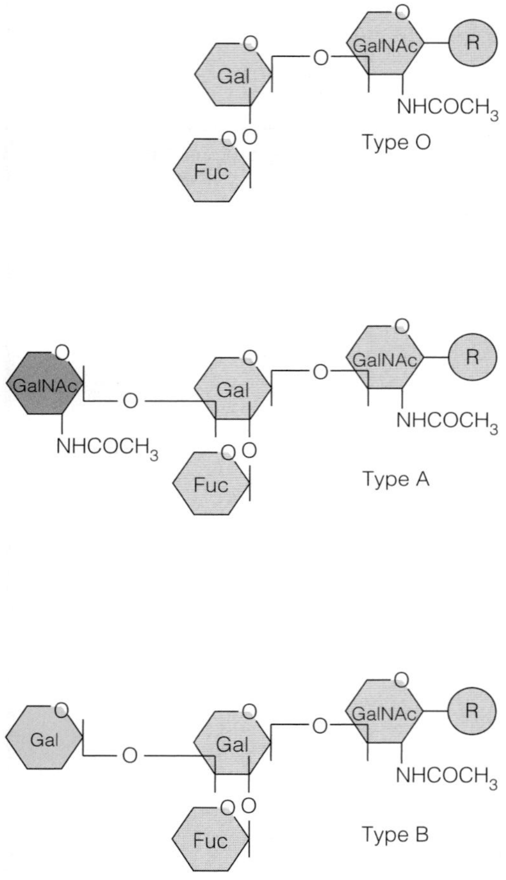

FIGURE 9.27

The ABO blood group antigens. The O oligosaccharide (top) does not elicit antibodies in most humans. The A and B antigens are formed by addition of GalNAc or Gal, respectively, to the O oligosaccharide. Each of the A and B antigens can elicit a specific antibody. In this figure, R can represent either a protein molecule or a lipid molecule.

Humans can produce antibodies against the A and B oligosaccharides, but the O type are nonantigenic. Normally, a person does not produce antibodies against his or her own antigen but does produce them against the other antigen type. Thus, an individual with type A blood carries antibodies directed against the B polysaccharide. If he or she accepts blood from a type B donor, these antibodies will cause clumping and precipitation of the donated blood cells. Nor can a type B individual safely accept type A blood. People with type O blood normally have antibodies against both A and B and thus can receive from neither. Those with AB type, because they carry both A and B antigens themselves, have antibodies against neither.

In donating blood, an inverse relationship holds. Those with type O blood, which carries no antigenic determinants, can safely donate to any other person— they are the "universal donors." Type AB individuals can donate *only* to other ABs; a person of any other type will carry antibodies to A, or B, or both. These relationships are summarized in Table 9.6.

> The blood group substances are a set of antigenic oligosaccharides attached to the surfaces of red cells.

Erythropoietin: A Glycoprotein with Both O- and N-Linked Oligosaccharides

The glycoprotein **erythropoietin** is a hormone synthesized in kidney, which stimulates the production of red blood cells. The hormone, often called EPO, is a 165-residue polypeptide with N-linked oligosaccharides (13-mers) at Asn 24, 38, and 83 and an O-linked trisaccharide at Ser-126. The carbohydrate stabilizes EPO within the blood, preventing rapid removal by the kidney. Anemia—low red blood cell count—is a common side effect of cancer chemotherapy, and EPO is often administered along with anticancer drugs to counteract this effect. EPO is also subject to misuse, particularly by athletes who wish to improve their aerobic status. Recombinant EPO, which is usually used for this purpose, can be identified by drug-testing laboratories on the basis of its abnormal glycosylation pattern.

Influenza Neuraminidase, a Target for Antiviral Drugs

The concept that virus infection induces enzymes that carry out new metabolic pathways within infected cells was established in the 1950s and 1960s, through studies with bacterial viruses, or bacteriophages (Chapter 22, page 950). However, long before that it was known that the influenza virus, an RNA virus, carries on its surface a virus-coded enzyme, **neuraminidase**. As shown in Figure 9.28, the spherical virus particle has two kinds of spikes on its exterior surface. One such spike, made up of a protein called *hemagglutinin*, binds to *N*-acetylneuraminic acid (also called sialic acid, see page 323). The virus attaches to host cells through binding of the hemagglutinin to sialic acid residues in cell surface glycoproteins or glycolipids. At the conclusion of the virus infection cycle, release of virus from infected cells requires cleavage of the sialic acid from the rest of each oligosaccharide chain, and this is carried out by the viral neuraminidase. The crystal structure of the neuraminidase complex with sialic acid was solved in the 1980s (see Figure 7.43, page 274), leading to synthesis of sialic acid analogs that might inhibit the enzyme, and, hence, block release of virus particles from infected cells. Two such analogs, *zanamivir* and *oseltamivir*, are shown in Figure 9.29, along with the partial structure of the neuraminidase. Oseltamivir, marketed as *Tamiflu*, assumed great importance during the H1N1 influenza epidemic

TABLE 9.6 Transfusion relationships among ABO blood types			
Person Has Blood Type:	Makes Antibodies Against:	Can Safely Receive Blood from:	Can Safely Donate Blood to:
O	A, B	O	O, A, B, AB
A	B	O, A	A, AB
B	A	O, B	B, AB
AB	None	O, A, B, AB[a]	AB

[a]In principle, this relationship is true. However, ABs are never given donations from other types because the donor's antibodies could react with the recipient's antigens.

FIGURE 9.28

The structure of the influenza virus. The 13,600-nucleotide RNA genome is packaged within the sphere, about 120 nm in diameter. The spikes on the virion exterior include the hemagglutinin molecule and a spike that terminates in four neuraminidase molecules.

CDC/Science Photo Library.

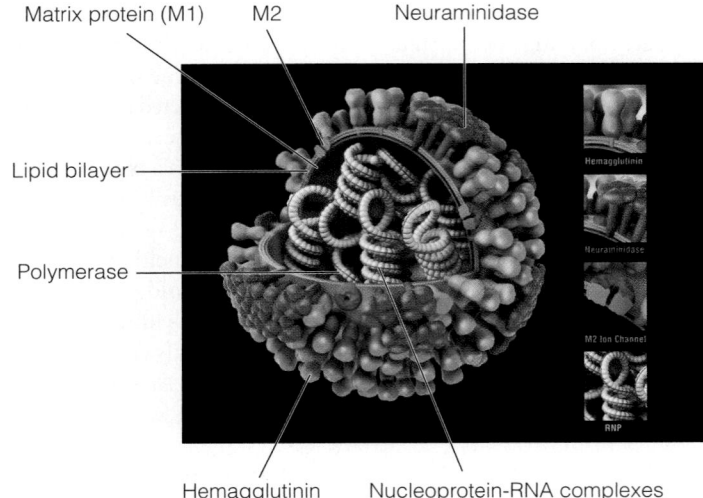

Matrix protein (M1)   M2   Neuraminidase

Lipid bilayer

Polymerase

Hemagglutinin   Nucleoprotein-RNA complexes

FIGURE 9.29

Rational design of neuraminidase inhibitors.
(a) Structures of sialic acid, zanamivir, and oseltamivir. **(b)** Partial model of the neuraminidase-zanamivir complex, showing amino acid residues that are close to the binding site for the inhibitor.

Fundamentals of Molecular Virology, Nicholas H. Acheson. © 2007 John Wiley & Sons, Inc. Reproduced with permission from John Wiley & Sons, Inc.

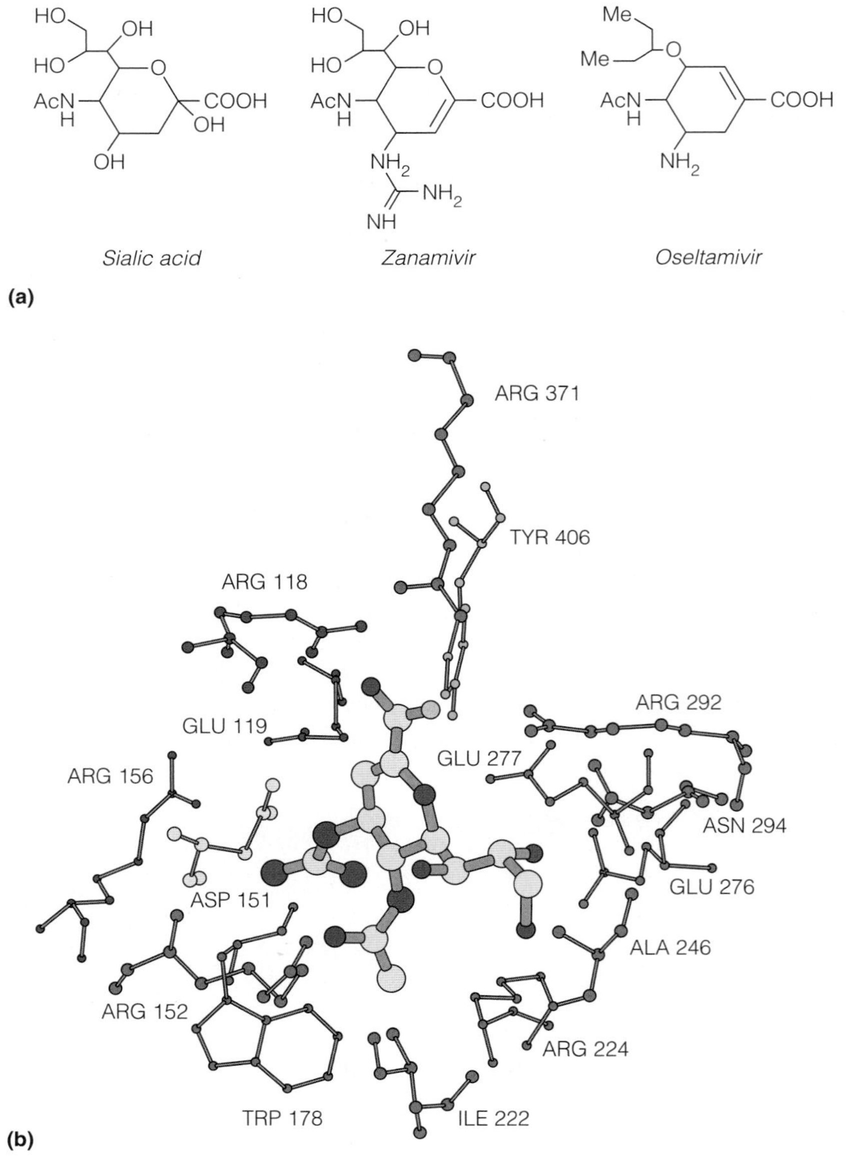

Sialic acid   *Zanamivir*   *Oseltamivir*

(a)

ARG 371

TYR 406

ARG 118

GLU 119

ARG 292

GLU 277

ARG 156

ASN 294

ASP 151

GLU 276

ALA 246

ARG 152

ARG 224

TRP 178   ILE 222

(b)

of 2009, before an effective vaccine had become available. Tamiflu acts, as predicted, by blocking release of newly formed virus particles from infected cells, but it must be administered very soon after the onset of flu symptoms in order to be effective.

Oligosaccharides as Cell Markers

Biologists realize that molecules such as the blood group antigens represent only a special case of a much more general phenomenon—cell marking by glycans. In a multicellular organism, it is essential that different kinds of cells be marked on their surfaces so that they can interact properly with other cells and molecules and so that an organism can recognize its own cells as immunologically distinct from foreign cells. In accord with this view is the growing appreciation that the surfaces of many cells are nearly covered with polysaccharides, which are attached to either proteins or lipids in the cell membrane (Figure 9.30a). Some animal cells have an extremely thick coating of polysaccharides called a **glycocalyx** (literally "sugar coat"); Figure 9.30b shows the glycocalyx of an intestinal cell. Glycocalyx oligosaccharides interact with other substances: with bacteria in the intestine and with collagen of the intercellular matrix in some other tissues.

For glycans to serve as recognition signals there must be proteins that bind to them specifically. One such class is the immunoglobulins. Another, very diverse group of saccharide-binding proteins is the **lectins**. The lectins were first recognized in plant tissues, where they appear to play defensive roles and to aid in adhering nitrogen-fixing bacteria to roots. We now know that lectins are widely distributed and play a great variety of roles in animals as well. For example, lectins seem to be involved in interactions between cells and proteins of the intercellular matrix, such as collagen, and help to maintain tissue and organ structure. Lectins in the walls of intestinal bacteria help to bond the bacteria to the glycocalyx of the intestinal epithelium. The fact that cell surface polysaccharides are important in determining cell–cell interactions (including both adhesion and avoidance) has much medical significance. For example, we know that the polysaccharides on the surfaces of many cancer cells are abnormal. This may account in part for the loss in tissue specificity that such cells commonly exhibit.

In animals, a related class of proteins is the **selectins**, which are mobilized as part of an inflammatory response. One such protein, *P-selectin*, is stored in platelets and epithelial cells. Early events in the inflammatory process release this protein from intracellular stores to the cell surface. There the carbohydrate moiety of P-selectin interacts with carbohydrates on the surface of leukocytes, a class of white blood cells. This slows movement of the leukocyte through the blood vessel, allowing it to leave the blood vessel and enter the site of an infection.

Why do oligosaccharides so often play the role of cellular markers? We don't know, but certain possibilities suggest themselves. First, oligosaccharides can

Many cells carry a complex layer of polysaccharides—the glycocalyx—on their surfaces.

FIGURE 9.30

Cell surface recognition factors. **(a)** Schematic view of a lipid membrane. Oligosaccharides are attached to the outer surface, through either membrane-embedded proteins or special lipid molecules. **(b)** Electron micrograph of the surface of an intestinal epithelial cell. The cellular projections, called microvilli, are covered on their outer surface by a layer of branched polysaccharide chains attached to proteins in the cell membrane. This carbohydrate layer, called the glycocalyx, is found on many animal cell surfaces.

(b) Steve Gschmeissner/Science Photo Library.

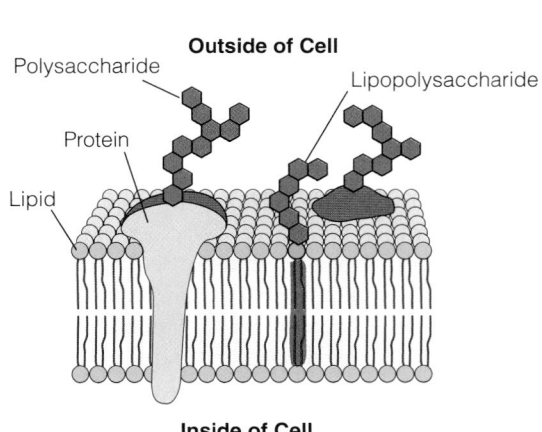

(a) Cell surface oligosaccharides

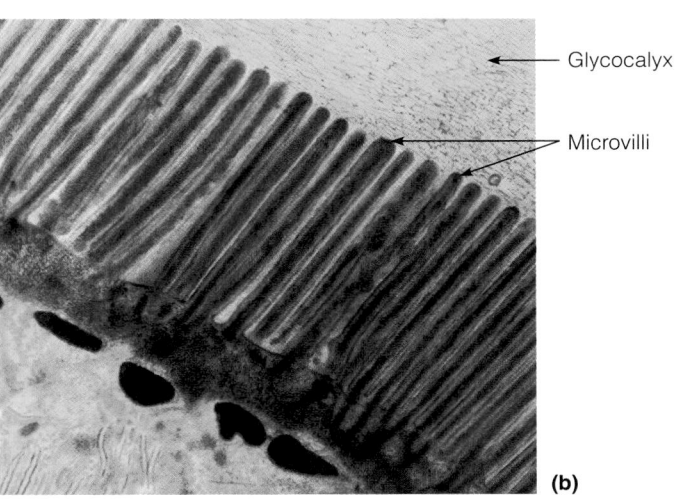

(b)

FIGURE 9.31

Imaging cell surface glycoproteins in living cells. Zebrafish embryos were allowed to incorporate an azido derivative of GalNAc and at fixed intervals the embryos were treated with a fluorescent reagent that couples to the azido group in the incorporated analog. In the example shown the embryos were treated with reagents of three different colors, each for a specific time interval, as described in the text. Scale, 100 μm.

From *Science* 320:664–667, S. T. Laughlin, J. M. Baskin, S. L. Amacher, and C. R. Bertozzi, In vivo imaging of membrane-associated glycans in developing zebrafish. © 2008. Reprinted with permission from AAAS.

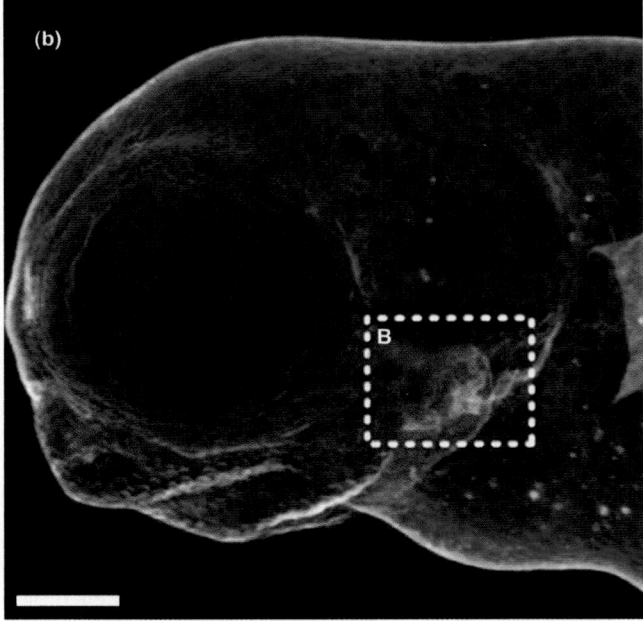

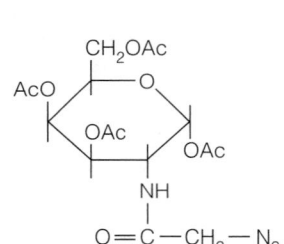

Peracetyl *N*-azidoacetylgalactosamine

present a seemingly limitless variety of structures in relatively short chains. The multiple choices of monomers (including modified sugars), linkages, and branching patterns allow a vast but specific vocabulary. Second, oligosaccharides are especially potent antigens, which means that specific antibodies can be elicited swiftly against them (see Chapter 7). Whether this interaction is the result of some intrinsic property of sugar molecules or of the antibody molecules is unclear. It is possible that antibodies evolved as a defense against bacteria, which have polysaccharide-rich walls, and thus have always favored glycans as targets.

With the realization that cell-surface glycoproteins are highly selective cell markers has come the desirability of imaging these markers in living cells, in much the same way that green fluorescent protein is used to image proteins in living cells (Chapter 6). An example of this emerging technology is shown in Figure 9.31. In this study, from the laboratory of Carolyn Bertozzi, living zebrafish embryos were incubated in the presence of peracetylated *N*-azidoacetylgalactosamine (Ac$_4$GalNAz), which is readily incorporated into mucin-type O-linked glycoproteins. At intervals after fertilization the embryos were treated with fluorescent reagents of different colors, each of which could react covalently with the azido group in the cell surface glycoproteins. In the figure shown, the intervals of fluorescent reagent treatment were 60–61, 62–63, and 72–73 hours post-fertilization, each period involving labeling with a dye of a different color. This experiment shows distinct spatial patterns of cell surface glycoprotein synthesis at different times during development of a live, developmentally normal embryo, thereby setting the stage for mechanistic analysis of glycoprotein metabolism in embryogenesis.

Biosynthesis of Glycoconjugates: Amino Sugars

A full appreciation of the excitement of contemporary *glycobiology* requires some understanding of the routes for biosynthesis of these diverse metabolites. Although enzymes and metabolic processes are dealt with in more detail in later chapters, we present some of the most significant processes here, with the realization that full appreciation of this material may come only after you have considered enzymes in Chapter 11 and carbohydrate metabolism in Chapter 13.

Because amino sugars such as glucosamine are major constituents of glycoconjugates, we first deal with their metabolic origins. Simple sugars, such as glucose and fructose, arise chiefly through photosynthesis, as was outlined in Figure 9.2. Sugar phosphates arise as intermediates in glycolysis, gluconeogenesis, and the pentose phosphate pathway, all of which are presented in Chapter 13. All amino sugars are derived metabolically from **glucosamine-6-phosphate**, which arises in turn from **fructose-6-phosphate** through an **amidotransferase** reaction, as shown below.

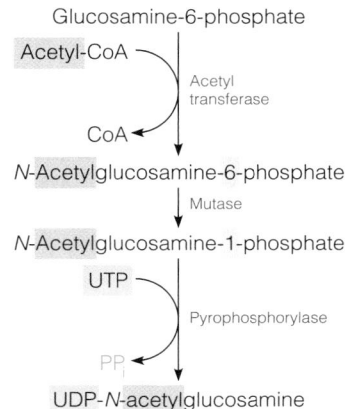

| Fructose-6-phosphate | Glutamine | Glucosamine-6-phosphate | Glutamate |

In this reaction the amide nitrogen of glutamine is transferred to C-2 of fructose-6-phosphate, with the glutamine being converted to glutamate; also the sugar undergoes an internal oxidoreduction, with the carbonyl moiety moving from C2 to C1. Several amidotransferases are involved in nucleotide metabolism and are discussed more fully in Chapter 22.

The further metabolism of glucosamine-6-phosphate involves acetylation of the amino group by **acetyl-coenzyme A**, or acetyl-CoA, as indicated in Figure 9.32. Acetyl-CoA is an active acetylating species, formed largely in mitochondria from oxidation of fatty acids and pyruvate (Chapter 12). The acetyl group is transferred to the nitrogen on glucosamine-6-phosphate. A **mutase** then isomerizes the sugar phosphate by transferring the phosphate from C6 to C1. Finally, a reaction with uridine triphosphate, like that shown on page 328, generates a nucleotide-linked sugar, prepared for further biosynthetic reactions.

The pathway outlined in simplified form in Figure 9.33 leads in five steps from UDP-N-acetylglucosamine to sialic acid, but in the process it generates several other intermediates in glycoconjugate synthesis. The **epimerase** enzyme in the first reaction inverts the sugar configuration at C4. Next, the uridine nucleotide is removed hydrolytically, and this is followed by an ATP-dependent phosphorylation at C6 coupled with another epimerization at C4. A complex reaction follows, in which the three carbons of **phosphoenolpyruvate** are incorporated into the sugar ring, giving a nine-carbon sugar phosphate. Hydrolytic removal of the phosphate gives N-acetylneuraminic acid, or sialic acid. Phosphoenolpyruvate, an intermediate in glycolysis (Chapter 13), was introduced in Chapter 3 as an ultra-high-energy compound.

As indicated earlier, the biosynthetic pathways leading to glycoconjugates involve glycosyltransferase reactions, which use a sugar activated for transfer to a growing glycan chain by formation of a nucleotide-linked sugar. Usually that nucleotide is UDP or another nucleoside diphosphate. Sialic acid presents an exception, in that the activating nucleotide is *cytidine* triphosphate and the activated product is an activated nucleoside *monophosphate* sugar.

$$CTP + \text{sialic acid} \rightarrow \text{CMP-sialic acid} + PP_i$$

For reasons unknown, this reaction occurs in the nucleus of animal cells, while all other nucleotide-linked sugars are synthesized in the cytosol.

Glycoconjugates of Interest

As noted throughout this chapter, the vast structural diversity possible among carbohydrates allows them to be used for both intracellular and extracellular molecular recognition in glycoproteins, even when present in very small amounts. The carbohydrate components of cell surface glycoproteins direct the interaction of the cells with other cells and with their environment. Such interactions involve processes as diverse as cell movements in development, cell adhesion, cell motility, cell growth control, oncogenic transformation, and **endocytosis** (internalization of material from the extracellular milieu—see Chapter 17).

FIGURE 9.32

Biosynthesis of UDP-N-acetylglucosamine from glucosamine-6-phosphate.

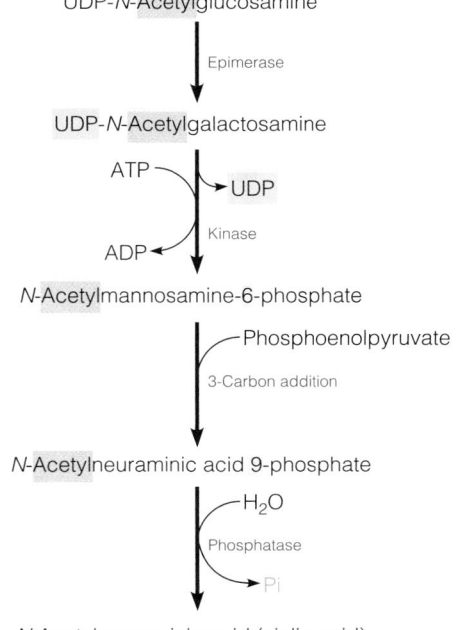

FIGURE 9.33

Biosynthesis of N-acetylneuraminic acid (sialic acid) from UDP-N-acetylglucosamine.

In this section we consider the biosynthesis of the carbohydrate components of some selected glycoconjugates—the glycan components of glycoproteins and the peptidoglycan and O-antigen portions of bacterial cell walls. The examples chosen have general biological interest, and they illustrate mechanisms used in synthesizing the vast known array of glycoconjugates. In general, as mentioned earlier, these mechanisms involve the action of specific glycosyltransferases that transfer a monosaccharide unit from a nucleotide-linked sugar to the nonreducing end of an oligosaccharide chain or to an appropriate functional group on the protein component. The glycosyltransferases are bound to membranes of the smooth or rough endoplasmic reticulum or the Golgi apparatus.

A particularly fascinating part of glycoprotein synthesis is the protein sorting, or traffic, that occurs as oligosaccharide chains are growing. In some cases the structure of the oligosaccharide chain or chains on a protein serves as a molecular recognition determinant, directing that protein to the proper intracellular location for the next step in oligosaccharide synthesis and, ultimately, to the site where the mature protein will reside. We introduce this topic here and return to it in Chapter 28.

O-Linked Oligosaccharides: Blood Group Antigens

Recall from page 340 that the blood group antigens are oligosaccharides bound to cell-surface proteins or lipids. As shown in Figure 9.34, the biosynthetic pathway begins with transfer of *N*-acetylgalactosamine from UDP-GalNAc to a serine or threonine hydroxyl group on a protein acceptor (R in the figure; the receptor can also be a lipid). Three successive glycosyltransferase reactions follow, leading to

> The presence or absence of one or two specific glycosyltransferases involved in O-linked oligosaccharide synthesis determines blood type A, B, AB, or O.

FIGURE 9.34

Biosynthesis of O-linked oligosaccharide units on glycoproteins of the O, A, and B blood group substances.

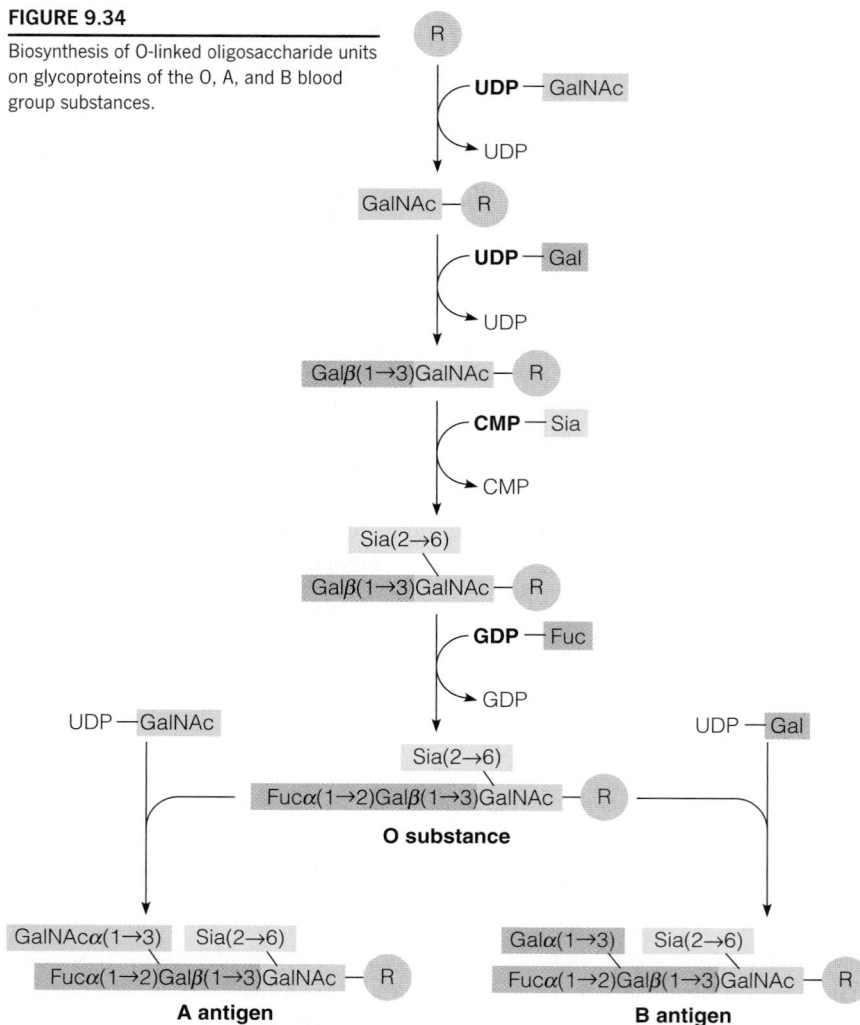

the tetrasaccharide O substance (not an antigen because it is not immunogenic). On the lower left of the figure, a glycosyltransferase transfers GalNAc from UDP-GalNAc to position 3 of the galactose residue in the O tetrasaccharide, giving the pentasaccharide A antigen. This is the enzyme missing in type B or type O individuals. Similarly, the enzyme shown at the lower right, which transfers Gal from UDP-galactose to the O substance, generates the B antigen, and this enzyme is missing in type A or O individuals. See page 340 for further details.

N-Linked Oligosaccharides: Glycoproteins

Quite a different process is involved in the synthesis of N-linked glycans, common to most glycoproteins. Here, oligosaccharide assembly occurs not on the polypeptide chain but on a *lipid-linked intermediate*. A precursor oligosaccharide is then transferred to a polypeptide chain, which is itself still in the midst of being synthesized. This type of reaction is called **cotranslational**. Finally, the transferred oligosaccharide is subject to further processing steps as it passes from the rough and smooth endoplasmic reticulum through the Golgi apparatus.

The many N-linked glycoproteins of known structure can be categorized according to three basic oligosaccharide structures, as summarized in Figure 9.35: **complex, hybrid**, and **high-mannose**. All of the known N-linked glycans have a common core pentasaccharide structure (the boxed areas in the figure; see also page 339).

That core is assembled as part of a larger oligosaccharide intermediate linked to the lipid compound **dolichol phosphate**. Dolichol phosphate is an *isoprenoid* compound, related metabolically to cholesterol (Chapter 10).

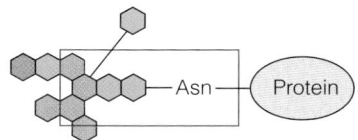

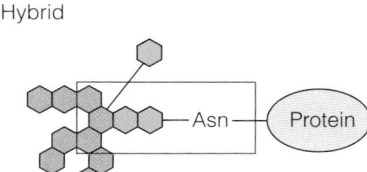

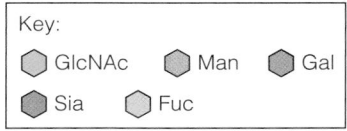

FIGURE 9.35

Structures of the major types of asparagine-linked (N-linked) oligosaccharides. The red boxes contain the pentasaccharide core common to all known N-linked structures. Here and in subsequent figures the carbohydrate rings are shown "fused" for simplicity.

The polysaccharide chains of N-linked glycoproteins are formed while linked to a lipid compound, dolichol phosphate.

Dolichol phosphate

In vertebrate tissues, dolichol contains 18 to 20 branched C_5 (isoprenoid) units with two *trans* double bonds, the remainder being *cis*, except for a saturated terminal isoprene unit. Dolichol is synthesized from the same pathway that yields cholesterol, other sterols, and other isoprenoid compounds (see Chapter 19) and is then phosphorylated.

Modifications of the carbohydrate chains of N-linked glycoproteins help target these proteins to their intra- or extracellular destinations.

Synthesis of the Lipid-Linked Intermediate

The first step in glycoprotein synthesis is assembly of a lipid-linked oligosaccharide intermediate, which serves as precursor to all known N-linked oligosaccharides. That process takes place in the endoplasmic reticulum (ER). Figure 9.36 summarizes the biosynthetic pathway leading to this intermediate. The first seven sugars are transferred to dolichol phosphate from nucleoside diphosphate sugars, UDP-GlcNAc and GDP-Man. Each reaction is catalyzed by a separate glycosyltransferase. The first of these enzymes is specifically inhibited by the antibiotic *tunicamycin*, which blocks the reaction of UDP-GlcNAc and Dol-P and hence inhibits the synthesis of all N-linked glycoproteins.

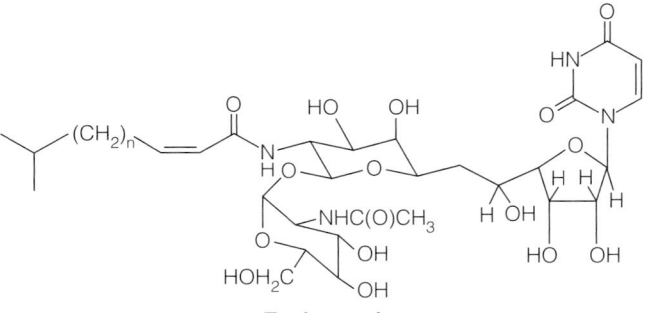

Tunicamycin

FIGURE 9.36

Biosynthesis of the lipid-linked oligosaccharide intermediate. The five sequential mannosyl transfer reactions from GDP-mannose are catalyzed by separate glycosyltransferases, as are the four mannosyl transfers from dolichol phosphate mannose. The latter is synthesized in turn from GDP-mannose. The acceptor site on the polypeptide chain is an asparagine residue two positions to the N side of a serine or threonine. The whole process occurs in the endoplasmic reticulum, with translocation to the lumen of the ER occurring after transfer of the fifth mannose residue.

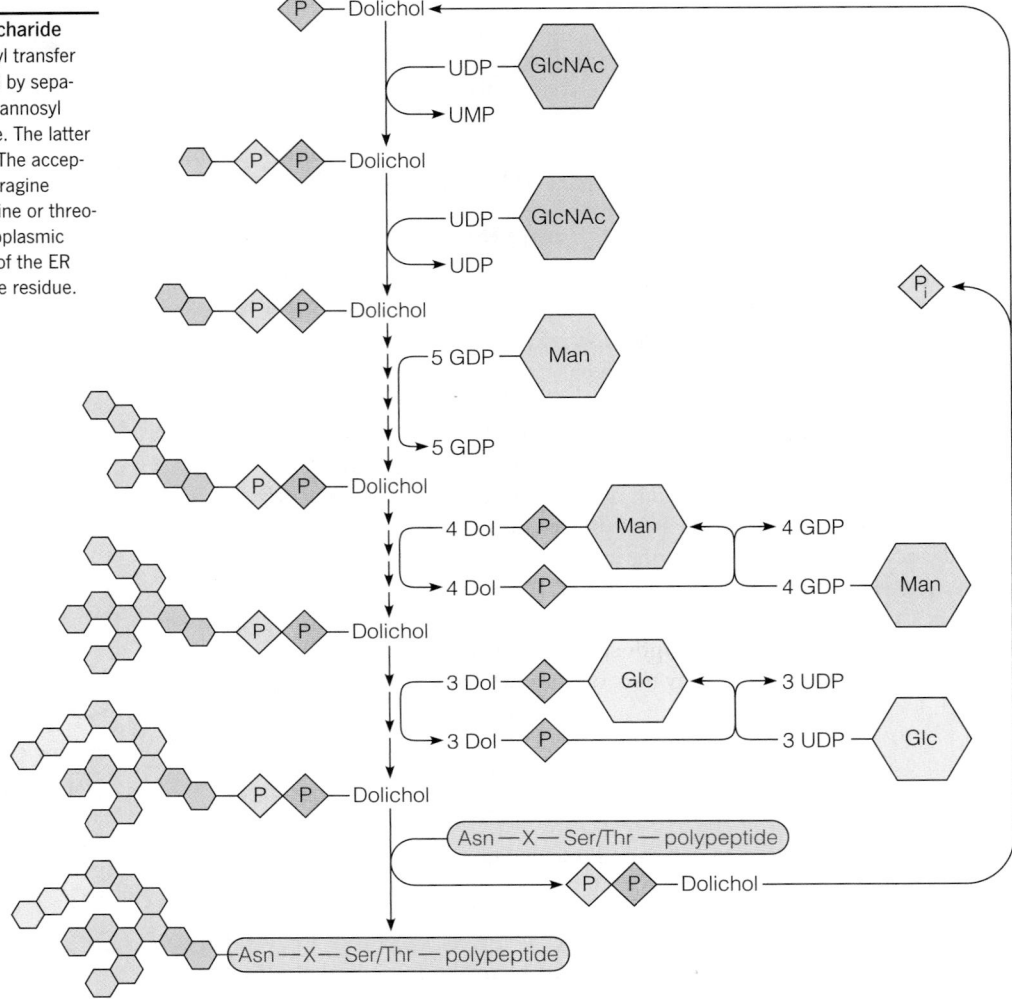

The next seven glycosyltransferases utilize as substrates not nucleotide-linked sugars but *dolichol-linked* sugars, Dol-P-Man and Dol-P-Glc. These in turn are synthesized from Dol-P plus GDP-Man or UDP-Glc, respectively. During this stage the lipid-linked intermediate ($Man_5GlcNAc_2$-P-P-Dol, the third intermediate in Figure 9.36) is somehow translocated from the exterior surface of the ER membrane to the luminal, or interior, side. This translocation explains the need for dolichol phosphate–linked sugars for the subsequent glycosylations because the nucleoside diphosphate–linked sugars cannot penetrate to the luminal side of the membrane.

The next step, transfer of the oligosaccharide unit to a polypeptide acceptor, also occurs in the lumen, catalyzed by a specific **oligosaccharyltransferase**. The acceptor site is an asparagine residue in the sequence Asn–X–Ser/Thr, where X is any amino acid. The acceptor site must be accessible in a loop or a bend in the polypeptide chain, which is probably why the transfer occurs simultaneously with translation of the acceptor polypeptide. The three glucosyl residues on the transferred oligosaccharyl unit somehow facilitate the transfer but are not absolutely required. For virtually all glycoproteins, these glucosyl residues are removed in subsequent processing steps.

Processing of the Oligosaccharides

Processing of the oligosaccharide-linked polypeptides begins in the lumen of the rough endoplasmic reticulum (ER) and continues as the nascent glycoprotein moves into the smooth ER and ultimately through various cisternae of the Golgi apparatus. A great variety of oligosaccharide structures is generated during this

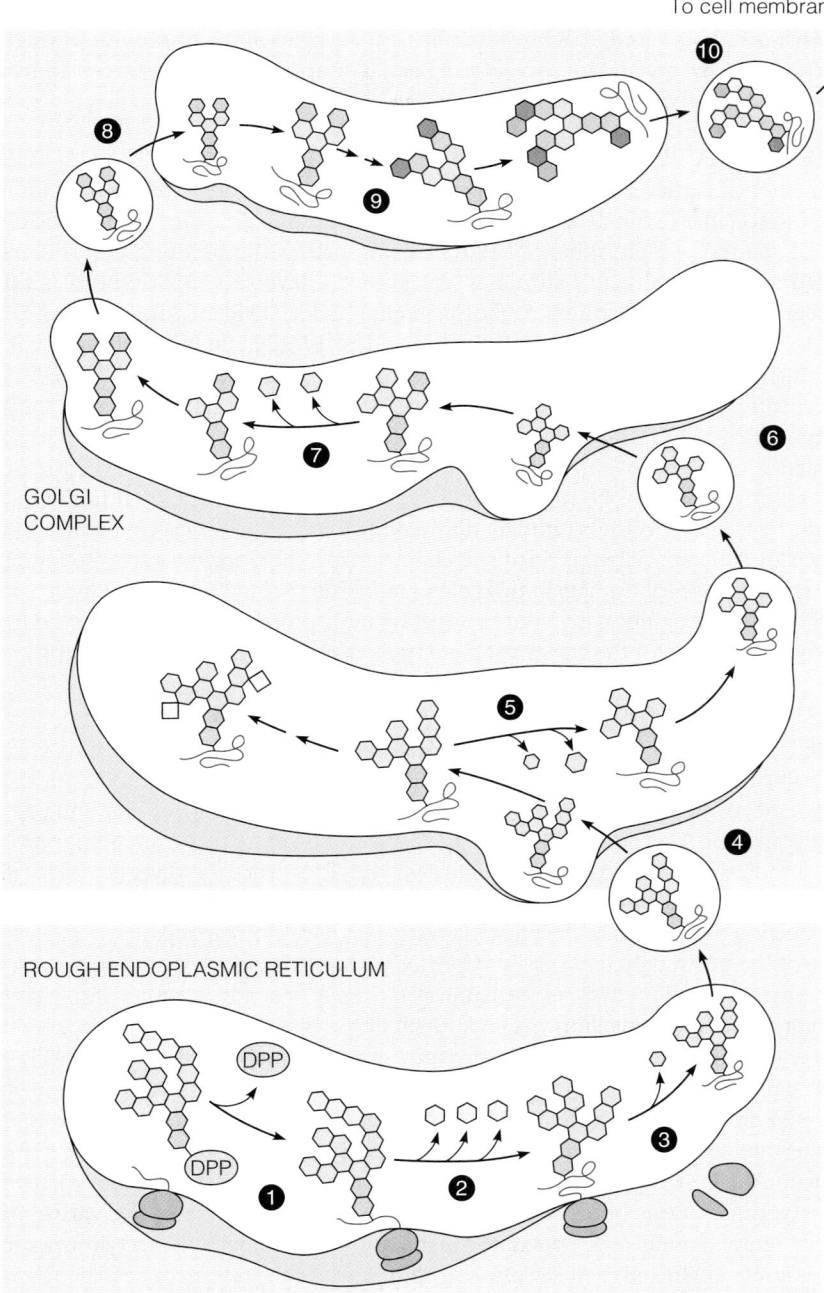

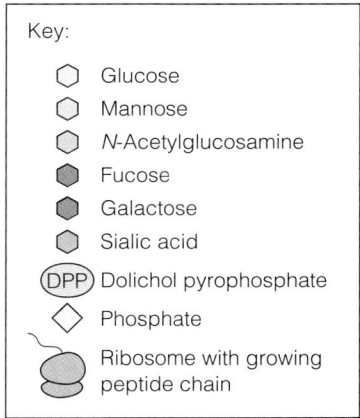

Key:
- ⬡ Glucose
- ⬡ Mannose
- ⬡ *N*-Acetylglucosamine
- ⬡ Fucose
- ⬡ Galactose
- ⬡ Sialic acid
- (DPP) Dolichol pyrophosphate
- ◇ Phosphate
- Ribosome with growing peptide chain

FIGURE 9.37

Schematic pathway of oligosaccharide processing on newly synthesized glycoproteins. Starting at the bottom of the figure, the oligosaccharide is first transferred from its dolichol phosphate carrier to a polypeptide chain (magenta) while the latter is still being synthesized on the ribosome (step 1). As the peptide chain grows within the endoplasmic reticulum, monosaccharides may be cleaved from the nonreducing ends of the oligosaccharide (step 2). After completion of the polypeptide chain synthesis (step 3), the nascent glycoprotein is carried in a transport vesicle (step 4) to the Golgi apparatus, where further modification of the oligosaccharide occurs. New monosaccharides may be added and others removed in a multistep process involving several transfers to different parts of the Golgi (steps 5–9). The completed glycoprotein is transported to its ultimate destination in a membrane (step 10) or is secreted.

processing. Part of the diversity arises from differences in conformation of the protein moiety in the vicinity of a carbohydrate chain, which affect the accessibility of the chain to glycosyltransferases and **glycosidases** (enzymes that hydrolytically remove sugars). Oligosaccharide chains probably generate recognition sites for targeting each processed compound to different sites both inside and later outside the Golgi, where specific membrane-bound processing enzymes are found (Figure 9.37). In virtually all cases, processing begins (after transfer to a polypeptide chain) with removal of the three glucosyl residues in the rough ER, followed by removal of some of the mannosyl residues in the Golgi apparatus. These processes are critical to the proper folding of some proteins in the ER. Those glycoproteins destined to be of the "complex" type are further processed by addition of *N*-acetylglucosamine, followed by further trimming of mannosyl residues. Fucosyl, galactosyl, and sialyl residues are added from appropriate nucleotide-linked sugars by specific glycosyltransferases. Comparable pathways lead to the other classes of glycoproteins.

Processing and Intracellular Protein Traffic

As mentioned earlier, oligosaccharide chains help direct glycoproteins to their ultimate intracellular destinations. A graphic demonstration is provided by the generation of mannose-6-phosphate residues during processing of the glycoproteins known as lysosomal acid hydrolases. All known enzymes of this type contain between one and five mannose-6-phosphate units, which evidently help target proteins to, and through, the lysosomal membrane. The role of mannose-6-phosphate in directing protein traffic is confirmed by the existence of a rare and fatal congenital abnormality called **I-cell disease.** In this condition, lysosomes cannot perform their normal function of intracellular digestion because a deficiency of the first glycosyltransferase that generates a mannose-6-phosphate residue causes the acid hydrolase content of lysosomes to be very low. Lysosomal enzymes, manufactured in the ER, cannot be targeted to their ultimate destinations but instead are secreted into the extracellular milieu. When purified and characterized, these enzymes are also found to lack mannose-6-phosphate in their oligosaccharide chains. Our understanding of the molecular recognition involved in targeting these proteins has been advanced by cloning of the gene for the lysosomal membrane mannose-6-phosphate receptor.

Microbial Cell Wall Polysaccharides: Peptidoglycan

Recall from Figure 9.24a that Gram-positive bacteria contain a rigid peptidoglycan cell wall surrounding the cytoplasmic membrane. Gram-negative bacteria, on the other hand, contain a third layer, in addition to the cytoplasmic membrane and peptidoglycan layer. This outer membrane is a complex structure that contains lipoproteins and lipopolysaccharides (see Figure 9.24b). The structural diversity of these macromolecules among different bacterial species is enormous, and space permits introduction of only two of the most interesting pathways. The first pathway, biosynthesis of the peptidoglycan layer in *Staphylococcus aureus,* is of interest for two reasons. First, as mentioned above, the pathway is the site of action of several important antibiotics, notably penicillins. Second, much of the biosynthetic process occurs on the outside of the cell, where there is no ready supply of energy in the form of ATP.

Recall from Figure 9.24 that bacterial peptidoglycans consist of polymeric chains of amino sugars cross-linked by oligopeptide chains to form a huge three-dimensional network in which the entire peptidoglycan layer of a cell is one giant macromolecule. The glycan, or polysaccharide, chain is an alternating polymer of *N*-acetylglucosamine and *N*-acetylmuramic acid, the latter a derivative of *N*-acetylglucosamine. In *S. aureus,* the carboxyl groups of all the *N*-acetylmuramic acid residues are linked to the terminal amino group of the tetrapeptide L-alanyl-D-γ-isoglutaminyl-L-lysyl-D-alanine. Each cross-link takes the form of a pentaglycine chain that joins the carboxyl group of a D-alanine residue to the ε-amino group of a lysine residue in an adjacent oligopeptide.

Biosynthesis of the *S. aureus* peptidoglycan can be considered in three distinct stages: (1) synthesis of *N*-acetylmuramylpeptide, (2) formation of the polysaccharide chain by polymerization of *N*-acetylglucosamine and *N*-acetylmuramylpentapeptide, and (3) cross-linking of individual peptidoglycan strands. Much of this pathway was elucidated in studies of the action of penicillin, which kills bacterial cells by blocking synthesis of the cell wall through inhibition of peptidoglycan synthesis.

Synthesis of *N*-Acetylmuramylpeptide

The first stage begins with synthesis of UDP-*N*-acetylmuramic acid from UDP-*N*-acetylglucosamine (see margin). One mole of phosphoenolpyruvate is transferred, giving a three-carbon side chain. This three-carbon group (shown in

UDP-*N*-acetylglucosamine

Phosphoenolpyruvate

UDP-*N*-acetylmuramic acid

magenta) is reduced by NADPH (a biological reductant; see Chapter 12) to give the 3-O-D-lactyl ether of N-acetylglucosamine, or N-acetylmuramic acid. Then a pentapeptide is built up in stepwise fashion, as shown in Figure 9.38. No messenger RNA template or ribosomes are involved here, as they are in the synthesis of polypeptide chains in protein synthesis. The specificity lies in the sequential actions of a series of ATP-dependent enzymes, which add first L-alanine, followed by D-glutamate (later amidated to D-*isoglutamine*), then L-lysine (linked to the γ-carboxyl group of glutamate), and finally the dipeptide D-alanyl-D-alanine. One of these two D-alanyl residues will be removed at a later step (see Figure 9.39).

Formation of the Peptidoglycan Chain

The next stage is polymerization of N-acetylglucosamine and N-acetylmuramylpentapeptide to give a linear peptidoglycan chain. This process involves a lipid carrier, **undecaprenol phosphate**, comparable to dolichol phosphate, which we just encountered in the synthesis of N-linked oligosaccharides.

$$H(CH_2-\underset{\underset{CH_3}{|}}{C}=CH-CH_2)_{10}-CH_2-\underset{\underset{CH_3}{|}}{C}=CH-CH_2-O-\underset{\underset{O^-}{\overset{O}{||}}}{P}-O^-$$

Undecaprenol phosphate

Undecaprenol phosphate is a 55-carbon compound containing 11 isoprenoid C_5 units, with phosphate linked at the terminus. To this phosphate is transferred the N-acetylmuramylpentapeptide moiety from UDP-N-acetylmuramylpentapeptide (Figure 9.39, step 1). This compound then accepts N-acetylglucosamine

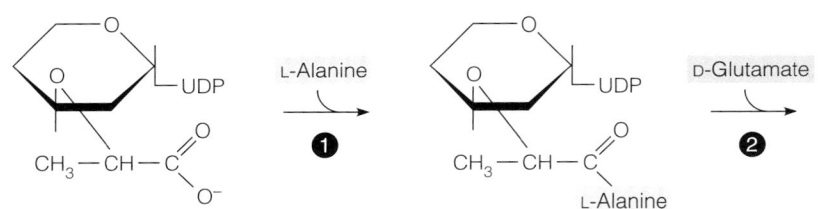

UDP-N-acetylmuramic acid

UDP-N-acetylmuramylpentapeptide

FIGURE 9.38

Biosynthesis of UDP-N-acetylmuramylpentapeptide from UDP-N-acetylmuramic acid. The sugar structure is shown in outline form.

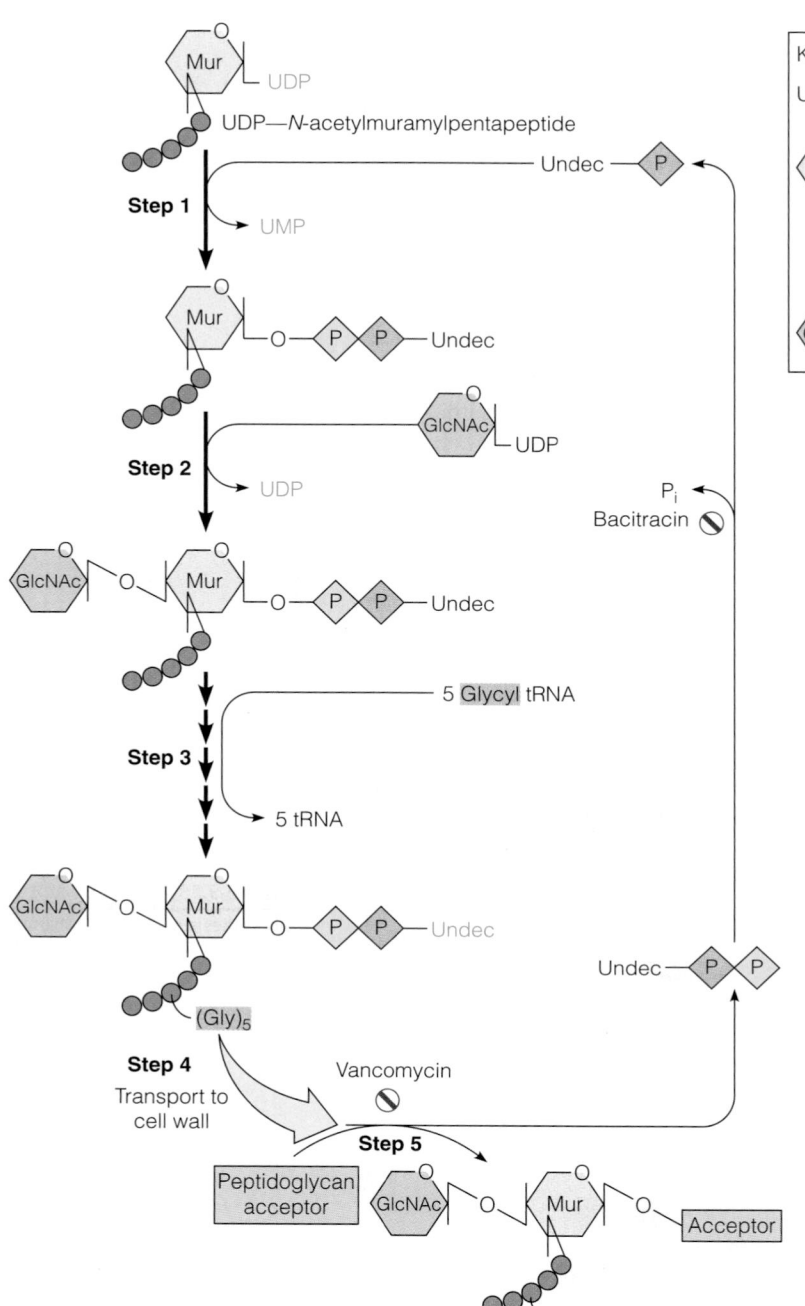

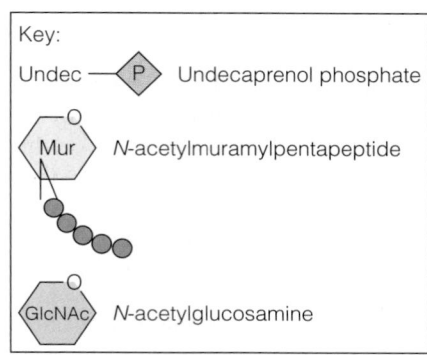

FIGURE 9.39

Synthesis of the linear peptidoglycan molecule of *Staphylococcus aureus*. The peptidoglycan molecule is synthesized by addition of *N*-acetylglucosamine and five glycyl residues to *N*-acetylmuramylpentapeptide, with undecaprenol phosphate acting as carrier. The figure does not show the ATP-dependent amidation of the D-glutamate residue of the pentapeptide, which occurs during this synthesis. Sites of inhibition by the antibiotics bacitracin and vancomycin are identified. Following synthesis, the peptidoglycan is transported through the cell membrane to the cell wall and added to the end of a chain in the peptidoglycan layer.

from UDP-*N*-acetylglucosamine (step 2), followed by the sequential addition of five glycyl residues, from glycyl transfer RNA (step 3). It seems likely that the phospholipid carrier then transports the peptidodisaccharide unit through the membrane at this stage (step 4) because polymerization—addition to the reducing end of a preexisting peptidoglycan chain—occurs on the outside of the cell wall (step 5). The antibiotics *bacitracin* and *vancomycin* inhibit specific steps in this process at the sites shown in Figure 9.39.

Cross-linking of Peptidoglycan Strands

Finally, the cross-linking occurs between adjacent chains, also outside the cell. This involves a **transpeptidation** reaction, with the cleavage of one peptide bond providing the energy needed to drive formation of another peptide bond. This means that the free

Much of the biosynthesis of peptidoglycan oligosaccharide chains occurs outside the cell wall, using activated intermediates synthesized inside the cell.

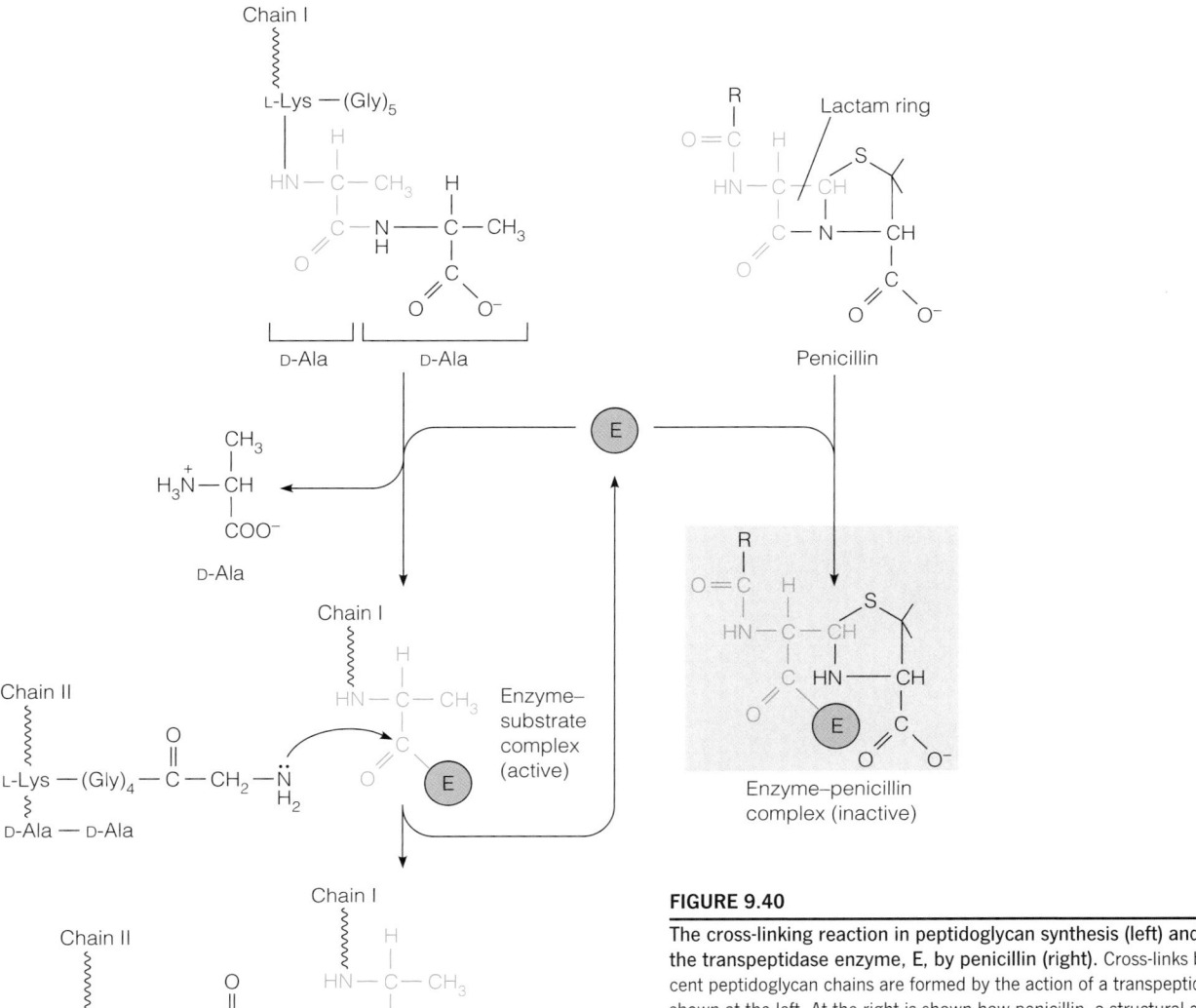

FIGURE 9.40

The cross-linking reaction in peptidoglycan synthesis (left) and inhibition of the transpeptidase enzyme, E, by penicillin (right). Cross-links between adjacent peptidoglycan chains are formed by the action of a transpeptidase enzyme, as shown at the left. At the right is shown how penicillin, a structural analog of the natural substrate, reacts with the active form of the enzyme to form an inactive covalent complex that resembles the enzyme–substrate complex.

energy needed to drive cross-link formation was built into the structures while they were still accessible to ATP, inside the cell. As shown in Figure 9.40, the transpeptidation involves nucleophilic attack by the free terminal amino group of the pentaglycine chain on the amide carbon linking the terminal D-alanines in an adjacent chain.

The cross-linking reaction is the target for the action of two important classes of antibiotics, the *penicillins* and the *cephalosporins*. Penicillin is thought to react irreversibly with the transpeptidase that catalyzes cross-linking. That enzyme normally forms an acyl-enzyme intermediate, via the penultimate D-alanine of the pentapeptide chain (Figure 9.40, right). Penicillin resembles the terminal dipeptide of this structure to the point that it can also react with the transpeptidase. The reaction is driven in part by the strain built into the four-membered **lactam ring** of penicillin, for that ring opens during the reaction. Penicillin has been widely studied as an "ideal" antibiotic because the cross-linking reaction has no counterpart in animal metabolism. Because the bacterial cell must continue to synthesize cell wall in order to grow and divide, inhibition of a step in this process provides a completely specific way to interfere with the growth of bacterial pathogens. Unfortunately, resistance to penicillin can be acquired. This resistance usually involves the synthesis, directed by an extrachromosomal gene, of **β-lactamase**, an enzyme that hydrolyzes the lactam ring of penicillin and destroys its ability to interfere with peptidoglycan synthesis.

The antibiotic activity of penicillin derives from its interference with extracellular peptidoglycan synthesis.

Microbial Cell Wall Polysaccharides: O Antigens

The final example we consider is the biosynthesis of the O antigen of the Gram-negative *Salmonella typhimurium*. The O antigen (not to be confused with the type O blood group substance of human erythrocytes) is the major lipopolysaccharide component of the outer membrane (see Figure 9.24b). Lipopolysaccharides contain repeating oligosaccharide units that are attached to a basal core polysaccharide. The latter is, in turn, attached to a complex called lipid A. The repeating oligosaccharide units protrude like minute fibers from the outer membrane surface. Because they represent the outer surface of the cell and are composed of specific carbohydrate structures, these fibers provoke strong immune reactions—hence the term *O antigen* that is applied to the fibers. Production of antibodies directed against O antigens represents a primary defense mechanism used by

FIGURE 9.41

Biosynthesis of the repeating oligosaccharide unit of the O antigen of *Salmonella typhimurium*. The first four reactions occur within the inner membrane. Transfer of the activated tetrasaccharide unit to the unactivated terminus of a growing polysaccharide unit occurs on the outside of the outer membrane.

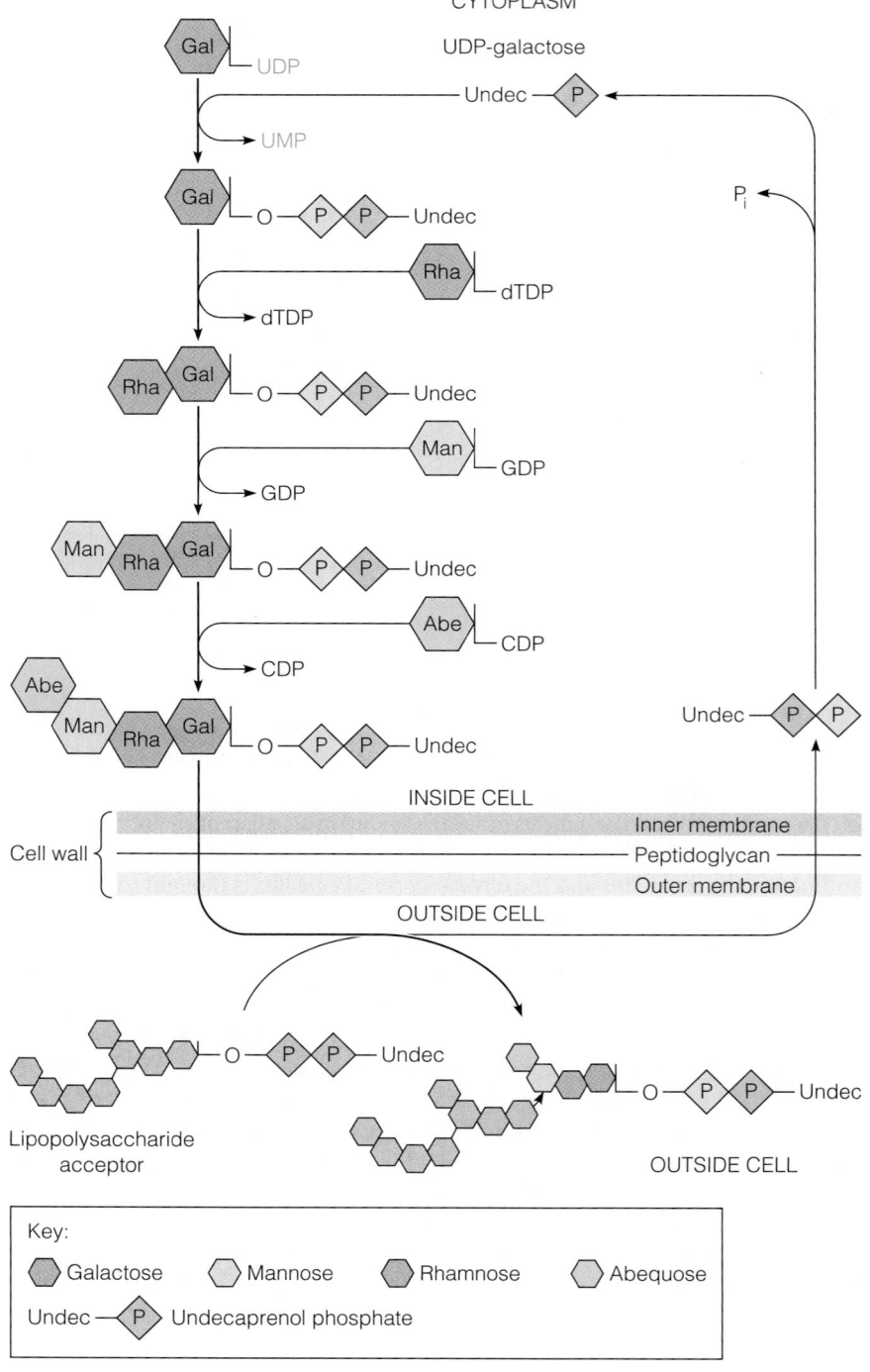

vertebrates against bacterial infection. Bacteria have responded evolutionarily by being able to change O-antigen structure through extremely rapid genetic change, the mechanism of which is not fully understood. Consequently, hundreds of different serotypes (immunologically distinct strains) of bacteria such as *S. typhimurium* exist, each with a different O-antigen repeating unit.

In wild strains of *S. typhimurium* the O-antigen repeating unit has the structure abequose α (1 → 3) mannose α (1 → 4) rhamnose β (1 → 3) galactose. Note from the margin that abequose and rhamnose are both deoxy sugars.

The oligosaccharide unit is assembled on a lipid carrier, undecaprenol phosphate, the same carrier we encountered in peptidoglycan synthesis. A lipid-linked tetrasaccharide is assembled within the inner membrane as shown in Figure 9.41 and then passes outward through the outer membrane, where the activated galactose of a lipid-linked polymer attacks the mannose of the activated tetrasaccharide. As noted earlier, this general scheme can lead to tremendous structural diversity among the lipopolysaccharides.

With development of new methods for structural analysis and synthesis of complex carbohydrates, the field has entered a new and exciting phase. As biochemists in the emerging field of glycobiology turn their attention more and more to the relationships between cells and tissues that make for an integrated organism, they are finding that carbohydrates are intimately involved in these processes. Understanding the roles of carbohydrates in these relationships is opening windows upon embryonic development, wound healing, inflammation, immunity, cell growth control, and fundamentally, all biological processes.

D-**Abequose**   L-**Rhamnose**

SUMMARY

Carbohydrates (or saccharides) are compounds with the stoichiometric formula $(CH_2O)_n$ or derivatives of such compounds. They are the primary product of photosynthesis, and their oxidation provides a major energy source for both plants and animals. Because most monosaccharides have multiple chiral centers, these saccharides exist as enantiomeric pairs (D and L mirror images) of multiple diastereomers. Monosaccharides may be either aldoses or ketoses. Those containing five or more carbons exist mainly in the form of rings of five (furanose) or six (pyranose) atoms, resulting from internal hemiacetal formation. Such rings exist as α or β anomers and exhibit multiple conformations (e.g., boat and chair) as well.

Important derivatives of monosaccharides include phosphate esters, acids and lactones, alditols, amino sugars, and glycosides. Phosphate esters are important as metabolic intermediates; glycosides represent a large class of compounds formed by elimination of water between a sugar and another hydroxy compound. Oligosaccharides and polysaccharides are formed by making glycosidic links between monosaccharides; the glycosidic linkage is metastable, so enzymes control its hydrolysis in vivo. Polysaccharides serve multiple functions—energy storage (starch and glycogen), structural roles (cellulose, xylans, chitin, glycosaminoglycans, cell wall polysaccharides), and identification tags (the oligosaccharides and polysaccharides on glycoproteins and cell surfaces). The blood group antigens are important examples of the identification function.

Complex glycan chains are assembled by stepwise transfer of monosaccharide units from nucleotide-linked sugars through the action of glycosyltransferases. In the assembly of N-linked glycans, which occurs cotranslationally with synthesis of the polypeptides, the activation of sugars involves not a nucleotide, but the isoprenoid compound dolichol phosphate. Processing of N-linked glycans occurs as nascent glycoproteins are guided through membranes of the endoplasmic reticulum and the Golgi apparatus, en route to their ultimate destinations. The biosynthesis of peptidoglycan components of bacterial cell walls also involves a lipid sugar carrier, undecaprenol phosphate. In Gram-positive bacteria, the cross-linking of peptidoglycan chains is the site of action of penicillin and related antibiotics. In Gram-negative bacteria, the peptidoglycan undergoes rapid changes in structure, to help the bacterium evade immune system detection.

REFERENCES

General

Binkley, R. W. (1988) *Modern Carbohydrate Chemistry*. Marcel Dekker, New York. A comprehensive survey.

Boraston, A., and B. Mulloy (2010) Structural glycobiology: Biosynthesis, recognition events, and new methods. *Curr. Opinion Struc. Biol.* 20:533–535. Introduction to a special issue of the journal concerned with current glycobiology.

Finklestein, J. (2007) Glycochemistry and glycobiology. *Nature* 446:999. Introduction to a special series of review articles on contemporary carbohydrate biochemistry.

Lindhorst, T. K. (2003) *Essentials of Carbohydrate Chemistry*, 2nd ed. Wiley-VCH, Weinheim, Germany. A valuable sourcebook for details of carbohydrate structure and chemistry.

Roseman, S. (2001) Reflections on glycobiology. *J. Biol. Chem.* 276: 41527–41542. A retrospective article by a longtime leader.

Carbohydrate Structure and Chemistry

Agard, N. J., and C. R. Bertozzi (2009) Chemical approaches to perturb, profile, and perceive glycans. *Acc. Chem. Res.* 42:788–797. Describes approaches for development of glycan microarray technology.

Barker, R., and A. S. Serianni (1986) Carbohydrates in solution: Studies with stable isotopes. *Acc. Chem. Res.* 19:307–313. A brief review of ^{13}C NMR work that establishes conformations.

Carver, J. P. (1991) Experimental structural determination of oligosaccharides. *Curr. Opin. Struct. Biol.* 1:716–720.

Laughlin, S. T., and C. R. Bertozzi (2009) Imaging the glycome. *Proc. Natl. Acad. Sci. USA* 106:12–17. Emerging technology for visualizing specific glycans in living cells.

Seeberger, P. H., and D. B. Werz (2007) Synthesis and medical applications of oligosaccharides. *Nature* 446:1046–1051. Sequence analysis and synthesis of complex carbohydrates both involve special problems.

Venkataraman, G., Z. Shriver, R. Raman, and R. Sasikekharan (1999) Sequencing complex polysaccharides. *Science* 286:537–542.

Bacterial and Plant Cell Walls

Goodwin, T. W., and E. I. Mercer (1983) *Introduction to Plant Biochemistry.* Pergamon, Oxford, UK.

Preiss, J., ed. (1988) *The Biochemistry of Plants: A Comprehensive Treatise.* Academic Press, New York.

Schockman, G. D., and J. F. Barnett (1983) Structure, function, and assembly of cell walls of Gram-positive bacteria. *Annu. Rev. Microbiol.* 37:501–527.

Glycoproteins

Freeze, H. H., and M. Aebi (2005) Altered glycan structures: The molecular basis of congenital disorders of glycosylation. *Curr. Opin. Struc. Biol.* 15:490–498.

Hart, G. W., M. P. Housley, and C. Slawson (2007) Cycling of O-linked β-N-acetylglucosamine on nucleoplasmic proteins. *Nature* 446: 1017–1022. Evidence for a function of reversible protein glycosylation in cell signaling.

Iozzo, R. V. (1998) Matrix proteoglycans: From molecular design to cellular function. *Ann. Rev. Biochem.* 67:609–652.

Rudd, P. M., and R. A. Dwek (1997) Glycosylation: Heterogeneity and the 3D structure of proteins. *Crit. Rev. Biochem. Mol. Biol.* 32:1–100.

van den Steen, P., P. M. Rudd, R. A. Dwek, and G. Opdenakker (1998) Concepts and principles of O-linked glycosylation. *Crit. Rev. Biochem. Mol. Biol.* 33:151–208.

Oligosaccharides and Cell Recognition

Fukuda, M., ed. (2006) *Functional Glycomics, Methods in Enzymology*, Vol. 417. Academic Press, New York. See also volumes 415 and 416, with the same editor.

Kilpatrick, D. C. (2002) Animal lectins: A historical introduction and overview. *Biochim. Biophys. Acta* 1572:187–197. Introduction to a special series of reviews on lectins in the same issue of the journal.

Labat-Robert, J., R. Timpl, and R. Ladiglas, eds. (1986) *Structural Glycoproteins in Cell–Matrix Interaction.* Karger, New York.

Taylor, M. E., and Drickamer, K. (2006) *Introduction to Glycobiology*, 2nd ed. Oxford University Press, Oxford, UK.

Varki, A. (2007) Glycan-based interactions involving vertebrate sialic-acid-recognizing proteins. *Nature* 446:1023–1029. Roles of sialic acid in cell recognition.

Glycoconjugate Synthesis

Alder, N. N., and A. E. Johnson (2004) Cotranslational membrane protein biosynthesis at the endoplasmic reticulum. *J. Biol. Chem.* 279:22787–22790.

Allan, B. B., and W. E. Balch (1999) Protein sorting by directed maturation of Golgi compartments. *Science* 285:63–66.

Gahmberg, C. G., and M. Tolvanen (1996) Why mammalian cell surface proteins are glycoproteins. *Trends Biochem. Sci.* 21:308–311. How the diversity of carbohydrate structures adapts them for use as recognition determinants on cell surfaces.

Kornfeld, S. (1992) Structure and function of the mannose 6-phosphate/insulinlike growth factor II receptors. *Annu. Rev. Biochem.* 61:307–330. A review describing the lysosomal membrane targeting system and the unexpected discovery that the receptor is identical to a cell-growth factor discovered independently.

Walsh, C. T. (1989) Enzymes in the D-alanine branch of bacterial cell wall peptidoglycan assembly. *J. Biol. Chem.* 264:2393–2396. A minireview describing mechanisms in this pathway, which is the site of action of penicillin and other antibiotics.

PROBLEMS

1. Draw Haworth projections for the following:
   (a) in α-furanose form. Name the sugar.

   ```
          CHO
           |
      H — C — OH
           |
     HO — C — H
           |
      H — C — OH
           |
         CH₂OH
   ```

   (b) The L isomer of (a)
   (c) α-D-GlcNAc
   (d) (d) α-D-Fructofuranose

2. α-D-Galactopyranose rotates the plane of polarized light, but the product of its reduction with sodium borohydride (galactitol) does not. Explain the difference.

3. Provide an explanation for the fact that α-D-mannose is more stable than β-D-mannose, whereas the opposite is true for glucose.

*4. Using data in Table 9.2, calculate the standard state free energy change for the conversion of D-glucose from the α to the β anomer at 40 °C. Can you provide a qualitative explanation for this? Do you think that ΔG arises primarily from an enthalpy or entropy contribution?

5. The disaccharide α,β-trehalose differs from the α,α structure in Figure 9.16a by having an (α1 → β1) linkage. Draw its structure.

6. A *reducing* sugar will undergo the Fehling reaction (see page 322), which requires a (potential) free aldehyde group. Which of the disaccharides shown in Figure 9.15 are reducing, and which are nonreducing?

7. *Dextrans* are polysaccharides produced by certain species of bacteria. They are glucans, with primarily α (1 → 6) linkages and with frequent α(1 → 3) branching. Draw a Haworth projection of a portion of a dextran, including one (1 → 3) branch point.

8. What is the natural polysaccharide whose repeating structure can be symbolized by GlcUAβ(1 → 3)GlcNAc, with these units connected by β(1 → 4) links?

*9. Indicate whether the structures shown are R or S in the absolute system.

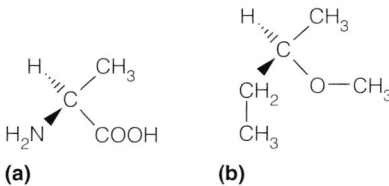

   (a)   (b)

*10. The reagent periodate (IO₄⁻) oxidatively cleaves the carbon–carbon bonds between two carbons carrying hydroxyl groups. Explain how periodate oxidation might be used to distinguish between methyl glycosides of glucose in the pyranose and furanose forms.

11. Draw (using Haworth projections) the fragments of xylan and glucomannan structure shown on page 333.

12. A research student is attempting to sequence an oligosaccharide attached to an orosomucoid (see Figure 9A.1) from a mutant cell line. She finds from analysis the presence of Sia, Gal, and GlcNAc. Neuraminidase cleavage succeeds, but β-galactosidase is without effect. Suggest an explanation.

13. One or more of the compounds shown below will satisfy each of the following statements. Not all compounds may be used; some may be used twice. Put the letter(s) in the blank.
   (a) Found in chitin. _____
   (b) An L-saccharide. _____
   (c) The first residue attached to asparagine in N-linked glycans. _____
   (d) A uronic acid. _____
   (e) A ketose. _____

   (a)   (b)   (c)

   ```
     CH₂OH                              
      |          CHOH        CH₂OH
      C = O        |            |
      |          CHOH         C = O
 H — C — OH        |            |
      |          CH₂OH    HO — C — H
    CH₂OH                      |
                            CH₂OH
   ```

   (d)   (e)   (f)

14. Why, do you suppose, is the influenza virus protein that binds the virus to an infected cell called hemagglutinin? Hemagglutination is the clumping together of red blood cells.

15. The diversity of functional groups on sugars that can form glycosidic bonds greatly increases the information content of glycans relative to oligopeptides. Consider three amino acids, A, B, and C. How many tripeptides can be formed from one molecule of each amino acid? Now consider three sugars—glucose, glucuronic acid, and N-acetylglucosamine. Use shorthand (e.g., Glc[1 → 4]GlcUA[1 → 4]GlcNAC) to represent ten trisaccharides with the sequence Glc-GlcUA-GluNAc. Is your list exhaustive?

16. Are mannose and galactose epimers? Allose and altrose? Gulose and talose? Ribose and arabinose? Explain your answers.

*17. Propose synthetic pathways for the following nucleotide sugar intermediates involved in the biosynthesis of *Salmonella typhimurium* O antigen: UDP-Gal, GDP-Man, CDP-abequose, and dTDP-L-rhamnose.

18. Write the structure of each nucleotide-linked sugar in the pathway shown in Figure 9.33.

TOOLS OF BIOCHEMISTRY 9A

SEQUENCING OLIGOSACCHARIDES

Determining oligosaccharide sequences presents problems similar to, but more difficult than, those encountered in protein sequencing. Because of the many types of monomers that may be encountered and the variety of linkages between them, no single method like the Edman degradation of polypeptides has been devised.

The first step in any sequence analysis is, as in polypeptide analysis, the determination of composition. The oligomer is hydrolyzed in acidic solution, yielding a mixture of monosaccharides. At present, these monosaccharides are almost always separated, identified, and quantified by gas or liquid chromatography.

Determination of the oligosaccharide sequence itself is much more difficult. In the past, chemical methods were extensively used, but they have been largely supplanted by enzymatic cleavage of the oligomer followed by sophisticated methods for identification of the fragments. Researchers are now familiar with a large number of enzymes (**glycosidases**) that catalyze the cleavage of glycosidic bonds between sugar moieties. Some of these enzymes are very specific in their action. They may be divided into two groups: *exoglycosidases,* which remove the terminal residue from an oligosaccharide chain, and *endoglycosidases,* which catalyze cleavage within the chain. A few examples of these enzymes are listed in Table 9A.1.

A simple example of the application of specific glycosidases to sequence determination is shown in Figure 9A.1. The oligosaccharide shown is a portion of one of several oligosaccharides attached to the blood serum protein *orosomucoid.* The residue at the nonreducing end of the chain can be removed by neuraminidase. According to Table 9A.1, the terminal residue must be sialic acid, attached to either Gal or GlcNAc. Sialic acid is found, and subsequent release of Gal by *Streptococcus* β-galactosidase indicates that the next residue is Gal, attached by a 1 → 4 linkage to GlcNAc. The latter residue is confirmed by its release by β-*N*-acetylglucosaminidase. Using modern techniques for analysis, sequences may be determined in this manner at the picomole level.

Although such procedures can often provide definitive information, they are laborious and not always effective. In recent years there has been remarkable development of high-resolution NMR and mass spectrometry as techniques for identification of complex oligosaccharides. The use of these methods is described in the References, but is too technical to be presented here.

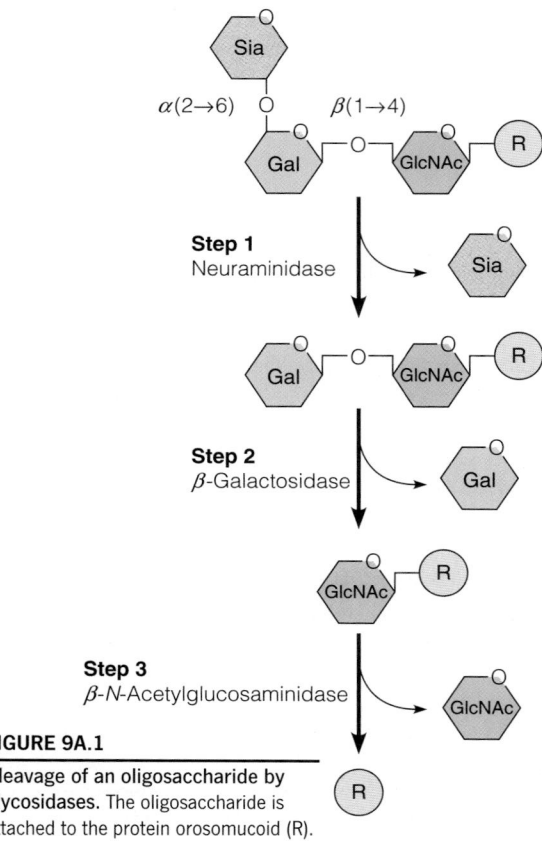

FIGURE 9A.1

Cleavage of an oligosaccharide by glycosidases. The oligosaccharide is attached to the protein orosomucoid (R).

References

Dell, A., and H. R. Morris (2001) Glycoprotein structure determination by mass spectrometry. *Science* 291:2351–2356. A relatively recent brief review.

Jones, C. (1991) Nuclear magnetic resonance spectroscopy methods for the analysis of polysaccharides and glycoprotein carbohydrate chains. *Adv. Carbohydr. Anal.* 1:145–184. Includes a survey of useful techniques.

McCleary, B. V., and N. K. Matheson (1986) Enzymatic analysis of polysaccharide structures. *Adv. Carbohydr. Chem. Biochem.* 44:147–276.

Paulson, J. C., O. Blixt, and B. E. Collins (2006) Sweet spots in functional genomics. *Nature Chem. Biol.* 2:238–248. Recent techniques in glycobiology.

TABLE 9A.1	Some specific glycosidases used in sequencing oligosaccharides	
Enzyme Name	Source	Specificity
Exoglycosidases		
Neuraminidase	*Streptococcus pneumoniae*	Siaα(2 → 3 or 6)Gal or Siaα(2 → 6)GlcNAc
β-Galactosidase	*Streptococcus pneumoniae*	Galβ(1 → 4)GlcNAc
α-Fucosidase	*Clostridium perfringens*	Fucα(1 → 2)Gal
Endoglycosidases		
Endo-β-galactosidase	*Escherichia freundii*	· · · GlcNAcβ(1 → 3)Galβ(1 → 4)Glc(GlcNAc)· · ·
Almond emulsion	Bitter almond seeds	Cleaves bond to Asn in many N-linked oligosaccharides

CHAPTER 10

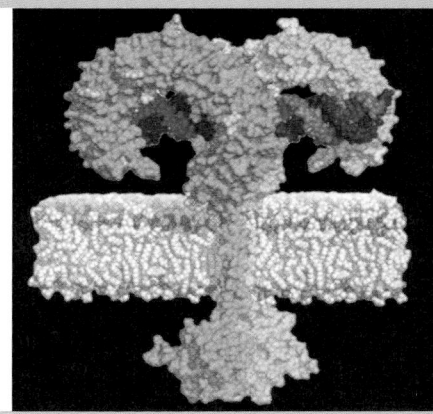

Lipids, Membranes, and Cellular Transport

The lipids are a structurally and functionally diverse group of molecules. They are generally hydrophobic, although many also possess polar or charged groups in addition to a hydrocarbon core. Given their structural diversity, lipids carry out multiple functions. Primary among these are energy storage, signaling, and formation of membrane structures. In this chapter we focus on the structures of lipid molecules, the general features of membranes formed from lipids, and selective transport across membranes. The roles of lipids as energy stores and signaling molecules are described in great detail in Chapters 17, 19, and 23. In those chapters, we also discuss lipid biosynthesis as well as lipid transport and the lipid–protein complexes, or *lipoproteins*, involved in that process.

The largest fraction of the lipids in most cells is used to form **membranes**, the partitions that divide compartments from one another and separate the cell from its surroundings. Cellular membranes are much more than passive walls, for they contain highly selective gates that control the passage of materials in specific directions. It is this property of *selective membrane permeability* that allows each part of the cell to carry out its specific operations. It is difficult to imagine the organization and evolution of complex multicellular organisms without membranes. Membranes establish order (by compartmentation) and allow free energy to be stored in the form of concentration gradients (e.g., proton gradients in mitochondria, or ion gradients in nerve cells). Thus, membranes have a vital role in maintaining the living state, which is essentially an ongoing campaign against entropy and the tendency to move toward chemical equilibrium.

The Molecular Structure and Behavior of Lipids

Compared to the other classes of biomolecules we have considered in earlier chapters, lipids are not generally water-soluble, due to the large portion of their structure that is hydrocarbon. This means that lipids are rarely found free in solution, rather, they are either in complex with soluble protein transporters, or part of

Lipid molecules tend to be insoluble in water, but they can associate to form water-soluble structures, such as micelles, vesicles, and bilayers.

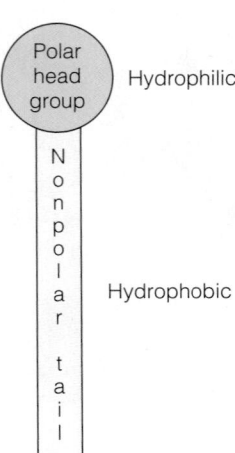

**A simplified representation
of an amphipathic lipid molecule**

higher order assemblies that sequester the hydrophobic surface area from the surrounding aqueous environment.

Unlike the amino acids in proteins, the nucleotides in nucleic acids, and the monosaccharides in complex carbohydrates, lipids do not form large covalent polymers. Rather, they tend to associate with each other through noncovalent forces. For example, the lipids that make up membranes are usually characterized by the kind of structure shown in the margin: a polar, hydrophilic "head" connected to a larger nonpolar, hydrophobic hydrocarbon "tail." Such lipids in an aqueous environment tend to associate together by noncovalent interactions for two fundamental reasons. Just as nonpolar groups in proteins associate via an entropy-driven hydrophobic effect, so do the nonpolar tails of lipids. A second stabilizing force comes from the van der Waals interactions between the hydrocarbon regions of the molecules.

The polar, hydrophilic head groups of membrane lipids tend to associate with water. Such lipids are prime examples of the kind of amphipathic substance described in Chapter 2. The amphipathic nature of membrane lipids has a number of consequences, including the formation of surface monolayers, bilayers, micelles, and vesicles by lipids in contact with water (see Figure 2.15, page 38). From a biological point of view, the most important of these consequences is the tendency of lipids to form micelles and membrane bilayers. Exactly what kind of structure is formed when a lipid is in contact with water depends on the specific molecular structure of the hydrophilic and hydrophobic parts of that lipid molecule. Thus, it is appropriate that we now examine the structure of some of the major types of lipids.

Fatty Acids

The simplest lipids are the **fatty acids**, which are also constituents of many more complex lipids. Their basic structure exemplifies the general lipid model described above: A hydrophilic carboxylate group is attached to one end of the hydrocarbon chain, which contains typically 12 to 24 carbons. An example is *stearic acid*, which is widely distributed in organisms. We show it in Figure 10.1 as the ionized form, the *stearate* ion. Stearic acid is an example of a **saturated** fatty acid, one in which the carbons of the tail are all saturated with hydrogen atoms (i.e., no C$=$C double bonds). A number of the more biologically important saturated fatty acids are listed in Table 10.1. Note that each has a common name (such as stearic acid) and a systematic name (in this case, octadecanoic acid).

Many important naturally occurring fatty acids are **unsaturated**—that is, they contain one or more double bonds (see Table 10.1). One such example is *oleic acid* (or *oleate*), which is found in many animal fats (Figure 10.1). In most of the

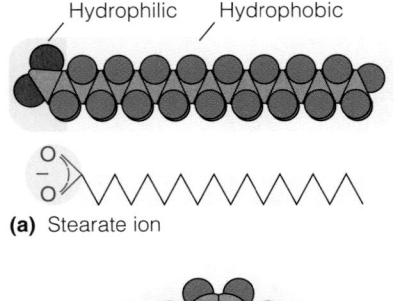

(a) Stearate ion

(b) Oleate ion

FIGURE 10.1

Structures of the ionized forms of some representative fatty acids. Hydrophilic portions (head groups) of the molecules are indicated by a pale blue background in the models, hydrophobic portions (tails) by yellow. **(a)** Stearate (the anion of stearic acid) is a saturated fatty acid. **(b)** Oleate is an unsaturated fatty acid with one *cis* double bond. **(c)** Alternate representations for (a) and (b).

CCH$_2$CH$_2$CH$_2$CH$_2$CH$_2$CH$_2$CH$_2$CH$_2$CH$_2$CH$_2$CH$_2$CH$_2$CH$_2$CH$_2$CH$_2$CH$_2$CH$_3$

Polar head group   Hydrocarbon tail

Stearate ion

CCH$_2$CH$_2$CH$_2$CH$_2$CH$_2$CH$_2$CH$_2$C$=$CCH$_2$CH$_2$CH$_2$CH$_2$CH$_2$CH$_2$CH$_2$CH$_3$

Oleate ion

(c) Formulas

TABLE 10.1   Some biologically important fatty acids

Common Name	Systematic Name	Abbreviation	Structure	Melting Point (°C)
Saturated Fatty Acids				
Capric	Decanoic	10:0	$CH_3(CH_2)_8COOH$	31.6
Lauric	Dodecanoic	12:0	$CH_3(CH_2)_{10}COOH$	44.2
Myristic	Tetradecanoic	14:0	$CH_3(CH_2)_{12}COOH$	53.9
Palmitic	Hexadecanoic	16:0	$CH_3(CH_2)_{14}COOH$	63.1
Stearic	Octadecanoic	18:0	$CH_3(CH_2)_{16}COOH$	69.6
Arachidic	Eicosanoic	20:0	$CH_3(CH_2)_{18}COOH$	76.5
Behenic	Docosanoic	22:0	$CH_3(CH_2)_{20}COOH$	81.5
Lignoceric	Tetracosanoic	24:0	$CH_3(CH_2)_{22}COOH$	86.0
Cerotic	Hexacosanoic	26:0	$CH_3(CH_2)_{24}COOH$	88.5
Unsaturated Fatty Acids				
Palmitoleic	cis-9-Hexadecenoic	16:1Δ9	$CH_3(CH_2)_5CH = CH(CH_2)_7COOH$	0
Oleic	cis-9-Octadecenoic	18:1Δ9	$CH_3(CH_2)_7CH = CH(CH_2)_7COOH$	16
Linoleic	cis,cis-9, 12-Octadecadienoic	18:2cΔ9,12	$CH_3(CH_2)_4CH =$ $CHCH_2CH = CH(CH_2)_7COOH$	5
Linolenic	all-cis-9,12, 15-Octadecatrienoic	18:3cΔ9,12,15	$CH_3(CH_2)_4CH = CHCH_2CH =$ $CHCH_2CH = CH(CH_2)_7COOH$	−11
Arachidonic	all-cis-5,8,11, 14-Eicosatetraenoic	20:4cΔ5,8,11,14	$CH_3(CH_2)_4CH =$ $CHCH_2CH = CHCH_2CH =$ $CHCH_2CH = CH(CH_2)_3COOH$	−50
Branched and Cyclic Acids				
Tuberculostearic	10-R-methyloctadecanoic		$CH_3(CH_2)_7\overset{\displaystyle CH_3}{\underset{\displaystyle \vert}{CH}}(CH_2)_8COOH$	13.2
Lactobacillic	2-hexyl-cyclopropyl -decanoic		$CH_3(CH_2)_5CH \overset{\displaystyle CH_2}{-} CH(CH_2)_9COOH$	29

naturally occurring unsaturated fatty acids, the orientation about double bonds is *cis* rather than *trans*. This orientation has an important effect on molecular structure, for each *cis* double bond inserts a bend into the hydrocarbon chain. Keep in mind, however, that although Figure 10.1 depicts the molecules as extended structures, there is freedom of rotation about each single bond in the hydrocarbon chain. Thus, many conformations are possible.

Most of the naturally occurring fatty acids have an even number of carbon atoms because they are synthesized by sequential additions of a two-carbon precursor (see Chapter 17). Although the hydrocarbon chains are linear in most fatty acids, some fatty acids (found primarily in bacteria) contain branches or even cyclic structures (Table 10.1).

To provide a more convenient and definitive way of referring to fatty acids, a system of abbreviations has been developed; it is illustrated in Table 10.1. The rules are as follows: The number before the colon gives the total number of carbons, and the number after the colon gives the count of double bonds. The configurations and positions of double bonds are indicated by c (*cis*) or t (*trans*) followed by Δ and one or more numbers. These numbers denote the carbon atom (the carboxylic carbon is designated as "1") where each double bond starts. Thus, oleic acid is designated by 18:1cΔ9 and linolenic by 18:3cΔ9, 12, 15. With several bonds, linolenic acid is an example of a **polyunsaturated fatty acid** or **PUFA**.

Animals generally do not produce via biosynthesis all the PUFAs they need for proper cellular function; thus, these fatty acids must be part of their diet. The nutritional value and consequences of ingesting various fats are discussed in Chapter 17 in the context of fuel metabolism. It should be mentioned here that in contrast to the nomenclature rules listed above, nutritionists use an alternative

Membrane lipids are amphipathic. They tend to form surface monolayers, bilayers, micelles, or vesicles when in contact with water.

Most naturally occurring fatty acids contain an even number of carbon atoms. If double bonds are present (unsaturation), they are usually *cis*.

Fats, or triacylglycerols, are triesters of fatty acids and glycerol. They are the major long-term energy storage molecules in many organisms.

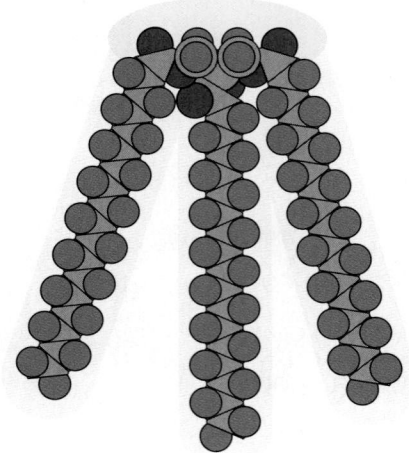

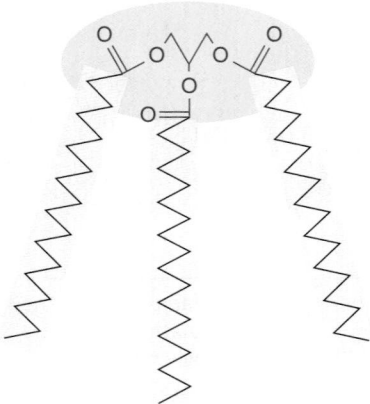

FIGURE 10.2

The structure of tristearin, a fat. Tristearin is a triacylglycerol (fat) composed of glycerol and three stearate molecules.

naming convention in which the locations of the C=C bonds are referenced to the "omega carbon" (ω-carbon; i.e., the last C atom in the chain), where the α-carbon is the —COO$^-$ carbon. In this system, the so-called ω-3 fatty acids have a C=C bond that includes the third C atom from the ω carbon. Linolenic acid is one example of an ω-3 fatty acid.

The fatty acids are weak acids, with pK_a values averaging about 4.5:

$$RCOOH \underset{pK_a \cong 4.5}{\rightleftharpoons} RCOO^- + H^+$$

Thus, these acids exist in the anionic form (RCOO$^-$) at physiological pH. Due to this hydrophilic charge and the long hydrophobic tail, fatty acids behave like amphipathic substances when dissolved in water. As shown in Figure 2.15, they tend to form **monolayers** at the air–water interface, with the carboxylate groups immersed in water and the hydrocarbon tails out of water.

If fatty acids are shaken with water, they will make spherical **micelles**, in which the hydrocarbon tails cluster together within the structure and the carboxylate heads are in contact with the surrounding water. If fatty acids are mixed with water *and* an oily or greasy substance (for example, a hydrocarbon), the micelles will form around the oil droplets, emulsifying them. In this way soaps and synthetic detergents solubilize grease.

Although the fatty acids play important roles in metabolism, large quantities of the free acids or their anions are not found in living cells. Instead, these compounds almost always occur as constituents of more complex lipids. We now turn to consideration of some of these classes of biologically important lipid molecules.

Triacylglycerols: Fats

The long hydrocarbon chains of fatty acids are extraordinarily efficient for energy storage because they contain carbon in a reduced form and will therefore yield a large amount of energy on oxidation. They are, in fact, much more efficient energy stores than are carbohydrates (quantitative analysis of this difference is made in Chapter 17). Because they are hydrophobic, fat stores are also anhydrous, whereas glycogen is hydrated. Thus, a gram of stored fat yields much more metabolic energy than a gram of stored carbohydrate. For these reasons, lipids are used by many organisms, including humans, for storage of metabolic energy. A typical human stores sufficient calories in fats to survive for several weeks (assuming access to water).

Storage of fatty acids in organisms is largely in the form of **triacylglycerols**, or **triglycerides**, or simply, **fats**. These substances are *triesters* of fatty acids and **glycerol**; the general formula is shown here.

$$\begin{array}{c} \quad\quad\quad\quad O \\ \quad\quad\quad\quad \| \\ H_2C-O-C-R_1 \\ | \quad\quad\quad\quad O \\ | \quad\quad\quad\quad \| \\ H-C-O-C-R_2 \\ | \quad\quad\quad\quad O \\ | \quad\quad\quad\quad \| \\ H_2C-O-C-R_3 \end{array}$$

Triacylglycerol

Here R$_1$, R$_2$, and R$_3$ correspond to the hydrocarbon tails of various fatty acids. We have depicted the structure with the hydrophobic chains to the right, according to our convention. This convention does not indicate stereochemical configuration (the correct stereochemical configuration around each C atom in the glycerol moiety is depicted in Figure 10.7a, page 366). As a particular example, if R$_1$=R$_2$=R$_3$=(CH$_2$)$_{16}$CH$_3$, the hydrocarbon tail of stearic acid, the molecule is *tristearin* (Figure 10.2). Triacylglycerols with the same fatty acid esterified at each position are called "simple fats." Most triacylglycerols, however, are "mixed fats" that contain a mixture of fatty acids, often including unsaturated ones. Table 10.2 lists the fatty acid composition of some naturally occurring fats.

Comparison of common experience with these fats and data in the table reveals an interesting correlation. Fats rich in unsaturated fatty acids (like olive oil) are liquid at room temperature, whereas those with a higher content of saturated fatty acids (like butter) are more solid. Indeed, a wholly saturated fat yields a firm solid, especially if the hydrocarbon chains are long. This is made clear from the melting point data in Table 10.1. The reason is simple: Long saturated chains can pack closely together, thereby increasing the number of van der Waals contacts to form regular, semicrystalline structures. In contrast, the kind of bend imposed by one or more *cis* double bonds (see Figure 10.1b) makes molecular packing less regular, and therefore more dynamic. Indeed, partial **hydrogenation** of unsaturated fat oils (like corn oil) is used commercially to convert them to firmer fats, which can be used as butter substitutes such as oleomargarine or to stabilize them against spoilage. Oxidative cleavage of the double bonds in fats yields volatile aldehydes and carboxylic acids, which contribute to the rancid odor of spoiled fats. Hydrogenation reduces such bonds to single bonds, but it also converts some *cis* double bonds to *trans* double bonds, yielding so-called trans fats. Because trans fats have been associated with an elevated risk of cardiovascular disease, the FDA requires food packaging to list the content of trans fats and several cities have banned use of trans fats for cooking in restaurants.

Esterification with glycerol diminishes the hydrophilic character of the head groups of the fatty acids. As a consequence, triacylglycerols are water-insoluble. Fats accumulated in plant and animal cells therefore form as oily droplets in the cytoplasm. In **adipocytes**, animal cells specialized for fat storage, almost the entire volume of each cell is filled by a fat droplet (Figure 10.3). Such cells make up most of the adipose (fatty) tissue of animals.

Fat storage in animals serves three distinct functions:

1. *Energy production.* As described in Chapter 17, most fat in most animals is oxidized for the generation of ATP, to drive metabolic processes.

2. *Heat production.* Some specialized cells (in "brown fat" of warm-blooded animals, for example) oxidize triacylglycerols for heat production, rather than to make ATP.

3. *Insulation.* In animals that live in a cold environment, layers of fat cells under the skin serve as thermal insulation. The blubber of whales is one obvious example.

Soaps and Detergents

If fats are hydrolyzed with strong bases such as NaOH or KOH (in earlier times, wood ashes were used), a *soap* is produced. This process is called **saponification**. The fatty acids are released as either sodium or potassium salts, which are fully ionized. However, as cleansers, soaps have the disadvantage that the fatty acids are precipitated by the calcium or magnesium ions present in "hard" water, forming a scum and destroying the emulsifying action. Synthetic detergents have been devised that do not have this defect. One class is exemplified by *sodium dodecyl sulfate (SDS)*:

$$Na^{+-}O_3SO(CH_2)_{11}CH_3$$

The salts of dodecyl sulfate with divalent cations (i.e., Ca^{2+} and Mg^{2+}) are more soluble. Recall that SDS is widely used in forming micelles about proteins for gel electrophoresis (see Tools of Biochemistry 6B, page 228). There are also synthetic nonionic detergents, like *Triton X-100*:

	Number of C Atoms in Chain	Percent Present in:		
		Olive Oil	Butter[a]	Beef Fat
Saturated	4–12	2	11	2
	14	2	10	2
	16	13	26	29
	18	3	11	21
Unsaturated	16–18	80	40	46

TABLE 10.2 Composition of some natural fats in percent of total fatty acids

[a]Numbers do not total 100% because the substance contains small amounts of other fatty acids.

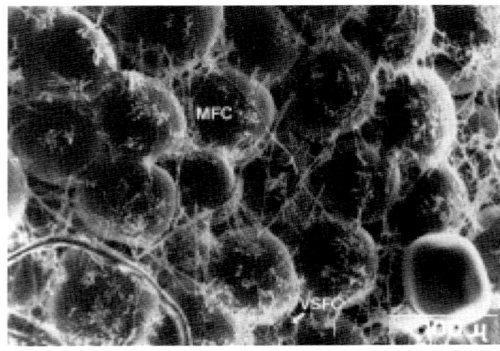

FIGURE 10.3

Adipocytes. Adipocytes, or animal fat storage cells, make up a large part of adipose tissue. The designations MFC and VSFC correspond to "mature fat cell" and "very small fat cell," respectively.

The Journal of Lipid Research 30:293–299, P. Julien, J.-P. Despres, and A. Angel, Scanning electron microscopy of very small fat cells and mature fat cells in human obesity. Reprinted with permission. © 1989 The American Society for Biochemistry and Molecular Biology. All rights reserved.

H(OCH₂CH₂)ₙ — O — [benzene ring] — C(CH₃)(CH₃) — CH₂ — C(CH₃)(CH₃) — CH₃

Oleic acid

Oleyl alcohol

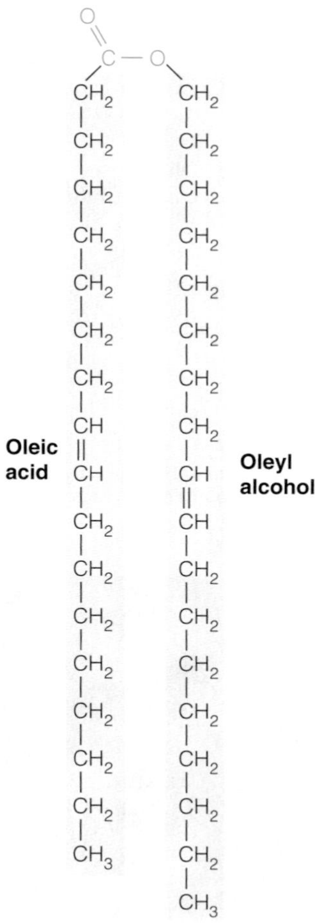

FIGURE 10.4

Structure of a typical wax. Waxes are formed by esterification of fatty acids and long-chain alcohols. The small head group can contribute little hydrophilicity, in contrast to the significant hydrophobic contribution of the two long tails.

The major lipid components of biological membranes are glycerophospholipids, sphingolipids, glycosphingolipids, and glycoglycerolipids.

The hydrophilic head group here is the polyoxyethylene group (shown in blue), which in the commercial product averages about 9.5 residues in length.

Waxes

In the natural **waxes**, a long-chain fatty acid is esterified to a long-chain alcohol (Figure 10.4). This yields a head group that is only weakly hydrophilic, attached to two long hydrocarbon chains. As a consequence, the waxes are completely water-insoluble. In fact, they are so hydrophobic that they often serve as water repellents, as in the feathers of some birds and the leaves of some plants. In some marine microorganisms, waxes are used instead of other lipids for energy storage. In beeswax, they serve a structural function. As with the triacylglycerols, the firmness of waxes increases with chain length and degree of hydrocarbon saturation.

The Lipid Constituents of Biological Membranes

All biological membranes contain lipids as major constituents. The molecules that play the dominant roles in membrane formation all have highly polar head groups and, in most cases, two hydrocarbon tails. This composition makes molecular sense: If a large head group is attached to a single hydrocarbon chain, the molecule is wedge-shaped and will tend to form spherical micelles (Figure 10.5a). A double tail yields a roughly cylindrical molecule (Figure 10.5b); such cylindrical molecules can easily pack in parallel to form extended sheets of **bilayer** membranes with the hydrophilic head groups facing outward into the aqueous regions on either side (Figure 10.5c). The four major classes of membrane-forming lipids—glycerophospholipids, sphingolipids, glycosphingolipids, and glycoglycerolipids—share this type of cylindrical molecular structure. They differ principally in the nature of the head group. We shall describe a few examples of each class.

The bilayer is roughly 6 nm thick, with ~1.5 nm of interface on either side of the ~3 nm hydrophobic core (Figure 10.5c). The interface region is composed of the lipid head groups and associated water molecules. Beyond the interface is the bulk water.

Glycerophospholipids

Glycerophospholipids (also called *phosphoglycerides*) are the major class of naturally occurring **phospholipids**, lipids with phosphate-containing head groups. These compounds make up a significant fraction of the membrane lipids throughout the bacterial, plant, and animal kingdoms. Like the other biomolecular building

FIGURE 10.5

Phospholipids and membrane structure. **(a)** Fatty acids are wedge-shaped and tend to form spherical micelles. **(b)** Phospholipids are more cylindrical and pack together to form a bilayer structure. **(c)** A computer simulation of a phospholipid bilayer showing approximate boundaries for the hydrophobic core, the interface regions, and the bulk water (above and below the interfacial regions). Water is shown with white (H) and red (O) space-filling atoms. The hydrocarbon portions of the bilayer lipids are shown in green and cyan.

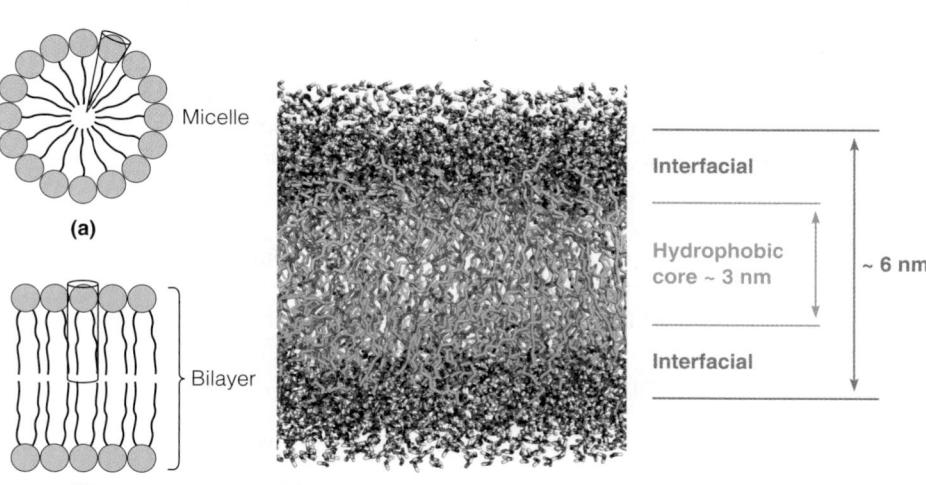

blocks we have considered, lipids possess a specific stereochemistry. Glycerol does not possess any stereocenters; however, derivatization of one or the other of the equivalent — CH$_2$OH groups bonded to the central carbon atom will generate an asymmetric center with a defined stereochemistry at C2 (Figure 10.6a). Thus, glycerol is an example of a **prochiral** molecule—it has no stereocenter until it is derivatized. As shown in Figure 10.6b, the two conventions for assigning stereochemistry that we have discussed so far, the D/L and the *R/S* systems, do not assign a uniform carbon numbering to glycerol derivatives. To avoid potential confusion in naming such compounds, the **sn** (stereospecific numbering) system has been developed (Figure 10.6c).

As shown in Figure 10.6a, we can define the carbon atoms in the two — CH$_2$OH groups as **pro-S** or **pro-R** based on the stereochemistry of the product that results from derivatization of one or the other of these groups. Consider *R*-glycerol phosphate, which can be described equally well as L-glycerol-3-phosphate or D-glycerol-1-phosphate, depending on which carbon is designated as #1 (Figure 10.6b). As we consider more complex lipids it is desirable to have a carbon numbering convention that is referenced to a single stereochemical representation. The *sn* system provides this by assigning C1 to the pro-*S* carbon of glycerol (Figure 10.6c). Following this convention, all glycerophospholipids can be considered to be derivatives of *sn*-glycerol-3-phosphate. The general structure and stereochemistry for a glycerophospholipid is shown in Figure 10.7a. Because the groups at C1 and C2 are generally hydrophobic and make up the interior of the

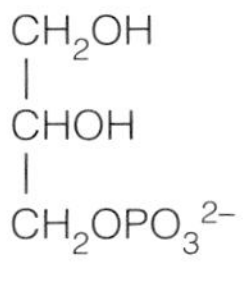

CH$_2$OH
|
CHOH
|
CH$_2$OPO$_3$$^{2-}$

Glycerol-3-phosphate

(a)

(b)

(c) pro-*S* carbon is assigned as carbon #1 in the **sn** labeling convention

sn-glycerol-3-phosphate

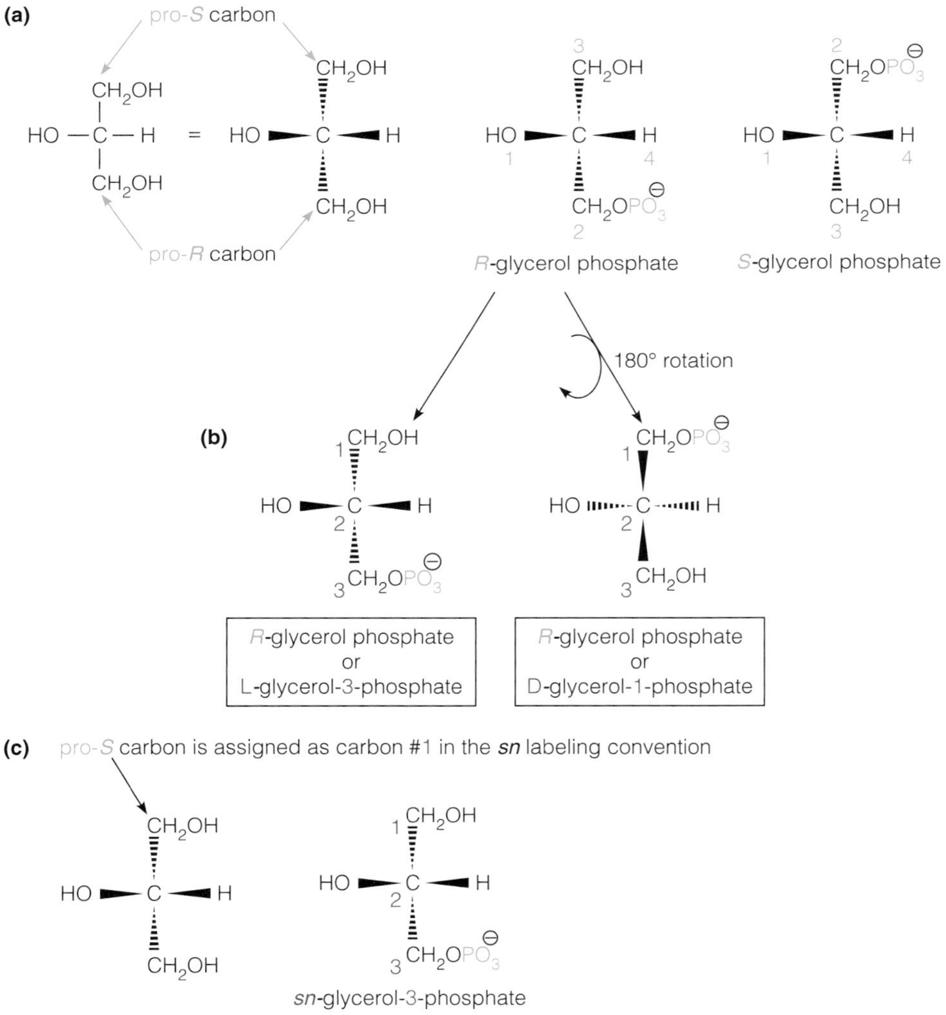

FIGURE 10.6

Stereochemistry of glycerophospholipids. **(a)** Glycerol is a prochiral molecule. Phosphorylation of one — CH$_2$OH group or the other gives the *R*- or *S*-enantiomer of glycerol phosphate. The numbers in blue text give the Cahn-Ingold-Prelog priority for these groups (see Figure 9.5, page 312). **(b)** The same molecule can be called L-glycerol-3-phosphate or D-glycerol-1-phosphate depending on the carbon numbering scheme. Carbon numbering is indicated in red text. **(c)** The *sn* (stereochemical numbering) system assigns the pro-*S* carbon as C1. All glycerophospholipids are derivatives of *sn*-glycerol-3-phosphate.

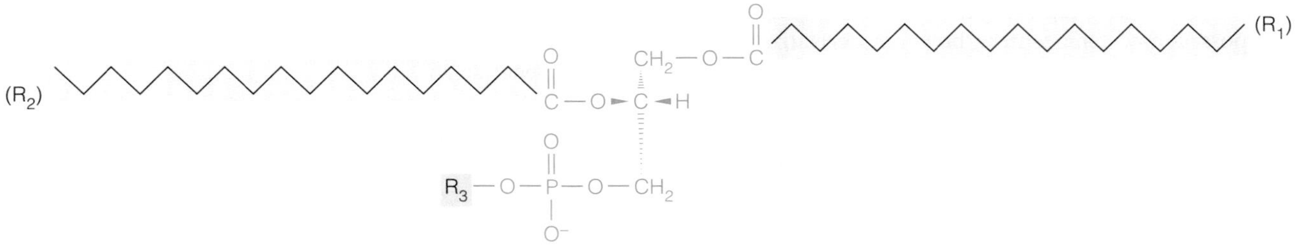

(a)

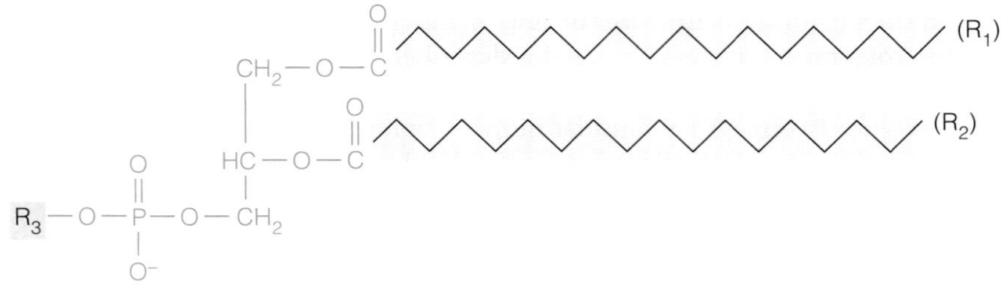

(b)

FIGURE 10.7

Glycerophospholipid structure. **(a)** Stereochemical view of a generalized glycerophospholipid. **(b)** The same structure represented in the convention used in this text, with hydrophobic groups to the right, hydrophilic to the left. R_3 is a hydrophilic group (see Table 10.3).

TABLE 10.3	The hydrophilic groups[a] that distinguish common glycerophospholipids

Name of Glycerophospholipid	R_3 (in Figure 10.7)
Phosphatidic acid	$H-$ (ionized at neutral pH)
Phosphatidyl-ethanolamine (PE)	$H_3\overset{+}{N}-CH_2-CH_2-$
Phosphatidylcholine (PC)	$(CH_3)_3\overset{+}{N}-CH_2-CH_2-$
Phosphatidylserine (PS)	$H_3\overset{+}{N}-\overset{\overset{H}{\mid}}{\underset{\underset{COO^-}{\mid}}{C}}-CH_2-$
Phosphatidyl inositol (PI)	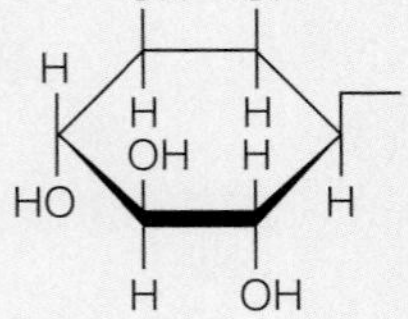

[a]These are the R_3 groups in Figure 10.7. In addition to this variation, there is also a great deal of variation in the hydrocarbon tails (R_1, R_2 groups).

bilayer, and the group at C3 is polar and on the outside face of the bilayer, we will draw these groups in the manner shown in Figure 10.7b, with the hydrophobic tails drawn to the right and the hydrophilic head group to the left.

Typically, the groups R_1 and R_2 are acyl side chains derived from the fatty acids; often R_1 is saturated, and R_2 is unsaturated. The hydrophilic R_3 head group varies greatly, and confers the greatest variation in properties among the glycerophospholipids. A gallery of the most common glycerophospholipid head groups is shown in Table 10.3, and their relative abundances in some membranes are given in Table 10.4. The simplest member of the group, **phosphatidic acid**, is only a minor membrane constituent; its principal role is as an intermediate in the synthesis of other glycerophospholipids or triglycerides (described in Chapter 19). The names of glycerophospholipids are derived from phosphatidic acid: *phosphatidylcholine*, *phosphatidylethanolamine*, and so on. As Table 10.3 shows, the glycerophospholipids have very polar head groups. The net charge on a phospholipid is

TABLE 10.4	Lipid composition of some biological membranes			
	Percentage of Total Composition in			
Lipid	Human Erythrocyte Plasma Membrane	Human Myelin	Bovine Heart Mitochondria	*E. coli* Cell Membrane
Phosphatidic acid	1.5	0.5	0	0
Phosphatidylcholine	19	10	39	0
Phosphatidylethanolamine	18	20	27	65
Phosphatidylglycerol	0	0	0	18
Phosphatidylinositol	1	1	7	0
Phosphatidylserine	8.0	8.0	0.5	0
Sphingomyelin	17.5	8.5	0	0
Glycolipids	10	26	0	0
Cholesterol	25	26	3	0
Others	0	0	23.5	17

Data from C. Tanford (1973) *The Hydrophobic Effect*. Wiley, New York.

a function of the charge (if any) on the R_3 group in combination with the negatively charged phosphodiester in the head group. Because the hydrocarbon tails are derived from the naturally occurring fatty acids in various combinations, an enormous variety of glycerophospholipids exists. For example, the erythrocyte membrane contains molecules with hydrocarbon chains of 16 to 24 carbons, with 0 to 6 double bonds. Such variation in membrane composition allows "fine-tuning" of membrane properties for the diverse functions that different membranes must perform.

Sphingolipids and Glycosphingolipids

A second major class of membrane constituents is built on the amino alcohol **sphingosine**, rather than on glycerol. The structure of sphingosine includes a long-chain hydrophobic tail, so it requires the addition of only one fatty acid to make it suitable as a membrane lipid. If a fatty acid is linked via an amide bond to the NH_2 group, the class of **sphingolipids** referred to as **ceramides** is obtained:

A wide variety of sphingolipids is built upon a sphingosine core.

Sphingosine = D-4-sphinganine

General structure of a ceramide (R = hydrocarbon)

Ceramides consist of sphingosine and a fatty acid. Further modification, by addition of groups to the C-1 hydroxyl of sphingosine, leads to a variety of other sphingolipids. An especially important example is *sphingomyelin*, in which a *phosphocholine* group is attached to the C-3 hydroxyl.

Phosphocholine   Ceramide

Sphingomyelin

In some of the membrane lipids built on sphingosine, the head group contains saccharides. Lipids containing saccharide groups go under the general name of **glycolipids**. The **glycosphingolipids** constitute the third major class of membrane lipids. Their general structure is shown in the margin. In addition to being constituents of the ABO blood group antigens (described in Chapter 9), they include such molecules as the **cerebrosides** (monoglycosyl ceramides) and **gangliosides**, anionic glycosphingolipids containing one or more sialic acid residues. Examples of these are shown in Figure 10.8. As the names of these compounds suggest, they are especially common in the membranes of brain and nerve cells.

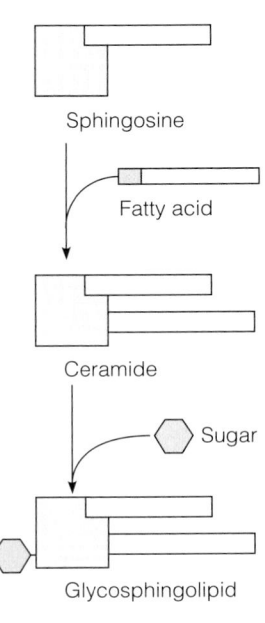

Composition of generalized glycosphingolipid

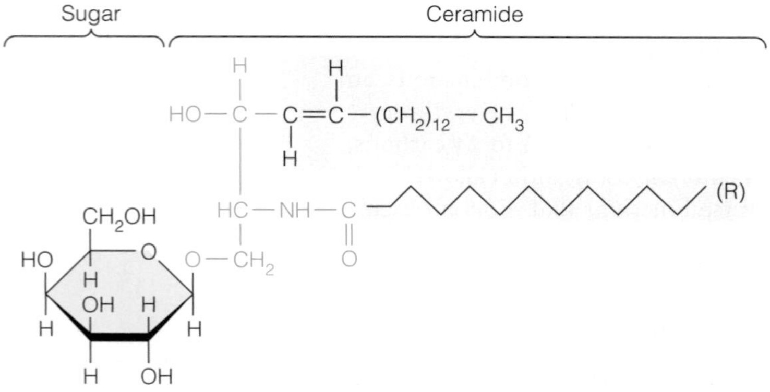

FIGURE 10.8

Examples of glycosphingolipids. **(a)** A cerebroside, an important constituent of brain cell membranes. **(b)** A ganglioside. This particular ganglioside, called GM_2 or the Tay-Sachs ganglioside, accumulates in neural tissue of infants with Tay-Sachs disease. The defect responsible for this inherited condition is the lack of an enzyme that normally cleaves the terminal GalNAc (see Chapter 19).

(a) Galactosylceramide

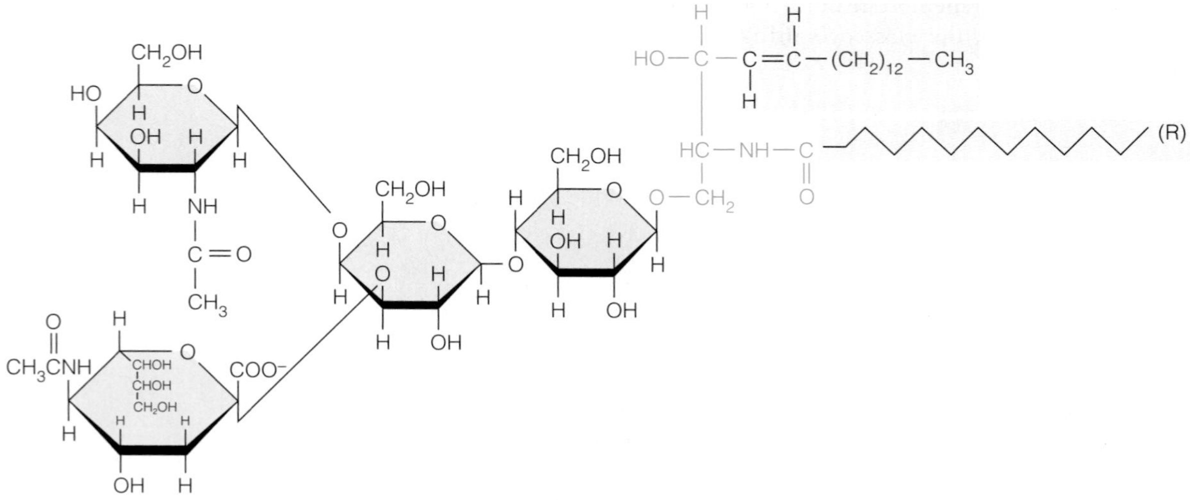

(b) GalNAcβ(1→4)Galβ(1→4)Glcβ(1→1)ceramide or, Ganglioside GM_2

$$\left(\begin{array}{c} 3 \\ \uparrow \\ \alpha 2 \end{array}\right)$$
Sia

Glycoglycerolipids

Another class of lipids, less common in animal membranes but widespread in plant and bacterial membranes, are the **glycoglycerolipids**, exemplified by **monogalactosyl diglyceride**:

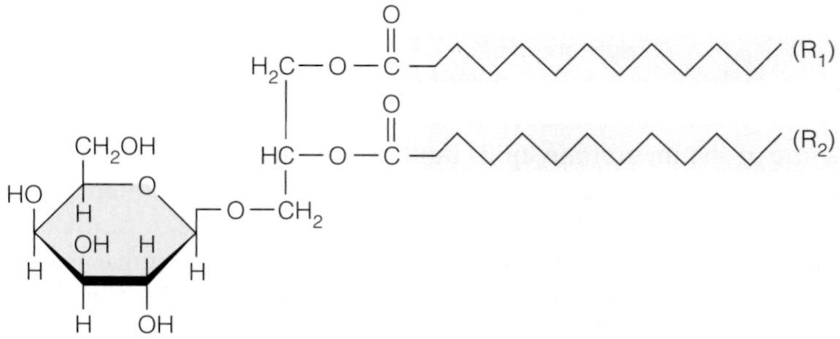

This compound may actually be the most abundant of all polar lipids, for it constitutes about half the lipid in chloroplast membranes (see Chapter 16). Such lipids are also abundant in archaea, where they are the major membrane components.

Cholesterol

One important lipid constituent of many membranes bears little superficial resemblance to the compounds we have studied so far. This substance is **cholesterol**, the structure of which is shown in Figure 10.9. Cholesterol is a member of a large group of substances called **steroids**, which include a number of important hormones, among them the sex hormones of higher animals. In fact, cholesterol is the precursor for many of these substances. Its role in these syntheses is discussed in Chapter 19, along with a detailed description of other steroids and their functions in cell signaling. Another important and diverse class of signaling molecules derived from lipids is the **eicosanoids**, which are derived from arachidonic acid (Table 10.1). These are potent activators of a wide range of physiological functions, including inflammation, blood clotting, blood pressure regulation, and reproduction. The biosynthesis and activities of the steroids and eicosanoids are described in Chapter 19.

Cholesterol is a weakly amphipathic substance because of the hydroxyl group at one end of the molecule. As the conformational structure in Figure 10.9b shows, the fused cyclohexane rings in cholesterol are all in the chair conformation. This makes cholesterol a bulky, rigid structure as compared with other hydrophobic membrane components such as the fatty acid tails; thus, the cholesterol molecule tends to disrupt regular packing of fatty acid tails in membrane structure. This property can have a major effect because cholesterol constitutes 25% or more of the lipid content in some membranes (see Table 10.4). Such changes in membrane structure can have, as we shall see, profound effects on such properties as membrane stiffness and permeability. Other steroids are also found in membranes; for example, *lanosterol* (page 798) is prominent in plant cell membranes.

The molecules described above constitute the major portion of membrane lipids in most organisms. However, one of the "three kingdoms" of organisms—the archaea—are unique in having glycoglycerolipids as their major membrane lipids.

The Structure and Properties of Membranes and Membrane Proteins

The membranes of living cells are remarkable bits of molecular architecture, with many and varied functions. To say that a membrane is essentially a phospholipid bilayer is a gross oversimplification. To be sure, the phospholipid bilayer, as depicted in Figure 10.5b and c, forms the basic structure, but there is much more to the membranes found in living cells. Some of the complex features of a typical eukaryotic cell membrane are shown in Figure 10.10. An important feature of cellular membranes is the wide variety of specific proteins contained within the lipid bilayer or bound to its surface. Many of these proteins carry oligosaccharide groups that project into the surrounding aqueous medium. Other oligosaccharides are carried by glycolipids, with the lipid portions inserted in the membrane. The two sides of the bilayer are usually different, both in lipid composition and in the placement and orientation of proteins and oligosaccharides.

The protein content varies greatly among different kinds of membranes (see Table 10.5) and appears to be directly related to the functions a particular membrane must carry out. Mitochondrial inner membranes and bacterial cell-wall membranes, which carry out many functions, are about 75% protein. The myelin of nerve fibers, which acts primarily as an electrical insulator, has a much lower protein content (~20%). As a rule of thumb, a typical membrane is roughly 60% protein and 40% lipid by mass.

Much of our current understanding concerning biological membranes is based upon the **fluid mosaic model** proposed by S. J. Singer and G. L. Nicolson in 1972. This is the model depicted in Figure 10.10. The fluid, asymmetric lipid bilayer carries within it a host of proteins. Some of them, called **peripheral membrane**

Cholesterol, a component of many animal membranes, influences membrane fluidity by its bulky structure.

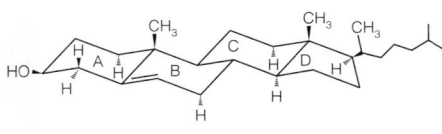

(a)

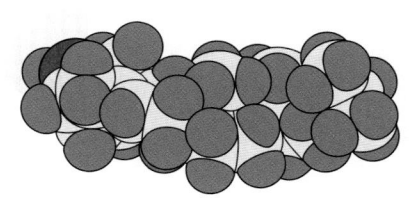

(b)

(c)

FIGURE 10.9

Cholesterol. **(a)** Structural formula. **(b)** Conformational model. **(c)** Space-filling model.

FIGURE 10.10

Structure of a typical cell membrane. In this schematic view, a strip of the plasma membrane of a eukaryotic cell has been peeled off. Proteins are embedded in and on the phospholipid bilayer; some of them are glycoproteins, carrying oligosaccharide chains. The membrane is about 6 nm thick. Most membranes are more densely packed with proteins than is shown here.

According to the fluid mosaic model, a membrane is a fluid mixture of lipids and proteins.

Peripheral membrane proteins are associated with one side of the bilayer and can be separated from the membrane without disrupting the bilayer. Integral membrane proteins are more deeply embedded in the bilayer and can only be extracted under conditions that disrupt membrane structure. Many integral membrane proteins extend through the bilayer.

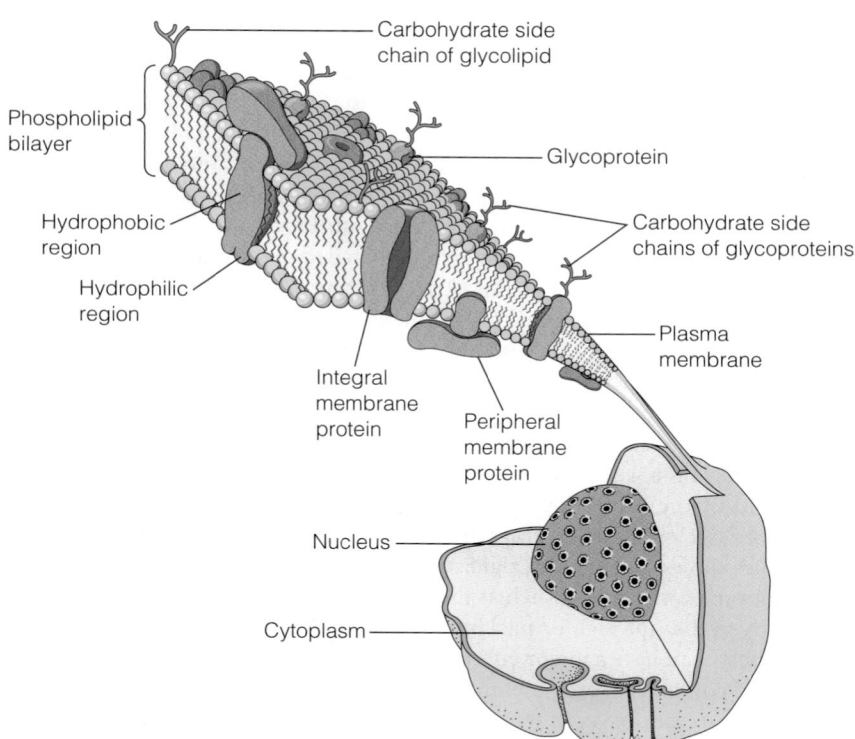

FIGURE 10.11

Experimental demonstration of membrane fluidity. When cells with surface membrane protein marked by fluorescent tags are induced to fuse, the proteins gradually mix over the fused surface.

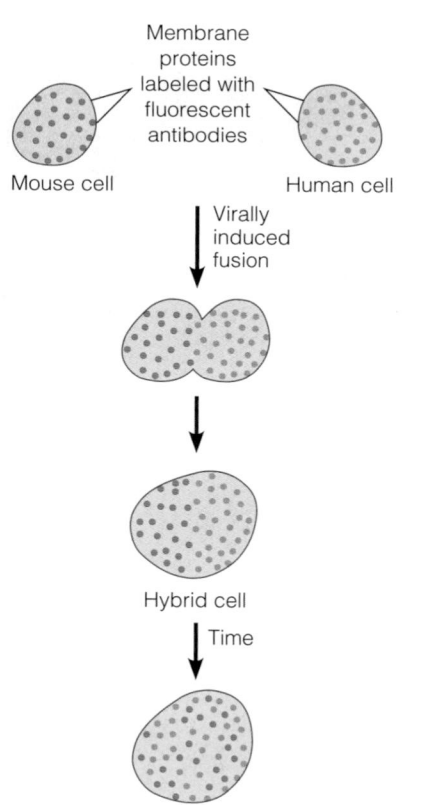

proteins, are exposed at only one membrane face or the other. They are held to the membrane by interaction with lipid heads or integral membrane proteins. The **integral membrane proteins** are largely buried within the membrane but are usually exposed on both faces. Integral proteins are frequently involved in transmitting either specific substances or transducing chemical signals through the membrane. The whole membrane is a mosaic of lipids and proteins. Current research suggests that the membrane surface is more crowded, and the distribution of proteins in membranes is more highly organized than is depicted in Figure 10.10.

Motion in Membranes

A functioning biological membrane is not a rigid, frozen structure. In fact, most of the lipid and protein components are in constant motion. This motion can be demonstrated in a direct and dramatic way. If human and mouse cells, each carrying a distinctive fluorescent marker in its plasma membrane, are fused together, the two kinds of markers gradually become intermixed (Figure 10.11). This

TABLE 10.5   Protein, lipid, and carbohydrate content of some membranes

Membrane	Percent by Weight		
	Protein	Lipid	Carbohydrate
Myelin	18	79	3
Human erythrocyte (plasma membrane)	49	43	8
Bovine retinal rod	51	49	0
Mitochondria (outer membrane)	52	48	0
Amoeba (plasma membrane)	54	42	4
Sarcoplasmic reticulum (muscle cells)	67	33	0
Chloroplast lamellae	70	30	0
Gram-positive bacteria	75	25	0
Mitochondria (inner membrane)	76	24	0

Adapted from *Annual Review of Biochemistry* 41:731, G. Guidotti, Membrane proteins. © 1972 Annual Reviews.

demonstrates that *lateral diffusion* (parallel to the membrane surface) can occur in the membrane. The rapidity with which such two-dimensional diffusion can occur depends on the membrane fluidity, which in turn depends on temperature and lipid composition. Under physiological conditions, the average time required for a phospholipid molecule to wander completely around a cell is on the order of seconds to minutes; membrane proteins also move, but more slowly, and their range may be constrained by other structural features of the membrane.

Motion in Synthetic Membranes

The effects of temperature and composition on fluidity can be most simply studied using artificial membranes containing only one or a few kinds of lipids and no proteins (see Tools of Biochemistry 10A). Figure 10.12a depicts the behavior of a membrane made entirely from phosphatidylcholine carrying two 16-carbon saturated chains (PC-16:0/16:0, in shorthand). At low temperatures the hydrocarbon tails pack together closely to form a nearly solid *gel* state. If the temperature is raised above 41 °C, a **phase change** occurs in which this regular order is lost, and the hydrocarbon tails become free to move about. The membrane "melts" to adopt a semifluid *liquid crystalline* state. The temperature at which this happens is called the *transition temperature*. This abrupt change in synthetic membrane properties can be detected by a number of the techniques described in Tools of Biochemistry 10A. Figure 10.12b shows the transition as detected by calorimetry (see Tools of Biochemistry 6C).

The transition temperature is very sensitive to the nature of the hydrocarbon tails. If a PC-14:0/14:0 membrane is used (with tails only two carbons shorter than those just described), the transition temperature drops to 23 °C. If a single *cis* double bond is incorporated into each 16-carbon tail (PC-16:1/16:1), melting occurs at 36 °C. As explained earlier, *cis* double bonds put bends in the chains, disrupting their close packing; thus, such chains must be cooled to a lower temperature to produce the gel phase. Changing the head group can also make a big difference: If phosphatidylethanolamine (Table 10.3) is substituted for phosphatidylcholine in PC-16:0/16:0, the thermal transition is raised to 63 °C. The sensitivity of the transition to lipid composition is shown dramatically by the fact that combining several of the changes described above can change the membrane transition temperature over a range of 100 °C.

Motion in Biological Membranes

Biological membranes, which contain complex mixtures of lipid components plus protein, exhibit much broader and more complex phase transitions than those observed for synthetic bilayers of the kind described above. Indeed, there is now evidence for the existence of quite stable "domains" of different composition in different parts of a cell membrane. This explains the broader transition observed. Because it is essential that the membranes in living cells be fluid, the membrane composition is regulated so as to keep the transition temperature below the body temperature of the organism. One example is found in bacteria, which will alter the saturated/unsaturated fatty acid ratio in their membranes in response to a change in the temperature at which they are grown. A remarkable case in the animal kingdom is that of the reindeer's leg. Its cell membranes show an increase in relative amount of unsaturated fatty acids near the hoof, which is usually cooler than the rest of the body.

Cholesterol has a specific and complex effect on membrane fluidity. As Figure 10.12b shows for a synthetic membrane, cholesterol does not influence the transition temperature markedly, but it does broaden the transition. It has been hypothesized that this broadening occurs because cholesterol can both stiffen the membrane above the transition temperature and inhibit regularity in structure formation below the transition temperature. Thus, it blurs the distinction between the gel and the fluid state. There is evidence that variations in cholesterol content are used to regulate membrane behavior in some organisms.

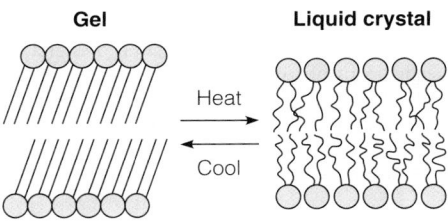

Gel **Liquid crystal**

Heat →
← Cool

Head groups tightly packed; tails regular; membrane thicker

Head groups loosely packed; tails disordered; membrane thinner

(a) Transition

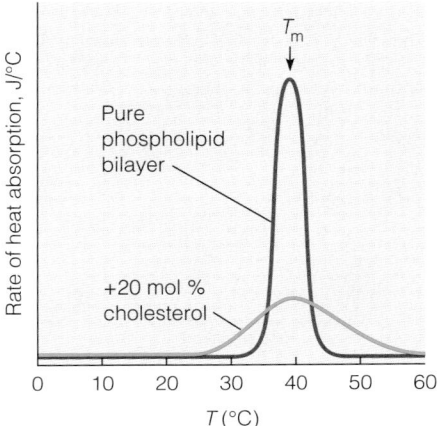

(b) Transition with and without cholesterol

FIGURE 10.12

The gel–liquid crystalline phase transition in a synthetic lipid bilayer. **(a)** A schematic view of the change at the transition temperature. Below this temperature the hydrocarbon tails are packed together in a nearly crystalline gel state (left). Above this temperature the movement of the chains becomes more dynamic, and the interior of the membrane resembles a liquid hydrocarbon (right). **(b)** Detection of the transition by calorimetry (see Tools of Biochemistry 6C). Measurement of the heat absorbed by a membrane as the temperature is raised each degree shows a sharp spike at the transition temperature (T_m) for a pure dipalmitoylphosphatidylcholine bilayer. This well-defined transition from gel to liquid is called melting of the membrane. When 20 mol % cholesterol is mixed into the bilayer, the transition temperature is not changed, but the transition is broadened.

The transition temperature for a membrane depends on its lipid composition. Lipids with longer, saturated tails tend to increase the transition temperature, whereas those with more *cis* double bonds and/or shorter tails will reduce the transition temperature.

Under physiological conditions, biological membranes exist in a semifluid liquid crystalline state.

FIGURE 10.13

A schematic model of the effects of cholesterol on synthetic membrane structure. The data indicate an initial effect of the cholesterol/phospholipid ratio, C:PL, on membrane width. At C:PL mode ratios at or below 0.8:1, bilayer width increases with increasing C:PL. At C:PL mole ratios above 0.9:1, two separate lamellar phases form, one representing a liquid crystalline lipid bilayer and another representing an immiscible cholesterol phase. The immiscible cholesterol phase forms by phase separation as the membrane "saturates" with cholesterol.

The Journal of Lipid Research 39:947–956, T. Tulenko, M. Chen, P. E. Mason, and R. P. Mason, Physical effects of cholesterol on arterial smooth muscle membranes: Evidence of immiscible cholesterol domains and alterations in bilayer width during atherogenesis. Adapted with permission. © 1998 The American Society for Biochemistry and Molecular Biology. All rights reserved.

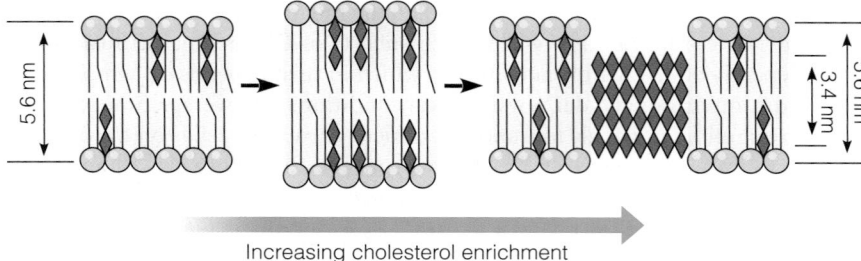

Increasing cholesterol enrichment

The effects of cholesterol on membrane structure are strongly dependent on its concentration in the membrane. X-ray scattering studies on reconstituted synthetic bilayers have shown that at moderate concentrations cholesterol fits into the bilayer and may thicken it (Figure 10.13), whereas at higher concentrations, "islands" of cholesterol bilayers form. These, it is speculated, may provide nuclei for the formation of cholesterol plaques in the circulatory system. The formation of such structures has been demonstrated to be an in vivo effect by experiments in which rats were fed high levels of cholesterol for many weeks. As we discuss later in this chapter, cholesterol also appears to play a role in organizing smaller regions of the bilayer into functional units referred to as "lipid rafts." The effect of cholesterol on modulating the thickness of biological membranes appears to be marginal, based on recent experiments in which cholesterol was removed from the membranes of rat liver cells and the associated organelles. In these membrane systems, protein content of the membrane had a much greater effect on regulating bilayer thickness than did cholesterol.

The Asymmetry of Membranes

In contrast to the ease of lateral movement, the "flip-flop" of lipid molecules *across* synthetic lipid bilayers, from one side to the other, is much slower. The reason is not hard to see: When a phospholipid molecule turns from one face to the other, it must pass its very hydrophilic head through the hydrophobic medium of the hydrocarbon tails. Such an event is very unfavorable from an energetic point of view, and hence the process is slow. There exist enzymes (*translocases* and *flippases*) that catalyze membrane lipid flipping. In contrast to phospholipids, the transport of fatty acids across membranes is much more rapid (less than a second), for a simple reason. Unlike the phosphate groups on phospholipids, the carboxylates on fatty acids have a pK_a that is sufficiently high to allow protonation in the nonpolar environment of the lipid bilayer. Thus, they lose their negative charge and can partition into the hydrophobic core of the membrane more favorably than an ionized molecule.

Every biological membrane has two distinct faces, each encountering a different environment. Examples can be seen in the illustrations of cell ultrastructure shown in Chapter 1. The plasma membrane of a cell faces the external environment on the outside and the cytoplasm on the inside, whereas the membrane around a chloroplast faces the photosynthetic apparatus on the inside and the cytoplasm on the outside. Because the two faces of a membrane must deal with different surroundings, the faces are usually quite different in composition and structure.

This difference extends even to the level of phospholipid composition. Recall that all phospholipid membranes are bilayers; the two individual layers are called *leaflets*. The compositions of the two leaflets in the plasma membranes of several kinds of cells are shown in Figure 10.14. Not only are the individual lipids distributed very asymmetrically, but the distribution also varies considerably among cell types.

The two leaflets of a membrane usually differ in lipid composition.

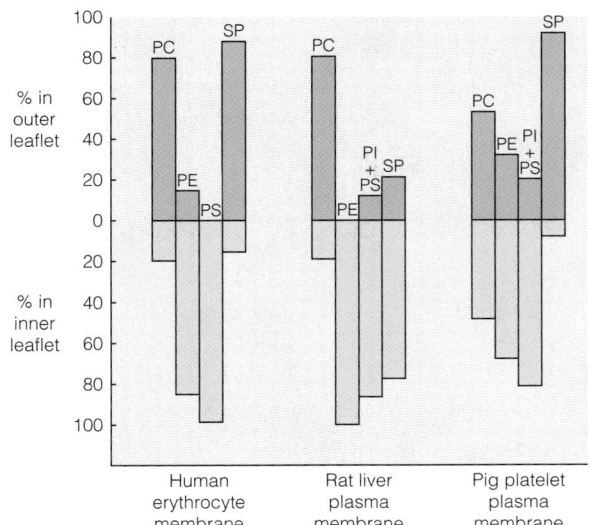

FIGURE 10.14

Phospholipid asymmetry in plasma membranes. Lipid composition in the outer leaflet (green) and inner leaflet (gold) of the plasma membrane is graphed for three cell types. PC = phosphatidylcholine; PE = phosphatidylethanolamine; PS = phosphatidylserine; PI = phosphatidylinositol; SP = sphingomyelin.

Membranes are highly specialized structures that serve diverse functions in different cells/tissues. The lipid and protein content of a given membrane is tailored to the specific function of that membrane. For example, differences in charged groups between bilayer leaflets lead to differences in the electrical potentials (discussed later in this chapter) across various membranes. Glycoproteins and glycolipids carried in the outer leaflet of a plasma membrane contribute, via their oligosaccharide chains, to identification of cells (see Chapter 9).

Much of our knowledge of membrane asymmetry comes from studies of **vesicles**, fragments of membrane that have resealed to form hollow shells, with an inside and an outside. Reagents can be either captured inside the vesicle or added only to the surrounding solution so that they can react specifically with either outward-facing or inward-facing proteins or lipids. A membrane protein in a vesicle may be reacted covalently with a radioactively labeled reagent, isolated, and cleaved into peptides by proteases. Identification of which peptides are labeled by "inside" or "outside" reactants can reveal which portions of the protein were on the inner face and which were on the outer face. In a similar way, lipids can be tested, using enzymes or other reagents that cleave off or otherwise modify the head groups. Experiments of this kind, performed inside or outside vesicles, have provided much of the information shown in Figure 10.14.

Biological membranes are highly dynamic structures. Not only must they continually expand as cells grow and divide but also, even in resting cells, a continual turnover and renewal of the membrane components apparently occurs. Indeed, this dynamic, nonequilibrium state is a necessity if asymmetry is to be maintained. From the second law of thermodynamics (see Chapter 3) we know that the equilibrium state of a two-layer membrane would require an equal distribution of every component on either side. Like so many other biological systems, membranes exist as they do because they are *not* in equilibrium but rather represent dynamic, steady-state structures.

Characteristics of Membrane Proteins

Membrane proteins possess special characteristics that distinguish them from other globular proteins. They often contain a high proportion of hydrophobic amino acids in the parts of the protein molecules that are embedded in the membrane (see Figure 10.15). The segments of proteins that span membranes are often α-helical, although β-barrels are also common membrane-spanning motifs. Figure 10.16 depicts bacteriorhodopsin, an integral membrane protein whose structure has been solved to high resolution. Like many such proteins, it contains a bundle of seven α-helical segments that pass back and forth through the

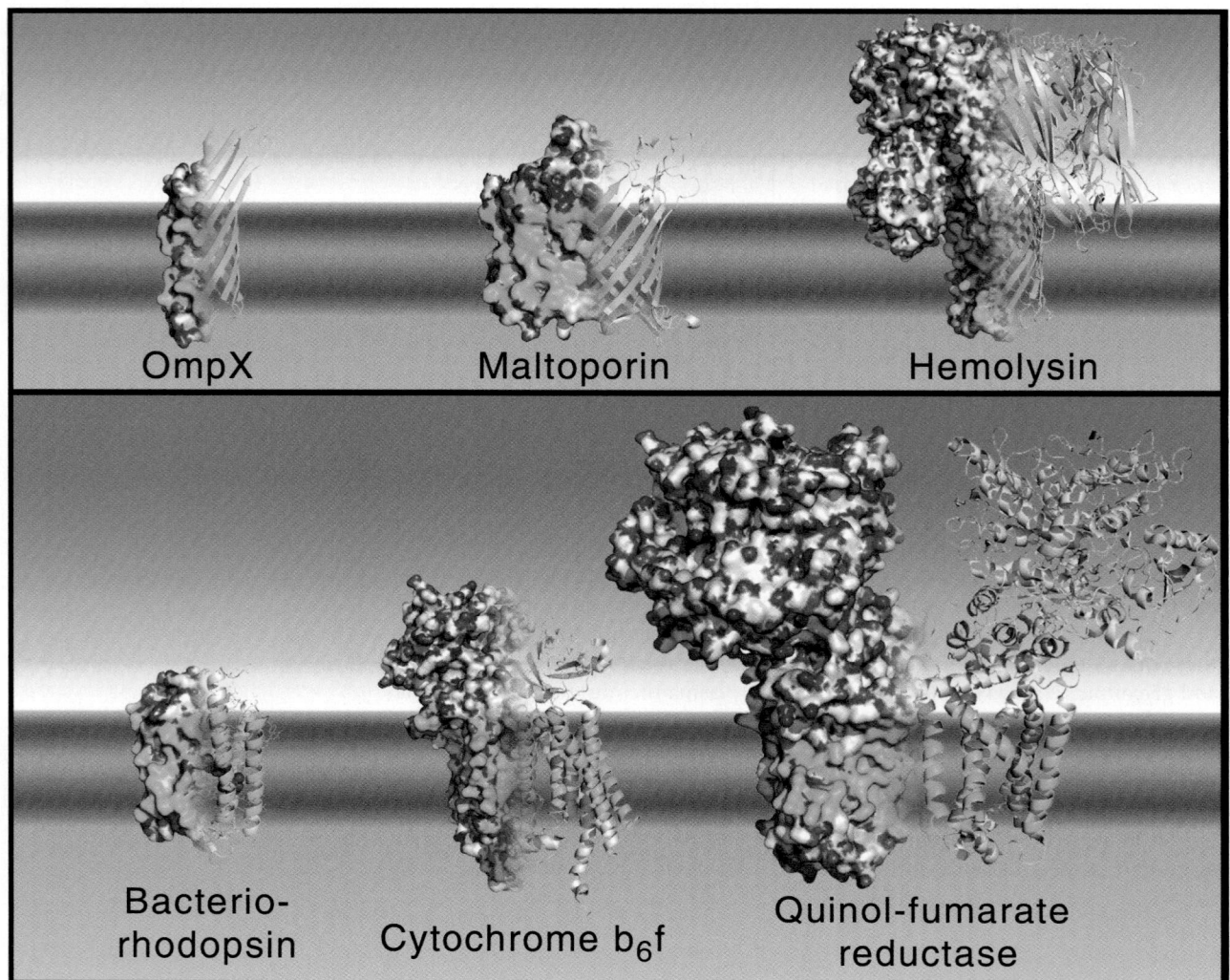

FIGURE 10.15

Examples of structures for several integral membrane proteins. Each protein is rendered half in a space-filling model (left half of each structure) and half in a cartoon (right half). The cartoon rendering shows the differences in membrane-spanning structure, whereas the space-filling models show similar distributions of hydrophilic (red and blue) and hydrophobic (green and gray) surface areas for both classes of proteins. The membrane-spanning regions are dominated by hydrophobic surface area. The bilayer is indicated by the brown band, and the cytosolic region is indicated by the aqua-colored region. **Top panel**: Examples of β-barrel transmembrane structures. **Bottom panel**: Examples of α-helical transmembrane proteins. *E. coli* outer membrane protein X (PDB ID: 1QJ9) is involved in biofilm formation. *S. typhimurium* maltoporin (PDB ID: 2MPR) facilitates diffusion of certain saccharides across the outer membrane of gram negative bacteria. *S. aureus* hemolysin (PDB ID: 7AHL) is a toxin that opens a pore in target cells and lyses them. *H. salinarum* bacteriorhodopsin (PDB ID: 1C3W) is a proton pump in photosynthetic bacteria. Cytochrome *b6f* from *C. reinhardtii* (PDB ID: 1Q90) is a photosynthetic proton pump and electron transporter. *E. coli* quinol-fumarate reductase (PDB ID: 1L0V) is an electron transport protein that catalyzes the reduction of fumarate to succinate.

Courtesy of Gary Carlton.

membrane. The presence of such transmembrane segments can sometimes be inferred from the kind of **hydrophobicity plot** shown in Figure 10.17. This plot has been calculated according to the hydrophobicity scale of Kyte and Doolittle given in Table 6.5. It reveals maxima in regions of the sequence corresponding to the transmembrane helices. Transmembrane helices typically contain 20–25 residues, which are predominantly hydrophobic.

Another class of membrane-associated proteins is covalently modified with lipids as shown in Figure 10.18. Many proteins involved in signaling are modified in this way, via reactions that are formally acyl group transfers, or **acylations**. Specific sequences at the N-terminus or C-terminus of the protein are required for the addition of **geranylgeranyl**, **farnesyl**, and **myristoyl** groups. For example, myristoylation occurs at N-terminal Gly residues following cleavage of the N-terminal Met, and farnesylation requires a C-terminal "CaaX" motif (where C = Cys, a = an aliphatic amino acid, and X = any amino acid except Pro). These processes are presented in greater detail in Chapters 19 and 23. Another common protein modification is the addition of one or more palmitoyl groups to a Cys sidechain (or to a lesser extent, the N-terminal amino group).

As discussed below, the attached lipid may serve primarily to target the modified protein to a particular subcellular organelle or to a specific location in the plasma membrane. For example, proteins linked to **glycosylphosphatidylinositol** (GPI) are frequently associated with regions of the outer leaflet rich in cholesterol and sphingolipids. Although the lipid moiety may insert directly into the bilayer, several membrane proteins with lipid-binding sites are known. Thus, the association of a lipid-linked protein with the membrane may be mediated by protein-lipid binding.

Membranes are complex structures, with specific compositions for each of the two leaflets. To make the picture more concrete, let us consider in some detail the structure of one example, the plasma membrane of the erythrocyte (red blood cell).

The Erythrocyte Membrane: An Example of Membrane Structure

The erythrocytes of mammals are among the simplest of all cells. In their mature state in the circulating blood, they lack mitochondria, internal membranes, and a nucleus. They are essentially bags containing a concentrated solution of hemoglobin. They can easily be lysed to release their contents and produce membrane *ghosts*, large vesicles that represent a nearly pure preparation of the plasma membrane of the cells. Erythrocyte ghosts have the lipid composition given in Table 10.4, distributed between inner and outer leaflets as shown in Figure 10.14. If the total protein content of the erythrocyte ghost is extracted with detergent and analyzed by SDS gel electrophoresis, the pattern shown in Figure 10.19a is obtained.

Separating Peripheral and Integral Proteins

The plasma membrane of erythrocytes contains far fewer proteins than are found in most other cell membranes, in keeping with the simple metabolism of these cells. What are these proteins, and what do they do? To begin, these proteins are classified as either peripheral or integral membrane proteins. The peripheral proteins can be washed off the ghosts by simple changes in ionic strength or pH. All the peripheral proteins in erythrocytes are attached to the inside (cytoplasmic face) of the erythrocyte membrane (Figure 10.20). Electron microscopy after freeze-etching (see Tools of Biochemistry 10A) reveals a nearly smooth outer surface, with only integral membrane proteins and attached carbohydrates decorating the surface, whereas the inside of the membrane and the inner surface are rich in protein particles.

The major integral proteins of this membrane are the anion channel (band 3), band 4.5, and the glycophorins. Note that glycophorin and band 4.5 do not appear in Figure 10.19 because of the particular dye used in the analysis. The integral proteins, along with much of the lipid material, can be extracted from intact membranes by use of the nonionic detergent Triton X-100.

The Protein Skeleton

Surprisingly, treatment with Triton X-100 leaves intact a protein skeleton, which retains the shape of the membrane ghost. The proteins of this membrane skeleton

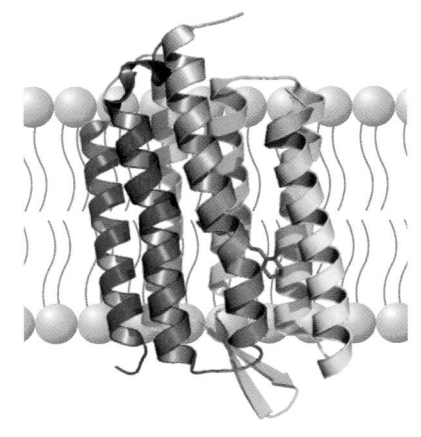

FIGURE 10.16

Bacteriorhodopsin—an integral membrane protein. Bacteriorhodopsin (PDB ID: 1C3W) functions as a light-driven proton pump in certain bacteria. Seven helices span the membrane and hold a molecule of the light-absorbing pigment **retinal** (magenta). The biosynthesis of retinal is described in Chapter 19.

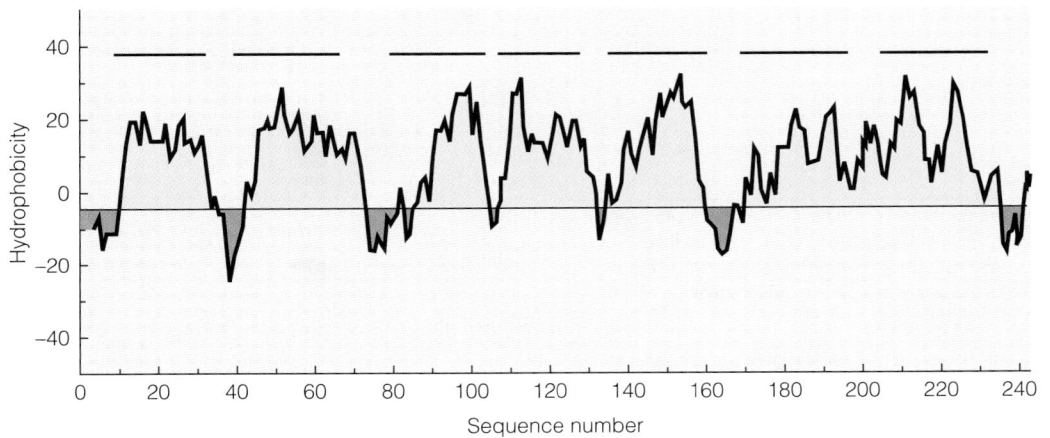

FIGURE 10.17

Hydrophobicity plot for the bacteriorhodopsin molecule depicted in Figure 10.16. The hydrophobicity index has been calculated at each residue by the method of Kyte and Doolittle. The black bars show the approximate positions of the transmembrane helices shown in Figure 10.16.

Adapted from *Journal of Molecular Biology* 157:105–132, J. Kyte and R. F. Doolittle, A simple method for displaying the hydropathic character of a protein. © 1982, with permission from Elsevier.

FIGURE 10.18

Protein lipidation. A given protein may be modified by more than one of these lipid acyl groups. Myristoylation, farnesylation, geranylgeranylation, and palmitoylation occur on the cytoplasmic side of the membrane. The glycosylphosphatidylinositol modification is found only on the outside of the bilayer. The glycan portion of GPI is composed of mannose (green), glucosamine (blue), and inositol (red).

Courtesy of Gary Carlton.

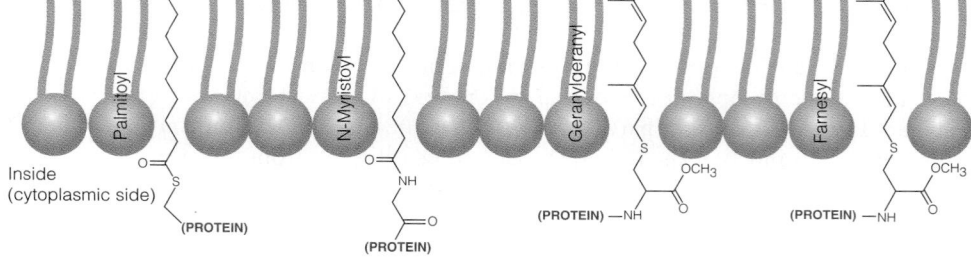

FIGURE 10.19

Gel electrophoretic analysis of erythrocyte membrane proteins. (a) Peripheral and integral proteins of the erythrocyte ghost. Glycophorin and band 4.5 protein do not show up here because they do not stain with the dye that was used. **(b)** Proteins of the erythrocyte skeleton (see Figure 10.20). The skeleton consists of the peripheral membrane proteins that remain after integral proteins have been extracted with the detergent Triton X-100.

Reprinted from *Cell* 24:24–32, D. Branton, C. M. Cohen, and J. Tyler, Interaction of cytoskeletal proteins on the human erythrocyte membrane. © 1981, with permission from Elsevier.

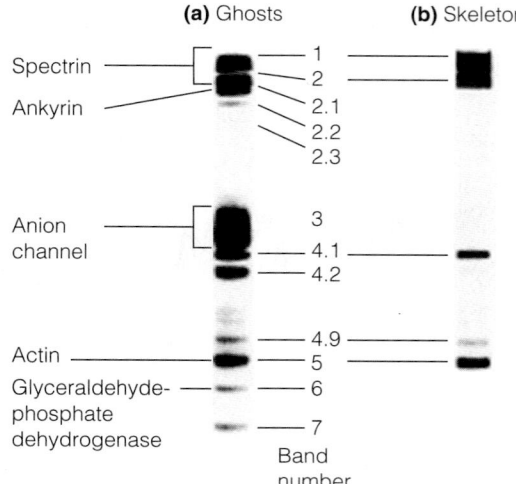

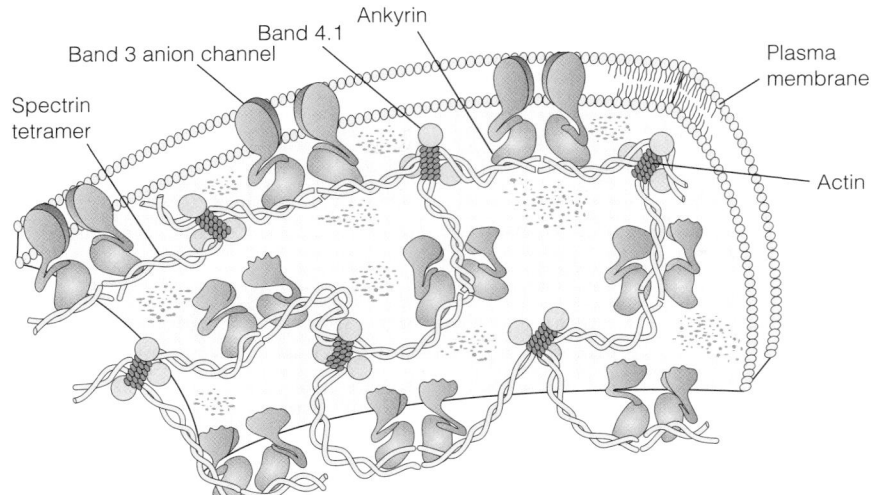

FIGURE 10.20

Model of the postulated structure of the erythrocyte membrane skeleton. The proteins are identified in Figure 10.19. Note that ankyrin anchors the membrane to the skeleton by interacting with both spectrin and the integral band 3 protein (anion channel).

are identified in Figure 10.19b, and a schematic picture is shown in Figure 10.20. The skeleton is a two-dimensional network of some of the peripheral proteins, the major constituents being **spectrins**, actin, and band 4.1. The 200-nm-long fibers in this network are made up of $\alpha_2\beta_2$ tetramers of spectrin molecules. These very elongated molecules contain a large fraction of α-helix and appear to be linked at their ends through short chains of actin molecules, together with the band 4.1 protein and adducin. The actin is also complexed with an erythrocyte-specific tropomyosin (see Chapter 8). The skeleton is anchored to the membrane itself by the protein **ankyrin**, which binds to both spectrin and the band 3 integral protein. Band 4.1 adds to the structure by also binding to glycophorin, another integral protein.

What purpose does this elaborate underpinning to the erythrocyte membrane serve? An obvious suggestion is that it helps to maintain the shape of the erythrocyte under the extreme distortions and shear stress the red cell suffers in passing through the circulatory system. The erythrocyte is durable, typically surviving for about 120 days, or 10 million heartbeats. The discoid shape of the cell allows efficient exchange of O_2 and CO_2 to the hemoglobin inside, and even if that shape is momentarily deformed, the skeleton helps to re-establish it. Indeed, deficiencies in these skeletal proteins result in easy lysis of erythrocytes, which is the cause of some forms of anemia. The kind of structure described here is not confined to erythrocytes; many other cell types contain similar but distinct membrane skeletons. It seems likely that connections exist between the membrane skeleton and the cytoskeleton (see Chapter 8) so that the membrane is linked to intracellular structure.

The Major Integral Membrane Proteins

The most abundant protein in the erythrocyte membrane is the band 3 protein, which is an anion channel, facilitating the exchange of HCO_3^- for Cl^- across the erythrocyte membrane. Recall from Chapter 7 the importance of bringing HCO_3^- into the erythrocytes to accomplish CO_2 transport; exchange for Cl^- keeps ionic balance. Band 3 protein functions as a complex of 2 or 4 subunits. Each subunit chain crosses and recrosses the membrane a number of times, to create the channel through which the ions can be exchanged. The N-terminal portion of the protein extends into the cytosol, where it makes a number of interesting interactions. As mentioned previously, it contacts the skeletal protein ankyrin, providing a major attachment between the skeleton and the membrane. In addition, band 3 protein contacts a number of cytosolic proteins, including some glycolytic enzymes (glyceraldehyde-3-phosphate dehydrogenase and aldolase, for example; see Chapter 13) and hemoglobin. The significance of these interactions is not wholly clear.

The erythrocyte, like many other cells, has a complex "skeleton" of proteins underlying and attached to its plasma membrane.

The other major integral red cell membrane proteins are the **glycophorins**, which apparently serve a variety of functions. Each of these proteins has an external carbohydrate-carrying domain, a single transmembrane helix, and a cytosolic C-terminal domain (see Figure 10.21). The O-linked carbohydrate groups on the external N-terminal domain carry sialic acid residues, giving a negative charge to the exterior of the red blood cell. This may minimize the chances of such cells sticking to capillary walls during circulation. The role of the cytosolic domain is less clear, but in some glycophorins it appears to interact with band 4.1, stabilizing skeleton–membrane adhesion.

The asymmetry of the orientation of integral proteins must be emphasized. Just as the erythrocyte membrane is asymmetric in its distribution of lipid components, it is equally asymmetric in its distribution of proteins because each kind of protein is oriented in a particular direction. There appear to be special components of the cellular protein synthesis machinery that direct the placement of proteins in membranes and ensure their asymmetric orientation. We discuss only one aspect of that now: The co-translational insertion of transmembrane helices into the bilayer.

Insertion of Proteins into Membranes

Greater than 30% of proteins must cross, or integrate into, a cellular membrane. How do hydrophilic, globular proteins make it across a hydrophobic membrane bilayer during, or following, their ribosomal synthesis? The ribosomes are either free in the cytosol or bound to the rough endoplasmic reticulum (see Chapter 28 for a discussion of protein **secretion**). Given the cytosolic location of protein synthesis and the asymmetric orientation of integral membrane proteins, two fundamental questions arise: (1) How do integral membrane proteins get inserted into the membrane? and (2) How are they inserted into the membrane in the correct orientation?

One solution to the problem of protein insertion into a membrane is illustrated in Figure 10.22. The membrane-spanning regions of several integral membrane proteins are inserted into the bilayer co-translationally (i.e., *during* ribosomal protein synthesis), where they then fold. This process is facilitated by a

FIGURE 10.21

The sequence and postulated structure of glycophorin A. This protein was the first integral membrane protein to be sequenced. The external (N-terminal) domain carries 15 O-linked and one N-linked oligosaccharides; together these constitute about 60% of the total protein mass. The single transmembrane helix is highly hydrophobic, whereas the cytosolic C-terminal domain is quite hydrophilic.

Adapted from *Seminars in Hematology* 16:3–20, V. T. Marchesi, Functional proteins of the human red blood cell membrane. © 1979, with permission from Elsevier.

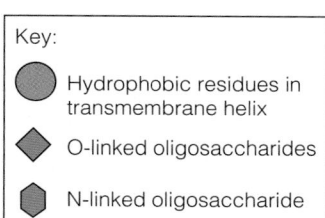

Key:
- Hydrophobic residues in transmembrane helix
- O-linked oligosaccharides
- N-linked oligosaccharide

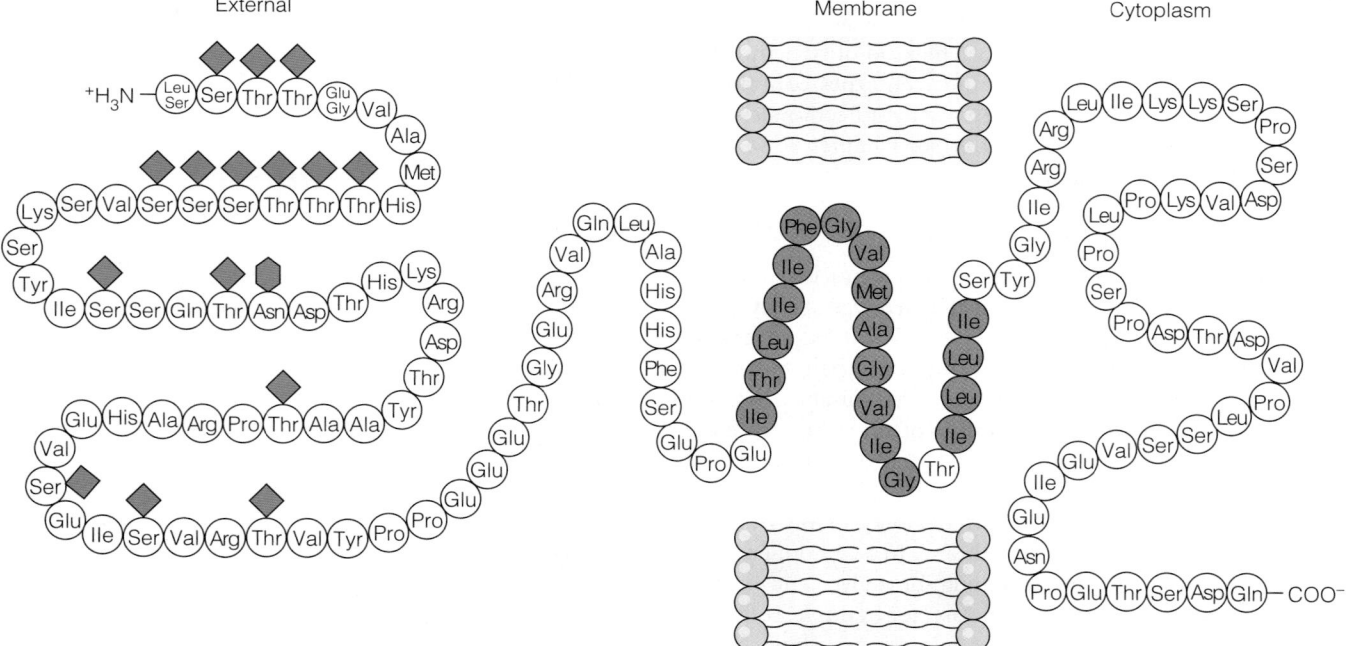

protein-conducting channel called a **translocon**. These protein channels are multisubunit complexes called **SecY** in prokaryotes and **Sec61** in eukaryotes, where "Sec" refers to proteins involved in secretion.

In the process of protein integration into the membrane, the translocon must perform three critical tasks. First, it must allow the hydrophobic transmembrane sequences of the emerging, or "nascent," protein to remain in the bilayer while conducting certain hydrophilic sequences across the membrane. Second, it must facilitate correct orientation, or "topology," of the transmembrane sequences. Membrane proteins must be properly oriented in the bilayer to ensure they can carry out their function. For example, a hormone receptor must display its ligand-binding site on the *extracellular* side of the membrane if it is to detect the presence of the hormone in circulation. Likewise, many transport proteins must move their substrates *in only one direction* across the bilayer to create electrochemical gradients that are critical for several cell functions. Thus, inserting transmembrane peptide sequences with the correct topology is a matter of survival. Finally, the translocon must do all of this while preventing the nonspecific transport of other molecules or ions across the bilayer, which would result in cell death due to collapse of electrochemical gradients.

A model for the function of the translocon is suggested by the crystal structure of the three subunit SecY complex from the bacterium *M. jannaschii* shown in Figure 10.23. The α subunit is largest, and it forms the protein channel from eight transmembrane helices and two others that do not completely traverse the bilayer. The β and γ subunits each have a single transmembrane helix. The α subunit is shaped like an hourglass, with a ring of six conserved Ile, Val, and Leu residues at the narrowest part of the channel. This ring is just large enough to accommodate the polypeptide chain as it emerges from the ribosome. As shown in Figures 10.23 and 10.24a, when the pore is not occupied by a translocating peptide, it is plugged with a small helix from the N-terminal sequence of the α subunit, thus explaining how SecY can seal itself and thereby prevent nonspecific transport of ions and other molecules.

How are the hydrophobic transmembrane segments of the inserted peptide recognized, and how do they get out of the channel and into the bilayer? One plausible explanation is that the channel opens along its length to allow the peptide to slide out laterally (Figure 10.24b and c). Hydrophobic peptides will be more likely to partition into the bilayer core, whereas polar sequences will remain in the channel and partition into the polar interface region of the membrane. In essence, the translocon is acting like a molecular separatory funnel, exposing the peptide sequence to polar and nonpolar phases. The peptide will move to the phase that best matches its polarity. This proposal is supported by the crystal

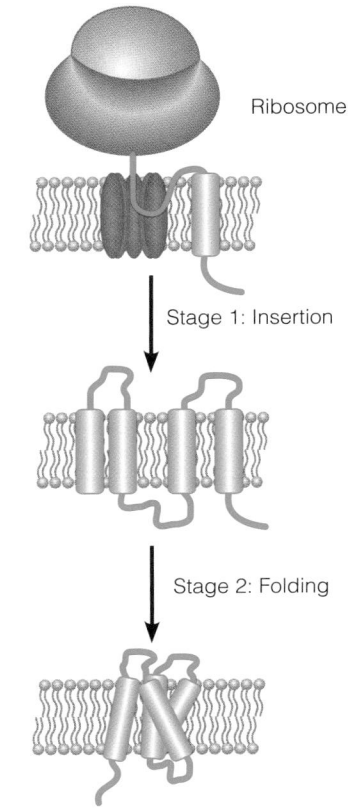

FIGURE 10.22

Co-translational insertion and folding of transmembrane helices in an integral membrane protein. The bilayer lipids are indicated in tan, the ribosome in brown, the translocon channel in purple, and the peptide in green. Some parts of the protein (in this case, the loops) are conducted across the bilayer through the translocon; transmembrane portions exit the translocon (see text) and remain embedded in the bilayer. The transmembrane helices fold after they are inserted.

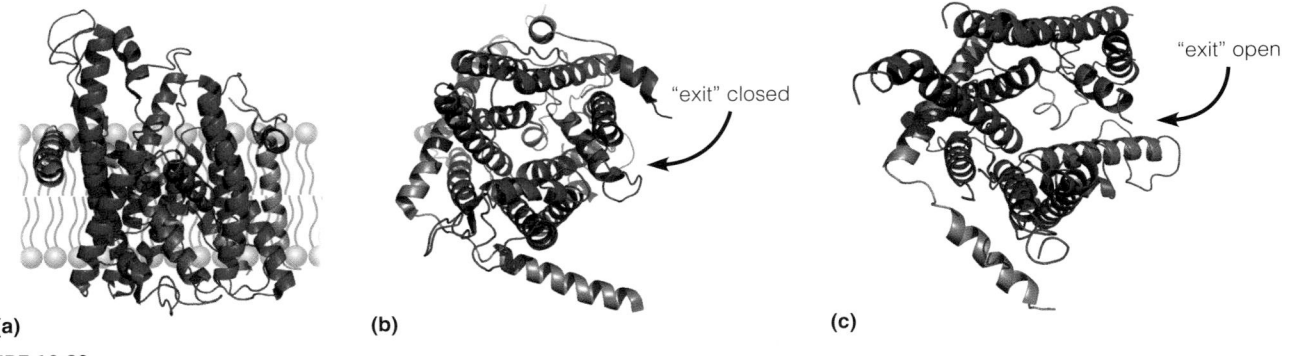

(a)   **(b)**   **(c)**

FIGURE 10.23

Crystal structures of SecY complex in "closed" and "open" conformations. The α subunit is highlighted in purple and red, with the β and γ subunits in gray. The small helix in the α subunit that plugs the pore is shown in blue. Panel **(a)**: Side view, looking toward the "exit," of SecY from *M. jannaschii* in the closed conformation (PDB ID: 1RHZ). The two helices on either side of the closed exit are highlighted in red. Note the location of the blue helix in the center of the channel. Panel **(b)**: Top view looking down the protein channel in the closed conformation. Panel **(c)**: Top view of SecY from *P. furiosus* in the open conformation (PDB ID: 3MP7). The two red helices have moved apart, opening a gap that runs the length of the channel and allows the nascent peptide sequence in the translocon access to the hydrophobic core of the membrane.

A protein channel called the "translocon" facilitates the insertion of integral membrane proteins into the membrane bilayer.

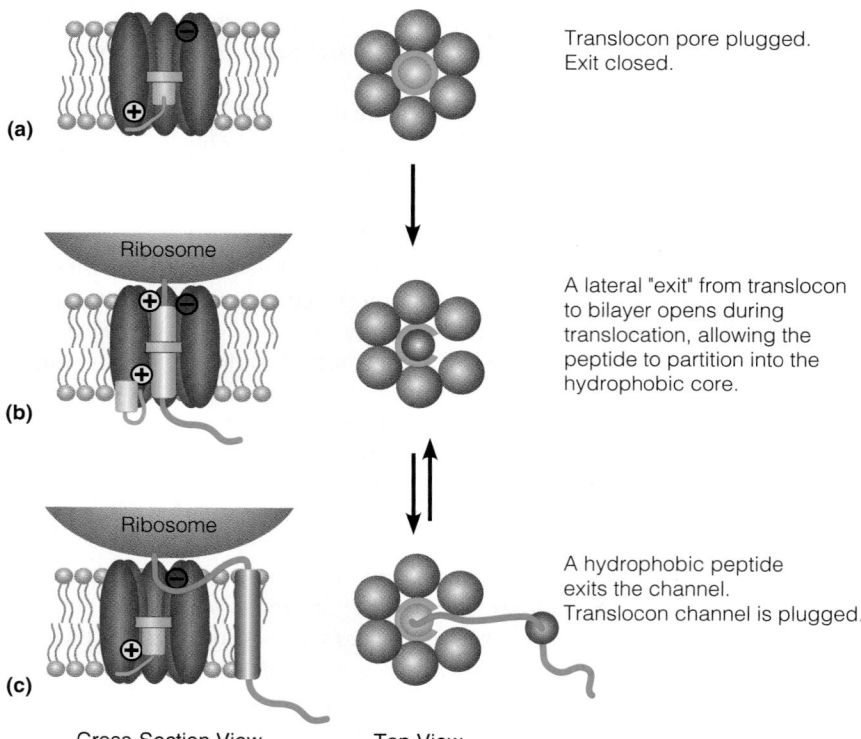

Translocon pore plugged. Exit closed.

A lateral "exit" from translocon to bilayer opens during translocation, allowing the peptide to partition into the hydrophobic core.

A hydrophobic peptide exits the channel. Translocon channel is plugged.

Cross-Section View   Top View

FIGURE 10.24

Model for translocon function. In all panels the cytoplasm is above the bilayer and the color coding matches that in Figures 10.22 and 10.23. **(a)** The "resting" translocon. The protein pore is plugged by a small helix (light blue). The ring of hydrophobic residues at the waist of the hourglass is shown in dark blue. **(b)** The "active" translocon. A nascent peptide (green) exits the ribosome and enters the translocon protein channel. The nascent peptide has access to the bilayer through a lateral opening in the translocon. In the yeast translocon, charged residues (yellow and red circles) in the channel facilitate the proper orientation of the nascent peptide in the bilayer. **(c)** If a segment of the nascent peptide is sufficiently hydrophobic it will partition into the lipid bilayer. Hydrophilic sequences (loops) partition to either side of the bilayer, depending on the orientation of the transmembrane segments.

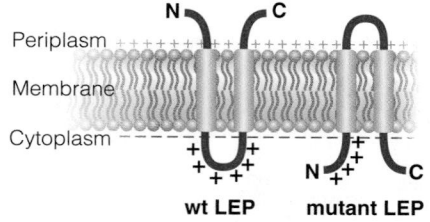

FIGURE 10.25

The "inside positive" rule. Wild-type leader peptidase (Lep) from *E. coli* orients its two transmembrane helices with the termini in the periplasm and the loop between the helices in the cytoplasm. Addition of four Lys to the N-terminus and removal of positive charge from the loop yield a mutant Lep that has the opposite membrane topology. The membrane potential is indicated by greater negative charge on the cyotplasmic side compared to the periplasmic side.

structure of SecY from *P. furiosus*, which shows a lateral opening in the translocon (Figure 10.23c).

Given this model for insertion of transmembrane helices, how are they inserted in the correct orientation? Several factors are likely at work to assure the proper topology for the inserted protein. Work by Gunnar von Heijne and colleagues has supported an "inside positive" rule that explains the enrichment in Lys and Arg residues in the cytoplasmic portions of proteins with multiple membrane-spanning regions. Due to the electrical potential across the membrane, $\Delta\psi$, the cytoplasmic side of the bilayer is more negatively charged compared to the outside; thus, a cluster of positively charged side chains would tend to migrate to the cytoplasmic side of the membrane. As a test of this hypothesis, von Heijne made a mutant of *E. coli* leader peptidase (Lep) in which four Lys residues were added to the N-terminus and several basic amino acids were removed from a cytoplasmic loop. The membrane topology of the mutant Lep was reversed compared to the wild-type (Figure 10.25). This "inside positive" orientation of transmembrane proteins is also aided by the preponderance of acidic (i.e., negatively charged) phospholipids on the inside of membranes. Finally, in some species, the translocon plays a role in directing the orientation of transmembrane peptides, via key charge–charge interactions. In the yeast translocon Sec61, basic residues on the periplasmic side and acidic residues on the cytoplasmic side of the channel (see Figure 10.24b) have been shown to be critical in determining the proper orientation of transmembrane segments. Reversing these charges in Sec61 results in a reversal of transmembrane protein topology.

Evolution of the Fluid Mosaic Model of Membrane Structure

Many features of the Singer–Nicolson fluid mosaic model have been confirmed since it was proposed in 1972; however, with improvements in imaging technology more details have emerged and prompted reinterpretation of the model. Specifically, most membranes appear to be very crowded with proteins and show significant variation in both thickness of the bilayer and distribution of proteins and lipids within a leaflet.

We mentioned previously that GPI-modified proteins are frequently localized in regions of the membrane that are enriched in cholesterol and sphingolipids. This observation led to the proposal that these three membrane components can coalesce to form separate membrane domains called **lipid rafts** (or **membrane rafts**). These membrane rafts are small (~10 nm), short-lived dynamic structures that, in response to certain stimuli, can transiently associate with each other to form larger "raft platforms" (Figure 10.26). Actin fibers may act to stabilize and/or initiate formation of the rafts. These raft platforms are thought to play significant roles in cell signaling and the sorting of proteins into specific organelles within a cell. GPI-anchored proteins are frequently involved in cell signaling, and the clustering of GPI proteins in raft platforms may accelerate signal transduction across the membrane—particularly in cases where dimerization of a signal receptor is required. More recently, rafts have been implicated in facilitating bacterial entry into host cells.

From studies in model membranes it is known that the bilayer thickness is a function of the lipid and protein composition in membrane domains (Figure 10.27). This may be due to the effect of cholesterol on the phase behavior of lipids (see Figure 10.13). At certain concentration ratios of sterols to lipids, membrane lipids form more ordered and elongated structures. For example, the membrane rafts described in Figure 10.26 are thicker than the surrounding

Membrane rafts are rich in cholesterol, sphingolipids, and GPI-linked proteins. The bilayer is thicker in the raft domains than in the surrounding membrane.

FIGURE 10.26

Membrane rafts. **(a)** Cholesterol, sphingolipids, and GPI-anchored proteins coalesce and form nanometer-sized dynamic raft domains, which may be stabilized by interactions with actin fibers. **(b)** Rafts can associate to form larger structures ("platforms"). Certain proteins interact preferentially with rafts (orange shading), while others do not (brown shading) or are excluded. Abbreviations: GPL, glycerophospholipid; GSL, glycosyl sphingolipid; GPI, glycosyl phosphatydylinositol.

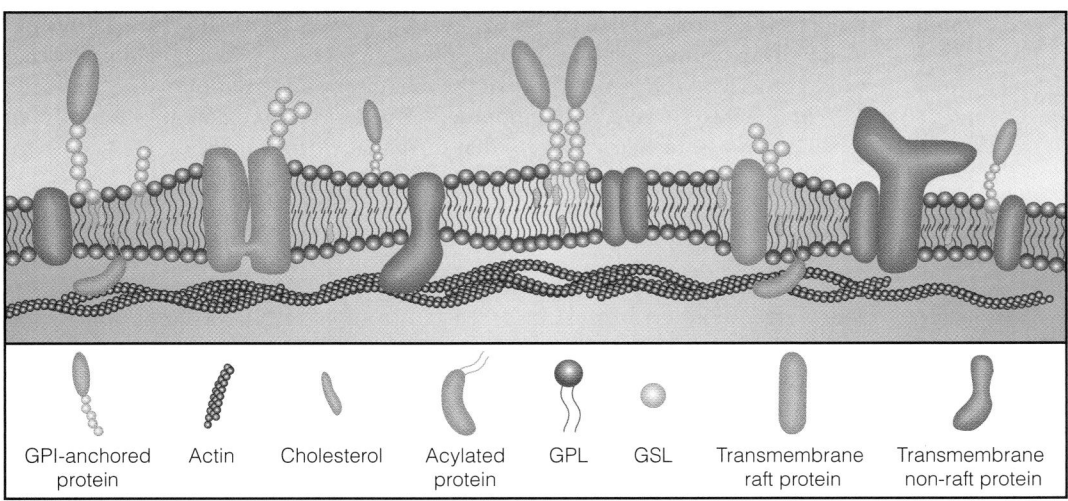

GPI-anchored protein | Actin | Cholesterol | Acylated protein | GPL | GSL | Transmembrane raft protein | Transmembrane non-raft protein

(a)

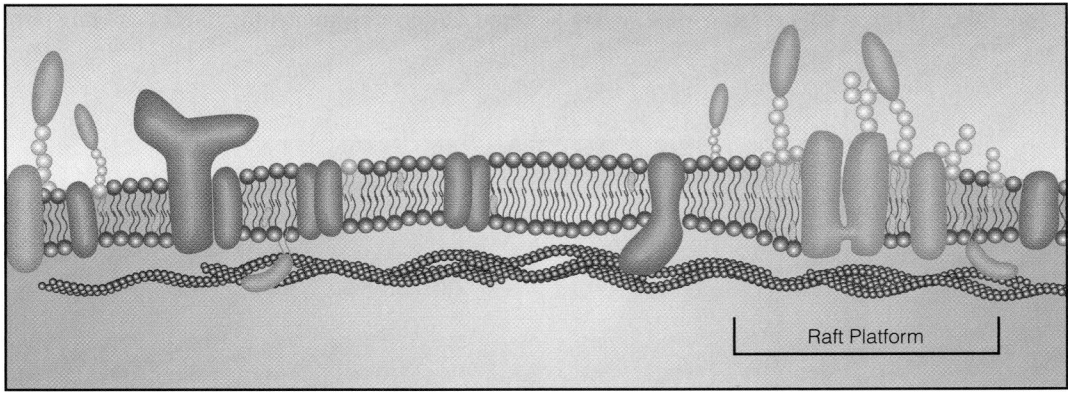

Raft Platform

(b)

FIGURE 10.27

Atomic force microscopy image of lipid domains in a model membrane. Top view of a bilayer composed of a mixture of 1,2-dilauroylphosphatidylcholine (DLPC) and 1,2-distearoylphosphatidylcholine (DSPC) separates into two phases as the temperature is lowered from 70 °C to 25 °C. The DSPC molecules coalesce and form thicker gel-like "islands" or domains (lighter gray spots) in a "sea" of fluid DLPC (darker gray background). Such simple systems have been used to model the more complex lipid rafts described in Figure 10.26.

Reprinted from *Biophysical Journal* 83:3380–3392, T. V. Ratto and M. L. Longo, Obstructed diffusion in phase-separated supported lipid bilayers: A combined atomic force microscopy and fluorescence recovery after photobleaching approach. © 2002, with permission from Elsevier.

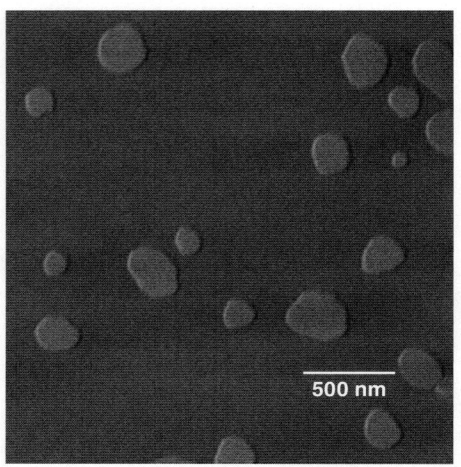

500 nm

Membrane curvature generates mechanical forces in the bilayer that can affect protein function, such as the opening and closing of membrane channels.

nonraft membrane. Membrane thickness is also influenced by the proteins in the bilayer. This raises the interesting question illustrated in Figure 10.28: Do proteins change conformation to optimize interactions with the bilayer as membrane thickness changes, or do lipids change their structure to accommodate proteins? Current evidence suggests that it is more common for lipids to adjust to proteins than vice versa; however, both adaptations are observed in cell membranes.

Lipid Curvature and Protein Function

As mentioned previously, X-ray scattering studies of membranes from various organelles have shown that protein composition, rather than lipid composition, is the major determinant of membrane thickness. Nonetheless, lipids are not passive constituents of the membrane. The lipid bilayer exerts significant lateral pressure on embedded proteins. This pressure can be up to 400 atm, and it arises due to the balancing of attractive and repulsive forces in the plane of the bilayer. The resulting pressures are experienced by proteins imbedded in the bilayer (Figure 10.29). If the membrane surrounding a cell or organelle expands or contracts, the resulting change in bilayer curvature will alter the magnitude and position of the forces acting on imbedded proteins. Thus, a *mechanical* stress on the bilayer is translated to these proteins. Such stress can alter protein conformation, as is observed in the opening and closing of the mechanosensitive channel of *Mycobacterium tuberculosis*, MscL (Figure 10.30). When a cell or bacterium is placed in a solution with a low solute concentration, or low **osmolarity**, the osmotic pressure on the cell changes and water will enter the cell. The resulting swelling changes the membrane curvature, which, in turn, alters the pressure on the MscL channel such that it opens (Figure 10.30b). The opening of the MscL channel allows the bacterium to rapidly expel internal fluid in response to changes in solution osmolarity that would otherwise result in rupture of the membrane and cell death.

For a cell to survive, its membrane(s) must tightly regulate the transport of materials across the bilayer. It is to that topic that we now turn.

Transport Across Membranes

A cell or an organelle can be neither wholly open nor wholly closed to its surroundings. Its interior must be protected from certain toxic compounds, and metabolites must be taken in and waste products removed. Because the cell must contend with thousands of substances, it is not surprising that much of the complex structure of membranes is devoted to the regulation of transport.

In this section we describe various ways in which molecules are transported across membranes. These include the actions of small molecules that act as ion carriers, larger proteins that are highly specific transporters, and proteins that promote the formation of membrane vesicles. Before considering specific mechanisms of transport, we review the underlying thermodynamics of transport processes.

FIGURE 10.28

Adaptation to hydrophobic mismatch in a membrane. If the thickness of the bilayer core and the hydrophobic surface area of an embedded protein do not match (middle image), either the protein will undergo conformational change (left) or the bilayer will change composition (right) until the dimensions of these hydrophobic regions match.

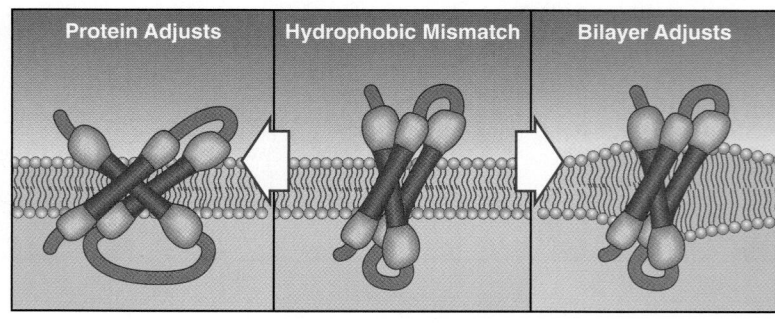

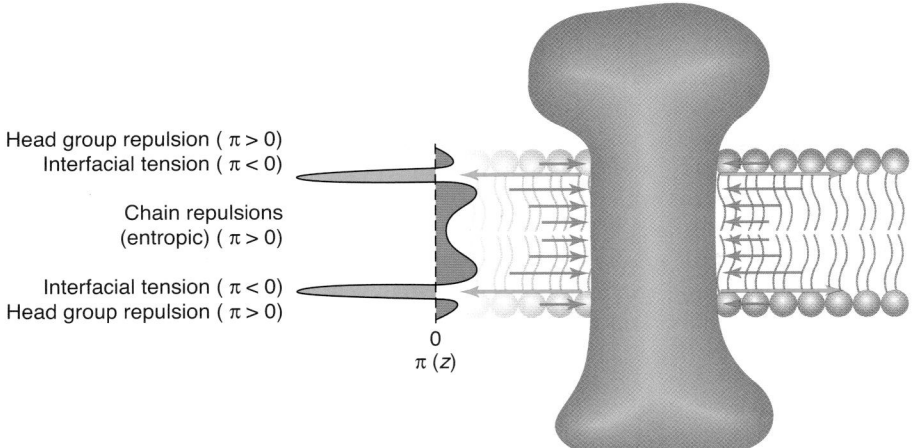

FIGURE 10.29

Lateral pressure in membranes. Formation of the bilayer results in opposing forces, of equal overall magnitudes, that exert lateral pressure, π, in the plane of the membrane. Head group and alkyl chain repulsions (green shading, left side of diagram) oppose interfacial tension (blue shading). Note that the surface area over which interfacial tension acts is much smaller than that over which the repulsive forces act. The repulsive forces exert a positive pressure (green arrows) on a protein imbedded in the membrane (red shading), whereas the interfacial tension exerts a negative pressure as indicated by the blue arrows. The length of the arrows is a rough indication of the magnitude of the pressure exerted at that point in the membrane.

Head group repulsion ($\pi > 0$)
Interfacial tension ($\pi < 0$)

Chain repulsions (entropic) ($\pi > 0$)

Interfacial tension ($\pi < 0$)
Head group repulsion ($\pi > 0$)

0
$\pi (z)$

The Thermodynamics of Transport

In Chapter 3 we discussed the general thermodynamic principles governing the transfer of substances across membranes. It was shown that the free energy change, ΔG, for transporting one mole of a substance from a location in which its concentration is C_1 to a different location where its concentration is C_2 is given by

$$\Delta G = \Delta G^{\circ\prime} + RT \ln Q = -RT \ln K_{eq} + RT \ln \frac{[C_2]}{[C_1]} \qquad (10.1a)$$

For a process that only involves transport of some substance across a membrane, the term $\Delta G^{\circ\prime}$ equals zero. Why is this so? Recall that $\Delta G^{\circ\prime}$ describes the *equilibrium state* for a process, and for the case of transport across a membrane, the equilibrium state is reached when the concentrations of the substance are the same on both sides of the membrane (i.e., at equilibrium $[C_2] = [C_1]$). In this case $K_{eq} = [C_2]/[C_1] = 1$. Since $\Delta G^{\circ\prime} = -RT \ln K_{eq}$, if $K_{eq} = 1$, $\Delta G^{\circ\prime} = 0$ for the process. Note that, when $[C_2] = [C_1]$, the term $Q = 1$; thus, $\Delta G = 0$, as expected for a system at equilibrium.

For a transport process that is not at equilibrium $[C_2] \neq [C_1]$; thus, $RT \ln Q \neq 0$, and ΔG for the process is given by

$$\Delta G = \Delta G^{\circ\prime} + RT \ln Q = 0 + RT \ln Q = RT \ln \frac{[C_2]}{[C_1]} \qquad (10.1b)$$

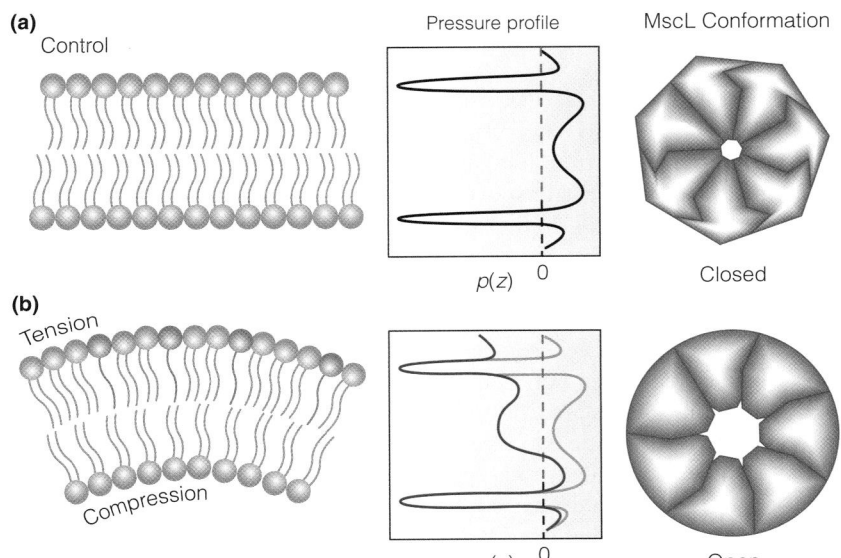

(a) Control

Pressure profile

MscL Conformation

$p(z)$ 0

Closed

(b) Tension

Compression

$p(z)$ 0

Open

FIGURE 10.30

Mechanical stress influences protein conformation. Changes in bilayer curvature (left panels) alter the lateral pressure profile in the membrane (middle panels). This results in a conformational switch in MscL that opens or closes the channel (right panels; the view is looking along the channel opening through the bilayer).

For a substance that can pass through a membrane, the normal state of equilibrium is achieved when the concentrations of the substance are equal on both sides of the membrane.

Equalization of the concentrations of some substance across a membrane can be circumvented (1) by binding of the substance to macromolecules, (2) by maintaining a membrane potential (if the substance is ionic), or (3) by coupling transport to an exergonic process.

A more detailed derivation of this equation is given in Chapter 3 (see Equation 3.19, pages 69–71). According to this equation, if $[C_2]$ is less than $[C_1]$, ΔG is negative, and the process is thermodynamically favorable. As more and more substance is transferred (between the two locations), $[C_1]$ decreases and $[C_2]$ increases, until $[C_2] = [C_1]$. At this point $\Delta G = 0$, and the system is at equilibrium. *Unless other factors are involved*, this equilibrium is the ultimate state approached by transport across any membrane. In short, a substance that can traverse the membrane will eventually reach the same concentration on both sides. We can describe the same process in kinetic terms. If the molecules are colliding with the membrane at random, the number entering from any side will be proportional to the concentration on that side. When the concentrations become equal, the rates of transport in the two directions will be the same, and no net transport will occur.

There are three circumstances under which this equilibrium state can be circumvented, and each is important in the behavior of real membranes:

1. A substance may be preferentially bound by macromolecules confined to one side of the membrane or it may be chemically modified once it crosses. We may find that compound A is more concentrated inside a cell (in terms of total moles of A per unit volume) than outside. But much of A may be bound to some cellular macromolecules or may have been modified; that portion does not really count in Equation 10.1, which simply states that the concentrations of *free* A on the two sides must be equal at equilibrium. An appropriate example is oxygen in erythrocytes. If we were to measure the *total* oxygen concentration in an erythrocyte, we would find it higher than the concentration of O_2 in the surrounding blood plasma. But the total concentration inside the cell includes oxygen bound to hemoglobin. The *free* oxygen concentration in the fluids inside and outside an erythrocyte is the same at equilibrium.

2. A **membrane electrical potential** may be maintained across a membrane that influences the distribution of ions. This tendency can be expressed quantitatively in the following way. For an ion of charge Z, the free energy change for transport across a cell or organelle membrane now involves two contributions: the normal concentration term, as given in Equation 10.1b, plus a second term describing the energy change (or work involved) in moving a mole of ions across the potential difference. We consider a process in which one mole of ions is transported from *outside* to inside.

$$\Delta G = RT \ln \frac{[C_{in}]}{[C_{out}]} + ZF\Delta\psi \tag{10.2}$$

Here F is the Faraday constant ($96.5 \text{ kJ mol}^{-1}\text{V}^{-1}$), and $\Delta\psi$ is the membrane potential in volts. We define $\Delta\psi$ in terms of the initial and final locations of the transported ion ($\Delta\psi = \psi_{final} - \psi_{initial}$: in this case, $\Delta\psi = \psi_{in} - \psi_{out}$). In this example, $\Delta\psi$ will be negative if the inside of the membrane is negatively charged compared to the outside. Under these conditions, if Z is positive, the $ZF\Delta\psi$ term in Equation 10.2 is negative and makes ΔG more exergonic (favorable). That is, the transport of cations *into* this hypothetical cell is favored. For anions, of course, the opposite is true; they will be driven out. In the presence of a non-zero membrane potential, the equilibrium state ($\Delta G = 0$) will *not* correspond to the same concentration of ions on the two sides of the membrane. However, energy must be expended continually to keep up the potential difference; otherwise migration of ions would neutralize it. Conversely, Equation 10.2 may be interpreted to mean that if a difference in ionic concentration is maintained, an electrical potential will be produced across the membranes (see Problem 9).

3. If some thermodynamically favored process is *coupled* to the transport, then the $\Delta G^{\circ\prime}$ and $RT \ln Q$ for this favorable process must be included in the free energy equation. This is the general case of *active transport*, for which we can write

$$\Delta G = \Delta G^{\circ\prime} + RT \ln \frac{Q[C_{in}]}{[C_{out}]} \tag{10.3a}$$

where the term Q includes activities for the species in the favorable reaction that is coupled to the transport of C. If the substance C being transported is an ion we must also include the $ZF\Delta\psi$ term

$$\Delta G = \Delta G^{\circ\prime} + RT \ln \frac{Q[C_{in}]}{[C_{out}]} + ZF\Delta\psi \tag{10.3b}$$

The quantities $\Delta G^{\circ\prime}$ and Q could correspond to a thermodynamically favored reaction (like hydrolysis of one mol ATP) that is coupled to the process of transport. This equation is a generalization of Equation 10.2, now allowing a variety of processes—not just those that maintain an electrical potential difference—to participate in the transport. In the case where hydrolysis of one mol of ATP results in transport of n mol of an ion C, we must modify Equation 10.3b to include the proper stoichiometry:

$$\Delta G = \Delta G^{\circ\prime} + RT \ln \frac{Q[C_{in}]^n}{[C_{out}]^n} + nZF\Delta\psi \tag{10.3c}$$

where $n = $ mol of C transported per mol of ATP. We will show examples of such calculations in the discussion that follows.

With this background, we turn now to the mechanisms whereby substances are passed through membranes. We introduce the problem by asking two questions: (1) Does the process approach a state in which there are equal concentrations of the free substance on both sides, or is it maintained far from equilibrium? (2) How fast does the transport occur? Some molecules that are not actively transported against a concentration gradient can still traverse some membranes very rapidly, whereas others are transported so slowly as to be effectively excluded.

Nonmediated Transport: Diffusion

Nonmediated diffusion across membranes is accomplished by the random wandering of molecules through membranes. The process is the same as the Brownian motion of molecules in any fluid, which is termed **molecular diffusion**. Nonmediated transport ultimately results in the free concentration of the diffusing substance being the same on both sides of the membrane. The net rate of transport, J (in moles per square centimeter per second), is, as you might expect, proportional to the concentration difference $(C_2 - C_1)$ across the membrane:

$$J = -\frac{KD_l(C_2 - C_1)}{l} \tag{10.4}$$

where l is the thickness of the membrane, D_l is the **diffusion coefficient** of the diffusing substance in the membrane, and K is the **partition coefficient** for the diffusing material between membrane lipid and water (i.e., the ratio of solubilities of the material in the lipid and water phases). For ions and other hydrophilic substances, K is a very small number, with the result that nonmediated diffusion of such substances through lipid membranes is extremely slow. There is simply not enough of the hydrophilic substance dissolved in the bilayer to provide rapid transport. In agreement with Equation 10.1, Equation 10.4 says that net transport will stop when $C_2 = C_1$. If the concentrations of C_1 and C_2 are expressed in mol/cm^3 and l in cm, then D_l has the units cm^2/s. D_l is not the same as the diffusion coefficient (D) that the same molecule would have in aqueous solution because it depends not only on the size and shape of the molecule but also on the viscosity of the membrane lipid.

The rate of nonmediated transport, as measured by membrane permeability, is proportional to the diffusion and partition coefficients and inversely proportional to membrane thickness.

We usually do not know either K, D_j, or the exact thickness of the membranes involved, so we often describe the rate of passive transport in terms of a **permeability coefficient**, P, which can be measured by direct experiment:

$$J = -P(C_2 - C_1) \qquad (10.5)$$

By comparing equations (10.5) and (10.4), we see that P is given by

$$P = \frac{KD_j}{l} \qquad (10.6)$$

with units of cm/s.

Table 10.6 lists permeability coefficients for a number of small molecules and ions in membranes. The low P values of the ions are as expected because ions have low values of K. However, the relatively large permeability value for water is surprising. Despite their hydrophobicity, biological membranes are not, in fact, very good barriers against water. Although the reasons for this are not entirely clear, it is probably fortunate for life, for it allows cells to exchange water, albeit slowly, with their surroundings. When water loss is to be strenuously avoided, as in the leaves of desert plants, waxy substances, with their much more hydrophobic structures, provide a nearly impermeable barrier. In some cells very rapid transport of water is required. As described below, such rapid transport of water is achieved by specific membrane-spanning channels called **aquaporins**.

Facilitated Transport: Accelerated Diffusion

For many substances, the slow transport provided by nonmediated diffusion is insufficient for the functional and metabolic needs of cells, and means must be found to increase transport rates. For example, exchange of Cl^- and HCO_3^- is essential to erythrocyte function. If we examine the permeability of erythrocyte membranes to chloride or bicarbonate ions, we find permeability coefficients of about 10^{-4} cm/s. This value is about 10 million times greater than the permeability coefficient for ions in pure lipid bilayers like the artificial phosphatidylserine membrane listed in Table 10.6. Clearly, some special mechanism is required to account for this difference. Three general types of **facilitated transport**, or **facilitated diffusion**, are known to occur: transport through pores or channels formed by transmembrane proteins (Figure 10.31a); transport by carrier molecules (Figure 10.31b); and transport by **permeases** (Figure 10.31c).

Carriers

Ionophores increase the permeability of a cell membrane to ions. Thus, many bacteria secrete ionophores that act as chemical warfare agents, or *antibiotics*, to kill other bacteria with which they compete for nutrients. The ionophores kill the

Facilitated transport, via pores, permeases, or carriers, can increase the rate of diffusion across a membrane by many orders of magnitude.

TABLE 10.6 Permeability coefficients for some ions and molecules through membranes		
Permeability Coefficient (cm/s) for	Synthetic Membrane (Phosphatidylserine)	Biological Membrane (Human Erythrocyte)
K^+	$<9 \times 10^{-13}$	2.4×10^{-10}
Na^+	$<1.6 \times 10^{-13}$	10^{-10}
Cl^-	1.5×10^{-11}	$1.4 \times 10^{-4*}$
Glucose	4×10^{-10}	$2 \times 10^{-5*}$
Water	5×10^{-3}	5×10^{-3}

Data from M. K. Jain and R. C. Wagner (1980) *Introduction to Biological Membranes*. Wiley, New York.
*Facilitated transport. Note that whenever facilitated transport is encountered, the permeability coefficient rises dramatically.

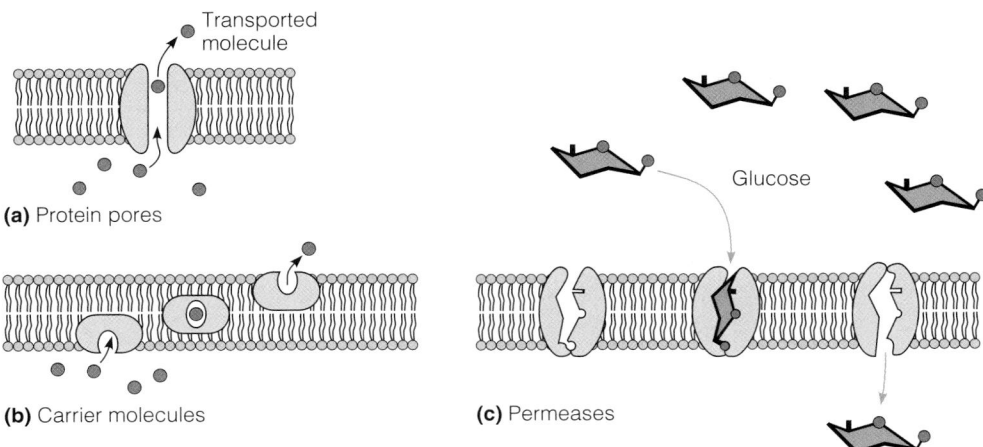

FIGURE 10.31

The three major mechanisms for facilitated transport. **(a)** Protein pores. **(b)** Carrier molecules. **(c)** Permeases.

(a) Protein pores

(b) Carrier molecules

(c) Permeases

Glucose

Transported molecule

neighboring bacteria because unregulated ion transport destroys the electrochemical gradients that store free energy needed to drive vital processes in living cells. Some ionophores create ion-conducting pores in the membrane, whereas others carry ions from one side of the bilayer to the other (see Figure 10.31b). For example, *valinomycin*, produced by a *Streptomyces*, is an ion carrier and has the structure shown in Figure 10.32. When complexed with K^+ it is a cyclic polypeptide-like molecule, involving three repeats of the sequence (D-valine)–(L-lactate)–(L-valine)–(D-hydroxyisovalerate). Its folded conformation presents an outside surface rich in —CH_3 groups and an interior cluster of nitrogen and oxygen atoms that is well suited to chelating cations. The dimensions of the interior cavity nicely accommodate a K^+ ion but do not fit other cations as well. This structure is exactly what is needed for a cation carrier: The outer surface is hydrophobic, making the molecule soluble in the lipid bilayer, whereas the inside mimics in some ways the hydration shell that the cation would have in aqueous solution. A number of other ion-carrier antibiotics have the same kind of structure. These molecules are either cyclic, or linear chains that can fold into cage-like structures. Their relative affinities for different ions vary greatly. For example, valinomycin has nearly a 20,000-fold preference for K^+ over Na^+, whereas the antibiotic *monensin* prefers Na^+ by only 10-fold.

A molecule like valinomycin can diffuse to one surface of a membrane, pick up an ion, and then diffuse to the other surface and release it. There is no *directed* flow, but the carrier in effect increases the solubility of the ion in the membrane. We could say that it increases the factor K in Equation 10.6. For such ion carriers, the net transport of ions will be in the direction that equalizes concentration of the ion on both sides of the membrane. Such facilitated transport is also called **passive transport** to distinguish it from **active transport**—a strictly directional process, which requires an input of free energy.

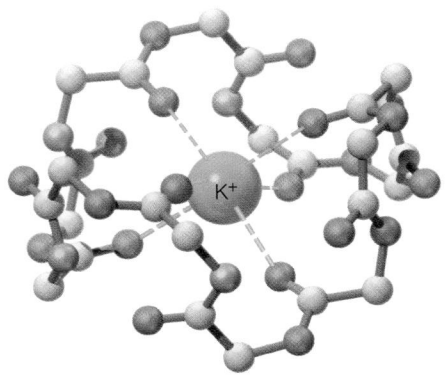

FIGURE 10.32

Valinomycin, an antibiotic that acts as an ion carrier. The outside of this roughly spherical cyclic polypeptide is hydrophobic. The central cavity surrounded by oxygens (red) complexes a K^+ ion. Nitrogens are shown in blue, oxygens in the polypeptide backbone are red, and carbons are gray. The surface is covered with CH_3 groups (not shown).

Facilitated transport can be passive or active. Passive transport can only achieve net transport of a substrate along its concentration gradient. Active transport can move a substrate against its concentration gradient, but this requires an input of free energy.

Permeases

Membrane-spanning proteins that recognize specific molecules for transport are called permeases or **transporters**. Like the carriers described above, some permeases act in a passive fashion—transporting their substrates in both directions, with a net flow toward the side of the membrane with the lower substrate concentration. The glucose transporter in erythrocytes (**GLUT1**, or band 4.5 protein) is thought to operate in this way. The small energy demands of an erythrocyte are met by glucose, which is readily available in the surrounding blood plasma. However, as Table 10.6 shows, the nonmediated transport of glucose through artificial phospholipid membranes is exceedingly slow: $P = 4 \times 10^{-10}$ cm/s. GLUT1, a 492-residue protein with 12 membrane-spanning helices, increases the glucose diffusion rate 50,000-fold. GLUT1 is quite discriminating; for example, D-glucose is transported

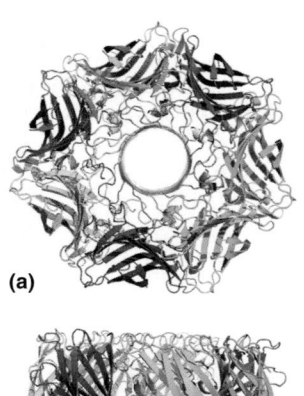

(a)

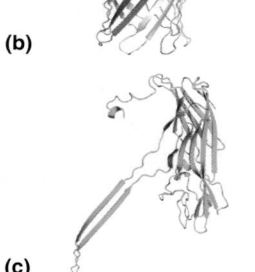

(b)

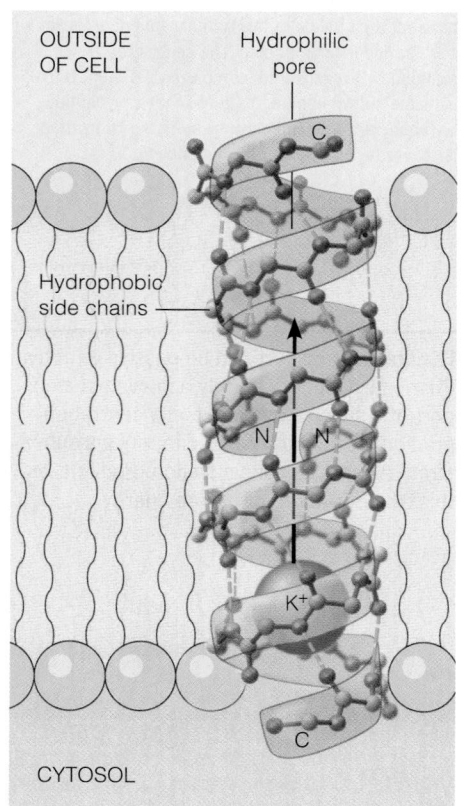

(c)

FIGURE 10.33

The channel-forming hemolysin from *Staphylococcus aureus.* Ribbon drawings of the α-hemolysin heptamer, viewed **(a)** looking down the sevenfold axis and **(b)** perpendicular to the sevenfold axis. **(c)** One protomer extracted from the heptamer structure. The heptamer is 10 nm in diameter and 10 nm in length, as measured along the sevenfold axis. The β-barrel stem, which penetrates the membrane, is about 6 nm long. PDB ID: 7AHL.

orders of magnitude more rapidly than L-glucose. Facilitated transport of metabolites like glucose appears to be a common feature in cells.

The key feature of permease function is shown schematically in Figure 10.31c, namely, the permease shifts between two conformations: one open only to the "outside," and the other open only to the "inside." Thus, the permease never forms a pore that allows unrestricted flow of the transported substrate. Transport requires both binding of the substrate and conformational change by the permease.

Some permeases couple the transport of more than one substrate or ion. When transport of the two molecules, or ions, is in the same direction, the transporter is referred to as a **symport**; when the substrates move in opposite directions, the transporter is called an **antiport**. This cotransport strategy allows the thermodynamically unfavorable transport of some substrate *against* its concentration gradient, when coupled to the favorable transport of the cosubstrate. We will return to this topic later, after we present the general features of active transport.

Pore-Facilitated Transport

Many pathogenic bacteria synthesize and secrete protein toxins that act as ionophores by creating pores in the plasma membranes of cells of the host organism. An example, shown in Figure 10.33 (and Figure 10.15), is α-hemolysin from *Staphylococcus aureus*. This protein is made up of seven subunits, which associate to produce a membrane-spanning ion channel. Likewise, the toxin *gramicidin A* produced by the bacterium *Bacillus brevis* acts as a cation-specific ion pore, allowing a breakdown in the unequal ratio of [K⁺] and [Na⁺] normally maintained between the inside and outside of living cells. Gramicidin A is a 15-residue polypeptide, containing alternating L- and D-amino acids (Figure 10.34). Gramicidin adopts an open helical conformation when dissolved in the membrane, but one molecule of the antibiotic is long enough to traverse only half the thickness of the membrane. An open pore forms only when two gramicidin molecules line up to form a head-to-head dimer (Figure 10.34). Potassium ions (and, to a lesser extent, sodium ions) can then pass through the channel.

In addition to channels that do damage to cells, many channels facilitate transport processes that are critical to cell survival. We have already mentioned the apparatus responsible for facilitated transport of Cl⁻ and HCO₃⁻ in erythrocyte membranes: it is the transmembrane protein called the band 3 protein, or anion channel. The band 3 protein forms a channel through which Cl⁻ and HCO₃⁻ can pass. As explained in Chapter 7, much of the CO₂ generated in tissues is transported as HCO₃⁻, which is formed in erythrocytes by the action of carbonic anhydrase on CO₂.

$$CO_2 + H_2O \underset{\text{anhydrase}}{\overset{\text{Carbonic}}{\rightleftharpoons}} HCO_3^- + H^+$$

Exit of the HCO₃⁻ is balanced by influx of Cl⁻; this both maintains charge balance and facilitates O₂ release (see Chapter 7). The band 3 protein doesn't just form a hole in the membrane for the passage of ions. Rather, the channel is a very selective antiport, exchanging HCO₃⁻ and Cl⁻ on a 1:1 basis. By contrast, such

OUTSIDE OF CELL

Hydrophilic pore

C

Hydrophobic side chains

N   N

K⁺

C

CYTOSOL

FIGURE 10.34

Gramicidin A, an antibiotic that acts as an ion pore. Two molecules of gramicidin A form a pore through the membrane by adopting a helical conformation, with their hydrophobic side chains in contact with the lipid. Note the N-termini are inside and the C-termini are outside the bilayer core. The inside of the helix forms the hydrophilic pore. The hydrogen bonding in this open helical structure resembles that in β-sheet polypeptides. This is possible because of the alternating D and L residues.

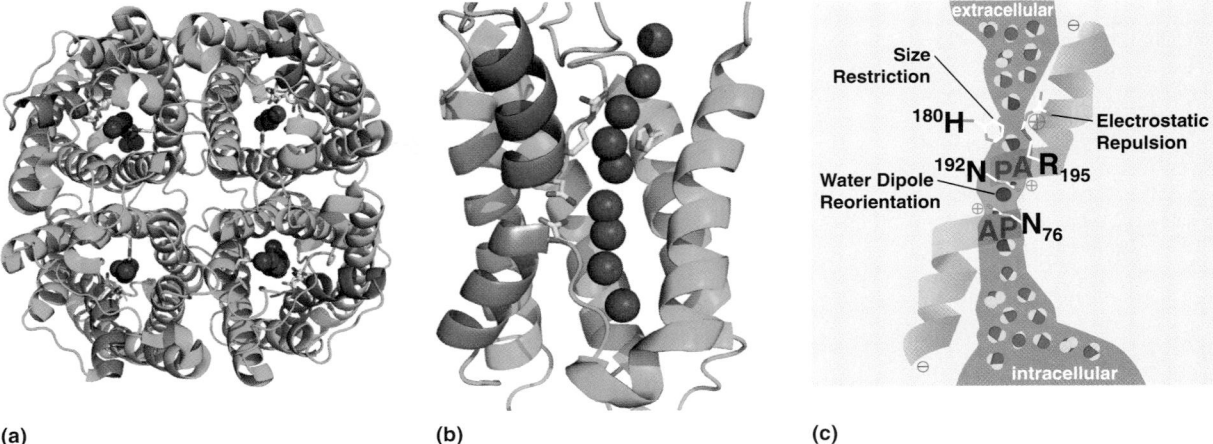

(a)　　　　　　　　　　(b)　　　　　　　　　　(c)

facilitated transport is not necessary for O_2. This tiny nonpolar molecule can move rapidly through the membrane by simple diffusion.

Many types of eukaryotic cells must move large amounts of water rapidly across their membranes as part of their physiological function. These include erythrocytes (which experience a wide range of solution osmolarity as they transit through lungs, capillaries, and kidneys), secretory cells in salivary glands, and epithelial cells in the kidney. Although water can cross membranes, it does so relatively slowly; thus, the inherent permeability of membranes toward water (see Table 10.6) is not sufficient to support the rapid transport observed in many cell types.

Such rapid transport is achieved by water-specific channels called **aquaporins**. The aquaporins function as tetramers of identical monomers. Each monomer contains six membrane-spanning helices and two shorter helices that contain a conserved N-terminal Asn-Pro-Ala (NPA) motif. Crystal structures of the aquaporins reveal that selectivity for water is achieved by three means (Figure 10.35). First, the channel is quite narrow (~0.28 nm) and excludes anything larger than a water molecule (including hydrated ions). Second, H_3O^+ is excluded by electrostatic repulsion. A conserved Arg places a (+) charge at this constriction, effectively repelling any cations. In addition, the two shorter helices are oriented with their N-termini pointing into the narrowest part of the channel; thus, the positive ends of the helical macrodipoles provide additional repulsion to H_3O^+. Third, water molecules can only pass through the channel in single file. As they do so, main chain carbonyl groups as well as the conserved Asn72 and Asn192 side chains in the conserved NPA motifs form H-bonds with the individual water molecules, thereby reorienting the water molecules and disrupting H-bonding between the water molecules in the channel. This is a critical feature of the transport mechanism because it prevents protons from traversing the membrane via an H-bonding network of water molecules. As will become clear later in this chapter, and in subsequent chapters, many membranes must maintain ion gradients to carry out critical processes (e.g., bacterial flagellar motion, ATP synthesis, firing of neurons, etc.). The aquaporins elegantly solve the problem of maintaining the osmotic balance in a cell while not destroying critical ion gradients.

Research that has increased our understanding of transport across membranes was recognized by the 2003 Nobel Prize in Chemistry, awarded to Peter Agre for the discovery of aquaporins and to Roderick MacKinnon for his work elucidating the structure and function of potassium ion channels. We turn now to the selective transport of ions across membranes.

Ion Selectivity and Gating

Two outstanding features of ion channels that we explore here are their selectivity for a particular ion and the control of ion transport through the channel. Many protein ion channels and transporters display high selectivity for the biologically

FIGURE 10.35

The aquaporin water channel. **(a)** Cartoon rendering of human aquaporin-5 tetramer (PDB ID: 3D9S), looking along the four water channels. Water molecules are shown as red spheres. The two short helices containing the NPA sequence are shown in blue. Side chains for Asn76, Asn192, His180, and Arg195 are highlighted in yellow. **(b)** A cutaway view of the water channel in one of the monomers. The narrowest part of the channel is where the two short helices meet. Note the location of Asn76 and Asn192 at this restriction. The two helical macrodipoles and Arg195 provide an electrostatic barrier to H_3O^+ passage. **(c)** Schematic view of the aquaporin channel, showing the electrostatic repulsion of H_3O^+ and the reorientation of the water molecules as they pass through the central restriction.

(c) Reprinted by permission of Federation of the European Biochemical Societies, *FEBS Letters* 555:72–78, P. Agre and D. Kozono, Water channels: Molecular mechanisms for human disease. © 2003.

Ion selectivity is achieved by optimal geometry of chelating groups in ion channels.

FIGURE 10.36

The structure of the potassium channel pore. The transmembrane pore region of the potassium channel KcsA (PDB ID 1BL8) from the bacterium *Streptomyces lividans* is shown in a "closed" conformation in panels **(a–c)**. The pore region of the potassium channel MthK (PDB ID 1LNQ) from the bacterium *M. thermautotrophicus* is shown in an "open" conformation in panels **(d–f)**. **(a)** A view looking along the pore axis from the extracellular face. The four transmembrane subunits are shown in different colors. The K$^+$ ions (purple spheres) are bound in the selectivity "filter" (highlighted in red) by amide carbonyl groups from each protein subunit. **(b)** A view from within the plane of the bilayer. Here, the periplasmic side is on top and the cytoplasmic side is on the bottom. Three K$^+$ ions (purple spheres) and one water molecule (smaller red sphere) are shown in the selectivity filter (see text and Figure 10.37b). **(c)** Here, two of the four subunits have been removed to show better the convergence of the pore helices that close the channel. The selectivity filter and the "hinge Gly" are highlighted in red. Bending of the helices at the hinge Gly residues opens the channel gate. The views in panels **(d–f)** are the same as those in **(a–c)**, except that K$^+$ ions were not crystallized in the selectivity filter.

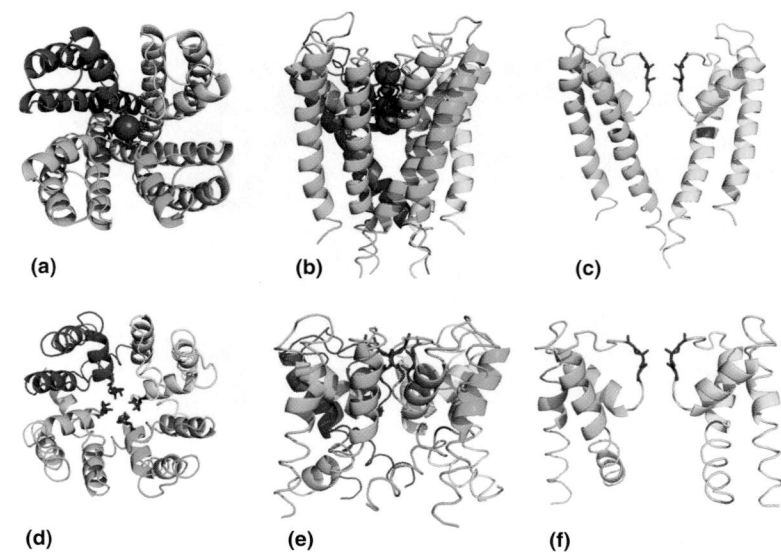

(a)   **(b)**   **(c)**

(d)   **(e)**   **(f)**

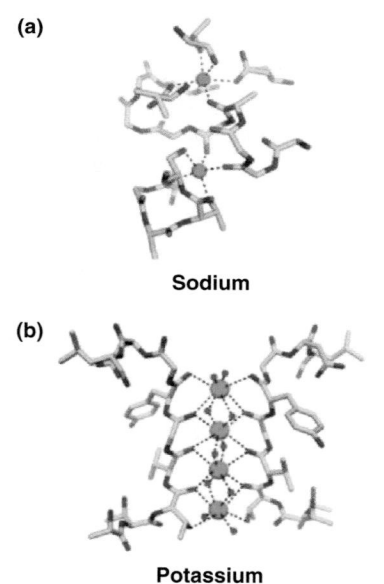

(a)

Sodium

(b)

Potassium

FIGURE 10.37

Selective binding of Na$^+$ and K$^+$ in ion channels. **(a)** Two Na$^+$ binding sites make up the selectivity filter in the transmembrane region of LeuT (PDB ID: 2A65). **(b)** Four K$^+$ are bound in the filter of the KcsA K$^+$ channel (PDB ID: 1K4C).

From *Science* 310:1461–1465, E. Gouaux and R. MacKinnon, Principles of selective ion transport in channels and pumps. © 2005. Reprinted with permission from AAAS.

relevant ions K$^+$, Na$^+$, Ca^{2+}, or Cl$^-$. These include the bacterial K$^+$ channel, KcsA (shown in Figure 10.36a–c), and the Na$^+$/Leu transporter, LeuT. It is instructive to compare and contrast the binding of K$^+$ and Na$^+$ by these two proteins to see how nature has achieved such exquisite selectivity. As shown in Figure 10.37, the Na$^+$ and K$^+$ bound in the **selectivity filter** are completely desolvated and chelated by multiple oxygen atoms. In LeuT, the two Na$^+$ binding sites provide either five or six chelating oxygen atoms (from main chain carbonyl or side chain carboxylate, hydroxyl, or amide groups) with a mean Na$^+$—O distance of 0.23 nm. In KsaA, each K$^+$ is bound by eight oxygen atoms (from main chain carbonyl or side chain —OH groups) at a mean K$^+$—O distance of 0.28 nm. These chelating groups replace solvating interactions the ions would make with water molecules; thus, there is no enthalpic penalty for desolvating the ions as they transit the channel. Based on the structures shown in Figure 10.37, the major determinant of discrimination between Na$^+$ and K$^+$ appears to be the geometry of the chelating groups in the selectivity filter.

It is critical that the activity of any ion channel is regulated to maintain proper cell function; thus, there should be "open" (i.e., ion-conducting) and "closed" conformations for the channel. The switching between conductive and nonconductive conformations is called **gating**. Figure 10.36 compares open and closed conformations for the K$^+$ channel. Note that the K$^+$ channel pore structure is highly conserved and is defined by two membrane-spanning helices and the shorter selectivity filter described above. In the closed conformation, one transmembrane helix from each subunit extends into the channel cavity on the cytoplasmic side of the channel. The convergence of these four helices occludes the channel and prevents K$^+$ transport (Figure 10.36b,c). In response to some gating stimulus (e.g., a change in pH, or membrane potential, or binding of some ligand to the extracellular portion of the channel protein, etc.), the conformations of these helices change. Near the middle of each helix is a so-called hinge Gly (highlighted in red in Figure 10.36c,f). Bending of the transmembrane helices around this Gly moves the C-terminal ends of the helices apart, thereby opening the channel.

To understand better the mechanism of channel gating, let us consider the current model for the opening and closing of voltage-gated K$^+$ channels. The voltage-gated channels include six helical transmembrane segments labeled S1–S6 (Figure 10.38), where the K$^+$ pore is formed by the S5 and S6 helices along with the intervening selectivity filter sequence. This pore is structurally similar to the channels shown in Figure 10.36. Sequences S1–S4 make up the voltage-sensing domain, where S4 contains a sequence in which every third residue is lysine or

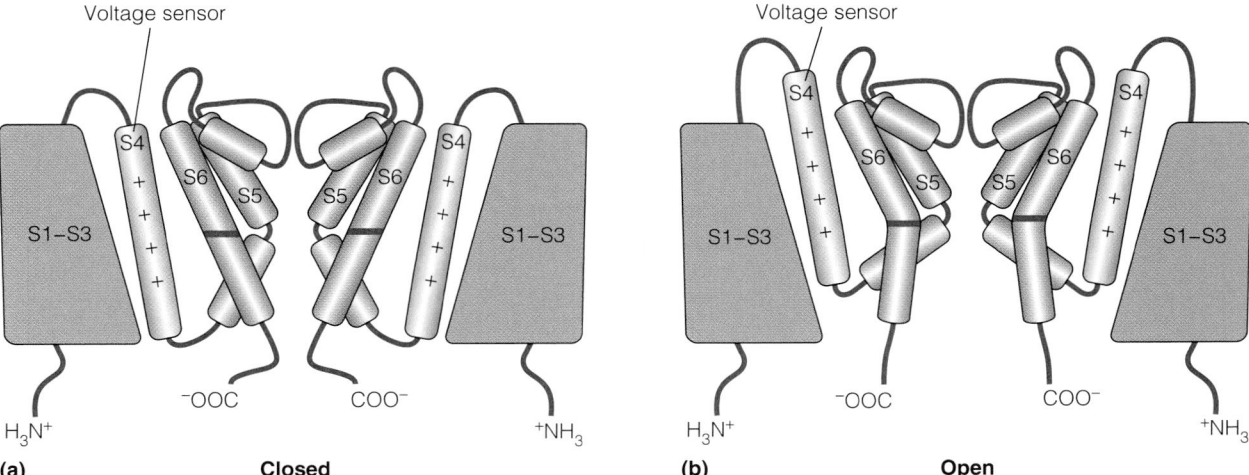

FIGURE 10.38

A model for voltage-gating in the K$^+$ channel. The channel portion of the voltage-gated K$^+$ channel is structurally homologous to the KcsA channel (compare to Figure 10.36). The Arg- and Lys-rich S4 helices are highlighted in blue. The depth of these helices in the bilayer changes as a function of the membrane potential. In the closed conformation **(a)**, the helices are closer to the cytosolic side of the bilayer and this position seals the channel. The channel opens when the charged helices move toward the extracellular side of the bilayer **(b)**. The red stripe in helix S6 indicates the location of the "hinge Gly" residue.

arginine, separated by two hydrophobic residues. The position of the S4 helix in the channel is thought to change as a function of the membrane potential ($\Delta\psi$). The cytoplasmic side of the membrane is more negatively charged; thus, in the resting state of the membrane, with the channel closed, the S4 helix is positioned closer to this side of the membrane. When the potential across the membrane changes, as in nerve signal conduction, the cytosolic side of the membrane becomes less negatively charged and the S4 helix moves toward the other side of the membrane. As it does so, it pulls S5 along, which in turn allows S6 to bend and open the channel. In this model S6 corresponds to the helix with the hinge Gly (Figure 10.37). The channel returns to the closed conformation upon restoration of the resting-state membrane potential by active ion transporters.

Now that we have described three different modes of facilitated transport, we will briefly contrast them to nonmediated transport, before discussing active transport.

Distinguishing Facilitated from Nonmediated Transport

How can facilitated transport be distinguished from nonmediated diffusion? Aside from the generally much higher transport rate, a simple test is that facilitated transport systems are *saturable*. Any membrane has a limited number of tranporters. Each carrier or permease can handle only one molecule or ion at a time; each pore can accommodate only one or a few ions or molecules at any moment. Thus, if we measure the rate of transport at increasingly higher concentration differences of the substance transported across the membrane, a limiting rate is approached when all transporters are occupied (Figure 10.39). The rate of nonmediated transport, on the other hand, increases linearly with the concentration difference, as predicted by Equation 10.4 or 10.5, because there are no sites to saturate.

There is also an easy way to distinguish between pore-facilitated and carrier-facilitated diffusion. The latter should be extremely sensitive to membrane fluidity because the carrier must actually move in the membrane. If the temperature of a membrane is lowered below its fluid–gel transition temperature, transport by a carrier like valinomycin virtually ceases. Transport by a pore structure like gramicidin A, on the other hand, is affected little by temperature changes. A simple analogy can be made to a ferry and a bridge: If the river freezes, the ferry is stopped, but the bridge can continue to transport.

In conclusion, we must emphasize that even though facilitated transport is sometimes very fast and very selective, it is still only a special form of diffusion. Transporters effectively increase the solubility of the substance in the membrane. The equilibrium state for a system exhibiting facilitated transport is the same as that for nonmediated transport—the substance will be transported down its

The rate of facilitated diffusion approaches a maximum value when all available transporters are saturated with substrate, whereas nonmediated diffusion shows a linear increase in rate as substrate concentration increases.

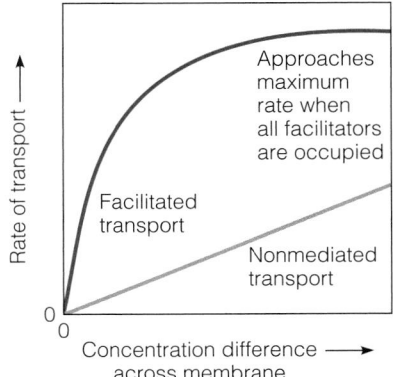

FIGURE 10.39

Facilitated and nonmediated transport. If the rate of transport is plotted against the concentration difference driving the transport, the graphs show that the rate of facilitated transport reaches a limit at a high concentration difference, whereas the rate of nonmediated transport increases linearly.

concentration gradient until the concentrations on both sides of the membrane are equal. In living cells it is necessary to transport many substances *against* a concentration gradient. The next section describes how this is achieved.

Active Transport: Transport Against a Concentration Gradient

It is imperative that some cells or cellular compartments are able to transport substances *against* concentration gradients, even very unfavorable ones. To take an extreme example, under some circumstances a $[Ca^{2+}]$ ratio of 30,000 must be established across membranes of the sarcoplasmic reticulum in muscle fibers (see Chapter 8). According to Equation 10.1, this ratio corresponds to $\Delta G = +26.6$ kJ/mol, a formidable barrier. Nevertheless, this ratio is built up and maintained in living cells. Such transport against a concentration gradient is called **active transport**. Clearly, to pump ions against a gradient requires a free energy source of some kind. In most cases, this energy comes from the hydrolysis of ATP. It is estimated that most cells spend 20–40% of total metabolic energy just on active transport. However, the hydrolysis of ATP can be coupled to transport in a number of different ways, some of them rather indirect. To give an idea of the range of these mechanisms, we shall now consider some specific examples.

Ion Pumps: Direct Coupling of ATP Hydrolysis to Transport

The best known physiological example of active transport is the maintenance of sodium and potassium gradients across the plasma membranes of cells. The fluid surrounding cells in most animals is about 145 mM in Na^+ and 4 mM in K^+. Yet animal cells maintain a Na^+ concentration of about 12 mM and a K^+ concentration of about 155 mM in their cytosol.

Even though Na^+ and K^+ pass very slowly through membranes by nonmediated diffusion, such inequalities would ultimately vanish unless something were done to keep K^+ moving in and Na^+ out. This movement is accomplished by the action of the **sodium–potassium pump** or **Na^+-K^+ ATPase** (Figure 10.40), first described by Jens Skou, who was awarded the 1997 Nobel Prize in Chemistry for this discovery. The sodium–potassium pump is only one member of a large class

In active transport, substances are moved across a membrane against a concentration gradient. Direct or indirect coupling of transport to ATP hydrolysis provides the required free energy.

The Na^+-K^+ pump acts in all cells to maintain higher concentrations of K^+ inside and Na^+ outside.

FIGURE 10.40

The structure of the Na^+-K^+ ATPase with K^+ bound. The α subunit is shown in cartoon rendering with the transmembrane domain in green, the ATP-binding domain in cyan, the phosphorylation domain in orange, and the actuator domain in purple. The actuator domain translates conformation changes in the cytoplasmic domains to the membrane-bound domain (see scheme in Figure 10.41). The β subunit is beige and the small regulatory protein (FXYD) is shown in gray. K^+ ions are shown as red spheres. Two K^+ are bound in the transmembrane helical bundle. These will be transported across the membrane. Another K^+ ion, thought to activate protein dephosphorylation, is bound in the phosphorylation domain. A phosphate ion analog, MgF_4^{2-}, is shown in green spheres at the site where reversible phosphorylation of the ATPase occurs. PDB ID: 2ZXE.

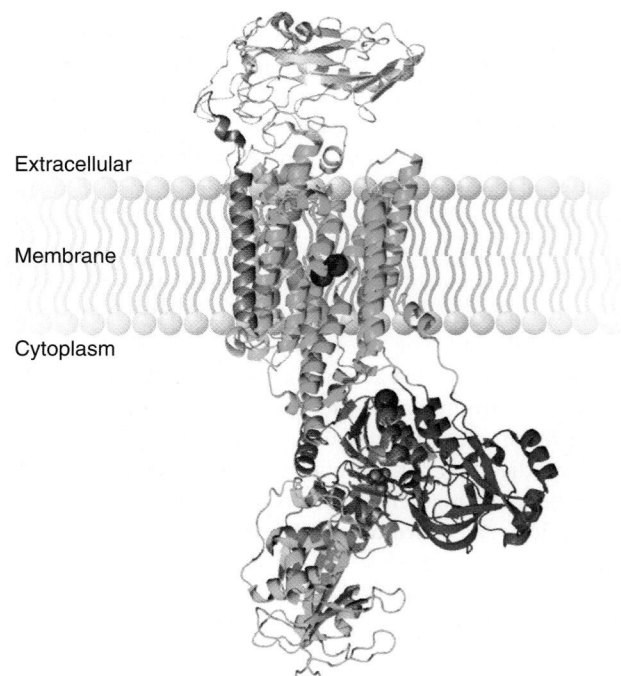

Extracellular

Membrane

Cytoplasm

of structurally related **P-type ATPases** that function in active transport across the plasma membrane. This molecular machine consists of a large 113 kDa α subunit, a 55 kDa β subunit, and frequently includes a much smaller regulatory subunit, γ. The α subunit is directly involved in the transport process and is an enzyme that hydrolyzes ATP. The free energy change in that reaction is used to drive the transport. The α subunit traverses the membrane 10 times, forming a multihelix channel. The site for ATP binding and phosphorylation lies on the cytoplasmic side (see Figure 10.40). Facing the outside are multiple sites for binding of **cardiac glycosides**, including ouabain and **digitoxin** (digitalis). The medical importance of these is discussed below. The β subunit has a single membrane-traversing helix and carries a large (20 kDa) polysaccharide on the outer surface. The β subunit acts as a chaperone and is required to target the α subunit to the plasma membrane. It also has a role in restricting K^+ movement in the conformational cycle that results in ion transport (see below).

Two K^+ ions are pumped into the cell and three Na^+ ions are pumped out for every ATP hydrolyzed. Is this estimate reasonable from a thermodynamic point of view? To answer this, we calculate the free energy required to take 3 moles of Na^+ from 12 mM to 145 mM and 2 moles of K^+ from 4 mM to 155 mM at 37 °C. First, let us use Equation 10.2 to calculate the free energy required to transport 3 moles of Na^+ from within the cell to outside. We must take into account the membrane potential of about 0.060 volt. The inside of the membrane is more negative than the outside, so this potential opposes the flow. Per mole of Na^+, we have

$$\Delta G = RT \ln \frac{[C_{Na^+}]_{out}}{[C_{Na^+}]_{in}} + Z_{Na^+}F\Delta\psi_{in \to out}$$

$$\Delta G = \left(0.008314 \frac{kJ}{mol\ K}\right)(310\ K)\left(\ln \frac{(0.145)}{(0.012)}\right) + (+1)\left(96.48 \frac{kJ}{mol\ V}\right)(+0.060\ V)$$

$$\Delta G = \left(6.4 \frac{kJ}{mol}\right) + \left(5.8 \frac{kJ}{mol}\right) = 12.2 \frac{kJ}{mol}$$

For 3 moles, then, we have $\Delta G = 3\ mol\ Na^+ \times +12.2\ kJ/(mol\ Na^+) = +36.6\ kJ$.

When K^+ is transported inward, the membrane potential is working in favor of the flow. Per mole of K^+, we have

$$\Delta G = \left(0.008314 \frac{kJ}{mol\ K}\right)(310\ K)\left(\ln \frac{(0.155)}{(0.004)}\right) + (+1)\left(96.48 \frac{kJ}{mol\ V}\right)(-0.060\ V)$$

$$\Delta G = \left(9.4 \frac{kJ}{mol}\right) + \left(-5.8 \frac{kJ}{mol}\right) = 3.6 \frac{kJ}{mol}$$

or, for 2 moles, $\Delta G = +7.2\ kJ$. The total free energy requirement for the outward transport of 3 moles of Na^+ and the inward transport of 2 moles of K^+ is then

$$\Delta G_{total} = 36.6\ kJ + 7.2\ kJ = +43.8\ kJ$$

At first glance, it would appear that the hydrolysis of 1 mole of ATP would not provide the necessary energy, for we have stated that $\Delta G°'$, the standard-state free energy change for ATP hydrolysis under physiological conditions, is about -30 kJ/mol. In most cells, however, ATP is in much higher concentration than ADP so the actual free energy change per mole is typically -45 to -50 kJ/mol (see Chapter 12 and Problem 5). Thus, ATP hydrolysis is sufficient to maintain these concentration gradients under the observed stoichiometry of transport, but it could not transport any more than 3 mol of Na^+ and 2 mol of K^+ per mol ATP hydrolyzed.

Despite the transport against strong electrochemical gradients, the sodium–potassium pump involves no violation of thermodynamic principles. The only requirement is that ATP hydrolysis and transport be *coupled*. This coupling is apparently accomplished in a multistep process. A schematic model for the entire process is diagrammed as a cycle in Figure 10.41. It is proposed that the pump can adopt two conformations, one open only to cytosol, the other open

FIGURE 10.41

A schematic diagram of the functional cycle of the Na⁺–K⁺ pump. The α subunit is believed to have two states, one open only to the outside (brown), the other open only to the inside (blue). A dot (·) between two symbols indicates noncovalent binding, and a line (|) indicates covalent attachment (as in phosphorylation).

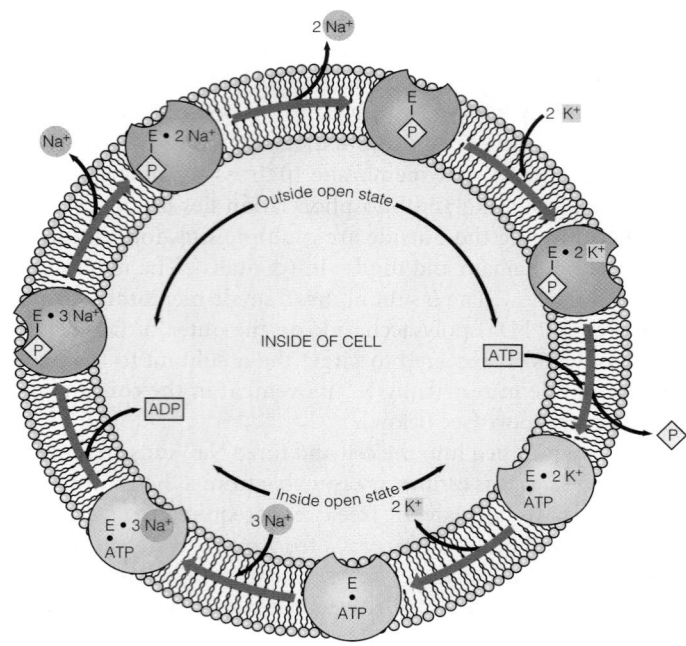

Cardiotonic steroids inhibit the Na⁺–K⁺ pump, resulting in increased Ca²⁺ ion concentration in heart muscle, which, in turn, leads to stronger contractions of the heart muscle.

In cotransport, the unfavorable movement of a substance through the membrane is coupled to the favorable transport of another substance.

only to the cell's surroundings. Completion of one conformational cycle requires hydrolysis of one ATP. Transition to the cytosol-open conformation, which allows K⁺ release and Na⁺ uptake, is triggered by binding of ATP and release of phosphate. Transition to the outside-open state, which permits Na⁺ release and K⁺ uptake, occurs upon phosphorylation of the α subunit and release of ADP.

The outside-open state designated E-P in Figure 10.41 has an especially high affinity for cardiotonic steroids, like digitoxin and ouabain. These agents inhibit the Na⁺–K⁺ pump by locking it in this conformation. Such inhibition has major effects on muscles, especially in the heart. The build-up of Na⁺ in cells leads to measures to reduce its concentration, including a Ca²⁺–Na⁺ exchange process catalyzed by another pump. The resulting increase in Ca²⁺ in the sarcoplasmic reticulum of heart muscle cells leads to much stronger contractions (see pages 293–294). This is why substances like digitoxin and ouabain are used as heart stimulants.

Because ATP hydrolysis and transport are tightly coupled, a pump driven *backward* can act as an ATP generator. That is, the same kind of molecular mechanism described above, if allowed to pass a substance down a gradient, can be used to *synthesize* ATP from ADP and P_i. In fact, this mechanism is the major way in which ATP is produced in living organisms (see Chapter 15).

Cotransport Systems

There are other kinds of active transport that do not depend directly on ATP as an energy source but employ ATP hydrolysis in an indirect way. You can imagine how this might occur, if you consider that the kind of ATP-driven ion pumps described above can generate large ion concentration gradients across membranes. These ion gradients are far from equilibrium and hence represent in themselves a potential source of free energy. The **sodium–glucose cotransport system** of the small intestine (Figure 10.42) is an example of how an ion gradient is used in driving transport of glucose from a region of low [glucose] (the intestinal lumen) to one where [glucose] is higher (inside epithelial cells of the intestine wall). The transport of each glucose molecule from within the intestinal lumen into the epithelial cells is accompanied by the simultaneous movement of one Na⁺ ion in the same direction. Because a favorable Na⁺ gradient is maintained by the ATP-driven Na⁺–K⁺ pump of these cells, glucose can be transported against an unfavorable

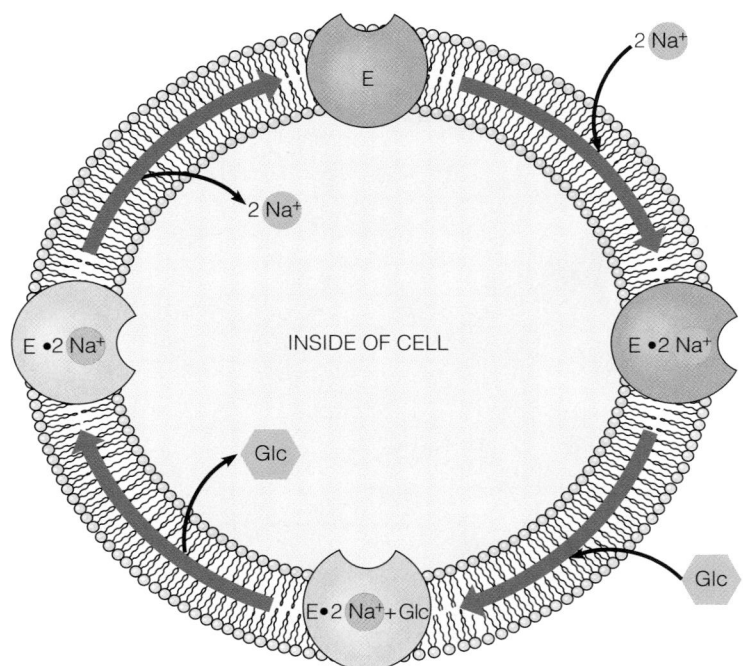

FIGURE 10.42

A schematic model for the sodium–glucose cotransport (symport) system. As in the case of the sodium–potassium pump, the sodium–glucose cotransport channel is presumed to have two possible states—one open only to the outside, the other open only to the inside of the cell. Transition to the inside-open state is stimulated by glucose binding to E · Na$^+$. Return to the outside-open state occurs upon Na$^+$ release to the inside of the cell. The sodium gradient from inside to outside provides the driving force for the unfavorable transport of glucose. That gradient must be maintained by the sodium–potassium pump.

gradient in glucose concentration. Glucose "piggybacks" on the thermodynamically favored Na$^+$ transport.

A large number of such cotransport systems are known, many of them utilized to move nutrients into cells. A few examples are listed in Table 10.7. Many use the Na$^+$ gradient as a driving force, but some, like the *lactose permease system* in *E. coli*, depend on an H$^+$ gradient. As we shall see in later chapters, generation of H$^+$ gradients is a central step in energy production by most cells.

Transport by Modification

One other method cells have of achieving transport against a gradient uses the following strategy: Suppose a molecule, on moving into a cell by nonmediated or facilitated transport, is chemically modified in such a way that it can no longer pass back through the membrane. The net result is that quantities of the modified molecule steadily accumulate within the cell. This method is used by many bacteria for the uptake of sugars. The sugars are phosphorylated, either during their diffusion through the membrane or as soon as they emerge into the cytosol. Membranes are impermeable to the charged, phosphorylated monosaccharides, and thus these products remain in the cell. In the best-studied example, the **phosphoenolpyruvate: glucose phosphotransferase system** of *E. coli*, transport is facilitated by a transmembrane protein. While in the pore of the transporter, the sugar molecule is apparently phosphorylated by phosphoenolpyruvate (PEP, a strong phosphoryl group donor—see Chapter 3). The process has the added convenience that phosphorylation of monosaccharides is, as discussed in Chapter 13, the first step in their metabolic utilization. Thus, the sugars taken up by *E. coli* are already primed for metabolism. Although this transport mechanism looks very different from the direct coupling in ion pumps, they are basically the same. In both cases a compound (ATP or PEP) with a very favorable free energy of phosphoryl group transfer has been hydrolyzed to accomplish the directed transport of a molecule across the membrane.

In this section we have described only a few examples of specific membrane transport. This phenomenon is encountered in later chapters on metabolism. To give an idea of how important specific transport is, Figure 10.43 illustrates some

In transport by modification, a substance that has diffused through a membrane is modified so that it cannot return.

TABLE 10.7	Some cotransport systems	
Molecule Transported	Ion Gradient Used	Organism or Tissue
Glucose	Na$^+$	Intestine, kidney of many animals
Amino acids	Na$^+$	Mouse tumor cells
Glycine	Na$^+$	Pigeon erythrocytes
Alanine	Na$^+$	Mouse intestine
Lactose	H$^+$	*E. coli*

FIGURE 10.43

Specific transport processes. This composite plant–animal cell illustrates some of the most important specific transport processes. All of the substances shown here, and many more, are transported in specific directions across cellular membranes. Magenta dots signify known transport proteins.

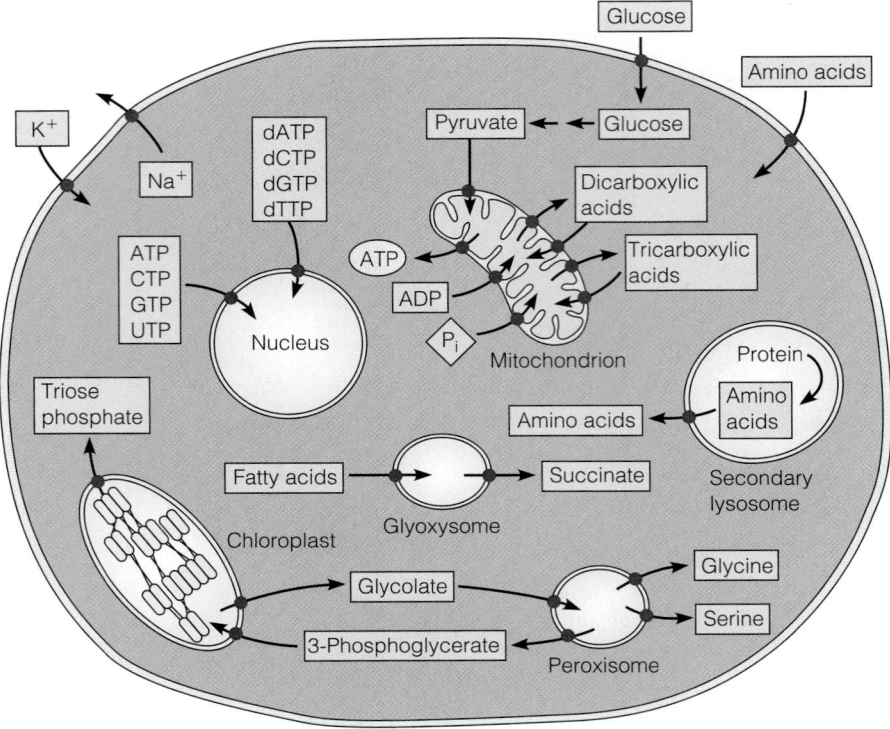

of the known examples, using a hypothetical cell containing some characteristics of both plant and animal cells.

The modes of transport we have described above for small molecules or ions typically involve the action of a single protein or carrier. Larger volumes of solution can be transported across various membranes in bulk as a result of remodeling of membranes. This is achieved by the actions of proteins that bind tightly to the bilayer and deform it, forming pits or sacs that enclose the volume to be transported.

Caveolae and Coated Vesicles

Bilayer deformation is required for the normal function of membranes in processes such as endocytosis (see Chapter 17), exocytosis, and membrane trafficking (Chapter 28). Endocytosis is a process in which extracellular substances are engulfed by a small portion of the membrane and imported into the cell (exocytosis is essentially the reverse process). Membrane trafficking involves the sorting of proteins into specific cellular organelles, via carrier vesicles. This requires that spherical portions of the membrane engulf some volume containing "cargo" and then bud from the membrane, in a process called **membrane fission**, so the cargo can be delivered from one site to another. The cargo may be soluble in the enclosed fluid, or it may be attached to a receptor in the membrane that encloses it. **Coated vesicles** (~0.1 microns in diameter) and smaller **caveolae**, shown in Figure 10.44, are two examples of such structures.

The formation of coated vesicles is mediated by several molecules of the protein **clathrin**, which forms a cage around a portion of the membrane and thereby distorts the bilayer into a so-called **coated pit** (top panels Figure 10.44a; see also Figure 17.10, page 718). Eventually the coated pit seals up and buds (bottom panel Figure 10.44a). This process is described in greater detail in Chapter 17 (pages 718–720).

In the case of caveola formation, the protein **caveolin** inserts into one leaflet of the bilayer and induces extreme curvature (Figure 10.44b), which leads to bud formation. Caveolin interacts with the components of membrane rafts discussed above (see page 381); thus, rafts appear to play a role in membrane trafficking.

Bulk transport across membranes involves the formation of clathrin-coated vesicles and/or caveolae.

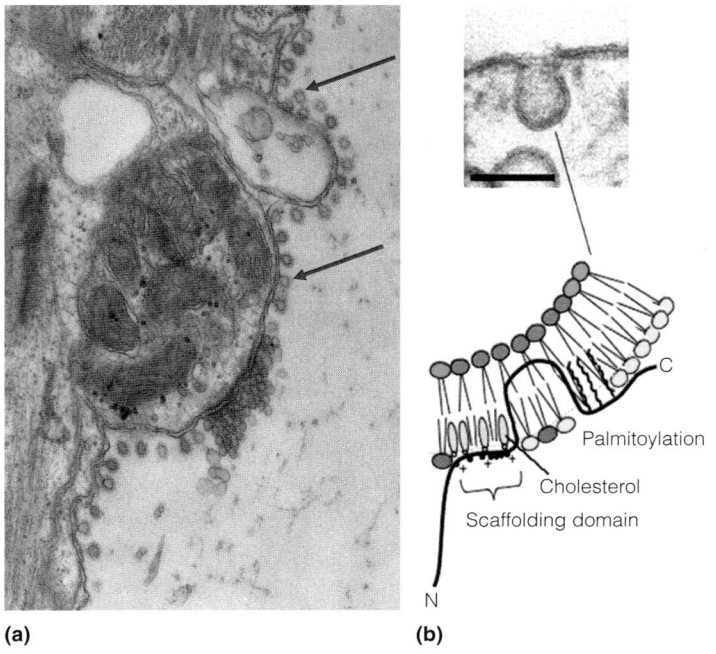

(a)   **(b)**

FIGURE 10.44

Bulk transport by coated vesicles and caveolae. Electron micrographs are shown in panel (a) and the top of panel (b). **(a)** Formation of a clathrin-coated vesicle initiates with formation of a coated pit (top two panels), which then buds (third panel from top), forming a free coated vesicle (bottom panel). **(b)** Caveola formation occurs at sites rich in cholesterol and sphingolipids ("lipid rafts," see Figure 10.26) due to insertion of caveolin into one leaflet of the membrane. Budding of the caveola yields a free vesicle.

(a) Adapted from *Science* 276:259–263, O. Shupliakov, P. Löw, D. Grabs, H. Gad, H. Chen, C. David, K. Takei, P. De Camilli, and L. Brodin, Synaptic vesicle endocytosis impaired by disruption of dynamin-SH3 domain interactions. © 2006. Reprinted with permission from AAAS, Pietro De Camilli and Lennart Brodin; (b) Adapted from *Journal of Cell Science* 119:787–796, R. G. Parton, M. Hanzal-Bayer, and J. F. Hancock, Biogenesis of caveolae: A structural model for caveolin-induced domain formation. © 1997 The Company of Biologists Ltd.

Once the cargo is enclosed it must be delivered to its target in the cell. This is achieved by recognition of proteins on the surface of the target membrane by the vesicle/caveola. The membranes of the vesicle and target must then fuse to allow the cargo to be transferred. This process, called **membrane fusion**, is mediated by proteins called **SNARE**s and is discussed in greater detail in Chapter 28 (see Figure 28.47, page 1220).

Excitable Membranes, Action Potentials, and Neurotransmission

We close with an example that demonstrates the enormous variety of properties that membranes can exhibit, through their ability to regulate ion transport. The conduction of neural impulses in animals is a remarkable process, but it depends on very simple physical principles.

Neurons, the cells responsible for conduction of electrical impulses and thus nervous system communication, have specialized thin cell projections called dendrites and axons that act as the "wires" of the nervous system (Figure 10.45). Neurons are truly remarkable cells that must meet unusual requirements. They

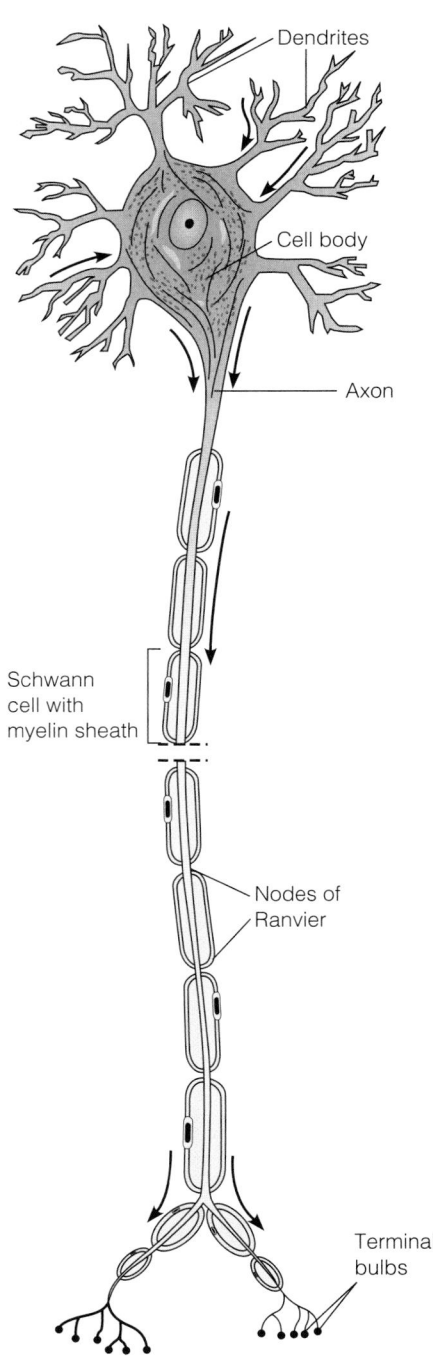

FIGURE 10.45

Structure of a typical mammalian motor neuron. A motor neuron transmits nerve impulses to muscles. The cell body contains the nucleus and most of the cellular machinery. The dendrites receive signals from the axons of other neurons; the axon transmits signals via the synaptic termini, which communicate to the dendrites of other neurons or to muscle cells. Along the axon are Schwann cells, which envelop the axon in layers of an insulating myelin membrane. The Schwann cells are separated by nonmyelinated regions called the nodes of Ranvier.

Neurons conduct electrical impulses by membrane potential changes in regions of the plasma membrane cell.

must be able to conduct impulses over relatively long distances without significant signal loss (e.g., from the spinal cord to the tip of the toe), and they must conduct impulses on millisecond timescales to control rapid and coordinated thought and behavior. Nerve conduction is accomplished not by electron flow, as in wires, but by waves in the membrane electrical potential on the surface of the membrane. To understand how nerve signaling works, we must examine how membrane potentials are generated and how they can be changed. In the limited space here, we can give only the briefest introduction to this vast and complex field.

The Resting Potential

To understand changes in membrane potentials, we must first understand the source and nature of the cell's resting membrane potential. We begin with a simplified model, which draws on our earlier discussion of the electrochemical potential difference across a semipermeable membrane (see Equation 10.2). Suppose we have an ion (M^Z) of charge Z that is present outside the membrane at concentration $[M^Z]_{out}$ and inside at concentration $[M^Z]_{in}$.

If the system is at equilibrium, ΔG for transport will be zero. Then, from equation (10.2), we find

$$\frac{RT}{ZF} \ln \frac{[M^Z]_{out}}{[M^Z]_{in}} = \Delta\psi \qquad (10.7)$$

where $\Delta\psi$ is defined as $\psi_{out} - \psi_{in}$. Equation 10.7 is one form of the **Nernst equation**. For monovalent ions ($Z = \pm1$) at 20 °C, this form of the Nernst equation reduces to

$$\Delta\psi = \pm59 \log_{10}\frac{[M]_{out}}{[M]_{in}} \qquad (10.8)$$

when $\Delta\psi$ is expressed in millivolts (mV).

According to Equation 10.8, if we maintain an ion concentration difference across a membrane, an electrical potential difference will be produced. For example, if an ion such as K^+ ($Z = +1$) is kept 10 times as concentrated inside as outside, the membrane will be polarized, with $\Delta\psi = -59$ mV. If the single ion that is unevenly distributed in this way is chloride, the potential will be $+59$ mV.

The major mechanisms creating ionic imbalance across cellular membranes are the specific ion pumps (e.g., the Na^+-K^+ ATPase) that continually act to concentrate certain ions on one side or the other. This imbalance gives rise to the resting potential across the membrane of a nerve axon. A much-studied example is the giant axon of the squid. This is a favorite experimental tool because squids are unusual in having axons as large as 1 mm in diameter. It is possible, as shown in Figure 10.46, to insert recording electrodes into such an axon and measure the

FIGURE 10.46

Use of squid giant axons for studies of neural transmission. Electrodes attached to a voltmeter record the potential across the axonal membrane. At the resting axon ion concentrations shown here, the voltmeter would read about −60 mV. If the axon is stimulated at point A by a depolarizing pulse, the traveling membrane potential will shortly pass point B, where it can be recorded.

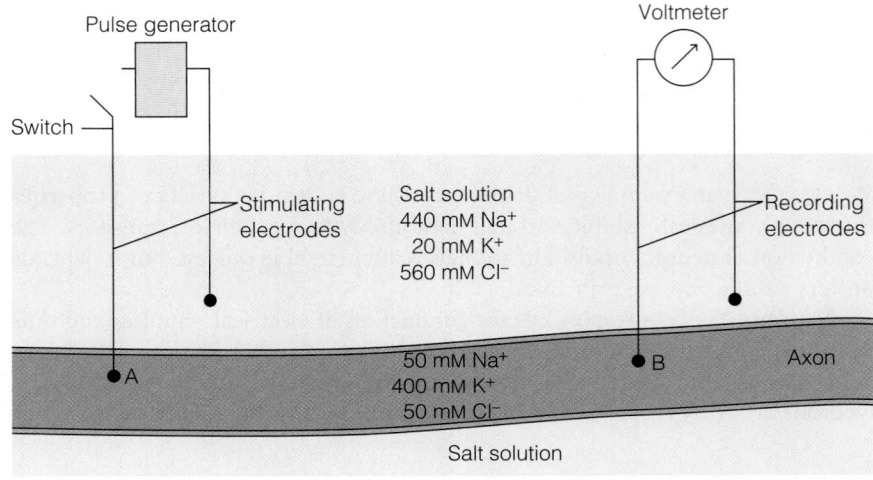

potential difference across the membrane. (The figure also shows stimulating electrodes, which are used in an experiment we will describe shortly.)

In the squid axon, we have a situation more complex than can be described by the Nernst equation. Several ions are involved, each of which can pass through the membrane, at least to some degree, and which are maintained in unequal concentrations on the two sides. If we used the Nernst equation to calculate the potential from the distribution of K^+ alone, we would predict a value of -75 mV (which we will call $\Delta\psi_{K^+}$). On the other hand, using the Nernst equation with the Na^+ concentration inside and outside, we would find $\Delta\psi_{Na^+} = +55$ mV. When we measure the potential across a resting squid axonal membrane, we find a value of about -60 mV. What determines the actual potential in this case?

The key is that when several kinds of ions can pass through the membrane, the *permeability* of the membrane to different ions is important in determining $\Delta\psi$. The various ions are not at true equilibrium across the axonal membrane but are in a steady state, the position of which is determined in part by the individual permeabilities. This steady state can be described quantitatively by the **Goldman equation**. For the membrane potential determined by a collection of monovalent anions and cations, the Goldman equation gives

$$\Delta\psi = \frac{RT}{F} \ln \frac{\sum_+ P_i[M_i^+]_{out} + \sum_- P_j[X_j^-]_{in}}{\sum_+ P_i[M_i^+]_{in} + \sum_- P_j[X_j^-]_{out}} \qquad (10.9)$$

Here, the sums (denoted by Σ) are taken over all cations (Σ_+) and anions (Σ_-) with significant permeability, and the P values are the relative membrane permeabilities for these ions. Note that if any one ion were to have a much greater permeability than any of the others, it would dominate in the Goldman equation, and Equation 10.9 would reduce to the Nernst equation for that ion.

The resting potential is largely determined by K^+ **leak channels**, which, unlike the gated K^+ channels discussed previously, are open under most conditions. The leak channels increase the permeability of cell membranes to K^+, resulting in a membrane potential that is close to that predicted by the Goldman equation. The only ions that contribute appreciably to the membrane potential in this case are K^+, Na^+, and Cl^-, and their relative permeabilities are $P_K^+ = 1.0$, $P_{Na}^+ = 0.04$, and $P_{Cl^-} = 0.45$. If these values are inserted into Equation 10.9, together with the ion concentrations given in Figure 10.46, we find $\Delta\psi = -61$ mV, in agreement with experimental observation.

The Action Potential

The action potential is a controlled and rapidly propagated change in membrane potential that is transmitted down the length of an axon. This process involves a unique set of voltage-gated potassium and voltage-gated sodium channels. Due to the biochemical properties of the ion channels, the action potential is self-propagating and travels in a single direction down an axon. We briefly describe how this unique membrane potential change is created and propagated.

An action potential is created if the changes in the membrane potential reach a critical threshold. The resting potential value of -61 mV has a vital significance. Because -61 mV lies much closer to -75 mV than to $+55$ mV, it means that at the potential existing across the squid axon membrane, K^+ is much closer to its equilibrium distribution than is Na^+. If the membrane were to become fully permeable to Na^+ ions, the major event would be a massive influx of Na^+ ions, with an accompanying shift in the membrane potentials toward $\Delta\psi_{Na^+}$. This is just what happens when an action potential is transmitted along a nerve (Figure 10.47).

Axonal fibers have specific, voltage-gated channels for the facilitated transport of Na^+ and K^+ through the membrane. As described previously, voltage-gated channels are open or closed, depending on the membrane potential. For example, they are closed in the resting state (Figure 10.47a). Suppose we perform an experiment using electrodes placed in a squid giant axon as shown in Figure 10.46. At some distance from the recording electrodes, we place stimulating electrodes

The resting potential of a nerve fiber is determined by the permeabilities of the membrane to different ions, particularly K^+, which has a high permeability due to K^+ leak channels.

The action potential is generated and propagated because a small depolarization of the nerve cell membrane opens voltage gated channels, allowing ions to flow through.

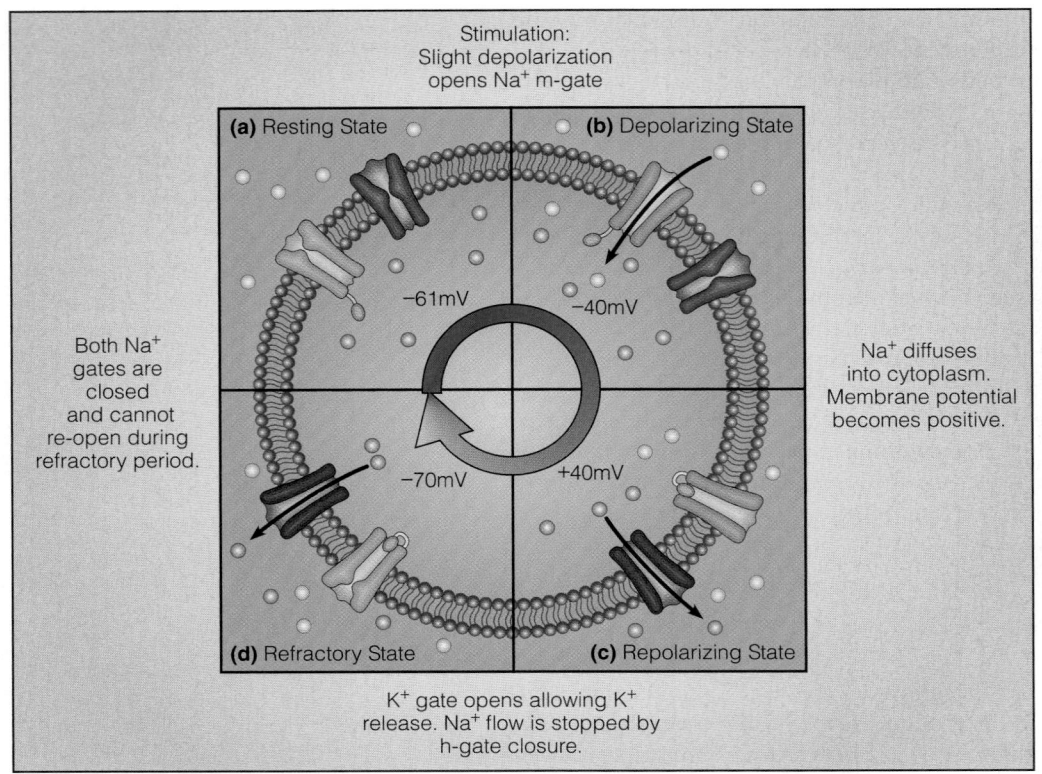

Stimulation:
Slight depolarization
opens Na⁺ m-gate

(a) Resting State   **(b)** Depolarizing State

−61mV   −40mV

Both Na⁺
gates are
closed
and cannot
re-open during
refractory period.

Na⁺ diffuses
into cytoplasm.
Membrane potential
becomes positive.

−70mV   +40mV

(d) Refractory State   **(c)** Repolarizing State

K⁺ gate opens allowing K⁺
release. Na⁺ flow is stopped by
h-gate closure.

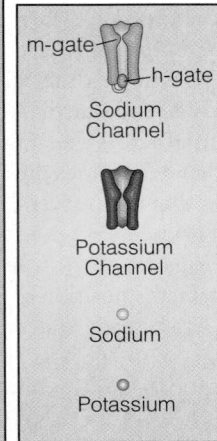

m-gate ── ── h-gate
Sodium
Channel

Potassium
Channel

○ Sodium

○ Potassium

FIGURE 10.47

The voltage-gated channel cycles during an action potential. **(a)** In the resting state, K⁺ is concentrated on the inside of the axon membrane and Na⁺ on the outside. **(b)** Electrical stimuli cause local, partial depolarization of the membrane, which in turn causes the voltage-gated sodium channel activation gates to open, allowing a large inward rush of Na⁺. The opening of these channels causes the membrane to be primarily permeable to Na⁺ and thus the membrane potential is determined primarily by the potential of Na⁺. As Na⁺ approaches its equilibrium, the inrush slows, and Na⁺ diffuses along inside the axon. **(c)** The increase in membrane potential closes the Na⁺ channel h-gates, and opens the K⁺ channels. K⁺ gates open, and K⁺ flows outward. **(d)** The outward flow of K⁺ causes an extreme decrease in the membrane potential, which becomes temporarily more negative than the resting potential. This causes the Na⁺ m-gates to close (in addition to the h-gates). Movement of K⁺ ions restores the resting potential, but the membrane is inactivated (refractory) during this time period. This refractory period is crucial for ensuring unidirectional propagation of the action potential down the axon.

Courtesy of Gary Carlton.

connected to a voltage source. If we apply at this electrode a pulse sufficient to depolarize the membrane locally by about 20 mV—that is, to a potential of −40 mV (the threshold for opening the Na⁺ channels)—the Na⁺ channel **activation gates** (m-gates) are opened (Figure 10.47b). The permeability to Na⁺ increases about 100-fold and a flood of sodium ions rushes in, bringing the membrane potential up to about +40 mV within less than a millisecond (Figure 10.47c). In terms of the Goldman equation, the sudden high permeability of Na⁺ ions makes them dominate, and the membrane potential *approaches* $\Delta\psi_{Na^+}$ (+55 mV). It does not reach this value, however, for further changes now occur. As seen in Figure 10.47b, the stimulus also causes K⁺ channel gates to open, but more slowly, and these allow the influx of K⁺ ions into the surrounding medium. This event reverses the potential again, overshooting to about −70 mV. This change closes the Na⁺ channel **inactivation gates** (h-gates, which are separate from the activation gates), temporarily making the Na⁺ channels resistant to opening (the refractory period— Figure 10.47d). The potential and permeability changes shown in Figure 10.47b–d occur in just a few milliseconds. The same changes are depicted graphically in Figure 10.48.

These steep changes in membrane potential would be a localized effect were it not for the sequential opening and closing of gates as shown in Figure 10.47c and d. As Na⁺ ions rush in, they diffuse away from the region of stimulus and trigger the same round of depolarization in an adjacent section of the fiber. Thus a wave of depolarization can spread through the axon. Following the wave front are the reverse polarization due to K⁺ efflux, the refractory zone where K⁺ channels are open and the Na⁺ channels are inactivated, and the final return to the resting condition (Figures 10.47a and 10.49). Thus, the recording electrode in Figure 10.46 will see traveling past it, at some time after the stimulating pulse, exactly the same pattern of depolarization and reversal depicted in Figure 10.47b–d and Figure 10.48. This traveling pulse is called the *action potential*. The movement is depicted in Figure 10.49. The time required for the impulse to pass from stimulating electrode to recording electrode is proportional to the distance between the two electrodes and is inversely proportional to the velocity of propagation of the pulse. Typical values for propagation of the action potential range from 1 to 100 meters per second.

Toxins and Neurotransmission

Many extremely toxic substances have their effect by blocking the action of the specific ion gates necessary for development of the action potential. These substances are often called **neurotoxins**. *Tetrodotoxin* is found in some organs of the puffer fish. This fish is considered a delicacy in Japan, where special chefs are trained and certified for their ability to remove the toxin-containing organs. Tetrodotoxin binds specifically to the Na^+ channel, blocking all ion movement. The same effect is produced by *saxitoxin*, contained in the marine dinoflagellates responsible for "red tide." These microscopic algae, along with their toxin, are ingested by shellfish and can in turn be consumed by humans. These two toxins, which attack a fundamental process of the nervous system, are among the most poisonous substances known, and their accidental ingestion leads to many deaths every year. A third very poisonous substance, *veratridine*, is found in the seeds of a plant of the lily family, *Schoenocaulon officinale*. This toxin also binds to the Na^+ channels but blocks them in the "open" configuration.

These toxins have proven to be useful in studies of axonal structure and conduction, for their tight binding makes them excellent affinity labels for the channel.

We have described here only one part of the whole phenomenon of the transmission of neural impulses—the conduction along a single nerve fiber. The

FIGURE 10.48

The action potential. **(a)** Changes in membrane conductance at a point on an axon as a neural impulse passes. The membrane first becomes permeable to sodium ions, allowing a large inward rush of Na^+. A decrease in the Na^+ permeability results and is in turn followed by an outward flow of K^+. **(b)** Changes in membrane potential accompanying the permeability changes shown in (a). As Na^+ rushes in, the potential increases and becomes positive. As K^+ influx increases, the potential decreases, undershooting the resting potential, $\Delta\psi_m$, before returning to the resting potential.

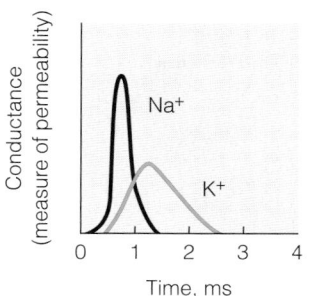

(a)

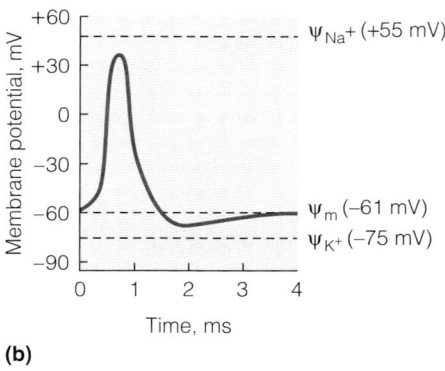

(b)

Neurotoxins can act by blocking gates in the axonal membrane in closed or open states.

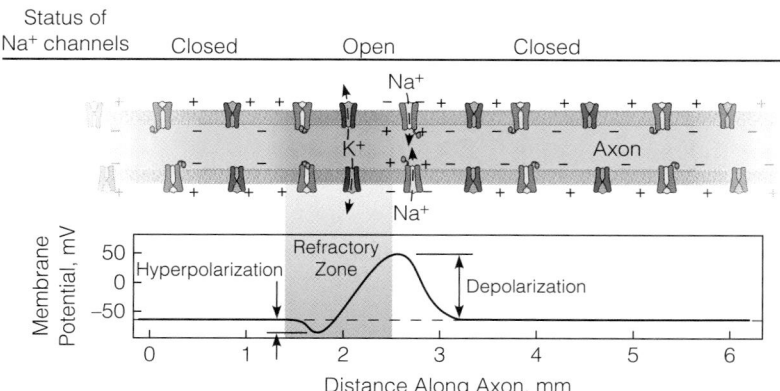

(a) Time = 0

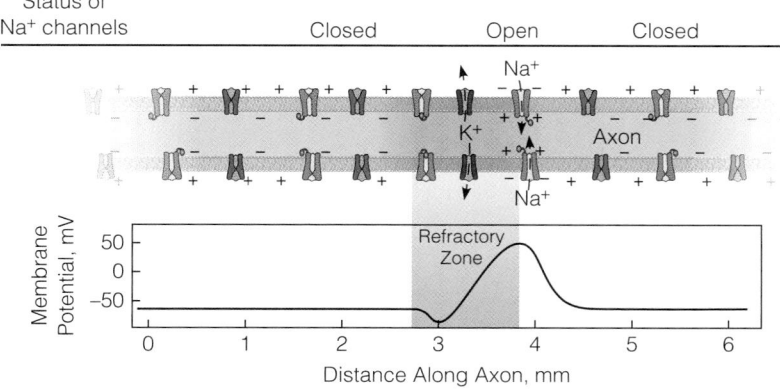

(b) Time = 1 ms

FIGURE 10.49

Transmission of the action potential. Shown are two "snapshots," taken 1 ms apart, of potential along the axon. Arrows show Na^+ influx and K^+ efflux. **(a)** At time = 0, an action potential is occurring at the 2.5-mm position. The depolarization spreads down the axon, triggering development of the action potential downstream (to the right in this diagram). **(b)** At time = 1 ms, the action potential peak has moved to the 3.8-mm position. The potential can move in only one direction because after it has passed, the region behind the potential becomes refractory for a few milliseconds due to the inactivation gate (see Figure 10.47).

Courtesy of Gary Carlton.

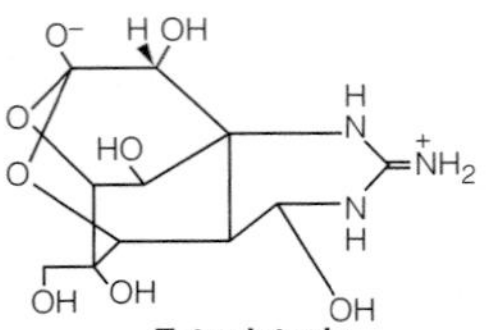

Tetrodotoxin

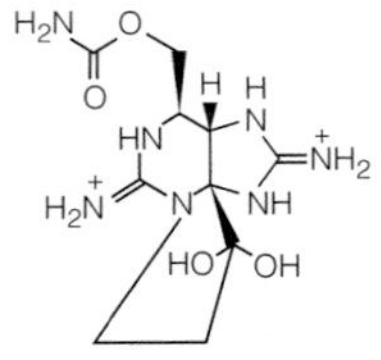

Saxitoxin

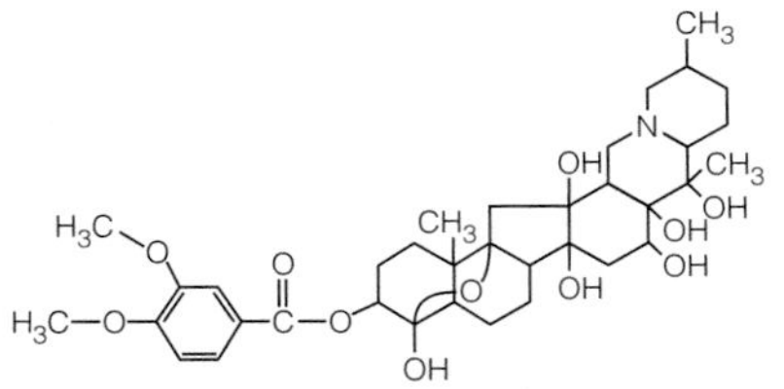

Veratridine

equally important problem of how these impulses are transmitted from one cell to another is discussed in Chapter 23, where the actions of *neurotransmitter* substances are described.

SUMMARY

Many of the important properties of lipids stem from the fact that these substances are largely hydrophobic. Some are amphipathic, containing both hydrophobic and hydrophilic regions. Most naturally occurring fatty acids contain an even number of carbon atoms. When they are unsaturated, the double bonds are usually *cis*. Fatty acids are present in fats (triacylglycerols), where they serve for energy storage and insulation, and in membranes, where they are constituents of phospholipids, sphingolipids, glycosphingolipids, and glycoglycerolipids. Membranes are bilayer structures, containing proteins and lipids in a fluid mosaic. The two leaflets differ in protein and lipid composition. Peripheral proteins are confined to one face or the other, whereas integral proteins extend through the membrane, with hydrophobic α helices common in the transmembrane region.

Transport through membranes may be achieved by nonmediated diffusion; may be facilitated by pores, permeases, or carriers; or may be actively driven by exergonic reactions. Only in the last case can transport against a concentration gradient occur. An example is the Na^+-K^+ pump, which maintains the ionic imbalance and membrane potential found between cells and their surroundings. Active transport may be indirect, as in cotransport or transport by modification.

Neural conduction of impulses depends upon a moving wave of depolarization (an action potential) in the membrane potential of a neural cell. This depolarization is produced by the flow of ions through voltage-gated channels in the membrane. The rate of neural transmission depends upon the dimensions of the axons and whether or not they are myelinated.

APPENDIX

Guidelines for Evaluating the Thermodynamics of Ion Transport

In biological systems the concentrations of solutes, especially ions, are usually significantly different across a membrane. Such concentration differences for ions are called "electrochemical gradients," or "electrochemical potentials." The free energy of ion movement through an electrochemical gradient is described by Equation 10a.1:

$$\Delta G = \Delta G^{\circ\prime} + RT \ln Q + nzF\Delta\psi \qquad (10a.1)$$

where R = the gas constant, T = temperature in Kelvin, Q = steady-state mass action expression, n = mol of ions, z = charge on the ion, F = the Faraday constant, and $\Delta\psi$ = the membrane potential. For the movement of a solute across a membrane by diffusion, $\Delta G^{\circ\prime} = 0$. Note that for *active* transport, $\Delta G^{\circ\prime}$ will not equal zero because $\Delta G^{\circ\prime}$ for ATP hydrolysis is not equal to zero, and $\Delta G^{\circ\prime}$ for the *overall transport process* is the sum of the values of $\Delta G^{\circ\prime}$ for the individual steps (ion transport + ATP hydrolysis).

The term $RT \ln Q$ describes the energy required to move an ion through a concentration gradient and $nzF\Delta\psi$ describes the energy required to move an ion through an electrical potential. Given the intra- and extra-cellular concentrations of ions: $[Na^+]_{in}$ = 12 mM, $[Na^+]_{out}$ = 145 mM, $[K^+]_{in}$ = 155 mM, $[K^+]_{out}$ = 4 mM and the potential across the membrane of 60 mV

(inside negative), one can apply basic thermodynamic principles to correctly predict the effects that the terms $RT \ln Q$ and $nzF\Delta\psi$ will have on the sign of ΔG (Equation 10a.1).

To begin, consider the simpler case of ion transport without ATP hydrolysis. Recall that changes in state functions are calculated as State$_{(final)}$ − State$_{(initial)}$. In this case "final" and "initial" states more appropriately refer to final and initial *conditions* of the transported ion (i.e., the ion is either inside or outside the cell). As described in Chapter 3, the steady-state distribution of species is described by the mass-action term Q, where

$$Q = \text{(activities of species in final condition)}$$
$$\div \text{ (activities of species in initial condition)}$$

Likewise $\Delta\psi$ is calculated as $\psi_{final} - \psi_{initial}$, where in this case "final" and "initial" refer to location:

$\Delta\psi = \psi_{final\ location} - \psi_{initial\ location}$ (in this case "location" = "in" or "out," depending which direction the ion is transported)

Putting these parameters back into Equation 10a.1:
For the transport of $[Na^+]_{in} \rightarrow [Na^+]_{out}$

$$\Delta G = \Delta G^{\circ\prime} + RT \ln Q + nzF\Delta\psi \text{ or}$$
$$\Delta G = RT \ln ([Na^+]_{out}/[Na^+]_{in}) + nzF (\psi_{out} - \psi_{in}) \qquad (10a.2)$$

Conversely, for the transport of $[Na^+]_{out} \rightarrow [Na^+]_{in}$

$$\Delta G = RT \ln ([Na^+]_{in}/[Na^+]_{out}) + nzF (\psi_{in} - \psi_{out}) \quad (10a.3)$$

The proper evaluation of the Q term is described in detail in Chapter 3 (see pages 79–80). Here, we describe two methods for correctly calculating the value of $\Delta\psi$.

Method 1: The conditions given above, where $\Delta\psi = 60$ mV (inside negative), indicate that the magnitude of the electrical potential difference between ψ_{in} and $\psi_{out} = 60$ mV, and that the inside is negative with respect to the outside. So ψ_{in} is a relatively negative number and ψ_{out} is a relatively positive number, thus:

For the transport of $[Na^+]_{out} \rightarrow [Na^+]_{in}$:

$\psi_{in} - \psi_{out} = [a\,(-)number]$ minus $[a\,(+)number] = (-)number$,

so $\Delta\psi$ will be

$$-60\,mV\,(or\,-0.060\,V)$$

and

$$nzF\Delta\psi = (1\,mol\,Na^+)(+1)(96.5\,kJ\,mol^{-1}\,V^{-1})\,(-0.060\,V)$$
$$= -5.79\,kJ \quad (10a.4)$$

For the transport of $[Na^+]_{in} \rightarrow [Na^+]_{out}$:

$\psi_{out} - \psi_{in} = [a\,(+)number]$ minus $[a\,(-)number] = (+)number$,

so $\Delta\psi$ will be

$$+60\,mV\,(or\,+0.060\,V)$$

and

$$nzF\Delta\psi = (1\,mol\,Na^+)(+1)(96.5\,kJ\,mol^{-1}\,V^{-1})(+0.060\,V)$$
$$= +5.79\,kJ \quad (10a.5)$$

Method 2: Check this calculation against the guiding principle given below. As seen from method 1, the *magnitude* of $nzF\Delta\psi$ is the same in either case (±5.79 kJ/mol). The sign is different depending on which direction the ion is transported. The most common error in calculating $nzF\Delta\psi$ is incorrectly calculating the sign of $\Delta\psi$. This is where understanding a guiding principle can help.

For the transport of $[Na^+]_{out} \rightarrow [Na^+]_{in}$:

A $(+)$ charged sodium ion is moving from an environment of relative $(+)$ charge outside the cell to one of relative $(-)$ charge inside the cell. *Such ion movement will decrease unfavorable repulsive interactions and increase favorable attractive interactions.* Overall, this should be a thermodynamically *favorable* process and the sign of $nzF\Delta\psi$ should be $(-)$ to reflect that fact (see Equation 10a.4).

For the transport of $[Na^+]_{in} \rightarrow [Na^+]_{out}$:

A $(+)$ charged sodium ion is moving from an environment of relative $(-)$ charge inside the cell to one of relative $(+)$ charge outside the cell. *Such ion movement will increase unfavorable repulsive interactions and decrease favorable attractive interactions.* Overall, this should be a thermodynamically *unfavorable* process and the sign of $nzF\Delta\psi$ should be $(+)$ to reflect that fact (see Equation 10a.5).

Given the ion concentrations and membrane potential above, Table 10a.1 summarizes predictions for the signs on the $RT \ln Q$ and $nzF\Delta\psi$ terms in the free energy calculation based on the guiding principle above (which applies to $nzF\Delta\psi$) and another given here (which applies to $RT \ln Q$): *Transport from a location of low concentration to one of higher concentration is unfavorable; transport from a location of high concentration to one of lower concentration is favorable.*

TABLE 10a.1

	Sign on $RT \ln Q$	Sign on $nzF\Delta\psi$
For $[Na^+]_{in} \rightarrow [Na^+]_{out}$	positive (unfavorable)	positive (unfavorable)
For $[Na^+]_{out} \rightarrow [Na^+]_{in}$	negative (favorable)	negative (favorable)
For $[K^+]_{in} \rightarrow [K^+]_{out}$	negative (favorable)	positive (unfavorable)
For $[K^+]_{out} \rightarrow [K^+]_{in}$	positive (unfavorable)	negative (favorable)

REFERENCES

General

Engleman, D. M. (2005) Membranes are more mosaic than fluid. *Nature* 438:578–580. This issue of *Nature* includes several review articles on membranes.

Gennis, R. B. (1989) *Biomembranes.* Springer-Verlag, New York.

Gurr, A. I., and J. L. Harwood (1991) *Lipid Biochemistry: An Introduction,* 4th ed. Chapman & Hall, New York. A valuable source for general information concerning lipids.

Membrane Asymmetry and Structure

Daleke, D. L. (2007) Phospholipid flippases. *J. Biol. Chem.* 282:821–825.

Hartlova, A., L. Cerveny, M. Hubalek, Z. Krocova, and J. Stulik (2010) Membrane rafts: A potential gateway for bacterial entry into host cells. *Microbiol. Immunol.* 54:237–245.

Laude, A. J., and I. A. Prior (2004) Plasma membrane microdomains: Organization function and trafficking. *Molec. Membr. Biol.* 21:193–205.

Lingwood, D., and K. Simons (2010) Lipid rafts as a membrane-organizing principle. *Science* 327:46–50.

Mitra, K., I. Ubarretxena-Belandia, T. Taguchi, G. Warren, and D. M. Engelman (2004) Modulation of the bilayer thickness of exocytic pathway membranes by membrane proteins rather than cholesterol. *Proc. Natl. Acad. Sci. USA* 101:4083–4088.

Phillips, R., T. Ursell, P. Wiggins, and P. Sens (2009) Emerging roles for lipids in shaping membrane-protein function. *Nature* 459:379–385.

Singer, S. J., and G. L. Nicolson (1972) The fluid mosaic model of the structure of membranes. *Science* 175:720–731. The classic paper presenting this model.

Unwin, N., and R. Henderson (1984) The structure of proteins in biological membranes. *Sci. Am.* 250(2):78–94. Describes a pioneering structural study.

Vereb, G., J. Szöllösi, J. Matkó, P. Nagy, T. Farkas, L. Vigh, L. Mátyus, T. A. Waldmann, and S. Damjanovich (2003) Dynamic, yet structured: The cell membrane three decades after the Singer-Nicolson model. *Proc. Natl. Acad. Sci. USA* 100:8053–8058.

Voelker, D. R. (1996) Lipid assembly into cell membranes. In *Biochemistry of Lipids, Lipoproteins, and Membranes*, D. E. Vance and J. E. Vance, eds. Elsevier Science, Amsterdam.

Membrane Proteins

Booth, P. J., and P. Curnow (2009) Folding scene investigation: Membrane proteins. *Curr. Op. Struct. Biol.* 19:8–13.

Bowie, J. U. (2005) Solving the membrane protein folding problem. *Nature* 438:581–589.

Marguet, D., P.-F. Lenne, H. Rigneault, and H.-T. He (2006) Dynamics in the plasma membrane: How to combine fluidity and order. *EMBO J.* 25:3446–3457.

Müller, D. J., N. Wu, and K. Palczewski (2008) Vertebrate membrane proteins: Structure, function, and insights from biophysical approaches. *Pharm. Rev.* 60:43–78.

van Klompenburg, W., I. M. Nilsson, G. von Heijne, and B. de Kruijff (1997) Anionic phospholipids are determinants of membrane protein topology. *EMBO J.* 16:4261–4266.

von Heijne, G. (1989) Control of topology and mode of assembly of a polytopic membrane protein by positively charged residues. *Nature* 341:456–458. A test of the "inside positive" rule.

White, S. H. (2007) Membrane protein insertion: The biology-physics nexus. *J. Gen. Physiol.* 129:363–369.

The Membrane Skeleton

Bennett, V. (1985) The membrane skeleton of human erythrocytes and its implication for more complex cells. *Annu. Rev. Biochem.* 54:273–304.

Coleman, T. R., D. J. Fishkind, M. E. Mooseker, and J. S. Morrow (1989) Functional diversity among spectrin isoforms. *Cell Motility Cytoskeleton* 12:225–247.

Liu, S.-C., and L. H. Derick (1992) Molecular anatomy of the red blood cell membrane skeleton: Structure–function relationships. *Semin. Hematol.* 29:231–243.

Translocon Structure and Function

Becker, T., S. Bhushan, A. Jarasch, J.-P. Armache, S. Funes, F. Jossinet, J. Gumbart, T. Mielke, O. Berninghausen, K. Schulten, E. Westhof, R. Gilmore, E. C. Mandon, and R. Beckmann (2009) Structure of monomeric yeast and mammalian Sec61 complexes interacting with the translating ribosome. *Science* 326:1369–1373.

Egea, P. F., and R. M. Stroud (2010) Lateral opening of a translocon upon entry of protein suggests the mechanism of insertion into membranes. *Proc. Natl. Acad. Sci. USA* 107:17182–17187.

Van den Berg, B., W. M. Clemons Jr., I. Collinson, Y. Modis, E. Hartmann, S. C. Harrison, and T. A. Rapoport (2004) X-ray structure of a protein-conducting channel. *Nature* 427:36–44.

Xie, K., and R. E. Dalby (2008) Inserting proteins into the bacterial cytoplasmic membrane using the Sec and YidC translocases. *Nature Rev. Microbiol.* 6:234–244.

Transport Across Membranes

Catterall, W. A. (2010) Ion channel voltage sensors: Structure, function and pathophysiology. *Neuron* 67:915–928.

Gouaux, E., and R. MacKinnon (2005) Principles of selective ion transport in channels and pumps. *Science* 310:1461–1465.

Jiang, Y., V. Ruta, J. Chen, A. Lee, and R. MacKinnon (2003) The principle of gating charge movement in a voltage-dependent K^+ channel. *Nature* 423:42–48.

King, L. S., D. Kozono, and P. Agre (2004) From structure to disease: The evolving tale of aquaporin biology. *Nature Rev. Mol. Cell Biol.* 5:687–698.

Lee, S.-Y., A. Lee, J. Chen, and R. MacKinnon (2005) Structure of the KvAP voltage-dependent K^+ channel and its dependence on the lipid membrane. *Proc. Natl. Acad. Sci. USA* 102:15441–15446.

Martinac, B., Y. Saimi, and C. Kung (2008) Ion channels in microbes. *Physiol. Rev.* 88:1149–1490.

Wang, W., S. S. Black, M. D. Edwards, S. Miller, E. L. Morrison, W. Bartlett, C. Dong, J. H. Naismith, and I. R. Booth (2008) The structure of an open form of an *E. coli* mechanosensitive channel at 3.45 Å resolution. *Science* 321:1179–1183.

Neural Transmission

Ben-Abu, Y., Y. Zhou, N. Zilberberg, and O. Yifrach (2009) Inverse coupling in leak and voltage-activated K^+ channel gates underlies distinct roles in electrical signaling. *Nature Struct. Mol. Biol.* 16:71–79.

Bradford, H. F. (1986) *Chemical Neurobiology*. Freeman, San Francisco.

Hille, B. (2001) *Ionic Channels of Excitable Membranes*. Sinauer Associates, Sunderland, Mass.

PROBLEMS

1. Give structures for the following, based on the data in Table 10.1.
   (a) *cis*-9-Dodecenoic acid
   (b) 18:1cΔ11
   (c) A saturated fatty acid that should melt below 30 °C

2. Given these molecular components—glycerol, fatty acid, phosphate, long-chain alcohol, and carbohydrate—answer the following:
   (a) Which two are present in both waxes and sphingomyelin?
   (b) Which two are present in both fats and phosphatidylcholine?
   (c) Which are present in a ganglioside but not in a fat?

3. The classic demonstration that cell plasma membranes are composed of bilayers depends on the following kinds of data:
   - The membrane lipids from 4.74×10^9 erythrocytes will form a monolayer of area 0.89 m^2 when spread on a water surface.
   - The surface of one erythrocyte is approximately $100 \, \mu\text{m}^2$ in area.

   Show that these data can be accounted for only if the erythrocyte membrane is a bilayer.

4. The lipid portion of a typical bilayer is about 3 nm thick.
   (a) Calculate the number of residues in an α helix that will just span this distance.
   (b) The epidermal growth factor receptor has a single transmembrane helix. Find it in this partial sequence:

   $$\cdots \text{RGPKIPSIATGMVGALLLLVVALGIGILFMRRRH} \cdots$$

5. In the situations described below, what is the free energy change if 1 mole of Na^+ is transported across a membrane from a region where the concentration is $1 \, \mu\text{M}$ to a region where it is 100 mM? (Assume $T = 37 \, °C.$)
   (a) In the absence of a membrane potential.
   (b) When the transport is opposed by a membrane potential of 70 mV.
   (c) In each case, will hydrolysis of 1 mole of ATP suffice to drive the transport of 1 mole of ion, assuming pH 7.4 and the following cytoplamic concentrations: ATP = 4.60 mM, P_i = 5.10 mM, ADP = $310 \, \mu\text{M}$?

*6. Consider passive diffusion of ions across the erythrocyte membrane, as measured by the permeability coefficients in Table 10.6. Calculate the number of moles of K^+ that would diffuse across a single erythrocyte membrane in 1 min, given the following data:

   $$C_{K^+}(\text{inside}) = 100 \text{ mM}$$
   $$C_{K^+}(\text{outside}) = 15 \text{ mM}$$

   The surface area of one erythrocyte $= 100 \, \mu\text{m}^2$.

7. If the volume of the erythrocyte in Problem 6 is about $100 \, \mu\text{m}^3$, what percentage of the K^+ ions would escape by passive diffusion in 1 minute?

*8. Assuming that the flow, J, in carrier-facilitated transport is proportional to the fraction of carriers occupied, derive an equation for J as a function of the concentration [A] of the substance transported.

9. Suppose calcium ion is maintained within an organelle at a concentration 1000 times greater than outside the organelle ($T = 37 \, °C$). What is the contribution of Ca^{2+} to the membrane potential? Which side of the organelle membrane is positive, and which is negative?

10. Calculate the equilibrium membrane potentials to be expected across a membrane at 37 °C, with a NaCl concentration of 0.10 M on the right and 0.01 M on the left, given the following conditions. In each case, state which side is (+) and which is (−).
    (a) Membrane permeable only to Na^+.
    (b) Membrane permeable only to Cl^-.
    (c) Membrane equally permeable to both ions.

11. In each of a, b, and c of Problem 10, will any appreciable transport of material take place in establishing the membrane potential?

*12. The rod cells in the retina contain membranes that are depolarized when a photon of light is absorbed. Suppose the following concentrations exist inside and outside a retinal rod cell:

	Concentration (mM)	
	Inside	Outside
K^+	100	5
Na^+	10	140
Cl^-	10	100

(a) If the relative permeabilities of K^+ and Cl^- are 1.0 and 0.45, respectively, what must be the relative permeability for Na^+ to give a potential of 30 mV, the value found across the resting rod cell membrane? Assume $T = 37 \, °C$. [Hint: You must rearrange and solve the Goldman equation.]
(b) If the Na^+ gates were to close completely on stimulation by a photon, what value of the membrane potential would be reached?

13. Many transmembrane proteins are oligomeric, with several identical subunits. The oligomers are usually found to have some form of C_n symmetry, rather than D_n or any higher order. Suggest a reason for this observation.

14. The average human generates approximately his or her weight in ATP every day. A resting person uses about 25% of this in ion transport—mostly via the $Na^+ - K^+$ ATPase. About how many grams of Na^+ and K^+ will a sedentary 70-kg person pump across membranes in a day?

*15. The concentration of glucose in your circulatory system is maintained near 5.0 mM by the actions of the pancreatic hormones glucagon and insulin. Glucose is imported into cells by protein transporters that are highly specific for binding glucose. Inside the liver cells the imported glucose is rapidly phosphorylated to give glucose-6-phosphate (G-6-P). This is an ATP-dependent process that consumes 1 mol ATP per mol of glucose. Given the steady-state intracellular concentrations below, calculate the theoretical maximum concentration of G-6-P inside a liver cell at 37 °C, pH = 7.2 when the glucose concentration outside the cell (i.e., [glucose]$_{\text{outside}}$) is 5.0 mM:

   $$\text{ATP} = 4.7 \text{ mM; ADP} = 0.15 \text{ mM; P}_i = 6.1 \text{ mM}$$

   For: ATP + $H_2O \rightarrow$ ADP + P_i + H^+ $\Delta G^{°\prime} = -30.5 \text{ kJ/mol}$ and

   $$\text{G-6-P} + H_2O \rightarrow \text{Glucose} + P_i \quad \Delta G^{°\prime} = -13.8 \text{ kJ/mol}$$

   The glucose phosphorylation reaction is

   $$\text{ATP} + \text{glucose}_{\text{inside}} \rightarrow \text{ADP} + \text{glucose-6-phosphate} + H^+$$

*16. ATP is synthesized from ADP, P_i, and a proton on the *matrix side* of the inner mitochondrial membrane. We will refer to the matrix side as the "inside" of the inner mitochondrial membrane (IMM).

   (a) H^+ transport from the outside of the IMM into the matrix drives this process. The pH inside the matrix is 8.2 and the outside is more acidic by 0.8 pH units. Assuming the IMM membrane potential is 168 mV (inside negative), calculate ΔG for the transport of 1 mol of H^+ across the IMM into the matrix at 37 °C: $H^+_{(\text{outside})} \rightarrow H^+_{(\text{inside})}$.

   (b) Assume three mol H^+ must be translocated to synthesize one mol ATP by coupling of the following reactions:

   $$\text{ADP} + P_i + H^+_{(\text{inside})} \rightarrow \text{ATP} + H_2O \, (\text{ATP synthesis})$$
   $$3H^+_{(\text{outside})} \rightarrow 3H^+_{(\text{inside})} \, (\text{proton transport})$$

   Write the overall reaction for ATP synthesis coupled to H^+ transport [and use this equation for part (c)]:
   (c) Assume three mol H^+ must be translocated to synthesize one mol ATP as described in part (b) above. Given the following steady-state concentrations: ATP = 2.70 mM and P_i = 5.20 mM, the membrane potential $\Delta\psi$ = 168 mV (inside negative), and the pH values in part (a), calculate the steady-state concentration of ADP at 37 °C when $\Delta G = -11.7 \text{ kJ/mol}$.

TOOLS OF BIOCHEMISTRY 10A

TECHNIQUES FOR THE STUDY OF MEMBRANES

Electron Microscopic Methods

Examination of membrane structure as it exists within cells depends heavily on electron microscopy (EM), and virtually all of the variants of that method mentioned in Tools of Biochemistry 1A have been employed at one time or another. For example, transmission EM of thin sections of cells embedded in a plastic matrix reveals cellular membranes in cross-section (see Figure 1.2 for examples), whereas scanning EM can show surface details. An especially useful variant of this method is the freeze fracture technique. If a membrane is frozen quickly and then broken by a sharp blow from a microtome knife, it frequently splits along the plane between the bilayer leaflets (Figure 10A.1). One layer is thus peeled back, revealing the internal structure. The sample can then be metal-shadowed and studied by scanning EM. In a variant called freeze-etching, some of the ice is sublimed off before shadowing, revealing surface and subsurface details.

Although electron microscopy has revealed much of the elaborate architecture of natural membranes, investigators often need to use simplified systems to study specific membrane properties. For these purposes, synthetic bilayers and vesicles are frequently used.

Preparation of Bilayers and Vesicles

Membranes from individual types of cells or purified organelles can usually be obtained by lysis of the cell or organelle, followed by differential centrifugation.

If, as shown in Figure 10A.2, a membrane is extracted with organic solvents (a chloroform–ethanol mixture, for example),

the soluble lipid constituents can be separated from insoluble protein and oligosaccharides. The lipid mixture can then be fractionated by methods such as high-performance liquid chromatography (HPLC) to yield pure lipid components and an analysis of the lipid content. Alternatively, the investigator may wish to use the entire lipid mixture from a membrane.

If the organic solvent in such a preparation is removed by evaporation, and the membrane lipids are dispersed in aqueous solution, they will form vesicles (also called *liposomes*)—small spherical bilayer structures. Alternatively, a bilayer may be spread across a small hole in a partition between two compartments. Such preparations are often used to study permeability across membranes.

The vesicles can be used for many kinds of studies. For example, it is possible to reconstitute specific transport systems by isolating the transport proteins through detergent solubilization and then adding them to a vesicle-forming system. In the presence of ATP, properly reconstituted systems will display active transport. Figure 10A.3 demonstrates reconstitution of the Ca^{2+} pump of muscle cells. Vesicles made with specific mixtures of lipids and other components are also excellent objects for the study of such processes as the phase transition or diffusion in membranes.

In other kinds of experiments, it is desirable to retain all of the natural components in the membrane. This can often be done by carefully lysing cells or organelles, isolating the intact membranes, and then dispersing them in a solution in which they will reseal to form vesicles. In some cases, as shown with an erythrocyte in Figure 10A.4, it is possible to adjust conditions so that the vesicles reseal preferentially either right side out or inside out.

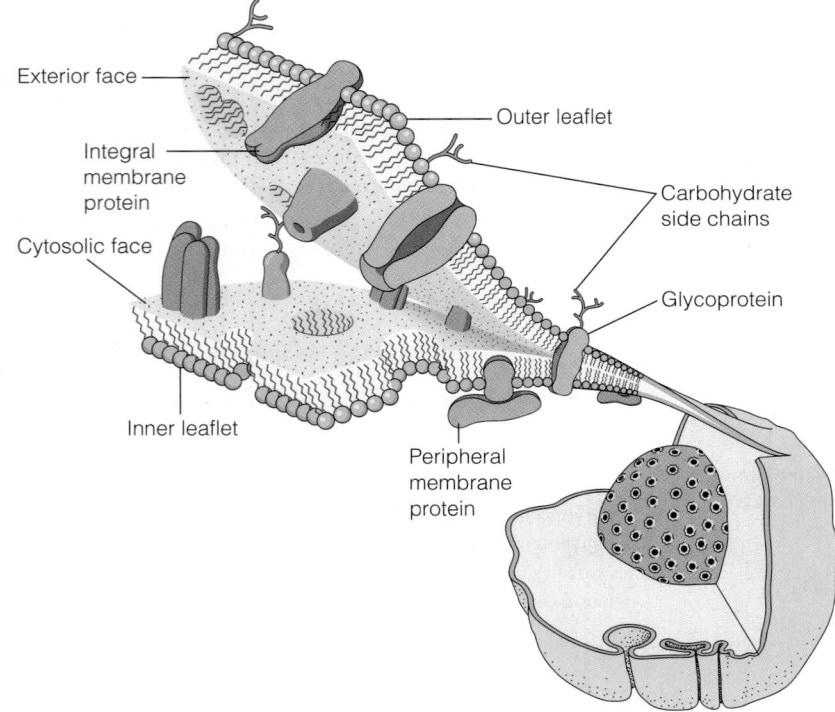

FIGURE 10A.1

Freeze fracture. A schematic view of a freeze-fractured membrane.

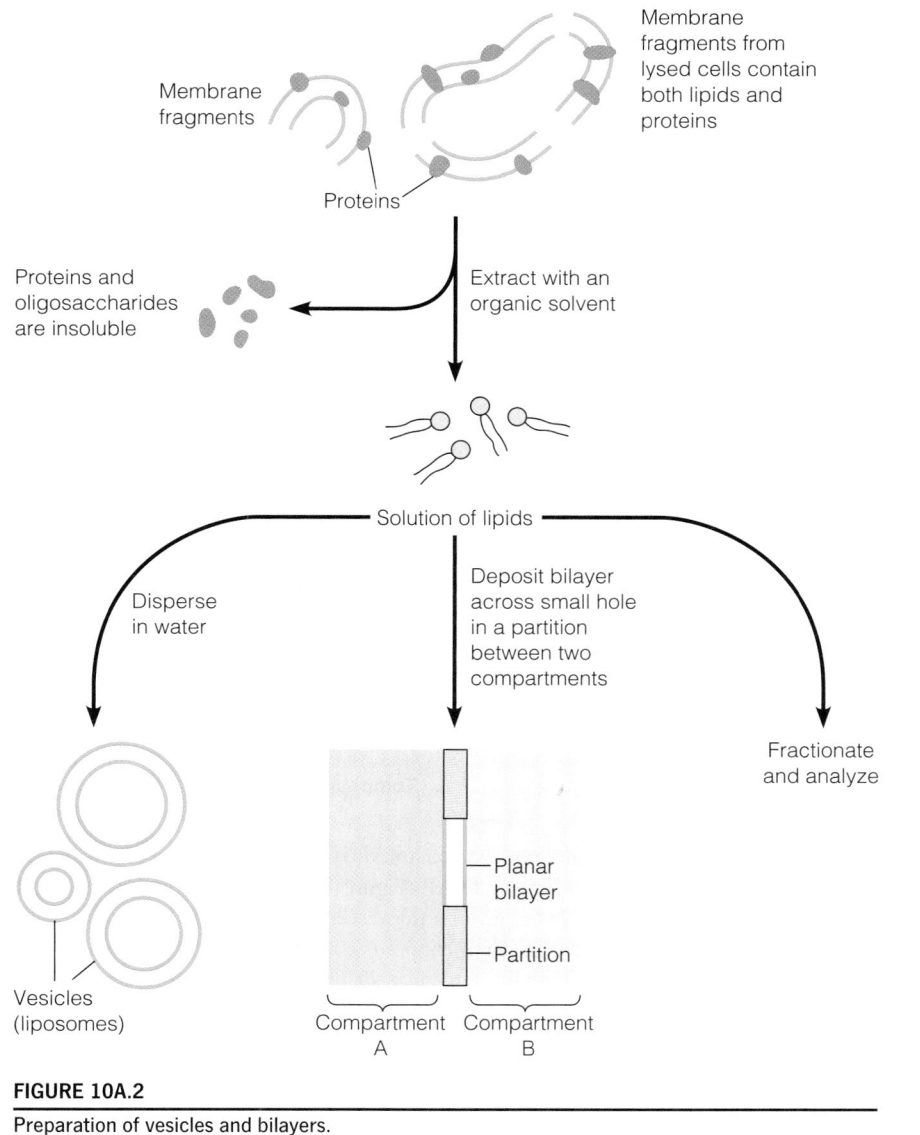

FIGURE 10A.2

Preparation of vesicles and bilayers.

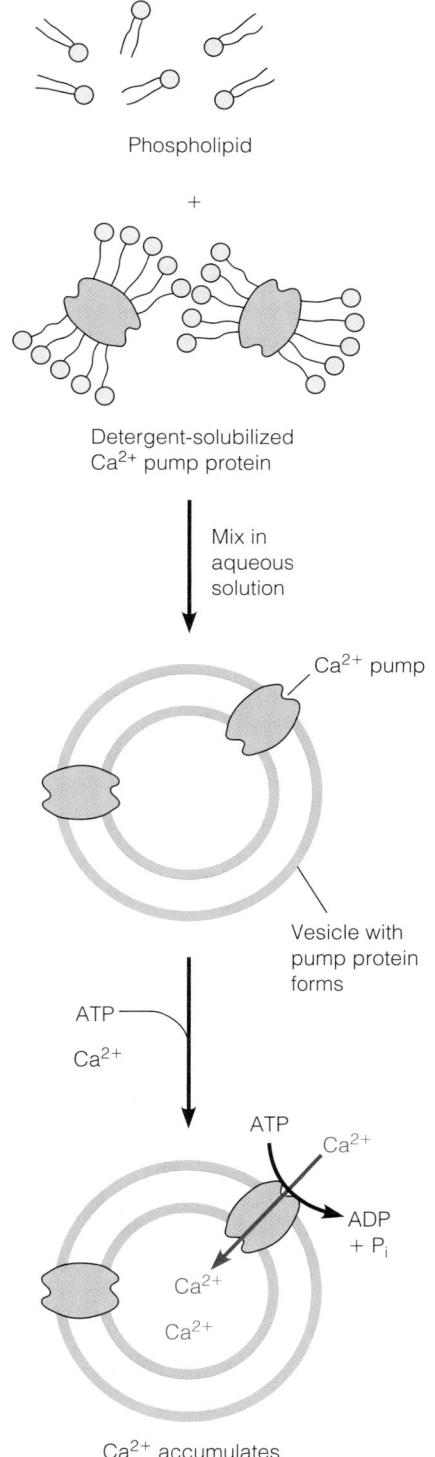

FIGURE 10A.3

Reconstitution of the Ca^{2+} pump.

Such preparations have provided much of our information concerning membrane asymmetry.

Physical Techniques

Much of our information concerning structural transitions in membranes has come from differential scanning calorimetry (DSC). This technique was described in Tools of Biochemistry 6C for the study of protein unfolding, which is a process of phase change. Likewise, DSC is a useful technique for the study of phase transitions (e.g., from semi-crystalline to fluid) in membranes.

A simplified diagram of a DSC instrument is shown in Figure 6C.1 (page 230). To carry out a DSC experiment, samples of a lipid vesicle suspension and a buffer blank are heated in parallel, and the difference in energy input required to keep them at the same temperature is carefully monitored. As the transition temperature (T_m) is passed, more heat has to be put into the lipid sample in order to melt the membrane structure. This transition shows up as the kind of spike illustrated in Figure 10.12b. The experiment reveals the transition temperature, the sharpness of the transition, and the total energy required.

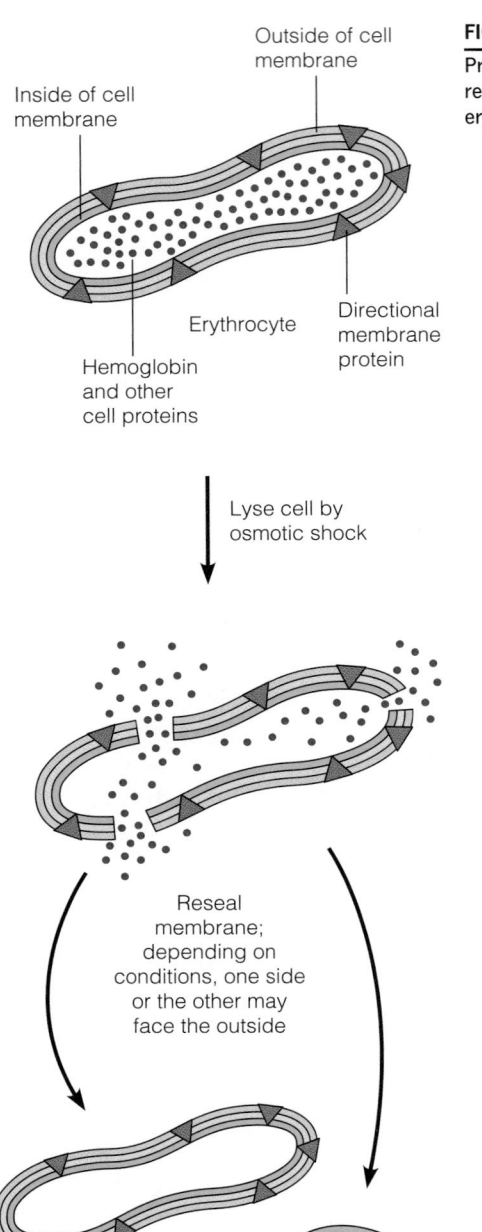

FIGURE 10A.4

Preparation and resealing of erythrocyte ghosts.

Outside of cell membrane

Inside of cell membrane

Erythrocyte

Directional membrane protein

Hemoglobin and other cell proteins

Lyse cell by osmotic shock

Reseal membrane; depending on conditions, one side or the other may face the outside

Right side out

Inside out

Erythrocyte ghosts

To study membrane fluidity more directly, and to examine the motion of individual kinds of molecules in the membrane, a number of other techniques are employed. Electron spin resonance (ESR) has been of major importance. Electron spin resonance bears similarities to nuclear magnetic resonance (see Tools of Biochemistry 6A) but involves changes in the spin of unpaired electrons rather than nuclei. The resonance spectrum is sensitive, in both absorption line spacing and sharpness, to the environment of the unpaired electron. As in NMR, narrow spectral lines are characteristic of a fluid environment with rapid molecular motion, and broadened lines are observed when molecular motion is sluggish.

Most compounds in natural membranes do not have unpaired electrons, but certain nitroxide compounds, such as *tempocholine*, contain an unpaired electron in an N—O bond. Tempocholine has been substituted for choline in phosphatidylcholine to act as a "reporter group," sensing the freedom of motion of the head groups in membranes. Other reporter groups or molecules can probe the fluidity of the inside of membranes, or the mobility of membrane proteins.

$$HO-CH_2-CH_2-N^+(CH_3)_2-\cdots-N-O\cdot$$

Tempocholine

Another powerful technique is called fluorescence recovery after photobleaching or **FRAP** (Figure 10A.5). Selected molecules in cell membranes are given a fluorescent label. Then a tiny spot on the membrane surface is exposed to a high-intensity laser beam. This "bleaches" the fluorescent label, forming a nonfluorescent spot on the cell surface. The cell is then observed under a fluorescence microscope. As bleached molecules diffuse out of the spot and unbleached molecules diffuse in, the spot gradually recovers its original fluorescence intensity. This procedure provides a direct way to measure the lateral movement of selected molecules in membranes.

References

Marguet, D., P.-F. Lenne, H. Rigneault, and H.-T. He (2006) Dynamics in the plasma membrane: How to combine fluidity and order. *EMBO J.* 25:3446–3457. Review of fluorescence-based methods of measuring membrane dynamics.

Prasad, R., ed. (1996) *Manual on Membrane Lipids.* Springer-Verlag, New York. This little manual describes techniques for a wide variety of lipid and membrane problems.

FIGURE 10A.5

Fluorescence recovery after photobleaching (FRAP).

Molecules on a cell surface are labeled with a fluorescent dye.

A spot on the surface is bleached by an intense, highly focused laser (∿∿▶).

As labeled molecules diffuse into the spot, the contrast begins to fade.

Eventually the spot is indistinguishable from the rest of the cell surface.

PART 3

Dynamics of Life: Catalysis and Control of Biochemical Reactions

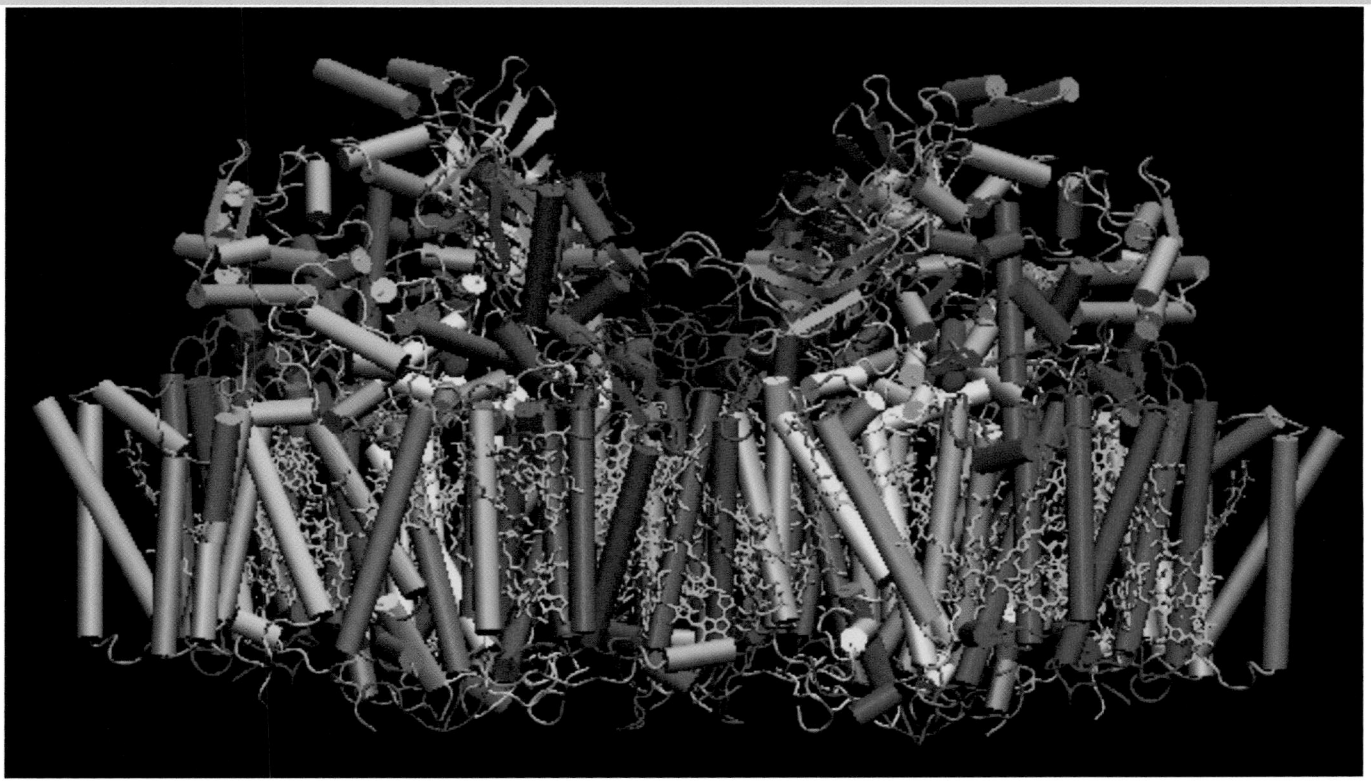

Photosystem II, part of the photosynthetic apparatus of a thermophilic cyanobacterium. Photosystem II, buried in the chloroplast thylakoid membrane, creates oxygen from water and excites electrons for reductive biosynthesis. Source: Dean Appling (Figure 16.14) (PDB ID 3BZ1, 3BZ2)

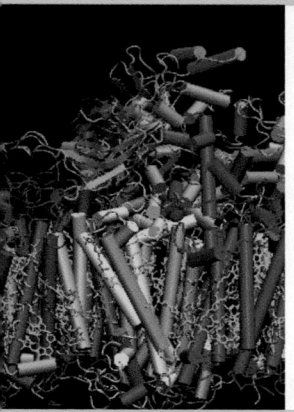

CHAPTER 11

Enzymes: Biological Catalysts

In earlier chapters we have alluded to the importance of specific catalysts called **enzymes** in regulating the chemistry of cells and organisms. Catalysis is essential to make most critical biochemical reactions proceed at useful rates under physiological conditions. A reaction that takes many hours to approach completion cannot be metabolically useful to a bacterium that must reproduce in 20 minutes or to a human nerve cell that must respond instantly to a stimulus.

In fact, most reactions in cells must be catalyzed to proceed at rates that can support the living state. In the complex milieu of the cell, countless thermodynamically favorable reactions are possible. The cell employs specific catalysis to direct reactive substances into useful pathways rather than into unproductive side reactions. Furthermore, enzymes are regulated—in almost all cases, the activities of enzymes can be controlled to modulate the production of different substances in response to cellular and organismal needs.

The rest of this book elaborates on this theme. Each of the hundreds of reactions we will encounter is catalyzed by a specific enzyme, optimized by evolution to perform its required task. What are the special properties of enzymes that make them such efficient catalysts? That is what we will explore in this chapter.

The Role of Enzymes

Catalysts increase the velocity of chemical reactions. Enzymes are biological catalysts.

In general terms, a **catalyst** is a substance that increases the rate, or velocity, of a chemical reaction without itself being changed in the overall process. Enzymes are biological catalysts, most of which are proteins. We have already encountered a few enzymes in earlier chapters. For example, the protein trypsin catalyzes hydrolysis of peptide bonds in proteins and polypeptides. The substance that is acted on by an enzyme is called the **substrate** of that enzyme. Thus, polypeptides are the natural substrates for trypsin.

We can see the power of enzyme catalysis in a familiar example—the decomposition of hydrogen peroxide into water and oxygen:

$$2H_2O_2 \rightarrow 2H_2O + O_2$$

This reaction, although strongly favored thermodynamically, is very slow unless catalyzed. A bottle of H_2O_2 (hydrogen peroxide) solution can be kept for many months before it breaks down. If, however, a bit of Fe^{3+} ion (as $FeCl_3$, for example) is added, the reaction rate increases about 1000-fold. The iron-containing protein hemoglobin is even better at increasing the rate of this reaction. When hydrogen peroxide solution is applied to a cut finger, bubbles from released O_2 appear immediately—the reaction is now proceeding about 1 million times faster than the uncatalyzed process. However, even higher rates can be achieved with an enzyme specific for this reaction; *catalase*, an enzyme present in many cells, increases the rate of H_2O_2 decomposition about 1 billion-fold over the uncatalyzed rate. Hydrogen peroxide is produced in some cellular reactions and is a dangerous oxidant (see Chapter 15); thus, selective pressures have resulted in the evolution of catalase to defend cells against the damaging effects of H_2O_2. This example shows that the rate of a favorable reaction depends greatly on whether a catalyst is present and upon the nature of the catalyst. Enzymes are among the most efficient and specific catalysts known.

Two facts deserve emphasis. First, although a true catalyst participates in the reaction process, it is unchanged by the process. For example, after catalyzing the decomposition of an H_2O_2 molecule, catalase is found again in exactly the same state as before, ready for another round. In contrast, although hemoglobin accelerates the rate of H_2O_2 decomposition, it is oxidized in the process, from the active Fe^{2+} to the inactive Fe^{3+} form; thus, hemoglobin is not a true catalyst for this reaction. Second, catalysts change *rates* of processes but do not affect the position of equilibrium for a reaction. A thermodynamically favorable process is not made more favorable, nor is an unfavorable process made favorable, by the presence of a catalyst. The equilibrium state is just approached more quickly.

Chemical Reaction Rates and the Effects of Catalysts: A Review

Reaction Rates, Rate Constants, and Reaction Order

First-Order Reactions: The Rate Constant

To understand what is meant by a reaction rate and how it might be measured, let us first consider the simplest possible reaction, the *irreversible* conversion of substance A to substance B:

$$A \rightarrow B$$

The single arrow here means that the reverse reaction ($B \rightarrow A$) proceeds to only an infinitesimal extent; that is, the equilibrium state lies far to the right.

We can define the **reaction rate**, or **velocity** (**v**), at any instant as the rate of formation of the product, in this case B:

$$v = \frac{d[B]}{dt} \tag{11.1}$$

The units of v are *concentration per unit time* (e.g., *molar* per second: $\text{M} \cdot \text{s}^{-1}$, where [B] symbolizes molar concentration of B). If we note that, for every B molecule formed, an A molecule must disappear, it is clear that v can equally well be written as

$$v = -\frac{d[A]}{dt} \tag{11.2}$$

where the negative sign indicates [A] is decreasing with time. The change of each molecule of A into B is an independent event. Therefore, as molecules of A are consumed, the number of molecules left to change is diminished, and the rate

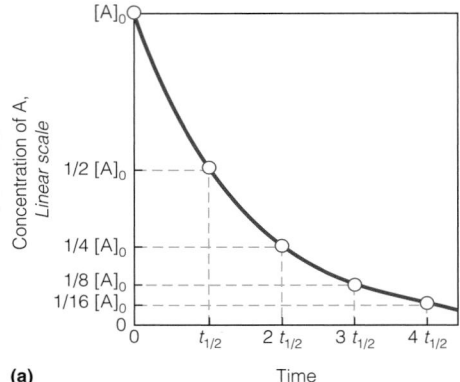

(a)   Time

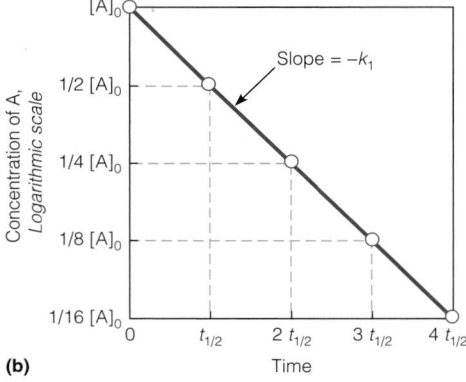

(b)   Time

FIGURE 11.1

Determining the order and rate constant of an irreversible first-order reaction. Graphs **(a)** and **(b)** analyze the rate of a single reaction, with time expressed as multiples of the half-life ($t_{1/2}$) of the reactant. Note that for each interval of $t_{1/2}$ the reactant concentration is halved. **(a)** A graph of [A] versus t shows that the rate, defined as the slope of the curve, decreases as the reaction continues. **(b)** A graph of ln[A] versus t, when linear, indicates that the reaction follows equation (11.4d) and is first-order. The slope of this line ($d\ln[A]/dt$) is equal to $-k_1$.

A first-order reaction is one whose rate is directly proportional to the first power of the reactant concentration.

A first-order reaction is characterized by single exponential decay of the reactant.

---

decreases as the reaction proceeds (see Figure 11.1a). Mathematically, we state this by saying that the rate is proportional to [A]:

$$v = \frac{d[B]}{dt} = -\frac{d[A]}{dt} = k_1[A]^n \tag{11.3}$$

The factor n in Equation 11.3 describes the dependence of the observed rate on the concentration of A. In the case where $n = 1$, the reaction rate depends on the first power of the reactant concentration and the reaction is called a **first-order reaction**. The constant k_1 is called the **rate constant** and for a first-order reaction has units of 1/(time) (typically, s^{-1} or min^{-1}). The rate constant provides a direct measure of how fast this reaction is. The larger the value of k_1, the more rapid the rate, and vice versa. The most common example of a first-order reaction is the decay of radioactive elements (see Tools of Biochemistry 12A).

Initially, the value of n is not known. It may be 1 (first-order), or 2 (second-order), or 3 (third-order), and so on. The order of a reaction is determined experimentally by comparing the kinetic data (i.e., [A] vs. time) to mathematical models, which describe the predicted change in [A] for each type of reaction. We illustrate this idea for a first-order reaction.

Let us begin with an equation that describes how the concentration of A changes with time during a first-order reaction. This description can be obtained by rearranging, then integrating Equation 11.3, where $n = 1$:

$$\frac{d[A]}{[A]} = -k_1 \, dt \tag{11.4a}$$

$$\int_{[A]_0}^{[A]_t} \frac{d[A]}{[A]} = -k_1 \int_0^t dt \tag{11.4b}$$

In equation 11.4b we integrate both sides of the equation over limits from time $= 0$ (where [A] equals the initial concentration "$[A]_0$") to time $= t$ (where $[A] = [A]_t$), yielding:

$$\ln \frac{[A]_t}{[A]_0} = -k_1 t \tag{11.4c}$$

or

$$\ln [A]_t = \ln [A]_0 - k_1 t \tag{11.4d}$$

or

$$[A]_t = [A]_0 e^{-k_1 t} \tag{11.4e}$$

Equation 11.4e tells us that the concentration of A decreases exponentially with time, as shown in Figure 11.1a. A characteristic of such an exponential decay is the **half-life** ($t_{1/2}$), which is the time needed for [A] to decrease by one half. For a first-order reaction, the half-life is inversely proportional to k_1 (see Problem 1 at the end of this chapter). To test whether a reaction is first-order, we need only make a graph of ln[A] versus t, as shown in Figure 11.1b. A straight line with a slope of $-k_1$ and an intercept of $[A]_0$, as predicted by Equation 11.4d, is consistent with a first-order reaction.

Most biochemical processes cannot be described over their full course by equations as simple as Equation 11.4c. One reason is that many of the reactions and processes we encounter are *reversible*, and as product accumulates, the reverse reaction becomes important. For example, we may have a reaction like the following:

$$A \underset{k_{-1}}{\overset{k_1}{\rightleftharpoons}} B$$

Because A is being consumed in the reaction to the right and formed by the reaction to the left, the corresponding rate equation is

$$v = -\frac{d[A]}{dt} = k_1[A]^n - k_{-1}[B]^m = \frac{d[B]}{dt} \tag{11.5}$$

In the case where n and m both equal 1, k_1 and k_{-1} are, respectively, the rate constants for the first-order forward and reverse reactions. Such a reaction approaches a state of equilibrium, at which point the rates of the forward and reverse reactions become equal, and so the observed rate becomes zero (i.e., there is no apparent change in [A] or [B] over time). Thus, at equilibrium

$$k_1[A] = k_{-1}[B] \tag{11.6a}$$

Recall from the discussion of reversible ligand binding in Chapter 7 (Equation 7.7, page 240), that the equilibrium constant, K, for a reversible reaction can be written as the ratio of the forward and reverse rate constants:

$$K = \frac{[B]}{[A]} = \frac{k_1}{k_{-1}} \tag{11.6b}$$

Second-Order Reactions

The reactions we have described so far are first-order—they involve changes happening in individual molecules; but, many biochemical reactions are more complex, involving encounters between molecules. A **second-order reaction** occurs typically when two molecules must come together to form products:

$$A + B \rightarrow C$$

This is illustrated by the binding of oxygen to myoglobin:

$$Mb + O_2 \xrightarrow{k_1} MbO_2$$

The rate for this process is given by

$$v = -\frac{d[Mb]}{dt} = -\frac{d[O_2]}{dt} = k_1[Mb]^n[O_2]^m \tag{11.7}$$

where n and m both equal 1, and k_1 is the **second-order rate constant**, with dimensions of $M^{-1}s^{-1}$. Equation 11.7 predicts that the rate of formation of the product, MbO_2, is dependent on both the concentration of free Mb *and* the concentration of free O_2. In this case we say the reaction is first order in [Mb] ($n = 1$) and first order in [O_2] ($m = 1$), and second order overall ($n + m = 2$). The binding of substrate (S) to an enzyme to form an **enzyme-substrate complex** is formally a second-order process:

$$Enz + S \rightarrow [Enz \cdot S] \rightarrow Enz + Product$$

> A first-order rate constant has units of (time)$^{-1}$, whereas a second-order rate constant has units of (concentration)$^{-1}$ (time)$^{-1}$.

Later in this chapter we will explore this further when we consider simple kinetic models for enzyme-catalyzed reactions.

Many more complicated kinds of reactions occur, including complex, multistep processes. We shall not be concerned with their kinetics at this point, although we shall see that enzyme-catalyzed reactions, when analyzed in detail, are generally more complicated than those described above. Often, however, the analysis of complex multistep reaction schemes can be simplified by the recognition of a **rate-limiting step**. The rate-limiting step is the slowest step in a multistep process. As such, it determines the experimentally observed rate for the entire process.

Transition States and Reaction Rates

What determines the rate of a chemical reaction? That is, what makes a rate constant large or small? The thermodynamic concepts we have presented in earlier chapters allow us to determine whether or not a reaction is favorable. However, such information, on its own, does not explain reaction rates. A free energy diagram for a favorable reaction, drawn on the basis of thermodynamics alone, will look like Figure 11.2a. It shows the free energy of the system versus the **reaction coordinate**, a generalized measure of the progress of the reaction through intermediate

FIGURE 11.2

Free energy diagrams for the simple reaction A → B. (a) Information provided by thermodynamic studies of the equilibrium: Only the free energy difference between the initial state and the final state is revealed. $G°_A$ and $G°_B$ represent the standard free energies per mole, respectively, of A and B molecules. $\Delta G°$ is the standard state free energy change for the reaction. **(b)** Free energy diagram filled in to include the transition state through which the molecule must pass to go from A to B or vice versa. $\Delta G°^{\ddagger}_1$ is the energy of activation for the A → B transition, and $\Delta G°^{\ddagger}_{-1}$ is for the B → A transition. **(c)** A reasonable path for the transition of a pyranose (such as glucose) from boat (1) to chair (3) conformation. The highest energy state—the transition state—will look something like (2).

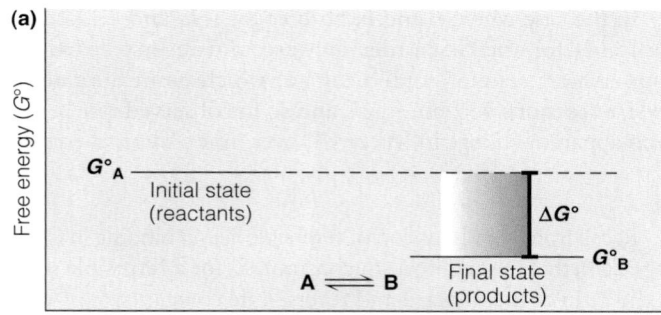

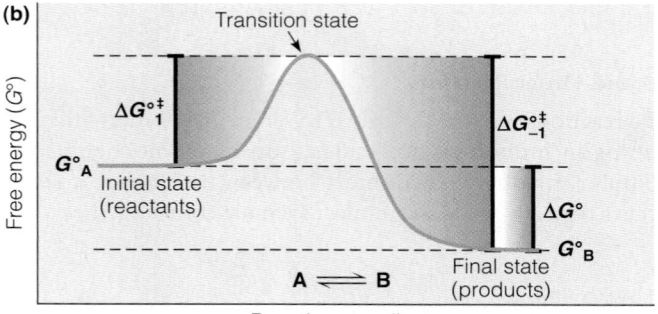

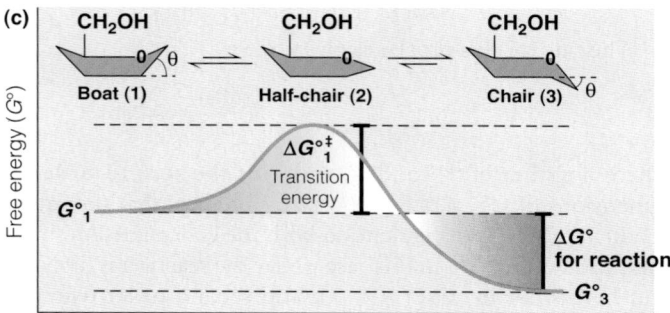

states. Its physical meaning will differ from one reaction to another (as discussed below). For a favorable reaction, the standard-state free energy of the products is lower than that of the reactants. However, what may be most important in determining the reaction rate is what happens in the *transition* from reactants to product. Equilibrium measurements, which pertain to final and initial states, do not reveal any information about the transition between these states or the energetic barrier that the transition presents.

A molecule in a first-order reaction must only *occasionally* reach an energy state in which the process can occur; otherwise, all molecules would already have reacted. This observation suggests that only a fraction of the molecules—those that are sufficiently energetic—can undergo reaction. Similarly, in a second-order reaction not all encounters between reactants can be productive because some collisions may not be sufficiently energetic, or the colliding molecules may not be properly oriented with respect to each other to react. Such considerations have given rise to the idea of a **free energy barrier** to reaction and the concept of an **activated state**, or **transition state** (symbolized by ‡). The transition state is thought of as a stage through which the reacting molecule or molecules must pass, often one in which a molecule is strained or distorted or has a particular electronic structure, or in which molecules collide productively. To make concrete the idea of a transition state, we diagram in Figure 11.2b the free energy as a function of the reaction coordinate. This concept is somewhat abstract, but we give a simple, concrete example in Figure 11.2c, the boat → chair conversion of a pyranose ring (see Chapter 9, page 319). The reaction coordinate in this case is just the angle θ. Both the initial state (boat) and the final state (chair) are of lower energy than the most strained, flattened state (the half-chair state). To make the conversion, the ring must go through the half-chair state, which represents the high-energy transition state in this example.

We now consider two strategies for increasing reaction rates: (1) raising the temperature of the reaction, and (2) lowering the free energy of the transition state. The **standard free energy of activation**, $\Delta G^{\circ\ddagger}$, represents the additional free energy (above the average free energy of reactant molecules) that molecules must have to attain the transition state. If the activation barrier to the reaction is high, only a small fraction of the molecules will have enough energy to surmount it, or only a small fraction of collisions will be energetic enough for the reaction to occur. We know that in any sample or solution, not all molecules have the same energy at any instant. As shown in Figure 11.3a, at a given temperature some molecules will have lower kinetic energies and others will have much higher kinetic energies. As temperature increases the average kinetic energy of the molecules in a sample increases; thus, as the temperature of a sample increases we expect more molecules to have sufficient energy to attain the transition state (compare the shaded portions of Figure 11.3a), and the rate of the reaction should increase. Except for the organisms that have adapted to extreme environments (e.g., hydrothermal vents in the ocean), most organisms are sensitive to even small increases in temperature; thus, raising the temperature inside a cell beyond 2–3 °C is not generally tolerated. For example, many laboratory strains of *E. coli* grow well at 37 °C but do not survive at 42 °C. In humans, a body temperature above 41 °C is considered a medical emergency. There is another reason that raising reaction temperature is not necessarily beneficial—exothermic reactions become less favorable at elevated temperature.

Nature's response to the need to increase reaction rates in cells has been not to increase the temperature inside the cells but rather to use enzymes to lower the activation energy for the reaction. We can use the thermodynamic principles discussed in Chapter 3 to describe this effect in qualitative terms.

If we consider the simple reaction $A \rightleftarrows B$, and assume that molecules of the reactant A are in equilibrium between the initial state and the activated state, the concentration of activated molecules at any instant will be given by Equation 3.28, introduced on page 73, for a reaction at equilibrium ($K = e^{-\Delta G^{\circ}/RT}$). It is important to recognize that a true thermal equilibrium is not possible between the reactant state and something as fleeting and dynamic as the transition state. Instead we will assume that $A^{\ddagger}$ represents an average structure that is a reasonable approximation of the transition state. With that caveat in mind, if we let $[A^{\ddagger}]$ represent the fraction of molecules with sufficient energy to attain the transition state and $[A]$ the fraction of A that remains in the reactant state, we can use Equation 3.28 to describe the ratio $[A^{\ddagger}]/[A]$ as a function of the activation energy, $\Delta G^{\circ\ddagger}$:

$$\frac{[A^{\ddagger}]}{[A]} = e^{\left(\frac{-\Delta G^{\ddagger}}{RT}\right)} \tag{11.8}$$

where T is temperature and R is the gas constant. Equation 11.8 and Figure 11.3b provide us with a framework to understand the most important feature of enzyme catalysis: Enzymes achieve faster rates by lowering the activation energy for a reaction. As $\Delta G^{\circ\ddagger}$ decreases, a larger fraction of molecules will possess sufficient energy to attain the transition state, and the rate of reaction will increase (compare the shaded portions of Figure 11.3b).

In 1921 Michael Polanyi proposed that a reaction catalyst preferentially binds the transition state structure and thereby stabilizes it, relative to the ground state, leading to a reduction in activation energy. Twenty-five years later, Linus Pauling

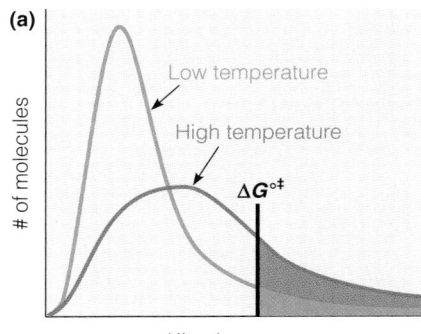

(a)

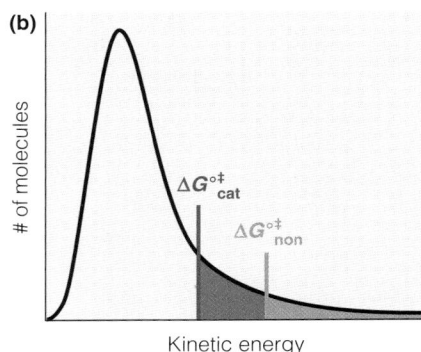

(b)

FIGURE 11.3

Effect of increasing temperature or lowering $\Delta G^{\circ\ddagger}$ on the rates of reactions. The rates of reactions are proportional to the number of molecules with sufficient energy to overcome the activation barrier $\Delta G^{\circ\ddagger}$. **(a)** At higher temperature more molecules have this energy (compare the shaded area under the light blue curve vs. the shaded area under the orange curve). **(b)** Lowering the value of $\Delta G^{\circ\ddagger}$ also increases the number of molecules with sufficient energy to attain the transition state (compare shaded areas under the curve for each value of $\Delta G^{\circ\ddagger}$). The subscripts "cat" and "non" indicate, respectively, the catalyzed and noncatalyzed processes.

Catalysts increase reaction rates by lowering the activation energy.

extended this idea to biological catalysts, suggesting that specific, complementary binding interactions between the transition-state structure and the enzyme active site would account for the extraordinary rate enhancements achieved by enzymes. Indeed, **transition state theory** has proved to be widely applicable to the study of enzyme catalysts, both in the laboratory and in computer simulations.

Transition State Theory Applied to Catalysis

In its simplest formulation, transition state theory assumes that a reactant molecule that attains the transition state rapidly decomposes to a lower energy state such as the product state, or to an intermediate state (see below). Because bonds are in the process of breaking and/or forming in the transition state, the lifetime of the transition state is similar to the vibrational frequencies of covalent bonds—on the order of a picosecond (10^{-12} s). These concepts are summarized in the following general expression for the rate constant, k:

$$k = \gamma \left(\frac{k_B T}{h} \right) e^{\left(\frac{-\Delta G^{\ddagger}}{RT} \right)} \tag{11.9}$$

where k_B is the Boltzmann constant, h is Planck's constant, T is temperature, R is the gas constant, and γ is called the "transmission coefficient." The term $(k_B T/h)$ is a factor that accounts for the vibrational frequency of bonds in the transition state. At 310 K (37 °C), $(k_B T/h)$ has a value of ~6.4×10^{12} s^{-1}. The transmission coefficient γ takes into account several factors that we will not discuss here because, in many cases, they make relatively minor contributions to the overall value of the rate constant. The value of γ for various enzyme reactions has been calculated from quantum mechanical principles and is often very close to 1 but may be as large as ~10^3 for some hydrogen transfer reactions where quantum tunneling* is significant. Let us define the term **rate enhancement** (or **rate acceleration**) before making comparisons between the different parameters in Equation 11.9.

The rate enhancement is the ratio of the rate constants for the catalyzed (k_{cat}) and the noncatalyzed (k_{non}) reactions for a given set of conditions (e.g., temperature, pH, etc.). The rate enhancement indicates how much faster the reaction occurs in the presence of the enzyme. For example, we can compare the rate of amide bond hydrolysis by the enzyme *carboxypeptidase A* vs. the noncatalyzed rate at pH = 8 and 23 °C:

$$\text{rate enhancement} = \frac{k_{cat}}{k_{non}} = \frac{238 \text{ s}^{-1}}{1.8 \times 10^{-11} \text{ s}^{-1}} = 1.3 \times 10^{13} \tag{11.10}$$

> The rate enhancement for an enzyme-catalyzed reaction is the ratio of the rate constants for the catalyzed (k_{cat}) and the noncatalyzed (k_{non}) reactions. The rate enhancement indicates how much faster the reaction occurs in the presence of the enzyme.

How significant is this rate enhancement? The noncatalyzed peptide-bond hydrolysis has a half-life of ~2500 years, whereas the enzyme-catalyzed reaction has a half-life of ~0.005 second! Without the catalyst, this reaction would not occur on a physiologically useful timescale. Figure 11.4 illustrates the enormous rate enhancements that are characteristic of enzyme-catalyzed reactions.

We can combine Equations 11.9 and 11.10 to evaluate by how much an enzyme must stabilize the transition state to achieve these observed rate enhancements:

$$\text{rate enhancement} = \frac{k_{cat}}{k_{non}} = \frac{\gamma_{cat} \left(\frac{k_B T}{h} \right) e^{\left(\frac{-\Delta G^{\circ\ddagger}_{cat}}{RT} \right)}}{\gamma_{non} \left(\frac{k_B T}{h} \right) e^{\left(\frac{-\Delta G^{\circ\ddagger}_{non}}{RT} \right)}} = \left(\frac{\gamma_{cat}}{\gamma_{non}} \right) e^{\left(\frac{\Delta\Delta G^{\circ\ddagger}}{RT} \right)} \tag{11.11}$$

---

*Some reactions occur much faster than would be predicted by the height of the activation barrier ($\Delta G^{\ddagger}$); thus, the reaction coordinate appears to "tunnel" though the barrier. This is allowed by quantum mechanics for particles, such as electrons and protons, with small masses.

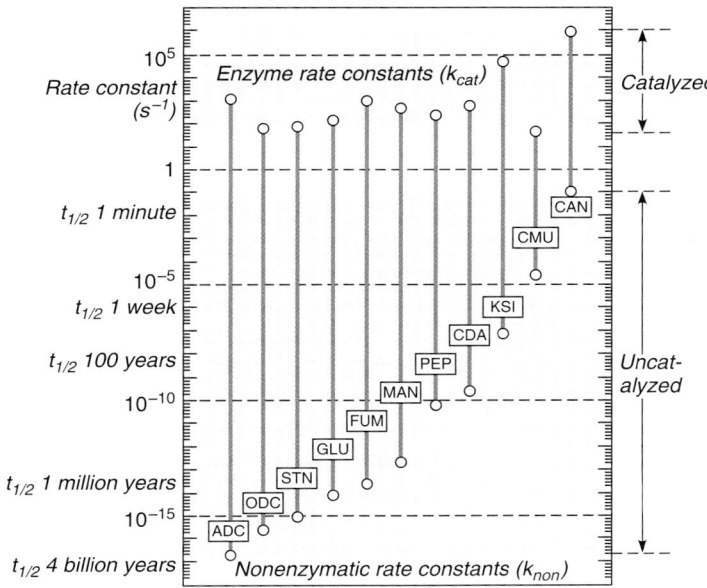

FIGURE 11.4

Enzymatic rate enhancements. Logarithmic scale of k_{cat} and k_{non} values (white circles) for some representative reactions at 25 °C. The length of each vertical bar represents the rate enhancement achieved by the enzyme. ADC = arginine decarboxylase; ODC = orotidine 5'-phosphate decarboxylase; STN = staphylococcal nuclease; GLU = sweet potato α-amylase; FUM = fumarase; MAN = mandelate racemase; PEP = carboxypeptidase B; CDA = *E. coli* cytidine deaminase; KSI = ketosteroid isomerase; CMU = chorismate mutase; CAN = carbonic anhydrase.

Reprinted with permission from *Accounts of Chemical Research* 34:938–945, R. Wolfenden and M. J. Snyder, The depth of chemical time and the power of enzymes as catalysts. © 2001 American Chemical Society.

where $\Delta\Delta G^{\circ\ddagger} = (\Delta G^{\circ\ddagger}_{non} - \Delta G^{\circ\ddagger}_{cat})$, or the difference in activation energies between the noncatalyzed and catalyzed reactions (see Figure 11.5). $\Delta\Delta G^{\circ\ddagger}$ indicates by how many kJ/mol the transition state is stabilized in the catalyzed reaction. As is described below, the value of $\Delta\Delta G^{\circ\ddagger}$ is often modest—equivalent to the energy gained by formation of a few noncovalent bonds ($\sim30-90$ kJ/mol).

Figure 11.5 illustrates an important point—that a catalyst lowers the activation energy barrier in *both* directions to the same extent. Thus, the rate accelerations for the forward and reverse reactions are identical (i.e., $\Delta\Delta G^{\circ\ddagger}$ is the same in both directions). This is the basis for the earlier statement that a catalyst accelerates the approach to equilibrium (from either direction) but not its position.

In summary, a catalyst lowers the energy barrier to a reaction, thereby increasing the fraction of molecules that have enough energy to attain the transition state, and it makes the reaction go faster in both directions. However, the presence of a catalyst has no effect on the position of equilibrium because ΔG° is the same whether a catalyst is present or not. Thus, *K*, which equals k_1/k_{-1}, is unchanged by a catalyst, even though k_1 and k_{-1} may each be thousands or millions of times larger than they were for the noncatalyzed process.

How does a catalyst lower the activation energy barrier for a reaction such that $\Delta G^{\circ\ddagger}_{cat} < \Delta G^{\circ\ddagger}_{non}$? To answer this question, let us examine $\Delta G^{\circ\ddagger}$ a bit more closely. Since $\Delta G^{\circ\ddagger} = \Delta H^{\circ\ddagger} - T\Delta S^{\circ\ddagger}$, we can imagine two possibilities for making $\Delta G^{\circ\ddagger}_{cat} < \Delta G^{\circ\ddagger}_{non}$: either reduce $\Delta H^{\circ\ddagger}$ (such that $\Delta H^{\circ\ddagger}_{cat} < \Delta H^{\circ\ddagger}_{non}$), or increase

> The presence of a catalyst increases forward and reverse rates for a reaction, but does not affect the equilibrium composition of reactants and products.

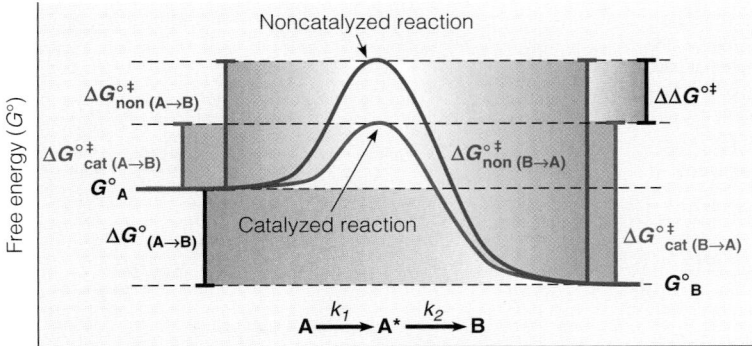

FIGURE 11.5

Effect of a catalyst on activation energy. A free energy diagram (blue curve) similar to that in Figure 11.2b is shown, along with an alternative *catalyzed* path (red curve) for the reaction. The catalyst lowers the standard free energy of activation, $\Delta G^{\circ\ddagger}$, and thereby accelerates the rate because more of the reactant molecules have the energy needed to reach this lowered transition state. The rate enhancement is related to $\Delta\Delta G^{\circ\ddagger}$ (see text). Note that the values of $\Delta G^{\circ}_{A\rightarrow B}$ for both the catalyzed and noncatalyzed reactions are the same; thus, the reaction equilibrium is not perturbed by the presence of the catalyst.

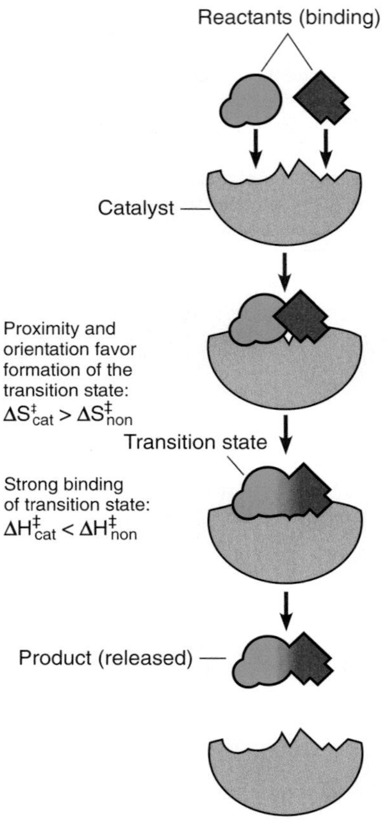

Reactants (binding)

Catalyst

Proximity and orientation favor formation of the transition state:
$\Delta S_{cat}^{\ddagger} > \Delta S_{non}^{\ddagger}$

Transition state

Strong binding of transition state:
$\Delta H_{cat}^{\ddagger} < \Delta H_{non}^{\ddagger}$

Product (released)

FIGURE 11.6

Entropic and enthalpic factors in catalysis. In this example, two reactants are bound to sites on the catalyst, which ensures their correct mutual orientation and proximity, and binds them most strongly when they are in the transition state conformation.

An enzyme active site is complementary in shape, charge, and polarity to the transition state for the reaction, and to a lesser extent, its substrate. Enzyme-substrate complementarity is the basis for the specificity of enzyme-catalyzed reactions.

$\Delta S^{\circ\ddagger}$ (such that $\Delta S_{cat}^{\circ\ddagger} > \Delta S_{non}^{\circ\ddagger}$). $\Delta H^{\circ\ddagger}$ can be lowered by increasing the number of bonding interactions between the catalyst and the transition state. As described below, this appears to be the dominant effect in most enzyme-catalyzed reactions. The $\Delta S^{\circ\ddagger}$ term reflects the fact that a particular *orientation* between reactants or parts of a molecule may be necessary to achieve the transition state. For example, when collisions between molecules occur (as in a second-order reaction), most encounters are unproductive just because the molecules happen to be pointed the wrong way when they hit. A catalyst that can bind two reacting molecules in proper mutual orientation will increase their reactivity by making the entropy of activation less negative (Figure 11.6). In a first-order reaction, which involves something happening *within* a single molecule, parts of the molecule must often be re-oriented properly to allow the transition state to be reached. Such re-orientation is sometimes referred to as *strain*. In the case of enzymes, which are structurally dynamic catalysts, it is not always clear whether conformational strain between the enzyme and substrate induces a larger change in the conformation of the enzyme or in the substrate. The thermodynamic cost of making $\Delta S^{\circ\ddagger}$ more positive is paid by favorable enthalpic interactions between the catalyst and substrate.

In some cases the catalyst can reduce the activation energy requirement by altering the reaction pathway and stabilizing an **intermediate** state that resembles the transition state but is of lower energy (Figure 11.7). The result is that two lower activation energy barriers replace the single higher barrier. We distinguish such an intermediate state from a transition state by the fact that the former corresponds to a local free energy minimum, and the latter to a free energy maximum.

How Enzymes Act as Catalysts: Principles and Examples

General Principles: The Induced Fit Model

We have seen that the role of a catalyst is to decrease $\Delta G^{\circ\ddagger}$ by facilitating the formation of the transition state. An enzyme binds a molecule of substrate (or in some cases several substrates) into a region of the enzyme called the **active site**, as shown schematically in Figure 11.8. The active site is often a pocket or cleft surrounded by amino acid residues that help bind the substrate and by other residues that play a role in catalysis. The extraordinary specificity of enzyme catalysis is due in part to the complex tertiary structure of an enzyme, which enables the active site to recognize the substrate through *complementary* binding interactions (like those we described for the binding between an antibody and its cognate antigen in Chapter 7). This possibility was realized as early as 1894 by the great German biochemist Emil Fischer, who proposed a *lock-and-key hypothesis* for enzyme action. According to this model, the enzyme accommodates the specific substrate as a lock does its specific key (Figure 11.8a). Although the lock-and-key

FIGURE 11.7

Importance of intermediate states. An enzyme may alter the reaction pathway to one that includes one or more intermediate states that resemble the transition state but have a lower free energy (red curve). In the case of a single intermediate, the activation energies for formation of the intermediate state and for conversion of the intermediate to product ($\Delta G_1^{\circ\ddagger}$ and $\Delta G_2^{\circ\ddagger}$, respectively) are lower than the activation energy for the uncatalyzed reaction (blue curve). In this figure, only activation energies for the enzyme-catalyzed forward reaction (A → B) are shown.

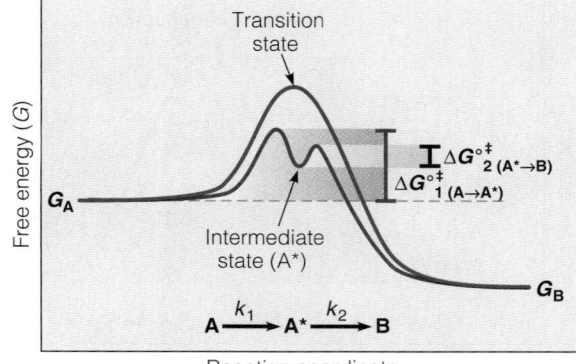

model explained enzyme specificity, it did not increase our understanding of the catalysis itself because a lock does nothing to its key. This understanding came from an elaboration of Fischer's idea: What fits the enzyme active site best is a substrate molecule induced to take up a configuration approximating the transition state. In other words, the enzyme does not simply accept the substrate—the enzyme also demands that the substrate be distorted into something close to the transition state. This *induced fit hypothesis*, proposed by Daniel Koshland in 1958, is still an important model for enzymatic catalysis; however, it has been extended to include the idea that the substrate can also induce conformational changes in the enzyme that lead to stabilization of the transition state.

The structural changes induced upon substrate binding may be local distortions or may involve major changes in enzyme conformation (Figure 11.8b). A conformational change of this kind can be seen when substrate binds the enzyme *hexokinase*, which catalyzes the phosphorylation of glucose to glucose-6-phosphate in the first step in the metabolic pathway called **glycolysis** (see Chapter 13). The structure of this enzyme has been determined by X-ray diffraction both in the presence and absence of bound glucose. As Figure 11.9 shows, the binding of glucose causes two domains of the enzyme to fold toward each other, closing the binding-site cleft about the substrate.

But enzymes do more than simply distort or position their substrates. Often, we find specific amino acid side chains poised in exactly the right places to aid in the catalytic process itself. In many cases, these side chains are acidic or basic groups that can promote the addition or removal of protons. In other instances, the enzyme holds a metal ion in exactly the right position to participate in catalysis. Thus, an enzyme (1) binds the substrate or substrates, (2) lowers the energy of the transition state, and (3) directly promotes the catalytic event. When the catalytic process has been completed, the enzyme must be able to release the product or products and return to its original state, ready for another round of catalysis. For an enzyme (E) that catalyzes the conversion of a single substrate (S) into a single product (P), the expression for the reaction includes three steps:

$$E + S \underset{k_{-1}}{\overset{k_1}{\rightleftharpoons}} ES \underset{k_{-2}}{\overset{k_2}{\rightleftharpoons}} EP \underset{k_{-3}}{\overset{k_3}{\rightleftharpoons}} E + P \qquad (11.12)$$

Here ES represents the **enzyme–substrate complex**, and EP represents the enzyme bound to the product. For many enzyme-catalyzed reactions the first step, binding of substrate, is reversible (i.e., k_1 and $k_{-1} \gg k_2$); the second step, conversion of ES to EP, lies far to the right (i.e., $k_2 \gg k_{-2}$); and the third step, release of product, is rapid compared to the catalytic step (i.e., $k_3 \gg k_2$). We shall see later that much more complex representations are possible, with one or more intermediate states, but the equations above illustrate the basic model of enzymatic catalysis. Later in this chapter, we will develop a rate expression based on this model and the three assumptions described above.

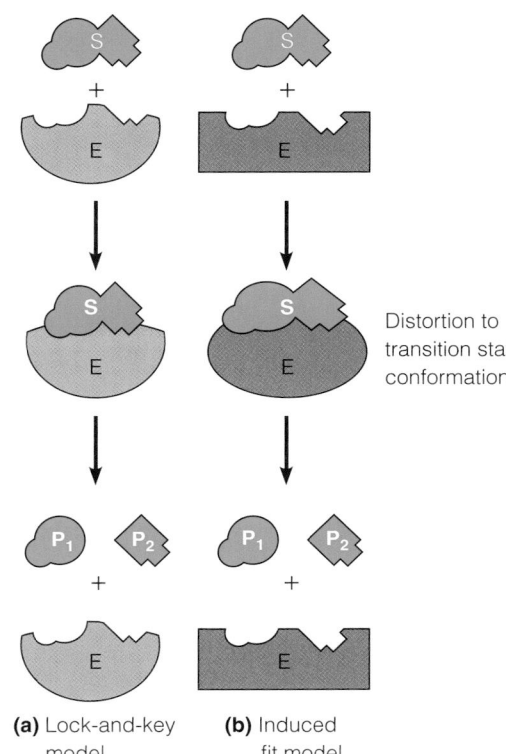

Distortion to transition state conformation

(a) Lock-and-key model   **(b)** Induced fit model

FIGURE 11.8

Two models for enzyme–substrate interaction. In this example, the enzyme catalyzes a cleavage reaction. **(a)** The lock-and-key model. In this early model, the active site of the enzyme fits the substrate as a lock does a key. **(b)** The induced fit model. In this elaboration of the lock-and-key model, both enzyme and substrate are distorted on binding. The substrate is forced into a conformation approximating the transition state; the enzyme keeps the substrate under strain.

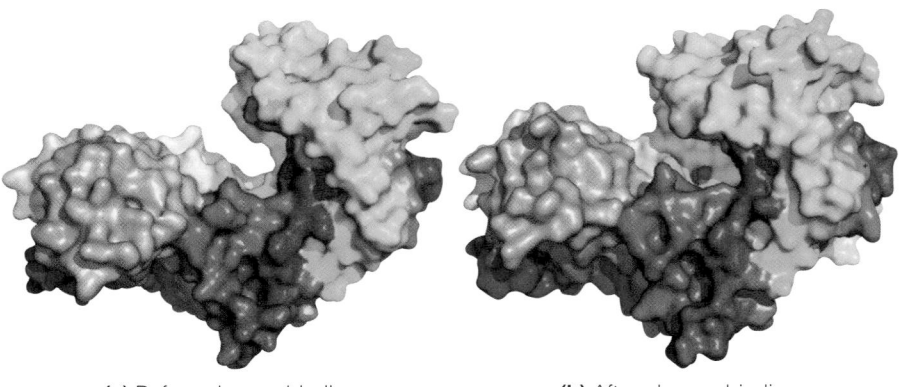

(a) Before glucose binding   **(b)** After glucose binding

FIGURE 11.9

The induced conformational change in hexokinase. The binding of glucose to hexokinase induces a significant conformational change in the enzyme. The enzyme is a single polypeptide chain, with two major domains. A surface representation is shown with rainbow coloring (N-terminus is blue; C-terminus is red). Notice how the obvious cleft between the domains (panel a) closes around the glucose molecule (magenta spheres in panel b). PDB IDs: panel **(a)** 2YHX; panel **(b)** 3B8A.

Enzyme binds preferentially to ‡ : $\Delta G^{\circ\ddagger}_{cat} < \Delta G^{\circ\ddagger}_{non}$

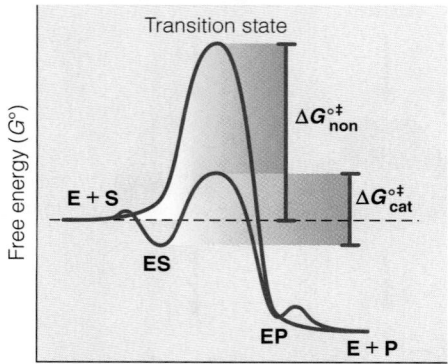

Enzyme binds substrate and ‡ with equal affinity: $\Delta G^{\circ\ddagger}_{cat} = \Delta G^{\circ\ddagger}_{non}$

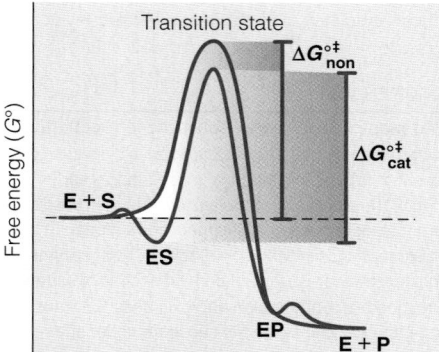

Enzyme binds preferentially to substrate: $\Delta G^{\circ\ddagger}_{cat} > \Delta G^{\circ\ddagger}_{non}$

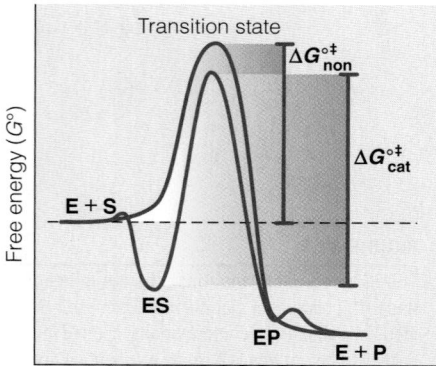

FIGURE 11.11

The effect on $\Delta G^{\circ\ddagger}_{cat}$ of differential binding to ES and the transition state. A rate enhancement is seen only for the case where $\Delta G^{\circ\ddagger}_{cat} < \Delta G^{\circ\ddagger}_{non}$ (top panel).

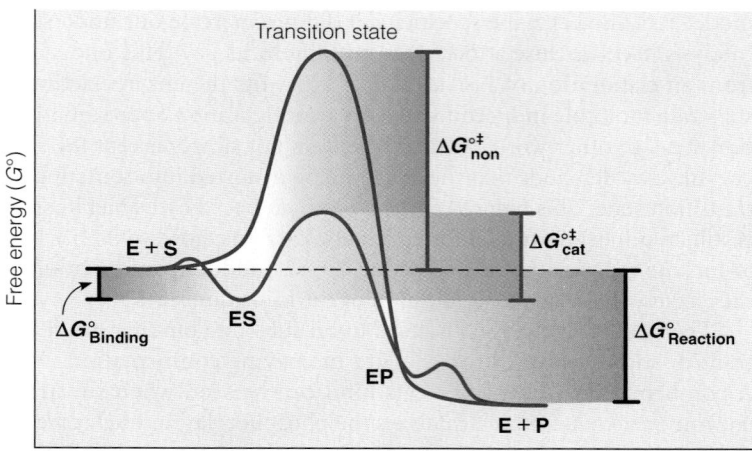

FIGURE 11.10

Reaction coordinate diagram for a simple enzyme-catalyzed reaction. The blue curve shows the free energy profile for the noncatalyzed reaction; the red curve shows the free energy profile for a simple enzyme-catalyzed reaction that follows the scheme shown in Equation 11.12.

A reaction coordinate diagram based on Equation 11.12 is shown by the red curve in Figure 11.10. This figure illustrates several important points. First, formation of the ES complex tends to be thermodynamically favorable, due to complementary binding interactions between the enzyme and substrate. Second, for maximum efficiency, binding to product should be less favorable than its release from the active site. Third, the enzyme active site must bind the transition state more favorably than it binds the substrate to achieve $\Delta G^{\circ\ddagger}_{cat} < \Delta G^{\circ\ddagger}_{non}$. It is critical to distinguish the ES complex (a stable intermediate state) from the transition state (an unstable state) in this analysis—they represent *different* states on the reaction coordinate.

As shown in the top panel of Figure 11.11, rate enhancement is achieved only when the enzyme binds the transition state with an affinity that is *greater than* its binding affinity for the substrate. In this case, $\Delta G^{\circ\ddagger}_{cat} < \Delta G^{\circ\ddagger}_{non}$. If the enzyme bound the reaction transition state with an affinity *equal to* its binding affinity for the substrate (Figure 11.11 middle panel), the activation barrier would be the same for both the catalyzed and noncatalyzed processes (i.e., $\Delta G^{\circ\ddagger}_{cat} = \Delta G^{\circ\ddagger}_{non}$). Finally, if some hypothetical enzyme bound the reaction transition state with an affinity *less than* its binding affinity for the substrate (Figure 11.11 bottom panel), the activation barrier would be the greater for the "catalyzed" processes (i.e., $\Delta G^{\circ\ddagger}_{cat} > \Delta G^{\circ\ddagger}_{non}$). Thus, an efficient catalyst will stabilize the transition state and thereby lower its free energy relative to the other states along the reaction coordinate.

As shown in Figure 11.4, rate enhancements reported for enzyme-catalyzed reactions range from 10^7 to 10^{19}. If we consider these observations in terms of transition state theory, assuming 10^3 as an upper limit for the value of ($\gamma_{cat}/\gamma_{non}$), we expect $e^{(\Delta\Delta G^{\circ\ddagger}/RT)}$ to contribute at least 10^4 to 10^{16} to the rate enhancement (see Equation 11.11). This means the enzyme must provide a reduction in the activation energy (i.e., $\Delta\Delta G^{\circ\ddagger}$) on the order of 24–95 kJ/mol at 37 °C (see Table 11.1). This is equivalent to the energy of a few noncovalent bonds; thus, it is reasonable to propose that specific noncovalent interactions between the enzyme active site and the transition state can account for the stabilization that gives rise to the observed rate enhancements.

We have discussed several means by which an enzyme might achieve such significant rate enhancements:

1. Preferential binding to the transition state through complementary noncovalent bonding interactions (H-bonds, charge–charge interactions, etc.).

Recall from Chapter 2 that noncovalent bonds are electrostatic in nature; thus, this is referred to as **electrostatic catalysis**.

2. Distortion of the substrate and/or active site, which promotes reduction of the activation energy (induced fit).

3. Binding of substrates to optimize proximity and orientation (making $\Delta S^{\circ\ddagger}$ more favorable).

4. Altering the reaction pathway to include intermediate states (see Figure 11.7). This is typical of **covalent catalysis**.

In addition to these, we will add two other prevalent mechanisms: **general acid/base catalysis** and **metal ion catalysis**.

General acid/base catalysis (**GABC**) is important in reactions involving proton transfer. An active-site amino acid residue is classified as a **general acid** if it donates an H^+ to an atom that develops negative charge in the transition state (Figure 11.12). A **general base** removes an H^+ from an atom that develops positive charge in the transition state. Thus, GABC may be viewed as a specialized case of electrostatic catalysis involving the transfer of a positive charge (H^+).

In general, the catalytically important residues in the enzyme active site are polar. Histidine is a very common GABC catalyst in active sites, due to its ability to accept or donate protons at physiological pH. Residues like Glu, Asp, Lys, and Arg are also common participants in proton transfers, as well as frequently serving to make electrostatic bonds with substrate molecules. A number of other residues, such as Ser, Tyr, and Cys, are also found to play important roles in the active sites of enzymes, either as H-bond donors/acceptors or as nucleophiles.

Over one third of enzymes characterized to-date contain metal ions in their active sites; thus, **metalloenzymes** are an important class of enzymes and much

TABLE 11.1	The relationship between transition-state stabilization ($\Delta\Delta G^{\circ\ddagger}$) and rate enhancement for reactions catalyzed at 37 °C.

$\Delta\Delta G^{\circ\ddagger}$ (kJ/mol)	Rate Enhancement
24	10^4
36	10^6
47	10^8
59	10^{10}
71	10^{12}
83	10^{14}
95	10^{16}

An enzyme binds substrate(s), preferentially stabilizes the transition state, and releases product(s).

Enzyme-catalyzed ester cleavage

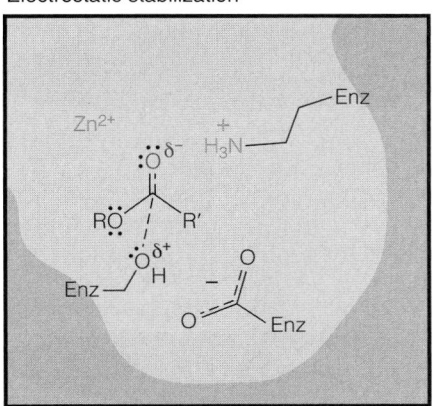

Transition state

Tetrahedral oxyanion intermediate

Stabilizing interactions in transition state

Electrostatic stabilization

General acid/base catalysis

FIGURE 11.12

Enthalpic stabilization of the transition state in an enzyme-catalyzed reaction. The upper panel shows the transition state and tetrahedral intermediate for an enzyme-catalyzed ester cleavage. This transition state might be stabilized by electrostatic interactions with active site amino acids and/or metal ions (lower left panel), or it might be stabilized by general acids or bases (lower right panel). The direction of proton transfer from the proton donor to the proton acceptor is indicated by the placement of the wide end of the dashed bond near the proton acceptor. The GAC is a proton donor and the GBC is a proton acceptor.

current research is devoted to understanding the roles of the metal ions in catalysis. If the metal ion behaves as a Lewis acid, by accepting electron density from an electron-rich atom (e.g., an atom that develops negative charge in the transition state), it is acting as an electrostatic catalyst. Metal ions can also promote the formation of hydroxide ion ($^-$OH) in the enzyme active site. This species is an important nucleophile in many hydrolytic reactions, such as the cleavage of peptide bonds in proteins or phosphodiester bonds in DNA and RNA.

To illustrate these general principles in a more concrete way, we consider the mechanisms of two specific enzyme-catalyzed reactions, for which details of the catalysis are well understood.

Lysozyme

Lysozyme was briefly described in Chapter 9 as an enzyme that cleaves the peptidoglycan layer of gram-positive bacteria (Figure 9.25), resulting in lysis of the bacterial cell. As such, it defends against bacterial infection, and it is found in the secretions of tissues that contact the external environment, such as tears, saliva, and mucus. The first X-ray crystal structure of an enzyme was that of lysozyme from hen egg white, reported in 1965 by the laboratory of David Phillips at the Royal Institution in London.

The active site of lysozyme is located in a deep cleft (Figure 11.13) where six glycosyl residues bind in subsites labeled A–F. The glycosidic bond cleavage occurs between residues D and E, with retention of configuration at the anomeric carbon (i.e., C1 of the pyranose in the D subsite). Based on the crystal structure of the enzyme, Phillips proposed a stereochemical mechanism in which glutamic acid 35 (E35) acts as a general acid and aspartate 52 (D52) acts as an electrostatic catalyst

FIGURE 11.13

The active site cleft of lysozyme. Top: The solvent-accessible surface of hen lysozyme (PDB ID: 2WAR) is shown in blue. The trisaccharide NAM-NAG-NAM is shown in stick representation bound to the active site. **Bottom:** A schematic drawing of (NAG-NAM)$_3$ bound to the A–F subsites in lysozyme. The site of glycosidic bond cleavage, between the D and E subsites, is shown by the red dashed line.

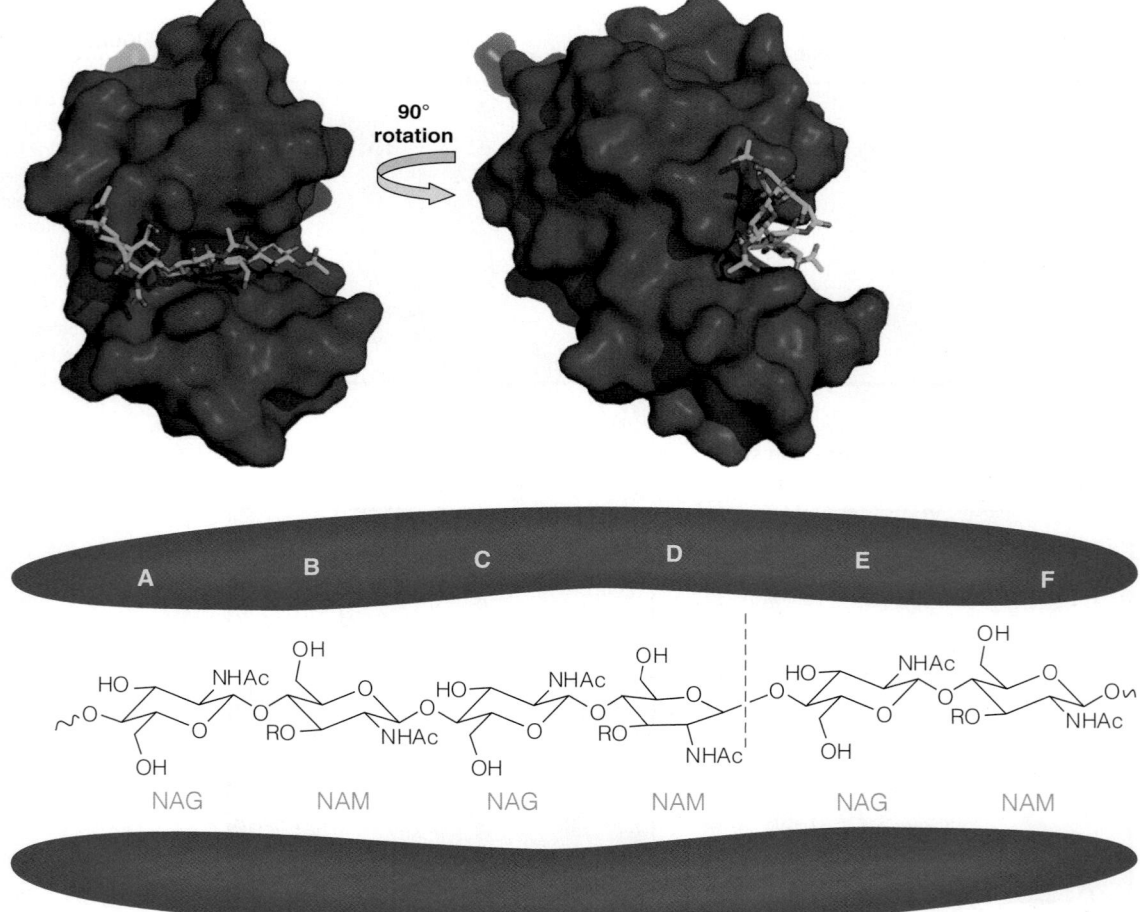

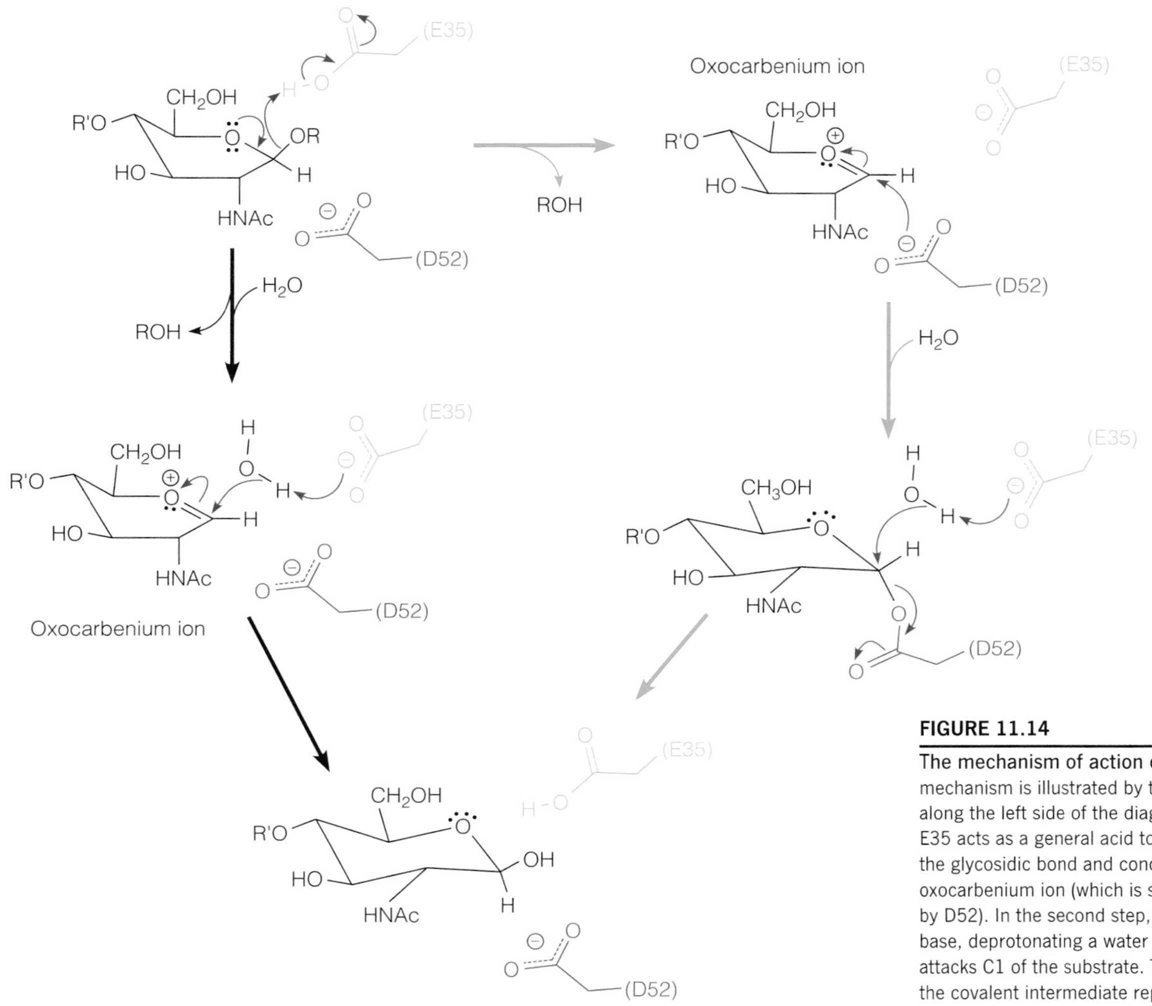

FIGURE 11.14

The mechanism of action of lysozyme. The Phillips mechanism is illustrated by the black reaction arrows along the left side of the diagram. In the first step, E35 acts as a general acid to promote cleavage of the glycosidic bond and concomitant formation of the oxocarbenium ion (which is stabilized electrostatically by D52). In the second step, E35 acts as a general base, deprotonating a water molecule, which then attacks C1 of the substrate. The pathway that includes the covalent intermediate reported by Steve Withers follows the green reaction arrows along the right side of the diagram. In this case, the second step involves covalent bond formation between C1 of the substrate and D52. Attack of the water displaces D52 in the subsequent step.

during the generation of an oxocarbenium ion in the transition state (Figure 11.14). Phillips also proposed that the residue in the D subsite is distorted into a half-chair conformation upon binding in the active site (see Figure 11.2). This conformation approximates that of the oxocarbenium ion and is an example of the substrate distortion ("strain") discussed above. Evidence for this distortion of the D site residue was reported in 1991 by Michael James based on the X-ray crystal structure of the trisaccharide NAM-NAG-NAM bound to the B-C-D subsites of lysozyme.

Additional experimental evidence for the proposed mechanism includes the observation of *kinetic isotope effects* (discussed in Tools of Biochemistry 11A) and the results of mutagenesis studies in which E35 and D52 were changed to nonionizing residues such as glutamine (Q), asparagine (N), and alanine (A). Amino acid mutations are often described using the single letter code and the following convention: [native amino acid][residue number][mutant amino acid]. For example, a mutant in which glutamine replaces glutamic acid at residue 35 would be "E35Q." In kinetics experiments using defined substrates, the E35Q and D52N mutants each showed <0.1% of wild-type activity, suggesting that both residues play critical roles in catalysis. As shown in Figure 11.15, lysozyme has optimum activity at a pH of ~5 and is dependent on a base with a pK_a of ~4 and an acid with a pK_a of ~6. The pK_as of all the Glu and Asp residues in lysozyme were determined by NMR titration (see Tools of Biochemistry 6A), and the pK_as of E35 and D52 were found to be, respectively, 6.2 and 3.7. Thus, at values of pH between 3.7 and 6.2, E35 would be predominantly protonated and D52 would be predominantly deprotonated, as required by the Phillips mechanism.

FIGURE 11.15

The effect of pH on the activity of lysozyme. E35 must be protonated to act as a general acid catalyst in the first step of the mechanism; thus, at pH values *below* 6.2 (blue line) the ratio of [COOH]/[COO⁻] is greatest, favoring catalysis. D52 must be deprotonated to interact with the oxocarbenium ion; thus, at pH values *above* 3.7 (red line) the ratio of [COO⁻]/[COOH] is greatest, favoring catalysis. These two boundary requirements give rise to the observed pH optimum (~5) where both protonated E35 and deprotonated D52 are abundant.

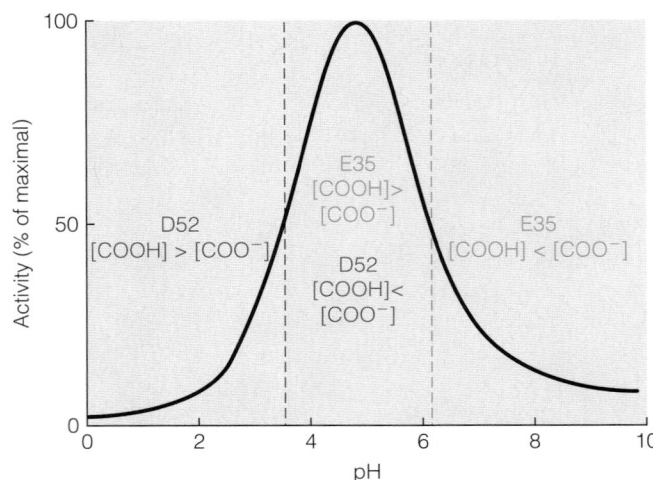

Lysozyme activity vs. pH

The Phillips mechanism proposes that D52 stabilizes the oxocarbenium ion via electrostatic stabilization and also shields one side of the ion from attack by water, thus explaining the retention of configuration at the anomeric carbon (C1). This aspect of the mechanism was challenged by (1) the prediction that the lifetime of the oxocarbenium ion is too short to allow water to diffuse into the active site after the E and F glycosyl residues vacate the active site and (2) the observation that several other related enzymes proceed through an intermediate in which a stabilizing bond is made at C1 of the oxocarbenium ion. Daniel Koshland had proposed in 1953 that the mechanism of lysozyme includes such a stabilizing covalent intermediate, and when the crystal structure was obtained, D52 appeared well-positioned to make such a bond. This proposal was controversial for many years; however, in 2001, Steve Withers' laboratory reported a crystal structure of an enzyme-glycosyl intermediate that showed the D52 side chain covalently bound to C1 (Figure 11.16 and right side of Figure 11.14). This result was obtained using the E35Q mutant with the synthetic substrate NAG-2FGlcF. This

NAG-2FGlcF

FIGURE 11.16

Evidence for the covalent intermediate in the mechanism of lysozyme. The covalent adduct between the synthetic substrate NAG-2FGlcF and D52 in the active site of the E35Q mutant of lysozyme is shown (PDB ID 1H6M). The enzyme backbone is shown in cartoon representation (purple). The side chains of D52 and Q35 are shown in stick form (C atoms green, O atoms red, N atoms blue). The NAG-2FGlc adduct is shown in stick representation, where C1 is colored black and the F atom at C2 is light blue. The red arrow points to C1 of NAG-2FGlcF, which is the site Asp52 attacks to form the covalent intermediate.

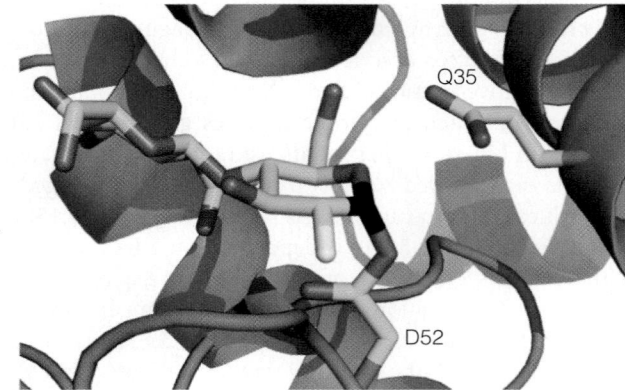

substrate readily forms the oxocarbenium ion; however, the subsequent addition of —OH to the C1 is slowed by the E35Q mutation (Q does not act as a general base). Thus, the covalent intermediate is sufficiently long-lived that it is possible to obtain a crystal structure.

For the reaction to be catalytic the enzyme active site must be restored to its initial state; in this case, the side chain of E35 must be protonated and that of D52 must be free and deprotonated. Both of these requirements are achieved in the second half of the mechanism shown in Figure 11.14, in which a water molecule attacks the intermediate. Note that in this step E35 acts as a general base, removing a proton from the attacking water molecule.

In summary, lysozyme employs GABC, substrate distortion, and covalent catalysis to achieve its rate enhancement.

Serine Proteases

As a second example, let us consider the catalysis of peptide-bond hydrolysis by one of the **serine proteases**. This important class of enzymes includes trypsin and chymotrypsin, which we first encountered in Chapter 5. These enzymes are called proteases because they catalyze the hydrolysis of peptide bonds in polypeptides and proteins. Many kinds of proteases exist, exhibiting a wide range of substrate specificities and utilizing a variety of catalytic mechanisms. The *serine* proteases are distinct because they all have a critical serine nucleophile in the active site. The serine proteases also hydrolyze a wide variety of esters—a fact of little physiological importance, but one that biochemists make use of in kinetic studies, as described below.

Catalysis of peptide-bond hydrolysis by a serine protease (chymotrypsin, in this example) proceeds as shown in Figure 11.17. First, the polypeptide chain to be cleaved is bound to the enzyme surface. Most of the polypeptide binds nonspecifically, but the side chain of the residue to the N-terminal side of the peptide bond to be cleaved must fit in the active site pocket. This pocket defines not only the position of the bond cleavage but also the *specificity* of serine proteases. Each of the serine proteases preferentially cleaves the amide bond immediately C-terminal to a specific kind of amino acid side chain. Examples of sites of preferential cleavage are shown in Table 5.4 (see page 149). For example, trypsin cleaves preferentially to the carboxylate side of basic amino acid residues like lysine or arginine, whereas chymotrypsin prefers a large hydrophobic residue like phenylalanine in this position. Chymotrypsin has a narrow specificity pocket lined with small glycine residues, which can accommodate bulky nonpolar side chains. Trypsin, on the other hand, has an aspartate side chain in the bottom of the specificity pocket. This negatively charged residue provides complementary binding interactions to the positive charge on an arginine or lysine side chain. This very specific binding of a particular type of amino acid also serves to place the active site serine very close to the carbonyl group of the bond to be cleaved—also called the **scissile bond**.

A common feature of serine proteases is the so-called **catalytic triad** of a *nucleophile*, a *general base*, and an *acid*. In many of the serine proteases that have been studied in detail, this catalytic triad is composed of serine, histidine, and aspartic acid residues presented in a similar 3-D orientation in the active site. As shown in Figure 11.18, in chymotrypsin the catalytic triad residues are Ser 195 (S195), His 57 (H57), and Asp 102 (D102). Other serine proteases have been discovered in which the triad pattern nucleophile–base–acid is conserved but the identities of the base and acid vary (e.g., Ser–His–His or Ser–Glu–Asp).

Due to the high pK_a of alcohol groups, serine side chains are usually in the protonated (i.e., —OH) form and are therefore not very reactive nucleophiles. However, S195 is in an environment that optimizes its reactivity. The S195 proton is transferred to the imidazole ring of H57, leaving a negative charge on the serine. Normally, this transfer would be unlikely (due to the

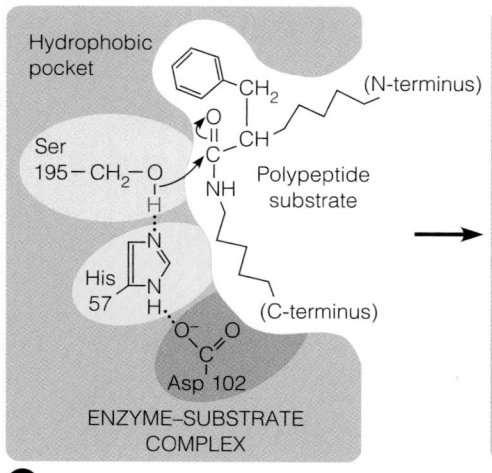

① Polypeptide substrate binds noncovalently with side chains of hydrophobic pocket.

② H⁺ is transferred from Ser to His. The substrate forms a tetrahedral oxyanion intermediate with the enzyme.

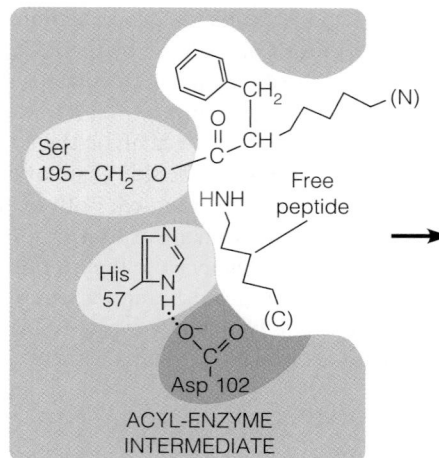

③ H⁺ is transferred to the C-terminal fragment, which is released by cleavage of the C—N bond. The N-terminal peptide is bound through acyl linkage to serine.

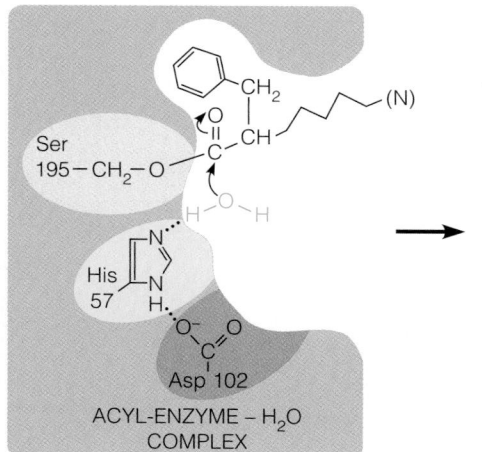

④ A water molecule binds to the enzyme in place of the departed polypeptide.

⑤ The water molecule transfers its proton to His 57 and its —OH to the remaining substrate fragment. Again a tetrahedral oxyanion intermediate is formed.

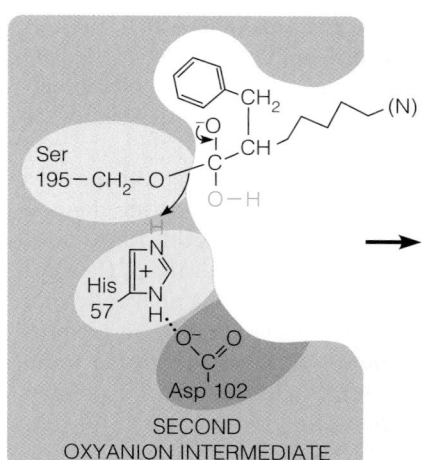

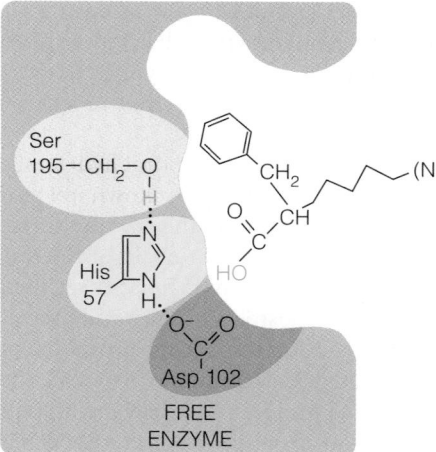

⑥ The second peptide fragment is released: The acyl bond is cleaved, the proton is transferred from His back to Ser, and the enzyme returns to its initial state.

FIGURE 11.17

Catalysis of peptide bond hydrolysis by chymotrypsin. The figure shows the steps in the cleavage of a polypeptide chain as catalyzed by chymotrypsin. The figure is highly schematic and does not represent the actual spatial arrangement of atoms (see Figure 11.18).

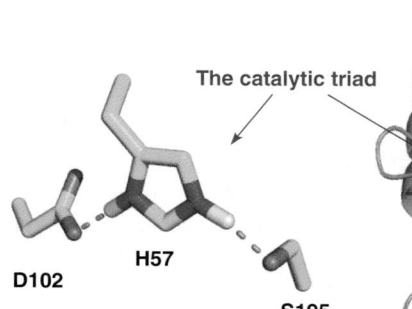

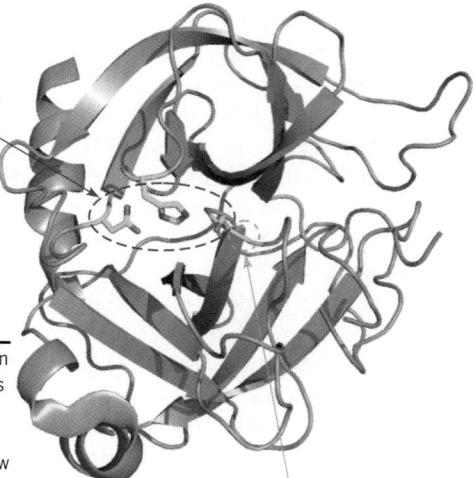

FIGURE 11.18

The structure of chymotrypsin and the serine protease catalytic triad. The backbone of bovine chymotrypsin detemined by X-ray crystallography (PDB ID: 4CHA) is shown in cartoon representation (blue) with the side chains of the catalytic triad, S195, H57, and D102, shown in stick form within the dotted red oval. The location of the oxyanion hole (see text) is indicated by the dotted orange circle. To the left is shown a high-resolution neutron diffraction structure of the catalytic triad from porcine elastase (PDB ID: 3HGN). The neutron diffraction data show the positions of H atoms with greater resolution than do the X-ray data. Here, H57 is protonated, S195 and D102 are deprotonated, and H-bonds are shown by orange dashes.

lower pK_a of a His side chain vs. a typical Ser side chain), but it appears to be facilitated by D102, which, by its negative charge, stabilizes the protonation of the adjacent H57 side chain (see Figure 11.18). These interactions make S195 an unusually reactive nucleophile that attacks the carbonyl amide of the scissile bond.

An alternative explanation for the unusual reactivity of S195 is based on X-ray crystal and solution-phase NMR evidence for what appears to be an unusually short and strong hydrogen bond between H57 and D102. When the pK_as of an H-bond donor and acceptor are nearly equal, the hydrogen is equally shared between them, and a so-called **low-barrier hydrogen bond** (LBHB) arises. Low-barrier hydrogen bonds are shorter (<0.25 nm), and 3–6 times stronger (~30–80 kJ/mol), with more covalent character than the typical hydrogen bonds described in Chapter 2. The proposed effect of the LBHB in chymotrypsin is to increase the pK_a of H57 to ~12 by stabilizing the protonated form of the imidazole side chain; thus, H57 would be better suited to deprotonate S195 (step 1 of Figure 11.17). The role of LBHBs in enzyme catalysis has been vigorously debated in the scientific literature because it remains challenging to measure directly the strength of such interactions in enzymes. Recent evidence from high-resolution crystal structure studies of the serine proteases α-lytic protease and elastin suggests that the hydrogen bond between H57 and D102 is shorter (~0.27 nm) than a typical hydrogen bond, but is not a LBHB. Further details of this debate can be found in the citations at the end of this chapter.

Attack by the serine nucleophile on the amide carbonyl results in the formation of a **tetrahedral oxyanion**, which then collapses to an **acyl–enzyme intermediate** (Figure 11.17). The formation of the oxyanion intermediate requires that the planar sp^2 hybridized amide carbon adopt a tetrahedral sp^3 configuration. Modeling of these species in the X-ray crystal structure of a serine protease active site led Jon Robertus and Joseph Kraut to propose the existence of an "**oxyanion hole**" that stabilizes the tetrahedral intermediate through specific H-bonding interactions to the negatively charged oxygen atom in the oxyanion (see Figure 11.19). In chymotrypsin, the H-bond donors are two backbone amido protons from residues S195 and G193. In the serine protease subtilisin, an asparagine side chain acts as an H-bond donor in the oxyanion hole. These H-bonds are stronger when the configuration of the amide carbon atom is tetrahedral rather than planar. Indeed, one of the H-bonds (from G193) appears to form only after the oxyanion is generated by attack of S195. Presumably the enzyme also stabilizes the transition state between the ES complex and this oxyanion intermediate because both states have significant negative charge on the oxygen atom and sp^3-like geometry. The oxyanion hole is a clear example of enthalpic interactions that preferentially stabilize high-energy states along the reaction coordinate and thereby lower the overall energies of those states.

The protonated H57 acts as a general acid in the collapse of the oxyanion that yields the acyl–enzyme intermediate. This leaves the N-terminal part of the polypeptide substrate covalently bound to the enzyme and allows the C-terminal portion to diffuse from the active site. A water molecule can now enter the active site to cleave the acyl–enzyme intermediate. H57 acts as a general base, deprotonating the H_2O to make it a more potent nucleophile, which attacks the acyl-enzyme ester to form a second tetrahedral oxyanion intermediate. Finally, H57, acting as a general acid, facilitates the collapse of the oxyanion to regenerate S195 and release the remainder of the cleaved polypeptide chain. The enzyme is back in its original state, ready to catalyze the hydrolysis of another amide bond.

Mutagenesis studies have been used to confirm the critical contributions of each residue in the catalytic triad toward the ~10^{10} rate enhancement observed for serine proteases. Working with a bacterial serine protease called subtilisin, Paul Carter and Jim Wells showed that S195A or H57A mutations reduced the rate

The catalysis of peptide-bond cleavage by serine proteases involves stabilization of transition states and tetrahedral intermediate states.

FIGURE 11.19

Binding in the oxyanion hole of a serine protease.
Left: The active site of α-lytic protease from the bacterium *Lysobacter enzymogenes* is shown (PDB ID: 1GBB). The backbone is shown in cartoon form (green) with the catalytic triad in stick form (C atoms yellow, N atoms blue, O atoms red) and the peptide boronic ester in stick form (C atoms light blue, B atom pink). H-bonds from the main chain amides of S195 and G193 in the oxyanion hole are shown by black dashed lines. **Right:** The attack of S195 changes the geometry of the boron atom. This places an O atom in the oxyanion hole, where H-bonds to the main chain form.

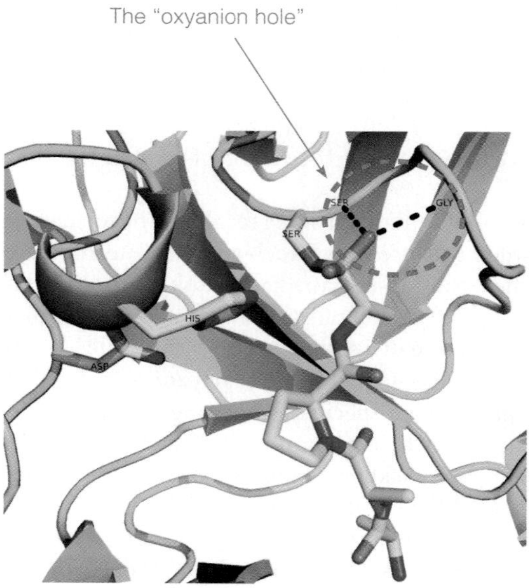

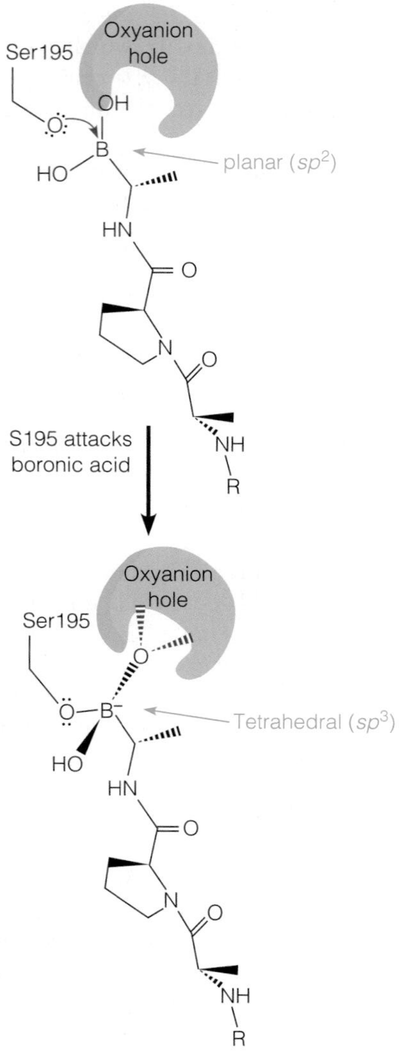

enhancement by ~10^6 compared to the wild-type enzyme, and the D102A mutant showed a reduction of ~10^4. These results indicate that S195 and H57 contribute 100-fold more to the rate enhancement than does D102. The triple mutant, S195A:H57A:D102A, also shows a ~10^6 reduction in rate enhancement, which suggests that other features of the enzyme active site, such as the oxyanion hole, contribute ~10^3–10^4 to the overall rate enhancement (see Problem 14 at the end of this chapter).

Additional support for the mechanism shown in Figure 11.17 comes from a large number of crystal structures of serine proteases bound to reaction intermediates as well as to **transition-state analogs**. Transition-state analogs are compounds that are designed to mimic the transition state for a reaction. As such, they are meant to possess greater shape and electrostatic complementarity to the enzyme active site than do the natural substrates. In fact, transition-state analogs typically bind to enzymes with affinities that are at least 10^2–10^4 greater than those of the natural substrates. In some cases the tight-binding analog is formed in the enzyme active site. An example is shown in Figure 11.19, where the tetrapeptide Ala–Ala–Pro–Ala bearing a C-terminal boronic acid is converted to a mixed borate ester following attack by S195 on the boron atom. This converts the planar boronic acid to the tetrahedral borate ester, which can then interact with the amides in the oxyanion hole.

TABLE 11.2	The strategies used by lysozyme and serine proteases to lower $\Delta G^{\circ\ddagger}$	
Catalytic Strategy	Lysozyme	Ser Proteases
GABC	E35	H57 (catalytic triad)
Covalent	D52	S195 (catalytic triad)
Electrostatic		oxyanion hole
Other	strain of D ring	low-barrier H-bond?

In summary, serine proteases employ covalent catalysis and electrostatic stabilization of the transition state to achieve rate enhancement (Table 11.2).

It is important to keep in mind that there are many enzyme-catalyzed reactions that are thought to work in ways we have not discussed here, and some of these, like the proposed low-barrier H-bond in chymotrypsin, are controversial. For example, elucidating the role of protein dynamics in rate acceleration by enzymes is a matter of active research and vigorous debate. We provide a brief introduction to this topic in the next section.

The Role of Dynamics in Catalysis

Proteins are dynamic structures—a fact that is difficult to represent with static figures in the pages of a textbook, but one that should be kept in mind while considering the functional properties of biomolecules. In previous chapters we have described several examples of protein conformational changes directly related to function, such as switching between T and R states in O_2 binding in hemoglobin, the movement of myosin along an actin filament, and active transport of ions by the Na^+-K^+ pump. Such large *directed* movements in proteins occur on a timescale of μs to s; however, as shown in Table 11.3, more subtle motions in proteins (e.g., "bending" or side chain rotation) occur more rapidly and, thus, appear random in nature. Recent work suggests that the rapid conformational dynamics of proteins may not be so random, and some researchers hypothesize that the dynamic behavior of proteins has therefore *evolved* to enhance protein function.

It is logical that enzymes must be dynamic to allow substrate binding and product release, to reorganize the active site for optimal stabilization of the transition state, or to allow allosteric regulation of activity (discussed later in this chapter). However, neither the relationship between dynamic motions and observed rates nor the evolution of the dynamic features of enzymes is well understood. Even so, it seems reasonable that active site dynamics can play a critical role in those hydrogen-transfer reactions that appear to proceed via a tunneling mechanism. Quantum tunneling is exquisitely sensitive to small changes in distance (i.e., 0.1 Å) between the hydrogen donor and acceptor atoms; thus, even rapid motions such as bond rotation in the active site could dramatically affect the rate of hydrogen transfer.

Recent NMR studies on the enzyme dihydrofolate reductase (see Chapter 20) have shown that dynamic motions of several amino acids in the protein are *correlated* and biased toward the handful of conformations that are relevant to the catalytic mechanism. Furthermore, the dynamic motions observed for each intermediate in the five-step catalytic cycle of dihydrofolate reductase appear to favor the formation of conformations which resemble that of the next intermediate in the cycle.

The five intermediates in the catalytic cycle of dihydrofolate reductase are shown in Figure 11.20. Starting on the left, the catalytic cycle begins with substrate binding to the enzyme complexed with the **cofactor** *reduced* nicotinamide adenine dinucleotide phosphate (NADPH). Cofactors are small molecules that

TABLE 11.3	Timescale of protein motions related to catalysis	
Motion	Approximate Time Scale (seconds)
Bond vibration	10^{-13} to 10^{-14}
Proton transfer	10^{-12}
Hydrogen bonding	10^{-11} to 10^{-12}
Elastic vibration of globular region	10^{-11} to 10^{-12}
Sugar repuckering	10^{-9} to 10^{-12}
Rotation of side chains at surface	10^{-10} to 10^{-11}
Torsional libration of buried groups	10^{-9} to 10^{-11}
Hinge bending at domain interfaces	10^{-7} to 10^{-11}
Water structure reorganization	10^{-8}
Helix-coil breakdown/ formation	10^{-7} to 10^{-8}
Allosteric transitions	1 to 10^{-5}
Local denaturation	10 to 10^{-5}
Rotation of medium-sized side chains in interior	1 to 10^{-4}

Data from *Science* 301:1196–1202 (2003), S. J. Benkovic and S. Hammes-Schiffer, A perspective on enzyme catalysis.

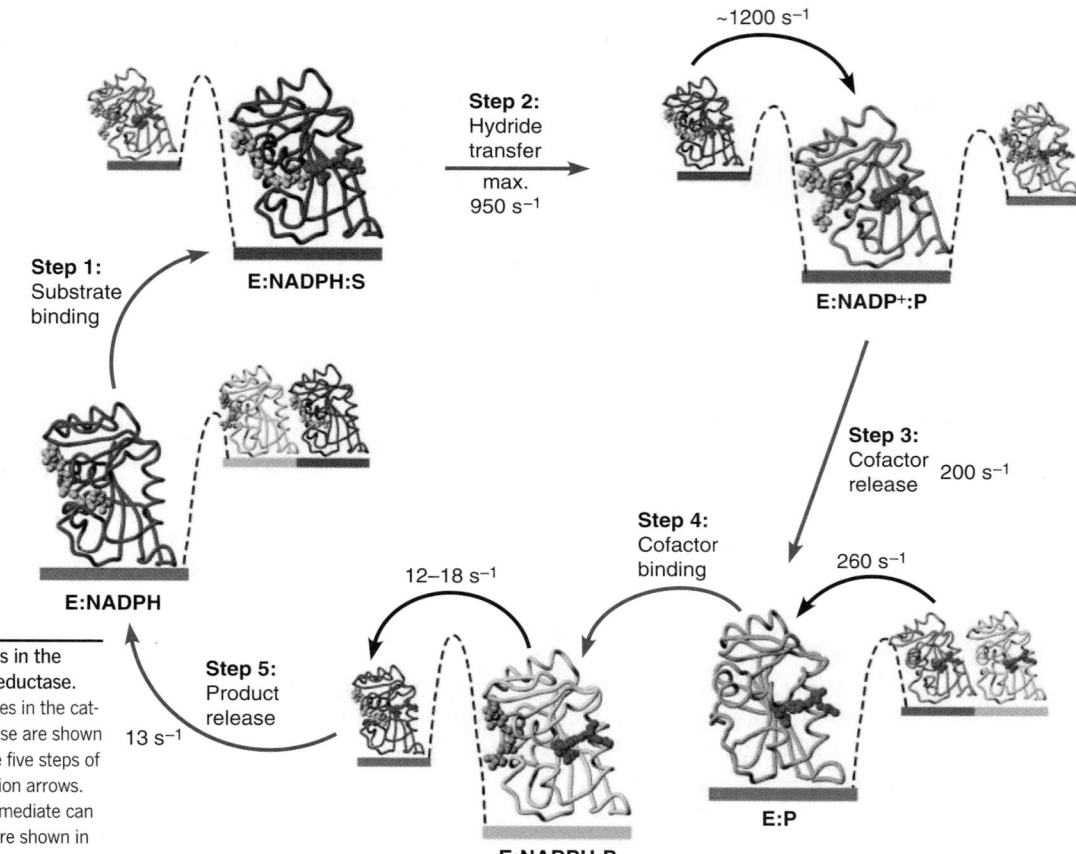

FIGURE 11.20

Dynamic conformational changes in the catalytic cycle of dihydrofolate reductase. Models of the five intermediate states in the catalytic cycle of dihydrofolate reductase are shown in larger scale and color coded. The five steps of the cycle are indicated by red reaction arrows. The other conformations each intermediate can sample through dynamic motions are shown in smaller scale with the same color coding. The rates of conformational exchange between intermediates are shown above the black reaction arrows. See the text for more detail.

From *Science* 313:1638–1642, D. Boehr, D. McElheny, H. J. Dyson and P. E. Wright, The dynamic energy landscape of dihydrofolate reductase catalysis. © 2006. Reprinted with permission from AAAS. Adapted with permission from Peter Wright.

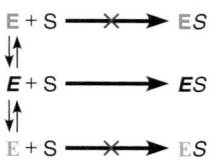

FIGURE 11.21

Conformational Selection vs. Induced Fit. Here *ES* represents the enzyme–substrate complex. Induced fit implies conformational homogeneity in the unbound enzyme, which is then distorted upon substrate binding. Conformational selection implies that the unbound enzyme is present in multiple conformations; but, the substrate can only bind to unbound enzyme that is in the same conformation as that of the *ES* complex. Substrate binding will perturb the conformational equilibrium of the unbound enzyme, driving more E into the *ES* conformation by Le Chatelier's principle (see Chapter 3).

expand the range of enzyme chemistry and are discussed in more detail later in this chapter. For example, NADPH is a source of hydride (H⁻), which is commonly used in biological redox reactions. When the E:NADPH complex (blue) binds substrate, the E:NADPH:S complex (red) is formed. Hydride transfer from the NADPH to the substrate yields the product complex E:NADP⁺:P, shown in purple (here NADP⁺ represents oxidized cofactor). Release of NADP⁺ gives the E:P complex (green), which can then bind to NADPH to give E:NADPH:P (yellow). Release of the product from this complex gives E:NADPH and the cycle is completed.

The important finding illustrated in Figure 11.20 is that each intermediate displays dynamic motions that allow it to adopt the conformation of a neighboring intermediate state. For example, the E:NADP⁺:P intermediate can adopt either the E:NADPH:S or the E:P conformation. The NMR experiment used in these studies also provides information on the rates of interconversion between the conformational states (black arrows in Figure 11.20). For three of the steps in the catalytic cycle, the rates of conformational dynamics closely match the rates determined in classical kinetics studies. For example, cofactor release occurs with a rate of 200 s⁻¹, and the conformational dynamics between the E:NADP⁺:P and E:P states occur at 260 s⁻¹. This observation suggests that cofactor release occurs from the E:P conformation (and not from the E:NADP⁺:P conformation), and that cofactor is released from E:NADP⁺:P nearly every time the E:NADP⁺:P intermediate adopts the E:P conformation.

Compared to the induced fit model, this **conformational selection** (or **selected fit**) model is a very different description of the binding interactions between an enzyme and its substrate/cofactors. As shown in Figure 11.21, conformational selection implies that the binding of a substrate doesn't impose a conformational change on the enzyme; rather, the substrate only binds to enzyme molecules that have conformations similar to that of the ES complex. A corollary is that substrate will be released when the ES complex samples conformations that are incompatible with substrate binding. This is the basis for the hypothesis that dynamic motions of enzymes, which determine the conformational state of the enzyme, have evolved due to selective pressure to optimize catalytic efficiency.

Finally, it has been suggested that dynamic conformational switching may require the *correlated* movements of several amino acid residues in the protein, and such concerted motions make up a dynamic "network." The existence of such networks in enzymes has been proposed to explain two observations: (1) some mutations distant from the active site have significant impacts on reaction rates even though these amino acids make no direct contact to the substrate and (2) such mutations tend to affect the dynamic behavior of a larger group of amino acids. The latter observation suggests that the effect of a distant mutation can be communicated to the active site through the coordinated motions of the residues that make up the dynamic network.

> The conformational selection model for enzyme catalysis suggests that free enzyme is present in multiple conformations; but, substrate binds only to an enzyme that is in a conformation similar to that of the enzyme in the ES complex.

The Kinetics of Enzymatic Catalysis

In the preceding section, we described details of the mechanisms for two well-studied enzyme-catalyzed reactions. Hundreds more are known, and such knowledge is essential in many aspects of modern medicine. A detailed understanding of the molecular basis for a disease guides the design of therapeutic drugs. For example, as discussed in Chapter 20, compounds that inhibit the activity of dihydrofolate reductase have found wide use in treating cancer and infectious diseases. Where does this mechanistic knowledge come from? X-ray crystallography and solution NMR studies have provided key structural insights into enzyme activity; however, much of our understanding of enzyme mechanisms comes from the careful mathematical analysis of enzyme kinetics. Thus, it is to that analysis that we now turn.

Reaction Rate for a Simple Enzyme-Catalyzed Reaction: Michaelis–Menten Kinetics

Earlier in this chapter we introduced Equation 11.12 as an expression for a simple reaction involving a single substrate and product:

$$E + S \underset{k_{-1}}{\overset{k_1}{\rightleftharpoons}} ES \underset{k_{-2}}{\overset{k_2}{\rightleftharpoons}} EP \underset{k_{-3}}{\overset{k_3}{\rightleftharpoons}} E + P \qquad (11.12)$$

If we analyze the **initial rate** of an enzyme-catalyzed reaction (i.e., before a significant concentration of P appears) and we assume that k_1, k_{-1}, and $k_3 \gg k_2$, Equation 11.12 simplifies to the following:

$$E + S \underset{k_{-1}}{\overset{k_1}{\rightleftharpoons}} ES \overset{k_{cat}}{\rightarrow} E + P \qquad (11.13)$$

where k_{cat} is the apparent rate constant for the rate-determining conversion of substrate to product.*

We have assumed that initial reaction conditions are such that the reverse reaction between E and P is negligible. The catalytic formation of the product, with enzyme regeneration, will then be a simple first-order reaction, and its rate will be determined solely by the concentration of ES and the value of k_{cat}. Therefore, the reaction rate, or velocity, defined as the observed rate of formation of products, can be expressed as

$$v = k_{cat}[ES] \qquad (11.14)$$

If v and [ES] can be measured for a specific enzyme and substrate, the rate constant, k_{cat}, for that particular reaction can be derived. In practice, [ES] is difficult to measure in kinetics experiments. The easily measured parameters are substrate

---

*k_{cat} is an aggregate rate constant $= k_2k_3/(k_2 + k_3)$. In the limiting case where $k_3 \gg k_2$, $k_{cat} \approx k_2$.

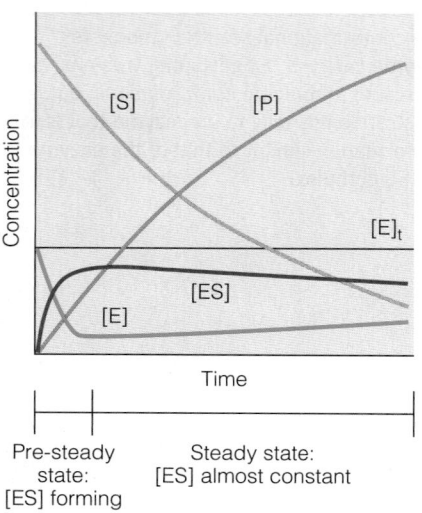

Time

Pre-steady
state:
[ES] forming

Steady state:
[ES] almost constant

FIGURE 11.22

The steady state in enzyme kinetics. The figure shows how the concentrations of substrate [S], free enzyme [E], enzyme–substrate complex [ES], and product [P] vary with time for a simple enzyme-catalyzed reaction described by $E + S \rightleftharpoons ES \rightarrow E + P$. After a very brief initial period, [ES] reaches a steady state in which ES is consumed approximately as rapidly as it is formed, so $d[ES]/dt \approx 0$. The amounts of E and ES are greatly exaggerated for clarity. Note that $[E] + [ES] = [E]_t$, or total enzyme concentration, and that [ES] actually falls very slowly as substrate is consumed, while [E] accordingly rises.

(or product) concentration and the total concentration of enzyme, which must be the sum of free enzyme and enzyme bound to substrate:

$$
\underset{\substack{\text{Total} \\ \text{enzyme}}}{[E]_t} = \underset{\substack{\text{Free} \\ \text{enzyme}}}{[E]} + \underset{\substack{\text{Enzyme in} \\ \text{ES complex}}}{[ES]} \tag{11.15}
$$

Thus, it is desirable to express the rate, v, in terms of the substrate concentration [S] and the total enzyme concentration $[E]_t$.

The way we have written Equation 11.13 suggests that E and S should be in equilibrium with ES, with an equilibrium dissociation constant K_S:

$$
K_S = \frac{k_{-1}}{k_1} = \frac{[E][S]}{[ES]} \tag{11.16}
$$

This is usually an incorrect assumption, but under certain circumstances (i.e., $k_{cat} \ll k_{-1}$), this approximation is valid. This assumption was used in early attempts to solve the problem of expressing the reaction rate. It doesn't apply in general because E, S, and ES are not truly in equilibrium; some ES is continually being drained off to make P. An analysis that avoids the assumption of equilibrium was presented by G. E. Briggs and J. B. S. Haldane in 1925. The Briggs–Haldane model is based on the following argument: The more ES that is present, the faster ES will dissociate either to products (k_{cat}) or back to reactants (k_{-1}). Therefore, when the reaction is started by mixing enzymes and substrates, the ES concentration builds up at first, but quickly reaches a **steady state**, in which it remains almost constant. This steady state will persist until almost all of the substrate has been consumed (Figure 11.22). Because the steady state accounts for nearly all the reaction time, we can calculate the reaction velocity by assuming steady-state conditions. Normally, we measure rates only after the steady state has been established and before [ES] has changed much. We can then express the velocity as follows.

In the steady state, the rates of formation and breakdown of ES are equal. Therefore,

$$
\underset{\substack{\text{Formation of} \\ \text{ES complex}}}{k_1[E][S]} = \underset{\substack{\text{Dissociation} \\ \text{of ES complex}}}{k_{-1}[ES]} + \underset{\substack{\text{Breakdown} \\ \text{to E + P}}}{k_{cat}[ES]} \tag{11.17}
$$

which can be rearranged to give

$$
[ES] = \left(\frac{k_1}{k_{-1} + k_{cat}}\right)[E][S] \tag{11.18}
$$

Combining the ratio of rate constants in Equation 11.17 gives a single constant, K_M:

$$
K_M = \frac{k_{-1} + k_{cat}}{k_1} \tag{11.19}
$$

Equation 11.18 can now be rewritten as

$$
K_M[ES] = [E][S] \tag{11.20}
$$

At this point, [ES] is expressed in terms of [E] and [S]. To get $[E]_t$ into the equation, rather than [E], recall from Equation 11.15 that $[E] = [E]_t - [ES]$. Putting this into Equation 11.20 yields

$$
K_M[ES] = [E]_t[S] - [ES][S] \tag{11.21}
$$

This rearranges to

$$
[ES] = \frac{[E]_t[S]}{K_M + [S]} \tag{11.22}
$$

Finally, inserting this result into Equation 11.14 gives an expression for v in terms of $[E]_t$ and $[S]$:

$$v = \frac{k_{cat}[E]_t[S]}{K_M + [S]} \quad (11.23)$$

Equation 11.23 is called the **Michaelis–Menten equation**, and K_M the **Michaelis constant**, honoring two pioneers in the analysis of enzyme kinetics, Leonor Michaelis and Maude Menten. We will discuss the meaning of K_M shortly; in the meantime, there are two important points to keep in mind. First, because K_M is a ratio of the rate constants for a specific reaction (see Equation 11.19), it is a characteristic of that reaction. Thus, a given enzyme acting upon a given substrate has a defined K_M. Second, you can see from Equations 11.19 and 11.20 that K_M has units of concentration.

Now consider the graph of v versus $[S]$ shown in Figure 11.23. At high substrate concentrations, where $[S]$ is much greater than K_M, the reaction approaches a **maximum velocity**, V_{max}, because the enzyme molecules are *saturated*; every enzyme molecule is bound by substrate; thus, $[ES] = [E]_t$ and Equation 11.14 will reach its maximum value. Recall from Chapter 10 that such saturation behavior is a characteristic of transport proteins in membranes. When $[S] \gg K_M$, $K_M + [S] \approx [S]$, and Equation 11.23 simplifies to the expression for V_{max}:

$$V_{max} = k_{cat}[E]_t \quad (11.24)$$

Thus, $k_{cat}[E]_t$ in Equation 11.23 is equivalent to V_{max}, and the Michaelis–Menten rate equation is

$$v = \frac{V_{max}[S]}{K_M + [S]} \quad (11.25)$$

This is the most familiar form of the Michaelis–Menten equation.

The Significance of K_M, k_{cat}, and k_{cat}/K_M

The two quantities that characterize an enzyme obeying Michaelis–Menten kinetics are K_M and k_{cat}. What do they signify? The Michaelis constant, K_M, is often associated with the affinity of enzyme for substrate. However, this relationship is only true in the limiting case: a two-step reaction in which $k_{cat} \ll k_{-1}$, where Equation 11.19 then yields $K_M \approx k_{-1}/k_1 = K_S$, the equilibrium constant defined in Equation 11.16. It is important to note that for more complex kinetic schemes, K_M is a ratio of several rate constants. For any reaction that follows the Michaelis–Menten equation, K_M is numerically equal to the substrate concentration at which the reaction velocity has attained *half* of its maximum value (see Figure 11.23). Thus, K_M is a measure of the substrate concentration required for effective catalysis to occur. *An enzyme with a high K_M requires a higher substrate concentration to achieve a given reaction velocity than an enzyme with a low K_M but the same k_{cat}.* Table 11.4 lists K_M values for a number of important enzymes.

The second constant, k_{cat}, gives a direct measure of the rate of product formation under optimum conditions (saturated enzyme). The units of k_{cat} are usually given as s^{-1}, so the reciprocal of k_{cat} can be thought of as a time—the time required by an enzyme molecule to "turn over" one substrate molecule. Alternatively, k_{cat} measures the number of substrate molecules turned over per enzyme molecule per second. Thus, k_{cat} is sometimes called the **turnover number**. Some typical turnover numbers are listed in Table 11.4.

The enzymes listed in Table 11.4 are arranged in increasing order of the ratio k_{cat}/K_M. This ratio is often thought of as a measure of enzyme efficiency. Note that either a large value of k_{cat} (i.e., rapid turnover) or a small value of K_M (i.e., $\frac{1}{2}V_{max}$ occurs at relatively low $[S]$) will make k_{cat}/K_M large. We can gain another insight into the meaning of k_{cat}/K_M by considering the situation at very low substrate

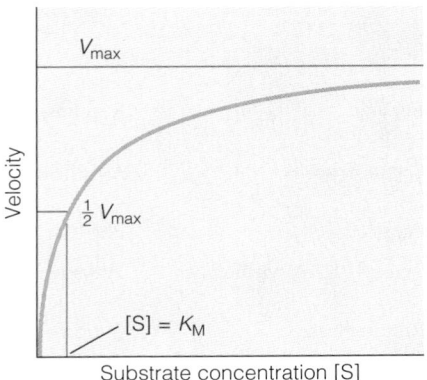

FIGURE 11.23

Reaction velocity as a function of substrate concentration. This graph, a plot of Equation 11.25, shows the variation of reaction velocity with substrate concentration according to the Michaelis–Menten model of enzyme kinetics. The values of v plotted here are determined from the *initial rates* of the reaction (see Figure 11.24). At the point where $[S] = K_M$, the reaction has exactly half its maximum velocity. Note that V_{max} is approached asymptotically.

The steady-state assumption proposes that the concentration of enzyme–substrate complex remains nearly constant through much of the reaction.

The Michaelis constant, K_M, indicates the substrate concentration at which the reaction rate is $\frac{1}{2}V_{max}$.

The turnover number, k_{cat}, measures the rate of the catalytic process.

TABLE 11.4 Michaelis–Menten parameters for selected enzymes, arranged in order of increasing efficiency as measured by k_{cat}/K_M

Enzyme	Reaction Catalyzed	K_M (mol/L)	$k_{cat}(s^{-1})$	k_{cat}/K_M [(mol/L)$^{-1}$ s^{-1}]
Chymotrypsin	Ac–Phe–Ala $\xrightarrow{H_2O}$ Ac–Phe + Ala	1.5×10^{-2}	0.14	9.3
Pepsin	Phe–Gly $\xrightarrow{H_2O}$ Phe + Gly	3×10^{-4}	0.5	1.7×10^3
Tyrosyl-tRNA synthetase	Tyrosine + tRNA $\longrightarrow$ tyrosyl-tRNA	9×10^{-4}	7.6	8.4×10^3
Ribonuclease	Cytidine 2′, 3′ cyclic phosphate $\xrightarrow{H_2O}$ cytidine 3′-phosphate	7.9×10^{-3}	7.9×10^2	1.0×10^5
Carbonic anhydrase	$HCO_3^- + H^+ \longrightarrow H_2O + CO_2$	2.6×10^{-2}	4×10^5	1.5×10^7
Fumarase	Fumarate $\xrightarrow{H_2O}$ malate	5×10^{-6}	8×10^2	1.6×10^8

concentrations. In this case, $[S] \ll K_M$, and most of the enzyme is free, so $[E]_t \approx [E]$. Then equation (11.23) becomes

$$v = \frac{k_{cat}}{K_M}[E][S] \tag{11.26}$$

Therefore, under these circumstances, the ratio k_{cat}/K_M behaves as a second-order rate constant for the reaction between substrate and free enzyme, and it provides a direct measure of enzyme efficiency and specificity. It shows what the enzyme and substrate can accomplish when abundant enzyme sites are available, and it allows direct comparison of the effectiveness of an enzyme toward different substrates. Suppose an enzyme has a choice of two substrates, A or B, present at equal concentrations. Then under conditions in which both substrates are dilute and are competing for the enzyme, we find

$$\frac{v_A}{v_B} = \frac{\left(\dfrac{k_{cat}}{K_M}\right)_A [E][A]}{\left(\dfrac{k_{cat}}{K_M}\right)_B [E][B]} = \frac{\left(\dfrac{k_{cat}}{K_M}\right)_A}{\left(\dfrac{k_{cat}}{K_M}\right)_B} \tag{11.27}$$

Table 11.5 lists values of k_{cat}/K_M for cleavage of various esters by chymotrypsin. Within the group shown, k_{cat}/K_M varies 1 million-fold, showing the range of preference the enzyme has for different peptide substrates. The preference to cleave next to the most hydrophobic residues is quite clear from these data.

We have just seen that the ratio k_{cat}/K_M corresponds to the second-order rate constant for enzyme–substrate combination under circumstances of low substrate concentration. Such a rate constant has a maximum possible value, which is determined

The ratio k_{cat}/K_M is a convenient measure of enzyme efficiency.

N-acetyl amino acid methyl ester

TABLE 11.5 Preferences of chymotrypsin in the hydrolysis of several N-acetyl amino acid methyl esters, as measured by k_{cat}/K_M

Amino Acid in Ester	Amino Acid Side Chain	k_{cat}/K_M [(mol/L)$^{-1}$ s^{-1}]
Glycine	—H	1.3×10^{-1}
Norvaline	—$CH_2CH_2CH_3$	3.6×10^2
Norleucine	—$CH_2CH_2CH_2CH_3$	3.0×10^3
Phenylalanine	—CH_2—	1.0×10^5

by the frequency with which enzyme and substrate molecules can collide in solution. A reaction that attains such a velocity is said to be "diffusion-limited"; every encounter leads to reaction, so nothing but the rate of molecular encounters limits the velocity. If *every* collision results in formation of an enzyme–substrate complex, diffusion theory predicts that k_{cat}/K_M will attain a maximum value of about 10^8 to $10^9 (mol/L)^{-1} s^{-1}$. Thus, an enzyme that approaches maximum possible efficiency will demonstrate this by having a value of k_{cat}/K_M in this range. As Table 11.4 shows, enzymes such as carbonic anhydrase and fumarase actually approach this limit. Triose phosphate isomerase (discussed in Chapter 13) is another example, having $k_{cat}/K_M = 2.4 \times 10^8 (mol/L)^{-1} s^{-1}$. In fact, it has been argued that triose phosphate isomerase is an almost perfect enzyme, having evolved to nearly maximum efficiency. In support of this idea is the observation that triose phosphate isomerases from organisms as evolutionarily distant as yeasts and vertebrates show very little change in structure. Apparently, this enzyme (which plays a vital role in cellular ATP production from carbohydrates) reached near perfection early in evolution and has changed little since that time.

Analysis of Kinetic Data: Testing the Michaelis–Menten Equation

The measurement of reaction velocity as a function of substrate concentration is used to determine whether an enzyme-catalyzed reaction follows the Michaelis–Menten model (Equation 11.25) and, if so, to determine the constants K_M and k_{cat}.

A few analytical methods for the measurement of rates are described in Tools of Biochemistry 11A. One general point should be noted: In principle, one could simply mix enzyme and substrate and follow the change in substrate concentration with time, as shown in Figure 11.1a. As substrate is consumed, the reaction velocity decreases, until equilibrium is eventually reached. But measuring the instantaneous velocity at specific times during the reaction is difficult and usually inaccurate. It is usually easier to set up a series of experiments, all at the same enzyme concentration but at different substrate concentrations, and measure the *initial* rates (Figure 11.24). Because we know the initial [S] precisely, and the change in [S] versus t is almost linear in the initial stages, accurate data for v as a function of [S] can be obtained. An enzyme that obeys the Michaelis–Menten kinetic model will yield a plot of initial velocity vs. substrate concentration that is hyperbolic—compare Figures 11.23 and 11.24b.

Given such data for concentrations and initial rates, how are K_M and k_{cat} calculated? In practice, modern nonlinear curve-fitting software can fit the data plotted in Figure 11.24b directly to provide these parameters. Before such data analysis became widely available, a different method was (and still is) used. Equation 11.25 can be rearranged to give the expression for a linear graph. Several kinds of graphs are possible, but it is most common to use a **double reciprocal plot**, also called a **Lineweaver–Burk plot** (Figure 11.25). A Lineweaver–Burk plot provides a quick test for adherence to Michaelis–Menten kinetics and allows easy evaluation of the critical constants. As described below, it also allows discrimination between different kinds of enzyme inhibition and regulation. Taking the reciprocal of both sides of Equation 11.25, we find

$$\frac{1}{v} = \frac{K_M + [S]}{V_{max}[S]} = \frac{K_M}{V_{max}[S]} + \frac{[S]}{V_{max}[S]} \tag{11.28a}$$

or

$$\frac{1}{v} = \left(\frac{K_M}{V_{max}}\right)\frac{1}{[S]} + \frac{1}{V_{max}} \tag{11.28b}$$

So, plotting $1/v$ versus $1/[S]$ yields a straight line. At $1/[S] = 0$, [S] is infinitely large and the reaction velocity is at its maximum. Therefore, the y-intercept at

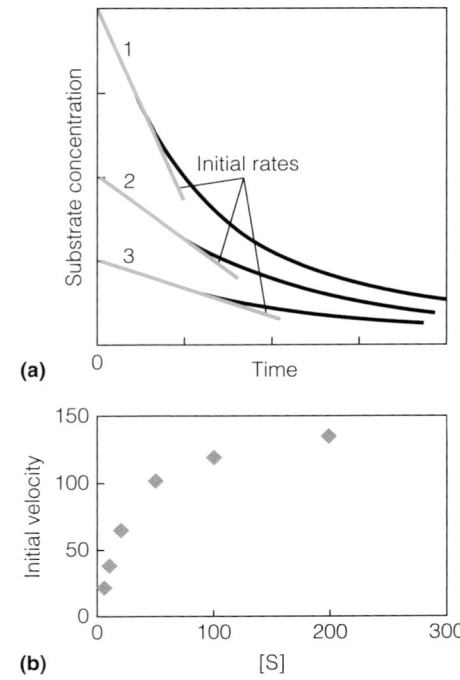

(a)

(b)

FIGURE 11.24

Analysis of intial rates. (a) Several reactions are performed with varying concentrations of substrate and the values of the initial rates are determined from the slopes of the curves in the early phase for each reaction. **(b)** Initial rate data, determined as described in (a), are plotted as a function of the substrate concentration. The enzyme appears to obey the Michaelis–Menten kinetic model (compare this plot to Figure 11.23).

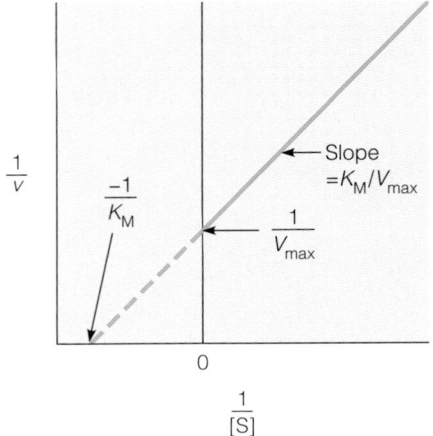

FIGURE 11.25

A Lineweaver–Burk plot. In this double reciprocal plot, $1/v$ is graphed versus $1/[S]$, according to Equation 11.28b. Note that a linear extrapolation of the data gives both V_{max} and K_M.

Lineweaver–Burk plots or Eadie–Hofstee plots provide convenient ways to determine K_M and k_{cat} from intial-rate data.

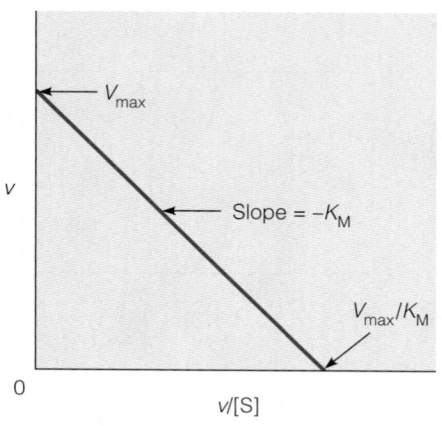

FIGURE 11.26

An Eadie–Hofstee plot. Graphing v versus $v/[S]$, we obtain V_{max} at $(v/[S]) = 0$ and K_M from the slope of the line.

The observed effects of an amino acid mutation in an enzyme active site on K_M and k_{cat} can be used to identify the role of the amino acid in substrate binding (K_M effects) and transition-state stabilization (k_{cat} effects).

Multisubstrate reactions fall into several classes, depending on the order of substrate binding: random, ordered, or ping-pong.

$1/[S] = 0$ is equal to $1/V_{max}$. Given V_{max} and $[E]_t$ (from the initial conditions of the kinetics experiment), k_{cat} can be calculated from $V_{max} = k_{cat}[E]_t$. In similar fashion, K_M is calculated using the slope of the plot, which gives K_M/V_{max} and the value of V_{max} obtained from the y-intercept. Alternatively, K_M can be calculated from the intercept of the Lineweaver–Burk plot with the x-axis where $1/v = 0$. Setting $1/v = 0$ gives

$$0 = \left(\frac{K_M}{V_{max}}\right)\frac{1}{[S]_0} + \frac{1}{V_{max}} \tag{11.29}$$

where $[S]_0$ denotes the value of $[S]$ at $1/v = 0$. We then obtain from Equation 11.29:

$$\frac{1}{[S]_0} = -\frac{1}{K_M} \tag{11.30}$$

Thus, the intercept of the Lineweaver–Burk plot with the x-axis gives $-1/K_M$ (see Figure 11.25).

A disadvantage of a Lineweaver–Burk plot is that a long extrapolation is often required to determine K_M, which introduces corresponding uncertainty in the result. Consequently, other ways of plotting the data are sometimes used. One alternative is to rearrange Equation 11.25 into the form

$$v = V_{max} - K_M \frac{v}{[S]} \tag{11.31}$$

and graph v versus $v/[S]$. This yields what is called an **Eadie–Hofstee plot** (Figure 11.26). These linear plots provide useful methods for recognizing modes of inhibition; however, they suffer from unequal weighting of the data. Performing nonlinear curve fitting to the raw data with readily available software is now preferred.

As discussed above for lysozyme and subtilisin, site-directed mutagenesis has been used extensively to test mechanistic models for most enzymes. A particular mutation may change the apparent value of k_{cat} or K_M or both. The simplest interpretation of such data is that a mutation that affects only k_{cat} changes an amino acid side chain involved solely in catalysis (e.g., the active site nucleophile, or a GABC, etc.) but not involved in binding to the substrate in the ground state. Conversely, a mutation that affects only K_M alters a side chain that binds to the substrate but is not involved in stabilizing the transition state. Frequently, both k_{cat} and K_M are affected by mutations to active site residues. This would occur for a residue that binds substrate and then makes stronger interactions with the transition state as a result of an altered geometry in the transition state that optimizes the interaction.

Multisubstrate Reactions

Our discussions of enzyme kinetics have, to this point, centered on simple reactions in which one substrate molecule is bound to an enzyme and undergoes reaction there. In fact, such reactions are in the minority. Most biochemical reactions involve two or more substrates. An example we have already discussed is proteolysis, which involves two substrates (the polypeptide and water) and two products (the two fragments of the cleaved polypeptide chain). Phosphorylation of glucose, as catalyzed by hexokinase, is another such case: The two substrates are glucose and ATP, and the products are glucose-6-phosphate and ADP.

When an enzyme binds two or more substrates and releases multiple products, the order of the steps becomes an important feature of the enzyme mechanism. Several major classes of mechanisms for multisubstrate reactions are recognized. We shall illustrate them with examples using two substrates, S1 and S2, and two products, P1 and P2.

Random Substrate Binding

In random substrate binding, either substrate can be bound first, although in many cases one substrate will be favored for initial binding, and its binding may promote the binding of the other. The general pathway is

$$\text{either} \quad S1 \nearrow \quad E \cdot S1 \searrow S2$$
$$E \qquad\qquad E \cdot S1 \cdot S2 \longrightarrow E + P1 + P2$$
$$\text{or} \quad S2 \searrow \quad E \cdot S2 \nearrow S1$$

The phosphorylation of glucose by ATP, with hexokinase as the enzyme, appears to follow such a mechanism, although there is some tendency for glucose to bind first.

Ordered Substrate Binding

In some cases, one substrate *must* bind before a second substrate can bind significantly:

$$E \underset{}{\overset{S1}{\rightleftharpoons}} E \cdot S1 \underset{}{\overset{S2}{\rightleftharpoons}} E \cdot S1 \cdot S2 \longrightarrow E + P1 + P2$$

This mechanism is often observed in oxidations of substrates by the cofactor NAD^+, which is related to the cofactor $NADP^+$ discussed above in the dihydrofolate reductase mechanism.

The Ping-Pong Mechanism

Sometimes the sequence of events in catalysis goes like this: One substrate is bound, one product is released, a second substrate comes in, and a second product is released. This is called a "ping-pong" reaction:

$$E \underset{}{\overset{S1}{\rightleftharpoons}} E \cdot S1 \overset{P1}{\longrightarrow} E^* \underset{}{\overset{S2}{\rightleftharpoons}} E^* \cdot S2 \overset{P2}{\longrightarrow} E$$

Here E^* is a modified form of the enzyme, often carrying a fragment of S1. A good example is the cleavage of a polypeptide chain by a serine protease such as trypsin or chymotrypsin. In that case, we describe the polypeptide as $S = B—A$, where A and B designate, respectively, the C-terminal and N-terminal portions of the peptide chain on either side of the scissile bond:

$$E \underset{}{\overset{S}{\rightleftharpoons}} E \cdot S \overset{A}{\rightleftharpoons} E^* \cdot B \overset{H_2O}{\rightleftharpoons} E^* \cdot B \cdot H_2O \overset{B}{\longrightarrow} E$$

Here $E^* \cdot B$ and $E^* \cdot B \cdot H_2O$ indicate the covalent intermediates described above (see Figure 11.17).

A Closer Look at Some Complex Reactions

How are the mechanism of a complex enzyme-catalyzed reaction and the rate constants for different steps determined? For example, consider the cleavage of a substrate by a serine protease such as chymotrypsin. Note that the step $E^* \cdot B + H_2O \longrightarrow E^* \cdot B \cdot H_2O$ cannot be analyzed because the concentration of water is essentially fixed in aqueous solution and is not a variable. Therefore, it will suffice to write the reaction as

$$E + S \underset{k_{-1}}{\overset{k_1}{\rightleftharpoons}} E \cdot S \overset{A}{\underset{k_2}{\longrightarrow}} E^* \cdot B \overset{}{\underset{k_3}{\longrightarrow}} E + B$$

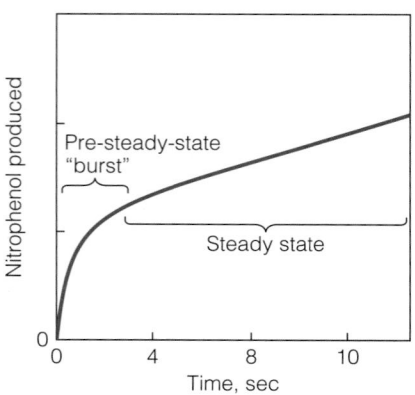

FIGURE 11.27

The pre-steady-state. This plot shows the kinetics of chymotrypsin-catalyzed hydrolysis of *p*-nitrophenyl acetate. Production of the first product (*p*-nitrophenol) is followed spectrophotometrically after the enzyme and substrate are mixed. The initial burst of product formation ceases when enzyme is almost all bound in the acyl–enzyme intermediate.

In this example there are several constants to determine. Steady-state measurements will not in themselves be sufficient. It can be shown that the steady-state velocity for this reaction is given by the equation

$$v = \frac{\left(\dfrac{k_2 k_3}{k_2 + k_3}\right)[E]_t[S]}{[S] + \left(\dfrac{K_S k_3}{k_2 + k_3}\right)} \tag{11.32}$$

In other words, the enzyme obeys Michaelis–Menten kinetics, but k_{cat} and K_M are now defined as follows:

$$k_{cat} = \frac{k_2 k_3}{k_2 + k_3} \tag{11.33a}$$

$$K_M = \frac{K_S k_3}{k_2 + k_3} \tag{11.33b}$$

$$K_S = \frac{k_{-1}}{k_1} \tag{11.33c}$$

These results emphasize that the appropriate expressions for k_{cat} and K_M depend upon the reaction mechanism, even in cases where the Michaelis–Menten equation describes the velocity. To obtain the individual rate constants in such a case, measurements outside the steady-state range must be employed. One of the first indications that an enzyme-bound intermediate might be involved in proteolysis came from rapid kinetic studies of the early stages in the hydrolysis of esters by chymotrypsin. Monitoring the release of product A (*p*-nitrophenol in this example) shows that its concentration increases quickly for a few seconds, until about 1 mole has been produced per mole of enzyme. After this time, the rate is nearly constant, as expected for a reaction at steady state (Figure 11.27).

The initial rapid production of the first product, called *pre-steady-state* production, or the *burst* phase, has been explained in the following way. For ester hydrolysis, k_3 is much smaller than k_2. Thus, the acyl intermediate forms quickly on each enzyme molecule, with accompanying release of 1 mole of product A. But after this period, more A can be formed only after each acyl intermediate breaks down and the enzyme becomes available again. The dissociation of the acyl intermediate is the rate-limiting step.

Still faster measurements, using stopped-flow techniques (see Tools of Biochemistry 11A), allow measurement of the rate of formation of the enzyme–substrate complex (ES). Measurements of the decay of the acyl intermediates after substrate is exhausted provide k_3. By using a combination of such methods, together with steady-state studies, we can obtain all of the constants in Equation 11.32. Two examples of such detailed kinetic data are given in Table 11.6 for hydrolysis of *N*-acyl amino acid esters. In example 1, we find $k_{cat} \cong k_2$, and $K_M \cong K_S$. These are the results expected when the acylation reaction (k_2) is rate-limiting, with $k_2 < k_3$. In the second example, deacylation is rate-limiting ($k_2 > k_3$) and $k_{cat} \cong k_3$. In this situation, $K_M = K_S(k_3/k_2)$. Each of these statements can be checked by examination of how Equations 11.33a–c behave in the specific cases.

This example demonstrates that a steady-state analysis is only a first step in the study of any enzyme and that a variety of techniques must be employed to unravel the details of catalytic mechanisms.

Single-Molecule Studies of Enzyme Activity

Our discussion of enzyme kinetics so far has been based on the typical conditions used for such studies—purified proteins present in large numbers in the reaction vessel. Thus, the kinetic behavior observed is the average behavior of a huge

TABLE 11.6 Rate constants for the hydrolysis of two *N*-acyl amino acid esters by chymotrypsin

Substrate	k_{cat} (s^{-1})	k_2 (s^{-1})	k_3 (s^{-1})	K_M (mM)	K_S (mM)
(structure 1)	0.069	0.069	0.6	5.87	5.97
(structure 2)	192	5000	200	0.663	17.2

ensemble of molecules. Recently it has been possible to record the kinetics of *single enzyme molecules* tethered in place on a fluorescence microscope. Sunney Xie and coworkers studied β-galactosidase in this way. They found that the kinetic behavior of single molecules was variable—a given molecule displayed multiple turnover rates, and these rates differed between different molecules. They interpreted these results in terms of multiple conformations for each enzyme molecule (Figure 11.28a), where each conformation is associated with a different value of k_{cat}. Figure 11.28b illustrates how we might imagine variation in k_{cat} for a single enzyme molecule if we plot the reaction coordinate (i.e., the red curve in Figure 11.10) in a second dimension that describes the multiple conformations that molecules can adopt as a consequence of dynamic motions. Each state of the enzyme (e.g., E, ES, and EP) has multiple conformations and each conformation is associated with a different free energy. Energy barriers along *both* dimensions could give rise to the variations in k_{cat} observed in these experiments.

This does not mean that all the enzyme kinetics presented above is invalid! The Michaelis–Menten model provides a good description of the kinetics of a large ensemble of molecules. Furthermore, Xie found that when single molecules were observed for many turnovers, their average kinetic behaviour could also be described by Michaelis–Menten kinetics. Thus, we will continue to interpret data from kinetics studies in terms of the Michaelis–Menten model.

FIGURE 11.28

Conformational dynamics and an expanded reaction coordinate landscape. **(a)** Studies of single molecules suggest that enzymes can adopt multiple conformations, not only along the reaction coordinate trajectory (red arrows) but also between similar states along the reaction coordinate (blue arrows between free enzyme states, ES complex, EP complex). **(b)** A model for the free energy landscape defined by the scheme in (a). For real enzymes this landscape is probably more rugged (i.e., higher peaks and valleys than shown here), with a few saddle points between the ES and EP states. The variability seen in observed rates for single molecules is attributed to the differences in barrier heights between the various states shown in (a).

Adapted with permission from *Biochemistry* 47: 3317–3321, S. J. Benkovic, G. G. Hammes, and S. Hammes-Schiffer, Free-energy landscape of enzyme catalysis. © 2008 American Chemical Society.

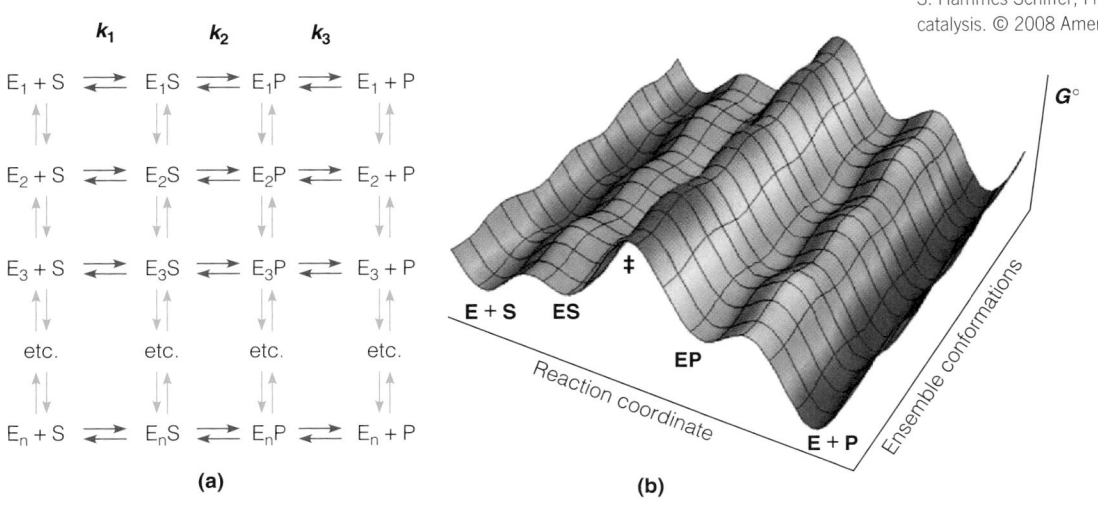

(a)   (b)

Enzyme Inhibition

Studies of enzyme inhibition can provide critical insights into the mechanisms of enzyme catalysis. Many different kinds of molecules inhibit enzymes, and they act in a variety of ways. A major distinction must be made between **reversible inhibitors** and **irreversible inhibitors**. The former involves *noncovalent* binding of the inhibitor and can always be reversed, at least in principle, by removal of the inhibitor. In some cases, noncovalent binding may be so strong as to appear irreversible under physiological conditions. Trypsin inhibitor binding to trypsin is one such example (see Figure 6.42). In irreversible inhibition, on the other hand, a molecule is *covalently* bound to the enzyme and truly incapacitates it. Irreversible inhibition is frequently encountered in the action of specific toxins and poisons, many of which kill by incapacitating key enzymes. On the other hand, the therapeutic action of many drugs depends on their acting as enzyme inhibitors, as we shall see in many examples.

Reversible Inhibition

The various modes of reversible inhibition all involve the noncovalent binding of an inhibitor to the enzyme, but they differ in the mechanisms by which they decrease the enzyme's activity and in how they affect the kinetics of the reaction.

Competitive Inhibition

Suppose a molecule exists that so closely resembles the substrate for an enzyme-catalyzed reaction that the enzyme will accept this molecule in its active site. If this molecule can also be processed by the enzyme, it is merely a competing alternative substrate. However, if the molecule binds to the active site but *cannot* undergo the catalytic step, it effectively reduces the enzyme's availability to carry out chemistry on true substrates. Such a molecule is called a **competitive inhibitor** because it competes with the substrate for binding to the same site on the enzyme (Figure 11.29).

For whatever fraction of time a competitive inhibitor molecule is occupying the active site, the enzyme is unavailable for catalysis. The overall effect is as if the enzyme cannot bind substrate as well when the inhibitor is present. Thus, the enzyme is predicted act as if its K_M were increased by the presence of the inhibitor. These ideas can be expressed by writing the reaction scheme as

$$E + S \underset{k_{-1}}{\overset{k_1}{\rightleftharpoons}} ES \xrightarrow{k_{cat}} E + P$$

$$+$$

$$K_I \updownarrow$$

$$EI$$

Here I stands for the inhibitory substance and K_I is a dissociation constant for inhibitor binding, defined as $K_I = [E][I]/[EI]$, where $[I]$ = concentration of free inhibitor. The rate equations can be solved as shown in the previous section, but note that now

$[E]_t$	=	$[E]$	+	$[ES]$	+	$[EI]$	
Total enzyme		Free enzyme		Enzyme bound to substrate		Enzyme bound to inhibitor	(11.34)

Upon analysis of this case, the expression for v is found to be

$$v = \frac{k_{cat}[E]_t[S]}{K_M\left(1 + \dfrac{[I]}{K_I}\right) + [S]} \tag{11.35a}$$

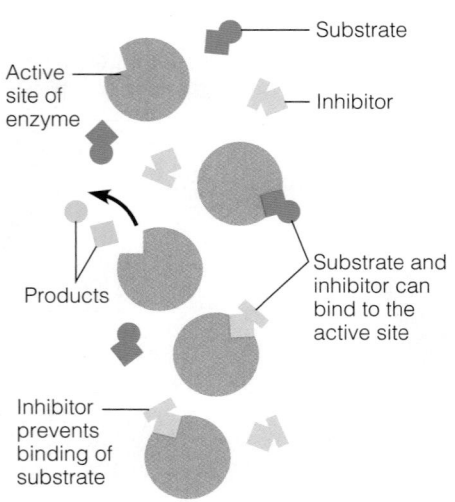

Active site of enzyme

Products

Inhibitor prevents binding of substrate

Substrate

Inhibitor

Substrate and inhibitor can bind to the active site

FIGURE 11.29

Competitive inhibition. Both substrate and inhibitor can fit the active site. Substrate can be processed by the enzyme, but inhibitor cannot.

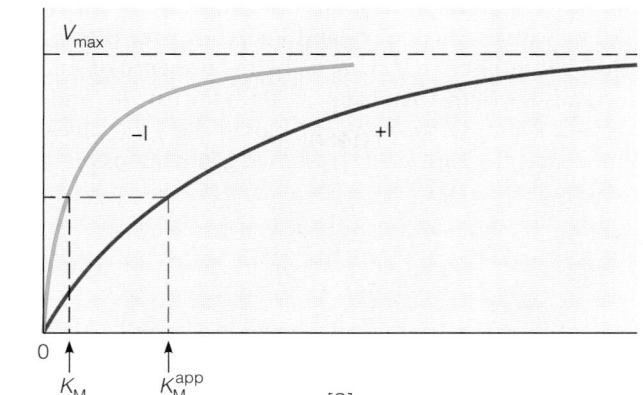

(a)

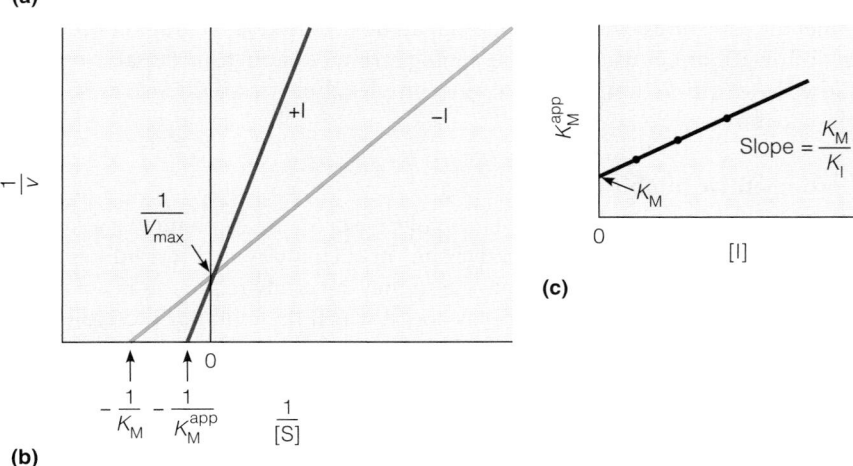

(b)

(c)

FIGURE 11.30

Effects of competitive inhibition on enzyme kinetics. (a) The effect of a competitive inhibitor (I) on reaction velocity at different substrate concentrations. Two sets of substrate-velocity experiments were carried out, one with (red line) and one without (blue line) inhibitor present. Addition of the inhibitor decreases the velocity but not the V_{max}. The apparent K_M is higher in the presence of inhibitor. **(b)** Lineweaver–Burk plots of the reactions shown in (a). The lines cross the $1/v$ axis at the same V_{max}, showing that I is a competitive inhibitor. **(c)** Determination of K_M and K_I. If the measurement of K_M^{app} is repeated at different concentrations of I, K_I can be determined from the slope of the line, and the true K_M from the line's intercept where [I] = 0 (see Equation 11.36).

or

$$v = \frac{V_{max}[S]}{\alpha K_M + [S]} \qquad (11.35b)$$

where $\alpha = (1 + [I]/K_I)$. The term αK_M may also be written as K_M^{app}, which gives

$$v = \frac{V_{max}[S]}{K_M^{app} + [S]} \qquad (11.35c)$$

This expression looks just like the Michaelis–Menten equation, with an "apparent" K_M given by $\alpha K_M = (1 + [I]/K_I)K_M$. As predicted, increasing [I] causes an increase in the apparent K_M. Because the formation of EI depends on [I] just as the formation of ES depends on [S], the rate of a competitively inhibited reaction is strictly dependent on the relative concentrations of I and S. It is important to note that K_M for the substrate is not changing per se; rather, the presence of a competitive inhibitor increases the value of [S] which gives half V_{max}. Thus, an *apparent* K_M is measured for the inhibited reaction, which is always greater than the actual K_M for the substrate because α is always >1.

For competitive inhibition the maximum velocity is unchanged because as [S] becomes very large, v approaches V_{max} (just as in the absence of inhibition), and $V_{max} = k_{cat}[E]_t$. Physically, this simply means that when [S] is very large at a given [I], the more numerous substrate molecules will outcompete the inhibitor. The effect of competitive inhibition on a graph of v versus [S] is shown in Figure 11.30a.

UpA: ribonuclease substrate

UpcA: competitive inhibitor of ribonuclease

FIGURE 11.31

A substrate and its competitive inhibitor. The substrate UpA and the structurally similar molecule UpcA are competitors for the enzyme ribonuclease. The single difference between the substrate and the inhibitor is shown in magenta.

Because the system, at a given [I], still obeys an equation of the Michaelis–Menten form, the Lineweaver–Burk plots and Eadie–Hofstee plots should be linear graphs, with K_M (but not V_{max}) changed by the presence of inhibitor. As Figure 11.30b shows, this is exactly what happens. Both the true K_M and K_I can be determined as shown in Figure 11.30c because a plot of the K_M^{app} values obtained at different inhibitor concentrations versus [I] should give a line as predicted by

$$K_M^{app} = \alpha K_M = \left(1 + \frac{[I]}{K_I}\right)K_M = K_M + \frac{K_M}{K_I}[I] \tag{11.36}$$

A clear example of a competitive inhibitor is shown in Figure 11.31. The dinucleotide UpA is an excellent substrate for the enzyme ribonuclease, which catalyzes hydrolysis of the phosphodiester bond between the two nucleotides. But if the oxygen atom at the cleavage site in UpA is replaced by a CH_2 group, the molecule becomes the *phosphonate* analog UpcA, a strong competitive inhibitor. Ribonuclease binds the analog strongly enough in the active site to allow X-ray diffraction studies of the complex, but it cannot cleave the phosphonate bond.

Uncompetitive Inhibition

This form of inhibition occurs when a molecule or an ion can bind to a *second* site on an enzyme surface (not the active site) in such a way that it modifies k_{cat}. It might, for example, distort the enzyme so that the catalytic process is not as efficient. The simplest case to consider is one in which the inhibitor molecule binds only to the ES complex and does not interfere in any way with substrate binding, but completely prevents the catalytic step (Figure 11.32).

$$E + S \underset{k_{-1}}{\overset{k_1}{\rightleftharpoons}} ES \xrightarrow{k_{cat}} E + P$$
$$+$$
$$I$$
$$\Big\updownarrow K_I'$$
$$EIS \xrightarrow{\quad\times\quad} \text{No reaction}$$

Here we distinguish the equilibrium constant for inhibitor binding to free enzyme, K_I, from K_I', the equilibrium constant for inhibitor binding to the ES complex. Because the inhibitor binding site is completely separate from the substrate binding site, such an **uncompetitive inhibitor** can be a molecule that does not resemble the substrate at all.

It can be shown (see Problem 28a) that the Michaelis–Menten equation for the uncompetitive inhibition scheme above is

$$v = \frac{V_{max}^{app}[S]}{K_M^{app} + [S]} = \frac{V_{max}[S]}{K_M + \alpha'[S]} \tag{11.37}$$

where $\alpha' = (1 + [I]/K_I')$, $V_{max}^{app} = V_{max}/\alpha'$, and $K_M^{app} = K_M/\alpha'$. In the presence of an uncompetitive inhibitor both V_{max} and K_M appear to be *reduced* by a factor of $1/\alpha'$ (Figure 11.33a). By binding only to the ES complex, an uncompetitive inhibitor increases the effective S binding, which reduces the apparent K_M. How can this observation be explained? At $[S] < K_M$ the effect of the inhibitor is minimal because as [S] decreases, v approaches $V_{max}[S]/K_M$. At $[S] > K_M$, the effect of the inhibitor on reducing V_{max} is apparent because as [S] increases, v approaches V_{max}/α'. Thus, as shown in Figure 11.33a, the v vs. [S] plots overlap at low [S] but diverge at higher [S].

Uncompetitive inhibition is distinguished from competitive inhibition by two observations: (1) uncompetitive inhibition cannot be reversed by increasing [S] and (2) as shown in Figure 11.33b, the Lineweaver–Burk plot yields parallel rather

An uncompetitive inhibitor does not compete for the active site but affects the catalytic event. It reduces both the apparent V_{max} and apparent K_M. These effects cannot be reversed by increasing [S].

FIGURE 11.32

Uncompetitive inhibition. The inhibitor binds at a site on the enzyme surface different from that of the substrate. In this simple example, the inhibitor binds only to the ES complex and inhibits the catalytic event.

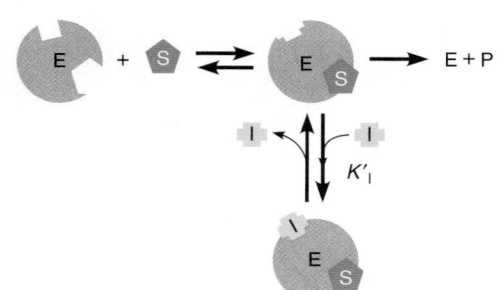

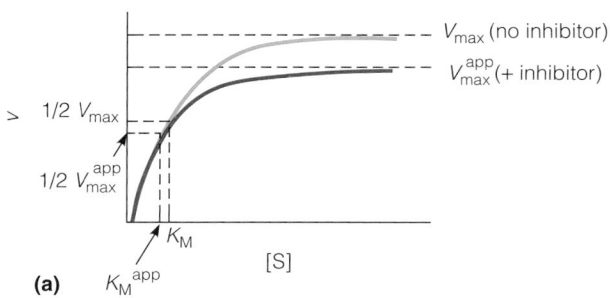

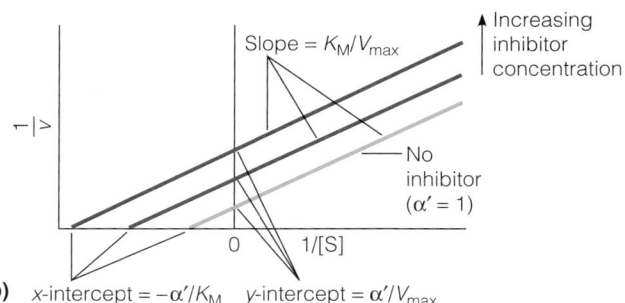

than intersecting lines (because the factor drops out of the ratio to give a slope $= K_M/V_{max}$ for all values of α').

This behavior is found in the inhibition of acetylcholinesterase (see Chapter 23) by tertiary amines (R_3N). Such compounds bind to the enzyme in its various forms, but the acyl–intermediate–amine complex cannot break down into enzyme plus product.

Mixed Inhibition

This form of inhibition occurs when a molecule or an ion can bind to both the free enzyme and the ES complex (see Figure 11.34):

$$E + S \underset{k_{-1}}{\overset{k_1}{\rightleftarrows}} ES \xrightarrow{k_{cat}} E + P$$

$$EI + S \rightleftharpoons EIS \xrightarrow{\hspace{0.3cm}\times\hspace{0.3cm}} No\ reaction$$

Again, we distinguish the equilibrium constants K_I and K'_I, respectively, for inhibitor binding to E and ES. In the case where $K_I = K'_I$, the mixed mode of inhibition is also called **noncompetitive inhibition**. Substrate binding to the EI complex (green arrows in the scheme above) is typically significantly reduced compared to that for binding to free enzyme. Thus, the process EI $+$ S $\rightleftarrows$ EIS will not be considered here, although it is part of a complete thermodynamic analysis; in doing so, the derivation of the Michaelis–Menten equation for mixed inhibition is greatly simplified without significantly altering the conclusions of the analysis. For this simplified case:

$$v = \frac{V_{max}^{app}[S]}{K_M^{app} + [S]} = \frac{V_{max}[S]}{\alpha K_M + \alpha'[S]} \tag{11.38}$$

where α and α' are defined as above, $V_{max}^{app} = V_{max}/\alpha'$, and $K_M^{app} = \alpha K_M/\alpha'$ (see Problem 28b for the derivation of this equation). In most cases, the inhibitor has a greater affinity for the free enzyme than for the ES complex; thus, α is typically greater than α'.

This mode of inhibition is "mixed" because the denominator of Equation 11.38 contains terms found in the equations for competitive inhibition (αK_M) and uncompetitive inhibition ($\alpha'[S]$). As for competitive inhibition, K_M appears to be *increased* (by a factor of α/α'), and as for uncompetitive inhibition, V_{max} appears to be *reduced* (by a factor of $1/\alpha'$). Figure 11.35 shows that the Lineweaver–Burk plot for mixed inhibition reflects this decreased V_{max} and increased K_M, and is distinct from similar plots for competitive and uncompetitive modes of inhibition.

Finally, mixed inhibitors effectively reduce v at low and high values of [S]. At [S] $>$ K_M, v approaches $V_{max}[S]/\alpha K_M$, and at [S] $>$ K_M, v approaches V_{max}/α'. Thus, V_{max}^{app} will be less than V_{max} for all values of [S].

FIGURE 11.33

Effects of uncompetitive inhibition on enzyme kinetics. **(a)** The effect of an uncompetitive inhibitor (I) on reaction velocity at different substrate concentrations. In this simple example, both K_M and V_{max} are decreased by a factor of $1/\alpha'$. **(b)** Lineweaver–Burk plots of the reactions shown in (a). The lines are parallel and cross the $1/v$ axis at different points, clearly distinguishing this situation from competitive inhibition (see Figure 11.30b).

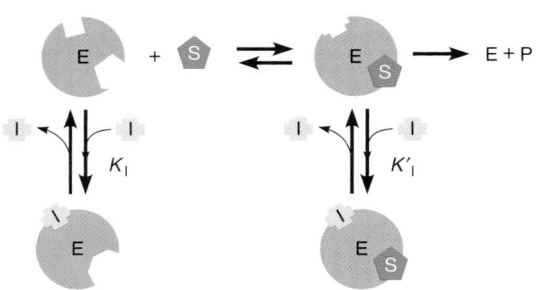

FIGURE 11.34

A model for mixed inhibition. The inhibitor binds at a site on the enzyme surface different from that of the substrate. In this simplified example, the inhibitor binds to both free enzyme and the ES complex. EI has reduced substrate binding affinity compared to free enzyme. The EIS complex cannot carry out the catalytic event. As noted in the text, the process EI + S $\rightleftarrows$ EIS is not considered here.

A mixed inhibitor does not compete for the active site but affects the catalytic event. It reduces the apparent V_{max} at all [S] and increases the apparent K_M.

FIGURE 11.35

Lineweaver–Burk plot for mixed inhibition kinetics. V_{max} is decreased by a factor of $1/\alpha'$ and K_M is increased by a factor of α/α'. Compare this plot with those in Figures 11.30b and 11.33b to find the features that distinguish competitive, uncompetitive, and mixed modes of inhibition.

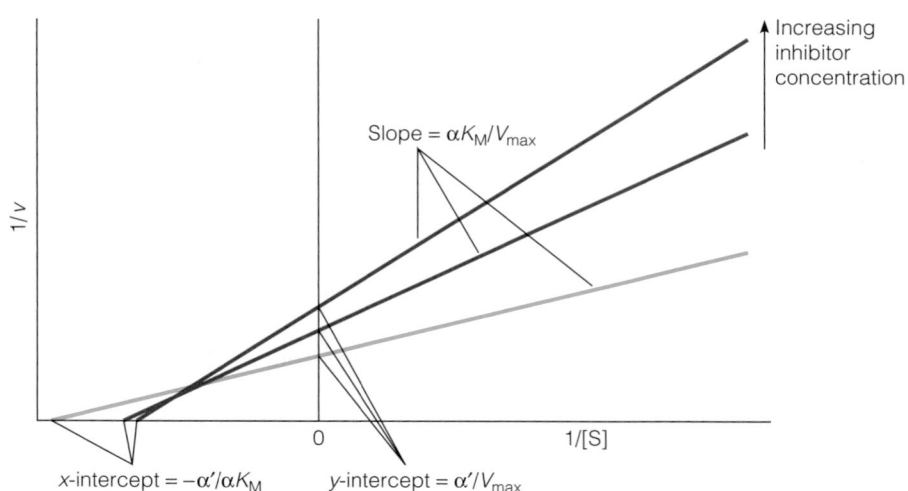

Slope = $\alpha K_M/V_{max}$

Increasing inhibitor concentration

x-intercept = $-\alpha'/\alpha K_M$ y-intercept = α'/V_{max}

Many irreversible inhibitors bind covalently to the active sites of enzymes.

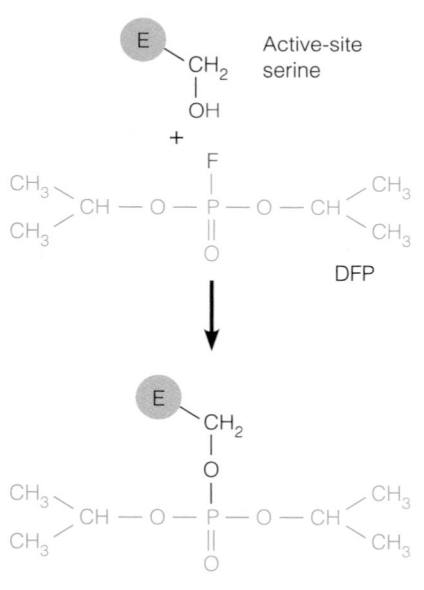

Active-site serine

DFP

+ HF

FIGURE 11.36

Irreversible inhibition by adduct formation. Diisopropyl fluorophosphate (DFP) reacts with a serine group on a protein to form a covalent adduct. The covalent bond renders the catalytically important serine ineffective in catalysis. The adduct also may block substrate binding to the active site.

Irreversible Inhibition

Some substances combine *covalently* with enzymes to inactivate them irreversibly. Almost all **irreversible enzyme inhibitors** are toxic substances, either natural or synthetic. Table 11.7 lists a number of them. In most cases, such substances react with some functional group in the active site to leave it catalytically inactive or to block substrate binding.

A typical example of an irreversible competitive inhibitor is found in *diisopropyl fluorophosphate (DFP)*. This compound reacts rapidly and irreversibly with serine hydroxyl groups to form a covalent *adduct*, as shown in Figure 11.36. Therefore, DFP acts as an irreversible inhibitor of enzymes that contain an essential serine in their active site. These enzymes include, among others, the serine proteases and the enzyme *acetylcholinesterase*. It is the inhibition of acetylcholinesterase that makes DFP such an exceedingly toxic substance to animals. Acetylcholinesterase is essential for nerve conduction (see Chapter 23), and its inhibition causes rapid paralysis of vital functions. Many insecticides and nerve gases resemble DFP and are potent acetylcholinesterase inhibitors (Table 11.7).

For such irreversible inhibitors to react selectively with a critical residue, they must bind strongly to the active site. Many do so because they are transition state analogs. Examples include DFP and the nerve gas *sarin*, which have a tetrahedral structure surrounding the phosphorus atom that is similar to the tetrahedral oxyanion transition state of the substrate in many hydrolytic enzymes.

Some selective irreversible inhibitors mimic the substrate of the target enzyme. An example is *N-tosyl-L-phenylalaninechloromethyl ketone (TPCK)* (see Table 11.7). TPCK is an excellent inhibitor for chymotrypsin because the phenyl group fits nicely into the active site pocket, positioning the chlorine for nucleophilic displacement by a nitrogen of the imidazole ring of His 57. The resulting covalent adduct blocks access to the enzyme active site—effectively rendering the enzyme completely inactive. A large number of such specific irreversible inhibitors have been synthesized to aid in the analysis of enzyme mechanisms and to control enzyme activity. For example, a biochemist who is using chymotrypsin to hydrolyze a protein can stop the reaction instantly at any point by simply adding TPCK. Another use for such substances is to label active site residues of an enzyme specifically to aid in their identification. When irreversible inhibitors are used in this way, they are often called **affinity labels**. In some cases an affinity label is unreactive until it is acted on by the enzyme, at which point it binds irreversibly. Such substances are called **suicide inhibitors** because the enzyme "kills" itself by processing the inhibitor from a benign to a reactive form.

Many natural toxins are irreversible enzyme inhibitors. The alkaloid *physostigmine* (see Table 11.7), which is contained in calabar beans, is toxic

TABLE 11.7　Irreversible enzyme inhibitors

Name	Formula[a]	Source	Mode of Action
Cyanide	CN^-	Bitter almonds	Reacts with enzyme metal ions (i.e., Fe, Zn, Cu); respiratory chain enzymes are primary targets (see Chapter 15)
Diisopropyl fluorophosphate (DFP, or DIFP)		Synthetic	Inhibits enzymes with active site serine, including acetylcholinesterase and serine proteases
Sarin		Synthetic (nerve gas)	Like DFP
Physostigmine		Calabar beans	Like DFP
Parathion		Synthetic (insecticide)	Like DFP, but especially inhibitory to insect acetylcholinesterase
N-Tosyl-L-phenyl-alaninechloro-methyl ketone (TPCK)		Synthetic	Reacts with His 57 of chymotrypsin
Penicillin		From *Penicillium* fungus	Inhibits enzymes in bacterial cell wall synthesis (see Chapter 9)

[a]R = variable group; differs on different penicillins.

because it is a potent inhibitor of acetylcholinesterase. The *penicillin* antibiotics also act as irreversible inhibitors of serine-containing enzymes used in bacterial cell wall synthesis (see Chapter 9).

Cofactors, Vitamins, and Essential Metals

The complexity of globular protein structure and the variety of side chain structures in a protein allow the formation of many kinds of catalytic sites. This variability allows enzymes to act as efficient catalysts for many reactions. However, for some kinds of biological processes, the chemical potential of amino acid side chains alone is not sufficient. A protein may require the help of some other small molecule or ion to carry out the reaction. The ions or molecules that are bound to enzymes for this

Many enzymes use ions or small bound molecules called cofactors or coenzymes to aid in catalysis.

purpose are called **cofactors** or **coenzymes**. Like enzymes, cofactors are not irreversibly changed during catalysis; they are either unmodified or regenerated, as discussed above in the case of dihydrofolate reductase, which uses the NADPH cofactor.

Cofactors and What They Do

Many essential vitamins are constituents of enzyme cofactors.

Cofactors often have complex organic structures that cannot be synthesized by some organisms—mammals in particular. The water-soluble vitamins, those usually referred to as the vitamin-B complex, are metabolic precursors of several cofactors. This is why such vitamins are so important in metabolism. Our approach in this book is to introduce the detailed biochemistry of each cofactor as we first encounter that cofactor in discussions of metabolic pathways. Table 11.8 lists a number of important enzyme cofactors, together with their related vitamins, the kinds of reactions they are associated with, and where in this textbook you will find detailed descriptions and discussion. At this point, just to give a more concrete idea of how cofactors function, we will describe one class in some detail. These are the nicotinamide nucleotides, a major example being **nicotinamide adenine dinucleotide** (**NAD$^+$**) derived from the vitamin **niacin**.

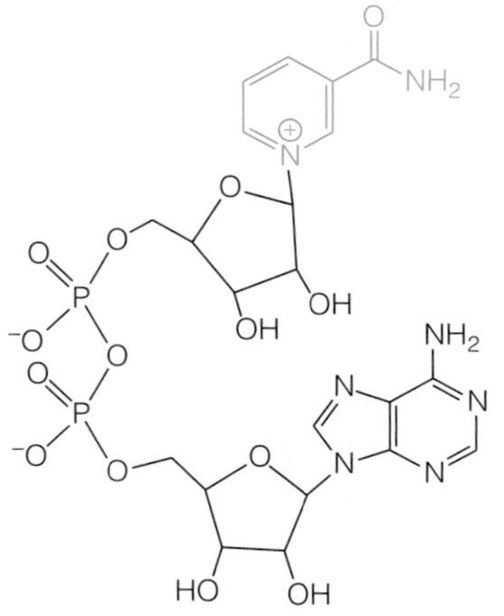

NAD$^+$

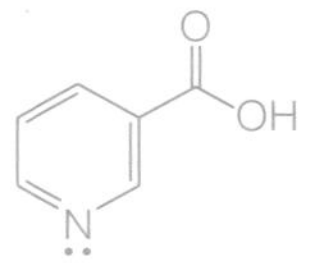

Niacin

TABLE 11.8   Some important enzyme cofactors and related vitamins

Vitamin	Cofactor	Reactions Involving These Cofactors	Page Where Cofactor Is Introduced
Thiamine (vitamin B$_1$)	Thiamine pyrophosphate	Activation and transfer of aldehydes	536–538
Riboflavin (vitamin B$_2$)	Flavin mononucleotide; flavin adenine dinucleotide	Oxidation–reduction	599
Niacin	Nicotinamide adenine dinucleotide; nicotinamide adenine dinucleotide phosphate	Oxidation–reduction	446, 485, 530
Pantothenic acid	Coenzyme A	Acyl group activation and transfer	601
Pyridoxine	Pyridoxal phosphate	Various reactions involving amino acid activation	845
Biotin	Biotin	CO$_2$ activation and transfer	617–619
Lipoic acid	Lipoamide	Acyl group activation; oxidation–reduction	598–599
Folic acid	Tetrahydrofolate	Activation and transfer of single-carbon functional groups	848
Vitamin B$_{12}$	Adenosyl cobalamin; methyl cobalamin	Isomerizations and methyl group transfers	853

The nicotinamide portion is the metabolically relevant part of NAD^+, because it is capable of being reduced and thus can serve as an *oxidizing* agent to which two electrons and a proton are added to the nicotinamide ring: $NAD^+ + 2e^- + H^+ \rightarrow NADH$. The reaction is reversible (i.e., NADH acts as a *reducing* agent in several reactions) and is formally a **hydride ion** transfer: $NAD^+ + H^- \rightleftharpoons NADH$.

Here, R stands for the remainder of the molecule.

A typical reaction in which NAD^+ acts as an oxidizing agent is the conversion of alcohols to aldehydes or ketones (for example, by the *alcohol dehydrogenase* of liver):

It is the C-linked H, not the O-linked H, that is transferred to NAD^+, as can be demonstrated by studies using deuterated compounds. Furthermore, these reactions are stereospecific. Even when the hydroxyl carbon has *two* hydrogens attached (as with ethanol), a particular one of the hydrogens is transferred to NAD^+. This specificity may seem surprising because the hydroxyl carbon of ethanol is not a chiral center. How can a particular hydrogen be favored when the substrate molecule has a plane of symmetry? The answer lies in the asymmetric nature of the enzyme surface, to which both NAD^+ and the alcohol are bound. If a symmetrical molecule like ethanol is bound by at least *three* points to an asymmetric object, the two H atoms are no longer equivalent; they are said to be *prochiral* (Figure 11.37). Furthermore, although the nicotinamide ring is planar, the transfer of hydrogen in a particular reaction is always to a specific face of the ring, as the two faces are *not* equivalent in the asymmetric active site of an enzyme. Such considerations lie behind the high stereospecificity of many enzyme-catalyzed reactions, in contrast to nonenzymatic catalysis.

Sometimes it is difficult to make a clear distinction between a true cofactor and a second substrate in a reaction. The reactions we have just discussed are good examples of this problem. The dehydrogenase enzymes, such as alcohol dehydrogenase, each have a strong binding site for the oxidized form of the cofactor, NAD^+. After oxidation of the substrate, the reduced form, NADH, leaves the enzyme and is reoxidized by other electron-acceptor systems in the cells. The NAD^+ so formed can now bind to another enzyme molecule and repeat the cycle. In such cases, NAD^+ is acting more like a second substrate than a true cofactor. Yet NAD^+ and NADH differ from most substrates in that they are continually recycled in the cell and are used over and over again. Because of this behavior, we consider them cofactors.

An example of NAD^+ behaving unambiguously as a cofactor is found in the reaction involving *UDP-galactose 4-epimerase* shown in Figure 11.38. This enzyme

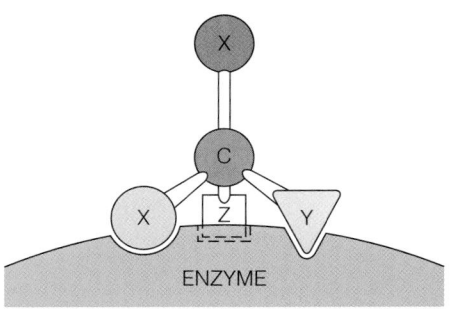

FIGURE 11.37

Stereospecificity conferred by an enzyme. This figure shows how the asymmetric surface of an enzyme can confer stereospecificity in the reaction of a symmetric substrate. If the substrate molecule X_2CYZ makes at least three contacts with unique complementary groups on the enzyme, its two X atoms are no longer equivalent. Only a specific one of the two X atoms can contact the surface properly. In many cases, a minimum of *four* contacts is required to distinguish prochiral functional groups. This is discussed in greater detail in Figure 14.11 (page 605; see also page 607).

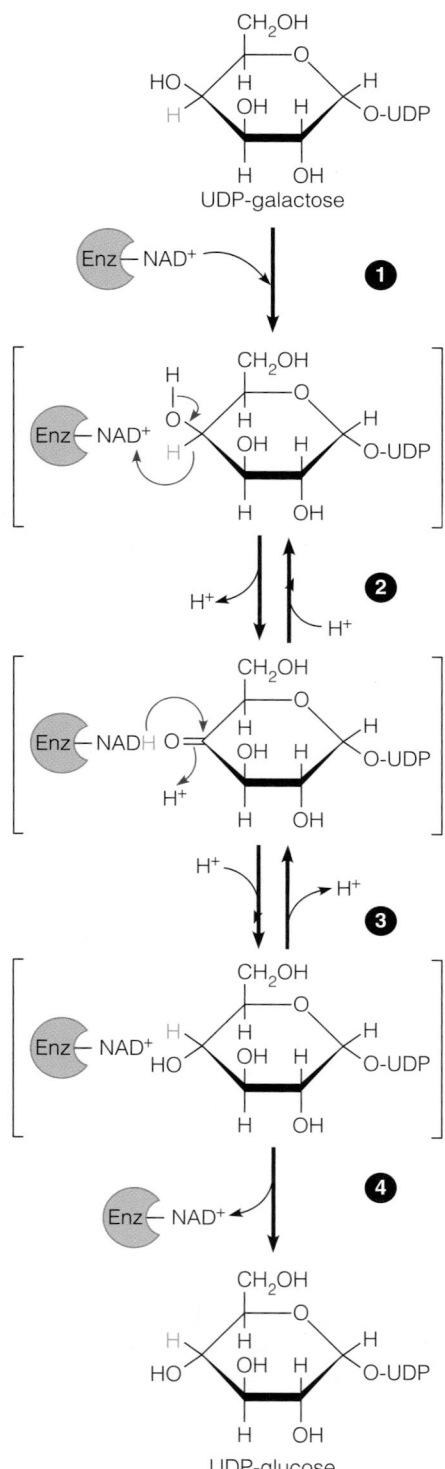

FIGURE 11.38

Proposed mechanism for UDP-galactose epimerase. UDP-galactose is bound to the enzyme, which carries the coenzyme NAD⁺ (step 1). Hydride is transferred to NAD⁺ from C4 of the galactose ring to produce the carbonyl intermediate (step 2) and then transferred back to C4 to give the opposite stereochemistry (step 3). The product, UDP-glucose, is then released (step 4).

facilitates synthesis of complex polysaccharides by interconverting UDP-glucose and UDP-galactose (see Chapters 9 and 13). The mechanism by which the stereochemistry of the hydroxyl at position 4 is changed involves oxidation of the hydroxyl to a carbonyl as an intermediate state. In this case, NAD^+ is reduced to NADH (step 2), which is, in turn, reoxidized (step 3) to regenerate the NAD^+ cofactor. This reaction provides a good example of what cofactors do and why they are necessary. The carbonyl intermediate provides an excellent intermediate state for interconversion of the sugars, but none of the normal amino acid side chains of a protein is well suited to promote this kind of redox reaction. By binding NAD^+, the enzyme can carry out this function.

Metal Ions in Enzymes

Many enzymes contain one or more metal ions, usually held by coordinate covalent bonds from amino acid side chains, but sometimes bound by a prosthetic group like heme. Such enzymes are called **metalloenzymes**. The bound ion acts in much the same way as a coenzyme, conferring on the metalloenzyme a property it would not possess in its absence. As Table 11.9 shows, the roles these ions play are diverse. For example, the zinc ion in *carboxypeptidase A* binds the water molecule that attacks the carbonyl of the scissile bond and also acts as an electrostatic catalyst (Figure 11.39). The zinc ion stabilizes the tetrahedral oxyanion in the transition state and the intermediate state, in much the same way as the oxyanion hole serves this function in chymotrypsin (see Figure 11.19 for comparison).

In other cases, the metal in a metalloenzyme serves as a redox reagent. We have mentioned the example of the heme-iron–containing enzyme *catalase*, which catalyzes the breakdown of hydrogen peroxide, a potentially destructive agent in cells. Because the reaction involves both reduction and oxidation of H_2O_2, the Fe^{2+} is reversibly oxidized and reduced, acting as an electron exchanger. As noted previously, catalase is a very efficient enzyme; its k_{cat}/K_M value of $4 \times 10^7 \, M^{-1} \, s^{-1}$ approaches the theoretical diffusion limit. Hemoglobin, which also contains Fe^{2+}, has weaker catalase activity because oxidation of Fe^{2+} to Fe^{3+} is largely irreversible in hemoglobin. Such redox activity requires metals like Fe or Cu with multiple oxidation states.

In many other enzymatic reactions, certain ions are necessary for catalytic efficiency, even though they may not remain permanently attached to the protein nor play a direct role in the catalytic process. For example, a number of enzymes that couple ATP hydrolysis to other processes require Mg^{2+} for efficient function. In most cases Mg^{2+} is necessary because the Mg–ATP complex (see Chapter 3) is a better substrate than ATP itself.

TABLE 11.9	Metals and trace elements important as enzymatic cofactors	
Metal	Example of Enzyme	Role of Metal
Fe	Cytochrome oxidase	Oxidation–reduction
Cu	Ascorbic acid oxidase	Oxidation–reduction
Zn	Alcohol dehydrogenase	Helps bind NAD^+
Mn	Histidine ammonia lyase	Aids in catalysis by electron withdrawal
Co	Glutamate mutase	Co is part of cobalamin coenzyme
Ni	Urease	Catalytic site
Mo	Xanthine oxidase	Oxidation–reduction
V	Nitrate reductase	Oxidation–reduction
Se	Glutathione peroxidase	Replaces S in one cysteine in active site
Mg^{2+}	Many kinases	Helps bind ATP

The Diversity of Enzymatic Function

Classification of Protein Enzymes

By this point, it should be clear that an enormous number of different proteins act as enzymes. Many of these enzymes were given common names, especially during the earlier years of enzymology. Some enzyme names, like *triose phosphate isomerase*, are descriptive of the enzyme's function; others, like *trypsin*, are not. To reduce confusion, a rational naming and numbering system has been devised by the Enzyme Commission of the International Union of Biochemistry and Molecular Biology (IUBMB). Enzymes are divided into six major classes, with subgroups and sub-subgroups to define their functions more precisely. The major classes are as follows:

1. *Oxidoreductases* catalyze oxidation–reduction reactions.
2. *Transferases* catalyze transfer of functional groups from one molecule to another.
3. *Hydrolases* catalyze hydrolytic cleavage.
4. *Lyases* catalyze removal of a group from or addition of a group to a double bond, or other cleavages involving electron rearrangement.
5. *Isomerases* catalyze intramolecular rearrangement.
6. *Ligases* catalyze reactions in which two molecules are joined.

The IUBMB Enzyme Commission (EC) has given each enzyme a number with four parts, such as: EC 3.4.21.5. The first three numbers define major class, subclass, and sub-subclass, respectively. The last is a serial number in the sub-subclass, indicating the order in which each enzyme is added to the list, which is continually growing. Listings for almost all currently recognized enzymes, together with information on each enzyme and literature references, can be found in online databases such as BRENDA (BRaunschweig ENzyme DAtabase; http://www.brenda-enzymes.info) or ExPASy (Expert Protein Analysis System; http://expasy.org/enzyme). The 5000+ entries in these databases do not include all enzymes; more are being discovered all the time. Indeed, it has been estimated that the typical cell contains many thousands of different kinds of enzymes. Table 11.10 lists one example enzyme and reaction from each of the major classes. We will discuss each of these reactions later in this book. The main point here is the enormous diversity of enzymatic functions and how their nomenclature has been rationalized.

Molecular Engineering of New and Modified Enzymes

Despite the variety of enzymatic functions available in nature, modern biotechnology continually faces needs for substances with new catalytic abilities, or enzymes that function with different specificities or under unusual conditions. These needs have generated a field of enzyme design and engineering, which has enormous potential in the design of industrial catalysts as well as biopharmaceuticals. Several approaches are being taken to achieve the goal of generating such tailor-made catalysts. These include site-directed mutagenesis, fusion of one or more functional domains, selection of a sequence with the desired activity from a large pool of randomly generated protein sequences, generation of "catalytic antibodies," and computational design. Tools of Biochemistry 11B provides a brief introduction to these techniques.

Nonprotein Biocatalysts: Catalytic Nucleic Acids

Throughout this chapter, we have described how the proteins called enzymes function as biocatalysts. Indeed, for many years, it was assumed that *all* biochemical catalysis was carried out by proteins. But biochemistry is full of surprises, and

Some enzymes require metal ions for their catalytic function.

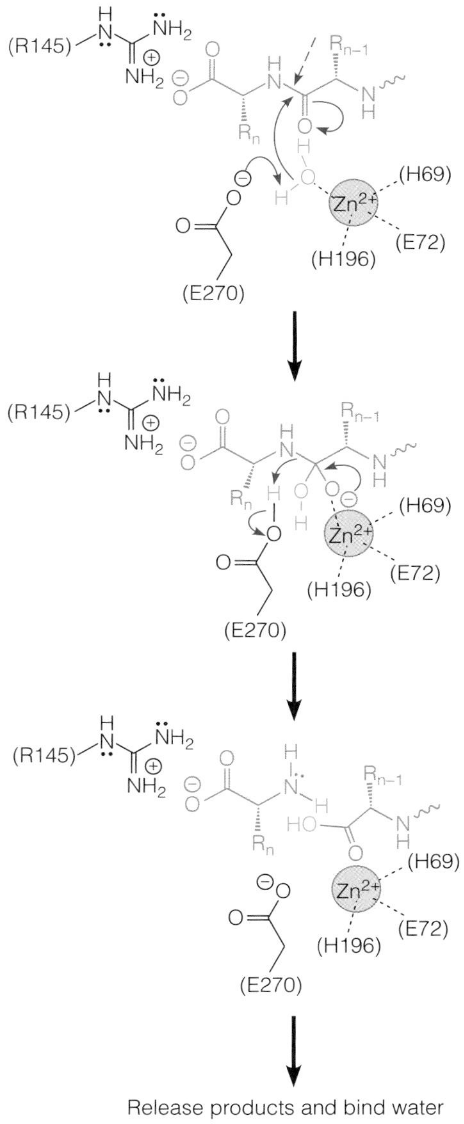

FIGURE 11.39

The mechanism of the protease carboxypeptidase A. The zinc ion (orange circle) binds a water molecule (blue) and serves as an electrostatic catalyst to promote hydrolysis of the C-terminal amino acid from a peptide substrate (green). It does so by stabilizing the negative charge on the oxygen in the tetrahedral transition state. Enzyme active site residues are indicated by black coloring. The bond cleaved is indicated by the dashed red arrow.

New or radically modified enzymes can be created by "protein engineering," which includes a number of techniques such as site-directed mutagenesis, protein domain fusion, selection from randomly generated libraries, and computational design.

TABLE 11.10   Examples of each of the major classes of enzymes

Class	Example (reaction type)	Reaction Catalyzed
1. Oxidoreductases	Alcohol dehydrogenase (EC 1.1.1.1) (oxidation with NAD^+)	
2. Transferases	Hexokinase (EC 2.7.1.2) (phosphorylation)	
3. Hydrolases	Carboxypeptidase A (EC 3.4.17.1) (peptide bond cleavage)	
4. Lyases	Pyruvate decarboxylase (EC 4.1.1.1) (decarboxylation)	
5. Isomerases	Maleate isomerase (EC 5.2.1.1) (cis–trans isomerization)	
6. Ligases	Pyruvate carboxylase (EC 6.4.1.1) (carboxylation)	

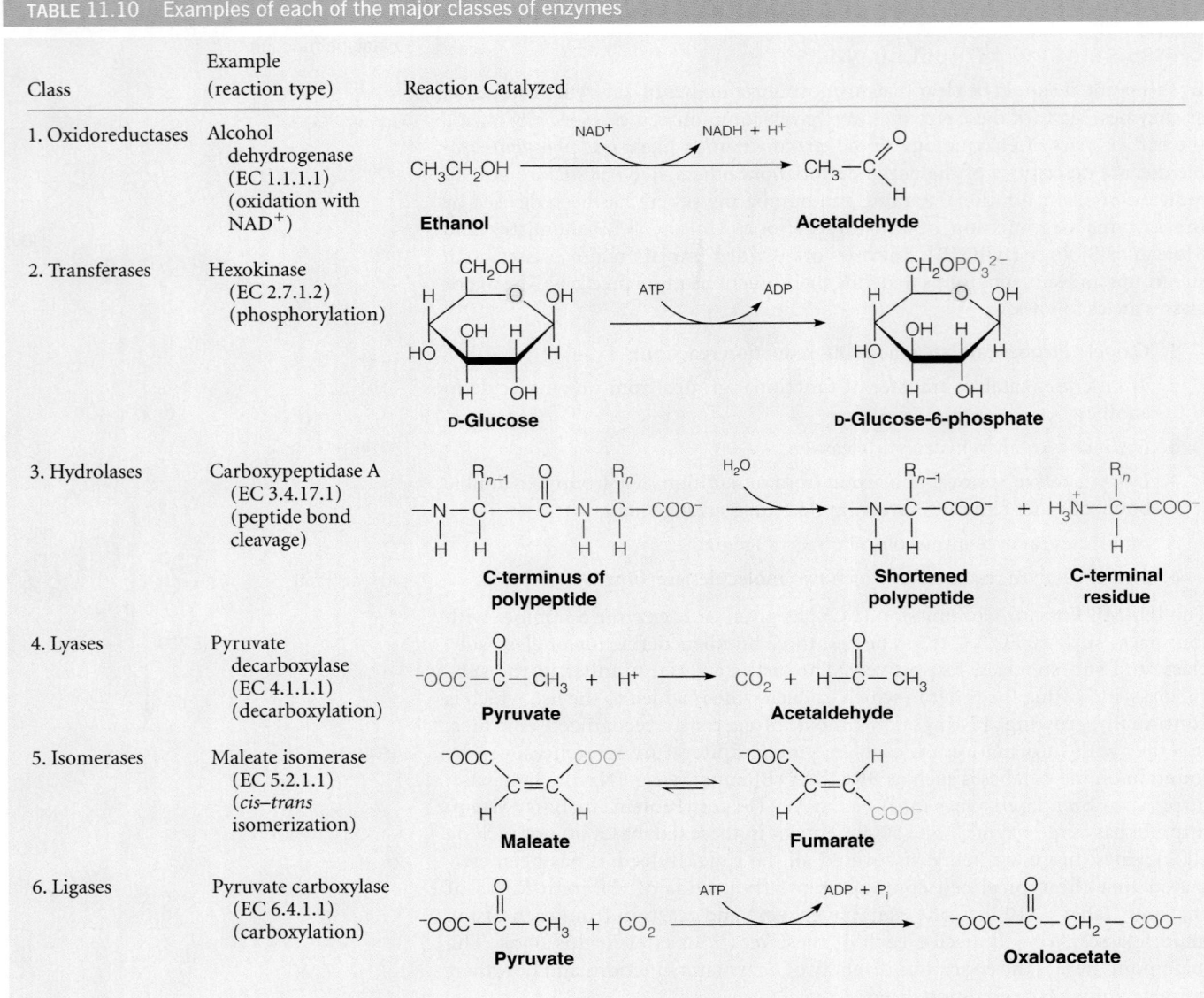

Ribozymes, a class of ribonucleic acids, function as biological catalysts.

research performed in the 1980s revealed something wholly unexpected: Some RNA molecules, called **ribozymes**, can act as enzymes.

The first hint that RNA might have catalytic activity came from studies of **ribonuclease P**, an enzyme that cleaves the precursors of tRNAs to yield the functional tRNAs (Figure 11.40 and Chapter 27). It had been known for some time that active ribonuclease P contained both a protein portion and an RNA "cofactor," but it was widely assumed that the active site resided on the protein portion. However, careful studies of the isolated components by Sidney Altman and coworkers in 1983 revealed an astonishing fact: Whereas the protein component alone was wholly inactive, the RNA by itself, if provided with either a sufficiently high concentration of magnesium ion or a small amount of magnesium ion plus the small basic molecule spermine, was capable of catalyzing the specific cleavage of pre-tRNAs. Furthermore, the RNA acted like a true enzyme, being unchanged in the process and obeying Michaelis–Menten kinetics. Addition of the protein portion of ribonuclease P does enhance the activity (k_{cat} is markedly increased) but is in no way essential for either substrate binding or cleavage. At high salt concentrations the RNA itself becomes a very efficient catalyst; K_M becomes very low, and k_{cat}/K_M approaches 10^7 $(mol/L)^{-1}$ s^{-1}.

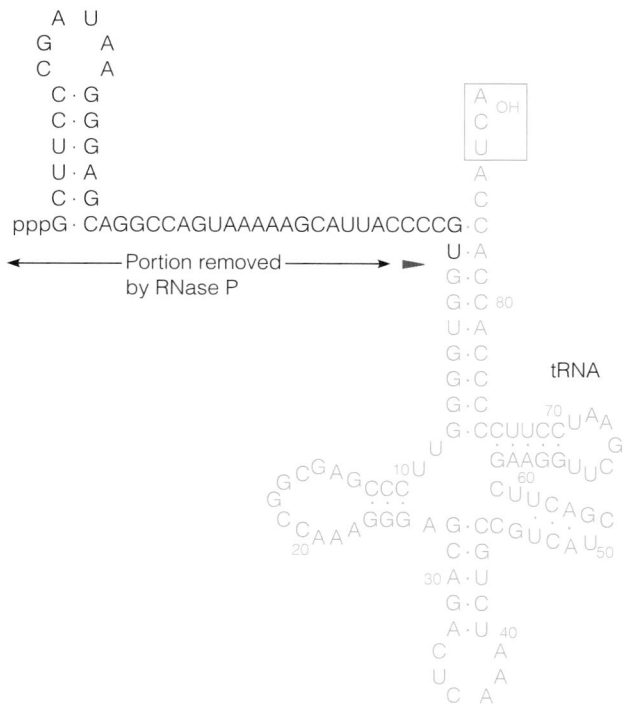

FIGURE 11.40

Cleavage of a typical pre-tRNA by ribonuclease P. The production of tRNA from pre-tRNA is catalyzed by an RNA–protein complex called ribonuclease P. The portion removed from tRNA is shown in black, and the resulting tRNA is in blue. The RNA portion of ribonuclease P can by itself catalyze the hydrolysis of the specific phosphodiester bond indicated by the megenta wedge. The 3′ terminal —OH group is shown as a subscript to the 3′-terminal adenosine.

At about the same time, another remarkable class of RNA-catalyzed reactions was discovered by Thomas Cech and his colleagues. Examining the removal of an intron (intervening sequence, or IVS) from the preribosomal RNAs of the protist *Tetrahymena*, they found that the rRNA *itself* carried out the excision of its 413-nucleotide intron and the necessary resplicing as well (Figure 11.41a). In addition, the excised IVS went through a further series of site-specific reactions. The final product is a molecule called L-19 IVS—the intervening sequence with 19 more nucleotides removed. This activity is not regarded as true catalysis because the "catalyst" itself is modified in the reaction. However, L-19 IVS does have true catalytic abilities: It is capable of either lengthening or shortening small oligonucleotides, in the manner shown in Figure 11.41b. This example is by no means unique; we discuss other RNA-catalyzed reactions in Chapters 27 and 28.

At that time many biochemists were astounded by these discoveries, although in retrospect we can see that there is no reason why RNA molecules should not have catalytic functions. As discussed in Chapter 4, RNA molecules can adopt complex tertiary structures, just as proteins do, and such complex structure appears to be essential for enzymatic activity.

As mentioned in Chapter 4, the fact that RNA molecules possess the potential for both self-replication and catalysis has led some scientists to suggest that these molecules may well have been the primordial substances in the evolution of life. Such theorists envision an "RNA world" before proteins and DNA had evolved, where only self-replicating RNA molecules existed, capable of catalyzing a simple metabolism. Support for the idea that ribozymes could be self-replicating comes from remarkable experiments. Wright and Joyce (see References) have shown that the ribozyme depicted in Figure 11.42 is capable of self-catalyzed replication (with a little help from some protein enzymes). The RNA is capable of catalyzing the ligation of another RNA segment (in this case carrying a bacteriophage T7 promoter) to itself. In the presence of the enzymes reverse transcriptase and RNA polymerase (see, respectively, Chapters 25 and 27), a new copy of the RNA is generated. If this is supplied with an excess of the T7 promoter fragment, ligation occurs and a new copy of the ribozyme is formed. In one such experiment, the

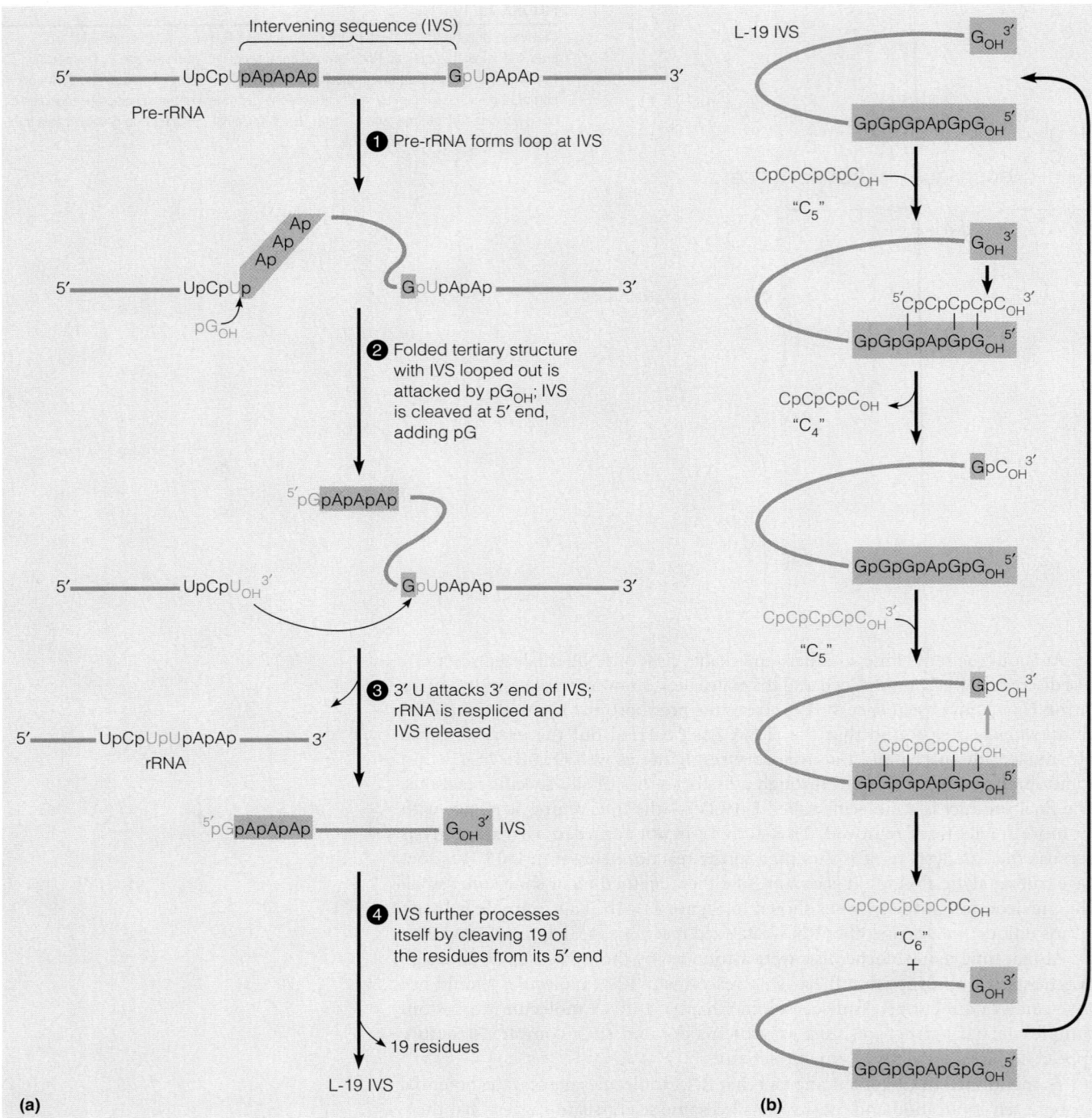

FIGURE 11.41

Catalysis by the intervening sequence in *Tetrahymena* preribosomal RNA. **(a)** Self-excision and splicing of the intervening sequence (IVS). Note that a pG_{OH} is added in the reaction. A series of further steps reduces the IVS to L-19 IVS. **(b)** Conversion of $2\,C_5$ to $C_4 + C_6$ by L-19 IVS. This oligonucleotide can itself either shorten or elongate small oligonucleotides, acting here as a true ribozyme catalyst.

system performed over 1000 doublings in a period of 2 days. During this time, the sequence evolved to a more efficient ribozyme. Such studies provide support for the "RNA-world" model for life's origin.

Like RNA, single-stranded DNA can form complex tertiary structures, which are required for specific binding to a ligand/substrate and catalytic activity. Given the chemical similarities between DNA and RNA, it is reasonable to ask whether DNA can carry out meaningful biological catalysis. So far, no *naturally occuring* catalytic DNA, or **DNAzyme**, has been discovered in cells. However, many DNAzymes have been developed in the laboratory since the first catalytic DNA molecule was

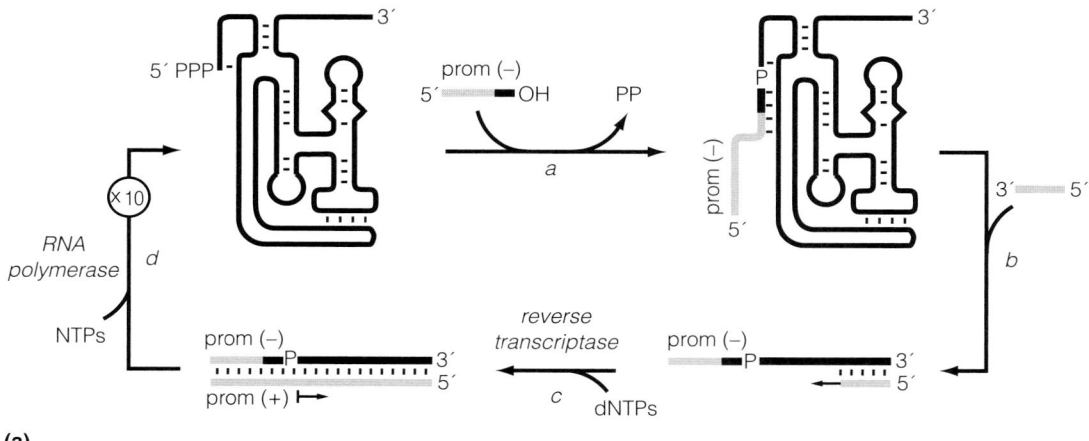

(a)

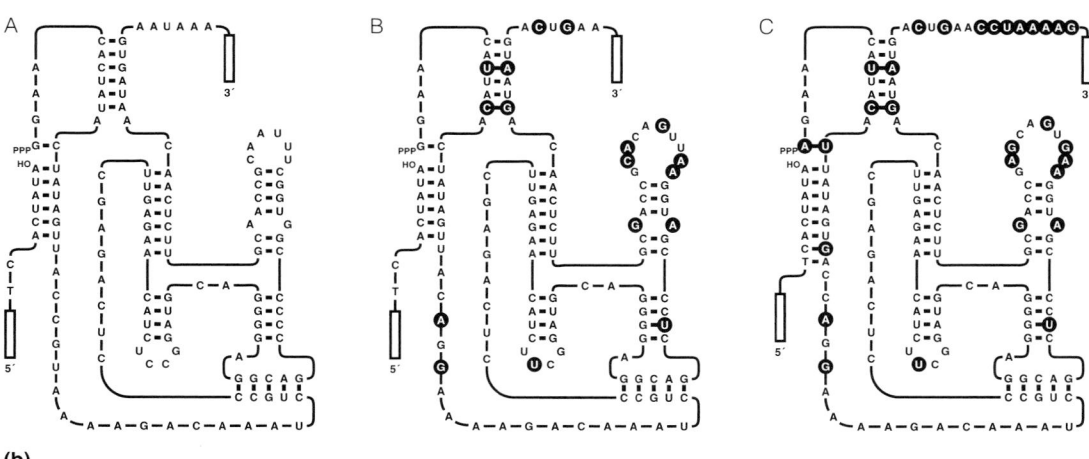

(b)

described in 1994 by Ronald Breaker and Gerald Joyce. The DNAzymes characterized to date have been obtained by selection of active sequences from large pools of randomized oligonucleotides and in vitro evolution (Figure 11.43).

Like ribozymes, DNAzymes have been shown to catalyze a diverse set of reactions, with significant rate enhancements (Table 11.11). For example, the types of reactions catalyzed by DNAzymes include hydrolysis (RNA and DNA cleavage), C—C bond cleavage, and photolytic repair of damaged DNA. Because DNA has greater chemical stability than do RNA and peptides, there is much current interest in developing DNAzymes as therapeutics, diagnostics, and biosensors.

The Regulation of Enzyme Activity: Allosteric Enzymes

So far we have described the basic features of enzyme function. We now turn our attention to an equally important feature of enzymes: the regulation of their activity. In Chapter 1 we made an analogy between a living cell and a factory. That analogy is especially appropriate when we consider the roles that enzymes play in living cells. We note that a cell has certain raw materials available to it and must produce specific products from them. Enzymes make up the majority of the machines that facilitate these transformations in the cell. Often, as we shall see, enzymes are arranged in "assembly lines" to carry out the necessary sequential steps in a metabolic pathway.

FIGURE 11.42

Self-replication and mutation of a ribozyme.
(a) The ribozyme replication cycle, which involves repeated replication of the ribozyme, when supplied with excess T7 promoter sequence and ribonucleoside triphosphates. Reverse transcriptase and RNA polymerase are also needed. **(b)** Mutations that arose during ribozyme evolution. Open rectangles indicate the 5′ portion of the substrate and the primer binding site at the 3′ end of the ribozyme (5-CCAAUCGCAGGCUCAGC-3), both of which are immutable during evolution. Highlighted residues are those that were mutated relative to the starting ribozyme. **A** Prototype ribozyme used to construct the initial pool. **B** An individual ribozyme isolated before beginning continuous evolution. **C** An individual ribozyme isolated after 52 hours of continuous evolution.

From *Science* 276:614–616, M. C. Wright and G. F. Joyce, Continuous in vitro evolution of catalytic function. © 1997. Reprinted with permission from AAAS.

Selection/*In vitro* evolution

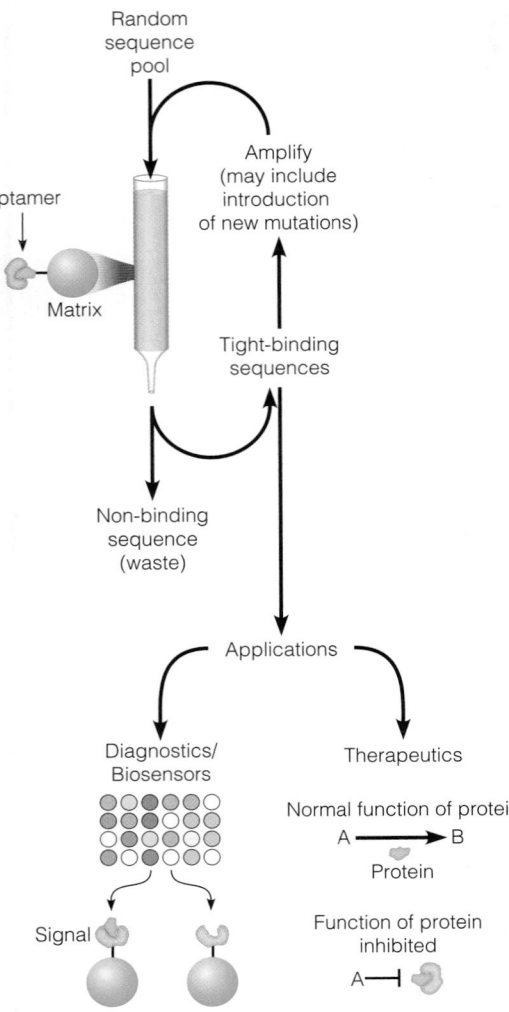

FIGURE 11.43

Affinity selection from a random sequence library.
A large library of random sequences is applied to a chromatography matrix that displays the target ligand. Those nucleic acid sequences, or "aptamers," that bind are selectively eluted from the column and amplified. The amplification may be performed in a manner that introduces greater variation in the selected sequences (i.e., in vitro evolution). The amplified sequences are re-applied to the column. This process is repeated until sequences with high affinity for the target are obtained. The clinical applications of the selected sequences include diagnostics and therapeutics.

Regulation of enzyme activity is essential for the efficient and ordered flow of metabolism.

TABLE 11.11   Examples of reaction types and rate enhancements for DNAzymes

Reaction Type	k_{cat} (min^{-1})	Rate Enhancement
Various RNA transesterifications	0.007–4.3	10^5–10^8
DNA cleavage	0.05–0.2	10^7–10^8
Porphyrin metallation	1.3	10^3
DNA ligation	0.0001–0.07	10^2–10^5
Adenylylation	0.005	10^{10}
N-Glycosyl cleavage	0.2	10^6
Phosphorylation	0.012	10^9

From *Cellular and Molecular Life Sciences* 59:596–607, G. M. Emilsson and R. R. Breaker, Deoxyribozymes: New activities and new applications, © 2002, with kind permission from Springer Science+Business Media B.V.

No factory operates efficiently if every machine is operating at its maximum rate. The capabilities of machines vary greatly, and if all of them were running at top speed, massive problems would soon arise. Intermediate products would pile up in some assembly lines, and certain parts of the finished product would be produced in vast excess. Different assembly lines might draw on the same raw material, and the faster ones could deplete the supplies so completely that other, equally important, lines would have to shut down. Obviously, *coordination* and *regulation* are required to run a large factory efficiently.

The same kinds of problems could occur if the enzymatic machinery of the cell were not regulated precisely. The efficiencies with which individual enzymes operate must be controlled in a manner that reflects the availability of substrates, the utilization of products, and the overall needs of the cell. In the following chapters we will see many examples of such regulation.

Substrate-Level Control

Some enzyme regulation occurs in a simple way, through direct interaction of the substrates and products of each enzyme-catalyzed reaction with the enzyme itself. This is referred to as **substrate-level control**. This is a consequence of the law of mass action (described in Chapters 2 and 3). As our analysis of kinetics has shown, the higher a substrate concentration is, the more rapidly a reaction occurs, at least until saturation of the enzyme is approached. Conversely, high levels of product, which can also bind to the enzyme, tend to inhibit the conversion of substrate to product. Insofar as the metabolically desired reaction is concerned, the product can act as an inhibitor. As an example, consider the first step in glycolysis (see Chapter 13)—the phosphorylation of glucose to yield glucose-6-phosphate (G6P):

$$\text{Glucose} + \text{ATP} \xrightarrow{\text{Hexokinase}} \text{glucose-6-phosphate} + \text{ADP}$$

The enzyme hexokinase, which catalyzes this reaction, is inhibited by its product, G6P. If subsequent steps in glycolysis are blocked for any reason, G6P will accumulate and bind to hexokinase. This results in inhibition of hexokinase and slows down further production of G6P from glucose. In many cases the reaction product binds the enzyme active site and therefore acts as a competitive inhibitor. Hexokinase is an interesting example because its product, G6P, can act both as a competitive inhibitor (by binding to the active site), as well as an uncompetitive inhibitor (by binding at another site).

Substrate-level control is not sufficient for the regulation of many metabolic pathways. In many instances, it is advantageous to have an enzyme regulated by some substance quite different from the substrate or immediate product. Such

regulation can be achieved with concentrations of the inhibitor that are significantly lower than those of the substrates.

Feedback Control

We have emphasized that most metabolic pathways resemble assembly lines. The
simplest metabolic assembly line looks like this:

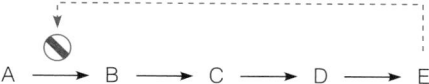

where A is the initial reactant or raw material; B, C, and D are intermediate
products; and E is the final product.

The final product of this pathway, E, will probably be used in some other pathway. Similarly, the "raw material," A, may also participate in some other set of
processes. Suppose the utilization of E suddenly slows down. If everything kept
going as before, E would accumulate, and consumption of A would continue. But
this process is inefficient. A more efficient process would solve this problem by
closely monitoring the concentration of E and, as E accumulated, sending a signal
back to inhibit its production. The cell can control generation of the final product
through activation (🔼) or inhibition (🚫) of a key step in the pathway. It would be
most efficient to slow the *first step*—the conversion of A to B. So the A → B
"machine" should be regulated by the concentration of E.

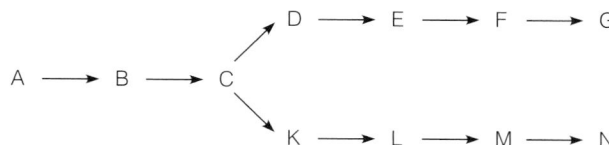

This type of **feedback control** is called **feedback inhibition** because an *increase*
in the concentration of E leads to a *decrease* in its rate of production. Note that by
inhibiting the first step, we prevent both unwanted utilization of A and accumulation
of E. Furthermore, because most biochemical processes are reversible to some extent,
generation of a large quantity of E will tend to build up the concentration of intermediate products. The feedback control mechanism visualized above prevents accumulation of any intermediates, which might have undesired effects on metabolism.

Other metabolic situations require more complicated patterns, in which
activation as well as inhibition may be useful. For example, consider a slightly
more complex case, in which A is fed into two pathways, which lead to two products needed in roughly equivalent amounts. Then a scheme like the following
emerges:

$$
\begin{array}{ccccccc}
 & & & & D \longrightarrow E \longrightarrow F \longrightarrow G \\
A \longrightarrow B \longrightarrow C & & & \nearrow \\
 & & & \searrow \\
 & & & & K \longrightarrow L \longrightarrow M \longrightarrow N
\end{array}
$$

To control the pathways so that G and N keep in balance, high concentrations of
G might *inhibit* the C → D enzyme and/or *activate* the C → K enzyme.
Conversely, N might inhibit the C → K enzyme and/or activate the C → D
enzyme. Finally, it might be useful to have G and N act *together* to inhibit the
A → B enzyme, to provide overall regulation. An example of this kind of control
is found in the synthesis of the purine and pyrimidine monomers that go into
making DNA because approximately equal quantities of all four deoxyribonucleotides are required for DNA replication.

It is important to note that both inhibition and activation of enzymes are
essential to regulate metabolism. Furthermore, control of pathways by their end
products means that the necessary inhibitions and activations *must* be produced
by molecules that come from far down the assembly line and therefore bear little
or no resemblance to either the substrates or the direct products of the enzymes to

Feedback control is important in the efficient
regulation of complex metabolic pathways.

FIGURE 11.44

Effect of cooperative substrate binding on enzyme kinetics. **(a)** Comparison of v vs. [S] curves for a non-cooperative enzyme and an allosteric enzyme with cooperative binding. The two enzymes are assumed to have the same V_{max}. Compare this plot with the myoglobin and hemoglobin oxygen binding curves shown in Figure 7.10 (page 242). **(b)** Lineweaver–Burk Plot. Corresponding to the cooperative binding curve shown in (a). The T state has a high K_M (V_{max} is achieved at higher [S]). As more S is bound, the T → R equilibrium shifts toward the R state, which has a lower K_M.

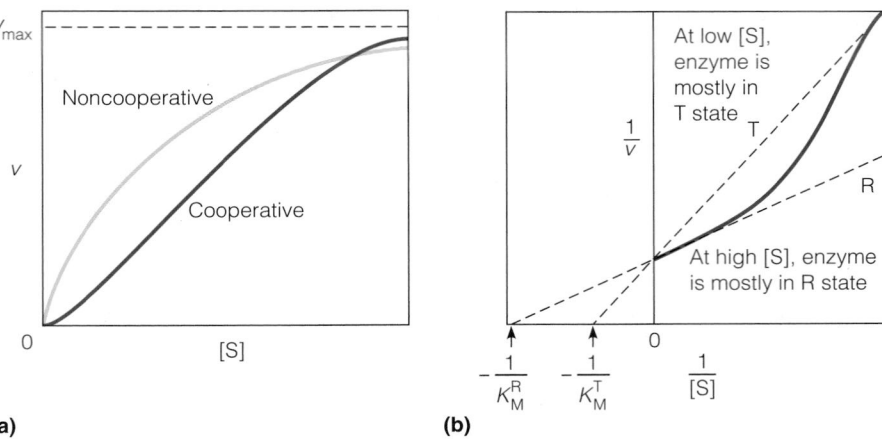

(a)　　　　　　　　　　　(b)

be regulated. None of the kinds of regulation we have discussed up to this point will satisfy these needs. To attain this kind of control, organisms have evolved a special class of enzymes, capable of **allosteric regulation**. The term *allosteric* is derived from Greek words meaning "other structure," emphasizing that structures of regulators need not resemble substrate or direct product.

Allosteric Enzymes

Allosteric enzymes are frequently multisubunit proteins, with multiple active sites. They exhibit cooperativity in substrate binding (**homoallostery**) and regulation of their activity by other, effector molecules (**heteroallostery**).

We have already studied an example of allosteric control of protein function. Hemoglobin (see Chapter 7) is a four-subunit protein that has four binding sites for its "substrate," oxygen. The binding of oxygen is cooperative and is influenced by other molecules and ions. The basic ideas that were presented in Chapter 7 for the analysis of hemoglobin function apply equally well to allosteric enzymes.

Homoallostery

Let us first consider the homoallosteric effects (cooperative substrate binding). In Chapter 7 we contrasted O_2 binding by the single-subunit protein myoglobin with binding by the multisubunit hemoglobin. Myoglobin gives a hyperbolic binding curve (Figure 7.7); hemoglobin, with its cooperative binding, gives a sigmoidal curve (Figure 7.10d). We find *exactly the same contrast* when we compare the v vs. [S] curve of a single-site enzyme obeying Michaelis–Menten kinetics with that of a multisite enzyme showing cooperative binding (Figure 11.44a). The same kind of reasoning applies: An enzyme that binds substrate cooperatively will behave, at low substrate concentration, as if it were poor at substrate binding (that is, as if it had a large K_M). But as the substrate levels are increased and more substrate is bound, the enzyme becomes more and more effective because it binds substrate more avidly in the last sites to be filled (see Figure 11.44b). We imagine this happening, as with hemoglobin, because as more substrate is bound, the enzyme undergoes a transition from a lower affinity state (T state) to a higher affinity state (R state). The kinds of models that have been used to describe O_2 binding by hemoglobin (see Figure 7.12, page 245) can account equally well for the kinetics exhibited by enzymes that show cooperative substrate binding.

What physiological function does sigmoidal kinetics fulfill? In extreme cases, enzymes obeying sigmoidal kinetics can regulate substrate levels to quite constant values. Consider a substrate that is being supplied constantly by other reactions and is acted on by an enzyme that exhibits the extreme cooperativity shown in Figure 11.45. Substrate can easily accumulate up to the critical level $[S]_c$; the enzyme is essentially inactive at lower [S], allowing [S] to increase up to $[S]_c$.

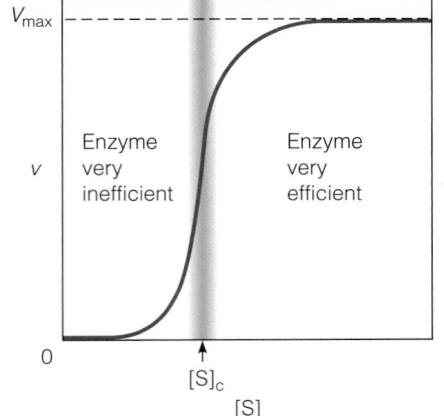

FIGURE 11.45

Effect of extreme homoallostery. The v vs. [S] curve is shown for a hypothetical enzyme with extreme positive cooperativity in substrate binding. At concentrations below $[S]_c$, the enzyme is almost inactive; above this concentration, it is very active. Substrate can easily accumulate to the level $[S]_c$, but at higher concentrations it will be processed rapidly. The vertical blue line represents the homeostatic concentration range for S.

However, any further increase leads to a greatly increased enzyme activity so that the substrate will be more rapidly consumed and its concentration will be maintained near the value $[S]_c$. Although real allosteric enzymes rarely if ever exhibit curves as extremely sigmoidal as that in Figure 11.45, the principle remains: Multisubunit enzymes may help to maintain the homeostasis of a dynamic system. In other words, homoallostery enhances substrate-level control.

Heteroallostery

The major advantage of allosteric control is found in the role of **heteroallosteric effectors**, which may be either inhibitors or activators. These effectors are the analogs, in enzyme kinetics, of the CO_2, BPG, and H^+ that so elegantly regulate O_2 binding by hemoglobin. The activation and inhibition of enzymes by allosteric effectors are the keys to the kind of complex feedback control described above. If an enzyme molecule can exist in two conformational states (T and R) that differ dramatically in the strength with which substrate is bound or in the catalytic rate, then its kinetics can be controlled by *any* other substance that, in binding to the protein, alters the T → R equilibrium. Allosteric *inhibitors* shift the equilibrium toward T, and *activators* shift it toward R (Figure 11.46). Some enzymes are regulated by multiple inhibitors and activators, allowing extremely subtle and complex patterns of metabolic control.

Aspartate Carbamoyltransferase: An Example of an Allosteric Enzyme

An excellent example of allosteric regulation is provided by the enzyme **aspartate carbamoyltransferase** (also known as aspartate transcarbamoylase, or ATCase), a key enzyme in pyrimidine synthesis (Chapter 22). As can be seen from Figure 11.47, ATCase stands at a crossroads in biosynthetic pathways. Glutamine, glutamate, and aspartate are also used in protein synthesis; but once aspartate has been carbamoylated to form **N-carbamoyl-L-aspartate** (CAA), the molecule is committed to pyrimidine synthesis. Thus, the enzyme that controls this step must be sensitive to pyrimidine need. In bacteria like *E. coli*, the activity of ATCase is regulated to respond to this need. This enzyme, as shown in Figure 11.48, is inhibited by cytidine triphosphate (CTP) and activated by ATP. Both responses make physiological sense; when CTP levels are already high, more pyrimidines are not needed. On the other hand, high ATP signals both a purine-rich state (signaling a need for increased pyrimidine synthesis) and an energy-rich cell condition under which DNA and RNA synthesis will be active.

Like most allosteric enzymes, ATCase is a multisubunit protein. Its quaternary structure has been examined in some detail and is depicted schematically in

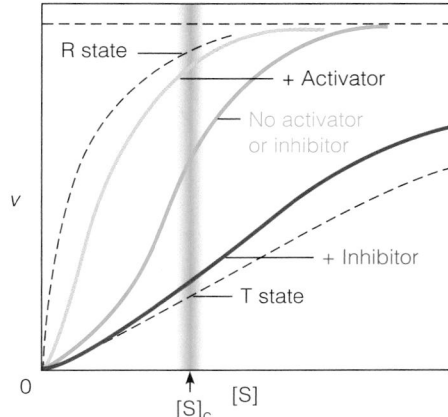

FIGURE 11.46

Heteroallosteric control of an enzyme. In the absence of activation or inhibitors, the *v* vs. [S] curve is sigmoidal. Activators shift the system toward the R state; inhibitors stabilize the T state. $[S]_c$ represents the homeostatic concentration range for S. Note that effectors significantly alter the activity of the enzyme over this range of [S].

Allosteric enzymes show cooperative substrate binding and can respond to a variety of inhibitors and activators.

FIGURE 11.47

Control points in pyrimidine synthesis. This figure shows the formation of N-carbamoyl-L-aspartate from carbamoyl phosphate and aspartate. This reaction is the first committed step in a series of reactions that leads to synthesis of pyrimidine nucleotides, so control at or near this point is essential. In prokaryotes, the aspartate carbamoyltransferase is regulated; in most eukaryotes, regulation is on the preceding step, catalyzed by carbamoyl phosphate synthetase II.

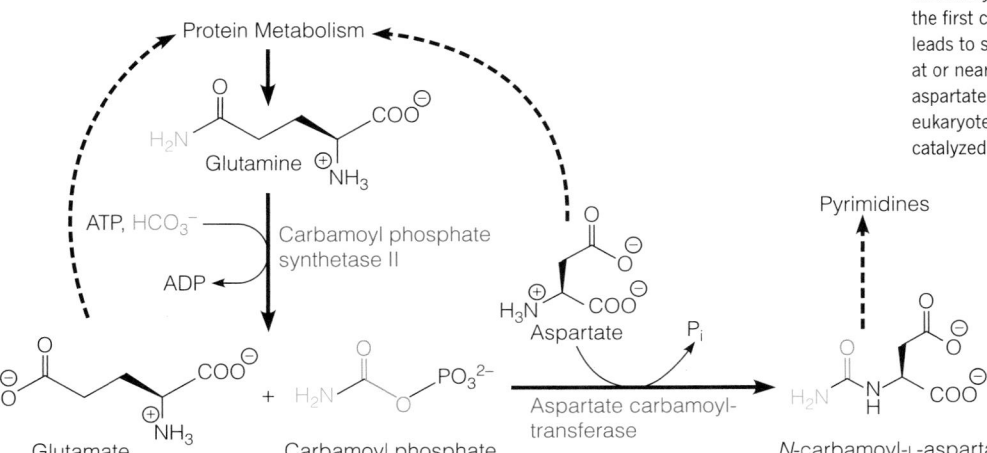

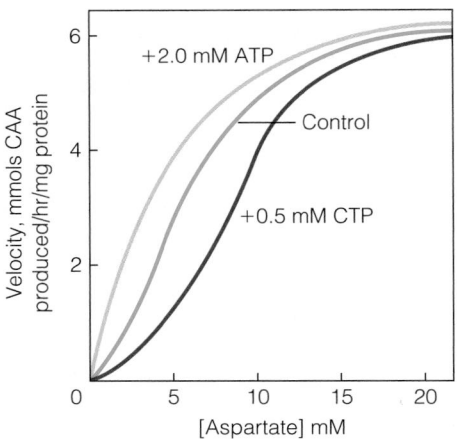

FIGURE 11.48

Regulation of aspartate carbamoyltransferase by ATP and CTP. ATP is an activator of aspartate carbamoyltransferase, and CTP is an inhibitor. The curve marked as "control" shows the behavior of the enzyme in the absence of both regulators. N-Carbamoyl-L-aspartate (CAA) is the product of the reaction.

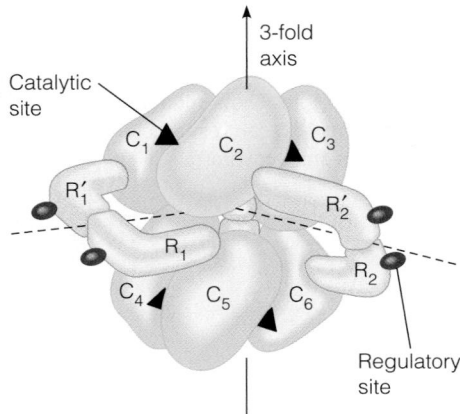

(a) ATCase: T state

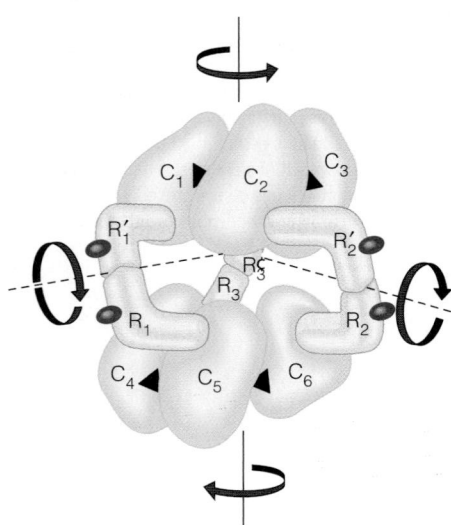

(b) ATCase: R state

Figure 11.49a. There are six *catalytic* subunits, in two tiers of three, held together by six *regulatory* subunits. Pairs of regulatory subunits appear to connect catalytic subunits in the two tiers. The three-dimensional structure of ATCase has been solved to high resolution, and a detailed representation of one catalytic subunit and one regulatory subunit is shown in Figure 11.50. The catalytic subunit comprises two domains, one binding aspartate and the other carbamoyl phosphate, and the active site lies between them. The regulatory subunit is likewise of two parts; the so-called zinc domain and the allosteric domain. The former binds a structurally necessary zinc atom; the latter contains the ATP/CTP binding site. ATP and CTP thus compete for the same site so that the activity of ATCase is regulated by the *ratio* of ATP to CTP in the cell.

As in the case of hemoglobin, the allosteric regulation of ATCase involves changes in the quaternary structure of the molecule. Conformations of the R and T states have been determined by X-ray diffraction. As Figure 11.49b shows, a major rearrangement of subunit positions occurs in the T → R transition.

Virtually every metabolic pathway we shall encounter in the following chapters is subject to complex feedback control, and in almost all cases multisubunit, allosteric enzymes are employed. The pattern of control, even in a given pathway, is not the same in every organism. To take a relevant example, whereas ATCase is the major control point in the pyrimidine pathway in bacteria, eukaryotes regulate at the preceding step—the synthesis of carbamoyl phosphate (see Figure 11.47). In mammals, the **carbamoyl phosphate synthetase II** is inhibited by UDP, UTP, CTP, dUDP, and UDP-glucose. These compounds all inhibit binding of the ATP substrate. In addition, glycine acts as a competitive inhibitor for glutamine.

Recently, examples of single subunit proteins under allosteric control have been described. Here, dynamics is thought to be of critical importance. In this model for allostery, the enzyme samples different conformational states, corresponding to higher ("R-like") and lower ("T-like") activities, and the effector binds to a particular conformation and stabilizes it. A positive effector binds the higher activity conformation, whereas a negative effector binds the lower activity conformation. This model suggests that any dynamic protein could, in principle, be subject to allosteric regulation.

It should be clear at this point that organisms can regulate metabolism in complex and subtle ways through allosteric enzymes; however, this kind of regulation is not sufficient for all needs. We turn now to covalent modification, an entirely different kind of regulatory mechanism.

Covalent Modifications Used to Regulate Enzyme Activity

In the factory analogy, allosteric regulation can be thought of as the feedback control of continuously running machines. But any large factory also has machinery that is used only from time to time and is left on standby until needed. The same is true for the cell. In this section, we discuss enzymes that are essentially inactive until they are changed by a **covalent modification** and then begin to function. In some cases such modification acts in the opposite direction, to inactivate otherwise active enzymes. Some such modifications can be reversed; others cannot.

FIGURE 11.49

Quaternary structure of aspartate carbamoyltransferase (ATCase). **(a)** Quaternary structure of ATCase in the T state. This schematic view of the enzyme shows the six catalytic subunits (C) and six regulatory subunits (R). Six catalytic sites (black triangles) lie in or near the grooves between the catalytic subunits. Regulatory sites (red ellipses) lie on the outer surfaces of the regulatory subunits. The molecule has 1 three-fold axis and 3 two-fold axes (D_3 symmetry). This is a side view of the molecule with the three-fold axis in the plane of the paper. **(b)** Transition of ATCase to the R state. The transition involves a rotation of the regulatory subunits, which pushes the two tiers of catalytic subunits apart and rotates them slightly about the three-fold axis.

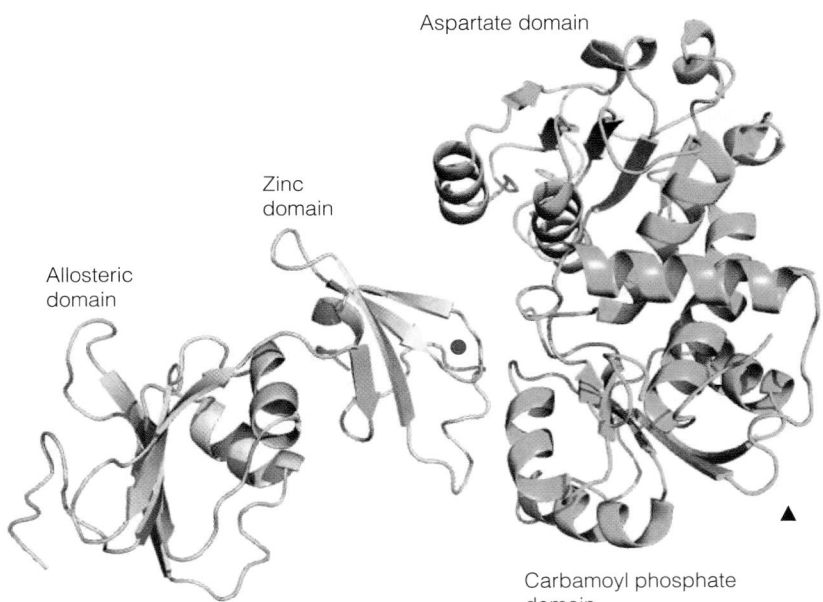

Aspartate domain

Zinc domain

Allosteric domain

Carbamoyl phosphate domain

FIGURE 11.50

The detailed structure of one catalytic subunit (green) and adjacent regulatory subunit (yellow) of ATCase. The view is down the three-fold axis, which is at the lower right (black triangle). The regulatory subunit lies mostly below the plane, the catalytic subunit mostly above. The location of the Zn^{2+} is shown by a red circle. PDB ID: 1R0C.

A number of kinds of covalent modification are commonly used to regulate enzyme activity (Figure 11.51). The most widespread is phosphorylation or dephosphorylation of various amino acid side chains (serine, threonine, tyrosine, and histidine, for example). Other covalent modifications include **adenylylation**, the transfer of an adenylate moiety from ATP; **ADP-ribosylation**, the transfer of an ADP-ribosyl moiety from NAD^+; and **acetylation**, the transfer of an acetyl group from acetyl-coenzyme A (see Table 11.5).

The majority of enzymes, and their associated metabolic and signaling pathways, are regulated by reversible phosphorylation. **Protein kinases** are ATP-dependent enzymes that add a phosphoryl group to the —OH group of a Tyr, Ser, or Thr on some target protein (Figure 11.52). This process is made reversible by a second class of enzymes, called **phosphatases**, which hydrolyze the resulting side chain phosphate esters, releasing P_i. In addition, several protein kinases have been found to be **oncogene** (cancer-causing gene) products. Aberrant activities of these kinases are involved in the transformation of a normal cell to a cancer cell. Much research activity has been devoted to understanding the roles of various kinases and phosphatases in cell signaling and regulation of metabolism.

Protein phosphorylation and acetylation are part of complex regulatory pathways, frequently under hormonal control. Several of these pathways are discussed in great detail in subsequent chapters, where their significance will be clearer. At this point, we shall concentrate instead on a third kind of covalent modification—the irreversible activation of certain enzymes by proteolytic cleavage of precursor forms.

Pancreatic Proteases: Activation by Cleavage

An important example of covalent enzyme activation, **proteolytic cleavage**, is found in the maturation of **pancreatic proteases**. These include a number of enzymes—for example, trypsin, chymotrypsin, elastase, and carboxypeptidase—some of which we have already discussed. All are synthesized in the pancreas. They are secreted through the pancreatic duct into the duodenum of the small intestine in response to a hormone signal generated when food passes from the stomach. They are not, however, synthesized in their final, active form because a battery of potent proteases free in the pancreas would digest the pancreatic tissue. Rather, they are made as slightly longer, catalytically inactive molecules, called **zymogens**. The names given to the zymogens of the enzymes mentioned above are *trypsinogen,*

FIGURE 11.51

Four types of covalent modifications that control the activities of enzymes. The target residue for phosphorylation or adenylylation is usually serine, threonine, or tyrosine, whereas ADP-ribosylation can involve arginine, glutamate, aspartate, or a modified histidine residue. N-acetylation involves reaction between a lysine side chain and acetyl-coenzyme A.

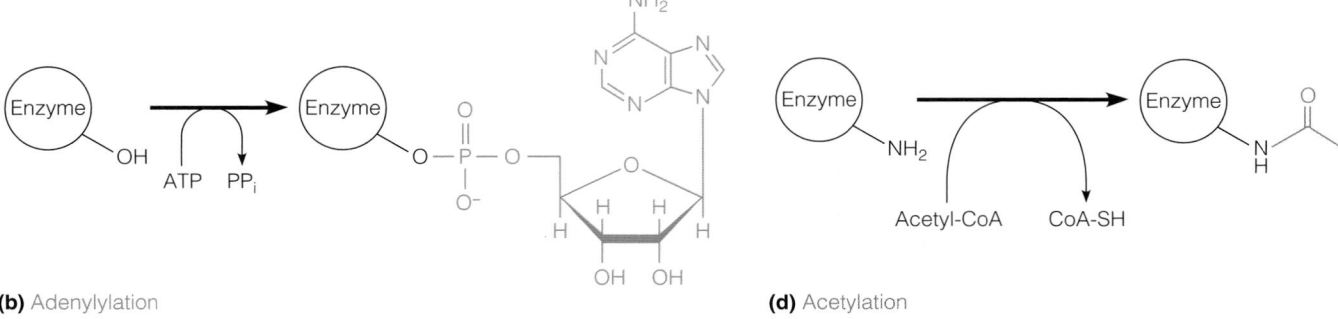

(a) Phosphorylation

(b) Adenylylation

(c) ADP-ribosylation

(d) Acetylation

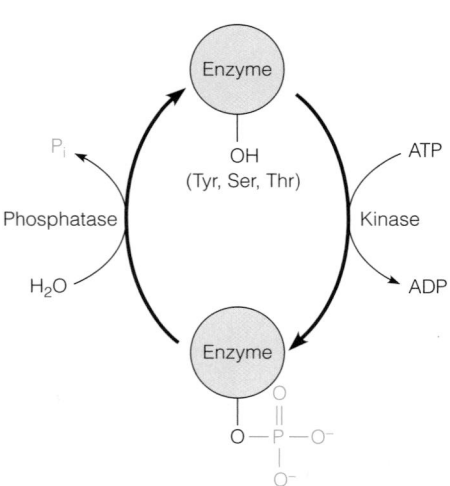

FIGURE 11.52

Reversible covalent modification by kinases/phosphatases. The target residues for ATP-dependent phosphorylation by kinases are serine, threonine, or tyrosine. The phosphoprotein is dephosphorylated by a phosphatase-catalyzed hydrolysis reaction.

chymotrypsinogen, proelastase, and *procarboxypeptidase,* respectively. The zymogens must be cleaved proteolytically in the intestine to yield the active enzymes. These enzymes are broken down after serving their purpose, so they do not endanger the intestinal tissue, which is also somewhat protected by its glycosylated surface. The cleavage of zymogens to active enzymes is diagrammed in Figure 11.53.

The first step is the activation of trypsin in the duodenum. A hexapeptide is removed from the N-terminal end of trypsinogen by *enteropeptidase,* a protease secreted by duodenal cells. This action yields the active trypsin, which then activates the other zymogens by specific proteolytic cleavages. In fact, once some active trypsin is present, it will activate other trypsinogen molecules to make more trypsin; thus its activation is *autocatalytic.* This is an example of the kind of **cascade** process frequently observed when enzymes are activated by covalent modification. The production of just a few trypsin molecules leads quickly to many more, as each enzyme molecule, when activated, can process many more every minute. These molecules can in turn activate the other zymogens. In effect, an enzyme cascade amplifies the original signal (e.g., hormone binding to the surface of a cell) and mounts a rapid, overwhelming response to that signal.

The activation of chymotrypsinogen to chymotrypsin is one of the most complex and best-studied examples of proteolytic activation of an enzyme; it is illustrated in Figure 11.54. In the first step, trypsin cleaves the bond between arginine 15 and isoleucine 16. The N-terminal peptide remains attached to the rest of the molecule because of the disulfide bond between residues 1 and 122. The product, called *π-chymotrypsin,* is an active enzyme.

Just how the cleavage of one peptide bond transforms an essentially inactive protein into an active one can now be understood as a result of detailed X-ray diffraction studies of the zymogen and the active enzyme. Cleavage of the peptide

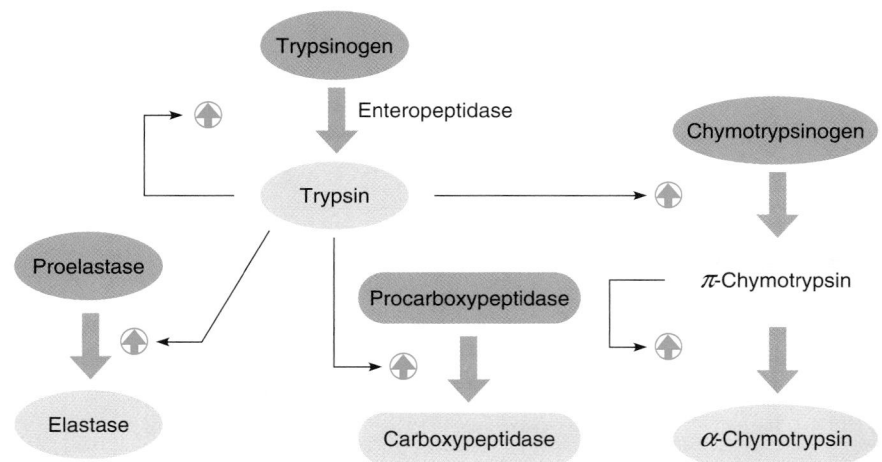

FIGURE 11.53

Zymogen activation by proteolytic cleavage. This schematic view shows the activation of pancreatic zymogens, molecules that become catalytically active when cleaved. Zymogens are shown in orange and active proteases in yellow or green. The difference between π-chymotrypsin and α-chymotrypsin is shown in Figure 11.54.

Some enzymes, such as pancreatic proteases, are irreversibly switched on by proteolytic cleavage.

bond between residues 15 and 16 creates a new, positively charged N-terminal residue at Ile 16. This residue shifts its position and forms a salt bridge with Asp 194, the neighbor of the active site Ser 195 (see Figure 11.18). This change in turn triggers further conformational rearrangements in the active site. These changes result in the formation of a catalytically competent active site pocket, including the movement of main chain amino groups of residues 193 and 195 to form the oxyanion hole. Thus, both the binding pocket and the catalytic site are formed correctly only after the peptide bond between Arg 15 and Ile 16 has been cleaved.

π-Chymotrypsin is not the most active form of chymotrypsin. More autocatalytic cleavages remove residues 14–15 and 147–148 from the molecule, to produce the final *α-chymotrypsin*, which is the principal and fully active form found in the digestive tract.

This battery of enzymes, trypsin, chymotrypsin, elastase, and carboxypeptidase, together with the *pepsin* of the stomach and other proteases secreted by the intestinal wall cells, is capable of ultimately digesting most ingested proteins into free amino acids, which can be absorbed by the intestinal epithelium. The enzymes themselves are continually subjected to mutual digestion and autodigestion so that high levels of these enzymes never accumulate in the intestine.

Even inactive zymogens are a potential source of danger to the pancreas. Because trypsin activation can be autocatalytic, the presence of even a single active trypsin molecule could set the activation cascade in motion prematurely. Therefore the pancreas protects itself further by synthesizing a protein called the *secretory pancreatic trypsin inhibitor* (to be distinguished from the pancreatic trypsin inhibitor shown in Figure 6.42, which is an intracellular protein found only in ruminants). This competitive inhibitor binds so tightly to the active site of trypsin that it effectively inactivates it even at very low concentration. The bonding between trypsin and its inhibitor is among the strongest noncovalent associations known in biochemistry. Only a tiny amount of trypsin inhibitor is present— far less than needed to inhibit all of the potential trypsin in the pancreas. Thus, only a fraction of the trypsin generated in the duodenum is inhibited, and the rest can be activated. Because protection is limited, zymogen activation can sometimes be triggered in the pancreas—for example, if the pancreatic duct is blocked. The active enzymes then begin to digest the pancreatic tissue itself. This condition, called *acute pancreatitis*, is extremely painful and sometimes fatal.

The first well-understood regulatory cascade was the one controlling glycogen breakdown in animal cells, a critical process that provides carbohydrate substrates for energy generation. This regulatory cascade, involving enzyme phosphorylation and dephosphorylation, is described in detail in Chapter 13. Another spectacular example of an enzyme cascade occurs in blood clotting, which we describe in the next section.

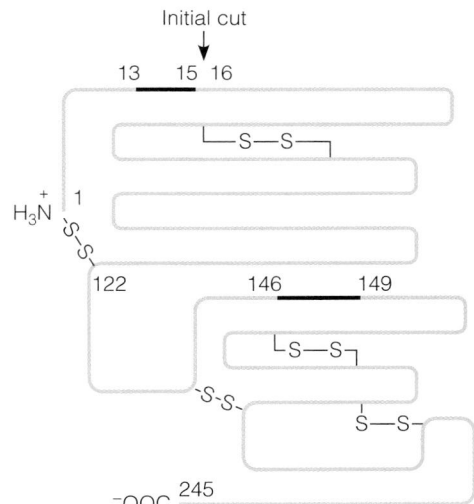

FIGURE 11.54

Activation of chymotrypsinogen. The figure is a schematic rendition of the chymotrypsinogen molecule. A series of cleavages produces the enzyme chymotrypsin, with the disulfide bonds continuing to hold the structure together. The initial cleavage between amino acids 15 and 16 (arrow) results in the formation of π-chymotrypsin. Subsequent removal of the segments shown in black yields α-chymotrypsin.

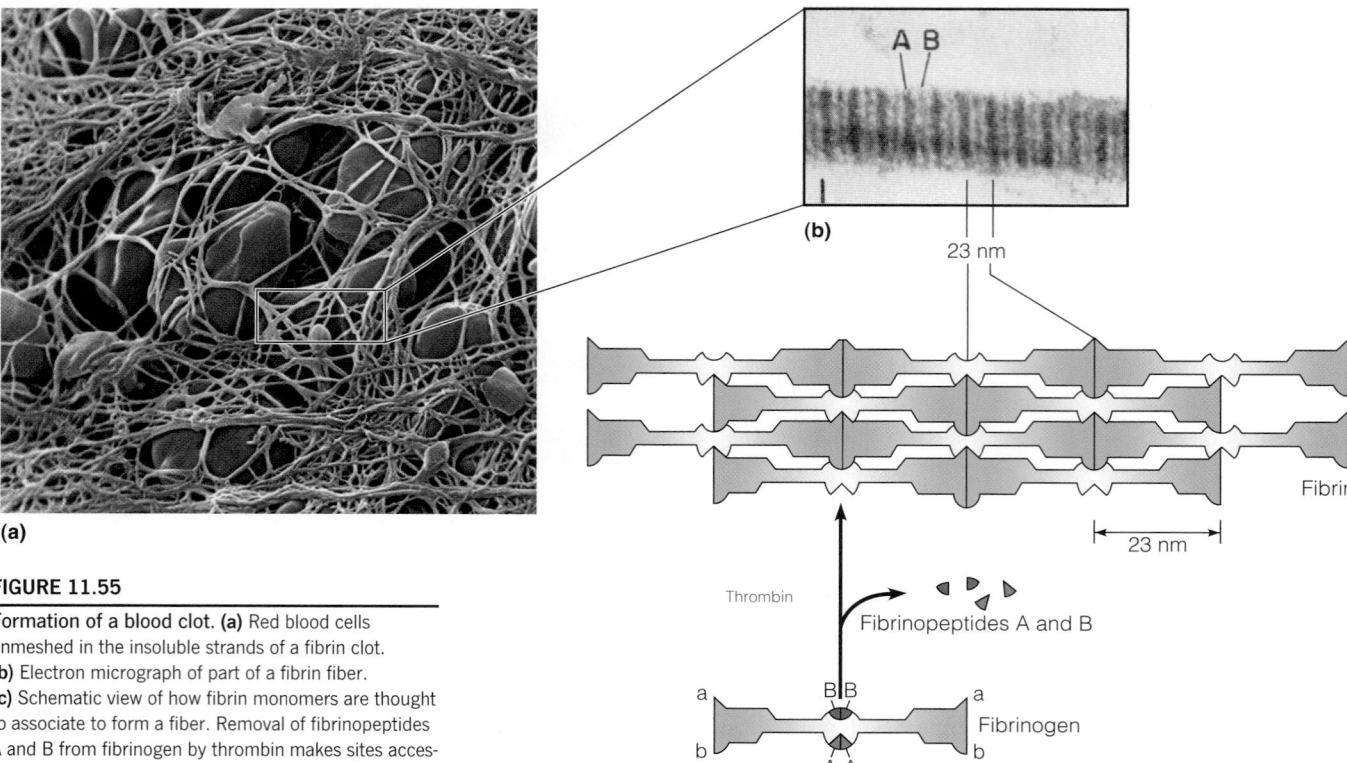

FIGURE 11.55

Formation of a blood clot. (a) Red blood cells enmeshed in the insoluble strands of a fibrin clot. **(b)** Electron micrograph of part of a fibrin fiber. **(c)** Schematic view of how fibrin monomers are thought to associate to form a fiber. Removal of fibrinopeptides A and B from fibrinogen by thrombin makes sites accessible for association with complementary sites a and b on adjacent monomers. The molecules are believed to overlap as shown because the striations seen in the fibers are 23 nm wide, exactly half the length of the fibrinogen molecule.

Blood clotting involves a cascade of proteolytic activation of specific proteases, culminating in the transition of fibrinogen to fibrin.

A Further Look at Activation by Cleavage: Blood Clotting

Activation of zymogens is the key to another biologically important process—the clotting of vertebrate blood. If a blood clot is examined in the electron microscope, it is found to be composed of striated fibers of a protein called **fibrin** (Figure 11.55a). The fibrin monomers are elongated molecules, about 46 nm long, that stick together in a staggered array as shown in Figure 11.55c. Fibrin monomers are derived from a precursor, **fibrinogen**, by proteolytic cleavages that release small *fibrinopeptides* (A and B in Figure 11.55c). Loss of these peptides uncovers positions at which the fibrin molecules can stick together. After the clot is formed, it is further stabilized by covalent cross-links between glutamine and lysine residues.

The proteolysis of fibrinogen to fibrin is catalyzed by the serine protease **thrombin**. Thrombin has sequence and structural similarities to trypsin, but as a protease with a very specific function it cleaves only a few types of bonds, mainly Arg–Gly. Thrombin itself is produced from **prothrombin** by another specific protease; in fact, as Figure 11.56 shows, a whole cascade of proteolytic activation reactions leads ultimately to the formation of a fibrin clot. Involved are a series of proteases referred to as *factors*. In damaged tissues, the proteins *kininogen* and *kallikrein* activate factor XII (also called *Hageman factor*), which in turn activates factor XI—and the cascade of reactions proceeds as shown. This set of initial reactions is called the **intrinsic pathway**. Alternatively, damage to blood vessels leads to the release of *tissue factor* and activation of factor VII, starting the **extrinsic pathway**. The two pathways merge in the activation of factor X, which will proteolyze and thereby activate prothrombin.

Some of the activation steps require auxiliary proteins. For example, activation of factor X in the intrinsic pathway by factor IX (*Christmas factor*) requires a 330-kilodalton protein called *antihemophilic factor* (factor VIII). The partial or complete absence of factor VIII activity is the cause of classic **hemophilia**. The gene for factor VIII is carried on the X chromosome, so women, who have two

FIGURE 11.56

The cascade process in blood clotting. Each factor (protease) in the pathway can exist in an inactive form (orange) or an active form (green). The cascade of proteolytic activations can start from exposure of blood at damaged tissue surfaces (intrinsic pathway) or from internal trauma to blood vessels (extrinsic pathway). The common result is activation of fibrinogen to clotting fibrin. Auxiliary factors that aid some steps are also shown. Asterisk (*) denotes serine proteases.

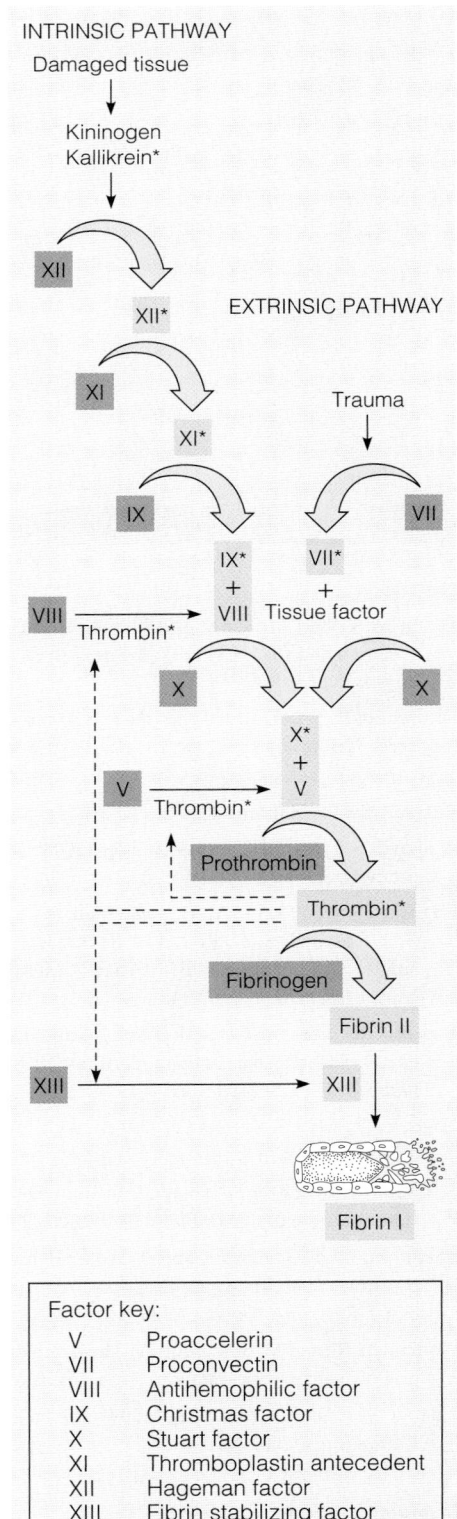

Factor key:

V	Proaccelerin
VII	Proconvectin
VIII	Antihemophilic factor
IX	Christmas factor
X	Stuart factor
XI	Thromboplastin antecedent
XII	Hageman factor
XIII	Fibrin stabilizing factor

copies of this chromosome, can be heterozygous carriers of the trait but will exhibit the symptoms only if they are homozygous. However, a male descendent who receives on his single X chromosome the damaged copy of the factor VIII gene will experience more or less severe difficulty in blood clotting. The condition can now be treated by frequent transfusions of a blood serum fraction concentrated in factor VIII. The gene for this protein has recently been cloned and expressed in bacteria, and the availability of this synthetic factor VIII may allow such patients to avoid the need for frequent transfusions.

As wounds heal, or tissue damage is repaired, it is essential that blood clots be dissolved. The principal agent for clot dissolution is an enzyme called **plasmin**, which cleaves fibrin. Plasmin itself is derived by proteolytic cleavage of an inactive precursor, **plasminogen**. Activation of plasminogen is catalyzed by a number of proteases, the most important being tissue-type **plasminogen activator (t-PA)**. In addition to its normal function, t-PA can be extremely effective in initiating the cascade to dissolve the unwanted blood clot involved in stroke or heart attack.

The mechanisms of regulation we have mentioned here by no means describe the cell's whole repertoire. In addition to regulation of enzyme *performance*, cells and organisms can regulate both the synthesis and degradation of enzymes, as well as the compartmentalization of enzymes within specific organelles or multi-enzyme complexes. However, description of these processes is more appropriate in the broader framework of Chapter 12.

SUMMARY

The rate of a chemical reaction is determined by reactant concentrations and by the rate constant. The rate constant in turn depends upon the activation energy needed to reach the transition state. All catalysts function by lowering the activation energy for a reaction. In doing so, they do not affect chemical equilibrium but only increase rates.

Enzymes are biological catalysts that increase the rates of biochemical processes but are themselves unchanged. Most (but not all) enzymes are proteins. In enzyme catalysis, one or more substrates are bound at the active site of an enzyme, to form the enzyme–substrate complex; products are then released.

The induced fit hypothesis states that enzymes induce bound substrates to adopt conformations close to the transition state, though binding may also bring about conformational change in the enzyme. The role of dynamic motions in enzyme catalysis is not well understood, but may be critical for achieving high rate accelerations—particularly in reactions involving hydrogen transfer.

Most simple enzymatic reactions can be described by the Michaelis–Menten equation, with two parameters, the Michaelis constant, K_M, and the turnover number, k_{cat}. Enzymes can be inhibited reversibly or irreversibly. Reversible inhibition can be competitive, uncompetitive, or mixed. Competitive inhibition increases the apparent K_M, uncompetitive inhibition reduces the apparent V_{max} and apparent K_M, and mixed inhibition reduces the apparent V_{max} and increases the apparent K_M. Irreversible inhibition usually involves covalent binding to the active site.

Many enzymes utilize cofactors in their function; others require specific metal ions. A number of enzyme cofactors are closely related to vitamins required in the human diet.

It is possible to create new or modified enzymes by molecular engineering or in vitro evolution. A common method involves selection of active molecules from a large library of candidate clones, as in the case of the generation of catalytic antibodies. Computational approaches allow for the evaluation of candidate enzymes from even larger libraries of possible structures. In addition, it has been found that some nucleic acid molecules function as enzymes; they are called ribozymes or DNAzymes.

Regulation of enzymatic activity takes many forms. Substrate-level regulation simply depends on ambient concentrations of reactants and products. Allosteric regulation provides sensitive feedback control for complex metabolic pathways. For more drastic changes in activity, some enzymes are switched on or off (or both) by covalent modification.

REFERENCES

General

Fersht, A. (1999) *Structure and Mechanism in Protein Science.* W. H. Freeman and Co., New York. A fine treatise on almost all aspects of enzymology.

Gutfreund, H. (1995) *Kinetics for the Life Sciences.* Cambridge University Press, Cambridge, UK.

Enzyme Mechanisms and Kinetics

Benkovic, S. J., and S. Hammes-Schiffer (2003) A perspective on enzyme catalysis. *Science* 301:1196–1202.

English, B. P., W. Min, A. M. van Oijen, K. T. Lee, G. Luo, H. Sun, B. J. Cherayil, S. C. Kou, and X. S. Xie (2006) Ever-fluctuating single enzyme molecules: Michaelis-Menten equation revisited. *Nature Chem. Biol.* 2:87–94.

Garcia-Viloca, M., J. Gao, M. Karplus, and D. Truhlar (2004) How enzymes work: Analysis by modern rate theory and computer simulation. *Science* 303:186–195.

Nagel, Z. D., and J. P. Klinman (2009) A 21st century revisionist's view at a turning point in enzymology. *Nature Chem. Biol.* 8:543–550.

Schramm, V. L. (2007) Enzymatic transition state theory and transition state analogue design. *J. Biol. Chem.* 282:28297–28300.

Snider, M. G., B. S. Temple, and R. Wolfenden (2004) The path to the transition state in enzyme reactions: A survey of catalytic efficiencies. *J. Phys. Org. Chem.* 17:586–591.

Wolfenden, R., and M. Snider (2001) The depth of chemical time and the power of enzymes as catalysts. *Accts. Chem. Res.* 34:938–945.

Zalatan, J. G., and D. Herschlag (2009) The far reaches of enzymology. *Nature Chem. Biol.* 8:516–520.

Lysozyme and Serine Proteases

Bartik, K., C. Redfield, and C. M. Dobson (1994) Measurement of the individual pK_a values of acidic residues of hen and turkey lysozymes by two-dimensional ^{1}H NMR. *Biophys. J.* 66:1180–1184.

Carter, P., and J. A. Wells (1988) Dissecting the catalytic triad of a serine protease. *Nature* 332:564–568.

Cleland, W. W., P. A. Frey, and J. A. Gerlt (1998) The low barrier hydrogen bond in enzymatic catalysis. *J. Biol. Chem.* 273:25529–25532.

Corey, D. R., and C. S. Craik (1992) An investigation into the minimum requirements for peptide hydrolysis by mutation of the catalytic triad of trypsin. *J. Am. Chem. Soc.* 114:1784–1790.

Frey, P. A., S. A. Whitt, and J. B. Tobin (1994) A low-barrier hydrogen bond in the catalytic triad of serine proteases. *Science* 264:1927–1930.

Fuhrmann, C. N., M. D. Daugherty, and D. A. Agard (2006) Subangstrom crystallography reveals that short ionic hydrogen bonds, and not a His-Asp low-barrier hydrogen bond, stabilize the transition state in serine protease catalysis. *J. Amer. Chem. Soc.* 128:9086–9102.

Matsumura, I., and J. F. Kirsch (1996) Is aspartate 52 essential for catalysis by chicken egg white lysozyme? The role of natural substrate-assisted hydrolysis. *Biochemistry* 35:1881–1889.

Perrin, C. L. (2010) Are short low-barrier hydrogen bonds unusually strong? *Accts. Chem. Res.* 43:1550–1557.

Polgár, L. (2005) The catalytic triad of serine peptidases. *Cell. Mol. Life Sci.* 62:2161–2172.

Robertus, J. D., J. Kraut, R. A. Alden, and J. J. Birktoft (1972) Subtilisin: A stereochemical mechanism involving transition state stabilization. *Biochemistry* 11:4293–4303.

Strynadka, N. C. J., and M. N. G. James (1991) Lysozyme revisited: Crystallographic evidence for distortion of an N-acetylmuramic acid residue bound in site D. *J. Mol. Biol.* 220:401–424.

Tamada, T., T. Kinoshita, K. Kurihara, M. Adachi, T. Ohhara, T. Imai, R. Kuroki, and T. Tada (2009) Combined high-resolution neutron and X-ray analysis of inhibited elastase confirms the active-site oxyanion hole but rules against a low-barrier hydrogen bond. *J. Am. Chem. Soc.* 131:11033–11040.

Vocadlo, D. J., G. J. Davies, R. Laine, and S. G. Withers (2001) Catalysis by hen egg-white lysozyme proceeds via a covalent intermediate. *Nature* 412:835–838.

Wilmouth, R. C., K. Edman, R. Neutze, P. A. Wright, I. J. Clifton, T. R. Schneider, C. J. Schofield, and J. Hajdu (2001) X-ray snapshots of serine protease catalysis reveal a tetrahedral intermediate. *Nature Struct. Biol.* 8:689–694.

Dynamic Motions of Enzymes and Catalysis

Agarwal, P. K. (2005) Role of protein dynamics in reaction rate enhancement by enzymes. *J. Am. Chem. Soc.* 127:15248–15256.

Benkovic, S. J., G. G. Hammes, and S. Hammes-Schiffer (2008) Free-energy landscape of enzyme catalysis. *Biochemistry* 47:3317–3321.

Boehr, D. D., D. McElheny, H. J. Dyson, and P. E. Wright (2006) The dynamic energy landscape of dihydrofolate reductase catalysis. *Science* 313:1638–1641.

Eisenmesser, E. Z., O. Millet, W. Labeikovsky, D. M. Korzhnev, M. Wolf-Watz, D. A. Bosco, J. J. Skalicky, L. E. Kay, and D. Kern (2005) Intrinsic dynamics of an enzyme underlies catalysis. *Nature* 438:117–121.

Kamerlin, S. C. L., and A. Warshel (2010) At the dawn of the 21st century: Is dynamics the missing link for understanding enzyme catalysis? *Proteins* 78:1339–1375.

Roca, M., B. Messer, D. Hilvert, and A. Warshel (2008) On the relationship between folding and chemical landscapes in enzyme catalysis. *Proc. Natl. Acad. Sci. USA* 105:13877–13882.

Schwartz, S. D., and V. Schramm (2009) Enzymatic transition states and dynamic motion in barrier crossing. *Nature Chem. Biol.* 8:551–558.

Ribozymes and DNAzymes

Breaker, R. R., and G. F. Joyce (1994) A DNA enzyme that cleaves RNA. *Chem. Biol.* 1:223–229.

Chandra, M., A. Sachdeva, and S. K. Silverman (2009) DNA-catalyzed sequence-specific hydrolysis of DNA. *Nature Chem. Biol.* 5:718–720.

Cech, T. R. (1987) The chemistry of self-splicing RNA and RNA enzymes. *Science* 236:1532–1539.

Emilsson, G. M., and R. R. Breaker (2002) Deoxyribozymes: New activities and new applications. *Cell. Mol. Life Sci.* 59:596–607.

Joyce, G. F. (2002) The antiquity of RNA-based evolution. *Nature* 418:214–221.

McCorkle, G. M., and S. Altman (1987) RNA's as catalysts. *Concepts Biochem.* 64:221–226.

Sen, D., and C. R. Geyer (1998) DNA enzymes. *Curr. Op. Chem. Biol.* 2:680–687.

Strobel, S. A., and J. C. Cochrane (2007) RNA catalysis: Ribozymes, ribosomes, and riboswitches. *Curr. Op. Chem. Biol.* 11:636–643.

Willner, I., B. Shylahovsky, M. Zayats, and B. Willner (2008) DNAzymes for sensing, nanobiotechnology and logic gate applications. *Chem. Soc. Rev.* 37:1153–1165.

Wright, M. C., and G. F. Joyce (1997) Continuous in vitro evolution of catalytic function. *Science* 276:614–616.

Allosteric Regulation

Goodey, N. M., and S. J. Benkovic (2008) Allosteric regulation and catalysis emerge via a common route. *Nature Chem. Biol.* 8:474–482.

Gunasekaran, K., B. Ma, and R. Nussinov (2004) Is allostery an intrinsic property of all dynamic proteins? *Proteins: Struct. Func. Bioinform.* 57:433–443.

Lipscomb, W. N. (1994) Aspartate transcarbamylase from *Escherichia coli*: Activity and regulation. *Adv. Enzymol.* 73:67–151.

Monod, J., J.-P. Changeux, and F. Jacob (1963) Allosteric proteins and cellular control systems. *J. Mol. Biol.* 6:306–329. This paper introduced the concept of allosteric control.

Swain, J. F., and L. M. Gierasch (2006) The changing landscape of protein allostery. *Curr. Op. Struct. Biol.* 16:102–108.

Zymogen Activation

Bode, W., and R. Huber (1986) Crystal structure of pancreatic serine endopeptidases. In: *Molecular and Cellular Basis of Digestion,* edited by P. Desnuelle, H. Sjorstrom, and O. Noren, pp. 213–234. Elsevier, New York.

Neurath, H. (1986) The versatility of proteolytic enzymes. *J. Cell. Biochem.* 32:35–49.

Blood Clotting

Davie, E. W. (1986) Introduction to the blood coagulation cascade and the cloning of blood coagulation factors. *J. Protein Chem.* 5:247–253.

Doolittle, R. F. (1984) Fibrinogen and fibrin. *Annu. Rev. Biochem.* 53:195–229.

PROBLEMS

1. Show that the half-life for a first-order reaction is inversely proportional to the rate constant, and determine the constant of proportionality.

2. A substance A is consumed by a reaction of unknown order. The initial concentration is 2 mM, and concentrations at later times are as shown:

Time (min)	[A] (mM)
1	1.66
2	1.44
4	1.12
8	0.76
16	0.48

Test whether or not a first-order reaction fits the data.

3. The enzyme urease catalyzes the hydrolysis of urea to ammonia plus carbon dioxide. At 21 °C the uncatalyzed reaction has an activation energy of about 125 kJ/mol, whereas in the presence of urease the activation energy is lowered to about 46 kJ/mol. By what factor does urease increase the velocity of the reaction?

4. An enzyme contains an active site aspartic acid with a $pK_a = 5.0$, which acts as a *general acid catalyst*. On the template below, draw the curve of enzyme activity (reaction rate) vs. pH for the enzyme (assume the protein is stably folded between pH 2–12 and that the active site Asp is the only ionizable residue involved in catalysis). Briefly explain the shape of your curve.

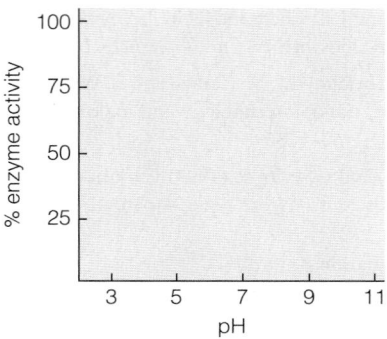

*5. In some reactions, in which a protein molecule is binding to a specific site on DNA, a rate *greater* than that predicted by the diffusion limit is observed. Suggest an explanation. [Hint: The protein molecule can also bind weakly and nonspecifically to *any* DNA site.]

*6. Derive an expression for the change of concentration of reactant with time for a reversible first-order reaction (Equation 11.5), assuming [B] = 0 at $t = 0$. [Warning: This is much more difficult than it looks.]

7. (a) Derive an expression for the concentration of reactant as a function of time for the second-order reaction $2A \rightarrow B$.
   (b) On the basis of your answer to (a), show how a linear graph could be obtained for a second-order reaction.
   (c) Test the data given in Problem 2 with your method.

8. The initial rate for an enzyme-catalyzed reaction has been determined at a number of substrate concentrations. Data are as follows:

[S] (μmol/L)	v [(μmol/L) min^{-1}]
5	22
10	39
20	65
50	102
100	120
200	135

   (a) Estimate V_{max} and K_M from a direct graph of v versus [S] (these data are plotted in Figure 11.24b). Do you find difficulties in getting clear answers?
   (b) Now use a Lineweaver–Burk plot to analyze the same data. Does this work better?
   (c) Finally, try an Eadie–Hofstee plot of the same data.

9. (a) If the total enzyme concentration in Problem 8 was 1 nmol/L, how many molecules of substrate can a molecule of enzyme process in each minute?
   (b) Calculate k_{cat}/K_M for the enzyme reaction in Problem 8. Is this a fairly efficient enzyme? (See Table 11.4.)

*10. (a) If we write the Michaelis–Menten equation as

$$\frac{d[S]}{dt} = -\frac{d[P]}{dt} = -\frac{V_{max}[S]}{K_M + [S]}$$

   it should be possible to integrate it and obtain an expression for substrate concentration as a function of time. Do so, calling [S]$_0$ the initial substrate concentration at $t = 0$.
   (b) Show from the result in (a) that under conditions where [S]$_0 \gg K_M$, the decrease in substrate concentration with time is approximately linear. Interpret this result.

11. Figure 11.39 shows a proposed mechanism for carboxypeptidase A.
   (a) What is the role of Glu 270 in catalysis?
   (b) What is the role of Arg 145 in catalysis?

12. The catalytic efficiency of many enzymes depends on pH. Chymotrypsin shows a maximum value of k_{cat}/K_M at pH 8. Detailed analysis shows that k_{cat} increases rapidly between pH 6 and 7 and remains constant at higher pH. K_M increases rapidly between pH 8 and 10. Suggest explanations for these observations.

13. The following data describe the catalysis of cleavage of peptide bonds in small peptides by the enzyme elastase.

Substrate	K_M (mM)	k_{cat} (s^{-1})
P A P A $\downarrow$ G	4.0	26
P A P A $\downarrow$ A	1.5	37
P A P A $\downarrow$ F	0.64	18

The arrow indicates the peptide bond cleaved in each case.
(a) If a mixture of these three substrates was presented to elastase with the concentration of each peptide equal to 0.5 mM, which would be digested most rapidly? Which most slowly? (Assume enzyme is present in excess.)
(b) On the basis of these data, suggest what features of amino acid sequence dictate the specificity of proteolytic cleavage by elastase.
(c) Elastase is closely related to chymotrypsin. Suggest two kinds of amino acid residues you might expect to find in or near the active site.

14. At 37 °C, the serine protease subtilisin has $k_{cat} = 50$ s^{-1} and $K_M = 1.4 \times 10^{-4}$ M. It is proposed that the N155 side chain contributes a hydrogen bond to the oxyanion hole of subtilisin. J. A. Wells and colleagues reported (1986, *Phil. Trans. R. Soc. Lond. A* 317:415–423) the following kinetic parameters for the N155T mutant of subtilisin: $k_{cat} = 0.02$ s^{-1} and $K_M = 2 \times 10^{-4}$ M.
(a) Subtilisin is used in some laundry detergents to help remove protein-type stains. What unusual kind of stability does this suggest for subtilisin?
(b) Subtilisin does have a problem, in that it becomes inactivated by oxidation of a methionine close to the active site. Suggest a way to make a better subtilisin.
(c) Is the effect of the N155T mutation what you would expect for a residue that makes up part of the oxyanion hole? How do the reported values of k_{cat} and K_M support your answer?
(d) Assuming that the T155 side chain cannot H-bond to the oxyanion intermediate, by how much (in kJ/mol) does N155 appear to stabilize the transition state at 37 °C?
(e) The value you calculated in part (d) represents the strength of the H-bond between N155 and the oxyanion in the transition state. This value is higher than typical H-bonds in water. How might this observation be rationalized? Hint: consider Equation 2.2 (Coulomb's Law).

15. The steady-state kinetics of an enzyme are studied in the absence and presence of an inhibitor (inhibitor A). The initial rate is given as a function of substrate concentration in the following table:

[S] (mmol/L)	v [(mmol/L)min^{-1}]	
	No inhibitor	Inhibitor A
1.25	1.72	0.98
1.67	2.04	1.17
2.50	2.63	1.47
5.00	3.33	1.96
10.00	4.17	2.38

(a) What kind of inhibition (competitive, uncompetitive, or mixed) is involved?
(b) Determine V_{max} and K_M in the absence and presence of inhibitor.

16. The same enzyme as in Problem 15 is studied in the presence of a different inhibitor (inhibitor B). In this case, two different concentrations of inhibitor are used. Data are as follows:

	$v\,[\,(\mathrm{mmol/L})\,\mathrm{min}^{-1}]$		
[S] (mmol/L)	No inhibitor	3 mM inhibitor B	5 mM inhibitor B
1.25	1.72	1.25	1.01
1.67	2.04	1.54	1.26
2.50	2.63	2.00	1.72
5.00	3.33	2.86	2.56
10.00	4.17	3.70	3.49

(a) What kind of inhibitor is inhibitor B?

(b) Determine the apparent V_{max} and K_M at each inhibitor concentration.

(c) Estimate K_I from these data.

17. We have mentioned Eadie–Hofstee plots as an alternative to Lineweaver–Burk plots for expression of kinetic data. Sketch what Eadie–Hofstee plots would look like for a series of experiments at different concentrations of

(a) a competitive inhibitor

(b) a mixed inhibitor

18. We have discussed the role of TPCK as an irreversible inhibitor of chymotrypsin. Design a comparable inhibitor for trypsin.

19. The folding and unfolding rate constants for a myoglobin mutant have been determined. The unfolding rate constant $k_{F \to U} = 3.62 \times 10^{-5}\,\mathrm{s}^{-1}$ and the folding rate constant $k_{U \to F} = 255\,\mathrm{s}^{-1}$, where F is the folded protein and U is the unfolded (denatured) protein. For wild-type myoglobin, $\Delta G^{\circ\prime}_{F \to U} = +37.4\,\mathrm{kJ/mol}$. Which myoglobin is more thermodynamically stable, the mutant or the wild-type?

20. The following data are found for the steady state kinetics of a multi-subunit enzyme:

[S] (mmol/L)	Initial Rate [(mmol/L)s^{-1}]
0.25	0.26
0.33	0.45
0.50	0.92
0.75	1.80
1.00	2.50
2.00	4.10
4.00	4.80

(a) Does this enzyme follow Michaelis–Menten kinetics?

(b) Estimate V_{max}.

*21. Derive an equation expressing $\left(\dfrac{v/V_{max}}{1 - v/V_{max}}\right)$ as a function of $\log[S]$ for an enzyme that obeys the Michaelis–Menten equation.

For an *allosteric* enzyme, show how you could use this equation to obtain K_M^T and K_M^R (the K_M values for T and R states) from a Hill plot of kinetic data (see Figure 7.9).

*22. In a few instances, multisubunit enzymes have been demonstrated to exhibit *negative cooperativity*; that is, the binding of the first substrate molecule markedly decreases the affinity for binding of subsequent molecules. Show what a Hill plot would look like for an enzyme exhibiting negative cooperativity. [Hint: Consult Problem 21a.]

23. The MWC theory (see pages 245–246 in Chapter 7) cannot account for negative cooperativity, but the KNF theory can. Explain.

*24. Although the MWC theory describing binding to a multisite protein is quite complicated, we can easily derive equations for a simple case. Assume a dimeric enzyme, with two active sites and two conforma-

tional states: T and R. For simplicity, assume that the R state can bind one or two molecules of substrate, but the T state cannot bind. We then have the equations

$$T \rightleftharpoons R \qquad L = \frac{[T]}{[R]}$$

$$R + S \rightleftharpoons RS \qquad K = \frac{2[R][S]}{[RS]}$$

$$RS + S \rightleftharpoons RS_2 \qquad 2K = \frac{[RS][S]}{[RS_2]}$$

The factor 2 appears because there are two equivalent sites on each R to bind S.

(a) From this information, show that the fraction of sites occupied at substrate concentration [S] is

$$Y = \frac{[S]}{K}\left(\frac{1 + \dfrac{[S]}{K}}{L + \left(1 + \dfrac{[S]}{K}\right)^2}\right) = \frac{v}{V_{max}}$$

(b) Picking a large value of L (say 10^5), and with $K = 10^{-6}$ M, show that a sigmoidal curve is obtained for v/V_{max} versus [S].

25. Suggest the effects of each of the following mutations on the physiological role of chymotrypsinogen:

(a) R15S

(b) C1S

(c) T147S

26. Suppose you had available a sample of the ribozyme IVS-19 from *Tetrahymena*. Describe the protocol for a simple experiment that would demonstrate the reaction shown in Figure 11.41b.

27. Read Tools of Biochemistry 11B before answering this question. From the example shown Figure 11.B2, suggest the design of an antigen that might produce a catalytic antibody capable of amide bond hydrolysis.

28. (a) Derive the Michaelis–Menten expression for uncompetitive inhibition under steady-state conditions.

(b) Derive the Michaelis–Menten expression for the simple case of mixed inhibition (described on pages 443–444) under steady-state conditions. [Hint: In each case, start with the appropriate expression for $[E]_T$, and recall that under steady-state conditions, $K_M = [E][S]/[ES]$.]

29. Enalapril is an anti-hypertension "prodrug" (i.e., a drug precursor) that is inactive until the ethyl ester (arrow in figure) is hydrolyzed by esterases present in blood plasma. The active drug is the dicarboxylic acid ("enalaprilat") that results from this hydrolysis reaction.

(a) Enalapril is administered in pill form, but enalaprilat must be administered intravenously. Why do you suppose enalapril works as a pill, but enalaprilat does not?

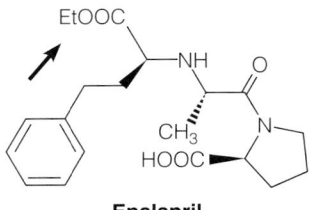

Enalapril

(b) Enalaprilat is a *competitive* inhibitor of the angiotensin-converting enzyme (ACE), which cleaves the blood-pressure regulating peptide angiotensin. ACE has a $K_M = 12\,\mu\mathrm{M}$ for angiotensin, which is present in plasma at a concentration of 75 μM. When enalaprilat is present at 2.4 nM, the activity of ACE in plasma is 10% of its uninhibited activity. What is the value of K_I for enalaprilat?

TOOLS OF BIOCHEMISTRY 11A

HOW TO MEASURE THE RATES OF ENZYME-CATALYZED REACTIONS

There are essentially two approaches to enzyme kinetic analysis. The first and simplest is to make measurements of rates under conditions in which the steady-state approximation holds (see page 432). Under these conditions, the Michaelis–Menten equation is often applicable, and determination of the reaction velocity as a function of substrate and enzyme concentrations will yield K_M and k_{cat}. Almost all enzymatic studies at least start in this way. But if the experimenter wishes to learn more of the details of the mechanism, it is often important to carry out studies before the steady state has been attained. Such *pre-steady-state* experiments require the use of special fast techniques. On pages 437–438 we described how a combination of such approaches can be used to dissect a complex enzymatic process and to understand it in detail. Here we describe some of the experimental techniques that can be employed.

Analysis in the Steady State

The steady state in most enzymatic reactions is established within seconds or a few minutes and persists for many minutes or even hours thereafter. Therefore, extreme rapidity of measurement is not important, and many techniques are available to the experimenter wishing to follow the reaction. Descriptions of the most commonly used techniques follow.

Spectrophotometry

Spectrophotometric methods are simple and accurate (see Tools of Biochemistry 6A). However, an obvious requirement is that either a substrate or a product of the reaction must absorb light in a spectral region where other substrates or products do not. Classic examples are reactions that generate or consume NADH. NADH absorbs quite strongly at 340 nm, but NAD^+ does not absorb in this region. Thus we could, for example, follow the oxidation of ethanol to acetaldehyde, as catalyzed by alcohol dehydrogenase, by measuring the formation of NADH spectrophotometrically. Even if the reaction being studied does not involve a light-absorbing substance, it may be possible to couple this reaction to another, very rapid reaction that does.

Fluorescence

The applications of fluorescence are similar to those of spectrophotometry, and the problems are similar: A substrate or a product must have a distinctive fluorescence emission spectrum (see Tools of Biochemistry 6A). However, fluorescence often has the advantage of high sensitivity, so extremely dilute solutions may be employed, enabling an experimenter to greatly extend the concentration range (i.e., [S]) over which studies are practicable.

Automatic Titration

If the reaction produces or consumes acid or base, it can be followed by using a device called a pH-*stat*. A glass electrode senses the pH of the solution, and its signal is used to actuate a motor-driven syringe that titrates acid or base into the reaction vessel to keep the pH constant. The time-based record of acid or base consumed is then a record of the progress of the enzymatically catalyzed reaction.

Radioactivity Assays

If a substrate is labeled with a radioactive isotope that will be lost or transferred during the reaction to be studied, measurement of changes in radioactivity can be an extremely sensitive kinetic method. This procedure requires that the labeled compound can be separated quickly at different, precisely defined times during the reaction. An example is a method often used with radioactive ATP. The ATP can be adsorbed on charcoal-impregnated filter disks by very fast filtration of aliquots from the reaction mixture. The radioactivity can then be measured in a scintillation counter (see Tools of Biochemistry 12A). Another example of the use of radioisotopes comes from measuring the rates of peptide-bond cleavage (by a protease), or protein biosynthesis (e.g., ribosomal protein synthesis). Peptides are most commonly labeled with radioactive amino acids that contain ^{3}H, ^{14}C, or [35]S. The rate of a peptide cleavage or synthesis reaction can be monitored by rapidly precipitating the peptide (or peptide fragments) from the reaction solution using cold trichloroacetic acid and collecting the precipitate on filter paper. As described above, the radioactivity present on the filter paper can be quantitated using a scintillation counter.

Analysis of Very Fast Reactions

Reactions that are extremely rapid require special techniques to investigate the pre-steady-state processes. Two major methods are currently employed to cover the rapid time scales shown in Figure 11A.1.

Stopped Flow

Figure 11A.2 shows a **stopped-flow apparatus**, first described by Quentin Gibson in the 1950s. Enzyme and substrate are initially in separate syringes. The syringes are driven, within a few milliseconds, to deliver their contents through a mixing chamber and into a third, "stopping" syringe. This step triggers a detector to begin observing (for example, by light absorption or fluorometry) the solution in the tube connecting the mixer to the stopping syringe. Flow rates can easily be made as high as 1000 cm/s. If the mixture was moving at this rate when the flow was stopped, and if the observation point is 1 cm from the mixer, the detection system first sees a mixture that is 1 ms "old." The reaction can then be followed for as long as desired—often for a period of only a few seconds. The limitations of the method are imposed only by the initial "dead time" (i.e., the time it takes the mixed solutions

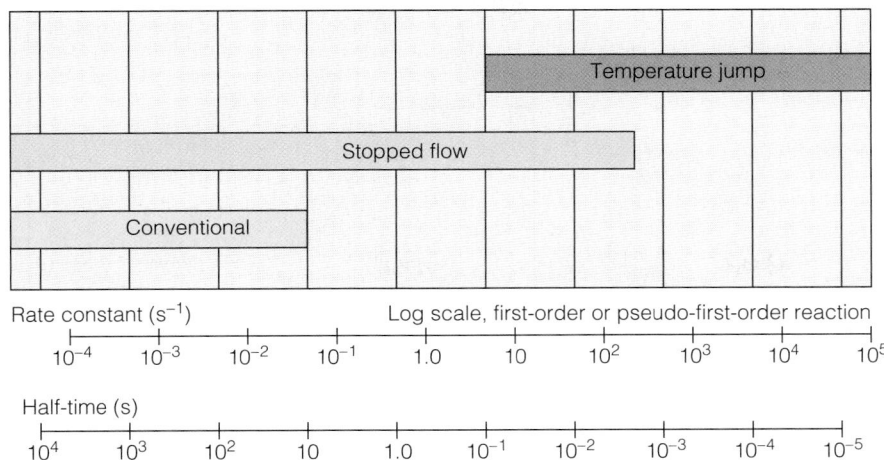

FIGURE 11A.1

Time scales for kinetic techniques described here.

Courtesy of Thermo Fisher Scientific.

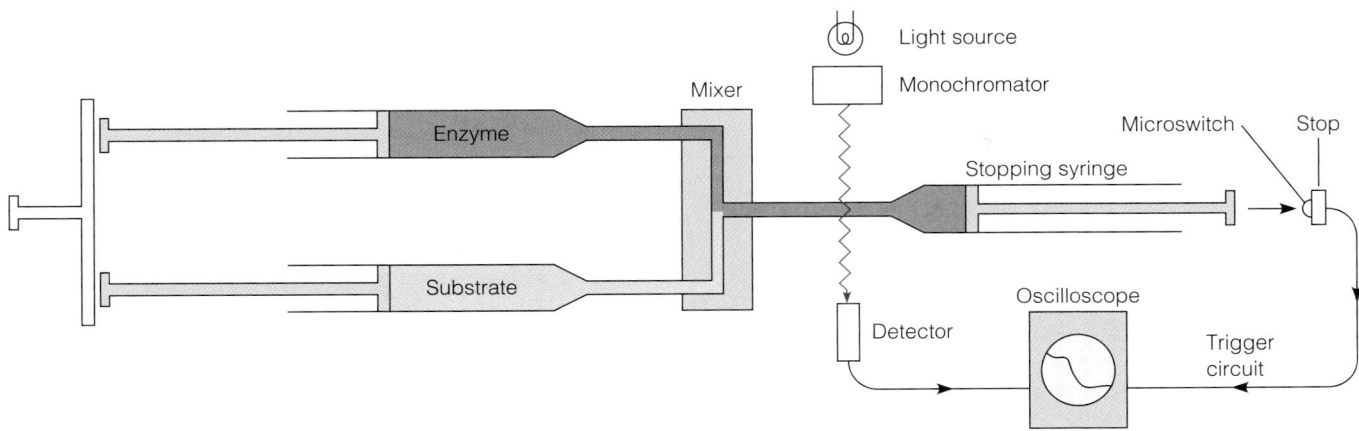

FIGURE 11A.2

Typical stopped-flow apparatus.

to arrive at the detector—in the example above, 1 ms) and the rapidity of the detection system.

Stopped-flow is used to measure rates of rapid enzymatic reactions as well as ligand binding events, such as O_2 binding to, or release from, hemoglobin (see Chapter 7).

Temperature Jump

Some processes are so fast that they are essentially completed in the dead time of a stopped-flow apparatus. The experimenter may then turn to temperature jump (T-jump) methods. The basic apparatus and principle of the method are shown in Figure 11A.3a and b, respectively. A reaction mixture that is at equilibrium at a temperature T_1 is suddenly jumped to a temperature T_2. Because chemical equilibria are typically temperature-dependent, the position of equilibrium will shift, and the system must now react to attain this new equilibrium. A rapid jump in temperature (5–10 °C in 1 μs) can be obtained by passing a large burst of electrical current between electrodes immersed in the reaction mixture. Even more rapid jumps (10–100 ns) can be obtained if a pulsed infrared laser is used to heat the mixture. The relaxation (approach) to a

new equilibrium, monitored by absorption or fluorescence measurements, is an exponential process. For a simple reaction, the change in reactant concentration is given by

$$\Delta[A] = (\Delta[A]_{total})e^{-t/\tau} \qquad (11.A1)$$

where τ is called the *relaxation time* and can be related to the rate constants for the reaction. For example, for the simple reversible reaction

$$A \underset{k_{-1}}{\overset{k_1}{\rightleftharpoons}} B$$

we have

$$\frac{1}{\tau} = k_1 + k_{-1} \qquad (11.A2)$$

More complex reactions involve multiple relaxation times and more complex curves than expressed by equation (11.A1). *T*-jump experiments are appropriate for reactions with τ values as low as about 10^{-5} s.

Although a number of other techniques are employed for even faster reactions, including some newly developed NMR

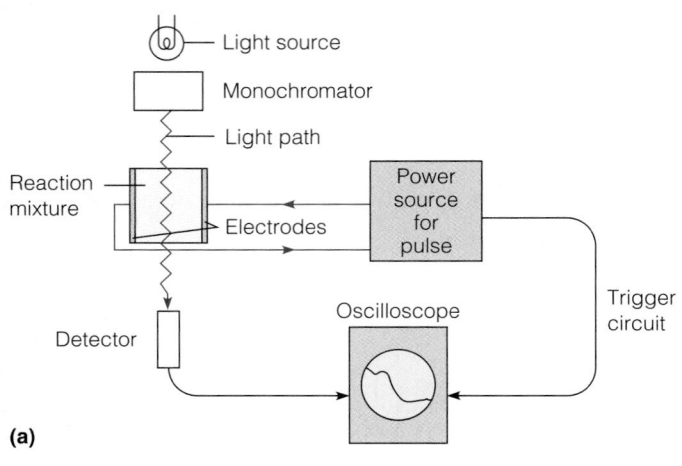

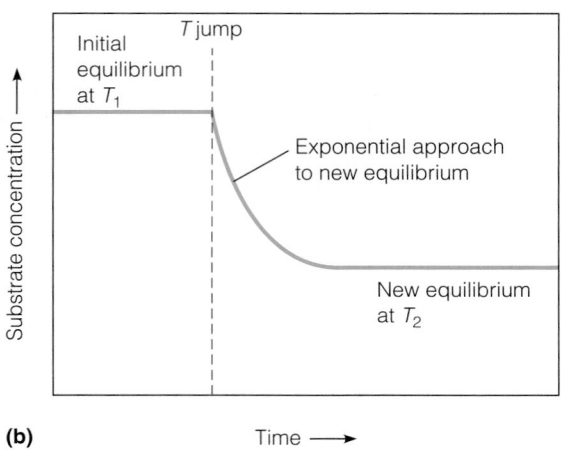

(a)   **(b)**

FIGURE 11A.3

The temperature jump method.

methods and pulsed laser techniques, the methods described here are widely used. If we consider the variety of techniques available to the experimenter, we can see that they cover a wide time range. Altogether, times from nanoseconds to hours can be studied.

Relating Kinetics to a Mechanism: Kinetic Isotope Effects

Kinetic data are the basis for proposing a detailed chemical mechanism such as those illustrated in Figures 11.14, 11.17, 11.38, and 11.39. The determinations of rates for the turnover of isotopically-labeled substrates are among the more useful data for distinguishing one possible mechanism from another (e.g., see the discussion of the lysozyme mechanism).

The rates of bond formation/cleavage depend on the masses of the atoms involved because the vibrational frequencies of bonds are sensitive to the masses of the bonded atoms. Bond

cleavage/formation reactions involving heavier isotopes proceed with slower rates, and this effect is known as a **kinetic isotope effect** (or KIE). Synthetic chemists have developed methods to synthesize substrates for enzyme-catalyzed reactions with atom-specific isotopic substitutions. In other words, at some specific location in the substrate molecule a deuterium (^2H or D) or a tritium (^3H or T) might be substituted for the common isotope of hydrogen, ^1H (sometimes referred to as "protium"). The KIE is largest for hydrogen isotopes because the change in mass is more significant for ^1H vs. ^2H (or ^3H), than it is for heavier elements (e.g., ^{12}C vs. ^{13}C, or ^{16}O vs. ^{18}O).

A so-called primary KIE is observed when the bond including the atom in question is broken/formed in the rate-determining step, whereas a secondary KIE is observed when the bond adjacent to the atom in question is broken/formed (Figure 11A.4). The KIE is recorded as the ratio of rates for the reactions of a substrate

Primary Kinetic Isotope Effect:
- the bond *to* the isotope is broken/formed
- k_H/k_D = 3–5 for hydride transfer

Secondary Kinetic Isotope Effect:
- the bond *adjacent* to the isotope is broken/formed
- k_H/k_D = 1.05–1.12

FIGURE 11A.4

Examples of primary and secondary kinetic isotope effects.

(a)   **(b)**

labeled with two different isotopes, for example k_H/k_D. For primary KIE, the range of observed values for hydrogen transfer reactions is 2–15. For secondary KIE the range is closer to 1 (1.05–1.12). These differences in magnitude allow us to distinguish primary and secondary KIEs.

If a researcher suspects that a particular bond is broken during the slow step of a reaction she/he can synthesize a labeled substrate and compare the kinetics of the reaction with unlabeled and labeled substrates. If the expected KIE is observed, the proposed mechanism may be correct. If the expected KIE is not observed, the proposed mechanism is likely incorrect, or the bond cleavage/formation is more rapid than some other rate-determining step in the mechanism. In this way, KIEs are useful tools for the elucidation of mechanistic detail in enzyme-catalyzed reactions.

References

Cleland, W. W. (2003) The use of isotope effects to determine enzyme mechanism. *J. Biol. Chem.* 278:51975–51984.

Fersht, A. (1999) *Structure and Mechanism in Protein Science.* W. H. Freeman and Co., New York.

Himori, K. (1979) *Kinetics of Fast Enzyme Reactions.* Halstead, New York.

Johnson, K. A. (2003) Introduction to kinetic analysis of enzyme systems. In *Kinetic Analysis of Macromolecules: A Practical Approach,* K. A. Johnson (ed.). Oxford University Press, New York.

Schramm, V. L. (2005) Enzymatic transition states: Thermodynamics, dynamics and analogue design. *Arch. Biochem. Biophys.* 433:13–26.

TOOLS OF BIOCHEMISTRY 11B

INTRODUCTION TO PROTEIN ENGINEERING OF ENZYMES

The search for catalysts with "tailor-made" function has led to the development of many strategies to generate novel protein structures that possess the desired function. These fall broadly into two general categories: "rational design" methods, which are based on application of fundamental principles of protein folding, stability, and function as described in Chapters 5–7, and "directed evolution" methods, which are based on the generation of large pools of protein variants ("libraries"), followed by stringent selection for the desired function. Often, these methods are combined, such that a candidate catalyst with modest activity will be used as a starting template to generate a large library of variants, which is then screened to identify those variants with improved activity. Although we illustrate these principles with examples of protein catalysts, these same strategies are applied in the design/selection of nucleic acid catalysts (see Figure 11.43).

We begin this brief survey with a reminder that the manipulation of gene sequence allows for changes in amino acid sequence in both site-specific ("rational") and random ("directed evolution") strategies.

Site-Directed Mutagenesis

It is now routine to clone the genes for enzymes and to make specific mutations at particular sites in the sequence (as described in Tools of Biochemistry 5A and 25E). The method has proved extremely powerful in the elucidation of enzyme mechanisms, as described previously for lysozyme and the serine proteases (see pages 425–429); however, it can also be employed to change enzyme specificities. For example, a study of the protease *subtilisin* by James Wells and coworkers focused on mutations at a specific site (residue 166) in the specificity pocket. This site is normally occupied by Gly, and the enzyme preferentially cleaves polypeptide chains next to a bulky hydrophobic residue. Activity toward polypeptides containing glutamic acid in the same position is very low. Replacing Gly 166 by Lys increases by 500-fold the frequency of cleavage next to glutamic acid.

Although some attempts at protein engineering for industrial purposes have aimed at tailoring enzymes for new reactions, more have focused on developing tolerance to extreme environmental conditions (heat, acidity, salinity) typical of industrial or specialized agricultural operations. However, to date, nature (evolution) appears to have been a better designer than humans; the most interesting enzymes capable of tolerating extreme environments have been found in bacteria that inhabit hot springs, deserts, and undersea thermal vents. It is now clear that there exist in nature enzymes capable of efficient function at temperatures as high as 100 °C.

Domain Fusion: Chimeric Enzymes

The rearrangement of genes that is possible with modern molecular biology techniques allows production of **fusion proteins**—proteins coded for by genes that have been spliced together in vitro from two or more sources. Such gene fusion makes it possible to join proteins, or specific protein domains, in novel ways. As a simple example of this approach, consider the problem of digesting cereal β-glucans, as found in barley. These polysaccharides involve both $\beta(1 \rightarrow 4)$ and $\beta(1 \rightarrow 3)$ linkages between glucose residues. In brewing practice, a glucanase and a microbial cellulase are used, but the latter is often ineffective in industrial conditions, leading to viscous products that interfere with beer manufacture. A hybrid enzyme has been constructed that possesses the capability of cleaving both kinds of bonds and thereby digesting the glucan to glucose in one step.

FIGURE 11B.1

Generating novel chemical structures by engineering of polyketide synthases. In this simplified example, a PKS containing 6 domains yields the erythromycin core structure (leftmost figure). Deletion of domain 4 or domain 5 (right side of figure) alters the structure of the final product (see areas highlighted in blue). See Figure 17.37 for a more detailed description of the organization and activities of the various domains in a PKS.

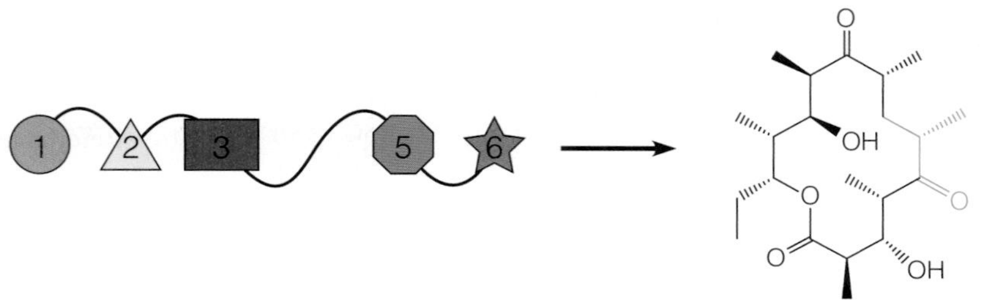

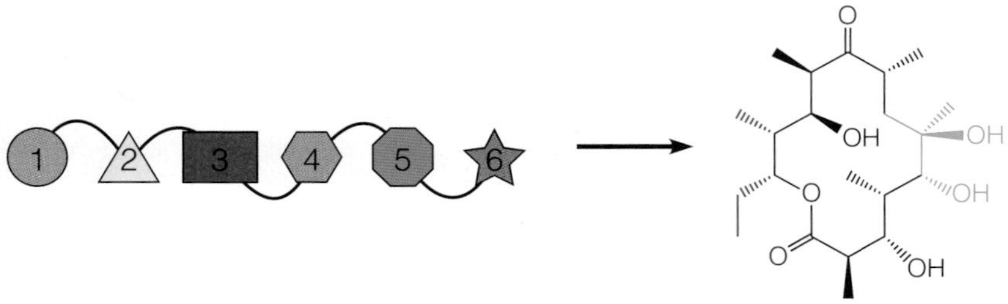

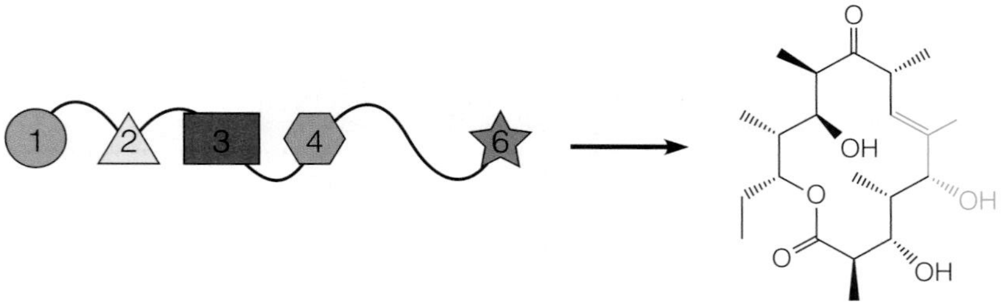

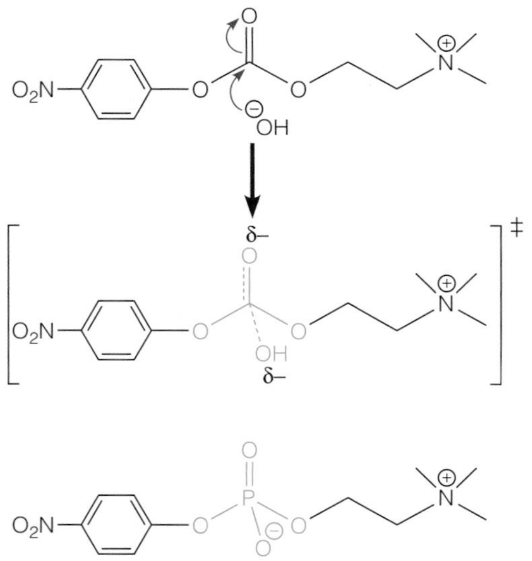

Transition state analog for carbonate hydrolysis

FIGURE 11B.2

Modeling the transition state for carbonate hydrolysis with a phosphonate ester. **Top:** The hydrolysis of a carbonate at alkaline pH involves the attack of a hydroxide ion on the carbonate carbon. **Middle:** A model of the transition state for this reaction. **Bottom:** A transition state analog for the alkaline hydrolysis reaction. This analog is relatively stable; thus, it can be used for affinity purification of candidate catalysts from a large pool of clones.

In addition to creating novel enzymes through domain fusion, it is also possible to modify the sequences of multidomain proteins to alter their function. For example, polyketide synthases (PKS) are very large multidomain enzymes that carry out the multistep synthesis of complex natural products, such as erythromycin. PKS are of great interest because the molecular engineering of these enzymes may lead to new antibiotics that could be effective in the treatment of patients infected with antibiotic-resistant strains of bacteria. Each domain in a PKS performs a specific chemical transformation on the substrate, and the *specific arrangement* of the domains within the primary sequence of the PKS dictates the chemical structure and stereochemistry of the final product (Figure 11B.1). Different products can be obtained by selectively rearranging, or deleting, gene sequences encoding PKS domains. The chemistry performed by PKS is described in greater detail in Chapter 17 (e.g., see Figure 17.37, page 747).

Selection from Random Libraries: Catalytic Antibodies

Recall from Chapter 7 that antibodies show remarkably high specificity in binding to their target antigens. Enzymes bind most strongly to the transition state in a reaction. What happens if we make antibodies against molecules that are transition-state analogs for a particular reaction? The answer is that these antibodies act like enzymes: They show specificity for a particular substrate and rate enhancement for the reaction of interest. One of the first examples of antibody-mediated catalysis was for the carbonate ester hydrolysis shown in Figure 11B.2. Just as in the hydrolysis of amides, a tetrahedral transition state is generated in ester hydrolysis. A tetrahedral phosphonate ester was used to mimic the transition state for the carbonate hydrolysis reaction. Because the phosphonate is stable, it was also used to make an affinity column (Tools of Biochemistry 5A) that was, in turn, used to purify antibodies that bind tightly to the phosphonate. The antibodies purified in this fashion bound the analogous carbonate and catalyzed its hydrolysis with a rate enhancement of $\sim 10^3$.

Using various kinds of molecules as antigens, it has been possible to produce catalytic antibodies for a number of the classes of reactions shown in Table 11.10. In some cases, rate enhancements as much as 10^7 have been obtained; however, the majority of catalytic antibodies reported to date show rate enhancements in the range of 10^4–10^5. Some researchers in this field have speculated that this value may represent the practical upper limit for rate enhancements achievable with catalytic antibodies.

For many years, the major difficulty in generating catalytic antibodies directed toward specific compounds or functional groups was the necessity of using the immune system of some animal to make the selection. Systems have been developed recently that circumvent this requirement. The basic idea is that libraries of F_{ab} or F_v fragments (see Chapter 7) with random complementarity determining regions are cloned and expressed on the surface of bacteriophage particles. The collection of phage particles is called a "library," which typically contains 10^6–10^8 different clones. This library is then subjected to affinity chromatography to select for those clones that display an F_{ab} or F_v that can bind the desired molecule or structure. Such techniques have allowed the development of catalytic antibodies directed toward synthetic molecule substrates that would be very difficult to present as antigens to an in vivo system because of toxic effects or chemical instability of the antigen.

The generation of such random sequence libraries allows for so-called **directed evolution** of protein (or nucleic acid) function in the laboratory. A diverse set of protein sequences can be generated by subjecting multiple sites in a protein gene to simultaneous random mutagenesis. The identification of the few clones within a library that possess the desired novel function is carried out using the appropriate selection or screen for that function. As we shall see in the next section, one or two rounds of directed evolution are often required to fine-tune the function of a designed protein.

Mutagenesis *in silico*: Computational Design of Enzymes

Recall from Chapter 6 that we described some modern computational methods for protein-structure prediction from amino acid sequence (see Figures 6.35 and 6.36). The reverse problem, predicting what sequence of amino acids will yield a desired tertiary structure, has also been addressed using computation. We can now bring several topics together to ask a fascinating question: Is it possible to design a protein catalyst with an active site that is complementary to the transition state for a

reaction? Protein chemists have pursued an answer to this difficult question for many years, and recent results from David Baker and colleagues suggest that the answer is "yes"—albeit, a qualified yes.

In this approach, the transition state for a reaction is modeled—we present an example for which the desired activity is that of a retro-Aldolase, which results in C—C bond cleavage (Figure 11B.3a; the retroaldol reaction is discussed in greater detail in Chapter 13). A library of model active sites is then built around the transition state (Figure 11B.3b/c). This library may have 10^{18} different models. In the next step, these model active sites are matched to the structures of known protein scaffolds that can display the critical catalytic residues in the same orientation required by the various active site models (Figure 11B.3d). This matching step is computationally intensive; however, the computer is able to evaluate many more potential active sites than is practical to clone and test in the laboratory. In the final computational design step, the specific mutations in the candidate scaffold proteins are identified and the genes for the candidate enzymes are cloned and expressed. In this example, 72 candidate sequences were cloned and evaluated and roughly half showed some catalytic activity. The best clones exhibited a rate enhancement of $\sim 10^4$—significant, but not as large as the values for naturally occurring enzymes (see Figure 11.4).

A similar approach produced a Kemp elimination catalyst with a 10^4 rate enhancement. This could be increased to 10^6 following directed evolution of the computationally designed sequence. Two important results emerged from these studies: (1) the computational approach produced catalysts with good specificity and modest rate accelerations and (2) the crystal structures of the resulting enzymes very closely matched the structures generated during the computational design. More recently, a Diels–Alder catalyst has been generated using this methodology (see citations in the References section).

An important question remains: Why do designed catalysts typically achieve rate accelerations far below those of naturally occurring enzymes? The answer may be revealed as we gain a greater quantitative understanding of the role of protein dynamics in the function of enzymes.

References

Janda, K. D., L.-C. Lo, C.-H. Lo, M.-M. Sim, R. Wang, C.-H. Wong, and R. A. Lerner (1997) Chemical selection for catalysis in combinatorial antibody libraries. *Science* 275:945–948.

Jiang, L., E. A. Althoff, F. R. Clemente, L. Doyle, D. Röthlisberger, A. Zanghellini, J. L. Gallaher, J. L. Betker, F. Tanaka, C. F. Barbas III, D. Hilvert, K. Houk, B. L. Stoddard, and D. Baker (2008) De novo computational design of retro-aldol enzymes. *Science* 319:1387–1391.

Khosla, C. (1997) Harnessing the biosynthetic potential of modular polyketide synthases. *Chem. Rev.* 97:2577–2590.

Lutz, S. (2010) Reengineering enzymes. *Science* 329:285–287.

Nanda, V., and R. L. Koder (2010) Designing artificial enzymes by intuition and computation. *Nature Chem.* 2:15–24.

Olsen, O., K. K. Thomsen, J. Weber, J. Duus, I. Svendsen, C. Wegener, and D. von Wettstein (1996) Transplanting two unique β-glucanase catalytic activities into one multienzyme which forms glucose. *Nat. Biotechnol.* 14:71–76.

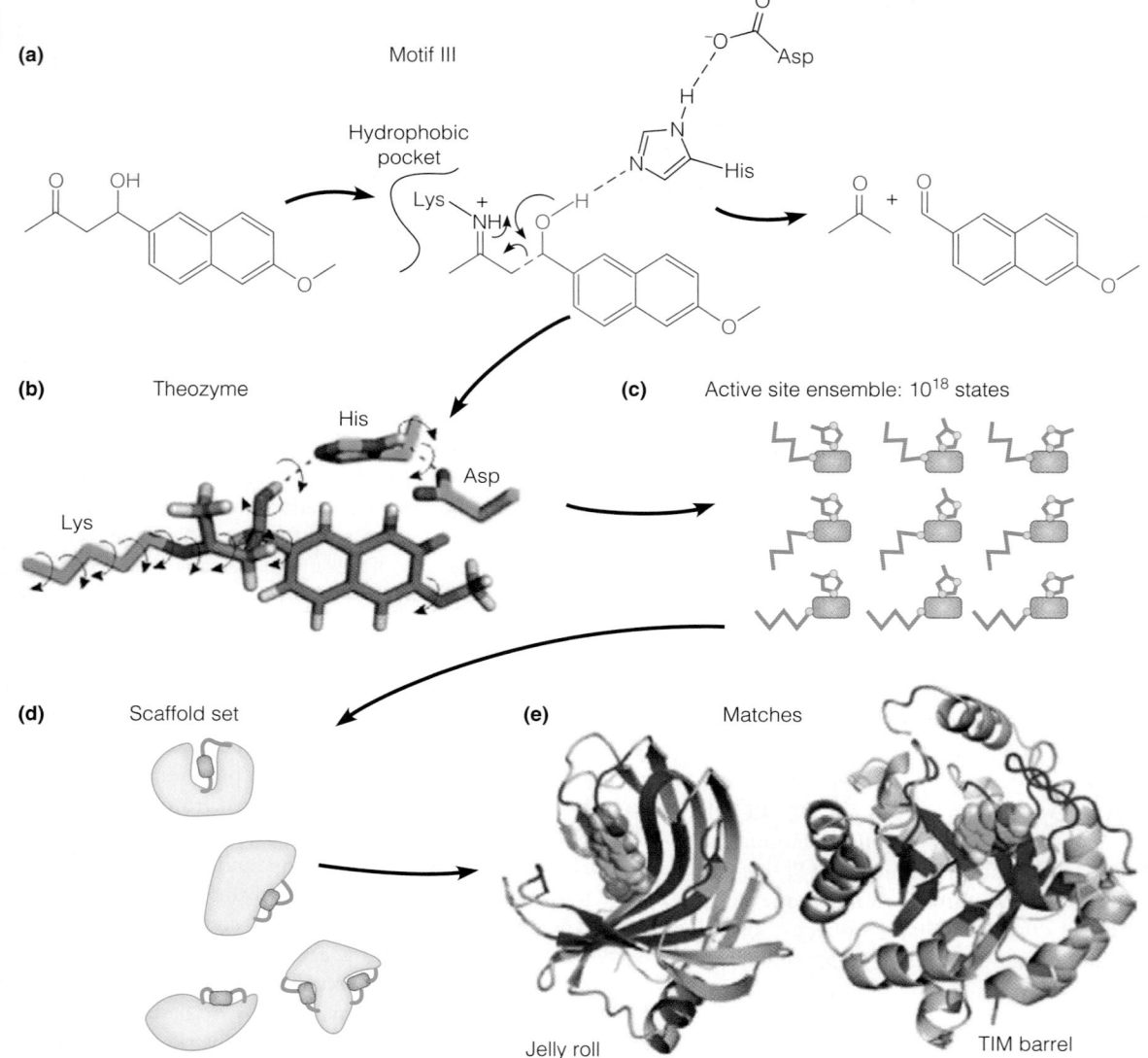

FIGURE 11B.3

Computational design of a retro-Aldolase. See the text for an explanation of the steps illustrated here.

Reprinted by permission from Macmillan Publishers Ltd. *Nature Chemical Biology* 4:273–275, V. Nanda, Do-it-yourself enzymes. © 2008.

Röthlisberger, D., O. Kersonsky, A. M. Wollacott, L. Jiang, J. DeChancie, J. Betker, J. L. Gallaher, E. A. Althoff, A. Zanghellini, O. Dym, S. Albeck, K. N. Houk, D. S. Tawfik, and D. Baker (2008) Kemp elimination catalysts by computational enzyme design. *Nature* 453:190–195.

Siegel, J. B., A. Zanghellini, H. M. Lovick, G. Kiss, A. R. Lambert, J. L. St.Clair, J. L. Gallaher, D. Hilvert, M. H. Gelb, B. L. Stoddard, K. N. Houk, F. E. Michael, and D. Baker (2010) Computational design of an enzyme catalyst for a stereoselective bimolecular Diels–Alder reaction. *Science* 329:309–313.

Turner, N. (2009) Directed evolution drives the next generation of biocatalysts. *Nature Chem. Biol.* 8:567–573.

Walsh, C. T. (2004) Polyketide and nonribosomal peptide antibiotics: Modularity and versatility. *Science* 303:1805–1810.

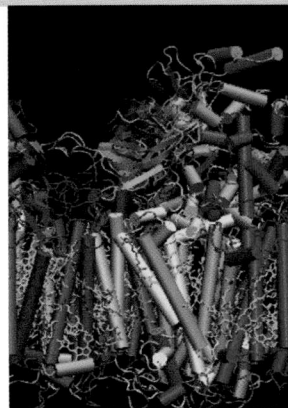

CHAPTER 12

Chemical Logic of Metabolism

A chemist carrying out an organic synthesis rarely runs more than one reaction in a single reaction vessel. This procedure is essential to prevent side reactions and to optimize the yield of the desired product. Yet a living cell carries out thousands of reactions simultaneously, with each reaction sequence controlled so that unwanted accumulations or deficiencies of intermediates and products do not occur. Reactions of great mechanistic complexity and stereochemical selectivity proceed smoothly under mild conditions—1 atm pressure, moderate temperature and osmotic pressure, and a pH near neutrality. How do cells avoid metabolic chaos? A goal of the next several chapters is to understand how cells carry out and regulate these complex reaction sequences.

A First Look at Metabolism

As noted above, a principal task of the biochemist is to understand how a cell regulates its myriad reaction sequences and, in so doing, controls its internal environment. In Chapter 11 we discussed the properties of individual enzymes and control mechanisms that affect their activity. Here we consider specific reaction sequences, or **pathways**; the relationship between each pathway and cellular architecture; the biological importance of each pathway; control mechanisms that regulate **flux**, or intracellular reaction rate; and experimental methods used to investigate metabolism. Figure 12.1, a simplified view of the processes we shall consider, illustrates two important principles. First, metabolism can be subdivided into two major categories—**catabolism**, those processes in which complex substances are degraded to simpler molecules, and **anabolism**, those processes concerned primarily with the synthesis of complex organic molecules (Figure 12.1a). Catabolism is generally accompanied by the net release of chemical energy, and anabolism requires a net input of chemical energy; we will see that these two sets of reactions are coupled together by ATP. Second, both catabolic and anabolic pathways occur in three stages of complexity—stage 1, the interconversion of

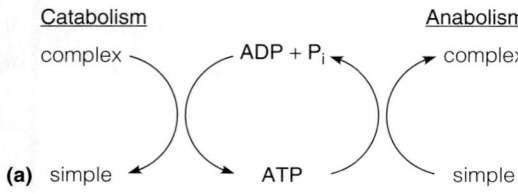

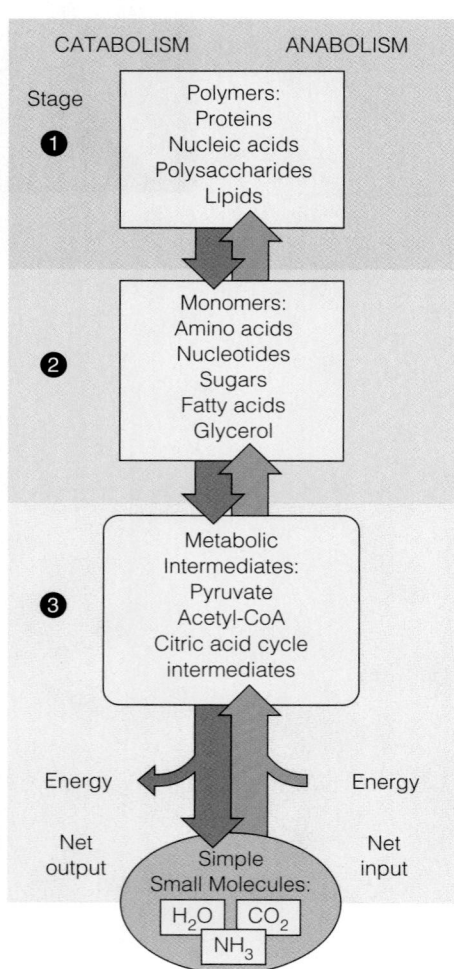

(b)

FIGURE 12.1

A brief overview of metabolism.

Intermediary metabolism refers primarily to the biosynthesis, utilization, and degradation of low-molecular-weight compounds (intermediates).

polymers and complex lipids with monomeric intermediates; stage 2, the interconversion of monomeric sugars, amino acids, and lipids with still simpler organic compounds; and stage 3, the ultimate degradation to, or synthesis from, inorganic compounds, including CO_2, H_2O, and NH_3 (Figure 12.1b). As we proceed through this chapter, we shall add detail to this figure, introducing you thereby to each major metabolic process and identifying the functions of each.

During our discussion we shall see that energy-yielding pathways also generate intermediates used in biosynthetic processes. Thus, although we will focus at first upon degradation of organic compounds to provide energy, you should be aware that metabolism is really a continuum, with many of the same reactions playing roles in both degradative and biosynthetic processes.

We shall also use the terms **intermediary metabolism**, **energy metabolism**, and **central pathways**. Intermediary metabolism comprises all reactions concerned with storing and generating metabolic energy and with using that energy in biosynthesis of low-molecular-weight compounds (intermediates) and energy-storage compounds. Not included are nucleic acid and protein biosynthesis from monomeric precursors. The reactions of intermediary metabolism can be thought of as those that do not involve a nucleic acid template because the information needed to specify each reaction is provided within the structure of the enzyme catalyzing that reaction. Energy metabolism is that part of intermediary metabolism consisting of pathways that store or generate metabolic energy. Chapters 13 through 22 present intermediary metabolism, and a major focus in Chapters 13 through 17 is on its energetic aspects. The central pathways of metabolism are substantially the same in many different organisms, and they account for relatively large amounts of mass transfer and energy generation within a cell; they are the quantitatively major pathways. Later we will identify and further discuss each central pathway, but for now let us focus on the overall process.

Most organisms derive both the raw materials and the energy for biosynthesis from organic fuel molecules such as glucose. The central pathways involve the oxidation of fuel molecules and the synthesis of small biomolecules from the resulting fragments; these pathways are found in all aerobic organisms. But a fundamental distinction among these organisms lies in the source of their fuel molecules.

Autotrophs (from Greek, "self-feeding") synthesize glucose and all of their other organic compounds from inorganic carbon, supplied as CO_2. In contrast, **heterotrophs** ("feeding on others") can synthesize their organic metabolites only from other organic compounds, which they must therefore consume. A primary difference between plants and animals is that plants are autotrophs and animals are heterotrophs. With the exception of rare insect-eating plants, such as the Venus flytrap, green plants obtain all of their organic carbon through photosynthetic fixation of carbon dioxide. Animals feed on plants or other animals and synthesize their metabolites by transforming the organic molecules they consume. Microorganisms exhibit a wide range of biosynthetic capabilities and sources of metabolic energy.

Microorganisms also show adaptability with respect to their ability to survive in the absence of oxygen. Virtually all multicellular organisms and many bacteria are strictly **aerobic** organisms; they depend absolutely upon **respiration**, the coupling of energy generation to the oxidation of nutrients by oxygen. By contrast, many microorganisms either can, or must, grow in **anaerobic** environments, deriving their metabolic energy from processes that do not involve molecular oxygen.

To the extent that biological molecules are synthesized ultimately from CO_2 that undergoes photosynthetic carbon fixation, the sun can be considered as the ultimate source of biological energy. However, this concept is not quite accurate because of the existence of a relatively large group of prokaryotic cells (both eubacteria and archaea) that derive their energy from other sources—for example, extremely **thermophilic** organisms, which live at temperatures as high as

100 °C or more, either in hydrothermal vents deep in the ocean or in geothermal vents in active volcanic regions. Although there is much to learn about metabolism in these organisms, it is clear that most of their metabolic energy does not come from sunlight.

Freeways on the Metabolic Road Map

Now let us return to the central pathways and their identification. Probably you have seen metabolic charts—those wall hangings like giant road maps that adorn biochemistry laboratories and offices. Figure 12.1 is a highly simplified metabolic chart. Figure 12.2 presents metabolism in more detail and is the basic road map for this section of the book. This figure will reappear in subsequent chapters, with the pathways presented in those chapters highlighted.

Faced with the thousands of individual reactions that constitute metabolism, how do we approach this vast topic? Our first concerns are with central pathways and with energy metabolism. Therefore, in the next several chapters we consider the degradative processes that are most important in energy generation—the catabolism of carbohydrates and lipids. We shall also consider how these substances are biosynthesized. These reactions are located in the middle of the metabolic charts and are illustrated with the biggest arrows—freeways, so to speak, on the metabolic road map.

The road map analogy is also useful when we consider directional flow in metabolism. Just as traffic flows from the suburbs to downtown in the morning and in the reverse direction in the evening, so also will we see that some conditions favor biosynthesis whereas others favor catabolism, and that parts of the same highways are used in both sets of processes.

Central Pathways of Energy Metabolism

Recall from our discussion of Figure 12.1 that metabolism can be subdivided into three stages of complexity of the metabolites involved. The first pathway that we present in detail (in Chapter 13) is **glycolysis**, a stage 2 pathway for degradation of carbohydrates, in either aerobic or anaerobic cells. As schematized in Figure 12.3, the major input to glycolysis is glucose, usually derived from either energy-storage polysaccharides or dietary carbohydrates. This pathway leads to pyruvate, a three-carbon α-keto acid. Anaerobic organisms reduce pyruvate to a variety of products—for example lactate, or ethanol plus carbon dioxide. These processes are called *fermentations* (see page 487). In oxidative metabolism (respiration), the major fate of pyruvate is its oxidation to a metabolically activated two-carbon fragment, **acetyl-coenzyme A**, or acetyl-CoA (see page 601). The two carbons in the acetyl group then undergo oxidation in the **citric acid cycle** (Figure 12.4). In aerobic organisms the citric acid cycle, presented in Chapter 14, is the principal stage 3 pathway. This cyclic pathway accepts simple carbon compounds, derived not only from carbohydrate but also from lipid or protein, and oxidizes them to CO_2. Using the freeway analogy again, we will see that numerous on-ramps from the highways and byways of stage 1 and stage 2 metabolism lead to the citric acid cycle. In fact, all catabolic pathways converge at this point.

Oxidative reactions of the citric acid cycle generate reduced electron carriers whose reoxidation drives ATP biosynthesis, primarily through processes in the mitochondrial respiratory chain—**electron transport** and **oxidative phosphorylation**, shown also in Figure 12.4. As described in Chapter 15, the mitochondrial membrane uses oxidative energy to maintain a transmembrane gradient of hydrogen ion concentration (the **protonmotive force**), and discharge of this electrochemical potential energy powers the synthesis of ATP from ADP + P_i.

Stage 2 pathways other than glycolysis also deliver fuel to the citric acid cycle. Acetyl-CoA comes not only from pyruvate oxidation but also from the

The central pathways account for most of the mass transformations in metabolism.

In aerobic organisms all catabolic pathways converge at the citric acid cycle.

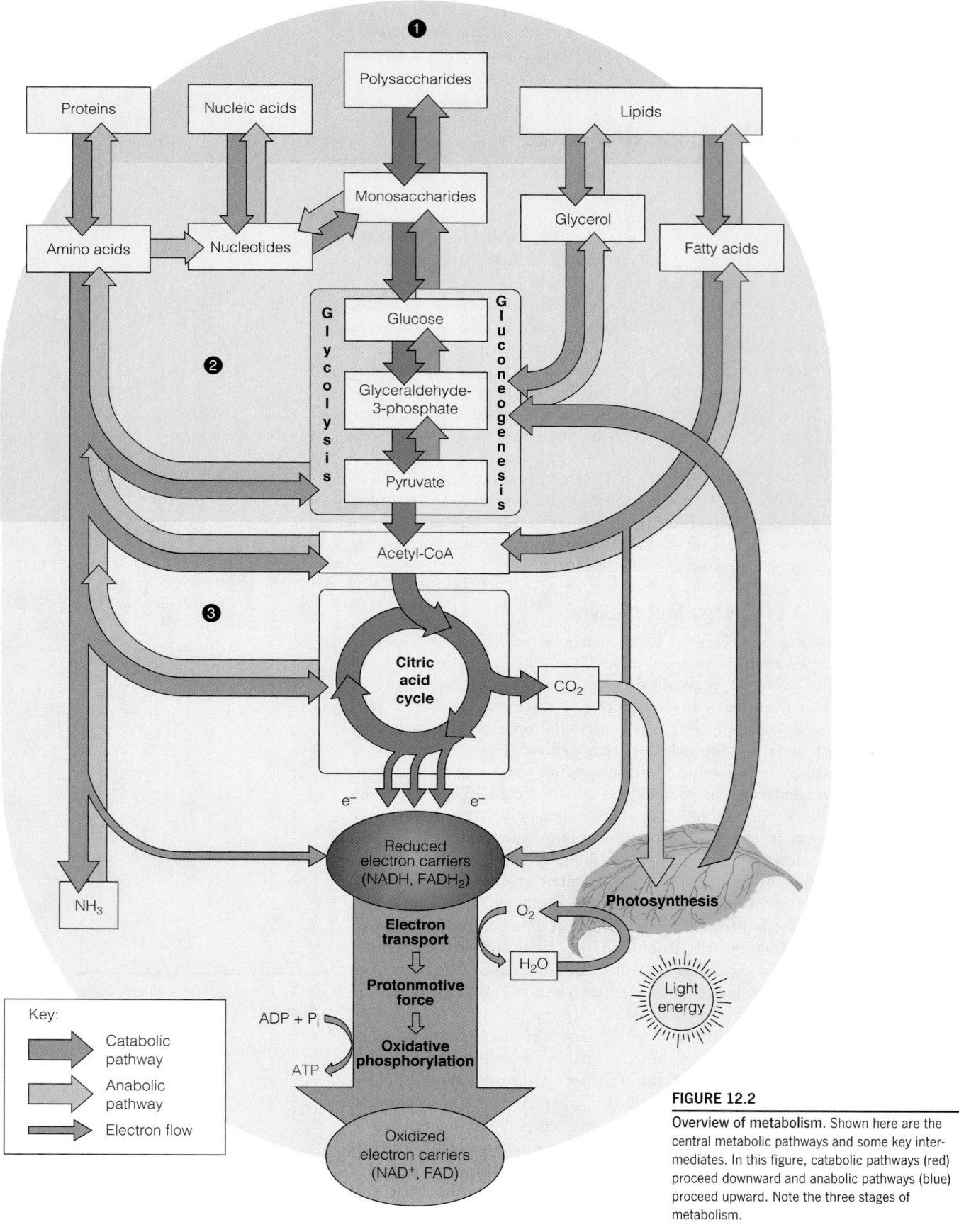

FIGURE 12.2

Overview of metabolism. Shown here are the central metabolic pathways and some key intermediates. In this figure, catabolic pathways (red) proceed downward and anabolic pathways (blue) proceed upward. Note the three stages of metabolism.

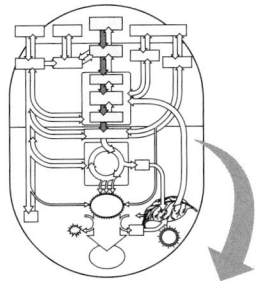

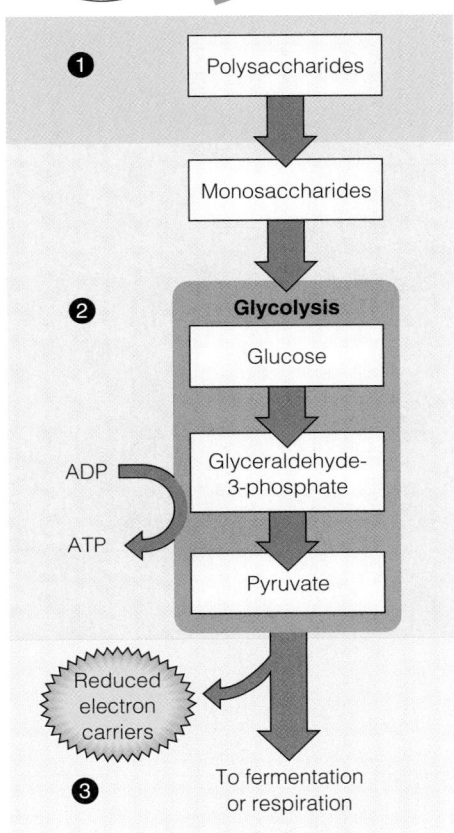

FIGURE 12.3

The initial phase of carbohydrate catabolism: glycolysis. Pyruvate either undergoes reduction in fermentation reactions or enters oxidative metabolism (respiration) via conversion to acetyl-CoA as shown in Figure 12.4.

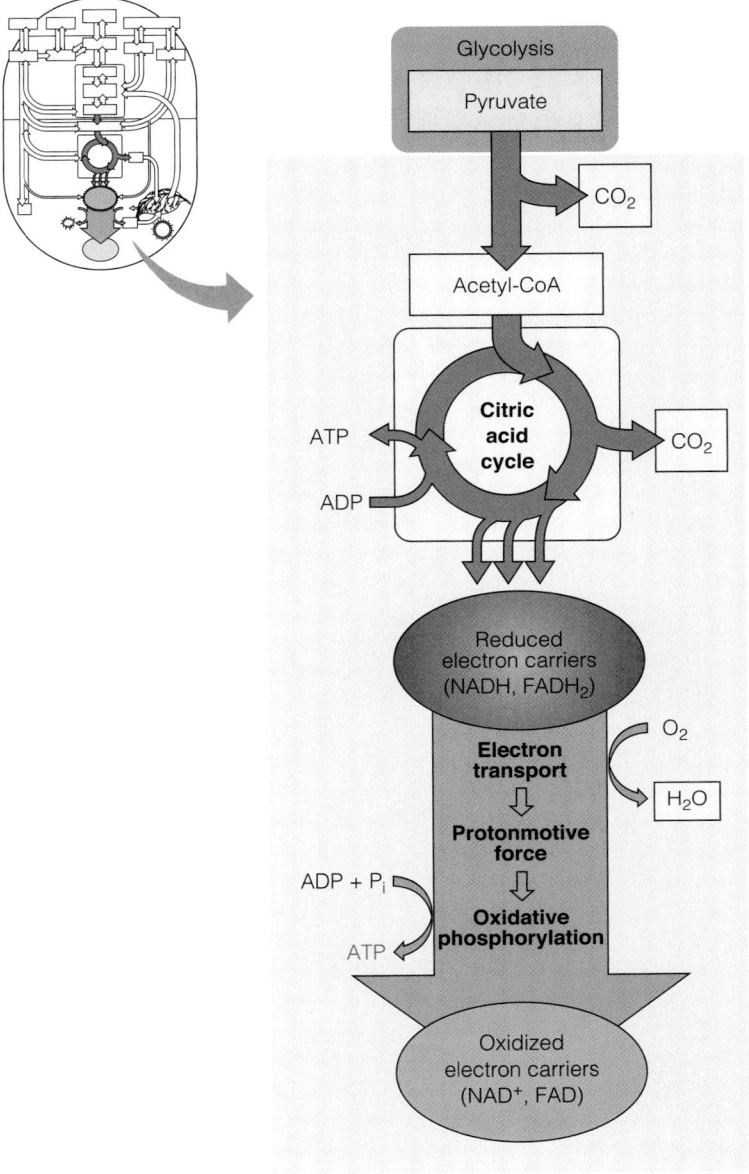

FIGURE 12.4

Oxidative metabolism. Oxidative metabolism includes pyruvate oxidation, the citric acid cycle, electron transport, and oxidative phosphorylation. Pyruvate oxidation supplies acetyl-CoA to the citric acid cycle.

breakdown of fatty acids by **β-oxidation** (presented in Chapter 17) and from some amino acid oxidation pathways (presented in Chapters 20 and 21). If the two carbons of acetyl-CoA are not oxidized in the citric acid cycle, they can go in the anabolic direction, providing substrates for synthesis of fatty acids and steroids (presented in Chapters 17 and 19). These and other biosynthetic processes use a reduced electron carrier, NADPH, which is structurally very similar to NADH (page 488).

Several important biosynthetic processes for carbohydrates will concern us in this book (Figure 12.5). Chapter 13 presents **gluconeogenesis**, the synthesis of glucose from noncarbohydrate precursors, and polysaccharide biosynthesis, particularly the biosynthesis of glycogen in animal cells. In Chapter 16 we

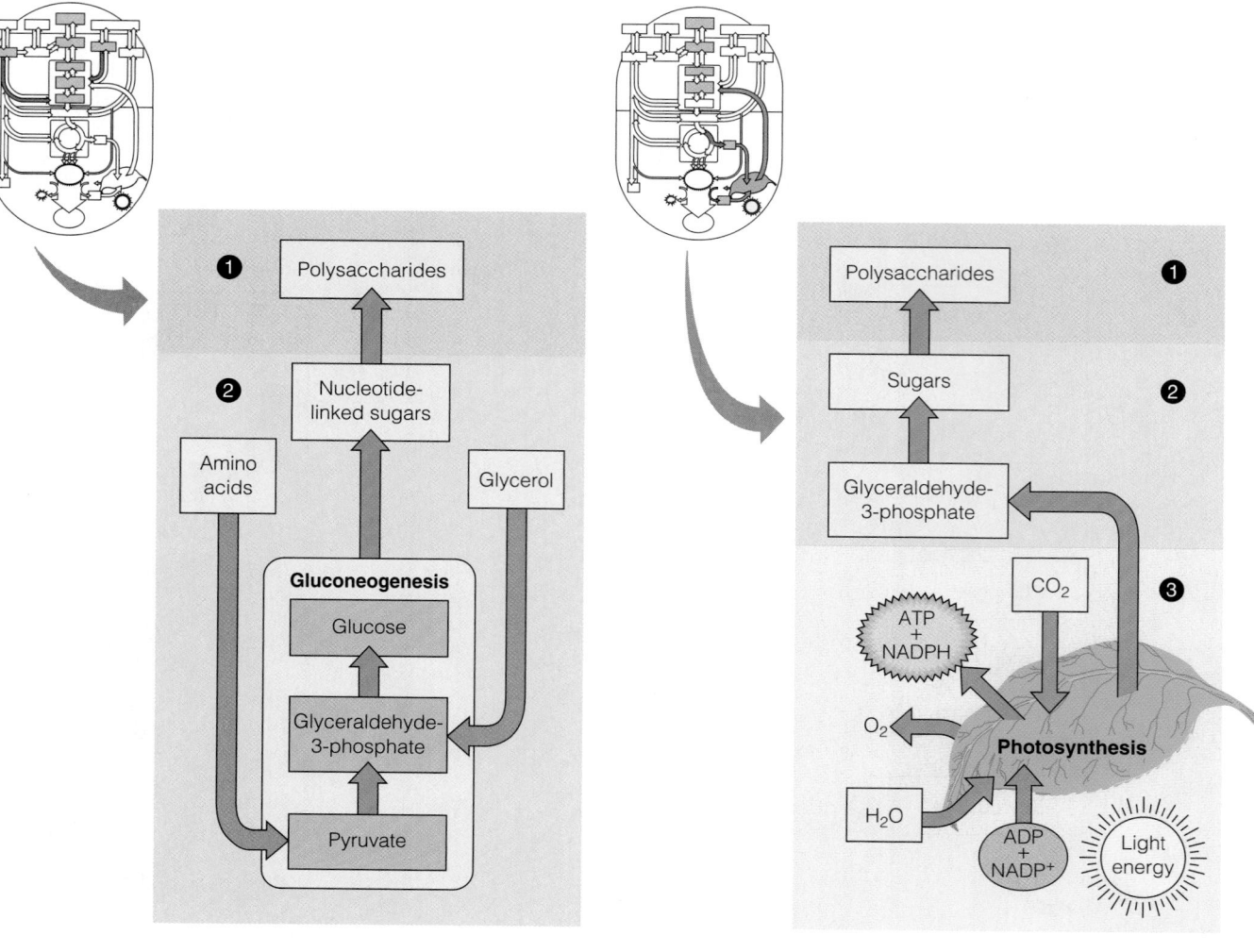

FIGURE 12.5

Carbohydrate anabolism. Biosynthesis of carbohydrates includes gluconeogenesis and polysaccharide synthesis.

FIGURE 12.6

Photosynthesis.

present **photosynthesis** (Figure 12.6), the supremely important process by which green plants capture light energy to drive the generation of energy (ATP) and reducing power (NADPH), both of which are used for carbohydrate synthesis.

Distinct Pathways for Biosynthesis and Degradation

It may appear from Figure 12.2 that some pathways operate simply as the reversal of other pathways. For example, fatty acids are synthesized from acetyl-CoA, but they are also converted to acetyl-CoA by β-oxidation. Similarly, glucose-6-phosphate is synthesized from pyruvate in gluconeogenesis, which looks at first glance like a simple reversal of glycolysis. It is important to realize that in these cases the opposed pathways are quite distinct from one another. They may share some common intermediates or enzymatic reactions, but they are separate reaction sequences, regulated by distinct mechanisms and with different enzymes catalyzing their regulated reactions. They may even occur in separate cellular compartments. For example, fatty acid synthesis takes place in cytosol, whereas fatty acid oxidation takes place in mitochondria.

Biosynthetic and degradative pathways are never simple reversals of one another, even though they often begin and end with the same metabolites. The existence of separate unidirectional pathways is important for two reasons. First, to proceed in a particular direction, a pathway must be exergonic in that direction. If a pathway is strongly exergonic, then reversal of that pathway is just as strongly endergonic, and thus impossible, under the same conditions. Thus, opposed biosynthetic and degradative pathways must both be exergonic, and thus unidirectional, in their respective directions.

Second and equally important is the need to control the flow of metabolites in relation to the bioenergetic status of a cell. When ATP levels are high, there is less need for carbon to be oxidized in the citric acid cycle. At such times the cell can store carbon as fats and carbohydrates, so fatty acid synthesis, gluconeogenesis, and related pathways come into play. When ATP levels are low, the cell must mobilize stored carbon to generate substrates for the citric acid cycle, so carbohydrate and fat breakdown must occur. Using separate pathways for the biosynthetic and degradative processes is crucial for control, so conditions that activate one pathway tend to inhibit the opposed pathway and vice versa.

Consider what would happen, for example, if fatty acid synthesis and oxidation took place in the same cell compartment and in an uncontrolled fashion. Two-carbon fragments released by oxidation would be immediately used for resynthesis, a situation called a **futile cycle**. No useful work is done, and the net result is simply consumption of more ATP in the endergonic reactions of fatty acid synthesis than produced in the oxidation reactions. A similar futile cycle could result from the interconversion of fructose-6-phosphate with fructose-1,6-bisphosphate in carbohydrate metabolism.

Fructose-6-phosphate + ATP $\longrightarrow$ fructose-1,6-bisphosphate + ADP

Fructose-1,6-bisphosphate + H_2O $\longrightarrow$ fructose-6-phosphate + P_i

Net: ATP + H_2O $\longrightarrow$ ADP + P_i

The first reaction occurs in glycolysis, and the second participates in a biosynthetic pathway, gluconeogenesis. Both processes occur in the cytosol. The net effect of carrying out both reactions simultaneously would be the wasteful hydrolysis of ATP to ADP and P_i. However, enzymes catalyzing both of the above reactions respond to allosteric effectors, such that one enzyme is inhibited by conditions that activate the other. This reciprocal control prevents the futile cycle from occurring, even though the two enzymes occupy the same cell compartment. Therefore, it is more appropriate to call this situation—two seemingly opposed cellular reactions that are independently controlled—a **substrate cycle**.

Studies on metabolic control suggest that a substrate cycle represents an efficient regulatory mechanism because a small change in the activity of either or both enzymes can have a much larger effect on the flux of metabolites in one direction or the other. You can test this idea in Problem 6 at the end of this chapter.

Biochemical Reaction Types

There is no metaphysics in biochemistry—the chemistry of living systems follows the same chemical and physical laws as the rest of nature. The complexity of these biochemical pathways may at first glance seem overwhelming, but only five general types of chemical transformations are commonly found in cells. Although all of these transformations are catalyzed by enzymes in cells, the reactions proceed by straightforward organic chemistry mechanisms. We will discuss these reaction types only briefly here because you previously learned about them in organic chemistry. However, you should consult your organic chemistry textbook if you need a more detailed review.

Degradative and biosynthetic pathways are distinct for two reasons: A pathway can be exergonic in only one direction, and pathways must be separately regulated to avoid futile cycles.

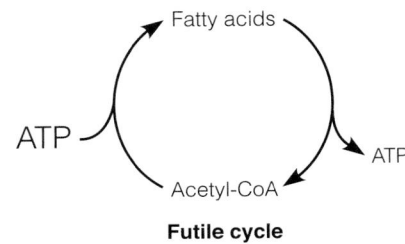

Futile cycle

Compartmentation and allosteric control of anabolic and catabolic processes prevent futile cycles, which simply waste energy.

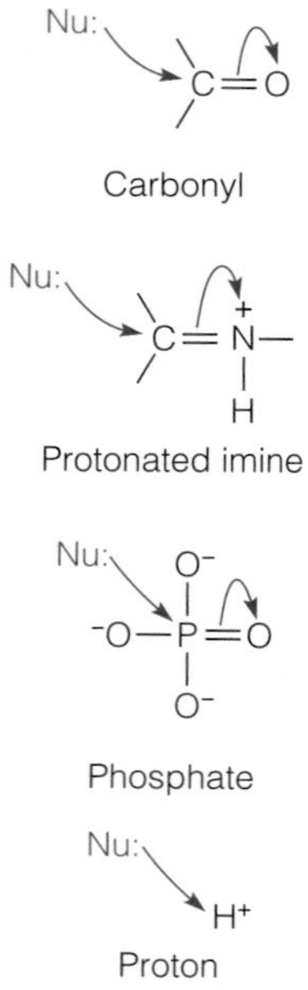

Carbonyl

Protonated imine

Phosphate

Proton

Electrophiles

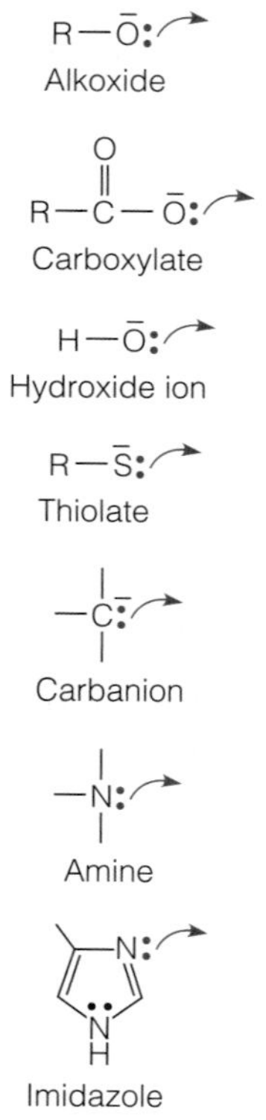

R—Ö:
Alkoxide

R—C—Ö:
Carboxylate

H—Ö:
Hydroxide ion

R—S:
Thiolate

Carbanion

Amine

Imidazole

Nucleophiles

Nucleophilic Substitutions

Much of the chemistry of biological molecules is the chemistry of the carbonyl group (C=O) because the vast majority of biological molecules contain them. And most of the chemistry of carbonyl groups involves **nucleophiles** (abbreviated "**Nu:**") and **electrophiles**. Recall that a nucleophile is a "nucleus-loving" substance with a negatively polarized, electron-rich atom that can form a bond by donating a pair of electrons to an electron-poor atom. An electrophile is an "electron-loving" substance with a positively polarized, electron-poor atom that can form a bond by accepting a pair of electrons from an electron-rich atom. Carbonyl groups are polar, with the electron-poor C atom bearing a partial positive charge and the electron-rich O atom bearing a partial negative charge.

Aldehyde   **Ketone**   **Carboxylic acid**   **Ester**

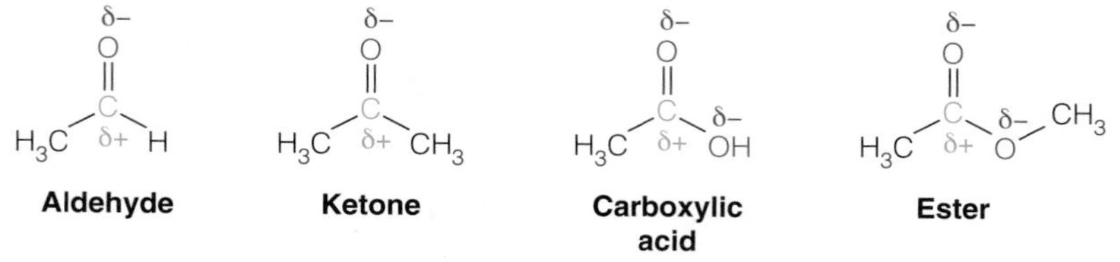

Carbonyl carbons are thus very common electrophiles in biochemical reactions. Other common electrophiles are protonated imines, phosphate groups, and protons.

Oxyanions (e.g., hydroxide ion, alkoxides, or ionized carboxylates), thiolates (deprotonated sulfhydryls), carbanions, deprotonated amines, and the imidazole side chain of histidine are common nucleophiles in biochemical reactions.

In a nucleophilic substitution reaction, one nucleophile replaces a second nucleophile (the leaving group) on an sp^3-hybridized carbon atom. The leaving group develops a partial negative charge in the transition state, and the best leaving groups are those that are stable as anions. Thus, halides and the conjugate bases of strong acids, such as the phosphate anion, are good leaving groups. Nucleophilic substitutions proceed by either S_N1 or S_N2 mechanisms. In the S_N1 (Substitution, Nucleophilic, Unimolecular) mechanism, the leaving group ($Y:^-$) departs with the bonding electrons, generating a carbocation intermediate, *before* the attacking nucleophile ($X:^-$) arrives. S_N1 reactions result in either retention of configuration or racemization at the reacting center.

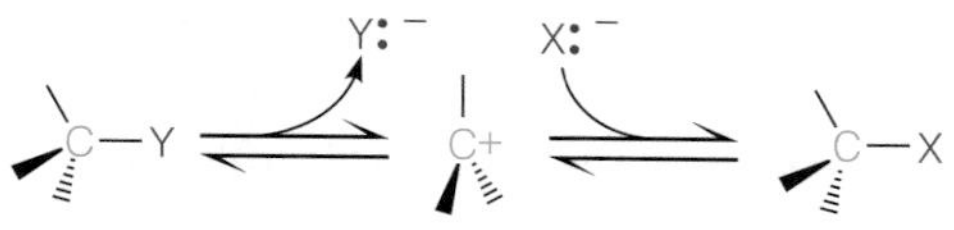

Carbocation intermediate

In the S_N2 (Substitution, Nucleophilic, Bimolecular) mechanism, the attacking nucleophile approaches one side of the electrophilic center while the leaving group remains partially bonded to other side, resulting in a transient pentavalent intermediate.

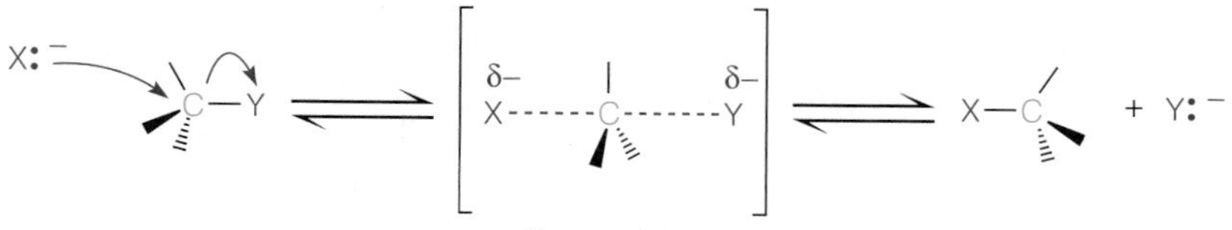

Pentavalent intermediate

In the S_N2 mechanism, departure of the leaving group from the opposite side results in a substituted product with inverted configuration.

A very important class of substitution reaction in biochemistry is the **nucleophilic *acyl* substitution** of carboxylic acid derivatives. Acyl substitutions occur most readily when the carbonyl carbon is bonded to an electronegative atom (such as O or N) or a highly polarizable atom (such as S) that can stabilize a negative charge and thereby act as a good leaving group. Thus, carboxylic acids and their derivatives (esters, amides, thioesters, acyl phosphates) are common substrates in acyl substitution reactions. In contrast to the S_N1 and S_N2 mechanisms of *alkyl* substitutions, *acyl* substitutions involve a tetrahedral oxyanion reaction intermediate.

Oxyanion intermediate

The planar carbonyl group is converted to a tetrahedral geometry as the carbonyl carbon rehybridizes from sp^2 to sp^3. As the electron pair moves from the oxyanion back toward the central carbon, the leaving group is expelled and the $C{=}O$ bond is regenerated. Many enzymes catalyze nucleophilic acyl substitutions, such as carboxypeptidase A, in which an activated water molecule serves as the attacking nucleophile (see Figure 11.39, page 449). This mechanism is also the basis of a variety of **group transfer reactions**, in which an acyl, glycosyl, or phosphoryl group is transferred from one nucleophile to another.

Nucleophilic Additions

Unlike carboxylic acids and their derivatives, the carbonyl carbon in aldehydes and ketones is bonded to atoms (C and H) that cannot stabilize a negative charge, and thus are not good leaving groups. These carbonyl groups typically undergo **nucleophilic addition reactions**, instead of substitution reactions. Like the nucleophilic acyl substitution mechanism, addition of a nucleophile leads to a tetrahedral oxyanion intermediate, as the electron pair from the $C{=}O$ bond moves onto the oxygen. Recall from Chapter 11 that an oxyanion intermediate is formed in the initial steps of peptide bond hydrolysis by the serine proteases. The oxyanion intermediate then has several fates, depending on the nucleophile (Figure 12.7). When the attacking nucleophile is a hydride ion (H:$^-$), the oxyanion intermediate undergoes protonation to form an alcohol. Alcohols also result when the attacking nucleophile is a carbanion (R_3C^-), and this is one of the mechanisms that yield new $C{-}C$ bonds. When an oxygen nucleophile adds, such as an alcohol (ROH), the oxyanion intermediate undergoes proton transfer to yield a hemiacetal. This reaction is the basis of ring formation in monosaccharides (Chapter 9). Reaction with a second equivalent of alcohol gives an acetal. When the attacking nucleophile is a primary amine ($R'NH_2$), the oxyanion intermediate picks up a proton from the amino group, giving a carbinolamine, which loses water to form an **imine** ($R_2C{=}NR'$). Imines (called **Schiff bases**) are common reaction intermediates in many biochemical reactions due to their ability to delocalize electrons.

Carbonyl Condensations

Formation of new $C{-}C$ bonds is a critical element of metabolism, and the condensation of two carbonyl compounds is a common strategy used in many biosynthetic pathways. A carbonyl condensation reaction relies on the weak acidity of the carbonyl α hydrogen, producing a carbanion, which is in resonance with a nucleophilic enolate ion.

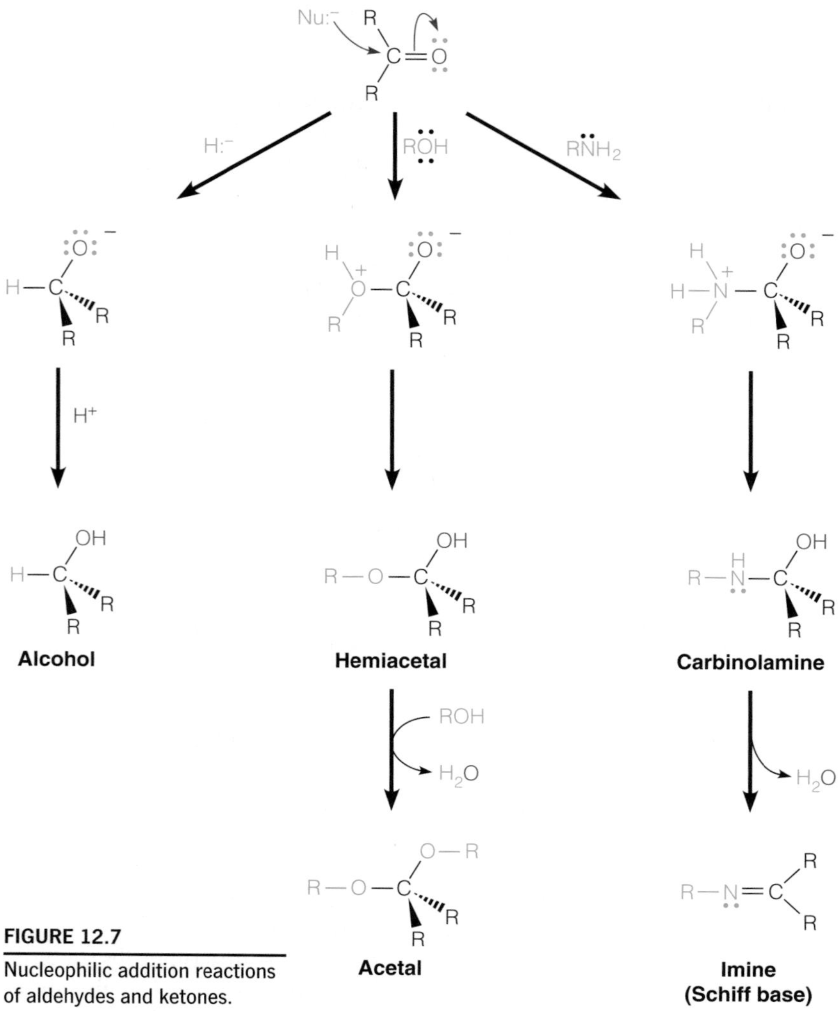

FIGURE 12.7

Nucleophilic addition reactions of aldehydes and ketones.

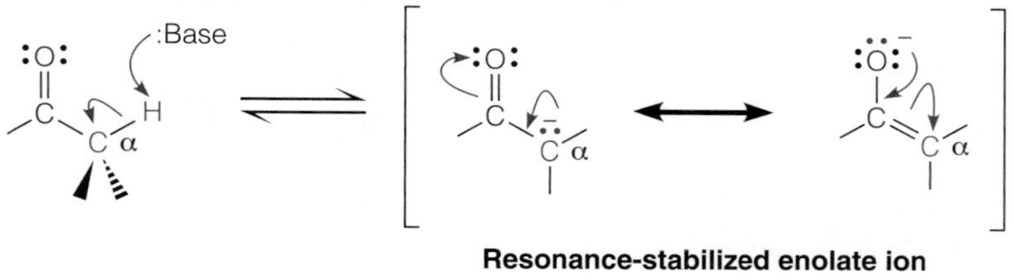

Resonance-stabilized enolate ion

The enolate ion, stabilized by resonance, nucleophilically adds to the electrophilic carbon of a second carbonyl, forming a new C—C bond (Figure 12.8). If the second carbonyl is an aldehyde or ketone (an **aldol condensation**), this nucleophilic addition produces an oxyanion intermediate, which is protonated to give a β-hydroxy carbonyl product. If the second carbonyl is an ester (a **Claisen condensation**), the intermediate oxyanion expels the ester alkoxide (RO—) as the leaving group, giving a β-keto product. Carbonyl condensations thus result in a new bond formed between the carbonyl carbon of one reactant with the α carbon of the other. Aldol and Claisen condensations are both reversible, and such "retro-aldol" and "retro-Claisen" reactions are frequently used to cleave C—C bonds. Indeed, β-keto compounds readily undergo cleavage or decarboxylation by a retro-aldol mechanism, in which the electron-accepting carbonyl group two carbons away from the carboxylate stabilizes the formal negative charge of the carbanionic transition state.

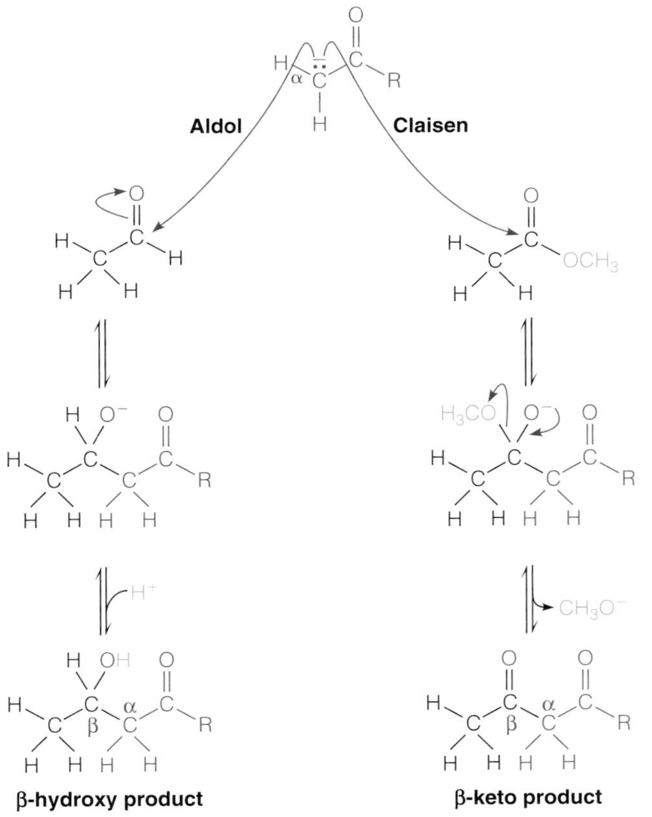

FIGURE 12.8

Carbonyl condensation reactions. These reactions are initiated by deprotonation of the weakly acidic α hydrogen to give a resonance-stabilized enolate ion (top). In an aldol condensation (left side), the enolate adds to an aldehyde or ketone, yielding a β-hydroxy carbonyl product. In a Claisen condensation (right side), the enolate adds to an ester, yielding a β-keto product.

β-hydroxy product

β-keto product

Eliminations

Eliminations of the type

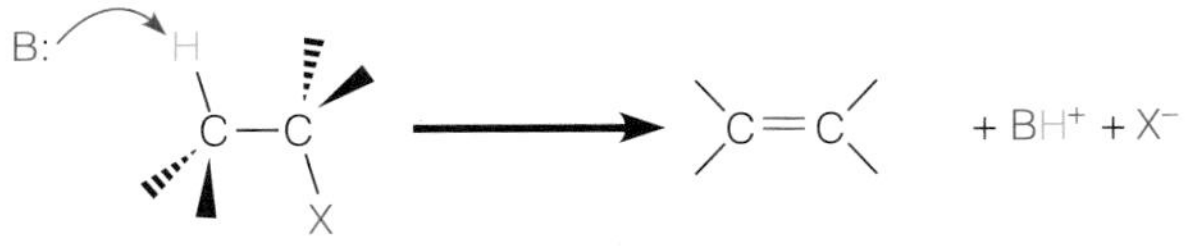

are also quite common in biochemical pathways. Elimination reactions can occur by several different mechanisms, but the most common one involves a carbanion intermediate. The reactant is often a β-hydroxy carbonyl (where X=OH) in which the H atom to be removed is made more acidic by being adjacent to a carbonyl group. A base abstracts the proton to give a carbanion intermediate (resonance-stabilized with the enolate) that loses OH^- to form the C=C double bond. β-hydroxy carbonyl compounds are readily dehydrated via these α,β-elimination reactions.

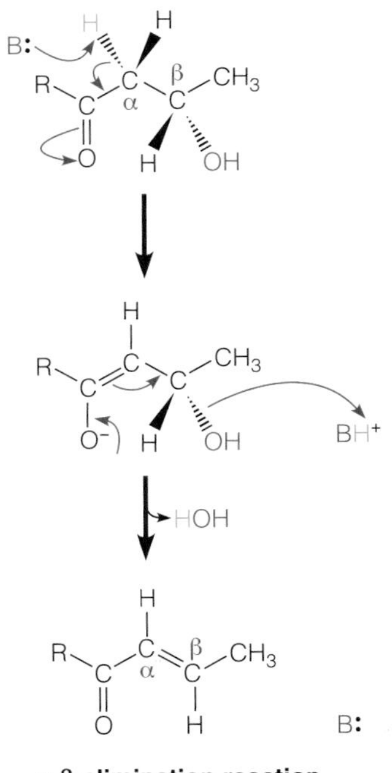

Oxidations and Reductions

Energy production in most cells involves the oxidation of fuel molecules such as glucose. Oxidation-reduction, or **redox**, chemistry thus lies at the core of metabolism. Redox reactions involve reversible electron transfer from a donor (the **reductant**) to an acceptor (the **oxidant**). Cells have evolved a number of electron carriers, such as the NAD^+ (nicotinamide adenine dinucleotide) coenzyme introduced in Chapter 11. Oxidations involving coenzymes such as NAD^+ occur by a reversible hydride (H^-) transfer mechanism, illustrated by the oxidation of an alcohol to a carbonyl compound.

α,β-elimination reaction

NAD⁺   **NADH**

A base abstracts the weakly acidic O—H proton, the electrons from that bond move to form a C=O bond, and the C—H bond is cleaved. The hydrogen *and the electron pair* (i.e., hydride, H^-) add to NAD^+ in a nucleophilic addition reaction, reducing it to NADH. The plus sign in NAD^+ reflects the charge on the pyridine ring nitrogen in the oxidized form; this charge is lost as the electron pair moves through the ring onto this nitrogen. Because the alcohol has lost a pair of electrons and two hydrogen atoms, this type of oxidation is called a **dehydrogenation**, and enzymes that catalyze this reaction are called **dehydrogenases**. Remember, however, that redox reactions are reversible, and dehydrogenases can catalyze the reductive direction as well. While two-electron oxidations exemplified by the NAD^+-dependent dehydrogenases are the most common redox reactions in metabolism, they are certainly not the only ones. A number of redox processes involve one-electron transfers, and various electron carriers exist to handle single electrons—you'll learn more about these in Chapter 15.

There are other, less common types of reactions in biochemical pathways, such as free radical reactions, but these five represent the basic toolkit that cells use to carry out the vast majority of their chemical transformations.

Some Bioenergetic Considerations

Oxidation as a Metabolic Energy Source

Because the primary focus of the next few chapters will be on energy metabolism, we will now consider briefly how metabolic energy is generated. As we saw in Chapter 3, a thermodynamically unfavorable, or endergonic, reaction will proceed readily in the unfavored direction if it can be coupled to a thermodynamically favorable, or exergonic, reaction. In principle, any exergonic reaction can serve this purpose, provided that it releases sufficient free energy. In living systems, most of the energy needed to drive biosynthetic reactions is derived from the *oxidation* of organic substrates. Oxygen, the ultimate electron acceptor for aerobic organisms, is a strong oxidant; it has a marked tendency to attract electrons, becoming reduced in the process. Given this tendency and the abundance of oxygen in our atmosphere, it is not surprising that living systems have gained the ability to derive energy from the oxidation of organic substrates.

Biological Oxidations: Energy Release in Small Increments

In a thermodynamic sense, the biological oxidation of organic substrates is comparable to nonbiological oxidations such as the burning of wood. The free energy release is the same whether we are talking about oxidation of the glucose polymer cellulose in a wood fire, combustion of glucose in a calorimeter, or the metabolic oxidation of glucose:

$$C_6H_{12}O_6 + 6O_2 \rightarrow 6CO_2 + 6H_2O \qquad \Delta G^{\circ\prime} = -2870 \, \text{kJ/mol} \qquad (12.1)$$

This equation reveals the conservation stoichiometry, or simple **reaction stoichiometry**, of glucose combustion. Biological oxidations, however, are far more complex processes than combustion. When wood is burned, all of the energy is released as heat; useful work cannot be performed, except through the

Most biological energy derives from oxidation of reduced metabolites in a series of reactions, with oxygen as the final electron acceptor.

action of a device such as a steam engine. In biological oxidations, by contrast, oxidation reactions occur without a large increase in temperature and with capture of some of the free energy as chemical energy. This energy capture occurs largely through the synthesis of ATP, and, as discussed in Chapter 3, the hydrolysis of ATP can be coupled to many processes to provide energy for biological work. In catabolism of glucose, about 40% of the released energy is used to drive the synthesis of ATP from ADP and P_i.

Unlike the oxidation of glucose by oxygen shown in the previous equation, most biological oxidations do not involve direct transfer of electrons from a reduced substrate to oxygen (Figure 12.4). Rather, a series of coupled oxidation–reduction reactions occurs, with the electrons passed to intermediate electron carriers such as NAD^+ and FAD, and finally transferred to oxygen. We have already briefly mentioned the role of NAD^+ as an electron carrier; FAD (flavin adenine dinucleotide) will be described in detail in Chapter 14. Thus, the biological oxidation of glucose might be more accurately represented by the following coupled reactions:

$$C_6H_{12}O_6 + 10NAD^+ + 2FAD + 6H_2O \rightarrow 6CO_2 + 10NADH + 10H^+ + 2FADH_2 \quad (12.2)$$

$$10NADH + 10H^+ + 2FADH_2 + 6O_2 \rightarrow 10NAD^+ + 2FAD + 12H_2O \quad (12.3)$$

$$\text{Net: } C_6H_{12}O_6 + 6O_2 \rightarrow 6CO_2 + 6H_2O \quad (12.4)$$

The net reaction of the biological oxidation process (reaction 12.4) is identical to that of direct combustion (reaction 12.1). Reactions 12.2 and 12.3 are examples of **obligate-coupling stoichiometry**. These are stoichiometric relationships fixed by the chemical nature of the process. Thus, the complete oxidation of glucose requires the transfer of 12 pairs of electrons from glucose to molecular oxygen, whether as a direct process, as in combustion, or via intermediate electron carriers, as in the biological process. In the biological process, 12 moles of electron carriers (NAD^+ and FAD) are obligately coupled to the oxidation of 1 mole of glucose to 6 moles of CO_2.

The transfer of electrons from these intermediate electron carriers to oxygen is catalyzed by the **electron transport chain**, or **respiratory chain**, and oxygen is called the **terminal electron acceptor**. Because the potential energy stored in the organic substrate is released in small increments, it is easier to control oxidation and capture some of the energy as it is released—small energy transfers waste less energy than a single large transfer does.

Not all metabolic energy comes from oxidation by oxygen. Substances other than oxygen can serve as terminal electron acceptors. Many microorganisms either can or must live **anaerobically** (in the absence of oxygen). For example, *Desulfovibrio* carry out anaerobic respiration using sulfate as the terminal electron acceptor:

$$SO_4^{2-} + 8e^- + 8H^+ \rightarrow S^{2-} + 4H_2O$$

Most anaerobic organisms, however, derive their energy from **fermentations**, which are energy-yielding catabolic pathways that proceed with no net change in the oxidation state of the products as compared with that of the substrates. A good example is the production of ethanol and CO_2 from glucose, presented in Chapter 13. Other anaerobic energy-yielding pathways are seen in some deep-sea hydrothermal vent bacteria, which reduce sulfur to sulfide as the terminal electron transfer reaction, and in other bacteria that reduce nitrite to ammonia. These organisms oxidize the substrates that sustain them, but they use terminal electron acceptors other than oxygen.

Energy Yields, Respiratory Quotients, and Reducing Equivalents

If metabolic energy comes primarily from oxidative reactions, it follows that the more highly reduced a substrate, the higher its potential for generating biological energy. We can use a calorimeter to measure the heat output (enthalpy) from oxidation of fat, carbohydrate, or protein. The combustion of fat provides more heat

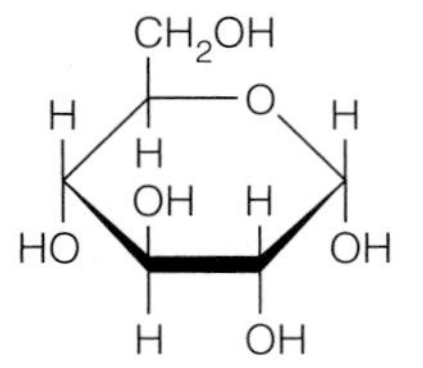

CH$_2$OH

Glucose

CH$_3$(CH$_2$)$_{14}$COOH
Palmitic acid

energy than the combustion of an equivalent mass of carbohydrate. In other words, fat has a higher **caloric content** than carbohydrate. For illustration, compare the oxidation of glucose with the oxidation of a typical saturated fatty acid, palmitic acid:

$$C_6H_{12}O_6 + 6O_2 \rightarrow 6CO_2 + 6H_2O \qquad \Delta G°' = -3.74 \, kcal/g$$

$$C_{16}H_{32}O_2 + 23O_2 \rightarrow 16CO_2 + 16H_2O \qquad \Delta G°' = -9.30 \, kcal/g$$

Converting calories (the units of nutrition) to joules (the modern units of thermodynamics), we see that the oxidation of glucose yields 15.64 kJ/g, and the oxidation of palmitic acid yields 38.90 kJ/g. The carbons in fat are in general more highly reduced than those in carbohydrate; thus, they contain more protons and electrons to combine with oxygen on the path to CO_2 than do the carbons in sugar. We can see this by counting oxygen atoms. Glucose has more oxygens per carbon than does palmitic acid; each carbon in glucose is linked to at least one oxygen atom.

We can also tell that glucose is the more highly oxidized substance because its oxidation produces more moles of CO_2 per mole of O_2 consumed during oxidation, a ratio called the **respiratory quotient**, or **RQ**. The above reaction stoichiometries reveal RQ for glucose to be 1.0 (6CO_2/6O_2), whereas that for palmitic acid is 0.70 (16CO_2/23O_2). In general, the lower the RQ for a substrate, the more oxygen consumed per carbon oxidized and the greater the potential per mole of substrate for generating ATP.

Another way to express the degree of substrate oxidation is to say that more **reducing equivalents** are derived from oxidation of fat than from oxidation of carbohydrate. A reducing equivalent can be defined as 1 mole of hydrogen atoms (one proton and one electron per H atom). For example, two moles of reducing equivalents are used in the reduction of one-half mole of oxygen to water:

$$\tfrac{1}{2}O_2 + 2e^- + 2H^+ \rightarrow H_2O$$

Whereas the breakdown of complex organic compounds yields both energy and reducing equivalents, the biosynthesis of such compounds utilizes both. For example, we know that both carbons of acetate are used for fatty acid biosynthesis:

$$8CH_3COO^- \rightarrow \rightarrow \rightarrow CH_3(CH_2)_{14}COO^-$$

Acetate   Palmitate

Fifteen of the 16 carbon atoms of palmitate are highly reduced—14 at the methylene level and 1 at the methyl level. Therefore, many reducing equivalents are required to complete this biosynthesis.

The major source of electrons for reductive biosynthesis is **NADPH, nicotinamide adenine dinucleotide phosphate (reduced)**. NADP$^+$ and NADPH are identical to NAD$^+$ and NADH, respectively, except that the former have an additional phosphate esterified at C-2 on the adenylate moiety. NAD$^+$ and NADP$^+$ are equivalent in their thermodynamic tendency to accept electrons; they have equal standard reduction potentials (Chapter 3). However, nicotinamide nucleotide–linked enzymes that act primarily in a catabolic direction usually use the NAD$^+$–NADH pair, whereas those acting primarily in anabolic pathways use NADP$^+$ and NADPH. In other words, as shown in Figure 12.9, nicotinamide nucleotide–linked enzymes that oxidize substrates (dehydrogenases) usually use NAD$^+$, and those enzymes that reduce substrates (reductases) usually use NADPH. Of course, both dehydrogenases and reductases catalyze reversible reactions—the directions are determined by the **redox state**, or ratio of the oxidized and reduced forms of each pair that prevail in the cell. Thus, the NAD$^+$/NADH pair is maintained at a more oxidized level than the NADP$^+$/NADPH pair in a healthy cell. These ratios generally drive NAD$^+$-linked

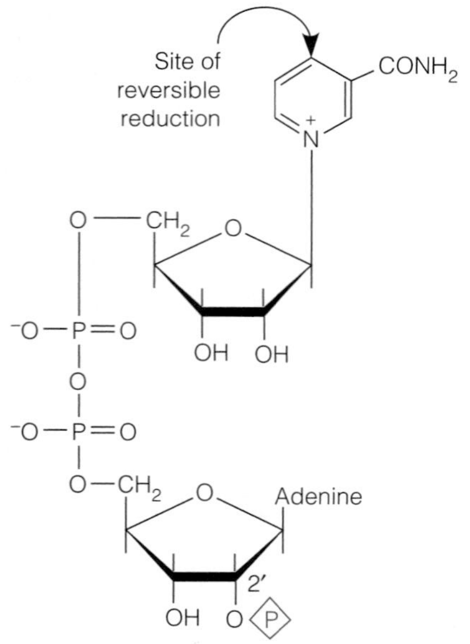

Site of reversible reduction
CONH$_2$

Adenine

2'

NADP$^+$
Nicotinamide adenine dinucleotide phosphate (oxidized)

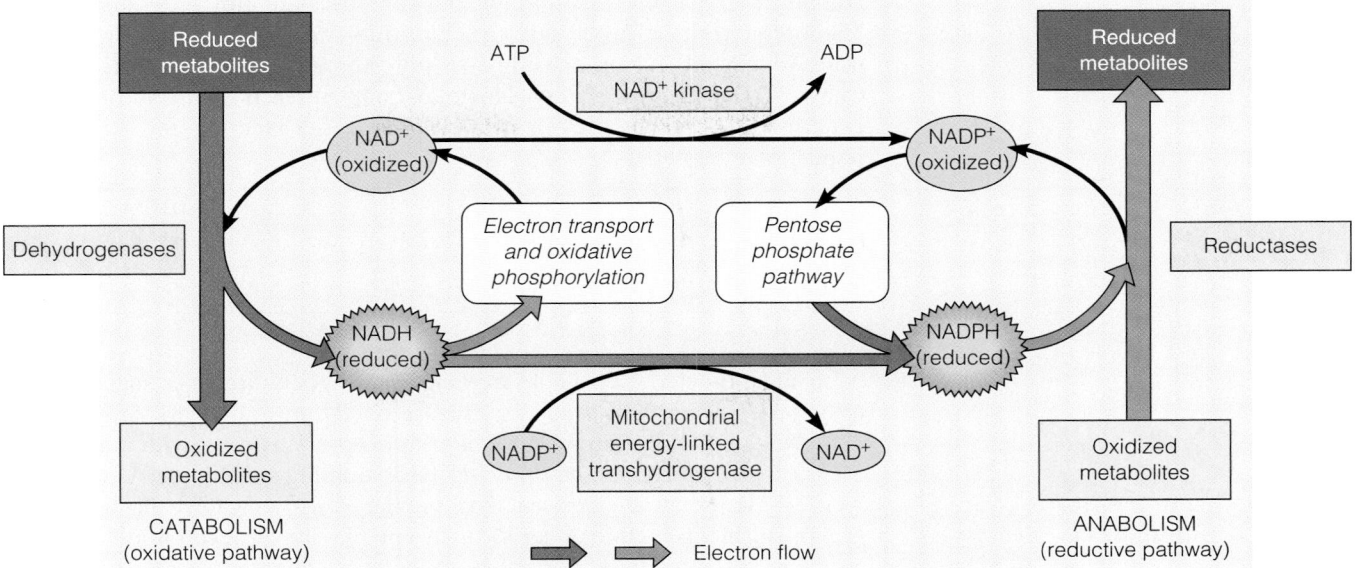

FIGURE 12.9

Nicotinamide nucleotides in catabolism and biosynthesis. NAD^+ is the cofactor for most enzymes that act in the direction of substrate oxidation (dehydrogenases), whereas NADPH usually functions as a cofactor for reductases, enzymes that catalyze substrate reduction. NADPH is regenerated either from $NADP^+$ in the pentose phosphate pathway (see Chapter 13) or from NADH through the action of mitochondrial energy-linked transhydrogenase (see Chapter 15). $NADP^+$ is synthesized from NAD^+ by an ATP-dependent kinase reaction.

reactions in the oxidative direction and $NADP^+$-linked reactions in the reductive direction. An exception is two $NADP^+$-linked dehydrogenases in the **pentose phosphate pathway** (see Chapter 13), which convert $NADP^+$ to NADPH and represent a major route for regeneration of the reduced nucleotide.

ATP as a Free Energy Currency

ATP is commonly referred to as a "free energy currency." What does this mean? Currency is a medium of exchange. A $20 bill has a generally recognized value that can be readily exchanged for various goods or services—for example, dinner at a moderately priced restaurant, or about a quarter-hour's labor by a skilled auto mechanic. In the same sense, cells exchange the energy released from the breakdown of ATP to carry out essential functions, often converting the chemical energy released in ATP hydrolysis to other forms of energy—mechanical energy, for example, in muscle contraction, or electrical energy in conducting nerve impulses, or osmotic energy in transporting substances across membranes against a concentration gradient. Thus ATP serves as an immediate donor of free energy—it is continuously formed and consumed. It is estimated that a resting human turns over as much as 65 kg ATP every 24 hours, about equal to the entire body weight! (The Na^+/K^+–ATPase you learned about in Chapter 10 accounts for approximately 25% of this resting ATP consumption.) During strenuous exercise, ATP turnover can be as high as 0.5 kg/min.

As summarized in Figures 12.3 and 12.4, regeneration of all this ATP is coupled to fermentative or oxidative processes. Thus, as we will see in Chapter 15, the oxidation of 1 mole of glucose during respiration is coupled to the phosphorylation of ~32 moles of ADP:

$$C_6H_{12}O_6 + 6O_2 + 32ADP + 32P_i \rightarrow 6CO_2 + 38H_2O + 32ATP \qquad (12.5)$$

The stoichiometries for ADP, P_i, and ATP in this equation are fundamentally different from the simple reaction and obligate-coupling stoichiometries we have encountered thus far. There is no chemical necessity for the production of 32 moles of ATP (in fact, there is not even a direct chemical connection between glucose oxidation and ATP synthesis). We could not predict a stoichiometry of 32 ATP from chemical considerations. This is instead an **evolved-coupling**

NAD^+ is the cofactor for most dehydrogenases that oxidize metabolites. NADPH is the cofactor for most reductases.

stoichiometry. Evolved-coupling stoichiometries are biological adaptations, phenotypic traits acquired during evolution—they are the result of compromise. So why did evolution settle on a stoichiometry of ~32 ATP? Recall from Chapter 3 that the free energy changes of coupled reactions are additive. Thus equation 12.5 can be broken into its free energy yielding and free energy requiring processes:

	ΔG (kJ/mol glucose)
$C_6H_{12}O_6 + 6O_2 \rightarrow 6CO_2 + 6H_2O$	-2900
$32ADP + 32P_i \rightarrow 32H_2O + 32ATP$ $\Delta G = +50\,kJ/mol \times 32\,mol\,ATP = +1600$	
Sum: $C_6H_{12}O_6 + 6O_2 + 32ADP + 32P_i \rightarrow 6CO_2 + 38H_2O + 32ATP$	-1300

Using estimates of the Gibbs free energy changes under physiological conditions (ΔG) of -2900 kJ/mol for glucose oxidation and -50 kJ/mol for the hydrolysis of ATP, we can see that the coupled process occurs with a net ΔG of -1300 kJ/mol glucose based on an evolved-coupling stoichiometry of 32 ATP. A ΔG of this magnitude represents such an immense driving force that respiration is emphatically favorable under virtually any physiological condition, and thus goes to completion. Because the ATP-coupling stoichiometry is an evolved trait, perhaps a different stoichiometry evolved in some ancient organism. Imagine a cell that mutated such that it ended up with an ATP-coupling stoichiometry for respiration of 58 instead of 32. Repeating the calculations above, the energy requiring process would have a ΔG of $+2900$ kJ/mol glucose (58 ATP $\times$ $+50$ kJ/mol), giving an overall ΔG of 0 for respiration. Thus, there would be no net driving force for the reaction. Although this hypothetical cell might have a higher yield of ATP per glucose oxidized, it would come to equilibrium before very much glucose was metabolized. This cell would probably be a very poor competitor, especially if glucose concentrations were limiting, and the mutation(s) that led to this stoichiometry would likely be selected against. The ATP-coupling stoichiometry that we actually find (~32) is thus an evolved compromise between maximizing ATP yield and ensuring that the overall process is unidirectional under any conceivable conditions the cell might encounter.

Of course, as emphasized in Figure 12.1, there are many biochemical pathways that consume ATP. The role of ATP in these anabolic processes is to convert a thermodynamically unfavorable process into a favorable process. Remember that biosynthetic and degradative pathways are never simple reversals of one another. In particular, opposed pathways will always have different ATP stoichiometries. That is, the number of ATP (or ATP equivalents) produced in the catabolic direction will always be different from the number of ATP (or ATP equivalents) required in the anabolic direction. We refer to these numbers as the **ATP-coupling coefficient** of the reaction or pathway. For example, in the fructose-6-phosphate/fructose-1,6-bisphosphate substrate cycle (page 481), the glycolytic reaction,

$$\text{Fructose-6-phosphate} + \text{ATP} \rightarrow \text{fructose-1,6-bisphosphate} + \text{ADP} \quad (12.6)$$

has an ATP-coupling coefficient of -1. The gluconeogenic reaction,

$$\text{Fructose-1,6-bisphosphate} + H_2O \rightarrow \text{fructose-6-phosphate} + P_i \quad (12.7)$$

has an ATP-coupling coefficient of 0. Likewise, as we will learn in Chapter 13, glycolysis has an overall ATP-coupling coefficient of $+2$, whereas gluconeogenesis has an ATP-coupling coefficient of -6. These differences in ATP-coupling coefficients between opposed reactions or pathways are essential for the unidirectionality of both processes.

What do we really mean when we say that ATP converts a thermodynamically unfavorable process into a favorable process? Coupling ATP hydrolysis to a pathway provides a new chemical route, using different reactions with different stoichiometries, resulting in a different overall equilibrium constant for the process.

The fundamental biological role of ATP as an energy-coupling compound is to convert thermodynamically unfavorable processes into favorable processes.

In fact, coupling ATP hydrolysis to a process changes the equilibrium ratio of certain [reactants] to [products] by a factor of 10^8! To illustrate this, let's consider how the equilibrium ratio of [fructose-1,6-bisphosphate] to [fructose-6-phosphate] changes when we compare a direct addition of P_i to fructose-6-phosphate vs. the reaction pathway that is coupled to ATP hydrolysis. The direct addition of P_i is simply reversal of reaction 12.7:

$$\text{Fructose-6-phosphate} + P_i \rightleftharpoons \text{fructose-1,6-bisphosphate} + H_2O \qquad (12.8)$$

In this direction, the standard free energy change, $\Delta G^{\circ\prime}$, is $+16.3$ kJ/mol (see Table 13.2, page 545). Let's calculate the equilibrium constant, using Equation 3.28 from Chapter 3:

$$K = e^{\left(\frac{-\Delta G^{\circ\prime}}{RT}\right)} = e^{\left(\frac{-16300\,\text{J/mol}}{(8.315\,\text{J/mol}\cdot\text{K})(298\,\text{K})}\right)} = 0.0014 = \frac{[\text{fructose-1,6-bisphosphate}]}{[\text{fructose-6-phosphate}][P_i]} \qquad (12.9)$$

In many cells, the normal intracellular concentration of P_i is ~1mM (10^{-3} M), so the equilibrium ratio of [fructose-1,6-bisphosphate]/[fructose-6-phosphate] achieved by this reaction would be $(0.0014)(10^{-3}) = 1.4 \times 10^{-6}$. Now let's compare that to a reaction that couples the phosphorylation of fructose-6-phosphate to ATP hydrolysis, as in reaction 12.6 above. $\Delta G^{\circ\prime}$ for this reaction is -14.2 kJ/mol (see Table 13.1, page 534), and thus the equilibrium constant, K, for this reaction is 308.

$$K = e^{\left(\frac{14200\,\text{J/mol}}{(8.315\,\text{J/mol}\cdot\text{K})(298\,\text{K})}\right)} = 308 = \frac{[\text{fructose-1,6-bisphosphate}][\text{ADP}]}{[\text{fructose-6-phosphate}][\text{ATP}]} \qquad (12.10)$$

To compare the resulting equilibrium ratio of [fructose-1,6-bisphosphate]/[fructose-6-phosphate], we need to estimate the intracellular concentrations of ADP and ATP. In most normal, healthy cells, [ATP] is several-fold (3–10 times) higher than [ADP]. For the purposes of this calculation, let's make the simplifying assumption that the intracellular concentrations of ADP and ATP are roughly equal, that is, their concentration ratio is ~1. Thus, [ADP] and [ATP] cancel out, and the equilibrium ratio of [fructose-1,6-bisphosphate]/[fructose-6-phosphate] for this reaction is 308. This equilibrium ratio is indeed 10^8 times higher than that achieved without ATP coupling $[308/(1.4 \times 10^{-6}) = 2.2 \times 10^8]$. This effect is completely general—it depends only on the existence of a chemical mechanism that couples ATP hydrolysis to the process at hand (e.g., an enzyme like phosphofructokinase that catalyzes reaction 12.6 in glycolysis). Note that ATP is not directly hydrolyzed in this reaction, but is instead converted to ADP as its phosphate is transferred to fructose-6-phosphate. However, any process that couples the conversion of ATP to ADP gains the thermodynamic equivalent of the free energy of hydrolysis of ATP.

Metabolite Concentrations and Solvent Capacity

The use of ATP as the fundamental coupling agent in biochemical systems has had far-reaching implications for the evolutionary design of cell metabolism. Evolution has produced a complex cell metabolism comprising thousands of enzymes and metabolic intermediates each at very low individual concentrations. There are at least a couple of reasons why a cell must maintain its components at very low concentrations. Even a simple bacterial cell contains several thousand different metabolites dissolved in the aqueous compartment. This aqueous compartment has a finite capacity for the amount of dissolved substances (metabolites and macromolecules)—this is its **solvent capacity**. Thus, individual metabolites must exist at low concentrations ($10^{-3} - 10^{-6}$ M, or even lower) to avoid exceeding the solvent capacity of the cell. Secondly, low metabolite concentrations minimize unwanted side reactions. By way of illustration, imagine two metabolites, A and B, that can react nonenzymatically to form C. Assume that this unwanted side reaction is first-order with respect to each metabolite so that the reaction rate is directly proportional to the product of [A] and [B] ($v = k$[A][B]). We can compare the amount of C that would be produced under two different metabolite concentrations: 1 M each vs. 10^{-5} M each. Whatever the rate constant is, the velocity of the unwanted side reaction at the two different concentrations will differ by a factor of 10^{10}. Thus, the same amount of C that would be produced in 1 sec if each metabolite is at 1 M would take ~317 years (10^{10} sec) to produce if each metabolite is at 10^{-5} M instead.

How does ATP coupling help avoid high metabolite concentrations? It does it by activating metabolic intermediates. Let's return to our previous example, comparing the phosphorylation of fructose-6-phosphate using inorganic phosphate vs. ATP. Assume that the cell needs to maintain a [fructose-1,6-bisphosphate]/[fructose-6-phosphate] ratio of 10 to thrive. What concentration of P_i is required to ensure this nonequilibrium ratio using reaction 12.8? We can solve for the required [P_i] by recognizing that ΔG for this reaction must be <0 and rearranging the relevant form of Equation 3.23:

$$\Delta G = \Delta G^{\circ\prime} + RT \ln\left(\frac{[\text{F-1,6-BP}][\text{H}_2\text{O}]}{[\text{F-6-P}][\text{P}_i]}\right) < 0$$

or

$$\ln\left(\frac{[\text{F-1,6-BP}][\text{H}_2\text{O}]}{[\text{F-6-P}][\text{P}_i]}\right) < \frac{-\Delta G^{\circ\prime}}{RT} \tag{12.11}$$

$$\ln\left(\frac{[10][1]}{[1][\text{P}_i]}\right) < \frac{-16.3\,\frac{\text{kJ}}{\text{mol}}}{\left(0.008314\,\frac{\text{kJ}}{\text{mol K}}\right)(298\,\text{K})} \tag{12.12}$$

Solving for [P_i]:

$$[\text{P}_i] > \frac{10}{0.0014} = 7143\,\text{M}!$$

Of course, there is no way to fit 7000 moles of phosphate into a liter of aqueous solution. Even if the cell could survive with a [fructose-1,6-bisphosphate]/[fructose-6-phosphate] ratio of 1, it would still need a [P_i] of > 714 M to make the reaction thermodynamically favorable (i.e., to make $\Delta G < 0$).

Now let us consider the process coupled to ATP hydrolysis, via the phosphofructokinase reaction (equation 12.6). Here, the relevant calculation is

$$\frac{[\text{F-1,6-BP}][\text{ADP}]}{[\text{F-6-P}][\text{ATP}]} < e^{\left(\frac{-\Delta G^{\circ\prime}}{RT}\right)} \tag{12.13}$$

with $\Delta G^{\circ\prime} = -14.2$ kJ/mol. If we assume an [ADP]/[ATP] ratio of ~1 under physiological conditions, we can calculate the maximum ratio of [fructose-1,6-bisphosphate]/[fructose-6-phosphate] that still makes the reaction favorable:

$$\frac{[\text{fructose-1,6-bisphosphate}]}{[\text{fructose-6-phosphate}]} < 308$$

Thus, by coupling the reaction to ATP hydrolysis the reaction will be thermodynamically favorable as long as the ratio of [F-1,6-BP] to [F-6-P] remains under 308—far above the desired 10. This scenario completely avoids the involvement of impossible concentrations of inorganic phosphate. In this example, ATP can be thought of as an activated form of phosphate. Cells have evolved a number of activated intermediates besides ATP, including acetyl-CoA, acyl-lipoate, and nucleoside diphosphate sugars (NDP-sugars). We will discuss each of these in later chapters, but suffice it to say here that they all function to allow reactions to occur under physiologically relevant concentrations of intermediates.

Activated intermediates, such as ATP, allow reactions to occur under physiologically relevant concentrations of metabolic intermediates.

Thermodynamic Properties of ATP

What factors equip ATP for its special role as energy currency? First, there is nothing unique about the chemistry of ATP; the phosphoanhydride bonds whose breakdown can be coupled to drive endergonic reactions are shared by all other nucleoside di- and triphosphates and by several other metabolites. When we call ATP a "high-energy compound," as we did in Chapter 3, we use that term within a defined context; i.e., a high-energy compound is one containing at least one bond with a sufficiently favorable $\Delta G^{\circ\prime}$ of hydrolysis. ATP has two phosphoanhydride bonds. Cleavage of one yields adenosine diphosphate (ADP) and inorganic phosphate (P_i), and cleavage of the other gives adenosine monophosphate (AMP) and pyrophosphate (PP_i). Either reaction proceeds with a large negative $\Delta G^{\circ\prime}$ of hydrolysis (Figure 3.7, page 76). Note, however, that calling a substance a high-energy compound does not mean that it is chemically unstable or unusually reactive. In fact, ATP is a *kinetically* stable compound; its spontaneous hydrolysis is slow, but when hydrolysis does occur, whether spontaneously or enzyme-catalyzed, substantial free energy is released. Be aware, however, that utilization of that energy to drive endergonic reactions usually does *not* involve hydrolysis. Rather, ATP breakdown is usually coupled with a thermodynamically unfavorable reaction, such as the synthesis of glucose-6-phosphate from glucose. In this case the phosphate released from ATP does not become P_i but instead is transferred directly to glucose, forming the esterified phosphate of glucose-6-phosphate. Thus, as discussed in Chapter 3, it is more accurate to say that ATP has a "high phosphoryl group transfer potential" than to call it a high-energy compound. However, so long as you understand the context in which the term is used, the concept of a high-energy compound is quite useful.

Recall from Chapter 3 that several factors contribute to the thermodynamic stability of a hydrolyzable bond and determine whether $\Delta G^{\circ\prime}$ for hydrolysis is highly favorable, as in the phosphoric anhydride bonds of ATP, ADP, or pyrophosphate, or less favorable, as in the phosphoric ester bonds of glucose-6-phosphate or AMP. These factors include electrostatic repulsion among the negative charges in the molecule before hydrolysis, resonance stabilization of the products of hydrolysis, and the tendency of the hydrolysis products to deprotonate (see page 77). By contrast, the hydrolysis of a phosphate ester, such as AMP, generates an alcohol (the sugar 5′ hydroxyl), which has almost no tendency to lose a proton.

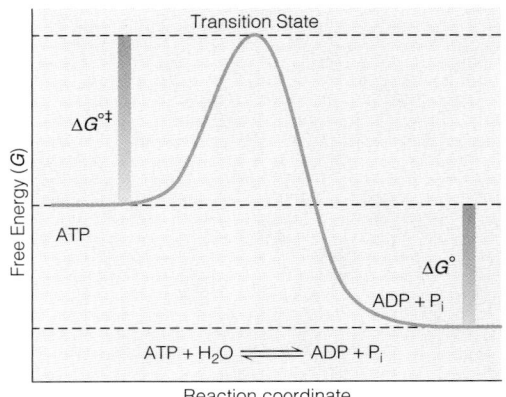

Phosphoanhydride bonds are *thermodynamically* unstable, but *kinetically* stable—large free energies of activation require enzymes to lower the activation barrier.

The above mentioned factors combine to give ATP hydrolysis a $\Delta G^{\circ\prime}$ of -30.5 kJ/mol, twice the phosphate transfer potential of phosphate esters such as AMP. On the other hand, several important metabolites have $\Delta G^{\circ\prime}$ values for hydrolysis that are much more negative than that of ATP. Examples include phosphoenolpyruvate, 1,3-bisphosphoglycerate, and creatine phosphate (see Figure 3.7, page 76), with $\Delta G^{\circ\prime}$ values of -61.9, -49.4, and -43.1 kJ/mol, respectively. This means that ATP is actually intermediate on the scale of "chemical potential." This is important as well because it means that the breakdown of a compound such as phosphoenolpyruvate can be coupled to drive the synthesis of ATP itself from ADP and P_i. In fact, such coupled reactions, called **substrate-level phosphorylation** reactions, represent the process by which ATP is synthesized in glycolysis, as we discuss in Chapter 13.

The Important Differences Between ΔG and $\Delta G°'$

But what provides the energy for synthesis of compounds with a much higher phosphate transfer potential than that of ATP itself? Much of the answer lies in the fact that ΔG values under intracellular conditions are quite different from standard ($\Delta G°'$) values. This is mainly because intracellular concentrations are far different from the 1 M concentrations used to compute standard free energies. If you work Problem 5 at the end of this chapter, you will see that ATP hydrolysis has a considerably more negative ΔG value at intracellular concentrations of ATP, ADP, AMP, and P_i than it has under standard conditions. In fact, the physiological ratio of [ATP]/[ADP] is $\sim 10^8$ times higher than the equilibrium ratio. This value is not coincidental—if the physiological [ATP]/[ADP] ratio is 10^8 times larger than its equilibrium ratio, then the equilibrium ratio of any other reaction that is coupled to ATP hydrolysis will be altered by this same order of magnitude (see page 491). As we shall see, maintenance of the physiological ratio of [ATP]/[ADP] so far away from equilibrium is accomplished by *kinetic* control, i.e., by regulation of enzymes. The key point is that maintaining the physiological ratio so far from equilibrium provides the thermodynamic driving force for nearly every biochemical event in the cell.

Kinetic Control of Substrate Cycles

Substrate cycles provide a beautiful illustration of the advantages of ATP as a coupling agent and the role of kinetic control over the direction, as well as the rate, of these opposing pathways. Let's return once again to the fructose-6-phosphate/fructose-1,6-bisphosphate substrate cycle:

	$\Delta G°'$	K	ATP coupling coefficient
Fructose-6-phosphate + ATP → fructose-1,6-bisphosphate + ADP (PFK)	−14.2	308	−1
Fructose-1,6-bisphosphate + H_2O → fructose-6-phosphate + P_i (FBPase)	−16.3	719	0

Net: ATP + H_2O → ADP + P_i

For the phosphofructokinase (PFK) reaction, under physiological conditions where [ATP] $\approx$ [ADP], we calculated a value of 308 for the equilibrium ratio of [fructose-1,6-bisphosphate]/[fructose-6-phosphate] (equation 12.10). Thus, this reaction is favorable in the direction of fructose-1,6-bisphosphate until its concentration approaches a level 308 times that of fructose-6-phosphate. For the fructose-1,6-bisphosphatase (FBPase) reaction, using the physiological concentration of P_i ($\sim 10^{-3}$ M), we obtain a value of 719,000 for the equilibrium ratio of [fructose-6-phosphate]/[fructose-1,6-bisphosphate]. Thus, under virtually any likely cellular condition, the FBPase reaction is thermodynamically favorable in the direction of fructose-6-phosphate. It follows then that *both* the PFK and FBPase reactions are thermodynamically favorable in their respective directions as long as [fructose-1,6-bisphosphate]/[fructose-6-phosphate] ratio is between 0.0000014 (1/719,000) and 308. In fact, the ratio will *always* be in this wide range in a healthy cell, and thus both reactions are *always* favorable. What then prevents these two reactions from occurring simultaneously, accomplishing nothing more than the net hydrolysis of ATP (i.e., a futile cycle)? The answer, of course, is the independent, but coordinated, kinetic control of the two opposed enzymes. As we will see in the next chapter, enzymes in substrate cycles are kinetically regulated by the levels of allosteric effectors.

Thus, substrate cycles illustrate the important metabolic design feature mentioned previously: ATP-coupling coefficients of opposing reactions or pathways always differ. This difference allows both sequences to be thermodynamically favorable at all times. The choice of which pathway operates, however, is determined entirely by the metabolic needs of the cell (via allosteric effectors), not by thermodynamics (because both pathways are favorable). Why is it so important for both pathways to be thermodynamically favorable at all times? *Because regulation can be imposed only on reactions that are displaced far from equilibrium.* Consider the following analogy:

A dam separates two bodies of water. If the water level is the same on both sides of the dam, we can say the system is at equilibrium. If we now open the floodgate, what happens? Water may move back and forth through the floodgate, but there will be no net movement of water or change in water levels. Thus, regulation (opening the floodgate) has no impact on a system at equilibrium. Now imagine a different system.

Here, the water level on the left side of the dam is much higher than that on the right side of the dam. We would say this system is far from equilibrium. If we now open the floodgate, water will rush from the left side to the right side and will continue flowing until it reaches equilibrium, or until we impose regulation (close the floodgate). This characteristic, that each pathway of an opposing pair is thermodynamically favorable (made possible by the different ATP-coupling coefficients for each), results in each pathway being unidirectional (because the reverse direction is highly thermodynamically unfavorable). These opposing pathways (e.g, glycolysis vs. gluconeogenesis) are like dams holding back high waters, each poised to flow downhill (thermodynamics), just waiting for the signal to open the floodgates (kinetics). The signal (e.g., allosteric effectors) will logically be different for each floodgate (regulatory enzyme) so that both pathways never flow at the same time (but both could if allowed).

The theoretical basis of this concept, which was developed in Chapter 3, is illustrated in Figure 12.10. It shows a plot of G (Gibbs free energy) vs. the log of the ratio of Q (the mass action ratio; page 69) to K (the equilibrium constant) for the reaction A $\rightleftharpoons$ B. This plot gives a parabola and the slope at any point on the curve corresponds to ΔG. The free energy is at a minimum when the Q/K ratio = 1 (i.e., when the actual concentrations of reactants and products is equal to their equilibrium concentrations). In other words, when the system is at equilibrium, $\Delta G = 0$ (verified by the slope = 0 at this point). The magnitude of ΔG is greatest when the mass action ratio is far from equilibrium concentrations, and the further the reaction is from equilibrium, the greater the driving force. ΔG will be large and negative when [A] is high and [B] is low (left side of parabola; $Q/K < 1$), and the reaction will proceed to the right due to the large driving force. Conversely, when [B] is high and [A] is low (right side of parabola; $Q/K > 1$), the reaction will proceed to the left due to an equally large driving force.

Earlier we estimated a ΔG of approximately −1300 kJ/mol for the oxidation of one mole of glucose during respiration (page 490). We can calculate the corresponding Q/K ratio at 298 K using Equation 3.33 from Chapter 3:

$$\Delta G = RT \ln\left(\frac{Q}{K}\right)$$

$$\frac{Q}{K} = e^{\left(\frac{\Delta G}{RT}\right)} = e^{\left(\frac{-1300 \times 10^3 \text{ J/mol}}{(8.315 \text{ J/mol}) \cdot (298 \text{ K})}\right)} \approx 10^{-228}$$

This Q/K ratio, 10^{-228}, is based on approximations of ΔG and might be off by a few orders of magnitude. Whatever the actual value, however, it lies far, far up the left arm of the parabola in Figure 12.10, and thus represents an immense driving force.

Regulation can be imposed only on reactions that are displaced far from equilibrium.

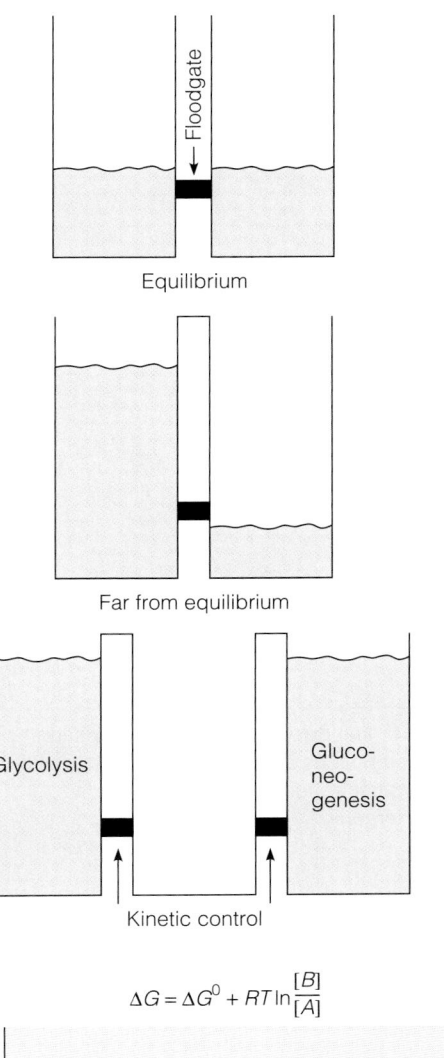

$$\Delta G = \Delta G^0 + RT \ln\frac{[B]}{[A]}$$

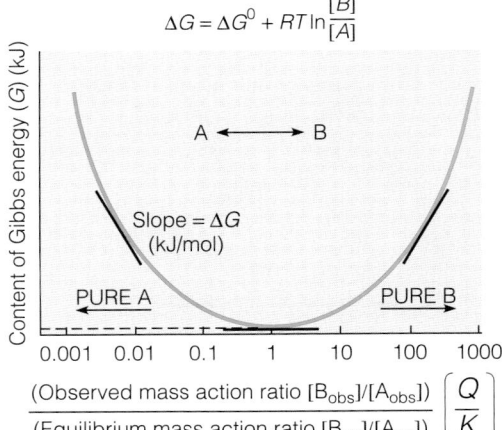

FIGURE 12.10

Gibbs free energy of a reaction as a function of its displacement from equilibrium.

Modified from *Bioenergetics*, 3rd ed., D. G. Nicholls and S. J. Ferguson, p. 35. © 2002, with permission from Elsevier.

Other High-Energy Phosphate Compounds

This same principle allows ATP to drive the synthesis of compounds of even higher phosphate transfer potential, such as creatine phosphate. This compound shuttles phosphate bond energy from ATP in mitochondria to myofibrils, where that bond energy is transduced to the mechanical energy of muscle contraction. Creatine phosphate (CrP) is produced from creatine by the enzyme **creatine kinase**:

$$H_2N-\overset{\overset{+NH_2}{\|}}{C}-\underset{\underset{CH_3}{|}}{N}-CH_2-COO^- \ + \ ATP \ \rightleftharpoons \ ^-O-\overset{\overset{O}{\|}}{\underset{\underset{O}{|}}{P}}-\overset{H}{\underset{}{N}}-\overset{\overset{+NH_2}{\|}}{C}-\underset{\underset{CH_3}{|}}{N}-CH_2-COO^- \ + \ ADP$$

Creatine   **Creatine phosphate**

Mammalian cells have several different creatine kinase isozymes, one of which is localized to the intermembrane space of mitochondria (mCK; see margin). From the respective $\Delta G^{\circ\prime}$ values for creatine phosphate and ATP hydrolysis (Figure 3.7, page 76), you can calculate that this reaction is endergonic under standard conditions, with a $\Delta G^{\circ\prime}$ value of $+12.6$ kJ/mol. However, because ATP levels are high within mitochondria, and creatine phosphate levels are relatively low ($Q/K << 1$), the reaction is exergonic as written and proceeds to the right in the mitochondrial intermembrane space. Creatine phosphate then diffuses from mitochondria to the myofibrils, where it provides the energy for muscle contraction. However, the direct energy source for that contraction is ATP hydrolysis once again. High levels of ADP formed during contraction favor the reverse reaction, catalyzed by a myofibril creatine kinase isozyme (myoCK). ATP is resynthesized at the expense of creatine phosphate cleavage to creatine, which can then return to mitochondria for resynthesis of creatine phosphate. The popularity of creatine as a dietary supplement for athletes, to increase muscle strength, suggests that the biosynthesis of creatine itself (see Chapter 21 for the pathway) may be a limiting factor in operating this intracellular energy shuttle.

In some invertebrate animals **arginine phosphate**, instead of creatine phosphate, plays a similar role in storing high-energy phosphate for rapid production, as needed, of ATP. In both instances, the varying concentrations of adenine nucleotides in the two environments—mitochondria and myofibrils—both at levels far from standard values, are critical in understanding how these high-energy compounds can be synthesized and utilized.

But factors in addition to concentration combine to make physiological ΔG values quite different from standard values. For example, as pH increases, the negative charge on the ATP molecule increases. This in turn increases the electrostatic repulsion between oxygen atoms linked to adjacent phosphorus atoms, and this in turn promotes hydrolysis, thus making ΔG more negative.

Also significant is the fact that most ATP is chelated within cells, as a complex with Mg^{2+}:

$$Adenosine-O-\overset{\overset{O}{\|}}{\underset{\underset{O^-}{|}}{P}}-O-\overset{\overset{O}{\|}}{\underset{\underset{O^-}{|}}{P}}-O-\overset{\overset{O}{\|}}{\underset{\underset{O^-}{|}}{P}}-O^-$$
$$Mg^{2+}$$

ADP is also complexed with Mg^{2+}, but ADP has a different affinity for Mg^{2+} from that of ATP. Varying levels of Mg^{2+} will change ΔG in complicated ways, depending upon relative affinities of reactants and products for the magnesium ion. (For a detailed discussion of the effects of pH, magnesium, and other ionic conditions, see the article by R. Alberty cited in the Chapter 3 references.)

ATP can drive the synthesis of higher-energy compounds, if nonequilibrium intracellular concentrations make such reactions exergonic.

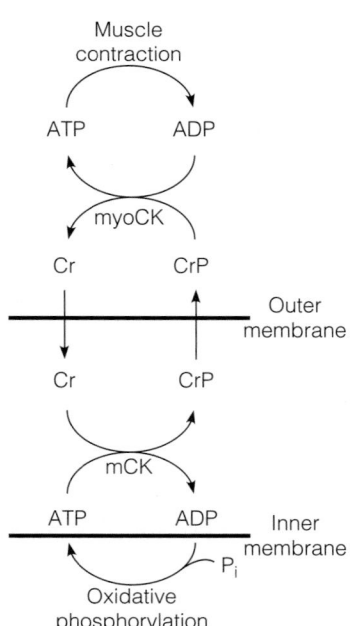

Creatine–creatine phosphate shuttle in muscle contraction

Other High-Energy Nucleotides

As discussed earlier, there is nothing unique about the properties endowing ATP with its special role as energy currency. All other nucleoside triphosphates, as well as more complex nucleotides, such as NAD^+, have $\Delta G^{\circ\prime}$ values for hydrolysis close to -31 kJ/mol and could have been selected for the role played by ATP. However, evolution has created an array of enzymes that preferentially bind ATP and use its chemical potential to drive endergonic reactions. There are exceptions, such as the use of GTP as the primary energy-providing nucleotide in protein synthesis. But phosphate-bond energy is created almost exclusively at the adenine nucleotide level, through oxidative phosphorylation in aerobic cells, photosynthesis in plants, and substrate-level phosphorylation during glycolysis in virtually all organisms. As a result, ATP is usually the most abundant nucleotide.

In most cells, ATP levels, at 2–8 mM, are severalfold higher than those of the other nucleoside triphosphates and also severalfold higher than the levels of ADP or AMP. These factors give ATP a strong tendency to distribute its γ (outermost) phosphate in the synthesis of other nucleoside triphosphates. This is accomplished through the action of **nucleoside diphosphate kinase**, which synthesizes CTP from CDP in the following example.

$$ATP + CDP \rightleftharpoons ADP + CTP$$

Nucleoside diphosphate kinase is active with a wide variety of phosphate donors and acceptors. Because its equilibrium constant is close to unity and because ATP is the most abundant nucleotide within cells, the enzyme normally uses ATP to drive the synthesis of the other common ribo- and deoxyribonucleoside triphosphates from their respective diphosphates.

Some metabolic reactions, such as the activation of amino acids for protein synthesis, cleave ATP, not to ADP and P_i, but to AMP and PP_i. The conversion of AMP to ATP, allowing reuse of the nucleotide, involves another enzyme, **adenylate kinase** (also called myokinase because of its abundance in muscle).

$$AMP + ATP \rightleftharpoons 2ADP$$

ADP is reconverted to ATP by substrate-level phosphorylation, oxidative phosphorylation, or (in plants) photosynthetic energy. Because the reaction is readily reversible, it can also be used for resynthesis of ATP when ADP levels rise; for example, after a burst of energy consumption. This function is particularly important in muscle metabolism.

Adenylate Energy Charge

As noted above, ATP levels are normally severalfold higher than the levels of ADP and AMP in well-nourished, energy-sufficient cells. Many enzymes that participate in regulating energy-generating or storage pathways are acutely sensitive to concentrations of adenine nucleotides. In general, energy-generating pathways, such as glycolysis and the citric acid cycle, are activated at low energy states, when levels of ATP are relatively low and those of ADP and AMP are relatively high. It is useful to be able to describe the energy status of a cell in quantitative terms. In his excellent book (cited in the references at the end of this chapter), Daniel Atkinson has likened the cell to a battery. When the cellular battery is fully charged, all of the adenine ribonucleotides are present in the form of ATP. When fully discharged, all of the ATP has been broken down to AMP. Atkinson has proposed the term **adenylate energy charge**, which is defined in terms of the intracellular concentrations of ATP, ADP, and AMP.

$$\text{Adenylate energy charge} = \frac{[ATP] + 0.5[ADP]}{[ATP] + [ADP] + [AMP]} \tag{12.14}$$

The term is the proportion of total energy-rich bonds that could be present in the adenine nucleotides in a cell to those that are actually present. Note that ADP has one energy-rich bond compared with two in ATP, so it carries one-half the weight of ATP in the numerator of equation 12.14. The value of the adenylate energy charge is probably never far from 1 in well-nourished aerobic cells.

Arginine phosphate

Major Metabolic Control Mechanisms

The living cell uses a marvelous array of regulatory devices to control its functions. These mechanisms include those that act primarily to control enzyme *activity*, such as substrate concentration and allosteric control, as discussed in Chapter 11. Control of enzyme *concentration*, through regulation of enzyme synthesis and degradation, is the major focus of Chapters 27–29 and 20, respectively. In eukaryotic cells, *compartmentation* represents another regulatory mechanism, with the fate of a metabolite being controlled by the flow of that metabolite through a membrane. Overlying all of these mechanisms are the actions of *hormones*, chemical messengers that act at all levels of regulation.

Control of Enzyme Levels

If you were to prepare a cell-free extract of a particular tissue and determine intracellular concentrations of several different enzymes, you would find tremendous variations. Enzymes of the central energy-generating pathways are present at many thousands of molecules per cell, whereas enzymes that have limited or specialized functions might be present at fewer than a dozen molecules per cell. Two-dimensional gel electrophoresis of a cell extract (see Figure 1.11, page 19) gives an impression, from the varying spot intensities, of the wide variations in amounts of individual proteins in a particular cell.

The level of a single enzyme can also vary widely under different environmental conditions, in large part because of variation in the enzyme's rate of synthesis. For example, when a usable substrate is added to a bacterial culture, the abundance of the enzymes needed to process the substrate may increase, through synthesis of new protein, from less than one molecule per cell to many thousands of molecules per cell. This phenomenon is called enzyme **induction**. Similarly, the presence of the end product of a pathway may turn off the synthesis of enzymes needed to generate that end product, a process called **repression**.

For some time it was thought that controlling the intracellular level of a protein was primarily a matter of controlling the *synthesis* of that protein—in other words, through genetic regulation. We now know that intracellular protein *degradation* is also important in determining enzyme levels (see Chapter 20).

Enzyme levels in a cell may change in response to changes in metabolic needs.

Control of Enzyme Activity

The *catalytic activity* of an enzyme molecule can be controlled in two ways: by reversible interaction with ligands (substrates, products, or allosteric modifiers) and by covalent modification of the protein molecule.

Enzyme activity is most commonly controlled by low-molecular-weight ligands, principally substrates and allosteric effectors. Substrates are usually present within cells at concentrations lower than the K_M values for the enzymes that act on them, but within an order of magnitude of these values. In other words, substrate concentrations usually lie within the first-order ranges of substrate concentration–velocity curves for the enzymes that act on them. Therefore, reaction velocities respond to small changes in substrate concentration. Ligands that control enzyme activity can also be polymers. For example, protein–protein interactions can affect enzyme activity, and several enzymes of nucleic acid metabolism are activated by binding to DNA.

We saw in Chapter 11 that allosteric activation or inhibition usually acts on committed steps of a metabolic pathway, often initial reactions. The effectors function by binding at specific regulatory sites, thereby affecting subunit–subunit interactions in the enzyme. This effect, in turn, either facilitates or hinders the binding of substrates. Such a mechanism operates in an obvious way to control product formation if a pathway is unidirectional and unbranched. However, some substrates are involved in numerous pathways, so many branch points exist. Therefore, some of the allosteric enzymes that we will describe display somewhat more complicated regulation than the examples presented in Chapter 11.

Enzyme activity is regulated by interaction with substrates, products, and allosteric effectors and by covalent modification of enzyme protein.

Covalent modification of enzyme structure represents another efficient way to control enzyme activity. Chapter 11 introduced several types of covalent modification that are used to regulate enzyme activity, including **phosphorylation**, **acetylation**, **methylation**, **adenylylation** (the transfer of an adenylate moiety from ATP), and **ADP-ribosylation** (the transfer of an ADP-ribosyl moiety from NAD^+). Many other less common covalent modifications are now known. However, phosphorylation is by far the most widespread covalent modification used to control enzyme activity.

Control through covalent modification is often associated with regulatory cascades. Modification activates an enzyme, which in turn acts on a second enzyme, which may activate yet a third enzyme, which finally acts on the substrate. Because enzymes act catalytically, this cascading provides an efficient way to *amplify* the original biological signal. Suppose that the original signal modifying enzyme A activates it 10-fold, that modified enzyme A then activates enzyme B 100-fold, and that B activates enzyme C 1000-fold. Thus, with the involvement of relatively few molecules of enzyme, a pathway can be activated by a million-fold ($10^1 \times 10^2 \times 10^3$).

The first well-understood regulatory cascade was the one controlling glycogen breakdown in animal cells, a critical process that provides carbohydrate substrates for energy generation. This regulatory cascade, involving enzyme phosphorylation and dephosphorylation, is described in detail in Chapter 13. Blood clotting, described in Chapter 11, is another well-understood regulatory cascade.

Compartmentation

We have already described the physical division of labor that exists in a eukaryotic cell, in the sense that enzymes participating in the same process are localized to a particular *compartment* within the cell. For example, RNA polymerases are found in the nucleus and nucleolus, where DNA transcription occurs, and the enzymes of the citric acid cycle are all found in mitochondria. Figure 12.11 presents the locations of a number of metabolic pathways within eukaryotic cells.

Compartmentation creates a division of labor within a cell, which increases the efficiency of cell function. The creatine–creatine phosphate shuttle (page 496) is a good example of this. In addition, compartmentation has an important

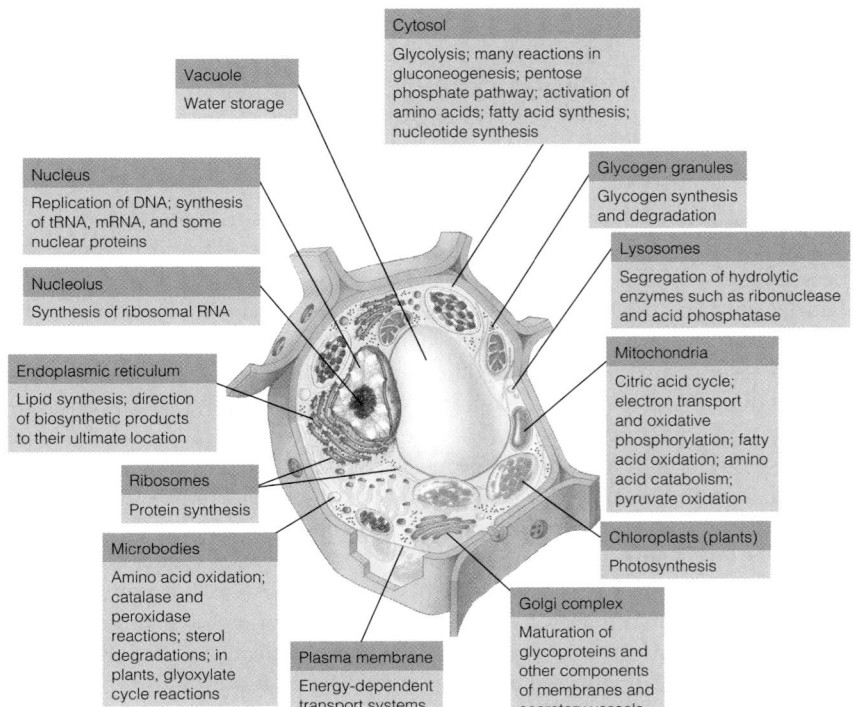

FIGURE 12.11

Locations of major metabolic pathways within a eukaryotic cell. This hypothetical cell combines features of a plant cell and an animal cell.

Biology, 5th ed., Neil A. Campbell, Jane B. Reece, and Lawrence A. Mitchell. © 1999. Reprinted by permission of Pearson Education Inc., Upper Saddle River, NJ.

regulatory function. This function derives largely from the selective permeability of membranes to different metabolites, thereby controlling the passage of intermediates from one compartment into another. Typically, intermediates of a pathway remain trapped within an organelle, while specific carriers allow substrates to enter and products to exit. The flux through a pathway, therefore, can be regulated by controlling the rate at which a substrate enters the compartment. For example, one of the ways in which the hormone insulin stimulates carbohydrate utilization is by moving glucose transporters into the plasma membrane so that glucose is more readily taken into cells for catabolism or for synthesis of glycogen.

Compartmentation is more than a matter of tucking enzymes into the proper organelles. Juxtaposition of enzymes that catalyze sequential reactions localizes substrates even in the absence of membrane-bound organelles. Opportunities for diffusion are reduced because the product of one reaction is released close to the active site of the next enzyme in a pathway. The enzymes may be bound to one another in a membrane, as are the enzymes of mitochondrial electron transport. Alternatively, they may be part of a highly organized multiprotein complex, such as the pyruvate dehydrogenase complex, a major entry point to the citric acid cycle.

Compartmentation can also result from weak interactions among enzymes that do not remain complexed when they are isolated. For example, conversion of glucose to pyruvate by glycolysis is catalyzed by enzymes that interact quite weakly in solution. However, there is evidence that these enzymes interact within the cytosol, forming a supramolecular structure that facilitates the multistep glycolytic pathway. The concept of intracellular interactions among readily solubilized enzymes developed as scientists began to realize that the cytosol is much more highly structured than was formerly thought. High-resolution electron micrographs of mammalian cytosol reveal the outlines of an organized structure that has been termed the **cytomatrix**, a model of which is shown in Figure 12.12. It is likely that such structures form as a result of the extremely high concentrations of proteins inside cells, which decrease the concentration of water and drive weakly interacting proteins to associate. It has been proposed that soluble enzymes are bound within the cell to the structural elements of the cytomatrix.

Whether highly structured or loosely associated, multienzyme complexes allow for efficient control of reaction pathways. Enzyme complexes restrict diffusion of intermediates, thereby keeping the average concentrations of intermediates low (but their local concentrations high, at enzyme catalytic sites). This complexing reduces *transient time*, the average time needed for a molecule to traverse a pathway. Thus, the flux through a pathway can change quickly in response to a change in the concentration of the first substrate for that pathway.

Hormonal Regulation

Overlaid and interspersed with the regulatory mechanisms operating within a eukaryotic cell are messages dispatched from other tissues and organs. The process of transmitting these messages and bringing about metabolic changes is called **signal transduction**. The extracellular messengers include hormones, growth factors, neurotransmitters, and pheromones, which interact with specific receptors, resulting in specific metabolic changes in the target cell.

Metabolic responses to hormones can involve changes in gene expression, leading to changes in enzyme levels. This type of response typically operates on a timescale of hours to days, resulting in a reprogramming of the metabolic capability of the cell. On a shorter timescale (seconds to hours), some hormones stimulate the synthesis of intracellular **second messengers** that control metabolic reactions. One of the most important second messengers is **adenosine 3′,5′-cyclic monophosphate**, more commonly called **cyclic AMP**, or simply **cAMP**. The hormone (the first messenger) acts extracellularly, by binding to receptors in the plasma membrane. Because the receptor protein traverses the membrane, it can, on the intracellular side, stimulate the formation of a second messenger in response to extracellular

Enzymes catalyzing sequential reactions are often associated, even in the cytosol, where organized structures are difficult to visualize.

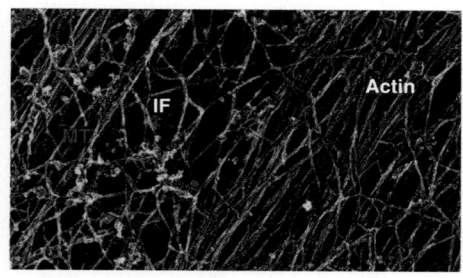

FIGURE 12.12

Structural organization of the cytomatrix. This electron micrograph of the cytoplasmic matrix of a cultured mammalian fibroblast reveals networks of filaments anchored to the plasma membrane. MT = microtubules; IF = intermediate filaments. Approximate magnification, 150,000.

© The Rockefeller University Press. *The Journal of Cell Biology* 1979, 82:114–139, J. J. Wolosewick and K. R. Porter, Microtrabecular lattice of the cytoplasmic ground substance: Artifact or reality.

binding of the first messenger. These signal transduction systems can effect precise control over metabolic pathways, often through reversible covalent modification of critical enzymes in the pathway.

Signal transduction systems are modular in nature, allowing for a diversity of metabolic responses based on the same operating principles. Thus, secretion of one hormone can have quite diverse effects in different tissues, depending upon the nature of the receptor and other components and second messenger systems in different target cells. Moreover, a single second messenger may have diverse effects within a single cell. For example, cyclic AMP, which activates glycogen degradation, also activates a cascade that inhibits the synthesis of glycogen. This dual effect is an example of a coordinated metabolic response: Glycogen synthesis is *inhibited* under the same physiological conditions that *promote* glycogen breakdown. These signal transduction systems will be described in great detail in Chapters 13, 18, and 23.

Distributive Control of Metabolism

In recent years, an important principle of metabolic regulation has emerged. With the discovery in the 1950s and 1960s of allosterically regulated enzymes, the concept arose that metabolic pathway flux is regulated primarily through control of the intracellular activity of one or a few key enzymes in that pathway. When it was realized that allosteric enzymes often catalyze committed reactions, i.e., the first reaction in a pathway that leads to an intermediate with no other known function, the concept arose that these enzymes catalyze the "rate-limiting reactions" in metabolic pathways. Taken literally, the term *rate-limiting reaction* implies that the flux rate of the pathway is identical to the intracellular activity of the rate-limiting step. However, because in a metabolic steady state all of the steps along a linear pathway are proceeding at the same rate, it is difficult or impossible to establish that the rate is controlled by a single enzymatic step. In other words, the concept of a rate-limiting enzyme is seriously flawed.

We now realize that metabolic regulation is more complex and that all of the enzymes in a pathway contribute toward control of pathway flux. An approach called **metabolic control analysis** assigns to each enzyme in a pathway a **flux control coefficient, C^J**, a value that can vary between zero and one. Consider a metabolic pathway where substrate A is converted to product D in several steps. The **flux** through the pathway, **J**, is equal to the rate of the forward process, v_f, minus the rate of the reverse process, v_r:

$$J = v_f - v_r$$

For a given enzyme, the flux control coefficient is the relative increase in flux, divided by the relative increase in enzyme activity that brought about that flux increase. For a true rate-limiting enzyme, the flux control coefficient is 1; a 20% increase in the activity of that enzyme would increase that flux rate by 20%. But metabolic control theory predicts that all enzymes in a pathway contribute toward regulation, meaning that all enzymes have flux-control coefficients greater than zero (but none has a value as high as 1, which would be the case if flux were truly controlled solely by one rate-limiting enzyme). Flux-control coefficients are properties of the pathway or metabolic system, and thus all the flux-control coefficients in the pathway must sum to 1:

$$C_1^J + C_2^J + \cdots + C_n^J = 1$$

In the hypothetical pathway in the margin, intermediate C has two alternative fates. Because reaction 4 draws C away from the pathway A → D, it decreases flux through the pathway, and the enzyme that catalyzes step 4 has a negative flux control coefficient. The sum of the four flux control coefficients in the pathway still equals 1.

The predictions of metabolic control theory can be tested, for example, by using mutations affecting a specific enzyme to bring about defined changes in the activity of that enzyme in vivo, and then measuring the change in flux rate of a pathway in which that enzyme is involved. Such analyses confirm that enzymes catalyzing committed

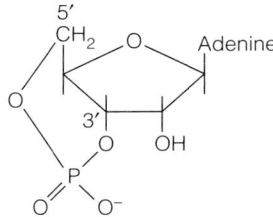

Adenosine-3′,5′-cyclic monophosphate (cyclic AMP)

Second messengers transmit information from hormones bound at the cell surface, thereby controlling intracellular metabolic processes.

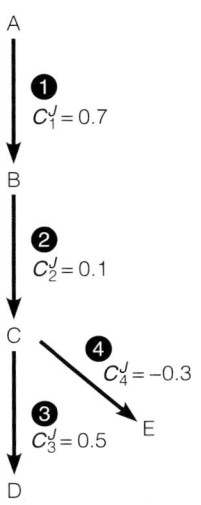

reactions do play large roles in regulation (i.e., they have high flux-control coefficients). More important, however, these analyses confirm that *every* enzyme in a pathway contributes toward control of that pathway; in other words, every enzyme has a flux-control coefficient greater than zero. Thus, regulation of a pathway is *distributed* among all of the enzymes involved in the pathway, giving rise to the concept of **distributive control of metabolism**. In retrospect, this concept should have been predicted simply from the long-understood complexity of metabolism. Many intermediates participate in more than one pathway, making different pathways interdependent and interlocking. Regulatory schemes that depend on control of just one or two enzymes in each pathway lack the flexibility and subtlety to account for the ability of cells to maintain homeostasis under widely varying nutritional and energetic conditions. To be sure, for each pathway or process, one or a few regulatory enzymes of primary importance have been identified, and these will be pointed out as we present the individual pathways involved.

Experimental Analysis of Metabolism

Goals of the Study of Metabolism

Given that metabolism consists of all the chemical reactions in living matter, how does a biochemist approach metabolism in the laboratory? To subdivide a particular metabolic process into experimentally attainable goals, the biochemist seeks (1) to identify reactants, products, and cofactors, plus the stoichiometry, for each reaction involved; (2) to understand how the rate of each reaction is controlled in the tissue of origin; and (3) to identify the physiological function of each reaction and control mechanism. These goals necessitate isolating and characterizing the enzyme catalyzing each reaction in a pathway. This latter task—extrapolating from test tube biochemistry to the intact cell—is especially challenging. For instance, given that most enzymes catalyze reactions that can proceed in either direction, what is the direction of an enzymatic reaction in vivo? Many reactions originally found to proceed one way in vitro have been shown to proceed in the opposite direction in vivo. The mitochondrial enzyme that synthesizes ATP from ADP, for example, was originally isolated as an ATPase—an enzyme that hydrolyzes ATP to ADP and P_i. Therefore, it is not enough to isolate an enzyme and to demonstrate that it catalyzes a particular reaction in the test tube. We must show also that the same enzyme catalyzes the same reaction in intact tissue, usually a more difficult task.

To achieve the stated goals, a biochemist must perform analyses at several levels of biological organization, from living organisms and intact cells to broken-cell preparations and, ultimately, to purified components. Cell-free, or in vitro, preparations can be manipulated in ways that intact cells cannot—for example, by addition of substrates and cofactors that will not pass through cell membranes. The researcher attempts to duplicate in vitro the process that is known to occur in vivo.

Preparation of cell-free components usually destroys biological organization. Cells can be broken open by sonication, shear forces, or enzymatic digestion of cell walls. These harsh treatments lead to mixing of components that were in separate compartments in the intact cell, making it possible to misinterpret data obtained from in vitro systems. An example comes from studies of protein biosynthesis. The existence of messenger RNA as a template for protein synthesis was predicted in 1961 from the behavior of bacterial mutants altered in genetic regulation. However, it was difficult to demonstrate the existence of messenger RNA in vitro because the putative template was present in very small amounts and was rapidly degraded by enzymes in cell-free extracts. Only when investigators learned how to prevent this degradation could they prove the existence of messenger RNA.

The foregoing discussion reveals the necessity of studying metabolism at various levels of biological organization, from intact organism to purified chemical components. Here we discuss what can be learned at each level.

Levels of Organization at Which Metabolism Is Studied

Whole Organism

Biochemists must investigate metabolism in whole organisms because our ultimate aim is to understand chemical processes in intact living systems. Radioisotopic tracers are widely used to characterize metabolic pathways, as described in Tools of Biochemistry 12A. A classic example, described in Chapter 19, is the elucidation of cholesterol synthesis in the 1940s. Konrad Bloch injected ^{14}C-labeled acetate into rats and followed the flow of label into intermediates by sacrificing rats at intervals and analyzing the radioactive compounds in their livers. In designing experiments of this type, the investigator must pay attention to the efficiency of transport of the labeled precursor to the organs of interest, uptake of the precursor into cells, and competition of exogenous precursors with preexisting pools of unlabeled intermediates.

Many diagnostic tests in clinical medicine are in vivo metabolic experiments. Instead of using radioisotopes, we sample tissue at intervals and carry out biochemical assays. In the **glucose tolerance test**, for example, a subject consumes a large oral dose of glucose, and its level in the blood is then determined at intervals over several hours. The glucose tolerance test is used to diagnose diabetes and other disorders of carbohydrate metabolism.

In recent years, *nuclear magnetic resonance* (NMR) spectroscopy has become widely available for noninvasive monitoring of intact cells and organs. As explained in Tools of Biochemistry 6A, compounds containing certain atomic nuclei can be identified from an NMR spectrum, which measures shifts in the frequency of absorbed electromagnetic radiation. A researcher can determine an NMR spectrum of whole cells, or of organs or tissues in an intact plant or animal. NMR has even become a powerful noninvasive diagnostic tool, referred to as magnetic resonance imaging (MRI) in the medical arena. For the most part, macromolecular components do not contribute to the spectrum, nor do compounds that are present at less than about 0.5 mM. The nuclei most commonly used in this in vivo technique are ^1H, ^{31}P, and ^{13}C. Figure 12.13 shows ^{31}P NMR spectra that represent components in the human forearm muscle. The five major peaks correspond to the phosphorus nuclei in orthophosphate (P_i), creatine phosphate, and the three phosphates of ATP. Because peak area is proportional to concentration, the energy status of intact cells can be determined. For example, an energy-rich muscle has lots of creatine phosphate, whereas a fatigued muscle uses up most of its creatine phosphate in order to maintain ATP levels (note also the accumulation of AMP—peak 6—in the third scan). NMR is finding wide applicability in monitoring recovery from heart attacks, in which cellular ischemia (insufficient oxygenation) damages cells by reducing ATP content. NMR can also be used to study metabolite compartmentation, flux rates through major metabolic pathways, and intracellular pH.

Isolated or Perfused Organ

Some of the difficulties in transporting a precursor or inhibitor to the desired organ can be circumvented by using an isolated organ. A researcher usually **perfuses** the isolated organ during the experimental manipulations. This procedure involves pumping a buffered isotonic solution containing nutrients, drugs, or hormones through the organ. The solution partly takes the place of the normal circulation, delivering nutrients and removing waste products. The researcher can also perfuse an organ within a living animal, following appropriate surgical procedures. Of course, perfusion is much less efficient than circulation; thus, experiments at this level must be of limited duration.

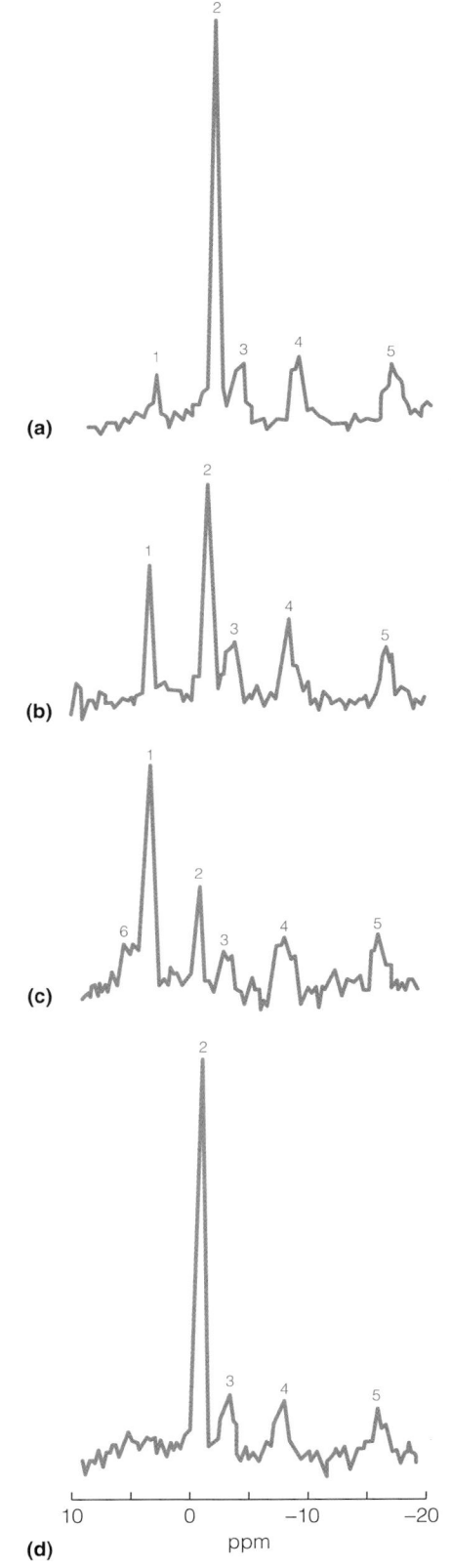

FIGURE 12.13

Effect of anaerobic exercise on ^{31}P NMR spectra of human forearm muscle. **(a)** Before exercise. **(b)** One minute into a 19-minute exercise period. **(c)** The 19th minute. **(d)** Ten minutes after exercise. Peak areas are proportional to intracellular concentrations. Peak 1 = P_i; 2 = creatine phosphate; 3 = ATP γ-phosphate; 4 = ATP α-phosphate; 5 = ATP β-phosphate; 6 = phosphomonoesters. See Tools of Biochemistry 6A for interpretation of NMR spectra.

From *Science* 233:640–645, G. K. Radda, The use of NMR spectroscopy for the understanding of disease. © 1986. Reprinted with permission from AAAS and G. K. Radda.

The circulation problem can be partly overcome by cutting the tissue into thin slices before the experimental manipulations begin. The structural integrity of the organ is lost, but most of the cells remain intact, and they are in better contact with the fluid that bathes the tissue. The cells may be better oxygenated and supplied with substrates than in a whole organ. The elucidation of the citric acid cycle resulted largely from experiments with slices of liver and heart.

Whole Cells

Tissue slices are not now widely used, partly because methods are now available for disaggregating an organ or a tissue into its component cells. Liver, kidney, and heart cells can be prepared by treatment of the organ with trypsin or collagenase, to break down the extracellular matrix that holds the organ together. For plant cells, enzymes such as cellulase or pectinase, which attack the cell wall, can be used to make comparable preparations.

Any plant or animal organ contains a complex mixture of different cell types. Several means of fractionating the cells after disaggregation of an organ are used to obtain preparations enriched in one cell type. One method is centrifugation to separate cells on the basis of size. In recent years the **fluorescence-activated cell sorter** has come into widespread use. In a typical application a cell suspension is treated with a fluorescent-tagged antibody to a cell surface antigen, which is present in varying amounts among different cell types. Cells pass in single file through a laser beam and are physically separated according to the amount of fluorescence recorded from each cell. Such machines, which can sort several thousand cells per second, result in fractionation based on the abundance of the selected surface antigen.

Uniformity in a cell population is often achieved by growth of cells in **tissue culture**. Disaggregated cells of an organ or a tissue can, with special care, be induced to grow in a medium containing cell nutrients and protein growth factors. The cells grow and divide independently of one another, much like the cells in a bacterial culture. Although animal cells usually cease growth after a certain number of divisions in culture, variant lines arise that are capable of indefinite growth, as long as they are adequately nourished. In such cultures, **clonal** cell lines can be generated in which all of the cells in a line are derived from a single cell so that they are genetically and metabolically uniform. This uniformity is a boon for many biochemical investigations. For example, much of our understanding of virus replication depends upon the ability to infect a large number of identical cells in culture simultaneously and then follow the metabolic changes by sampling the cell culture at various times after infection.

One problem with tissue culture is that cells adapted to long-term growth in culture take on characteristics different from those of their parent cells, which were originally embedded in plant or animal tissue. Maintaining specialized cell characteristics in culture always presents a challenge.

Cell-Free Systems

Problems of transport through membranes are obviated by working with broken-cell preparations. Animal cells are easily lysed (ruptured) by mild shear forces, suspension in hypotonic medium, or freezing and thawing. Bacterial cells have a rigid cell wall that requires vigorous treatment such as sonication (using ultrasound energy to agitate particles). Enzymatic digestion with lysozyme is often used to open bacterial cells under relatively mild conditions. Breaking the especially tough cell walls of yeasts and plants usually requires combinations of enzymatic and mechanical treatments.

Initial metabolic experiments are usually carried out in unfractionated cell-free homogenates. However, localizing a metabolic pathway within a particular cell compartment requires fractionating the homogenate to separate the organelles, usually by **differential centrifugation**. Lysis is carried out in isotonic sucrose solutions and generally yields morphologically intact organelles. These components can be pelleted by centrifugation at different speeds and for different lengths of time. Typically, nuclei, mitochondria, chloroplasts, lysosomes, and **microsomes** (artifactual membrane vesicles formed from disrupted endoplasmic reticulum) can all be at least partially separated from each other. The contents of the cytosol remain in the supernatant after the final centrifugation step. Much of our understanding of DNA replication and transcription in eukaryotic cells comes from investigations with isolated nuclei, whereas purified mitochondria have yielded much of our understanding of respiratory electron transport and oxidative phosphorylation.

Purified Components

To understand a biological process at the molecular level, the investigator must purify to homogeneity all of the factors thought to be involved and determine their interactions. Often, as with the citric acid cycle, this process is simply a matter of purifying the individual enzymes involved, determining the substrate and cofactor requirements of each, recombining the purified enzymes, and showing that the entire process can be catalyzed by purified components. This process is called **reconstitution**. Some pathways require cell constituents other than enzymes, such as the ribosomes and transfer RNAs needed for protein synthesis.

In purifying individual components, the biochemist continually risks losing factors that are essential for normal control or for some other aspect of the process under study. Avoiding such pitfalls requires painstaking experiments in which the researcher defines criteria for biological activity and continually examines fractions to ensure that each activity is retained through each fractionation step. A good example of this approach is presented in Chapter 25, in which we discuss the several enzymes and proteins that must function at a DNA replication fork.

Metabolic Probes

An invaluable biochemical aid is the use of metabolic probes, agents that allow a researcher to interfere specifically with one or a small number of reactions in a pathway. The consequences of such interference can be extremely informative. Two kinds of probes are most widely used—*metabolic inhibitors* and *mutations*. By blocking a specific reaction in vivo and determining the results of the blockade, such probes help identify the metabolic role of a reaction. For example, respiratory poisons such as carbon monoxide and cyanide block specific steps in respiration, and metabolic inhibitors helped identify the order of electron carriers in the respiratory electron transport chain (see Chapter 15).

Inhibitors can present difficulties for the researcher, such as poor transport into cells or multiple sites of action. It is often easier to interfere with a pathway by selecting mutant strains deficient in the enzyme of interest. In the 1940s George Beadle and Edward Tatum were the first to use mutations as biochemical probes, in their work with the bread mold *Neurospora crassa*. Beadle and Tatum isolated a number of X-ray-induced mutants that required arginine for growth, in addition to the constituents of minimal medium. Furthermore, different mutations affected different enzymes in the arginine biosynthetic pathway, with each mutant accumulating in the culture medium the intermediate that was the substrate for the deficient enzyme. This observation allowed Beadle and Tatum to order the enzymes according to the reactions they controlled, by the rationale illustrated in Figure 12.14. If a culture filtrate (containing an accumulated intermediate) from one mutant allowed a second mutant to grow without arginine, the researchers concluded that the first mutation blocked an enzymatic step *later* in the pathway than the step blocked by the second mutation. Ultimately, the accumulating intermediates were identified and the pathway was elucidated. The impact of this "biochemical genetics" approach was much wider than simply working out the details of a metabolic pathway. Beadle and Tatum recognized that there was a one-to-one correspondence between a genetic mutation and the loss of a specific enzyme, leading them to propose the **one gene–one enzyme hypothesis**, years before the chemical nature of the gene was known.

In addition to identifying pathways, mutants have been used to elucidate regulatory mechanisms. The earliest successes came from studies in Paris by François Jacob and Jacques Monod, who isolated dozens of *Escherichia coli* mutants with

By inactivating individual enzymes, mutations and enzyme inhibitors help identify the metabolic roles of enzymes.

A mutant defective in enzyme:	Accumulates metabolite in culture medium:	Requires an external source of:	Culture filtrate allows the growth of another mutant, defective in enzyme:
I	A	B, C, D, or E	—
II	B	C, D, or E	I
III	C	D or E	I or II
IV	D	E	I, II, or III

Analysis of mutants

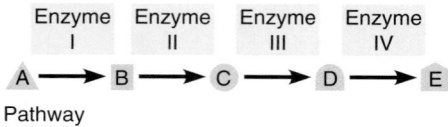

Pathway

FIGURE 12.14

Using mutations as biochemical probes. The steps of a hypothetical metabolic pathway are identified by analysis of mutants defective in individual steps of the pathway. For example, we can identify metabolite C as the substrate for enzyme III by the absence of this enzyme in mutants that accumulate C. We know that D and E follow C in the pathway because feeding either D or E to mutants defective in enzyme III bypasses the genetic block and allows the cells to grow.

defects in regulation of lactose catabolism or with abnormalities in virus–host relationships. These data led ultimately to the discovery of mRNA and the repressor–operator mechanism of genetic regulation, which we discuss in Chapter 27. Recombinant DNA technologies (see Tools of Biochemistry 4B) have allowed even more sophisticated perturbations of metabolic regulatory systems. For example, a researcher might use site-directed mutagenesis to disable the allosteric regulatory sites on an enzyme in a particular substrate cycle. The mutated enzyme is then introduced back into the cell, replacing the wild-type version, and the effects of loss of regulation can be examined in vivo.

A single investigation can use both metabolic inhibitors and mutations. This combined approach helped to elucidate the function of DNA gyrase, one of the DNA topoisomerases mentioned in Chapter 4. This enzyme is inhibited by nalidixic acid. When nalidixic acid is administered to bacteria, DNA replication is inhibited, suggesting that DNA gyrase plays an essential role in DNA replication. However, because nalidixic acid *could* inhibit DNA replication by blocking some other enzyme, stronger evidence is needed. Such evidence was obtained when mutants resistant to nalidixic acid were found to contain an altered form of DNA gyrase that was resistant to nalidixic acid. Thus, a single mutation abolished nalidixic acid sensitivity for both the enzyme and the ability of the cells to replicate their DNA, strongly supporting an essential role for DNA gyrase in DNA replication.

Finally, **systems biology** approaches are now being applied to the study of metabolism. Systems biologists try to catalog the full set of components in a system, and then study the interactions between these components and how these interactions give rise to the specific function and behavior of that system. Metabolic profiling (see Tools of Biochemistry 12B) is an example of this kind of approach.

In this chapter we have described the general strategy of metabolism, identified the major pathways, outlined how pathways are regulated, and identified experimental approaches to understanding metabolism. We are now prepared for detailed descriptions of metabolic pathways, which we begin in Chapter 13 with carbohydrates.

SUMMARY

Metabolism is the totality of chemical reactions occurring within a cell. Catabolic pathways break down substrates to provide energy, largely through oxidative reactions, whereas anabolic pathways synthesize complex biomolecules from small molecules, often from intermediates in catabolic pathways. Catabolic and anabolic pathways with the same end points are actually different pathways, not simple reversals of each other, so that both pathways are thermodynamically favorable. Regulation can be imposed only on reactions that are displaced far from equilibrium. Most metabolic energy comes from oxidation of substrates, with energy release coming in a series of small steps as the electrons released are transferred ultimately to oxygen. The more highly reduced a substrate, the more energy is released through its catabolism.

Flux through metabolic pathways is controlled by regulation of enzyme concentration (through control of enzyme synthesis and degradation), enzyme activity (through concentrations of substrates, products, and allosteric effectors, and covalent modification of enzymes), compartmentation, and hormonal control. Hormonal regulation may involve control of enzyme synthesis at the genetic level or regulation of enzyme activity. In the latter case, intracellular second messengers are formed in response to hormonal signals.

The understanding of metabolic processes requires identification of each reaction in a pathway and a knowledge of the reaction's function and control. This understanding requires experimentation at all levels of biological organization, from living organism to purified enzyme. The ability to block specific enzymes, either with inhibitors or mutations, is of great value in identifying the functions of those enzymes.

REFERENCES

Metabolic Design Principles

Atkinson, D. E. (1977) *Cellular Energy Metabolism and Its Regulation.* Academic Press, New York. This excellent book lays out the bioenergetic foundations of metabolic processes.

Brosnan, J. T. (2005) Metabolic design principles: Chemical and physical determinants of cell chemistry. *Adv. Enzyme Regul.* 45:27–36. This essay, based largely on concepts in Atkinson's book, discusses how the design of metabolic systems follows logically from chemical and physical constraints.

Experimental Techniques in the Study of Metabolism

Cunningham, R. E. (2010) Overview of flow cytometry and fluorescent probes for flow cytometry. *Methods Mol. Biol.* 588:319–326. The power of fluorescence-activated cell sorting.

Shulman, R. G., and D. L. Rothman (2001) ^{13}C NMR of intermediary metabolism: Implications for systemic physiology. *Annu. Rev. Physiol.* 63:15–48.

Tsien, R. Y. (2009) Constructing and exploiting the fluorescent protein paintbox (Nobel Lecture). *Angew. Chem. Int. Ed. Engl.* 48:5612–5626. This review by the 2008 Chemistry Nobel Laureate describes another powerful technique for noninvasive metabolic monitoring of individual cells.

Compartmentation and Intracellular Enzyme Organization

Dzeja, P. P., and A. Terzic (2003) Phosphotransfer networks and cellular energetics. *J. Exp. Biol.* 206:2039–2047. A contemporary account of the bioenergetic role of creatine phosphate, emphasizing the importance of compartmentation as a metabolic control phenomenon.

Goodsell, D. S. (1991) Inside a living cell. *Trends Biochem. Sci.* 16:203–206. Classic drawings of the interior of a bacterial cell, based upon physical information about the sizes, shapes, and distribution of cellular constituents.

Ovádi, J., and V. Saks (2004) On the origin of intracellular compartmentation and organized metabolic systems. *Mol. Cell. Biochem.* 256–257:5–12. A concise review of evidence for the organization of sequential metabolic pathways, including both membranous complexes and complexes involving soluble enzymes.

Enzyme Control and Metabolic Regulation

Fell, D. (1997) *Understanding the Control of Metabolism.* Portland Press Ltd., London. This book goes into depth on most of the topics of this chapter, including metabolic control analysis.

Newsholme, E. A., R. A. J. Challiss, and B. Crabtree (1984) Substrate cycles: Their role in improving sensitivity in metabolic control. *Trends Biochem. Sci.* 9:277–280. A brief but lucid discussion of substrate cycle control, with several examples.

PROBLEMS

1. Write a balanced equation for the complete oxidation of each of the following, and calculate the respiratory quotient for each substance.
   (a) Ethanol
   (b) Acetic acid
   (c) Stearic acid
   (d) Oleic acid
   (e) Linoleic acid

2. Given what you know about the involvement of nicotinamide nucleotides in oxidative and reductive metabolic reactions, predict whether the following intracellular concentration ratios should be (1) unity, (2) greater than unity, or (3) less than unity. Explain your answers.
   (a) [NAD$^+$]/[NADH]
   (b) [NADP$^+$]/[NADPH]

   Because NAD$^+$ and NADP$^+$ are essentially equivalent in their tendency to attract electrons, discuss how the two concentration ratios might be maintained inside cells at greatly differing values.

3. (a) NAD$^+$ kinase catalyzes the ATP-dependent conversion of NAD$^+$ to NADP$^+$. How many reducing equivalents are involved in this reaction?
   (b) How many reducing equivalents are involved in the conversion of ferric to ferrous ion?
   (c) How many reducing equivalents are involved in reducing one molecule of oxygen gas to water?

4. On page 488 we showed that the oxidation of glucose and palmitic acid yields 15.64 kJ/g and 38.90 kJ/g, respectively. Calculate these values in terms of kJ/mol and kJ per carbon atom oxidized for both glucose and palmitic acid.

5. Free energy changes under intracellular conditions differ markedly from those determined under standard conditions. $\Delta G^{\circ\prime}$ for ATP hydrolysis to ADP and P$_i$ is -30.5 kJ/mol. Calculate ΔG for ATP hydrolysis in a cell at 37 °C that contains ATP at 3 mM, ADP at 1 mM, and P$_i$ at 1 mM.

6. Consider the following hypothetical metabolic pathway:

$$A \rightleftharpoons B \xrightarrow{\text{X}} C \rightleftharpoons D$$

   (a) Under intracellular conditions, the activity of enzyme X is 100 pmol/10^6 cells/sec. Calculate the effect on metabolic flux rate (pmol/10^6 cells/sec) of B $\rightarrow$ C of the following treatments. Calculate as % change (increase or decrease).

Treatment	% Change in flux
Inhibitor that reduces activity of X by 10%	
Activator that increases activity of X by 10%	

   (b) Now consider a substrate cycle operating with enzymes X and Y in the same hypothetical metabolic pathway:

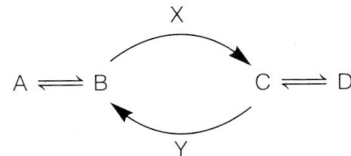

Under intracellular conditions, the activity of enzyme X is 100 pmol/10^6 cells/sec, and that of enzyme Y is 80 pmol/10^6 cells/sec. What are the direction and rate (pmol/10^6 cells/sec) of metabolic flux between B and C?

(c) Calculate the effect on direction and metabolic flux rate of the following treatments. Calculate as % change (increase or decrease).

Treatment	Direction of flux (B → C or C → B)	% Change in flux
Inhibitor that reduces activity of X by 10%		
Activator that increases activity of X by 10%		
Doubling the activity of enzyme Y		

(d) Briefly summarize the regulatory advantage(s) of a substrate cycle in a pathway.

7. The glucose/glucose-6-phosphate substrate cycle involves distinct reactions of glycolysis and gluconeogenesis that interconvert these two metabolites. Assume that under physiological conditions, [ATP] = [ADP]; [P_i] = 1 mM.

Consider the glycolytic reaction catalyzed by hexokinase:

$$\text{ATP} + \text{glucose} \rightleftharpoons \text{ADP} + \text{glucose-6-phosphate}$$
$$\Delta G^{\circ\prime} = -16.7 \text{ kJ/mol}$$

(a) Calculate the equilibrium constant (K) for this reaction at 298 °K, and from that, calculate the maximum [glucose-6-phosphate]/[glucose] ratio that would exist under conditions where the reaction is still thermodynamically favorable.

(b) Reversal of this interconversion in gluconeogenesis is catalyzed by glucose-6-phosphatase:

$$\text{glucose-6-phosphate} + \text{H}_2\text{O} \rightleftharpoons \text{glucose} + \text{P}_i$$
$$\Delta G^{\circ\prime} = -13.8 \text{ kJ/mol}$$

K for this reaction is 262. Calculate the maximum ratio of [glucose]/[glucose-6-phosphate] that would exist under conditions where the reaction is still thermodynamically favorable.

(c) Under what cellular conditions would both directions in the substrate cycle be strongly favored?

(d) What ultimately controls the direction of net conversion of a substrate cycle such as this in the cell?

Read Tools of Biochemistry 12A before attempting to work Problems 8–10.

*8. Two-dimensional gel electrophoresis of proteins in a cell extract provides a qualitative way to compare proteins with respect to intracellular abundance. Describe a quantitative approach to the determination of number of molecules of an enzyme per cell.

*9. Mammalian cells growing in culture were labeled for a long time in [^{3}H]thymidine to estimate the rate of DNA synthesis. The thymidine administered had a specific activity of 3000 cpm/pmol. At intervals, samples of culture were taken and acidified to precipitate nucleic acids. The rate of incorporation of isotope into DNA was 1500 cpm/10^6 cells/min. A portion of culture was taken for determination of specific activity of the intracellular dTTP pool, which was found to be 600 cpm/pmol.

(a) What fraction of the intracellular dTTP is synthesized from the exogenous precursor?

(b) What is the rate of DNA synthesis, in molecules per minute per cell of thymine nucleotides incorporated into DNA?

(c) How might you determine the specific activity of the dTTP pool?

*10. Suppliers of radioisotopically labeled compounds usually provide each product as a mixture of labeled and unlabeled material. Unlabeled material is added deliberately as a **carrier**, partly because the specific activity of the carrier-free product is too high to be useful and partly because the product is more stable at lower specific activities. Using the radioactive decay law, calculate the following.

(a) The specific activity of carrier-free [^{32}P]orthophosphate, in mCi/mmol.

(b) In a preparation of uniform-label [^{3}H]leucine, provided at 10 mCi/mmol, the fraction of H atoms that are radioactive.

TOOLS OF BIOCHEMISTRY 12A

RADIOISOTOPES AND THE LIQUID SCINTILLATION COUNTER

Radioisotopes revolutionized biochemistry when they became available to investigators shortly after World War II. Radioisotopes extend by orders of magnitude the sensitivity with which chemical species can be detected. Traditional chemical analysis can detect and quantify molecules in the micromole or nanomole range (i.e., 10^{-6} to 10^{-9} mole). A compound that is "labeled," containing one or more atoms of a radioisotope, can be detected in picomole or even femtomole amounts (i.e., 10^{-12} or 10^{-15} mole). Radiolabeled compounds are called **tracers** because they allow an investigator to follow specific chemical or

biochemical transformations in the presence of a huge excess of nonradioactive material.

Recall that isotopes are different forms of the same element. They have different atomic weights but the same atomic number. Thus, the chemical properties of the different isotopes of a particular element are virtually identical. Isotopic forms of an element exist naturally, and substances enriched for rare isotopes can be isolated and purified from natural sources. Most of the isotopes used in biochemistry, however, are produced in nuclear reactors. Simple chemical compounds produced in such reactors are then converted to radiolabeled biochemicals by chemical and enzymatic synthesis.

Stable Isotopes

Although radioisotopes are widely used in biochemistry, stable isotopes are also used as tracers, particularly when convenient radioisotopes are not available. For example, the two rare isotopes of hydrogen include a stable isotope ($^2H_1$, or **deuterium**) and a radioactive isotope ($^3H_1$, or **tritium**). Of the many uses of stable isotopes in biochemical research, we mention three applications here. First, incorporation of a stable isotope often increases the density of a material because the rare isotopes usually have higher atomic weights than their more abundant counterparts. This difference presents a way to separate labeled from nonlabeled compounds physically, as in the Meselson–Stahl experiment on DNA replication (see Chapter 4). Second, compounds labeled with stable isotopes, particularly ^{13}C, are widely used in nuclear magnetic resonance studies of molecular structure and dynamics (see Tools of Biochemistry 6A). Third, stable isotopes are used to study reaction mechanisms. The "kinetic isotope effect" refers to the effect on reaction rate of substitution of an atom by a heavy isotope. As discussed in Chapter 11, such an effect helps to identify rate-limiting steps in enzyme-catalyzed reactions. Table 12A.1 gives information about the isotopes, both stable and radioactive, that have found the greatest use in biochemistry.

The Nature of Radioactive Decay

The atomic nucleus of an unstable element can decay, giving rise to one or more of the three types of ionizing radiation: α-, β-, or γ-rays. Only β- and γ-emitting radioisotopes are used in biochemical research; the most useful are shown in Table 12A.1. A β-ray is an emitted electron, and a γ-ray is a high-energy photon.

TABLE 12A.1 Some useful isotopes in biochemistry

Isotope	Stable or Radioactive	Emission	Half-Life	Maximum Energy (MeV[a])
^2H	Stable			
^3H	Radioactive	β	12.1 years	0.018
^{13}C	Stable			
^{14}C	Radioactive	β	5568 years	0.155
^{15}N	Stable			
^{18}O	Stable			
^{24}Na	Radioactive	β (and γ)	15 hours	1.39
^{32}P	Radioactive	β	14.2 days	1.71
^{35}S	Radioactive	β	87 days	0.167
^{45}Ca	Radioactive	β	164 days	0.254
^{59}Fe	Radioactive	β (and γ)	45 days	0.46, 0.27
^{131}I	Radioactive	β (and γ)	8.1 days	0.335, 0.608

[a]MeV = million electron volts.

γ-Ray detectors have found wide use in immunological research because γ-emitting isotopes of iodine are available, and antibodies, like many proteins, can easily be iodinated without substantial changes in their biological properties. Most biochemical uses of radioisotopes, however, involve β-emitters.

Radioactive decay is a first-order kinetic process. The probability that a given atomic nucleus will decay is affected neither by the number of preceding decay events that have occurred nor by interaction with other radioactive nuclei. Rather, it is an intrinsic property of that nucleus. Thus, the number of decay events occurring in a given time interval is related only to the number of radioactive atoms present. This phenomenon gives rise to the **law of radioactive decay**:

$$N = N_0 e^{-\lambda t}$$

where N_0 is the number of radioactive atoms at time zero, N is the number remaining at time t, and λ is a radioactive decay constant for a particular isotope, related to the intrinsic instability of that isotope. This equation states that the *fraction* of nuclei in a population that decays within a given time interval is constant. For this reason, a more convenient parameter than the decay constant λ is the **half-life**, $t_{1/2}$, the time required for half of the nuclei in a sample to decay. The half-life is equal to $-\ln 0.5/\lambda$, or $+0.693/\lambda$. The half-life, like λ, is an intrinsic property of a given radioisotope (see Table 12A.1).

The basic unit of radioactive decay is the **curie** (Ci). This unit is defined as an amount of radioactivity equivalent to that in 1 g of radium, namely 2.22×10^{12} disintegrations per minute (dpm). Biochemists usually work with far smaller amounts of radioactive material—the **millicurie** (mCi) and the **microcurie** (μCi), which correspond, respectively, to 2.2×10^9 and 2.2×10^6 dpm. Because detectors of radioactivity rarely record every decay event in a sample—that is, they are not 100% efficient—we often speak of radioactivity in terms of the decay events actually recorded: counts per minute, or cpm. Counting efficiency is the percentage of decay events actually recorded, as determined, for example, by reference to standards. A counter that is 50% efficient for a given isotope would show a count rate of 1.1×10^5 cpm for a 0.1μCi sample.

Detection of Radioactivity: The Liquid Scintillation Counter

In scintillation counting, the most widely used method for measuring β-emissions, the sample is dissolved or suspended in an organic solvent, although aqueous mixtures are available as well. Also present are one or two fluorescent organic compounds, or **fluors**. A β-particle emitted from the sample has a high probability of hitting a molecule of the solvent; this contact excites the solvent molecule, driving an electron into a higher-energy orbital. When that electron returns to the ground state, a photon of light is emitted. The photon is absorbed by a molecule of the fluor, which in turn becomes excited. Fluorescence involves the absorption of light at a given energy, followed by emission of that light at lower energy, or longer wavelength. A photomultiplier detects the fluorescence and for each disintegration converts it to an electrical signal, which is recorded and counted.

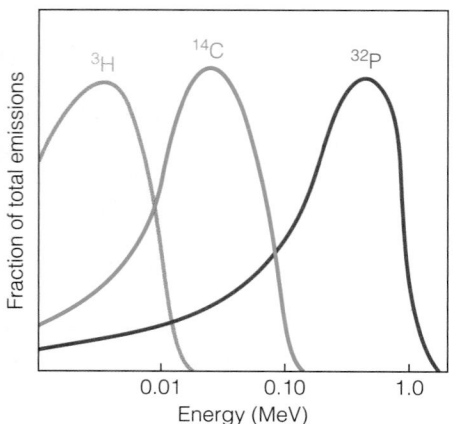

FIGURE 12A.1

Energy spectra for some β-emitting isotopes. Shown here are spectra for the three β-emitters most widely used in biochemistry—^3H, ^{14}C, and ^{32}P.

Whereas γ-rays are emitted at distinct and characteristic energy values, a β-emitting isotope will display a range of energy values among its emissions. Each β-emitter shows a characteristic **energy spectrum**—that is, a plot of energy (in millions of electron volts, or MeV) against the probability of an individual emission having that energy. Figure 12A.1 shows the energy spectra for three widely used radioisotopes—^3H, ^{14}C, and ^{32}P.

These differences in emission energies are exploited in the liquid scintillation counter so that two isotopes in the same sample can be simultaneously quantified (a dual-label experiment). A strong β-emission excites more fluor molecules than a weak emission does, so a brighter light flash is produced. This flash level can be detected by setting electronic discriminators so that energy values falling only within a desired range are recorded—that is, setting a "window" that is optimized for detection of a particular isotope. This restriction reduces counting efficiency because only a fraction of the emissions from the higher-energy isotope are counted, but it does permit great selectivity. Most scintillation counters have three different sets of discriminators, or **counting channels**. Thus, isotopes can be counted simultaneously in three different windows.

Because radioactive decay is a random process, the more counts observed during the analysis of a sample, the more closely the measured radioactivity will approach the true decay rate. In practice, therefore, a researcher wants to count a sample for enough time to accumulate several thousand counts. Also, the researcher must consider the signal-to-noise ratio and design an experiment so that the measured radioactivity of the samples of interest will be many times higher than **background radioactivity**, or the counts recorded in the absence of a radioactive sample.

Some Uses of Radioisotopes in Biochemistry

Radioisotopes have many uses in metabolic investigations. Here we describe a few examples. Tracer experiments can be categorized in terms of the time of exposure of the biological system to the radioisotope and include (1) equilibrium labeling, (2) pulse labeling, and (3) pulse-chase labeling.

In **equilibrium labeling** the exposure to the tracer is relatively long so that each labeled species reaches a constant **specific radioactivity**, or **specific activity**. The specific activity is a measure of the relative abundance of radioactive molecules in a labeled sample and is reported as radioactivity per unit mass—for example, cpm/mol. Experiments to identify metabolic precursors often involve equilibrium labeling conditions. The investigator administers a radiolabeled precursor so as to maximize the chance of detecting label in the product. A good example, mentioned in Chapter 19, is Konrad Bloch's study of cholesterol biosynthesis, in which ^{14}C-labeled acetate was administered to rats. Bloch then isolated cholesterol from the liver and determined, by chemical degradation, the specific carbon atoms in cholesterol that had incorporated radioactivity.

Another use of equilibrium labeling is to measure rates of biological processes from the rates of labeling by low-molecular-weight precursors. An example is the widespread use of radiolabeled thymidine to follow rates of DNA synthesis, either in cell cultures or in intact organisms. Incorporation of thymidine into macromolecules other than DNA is negligible, so the investigator merely samples the population with time and observes the incorporation into acid-insoluble material. However, this measurement alone does not give the true rate of DNA replication. The investigator must also consider the metabolism of the labeled precursor en route to its ultimate destination. If intracellular metabolism is producing dTTP, that nucleotide will be nonradioactive, and its mixing with the labeled pool will dilute the specific radioactivity incorporated into DNA. The labeling experiment gives only a rate of DNA labeling, in cpm incorporated per cell per unit time. To get a true rate, in molecules incorporated per cell per unit time, the initial result must be divided by the specific activity of the immediate precursor, in this case, dTTP. Calculating that specific activity requires isolation of dTTP in sufficient amount and purity to allow determination of both its mass and radioactivity.

Pulse labeling involves administration of an isotopic precursor for an interval that is short relative to the process under study. Radiolabel accumulates preferentially in the shortest-lived species—that is, the earliest intermediates in a metabolic pathway—because the shorter the labeling interval is, the less time is available for loss of radioactivity from a labeled pool through breakdown of rapidly labeled and short-lived metabolites. Melvin Calvin identified the pathway of photosynthetic carbon fixation by labeling green algae with $^{14}CO_2$ for just a few seconds. If labeling was carried out for 10 s, a dozen or more radioactive compounds could be detected. After a 5-s pulse of radioactivity, only a single compound was labeled, namely 3-phosphoglycerate. This finding led ultimately to the discovery of ribulose-1,5-bisphosphate carboxylase as the first enzyme in the photosynthetic carbon fixation pathway (see page 695 in Chapter 16).

In a **pulse-chase** experiment, the investigator administers label for a short time (the pulse) and then rapidly reduces the specific activity of the isotopic precursor to prevent further incorporation. This reduction can be done by adding unlabeled precursor at a molar excess of about 1000-fold, which greatly dilutes the radioisotope still present (the chase). The investigator samples at various times afterward to determine the metabolic fate of the material labeled during the pulse. Messenger RNA was originally detected by pulse-labeling bacterial cultures (see Chapter 27). Traditional analytical methods could not detect mRNA because of its low abundance and its metabolic instability in bacterial cells. When label incorporated into mRNA by a pulse of [^{32}P]orthophosphate or labeled uridine was chased out, the label was ultimately found to be distributed uniformly in all cellular RNA species. This finding showed that the metabolic instability of mRNA involves its degradation to nucleotides, which can then be used for synthesis of other RNA species.

A final application of radioisotopes is **radioautography**, in which the investigator incorporates an isotopic precursor into a biomolecule and prepares an image of the radiolabeled molecule on a sheet of photographic film. Chapter 25 shows several radioautographs of individual DNA molecules made radioactive by growth of *E. coli* in the presence of [^{3}H]thymidine. The two-dimensional gel shown in Figure 1.11 (page 19) is a radioautograph because the radiolabeled proteins were detected by their ability to darken photosensitive film. Film radioautography has been largely replaced by **phosphorimaging**, in which the photographic film is replaced by an imaging plate coated with photostimulable phosphors. Phosphorimaging is much more accurate than film radioautography in quantifying the amount of radioactivity in the sample because its response to radioactivity is far more linear than that of X-ray film. Moreover, the imaging plate can be erased and reused indefinitely.

Reference

Freifelder, D. (1982) *Physical Biochemistry*, 2nd ed. W. H. Freeman, San Francisco. Chapter 5 of this book presents a clear description of techniques in radioactive labeling and counting.

TOOLS OF BIOCHEMISTRY 12B

METABOLOMICS

Virtually all the facts in this textbook came from experiments in which a biochemist measured something in a cell extract: the level of a particular mRNA or protein, the activity of a particular enzyme, or the concentration of a particular metabolite. Indeed, a large part of experimental biochemistry involves the development of specific and sensitive *assays* for a particular cellular component. For example, Hans Krebs's elucidation of the citric acid cycle (see Chapter 14) depended on his ability to accurately measure the concentrations of potential substrates such as pyruvate, citrate, succinate, or oxaloacetate in muscle or liver slices. A specific assay had to be developed to measure each metabolite. To fully analyze a pathway such as the citric acid cycle or glycolysis might require 10 or more separate metabolite assays on each sample. Furthermore, these assays could only measure already known metabolites—they were not very useful for discovering new or missing metabolic intermediates.

In recent years, however, new technologies have made it possible to move beyond the measurement of single mRNAs, single proteins, or single metabolites. The development of these new technologies has driven the "-omics" revolution—genomics, transcriptomics, proteomics, and metabolomics—where hundreds or even thousands of specific components are measured simultaneously in a biological sample. Thus, it is now feasible to measure the full set of transcripts (transcriptome), proteins (proteome), or metabolites (metabolome) in a particular cell or tissue. The metabolome represents the ultimate molecular phenotype of a cell under a given set of conditions because all the changes in gene expression and enzyme activity eventually lead to changes in cellular metabolite levels (the metabolic state or profile). The metabolic state of a cell or a whole organism is a sensitive indicator of the physiological status of the organism. Changes in metabolic state can be used to understand and diagnose disease, study effects of drugs, and even predict the effectiveness of a drug in a particular patient. There are many analytical approaches to determining the metabolic state of a cell (metabolic profiling), but they all follow the same basic process (Figure 12B.1): sample extraction; metabolite identification and quantitation; and data analysis (informatics).

Metabolic Profiling

Sample Collection and Extraction

Samples of interest are collected—for example drug-treated vs. untreated cells, serum from a patient, or a tissue biopsy from a tumor. The small molecules are extracted from the samples by methods that are matched to the analytical technique to be used. No single method will extract all metabolites from a sample. Different extraction procedures can be used to select specific subsets of the metabolome (e.g., lipophilic metabolites, amino acids, or carbohydrates).

Metabolite Identification and Quantitation

While the number of distinct metabolites in a cell is not known, it is certainly on the order of a few thousand. Therefore, the analytical method must have both a separation component and a detection

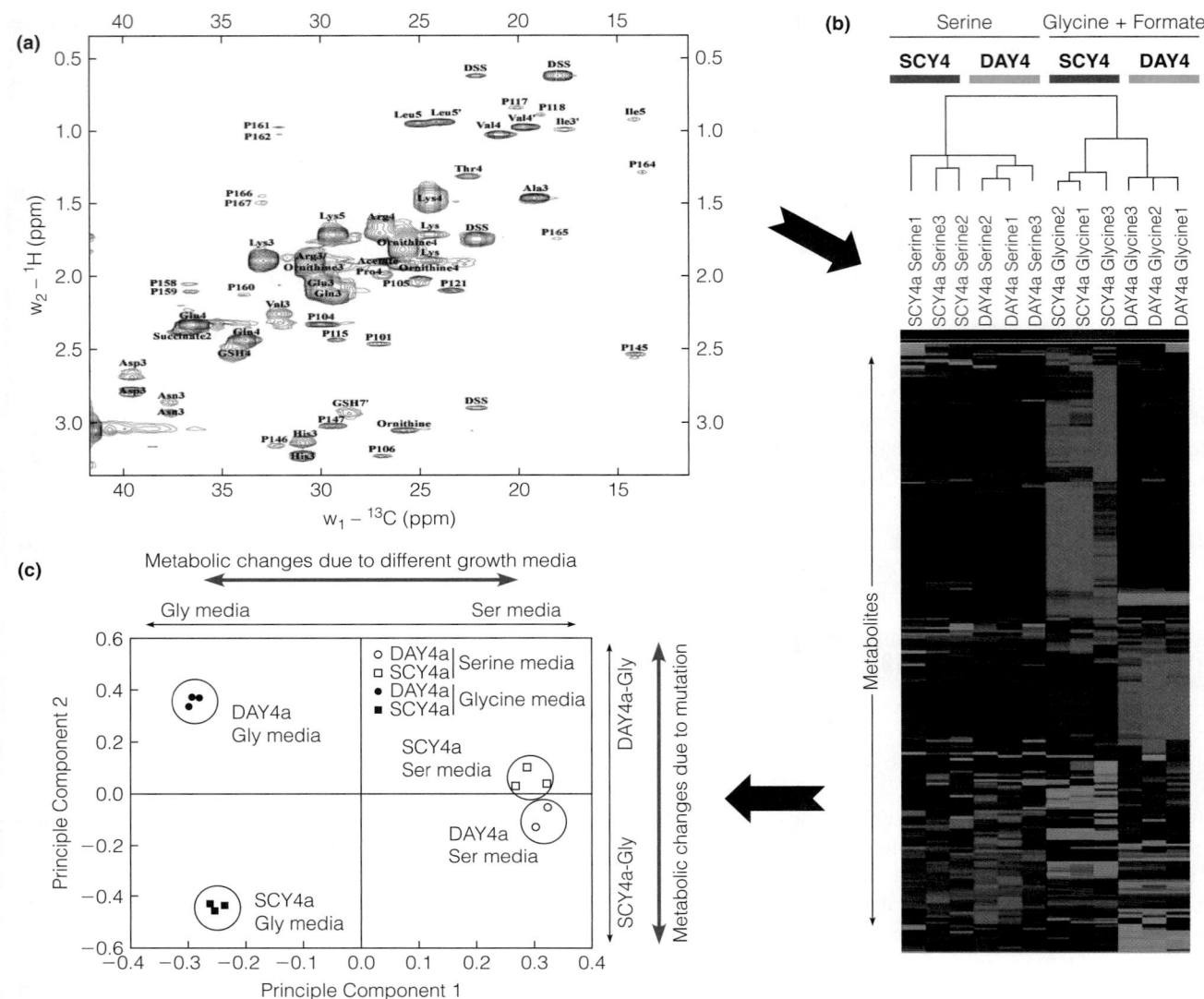

FIGURE 12B.1

Basic process of metabolic profiling. Metabolites are identified and quantified by an analytical method. Panel **(a)** shows a section of a $^1H-^{13}C$ 2D-NMR spectrum. Data are collected and visualized by informatics approaches. In the "heat map" shown in panel **(b)**, each row corresponds to a single 2D-NMR metabolite peak, with columns representing different experimental conditions or cell type. The normalized magnitude of each 2D-NMR peak is indicated by color, with red indicating an increase in metabolite abundance relative to the control, green indicating a decrease in abundance, and black for the median level of abundance. Informatics approaches are then used to reveal relationships and patterns among the samples (panel **(c)**).

Modified from *Metabolic Engineering* 9:8–20, P. Lu, A. Rangan, S. Y. Chan, D. R. Appling, D. W. Hoffman, and E. M. Marcotte, Global metabolic changes following loss of a feedback loop reveal dynamic steady states of the yeast metabolome. © 2007, with permission from Elsevier.

component. Mass spectrometry (MS) and nuclear magnetic resonance (NMR) are the most widely used detection methods. The MS-based techniques use liquid chromatography (LC), gas chromatography (GC), or capillary electrophoresis (CE) to separate the metabolites based on some chemical or physical property (size, charge, hydrophobicity, etc.). Effluent from the LC, GC, or CE is introduced into the mass spectrometer, where the metabolites are detected and quantified (see Tools of Biochemistry 5B). NMR (see Tools of Biochemistry 6A) can be applied directly to samples without a separation step by obtaining spectra in two dimensions (e.g., ^1H vs. ^{13}C), termed 2D-NMR (Figure 12B.1a). Both MS and NMR methods are capable of detecting and quantifying hundreds of distinct metabolites in a single run, and both have advantages and disadvantages. NMR can detect many different classes of metabolites and offers great power in metabolite identification and quantitation. MS-based methods are much more sensitive than NMR, but quantitation is not as straightforward. Note, however, that both methods can detect and quantify unidentified metabolites, which can facilitate the discovery of new intermediates or even new pathways.

Data Analysis (Informatics)

The third critical step in metabolic profiling is data analysis. Because the datasets are generally quite large, and sophisticated statistical methods are required to visualize and compare metabolic profiles, data analysis uses modern informatics tools. These informatics algorithms attempt to identify patterns in the data that are reproducibly characteristic of a particular metabolic state. These patterns, or fingerprints, can then be compared between samples. The most common informatics methods for analyzing and comparing metabolome data are hierarchical clustering (the partitioning of a data set into subsets or clusters) (Figure 12B.1b) and principal component analysis (PCA) (Figure 12B.1c). PCA is used to reduce multidimensional data sets to lower dimensions (principal components). PCA can often reveal the internal structure of the data in a way which best explains the variance in the data.

Applications

Biomarker Discovery

The identification of new biomarkers to predict or diagnose disease is a major goal of biomedical research. Although there are some useful single molecule biomarkers known, for many diseases a pattern of several metabolites will be a more informative biomarker. The following example illustrates the power of metabolic profiling to discover new biomarkers. Acetaminophen (e.g., Tylenol) is a widely used over-the-counter analgesic. Excess acetaminophen is toxic to the liver because it causes oxidative stress, but individuals vary in their

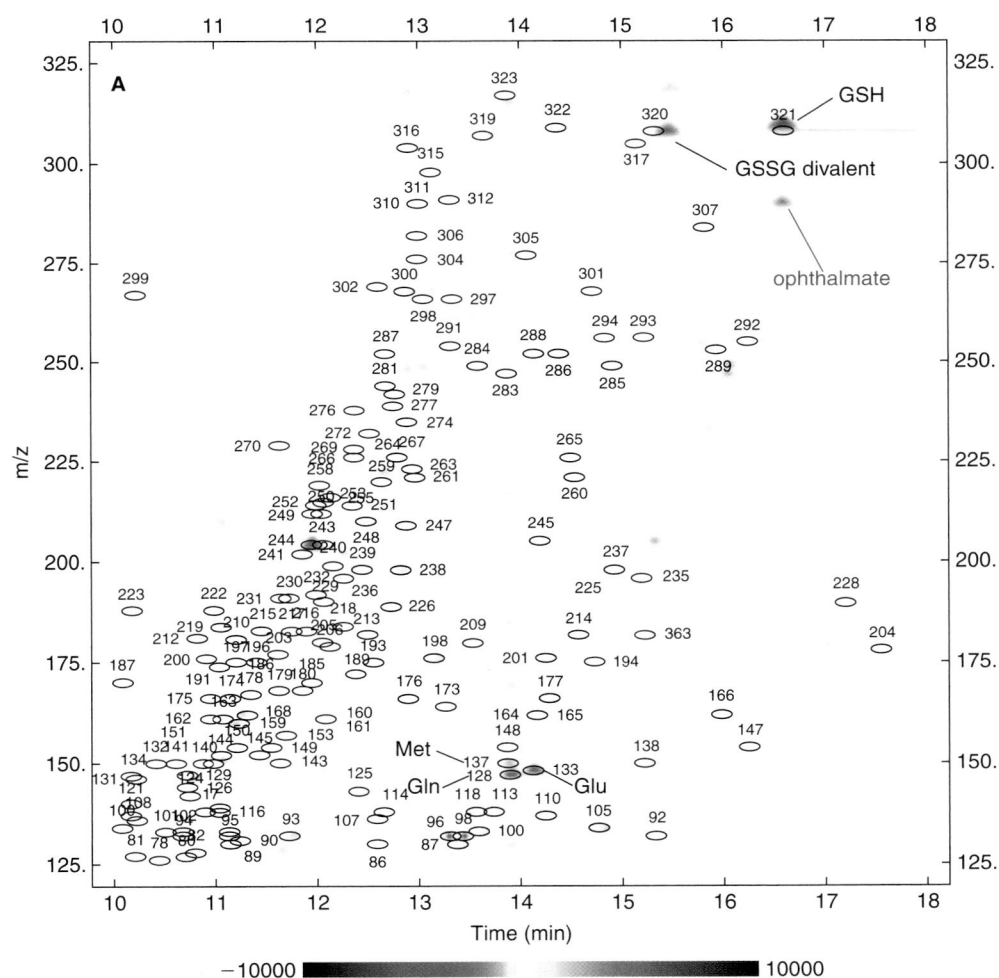

FIGURE 12B.2

Differential metabolic profile from livers of control mice and livers from mice 2 h after treatment with acetaminophen. The 2D plot (m/z vs. elution time) shows cations detected by CE-MS analysis. The arrow points to the metabolite subsequently identified as ophthalmate whose level significantly increased in treated mice. The color bar indicates the increase (red) or the decrease (blue) of metabolite level after drug treatment.

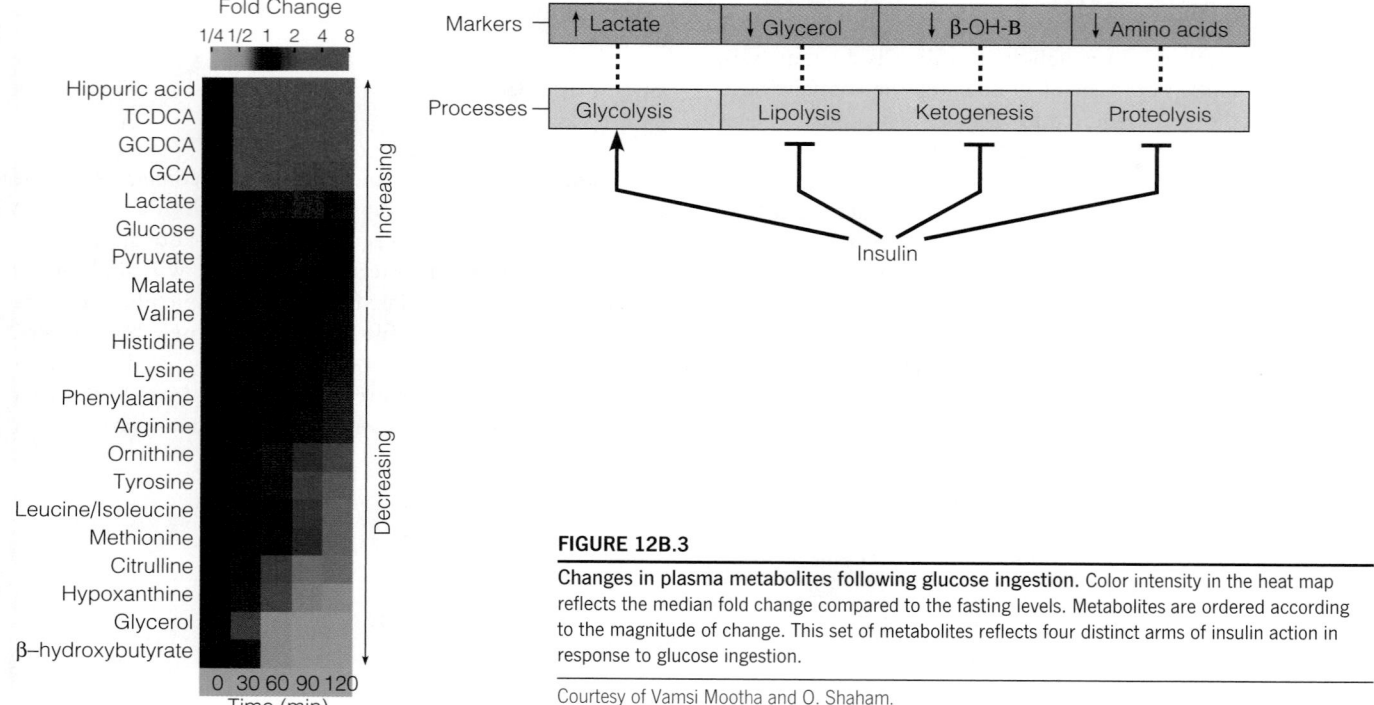

FIGURE 12B.3

Changes in plasma metabolites following glucose ingestion. Color intensity in the heat map reflects the median fold change compared to the fasting levels. Metabolites are ordered according to the magnitude of change. This set of metabolites reflects four distinct arms of insulin action in response to glucose ingestion.

Courtesy of Vamsi Mootha and O. Shaham.

susceptibility to liver damage. To identify potential biomarkers for the oxidative stress caused by acetaminophen, CE-MS was used to study changes in liver metabolites in mice treated with excess acetaminophen. Metabolic profiling identified 132 compounds, including a new metabolite (ophthalmate) (Figure 12B.2). This compound is not a metabolite of acetaminophen, but its levels rise in response to the drug. This new metabolite may serve as a useful biomarker for acetaminophen-induced hepatotoxicity in humans.

Disease Diagnosis

The incidence of diabetes has increased so rapidly in the United States that it is now considered an epidemic. Today, more than 21 million Americans have diabetes, and it is estimated that 54 million have prediabetes—they have elevated blood sugar levels, but not high enough to be classified as diabetes. Prediabetics are at high risk of developing type 2 diabetes. If detected early enough, diabetes is a manageable disease. Insulin resistance, traditionally defined as the reduced ability of insulin to promote glucose uptake, is a characteristic feature of type 2 diabetes. However, insulin regulates many other metabolic processes, and metabolic profiling can be used to look beyond glucose metabolism. LC-MS has been used to quantify changes in 191 metabolites in human plasma following glucose ingestion. Eighteen of these were observed to change reproducibly in healthy individuals, reflecting four distinct arms of insulin function (Figure 12B.3). In prediabetic patients, however, a subset of these metabolites showed blunted responses to glucose ingestion, providing a strong indicator of loss of insulin sensitivity. This application of metabolic profiling holds great promise in early diagnosis of diabetes.

Predicting Drug Response

One of the most serious problems with therapeutic drugs is that patients can vary considerably in their response to the drug and in their susceptibility to side effects. For example, the tyrosine kinase inhibitor imatinib is a highly effective drug for treatment of chronic myeloid leukemia (CML). Unfortunately, some patients develop resistance to this drug, and treatment fails. To try to understand the mechanisms underlying this individual variation in drug response, blood plasma metabolites were profiled in CML patients previously determined to be either sensitive or resistant to imatinib. Principal component analysis (PCA) revealed metabolic profiles that differed between those patients who were sensitive to drug treatment from those who were resistant (Figure 12B.4). Thus metabolic profiling can be used to predict whether a drug will be effective in a particular patient.

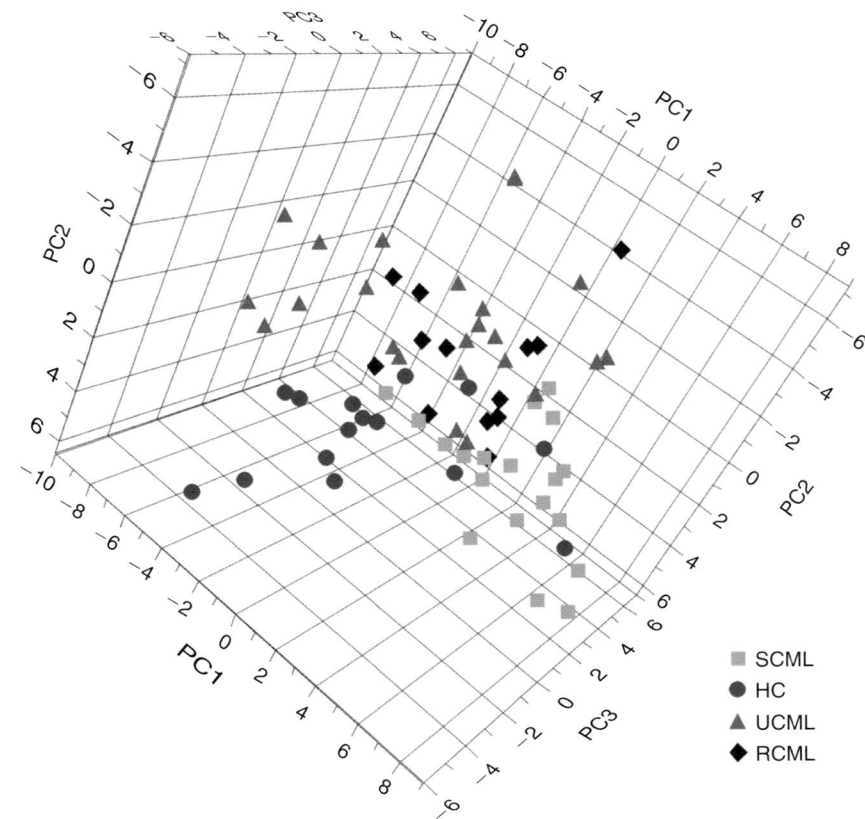

FIGURE 12B.4

Principal component analysis of blood plasma metabolites in chronic myeloid leukemia (CML) patients. 3D scatter plot of the first three principal components (PC1 vs. PC2 vs. PC3). Green squares are patients who are sensitive to drug treatment (SCML). Black diamonds are patients who are resistant to drug treatment (RCML). Red triangles are untreated CML patients (UCML). Blue circles are healthy controls (HC).

A. J. S. Qian, G. Wang, B. Yan, S. Zhang, et al. (2010) Chronic myeloid leukemia patients sensitive and resistant to imatinib treatment show different metabolic responses. PLoS ONE 5(10):e13186. doi:10.1371/journal.pone. 0013186.

- ■ SCML
- ● HC
- ▲ UCML
- ◆ RCML

Reference

Kaddurah-Daouk, R., B. S. Kristal, and R. M. Weinshilboum, (2008) Metabolomics: A global biochemical approach to drug response and disease. *Ann. Rev. Pharmacol. Tox.* 48:653–683.

PART 4

Dynamics of Life: Energy, Biosynthesis, and Utilization of Precursors

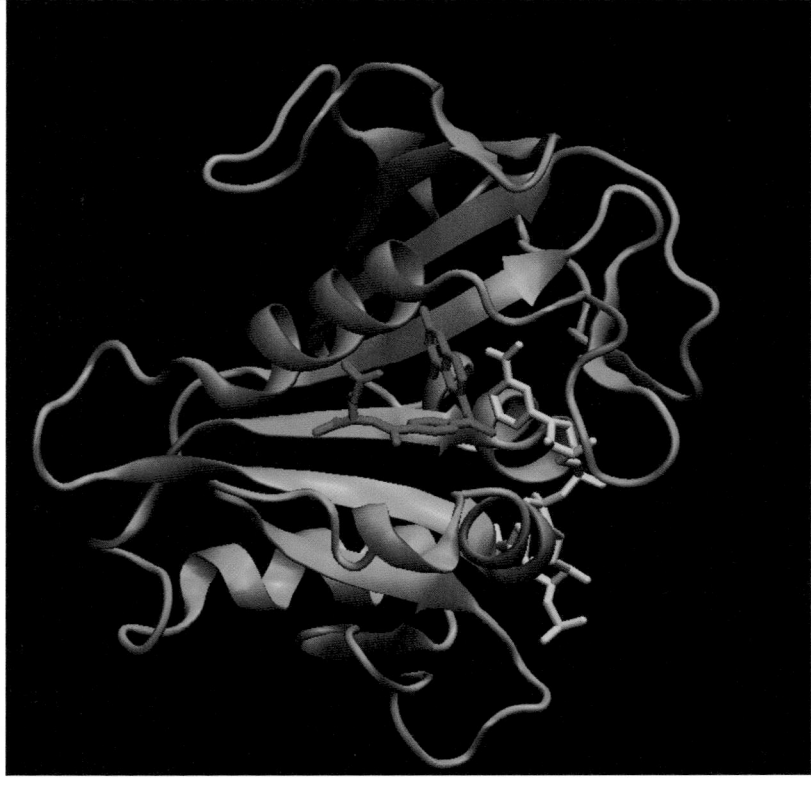

Dihydrofolate reductase, a target for antimicrobial and anticancer drugs. The anticancer drug methotrexate (in red) binds tightly to the complex of the enzyme and its coenzyme, NADPH (in yellow), leading to interference with nucleotide and amino acid biosynthesis. Source: Dean Appling (Figure 20.16) from PDB coordinates.

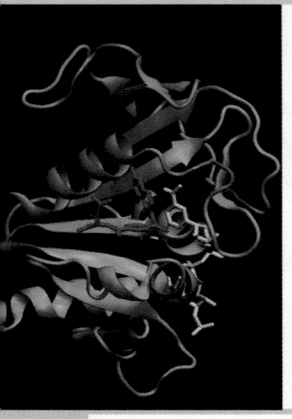

CHAPTER 13

Carbohydrate Metabolism: Glycolysis, Gluconeogenesis, Glycogen Metabolism, and the Pentose Phosphate Pathway

Our detailed study of metabolism begins with the anaerobic phases of carbohydrate metabolism (Figure 13.1). Much of this chapter is devoted to glycolysis, the initial pathway in the catabolism of carbohydrates. The term *glycolysis* is derived from Greek words meaning "sweet" and "splitting." These words are literally correct terms, for glycolysis is the pathway by which six-carbon sugars are split, yielding a three-carbon compound, pyruvate. During glycolysis, some of the potential energy stored in the hexose structure is used to drive the synthesis of ATP from ADP. Glycolysis can proceed under anaerobic conditions, with no net oxidation of the sugar substrates taking place. **Anaerobes**, microorganisms that live in oxygen-free environments, can derive all of their metabolic energy from this process. Indeed, carbohydrate is the only fuel whose catabolism can produce ATP in the absence of oxygen. However, aerobic cells also use glycolysis. In these cells, glycolysis is the initial, anaerobic part of an overall degradation pathway that ultimately involves considerable oxygen consumption and the complete oxidation of carbohydrates to CO_2 and H_2O.

This chapter will also introduce *gluconeogenesis*, the synthesis of glucose from noncarbohydrate precursors, as well as the synthesis of storage polysaccharides, primarily glycogen in animals. We discuss these two energy-requiring biosynthetic processes here because their regulation is so intimately coordinated with that of glycolysis. We will end the chapter with the *pentose phosphate pathway*, a multipurpose pathway that represents an alternative process for the catabolism of glucose.

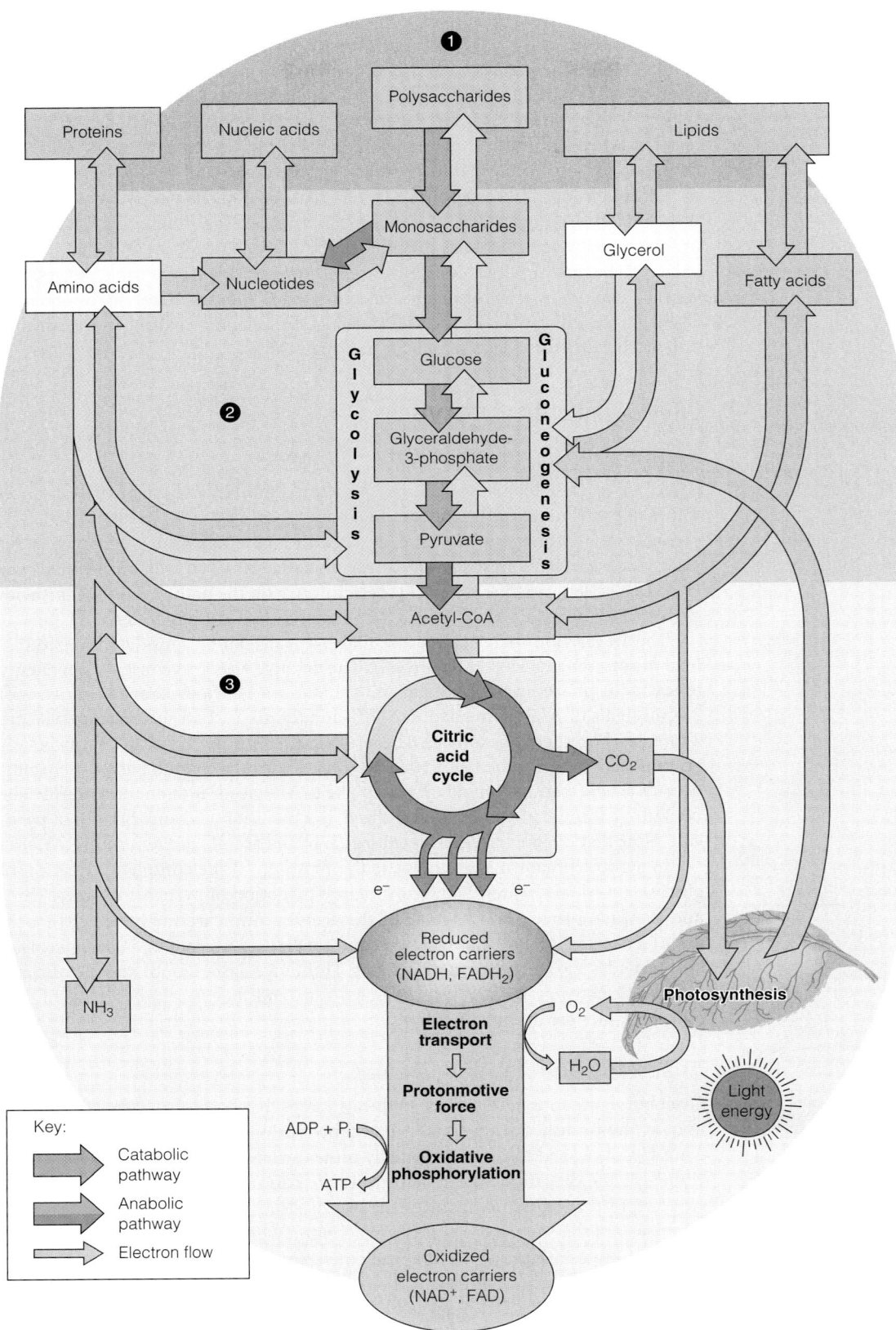

FIGURE 13.1

Catabolic and anabolic processes in anaerobic carbohydrate metabolism. The gold arrows show the glycolytic pathway and the breakdown of polysaccharides that supply this pathway. Glycolysis generates ATP anaerobically and provides fuel for the aerobic energy-generating pathways. The green arrows show the gluconeogenesis pathway, the synthesis of polysaccharides such as glycogen. The blue arrow shows the pentose phosphate pathway, an alternative carbohydrate oxidation pathway needed for nucleotide synthesis. The numbers 1, 2, and 3 identify the three stages of metabolism (see Chapter 12).

Glycolysis is an appropriate point to begin a detailed study of metabolism, for several reasons. First, it was the earliest metabolic pathway to be understood in detail. Second, the pathway is nearly universal in living cells. Third, the regulation of glycolysis is particularly well understood. Last, but not least, is the central metabolic role this pathway plays in generating both energy and metabolic intermediates for other pathways. It is one of the busiest freeways on the metabolic road map, but it is also connected to many less traveled roads.

Although cells can metabolize a variety of hexose sugars via glycolysis, glucose is the major carbohydrate fuel for many cells. Indeed, some animal tissues, such as brain, normally use glucose as the sole energy source, and all energy generation in such cells begins with glycolysis. Most cells, however, can utilize other sugars, and we shall explore how those sugars are converted to intermediates in glycolysis. In addition, we will consider processes by which stored carbohydrate in the form of polysaccharides is made available for use in glycolysis.

Glycolysis: An Overview

Relation of Glycolysis to Other Pathways

Glycolysis is a 10-step pathway that converts one molecule of glucose to two molecules of pyruvate, with the concomitant generation of two molecules of ATP. The breakdown of storage polysaccharides and the metabolism of oligosaccharides yield glucose, related hexoses, and sugar phosphates, all of which find their way into the glycolytic pathway. We will focus initially on the pathway as it begins with glucose and then discuss the routes for entry of other carbohydrates.

The 10 reactions between glucose and pyruvate can be considered as two distinct phases, schematized in Figure 13.2. The first five reactions constitute an **energy investment phase**, in which sugar phosphates are synthesized at the expense of 2 equivalents of ATP (converted to ADP), and the six-carbon substrate is split into 2 three-carbon sugar phosphates. The last five reactions represent an **energy generation phase**, in which the two triose phosphates are converted to energy-rich compounds. These transfer 4 moles of phosphate to ADP, leading to 4 moles of ATP. The net yield, per mole of glucose metabolized, is 2 moles of ATP and 2 moles of pyruvate. Note that 2 reducing equivalents are generated as well, in the form of NADH.

In aerobic organisms, glycolysis is the first step in the complete oxidation of glucose to CO_2 and water. The second step is oxidation of pyruvate to acetyl-CoA, and the final process is oxidation of the acetyl group carbons in the citric acid cycle (see Figure 13.1). Chapter 14 presents the latter processes in detail. Glycolysis also provides biosynthetic intermediates. Thus, glycolysis is both an anabolic and a catabolic pathway, with an importance that extends beyond the synthesis of ATP and substrates for the citric acid cycle.

Anaerobic and Aerobic Glycolysis

Glycolysis is an ancient metabolic pathway that probably evolved before the earliest known photosynthetic organisms began contributing O_2 to the earth's atmosphere. Thus glycolysis had to function initially under anaerobic conditions—with no net change in the oxidation state as substrates are converted to products. Recall from Chapter 12 that oxidation involves the loss of electrons from a substrate, and the electrons are transferred to an electron acceptor, which thereby becomes reduced. However, note in Figure 13.2 that the conversion of glucose to pyruvate, which does oxidize the carbons of glucose, involves the concomitant reduction of 2 equivalents of NAD^+ to NADH. For the pathway to operate anaerobically, NADH must be reoxidized to NAD^+ by transferring its electrons to an **electron acceptor** so that *a steady-state concentration of* NAD^+ *is maintained.* Some microorganisms growing anaerobically can generate additional energy by transferring the electrons to inorganic substances such as sulfate ion or nitrate ion, and some microorganisms

The 10 reactions of glycolysis occur in two phases: energy investment (first five reactions) and energy generation (last five reactions).

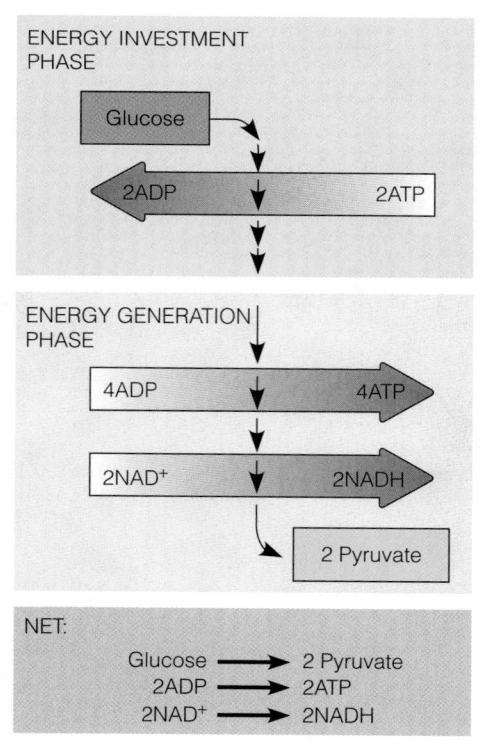

FIGURE 13.2

The two phases of glycolysis and the products of glycolysis.

reduce organic substrates. Most straightforward is the route used by lactic acid bacteria, which simply use NADH to reduce pyruvate to lactate, via the enzyme **lactate dehydrogenase**.

$$
\begin{array}{c}
\text{COO}^- \\
| \\
\text{C}=\text{O} \\
| \\
\text{CH}_3
\end{array}
+ \text{NADH} + \text{H}^+
\rightleftharpoons
\begin{array}{c}
\text{COO}^- \\
| \\
\text{HO}-\text{C}-\text{H} \\
| \\
\text{CH}_3
\end{array}
+ \text{NAD}^+ \qquad \Delta G^{\circ\prime} = -25.1 \text{ kJ/mol}
$$

Pyruvate   L-**Lactate**

Lactate formation is strongly favored under standard conditions, as indicated by the large negative standard free energy change.

Glycolysis is therefore part of a *fermentation*, which we defined as an energy-yielding metabolic pathway that involves no net change in oxidation state of the carbon substrate (page 487). This **homolactic fermentation** (conversion of all six carbons of glucose to lactate*) is also important in the manufacture of cheese. **Alcoholic fermentation** involves cleavage of pyruvate to acetaldehyde and CO_2 (see page 536), with the acetaldehyde then reduced to ethanol by **alcohol dehydrogenase**:

$$
\begin{array}{c}
\text{H} \quad\; \text{O} \\
\diagdown\;\; \diagup \\
\text{C} \\
| \\
\text{CH}_3
\end{array}
+ \text{NADH} + \text{H}^+
\rightleftharpoons
\begin{array}{c}
\text{OH} \\
| \\
\text{CH}_2 \\
| \\
\text{CH}_3
\end{array}
+ \text{NAD}^+ \qquad \Delta G^{\circ\prime} = -23.7 \text{ kJ/mol}
$$

Acetaldehyde   **Ethanol**

As carried out by yeasts, this fermentation generates the ethanol in alcoholic beverages. Yeasts used in baking also carry out the alcoholic fermentation; the CO_2 produced by pyruvate decarboxylation causes bread to rise, and the ethanol produced evaporates during baking. Among the dozens of other useful fermentations are those leading to acetic acid (manufacture of vinegar) and propionic acid (manufacture of Swiss cheese). All of these fermentations are simply variations on the same theme: The NADH produced earlier in the pathway must be reoxidized to NAD^+ by transferring its electrons to some abundant electron acceptor. The simplest way to ensure that there is enough electron acceptor available is for the cell to use the same carbon skeleton that the electrons came from in the first place, that is, pyruvate, or a derivative of pyruvate. Thus, the large diversity of fermentations found in nature are distinguished really only by the final fermentation product. Figure 13.3 illustrates the common strategy used by virtually all fermentations.

Animal cells, like lactic acid bacteria, can reduce pyruvate to lactate, and they do so when pyruvate is produced faster than it can be oxidized through the citric acid cycle. During strenuous exertion, skeletal muscle cells derive most of their energy from **anaerobic glycolysis**—glycolysis occurring under anaerobic conditions.

By contrast, consider a cell undergoing active *respiration*, the oxidative breakdown and release of energy from nutrient molecules by reaction with oxygen. In these cells, pyruvate is oxidized to acetyl-CoA, which enters the citric acid cycle. The NADH produced during glycolysis is reoxidized through the mitochondrial electron transport chain for additional energy production (see Chapter 15), with the electrons transferred ultimately to O_2, the terminal electron acceptor. The conversion of glucose to pyruvate in a respiring cell is called **aerobic glycolysis**.

A fermentation is an energy-yielding metabolic pathway with no net change in the oxidation state of products compared to substrates.

Anaerobic glycolysis (like aerobic glycolysis) leads to pyruvate, but the pyruvate is then reduced, so no net oxidation of glucose occurs.

---

*Some organisms carry out a **heterolactic fermentation**, in which only one lactate is produced, with the other three carbons converted to one ethanol plus one CO_2.

FIGURE 13.3

Fermentations use a common strategy to regenerate oxidized NAD$^+$. **(a)** Homolactic **(b)** Alcoholic **(c)** Butanediol.

The Microbial World, 3rd ed. Roger Y. Stanier, Michael Doudoroff, and Edward A. Adelberg. © 1970. Modified with permission of Pearson Education Inc., Upper Saddle River, NJ.

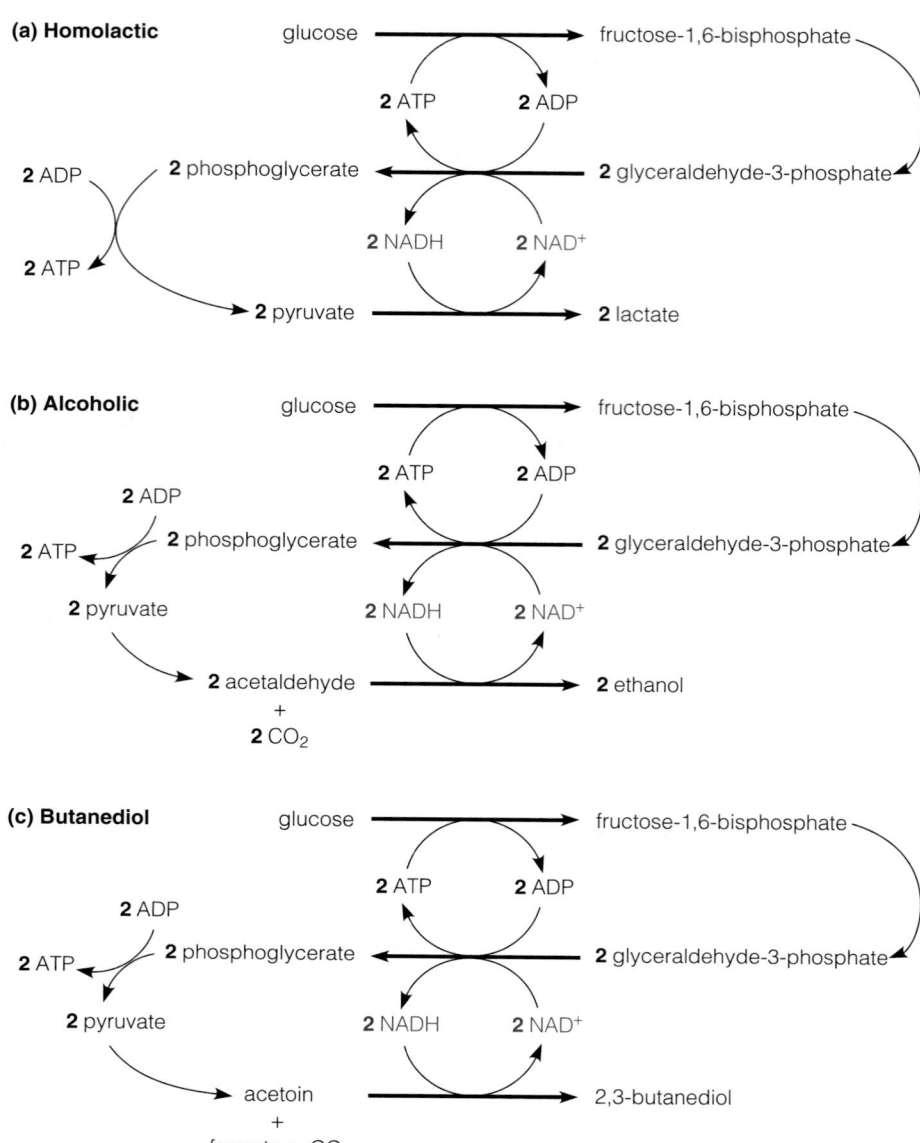

The Crucial Early Experiments

For as long as humans have used yeasts in baking and in brewing, fermentation has been exploited, even though the biochemical process was not understood until the last century. (The early definition of *fermentation* was "a chemical change with effervescence.") Louis Pasteur's demonstration in 1856 that fermentations are carried out by microorganisms ranks as a milestone in the history of science. The dominant viewpoint of the time was that a process such as the fermentation of glucose to ethanol was so complex that it could not be reconstituted outside a living cell. As we saw in Chapter 1, however, Eduard and Hans Büchner showed in 1897 that fermentation could occur under cell-free conditions.

In 1905 Arthur Harden and William Young found that inorganic phosphate, when added to yeast extract, stimulated and prolonged the fermentation of glucose. During fermentation the inorganic phosphate disappeared from the reaction medium, which led Harden and Young to suggest that fermentation was functioning via the formation of one or more sugar phosphate esters.

This observation opened the door to dissection of the individual chemical reactions involved in fermentation, a feat accomplished in Germany in the 1930s, largely by G. Embden, O. Meyerhof, Jacob Parnas, and O. Warburg. In fact, glycolysis is often referred to as the **Embden–Meyerhof–Parnas pathway**. These scientists identified 10 different reactions, virtually identical in a wide range of organisms, leading from glucose to pyruvate. Glycolysis is the first metabolic pathway to have been elucidated as a series of defined chemical reactions. Extensive information is now available about the structure and mechanism of action of each enzyme involved.

Strategy of Glycolysis

Glycolysis is such an important pathway that we shall examine each of its 10 reactions in some detail. Before doing so, let us look at the pathway as a whole. First, recall from Chapter 12 that in eukaryotic cells glycolysis occurs in the cytosol, and the further oxidation of pyruvate occurs in mitochondria. (Certain trypanosomes, the parasitic protozoans that cause African sleeping sickness, present an interesting exception. They carry out the first seven reactions of glycolysis in an organized cytoplasmic organelle called the **glycosome**.)

The chemical strategy of glycolysis can be condensed into three processes:

1. Add phosphoryl groups to glucose, yielding compounds with low phosphate group–transfer potential = priming.

2. Chemically convert these low phosphate group–transfer potential intermediates into compounds with high phosphate group–transfer potential.

3. Chemically couple the energy–yielding hydrolysis of these high phosphate group–transfer potential compounds to the synthesis of ATP by transfer of the phosphate group to ADP.

Figure 13.4 presents an abbreviated look at the conversion of glucose to pyruvate. In the energy investment phase (the first five reactions), the sugar is metabolically activated by phosphorylation. This priming process yields a six-carbon doubly phosphorylated sugar, **fructose-1,6-bisphosphate**, which undergoes cleavage to yield 2 equivalents of triose phosphate: **glyceraldehyde-3-phosphate** and **dihydroxyacetone phosphate**. Both of these compounds have a lower phosphate transfer potential than that of ATP.

In the energy generation phase (reactions 6 through 10), the triose phosphates undergo further activation to yield two compounds containing high phosphate transfer potential—first **1,3-bisphosphoglycerate** and then **phosphoenolpyruvate**. Recall from Figure 3.7 (page 76) that each of these compounds has a higher $\Delta G^{\circ\prime}$ of hydrolysis than ATP. The energy used to boost the phosphate transfer potential of these compounds comes from oxidation reactions. During the energy generation phase, the high phosphate transfer potential of these compounds drives the phosphorylation of ADP, yielding ATP. This process is called **substrate-level phosphorylation**—the direct transfer of a phosphoryl group from a donor compound to ADP, yielding ATP. Substrate-level phosphorylation is distinguished from **oxidative phosphorylation**, the indirect synthesis of ATP driven by the transmembrane gradient of hydrogen ion concentration (the **protonmotive force**, see Chapter 15), and **photophosphorylation**, the utilization of photosynthetic light energy to generate a protonmotive force that drives ATP synthesis (see Chapter 16).

Because 2 moles of triose phosphate are metabolized per mole of glucose, the yield from the two substrate-level phosphorylations of glycolysis is 4 moles of ATP per mole of glucose. Subtracting the 2 moles of ATP invested in the first, priming phase (reactions 1–5), we see a net gain of two ATP molecules synthesized per molecule of glucose converted to pyruvate (see Figures 13.2 and 13.3).

ATP is synthesized by three major routes— substrate-level phosphorylation, oxidative phosphorylation, and photophosphorylation.

Reactions of Glycolysis

Now let us consider in sequence the 10 reactions leading from glucose to pyruvate, numbering each reaction as indicated in Figure 13.4. The complete names of substrates and products are given when each reaction is presented, but in the text

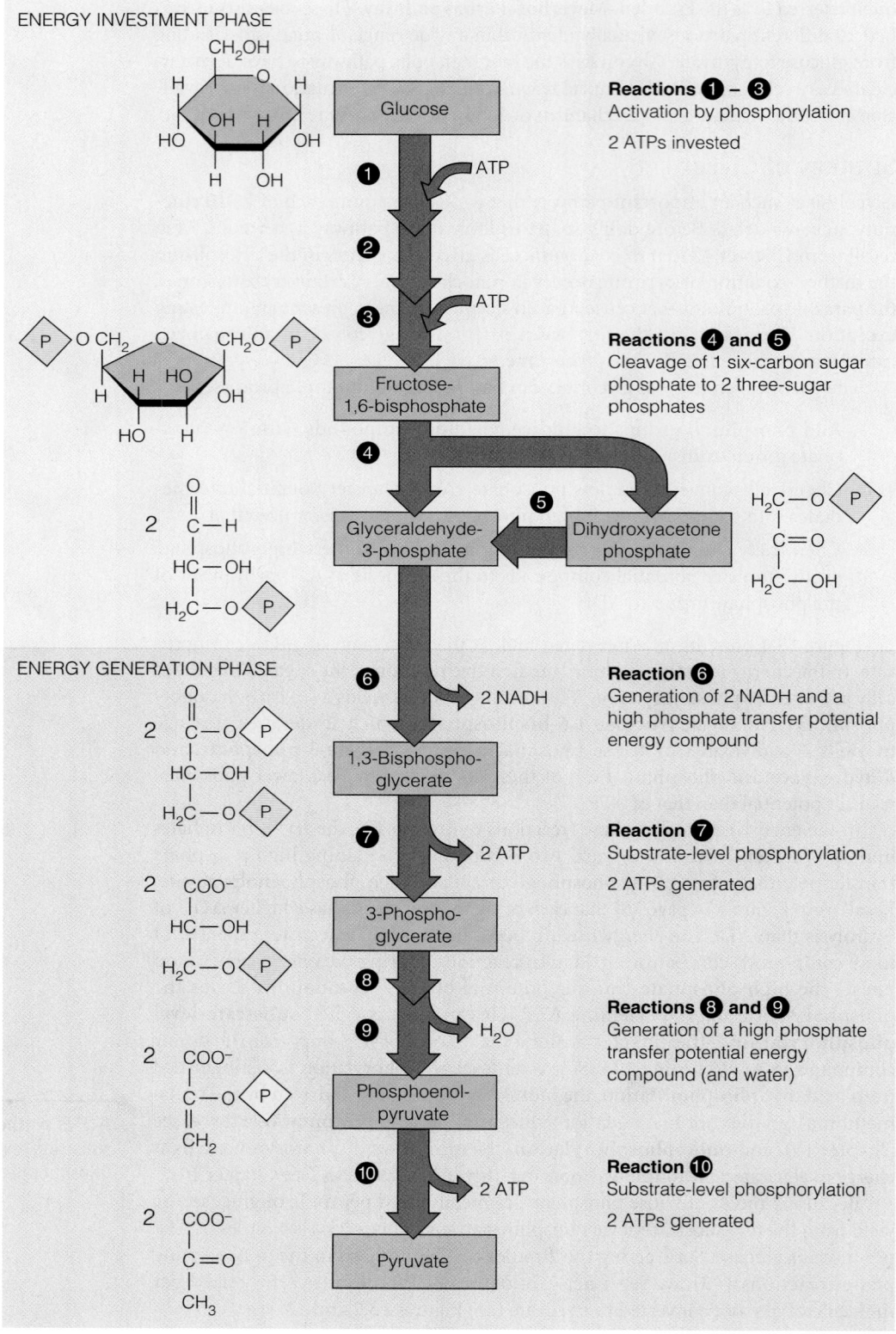

FIGURE 13.4

An overview of glycolysis. This condensed view of glycolysis shows the key intermediates and reactions in each of the two major phases. In the energy-generating phase, two ATPs are produced for each ATP utilized in the energy-investment phase. Here and elsewhere, the phosphate groups of high phosphate transfer potential compounds are highlighted in yellow.

these names are shortened for simplicity. Thus, glucose-6-phosphate is the same as α-D-glucose-6-phosphate.

Reactions 1–5: The Energy Investment Phase

The first five reactions, which constitute the energy investment phase, are summarized in the margin.

Reaction 1: The First ATP Investment

We begin with the ATP-dependent phosphorylation of glucose, catalyzed by **hexokinase**. (To simplify the ring structures used throughout this and subsequent chapters, hydrogens have been omitted. An empty vertical bond in a furanose or pyranose ring implies an H.)

$$\alpha\text{-}\text{D-Glucose} + \text{ATP} \xrightarrow{\text{Mg}^{2+}} \alpha\text{-}\text{D-Glucose-6-phosphate} + \text{ADP} + \text{H}^+ \quad \Delta G^{\circ\prime} = -16.7 \text{ kJ/mol}$$

The reaction involves a nucleophilic attack of the C6—OH of glucose on the electrophilic γ-phosphate of ATP. Magnesium ion is required because the reactive form of ATP is its chelated complex with Mg^{2+} (see page 496). This is true for virtually all ATP-requiring enzymes. Mg^{2+} partially neutralizes the negative charges on the oxygen atoms, making $\text{Mg}^{2+} \cdot \text{ATP}$ more susceptible to nucleophilic attack, that is, making it a better electrophile. Hexokinase exists in various forms in different organisms but is generally characterized by broad specificity for sugars and low K_M for the sugar substrate (0.01 to 0.1 mM). The broad specificity allows phosphorylation of various hexose sugars, including fructose and mannose, leading to their utilization via glycolysis. The low K_M allows glycolysis to proceed even when the blood glucose concentration is in the lowest part of its normal range (4 mM or less; see margin). As noted in Chapter 11, hexokinase is inhibited by its product, glucose-6-phosphate, a mechanism that controls the influx of substrates into the glycolytic pathway. Besides serving the priming function, phosphorylation of glucose aids in its retention in the cell—phosphorylated compounds cross the plasma membrane very poorly. Recall also that the structure of hexokinase provides striking evidence for the induced fit model of enzyme catalysis (see page 419).

Mammals possess several molecular forms of hexokinase. Different molecular forms of an enzyme catalyzing the same reaction are called **isoenzymes**, **isozymes**, or **isoforms**. Most tissues express hexokinase I, II, or III, all low K_M enzymes. Because intracellular glucose levels (2–15 mM) are usually far higher than the K_M value for hexokinase, the enzyme often functions in vivo at saturating substrate concentrations. Vertebrate liver expresses a distinctive isozyme, hexokinase IV, characterized by a much lower affinity for glucose, and an insensitivity to inhibition by physiological concentrations of glucose-6-phosphate. More important, hexokinase IV exhibits a sigmoidal concentration dependence on glucose (see margin), requiring 5–10 mM glucose for half-saturation. This special hexokinase isozyme, previously called glucokinase, allows the liver to adjust its rate of glucose utilization in response to variations in blood glucose levels. In fact, as discussed in Chapter 18, a major role of liver is to regulate blood glucose levels, and this enzyme represents one of the principal mechanisms by which it does so.

Although most bacteria contain a hexokinase, the usual route for glucose uptake and phosphorylation is the phosphenolpyruvate:glucose phosphotransferase system. This system, introduced in Chapter 10 as an example of transport by modification

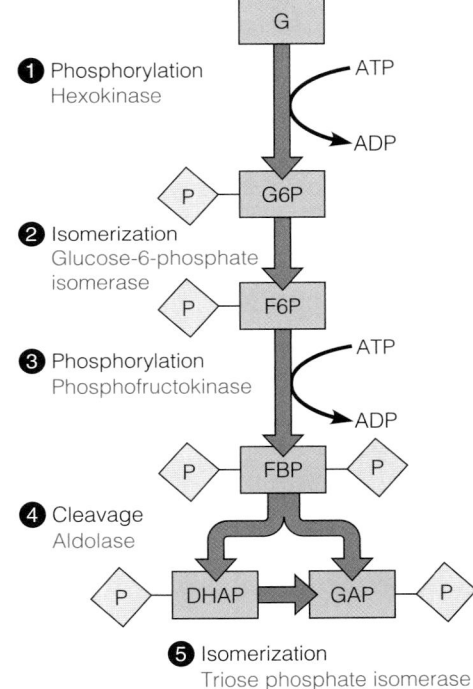

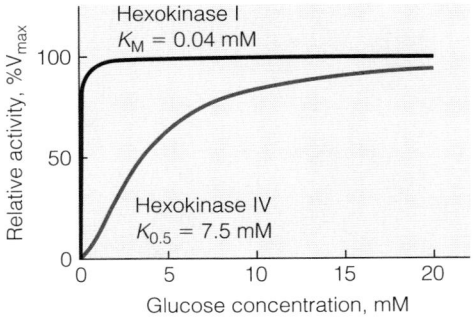

A low affinity isozyme of hexokinase in liver with a sigmoidal dependence on glucose concentration permits that organ to adjust glucose utilization to glucose supply at high blood glucose levels.

(page 395), couples the phosphorylation of glucose to its translocation across the cytoplasmic membrane. The energy to drive these two coupled processes is provided by phosphoenolpyruvate (PEP), one of the high energy compounds produced in the energy generation phase of glycolysis. The phosphenolpyruvate:glucose phosphotransferase system catalyzes the following overall process:

$$PEP_{in} + glucose_{out} \rightarrow pyruvate_{in} + glucose\text{-}6\text{-}phosphate_{in}$$

Glucose-6-phosphate, now in the cytoplasm, directly enters glycolysis at reaction 2. Bacteria possess a parallel set of phosphotransferase systems to allow transport and phosphorylation of numerous monosaccharides, disaccharides, and other sugar derivatives.

Reaction 2: Isomerization of Glucose-6-phosphate

The next reaction, catalyzed by **glucose-6-phosphate isomerase** (also called phosphoglucoisomerase), is the readily reversible isomerization of the aldose, glucose-6-phosphate (G6P), to the corresponding ketose, **fructose-6-phosphate** (F6P).

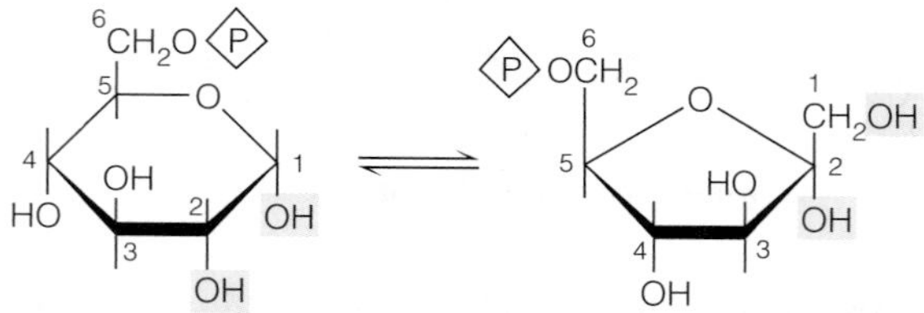

$$\Delta G^{\circ\prime} = +1.7 \text{ kJ/mol}$$

α-D-Glucose-6-phosphate   **D-Fructose-6-phosphate**

This reaction proceeds via an enediolate intermediate: B: and B-H represent active-site amino acid residues acting as bases and acids, respectively.

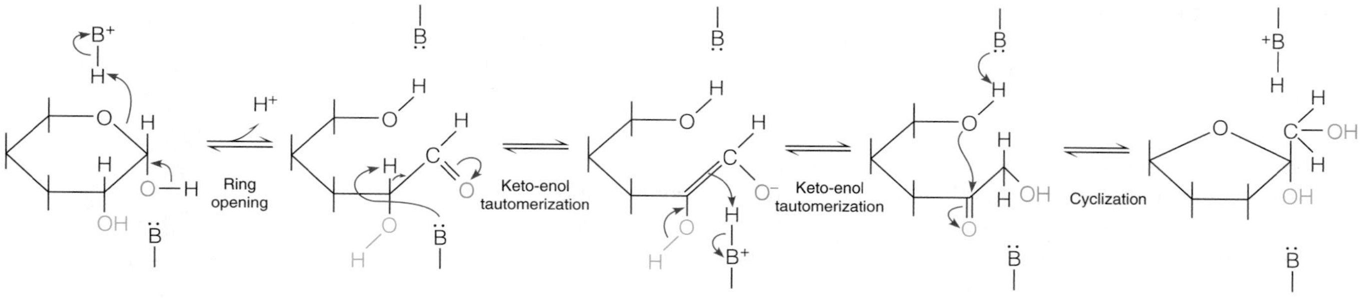

Transferring the carbonyl oxygen from carbon 1 to carbon 2 has two important effects. The hydroxyl group generated at carbon 1 can be readily phosphorylated in the reaction 3. It also sets up the sugar for a symmetric aldol cleavage in reaction 4. We shall encounter other aldose–ketose isomerizations that proceed by a similar mechanism.

Reaction 3: The Second Investment of ATP

In reaction 3, **phosphofructokinase** catalyzes a second ATP-dependent phosphorylation, to give a hexose derivative, fructose-1,6-bisphosphate (FBP), phosphorylated at both carbons 1 and 6.

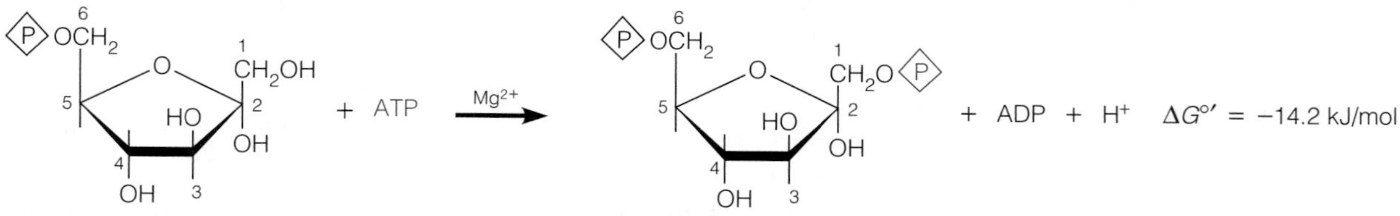

D-Fructose-6-phosphate   **D-Fructose-1,6-bisphosphate**

$$+ \text{ ADP } + \text{ H}^+ \quad \Delta G^{\circ\prime} = -14.2 \text{ kJ/mol}$$

The reaction involves the same chemistry we saw in step 1—nucleophilic substitution. Here, however, the C1—OH of fructose-6-phosphate is the nucleophile that attacks the electrophilic γ-phosphate of ATP. Like phosphorylation at the 6 position, this reaction is sufficiently exergonic to be essentially irreversible in vivo. Irreversibility is important because phosphofructokinase (PFK) represents the primary site for regulation of the flow of carbon through glycolysis. PFK is an allosteric enzyme whose activity is acutely sensitive to the energy status of the cell, as well as to the levels of various other intermediates, particularly citrate and fatty acids. Interactions with allosteric effectors, which are discussed later in this chapter, activate or inhibit PFK. This allosteric regulation increases carbon flux through glycolysis when there is a need to generate more ATP and inhibits it when the cell contains ample stores of ATP or oxidizable substrates.

Higher plants and a number of prokaryotes contain two different PFKs—the ATP-dependent enzyme and a unique form, which uses pyrophosphate (PP_i) instead of ATP as the phosphorylating agent.

Fructose-6-phosphate + PP_i $\rightleftharpoons$ fructose-1,-6-bisphosphate + P_i $\Delta G^{\circ\prime} = -2.9\,\text{kJ/mol}$

Whereas phosphorylation of fructose-6-phosphate by the ATP-dependent PFK is virtually irreversible in vivo, the pyrophosphate-dependent enzyme catalyzes a near-equilibrium reversible reaction. The pyrophosphate-dependent enzyme has been proposed to be involved in adaptation of plants to nonoptimal conditions, such as phosphate deficiency or low oxygen stress.

Reaction 4: Cleavage to Two Triose Phosphates

Reaction 4 is catalyzed by **fructose-1,6-bisphosphate aldolase**, usually called **aldolase**, because its reaction is similar to the reverse of an aldol condensation. In this reaction the "splitting of sugar" that is connoted by the term *glycolysis* occurs because the six-carbon compound fructose-1,6-bisphosphate is cleaved to give 2 three-carbon intermediates, glyceraldehyde-3-phosphate and dihydroxyacetone phosphate (DHAP).

D-Fructose-1,6-bisphosphate   **Dihydroxyacetone phosphate**   **D-Glyceraldehyde-3-phosphate**

$\Delta G^{\circ\prime} = +23.9\,\text{kJ/mol}$

This reaction illustrates an important metabolic principle. Note that the reaction is strongly endergonic under standard conditions, such that formation of fructose-1,6-bisphosphate is highly favored. However, from the actual intracellular concentrations of the reactant and products, as determined in rabbit skeletal muscle, a ΔG of $-1.3\,\text{kJ/mol}$ is calculated, consistent with the observation that the reaction proceeds as written in vivo. As was emphasized in Chapter 3, this example illustrates the importance of considering conditions *in the cell* (ΔG), rather than standard state conditions ($\Delta G^{\circ\prime}$) in deciding which direction of a reaction is favored.

Aldolase from most vertebrate sources is a tetrameric protein. The enzyme activates the substrate for cleavage by nucleophilic attack on the keto carbon at position 2 with a lysine ε-amino group in the active site, as shown in Figure 13.5.

The phosphofructokinase reaction is the primary step at which glycolysis is regulated.

Aldolase cleaves fructose-1,6-bisphosphate under intracellular conditions, even though the equilibrium lies far toward fructose-1,6-bisphosphate under standard conditions.

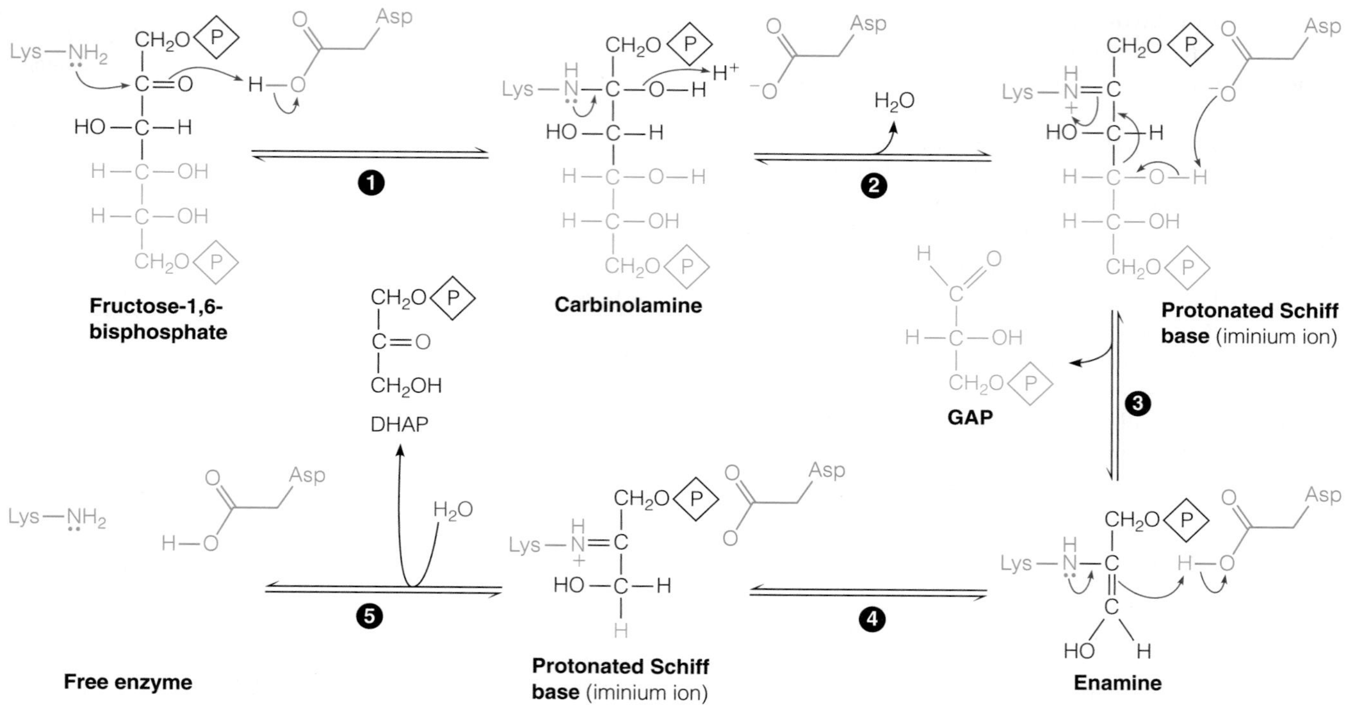

FIGURE 13.5

Reaction mechanism for fructose-1,6-bisphosphate aldolase. The figure shows the protonated Schiff base intermediate (iminium ion) between the substrate and an active site lysine residue. An aspartate residue facilitates the reaction via general acid–base catalysis. See text for details.

This is facilitated by protonation of the carbonyl oxygen by an active site acid (aspartate) (step 1). The resulting carbinolamine undergoes dehydration to give a protonated **Schiff base** (iminium ion) (step 2). A Schiff base is a nucleophilic addition product between an amino group and a carbonyl group. A retro-aldol reaction then cleaves the protonated Schiff base into an enamine plus glyceraldehyde-3-phosphate (step 3). The enamine is protonated to give another iminium ion (protonated Schiff base) (step 4), which is then hydrolyzed off the enzyme to give the second product, DHAP (step 5).

The advantage of the Schiff base intermediate derives from its ability to delocalize electrons. The positively charged iminium ion is thus a better electron acceptor than a ketone carbonyl, facilitating retro-aldol reactions like this one and, as we shall see, many other biological conversions. This mechanism also reveals the importance of the earlier glucose-6-phosphate isomerase reaction in setting the symmetry of the aldolase reaction. If glucose had not been isomerized to fructose (moving the carbonyl to C-2 instead of C-1), the aldolase reaction would have given two- and four-carbon fragments, instead of the metabolically equivalent three-carbon fragments.

Reaction 5: Isomerization of Dihydroxyacetone Phosphate

As noted above, the symmetrical aldolase reaction yields 2 three-carbon sugar phosphates. The function of reaction 5, catalyzed by **triose phosphate isomerase**, is conversion of one of these products, dihydroxyacetone phosphate (DHAP), to the other, glyceraldehyde-3-phosphate (GAP). Because GAP is the substrate for the next glycolytic reaction, this reaction permits use of all six carbon atoms of glucose: C-1 and C-6 of glucose become C-3 of glyceraldehyde-3-phosphate; C-2 and C-5 become C-2; and C-3 and C-4 become C-1 of glyceraldehyde-3-phosphate.

Dihydroxyacetone phosphate　　　　**Enediol intermediate**　　　　**D-Glyceraldehyde-3-phosphate**

$\Delta G^{\circ\prime} = +7.6$ kJ/mol

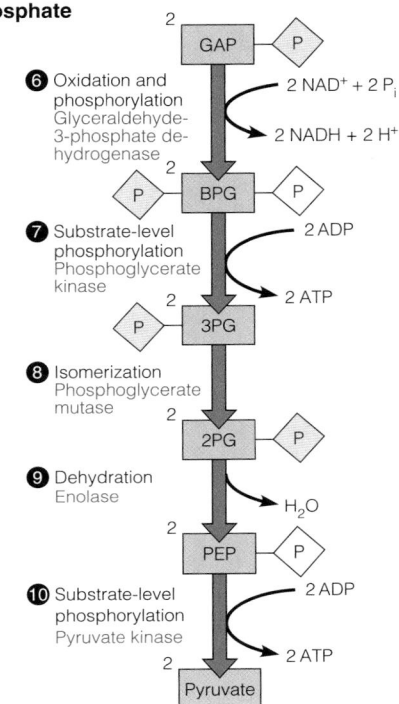

This reaction is also weakly endergonic under standard conditions, but the intracellular concentration of glyceraldehyde-3-phosphate is low (because it is consumed in subsequent reactions), drawing the reaction toward the right. The isomerization of dihydroxyacetone phosphate proceeds via an enediol intermediate.

At this point, glycolysis has expended two ATP molecules and converted one hexose sugar to two molecules of glyceraldehyde-3-phosphate, each of which will next be metabolized to give compounds with high phosphate transfer potential that can drive the synthesis of ATP. The energy investment phase of the cycle is complete, and the energy generation phase is about to begin.

Reactions 6–10: The Energy Generation Phase

The five reactions of the energy generation phase are summarized in the margin.

Reaction 6: Generation of the First Energy-Rich Compound

This reaction, catalyzed by **glyceraldehyde-3-phosphate dehydrogenase**, is among the most important in glycolysis, partly because it generates the first intermediate with high phosphate transfer potential and partly because it generates a pair of reducing equivalents (see Figure 13.6). The overall reaction is as follows:

D-Glyceraldehyde-3-phosphate　　　　**1,3-Bisphosphoglycerate**

$+ \text{ NAD}^+ + \text{P}_i \rightleftharpoons \quad + \text{ NADH } + \text{ H}^+ \quad \Delta G^{\circ\prime} = +6.3$ kJ/mol

Reaction 6 involves a two-electron oxidation of the carbonyl carbon of glyceraldehyde-3-phosphate to the carboxyl level, a reaction that is normally quite exergonic. However, the overall reaction is slightly *endergonic* (under *standard* conditions) because the enzyme utilizes most of the energy released by the oxidation to drive the synthesis of a compound with high phosphate transfer potential, 1,3-bisphosphoglycerate (BPG). This compound contains a carboxylic–phosphoric acid mixed anhydride, or an **acyl-phosphate group**, at position 1, a functional group with a very high standard free energy of hydrolysis, -49.4 kJ/mol. This enzyme also requires a coenzyme, NAD^+, to accept electrons from the substrate being oxidized.

Because the acyl-phosphate group is much more energy rich than the phosphate anhydride of ATP, 1,3-bisphosphoglycerate can drive the synthesis of ATP

Glyceraldehyde-3-phosphate dehydrogenase creates a high-energy compound and generates a pair of reducing equivalents.

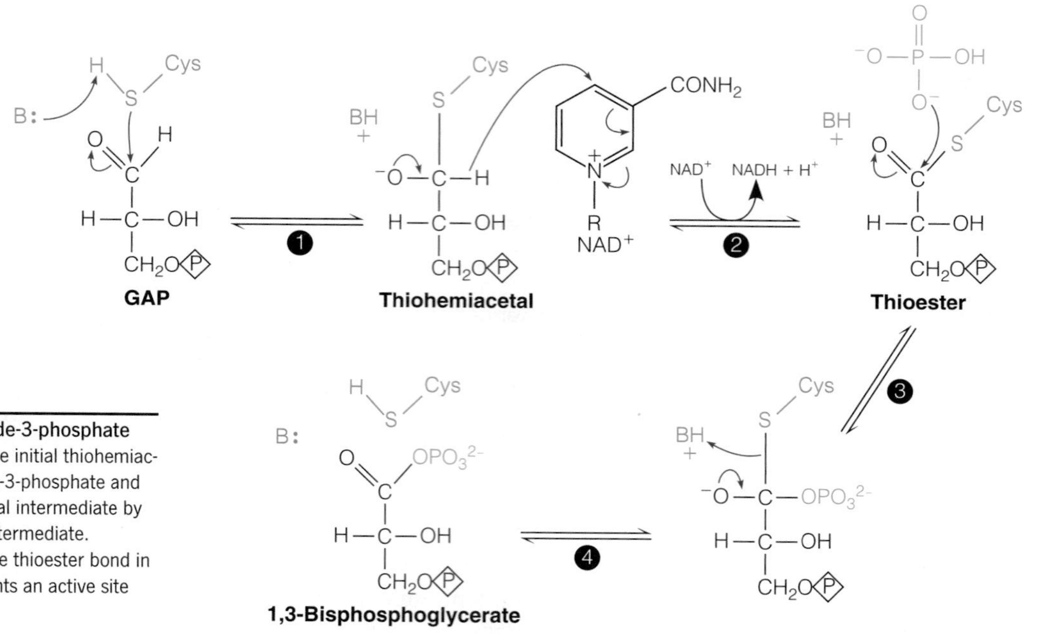

FIGURE 13.6

Reaction mechanism for glyceraldehyde-3-phosphate dehydrogenase. **Step 1:** Formation of the initial thiohemiacetal intermediate between glyceraldehyde-3-phosphate and the enzyme. **Step 2:** Oxidation of the initial intermediate by NAD^+ to give an acyl-thioester enzyme intermediate. **Steps 3, 4:** Phosphorolytic cleavage of the thioester bond in the acyl-enzyme intermediate. B: represents an active site histidine acting as a general base.

from ADP. Indeed, it does so in the next reaction in the sequence, the first of two substrate-level phosphorylations in glycolysis. Because of the importance of understanding how ATP is synthesized, much attention has focused on understanding how the high phosphate transfer potential compounds in substrate-level phosphorylation are synthesized.

For glyceraldehyde-3-phosphate dehydrogenase, that understanding derived in large part from an old observation that glycolysis is inhibited by iodoacetate and by heavy metals such as mercury. Both compounds react with free sulfhydryl groups, as the following shows for iodoacetate:

$$RSH + ICH_2COO^- \longrightarrow RS{-}CH_2COO^- + HI$$

The finding that these compounds inhibit glycolysis specifically by inhibition of glyceraldehyde-3-phosphate dehydrogenase strongly implied that the enzyme contains one or more essential thiol groups. We now know that the reaction proceeds as outlined in Figure 13.6, starting with formation of a **thiohemiacetal** group involving the substrate carbonyl group and a cysteine thiol group on the enzyme. The thiohemiacetal is next oxidized by NAD^+ to give an acyl-enzyme intermediate, or thioester. Thioesters are high-energy compounds (see Figure 14.8, page 601); cleavage of this thioester by P_i preserves much of the energy as the acyl phosphate, which is the product.

This mechanism conserves the energy of oxidation by coupling an exergonic reaction to an endergonic reaction:

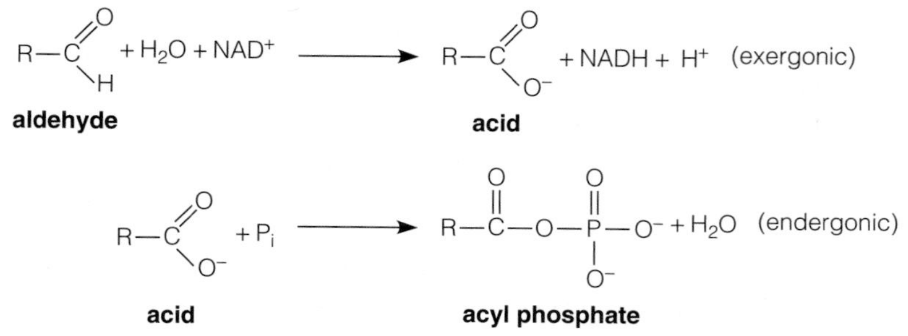

An aldehyde is oxidized to the level of an acid (a two-electron process). But rather than releasing the acid, the enzyme incorporates P_i using the oxidation energy to create a "high-energy" acyl-phosphate compound, possessing higher phosphate transfer potential than ATP. This reaction also explains the stimulation of glycolysis observed by Harden and Young when they added inorganic phosphate to yeast extracts (page 522).

The overall stoichiometry of the reaction involves reduction of 1 mole of NAD^+ to $NADH + H^+$ per mole of glyceraldehyde-3-phosphate. This reaction is the source of the NADH formed in glycolysis, which was first identified in Figures 13.2 and 13.3.

Reaction 7: The First Substrate-Level Phosphorylation

As noted previously, 1,3-bisphosphoglycerate, because of its high phosphate transfer potential, has a strong tendency to transfer its acyl-phosphate group to ADP, with resultant formation of ATP. This substrate-level phosphorylation reaction is catalyzed by **phosphoglycerate kinase**, as follows:

1,3-Bisphosphoglycerate   **3-Phosphoglycerate** $+$ ATP $\Delta G^{\circ\prime} = -18.8$ kJ/mol

The glyceraldehyde-3-phosphate dehydrogenase and phosphoglycerate kinase are thermodynamically coupled:

	$\Delta G^{\circ\prime}$ (kJ/mol)
glyceraldehyde-3-P $+$ P_i $+$ NAD$^+$ $\rightarrow$ 1,3-bisphosphoglycerate $+$ NADH $+$ H$^+$	$+6.3$
1,3-bisphosphoglycerate $+$ ADP $\rightarrow$ 3-phosphoglycerate $+$ ATP	-18.8
glyceraldehyde- 3-P $+$ P_i $+$ ADP $+$ NAD$^+$ $\rightarrow$	
3-phosphoglycerate $+$ ATP $+$ NADH $+$ H$^+$ $\Delta G^{\circ\prime}_{Sum} = -12.5$	

Thus, through two consecutive reactions, the energy of oxidation of an aldehyde to a carboxylic acid is conserved in the form of ATP. At this stage the net ATP yield from the glycolytic pathway is zero. Recall that 2 moles of ATP per mole of glucose were invested to generate 2 moles of triose phosphate. The phosphoglycerate kinase reaction generates 1 mole of ATP from each mole of triose phosphate, or 2 moles of ATP per mole of glucose (see Figure 13.3). The pathway as a whole becomes exergonic in the remaining three reactions. This involves activation of the remaining phosphate, which, in 3-phosphoglycerate (3PG), has a relatively low phosphate transfer potential.

Phosphoglycerate kinase catalyzes the first glycolytic reaction that forms ATP.

Reaction 8: Preparing for Synthesis of the Next High-Energy Compound

Activation of 3-phosphoglycerate begins with an isomerization catalyzed by **phosphoglycerate mutase**. The enzyme transfers phosphate from position 3 to position 2 of the substrate to yield 2-phosphoglycerate. Mg^{2+} is required.

3-Phosphoglycerate   **2-Phosphoglycerate** $\Delta G^{\circ\prime} = +4.4$ kJ/mol

N-Phosphohistidine residue

The reaction is slightly endergonic under standard conditions. However, the intracellular level of 3-phosphoglycerate is high relative to that of 2-phosphoglycerate (2PG), so that in vivo the reaction proceeds to the right without difficulty. The enzyme contains a phosphohistidine residue in the active site. In the first step of the reaction, the phosphate is transferred from the enzyme to the substrate to give an intermediate, 2,3-bisphosphoglycerate. Transfer of the phosphate from C-3 to the active site of the enzyme regenerates the phosphorylated enzyme and forms the product, which is released.

$$\text{Enzyme-P + 3-P-glycerate} \rightleftharpoons \left[\begin{array}{c} \text{Enzyme-2,3-bis-P-glycerate} \end{array} \right] \rightleftharpoons \text{Enzyme-P + 2-P-glycerate}$$

Enzyme-2,3-bis-P-glycerate

Moving the phosphate from the 3 position to the 2 position sets up the next energy-conserving step in glycolysis, again by conversion of a compound with low phosphate transfer potential to one with very high phosphate transfer potential.

Reaction 9: Synthesis of the Second High-Energy Compound

Reaction 9, catalyzed by **enolase**, generates phosphoenolpyruvate (PEP), another compound with very high phosphate transfer potential. PEP participates in the second substrate-level phosphorylation of glycolysis.

$\Delta G^{\circ\prime} = -3.2 \text{ kJ/mol}$

2-Phosphoglycerate   **Phosphoenolpyruvate**

Like most β-hydroxy carbonyl compounds, 2-phosphoglycerate is readily dehydrated via an α,β-elimination reaction. An active site Lys, acting as a general base, abstracts a proton from C-2 (α carbon) of 2PG, generating a carbanionic intermediate. An active site Glu facilitates the elimination, protonating the leaving group (OH^-) by general acid catalysis. Product formation is also favored by the stability of the conjugated double bond system in PEP. Enolase requires two Mg^{2+} ions, which neutralize the negative charges on the substrate and lower the pK_a of the C-2 proton, making it easier to abstract. Despite its small overall free energy change, this seemingly simple transformation increases enormously the standard free energy of hydrolysis of the phosphate bond—from -15.6 kJ/mol for 2-phosphoglycerate to -61.9 kJ/mol for phosphoenolpyruvate. Phosphorylation of the C2—OH prevents the otherwise highly favorable tautomerization to the keto form. When the phosphate is transferred to ADP, the enol to keto tautomerization can take place, driving the reaction forward. As discussed in Chapter 3 (page 79), the great thermodynamic instability of enolpyruvate is chiefly responsible for the large negative free energy of hydrolysis of phosphoenolpyruvate.

Reaction 10: The Second Substrate-Level Phosphorylation

In the last reaction, catalyzed by **pyruvate kinase**, phosphoenolpyruvate transfers its phosphoryl group to ADP in another substrate-level phosphorylation. Note that the enzyme is named as if it were acting leftward in the reaction shown below, even though it is strongly exergonic in the rightward direction. Many enzymes were named before the function or direction of intracellular catalysis had been identified.

Phosphoenolpyruvate   **Pyruvate**

The enzyme requires Mg^{2+} and K^+. Even though the reaction involves the endergonic synthesis of ATP, the overall reaction is strongly exergonic because, as noted in Chapter 3, the spontaneous tautomerization of the product, enolpyruvate, to the highly favored keto form provides a strong thermodynamic drive ($\Delta G^{\circ\prime} = -46\,\text{kJ/mol}$) in the forward direction.

The pyruvate kinase (PK) reaction is another site for metabolic regulation. PK, like many enzymes, exists in animal tissues as multiple isozymes. Mammals carry two paralogous pyruvate kinase genes, which, through alternative splicing, produce four different PK isozymes. As we will see in Chapter 27, alternative splicing is a common mechanism used by eukaryotes to greatly expand the number of different protein products produced from a single gene. The PK-L and PK-R isozymes, both encoded by the *PKLR* gene, are expressed specifically in liver and red blood cells, respectively. The M1 and M2 isozymes are encoded by the *PKM2* gene. PK-M1 is expressed in muscle and brain and other terminally differentiated tissues. PK-M2 is expressed during embryonic development, but also in tumor cells, where it appears to confer a selective proliferation advantage to the tumor cells—more about this later. All four PK isozymes are homotetramers of ~60 kDa subunits, and except for the M1 isozyme, require allosteric activation by fructose-1,6-bisphosphate for full activity. The L and R isozymes are allosterically inhibited at high ATP concentrations. The M1 isozyme, on the other hand, is constitutively active. On top of these allosteric mechanisms, the synthesis of the liver enzyme is also under dietary control; intracellular activity may increase as much as 10-fold from increased enzyme synthesis, or induction, as a result of high carbohydrate ingestion.

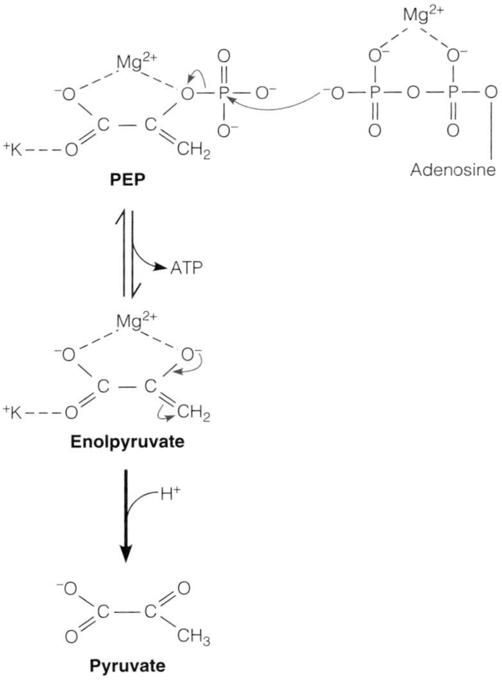

The liver pyruvate kinase isozyme (PK-L) is also regulated by phosphorylation and dephosphorylation. The dephosphorylated form is far more active than the phosphorylated form. Phosphorylation, which is under hormonal control, diverts phosphoenolpyruvate to gluconeogenesis (see below) when fatty acid oxidation and the citric acid cycle are already operating at rates sufficient to meet the energy needs of the cell. By contrast, virtually all of the phosphoenolpyruvate produced in muscle is converted to pyruvate.

Human genetic deficiencies of erythrocyte pyruvate kinase (PK-R) reveal an interesting link between glycolysis and hemoglobin function. Accumulation of phosphoenolpyruvate leads to excessive levels in blood of other glycolytic intermediates and three-carbon metabolites, including 2,3-bisphosphoglycerate (2,3-BPG). 2,3-BPG was introduced in Chapter 7 as an allosteric inhibitor of oxygen binding to hemoglobin, and its accumulation leads to decreased affinity of hemoglobin for oxygen in PK-R–deficient patients. However, the anemia associated

Pyruvate kinase catalyzes the second glycolytic reaction that forms ATP.

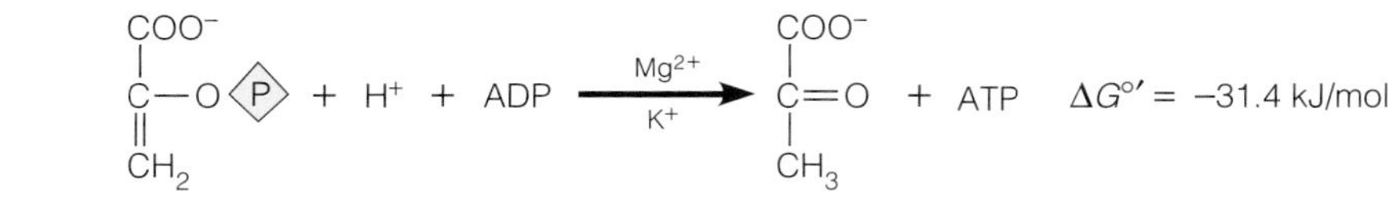

PEP

Enolpyruvate

Pyruvate

Dietary carbohydrate induces the biosynthesis of pyruvate kinase and increases the ability of the body to obtain energy from glycolysis.

TABLE 13.1   Summary of glycolysis

Reaction	Enzyme	ATP yield	$\Delta G^{\circ\prime}$	ΔG
Glucose (G) $\xrightarrow[\text{ADP}]{\text{ATP}}$ ❶	HK	−1	−16.7	−33.5
Glucose-6-phosphate (G6P) ❷	PGI		+1.7	−2.5
Fructose-6-phosphate (F6P) $\xrightarrow[\text{ADP}]{\text{ATP}}$ ❸	PFK	−1	−14.2	−22.2
Fructose-1,6-bisphosphate (FBP) ❹	ALD		+23.9	−1.3
Glyceraldehyde-3-phosphate (GAP) + dihydroxyacetone phosphate (DHAP) ❺	TPI		+7.6	~0
Two glyceraldehyde-3-phosphate (GAP) $\xrightarrow[\text{NADH} + 2H^+]{\text{NAD}^+ + 2P_i}$ ❻	GAPDH		+6.3 (+12.6)	−1.7 (−3.4)
1,3-Bisphosphoglycerate (BPG) $\xrightarrow[\text{ATP}]{\text{ADP}}$ ❼	PGK	+1 (+2)	−18.8 (−37.6)	~0
3-Phosphoglycerate (3PG) ❽	PGM		+4.4 (+8.8)	~0
2-Phosphoglycerate (2PG) ❾ $\xrightarrow{\text{H}_2\text{O}}$	ENO		−3.2 (−6.4)	−3.3 (−6.6)
Phosphoenolpyruvate (PEP) $\xrightarrow[\text{ATP}]{\text{ADP}}$ ❿	PK	+1 (+2)	−31.4 (−62.8)	−16.7 (−33.4)
Pyruvate (Pyr)				

Net: Glucose + 2ADP + 2P$_i$ + 2NAD$^+$ → 2 pyruvate + 2ATP + 2NADH + 2H$^+$ + 2H$_2$O   +2   −83.1   −102.9

Note: $\Delta G^{\circ\prime}$ and ΔG values in kJ/mol. The values in parentheses are based on doubling all the reactions past reaction 5, since the energy generation phase involves 2 three-carbon substrates per glucose molecule. ΔG values are estimated from the approximate intracellular concentrations of glycolytic intermediates in rabbit skeletal muscle.

with this enzyme deficiency is generally well tolerated because reduced O_2 binding in the lungs is compensated by increased O_2 release in the tissues (see Figures 7.10 and 7.24).

The pyruvate kinase reaction converts the overall glycolytic pathway from an energy-neutral process to one that involves net synthesis of ATP. Two high-energy phosphates (i.e., ATP) per hexose are generated here, to go with the two generated by phosphoglycerate kinase (see Figure 13.3). Subtracting the two ATPs invested at hexokinase and phosphofructokinase gives a net yield of 2 moles of high-energy phosphates per mole of glucose—not a high yield, to be sure, but the process is fast and can meet the energy requirements of many anaerobes. Moreover, subsequent metabolism of pyruvate through aerobic pathways generates much additional high-energy phosphate.

Table 13.1 summarizes the reactions of glycolysis, showing free energy changes and ATP yields at each step. Note the difference between standard ($\Delta G^{\circ\prime}$) and estimated (ΔG) Gibbs free energy changes, and the three values in red—we will discuss the importance of these shortly.

Metabolic Fates of Pyruvate

Pyruvate represents a central metabolic branch point. Its fate depends crucially on the oxidation state of the cell, which is related to the reaction catalyzed by glyceraldehyde-3-phosphate dehydrogenase (reaction 6). Recall that this reaction converts 1 mole of NAD^+ per mole of triose phosphate to NADH. The cytoplasm has a finite supply of NAD^+, *so this NADH must be reoxidized to NAD^+* for glycolysis to continue. During aerobic glycolysis, as noted earlier, this NADH is oxidized by the mitochondrial electron transport chain, with the electrons transferred ultimately to molecular oxygen. This oxidation of NADH, which we consider in detail in Chapter 15, yields additional energy, with about 2.5 moles of ATP synthesized from ADP per mole of NADH oxidized. Because 2 moles of NADH are produced per mole of glucose entering the pathway, aerobic glycolysis yields considerably more ATP than anaerobic glycolysis. In addition, the oxidation of pyruvate through the citric acid cycle generates much more energy, via respiration.

Lactate Metabolism

In aerobic cells that are undergoing very high rates of glycolysis, the NADH generated in glycolysis cannot all be reoxidized at comparable rates in the mitochondrion. In such cases, or in anaerobic cells, which lack mitochondria, NADH must be used to drive the reduction of an organic substrate in order to maintain redox homeostasis. As noted earlier, that substrate is pyruvate itself, both in eukaryotic cells and in lactic acid bacteria, and the product is lactate. The enzyme catalyzing this reaction is lactate dehydrogenase (see page 521). The equilibrium for this reaction lies far toward lactate. As illustrated in Figure 13.3a, the NADH produced in the oxidation of glyceraldehyde-3-phosphate is used to reduce pyruvate to lactate. Thus, during anaerobic glycolysis, an overall electron balance is maintained.

Even in aerobic vertebrates, some tissues such as red blood cells derive most of their energy from anaerobic metabolism. Skeletal muscle, which derives most of its energy from respiration when at rest, relies heavily on anaerobic glycolysis during exertion, when glycogen stores are rapidly broken down, or mobilized, to provide glucose for glycolysis. Normally the lactate produced diffuses from the tissue and is transported through the bloodstream to highly aerobic tissues, such as heart and liver. The aerobic tissue can catabolize lactate further, through respiration, or can convert it back to glucose, through gluconeogenesis. However, if lactate is produced in large quantities, it cannot be readily consumed. Then, as we

Pyruvate must be reduced to lactate when tissues are insufficiently aerobic to oxidize all of the NADH formed in glycolysis.

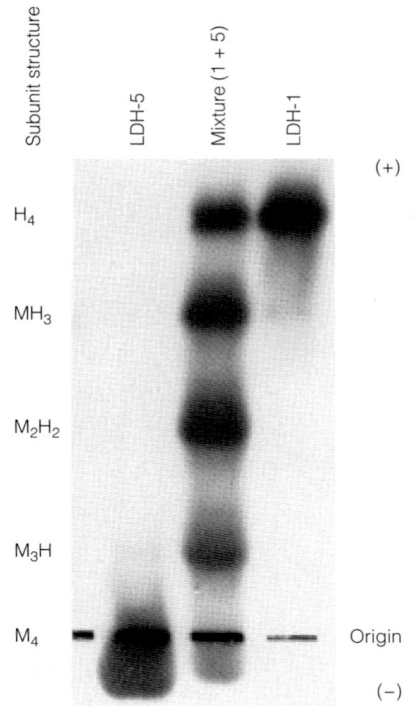

FIGURE 13.7

Structural basis for the existence of isozymes of lactate dehydrogenase. Preparations were subjected to electrophoresis in a starch gel, which was then treated to reveal bands containing enzymatically active protein. LDH-1 is a tetramer containing only the H subunit, whereas LDH-5 contains only M subunits. The middle lane depicts an experiment in which equal amounts of LDH-1 and LDH-5 were mixed. The subunits were dissociated and then allowed to reassociate. The presence of five different enzyme forms and their relative amounts show that individual M and H subunits can associate randomly to form tetramers of mixed-subunit composition.

From *Science* 140:1329, C. L. Markert, Lactate dehydrogenase isozymes: Dissociation and recombination of subunits. © 1963. Reprinted by permission from AAAS.

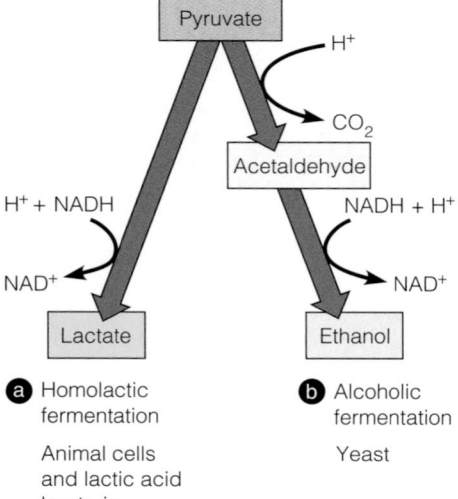

ⓐ Homolactic fermentation

Animal cells and lactic acid bacteria

ⓑ Alcoholic fermentation

Yeast

discussed in Chapter 7, the blood pH falls and the Bohr effect functions to increase oxygen supplies to the tissues.

Until recently it was thought that lactate accumulation in skeletal muscle was largely a consequence of anaerobic metabolism, which occurs when the need for tissues to generate energy exceeds their capacity to oxidize the pyruvate produced in glycolysis. Metabolic studies, including ^{31}P NMR analyses of the levels of phosphorylated intermediates in living muscle cells during exercise, suggest that lactate is actually an intermediate and not a metabolic "dead end," whose only fate is reconversion to pyruvate. These studies show that even in fully oxygenated muscle tissue, lactate is formed and utilized continuously. Lactate is actively oxidized by skeletal muscle mitochondria, and during exercise oxidation can account for up to 75% of lactate removal, with the remainder being used for gluconeogenesis.

Isozymes of Lactate Dehydrogenase

Lactate dehydrogenase was the first enzyme that established the structural basis for the existence of isozymes. Most tissues contain five isozymes of lactate dehydrogenase. They can be resolved electrophoretically, as shown in Figure 13.7.

Lactate dehydrogenase (LDH) is a tetrameric protein consisting of two types of subunits, called M and H, which have small differences in amino acid sequence. M subunits predominate in skeletal muscle and liver, and H subunits predominate in heart. M and H subunits combine randomly with each other, so that the five major isozymes have the compositions M_4, M_3H, M_2H_2, MH_3, and H_4. Because of random subunit reassortment, the isozymic composition of a tissue is determined primarily by the expression levels of the genes specifying the two subunits.

The physiological need for the existence of different forms of this enzyme is not yet clear. However, the tissue specificity of isozyme patterns is useful in clinical medicine. Such pathological conditions as myocardial infarction, infectious hepatitis, and muscle diseases involve cell death of affected tissue, with release of cell contents to the blood. The pattern of LDH isozymes in the blood serum is representative of the tissue that released the isozymes. This information can be used to diagnose such conditions and to monitor the progress of treatment.

Ethanol Metabolism

Pyruvate has numerous alternative fates in anaerobic microorganisms. As we have seen, lactic acid bacteria reduce pyruvate to lactate in a single step (see margin). By contrast, yeasts convert pyruvate to ethanol in a two-step pathway. This alcoholic fermentation starts with the nonoxidative decarboxylation of pyruvate to acetaldehyde, catalyzed by **pyruvate decarboxylase**. NAD^+ is regenerated in the next reaction, the NADH-dependent reduction of acetaldehyde to ethanol, catalyzed by alcohol dehydrogenase.

Pyruvate → **Acetaldehyde** → **Ethanol**

The first reaction requires **thiamine pyrophosphate** as a coenzyme. This coenzyme is derived from thiamine, the first of the B vitamins to be identified, and so it is also called vitamin B_1. The vitamin is structurally complex, but its conversion to the coenzyme form, thiamine pyrophosphate, or TPP, involves simply an ATP-dependent pyrophosphorylation.

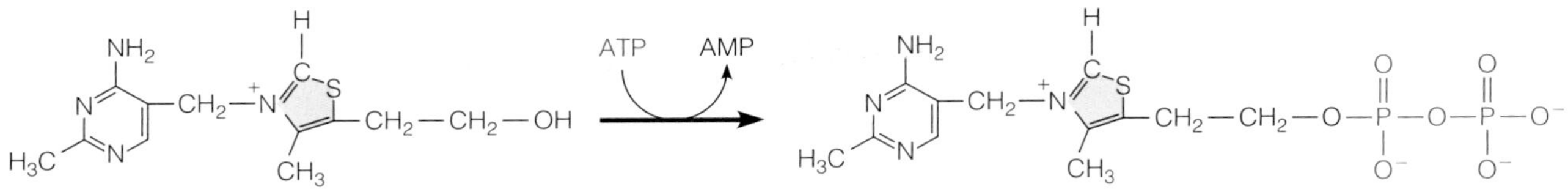

Thiamine   **Thiamine pyrophosphate**

Thiamine pyrophosphate is the coenzyme for all decarboxylations of α-keto acids. Unlike β-keto acids, α-keto acids cannot stabilize the carbanion transition state that develops during decarboxylation, and they thus require the aid of a cofactor (TPP). The mechanism shown in Figure 13.8 for pyruvate decarboxylation is involved in all of these reactions. Note that TPP contains two heterocyclic rings: a substituted pyrimidine and a thiazole. Recent NMR studies have shown that both rings participate in the formation of a reactive carbanion at C-2 of the thiazole ring—the carbon atom between the nitrogen and the sulfur (step 1 in Figure 13.8). Because the thiazole ring is weakly acidic ($pK_a \sim 18$ for the ring hydrogen), a general base is required in its ionization. As shown in the following diagram, a glutamate carboxyl group in the enzyme deprotonates N-1 of the pyrimidine, which in turn increases the basicity of the amino group, facilitating the deprotonation of C-2 of the thiazole ring to give a dipolar thiazolium carbanion (or **ylid**).

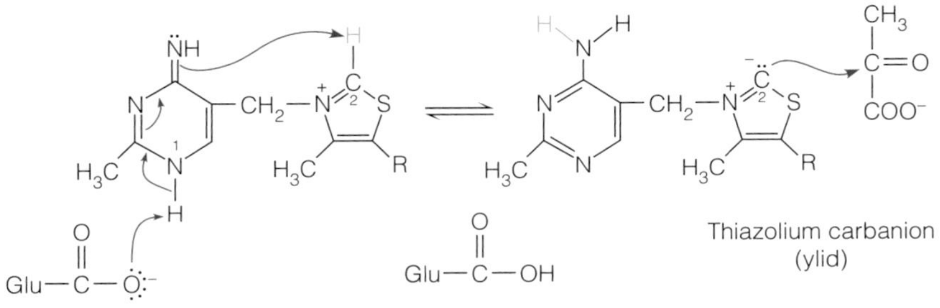

> The thiazole ring of TPP is the functional part of the coenzyme, allowing it to bind and transfer activated aldehydes.

Thiazolium carbanion (ylid)

This carbanion can attack the carbonyl carbon of α-keto acids, such as pyruvate, giving an addition compound (step 2 in Figure 13.8). The addition compound undergoes nonoxidative decarboxylation (step 3), with the thiazole ring acting as an electron sink in forming a resonance-stabilized carbanion. Protonation (step 4) gives a species called **active acetaldehyde** or, more accurately, **hydroxyethyl-TPP**. In pyruvate decarboxylase, this intermediate undergoes an elimination reaction to yield acetaldehyde and the TPP carbanion (step 5). In the pyruvate dehydrogenase reaction (not shown here), the activated two-carbon fragment is simultaneously oxidized and transferred to another enzyme, as discussed in Chapter 14. Thus, in general terms, TPP functions in the generation of an activated aldehyde species, which may or may not undergo oxidation as it is transferred to an acceptor. We will come across this coenzyme several times in other pathways.

Industrial production of ethanol has assumed immense importance as humanity attempts to deal with two serious problems—(1) replacement of nonrenewable petroleum by renewable fuel sources and (2) utilization of biological waste materials. Efforts in these arenas involve bioengineering, to generate bacterial strains that can convert materials such as the cellulose in wood waste or straw or more complex materials in human and animal wastes to hexose sugars, and analysis of glycolytic regulation, aimed at maximizing ethanol production from such sugars. Systems-level proteomics (see Tools of Biochemistry 5D) and metabolomics (see Tools of Biochemistry 12B) approaches are revealing new engineering targets for optimizing industrial cellulosic fermentation processes. However, there are some significant downsides to the use of croplands for the production of biofuels. For example, "first generation" biofuels produced from food crops such as corn may lead to increases in the price of corn and other food

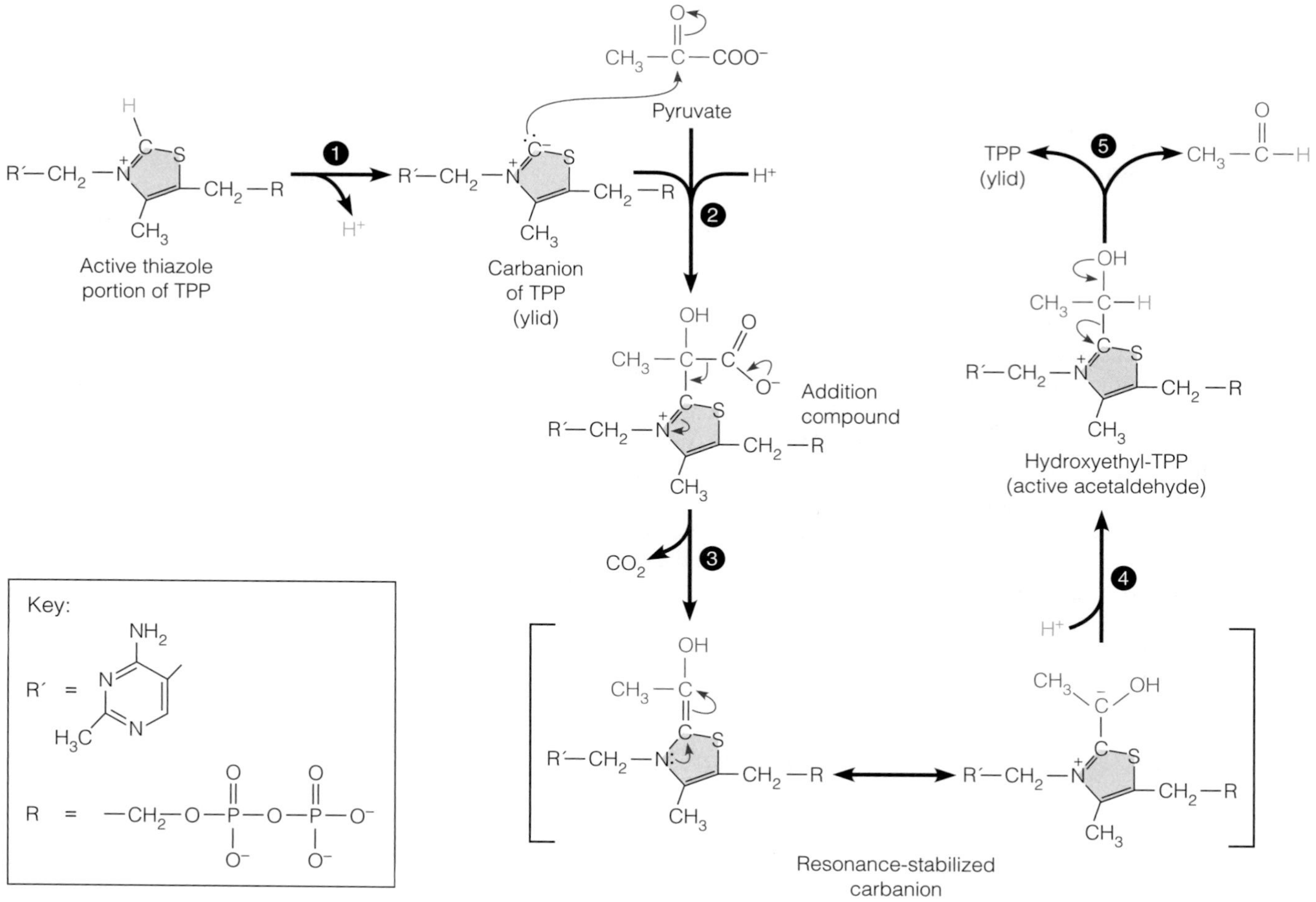

FIGURE 13.8

Thiamine pyrophosphate in the pyruvate decarboxylase reaction. Thiamine pyrophosphate (TPP) is the coenzyme for all decarboxylations of α-keto acids. The key reaction (step 2) is attack by the carbanion of TPP on the carbonyl carbon of pyruvate and is followed by nonoxidative decarboxylation of the coenzyme-bound pyruvate to give another carbanion (step 3), which is resonance-stabilized. In step 4, the two-carbon fragment (red) bound to TPP extracts a proton from the enzyme, generating a hydroxyethyl group. This fragment remains at the aldehyde oxidation level. In the final step, acetaldehyde is released and TPP carbanion is regenerated (step 5).

staples, excess fertilizer use, and soil depletion. These concerns have stimulated the development of "second generation" biofuels produced from nonfood biomass (ligno-cellulosic feedstocks), such as agricultural by-products and switchgrass. However, all of these approaches will result in increased water demand for irrigation, increased water pollution, and even increased greenhouse emissions as farmers respond to higher prices by converting forest and grassland to new cropland. Thus, many technical hurdles remain before biofuels can become commercially competitive with oil products and environmentally friendly at the same time.

Animal tissues also contain alcohol dehydrogenase, even though ethanol is not a major metabolic product in animal cells. Some of the major metabolic consequences of ethanol intoxication result from ethanol oxidation by this enzyme in the liver. First, there is massive reduction of NAD^+ to NADH, which depletes NAD^+ levels, thereby decreasing flux through glyceraldehyde-3-phosphate dehydrogenase, with consequent inhibition of energy generation. Second, acetaldehyde is quite toxic, and many of the unpleasant effects of hangovers result from actions of acetaldehyde and its metabolites.

Energy and Electron Balance Sheets

By writing a balanced chemical equation for glycolysis, we can compute the energy yield accompanying conversion of 1 mole of glucose. For homolactic fermentation, we can write the following balanced equation:

$$\text{Glucose} + 2ADP + 2P_i + 2H^+ \longrightarrow 2\,\text{lactate} + 2ATP + 2H_2O$$

Similarly, we can write a balanced equation for alcoholic fermentation:

$$\text{Glucose} + 2ADP + 2P_i + 4H^+ \longrightarrow 2\,\text{ethanol} + 2CO_2 + 2ATP + 2H_2O$$

Note first that both processes involve no net change in oxidation state; NAD^+ and NADH, both of which participate in the reaction pathways, do not appear in the overall reactions. This is an example of the obligate-coupling stoichiometry that was first introduced in Chapter 12. You can see that metabolism of glucose to either lactate or ethanol represents a nonoxidative process, by comparing the empirical formulas for glucose ($C_6H_{12}O_6$) and lactate ($C_3H_6O_3$). Clearly, there is no change in the overall oxidation state of the carbons because the numbers of hydrogens and oxygens bound per carbon atom are identical for glucose and lactate. The same is true for ethanol plus CO_2, when one counts the atoms in both. However, some individual carbon atoms of lactate, and of ethanol plus CO_2, undergo oxidation, and some become reduced. By contrast, pyruvate is more highly oxidized than glucose, as seen from its empirical formula ($C_3H_4O_3$).

Next, note that both of these balanced equations represent an exergonic process coupled to an endergonic process. In alcoholic fermentation, for example:

Exergonic: $\text{Glucose} \longrightarrow 2\,\text{ethanol} + 2CO_2 \qquad \Delta G^{\circ\prime} = \qquad -228\,\text{kJ/mol glucose}$

Endergonic: $2ADP + 2P_i \longrightarrow 2ATP + 2H_2O \qquad \underline{\Delta G^{\circ\prime} = 2 \times 30.5 = 61\,\text{kJ/mol glucose}}$

$\Delta G^{\circ\prime}_{\text{Sum}} = \qquad -167\,\text{kJ/mol glucose}$

Thus, the efficiency of the process, the amount of released free energy actually captured in ATP, is $61/228 = 27\%$ under standard conditions. The rest of the free energy is used to ensure that the process goes to completion. The equilibrium constant K for this process can be calculated from Equation 3.28 in Chapter 3:

$$K = e^{\left(\frac{-\Delta G^{\circ\prime}}{RT}\right)} = 1.9 \times 10^{29}$$

Clearly, glycolysis is highly favorable, and does indeed go to completion under standard conditions. What about physiological conditions? Table 13.1 lists ΔG values for the individual steps, estimated from the approximate intracellular concentrations of glycolytic intermediates in rabbit skeletal muscle. These estimated ΔG values are plotted in Figure 13.9 to illustrate the actual free energy changes in the pathway. All but three of the reactions function at or near equilibrium, and are freely reversible in vivo. The three exceptions are the reactions catalyzed by hexokinase, phosphofructokinase, and pyruvate kinase, which take place with large decreases in free energy. These nonequilibrium reactions are irreversible in vivo, and they make the entire glycolytic pathway unidirectional. As we will see in the next section, these are the steps that must be bypassed in gluconeogenesis. Not coincidentally, these three nonequilibrium reactions are also the sites of regulation of glycolysis—recall from Chapter 12 that regulation can be imposed *only* on reactions displaced far from equilibrium.

This 10-step pathway is fast—anaerobic glycolysis can produce ATP at rates 100 times higher than aerobic oxidative phosphorylation. Muscle cells take advantage of this during strenuous exercise, when their oxygen supply cannot keep up with their demand for ATP. Cancer cells also take advantage of the high rate of ATP production afforded by glycolysis, supporting their high proliferation rates. Indeed, most rapidly dividing cancer cells metabolize glucose by glycolysis, producing lactate even though oxygen is abundant. This phenomenon was first noted

Glycolysis, which yields 2 ATP per glucose, is fast but releases only a small fraction of the energy available from glucose.

FIGURE 13.9

Energy profile of anaerobic glycolysis. The graph shows the change in actual free energy for each reaction in the pathway, based on estimated ΔG values calculated in Table 13.1. Metabolite and enzyme abbreviations are as defined in Table 13.1. Most of the reactions function at or near equilibrium and are freely reversible in vivo. The three enzymes that catalyze reactions so highly exergonic as to be virtually irreversible (arrows) are subject to allosteric control.

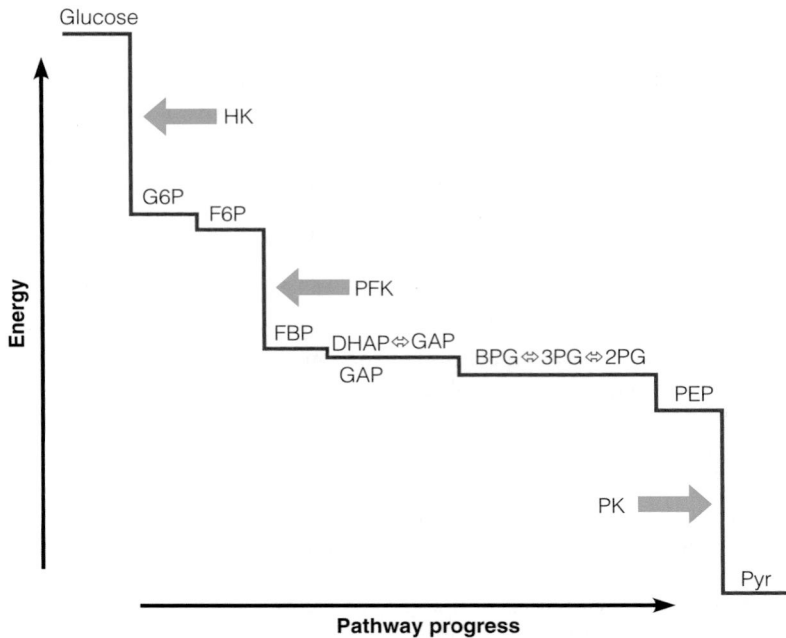

by Otto Warburg in 1925, and it is known as the "Warburg effect"—more about this in Chapter 18. Of course, this high rate of ATP production via glycolysis requires a high rate of glucose utilization, since only two ATPs are produced per glucose. This brings us to a final point about the bioenergetics of glycolysis.

Glycolysis releases but a small fraction of the potential energy stored in the glucose molecule. As noted earlier (see Chapters 3 and 12), the complete combustion of glucose to CO_2 and H_2O releases 2870 kJ/mol of free energy under standard conditions. Complete combustion of 2 moles of lactate to CO_2 and H_2O releases 2×1379 kJ/mol = 2757 kJ/mol of free energy under standard conditions. Thus, in homolactic fermentation, 2757/2870, or 96%, of the free energy available in the original glucose molecule is still present in the fermentation product, lactate. Alcoholic fermentation is similarly low yield.

As we shall see in Chapter 15, about 30–32 moles of ATP are synthesized from ADP per mole of glucose carried completely through glycolysis and the citric acid cycle. Aerobic metabolism yields more energy from glucose; therefore, aerobic organisms in general are more successful and widespread than anaerobic organisms. The early evolution of aerobic metabolism made possible the large, active animals that exist today. Nevertheless, many large animals still derive a large fraction of their metabolic energy from glycolysis, under certain physiological circumstances. A good example is the crocodile—torpid (and aerobic) for much of its life, yet capable of short bursts of intensely rapid movement. In the latter circumstance glycolysis, coupled with the breakdown of carbohydrate energy stores, represents a rapid, though inefficient, way to mobilize energy.

Gluconeogenesis

Gluconeogenesis is the synthesis of glucose from noncarbohydrate precursors. Here we encounter the first instance of a principle enunciated in Chapter 12: *Biosynthetic processes are never simply the reversal of corresponding catabolic pathways.* Superficially, gluconeogenesis looks very much like glycolysis in reverse, but different enzymatic reactions are used at crucial sites. These sites are strongly exergonic reactions that are controlled largely in reciprocal fashion, so that physiological conditions that activate glycolysis inhibit gluconeogenesis and vice versa. Much the same picture will emerge from our discussion later in this chapter of glycogen synthesis as compared with glycogen mobilization.

Physiological Need for Glucose Synthesis in Animals

Most animal organs can metabolize a variety of carbon sources to generate energy—triacylglycerols, various sugars, pyruvate, amino acids, and so forth. However, the brain and central nervous system require glucose as the sole or primary carbon source. The same is true for some other tissues, such as kidney medulla, testes, and erythrocytes (Figure 13.10). Consequently, animal cells must be able to synthesize glucose from other precursors and also to maintain blood glucose levels within narrow limits—both for proper functioning of the brain and central nervous system and also for providing precursors for glycogen storage in other tissues. The glucose requirements of the human brain are relatively enormous— 120 grams per day, out of about 160 grams needed by the entire body. The amount of glucose that can be generated from the body's glycogen reserves at any time is about 190 grams, and the total amount of glucose in body fluids is little more than 20 grams. Thus, the readily available glucose reserves amount to about one day's supply. During periods of fasting for more than one day, glucose must be formed from other precursors. In the overnight post-absorptive state, glycogenolysis and gluconeogenesis both contribute to overall glucose production. However, if you skip breakfast your glycogen stores become depleted and gluconeogenesis becomes the predominant process for maintaining blood glucose levels. The same occurs during intense exertion, for example, during a marathon run. Initially, hepatic glycogenolysis is the primary source of extra glucose for skeletal muscle, but hepatic gluconeogenesis becomes gradually more important as the glycogen stores deplete.

The biosynthetic process is called **gluconeogenesis**—literally, the production of new glucose. Gluconeogenesis is defined as the biosynthesis of carbohydrate from three-carbon and four-carbon precursors, generally noncarbohydrate in nature. The principal substrates for gluconeogenesis are *lactate*, produced primarily from glycolysis in skeletal muscle and erythrocytes; *amino acids*, generated from dietary protein or from the breakdown of muscle protein during starvation; the specific amino acid *alanine*, produced in muscle through the glucose–alanine cycle (see Figure 20.13, page 845); *propionate*, derived from the breakdown of fatty acids with odd-numbered carbon chains and some amino acids; and *glycerol*, derived from the catabolism of fats. The fatty acids released during lipid breakdown are mostly converted to acetyl-CoA and cannot be used for carbohydrate synthesis, except in organisms that have a functioning glyoxylate cycle (see Figure 14.21, page 620).

Gluconeogenesis occurs primarily in the cytosol, although some precursors are generated in mitochondria and must be transported to the cytosol to be utilized. The primary gluconeogenic organ in animals is the liver, with the kidney cortex contributing in a lesser but still significant way (see Figure 13.10). The major fates of glucose formed by gluconeogenesis are catabolism by nervous tissue and utilization by skeletal muscle. In addition, glucose is the primary precursor for all other carbohydrates, including amino sugars, complex polysaccharides, and the carbohydrate components of glycoproteins and glycolipids. The need for glucose as a biosynthetic intermediate means that gluconeogenesis is an important pathway in plants and microorganisms, as well as in animals, and the pathway is essentially identical in all organisms. However, the wealth of information on control of gluconeogenesis in animals leads us to concentrate on animal metabolism for the first part of this chapter.

Enzymatic Relationship of Gluconeogenesis to Glycolysis

Gluconeogenesis should be an easy pathway to learn because it closely resembles glycolysis in reverse. However, there are some important differences, which allow the pathway to run in the direction of glucose *synthesis* in the cell.

Recall that a metabolic pathway can proceed smoothly only if ΔG is strongly negative for the overall pathway in the direction written. We just saw that glycolysis from glucose to pyruvate is strongly exergonic; under typical intracellular conditions ΔG is about -103 kJ/mol (Table 13.1). How, then, can the conversion of pyruvate to

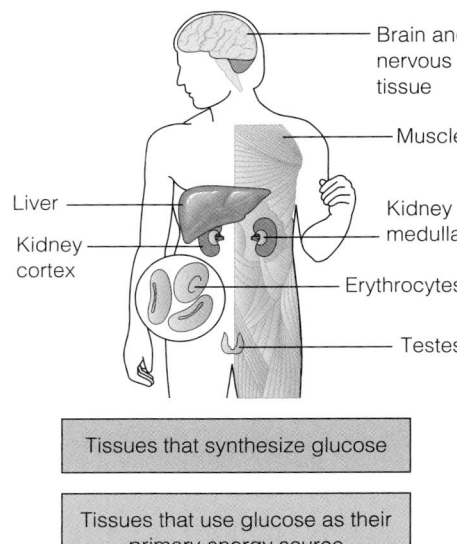

FIGURE 13.10

Synthesis and use of glucose in the human body. Liver and kidney cortex are the primary gluconeogenic tissues. Brain, skeletal muscle, kidney medulla, erythrocytes, and testes use glucose as their sole or primary energy source, but they lack the enzymatic machinery to synthesize it.

Synthesis of glucose from noncarbohydrate precursors is essential for maintenance of blood glucose levels within acceptable limits.

Gluconeogenesis uses specific enzymes to bypass three irreversible reactions of glycolysis.

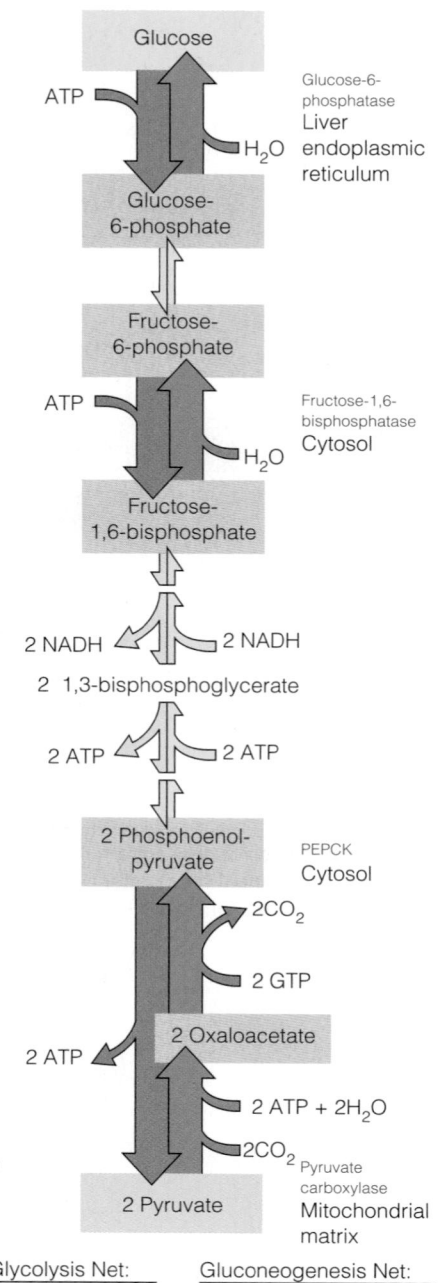

Glycolysis Net:
+ 2ATP + 2NADH

Gluconeogenesis Net:
− 4ATP − 2GTP − 2NADH

FIGURE 13.11

Reactions of glycolysis and gluconeogenesis. Irreversible reactions of glycolysis are shown in dark purple. The opposed reactions in gluconeogenesis, which bypass these steps, are shown in dark blue. Pale arrows identify reversible reactions used in both pathways.

glucose be made exergonic? Recall that three reactions of the glycolytic pathway are so strongly exergonic as to be essentially irreversible—those catalyzed by hexokinase, phosphofructokinase, and pyruvate kinase (Figure 13.9). In gluconeogenesis, different chemistry and enzymes are used at each of these steps, so that, for example, the conversion of fructose-1,6-bisphosphate to fructose-6-phosphate is not simply a reversal of the phosphofructokinase reaction. In essence, the three irreversible reactions of glycolysis are bypassed by enzymes specific to gluconeogenesis, which catalyze quite different reactions that run strongly in the direction of glucose synthesis. This biosynthetic process involves a substantial energy cost, which must be paid if the overall process is to be thermodynamically favored. The remaining seven reactions of gluconeogenesis are catalyzed by the glycolytic enzymes that catalyze reversible reactions and are driven in either direction by mass action (near-equilibrium reactions; see Figure 13.9). Another way to relate glycolysis to gluconeogenesis is to say that they differ only at three steps controlled by *substrate cycles* (see page 481).

The entire gluconeogenic pathway, from pyruvate to glucose, is summarized in Figure 13.11. We focus here on the reactions that bypass the three irreversible steps in glycolysis.

Bypass 1: Conversion of Pyruvate to Phosphoenolpyruvate

The bypass of pyruvate kinase begins in the mitochondrion and involves two reactions. **Pyruvate carboxylase** catalyzes the ATP- and biotin-dependent conversion of pyruvate to oxaloacetate. The structure and mechanism of this enzyme will be discussed in more detail in Chapter 14 (see page 616). The enzyme requires acetyl-CoA as an allosteric activator:

$$\begin{array}{c} CH_3 \\ | \\ C{=}O \\ | \\ COO^- \end{array} + HCO_3^- + ATP \;\rightleftharpoons\; \begin{array}{c} COO^- \\ | \\ CH_2 \\ | \\ C{=}O \\ | \\ COO^- \end{array} + ADP + P_i + 2H^+ \qquad \Delta G^{\circ\prime} = -2.1\ \text{kJ/mol}$$

Pyruvate   **Oxaloacetate**

As we shall see in the next chapter, this is one of the **anaplerotic** ("filling up") reactions used to maintain levels of citric acid cycle intermediates (see page 616). Pyruvate carboxylase generates oxaloacetate in the mitochondrial matrix, where it can be oxidized in the citric acid cycle. To be used for gluconeogenesis, oxaloacetate must move out of the mitochondrion to the cytosol, where the remainder of the pathway occurs. However, the mitochondrial membrane does not have an effective transporter for oxaloacetate. Therefore, oxaloacetate is reduced by mitochondrial malate dehydrogenase to malate, which is transported into the cytosol by exchange for orthophosphate and then reoxidized by cytosolic malate dehydrogenase.

Once in the cytosol, oxaloacetate is acted on by **phosphoenolpyruvate carboxykinase** (abbreviated **PEPCK**) to give phosphoenolpyruvate:

Oxaloacetate + GTP $\rightleftharpoons$ phosphoenolpyruvate + CO_2 + GDP $\Delta G^{\circ\prime} = +2.9\ \text{kJ/mol}$

Note the use of GTP, rather than ATP, as an energy donor. Note also that the same CO_2 that was fixed by pyruvate carboxylase is released in this reaction, so that no net fixation of CO_2 occurs. The PEPCK reaction requires Mg^{2+} or Mn^{2+} and is readily reversible. In the reaction the carboxyl group formed from the transferred CO_2 provides electrons to facilitate O — P bond formation:

Eukaryotes express two distinct isozymes of PEPCK: one in the cytosol and one in mitochondria. The cytosolic form appears to be the more important player in gluconeogenesis, as shown by the fact that its intracellular concentration is regulated by hormonal conditions known to control gluconeogenesis. However, in the livers of most mammals (including humans), at least half of the total PEPCK activity is mitochondrial. Consideration of the redox requirements and the compartmentation of gluconeogenesis suggests a metabolic explanation for the two isozymes. As shown in Figure 13.11, gluconeogenesis requires electrons in the form of NADH to reduce 1,3-bisphosphoglycerate to glyceraldehyde-3-phosphate. However, the cytosolic $[\text{NADH}]/[\text{NAD}^+]$ ratio is about 10^5 times lower than in mitochondria. In those tissues where the cytosolic PEPCK predominates, transport of malate from mitochondria to cytosol and the operation of the two malate dehydrogenase isozymes results in the transport of mitochondrial reducing equivalents to the cytosol for use in the glyceraldehyde-3-phosphate dehydrogenase reaction (Figure 13.12). However, if lactate is available as a gluconeogenic precursor, its oxidation to pyruvate by the cytoplasmic lactate dehydrogenase provides the necessary cytoplasmic reducing equivalents. Oxaloacetate produced by pyruvate carboxylase can now be converted directly to PEP by the mitochondrial PEPCK, since mitochondria possess a transporter for PEP.

The overall reaction for the bypass of pyruvate kinase is as follows:

$$\text{Pyruvate} + \text{ATP} + \text{GTP} \longrightarrow \text{phosphoenolpyruvate} + \text{ADP} + \text{GDP} + \text{P}_i + \text{H}^+ \qquad \Delta G^{\circ\prime} = +0.8\,\text{kJ/mol}$$

$\Delta G^{\circ\prime}$ for the two reactions combined is slightly positive. However, under intracellular conditions the sequence is quite exergonic, with ΔG of about $-25\,\text{kJ/mol}$. As shown in the summary reaction, two high-energy phosphates must be invested for the synthesis of one phosphoenolpyruvate. After this bypass, phosphoenolpyruvate is converted to fructose-1,6-bisphosphate by glycolytic enzymes acting in reverse. Note that the glyceraldehyde-3-phosphate dehydrogenase reaction requires NADH (Figure 13.11), provided by one of the alternative pathways just described (Figure 13.12).

Bypass 2: Conversion of Fructose-1,6-bisphosphate to Fructose-6-phosphate

The phosphofructokinase reaction of glycolysis is essentially irreversible, but only because it is driven by phosphate transfer from ATP. A bypass reaction in gluconeogenesis involves a simple hydrolytic reaction, catalyzed by **fructose-1,6- bisphosphatase**.

$$\text{Fructose-1,6-bisphosphate} + \text{H}_2\text{O} \xrightarrow{\text{Mg}^{2+}} \text{fructose-6-phosphate} + \text{P}_i \qquad \Delta G^{\circ\prime} = -16.3\,\text{kJ/mol}$$

The negative $\Delta G^{\circ\prime}$ for this reaction favors the rightward reaction. The multisubunit enzyme requires Mg^{2+} for activity and represents one of the major control sites regulating the overall gluconeogenic pathway. Fructose-6-phosphate formed in this reaction then undergoes isomerization by phosphoglucoisomerase to glucose-6-phosphate.

Bypass 3: Conversion of Glucose-6-phosphate to Glucose

Glucose-6-phosphate cannot be converted to glucose by reverse action of hexokinase or glucokinase because of the high positive $\Delta G^{\circ\prime}$ of that reaction; phosphate transfer from ATP makes that reaction virtually irreversible. Another enzyme specific to gluconeogenesis, **glucose-6-phosphatase**, comes into play instead. This bypass reaction also involves a simple hydrolysis.

$$\text{Glucose-6-phosphate} + \text{H}_2\text{O} \xrightarrow{\text{Mg}^{2+}} \text{glucose} + \text{P}_i \qquad \Delta G^{\circ\prime} = -13.8\,\text{kJ/mol}$$

Glucose-6-phosphatase, which also requires Mg^{2+}, is found in the endoplasmic reticulum (ER) membrane, with its active site facing the interior space, or **lumen**, of the ER. The enzyme is part of a multiprotein complex that includes transporters to import the substrate, glucose-6-phosphate, into the ER lumen and to export the products, glucose and inorganic phosphate, back to the cytoplasm.

FIGURE 13.12

PEPCK isozymes provide alternative routes to PEP and cytoplasmic reducing equivalents. MDH, malate dehydrogenase; LDH, lactate dehydrogenase. Blue ovals indicate inner membrane transporters.

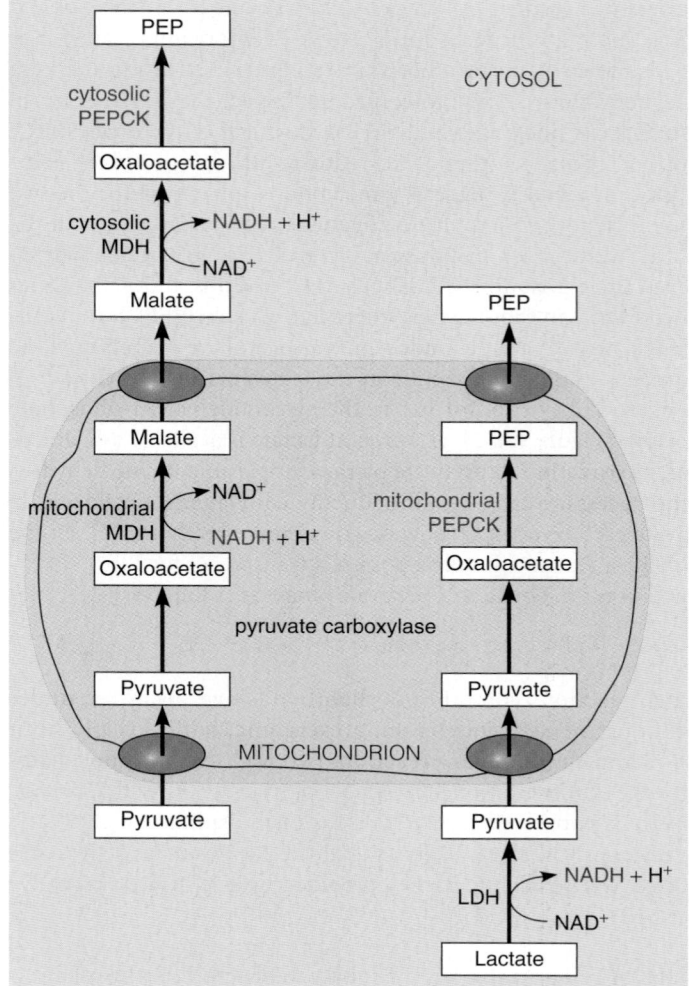

Mammals express three glucose-6-phosphatase isozymes. The major isozyme is expressed predominantly in liver and kidney, and the minor isozymes are expressed at lower levels in most other tissues, including brain, muscle, intestine, and pancreatic islets. The significance of its location in liver is that a unique function of liver is to synthesize glucose for export to the tissues via the bloodstream. The role of glucose-6-phosphatase in other tissues, especially brain and skeletal muscle, is not yet understood.

Stoichiometry and Energy Balance of Gluconeogenesis

We have emphasized that catabolic pathways generate energy, whereas anabolic pathways carry an energy cost. Let's now estimate that cost for gluconeogenesis. The overall conversion of 2 moles of pyruvate to 1 mole of glucose is quite exergonic, as shown in Table 13.2. $\Delta G^{\circ\prime}$ for the overall process is about -33 kJ/mol.

Gluconeogenesis:

$$2\,\text{Pyruvate} + 4\text{ATP} + 2\text{GTP} + 2\text{NADH} + 2\text{H}^+ + 4\text{H}_2\text{O} \longrightarrow \text{glucose} + 4\text{ADP}$$
$$+2\text{GDP} + 6\text{P}_\text{i} + 2\text{NAD}^+ \qquad\qquad \Delta G^{\circ\prime} = -33\,\text{kJ/mol}$$

The equivalent of 11 high-energy phosphates are consumed per mole of glucose synthesized by gluconeogenesis.

But the synthesis of glucose is energetically expensive. Six high-energy phosphate groups are consumed (four ATPs and two GTPs), as well as 2 moles of NADH, which is the energetic equivalent of five more ATPs (because mitochondrial oxidation of 1 mole of NADH generates ~2.5 moles of ATP).

TABLE 13.2   Summary of gluconeogenesis, from pyruvate to glucose

Reaction	$\Delta G^{\circ\prime}$ (kJ/mol)
Pyruvate + HCO_3^- + ATP $\longrightarrow$ oxaloacetate + ADP + P_i	$-2.1\ (-4.2)$
Oxaloacetate + GTP $\rightleftharpoons$ phosphoenolpyruvate + CO_2 + GDP	$+2.9\ (+5.8)$
Phosphoenolpyruvate + H_2O $\rightleftharpoons$ 2-phosphoglycerate	$+6.4\ (+12.8)$
2-Phosphoglycerate $\rightleftharpoons$ 3-phosphoglycerate	$-4.4\ (-8.8)$
3-Phosphoglycerate + ATP $\rightleftharpoons$ 1,3-bisphosphoglycerate + ADP	$+18.8\ (+37.6)$
1,3-Bisphosphoglycerate + NADH + H^+ $\rightleftharpoons$ glyceraldehyde-3-phosphate + NAD^+ + P_i	$-6.3\ (-12.6)$
Glyceraldehyde-3-phosphate $\rightleftharpoons$ dihydroxyacetone phosphate	-7.6
Glyceraldehyde-3-phosphate + dihydroxyacetone phosphate $\rightleftharpoons$ fructose-1,6-bisphosphate	-23.9
Fructose-1,6-bisphosphate + H_2O $\longrightarrow$ fructose-6-phosphate + P_i	-16.3
Fructose-6-phosphate $\rightleftharpoons$ glucose-6-phosphate	-1.7
Glucose-6-phosphate + H_2O $\longrightarrow$ glucose + P_i	-13.8
Net: 2 Pyruvate + 4ATP + 2GTP + 2NADH + $2H^+$ + $4H_2O$ $\longrightarrow$ glucose + 4ADP + 2GDP + $6P_i$ + $2NAD^+$	-32.7

Note: The reactions in purple type are those that bypass irreversible glycolytic reactions; the remaining reactions are reversible reactions of glycolysis. The $\Delta G^{\circ\prime}$ values in parentheses are based on doubling the first six reactions because 2 three-carbon precursors are required to make one molecule of glucose. The individual reactions are not necessarily balanced for H^+ and charge.

By contrast, if glycolysis could operate in reverse, the net equation would show far less energy input—2 moles of NADH and 2 moles of high-energy phosphate:

Reversal of Glycolysis:

$$2\,\text{Pyruvate} + 2\text{ATP} + 2\text{NADH} + 2H^+ + 2H_2O \longrightarrow \text{glucose} + 2\text{ADP} + 2P_i + 2NAD^+$$

$$\Delta G^{\circ\prime} = +83.1\ \text{kJ/mol}$$

This process would be highly endergonic, however, with a $\Delta G^{\circ\prime}$ of $+83.1$ kJ/mol. These are of course standard free energy changes, but it is clear that the investment of four extra high-energy phosphate bonds is essential if the net synthesis of glucose is to occur as an irreversible process in vivo.

Substrates for Gluconeogenesis

As indicated earlier, gluconeogenesis draws precursors from diverse sources, including lactate, amino acids, glycerol, and propionate. The pathways by which these substrates enter gluconeogenesis are shown in Figure 13.13 and are discussed in this section.

Lactate

In quantitative terms, lactate is the most significant gluconeogenic precursor. Recall that skeletal muscle derives much of its energy from glycolysis, particularly during intense exertion, when respiration cannot deliver sufficient oxygen to the tissues for complete oxidation of glucose. Under these conditions, glycolysis produces pyruvate more rapidly than it can be further metabolized via the citric acid cycle. Lactate dehydrogenase is abundant in muscle, and the equilibrium strongly favors pyruvate reduction to lactate. Thus, lactate from working muscle is released to the blood, whence it is readily taken up by the heart and oxidized as fuel. The accumulation of lactate during prolonged exertion is a significant factor limiting athletic performance.

Some of the lactate produced in muscle enters the liver and is reoxidized to pyruvate by liver LDH. This pyruvate can then undergo gluconeogenesis to give glucose, which is returned to the bloodstream and taken up by muscle to regenerate the glycogen stores. This process, described originally by Carl and Gerti Cori and appropriately called the **Cori cycle**, is schematized in Figure 13.14. The pathway is particularly

The most important gluconeogenic precursors are lactate, alanine, glycerol, and propionate.

FIGURE 13.13

Outline of pathways for glucose synthesis from the major gluconeogenic precursors. Note that both glucose and lactate are carried in the blood.

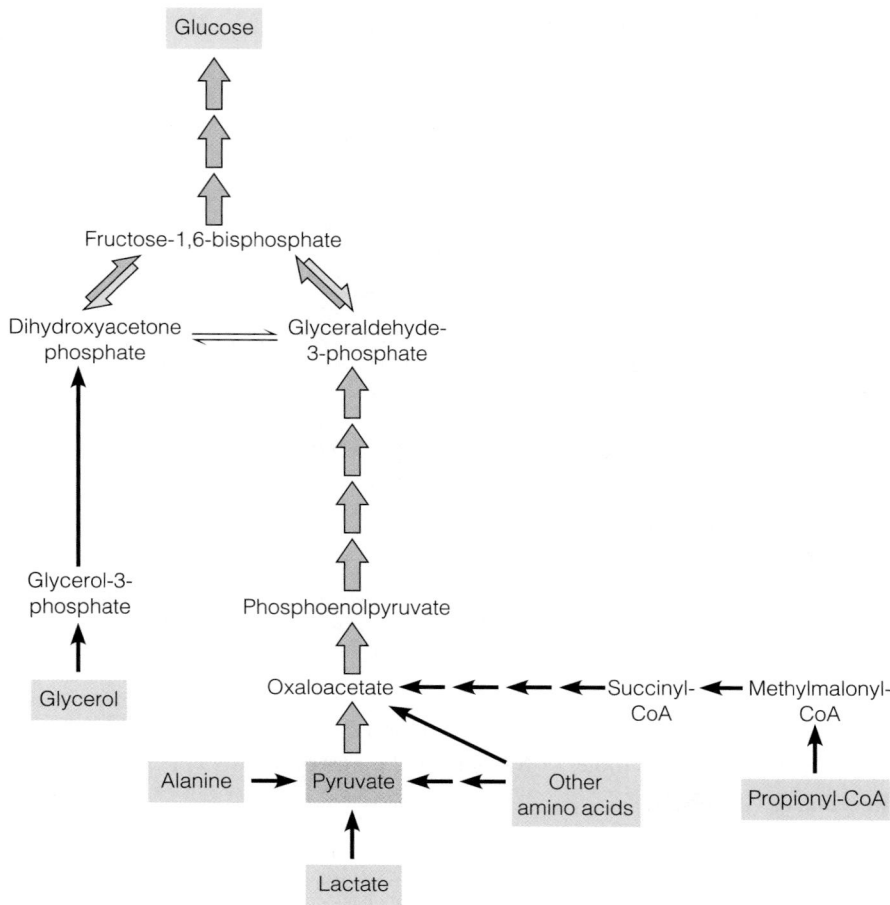

active during recovery from intense muscular exercise. During this time the breathing rate is elevated, and the increased oxidative metabolism generates more ATP, much of which is used to rebuild glycogen stores via gluconeogenesis.

In a parallel process, the **glucose–alanine cycle**, pyruvate in peripheral tissues undergoes transamination to alanine, which is returned to the liver and used for gluconeogenesis. This pathway, which is presented in detail in Chapter 20, helps tissues to dispose of toxic ammonia formed during protein degradation.

Amino Acids

Like alanine, many other amino acids can readily be converted to glucose, primarily through degradative pathways that generate citric acid cycle intermediates, which can be converted to oxaloacetate (see Figure 13.13). As we will learn in Chapter 21, such amino acids are called **glucogenic** (that is, able to be converted to glucose), although *gluconeogenic* is probably a more accurate term. Among the 20 amino acids found in proteins, only the catabolic pathways for leucine and lysine do not generate gluconeogenic precursors. During fasting, when insufficient carbohydrate is ingested, the catabolism of muscle proteins is the major source of intermediates needed to maintain normal blood glucose concentrations. The same is true in the disease diabetes mellitus, as discussed further in Chapter 18.

Glycerol

In general, lipids are poor gluconeogenic precursors. Catabolism of triacylglycerols yields fatty acids and glycerol. Fatty acids undergo β-oxidation to yield acetyl-CoA (Chapter 17). In animals, acetyl-CoA cannot be converted to pyruvate or to any other gluconeogenic precursor. Hence, *fatty acids cannot undergo net conversion to carbohydrates*. Although it is true that two-carbon units from acetyl-CoA can proceed to oxaloacetate in the citric acid cycle, there is no net conversion because two carbons

Animals cannot undergo net conversion of fat to carbohydrate.

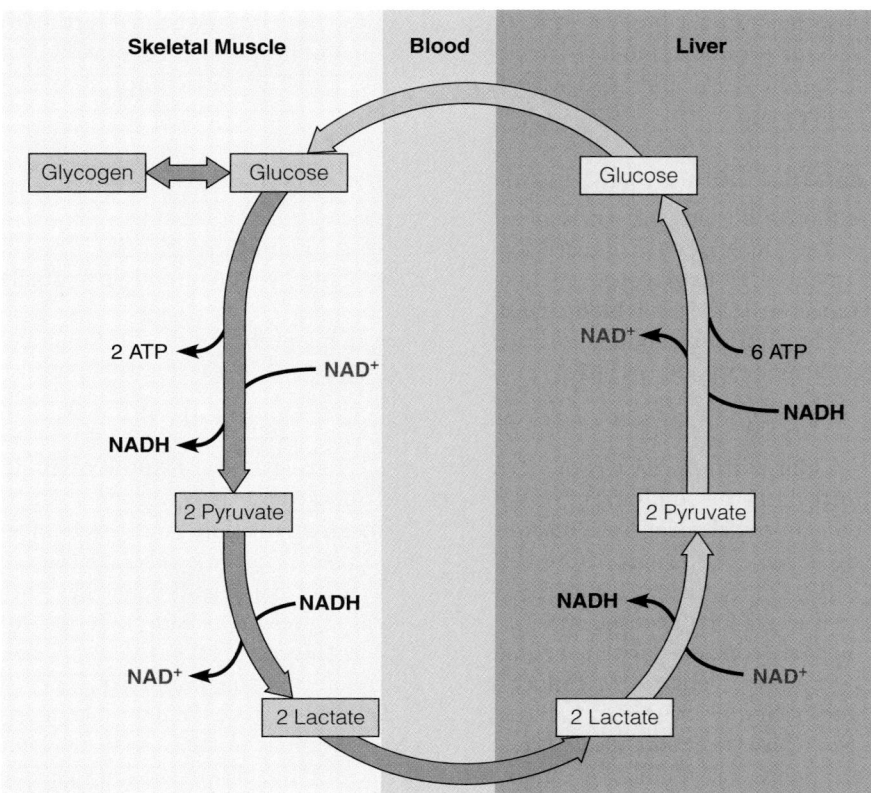

FIGURE 13.14

The Cori cycle. Lactate produced in glycolysis during muscle exertion is transported to the liver, for resynthesis of glucose by gluconeogenesis. Transport of glucose back to muscle for synthesis of glycogen, and its reutilization in glycolysis, completes the cycle.

are lost in each turn of the cycle. Therefore, aside from a minor contribution by odd-chain fatty acids (explained in the next section), the only fat breakdown product that can enter gluconeogenesis is the 3-carbon glycerol. Its utilization involves phosphorylation, followed by dehydrogenation to dihydroxyacetone phosphate (see Figure 13.13). Note that plants and bacteria *can* incorporate acetyl-CoA into carbohydrate via the glyoxylate cycle, which we will discuss in Chapter 14.

Propionate

In all organisms a three-carbon acyl-CoA, **propionyl-CoA**, is generated either from the breakdown of some amino acids or from the oxidation of fatty acids with odd numbers of carbon atoms. Propionyl-CoA enters gluconeogenesis via its conversion to succinyl-CoA and then to oxaloacetate. The process, which is detailed in Chapter 17, involves a coenzyme derived from vitamin B_{12}.

Propionyl-CoA → (CO_2) → **Methylmalonyl-CoA** → **Succinyl-CoA**

Although all organisms use propionate as a gluconeogenic substrate, it is particularly important in the metabolism of ruminant animals such as cattle. These animals display enormous rates of gluconeogenesis from a variety of substrates, owing to the great amount of bacterial fermentation occurring in their several stomach chambers. In cattle the four chambers of the stomach have a total volume as high as 70 liters. The action of various bacteria degrades plant materials, particularly cellulose, to glucose. But before the glucose can be absorbed into the bloodstream, as in human digestion, it is fermented further to various products, notably lactate and propionate. Propionate is converted to propionyl-CoA and thence to succinyl-CoA, and lactate is simply reduced to pyruvate.

Ethanol Consumption and Gluconeogenesis

Although it is possible to visualize pathways by which ethanol could be converted to glucose, ethanol is actually a poor gluconeogenic precursor. In fact, ethanol strongly inhibits gluconeogenesis and can bring about **hypoglycemia**, a potentially dangerous decrease in blood glucose levels.

Ethanol is metabolized primarily in the liver, by reversal of the alcohol dehydrogenase reaction (page 521):

$$\text{Ethanol} + \text{NAD}^+ \rightleftharpoons \text{acetaldehyde} + \text{NADH} + \text{H}^+$$

This reaction elevates the $[\text{NADH}]/[\text{NAD}^+]$ ratio in liver cytosol, which in turn shifts the equilibrium of the lactate dehydrogenase and glyceraldehyde 3-phosphate dehydrogenase reactions, inhibiting glycolysis. The same mechanism shifts the equilibrium of the cytosolic malate dehydrogenase reaction (Figure 13.12), so that oxaloacetate tends to be reduced to malate and hence becomes unavailable for gluconeogenesis. The resultant hypoglycemia can affect the parts of the brain concerned with temperature regulation. This response, in turn, can lower the body temperature by as much as 2 °C. Therefore, the time-honored practice of feeding brandy or whiskey to those rescued from cold or wet conditions is counterproductive. To be sure, alcohol creates a sense of warming through vasodilation, but this peripheral vasodilation causes further heat loss. Metabolically speaking, glucose would be far more effective in raising body temperature.

Roles of Extrahepatic Phosphoenolpyruvate Carboxykinase

In addition to its role in liver, the predominant gluconeogenic tissue, PEPCK is important in the metabolism of two other tissues. First, in kidney cortex, the other major gluconeogenic tissue, PEPCK participates in acid–base regulation. The kidney is responsible for regulating acid–base balance through the synthesis and excretion of ammonium (NH_4^+) in the urine. The primary source of ammonia is glutamine, which generates 2 moles of ammonia via the sequential action of **glutaminase** and **glutamate dehydrogenase**:

$$\text{Glutamine} + \text{H}_2\text{O} \longrightarrow \text{glutamate} + \text{NH}_4^+$$

$$\text{Glutamate} + \text{NAD}^+ + \text{H}_2\text{O} \longrightarrow \alpha\text{-ketoglutarate} + \text{NH}_4^+ + \text{NADH}$$

During metabolic acidosis, such as occurs in diabetes, excretion of ammonium ions into the urine facilitates the excretion of metabolically produced acids to compensate for the acidosis. The α-ketoglutarate produced in the second reaction is converted via the citric acid cycle to oxaloacetate, which generates phosphoenolpyruvate via the PEPCK reaction, for ultimate synthesis of glucose.

Second, in adipose (fatty) tissue, PEPCK participates in a process called **glyceroneogenesis**, which acts to produce sufficient glycerol-3-phosphate for triacylglycerol formation. Reesterification of free fatty acids to glycerol-3-phosphate is essential in both the fed and fasting states, in order to maintain balance between triacylglycerol breakdown and resynthesis. In adipose tissue, the phosphoenolpyruvate produced by pyruvate carboxylase and PEPCK is *not* converted to glucose. Instead, it is diverted at dihydroxyacetone phosphate for reduction to glycerol-3-phosphate, which combines with the coenzyme A derivatives of free fatty acids to produce triacylglyerols. These pathways will be discussed in more detail in Chapter 17.

Evolution of Carbohydrate Metabolic Pathways

Genome comparisons suggest that the Embden–Meyerhof–Parnas (glycolytic) pathway probably first evolved in the direction of gluconeogenesis. The earliest organisms were most likely thermophilic chemoautotrophs, living in the volcanic flows of submarine hydrothermal vents under anoxic or microaerobic conditions. It is proposed that these organisms used dissolved gases and inorganic substrates for energy production, which probably contained few carbohydrates for use as organic growth substrates. They reduced CO_2 or CO (carbon monoxide) into an activated form of acetic acid, acetyl-CoA, using electrons from H_2 as the main electron donor. This served as the starting material for the biosynthesis of more complex carbohydrates, using the primordial gluconeogenesis pathway. Only after large quantities of sugar-containing cyanobacterial and plant cell walls became available did organisms begin to use sugars as growth substrates, via the glycolytic direction of the pathway.

Coordinated Regulation of Glycolysis and Gluconeogenesis

Glycolysis and gluconeogenesis are closely coordinated with other major pathways of energy generation and utilization, notably synthesis and breakdown of glycogen (or starch), the pentose phosphate pathway (described later in this chapter), and the citric acid cycle (Chapter 14). Metabolic factors that control glycolysis tend to regulate other processes in a coordinated fashion. Thus, it is difficult to consider regulation of glycolysis in isolation from these other processes, and we return to this topic after we have presented the other major pathways in energy metabolism (see Chapter 18). However, glycolysis and gluconeogenesis provide a very useful introduction to the principles of coordinated metabolic regulation, and so we will describe here the key enzymes that serve as regulatory targets in these two opposing pathways.

The Pasteur Effect

Long before anything was known about pathways of glucose utilization, much less control mechanisms, Louis Pasteur observed that when anaerobic yeast cultures metabolizing glucose were exposed to air, the rate of glucose utilization decreased dramatically. It became clear that this phenomenon, known as the **Pasteur effect**, involves the inhibition of glycolysis by oxygen. This effect makes biological sense because far more energy is derived from complete oxidation of glucose than from glycolysis alone. What is the mechanism of this effect if oxygen is not an active participant in glycolysis? The needed insight came much later from analyses of the intracellular contents of glycolytic intermediates in aerobic and anaerobic cells. These analyses required techniques for the rapid interruption of metabolism and extraction of metabolites. One such technique is **freeze-clamping**, in which tissue is rapidly compressed between metal plates cooled to liquid nitrogen temperatures. The solid tissue can then be powdered and extracted for analysis.

Experiments of this type revealed that when oxygen is introduced to anaerobic cells, the levels of all the glycolytic intermediates from fructose-1,6-bisphosphate onward *decrease*, while all of the *earlier* intermediates accumulate at higher levels (see margin). This finding is consistent with the idea that the metabolic flux through phosphofructokinase is specifically decreased in the presence of O_2, probably due to changes in allosteric effector levels.

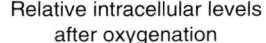
Relative intracellular levels after oxygenation

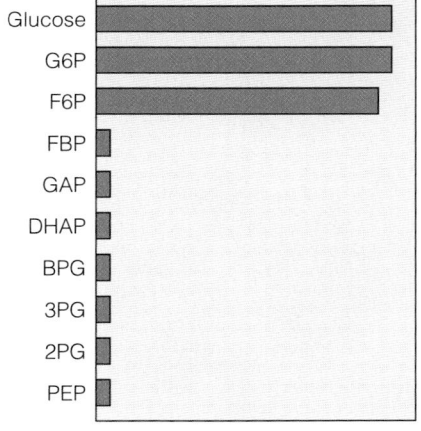

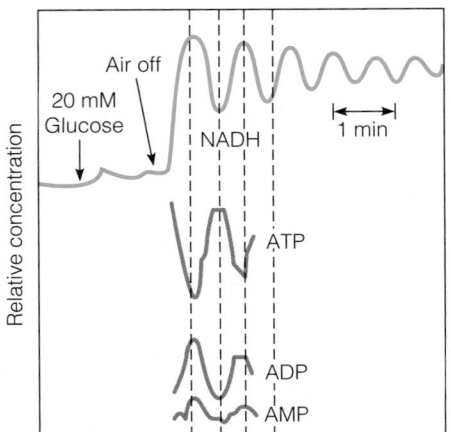

FIGURE 13.15

Periodic oscillations of the levels of glycolytic intermediates in yeast cells undergoing glycolysis. The culture was shifted from aerobic to anaerobic conditions at the point labeled "Air off." The upper tracing (blue) depicts continuous monitoring of fluorescence of the cell suspension, which is related to the intracellular concentration of NADH. The nucleotide levels (orange) were determined in a parallel experiment in which samples of culture were removed at various times, extracted, and assayed for their contents of ATP, ADP, and AMP.

Adapted with permission from *Archives of Biochemistry and Biophysics* 109:586, A. Betz and B. Chance, Influence of inhibitors and temperature on the oscillation of reduced pyridine nucleotides in yeast cells. © 1964, with permission from Elsevier.

Gluconeogenic flux rates are inversely related to the carbohydrate content of the diet. This effect is mediated hormonally.

Oscillations of Glycolytic Intermediates

Other important conclusions emerged from the discovery that the intracellular levels of glycolytic intermediates are not constant under many conditions but undergo periodic variations, or oscillations, as shown in Figure 13.15. These variations can be visualized most readily by following the fluorescence of a yeast cell suspension at 450 nm. The major contributor to this fluorescence is NADH, so this type of experiment monitors changes in the intracellular NADH pool with time. Oscillations are a common feature of feedback-controlled systems, and the cyclic variations in levels of glycolytic intermediates provide important clues to regulatory mechanisms affecting glycolysis.

During the time that the fluorescence of a yeast cell suspension is increasing, NADH is accumulating, through reduction of intracellular NAD^+. Under these conditions, glycolysis is turned on and NADH is being produced by glyceraldehyde-3-phosphate dehydrogenase faster than it can be used to reduce pyruvate. Presumably, during this period one or more regulatory substances are also accumulating. Once they have accumulated sufficiently to inhibit glycolysis, the NADH level falls—until the supply of regulators is depleted to the point that the pathway is inhibited again. This cycle occurs repeatedly.

Once investigators realized that the intracellular levels of NADH were varying periodically, they began sampling extracts of oscillating cells to analyze the intracellular levels of other substrates for glycolytic enzymes. These intermediates were found to rise and fall periodically as well. Note from Figure 13.15 that the levels of ADP and AMP rise and fall precisely in phase with NADH, whereas the level of ATP is 180° out of phase. This pattern suggests that the activity of glycolysis depends in some way on the adenylate energy charge (see Chapter 12): When the charge is high, the pathway is turned off; when it is low, the pathway is activated. These and other observations suggested an enzyme regulated by energy charge as the major point of regulation. Phosphofructokinase is such an enzyme.

Regulation of glycolysis is crucial for many physiological functions. It is important to recognize that glycolysis not only generates ATP and provides pyruvate for oxidation via the citric acid cycle but also is a biosynthetic pathway. Intermediates in glycolysis are precursors to a number of compounds, particularly lipids and amino acids. These processes are discussed throughout the book, but Figure 13.16 summarizes some of the major biosynthetic roles of glycolytic intermediates. This figure illustrates why glycolysis is considered a major metabolic thoroughfare. Many pathways lead into glycolysis, and many pathways diverge from it, creating a substantial flux through the pathway.

Regulation of gluconeogenesis is also crucial for many physiological functions, but particularly so for proper functioning of nervous tissue. Although other organs can use a variety of energy sources, the well-being of the central nervous system demands maintenance of blood glucose levels within narrow limits. Gluconeogenic control is important also as an animal adjusts to muscular exertion or to cycles of feeding and fasting. Flux through the pathway must increase and decrease, based on the availability of lactate produced by the muscles, of glucose from the diet, or of other gluconeogenic precursors.

Gluconeogenesis is controlled in large part by the diet. Animals fed a high-carbohydrate diet show low rates of gluconeogenesis, whereas fasted animals or those fed carbohydrate-poor diets show high flux through this pathway. As we introduced in Chapter 12, these hormonal effects, mediated primarily through insulin and glucagon, involve both control of the synthesis of phosphoenolpyruvate carboxykinase and regulation effected through control of cyclic AMP levels. Our discussion here focuses upon these cAMP-mediated effects as well as other mechanisms affecting enzyme activities. We return to discussion of hormonal effects upon enzyme synthesis in Chapter 23, when we present hormone action in detail.

The regulatory demands placed on the cell are too complex to meet with a single rate-controlling reaction, and thus both glycolysis and gluconeogenesis are controlled at multiple points.

Reciprocal Regulation of Glycolysis and Gluconeogenesis

Gluconeogenesis and glycolysis both proceed largely in the cytosol. Because gluconeogenesis synthesizes glucose and glycolysis catabolizes glucose, it is evident that *gluconeogenesis and glycolysis must be controlled in reciprocal fashion*. In other words, intracellular conditions that activate one pathway tend to inhibit the other. If it were not for reciprocal control, glycolysis and gluconeogenesis would operate together as a giant futile cycle. Reciprocal regulation is related in large part to the adenylate energy charge. Conditions of low energy charge tend to activate the rate-controlling steps in glycolysis while inhibiting carbon flux through gluconeogenesis. Conversely, gluconeogenesis is stimulated at high energy charge, under conditions where catabolic flux rates are low, but are adequate to maintain sufficient ATP levels.

Glycolysis is controlled primarily by regulation of the three strongly exergonic, nonequilibrium reactions of the pathway—those catalyzed by hexokinase, phosphofructokinase, and pyruvate kinase (see Figure 13.9). The opposed reactions in gluconeogenesis—those catalyzed by glucose-6-phosphatase, fructose-1,6-bisphosphatase, and the combination of pyruvate carboxylase and phosphoenolpyruvate carboxykinase—are also strongly exergonic and represent the chief targets for control of this pathway. In other words, the three substrate cycles that differentiate glycolysis from gluconeogenesis (Figure 13.17) represent the primary sites for reciprocal regulation of these pathways. This illustrates the principles introduced in Chapter 12: Regulation can be imposed *only* on reactions displaced far from equilibrium, such as those that comprise these three substrate cycles. Figure 13.18 identifies the major allosteric activators and inhibitors of the key exergonic reactions in glycolysis and gluconeogenesis.

Regulation at the Phosphofructokinase/ Fructose-1,6-Bisphosphatase Substrate Cycle

Phosphofructokinase is the primary flux-controlling enzyme in glycolysis. Energy charge affects the control of glycolysis and gluconeogenesis by regulating the interconversion of fructose-6-phosphate and fructose-1,6-bisphosphate. In mammals, PFK is activated by AMP and ADP, whereas the opposing enzyme, fructose-1,6-bisphosphatase, is inhibited by AMP (Figure 13.18). Thus, as energy charge decreases, glycolysis is activated and gluconeogenesis inhibited by the opposing effects of AMP on phosphofructokinase and fructose-1,6-bisphosphatase. Although intracellular adenine nucleotide levels do vary in parallel with changes in flux through glycolysis and gluconeogenesis, as expected if energy charge is a major regulatory factor, the correspondence is not absolute. These observations suggested additional control mechanisms and led to the discovery of the much more important physiological regulator, **fructose-2,6-bisphosphate.**

Fructose-2,6-bisphosphate and the Control of Glycolysis and Gluconeogenesis

Phosphofructokinase is a homotetramer that interconverts between two conformational states, R and T (recall the discussion of R and T states in Chapter 11, page 456). In addition to the catalytic sites that bind substrates (ATP, fructose-6-phosphate), mammalian PFK has binding sites for several allosteric effectors, including AMP, ADP, ATP, citrate, and fructose-2,6-bisphosphate. Bacterial PFKs have only a single allosteric site that can bind either an inhibitor (phosphoenolpyruvate) or an activator (ADP). Fructose-2,6-bisphosphate is considered to

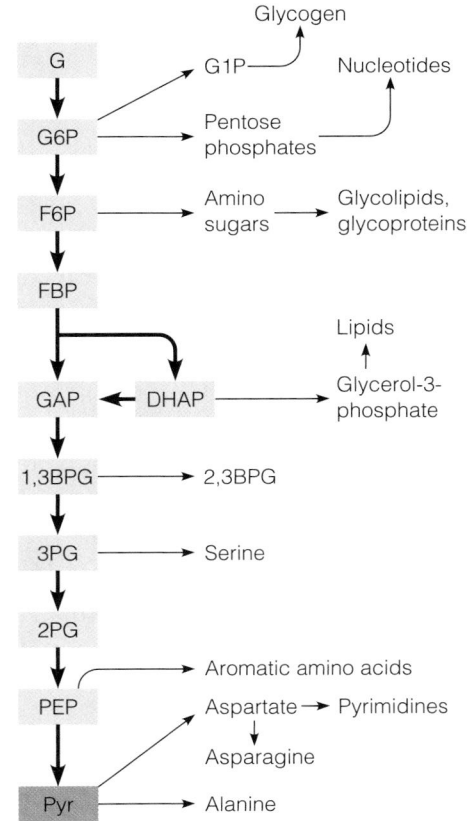

FIGURE 13.16

Alternative fates of glycolytic intermediates in biosynthetic pathways. G1P is glucose-1-phosphate.

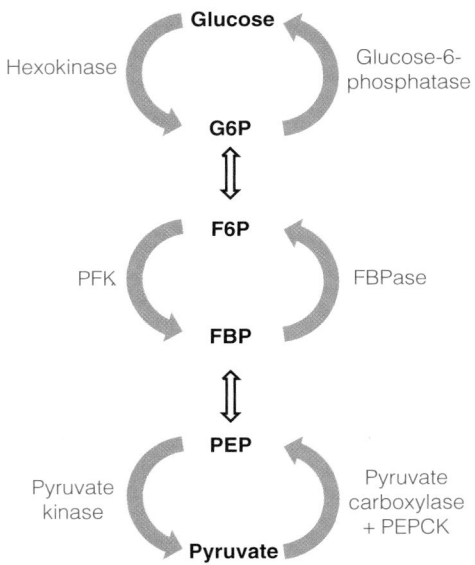

FIGURE 13.17

Substrate cycles in glycolysis/gluconeogenesis.

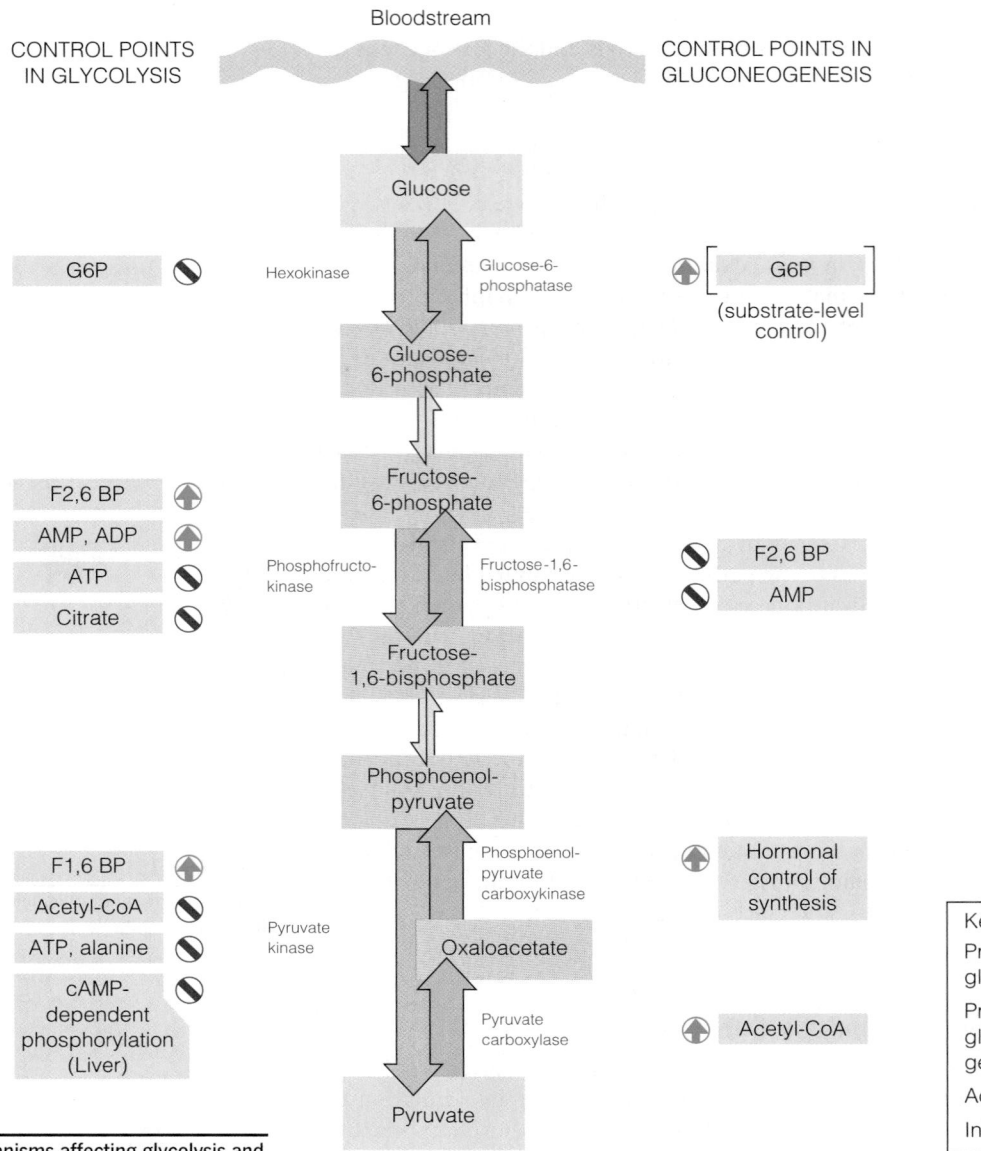

FIGURE 13.18

Major control mechanisms affecting glycolysis and gluconeogenesis. The figure shows the strongly exergonic reactions of glycolysis and gluconeogenesis and the major activators and inhibitors of these reactions.

Conditions that promote glycolysis inhibit gluconeogenesis, and vice versa.

be the major regulator controlling carbon flux through glycolysis and gluconeogenesis in mammalian liver—it is active at much lower concentrations than the other physiological regulators we have discussed. As shown in Figure 13.19a, a very low concentration of fructose-2,6-bisphosphate *activates* phosphofructokinase. AMP and ADP also activate PFK. The most significant *inhibitors* of mammalian phosphofructokinase, from a biological standpoint, are ATP (Figure 13.19b) and citrate. The effect of ATP may seem anomalous because ATP is a substrate and hence essential for the reaction. As an inhibitor, ATP binds to a site on the enzyme separate from the catalytic site, and with lower affinity (Figure 13.19c). At low ATP concentrations, the substrate saturation curve for fructose-6-phosphate is nearly hyperbolic because the regulatory site is not occupied, and the enzyme is almost all in the R state. At high ATP levels the T state predominates, causing the curve to become sigmoidal and shift far to the right (Figure 13.19b). Thus, inhibition is achieved because the apparent affinity for fructose-6-phosphate is greatly reduced. Activators such as AMP, ADP, and fructose-2,6-bisphosphate stabilize the R state, thus increasing the apparent affinity for the substrate fructose-6-phosphate.

The control of PFK by adenine nucleotides represents a way in which energy metabolism responds to the adenylate energy charge. At high energy charge, the relative abundance of ATP signals that the energy-yielding glycolytic pathway should diminish in activity; the signal effects inhibition of PFK. Conversely, a high AMP or ADP level signals that energy charge is low and that flux through glycolysis should increase. Inhibition by citrate represents another energy level sensor. At high energy charge, flux through the citric acid cycle diminishes, via mechanisms that are discussed in Chapter 14. Under these conditions, citrate accumulates and is transported out of mitochondria. Interaction with PFK in the cytosol can signal that energy generation is adequate, and hence the production of citric acid cycle precursors via glycolysis can be diminished.

Fructose-2,6-bisphosphate also regulates the gluconeogenic side of this substrate cycle. But here it is a potent inhibitor of fructose-1,6-bisphosphatase, at least in vitro. Thus, accumulation of the same regulatory molecule has the effect of simultaneously activating glycolysis and inhibiting gluconeogenesis (Figure 13.18).

Fructose-2,6-bisphosphate is formed from fructose-6-phosphate by a 6-phosphofructo-2-kinase, called PFK-2 to distinguish it from the well-known PFK of glycolysis, which we can call PFK-1 for clarity. Another enzyme activity, called **fructose-2,6-bisphosphatase**, cleaves fructose-2,6-bisphosphate back to fructose-6-phosphate. This activity is abbreviated FBPase-2 to distinguish it from the FBPase of gluconeogenesis.

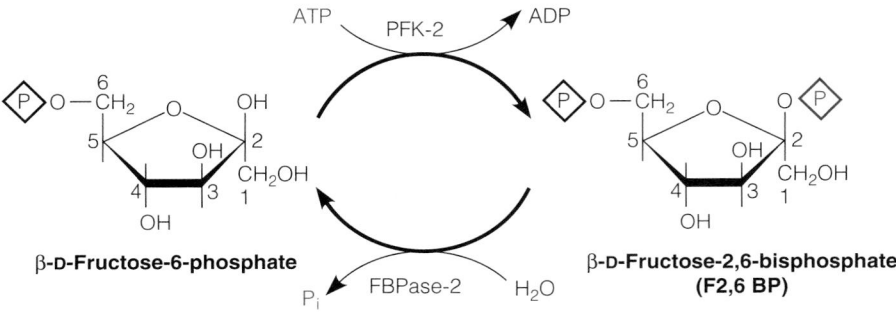

β-D-**Fructose-6-phosphate**

β-D-**Fructose-2,6-bisphosphate**
(F2,6 BP)

These two activities are contained on separate domains of a 100-kilodalton enzyme, PFK-2/FBPase-2 (Figure 13.20). This bifunctional enzyme thus catalyzes the opposing reactions of a substrate cycle that determines the level of this important regulatory molecule. The velocities of the two reactions of this substrate cycle are in turn controlled by phosphorylation/dephosphorylation of the bifunctional enzyme. Conformational changes caused by phosphorylation increase the activity of one domain while decreasing the activity of the other domain, thereby altering the ratio of kinase to bisphosphatase activity. Mammals express several different tissue-specific PFK-2/FBPase-2 isozymes that are generated by alternative splicing from four distinct genes. Each isozyme has a characteristic kinase/bisphosphatase activity ratio, and different regulatory properties that fit the metabolic needs of that tissue.

The activities of the liver isozyme are controlled by the pancreatic hormones insulin and glucagon and by glucose. All of these regulatory molecules act through signaling cascades that lead to the reversible phosphorylation of a specific serine residue, which results in conformational changes in the protein. Phosphorylation *decreases* the activity of PFK-2 and *increases* the activity of FBPase-2 (Figure 13.21). Dephosphorylation reverses this effect.

Phosphorylation of PFK-2/FBPase-2 is catalyzed by cyclic AMP–dependent protein kinase. This important regulatory kinase, also known as protein kinase A (PKA), will be discussed in greater detail later in this chapter. As pointed out in Chapter 12, cyclic AMP (cAMP) plays numerous roles in regulating metabolism, both in eukaryotes and in prokaryotes. In eukaryotes its role is as a second messenger, receiving hormonal messages originating outside the cell and transmitting them within the cell. This transmission involves the activation of some metabolic processes and the inhibition of others. Glucagon is the primary hormone whose action raises cAMP levels in liver. Dephosphorylation of PFK-2/FBPase-2 is catalyzed by one or more specific protein phosphatases (Figure 13.21). Insulin stimulates dephosphorylation, and thus

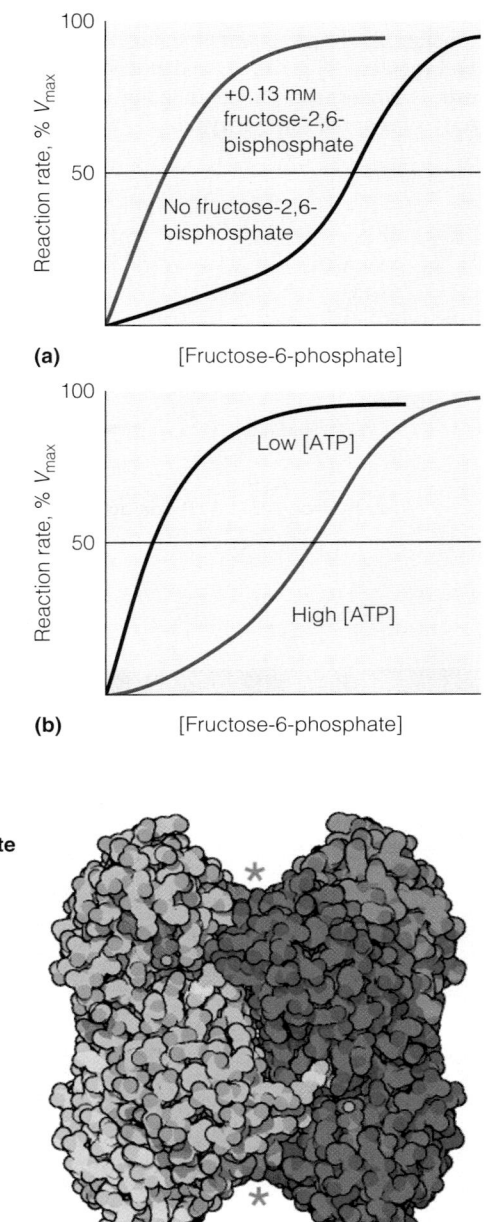

(c)

FIGURE 13.19

Allosteric control of liver phosphofructokinase. **(a)** Activation by fructose-2,6-bisphosphate. **(b)** How ATP increases the apparent K_M for substrate fructose-6-phosphate. **(c)** Model of *E. coli* PFK homotetramer, based on PDB ID 4PFK. The catalytic sites are indicated by the bound sugar (orange) and Mg · ADP (ADP in red; magnesium ion in green). The regulatory sites at the top and bottom of the tetramer, marked by asterisks, contain barely visible ADP molecules (red).

(c) Courtesy of David S. Goodsell, RCSB Protein Data Bank.

Fructose-2,6-bisphosphate, the most important regulator of glycolysis and gluconeogenesis, is synthesized and degraded by different forms of the same enzyme.

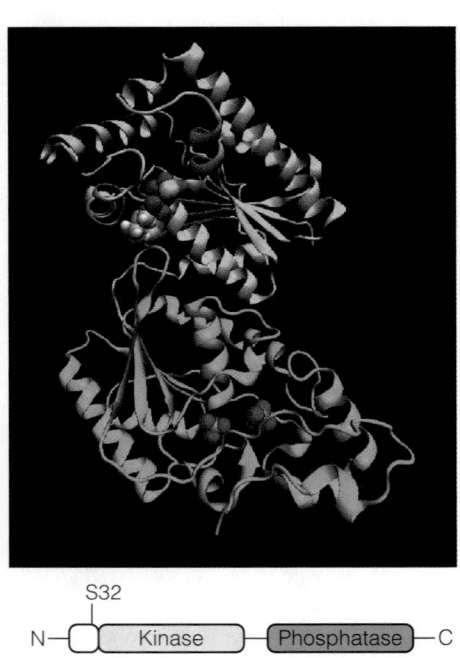

ATPγS
(non-hydrolyzable ATP analog)

FIGURE 13.20

Bifunctional PFK-2/FBPase-2. This figure shows the X-ray crystal structure of one subunit of the homodimeric human liver enzyme (PDB ID 1K6M). The N-terminal 6-phosphofructo-2-kinase domain is shown in yellow, and the C-terminal fructose-2,6-bisphosphatase domain is shown in green. The kinase active site is marked by a bound nonhydrolyzable ATP analog (ATPγS). The bisphosphatase active site is marked by bound phosphates. The diagram of the primary structure of the 470 amino acid liver isozyme shows the location of the phosphorylation site (serine-32). Residues 1–38, including Ser-32, are not visible in the X-ray structure, suggesting that this segment is highly flexible.

activation of the 6-phosphofructo-2-kinase domain, but the signaling pathway is not known. The glucose signal is mediated by protein phosphatase 2A (PP2A). An increase in liver glucose concentration leads to increased synthesis of intermediates in the pentose phosphate pathway (discussed later in this chapter). One of these metabolites, xylulose-5-phosphate, is a specific activator of PP2A.

We will discuss the hormonal control of energy metabolism in much greater detail in Chapter 18, but we can begin to grasp some of the mechanistic details here. Glucagon, released by the pancreas in response to low blood glucose levels, binds to its plasma membrane receptors on liver cells. This activates the receptor, which then initiates the cAMP cascade and results in an active protein kinase A that catalyzes the phosphorylation of PFK-2/FBPase-2, stimulating its fructose-2,6-bisphosphatase activity. The resultant drop in fructose-2,6-bisphosphate levels causes this regulatory molecule to dissociate from PFK-1, thereby increasing the sensitivity of PFK-1 to allosteric inhibitors—citrate and ATP. This in turn *reduces* flux through glycolysis and *stimulates* gluconeogenesis by relieving the inhibition of fructose-1,6-bisphosphatase. This is one mechanism by which glucagon increases blood glucose concentration. Insulin, released by the pancreas

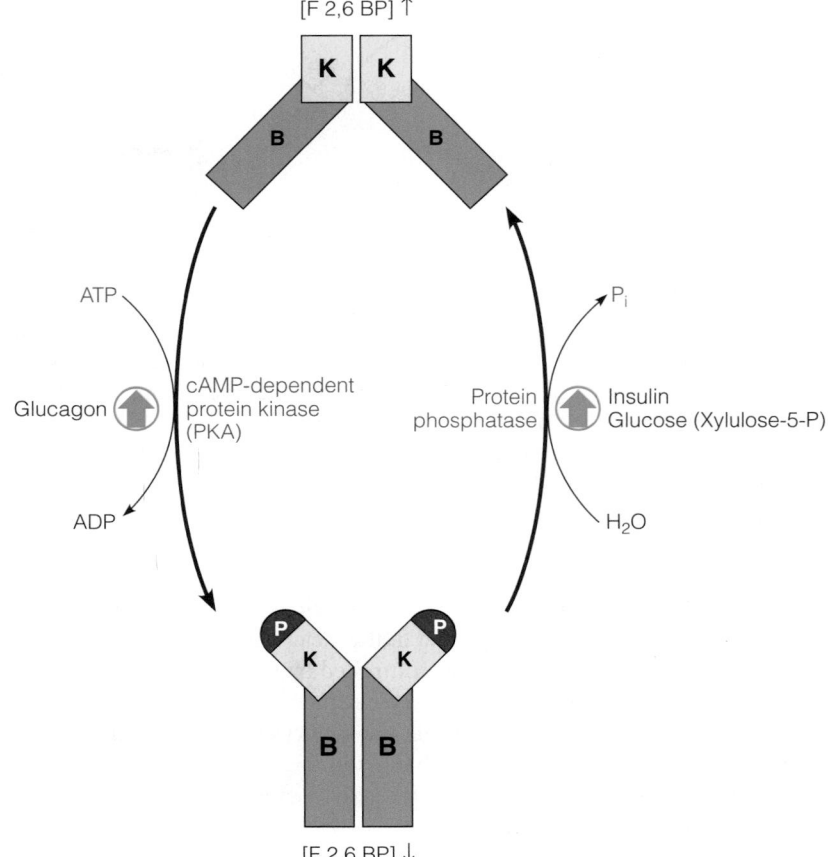

FIGURE 13.21

Regulation of the synthesis and degradation of fructose-2,6-bisphosphate in liver. The bifunctional PFK-2/FBPase-2 is controlled by reversible phosphorylation of a specific serine residue near the N-terminus of each subunit of the homodimeric protein. In the unphosphorylated form, the 6-phosphofructo-2-kinase domain (K) is active, and fructose-2,6-bisphosphate (F2,6BP) is synthesized. In the phosphorylated form, the fructose-2,6-bisphosphatase domain (B) is active, and F2,6BP is degraded. Glucagon stimulates phosphorylation by activating cAMP-dependent protein kinase (PKA). Insulin and glucose (via xyulose-5-phosphate) stimulate dephosphorylation by activating a protein phosphatase.

in response to high blood glucose levels, or glucose itself, stimulates dephosphorylation of PFK-2/FBPase-2, activating its 6-phosphofructo-2-kinase activity. The resultant rise in fructose-2,6-bisphosphate levels *stimulates* glycolysis and *inhibits* gluconeogenesis.

Regulation of PFK-2/FBPase-2 in Heart and Skeletal Muscle

Heart expresses a different isozyme of this bifunctional regulatory enzyme, and unlike in liver, phosphorylation of the heart isozyme *increases* its kinase-to-bisphosphatase activity ratio. Not surprisingly, the site of phosphorylation is different, a serine residue near the C-terminus of the heart isozyme. At least two protein kinases are known to phosphorylate heart PFK-2/FBPase-2: AMP-activated protein kinase (AMPK) and protein kinase B (PKB, also called Akt). As we will see in Chapter 18, AMPK is a master sensor and integrator of signals that control cellular energy balance in all eukaryotes. When the oxygen supply is restricted (**ischemia**), heart cells must switch to anaerobic glycolysis to supply ATP for continued muscle contraction. AMPK is activated by ischemia, whereupon it phosphorylates and activates the 6-phosphofructo-2-kinase activity of PFK-2/FBPase-2. Fructose-2,6-bisphosphate levels rise, stimulating glycolysis.

The PFK-2/FBPase-2 isozyme found in skeletal muscle is not subject to control by phosphorylation/dephosphorylation. Instead, skeletal muscle PFK-2/FBPase-2 is controlled by the intracellular concentration of fructose-6-phosphate, which is both an allosteric modifier and a substrate. The kinase-to-bisphosphatase activity ratio is increased by high fructose-6-phosphate, which signals the availability of substrate for glycolysis. Fructose-2,6-bisphosphate levels rise as a result, stimulating PFK-1 and thus glycolysis.

Fructose-2,6-bisphosphate even plays a regulatory role in plants. By inhibiting a cytosolic fructose-1,6-bisphosphatase, it controls the flow of three-carbon sugars, produced by photosynthesis, out of chloroplasts and into the pathway for sucrose synthesis in the cytosol.

Regulation at the Pyruvate Kinase/Pyruvate Carboxylase + PEPCK Substrate Cycle

Earlier, we identified pyruvate kinase as a control point for glycolysis. Like PFK, the L (liver) and R (erythrocyte) isozymes of pyruvate kinase are allosterically inhibited at high ATP concentrations (Figure 13.18), in a kinetically similar fashion: High ATP levels reduce the apparent affinity of pyruvate kinase for its other substrate, phosphoenolpyruvate. A second allosteric effect is the **feedforward activation** of pyruvate kinase by fructose-1,6-bisphosphate. This effect, the converse of feedback inhibition, ensures that carbon passing the first regulated step in the pathway (PFK) will be able to complete its passage through glycolysis and that undesirable accumulation of intermediates will not occur. A third feedback control effect is inhibition of pyruvate kinase by acetyl-CoA, the major product of fatty acid oxidation. This inhibition allows the cell to reduce glycolytic flux when ample substrates are available from fat breakdown. Finally, pyruvate kinase is inhibited by some amino acids, particularly alanine, the major gluconeogenic precursor among the amino acids. This relationship allows inhibition of glycolysis, with consequent activation of gluconeogenesis, specifically in gluconeogenic tissues, when ample energy and substrates are available. Control at the pyruvate kinase step allows conservation of high-energy phosphate in the phosphoenolpyruvate molecule.

Recall that the liver pyruvate kinase isozyme is also regulated by reversible phosphorylation/dephosphorylation, with the dephosphorylated form far more active than the phosphorylated form (page 533). Phosphorylation is stimulated by glucagon, acting through the same cAMP-dependent protein kinase pathway that phosphorylates PFK-2/FBPase-2. Thus, when blood glucose is low, glucagon secretion leads to inactivation of pyruvate kinase in liver and inhibition of glycolysis. Phosphoenolpyruvate is diverted instead to gluconeogenesis. Muscle pyruvate kinase, by contrast, is not regulated by covalent modification, and virtually all of the phosphoenolpyruvate produced in muscle is converted to pyruvate.

Acetyl-CoA can also be seen as a reciprocal regulator of glycolysis and gluconeogenesis, acting on the enzymes that interconvert pyruvate and phosphoenolpyruvate (Figure 13.18). Acetyl-CoA is a required activator of pyruvate carboxylase and an inhibitor of pyruvate kinase. It can thus signal, when its levels rise, that adequate substrates are available to provide energy through the citric acid cycle and that more carbon can be shuttled into gluconeogenesis and ultimately stored as glycogen. However, some biochemists have questioned whether the activation of pyruvate carboxylase by acetyl-CoA represents an important regulatory mechanism because its intramitochondrial levels under most conditions are far higher than the concentration giving half-maximal stimulation. Thus, the activity of pyruvate carboxylase in vivo might not vary in response to changing acetyl-CoA levels.

Finally, glucagon controls levels of the key gluconeogenic enzyme phosphoenolpyruvate carboxykinase (PEPCK), by activating transcription of the structural gene for PEPCK (Figure 13.18). Insulin has the converse effect. By inhibiting PEPCK gene transcription, insulin tends to depress gluconeogenic flux rates. Glucagon has an additional action at the genetic level: It represses synthesis of pyruvate kinase, thereby contributing to increased gluconeogenic flux from pyruvate to phosphoenolpyruvate.

Regulation at the Hexokinase/Glucose-6-Phosphatase Substrate Cycle

Recall that mammals possess several different isozymes of hexokinase that differ in their kinetic and regulatory properties (page 525). The hexokinase isoenzymes expressed in most tissues (HK-I, -II, -III) are inhibited by their product, glucose-6-phosphate, a mechanism that controls the influx of substrates into the glycolytic pathway. The liver isozyme, hexokinase IV, is not subject to feedback inhibition by glucose-6-phosphate. HK-IV is instead regulated indirectly by glucose through a mechanism that involves protein–protein interactions (Figure 13.22). When intracellular glucose concentrations are low, HK-IV is sequestered in the nucleus bound to a 68-kDa glucokinase regulatory protein (GKRP), thereby preventing hepatic glycolysis under these conditions. Elevated concentrations of glucose favor dissociation of HK-IV from GKRP, and the hexokinase translocates from the nucleus into the cytoplasm, where it can initiate glycolysis. The effect of glucose may not be direct, since fructose-1-phosphate and its precursors (such as fructose and sorbitol) also induce dissociation. Fructose-6-phosphate favors HK-IV binding to GKRP (F-6-P and F-1-P compete for a common binding site

FIGURE 13.22

Regulation of liver hexokinase by protein–protein interactions. Activity of the liver hexokinase isozyme IV (HKIV) is regulated by alternative protein–protein interactions with the glucokinase regulatory protein (GKRP) in the nucleus and dephosphorylated PFK-2/FBPase-2 in the cytoplasm. Insulin binding to its plasma membrane receptor stimulates dephosphorylation of PFK-2/FBPase-2. Glut2, plasma membrane glucose transporter. See legend to Figure 13.21 and text for details.

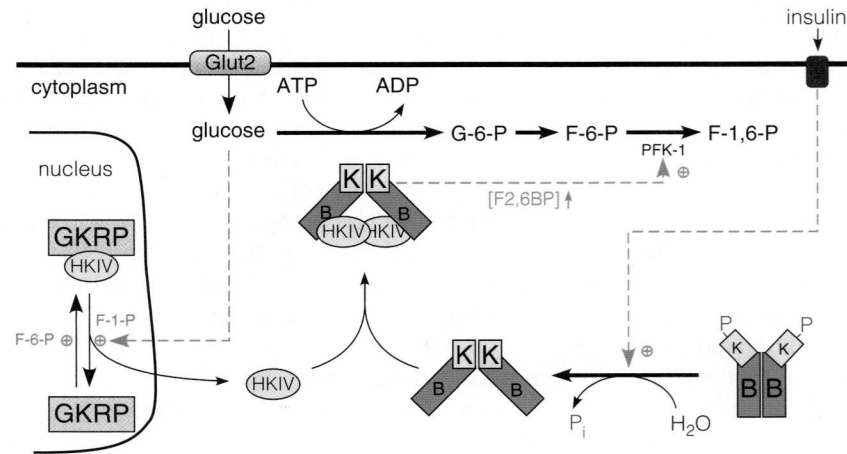

on GKRP). When HK-IV translocates to the cytoplasm, it forms a complex with the dephosphorylated form of PFK-2/FBPase-2. HK-IV is activated in this complex, further stimulating glycolytic flux. Note that activation of HK-IV is mediated by a protein–protein interaction rather than by fructose-2,6-bisphosphate. Recall from Figure 13.21 that dephosphorylation of PFK-2/FBPase-2 is stimulated by insulin or high glucose, increasing its kinase-to-bisphosphatase activity ratio. Moreover, binding of HK-IV increases this ratio, resulting in even higher fructose-2,6-bisphosphate levels. Thus, in response to a high carbohydrate meal (containing, for example, glucose or fructose), the liver PFK-2/FBPase-2 system appears to coordinate the rapid up-regulation of glucose phosphorylation (by hexokinase) with subsequent flux through glycolysis (at the PFK-1 step).

Glucose-6-phosphatase is not known to be allosterically controlled, but its K_M for glucose-6-phosphate is far higher than intracellular concentrations of this metabolite. Thus, intracellular activity is largely controlled in first-order fashion by the concentration of this substrate.

In summary, glycolysis and gluconeogenesis are controlled in large part by the energy charge of the cell, and by fuel status. Regulation is distributed over multiple steps, and is highly coordinated so that the two pathways never operate simultaneously in the same cell. Liver has additional control systems that reflect its special role in maintaining glucose homeostasis for the entire animal. The other major control point of glucose metabolism, at least in animals, is the breakdown and synthesis of glycogen. This extremely important process will be discussed shortly.

Entry of Other Sugars into the Glycolytic Pathway

Thus far our discussion of glycolysis has focused on glucose as a source of carbon for this pathway. Many other sources of carbohydrate energy are available, whether through digestion of foodstuffs or utilization of endogenous metabolites. This section focuses on the utilization of monosaccharides other than glucose, of disaccharides, and of glycerol derived from fat metabolism. These pathways are summarized in Figure 13.23.

Monosaccharide Metabolism

As stated earlier, hexokinases I, II, and III have broad substrate specificities. Thus, they can participate in utilization of hexoses other than glucose, particularly fructose and mannose. A separate enzyme, galactokinase, converts galactose to galactose-1-phosphate.

Galactose Utilization

D-Galactose is derived principally from hydrolysis of the disaccharide lactose [Galβ(1 → 4)Glc], which is particularly abundant in milk. The main route for galactose utilization is conversion to glucose-6-phosphate, via the Leloir pathway, as shown in Figure 13.24. This pathway, named after Argentine biochemist Luis Leloir, begins with epimerization of β-D-galactose, released from lactose, to the α anomer (Figure 13.24, reaction 1). This step is necessary because the next reaction, an ATP-dependent phosphorylation of the hydroxyl at C-1, is catalyzed by an enzyme (**galactokinase**) that is specific for α-D-galactose (reaction 2). Transformation of galactose-1-phosphate to glucose-1-phosphate involves epimerization at carbon 4. However, before epimerization can occur, galactose-1-phosphate must be metabolically activated (reaction 3), by a transferase reaction with a nucleotide-linked sugar, **uridine diphosphate glucose**, also called UDP-glucose, or UDP-Glc. This reaction produces another nucleotide-linked sugar, **uridine diphosphate galactose**, abbreviated UDP-galactose or UDP-Gal. The UDP donor in reaction 3, UDP-Glc, is formed from glucose-1-phosphate and UTP by

FIGURE 13.23

Routes for utilizing substrates other than glucose in glycolysis. In animals, most of the carbohydrate other than glucose and glycogen comes from the diet, and most of the glycerol is derived from lipid catabolism.

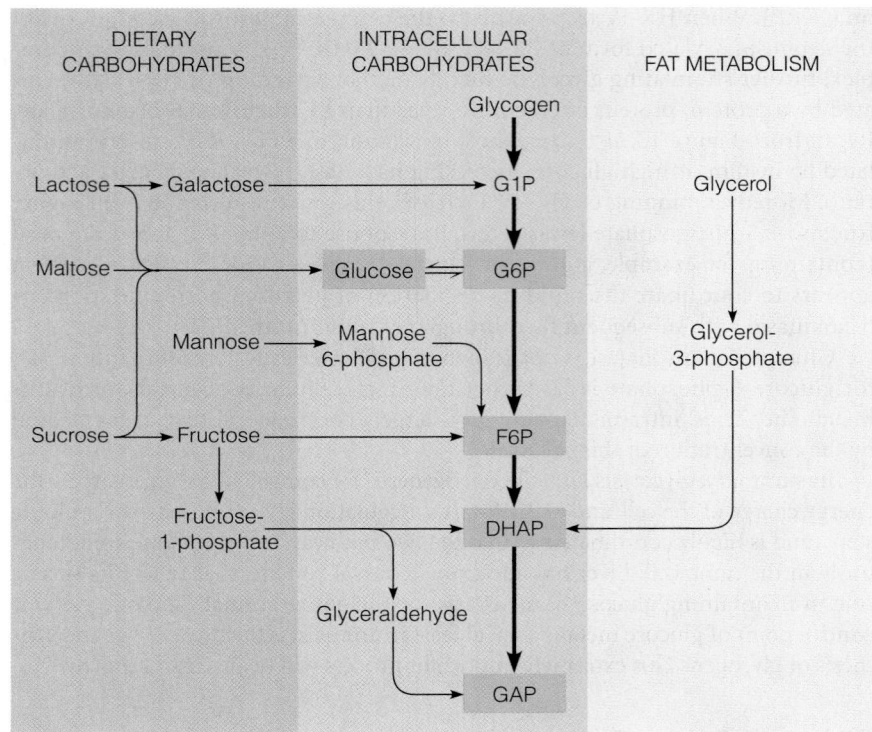

UDP-Glc pyrophosphorylase. This enzyme is named from its reverse reaction, involving cleavage of the phosphoric acid anhydride bond in UDP-Glc by addition across that bond of the elements of pyrophosphoric acid.

Uridine diphosphate glucose

The glucose-1-phosphate formed in reaction 3 is then converted to glucose-6-phosphate by **phosphoglucomutase** (reaction 4), an enzyme involved also in glycogen synthesis. The final step regenerates UDP-glucose from UDP-Gal, by epimerization at C-4 (reaction 5). Details of this NAD^+-linked reaction, catalyzed by **UDP-galactose 4-epimerase**, were shown previously in Figure 11.38 (page 448). The net stoichiometry of reactions 1–5 is:

$$\text{Galactose } + \text{ ATP} \rightarrow \text{glucose-6-phosphate } + \text{ ADP}$$

This overlooks the critical involvement of the nucleoside diphosphate sugars UDP-Glc and UDP-Gal. As we will soon see in our discussion of glycogen biosynthesis, nucleoside diphosphate sugars are widely used intermediates in polysaccharide biosynthesis (see Chapter 9, page 328).

The enzymes on the left side of Figure 13.24 (UDP-Glc pyrophosphorylase and UDP-galactose 4-epimerase) also participate in mammary gland in the synthesis of lactose in milk. Lactose is formed from UDP-Gal plus glucose by **lactose synthase**, in the presence of the protein α-**lactalbumin** (see Figure 9.16, page 329). The endergonic synthesis of UDP-Gal in this tissue proceeds smoothly because of its high rate of conversion to lactose. As discussed in Chapters 9 and 19, UDP-Gal is also used in glycoprotein and glycolipid biosynthesis.

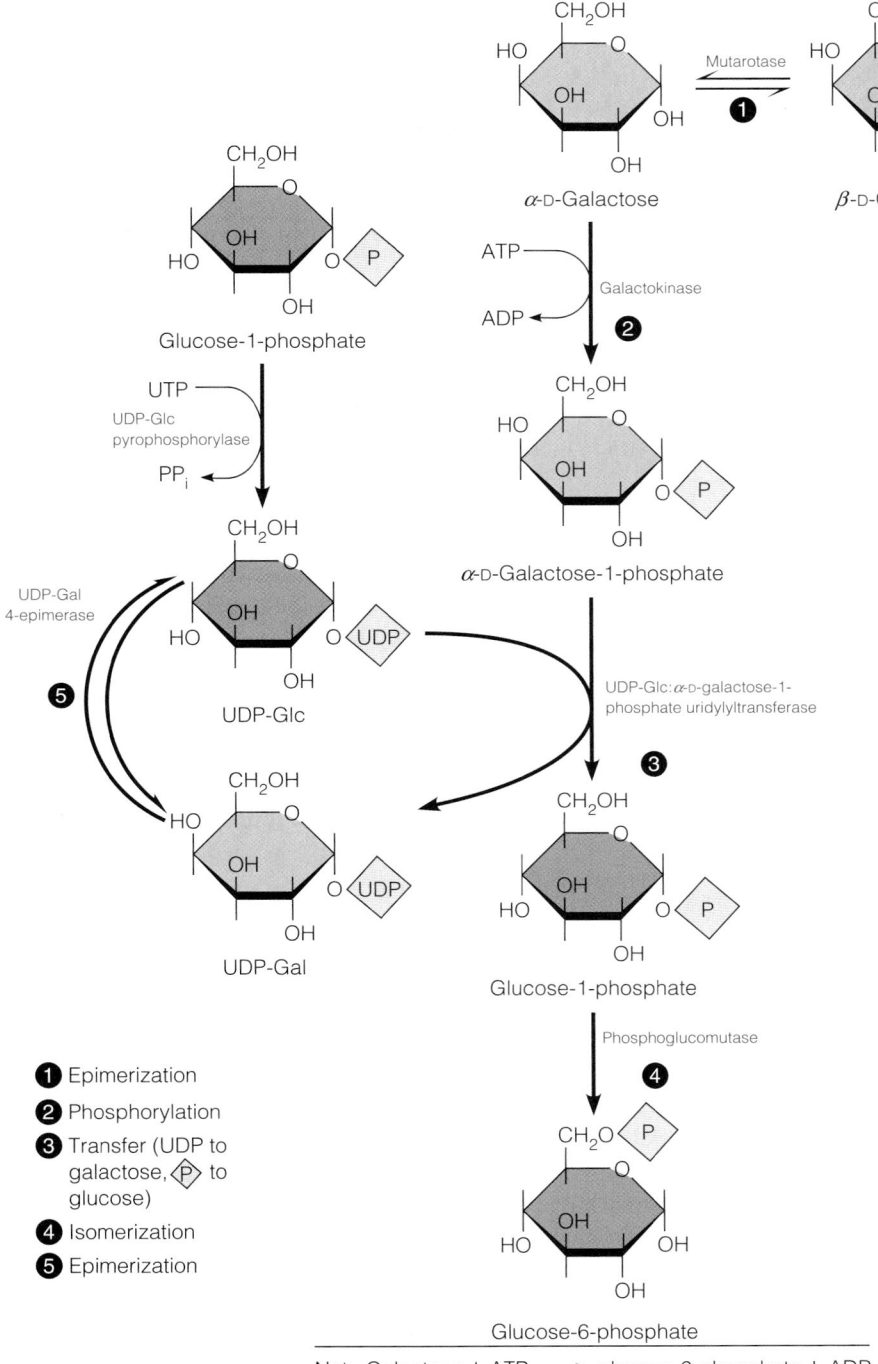

FIGURE 13.24

Leloir pathway for utilizing galactose by converting it to glucose-6-phosphate. The galactose and glucose rings are colored blue and orange, respectively, to emphasize that reaction 3 is a group transfer, not an epimerization. Recall that glucose and galactose are epimers at C-4 (see Figure 9.11).

Glucose-1-phosphate

α-D-Galactose   β-D-Galactose

Mutarotase

ATP

ADP

Galactokinase

UTP

UDP-Glc pyrophosphorylase

PPᵢ

α-D-Galactose-1-phosphate

UDP-Gal 4-epimerase

UDP-Glc

UDP-Glc:α-D-galactose-1-phosphate uridylyltransferase

UDP-Gal

Glucose-1-phosphate

Phosphoglucomutase

❶ Epimerization

❷ Phosphorylation

❸ Transfer (UDP to galactose, P to glucose)

❹ Isomerization

❺ Epimerization

Glucose-6-phosphate

Net: Galactose + ATP ⟶ glucose-6-phosphate + ADP

A variety of genetic disorders in humans go by the generic name **galactosemia**. They all involve a failure to metabolize galactose, so that galactose, galactose-1-phosphate, or both accumulate in the blood and tissues. Clinical consequences include mental retardation, visual cataracts, and enlargement of the liver and other organs. These disorders result from hereditary deficiency of any one of three enzymes involved in galactose utilization. The most common form is caused by a deficiency of UDP-glucose: α-D-galactose-1-phosphate uridylyltransferase (see Figure 13.24, reaction 3). Rarer forms involve deficiencies of galactokinase or UDP-galactose 4-epimerase. Because the major dietary source of galactose is lactose in milk, the symptoms usually occur in infants. The condition can be alleviated by withholding milk and milk products from the diet.

Fructose Utilization

Fructose is present as the free sugar in many fruits, and it is also derived from hydrolysis of sucrose (see Figure 13.23). Phosphorylation of fructose in most tissues yields fructose-6-phosphate, a glycolytic intermediate. A different pathway is involved in vertebrate liver, where the enzyme **fructokinase** phosphorylates fructose to **fructose-1-phosphate** (F1P). This intermediate is then cleaved by a specific enzyme, **aldolase B**. Cleavage products are dihydroxyacetone phosphate, a glycolytic intermediate, and D-glyceraldehyde. The latter is then phosphorylated in an ATP-dependent reaction to give the glycolytic intermediate glyceraldehyde-3-phosphate. This pathway of utilization bypasses phosphofructokinase regulation and may account for the ease with which dietary sucrose is converted to fat (i.e., F1P → GAP + DHAP → glycerol-3 phosphate → triacylglycerols; see Chapter 17).

Fructose is also the major component in the "high fructose corn syrup" used to sweeten soft drinks, and there is mounting evidence linking consumption of high fructose corn syrup to increased risk of obesity and type 2 diabetes.

Mannose Utilization

Finally, among the major hexoses, mannose arises through digestion of foods containing certain polysaccharides or glycoproteins. The hexokinase-catalyzed phosphorylation of mannose to mannose-6-phosphate is followed by isomerization of the latter to fructose-6-phosphate (see Figure 13.23).

Disaccharide Metabolism

The three disaccharides most abundant in foods are maltose, lactose, and sucrose. Maltose is available primarily as an artificial sweetener, derived from starch, while lactose and sucrose are abundant natural products. In animal metabolism they are hydrolyzed in cells lining the small intestine, to give the constituent hexose sugars:

$$\text{Maltose} + H_2O \xrightarrow{\text{Maltase}} 2 \text{ D-glucose}$$

$$\text{Lactose} + H_2O \xrightarrow{\text{Lactase}} \text{D-galactose} + \text{D-glucose}$$

$$\text{Sucrose} + H_2O \xrightarrow{\text{Sucrase}} \text{D-fructose} + \text{D-glucose}$$

The hexose sugars pass via the portal vein to the liver, where they are catabolized as described in the previous section.

Lactase is secreted in the intestines of infants to digest the lactose in their mothers' milk. Because milk is not ingested by most mammals after weaning, lactase secretion decreases in adults as part of a normal developmental program. Humans are unusual in the animal kingdom in that many of us continue to drink milk into adulthood. Lactase deficiency in adult humans ranges from 5–20% in whites, to 75% in blacks, to almost 90% in Asians. This causes **lactose intolerance**, a condition in which ingestion of milk or lactose-containing milk products causes intestinal distress because the gut bacteria ferment the lactose that accumulates.

Plants and microorganisms have different pathways for metabolizing disaccharides. Bacteria metabolize sucrose through the action of **sucrose phosphorylase**:

$$\text{Sucrose} + P_i \rightleftharpoons \text{D-glucose-1-phosphate} + \text{D-fructose}$$

The digestion of neutral fat (triacylglycerols) and most phospholipids generates glycerol as one product. In animals, glycerol first enters the glycolytic pathway by the action in liver of **glycerol kinase**.

Glycerol   **Glycerol-3-phosphate**

The product is then oxidized by **glycerol-3-phosphate dehydrogenase** to yield dihydroxyacetone phosphate, which is catabolized by glycolysis (see Figure 13.23).

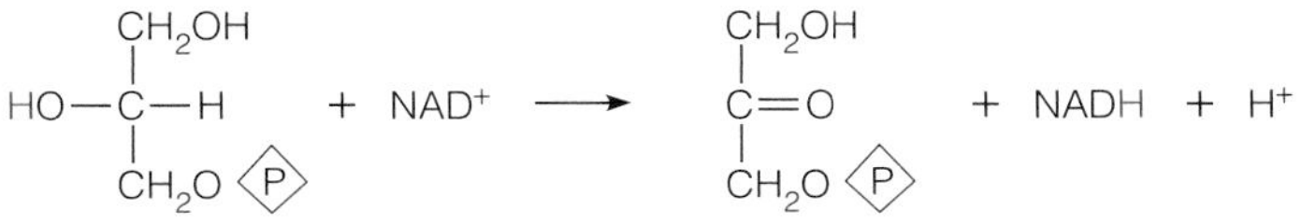

$$\underset{\substack{\text{Glycerol-}\\\text{3-phosphate}}}{\text{HO}-\overset{\overset{\displaystyle CH_2OH}{|}}{\underset{\underset{\displaystyle CH_2O-\text{\textcircled{P}}}{|}}{C}}-H} + NAD^+ \longrightarrow \underset{\substack{\text{Dihydroxyacetone}\\\text{phosphate}}}{\overset{\overset{\displaystyle CH_2OH}{|}}{\underset{\underset{\displaystyle CH_2O-\text{\textcircled{P}}}{|}}{C}}=O} + NADH + H^+$$

Polysaccharide Metabolism

In animal metabolism, two primary sources of glucose are derived from polysaccharides: (1) digestion of dietary polysaccharides, chiefly starch from plant foodstuffs and glycogen from meat; and (2) mobilization of the animal's own glycogen reserves. Recall from Chapter 9 that starch, the major nutrient polysaccharide of plants, consists of the unbranched glucose polymer amylose and the branched polymer amylopectin. Glucose residues in both polymers are linked by $\alpha(1 \rightarrow 4)$ glycosidic bonds, but amylopectin also has $\alpha(1 \rightarrow 6)$ linkages, which provide branch points in the otherwise linear polymer. Glycogen is chemically similar to amylopectin, except that it is more highly branched and is of higher molecular weight. Many microorganisms, like animals, store carbohydrate as glycogen.

Hydrolytic and Phosphorolytic Cleavages

Polysaccharide digestion and glycogen mobilization both involve sequential cleavage of monosaccharide units from nonreducing ends of glucose polymers. The first of these processes occurs via *hydrolysis* and the second via *phosphorolysis.* These processes are chemically similar, involving either water or inorganic phosphate as the nucleophile (Figure 13.25). Hydrolysis is the cleavage of a bond by addition across that bond of the elements of water, and a phosphorolytic cleavage occurs by addition of the elements of phosphoric acid. The reaction catalyzed by glyceraldehyde-3-phosphate dehydrogenase (Figure 13.6, page 530) is a good example of a phosphorolysis. An enzyme catalyzing a phosphorolysis is often called a **phosphorylase**, to be distinguished from a *phosphatase* (or, more precisely, a *phosphohydrolase*), which catalyzes the hydrolytic cleavage of a phosphate ester bond.

Energetically speaking, the advantage of a phosphorolytic mechanism is that mobilization of glycogen yields most of its monosaccharide units in the form of sugar phosphates. These units can be converted to glycolytic intermediates directly, without the investment of additional ATP. By contrast, starch digestion yields glucose plus some maltose; ATP and the hexokinase reaction are necessary to initiate glycolytic breakdown of these sugars.

The hydrolytic mechanism is useful, however, for the digestion of dietary carbohydrate, which occurs largely in the intestine. Digestion products must be absorbed and transported to the liver, where they are converted into glucose. Because sugar phosphates, like other charged compounds, are inefficiently transported across cell membranes, the hydrolytic digestion of polysaccharides to yield hexose sugars facilitates their uptake by tissues.

In plant metabolism we are not concerned with digestion because with few exceptions plants synthesize both monosaccharides and energy-storage polysaccharides via photosynthesis. However, the same enzymatic mechanisms are used to mobilize stored carbohydrate in plant metabolism as in animals—both hydrolysis and phosphorolysis of starch, with hydrolysis predominating. In the brewing of beer, the controlled germination of cereal seeds such as barley develops hydrolytic enzymes that break starch down to mono- and disaccharides for later fermentation by yeast. This process is called malting.

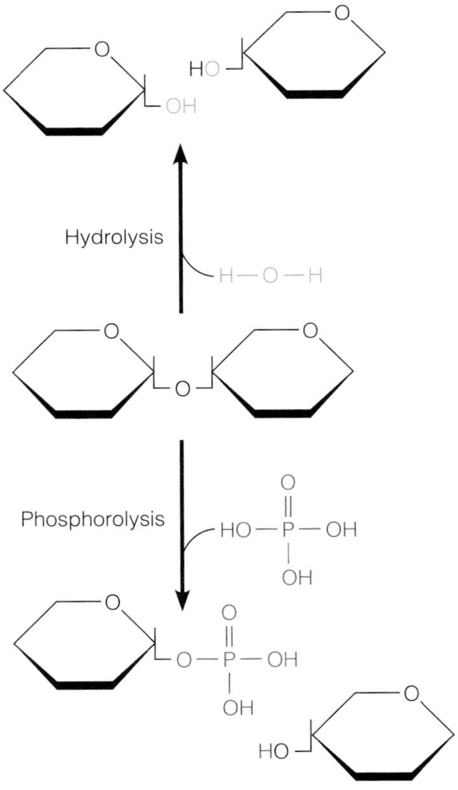

FIGURE 13.25

Cleavage of a glycosidic bond by hydrolysis or phosphorolysis. This formal diagram shows how the elements of water or phosphoric acid, respectively, are added across a glycosidic bond.

Dietary polysaccharides are metabolized by hydrolysis to monosaccharides. Intracellular carbohydrate stores, as glycogen, are mobilized as phosphorylated monosaccharides by phosphorolysis.

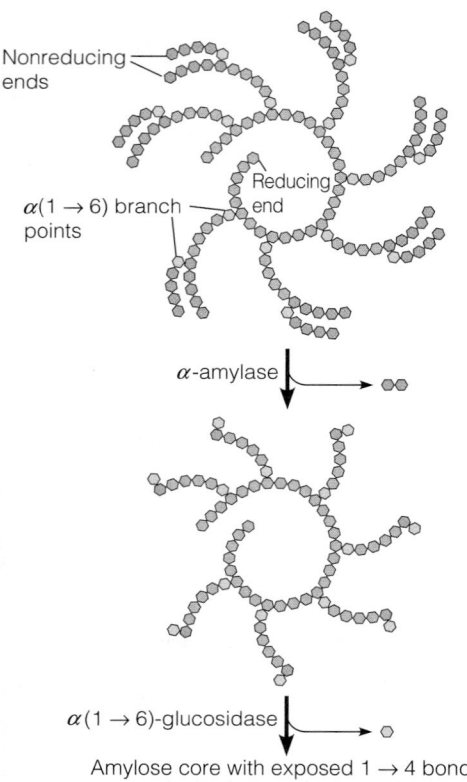

Nonreducing ends

$\alpha(1 \rightarrow 6)$ branch points

Reducing end

α-amylase

$\alpha(1 \rightarrow 6)$-glucosidase

Amylose core with exposed $1 \rightarrow 4$ bonds for further amylase attack

FIGURE 13.26

Sequential digestion of amylopectin or glycogen by α-amylase and $\alpha(1 \rightarrow 6)$-glucosidase. **(Top)** α-Amylase in saliva cleaves $1 \rightarrow 4$ bonds between the maltose units of amylopectin (or glycogen). However, it cannot cleave $1 \rightarrow 6$ glycosidic bonds in the branched polymer, and a limit dextrin (gray) accumulates unless $\alpha(1 \rightarrow 6)$-glucosidase (debranching enzyme) is present. **(Bottom)** $\alpha(1 \rightarrow 6)$-Glucosidase in the intestine cleaves the branch points, exposing the amylose core to further digestion by amylase.

Starch and Glycogen Digestion

In animals, the digestion of starch and glycogen begins in the mouth, with the action of **α-amylase** secreted in saliva. This enzyme cleaves internal $\alpha(1 \rightarrow 4)$ linkages of both polymers. In the intestine, digestion continues, aided by α-amylase secreted by the pancreas. α-Amylase degrades amylose to maltose and a little glucose. However, it only partially degrades amylopectin and glycogen, as shown in Figure 13.26, because it cannot cleave the $\alpha(1 \rightarrow 6)$ linkages found at branch points. The product of exhaustive digestion of amylopectin or glycogen by α-amylase is called a **limit dextrin**; its continued degradation requires the action of a "debranching enzyme," $\alpha(1 \rightarrow 6)$-**glucosidase** (also called isomaltase). This action exposes a new group of $\alpha(1 \rightarrow 4)$-linked branches, which can be attacked by α-amylase until a new set of $\alpha(1 \rightarrow 6)$-linked branches is reached. The end result of the sequential action of these two enzymes is the complete breakdown of starch or glycogen to maltose and some glucose. Maltose is cleaved hydrolytically by **maltase**, yielding 2 molecules of glucose, which is then absorbed into the bloodstream and transported to various tissues for utilization.

Glycogen Metabolism in Muscle and Liver

Before describing the enzymology and regulation of animal glycogen metabolism, we should have some idea of the different functions of the glycogen stores in muscle and liver. Glycogen is the major energy source for contraction of skeletal muscle. Because liver derives most of its own metabolic energy from fatty acid oxidation, liver glycogen has quite a different function—as a source for blood glucose, to be transported to other tissues for catabolism. Liver serves primarily as a "glucostat," adjusting synthesis and breakdown of glycogen to maintain appropriate blood glucose levels. As befits this role, the liver contains relatively large glycogen stores, from 2% to 8% of the weight of the organ. In liver the maximal rates of glycogen synthesis and degradation are about equal, whereas in muscle the maximal rate of glycogenolysis exceeds that of glycogen synthesis by about 300-fold. Although the enzymology of glycogen synthesis and breakdown is similar in liver and muscle, the endocrine control in liver is quite different, as we discuss here and in Chapter 18. The enzymes differ structurally as well.

Glycogen Breakdown

The principal glycogen stores in vertebrates are in skeletal muscle and liver. Breakdown of these stores into usable energy, or **mobilization** of glycogen, involves sequential phosphorolytic cleavages of $\alpha(1 \rightarrow 4)$ bonds, catalyzed by **glycogen phosphorylase**. In plants, starch is similarly mobilized by the action of **starch phosphorylase**. Both reactions release glucose-1-phosphate from nonreducing ends of the glucose polymer:

$$\text{Glucose}\,\alpha(1 \longrightarrow 4)\text{glucose}\,\alpha(1 \longrightarrow 4)\text{glucose}\,\alpha(1 \longrightarrow 4)\text{glucose} \cdots$$

$$\text{P}_i \searrow \Big\downarrow \text{Phosphorylase}$$

$$\alpha\text{-D-Glucose-1-}\langle\text{P}\rangle \; + \; \text{glucose}\,\alpha(1 \longrightarrow 4)\text{glucose}\,\alpha(1 \longrightarrow 4)\text{glucose}$$

The cleavage reaction is slightly disfavored under standard conditions ($\Delta G^{\circ\prime} = +3.1$ kJ/mol), but the relatively high intracellular levels of inorganic phosphate cause this reaction to operate in vivo almost exclusively in the degradative, rather than the synthetic, direction. Note that the reaction proceeds with retention of configuration at carbon 1, i.e., the phosphate is in α linkage in the glucose-1-phosphate product.

Like α-amylase, phosphorylases cannot cleave past $\alpha(1 \rightarrow 6)$ branch points. In fact, cleavage stops four glucose residues from a branch point. The debranching process involves the action of a second enzyme, as shown in Figure 13.27. This glycogen "debranching enzyme," $(\alpha 1,4 \rightarrow \alpha 1,4)$**glucantransferase**, catalyzes two reactions. First is the transferase activity, in which the enzyme removes three of the remaining glucose residues and transfers this trisaccharide moiety intact to the end of some other outer branch via a new $\alpha(1 \rightarrow 4)$ linkage. Next, the remaining glucose residue, which is still attached to the chain by an $\alpha(1 \rightarrow 6)$ bond, is cleaved by the $\alpha(1 \rightarrow 6)$-glucosidase activity of the same debranching enzyme. This yields one molecule of free glucose and a branch extended by three $\alpha(1 \rightarrow 4)$-linked glucose residues. This newly exposed branch is now available for further attack by phosphorylase. The end result of the action of these two enzymes is the complete breakdown of glycogen to glucose-1-phosphate (~90%) and glucose (~10%).

At this point you might wonder why the glycogen breakdown scheme has evolved to include this complex debranching process. The importance of storing carbohydrate energy in the form of a highly branched polymer may well lie in an animal's need to generate energy very quickly following appropriate stimuli. Glycogen phosphorylase attacks *exoglycosidic* bonds—it cleaves sequentially from nonreducing ends. The more such ends that exist in a polymer, the faster the polymer can be mobilized.

To be metabolized via glycolysis, the glucose-1-phosphate produced by phosphorylase action must be converted to glucose-6-phosphate. This isomerization is accomplished by phosphoglucomutase. This reaction is also important in glycogen synthesis. The reaction is mechanistically similar to that of phosphoglycerate mutase (page 532), except that in phosphoglucomutase a phosphoserine residue on the enzyme reacts with substrate, instead of phosphohistidine:

Enzyme-ser-phosphate + glucose-1-phosphate $\rightleftharpoons$ enzyme-ser + glucose-1,6-bisphosphate

Glucose-1,6-bisphosphate + enzyme-ser $\rightleftharpoons$ enzyme-ser-phosphate + glucose-6-phosphate

Net: Glucose-1-phosphate $\rightleftharpoons$ glucose-6-phosphate $\Delta G^{\circ\prime} = -7.3$ kJ/mol

The serine residue that carries the phosphate group is unusually reactive, as shown by the fact that phosphoglucomutase, like chymotrypsin and other serine proteases, is irreversibly inhibited by diisopropylfluorophosphate. The inhibition, like that of chymotrypsin (see Chapter 11), involves acylation of only the active site serine.

Most of the glycogen in vertebrate animals is stored as granules in cells of liver and skeletal muscle. As we learned from our discussion of gluconeogenesis, a major function of the liver is to provide glucose to other tissues for metabolism. This function is accomplished both through glycogen mobilization and through gluconeogenesis. Both processes yield phosphorylated forms of glucose, which cannot exit from liver cells. Conversion to free glucose requires the action of glucose-6-phosphatase, the same enzyme used in gluconeogenesis.

Glycogen Biosynthesis

A major fate of glucose in animals is the synthesis of glycogen. The mechanisms that are used to form glycosidic bonds in glycogen are general mechanisms used in the synthesis of all polysaccharides. For many years it was thought that reversal of the glycogen phosphorylase reaction was the major route for glycogen synthesis. However, three observations could not be reconciled with this notion. First, intracellular levels of orthophosphate are relatively high, which would make it difficult on equilibrium grounds for phosphorylase to catalyze glycogen synthesis in vivo. Second, although phosphorylase can synthesize glycogen in vitro, the product is of much lower molecular weight than natural glycogen. Third, epinephrine secretion activates glycogen metabolism only in the direction of breakdown; in

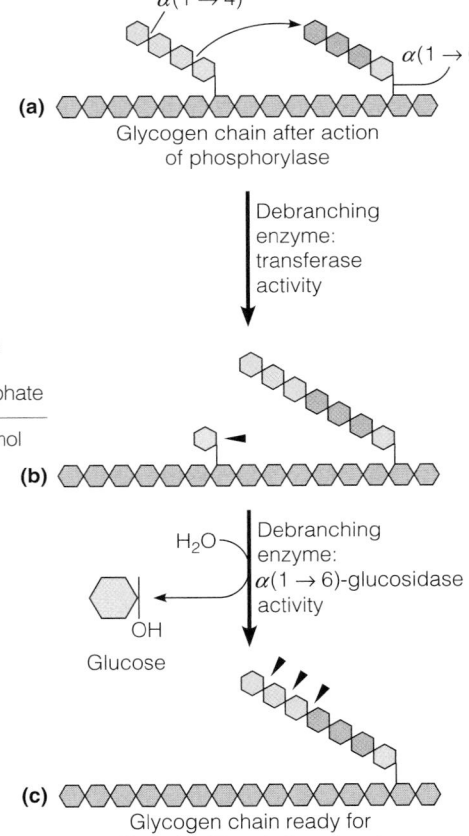

FIGURE 13.27

The debranching process in glycogen catabolism. **(a)** A glycogen chain following activity by phosphorylase, which cleaves off glucose residues to within four residues of the branch point. **(b)** The glycogen chain following transferase activity by the debranching enzyme. The three remaining glucose residues with $\alpha(1 \rightarrow 4)$ linkage have been transferred to a nearby nonreducing end. **(c)** The glycogen chain following $\alpha(1 \rightarrow 6)$-glucosidase activity by the debranching enzyme, which has removed the last remaining glucose residue of the branch. Phosphorylase will cleave off all but four glucose units of the newly elongated branch, beginning the debranching process again. The new cleavage points are indicated by wedges.

fact, epinephrine inhibits glycogen biosynthesis. All of these factors suggested that glycogen is synthesized by a different enzyme.

Glycogen Synthase and the Branching Process

In the late 1950s, Luis Leloir discovered that the substrate for glycogen biosynthesis is *uridine diphosphate glucose* (UDP-Glc, page 558). The enzyme involved, **glycogen synthase**, is bound tightly to intracellular glycogen granules.

Biosynthesis of UDP-Glucose

Let us first review how the substrate UDP-Glc is synthesized from blood glucose. Glucose is transported into cells by a plasma membrane **glucose transporter**. As shown in Figure 13.28, it is then phosphorylated by hexokinase to give glucose-6-phosphate, which is isomerized to glucose-1-phosphate by phosphoglucomutase. The enzyme **UDP-glucose pyrophosphorylase** then catalyzes the synthesis of UDP-glucose. The free energy change of this phosphoanhydride exchange reaction is ~0, but it is drawn forward by rapid enzymatic cleavage of pyrophosphate to orthophosphate, catalyzed by pyrophosphatase. Hydrolysis of pyrophosphate is exergonic by ~19 kJ/mol under standard conditions.

> UDP-glucose is the metabolically activated form of glucose for glycogen synthesis.

The Glycogen Synthase Reaction

UDP-glucose is the immediate donor of a glucosyl residue to the nonreducing end of a glycogen branch, which must be at least four glucose residues in length. Glycogen synthase is a **glycosyltransferase**—an enzyme that transfers an activated sugar unit to a nonreducing sugar hydroxyl group. The reaction, depicted in Figure 13.29, generates an $\alpha(1 \rightarrow 4)$ glycosidic linkage between C-1 of the incoming glucosyl moiety and C-4 of the glucose residue at the terminus of the glycogen chain. The reaction involves nucleophilic attack by the 4—OH of the incoming glucosyl residue on C-1 of UDP-glucose. C-1 is rendered electrophilic by elimination of UDP, an excellent leaving group, but the precise mechanism remains unsettled. The enzyme continues to add glucose residues successively to the 4-hydroxyl group at the nonreducing end. Because UDP-Glc is a high-energy compound, the glycogen synthase reaction is exergonic, with a $\Delta G^{\circ\prime}$ of about -13.4 kJ/mol. Glycogen synthase catalyzes the rate-limiting step of glycogen biosynthesis, and is the site of regulation of this anabolic pathway.

The primer for glycogen synthase is a short chain of glucose residues assembled by a small ($M_R = 37,000$) protein called **glycogenin**, which transfers glucose from UDP-Glc to a tyrosine residue on the protein itself. Glycogenin then transfers additional glucosyl units from UDP-Glc, to give $\alpha(1 \rightarrow 4)$-linked primers up to eight residues long. These primers are then extended by glycogen synthase. Glycogenin thus forms the core of the mature glycogen particle that eventually consists of up to 60,000 glucose residues. These particles are stored in liver and muscle cells in granules (see Figure 9.17, page 330), which contain all of the enzymes that metabolize glycogen as well.

Formation of Branches

> Glycogen biosynthesis requires glycogen synthase for polymerization and a transglycosylase to create branches.

Glycogen synthesis involves both polymerization of glucose units and branching from $\alpha(1 \rightarrow 6)$ linkages. These branches are important because they increase the solubility of the polymer and also increase the number of nonreducing ends from which glucose-1-phosphate can be derived during glycogen mobilization. However, these branches cannot be introduced by glycogen synthase. Another enzyme, called **branching enzyme** but more accurately termed **amylo-(1,4 $\rightarrow$ 1,6)-transglycosylase**, comes into play, as shown in Figure 13.30. This branching enzyme transfers a terminal fragment, some 6 or 7 residues long, from a branch terminus at least 11 residues in length to a hydroxyl group at the 6-position of a glucose residue in the interior of the polymer. The reaction involves nucleophilic attack of the C-6 hydroxyl on C-1 of the oligosaccharide that will form the

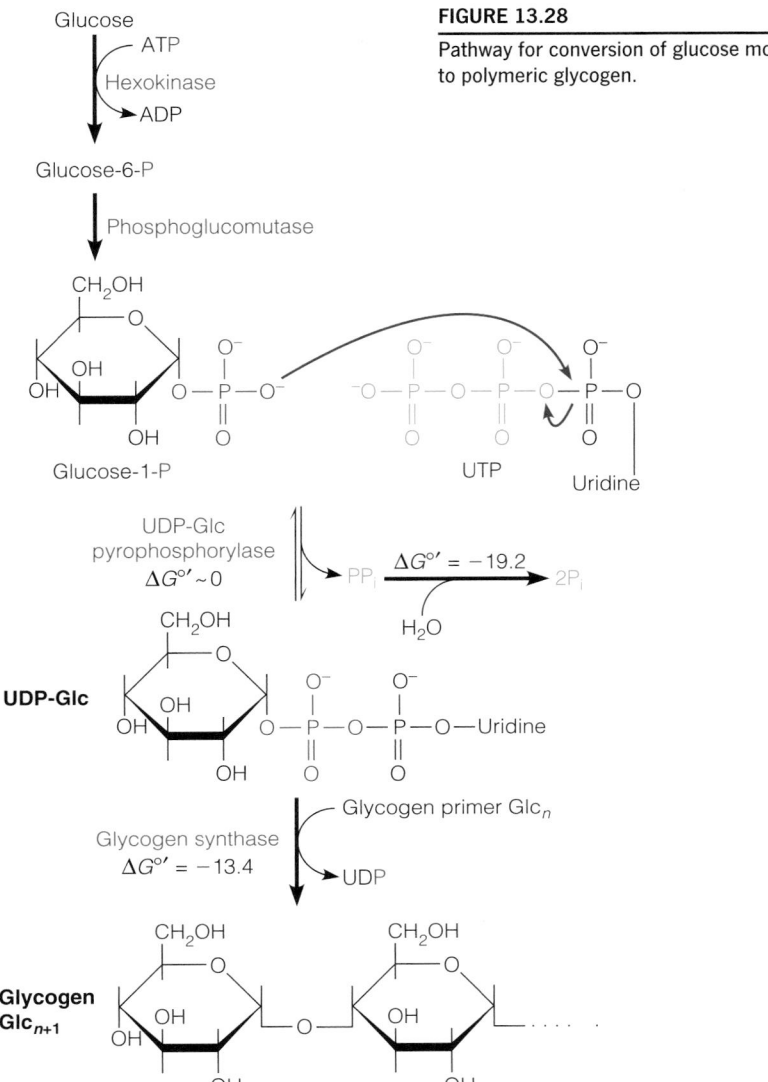

FIGURE 13.28

Pathway for conversion of glucose monomers to polymeric glycogen.

branch. The reaction thus creates two nonreducing termini for continued action by glycogen synthase, whereas just one existed before. The branching process does not involve a large free energy change because of the chemical similarity of $(1 \rightarrow 4)$ and $(1 \rightarrow 6)$ linkages.

Coordinated Regulation of Glycogen Metabolism

In Chapter 11 we mentioned the control of glycogen breakdown, or glycogenolysis, as a particularly well-understood example of a regulatory cascade, a process in which the intensity of an initial regulatory signal is amplified manyfold through a series of enzyme activations. This amplification is particularly important in the case of glycogenolysis because fright, for example, or the need to catch prey, can trigger an instantaneous requirement for increased energy generation and utilization. Glycogen represents the most immediately available *large-scale* source of metabolic energy, and hence it is important that animals be able to activate glycogen mobilization rapidly. Moreover, glycogen breakdown is the first hormone-controlled process for which the molecular action of the hormone was understood in detail (Figure 13.31).

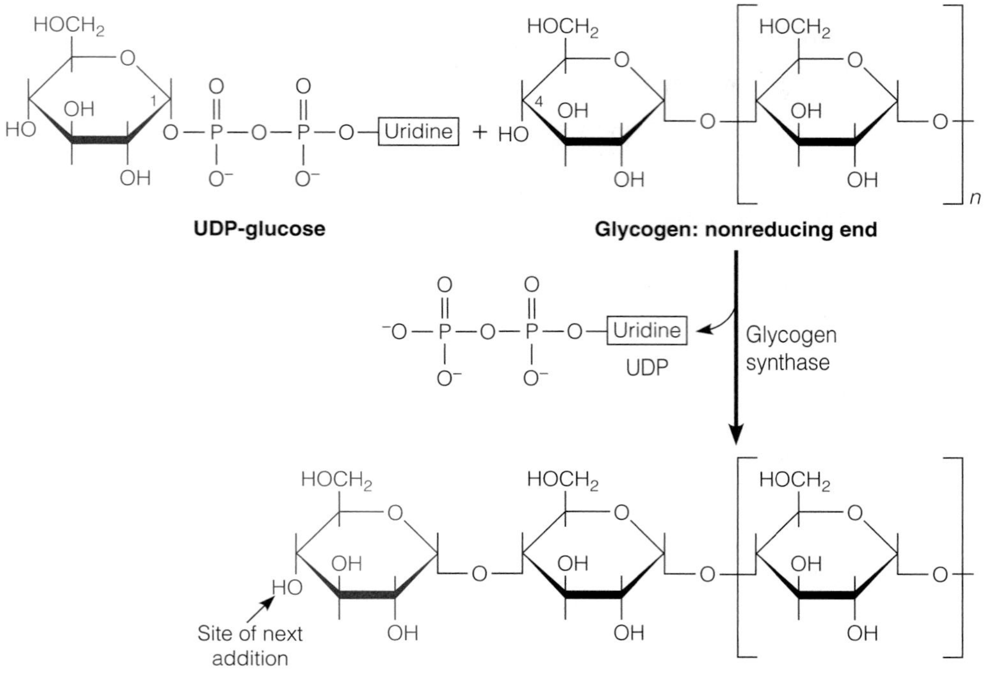

FIGURE 13.29

The glycogen synthase reaction.

UDP-glucose   **Glycogen: nonreducing end**

Glycogen synthase

UDP

Site of next addition

Structure of Glycogen Phosphorylase

In skeletal muscle, glycogen phosphorylase is a dimer containing two identical polypeptide chains, each of 97,400 daltons. The enzyme exists in two interconvertible forms—the relatively *active* phosphorylase *a* and the relatively *inactive* phosphorylase *b*.* Phosphorylation of serine 14 on each subunit induces a conformational change that converts the relatively inactive phosphorylase *b* to the active phosphorylase *a*. Thus, phosphorylation shifts the conformational equilibrium from the less active T state to the more active R state. As shown in Figure 13.31, activation is catalyzed by a specific **phosphorylase *b* kinase**, which transfers phosphate from ATP to the two serine residues. Deactivation is brought about by a specific phosphorylase phosphatase, also called **phosphoprotein phosphatase 1** (PP1). As we shall see, PP1 is a ubiquitous serine/threonine protein phosphatase in eukaryotes that regulates many cellular processes through the dephosphorylation of dozens of substrates. PP1 is directed to these various functions through its association with a diverse set of targeting proteins.

FIGURE 13.30

The branching process in glycogen synthesis. Branching is brought about by the action of amylo-$(1,4 \rightarrow 1,6)$-transglycosylase.

$(1 \rightarrow 4)$

Amylo-$(1,4 \rightarrow 1,6)$-transglycosylase (branching enzyme)

1,6 Branch point

Glycogen core

Nonreducing termini

Glycogen core

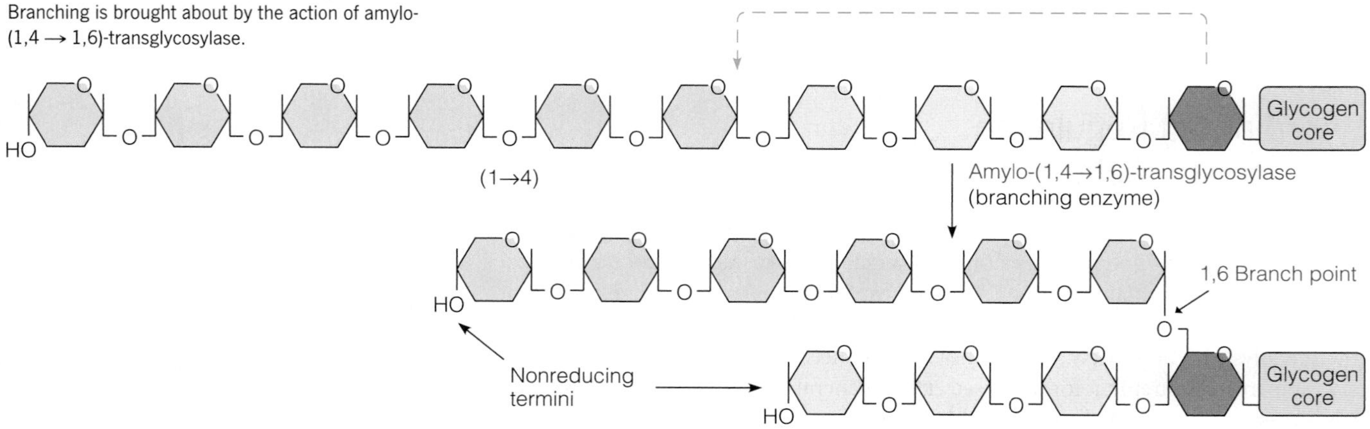

*For enzymatically interconvertible enzyme systems such as glycogen phosphorylase, *a* and *b* refer to the more active and less active forms, respectively.

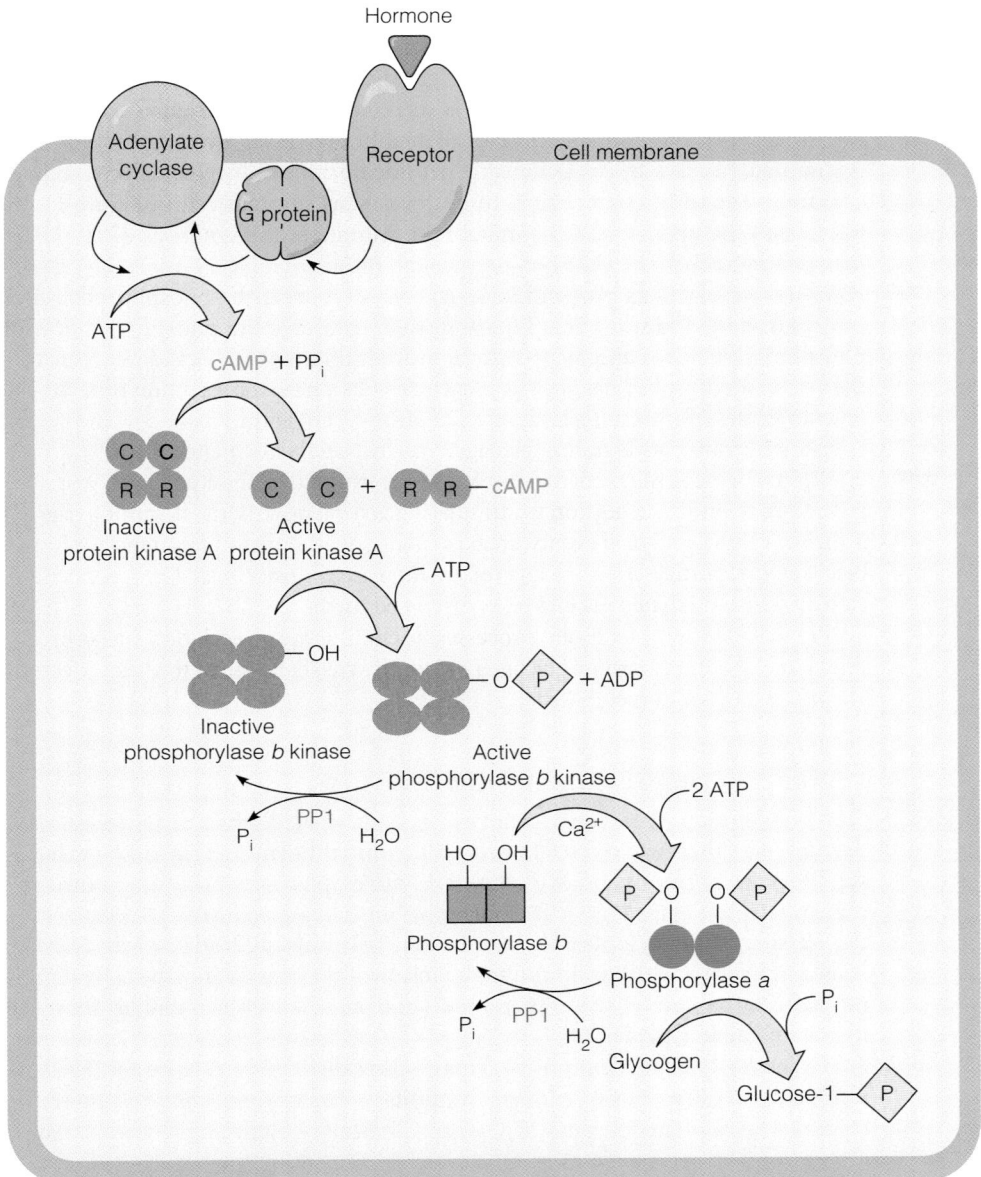

FIGURE 13.31

The regulatory cascade controlling glycogen breakdown. The hormonal regulation of glycogen breakdown is shown here as it might occur in a muscle cell after epinephrine stimulation or a liver cell after glucagon or epinephrine stimulation. Hormone binding to its plasma membrane receptor triggers an interaction with a G protein, which in turn interacts with and activates adenylate cyclase (see Chapter 23). Binding of cyclic AMP to the R (regulatory) subunits of the protein kinase A (PKA) R_2C_2 tetramer causes their dissociation from the C (catalytic) subunits. The active C monomer catalyzes phosphorylation of specific serine residues on inactive phosphorylase b kinase, activating phosphorylase b kinase in the process. This active kinase phosphorylates a serine residue on each of the two subunits of the homodimeric phosphorylase b. This converts the inactive phosphorylase b to the active phosphorylase a, which then catalyzes glycogen breakdown. Each reaction in the regulatory cascade amplifies the hormonal signal, so that binding of very few hormone molecules at the cell surface triggers enormous release of glucose-1-phosphate from intracellular glycogen stores. Inactivation of the pathway involves the action of protein phosphatase 1 (PP1), which removes the phosphates from phosphorylase b kinase and phosphorylase a. The activity of PP1 is also subject to hormonal control.

Control of Phosphorylase Activity

Phosphorylase b kinase is also activated by phosphorylation from an inactive to an active form (Figure 13.31). This reaction is catalyzed by the same cyclic AMP–dependent protein kinase that phosphorylates the bifunctional PFK-2/FBPase-2 in glycolysis and gluconeogenesis. In glycogenolysis, cyclic AMP exerts

Glycogen mobilization is controlled hormonally by a metabolic cascade that is activated by cAMP formation and involves successive phosphorylations of enzyme proteins.

The rapid mobilization of muscle glycogen triggered by epinephrine is one of several components of the "fight or flight" response.

a rapid and efficient activation. At the same time, it inhibits glycogen synthesis through a separate regulatory cascade, discussed below.

The primary hormone promoting glycogenolysis in muscle is epinephrine (formerly called adrenaline), which is secreted from the adrenal medulla and binds to specific receptors on muscle cell membranes. Liver glycogen mobilization is stimulated largely by the pancreatic peptide hormone glucagon, although liver can also respond to epinephrine. In both cases, as summarized in Figure 13.31, binding of the hormone at the membrane stimulates the synthesis of cAMP by membrane-bound adenylate cyclase, through the action of a G protein, $G_{s\alpha}$. Cyclic AMP in turn activates protein kinase A, which catalyzes the phosphorylation of phosphorylase b kinase. This kinase in turn catalyzes the phosphorylation of phosphorylase b to a and, hence, the activation of glycogen breakdown, through the action of phosphorylase a. These events explain how the secretion of relatively few molecules of hormone, such as epinephrine, can, within just a few moments, trigger a massive conversion of glycogen to glucose-1-phosphate.

Epinephrine is the principal hormone governing the "fight or flight" response to various stimuli. In addition to stimulating glycogenolysis, the hormone triggers a variety of physiological events, such as increasing depth and frequency of heartbeats. These cardiac effects, triggered by increased intracellular Ca^{2+} concentrations, are also mediated via cAMP, as discussed further in Chapter 23. Cyclic AMP also regulates other metabolic processes, including the stimulation of fat breakdown and the inhibition of glycogen synthesis. We shall return to these effects as we proceed through metabolism.

Proteins in the Glycogenolytic Cascade

Our presentation of the glycogenolytic cascade started with the phosphorylase reaction and then worked backward, to the initial hormonal signal. Now let us start with the hormone and work forward, with emphasis on the proteins involved (again, refer to Figure 13.31). The hormone binds to a specific receptor located on the outside of the cytoplasmic membrane. This binding leads to activation of adenylate cyclase, which is bound to the inside of the membrane; the activation is mediated by a G protein, $G_{s\alpha}$. (G proteins and adenylate cyclase will be discussed in great detail in Chapter 23.)

Cyclic AMP–dependent protein kinase, also called protein kinase A (PKA), is a tetramer consisting of two *catalytic* subunits, C, and two *regulatory* subunits, R (Figure 13.32a). The tetramer, R_2C_2, is catalytically inactive. Structural analysis revealed that the catalytic subunits possess a core structure that is similar in all known protein kinases. The catalytic mechanism involves nucleophilic attack by a serine residue in the protein substrate upon the γ-phosphate of ATP. In the R_2C_2 tetramer, however, the catalytic subunits are competitively inhibited by a short inhibitor segment (IS) from the R subunit that binds in the C subunit's active site (Figure 13.32b). Each R subunit possesses two cAMP-binding sites (domains A and B). When bound to the C subunit, the R subunit has an extended dumbbell shape, with the two cAMP-binding domains wrapped around the large lobe of the catalytic subunit (Figure 13.32b). Binding of cAMP to the R subunits causes a dramatic conformational change in the R subunits, with the two cAMP-binding domains packed together in a compact globular structure (Figure 13.32c). This conformational change causes the tetramer to dissociate, releasing the catalytically active C subunits to catalyze the phosphorylation of target proteins, including phosphorylase b kinase.

Phosphorylase b kinase is a complex multisubunit protein of ~1.3 MDa, composed of four copies each of α, β, γ, and δ subunits. The γ subunit contains the catalytic site, and the regulatory α and β subunits contain the sites of phosphorylation by PKA. The δ subunit is a protein called **calmodulin**, or <u>cal</u>cium-<u>modul</u>ating prote<u>in</u>. Calcium ion has long been known as an important physiological regulator, particularly of processes related to nerve conduction and muscle contraction.

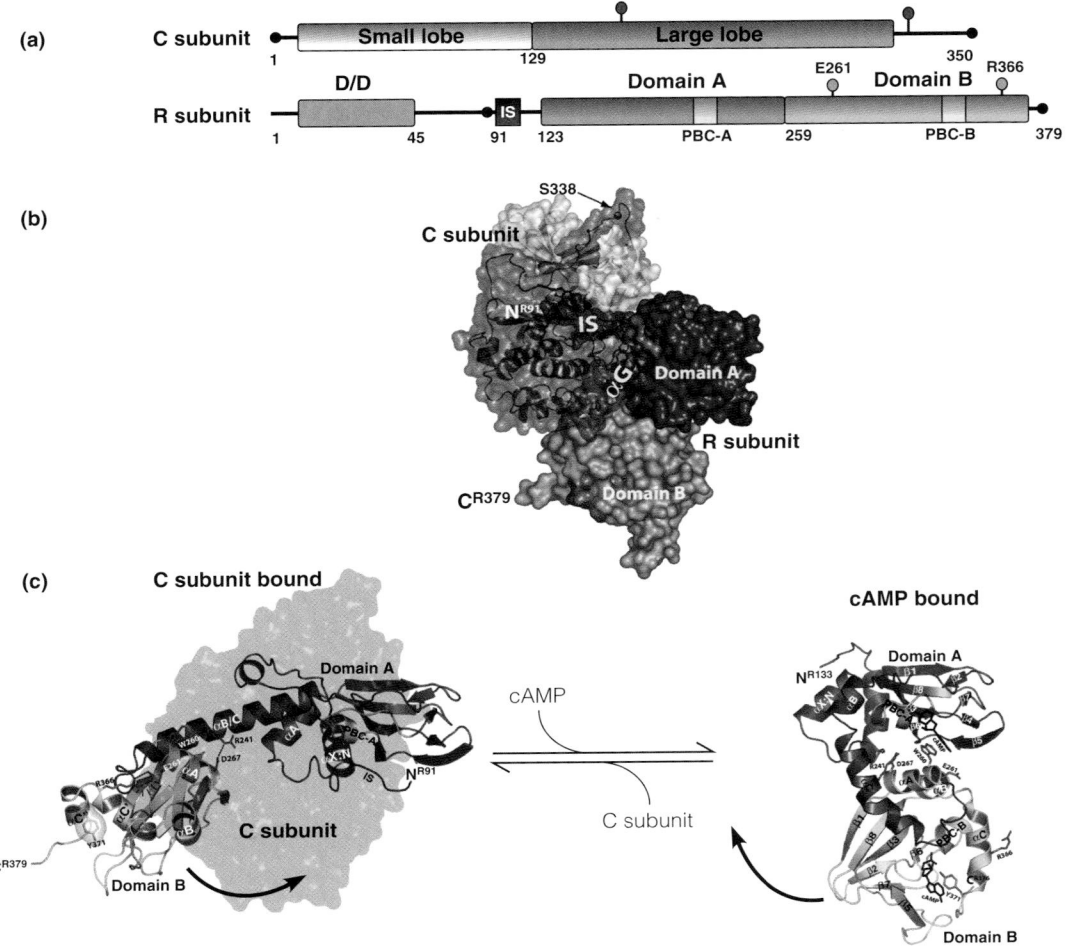

FIGURE 13.32

Structure and activation of cyclic AMP-dependent protein kinase (protein kinase A). **(a)** Domain structure of the catalytic (C) and regulatory (R) subunits. The two red spheres in the catalytic subunit indicate threonine or serine residues that must be phosphorylated for full activity. In the R subunit, IS indicates the inhibitor segment; PBC-A and -B indicate the locations of the cAMP binding sites. **(b)** Surface representation of the holoenzyme complex containing both C and R subunits. Each domain is colored with the same scheme used in panel (a). The active site of the catalytic subunit is marked by the inhibitor segment (red). **(c)** Conformational change of regulatory subunit upon binding cAMP. In the C subunit-bound structure (left), the R subunit has an extended dumbbell shape, with the two cAMP-binding domains wrapped around the catalytic subunit (gray). Upon binding cAMP, Domain B rotates ~125° toward Domain A (indicated by arrow) to adopt a compact globular structure (right), releasing the active catalytic subunits.

Modified from *Cell* 130:1032–1043, C. Kim, C. Y. Cheng, S. A. Saldanha, and S. S. Taylor, PKA-I holoenzyme structure reveals a mechanism for cAMP-dependent activation. © 2007, with permission from Elsevier.

Most of these effects are mediated through the binding of Ca^{2+} to calmodulin, which amplifies small changes in intracellular Ca^{2+} concentration.

Calmodulin is a small protein ($M_r \sim 17,000$) of highly conserved amino acid sequence. It contains four calcium ion binding sites (Figure 13.33a). Each Ca^{2+}-binding site (Figure 13.33b) is composed of a helix-loop-helix motif known as an EF hand (Figure 13.34). This motif is found in a large number of Ca^{2+}-binding proteins. EF hand domains bind Ca^{2+} with a K_D of about 10^{-6} M, consistent with observations that calcium can effect intracellular metabolic changes in concentrations as low as $1 \mu M$. Binding stimulates a major conformational change in the protein, leading to a more compact and more highly helical structure, which augments the affinity of calmodulin for a number of regulatory target proteins (Figure 13.33c). In the case of phosphorylase b kinase, calmodulin plays a special role as an integral subunit of the enzyme. Hence, the glycogenolysis cascade depends on intracellular calcium concentration as well as on cyclic AMP levels. This dependence is particularly important in muscle, where contraction is stimulated by calcium release. Thus, Ca^{2+} plays a dual role, in provision of the energy substrates needed to support muscle contraction and in contraction itself.

Nonhormonal Control of Glycogenolysis

Glycogen breakdown is under nonhormonal, as well as hormonal, control. Muscle and liver possess distinct isozymes of glycogen phosphorylase that differ somewhat in their allosteric properties. Recall that phosphorylase b is relatively inactive, existing primarily in its T state. This form of the enzyme is activated allosterically by 5'-AMP (but not by cyclic AMP). Usually this activation does not occur in the cell because

FIGURE 13.33

Calmodulin. (a) This model shows a backbone representation of bovine brain calmodulin, as determined by X-ray crystallography (PDB ID 1CLL). Ca^{2+} ions are depicted as yellow balls. The four Ca^{2+} binding domains are colored orange, violet, red, and blue, and the two ends are connected by a long central α-helix (green). **(b)** Each Ca^{2+}-binding domain is composed of an α-helix motif known as an EF hand. The dotted lines show the interaction between Ca^{2+} and oxygen atoms (red) on the side chains of Asp, Thr, and Glu residues. **(c)** The long central α-helix (green) undergoes conformational changes as a result of calcium binding. These conformational changes are responsible for changes in calmodulin's affinity for calcium-regulated targets. The structure on the left shows calmodulin (CaM) bound to the target peptide (cyan) of myosin light chain kinase (MLCK) in a "wraparound" conformation (PDB ID 1CDL). The structure on the right shows a calmodulin dimer bound to the target peptides (cyan) of a calcineurin A in an "extended" conformation (PDB ID 2W73). This illustrates the phenomenon of "domain swapping," in which each peptide is bound between the N-terminal lobe from one calmodulin monomer and the C-terminal lobe from the other calmodulin monomer.

(a, b) Modified *from BMC Systems Biology* 2:48, N. V. Valeyev, D. G. Bates, P. Heslop-Harrison, I. Postlethwaite, and N. V. Kotov, Elucidating the mechanisms of cooperative calcium-calmodulin interactions: A structural systems biology approach. © 2008 Valeyev et al. licensee BioMed.

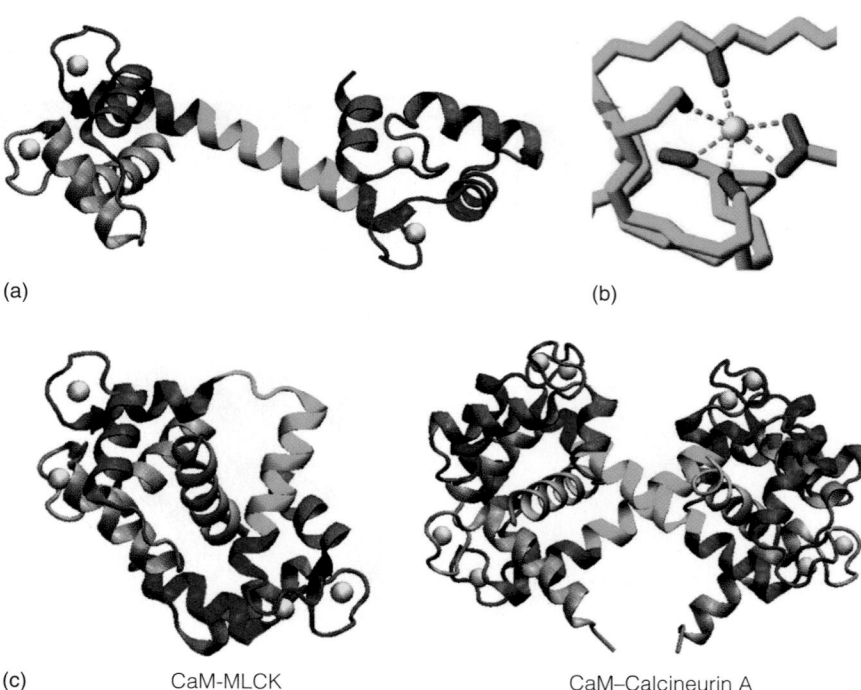

(a)

(b)

(c)   CaM-MLCK   CaM–Calcineurin A

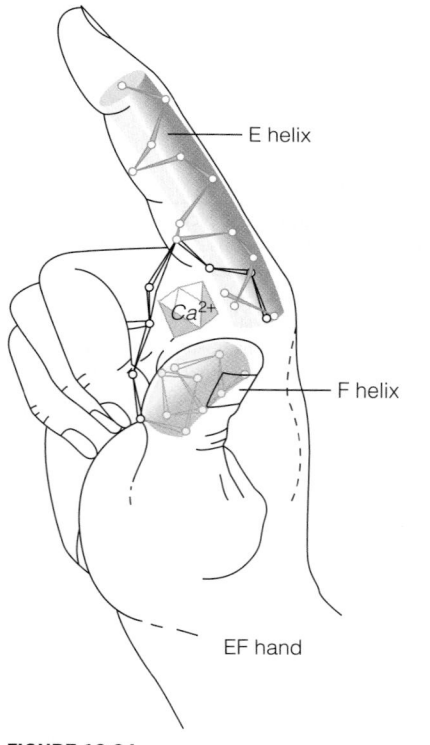

E helix

Ca^{2+}

F helix

EF hand

FIGURE 13.34

The EF hand Ca^{2+}-binding domain. This common helix-loop-helix motif is found in many Ca^{2+}-binding proteins.

Fundamentals of Biochemistry: Life at the Molecular Level, 3rd ed. Donald Voet, Judith G. Voet, and Charlotte W. Pratt. © 2008 John Wiley & Son, Inc. Reproduced with the permission of John Wiley & Sons, Inc. Modified with permission from *The Annual Review of Biochemistry* 45:241, R. H. Kretsinger, Calcium-binding proteins. © 1976 Annual Reviews.

ATP, which is far more abundant and does not activate phosphorylase *b*, competes with AMP for binding to the enzyme. However, under energy-deprived conditions, AMP may accumulate at the expense of ATP breakdown, to the extent that phosphorylase *b* and hence glycogenolysis are activated. Recent crystallographic studies show that the structural changes induced in phosphorylase *b* by AMP are remarkably similar to those induced by phosphorylation of phosphorylase *b* to *a*, even though AMP binding and phosphorylation occur at quite distant sites. Thus, AMP binding shifts the conformational equilibrium of phosphorylase *b* to the more active R state (Figure 13.35). ATP and glucose-6-phosphate, signs of adequate energy status, shift the equilibrium of phosphorylase *b* back to the less active T state. Once glycogen phosphorylase is phosphorylated, it exists primarily in the R form and is unresponsive to most metabolite effectors. However, glucose and glucose-6-phosphate act synergistically on phosphorylase *a*, shifting its equilibrium slightly back toward the T state. In the T state, the phosphoserine side chains are more accessible to phosphoprotein phosphatase 1, and the T state is more readily dephosphorylated than the R state.

Thus, the mobilization of energy reserves from glycogen can be brought about either by hormonal stimulation, reflecting a physiological need for increased ATP production, or by an allosteric mechanism triggered when the energy level is deficient for maintenance of normal functions. The nonhormonal mechanism, which does not involve a metabolic cascade, stimulates glycogenolysis in response to a low energy charge, whereas the hormonally induced cascade predominates when the need is to rapidly augment energy generation. In both cases the phosphorolysis of glycogen to glucose-1-phosphate is enhanced. If, on the other hand, the cell has a high energy charge, signaled by high ATP and/or glucose-6-phosphate levels, glycogenolysis is turned off.

Control of Glycogen Synthase Activity

Earlier we noted that epinephrine secretion inhibits glycogen synthesis in muscle at the same time that it promotes glycogen mobilization. Glucagon has similar effects in liver. Control of both synthesis and degradation of glycogen is mediated by distinct

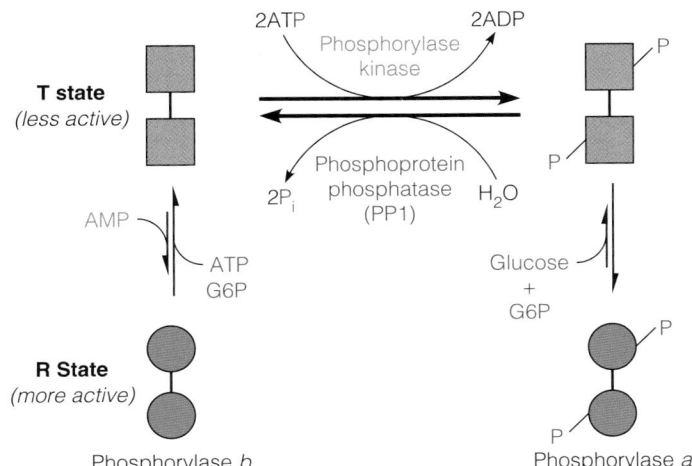

FIGURE 13.35

Control of glycogen phosphorylase activity. The enzyme exists in an equilibrium between a less active T state and a more active R state. The enzyme can be phosphorylated, in response to hormonal signals, by phosphorylase *b* kinase. Phosphorylation shifts the equilibrium largely to the R state. Dephosphorylation is catalyzed by phosphoprotein phosphatase 1 (PP1). Both T and R states can be dephosphorylated by PP1, but the T state is a better substrate because the phosphoserine side chains are more accessible in this conformation. The unphosphorylated form, phosphorylase *b*, exists largely in the T state, and its T $\Leftrightarrow$ R equilibrium is controlled by allosteric effectors AMP, ATP, and glucose-6-phosphate (G6P). Muscle phosphorylase *b* is much more sensitive to AMP and G6P than liver phosphorylase *b*. Glucose and G6P synergistically inhibit phosphorylase *a* by shifting its equilibrium to the T state, which is then rapidly dephosphorylated by PP1.

regulatory cascades involving cyclic AMP–dependent protein kinase and reversible protein phosphorylations. However, whereas the cascade controlling glycogenolysis leads to *activation* of glycogen phosphorylase (see Figure 13.31), the cascade controlling glycogen synthesis leads to *inhibition* of glycogen synthase (Figure 13.36).

> Conditions that activate glycogen breakdown inhibit glycogen synthesis, and vice versa.

Phosphorylation of Glycogen Synthase

Glycogen synthase from vertebrate tissues is a tetrameric protein comprising four identical subunits and having a total molecular weight of about 340 kilodaltons. Glycogen synthase activity is controlled by covalent modification, allosteric activation, and intracellular location. Like phosphorylase, glycogen synthase exists in phosphorylated and dephosphorylated states, with up to nine serine residues on each subunit subject to this modification. Several different protein kinases are known to act on glycogen synthase (Figure 13.36). Dephosphorylation is catalyzed by PP1, the same phosphatase that acts on glycogen phosphorylase and phosphorylase *b* kinase.

In contrast to glycogen phosphorylase, it is the unphosphorylated form, glycogen synthase *a*, that is the active form. Glycogen synthase *a* is active even in the absence of G6P, whereas the phosphorylated forms (glycogen synthase *b*) are dependent on allosteric activation by G6P. Binding of this effector overrides inhibition caused by phosphorylation (Figure 13.36). In addition, G6P binding induces a conformational change that makes the enzyme a better substrate for dephosphorylation by PP1.

> Glycogen synthase activity is controlled by phosphorylation, through mechanisms comparable to those controlling glycogen breakdown by phosphorylase but having reciprocal effects on enzyme activity.

Let's examine the consequences of hormone release upon glycogen synthase (see Figure 13.36). Just as depicted in Figure 13.31, the activation of adenylate cyclase by epinephrine (in muscle) or glucagon (in liver) promotes the dissociation of cAMP-dependent protein kinase (PKA) to give free catalytic C subunits. These C subunits phosphorylate active glycogen synthase *a* to inactive glycogen synthase *b*. To complicate the picture somewhat, several additional protein kinases, including AMP-activated protein kinase (AMPK), glycogen synthase kinase 3 (GSK3), and casein kinase II, can act on glycogen synthase *a*. The most important of these is GSK3. Each of these kinases phosphorylate different serine residues, but in a hierarchical manner, so that there are several different forms of glycogen synthase *b*, and it is an oversimplification to speak of just two forms. In general, as more sites are phosphorylated, the activity of the enzyme progressively decreases because of the following changes: (1) decreased affinity for the substrate, UDP-glucose; (2) decreased affinity for the allosteric activator, glucose-6-phosphate; and (3) increased affinity for ATP and P_i, both of which tend to antagonize the activation by glucose-6-phosphate. Thus, there is a graded series of responses to changing metabolic conditions, involving a series of different protein kinases. Whichever

FIGURE 13.36

Control of glycogen synthase activity. The enzyme can be phosphory-lated, in response to hormonal signals, by several different protein kinases, including the catalytic subunit of cAMP-dependent protein kinase (PKA), AMP-activated protein kinase (AMPK), glycogen synthase kinase 3 (GSK3), and casein kinase II (CKII). Each subunit of the homote-tramer can be phosphorylated on as many as nine serine residues. Dephosphorylation is catalyzed by phosphoprotein phosphatase 1 (PP1). Glucose-6-phosphate (G6P) can allosterically activate the phosphorylated enzyme. The conformational change also makes the enzyme a better substrate for dephosphorylation by PP1.

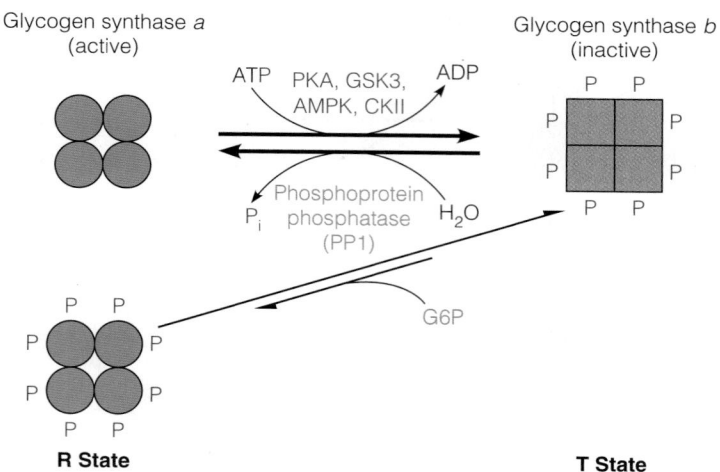

kinase is used, the net effect of phosphorylating glycogen synthase is inhibition of the enzyme, with consequent inhibition of glycogen synthesis.

Note that the glycogen *synthesis* cascade has one less stage than the glycogen *breakdown* cascade because PKA phosphorylates glycogen synthase directly, whereas it can act on glycogen phosphorylase only through its action on phosphorylase *b* kinase. The extra stage allows for more sensitive regulation of glycogen breakdown than of its synthesis, which is consonant with needs of animals for exceedingly rapid changes in demand for energy generation in muscle. In fact, experimental observations show that the maximum rate of muscle glycogen breakdown is some 300-fold higher than that of glycogen synthesis.

A Closer Look at Glycogen Synthase Regulation: Dephosphorylation of Glycogen Synthase *b*

As mentioned above, phosphoprotein phosphatase 1 (PP1) is the primary phosphatase that acts on glycogen synthase *b*. PP1 also dephosphorylates glycogen phosphorylase and phosphorylase *b* kinase. How is the activity and specificity of this enzyme controlled? Recall that PP1 is involved in the regulation of many cellular processes. The catalytic subunit of PP1 (PP1c) does not exist freely in the cell, but rather is associated with a host of different regulatory subunits. Indeed, PP1c has been shown to interact with as many as 100 different proteins and peptides. These regulatory subunits determine the intracellular localization, substrate specificity, and activity of PP1. One family of regulatory proteins is the glycogen-targeting subunits (G subunits). Mammals express at least 7 different G subunits, including G_M in muscle and G_L in liver, and these G subunits serve to anchor PP1 to the glycogen particle. Glycogen synthase, glycogen phosphorylase, and phosphorylase *b* kinase are also bound to the glycogen particles, and their dephosphorylation is mediated by PP1 bound to its G_M or G_L regulatory subunit. All G subunits possess a PP1-binding motif and a targeting domain with binding sites for glycogen and the PP1 substrates. G_M and G_L confer different regulatory properties onto PP1 in muscle and liver, respectively.

Let's start with muscle (see Figure 13.37). PP1c-G_M is regulated by phosphorylation of the G_M subunit, catalyzed by cAMP-dependent protein kinase (PKA). Phosphorylation of G_M leads to dissociation of PP1c, releasing it into the cytoplasm, where it is bound by a small protein called **phosphoprotein phosphatase inhibitor 1**, or **inhibitor 1**. When phosphorylated on a threonine residue, inhibitor 1 potently inhibits PP1c activity. Phosphorylation of inhibitor 1 is also carried out by PKA. Thus, as shown in Figure 13.37, cAMP exerts two effects in inhibiting glycogen synthesis: (1) phosphorylation of glycogen synthase, causing its inactivation; and (2) inhibition of phosphoprotein phosphatase 1, whose activity would tend to restore activity of glycogen synthase by catalyzing its dephosphorylation. Inhibitor 1 is

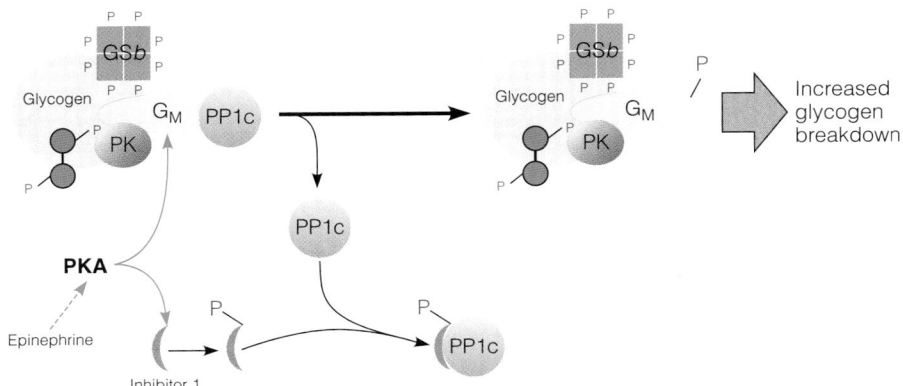

FIGURE 13.37

Regulation of phosphoprotein phosphatase 1 (PP1) in muscle. The active catalytic subunit of PP1 (PP1c) is anchored to the glycogen particle by its G subunit, G_M. Glycogen synthase b (GSb), phosphorylase b kinase (PK), and glycogen phosphorylase a (green) are also bound to the glycogen particle. Phosphorylation of G_M by cAMP-dependent protein kinase (PKA) leads to dissociation of PP1c, releasing it into the cytoplasm, where it is bound by inhibitor 1. Inhibitor 1 is also phosphorylated by PKA, forming a potent PP1c inhibitor. Sequestration and inhibition of PP1c increases the phosphorylation state of all the enzymes bound to the glycogen particle, resulting in inhibition of glycogen synthesis and stimulation of glycogen breakdown.

widely expressed in mammalian tissues, but is a particularly critical regulator of PP1 function in heart and in neuronal signaling. Thus, the phosphorylation cascade initiated by epinephrine binding to its receptor on the muscle cell leads to *inhibition* of glycogen synthesis (by phosphorylation of glycogen synthase and by inhibition of PP1 phosphatase activity) and *activation* of glycogen breakdown (by phosphorylation of glycogen phosphorylase and by inhibition of PP1). The net result is the flow of glucose from glycogen into glycolysis to provide ATP for muscle contraction.

Insulin has the opposite effect on muscle glycogen metabolism. In humans, skeletal muscle is the primary site for glucose storage, where up to 90% of glucose from a meal is converted to glycogen. During rest, insulin stimulates the storage of glucose as glycogen in muscle by enhancing glucose uptake and activating glycogen synthase. Glucose is imported in muscle mainly by GLUT4, a high-affinity, low-capacity glucose transporter. GLUT4 is initially localized in internal vesicles, which translocate to the plasma membrane of the muscle cell in response to insulin (see Figure 18.5, page 762). Entering glucose is phosphorylated to glucose-6-phosphate by hexokinase and then converted to glucose-1-phosphate for glycogen synthesis. Insulin binding also initiates a signaling pathway that leads to phosphorylation and inhibition of GSK3. Phosphorylation of glycogen synthase ceases, but its dephosphorylation continues. Recall that glucose-6-phosphate induces a conformational change in glycogen synthase *b* that makes it a better substrate for dephosphorylation by PP1. The net result is conversion of inactive phosphorylated glycogen synthase *b* to active dephosphorylated glycogen synthase *a*, and activation of glycogen synthesis.

Phosphoprotein phosphatase 1 is regulated differently in liver, reflecting the different roles of glycogen metabolism in liver and muscle (see Figure 13.38). Although the liver-specific G_L regulatory subunit anchors PP1 to glycogen granules in that tissue as well, G_L is not subject to reversible phosphorylation. Instead, PP1c-G_L possesses an allosteric binding site for glycogen phosphorylase *a* (the phosphorylated form), which *inhibits* the phosphatase activity of PP1c. Recall that PP1 binds both the T and R states of phosphorylase *a*, but only in the T state are the phosphoserine side chains accessible to hydrolytic activity. Thus, as long as phosphorylase *a* remains in its active R state, it sequesters PP1 from other potential substrates. If, however, glucose and glucose-6-phosphate levels rise, their binding to phosphorylase *a* shifts its equilibrium back toward the T state (Figure 13.35), exposing its phosphoserine side chains. PP1c now readily dephosphorylates the enzyme, converting it to phosphorylase *b*, which has very low affinity for PP1c-G_L. PP1c-G_L is released from the inhibited complex, where it is free to dephosphorylate other substrates in the glycogen granule, including glycogen synthase (activating it) and phosphorylase *b* kinase (inactivating it). This mechanism explains the experimental observation that activation of glycogen synthase occurs *after* inactivation of glycogen phosphorylase.

These regulatory mechanisms explain the observation that when plasma glucose levels rise after a meal, the liver gradually takes up glucose and stores it as glycogen.

The liver regulates blood glucose levels partly by control of its glycogen synthase and phosphorylase.

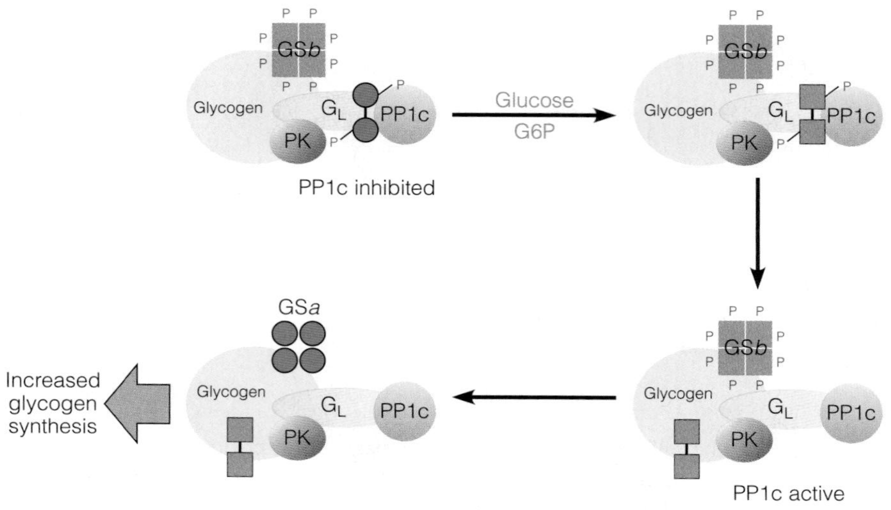

FIGURE 13.38

Regulation of phosphoprotein phosphatase 1 (PP1) in liver. The active catalytic subunit of PP1 (PP1c) is anchored to the glycogen particle by its G subunit, G_L. Glycogen synthase b (GSb), phosphorylase b kinase (PK), and glycogen phosphorylase a (green) are also bound to the glycogen particle. Glucose and glucose-6-phosphate (G6P) favor the T state of glycogen phosphorylase a (pink), which is readily dephosphorylated by PP1c, giving phosphorylase b. PP1c-G_L is released from the inhibited complex, where it is free to dephosphorylate other substrates in the glycogen granule, including glycogen synthase b, resulting in inhibition of glycogen breakdown and stimulation of glycogen synthesis.

Human mutations affecting enzymes of glycogen metabolism can have mild or profound clinical consequences.

Insulin is released from the pancreas under these conditions, stimulating glycogen synthesis. Recall from Figure 13.22 that insulin indirectly stimulates the liver isozyme of hexokinase (HK-IV), resulting in increased production of glucose-6-phosphate for incorporation into glycogen. Conversely, when blood glucose concentrations fall during fasting or exercise, glucagon secretion inhibits hepatic glycolysis (Figure 13.21) and stimulates glycogen breakdown (Figure 13.31). The liver thus maintains circulating blood glucose levels by mobilizing its glycogen stores.

Skeletal muscle, which does not carry out gluconeogenesis, relies on blood glucose supplied from the liver, as well as its own glycogen stores for fuel. Because muscle cells lack receptors for glucagon and express an isoform of pyruvate kinase (PK-M1) that is not regulated by covalent modification, muscle glycolysis is not inhibited when blood glucose concentrations are low. Muscle cells respond instead to epinephrine as part of the "fight or flight" response. Epinephrine stimulates glycogen breakdown (Figure 13.31), producing glucose-6-phosphate for ATP generation via glycolysis.

Congenital Defects of Glycogen Metabolism in Humans

A number of inherited human diseases involve mutations in genes encoding enzymes of glycogen metabolism. The clinical symptoms of these conditions, called **glycogen storage diseases**, can be quite severe and usually result from storage of abnormal quantities of glycogen or storage of glycogen with abnormal properties. Accumulation of abnormal glycogen results from its failure to be broken down. Studies on these conditions have helped identify the roles of the enzymes involved in glycogen metabolism.

Among the earliest glycogen storage diseases to be described was *von Gierke disease*, named for a German physician who studied an 8-year-old girl with a chronically enlarged liver. After her death in 1929 from influenza, her liver was found to contain 40% glycogen. The glycogen appeared normal but could not be degraded by extracts of the girl's liver, only by extracts of other livers. Today it is recognized that these symptoms can result from deficiency of either glucose-6-phosphatase or the debranching enzyme. When the debranching enzyme is deficient, phosphorylase can degrade glycogen only until branch points are reached and no farther.

Table 13.3 provides information on several of the glycogen storage diseases that have been characterized. Among the most serious clinically is the type I disease, resulting from functional lack of glucose-6-phosphatase. Individuals with this condition can break down glycogen normally, but because they cannot cleave G6P to glucose for release from the liver to the bloodstream, they are chronically hypoglycemic. In a less severe form of this disease, blood glucose levels are normal except after stress, when the normal hyperglycemic response is inhibited. One form of this disease (type Ia) results from deficiency of glucose-6-phosphatase

TABLE 13.3 Human congenital defects of glycogen metabolism

Type	Common Name	Enzyme Deficiency	Glycogen Structure	Organ Affected
Ia	von Gierke Disease	Glucose-6-phosphatase (ER)	Normal	Liver, kidney, intestine
Ib		Glucose-6-phosphate transporter (ER)	Normal	Liver
III	Cori or Forbes Disease	Debranching enzyme	Short outer chains	Liver, heart, muscle
IV	Andersen Disease	Branching enzyme	Abnormally long unbranched chains	Liver and other organs
V	McArdle Disease	Muscle glycogen phosphorylase	Normal	Skeletal muscle
VI	Hers Disease	Liver glycogen phosphorylase	Normal	Liver, leukocytes
VII	Tarui Disease	Muscle phosphofructokinase	Normal	Muscle
IX		Liver phosphorylase kinase	Normal	Liver
-		Glycogen synthase	Normal	Liver

itself. The type Ib disease involves deficiency of the specific transporter for glucose-6-phosphate into the lumen of the endoplasmic reticulum (ER). Recall that this transporter is part of a multiprotein complex, which includes glucose-6-phosphatase itself, located on the lumenal face of the ER (page 543).

Other forms of glycogen storage diseases involve abnormalities that can be easily understood in terms of the known enzymatic defect. In type III individuals, who have a defective debranching enzyme, glycogen with very short outer branches accumulates, which leads to enlargement of the liver. By contrast, type IV disease, which is associated with a defective branching enzyme, involves accumulation of glycogen with very long outer branches. Early death from liver failure is often observed in type IV individuals. Type III, V, VI, VII, and IX diseases have less severe symptoms. For instance, individuals with type V disease, with a deficiency of muscle glycogen phosphorylase, usually show no symptoms until about age 20. Once symptoms have appeared, the principal ones are severe muscle cramps upon exercising and failure of lactate to accumulate in blood after exercise. There are even some rare cases of hepatic glycogen synthase deficiency, in which affected patients have severely decreased liver glycogen stores.

Biosynthesis of Other Polysaccharides

As first discussed in Chapter 9, the synthesis of other polysaccharides involves the same mechanisms just presented for glycogen, particularly the use of nucleotide-linked sugars as activated biosynthetic intermediates and glycosyltransferase enzymes. In this section we consider briefly the synthesis of two of the most abundant and widely distributed polysaccharides—cellulose and starch.

UDP-glucose is used in plants and some bacterial species for the synthesis of cellulose, a straight-chain glucose homopolymer with $\beta(1 \rightarrow 4)$ linkages (page 332). The mechanism is identical to that of glycogen synthesis, except for the stereochemistry of glycosidic bond formation. Other nucleotide-linked sugars are also active in polysaccharide synthesis. *Adenosine* diphosphate glucose and *cytidine* diphosphate glucose are the substrates for cellulose biosynthesis in some plants. Because of the importance of cellulose in textiles and other fiber-based products, and as a potential feedstock for biofuels, there is great interest in the mechanism of cellulose biosynthesis.

Polysaccharide biosynthesis in general involves nucleotide-activated sugar intermediates and glycosyltransferase enzymes.

A Biosynthetic Pathway That Oxidizes Glucose: The Pentose Phosphate Pathway

The predominant pathway for glucose catabolism is glycolysis to yield pyruvate, followed by oxidation to CO_2 in the citric acid cycle (Chapter 14). An alternative process, the **pentose phosphate pathway**, is a remarkable, multipurpose pathway

FIGURE 13.39

Overall strategy of the pentose phosphate pathway. The pentose phosphate pathway converts glucose to various other sugars, which can be used for energy. Its most important products, however, are NADPH and ribose-5-phosphate. In **stage 1**, the oxidative phase, glucose-6-phosphate is oxidized to ribulose-5-phosphate and CO_2, with production of NADPH. (The three reactions involved in this stage are shown in Figure 13.40.) The remaining stages constitute the nonoxidative phase of the pathway. In **stage 2**, some ribulose-5-phosphate is converted to other five-carbon sugars, including ribose-5-phosphate. The ribose phosphate may be used in nucleotide synthesis (its primary use) or in the next stage of the pentose phosphate pathway. In **stage 3**, a series of reactions converts three molecules of five-carbon sugar to two molecules of six-carbon sugar and one of three-carbon sugar. In **stage 4**, some of these sugars are converted to glucose 6-phosphate, and the cycle repeats. Figure 13.43 presents a more detailed overview of the pathway and shows how it varies under different metabolic conditions.

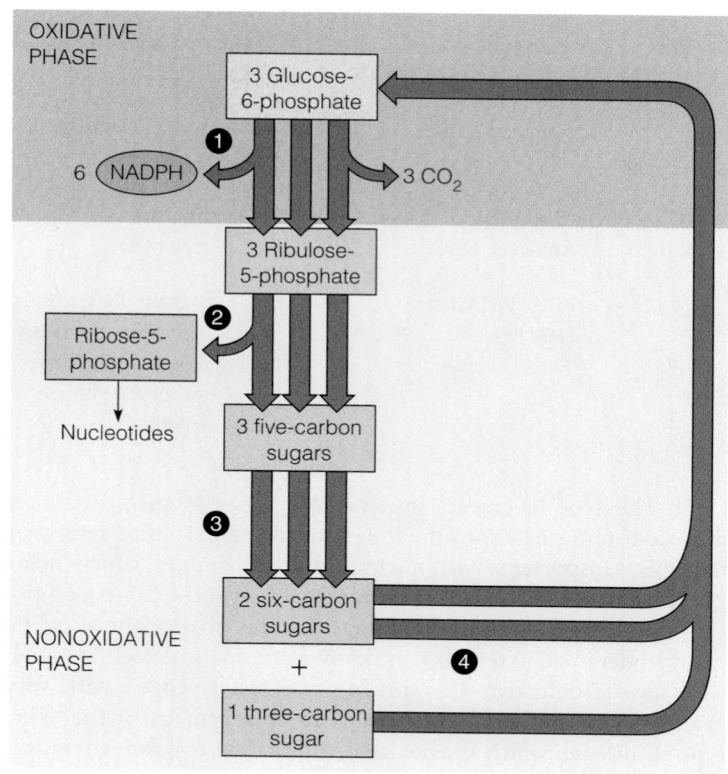

The pentose phosphate pathway primarily generates NADPH for reductive biosynthesis and ribose-5-phosphate for nucleotide biosynthesis.

that operates to varying extents in different cells and tissues. The role of this pathway is primarily anabolic rather than catabolic, but we present the pathway in this chapter because it does involve catabolism of glucose. The pathway, which operates exclusively in the cytosol, is summarized in Figure 13.39.

The pentose phosphate pathway has two primary functions: (1) to provide reducing equivalents (in the form of NADPH) for reductive biosynthesis and dealing with oxidative stress; and (2) to provide ribose-5-phosphate for nucleotide and nucleic acid biosynthesis. In addition, the pathway operates to metabolize dietary pentose sugars, derived primarily from the digestion of nucleic acids. In plants, a variant of the pentose phosphate pathway operates in reverse as part of the carbon fixation process of photosynthesis (see Chapter 16).

Recall from Chapter 12 that $NADP^+$ is identical to NAD^+ except for the additional $2'$ phosphate on one of the ribose moieties of $NADP^+$. Metabolically, the difference between NAD^+ and $NADP^+$ is that nicotinamide nucleotide–linked enzymes whose primary function is to *oxidize* substrates use the $NAD^+/NADH$ pair, whereas enzymes functioning primarily in a *reductive* direction use $NADP^+$ and NADPH. Because NADPH is used for fatty acid and steroid biosynthesis, tissues such as adrenal gland, liver, adipose, and mammary gland are rich in enzymes of the pentose phosphate pathway. NADPH is also the ultimate electron source for reduction of ribonucleotides to deoxyribonucleotides for DNA synthesis, so rapidly proliferating cells generally have high activity of pentose phosphate pathway enzymes, for production of both NADPH and ribose-5-phosphate.

The Oxidative Phase: Generation of Reducing Power as NADPH

It is convenient to think of the pentose phosphate pathway as operating in two phases—oxidative and nonoxidative. Two of the first three reactions in this pathway are oxidative, each involving reduction of one $NADP^+$ to NADPH. As shown in Figure 13.40, the first reaction, catalyzed by **glucose-6-phosphate dehydrogenase**,

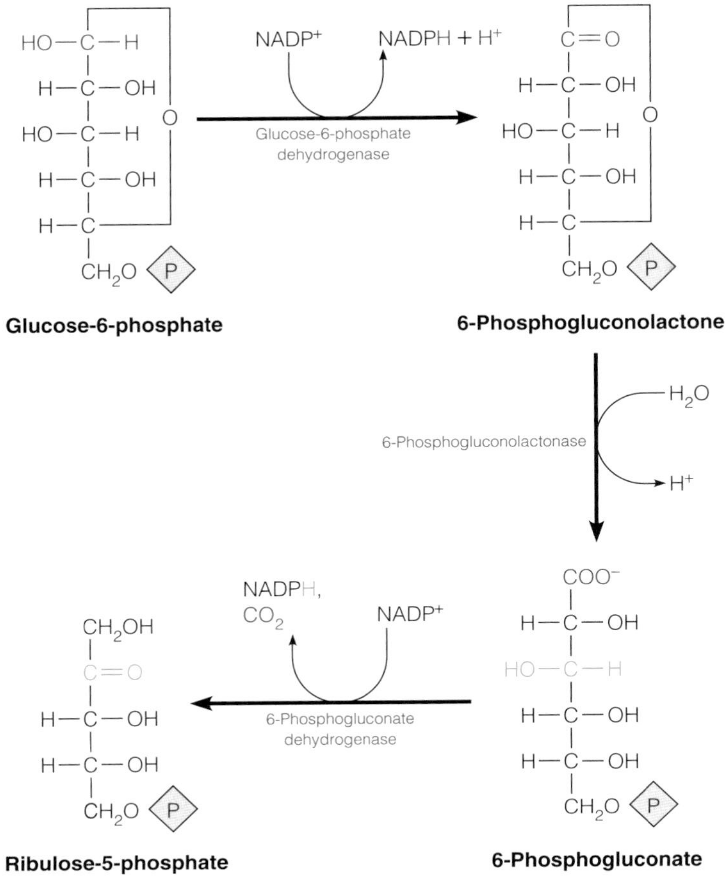

FIGURE 13.40

Oxidative phase of the pentose phosphate pathway. The three reactions of the oxidative phase include two oxidations, which produce NADPH.

oxidizes glucose-6-phosphate to 6-phosphogluconolactone, the corresponding **lactone** (an internal ester linking carbons 1 and 5). Phosphogluconolactone is hydrolyzed by **6-phosphogluconolactonase** to **6-phosphogluconate**, which undergoes an oxidative decarboxylation to yield CO_2, another NADPH, and **ribulose-5-phosphate** (a pentose phosphate). The net result of the oxidative phase is generation of 2 molecules of NADPH, oxidation of one carbon to CO_2, and synthesis of 1 molecule of pentose phosphate per G6P.

The Nonoxidative Phase: Alternative Fates of Pentose Phosphates

In the nonoxidative phase, some of the ribulose-5-phosphate produced in the oxidative phase is converted to ribose-5-phosphate by **phosphopentose isomerase**.

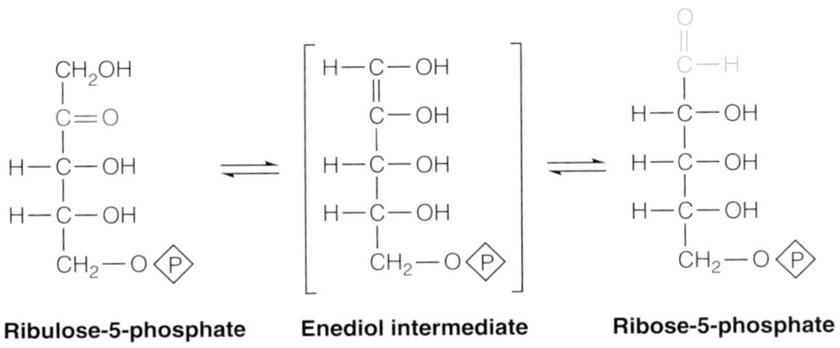

The reaction proceeds via an enediol intermediate, just as in two different reactions of glycolysis—those catalyzed by triose phosphate isomerase (see page 529) and phosphoglucoisomerase (see page 526).

Production of Six-Carbon and Three-Carbon Sugar Phosphates

At this stage the primary functions of the pathway have been fulfilled, namely the generation of NADPH and ribose-5-phosphate. We can write a balanced equation for what has transpired thus far:

$$\text{Glucose-5-phosphate} + 2NADP^+ \longrightarrow \text{ribose-5-phosphate} + CO_2 + 2NADPH + 2H^+$$

Many cells need the NADPH for reductive biosynthesis but do not need the ribose-5-phosphate in such large quantities. How, then, is this ribose-5-phosphate catabolized? The process involves a series of sugar phosphate transformations that may look complicated but that have a simple result. *The reaction sequence converts 3 five-carbon sugar phosphates to 2 six-carbon sugar phosphates and 1 three-carbon sugar phosphate.* The hexose phosphates formed can be catabolized either by recycling through the pentose phosphate pathway or by glycolysis. The triose phosphate is glyceraldehyde-3-phosphate, a glycolytic intermediate. Three enzymes are involved: **phosphopentose epimerase**, **transketolase**, and **transaldolase**.

The pathway begins with both ribulose-5-phosphate and ribose-5-phosphate, the latter having been formed by phosphopentose isomerase. Phosphopentose epimerase converts ribulose-5-phosphate to its epimer, xylulose-5-phosphate.

Ribulose-5-phosphate　　　　　　　　**Xylulose-5-phosphate**

Recall that xylulose-5-phosphate also plays a regulatory role—it is a specific activator of protein phosphatase 2A (PP2A) in liver. PP2A dephosphorylates PFK-2/FBPase-2 (see Figure 13.21, page 554), leading to increased levels of fructose-2,6-bisphosphate, which activates the glycolytic enzyme phosphofructokinase (PFK-1). Thus, as xylulose-5-phosphate levels rise in response to excess glucose via the pentose phosphate pathway, flux through glycolysis also increases.

One mole of xylulose-5-phosphate then reacts with one mole of ribose-5-phosphate. The reaction is catalyzed by **transketolase**, which transfers a two-carbon fragment from xylulose-5-phosphate to ribose-5-phosphate to give a triose phosphate, glyceraldehyde-3-phosphate, and a seven-carbon sugar, **sedoheptulose-7-phosphate**.

Xylulose-5-phosphate　　**Ribose-5-phosphate**　　　　　　**Glyceraldehyde-3-phosphate**　　**Sedoheptulose-7-phosphate**

The two-carbon fragment transferred is an activated **glycolaldehyde** fragment (see margin, next page). Recall that pyruvate decarboxylase transfers an activated *acetaldehyde* fragment (see Figure 13.8, page 538), with the aid of thiamine

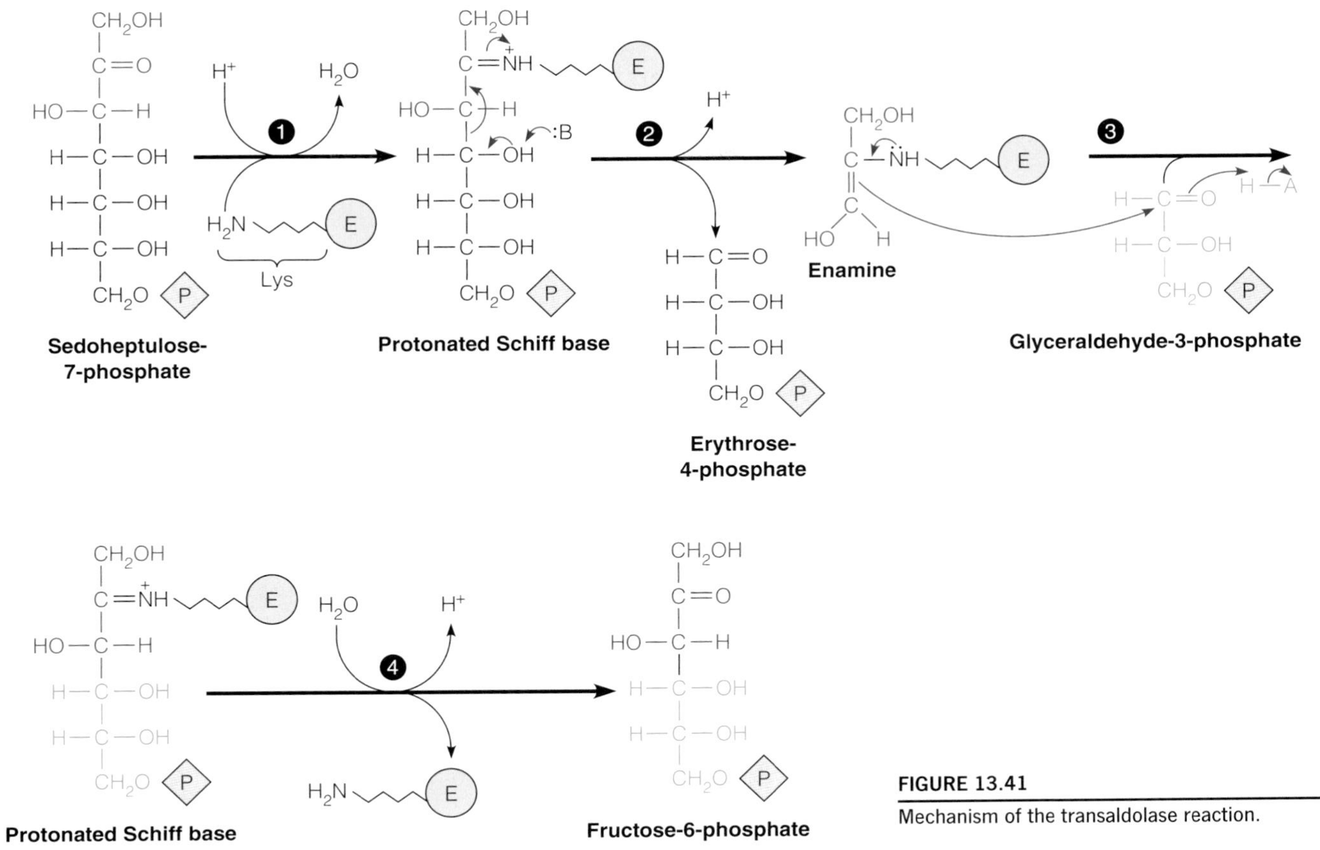

FIGURE 13.41

Mechanism of the transaldolase reaction.

pyrophosphate (TPP). Transketolase also requires TPP as a cofactor, with the two-carbon fragment bound transiently to carbon 2 of the thiazole ring of TPP. The mechanism of activation and transfer of two-carbon fragments is very similar in the reactions catalyzed by these two enzymes.

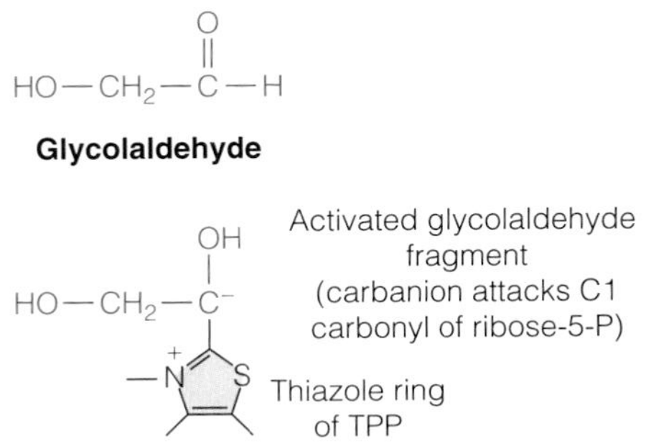

Next, transaldolase acts on the two products of the transketolase reaction, with transfer of a three-carbon **dihydroxyacetone** unit from the seven-carbon substrate to the three-carbon substrate. The products are a four-carbon sugar phosphate and a six-carbon sugar phosphate: **erythrose-4-phosphate** and fructose-6-phosphate, respectively. The combined actions of transketolase and transaldolase convert 2 five-carbon sugar phosphates to a four-carbon sugar phosphate and a six-carbon sugar phosphate.

Figure 13.41 shows the transaldolase reaction in more detail. The enzyme activates the ketose substrate by forming a Schiff base with a lysine residue on the enzyme (step 1). Protonation of the Schiff base leads to carbon–carbon bond cleavage, much as occurs in the fructose-bisphosphate aldolase reaction of glycolysis (see Figure 13.5, page 528), with release of a four-carbon aldose phosphate (step 2). The dihydroxyacetone unit remains bound as an enamine, which then adds to the carbonyl carbon of glyceraldehyde-3-phosphate, in an aldol condensation reaction (step 3). Hydrolysis of the protonated Schiff base yields the six-carbon product, fructose-6-phosphate (step 4).

The transketolase and transaldolase mechanisms represent two different chemical strategies to effect C—C bond cleavage (Figure 13.42). Both mechanisms use an iminium ion as an electron sink. In addition, both mechanisms involve carbanion intermediates that must be stabilized by resonance. Transketolase uses a cofactor (TPP), while transaldolase uses a protonated Schiff base with an active site lysine to solve the same problem.

Glycolaldehyde

Activated glycolaldehyde fragment (carbanion attacks C1 carbonyl of ribose-5-P)

Thiazole ring of TPP

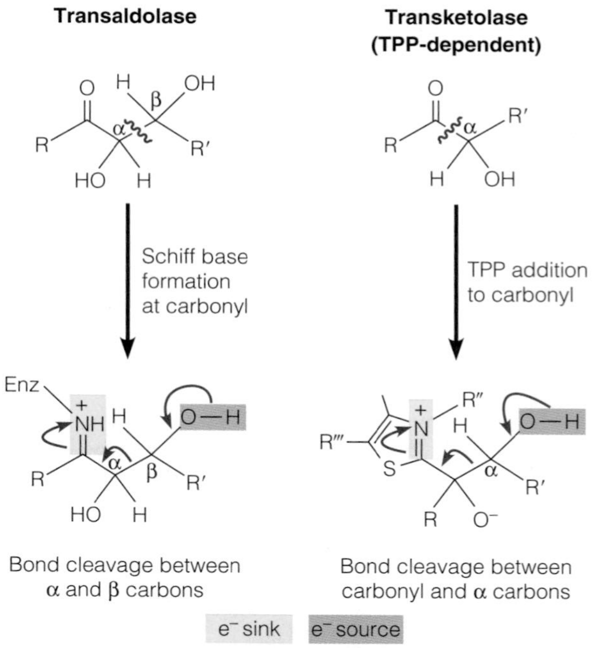

FIGURE 13.42

C—C bond cleavage in transketolase and transaldolase reactions. Yellow, electron sink; green, electron source.

Transaldolase   **Transketolase (TPP-dependent)**

Bond cleavage between α and β carbons

Bond cleavage between carbonyl and α carbons

e⁻ sink   e⁻ source

In the final reaction of pentose phosphate catabolism, transketolase acts on another molecule of xylulose-5-phosphate, transferring a glycolaldehyde fragment to erythrose-4-phosphate and generating a three-carbon product and a six-carbon product—glyceraldehyde-3-phosphate and fructose-6-phosphate, respectively.

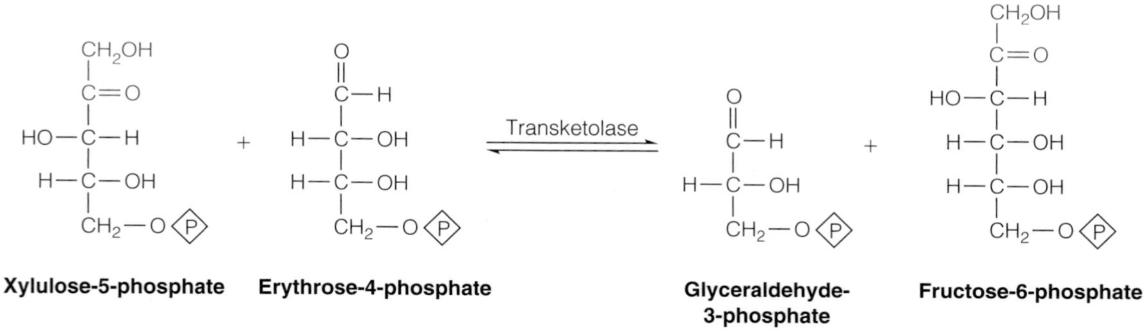

Xylulose-5-phosphate   **Erythrose-4-phosphate**   **Glyceraldehyde-3-phosphate**   **Fructose-6-phosphate**

So far, the pathway has required input of three pentose phosphate molecules—two for the first transketolase reaction and one for the second. Thus, to summarize the pathway to this point, we must consider three molecules of glucose-6-phosphate passing through the oxidative phase.

$$3\,\text{Glucose-6-phosphate} + 6\text{NADP}^+ + 3\text{H}_2\text{O} \longrightarrow 3\,\text{pentose-5-phosphate} + 6\text{NADPH} + 6\text{H}^+ + 3\text{CO}_2$$

Now the rearrangements of the nonoxidative phase result in conversion of three pentose phosphates to 2 six-carbon and 1 three-carbon sugar phosphates.

$$2\,\text{Xylulose-5-phosphate} + \text{ribose-5-phosphate} \longrightarrow 2\,\text{fructose-6-phosphate} + \text{glyceraldehyde-3-phosphate}$$

Thus, we can write a balanced equation for the entire pathway as follows.

$$3\,\text{Glucose-6-phosphate} + 6\text{NADP}^+ + 3\text{H}_2\text{O} \longrightarrow 2\,\text{fructose-6-phosphate} + \text{glyceraldehyde-3-phosphate} + 6\text{NADPH} + 6\text{H}^+ + 3\text{CO}_2$$

Tailoring the Pentose Phosphate Pathway to Specific Needs

In the equation for the overall reaction, three hexose phosphates yield two hexose phosphates, one triose phosphate, and three molecules of CO_2. In a formal sense, therefore, the pathway can be seen as a means to oxidize the six carbons of glucose-6-phosphate to CO_2, just as occurs in glycolysis plus the citric acid cycle.

However, as noted above, the pentose phosphate pathway is not primarily an energy-generating pathway. The actual fate of the sugar phosphates depends on the metabolic needs of the cell in which the pathway is occurring. If the primary need is for nucleotide and nucleic acid synthesis, the major product is ribose-5-phosphate, and most of the rearrangements of the nonoxidative phase do not take place (Figure 13.43a). If the primary need is for NADPH generation (for fatty acid or steroid synthesis), the nonoxidative phase generates compounds that can easily be reconverted to glucose-6-phosphate, for subsequent passage through the oxidative phase (Figure 13.43b). In this mode, repeated turns of the cycle result ultimately in the complete oxidation of glucose-6-phosphate to CO_2 and water, with maximal generation of reducing equivalents.

Finally, in a cell with moderate needs for both NADPH and pentose phosphates, the fructose-6-phosphate and glyceraldehyde-3-phosphate produced in the nonoxidative phase can be further catabolized by glycolysis and the citric acid cycle (Figure 13.43c). Because of the multiple metabolic needs of a cell for biosynthesis, it is unlikely that any one of these three modes operates exclusively in any one cell.

Regulation of the Pentose Phosphate Pathway

The pentose phosphate pathway competes with glycolysis for glucose-6-phosphate. Whereas glycolysis is regulated primarily by energy charge and fuel availability, flux through the pentose phosphate pathway is sensitive to the $NADP^+$/NADPH ratio of the cell. The first enzyme of the pathway, glucose-6-phosphate dehydrogenase, represents the committed step, and its activity controls flux through the entire pentose phosphate pathway. Glucose-6-phosphate dehydrogenase is regulated by the availability of $NADP^+$. If the $NADP^+$/NADPH ratio is low, indicating that the cell has plenty of reducing power, glucose-6-phosphate dehydrogenase activity will be low and the pathway will not divert glucose-6-phosphate from glycolysis. If, however, the cell needs more reducing equivalents, the high $NADP^+$/NADPH ratio will stimulate flux through glucose-6-phosphate dehydrogenase, regenerating the necessary NADPH.

Human Genetic Disorders Involving Pentose Phosphate Pathway Enzymes

The pentose phosphate pathway is particularly active in the generation of reducing power in the red blood cells of vertebrates. The importance of this activity became apparent through investigation of a fairly widespread human genetic disorder, a deficiency of glucose-6-phosphate dehydrogenase.

During World War II the antimalarial drug primaquine was prophylactically administered to members of the armed forces. As a result, a significant proportion of servicemen suffered a severe hemolytic anemia (massive destruction of red blood cells). They were also sensitive to a variety of compounds that, like primaquine, generate oxidative stress, as manifested by the appearance of hydrogen peroxide and organic peroxides in red cells. These individuals were later found to be deficient in glucose-6-phosphate dehydrogenase.

Primaquine

Abnormal sensitivity to antimalarial drugs was shown to result from mutations affecting glucose-6-phosphate dehydrogenase.

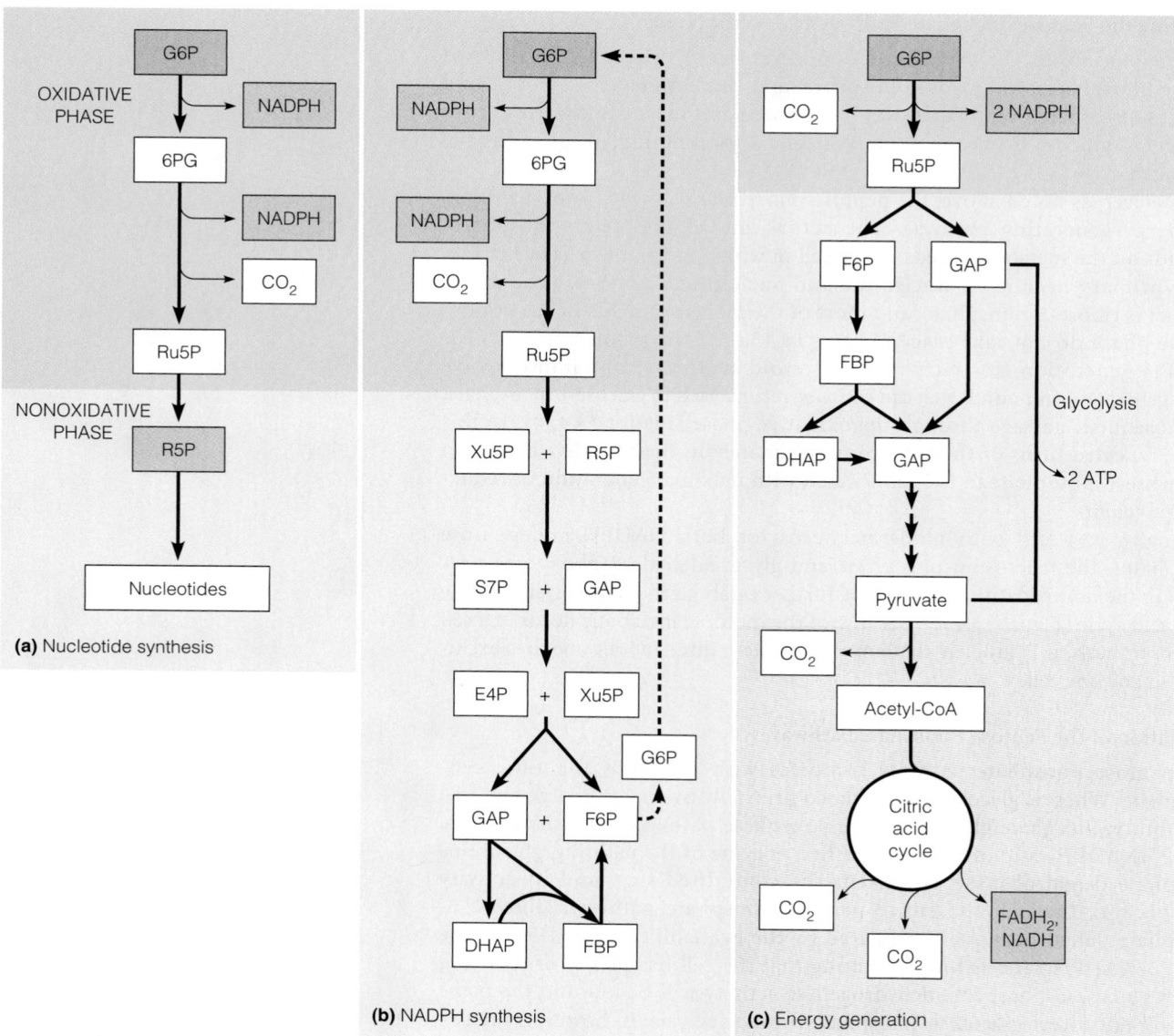

FIGURE 13.43

Alternative pentose phosphate pathway modes. The pentose phosphate pathway has different modes of operation to meet varying metabolic needs. **(a)** When the primary need is for nucleotide biosynthesis, the primary product is ribose-5-phosphate. Reducing equivalents from NADPH are used to reduce ribonucleotides to deoxyribonucleotides (Chapter 22). **(b)** When the primary need is for reducing power (NADPH), fructose phosphates are reconverted to glucose-6-phosphate for reoxidation in the oxidative phase. **(c)** When only moderate quantities of pentose phosphates and NADPH are needed, the pathway can also be used to supply energy, with the reaction products being oxidized through glycolysis and the citric acid cycle. Abbreviations: G6P = glucose-6-phosphate; 6PG = 6-phosphogluconate; R5P = ribose-5-phosphate; Ru5P=ribulose-5-phosphate; Xu5P = xylulose-5-phosphate; S7P = sedoheptulose-7-phosphate; GAP = glyceraldehyde-3-phosphate; E4P = erythrose-4-phosphate; F6P = fructose-6-phosphate; DHAP = dihydroxyacetone phosphate; FBP = fructose-1,6-bisphosphate.

Normally, peroxides are inactivated via reduction by **glutathione**, which is the tripeptide γ-glutamylcysteinylglycine.

$$H_3\overset{+}{N}-CH-CH_2-CH_2-\overset{\overset{\displaystyle O}{\displaystyle\|}}{C}-NH-CH-\overset{\overset{\displaystyle O}{\displaystyle\|}}{C}-NH-CH_2-COO^-$$

γ-Glu   Cys   Gly

Glutathione

Glutathione is abundant in most cells and, because of its free thiol group, it represents a major protective mechanism against oxidative stress. For example, it helps keep cysteine thiol groups in proteins in the reduced state. If two thiol groups become oxidized, they can be reduced nonenzymatically by glutathione.

Glutathione, an abundant thiol-containing tripeptide, is a major intracellular reductant.

$$\text{Protein} \begin{array}{c} S \\ \\ S \end{array} + 2\gamma\text{-Glu}-\text{Cys}-\text{Gly} \rightleftharpoons \text{Protein} \begin{array}{c} SH \\ \\ SH \end{array} + \begin{array}{c} \gamma\text{-Glu}-\text{Cys}-\text{Gly} \\ | \\ S \\ | \\ S \\ | \\ \gamma\text{-Glu}-\text{Cys}-\text{Gly} \end{array}$$

And, as noted, glutathione also carries out the reduction of peroxides; this is an enzymatic reaction, catalyzed by **glutathione peroxidase** (see page 667).

$$R-OOH + 2\gamma\text{-Glu}-\text{Cys}-\text{Gly} \rightleftharpoons \gamma\text{-Glu}-\text{Cys}-\text{Gly} + H_2O + ROH$$

Oxidized glutathione (GSSG) is reduced by the NADPH-dependent enzyme **glutathione reductase**.

$$GSSG + NADPH + H^+ \rightarrow 2GSH + NADP^+$$

This enzyme acts essentially unidirectionally, so that the ratio of reduced glutathione (GSH) to oxidized glutathione in most cells is about 500 to 1.

In the erythrocyte, a particularly important role of glutathione is to maintain hemoglobin in the reduced (Fe^{2+}) state; recall that methemoglobin (Fe^{3+}) cannot bind O_2 (see page 238). Therefore, the erythrocyte is especially sensitive to depletion of reduced glutathione. And because the pentose phosphate pathway is the major pathway for generation of NADPH, the erythrocyte is especially vulnerable to conditions that impair flux through this pathway and thereby lower intracellular NADPH levels. Thus, the individuals who were deficient in glucose-6-phosphate dehydrogenase were the ones most sensitive to oxidative stress caused by primaquine.

In most cases of glucose-6-phosphate dehydrogenase deficiency, the enzyme in red cells is not totally inactive but instead is decreased in activity by about 10-fold. Individuals with this deficiency are asymptomatic until stressed. That is, they are asymptomatic until primaquine or a related agent generates enough peroxides that the available GSH becomes depleted. Reduction of the resultant GSSG back to GSH is impaired because NADPH levels are inadequate to allow glutathione reductase to function. This causes methemoglobin (Fe^{3+}) to accumulate at the expense of hemoglobin, which in turn changes the structure of the cell, weakening the membrane and rendering it sensitive to rupture, or hemolysis.

Interestingly, glucose-6-phosphate dehydrogenase deficiency, like sickle-cell trait, confers resistance to malaria caused by *Plasmodium falciparum* (see Chapter 7). Thus, the deficiency has a positive survival value in tropical and subtropical regions of the world, where malaria is common. This explains the observation that glucose-6-phosphate dehydrogenase deficiency is seen most frequently among individuals of African or Mediterranean extraction.

Another genetic disorder related to the pentose phosphate pathway is the **Wernicke–Korsakoff syndrome**. This mental disorder is coupled with loss of memory and partial paralysis and develops when affected individuals suffer a moderate thiamine deficiency. The symptoms often become manifest in alcoholics, whose diets are apt to be vitamin deficient.

A mutation in transketolase that increases its K_M for TPP is responsible for the neurological symptoms of Wernicke–Korsakoff syndrome.

The basis for the Wernicke–Korsakoff syndrome is an alteration of transketolase that reduces its affinity for thiamine pyrophosphate by about 10-fold. Other TPP-dependent enzymes are not affected. Symptoms of the disease become manifest when TPP levels drop below the values needed to saturate the abnormal transketolase. Normal individuals contain a transketolase that binds TPP strongly enough that no change in enzyme function occurs as a result of these slight to moderate thiamine deficiencies.

Both glucose-6-phosphate dehydrogenase deficiency and the Wernicke–Korsakoff syndrome, like sickle-cell disease (Chapter 7), illustrate the interdependence of genetic and environmental factors in the onset of clinical disease. Symptoms of the hereditary change express themselves only after some kind of moderate stress that does not affect normal individuals.

SUMMARY

Glycolysis is the central pathway by which energy is extracted from carbohydrates. A 10-step pathway leads from glucose to pyruvate in respiring cells. In anaerobic microorganisms or in cells with impaired respiration, pyruvate undergoes reductive reactions, so that the overall pathway can proceed with no net change in oxidation state. Glycolysis can be viewed as occurring in two phases—first, an energy investment phase in which ATP is used to synthesize a six-carbon sugar phosphate that is split to yield two triose phosphates, and second, an energy generation phase, in which the energy of two high-energy compounds is used to drive ATP synthesis from ADP. Phosphofructokinase, pyruvate kinase, and hexokinase are the major sites for control of the pathway. Much of the control is related to the energy needs of a cell, with conditions of low energy charge stimulating the pathway and conditions of energy abundance retarding the pathway. All organisms also carry out gluconeogenesis, the synthesis of carbohydrate from noncarbohydrate three-carbon and four-carbon compounds. Gluconeogenesis uses seven glycolytic enzymes and four specific gluconeogenic enzymes, the latter to bypass the three irreversible steps in glycolysis. The four enzymes specific to gluconeogenesis are pyruvate carboxylase plus phosphoenolpyruvate carboxykinase, fructose-1,6-bisphosphatase, and glucose-6-phosphatase. Regulation occurs at the sites of these three substrate cycles. Control is supremely important in animal metabolism, requiring that blood glucose levels be maintained within narrow limits. Hormonal and allosteric mechanisms are involved, with fructuose-2,6-bisphosphate being a key regulator.

Intracellular polysaccharide stores in animals are mobilized by a hormonally controlled metabolic cascade, in which cyclic AMP transmits the hormonal signal and sets in motion events that activate the breakdown of glycogen to glucose-1-phosphate. Glycogen phosphorylase is the rate-limiting step of this process. Synthesis of polysaccharides such as glycogen involves glycosyltransferases, enzymes that transfer the sugar unit from a nucleotide-linked or otherwise activated sugar to an acceptor sugar, at a nonreducing end. Glycogen synthase uses uridine diphosphate glucose as its glucosyl donor. The enzyme is regulated by hormonal and nonhormonal processes that are complementary and opposed to those that regulate glycogen breakdown by phosphorylase.

An alternative glucose oxidative pathway, the pentose phosphate pathway, generates NADPH for reductive biosynthesis and pentose phosphates for nucleotide biosynthesis.

REFERENCES

Intracellular Organization of Glycolytic Enzymes

Ovadi, J., and P. A. Srere (2000) Macromolecular compartmentation and channeling. *Int. Rev. Cytol.* 192:255–280. This review was written by two of the pioneers of metabolic compartmentation.

Glycolytic and Gluconeogenic Enzymes

Hanson, R. W., and L. Reshef (1997) Regulation of phosphoenolpyruvate carboxykinase (GTP) gene expression. *Annu. Rev. Biochem.* 66:581–611.

A great deal is known about hormonal and dietary regulation of the synthesis of PEPCK.

Hutton, J. C., and R. M. O'Brien (2009) Glucose-6-phosphatase catalytic subunit gene family. *J. Biol. Chem.* 284:29241–29245.

Jitrapakdee, S., M. St Maurice, I. Rayment, W. W. Cleland, J. C. Wallace, and P. V. Attwood (2008) Structure, mechanism and regulation of pyruvate carboxylase. *Biochem. J.* 413:369–387.

Kim, J. W., and C. V. Dang (2005) Multifaceted roles of glycolytic enzymes. *Trends Biochem. Sci.* 30:142–150. Several glycolytic enzymes have a surprising range of functions in addition to catalytic roles in glycolysis.

Kresge, N., R. D. Simoni, and R. L. Hill (2005) Otto Fritz Meyerhof and the elucidation of the glycolytic pathway. *J. Biol. Chem.* 280:e3. This review summarizes the classic papers from Meyerhof and coworkers.

Glycogen Metabolism

Aiston, S., L. Hampson, A. M. Gomez-Foix, J. J. Guinovart, and L. Agius (2001) Hepatic glycogen synthesis is highly sensitive to phosphorylase activity: Evidence from metabolic control analysis. *J. Biol. Chem.* 276:23858–23866. This paper applies metabolic control analysis (see Chapter 12) to the complexities of controlling glycogen synthesis.

Chen, Y.-T. (2001) Glycogen storage diseases. In: *The Metabolic and Molecular Bases of Inherited Disease*, edited by C. R. Scriver, A. L. Beaudet, W. S. Sly, D. Valle, B. Childs, K. W. Kinzler, and B. Vogelstein, Vol. I, Ch. 71, pp. 1521–1551. McGraw-Hill, New York. A chapter in the four-volume treatise considered the most authoritative reference on heritable metabolic human diseases.

Greenberg, C. C., M. J. Jurczak, A. M. Danos, and M. J. Brady (2006) Glycogen branches out: New perspectives on the role of glycogen metabolism in the integration of metabolic pathways. *Am. J. Physiol. Endocrinol. Metab.* 291:E1–E8.

Holton, J. B., J. H. Walter, and L. A. Tyfield (2001) Galactosemia. In *The Metabolic and Molecular Bases of Inherited Disease*, edited by C. R. Scriver, A. L. Beaudet, W. S. Sly, D. Valle, B. Childs, K. W. Kinzler, and B. Vogelstein, Vol. I, Ch. 72, pp. 1553–1587. McGraw-Hill, New York. A comprehensive review of galactosemias and related disorders.

Johnson, L. N. (2009) The regulation of protein phosphorylation. *Biochem. Soc. Trans.* 37:627–641. This excellent review describes recent as well as historical work on protein phosphorylation, including the classical studies on glycogen phosphorylase.

Leloir, L. F. (1983) Long ago and far away. *Annu. Rev. Biochem.* 52:1–16. A personal reminiscence, describing the author's Nobel Prize–winning role in the discovery of nucleotide-linked sugars and the mechanism of glycogen synthesis.

Millward, T. A., S. Zolnierowicz, and B. A. Hemmings (1999) Regulation of protein kinase cascades by protein phosphatase 2A. *Trends Biochem. Sci.* 24:186–191. A classic minireview about the control of protein phosphorylation and dephosphorylation.

Toole, B. J., and P. T. W. Cohen (2007) The skeletal muscle-specific glycogen-targeted protein phosphatase 1 plays a major role in the regulation of glycogen metabolism by adrenaline in vivo. *Cell. Signal.* 19:1044–1055.

Regulation of Carbohydrate Metabolism

Agius, L. (2008) Glucokinase and molecular aspects of liver glycogen metabolism. *Biochem. J.* 414:1–18.

Bocarsly, M. E., E. S. Powell, N. M. Avena, and B. G. Hoebel (2010) High-fructose corn syrup causes characteristics of obesity in rats: Increased body weight, body fat and triglyceride levels. *Pharmacol. Biochem. Behav.* 97:101–106. This paper presents experimental evidence supporting the link between high-fructose corn syrup and obesity and type 2 diabetes.

Brosnan, J. T. (1999) Comments on metabolic needs for glucose and the role of gluconeogenesis. *Eur. J. Clin. Nutr.* 53 Suppl 1: S107–S111.

Ceulemans, H., and M. Bollen (2004) Functional diversity of protein phosphatase-1, a cellular economizer and reset button. *Physiol. Rev.* 84:1–39.

Kim, C., C. Y. Cheng, S. A. Saldanha, and S. S. Taylor (2007) PKA-I holoenzyme structure reveals a mechanism for cAMP-dependent activation. *Cell* 130:1032–1043. Gives the structural basis for the activation of protein kinase upon dissociation into its subunits.

Lee, Y. H., Y. Li, K. Uyeda, and C. A. Hasemann (2003) Tissue-specific structure/function differentiation of the liver isoform of 6-phosphofructo-2-kinase/fructose-2,6-bisphosphatase. *J. Biol. Chem.* 278:523–530.

El-Maghrabi, M. R., F. Noto, N. Wu, and N. Manes (2001) 6-Phosphofructo-2-kinase/fructose-2,6-bisphosphatase: Suiting structure to need, in a family of tissue-specific enzymes. *Curr. Opin. Clin. Nutr. Metab. Care* 4:411–418.

Nordlie, R. C., J. D. Foster, and A. J. Lange (1999) Regulation of glucose production by the liver. *Annu. Rev. Nutr.* 19:379–406. A comprehensive review of the role of the glucose-6-phosphatase/glucokinase substrate cycle in glucose homeostasis.

Okar, D. A., C. Wu, and A. J. Lange (2004) Regulation of the regulatory enzyme, 6-phosphofructo-2-kinase/fructose-2,6-bisphosphatase. *Adv. Enzyme Regul.* 44:123–154.

Sims, R. E., W. Mabee, J. N. Saddler, and M. Taylor (2010) An overview of second generation biofuel technologies. *Bioresour. Technol.* 101: 1570–1580. A review article that describes processes under development for converting cellulosic feedstocks to usable ethanol.

Smith, W. E., S. Langer, C. Wu, S. Baltrusch, and D. A. Okar (2007) Molecular coordination of hepatic glucose metabolism by the 6-phosphofructo-2-kinase/fructose-2,6-bisphosphatase:glucokinase complex. *Mol. Endocrinol.* 21:1478–1487.

Sprang, S. R., S. G. Withers, E. J. Goldsmith, R. J. Fletterick, and N. B. Madsen (1991) Structural basis for the activation of glycogen phosphorylase *b* by adenosine monophosphate. *Science* 254:1367–1371. One of a series of reports describing the crystal structure of glycogen phosphorylase in activated and inactivated states.

Tolonen, A. C., W. Haas, A. C. Chilaka, J. Aach, S. P. Gygi, and G. M. Church (2011) Proteome-wide systems analysis of a cellulosic biofuel-producing microbe. *Mol. Syst. Biol.* 7:461. This proteomics analysis of a cellulose fermenting bacterium reveals new engineering targets for industrial biofuels production.

Valeyev, N. V., D. G. Bates, P. Heslop-Harrison, I. Postlethwaite, and N. V. Kotov (2008) Elucidating the mechanisms of cooperative calcium-calmodulin interactions: A structural systems biology approach. *BMC Syst. Biol.* 2:48. A discussion of the structural basis for the interactions of target proteins with calmodulin.

Vander Heiden, M. G., L. C. Cantley, and C. B. Thompson (2009) Understanding the Warburg effect: The metabolic requirements of cell proliferation. *Science* 324:1029–1033. This short review describes the role of aerobic glycolysis in cancer.

Welch, E. J., B. W. Jones, and J. D. Scott (2010) Networking with AKAPs: Context-dependent regulation of anchored enzymes. *Mol. Interv.* 10:86–97. Many hormonal signals involve cAMP-dependent protein kinase A activation, and location of protein kinase within the cell helps to establish specificity of particular signaling pathways.

Zhao, S., W. Xu, W. Jiang, W. Yu, Y. Lin, T. Zhang, J. Yao, L. Zhou, Y. Zeng, H. Li, Y. Li, J. Shi, W. An, S. M. Hancock, F. He, L. Qin, J. Chin, P. Yang, X. Chen, Q. Lei, Y. Xiong, and K. L. Guan (2010) Regulation of cellular metabolism by protein lysine acetylation. *Science* 327:1000–1004. This recent work suggests yet another covalent modification (acetylation) involved in regulation of glycolysis and gluconeogenesis.

Analysis of Carbohydrate Metabolism by In Vivo NMR

Shulman, R. G., and D. L. Rothman (2001) ^{13}C NMR of intermediary metabolism: Implications for systemic physiology. *Annu. Rev. Physiol.* 63:15–48. This review summarizes how NMR studies have changed some of our concepts about energy metabolism and its regulation.

van Zijl, P. C., C. K. Jones, J. Ren, C. R. Malloy, and A. D. Sherry (2007) MRI detection of glycogen in vivo by using chemical exchange saturation transfer imaging (glycoCEST). *Proc. Natl. Acad. Sci. USA* 104:4359–4364.

Oscillations of Glycolytic Intermediates

O'Neill, J. S., and A. B. Reddy (2011) Circadian clocks in human red blood cells. *Nature* 469:498–503. NADH and ATP levels oscillate with a 24-hour rhythm.

Richard, P. (2003) The rhythm of yeast. *FEMS Microbiol. Rev.* 27:547–557. A review of the oscillatory behavior of the glycolytic pathway in yeast.

Richter, P. H., and J. Ross (1981) Concentration oscillations and efficiency: Glycolysis. *Science* 211:715–716. A theoretical discussion of the energetic advantages to a living system of the oscillations observed in levels of glycolytic intermediates.

Tu, B. P., R. E. Mohler, J. C. Liu, K. M. Dombek, E. T. Young, R. E. Synovec, and S. L. McKnight (2007) Cyclic changes in metabolic state during the life of a yeast cell. *Proc. Natl. Acad. Sci. USA* 104:16886–16891. This paper nicely illustrates the power of metabolomics.

Evolution of Carbohydrate Metabolic Pathways

Martin, W., J. Baross, D. Kelley, and M. J. Russell (2008) Hydrothermal vents and the origin of life. *Nat. Rev. Microbiol.* 6:805–814.

Pentose Phosphate Pathway and Oxidative Stress

Sies, H. (1999) Glutathione and its role in cellular functions. *Free Radic. Biol. Med.* 27:916–921. Reviews the chemistry and biochemistry of this important biological reductant.

Wamelink, M. M., E. A. Struys, and C. Jakobs (2008) The biochemistry, metabolism, and inherited defects of the pentose phosphate pathway: A review. *J. Inherit. Metab. Dis.* 31:703–717.

PROBLEMS

1. Intracellular concentrations in resting muscle are as follows: fructose-6-phosphate, 1.0 mM; fructose-1,6-bisphosphate, 10 mM; AMP, 0.1 mM; ADP, 0.5 mM; ATP, 5 mM; and P_i, 10 mM. Is the phosphofructokinase reaction in muscle *more* or *less* exergonic than under standard conditions? By how much?

2. Methanol is highly toxic, not because of its own biological activity but because it is converted metabolically to formaldehyde, through action of alcohol dehydrogenase. Part of the medical treatment for methanol poisoning involves administration of large doses of ethanol. Explain why this treatment is effective.

3. Refer to Figure 13.9, which indicates ΔG for each glycolytic reaction under intracellular conditions. Assume that glyceraldehyde-3-phosphate dehydrogenase was inhibited with iodoacetic acid. Which glycolytic intermediate would you expect to accumulate most rapidly, and why?

4. In different organisms sucrose can be cleaved either by hydrolysis or by phosphorolysis. Calculate the ATP yield per mole of sucrose metabolized by anaerobic glycolysis starting with (a) hydrolytic cleavage and (b) phosphorolytic cleavage.

5. Suppose it were possible to label glucose with ^{14}C at any position or combination of positions. For yeast fermenting glucose to ethanol, which form or forms of labeled glucose would give the *most* radioactivity in CO_2 and the *least* in ethanol?

6. Write balanced chemical equations for each of the following: (a) anaerobic glycolysis of 1 mole of sucrose, cleaved initially by sucrose phosphorylase; (b) aerobic glycolysis of 1 mole of maltose; (c) fermentation of one glucose residue in starch to ethanol, with the initial cleavage involving α-amylase.

7. Because of the position of arsenic in the periodic table, arsenate (AsO_4^{3-}) is chemically similar to inorganic phosphate and is used by phosphate-requiring enzymes as an alternative substrate. However, organic arsenates are quite unstable and spontaneously hydrolyze. Arsenate is known to inhibit ATP production in glycolysis. Identify the target enzyme, and explain the mechanism of inhibition.

8. Suppose that you made some wine whose alcohol content was 10% w/v (i.e., 10 g of ethanol per 100 mL of wine). The initial fermentation mixture would have had to contain what molar concentration of glucose or its equivalent to generate this much ethanol? Is it likely that an initial fermentation mixture would contain that much glucose? In what other forms might the fermentable carbon appear?

9. Briefly discuss why each of the three common forms of galactosemia involves impaired utilization of galactose. Which metabolic process is blocked in each condition?

*10. Some anaerobic bacteria use an alternative pathway for glucose catabolism, shown below, that converts glucose to acetate rather than to pyruvate. The first part of this pathway (glucose to fructose-1,6-bisphosphate) is identical to the glycolytic pathway. In the second part of the alternative pathway, Enzymes 1-6 all have mechanisms/activities analogous to enzymes in glycolysis. Note that there are two C—C bond cleavage reactions in this new pathway: A → B + C (Enzyme 1) and C → B + D. All the steps where ATP is consumed or generated have been shown; however, the addition or loss of $NAD^+/NADH$, P_i, H_2O, or H^+ has not been shown explicitly. Draw the structures for the intermediates B, F, G, H, and I, and include other reaction participants as needed.

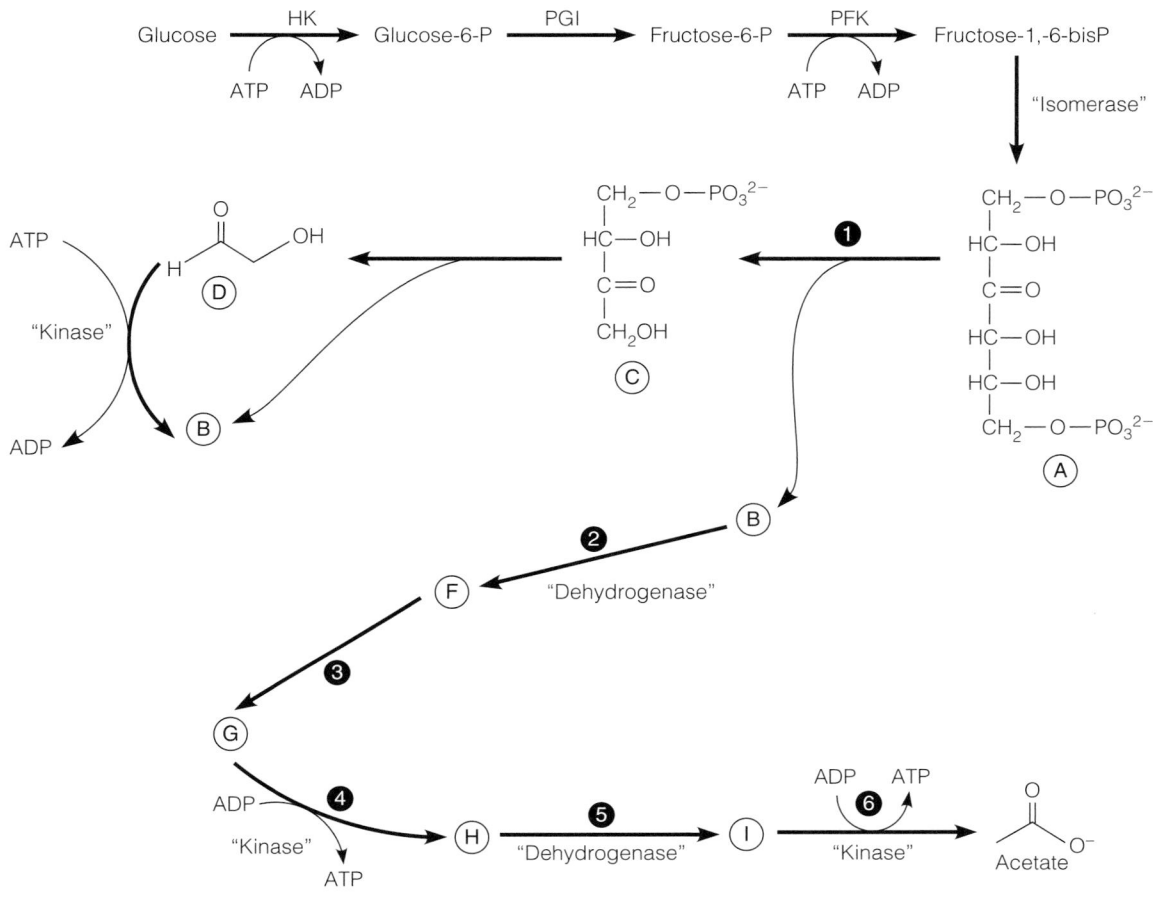

11. Write a pathway leading from glucose to lactose in mammary gland, and write a balanced equation for the overall pathway.

12. Sketch a curve that would describe the expected behavior of phosphofructokinase activity as a function of the adenylate energy charge.

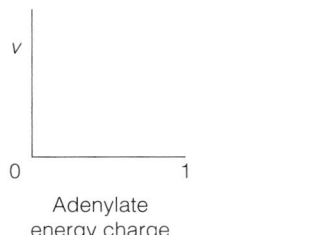

13. Explain the basis for the following statement. For efficient conversion of galactose to glucose-1-phosphate, UDP-glucose need be present in catalytic amounts only.

14. The muscle isozyme of lactate dehydrogenase is inhibited by lactate. Steady state kinetic analysis yielded the following data, with lactate either absent or present at a fixed concentration.

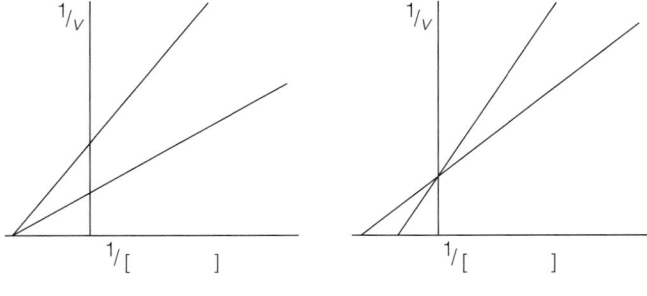

(a) Pyruvate is the substrate whose concentration is varied in one plot, NADH in the other. Identify each. Use an arrow and the appropriate letter (b, c, d, or e) to identify each of the following.
(b) Reciprocal of V_{max} for the uninhibited enzyme.
(c) The line representing data obtained in the presence of lactate acting as a competitive inhibitor with respect to the variable substrate.
(d) The line representing data obtained in the presence of lactate acting as a mixed inhibitor with respect to the variable substrate.

(e) Reciprocal of K_M in the presence of lactate acting as a competitive inhibitor.

(f) If K_M for NADH is 2×10^{-5} M, which is the most appropriate NADH concentration to use when determining K_M for pyruvate? 10^{-7} M, 10^{-6} M, 10^{-5} M, 10^{-4} M, or 10^{-3} M

15. How many ATP equivalents are consumed in the conversion of each of the following to a glucosyl residue in glycogen?
(a) Dihydroxyacetone phosphate
(b) Fructose-1,6-bisphosphate
(c) Pyruvate
(d) Glucose-6-phosphate

16. How many high-energy phosphates are generated or consumed in (a) converting 1 mole of glucose to lactate? (b) converting 2 moles of lactate to glucose?

17. Avidin is a protein that binds extremely tightly to biotin. Therefore, it is a potent inhibitor of biotin-requiring enzyme reactions. Consider glucose biosynthesis from each of the following substrates and predict which of these pathways would be inhibited by avidin.
(a) Lactate
(b) Oxaloacetate
(c) Malate
(d) Fructose-6-phosphate
(e) Phosphoenolpyruvate

18. $^{14}CO_2$ was bubbled through a suspension of liver cells that was undergoing gluconeogenesis from lactate to glucose. Which carbons in the glucose molecule would become radioactive?

19. Write a balanced equation for each of the following reactions or reaction sequences.
(a) The reaction catalyzed by PFK-2
(b) The conversion of 2 moles of oxaloacetate to glucose
(c) The conversion of glucose to UDP-Glc
(d) The conversion of 2 moles of glycerol to glucose
(e) The conversion of 2 moles of malate to glucose-6-phosphate

20. Sketch curves for reaction velocity vs. [fructose-6-phosphate] for the phosphorylated *and* nonphosphorylated forms of PFK-2 in liver.

21. Based on information presented on pages 570–574, sketch curves relating glycogen synthase reaction velocity to [UDP-glucose], for both the *a* and *b* forms of the enzyme, in the presence and absence of glucose-6-phosphate.

22. Glycogen synthesis and breakdown are regulated primarily at the hormonal level. However, important *nonhormonal* mechanisms also control the rates of synthesis and mobilization. Describe these nonhormonal regulatory processes.

23. Why does it make good metabolic sense for phosphoenolpyruvate carboxykinase, rather than pyruvate carboxylase, to be the primary target for regulation of gluconeogenesis at the level of control of enzyme synthesis?

24. What is the metabolic significance of the following observations? (1) Only the liver form of pyruvate kinase is inhibited by alanine, and (2) only gluconeogenic tissues contain appreciable levels of glucose-6-phosphatase.

*25. Predict how phosphoprotein phosphatase 1 (PP1) and phosphoprotein phosphatase inhibitor 1 (inhibitor 1) might interact with components of the glycogenolytic cascade to effect regulation reciprocal to their effects upon glycogen synthesis.

26. Write a one-sentence explanation for each of the following statements.
(a) In liver, glucagon stimulates glycogen breakdown via cyclic AMP. Although one might expect glucagon also to stimulate catabolism of the glucose formed, glucagon *inhibits* glycolysis and stimulates gluconeogenesis in liver.
(b) An individual with a glucose-6-phosphatase deficiency suffers from chronic hypoglycemia.
(c) The action of phosphorylase kinase simultaneously activates glycogen breakdown and inhibits glycogen synthesis.
(d) The presence in liver of glucose-6-phosphatase is essential to the function of the liver in synthesizing glucose for use by other tissues.

27. Write a balanced chemical equation for the pentose phosphate pathway in the first two modes depicted in Figure 13.43, where (a) ribose-5-phosphate synthesis is maximized and (b) NADPH production is maximized, by conversion of the sugar phosphate products to glucose-6-phosphate for repeated operations of the pathway.

*28. [1-^{14}C]Ribose-5-phosphate is incubated with a mixture of purified transketolase, transaldolase, phosphopentose isomerase, phosphopentose epimerase, and glyceraldehyde-3-phosphate. Predict the distribution of radioactivity in the erythrose-4-phosphate and fructose-6-phosphate that are formed in this mixture.

29. Pyruvate carboxylase is thought to activate CO_2 by ATP, through formation of carboxyphosphate as an intermediate. Propose a mechanism for formation of this intermediate.

TOOLS OF BIOCHEMISTRY 13A

DETECTING AND ANALYZING PROTEIN–PROTEIN INTERACTIONS

The enzymes of glycolysis are readily isolated as soluble proteins. However, several lines of evidence support the notion that they are physically associated within living cells. For years, biochemistry students have been told, "A cell is not a bag of enzymes," implying that enzymes are organized into functional supramolecular units in intact cells. Often these organized units are stabilized by weak, noncovalent forces that are easily disrupted when cells are broken open, as must occur if the enzymes within are to be isolated and characterized. Even when cells are gently lysed, most protein extraction processes dilute intracellular contents by several orders of magnitude, and that alone

can disrupt associations that are highly concentration-dependent. Biochemists are trying to define how the organization of functionally related enzymes facilitates the flow of metabolites and the control and coordination of metabolic pathways.

An early indication that glycolytic enzymes might interact within cells came from observations that the intracellular molar concentrations of glycolytic intermediates are actually lower than the concentrations of the enzymes that act upon those intermediates. This finding suggested that most of the supply of an intermediate is enzyme-bound within cells and this, in turn, led to the idea that intermediates are passed directly from enzyme to enzyme, without release to the surrounding milieu, as if the glycolytic enzymes functioned as part of a multienzyme complex.

If such a complex could be isolated intact, its properties could be explored by the methods for molecular weight determination, described in Tools of Biochemistry 6B. However, because of difficulties often encountered in isolating enzyme complexes held together by ephemeral forces, scientists usually use multiple approaches to demonstrate and characterize the protein–protein interactions involved. A few of those techniques are described here.

Bifunctional Cross-Linking Reagents

These are reagents containing two functional groups capable of forming covalent bonds with specific amino acid residues in closely associated proteins. For instance, *dimethylsuberimidate* (DMS) reacts with lysine ε-amino groups and N-terminal amino groups, cross-linking two proteins in a form that can be detected by gel electrophoresis because of the increase in molecular weight.

Some reagents have cleavable cross-links, such as a disulfide bond that can be reductively cleaved, allowing analysis of the separate cross-linked partners. Although the technique can be very informative, experimentation with many reagents is often required to find the right combination of functional groups and distance between the reactive partners for the cross-linking reaction to proceed to a measurable extent. Also, care must be taken not to overinterpret results, for even transient contacts between molecules, which may occur nonspecifically, sometimes lead to crosslinking. Chemical cross-linking is described in more detail in Tools of Biochemistry 28A.

Affinity Chromatography

In this technique, described in Tools of Biochemistry 5A, one protein is immobilized on a chromatographic support, and a mixture of proteins is passed through a column of this material. Proteins that are retained can be identified after elution, by biological activity or by electrophoretic techniques, such as immunoblotting or two-dimensional electrophoretic analysis. The chief limitations of this technique are the need to have one of the test proteins available in pure form for immobilization and the fact that interactions occur in a rather artificial environment. Again, controls are essential because of nonspecific retention of some proteins on affinity columns.

Immunoprecipitation

Antibody to a purified protein can be added to a protein mixture, often with immunoprecipitation (see Tools of Biochemistry 7A) of both the antigenic protein and any interactive proteins bound to it (co-immunoprecipitation). Although this technique is qualitative, like the approaches described above, it is simple to do, and it needs only small amounts of material. Because multiple assays can be run simultaneously, co-immunoprecipitation can be used, for example, to study the effects of the binding of small molecules (substrates or effectors) upon protein associations.

Kinetic Analysis

If enzymes catalyzing sequential reactions interact, the interactions can facilitate the flow of metabolites through multistep pathways (metabolic channeling), and this can be detected in vitro in several ways. Generally, a channeled pathway will display one or more of the following characteristics: (1) reduced *transient time*, the interval after initiating a multistep pathway needed for the formation of final product to reach its maximal rate; (2) steady-state levels of intermediates much lower than expected if they must seek the next enzyme acting on them by diffusion rather than by direct or facilitated transfer to a nearby enzyme molecule; and (3) restricted ability of an exogenous intermediate to equilibrate with the same intermediate in a channeled pathway, as determined usually by radioisotope experiments.

Library-Based Methods

These are methods that allow the screening of a large number, or library, of cloned genes. These methods allow tentative identification of interacting partners without a prior requirement for purification and identification of one of the partners. A popular method called the two-hybrid system uses a transcriptional activation system in yeast that requires two proteins to interact in order for transcription at an appropriate gene site to be initiated (see Chapters 27 and 29). One of these proteins binds at the DNA site, and the other activates transcription. Two hybrid, or fusion, proteins are generated by recombinant DNA techniques (see Tools of Biochemistry 4B); the gene for one test protein (X) is fused to the DNA-binding protein gene, and the gene for another test protein (Y), or a library of cloned genes, is fused to the gene for the

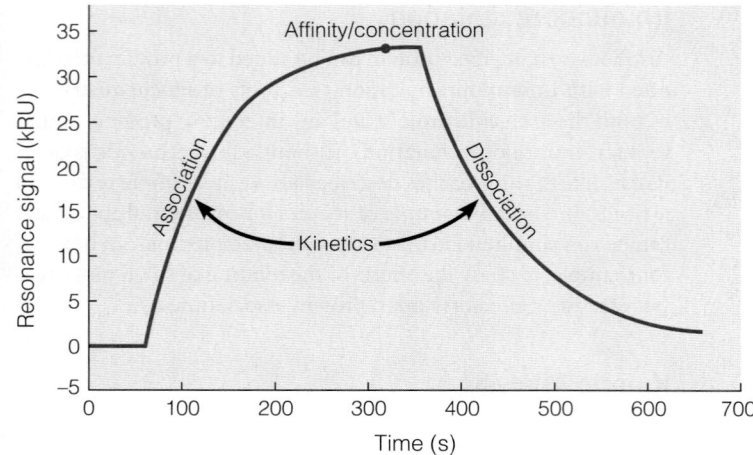

FIGURE 13A.1

Biacore analysis of a protein–protein association. Test protein flows past the immobilized protein in the association phase and is replaced by buffer in the dissociation phase. The height of the plateau response, as compared with standards, is related to the stoichiometry of the association.

Courtesy of Biacore Life Sciences.

transcriptional activation domain. The recombinant genes are transferred into yeast cells, where the interaction of proteins X and Y can form a fully functional transcriptional activator (assuming that the functional domains of the fusion proteins fold as they do in their native state). Transcription of the target gene is then monitored by assays for the activity of a reporter gene, a gene cloned downstream of the promoter and whose biological activity is easily assayed. Once a specific protein association has been detected, it becomes essential to isolate the interactive partners as full-length proteins and to ascertain that the interactions detected by this somewhat qualitative method are indeed biologically significant.

Biosensor Analysis

In recent years, a new kind of instrumentation has been developed that allows both qualitative and quantitative analysis of protein–protein interactions, using rather small quantities of purified proteins. One such instrument, a BIACORE, measures an optical property called *surface plasmon resonance*, which is related to minute changes in refractive index that occur when a protein in solution interacts with a protein immobilized on a chip. The signal measured is proportional to total protein concentration, over a wide range. Thus, the kinetics of a protein association reaction can

be monitored by following the increase in signal as a protein in solution passes over a chip containing immobilized protein. The amount of protein bound at equilibrium gives the affinity constant for the interaction, and the kinetics of dissociation can then be followed by passing buffer over the chip and following the decrease in signal, as indicated in Figure 13A.1. Limitations of this useful technique, which can be controlled for, include the possibility that immobilization alters the protein in a way that affects the interaction and the fact that the two interacting proteins are in different phases (solid, or immobilized, and liquid).

Finally, protein–protein interactions can be investigated using fluorescence spectroscopy methods, including fluorescence polarization and Förster Resonance Energy Transfer (FRET), as described in Tools of Biochemistry 6A.

References

Bruckner, A., C. Polge, N. Lentze, D. Auerbach, and U. Schlattner (2009) Yeast two-hybrid, a powerful tool for systems biology. *Int. J. Mol. Sci.* 10:2763–2788. A comprehensive review of the widely used two-hybrid technique.

Scarano, S., M. Mascini, A. P. Turner, and M. Minunni (2010) Surface plasmon resonance imaging for affinity-based biosensors. *Biosens. Bioelectron.* 25:957–966.

CHAPTER 14

Citric Acid Cycle
and Glyoxylate Cycle

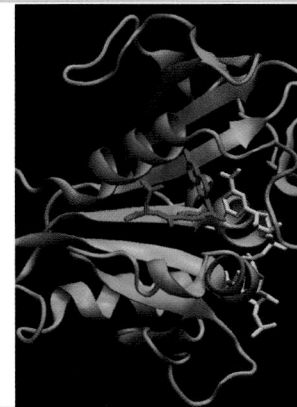

In Chapter 13 we explored the initial, fermentative, phase of carbohydrate degradation. Here we follow the subsequent aerobic reactions by which carbohydrates are ultimately oxidized to carbon dioxide and water via the citric acid cycle (Figure 14.1). As we shall see, the citric acid cycle is the central oxidative pathway in respiration, the process by which *all* metabolic fuels—carbohydrate, lipid, and protein—are catabolized in aerobic organisms and tissues.

We saw in Chapter 13 (page 540) that a fermentation process such as glycolysis releases only a fraction of the energy available in glucose. Complete combustion of glucose to CO_2 and H_2O releases 2870 kJ/mol of free energy under standard conditions. Ethanol and lactate, like other fermentation products of carbohydrate catabolism, are at the same oxidation level as the starting material, glucose. In fact, complete combustion of ethanol to CO_2 and H_2O releases 1326 kJ/mol of free energy under standard conditions. Because 2 moles of ethanol and 2 moles of CO_2 are produced per mole of glucose, $2 \times 1326 = 2652$ kJ/mol would be released. Thus, 92% (2652/2870) of the free energy available in glucose is retained in the fermentation product (ethanol), meaning very little of the potential energy stored in the original glucose molecule is released by its conversion to ethanol in the fermentation process. Indeed, all fermentations are characterized by a similarly low energy yield.

Far more energy is generated if organic fuels are completely oxidized to CO_2 and H_2O by molecular oxygen in the process termed **cellular respiration**. That energy release involves dehydrogenation reactions that generate reduced electron carriers, primarily NADH. These carriers are next reoxidized in the mitochondrial respiratory (electron transport) chain. These reactions provide the energy that drives ATP synthesis, through oxidative phosphorylation. The electrons released are ultimately transferred to oxygen, which becomes reduced to water. This chapter focuses on the fates of the oxidizable substrates, and the next chapter focuses on the chain of electron carriers and the synthesis of ATP.

The citric acid cycle is a pathway for oxidizing all metabolic fuels.

Most of the energy yield from substrate oxidation in the citric acid cycle comes from subsequent reoxidation of reduced electron carriers.

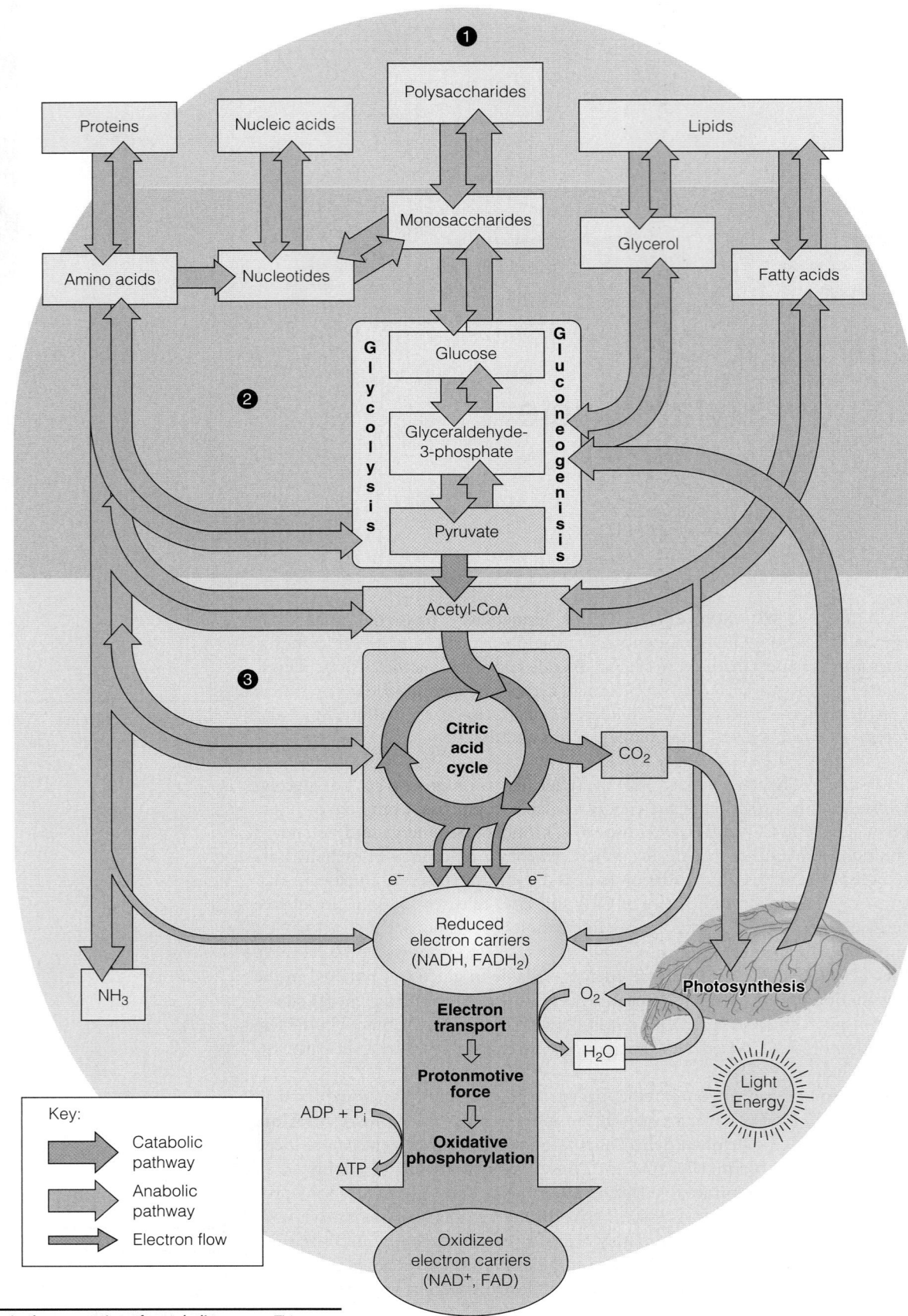

FIGURE 14.1

Oxidative processes in the generation of metabolic energy. This overview of intermediary metabolism highlights the citric acid cycle and the pathways that deliver fuel to the cycle for oxidation.

Overview of Pyruvate Oxidation and the Citric Acid Cycle

The Three Stages of Respiration

It is convenient to think of the metabolic oxidation of organic substrates as a three-stage process, schematized in Figure 14.2. Stage 1 is the generation of an activated two-carbon fragment—the acetyl group of **acetyl-coenzyme A**, or **acetyl-CoA**. (Recall from Chapter 12 that coenzyme A activates and transfers acyl groups; more about this on pages 601–602.) Stage 2 is oxidation of those two carbon atoms in the citric acid cycle to form 2 CO_2 and 4 pairs of electrons. Stage 3 is electron transport

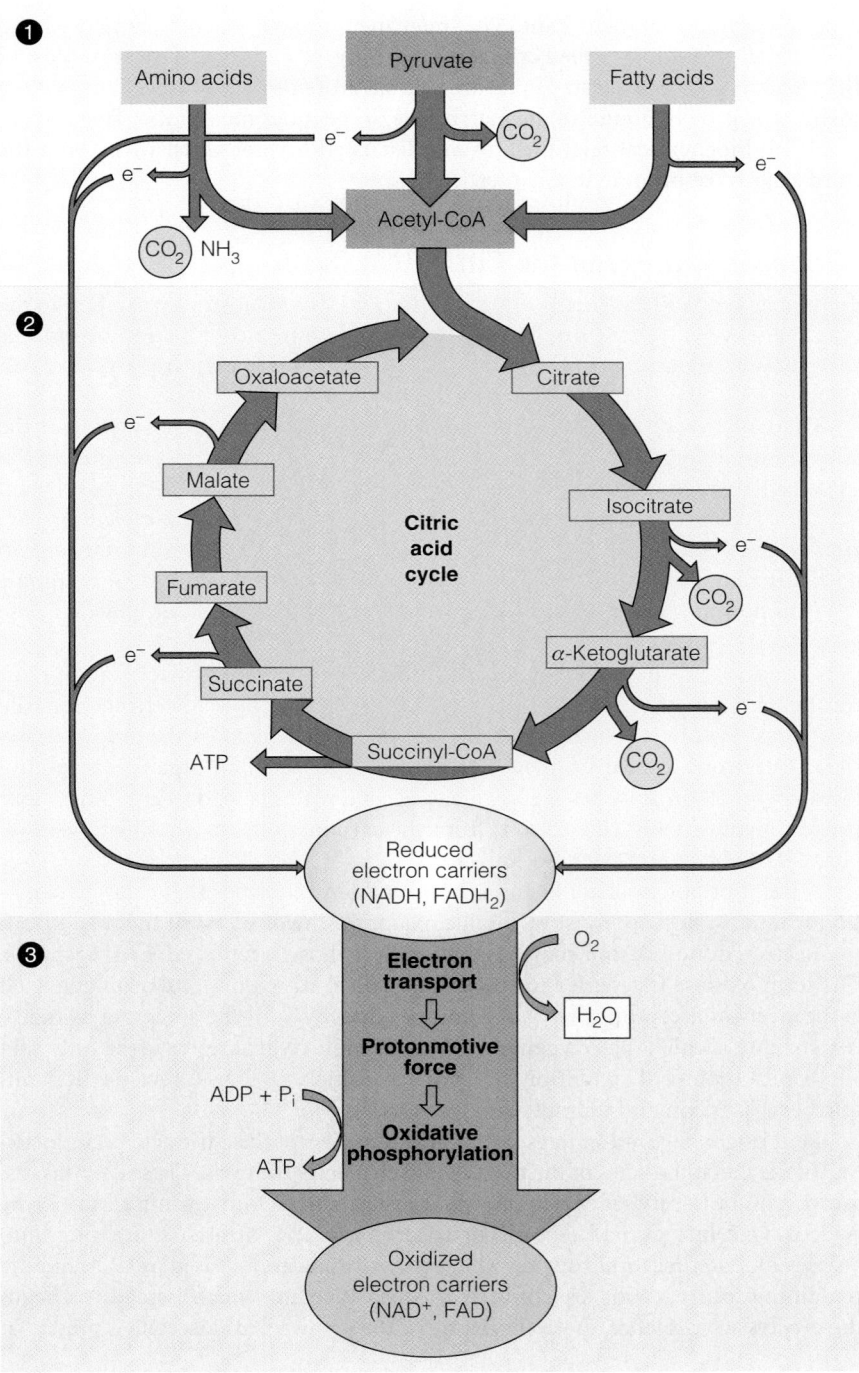

FIGURE 14.2

The three stages of respiration. In stage 1, carbon from metabolic fuels is incorporated into acetyl-CoA. In stage 2, the citric acid cycle oxidizes acetyl-CoA to produce CO_2, reduced electron carriers, and a small amount of ATP. In stage 3, the reduced electron carriers are reoxidized, providing energy for the synthesis of additional ATP.

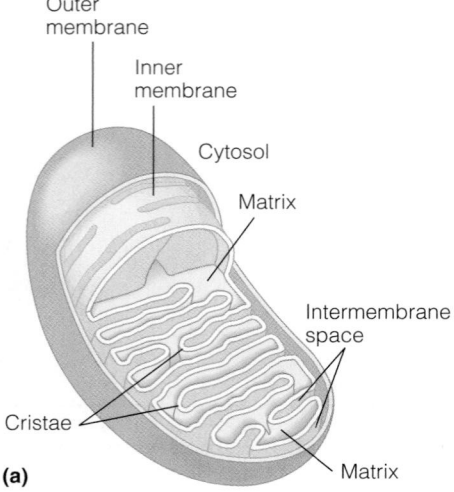

Outer membrane

Inner membrane

Cytosol

Matrix

Intermembrane space

Cristae

(a)

Matrix

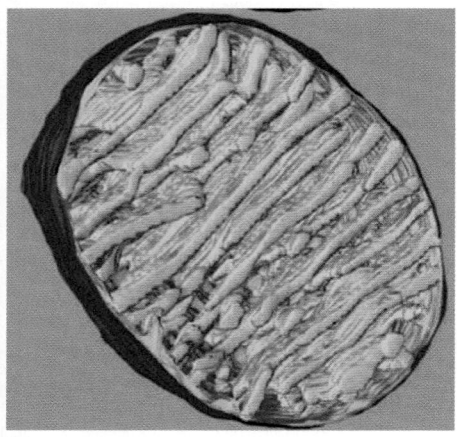

(b)

(a) Schematic of a mitochondrion. **(b)** Computer model generated from electron tomograms of a mitochondrion. Cristae are yellow, and the outer membrane is dark blue.

(b) Reprinted from *Trends in Biochemical Sciences* 25:319–324, T. G. Frey and C. A. Mannella, The internal structure of mitochondria. © 2000, with permission from Elsevier.

Dehydrogenases catalyze substrate oxidations. Oxidases catalyze the subset of oxidations in which O_2 is the direct electron acceptor.

and oxidative phosphorylation, in which the reduced electron carriers generated in the citric acid cycle become reoxidized, with concomitant synthesis of ATP. Stage 1 is a family of pathways, which operate separately on carbohydrate, fat, and protein. Carbon from carbohydrate enters stage 1 as pyruvate; oxidation of pyruvate to acetyl-CoA is described in this chapter. Fat breakdown generates acetyl-CoA primarily through β-oxidation of fatty acids (see Chapter 17), whereas several different pathways generate acetyl-CoA and citric acid cycle intermediates from amino acid catabolism (see Chapter 21).

In bacteria, the enzymes of pyruvate oxidation and the citric acid cycle are located in the cytoplasm and on the plasma membrane. In eukaryotic cells, respiration takes place in mitochondria. Reactions of the first two stages occur within the interior, **matrix** compartment of the mitochondrion, and electron transport and oxidative phosphorylation are catalyzed by membrane-bound enzymes in the inner mitochondrial membrane. The inner membrane is extensively stacked and folded into projections, called **cristae**, that greatly expand its surface area. Most of the enzymes of the citric acid cycle are soluble proteins in the matrix but one is a membrane protein bound to the matrix side of the inner membrane. These structural and biochemical relationships are discussed further when we explore the third stage of respiration in Chapter 15.

Chemical Strategy of the Citric Acid Cycle

To fully understand the chemical strategy that underlies substrate oxidation in the citric acid cycle, we first briefly review the oxidation and reduction of organic compounds. Quantitative aspects of biological oxidations are presented in Chapter 15.

Oxidation involves the loss of electrons from a substrate; that substrate is the electron donor, and the electrons are transferred to an electron acceptor, which thereby becomes reduced. Free electrons cannot exist in the cell; electrons released in an enzyme-catalyzed oxidation must be transferred to specialized electron carriers (e.g., NAD^+ or FAD). The oxidized substrate and the electron acceptor will have different affinities for electrons—this affinity difference drives an exergonic electron flow, releasing energy that can be captured by the cell, ultimately in the form of ATP.

Carbon atoms become oxidized either through loss of a hydride ion (H^-) or through combination with oxygen. The latter process removes electrons from the shell around a carbon nucleus because the electronegativity of the oxygen draws shared electrons toward its own nucleus. Similarly, when an organic compound loses a hydride ion, it loses the electron associated with that hydrogen. Thus, either process involves a loss of electrons from the carbon atom undergoing oxidation. Formally, the two processes are equivalent.

A point of potential confusion arises in naming enzymes that catalyze oxidation reactions. Because most metabolic oxidations involve loss of hydrogen from the electron donor, we call enzymes that catalyze those reactions **dehydrogenases**. The term **oxidase** is reserved for those enzymes in which molecular oxygen itself is the electron acceptor. If oxygen combines directly with the substrate oxidized, the enzyme is called an **oxygenase**. Oxidases and oxygenases catalyze only that small proportion of oxidation reactions in which O_2 is a direct participant. Further discussion and examples are given in Chapter 15.

Referring to Figure 14.3, let's take a bird's-eye view of the citric acid cycle, focusing on the metabolic fates of the two carbons that enter the cycle. These carbons, the acetyl group of acetyl-coenzyme A, are transferred to a four-carbon dicarboxylic acid, **oxaloacetate**, to yield a six-carbon tricarboxylic acid, citrate. Citrate enters into a series of seven reactions during which two carbons are released as CO_2 and the remaining four carbons are converted to oxaloacetate, which is ready to begin the process again. Hence, the cyclic nature of the pathway: oxaloacetate is present at

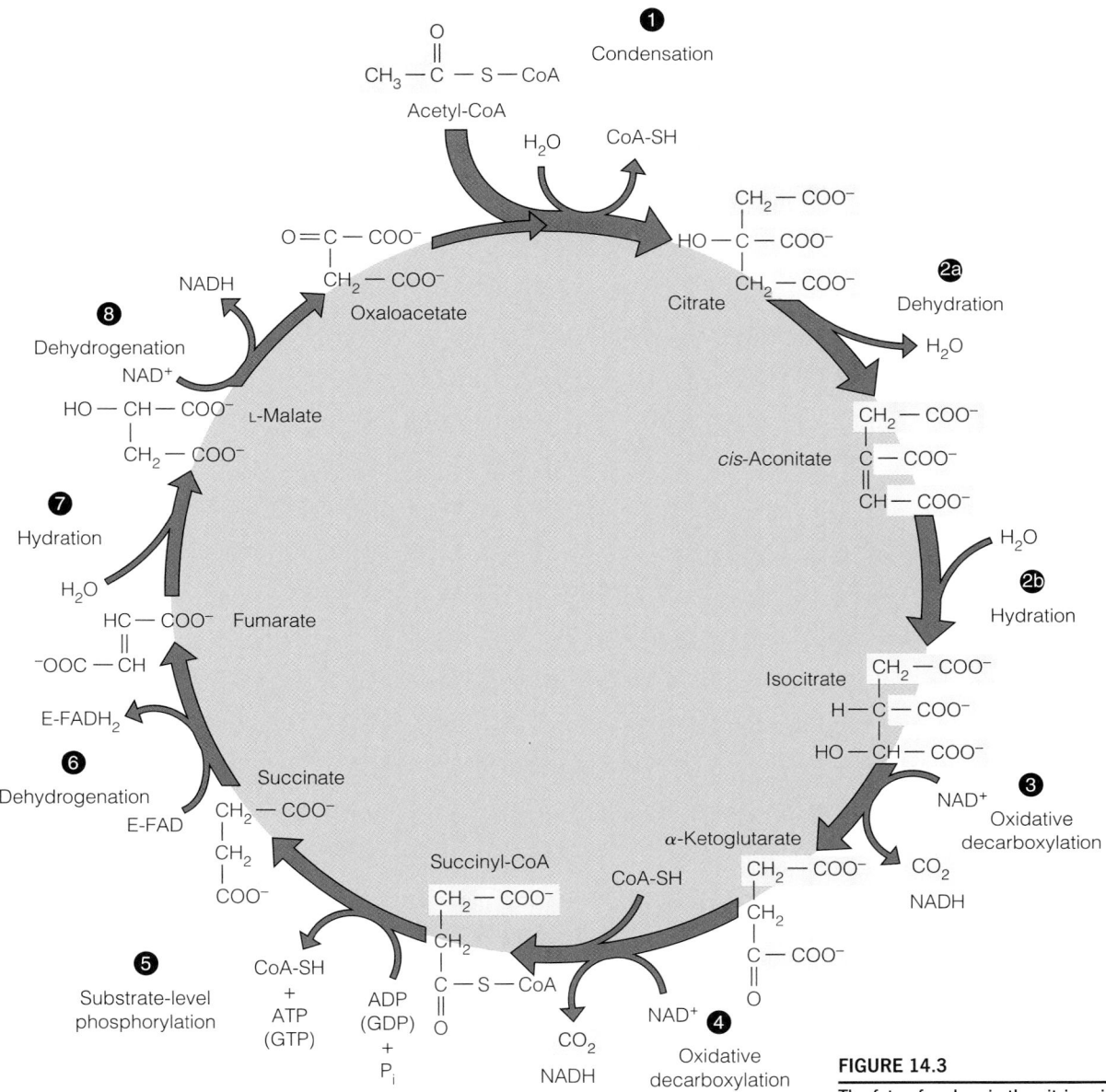

FIGURE 14.3

The fate of carbon in the citric acid cycle. Acetyl-CoA entering the citric acid cycle is highlighted (in blue) to show the fate of its two carbons through reaction 4. After reaction 5, the carbon atoms most recently entered are no longer highlighted because succinate and fumarate are symmetrical molecules. Thus C1 and C2 become indistinguishable from C3 and C4 beyond this point in the cycle. Carboxyl groups that leave the cycle as CO_2 in reactions 3 and 4 are shown in green. Note that these departing CO_2 groups derive from the two oxaloacetate carboxyl groups that were incorporated as acetyl-CoA in earlier turns of the cycle.

the beginning, to react with an activated two-carbon fragment, and it is present at the end, after two carbons have been oxidized to CO_2. Thus oxaloacetate, and indeed the entire cycle, acts as a catalyst for the oxidation of acetyl-CoA to CO_2. Note also that of the eight reactions shown in Figure 14.3, four are dehydrogenations, which together generate eight reducing equivalents (all derived from the acetyl-CoA) in the form of three $NADH/H^+$ and one $FADH_2$.

The oxidation of acetyl-CoA to 2 CO_2 would seem to be a relatively simple transformation. Why then, is such a complicated pathway used by cells? The answer lies in the chemistry: oxidation of acetyl-CoA requires $C—C$ bond cleavage, a difficult reaction for the two-carbon acetyl group. As discussed in Chapter 12, $C—C$ bond cleavage is much more facile if a carbonyl group is nearby to stabilize the carbanionic transition state, for example in the retro-aldol cleavage of fructose-1,6-bisphosphate in glycolysis (Figure 13.5, page 528). The strategy used in the citric acid cycle is to metabolize acetyl-CoA through a series of β-keto acid and α-keto acid intermediates, which are readily decarboxylated in enzyme-catalyzed reactions, thereby achieving the necessary $C—C$ bond cleavage.

Discovery of the Citric Acid Cycle

The idea that organic fuels are oxidized via a cyclic pathway was proposed in 1937 by Hans Krebs, on the basis of studies for which he later shared a Nobel Prize with Fritz Lipmann (the discoverer of coenzyme A). Beginning in 1932, Krebs tested the ability of various organic acids to be oxidized by following the rate of oxygen consumption in liver and kidney slices. He found that citrate, succinate, fumarate, malate, and acetate were readily oxidized in these tissues.

$$
\begin{array}{ccccc}
& & COO^- & COO^- & COO^- \\
& & | & | & | \\
H_2C-COO^- & CH_2 & CH & HO-CH \\
| & | & || & | & COO^- \\
HO-C-COO^- & CH_2 & HC & CH_2 & | \\
| & | & | & | & CH_3 \\
H_2C-COO^- & COO^- & COO^- & COO^- & \\
\textbf{Citrate} & \textbf{Succinate} & \textbf{Fumarate} & \textbf{Malate} & \textbf{Acetate}
\end{array}
$$

In 1935, the Hungarian biochemist Albert Szent-Györgyi, using minced pigeon breast muscle, a tissue with a very high rate of respiration, found that oxygen consumption was much greater than expected when he added small amounts of the dicarboxylic acids succinate, fumarate, or malate. Krebs recognized that these acids were somehow acting *catalytically*, rather than being used up as substrates in a linear pathway.

Then in 1937, Carl Martius and Franz Knoop discovered that citrate is converted to isocitrate and α-ketoglutarate, which was already known to undergo dehydrogenation to succinate. This linked the oxidation of citrate to that of succinate, and helped Krebs explain his earlier observation that citrate catalytically stimulated oxygen consumption. Krebs then found that malonate, an analog of succinate and a known inhibitor of succinate dehydrogenase, blocked the oxidation of pyruvate, pointing to a role for succinate dehydrogenase in pyruvate oxidation. Moreover, malonate-inhibited cells accumulated citrate, α-ketoglutarate, and succinate, suggesting that citrate and α-ketoglutarate are both normal precursors of succinate. The final piece to the puzzle was Krebs' discovery that addition of pyruvate plus oxaloacetate to this minced muscle preparation led to accumulation of citrate in the medium, suggesting that these two acids are precursors to citrate.

From these observations and a knowledge of the structures and reactivities of the organic acids that could stimulate respiration, Krebs proposed both the sequence of reactions involved and the cyclic nature of the pathway. It is worth noting that Krebs was already primed to recognize the cyclic organization of this pathway because he and Kurt Henseleit had just discovered the urea cycle (Chapter 20) in 1932.

Krebs postulated that carbohydrates entered the cycle via pyruvate, which reacted with oxaloacetate to give citrate plus CO_2. We now know that pyruvate must first be oxidized (releasing CO_2) and that acetyl-CoA is the species that subsequently reacts with oxaloacetate to form citrate. We know also of a CoA derivative of succinate, **succinyl-CoA**. Recognition of these activated intermediates in the cycle had to await the discovery of coenzyme A by Fritz Lipmann in 1947. Except for these changes, the pathway as proposed by Krebs was correct. The reactions are outlined in Figure 14.3.

The citric acid cycle is also known by other names: the Krebs cycle, after its discoverer, and the tricarboxylic acid (TCA) cycle because it was apparent from the outset that tricarboxylic acids are involved as intermediates. It became clear only later that citrate was one of those intermediates. The following discussion of the citric acid cycle will focus on mammalian biochemistry, but essentially identical versions of this pathway are found in nearly all organisms, and it is clearly an evolutionarily ancient pathway.

$$
\begin{array}{c}
COO^- \\
| \\
CH_2 \\
| \\
COO^-
\end{array}
$$

Malonate

Pyruvate Oxidation: A Major Entry Route for Carbon into the Citric Acid Cycle

As noted earlier, pyruvate derived from carbohydrate oxidation is but one of the major suppliers of acetyl-CoA for oxidation in the citric acid cycle. Chapters 17 and 20 focus upon oxidation of fatty acids and amino acids, respectively, to give this central metabolite. The conversion of pyruvate to acetyl-CoA, catalyzed by **pyruvate dehydrogenase complex** (**PDH complex**), is an oxidative decarboxylation. In the overall reaction the carboxyl group of pyruvate is released as CO_2, while the remaining two carbons form the acetyl moiety of acetyl-CoA.

$$CH_3-\overset{\overset{O}{\|}}{C}-COO^- + NAD^+ + CoA\text{-}SH \longrightarrow CH_3-\overset{\overset{O}{\|}}{C}-S\text{-}CoA + NADH + CO_2$$

Pyruvate   Acetyl-CoA $\Delta G^{\circ\prime} = -33.5$ kJ/mol

This reaction involves the decarboxylation of an α-keto acid (pyruvate). Unlike β-keto acids, α-keto acids cannot stabilize the carbanion transition state that develops during decarboxylation. To solve this problem, enzymatic decarboxylations of α-keto acid substrates all utilize the coenzyme **thiamine pyrophosphate** (**TPP**) to form a covalent adduct with the substrate and provide the electron delocalization required to stabilize the carbanion intermediate. We saw this same strategy used in the pyruvate decarboxylase reaction of alcohol fermentation (Chapter 13, page 538), and, indeed, the chemistry of these two decarboxylations is nearly identical.

Although the overall reaction may look straightforward, it is in fact rather complicated, involving generation of a reduced electron carrier (NADH), decarboxylation of pyruvate, and metabolic activation of the remaining two carbons of pyruvate. The reaction is highly exergonic and is essentially irreversible in vivo. Three enzymes are involved—**pyruvate dehydrogenase** (E_1), **dihydrolipoamide transacetylase** (E_2), and **dihydrolipoamide dehydrogenase** (E_3)—in the five-step reaction. In addition, five coenzymes are required, including the two coenzymes—NAD^+ and coenzyme A—that appear in the overall reaction. The three enzymes (E_1, E_2, and E_3) involved are assembled into a highly organized multienzyme complex called the **pyruvate dehydrogenase complex**.

Figure 14.4 shows an electron micrograph of the complex as purified from *E. coli* and a model of the eukaryotic PDH complex, based upon cryoelectron

Pyruvate oxidation to acetyl-CoA is a virtually irreversible reaction that involves three enzymes and five coenzymes.

FIGURE 14.4

Structure of the pyruvate dehydrogenase complex. **(a)** Electron micrograph of the purified pyruvate dehydrogenase complex from *E. coli*. **(b–e)** A model for the eukaryotic complex, based upon cryoelectron microscopy of the yeast (*Saccharomyces cerevisiae*) PDH complex and its subcomplexes. The complex is viewed along a 3-fold axis. **(b)** E_2 core subcomplex (60 E_2 monomers). **(c)** E_2-E_3 subcomplex (E_2 core + 12E_3 dimers). **(d)** Full PDH complex (E_2-E_3 subcomplex + ~30 E_1 tetramers). **(e)** Cutaway reconstruction of PDH complex. E_1, E_2, and E_3 are shown in yellow, green, and red, respectively.

(a) Reprinted from *Electron Microscopy of Proteins, Vol. 2*, R. M. Oliver and L. J. Reed; J. R. Harris, ed. © 1982, with permission from Elsevier; (b–e) *The Journal of Biological Chemistry* 276:38329–38336, L. Reed, A trail of research from lipoic acid to α-keto acid dehydrogenase complexes. Reprinted with permission. © 2001 The American Society for Biochemistry and Molecular Biology. All rights reserved.

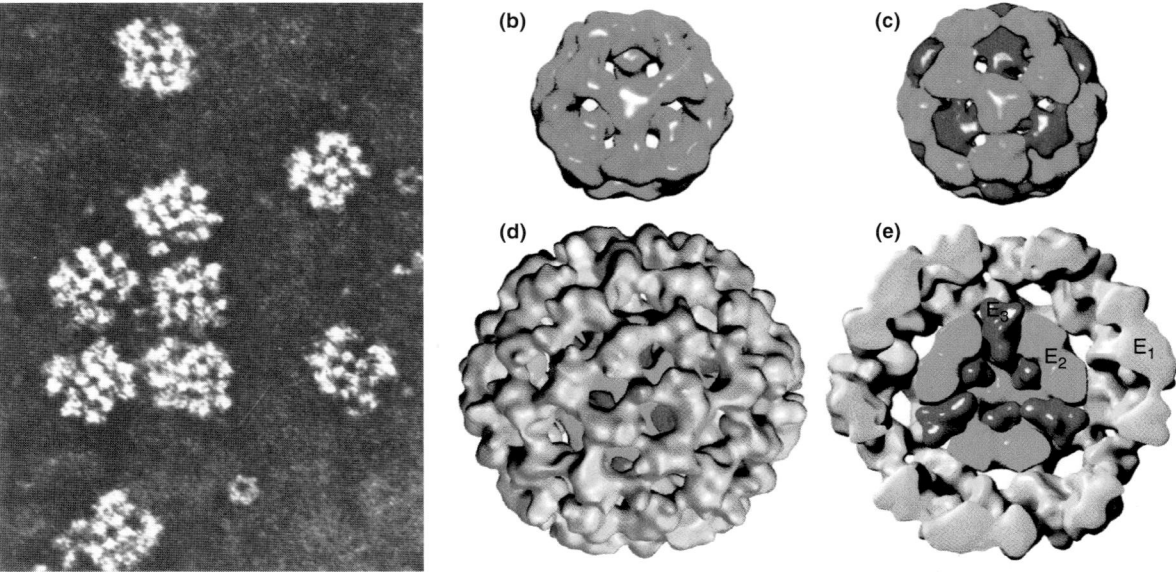

(a)   (b)   (c)   (d)   (e) E_3 E_2 E_1

TABLE 14.1   Coenzymes of the pyruvate dehydrogenase reaction

Cofactor	Location	Function
Thiamine pyrophosphate (TPP)	Tightly bound to E_1	Decarboxylates pyruvate, yielding hydroxyethyl-TPP
Lipoic acid (lipoamide)	Covalently bound to E_2 via lysine ("swinging arm")	Accepts hydroxyethyl carbanion from TPP as acetyl group
Coenzyme A (CoA)	Dissociable substrate for E_2	Accepts acetyl group from lipoamide
Flavin adenine dinucleotide (FAD)	Tightly bound to E_3	Accepts pair of electrons from reduced lipoamide
Nicotinamide adenine dinucleotide (NAD^+)	Dissociable substrate for E_3	Accepts pair of electrons from reduced $FADH_2$

microscopy of the yeast (*Saccharomyces cerevisiae*) complex. The *E. coli* PDH complex contains 24 polypeptide chains of E_1, 24 chains of E_2, and 12 chains of E_3. The pyruvate dehydrogenase complex as isolated from *E. coli* has a mass of about 4.6 million daltons, slightly larger than a ribosome. In mammalian mitochondria the complex is about twice that size. The eukaryotic complex (Figure 14.4b–e) is composed of a core of 60 E_2 monomers arranged as a pentagonal dodecahedron (icosahedral symmetry, see Figure 6.38, page 213). This core structure contains 12 E_3 homodimers and is surrounded by ~30–45 $E_1\alpha_2\beta_2$ heterotetramers. In addition, the mammalian PDH complex contains ~12 copies of E_3 binding protein (E3BP) and small amounts of two regulatory enzymes as well—a kinase that phosphorylates three serine residues in E_1 and a phosphatase that removes those phosphates. We will discuss regulation of the activities of the PDH complex later in this chapter.

To understand how the three enzymes (E_1, E_2, and E_3) in the pyruvate dehydrogenase complex interact, we must understand the functions of the five coenzymes that participate in the reaction: *thiamine pyrophosphate (TPP)*, *lipoic acid*, *coenzyme A*, *flavin adenine dinucleotide (FAD)*, and NAD^+ (Table 14.1). TPP is tightly bound to E_1, lipoic acid is covalently bound to E_2, and FAD is tightly bound to E_3. Although we briefly introduced all five coenzymes in Chapter 11, we have discussed only NAD^+ and TPP in detail (Chapter 13; glyceraldehyde-3-phosphate dehydrogenase and pyruvate decarboxylase reactions, respectively). Therefore, we digress here to describe the chemistry of the other three.

Coenzymes Involved in Pyruvate Oxidation and the Citric Acid Cycle

Thiamine Pyrophosphate (TPP)

The chemistry of **thiamine pyrophosphate (TPP)** was discussed in the context of its role in the pyruvate decarboxylase reaction of alcohol fermentation (Figure 13.8, page 538). As we shall see, it functions in exactly the same manner in the pyruvate dehydrogenase reaction, stabilizing a carbanion intermediate.

Lipoic Acid (Lipoamide)

In pyruvate oxidation, the acceptor of the active aldehyde (hydroxyethyl group) generated by TPP is **lipoic acid**, which is the internal disulfide of 6,8-dithiooctanoic acid. Lipoic acid was identified in 1951 by Lester J. Reed at the University of Texas and I. C. Gunsalus at the University of Illinois. With the help of Eli Lilly and Co., they processed nearly 10 tons of pork and beef liver residue to isolate ~30 mg of crystalline substance! The coenzyme is joined to E_2 via an amide bond linking the carboxyl group of lipoic acid to a lysine ε-amino group. Thus, the reactive species is an amide, called **lipoamide**, or **lipoyllysine**. Each lipoyllysine side chain is ~14 Å long, and is located within a flexible lipoyl domain of E_2 (and E_3BP),

Lipoamide is a carrier of both electrons and acyl groups.

allowing it to function as a "swinging arm" that can interact with the active sites of both the E_1 and E_3 components of the PDH complex (see below).

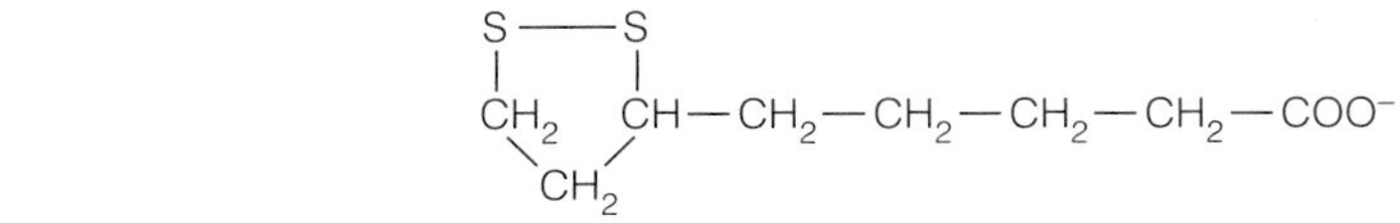

Lipoic acid

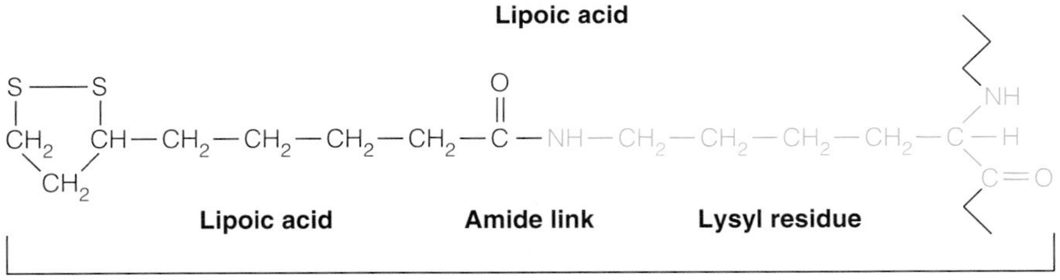

| Lipoic acid | Amide link | Lysyl residue |

Lipoamide (lipoyllysine)

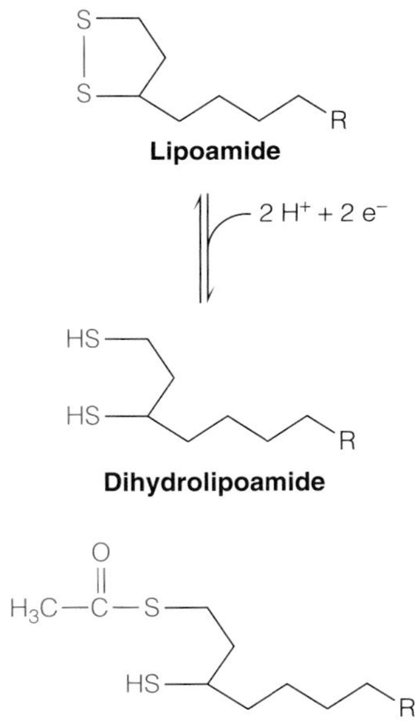

Transfer of the active aldehyde moiety (hydroxyethyl group) from TPP to the disulfide of lipoamide involves simultaneous oxidation of the aldehyde, coupled to reduction of the disulfide of lipoamide. This generates an acyl group, which, in pyruvate dehydrogenase, is transferred next to coenzyme A. The pair of electrons is transferred to lipoamide to form dihydrolipoamide (Figure 14.5). Thus, lipoamide is both an electron carrier and an acyl group carrier.

Flavin Adenine Dinucleotide (FAD)

Flavin adenine dinucleotide, or **FAD**, is one of two coenzymes derived from vitamin B_2, or **riboflavin**. The other is the simpler **flavin mononucleotide (FMN)**, or riboflavin phosphate (Figure 14.6). The functional part of both coenzymes is the

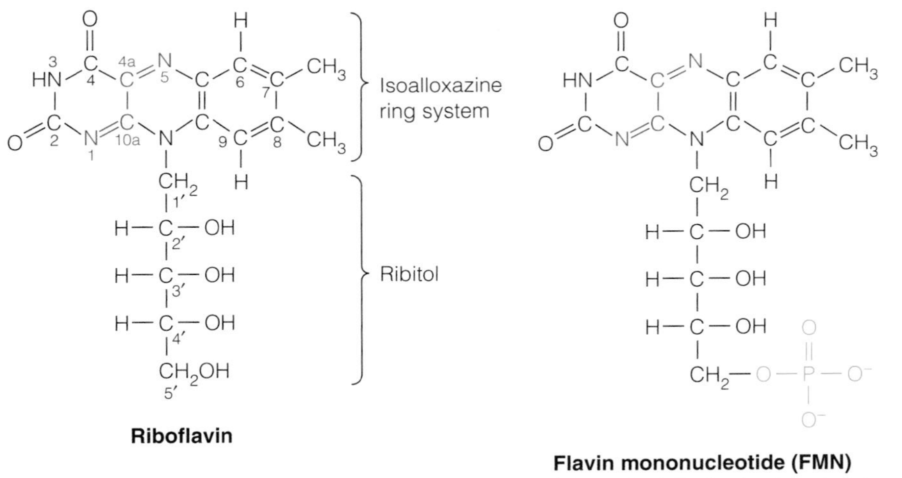

Lipoamide

$2 H^+ + 2 e^-$

Dihydrolipoamide

Acetyl-dihydrolipoamide

FIGURE 14.5

Oxidized and reduced forms of lipoamide. The cyclic disulfide of lipoamide can undergo a reversible two-electron reduction to form the dithiol, dihydrolipoamide. In pyruvate dehydrogenase, this reduction is coupled to the transfer of the hydroxyethyl group moiety from TPP, giving an acetyl thioester of the reduced dihydrolipoamide.

Riboflavin

Flavin mononucleotide (FMN)
(also called riboflavin phosphate)

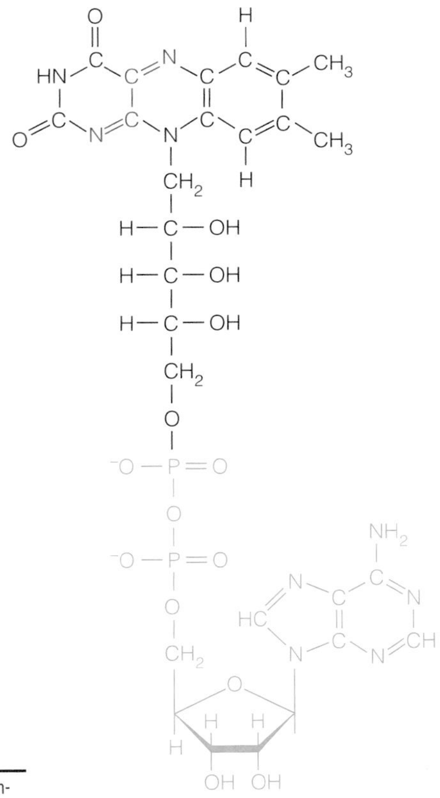

Flavin adenine dinucleotide (FAD)

FIGURE 14.6

Structures of riboflavin and the flavin coenzymes. Riboflavin and its coenzyme derivatives, FMN and FAD, all contain an isoalloxazine ring system and ribitol. The figure identifies (in red) the cluster of two carbon and two nitrogen atoms within the ring system that participates in the oxidation–reduction reactions of the flavin coenzymes.

isoalloxazine ring system, which serves as a two-electron acceptor. Compounds containing such a ring system are called **flavins**. In riboflavin and its derivatives the ring system is attached to **ribitol**, an open-chain version of ribose with the aldehyde carbon reduced to the alcohol level. The five-carbon of ribitol is linked to phosphate in FMN, and FAD is the adenylylated dinucleotide derivative of FMN. Thus, these compounds are somewhat analogous to nicotinamide mononucleotide and nicotinamide adenine dinucleotide (NAD^+), respectively.

Enzymes that use a flavin coenzyme are called **flavoproteins**, or **flavin dehydrogenases**. FMN and FAD undergo virtually identical electron transfer reactions. Flavoprotein enzymes preferentially bind either FMN or FAD. In a few cases, that binding is covalent. In most cases, however, the flavin is bound tightly but noncovalently, such that the coenzyme cannot easily dissociate from the enzyme. Thus flavins do not transfer electrons by diffusing from one enzyme to another, like the nicotinamide coenzymes. Instead, flavin dehydrogenases temporarily hold the electrons obtained from a reduced substrate and transfer them to a different electron acceptor. As we will see in the next chapter, another important feature of flavoproteins is that the tight binding of the flavin coenzyme to the protein (either covalently or noncovalently) confers a unique standard reduction potential ($E'°$) on the flavin ring.

Like the nicotinamide coenzymes, the flavins undergo two-electron oxidation and reduction reactions. The flavins, however, are distinctive in having a stable one-electron-reduced species, a **semiquinone** free radical, as shown in Figure 14.7. This free radical can be detected spectrophotometrically; whereas oxidized FAD and FMN are bright yellow, and fully reduced flavins are colorless, the semiquinone intermediate is either red or blue, depending on pH. The stability of the semiquinone intermediate gives flavins a catalytic versatility not shared by nicotinamide coenzymes, in that flavins can interact with either two-electron or one-electron donor–acceptor pairs. Also, flavoproteins can interact directly with oxygen. Thus, some, but not all, flavoproteins are oxidases.

Flavin coenzymes participate in two-electron oxidoreduction reactions that can proceed in 2 one-electron steps.

FIGURE 14.7

Oxidation and reduction reactions involving flavin coenzymes. Flavins participate in two-electron reactions, but the existence of the stable semiquinone free radical intermediate allows these reactions to proceed one electron at a time. Thus, reduced flavins can readily be oxidized by one-electron acceptors. Spectral maxima (λ_{max}) are indicated for the oxidized flavin and the protonated and deprotonated forms of the semiquinone intermediate. In both semiquinone forms the unpaired electron is delocalized between N-5 and C-4a.

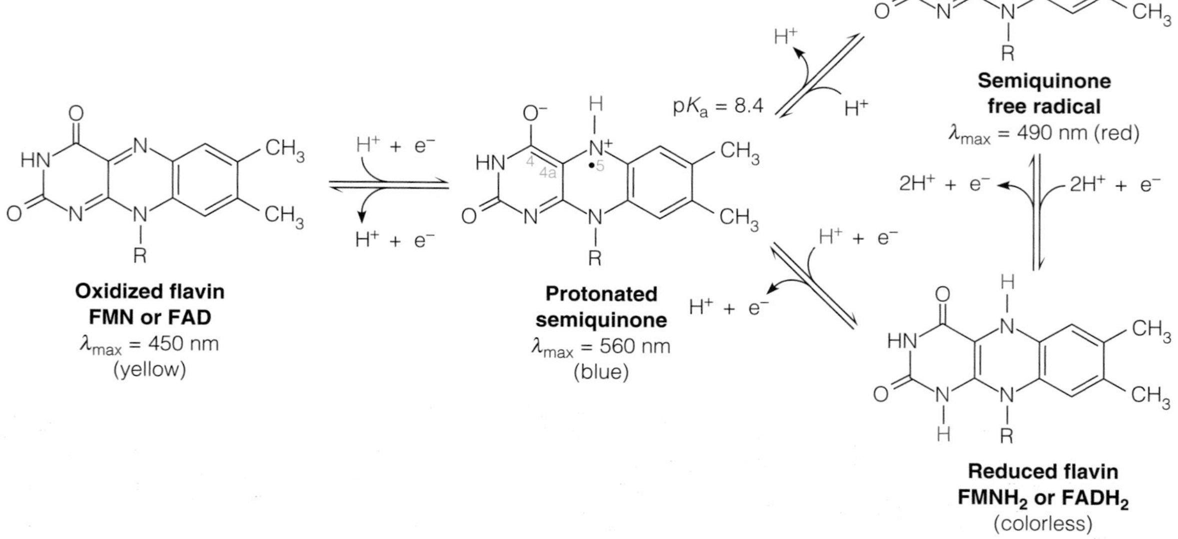

Oxidized flavin
FMN or FAD
λ_{max} = 450 nm
(yellow)

Protonated semiquinone
λ_{max} = 560 nm
(blue)

pK_a = 8.4

Semiquinone free radical
λ_{max} = 490 nm (red)

Reduced flavin
FMNH$_2$ or FADH$_2$
(colorless)

Coenzyme A: Activation of Acyl Groups

Coenzyme A (A for acyl) participates in activation of acyl groups in general, including the acetyl group derived from pyruvate. The coenzyme is derived metabolically from ATP, the vitamin **pantothenic acid**, and β-mercaptoethylamine.

β-Mercaptoethylamine   **Pantothenic acid**   **Adenosine 3′-phosphate 5′-diphosphate**

A free thiol on the β-mercaptoethylamine moiety is the business end of the coenzyme molecule; the rest of the molecule provides enzyme-binding sites. In acylated derivatives, such as acetyl-coenzyme A, the acyl group is linked to the thiol group to form an energy-rich thioester.

Coenzyme A   **Acetic acid**   **Acetyl-CoA**

We designate the acylated forms of coenzyme as acyl-CoA, and the unacylated form as CoA-SH.

The energy-rich nature of thioesters, as compared with ordinary esters, is related primarily to resonance stabilization (Figure 14.8). Most esters have two resonance forms. Stabilization involves π-electron orbital overlap, giving partial double-bond character to the C—OR link. In thioesters, the larger atomic size of S (as compared with O) reduces the π-electron overlap between C and S so that the C—SR structure does not contribute significantly to resonance stabilization. Thus, the thioester is *destabilized* relative to an ester so that its ΔG of hydrolysis is increased.

> Thioesters such as acetyl-CoA are energy rich because thioesters are destabilized relative to ordinary oxygen esters.

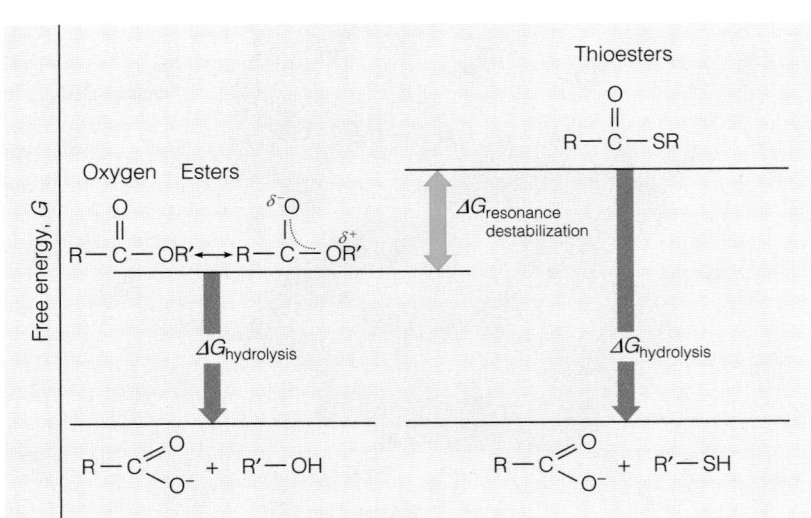

FIGURE 14.8

Comparison of free energies of hydrolysis of thioesters and oxygen esters. Lack of resonance stabilization in thioesters is the basis for the higher ΔG of hydrolysis of thioesters, relative to that of ordinary oxygen esters. The free energies of the hydrolysis products are similar for the two classes of compounds.

The lack of double-bond character in the C—SR bond of acyl-CoAs makes this bond weaker than the corresponding C—OR bond in ordinary esters, in turn making the thioalkoxide ion (R—S⁻) a good leaving group in nucleophilic displacement reactions. Thus, the acyl group is readily transferred to other metabolites, which is what occurs in the first reaction of the citric acid cycle.

Action of the Pyruvate Dehydrogenase Complex

As noted above, the oxidation of pyruvate to acetyl-CoA involves the coenzymes TPP, lipoic acid, CoA-SH, FAD, and NAD^+, acting in concert with three enzymes in the pyruvate dehydrogenase complex. Now we are ready to see how all of these components function together to effect the conversion of pyruvate to acetyl-CoA. The overall process is summarized in Figure 14.9.

The entire process, beginning with the decarboxylation of pyruvate, and ending with the transfer of a pair of electrons to NAD^+, requires five steps. A central feature of this reaction sequence is that in the three middle steps, reaction intermediates are formed and transferred between active sites by covalent attachment to the lipoamide moieties on the flexible lipoyl domains of E_2 (and E3BP). The reaction sequence begins at the active site of E_1 (**pyruvate dehydrogenase**), which catalyzes a nucleophilic addition of the TPP carbanion (ylid form of thiazolium moiety) to the ketone carbonyl group of pyruvate to form an addition product that undergoes decarboxylation to give hydroxyethyl-TPP (reaction 1, Figure 14.9). The chemistry of this step is identical to that of the pyruvate decarboxylase reaction (Figure 13.8, page 538), except that the hydroxyethyl group is not released as acetaldehyde. Instead, it is transferred to a lipoamide moiety on E_2. This reaction, also catalyzed by E_1, occurs by an S_N2-like attack of the hydroxyethyl carbanion on the lipoamide disulfide, followed by elimination of TPP to form an acetyl thioester on dihydrolipoamide and regenerate E_1 (reaction 2, Figure 14.9).

FIGURE 14.9

Mechanisms of the pyruvate dehydrogenase complex. This diagram of the oxidation of pyruvate to acetyl-CoA shows the role of the swinging arms of lipoamide in the functioning of the pyruvate dehydrogenase complex. E_1: pyruvate dehydrogenase. E_2: dihydrolipoamide transacetylase with its lipoyl domains (LD). E_3: dihydrolipoamide dehydrogenase. The subunit colors correspond to those in the structural model in Figure 14.4e (page 597). **Reaction 1:** Pyruvate reacts with the TPP carbanion of E_1 to form an addition product that undergoes decarboxylation, giving hydroxyethyl-TPP. **Reaction 2:** The hydroxyethyl group is transferred by E_1 to a lipoamide swinging arm on E_2, resulting in oxidation of the 2-carbon fragment to an acetyl group, with concomitant reduction of the lipoamide disulfide to dihydrolipoamide. **Reaction 3:** The acetyl group is transferred to CoA-SH, producing acetyl-CoA and dihydrolipoamide. **Reaction 4:** E_3 reoxidizes the reduced lipoamide swinging arm by transferring two electrons to an E_3 Cys-Cys disulfide bond. **Reaction 5:** E_3 catalyzes transfer of the electrons from the Cys sulfhydryl groups to NAD^+, regenerating the oxidized form of E_3 and releasing reduced NADH. Tightly bound FAD is used as an intermediate electron carrier in this step.

Modified from *Cellular and Molecular Life Sciences* 64:830–849, T. E.Roche, and Y. Hiromasa, Pyruvate dehydrogenase kinase regulatory mechanisms and inhibition in treating diabetes, heart ischemia, and cancer. © 2007, with kind permission from Springer Science+Business Media B.V.

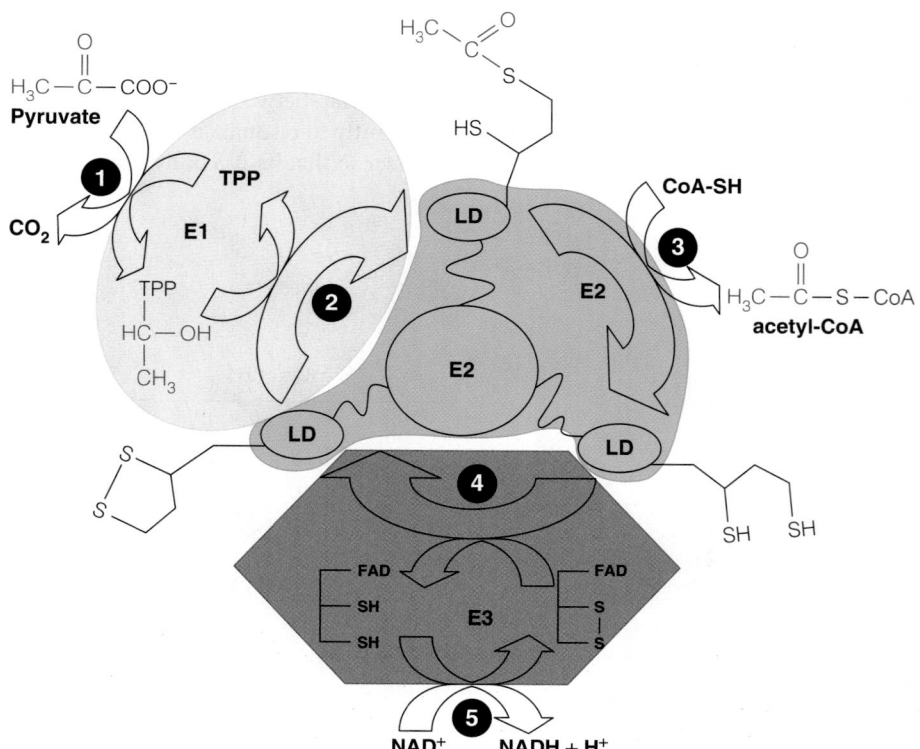

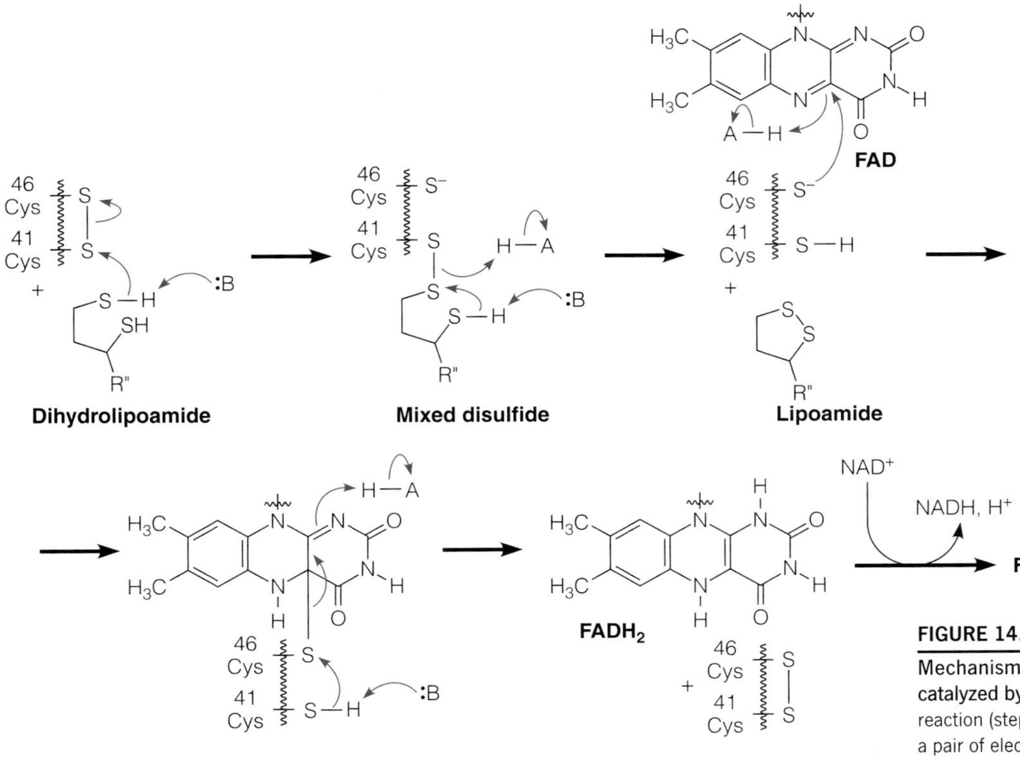

FIGURE 14.10

Mechanism of the reoxidation of dihydrolipoamide catalyzed by dihydrolipoamide dehydrogenase (E_3). This reaction (steps 4 and 5 in Figure 14.9) results in transfer of a pair of electrons from dihydrolipoamide to NAD^+, regenerating the oxidized form of E_3 and releasing reduced NADH.

Through the combination of steps 1 and 2, pyruvate undergoes a two-electron oxidation to an acetyl group, with concomitant two-electron reduction of the lipoamide disulfide to dihydrolipoamide.

The acetyl group is now covalently bound to the active site of E_2, via the flexible lipoyllysine "swinging arm." E_2 (**dihydrolipoamide transacetylase**) next catalyzes transfer of the acetyl group to CoA. This nucleophilic acyl substitution reaction simply exchanges one thioester for another, giving acetyl-CoA and dihydrolipoamide (reaction 3, Figure 14.9).

The last two steps of the process (reaction 4 and 5, Figure 14.9) are required to reoxidize the dihydrolipoamide of E_2, and transfer the pair of electrons to a dissociable carrier (NAD^+). E_3 (**dihydrolipoamide dehydrogenase**) catalyzes the transfer of the electron pair from dihydrolipoamide to NAD^+. The active site of E_3 contains a redox active Cys–Cys disulfide bond and a tightly bound FAD. The reaction involves initial formation of a mixed disulfide between dihydrolipoamide and Cys-41 of E_3 (Figure 14.10). The E_2 lipoamide disulfide is reformed by nucleophilic addition of the Cys-46 thiol to FAD to give reduced $FADH_2$ and reoxidized Cys–Cys disulfide on E_3. The final step (reaction 5, Figure 14.9) is transfer of the electron pair from $FADH_2$ to NAD^+, regenerating the oxidized form of E_3 and releasing reduced NADH.

Physical juxtaposition of the enzymes of the complex, and covalent attachment of reaction intermediates via the lipoamide swinging arm, provides a number of advantages. The five-step reaction sequence in Figure 14.9 nicely illustrates the concept of **substrate channeling**: intermediates of a multistep pathway are "handed off" from one active site to the next without diffusing from the complex. Channeling allows the overall reaction to proceed smoothly, without unwanted side reactions or diffusion of intermediates from catalytic sites. The local concentration of substrates can be very high, allowing greater flux through the pathway. The pyruvate dehydrogenase complex represents one of the best-understood

Lipoamide is tethered to one enzyme (E_2) in the pyruvate dehydrogenase complex, but it interacts with all three enzymes via a flexible swinging arm.

examples of how cells can achieve economy of function by formation of **multienzyme complexes** catalyzing sequential reactions in a pathway. In fact, this same E_1–E_2–E_3 multienzyme structure is used to oxidize several other α-keto acids. Other examples include the branched-chain α-ketoacid dehydrogenase complex (Chapter 21, page 870) and the α-ketoglutarate dehydrogenase complex, which we will discuss shortly. Indeed, the E_3 subunits are identical in all three complexes from a given species. It is clear from Figure 14.9 that each complex must have a unique E_1 subunit specific to its particular α-ketoacid substrate. But once the acyl group has been transferred to CoA-SH by E_2, the resulting dihydrolipoamide, the substrate of the E_3 subunits, is identical in each complex. Thus, as organisms evolved new α-ketoacid dehydrogenase complexes, the same E_3 subunit could be reused.

Arsenic poisoning, both intentional and unintentional, has had a long history, dating back to at least the eighth century. Trivalent As(III) compounds such as arsenite (AsO_3^{3-}) and organic arsenicals react readily with thiols, and they are especially reactive with dithiols, such as dihydrolipoamide, forming bidentate adducts:

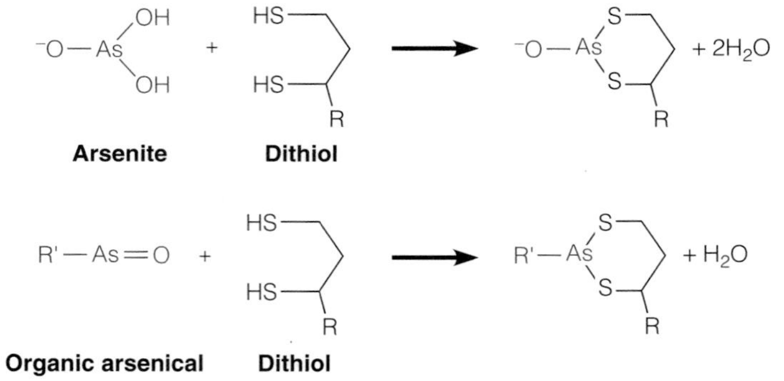

This covalent modification of lipoamide groups inactivates E_1—E_2—E_3 multienzyme complexes, including the pyruvate dehydrogenase and α-ketoglutarate dehydrogenase complexes of the citric acid cycle, thereby inhibiting respiration and accounting for the toxicity of these compounds. Arsenic became a favorite poison during the Middle Ages and the Renaissance, often being employed by impatient heirs to "get a jump" on their inheritances! Several prominent historical figures, including Francisco de' Medici, George III, and Napoleon Bonaparte may have been poisoned by arsenic. Arsenic was also an ingredient in many "tonics" that were popular during the Victorian Era. Charles Darwin may in fact have suffered from chronic arsenic poisoning from his use of these tonics. Organic arsenicals were even used in the early twentieth century to treat syphilis and trypanosomiasis because these pathogens' lipoamide-containing enzymes are often more sensitive than the host enzymes. The toxicity of these arsenic compounds led to their discontinued use once penicillin and other antibiotics were developed.

The Citric Acid Cycle

Figure 14.3 presented the entire citric acid cycle, showing the structure of each intermediate. We will now discuss the chemistry and enzymology of each reaction. The cycle is composed of eight steps, beginning with addition of a two-carbon moiety (acetyl-CoA) to a four-carbon compound (oxaloacetate) to give a six-carbon tricarboxylic acid, citrate, followed by loss of two carbons as CO_2, and finally the regeneration of oxaloacetate. During this process, four pairs of electrons are removed from the substrates and passed to carriers (NAD$^+$ and FAD) on their way to the respiratory chain.

Step 1: Introduction of Two Carbon Atoms as Acetyl-CoA

The initial reaction, catalyzed by **citrate synthase**, is akin to an aldol condensation.

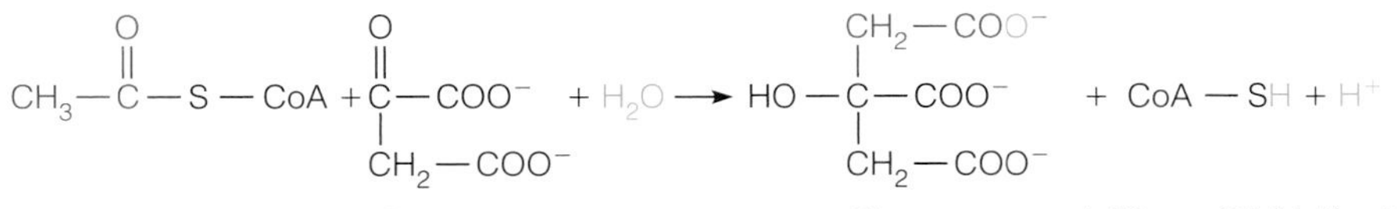

Acetyl-CoA   **Oxaloacetate**   **Citrate** $\Delta G^{\circ\prime} = -32.2$ kJ/mol

As shown in Figure 14.11, the acetyl moiety is activated by removal of an acidic α proton from the methyl carbon, and donation of a proton to the carbonyl oxygen, to give the enol (or enolate) form. A general base (side chain carboxylate of aspartate 375) abstracts the α proton and a general acid (imidazole of histidine 274) protonates the carbonyl oxygen of acetyl-CoA. The enolate then nucleophilically attacks the carbonyl carbon of oxaloacetate to give the enzyme-bound intermediate (**S**)-**citroyl-CoA**. This step is facilitated by a second active site histidine (His 320), acting as a general acid to protonate the aldol product. Citroyl-CoA is highly unstable, and spontaneously hydrolyzes to yield the products citrate + CoA-SH. This thioester hydrolysis makes the forward reaction highly exergonic—the K_{eq} for this reaction is about 3×10^5, which ensures continued operation of the cycle even when the oxaloacetate concentration is low. As expected for the first committed step in a pathway, this reaction is an important site of regulation for the overall pathway (see page 612). Finally, crystallographic analysis of citrate synthase (Figure 14.12) gives excellent evidence for the induced fit model of enzyme catalysis (see Chapter 11).

Step 2: Isomerization of Citrate

The tertiary alcohol of citrate presents yet another chemical problem: Tertiary alcohols cannot be oxidized without breaking a carbon–carbon bond because the carbon atom bearing the hydroxyl group is already bonded to three other carbons and cannot form a carbon–oxygen bond. To set up the next oxidation in the pathway, citrate is converted to isocitrate, a chiral secondary alcohol, which can be more readily oxidized. This isomerization reaction, catalyzed by **aconitase**, involves successive

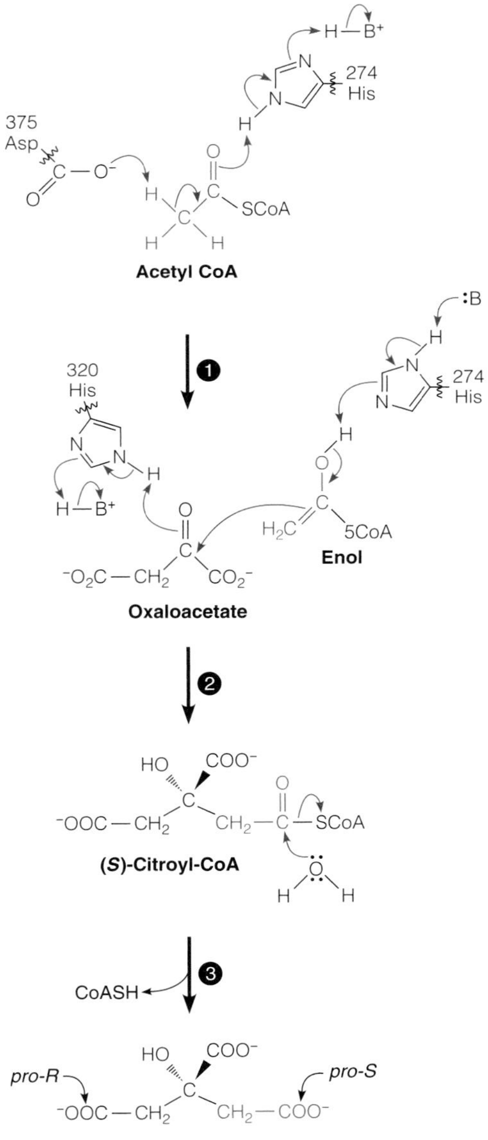

FIGURE 14.11

Mechanism of the citrate synthase reaction. **Step 1:** Asp 375 extracts a proton from the methyl group, and His 274 donates a proton to the carbonyl oxygen of acetyl-CoA, creating an enol. **Step 2:** His 274 deprotonates the acetyl-CoA enol, stabilizing the nucleophilic enolate that attacks the keto carbon of oxaloacetate. His 320 protonates the aldol product (S)-citroyl-CoA. **Step 3:** The citroyl-CoA intermediate spontaneously hydrolyzes to citrate by a nucleophilic acyl substitution reaction. Although citrate is a symmetrical molecule, the two carboxymethyl (—CH$_2$COO$^-$) groups occupy different positions relative to the OH and COO$^-$ groups on C3. Thus, citrate is **prochiral**—it can become chiral by substitution of one of its carboxymethyl groups—and these two chemically equivalent groups are designated pro-R and pro-S. The two carbons derived from acetyl-CoA form the pro-S arm in the final product citrate. This stereospecificity arises from the fact that citrate synthase catalyzes the attack of the acetyl-CoA enolate only on the si face of oxaloacetate's carbonyl carbon, giving exclusively the S stereoisomer of citroyl-CoA (recall that a carbonyl group is trigonal planar so that it has two distinct faces, referred to as si and re, at which a nucleophile can attack). Residue numbering is based on the pig heart enzyme.

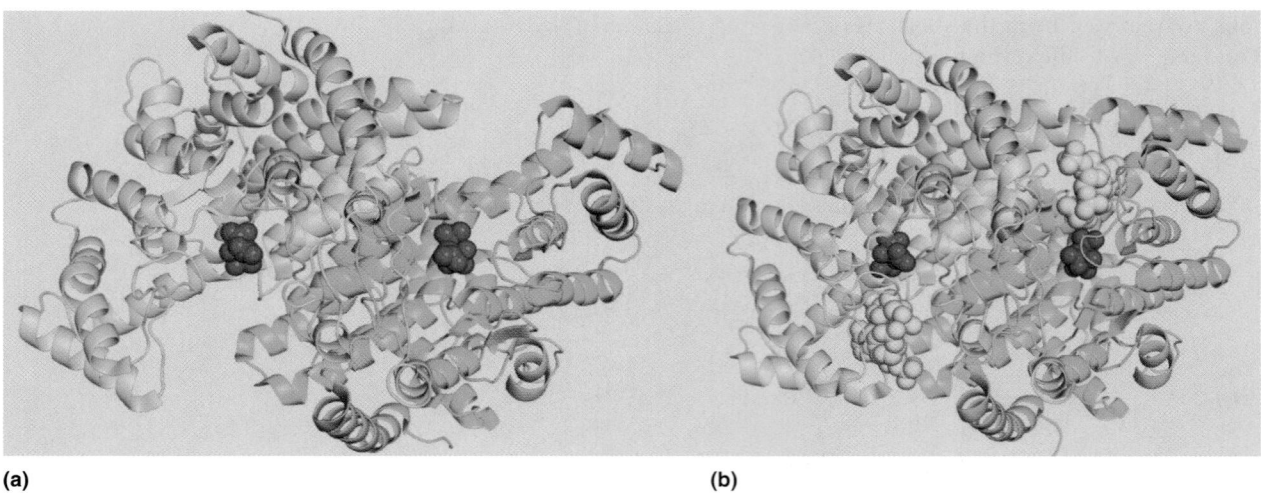

(a)   **(b)**

FIGURE 14.12

Three-dimensional structure of citrate synthase. The two forms of pig heart citrate synthase homodimer shown here were determined by crystallographic methods and support the induced fit model of enzyme catalysis (see Chapter 11). **(a)** In the absence of CoA-SH the enzyme crystallizes in an "open" form (PDB ID 1CTS). Citrate (red) binds at the base of large clefts in both catalytic domains of the homodimeric protein. **(b)** Binding of CoA-SH (yellow) causes the enzyme to adopt a "closed" conformation, with the clefts essentially filled (PDB ID 2CTS).

dehydration and hydration, through *cis*-aconitate as a dehydrated intermediate, which remains enzyme-bound.

$$\Delta G^{\circ\prime} = +6.3 \text{ kJ/mol}$$

Citrate   **cis-Aconitate**   **D-Isocitrate**
(3° alcohol)   **(2° alcohol)**

The enzyme contains nonheme iron and acid-labile sulfur in a cluster called a 4Fe–4S *iron–sulfur center*, normally associated with oxidoreductases (see Chapter 15). In aconitase this iron–sulfur cluster coordinates the hydroxyl group and one of the carboxyl groups on the citrate molecule. The dehydration occurs specifically on the *pro-R* arm, and the *pro-R* proton is removed by an active site serine acting as a base. Labeling studies showed that the —OH group added back in the hydration step is different from the one eliminated at the dehydration step, but the proton added back is the same—it is held in the active site by the serine. Furthermore, the proton removed during dehydration is added back to the *opposite* face of *cis*-aconitate during hydration. This implies that the *cis*-aconitate intermediate must flip 180° during the reaction, presumably by releasing from the enzyme, then rebinding with the iron–sulfur cluster, but in the opposite orientation. Thus, of the four possible diastereomers of isocitrate, only one, the *2R,3S* diastereomer, is produced.

Citrate   **cis-Aconitate**   **(2R, 3S)-Isocitrate**

How can citrate, a symmetric molecule, react asymmetrically with aconitase? Andrew Mesecar and Dan Koshland Jr. pointed out that if the enzyme bound the substrate at four points, then the binding site itself was asymmetric and could bind the substrate in only one way. Thus, citrate was the first substance to be recognized as **prochiral**, or symmetrical, but that becomes asymmetric upon binding to the asymmetric surface of an enzyme, or upon some similar change in one of two equivalent groups.

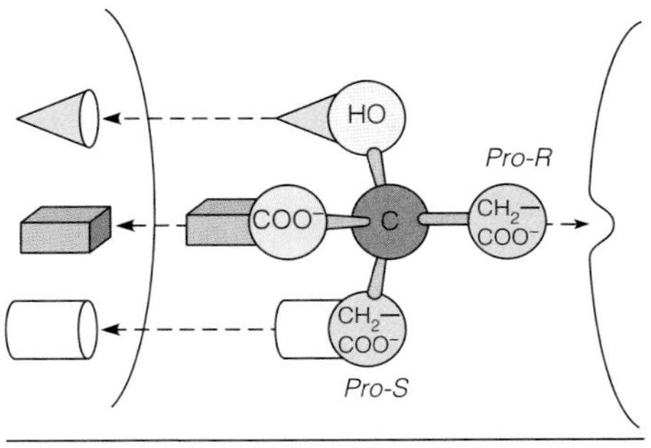

The aconitase reaction is freely reversible, and an equilibrium mixture of these three acids at 25 °C contains about 90% citrate, 4% *cis*-aconitate, and 6% isocitrate, but the exergonic nature of the next reaction draws the reaction to the right as written.

Aconitase is the target site for the toxic action of **fluoroacetate**, a plant product that was originally used as a rodenticide. Its use by ranchers in the West to control coyote populations also led to the death of eagles and other endangered animals. Fluoroacetate blocks the citric acid cycle by its metabolic conversion to **2-fluorocitrate**, which is a potent **mechanism-based**, or **suicide**, **inhibitor** of aconitase. As described in Chapter 11, a mechanism-based inhibitor undergoes the first few chemical steps of the enzymatic reaction but is then converted into a compound that binds tightly, often irreversibly, thereby inactivating the enzyme. In other words, the inhibitor requires the normal enzyme mechanism to inactivate the enzyme. Similarly, fluoroacetate can be considered a **suicide substrate**: not toxic to cells by itself, but it resembles a normal metabolite closely enough that it undergoes metabolic transformation to a product that does inhibit a crucial enzyme. The cell "commits suicide" by transforming the analog to a toxic product. In this case fluoroacetate is first converted to fluoroacetyl-CoA by acetate thiokinase (see page 621), and then to 2-fluorocitrate by citrate synthase. 2-Fluorocitrate then inhibits aconitase, halting the entire citric acid cycle.

Fundamentals of Biochemistry: Life at the Molecular Level, 3rd ed., Donald Voet, Judith G. Voet, and Charlotte W. Pratt. © 2008 John Wiley & Sons, Inc. Modified with the permission of John Wiley & Sons, Inc.

Fluoroacetate is an example of a mechanism-based, or suicide, inhibitor of aconitase.

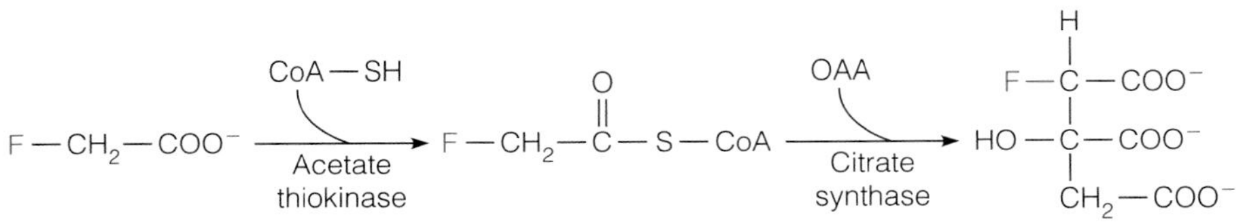

Fluoroacetate   **Fluoroacetyl-CoA**   **2-Fluorocitrate**

Step 3: Generation of CO_2 by an NAD^+-Linked Dehydrogenase

The first of two oxidative decarboxylations in the cycle is catalyzed by **isocitrate dehydrogenase**. Isocitrate is oxidized to a ketone, **oxalosuccinate**, an unstable enzyme-bound intermediate that spontaneously decarboxylates to give the product, α-ketoglutarate. The strategy here is to oxidize isocitrate's secondary alcohol to a keto group β to the carboxyl group to be removed. The β-keto group acts as an electron sink to stabilize the carbanionic transition state, facilitating decarboxylation.

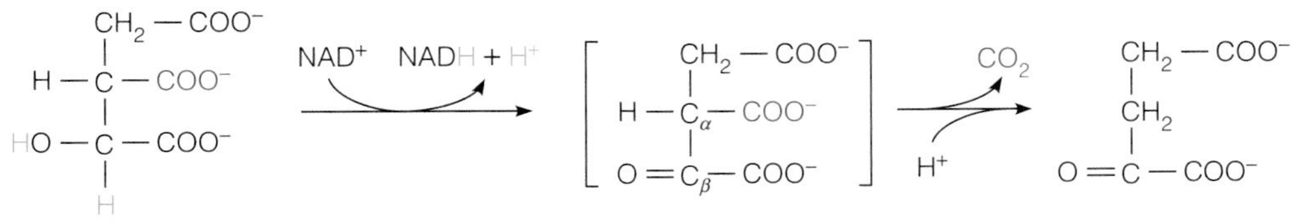

Isocitrate   **Oxalosuccinate**   **α-Ketoglutarate**

This reaction occurs with a $\Delta G^{\circ\prime}$ of -11.6 kJ/mol, and under physiological conditions it is sufficiently exergonic to pull the aconitase reaction forward. A mitochondrial form of isocitrate dehydrogenase that is specific for NAD^+ is probably the chief participant in the citric acid cycle in most cells. The NADH produced in this reaction, carrying two reducing equivalents, is the first link between

Two carbon atoms enter the citric acid cycle as acetyl-CoA, and two are lost as CO_2 in the oxidative decarboxylations of steps 3 and 4.

the citric acid cycle and the electron transport process of respiration. As such, this enzyme is an important regulatory site for controlling flux through the cycle (see page 612). Most cells also contain an $NADP^+$-specific form of isocitrate dehydrogenase, found in both cytosol and mitochondria; its likely role is the generation of NADPH for reductive biosynthetic processes.

Step 4: Generation of a Second CO_2 by an Oxidative Decarboxylation

The fourth reaction of the citric acid cycle is a multistep reaction entirely analogous to the pyruvate dehydrogenase reaction. An α-keto acid substrate undergoes oxidative decarboxylation, with concomitant formation of an acyl-CoA thioester.

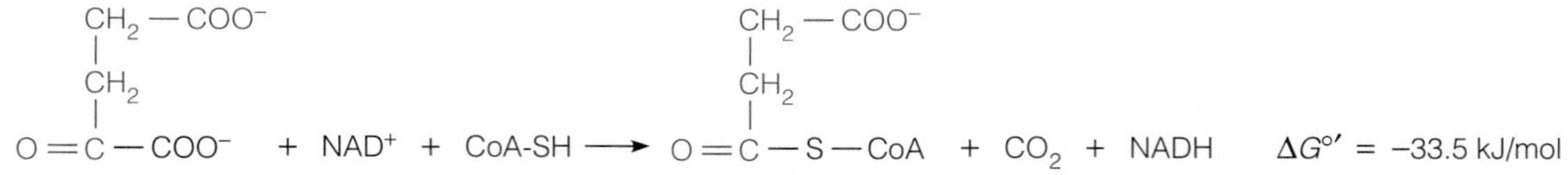

$$
\begin{array}{c}
CH_2-COO^- \\
|\\
CH_2 \\
|\\
O=C-COO^-
\end{array}
+ NAD^+ + CoA\text{-}SH \longrightarrow
\begin{array}{c}
CH_2-COO^- \\
|\\
CH_2 \\
|\\
O=C-S-CoA
\end{array}
+ CO_2 + NADH \qquad \Delta G^{\circ\prime} = -33.5 \text{ kJ/mol}
$$

α-Ketoglutarate   **Succinyl-CoA**

This reaction is catalyzed by the **α-ketoglutarate dehydrogenase complex**, an enzyme cluster similar to the pyruvate dehydrogenase complex, with three analogous enzyme activities and the same five coenzymes—TPP, lipoic acid, CoA-SH, FAD, and NAD^+. Indeed, the E_3 subunits are shared by both complexes. TPP is required for the same reason we learned with the PDH reaction: α-keto acids cannot stabilize the carbanion transition state that develops during decarboxylation. Thus the first step of the reaction decarboxylates α-ketoglutarate, producing a four-carbon TPP derivative (Figure 14.13). Subsequent transfer of the four-carbon unit to lipoic acid, transesterification of the dihydrolipoamide thioester with CoA-SH, and oxidation by FAD and NAD^+ are analogous to the reactions shown in Figure 14.9 for the pyruvate dehydrogenase complex. The succinyl-CoA product is a high-energy thioester of succinic acid.

At this point in the cycle, two carbon atoms have been introduced as acetyl-CoA (at the citrate synthase step), and two have been lost as CO_2. Because of the stereochemistry of the aconitase reaction, the two carbon atoms lost are *not* the same as the two carbons introduced at the beginning of the cycle. In the remaining reactions the four-carbon oxaloacetate is regenerated from the four-carbon succinyl-CoA, with two of the four steps involving oxidations.

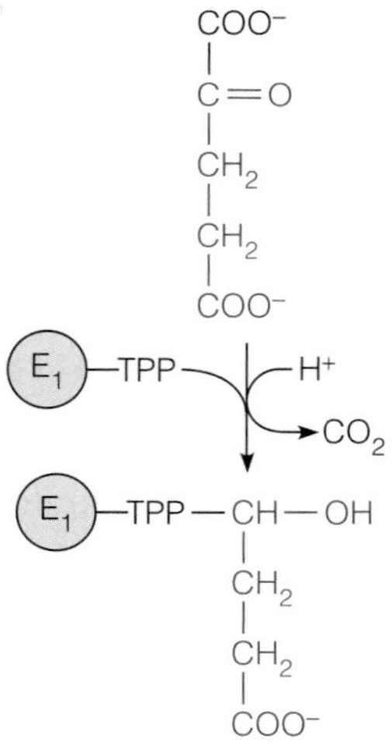

Step 5: A Substrate-Level Phosphorylation

Succinyl-CoA is an energy-rich thioester compound ($\Delta G^{\circ\prime}$ for hydrolysis $\approx$ -36 kJ/mol), and its potential energy is used to drive the formation of a nucleoside triphosphate ($\Delta G^{\circ\prime} = +30.5$ kJ/mol). This reaction, catalyzed by **succinyl-CoA synthetase**, is comparable to the two substrate-level phosphorylation reactions that we encountered in glycolysis, except that in animal cells the energy-rich nucleotide product is not always ATP but, in some tissues, GTP.

Succinyl-CoA + P_i + ADP (GDP) $\rightleftharpoons$ succinate + ATP (GTP) + CoA-SH

$$\Delta G^{\circ\prime} = -2.9 \text{ kJ/mol}$$

Succinyl-CoA synthetase is a heterodimer of α and β subunits, with the β subunit determining the substrate specificity (ADP or GDP). In animals, tissues that are dependent on oxidative metabolism, such as brain, heart, and skeletal muscle, contain the ATP-linked enzyme, while in kidney and liver ("biosynthetic," or anabolic, tissues) the GTP-linked succinyl-CoA synthetase predominates. The two isozymes of succinyl-CoA synthetase presumably serve different roles in the various tissues. The ratio of GTP to GDP in mammalian mitochondria is approximately 100:1, so the equilibrium of the GTP-linked succinyl-CoA synthetase reaction will lie further toward succinyl-CoA, ensuring adequate succinyl-CoA for ketone body production

$$
\begin{array}{c}
COO^- \\
|\\
C=O \\
|\\
CH_2 \\
|\\
CH_2 \\
|\\
COO^-
\end{array}
$$

E_1—TPP $\quad$ H$^+$

$\searrow$ CO$_2$

E_1—TPP—CH—OH
$\qquad\quad$ |
$\qquad\quad$ CH$_2$
$\qquad\quad$ |
$\qquad\quad$ CH$_2$
$\qquad\quad$ |
$\qquad\quad$ COO$^-$

FIGURE 14.13

Decarboxylation of α-ketoglutarate. The first step catalyzed out by the α-ketoglutarate dehydrogenase complex is a decarboxylation catalyzed by α-ketoglutarate decarboxylase (E_1 of the complex), producing a four-carbon TPP derivative.

FIGURE 14.14

Covalent catalysis by the succinyl-CoA synthetase reaction. Three successive nucleophilic substitution reactions conserve the energy of the thioester of succinyl-CoA in the phosphoanhydride bond of ATP (or GTP). An active site histidine side chain is transiently phosphorylated (*N*-phosphohistidine) during the reaction.

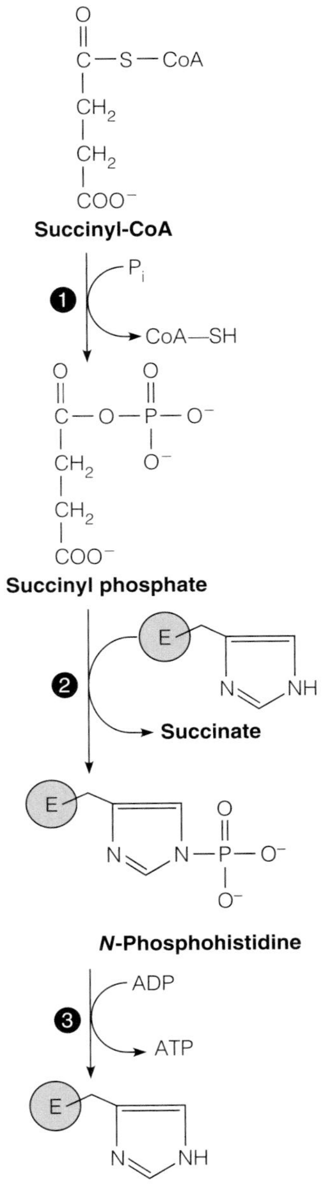

in liver (see Chapter 17, page 735). The ATP to ADP ratio in mitochondria, on the other hand, is closer to 1, and the equilibrium of the ATP-linked reaction will lie toward succinate. An ATP-linked succinyl-CoA synthetase is thus better suited for tissues dependent on oxidative metabolism for energy production. It is also possible that some of the GTP formed by the GDP-specific isozyme could be used to drive the synthesis of ATP, through the action of **nucleoside diphosphate kinase**.

$$GTP + ADP \rightleftharpoons ATP + GDP \quad \Delta G^{\circ\prime} = 0 \text{ kJ/mol}$$

Whichever nucleotide is used, the reaction occurs via a phosphorylated enzyme intermediate (covalent catalysis). As shown in Figure 14.14, an initial mixed anhydride between phosphate and a carboxyl group of succinate is formed in a nucleophilic acyl substitution reaction, displacing CoA-SH (step 1). In a second nucleophilic substitution, an active site histidine attacks the phosphorus atom of succinyl phosphate, displacing succinate (step 2). The resulting *N*-phosphohistidine residue then transfers its phosphate to the nucleoside diphosphate substrate (ADP or GDP) in a final nucleophilic substitution reaction (step 3).

X-ray crystal structures of the *E. coli* and pig heart succinyl-CoA synthetases revealed an interesting mechanism by which the enzyme stabilizes the *N*-phosphohistidine catalytic intermediate. These enzymes are composed of α and β subunits, with the active site formed at the interface of the two subunits. Each subunit of the $\alpha\beta$-dimer contributes an α-helix whose N-terminal end points toward the active site. Recall from Chapter 6 that α-helices possess a helical dipole moment with a partial (+) charge at their N-terminus. The positive ends of the electric dipoles of these two α-helices (termed "power" helices) stabilize the transient negatively charged phosphohistidine (Figure 14.15).

Step 6: A Flavin-Dependent Dehydrogenation

Completion of the cycle involves conversion of the four-carbon succinate to the four-carbon oxaloacetate. The first of the three reactions, catalyzed by **succinate dehydrogenase**, is the FAD-dependent dehydrogenation of two saturated carbons to a double bond.

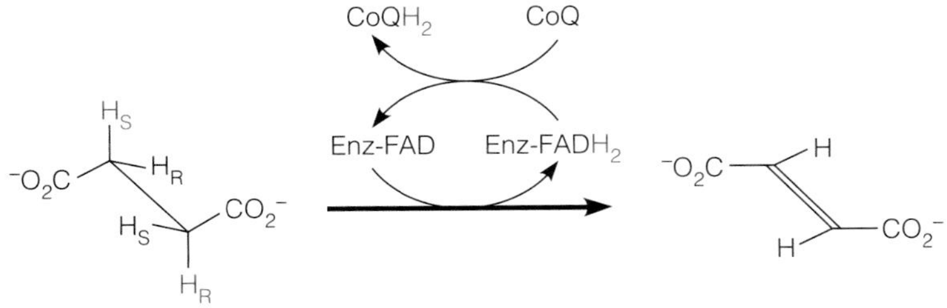

Succinate → **Fumarate**

Succinate dehydrogenase is competitively inhibited by malonate, a structural analog of succinate. Malonate inhibition of pyruvate oxidation was one of the clues that led Krebs to propose the cyclic nature of this pathway. Note that succinate

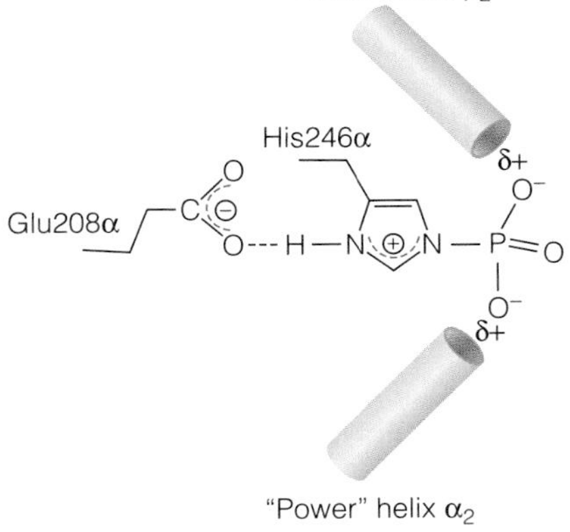

FIGURE 14.15

A charge–dipole interaction stabilizes the phosphohistidine intermediate in the succinyl-CoA synthetase reaction. In this schematic, based on the *E. coli* enzyme structure, the permanent dipoles of two α-helices (the "power" helices) are oriented such that the partial positive charges at their N-termini interact with the negative charges on the phosphate group of the active site *N*-phosphohistidine, stabilizing this transient reaction intermediate.

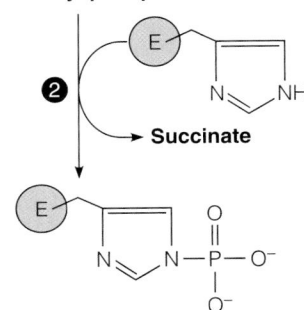

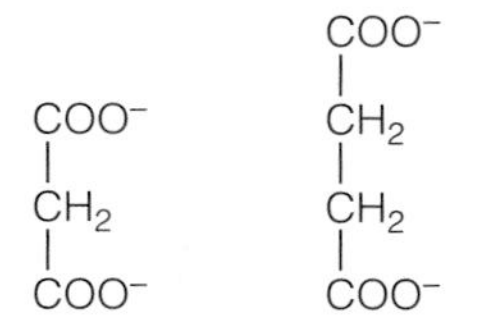

Malonate   **Succinate**

dehydrogenase is stereoselective, removing the *pro-S* hydrogen from one carbon and the *pro-R* hydrogen from the other, producing only the *trans* isomer, fumarate. The *cis* isomer, maleate, is not formed.

A C—C single bond is more difficult to oxidize than a C—O bond. Therefore, the redox coenzyme for succinate dehydrogenase is not NAD^+ but the more powerful oxidant FAD. The flavin is bound covalently to the enzyme protein, designated E, through a specific histidine residue.

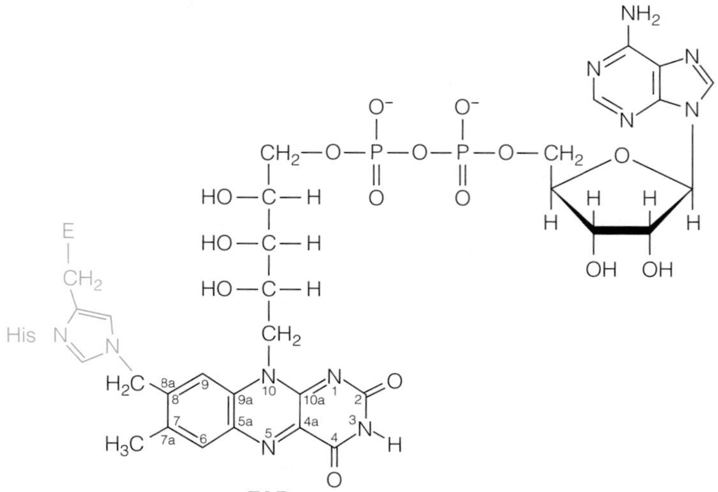

FAD

The importance of this binding is that the reduced flavin must be reoxidized for the enzyme to act again. The two electrons from the reduced flavin are transferred, through three iron–sulfur centers in the enzyme molecule, to an electron carrier (Coenzyme Q) in the mitochondrial electron transport system. Thus the reaction catalyzed by succinate dehydrogenase can be summarized as:

$$\text{succinate} + Q \rightleftharpoons \text{fumarate} + QH_2$$

In fact, succinate dehydrogenase is so tightly coupled to the other electron carriers that it is referred to as Complex II of the respiratory chain (more about this in Chapter 15). This also explains the fact that, unlike the other enzymes of the citric acid cycle, succinate dehydrogenase is an integral membrane protein, tightly bound to the mitochondrial inner membrane.

The succinate dehydrogenase reaction is followed by hydration of the fumarate double bond and dehydrogenation of the resulting α-hydroxy acid to give the α-keto acid oxaloacetate.

Step 7: Hydration of a Carbon–Carbon Double Bond

The stereospecific *trans* hydration of the carbon–carbon double bond is catalyzed by **fumarate hydratase**, more commonly called **fumarase**.

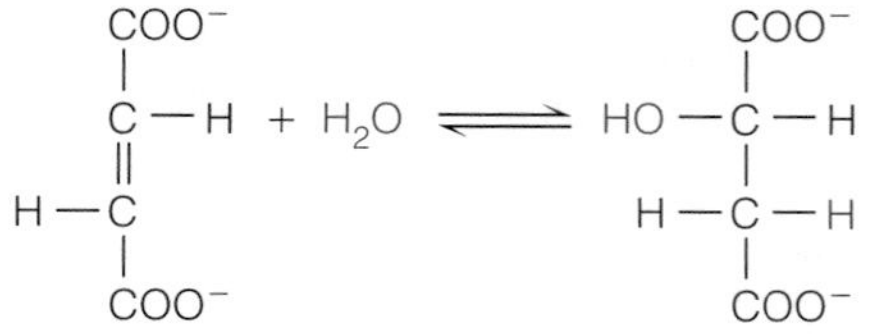

Fumarate   **L-Malate**

The reaction is mechanistically similar to the addition of water to *cis*-aconitate in the aconitase reaction, producing exclusively the *S* enantiomer (L-malate). The *cis* isomer of fumarate, namely maleate, is not a substrate for the forward reaction, nor can the enzyme act on D-malate in the reverse direction.

Step 8: A Dehydrogenation That Regenerates Oxaloacetate

Finally, the cycle is completed with the NAD^+-dependent dehydrogenation of malate to oxaloacetate, catalyzed by **malate dehydrogenase**.

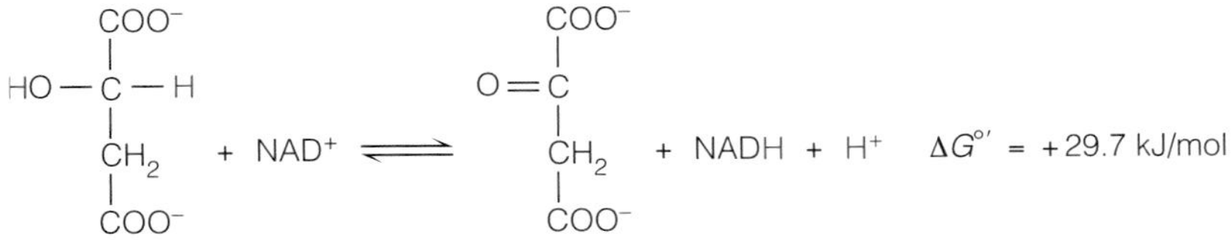

$$
\begin{array}{c}
COO^- \\
| \\
HO-C-H \\
| \\
CH_2 \\
| \\
COO^- \\
\text{L-Malate}
\end{array}
+ NAD^+ \rightleftharpoons
\begin{array}{c}
COO^- \\
| \\
O=C \\
| \\
CH_2 \\
| \\
COO^- \\
\text{Oxaloacetate}
\end{array}
+ NADH + H^+ \quad \Delta G^{\circ\prime} = +29.7 \text{ kJ/mol}
$$

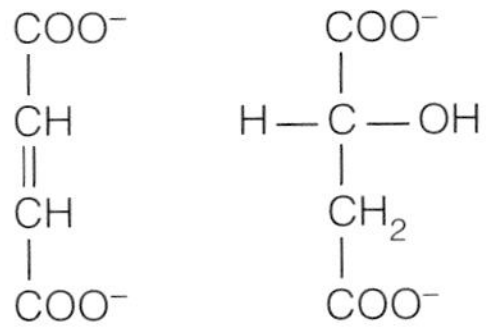

$$
\begin{array}{cc}
COO^- & COO^- \\
| & | \\
CH & H-C-OH \\
\| & | \\
CH & CH_2 \\
| & | \\
COO^- & COO^- \\
\textbf{Maleate} & \textbf{D-Malate}
\end{array}
$$

Despite the large positive standard free energy change ($\Delta G^{\circ\prime} = +29.7 \text{ kJ/mol}$), this reaction proceeds to the right in mitochondria because the highly exergonic citrate synthase reaction (the next reaction in the cycle) keeps intramitochondrial oxaloacetate levels exceedingly low (below 10^{-6} M).

Stoichiometry and Energetics of the Citric Acid Cycle

Now let us review what has been accomplished in one turn of the citric acid cycle, summarized in Table 14.2. The cycle started when a two-carbon fragment (acetyl-CoA) combined with a four-carbon acceptor (oxaloacetate). Then, two carbons were removed as CO_2 as the resultant citrate was further metabolized (but not the same two carbons that entered as acetyl-CoA). Four oxidation reactions occurred during the cycle, with NAD^+ serving as electron acceptor for three and FAD as electron acceptor for the fourth. Together, these dehydrogenations accomplish the six-electron oxidation of the methyl group and the two-electron oxidation of the carbonyl carbon of acetyl-CoA (all eight electrons derived from acetyl-CoA). High-energy phosphate was generated directly at only one reaction (catalyzed by succinyl-CoA synthetase). Finally, oxaloacetate was regenerated; it is now ready to start the cycle again by condensation with another molecule of acetyl-CoA.

TABLE 14.2   Reactions of the citric acid cycle

Reaction	Enzyme	$\Delta G^{\circ\prime}$ (kJ/mol)	ΔG (kJ/mol)
1. Acetyl-CoA + oxaloacetate + H_2O $\longrightarrow$ citrate + CoA–SH + H^+	Citrate synthase	−32.2	~ −55
2a. Citrate $\rightleftharpoons$ cis-aconitate + H_2O	Aconitase	+6.3	~ 0
2b. cis-Aconitase + H_2O $\rightleftharpoons$ isocitrate	Aconitase		
3. Isocitrate + NAD^+ $\rightleftharpoons$ α-ketoglutarate + CO_2 + NADH	Isocitrate dehydrogenase	−11.6	~ −20
4. α-Ketoglutarate + NAD^+ + CoA–SH $\rightleftharpoons$ succinyl-CoA + CO_2 + NADH	α-Ketoglutarate dehydrogenase complex	−33.5	~ −40
5. Succinyl-CoA + P_i + ADP(GDP) $\rightleftharpoons$ succinate + ATP(GDP) + CoA–SH	Succinyl-CoA synthetase	−2.9	~ 0
6. Succinate + FAD (enzyme-bound) $\rightleftharpoons$ fumarate + $FADH_2$ (enzyme-bound)	Succinate dehydrogenase	0	~ 0
7. Fumarate + H_2O $\rightleftharpoons$ L-malate	Fumarase	−3.8	~ 0
8. L-Malate + NAD^+ $\rightleftharpoons$ oxaloacetate + NADH + H^+	Malate dehydrogenase	+29.7	~ 0
	Net	−48.0	~ −115

Note: $\Delta G^{\circ\prime}$ value for reaction 3 was calculated from the $E^{\circ\prime}$ values for α-ketoglutarate/isocitrate (-0.38 V) and NAD/NADH (-0.32 V).

We can write a chemical equation representing the sum of the eight reactions involved in one turn of the cycle:

$$Acetyl\text{-}CoA + 2H_2O + 3NAD^+ + FAD + ADP + P_i \rightarrow 2CO_2 + 3NADH/H^+ + FADH_2 + CoA\text{-}SH + ATP$$

In those tissues that use a GTP-dependent isozyme of succinyl-CoA synthetase, the GTP formed in the reaction is energetically equivalent to ATP because nucleoside diphosphate kinase can convert the GTP that is formed to ATP at no net free energy cost.

Now if we take into account the pyruvate dehydrogenase reaction, and if we recall that each molecule of glucose generates two molecules of pyruvate, we can write the following equation for catabolism of glucose through glycolysis and the citric acid cycle.

$$Glucose + 2H_2O + 10NAD^+ + 2FAD + 4ADP + 4P_i \rightarrow 6CO_2 + 10NADH/H^+ + 2FADH_2 + 4ATP$$

Of the 10 moles of NADH produced per glucose, 2 moles are generated in the cytoplasm at the glyceraldehyde-3-phosphate dehydrogenase step. At this point the ATP yield per mole of glucose metabolized has not increased greatly over the yield from glycolysis: 2 moles of ATP per glucose in glycolysis alone to 4 moles here. Most of the ATP generated during glucose oxidation is not formed directly from reactions of glycolysis and the citric acid cycle (substrate-level phosphorylations) but is formed from the reoxidation of reduced electron carriers in the respiratory chain. These compounds, NADH and $FADH_2$, are themselves energy rich, in the sense that their oxidation is highly exergonic. As electrons are transferred from these reduced carriers to molecular oxygen, in a stepwise fashion, the coupled synthesis of ATP from ADP takes place, producing about 2.5 moles of ATP per mole of NADH reoxidized and about 1.5 moles of ATP per mole of $FADH_2$ reoxidized. As we see in Chapter 15, this coupled synthesis generates about 30–32 moles of ATP per mole of glucose oxidized to CO_2 and water.

Regulation of Pyruvate Dehydrogenase and the Citric Acid Cycle

Because the citric acid cycle is a source of biosynthetic intermediates, as well as a route for generating metabolic energy, regulation of the cycle is somewhat more complex than if it were just an energy-generating pathway. As in glycolysis, regulation occurs both at the level of entry of fuel into the cycle (pyruvate dehydrogenase complex and citrate synthase) and at the level of control of key reactions within the cycle (isocitrate dehydrogenase and α-ketoglutarate dehydrogenase). Figure 14.16 summarizes the major factors involved in regulation at both levels.

Control of Pyruvate Oxidation

Fuel enters the cycle primarily as acetyl-CoA, which arises from carbohydrate via pyruvate dehydrogenase and from the β-oxidation of fatty acids. Because fatty acid oxidation is discussed in Chapter 17, we concentrate here on control of pyruvate dehydrogenase. The activity of this complex is controlled by feedback inhibition and, as noted earlier, by a covalent modification that is in turn controlled by the energy state of the cell. E_2, the transacetylase component (refer to Figure 14.9), is competitively inhibited by acetyl-CoA. E_3, the dihydrolipoamide dehydrogenase component, is competitively inhibited by NADH. Thus, if the products of the reaction (acetyl-CoA and NADH) are not being continuously removed by subsequent metabolic processes, feedback inhibition by these products will shut down further pyruvate oxidation (Figure 14.17).

In the mammalian pyruvate dehydrogenase complex, however, the primary mechanism by which enzyme activity is controlled is by covalent modification of E_1, the pyruvate decarboxylase component of the complex. As shown schematically

One turn of the citric acid cycle generates one high-energy phosphate through substrate-level phosphorylation, plus three NADH and one $FADH_2$ for subsequent reoxidation in the electron transport chain.

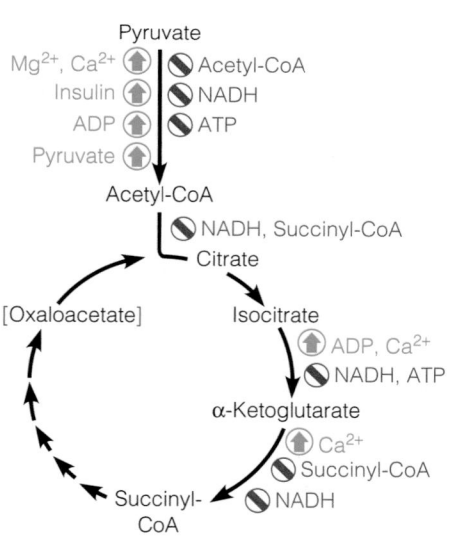

FIGURE 14.16

Major regulatory factors controlling pyruvate dehydrogenase and the citric acid cycle. Red brackets indicate concentration dependence. NADH can inhibit through allosteric interactions, but apparent NADH inhibition can also be a reflection of reduced NAD^+ availability.

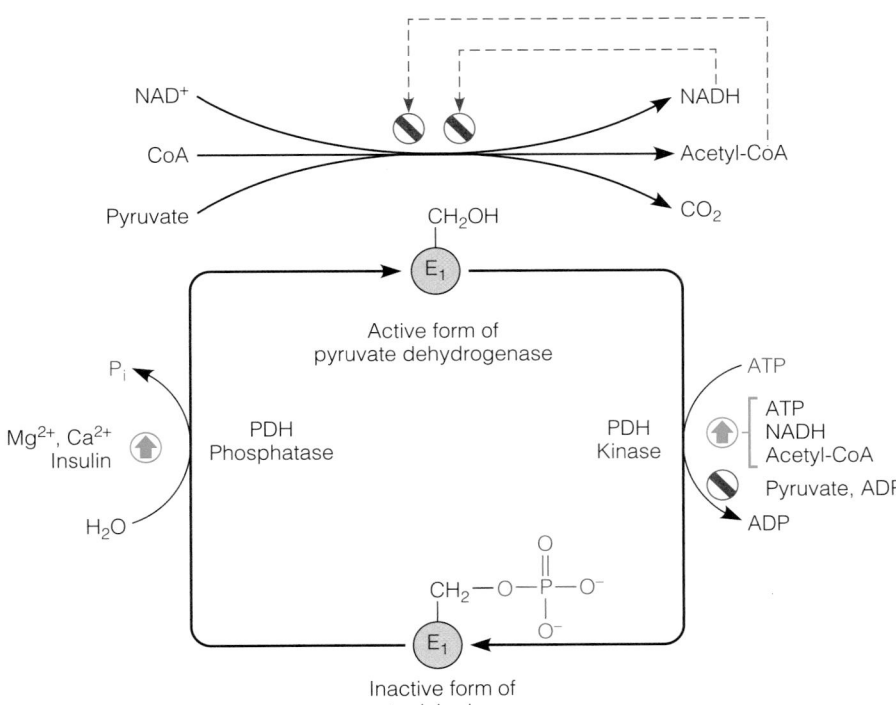

FIGURE 14.17

Regulation of the mammalian pyruvate dehydrogenase complex by feedback inhibition and by covalent modification of E_1. A kinase and a phosphatase inactivate and activate the first component (E_1) of the pyruvate dehydrogenase complex by phosphorylating and dephosphorylating, respectively, three specific serine residues (depicted as —CH_2OH). The active form of the pyruvate dehydrogenase complex is feedback inhibited by acetyl-CoA and NADH.

in Figure 14.17, this involves phosphorylation and dephosphorylation of serine residues in E_1. Mammals express four different isozymes of **pyruvate dehydrogenase kinase**, which phosphorylate several specific E_1 serine residues, resulting in loss of activity of pyruvate dehydrogenase. Two isozymes of **pyruvate dehydrogenase phosphatase** are known that hydrolytically remove the bound phosphate and reactivate the PDH complex. These protein kinase and phosphatase isozymes are expressed in different tissues and mediate tissue-specific regulation of the PDH complex. Recall that these regulatory enzymes are both integrated within the PDH complex, and PDH activity in different tissues is a subtle balance between the relative activities of the kinases and the phosphatases. PDH kinase is *activated* by both NADH and acetyl-CoA (products of the PDH reaction) and by ATP. Thus, the kinase turns off PDH activity when products of the reaction accumulate. PDH kinase is *inhibited* by ADP and pyruvate. These inhibitors of the kinase, which indicate low energy and available substrate, respectively, result in an increase in the ratio of active to inactive PDH, increasing flux through the complex. PDH phosphatase is *activated* by Ca^{2+} and Mg^{2+} and also as a result of insulin secretion. The Ca^{2+} effect on PDH phosphatase mediates stimulation of PDH activity during muscle contraction and in response to epinephrine. Recall from Chapter 8 that Ca^{2+} is a critical signaling molecule for contraction in vertebrate muscle. Using this same signaling molecule to regulate PDH flux provides an elegant mechanism to match ATP demand to its production by the citric acid cycle and subsequent oxidative phosphorylation. The Mg^{2+} effect on PDH phosphatase regulates flux through PDH in response to the energy charge ([ATP]/[ADP] ratio). Because ATP binds Mg^{2+} more tightly than does ADP, the concentration of free Mg^{2+} reflects the ATP/ADP ratio within the mitochondrion. That is, free Mg^{2+} accumulates at low [ATP]/[ADP] ratios, and it increases PDH activity by stimulating PDH phosphatase activity (and thus dephosphorylation of the complex). On the other hand, when ATP is abundant and further energy production is unneeded, pyruvate dehydrogenase is turned off due to activation of PDH kinase (which phosphorylates the complex).

Together, these short-term regulatory mechanisms allow the cell to manage the utilization of fuels for entry into the citric acid cycle. As will be seen in Chapter 17,

Activity of the pyruvate dehydrogenase complex is regulated by reversible phosphorylation of the E_1 subunit.

oxidation of fatty acids is another important source of acetyl-CoA and NADH, and both of these activate the PDH kinase. Thus, when metabolic conditions favor fatty acid oxidation as the primary fuel source (e.g., during fasting or long-term exercise), carbohydrate reserves are conserved by turning off PDH activity. Once the carbohydrate supply is replenished, PDH can be quickly turned back on.

Control of the Citric Acid Cycle

Flux through the citric acid cycle is controlled by allosteric interactions, but the concentrations of substrates also play a critical role. Though details of regulation vary among different cells and tissues, the major effects are as summarized in Figure 14.16. As we saw in glycolysis, most of the reactions of a pathway operate near equilibrium under cellular conditions ($\Delta G \sim 0$; Table 14.2). Key sites for allosteric regulation are thus those enzymes that catalyze reactions that occur with large free energy decreases: citrate synthase, isocitrate dehydrogenase, and α-ketoglutarate dehydrogenase (Table 14.2).

The most important factor controlling citric acid cycle activity is the intramitochondrial ratio of [NAD$^+$] to [NADH]. NAD$^+$ is a substrate for three cycle enzymes (Figure 14.3) as well as for pyruvate dehydrogenase. Under conditions that *decrease* the [NAD$^+$]/[NADH] ratio, such as limitation of the oxygen supply, the low concentration of NAD$^+$ can limit the activities of these dehydrogenases.

In some mammalian tissues, notably liver, the levels of citrate vary as much as 10-fold, and at low levels flux through the citrate synthase reaction is limited by substrate availability. Recall that in some animal tissues citrate is also a prime regulator of flux through *glycolysis* via allosteric regulation of phosphofructokinase (PFK), helping to match the rate of glycolysis to that of the citric acid cycle. This is not true for all tissues. Heart cells, for example, cannot transport citrate out of mitochondria, so interaction with cytosolic PFK probably does not occur to a significant extent. However, citrate levels still control the citric acid cycle in heart.

The other important regulatory sites are the reactions catalyzed by isocitrate dehydrogenase and α-ketoglutarate dehydrogenase. In many cells, isocitrate dehydrogenase is allosterically activated by ADP and is allosterically inhibited by NADH and ATP; this control is in addition to the indirect reduction of activity seen at low [NAD$^+$]/[NADH] ratios. In bacteria, isocitrate dehydrogenase is also inactivated by phosphorylation at one serine residue, which prevents binding of isocitrate. α-Ketoglutarate dehydrogenase activity is inhibited by accumulation of its products succinyl-CoA and NADH. The mechanisms are comparable to the mechanisms by which levels of acetyl-CoA and NADH control pyruvate dehydrogenase activity (Figure 14.17). Finally, in vertebrates, Ca^{2+} allosterically stimulates both isocitrate dehydrogenase and α-ketoglutarate dehydrogenase. Ca^{2+} can be thought of as a second messenger in a signal transduction pathway, but one that can cross the mitochondrial inner membrane. Ca^{2+} thus allows the rate of substrate oxidation by the citric acid cycle to respond to increased ATP demand during muscle contraction.

For some time it was thought that citrate synthase represented another site for allosteric control. The enzyme is subject to inhibition by NADH, NADPH, or succinyl-CoA. However, measurements of intramitochondrial levels of oxaloacetate, acetyl-CoA, and citrate show that the enzyme is operating close to equilibrium conditions; that is, the ratio of [citrate] to ([acetyl-CoA] × [oxaloacetate]) is close to K_{eq} for the citrate synthase reaction. On the other hand, it is evident that intramitochondrial levels of oxaloacetate, which are quite low, can exercise substrate-level control over flux through citrate synthase.

To summarize, citric acid cycle flux is responsive to the energy state of the cell, through allosteric activation of isocitrate dehydrogenase by ADP; to the redox state of the cell, through flux rate limitation caused when intramitochondrial [NAD$^+$] decreases; and to the availability of energy-rich compounds, through inhibition of relevant enzymes by acetyl-CoA or succinyl-CoA.

The citric acid cycle is controlled primarily by the relative intramitochondrial concentrations of NAD$^+$ and NADH.

Organization of the Citric Acid Cycle Enzymes

The matrix of mitochondria, where the citric acid cycle enzymes are all located, is not the simple aqueous solution we might imagine from the cartoons of mitochondria found in textbooks (including this one, see page 594). Indeed, it is estimated that the protein concentration of the mitochondrial matrix approaches *500 mg/mL or more*, and thus the matrix is more like a viscous gel. Consistent with this incredibly high protein concentration, there is now considerable evidence that the enzymes of the citric acid cycle are organized in a supramolecular multienzyme complex, or **metabolon**, that is associated with the matrix side of the inner membrane where succinate dehydrogenase is anchored. Physical association of enzymes catalyzing sequential steps in a metabolic pathway could provide significant kinetic advantages through substrate channeling, as was discussed for the pyruvate dehydrogenase complex. In fact, ^{13}C-NMR studies in the yeast *Saccharomyces cerevisiae* provide evidence for substrate channeling in the citric acid cycle. It is likely that many, if not most, multistep metabolic pathways are organized into metabolons in intact cells.

Evolution of the Citric Acid Cycle

As we discussed in Chapter 12, metabolic pathways are the products of evolution, being built from enzymes and pathways that may have had other functions initially. Indeed, as we shall see below, even the citric acid cycle as we know it in extant (currently existing) aerobic organisms is used for more than just oxidation of acetyl-CoA. Genomic analyses have revealed the existence of genes for citric acid cycle enzymes in organisms from all three domains of life (Bacteria, Archaea, and Eukarya), including anaerobic chemotrophs. These latter organisms do not harvest energy by oxidizing glucose, but rather possess an incomplete citric acid cycle as a fermentative pathway and for producing precursors for other biosynthetic processes. In the reductive branch (see margin), reversing the last four enzymes of the cycle (oxaloacetate to succinate) regenerates the oxidized cofactor NAD$^+$ from the NADH produced in the glyceraldehyde-3-phosphate dehydrogenase step of glycolysis (Chapter 13). In the oxidative branch, the first three steps of the citric acid cycle produce α-ketoglutarate, an important biosynthetic precursor. However, these anaerobic chemotrophs lack the enzymes necessary for conversion of α-ketoglutarate to succinate. We can imagine that these reductive and oxidative branches of an incomplete citric acid cycle existed in organisms that evolved before the appearance of atmospheric oxygen some 2.5 billion years ago. Once oxygen levels rose to a level that would support the much more efficient aerobic energy metabolism, it is not difficult to see how a complete citric acid cycle could evolve by recruitment of just a couple of new enzymes (α-ketoglutarate dehydrogenase and succinyl-CoA synthetase). Finally, phylogenetic analyses suggest that the citric acid cycle arose initially as a reductive pathway, used by early autotrophs to fix CO_2 (at the α-ketoglutarate dehydrogenase, isocitrate dehydrogenase, and pyruvate dehydrogenase steps).

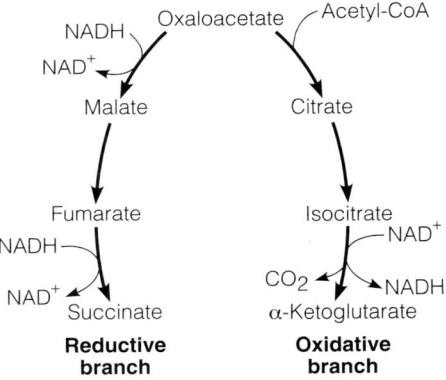

Citric Acid Cycle Malfunction as a Cause of Human Disease

Given the ubiquitous nature of the citric acid cycle, and the critical role it plays in the energy metabolism of a cell, one might expect that a deficiency in any of the enzymes in the cycle would be lethal. However, we now know that defects in certain citric acid cycle enzymes, while not lethal, are linked to a number of rare neurodegenerative diseases and tumors in humans. For example, inherited deficiency of succinate dehydrogenase leads to formation of paraganglioma or pheochromocytoma tumors, depending on which subunit is defective. Mutations in the fumarase gene are linked to uterine and/or renal cell cancer. Defects in α-ketoglutarate dehydrogenase, succinate dehydrogenase, and fumarase cause neurodegenerative diseases (Leigh Syndrome or other encephalopathies). Mutations in isocitrate dehydrogenase are found in a majority of several types of malignant gliomas, the most common type of brain tumors in humans. Although we do not fully understand the mechanisms underlying these processes, a link between organic acid accumulation and abnormal cell proliferation, resulting in tumor formation, is suspected. It is hypothesized that accumulation of certain citric acid cycle metabolites leads to activation of hypoxia-inducible factor 1 (HIF-1), a transcription factor that regulates tumor angiogenesis and tumor-cell energy metabolism. Despite the wealth of knowledge we have acquired since Krebs first elucidated the citric acid cycle, we still have a lot to learn about this pathway and its role in cellular physiology.

Anaplerotic Sequences: The Need to Replace Cycle Intermediates

Citric acid cycle intermediates used in biosynthetic pathways must be replenished to maintain flux through the cycle. Anaplerotic pathways serve this purpose.

So far, our discussion of the citric acid cycle has focused on its role in catabolism and energy generation. The cycle also serves as an important source of biosynthetic intermediates, and is thus considered an **amphibolic** pathway. Figure 14.18 summarizes the most important anabolic pathways involved. These pathways tend to draw carbon from the cycle by utilizing intermediates in the cycle. Succinyl-CoA is used in the synthesis of heme and other porphyrins. Oxaloacetate and α-ketoglutarate are the α-keto acid analogs of the amino acids aspartate and glutamate, respectively, and are used in the synthesis of these and other amino acids by transamination (see pages 618–619). In some tissues citrate is transported from mitochondria to the cytosol, where it is cleaved to provide acetyl-CoA for fatty acid biosynthesis. Because these and other reactions tend to deplete citric acid cycle intermediates by drawing carbon away, operation of the cycle would be impaired were it not for other processes that replenish the stores of citric acid cycle intermediates. These processes are called **anaplerotic** pathways, from a Greek word that means "filling up." In most cells the flow of carbon out of the cycle is balanced by these anaplerotic reactions so that the intramitochondrial concentrations of citric acid cycle intermediates remain constant with time. The anaplerotic processes are summarized by the red arrows in Figure 14.18.

Reactions That Replenish Oxaloacetate

In animals the most important anaplerotic reaction, particularly in liver and kidney, is the reversible, ATP-dependent carboxylation of pyruvate to give oxaloacetate. This reaction is catalyzed by **pyruvate carboxylase**, which we first saw in gluconeogenesis (Chapter 13).

$$\begin{array}{c} CH_3 \\ | \\ C=O \\ | \\ COO^- \end{array} + HCO_3^- + ATP \rightleftharpoons \begin{array}{c} COO^- \\ | \\ CH_2 \\ | \\ C=O \\ | \\ COO^- \end{array} + ADP + P_i + 2H^+$$

Pyruvate **Oxaloacetate**

Recall that the enzyme is activated allosterically by acetyl-CoA; in fact, it is all but inactive in the absence of this effector (see Figure 13.18, page 552). This process represents a *feedforward* activation because the effect of acetyl-CoA accumulation is to promote its own utilization by stimulating the synthesis of oxaloacetate. The latter compound in turn reacts with acetyl-CoA, via the citrate synthase reaction. Alternatively, oxaloacetate can be used for carbohydrate synthesis, via gluconeogenesis, and acetyl-CoA accumulation can be seen as a signal that adequate carbon is available for some of it to be stored as carbohydrate.

In Chapter 11 we identified biotin as a cofactor in most carboxylation reactions involving CO_2. Pyruvate carboxylase is a tetrameric protein with each

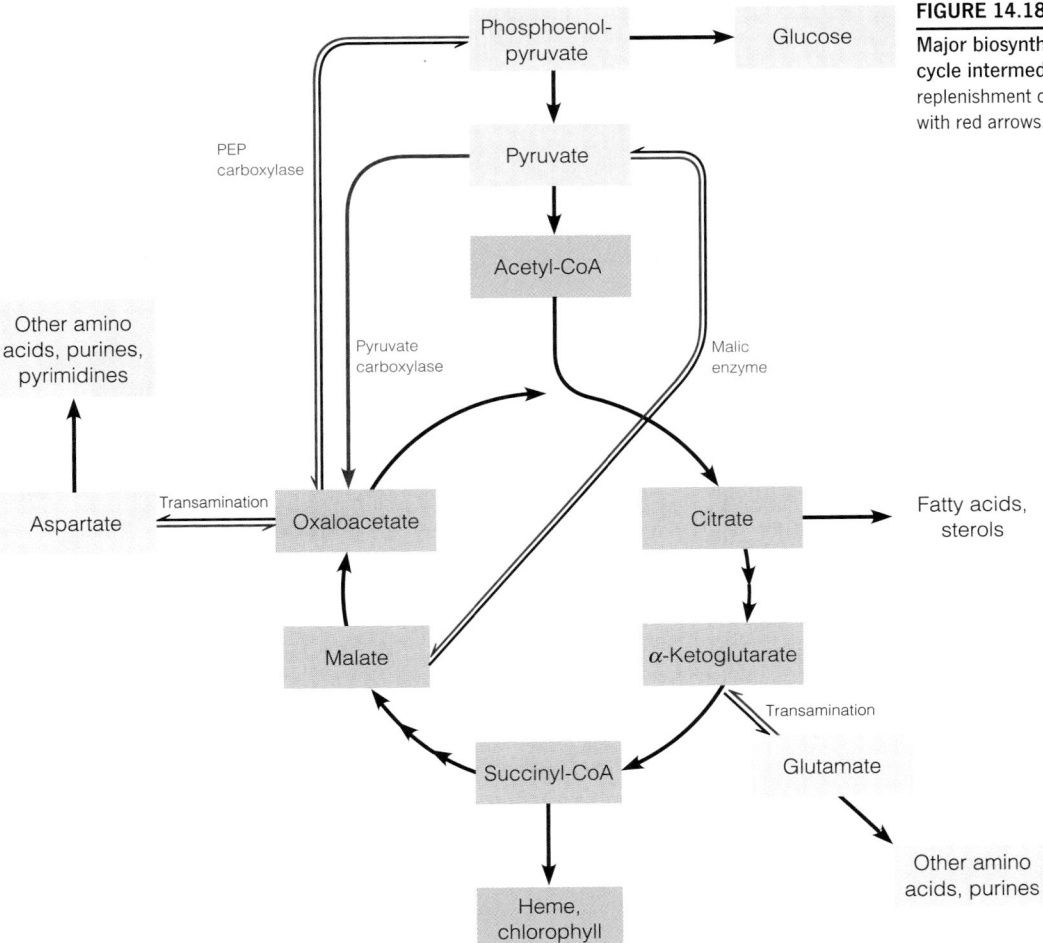

FIGURE 14.18

Major biosynthetic roles of some citric acid cycle intermediates. Anaplerotic pathways for replenishment of these intermediates are shown with red arrows.

identical subunit composed of four domains with distinct functions (Figure 14.19). A biotin carrier (BCCP) domain contains the biotin cofactor covalently bound through an amide linkage involving the ε-amino group of a lysine residue. The biotin carboxylation (BC) domain catalyzes an ATP-dependent carboxylation of the cofactor to give *N*-carboxybiotin (Figure 14.20). The carboxyltransferase (CT) domain catalyzes transfer of the carboxyl group from *N*-carboxybiotin to pyruvate to form the product oxaloacetate. During the reaction cycle, the BCCP domain acts as a swinging arm to transfer the biotin-bound carboxyl group between the two catalytic domains, much like the lipoamide swinging arms of α-keto acid dehydrogenase complexes we discussed earlier. The fourth domain is the regulatory domain, where the allosteric activator acetyl-CoA binds. This type of enzyme, with multiple domains carrying out multiple catalytic functions, is called a **multifunctional enzyme**.

In plants and bacteria an alternative route leads directly from phosphoenolpyruvate to oxaloacetate. Because phosphoenolpyruvate is such an energy-rich compound, this reaction, catalyzed by **phosphoenolpyruvate carboxylase**, requires neither an energy cofactor nor biotin. This reaction is important in the C_4 pathway of photosynthetic CO_2 fixation (Chapter 16).

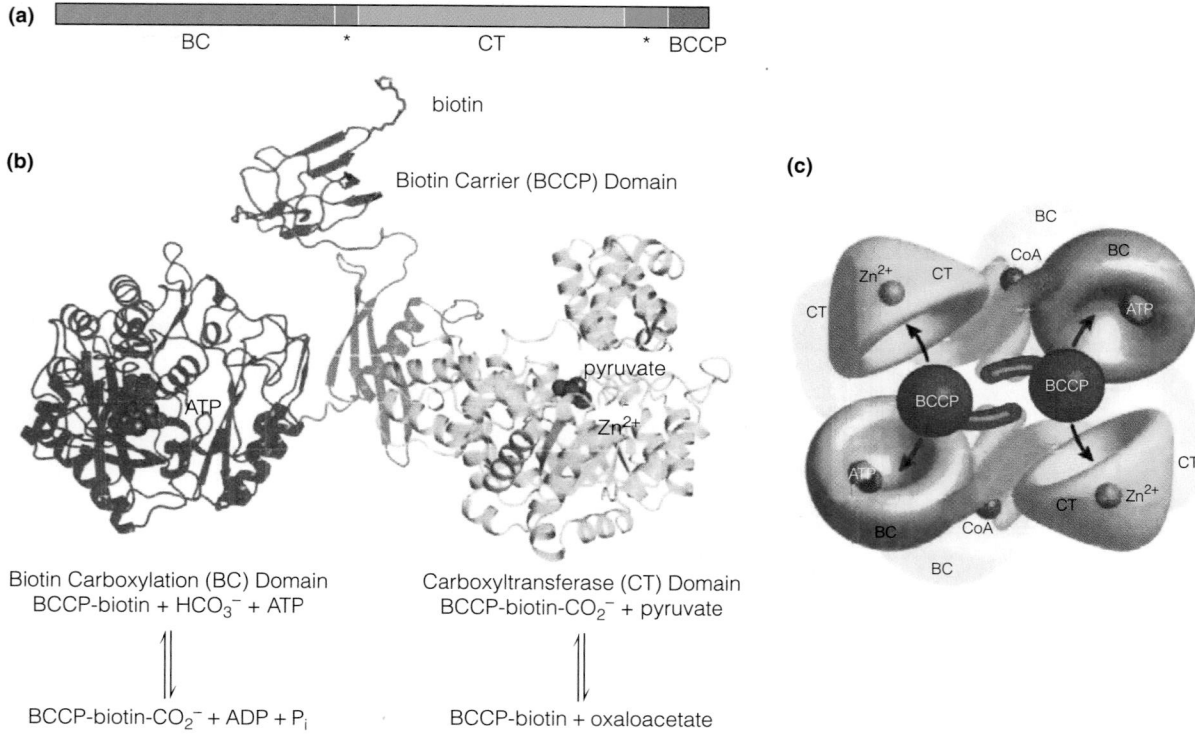

FIGURE 14.19

Structure of the multifunctional biotin-dependent pyruvate carboxylase. **(a)** This schematic illustrates the linear arrangement of the four domains within the primary structure of each subunit. The biotin cofactor is covalently linked to a lysine residue in the BCCP domain (red) near the C-terminus of each subunit. The dark green regions fold into the allosteric regulatory domain. **(b)** Shows the X-ray crystal structure of one subunit of pyruvate carboxylase from *Staphylococcus aureus*, along with the reactions catalyzed by the BC (blue) and CT (yellow) domains. Note the central allosteric domain connecting the two catalytic domains with the BCCP domain. **(c)** A model for the tetrameric enzyme shows how the BCCP domain swings between BC and CT active sites located on opposing polypeptide chains.

(a) From *Science* 317:1076–1079, M. St. Maurice, L. Reinhardt, K. H. Surinya, P. V. Attwood, J. C. Wallace, W. W. Cleland, and I. Rayment, Domain architecture of pyruvate carboxylase, a biotin-dependent multifunctional enzyme. © 2007. Reprinted with permission from AAAS; (b) Reproduced with permission from *Biochemical Journal* 413:369–387, S. Jitrapakdee, M. St. Maurice, I. Rayment, W. W. Cleland, J. C. Wallace, and P. V. Attwood, Structure, mechanism and regulation of pyruvate carboxylase. © 2008, The Biochemical Society.

A related enzyme, phosphoenolpyruvate carboxykinase, also converts phosphoenolpyruvate to oxaloacetate. The primary role of this enzyme is in gluconeogenesis, as we discussed in Chapter 13. However, in heart and skeletal muscle, this same enzyme can be used in the anaplerotic direction to replenish oxaloacetate.

The Malic Enzyme

In addition to pyruvate carboxylase and phosphoenolpyruvate carboxylase, a third anaplerotic process is provided by an enzyme commonly known as **malic enzyme** but more officially as **malate dehydrogenase (decarboxylating:NADP$^+$)**. The malic enzyme catalyzes the reductive carboxylation of pyruvate to give malate.

$$\underset{\textbf{Pyruvate}}{\begin{array}{c} CH_3 \\ | \\ C=O \\ | \\ COO^- \end{array}} + HCO_3^- + NADPH + H^+ \rightleftharpoons \underset{\textbf{L-Malate}}{\begin{array}{c} COO^- \\ | \\ CH_2 \\ | \\ H-C-OH \\ | \\ COO^- \end{array}} + NADP^+ + H_2O$$

Note that this enzyme uses NADPH, rather than NADH, as the electron donor in this reduction. As we shall see in Chapter 17, the malic enzyme reaction, running in the opposite direction, is an important source of NADPH for fatty acid synthesis.

Reactions Involving Amino Acids

Although not usually classified as anaplerotic pathways, **transamination** reactions can be regarded that way also because they are reversible reactions that can yield citric acid cycle intermediates. In transamination an amino acid transfers its

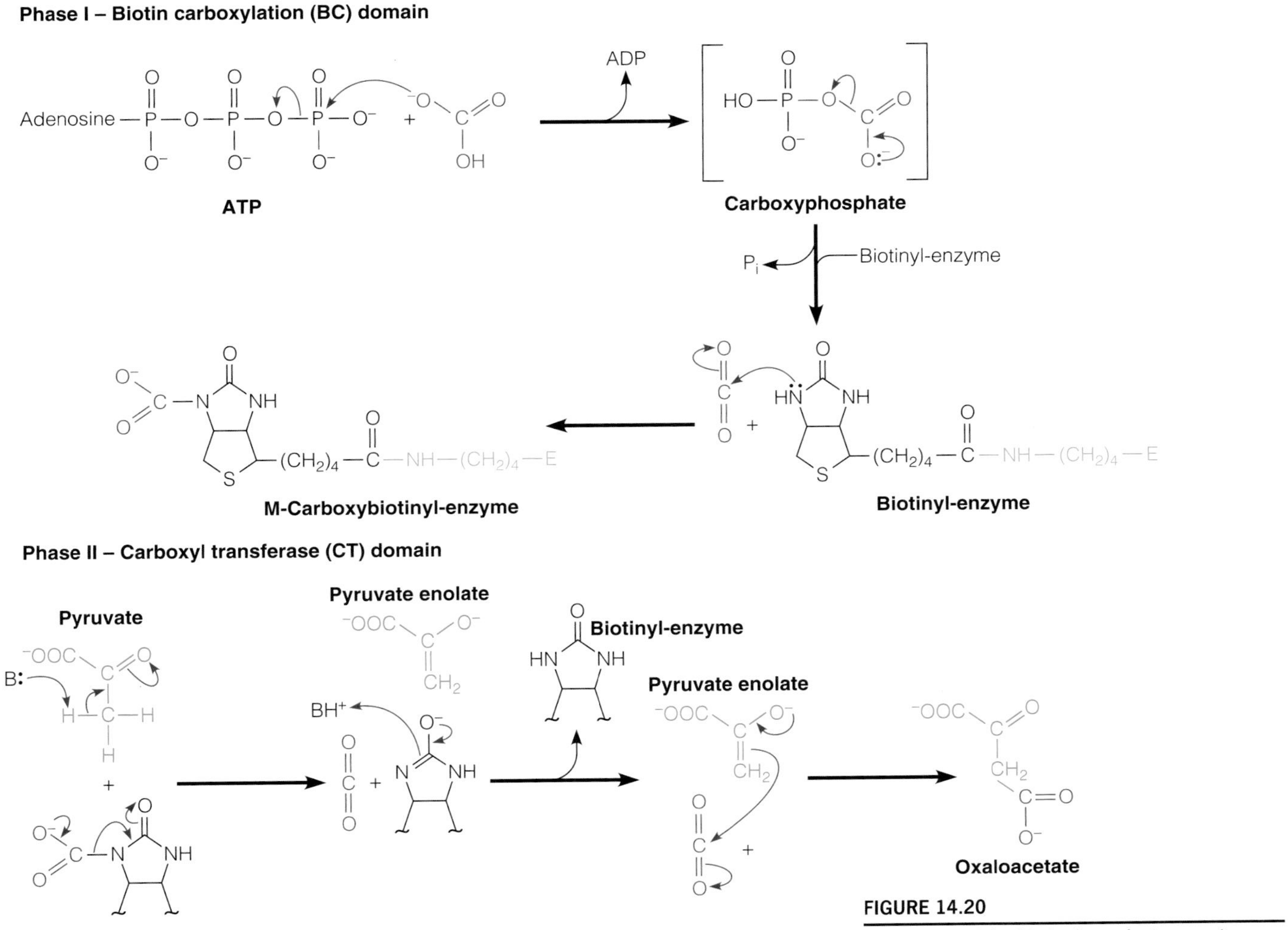

Phase I – Biotin carboxylation (BC) domain

ATP

Carboxyphosphate

M-Carboxybiotinyl-enzyme

Biotinyl-enzyme

Phase II – Carboxyl transferase (CT) domain

Pyruvate

Pyruvate enolate

Biotinyl-enzyme

Pyruvate enolate

Oxaloacetate

FIGURE 14.20

Mechanism of the biotin-dependent pyruvate carboxylase reaction. The reaction occurs in two phases, with the first phase catalyzed by the biotin carboxylation (BC) domain and the second phase catalyzed by the carboxyltransferase (CT) domain. In phase I, bicarbonate is dehydrated via the ATP-dependent formation of carboxyphosphate. This intermediate decomposes to give free CO_2 containing sufficient free energy to carboxylate biotin, forming N-carboxybiotin. In phase II, the BCCP domain swings the N-carboxybiotin to the CT domain, where pyruvate is bound. A base (B:) in the CT active site generates the pyruvate enolate, which attacks the free CO_2 produced by elimination of biotin, to give the final product oxaloacetate.

Biochemistry, 3rd ed., Donald Voet and Judith G. Voet. © 2005 John Wiley & Sons, Inc. Modified with permission from John Wiley & Sons, Inc.

amino group to an α-keto acid and is thereby converted itself to a keto acid. The mechanism is discussed in Chapter 20.

Glutamate and aspartate undergo transamination to generate the citric acid cycle intermediates α-ketoglutarate and oxaloacetate, respectively. Hence, cells containing amino acids in abundance can convert them via transamination to citric acid cycle intermediates. Another enzyme, **glutamate dehydrogenase**, presents an additional route for synthesis of α-ketoglutarate from glutamate.

$$\text{Glutamate} + \text{NAD(P)}^+ + H_2O \rightleftharpoons \alpha\text{-ketoglutarate} + \text{NAD(P)H} + NH_4^+$$

Glutamate dehydrogenase, which we discuss in more detail in Chapter 20, uses either NAD^+ or $NADP^+$. Being reversible, transaminations and the glutamate dehydrogenase reaction can be used either for amino acid synthesis or for replenishment of citric acid cycle intermediates, depending upon the needs of the cell.

Finally, many plants and bacteria can convert two-carbon fragments to four-carbon citric acid cycle intermediates via the glyoxylate cycle, as described in the next section.

Glutamate

α-Ketoglutarate

Aspartate

Oxaloacetate

Glyoxylate

The glyoxylate cycle allows plants and bacteria to carry out net conversion of fat to carbohydrate, bypassing CO$_2$-generating reactions of the citric acid cycle.

Glyoxylate Cycle: An Anabolic Variant of the Citric Acid Cycle

Metabolically, plant and animal cells differ in many important respects. Of particular concern here is that plant cells, along with some microorganisms, can carry out the net synthesis of carbohydrate from fat. This conversion is crucial to the development of seeds, in which a great deal of energy is stored in the form of triacylglycerols. (In fact, most vegetable oils available in grocery stores are mixtures of triacylglycerols derived from seeds.) When the seeds germinate, triacylglycerol is broken down and converted to sugars, which provide energy and raw material needed for growth of the plant. By contrast, animal cells cannot carry out the net synthesis of carbohydrate from fat.

Plants synthesize sugars by using the **glyoxylate cycle**, which can be considered an anabolic variant of the citric acid cycle. To understand the importance of this cycle, consider first the two primary fates of acetyl-CoA in animal metabolism—oxidation through the citric acid cycle, and the synthesis of fatty acids. Because of the virtual irreversibility of the pyruvate dehydrogenase reaction, acetyl-CoA cannot undergo net conversion to pyruvate and hence cannot participate in the *net* synthesis of carbohydrate. To be sure, the two carbons of acetyl-CoA can be incorporated into oxaloacetate, which is an efficient gluconeogenic precursor. However, because two carbons are lost in this part of the citric acid cycle, there is no *net* accumulation of carbon in carbohydrate. The glyoxylate cycle, however, permits the net synthesis of oxaloacetate by bypassing the reactions in which CO$_2$ is lost.

The glyoxylate cycle (Figure 14.21) is a cyclic pathway that results in the net conversion of two acetyl units, as acetyl-CoA, to one molecule of succinate. The

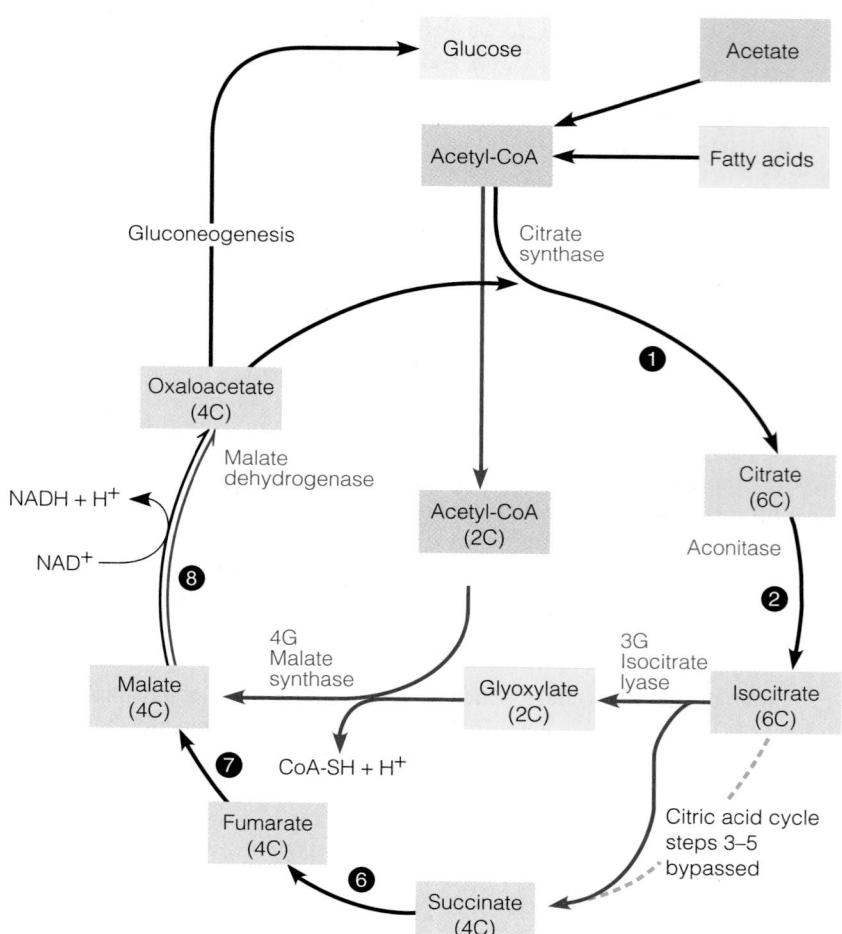

FIGURE 14.21

Reactions of the glyoxylate cycle. Two acetyl-CoA molecules enter the cycle, one at the citrate synthase step, and the second at the malate synthase step. The reactions catalyzed by isocitrate lyase and malate synthase (red arrows) bypass the three citric acid cycle steps between isocitrate and succinate (blue dashes) so that the two carbons lost in the citric acid cycle are saved, resulting in the net synthesis of oxaloacetate. The numbered reactions are identical to those in the citric acid cycle; however, reactions 1, 2, 3G, 4G, and 8 are catalyzed by unique isozymes located in the glyoxysome.

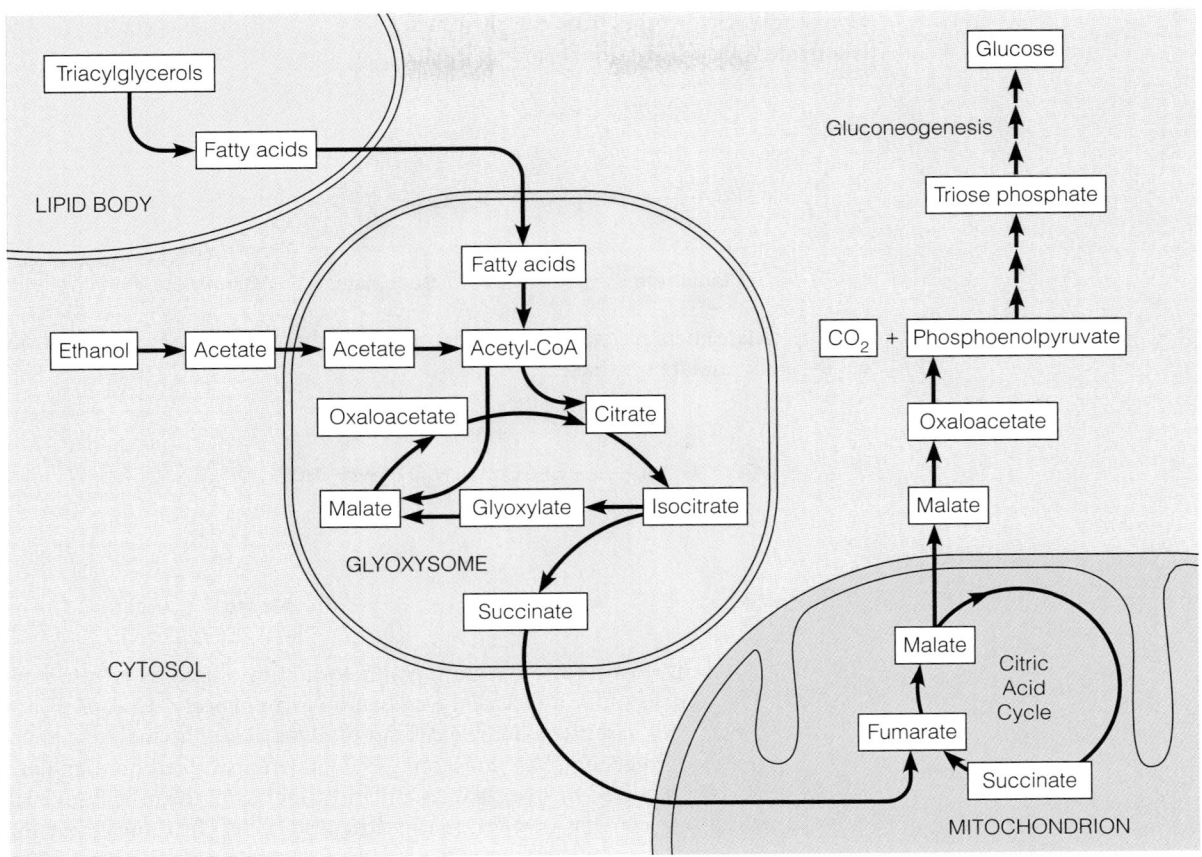

FIGURE 14.22

Intracellular relationships involving the glyoxylate cycle in plant cells. Fatty acids released in lipid bodies are oxidized in glyoxysomes to acetyl-CoA, which can also come directly from acetate. Acetyl-CoA is then converted to succinate in the glyoxylate cycle, and the succinate is transported to mitochondria. There it is converted in the citric acid cycle to oxaloacetate, which is readily converted to sugars by gluconeogenesis.

pathway uses some of the same enzymes as the citric acid cycle, but it bypasses the reactions in which carbon is lost during the citric acid cycle. The second mole of acetyl-CoA is brought in during this bypass (Figure 14.21). Thus, each turn of the cycle involves incorporation of 2 two-carbon fragments and results in the net synthesis of a four-carbon molecule. This process occurs in the **glyoxysome**, a specialized organelle that carries out both β-oxidation of fatty acids to acetyl-CoA and utilization of that acetyl-CoA in the glyoxylate cycle. Figure 14.22 illustrates the relationships of the intracellular compartments in a plant cell, and the exchange of metabolites between those compartments. The succinate generated is transported from the glyoxysome to the mitochondrion, where it is converted, via reactions 6, 7, and 8 of the citric acid cycle (see Figure 14.3), to oxaloacetate. The oxaloacetate is readily utilized for carbohydrate synthesis via gluconeogenesis.

The glyoxylate cycle also allows many microorganisms to metabolize two-carbon substrates, such as acetate. *E. coli*, for example, can grow in a medium that provides acetate as the sole carbon source, as can many fungi, protozoans, and algae. These cells synthesize acetyl-CoA, which is used both for energy production, via the citric acid cycle, and for synthesis of gluconeogenic precursors, via the glyoxylate cycle.

Now let us examine the individual reactions of the glyoxylate cycle. As noted, acetyl-CoA is provided from fatty acid oxidation. Alternatively, acetate itself is converted to acetyl-CoA by **acetate thiokinase**, an enzyme found in nearly all organisms, including those lacking the glyoxylate cycle.

$$\text{Acetate} + \text{CoA-SH} + \text{ATP} \rightleftharpoons \text{acetyl-CoA} + \text{AMP} + \text{PP}_i$$

Next, acetyl-CoA condenses with oxaloacetate to give citrate, just as in the citric acid cycle, and citrate reacts with aconitase to give isocitrate. At this point the glyoxylate cycle diverges from the citric acid cycle. The next reaction, catalyzed by **isocitrate lyase**, cleaves the 6-carbon isocitrate to glyoxylate and succinate.

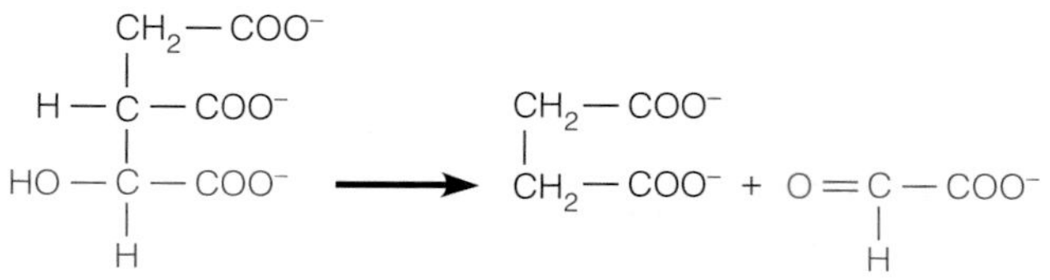

Isocitrate   **Succinate**   **Glyoxylate**

Glyoxylate then accepts two carbons from another acetyl-CoA, in a reaction catalyzed by **malate synthase**.

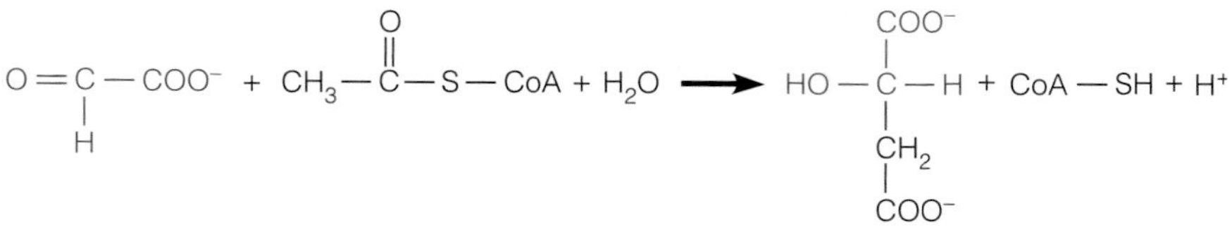

Glyoxylate   **Acetyl-CoA**   **Malate**

Mechanistically, this reaction is comparable to that catalyzed by citrate synthase, involving nucleophilic attack of the carbanion form of acetyl-CoA on a carbonyl carbon, in this case the aldehyde carbon of glyoxylate. The malate is then dehydrogenated to regenerate oxaloacetate. The enzyme involved here, malate dehydrogenase, is localized in glyoxysomes and is distinct from the mitochondrial form of the enzyme, which is involved in the citric acid cycle. The same is true for citrate synthase and aconitase isozymes used in the glyoxylate cycle (Figure 14.22).

As noted earlier, the glyoxylate cycle results in the net conversion of 2 two-carbon fragments, acetyl-CoA, to a four-carbon compound, succinate, as shown by the following balanced equation.

$$2 \text{ Acetyl-CoA} + NAD^+ + 2H_2O \longrightarrow \text{succinate} + NADH + H^+ + 2CoA\text{-}SH$$

The primary fate of succinate is its entry into gluconeogenesis via its conversion to oxaloacetate.

SUMMARY

The citric acid cycle is a central pathway for oxidation of carbohydrates, lipids, and proteins. A principal entrant to this cyclic pathway is pyruvate produced in glycolysis, which undergoes oxidation to acetyl-CoA by the pyruvate dehydrogenase complex. This three-enzyme system uses five coenzymes: NAD^+, CoA-SH, FAD, lipoamide, and thiamine pyrophosphate (TPP). Each turn of the citric acid cycle involves entry of two carbons as the acetyl group of acetyl-CoA and loss of two carbons as CO_2. Acetyl-CoA condenses with oxaloacetate to form citrate. After one turn of the cycle, oxaloacetate is regenerated, to begin the process anew. During the cycle, reduced electron carriers, primarily NADH, are generated, and their reoxidation in mitochondria provides the energy for ATP synthesis. Regulation of the citric acid cycle occurs both at the level of entry of fuel into the cycle (at the pyruvate dehydrogenase and citrate synthase steps) and at the level of control of key reactions within the cycle (isocitrate dehydrogenase and α-ketoglutarate dehydrogenase). In mammals, activity of the pyruvate dehydrogenase complex is regulated by phosphorylation/dephosphorylation of its E_1 subunit, catalyzed by specific protein kinases and phosphatases. Anaplerotic reactions replace citric acid cycle intermediates that are consumed in biosynthetic pathways. In plants and bacteria the glyoxylate cycle bypasses the two decarboxylation reactions of the citric acid cycle, allowing acetyl-CoA to undergo net conversion to carbohydrate.

REFERENCES

Regulation of the Citric Acid Cycle

Atkinson, D. E. (1977) *Cellular Energy Metabolism and Its Regulation.* Academic Press, New York. Provocative remarks by the person who originated the concept of adenylate energy charge.

Denton, R. M., and J. G. McCormack (1990) Ca^{2+} as a second messenger within mitochondria of the heart and other tissues. *Annu. Rev. Physiol.* 52:451–466. Detailed review of the relationship between intramitochondrial calcium levels and the demand for energy generation.

Maj, M. C., J. M. Cameron, and B. H. Robinson (2006) Pyruvate dehydrogenase phosphatase deficiency: Orphan disease or an underdiagnosed condition? *Mol. Cell. Endocrinol.* 249:1–9.

Nichols, B. J., M. Rigoulet, and R. M. Denton (1994) Comparison of the effects of Ca^{2+}, adenine nucleotides and pH on the kinetic properties of mitochondrial NAD(+)-isocitrate dehydrogenase and oxoglutarate dehydrogenase from the yeast *Saccharomyces cerevisiae* and rat heart. *Biochem. J.* 303:461–465.

Ottaway, J. H., J. A. McClellan, and C. L. Saunderson (1981) Succinic thiokinase and metabolic control. *Int. J. Biochem.* 13:401–410. Discusses equilibrium considerations of ATP- vs. GTP-dependent succinyl-CoA synthetase.

Roche, T. E., and Y. Hiromasa (2007) Pyruvate dehydrogenase kinase regulatory mechanisms and inhibition in treating diabetes, heart ischemia, and cancer. *Cell. Mol. Life Sci.* 64:830–849.

Sugden, M. C., and M. J. Holness (2006) Mechanisms underlying regulation of the expression and activities of the mammalian pyruvate dehydrogenase kinases. *Arch. Physiol. Biochem.* 112:139–149. These two reviews discuss the role of phosphorylation in the regulation of the pyruvate dehydrogenase complex in mammals.

Enzymes of the Citric Acid Cycle and Related Pathways

Briere, J.-J., J. Favier, A.-P. Gimenez-Roqueplo, and P. Rustin (2006) Tricarboxylic acid cycle dysfunction as a cause of human diseases and tumor formation. *Am. J. Physiol. Cell. Physiol.* 291:C1114–C1120. Reviews the linkage between defects in citric acid cycle enzymes and human disease.

Jitrapakdee, S., St. M. Maurice, I. Rayment, W. W. Cleland, J. C. Wallace, and P. V. Attwood (2008) Structure, mechanism and regulation of pyruvate carboxylase. *Biochem. J.* 413:369–387. An excellent review on the structure and function of this multifunctional biotin-dependent enzyme.

Kaelin, W. G., Jr., and C. B. Thompson (2010) Q&A: Cancer: Clues from cell metabolism. *Nature* 465:562–564. This short article describes how mutations in metabolic enzymes, including citric acid cycle enzymes, can lead to cancer.

Kern, D., G. Kern, H. Neef, K. Tittmann, M. Killenberg-Jabs, C. Wilkner, G. Schneider, and G. Hübner (1997) How thiamine diphosphate is activated in enzymes. *Science* 275:67–70. This paper describes NMR experiments that establish how the reactive carbanion is formed in thiamine pyrophosphate–dependent reactions.

Lambeth, D. O. (2006) Reconsideration of the significance of substrate-level phosphorylation in the citric acid cycle. *Biochem. Mol. Biol. Educ.* 34:21–29. Discusses the existence and roles of ATP- and GTP-dependent succinyl-CoA synthetase isozymes in animals.

Lauble, H., M. C. Kennedy, M. H. Emptage, H. Beinert, and C. D. Stout (1996) The reaction of fluorocitrate with aconitase and the crystal structure of the enzyme-inhibitor complex. *Proc. Natl. Acad. Sci. USA* 93:13699–13703.

Mesecar, A. D., and D. E. Koshland Jr. (2000) A new model for protein stereospecificity. *Nature* 403:614–615. A concise illustration of the four-point location model to explain stereospecific binding to enzymes.

Perham, R. N. (2000) Swinging arms and swinging domains in multi-functional enzymes: Catalytic machines for multistep reactions. *Annu. Rev. Biochem.* 69:961–1004. An excellent review of multienzyme complexes, including those that use lipoic acid and biotin.

Rutter, J., D. R. Winge, and J. D. Schiffman (2010) Succinate dehydrogenase—Assembly, regulation and role in human disease. *Mitochondrion* 10:393–401. A concise review on the biochemistry and medical connections of this key respiratory enzyme.

Srere, P. A., A. D. Sherry, C. R. Malloy, and B. Sumegi (1997) Channelling in the Krebs tricarboxylic acid cycle. In *Channelling in Intermediary Metabolism*, L. Agius, and H. S. A. Sherratt, eds., Vol. IX, pp. 201–217. Portland Press Ltd., London. A general review of the concept of metabolons and channeling in metabolic pathways.

Velot, C., M. B. Mixon, M. Teige, and P. A. Srere (1997) Model of a quinary structure between Krebs TCA cycle enzymes: A model for the metabolon. *Biochemistry* 36:14271–14276. This paper describes a novel experimental approach to understanding how enzymes that catalyze sequential reactions interact with each other to facilitate catalysis of multistep pathways.

Wolodko, W. T., M. E. Fraser, M. N. James, and W. A. Bridger (1994) The crystal structure of succinyl-CoA synthetase from *Escherichia coli* at 2.5-Å resolution. *J. Biol. Chem.* 269:10883–10890. This paper describes the role of helix dipoles in stabilizing reaction intermediates.

Zhou, Z. H., D. B. McCarthy, C. M. O'Connor, L. J. Reed, and J. K. Stoops (2001) The remarkable structural and functional organization of the eukaryotic pyruvate dehydrogenase complexes. *Proc. Natl. Acad. Sci. USA* 98:14802–14807.

Experimental Background of the Citric Acid Cycle

Krebs, H. A. (1970) The history of the tricarboxylic acid cycle. *Perspect. Biol. Med.* 14:154–170. A historical account by the man responsible for most of the history.

Reed, L. J. (2001) A trail of research from lipoic acid to alpha-keto acid dehydrogenase complexes. *J. Biol. Chem.* 276:38329–38336. A historical account by the discoverer of lipoic acid and its role in pyruvate dehydrogenase.

Snell, E. E. (1993) From bacterial nutrition to enzyme structure: A personal odyssey. *Annu. Rev. Biochem.* 62:1–28. A memoir by one of the scientists most intimately involved in discoveries of vitamins and coenzymes.

Sumegi, B., A. D. Sherry, and C. R. Malloy (1990) Channeling of TCA cycle intermediates in cultured *Saccharomyces cerevisiae*. *Biochemistry* 29:9106–9110. This paper describes ^{13}C-NMR studies that revealed nonrandom labeling of the symmetrical succinate and fumarate intermediates, suggesting substrate channeling in the cycle.

The Glyoxylate Cycle

Eastmond, P. J., and I. A. Graham (2001) Re-examining the role of the glyoxylate cycle in oilseeds. *Trends Plant Sci.* 6:72–78.

PROBLEMS

1. Design a radiotracer experiment that would allow you to determine which proportion of glucose catabolism in a given tissue preparation occurs through the pentose phosphate pathway and which proportion through glycolysis and the citric acid cycle. Assume that you can synthesize glucose labeled with ^{14}C in any desired position or combination of positions. Assume also that you can trap CO_2 after administration of labeled glucose and determine its radioactivity.

2. Consider the fate of pyruvate labeled with ^{14}C in each of the following positions: carbon 1 (carboxyl), carbon 2 (carbonyl), and carbon 3 (methyl). Predict the fate of each labeled carbon during one turn of the citric acid cycle.

3. Suppose that aconitase did *not* bind its substrate asymmetrically. What fraction of the carbon atoms introduced in one cycle as acetyl-CoA would be released in the first turn of the cycle? What fraction of the carbon atoms that entered in the first cycle would be released in the second turn?

4. [methyl-^{14}C]Pyruvate was administered to isolated liver cells in the presence of sufficient malonate to block succinate dehydrogenase completely. After a time, isocitrate was isolated and found to contain label in both carbon 2 and carbon 5:

$$\begin{array}{l} ^{14}CH_2-COO^- \\ \quad | \\ HC-COO^- \\ \quad | \\ H^{14}C-COO^- \\ \quad | \\ \quad OH \end{array}$$

How do you explain this result?

5. Considering the evidence that led Krebs to propose a cyclic pathway for oxidation of pyruvate, discuss the type of experimental evidence that might have led to realization of the cyclic nature of the glyoxylate pathway.

6. Which carbon or carbons of glucose, if metabolized via glycolysis and the citric acid cycle, would be most rapidly lost as CO_2?

7. In deciding which form of isocitrate dehydrogenase plays the more important role in the citric acid cycle—the NAD^+-dependent or the $NADP^+$-dependent form—what kinds of information would help you?

8. Referring to Figure 14.9, write a balanced equation for the overall reaction catalyzed by each of the three enzymes in the pyruvate dehydrogenase complex—E_1, E_2, and E_3.

9. Briefly describe the biological rationale for each of the following allosteric phenomena: (a) activation of pyruvate carboxylase by acetyl-CoA; (b) activation of pyruvate dehydrogenase kinase by NADH; (c) inhibition of isocitrate dehydrogenase by NADH; (d) activation of isocitrate dehydrogenase by ADP; (e) inhibition of α-ketoglutarate dehydrogenase by succinyl-CoA; (f) activation of pyruvate dehydrogenase phosphatase by Ca^{2+}.

10. Certain microorganisms with a modified citric acid cycle decarboxylate α-ketoglutarate to produce succinate semialdehyde:

$$\begin{array}{l} COO^- \\ | \\ CH_2 \\ | \\ CH_2 \\ | \\ C=O \\ | \\ COO^- \end{array} \xrightarrow{\quad CO_2 \quad} \begin{array}{l} COO^- \\ | \\ CH_2 \\ | \\ CH_2 \\ | \\ C=O \\ | \\ H \end{array}$$

α-Ketoglutarate　　　　**Succinate semialdehyde**

(a) Succinate semialdehyde is then converted to succinate, which is further metabolized by standard citric acid cycle enzymes. What kind of reaction is required to convert succinate semialdehyde to succinate? Show any coenzymes that might be involved.

(b) Based on your answer in part (a), how does this pathway compare to the standard citric acid cycle in energy yield?

11. Given what you know about the function of the glyoxylate cycle and the regulation of the citric acid cycle, propose control mechanisms that might regulate the glyoxylate cycle.

12. Write a balanced equation for the conversion in the glyoxylate cycle of two acetyl units, as acetyl-CoA, to oxaloacetate.

13. As discussed in Chapter 15, FAD is a stronger oxidant than NAD^+; FAD has a higher standard reduction potential than NAD^+. Yet in the last reaction of the pyruvate dehydrogenase complex, $FADH_2$ bound to E_3 is oxidized by NAD^+. Explain this apparent paradox.

14. Given the roles of $NAD^+/NADH$ in dehydrogenation reactions and $NADPH/NADP^+$ in reductions, as discussed on page 488, would you expect the intracellular ratio of NAD^+ to NADH to be high or low? What about the ratio of $NADP^+$ to NADPH? Explain your answers.

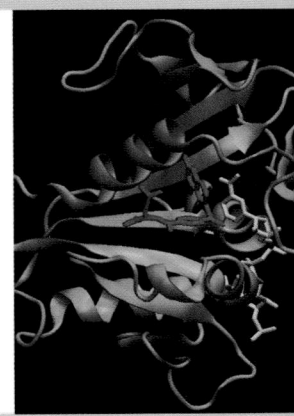

CHAPTER 15

Electron Transport, Oxidative Phosphorylation, and Oxygen Metabolism

The average adult human synthesizes ATP at a rate of nearly 10^{21} molecules per second, equivalent to producing his or her own weight in ATP *every day*. How is this massive amount of energy converted? As we saw in Chapters 13 and 14, glycolysis and the citric acid cycle by themselves generate relatively little ATP directly. However, under aerobic conditions, six dehydrogenation steps—one in glycolysis, another in the pyruvate dehydrogenase reaction, and four more in the citric acid cycle—collectively reduce 10 moles of NAD^+ to NADH and 2 moles of FAD to $FADH_2$ per mole of glucose. Reoxidation of these reduced electron carriers in the process termed **cellular respiration** generates most of the energy needed for ATP synthesis. This represents the third stage of the metabolic oxidation of substrates (Figure 15.1). In eukaryotic cells NADH and $FADH_2$ are reoxidized by electron transport proteins bound to the inner mitochondrial membrane. A series of linked oxidation and reduction reactions occurs, with electrons being passed along a series of electron carriers—the electron transport chain, or **respiratory chain** (Figure 15.1). The final step is reduction of O_2 to water. The overall electron transport sequence is quite exergonic. One pair of reducing equivalents, generated from 1 mole of NADH, suffices to drive the synthesis of between 2 and 3 moles of ATP from ADP and P_i by a process termed **oxidative phosphorylation**. How is the energy released from the oxidative reactions of the respiratory chain harnessed, or **coupled**, to drive the synthesis of ATP? The mechanism of this coupling will concern us throughout this chapter. In addition, we will consider a number of other important metabolic roles oxygen plays in aerobic cells.

Oxidation of 1 mole of NADH by the respiratory chain provides sufficient energy for synthesis of ~2.5 moles of ATP from ADP.

We saw in glycolysis that the energy released during the breakdown of glucose is used to convert low-energy phosphorylated intermediates into high-energy phosphorylated intermediates, which then transfer their phosphates to form ATP. It was thus assumed that a similar mode of direct chemical coupling would underlie the formation of ATP in respiration. For many years, researchers around the world searched for a high-energy intermediate that could link electron transport to ATP synthesis. There were, however, several nagging facts that argued against a

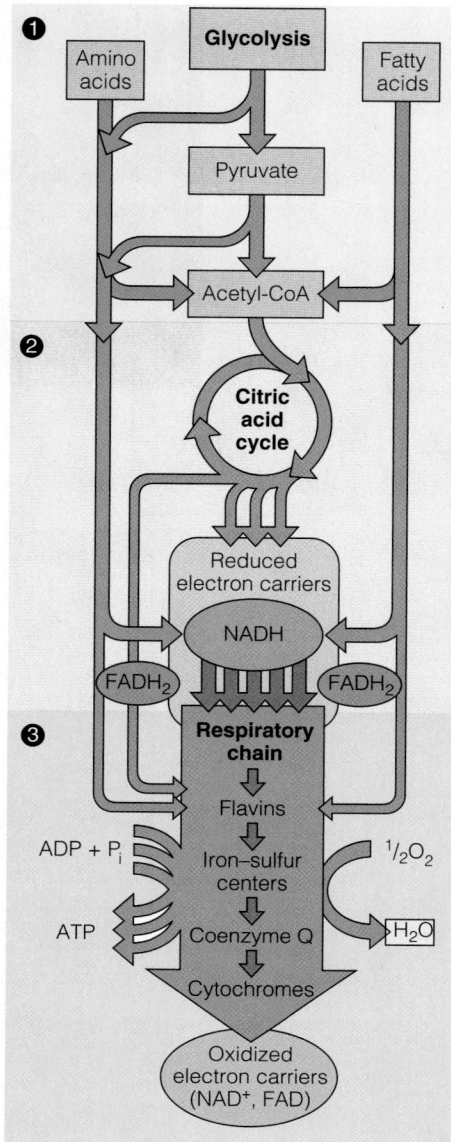

FIGURE 15.1

Overview of oxidative energy generation.

direct chemical coupling process like the phosphoglycerate kinase and pyruvate kinase reactions in glycolysis. Perhaps most troubling was the realization that the number of ATP molecules produced per glucose during respiration was not a constant: It tended to vary between 28 and 38. That is, the passage of a pair of electrons down the respiratory chain generates approximately 2.5 ATPs, not an integer, as would be predicted by a direct chemical coupling mechanism. Furthermore, respiration was shown to require a membrane. If the membrane is disrupted, the passage of electrons down the respiratory chain becomes *uncoupled* from ATP synthesis: Electrons still flow, but ATP production ceases. As we shall soon see, cells long ago evolved an elegantly simple mechanism that explains all of these observations, a mechanism that turns out to have much broader significance than just the synthesis of ATP. In short, the passage of electrons down the respiratory chain creates a proton gradient across the inner mitochondrial membrane, and the energy in this proton gradient provides the driving force for ATP synthesis.

The Mitochondrion: Scene of the Action

Our comprehension of biological oxidations requires an understanding of both the chemistry of oxidation–reduction reactions and the cell biology of the mitochondrion. Before reviewing the chemistry, let us describe the intracellular sites where these reactions occur. Cellular metabolism generates reduced compounds in all of the major compartments of a eukaryotic cell. As noted earlier, glycolysis takes place in the cytosol of eukaryotic cells, whereas pyruvate oxidation, fatty acid β-oxidation, amino acid oxidation, and the citric acid cycle occur within the mitochondrial matrix. Individual cells vary widely in the abundance and structure of their mitochondria. Most vertebrate cells contain from a few hundred to a few thousand mitochondria, but the number can be as low as 1 or as high as 100,000.

The mitochondrion consists of four distinct subregions, shown in Figure 15.2a and b—the outer membrane, the inner membrane, the intermembrane space, and the matrix, located within the inner membrane. The inner membrane is highly folded into **cristae** that project into, and often nearly through, the interior of the mitochondrion. Because respiratory proteins are embedded in the inner membrane, the density of cristae is related to the respiratory activity of a cell. For example, heart muscle cells, which have high rates of respiration, contain mitochondria with densely packed cristae. By contrast, liver cells have much lower respiration rates and mitochondria with more sparsely distributed cristae.

Whatever the compartment in which biological oxidations occur, all of these processes generate reduced electron carriers, primarily NADH. Most of this NADH is reoxidized, with concomitant ATP production, by the enzymes of the respiratory chain, firmly embedded in the inner membrane. The inner membrane itself consists of about 70% protein and 30% lipid, making it perhaps the most protein rich of all biological membranes. About half of the inner membrane protein in bovine heart mitochondria consists of proteins directly involved in electron transport and oxidative phosphorylation. Most of the remaining proteins are involved in transport of substances into and out of mitochondria. By contrast, a completely different set of proteins is bound to the outer membrane, including

FIGURE 15.2

Localization of respiratory processes in the mitochondrion. **(a)** A mitochondrion from a pancreatic cell, shown as a thin section in the electron micrograph. The major intramitochondrial compartments are shown, along with principal enzymes and pathways localized to each compartment. Magnification, 155,000. **(b)** Overview of oxidative phosphorylation. Reduced electron carriers, produced by cytosolic dehydrogenases and mitochondrial oxidative pathways, become reoxidized by enzyme complexes bound in the inner membrane. These complexes actively pump protons outward, creating an energy gradient whose discharge through complex V drives ATP synthesis.

(a) Springer and Plenum/*Mitchondria*, 1982, A. Tzagoloff, with kind permission from Springer Science+ Business Media B.V.

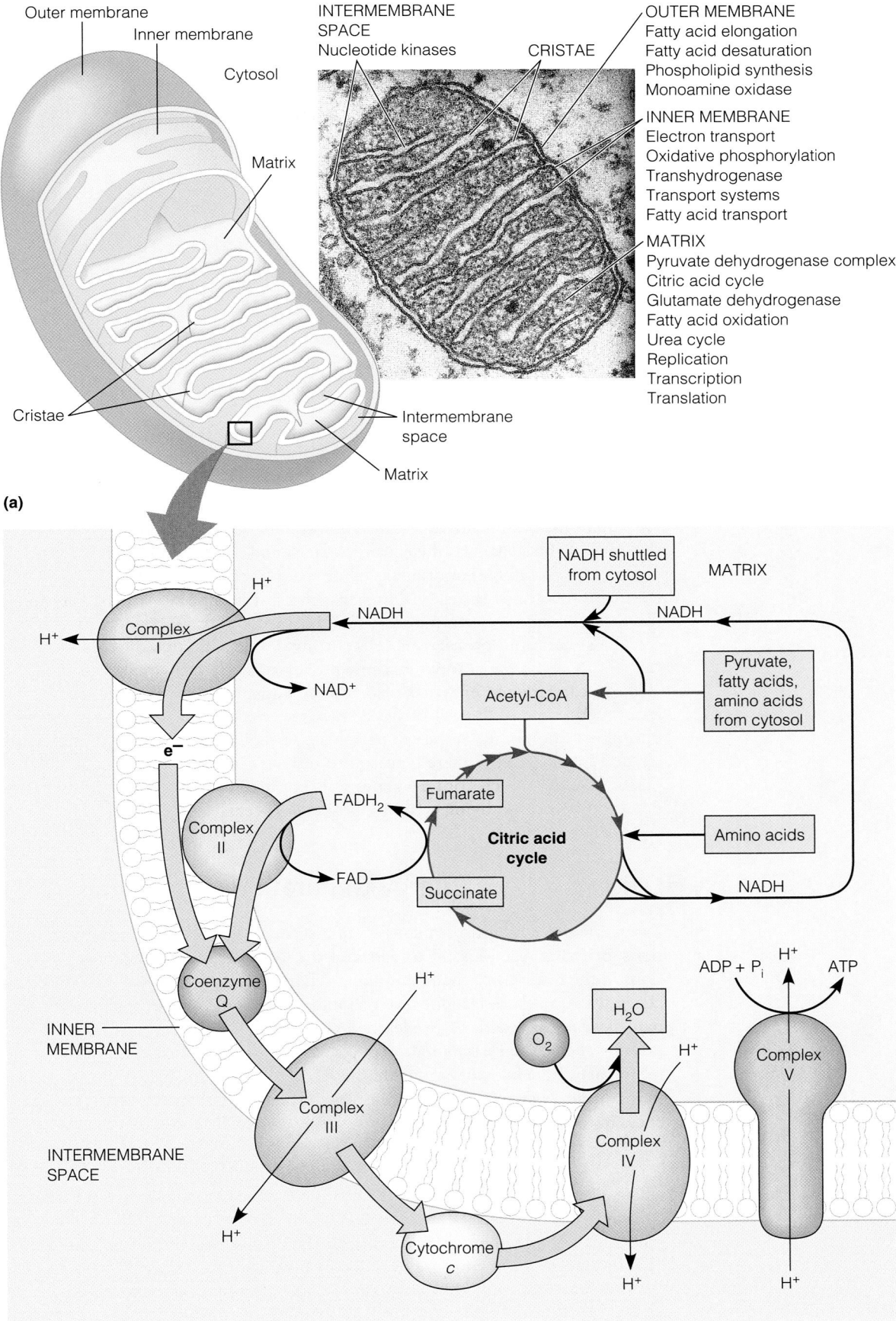

(a)

(b)

enzymes of amino acid oxidation, fatty acid elongation, membrane phospholipid biosynthesis, and enzymatic hydroxylations.

Embedded within the inner membrane are the protein carriers that constitute the respiratory chain. They are assembled in the form of five multiprotein enzyme complexes, named I, II, III, IV, and V (Figure 15.2b). Complex I and complex II receive electrons from the oxidation of NADH and succinate, respectively, and pass them along to a lipid-soluble electron carrier, coenzyme Q (see page 631), which moves freely in the membrane. Complex III catalyzes the transfer of electrons from the reduced form of coenzyme Q to cytochrome c, a protein electron carrier that is also mobile within the intermembrane space. Finally, complex IV catalyzes the oxidation of cytochrome c, reducing O_2 to water. The energy released by these exergonic reactions creates a proton gradient across the inner membrane, with protons being pumped from the matrix into the intermembrane space. Protons then re-enter the matrix through a specific channel in complex V. The energy released by this exergonic process drives the endergonic synthesis of ATP from ADP and inorganic phosphate. Throughout this chapter we develop the structural and functional basis for our understanding of these energy coupling processes.

Critical to comprehension of these processes was the isolation of physiologically intact mitochondria, using differential centrifugation of cell homogenates. This feat was accomplished in the late 1940s by Eugene Kennedy and Albert Lehninger, who demonstrated that isolated mitochondria could synthesize ATP from ADP and P_i in vitro, but only if an oxidizable substrate was present as well. Much of our understanding of the sequence of electron carriers, and of the localization of specific enzymes within mitochondria, has come from fractionation and analysis of the complexes described earlier, yielding snapshots of individual parts of the overall reaction sequence of electron transport and oxidative phosphorylation.

The situation in prokaryotic cells is comparable, although different electron carriers are involved. However, because prokaryotic cells lack organelles, all of the electron carriers and enzymes of oxidative phosphorylation are bound to the inner membrane of the cell surface. Therefore, electron transport and oxidative phosphorylation occur at the cell periphery. As discussed at the end of this chapter, and in Chapter 16, there is reason to believe that mitochondria and chloroplasts, both of which contain genes and the machinery to express them, are descended from free-living primitive prokaryotic cells.

Most electron carriers in the respiratory chain are embedded in the mitochondrial inner membrane.

Oxidations and Energy Generation

Biological electron transport consists of a series of linked oxidations and reductions, or redox reactions. To understand the logic behind the sequence of reactions in the respiratory chain, and the mechanisms by which metabolic energy is generated from these reactions, you should recall from Chapter 3 the thermodynamics of redox reactions. We learned that the higher the value of the *standard reduction potential*, E_0', for a redox couple, the greater is the tendency for that couple to participate in oxidation of another substrate. We can describe this tendency in quantitative terms because free energy changes are directly related to differences in reduction potential:

$$\Delta G^{\circ\prime} = -nF\Delta E^{0\prime} = -nF(E^{0\prime}_{acceptor} - E^{0\prime}_{donor}) \tag{15.1}$$

where n is the number of electrons transferred in the half-reactions, F is Faraday's constant ($96485\ J\ mol^{-1}V^{-1}$), and $\Delta E^{0\prime}$ is the difference in standard reduction potentials between the two redox couples. $E^{0\prime}$ values for a number of biochemically important redox pairs are recorded in Table 15.1.

The values given in Table 15.1 allow calculation of free energy changes only under standard conditions (including, by convention, a pH of 7.0). For nonstandard

TABLE 15.1 Standard reduction potentials of interest in biochemistry

Oxidant		Reductant	n	$E^{0'}(V)$
Acetate + CO_2 + $2H^+$ + $2e^-$	$\rightleftharpoons$	Pyruvate + H_2O	2	−0.70
Succinate + CO_2 + $2H^+$ + $2e^-$	$\rightleftharpoons$	α-Ketoglutarate + H_2O	2	−0.67
Acetate + $3H^+$ + $2e^-$	$\rightleftharpoons$	Acetaldehyde + H_2O	2	−0.60
Ferredoxin (oxidized) + e^-	$\rightleftharpoons$	Ferredoxin (reduced)	1	−0.43
$2H^+$ + $2e^-$	$\rightleftharpoons$	H_2	2	−0.42
α-Ketoglutarate + CO_2 + $2H^+$ + $2e^-$	$\rightleftharpoons$	Isocitrate	2	−0.38
Acetoacetate + $2H^+$ + $2e^-$	$\rightleftharpoons$	β-Hydroxybutyrate	2	−0.35
Pyruvate + CO_2 + H^+ + $2e^-$	$\rightleftharpoons$	Malate	2	−0.33
NAD^+ + H^+ + $2e^-$	$\rightleftharpoons$	NADH	2	−0.32
$NADP^+$ + H^+ + $2e^-$	$\rightleftharpoons$	NADPH	2	−0.32
Lipoate (oxidized) + $2H^+$ + $2e^-$	$\rightleftharpoons$	Lipoate (reduced)	2	−0.29
1,3-Bisphosphoglycerate + $2H^+$ + $2e^-$	$\rightleftharpoons$	Glyceraldehyde-3-phosphate + P_i	2	−0.29
Glutathione (oxidized) + $2H^+$ + $2e^-$	$\rightleftharpoons$	2 Glutathione (reduced)	2	−0.23
FAD + $2H^+$ (free coenzyme) + $2e^-$	$\rightleftharpoons$	$FADH_2$	2	−0.22
Acetaldehyde + $2H^+$ + $2e^-$	$\rightleftharpoons$	Ethanol	2	−0.20
Pyruvate + $2H^+$ + $2e^-$	$\rightleftharpoons$	Lactate	2	−0.19
Oxaloacetate + $2H^+$ + $2e^-$	$\rightleftharpoons$	Malate	2	−0.17
O_2 + e^-	$\rightleftharpoons$	$O_2^{\cdot-}$ (superoxide)	1	−0.16
α-Ketoglutarate + NH_4^+ + $2H^+$ + $2e^-$	$\rightleftharpoons$	Glutamate + H_2O	2	−0.14
FAD (enzyme-bound) + $2H^+$ + $2e^-$	$\rightleftharpoons$	$FADH_2$ (enzyme-bound)	2	~0 to −0.30
Methylene blue (oxidized) + $2H^+$ + $2e^-$	$\rightleftharpoons$	Methylene blue (reduced)	2	0.01
Fumarate + $2H^+$ + $2e^-$	$\rightleftharpoons$	Succinate	2	0.03
Q + $2H^+$ + $2e^-$	$\rightleftharpoons$	QH_2	2	0.04
Dehydroascorbate + $2H^+$ + $2e^-$	$\rightleftharpoons$	Ascorbate	2	0.06
Cytochrome b (+3) + e^-	$\rightleftharpoons$	Cytochrome b (+2)	1	0.07
Cytochrome c_1 (+3) + e^-	$\rightleftharpoons$	Cytochrome c_1 (+2)	1	0.23
Cytochrome c (+3) + e^-	$\rightleftharpoons$	Cytochrome c (+2)	1	0.25
Cytochrome a (+3) + e^-	$\rightleftharpoons$	Cytochrome a (+2)	1	0.29
O_2 + $2H^+$ + $2e^-$	$\rightleftharpoons$	H_2O_2	2	0.30
Ferricyanide + e^-	$\rightleftharpoons$	Ferrocyanide	1	0.36
NO_3^- (Nitrate) + $2H^+$ + $2e^-$	$\rightleftharpoons$	NO_2^- (Nitrite) + H_2O	2	0.42
Cytochrome a_3(+3) + e^-	$\rightleftharpoons$	Cytochrome a_3(+2)	1	0.55
Fe (+3) + e^-	$\rightleftharpoons$	Fe (+2)	1	0.77
$\frac{1}{2}O_2$ + $2H^+$ + $2e^-$	$\rightleftharpoons$	H_2O	2	0.82

Note: $E^{0'}$ is the standard reduction potential at pH 7 and 25 °C, n is the number of electrons transferred, and each potential is for the partial reaction written as follows: Oxidant + $ne^- \rightleftharpoons$ reductant.

conditions such as those that might exist in a cell, recall that we learned in Chapter 3 to use Equation 3.23 to evaluate ΔG for a redox reaction such as

$$A_{ox} + B_{red} \rightarrow A_{red} + B_{ox}$$

$$\Delta G = \Delta G^{\circ\prime} + RT \ln\left(\frac{[A_{red}][B_{ox}]}{[A_{ox}][B_{red}]}\right) \qquad (15.2)$$

In this example A is the e^- acceptor and B is the donor.

Free Energy Changes in Biological Oxidations

Each of the coupled redox reactions in biological electron transport involves the transfer of electrons from one redox couple to another couple of higher

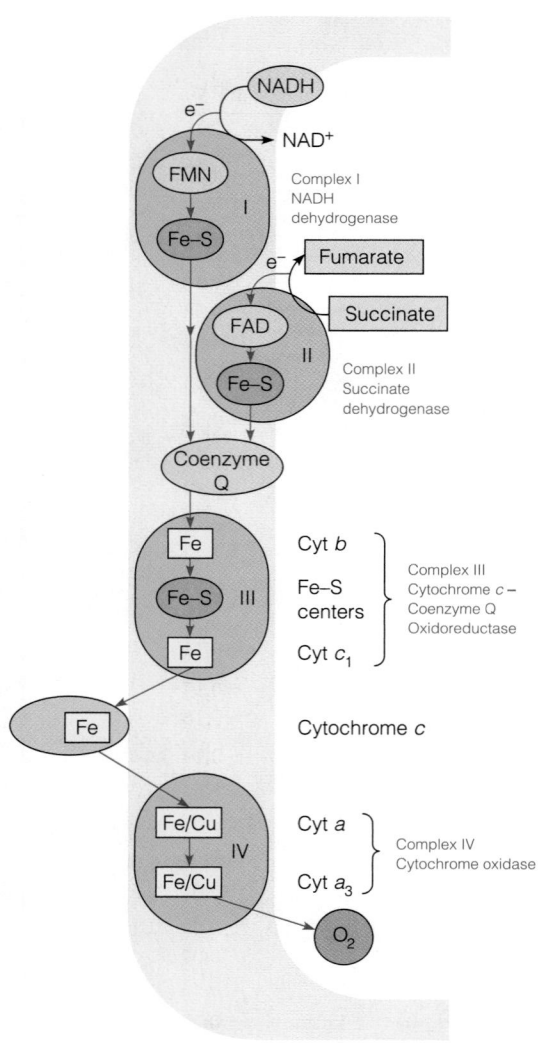

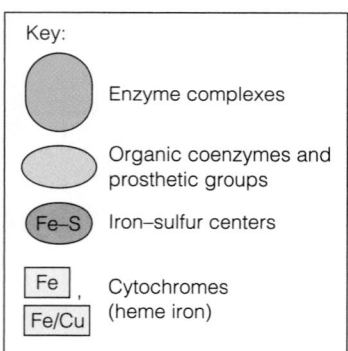

FIGURE 15.3

Respiratory electron carriers in the mitochondrion. This figure shows the sequence of electron carriers that oxidize succinate and NAD⁺-linked substrates in the inner membrane.

The respiratory chain catalyzes the transport of electrons from low-potential carriers to high-potential carriers.

(more positive) reduction potential. Thus, each individual redox reaction in the sequence is exergonic under standard conditions. For electrons entering the respiratory chain as NADH, the overall reaction sequence is given by the following equation:

$$NADH + H^+ + \tfrac{1}{2}O_2 \rightleftharpoons NAD^+ + H_2O$$

This sequence is strongly exergonic under standard conditions:

$$\Delta G°' = -nF\,\Delta E^{0'} = -2(96485\ \text{J mol}^{-1}\text{V}^{-1})(0.82\ \text{V} - (-0.32\ \text{V})) = -220\ \text{kJ/mol} \quad (15.3)$$

As discussed later in this chapter, the oxidation of 1 mole of NADH in the respiratory chain proceeds concomitantly with synthesis of about 2.5 moles of ATP from ADP and P_i. Because $\Delta G°'$ for ATP hydrolysis is -30.5 kJ/mol, synthesis of 2.5 ATPs requires 76 kJ under standard conditions, giving an efficiency for oxidative phosphorylation of about 35%. In Chapter 3 (pages 84–85), we calculated the ΔG for this process using estimates of substrate and product concentrations that might exist in the mitochondrion. We arrived at a value for $\Delta G = -381$ kJ/mol O_2, or -190 kJ per pair of electrons. Because ΔG for ATP hydrolysis under intracellular conditions is -50 kJ/mol or more, the efficiency of oxidative phosphorylation probably approaches 60 or 70% in vivo.

Electron Transport

Electron Carriers in the Respiratory Chain

If you compare the sequence of respiratory electron carriers (Figure 15.3) with the standard reduction potentials of those carriers (Table 15.1), you will see that E_0' for each carrier increases in the same order as the sequence of their use in electron transport. This order suggests that each individual oxidoreduction reaction in electron transport is exergonic under standard conditions and that electrons flow in continuous fashion from low-potential to high-potential carriers. Very neat, but is it real? After all, we have seen that glycolysis and the citric acid cycle both proceed smoothly despite the involvement of some reactions with large positive $\Delta G°'$ values. We shall explore some of the lines of evidence by which the currently accepted pathway of electron transport was determined. First, however, let us become better acquainted with the participants—the electron carriers involved.

Flavoproteins

Flavoproteins, introduced in Chapter 14 (page 600), contain tightly bound FMN or FAD as redox cofactors. Each flavoprotein provides a different microenvironment for the isoalloxazine ring, conferring a unique standard reduction potential on the flavin. Flavin nucleotides can act as transformers between two-electron and one-electron processes due to their ability to exist as stable one-electron reduced semiquinone intermediates.

Iron–Sulfur Proteins

Iron–sulfur clusters consist of nonheme iron complexed to sulfur in four known ways (Figure 15.4). These clusters function as the prosthetic groups of **iron–sulfur proteins**. The simplest form, designated **FeS**, involves one iron tetrahedrally complexed with the thiol sulfurs of four cysteine residues. The second form, denoted Fe_2S_2, contains two irons, each complexed with two cysteine residues and two inorganic sulfides. The most complicated form (Fe_4S_4) contains four irons, four sulfides, and four cysteine residues. Not all iron–sulfur clusters contain equal numbers of iron and sulfide ions—the Fe_3S_4 cluster has the same geometry as the Fe_4S_4 cluster but lacks one Fe atom and one Cys ligand. NADH dehydrogenase contains both Fe_2S_2 and Fe_4S_4 centers. In all of these centers the iron can undergo cyclic one-electron oxidoreduction between ferrous and ferric states. The standard reduction potential of the iron in these iron–sulfur clusters varies dramatically depending on the type of cluster and the microenvironment provided by the protein to which it is attached.

Coenzyme Q

The respiratory electron carrier coenzyme Q was discovered when treatment of isolated mitochondria with an organic solvent such as isooctane was observed to completely abolish the ability of the mitochondria to oxidize substrates. Addition of the material extracted by isooctane completely restored the oxidative capacities of the mitochondria, suggesting the presence of an extremely lipophilic electron carrier that was only loosely linked to protein. This carrier was found to be a benzoquinone linked to a number of isoprene units, usually 10 in mammalian cells and 6 in bacteria. Because the substance is ubiquitous in living cells, one group of researchers named it *ubiquinone*, while another called it **coenzyme Q**, or **Q**. The term Q_{10} is used to specify the form of Q containing 10 isoprene units. The isoprenoid tail gives the molecule its hydrophobic character, which allows Q to diffuse rapidly through the inner mitochondrial membrane. Like the flavins (FMN and FAD), Q oxidoreduction proceeds one electron at a time through a stable semiquinone intermediate so that Q provides another interface between two-electron carriers such as NADH and the one-electron carriers such as the cytochromes.

Oxidized coenzyme Q_{10} (CoQ)

$H^+ + e^-$

Semiquinone form of coenzyme Q

$H^+ + e^-$

Reduced coenzyme Q_{10} (CoQH$_2$)

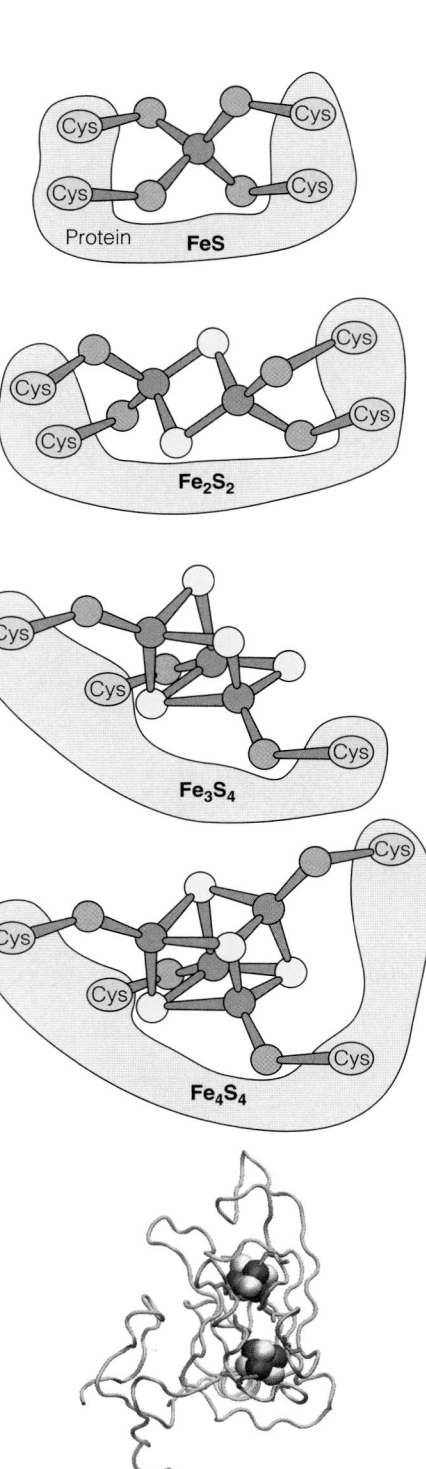

Key:
- ● Iron   ○ Inorganic sulfur
- ● Cysteine sulfur

FIGURE 15.4

Structures of iron–sulfur clusters. The bottom panel illustrates the covalent attachment of a pair of Fe_4S_4 clusters in a subunit of complex I from *Thermus thermophilus* (PDB ID 3IAS).

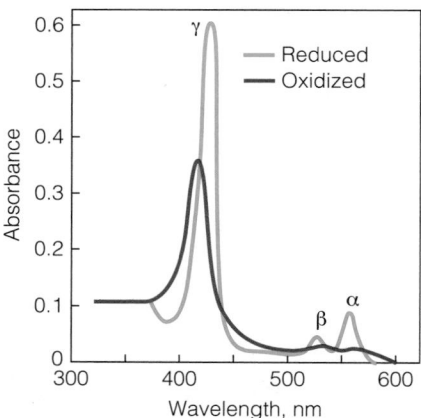

(a) Cytochrome *b*

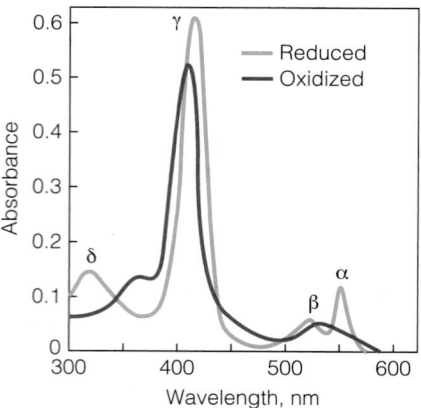

(b) Cytochrome *c*

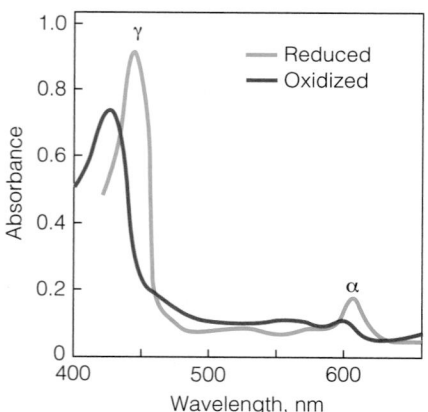

(c) Cytochromes *a* and *a₃*

FIGURE 15.5

Absorption spectra of cytochromes. The plots show the absorption spectra of cytochromes *b*, *c*, and *a* in their oxidized (red) and reduced (blue) states.
(a) Cytochrome *b* from *Neurospora*; **(b)** cytochrome *c* from horse heart; **(c)** bovine heart cytochrome oxidase (which contains both cytochromes *a* and *a₃*).

Springer and Plenum/*Mitchondria*, 1982, A. Tzagoloff, adapted with kind permission from Springer Science+ Business Media B.V.

Cytochromes

Finally, we come to the cytochromes, a group of red or brown heme proteins having distinctive visible-light spectra. These proteins were first characterized and their role in respiration demonstrated by the Englishman David Keilin in 1925. Using a hand spectroscope, Keilin observed red-brown pigments in the flight muscles of insects. During muscular exertion (when an immobilized fly tried to free itself), the spectra of these pigments underwent marked changes. This observation led Keilin to postulate a role for these substances in carrying electrons from biological fuels to oxygen.

The major respiratory cytochromes are classified as *b*, *c*, or *a*, depending on the wavelengths of the spectral absorption peaks. Figure 15.5 shows the spectral characteristics of typical *b*-, *c*-, and *a*-type cytochromes. The unique spectra of each class result from differences in heme substitution and ligation. Within each class (*b*, *c*, or *a*), the cytochromes are distinguished by smaller spectral differences. For example, cytochrome c_1 has a spectrum similar to that of cytochrome *c*, but the α and γ absorption peaks are slightly red-shifted (i.e., shifted to longer wavelengths).

Among the respiratory electron carriers are three *b*-type cytochromes, cytochromes *c* and c_1, and cytochromes *a* and a_3. Cytochromes *b*, *c*, and c_1 all contain the same heme found in hemoglobin and myoglobin–iron complexed with protoporphyrin IX. In cytochromes *c* and c_1, but not *b*, this heme is covalently linked to the protein component, via thioether bonds formed between two of the vinyl side chains and two cysteine residues (Figure 15.6a). Cytochromes *a* and a_3 contain a modified form of heme, called heme A, in which one of the side chains is modified with a hydrophobic tail composed of three isoprene units (Figure 15.6b). Cytochromes *a* and a_3 represent two identical heme A moieties, attached to the same polypeptide chain, but within different environments in the inner membrane and, hence, having different reduction potentials. In cytochromes *a* and a_3, each of the hemes is associated with a copper ion, located close to the heme iron. The axial ligands of the heme iron also vary with the cytochrome type. Figure 15.6c shows a *b*-type heme bound in a four-helix bundle, a common structural motif for many cytochromes. Nature uses these differences in heme environments to access a wide range of standard reduction potentials for the cytochromes. Cytochromes undergo oxidoreduction through the complexed metal, which cycles between +2 and +3 states of heme iron and +1 and +2 states for the copper in cytochromes *a* and a_3. Thus, the cytochromes are one-electron carriers.

Cytochrome *c* is a small protein (~100 amino acids), which is associated with the inner membrane but is readily extracted in soluble form. Because it is small and relatively abundant, detailed structural studies have been carried out with this protein. Recall from Chapter 5 that the amino acid sequence of cytochrome *c* has been highly conserved in evolution, with nearly 50% identity between residues at corresponding positions of cytochromes *c* in organisms as diverse as yeast and human. The other cytochromes are integral membrane proteins and are exceedingly difficult to dissociate from the membrane. Accordingly, we know less about their structures.

Determining the Sequence of Respiratory Electron Carriers

To comprehend the mechanism by which energy from biological oxidations is captured to drive ATP synthesis, we must understand the oxidation reactions of electron transport, both the sequence in which electrons are carried from reduced substrates to oxygen and the energetics of individual reactions. Figure 15.7 shows E_0' values for the major respiratory electron carriers. If this figure accurately represents the sequence, then we can visualize respiratory electron transport as a series of coupled exergonic reactions in which the total energy available from oxidation of NADH by O_2 is released in a series of small steps, some of which are sufficiently exergonic to generate the 30.5 kJ/mol needed to drive ATP synthesis.

FIGURE 15.6

The hemes found in cytochromes. **(a)** The covalent bond formed between heme and the protein component in cytochromes c and c_1. The vinyl groups on heme are linked to the thiol groups of two cysteine residues (red). **(b)** Heme A, the form found in cytochromes a and a_3. Note the modified side chains—a formyl group (red) and an isoprenoid side chain (blue). **(c)** Four-helix bundle structure of cytochrome b_{562} from *E. coli* (PDB ID 1LM3).

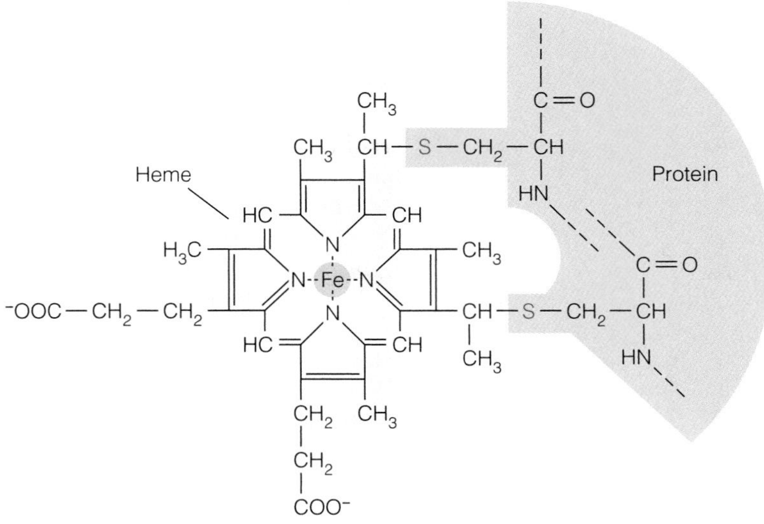

(a) General structure of cytochromes c and c_1

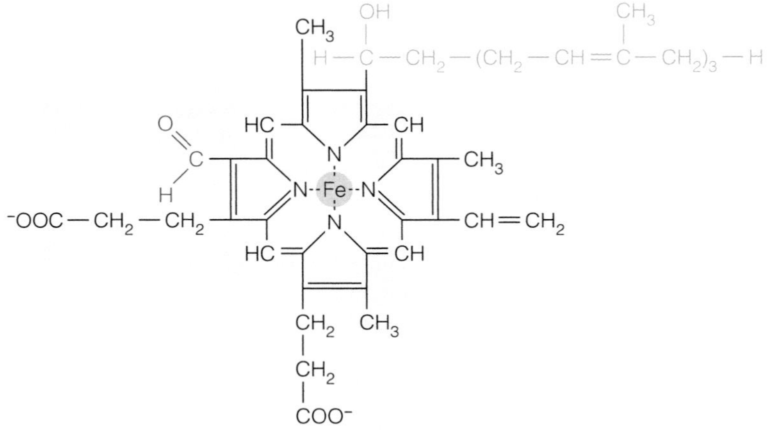

(b) Heme A in cytochromes a and a_3

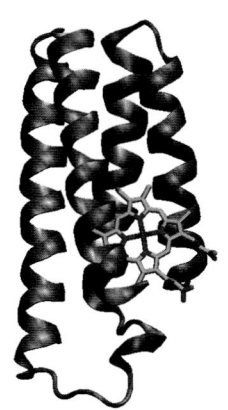

(c) 4-Helix bundle of cyt b_{562}

However, the data of Figure 15.7 involve standard conditions, and conditions are much different within the mitochondrion. In particular, the hydrophobic nature of the membrane environment changes E values in ways that are difficult to predict or to measure. We shall outline here three experimental techniques that were used to identify both the actual order of electron carriers and the specific reactions that drive ATP synthesis. We focus first on three approaches used to identify the order of electron carriers: (1) spectrophotometric techniques to measure the redox status of electron carriers in intact mitochondria; (2) use of specific respiratory inhibitors and artificial electron acceptors; and (3) fractionation of mitochondria into respiratory subassemblies, each capable of catalyzing specific portions of the overall sequence.

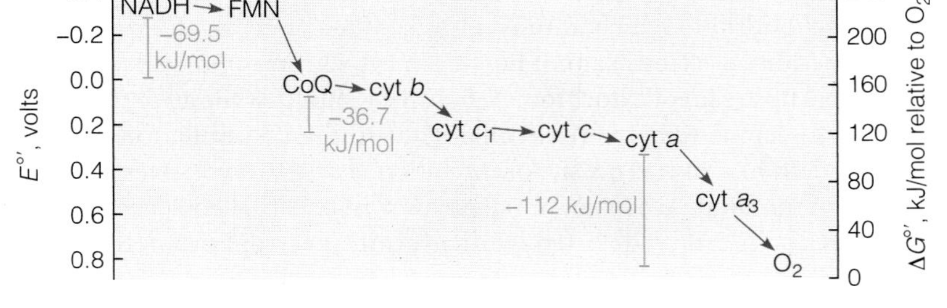

FIGURE 15.7

Standard reduction potentials of the major respiratory electron carriers. Three reactions in the respiratory chain have $\Delta G^{\circ\prime}$ values greater than -30.5 kJ/mol, the $\Delta G^{\circ\prime}$ for ATP hydrolysis: FMN $\rightarrow$ Q; cyt $b \rightarrow$ cyt c_1; and cyt $a \rightarrow O_2$.

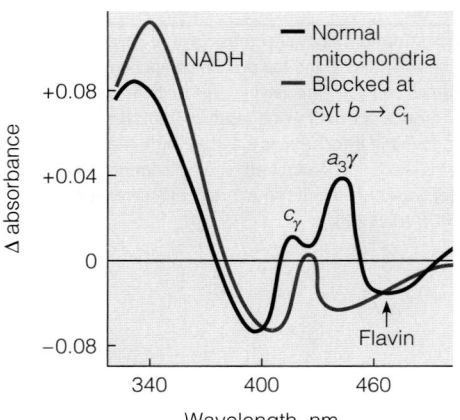

(a) Difference spectra for wavelengths below 500 nm

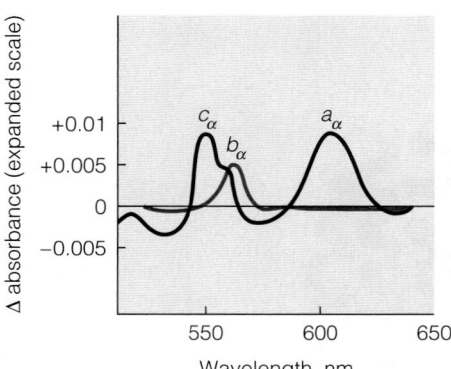

(b) Difference spectra continued with extended scale

FIGURE 15.8

Difference spectra of mitochondria. These difference spectra of rat liver mitochondria were recorded in a double-beam spectrophotometer. The black line shows the difference spectrum of fully reduced versus fully oxidized mitochondria. Mitochondria reduced with substrate under anaerobic conditions were in the sample chamber, and oxygenated mitochondria in the reference chamber. The peaks and shoulders refer to NADH, flavin, and the α and γ absorption bands, as indicated, for cytochromes a, a_3, b, and c. The red line shows the effect of adding a respiratory inhibitor, antimycin A, which blocks electron flow from cytochrome b to c_1. The inhibitor causes all of the carriers beyond cytochrome b to become fully oxidized, while NADH, flavin, and cytochrome b are in the reduced state. Note the expanded absorbance scale beyond 500 nm.

Springer and Plenum/*Mitchondria*, 1982, A. Tzagoloff, adapted with kind permission from Springer Science+Business Media B.V.

Difference Spectra

For nicotinamide nucleotides, flavin nucleotides, and cytochromes, the absorption spectrum for the reduced carrier differs from that of its oxidized counterpart. We should, therefore, be able to scan the absorption spectrum of a mixture of these carriers and ascertain the proportions of each in the oxidized and reduced states. The sensitivity of the technique is increased if a **difference spectrum** is obtained. Here the sample cuvette contains the mixture of electron carriers under study, and the reference cuvette contains not a blank but an equimolar mixture of carriers in a known state, for instance, entirely oxidized. Thus, any small absorbance changes, either positive or negative, result from reduction of a portion of the carriers in the test sample.

When these carriers are embedded in the mitochondrion, however, the task becomes quite difficult. Mitochondrial suspensions are turbid, and the resultant light scattering makes it impossible to measure difference spectra with ordinary spectrophotometers. Britton Chance greatly improved this technique in the mid-1950s when he developed a dual-wavelength, double-beam spectrophotometer that allowed him to obtain difference spectra with intact mitochondria. In the example shown by the black lines in Figure 15.8, the reference cuvette contains mitochondria saturated with oxygen so that all carriers are oxidized, while the sample cuvette contains anaerobic mitochondria plus an oxidizable substrate so that all of the carriers are reduced. The difference spectrum identifies wavelengths of maximal and minimal absorbance differences. Absorbance readings at these wavelengths allow determinations of the concentrations of the reduced and oxidized forms of an electron carrier that absorbs light in these ranges. For example, the higher the absorbance at 340 nm, the greater the proportion of the $NAD^+/NADH$ couple that is present as NADH; and the lower the negative absorbance at 460 nm, the greater the proportion of flavin nucleotide that is in the oxidized form.

Two important observations were made soon after the introduction of this technique. First, in actively respiring mitochondria, NADH predominated over NAD^+, whereas cytochrome a_3 was largely oxidized. For the intermediate carriers, the proportion in the oxidized state increased in the same order as the order of their presumed function in respiration (Figure 15.7). Second, when oxygen was added to anaerobic mitochondria, and difference spectra were obtained at intervals after oxygen addition, it was possible to determine the order in which each carrier went from fully reduced to partially oxidized. That order was the same as the presumed order of function in respiration.

Inhibitors and Artificial Electron Acceptors

Further information was obtained from difference spectrophotometry in conjunction with exogenous compounds that functioned either as respiratory inhibitors or as artificial electron donors or acceptors. The sites of action of several important inhibitors are shown in Figure 15.9. These inhibitors include (1) **rotenone**, a plant product from South America that is used as an insecticide and that blocks electron flow from NADH to coenzyme Q; (2) **amytal**, a barbiturate drug that acts at the same site; (3) **antimycin A**, a *Streptomyces* antibiotic that blocks electron flow from cytochrome b to c_1; and (4) **cyanide**, **azide**, and **carbon monoxide**. Cyanide and azide react with the oxidized form of cytochrome a_3, and CO reacts with the reduced form.

To see the utility of respiratory inhibitors, consider what happens in actively respiring mitochondria to which antimycin A is added (refer to the red lines in Figure 15.8). Because electrons cannot flow from cytochrome b to cytochrome c_1, all of the carriers upstream of cytochrome b become reduced, while all of the downstream carriers become fully oxidized, analogous to the accumulation of water upstream of a dam in a river. This site of inhibition is called a **crossover point**—a specific target of inhibition when an overall pathway is blocked. Crossover points in the respiratory chain can be detected by dual-beam spectrophotometry, which

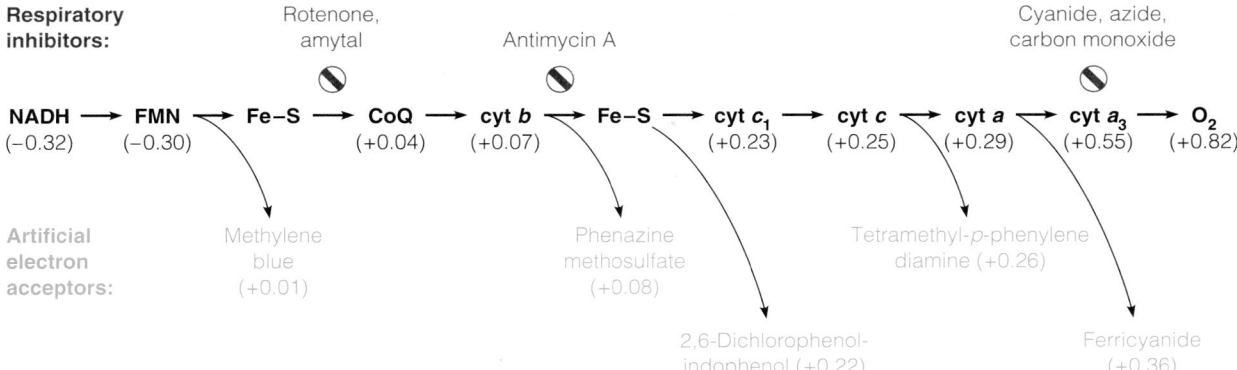

allows identification of all the carriers preceding a site of inhibition (NAD, flavin, Q, and cytochrome b after antimycin treatment) and all those following that site (cytochromes c_1, c, a, and a_3).

Artificial electron donors and acceptors are compounds that can either feed in or draw off electrons from the respiratory chain in spontaneous nonenzymatic redox reactions. For example, 2,6-dichlorophenol-indophenol (DCIP) can spontaneously oxidize cytochrome b, but probably not c_1, because of the $E^{0'}$ values involved (see Figure 15.9). This demonstrated that cytochrome b lies beyond the entry point for electrons from succinate, as well as from NADH. The key observation was that addition of DCIP to cyanide-inhibited mitochondria allows these mitochondria to oxidize both NAD^+-linked substrates and succinate. In this system, electrons flow from substrate to cytochrome b and then to the exogenous electron carrier, DCIP, which becomes reduced. Ferricyanide, with its large positive $E^{0'}$, can accept electrons from any of the carriers up to cyt a_3.

The order of action of electron carriers in the respiratory chain was elucidated by difference spectrophotometry, analysis of respiratory inhibitors, and properties of membrane complexes.

Respiratory Complexes

Mitochondria can be disrupted by mechanical treatment, such as sonication, or by low concentrations of nonionic detergents such as *digitonin*, which preferentially solubilizes the outer membrane but leaves many protein–protein associations intact. By combinations of these techniques, one can fractionate the mitochondrial respiratory chain into four separate enzyme complexes (complexes I, II, III, and IV—see Figure 15.2b), each of which contains part of the entire respiratory sequence, plus a fifth (complex V), which catalyzes ATP synthesis from ADP. The electron-transfer activity of these multisubunit complexes is retained during solubilization and fractionation, revealing that complexes I, II, III, and IV are membrane-embedded enzymes that catalyze the transfer of electrons from a relatively mobile electron carrier (one that is not tightly membrane-bound) to another mobile carrier. These mobile carriers are NADH, succinate, coenzyme Q, cytochrome c, and oxygen. Analysis of each complex for the presence of electron carriers, as well as for reactions catalyzed, has helped to establish the currently accepted sequence of carriers. Figure 15.10 provides a summary of the protein composition and catalytic activities of each complex. We will now examine in more detail the structure and function of each complex and how they are organized in a respiratory chain.

NADH Dehydrogenase (Complex I)

The main donor of electrons into the respiratory chain is the reduced nicotinamide nucleotide NADH. Numerous dehydrogenases in the cell catalyze the oxidation of substrates using NAD^+ as the electron acceptor:

$$\text{reduced substrate} + NAD^+ \rightleftharpoons \text{oxidized substrate} + NADH + H^+$$

As described in Chapter 11, these reversible dehydrogenations involve the transfer of two electrons, in the form of a hydride ion (H^-), from the substrate to

FIGURE 15.10

FIGURE 15.10

Multiprotein complexes in the mitochondrial respiratory assembly. The subscripts for the *b* cytochromes denote their spectral maxima. The two *b* hemes in complex III, identified as cyt b_{562} and cyt b_{566}, are bound to the same polypeptide chain. The yellow arrows denote the energy released by the actions of complexes I, II, and IV used to drive the synthesis of ATP by complex V (ATP synthase).

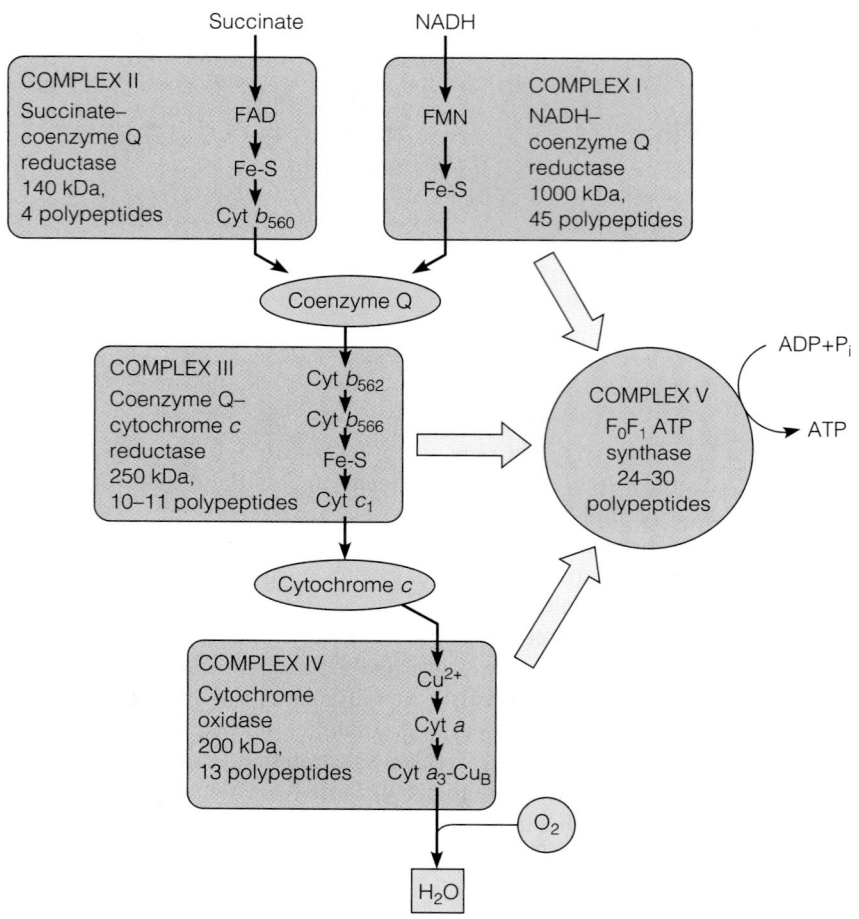

NAD^+ to give NADH. The nicotinamide nucleotides readily dissociate from their enzymes and thus act as soluble redox cofactors that carry electrons between different enzymes and pathways. In mitochondrial respiration, NADH from many different dehydrogenases becomes oxidized in the first step of electron transport by complex I, or **NADH dehydrogenase**, which catalyzes the following reaction.

$$NADH + H^+ + Q \rightleftharpoons NAD^+ + QH_2$$

Mitochondrial NADH dehydrogenase is a large, membrane-embedded multisubunit complex (~1000 KDa) with about 45 separate polypeptide chains. The bacterial complex is much smaller, composed of just 14 "core" subunits which are conserved from bacteria to humans. The larger mitochondrial enzyme evolved by gradual recruitment of additional subunits to this core. The complex contains **flavin mononucleotide (FMN)** as a tightly bound prosthetic group and eight **iron–sulfur clusters**, which transfer electrons from reduced flavin to coenzyme Q (Q).

The overall reaction catalyzed by the NADH dehydrogenase complex is shown below. Because the electrons are subsequently used to reduce coenzyme Q, a more descriptive name for this complex is **NADH–coenzyme Q reductase**.

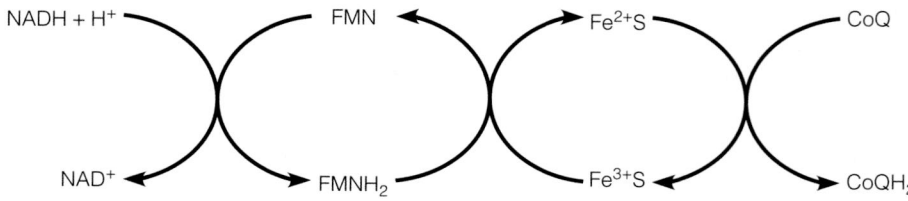

Notice that this process begins with a two-electron donor (NADH), but uses several iron–sulfur clusters, which can carry only single electrons. FMN acts as a transformer between the two-electron and one-electron carriers.

Figure 15.11 illustrates our current understanding of how these various electron carriers function in complex I. Electron microscopy of complex I from several different eukaryotic and bacterial sources reveals an L-shaped structure with two arms: a hydrophobic membrane arm embedded in the mitochondrial inner membrane (or bacterial plasma membrane), and a hydrophilic peripheral arm extending into the mitochondrial matrix (or bacterial cytoplasm) (Figure 15.11a). An X-ray crystal structure of the entire complex I from the archaea *Thermus thermophilus* reveals a 15-subunit complex with FMN and nine iron–sulfur clusters in the peripheral arm (Figure 15.11b, c). Complex I catalyzes a hydride transfer (two electrons) from NADH to FMN, which then transfers the electrons, one at a time, to the series of iron–sulfur clusters. In the final step of the reaction, Q is reduced, one electron at a time, to QH_2.

Complex I catalyzes a second important process, which is tightly coupled to the electron flow through the complex: the transfer of four protons (H^+) from the matrix side to the intermembrane space side of the inner membrane (Figure 15.11). The X-ray structure of the *T. thermophilus* enzyme suggests a possible mechanism to explain how electron transfer drives proton pumping (Figure 15.11d). The NuoL subunit of the membrane arm possesses a long α-helix (110 Å) that projects through the adjacent membrane subunits. As electrons pass from FMN to coenzyme Q, the NuoA/J/K and H subunits undergo conformational changes that push the long α-helix toward the other membrane arm subunits. This tilts α-helices (red in Figure 15.11 c and d) in those subunits, allowing protons to pass through the channels across the membrane.

Thus complex I acts as a proton pump, using the energy released in the exergonic transfer of two electrons from NADH to Q to transport four H^+ *against* their electrochemical gradient. To account for these four translocated protons, we can rewrite the reaction catalyzed by complex I as:

$$NADH + 5H^+_{matrix} + Q \rightleftharpoons NAD^+ + QH_2 + 4H^+_{intermembrance\ space}$$

As we will see shortly, this obligately coupled process is the essence of the chemiosmotic mechanism that drives ATP synthesis.

Succinate–Coenzyme Q Reductase (Complex II)

Coenzyme Q (Q) draws electrons into the respiratory chain, not only from NADH but also from succinate, as shown in Figure 15.3, and from intermediates in fatty acid oxidation. Succinate dehydrogenase uses an FAD coenzyme, as noted in Chapter 14. Unlike the other citric acid cycle enzymes, succinate dehydrogenase is an inner membrane protein (Figure 15.12). The enzyme thus transfers electrons directly from its bound $FADH_2$ to the other membrane-bound respiratory carriers. Like NADH dehydrogenase (complex I), succinate dehydrogenase transfers electrons via a series of iron–sulfur centers to coenzyme Q, and it is more completely named **succinate–coenzyme Q reductase** (it is also called complex II—see Figures 15.2 and 15.3). Complex II does not pump protons, so the reaction it catalyzes can be summarized as:

$$succinate + Q \rightleftharpoons fumarate + QH_2$$

At least two other flavoprotein dehydrogenases, including the **ETF:ubiquinone oxidoreductase** and **glycerol-3-phosphate dehydrogenase** also deliver electrons to Q (Figure 15.13). These enzymes transfer electrons not from citric acid cycle intermediates, but rather from other oxidation pathways. Like succinate dehydrogenase, ETF:ubiquinone oxidoreductase is bound to the matrix side of the inner membrane, and it uses FAD and an Fe_4S_4 cluster as electron carriers. **ETF** (**electron-transferring flavoprotein**) is a small, soluble protein in the mitochondrial matrix

Electron transport through complex I is coupled to the pumping of protons from the matrix to the intermembrane space.

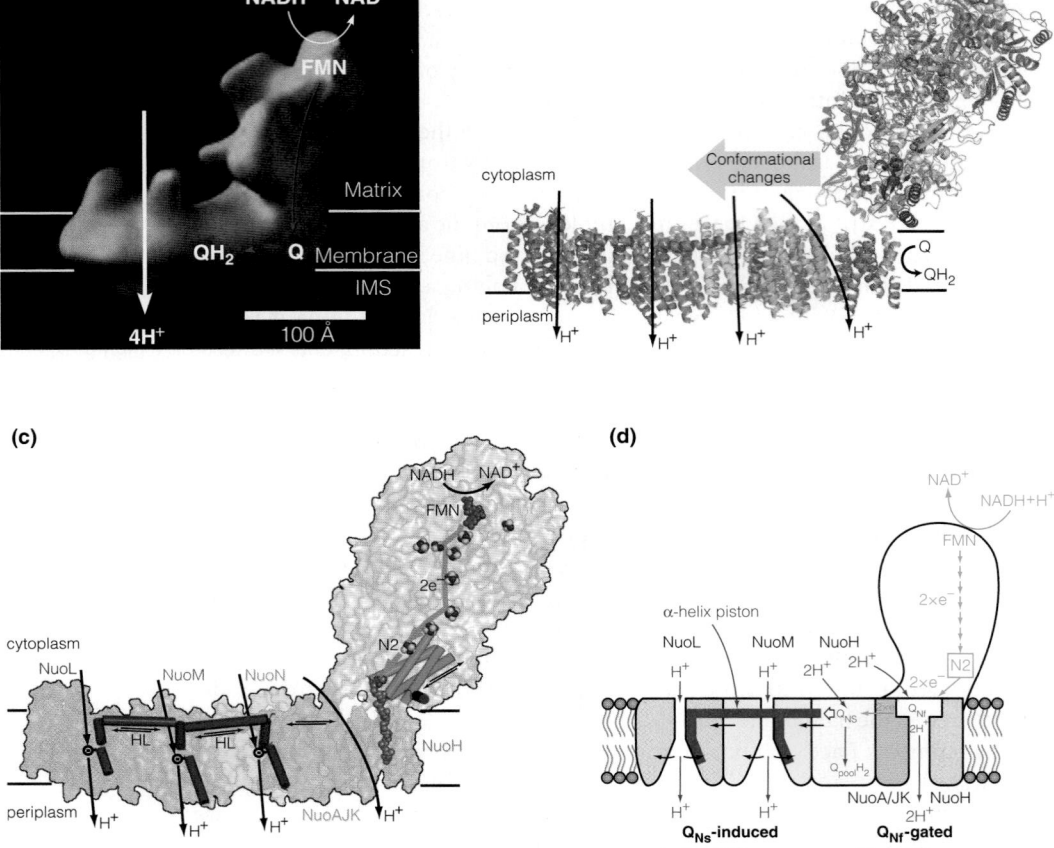

FIGURE 15.11

Structure and function of complex I (NADH–coenzyme Q reductase). **(a)** Model of complex I from the yeast *Yarrowia lipolytica* determined by three-dimensional electron microscopy. The membrane arm is composed of ~30 subunits, and contains the Q binding site and the proton pumping machinery. The peripheral arm is composed of eight subunits. The path of electron transport from NADH to Q, and the direction of H^+ pumping, is shown schematically. **(b)** X-ray structure of the entire complex I from the archaea *Thermus thermophilus*. The long α-helix of subunit NuoL that projects onto subunits NuoM (blue) and N (yellow) is dark purple. Regarding proton movement, the cytoplasm and periplasm in bacteria are equivalent to the matrix and intermembrane space (IMS), respectively, in mitochondria. **(c)** Proposed model of proton translocation by complex I. The subunits are named according to the *E. coli* structure, and are colored as in panel b. Locations of the FMN (pink), iron–sulfur clusters (red/yellow), and coenzyme Q (dark blue) are highlighted. The path of electron transport from FMN through the iron–sulfur clusters to Q is shown by the blue arrows. NADH binds near FMN. **(d)** Proposed model for coupling of electron transport to proton pumping by complex I. The long α-helix from NuoL is represented by the red horizontal bar. The vertical red bars represent proton channel α-helices that tilt in response to the conformational change in the green and orange subunits. Q_{Nf} and Q_{NS} represent distinct species of coenzyme Q.

(a) Adapted from *The Annual Review of Biochemistry* 75:69–92, U. Brandt, Energy converting NADH: Quinone oxidoreductase (complex I). © 2006 Annual Reviews; (b, c) Reprinted from *Current Opinion in Structural Biology* 21:532–540, R. G. Efremov and L. A. Sazanov, Respiratory complex I: "Steam engine" of the cell? © 2011, with permission from Elsevier. (d) Modified with permission of Federation of the European Biochemical Societies from *FEBS Letters* 584:4131–4137, T. Ohnishi, E. Nakamaru-Ogiso, and S. T. Ohnishi, A new hypothesis on the simultaneous direct and indirect proton pump mechanisms in NADH-quinone oxidoreductase (complex I). © 2010.

that receives electrons from at least 12 different mitochondrial FAD-containing dehydrogenases involved in fatty acid and amino acid oxidations. For example, in the first step of mitochondrial fatty acid β-oxidation (see Chapter 17), **acyl-CoA dehydrogenase** catalyzes the transfer of two electrons from the fatty acyl-CoA

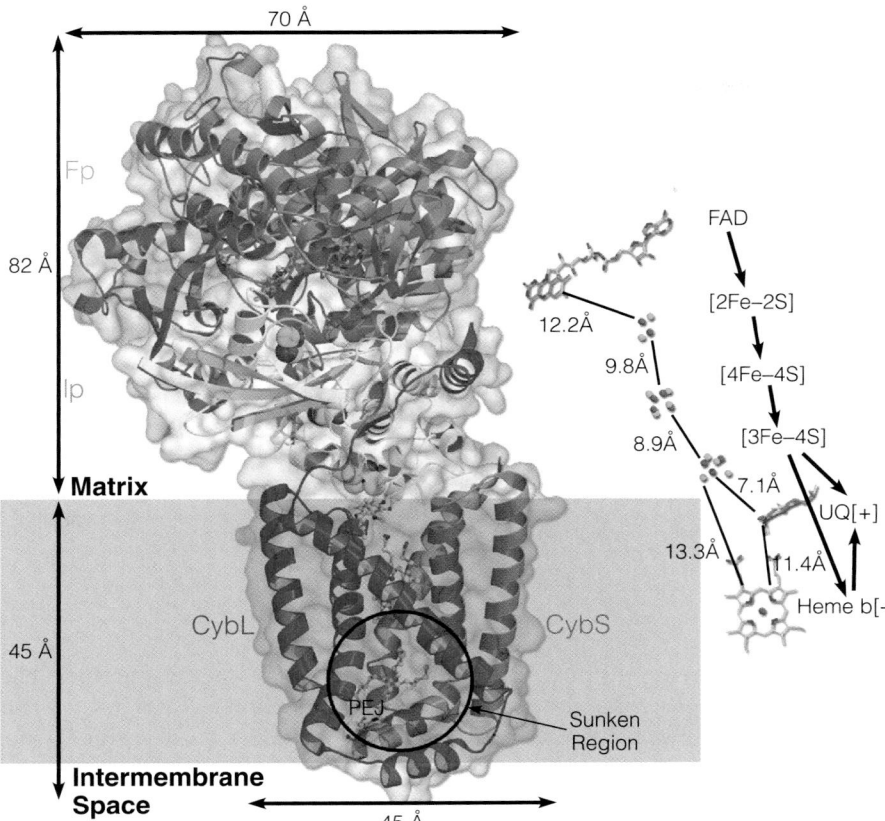

FIGURE 15.12

Structure of complex II (succinate dehydrogenase) from pig heart mitochondria (PDB ID 1ZOY). The enzyme is composed of two hydrophilic subunits extending into the matrix, the FAD-binding subunit (blue) and the iron–sulfur subunit (yellow), and two transmembrane subunits (pink and gold). The path of electron transport from FAD through the three iron–sulfur clusters to coenzyme Q (UQ) is shown on the right. The enzyme also has a b-type heme bound by the transmembrane subunits, but its role in the electron transfer process is not yet understood. Succinate binds near FAD in the blue subunit.

Adapted from *Cell* 121:1043–1057, F. Sun, X. Huo, Y. Zhai, A. Wang, J. Xu, D. Su, M. Bartlam, Z. Rao, Crystal structure of mitochondrial respiratory membrane protein complex II. © 2005, with permission from Elsevier.

substrate to ETF. ETF, with its bound $FADH_2$, is then reoxidized by ETF:ubiquinone oxidoreductase, transferring the electrons to Q in the respiratory chain. Glycerol-3-phosphate dehydrogenase, located on the *intermembrane face* of the inner membrane, catalyzes the oxidation of glycerol-3-phosphate to dihydroxyacetone phosphate (DHAP), reducing Q to QH_2 (Figure 15.13). We shall see later that this reaction plays an important role in shuttling electrons from cytoplasmic NADH into the mitochondrial matrix (see Figures 15.2 and 15.30). Because Q is subsequently oxidized by complex III, it can be seen as a collection point, gathering electrons from several flavoprotein dehydrogenases and passing them along the respiratory chain, ultimately to O_2 (Figure 15.10).

Coenzyme Q:cytochrome *c* Oxidoreductase (Complex III)

The oxidation of reduced coenzyme Q is mediated by another large multisubunit complex embedded in the inner membrane, complex III of the respiratory chain. This enzyme catalyzes the transfer of electrons from QH_2 to cytochrome *c*, and is thus called **coenzyme Q:cytochrome *c* oxidoreductase**. Mammalian complex III functions as a dimer, with each monomer composed of 10 or 11 protein chains (~250 kDa), including cytochrome *b*, cytochrome c_1, and a protein called the Rieske iron–sulfur protein. Electron transfer through this complex involves, per monomer, two *b*-type hemes, one *c*-type heme, and one Fe_2S_2 cluster, thus complex III is also referred to as **cytochrome bc_1 complex**. Figure 15.14 shows the X-ray crystal structure of the bovine mitochondrial complex III and the locations of the redox centers.

The path of electrons from QH_2 to cyt *c* through complex III is more complicated than indicated in Figure 15.10 because at this point, a two-electron donor,

FIGURE 15.13

Coenzyme Q collects electrons from multiple flavoproteins. This lipid-soluble mobile redox cofactor serves as the electron acceptor for at least four mitochondrial flavoprotein dehydrogenases. Electrons from NADH and succinate are delivered via complex I and complex II (succinate dehydrogenase), respectively. Electron-transferring flavoprotein (ETF):ubiquinone oxidoreductase catalyzes the transfer of electrons from reduced ETF to Q. These electrons originate in the acyl-CoA dehydrogenase step of fatty acid β oxidation. Glycerol-3-phosphate dehydrogenase, located on the *intermembrane face* of the inner membrane, delivers electrons from glycerol-3-phosphate to coenzyme Q. Electrons from reduced QH_2 eventually pass to complex III of the respiratory chain.

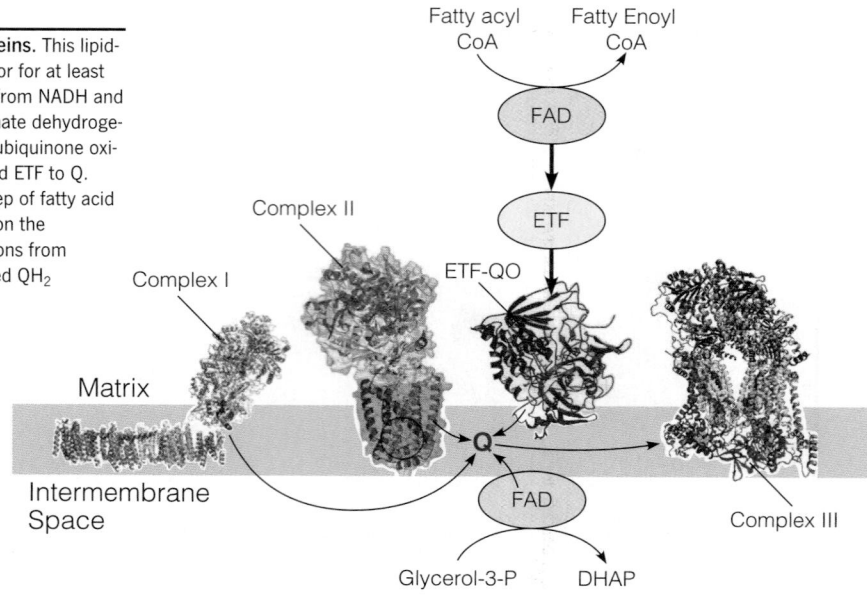

FIGURE 15.14

Structure of Complex III (coenzyme Q:cytochrome c oxidoreductase). (a) X-ray structure of the dimeric complex from bovine mitochondria (PDB ID 1PPJ). Cytochrome *b* subunits are yellow; each binds two *b*-type hemes (b_H and b_L), shown in red. The Rieske iron–sulfur protein (ISP) subunits are magenta, with the Fe_2S_2 iron–sulfur center shown in orange (the second ISP subunit and its iron–sulfur center are hidden behind the structure). Cytochrome c_1 subunits are blue, with their *c*-type heme in shown in green. The remaining non-core subunits are shown in silver.
(b) Cartoon of the complex III dimer, with the same color scheme as in the X-ray structure, illustrating the arrangement of the subunits and redox carriers. The approximate location of the complex in the inner membrane is indicated.

QH_2, is transferring electrons to one-electron acceptors, the cytochromes. The **Q cycle** has been proposed to account for this stoichiometry (Figure 15.15). Complex III has two binding sites for Q: Q_0 and Q_1. QH_2 is oxidized at the Q_0 site, where the two electrons take two different paths. The first electron is passed to the iron–sulfur cluster of the Rieske iron–sulfur protein, to cytochrome c_1 and then to cytochrome *c*. The resultant $QH^\cdot$ semiquinone then transfers its second electron to the low-potential b_L heme component of cytochrome *b* (also known as b_{566} because this heme has a light absorption maximum at 566 nm), and this electron next passes to the high-potential b_H heme (b_{562}, light absorption maximum at 562 nm) component. The b_H heme is located at the Q_1 site near the matrix side of the membrane, where it reduces a molecule of oxidized Q to the $QH^\cdot$ semiquinone. This process is repeated, with a second molecule of QH_2 being oxidized at the Q_0 site, and one electron passing to the Rieske iron–sulfur protein and onto

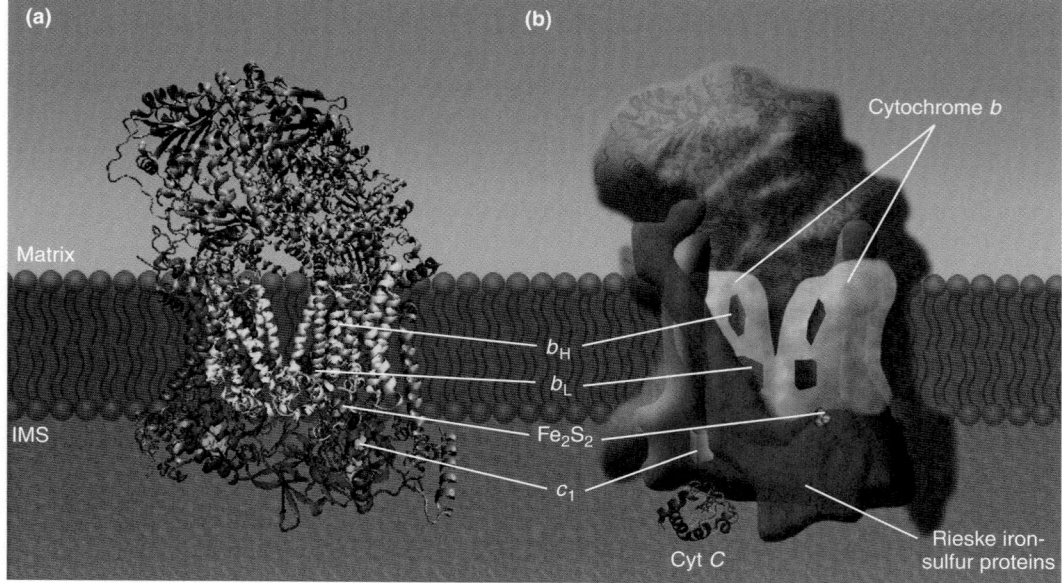

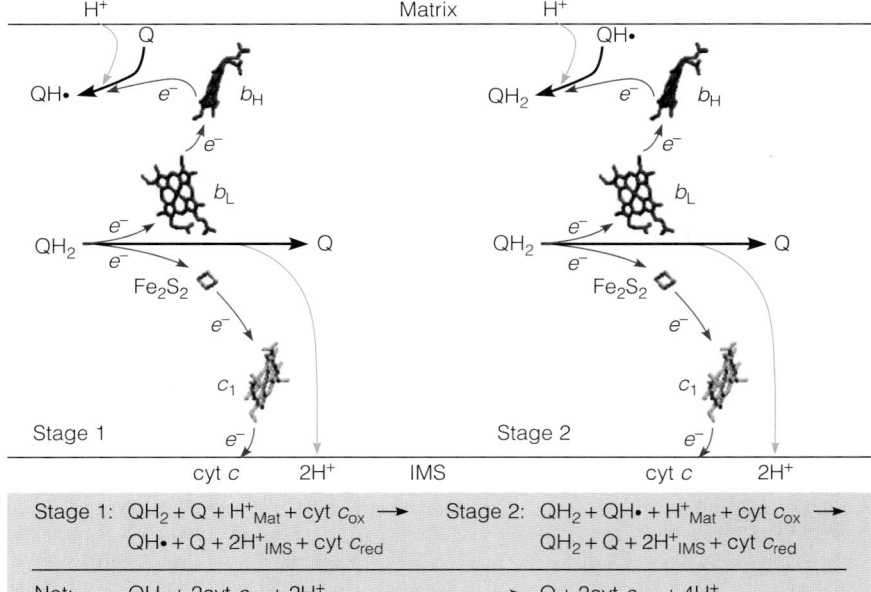

FIGURE 15.15

Q cycle. The spatial arrangement of the redox centers in one monomer of the dimeric bovine mitochondrial complex III (see Figure 15.14) is shown without the protein components (PDB ID 1PPJ). The path of electrons in each stage is indicated by red arrows, and the path of protons is indicated by blue arrows. The Q_0 site is in the middle of the diagram; the Q_1 site is at the top (matrix side of the inner membrane). The stoichiometry of each stage of the cycle is shown, as well as the net stoichiometry of the full cycle.

Stage 1:	$QH_2 + Q + H^+_{Mat} + cyt\ c_{ox} \longrightarrow$	Stage 2:	$QH_2 + QH\bullet + H^+_{Mat} + cyt\ c_{ox} \longrightarrow$
	$QH\bullet + Q + 2H^+_{IMS} + cyt\ c_{red}$		$QH_2 + Q + 2H^+_{IMS} + cyt\ c_{red}$

Net:	$QH_2 + 2cyt\ c_{ox} + 2H^+_{Mat} \longrightarrow Q + 2cyt\ c_{red} + 4H^+_{IMS}$

cytochrome c as before. However, this time the electron from the b_H heme reduces the QH$\bullet$ semiquinone to QH_2 at the Q_1 site (Figure 15.15). The result is that two molecules of QH_2 become oxidized and one molecule of Q becomes reduced, for a net transfer of two electrons to reduce two molecules of cytochrome c. Because the proton-consuming reactions occur within the matrix, while proton release takes place in the intermembrane space, the Q cycle contributes to the proton gradient needed to drive ATP synthesis (see page 645).

Note the location of cytochrome c: It is on the intermembrane space side of the inner membrane. Cytochrome c plays another important role in the cell besides as an electron carrier in respiration. Cytochrome c is a central signaling component in the **intrinsic pathway of apoptosis**, or programmed cell death. This pathway is initiated in response to oxidative damage, DNA damage, and many other chemical and physical insults that damage the cell directly. These diverse *intrinsic* stresses all trigger the same response: The mitochondrial outer membrane becomes more permeable and cytochrome c is released from the intermembrane space into the cytoplasm, where it binds with several other proteins in a complex (the **apoptosome**), which leads to activation of a proteolytic cascade, eventually resulting in cell death (apoptosis will be discussed in more detail in Chapter 28). Thus, this integral component of the respiratory chain is also an integral component of a cell death pathway, hinting at the relationship between mitochondria and the evolution of the eukaryotic cell (more about this at the end of this chapter).

Electron transport through complex III is coupled to the pumping of protons from the matrix to the intermembrane space.

Cytochrome c Oxidase (Complex IV)

The final stage of electron transport is carried out by cytochrome c oxidase (complex IV). The mitochondrial enzyme exists as a large homodimeric complex in the inner membrane. Each monomer is composed of 13 subunits, with 28 transmembrane helices separating hydrophilic domains facing the matrix and intermembrane space compartments (Figure 15.16). Bacterial cytochrome c oxidase is much simpler, composed of just four subunits. However, homologs of three of the bacterial subunits form the core of the mitochondrial complex, revealing the evolutionary origin of this enzyme. Indeed, subunits I, II, and III of cytochrome c oxidase are encoded by the mitochondrial genome and synthesized on mitochondrial ribosomes in the matrix. Cytochrome c oxidase contains two hemes, a and a_3,

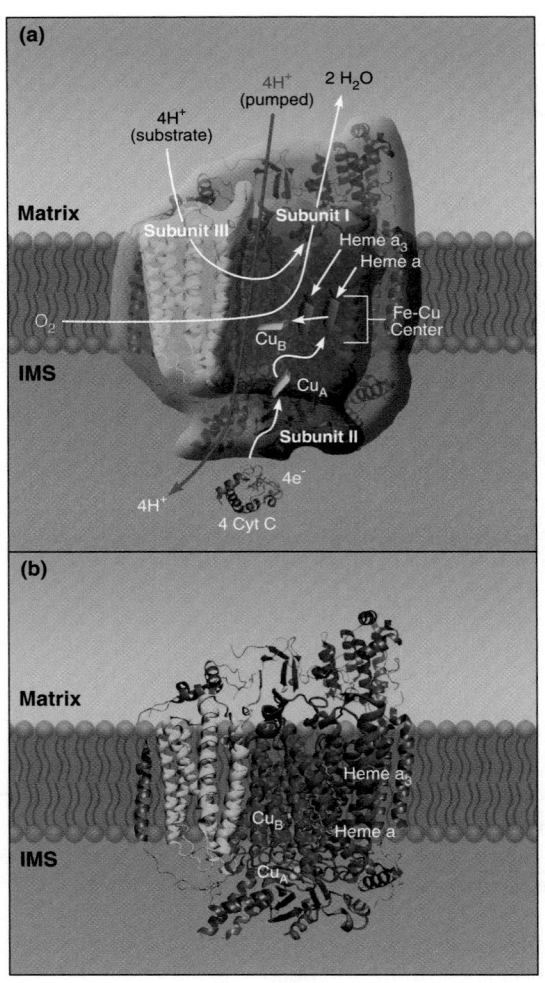

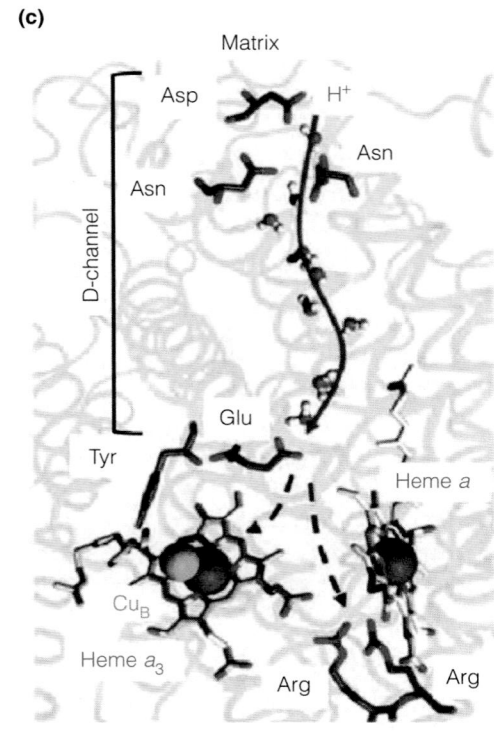

FIGURE 15.16

Structure of cytochrome c oxidase (complex IV).
(a) Cartoon of reaction catalyzed by complex IV. Four electrons are donated, one at a time, by reduced cyt c, to the Cu_A center, through heme a, and on to the catalytic site (binuclear a_3–Cu_B site) where one molecule of O_2 is reduced, yielding two molecules of H_2O. Protons are pumped from the matrix side to the intermembrane space (IMS) side of the membrane. **(b)** X-ray structure of one monomer of the bovine mitochondrial enzyme (PDB ID 2EIJ). Subunits I, II, and III are colored as in the cartoon in panel (a); the remaining subunits are gray. The core two hemes are orange, and the two Cu centers are green. The approximate location of the complex in the inner membrane is indicated. Cyt c binds near the Cu_A center of subunit II. **(c)** The D-channel proton-transfer pathway in subunit I of the *R. sphaeroides* cytochrome c oxidase (PDB ID 1M57) is indicated by the red arrow and the water molecules. The dashed black arrows indicate the branching paths of protons destined for the catalytic site and the pump site.

(c) Modified with permission from *Biochemical Society Transactions* 36:1169–1174, P. Brzezinski, J. Reimann, and P. Adelroth, Molecular architecture of the proton diode of cytochrome c oxidase. © 2008 The Biochemical Society.

bound by subunit I in the interior of the membrane, and two copper centers (Cu_A and Cu_B). The Cu_B atom is ligated by three histidine residues and sits within 5 Å of heme a_3 on subunit I. The iron of heme a_3 and Cu_B thus constitute a "binuclear center," functioning as a single unit in electron transfer. The Cu_A center is composed of two Cu atoms bound by subunit II at the intermembrane space face of the membrane.

Complex IV catalyzes the transfer of electrons from reduced cytochrome c to oxygen. The initial oxidation of cytochrome c is carried out by Cu_A, with the electron transferred to heme a, and then to the binuclear heme a_3–Cu_B site. The binuclear center is the catalytic site where O_2 undergoes its four-electron reduction to water. The redox reaction catalyzed by complex IV is as follows:

$$4 \text{ cyt } c_{red} (\text{Fe}^{2+}) + O_2 + 4H^+_{matrix} \rightleftharpoons 4 \text{ cyt } c_{ox} (\text{Fe}^{3+}) + 2H_2O$$

For each electron transferred, complex IV pumps approximately one proton from the matrix side to the intermembrane space side of the membrane. Thus, the net reaction catalyzed by complex IV can be summarized as follows:

$$4 \text{ cyt } c_{red} (\text{Fe}^{2+}) + O_2 + 8H^+_{matrix} \rightleftharpoons 4 \text{ cyt } c_{ox} (\text{Fe}^{3+}) + 2H_2O + 4H^+_{intermembrance\ space}$$

How are these four protons per molecule of oxygen reduced pumped from the matrix to the intermembrane space? X-ray structures of cytochrome c oxidase, such as that in Figure 15.16, reveal the existence of two distinct proton channels leading from the matrix to the catalytic center in subunit I, called the D-channel and the K-channel. The K-channel (named for an essential lysine residue) provides two of the protons required to form water. The D-channel (named for an essential aspartate residue at the entrance to the channel; top of Figure 15.16c) provides the other two protons in the O_2 reduction to $2H_2O$, as well as all four of the pumped protons. A series of ~10 water molecules can be seen in the D-channel in the X-ray structures (Figure 15.16c), providing a mechanism to transfer protons through a hydrogen-bonded network (recall the proton-hopping described in Chapter 2) from the matrix to the catalytic site. How do the pumped protons move from the catalytic site, buried in the middle of the inner membrane, to the intermembrane space side? The mechanism is incompletely understood, but it probably involves conformational changes in the protein coupled to changes in redox state of the binuclear heme $a_3 - Cu_B$ center. These conformational changes alter the pK_as of proton-binding residues (e.g., Glu or Asp), as described for the Bohr effect in hemoglobin (see Chapter 7). One model proposes that following electron transfer to the catalytic site, a proton is transferred from a conserved glutamate residue (located at the bottom of the D-channel, near the binuclear center) to the catalytic site. Deprotonation of this Glu triggers a conformational change that raises the pK_a of a second proton acceptor such that it now has a greater affinity for protons and binds a proton from the D-channel. Reprotonation of Glu with another proton from the D-channel reverses the conformational change back to the original state, lowering the pK_a of the second proton acceptor, triggering release of the pumped proton on the other side of the membrane.

Supramolecular Organization of Mitochondrial Respiratory Chain Complexes

Recent kinetic, biochemical, and electron microscopy studies suggest that the respiratory chain complexes exist as highly organized "supercomplexes" in the inner membrane of intact mitochondria. For example, supercomplexes containing complex I, complex III, and complex IV in a 1:2:1 ratio can be isolated from rat and bovine mitochondria after solubilization with mild detergents like digitonin. Electron microscopy and 3D reconstructions show that the Q and cytochrome c binding sites of the individual complexes are facing each other, suggesting the possibility of substrate channeling (Figure 15.17). Indeed, kinetic studies suggest that the respiratory complexes form a series of tightly linked redox reactions, behaving much like a tiny electrical wire.

Oxidative Phosphorylation

Having discussed how the free energy of electron transport down the respiratory chain is used to pump protons out of the matrix, we turn now to the question of how that energy is made available for ATP synthesis—in short, the mechanism of oxidative phosphorylation. Mechanistically, oxidative phosphorylation is far more complex than the substrate-level phosphorylation reactions in glycolysis and the citric acid cycle, and this field of inquiry was one of the most contentious biochemical research arenas for many years. As the late Efraim Racker stated, "Anyone who is not confused about oxidative phosphorylation just doesn't understand the situation." As we shall see, a great deal of the needed understanding has developed in the three decades since Racker wrote these words.

The P/O Ratio: Efficiency of Oxidative Phosphorylation

Before considering the mechanism of oxidative phosphorylation, it is useful to recall the energetics of the process. We previously calculated (page 630) that for a pair of electrons entering the respiratory chain as NADH, and traversing the entire chain to O_2, the standard free energy change, $\Delta G^{\circ\prime}$, is -220 kJ / mol:

$$NADH + H^+ + \tfrac{1}{2}O_2 \rightleftharpoons NAD^+ + H_2O$$
$$\Delta G^{\circ\prime} = -nF \Delta E^{0\prime} = -2(96485)(-0.82 - (-0.32)) = -220 \, kJ/mol$$

Electron transport through Complex IV is coupled to the pumping of protons from the matrix to the intermembrane space.

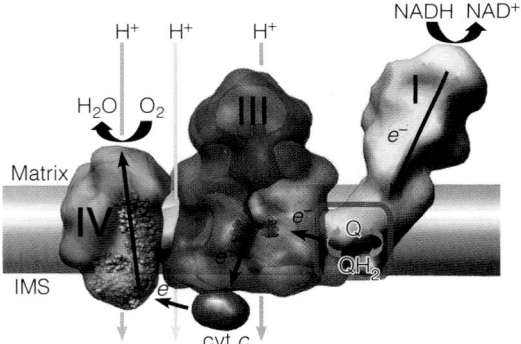

FIGURE 15.17

Model of proposed supercomplex of complexes I, III, and IV. This model is based on biochemical, X-ray structural, and electron microscopy results. Complex I is yellow, the dimer of complex III is red, and complex IV is green. Cytochrome c is gray. The path of electrons from NADH to O_2 is indicated, and the putative Q binding sites in complexes I and III are shown in the gray box.

Adapted from *Biochimica et Biophysica Acta* 1793:117–124, J. Vonck and E. Schafer, Supramolecular organization of protein complexes in the mitochondrial inner membrane. © 2009, with permission from Elsevier.

How much of this free energy is actually captured as ATP in oxidative phosphorylation? This question could be answered once we could measure the quantity of ATP synthesized per mole of substrate oxidized in isolated mitochondria. What we usually measure is the **P/O ratio**, which is the number of molecules of ATP synthesized per pair of electrons carried through electron transport. ATP synthesis is quantitated as phosphate incorporation into ATP, and electron pairs are quantitated as oxygen uptake, in μmol of O atoms (not O_2 molecules) reduced to water. Oxygen uptake is determined with an oxygen electrode.

As you might imagine, precise measurements of oxygen consumption and ATP synthesis in a mitochondrial preparation are difficult to obtain, and many experimental pitfalls can lead to inaccurate estimates of the P/O ratio. Early experiments suggested that the mitochondrial oxidation of NADH proceeds with a P/O ratio of 3 and oxidation of succinate proceeds with a P/O ratio of 2. These values were consistent with older theories of oxidative phosphorylation, which postulated the existence of a high-energy intermediate directly coupled to ATP synthesis, as seen in the substrate-level phosphorylation reactions of glycolysis. However, as researchers got better at preparing intact mitochondria and measuring oxygen consumption and ATP synthesis, it became clear that the P/O ratios were *not integers*. The general consensus now is that the P/O ratio is ~2.5 for oxidation of NADH and ~1.5 for oxidation of succinate. Indeed, these noninteger P/O values contributed to the realization that phosphorylation and oxidation are *not* directly coupled. As Figure 15.2 suggests, coupling occurs indirectly. As we will see next, this mechanism does not require an integral stoichiometric relationship between reducing equivalents consumed and ATP synthesized. With this in mind, we can write a balanced equation for the mitochondrial oxidation of NADH coupled to the synthesis of ATP:

$$\text{NADH} + \text{H}^+ + \tfrac{1}{2}\text{O}_2 + 2.5\text{ADP} + 2.5\text{P}_i \rightleftharpoons \text{NAD}^+ + \text{H}_2\text{O} + 2.5\text{ATP}$$

As discussed above, the oxidation of NADH by O_2 has a $\Delta G^{\circ\prime}$ of -220 kJ/mol. Because $\Delta G^{\circ\prime}$ for ATP hydrolysis is -30.5 kJ/mol, synthesis of 2.5 ATPs requires 76 kJ under standard conditions, giving an efficiency for oxidative phosphorylation of about 35% under standard conditions. Because ΔG for ATP hydrolysis under intracellular conditions is -50 kJ/mol or more, the efficiency of oxidative phosphorylation probably approaches 60% or 70% in vivo.

Oxidative Reactions That Drive ATP Synthesis

A glance at Figure 15.10 reveals that the transfer of reducing equivalents from an NAD^+-oxidizable substrate to O_2 involves about a dozen consecutive, linked oxidoreduction reactions. Which of these reactions actually drive ATP synthesis? This question was of paramount concern in the early days of bioenergetics research, when it was thought that ATP synthesis was directly coupled to individual exergonic reactions, as it is in substrate-level phosphorylation. The most straightforward interpretation of the P/O ratio of 3 initially estimated for NADH oxidation was that three of the individual reactions of the respiratory chain are sufficiently exergonic to drive the synthesis of one ATP molecule each. Indeed, three of these reactions do have $\Delta G^{\circ\prime}$ values exceeding 30.5 kJ/mol, the minimum barrier that must be overcome (under standard conditions) to make the synthesis of each ATP exergonic (see Figure 15.7). Those three reactions are the oxidation of NADH by coenzyme Q ($\Delta G^{\circ\prime} = -69.5$ kJ/mol; catalyzed by complex I), the oxidation of QH_2 by cytochrome c ($\Delta G^{\circ\prime} = -36.7$ kJ/mol; catalyzed by complex III), and the oxidation of reduced cytochrome c by O_2 ($\Delta G^{\circ\prime} = -112$ kJ/mol; catalyzed by cytochrome c oxidase). Each of these reactions was thus considered to be a "**coupling site**" for ATP synthesis; that is, each was considered to be a reaction in which ATP synthesis was driven directly by the energy released from that reaction.

Because we now know that coupling between oxidation and ATP synthesis is indirect, the concept of coupling sites is an oversimplification. Nevertheless, the concept was useful because it provided a framework for experiments identifying each of the above three reactions as individually capable of energizing the membrane for

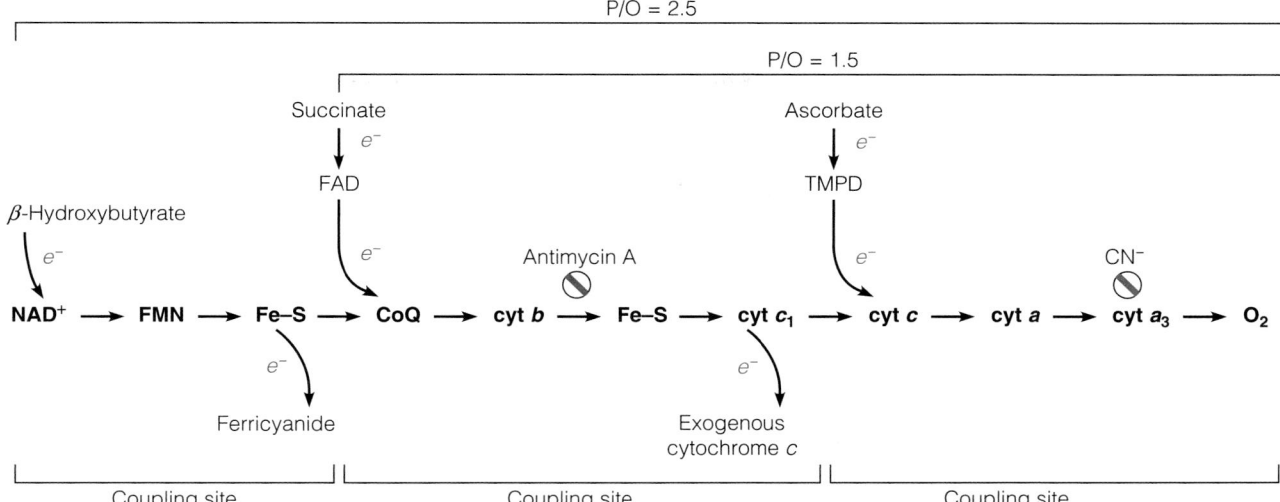

FIGURE 15.18

Experimental identification of "coupling sites." Electron transport is restricted to particular parts of the chain by use of selected electron donors, electron acceptors, and respiratory inhibitors, as indicated. For each segment of the chain that is thus isolated, P/O ratios are determined, allowing identification of coupling sites.

ATP synthesis, even when the entire electron transport chain is not operating. With that in mind, let us examine this evidence, which is summarized in Figure 15.18.

First, succinate is oxidized with a P/O ratio of ~1.5, not 2.5, suggesting the existence of a coupling site associated with NADH dehydrogenase (complex I). This was confirmed by blocking electron transport past cytochrome b with antimycin A. Ferricyanide was added as an artificial electron acceptor so that electrons could continue to flow. Under these conditions NAD^+-linked substrates, such as β hydroxybutyrate, were oxidized with a P/O ratio of ~1, confirming the existence of one site before cytochrome b.

Another approach involved an artificial electron donor, ascorbate. In the presence of another electron acceptor, tetramethyl-p-phenylene diamine (TMPD—see Figure 15.9), electrons could be supplied to the respiratory chain at cytochrome c. These electron carriers reduced cytochrome c nonenzymatically, and its oxidation via cytochrome oxidase proceeded with a P/O ratio of ~1, thus localizing one coupling site beyond cytochrome c. Finally, purified oxidized cytochrome c, when added to mitochondria, can act as an electron acceptor, withdrawing electrons from the electron transport chain. The further addition of a cytochrome oxidase inhibitor, such as cyanide, forces electrons to exit from the respiratory chain at cytochrome c. Under these conditions succinate is oxidized with a P/O ratio of ~1, which localizes a site between cytochromes b and c. However, if antimycin A is then added, no ATP is synthesized, showing that complex II (succinate dehydrogenase) is not a coupling site. In sum, these experiments demonstrate that complexes I, III, and IV are each capable of driving ATP synthesis, but that complex II is not (Figure 15.18). We now know that these three "coupling sites" indirectly drive ATP synthesis by pumping protons across the membrane as electrons flow through the complexes.

Mechanism of Oxidative Phosphorylation: Chemiosmotic Coupling

What is the actual mechanism by which energy released from electron transport through the respiratory chain is harnessed to drive the synthesis of ATP? For many years, researchers around the world searched for a high-energy intermediate that could link electron transport to ATP synthesis. However, such activated intermediates have never been demonstrated in oxidative phosphorylation.

Although other models have been considered, there is now widespread acceptance of a model involving **chemiosmotic coupling**, proposed in 1961 by British biochemist Peter Mitchell. Although this model was resisted at first, overwhelming evidence has now accumulated in its support, and Mitchell's achievements were recognized in 1978 with a Nobel Prize. In its most basic form, this model proposes that *the free energy from electron transport drives an active transport system, which pumps protons out of the*

FIGURE 15.19

Chemiosmotic coupling of electron transport and ATP synthesis. This depiction of protein complexes in the inner membrane shows the sequence of electron carriers from NAD^+ to O_2. Complex II is not included because it does not contribute to the proton gradient. Protons are pumped by complexes I, III, and IV as electrons flow through the complexes, generating an electrochemical gradient across the membrane (protonmotive force, pmf). Proton re-entry to the matrix, through the F_0 channel of ATP synthase (complex V), provides the energy to drive ATP synthesis.

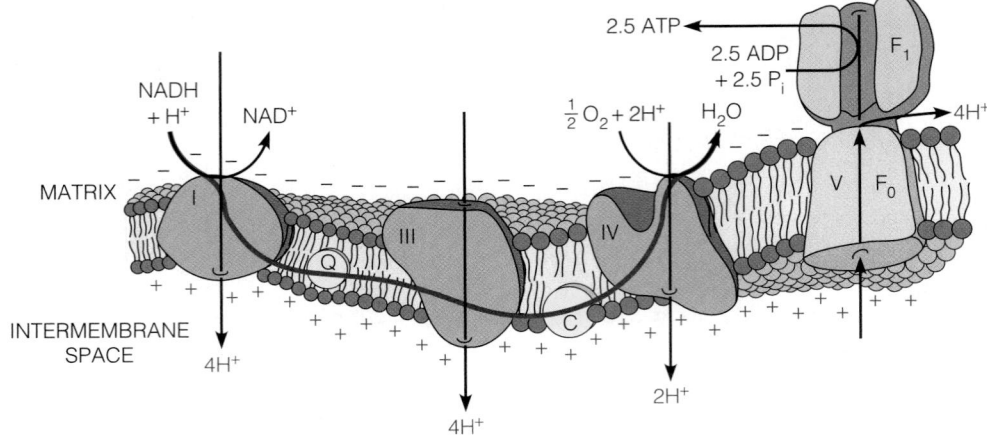

mitochondrial matrix into the intermembrane space. This action generates an electrochemical gradient for protons. The protons on the outside have a thermodynamic tendency to flow back in, down their electrochemical gradient, and this provides the driving force for ATP synthesis. Another way to state this is that free energy must be expended to maintain the proton gradient. When protons do flow back into the matrix, that energy is dissipated, some of it being harnessed to drive the synthesis of ATP.

To understand the chemiosmotic theory in more detail, recall that some, but not all, of the reactions of electron transport transfer hydrogen ions (protons) as well as electrons. These reactions include the dehydrogenations of NADH, $FMNH_2$, $FADH_2$, and reduced coenzyme Q. Mitchell proposed that the enzymes catalyzing these dehydrogenations are asymmetrically oriented in the inner membrane so that protons are always taken up from inside the matrix and released in the intermembrane space. Figure 15.19 shows how this process occurs. This **proton pumping** by respiratory proteins results in conversion of the energy of respiration to osmotic energy, in the form of an **electrochemical gradient**—or a gradient of chemical concentration that establishes an electrical potential (as discussed in detail in Chapter 10). The energy released from discharging this gradient can be coupled with the phosphorylation of ADP to ATP, with no isolatable "high energy" intermediates being formed. This process involves the ATP synthase (complex V). The F_0 portion of the complex spans the inner membrane and contains a specific channel for return of protons to the mitochondrial matrix. The free energy released as protons traverse this channel to return to the matrix is harnessed to drive the synthesis of ATP, catalyzed by the F_1 component of the complex. Before we describe the structure and function of ATP synthase, we will take a closer look at the evidence for the chemiosmotic coupling mechanism.

A Closer Look at Chemiosmotic Coupling: The Experimental Evidence

Let us now explore the experimental evidence for chemiosmotic coupling in some detail—partly because of its importance to an understanding of oxidative phosphorylation and partly because it provides insight into other biological processes, including active transport across membranes and photosynthesis. Recall from the introduction to this chapter that there were several nagging facts that argued against ATP synthesis by a direct chemical coupling process like that in glycolysis. The first was the realization that the number of ATP molecules produced per pair of electrons generates approximately 2.5 ATPs, not an integer as would be predicted by a direct chemical coupling mechanism. Secondly, oxidative phosphorylation was shown to require a membrane. If the membrane is disrupted, the passage of electrons down the respiratory chain becomes *uncoupled* from ATP synthesis: Electrons still flow, but ATP production ceases. These observations are nicely explained by the chemiosmotic coupling mechanism.

Chemiosmotic coupling refers to the use of a transmembrane proton gradient to drive endergonic processes like ATP synthesis.

Membranes Can Establish Proton Gradients

When it became possible to measure changes in pH and electrical potential across mitochondrial membranes, it became clear that mitochondria can pump protons from the matrix to the intermembrane space. In fact, the pH value outside an actively respiring mitochondrion is about 0.75 units lower than in the matrix. The pH gradient also generates an electrical potential of 150 to 200 millivolts (mV) across the membrane because of the net movement of positively charged protons outward across the inner membrane. The pH gradient and the membrane potential both contribute to an *electrochemical H⁺ gradient*, or **protonmotive force (pmf)**, although the electrical component is by far the major contributor. An electrical charge of 150 mV across a membrane may not seem like much (1/10 the voltage of a 1.5 V flashlight battery). However, as the British biochemist Nick Lane has pointed out, considering the thickness of a biological membrane ($\sim 5 \times 10^{-9}$ m), this corresponds to 30 million volts/m, similar to the voltage of a lightning bolt! We can calculate the free energy change of this electrochemical gradient using Equation 10.2 from Chapter 10:

$$\Delta G = RT \ln\left(\frac{C_2}{C_1}\right) + ZF \Delta\psi \qquad (15.4)$$

C_2 and C_1 are the concentrations of the ion in the two compartments, Z is the charge on the ion (+1 for H⁺), and $\Delta\psi$ is the membrane potential, in volts. ΔG in this equation is equivalent to $\Delta\mu_H$, the proton electrochemical gradient. For a proton electrochemical gradient, because pH is a logarithmic function of [H⁺], Equation 15.4 can be simplified to:

$$\Delta\mu_H = 2.3 \, RT \, \Delta\text{pH} + F\Delta\psi \qquad (15.5)$$

$\Delta\mu_H$ is also called Δp, the protonmotive force, or pmf. ΔpH has a positive value (+0.75) because it is defined as the pH in the matrix minus the pH in the intermembrane space (recall that pH $= -\log[\text{H}^+]$). Thus, the contribution of the pH gradient is $2.3RT(0.75) = +4.5$ kJ/mol at 37 °C (310 K). The membrane potential ($\Delta\psi$) across the inner membrane of an actively respiring mitochondrion is 0.15 to 0.20 V, so the contribution of the electrical component is +14.5 to +19.3 kJ/mol. Thus the total free energy change of transporting a proton from the matrix to the intermembrane space is on the order of +21 kJ per mole of protons. Because the formation of this protonmotive force is an endergonic process (positive ΔG), discharge of the gradient is an exergonic process: It is this free energy that is used to drive the phosphorylation of ADP. As summarized in Figure 15.19, approximately 10 protons are pumped per pair of electrons transferred from NADH to O_2. Thus, the protonmotive force conserves approximately 210 kJ of free energy for ATP synthesis. Recall that the free energy change for the oxidation of NADH by O_2 is -220 kJ/mol under standard conditions, most of which is conserved in the electrochemical proton gradient. If we make some reasonable assumptions regarding the concentrations of reactants and products in the matrix, we can estimate that in vivo ΔG for NADH oxidation is roughly -200 kJ/mol (page 630), and ΔG for the synthesis of ATP from ADP and P_i is roughly +50 kJ/mol. Thus, electron transport provides sufficient free energy to synthesize approximately 4 moles of ATP per mole of NADH oxidized, but only about 2.5 are synthesized, reflecting the indirect, evolved coupling stoichiometry of oxidative phosphorylation.

Comparable experiments have shown that electrochemical proton gradients are used in energy transactions other than oxidative phosphorylation. Bacterial membranes use proton pumping to transduce energy both for oxidative phosphorylation and for driving flagellar motors that allow movement of the cell. Bacteria also have many membrane solute transporters that use the protonmotive force to pump solutes into or out of their cells. Proton pumping across the chloroplast thylakoid membrane drives ATP synthesis in photophosphorylation (see Chapter 17). Proton gradients also drive active transport (see Chapter 10) and heat production (see below). The existence of chemiosmotic coupling in all three domains of life, and its myriad uses, suggests an ancient evolutionary history for this energy conserving mechanism.

2,4-Dinitrophenol (DNP)

Trifluorocarbonylcyanide phenylhydrazone (FCCP)

An Intact Inner Membrane Is Required for Oxidative Phosphorylation

When the physical continuity of the membrane is interrupted—for example, by sonication—the resultant particles can carry out electron transport but not ATP synthesis. The necessity of a structurally intact membrane for maintenance of a membrane potential is consistent with the idea that a proton gradient is essential for oxidative phosphorylation.

Key Electron Transport Proteins Span the Inner Membrane

If the respiratory proteins are to serve as proton pumps, then the electron carriers that pump protons should be in contact with both the inner and the outer sides of the membrane. Moreover, these carriers should be asymmetrically oriented in the membrane to account for proton transport in one direction—outward. Asymmetric orientation has been demonstrated by the use of agents that react with respiratory proteins but cannot themselves traverse the membrane, such as antibodies, proteolytic enzymes, or labeling reagents. Treatment of intact mitochondria with such reagents allows detection of proteins located at the outer surface of the inner membrane, whereas reaction with membrane vesicles allows access to the inner, or matrix, side. Such approaches have shown, for example, that the cytochrome oxidase complex (complex IV) binds to cytochrome *c* only on the intermembrane space side (see Figure 15.16). Moreover, 9 of the 13 subunits of the complex can be labeled from only one side or the other, indicating asymmetric placement of the complex in the membrane. Similar findings have now been made for the subunits of complexes I, II, and III.

Uncouplers Act by Dissipating the Proton Gradient

A class of compounds, exemplified by the lipophilic weak acids **2,4-dinitrophenol (DNP)** and **trifluorocarbonylcyanide phenylhydrazone (FCCP)** are called uncoupling agents, or **uncouplers**. Uncoupling agents, when added to mitochondria, permit electron transport along the respiratory chain to O_2 to occur in the absence of ATP synthesis. That is, they *uncouple* the process of electron transport from the process of ATP synthesis. The pK_a of the phenolic hydroxyl group in DNP is such that it is normally dissociated at intracellular pH. However, a DNP molecule that approaches the inner membrane from the outside becomes protonated because of the lower pH value in this vicinity. The protonated DNP diffuses into and through the membrane by mass action. Once inside the matrix, the higher pH causes the phenolic hydroxyl to deprotonate. The deprotonated dinitrophenolate ion is still lipophilic enough to pass back across the membrane, where it can bind another proton and repeat the cycle. Thus, the uncoupler has the effect of transporting H^+ into the matrix, bypassing the F_0 proton channel and thereby preventing ATP synthesis.

Extensive data on the transport of ions other than H^+ confirm that a functionally intact membrane is essential to oxidative phosphorylation. The antibiotic *valinomycin* (see page 387 in Chapter 10) is an example of an **ionophore** ("ion carrier"). This lipid-soluble compound forms a specific complex with potassium ion. Because the complex is lipophilic and can diffuse into the membrane, just as protonated DNP does, valinomycin brings about the transport of K^+ through the inner membrane in much the same sense that DNP transports protons. Valinomycin acts by decreasing the $\Delta\psi$ (membrane potential) component of the pmf, without a direct effect on the pH gradient. Another antibiotic, **nigericin**, acts as a K^+/H^+ **antiporter**; it carries H^+ in one direction, coupled with the reverse transport of K^+. Thus, nigericin dissipates the pH component of the pmf, with little effect on $\Delta\psi$. Neither compound alone is a particularly effective uncoupler of oxidative phosphorylation, but in combination they collapse both elements of the pmf, and ATP synthesis is effectively inhibited.

Generation of a Proton Gradient Permits ATP Synthesis Without Electron Transport

Andre Jagendorf, while studying photosynthetic ATP production, provided important evidence for chemiosmotic coupling in the chloroplast. As presented in Chapter 16, the chloroplast couples light energy to ATP synthesis. Jagendorf showed that ATP synthesis can proceed in the chloroplast in the absence of electron transport, so long as a proton gradient is present. Chloroplasts were incubated at pH 4 for several hours and then quickly transferred to a buffer at pH 8. Thus, like the situation in intact cells, the inside of the organelle was at a lower pH than the outside (chloroplast membranes pump protons *inward*, not outward). Addition of ADP and P_i to these chloroplasts generated a burst of ATP synthesis, simultaneous with dissipation of the pH gradient. Similar results have now been observed with mitochondria. These experiments show that the establishment of a proton gradient, even without a corresponding energy input, suffices to drive the synthesis of ATP.

A dramatic variation on Jagendorf's experiment involved a membrane protein, *bacteriorhodopsin* (Chapter 10, page 374), from the photosynthetic bacterium *Halobacterium halobium*. Bacteriorhodopsin pumps protons when the bacteria are supplied with light. Bacteriorhodopsin was isolated in a membrane-free form and then incorporated into synthetic vesicles, along with isolated liver mitochondrial F_0F_1 ATPase in the right orientation. When these vesicles were illuminated, ATP synthesis occurred, showing that phosphorylation can occur as a direct consequence of the formation of a proton gradient. This completely artifical reconstituted system, composed of bacterial and mammalian components, elegantly demonstrates the essence of chemiosmotic coupling.

Complex V: The Enzyme System for ATP Synthesis

That complexes I, III, and IV could individually drive ATP synthesis was confirmed in elegant reconstitution studies involving complex V, which catalyzes the actual synthesis of ATP from ADP. Let us review the discovery and nature of complex V and then discuss the reconstitution experiments.

When mitochondria are negatively stained with phosphotungstate, electron microscopy reveals that the cristae are covered with knoblike projections on the matrix side, each attached to the inner membrane by a short stalk (Figure 15.20). The knobs are known as F_1 **spheres**. Disruption of mitochondria by sonication generates fragments of inner membrane, which reseal in the form of closed vesicles. The membrane closes on itself inside out, so the knoblike projections are on the outside. These **submitochondrial particles** respire and synthesize ATP, just as intact mitochondria do. Efraim Racker and his colleagues showed that treatment of these vesicles with trypsin or urea caused the knobs to dissociate from the vesicles. After centrifugation to separate the "stripped" vesicles from the knobs, the vesicles could still oxidize substrates and reduce oxygen, but no ATP was synthesized. When knobs were added back to the vesicles, there was substantial reconstitution of particles that could then catalyze ATP synthesis as a result of the oxidation of exogenous substrates.

Because readdition of the knobs recoupled ATP synthesis to electron transport, the knobs were originally called **coupling factors**. Purification of the knobs attached to the underlying stalks revealed a large multiprotein aggregate consisting of more than a dozen polypeptide chains, as schematized in Figure 15.21. The entire structure, called the F_0F_1 **complex**, consists of the knob, the stalk to which it is attached, and an attached complex that is embedded in the inner membrane. The stalked knob is called F_1, and the base is called F_0. The F_1 complex consists of five proteins, designated $\alpha, \beta, \gamma, \delta$, and ε, with a subunit stoichiometry of $\alpha_3\beta_3\gamma\delta\varepsilon$. The F_0 complex consists of an oligomer of c subunits (10 copies in yeast mitochondrial ATP synthase; 8 copies in higher eukaryotes) plus one subunit a, and single copies of subunits b, d, F6 and OSCP (oligomycin sensitivity conferral protein) that form a peripheral stalk.

The entire F_0F_1 complex has an ATP *hydrolysis* activity in vitro, as does factor F_1 alone; this ATPase activity was assumed to represent the reverse of the true physiological reaction, namely, ATP synthesis. The OSCP subunit in the F_0 complex is the site of binding of the ATP synthesis inhibitor **oligomycin** (discussed further on page 655). For this reason, some workers call this structure F_o (the letter o stands for oligomycin), rather than F_0. The ATPase activity, sensitivity to oligomycin, and results of the reconstitution experiments confirm that the role of the F_0F_1 complex (also called complex V) is to synthesize ATP. The action of this

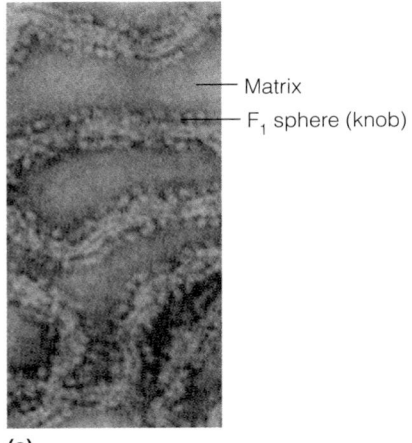

— Matrix
— F_1 sphere (knob)

(a)

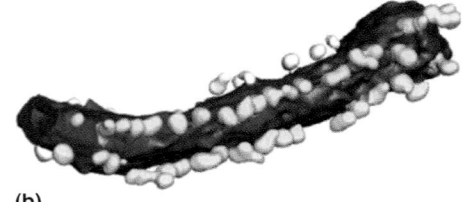

(b)

FIGURE 15.20

Fine structure of mitochondrial cristae. Preparations of the inner membrane show the F_1 spheres as "knobs" projecting from the cristae. **(a)** A negatively stained portion of bovine heart mitochondrial inner membrane, showing knoblike projections along the matrix side of the membrane. The knob is attached by a short stalk to a base, which is embedded in the inner membrane of intact mitochondria. **(b)** Surface rendering of tubular vesicles from rat liver mitochondria imaged by cryo-electron tomography. F_1 knobs are yellow, and the membrane is gray. The length of the tube is 280 nm. The F_1 knobs appear to exist in double rows, suggesting that ATP synthase is organized as dimers.

(a) Springer and Plenum/*Mitchondria*, 1982, A. Tzagoloff, with kind permission from Springer Science+Business Media B. V.; (b) Courtesy of K. M. Davies, M. Strauss, B. Daum, J. H. Kief, H. D. Osiewacz, A. Rycovska, V. Zickermann, and W. Küehlbrandt. Reprinted with permission from Werner Küehlbrandt.

Antibiotics such as valinomycin and nigericin interfere with maintenance of the protonmotive force.

Chloroplasts use light energy to drive protons inward, establishing a proton gradient for ATP synthesis.

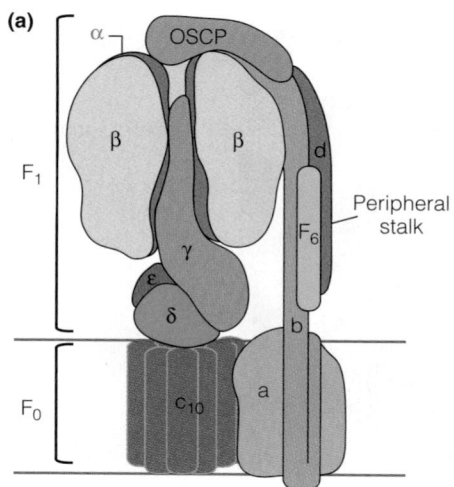

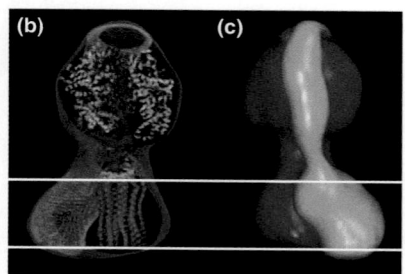

FIGURE 15.21

Structure of the F_0F_1 complex. (a) The F_0F_1 complex, also called ATP synthase or complex V, contains an F_1 knob projecting into the mitochondrial matrix and connected by a central stalk to the F_0 base. The globular F_1 knob contains three $\alpha\beta$ dimers, arranged about the central stalk, which is made up of γ, δ, and ε subunits. The F_0 base is composed of 10–12 c subunits (the c-ring) and one subunit a. The peripheral stalk (subunits b, d, F6, and OSCP) is attached to the F_0 base via subunit a and at least four other minor membrane-embedded subunits that are not shown. The central stalk and the c-ring compose the "rotor" of ATP synthase. The remainder of the subunits make up the "stator," a structure that prevents the rotation of the three $\alpha\beta$ dimers of F_1. This model is based upon the X-ray crystal structures of the yeast and bovine mitochondrial F_0F_1 complex. (b) The yeast mitochondrial F_1F_0 X-ray structure has been docked onto cyroelectron microscopy reconstructions of the bovine complex. The 10 c subunits of the c-ring are magenta, and the γ subunit and $\alpha\beta$ dimers are colored as in panel (a). The green density represents the peripheral stalk subunits. (c) The image in (b) is rotated to highlight the peripheral stalk and its function as a stator. The approximate location of the complex in the inner membrane is indicated by the horizontal lines in each panel.

Reprinted from *Biochimica et Biophysica Acta* 1757:286–296, J. E. Walker and V. K. Dickson, The peripheral stalk of the mitochondrial ATP synthase. © 2006, with permission from Elsevier.

amazing structure, a molecular "rotary engine," will be discussed after we describe its structure in more detail.

Continuing with his reconstitution experiments, Racker found that isolated submitochondrial respiratory complexes (I, II, and so on) could be reconstituted by sonication into artificial membranes (liposomes) containing purified phospholipids. When F_0F_1 complex was included in the sonication mixture, it was also incorporated into the vesicles. In this case each preparation had the electron transport properties of the original complex, plus a phosphorylation activity. For reconstituted complexes I, III, and IV, the P/O ratio was ~1 in each case. For example, reconstituted complex III could transfer electrons from coenzyme Q to cytochrome *c*, with concomitant synthesis of ~1 mole of ATP per pair of reducing equivalents. This type of evidence showed that respiratory complexes I, III, and IV each contain one coupling site. By contrast, complex II (succinate dehydrogenase) showed no ATP synthesis, confirming the absence of a coupling site in this complex.

FIGURE 15.22

Structure of the mitochondrial F_1 ATP synthase complex. (a) X-ray structure of the bovine mitochondrial F_1 complex (PDB ID 1BMF). α subunits (red) and β subunits (yellow) are arranged in a hexameric ring around the central γ subunit (blue), in the manner shown schematically in (d). The membrane would be at the bottom of the structure in this view. In (b), only one α subunit and one β subunit, from opposite sides of the hexameric ring, and the γ subunit are shown to illustrate their nearly identical tertiary structures. An ATP analog (AMP-PNP; green) can be seen bound to each subunit. (c) In this view from the matrix side of the membrane looking up, the arrangement of the α and β subunits as three $\alpha\beta$ dimers surrounding the central γ subunit is emphasized. Only the central nucleotide binding domain of each subunit is shown. The γ subunit is asymmetric, and makes unique contacts with each $\alpha\beta$ dimer. Although not obvious from this figure, each $\alpha\beta$ dimer has a slightly different conformation, especially in the central nucleotide-binding domain of each β subunit. The ATP analog AMP-PNP (green) can be seen bound in the "T" β subunit. ADP (purple) can be seen bound in the "L" β subunit and the "O" β subunit is empty.

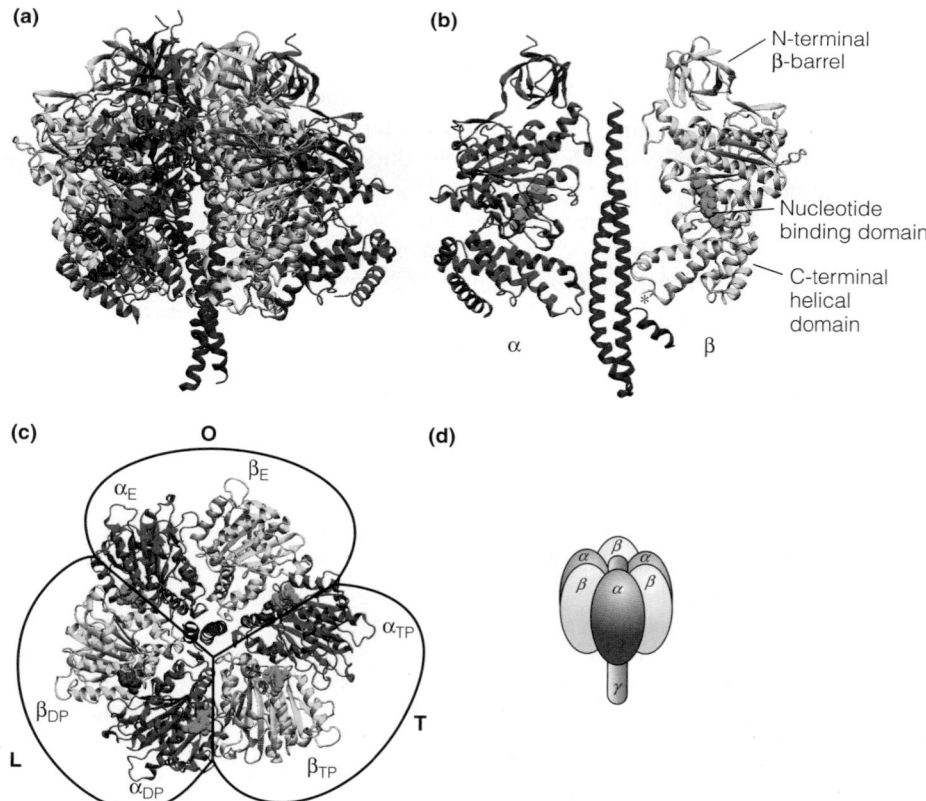

In 1994 the publication by John Walker's group of the crystal structure of the 371-kDa F_1 component of complex V helped to clarify the mechanism by which the passage of protons through the F_0F_1 complex drives the synthesis of ATP. As was suggested in Figure 15.21, the knob of the F_1 complex contains three identical $\alpha\beta$ dimers forming a flattened sphere 90–100 Å in diameter (Figure 15.22a). Each monomer can bind an adenine nucleotide at the interfaces between the α and β subunits, but the catalytic sites are located in the β subunits. The α and β subunits have nearly identical tertiary structures, composed of an N-terminal 6-stranded β barrel, a central domain containing the nucleotide-binding site, and a C-terminal bundle of helices (Figure 15.22b). An α-helical coiled-coil domain of the γ subunit of the central stalk penetrates the interior of the three $\alpha\beta$ dimers, where it interacts with the central and C-terminal domains of the α and β subunits. Careful analysis of the crystal structure showed important structural differences among the three dimers, especially in the central nucleotide-binding domain of each β subunit. Most significant, the nucleotide-binding site of one contained ADP, while another contained AMP-PNP (a nonhydrolyzable ATP analog added during crystallization), and the third was empty (Figure 15.22c). Thus, each $\alpha\beta$ dimer exists in one of three alternative conformations. What would cause otherwise identical $\alpha\beta$ dimers to adopt different conformations? They key is the interaction of the γ subunit in the interior of the three $\alpha\beta$ dimers. Because the γ subunit is itself asymmetric, it makes unique contacts with each $\alpha\beta$ dimer. Consequently, each $\alpha\beta$ dimer assumes a different conformation.

Kinetic experiments with purified F_1 component have revealed that the equilibrium constant for formation of ATP from ADP $+$ P_i *on the enzyme* is close to 1. Recall that the apparent K'_{eq} for ATP hydrolysis free in solution is approximately 10^5 M (Chapter 12). Is ATP synthase violating the first law of thermodynamics, making ATP with no energy input? The key is that the enzyme is not making *free* ATP, but, rather, *bound* ATP. In fact, ATP synthase binds ATP with much, much higher affinity than ADP (~7 orders of magnitude difference). This amounts to about 40 kJ/mol of binding energy, which drives the equilibrium toward ATP on the enzyme. ATP is so tightly bound that considerable energy must be used to release it from the enzyme. The protonmotive force lowers the binding affinity of ATP by a million-fold or more, allowing its dissociation from the enzyme. The energy-dependent step is thus not ATP *synthesis* but rather its *release* from a tight-binding site. This release is brought about by the energy-dependent rotation of γ, which drives conformational changes in the $\alpha\beta$ assemblies.

All of this supported a mechanism proposed by Paul Boyer some years earlier, in which rotation of the γ subunit causes sequential conformational changes in the $\alpha\beta$ dimer assemblies. In this model, called the binding-change model, the F_1 motor functions essentially as a three-cylinder engine. As shown in Figure 15.23, the three conformations of the nucleotide-binding site are termed loose (L), tight (T), and open (O). In step 1 of Figure 15.23, rotation of γ by 120°, fueled by proton passage through channels in F_0, opens site T, leading to ATP release, and an open O site changes to an L site, binding ADP and P_i. This same rotation causes the third site to change from an L conformation, with loosely bound ADP and P_i, to a T conformation, where the substrates are tightly bound, leading to ATP formation in step 2. Steps 3 and 4, and 5 and 6, are just repeats of steps 1 and 2, releasing two additional ATP molecules, except that the three conformations have migrated around the three $\alpha\beta$ dimers. There is now good evidence for additional conformational intermediates. For example, P_i binds to the O site before ADP. This prevents the unwanted binding of ATP instead of P_i + ADP and provides an explanation for how the enzyme is able to make ATP under cellular conditions where the ATP concentration is 10 to 50 times that of ADP. The structure determination of the F_0F_1 complex and the mechanism of ATP synthesis suggested by that structure led to the recognition of Walker and Boyer as recipients of the 1997 Nobel Prize in Chemistry.

Is F_1 really a rotating engine, as implied by the model of Figure 15.23? Several experimental approaches indicate that it is. The most graphic evidence comes from

The F_0F_1 complex contains a proton channel and the enzyme that synthesizes ATP.

AMP-PNP (5'-Adenylyl imidodiphosphate)

FIGURE 15.23

Binding change model for ATP synthase. The nucleotide binding site (catalytic site) of the three $\alpha\beta$ dimers exists in three different conformations, termed loose (L), tight (T), and open (O). In this scheme, the γ subunit rotates counterclockwise, driven by the passage of protons through channels in F_0, while the $\alpha\beta$ dimer assemblies are held stationary by a stator. Each $\alpha\beta$ dimer is distinguished by a different color, with the catalytic sites shown at the interface of α and β subunits. Step 1 represents 120° rotation of the γ subunit, leading to a conformational change in all three $\alpha\beta$ dimers, such that the T site changes to an O site, leading to ATP release, and an O site changes to an L site, binding ADP and P_i. The third site changes from an L conformation, with loosely bound ADP and P_i, to a T conformation, where the substrates are tightly bound, leading to ATP formation in step 2. Steps 3 and 4, and steps 5 and 6, are just repeats of steps 1 and 2, except that the three conformations have migrated around the three $\alpha\beta$ dimers.

Modified from *Bioenergetics*, 3rd ed., D. G. Nicholls and S. J. Ferguson, p. 209. © 2002, with permission from Elsevier.

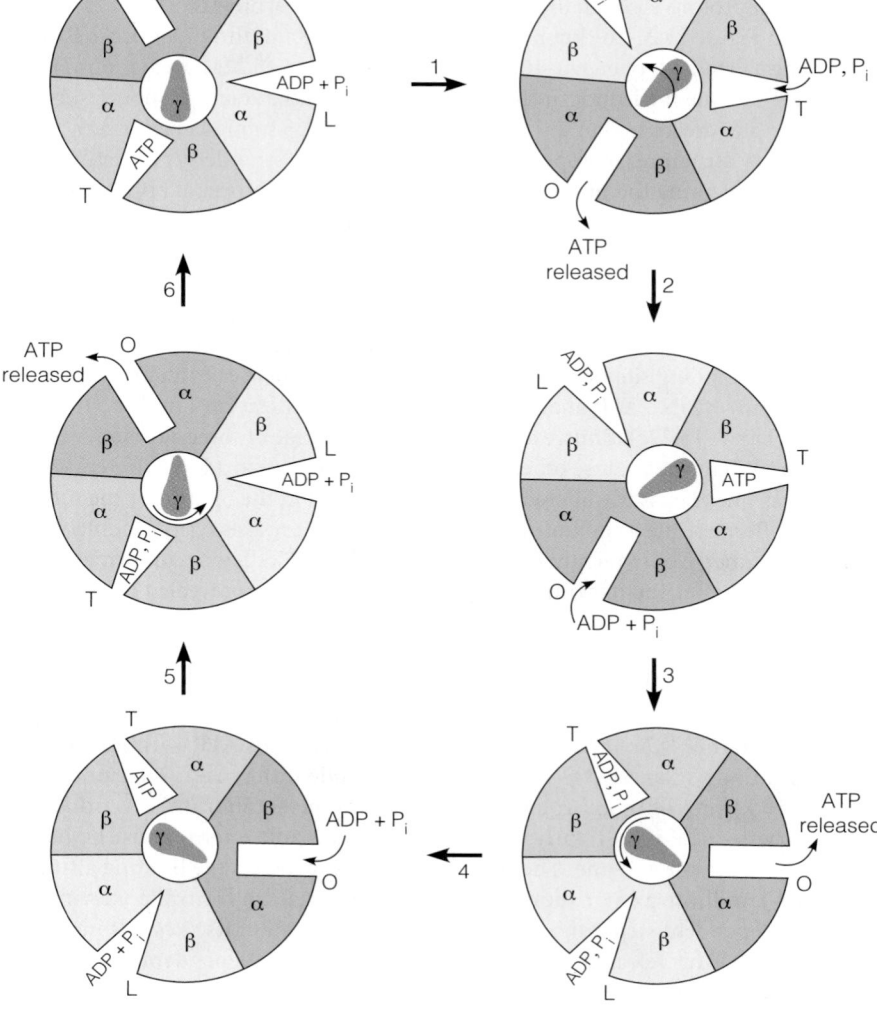

F_1 ATP synthase functions as a three-cylinder rotary engine driven by driven by the passage of protons through channels in F_0.

experiments in which the β subunits were immobilized to a glass coverslip and a fluorescent actin filament was attached to the γ protein (see Figure 15.24). α subunits were added, and F_1 complexes were allowed to assemble on the glass slide. When this structure was examined by fluorescence microscopy, the addition of ATP could be seen to stimulate the rotation of the fluorescent probe. Analysis showed the rotation to be counterclockwise (when viewed from the membrane) and to occur in discrete steps of 120°, just as predicted from the model. Rotation in a fully coupled F_1F_0 ATP synthase has now been observed in both synthesis and hydrolysis modes. The rotor turns counterclockwise in the ATP hydrolysis direction and clockwise in the ATP synthesis direction, when viewed from the membrane, and performs up to 700 revolutions per second (~100 revolutions per second in vivo).

How do protons flowing through the F_0 membrane component drive rotation of the γ subunit? As illustrated in Figure 15.21, the γ subunit is attached to the oligomer of c subunits (the c-ring) via the δ subunit. The a subunit is stationary in the membrane, anchoring the peripheral stalk stator subunits. The mechanism requires that the c-ring rotates relative to the stationary a subunit as protons flow from the intermembrane space through the complex into the matrix. Biochemical and crystallographic studies have identified putative access channels for proton movement through the a and c subunits of F_0 (Figure 15.25). Several charged residues lying in the center of the membrane are key players in the proton-transport mechanism. It was found that upon protonation of a specific acidic residue

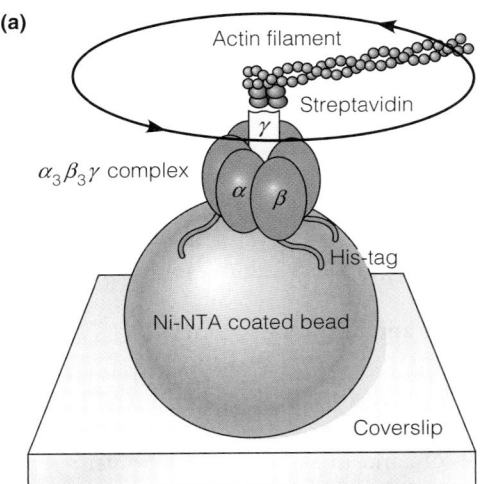

(a)
Actin filament
Streptavidin
γ
$\alpha_3\beta_3\gamma$ complex
α β
His-tag
Ni-NTA coated bead
Coverslip

FIGURE 15.24

The experimental system that permits observation of rotation in the F_1 component of F_0F_1 ATP synthase. **(a)** The cloned gene encoding the F_1 β subunit was modified by adding a sequence that encodes an oligohistidine sequence that binds to a nickel-coated bead (Ni-NTA coated bead) in the conformation shown. Thus, the F_1 complex was immobilized on the bead and attached to a glass coverslip. Streptavidin is a protein used to couple fluorescent-tagged actin to the γ subunit. **(b)** Fluorescence microscopic examination showed that, following addition of ATP and its subsequent hydrolysis by the $\alpha\beta$ catalytic subunits, the actin molecule was rotating, which proved that the γ subunit itself was rotating. The time interval between images is 133 ms.

(a) Reprinted from *Cell* 93:1117–1124, R. Yasuda, H. Noji, K. Kinosita Jr., and Y. Masasuke, F1-ATPase is a highly efficient molecular motor that rotates with discrete 120° steps. © 1998, with permission from Elsevier; (b) *Journal of Biological Chemistry* 276:1665–1668, H. Noji and M. Yoshida, The rotary machine in the cell, ATP synthase. © 2001 The American Society for Biochemistry and Molecular Biology. All rights reserved.

(b)

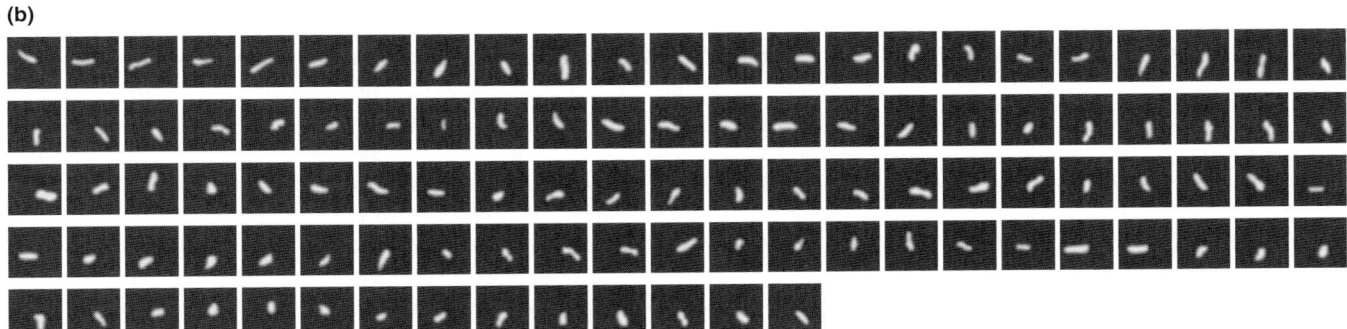

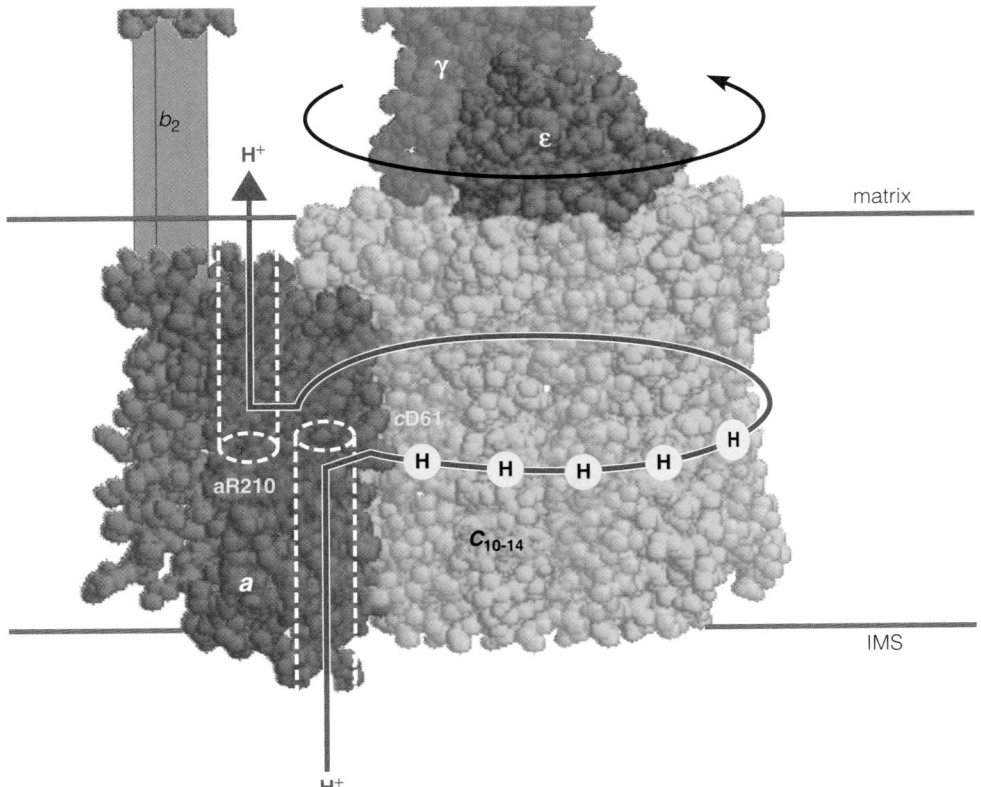

b_2
H+
matrix
γ
ε
cD61
aR210
C_{10-14}
a
IMS
H+

FIGURE 15.25

Proton-driven rotation of the c-ring of the F_0 component of F_0F_1 ATP synthase. This model is based on the X-ray structure of the *E. coli* ATP synthase. The membrane-spanning α-helical *c* subunits of the c-ring (10 in *E. coli*) are indicated in alternating gray and light blue. The *a* subunit (brown) and *b* subunit (magenta) are stationary in the membrane. The γ subunit (dark blue) and ε subunit (green) are attached to the matrix side of the c-ring. Proposed proton channels in and out of the *a* subunit, and proton-driven rotation of the c-ring, is indicated by the red arrow. Critical residues involved in protonation/deprotonation reactions are shown (Arg210 on the *a* subunit and Asp61 on the *c* subunits). In mitochondrial ATP synthase, the proton-binding amino acid on the *c* subunits is a glutamate.

Reprinted by permission of Federation of the European Biochemical Societies from *FEBS Letters* 545:61–70, J. Weber and A. E. Senior, ATP synthesis is driven by proton transport in F_1F_0-ATP synthase. © 2003.

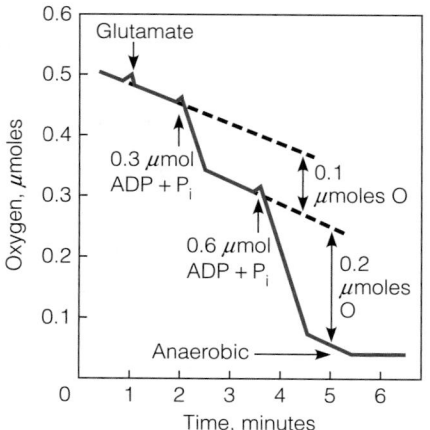

FIGURE 15.26

Experimental demonstration of respiratory control. Oxygen uptake is monitored in carefully prepared coupled mitochondria. The addition of an exogenous oxidizable substrate (glutamate) stimulates respiration only slightly, unless ADP + P_i is added as well. Both ADP additions represent limiting amounts; the second addition is twice the amount of the first, to show that the magnitude of oxygen uptake is stoichiometric. The slow oxygen uptake at the beginning results from endogenous substrates in the mitochondria. ADP stimulates respiration only until all of the ADP + P_i has been converted to ATP. Oxygen uptake is recorded in μmoles O, because one pair of electrons reduces one atom of O, not one molecule of O_2.

FIGURE 15.27

Effects of an inhibitor and an uncoupler on oxygen uptake and ATP synthesis. The plot shows the results of an experiment in which an inhibitor and an uncoupler of oxidative phosphorylation were added to a mixture of isolated mitochondria, an oxidizable substrate (glutamate), and excess ADP + P_i. The red trace shows oxygen uptake; the blue trace represents ATP synthesis. The addition of oligomycin inhibits phosphorylation and consequently slows respiration. Dinitrophenol (DNP) uncouples respiration from phosphorylation so that O_2 uptake is stimulated even in the presence of oligomycin, but ATP synthesis remains blocked.

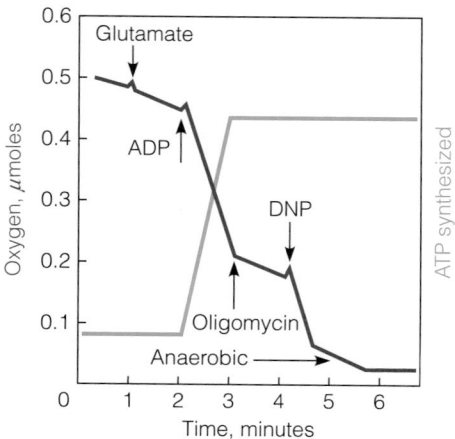

(glutamate in mitochondrial ATP synthase; aspartate in the bacterial enzymes), each c subunit undergoes a large rotation of its C-terminal helix, moving the protonated carboxylate to a more hydrophobic environment and generating rotation of the c-ring. As successive c subunits bind a proton, each protonated carboxylate interacts with subunit a, causing it to reionize, releasing the proton into the mitochondrial matrix. Thus, protonation/deprotonation reactions cause angular displacement of c versus a, resulting in net proton-driven c-ring rotation. Each 360° rotation produces three ATP molecules by F_1 and requires the translocation of one proton per each c subunit in the ring of F_0.

Respiratory States and Respiratory Control

Like any metabolic process, oxidative phosphorylation can occur only in the presence of adequate quantities of its substrates. It is controlled not by allosteric mechanisms but by substrate availability and thermodynamics. Those substrates include ADP, P_i, O_2, and an oxidizable metabolite that can generate reduced electron carriers—NADH and/or $FADH_2$. Under different metabolic conditions any one of these four substrates can limit the rate of oxidative phosphorylation.

The dependence of oxidative phosphorylation on ADP reveals an important general feature of this process: *Respiration is tightly coupled to the synthesis of ATP.* Not only is ATP synthesis absolutely dependent on continued electron flow from substrates to oxygen, but electron flow in normal mitochondria occurs only when ATP is being synthesized as well. This regulatory phenomenon, called **respiratory control**, makes biological sense because it ensures that substrates will not be oxidized wastefully. Their utilization is controlled by the physiological need for ATP.

In most aerobic cells the level of ATP exceeds that of ADP by 4- to 10-fold. Thus, it is convenient to think of respiratory control as a dependence of respiration on ADP as a substrate for phosphorylation. If the energy demands on a cell cause ATP to be consumed at high rates, the resultant accumulation of ADP will stimulate respiration, with concomitant activation of ATP resynthesis. Conversely, in a relaxed and well-nourished cell, ATP accumulates at the expense of ADP, and the depletion of ADP limits the rate of both electron transport and its own phosphorylation to ATP. Thus, the energy-generating capacity of the cell is closely attuned to its energy demands.

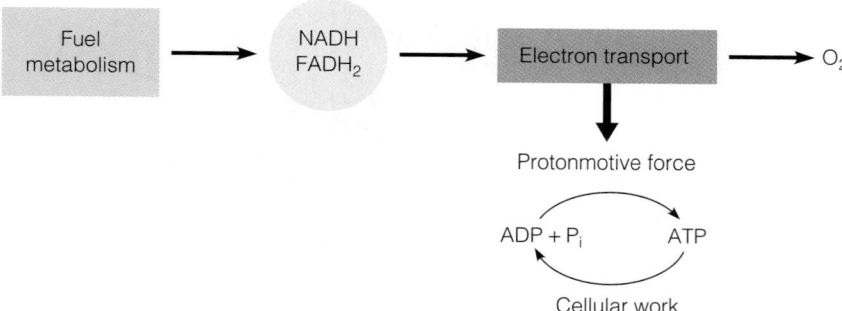

Experimentally, respiratory control is demonstrated by following oxygen utilization in isolated mitochondria (Figure 15.26). In the absence of added substrate or ADP, oxygen uptake, caused by oxidation of endogenous substrates, is slow. Addition of an oxidizable substrate, such as glutamate or malate, has but a small effect on the respiration rate. If ADP is then added, however, oxygen uptake proceeds at an enhanced rate until all of the added ADP has been converted to ATP, and then oxygen uptake returns to the basal rate. This stimulation of respiration is stoichiometric; that is, addition of twice as much ADP causes twice the amount of oxygen uptake at the enhanced rate. If excess ADP is present instead of oxidizable substrate, the addition of substrate in limiting amounts will stimulate oxygen uptake until the substrate is exhausted.

Maintenance of respiratory control depends on the structural integrity of the mitochondrion. Disruption of the organelle causes electron transport to become uncoupled from ATP synthesis. Under these conditions, oxygen uptake proceeds at high rates even in the absence of added ADP. ATP synthesis is inhibited, even though electrons are being passed along the respiratory chain and used to reduce O_2 to water. Before carrying out experiments with freshly isolated mitochondria, biochemists usually ascertain that their mitochondria are tightly **coupled**, by determining rates of oxygen uptake in the presence and absence of added ADP. In carefully prepared mitochondria the ratio of these two O_2 uptake rates may be as high as 10. By contrast, aged or disrupted mitochondria may yield ratios as low as 1, showing an absence of coupling. Uncoupling of respiration from phosphorylation can also be achieved chemically. As noted earlier, uncouplers such as DNP or FCCP act by dissipating the proton gradient. Addition of an uncoupler to mitochondria stimulates oxygen utilization even in the absence of added ADP (Figure 15.27).

Another group of compounds, exemplified by the antibiotic oligomycin, act as inhibitors of oxidative phosphorylation. Addition of oligomycin to actively respiring, well-coupled mitochondria inhibits both oxygen uptake and ATP synthesis, as shown in Figure 15.27. However, no direct inhibition of electron transport occurs, as shown by the fact that subsequent addition of an uncoupler such as DNP greatly stimulates oxygen uptake.

Figures 15.26 and 15.27 show that blocking ATP synthesis either by depletion of substrate (ADP) or inhibition by oligomycin simultaneously blocks electron transport in coupled mitochondria. It is easy to visualize how inhibition of electron transport inhibits ATP synthesis because there is no proton gradient being formed. But how does inhibition of ATP synthesis block electron transport? Oligomycin inhibits ADP phosphorylation by binding to specific sites in the F_0 complex, blocking the flow of protons through the F_0 proton channel and thus blocking rotation of the c-ring. Likewise, depletion of ADP inhibits ATP synthesis by preventing the $\alpha\beta$ dimers from going through their catalytic cycles and the associated conformational changes. If these conformational changes are prevented, neither the γ subunit, nor the c-ring it is attached to, can rotate. Without rotation of these components, proton flow through F_0 is blocked. But why would blocking proton flow through the F_0 channel inhibit electron transport through the respiratory chain?

The answer is that these processes are *mechanically* coupled. Consider a simple hand-operated pump, like the ones our ancestors used to draw water from a well. You used your muscles to pump the handle up and down to pump water up out of the ground and out the faucet. Now what would happen if we attached a fire hose to the faucet and we used the diesel motor on the fire truck to pump water through the pump and down into the ground? What would happen to the pump handle? It would move (rapidly!) up and down because water flow (in either direction) is coupled to the mechanical movement of the pump handle. (Like ATP synthase, this pump is reversible—as mentioned earlier, ATP synthase was originally discovered as an ATPase.) What if we now held onto the pump handle, preventing it from moving? What would happen to the flow of water from the fire truck through the pump? It would stop, unless its pressure was greater than the force holding the handle (or the hose bursts!).

Using an inhibitor of ATP synthase (e.g., oligomycin) is like holding the pump handle. If we block the ATP synthase from going through its catalytic cycle, there can be no rotation of the c subunits in the F_0 complex in the membrane, and thus no flow of protons through the F_0 due to the mechanical coupling of these two processes. If there is no flow of H^+ from outside back into the matrix to dissipate the protonmotive force, the proton pressure ($\Delta\psi$) builds up on the outside of the membrane to the point where pumping protons out of the matrix is no longer favorable. That is, the thermodynamically favorable electron transport through these complexes cannot provide sufficient free energy to overcome the high opposing protonmotive force (i.e., $\Delta G_{e^-\ \text{transport}} + \Delta G_{H^+\ \text{pumping}} = 0$). So

Oligomycin

Dean Appling

The rate of electron transport is limited by the availability of ADP for conversion to ATP.

FIGURE 15.28

Reversibility of F_1F_0 ATP synthase. (a) In normally respiring mitochondria, the pmf is high and ATP synthase operates in the direction of ATP synthesis. The ATP is exchanged for cytoplasmic ADP via the adenine nucleotide translocase (ant). **(b)** During hypoxia, the pmf falls, and the cell relies mainly on homolactic (or alcoholic) fermentation for ATP production. This ATP enters the matrix in exchange for ADP. F_1F_0 ATP synthase operates as a proton-translocating ATPase, pumping protons out of the matrix, temporarily sustaining the pmf to support other processes, such as metabolite transport.

Reprinted from *Trends in Biochemical Sciences* 34: 343–350, M. Campanella, N. Parker, C. H. Tan, A. M. Hall, and M. R. Duchen, IF1: Setting the pace of the F1Fo-ATP synthase. © 2009, with permission from Elsevier.

electron transport stops because it too is *mechanically* coupled to proton pumping through complexes I, III, and IV. Thus, the fundamental factor controlling the rate of respiration is the balance between ΔG for the phosphorylation of ADP (the ATP synthase reaction), ΔG for electron transport through the respiratory complexes, and ΔG for H^+ pumping (defined by the protonmotive force). If the free energy required to synthesize ATP is exactly balanced by the free energy available from the proton gradient, there will be no proton flow and no ATP synthesis (i.e., $\Delta G_{\text{ATP synthesis}} + \Delta G_{H^+ \text{re-entry to matrix}} = 0$). Now let's consider what happens when the cell begins to consume ATP to drive biosynthetic processes. As ATP is consumed, levels of ATP in the matrix will fall, thereby making ΔG for phosphorylation of ADP more favorable (i.e., $\Delta G_{\text{ATP synthesis}} + \Delta G_{H^+ \text{re-entry to matrix}} < 0$). ATP synthesis will then proceed, as will the coupled flow of protons through F_0. This increased proton re-entry will reduce the thermodynamic barrier to electron transport (i.e., $\Delta G_{e^- \text{transport}} + \Delta G_{H^+ \text{pumping}} < 0$). This in turn leads to increased electron transport, and hence, respiration rate.

The reversibility of the ATP synthase poses a potential problem for the cell. In normally respiring mitochondria, the pmf is high, while the exchange of matrix ATP for ADP ensures that the matrix ATP/ADP ratio is held relatively low, favoring ATP synthesis (Figure 15.28a). However, when mitochondrial respiration is disturbed—for example, by hypoxia, or by proton leak across the inner membrane—the energetics can reverse. If the pmf falls below a threshold, and cytosolic ATP is available (e.g., from upregulated glycolysis), the adenine nucleotide translocase will import ATP into the matrix, increasing the ATP/ADP ratio. These conditions will favor ATP *hydrolysis*, and the ATP synthase can reverse, hydrolyzing ATP to pump protons out of the mitochondrial matrix (Figure 15.28b). If the respiratory disturbance is mild, or temporary, this reverse mode can sustain the pmf to support other processes, such as metabolite transport (see below) that require a proton gradient. However, if the loss of membrane potential is extreme, this reversal of ATP synthase will eventually deplete all cellular ATP, leading to cell death.

To prevent this potentially catastrophic outcome, operation of the F_1F_0 ATP synthase as a proton-translocating ATPase is limited by an endogenous mitochondrial F_1F_0-ATPase inhibitor protein, IF_1 (inhibitor of F_1). IF_1 is a small (84 amino acids), highly conserved protein that inhibits the F_1F_0-ATPase activity by binding directly to the α and β subunits of the F_1 component. IF_1's inhibitory activity appears to be regulated by pH. During hypoxia, the cell relies mainly on homolactic fermentation for ATP production. The resulting lactic acid lowers the pH in the cytosol and mitochondria, activating the inhibitory function of IF_1. Thus, under conditions of mild hypoxia, inhibition of F_1F_0-ATPase by IF_1 preserves ATP at the expense of the membrane potential.

Under some natural conditions the ability to uncouple respiration from phosphorylation is highly desirable. Many mammals, particularly those that are born

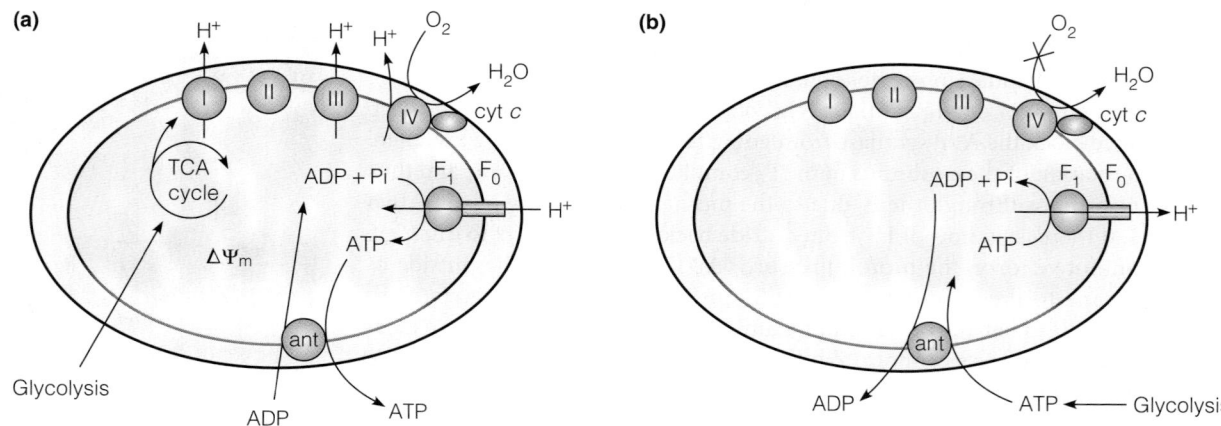

hairless (including human infants), those that hibernate, and those that are cold-adapted, have special needs for maintenance of core body temperature. Such animals have a special tissue, called brown adipose tissue (BAT), in the neck and upper back. Mitochondria in this tissue are especially rich in respiratory electron carriers, particularly cytochromes, which give BAT its brown color. These mitochondria are specialized to generate heat from fat oxidation, metabolizing the acetyl-CoA in the citric acid cycle. The inner membranes of BAT mitochondria are rich in **uncoupling protein 1 (UCP1)**, a 33-kDa channel that allows protons to return to the matrix in a process that bypasses ATP synthase, thereby uncoupling electron transport from ATP synthesis. Thus, the energy that is derived from fat oxidation is dissipated as heat. Mammals express at least five different uncoupling proteins (UCP1–UCP5). Whereas the function and regulation of UCP1 in response to cold exposure by the sympathetic nervous system is well-understood, the physiological function of the other UCPs in other tissues, and the role of BAT in adults, remains an area of intense investigation.

A comparable phenomenon is seen in the plant world among species that emerge in early spring, often when the ground is still covered with snow. The floral spike of the skunk cabbage is a particularly dramatic example; this tissue can maintain a temperature some 10° to 25° above ambient temperature, by uncoupling oxidation from phosphorylation.

Mitochondrial Transport Systems

Whereas the mitochondrial outer membrane is freely permeable to molecules up to ~5000 Da, the permeability of the inner membrane is severely limited. The importance of this selective permeability can be seen from our discussions of electrochemical gradients and the shuttle systems used to transport reducing equivalents into the mitochondrion. We must also consider substrate transport, including the inward transport of intermediates for oxidation in the citric acid cycle, the export of intermediates used for biosynthesis in other cell compartments, and the exit of newly synthesized ATP. Properties of the principal mitochondrial transport systems are outlined in Figure 15.29. Because the matrix is negatively charged relative to the cytoplasm, it is energetically unfavorable for a negatively charged solute to enter the matrix. Consequently, mitochondrial carriers use cotransport with protons, or exchange with OH^-, to import these metabolites. (Note: Additional systems transport fatty acids and amino acids into mitochondria for subsequent oxidation.) Most of these transporters are members of the mitochondrial carrier family (MCF).

First let us consider ATP, ADP, and P_i, the participants in oxidative phosphorylation. Two systems are involved, an **adenine nucleotide translocase (ADP/ATP carrier, or ANT)** and a **phosphate translocase**. The ADP/ATP carrier spans the inner membrane, and it binds ADP to a specific site on the outer surface of the inner membrane. The protein couples the efflux of free ATP from the matrix to the influx of an equivalent amount of free ADP from the intermembrane space (Mg-ADP and Mg-ATP are not substrates of the carrier). Because this antiporter exchanges ATP, with a charge of -4, for ADP, with a charge of -3, its action is driven by the membrane potential (outside positive). It is generally true of the mitochondrial transport systems that at least one of the participants is moving down a concentration gradient, or is coupled to the proton gradient so that no further energy source is required.

The phosphate translocase acts in either the *antiport* or the *symport* mode (see Chapter 10). As an antiporter, it transports $H_2PO_4^-$ into the matrix, coupled with the efflux of a hydroxide ion. In the alternative symport mode, it transports HPO_4^{2-} into the matrix, along with two protons. Both modes of transport maintain electrical neutrality and are driven by the ΔpH component of the protonmotive force generated by electron transport. The net effect of the adenine nucleotide and phosphate transport systems is to couple the inward transport of the substrates of oxidative phosphorylation, ADP and P_i, to the efflux of the product, ATP.

FIGURE 15.29

Major inner membrane transport systems for respiratory substrates and products. The ADP/ATP carrier and the phosphate translocase move substrates for oxidative phosphorylation (ADP and P_i) into the mitochondrion and the product (ATP) out. Other transport systems move substrates and products for citric acid cycle oxidation into or out of the matrix, as dictated by the metabolic needs of the cell.

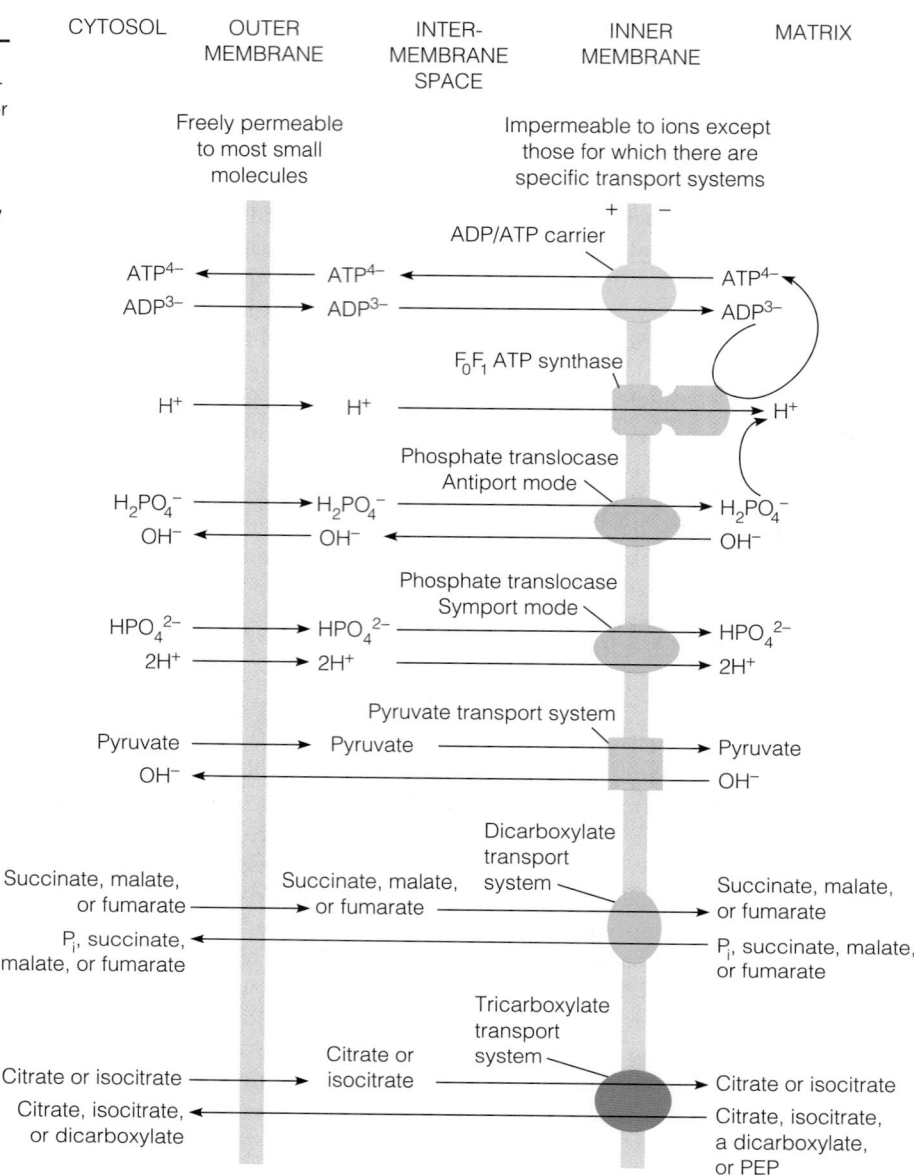

As might be expected by the critical role the ADP/ATP and phosphate carriers play in providing substrates for the ATP synthase, their deficiency in humans is usually characterized by exercise intolerance, muscular hypotonia, hypertrophic cardiomyopathy, elevated plasma lactate levels, and lactic acidosis.

Next, let us consider the substrates for oxidation. The major substrate from carbohydrate catabolism is pyruvate, which, like phosphate, is exchanged for OH^-. Dicarboxylic acid substrates—namely succinate, fumarate, and malate—can be exchanged for each other or for orthophosphate in the dicarboxylate transport system. Similarly, the tricarboxylate transport system carries either citrate or isocitrate, coupled each with the other or with a dicarboxylic acid or phosphoenolpyruvate. The influx of fatty acids for β-oxidation involves an additional transport system, which is presented in Chapter 17.

In addition to the transport systems shown in Figure 15.29, there is an important transport system for calcium ion. Ca^{2+} regulates a number of metabolic processes in the cytosol, and its release from mitochondrial stores may represent one way to change the cytosolic concentration of Ca^{2+} (see Chapter 23). Calcium is transported inward by a uniporter driven by the membrane potential, which is more negative inside. It is effluxed by exchange with Na^+ from the outside.

Shuttling Cytoplasmic Reducing Equivalents into Mitochondria

Another important role of mitochondrial carriers is to shuttle cytoplasmic reducing equivalents into the matrix for subsequent reoxidation by the respiratory chain. Recall that during aerobic glycolysis the NADH produced at the glyceraldehyde-3-phosphate dehydrogenase step is not reoxidized because pyruvate can be further oxidized in the citric acid cycle. To extract the energy from this NADH, and to regenerate oxidized NAD$^+$ for continued glycolysis, the reducing equivalents must be transferred into the mitochondrion. However, the NADH generated by a cytosolic NAD-linked dehydrogenase cannot itself traverse the mitochondrial membrane to be oxidized by the respiratory chain. Therefore, the reducing equivalents must be *shuttled* to respiratory assemblies in the inner mitochondrial membrane, without physical movement of the coenzyme. This process involves the reduction of a substrate by NADH in the cytoplasm, passage of the reduced substrate into the mitochondrial matrix via a specific transport system, reoxidation of that compound inside the matrix, and passage of the oxidized substrate back to the cytoplasm, where it can undergo the same cycle again.

The earliest known shuttle system is the dihydroxyacetone phosphate/glycerol-3-phosphate shuttle, which is particularly active in brain and skeletal muscle (and in the flight muscle of insects). As shown in Figure 15.30a, dihydroxyacetone phosphate (DHAP) is reduced by NADH in the cytosol, followed by reoxidation of the resultant glycerol-3-phosphate by a flavin-dependent glycerol-3-phosphate dehydrogenase, bound at the outer face of the inner mitochondrial membrane. This process involves reduction of FAD, followed by transfer of an electron pair from FADH$_2$ to coenzyme Q, just as intramitochondrial NADH transfers electrons to Q (see Figure 15.13). Once dihydroxyacetone phosphate has returned to the cytosol, the net effect has been to transfer two reducing equivalents from cytosolic NADH to mitochondrial FADH$_2$ and from there down the respiratory chain.

Electrons are transported into mitochondria by metabolic shuttles.

FIGURE 15.30

Shuttles for transfer of reducing equivalents from cytosol into mitochondria. **(a)** The dihydroxyacetone phosphate/glycerol-3-phosphate shuttle. **(b)** The malate/aspartate shuttle. Red arrows indicate flow of reducing equivalents. Glu = glutamate; Asp = aspartate; αKG = α-ketoglutarate.

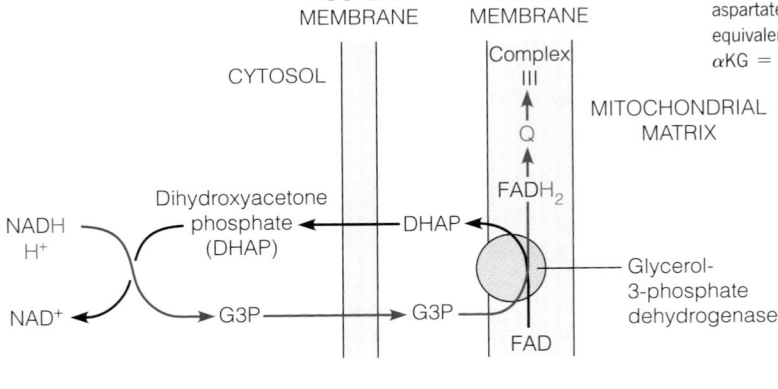

(a)

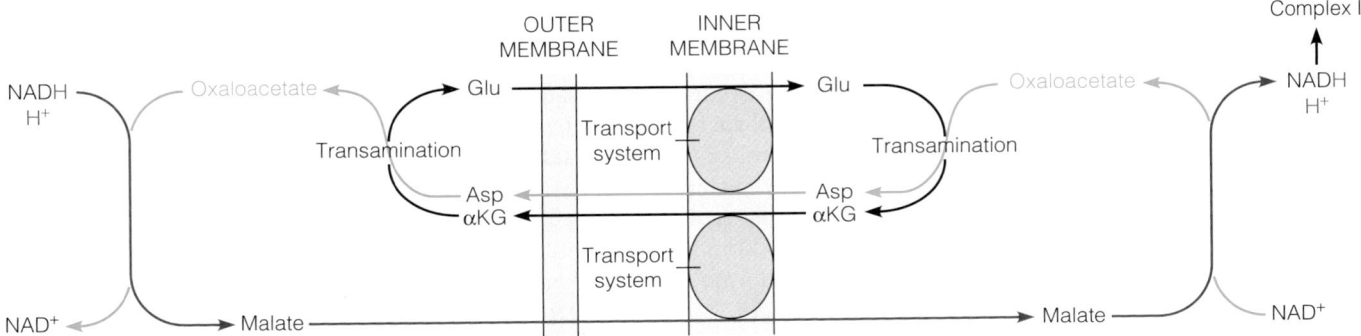

(b)

A different shuttle system, particularly active in liver, kidney, and heart, is the malate/aspartate shuttle, shown in Figure 15.30b. Here, a cytosolic isozyme of malate dehydrogenase, together with NADH, reduces oxaloacetate to malate, which passes into the matrix via a specific α-ketoglutarate/malate exchanger in the inner mitochondrial membrane. The malate is reoxidized by the malate dehydrogenase of the citric acid cycle, which also uses NAD^+. The resulting matrix NADH is then oxidized by complex I. Because oxaloacetate cannot cross the inner membrane, it is transaminated to aspartate, which is then transported out via a specific aspartate/glutamate exchanger. Once in the cytoplasm, aspartate is reconverted to oxaloacetate by transamination, to begin the cycle anew. Because of the transaminations involved, this process requires that α-ketoglutarate be continuously transported out of mitochondria and that glutamate be continuously transported in. This balance is ensured by the substrate specificity of the two exchangers.

Thus, in tissues that use the malate/aspartate shuttle, approximately 2.5 moles of ATP are generated per mole of cytoplasmic NADH. In tissues that use the dihydroxyacetone phosphate/glycerol-3-phosphate shuttle, only about 1.5 moles of ATP are generated per mole of NADH because those cytoplasmic reducing equivalents enter the respiratory chain at complex III (via QH_2), rather than at complex I.

Energy Yields from Oxidative Metabolism

Much of the past three chapters has been devoted to the pathways by which carbohydrates are oxidized to CO_2 and water. Finally, we are in a position to calculate the total energy yield and metabolic efficiency of these combined pathways. Let us review how much energy is recovered in the form of ATP from the entire oxidative catabolism of glucose. First, we present a balanced equation for each of the three pathways involved, and then we estimate the amount of ATP that can be derived through oxidative phosphorylation from the reduced electron carriers.

Glycolysis:
Glucose + 2ADP + $2P_i$ + $2NAD^+ \rightarrow$ 2 pyruvate + 2ATP + 2NADH + $2H_2O$ + $2H^+$

Pyruvate dehydrogenase complex:
2 pyruvate + $2NAD^+$ + 2 CoA-SH $\rightarrow$ 2 acetyl-CoA + 2NADH + $2CO_2$

Citric acid cycle (including conversion of GTP to ATP):
2 acetyl-CoA + $4H_2O$ + $6NAD^+$ + 2FAD + 2ADP + $2P_i \rightarrow 4CO_2$ + 6NADH + 2FADH$_2$
+ 2 CoA-SH + 2ATP + $4H^+$

Net:
Glucose + $10NAD^+$ + 2FAD + $2H_2O$ + 4ADP + $4P_i \rightarrow 6CO_2$ + 10NADH
+ $6H^+$ + 2FADH$_2$ + 4ATP

These three processes generate 4 moles of ATP directly, plus 10 moles of NADH and 2 moles of FADH$_2$. Using P/O ratios for oxidation of NADH and FADH$_2$ of 2.5 and 1.5, respectively, the total realizable ATP yield is about 32 per mole of glucose oxidized [4 + (2.5 × 10) + (1.5 × 2)]. In prokaryotes and in cells using the malate/aspartate shuttle (see Figure 15.30b), the reducing equivalents from cytoplasmic glycolysis are carried into the mitochondrion with no energy cost. However, cells using the glycerol phosphate shuttle (see Figure 15.30a) must pay an energy cost because the electrons from cytosolic NADH enter the respiratory chain as FADH$_2$. Therefore, the ATP yield from each of these two NADHs is ~1.5, not 2.5. This decreases the overall ATP yield to 30 per mole of glucose. For the following discussion, we shall use 32 as the best estimate of total moles of ATP derived per mole of glucose oxidized. Recalling that $\Delta G^{\circ\prime}$ for glucose oxidation is −2870 kJ/mol (see Chapter 12, page 486) and that $\Delta G^{\circ\prime}$ for ATP hydrolysis is −30.5 kJ/mol, we can calculate the efficiency for operation of this biochemical machine: (32 × 30.5)/2870, or ~34%, under standard conditions. As noted earlier, ATP hydrolysis is worth 50–60 kJ/mol in vivo, and thus the efficiency in the cell is considerably higher.

The complete oxidation of 1 mole of glucose generates about 30–32 moles of ATP synthesized from ADP.

The Mitochondrial Genome and Disease

Mitochondria possess a double-stranded circular genome of approximately 16,500 bp in length. In humans, mitochondrial DNA (mtDNA) contains 37 genes that encode for 13 proteins, all of which are subunits of the respiratory chain complexes (Figure 15.31a). Complexes I, III, IV, and V all contain some subunits encoded by mtDNA, whereas complex II and cytochrome *c* are encoded by nuclear DNA. The remaining mtDNA genes encode 22 tRNAs and 2 rRNAs. These tRNAs and rRNAs are part of the mitochondrial protein synthesis machinery, required to translate the 13 proteins encoded in the mtDNA. Of course, most of the proteins found in mitochondria (more than 900) are encoded by the nucleus, translated on cytoplasmic ribosomes, and then transported into the mitochondria.

FIGURE 15.31

Mitochondrial DNA and mitochondrial diseases. **(a)** The five multisubunit complexes of the respiratory chain are shown with mtDNA-encoded subunits in red and nuclear-encoded subunits in blue. **(b)** A morbidity map of the human mitochondrial genome. The 16,569-bp mtDNA shows the locations of the 13 protein-coding genes in red, the 12S and 16S ribosomal RNAs in dark blue, and the 22 transfer RNAs (identified by one-letter codes for the corresponding amino acids) in light blue. Diseases due to mutations in protein-coding genes are boxed in red; diseases due to mutations in mitochondrial protein synthesis components are boxed in blue.

Reprinted from *Biochimica et Biophysica Acta* 1658: 80–88, S. DiMauro, Mitochondrial diseases. © 2004, with permission from Elsevier.

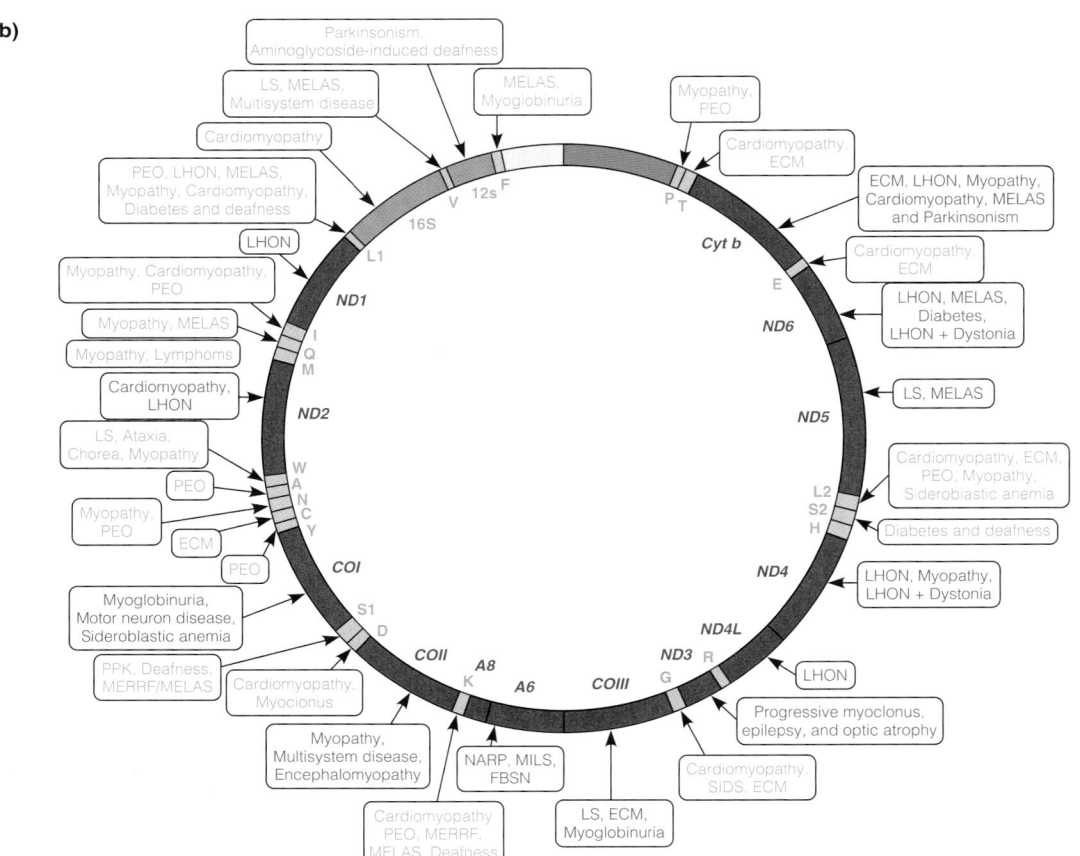

Diseases that affect mitochondrial function (mitochondrial diseases) are probably quite common in humans—perhaps as much as 2% of the population has some mitochondrial defect. However, only a small number of these disorders are caused by mutations in mtDNA (~1 in 10,000 individuals). mtDNA mutations cause defects in the respiratory chain, either directly through mutations in one of the 13 structural genes (Figure 15.31a), or indirectly through mutations in a tRNA or rRNA gene. Many of these diseases involve brain and skeletal muscle, and are thus known as *mitochondrial encephalomyopathies*. A key feature of mtDNA diseases is that the tissues affected and the severity of the disease vary considerably among patients. This variability is due to the fact that each cell contains hundreds or thousands of mtDNA copies, which, at cell division, distribute randomly among daughter cells. If one of these mtDNAs becomes mutated, some daughter cells will inherit mutant mtDNA along with normal mtDNA. When a cell, tissue, or an individual has both normal and mutant mtDNA, it is said to be *heteroplasmic*. The clinical phenotype of a mtDNA mutation is largely determined by the relative proportion of normal and mutant mtDNA genomes in different tissues. A minimum number of mutant mtDNAs in a particular organ or tissue is generally required to cause mitochondrial dysfunction that is severe enough to result in disease. This is referred to as a *threshold effect*.

The "morbidity map" in Figure 15.31b shows the locations of mtDNA mutations and the mitochondrial encephalomyopathies they cause. These include LHON (Leber's hereditary optic neuropathy), MELAS (mitochondrial encephalomyopathy, lactic acidosis, and stroke-like episodes), and MERRF (myoclonic epilepsy with ragged-red fibers). Although the clinical features of these disorders are quite variable, patients with mtDNA disease often exhibit progressive and disabling neurological problems, muscle weakness, chronic progressive ophthalmoplegia, and exercise-induced fatigue, all of which stem from defective respiration and ATP production.

As mentioned above, the majority of inherited mitochondrial diseases are due to mutations in nuclear genes. This is because efficient assembly and functioning of the respiratory chain requires a large number of nuclear-encoded proteins. For example, both the mitochondrial DNA polymerase and RNA polymerase are encoded by nuclear genes, as are all of the protein subunits of the mitochondrial ribosomes. More than 150 point mutations in the gene for the catalytic subunit of mitochondrial DNA polymerase (POLG) are linked with a wide variety of mitochondrial diseases, including progressive external ophthalmoplegia, Alpers' syndrome, Parkinsonism, ataxia-neuropathy syndromes, and male infertility. In addition, there are a number of nuclear-encoded proteins required for mitochondrial dynamics (biogenesis, fission, fusion, degradation). Charcot-Marie-Tooth disease, Friedreich's ataxia, and hereditary spastic paraplegia are examples of inherited diseases caused by mutations in genes required for these processes. Finally, there is a group of late-onset neurodegenerative diseases that all involve mitochondrial dysfunction, including Huntington's disease, Alzheimer's disease, and amyotrophic lateral sclerosis (ALS, also known as Lou Gehrig's disease). The precise role of mitochondria in the pathogenesis of these hereditary neurodegenerative diseases is not yet understood, but is an area of intense investigation.

Mitochondria and Evolution

One of the biggest questions in evolutionary biology concerns the origin of mitochondria. Where did they come from? Although there are many theories, probably the most widely accepted one is the *endosymbiont hypothesis*. In this scenario, over a billion years ago an Archaea, perhaps a methanogen that used H_2, entered into a symbiotic relationship with a facultative anaerobic α-proteobacterium, perhaps one that produced H_2. Eventually, the Archaeal host completely engulfed the α-proteobacterium, without killing it. In this endosymbiotic relationship, the host supplied the endosymbiont with oxidizable substrates, and the endosymbiont in turn supplied the host with energy (ATP). This relationship was sealed forever as most of the genes of the endosymbiont were transferred to the host genome, giving rise to the first eukaryotic cell, with mitochondria representing the modern descendants of the original α-proteobacterium.

Oxygen as a Substrate for Other Metabolic Reactions

In most cells, at least 90% of the molecular oxygen consumed is utilized for oxidative phosphorylation. The remaining O_2 is used in a wide variety of specialized metabolic reactions. At least 200 known enzymes use O_2 as a substrate. Because O_2 is rather unreactive, virtually all of these 200 enzymes use a metal ion to enhance the reactivity of oxygen, just as cytochrome oxidase does. In this section we briefly categorize these enzymes, and we consider the metabolism of partially reduced forms of oxygen, which arise continually in all cells and are highly toxic because of their great reactivity.

Oxidases and Oxygenases

The term **oxidase** is applied to enzymes that catalyze the oxidation of a substrate *without* incorporation of oxygen from O_2 into the product. A two-electron oxidation is usually involved, so the oxygen is converted to H_2O_2. Most oxidases utilize either a metal or a flavin coenzyme. D-Amino acid oxidases, for example, use FAD as a cofactor.

> Oxidases and oxygenases are enzymes that use O_2 as a substrate.

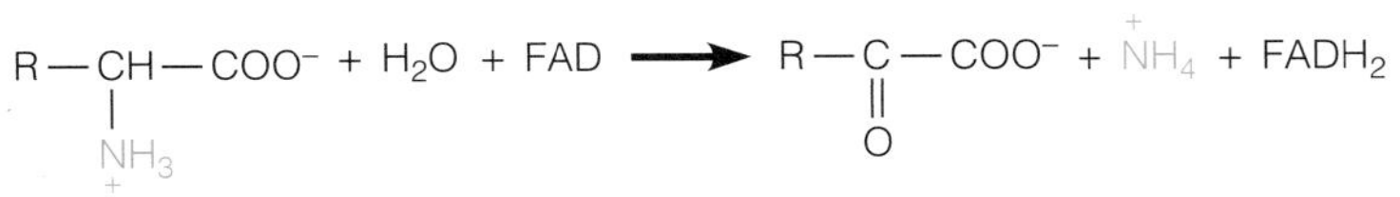

$$R-\underset{\underset{+}{NH_3}}{\underset{|}{CH}}-COO^- + H_2O + FAD \longrightarrow R-\underset{\underset{O}{\parallel}}{C}-COO^- + \overset{+}{NH_4} + FADH_2$$

$$FADH_2 + O_2 \longrightarrow FAD + H_2O_2$$

Oxygenases are enzymes that incorporate oxygen atoms from O_2 into the oxidized products; there are two classes—monooxygenases and dioxygenases. **Dioxygenases**, which incorporate both atoms of O_2 into one substrate, are of limited distribution. An example is tryptophan 2,3-dioxygenase, which contains a heme cofactor and catalyzes the first reaction in tryptophan catabolism:

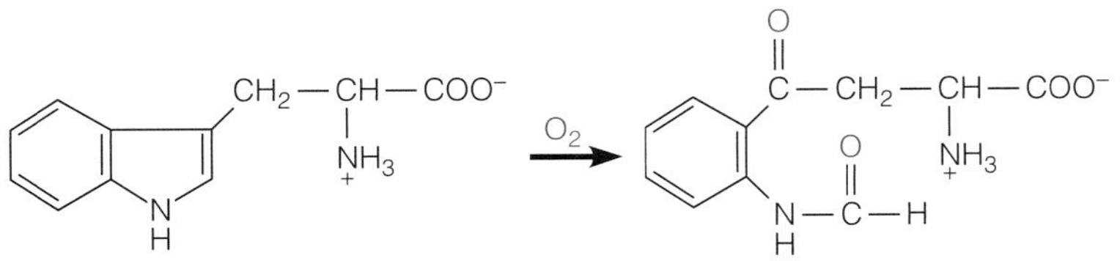

Tryptophan   ***N*-Formylkynurenine**

Far more widely distributed are **monooxygenases**, which incorporate one atom from O_2 into a product and reduce the other atom to water. A monooxygenase has one substrate that accepts oxygen and another that furnishes the two H atoms that reduce the other oxygen to water. Because two substrates are oxidized, enzymes of this class are also called **mixed-function oxidases**. The general reaction catalyzed by monooxygenases is as follows:

$$AH + BH_2 + O{=}O \longrightarrow A-OH + B + H_2O$$

Because the substrate AH usually becomes hydroxylated by this class of enzymes, the term **hydroxylase** is also used. An example of this type of reaction is the hydroxylation of steroids. Here the reductive cofactor, NADPH, appears as BH_2.

$$RH + NADPH + H^+ + O_2 \longrightarrow R-OH + NADP^+ + H_2O$$

Several compounds other than NADPH/H^+ function as BH_2 in monooxygenase reactions, including α-ketoglutarate, as in the hydroxylation of phytanoyl-CoA (Chapter 17) or collagen proline residues (Chapter 21).

Cytochrome P450

Cytochromes P450 catalyze hydroxylations of numerous unreactive substrates, making them easier to metabolize.

The most numerous hydroxylation reactions involve a superfamily of heme proteins with the collective name **cytochrome P450**, found in nearly all organisms, from bacteria to mammals. The human genome contains 57 different structural genes for cytochromes P450, making this a large and diverse protein family. These proteins resemble hemoglobin and mitochondrial cytochrome oxidase in being able to bind both O_2 and carbon monoxide. Cytochromes P450 are distinctive, however, in that the reduced form of the heme, when complexed with carbon monoxide, absorbs light strongly at 450 nm. A common structural feature of cytochromes P450 is a cysteine thiolate ion, in which the sulfur occupies one of the six coordination positions with the heme iron (Figure 15.32a). The other ligands are the four pyrrole nitrogens of the protoporphyrin IX in heme and the bound oxygen. Not all proteins with the cysteinyl–heme configuration, however, are classified as cytochromes P450. This same structural motif is used, for example, by thromboxane synthase (page 814)

(a)

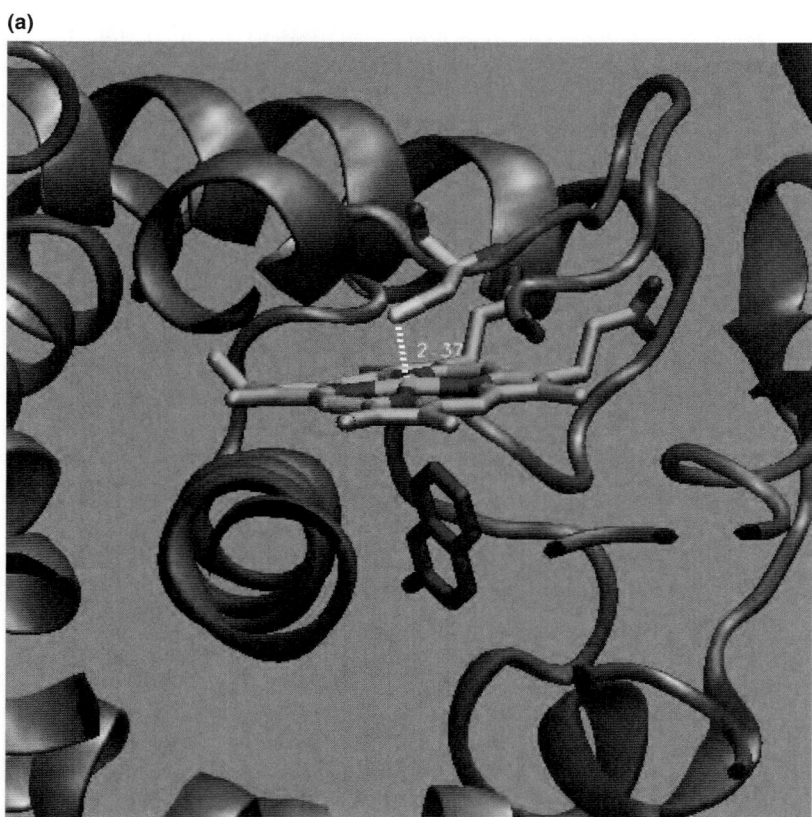

FIGURE 15.32

Cytochrome P450. (a) Active site of human cytochrome P450 2A6 in complex with the anticoagulant coumarin (PDB ID 1Z10). The Fe atom of the heme (shown edgewise) is axially liganded to Cys349. The coumarin substrate (purple) is on the opposite face of the heme. **(b)** The flavoprotein enzyme P450 oxidoreductase (POR) delivers electrons one at a time from NADPH to cytochrome P450.

(b) Courtesy of W. L. Miller. Published in *Proceedings of the National Academy of Sciences of the United States of America*, 2008, 105:1733–1738 (2008), N. Huang, V. Agrawal, K. M. Giacomini, and W. L. Miller, Genetics of P450 oxidoreductase: Sequence variation in 842 individuals of four ethnicities and activities of 15 missense mutants.

(b)

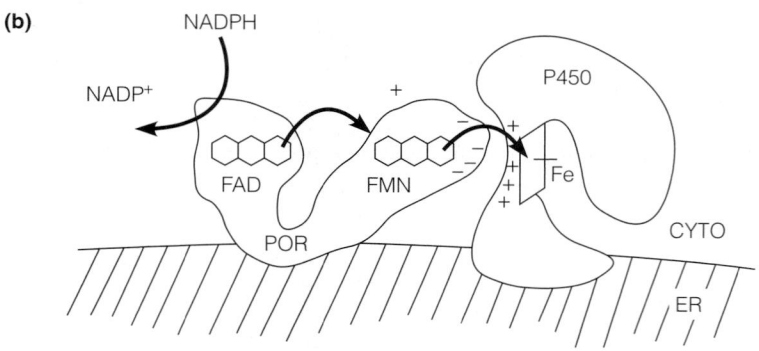

and nitric oxide synthase (page 884). Type I cytochromes P450 are found in bacteria and mitochondria. The more common type II enzymes are found embedded in the endoplasmic reticulum membrane of eukaryotic cells.

Cytochromes P450 are involved in hydroxylating a large variety of compounds. These reactions include the hydroxylations in steroid hormone biosynthesis (see Chapter 19) and the synthesis of hydroxylated fatty acids and fatty acid epoxides. In addition, cytochromes P450 act upon thousands of **xenobiotics** (foreign compounds), including drugs such as phenobarbital and environmental carcinogens such as benzo[*a*]pyrene, a constituent of tobacco smoke, or aflatoxin B, a carcinogenic compound produced by a mold and found in peanuts that have not been properly screened. Hydroxylation of foreign substances usually increases their solubility and is an important step in their detoxification, or metabolism and excretion. However, some of these reactions result in activation of potentially carcinogenic substances to more reactive species, as shown for aflatoxin B_1, which is converted to more reactive species either by hydroxylation or epoxidation (see margin).

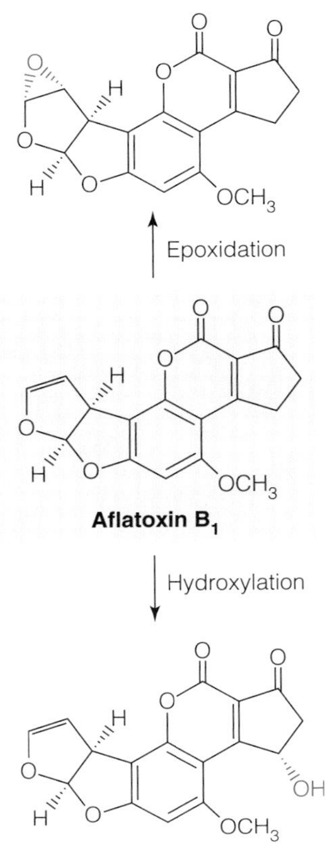

A key to the reactivity of cytochrome P450 is its ability to split O_2, with one oxygen atom binding to the cytochrome's heme iron. This bond forms a **perferryl iron–oxygen complex** (FeO^{3+}). This highly reactive group can abstract a hydrogen atom, even from an unreactive substrate such as a hydrocarbon. In such a hydroxylation, reducing equivalents are typically transferred to the cytochrome from NADPH, the usual electron donor in hydroxylation reactions, via a flavoprotein, P450 oxidoreductase (POR) (Figure 15.32b).

Cytochrome P450 systems participate in a wide variety of additional reactions, including epoxidation, peroxygenation, desulfuration, dealkylation, deamination, and dehalogenation. These reactions are particularly active in liver, where a number of cytochromes P450 are inducible; that is, their synthesis is stimulated by substrates that are metabolized by these enzymes. Inducers include drugs such as phenobarbital and other barbiturates.

Reactive Oxygen Species, Antioxidant Defenses, and Human Disease

Formation of Reactive Oxygen Species

As we have seen, the terminal step in electron transport is the four-electron reduction of O_2 to water. Cytochrome oxidase, like most oxidases, transfers electrons to oxygen from metal ions that change their valence states by one electron at a time—such as the heme iron and copper in cytochrome oxidase. Because the interactions of one-electron carriers with two-electron carriers are rarely 100% efficient, oxidases often generate incompletely reduced oxygen species—**superoxide** ($O_2^{\cdot-}$), formed from a one-electron reduction of O_2; hydrogen peroxide (H_2O_2), formed from a two-electron reduction; and **hydroxyl radical** ($OH^{\cdot}$), formed via a three-electron reduction (Figure 15.33). In addition, some enzymes, such as xanthine oxidase (Chapter 22) and amino acid oxidases (page 840), generate hydrogen peroxide as their ordinary products. Superoxide, peroxide, and hydroxyl radical are all more reactive than oxygen and are referred to collectively as **reactive oxygen species** (ROS). Hydroxyl radical is particularly reactive and is responsible for damage to other biological molecules. Hydroxyl radical damages proteins in various ways and damages membranes by initiating the oxidation of fatty acids in membrane lipids, a process termed **lipid peroxidation**. As shown in the margin on the following page, lipid peroxidation is a chain reaction because each fatty acyl moiety that undergoes peroxidation generates a radical that can initiate another peroxidation reaction. Hydroxyl radical also damages nucleic acids, both by causing polynucleotide strand breakage (double-stranded DNA breaks are lethal) and by changing the structure of DNA bases. About 20 different base changes, or DNA lesions, are known to result from reactions of hydroxyl radical with DNA. Some lesions are

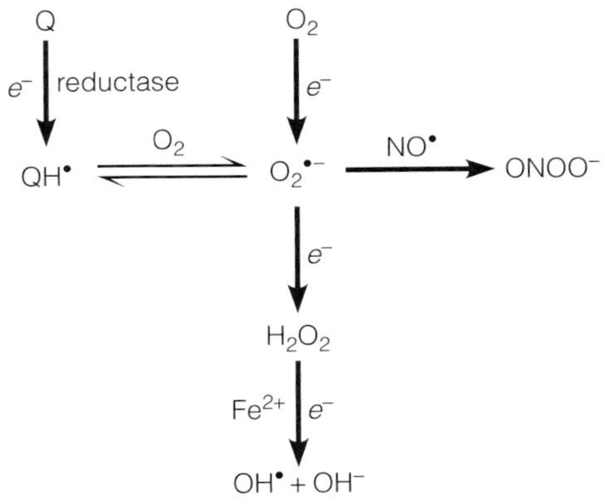

Epoxidation

Aflatoxin B₁

Hydroxylation

Partially reduced oxygen species—superoxide, peroxide, and hydroxyl radical—are extremely toxic. Their toxicity is counteracted by both enzymatic and nonenzymatic mechanisms.

FIGURE 15.33

Reactive oxygen species (ROS). The generation and interconversion of the most common reactive oxygen species are shown. $O_2^{\cdot-}$, superoxide; $OH^{\cdot}$; hydroxyl radical; $NO^{\cdot}$, nitric oxide; $OONO^{-}$, peroxynitrite; Q, oxidized coenzyme Q; $QH^{\cdot}$, semiquinone radical; H_2O_2 hydrogen peroxide.

mutagenic because the altered base created (such as **8-oxoguanine**, see margin) forms non-Watson–Crick base pairs during DNA replication. Other lesions, such as **thymine glycol**, are potentially lethal because, unless the lesion is repaired (Chapter 26), their occurrence in DNA blocks replication past that site. Hydroxyl radical is the most active mutagen derived from ionizing radiation. Hydroxyl radical is also produced from H_2O_2 in the Fenton reaction:

$$H_2O_2 + Fe^{2+} (\text{or } Cu^+) \rightarrow Fe^{3+} (\text{or } Cu^{2+}) + OH^\cdot + OH^-$$

Superoxide per se is relatively nontoxic. However, because it contains an unpaired electron, it is a free radical, and it combines readily with another free radical, **nitric oxide** ($NO^\cdot$) (Figure 15.33), a biological signaling agent that is produced in many animal tissues (Chapters 7, 21, and 23). The product is **peroxynitrite** ($OONO^-$), also considered a reactive oxygen species. Peroxynitrite causes lipid peroxidation and also causes nitration of tyrosyl hydroxyl groups in proteins, a reaction particularly damaging to membrane proteins.

Normal cellular metabolism produces ROS, and mitochondria are an important source of these ROS. It was initially estimated that as much as 1–2% of all the electrons that start down the respiratory chain never make it to cytochrome oxidase but leak from complexes I, II, and III, and bring about one-electron reductions of oxygen to superoxide. It is now generally accepted that these in vitro experiments overestimated the in vivo production of superoxide and hydrogen peroxide by at least two orders of magnitude. Indeed, there is growing evidence that ROS constitute a second messenger system that fine tunes cellular energy metabolism in response to changes in redox status and mitochondrial function. There are even some cases where overproduction of reactive oxygen species is a normal part of the functioning of a cell. For example, certain white blood cells contribute to defense against infectious agents by **phagocytosis** (Greek, "cell eating"). Such cells can engulf a bacterial cell. This event is followed by a **respiratory burst**, a rapid increase in oxygen uptake. In a deliberate and controlled process, much of this oxygen is reduced to superoxide ion and to H_2O_2. The hydrogen peroxide is then converted to more reactive oxidants such as hypochlorous acid (HOCl), which help to kill the engulfed bacterium.

However, uncontrolled overproduction of reactive oxygen species has the potential to inflict considerable damage on the tissues in which they are produced, a situation called **oxidative stress**. A series of elaborate mechanisms has evolved to minimize its harmful consequences.

Dealing with Oxidative Stress

As discussed in Chapter 16, the earth had an anaerobic atmosphere for its first billion years, and oxygen was intensely toxic to all life forms existing at that time. With the evolution of oxygen in our atmosphere, life forms developed both enzymatic and nonenzymatic defenses against oxidative stress. The nonenzymatic protection is afforded by **antioxidant** compounds, including glutathione (Chapter 13), vitamins C and E, and **uric acid**, an end product of purine metabolism (Chapter 22). These compounds can scavenge ROS before they can cause damage, or they can prevent oxidative damage from spreading. For example, trapping one radical can break a chain in which multiple fatty acyl moieties would otherwise be damaged in lipid peroxidation. Vitamin E, a family of compounds in which α-tocopherol is most common, is the principal lipid-soluble antioxidant compound and plays an important role in preventing membrane damage. β-Carotene and other carotenoid compounds related to vitamin A (page 808) are lipid-soluble antioxidants that also play roles in free radical trapping. Glutathione, as noted in Chapter 13, is abundant within cells and plays a particularly important role in cellular antioxidant protection. Vitamin C, or **ascorbic acid**, is an important antioxidant by virtue of its water solubility and its ready oxidation to dehydroascorbic acid. In extracellular fluids, ascorbate levels are far higher than those of glutathione,

Lipid peroxidation

8-Oxoguanine

Thymine glycol

and ascorbate probably plays the predominant role in extracellular antioxidant protection. Recent evidence suggests that a major antioxidant role of uric acid is its ability to bind and inactivate peroxynitrite.

Among enzymatic mechanisms the first line of defense is **superoxide dismutase (SOD)**, a family of metalloenzymes that catalyze a **dismutation** (a reaction in which two identical substrate molecules have different fates). Here, one molecule of superoxide is oxidized and one is reduced.

$$O_2^{\cdot -} + O_2^{\cdot -} + 2H^+ \rightarrow H_2O_2 + O_2$$

A copper- and zinc-containing form of this enzyme (SOD1) is found in the cytosol of eukaryotic cells; a manganese-containing form (MnSOD) is found in both mitochondria and bacterial cells; and a related iron-containing form is found in bacteria, cyanobacteria, and some plants. A nickel-containing bacterial SOD has recently been described.

Hydrogen peroxide is metabolized either by **peroxiredoxins**, a ubiquitous family of thiol proteins, **catalase**, another widely distributed enzyme, or by a more limited family of **peroxidases**. Peroxiredoxins are found in all domains of life, and these abundant enzymes can be located in the cytosol, in mitochondria, in chloroplasts and peroxisomes, and associated with nuclei and membranes. Peroxiredoxins play an antioxidant role through their peroxidase activity:

$$ROOH + 2H^+ + 2e^- \rightarrow ROH + H_2O$$

These enzymes can reduce and detoxify not only hydrogen peroxide, but also peroxynitrite and a wide range of organic hydroperoxides (ROOH). The catalytic mechanism involves oxidation of an active site cysteine to a sulfenic acid (S-OH) by the peroxide substrate. The sulfenic acid must then be reduced back to the thiol state by a disulfide reductase, with electrons provided by a dithiol such as thioredoxin.

Catalase is a heme protein with an extremely high turnover rate (>40,000 molecules per second). It catalyzes the following reaction:

$$2H_2O_2 \rightarrow 2H_2O + O_2$$

Peroxidases, which are widely distributed in plants, reduce H_2O_2 to water at the expense of oxidation of an organic substrate. An example of a peroxidase is found in erythrocytes, which are especially sensitive to peroxide accumulation. (See pages 581–584 in Chapter 13 for a discussion of the consequences of peroxide accumulation in glucose-6-phosphate dehydrogenase deficiency.) Within erythrocytes is *glutathione peroxidase*, an enzyme that reduces H_2O_2 to water, along with the oxidation of glutathione.

$$2GSH + H_2O_2 \rightarrow GSSG + 2H_2O$$

Glutathione peroxidase is interesting in that it contains one residue per mole of an unusual amino acid, *selenocysteine* (Chapter 5), an analog of cysteine that contains selenium in place of sulfur (more on this in Chapter 27).

Oxygen Metabolism and Human Disease

Because oxidative stress can damage many biomolecules—lipids, proteins, and nucleic acids—the tissue injury that results can, in principle, lead to a variety of disease states. Oxidative damage has been implicated in many different human disorders, including cardiovascular disease, cancer, stroke, neurodegenerative diseases, chronic inflammatory diseases, and even aging. Determining precise cause-and-effect relationships is difficult at this stage, and it is clear that in some states, oxidative stress is not a cause but rather a result of tissue injury from some other factor, which exacerbates the original problem. However, epidemiological evidence points strongly to the value of adequate dietary intake of natural antioxidant compounds in preventing many of these diseases. The evident health-promoting effects of diets rich in fresh fruits and vegetables, particularly with respect to incidence of cardiovascular disease and cancer, probably result in large part from their high content of antioxidant compounds, particularly

γ-Glu - Cys - Gly
Glutathione

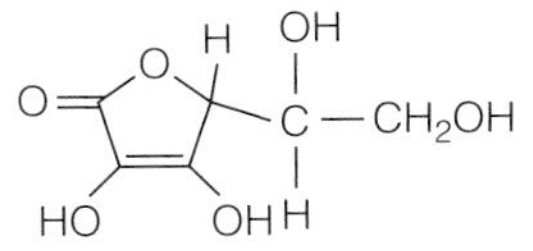

**L-Ascorbic acid
(vitamin C)**

$2e^-$

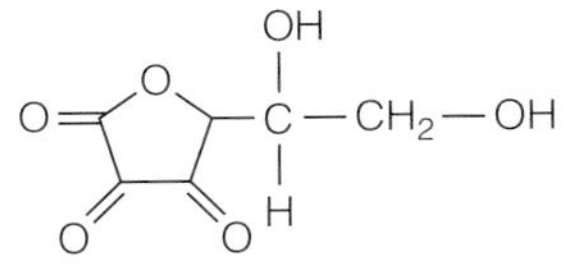

Dehydroascorbic acid

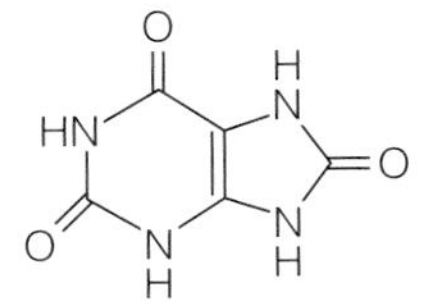

Uric acid

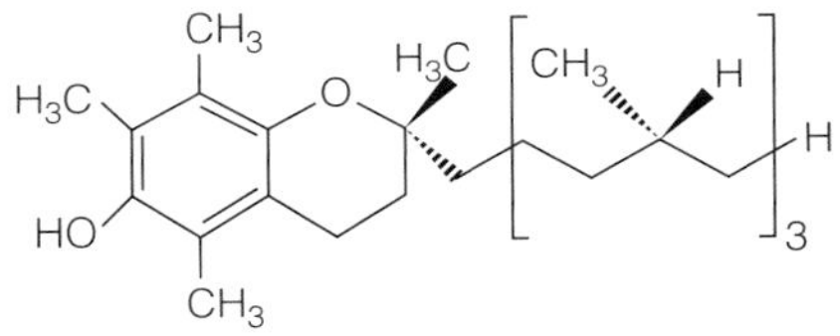

**α-Tocopherol
(vitamin E)**

vitamins C and E. Many people now take dietary supplements of vitamins C and E as a preventive measure.

Whatever the cause-and-effect relationship between oxidative damage and disease, there is little doubt that the DNA damage caused by ionizing radiation, which is known to be carcinogenic, is mediated through the mutagenic effects of hydroxyl radicals. In addition, hydroxyl radicals are generated by oxidizing agents independent of radiation effects. As discussed in Chapters 23 and 25, cancer is clearly a genetic disease, resulting from an accumulation within a precancerous cell of mutations that ultimately destroy the cell's ability to regulate its own growth and its programmed death (apoptosis). The generation of altered bases in DNA, such as 8-oxoguanine or 5-hydroxycytosine, is intensely mutagenic. Even though such lesions are usually removed by DNA repair systems (Chapter 26), the repair systems are not 100% effective, and with time the accumulation of unrepaired DNA damage can contribute to the burden of mutations that ultimately transform a normal cell to a cancer cell. Partly because cancer incidence is strongly correlated with age, many scientists also attribute normal aging to the accumulation of unrepaired mutagenic DNA lesions, and oxidative stress is implicated in what has been called the "free radical theory of aging."

SUMMARY

Most of the energy captured for ATP synthesis from oxidative reactions in cells is generated in mitochondrial oxidative phosphorylation. Reduced electron carriers, both NADH and $FADH_2$, shuttle reducing equivalents in the mitochondrial matrix. Enzyme complexes bound to the inner mitochondrial membrane pass these electrons through the respiratory chain, a series of electron carriers of ever-increasing reduction potential. The complexes are numbered I (NADH dehydrogenase), II (succinate–coenzyme Q oxidoreductase), III (coenzyme Q–cytochrome c oxidoreductase), IV (cytochrome c oxidase), and V (ATP synthase). Electrons are eventually transferred to O_2, which is reduced to water. The redox reactions of complexes I, III, and IV provide energy to pump protons from the matrix across the inner membrane, generating an electrochemical gradient termed the protonmotive force. Discharge of the resultant proton gradient, when protons pass back into the matrix through a specific ion channel, generates energy that is used to drive ATP synthesis. Although respiration accounts for about 90% of the total oxygen uptake in most cells, in numerous other reactions dozens of enzymes use O_2 as a substrate—oxygenases, oxidases, and hydroxylases. Some reactions generate partially reduced oxygen species—as hydroxyl radical, superoxide, and peroxide—termed reactive oxygen species, which are toxic and mutagenic. Cells possess numerous mechanisms for detoxification of these reactive oxygen species.

REFERENCES

Historical Background

Lehninger, A. L. (1965) *The Mitochondrion: Molecular Basis of Structure and Function.* Benjamin, New York. An account of the earlier work by one who contributed much to it.

Saier, M. H., Jr. (1997) Peter Mitchell and his chemiosmotic theories. *ASM News* 63:13–21. A short scientific biography of the biochemist who proposed the proton gradient as the driving force for ATP synthesis.

Mitochondrial Structure and Function

Kiberstis, P. A. (1999) Mitochondria make a comeback. *Science* 283:1475. An introductory essay to a special section of *Science*, with four contemporary reviews—of mitochondrial evolution, mitochondrial diseases, oxidative phosphorylation, and mitochondrial genetics.

Palmieri, F. (2004) The mitochondrial transporter family (SLC25): Physiological and pathological implications. *Pflugers Arch.* 447:689–709. A review on the carriers that transport metabolites across the inner membrane.

Scheffler, I. E. (2008) *Mitochondria.* Wiley-Liss, Hoboken, New Jersey. An up-to-date compendium of mitochondrial structure, function, genetics, and evolution.

Tzagoloff, A. (1982) *Mitochondria.* Plenum, New York. A concise, well-illustrated book-length review of mitochondrial structure and function.

Mechanisms in Electron Transport

Beinert, H., R. H. Holm, and E. Münck (1997) Iron–sulfur clusters: Nature's modular, multipurpose structures. *Science* 277:653–659. A review of the roles of these structures in oxidative enzymes and their numerous other roles.

Brandt, U. (2006) Energy converting NADH:quinone oxidoreductase (complex I). *Annu. Rev. Biochem.* 75:69–92. A review of the structure and function of complex I.

Brzezinski, P., J. Reimann, and P. Adelroth (2008) Molecular architecture of the proton diode of cytochrome *c* oxidase. *Biochem. Soc. Trans.* 36:1169–1174. A review of the structure and proton pumping mechanism of complex IV.

Crofts, A. R., J. T. Holland, D. Victoria, D. R. Kolling, S. A. Dikanov, R. Gilbreth, S. Lhee, R. Kuras, and M. G. Kuras (2008) The Q-cycle reviewed: How well does a monomeric mechanism of the bc_1 complex account for the function of a dimeric complex? *Biochim. Biophys. Acta* 1777:1001–1019.

Huang, L. S., D. Cobessi, E. Y. Tung, and E. A. Berry (2005) Binding of the respiratory chain inhibitor antimycin to the mitochondrial bc_1 complex: A new crystal structure reveals an altered intramolecular hydrogen-bonding pattern. *J. Mol. Biol.* 351:573–597. Insights into the Q cycle from the structure of complex III.

Sazanov, L. A., and P. Hinchliffe (2006) Structure of the hydrophilic domain of respiratory complex I from *Thermus thermophilus*. *Science* 311:1430–1436.

Sun, F., X. Huo, Y. Zhai, A. Wang, J. Xu, D. Su, M. Bartlam, and Z. Rao (2005) Crystal structure of mitochondrial respiratory membrane protein complex II. *Cell* 121:1043–1057.

Vonck, J., and E. Schafer (2009) Supramolecular organization of protein complexes in the mitochondrial inner membrane. *Biochim. Biophys. Acta* 1793:117–124.

Mechanisms in Oxidative Phosphorylation

Abrahams, J. P., A. G. Leslie, R. Lutter, and J. E. Walker (1994) Structure at 2.8 Å resolution of F_1 ATPase from bovine heart mitochondria. *Nature* 370:621–626. The X-ray structure that confirms essential features of the mechanism of oxidative phosphorylation.

Boyer, P. D. (1997) The ATP synthase—A splendid molecular machine. *Annu. Rev. Biochem.* 66:717–750. A mechanistic analysis of the function of F_0F_1 ATP synthase by the person who predicted the correct mechanism of ATP synthesis and did the crucial early experiments.

Campanella, M., N. Parker, C. H. Tan, A. M. Hall, and M. R. Duchen (2009) IF_1: Setting the pace of the F_1F_0-ATP synthase. *Trends Biochem. Sci.* 34:343–350. A discussion of the role of an F_1F_0-ATP synthase inhibitor protein.

Hinkle, P. C., M. A. Kumar, A. Resetar, and D. L. Harris (1991) Mechanistic stoichiometry of mitochondrial oxidative phosphorylation. *Biochemistry* 30:3576–3582. A careful conceptual and experimental analysis that questions whether P/O ratios need be integral.

Nicholls, D. G., and S. J. Ferguson, (2002) *Bioenergetics 3*, Academic Press, London, UK. An excellent source on the thermodynamics and mechanisms of chemiosmosis and redox chemistry.

Noji, H., and M. Yoshida (2001) The rotary machine in the cell, ATP synthase. *J. Biol. Chem.* 276:1665–1668. A minireview on the experimental evidence for rotation of the complex in the membrane.

Strauss, M., G. Hofhaus, R. R. Schröder, and W. Kühlbrandt (2008) Dimer ribbons of ATP synthase shape the inner mitochondrial membrane. *EMBO J.* 27:1154–1160.

Walker, J. E., and V. K. Dickson (2006) The peripheral stalk of the mitochondrial ATP synthase. *Biochim. Biophys. Acta* 1757:286–296. A review of the structure of the F_1F_0 ATP synthase.

Watt, I. N., M. G. Montgomery, M. J. Runswick, A. G. Leslie, and J. E. Walker (2010) Bioenergetic cost of making an adenosine triphosphate molecule in animal mitochondria. *Proc. Natl. Acad. Sci. USA* 107:16823–16827. The structure of the bovine mitochondrial c-ring reveals the stoichiometry of ATP synthesis in higher eukaryotes.

Weber, J., and A. E. Senior (2003) ATP synthesis driven by proton transport in F_1F_0-ATP synthase. *FEBS Lett.* 545:61–70. A review on the path of protons through the F_0 component.

Yasuda, R., H. Noji, K. Kinosita, Jr., and M. Yoshida (1998) F_1 ATPase is a highly efficient molecular motor that rotates with discrete 120° steps. *Cell* 93:1117–1124. Direct evidence for rotation from fluorescence microscopy.

Mitochondrial Genetics, Diseases, and Evolution

DiMauro, S. (2004) Mitochondrial diseases. *Biochim. Biophys. Acta* 1658:80–88.

Embley, T. M., and W. Martin (2006) Eukaryotic evolution, changes and challenges. *Nature* 440:623–630. Reviews current thinking on the origins of mitochondria and eukaryotic cells.

Lane, N. (2005) *Power, Sex, Suicide. Mitochondria and the Meaning of Life.* Oxford University Press, Oxford. This excellent book presents the case for the central role played by mitochondria in the evolution of eukaryotic cells, and the consequences for human disease and aging.

Palmieri, F. (2008) Diseases caused by defects of mitochondrial carriers: A review. *Biochim. Biophys. Acta* 1777:564–578.

Oxygen Metabolism

Addabbo, F., M. Montagnani, and M. S. Goligorsky (2009) Mitochondria and reactive oxygen species. *Hypertension* 53:885–892.

Beckman, K. B., and B. N. Ames (1998) The free radical theory of aging. *Physiol. Rev.* 78:547–581. A comprehensive review from Bruce Ames, an early proponent of the idea that oxidative damage causes cancer and aging.

Dickinson, B. C., and C. J. Chang (2011) Chemistry and biology of reactive oxygen species in signaling or stress responses. *Nat. Chem. Biol.* 7:504–511.

Fridovich, I. (1995) Superoxide radical and superoxide dismutases. *Annu. Rev. Biochem.* 64:97–112. A review by the discoverer of superoxide dismutase.

Guengerich, F. P. (2008) Cytochrome P450 and chemical toxicology. *Chem. Res. Toxicol.* 21:70–83.

Hall, A., P. A. Karplus, and L. B. Poole (2009) Typical 2-Cys peroxiredoxins—structures, mechanisms and functions. *FEBS J.* 276:2469–2477.

Murphy, M. P. (2009) How mitochondria produce reactive oxygen species. *Biochem. J.* 417:1–13.

Wallace, D. C. (2005) A mitochondrial paradigm of metabolic and degenerative diseases, aging, and cancer: A dawn for evolutionary medicine. *Annu. Rev. Genet.* 39:359–407.

Winterbourn, C. C. (2008) Reconciling the chemistry and biology of reactive oxygen species. *Nat. Chem. Biol.* 4:278–286.

PROBLEMS

1. Referring to Table 15.1 for E_0' values, calculate $\Delta G^{\circ\prime}$ for oxidation of malate by malate dehydrogenase.

2. When pure reduced cytochrome c is added to carefully prepared mitochondria along with ADP, P_i, antimycin A, and oxygen, the cytochrome c becomes oxidized, and ATP is formed, with a P/O ratio approaching 1.0.
   (a) Indicate the probable flow of electrons in this system.
   (b) Why was antimycin A added?
   (c) What does this experiment tell you about the location of coupling sites for oxidative phosphorylation?
   (d) Write a balanced equation for the overall reaction (including cyt c oxidation and ATP synthesis).
   (e) Calculate ΔG°, for the above reaction, using E_0' values from Table 15.1 and a ΔG° value for ATP hydrolysis of -30.5 kJ/mol.

3. Freshly prepared mitochondria were incubated with β-hydroxybutyrate, oxidized cytochrome c, ADP, P_i, and cyanide. β-Hydroxybutyrate is oxidized by an NAD^+-dependent dehydrogenase.

$$
\begin{array}{ccc}
COO^- & & COO^- \\
| & & | \\
CH_2 & & CH_2 \\
| & \longrightarrow & | \\
H-C-OH & & C=O \\
| & NAD^+ \quad NADH & | \\
CH_3 & \quad\quad +H^+ & CH_3
\end{array}
$$

The experimenter measured the rate of oxidation of β-hydroxybutyrate and the rate of formation of ATP.
   (a) Indicate the probable flow of electrons in this system.
   (b) How many moles of ATP would you expect to be formed per mole of β-hydroxybutyrate oxidized in this system?
   (c) Why is β-hydroxybutyrate added rather than NADH?
   (d) What is the function of the cyanide?
   (e) Write a balanced equation for the overall reaction occurring in this system (electron transport and ATP synthesis).
   (f) Calculate the net standard free energy change ($\Delta G^{\circ\prime}$) in this system, using E_0' values from Table 15.1 and a $\Delta G^{\circ\prime}$ value for ATP hydrolysis of -30.5 kJ/mol.

4. If you were to determine the P/O ratio for oxidation of α-ketoglutarate, you would probably include some malonate in your reaction system. Why? Under these conditions, what P/O ratio would you expect to observe?

5. Of the various oxidation reactions in glycolysis and the citric acid cycle, the only one that does not involve NAD^+ is the succinate dehydrogenase reaction. What would $\Delta G^{\circ\prime}$ be for an enzyme that oxidizes succinate with NAD^+ instead of FAD? If the intramitochondrial concentration of succinate was 10-fold higher than that of fumarate, what minimum $[NAD^+]/[NADH]$ ratio in mitochondria would be needed to make this reaction exergonic at 37 °C?

6. Intramitochondrial ATP concentrations are about 5 mM, and phosphate concentration is about 10 mM. If ADP is five times more abundant than AMP, calculate the molar concentrations of ADP and AMP at an energy charge of 0.85. Calculate ΔG for ATP hydrolysis at 37 °C under these conditions. The energy charge is the concentration of ATP plus half the concentration of ADP divided by the total adenine nucleotide concentration:

$$
\frac{[ATP] + 1/2[ADP]}{[ATP] + [ADP] + [AMP]}
$$

7. From E_0' values in Table 15.1, calculate the equilibrium constant for the glutathione peroxidase reaction at 37 °C.

*8. In the early days of "mitochondriology," P/O ratios were determined from measurements of volume of O_2 taken up by respiring mitochondria and chemical assays for disappearance of inorganic phosphate. Now, however, it is possible to measure P/O ratios simply with a recording oxygen electrode. How might this be done?

9. Years ago there was interest in using uncouplers such as dinitrophenol as weight control agents. Presumably, fat could be oxidized without concomitant ATP synthesis for re-formation of fat or carbohydrate. Why was this a bad idea?

10. Referring to Figure 15.18, predict the P/O ratio for oxidation of ascorbate by isolated mitochondria.

11. As a representation of the respiratory chain, what is wrong with this picture? There are four deliberate errors.

$$
Malate \begin{array}{c} \nearrow \\ \searrow \end{array} \begin{array}{c} NAD^+ \\ NADH \end{array} \begin{array}{c} \searrow \\ \nearrow \end{array} \begin{array}{c} FADH_2 \\ FAD \end{array} \begin{array}{c} \nearrow \\ \searrow \end{array} \begin{array}{c} CoQH_2 \\ CoQ \end{array} \begin{array}{c} \searrow \\ \nearrow \end{array} \begin{array}{c} cyt\ b^{2+} \\ cyt\ b^{3+} \end{array}
$$

Oxaloacetate

$$
\begin{array}{c} cyt\ c^{3+} \\ cyt\ c^{2+} \end{array} \begin{array}{c} cyt\ c_1^{2+} \\ cyt\ c_1^{3+} \end{array} \begin{array}{c} cyt\ a\text{-}a_3^{3+} \\ cyt\ a\text{-}a_3^{2+} \end{array} \begin{array}{c} H_2O_2 \\ 1/2 O_2 \end{array}
$$

12. $GSSG + NADPH + H^+ \rightarrow 2GSH + NADP^+$
   (a) Calculate $\Delta G^{\circ\prime}$ for the glutathione reductase reaction in the direction shown, using E_0' values from Table 15.1.
   (b) Suppose that a cell contained an isoform of glutathione reductase that used NADH instead of NADPH as the reductive coenzyme. Would you expect $\Delta G^{\circ\prime}$ for this enzyme to be higher, lower, or the same as the corresponding value for the real glutathione reductase? Briefly explain your answer.
   (c) Given what you know about the metabolic roles and/or intracellular concentration ratios of $NAD^+ / NADH$ and $NADP^+ / NADPH$, would you expect ΔG (not $\Delta G^{\circ\prime}$) for this enzyme to be higher, lower, or the same as ΔG for the real enzyme under intracellular conditions? Briefly explain your answer.

13. To carefully prepared mitochondria were added succinate, oxidized cytochrome c, ADP, orthophosphate, and sodium cyanide. Referring to Figures 15.9 and 15.18, answer the following.
   (a) List the sequence of electron carriers in this system.
   (b) Write a balanced equation for the overall reaction occurring in this system, showing oxidation of the initial electron donor, reduction of the final acceptor, and synthesis of ATP.
   (c) Calculate $\Delta G^{\circ\prime}$ for the overall reaction. $\Delta G^{\circ\prime}$ for ATP hydrolysis is -30.5 kJ/mol.
   (d) Why was cyanide added in this experiment?
   (e) What would the P/O ratio be if the same experiment were run with addition to the mitochondria of 2,4-dinitrophenol?

14. In order to function as an oxidative phosphorylation uncoupler (page 648), 2,4-dinitrophenol must act catalytically, not stoichiometrically. What does this mean? Identify and discuss an important implication of this conclusion.

15. (a) Calculate the standard free energy change as a pair of electrons is transferred from succinate to molecular oxygen in the mitochondrial respiratory chain.

(b) Based on your answer in part (a), calculate the maximum number of protons that could be pumped out of the matrix into the intermembrane space as these electrons are passed to oxygen. Assume 25 °C, $\Delta pH = 1.4$; $\Delta\psi = 0.175$ V (matrix negative).

(c) At which site(s) are these protons pumped?

16. Four electron carriers, a, b, c, and d, whose reduced and oxidized forms can be distinguished spectrophotometrically, are required for respiration in a bacterial electron-transport system. In the presence of substrates and oxygen, three different inhibitors block respiration, yielding the patterns of oxidation states shown below. What is the order of the carriers in the chain from substrates to O_2, and where do the three inhibitors act?

	Carriers			
Inhibitor	a	b	c	d
1	O	O	R	O
2	R	R	R	O
3	O	R	R	O

O and R indicate fully oxidized and fully reduced, respectively.

17. Biochemists working with isolated mitochondria recognize five energy "states" of mitochondria, depending on the presence or absence of essential substrates for respiration—O_2, ADP, oxidizable substrates, and so forth. The characteristics of each state are:

state 1: mitochondria alone (in buffer containing P_i)
state 2: mitochondria + substrate, but respiration low due to lack of ADP
state 3: mitochondria + substrate + limited amount of ADP, allowing rapid respiration
state 4: mitochondria + substrate, but all ADP converted to ATP, so respiration slows
state 5: mitochondria + substrate + ADP, but all O_2 used up (anoxia), so respiration stops

(a) On the graph, identify the state that might predominate in each stage of the trace indicated with a letter.

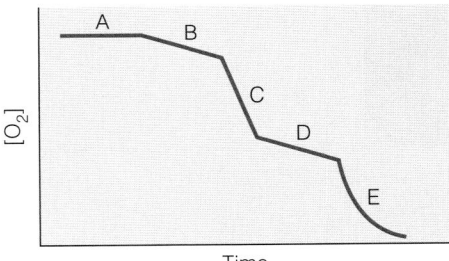

(b) To determine whether isolated mitochondria exhibit respiratory control, one determines the ratio of rates of oxygen uptake in two different states. Which states?

(c) Which state probably predominates in vivo in skeletal muscle fatigued from a long and strenuous workout?

(d) Which state probably predominates in resting skeletal muscle of a well-nourished animal?

(e) Which state probably predominates in heart muscle most of the time?

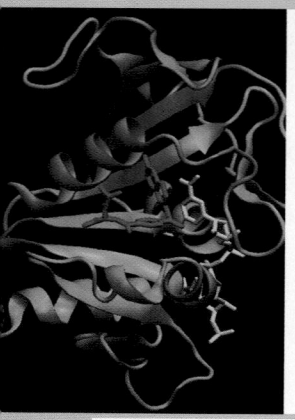

CHAPTER 16

Photosynthesis

In earlier chapters, we described in considerable detail the ways in which organisms extract a substantial portion of the energy available from the oxidation of carbohydrates. Using glucose as an example, we wrote for the overall reaction:

$$C_6H_{12}O_6 + 6O_2 \longrightarrow 6CO_2 + 6H_2O \qquad \Delta G^{\circ\prime} = -2870\,\text{kJ/mol}$$

We noted that as much as 40% of this energy could be recovered for useful biochemical work.

But life cannot depend on oxidative metabolism as its ultimate source of energy, and it cannot continue indefinitely returning organic carbon to the atmosphere as CO_2. The reaction above is only half of the great energy–carbon cycle of nature (Figure 16.1). The reverse of the carbohydrate oxidation reaction is accomplished by plants, algae, and some microorganisms, using the energy from sunlight to provide the enormous amount of free energy required.

$$6CO_2 + 6H_2O \xrightarrow{\text{Light energy}} C_6H_{12}O_6 + 6O_2 \qquad \Delta G^{\circ} = +2870\,\text{kJ/mol}$$

Photosynthesis provides carbohydrates for energy production, fixes CO_2, and is the major source of atmospheric O_2.

This process is called **photosynthesis**. Not only does it provide carbohydrates for energy production in plants and animals, but it is also the major path through which carbon re-enters the biosphere—that is, the principal means of carbon fixation. Furthermore, photosynthesis is the major source of oxygen in the earth's atmosphere.

Prior to the evolution of photosynthetic organisms, the earth's atmosphere was probably devoid of oxygen (though rich in carbon dioxide). Prephotosynthetic organisms must have used hydrogen/electron donors, such as H_2S, NH_3, Fe^{2+}, and organic acids, which were in limited supply compared to water. Without the advent of photosynthesis, these energy sources would have eventually been wholly consumed, and life would have perished. Photosynthesis provided life with an unlimited supply of reducing equivalents (the oceans of water) needed to convert carbon dioxide into carbohydrates and the other organic molecules necessary for life.

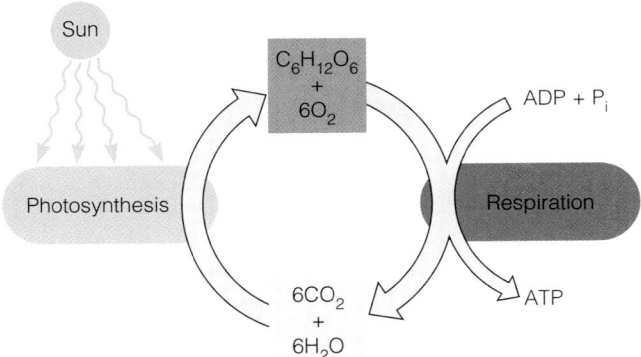

FIGURE 16.1

The carbon cycle in nature. Carbon dioxide and water are combined through photosynthesis to form carbohydrates. In both photosynthetic and nonphotosynthetic organisms these carbohydrates can be reoxidized to regenerate CO_2 and H_2O. Part of the energy obtained from both photosynthesis and fuel oxidation is captured in ATP.

The fossil and geochemical records suggest that photosynthetic organisms first appeared approximately 3.4 billion to 2.3 billion years ago. Their gradual conversion of the primitive, nonoxidizing atmosphere of the earth to an oxidizing atmosphere paved the way for aerobic metabolism and the evolution of animals. Today, photosynthesis represents the ultimate source of energy for almost all life.* It is used by plants, algae, and a wide variety of prokaryotes—all of which are food sources for other organisms. A comprehensive view of the relationship of photosynthesis to other pathways we have studied is shown in Figure 16.2.

The Basic Processes of Photosynthesis

The equation we have just shown for the photosynthetic reaction is, of course, a great oversimplification. As you might expect, the actual process of photosynthesis involves many intermediate steps. Furthermore, a hexose itself is not the primary carbohydrate product. Therefore, the photosynthetic reaction is usually written in this more general form:

$$CO_2 + H_2O \xrightarrow{\text{Light energy}} [CH_2O] + O_2$$

where $[CH_2O]$ represents a general carbohydrate.

Because the catabolism of carbohydrates to form CO_2 is an oxidative process, converting CO_2 into carbohydrate must involve a *reduction* of the carbon. The preceding reaction statement shows H_2O as the ultimate reducing agent, which is the case in plants, most algae, and cyanobacteria. However, there are photosynthetic processes in many bacteria that use other reductants. Thus, an even more general reaction can be written:

$$CO_2 + 2H_2A \xrightarrow{\text{Light energy}} [CH_2O] + H_2O + 2A$$

where H_2A is a general reductant and A is the oxidized product. Examples of photosynthetic reactions are given in Table 16.1. Comparison of the reactions shown in the table suggests that the source of the oxygen released in photosynthesis by plants, algae, and cyanobacteria must be H_2O, rather than CO_2. This source was predicted in the early 1930s by C. B. van Niel, one of the pioneers in photosynthesis studies, and was confirmed in 1941 by Samuel Ruben and Martin

Photosynthesis requires a reductant, usually H_2O, to reduce CO_2 to the carbohydrate level.

---

*Recent developments make it necessary to qualify the statement that all life derives its energy, directly, from photosynthesis. It has been found that some bacteria, such as those associated with submarine "black smoker" hydrothermal vents, use the oxidation of substances like H_2S, formate, or H_2 as an alternate energy source, in the complete absence of light. This energy cycle represents, however, only a small fraction of the energy flow in the biosphere.

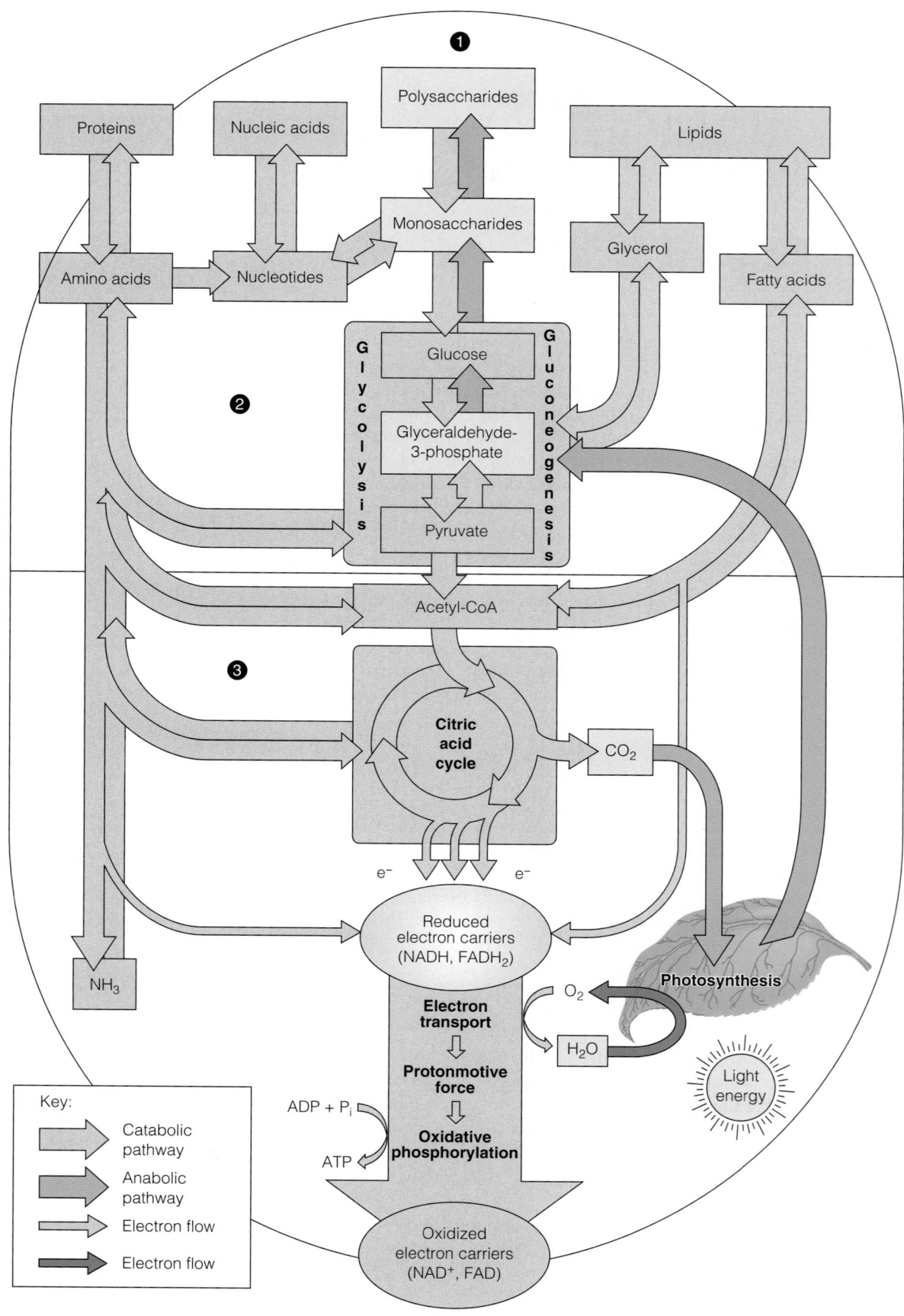

FIGURE 16.2

The role of photosynthesis in metabolism. The major biosynthetic pathways leading from carbon dioxide and water to polysaccharides are highlighted in green. Oxygen derived from the water is released as a by-product of photosynthesis.

TABLE 16.1 Examples of some photosynthetic reactions

Organisms	Reductant	Carbon Assimilation Reaction
Plants, algae, cyanobacteria	H_2O	$CO_2 + 2H_2O \rightarrow [CH_2O] + H_2O + O_2$
Green sulfur bacteria	H_2S	$CO_2 + 2H_2S \rightarrow [CH_2O] + H_2O + 2S$
Purple bacteria	$[HSO_3^-]$	$CO_2 + H_2O + 2[HSO_3^-] \rightarrow [CH_2O] + 2[HSO_4^-]$
Nonsulfur photosynthetic bacteria	H_2 or many other reductants such as lactate	$CO_2 + 2H_2 \rightarrow [CH_2O] + H_2O$ $CO_2 + 2(HC\text{—}OH) \rightarrow [CH_2O] + H_2O + 2(C\equiv O)$

$$CO_2 + 2\left(\begin{array}{c} CH_3 \\ | \\ HC\text{—}OH \\ | \\ COO^- \end{array}\right) \longrightarrow [CH_2O] + H_2O + 2\left(\begin{array}{c} CH_3 \\ | \\ C\text{=}O \\ | \\ COO^- \end{array}\right)$$

Lactate   Pyruvate

Kamen in isotope labeling experiments using ^{18}O-labeled water and unlabeled CO_2. These experiments showed that neither of the oxygen atoms in O_2 comes from CO_2. Therefore, it is more correct to write the photosynthetic reactions in the fashion shown in Table 16.1, which makes it clear that one of the oxygens from CO_2 ends up in carbohydrate, the other in water:

$$CO_2 + 2H_2O \xrightarrow{\text{Light energy}} [CH_2O] + H_2O + O_2$$

Light energy cannot be used *directly* to drive this reaction; in fact, H_2O does not reduce CO_2 *directly* under any known circumstances. The overall process we have just described is actually separated, both chemically and physically, into two subprocesses in all photosynthetic organisms. A slightly more sophisticated version of what actually happens is shown in Figure 16.3. In the first subprocess, in a series of steps called the **light reactions**, energy from sunlight is used to carry out the photochemical oxidation of H_2O. Two things are accomplished by this oxidation. First, the oxidizing agent $NADP^+$ is reduced to NADPH and O_2 is released. Second, part of the energy from sunlight is captured by phosphorylating ADP to produce ATP. This is called **photophosphorylation**. In the second subprocess, the so-called **dark reactions** of photosynthesis, the NADPH and ATP produced by the light reactions are used in the reductive synthesis of carbohydrate from CO_2 and water. These reactions were originally termed *dark* to emphasize that they do not require the direct participation of light energy. Though this is true, the term carries the unfortunate implication that these synthetic reactions occur only in the dark. Nothing could be further from the truth. Rather, these reactions occur at all times and are actually accelerated by light. Because the term *dark reactions* is well established, we shall retain it, but you should not be misled by it.

Before considering details of either the light reactions or the dark reactions, it is appropriate to consider the sites of photosynthesis. Just as all eukaryotic cells have organelles (mitochondria) specialized for oxidative metabolism, plants and algae have organelles specialized for photosynthesis.

The Chloroplast

In all higher plants and algae, photosynthetic processes are localized in **chloroplasts**. In plants, most of these organelles are found in cells just under the leaf surface (mesophyll cells). Each cell may contain 20 to 50 chloroplasts (Figure 16.4). Eukaryotic algae also have chloroplasts, but often only one very large one is found in each cell.

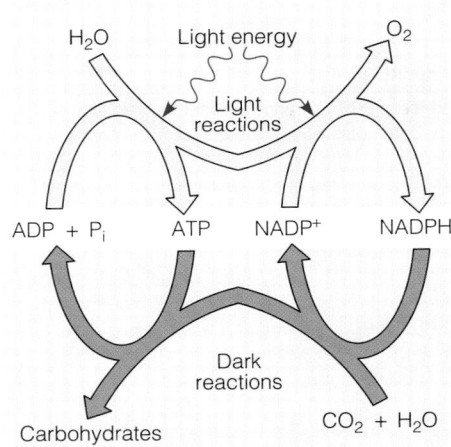

FIGURE 16.3

The two subprocesses of photosynthesis. The overall process of photosynthesis is divided into light reactions and dark reactions. The light reactions, which require visible light as an energy source, produce reducing power (in the form of NADPH), ATP, and O_2. The NADPH and ATP drive the so-called dark reactions, which occur in both the presence and the absence of light and fix CO_2 into carbohydrates.

Photosynthesis can be divided into light reactions, which use sunlight energy to produce NADPH and ATP, releasing O_2 in the process, and dark reactions, which use NADPH and ATP to fix CO_2.

Photosynthesis in plants and algae occurs in chloroplasts.

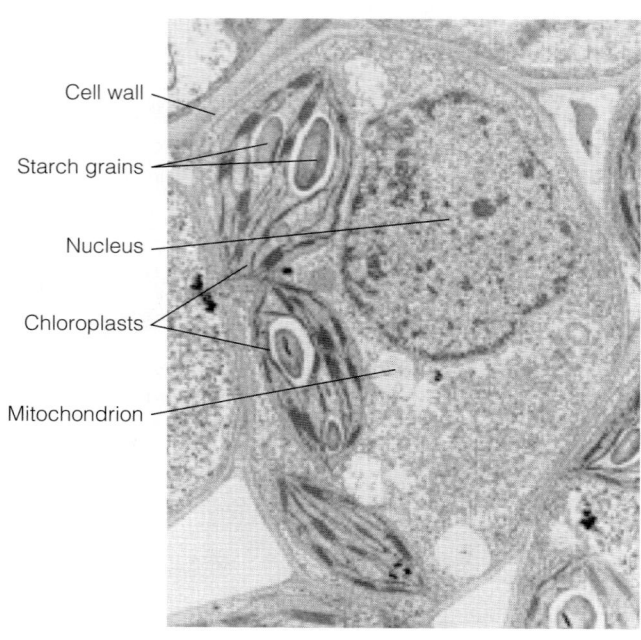

Cell wall

Starch grains

Nucleus

Chloroplasts

Mitochondrion

(a)

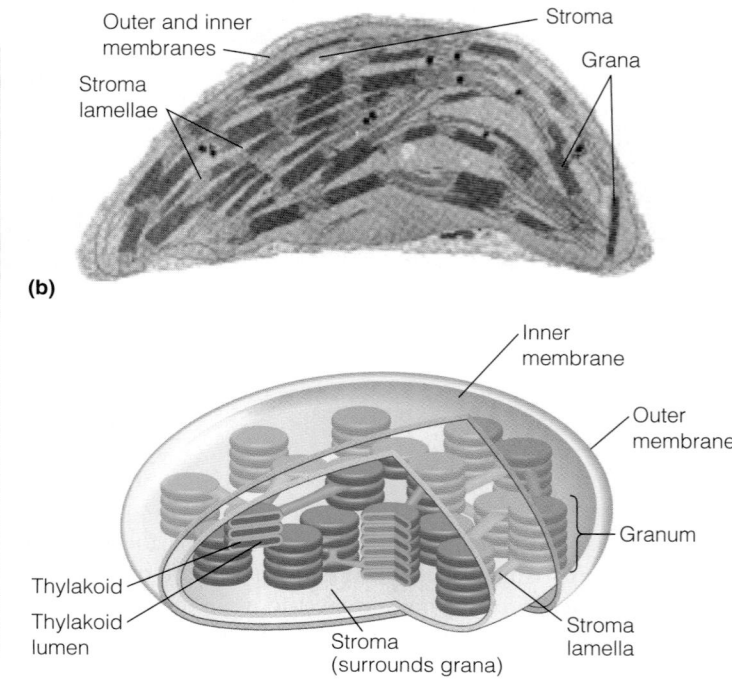

Outer and inner membranes

Stroma lamellae

Stroma

Grana

(b)

Inner membrane

Outer membrane

Granum

Thylakoid

Thylakoid lumen

Stroma (surrounds grana)

Stroma lamella

(c)

FIGURE 16.4

Chloroplasts, the photosynthetic organelles of green plants and algae. **(a)** Several chloroplasts are shown in a cross-section of a cell from a *Coleus* leaf. **(b)** Enlarged view of a single chloroplast from a leaf of timothy grass. **(c)** Schematic rendering of a chloroplast.

(a) Micrograph by M. W. Steer, photo provided by E. H. Newcomb; (b) micrograph by K. P. Wergin, photo provided by E. H. Newcomb/BPS; (c) *Biology*, 5th ed., Neil A. Campbell, Jane B. Reece, and Lawrence A. Mitchell. © 1999. Reprinted by permission of Pearson Education Inc., Upper Saddle River, NJ.

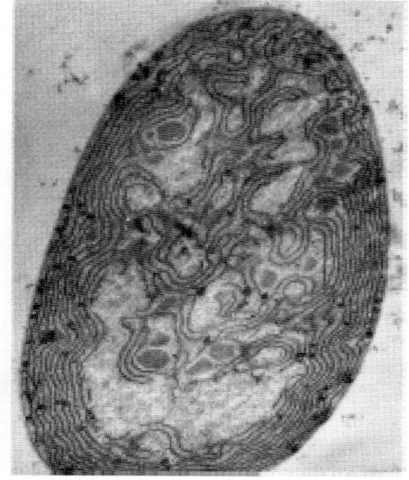

FIGURE 16.5

A photosynthetic prokaryote. This electron micrograph of a thin section of the cyanobacterium *Anabaena azollae* shows the folded membranes, which resemble the thylakoids of eukaryotic chloroplasts.

Courtesy of N. Lang, University of California, Davis/BPS.

Like mitochondria, chloroplasts are semiautonomous, carrying their own DNA to code for some of their proteins, as well as the ribosomes necessary for translation of the appropriate messenger RNAs. There is now much evidence that chloroplasts evolved from unicellular organisms similar to cyanobacteria (blue-green algae). Such prokaryotic photosynthesizers do not contain chloroplasts but have membrane structures that play the same roles as chloroplast membranes (Figure 16.5). To a certain extent, the cyanobacteria resemble free-living chloroplasts. It is believed that, early in evolution, primitive unicellular organisms took up cyanobacteria-like prokaryotes and that eventually the relationship became symbiotic: The photosynthetic organelles were no longer capable of independent life, and the algae depended upon them as energy sources. Today, some chloroplast genes are coded in the organelle genome, and some are in the cell nucleus.

The internal structure of a chloroplast, as shown in Figure 16.4b and c, bears some resemblance to that of a mitochondrion (see Figure 15.2a), except that chloroplasts have a third set of membranes. There is an outer, freely permeable membrane and an inner membrane that is selectively permeable. The inner membrane encloses a compartment called the **stroma** that is analogous to the mitochondrial matrix. Immersed in the stroma are many flat, saclike membrane structures called **thylakoids**, which are often stacked like coins to form units called *grana* (see Figure 16.4c). Individual grana are irregularly interconnected by thylakoid extensions called *stroma lamellae*. This third membrane, the thylakoid membrane, encloses an interior space, the **lumen** of the thylakoid.

The division of labor within a chloroplast is simple. Absorption of light and all of the light reactions occur within or on the thylakoid membranes. The ATP and NADPH produced by these reactions are released into the surrounding stroma where all of the synthetic dark reactions occur. Thus, there are analogies in structure and role between mitochondrial matrix and chloroplast stroma and between the inner membrane of the mitochondrion and the thylakoid membrane of the chloroplast. Indeed, we shall find that a very similar kind of chemiosmotic ATP generation is carried out across these membranes in both mitochondria and chloroplasts. To see how this ATP generation occurs, we must first examine the light reactions in detail, beginning with the process of light absorption.

The Light Reactions

Absorption of Light: The Light-Harvesting System

The Energy of Light

Absorption of light and the light reactions occur in the chloroplast membranes. The dark reactions occur in the stroma.

To understand how energy from sunlight can be captured and utilized, we must first review the nature of electromagnetic radiation. The quantum mechanical theory of radiation states that light (and all other electromagnetic radiation) has two aspects: wavelike and particle-like. We can characterize a particular kind of radiation by its wavelength (λ) or frequency (ν); these parameters characterize the *wave* aspects of the light. If waves with a length of λ are passing an observer at a velocity c, the number of waves passing per second is the frequency, ν. Thus,

$$\nu = \frac{c}{\lambda} \qquad (16.1)$$

where c is the velocity of light, 2.998×10^8 m/s. The red light from a neon laser has a wavelength of 632.8 nm, or 6.328×10^{-7} m. Thus, its frequency is $4.74 \times 10^{14}\ s^{-1}$. But to see how *energy* might be obtained from light, it is necessary to consider the particulate aspect of radiation. We must think of a light beam as a stream of light particles, or **photons**. Each photon has an associated unit of energy called a **quantum**. The energy value of a quantum—that is, the energy per photon—is related to the frequency of the light by one of the most basic equations in physics, Planck's law (see Tools of Biochemistry 6A)

$$E = h\nu = \frac{hc}{\lambda} \qquad (16.2)$$

where h is Planck's constant, 6.626×10^{-34} J s. Thus, the neon laser in our example can deliver light energy only in packets, or quanta, of 3.14×10^{-19} J (or $[6.626 \times 10^{-34}\ \text{J s}] \times [4.74 \times 10^{14}\ s^{-1}]$). However, biochemists rarely deal with single photons. Because we are interested in how radiation can promote chemical or biochemical processes, which are usually expressed on a molar basis, the more appropriate quantity for our purposes is the energy of a *mole* (6.02×10^{23}) of photons. For the neon laser light, multiplying the energy per photon by 6.02×10^{23} gives 189 kJ. A mole of photons is called one **einstein**.

Figure 16.6 shows a graph of energy per mole of photons as a function of wavelength, through the infrared, visible, and ultraviolet parts of the spectrum. For comparison, the energies associated with molecular vibrations and various covalent bonds are indicated. When photons of infrared radiation are absorbed by a molecule, they can do little except stimulate molecular vibrations, which we perceive as heat. Photons of far-ultraviolet radiation, on the other hand, have energies quite capable of breaking covalent bonds. Far-ultraviolet radiation is chemically

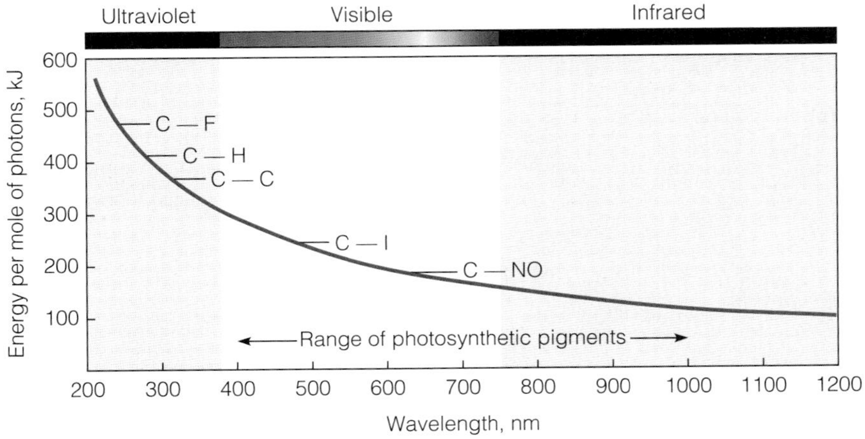

FIGURE 16.6

The energy of photons. The graph shows energy per mole of photons as a function of wavelength, compared with energies of several chemical bonds. Light in the ultraviolet range has enough energy to break many chemical bonds directly. Visible light can break some weak bonds. Light in the long-wavelength portion of infrared region of the spectrum causes only heat-producing molecular vibrations.

destructive to humans and to other organisms, but fortunately most of it is screened from the earth's surface by the ozone layer. This is one of the reasons why depletion of the ozone layer is of such serious concern.

Photosynthesis depends primarily on light in the visible and near-infrared regions of the spectrum, lying between the extremes of covalent bond–breaking and stimulating molecular vibrations. Photons in the visible and near infrared are not highly destructive, but can cause transitions in the electronic states of organic molecules that can drive reactions and thus capture the energy in a chemical form. The ability to use radiation in this range has had clear evolutionary advantages for photosynthetic organisms. Most of the sun's energy that reaches the earth's surface lies in this spectral range. The small amount of ultraviolet radiation that does get through can penetrate only a very short distance into water and thus would have been unavailable to primitive photosynthetic organisms living in the sea. The photons of far-infrared radiation have energies too low to be useful for any photochemical processes.

The Light-Absorbing Pigments

To capture the useful portion of the light energy, photosynthetic organisms have evolved a set of pigments that efficiently absorb visible and near-infrared light. The light-absorbing portions of these pigments are referred to as **chromophores**. Structures of a few of the most important photosynthetic chromophores are shown in Figure 16.7. In Figure 16.8 the absorption spectra of these photosynthetic pigments are compared with the distribution of solar radiation in the spectrum. Together, the chromophores "blanket" the visible spectrum; scarcely a photon can

FIGURE 16.7

Some photosynthetic pigments. Chlorophylls *a* and *b* are the most abundant plant and algal pigments, whereas β-carotene and phycocyanin are examples of accessory pigments. Phycocyanin and the related phycoerythrin are open-chain tetrapyrroles that are covalently attached to **phycobiloproteins** via a sulfhydryl group, and they are abundant in aquatic photosynthetic organisms. These pigments absorb strongly in the 500–600 nm range, wavelengths that can efficiently pass through water. There are also bacteriochlorophylls, which differ slightly in structure. The colored highlights indicate the extended conjugated double-bond systems.

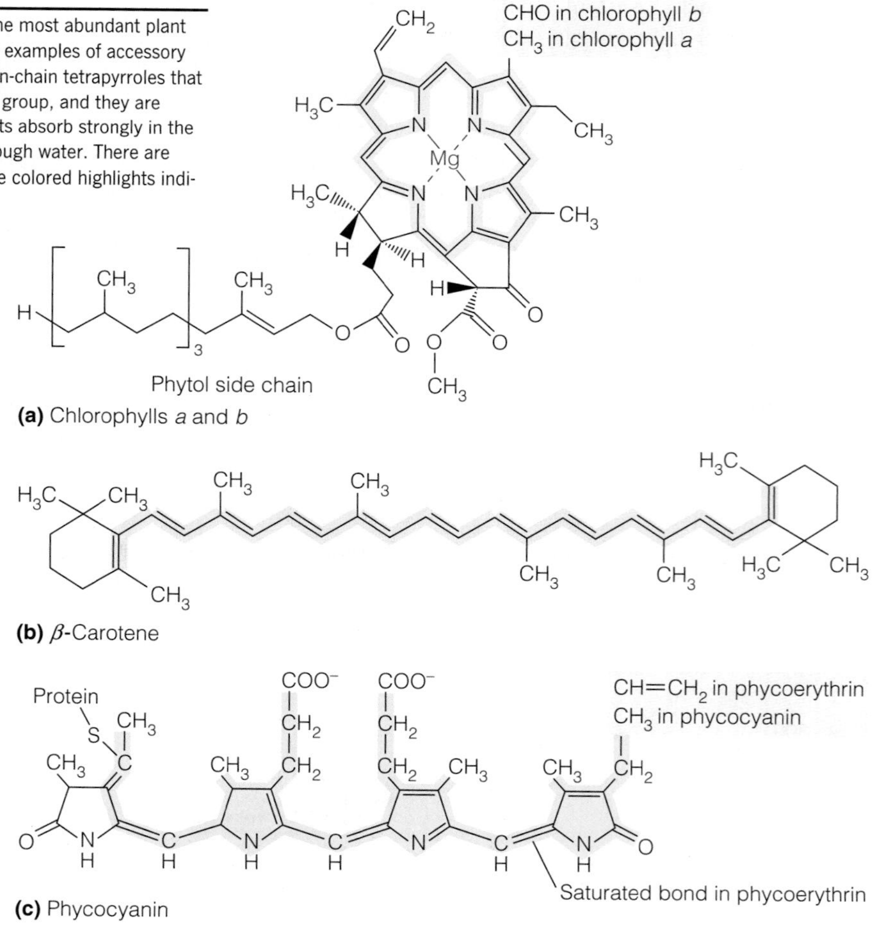

(a) Chlorophylls *a* and *b*

(b) β-Carotene

(c) Phycocyanin

come through that cannot be absorbed by one chromophore or another. The most abundant pigments in higher plants are chlorophyll *a* and chlorophyll *b*. As you can see by comparing Figure 16.7a with Figure 7.4b (page 236), these molecules are related to the protoporphyrin IX found in hemoglobin and myoglobin. However, the bound metal in the chlorophylls is Mg^{2+} rather than Fe^{2+}. In Figure 16.7b and c, two accessory pigments are also shown. All of these molecules absorb light in the visible region of the spectrum because they have large conjugated double-bond systems. Because chlorophylls *a* and *b* absorb strongly in both the deep blue and red, the light that is *not* absorbed but *reflected* from chloroplasts is green, the color we associate with most growing plants. The other observed colors, such as the red, brown, or purple of algae and photosynthetic bacteria, are accounted for by differing amounts of *accessory pigments*. Loss of chlorophylls in autumn leaves allows the colors of the accessory pigments, as well as nonphotosynthetic pigments, to become evident. Carotenoids are the most abundant accessory pigments in plants. These include the red-orange β-carotene (Figure 16.7b) and xanthophylls (oxygen-containing carotenoids) such as the yellow lutein. Some photosynthetic bacteria use pigments that absorb wavelengths up to about 1000 nm, in the near infrared.

The Light-Gathering Structures

Chlorophyll and some of the accessory pigments are contained in the **thylakoid membranes** of the chloroplast. The composition of these membranes is rather unusual. They contain only a small fraction of the common phospholipids but are rich in glycolipids. They also contain much protein, and some of the photosynthetic pigments are attached to certain of these proteins. Other photosynthetic pigments, including chlorophylls *a* and *b*, are not covalently bound but interact with both proteins and membrane lipids. These pigments interact with membrane lipids through their hydrophobic phytol tails (see Figure 16.7a).

The assemblies of light-harvesting pigments in the thylakoid membrane, together with their associated proteins, are organized into well-defined **photosystems**, structural units dedicated to the task of absorbing light photons and recovering some of their energy in a chemical form. As we shall see, plants use two distinct photosystems (I and II). The first part of photosynthesis takes place in what are referred to as **light-harvesting complexes (LHCs)**. Each is a multisubunit protein complex containing multiple **antenna** pigment molecules (chlorophylls and some accessory pigments) and a pair of chlorophyll molecules that act as the **reaction center**, trapping energy quanta excited by the absorption of light.

To understand how this system functions, we must look a bit more closely into what can happen when a molecule absorbs a quantum of radiant energy. Recall from Tools of Biochemistry 6A that absorption in the visible region of the spectrum excites the molecule from the ground state to a higher electronic state. In the case of the photosynthetic pigments, the excited electron occupies a π orbital in the conjugated bond system. In Tools of Biochemistry 6A we described two ways in which the energy could be lost to return the molecule to its ground state: radiationless dissipation of the energy as heat, or re-radiation as fluorescence. However, when similar absorbing molecules are packed tightly together, as in a photosystem, two other possibilities arise. First, the excitation energy may be passed from one molecule to an adjacent one—a process called *resonance transfer* or *exciton transfer* (Figure 16.9a). Alternatively, the excited electron itself may be

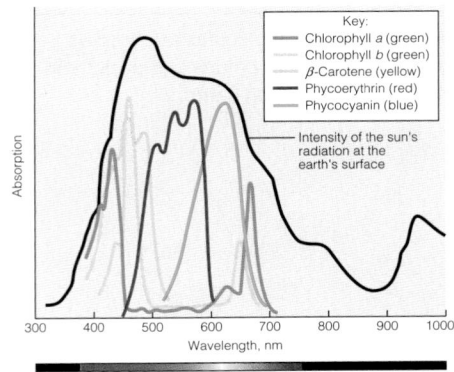

FIGURE 16.8

Absorption spectra and light energy. The absorption spectra of various plant pigments are compared with the spectral distribution of the sunlight that reaches the earth's surface.

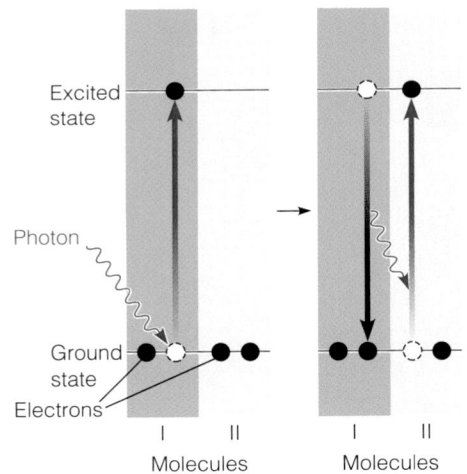

(a) Resonance transfer

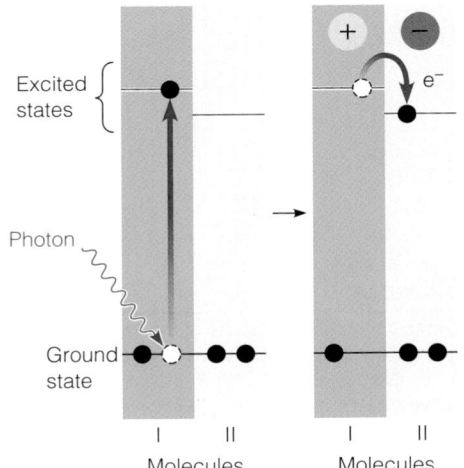

(b) Electron transfer

FIGURE 16.9

Two modes of energy transfer following photoexcitation. For each of the two types of energy transfer that occur in a photosystem, the left-hand illustration shows a molecule being excited to a higher energy state by absorption of a photon of radiation. The right-hand illustrations show how the energy is transferred to an adjacent molecule. **(a)** In *resonance transfer* molecule I transfers its excitation energy to an identical molecule II, which rises to its higher energy state as molecule I falls back to the ground state. Resonance transfer is extraordinarily fast. **(b)** In *electron transfer* an excited electron in molecule I is transferred to the slightly lower excited state of molecule II, making molecule I a cation and molecule II an anion.

Most chlorophyll molecules are used as antennae to catch photons and pass their energy on to reaction centers.

passed to a nearby molecule with a slightly lower excited state—an *electron transfer* reaction (Figure 16.9b). Both of these processes are important in photosynthesis.

The clue that eventually led to the recognition that resonance transfer played a role in photosynthesis came from measurements by Robert Emerson and William Arnold in the 1930s. They showed that even when the photosynthetic system of the alga *Chlorella* was operating at maximum efficiency, only one O_2 molecule was produced for every 2500 chlorophyll molecules. As we now realize, most of the chlorophyll molecules are not directly engaged in the photochemical process itself but act, instead, as antenna molecules of the light-harvesting complexes. The structure of a light-harvesting complex (LHCII) is shown in Figure 16.10. Antenna molecules absorb photons, and the energy is passed by resonance transfer to specific chlorophyll molecules in a relatively few reaction centers. In other words, the energy of a photon absorbed by any antenna molecule in a photosystem wanders about the system randomly (Figure 16.11). Eventually (meaning in about 10^{-10} s), the energy finds its way to a chlorophyll molecule in the reaction center. This molecule is like the other chlorophylls, except it is in a different protein microenvironment so that its excited state energy level is a bit lower. Thus, it acts as a trap for quanta of energy absorbed by any of the other pigment molecules. It is the excitation of this reaction center that begins the actual photochemistry of the light reactions, for it starts a series of electron transfers.

Photochemistry in Plants and Algae: Two Photosystems in Series

Our understanding of the photochemical light reactions has developed from many elegant experiments in many different laboratories. A pioneering study in 1939, by Robert Hill at the University of Cambridge, made the seminal observation that isolated chloroplasts could promote redox chemistry when illuminated

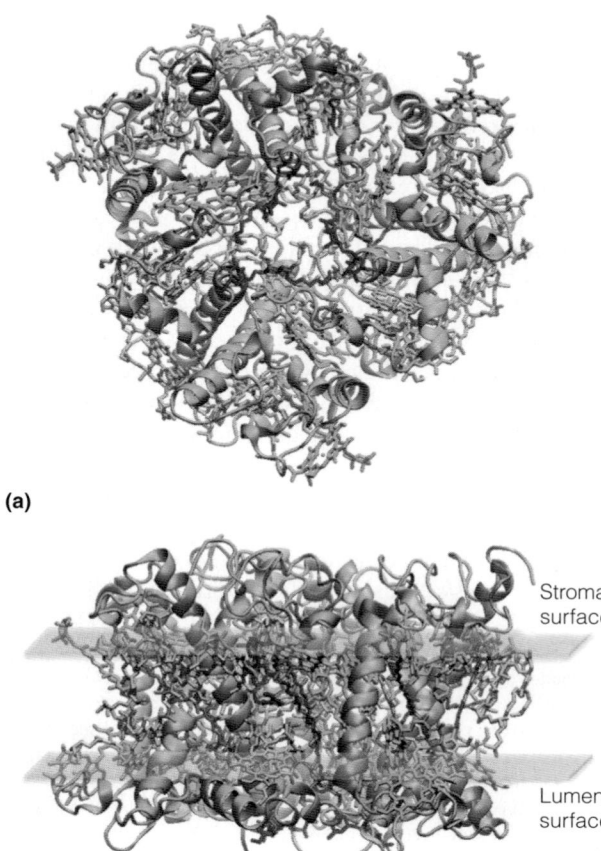

(a)

(b)

Stromal surface

Lumenal surface

FIGURE 16.10

Three-dimensional structure of the trimeric light-harvesting complex II of plants. The X-ray structure of the pea LHCII (PDB ID 2BHW) shows that the protein exists as a homotrimer buried in the thylakoid membrane. **(a)** A top view from the lumenal side of the membrane. The three monomers are in different shades of cyan. **(b)** A side view of the trimer (approximate locations of the stromal and lumenal faces of the membrane are indicated). Each trimer contains 24 chlorophyll *a* molecules (light green), 18 chlorophyll *b* molecules (green), and 12 carotenoids (luteins and xanthophylls; orange), all serving as antenna molecules. Bound lipids are shown in purple.

in the presence of any of a variety of electron acceptors. For example, when ferricyanide was used, the following reaction proceeded efficiently:

$$4\text{Ferricyanide } [\text{Fe(CN)}_6{}^{3-}] + 2H_2O \xrightarrow{\text{Light energy}} 4\text{Ferrocyanide } [\text{Fe(CN)}_6{}^{4-}] + 4H^+ + O_2$$

A number of such reactions involving different inorganic oxidants are known and are now referred to collectively as *Hill reactions*. Such reactions, in the absence of photochemical activation, are very unfavorable. Ferricyanide, for example, is a much weaker oxidant than O_2 (refer to Table 15.1, page 629); $\Delta G^{\circ\prime}$ for the reaction as written is about $+178 \text{ kJ/mol } O_2$ [from eqn. 15.1; $\Delta G^{\circ\prime} = -4(96485 \text{ J mol}^{-1}\text{V}^{-1}) \times (0.36 \text{ V} - 0.82 \text{ V})$]; so the equilibrium should lie far to the left. Hill's discoveries showed that *chloroplasts irradiated by light are capable of driving thermodynamically unfavorable reactions*. The Hill reactions also demonstrated that the photosynthetic system can oxidize water to O_2 without any involvement of CO_2 (see Figure 16.3). This observation was the first clear indication that the light and dark reactions are separate processes, and it led ultimately to the discovery that the final electron acceptor of the light reactions in vivo is $NADP^+$, yielding NADPH.

Further studies revealed that *two* kinds of photosystems must be involved in photosynthesis in plants. The first hint came from experiments that measured the quantum efficiency of photosynthesis in algae, using light of different wavelengths. The *quantum efficiency* (Q) is the ratio of oxygen molecules released to photons absorbed. As the wavelength of the monochromatic light used was raised above 680 nm (far red), an abrupt drop in Q was noted. This "red drop" was a strange observation, for chlorophylls in plants still show appreciable absorbance even at higher wavelengths. Somehow the energy was just not being used as efficiently above 680 nm. Even more remarkable was the observation that simultaneous illumination with yellow light (650 nm) produced a marked increase in the quantum efficiency from light at 700 nm. Even if the yellow light was switched off a few minutes before the measurement, the quantum efficiency remained high. The only reasonable explanation for these results is that two complementary photosystems exist, one absorbing most strongly at wavelengths around 700 nm and the other at shorter wavelengths. The action of *both* must be required for photosynthesis to proceed with maximal efficiency.

The two photosystems predicted by early experimenters have now been identified and characterized. They are both localized in the thylakoid membrane. Each photosystem is a multisubunit, transmembrane protein complex, carrying antenna and reaction center chlorophyll molecules and electron transport agents. The photosystems have been named according to the order in which they were discovered. The one showing absorbance up to 700 nm is called **photosystem I (PSI)**, and the one that absorbs only to a wavelength of about 680 nm is called **photosystem II (PSII)**. In algae, cyanobacteria, and all higher plants, these two photosystems are linked in series to carry out the complete sequence of the light reactions. The basic sequence is illustrated in Figure 16.12a, which depicts the

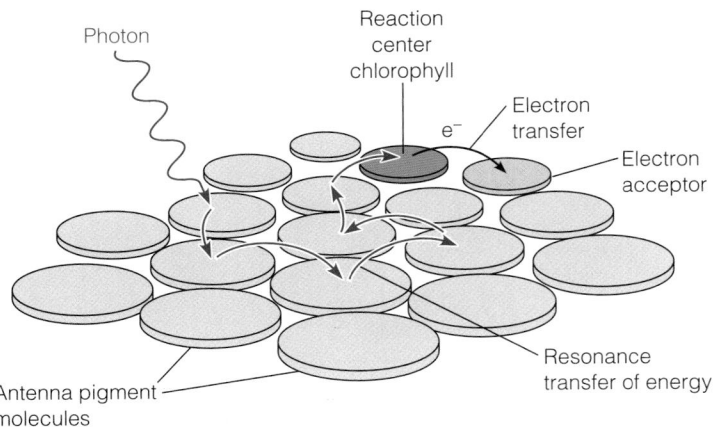

Photon

Reaction center chlorophyll

e^-

Electron transfer

Electron acceptor

Resonance transfer of energy

Antenna pigment molecules

FIGURE 16.11

Resonance transfer of energy in a light-harvesting complex. The excitation energy originating in a photon of light wanders from one antenna molecule to another until it reaches a reaction center. There an electron is transferred to a primary electron acceptor molecule, and the energy is trapped.

FIGURE 16.12

The two-photosystem light reactions. In the two-photosystem mode of photosynthesis, the light reactions are carried out by two photosystems linked in series. **(a)** A schematic view of the path of electrons through the two photosystems. The two systems and the cytochrome complex are embedded in the thylakoid membrane. Electrons taken from water in photosystem II are transferred to photosystem I via plastoquinones (Q), the cytochrome b_6f complex, and plastocyanin (PC). In photosystem I the electrons are excited by light again, for transfer via a series of intermediates to ferredoxin. Reduced ferredoxin reduces $NADP^+$. **(b)** Energetics of the two-photosystem light reactions. In each of the two reaction centers, P680 and P700, electrons are raised to an excited state by absorption of photons. In each photosystem the excited electrons are passed through an electron transport chain, which drives the pumping of protons into the thylakoid lumen. The two-photosystem mode has historically been called the Z-scheme because of the pattern of energy changes shown here, but N-scheme would be more accurate. The estimate of eight protons pumped by the cytochrome b_6f complex is based on the Q cycle translocating two protons for each electron transported. **Key:** OEC = oxygen-evolving complex; Y_Z = donor to P680; P680 = photosystem II reaction center chlorophyll; Ph = pheophytin acceptor; Q_A, Q_B = protein-bound plastoquinones; QH_2 = plastoquinol (reduced plastoquinone) in membrane; Cyt b_6f = cytochrome b_6f complex; PC = plastocyanin; P700 = photosystem I reaction center chlorophyll; A_0 = chlorophyll acceptor; A_1 = protein-bound phylloquinone; F_A, F_B, F_X = iron–sulfur clusters; Fd = ferredoxin; FNR = ferredoxin:$NADP^+$ oxidoreductase.

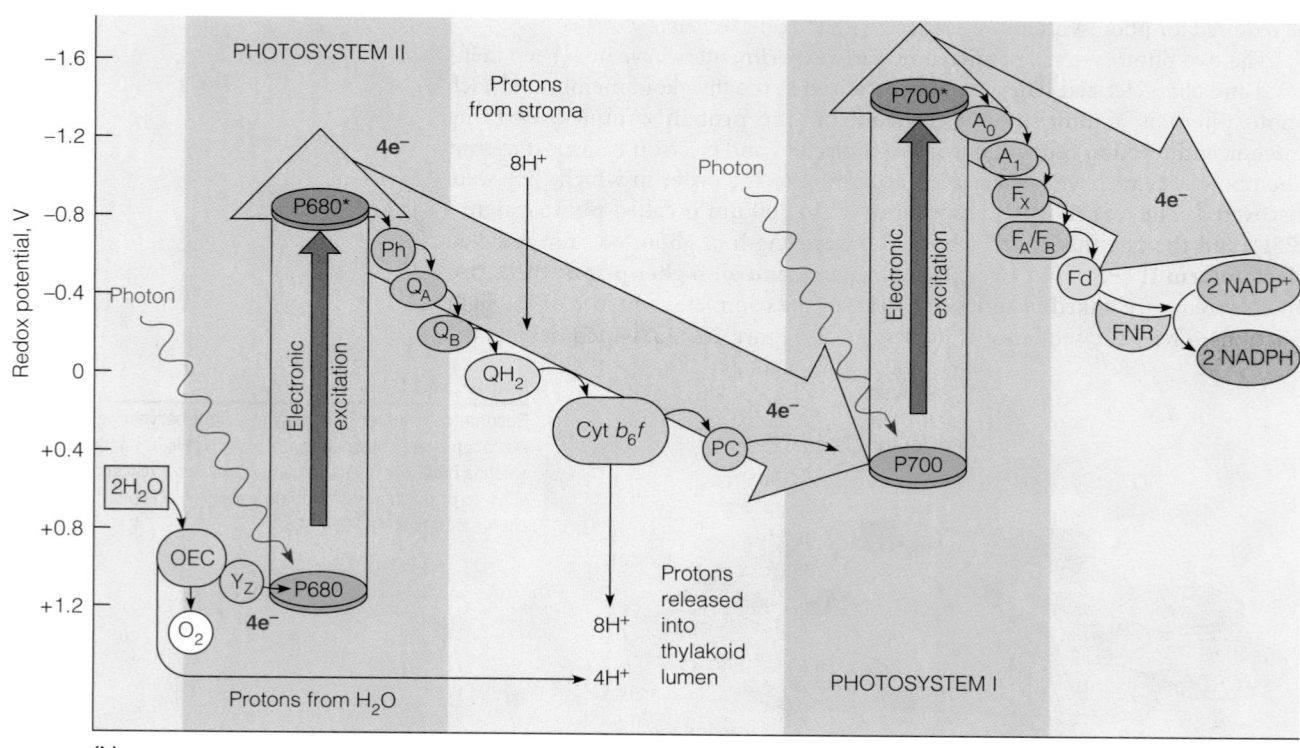

path of electrons through the two systems. Figure 16.12b emphasizes the energetics of the electron flow and places the major participants in the light reactions on a scale of reduction potential.

In each of the two photosystems, the primary step is transfer of a light-excited electron from a reaction center (P680 or P700) into an electron transport chain. The ultimate source of the electrons is the water molecules shown at the left in both parts of Figure 16.12. The final destination of the electrons is the molecule of $NADP^+$ at the right, which is thereby reduced to NADPH. At two stages in the electron transport process, protons are released into the thylakoid lumen. Some of the protons come from the H_2O that is broken down, some come from the stroma. This transfer of protons into the lumen produces a pH gradient across the thylakoid membrane. It should not surprise you to find that this proton gradient drives ATP production, just as a proton gradient drives ATP synthesis in mitochondria (see Chapter 15). Thus, ATP and reducing power in the form of NADPH are the products of the light reactions. These compounds are exactly what is needed to drive the syntheses carried out in the dark reactions. To examine their generation in detail, we begin with photosystem II, for it is where electrons enter the scheme.

Two photosystems, linked in series, are involved in the photosynthetic light reactions in algae, cyanobacteria, and higher plants.

Photosystem II: The Splitting of Water

Each of the photosystems contains an electron transport chain, which extracts energy when an excited electron loses its energy of excitation in a stepwise fashion. The photosystem carries out a series of oxidation–reduction reactions. It is easiest to follow the events in a photosystem by starting with the absorption of a photon picked up by the light-harvesting system of photosystem II. The photon is funneled to a reaction center chlorophyll, designated P680 in Figure 16.12. Excitation of P680 raises the molecule from the ground state to an excited state at -0.8 volts. Thus, the excited P680 has become an excellent reducing agent, able to quickly transfer an electron from P680 to a lower-energy primary electron acceptor—*pheophytin a* (*Ph*), as shown in Figure 16.12b. The pheophytins are molecules identical to chlorophylls, except that two protons substitute for the centrally bound magnesium ion. We can consider this excited electron as a low-redox-potential electron.

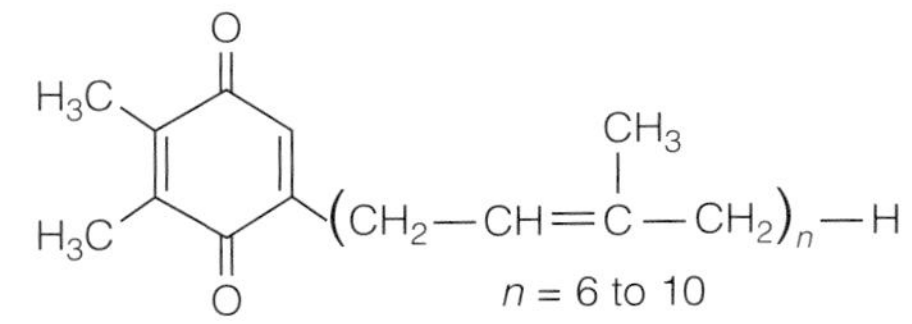

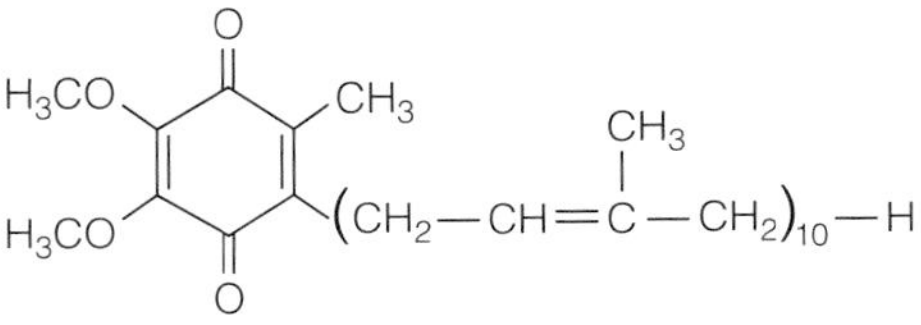

The electron is then transferred to a series of *plastoquinone* molecules (Q_A and Q_B) associated with PSII proteins. Ultimately, two electrons and two protons are picked up by the plastoquinone Q_B; the protons come from the stroma. The reduced plastoquinone, QH_2 (*plastoquinol*), is then released into the lipid portion of the thylakoid membrane. The overall reduction of plastoquinone can be written as follows:

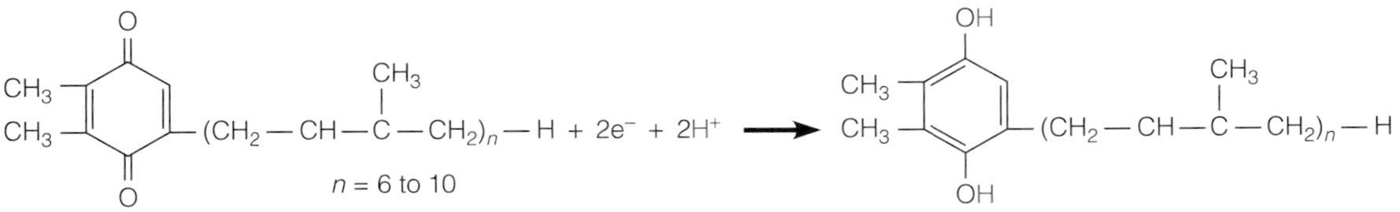

Plastoquinone

Plastoquinol

Note the structural similarity between plastoquinone and the ubiquinone (coenzyme Q) found in respiratory chains (Chapter 15). Plastoquinol then interacts with a membrane-bound complex of cytochromes and iron–sulfur proteins, the cytochrome $b_6 f$ complex. This complex catalyzes the transfer of the electrons to a copper protein, *plastocyanin* (PC). In doing so, the $b_6 f$ complex serves two purposes. First, it transmits activated electrons from photosystem II to photosystem I. At the same time, it pumps protons from the stroma into the thylakoid lumen (2 H^+ per electron). The major components of this complex are cytochrome f (which contains one c-type heme), cytochrome b_6 (which contains two b-type hemes), and a Rieske iron–sulfur protein (Figure 16.13a). The cytochrome $b_6 f$

FIGURE 16.13

Structure of the cytochrome b_6f complex. **(a)** The X-ray structure (PDB ID 2D2C) of the complex from the thermophilic cyanobacterium, *Mastigocladus laminosus*, shows that the protein exists as a homodimer buried in the thylakoid membrane. This view is in the plane of the membrane, with the thylakoid lumen at the top. Cytochrome *f* subunits are red, with the *c*-type hemes shown in white; the Rieske iron–sulfur protein subunits are yellow, with the Fe_2S_2 iron–sulfur centers shown in orange; Cytochrome b_6 subunits are blue, with the b_H and b_L hemes shown in white; and subunits IV are purple. Four additional small subunits per monomer are shown in green. The lower panels compare the structures of the photosynthetic cytochrome b_6f complex **(b)** (PDB ID 2E74) and the mitochondrial complex III **(c)** (PDB ID 1L0L). Homologous or analogous subunits have the same colors, based on the color scheme in panel (a).

(b, c) *Photochemistry and Photobiology* 84:1349–1358, D. Baniulis, E. Yamashita, H. Zhang, S. S. Hasan, and W. A. Cramer, Structure–function of the cytochrome b_6f complex. © 2008 John Wiley & Sons, Inc. Reproduced with permission from John Wiley & Sons, Inc.

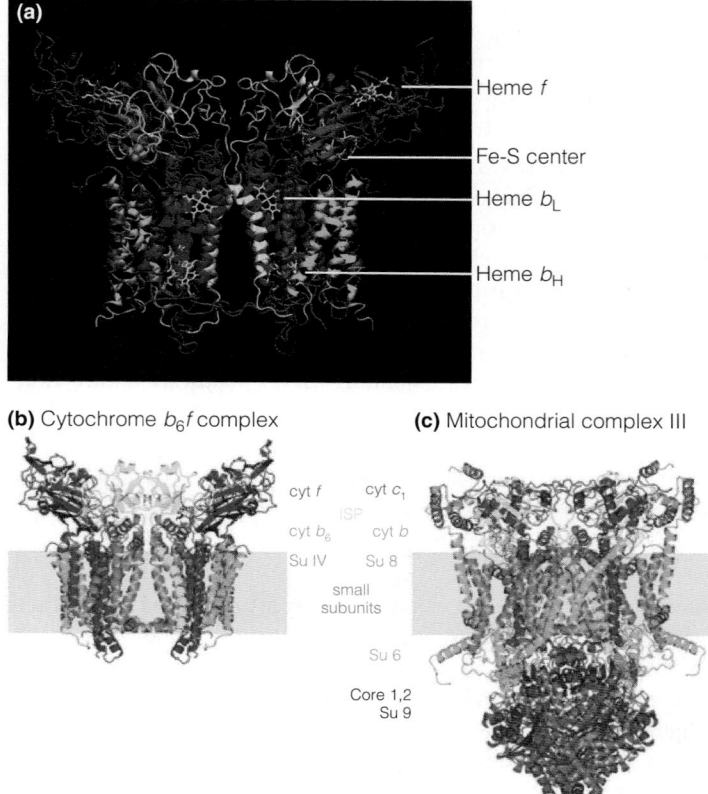

(a)

Heme *f*

Fe-S center

Heme b_L

Heme b_H

(b) Cytochrome b_6f complex

cyt *f*　　cyt c_1
ISP
cyt b_6　　cyt *b*
Su IV　　Su 8
small subunits

(c) Mitochondrial complex III

Su 6

Core 1,2
Su 9

Photosystem II extracts electrons from water, passing them to photosystem I and releasing O_2.

complex is thus analogous to complex III of the mitochondrial respiratory chain (Figure 16.13b, c), and catalyzes a similar Q cycle (as shown in Figure 15.15, page 641). As plastoquinol is oxidized back to plastoquinone, the two protons it has taken from the stroma are released into the thylakoid lumen. Plastocyanin, a mobile protein in the thylakoid lumen, passes the electrons on to P700 reaction centers. In this process the copper in plastocyanin is first reduced to Cu(I) and then reoxidized to Cu(II). We shall consider the fate of the electrons passed to P700 when we discuss photosystem I.

Note that the processes we have described so far have left P680 reaction centers deficient in electrons—in other words, oxidized to a strong oxidant, $P680^+$. These electrons are regained from water, which is split in the presence of an electron acceptor, releasing oxygen in the process. New insight into the mechanism of this water-splitting process has recently been provided by X-ray structures of photosystem II complexes from several cyanobacterium. Figure 16.14 shows PSII as a large multisubunit complex embedded in the thylakoid membrane. Each monomer of the homodimer comprises 20 subunits, including antenna proteins with their numerous antenna chlorophylls surrounding two subunits (D1 and D2) that contain the P680 reaction center chlorophylls and the water-splitting catalytic components. As indicated in Figure 16.12b (left), the electron acceptor is a subunit of PSII referred to as the oxygen-evolving complex (OEC). The OEC contains a cube-shaped cluster of four oxygen-bridged manganese ions and one calcium ion. This metal cluster can exist in a series of oxidation states, as indicated in Figure 16.15a; light-driven cycling through these oxidation states allows the cluster to dismantle two water molecules, passing four electrons back to P680 and releasing the four accompanying protons into the thylakoid lumen. Exactly at which points in the cycle individual electrons and protons are released is still a matter of

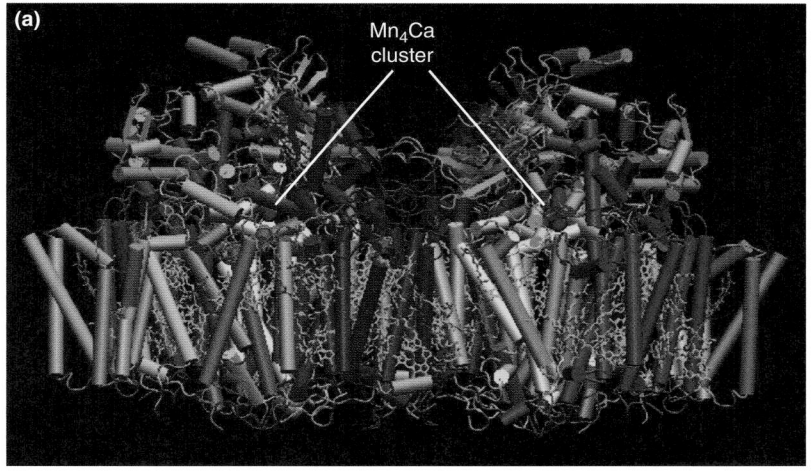

FIGURE 16.14

Structure of photosystem II. (a) The X-ray structure (PDB ID 3BZ1, 3BZ2) of PSII from the thermophilic cyanobacterium, *Thermosynechococcus elongatus*, shows that the protein exists as a homodimer buried in the thylakoid membrane. This view is in the plane of the membrane, with the thylakoid lumen at the top. Each monomer is composed of 20 subunits, including the antenna proteins CP47 (red) and CP43 (magenta), the reaction center subunits D1 (yellow) and D2 (orange), and two cytochrome b559 subunits (lime). Each monomer contains 35 chlorophyll a molecules (green) and 12 carotenoids (orange) bound to CP43 and CP47. The Mn_4Ca cluster of the oxygen-evolving center is shown in blue, bound by the D1 subunit of each monomer. **(b)** The structure of the Mn_4Ca cluster from the PSII of *Thermosynechococcus vulcanus* is shown in greater detail. Distances (in angstroms) between the metal atoms and oxygen atoms or water molecules are indicated.

(b) Reprinted by permission of Macmillan Publishers Ltd. *Nature* 473:55–60, Y. Umena, K. Kawakami, J.-R. Shen and N. Kamiya, Crystal structure of oxygen-evolving photosystem II at a resolution of 1.9Å. © 2011.

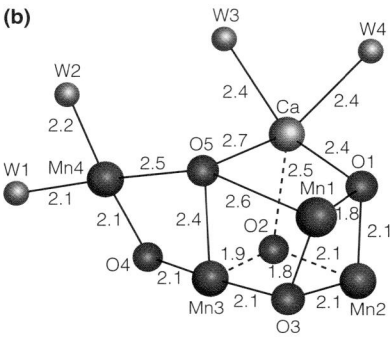

debate, with several models proposed. In the version shown in Figure 16.15b, electrons and protons are released in pairs, which amounts to a hydrogen extraction. This idea is consistent with the observation that the electron donor returning electrons to oxidized P680 is a redox-active tyrosine in the D1 subunit of PSII. This yields a tyrosine radical ($Y_Z\cdot$) which has been observed by electron spin resonance (see Tools of Biochemistry 10A), and releases a proton. It is proposed, then,

FIGURE 16.15a

A model for the catalytic function of the oxygen-evolving complex (OEC) cluster in PSII. **(a)** S_0–S_4 represent the different oxidation states that the ligated metal cluster cycles through as e^- and H^+ are abstracted from the H_2O molecules. In the first four transitions, light energy is used to oxidize P_{680} to P_{680}^+, which in turn oxidizes the metal-oxo cluster of the OEC. The $S_4 \rightarrow S_0$ transition is light-independent and releases O_2. Thus, four photons are required to oxidize two H_2O to one O_2.

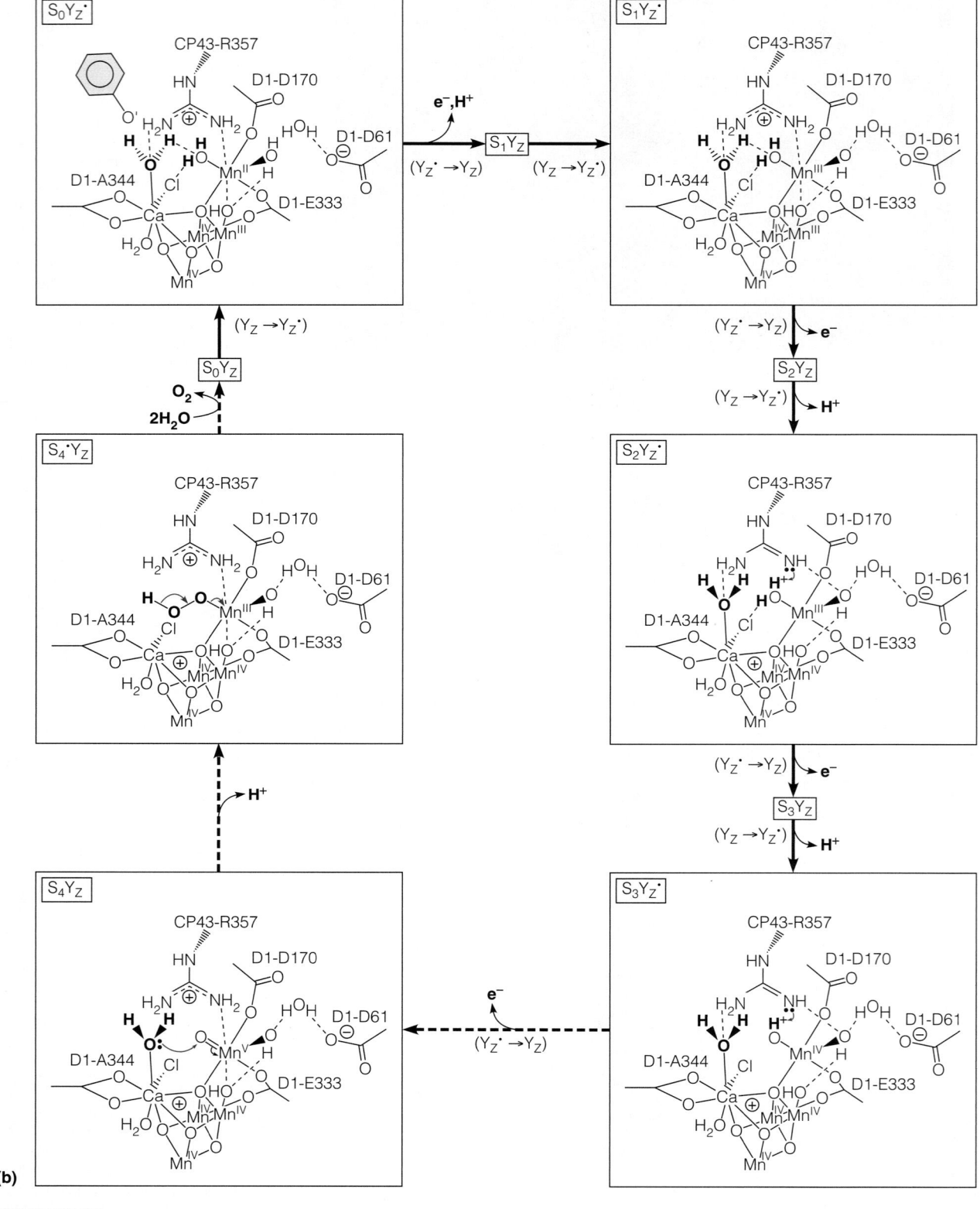

(b)

FIGURE 16.15b

(b) In this proposed mechanism, Y_Z represents the redox-active tyrosine residue of the D1 subunit; the oxidized form of Y_Z is the neutral, deprotonated radical species, $Y_Z\cdot$. The two H_2O molecules are bound in the S_0Y_Z state by the Ca^{2+} ion and the "dangler" Mn^{III} ion. Each transition passes through an intermediate $S_nY_Z\cdot$, in which the oxidized tyrosine is poised to oxidize the metal cluster of the OEC, but for clarity, the tyrosine radical is shown only once. In this model, only two of the Mn ions change their oxidation states. The $S_4 \rightarrow S_4'$ transition is the crucial O—O bond-forming step and involves the attack by a calcium-bound water upon the electrophilic Mn^V=O oxo group. O_2 is released, and two new H_2O are bound to initiate another cycle.

(a, b) Modified with permission from *Chemical Reviews* 106:4455–4483, J. P. McEvoy and G. W. Brudvig, Water-splitting chemistry of photosystem II. © 2006 American Chemical Society.

that each of the steps in Figure 16.15 in which a hydrogen atom $(H^+ + e^-)$ is extracted involves the following cycle:

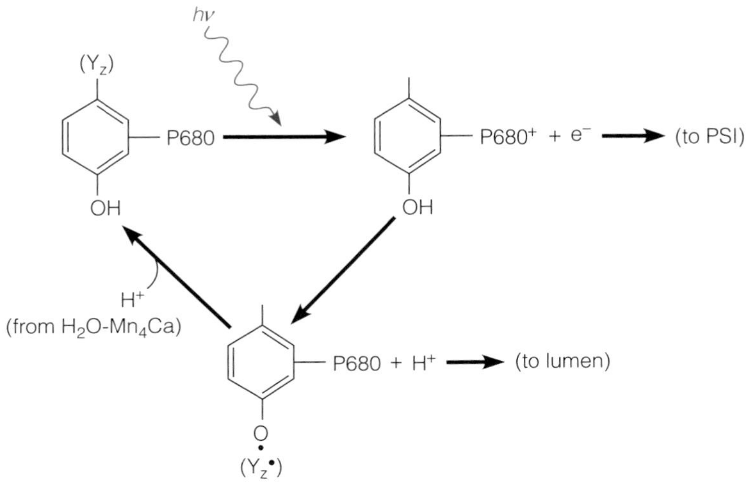

The system has in effect stripped four electrons from the four hydrogen atoms in two water molecules. The oxygen produced diffuses out of the chloroplast. The four protons that are produced from the two water molecules are released into the thylakoid lumen, helping to generate a pH difference between the lumen and stroma. We may summarize the reaction carried out by photosystem II as follows:

$$2H_2O \xrightarrow{4h\nu} 4H^+ + 4e^- + O_2$$

The electrons produced travel through the transport chain of photosystem II and are passed on to photosystem I through the b_6f complex.

Photosystem I: Production of NADPH

We have seen that in plants, which utilize two photosystems, photosystem II accomplishes the splitting of water with evolution of O_2 and helps generate a proton gradient across the thylakoid membrane. However, the electrons from the water molecules have not yet reached their final destination in NADPH. This process is the task of photosystem I, in which electrons are again released from a reaction center by light excitation and passed through a second electron transport chain. These electrons are replaced by those passed on from photosystem II.

Photosystem I is a multiprotein complex, containing at least 19 polypeptide chains. It also contains many antenna chlorophylls and a reaction center chlorophyll, P700, which can absorb light of up to 700 nm. As shown in Figure 16.12b, excitation by a photon absorbed by antenna chlorophylls raises electrons in P700 from a ground state to an excited state at about -1.3 V—probably the most powerful reductant in nature. Each excited electron then passes through a transport chain. It is first taken up by a special chlorophyll acceptor (designated A_0), then transferred to a molecule of *phylloquinone* (A_1, also known as vitamin K_1— see page 811), and finally passed through a series of three iron–sulfur proteins (F_X, F_B, and F_A). These proteins contain Fe_4S_4 clusters of the kinds depicted in Figure 15.4. Finally, the electron is transferred to another iron–sulfur protein, *soluble ferredoxin* (Fd), which is present in the stroma. The enzyme ferredoxin:NADP$^+$ oxidoreductase catalyzes the transfer of electrons to NADP$^+$, after ferredoxin has been reduced by photosystem I:

$$2Fd\ (red) + H^+ + NADP^+ \xrightarrow{\text{Fd-NADP}^+ \text{ Reductase}} 2Fd\ (ox) + NADPH$$

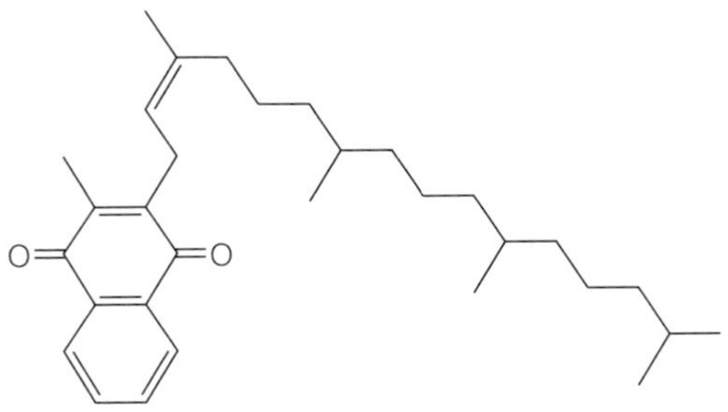

In a sense, ferredoxin, rather than NADP$^+$, can be considered the *direct* recipient of electrons from the pathway. Although much of the reduced ferredoxin is used

Phylloquinone

Photosystem I receives electrons from photosystem II and transfers them to NADP$^+$ to make NADPH.

to reduce NADP$^+$, some is used for other reductive reactions, which we discuss later. In fact, we may consider reduced ferredoxin a source of low-potential electrons for many reductive processes. The NADPH produced by ferredoxin oxidation is released into the stroma, where it will be used in dark reactions.

The electrons that have been driven through photosystem I originated in electron transfer from P700 reaction centers. The electron-deficient, oxidized reaction centers (P700$^+$) so produced must be resupplied with electrons for photosynthesis to continue. In two-system photosynthesis, these electrons are provided from photosystem II via plastocyanin.

Recently, X-ray diffraction studies have revealed structures of the entire photosystem I complex from cyanobacteria and plants (see References). The cyanobacterial complex exists as a trimer, whereas PSI in plants is a monomer. Both types of PSI consist of multiple polypeptide chains, several hundred chlorophylls (most of which are antenna molecules), and the other components of the electron transport chain in a complex that spans the thylakoid membrane. Plant PSI consists of two membrane complexes, the core complex and the light-harvesting complex I (LHCI) (Figure 16.16a). The heart of the core complex is formed by the PsaA and PsaB subunits, which bind all of the electron transport chain components, including the P700 reaction center chlorophylls, A$_0$, A$_1$, and the iron–sulfur clusters, as well as 80 chlorophylls that function as light-harvesting antennae. Three stromal subunits (PsaC, D, E) and one lumenal subunit (PsaF) complete the core complex. LHCI is composed of four subunits (Lhca1–4) located in the

FIGURE 16.16

The structure of plant photosystem I. **(a)** A view of the pea PSI structure (PDB ID 2O01) as seen from the plane of the membrane, with the thylakoid lumen at the bottom. The four LHCI subunits (Lhca1–4) are in front, and are colored green, light blue, magenta, and yellow, respectively. The core subunits PsaA (pink) and PsaB (silver) are hidden behind the LHCI subunits. Antenna chlorophylls of LHCI are blue and cyan, and antenna chlorophylls of the core are green. P700 reaction center chlorophylls are red, and the Fe$_4$S$_4$ clusters are shown as yellow-red spheres. Dimensions in Å are indicated. **(b)** Model for the electron transport chain and binding sites for plastocyanin and soluble ferredoxin. The A$_0$ chlorophylls are green and the A$_1$ phylloquinones are blue. The path of electron transport is indicated by red arrows. Amino acids involved in electrostatic and hydrophobic interactions between plastocyanin and PsaA/PsaB are also shown.

Modified from *Structure* 17:637–650, A. Amunts and N. Nelson, Plant photosystem I design in the light of evolution. © 2009, with permission from Elsevier.

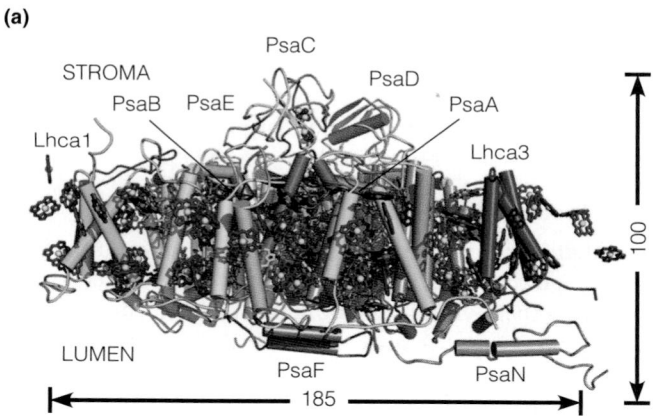

(a)

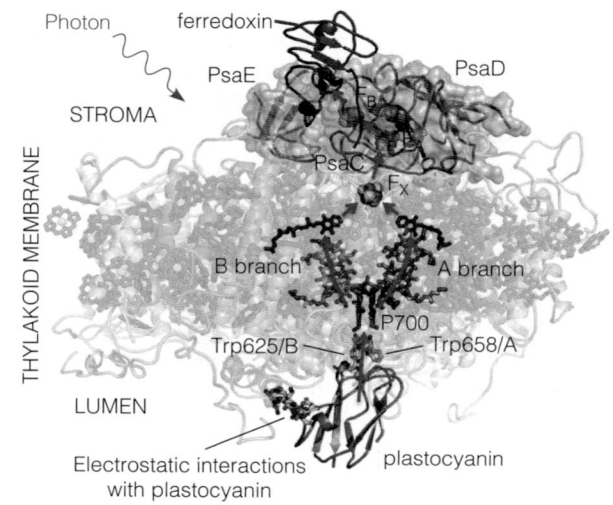

(b)

membrane on one side of the core complex. Like the LHCII described earlier, LHCI serves as an additional antenna system that collects photons and transmits the energy to the core complex. The entire plant PSI-LHCI supercomplex is ~600 kDa and contains 45 transmembrane helices and 168 antenna chlorophylls. The center-to-center distances of these antenna chlorophylls to their nearest neighbor are in the range of 7–16 Å, a favorable distance for fast excitation energy transfer.

PSI catalyzes the light-driven electron transfer from the soluble electron carrier plastocyanin, located at the lumenal side of the thylakoid membrane, to ferredoxin, located at the stromal side of the membrane. Figure 16.16b shows the proposed binding sites for plastocyanin and ferredoxin on PSI and the locations of the electron transport chain components. The path of electrons from the first P700 iron–sulfur cluster (F_x) is branched, with the A_0 chlorophylls and A_1 phylloquinones arranged in symmetric pairs. Almost every photon absorbed by the PSI complex is used to drive electron transport. PSI is remarkably efficient, with a quantum yield of nearly 1, meaning that almost every photon absorbed by the PSI complex is captured and used to drive electron transport.

Summation of the Two Systems: The Overall Reaction and ATP Generation

We can now summarize the electron flow through the two-system light reactions. As shown in Figure 16.12, electrons are taken from water and end up in NADPH. We wrote for the overall reaction in photosystem II:

$$2H_2O \xrightarrow{4h\nu} 4H^+ + 4e^- + O_2$$

The reactions of photosystem I, if written for four electrons and with all intermediates eliminated, are

$$4e^- + 2H^+ + 2NADP^+ \xrightarrow{4h\nu} 2NADPH$$

Adding these two reactions gives us the following summation of the light reactions:

$$2H_2O + 2NADP^+ \xrightarrow{8h\nu} 2H^+ + O_2 + 2NADPH$$

The key to ATP generation is that additional protons have been pumped from the stroma into the thylakoid lumen during the passage of each electron through the electron transport chain. Current estimates of the total number of protons are somewhat uncertain because the number transported per electron by the $b_6 f$ complex is not known exactly. However, it is estimated that ~12 protons are translocated per O_2 released (corresponding to ~3 protons per electron passing from H_2O to $NADP^+$). The net result from the combined function of photosystem I and II is the reduction of $NADP^+$ and the generation of a proton gradient across the thylakoid membrane, with the lumen becoming more acidic than the stroma. Recall from Chapter 15 that the free energy available in a proton gradient ($\Delta\mu_H$, the protonmotive force) is composed of both a chemical component (the proton concentration gradient, ΔpH) and an electrical component (the membrane potential, $\Delta\Psi$):

$$\Delta\mu_H = 2.3RT\Delta pH + F\Delta\Psi \qquad (16.3)$$

Unlike the mitochondrial inner membrane, the thylakoid membrane of chloroplasts is permeable to ions such as Mg^{2+} and Cl^-. Movement of these ions across the thylakoid membrane maintains electrical neutrality, dissipating most of the membrane potential. (Light-induced translocation of Mg^{2+} into the stroma also has a regulatory role, as we shall see later.) Thus, in illuminated chloroplasts, the protonmotive force is dominated by the $[H^+]$ gradient. The pH difference produced across the thylakoid membrane can become very large—as much as 3.5 pH units in brightly illuminated chloroplasts. This pH gradient across the membrane corresponds to more than a 3000-fold difference in $[H^+]$ and a free energy change of about -20 kJ per mol of protons at 25 °C. Based on the estimated stoichiometry of 12 mol H^+ per mol O_2 produced, this corresponds to roughly 240 kJ/mol O_2 energy available to drive the synthesis of ATP.

As in the case of ATP generation in mitochondria, these protons can pass back through the thylakoid membrane only through membrane-bound ATP synthase complexes. In chloroplasts these complexes are called CF_0-CF_1 complexes, and they exhibit considerable resemblance to the F_0-F_1 complexes of mitochondria (see Chapter 15). It has been estimated that one ATP is produced for each three protons passing through the CF_0-CF_1 complex. Three moles of protons would supply ~60 kJ to drive the synthesis of 1 mole of ATP, a thermodynamically reasonable result. Because ~12 moles of H^+ are transported per mole of O_2 produced, $\sim 12/3 \approx 4$ moles of ATP are generated for each mole of O_2 evolved.

A summary view of the whole set of light reactions is shown in Figure 16.17. It should be noted that photosystems I and II, the cytochrome b_6f complex, and ATP synthase (CF_0-CF_1) are all individual entities embedded in the thylakoid membrane but are not necessarily contiguous. The components that link the photosystems and the b_6f complex are mobile—plastoquinone in the lipid phase of the membrane and plastocyanin in the thylakoid lumen. Thus, electrons can be moved over long distances in this system. Such long-range transport is necessary because of the arrangement of components in the thylakoid membrane. Careful analysis of the composition of grana indicates that the interior membrane layers of the grana are rich in photosystem II; by contrast, the stroma lamellae are rich in photosystem I (Figure 16.18). This physical segregation of photosystems is possible because of the high mobilities of plastoquinone and plastocyanin. Physical separation of PSI and PSII also provides a way to optimize the quantum yield of the photosynthetic process. For maximum efficiency, the energy absorbed by each

Both photosystems transport protons from the stroma into the thylakoid lumen. The return of protons, through CF_0-CF_1 complexes, is used to generate ATP.

FIGURE 16.17

Summary view of the light reactions as they occur in the thylakoid. Photosystems I and II and the cytochrome b_6f complex are physically separate protein complexes embedded in the thylakoid membrane. Electron transfer from PSII to cytochrome b_6f is by diffusion of reduced plastoquinone (QH_2) in the membrane lipid. Transfer from b_6f to PSI is mediated by plastocyanin (PC), soluble in the lumen. The protons added to the thylakoid lumen during the light reactions pass back across the thylakoid membrane via ATP synthase complexes (CF_0-CF_1). The ATP synthase catalytic subunits are in the CF_1 component and face into the stroma so that ATP is generated in that compartment. Reduction of $NADP^+$ by Fd–$NADP^+$ reductase (FNR) also occurs on or near the stromal surface of the membrane.

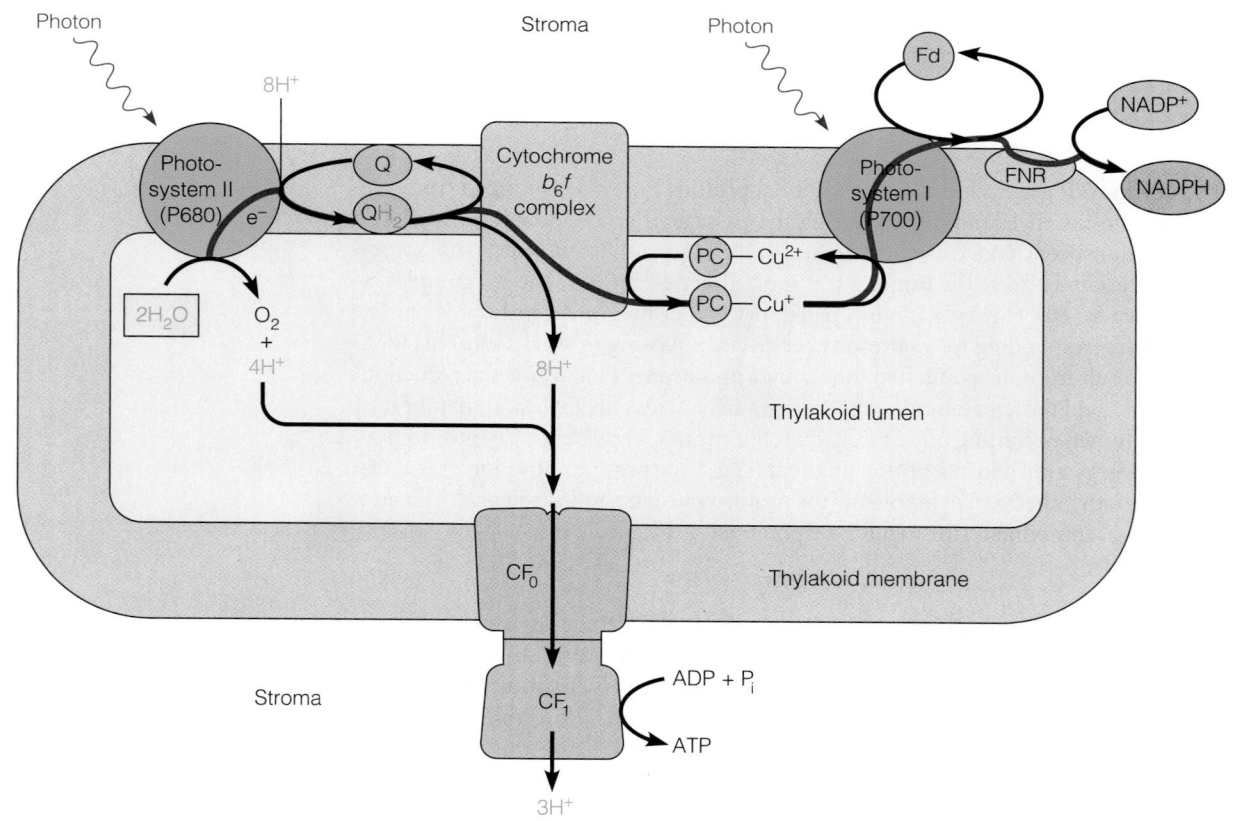

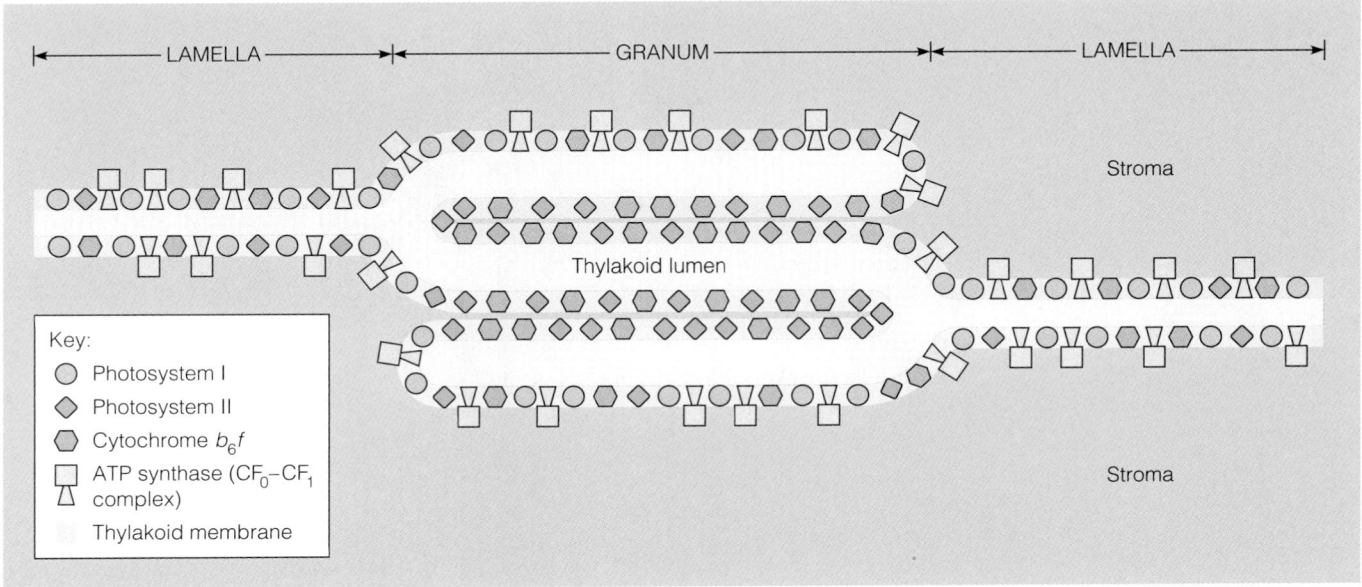

FIGURE 16.18

Arrangement of components of the two photosystems on the thylakoid membrane. The membrane layers in the interior of the granum are rich in photosystem II. The stroma lamellae and the top and bottom surfaces of the granum are rich in photosystem I and ATP synthase particles, allowing $NADP^+$ reduction and ATP generation to occur at or near these stroma-facing surfaces.

photosystem should be balanced. Because the intensity and spectral quality of light can change rapidly, higher plants and green algae have evolved an elegant mechanism to modulate the excitation energy of the two photosystems. Recall that PSII and PSI have distinct maximum absorption at 680 nm (blue–green light) and 700 nm (red light), respectively. If PSII is excited in excess over PSI (referred to as State 2), the plastoquinone pool becomes over-reduced because PSII reduces Q to QH_2 faster than the cytochrome b_6f complex and PSI can oxidize it (refer to Figure 16.17). In State 1, PSI is excited in excess over PSII, and the plastoquinone pool becomes over-oxidized. To maintain an optimal excitation balance, the chloroplast can switch the light-harvesting efficiency of the two photosystems in a process called **state transitions**. This is accomplished by redistributing the LHCII antenna complexes (Figure 16.10) between PSII and PSI within the thylakoid membrane. The trigger for this switch is phosphorylation of LHCII, catalyzed by specific thylakoid-bound kinases. The kinase is activated by accumulation of reduced plastoquinone. Phosphorylated LHCII dissociates from PSII and binds to PSI, thereby redirecting the absorbed excitation energy to PSI at the expense of PSII. Once PSI activity catches up with PSII, the plastoquinone pool will become oxidized, leading to deactivation of the LHCII kinase. A constitutively active protein phosphatase catalyzes dephosphorylation of LHCII, which then redistributes back to PSII. These state transitions occur on a timescale of seconds to minutes in response to changing light conditions.

An Alternative Light Reaction Mechanism: Cyclic Electron Flow

In the two-system light reactions just described, the electrons displaced from photosystem I by excitation are replaced by photosystem II, which receives them from water. The entire process is called **noncyclic electron flow**, and generation of ATP by this process is called **noncyclic photophosphorylation**. An alternative pathway for the light reactions, called **cyclic electron flow**, utilizes the components of photosystem I, plus plastocyanin and the cytochrome b_6f complex (Figure 16.19). Whether this pathway is used or not depends on the levels of $NADP^+$ in the chloroplast stroma. When $NADP^+$ is present in only small amounts, electrons excited in the P700 center are not transferred to $NADP^+$. Instead, they are passed from ferredoxin to the cytochrome b_6f complex and from there returned via plastocyanin to the P700 ground state. One way to look at this cyclic electron flow is to consider the b_6f

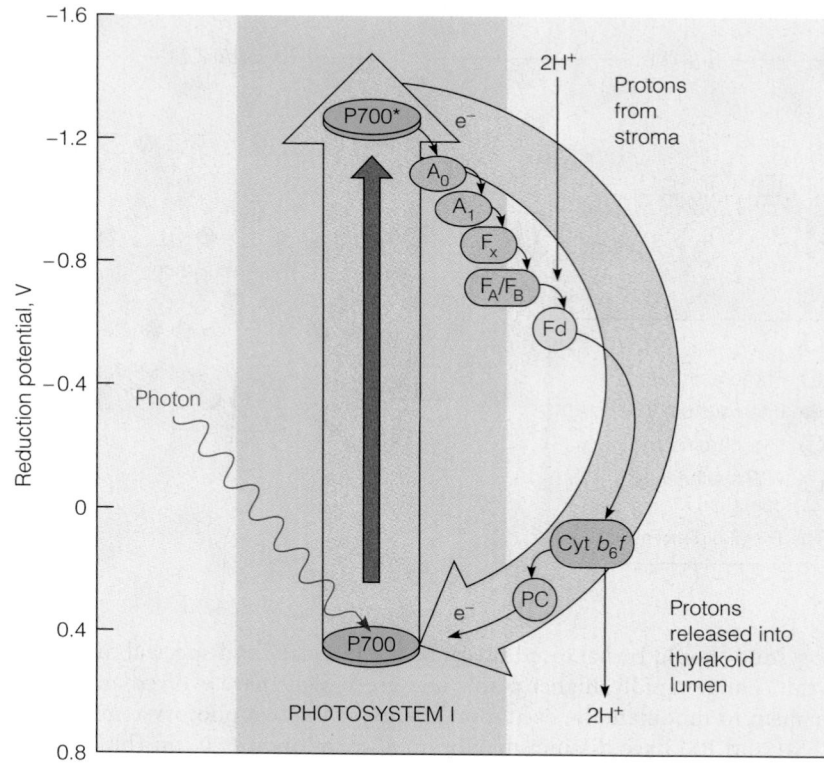

Cyclic electron flow, an alternative to two-system (noncyclic) electron flow, generates extra ATP when NADPH is plentiful.

complex and NADP$^+$ as competitors for the electrons from Fd. The b_6f complex pumps protons across the thylakoid membrane during this cyclic process, thereby ensuring the generation of ATP. Approximately one ATP is generated for every two electrons that complete the cycle, a process called **cyclic photophosphorylation**. However, in this process, no O_2 is released and no NADP$^+$ is reduced.

Cyclic electron flow apparently serves to generate ATP in situations when the reductant NADPH is abundant and little NADP$^+$ is available as an electron acceptor. It may also play a more fundamental role. As we shall see, the requirements for ATP in the photosynthetic dark reactions are substantial and may not always be fully met by noncyclic electron flow. Cyclic photophosphorylation, which produces ATP but no NADPH, helps maintain the necessary balance between ATP and NADPH production.

Reaction Center Complexes in Photosynthetic Bacteria

The light reactions just described are those that occur in plants and algae. However, some of our most precise information concerning functioning of the light reactions has come from studies of photosynthesizing bacteria, such as the purple bacteria (named for the colored bacteriochlorophyll and carotenoid pigments in their membranes). These bacteria are versatile organisms, able to obtain energy in the dark using cellular respiration, or in the light by switching on photosynthesis. Pioneering studies by Roderick Clayton showed that these organisms use only a single photosystem and that the reaction centers from these organisms could be isolated in pure form. Later, the entire molecular structure of a crystallized reaction center complex from the purple bacterium *Rhodopseudomonas viridis* was determined by Johann Deisenhofer, Hartmut Michel, and Robert Huber using X-ray diffraction. This work earned the Nobel Prize in Chemistry in 1988.

A structural model of a purple bacterial reaction center complex is shown in Figure 16.20. It is a transmembrane protein, consisting of four polypeptides. The

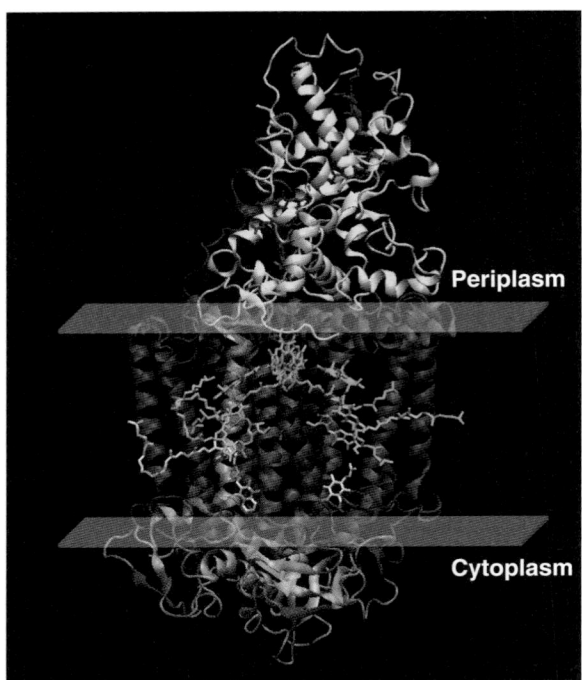

FIGURE 16.20

Model of a purple bacterial reaction center complex. A view of the reaction center from *Blastochloris viridis* (formally *Rhodopseudomonas viridis*) (PDB ID 3D38) as seen from the plane of the membrane (approximate locations of the periplasmic and cytoplasmic faces of the membrane are indicated). The cytochrome (yellow), carrying four heme groups (red), lies in the periplasmic space between the bacterial inner and outer membranes. Subunits M (silver) and L (tan), each with five transmembrane α-helices, span the membrane. These subunits carry four bacteriochlorophylls (green), two bacteriopheophytins (cyan), two quinones (white), and an iron atom (red)—all involved in photon harvesting and electron transfer. Two of the bacteriochlorophylls constitute the reaction center. Subunit H (orange) lies mostly on the cytosolic side of the membrane but has one membrane-spanning α-helix.

portion of the complex that lies outside the bacterial plasma membrane, in the periplasmic space, is a *c*-type cytochrome carrying four heme groups. Subunit H, on the other hand, lies largely on the cytosolic face of the membrane. Buried within the membrane are two subunits (L and M) that are largely α-helical. They carry four bacteriochlorophyll *b* molecules, two bacteriopheophytins, two quinones (designated Q_A and Q_B), and a bound iron atom. Two of the bacteriochlorophylls (the "special pair") lie very close together; they constitute the reaction center itself. Light absorbance is maximal in the near infrared, at about 870 nm, so the center is referred to as P870.

Chemically, the purple bacterial reaction center complex most closely resembles photosystem II in plants, given that it contains pheophytins (bacteriopheophytin, or BPh) and quinones. Studies of the kinetics of the reactions in isolated type-II centers have elucidated the electron pathway (Figure 16.21). Excitation of the reaction center leads very quickly (in about 10^{-12} s) to transfer of an electron to one of the two pheophytins. The electron is then passed on to Q_A and then to Q_B. These quinones are normally bound in the complex, but on receiving a second electron (and two protons), Q_B dissociates. QH_2 is believed to then move to and reduce a cytochrome bc_1 complex (homologous to the cytochrome bc_1 complex in mitochondria). The electron is returned from the bc_1 complex to the reaction center via the cytochrome in the reaction center complex. Because the electron flow is cyclic, there is no net oxidation–reduction, and thus no need for an external reductant, such as water. Consequently, this process does not produce O_2, and is referred to as *anoxygenic* photosynthesis.

However, the net result of this cyclic electron flow is to pump protons from the bacterial cytosol into the periplasmic space as QH_2 is oxidized by the bc_1 complex. The cytosols of such bacteria become quite alkaline as photosynthesis continues. Return of protons is through ATP synthase complexes spanning the plasma membrane, resulting in generation of ATP (Figure 16.21). The cyclic electron flow in these bacteria must be carefully distinguished from the cyclic flow that can occur via photosystem I in plants. The bacterial system is, with respect to its electron carriers, much more like photosystem II. Though no reducing power has been directly generated by the light reaction in this case, these bacteria can

Some photosynthetic bacteria use a photosystem that generates ATP in a manner analogous to that of photosystem II.

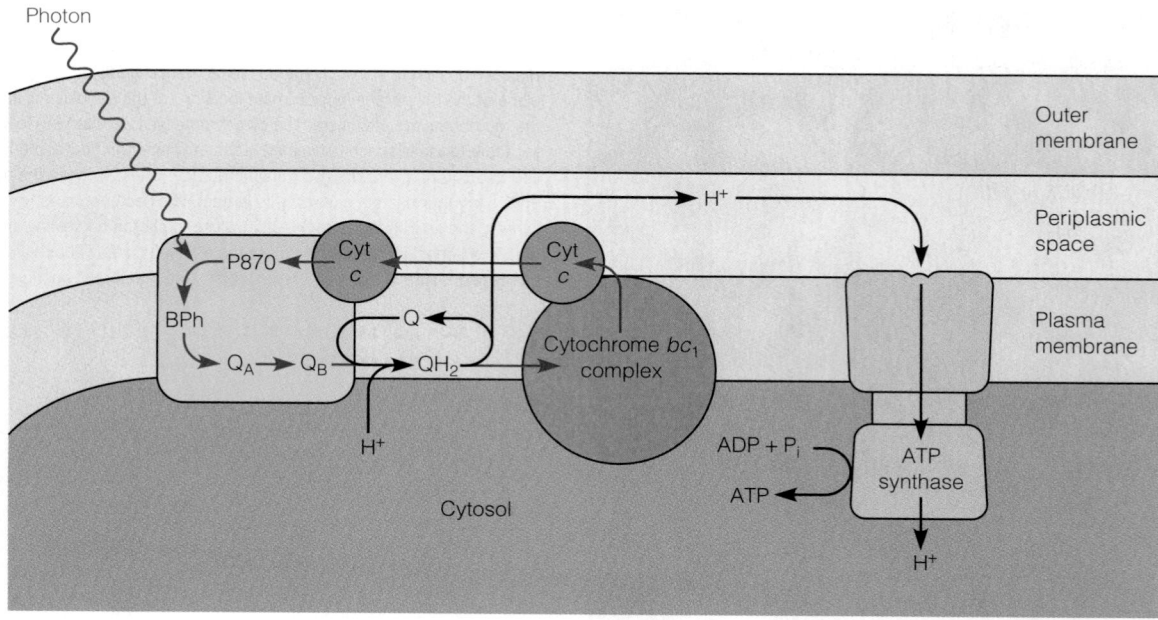

FIGURE 16.21

Postulated mechanism for purple bacterial photosynthesis. This process somewhat resembles the photosystem II reactions in the thylakoid (see Figure 16.17), with a reaction center and a membrane-bound cytochrome complex. However, there is only one kind of reaction center, water is not split, and electron flow is cyclic.

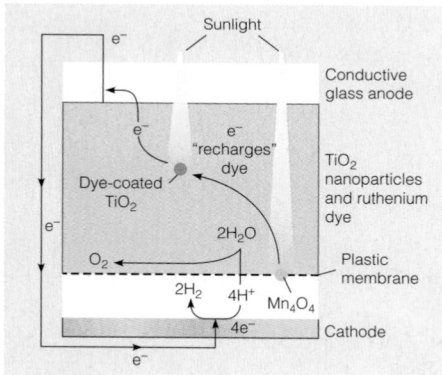

FIGURE 16.22

Artificial photosynthesis. In this solar cell, absorbed light energy drives the splitting of water into oxygen and protons, forming hydrogen (H_2) fuel.

From *Science* 325:1200–1201, Robert F. Service, New trick for splitting water with sunlight. © 2009. Reprinted with permission from AAAS and Gerhard Swiegers (University of Wollongong, Australia), G. Charles Dismukes (Princeton University, USA) and Leone Spiccia (Monash University, Australia).

carry out the dark reactions of photosynthesis (CO_2 assimilation) by using ATP energy to transfer electrons from various substrates such as H_2S, S, $S_2O_3^{2-}$ (thiosulfate), or H_2 to $NADP^+$ to produce the necessary NADPH.

Other anoxygenic photosynthetic bacteria, such as the anaerobic green bacterium *Chlorobium*, use a photosystem that more closely resembles photosystem I in plants. These type-I systems catalyze a linear light-driven electron transport process, in which electrons are extracted from H_2S to replenish the oxidized reaction center, and S_2 is released.

Artificial Photosynthesis

The great incentive to devise new, environmentally safe energy sources has led a number of scientists to attempt to mimic the most efficient solar energy converters known—photosynthesizing plants. Although many kinds of molecules can capture photons, in most cases the energy is simply degraded to heat at low temperatures and cannot produce useful work. The trick is to produce, by capture of radiation, a long-lived excited state that can be coupled to a desired energy-saving process. Recently, researchers succeeded in incorporating a bioinspired manganese–oxygen catalyst (Mn_4O_4) into a plastic membrane that allows protons to pass in one direction (Figure 16.22). The membrane separates two electrodes. The anode is immersed in a solution containing a ruthenium dye; sunlight excites electrons in the dye that are picked up by the anode and passed into an external circuit. The Mn_4O_4 complex also absorbs photons, whereupon it catalyzes the splitting (oxidation) of H_2O to O_2 and protons (H^+). The electrons from water are passed to the electron-deficient dye molecules, and the protons pass through the membrane to the cathode. These protons then combine with electrons from the external circuit, producing hydrogen gas (H_2). Thus, like natural photosynthesis, this artificial solar cell converts sunlight into chemical fuel, although with much lower efficiency.

In what is a rapidly progressing field, a wide variety of other model systems have been developed or proposed. It may well turn out that a major energy source in the future will be patterned on that developed by nature several billion years ago.

The Dark Reactions: The Calvin Cycle

The dark reactions occur in the stroma of the chloroplast (or in the cytoplasm of photosynthetic bacteria). Their function is to fix atmospheric carbon dioxide into carbohydrates, utilizing ATP energy and reducing power (NADPH) generated by the light reactions. As noted earlier, the dark reactions can occur without light but are accelerated in the presence of light.

Carbon dioxide fixation is accomplished by adding one CO_2 at a time to an acceptor molecule and passing the molecule through a cyclic series of reactions, shown schematically in Figure 16.23. The whole series is called the **Calvin cycle**, after the biochemist Melvin Calvin, who in 1961 received the Nobel Prize for his work in this field. Calvin, working with James Bassham and Andrew Benson in the 1940s and 50s, exposed cultures of unicellular algae such as *Chlorella* and *Scenedesmus* to ^{14}C-labeled CO_2 for a few seconds, and then rapidly killed the cells to stop all metabolism. Extracts were analyzed by the newly developed technique of two-dimensional paper chromatography and radioactive compounds were detected by exposing the chromatogram to X-ray film (autoradiography, see Tools of Biochemistry 12A). These methods allowed the researchers to discover the first product of CO_2 fixation (3-phosphoglycerate) and eventually to elucidate the entire cyclic pathway. The cycle results in the formation of hexoses and in the regeneration of the acceptor molecule. The Calvin cycle can be envisioned as divided into two stages. In stage I, the carbon dioxide is trapped as a carboxylate and reduced to the carbonyl level found in sugars, resulting in net carbohydrate synthesis. Stage II is dedicated to regenerating the acceptor molecule. Let us examine each stage in turn.

Stage I: Carbon Dioxide Fixation and Sugar Production

Incorporation of CO_2 into a Three-Carbon Sugar

Carbon dioxide is incorporated into glyceraldehyde-3-phosphate (GAP) via the intermediates shown in Figure 16.23. The acceptor molecule for CO_2 is **ribulose-1,5-bisphosphate (RuBP)**. Carbon dioxide from the air diffuses into the stroma of the chloroplast, where it is added at the carbonyl carbon of RuBP. The reaction is catalyzed by the enzyme **ribulose-1,5-bisphosphate carboxylase**, also known as ribulose-1,5-bisphosphate carboxylase/oxygenase (or **rubisco**). This enzyme is one of the most important in the biosphere and certainly the most abundant. It makes up about 15% of all chloroplast proteins, and there are an estimated 40 million tons of it in the world—about 20 pounds for every living person. There are four different forms of rubisco found in nature. Form I, found in higher plants, eukaryotic algae, and many cyanobacteria and proteobacteria, is composed of eight large (~50 kDa) catalytic subunits and eight small (~15 kDa) non-catalytic subunits. This L_8S_8 complex is arranged as a core of four homodimers of catalytic subunits capped on the top and bottom by four small subunits. The large, catalytic subunits are encoded by the chloroplast genome, whereas the small subunits are encoded by the nuclear genome. Thus, production of active rubisco in the chloroplast stroma requires coordination of nuclear and chloroplast gene expression and translocation of the cytoplasmically synthesized small subunits into the chloroplast for assembly of the holoenzyme. The other three known forms of rubisco are less widely distributed and are composed of various arrangements of large subunits only.

As its full name implies, rubisco also has an alternative oxygenase activity. We shall see the consequences of this other activity later. For the moment we will concentrate on its CO_2-fixing (carboxylase) function. The carboxylase reaction is a

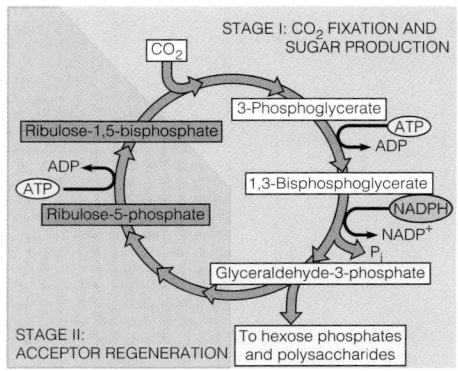

FIGURE 16.23

Schematic view of the Calvin cycle. The cycle may be divided into two stages. In stage I, CO_2 is fixed and glyceraldehyde-3-phosphate (GAP) is produced. Part of this GAP is used to make hexose phosphates and eventually polysaccharides. Another fraction of the GAP is used in stage II to regenerate the acceptor molecule, ribulose-1,5-bisphosphate.

The Calvin cycle uses the ATP and NADPH generated in the light reactions to fix atmospheric CO_2 into carbohydrate.

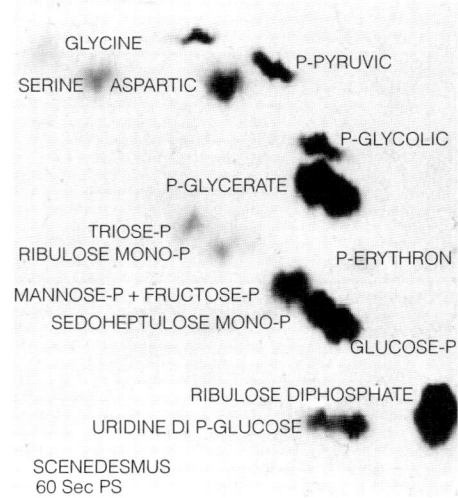

Autoradiogram of a 2D paper chromatogram showing labeled products after a 60-sec exposure of *Scenedesmus* to $^{14}CO_2$.

Springer/*Photosynthesis Research* 73:29–49, 2002, A. A. Benson, Following the path of carbon in photosynthesis: A personal story, with kind permission from Springer Science+Business Media B.V.

The Calvin cycle has two stages. First, CO_2 is fixed by addition to ribulose-1,5-bisphosphate (RuBP) and hexoses are formed. In the second stage RuBP is regenerated.

complex series of steps in which the actual CO_2 acceptor is proposed to be the five-carbon enediolate intermediate:

Ribulose-1,5-bisphosphate   **Enediolate intermediate**   **Carboxy-β-keto intermediate**

3-phosphoglycerate

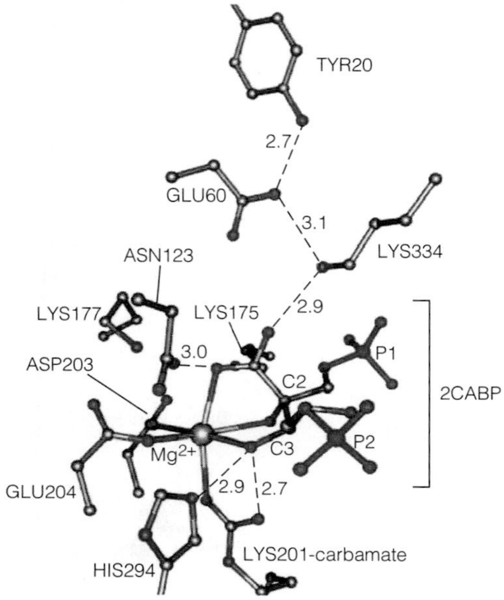

3-phosphoglycerate

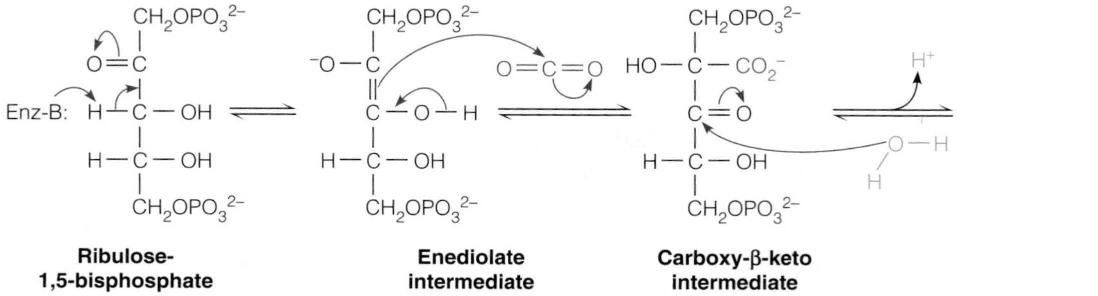

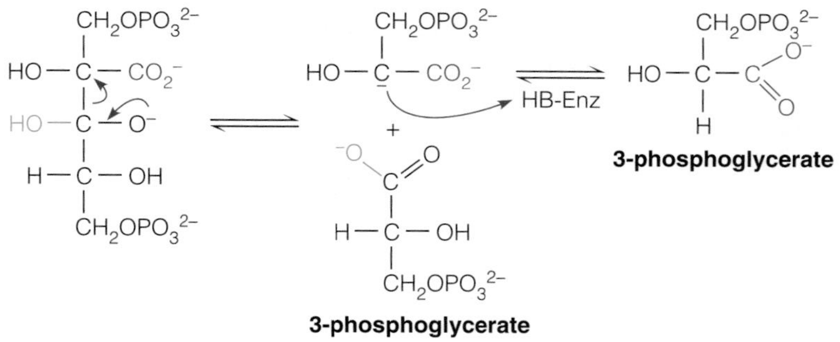

FIGURE 16.24

Active site of rubisco. This model is based on the X-ray structure (PDB ID 8RUC) of spinach rubisco complexed with Mg^{2+} and the transition-state analog 2-carboxy-D-arabinitol-1,5-bisphosphate (2CABP).

Modified with permission from *Journal of the American Chemical Society* 130:15063–15080, B. Kannappan and J. E. Gready, Redefinition of rubisco carboxylase reaction reveals origin of water for hydration and new roles for active-site residues. © 2008 American Chemical Society.

The chemistry of this reaction is the subject of intense research, and several mechanisms have been proposed. The enzyme is first activated by *carbamation* of an active site lysine residue. This carbamate, formed by nonenzymatic reaction of a nonsubstrate CO_2 with the ε-amino group of lysine, is analogous to the carbamates on the N-termini of hemoglobin (Chapter 7). The negatively charged carbamate forms part of the binding site for an essential Mg^{2+} ion (Figure 16.24). The Mg^{2+} ion is involved in binding RuBP and activating the H_2O molecule that hydrates the carboxy-β-keto intermediate. Once the enediolate intermediate is carboxylated, the product is hydrated and then cleaved to yield two molecules of 3-phosphoglycerate (3PG). Cleavage is facilitated by protonation by an active site acid (HB-Enz). The reaction is essentially irreversible, with $\Delta G^{\circ\prime} = -35.1$ kJ/mol. At this point CO_2 has already been fixed into a carbohydrate. The remainder of the Calvin cycle reactions are dedicated to producing hexoses from the triose and regenerating RuBP.

Each molecule of 3PG is phosphorylated by ATP, in a reaction catalyzed by *phosphoglycerate kinase*. The 1,3-bisphosphoglycerate so produced is then reduced to glyceraldehyde-3-phosphate (GAP), with accompanying loss of one phosphate. The reducing agent is NADPH, produced in the light reaction, and the reaction is catalyzed by the enzyme *glyceraldehyde-3-phosphate dehydrogenase*:

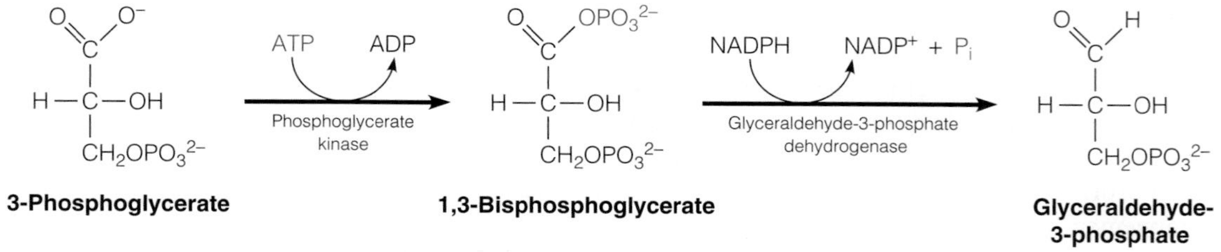

3-Phosphoglycerate   **1,3-Bisphosphoglycerate**   **Glyceraldehyde-3-phosphate**

We have encountered similar enzymes earlier, in connection with their roles in glycolysis (see Chapter 13).

At this stage of the cycle a molecule of CO_2 has been fixed into a simple (three-carbon) monosaccharide. It is useful to note the requirements in ATP and

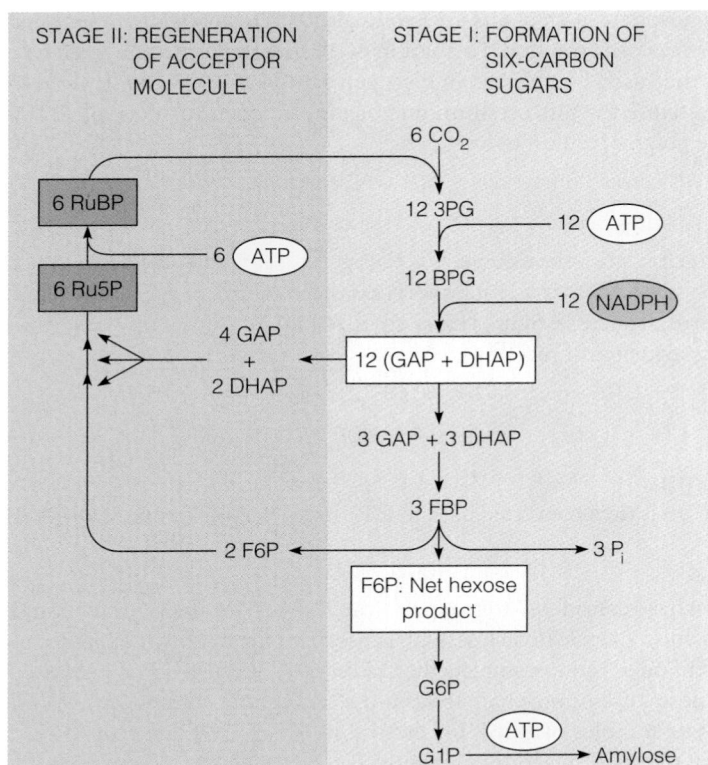

STAGE II: REGENERATION OF ACCEPTOR MOLECULE	STAGE I: FORMATION OF SIX-CARBON SUGARS

FIGURE 16.25

Stoichiometry of the Calvin cycle. In six turns of the Calvin cycle, six CO_2 molecules will have entered and bound to six molecules of ribulose-1,5-bisphosphate (RuBP) to yield 12 molecules of glyceraldehyde-3-phosphate (GAP). Because GAP is in isomeric equilibrium with dihydroxyacetone phosphate (DHAP), the 12 GAP may be considered an interconvertible stock of 12 molecules of (GAP + DHAP). Of these, six are used to make three molecules of fructose-1,6-bisphosphate (FBP), of which *one* constitutes the net hexose product of the six turns. The other two FBPs are used, together with the six remaining molecules of (GAP + DHAP), to form six molecules of ribulose-5-phosphate (Ru5P), which are then phosphorylated to regenerate the required six molecules of RuBP.

NADPH up to this point. For each CO_2 molecule that has passed through these steps, two molecules of ATP have been hydrolyzed and two molecules of NADPH have been oxidized. However, it is more appropriate to keep accounts on a "per glucose" basis because we want to see what must happen to account for the generation of one hexose molecule from CO_2. Figure 16.25 provides a schematic view of the stoichiometry of the whole Calvin cycle. Six molecules of CO_2 will have to enter the cycle to provide the six carbons needed for every new molecule of hexose produced. That will require formation of 12 GAP, and therefore 12 ATP and 12 NADPH will be needed.

At this point the pathway splits, so as to satisfy the two essential goals—to make hexoses and to regenerate the acceptor. Of the 12 molecules of GAP that have been produced, two will be used to make a molecule of a hexose. The remaining 10 will be utilized to regenerate the 6 molecules of ribulose bisphosphate necessary to maintain the cycle. That is, 10 three-carbon molecules will be converted to 6 five-carbon molecules.

Formation of Hexose Sugars

This is actually familiar ground, for it follows a portion of the gluconeogenic pathway described in Chapter 13. The reactions are shown schematically in Figure 16.25. Recall that glyceraldehyde-3-phosphate can be isomerized to dihydroxyacetone phosphate (DHAP) by triose phosphate isomerase (pages 528–529). Thus, the 12 molecules of GAP produced can be considered to be an interconvertible equilibrium pool of GAP and DHAP. A molecule of GAP and a molecule of DHAP can be combined, via the enzyme *fructose bisphosphate aldolase*, to yield fructose-1,6-bisphosphate (FBP). As Figure 16.25 shows, six of the GAP molecules follow this path, to yield three molecules of FBP. The FBP is dephosphorylated by fructose-1,6-bisphosphatase to yield three molecules of fructose-6-phosphate (F6P). Of these, two will be employed in the regeneration pathway, but one is available as a net product of the Calvin cycle; it is then isomerized to glucose-6-phosphate (G6P) and finally to glucose-1-phosphate (G1P).

Glucose-1-phosphate is, in plants as in animals, the precursor to oligosaccharide and polysaccharide formation. Formation of plant starch (amylose) follows a path similar to that used by animals in glycogen synthesis. However, instead of using UTP to activate the glucose monomer, as in glycogen formation, ATP is employed in the polymerization of amylose:

$$\text{Glucose-1-phosphate} + \text{ATP} \rightarrow \text{ADP-glucose} + PP_i$$

$$\text{ADP-glucose} + (\text{glucose})_n \rightarrow (\text{glucose})_{n+1} + \text{ADP}$$

Amylose, which is not very soluble, is a storage carbohydrate. However, much of the saccharide synthesized in plant leaves is exported to other parts of the plant, mostly in the form of sucrose. Sucrose is synthesized in the cytosol of plant leaves by the following sequence of reactions:

$$\text{UTP} + \text{glucose-1-P} \rightarrow \text{UDP-glucose} + PP_i$$

$$\text{UDP-glucose} + \text{fructose-6-P} \rightarrow \text{UDP} + \text{sucrose-6-P}$$

$$\text{Sucrose-6-P} + H_2O \rightarrow \text{sucrose} + P_i$$

The UDP produced is then converted back to UTP by phosphate transfer from ATP.

Stage II: Regeneration of the Acceptor

The reactions we have considered to this point can account for the introduction of one carbon into one molecule of hexose, with subsequent formation of oligosaccharides or polysaccharides. But to complete the Calvin cycle, it is necessary to regenerate enough ribulose-1,5-bisphosphate to keep the cycle going. This means we will need to regenerate 6 moles of RuBP for every 6 moles of CO_2 taken up. This is accomplished by the set of reactions shown in Figure 16.26, which constitute the

FIGURE 16.26

Regeneration phase of the Calvin cycle. The stoichiometry here follows that in Figure 16.25. The two molecules of fructose-6-phosphate entering at the top are combined with four molecules of GAP and two molecules of DHAP to yield the required six molecules of ribulose-5-phosphate. These are then phosphorylated to produce the required RuBP. Note the similarity of this pathway to parts of the pentose phosphate pathway running in reverse (see Figure 13.39, page 576).

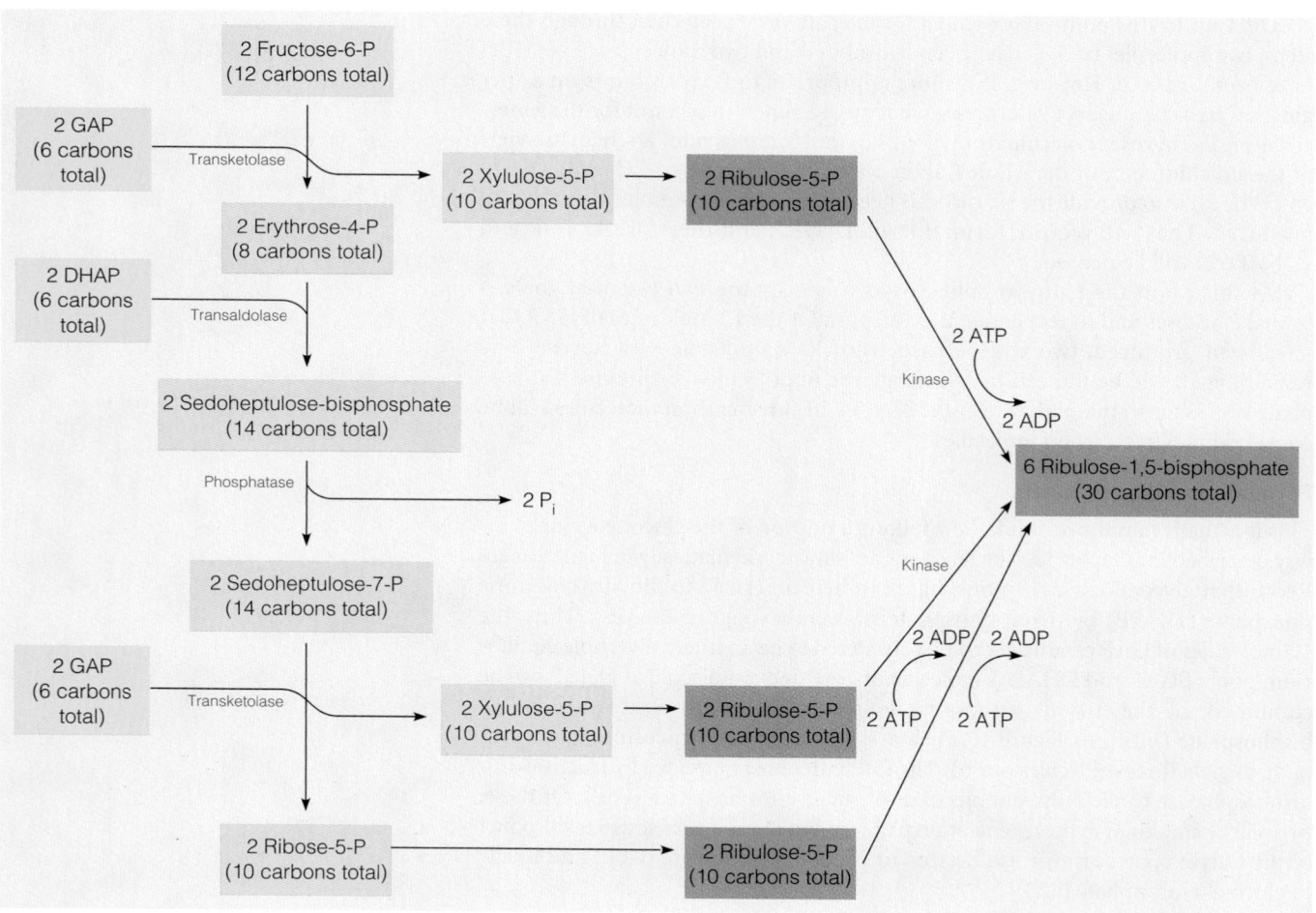

regenerative phase of the cycle schematized in Figures 16.23 and 16.25. Note that the *input* molecules in this somewhat complex reaction pathway are as follows:

1. Two molecules of DHAP and four molecules of GAP, from the six GAP that were diverted to the regeneration pathway in Figure 16.25.

2. Two of the three molecules of fructose-6-phosphate (F6P) that were produced from the remaining three GAP and three DHAP.

In order to make five-carbon molecules from six-carbon and three-carbon molecules, several rearrangements are required. They are accomplished by *transketolases* and *transaldolases*. The structures of the sugars involved in these reactions are all given in Chapter 9 and the chemistry of these reactions is discussed in Chapter 13 (pages 578–580). What is important here is the way in which two hexoses and six trioses have been rearranged and recombined to form six pentoses.

The final step in the regeneration of ribulose-1,5-bisphosphate is a phosphorylation, catalyzed by the enzyme *ribulose-5-phosphate kinase* and utilizing ATP. For six rounds of the cycle, this step will require 6 ATPs in addition to the 12 already accounted for. Therefore, the requirements for synthesizing 1 mole of hexose from CO_2 are 12 moles of NADPH and 18 moles of ATP:

$$6CO_2 + 12NADPH + 12H^+ \rightarrow C_6H_{12}O_6 + 12NADP^+ + 6H_2O$$

$$18ATP + 18H_2O \rightarrow 18ADP + 18P_i + 18H^+$$

[H is balanced in the second equation when the H in P_i (HPO_4^{2-}) is included.] Summing these two equations, we can write the overall dark reaction as

$$6CO_2 + 18ATP + 12NADPH + 12H_2O \rightarrow C_6H_{12}O_6 + 18ADP + 18P_i + 12NADP^+ + 6H^+$$

A Summary of the Light and Dark Reactions in Two-System Photosynthesis

The Overall Reaction and the Efficiency of Photosynthesis

The ATP and NADPH needed for the dark reactions are released into the stroma by the light reactions of photosynthesis. If we recall that two photons are required for every electron to pass through photosystems I and II, and that two electrons are required to reduce each $NADP^+$, then four photons are necessary for the production of each NADPH molecule. This corresponds to eight photons per O_2, a number in agreement with the quantum efficiency experimentally observed when both photosystems are operating—about 0.12 O_2 per photon. For the 12 NADPH needed in the dark reaction, as summarized in the previous section, 48 photons must be absorbed. If we assume that these photons will also pump enough protons across the thylakoid membrane to yield the 18 ATP required, we may, as an approximation, write the light reactions as

$$12H_2O + 12NADP^+ \rightarrow 12H^+ + 12NADPH + 6O_2$$

$$18ADP + 18P_i + 18H^+ \rightarrow 18ATP + 18H_2O$$

$$\text{Sum: } 12NADP^+ + 18ADP + 18P_i + 6H^+ \rightarrow 18ATP + 6H_2O + 12NADPH + 6O_2$$

This equation differs from the third equation on page 689 because we now include the ATP generation from the proton gradient. Adding this equation for the light reaction to the overall dark reaction, we obtain

$$6H_2O + 6CO_2 \xrightarrow{\text{48 photons}} C_6H_{12}O_6 + 6O_2$$

This estimate of 48 photons assumes that noncyclic photophosphorylation provides enough ATP for the dark reactions. If, as many workers in the field believe, additional ATP from cyclic photophosphorylation is required, the number of photons needed will be greater.

The overall energy efficiency of photosynthesis can approach 35%.

We can, on the basis of these calculations, estimate the energy efficiency of photosynthesis. Forming a mole of hexose from CO_2 and water requires, as we have seen, 2870 kJ. The energy input per photon depends on the wavelength of light used. Assuming that light of 650 nm wavelength is used, 48 einsteins of such light correspond to about 8000 kJ (see Figure 16.6). From this value, we estimate a theoretical efficiency of approximately 35%. Direct experimental measurements of the efficiency under optimal conditions give results in the same range or slightly lower. At high levels of illumination, when not all photons absorbed by the chloroplasts can be used for reaction center excitation, efficiency is much lower.

Regulation of Photosynthesis

It should be evident that the so-called dark reactions of photosynthesis, which result in the production of sugars, require careful regulation. Because the dark reactions depend on the reductive power and ATP supplied by the light reactions, it is not surprising that they are stimulated by the light reactions. There are three major ways in which this stimulation is accomplished. First, the central enzyme in the dark reactions, ribulose-1,5-bisphosphate carboxylase, is stimulated by high pH and by both CO_2 and Mg^{2+}. Recall from page 690 that the pumping of protons from the stroma into the thylakoid lumen by the light reactions increases the stromal pH; at the same time, Mg^{2+} ions enter the stroma to compensate for the positive charge of the H^+ ions that have been lost. Second, rubisco is sensitive to a number of naturally occurring sugar phosphates that act as tight-binding inhibitors, including xylulose-1,5-bisphosphate, 2-carboxy-D-arabinitol-1-phosphate (CA1P), as well as the ribulose-1,5-bisphosphate substrate itself. These inhibitors resemble transition state intermediates and cause the active site of rubisco to adopt a closed conformation, preventing carbamation and/or substrate binding. Genetic experiments with *Arabidopsis* led to the discovery of a regulatory enzyme, termed **rubisco activase**, that is required for activation and maintenance of Rubisco activity in vivo. Rubisco activase promotes the ATP-dependent dissociation of the inhibitor, facilitating the carbamation reaction and the binding of Mg^{2+} and substrate.

The photosynthetic dark reactions are regulated by the amount of light available to the organism through the activation of key enzymes.

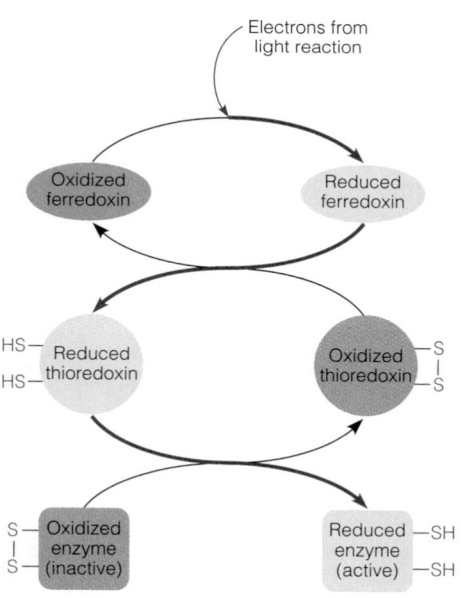

2-carboxy-D-arabinitol-1-phosphate (CA1P)

One of the most potent rubisco inhibitors, CA1P, is synthesized in chloroplasts in the dark, and is responsible for the characteristic decline in rubisco activity during darkness. When the chloroplasts are again illuminated, rubisco activase removes CA1P and reactivates rubisco. Rubisco activase is itself activated in the light by another light-dependent mechanism that will be described next.

The third way by which the dark reactions are stimulated by the light reactions depends on the redox state of critical disulfide bonds in several enzymes of the Calvin cycle. Rubisco activase, fructose-1,6-bisphosphatase (see page 697), sedoheptulose-1,7-bisphosphatase (the phosphatase shown in Figure 16.26), glyceraldehyde-3-phosphate dehydrogenase (see page 696), and ribulose-5-phosphate kinase (see Figure 16.26) are all activated by reduction of disulfides to sulfhydryls in the enzymes. Reduction is promoted by a disulfide exchange reaction with the protein *thioredoxin* (Figure 16.27). Thioredoxin, a small protein carrying two reversibly oxidizable Cys-SH groups, is used in a wide variety of redox reactions (see page 940). The reduction of thioredoxin, in turn, is promoted by oxidation of the reduced form of ferredoxin via a reaction catalyzed by the enzyme *ferredoxin–thioredoxin reductase*. In strongly irradiated chloroplasts, in which $NADP^+$ stores are depleted, reduced ferredoxin accumulates. High levels of reduced ferredoxin thereby lead to activation of the Calvin cycle enzymes, stimulating the Calvin cycle reactions when the light reactions are very active. The same compound, reduced thioredoxin, also stimulates the CF_0-CF_1 complexes, ensuring a high rate of ATP generation when illumination is intense.

In the dark, the plant "turns into an animal" in terms of its biochemistry. Although the dark reactions of photosynthesis may continue for some time, using photosynthesized ATP and NADPH, the plant must ultimately begin to draw on

FIGURE 16.27

Light-dependent activation of dark-reaction enzymes. Some enzymes of the Calvin cycle are activated by disulfide reduction, which is mediated by reduced thioredoxin. Thioredoxin is reduced by reduced ferredoxin, which accumulates in irradiated chloroplasts. The flow of electrons from the light reactions (photosystem I) is indicated by the magenta arrows. In the dark, the enzymes are deactivated by reoxidation (reversal of the pathway). Additional functions of thioredoxin are presented in Chapter 22.

its energy reserves, using pathways familiar from our studies of animal catabolism: glycolysis, the citric acid cycle, and the pentose phosphate pathway. In general, these pathways are inhibited in plants in the presence of sunlight and become more active in the dark. The key light-inhibited enzymes are phosphofructokinase (in glycolysis) and glucose-6-phosphate dehydrogenase (in the pentose phosphate pathway). The latter is inhibited by the same reduced form of thioredoxin that *activates* Calvin cycle enzymes.

Finally, it should be mentioned that there is now evidence for regulation of chloroplast genes at the transcriptional level. Sucrose and glucose can act as corepressors, turning off expression of the photosynthetic machinery.

Photorespiration and the C$_4$ Cycle

Ribulose bisphosphate carboxylase is a peculiar enzyme. Under normal environmental conditions, it can behave as an *oxygenase* as well as a carboxylase. In the oxygenase reaction, it is proposed that the enediolate intermediate nucleophilically attacks O_2 instead of CO_2 (compare with the carboxylase reaction on page 696):

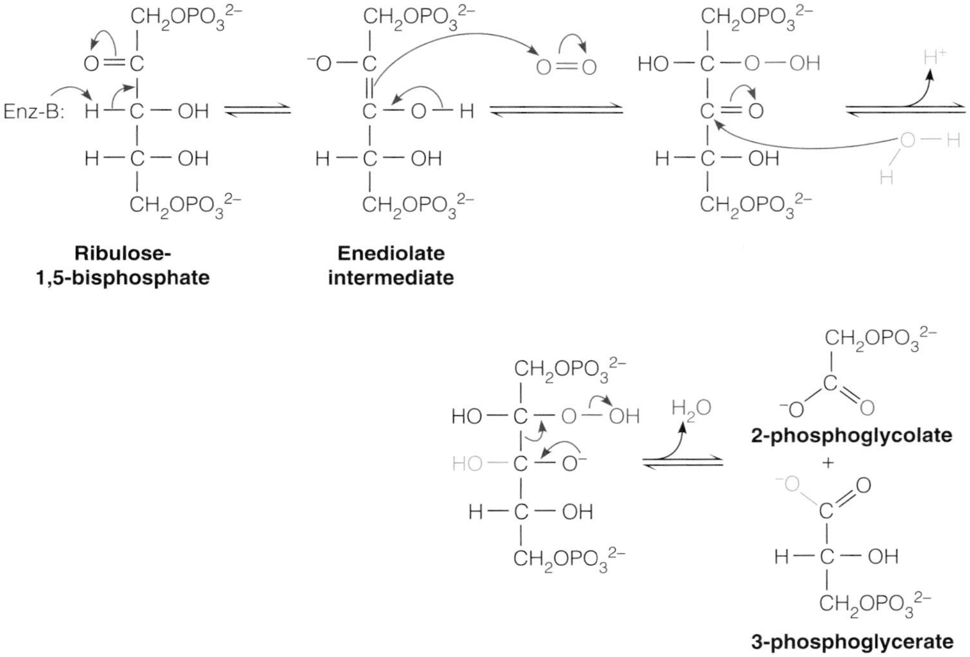

Ribulose-
1,5-bisphosphate

Enediolate
intermediate

2-phosphoglycolate

3-phosphoglycerate

The relative rates of the carboxylase and oxygenase reactions are determined by the concentrations of the two gases at the active site of the enzyme and its K_M values for CO_2 and O_2. Oxygenation occurs under conditions of high O_2 and low CO_2 concentrations (such as found in normal air) because the K_M for O_2 is about 10-fold higher than for CO_2. When the oxygenase reaction becomes significant, it initiates a reaction pathway known as **photorespiration**, with production of 3-phosphoglycerate and *phosphoglycolate* in the chloroplast. As Figure 16.28 shows, the phosphoglycolate is then dephosphorylated and passed to organelles called **peroxisomes**. Here, it is further oxidized, yielding glyoxylate and hydrogen peroxide. The toxic H_2O_2 is broken down by catalase, and the glyoxylate is transaminated, producing glycine. The glycine enters mitochondria, where *two* molecules are converted into *one* molecule of serine, plus one molecule each of CO_2 and NH_3. This process, involving the multiprotein glycine cleavage system and serine hydroxymethyltransferase, is described in detail in Chapter 20. The gases CO_2 and ammonia are released. The serine passes back into the peroxisome, where a series of reactions convert it to glycerate. Returning to the chloroplast, the glycerate is rephosphorylated (using ATP) to yield 3-phosphoglycerate (3PG).

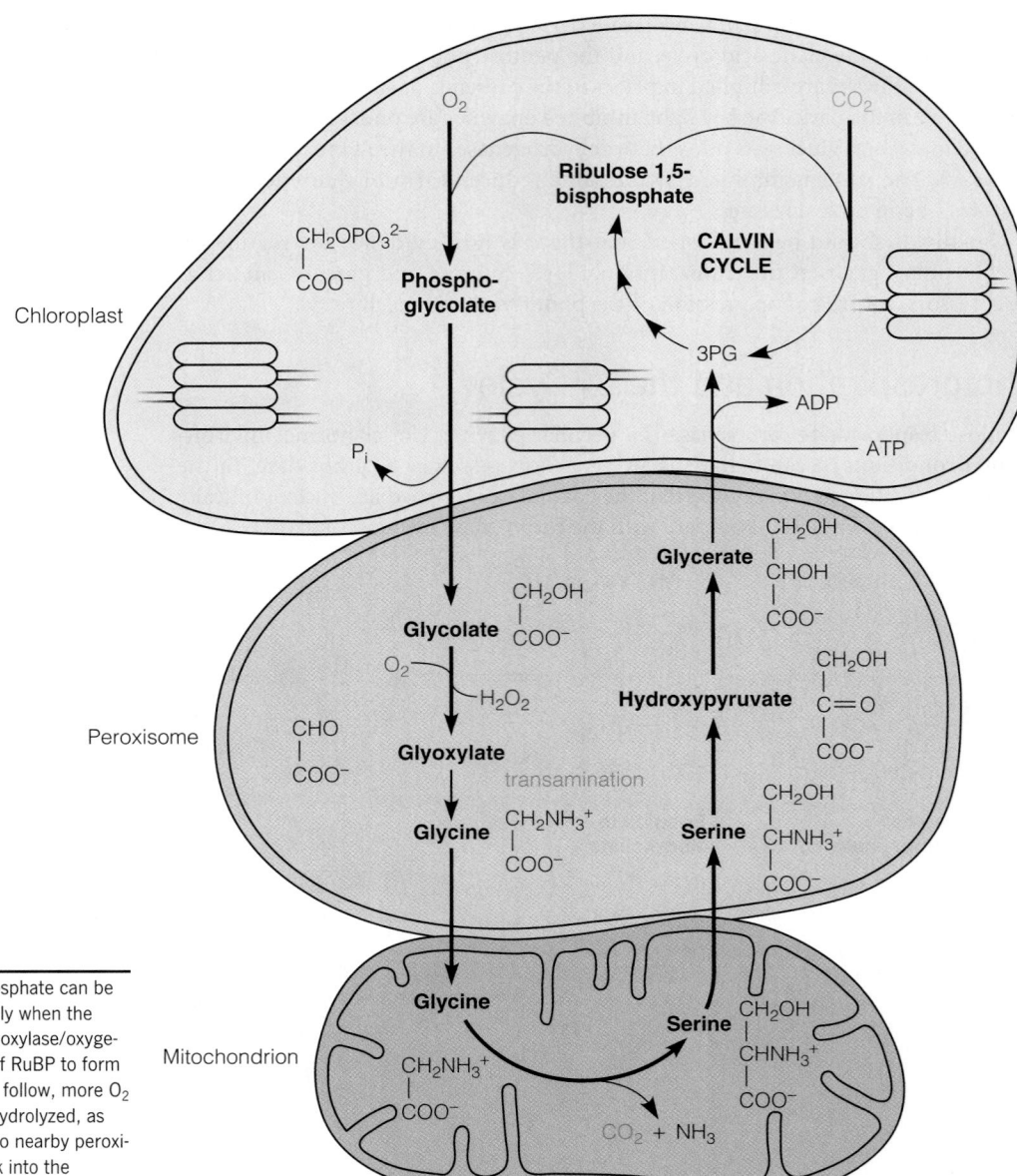

FIGURE 16.28

Photorespiration. Ribulose-1,5-bisphosphate can be diverted from the Calvin cycle, especially when the concentration of CO_2 is low. RuBP carboxylase/oxygenase (rubisco) catalyzes the oxidation of RuBP to form phosphoglycolate. In the reactions that follow, more O_2 is used, CO_2 is generated, and ATP is hydrolyzed, as metabolites pass from the chloroplast to nearby peroxisomes and mitochondria and then back into the chloroplast.

Under conditions of low CO_2 and high O_2, plants exhibit photorespiration, in which O_2 is consumed and CO_2 released.

Photorespiration appears to be a wasteful process that costs considerable energy and limits the carbon assimilation by the plant. Note the following:

1. Ribulose-1,5-bisphosphate is lost from the Calvin cycle.
2. The fixation of CO_2 is reversed: O_2 is consumed and CO_2 is released.
3. Only a part (75%) of the carbon is returned to the chloroplast.
4. ATP is expended.

So why does photorespiration exist? It is probably a result of the fact that evolution can only work with what it has at hand. Rubisco evolved as a carboxylase ~3 billion years ago, when the atmosphere was characterized by high CO_2 and low O_2 levels. Under these conditions, the oxygenation reaction was negligible. However, as O_2 levels increased in the atmosphere, its oxygenase activity became significant. Plants then had to evolve strategies to deal with the inefficiency introduced by the oxygenase activity of rubisco. You might expect they would modify the ribulose bisphosphate carboxylase/oxygenase enzyme to suppress the oxygenase function. Surprisingly, that is not the case. This enzyme, despite (or perhaps because of) its vital importance, has changed little over long ages. It remains a relatively inefficient catalyst (with $k_{cat} \cong 2 \, s^{-1}$) and has never lost its oxygenase function. Stuck with this "design flaw" in rubisco, plants instead evolved a pathway to reclaim the two-carbon fragment (phosphoglycolate) released by the oxygenase activity. This pathway, photorespiration, can thus be thought of as an evolutionary compromise, albeit an inelegant one.

Certain plants, which are called **C_4 plants**, have evolved an additional photosynthetic pathway that helps conserve CO_2 released by photorespiration. This pathway is called the **C_4 cycle** because it involves incorporation of CO_2 into a C_4 intermediate (oxaloacetate). This cycle is distinguished from the Calvin cycle, which utilizes a three-carbon intermediate, and is hence sometimes called the C_3 cycle. The C_4 cycle is found in several crop species (maize and sugarcane, for example) and is important in tropical plants, which are exposed to intense sunlight and high temperatures. Although photorespiration occurs to some extent at all times in all plants, it is most active under conditions of high illumination, high temperature, and CO_2 depletion.

C_4 photosynthesis is really just a CO_2 concentrating mechanism, providing a higher CO_2/O_2 ratio at the active site of rubisco that favors carboxylation. C_4 plants concentrate their Calvin cycle (C_3) photosynthesis in specialized *bundle sheath cells*, which lie below a layer of mesophyll cells (Figure 16.29a). The mesophyll cells, on the other hand, which are most directly exposed to external CO_2, contain the enzymes for the C_4 cycle. This pathway, as it operates in most C_4 plants, is shown in Figure 16.29b. It is essentially a mechanism for trapping CO_2 into a four-carbon compound and passing it on to bundle sheath cells for decarboxylation and use of the resulting CO_2 in their Calvin (C_3) cycle.

The key to the efficiency of C_4 plants is that the CO_2-fixing enzyme used in this pathway, *phosphoenolpyruvate carboxylase*, lacks the oxygenase activity shown by ribulose bisphosphate carboxylase, and has a much lower K_M for CO_2. Thus, even under conditions of high O_2 concentration and low CO_2 concentration in the atmosphere, the mesophyll cells continue to pump CO_2 to the photosynthesizing bundle sheath cells, where rubisco is localized. This process helps maintain high enough CO_2 levels in the bundle sheath cells that CO_2 fixation, rather than photorespiration, is favored. Furthermore, if photorespiration *does* occur, the CO_2 that is released in that process can be largely salvaged in the surrounding mesophyll cells and returned to the Calvin cycle.

As Figure 16.29 shows, the C_4 cycle costs the plant energy in the form of ATP. In fact, because ATP is hydrolyzed to AMP and inorganic phosphate in regenerating phosphoenolpyruvate, the expense is equivalent to *two* extra ATPs for every CO_2 molecule fixed.

The inefficiency of rubisco as an enzyme and its participation in photorespiration greatly reduce the efficiency of plants as food producers. Not only the very large amounts of rubisco that must be synthesized but also the energy expended in photorespiration place seemingly unnecessary demands on plant metabolism. If a more efficient enzyme could be developed, crop yields could be significantly increased and nitrogen demands reduced. There were intensive attempts to engineer a more efficient rubisco into crop plants, but these efforts were unsuccessful, and have been largely abandoned. This failure is perhaps not surprising, since evolution has had a few billion years to eliminate the oxygenase activity and has not managed to do so. An alternative approach that is currently being pursued is to introduce C4 photosynthesis into C3 crop plants.

Some plants (called C_4 plants) minimize the wastefulness of photorespiration by utilizing an adjunct to the Calvin cycle.

Evolution of Photosynthesis

How did photosynthesis arise? It is generally accepted that anoxygenic photosynthesis appeared first, probably soon after life began more than 3.8 billion years ago. These earliest photosynthetic bacteria possessed a single photosystem, the predecessor of modern type-I and type-II photosystems, and relied on substances like

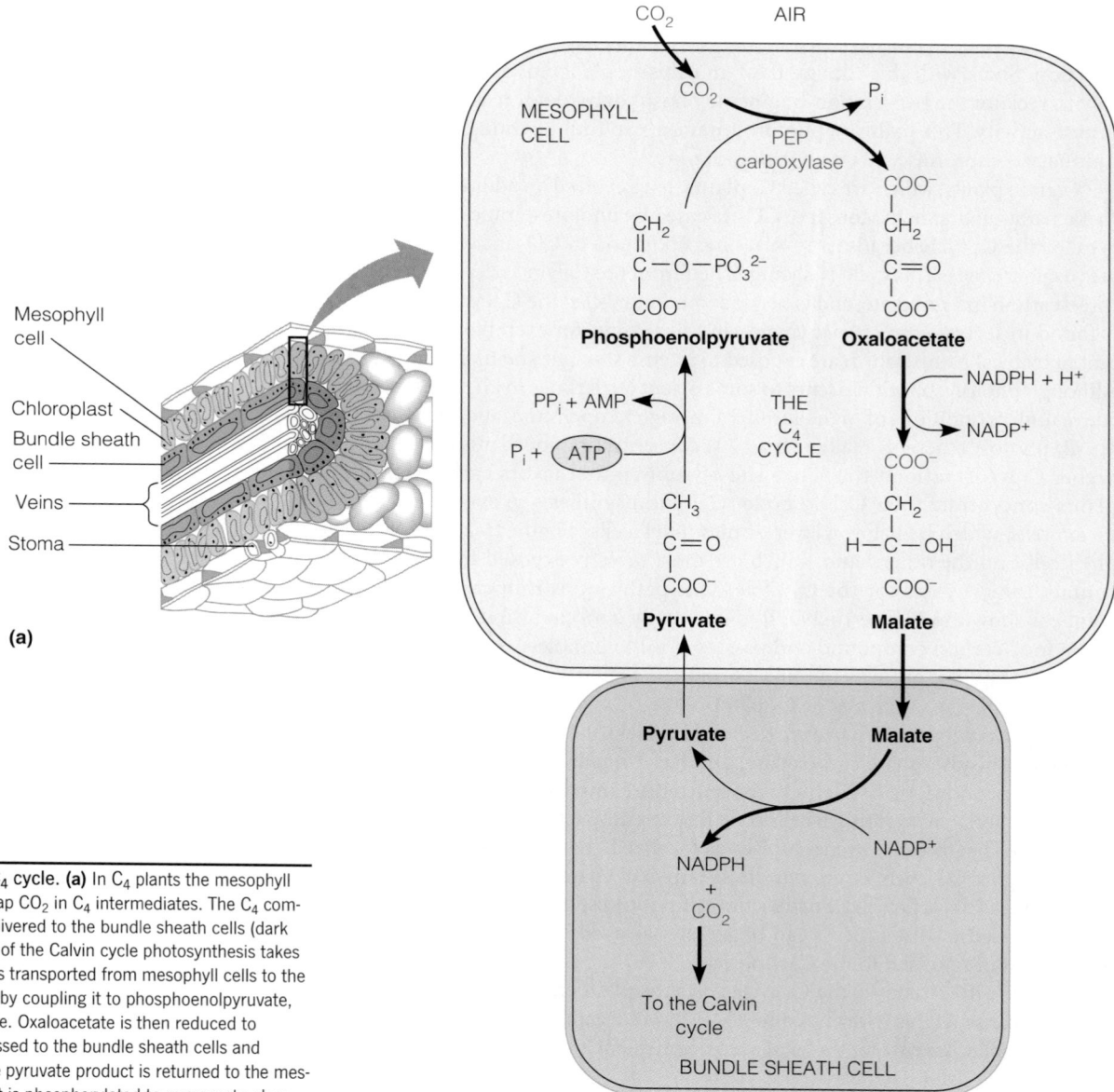

FIGURE 16.29

Reactions of the C$_4$ cycle. **(a)** In C$_4$ plants the mesophyll cells (light green) trap CO$_2$ in C$_4$ intermediates. The C$_4$ compounds are then delivered to the bundle sheath cells (dark green), where most of the Calvin cycle photosynthesis takes place (C$_3$). **(b)** CO$_2$ is transported from mesophyll cells to the bundle sheath cells by coupling it to phosphoenolpyruvate, forming oxaloacetate. Oxaloacetate is then reduced to malate, which is passed to the bundle sheath cells and decarboxylated. The pyruvate product is returned to the mesophyll cells, where it is phosphorylated to regenerate phosphoenolpyruvate.

hydrogen sulfide as an electron source. The modern descendants of these anoxygenic photosynthetic bacteria possess either a type-I or a type-II photosystem, but never both types in the same organism. Oxygenic photosynthesis, such as that seen in modern cyanobacteria, requires two photosystems, working in series, and a catalyst that splits water, producing O$_2$ and electrons—the manganese-dependent oxygen-evolving complex (OEC) of photosystem II (Figure 16.12b).

How all these components came together in one ancestral cell is the subject of much speculation. One theory, based on genomics analyses, argues that photosystem I was the ancestral prototype, from which photosystem II evolved, perhaps by a gene duplication event. An alternative theory holds that lateral gene transfer or fusion of a type-I cell with a type-II cell led to a "protocyanobacterium" possessing both photosystems (Figure 16.30). Either way, our ancient protocyanobacterium has a new problem. Recall that cyclic (type-II) photosystems supply electrons, via reduced quinones, into a proton-pumping complex (the cytochrome bc_1 complex) (Figure 16.21). A type-I photosystem operating in a linear fashion that uses electrons extracted from a donor such as H$_2$S supplies those electrons via ferredoxin to soluble electron acceptors such as NADP$^+$ for net reduction of CO$_2$. These two

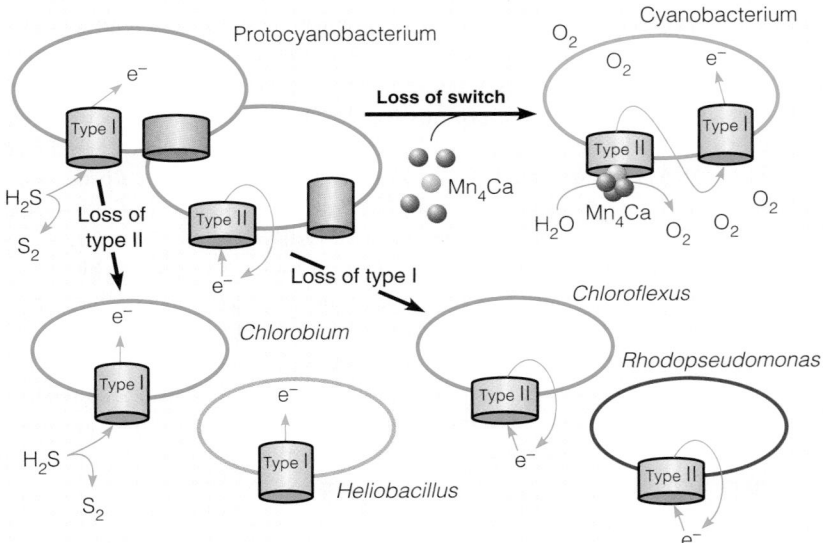

FIGURE 16.30

Proposed evolution of oxygenic photosynthesis. An ancestral anoxygenic "protocyanobacterium," possessing genes for both type-I and type-II photosystems, expresses only one of the two photosystems at any time, depending on its environment and metabolic needs. Modern anoxygenic descendants possessing only a type-I photosystem, such as *Chlorobium* or *Heliobacillus*, evolved by loss of the type-II genes. Modern anoxygenic descendants possessing only a type-II photosystem, such as *Chloroflexus* or *Rhodopseudomonas* (renamed *Blastochloris*), evolved by loss of the type-I genes. Acquisition of a manganese–metal cluster, and loss of the regulatory switch that prevented simultaneous expression of the two photosystems, led to a cell with two photosystems linked in series to a Mn_4Ca catalyst that could extract electrons from H_2O and produce O_2. This cell represents the ancestor of modern oxygenic Cyanobacteria.

Reprinted by permission from Macmillan Publishers Ltd. *Nature* 445:610–612, J. F. Allen and W. Martin, Evolutionary biology: Out of thin air. © 2007.

functions would be favored under different growth conditions, but would compete with each other if expressed in the same membrane at the same time. Thus, our protocyanobacterium must have had a mechanism to switch between expression of the two photosystems, depending on its environment and metabolic state. Modern single-photosystem anoxygenic photosynthetic bacteria could have evolved from this two-system ancestor through loss of the genes encoding one photosystem (Figure 16.30). This scenario may also explain the inefficiency of modern rubisco—as mentioned above, this enzyme evolved at a time when there was little or no oxygen in the atmosphere and CO_2 levels were much higher.

Now we come to the question of the invention of oxygenic photosynthesis. Manganese ions are readily photooxidized by ultraviolet light, which would have been abundant before the ozone layer formed in the early atmosphere. Upon oxidation, a Mn ion ejects an electron, an electron that could be used to replenish an oxidized photosystem. The oxidized Mn would then grab an electron from the most abundant source available—water. If our ancestral protocyanobacterium bound a few Mn ions to the surface of its photosystem II, it could extract electrons from H_2O and pass them through the proton-pumping complex, no longer requiring a cyclic pathway. Photosystem I could now be coupled to photosystem II, taking electrons from its cytochrome bc_1 complex and passing them via ferredoxin to soluble electron acceptors (Figure 16.30). Over time, the soluble manganese ions were replaced by the more stable Mn_4Ca cluster, and the photosystems became more efficient, eventually leading to the oxygenated, green earth that exists today. There would have been substantial selective pressure for this evolutionary scenario, since oxygenic photosynthesis released our ancestral protocyanobacterium from scarce inorganic or organic electron donors. As biochemist John F. Allen of University of London has said, "Water is everywhere, so the organisms never ran out of electrons. They were unstoppable."

SUMMARY

Photosynthesis is the source of most of the energy in the biosphere and accounts for fixation of atmospheric CO_2 and the production of most or all of the O_2 in the atmosphere. The whole process can be divided into light reactions and dark reactions. The light reactions use the energy of sunlight to extract electrons from water, producing O_2, reductive potential, and a proton gradient that drives ATP formation. The dark reactions reduce CO_2 into carbohydrates. In plants and higher algae, both types of reactions take place in chloroplasts. Photons for the light reaction are absorbed by antenna pigments, and the energy is transferred to reaction centers, where it enters either photosystem I or photosystem II. These two systems, working in conjunction, carry out the light reactions. Photosystem II oxidizes water, and photosystem I reduces $NADP^+$. Together, the systems drive the transport of protons across chloroplast membranes to provide a pH gradient to drive ATP production. Under conditions of high NADPH concentration, photosystem I can operate independently of photosystem II in a process called cyclic photophosphorylation, where only ATP is produced. Some photosynthetic bacteria, on the other hand, use a cyclic version of photosystem II to generate ATP.

The dark reactions are largely summarized in the Calvin cycle, which may be divided into two stages. In the first stage, CO_2 is added to ribulose-1,5-bisphosphate (RuBP), which is then cleaved and reduced to form trioses that can then be combined to form hexose. The second stage of the cycle uses most of the trioses and hexoses to regenerate RuBP. A number of these dark reactions are regulated by light intensity.

Under conditions of low CO_2 and high O_2, plants undergo an oxidative process called photorespiration. This process is essentially inefficient, and some tropical plants compensate for it via the C_4 cycle, which is less sensitive to high O_2 levels.

REFERENCES

General

Blankenship, R. E. (2007) *Molecular Mechanisms of Photosynthesis.* John Wiley & Sons, Ltd., Chichester. A concise, but complete, introduction to the history, chemistry, mechanisms, physiology, and evolution of photosynthetic systems.

Bowsher, C., M. W. Steer, and A. K. Tobin (2008) *Plant Biochemistry.* Garland Science, New York. An introduction to all aspects of the biochemistry of plants.

Clayton, R. K. (1980) *Photosynthesis: Physical Mechanisms and Chemical Patterns.* Cambridge University Press, Cambridge. Although somewhat superseded by more recent work, this remains an excellent summary of the more physical aspects of photosynthesis, written by one of the pioneers.

Evolution of Photosynthesis

Allen, J. F., and W. Martin (2007) Evolutionary biology: Out of thin air. *Nature* 445:610–612.

Leslie, M. (2009) On the origin of photosynthesis. *Science* 323:1286–1287.

Mulkidjanian, A. Y., E. V. Koonin, K. S. Makarova, S. L. Mekhedov, A. Sorokin, Y. I. Wolf, A. Dufresne, F. Partensky, H. Burd, D. Kaznadzey, R. Haselkorn, and M. Y. Galperin (2006) The cyanobacterial genome core and the origin of photosynthesis. *Proc. Natl. Acad. Sci. USA* 103:13126–13131. Describes a genomic analysis of cyanobacterial photosystems.

Raymond, J., O. Zhaxybayeva, J. P. Gogarten, S. Y. Gerdes, and R. E. Blankenship (2002) Whole-genome analysis of photosynthetic prokaryotes. *Science* 298:1616–1620. Provides evidence for lateral transfer of photosystem genes.

Light Reactions

Barber, J. (2008) Photosynthetic generation of oxygen. *Philos. Trans. R. Soc. Lond. B. Biol. Sci.* 363:2665–2674. A review of the structure and function of photosystem II.

Kargul, J., and J. Barber (2008) Photosynthetic acclimation: Structural reorganisation of light harvesting antenna—role of redox-dependent phosphorylation of major and minor chlorophyll a/b binding proteins. *FEBS J.* 275:1056–1068. A mini-review of state transitions in chloroplasts.

McEvoy, J. P., and G. W. Brudvig (2006) Water-splitting chemistry of photosystem II. *Chem. Rev.* 106:4455–4483. A detailed review of the chemistry and structure of the manganese cluster of the oxygen-evolving complex.

Rochaix, J.-D. (2011) Regulation of photosynthetic electron transport. *Biochim. Biophys. Acta-Bioenergetics* 1807:375–383.

Yano, J., J. Kern, Y. Pushkar, K. Sauer, P. Glatzel, U. Bergmann, J. Messinger, A. Zouni, and V. K. Yachandra (2008) High-resolution structure of the photosynthetic Mn_4Ca catalyst from X-ray spectroscopy. *Philos. Trans. R. Soc. Lond. B. Biol. Sci.* 363:1139–1147.

Structures

Amunts, A., and N. Nelson (2009) Plant photosystem I design in the light of evolution. *Structure* 17:637–650.

Cramer, W. A., H. Zhang, J. Yan, G. Kurisu, and J. L. Smith (2006) Transmembrane traffic in the cytochrome $b_6 f$ complex. *Annu. Rev. Biochem.* 75:769–790.

Deisenhofer, J., and H. Michel (2004) The photosynthetic reaction centre from the purple bacterium *Rhodopseudomonas viridis. Biosci. Rep.* 24:323–361. This is a republication of Deisenhofer's and Michel's 1988 Nobel Prize lecture—it describes the history, methods, struggles, and ultimate triumph of solving the X-ray structure of a membrane protein.

Guskov, A., J. Kern, A. Gabdulkhakov, M. Broser, A. Zouni, and W. Saenger (2009) Cyanobacterial photosystem II at 2.9-A resolution and the role of quinones, lipids, channels and chloride. *Nat. Struct. Mol. Biol.* 16:334–342. Describes the structure of the *Thermosynechococcus elongatus* PSII complex illustrated in Figure 16.14.

Li, L., S. Nachtergaele, A. M. Seddon, V. Tereshko, N. Ponomarenko, and R. F. Ismagilov (2008) Simple host-guest chemistry to modulate the process of concentration and crystallization of membrane proteins by detergent capture in a microfluidic device. *J. Am. Chem. Soc.* 130:14324–14328. Describes the structure of the purple bacterial reaction center complex illustrated in Figure 16.20.

Standfuss, J., A. C. Terwisscha van Scheltinga, M. Lamborghini, and W. Kuhlbrandt (2005) Mechanisms of photoprotection and nonphotochemical quenching in pea light-harvesting complex at 2.5 Å resolution. *EMBO J.* 24:919–928. The structure of LHCII (Figure 16.10).

Dark Reactions and Photorespiration

Benson, A. A. (2002) Following the path of carbon in photosynthesis: A personal story. *Photosyn. Res.* 73:29–49.

Foyer, C. H., A. J. Bloom, G. Queval, and G. Noctor (2009) Photorespiratory metabolism: Genes, mutants, energetics, and redox signaling. *Annu. Rev. Plant Biol.* 60:455–484.

Lemaire, S. D., L. Michelet, M. Zaffagnini, V. Massot, and E. Issakidis-Bourguet (2007) Thioredoxins in chloroplasts. *Curr. Genet.* 51: 343–365. A review of the role of thioredoxins in controlling the dark reactions.

Parry, M. A., A. J. Keys, P. J. Madgwick, A. E. Carmo-Silva, and P. J. Andralojc (2008) Rubisco regulation: A role for inhibitors. *J. Exp. Bot.* 59:1569–1580; A. R. Portis, Jr., C. Li, D. Wang, and M. E. Salvucci

(2008) Regulation of rubisco activase and its interaction with rubisco. *J. Exp. Bot.* 59:1597–1604. Recent reviews of the regulation of rubisco.

Tabita, F. R., T. E. Hanson, S. Satagopan, B. H. Witte, and N. E. Kreel (2008) Phylogenetic and evolutionary relationships of RubisCO and the RubisCO-like proteins and the functional lessons provided by diverse molecular forms. *Philos. Trans. R. Soc. Lond. B. Biol. Sci.* 363:2629–2640.

Artificial Photosynthesis

Brimblecombe, R., D. R. Kolling, A. M. Bond, G. C. Dismukes, G. F. Swiegers, and L. Spiccia (2009) Sustained water oxidation by $[Mn_4O_4]_7^+$ core complexes inspired by oxygenic photosynthesis. *Inorg. Chem.* 48:7269–7279. Describes development of the solar cell illustrated in Figure 16.22.

Service, R. F. (2009) New trick for splitting water with sunlight. *Science* 325: 1200–1201. Discusses recent advances in artificial photosynthesis.

PROBLEMS

1. According to Figure 16.12b, upon excitation the P700 reaction center is raised in potential from about +0.4 to −1.3 volts. To what value of $\Delta G^{\circ\prime}$ does this correspond? How does it compare with the energy in an einstein of 700 nm photons?

2. In cyclic photophosphorylation, it is estimated that two electrons must be passed through the cycle to pump enough protons to generate one ATP. Assuming that the ΔG for hydrolysis of ATP under conditions existing in the chloroplast is about −50 kJ/mol, what is the corresponding percent efficiency of cyclic photophosphorylation, using light of 700 nm?

3. Assume a pH gradient of 4.0 units across a thylakoid membrane, with the lumen more acidic than the stroma. What is the *longest* wavelength of light that could provide enough energy per photon to pump one proton against this gradient, assuming a 20% efficiency in photosynthesis and $T = 25\ °C$?

*4. Suppose a brief pulse of $^{14}CO_2$ is taken up by a green plant.
(a) Trace the ^{14}C label through the steps leading to fructose-1,6-bisphosphate synthesis, showing which carbon atoms in each compound should carry the label during the first cycle.
(b) Will all molecules of fructose-1,6-bisphosphate carry two ^{14}C atoms? Explain.

5. The flux of solar energy reaching the earth's surface is approximately 7 J/cm² s. Assume that *all* of this energy is used by a green leaf (10 cm² in area), with the maximal efficiency of 35%. How many moles of hexose could the leaf theoretically generate in an hour? You may use 600 nm for an average wavelength.

*6. The substance dichlorophenyldimethylurea (DCMU) is an herbicide that inhibits photosynthesis by blocking electron transfer between plastoquinones in photosystem II.
(a) Would you expect DCMU to interfere with cyclic photophosphorylation?
(b) Normally, DCMU blocks O_2 evolution, but addition of ferricyanide to chloroplasts allows O_2 evolution in the presence of DCMU. Explain.

7. Suppose a researcher is carrying out studies in which she adds a nonphysiological electron donor to a suspension of chloroplasts. Illumination of the chloroplasts yields oxidation of the donor. How could she tell whether photosystem I, II, or both are involved?

8. Suppose ribulose-5-phosphate, labeled with ^{14}C in carbon 1, is used as the substrate in dark reactions. In which carbon of 3PG will the label appear?

9. The following data, presented by G. Bowes and W. L. Ogre in *J. Biol. Chem.* (1972) 247:2171–2176, describe the relative rates of incorporation of CO_2 by rubisco under N_2 and under pure O_2. Decide whether O_2 is a competitive or uncompetitive inhibitor.

[CO_2] (mM)	Under N_2	Under O_2
0.20	16.7	10
0.10	12.5	5.6
0.067	8.3	4.2
0.050	7.1	3.2

*10. J. C. Servaites, in *Plant Physiol.* (1985) 78:839–843, observed that rubisco from tobacco leaves collected before dawn had a much lower specific activity than the enzyme collected at noon. This difference persisted despite extensive dialysis, gel filtration, or heat treatment. However, precipitation of the predawn enzyme by 50% $(NH_4)_2SO_4$ restored the specific activity to the level of noon-collected enzyme. Suggest an explanation.

11. It is believed that the ratio of cyclic photophosphorylation to noncyclic photophosphorylation changes in response to metabolic demands. In each of the following situations, would you expect the ratio to increase, decrease, or remain unchanged?
(a) Chloroplasts carrying out both the Calvin cycle and the reduction of nitrite (NO_2^-) to ammonia (This process does not require ATP.)
(b) Chloroplasts carrying out not only the Calvin cycle but also extensive active transport
(c) Chloroplasts using both the Calvin cycle and the C_4 pathway

12. If a photosynthetic organism is illuminated in a closed, sealed environment, it is observed that the CO_2 and O_2 levels in the surrounding atmosphere reach a constant ratio.
(a) Suggest an explanation.
(b) What factor would you think primarily determines the value of this ratio?

13. If algae are exposed to $^{14}CO_2$ for a brief period while illuminated, the labeled carbon is initially found almost entirely in the carboxyl group of 3-phosphoglycerate. However, if illumination is continued after the label pulse, other carbon atoms become labeled. Explain.

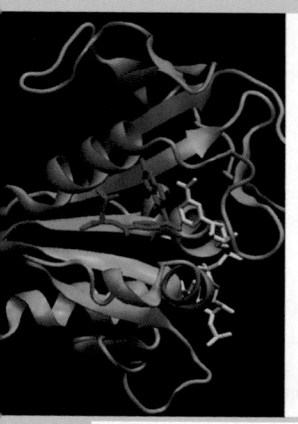

CHAPTER 17

Lipid Metabolism I: Fatty Acids, Triacylglycerols, and Lipoproteins

Like the carbohydrates we have discussed in previous chapters, lipids play roles in energy metabolism as well as in a variety of other processes. For lipids, those other processes include their roles as membrane constituents, hormones, fat-soluble vitamins, thermal insulators, and signaling molecules. This chapter focuses on bioenergetic aspects of lipid metabolism. We discuss the synthesis and breakdown of energy storage lipids, as well as fatty acid oxidation and biosynthesis (Figure 17.1), processes that are quite similar among plants, animals, and microorganisms. We also present topics related more directly to animal metabolism—fat digestion, absorption, storage, and mobilization. Metabolism of membrane lipids and lipids of more specialized metabolic functions is covered in Chapter 19.

Utilization and Transport of Fat and Cholesterol

As discussed in Chapter 10, the great bulk of the lipid in most organisms is in the form of *triacylglycerols* (formerly called *triglycerides*). The term *fat*, or *neutral fat*, refers to this most abundant class of lipids. Fat utilization in animals is intertwined with lipoprotein metabolism, as is the metabolism of cholesterol. Therefore, we will consider fat and cholesterol metabolism together although cholesterol biosynthesis is presented in Chapter 19.

A mammal contains 5% to 25% or more of its body weight as lipid, with as much as 90% of this lipid in the form of triacylglycerols. Most of this fat is stored in adipose tissue and constitutes the primary energy reserve. Mammals (including humans) are "fat burners"—we eat in pulses (i.e., meals), convert the excess carbohydrate into fat, and store it. The fat is then burned at a later time as needed. Indeed, certain tissues, such as heart and liver, obtain as much as 80% of their energy needs from fat oxidation. In animal systems, fat is stored in specialized cells called *adipocytes*, where giant fat globules occupy most of the intracellular space (see Figure 10.3, page 363). Plant seeds store great quantities of fat to provide energy to the developing plant embryo

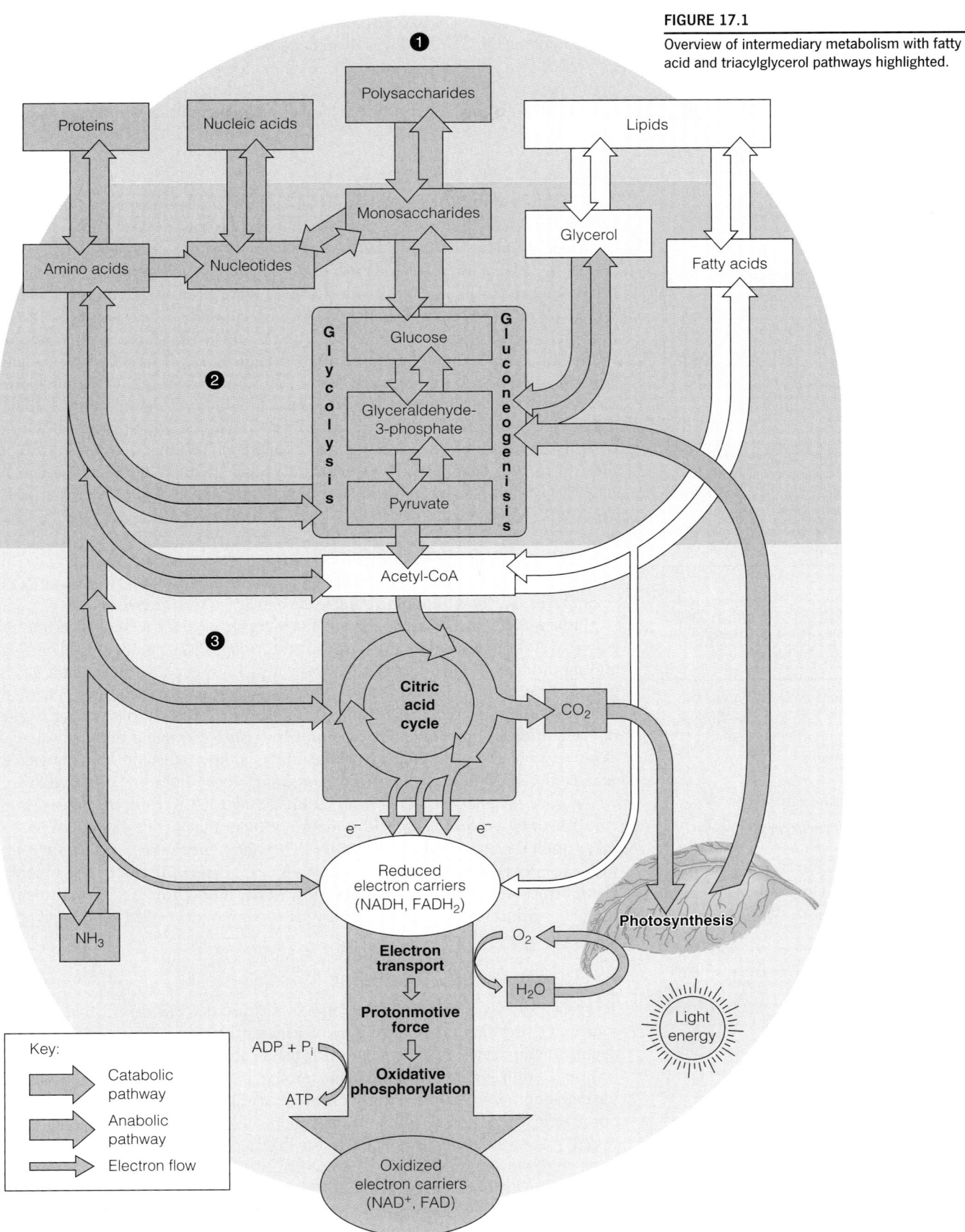

FIGURE 17.1

Overview of intermediary metabolism with fatty acid and triacylglycerol pathways highlighted.

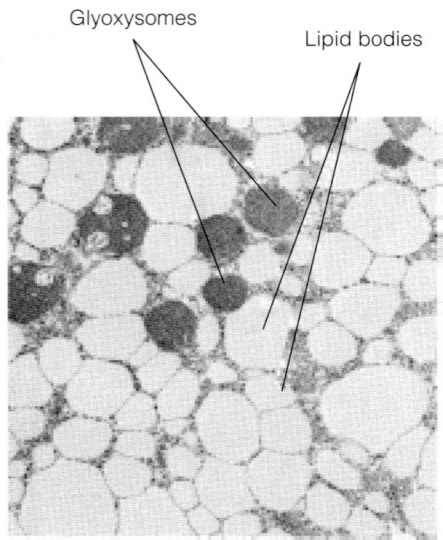

FIGURE 17.2

Fat storage in a plant seedling. The electron micrograph (×6500) shows a cell from a cucumber cotyledon (seed leaf) a few days after germination. Fat stored in lipid bodies is degraded, oxidized, and converted to carbohydrate in neighboring glyoxysomes (or microbodies) to support the growth of the plant.

Courtesy of R. N. Trelease, P. J. Gruber, W. M. Becker, and E. H. Newcomb, *Plant Physiology* 48:461 (1971), Microbodies (glyoxysomes and peroxisomes) in cucumber cotyledons: Correlative biochemical and ultrastructural study in light- and dark-grown seedlings.

Fat has six times more caloric content by weight than carbohydrate because fat is more highly reduced and is anhydrous.

Tissue Fuel Stores for Average 70-kg Human

Fuel	Weight (g)	Energy Content (kJ/g)	Total Energy (kJ)
Triacylglycerols	~15,000	37	555,000
Protein	~6,000	17	100,000
Glycogen	~400	17	6,800
Glucose	~20	17	340
Total fuel stores:			662,140

(Figure 17.2). Because plant lipids contain mostly unsaturated fatty acids, the triacylglycerols of seeds are largely in the form of liquid oils.

Triacylglycerols play roles other than in energy storage. Fat serves to cushion organs against shock, and it provides an efficient thermal insulator, particularly in marine mammals, which must maintain a body temperature far higher than that of the seawater in which they live.

Fats as Energy Reserves

Recall that most of the carbon in triacylglycerols is more highly reduced than the carbon in carbohydrates. To be sure, the carboxyl carbons of fatty acids are highly oxidized, but most of the fatty acid carbons are at the reduced methyl or methylene level. Thus, metabolic oxidation of fat consumes more oxygen, on a weight basis, than oxidation of carbohydrate, with correspondingly larger metabolic energy release. The complete metabolic oxidation of triacylglycerols yields 37 kJ/g or more, whereas that of carbohydrates and proteins yields about 17 kJ/g. Adding to this difference between fat and carbohydrate is the hydrophilic nature of glucose polymers. Glycogen binds about 2 grams of water per gram of carbohydrate. Fat, being extremely nonpolar, is anhydrous. Thus, because 1 gram of intracellular glycogen contains but $\frac{1}{3}$ gram of anhydrous glucose polymer, intracellular fat contains about six times as much potential metabolic energy, on a mass basis, as intracellular glycogen. This is an obvious advantage in many situations, such as in hibernating animals, which must store several months' worth of food, or in the flight muscles of small birds, in which weight is at a premium. Incredibly, some small land birds prepare for migration by increasing their body weight about 15% per day, with all of the weight gain being triacylglycerols. Such obese birds can then fly nonstop for 60 hours or more. In addition, the insolubility of fat allows it to be stored in cells without affecting intracellular osmotic pressure.

Little wonder, then, that fat is the major energy storage form in most organisms. A typical 70-kg human may have fuel reserves of 500,000 kJ or more in total body fat and about 100,000 kJ in total protein (mostly muscle protein). By contrast, the glycogen stores amount to just 6800 kJ of available energy, and the total free glucose to about 300 kJ. Fat stores are maintained from the diet; 35%–50% of the caloric value of Western diets comes from fat. Most nutritionists recommend that this value be closer to 25%–35% for cardiovascular health. In addition, carbohydrate ingested in excess of its ability to be catabolized or stored as glycogen is readily converted to fat.

Most of the energy derived from fat breakdown comes from oxidation of the constituent fatty acids. Fatty acid oxidation provides the major energy source for many animal tissues. Brain is distinctive in being unable to use fatty acids as a significant energy supply; brain has a highly specific requirement for glucose. However, under conditions of starvation, when blood glucose levels decrease, brain can adjust to use a class of lipid-related compounds called *ketone bodies*, as discussed later in this chapter.

Fat Digestion and Absorption

The triacylglycerols that mammals use as fuel are derived from three primary sources: (1) the diet; (2) de novo biosynthesis, particularly in liver; and (3) storage depots in adipocytes. Processes by which these sources are utilized in animals are summarized in Figure 17.3. The major problem that animals must cope with in the digestion, absorption, and transport of dietary lipids is their insolubility in aqueous media. The action of **bile salts**, detergent substances synthesized in liver and stored in the gallbladder, is essential to the digestion of lipids and their absorption through the intestinal mucosa. The problem of transport through the blood and lymph is dealt with in part by the complexing of the lipids with proteins to form soluble aggregates called **lipoproteins**.

A bile salt molecule is made up of a **bile acid**, such as cholic acid, and an associated cation. Bile acids (discussed further in Chapter 19) are derived from cholesterol.

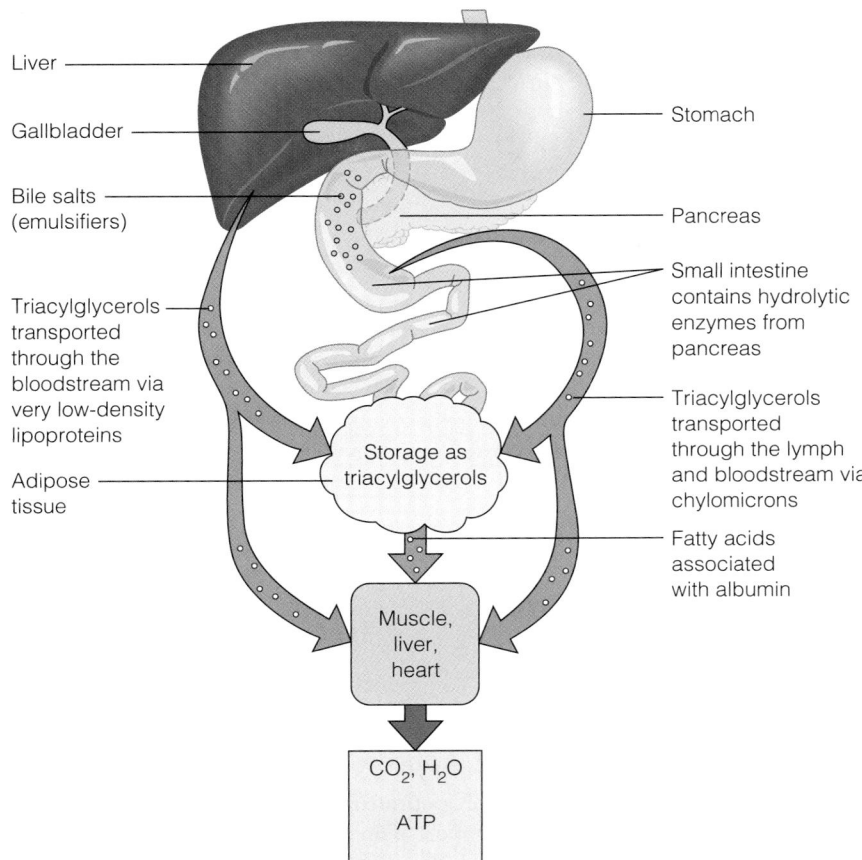

Liver

Gallbladder

Bile salts
(emulsifiers)

Triacylglycerols
transported
through the
bloodstream via
very low-density
lipoproteins

Adipose
tissue

Stomach

Pancreas

Small intestine
contains hydrolytic
enzymes from
pancreas

Triacylglycerols
transported
through the lymph
and bloodstream via
chylomicrons

Fatty acids
associated
with albumin

Storage as
triacylglycerols

Muscle,
liver,
heart

CO$_2$, H$_2$O

ATP

FIGURE 17.3

Overview of fat digestion, absorption, storage, and mobilization in the human. Triacylglycerols (fats) are ingested, synthesized in the liver, or mobilized from storage. Ingested triacylglycerols are hydrolyzed in the lumen of the small intestine by pancreatic lipase and other enzymes. Hydrolysis products absorbed by the intestinal mucosa are recombined into triacylglycerols, which combine with apoproteins to form the lipoproteins called chylomicrons. This process solubilizes the lipids and permits their transport through lymph and blood. Triacylglycerols synthesized in liver are combined with other apoproteins to form very low-density lipoproteins (VLDLs) for transport. Lipoproteins transported to peripheral tissues are hydrolyzed at the inner surfaces of capillaries. Hydrolysis products entering cells are either catabolized for energy or recombined into triacylglycerols for storage. Mobilization of stored triacylglycerols is hormonally regulated.

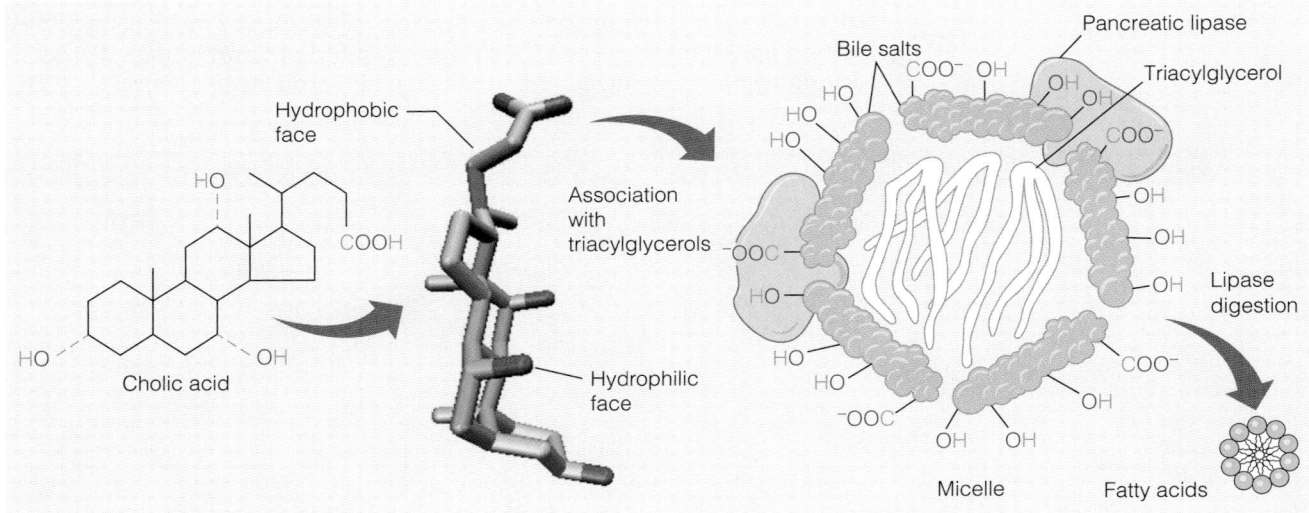

Hydrophobic
face

Cholic acid

Hydrophilic
face

Association
with
triacylglycerols

Bile salts

Pancreatic lipase

Triacylglycerol

Lipase
digestion

Micelle

Fatty acids

FIGURE 17.4

Action of bile salts in emulsifying fats in the intestine. Cholic acid, a typical bile acid, ionizes to give its cognate bile salt. The hydrophobic surface of the bile salt molecule associates with triacylglycerol, and several such complexes aggregate to form a micelle. The polar surface of the bile salts faces outward, allowing the micelle to associate with pancreatic lipase/colipase. Hydrolytic action of this enzyme frees the fatty acids to associate in a much smaller micelle that can be absorbed through the intestinal mucosa.

As shown in Figure 17.4, the bile salt molecule has both hydrophobic and hydrophilic surfaces. This amphipathic character allows bile salts to orient at an oil–water interface, with the hydrophobic surface in contact with the apolar phase and the hydrophilic surface in contact with the aqueous phase. This detergent action emulsifies lipids and yields micelles (see Chapter 10), allowing digestive attack by water-soluble enzymes and facilitating the absorption of lipid through intestinal mucosal cells. Most of the intestinal digestion occurs through action of **pancreatic lipase**, an unusual calcium-requiring enzyme that catalyzes a reaction at an oil–water interface. The substrate being cleaved is in an apolar phase, and the other substrate, of course, is water. Structural studies on pancreatic lipase show that the catalytic center is exposed as the result of a conformational change that occurs only at an oil–water interface. Pancreatic lipase also functions in a 1:1 complex with **colipase**, a 90-amino acid protein that aids in the binding to the lipid surface.

> Bile salts emulsify fats, thereby promoting their hydrolysis in digestion.

The products of fat digestion comprise a mixture of glycerol, free fatty acids, monoacylglycerols, and diacylglycerols. Less than 10% of the original triacylglycerol remains unhydrolyzed. During absorption through intestinal mucosal cells, much resynthesis of triacylglycerols occurs from the hydrolysis products. This resynthesis occurs in the endoplasmic reticulum and Golgi complex of mucosal cells. Triacylglycerols emerge into the lymph system complexed with protein to form the lipoproteins called **chylomicrons**. The chylomicron is essentially an oil droplet coated with more polar lipids and a skin of protein, which help disperse and partially solubilize the fat for transport to tissues. The chylomicron is also a transport vessel for dietary cholesterol.

Transport of Fat to Tissues: Lipoproteins

Chylomicrons constitute just one class of lipoproteins found in the bloodstream. These complexes play essential roles in the transport of lipids to tissues, either for energy storage or for oxidation. Free lipids are all but undetectable in blood. The polypeptide components of lipoproteins are called **apoproteins** or **apolipoproteins**. These are synthesized mainly in the liver, though about 20% are produced in intestinal mucosal cells.

Classification and Functions of Lipoproteins

Distinct families of lipoproteins have been described, each of which plays defined roles in lipid transport. These families are classified in terms of their density, as determined by centrifugation (Table 17.1). Lipoproteins in each class contain

TABLE 17.1   Properties of major human plasma lipoprotein classes

	Chylomicron	VLDL	IDL	LDL	HDL
Density (g/mL)	<0.95	0.950–1.006	1.006–1.019	1.019–1.063	1.063–1.210
Diameter (Å)	10^3–10^4	300–800	250–350	180–250	50–120
Components (% dry weight)					
Protein	2	8	15	22	40–55
Triacylglycerol	86	55	31	6	4
Free cholesterol	2	7	7	8	4
Cholesterol esters	3	12	23	42	12–20
Phospholipids	7	18	23	22	25–30
Apoprotein composition	A-I, A-II, A-IV, B-48, C-I, C-II, C-III, E	B-100, C-I, C-II, C-III, E	B-100, C-I, C-II, C-III, E	B-100, E	A-I, A-II, C-I, C-II, C-III, D, E

Data from A. Jonas (2002) Lipoprotein structure. In *Biochemistry of lipids, lipoproteins and membranes*, 4th ed., D. E. Vance and J. E. Vance, eds., Ch. 18, pp. 483–504, Elsevier, Amsterdam; and R. J. Havel and J. P. Kane (2001) Introduction: Structure and metabolism of plasma lipoproteins. In *The Metabolic and Molecular Bases of Inherited Disease*, C. R. Scriver, A. L. Beaudet, W. S. Sly, D. Valle, B. Childs, K. W. Kinzler, and B. Vogelstein, eds., Vol. II, Ch. 114, pp. 2705–2716, McGraw-Hill, New York.

TABLE 17.2 Apoproteins of the human plasma lipoproteins

Apoprotein	Molecular Weight	Characteristics
A-I	28,300	Major protein in HDL; activates LCAT
A-II	17,400	Major protein in HDL
A-IV	44,000	Found in chylomicrons
B-48	241,100	Found exclusively in chylomicrons
B-100	513,000	Major protein in LDL
C-I	6,600	Found in chylomicrons; activates LCAT and LPL
C-II	8,900	Found primarily in VLDL and chylomicrons; activates LPL
C-III	8,800	Found primarily in chylomicrons, VLDL, and HDL; inhibits LPL
D	33,000	HDL protein, also called cholesterol ester transfer protein
E	34,000	Found in VLDL, LDL, IDL, and HDL

Data from A. Jonas, (2002) Lipoprotein structure. In *Biochemistry of Lipids, Lipoproteins and Membranes*, 4th ed., D. E. Vance and J. E. Vance, eds. Ch. 18, pp. 483–504, Elsevier, Amsterdam; and R. J. Havel and J. P. Kane (2001) Introduction: Structure and metabolism of plasma lipoproteins in *The Metabolic and Molecular Bases of Inherited Disease*, Vol. II, C. R. Scriver, A. L. Beaudet, W. S. Sly, D. Valle, B. Childs, K. W. Kinzler, and B. Vogelstein, eds., Ch. 114, pp. 2705–2716, McGraw-Hill, New York. *Note:* LCAT = lecithin:cholesterol acyltransferase, LPL = lipoprotein lipase.

characteristic apoproteins and have distinctive lipid compositions. A total of 10 major apolipoproteins are found in human lipoproteins. Their properties are summarized in Table 17.2. Each of these is encoded by a unique nuclear gene, with the interesting exception of apolipoproteins B-48 and B-100. Sequence analysis of these two proteins revealed that Apo B-48 (241,000 Da) is identical to the N-terminal portion of apo B-100. A single structural gene encodes both apo B-48 and apo B-100, which is transcribed into a 14,000-nucleotide mRNA (Figure 17.5).

FIGURE 17.5

RNA editing of the apolipoprotein B gene transcript. The APOB gene, composed of 29 exons, is transcribed to produce a ~14,000-nucleotide transcript. In liver, this mRNA is translated to give the 4536-amino acid apo B-100 product. In intestine, a cytidine deaminase converts the C residue in codon 2153 to U, changing the Gln codon to a stop codon. This edited mRNA is translated to give the 2152-amino acid apo B-48 product.

Reprinted from *Biochimica et Biophysica Acta (BBA) - Gene Structure and Expression* 1494:1–13, A. Chester, J. Scott, S. Anant, N. Navaratnam, RNA editing: Cytidine to uridine conversion in apolipoprotein B mRNA. © 2000, with permission from Elsevier.

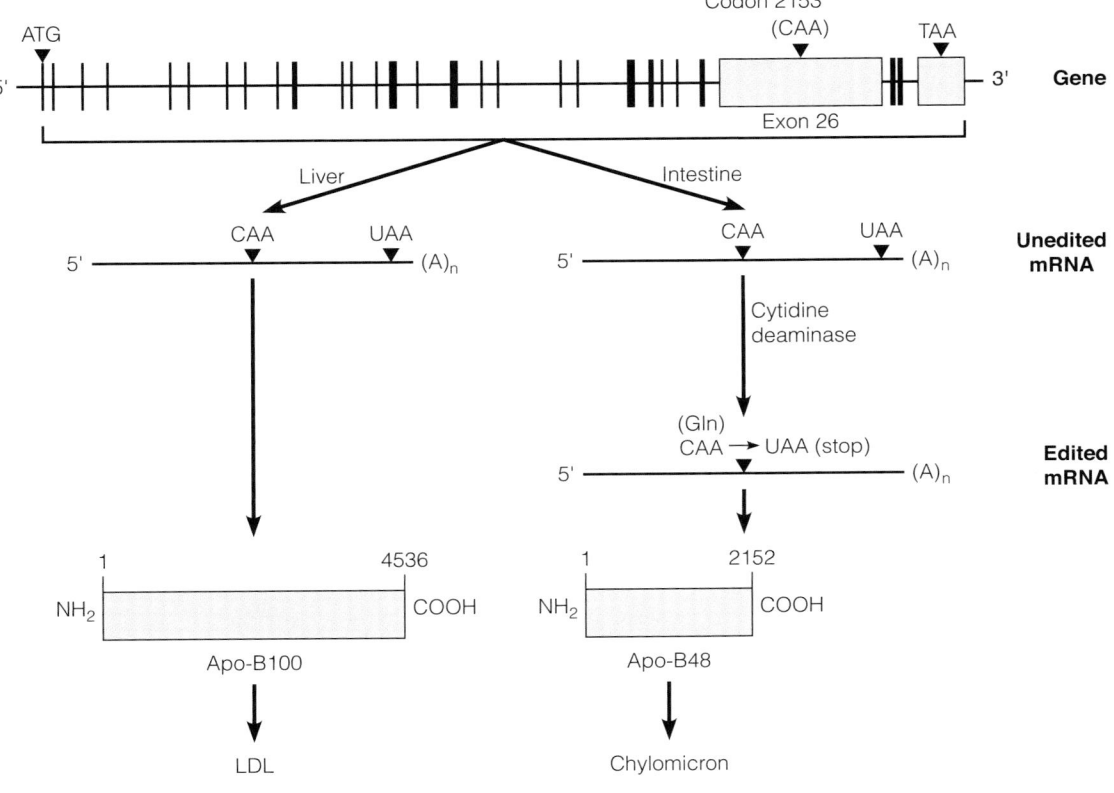

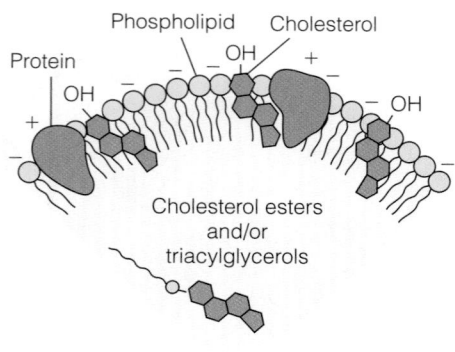

FIGURE 17.6

Generalized structure of a plasma lipoprotein. The spherical particle, part of which is shown, contains neutral lipids in the interior and phospholipids, cholesterol, and protein at the surface.

Lipoproteins are lipid–protein complexes that allow movement of apolar lipids through aqueous environments.

In liver, translation of the full-length open reading frame gives the 4536 amino acid apo B-100 (513,000 Da), which is secreted and assembled into **low-density lipoprotein (LDL)** particles. In intestine, where apo B-48 is expressed for chylomicron assembly, the apo B mRNA undergoes **RNA editing**: a cytidine deaminase found only in intestine specifically deaminates a single cytidine in the mRNA, in codon 2153, converting it to uridine. This changes the codon from CAA (Gln) to UAA, a termination codon. Translation of this edited mRNA gives the shorter apo B-48. RNA editing is fairly widespread—its mechanism and control will be discussed in Chapter 29.

Because lipids are of much lower density than proteins, the lipid content of a lipoprotein class is inversely related to its density: The higher the lipid abundance, the lower the density. The standard lipoprotein classification includes, in increasing order of density: *chylomicrons*, **very low-density lipoprotein (VLDL)**, **intermediate-density lipoprotein (IDL)**, **low-density lipoprotein (LDL)**, and **high-density lipoprotein (HDL)**. Some classification schemes recognize two classes of HDL, and in addition there is a quantitatively minor lipoprotein called very high-density lipoprotein (VHDL).

Despite their differences in lipid and protein composition, all lipoproteins share common structural features, notably a spherical shape that can be detected by electron microscopy. As shown in Figure 17.6, the hydrophobic parts, both lipid and apolar amino acid residues, form an inner core, and hydrophilic protein structures and polar head groups of phospholipids are on the outside.

Some apolipoproteins have specific biochemical activities other than their roles as passive carriers of lipid from one tissue to another. For instance, apo C-II is an activator of triacylglycerol hydrolysis by **lipoprotein lipase**, a cell surface glycoprotein that hydrolyzes triacylglycerols in lipoproteins. A human deficiency of apo C-II is associated with massive accumulation of chylomicrons and elevated triacylglycerol levels in blood. Other apoproteins target specific lipoproteins to specific cells by being recognized by receptors in the plasma membranes of these cells. Of great interest is an association of a variant form of apo E with increased risk for developing Alzheimer's disease. The mechanism underlying this association is not yet understood, but there is a solid epidemiological link between high serum cholesterol at midlife and Alzheimer's disease in later life, and apo E is the most abundant cholesterol transport protein in the central nervous system. There are three common allelic forms of apo E (E2, E3, and E4), and possessing at least one E4 allele is the major known genetic risk factor for Alzheimer's disease.

Following digestion and absorption of a meal, the lipoproteins help maintain in emulsified form some 500 mg of total lipid per 100 mL of human blood in the postabsorptive state. Of this 500 mg, typically about 120 mg is triacylglycerol, 220 mg is cholesterol (two-thirds esterified with fatty acids, one-third free), and 160 mg is phospholipids, principally phosphatidylcholine and phosphatidylethanolamine. Indeed, following a high-fat meal, chylomicrons are so abundant in blood that they give the plasma a milky appearance.

Transport and Utilization of Lipoproteins

As noted previously, chylomicrons represent the form in which dietary fat is transported from the intestine to peripheral tissues, notably heart, muscle, and adipose tissue (see Figure 17.3). VLDL plays a comparable role for triacylglycerols synthesized in liver. The triacylglycerols in both of these lipoproteins are hydrolyzed to glycerol and fatty acids at the inner surfaces of capillaries in the peripheral tissues. This hydrolysis involves activation of the extracellular enzyme lipoprotein lipase by apoprotein C-II, a component of both chylomicron and

CAPILLARY

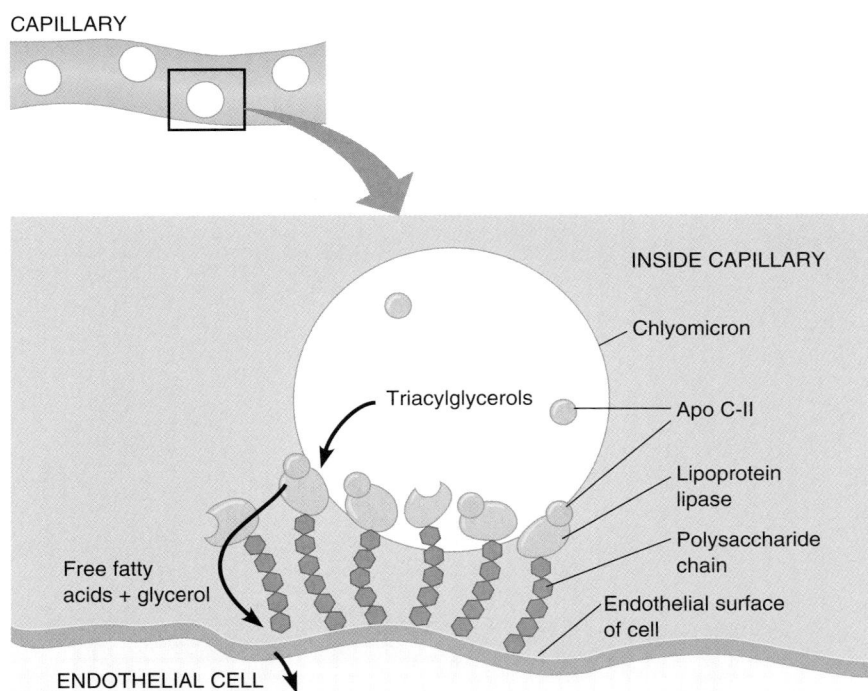

ENDOTHELIAL CELL

FIGURE 17.7

Binding of a chylomicron to lipoprotein lipase on the inner surface of a capillary. The chylomicron is anchored by lipoprotein lipase, which is linked by a polysaccharide chain to the lumenal surface of the endothelial cell. When activated by apoprotein C-II, the lipase hydrolyzes the triacylglycerols in the chylomicron, allowing uptake into the cell of the glycerol and the free fatty acids.

VLDL (Figure 17.7). Lipoprotein lipase is a member of the serine esterase family, which includes pancreatic lipase and **hormone sensitive lipase** (HSL, an enzyme involved in the regulated mobilization of stored fat from adipose tissue—see page 721). This family is characterized by use of a catalytic triad of serine, histidine, and aspartate and an acyl-enzyme intermediate, similar to the serine proteases described in Chapter 11. Some of the released fatty acids are absorbed by nearby cells, while others, still rather insoluble, become complexed with serum albumin for transport to more distant cells. After absorption into the cell, the fatty acids derived from lipoprotein lipase action can be either catabolized to generate energy or, in adipose cells, used to resynthesize triacylglycerols. However, because adipocytes lack glycerol kinase, glycerol-3-phosphate for resynthesis of triacylglycerols must come from glycolysis. Glycerol is returned from adipocytes to the liver, for resynthesis of glucose by gluconeogenesis. Figure 17.8 summarizes overall aspects of lipoprotein metabolism and transport.

As a consequence of triacylglycerol hydrolysis in the capillaries, both chylomicrons and VLDL are degraded to protein-rich remnants. The IDL class of lipoprotein is derived from VLDL, and chylomicrons are degraded to what are simply called chylomicron remnants. Both classes of remnants are taken up by the liver through interaction with specific receptors and further degraded in liver lysosomes. Apoprotein B-100 is reused for synthesis of LDL (via IDL). As described in the next section, LDL is the principal form in which cholesterol is transported to tissues, and HDL plays the primary role in returning excess cholesterol from tissues to the liver for metabolism or excretion. The importance of lipoproteins as transport vehicles is evident from the fact that a major consequence of chronic liver cirrhosis is fatty liver degeneration, where the liver becomes engorged with fat. Because the liver is the major site of apolipoprotein synthesis, damage to this organ causes endogenously

A major consequence of liver dysfunction is an inability to synthesize apolipoproteins and, hence, to transport fat out of the liver.

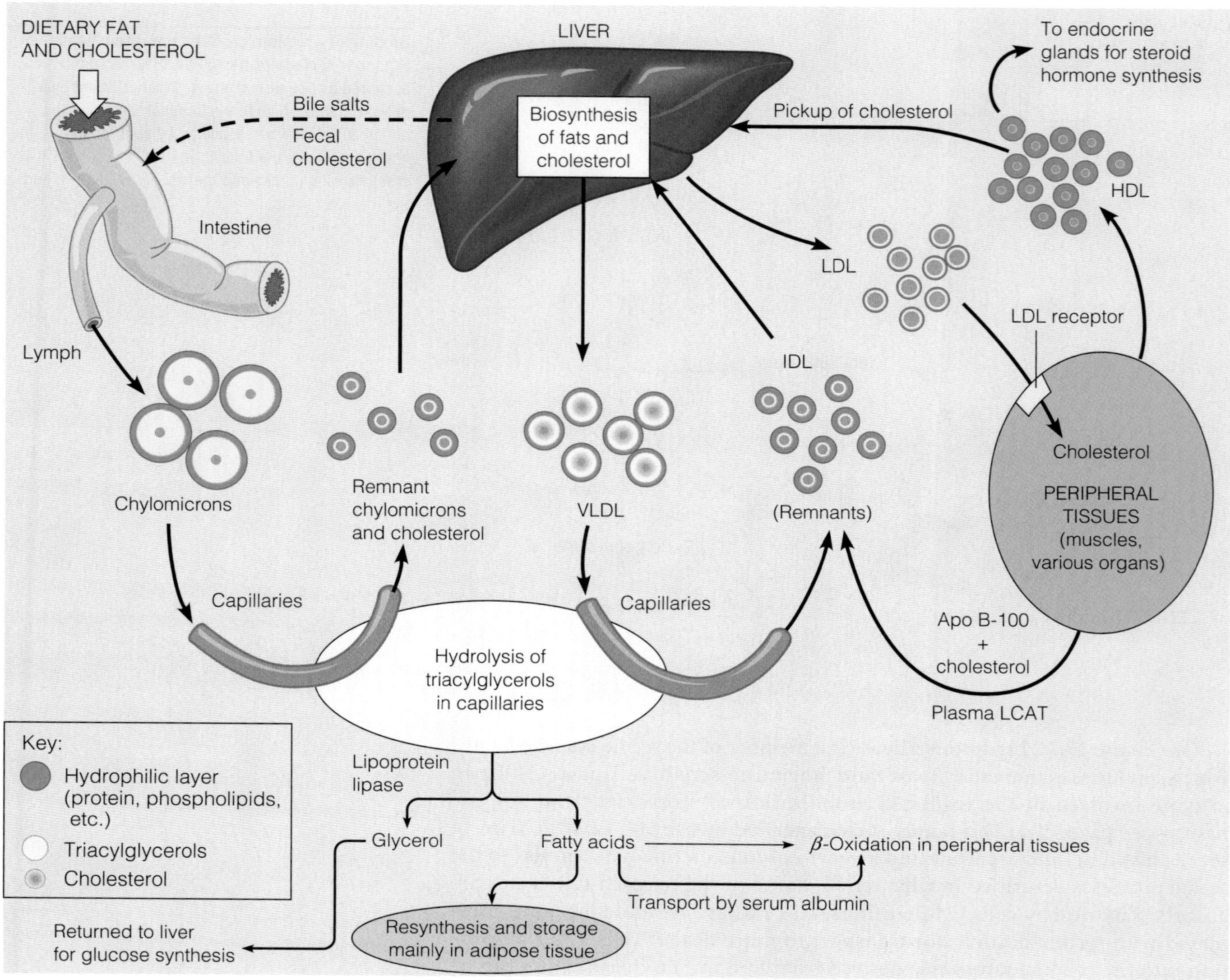

FIGURE 17.8

Overview of lipoprotein transport pathways and fates.

synthesized fat to accumulate there because it cannot be transported to peripheral tissues.

Cholesterol Transport and Utilization in Animals

Cholesterol accumulation in the blood is correlated with development of atherosclerotic plaque.

As you undoubtedly know, a primary risk factor predisposing to heart disease is an abnormally elevated level of cholesterol in the blood. Prolonged cholesterol accumulation contributes to the development of **atherosclerotic plaques**, fatty deposits that line the inner surfaces of coronary arteries.

Recall from Table 17.1 that cholesterol in plasma lipoproteins exists both as the free sterol and as cholesterol esters. Esterification occurs at the cholesterol hydroxyl position with a long-chain fatty acid, usually unsaturated. Cholesterol esters are synthesized in plasma from cholesterol and an acyl chain on phosphatidylcholine (lecithin), through the action of **lecithin:cholesterol acyltransferase (LCAT)**, an enzyme that is secreted from liver into the bloodstream, bound to HDL and LDL:

Phosphatidylcholine + cholesterol $\rightleftharpoons$ lysolecithin + cholesterol ester

Cholesterol esters are considerably more hydrophobic than cholesterol itself.

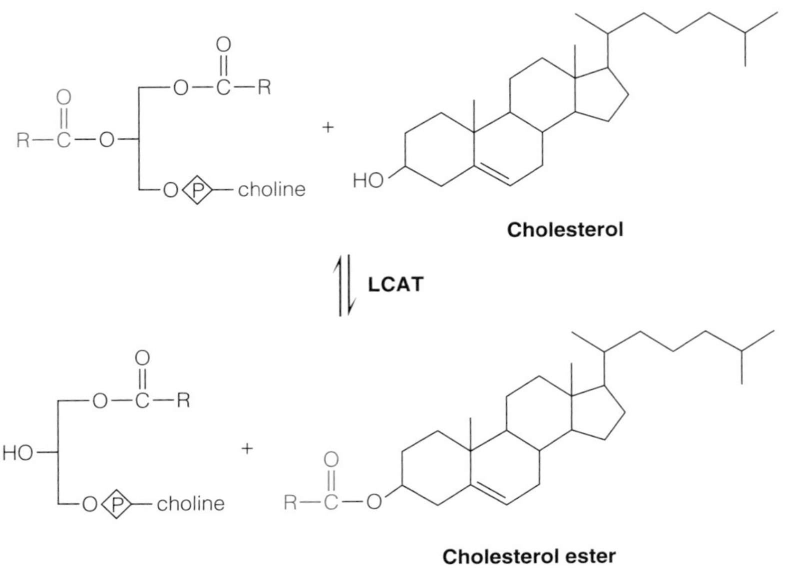

Cholesterol

LCAT

Cholesterol ester

Of the five lipoprotein classes, LDL is by far the richest in cholesterol. The amounts of cholesterol and cholesterol esters associated with LDL are typically about two-thirds of the total plasma cholesterol. In normal adults, total plasma cholesterol levels range from 3.5–6.7 mM (equivalent to 130–260 mg/100 mL of human plasma; total plasma cholesterol above 200 mg/100 mL is a major risk factor for heart disease). Approximately 40% of the weight of the LDL particle is cholesterol esters, and the total of esterified and free cholesterol approaches half the total weight. The LDL particle contains a single molecule of apoprotein B-100 as its primary protein component. Because cholesterol biosynthesis is confined primarily to the liver with some occurring also in intestine, LDL plays an important role in delivering cholesterol to other tissues.

The LDL Receptor and Cholesterol Homeostasis

The importance of understanding cholesterol homeostasis can be seen by reviewing the consequences of prolonged high plasma cholesterol levels. Excess LDL cholesterol accumulates in the inner arterial walls, forming fatty streaks, which attract white blood cells (macrophages). If cholesterol levels are too high for its subsequent removal into the bloodstream, these macrophages become engorged with fatty deposits, which then harden into plaque; this condition, called **atherosclerosis**, ultimately blocks key blood vessels and causes myocardial infarctions, or heart attacks.

To understand the relationship between elevated cholesterol levels and atherogenesis, we must know how cholesterol is taken up from LDL into cells because cholesterol esters are too hydrophobic to traverse cell membranes by themselves. The answer to this question came from the research of Michael Brown and Joseph Goldstein, who showed in the mid-1970s that cholesterol uptake by cells is a receptor-mediated process and that the quantity of receptors themselves is subject to regulation.

In 1972 Brown and Goldstein began to study a hereditary condition called **familial hypercholesterolemia**, or FH. Individuals with the rare homozygous form of this disease (about 1 in a million) have grossly elevated levels of serum cholesterol, from 650 to 1000 mg/100 mL (about five-fold over normal levels). They develop atherosclerosis early in life and usually die of heart disease before age 20. The more common heterozygous condition, characterized by one defective allele instead of two, affects about one individual in 500. These individuals have less severely elevated cholesterol levels, in the range of 350 to 500 mg/100 mL. They are at high risk to have heart attacks in their thirties and forties, although many enjoy a normal life span.

FIGURE 17.9

Feedback regulation of HMG-CoA reductase activity. Fibroblasts obtained from a normal subject (closed symbols) or from a patient homozygous for familial hypercholesterolemia (FH Homozygote) (open symbols) were grown in monolayer cultures. **(a)** At time zero, the medium was replaced with fresh medium depleted of lipoproteins, and HMG-CoA reductase activity was measured in extracts prepared at the indicated times. **(b)** Twenty-four hours after addition of the lipoprotein-deficient medium, human LDL was added to the cells at the indicated levels, and HMG-CoA reductase activity was measured at the indicated time.

Courtesy of Joseph L. Goldstein and Mike S. Brown.

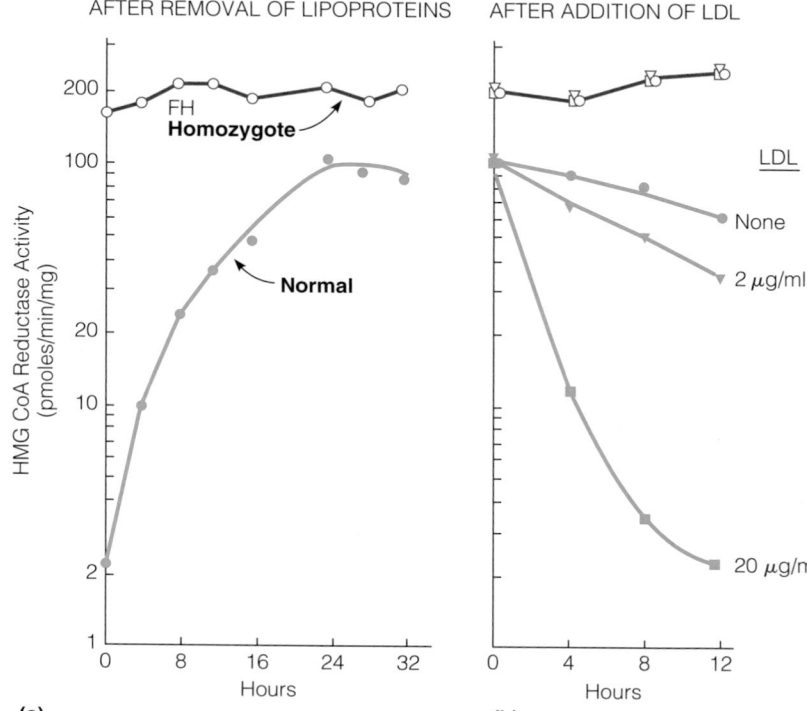

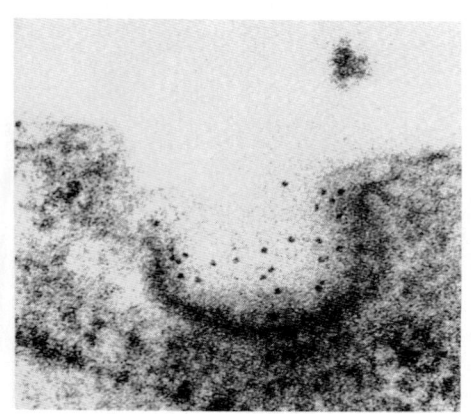

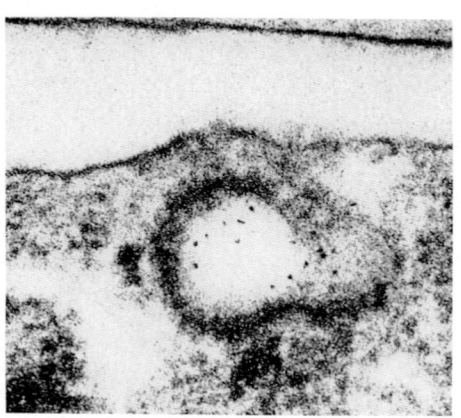

Crucial to the success of Brown and Goldstein was their ability to demonstrate the defective FH phenotype in cell culture. Fibroblasts from FH patients synthesized cholesterol at abnormally high rates in culture, whereas normal cells showed low rates of synthesis. Figure 17.9 reproduces the results of this seminal experiment. When cultured in the presence of LDL, normal cells showed low activity of hydroxymethylglutaryl-CoA reductase (HMG-CoA reductase, the rate-limiting and major regulated enzyme in cholesterol synthesis—see page 801, Chapter 19). In the absence of LDL the same cells showed reductase activities some 50- to 100-fold higher (Figure 17.9a). This high level of enzyme activity was rapidly suppressed upon addition of LDL to normal cells (Figure 17.9b). These results suggested that cholesterol is normally transported into the cell, where it regulates its own synthesis by suppressing the activity of the rate-limiting enzyme (feedback regulation). By contrast, cells from FH individuals showed high levels of reductase activity, whether cultured in the presence or absence of LDL, suggesting that they were deficient in ability to take up cholesterol from the medium.

These observations suggested that cholesterol is taken into cells through the action of a specific receptor, which is deficient or defective in FH patients. In short order Brown and Goldstein and their colleagues demonstrated the existence of this receptor, the **LDL receptor** (see Figure 17.8), and demonstrated a new mechanism by which cells can interact with their environment—**receptor-mediated endocytosis**. By conjugating LDL with an electron-dense material and binding the conjugate to cells, the investigators were able to visualize the LDL receptor on cell surfaces (Figure 17.10). These experiments showed that the

FIGURE 17.10

Receptor-mediated endocytosis of LDL. Low-density lipoprotein (LDL) was conjugated with ferritin to permit electron microscopic visualization. **(a)** The LDL–ferritin (dark dots) binds to a coated pit on the surface of a cultured human fibroblast (a type of connective tissue cell). **(b)** The plasma membrane closes over the coated pit, forming an endocytotic vesicle.

Courtesy of R. G. W. Anderson, M. S. Brown, and J. L. Goldstein.

receptors are clustered in a structure called a **coated pit**, an invagination whose most abundant protein is **clathrin**, a self-interacting protein capable of forming a cagelike structure (Figure 17.11).

Endocytosis is a process by which cells take up large molecules from the extracellular environment. Although LDL uptake involves a cell surface receptor, the interaction of LDL with its receptor is unlike the interaction of hormones such as epinephrine with their receptors. As discussed in Chapters 12 and 13, the binding of epinephrine at its receptor in the plasma membrane triggers intracellular metabolic changes, but the hormone itself does not enter the cell. By contrast, when LDL binds to its receptor, through recognition of the B-100 apoprotein by the receptor, the entire LDL molecule is engulfed and taken into the cell, as schematized in Figure 17.12. The plasma membrane fuses in the vicinity of the LDL–receptor complex, and the coated pit becomes an endocytotic vesicle. Several of these clathrin-lined vesicles fuse to form an **endosome**. The endosome then fuses with a lysosome, putting the LDL–receptor complex in contact with the hydrolytic enzymes of the lysosome. The LDL apoprotein is hydrolyzed to amino acids, and the cholesterol esters are hydrolyzed to give free cholesterol. The receptor itself is recycled, moving back to the plasma membrane to pick up more LDL. About 10 minutes is required for each round trip. The discovery of the LDL receptor and receptor-mediated endocytosis earned Brown and Goldstein the Nobel Prize for Physiology or Medicine in 1985.

Much of the cholesterol released moves to the endoplasmic reticulum, where it is used for membrane synthesis. The internalized cholesterol exerts three regulatory effects. (1) As mentioned earlier, it suppresses endogenous cholesterol synthesis, by inhibiting HMG-CoA reductase and also by suppressing transcription of the gene for this enzyme and accelerating degradation of the enzyme protein. (2) It activates **acyl-CoA:cholesterol acyltransferase (ACAT)**, an intracellular enzyme that synthesizes cholesterol esters from cholesterol and a long-chain acyl-CoA. This promotes the storage of excess cholesterol in the form of droplets of cholesterol esters. (3) It regulates the synthesis of the LDL receptor itself, by decreasing transcription of the receptor gene. Decreased synthesis of the receptor ensures that cholesterol will not be taken into the cell in excess of the cell's needs, even when extracellular levels are very high. This regulatory mechanism explains why excessive dietary cholesterol leads directly to elevations of blood cholesterol levels. With intracellular cholesterol levels so well regulated, the extracellular cholesterol accumulates because it has nowhere else to go.

This regulatory structure led to the development of inhibitors of HMG-CoA reductase, on the assumption that they would depress de novo cholesterol biosynthesis and, hence, intracellular cholesterol levels; the production of LDL receptors was expected to increase as a consequence, thereby leading to more rapid clearance of extracellular cholesterol from the blood. Indeed, HMG-CoA reductase inhibitors, called **statins**, work in exactly this way and are currently the gold standard for therapeutic approaches to lowering cholesterol levels (more about this in Chapter 19).

Studies of many individuals with FH have given a remarkably detailed picture of the LDL receptor and its action. The receptor is a glycoprotein with an 839-residue polypeptide chain and 18 O-linked oligosaccharide chains. The LDL receptor binds specifically the apolipoproteins B-100 and E, and also participates in the uptake of chylomicron remnants and VLDL remnants by the liver.

Gene cloning and DNA sequence analysis have allowed identification of five classes of mutations affecting the receptor and its metabolism in humans. First are mutations that lead to insufficient receptor synthesis. Second, and most common, are mutations in which the receptor is synthesized but fails to migrate from the endoplasmic reticulum to the Golgi complex, for transport to the cytoplasmic membrane. Third are mutations in which receptor is synthesized and processed normally and reaches the cell surface, but fails to bind LDL. Fourth are mutations

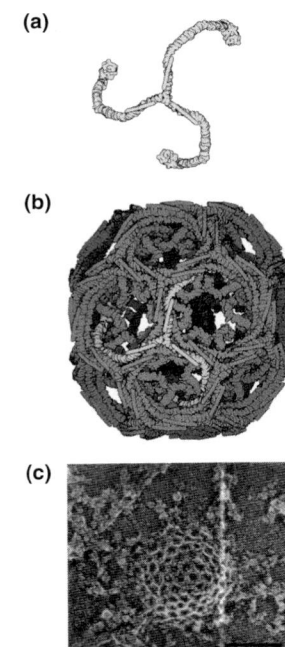

(a)

(b)

(c)

FIGURE 17.11

Structure of a clathrin-coated pit. **(a)** Clathrin, the major protein in coated pits, forms triskelions (named after the symbol of three legs radiating from the center), which assemble into polyhedral lattices composed of hexagons and pentagons, such as the barrel shown in the next panel. **(b)** Image reconstruction from electron cryomicroscopy of a clathrin barrel formed from 36 triskelions. A single clathrin triskelion is highlighted in light blue. **(c)** A coated pit on the inner surface of the plasma membrane of a cultured mammalian cell is visualized by freeze-fracture electron microscopy. The cagelike structure of the pit is due to the clathrin lattice.

(a, b) David S. Goodsell, RCSB Protein Data Bank; (c) Produced by John Heuser, Washington University School of Medicine.

Uptake of cholesterol from the blood occurs at the LDL receptor via receptor-mediated endocytosis.

Intracellular cholesterol regulates its own level by controlling (1) de novo cholesterol biosynthesis, (2) formation and storage of cholesterol esters, and (3) LDL receptor density.

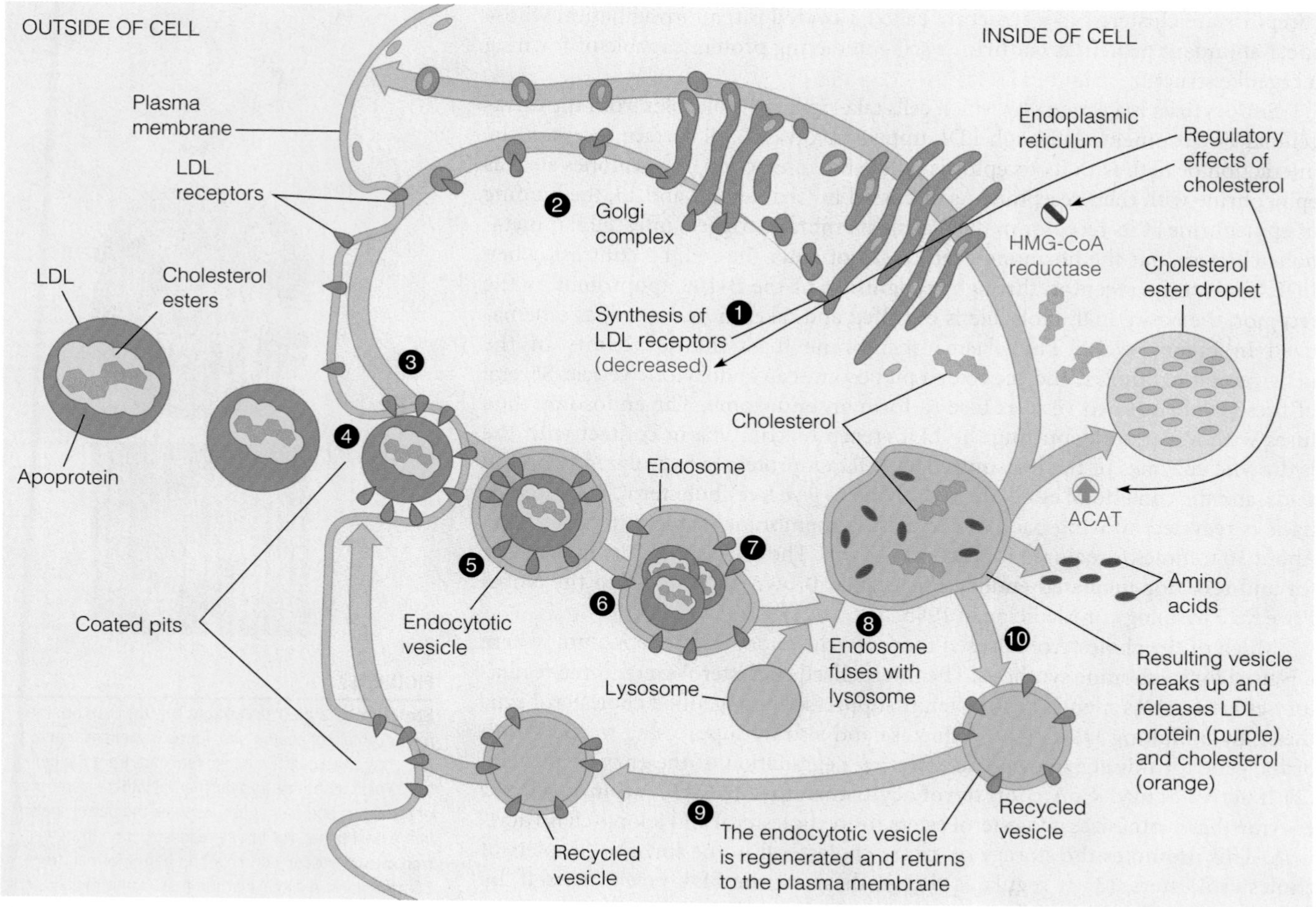

FIGURE 17.12

Involvement of LDL receptors in cholesterol uptake and metabolism. LDL receptors are synthesized in the endoplasmic reticulum ❶ and mature in the Golgi complex ❷. They then migrate to the cell surface, where they cluster in clathrin-coated pits ❸. LDL, made up of cholesterol esters and apoprotein, binds to the LDL receptors ❹ and is internalized in endocytotic vesicles ❺. Several such vesicles fuse to form an organelle called an endosome ❻. Proton pumping in the endosome membrane causes the pH to drop, which in turn causes LDL to dissociate from the receptors ❼. The endosome fuses with a lysosome ❽ and the receptor-bearing clathrin coat dissociates and returns to the membrane ❾. The receptor–LDL complex is degraded in the lysosomes ❿ and cholesterol has various fates. Regulatory actions of cholesterol are shown in red. ACAT, acyl-CoA:cholesterol acyltransferase.

Polyunsaturated fat (PUFA) ingestion is correlated with low plasma cholesterol levels. The mechanisms involved are not completely understood.

that reach the cell surface and bind LDL but fail to cluster in clathrin-coated pits and thus do not internalize LDL. Finally, there is a class of mutant receptors that bind and internalize LDL in coated pits, but fail to release LDL in the endosome, and do not recycle to the cell surface.

Receptor-mediated endocytosis is now known to be a widely used pathway for internalization of extracellular substances, including other lipoproteins, cell growth factors, the iron-binding protein transferrin, some vitamins, and even viruses.

Cholesterol, LDL, and Atherosclerosis

Thanks principally to the work of Brown and Goldstein, we now know a great deal about the genetic and biochemical factors that control serum cholesterol levels, and overwhelming epidemiological evidence now links prolonged hypercholesterolemia to the development of atherosclerotic plaque. Much, however, remains to be learned. We don't know, for example, why diets rich in saturated fatty acids tend to elevate serum cholesterol levels. Nor do we know why a particular class of polyunsaturated fatty acids (PUFAs) called ω-3 (**omega-3**) **fatty acids** tends to depress levels of both serum cholesterol and triacylglycerols and slow the build-up of atherosclerotic plaques. But nutritionists have found that adding to a Western diet fish or fish oils, which are abundant in this class of fatty acids, does indeed have this effect. That is why we are being urged to substitute fish for red meat, which tends to be rich in both saturated fatty acids and cholesterol. The most prominent ω-3 fatty acid is linolenic acid, which is an 18:3cΔ9,12,15 fatty acid (nutritionists number fatty acids backward from biochemists, so the term ω-3 refers to a double bond on the third carbon from the terminal methyl group, what we would call the bond between C-15 and C-16 in this 18-carbon molecule; see margin on next page).

Linolenic acid is an essential fatty acid for humans, but we lack the enzymes to synthesize it, so it must be obtained from the diet. Two other important ω-3 PUFAs, eicosapentaenoic acid (EPA) and docosahexaenoic acid (DHA) can be synthesized by humans from dietary linolenic acid.

Progress is being made, however, in learning how elevated cholesterol levels lead to atherogenesis (plaque formation). LDL undergoes rather ready oxidation, both in cells and in plasma, to a mixture of molecules collectively called **oxidized LDL**. Although the specific oxidation reactions are not well defined, they include peroxidation of unsaturated fatty acids (page 666), hydroxylation of cholesterol itself, and oxidation of amino acid residues in the apoprotein. Accumulation of oxidized LDL on the vessel wall triggers an inflammation response in which the endothelial cells express adhesion molecules that recruit monocytes and T lymphocytes. Some of these cells differentiate into macrophages that take up the lipids that accumulate at sites of arterial injury. Uptake occurs through one of a family of **scavenger receptors**; these receptors take up many substances in addition to oxidized LDL. Unlike the LDL receptor, the scavenger receptor is not down-regulated by cholesterol, so cholesterol uptake into these cells is virtually unlimited, which converts these cells to a cholesterol-engorged species called a **foam cell**. These events have a chemotactic effect, causing more white cells to migrate to the site and leading them to accumulate more cholesterol, which ultimately becomes one of the chief chemical constituents of the plaque that forms at such a site.

TV advertisements talk about "bad cholesterol" and "good cholesterol." These are actually inappropriate terms because cholesterol itself is a natural metabolite, an essential component of all membranes, and the precursor to all steroid hormones and bile acids (see Chapter 19). However, cholesterol present in LDL is considered "bad" because prolonged elevation of LDL levels is what leads to atherosclerosis. By contrast, cholesterol in HDL is called "good" because high levels of HDL counteract atherogenesis. Cholesterol cannot be metabolically degraded, and excess cholesterol is returned from peripheral cells to the liver, for passage through the bile to the intestine, for ultimate excretion (Figure 17.8). As the agent for this transport back to the liver, HDL plays a role in lowering total serum cholesterol levels, which is "good." Recent studies have shown that cholesterol from HDL is taken up into cells not by endocytosis, as with LDL, but instead by a "docking" reaction, in which HDL interacts with another member of the scavenger cell surface receptor family (SR-BI), deposits its cholesterol for uptake, and departs as a remnant without itself being incorporated in the cell's interior.

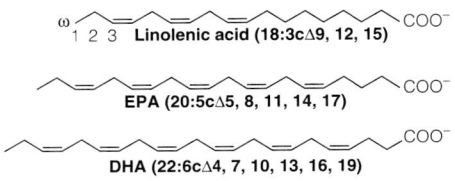

Uptake of oxidized LDL by a scavenger receptor is a key event in atherogenesis.

Mobilization of Stored Fat

In general, the capacity of animal storage depots to store fat is virtually unlimited. Whatever appears in the body from the diet is absorbed, and most of it is transported to adipose tissue for storage. The lack of control of this process is sadly evident from the prevalence of obesity among humans, in whom fat is stored in excess of its need to supply energy. By contrast, the release of fat from storage depots in adipose tissue is controlled hormonally, to meet the needs of the organism for energy generation.

The catabolism of fat (**lipolysis**) begins with the hydrolysis of triacylglycerol to yield glycerol plus free fatty acid (often abbreviated FFA). About 95% of the energy derived from subsequent oxidation of the fat comes from the fatty acids, with only 5% coming from glycerol. All of the carbons from fatty acids are catabolized to two-carbon fragments, as acetyl-coenzyme A, except for the small proportion of fatty acids that contain odd-numbered chains.

The release of metabolic energy stored in triacylglycerols is comparable to the mobilization of carbohydrate energy stored in animal glycogen, in that the first step of fat breakdown—its hydrolysis to glycerol and fatty acids—is hormonally regulated. Three lipolytic enzymes are now known to participate in this process: **triacylglycerol lipase**, also called hormone-sensitive lipase (**HSL**), **adipose triglyceride lipase** (**ATGL**), and **monoacylglycerol lipase** (**MGL**). All three of these

Fat mobilization in adipose cells is hormonally controlled, via the cyclic AMP–dependent phosphorylation of lipolytic enzymes and lipid droplet-associated proteins.

FIGURE 17.13

Mobilization of adipose cell triacylglycerols by lipolysis. Three lipases act sequentially to hydrolyze triacylglycerol (TG) to glycerol and free fatty acids (FFA). These enzymes act at the oil–water interface of the lipid droplet. FFA are exported to the blood plasma, where they are bound to albumin for transport to liver and other tissues for subsequent oxidation. Glycerol is released to the blood to be taken up by liver cells, where it serves as a gluconeogenic substrate. DG, diacylglycerol; MG, monoacylglycerol; ATGL, adipose triglyceride lipase; HSL, hormone-sensitive lipase; MGL, monoacylglycerol lipase.

The Journal of Lipid Research 50:3–21, R. Zechner, P. C. Kienesberger, G. Haemmerle, R. Zimmermann, and A. Lass, Adipose triglyceride lipase and the lipolytic catabolism of cellular fat stores. Modified with permission. © 2009 The American Society for Biochemistry and Molecular Biology. All rights reserved.

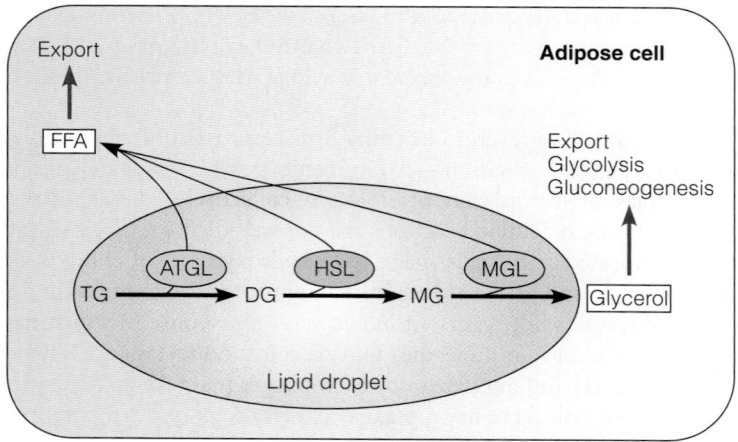

enzymes are serine esterases, catalyzing the hydrolysis of the ester linkage between the glycerol backbone and a fatty acid using a serine nucleophile to form the acyl-enzyme intermediate. HSL and MGL each possess the classic catalytic triad of serine, aspartate, and histidine, whereas ATGL has a catalytic dyad of serine and aspartate. Although these enzymes catalyze the same chemical reaction, their substrate specificities differ. HSL is capable of hydrolyzing triacylglycerol (TG), diacylglycerol (DG), monoacylglycerol (MG), and cholesteryl esters (CEs), but is 10 times more active with DG than with TG or MG. Indeed, HSL was long considered the only enzyme involved in this process. This view was overturned when HSL-knockout mice were created (see Tools of Biochemistry 26A), which totally lack HSL. To everyone's surprise, these mice did not become overweight or obese and could still release free fatty acids in response to hormonal stimulation. This result suggested that another TG hydrolase must exist, and led to the discovery of ATGL in 2004. ATGL has 10-fold higher substrate specificity for TG than for DG and does not hydrolyze MG. Recently, ATGL-knockout mice have been created. In contrast to HSL-deficient mice, ATGL-knockout mice accumulate excessive TG in all organs, whereas DG levels are normal. Thus, it appears that these three enzymes work together to convert triacylglycerol to free fatty acids and glycerol. ATGL catalyzes the first step in TG mobilization, generating DG and FFA. HSL catalyzes hydrolysis of DG, generating MG and FFA, and MGL releases the third FFA from the glycerol backbone (Figure 17.13).

How is lipolysis regulated? HSL activity is controlled by a cascade involving cyclic AMP (Figure 17.14). Depending on the physiological state, glucagon, epinephrine, parathyroid hormone, thyrotropin, or adrenocorticotropin binds to a β-adrenergic receptor on the plasma membrane, leading to activation of adenylate cyclase, as described in Chapter 13. This, in turn, activates protein kinase A (PKA), by releasing active catalytic subunits (C) from the inactive R_2C_2 tetramer. PKA phosphorylates several target proteins in the adipose cell to stimulate lipolysis. HSL activity is stimulated by phosphorylation, but only about two-fold. The major stimulatory effect is on **perilipin**, a protein that coats the cytosolic surface of adipocyte lipid droplets. Normally, perilipin blocks access of HSL to the lipid droplet. However, upon phosphorylation of perilipin by PKA, phosphorylated HSL is recruited to the lipid droplet where it can now access its DG substrate. ATGL activity also appears to be controlled by perilipin, albeit indirectly. In addition, ATGL requires a cofactor for full activity, a protein named CGI-58. In unstimulated adipocytes, CGI-58 is sequestered by perilipin A on the lipid droplet and is unavailable to activate ATGL. Following hormonal stimulation and PKA-dependent phosphorylation of perilipin, CGI-58 is released from perilipin to activate ATGL and deliver it to the lipid droplet (Figure 17.14). In adipose tissue the primary hormonal effects are mediated by epinephrine in stress situations and by glucagon during fasting.

The involvement of protein cofactors for lipases that act at an oil–water interface is quite common. We learned previously that pancreatic lipase requires colipase

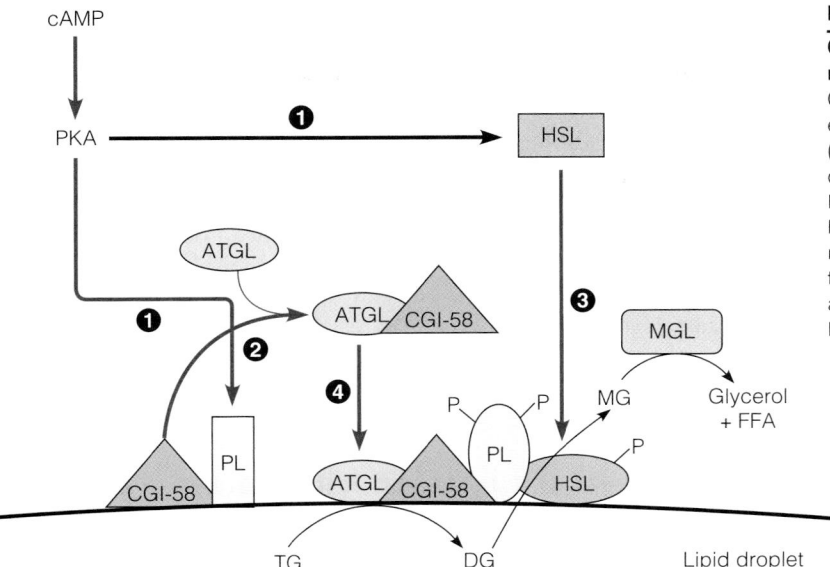

FIGURE 17.14

Control of lipolysis in adipose cells by a cyclic AMP-mediated cascade system. Hormonal activation of a β-adrenergic G-protein coupled receptor on the plasma membrane leads to elevation of cAMP levels, which in turn, activates protein kinase A (PKA). PKA phosphorylates perilipin (PL) and HSL ❶. CGI-58 dissociates from phosphorylated perilipin, and binds ATGL ❷. Phosphorylated HSL is recruited to the lipid droplet and activated by phosphorylated perilipin ❸. Phosphorylated perilipin also recruits the ATGL/CGI-58 complex to the lipid droplet, activating this lipase ❹. Activated ATGL hydrolyzes TG to DG+FFA; activated HSL hydrolyzes DG to MG+FFA; cytoplasmic MGL hydrolyzes MG to free glycerol + FFA.

(page 712) and lipoprotein lipase requires apoC-II (page 714) for full activity. In a similar manner, the action of both HSL and ATGL depends on cofactor proteins (perilipin and CGI-58, respectively). MGL, on the other hand, appears to be a cytoplasmic lipase that hydrolyzes MG released from the lipid droplet without the requirement of any cofactor.

The free fatty acid hydrolysis products exit the adipocyte by passive diffusion and find their way to the blood plasma, where the fatty acids become bound to **albumin**. This is the most abundant plasma protein, about 50% of the total plasma protein in humans. The protein, with an M_r of 66,200, contains 17 disulfide bridges. Each molecule of albumin can bind up to 10 molecules of free fatty acid, although the actual amount bound is usually far lower. Fatty acids are released from albumin and taken up by tissues largely by passive diffusion so that fatty acid uptake into cells is driven primarily by concentration. Most of the glycerol released to the bloodstream is taken up by liver cells, where it serves as a gluconeogenic substrate, leading to the production of glucose (see Figure 17.8).

Fatty Acid Oxidation

Early Experiments

The nature of the pathway by which fatty acids are oxidized was revealed as early as 1904, in a brilliant series of experiments by a German chemist named Franz Knoop. The experiments were inspired because they involved the first known use of metabolic tracers, more than 40 years before radioactive tracers became available. Knoop fed dogs a series of fatty acids in which the terminal methyl group was derivatized with a phenyl group. The expectation was that these analogs would follow metabolic pathways similar to those used for oxidizing normal fatty acids. Knoop found that when the fed fatty acid had an even-numbered carbon chain, the final breakdown product, recovered from urine, was phenylacetic acid. When the fed fatty acid had an odd-numbered chain, the product was benzoic acid (Figure 17.15).

These results led Knoop to propose that fatty acids are oxidized in a stepwise fashion, with initial attack on carbon 3 (the β-carbon with respect to the carboxyl group). This attack would release the terminal two carbons, and the remainder of the fatty acid molecule could undergo another oxidation. Release of a two-carbon fragment would occur at each step in the oxidation. With the analogs, the process would be repeated until the remaining acid, either phenylacetic or benzoic acid, could not be further metabolized and would be excreted in the urine.

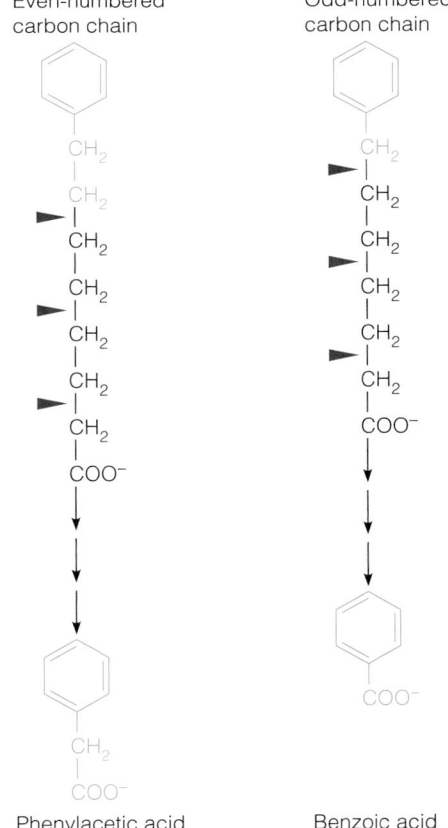

Even-numbered carbon chain   Odd-numbered carbon chain

Phenylacetic acid   Benzoic acid

FIGURE 17.15

Oxidation of phenyl derivatives of fatty acids in Knoop's experiment. Red triangles represent presumed sites of cleavage of these model fatty acids.

The next major development came in the 1940s, when Luis Leloir and Albert Lehninger independently demonstrated fatty acid oxidation in cell-free liver homogenates. Lehninger showed that ATP was essential for this process, suggesting that ATP somehow activates the carboxyl group of the fatty acid. Working with Eugene Kennedy, Lehninger also showed that the process occurs in mitochondria and that it releases two-carbon fragments that are oxidized in the citric acid cycle. In Munich, Feodor Lynen demonstrated that the ATP-dependent activation esterifies the fatty acid carboxyl group with the thiol group of coenzyme A, and it was later shown that all of the intermediates in the subsequent oxidative reactions are fatty acyl-CoA thioesters. Thus, by the mid-1950s the basic outlines of the fatty acid oxidation pathway were clear. As shown in Figure 17.16, the pathway consists of activation of the carboxyl group, transport into the mitochondrial matrix, and stepwise oxidation of the carbon chain, two carbons at a time, from the end containing the carboxyl group.

Fatty Acid Activation and Transport into Mitochondria

Fatty acids are activated for oxidation by ATP-dependent acylation of coenzyme A.

For the most part, fatty acids arise in the cytosol, either through biosynthesis (as discussed later in this chapter) or through triacylglycerol or fatty acid transport from fat depots outside the cell. These fatty acids must be transported into the mitochondrial matrix for oxidation. Because the inner membrane is impermeable to free long-chain fatty acids and acyl-CoAs, a specific transport system comes into play. That transport system operates hand in hand with the metabolic activation needed to initiate the β-oxidation pathway.

A series of **acyl-CoA synthetases**, specific for short-chain, medium-chain, or long-chain fatty acids, catalyzes formation of the fatty acyl thioester conjugate with coenzyme A (steps 1 and 1′ in the upper part of Figure 17.16):

$$R-COO^- + ATP + CoA-SH \rightleftharpoons R-\overset{O}{\overset{\|}{C}}-S-CoA + AMP + PP_i \qquad \Delta G^{\circ\prime} \approx -15 \text{ kJ/mol}$$

The synthetase specific for long-chain acids is a membrane-bound enzyme, found in both the endoplasmic reticulum and the outer mitochondrial membrane; the short-chain and medium-chain enzymes are found primarily in the mitochondrial matrix. The long-chain enzyme, which plays the predominant role in initiating fatty acid oxidation, acts on fatty acids with chain lengths of 10 to 20 carbons; the medium-chain enzyme acts on 4- to 12-carbon chains; and the short-chain enzyme prefers acetate and propionate.

Chemically, the energy-rich thioester link in long-chain fatty acyl-CoAs is identical to that of acetyl-CoA (see Chapter 14). Recall that pyruvate oxidation provides the energy to drive acetyl-CoA formation in the pyruvate dehydrogenase reaction. The acyl-CoA synthetases, on the other hand, use a two-step mechanism involving cleavage of ATP to drive the endergonic thioester formation (Figure 17.17). First comes activation of the carboxyl group by ATP to give a **fatty acyl adenylate**, with concomitant release of pyrophosphate. Next, the activated carboxyl group is attacked by the nucleophilic thiol group of CoA, thereby displacing AMP and forming the fatty acyl-CoA derivative. Carboxyl groups of amino acids are activated for protein synthesis in very similar fashion.

Although each fatty acyl-CoA, like ATP itself, is an energy-rich compound ($\Delta G^{\circ\prime}$ of hydrolysis ~−30 kJ/mol), cleavage of ATP to AMP ($\Delta G^{\circ\prime} = -45.6$ kJ/mol: Figure 3.7) provides the driving force for the formation of the fatty acyl-CoA. The reaction is made essentially irreversible because of the active pyrophosphatase present in most cells:

$$PP_i + H_2O \rightleftharpoons 2P_i \qquad \Delta G^{\circ\prime} = -19.2 \text{ kJ/mol}$$

Thus, the overall reaction (sum of the two previous reactions) proceeds far in the direction of completion, with a net $\Delta G^{\circ\prime}$ of about −35 kJ/mol.

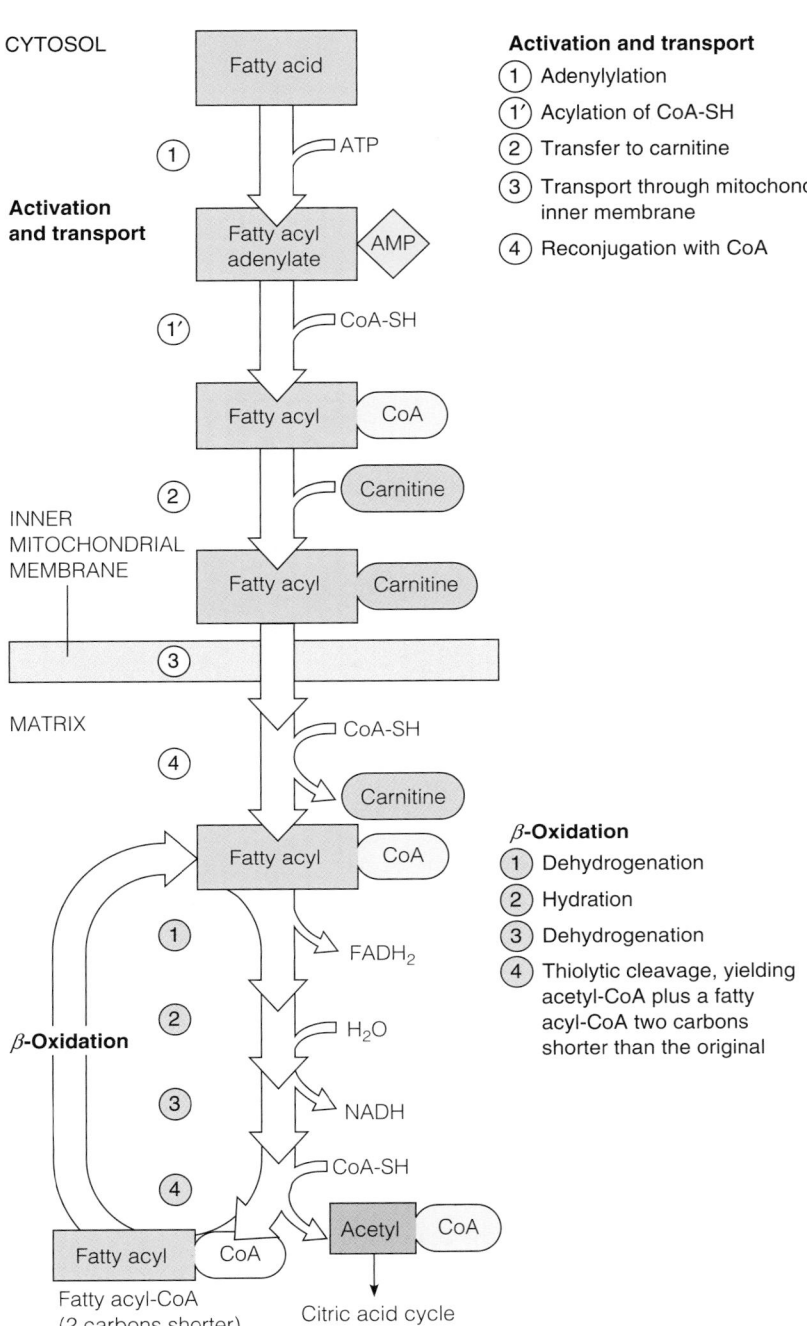

FIGURE 17.16

Overview of the fatty acid oxidation pathway.

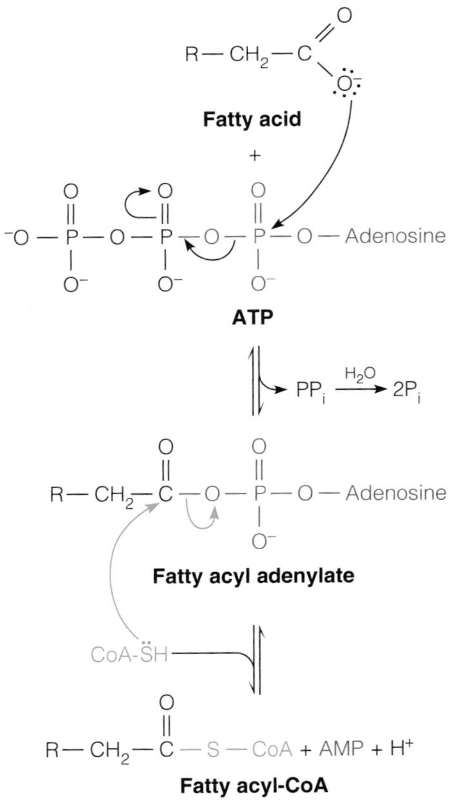

Fatty acyl-CoAs are formed on the outer mitochondrial membrane. Hence, they must move through the inner mitochondrial membrane to become oxidized. This movement involves transfer of the fatty acyl moiety to a carrier called **carnitine** (step 2, upper part of Figure 17.16). The reaction is catalyzed by **carnitine acyltransferase I** (also called **carnitine palmitoyltransferase I**, or **CPT I**), embedded in the outer mitochondrial membrane with its active site facing the cytoplasm. CPT I yields a derivative, **fatty acyl-carnitine**, that traverses the inner membrane via a specific carrier, the carnitine-acylcarnitine translocase (Figure 17.18 and step 3, upper part of Figure 17.16). A second enzyme, **carnitine acyltransferase II** (also called **carnitine palmitoyltransferase II**, or **CPT II**), loosely associated with the matrix side of the inner membrane, completes the transfer process by exchanging fatty acyl-carnitine for free carnitine and producing fatty acyl-CoA within the matrix (step 4, upper part of

FIGURE 17.17

Mechanism of acyl-CoA synthetase reactions.
The figure shows reversible formation of the activated fatty acyl adenylate, nucleophilic attack by the thiol sulfur of CoA-SH on the activated carboxyl group, and the quasi-irreversible pyrophosphatase reaction, which draws the overall reaction toward fatty acyl-CoA.

FIGURE 17.18

The carnitine acyltransferase system, for transport of fatty acyl-CoAs into mitochondria.

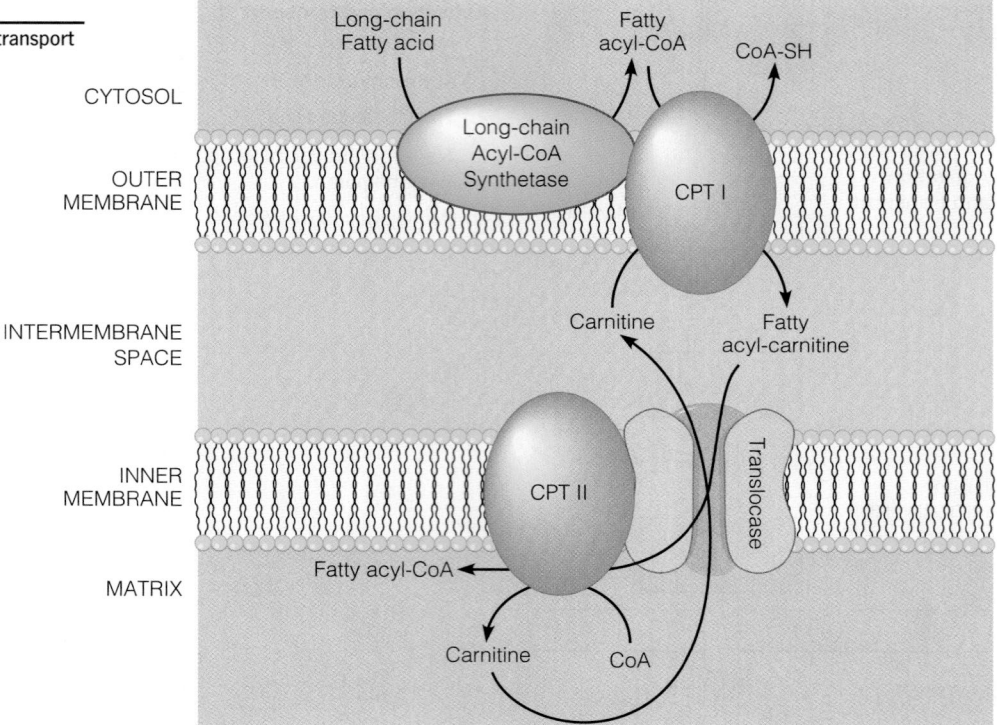

Figure 17.16). The inner membrane translocase is an antiporter, catalyzing the reversible exchange of acylcarnitine for free carnitine. Thus, the free carnitine formed in the matrix returns to the intermembrane space via the translocase and then moves to the cytoplasm through pores in the outer membrane. Although fatty acyl-carnitines are ordinary esters, the ester bond in these compounds is somewhat activated, as shown by the ready reversibility of the carnitine acyltransferase reactions.

What is the point of this rather complex shuttling process? It exists to regulate fatty acid oxidation, preventing the futile cycle that would occur if oxidation and resynthesis were taking place at the same time. Carnitine acyltransferase I is strongly inhibited by **malonyl-CoA**, the first committed intermediate in fatty acid synthesis (page 737) (CPT II is insensitive to malonyl-CoA). Indeed, entry of fatty acids into mitochondria is rate-limiting for β-oxidation and is the primary point of regulation. Thus, conditions in the cell that favor fatty acid synthesis prevent the transfer of fatty acyl moieties to their intracellular sites of oxidation and, hence, prevent that oxidation.

The β-Oxidation Pathway

Once inside the mitochondrial matrix, fatty acyl-CoAs are oxidized as predicted by Knoop, with oxidation of the β-carbon and a series of steps in which the fatty acyl chain is shortened by two carbons at a time. The two-carbon fragment is released in the form of acetyl-CoA. Each step involves four reactions (Figure 17.19 and steps 1–4, lower part of Figure 17.16). The pathway is cyclic in that each step ends with formation of an acyl-CoA, shortened by two carbons, which undergoes the same process in the next step, or cycle. For example, 1 mole of palmitoyl-CoA, derived from a 16-carbon fatty acid, undergoes seven cycles of oxidation to give 8 moles of acetyl-CoA. Each cycle releases 1 two-carbon unit, concomitant with 2 two-electron oxidation–reduction reactions. Because each cycle results in oxidation of the β-carbon (the α-carbon remains at the methylene oxidation state), the pathway is called **β-oxidation**.

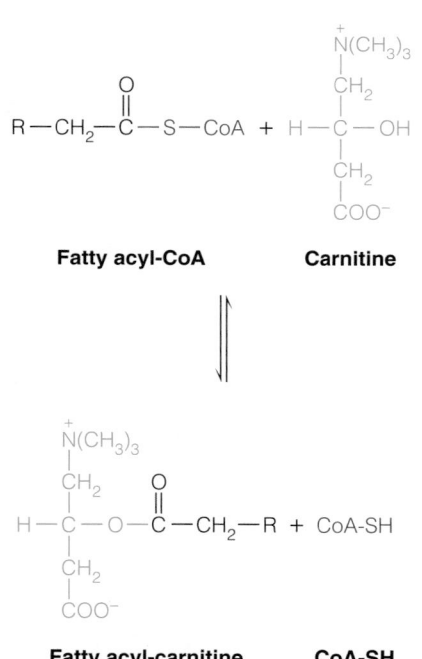

Carnitine transports acyl-CoAs into mitochondria for oxidation.

FIGURE 17.19

Outline of the β-oxidation of fatty acids. In the diagram a 16-carbon saturated fatty acyl-CoA (palmitoyl-CoA) undergoes seven cycles of oxidation to yield eight molecules of acetyl-CoA. These reactions correspond to steps 1–4, lower part of Figure 17.16.

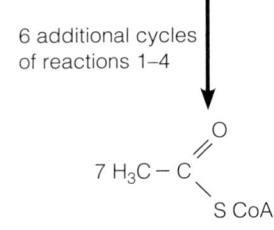

Mechanistically, this pathway is remarkably similar to that used to oxidize succinate in the citric acid cycle (compare with Figure 14.3, page 595). As shown in Figure 17.19, each cycle in the oxidation of a saturated fatty acyl-CoA involves the following reactions: (1) dehydrogenation to give an enoyl derivative; (2) hydration of the resultant double bond, with the β-carbon undergoing hydroxylation; (3) dehydrogenation of the hydroxyl group (oxidation to a ketone); and (4) cleavage by attack of a second molecule of coenzyme A on the β-carbon, to release acetyl-CoA and a fatty acyl-CoA two carbons shorter than the original substrate. Oxidation of unsaturated fatty acyl-CoAs is slightly different, as discussed on page 730.

Acetyl-CoA from β-oxidation enters the citric acid cycle, where it is oxidized to CO_2 in the same fashion as the acetyl-CoA derived from the oxidation of pyruvate. Like the citric acid cycle, β-oxidation generates reduced electron carriers, whose reoxidation in the mitochondria generates ATP via oxidative phosphorylation from ADP. Now let us describe the individual reactions in detail (refer to Figure 17.19).

Reaction 1: The Initial Dehydrogenation

The first reaction is catalyzed by an **acyl-CoA dehydrogenase**, which catalyzes the removal of two hydrogen atoms from the α- and β-carbons to give a *trans α,β*-unsaturated acyl-CoA (*trans*-2-enoyl-CoA) as the product. This oxidation introduces a conjugated double bond into a carbonyl compound, and like most enzymes that catalyze this type of oxidation, acyl-CoA dehydrogenase uses a tightly bound FAD prosthetic group. The enzyme first abstracts the acidic *pro-R* hydrogen from the α-carbon to give a thioester enolate. The *pro-R* hydrogen on the β-carbon is then transferred as a hydride equivalent to FAD to give the *trans* double bond and enzyme-bound $FADH_2$.

Thioester enolate

As shown in Figure 17.20, the enzyme-bound $FADH_2$ contributes a pair of electrons to a shuttle protein, the **electron-transferring flavoprotein (ETF)**. These electrons are passed in turn to coenzyme Q via ETF-Q oxidoreductase, an integral membrane protein, and are then shuttled along the respiratory chain, yielding ATP via oxidative phosphorylation. In this respect, ETF-Q oxidoreductase is comparable to NADH dehydrogenase and succinate dehydrogenase. All three are flavoproteins that transfer electrons to the mobile electron carrier coenzyme Q (see Figure 15.13, page 640).

Reactions 2 and 3: Hydration and Dehydrogenation

Like succinate oxidation, an initial FAD-dependent fatty acyl-CoA oxidation is followed by hydration and an NAD^+-dependent dehydrogenation. In β-oxidation the latter two reactions are catalyzed by **enoyl-CoA hydratase** and **3-L-hydroxyacyl-CoA dehydrogenase**, respectively (Figure 17.19). Both reactions are stereospecific. Because carbon 3 is β with respect to the carboxyl carbon, the products of these two reactions are sometimes called L-β-hydroxyacyl-CoA and β-ketoacyl-CoA, respectively. Hence the term *β-oxidation*.

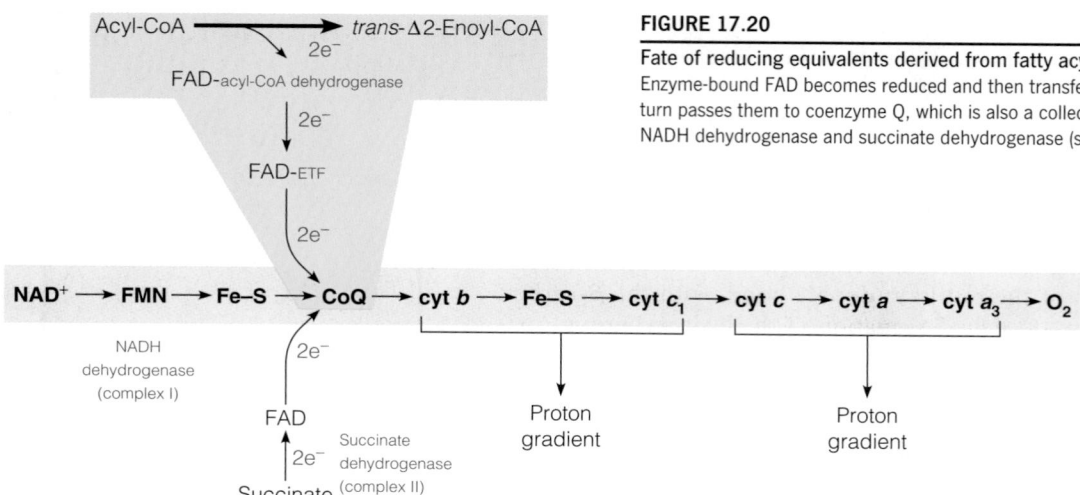

FIGURE 17.20

Fate of reducing equivalents derived from fatty acyl-CoA dehydrogenation. Enzyme-bound FAD becomes reduced and then transfers its electrons to ETF, which in turn passes them to coenzyme Q, which is also a collection point for electrons from NADH dehydrogenase and succinate dehydrogenase (see Figure 15.13, page 640).

Reaction 4: Thiolytic Cleavage

The fourth and last reaction in each cycle of the β-oxidation pathway involves attack of the nucleophilic thiol sulfur of coenzyme A on the electron-poor keto (β) carbon of 3-ketoacyl-CoA, with cleavage of the α—β bond and release of acetyl-CoA. The other product is a shortened fatty acyl-CoA, ready to begin a new cycle of oxidation:

$$\underset{\textbf{3-Ketoacyl-CoA}}{R-\overset{\overset{\text{O}}{\|}}{C}-CH_2-\overset{\overset{\text{O}}{\|}}{C}-S-CoA} + \text{CoA-SH} \longrightarrow \underset{\textbf{Acyl-CoA}}{R-\overset{\overset{\text{O}}{\|}}{C}-S-CoA} + \underset{\textbf{Acetyl-CoA}}{CH_3-\overset{\overset{\text{O}}{\|}}{C}-S-CoA}$$

Because this reaction involves cleavage by a thiol, it is referred to as a **thiolytic cleavage**, by analogy with hydrolysis, which involves cleavage by water. The enzyme is commonly called β-**ketothiolase**, or simply **thiolase**. An essential nucleophilic cysteine thiol group on the enzyme (E-SH) attacks the substrate, with formation of an acyl-enzyme intermediate and acetyl-CoA in a retro-Claisen reaction (see Chapter 12, page 484). Free CoA-SH then attacks the intermediate in a nucleophilic acyl substitution reaction (see margin).

As noted earlier, the overall oxidation pathway as just described is applicable to the most abundant fatty acids—those that contain even numbers of carbon atoms and are fully saturated. Presently, we shall describe the variations in this pathway that permit oxidation of other fatty acids. For the saturated, even-chain fatty acyl-CoAs, oxidation simply proceeds stepwise, with two carbons lost as acetyl-CoA after each cycle. For the C_{16} palmitoyl-CoA, the example shown in Figure 17.19, the first cycle yields acetyl-CoA plus the C_{14} myristoyl-CoA. A second cycle, acting on the latter substrate, yields acetyl-CoA plus the C_{12} lauroyl-CoA. In the seventh and last cycle, the 3-hydroxyacyl-CoA dehydrogenase reaction yields acetoacetyl-CoA. Thiolytic cleavage of this substrate yields 2 moles of acetyl-CoA (see margin on next page). Thus, the oxidation of 1 mole of palmitic acid involves six successive cycles, each of which yields 1 mole of acetyl-CoA, and a seventh cycle, which yields 2 moles. Other saturated even-chain fatty acids are degraded identically. For example, stearic acid oxidation involves eight steps, with two acetyl-CoAs resulting from the last cycle.

Mitochondrial β-Oxidation Involves Multiple Isozymes

Mammalian mitochondria possess multiple isozymes for several of the steps of the β-oxidation pathway. Four isozymes of acyl-CoA dehydrogenase exist with overlapping chain length specificities for short-, medium-, long-, or very long-chain fatty acyl-CoAs (referred to as SCAD, MCAD, LCAD, and VLCAD, respectively). All of these are soluble matrix proteins, except VLCAD (very long-chain acyl-CoA

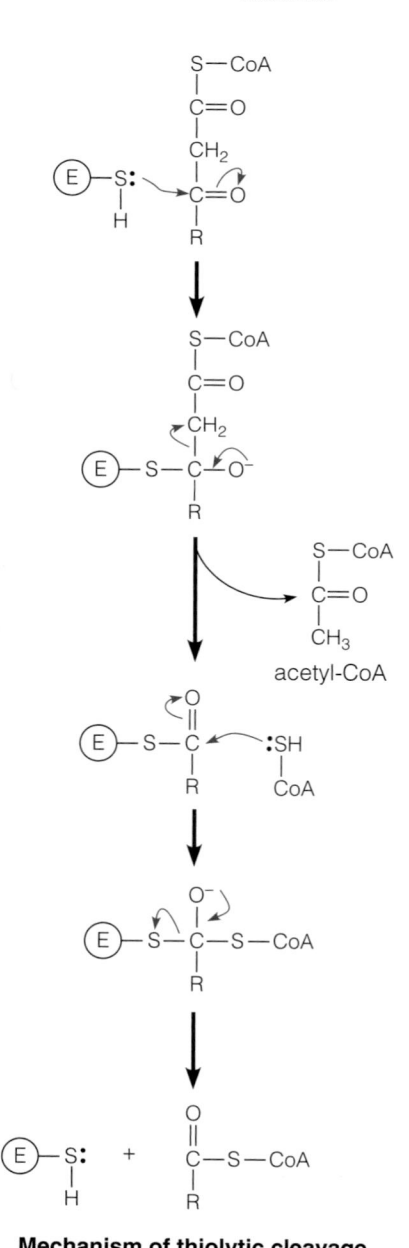

Mechanism of thiolytic cleavage

dehydrogenase), which is bound to the matrix side of the inner membrane. There are two isozymes for each of the next three enzymes in the pathway. One member of each pair is a soluble matrix enzyme, specific for shorter chain acyl-CoAs. The other isozyme of each pair is specific for longer chain substrates and is known to be an inner membrane protein. In 1992, a large protein was purified from rat and human mitochondrial inner membranes that possessed long-chain enoyl-CoA dehydratase, long-chain hydroxyacyl-CoA dehydrogenase, and long-chain ketoacyl-CoA thiolase activities. This 460,000-dalton protein, an $\alpha_4\beta_4$ octamer, catalyzes the last three steps of the β-oxidation cycle and was thus named mitochondrial trifunctional protein (MTP). One hypothesis, not yet proven, is that upon entry into the mitochondrial matrix, long-chain fatty acyl-CoAs are initially metabolized by VLCAD and MTP; after several rounds of β-oxidation by the membrane-bound isozymes, the shortened acyl-CoA substrates are handed off to the soluble isozymes for completion of the pathway.

Many disorders that affect mitochondrial fatty acid oxidation in humans have been described, including inherited defects in proteins of the carnitine acyltransferase system and the β-oxidation enzymes. Deficiency of medium-chain acyl-CoA dehydrogenase (MCAD) is the most common disorder of fatty acid metabolism, with a frequency of 1 in 15,000, and has been associated with some cases of sudden infant death syndrome (SIDS). Two different disorders affect the mitochondrial trifunctional protein: either isolated long-chain hydroxyacyl-CoA dehydrogenase (LCHAD) deficiency, or complete MTP deficiency with reduced activity of all three enzymes, depending on the nature of the mutation. Children with fatty acid oxidation disorders usually present within the first year of life with recurring episodes of fatty liver (steatosis), high blood levels of fatty acid intermediates, hypoglycemia, and may also exhibit skeletal and cardiac myopathies. The episodes are induced by fasting as a result of the regulatory mechanisms that control the switch from carbohydrate fuels to lipid fuels (more about this later), and thus treatment includes avoiding fasting and consuming a low-fat, high-carbohydrate diet. Improvements in mass spectrometry (see Tools of Biochemistry 12B) have facilitated the screening of newborn infants using blood spots, allowing affected infants to be detected before the onset of symptoms. This is important because many of these disorders can be managed by relatively simple dietary means.

Energy Yield from Fatty Acid Oxidation

We can now write a balanced equation for the overall degradation of palmitoyl-CoA to 8 moles of acetyl-CoA:

Palmitoyl-CoA + 7CoA-SH + 7FAD + 7NAD$^+$ + 7H$_2$O $\rightarrow$ 8 acetyl-CoA + 7FADH$_2$ + 7NADH + 7H$^+$

Each of the products is metabolized exactly as described earlier for oxidation of carbohydrates. Acetyl-CoA is catabolized via the citric acid cycle, and FADH$_2$ and NADH transfer electrons to the respiratory chain through ETF (page 727) and complex I, respectively. Thus, we can easily compute the metabolic energy yield from fatty acid oxidation in terms of moles of ATP synthesized from ADP. Recall from Chapter 15 that the P/O ratios for oxidation of flavoproteins and NADH are ~1.5 and ~2.5, respectively, and oxidation of acetyl-CoA in one turn of the citric acid cycle yields 10 ATPs. The following summation, using palmitate as an example, gives the total energy yield:

Reaction	ATP Yield
Activation of palmitate to palmitoyl-CoA	-2
Oxidation of 8 acetyl-CoA	$8 \times 10 = 80$
Oxidation of 7 FADH$_2$	$7 \times 1.5 = 10.5$
Oxidation of 7 NADH	$7 \times 2.5 = 17.5$
Net: Palmitate $\longrightarrow$ CO$_2$ + H$_2$O	106

Fatty acids are oxidized by repeated cycles of dehydrogenation, hydration, dehydrogenation, and thiolytic cleavage, with each cycle yielding acetyl-CoA and a fatty acyl-CoA shorter by two carbons than the input acyl-CoA.

The final thiolytic cleavage

From this, you can calculate the ATP yield per carbon oxidized to CO_2 as 106/16, or about 6.6. The corresponding value for glucose is 5–5.3 (30–32 ATPs formed per 6 carbons oxidized). Thus, the energy yield from fat oxidation is higher than that from oxidation of the less highly reduced carbohydrate, whether measured on a weight basis (see page 488) or a molar basis.

Oxidation of Unsaturated Fatty Acids

Recall from Chapter 10 that many fatty acids in natural lipids are unsaturated; that is, they contain one or more double bonds (see Table 10.1, page 361). Because these bonds are in the *cis* configuration, they cannot be hydrated by enoyl-CoA hydratase, which acts only on *trans* compounds. Two additional enzymes, **enoyl-CoA isomerase** and **2,4-dienoyl-CoA reductase**, must come into play for unsaturated fatty acids to be oxidized. The isomerase acts upon monounsaturated fatty acids, such as the 18-carbon $\Delta 9$ compound, oleic acid, which contains a *cis* double bond between carbons 9 and 10. Oleic acid is activated, transported into mitochondria, and carried through three cycles of β-oxidation just as are the saturated fatty acids. The product of the third cycle is the CoA ester of a 12-carbon fatty acid with a *cis* double bond between carbons 3 and 4. Not only is the double bond in the wrong configuration to be hydrated, but it is also in the wrong position. The enoyl-CoA isomerase enzyme converts this *cis*-3-enoyl-CoA to the corresponding *trans*-2-enoyl-CoA, which can then be acted on by enoyl-CoA hydratase (see margin). This hydration and all subsequent reactions are identical to those already described for saturated fatty acids.

The other auxiliary enzyme, 2,4-dienoyl-CoA reductase, comes into play during the oxidation of polyunsaturated fatty acids, such as linoleic acid (18:2cΔ9,12). This 18-carbon fatty acid contains *cis* double bonds between carbons 9 and 10 and between carbons 12 and 13. As shown in Figure 17.21, linoleyl-CoA undergoes three cycles of β-oxidation, just as does oleyl-CoA, to give a C_{12} acyl-CoA with *cis* double bonds between carbons 3 and 4 and between 6 and 7. Enoyl-CoA isomerase converts the $\Delta 3$ *cis* double bond to a $\Delta 2$ *trans* double bond. There follow hydration, dehydrogenation, and thiolytic cleavage to give acetyl-CoA plus a 10-carbon enoyl-CoA, unsaturated between carbons 4 and 5. Action of acyl-CoA dehydrogenase yields a dienoyl-CoA, unsaturated at C4—C5 and at C2—C3. The NADPH-dependent 2,4-dienoyl-CoA reductase converts this to a C_{10} *cis*-Δ3-enoyl-CoA. The isomerase comes into play once more, generating a *trans*-Δ2-enoyl-CoA, which undergoes the remaining cycles of β-oxidation normally.

By the pathways described here, both monounsaturated and diunsaturated 18-carbon fatty acids can undergo degradation to 9 moles of acetyl-CoA. There is, of course, a reduction in the overall energy yield because each double bond at an odd-numbered carbon in the original fatty acid means one less FAD reduction step in the overall process. A double bond at an even-numbered carbon must be reduced at the expense of NADPH, equivalent to the cost of ~2.5 ATP. The two auxiliary enzymes allow all of the even-chain polyunsaturated fatty acids to be similarly degraded, with the following exception. A significant portion of dietary lipid contains unsaturated fatty acids with double bonds in the *trans* configuration. These fatty acids arise through microbial action in the digestive systems of ruminant mammals and also chemically, through partial hydrogenation of fats and oils. Commercial vegetable oils are subjected to partial hydrogenation to protect them against oxidation by converting the double bonds to single bonds (more saturated). Unfortunately, some *cis* double bonds are isomerised to *trans* double bonds during this process, and *trans* fatty acids can thus be quite abundant in dairy products and in margarine and cooking oils. Because there is growing evidence implicating *trans* fatty acids in coronary artery disease, margarine manufacturers are beginning to remove them from their product and many cities such as New York have banned the use of *trans* fatty acids in restaurants.

Two enzymes, enoyl-CoA isomerase and 2,4-dienoyl-CoA reductase, play essential roles in the oxidation of unsaturated fatty acids.

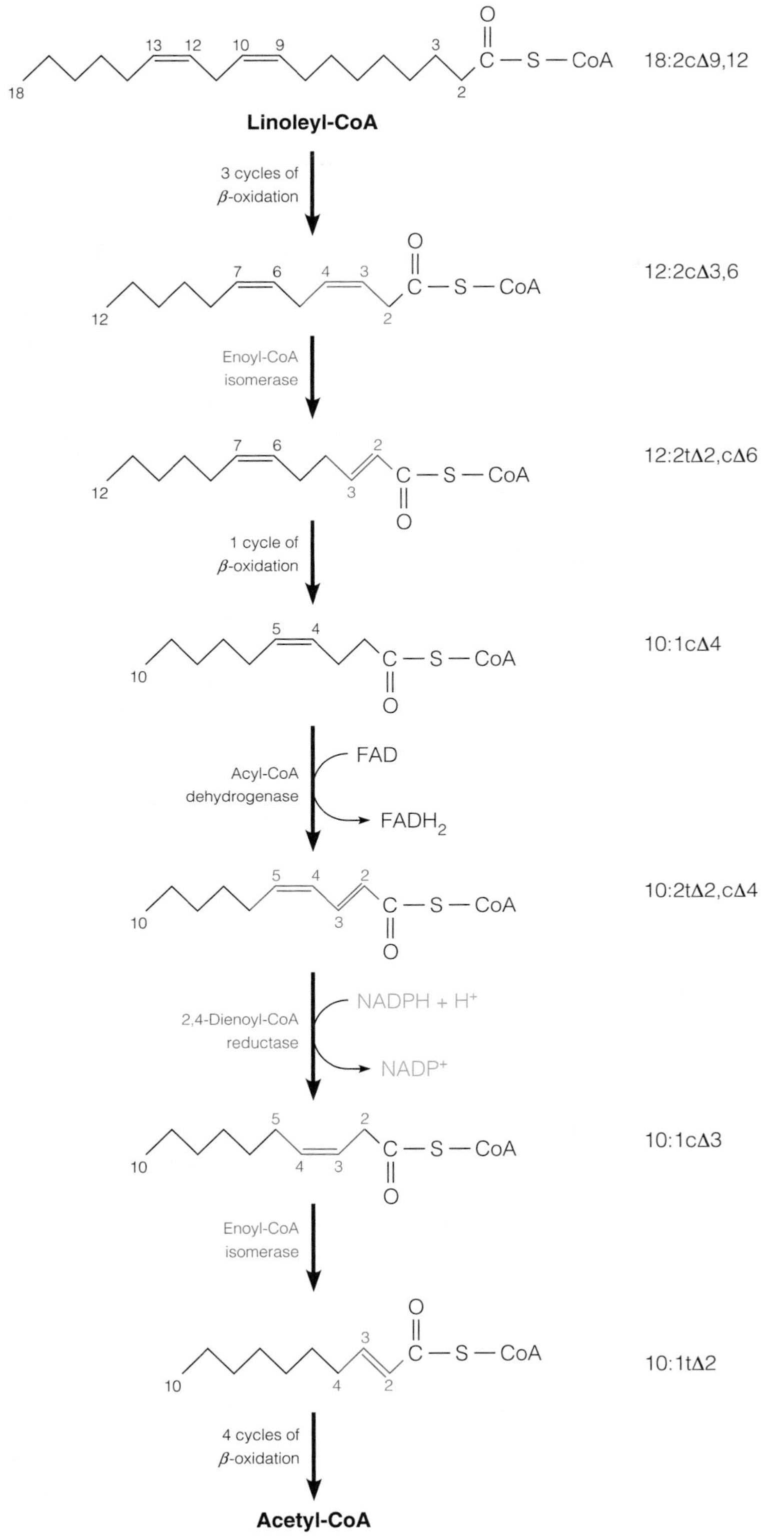

FIGURE 17.21

β-Oxidation pathway for polyunsaturated fatty acids. This example, using linoleyl-CoA, shows sites of action of enoyl-CoA isomerase and 2,4-dienoyl-CoA reductase, enzymes specific to unsaturated fatty acid oxidation (identified with magenta type).

Oxidation of Fatty Acids with Odd-Numbered Carbon Chains

Though most of the fatty acids in natural lipids contain even-numbered carbon chains, a small proportion have odd-numbered carbon chains. The latter group presents a special metabolic problem, which is solved in a novel way. The substrate for the last cycle of

Odd-numbered fatty acid chains yield upon oxidation 1 mole of propionyl-CoA, whose conversion to succinyl-CoA involves a biotin-dependent carboxylation and a coenzyme B_{12}–dependent rearrangement.

β-oxidation of an odd-chain acyl-CoA is a five-carbon acyl-CoA. Thiolytic cleavage of this substrate yields 1 mole each of acetyl-CoA and **propionyl-CoA**.

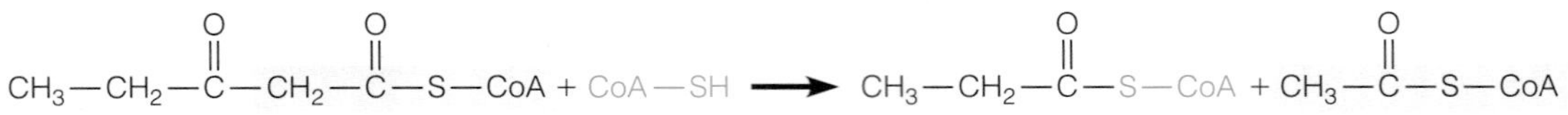

$$CH_3-CH_2-\overset{\overset{\displaystyle O}{||}}{C}-CH_2-\overset{\overset{\displaystyle O}{||}}{C}-S-CoA + CoA-SH \longrightarrow CH_3-CH_2-\overset{\overset{\displaystyle O}{||}}{C}-S-CoA + CH_3-\overset{\overset{\displaystyle O}{||}}{C}-S-CoA$$

Propionyl-CoA   **Acetyl-CoA**

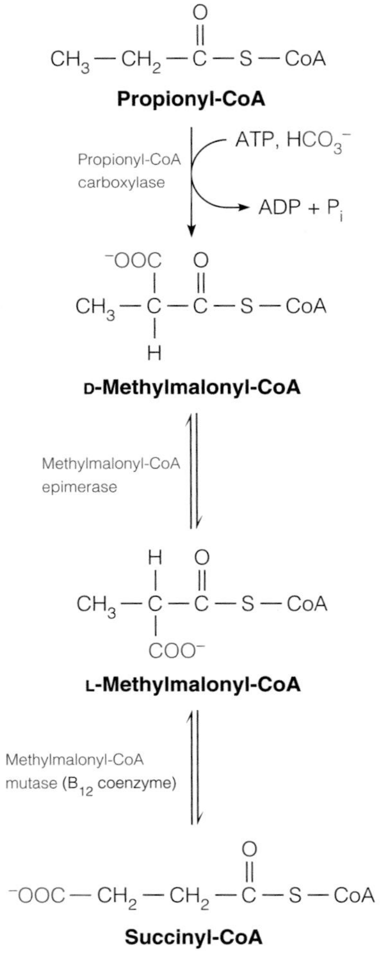

Unlike acetyl-CoA, which is catabolized via the citric acid cycle, propionyl-CoA must be further metabolized before its carbon atoms can enter the citric acid cycle for complete oxidation to CO_2. That further metabolism (Figure 17.22) involves first the ATP-dependent carboxylation of propionyl-CoA, catalyzed by the biotin-containing enzyme **propionyl-CoA carboxylase**. This enzyme is similar in domain structure and mechanism to pyruvate carboxylase, described in Chapter 14 (see Figures 14.19 and 14.20, pages 618–619). The product, D-methylmalonyl-CoA, then undergoes epimerization to its L stereoisomer by action of methylmalonyl-CoA epimerase. Next, this branched-chain acyl-CoA derivative is converted to the corresponding straight-chain compound, which happens to be succinyl-CoA, by an unusual reaction. The enzyme, L-methylmalonyl-CoA mutase, requires a cofactor called **5′-deoxyadenosylcobalamin**, derived from vitamin B_{12}. Mechanistically, the side-chain migration makes this reaction quite interesting. However, because B_{12} coenzymes are also involved in amino acid metabolism, we reserve study of this and other B_{12}-dependent reactions for Chapter 20.

Inability to catabolize propionyl-CoA properly has severe consequences in humans. If there is defective activity of L-methylmalonyl-CoA mutase or of the synthesis of the adenosylcobalamin coenzyme, L-methylmalonyl-CoA accumulates and exits from cells as methylmalonic acid. This process causes a severe acidosis (lowering of blood pH) and also damages the central nervous system. This rare condition, called **methylmalonic acidemia**, is usually fatal in early life. The disease can sometimes be treated by administering large doses of vitamin B_{12}. In these cases the mutation decreases affinity of the mutase for its B_{12} coenzyme, and the enzyme can be induced to function if the coenzyme concentration can be increased substantially.

Control of Fatty Acid Oxidation

In most cells, fatty acid oxidation is controlled by availability of substrates for oxidation, the fatty acids themselves. In animals this availability is controlled in turn by the hormonal control of fat mobilization in adipocytes. Because the function of adipose tissue is to store fat for use in other cells, it makes good metabolic sense for breakdown and release of this stored fat to be regulated by hormones, which are extracellular messengers. Recall from page 722 that triacylglycerol lipase activity is regulated by hormonally initiated regulatory cascades involving cyclic AMP. The action of glucagon or epinephrine causes fat breakdown and release, which leads ultimately to fatty acid accumulation in other cells. Also, as noted on page 726, malonyl-CoA provides another important regulatory mechanism, by inhibiting fatty acyl-CoA movement into mitochondria by the acylcarnitine shuttle. As we shall soon see (page 746), malonyl-CoA levels are also hormonally controlled, providing tight coordination between fatty acid oxidation and fatty acid synthesis.

Peroxisomal β-Oxidation of Fatty Acids

A modified version of the β-oxidation pathway occurs in **peroxisomes**, organelles that are present in most eukaryotic cells. Peroxisomes are quite similar to plant cell glyoxysomes, except that peroxisomes lack the enzymes of the glyoxylate pathway (see Chapter 14). The peroxisomal and glyoxysomal pathways use isozymes distinct from those of the mitochondrial pathway. Both organelles carry out a β-oxidation pathway in which an FAD-linked acyl-CoA dehydrogenase transfers

$$CH_3-CH_2-\overset{\overset{\displaystyle O}{||}}{C}-S-CoA$$

Propionyl-CoA

Propionyl-CoA carboxylase → ATP, HCO_3^- → ADP + P_i

$$CH_3-\overset{\overset{\displaystyle ^-OOC}{|}}{\underset{\underset{\displaystyle H}{|}}{C}}-\overset{\overset{\displaystyle O}{||}}{C}-S-CoA$$

D-Methylmalonyl-CoA

Methylmalonyl-CoA epimerase

$$CH_3-\overset{\overset{\displaystyle H}{|}}{\underset{\underset{\displaystyle COO^-}{|}}{C}}-\overset{\overset{\displaystyle O}{||}}{C}-S-CoA$$

L-Methylmalonyl-CoA

Methylmalonyl-CoA mutase (B_{12} coenzyme)

$$^-OOC-CH_2-CH_2-\overset{\overset{\displaystyle O}{||}}{C}-S-CoA$$

Succinyl-CoA

FIGURE 17.22

Pathway for catabolism of propionyl-CoA.

electrons not to the respiratory electron transport chain but directly to oxygen (these enzymes are thus more properly called **acyl-CoA oxidases**; see margin). The latter is reduced to hydrogen peroxide, which in turn is acted on by catalase.

The next two steps, hydration of the enoyl-CoA and oxidation of the hydroxyacyl-CoA, are catalyzed by a multifunctional enzyme that carries two additional activities required for the processing of unsaturated fatty acids (an epimerase and isomerase). The fourth step of the cycle is catalyzed by a monofunctional thiolase. Because electrons are not shuttled into the respiratory chain, the peroxisomal pathway is not coupled to energy production, but it does generate heat. In animal peroxisomes the pathway proceeds only as far as C_4 and C_6 acyl-CoAs. However, these acyl groups can be transferred to carnitine for transport into mitochondria, where β-oxidation can be completed. By contrast, plant glyoxysomes carry the oxidation all the way to acetyl-CoA, which is used for carbohydrate synthesis via the glyoxylate pathway. The function of the peroxisomal pathway is not yet clear, but it involves initial stages in oxidizing very long-chain fatty acids (VLCFA) and other lipids. The importance of the peroxisomal pathway is revealed by two human genetic diseases. Infants born with Zellweger syndrome are unable to import any proteins into the peroxisome, including those of the β-oxidation pathway, and they accumulate high levels of VLCFA in their blood. These infants have profound neurological impairment, and they die within their first year. Individuals with X-linked adrenoleukodystrophy (X-ALD) are also unable to metabolize VLCFA (≥ 22 C). The disease is characterized by elevated plasma and tissue levels of VLCFA. X-ALD is caused by mutations in the *ABCD1* gene, which encodes an ATP-binding-cassette (ABC) transporter located in the peroxisomal membrane. Patients with mutations in this gene are unable transport the CoA derivatives of these very long-chain fatty acids into peroxisomes, and thus their oxidation is blocked. Boys born with X-ALD develop normally for the first few years, but they then suffer progressive neurological impairment and death by age 10–15 years. Dietary treatment with a mixture of glyceryl trioleate and glyceryl trierucate (referred to as Lorenzo's Oil) normalizes plasma VLCFA levels in patients, presumably by competing with the enzyme system that elongates saturated fatty acids (see page 744). This treatment was the subject of a popular 1992 film, but the clinical efficacy of Lorenzo's Oil in preventing the neurological damage of X-ALD remains controversial.

α-Oxidation of Fatty Acids

Although β-oxidation is the major pathway for fatty acid degradation, a minor pathway, also localized to peroxisomes, oxidizes certain fatty acids with initial oxidation occurring on the α-carbon rather than the β-carbon. This pathway came to light through analysis of a rare and severe congenital neurological disorder called **Refsum's disease**. Patients with this condition accumulate large amounts of an unusual fatty acid, **phytanic acid**, which is derived from **phytol**, a constituent of chlorophyll (Figure 17.23). The methyl group on carbon 3 of phytol prevents β-oxidation of this substrate. However, the α-carbon can undergo oxidation, to give **pristanic acid**, a substrate that can be further degraded by β-oxidation (Figure 17.23). In Refsum's disease the α-oxidation pathway is defective, and the phytanic acid is not converted to a compound that can be degraded. The only known treatment is to feed a diet containing little or no chlorophyll. This treatment is difficult because it rules out both leafy green vegetables and meat and milk that come from herbivorous animals—all of these foods contain substantial amounts of phytanic acid.

Ketogenesis

Thus far we have written about acetyl-CoA as though it had only two major metabolic fates, either oxidation to CO_2 in the citric acid cycle or biosynthesis of fatty acids. Another major pathway comes into play in mitochondria (primarily in

Glyceryl trierucate

FIGURE 17.23

The α-oxidation pathway for phytanic acid oxidation. The methyl group on the β-carbon of phytanic acid prevents β-oxidation of this compound, so an additional pathway, involving α-oxidation, comes into play. Hydroxylation of phytanoyl-CoA at the α-position is catalyzed by an α-ketoglutarate-dependent monooxygenase (Chapter 15, page 663). After that, the rest of the molecule can be degraded by β-oxidation, producing propionyl and acetyl-CoA. The final oxidation product (isobutyryl-CoA is shown) remains speculative.

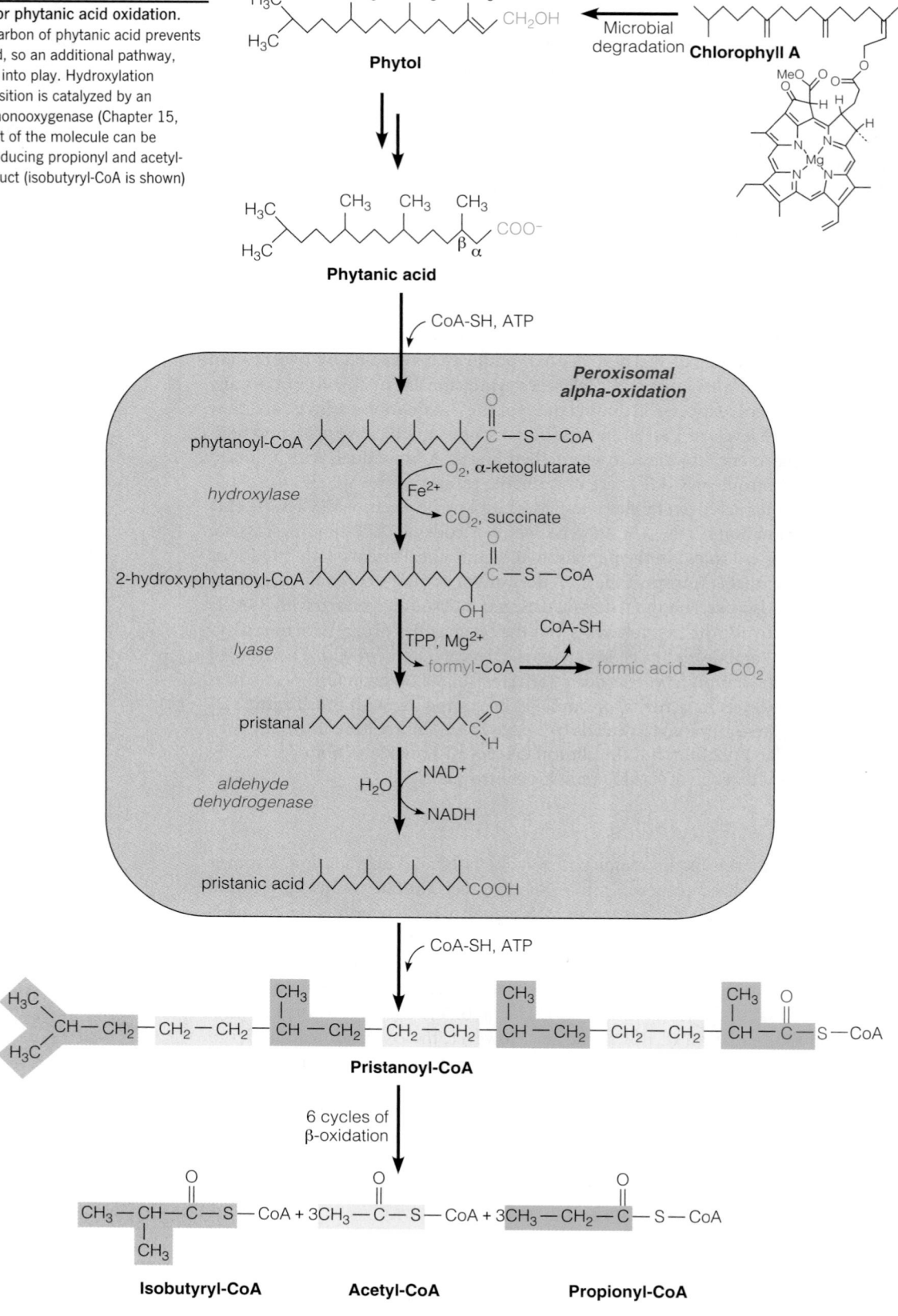

liver) when acetyl-CoA accumulates beyond its capacity to be oxidized or used for fatty acid synthesis. That pathway is called **ketogenesis**, and it leads to a class of compounds called **ketone bodies**.

During fasting or starvation, when carbohydrate intake is too low, oxaloacetate levels fall so that flux through citrate synthase is impaired, causing acetyl-CoA levels to rise. Under these conditions, 2 moles of acetyl-CoA undergo a reversal of the thiolase reaction to give acetoacetyl-CoA (Figure 17.24). Acetoacetyl-CoA can react in turn with a third mole of acetyl-CoA to give **3-hydroxy-3-methylglutaryl-CoA (HMG-CoA)**, catalyzed by **HMG-CoA synthase**. When formed in cytosol, HMG-CoA is an early intermediate in cholesterol biosynthesis (see Chapter 19). In mitochondria, however, HMG-CoA is acted on by **HMG-CoA lyase** to yield **acetoacetate** plus acetyl-CoA. Acetoacetate undergoes either NADH-dependent reduction to give D-**β-hydroxybutyrate** or, in very small amounts, spontaneous decarboxylation to acetone. Collectively, acetoacetate, acetone, and β-hydroxybutyrate are called ketone bodies, even though the last compound does not contain a keto carbonyl group. Liver also produces free acetate, by direct hydrolysis of acetyl-CoA. All of these compounds can be utilized by peripheral tissues as an alternative fuel under ketogenic conditions. Extrahepatic tissues take up acetate from the blood and transport it into mitochondria, where it is converted back to acetyl-CoA in an ATP-dependent reaction catalyzed by **acetyl-CoA synthetase**:

$$\text{acetate} + \text{ATP} + \text{CoA-SH} \rightleftharpoons \text{acetyl-CoA} + \text{AMP} + \text{PP}_i$$

Acetyl-CoA is then oxidized through the citric acid cycle for ATP production.

In some circumstances, ketogenesis can be considered an "overflow pathway." As noted above, it is stimulated when acetyl-CoA accumulates because of deficient carbohydrate intake. Ketogenesis occurs primarily in liver because of the high levels of HMG-CoA synthase in that tissue. Ketone bodies are transported from liver to other tissues, where acetoacetate and β-hydroxybutyrate can be reconverted to acetyl-CoA for energy generation. The reconversion involves enzymatic transfer of a CoA moiety from succinyl-CoA to acetoacetate, yielding acetoacetyl-CoA and succinate.

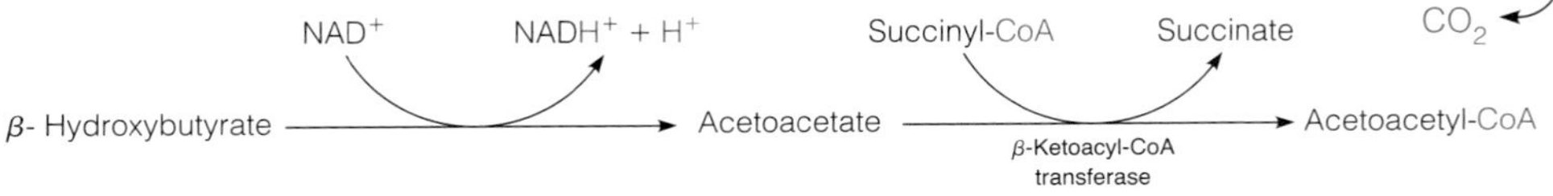

This enzyme, β-ketoacyl-CoA transferase, is present in all tissues except liver so that liver does not compete with the peripheral tissues for use of ketone bodies as fuel. The acetoacetyl-CoA is then converted to two acetyl-CoA by thiolase.

As discussed further in Chapter 18, ketogenesis becomes extremely important in fasting and starvation, when the brain, which normally uses glucose as its main fuel, undergoes metabolic adaptation to the use of ketone bodies. Excess ketone bodies are excreted in the urine, but because of its high volatility, acetone can be detected on the breath. Under normal conditions, certain other tissues, particularly heart, derive much of their energy by metabolizing ketone bodies produced in the liver. In untreated diabetes, where tissues are unable to efficiently use glucose, ketone bodies are produced in excess of the capacity of peripheral tissues to use them, a condition referred to as **ketosis**. Blood levels of ketone bodies in these patients can exceed 100 mg/dL and high levels of the weak acids acetoacetate and β-hydroxybutyrate lower blood pH (**acidosis**). This **ketoacidosis** is a diagnostic feature of diabetes.

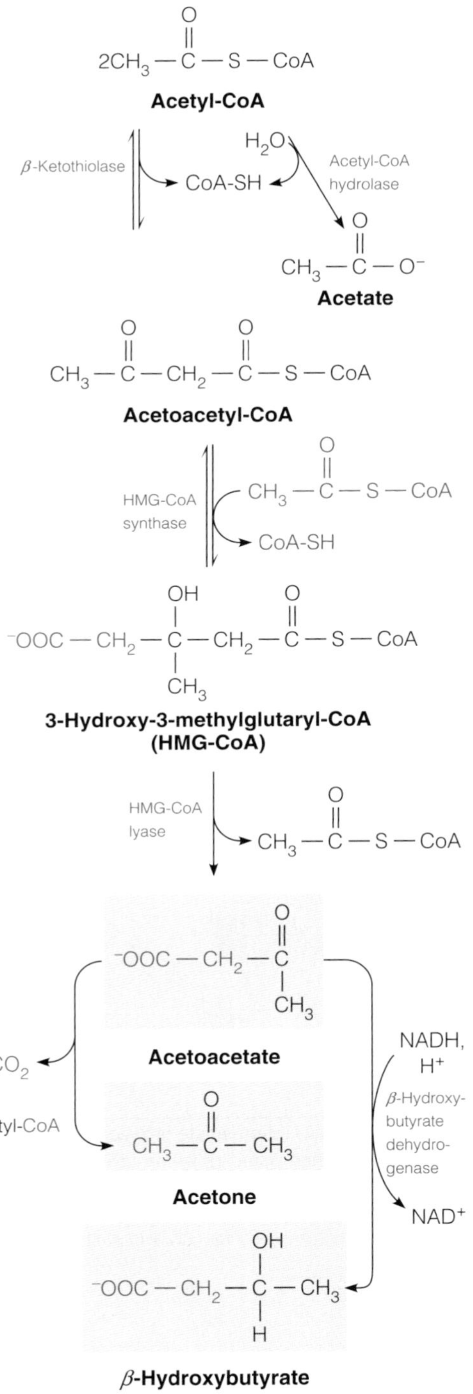

FIGURE 17.24

Biosynthesis of ketone bodies in the liver. The three water-soluble compounds commonly called ketone bodies are boxed. Acetone is formed in very small quantities, by nonenzymatic decarboxylation of acetoacetate. Acetate is also produced and released by liver for utilization by peripheral tissues.

Fatty Acid Biosynthesis

Relationship of Fatty Acid Synthesis to Carbohydrate Metabolism

We have noted that the vast majority of the stored fuel in most animal cells is in the form of fat. However, a large proportion of the caloric intake of many animal diets—certainly most human diets—is carbohydrate. Because carbohydrate storage reserves are strictly limited, there must be efficient mechanisms for conversion of carbohydrate to fat. In this section our primary focus is on fatty acid synthesis.

As schematized in Figure 17.25, a central metabolite is acetyl-CoA, which comes both from pyruvate in the pyruvate dehydrogenase reaction and from fatty acid β-oxidation. Acetyl-CoA is, in turn, converted in the cytosol to fatty acids. Thus, acetyl-CoA is derived from both fat breakdown and carbohydrate breakdown and is also the major fat precursor. However, in animals *acetyl-CoA cannot undergo net conversion to carbohydrate.* This is because of the virtual irreversibility of the pyruvate dehydrogenase reaction. As noted in Chapter 14, the glyoxylate cycle in plants and some microorganisms permits a bypass of this step, with net conversion of acetyl-CoA to gluconeogenic precursors. However, *in animals the conversion of carbohydrate to fat is unidirectional.* Moreover, although fatty acid synthesis is regulated, the total capacity for fat storage is not.

Early Studies of Fatty Acid Synthesis

Early in the twentieth century, when it became evident that most fatty acids in lipids contain even-numbered chains, it was reasonable to expect that the biosynthetic process would involve some stepwise addition of activated two-carbon fragments, in the same sense that oxidation proceeds two carbons at a time. Indeed, this process was demonstrated experimentally in the 1940s, in one of the first metabolic experiments using isotopic tracers. David Rittenberg and Konrad Bloch fed mice acetate labeled with the stable isotopes ^{13}C and deuterium ($^{13}C^{2}H_{3}^{13}COO^{-}$) and found both isotopes incorporated into fatty acids.

Once the β-oxidation pathway had been discovered, it was generally thought that fatty acid synthesis would proceed simply by a reversal of its degradation pathway. However, when biochemists began to fractionate enzyme systems capable of

---

When carbohydrate catabolism is limited, acetyl-CoA is converted to ketone bodies, mainly acetoacetate and β-hydroxybutyrate—important metabolic fuels in some circumstances.

Animals readily convert carbohydrate to fat, but cannot carry out net conversion of fat to carbohydrate.

Fatty acid synthesis occurs through intermediates similar to those of fatty acid oxidation, but with differences in electron carriers, carboxyl group activation, stereochemistry, and cellular location.

FIGURE 17.25

Acetyl-CoA as a key intermediate between fat and carbohydrate metabolism. Arrows identify major routes of formation or utilization of acetyl-CoA. Citrate serves as a carrier to transport acetyl units from the mitochondrion to the cytosol for fatty acid synthesis.

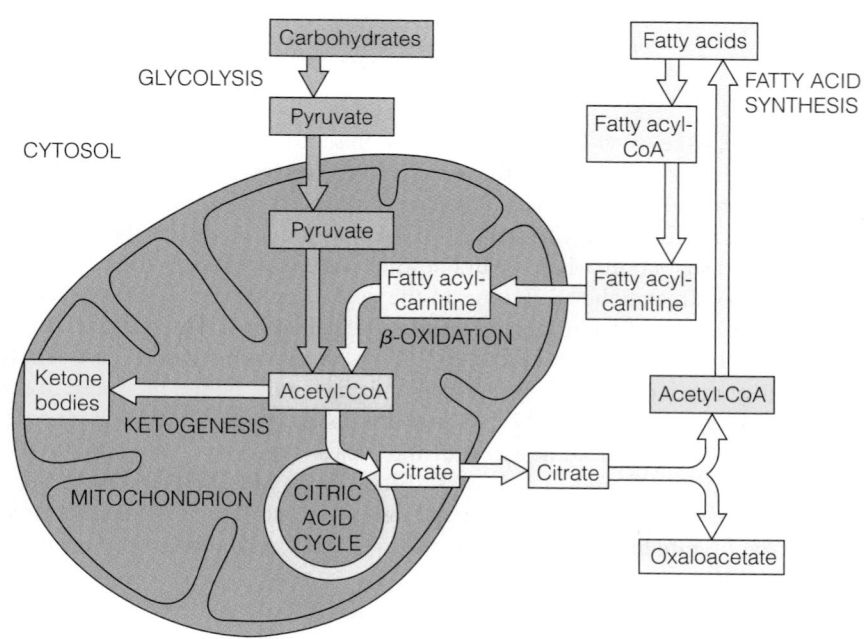

synthesizing fatty acids, they found that the activities of β-oxidation were lacking from their purified fractions. The key discovery that established fatty acid synthesis as an entirely different pathway was Salih Wakil's observation in the late 1950s that fatty acid synthesis has an absolute requirement for bicarbonate. The carbon from that bicarbonate was not incorporated in the final product, however. These observations led to the discovery of a three-carbon compound, *malonyl-CoA*, as the first committed intermediate in fatty acid biosynthesis. Today we know that, although the chemistries of fatty acid synthesis and degradation are similar, the pathways differ in the enzymes involved, acyl group carriers, stereochemistry of the intermediates, electron carriers, intracellular location, and regulation. Indeed, fatty acid metabolism is one of the best examples of the statement that anabolic pathways are never the simple reversal of catabolic pathways.

The overall process of fatty acid synthesis is similar in all prokaryotic and eukaryotic systems analyzed to date. Three separate enzyme systems catalyze, respectively, (1) biosynthesis of palmitate from acetyl-CoA, (2) chain elongation starting from palmitate, and (3) desaturation. In eukaryotic cells the first pathway occurs in the cytosol, chain elongation occurs both in mitochondria and in the endoplasmic reticulum, and desaturation occurs in the endoplasmic reticulum.

$$^-OOC-CH_2-\overset{\overset{\displaystyle O}{\|}}{C}-S-CoA$$

Malonyl-CoA

Biosynthesis of Palmitate from Acetyl-CoA

As outlined in Figure 17.26, the chemistry of palmitate synthesis is remarkably similar to that of palmitate oxidation run in reverse. The synthetic process comprises stepwise additions of two-carbon units, with each step proceeding via condensation, reduction, dehydration, and another reduction. The major distinctions are (1) the need for an activated intermediate, malonyl-CoA, at each two-carbon addition step, (2) the nature of the acyl group carrier, and (3) the use of NADPH-requiring enzymes in the reductive reactions. Details of these and the other reactions follow.

Synthesis of Malonyl-CoA

The first committed step in fatty acid biosynthesis is the formation of malonyl-CoA from acetyl-CoA and bicarbonate, catalyzed by **acetyl-CoA carboxylase (ACC)**.

$$CH_3-\overset{\overset{\displaystyle O}{\|}}{C}-S-CoA + ATP + HCO_3^- \longrightarrow {}^-OOC-CH_2-\overset{\overset{\displaystyle O}{\|}}{C}-S-CoA + ADP + P_i + H^+$$

Acetyl-CoA   **Malonyl-CoA**

Like other committed steps in biosynthetic pathways, this reaction is so exergonic as to be virtually irreversible. Similar to other enzymes catalyzing carboxylation reactions (see Chapter 14, page 618), acetyl-CoA carboxylase has a biotin cofactor, covalently bound via a lysine ε-amino group. The reaction proceeds via a covalently bound N-carboxybiotin intermediate.

$$E\text{-biotin} + ATP + HCO_3^- \longrightarrow E\text{-}N\text{-carboxybiotin} + ADP + P_i$$

$$E\text{-}N\text{-carboxybiotin} + acetyl\text{-}CoA \longrightarrow malonyl\text{-}CoA + E\text{-biotin}$$

The prokaryotic form of this enzyme, exemplified by the enzyme purified from *E. coli*, consists of three separate proteins: (1) a small carrier protein that contains the bound biotin, (2) a **biotin carboxylase**, which catalyzes the ATP-dependent formation of N-carboxybiotin, and (3) a **transcarboxylase**, which transfers the activated carboxyl group from N-carboxybiotin to acetyl-CoA. The hydrocarbon chains in both biotin and its associated lysine residue act as a flexible swinging arm, which allows the biotin to interact with the catalytic sites of both catalytic subunits, just as we saw for pyruvate carboxylase (Figure 14.19, page 618).

Oxidative degradation
Synthesis
R — CH2 — CH2 — CH2 — C — S — carrier
+ FAD
Dehydrogenation
Reduction
+ NADPH
+ H+
R — CH2 — CH=CH — C — S — carrier
+ H2O
Hydration
Dehydration
– H2O
OH
R — CH2 — CH — CH2 — C — S — carrier
(L configuration)
(D configuration)
+ NAD+
Dehydrogenation
Reduction
+ NADPH
+ H+
R — CH2 — C — CH2 — C — S — carrier
CO2
Thiolytic cleavage
Condensation
Acetyl-CoA
R — CH2 — C — S — carrier
(CoA or ACP)
Malonyl-CoA
Acetyl-CoA

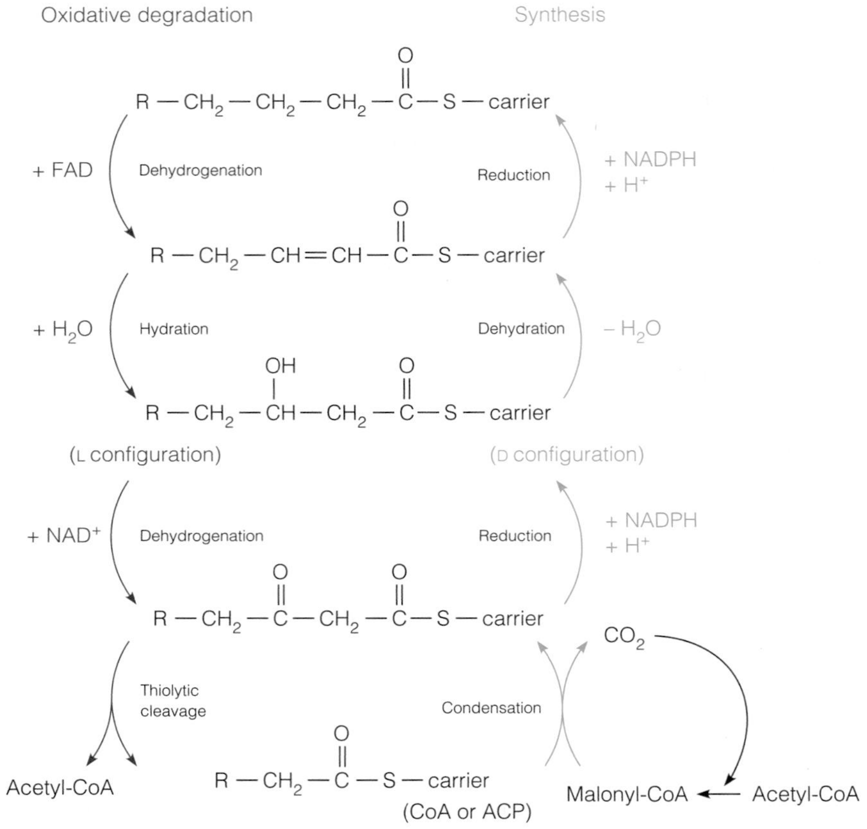

FIGURE 17.26

Chemical similarities between oxidation and synthesis of a fatty acid. The figure shows a single cycle of oxidation (down) or addition (up) of a two-carbon fragment. Coenzyme A is the acyl group carrier for oxidation, and acyl carrier protein (ACP) is the carrier for synthesis.

400 Å

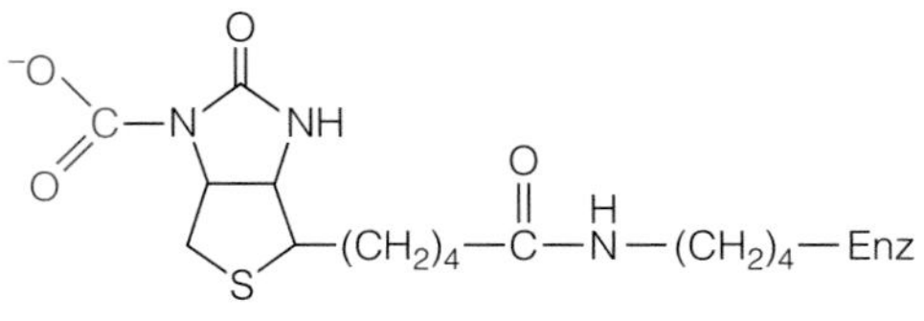

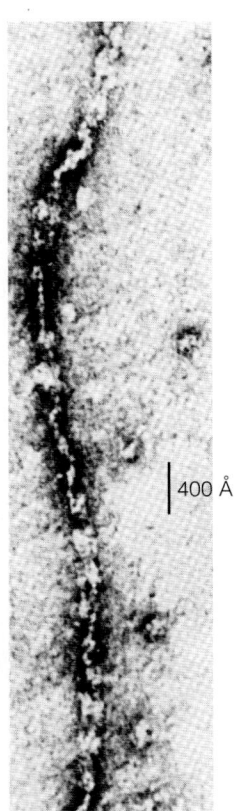

FIGURE 17.27

The active filamentous form of eukaryotic acetyl-CoA carboxylase.

Reprinted from *Current Topics in Cellular Regulation* 8:139–195, M. D. Lane, J. Moss, and S. E. Polakis. © 1974, with permission from Elsevier.

By contrast, ACC in eukaryotes consists of a single protein containing two identical polypeptide chains, each with M_r of about 265,000. The dimeric protein itself has low activity, but in the presence of citrate it polymerizes to a novel filamentous form, with M_r of $4-8 \times 10^6$, that can readily be visualized in the electron microscope (Figure 17.27). The equilibrium between inactive protein dimers and the active filamentous form, and its control by metabolic intermediates, represents an important mechanism for regulating fatty acid biosynthesis. The regulation of ACC and fatty acid synthesis will be discussed after we describe the rest of the pathway (page 745).

Malonyl-CoA to Palmitate

Recall that all of the intermediates in fatty acid oxidation are activated via their linkage to a carrier molecule, coenzyme A. A similar activation is involved in fatty acid synthesis, but the carrier is different. It is a small protein (77 residues in *E. coli*) called **acyl carrier protein (ACP)**. The chemistry of activation is identical to that in acyl-CoAs. Indeed, ACP uses an identical phosphopantetheine moiety with its reactive sulfhydryl group as that found in CoA. In ACP, the phosphopantetheine moiety is covalently linked to a serine group in the polypeptide (Figure 17.28). All of the chemistry catalyzed by fatty acid synthase occurs on substrates attached to ACP via a thioester linkage. As we shall see, the phosphopantetheine moiety acts as a swinging arm to transfer the acyl group between the active sites of the complex.

NH
(CH2)4—C—N—(CH2)4—Enz

***N*-Carboxybiotinyl-enzyme**

To begin the synthesis of a new palmitic acid molecule, the fatty acid synthase must first be charged with starter substrates. The acetyl moiety of acetyl-CoA is loaded onto ACP in a reaction catalyzed by **malonyl/acetyl-CoA-ACP transacylase (MAT)** (Figure 17.29, reaction 1a). The acetyl group is then transferred to a Cys-SH in the active site of the **β-ketoacyl-ACP synthase** (**KS**), giving acetyl-KS (reaction 1b). The phosphopantetheine moiety of ACP is now available to be charged with the second substrate, malonyl-CoA, also catalyzed by MAT (Figure 17.29,

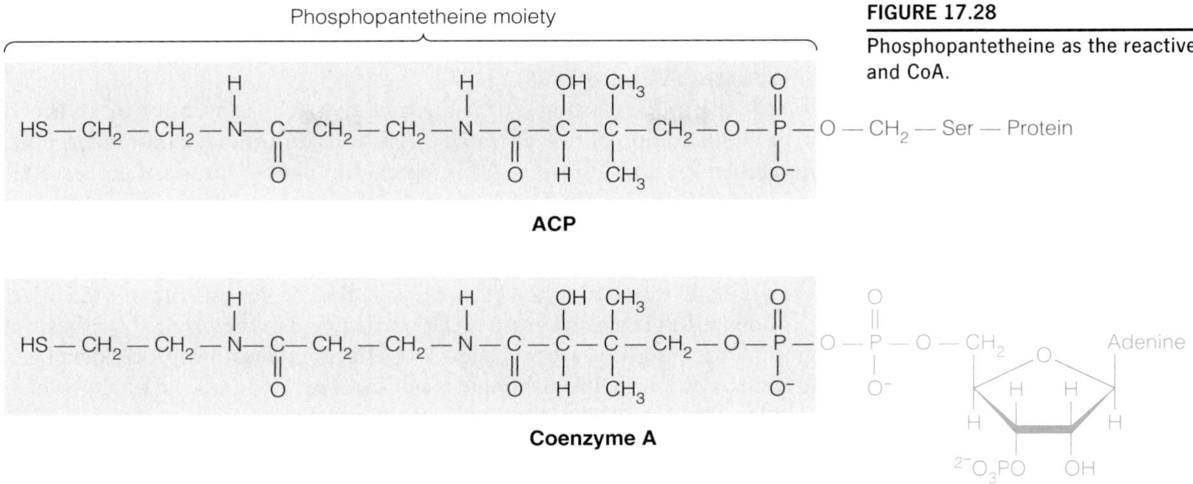

FIGURE 17.28

Phosphopantetheine moiety

ACP

Coenzyme A

FIGURE 17.28

Phosphopantetheine as the reactive unit in ACP and CoA.

reaction 2). Because the energy-rich thioester bonds in acyl-CoAs and acyl-ACPs are identical, these transacylase reactions are readily reversible. The fatty acid synthase is now activated for the first cycle of chain elongation. In each cycle, a primer substrate (an acyl group on KS) is condensed with an extender molecule (a malonyl group on ACP). The synthetic cycle proceeds via *condensation, reduction, dehydration*, and *reduction*.

For the first cycle of synthesis (Figure 17.30, reactions 1–4), we start with 1 mole each of malonyl-ACP and acetyl-KS, and in four reactions we generate 1 mole of butyryl-ACP. These are the reactions that resemble the reactions (in reverse) of fatty acid oxidation (see Figure 17.26). The key carbon–carbon bond forming reaction is a Claisen-type condensation (see Figure 12.8, page 485) between acetyl-KS and malonyl-ACP (Figure 17.30, reaction 1). This reaction, catalyzed by KS, involves decarboxylation of the malonyl moiety to give a nucleophilic enolate ion, which attacks the electrophilic acetyl group on KS. Breakdown of the tetrahedral intermediate involves elimination of the KS cysteine thiol, freeing it for attachment of the elongated acyl group at the end of the cycle (reaction 5). The condensation product, a β-ketoacyl-ACP thioester, is next reduced to a D-β-hydroxyacyl-ACP in an NADPH-dependent reaction catalyzed by **β-ketoacyl-ACP reductase (KR)** (reaction 2). Dehydration of the D-β-hydroxyacyl-ACP, catalyzed by **β-hydroxyacyl-ACP dehydrase (DH)** (reaction 3), yields a *trans*-2-enoyl-ACP, which undergoes a second NADPH-dependent reduction, catalyzed by **enoyl-ACP reductase (ER)** (reaction 4), to yield a fatty acyl-ACP—butyryl-ACP in the first cycle of synthesis. To end the first cycle, the butyryl group is translocated from ACP to the Cys-SH of KS (reaction 5), and ACP is charged with a second malonyl-CoA. To start the second cycle, butyryl-KS reacts with another molecule of malonyl-ACP, and the product of the second cycle is hexanoyl-ACP.

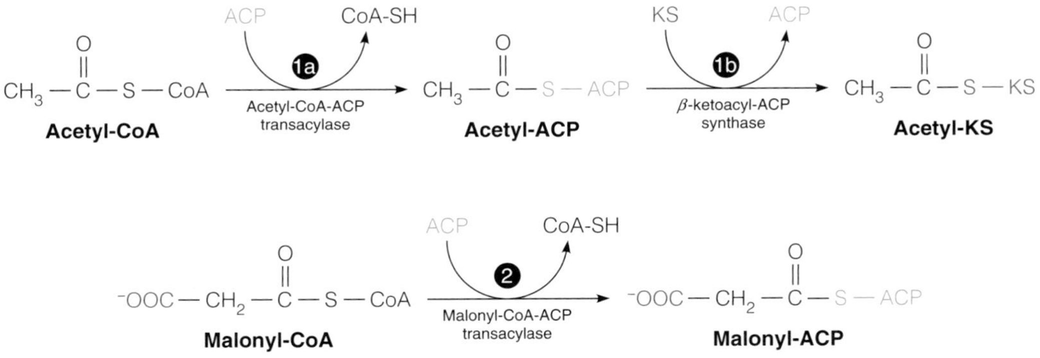

FIGURE 17.29

Malonyl/acetyl-CoA-ACP transacylase (MAT) loads fatty acid synthase with substrates. The acyl groups (acetyl or malonyl) are transferred from the SH group of CoA to the SH group of the phosphopantetheine moiety of acyl carrier protein (ACP). The reactions shown here produce acetyl-KS and malonyl-ACP, which are used in the remaining reactions of the cycle.

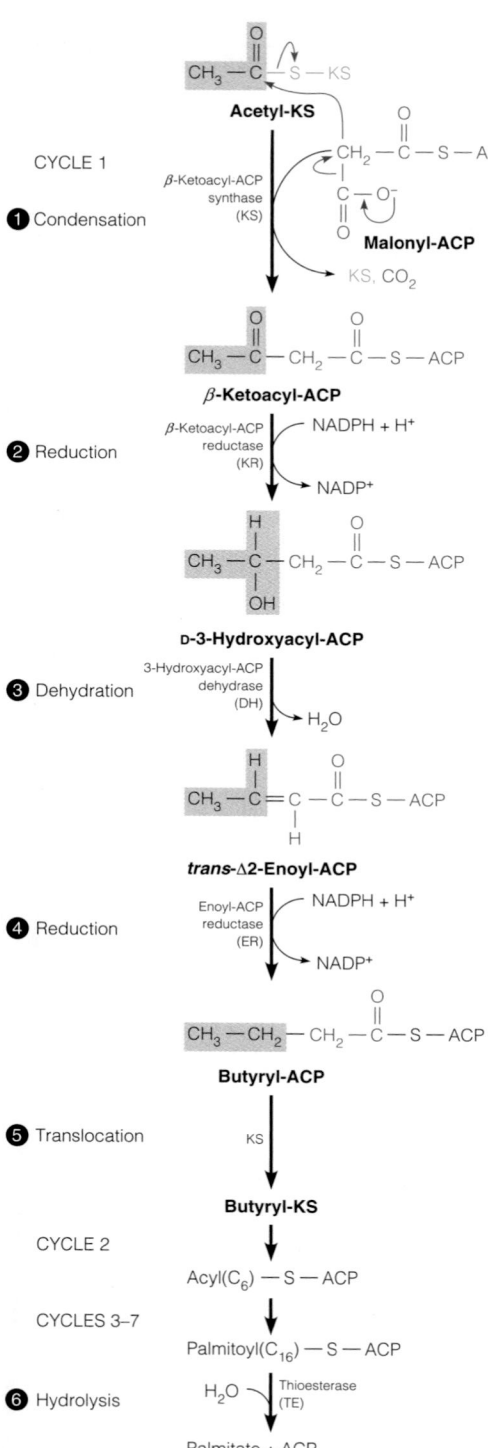

FIGURE 17.30

Synthesis of palmitate, starting with malonyl-ACP and acetyl-KS. The first cycle of four reactions generates butyryl-ACP. Following translocation from ACP, butyryl-KS reacts with a second molecule of malonyl-ACP, leading to a second cycle of two-carbon addition. A total of seven such cycles generates palmitoyl-ACP. Hydrolysis of this product releases palmitate.

The same pattern continues until the product of the seventh cycle, palmitoyl-ACP, undergoes hydrolysis to yield palmitate and free ACP. This final step is catalyzed by **thioesterase (TE)** (reaction 6).

What is the molecular logic of using malonyl-ACP as a donor of an acetyl unit? The condensation of two activated acetyl units is normally quite endergonic. A comparable reaction in reverse—namely the thiolytic cleavage of acetoacetyl-CoA—is strongly exergonic. However, the carboxyl group of malonyl-ACP is a good leaving group because the β-carbonyl group can act as an electron acceptor during the decarboxylation. This decarboxylation makes the condensation reaction exergonic. Ultimately, it is ATP hydrolysis that drives this otherwise endergonic condensation reaction because ATP participated in the original synthesis of malonyl-CoA from acetyl-CoA (page 737). This condensation process explains the early observation that bicarbonate is not incorporated into the final product. Rather, all of the carbons in fatty acids come from acetate.

Like most biosynthetic pathways, this one requires both *energy* (as ATP) and *reducing equivalents* (as NADPH). The quantitative requirements can be seen from the stoichiometry of the complete seven-cycle process:

$$\text{Acetyl-CoA} + 7\,\text{malonyl-CoA} + 14\,\text{NADPH} + 14\text{H}^+ \longrightarrow \text{palmitate} + 7\text{CO}_2$$
$$+ \ 14\text{NADP}^+ + 8\text{CoA-SH} + 6\text{H}_2\text{O}$$

Although one H_2O is released in each cycle, a net of only six H_2O is produced in the complete process because one H_2O is used to hydrolyze the thioester linkage to release free palmitate (Figure 17.30, reaction 6). To see the ATP requirement, we must consider the synthesis of the 7 moles of malonyl-CoA:

$$7\,\text{Acetyl-CoA} + 7\text{CO}_2 + 7\text{ATP} \longrightarrow 7\,\text{malonyl-CoA} + 7\text{ADP} + 7\text{P}_i + 7\text{H}^+$$

Hence, the following equation describes the overall process.

$$8\,\text{Acetyl-CoA} + 7\text{ATP} + 14\text{NADPH} + 7\text{H}^+ \longrightarrow \text{palmitate} + 14\text{NADP}^+$$
$$+ \ 8\text{CoA-SH} + 7\text{ADP} + 7\text{P}_i + 6\text{H}_2\text{O}$$

Multifunctional Proteins in Fatty Acid Synthesis

The fatty acid synthesis reaction pathway is essentially identical in all known organisms, but the enzymology involved is startlingly variable. Fatty acid synthesis was first worked out in *E. coli*, and it was discovered that all the reactions are catalyzed by seven distinct monofunctional enzymes, which can be separately purified. This same organization is found in all bacteria and plants, and is referred to as **type II fatty acid synthesis**. The earliest biochemical studies of fatty acid synthesis in animals and yeast revealed seven or eight individual polypeptide species in purified fatty acid synthase (FAS) preparations. It was eventually demonstrated that the eukaryotic enzymes were exquisitely sensitive to proteolysis, and trace amounts of proteases in FAS preparations resulted in proteolysis of the native polypeptide chains into smaller species. Once this was recognized, steps were taken to avoid proteolysis, and FAS could be purified intact from animals and yeast. These studies revealed that in animals and in lower eukaryotes all of the FAS activities are associated in a multifunctional enzyme referred to as a **megasynthase**, or **type I fatty acid synthase**. Two basic megasynthase architectures have evolved. The animal enzyme is a homodimer of 273,000 Da subunits (0.54 MDa total). Each subunit of the dimeric mammalian FAS is folded into seven distinct domains, and each domain carries out a specific function in the reaction sequence. Six of these domains carry active sites for the chemical steps and the seventh is the ACP domain with its phosphopantetheine moiety—truly, a *multifunctional* protein. The order of these domains along the polypeptide chain

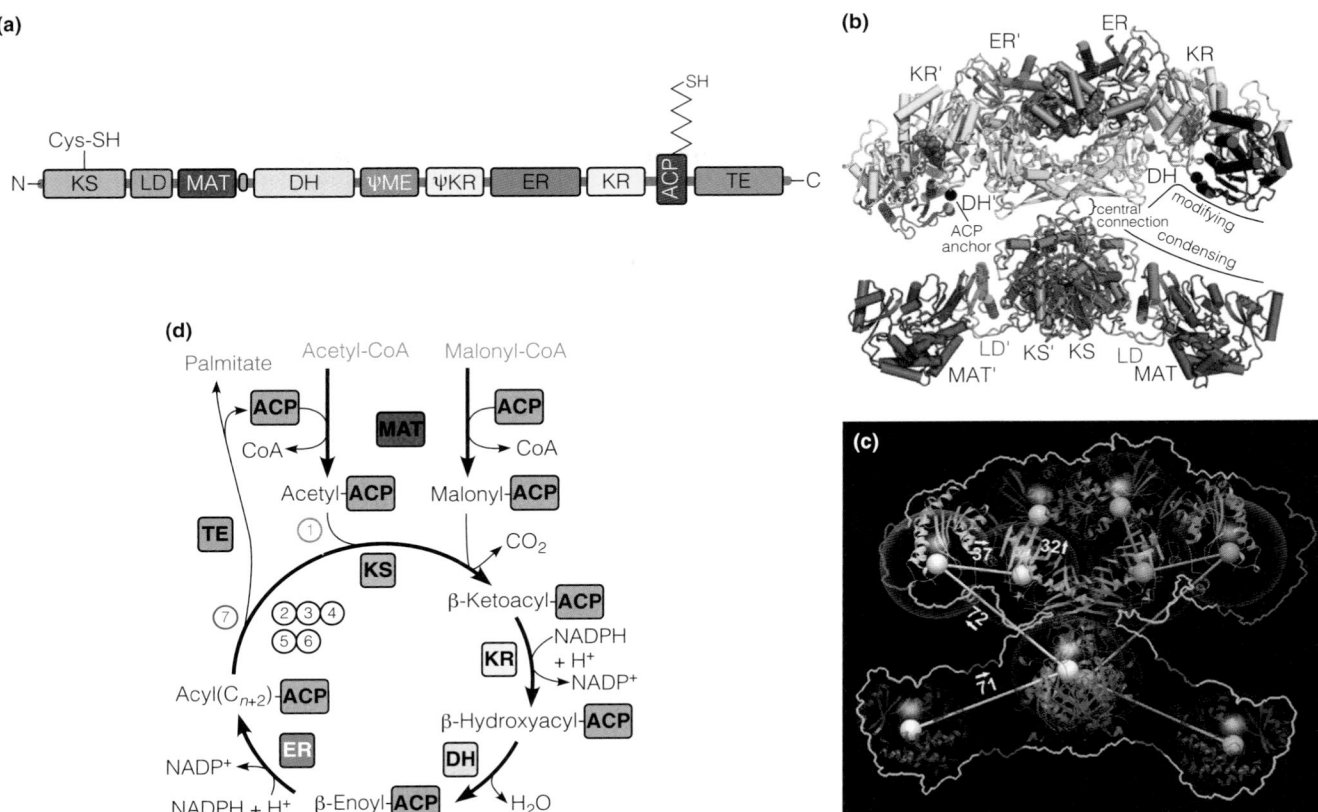

FIGURE 17.31

Structure and swinging arm mechanism in the mammalian fatty acid synthase complex. **(a)** Arrangement of domains on the 273,000-dalton polypeptide chain. KS = ketoacyl-ACP synthase; MAT = malonyl/acetyl transacylase; DH = hydroxyacyl-ACP dehydrase; ER = enoyl-ACP reductase; KR = ketoacyl-ACP reductase; ACP = acyl carrier protein; TE = thioesterase. The KR domain is actually composed of two discontinuous regions (ψKR + KR) split by the ER domain. LD = linker domain; ψME = noncatalytic structural domain. The cysteine thiol and phosphopantetheine thiol groups are shown attached to KS and ACP, respectively. **(b)** A cartoon representation of the X-ray structure of the porcine FAS homodimer (PDB ID 2VZ9). Each domain is abbreviated and colored as in panel a; domains of the second subunit are indicated by a prime ('). The two subunits are approximately C-shaped and intertwine at the central connection of the X-structure. The condensing and modifying domains are located in the lower and upper arms, respectively. The ACP anchor site is indicated by a black sphere, and NADP$^+$ bound at the KR active sites is shown in blue. **(c)** The ACP-phosphopantetheine swinging arm serves as a mechanism for bringing acyl groups into contact with all of the active sites (indicated by solid white and blue spheres). The hollow spheres that surround each active site show the length of the phosphopantetheine arm, indicating how close ACP must approach the individual domains during the catalytic cycle. The active sites are connected in order of the reaction sequence. The distances between the active sites, in Å, are indicated for the left subunit. **(d)** Summary of catalytic cycle. The first cycle ① begins with transfer of the acetyl group from acetyl-CoA to the ACP swinging arm. The acetyl group is then transferred to the cysteine thiol group on KS, and a malonyl group is transferred to the ACP swinging arm. KS catalyzes the condensation, and KR, DH, and ER catalyze the modifying steps. At the end of the first six cycles ②–⑥, KS translocates the reduced acyl group from ACP to its own Cys–SH to begin another cycle; after the last cycle ⑦, the palmitoyl group is hydrolyzed from ACP by the TE activity.

From *Science* 311:1258–1262, T. Maier, S. Jenni, and N. Ban, Architecture of mammalian fatty acid synthase at 4.5 Å resolution; and *Science* 321:1315–1322, T. Maier, M. Leibundgut, and N. Ban, The crystal structure of a mammalian fatty acid synthase. © 2006 and 2008. Modified with permission from AAAS and Nenad Ban.

is illustrated in Figure 17.31a. Note the locations of the ACP and KS domains—at opposite ends of the polypeptide chain. The sulfhydryl groups of these two domains must come into close proximity for the condensation step (reaction 1 in Figure 17.30). The X-ray structure of the porcine FAS (Figure 17.31b) reveals how this might occur. The two subunits form an intertwined X-structure on which two fatty acids can be synthesized simultaneously. The domains that catalyze the condensing reaction are found in the lower arms of the structure, whereas the domains catalyzing the next three reactions (the modifying reactions) are found in the upper arms. The ACP and thioesterase (TE) domains are not visible in this structure, but their attachment site (C-terminus of the KR domain) is known. During the reaction cycle, the phosphopantetheine moiety attached to the ACP domain serves as a swinging arm for bringing acyl groups into contact with all of the active sites of the complex. The order of the reaction sequence as the ACP

Malonyl-CoA represents an activated source of two-carbon fragments for fatty acid biosynthesis, with the loss of CO_2 driving C—C bond formation.

In eukaryotes, fatty acid synthesis is carried out by a megasynthase, an organized multienzyme complex that contains multifunctional proteins.

domain moves between the active sites is shown in Figure 17.31c. Figure 17.31d summarizes the catalytic cycle and the step at which each domain participates. The cycle begins with transfer of the acetyl group from acetyl-CoA to the ACP swinging arm. The acetyl group is then transferred to the cysteine thiol group on KS, and a malonyl group is transferred to the ACP swinging arm. KS catalyzes the condensation, and KR, DH, and ER catalyze the modifying steps. At the end of the first six cycles, KS translocates the reduced acyl group from ACP to its own Cys-SH to begin another cycle; after the last (seventh) cycle, the palmitoyl group is hydrolyzed from ACP by the TE activity.

Lower eukaryotes (yeast and fungi) possess a type I fatty acid synthase composed of two multifunctional polypeptide chains. This megasynthase has a mass of about 2.6 million daltons and is composed of six α subunits and six β subunits. The α subunit contains the ACP domain, the condensing enzyme (KS), and the β-ketothioester reductase (KR). The β subunit contains the remaining three activities (MAT, DH, and ER). The yeast complex lacks the thioesterase (TE) activity because the final product of fatty acid synthase activity in yeast is not palmitate but palmitoyl-CoA.

In 1988, it was discovered that mitochondria can also synthesize fatty acids. When the fungus *Neurospora crassa* was given [^{14}C]pantothenic acid, a small (125 amino acid) mitochondrial protein became labeled. This protein turned out to be a homolog of the bacterial acyl carrier protein (ACP), and this suggested that mitochondria might possess a fatty acid synthesis system. Subsequent studies have revealed that mitochondria from almost all species contain a complete type II FAS system of soluble monofunctional enzymes. One of the roles of this mitochondrial pathway is synthesis of **octanoic acid**, the precursor for the lipoic acid cofactor essential for pyruvate dehydrogenase and α-ketoglutarate dehydrogenase complexes of the citric acid cycle (Chapter 14). The mitochondrial system can synthesize fatty acids up to 14 carbons in length (myristic acid), but the physiological function for these longer fatty acids is not yet known.

Although all of these fatty acid synthase systems use essentially the same chemistry to build fatty acids, differences in their enzymology provide the possibility that specific inhibitors might be developed as antimicrobial drugs. For example, triclosan is widely used as an antibacterial agent. Triclosan is a potent inhibitor of bacterial enoyl-ACP reductase (ER), closely mimicking the structure of the enolate intermediate of the substrate (Figure 17.32). Unfortunately, the widespread use of triclosan in toothpastes, underarm deodorants, antiseptic soaps, plastic kitchenware, and many other household products has led to the development of triclosan resistance in several bacterial species, including *E. coli, Staphylococcus aureus, Pseudomonas aeruginosa,* and *Salmonella enterica.* Drug resistance is an increasingly common problem associated with the overuse of antibiotics.

It has long been known that intermediates between acetyl-CoA and palmitate do not accumulate in cells that are synthesizing fatty acids—the process is extremely fast. The basis for this is clear because all the intermediates are covalently bound to the ACP domain of a multifunctional protein. This arrangement ensures that substrates need not seek catalytic sites by simple diffusion. Indeed, starting from acetyl-CoA and malonyl-CoA, eukaryotic megasynthases can synthesize palmitate in less than a second. Moreover, because the genetic apparatus needs to regulate the synthesis of only one or two polypeptide chains instead of seven, this type of organization is attractive from the standpoint of coordinating the activities involved. Multifunctional proteins have now been described in most major metabolic processes, although few other enzymes carry as many as four activities on one polypeptide chain.

Transport of Acetyl Units and Reducing Equivalents into the Cytosol

Because acetyl-CoA is generated in the mitochondrial matrix, it must be transported to the cytosol for use in fatty acid synthesis. Like longer-chain acyl-CoAs,

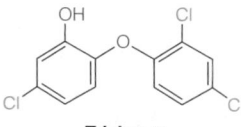

Triclosan

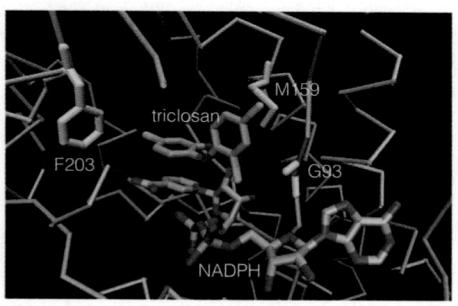

FIGURE 17.32

Triclosan is a specific inhibitor of bacterial enoyl-ACP reductase (ER). The X-ray structure of *E. coli* ER complexed with triclosan and NADPH (PDB ID 1D8A) shows that the phenol ring of triclosan stacks on top of the nicotinamide ring of NADPH. The three residues highlighted in yellow (phenylalanine-203, methionine-159, and glycine-93) make hydrophobic and van der Waals contacts with triclosan. Single amino acid substitutions of any of these three residues confers triclosan resistance to *E. coli.*

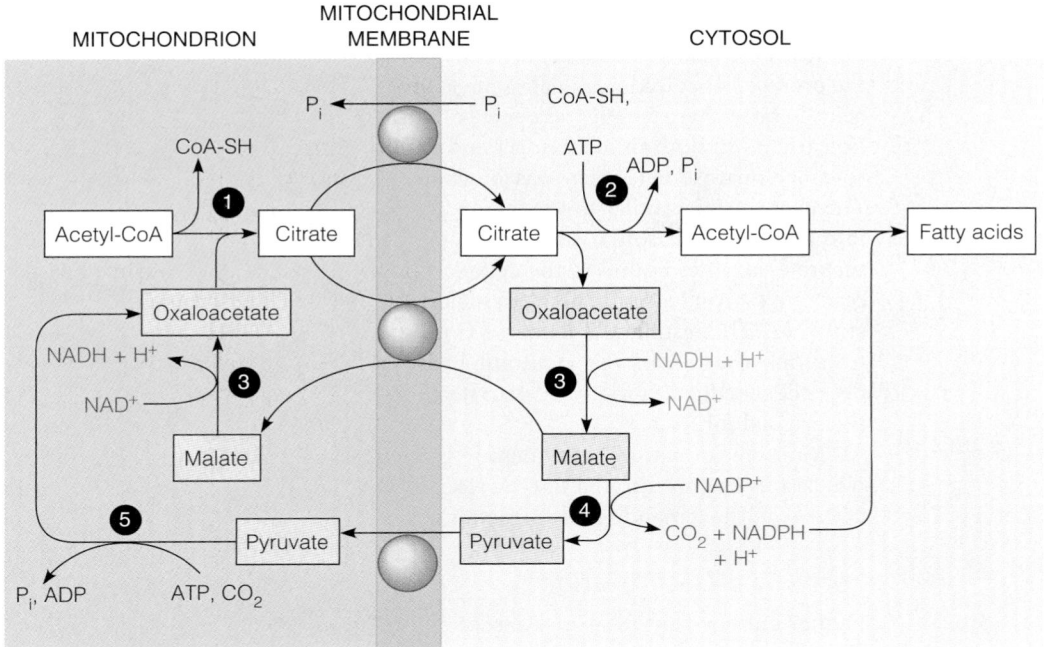

acetyl-CoA cannot penetrate the inner membrane. A shuttle system is used, which is interesting both because it provides a control mechanism for fatty acid synthesis and because it generates much of the NADPH needed for the process. The shuttle involves citrate, which is formed in mitochondria from acetyl-CoA and oxaloacetate in the first step of the citric acid cycle (Figure 17.33, step 1). When citrate is being generated in excess of the amount needed for oxidation in the citric acid cycle, it is transported through the mitochondrial membrane to the cytosol. There it is acted on by **citrate lyase** to regenerate acetyl-CoA and oxaloacetate at the expense of one ATP (step 2):

$$Citrate + ATP + CoA\text{-}SH \rightarrow acetyl\text{-}CoA + ADP + P_i + oxaloacetate$$

Oxaloacetate cannot directly return to the mitochondrial matrix because the inner membrane lacks a transporter for this compound. First it is reduced by a cytosolic malate dehydrogenase to malate (step 3), and some malate is oxidatively decarboxylated by the malic enzyme to give pyruvate (step 4; the malic enzyme is working here opposite to the direction shown in Chapter 14, page 618). Some of the malate formed returns to the mitochondrion in exchange for citrate, however.

$$Oxaloacetate + NADH + H^+ \longrightarrow malate + NAD^+$$

$$Malate + NADP^+ + H_2O \longrightarrow pyruvate + HCO_3^- + NADPH + H^+$$

The resultant pyruvate is transported back into mitochondria, where it is reconverted to oxaloacetate by pyruvate carboxylase (step 5; see Chapter 14, page 618).

$$Pyruvate + HCO_3^- + ATP \longrightarrow oxaloacetate + ADP + P_i + 2H^+$$

The net reaction catalyzed by these three enzymes is as follows:

$$NADP^+ + NADH + ATP + H_2O \longrightarrow NADPH + NAD^+ + ADP + P_i + 2H^+$$

For each mole of malate remaining in the cytosol, 1 mole of NADPH is generated. Much of the remainder of the 14 moles of NADPH required to synthesize 1 mole of palmitate is generated in the cytosol via the pentose phosphate pathway.

FIGURE 17.33

Transport of acetyl units and reducing equivalents used in fatty acid synthesis. This diagram shows the shuttle mechanism for transferring acetyl units and reducing equivalents from mitochondria to the cytosol, for use in fatty acid synthesis. Citrate must be exchanged for a carrier as it moves out of the mitochondrion. Some citrate is evidently exchanged for orthophosphate and some for malate. The malate that is not exchanged generates some of the NADPH for fatty acid synthesis, through action of the malic enzyme. Purple circles represent transport systems located in the mitochondrial membrane. 1 = citrate synthase; 2 = citrate lyase; 3 = malate dehydrogenase; 4 = malic enzyme; 5 = pyruvate carboxylase.

Citrate serves as a carrier of two-carbon fragments from mitochondria to cytosol for fatty acid biosynthesis.

Elongation of Fatty Acid Chains

Because fatty acid synthase action leads primarily to palmitate, we must consider the processes that lead from palmitate to give the variations observed among fatty acids in both chain length and degree of unsaturation. In eukaryotic cells, elongation occurs in both mitochondria and endoplasmic reticulum (ER). The latter, so-called microsomal, system has far greater activity and is the one described here. The chemistry is similar to the fatty acid synthase sequence that leads to palmitate, but the microsomal elongation system involves acyl-CoA derivatives and separate enzymes bound to the cytoplasmic face of the ER membrane. The first reaction is a condensation between malonyl-CoA and a long-chain fatty acyl-CoA substrate. The resultant β-ketoacyl-CoA undergoes NADPH-dependent reduction, dehydration of the resultant hydroxyacyl-CoA, and another NADPH-dependent reduction to give a saturated fatty acyl-CoA two carbons longer than the original substrate.

Multiple condensing enzymes are present in endoplasmic reticulum, one of which acts on unsaturated fatty acyl-CoAs. Apparently, a single set of enzymes carries out the remaining three reactions.

Fatty Acid Desaturation

Higher animals and fungi possess several endoplasmic reticulum–bound acyl-CoA desaturases that catalyze the production of mono- and polyunsaturated fatty acids. The first cis-double bond is always introduced between carbons 9 and 10, counting from the carboxyl end of the fatty acid. Additional double bonds are introduced toward the carboxyl end at three-carbon intervals so that the double bonds are separated by a methylene group, and therefore not conjugated. The most common monounsaturated fatty acids in animal lipids are oleic acid, an 18:1cΔ9 acid, and palmitoleic acid, a 16:1cΔ9 compound (see Table 10.1, page 361). These compounds are synthesized from stearate and palmitate, respectively, by a microsomal Δ9 desaturating system called **stearoyl-CoA desaturase**. The overall reaction for stearoyl-CoA desaturation is as follows:

$$\text{Stearoyl-CoA (18:0)} + \text{NAD(P)H} + \text{H}^+ + \text{O}_2 \longrightarrow \text{oleyl-CoA (18:1c}\Delta\text{9)} + \text{NAD(P)}^+ + 2\text{H}_2\text{0}$$

Note that both substrates, the fatty acid and the NAD(P)H, undergo two-electron oxidations in this reaction. The overall electron transfer in this reaction involves cytochrome b_5 and another enzyme, the flavin-dependent **cytochrome b_5 reductase** (Figure 17.34). The cytochrome b_5 component is found as an N-terminal domain of the desaturase. Mammalian endoplasmic reticulum also contains Δ5 and Δ6 desaturases which function similarly to the Δ9 desaturating system.

Whereas plants possess Δ12 and Δ15 desaturases (Figure 17.35), animals are unable to introduce double bonds beyond Δ9 (i.e., between C10 and the methyl carbon) in the fatty acid chain. Hence, they cannot synthesize either linoleic acid (18:2cΔ9,12) or linolenic acid (18:3cΔ9,12,15). These are called **essential fatty acids** because they are required lipid components that must be provided in the diet from plants.[*] After ingestion in mammals, they are in turn substrates for further desaturation and elongation reactions, as summarized in Figure 17.35. Linolenic acid is the precursor for the omega-3 PUFAs EPA (20:5cΔ5,8,11,14,17) and DHA (22:6cΔ4,7,10,13,16,19) (see margin, page 721). Dietary linoleic acid is the precursor for **arachidonic acid** (20:4cΔ5,8,11,14), which serves as precursor to a class of compounds called the eicosanoids. As discussed in Chapter 19, eicosanoids include two important classes of metabolic regulators, the prostaglandins and the thromboxanes.

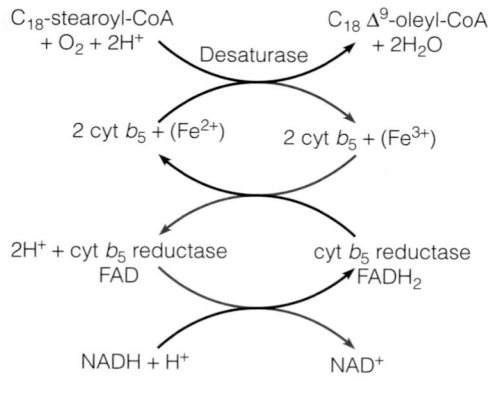

FIGURE 17.34

Fatty acid desaturation system. The black arrows indicate the path of electron flow as the two substrates are oxidized. Δ5 and Δ6 desaturases use the same mechanism.

[*]Fish are also good sources of linoleic acid and linolenic acid, but they cannot synthesize these essential fatty acids—like all animals, fish must obtain them from their diet.

Mammalian cells require specific amounts of polyunsaturated fatty acids for these signaling functions and to maintain proper membrane fluidity. Dietary PUFAs repress the expression of all three desaturase systems so that de novo production is turned off if adequate levels of these unsaturated fatty acids can be obtained from the diet.

Control of Fatty Acid Synthesis

To a large extent, fatty acid biosynthesis is controlled by hormonal mechanisms. Much of the fatty acid synthesis in animals takes place in adipose tissue, where fat is being stored for release and transport to other tissues on demand, to help meet their energy needs. As extracellular messengers, hormones are well suited to these interorgan regulatory roles.

Figure 17.36 summarizes the major effects in regulation of fatty acid synthesis in animal cells. Insulin acts in several ways to stimulate synthesis and storage of fatty acids in adipose tissue and liver. One of its effects is to increase glucose entry into cells by stimulating translocation of the glucose transporter to the plasma membrane. This effect increases flux through glycolysis and the pyruvate dehydrogenase reaction, which provides acetyl-CoA for fatty acid synthesis. Insulin also activates the pyruvate dehydrogenase complex, by stimulating its dephosphorylation to the active form (see Chapter 14, page 613).

Another site for regulation (not shown in the figure) is the transfer of acetyl units from the mitochondrial matrix to the cytosol, where fatty acid synthesis occurs. Citrate lyase, a key enzyme in this process (page 743), is activated by phosphorylation. Insulin and other growth factors stimulate this activation through the PI3K/Akt pathway (more about this in Chapter 18).

The first enzyme whose action is committed to fatty acid synthesis is acetyl-CoA carboxylase (ACC, page 737). Activities of this enzyme are quite low in starved animals, reflecting its regulation by allosteric and covalent modification mechanisms. Phosphorylation of acetyl-CoA carboxylase causes its inactivation. Two protein kinases are known to phosphorylate ACC: AMP-activated protein kinase (AMPK) and cyclic AMP–dependent protein kinase (PKA). The AMPK system acts as a sensor of cellular energy status—it is activated by increases in the cellular AMP:ATP ratio (more about this in Chapter 18). Thus, under conditions of low energy charge, such as would occur during fasting or starvation, activated AMPK switches off fatty acid synthesis by inhibiting ACC. ACC activity is also under hormonal control. Glucagon and epinephrine activate PKA, which phosphorylates and inhibits the enzyme.

Citrate and long-chain fatty acyl-CoAs are allosteric modulators of acetyl-CoA carboxylase. As noted earlier, ACC must undergo a reversible polymerization in order to be active. Long-chain fatty acyl-CoAs at low levels prevent polymerization, thereby inactivating the enzyme and providing feedback inhibition of the pathway. Citrate is an allosteric activator of ACC, stimulating its polymerization. As the carrier of acetyl units from mitochondria (see Figure 17.25, page 736), cytoplasmic citrate levels rise when mitochondrial acetyl-CoA and ATP concentrations increase. Thus the same molecule (citrate) serves as the precursor of acetyl-CoA for fatty acid biosynthesis and as an activator of the rate-limiting step for this process. The levels of fatty acyl-CoAs are lowered by insulin, another mechanism by which insulin stimulates fatty acid synthesis.

Fatty acid degradation is also regulated. Recall that malonyl-CoA, the product of the acetyl-CoA carboxylase reaction, inhibits fatty acid oxidation by blocking carnitine acyltransferase I, the first committed intermediate in fatty acid synthesis (page 726). This provides an elegant mechanism to coordinately regulate the synthesis and degradation of fatty acids.

Finally, there is evidence that fatty acid synthesis is controlled by the availability of reducing equivalents. Recall that NADPH comes from both the transport of citrate out of mitochondria (see page 743) and the pentose phosphate pathway

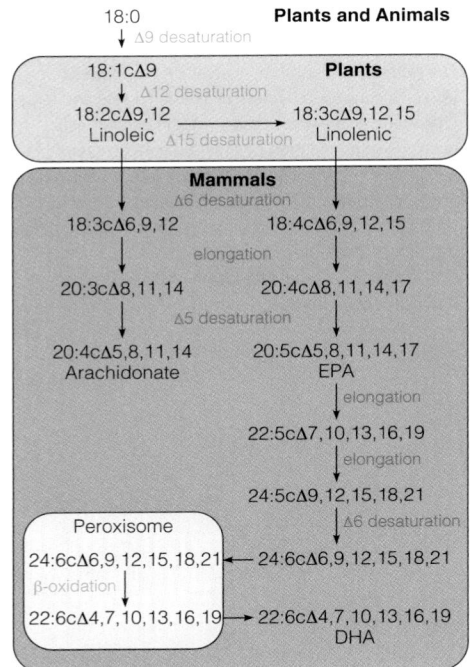

FIGURE 17.35

Pathway for synthesis of polyunsaturated fatty acids (PUFAs) in plants and animals. The essential fatty acids, linoleic and linolenic, are obtained in the diet. Mammals alternate Δ5 or Δ6 desaturation with two-carbon elongation cycles to produce arachidonic acid, eicosapentaenoic acid (EPA), and other PUFAs. Synthesis of docosahexaenoic acid (DHA) involves chain shortening of the 24:6 intermediate by one cycle of peroxisomal β-oxidation to give the 22:6 DHA. Although not shown, the mammalian desaturation and elongation reactions all operate on acyl-CoA derivatives, whereas the plant enzymes are specific for acyl-ACP substrates.

Adapted from *Trends in Biochemical Sciences* 27:467–473, J. G. Wallis, J. L. Watts, and J. Browse, Polyunsaturated fatty acid synthesis: What will they think of next? © 2002, with permission from Elsevier.

Acetyl-CoA carboxylase is the committed enzyme and major control point for fatty acid synthesis.

FIGURE 17.36

Regulation of fatty acid synthesis in animal cells. The rate-limiting enzyme, acetyl-CoA carboxylase (ACC), is controlled by both allosteric (citrate and long-chain fatty acids) and covalent modification mechanisms. Phosphorylation by AMP-activated protein kinase (AMPK) or cyclic AMP–dependent protein kinase (PKA) inactivates ACC. Insulin stimulates fatty acid synthesis by increasing glucose uptake and increasing flux through pyruvate dehydrogenase to produce acetyl-CoA. The dephosphorylated form of PDH is the enzymatically active form.

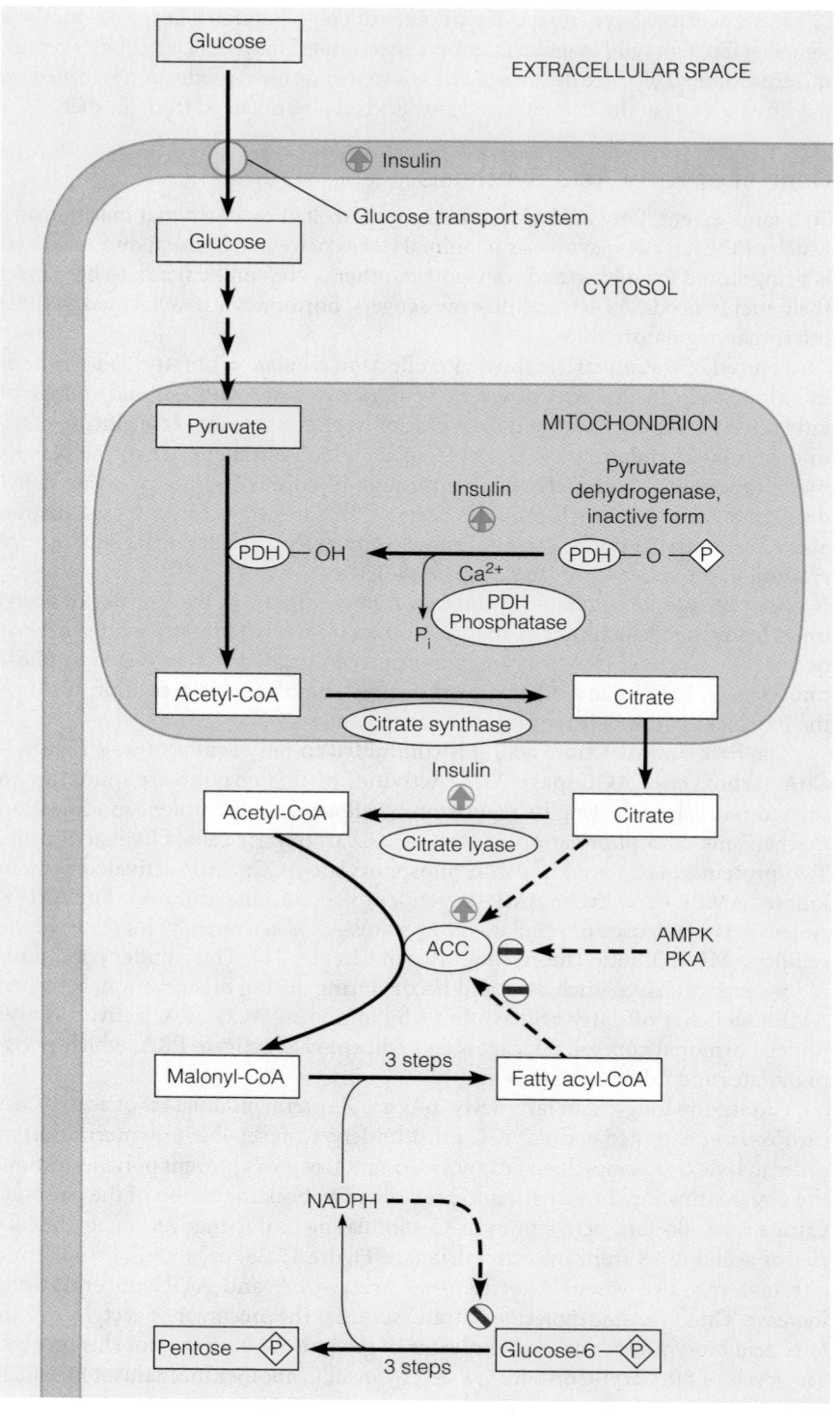

(see Chapter 13). The pentose phosphate pathway in turn is controlled through inhibition by NADPH of glucose-6-phosphate dehydrogenase and 6-phosphoglu-conate dehydrogenase. Typically, about 60% of the NADPH for fatty acid synthesis comes from the pentose phosphate pathway. When acetate is added to an experimental tissue preparation, that proportion can rise to 80%, indicating that flux through the pentose phosphate pathway can rise when the demand for NADPH increases.

Variant Fatty Acid Synthesis Pathways
That Lead to Antibiotics

We digress from lipid metabolism to introduce a related series of pathways in bacteria and fungi, involved in the biosynthesis of a class of antibiotics called **polyketides**. Erythromycin, which is synthesized by the bacterium *Saccharopolyspora erythraea*, is an example of this class. Another example, shown in the margin, is oxytetracycline, produced by *Streptomyces rimosus*. These polyketide antibiotics are potent inhibitors of bacterial protein synthesis (Chapter 28). Other polyketides, such as lovastatin and simvastatin (Chapter 19, page 803) have found clinical use as cholesterol-lowering drugs. Polyketides are synthesized in assembly-line fashion by giant enzyme megasynthases that consist of individual modules for rounds of carbon addition, with each module closely resembling the process whereby two carbons are added in a cycle of the fatty acid synthesis pathway. However, one or more of the modifying activities of fatty acid synthesis are missing in some of the modules, a factor that leads to great structural diversity among the products. For example, the synthesis of 6-deoxyerythronolide B, the aglycone precursor of erythromycin A, is shown in Figure 17.37a. 6-Deoxyerythronolide B synthase (DEBS) is composed of three large subunits that carry 28 active sites. Each subunit contains two modules, with each module responsible for one round of chain extension and modification. Figure 17.37b shows a model of the DEBS3 subunit, containing modules

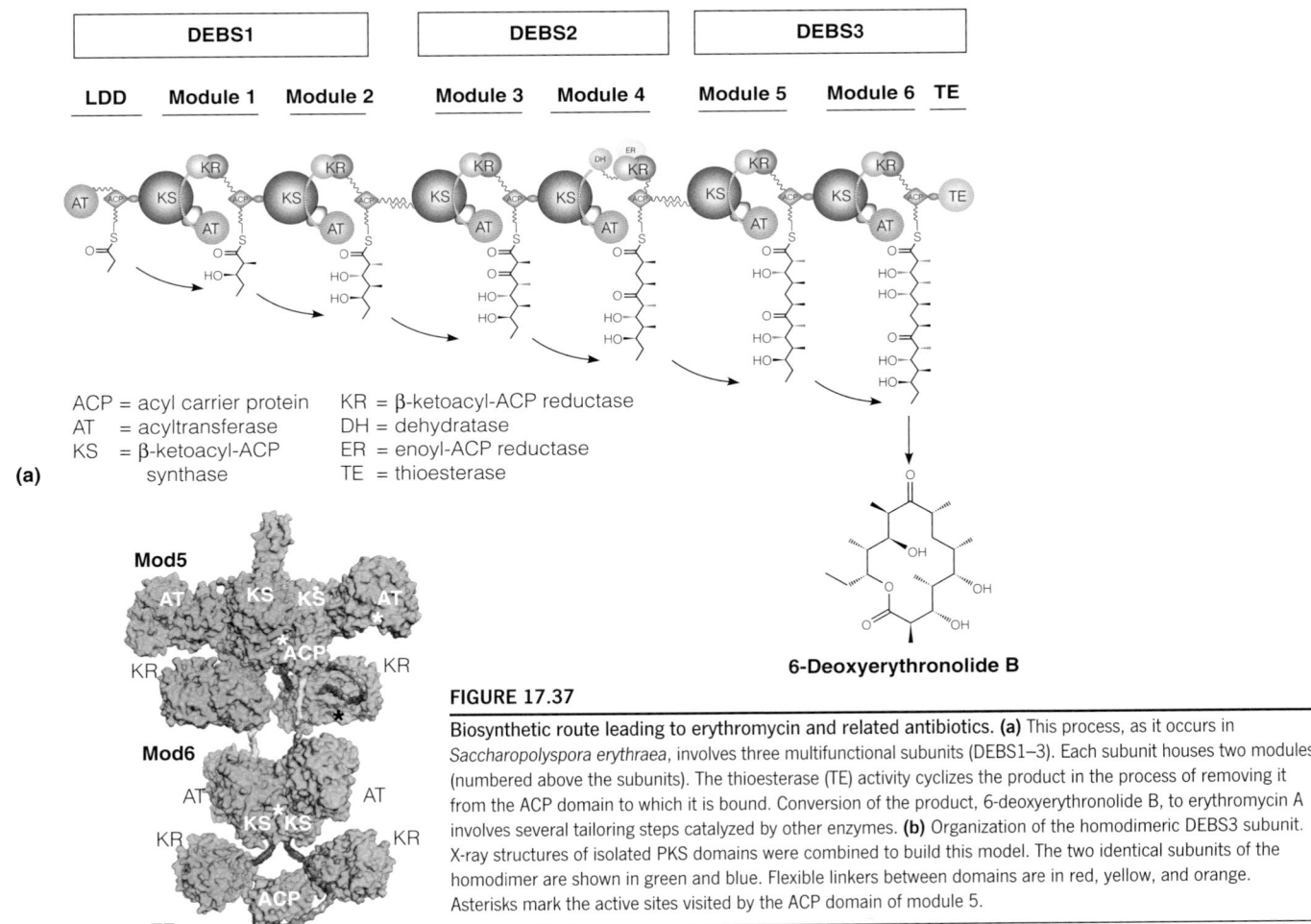

6-Deoxyerythronolide B

FIGURE 17.37

Biosynthetic route leading to erythromycin and related antibiotics. **(a)** This process, as it occurs in *Saccharopolyspora erythraea*, involves three multifunctional subunits (DEBS1–3). Each subunit houses two modules (numbered above the subunits). The thioesterase (TE) activity cyclizes the product in the process of removing it from the ACP domain to which it is bound. Conversion of the product, 6-deoxyerythronolide B, to erythromycin A involves several tailoring steps catalyzed by other enzymes. **(b)** Organization of the homodimeric DEBS3 subunit. X-ray structures of isolated PKS domains were combined to build this model. The two identical subunits of the homodimer are shown in green and blue. Flexible linkers between domains are in red, yellow, and orange. Asterisks mark the active sites visited by the ACP domain of module 5.

(a) Modified from *The Annual Review of Biochemistry* 76:195–221, C. Khosla, Y. Tang, A. Y. Chen, N. A. Schnarr, and D. E. Cane, Structure and mechanism of the 6-deoxyerythronolide B synthase. © 2007 Annual Reviews; (b) Courtesy of A. Keatinge-Clay, The University of Texas at Austin.

Erythromycin A

Oxytetracycline

5 and 6 plus the terminal thioesterase domain. Beginning with propionyl-CoA (instead of acetyl-CoA) and malonyl-CoA, this pathway involves seven two-carbon addition cycles, just like the synthesis of palmitoyl-ACP. However, only one of the seven modules contains a dehydratase domain, and two of the modules lack the ketoacyl reductase domain. Thus, the product, unlike fatty acids, contains both keto and hydroxyl oxygen atoms. Much current excitement involves protein engineering to modify and rearrange the individual modules, to further diversify the pathways involved, possibly leading to novel and useful antibiotics.

Biosynthesis of Triacylglycerols

Fatty acyl-CoAs and glycerol-3-phosphate serve as the major precursors to triacylglycerols and, as we shall see in Chapter 19, to the glycerophospholipids used to build membranes. In many cases, the glycerol backbone is the limiting substrate. Glycerol-3-phosphate is derived either from the reduction of the glycolytic intermediate dihydroxyacetone phosphate (DHAP), catalyzed by **glycerol-3-phosphate dehydrogenase**, or from the ATP-dependent phosphorylation of glycerol by **glycerol kinase** (see margin, page 749). The pathway involving DHAP predominates in adipose tissue because adipocytes lack glycerol kinase. During fasting or starvation conditions when glycolysis is reduced, adipocytes, hepatocytes, and cancer cells can synthesize glycerol-3-phosphate from pyruvate. This pathway, termed **glyceroneogenesis**, is an abbreviated version of gluconeogenesis (Chapter 13) that uses pyruvate carboxylase and phosphoenolpyruvate carboxykinase (PEPCK) to produce DHAP (Figure 17.38), which is then reduced to glycerol-3-phosphate by glycerol-3-phosphate dehydrogenase.

FIGURE 17.38

Glycerolipid/free fatty acid cycle and glyceroneogenesis. Mammals hydrolyze and resynthesize triacylglycerols (TG) in a glycerolipid/free fatty acid (GL/FFA) cycle. FFA are released from TG by lipases acting on lipid droplets. ATGL, adipose triglyceride lipase; HSL, hormone-sensitive lipase; MGL, monoacylglycerol lipase (see Figure 17.13). Some of the FFAs are released into the blood for transport and oxidation, but ~75% are reesterified back to TG. ATP hydrolysis in this futile cycle is shown in red. The glycerol backbone is produced via glyceroneogenesis, involving reactions catalyzed by pyruvate carboxylase and phosphoenolpyruvate carboxykinase (PEPCK), and reversal of glycolytic steps to give dihydroxyacetone phosphate (DHAP). DHAP is reduced to glycerol-3-phosphate, as shown in the margin on page 749. AMP-activated protein kinase (AMPK)-dependent phosphorylation of GPAT and HSL inhibits these steps of the cycle. The lysophosphatidic acid (LPA), phosphatidic acid (PA), and sn-1,2-diacylglycerol (DAG) intermediates in the TG resynthesis pathway are also important lipid signaling molecules.

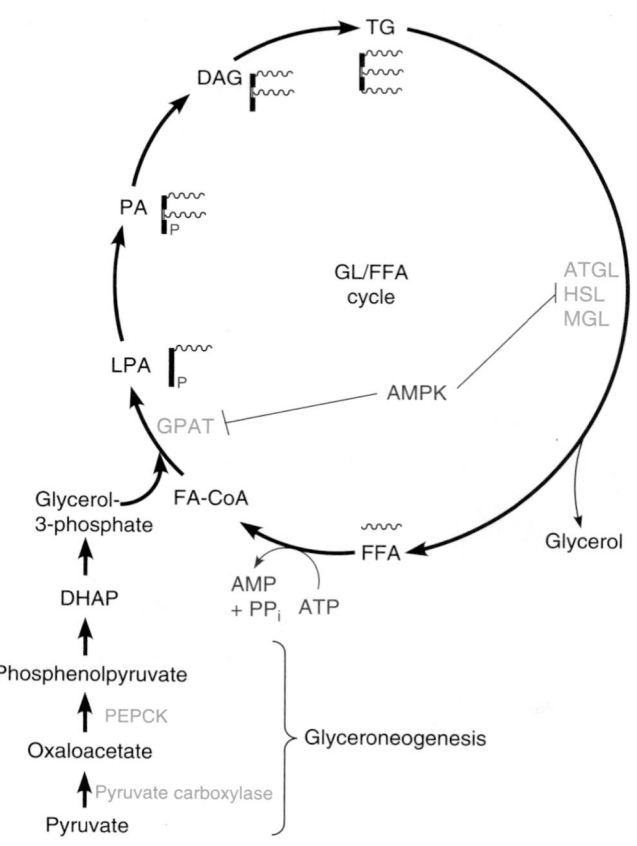

Whatever the source, glycerol-3-phosphate undergoes two successive enzymatic esterifications with fatty acyl-CoAs to yield **diacylglycerol-3-phosphate**. The first esterification is catalyzed by **glycerophosphate acyltransferase** (GPAT), and as the first committed step of the cycle, is an important site of regulation (see below). Diacylglycerol-3-phosphate, also called **phosphatidic acid**, is a precursor both to phospholipids and to triacylglycerols. The pathway to triacylglycerols involves hydrolytic removal of the phosphate, followed by transfer of another fatty acyl moiety from an acyl-CoA.

Careful studies on the release of glycerol and free fatty acids (FFA) during hormone-stimulated lipolysis of adipose tissue triacylglycerols (TG) in both rats and humans showed that nearly 40% of the FFAs were rapidly recycled back to TG. Some of the FFA and glycerol were released into the circulation and taken up by tissues such as heart and skeletal muscle for oxidation, but a substantial portion was taken up again and reesterified to TG. This recycling is now known to occur not only in adipose tissue but also in liver and skeletal muscle, and it is part of the **glycerolipid/free fatty acid cycle (GL/FFA cycle)**. It is estimated that about 75% of the total FFA released during adipose lipolysis is reesterified back to TG either in the adipose tissue itself or in other tissues. GL/FFA cycling can occur within a single cell, or between different organs in the body.

As can be seen from Figure 17.38, the glycerolipid/free fatty acid cycle is a classic example of a futile cycle (Chapter 12), continuously forming and hydrolyzing triacylglycerols at the expense of ATP and releasing the energy as heat. Indeed, one of the important roles of the GL/FFA cycle is heat production (thermogenesis) and body temperature maintenance. Excessive futile cycling can be suppressed by inhibition of two enzymes in the cycle. Both GPAT, catalyzing the first committed step of the cycle, and hormone-sensitive lipase (HSL), catalyzing triacylglycerol hydrolysis, are inhibited by AMPK-dependent phosphorylation in response to low energy charge (more about this in Chapter 18).

It is likely that another important function of this cycle is the production of lipid-signaling molecules, including **sn-1,2-diacylglycerol (DAG)**, **lysophosphatidic acid (LPA)**, and **phosphatidic acid (PA)** (more about these in Chapter 23). Finally, recent work suggests a role for the GL/FFA cycle in diabetes. Insulin resistance, the fundamental defect in type 2 diabetes, is associated with elevated levels of FFAs in the blood. These elevated FFAs lead to increased accumulation of fat inside muscle cells. GL/FFA cycling within the muscle cell leads to a buildup of intracellular DAG, which somehow shuts down the insulin-signaling pathway, making the muscle cells resistant to insulin. One class of antidiabetic drugs called thiazolidinediones (see margin, page 750) increases insulin sensitivity, in part by reducing FFA levels in the blood. The mechanism by which these drugs reduce circulating FFAs involves stimulation of the expression of genes involved in fatty acid uptake, transport, and storage in adipose tissue. Up-regulated genes include lipoprotein lipase, acyl-CoA synthetase, PEPCK, and glycerol kinase. As we will see in Chapter 18, these drugs act through **peroxisome proliferator-activated receptor-γ** (PPAR-γ), a transcription factor essential for fat cell differentiation and maintenance of normal adipocyte function and a master regulator of cellular metabolism.

As we have noted, triacylglycerols represent the major form in which energy can be stored. Normally in an adult animal, synthesis and degradation are balanced so that there is no net change in the total body amount of triacylglycerols. If dietary intake exceeds caloric needs, then proteins, carbohydrate, or fat can each readily provide acetyl-CoA to drive the synthesis of fatty acids and triacylglycerols. On the other hand, fat reserves allow animals to go for rather long times without eating and still maintain adequate energy levels. Such fasting does generate some metabolic stresses, though, as we describe further in Chapter 18.

Thiazolidinediones

Rosiglitazone (Avandia)

Pioglitazone (Actos)

Hibernating animals have adapted remarkably well to cope with such stresses. For example, bears store huge amounts of fat just before beginning a hibernation that may last as long as seven months. During this period all of the bear's energy comes from breakdown of the stored fat. Moreover, the bear excretes so little water that the water released from fat oxidation meets the animal's needs. Similarly, the glycerol released from triacylglycerols provides a source of gluconeogenic precursors.

Biochemical Insights into Obesity

About one-third of Americans are classified as seriously overweight, making obesity one of our most significant public health problems. Until recently, obesity was usually considered simply a consequence of overeating, leading to excessive triacylglycerol deposition in adipocytes. It is becoming clear that biochemical factors lead some individuals to be far more prone to obesity than others. In 1995 the product of the *OB* gene in mice (*OB*, for *obese*) was identified. Mice bearing two defective alleles of the *OB* gene (*ob/ob*) grow to body weights as much as three times normal. The *OB* gene specifies a 16-kDa protein called **leptin**, an **adipokine** synthesized in adipocytes which acts as a hormone, binding to a specific receptor (product of the *DB* gene) in neurons in the **arcuate nucleus** of the hypothalamus, where food intake is controlled. Mice with two defective alleles of the *DB* gene (*db/db*) are obese and develop diabetes. Leptin evidently functions as a "lipostat," sensing the amount of fat stored in the adipocytes. When fat stores are adequate, leptin levels are high, and the signaling system controls feeding behavior to limit fat deposition. Leptin also stimulates mitochondrial uncoupling in adipose tissue by increasing the synthesis of uncoupling proteins (UCP 1; Chapter 15, page 657). During starvation, leptin levels decline, which promotes feeding and fat storage within the adipocyte. *ob/ob* mice, lacking functional leptin, act as if perpetually starved, and their overeating makes them obese; injections of leptin lower their feeding rates and cause them to lose weight dramatically. Obese humans are different from obese mice, however, in that they contain high levels of leptin. Current research is aimed at the premise that these individuals are somehow unresponsive to normal leptin signaling. We will discuss the pathways through which leptin and other adipokines act in more detail in Chapter 18.

Evidence is accumulating for other biochemical factors as significant elements in weight control. A number of hormones, including **serotonin** (page 895), control satiety, the feeling of fullness after eating. The antiobesity drug **fenfluramine**, which acts by increasing serotonin levels and affecting the appetite, was becoming quite popular until it was found to do serious damage to the heart and was withdrawn from the market. Current biochemical attention is focused upon UCPs, which by uncoupling oxidative phosphorylation, decrease the amount of ATP synthesized during oxidative metabolism. Attention is also focused upon other hormonal factors and upon fatty acid–binding proteins, which evidently participate in the transport of fatty acids to different locations within the cell. Biochemical research on obesity is one of the most active current research frontiers.

SUMMARY

Triacylglycerols are the main form for storage of biological energy. In animals, dietary triacylglycerols are digested and then resynthesized as they are complexed with proteins to form chylomicrons, for transport to tissues. Triacylglycerols synthesized in liver are transported to peripheral tissues as very low-density lipoproteins. Low-density lipoproteins represent the major vehicle for transport of cholesterol to peripheral tissues. Cholesterol levels in blood are regulated through control of synthesis of LDL receptors, involved in cellular uptake of LDL by endocytosis. Faulty control of LDL levels contributes toward the development of atherosclerotic plaque.

Fat depots are mobilized by enzymatic hydrolysis of triacylglycerols to fatty acids plus glycerol. The process is hormonally controlled via cyclic AMP. Most fatty acid degradation occurs through β-oxidation, a mitochondrial process that involves stepwise oxidation and removal of two-carbon fragments as acetyl-CoA. Processing unsaturated fatty acids is a bit more complicated because of the stereochemistry involved, but the pathways are straightforward. Under conditions in which further oxidation of acetyl-CoA through the citric acid cycle is limited, acetyl-CoA is used to synthesize ketone bodies, which are excellent energy substrates for brain and heart.

Fatty acid biosynthesis occurs via the stepwise addition of two-carbon fragments, in a process that superficially resembles a reversal of β-oxidation. Metabolic activation involves acyl carrier protein, and the reductive power comes from NADPH. In eukaryotic cells the seven enzyme activities are linked covalently on multifunctional enzymes or multienzyme complexes called megasynthases. Fatty acid elongation beyond the C_{16} stage is mechanistically similar, but CoA derivatives are involved instead of acyl carrier protein. Elegant mechanisms have evolved to coordinately regulate the synthesis and degradation of fatty acids. Unsaturated fatty acids are formed primarily by an ER-associated desaturating system. Triacylglycerols are synthesized by straightforward pathways in which acyl groups in fatty acyl-CoAs are transferred to the hydroxyl groups of glycerol-3-phosphate and diacylglycerol. Faulty hormone control of triacylglycerol deposition, along with excess dietary intake, can be responsible for obesity.

REFERENCES

Lipid and Lipoprotein Metabolism in Animals

Brasaemle, D. L. (2007) Thematic review series: Adipocyte biology. The perilipin family of structural lipid droplet proteins: Stabilization of lipid droplets and control of lipolysis. *J. Lipid Res.* 48:2547–2559.

Chester, A., J. Scott, S. Anant, and N. Navaratnam (2000) RNA editing: Cytidine to uridine conversion in apolipoprotein B mRNA. *Biochim. Biophys. Acta* 1494:1–13.

Goldstein, J. L., and M. S. Brown (2009) The LDL receptor. *Arterioscler. Thromb. Vasc. Biol.* 29:431–438. A short account of the discovery and actions of the LDL receptor, written by its discoverers.

Havel, R. J., and J. P. Kane (2001) Introduction: Structure and metabolism of plasma lipoproteins. In *The Metabolic and Molecular Bases of Inherited Disease*, C. R. Scriver, A. L. Beaudet, W. S. Sly, D. Valle, B. Childs, K. W. Kinzler, and B. Vogelstein, eds., Vol. II, Ch. 114, pp. 2705–2716, McGraw-Hill, New York. The first in a series of 10 chapters dealing with clinical disorders of lipid and lipoprotein metabolism.

Schmid, S. L. (1997) Clathrin-coated vesicle formation and protein sorting: An integrated process. *Annu. Rev. Biochem.* 66:511–548. A review of the biochemistry of endocytosis and protein sorting.

Steinberg, D. (2009) The LDL modification hypothesis of atherogenesis: An update. *J. Lipid Res.* 50 Suppl:S376–S381. An update on the role of LDL oxidation in atherogenesis.

Zechner, R., P. C. Kienesberger, G. Haemmerle, R. Zimmermann, and A. Lass (2009) Adipose triglyceride lipase and the lipolytic catabolism of cellular fat stores. *J. Lipid Res.* 50:3–21. An excellent review on the enzymology and hormonal control of fat mobilization.

Fatty Acid Metabolism

Hiltunen, J. K., M. S. Schonauer, K. J. Autio, T. M. Mittelmeier, A. J. Kastaniotis, and C. L. Dieckmann (2009) Mitochondrial fatty acid synthesis type II: More than just fatty acids. *J. Biol. Chem.* 284: 9011–9015. A short review on the enzymes and physiological roles of mitochondrial fatty acid synthesis.

Jansen, G. A., and R. J. Wanders (2006) Alpha-oxidation. *Biochim. Biophys. Acta* 1763:1403–1412.

Kemp, S., and R. Wanders (2010) Biochemical aspects of X-linked adrenoleukodystrophy. *Brain Pathol.* 20:831–837.

Khosla, C., Y. Tang, A. Y. Chen, N. A. Schnarr, and D. E. Cane (2007) Structure and mechanism of the 6-deoxyerythronolide B synthase. *Annu. Rev. Biochem.* 76:195–221.

Kim, J. J., and K. P. Battaile (2002) Burning fat: The structural basis of fatty acid beta-oxidation. *Curr. Opin. Struct. Biol.* 12:721–728. Summarizes the different isozymes that participate in the mammalian mitochondrial pathway.

Maier, T., M. Leibundgut, and N. Ban (2008) The crystal structure of a mammalian fatty acid synthase. *Science* 321:1315–1322. Presents a model for the structure and mechanism of a type I fatty acid synthase.

McGarry, J. D., and N. F. Brown (1997) The mitochondrial carnitine palmitoyltransferase system. From concept to molecular analysis. *Eur. J. Biochem.* 244:1–14.

Prentki, M., and S. R. Madiraju (2008) Glycerolipid metabolism and signaling in health and disease. *Endocr. Rev.* 29:647–676. A detailed discussion of triacylglycerol/free fatty acid cycling and its metabolic roles.

Rector, R. S., R. M. Payne, and J. A. Ibdah (2008) Mitochondrial trifunctional protein defects: Clinical implications and therapeutic approaches. *Adv. Drug Deliv. Rev.* 60:1488–1496.

Saggerson, D. (2008) Malonyl-CoA, a key signaling molecule in mammalian cells. *Ann. Rev. Nutr.* 28:253–272. Reviews the regulation of fatty acid metabolism.

Smith, S., and S. C. Tsai (2007) The type I fatty acid and polyketide synthases: A tale of two megasynthases. *Nat. Prod. Rep.* 24: 1041–1072. A detailed discussion of the protein chemistry of the eukaryotic multifunctional proteins involved in fatty acid and polyketide synthesis.

Biochemical Insights into Obesity

Muoio, D. M., and C. B. Newgard (2006) Obesity-related derangements in metabolic regulation. *Annu. Rev. Biochem.* 75:367–401.

Savage, D. B., K. F. Petersen, and G. I. Shulman (2007) Disordered lipid metabolism and the pathogenesis of insulin resistance. *Physiol. Rev.* 87:507–520.

PROBLEMS

1. Calculate the ATP yield from oxidation of palmitic acid, taking into account the energy needed to activate the fatty acid and transport it into mitochondria. Do the same for stearic acid, linoleic acid, and oleic acid.

2. If palmitic acid is subjected to complete combustion in a bomb calorimeter, one can calculate a standard free energy of combustion of 9788 kJ/mol. From the ATP yield of palmitate oxidation, what is the metabolic efficiency of the biological oxidation, in terms of kilojoules saved as ATP per kilojoule released? (Ignore the cost of fatty acid activation.)

3. Calculate the number of ATPs generated by the complete metabolic oxidation of tripalmitin (tripalmitoylglycerol). Hydrolysis of the triacylglycerol occurs at the cell surface. Consider the energy yield from catabolism of glycerol, as well as from the fatty acids. Calculate the ATP yield per carbon atom oxidized, and compare it with the energy yield from glucose.

4. Write a balanced equation for the *complete* metabolic oxidation of each of the following. Include O_2, ADP, and P_i as reactants and ATP, CO_2, and H_2O as products.
   (a) Stearic acid
   (b) Oleic acid
   (c) Palmitic acid
   (d) Linoleic acid

5. Calculate the number of ATPs generated from the metabolic oxidation of the four carbons of acetoacetyl-CoA to CO_2. Now consider the homolog derived from oxidation of an odd-numbered carbon chain, namely propionoacetyl-CoA. Calculate the net ATP yield from oxidation of the five carbons of this compound to CO_2.

6. 2-Bromopalmitoyl-CoA inhibits the oxidation of palmitoyl-CoA by isolated mitochondria but has no effect on the oxidation of palmitoylcarnitine. What is the most likely site of inhibition by 2-bromopalmitoyl-CoA?

*7. When the identical subunits of chicken liver fatty acid synthase are dissociated in vitro, all of the activities can be detected in the separated subunits except for the β-ketoacyl synthase reaction and the overall synthesis of palmitate. Explain these observations.

8. Mammals cannot undergo *net* synthesis of carbohydrate from acetyl-CoA, but the carbons of acetyl-CoA can be incorporated into glucose and amino acids. Present pathways by which this could come about.

9. Describe a pathway whereby some of the carbon from a fatty acid with an odd-numbered carbon chain could undergo a net conversion to carbohydrate.

10. How many tritium atoms (^{3}H) are incorporated into palmitate when fatty acid synthesis is carried out in vitro with the following labeled substrate?

$$^-OOC - C^3H_2 - \overset{\overset{\textstyle O}{\|}}{C} - S - CoA$$

11. What would be the effect on fatty acid synthesis of an increase in intramitochondrial oxaloacetate level? Briefly explain your answer.

*12. Glucagon secretion causes inhibition of intracellular acetyl-CoA carboxylase activity by several mechanisms. Name all you can think of.

13. Identify and briefly discuss each mechanism ensuring against simultaneous fatty acid synthesis and oxidation in the same cell.

14. Discuss the metabolic rationale for phosphorylation of acetyl-CoA carboxylase by AMP-activated protein kinase (AMPK) and cyclic AMP–dependent protein kinase (PKA).

15. Describe the probable effect in adipocytes of insulin-stimulated uptake of glucose into these cells.

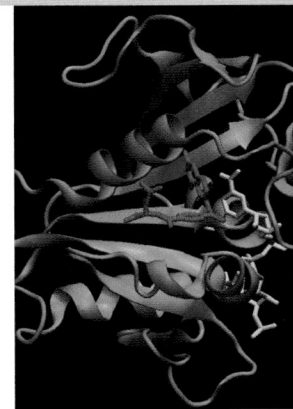

CHAPTER 18

Interorgan and Intracellular Coordination of Energy Metabolism in Vertebrates

Our presentation of intermediary metabolism in earlier chapters has placed primary emphasis on the cell, its reactions, its individual enzymes, and its other components. In this chapter we integrate these individual pathways in two ways. First, we review the metabolic profiles of the major organs in vertebrates: which fuels they use, which fuels they generate, and how the organs interact under stress to maintain appropriate energy balance. Second, we describe how these interactions are controlled in large part by hormonal signals, some of which have already been introduced. The molecular mechanisms of hormonal signal transduction will be discussed in detail in Chapter 23.

As we consider metabolic integration, keep in mind that metabolism is controlled to a great extent by the availability of substrates for specific metabolic pathways. In general, substrate concentrations within cells fall below saturating levels for the enzymes that metabolize them. Therefore, fluxes through particular enzymes vary as the concentrations of the substrate vary. A good example is the metabolic adaptation that occurs during a marathon run. Once the glycogen stores in liver and muscle are exhausted, flux through glycolysis decreases in muscle, not for any hormonal reason but simply because glucose phosphates are less available. Hormonal adjustments then occur to allow increased use of fatty acids in muscle, but the primary factor determining which metabolic pathway becomes functional in the cell—and thus which substrates are catabolized for energy—is the concentration of each of the usable substrates.

Metabolite concentrations represent a significant intracellular control mechanism.

Interdependence of the Major Organs in Vertebrate Fuel Metabolism

In this section we look at metabolism not as the activities in just one cell but as the totality of chemical reactions in a complex multicellular animal. We emphasize the specialized roles that each of the major organs plays in fuel metabolism—brain, muscle, liver, adipose tissue, and heart—and we describe the varying relationships among these organs as the animal encounters different physiological conditions.

Fuel Inputs and Outputs

In a differentiated organism, each tissue must be provided with fuels that it can use, in amounts sufficient to meet its own energy needs and to perform its specialized roles. The kidney, for example, must generate ATP for the osmotic work of transporting solutes against a concentration gradient for excretion. Muscle must generate ATP for the mechanical work of contraction, and in heart muscle the energy supply must be continuous. The liver generates ATP for biosynthetic purposes, whether for plasma protein synthesis, cholesterol generation, fatty acid synthesis, gluconeogenesis, or the production of urea for nitrogen excretion. Energy production must meet needs that vary widely, depending on level of exertion, composition of fuel molecules in the diet, time since last feeding, and so forth. For example, in humans the daily caloric intake may vary by four-fold, depending in part on the level of exertion—from 1500 to 6000 kcal/day in an average-sized human (1 kcal is equal to 1 **large calorie**, or **Cal**, the unit favored by nutritionists and found on our food labels). In the SI units used in this book, caloric intake ranges from 6000 to 25,000 kJ/day.

The major organs involved in fuel metabolism vary in their levels of specific enzymes so that each organ is specialized for the storage, use, and generation of different fuels. The major fuel depots are *triacylglycerols*, stored primarily in adipose tissue; *protein*, most of which exists in skeletal muscle; and *glycogen*, which is stored in both liver and muscle. In general, an organ specialized to *produce* a particular fuel lacks the enzymes to *use* that fuel. For example, liver is a major producer of ketone bodies, but little catabolism of ketone bodies occurs in the liver. Now let us review how the mobilization of each depot is controlled, and how the organs involved communicate with each other, to meet the energy needs of the animal. This information is summarized in Figure 18.1 and Table 18.1.

> The major fuel depots are triacylglycerols (adipose tissue), protein (muscle), and glycogen (muscle and liver).

Metabolic Division of Labor Among the Major Organs

Brain

The brain is the most fastidious, and one of the most voracious, of the organs. It must generate ATP in large quantities to maintain the membrane potentials essential for transmission of nerve impulses. Under normal conditions the brain uses only glucose to meet its prodigious energy requirement, which amounts to about 60% of the glucose utilization of a human at rest. The brain's need for about 120 grams of glucose per day is equivalent to 1760 kJ—about 15% of the total energy consumed by one person. The brain's quantitative requirement for glucose remains quite constant, even when an animal is at rest or asleep. The brain is a highly aerobic organ, and its metabolism demands some 20% of the total oxygen consumed by a human. Because the brain has no significant glycogen or other fuel reserves, the supply of both oxygen and glucose cannot be interrupted, even for a short time. Otherwise, irreversible brain damage results. However, the brain can adapt during fasting or starvation to use ketone bodies (see Chapter 17) instead of glucose as a major fuel.

Muscle

Muscle can utilize a variety of fuels—glucose, fatty acids, and ketone bodies. Skeletal muscle varies widely in its energy demands and the fuels it consumes, in line with its wide variations in activity. In resting muscle, fatty acids represent the major energy source; during exertion, glucose is the primary source. Early in a period of exertion, that glucose comes from mobilization of the muscle's glycogen reserves. Later, as glycogen reserves are depleted, fatty acids become the dominant fuel. Skeletal muscle stores about three-fourths of the total glycogen in humans, with most of the rest being stored in the liver. However, glucose produced from muscle glycogen cannot be released for use by other tissues. Muscle lacks the

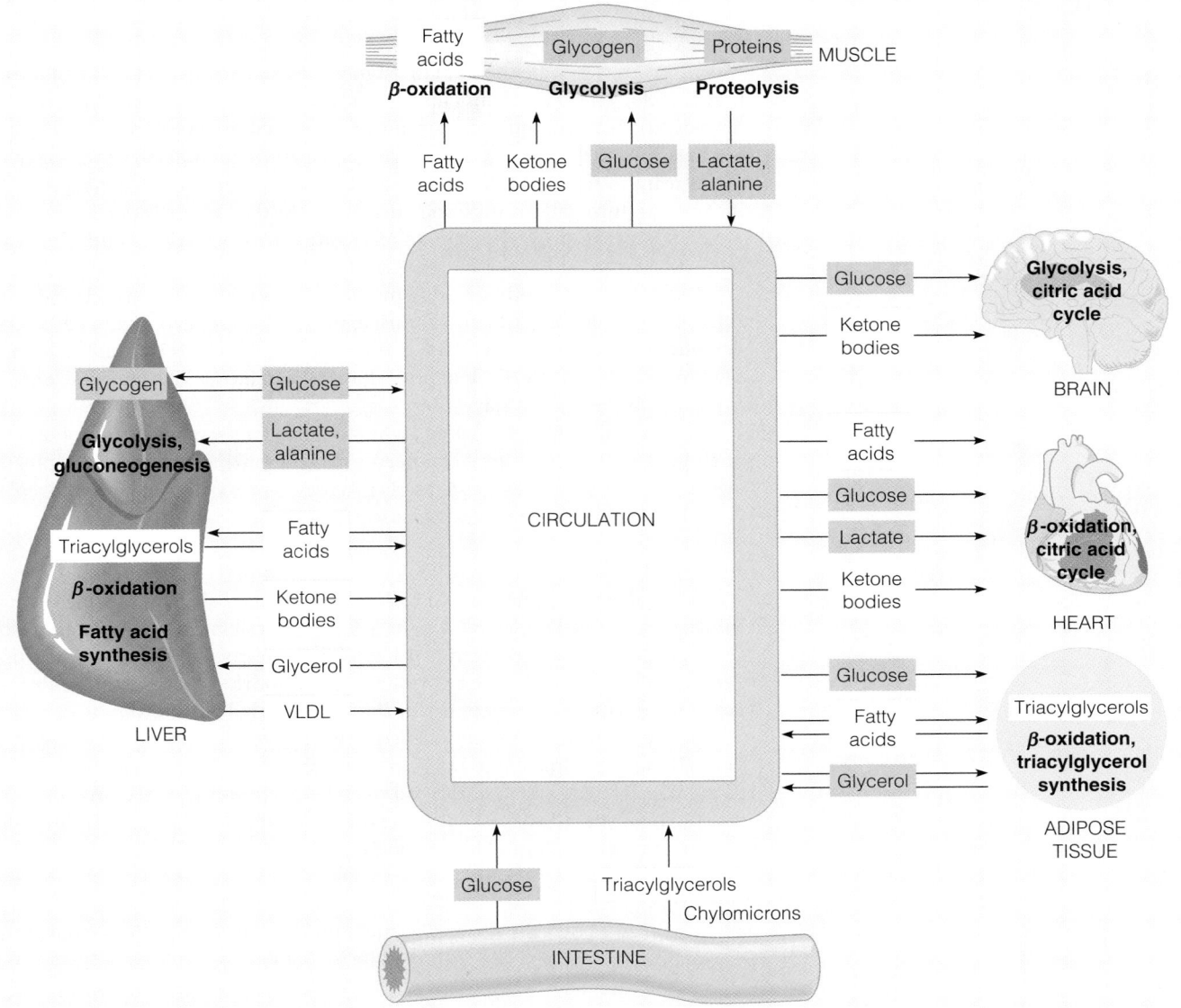

FIGURE 18.1

Metabolic interactions among the major fuel-metabolizing organs. The major fuel metabolites imported and exported by each organ are shown, along with the major energy pathways and fuel reserves in each organ. Lipid-derived metabolites are highlighted in yellow; carbohydrate-derived metabolites are highlighted in blue. VLDL, very low density lipoprotein.

TABLE 18.1   Profiles of the major vertebrate organs in fuel metabolism

Tissue	Fuel Store	Preferred Fuel	Fuel Sources Exported
Brain	None	Glucose (ketone bodies during starvation)	None
Skeletal muscle (resting)	Glycogen	Fatty acids	None
Skeletal muscle (during exertion)	None	Glucose	Lactate, alanine
Heart muscle	None	Fatty acids	None
Adipose tissue	Triacylglycerols	Fatty acids	Fatty acids, glycerol
Liver	Glycogen, triacylglycerols	Amino acids, glucose, fatty acids	Fatty acids, glucose, ketone bodies

enzyme glucose-6-phosphatase, so glucose phosphates derived from glycogen cannot be converted to free glucose for export—they are retained for use by the muscle cells.

During exertion, the flux rate through glycolysis exceeds that through the citric acid cycle, so lactate accumulates and is released. Another metabolic product is alanine, produced via transamination from pyruvate in the glucose–alanine cycle (see page 844 in Chapter 20). Both lactate and alanine are transported through the bloodstream to the liver, where they are reconverted through gluconeogenesis to glucose, for return to the muscle and other tissues by the Cori cycle (see Figure 13.14, page 547). However, the major fate of lactate during exertion is uptake by the heart for use as fuel, via oxidation to CO_2 (Chapter 13).

Muscle contains another readily mobilizable source of energy—its own protein. However, the breakdown of muscle protein to meet energy needs is both energetically wasteful and harmful to an animal, which needs its muscles to move about in order to survive. Protein breakdown is regulated so as to minimize amino acid catabolism except in starvation.

Finally, recall that muscle has an additional energy reserve in creatine phosphate, which generates ATP without the need for metabolizing fuels (see Chapter 12, page 496). This reserve is exhausted early in a period of exertion and must be replenished, along with glycogen stores, as muscle rests after prolonged exertion.

Heart

The metabolism of heart muscle differs from that of skeletal muscle in three important respects. First, work output is far less variable than in skeletal muscle. Second, the heart is a completely aerobic tissue, whereas skeletal muscle can function anaerobically for limited periods. Mitochondria are much more densely packed in heart than in other cells, making up nearly half the volume of a heart cell. Third, the heart contains negligible energy reserves as glycogen or lipid, although there is a small amount of creatine phosphate. Therefore, the supply of both oxygen and fuels from the blood must be continuous to meet the unending energy demands of the heart. The heart uses a variety of fuels—mainly fatty acids but also glucose, lactate, and ketone bodies.

Adipose Tissue

Adipose tissue represents the major fuel depot for an animal. The total stored triacylglycerols amount to some 555,000 kJ (133,000 Cal) in an average-sized human (see Chapter 17, page 710). This is enough fuel, metabolic complications aside, to sustain life for a couple of months in the absence of further caloric intake.

The adipocyte, or fat cell, is designed for continuous synthesis and breakdown of triacylglycerols, with breakdown controlled largely via the activation of hormone-sensitive triglyceride lipases (see Figure 17.13, page 722). Because adipocytes lack the enzyme glycerol kinase, some glucose catabolism must occur for triacylglycerol synthesis to take place—specifically, the formation of dihydroxyacetone phosphate, for reduction to glycerol-3-phosphate (see Chapter 17, page 749). Glucose acts as a sensor in adipose tissue metabolism. When glucose levels are adequate, continuing production of dihydroxyacetone phosphate generates enough glycerol-3-phosphate for resynthesis of triacylglycerols from the released fatty acids. When intracellular glucose levels fall, the concentration of glycerol-3-phosphate falls also, and fatty acids are released from the adipocyte for export as the albumin complex to other tissues.

Liver

A primary role of liver is the synthesis of fuel components for use by other organs. In fact, most of the low-molecular-weight metabolites that appear in the blood through digestion are taken up by the liver for this metabolic processing. The liver is a major site for fatty acid synthesis. It also produces glucose, both from its own glycogen stores and from gluconeogenesis, the latter using lactate and alanine from muscle, glycerol from adipose tissue, and the amino acids not needed for protein synthesis. Ketone bodies are also manufactured largely in the liver. In liver the level of malonyl-CoA, which is related to the energy status of the cell, is a determinant of the fate of fatty acyl-CoAs. When fuel is abundant, malonyl-CoA accumulates and inhibits carnitine acyltransferase I, preventing the transport of fatty acyl-CoAs into mitochondria for β-oxidation and ketogenesis. On the other hand, shrinking malonyl-CoA pools signal the cells to transport fatty acids into the mitochondria, for generation of energy and fuels.

An important role of liver is to buffer the level of blood glucose. It does this largely through the action of hexokinase IV (previously called glucokinase), an enzyme specific to liver, with a high $K_{0.5}$ (about 7.5 mM) for glucose, and partly through a high-K_M transport protein, the **glucose transporter (GLUT2)**, one member of a family of membrane proteins that carry out facilitated diffusion of glucose. Thus, liver is unique in being able to respond to high blood glucose levels by increasing the uptake and phosphorylation of glucose, which results eventually in its deposition as glycogen. Glucose-6-phosphate accumulation activates the *b* form of glycogen synthase (see Figure 13.36, page 572). In addition, glucose itself binds to glycogen phosphorylase *a*, increasing the susceptibility of phosphorylase *a* to dephosphorylation (see Figure 13.35, page 571), with consequent inactivation. Thus, in addition to hormonal effects, described shortly, liver senses the fed state and acts to store fuel derived from glucose. Liver also senses the fasted state and increases the synthesis and export of glucose when blood glucose levels are low. (Other organs also sense the fed state, notably the pancreas, which adjusts its glucagon and insulin outputs accordingly.)

To meet its internal energy needs, the liver can use a variety of fuel sources, including glucose, fatty acids, and amino acids.

> One of the most important roles of liver is to serve as a "glucostat," monitoring and stabilizing blood glucose levels.

Blood

All of the organs we have discussed are connected by the bloodstream, which transports what may be one organ's waste product but another organ's fuel (for example, alanine from muscle to liver). Blood also transports oxygen from lungs to tissues, enabling exergonic oxidative pathways to occur, followed by transport of the resultant CO_2 back to the lungs for exhalation, as described in Chapter 7. And, as described in Chapter 17, the lipoprotein components of blood plasma (e.g., chylomicrons and very low-density lipoproteins) play indispensable roles in transporting lipids. Of course, blood is also the medium of transport of hormonal signals from one tissue to another, and of exit for metabolic end products, such as urea, via the kidneys.

In terms of the blood's own energy metabolism, the most prominent pathway is glycolysis in the erythrocyte. Blood cells constitute nearly half the volume of blood, and erythrocytes constitute more than 99% of blood cells. Mammalian erythrocytes contain no mitochondria and depend exclusively upon anaerobic glycolysis to meet their limited energy needs.

Hormonal Regulation of Fuel Metabolism

In animals, it is supremely important to maintain blood glucose levels within rather narrow limits, particularly for proper functioning of the nervous system. Of course, blood glucose levels vary, depending on nutritional status. Several hours after a meal, the normal level in humans is about 80 mg per 100 mL of blood, or 4.4 mM. Shortly after a meal, that level might rise to 120 mg per 100 mL (6.6 mM). In response, homeostatic mechanisms come into play to promote uptake of glucose into cells and its use by tissues. Similarly, when glucose levels

> Maintenance of blood glucose within narrow limits is critical to brain function.

fall, several hours after a meal, other mechanisms promote both glucose release, from liver glycogen stores, and gluconeogenesis so that the normal level is maintained. Some of the homeostatic mechanisms were mentioned in the previous section; others involve hormonal regulation. Though we discuss molecular mechanisms of hormone action in detail in Chapter 23, it is appropriate here to discuss at the physiological level some of the hormones involved in fuel metabolism.

Actions of the Major Hormones

The most important hormone promoting glucose uptake and use is insulin, whereas both glucagon and epinephrine have the opposite effect, to increase blood glucose levels. The major effects of these agents are summarized in Table 18.2. Figure 18.2 illustrates the interplay between the two pancreatic hormones, insulin and glucagon.

Insulin

Insulin is a 5.8-kilodalton protein (see page 151) that is synthesized in the pancreas. The pancreas has both **endocrine cells**, which secrete hormones directly into the bloodstream, and **exocrine cells**, which secrete zymogen precursors of digestive enzymes into the upper small intestine. The endocrine tissue, which takes the form of cell clusters known as islets of Langerhans, contains at least five

> The key hormones regulating fuel metabolism are insulin, which promotes glucose use, and glucagon and epinephrine, which increase blood glucose.

TABLE 18.2   Major hormones controlling fuel metabolism in mammals

Hormone	Biochemical Actions	Enzyme Target	Physiological Actions
Insulin	↑ Glucose uptake (muscle, adipose tissue)	GLUT4	Signals fed state: ↓ Blood glucose level ↑ Fuel storage ↑ Cell growth and differentiation
	↑ Glycolysis (liver, muscle)	PFK-1 (via PFK-2/FBPase-2)	
	↑ Acetyl-CoA production (liver, muscle)	Pyruvate dehydrogenase complex	
	↑ Glycogen synthesis (liver, muscle)	Glycogen synthase	
	↑ Triacylglycerol synthesis (liver)	Acetyl-CoA carboxylase	
	↓ Gluconeogenesis (liver)	FBPase-1 (via PFK-2/FBPase-2)	
	↓ Lipolysis		
	↓ Protein degradation		
	↑ Protein, DNA, RNA synthesis		
Glucagon	↑ cAMP level (liver, adipose tissue)		Signals fasting state: ↑ Glucose release from liver ↑ Blood glucose level ↑ Ketone bodies as alternative fuel for brain
	↑ Glycogenolysis (liver)	Glycogen phosphorylase	
	↓ Glycogen synthesis (liver)	Glycogen synthase	
	↑ Triacylglycerol hydrolysis and mobilization (adipose tissue)	Hormone-sensitive lipase, perilipin, adipose triglyceride lipase	
	↑ Gluconeogenesis (liver)	FBPase-1 (via PFK-2/FBPase-2), pyruvate kinase, PEPCK	
	↓ Glycolysis (liver)	PFK-1 (via PFK-2/FBPase-2)	
	↑ Ketogenesis (liver)	Acetyl-CoA carboxylase	
Epinephrine	↑ cAMP level (muscle)		Signals stress: ↑ Glucose release from liver ↑ Blood glucose level
	↑ Triacylglycerol mobilization (adipose tissue)	Hormone-sensitive lipase, perilipin, adipose triglyceride lipase	
	↑ Glycogenolysis (liver, muscle)	Glycogen phosphorylase	
	↓ Glycogen synthesis (liver, muscle)	Glycogen synthase	
	↑ Glycolysis (muscle)	Glycogen phosphorylase, providing increased glucose	

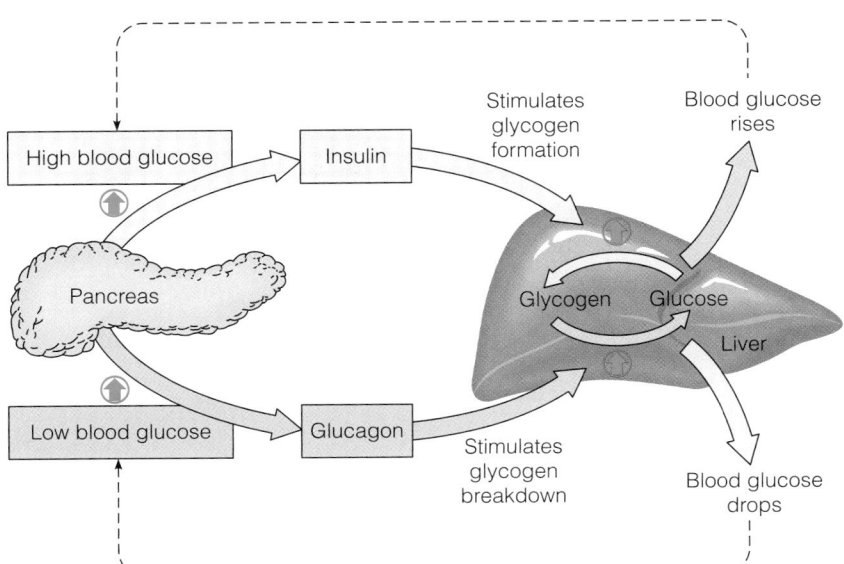

FIGURE 18.2

Aspects of the control of blood glucose levels by pancreatic secretion of insulin and glucagon. Conditions resulting from high glucose levels are shown in blue, and those from low glucose levels in pink. Green arrows indicate stimulatory responses.

different cell types, each specialized for synthesis of one hormone. The α (or A) cells produce glucagon; the δ (or D) cells, **somatostatin**; the ε cells, **ghrelin**; and the F cells, **pancreatic polypeptide**. Insulin is synthesized in the β (or B) cells, which sense glucose levels and secrete insulin in response to increased levels of blood glucose. The β cells take up and catabolize glucose, resulting in increased intracellular ATP levels, which causes closure of ATP-gated K^+ channels and depolarization of the plasma membrane (see Chapter 10). Voltage-gated Ca^{2+} channels open in response to this membrane depolarization, leading to increased cytosolic $[Ca^{2+}]$, which finally triggers exocytosis of insulin granules.

The simplest way to describe the several actions of insulin is to say that *insulin signals the fed state* and thereby promotes (1) uptake of fuel substrates into some cells, (2) storage of fuels (lipids and glycogen), and (3) biosynthesis of macromolecules (nucleic acids and protein). Specific effects include increased uptake of glucose in muscle and adipose tissue; activation of glycolysis in liver; increased synthesis of fatty acids and triacylglycerols in liver and adipose tissue; inhibition of gluconeogenesis in liver; increased glycogen synthesis in liver and muscle; increased uptake of amino acids into muscle with consequent activation of muscle protein synthesis; and inhibition of protein degradation. Because of its promotion of biosynthesis, it is appropriate to consider insulin a growth hormone.

The mechanism by which insulin stimulates glucose uptake into muscle and adipose cells is an area of intense investigation. One important action involves the glucose transporter, GLUT4 (see Figure 18.5). Before stimulation by insulin, this transporter is not present on the cell surface; rather, it is localized in vesicles in the cytosol. The protein is translocated to the cell surface in response to insulin, where it facilitates glucose uptake. An important consequence of glucose uptake in adipocytes is its conversion to glycerol-3-phosphate, which combines with fatty acids to stimulate triacylglycerol synthesis.

Glucagon

A 3.5-kilodalton polypeptide, glucagon is synthesized by α cells of the islets of Langerhans in the pancreas. These endocrine cells sense the blood glucose concentration and release the hormone in response to low levels (see Figure 18.2). Both synthesis and release of glucagon are controlled by insulin.

The primary target of glucagon is the liver, and its principal effect is to increase cyclic AMP levels in liver cells, as schematized in Figure 18.3. The resultant metabolic cascades, discussed in Chapter 13, promote glycogenolysis and

FIGURE 18.3

Actions of glucagon in liver that lead to a rise in blood glucose. Brackets indicate concentration; ↑ and ↓ indicate increase or decrease, respectively, in enzyme activity, pathway flux, or metabolite level.

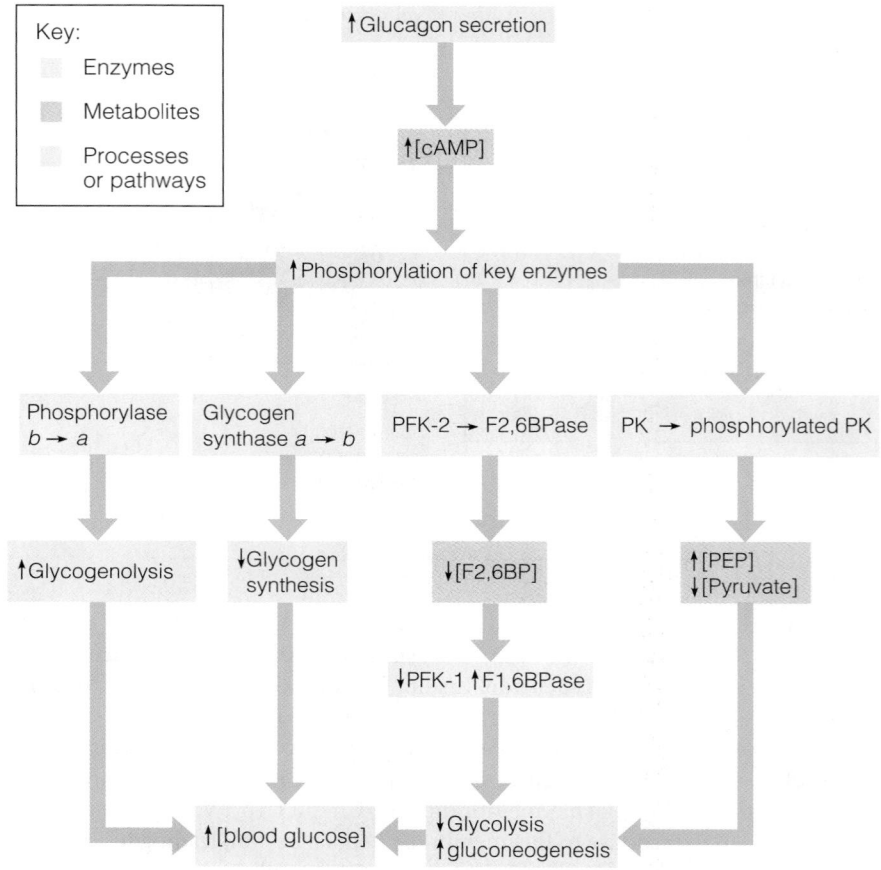

inhibit glycogen synthesis. In addition, by activating the hydrolysis of fructose-2, 6-bisphosphate, cAMP inhibits glycolysis and activates gluconeogenesis. Glucagon also brings about inhibition of pyruvate kinase (PK) in the liver, causing phosphoenolpyruvate (PEP) to accumulate. The level of pyruvate decreases, both because its synthesis from PEP is blocked and because it continues to be converted to PEP, via the pyruvate carboxylase and phosphoenolpyruvate carboxykinase reactions. Although accumulation of PEP is slight, it suffices to promote gluconeogenesis, while inhibition of pyruvate kinase diminishes the glycolytic flux rate.

Glucagon also raises cAMP levels in adipose tissue. There the chief effect of cAMP is to promote triacylglycerol mobilization via phosphorylation of hormone-sensitive lipase and perilipin, yielding glycerol and fatty acids (see Figure 17.14, page 723).

Epinephrine

The catecholamines epinephrine and norepinephrine, when released from presynaptic nerve endings, function as neurotransmitters (see Chapter 23). When released from adrenal medulla in response to low blood glucose levels, epinephrine interacts with second-messenger systems in many tissues, with varied effects. In muscle, epinephrine activates adenylate cyclase, with concomitant activation of glycogenolysis and inhibition of glycogen synthesis (see Chapter 13). Triacylglycerol breakdown in adipose tissue is also stimulated, providing fuel for the muscle tissue. Epinephrine also inhibits insulin secretion and stimulates glucagon secretion. These effects tend to increase glucose production and release by the liver. The net result is to increase blood glucose levels. Unlike glucagon, the catecholamines have short-lived metabolic effects. As discussed in Chapter 13, epinephrine action on skeletal and heart muscle cells is a crucial part of the "fight or flight" response.

Coordination of Energy Homeostasis

All organisms, indeed all cells, must balance the ingestion and absorption of fuel molecules with the metabolism and storage of these nutrients to meet immediate, as well as long-term, energy needs. Maintenance of this balance is termed **energy homeostasis**, and a complex regulatory system has evolved to coordinate and integrate these processes. This regulatory system comprises a bewildering number of components, but two protein kinases, AMPK and **mTOR**, play central roles in orchestrating the metabolic activity of mammalian cells. A third highly conserved family of enzymes, the **sirtuins**, are also emerging as important regulators of energy metabolism.

AMP-Activated Protein Kinase

We have seen AMPK (AMP-activated protein kinase) several times in the preceding chapters. This serine/threonine protein kinase, found in all eukaryotes, is activated when the energy charge of the cell is low (i.e., high AMP:ATP ratio), such as occurs during nutrient starvation or hypoxia. Once activated, AMPK initiates a signaling process that conserves cellular energy by stimulating pathways that lead to ATP production while inhibiting pathways that utilize ATP. AMPK is a heterotrimer, composed of a catalytic subunit (α) and two regulatory subunits (β and γ). Activation of the catalytic α subunit involves the binding of AMP to four nucleotide-binding sites in the γ subunit (Figure 18.4). Activation of AMPK also requires phosphorylation of a specific threonine residue in the α subunit, and AMP binding to the γ subunit allosterically protects against dephosphorylation of this residue in the α subunit. Two upstream kinases, **LKB1** and Ca^{2+}/calmodulin-dependent protein kinase kinase β (**CaMKKβ**), are known to phosphorylate the critical Thr residue.

AMPK phosphorylates multiple substrates that enhance energy-producing pathways, including targets that stimulate glucose uptake (via translocation of GLUT4 to the plasma membrane), glycolysis in heart (via stimulation of the kinase activity of the heart-specific PFK-2/FBPase-2; see Chapter 13, page 555), and mitochondrial biogenesis. At the same time, other AMPK targets inhibit energy-requiring pathways, including hepatic gluconeogenesis (via decreased transcription of gluconeogenic enzymes), fatty acid synthesis (via acetyl-CoA carboxylase; see Figure 17.36, page 746), triacylglycerol synthesis (via glycerophosphate acyltransferase and hormone-sensitive lipase; see Figure 17.38, page 748), and cholesterol synthesis (via HMG-CoA reductase; Chapter 19, page 802). These effects are summarized in Figure 18.5.

Mammalian Target of Rapamycin

mTOR (mammalian target of rapamycin) is the other main player in the regulation of energy homeostasis. Like AMPK, mTOR is a highly conserved serine/threonine protein kinase found in all eukaryotes. In contrast to AMPK, mTOR is active under nutrient-rich conditions and inactive under nutrient-poor conditions. Activated mTOR promotes anabolic processes, including cell proliferation, protein synthesis, and biosynthesis, but inhibits catabolic processes. mTOR exists as two distinct multiprotein complexes: a rapamycin-sensitive complex (mTORC1) composed of mTOR, mLST8, raptor, and PRAS40; and a rapamycin-insensitive complex (mTORC2) composed of mTOR, mLST8, rictor, Sin1, and

AMPK and mTOR protein kinases play central roles in orchestrating the metabolic activity of mammalian cells.

AMPK is activated when the energy charge of the cell is low.

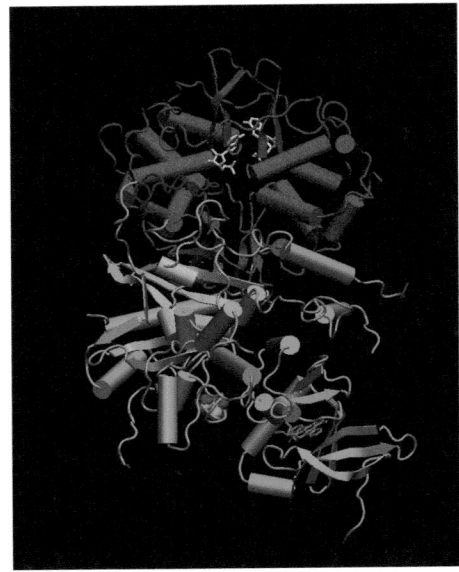

FIGURE 18.4

Mammalian AMP-activated protein kinase (AMPK). This figure shows the X-ray crystal structure of the heterotrimeric enzyme (PDB ID 2Y94). The α subunit, shown in cyan, is composed of two domains, an N-terminal kinase domain (dark cyan) and a C-terminal regulatory domain (light cyan). The kinase active site is marked by a bound inhibitor, staurosporine (orange). Phosphorylated threonine 172 is shown in space-filling representation. The regulatory β subunit (green) is normally 272 amino acids long, but only its C-terminal 85 residues are included in this structure. The regulatory γ subunit (red) is composed of four nucleotide-binding sites (CBS motifs, named after related sequences in cystathionine β synthase). Two of these sites are filled with AMP molecules (yellow); two are empty. In this structure of active AMPK, a loop of the α-subunit (blue) interacts with bound AMP in one of the CBS motifs of the γ-subunit. This interaction stabilizes the binding of the catalytic domain to the β-subunit, thereby inhibiting dephosphorylation of Thr 172 by phosphatase. This explains how AMP binding to the γ subunit allosterically protects against dephosphorylation of this residue in the catalytic subunit.

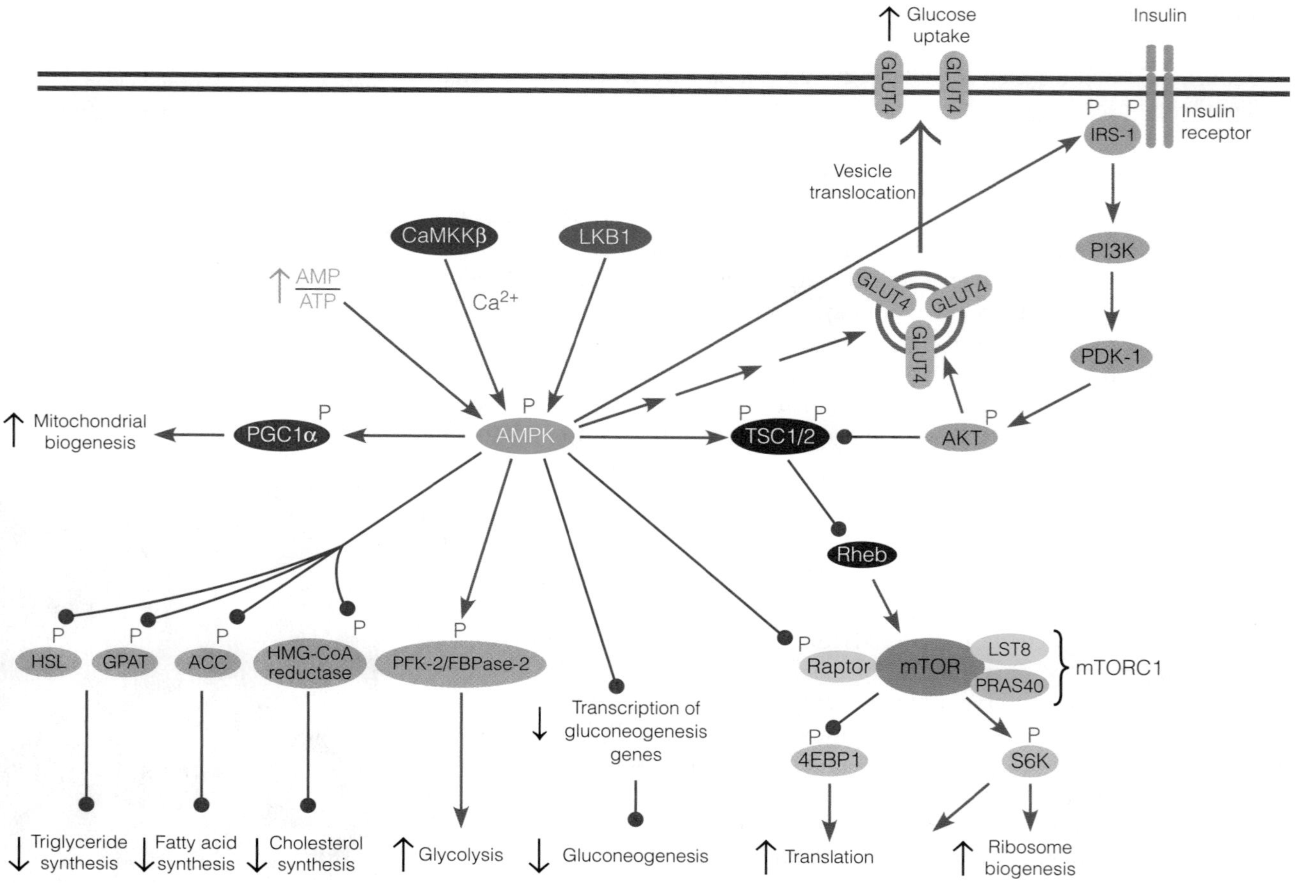

FIGURE 18.5

AMPK and mTOR signaling pathways. Green arrows indicate activation; red balls indicate inhibition. P indicates phosphorylation, causing either activation (P) or inhibition (P) of the phosphorylated target protein. For example, phosphorylation of TSC1/2 by AKT inhibits TSC1/2; phosphorylation by AMPK, at a different site, activates TSC1/2. Some of the metabolic responses mediated by AMPK and mTOR are tissue-specific. Refer to text for details. 4EBP1, eIF4E-binding protein 1; ACC, acetyl-CoA carboxylase; CaMKKβ, Ca²⁺/calmodulin-dependent protein kinase kinase β; GLUT4, glucose transporter; GPAT, glycerophosphate acyltransferase; HSL, hormone-sensitive lipase; IRS, insulin receptor substrate; PI3K, phosphoinositide 3-kinase; PGC-1α, PPARγ coactivator 1α; S6K, ribosomal protein S6 kinase; TSC1/2, tuberous sclerosis complex.

mTOR, in contrast to AMPK, is active under nutrient-rich conditions, and inactive under nutrient-poor conditions.

PRR5/Protor. mTOR was discovered during biochemical studies with the bacterial macrolide, **rapamycin**, a potent immunosuppressant. Rapamycin binds to a small cellular protein, FKBP12, and this FKBP12-rapamycin complex allosterically inhibits mTORC1, but has no effect on mTORC2.

The two best-characterized substrates of mTORC1 are **4EBP1** (eukaryotic initiation factor 4E-binding protein 1) and **S6K** (ribosomal protein S6 kinase) (Figure 18.5). Raptor acts as a scaffold to recruit these downstream substrates to the mTORC1 complex. 4EBP1 normally inhibits the translation initiation factor eIF4E (see Chapter 28); phosphorylation of 4EBP1 by mTORC1 suppresses its ability to bind eIF4E and inhibit translation. Phosphorylation of S6K activates this downstream protein kinase, which phosphorylates a number of proteins involved in translational control. The net effect of phosphorylation of 4EBP1 and S6K by mTORC1 is thus stimulation of protein synthesis and ribosome biogenesis.

mTORC1 activity is regulated by a number of upstream inputs, most of which are transmitted through the tuberous sclerosis complex (**TSC**) (Figure 18.5). This complex, first discovered as a tumor suppressor (see Chapter 23), is composed of TSC1 (hamartin) and TSC2 (tuberin). TSC2 contains a GTPase activating protein (GAP) domain that inactivates the small Ras-like GTPase **Rheb**. Rheb normally activates mTORC1, thus loss of TSC1 or TSC2 leads to hyperactivation of mTORC1. TSC serves as an integration point for regulation of mTORC1, receiving inputs from a wide variety of environmental signals, including growth factors (e.g., insulin and epidermal growth factor), energy status, and nutrient availability. Insulin signaling begins with phosphorylation of an **insulin receptor substrate** (IRS) protein by the tyrosine kinase activity of the plasma membrane insulin receptor. IRS proteins (humans express three isoforms, IRS-1, -2, and -4) integrate the activated insulin receptor to downstream adaptor proteins and enzymes. Thus, phosphorylated IRS activates **phosphoinositide 3-kinase** (PI3K), which converts membrane phosphatidylinositol 4,5-bisphosphate (PIP_2) to phosphatidylinositol 3,4,5-trisphosphate (PIP_3), a second messenger with a variety of intracellular targets. PIP_3 activates yet another protein kinase (PDK-1), which then activates a third protein kinase, Akt (also called PKB). Akt then phosphorylates TSC, inactivating the complex, resulting in activation of mTORC1. Activated Akt also stimulates GLUT4 translocation to the plasma membrane, accounting for insulin-stimulated glucose transport in muscle cells. We will discuss the insulin signal transduction pathway in more detail in Chapter 23.

Let's look more closely at the mechanism by which mTORC1 senses energy status because it involves AMPK. Recall that AMPK is activated during nutrient starvation, directly sensing the adenylate energy charge (AMP/ATP ratio). Activated AMPK then phosphorylates the TSC complex, which suppresses mTORC1 function by inhibiting Rheb. Activated AMPK can directly inhibit mTORC1 by phosphorylation of the raptor subunit of the complex. AMPK also modulates the insulin signaling pathway by phosphorylating IRS, thereby increasing the sensitivity of the cell to insulin (Figure 18.5).

In summary, AMPK and mTOR play opposing roles in controlling the metabolic activity of cells in response to intracellular and extracellular signals that report on the energy status of the individual cell and the organism as a whole. These two protein kinases, and the nutrient signaling pathways they control, are evolutionarily conserved from yeast to humans.

Rapamycin (Sirolimus)

Sirtuins

Sirtuins are a highly conserved family of protein deacetylases. These enzymes catalyze the deactylation of acetylated lysine residues in target proteins. Recall from Chapter 11 that acetylation of lysine residues is a common covalent modification of proteins, catalyzed by a family of protein acetyltransferases. Reversal of this modification requires an enzyme-catalyzed deacetylation. Of the three types of protein deacetylases found in nature, sirtuins are unique in that they require NAD^+ for their deacetylation activity. NAD^+ functions not as a redox cofactor in this complex reaction, but rather as a substrate that is cleaved to nicotinamide and 2'-O-acetyl-ADP-ribose (OAADPr).

> Sirtuins are a highly conserved family of NAD^+-dependent protein deacetylases.

Sirtuins are named after the founding member of the family, yeast Sir2 (silent information regulator 2), which deacetylates histones to regulate gene expression (silencing) of the mating type locus in yeast. Sirtuins are now known to act on many proteins besides histones, and they are structurally conserved from bacteria to humans (Figure 18.6). Mammals possess seven sirtuins (SIRT1-7), which differ in their cellular localization (nucleus, mitochondria, cytoplasm) and protein targets. For most protein targets, deacetylation increases the activity of the target protein. The deacetylase activity of sirtuins is sensitive to changes in the cellular NAD^+ levels, being enhanced at high NAD^+/NADH ratios. Thus, sirtuins act as metabolic sensors of the cellular redox state.

FIGURE 18.6

Catalytic core of sirtuins is evolutionarily conserved. X-ray crystal structures of the catalytic cores of sirtuins from **(a)** yeast (PDB ID 1Q14), **(b)** an archaea (PDB ID 1ICI), **(c)** a eubacterium (PDB ID 1S5P), and **(d)** human (PDB ID 2B4Y) show a high degree of structural similarity. The Rossmann-fold domain, characteristic of NAD^+/NADP-binding proteins, is magenta, the small zinc-binding domain is blue. The Zn ion is red. Disordered regions are indicated with dashed lines. Panel e shows the yeast Hst2 sirtuin with bound acetylated peptide substrate (green) and carba-NAD^+, a substrate analog (cyan).

Modified from *Biochimica et Biophysica Acta* 1804:1604–1616, B. D. Sanders, B. Jackson, and R. Marmorstein, Structural basis for sirtuin function: What we know and what we don't. © 2010, with permission from Elsevier.

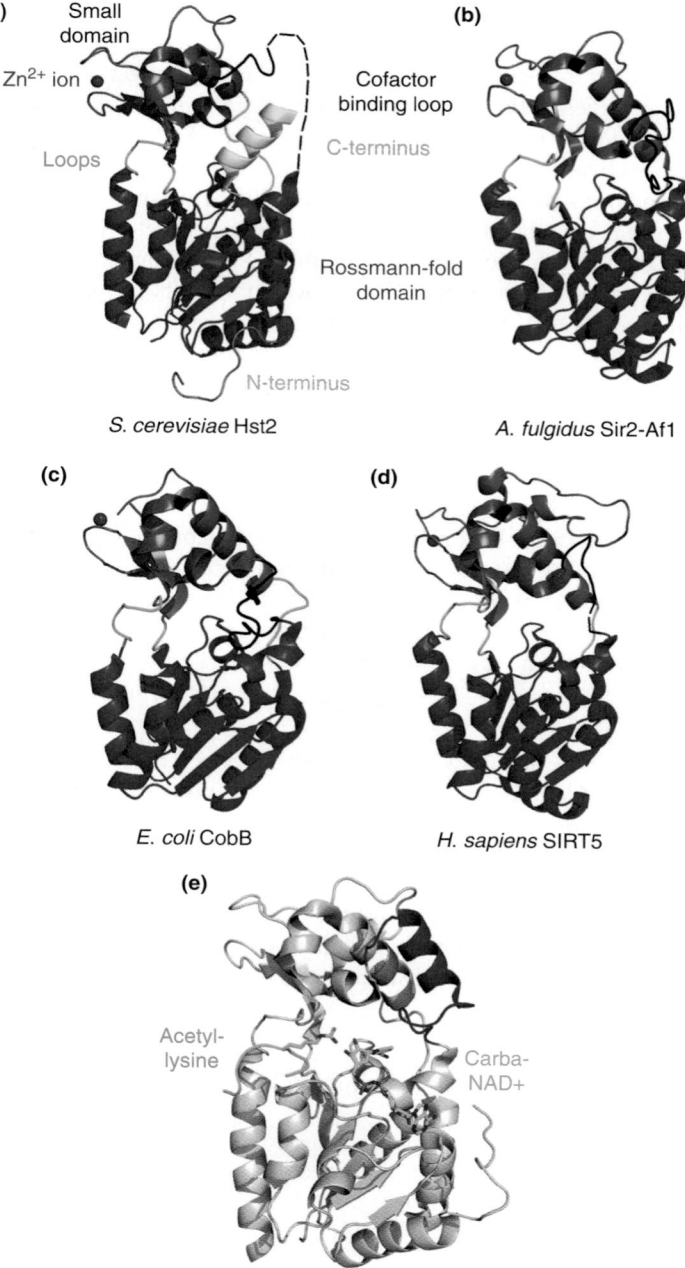

(a) *S. cerevisiae* Hst2

(b) *A. fulgidus* Sir2-Af1

(c) *E. coli* CobB

(d) *H. sapiens* SIRT5

(e)

Sirtuins are part of an intricate regulatory system that controls flux through fuel utilization pathways in response to the dietary availability of alternative fuels. For example, we will see in the next section that under fasting conditions, mammals initiate a reprogramming of their metabolic systems across several tissue types. This response includes increasing the rate of glucose synthesis in the liver and kidney and increasing the utilization of fatty acids as a fuel source in peripheral tissues. One of the most important participants in this metabolic reprogramming is the **peroxisome proliferator-activated receptor-γ coactivator 1α (PGC-1α)**. PGC-1α was initially discovered as a co-activator of the transcription factor peroxisome proliferator-activated receptor-γ (PPARγ), but PGC-1α binds to and stimulates the transcriptional activity of several transcription factors, including p53, nuclear respiratory factors 1 and 2 (NRF-1, NRF-2), and forkhead box O (FOXO).

The transcriptional coactivator function of PGC-1α is sensitive to its acetylation status, and PGC-1α can be deacetylated by SIRT1. Fasting (low nutrients) results in a higher NAD$^+$/NADH ratio, activating the sirtuins. Deacetylation of PGC-1α causes upregulation of its coactivator function (Figure 18.7). In liver, upregulation of PGC-1α function stimulates gluconeogenesis by activating the transcription of the phosphoenolpyruvate carboxykinase (PEPCK) and glucose-6-phosphatase genes. In skeletal muscle and heart, the transcriptional responses mediated by PGC-1α lead to increased oxidation of fatty acids and decreased utilization of glucose. Deacetylated PGC-1α also coactivates the transcription of nuclear-encoded genes that encode subunits of the mitochondrial respiratory chain (Chapter 15) as well as genes that encode components of the mitochondrial gene expression machinery. This enhanced mitochondrial biogenesis increases the capacity of the cell for fatty acid oxidation and is a critical part of the PGC-1α-dependent metabolic reprogramming that occurs in heart and skeletal muscle.

PGC-1α is capable of integrating multiple signals that monitor the cellular energy state. Recall that AMPK also stimulates mitochondrial biogenesis. AMPK phosphorylates PGC-1α, which also causes its activation (see Figure 18.5). It appears that phosphorylation primes PGC-1α for subsequent deacetylation by SIRT1. Thus, PGC-1α senses both the AMP/ATP ratio (via AMPK) and the NAD$^+$/NADH ratio (via SIRT1).

More than 2000 proteins have been found to be acetylated in mammalian cells, and the list of pathways known to be regulated by SIRT-mediated deacetylation is expanding rapidly. Table 18.3 summarizes our current understanding of this important family of regulatory enzymes.

The three mitochondrial sirtuins (SIRT3–5) are involved in the regulation of several metabolic processes, including the urea cycle and respiration. Deacetylation of their target proteins (Table 18.3) activates these pathways.

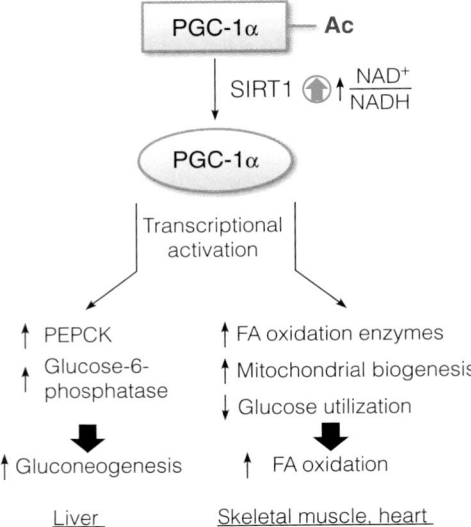

FIGURE 18.7

PGC-1α and SIRT1 control the reprogramming of fuel utilization pathways in response to fasting. A high NAD$^+$/NADH ratio, in response to low nutrients (fasting), activates SIRT1 to deacetylate PGC-1α, upregulating its transcriptional coactivator function. Tissue-specific transcriptional activation programs result in increased gluconeogenesis (liver) and increased fatty acid oxidation (skeletal and heart muscle).

	Subcellular Location	Deacetylation Substrates
TABLE 18.3	Substrates and cellular locations of mammalian sirtuins	
SIRT1	Nucleus, cytoplasm	Histones PGC-1α, FOXO, and many other transcription factors IRS-2
SIRT2	Nucleus, cytoplasm	Histones FOXO and other transcription factors
SIRT3	Mitochondrial matrix	Acetyl-CoA synthetase 2 (Chapter 17) Long-chain acyl-CoA dehydrogenase (Chapter 17) Complex I of electron transport chain (Chapter 15) Glutamate dehydrogenase (Chapter 20) Ornithine transcarbamoylase (Chapter 20) NADP$^+$-dependent isocitrate dehydrogenase (Chapter 14) Superoxide dismutase (Chapter 15)
SIRT4[a]	Mitochondrial matrix	Glutamate dehydrogenase[b] (Chapter 20)
SIRT5	Mitochondrial matrix	Carbamoyl phosphate synthetase I (Chapter 20)
SIRT6	Nucleus	Histones
SIRT7	Nucleolus	RNA polymerase I[c]

[a]SIRT4 catalyzes ADP ribosylation, rather than deacetylation, of its targets.

[b]ADP ribosylation of glutamate dehydrogenase decreases its activity.

[c]The specific RNA polymerase I target has yet to be identified.

SIRT4 does not appear to be a true protein deacetylase because it does not exhibit NAD^+-dependent deacetylation activity in vitro. Rather, SIRT4 catalyzes **ADP-ribosylation**, the transfer of ADP-ribose from NAD^+ to target proteins (see Figure 11.51, page 460).

Finally, in addition to regulating energy homeostasis, sirtuins play important roles in cellular stress responses, genomic stability, and tumorigenesis. In fact, sirtuins control lifespan in lower organisms, and evidence is accumulating that these evolutionarily conserved enzymes regulate the aging process in mammals as well. The first clue that sirtuins might influence the aging process came from studies on calorie restriction in laboratory rodents. Calorie restriction (CR), a reduction in food intake of 30–40%, can extend lifespan by up to 50% and delays the onset of age-associated disease. Moreover, CR extends lifespan in organisms ranging from yeast to rodents to primates. In a 20-year-long study of rhesus monkeys, CR reduced the incidence of diabetes, cancer, cardiovascular disease, and brain atrophy. Genetic studies in yeast led to the realization that sirtuins are important mediators of the CR response in most organisms, including mammals. For example, SIRT1 knockout mice do not exhibit the longevity normally associated with calorie restriction. Conversely, transgenic mice that overexpress SIRT1 exhibit many CR responses even when fed *ad libitum* (no food restriction). Enticed by the possibility that sirtuins might extend life, a virtual Gold Rush took place to find small molecules that could activate SIRT1. Resveratrol was the first of these compounds to be discovered by screening for activators of the NAD^+-dependent deacetylation activity of SIRT1. Resveratrol is a polyphenolic compound found in red grapes (and thus red wine), berries, and many other plants and has previously been associated with health benefits in humans. Although resveratrol and newer classes of SIRT1 activators do indeed confer many phenotypes of CR on fed animals, they are not likely to represent a simple "fountain of youth." Nonetheless, pharmacological manipulation of sirtuin function holds great promise for treatment of aging-related diseases.

Endocrine Regulation of Energy Homeostasis

AMPK, mTOR, and sirtuins all evolved in unicellular organisms as part of a control system that sensed the nutrient supply and initiated appropriate metabolic responses. If nutrients are abundant, the cells take up the fuels and metabolize them via glycolysis, a rapid, but relatively inefficient process (Figure 18.8, left panel). This *proliferative metabolism* provides the building blocks and free energy needed to produce biomass (new cells) during exponential growth. When nutrients are scarce, the cells adapt to a *starvation metabolism*. Biomass production ceases, and the cells switch to a slower, but more efficient, oxidative metabolism in order to extract maximum energy from the limiting nutrients. During the evolution of metazoans (multicellular organisms), this control system grew in complexity, responding to new inputs and developing new outputs. In contrast to unicellular organisms, most of the cells in multicellular organisms are bathed in a relatively constant supply of nutrients via the circulatory system, and this supply often exceeds the levels needed to support cell growth and replication. Growth control in metazoans thus evolved to occur at the levels of nutrient intake, transport, and utilization (metabolism). Thus, most mammalian cells exhibit a strict dependence on growth factors (hormones) to switch from a quiescent, differentiated state to a proliferative state. Nondividing, differentiated mammalian cells typically use oxidative metabolism (aerobic glycolysis + citric acid cycle) to metabolize glucose, as in the starvation metabolism of unicellular organisms (Figure 18.8, right panel). Upon stimulation by growth factors (e.g., insulin), differentiated cells switch to the faster glycolysis.

In 1925, Otto Warburg noted that, unlike nondividing, differentiated mammalian cells, most rapidly dividing cancer cells metabolize glucose by aerobic glycolysis, producing lactate even when oxygen is abundant. In this phenomenon, known as the "Warburg effect," the cancer cells have overcome their normal strict

Calorie restriction extends lifespan in organisms ranging from yeast to primates. The mechanisms are not fully understood, but sirtuins are important mediators of this response.

Resveratrol

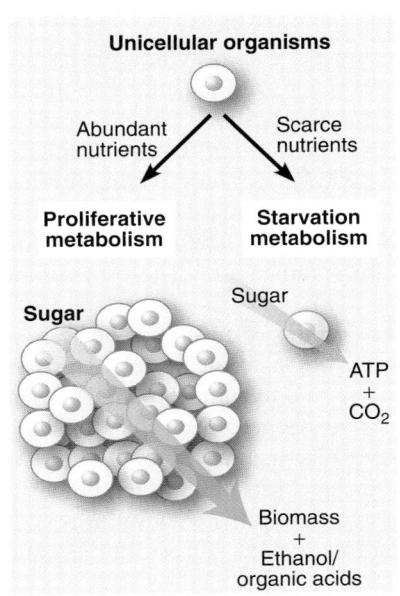
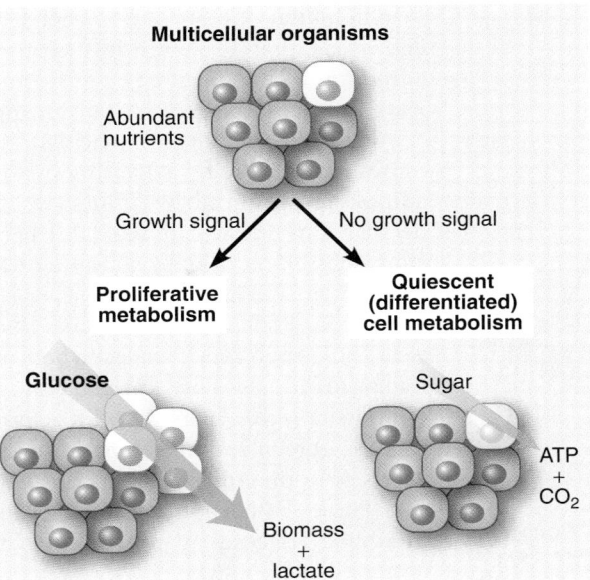

FIGURE 18.8

Proliferating and nonproliferating cells use different metabolic strategies to generate energy. Proliferative metabolism, in both unicellular and multicellular organisms, relies on glycolysis, a rapid but relatively inefficient process for generating ATP. This "fermentative" metabolism requires abundant nutrients. When nutrients are scarce, unicellular organisms adapt to a starvation metabolism, characterized by the slower, but more efficient, oxidative metabolism. This same efficient oxidative metabolism is used by nondividing, differentiated mammalian cells. Nutrient abundance is rarely an issue in multicellular organisms so that the switch between proliferative and quiescent metabolism is determined by the presence or absence of appropriate growth factors, rather than by nutrient availability.

From *Science* 324:1029–1033, M. G. Vander Heiden, L. C. Cantley, and C. B. Thompson, Understanding the Warburg effect: The metabolic requirements of cell proliferation. © 2009. Reprinted with permission from AAAS.

dependence on growth factors, and they switch to a proliferative metabolism. We now know that in most cases, cancer cells have acquired genetic mutations in components of the growth factor signaling pathways that normally control proliferation. For example, mutations that alter the signaling activity of PI3K are the most common genetic mutations found in cancer cells. We will see several other examples in Chapter 23.

In mammals, the brain coordinates whole body energy homeostasis. The brain receives information about the quality and quantity of nutrients being consumed, the levels of fuels already present in the blood, and the amounts of energy present in various storage depots in the body. These fuel and hormonal signals converge in neurons in the **arcuate nucleus** of the hypothalamus, where food intake is controlled. Not surprisingly, AMPK and mTOR play central roles in integrating these signals in the hypothalamus, where activation of AMPK promotes food intake (Figure 18.9). Hypothalamic AMPK is activated in response to low levels of nutrients such as glucose, branched-chain amino acids, and free fatty acids. The most important endocrine regulators of food intake are insulin and leptin (inhibit intake), and **ghrelin** and **adiponectin** (promote intake). Each of these act via specific receptors on particular cells in the arcuate nucleus to initiate signaling pathways that converge on AMPK and mTOR.

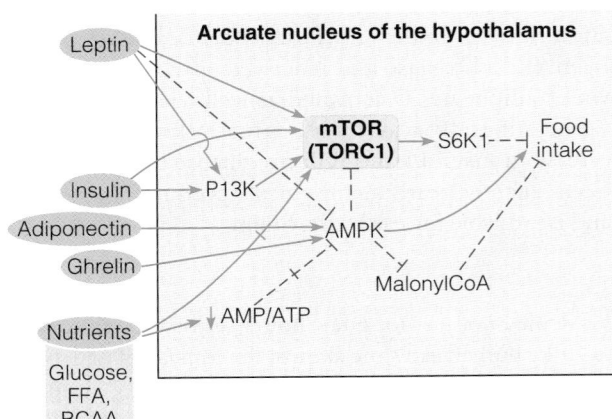

FIGURE 18.9

Fuel and hormonal control of food intake in the arcuate nucleus of the hypothalamus. Activation of AMPK promotes food intake; inhibition suppresses food intake. Green arrows indicate stimulatory effects; red bars indicate inhibitory effects. Insulin acts through the phosphoinositide 3-kinase (PI3K) cascade to inhibit food intake. Leptin has both direct and indirect effects (through PI3K and AMPK) on mTOR. Adiponectin and ghrelin stimulate food intake by activating AMPK. S6K, ribosomal protein S6 kinase; BCAA, branched-chain amino acids; FFA, free fatty acids.

Modified from *Annual Review of Nutrition* 28:295–311, S. C. Woods, R. J. Seeley, and D. Cota, Regulation of food intake through hypothalamic signaling networks involving mTOR. © 2008 Annual Reviews.

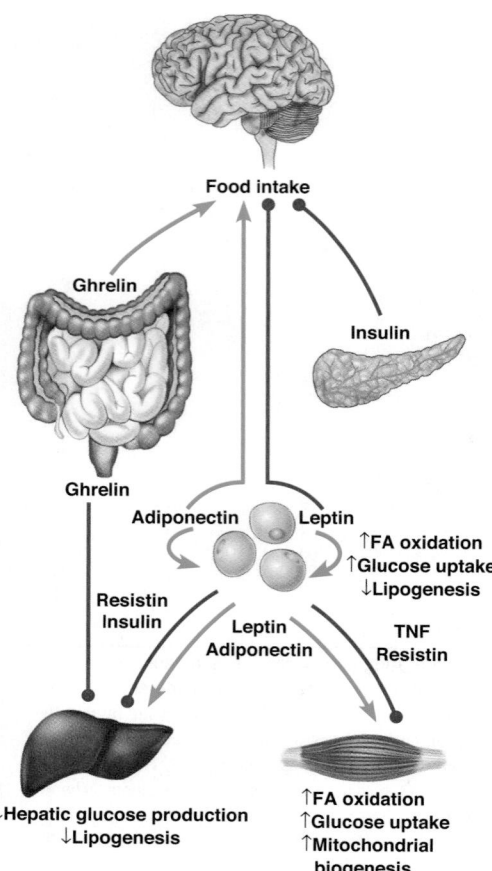

FIGURE 18.10

Endocrine regulation of food intake and energy homeostasis in mammals. This figure illustrates the effects of the major endocrine regulators on AMPK signaling in brain, adipose tissue, liver, and skeletal muscle. Green arrows indicate activation of AMPK; red circles indicate inhibition of AMPK. Resistin and TNF (tumor necrosis factor) are other adipokines not discussed here.

Modified with permission from *Physiological Reviews* 89:1025–1078, G. R. Steinberg and B. E. Kemp, AMPK in health and disease. © 2009 The American Physiological Society.

Insulin inhibits food intake through its activation of mTOR. As discussed at the end of Chapter 17, leptin is a peptide hormone released by adipocytes (an **adipokine**) when fat stores are adequate. Leptin activates mTOR and inhibits AMPK function in hypothalamus, suppressing food intake. Ghrelin, a small peptide hormone (28 amino acids) produced in cells lining the stomach, promotes food intake by activating hypothalamic AMPK. Adiponectin, another adipokine produced by adipocytes, is normally abundant in plasma, but decreases in obese individuals or people with type 2 diabetes. Adiponectin stimulates food intake by activating AMPK in hypothalamus. These effects are summarized in Figure 18.9.

Of course, all of these hormones have effects on tissues outside of the brain as well. Thus, leptin and adiponectin both activate AMPK signaling in skeletal muscle, leading to increased fatty acid oxidation, glucose uptake, and mitochondrial biogenesis (Figure 18.10). In liver, AMPK activation suppresses gluconeogenesis and lipogenesis. Ghrelin and insulin both suppress AMPK activity in liver, opposing the effects of leptin. Adiponectin has insulin-sensitizing effects on liver and other tissues. Finally, both leptin and adiponectin have autocrine activity, activating AMPK in adipocytes, leading to increased fatty acid oxidation, glucose uptake, and decreased lipogenesis.

Responses to Metabolic Stress: Starvation, Diabetes

An excellent way to understand how the interorgan and hormonal relationships we have discussed actually integrate fuel metabolism is to examine the effects of metabolic stress. In this section we consider two examples—prolonged fasting, in which the intake of fuel substrates is inadequate; and **diabetes mellitus**, in which a functional insufficiency of insulin impairs the ability of the body to use glucose, even when the sugar is present in abundance.

First let us review how glucose levels are maintained during normal feeding cycles (Figure 18.11). The blood glucose elevation occurring shortly after a carbohydrate-containing meal stimulates the secretion of insulin and suppresses the secretion of glucagon. Together these effects promote uptake of glucose into the liver, stimulate glycogen synthesis, and suppress glycogen breakdown. Flux through hexokinase IV increases in response to elevated glucose levels, providing substrates for glycogen synthesis. In addition, activation of acetyl-CoA carboxylase in the liver stimulates fatty acid synthesis, with subsequent transport to adipose tissue as triacylglycerols in very low-density lipoproteins. There, the increased levels of glycolytic intermediates and fatty acids stimulate triacylglycerol synthesis. Finally, increased glucose uptake into muscle increases levels of substrates for glycogen synthesis in that tissue as well.

Several hours later, when blood glucose levels begin to fall, the above events are reversed. Insulin secretion slows and glucagon secretion increases. This promotes glycogen mobilization in liver via the cAMP-dependent cascade mechanisms that activate glycogen phosphorylase and inactivate glycogen synthase. Triacylglycerol breakdown in adipocytes is activated as well, via the action of hormone-sensitive lipase, generating fatty acids for use as fuel by liver and muscle. At the same time, the decrease in insulin levels reduces glucose use by muscle, liver, and adipose tissue. Consequently, nearly all the glucose produced in the liver is exported to the blood and is available for use by the brain.

Starvation

Suppose that food intake is denied not just for a few hours, as just described, but for many days. Given that a 70-kg human can store at most the equivalent of 6700 kJ of energy as glycogen, this source of blood glucose will be exhausted in just a few hours. Because it is critical for brain function that blood glucose levels be

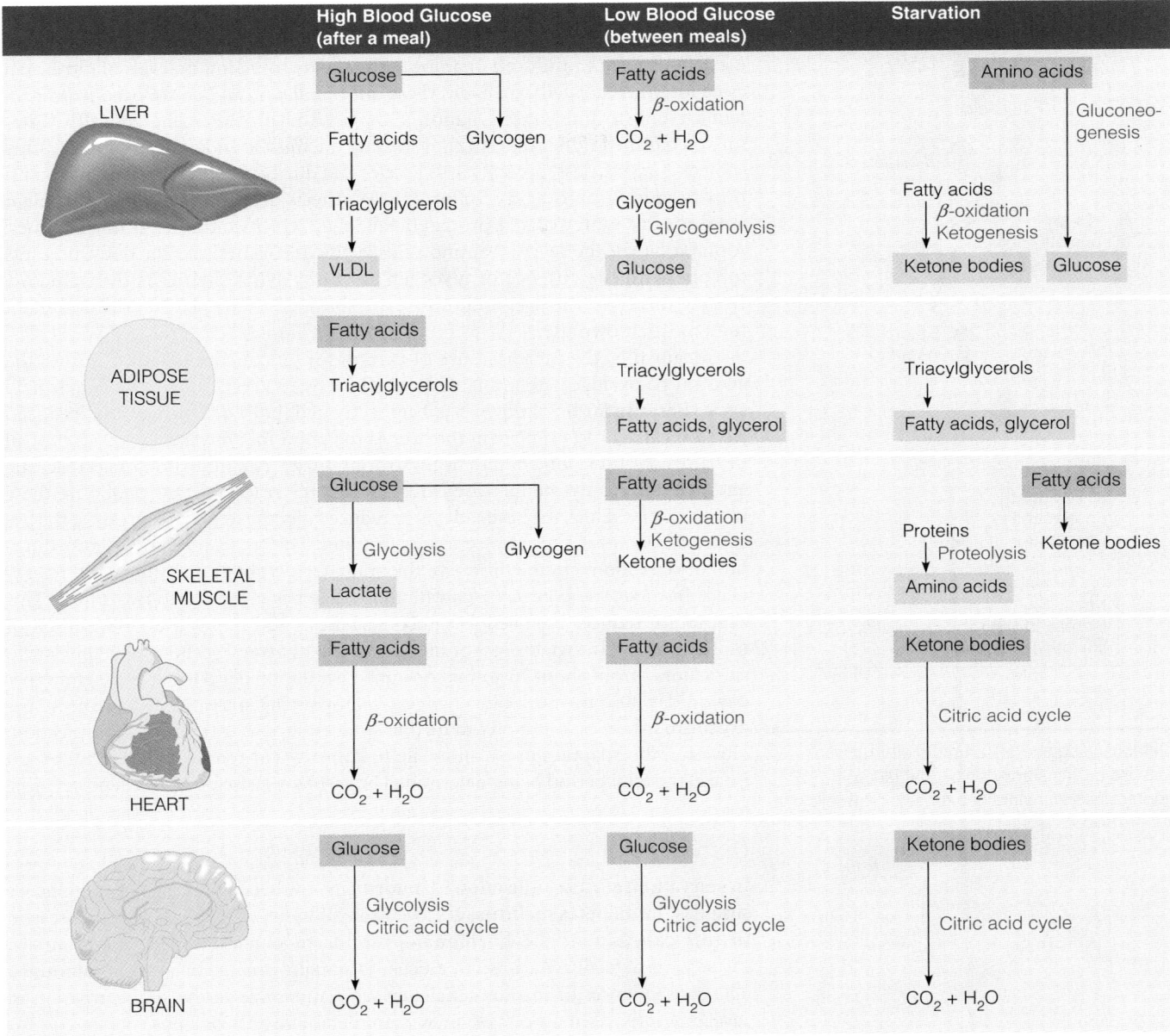

FIGURE 18.11

Major events in the storage, retrieval, and use of fuels in the fed and unfed states and in early starvation. Purple indicates fuels imported into the tissue; green indicates fuels exported from tissue.

maintained near 4.4 mM, the organism adapts metabolically to increase the use of fuels other than carbohydrate, primarily fat.

Before we discuss the metabolic adjustments involved, let us consider the other major energy stores: about 565,000 kJ as triacylglycerol, largely in adipose tissue, and 100,000 kJ as mobilizable proteins, largely in muscle. These stores provide sufficient energy to permit survival for up to several months. However, use of these stores presents problems. Triacylglycerol mobilization generates metabolic fuel largely in the form of acetyl-CoA, whose further oxidation in the citric acid cycle requires oxaloacetate. Recall from Chapter 14 that oxaloacetate and other citric acid cycle intermediates are used in other metabolic reactions and must be replenished via anaplerotic pathways. The most important of these processes is the pyruvate carboxylase reaction, with most of the pyruvate coming from carbohydrate catabolism. When carbohydrate availability is limited, the resupply of citric acid cycle intermediates is limited, and flux through the cycle may be reduced.

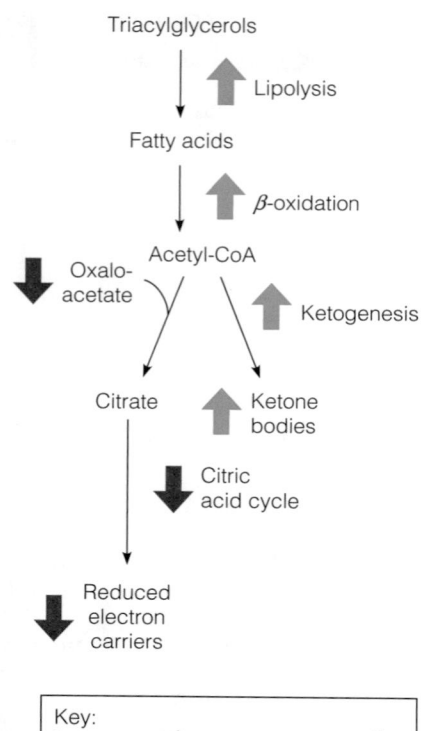

Key:
Increased (⬆) and decreased (⬇) flux during starvation.

——————————————•

Metabolic adaptations promote alternative fuel use during starvation so that glucose homeostasis is maintained for several weeks.

During carbohydrate limitation, citric acid cycle intermediates can be provided from other sources. For example, the glycerol released from lipolysis can be used, but it is not produced in amounts adequate to maintain levels of citric acid cycle intermediates. Alternatively, these intermediates can be produced from protein catabolism and transamination. However, this process is energetically wasteful and has the undesirable effect of wasting the muscle and weakening the fasting subject. Nevertheless, proteolysis is accelerated during the first few days of starvation because amino acids for protein synthesis are not present in sufficient amounts to counterbalance protein breakdown, which continues at normal rates. A major fate of the released amino acids is gluconeogenesis, as the body attempts to cope with the absence of glycogen stores by synthesizing its own glucose. During this time, the liver and muscle are shifting to fatty acids as the dominant fuels for their own use.

Meanwhile, the increased use of carbon for gluconeogenesis diminishes the amount of oxaloacetate available to combine with acetyl-CoA in the citric acid cycle. Because fat breakdown has been activated, both acetyl-CoA and reduced electron carriers accumulate in the liver to the point that the acetyl-CoA cannot all be oxidized, and ketone bodies begin to accumulate. Accumulation of acetoacetate and β-hydroxybutyrate increases flux through the reactions that catabolize these ketone bodies. Thus, the brain adapts to reduced glucose levels by increasing the use of ketone bodies as alternative energy substrates. This trend continues for the duration of starvation. On the third day, the brain derives about one-third of its energy needs from ketone bodies; by day 40, that usage has increased to two-thirds. This adaptation reduces the need for gluconeogenesis and spares the mobilization of muscle protein. In fact, the loss of muscle protein *decreases* by about four-fold late in starvation—from about 75 grams consumed per day on day 3 to about 20 grams per day on day 40. The metabolic changes accompanying starvation compromise the organism's abilities to respond to further stresses, such as extreme cold or infection. However, the adaptations do allow life to continue for many weeks without food intake, the total period being determined largely by the size of the fat deposits.

Diabetes

In starvation, glucose utilization is abnormally low because of inadequate glucose supplies. In **diabetes mellitus**, glucose utilization is similarly low, but the reason in this case is that the hormonal stimulus to glucose utilization—namely, insulin—is defective. As a result, glucose is actually present in excessive amounts. The consequences of insulin deficiency are comparable to those of starvation in revealing important aspects of interorgan metabolic relationships.

Diabetes is a major public health issue, having reached epidemic proportions in the United States and around the world. It is estimated that more than 12% of the adult population in the United States is afflicted with this disease. Diabetes is not a single disease but rather a family of diseases. **Type 1 diabetes**, formerly called insulin-dependent diabetes, or juvenile diabetes because of its typical early onset, often involves autoimmune destruction of the β cells of the pancreas, which can be caused by various factors, including viral infection. Some forms of type 1 diabetes have a genetic origin. Mutations in insulin structure can render the hormone inactive, and other mutations cause defects in the conversion of preproinsulin or proinsulin to the active hormone (see Chapter 5). Either way, type 1 diabetes is characterized by insulin deficiency, and can be treated by administration of insulin. **Type 2 diabetes**, formerly called adult-onset diabetes, obesity-related diabetes, or non-insulin-dependent diabetes mellitus, is characterized by insulin resistance—patients cannot respond to therapeutic doses of insulin. Type 2 diabetes accounts for more than 95% of people with diabetes.

The specific defects that lead to insulin resistance in type 2 diabetes are not known, but several clues are beginning to shed some light on the disease. The first

——————————————•

Diabetes results either from insulin deficiency or from defects in the insulin response mechanism.

is that most people with type 2 diabetes are also obese. In fact, obesity is so closely associated with insulin resistance that there must be a mechanistic link. Like diabetes, the prevalence of obesity has increased dramatically in the United States since the 1970s, with 34% of adults classified as obese in 2010. The second clue comes from the close relationship between type 2 diabetes and **metabolic syndrome**. Metabolic syndrome, which afflicts some 50 million Americans, is defined by abdominal obesity, hypertension, high blood sugar, and, most importantly, insulin resistance. These metabolic abnormalities often precede both cardiovascular disease and diabetes. Common to both obesity and metabolic syndrome is excess fuel intake and abnormal accumulation of lipid in "ectopic sites," primarily liver and skeletal muscle. Excess lipids initially accumulate in adipose cells, increasing their size. Eventually, fuel intake exceeds the storage capacity of adipose tissue, and excess lipids are shunted to ectopic sites. There is growing evidence that this abnormal lipid accumulation causes the insulin resistance by affecting downstream signaling pathways. Two related mechanisms have gained wide support as potential causes of the disease. The **lipid overload** hypothesis states that when fat accumulates in muscle cells, it blocks the insulin signaling pathway that normally stimulates translocation of GLUT4 (the major glucose transporter in muscle) to the plasma membrane (see Figure 18.5). Thus, insulin no longer efficiently stimulates glucose transport, that is, the cell is insulin resistant. The **inflammation** hypothesis states that as adipose cells increase in size with excess lipids, they secrete inflammatory adipokines and cytokines, including TNF-α, interleukins, and resistin. These cytokines bind their receptors in peripheral tissues such as muscle and stimulate an inhibitory phosphorylation of IRS, making it a poor substrate for the insulin receptor tyrosine kinase (see Figure 18.5). Once again, insulin signaling is shut down, and the cell becomes insulin resistant. Thus, although type 2 diabetes has long been characterized by defects in carbohydrate metabolism, abnormal lipid metabolism may be at the root of the disease.

> Two related mechanisms have been proposed as causes of type 2 diabetes: the lipid overload hypothesis and the inflammation hypothesis.

The antidiabetic drugs mentioned in Chapter 17 (the thiazolidinediones; page 750) increase the insulin sensitivity of people with type 2 diabetes by decreasing the expression of inflammatory adipokines that induce insulin resistance and by increasing production of adiponectin, which has insulin-sensitizing effects. These drugs act through PPARγ, the same transcription factor that PGC-1α coactivates. PPARγ is essential for fat cell differentiation and maintenance of normal adipocyte function and is a master regulator of cellular metabolism. Unfortunately, due to the pleiotropic roles of PPARγ, this class of drugs has serious side effects, including weight gain, osteoporosis, and heart failure. Recent studies have led to a better understanding of the mechanism by which these drugs activate PPARγ and provide an opportunity to develop new antidiabetic drugs with fewer side effects.

Whatever the cause of the functional insulin deficiency, diabetes can truly be called "starvation in the midst of plenty." The insufficient production of insulin or the failure of insulin to act normally in promoting glucose utilization, with resultant glucose accumulation in the blood, starves the cells of nutrients and promotes metabolic responses similar to those of fasting (Figure 18.12). Liver cells attempt to generate more glucose by stimulating gluconeogenesis. Most of the substrates come from amino acids, which in turn come largely from degradation of muscle proteins. Glucose cannot be reused for resynthesis of amino acids or of fatty acids, so a person with diabetes may lose weight even while consuming what would normally be adequate calories in the diet.

> Diabetes can be thought of as "starvation in the midst of plenty" because cells are unable to utilize the glucose that accumulates in the blood.

As cells attempt to generate usable energy sources, triacylglycerol depots are mobilized in response to the abnormally low insulin-to-glucagon ratio. Fatty acid oxidation is elevated, with concomitant generation of acetyl-CoA. Flux through the citric acid cycle may decrease because of the accumulation of reduced electron carriers and/or oxaloacetate limitation. In liver, both effects accelerate ketone body formation, generating increased levels of organic acids in the blood

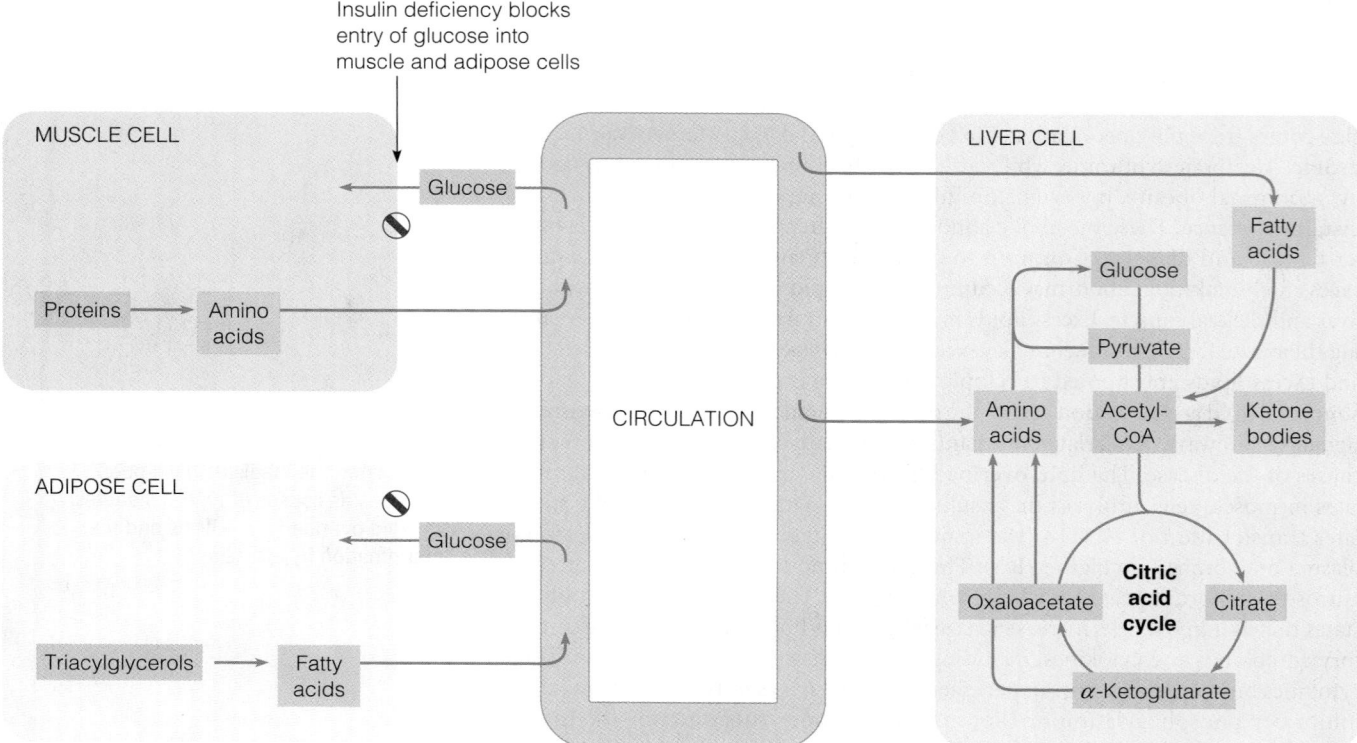

Insulin deficiency blocks entry of glucose into muscle and adipose cells

FIGURE 18.12

The metabolic abnormalities in diabetes. The insulin deficiency blocks the uptake of glucose into muscle and adipose tissue and reduces glucose catabolism in all tissues. Proteolysis in muscle and lipolysis in adipose tissue are enhanced. In the liver, gluconeogenesis from amino acids and citric acid cycle intermediates is stimulated as the cells attempt to remedy the perceived lack of usable glucose, and fatty acid oxidation and ketogenesis are also increased. Green indicates pathways activated; pink indicates pathways diminished.

(ketosis). These acids can lower the blood pH from the normal value of 7.4 to 6.8 or lower (ketoacidosis). Decarboxylation of acetoacetate, which is stimulated at low pH, generates acetone, which can be smelled on the breath of patients with severe ketoacidosis. A special danger is that such people may lose consciousness, and this, coupled with a sweet organic odor on the breath, may give the impression that they are intoxicated, when in fact their lives are in jeopardy.

The excessive concentrations of glucose in body fluids generate other metabolic problems, quite different from anything seen in starvation. At blood glucose levels above 10 mM, the kidney can no longer reabsorb all of the glucose out of the blood filtrate, and glucose is spilled into the urine, sometimes in amounts approaching 100 grams per day. In fact, the Latin name *diabetes mellitus* literally means "honey-sweet urine." Glucose excretion creates an osmotic load, which causes large amounts of water to be excreted as well, and under these conditions the kidney cannot reabsorb most of this water. In fact, the earliest indications of diabetes are often frequent and excessive urination, coupled with excessive thirst. Long before biochemistry was a science, the loss of nutrients, excessive urination, and breakdown of fat and protein were recognized as hallmarks of diabetes. As early as the first century A.D., diabetes was described as "the flesh and bones running together into urine." It took another 1800 years or so before Israel Kleiner in New York, and Frederick Banting, Charles Best, James Collip, and John Macleod in Toronto discovered that extracts of dog pancreas possessed the ability to lower glucose levels and restore health to children and young adults suffering from diabetes. These studies culminated in the identification of insulin as the active component in 1922.

In type 1 diabetes, the metabolic imbalance is usually more severe and difficult to control than in the milder and more common type 2 diabetes. The latter can often be controlled by exercise and dietary restriction of carbohydrate, whereas treatment for type 1 diabetes involves daily self-injection of insulin. For many years this insulin was purified from bovine pancreas, and its high cost, coupled with occasional problems resulting from the minor structural differences between human and bovine insulin, led the fledgling biotechnology industry to attempt to produce human insulin through recombinant DNA techniques. In the late 1970s the gene for human insulin was cloned into *E. coli* in a form that allowed it to be expressed, and in 1982 cloned human insulin became the first recombinant DNA product to be approved for human use.

SUMMARY

Each organ or tissue of a multicellular organism has a distinctive profile of metabolic activities that allow it to serve its specialized functions. Remote tissues must remain in constant communication to maintain homeostasis. In vertebrates the most essential element of this homeostasis is maintenance of constant blood glucose levels, primarily for proper brain function. The actions of three hormones—insulin, glucagon, and epinephrine—play the dominant roles in contributing to glucose homeostasis. Insulin signals the fed state and promotes glucose utilization and synthesis of energy storage compounds. Glucagon acts primarily upon liver cells, increasing blood glucose by several mechanisms involving cyclic AMP. Epinephrine has similar effects on muscle cells. Energy homeostasis, maintaining the balance of fuel intake with the metabolism and storage of nutrients to meet energy needs, is coordinated by a complex intracellular regulatory system. Two protein kinases, AMPK and mTOR, play central roles in orchestrating the metabolic activity of mammalian cells. The sirtuins, a highly conserved family of NAD^+-dependent protein deacetylases, act as metabolic sensors of the cellular redox state. The response to metabolic stresses such as starvation and diabetes reveals the interorgan and hormonal relationships that integrate fuel metabolism in mammals.

REFERENCES

Hormonal Regulation of Fuel Metabolism

Cheng, Z., Y. Tseng, and M. F. White (2010) Insulin signaling meets mitochondria in metabolism. *Trends Endocrinol. Metab.* 21:589–598. This review summarizes recent studies on insulin and mitochondrial metabolism, including links to the SIRT1/PGC1α pathway.

Fernandez-Marcos, P. J., and J. Auwerx (2011) Regulation of PGC-1α, a nodal regulator of mitochondrial biogenesis. *Am. J. Clin. Nutr.* 93:884S–890S.

Saggerson, D. (2008) Malonyl-CoA, a key signaling molecule in mammalian cells. *Ann. Rev. Nutr.* 28:253–272. This article reviews the diverse range of metabolic functions for this fatty acid intermediate.

Sugden, M. C., M. G. Zariwala, and M. J. Holness (2009) PPARs and the orchestration of metabolic fuel selection. *Pharmacol. Res.* 60:141–150. This review describes recent advances in our understanding of the role of the peroxisome proliferator-activated receptors in the control of fuel selection.

Wahren, J., and K. Ekberg (2007) Splanchnic regulation of glucose production. *Annu. Rev. Nutr.* 27:329–345. This article reviews the contributions of hepatic gluconeogenesis and glycogenolysis to the supply of glucose for the peripheral organs.

Xiao, B., M. J. Sanders, E. Underwood, R. Heath, F. V. Mayer, D. Carmena, C. Jing, P. A. Walker, J. F. Eccleston, L. F. Haire, P. Saiu, S. A. Howell, R. Aasland, S. R. Martin, D. Carling, and S. J. Gamblin, (2011) Structure of mammalian AMPK and its regulation by ADP. *Nature* 472:230–233.

AMPK, mTOR, and Sirtuins

Bao, J., and M. N. Sack (2010) Protein deacetylation by sirtuins: Delineating a post-translational regulatory program responsive to nutrient and redox stressors. *Cell. Mol. Life Sci.* 67:3073–3087.

Donmez, G., and L. Guarente (2010) Aging and disease: Connections to sirtuins. *Aging Cell* 9:285–290. This article reviews the connections between calorie restriction, sirtuins, and aging in mammals.

Finkel, T., C. X. Deng, and R. Mostoslavsky (2009) Recent progress in the biology and physiology of sirtuins. *Nature* 460:587–591.

Guan, K.-L., and Y. Xiong (2011) Regulation of intermediary metabolism by protein acetylation. *Trends Biochem. Sci.* 36:108–116. This review highlights new studies into the role of reversible acetylation in the regulation of glycolysis and gluconeogenesis, citric acid cycle, glycogen metabolism, fatty acid metabolism, and the urea cycle and nitrogen metabolism.

Hallows, W. C., B. C. Smith, S. Lee, and J. M. Denu (2009) Ure(k)a! Sirtuins regulate mitochondria. *Cell* 137:404–406 and Huang, J. Y., M. D. Hirschey, T. Shimazu, L. Ho, and E. Verdin (2010) Mitochondrial sirtuins. *Biochim. Biophys. Acta* 1804:1645–1651. These two articles summarize new results on the roles of sirtuins in regulating mitochondrial processes such as the urea cycle.

Hardie, D. G., S. A. Hawley, and J. W. Scott (2006) AMP-activated protein kinase—development of the energy sensor concept. *J. Physiol.* 574:7–15.

Steinberg, G. R., and B. E. Kemp (2009) AMPK in health and disease. *Physiol. Rev.* 89:1025–1078. A comprehensive review of AMPK function.

Woods, S. C., R. J. Seeley, and D. Cota (2008) Regulation of food intake through hypothalamic signaling networks involving mTOR. *Annu. Rev. Nutr.* 28:295–311. A recent review on the control of food intake in mammals.

Yang, Q., and K. L. Guan (2007) Expanding mTOR signaling. *Cell Res.* 17:666–681. This article reviews the many facets of mTOR function in energy homeostasis.

Diabetes and Obesity

Colman, R. J., R. M. Anderson, S. C. Johnson, E. K. Kastman, K. J. Kosmatka, T. M. Beasley, D. B. Allison, C. Cruzen, H. A. Simmons, J. W. Kemnitz, and R. Weindruch (2009) Caloric restriction delays disease onset and mortality in rhesus monkeys. *Science* 325:201–204.

Erion, D. M., and G. I. Shulman (2010) Diacylglycerol-mediated insulin resistance. *Nat. Med.* 16:400–402. A short review on the role of diacylglycerol in the lipid overload hypothesis of diabetes.

Friedman, J. M. (2010) A tale of two hormones. *Nat. Med.* 16:1100–1106. A brief history of the discoveries of insulin and leptin, written by the discoverer of leptin.

Houtkooper, R. H., and J. Auwerx (2010) Obesity: New life for antidiabetic drugs. *Nature* 466:443–444. This commentary discusses a new result that changes our view of how the thiazolidinediones work in the treatment of diabetes.

Hummasti, S., and G. S. Hotamisligil (2010) Endoplasmic reticulum stress and inflammation in obesity and diabetes. *Circ. Res.* 107:579–591. A detailed review on the inflammation hypothesis of diabetes.

Savage, D. B., K. F. Petersen, and G. I. Shulman (2007) Disordered lipid metabolism and the pathogenesis of insulin resistance. *Physiol. Rev.* 87:507–520. A detailed review of the lipid overload hypothesis.

PROBLEMS

1. Marathon runners preparing for a race engage in "carb loading" to maximize their carbohydrate reserves. This involves eating large quantities of starchy foods. Why is starch preferable to candy or sugar-rich foods?

2. Supposing that an average human consumes energy at the rate of 1500 kcal/day at rest and that long-distance running consumes energy at 10 times that rate, how long would the glycogen reserves last during a marathon run? Recall that 1 kcal is equivalent to 4.183 kJ.

3. What proportion of the total energy consumption supports brain function in an average resting human? What proportion in a human running in a marathon?

4. Proteolysis increases during the early phases of fasting, but later it decreases as the body adapts to using alternative energy sources. Given that feedback control mechanisms have not been described for intracellular proteases, how might you explain these apparent changes in protease activity?

5. Glucose has been found to react nonenzymatically with hemoglobin, through Schiff base formation between C-1 of glucose and the amino termini of the β chains. How might this finding be applied in monitoring diabetic patients?

6. Ketone bodies are exported from liver for use by other tissues. Because many tissues can synthesize ketone bodies, what enzymatic property of liver might contribute to its special ability to export these compounds?

7. Adipose tissue cannot resynthesize triacylglycerols from glycerol released during lipolysis (fat breakdown). Why not? Describe the metabolic route that is used to generate a glycerol compound for triacylglycerol synthesis.

8. (a) Briefly describe the relationship between intracellular malonyl-CoA levels in the liver and the control of ketogenesis.
   (b) Describe how the action of glucokinase helps the liver to buffer the level of blood glucose.

9. The action of glucagon on liver cells leads to inhibition of pyruvate kinase. What is the most probable mechanism for this effect?

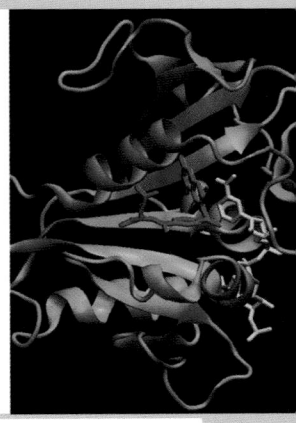

CHAPTER 19

Lipid Metabolism II: Membrane Lipids, Steroids, Isoprenoids, and Eicosanoids

Chapter 17 was concerned primarily with energetic aspects of lipid metabolism—synthesis and oxidation of fatty acids, interorgan transport of lipoproteins, and metabolism of triacylglycerols. In addition to their roles in energy storage, lipids function as membrane components and as biological regulators. Our attention now shifts to the roles played by these more complex lipids, as well as pathways for their synthesis and degradation. The cell biology of some of these processes is schematized in Figure 19.1.

This chapter focuses on the following major classes: *glycerophospholipids* (also called *phosphoglycerides*), which are primarily membrane components but which also play some specialized regulatory roles; *sphingolipids*, which in animals are found abundantly in nervous tissue; *steroids* and other *isoprenoid* compounds, which function as hormones, vitamins, and membrane constituents; and the *eicosanoids*, a class of biological regulators synthesized from arachidonic acid. Two major topics involving lipids and regulation are discussed later in this book: the actions of steroid hormones (in Chapters 23 and 27) and the second-messenger regulatory role of inositol phospholipids (in Chapter 23).

Metabolism of Glycerophospholipids

The most abundant phospholipids are those derived from glycerol. These **glycerophospholipids** are found primarily as components of membranes. Membrane phospholipids are also metabolic precursors to various regulatory elements of signal transduction pathways. In animals, phospholipids also participate in the transport of triacylglycerols and cholesterol, as discussed in Chapter 17, by forming the surface of lipoproteins. In addition, phospholipids play specific roles in processes as diverse as blood clotting and lung functioning. Pathways of glycerophospholipid synthesis are outlined in Figure 19.2 and are presented in more detail in the following sections. The major membrane lipids other than glycerophospholipids are the sphingolipids.

FIGURE 19.1

Intracellular synthesis and transport of membrane phospholipids.

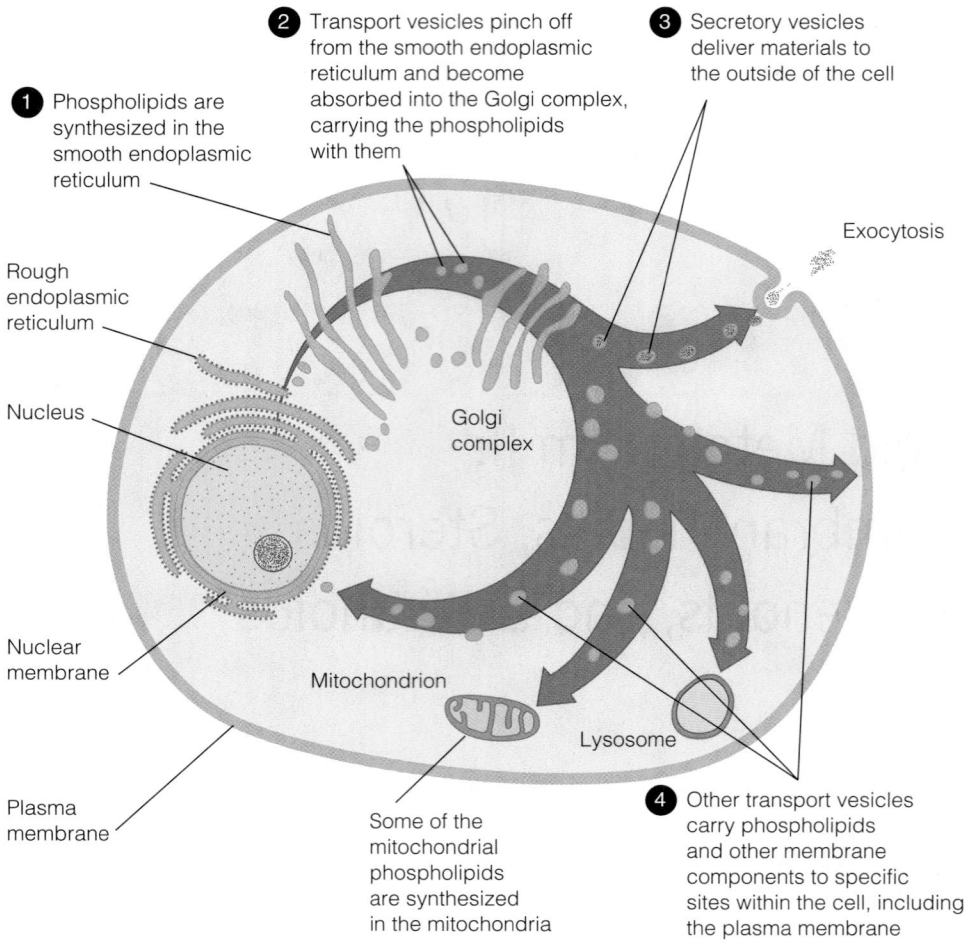

1 Phospholipids are synthesized in the smooth endoplasmic reticulum

2 Transport vesicles pinch off from the smooth endoplasmic reticulum and become absorbed into the Golgi complex, carrying the phospholipids with them

3 Secretory vesicles deliver materials to the outside of the cell

Exocytosis

Rough endoplasmic reticulum

Nucleus

Golgi complex

Nuclear membrane

Mitochondrion

Lysosome

Plasma membrane

Some of the mitochondrial phospholipids are synthesized in the mitochondria

4 Other transport vesicles carry phospholipids and other membrane components to specific sites within the cell, including the plasma membrane

E. coli membranes contain just three different phospholipids, which contain predominantly three different fatty acids.

Biosynthesis of Glycerophospholipids in Bacteria

We begin our discussion of phospholipid metabolism in the prokaryotic kingdom, partly because of the relative simplicity of the biological systems involved. In bacteria, phospholipids may constitute 10% of the dry weight of the cell, yet their only known role is as components of membranes. The membranes of *E. coli* contain only three phospholipids in significant amounts. Phosphatidylethanolamine (PE) constitutes 70%–80% of the total phospholipid, with phosphatidylglycerol (PG) and cardiolipin (CL) making up the remaining 20%–30%. The relative amounts of PG and CL depend on growth conditions, with PG more abundant in logarithmically growing cells and CL more abundant in stationary phase cells. The fatty acid content of these lipids is also simple, with three dominating— palmitate (16:0), palmitoleate (16:1cΔ9), and *cis*-vaccenate (18:1cΔ11).

Because they are easy to grow in large quantities, bacteria provide abundant sources for large-scale isolation of the enzymes involved in lipid metabolism. Much of our earliest information on both phospholipid synthesis and fatty acid synthesis came from studies with *E. coli*. Whereas the fatty acid chains are synthesized by a soluble type II fatty acid synthase (Chapter 17), all of the enzymes involved in phospholipid synthesis are bound to the plasma (cytoplasmic) membrane. More recently, studies on bacterial mutants have offered insight into control of membrane lipid synthesis and, in particular, the mechanisms used for temperature regulation of the fatty acid content of these lipids. Recall from Chapter 10 that, when grown at low temperature, bacteria increase the unsaturation of their membrane fatty acids to maintain optimal fluidity. One of the most intriguing areas of contemporary lipid biochemistry involves genetic analysis of mechanisms that maintain the optimal pattern of unsaturation at a given temperature.

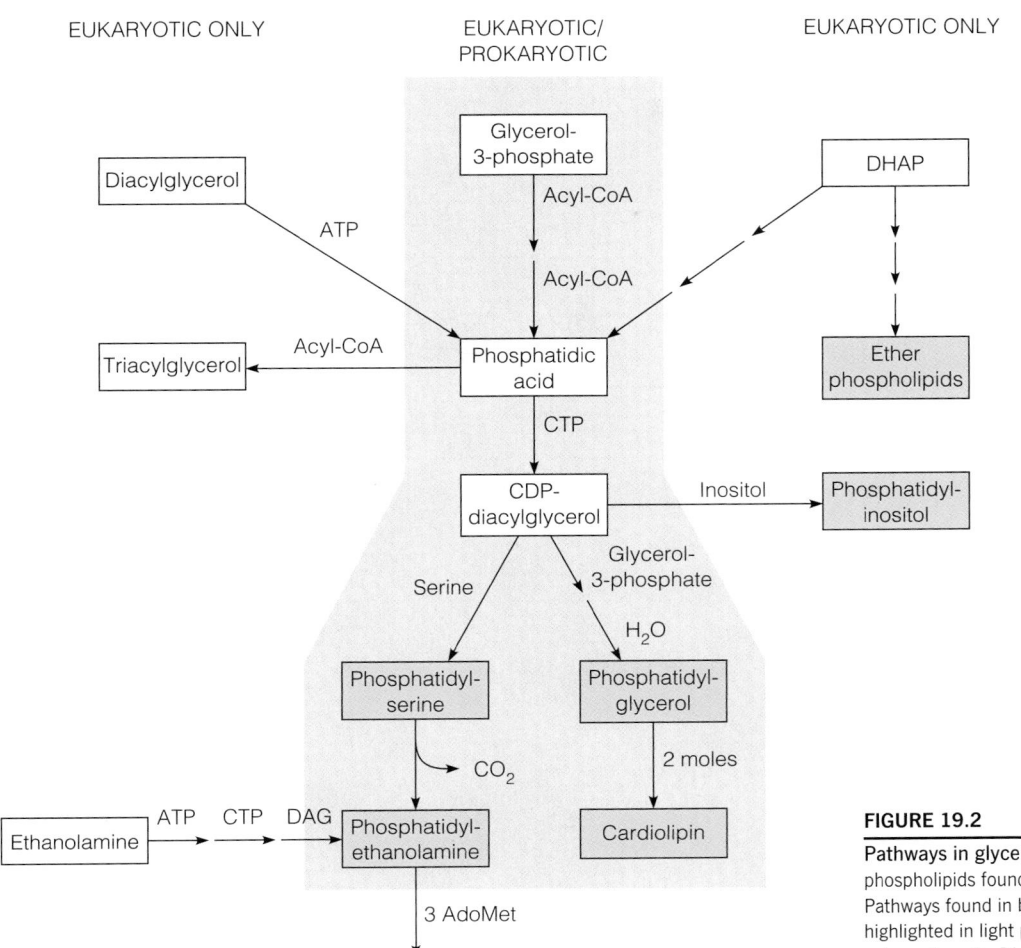

FIGURE 19.2

Pathways in glycerophospholipid biosynthesis. The major phospholipids found in membranes are shown in orange. Pathways found in both prokaryotic and eukaryotic cells are highlighted in light purple. Other reactions are confined to eukaryotic cells. DHAP = dihydroxyacetone phosphate; DAG = diacylglycerol; AdoMet = S-adenosylmethionine (see page 785).

Biosynthesis of Phosphatidic Acid and Polar Head Groups

In Chapter 17 we described the synthesis of phosphatidic acid (diacylglycerol-3-phosphate), starting from L-sn-glycerol-3-phosphate (refer to Chapter 10, page 365 for description of the stereospecific numbering system). As shown in Figure 19.2, phosphatidic acid represents a branch point between the syntheses of triacylglycerols and phospholipids. The energy cofactor for phospholipid biosynthesis is cytidine triphosphate (CTP), whose role is similar to that of UTP in polysaccharide synthesis (Chapter 9).

Recall that phosphatidic acid is synthesized by two successive acylations of glycerol-3-phosphate. In bacteria these 16- to 18-carbon fatty acyl groups are borne on acyl carrier protein (ACP) (Figure 19.3), the products of fatty acid synthase (Chapter 17). Two different glycerol phosphate acyltransferase enzymes are involved, as revealed in part by their differences in specificity for fatty acyl-ACPs: About 90% of the acyl groups esterified at position 1 are saturated, whereas 90% at position 2 are unsaturated. Phosphatidic acid next becomes metabolically activated by reaction with CTP. Mechanistically, this reaction is reminiscent of the activation of glucose-1-phosphate by UTP to yield UDP-glucose (Chapter 13, page 565). The phosphoryl oxygen of phosphatidic acid attacks the α phosphorus atom of CTP to form **CDP-diacylglycerol** and pyrophosphate. The $\Delta G^{\circ\prime}$ of this phosphoanhydride exchange is nearly zero. However, the reaction is pulled to completion by the highly exergonic hydrolysis of the PP$_i$.

CDP-diacylglycerol is now activated (CMP is a good leaving group) for nucleophilic attack by the various polar head groups. In one reaction sequence

Phosphatidic acid is a branch metabolite between triacylglycerol biosynthesis and phospholipid biosynthesis.

Metabolic activation of phospholipid precursors is carried out by reaction with CTP.

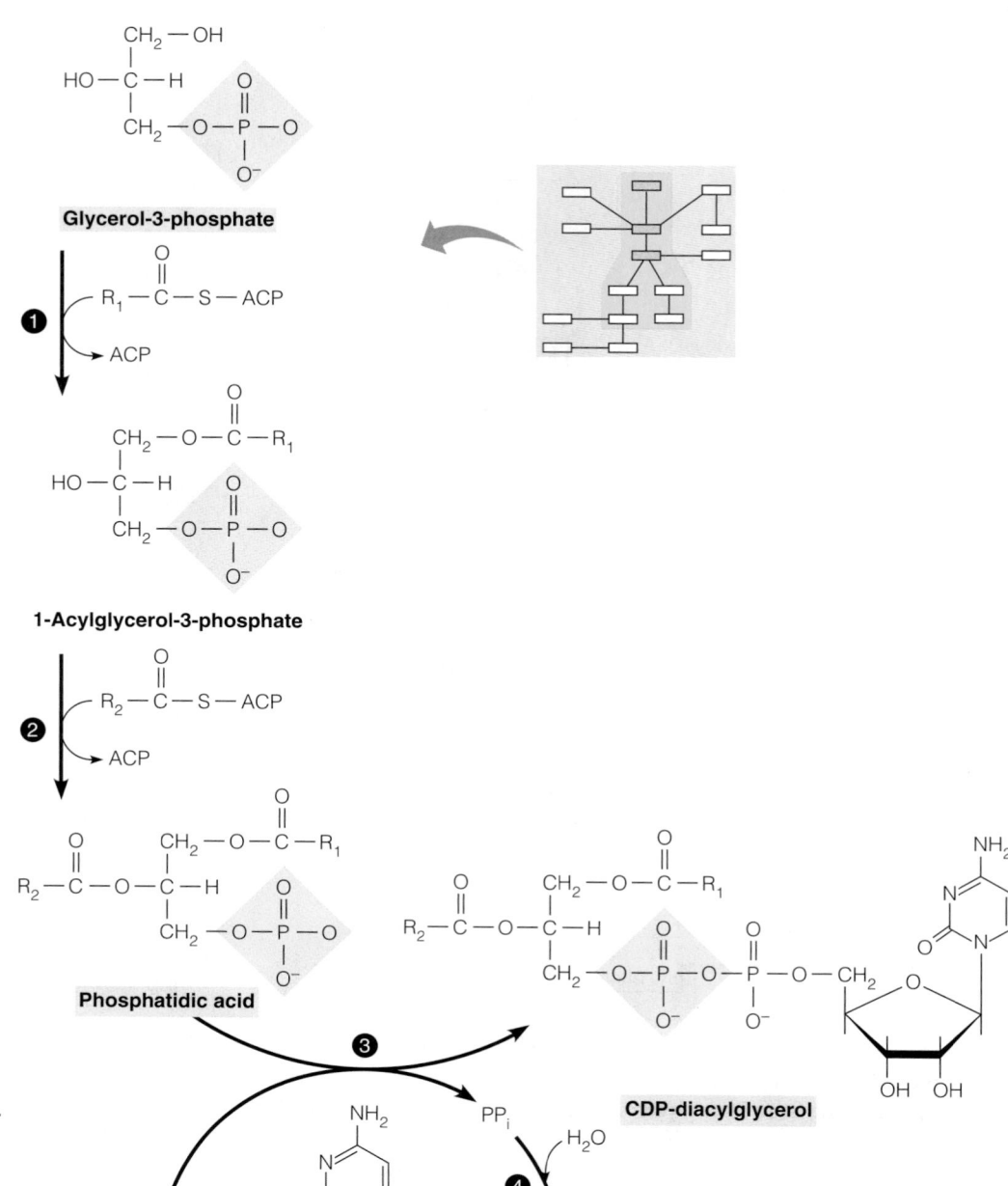

FIGURE 19.3

Synthesis of phosphatidic acid and CDP-diacylglycerol in bacteria. Two separate acyltransferases are involved in phosphatidic acid synthesis (reactions ❶ and ❷). The reaction of CTP with phosphatidic acid is catalyzed by CDP-diacylglycerol synthase (reaction ❸). This reaction is drawn to the right by the enzymatic hydrolysis of pyrophosphate, catalyzed by the ubiquitous pyrophosphatase (reaction ❹).

(Figure 19.4, left side), CMP is exchanged for serine, giving **phosphatidylserine**, which immediately undergoes decarboxylation to **phosphatidylethanolamine**. Consequently, phosphatidylserine does not accumulate in bacteria. In the other pathway (Figure 19.4, right side) the C1 hydroxyl of glycerol-3-phosphate attacks the highlighted phosphorus atom of CDP-diacylglycerol, eliminating CMP to form phosphatidylglycerol-3-phosphate, followed by a phosphatase reaction to give phosphatidylglycerol. Reaction with another mole of phosphatidylglycerol

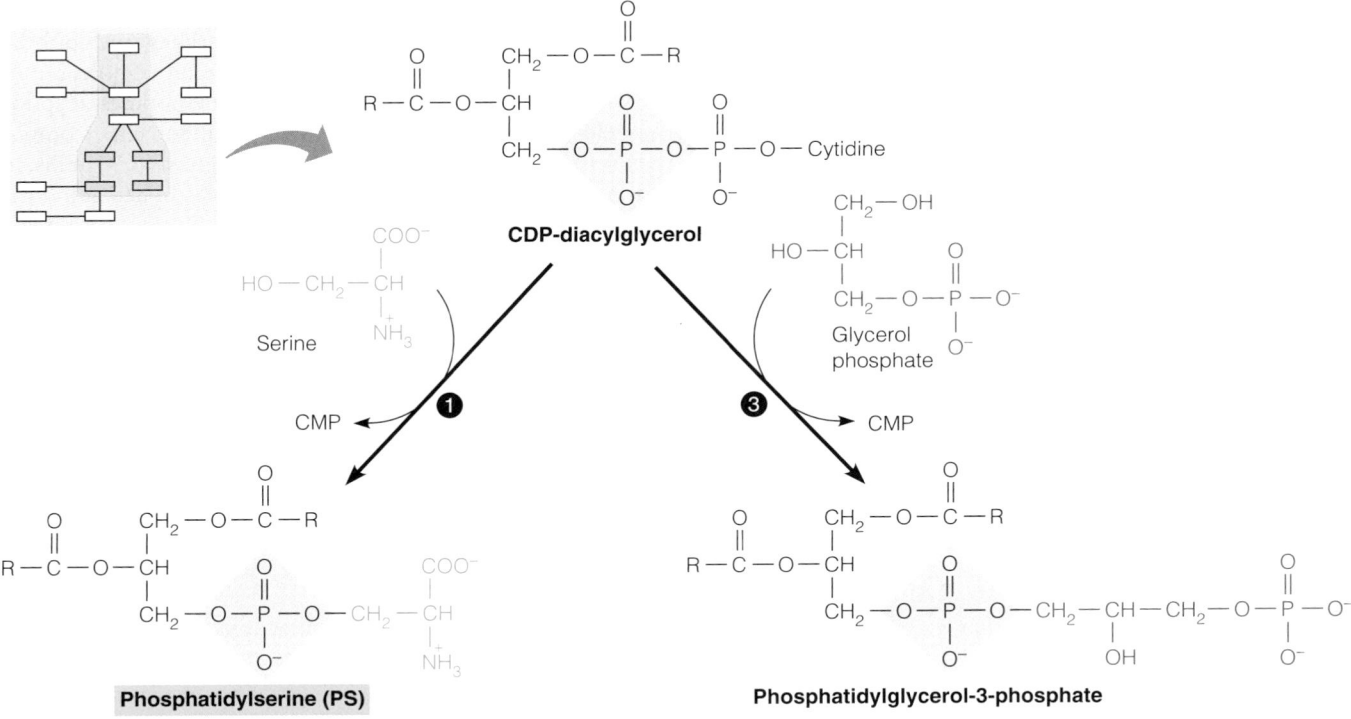

FIGURE 19.4

Synthesis of polar head groups of bacterial phospholipids. CDP-diacylglycerol is the precursor for both branches of the pathway. The left branch produces the zwitterionic PS and PE; the right branch produces the anionic PG and CL. ❶ Phosphatidylserine synthase; ❷ Phosphatidylserine decarboxylase; ❸ Phosphatidylglycerol-3-phosphate synthase; ❹ Phosphatidylglycerol-3-phosphate phosphatase; ❺ Cardiolipin synthase.

gives **diphosphatidylglycerol**, or **cardiolipin**. In this reaction, the phosphoryl group of one of the phosphatidylglycerol molecules undergoes nucleophilic attack by the glycerol C1 OH group of the other, displacing a molecule of glycerol. Cardiolipin, which is particularly abundant in the membranes of spirochetes (motile chemoheterotrophic bacteria), is the principal antigenic component measured in the Wassermann test, formerly used for diagnosis of syphilis. In *E. coli*, phosphatidylglycerol and cardiolipin play specific roles in activating the protein product of the *dnaA* gene, involved in initiation of DNA replication at a membrane site (Chapter 25, page 1063).

Not shown in Figure 19.2 is the fact that phosphatidylethanolamine and phosphatidylglycerol both turn over relatively rapidly in bacterial membranes. The glycerol-1-P head group of phosphatidylglycerol is transferred to membrane-derived oligosaccharides, which function in the periplasmic space to regulate osmotic pressure. The other product is diacylglycerol, which can be converted to phosphatidic acid by diacylglycerol kinase. Phosphatidylethanolamine turns over by transfer of the fatty acid at its 1-position to membrane lipoproteins in the periplasm. The other product of this reaction is 2-acylglycerolphosphoethanolamine, which can be re-acylated to form phosphatidylethanolamine.

Control of Phospholipid Synthesis in Prokaryotes

The genetic analysis of *E. coli* phospholipid metabolism is fairly advanced, in the sense that the structural genes for most of the enzymes involved have been identified, mutant phenotypes have been analyzed in detail, and all of the genes have been cloned and sequenced. However, we still know rather little about how phospholipid synthesis is regulated. Current evidence suggests that the rate of phospholipid synthesis is controlled primarily at the level of fatty acid synthesis by the long-chain acyl-ACP end products. Acyl-ACPs inhibit both the production of malonyl-CoA by acetyl-CoA carboxylase and the elongation of fatty acyl-ACPs by fatty acid synthase. Because fatty acids in bacteria are used primarily for membrane synthesis, and not as energy substrates, it makes good metabolic sense to limit the synthesis of phospholipids at the earliest steps committed to membrane formation. Furthermore, because the biophysical properties of membranes are determined in large part by the composition of the fatty acids that are produced de novo, control at the level of fatty acid synthesis is important for maintaining membrane homeostasis. Recall that the two branches of phospholipid synthesis compete for a common pool of CDP-diacylglycerol (Figure 19.4). Control of the phospholipid polar head group composition is determined largely by the activity of phosphatidylserine synthase, catalyzing the first reaction of the left branch. This enzyme (PssA in Figure 19.5) is a peripheral membrane protein that binds to and is activated by the anionic phospholipids phosphatidylglycerol and cardiolipin, both products of the right branch. Activated phosphatidylserine synthase increases the synthesis of

FIGURE 19.5

Regulation of membrane phospholipid composition in bacteria. Refer to text for details. PssA, Phosphatidylserine synthase; Psd, phosphatidylserine decarboxylase; PgsA, phosphatidylglycerol-3-phosphate synthase; PgpP, phosphatidylglycerol-3-phosphate phosphatase; CDP-DAG, CDP-diacylglycerol; PtdEtn, phosphatidylethanolamine; PtdGro, phosphatidylglycerol.

Modified with permission from Macmillan Publishers Ltd. *Nature Reviews Microbiology* 6:222–233, Y. M. Zhang and C. O. Rock, Membrane lipid homeostasis in bacteria. © 2008.

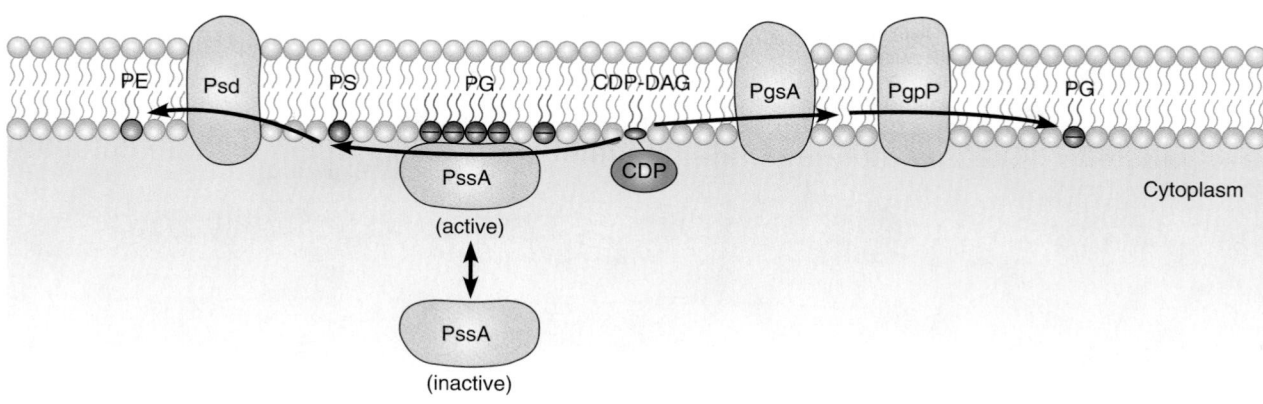

phosphatidylserine, which is converted to phosphatidylethanolamine by phosphatidylserine decarboxylase (Psd in Figure 19.5). Once the proportion of phosphatidylethanolamine in the bilayer reaches a certain threshold, PssA dissociates from the membrane and its activity decreases, allowing the right branch of the pathway to catch up [catalyzed by phosphatidylglycerol-3-phosphate synthase (PgsA), phosphatidylglycerol-3-phosphate phosphatase (PgpP), and cardiolipin synthase].

Glycerophospholipid Metabolism in Eukaryotes

Most eukaryotic cells contain six classes of glycerophospholipids—the same PE, PG, and CL found in bacteria, plus phosphatidylserine (PS), phosphatidylcholine (PC), and phosphatidylinositol (PI). As outlined in Figure 19.2, phosphatidic acid serves as a major precursor to all six compounds, and the pathways to PE, PS, PG, and CL are virtually identical to those already presented for bacteria. However, eukaryotic cells possess additional pathways that start with the free base—choline and ethanolamine, respectively, leading to PC and PE. These pathways are also outlined in Figure 19.2 and elaborated upon in a following section.

Synthesis of Phosphatidic Acid

Eukaryotes display three biosynthetic routes to phosphatidic acid (Figure 19.6). The major pathway, starting with glycerol-3-phosphate, is similar to that used

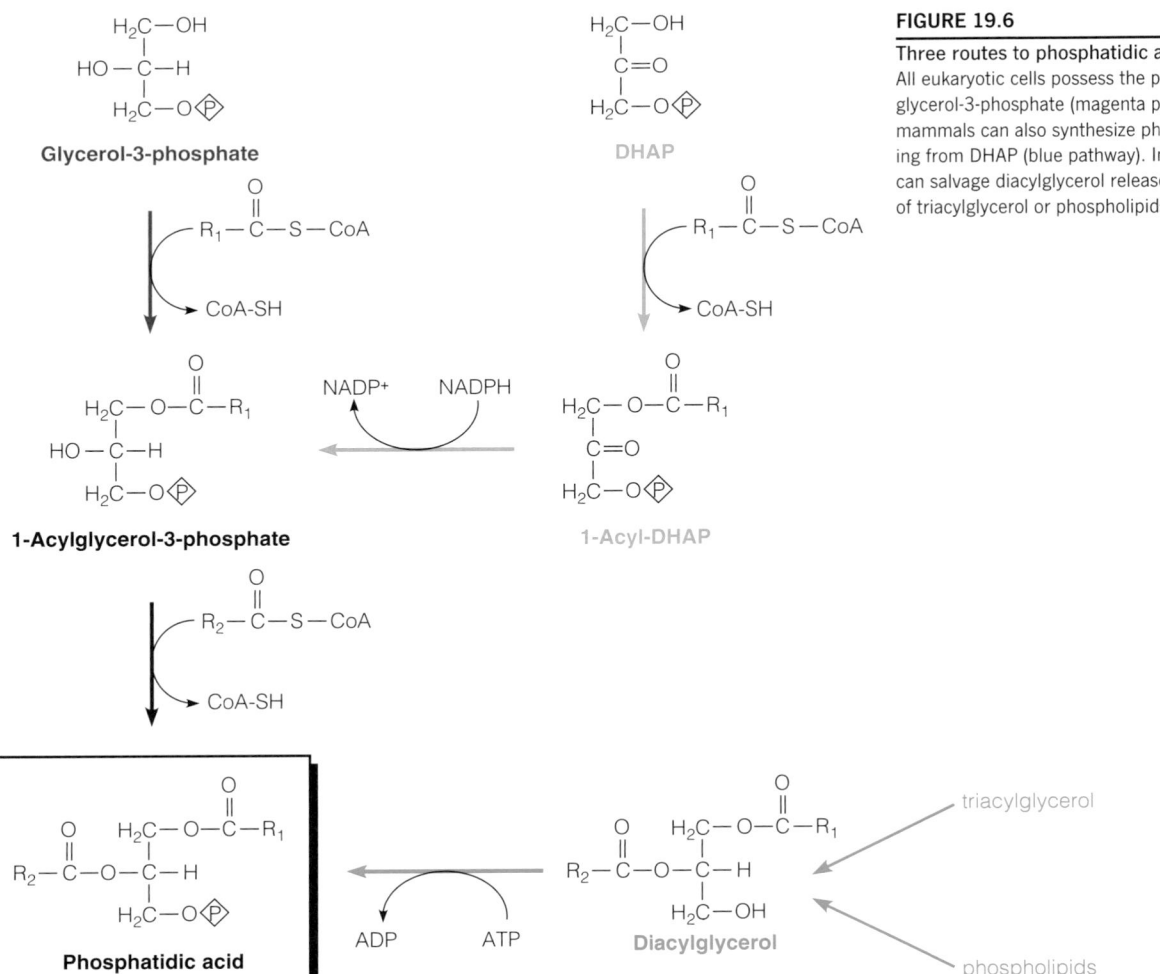

FIGURE 19.6

Three routes to phosphatidic acid in eukaryotes. All eukaryotic cells possess the pathway starting from glycerol-3-phosphate (magenta pathway). Yeast and mammals can also synthesize phosphatidic acid starting from DHAP (blue pathway). In addition, eukaryotes can salvage diacylglycerol released from turnover of triacylglycerol or phospholipids (green pathway).

In eukaryotes, phosphatidic acid has three different origins: glycerol-3-phosphate, dihydroxyacetone phosphate, and diacylglycerol.

by bacteria (see Figure 19.3), except that the acyltransferases use acyl-CoAs as substrates instead of acyl-ACPs. These acyltransferases are located in the endoplasmic reticulum. Yeast and mammals also possess a second pathway that starts with dihydroxyacetone phosphate (DHAP), which accepts a fatty acyl moiety at position 1 from an acyl-CoA, followed by reduction to 1-acylglycerol-3-phosphate and a second acylation. The enzymes that catalyze this pathway are located in mitochondria and peroxisomes, as well as endoplasmic reticulum.

The third route simply involves phosphorylation by a specific kinase of diacylglycerol, which arises from metabolic turnover of phospholipids (see page 748).

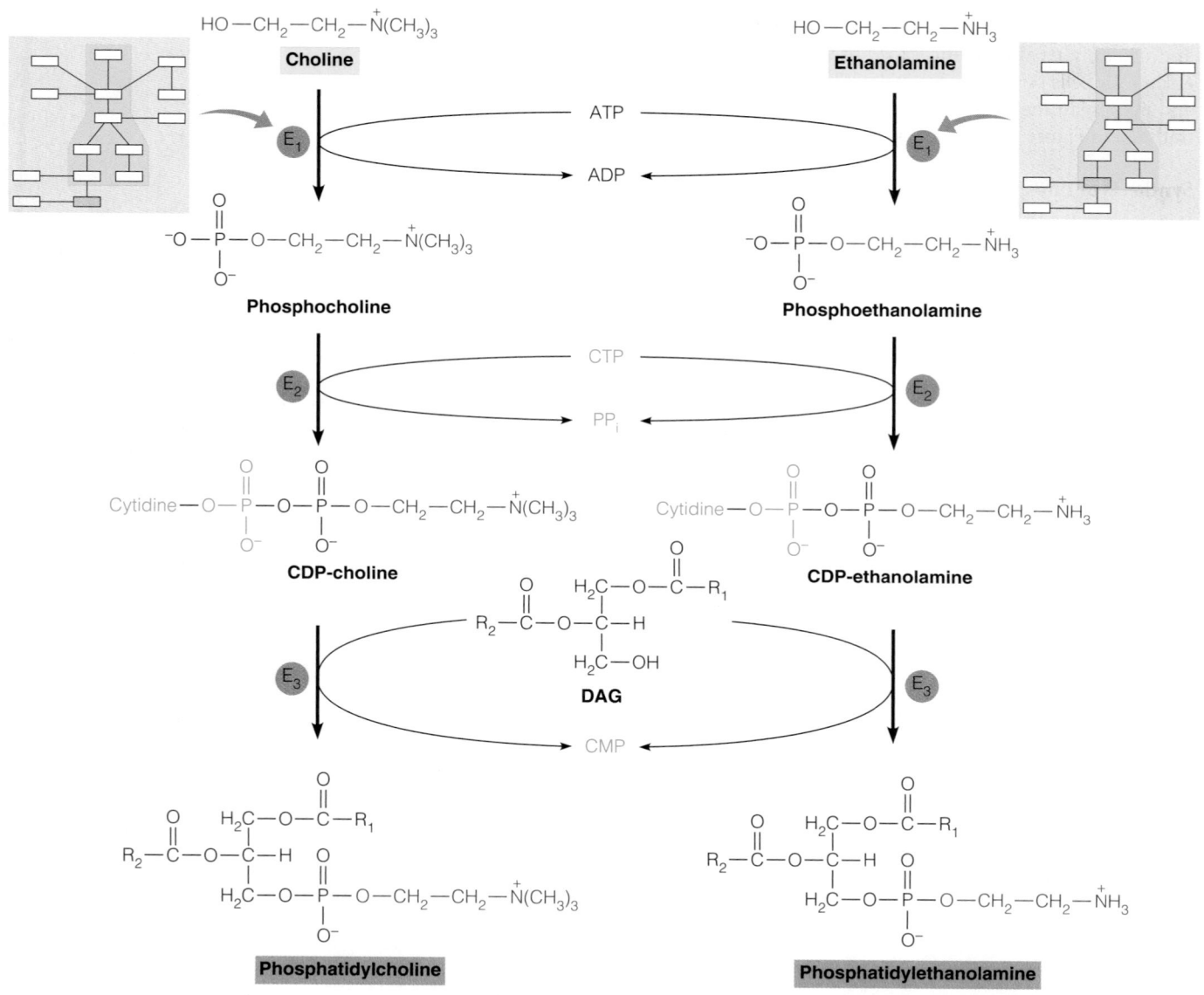

FIGURE 19.7

Synthesis of phosphatidylcholine and phosphatidylethanolamine in mammals. E_1 = choline kinase or ethanolamine kinase; E_2 = CTP:phosphocholine cytidylyltransferase or CTP:phosphoethanolamine cytidylyltransferase; E_3 = CDP-choline:1,2-diacylglycerol choline phosphotransferase or CDP-ethanolamine:1,2-diacylglycerol ethanolamine phosphotransferase.

Regardless of which route is used, phosphatidic acid is converted to CDP-diacylglycerol as previously described for bacteria. CDP-diacylglycerol in turn serves as precursor to PS, PE, PG, and CL. Another set of pathways starts from free bases, as described in the next section.

Pathways to Phosphatidylcholine and Phosphatidylethanolamine

The most abundant phospholipids in most eukaryotic cells are phosphatidylcholine and phosphatidylethanolamine. Both can be synthesized from phosphatidylserine or through alternative pathways that start with free choline or ethanolamine, respectively. Because choline and ethanolamine arise largely through the turnover of preexisting phospholipids, the latter pathways can be considered "salvage pathways" for reutilization of these breakdown products. The significance of reutilizing choline lies in the fact that the three methyl groups of choline are derived from the amino acid methionine. As we discuss in Chapter 20, methionine is nutritionally essential for many animals, also making essential the reuse of scarce metabolites such as choline. In fact, choline itself is an essential component of most animal diets.

The pathway for use of preformed choline or ethanolamine, which predominates in most animal cells, is summarized in Figure 19.7. The chemical strategy for forming the phosphodiester linkage between the polar head group and the diacylglycerol backbone is similar to that described for bacterial phospholipid synthesis (Figure 19.4), except that it is the head group that is activated rather than the diacylglycerol moiety. Choline or ethanolamine is phosphorylated, and the resultant **phosphocholine** or **phosphoethanolamine** is metabolically activated by reaction with CTP, to give the corresponding CDP derivatives (Figure 19.8). The phosphoryl group of the activated CDP-choline or CDP-ethanolamine then undergoes nucleophilic attack by the C3 OH group of diacylglycerol, displacing CMP to yield the corresponding phospholipid. The first enzyme in the pathway, **choline kinase** (E_1 in Figure 19.7) can phosphorylate both choline and ethanolamine. However, mammals also possess a specific **ethanolamine kinase**. The kinases are cytosolic enzymes, whereas the enzymes catalyzing the second step, **CTP:phosphocholine cytidylyltransferase** and **CTP:phosphoethanolamine cytidylyltransferase** (E_2), are distributed between cytosolic and endoplasmic reticulum fractions. These cytidylyltransferases catalyze the rate-limiting step of this pathway. The last enzyme, **CDP-choline:1, 2-diacylglycerol choline phosphotransferase** or **CDP-ethanolamine:1, 2-diacylglycerol ethanolamine phosphotransferase** (E_3), is membrane-bound in the endoplasmic reticulum. Recent evidence suggests that only the membrane-bound form of the phosphocholine cytidylyltransferase (E_2) is active and that the rate of phosphatidylcholine synthesis is controlled in part by translocation of this enzyme between cytosolic and membranous forms analogous to the regulation of phosphatidylserine synthase in bacteria (Figure 19.5). The enzyme has higher affinity for membranes with anionic head groups, which are thus depleted of phosphatidylcholine—membrane binding activates this rate-limiting enzyme to increase PC production. The physiological signals that regulate translocation of CTP:phosphocholine cytidylyltransferase to cell membranes are not yet known. This enzyme is an example of an **amphitropic** protein, one that can exist in two alternative states: soluble and membrane-bound.

The second pathway to PE and PC begins with the conversion of phosphatidylserine (PS) to PE, which is catalyzed by either of two different enzymes. **Phosphatidylserine decarboxylase** is a mitochondrial enzyme that, like the corresponding bacterial enzyme, decarboxylates phosphatidylserine to phosphatidylethanolamine. The second enzyme is a calcium-activated transferase, **phosphatidylethanolamine serinetransferase**, which exchanges free ethanolamine

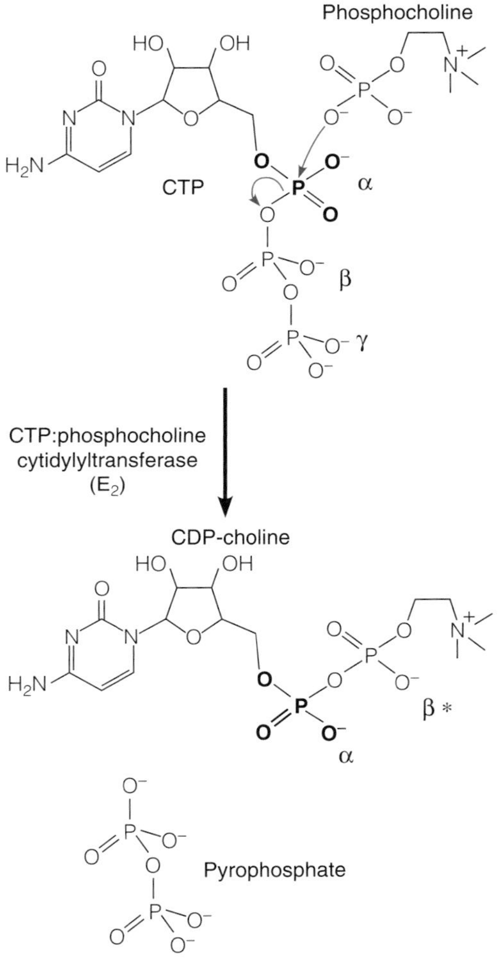

Salvage pathways to phospholipids, starting with choline or ethanolamine, are quantitatively important in eukaryotic cells.

FIGURE 19.8

Metabolic activation of phosphocholine. CTP: phosphocholine cytidylyltransferase (E_2 in Figure 19.7) catalyzes nucleophilic attack at the α-phosphate of CTP by a phosphoryl oxygen of phosphocholine, displacing the β,γ-phosphates (pyrophosphate) of CTP, forming CDP-choline.

for the serine moiety of phosphatidylserine, yielding phosphatidylethanolamine and serine.

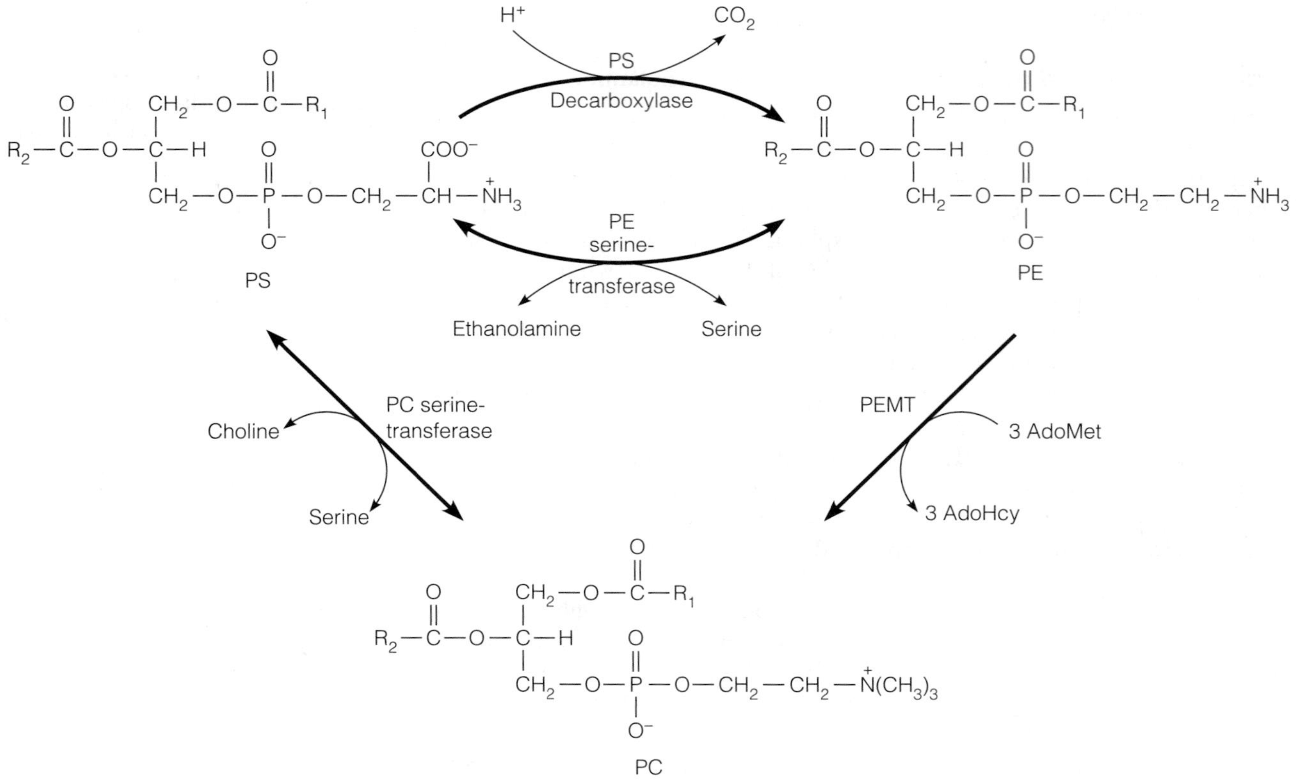

PE serinetransferase is found in endoplasmic reticulum and the Golgi complex. Mammals also possess a **phosphatidylcholine serinetransferase**, which exchanges free serine for the choline moiety of PC, yielding phosphatidylserine and choline. Both of these serinetransferases are readily reversible.

However formed, phosphatidylethanolamine then undergoes three successive methylations, all catalyzed by phosphatidylethanolamine N-methyltransferase (PEMT), to give phosphatidylcholine. In animals this pathway occurs primarily in the liver. The methyl group donor for these reactions is an activated derivative of methionine, **S-adenosyl-L-methionine** (AdoMet). The product of methyl group transfer is **S-adenosyl-L-homocysteine** (AdoHcy). As we shall see throughout the remainder of this book, AdoMet is a "universal" methyl donor, involved in numerous methyl group transfer reactions in lipid, protein, amino acid, and nucleic acid metabolism. AdoMet has an unstable **sulfonium ion**, which has a high thermodynamic tendency to transfer its strongly electrophilic methyl group to nucleophiles and lose its charge. Note that the sulfonium ion is a chiral center—the sulfur has a pyramidal configuration, with the lone electron pair forming the absent 4th ligand. All known AdoMet-dependent methyltransferases are specific for the S isomer (shown below).

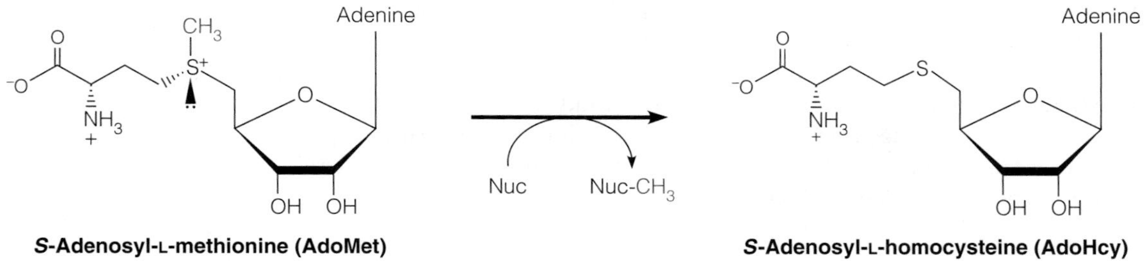

S-Adenosyl-L-methionine is the methyl group donor in synthesis of phosphatidylcholine and numerous other methylated metabolites.

AdoMet is formed from methionine and ATP in an unusual reaction catalyzed by methionine adenosyltransferase, in which ATP is cleaved to yield inorganic triphosphate (PPP_i) plus an adenosyl moiety linked directly via the ribose C5 to the methionine sulfur. The triphosphate is hydrolyzed by the enzyme to pyrophosphate (PP_i) and orthophosphate (P_i), drawing the reaction to completion.

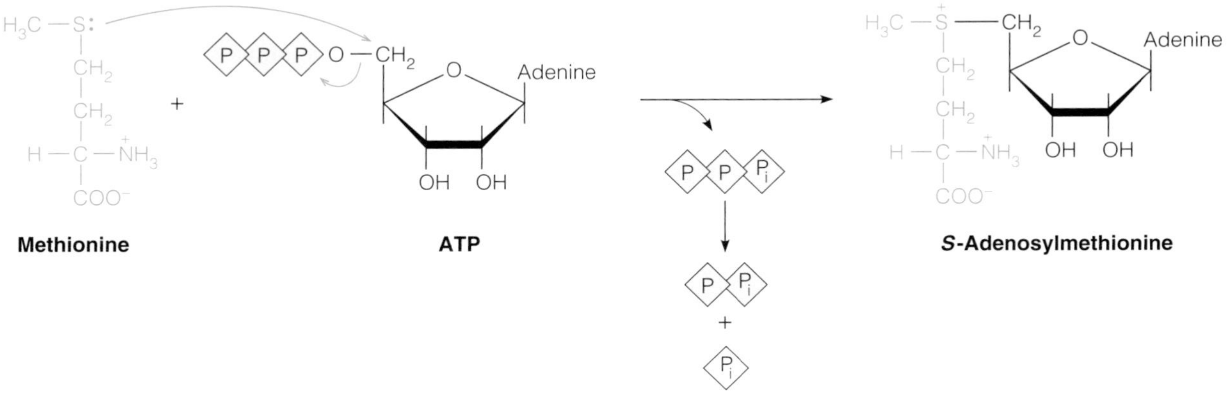

As might be expected from the critical role glycerophospholipids play in membrane synthesis and integrity, these biosynthetic pathways are essential in mammals. Disruption of the genes encoding these enzymes in mice is usually lethal during embryogenesis. The exceptions are for those enzymes that are encoded by multiple genes, or where alternative pathways exist. Thus, deficiency of PEMT can be compensated by the CDP-choline pathway (Figure 19.7). On the other hand, CTP:phosphoethanolamine cytidylyltransferase (Figure 19.7, E_2) and phosphatidylserine decarboxylase (page 784) cannot substitute for one another, suggesting that the PE synthesized from these two pathways is compartmentalized differently. In fact, the serinetransferases, PS decarboxylase, and PEMT enzymes are all localized to a specialized domain of the endoplasmic reticulum that is closely associated with the outer mitochondrial membrane, providing a mechanism for the compartmentalization of this pathway from serine to phosphatidylcholine.

Redistribution of Phospholipid Fatty Acids: Lung Surfactant and Phospholipases

Having described the biosynthesis of polar head groups, let us now focus on the fatty acid constituents of phospholipids, after which we will return to one more important polar head group—the inositol of phosphoinositides. A variety of isotope labeling experiments show that phospholipids, even after insertion into membranes, are not metabolically inert. Specifically, the fatty acyl chains can change in response to varying environmental conditions or needs. Recall that saturated and monounsaturated fatty acids are usually found at the sn-1 position, whereas polyunsaturated acyl groups are usually esterified at the sn-2 position. This fatty acyl asymmetry is generated by rapid turnover of the sn-2-acyl moiety in a remodeling pathway originally described by William Lands in 1958. This process (termed the Lands' cycle) involves the concerted and coordinated actions of **phospholipase A_2** and a family of acyltransferases specific for **lysophospholipids**, phospholipids lacking an acyl chain at the sn-2 position. Lysophospholipids derive their name from the fact that they are powerful detergents, able to solubilize membranes and hence cause cells to lyse; erythrocytes are particularly susceptible to this action (hemolysis). Phospholipase A_2 is one of a class of four enzymes that hydrolyze specific bonds in phospholipids; the others are phospholipases A_1, C, and D (Figure 19.9). One form of phospholipase A_2 comes from snake venom, which has the ability to cause hemolysis. The release of one fatty acyl chain from phosphatidylcholine yields 1-acylglycerophosphorylcholine, more commonly known as **lysolecithin** (phosphatidylcholine itself is commonly called lecithin).

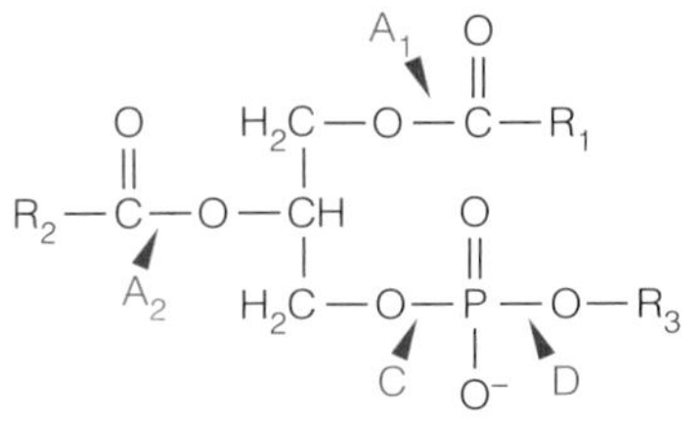

Fatty acid chains in phospholipids are remodeled through the concerted actions of phospholipase A_2 and specific lysophospholipid acyltransferases to meet the needs of the organism.

FIGURE 19.9

Specificities of phospholipases A_1, A_2, C, and D.

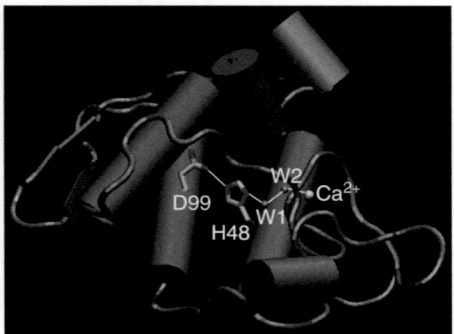

FIGURE 19.10

Structure of phospholipase A₂, an enzyme that metabolizes membrane phospholipids. Catabolism of phospholipids at membrane–water interfaces is important both in modification of membrane structure and as a source of second messengers and other regulators. The crystal structure of the porcine pancreas phospholipase A₂ (PDB ID 2B01) reveals the catalytic triad (D99-H48-water) and the active site Ca²⁺ ion, liganded to a second water molecule (W2). Hydrogen bonds and Ca²⁺ liganding interactions are indicated by thin white lines.

Phospholipases play roles in signal transduction and in membrane repair.

Several remodeling acyltransferases are now known that are specific for either lysolecithin, lysophosphatidylethanolamine, lysophosphatidylcardiolipin, or lysophosphatidylglycerol. Each enzyme also has a unique selectivity for the acyl-CoA chain it transfers to the *sn*-2 position of the lysophospholipid. This remodeling process is particularly important during biosynthesis of the phospholipid component of **pulmonary surfactant**, a lipid- and protein-containing substance that is secreted by alveolar cells in the lung and, by maintaining high surface tension, prevents collapse of the alveoli when air is expelled. Pulmonary surfactant contains some 50% to 60% **dipalmitoylphosphatidylcholine**, an unusual form of phosphatidylcholine in which saturated palmitoyl chains occupy both positions *sn*-1 and *sn*-2. Infants afflicted with **respiratory distress syndrome** show defects in the metabolism of pulmonary surfactant—deficiencies in either synthesis or secretion of this substance. Phospholipase A₂ first hydrolyzes the original acyl group at the *sn*-2 position to give a lysophosphatidylcholine. Alveolar cells express a specific lysophosphatidylcholine acyltransferase that has a clear preference for saturated fatty acyl-CoAs, especially palmitoyl-CoA, as acyl donor for replacement of the *sn*-2 chain. Synthesis of dipalmitoylphosphatidylcholine in lung tissue is greatly increased after birth, partly as a result of translocation of CTP:phosphocholine cytidylyltransferase (E₂ in Figure 19.7) to the endoplasmic reticulum and its consequent activation. Expression of the lysophosphatidylcholine acyltransferase is also increased during the perinatal period to provide the necessary remodeling activity. Because pulmonary surfactant production is not activated until babies reach full term, premature infants are often treated with a surfactant to prevent respiratory distress.

Phospholipases have been useful reagents in studies of both lipid and membrane structures. Structural studies on phospholipase A₂ have been of particular interest, partly because the enzyme is active at a membrane–water interface and serves as a probe of membrane structure, and partly because phospholipase actions lead to prostaglandins (page 813) and to second messengers (see Chapter 23). Phospholipase A₂ uses a catalytic triad similar to that in the serine proteases (Chapter 11), except that a water molecule replaces serine as the attacking nucleophile, and thus no acyl-enzyme intermediate forms in the reaction. As shown for the phospholipase A₂ from pig pancreas (Figure 19.10), the catalytic Asp99 and His 48 are located on adjacent α-helices, in a hydrogen bond network with a water molecule (W1). The active site also contains a catalytically important Ca²⁺ ion, which is liganded to a second water molecule (W2). The structure suggests that W2 is the catalytic water molecule, which is activated by both the Asp–His–W1 catalytic triad and the Ca²⁺ ion for nucleophilic attack on the ester linkage at the *sn*-2 position of the phospholipid. The Ca²⁺ ion is also proposed to stabilize the oxyanion of the tetrahedral reaction intermediate.

Another postulated role for phospholipase A₂ is in repair of damaged membrane phospholipids. As discussed in Chapter 15, fatty acids are susceptible to nonenzymatic attack by oxygen or reactive oxygen species such as superoxide to give fatty acid hydroperoxides. When a membrane phospholipid undergoes peroxidation of a fatty acyl chain, the structure of the membrane is distorted, and the function of the membrane can be affected. Current evidence suggests that phospholipase A₂ can remove these abnormal fatty acids from phospholipids still resident in a lipid bilayer, which leads to replacement of the damaged acyl chains by normal fatty acids.

The simplest possible lysophospholipid is **lysophosphatidic acid**, formed by action of phospholipase A₂ on phosphatidic acid. Lysophosphatidic acid has been identified as a biological signaling agent (see Chapter 23). Its release from activated cells such as platelets stimulates the growth of other cell types, through interaction with specific receptors, evidently as part of the wound-healing process.

Biosynthesis of Other Acylated Glycerophospholipids

The remaining major pathways outlined in Figure 19.2 are for synthesis of phosphatidylserine, phosphatidylglycerol, cardiolipin, and phosphatidylinositol. In yeast the CDP-diacylglycerol pathway to phosphatidylserine predominates. In animals, phosphatidylserine is synthesized primarily by the calcium-activated exchange between phosphatidylethanolamine and serine (page 784). However, the ultimate source of ethanolamine for this process is not yet known. In animals the synthesis of phosphatidylglycerol, which is largely confined to mitochondria, is identical to the route used in bacteria. However, its conversion to cardiolipin involves CDP-diacylglycerol, rather than a second mole of phosphatidylglycerol, as the second substrate (Figure 19.11).

The biosynthesis of phosphatidylinositol, catalyzed by **phosphatidylinositol synthase**, involves CDP-diacylglycerol and L-*myo*-inositol (Figure 19.11). The latter is one of nine possible stereoisomers of hexahydroxycyclohexane; it is synthesized

FIGURE 19.11

Biosynthesis of cardiolipin and phosphoinositides in eukaryotes.

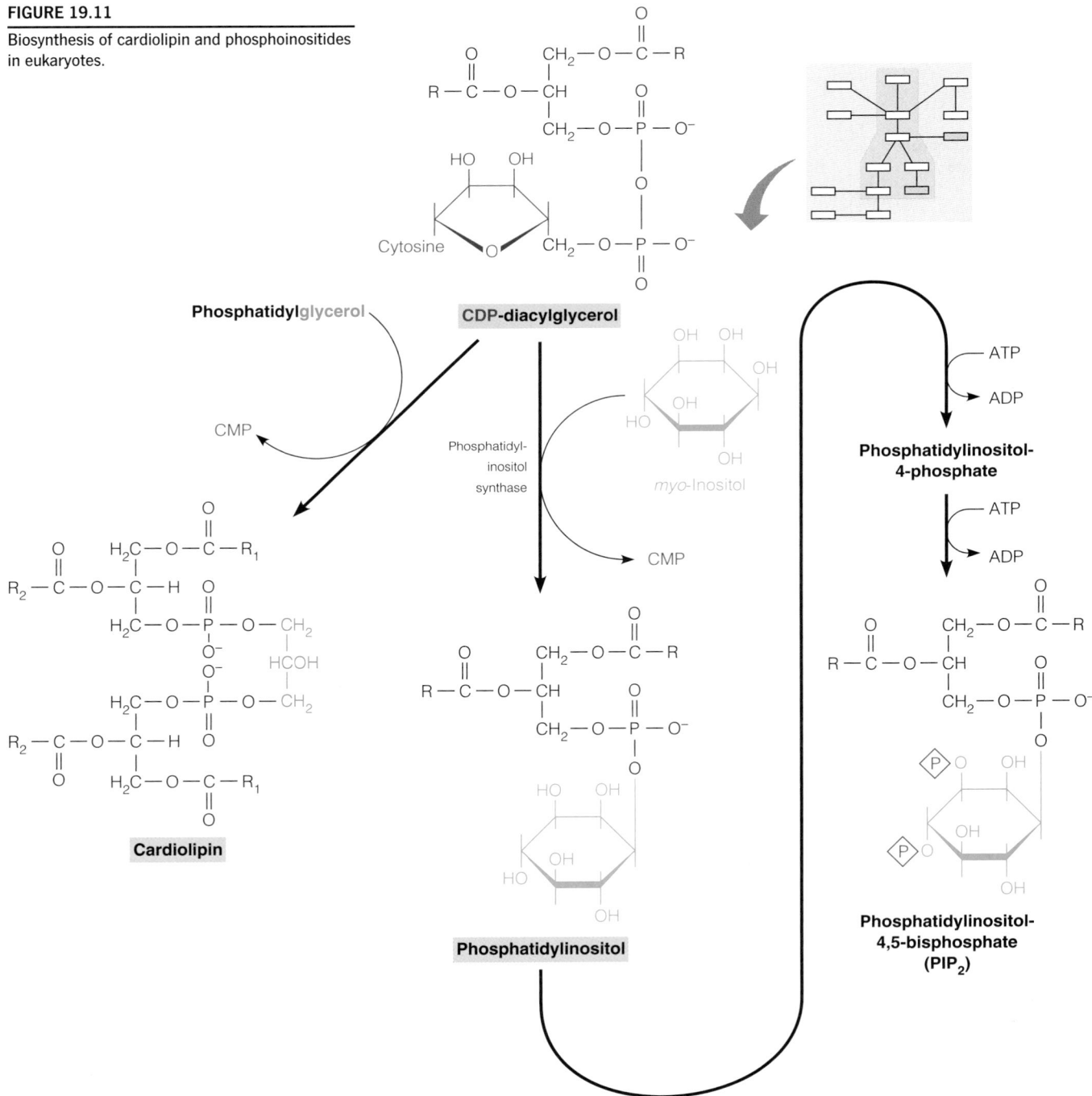

Phosphatidylinositol and its phosphorylated derivatives play important roles as precursors of second messengers.

FIGURE 19.12

Biosynthetic route to alkyl ether phospholipids.

FIGURE 19.12

Biosynthetic route to alkyl ether phospholipids.

from D-glucose-6-phosphate. Phosphatidylinositol undergoes two successive phosphorylations to yield phosphatidylinositol-4-phosphate and phosphatidylinositol-4,5-bisphosphate (PIP_2), both of which are present in small but appreciable amounts. All three of these lipids, which are collectively termed **phosphoinositides**, are enriched in arachidonic acid at position 2. This enrichment evidently occurs via the deacylation–reacylation process we discussed earlier (see page 786).

It has long been known from ^{32}P labeling studies that the phosphoinositides are in a state of active metabolic flux. They are synthesized and degraded rapidly, particularly in nervous tissue in response to the binding of neurotransmitters. As described in Chapter 18, phosphatidylinositol 3,4,5-trisphosphate (PIP_3) plays an important role as a second messenger in **transmembrane signaling**, the transmission of an extracellular signal to some element of the intracellular metabolic apparatus. Our current understanding of these events is presented in Chapter 23. Another important metabolic role for phosphatidylinositol is its participation in the glycosylphosphatidylinositol (GPI) linkages used to anchor some glycoproteins to the extracellular face of the plasma membrane (Chapter 10).

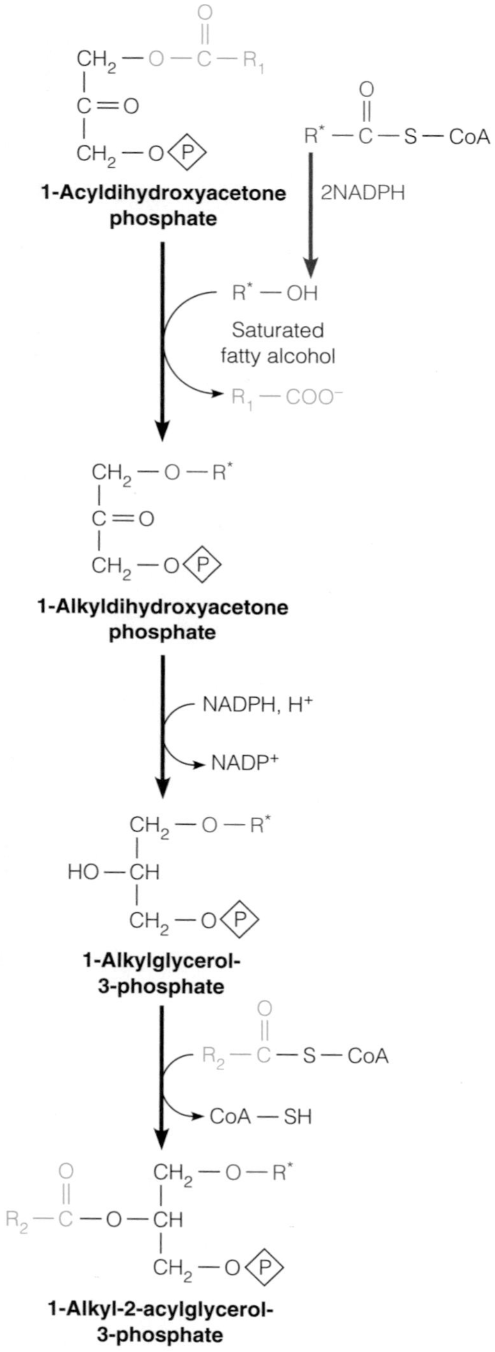

Ether Phospholipids

Ether lipids contain an alkyl group, rather than an acyl group, linked by an ether bond to one of the oxygen atoms of glycerol. Alkyl and alkenyl phospholipids are widely distributed, but their abundance in tissues varies greatly. For example, consider the **plasmalogens**, or **vinyl ethers**, phospholipids that contain an alkenyl ether at position sn-1 of glycerol. These compounds are found in most mammalian membranes, but are particularly abundant in heart, brain, and spermatozoa, where they can account for more than 50% of the membrane phospholipids. Ethanolamine plasmalogens are much more abundant than choline plasmalogens, except in muscle. So far, little is known about the functional significance of this class of lipids. The acid-labile vinyl ether bond is quite susceptible to oxidative damage, and plasmalogens are important targets of reactive oxygen species (ROS; see Chapter 15, page 665). Thus, it has been proposed that plasmalogens protect against oxidative damage, such as lipid peroxidation, by "scavenging" ROS. Plasmalogens may also be important sources of arachidonic acid and docosahexaenoic acid (DHA), released from the sn-2 position by the action of phospholipase A_2. Arachidonic acid is the precursor to the eicosanoids (more about these later) and DHA is a polyunsaturated fatty acid (PUFA) with the ability to depress levels of both serum cholesterol and triacylglycerols (Chapter 17). A genetic deficiency of plasmalogen synthesis has serious consequences. Biosynthesis of ether lipids occurs in peroxisomes, organelles that also carry out both β- and α-oxidation pathways (see Chapter 17, page 732). In a rare autosomal recessive disorder called **Zellweger syndrome**, peroxisomes are absent, and plasmalogen synthesis is severely deficient. Individuals with this condition suffer damage to brain, liver, and kidney, before reaching an early death.

The biosynthesis of ether phospholipids (Figure 19.12) begins in the peroxisome with 1-acyldihydroxyacetone phosphate (see Figure 19.6). This undergoes exchange of an alkyl group for the acyl group; the saturated fatty alcohol used in this reaction is derived from NADPH-dependent reduction of the corresponding fatty acyl-CoA. Carbon 2 is then reduced from the keto to the hydroxyl level to give 1-alkylglycerol-3-phosphate. This intermediate moves from the peroxisome to the endoplasmic reticulum, where C2 is acylated. This gives the 1-alkyl analog of phosphatidic acid (last structure in the figure), which is converted to saturated ether phospholipids, or **glyceryl ethers**, by the pathways already presented in Figure 19.2 for phospholipid biosynthesis. The primary route leads to the glyceryl ether of phosphatidylethanolamine; the serine and choline analogs arise from the ethanolamine analog (as on page 784) by base exchange and methylation, respectively. The synthesis of plasmalogens from glyceryl ethers then involves desaturation of the alkyl group at position *sn*-1 (Figure 19.13). The microsomal enzyme system involved, like that used for desaturation of stearoyl-CoA (see Figure 17.34, page 744), requires O_2, NADH, and cytochrome b_5.

An unusual ether lipid, called **platelet-activating factor**, has the structure **1-alkyl-2-acetylglycerophosphocholine**. Physiologically, this compound is perhaps the most potent compound known. At concentrations as low as 1 *picomolar* (10^{-12} M), it has numerous effects, both in normal physiology and in inflammation reactions, including stimulation of blood platelet aggregation, reduction of blood pressure, activation of several white blood cell classes, decreased cardiac output, stimulation of glycogenolysis, and stimulation of uterine contraction. This lipid is synthesized via acetylation of the corresponding 1-alkylglycerophosphocholine by acetyl-CoA.

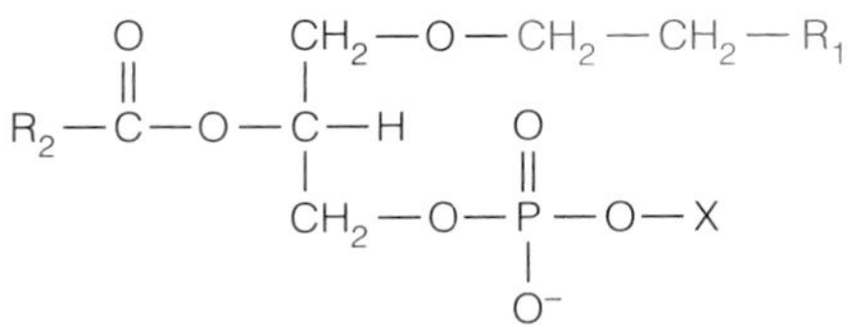

Phospholipid with an alkyl ether at position *sn*-1

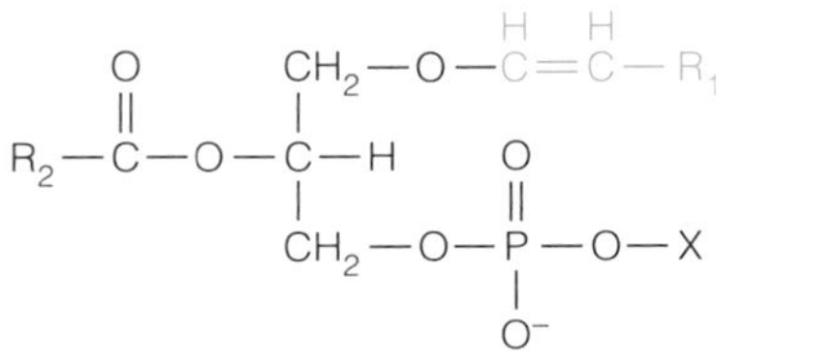

Phospholipid with an alkenyl ether at position *sn*-1 (Plasmalogen)

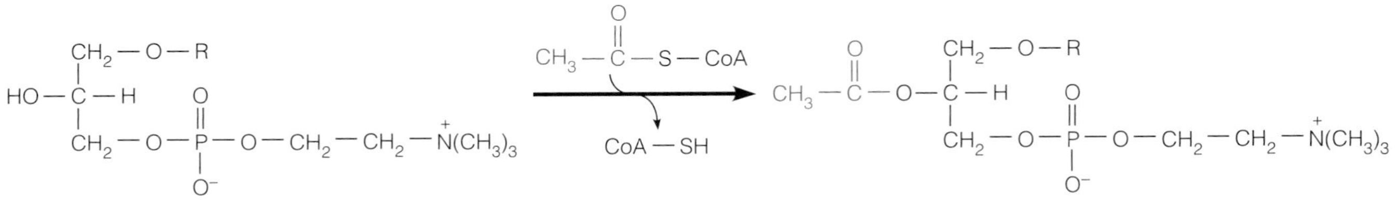

1-Alkylglycerophosphocholine

1-Alkyl-2-acetylglycerophosphocholine

The discovery of a phospholipid with such a striking biological activity was without precedent, and it has opened a fascinating new realm of biochemistry. The factor acts through binding to a high-affinity receptor in the membrane of susceptible cells. The receptors interact with signal transduction systems through G proteins (see Chapter 23).

Ether-containing lipids are quite abundant in the membranes of halophilic ("salt-loving") microorganisms. These bacteria and protozoans grow in media with NaCl concentrations as high as 4 M. Although we don't know the relationship between ether lipids and the ability to grow in a high-salt environment, the greater stability of alkyl ethers against hydrolysis, as compared with acyl esters, may be a factor.

Intracellular Transport of Membrane Phospholipids

Of the six major classes of glycerophospholipids in eukaryotic membrane lipids, phosphatidylglycerol and cardiolipin are found primarily in mitochondrial membranes and are synthesized in mitochondria. The remaining four classes are synthesized simultaneously with their insertion into the cytosolic side of membranes of the endoplasmic reticulum. From there they undergo translocation to the luminal side of the membrane and ultimately are transported to other membranes—the nuclear envelope, mitochondrial membranes, and the plasma membrane. Just

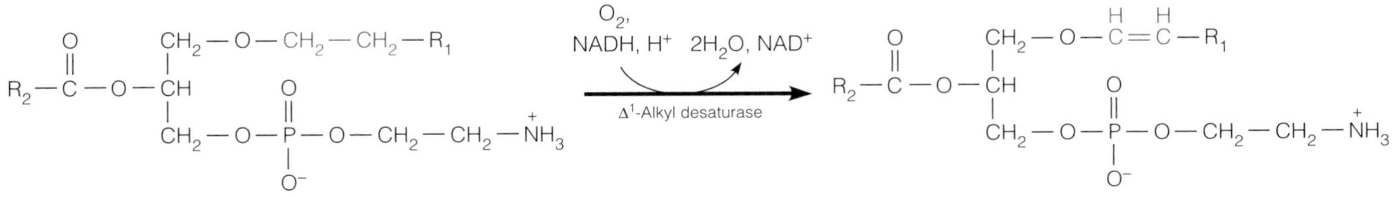

FIGURE 19.13

Synthesis of a plasmalogen from a glyceryl ether. Desaturation of 1-alkyl-2-acylglycerophosphoethanolamine (the alkyl analog of phosphatidylethanolamine) yields the corresponding vinyl ether, or plasmalogen.

1-Alkyl-2-acylglycerophospho-ethanolamine

Plasmalogens

how these events occur is the subject of one of the most active areas of contemporary cell biology. The three major questions are the following: (1) How do phospholipid molecules move from one side of a membrane to the other? (2) How do phospholipid molecules move from one site to another within the cell? (3) How does phospholipid transport directed to specific organelles account for the differences in phospholipid composition of membranes within a single cell?

Investigations of transmembrane movement of phospholipids (question 1) use specific lipid probes that allow detection of a lipid on only one side of a bilayer. As mentioned in Tools of Biochemistry 10A, one such approach involves the use of a **spin label**, a lipid analog that is detectable from its electron paramagnetic resonance spectrum. Such measurements show that transbilayer movement, or "flip-flop," does occur spontaneously but is quite slow. Measurements in vivo show much faster transbilayer movement, catalyzed by flippases and floppases described in Chapter 10.

Transport of phospholipids within the cell (question 2) involves largely the transfer of fragments of membranes of the ER into the Golgi complex, as was shown in Figure 19.1. Membrane vesicles are constantly pinched off from the Golgi, and these vesicles, containing secretory products, fuse with the plasma membrane for secretion of their contents via **exocytosis** (transport out of the cell). It seems likely that this route is used not only for extracellular secretion but also for transport of membrane lipids to the plasma membrane. Probably comparable processes transport membrane lipids to mitochondria, plant chloroplasts, and nuclei, although these processes are not as well understood.

To explain the variability of membrane lipid composition within a given cell (question 3), we can postulate the existence in Golgi membranes of specific targeting proteins—proteins that preferentially associate with certain lipids and have an affinity for certain organelles. Another mechanism involves the action of **phospholipid exchange proteins**—cytosolic proteins that bind a phospholipid and can catalyze its exchange with a corresponding membrane lipid. The protein-bound lipid moves into the membrane, and the membrane lipid becomes bound to the cytosolic protein. This mechanism does not provide for net transfer of lipid to a membrane, but it does allow for modulation of the lipid composition of a particular membrane.

Membranes are assembled by membrane vesicles moving from synthesis sites in the endoplasmic reticulum and Golgi complex to existing membranes and fusing with them.

$$CH_3(CH_2)_{12}\overset{H}{\underset{H}{C}}=C-\overset{H}{\underset{OH}{C}}-\overset{H}{\underset{\overset{+}{NH_3}}{C}}-CH_2OH$$

Sphingosine

$$CH_3(CH_2)_{12}CH_2-\overset{H}{\underset{OH}{C}}-\overset{H}{\underset{OH}{C}}-\overset{H}{\underset{\overset{+}{NH_3}}{C}}-CH_2OH$$

Phytosphingosine

Metabolism of Sphingolipids

Interest in sphingolipids focuses largely on their important role in nervous tissue and, related to this role, a number of human genetic defects of sphingolipid metabolism. Sphingolipids are also widely distributed in the membranes of plant cells and in lower eukaryotes such as yeast.

Recall from Chapter 10 that sphingolipids are derivatives of the base *sphingosine*. Plant sphingolipids contain a slightly different form of this compound, called **phytosphingosine**. The sphingolipids include *ceramide* (*N*-acylsphingosine), *sphingomyelin* (*N*-acylsphingosine phosphorylcholine), and a family of carbohydrate-containing sphingolipids called neutral and acidic **glycosphingolipids**; the latter substances include *cerebrosides* and *gangliosides* (which also contain sialic acid). Ceramide serves as the precursor to both sphingomyelin and the glycosphingolipids.

In animals the pathway to ceramide starts with the synthesis of a sphingosine derivative, **sphinganine**, from palmitoyl-CoA and serine (Figure 19.14). After reduction of the palmitoyl keto group, the amino group of sphinganine is acylated to give a ceramide. After transport to the Golgi complex, the sphinganine unit of this compound is desaturated to give a ceramide with a sphingosine base. Transfer of a phosphocholine unit from phosphatidylcholine yields sphingomyelin plus diacylglycerol.

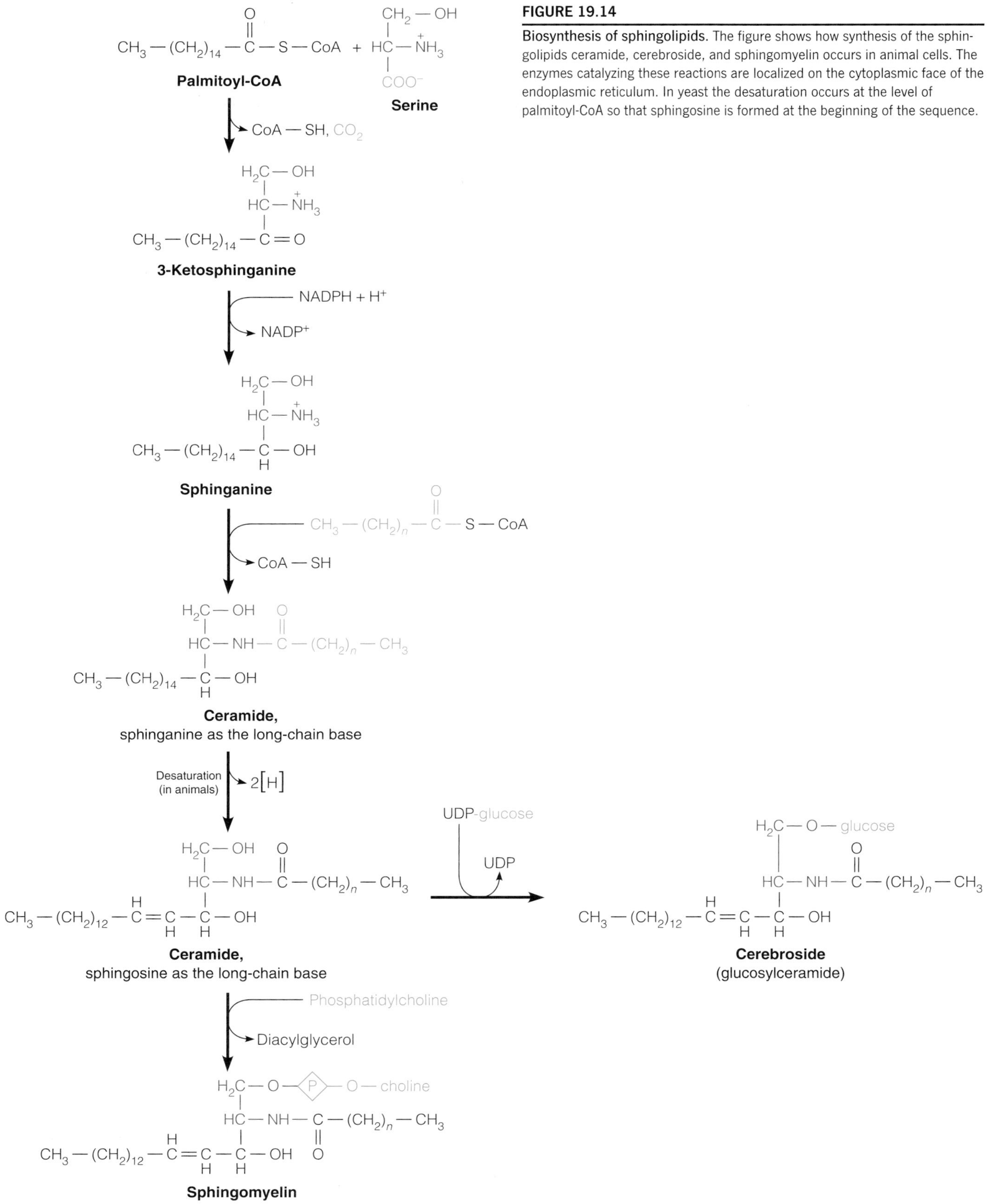

FIGURE 19.14

Biosynthesis of sphingolipids. The figure shows how synthesis of the sphingolipids ceramide, cerebroside, and sphingomyelin occurs in animal cells. The enzymes catalyzing these reactions are localized on the cytoplasmic face of the endoplasmic reticulum. In yeast the desaturation occurs at the level of palmitoyl-CoA so that sphingosine is formed at the beginning of the sequence.

The pathways leading to glycosphingolipids are more numerous, but the metabolic strategies are comparable to those we have encountered before in synthesis of the oligosaccharide chains of glycoproteins (see Chapter 9). The pathways involve the stepwise addition of monosaccharide units, using nucleotide-linked sugars as the activated biosynthetic substrates and with ceramide as the initial monosaccharide acceptor (Figure 19.14). The sugar nucleotides involved in glycosphingolipid synthesis include UDP-glucose (UDP-Glc), UDP-galactose (UDP-Gal), UDP-N-acetylgalactosamine (UDP-GalNAc), and CMP-N-acetylneuraminic acid (CMP-Sia, or CMP-sialic acid). Figure 19.15 shows pathways leading to some of the most abundant glycosphingolipids.

Sphingolipids, especially sphingomyelin, are abundant components of the *myelin sheath*, a multilayered structure that protects and insulates cells of the central nervous system (Figure 19.16; see also Figure 10.45, page 397). In human myelin, sphingolipids constitute some 25% of the total lipid. Sphingolipids are in a continuous state of metabolic turnover, both synthesis and degradation. Degradation occurs in the lysosomes, by a family of hydrolytic enzymes. These

Nucleotide-linked sugars and glycosyltransferases are involved in glycosphingolipid biosynthesis.

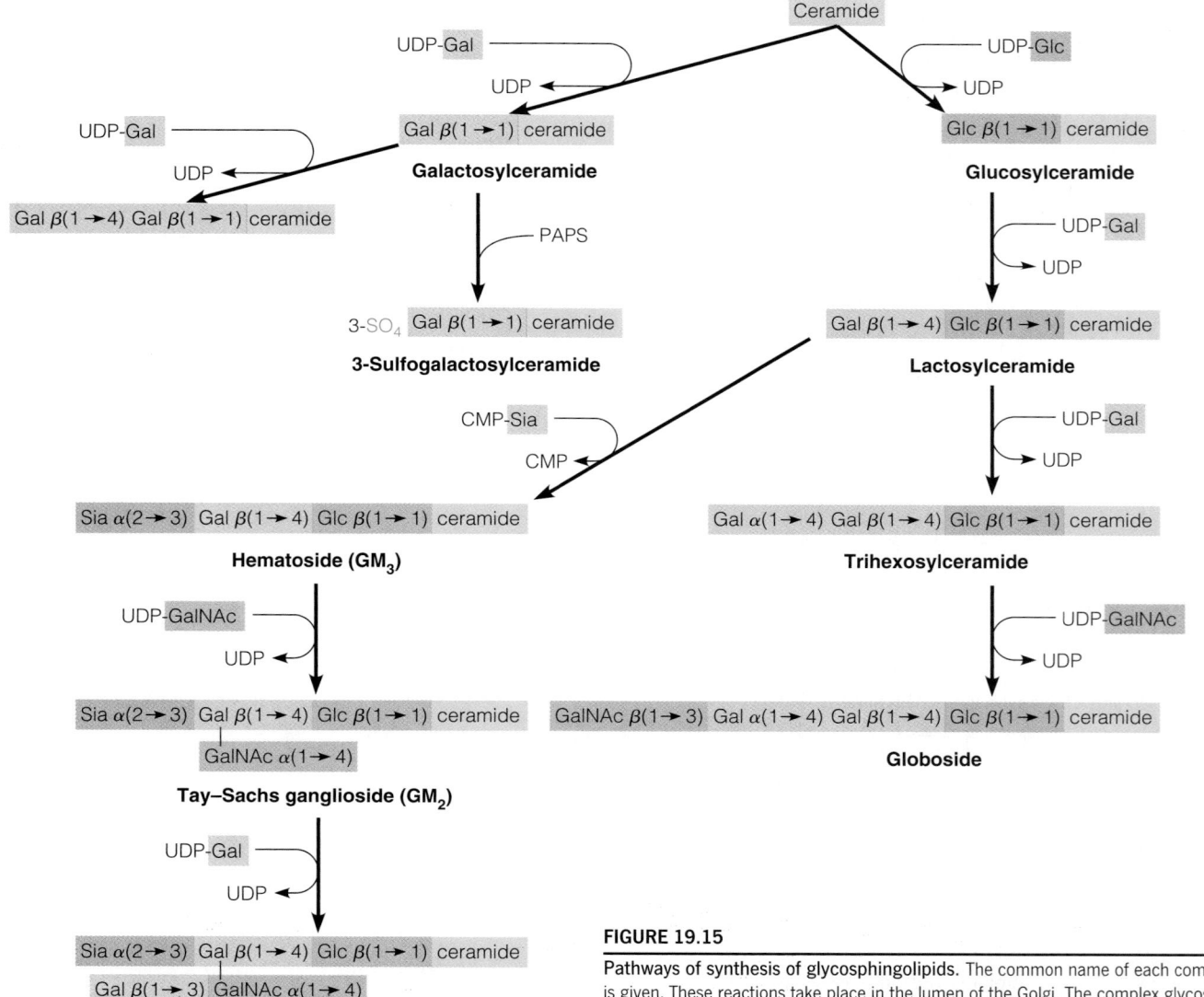

FIGURE 19.15

Pathways of synthesis of glycosphingolipids. The common name of each compound is given. These reactions take place in the lumen of the Golgi. The complex glycosphingolipids are subsequently transported to the plasma membrane via vesicular trafficking. PAPS is a sulfate group donor (see Chapter 21, page 901).

Axon   Myelin

FIGURE 19.16

A myelinated axon from the spinal cord. Myelin, an insulating layer wrapping about the axon, is rich in sphingomyelin.

Courtesy of Dr. Cedric Raine, New York.

TABLE 19.1   Inherited diseases of sphingolipid catabolism

Disease	Defective Enzyme[a]	Accumulated Intermediate
GM$_1$ gangliosidosis	❶ β-Galactosidase	GM$_1$ ganglioside
Tay–Sachs disease	❷ β-N-Acetylhexosaminidase A	GM$_2$ (Tay–Sachs) ganglioside
Fabry's disease	❸ α-Galactosidase A	Trihexosylceramide
Gaucher's disease	❹ β-Glucosidase	Glucosylceramide
Niemann–Pick disease (Types A and B)	❺ Sphingomyelinase	Sphingomyelin
Farber's lipogranulomatosis	❻ Ceramidase	Ceramide
Globoid cell leukodystrophy (Krabbe's disease)	❼ β-Galactosidase	Galactosylceramide
Metachromatic leukodystrophy	❽ Arylsulfatase A	3-Sulfogalactosyl-ceramide
Sandhoff disease	❾ N-Acetylhexosaminidases A and B	GM$_1$ ganglioside and globoside

[a]Numbers refer to enzymes shown in Figure 19.17.

pathways are of great medical interest because of their relationship to a group of congenital diseases called **sphingolipidoses** (also known as **lipid storage diseases**). Each condition is characterized by deficiency of one of the degradative enzymes, with concomitant accumulation within the lysosome of the substrate for the deficient enzyme (Table 19.1). In fact, structural analysis of the abnormal metabolites that accumulate helped to establish the degradative pathways, which are depicted in Figure 19.17. Most of these diseases are autosomal recessive, which means that two defective alleles of the gene encoding a particular enzyme must be present in an individual for disease symptoms to be manifest. Because of the large amounts of sphingolipids in nervous tissue, it is perhaps not surprising that most of the sphingolipidoses involve severely impaired central nervous system function.

The best known of the sphingolipidoses is **Tay–Sachs disease**, originally described in 1881, which is a deficiency of the lysosomal *N*-acetylhexosaminidase A. The enzyme deficiency causes accumulation of the ganglioside called GM$_2$, particularly in the brain (see Figure 19.17; structure shown in Figure 10.8b, page 368). The disease is devastating, causing nervous system degeneration, mental retardation, blindness, and death, usually by the age of four.

Although Tay–Sachs disease is rare in the general population, the defective gene is relatively common among Ashkenazic Jews (those of middle and eastern European extraction). Among American Jews, about 1 in 30 individuals carries the defective gene. Thus, two Jewish parents carry an appreciable risk of bearing a Tay–Sachs child. Because there is no known cure for the disease, attention has focused on prenatal detection. In fact, this was one of the first genetic diseases to be successfully diagnosed in conjunction with amniocentesis. Two prospective parents both shown to be heterozygous for this condition can be counseled that they have a 25% chance of conceiving a child with Tay–Sachs.

Sphingolipids function not only as membrane components but also as signaling molecules. Among the sphingolipids, ceramide signaling has been implicated in the regulation of cell growth and migration, differentiation, senescence, and **apoptosis**, or programmed cell death. This process, which we present in more

Genetic defects in glycosphingolipid catabolism cause breakdown intermediates to accumulate in nervous tissue, with severe consequences.

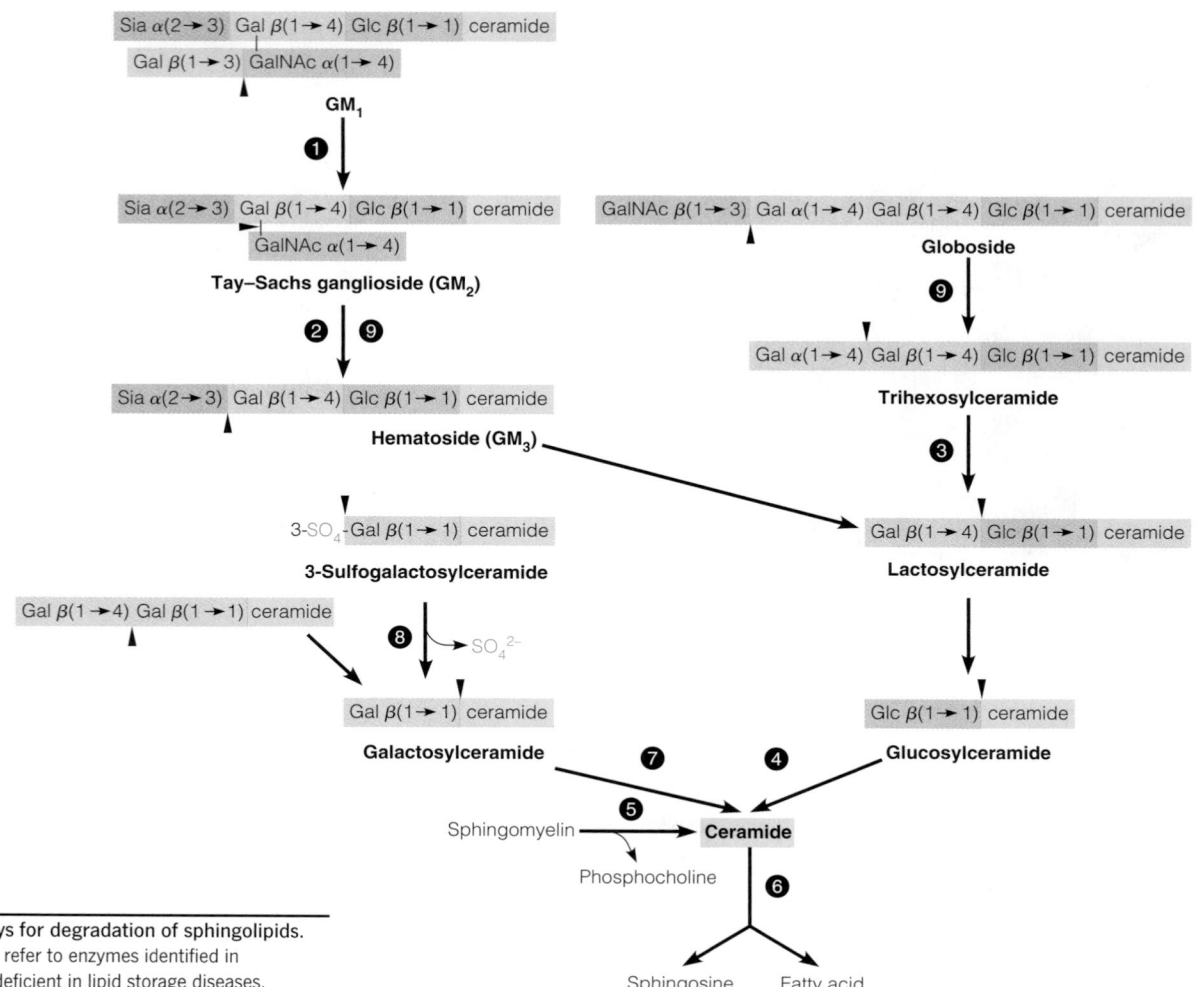

FIGURE 19.17

Lysosomal pathways for degradation of sphingolipids.
The circled numbers refer to enzymes identified in
Table 19.1 that are deficient in lipid storage diseases.

detail in Chapter 28, involves the death of certain cells as part of normal development or after environmental damage to a cell so severe that the cell's survival would be harmful to the organism. Extracellular factors activate the enzymatic cleavage of sphingomyelin within membranes, and the ceramide that is released activates specific protein phosphatases and protein kinases. Many details of this signal transduction pathway remain to be discovered.

Gangliosides act as receptors for specific agents, such as cholera toxin, which binds to ganglioside GM_1, or influenza virus, which recognizes the sialic acid portion of certain gangliosides. Influenza virus encodes a neuraminidase, which cleaves these gangliosides as part of the viral entry process into cells. Inhibitors of this enzyme, such as oseltamivir (Tamiflu), have found wide use as antiviral drugs to treat influenza. Also of great interest is the observation that some gangliosides promote the growth of neural tissue in cell culture, suggesting that they might be used to promote regeneration of nerve tissue after spinal cord injury. In fact, New York Jets lineman Dennis Byrd was treated with GM_1 ganglioside, as well as with surgery, after breaking his neck in a 1992 football injury, and this may have been partially responsible for his remarkable recovery.

Steroid Metabolism

We turn now to an extraordinarily large and diverse group of lipids, the **isoprenoids**, or **terpenes**. These compounds are built up from one or more five-carbon activated derivatives of **isoprene**. The family includes steroids and bile acids; the lipid-soluble

Isoprene

vitamins; the dolichol and undecaprenol phosphates we encountered in glycoprotein synthesis; phytol, the long-chain alcohol in chlorophyll; **gibberellins**, a family of plant growth hormones; insect juvenile hormones; the major components of rubber; coenzyme Q; and many more compounds.

Much of our discussion of isoprenoids focuses on a single steroid compound, cholesterol. As discussed in Chapter 10, this lipid is a major component of animal cell membranes, where it participates in modulation of membrane fluidity. In animals it also serves as precursor to all of the steroid hormones, to vitamin D, and to the bile acids, which aid in fat digestion. And as we discussed in Chapter 17, there is intense medical interest in cholesterol because of the relationships among diet, blood cholesterol levels, atherosclerosis, and heart disease. These biological relationships, coupled with the complex stereochemistry of its structure and the elegance of its biosynthetic pathway from a single low-molecular-weight precursor, have focused attention on this compound ever since its first isolation from gallstones, in 1784. Michael Brown and Joseph Goldstein have proclaimed cholesterol "the most highly decorated small molecule in biology," with 13 Nobel Prizes having been awarded to scientists, including themselves, who devoted major parts of their careers to cholesterol.

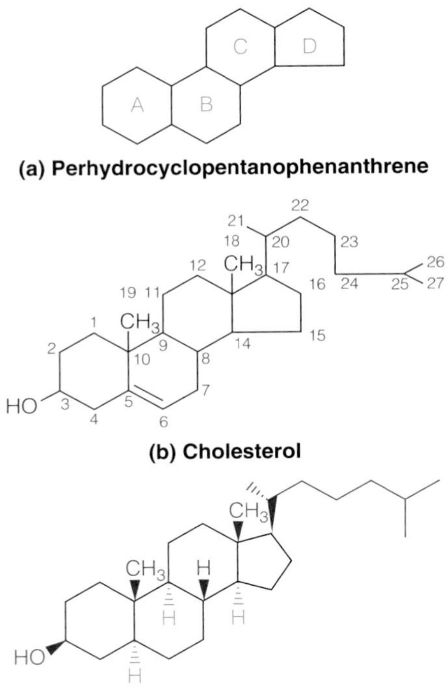

Some Structural Considerations

Steroids constitute a class of lipids that are derivatives of the saturated tetracyclic hydrocarbon **perhydrocyclopentanophenanthrene** (Figure 19.18). Note the letters used to denote the four rings—A, B, C, and D, with D being the five-membered ring—and the carbon numbering system. Cholesterol differs from the basic ring system in having an aliphatic chain at C-17, axial methyl groups at C-10 and C-13, a double bond in ring B, and a hydroxyl group in ring A. The alcoholic functional group and the carbon chain at C-17 make cholesterol a **sterol**, which is the generic term used to identify steroid alcohols.

The cyclohexane rings of the steroids adopt puckered conformations, of which the more stable chair form (see Figure 10.9, page 369) predominates over the boat. This gives cholesterol a rigid molecular structure, with only the hydroxyl group generating a little polarity at one end. Much of the cholesterol in lipoproteins and intracellular storage droplets is esterified at this position with a long-chain fatty acid, which makes the resultant cholesterol ester much more hydrophobic than cholesterol itself. The structure makes apparent how increasing concentrations of cholesterol in a membrane can alter the fluidity of that membrane, by reducing the proportion of total lipid that can undergo a phase transition and by reducing the lateral mobility of polar lipids within the membrane (see Figure 10.12b, page 371).

During most of our treatment of steroid metabolism, we shall use structural representations, as in Figure 19.18c, instead of three-dimensional configurational models. By convention, the methyl group at position 10 projects *above* the plane of the rings. This and all other substituents that project above the plane are denoted β and are drawn with a solid wedge. Substituents that project *below* the plane of the ring are called α and are denoted by a dashed wedge. These conventions are shown in Figure 19.18c for **cholestanol**, one of the two fully saturated derivatives of cholesterol.

Biosynthesis of Cholesterol

The pathway by which cholesterol is synthesized is worthy of study because of the diversity of metabolites synthesized by the pathway and the elegance of the pathway itself. Isotopic tracer studies have shown that all 27 carbons of cholesterol come from a two-carbon precursor—acetate. How could such a simple compound be built up to give a structure of the great complexity of cholesterol? That is what concerns us next.

(a) Perhydrocyclopentanophenanthrene

(b) Cholesterol

(c) Stereochemistry of cholestanol

FIGURE 19.18

Ring identification system (a) and carbon numbering system (b) used for steroids. (c) Structural conventions, with cholestanol as the example. α substituents project below the plane of the steroid ring system (blue dashed wedge), and β substituents project above that plane (magenta solid wedge). The hydrogens at positions 5, 9, and 14 have the α configuration, whereas the hydroxyl, the two methyl groups, the hydrogen at C-8, and the aliphatic side chain at C-17 are all β substituents.

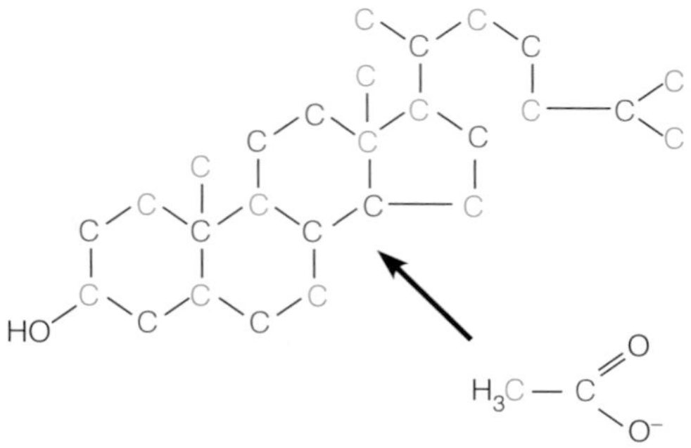

Cholesterol, the precursor to all steroids, derives all of its carbon atoms from acetate.

Early Studies of Cholesterol Biosynthesis

Most of our early insights into cholesterol biosynthesis came out of Konrad Bloch's laboratory in the 1940s. Taking note that cholesterol biosynthesis in vertebrates is confined largely to the liver, Bloch fed rats with acetate having ^{14}C either in the methyl group position or in the carboxyl group. After each administration, cholesterol was isolated from the liver and subjected to chemical degradation, with radioactive counting of the fragments. This procedure established the pattern shown in the margin, in which each carbon of cholesterol was found to originate from either the methyl carbon (blue) or the carboxyl carbon (magenta) of acetate (actually, acetyl-CoA).

Other early insights came from the realization that the five carbons of isoprene could be derived metabolically from three molecules of acetate, and the prediction that cholesterol was a product of the cyclization of the linear C_{30} hydrocarbon **squalene**. Squalene contains six isoprene units (delineated by magenta marks on the structures below), and its configuration makes it a plausible steroid precursor.

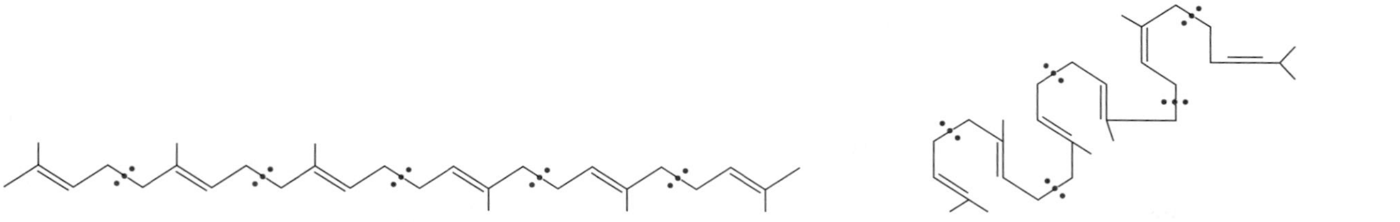

Squalene

Postulated precyclization configuration of squalene

In 1956 another important development occurred, when Karl Folkers discovered that a C_6 organic acid, **mevalonic acid**, could permit the growth of certain acetate-requiring strains of *Lactobacillus*. Folkers showed that mevalonic acid was readily converted to an activated C_5 isoprenoid compound, **isopentenyl pyrophosphate**. Interestingly, *Lactobacillus* does not synthesize steroids, but it uses the first several steps of the pathway to synthesize other isoprenoid compounds. In animals, mevalonate is readily converted to squalene. Once this had been established, the stage was set for considering cholesterol biosynthesis as three distinct processes.

1. Conversion of C_2 fragments (acetate) to a C_6 isoprenoid precursor (mevalonate)

2. Conversion of six C_6 mevalonates, via activated C_5 intermediates, to the C_{30} squalene

3. Cyclization of squalene and its transformation to the C_{27} cholesterol

Now let us consider these three processes in detail.

Stage 1: Formation of Mevalonate

The first part of the pathway is identical to reactions used in ketogenesis (Chapter 17), although it occurs in a different cell compartment. Ketogenesis occurs in mitochondria, whereas cholesterol biosynthesis occurs in the cytosol and the endoplasmic reticulum (ER).

Stage 1 begins with condensation of two molecules of acetyl-CoA to give acetoacetyl-CoA. Figure 19.19 shows the rest of this stage. Acetoacetyl-CoA reacts with a third molecule of acetyl-CoA to give 3-hydroxy-3-methylglutaryl-CoA (HMG-CoA). Recall from Figure 17.24 (page 735) that during ketogenesis, HMG-CoA cleaves to give acetoacetate plus acetyl-CoA in the mitochondrial matrix. However, the HMG-CoA lyase that accomplishes this cleavage is missing from the ER, where cholesterol biosynthesis begins. Instead, **HMG-CoA reductase**, an integral membrane protein in the ER, catalyzes the reduction of HMG-CoA to mevalonate. This multistep reaction requires two equivalents of NADPH (four electrons) to reduce the thioester to an alcohol. This is the major step that regulates the overall pathway of cholesterol biosynthesis.

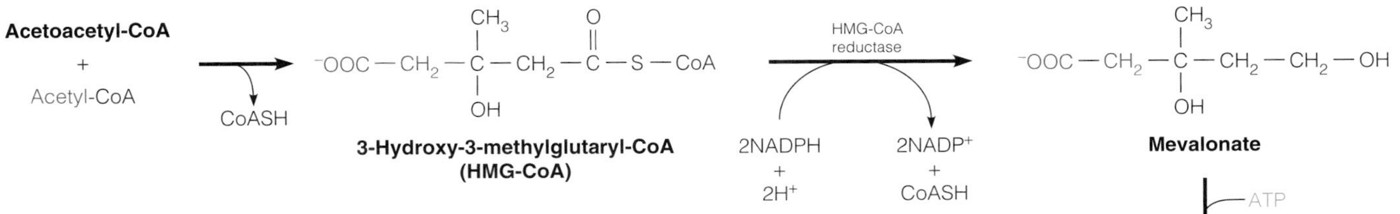

Stage 2: Synthesis of Squalene from Mevalonate

The next several reactions, shown in Figures 19.19 and 19.20, occur in the cytosol. First, mevalonate is activated by three successive phosphorylations (Figure 19.19). The first two are simple nucleophilic substitutions on the γ-phosphorous of ATP. The third phosphorylation, at position 3, sets the stage for a decarboxylation to give the five-carbon isopentenyl pyrophosphate (IPP). Recall from Chapter 12 (pages 484–485) that decarboxylations require the presence of an electron-accepting group two carbons away from the carboxylate in order to stabilize the formal negative charge resulting from loss of CO_2. β-keto acids, which readily undergo decarboxylation, provide this function by forming an enolate ion. However, 5-pyrophosphomevalonate has no carbonyl β to the departing carboxyl group. Instead the enzyme phosphorylates the tertiary hydroxyl group to give a tertiary phosphate, leading to spontaneous dissociation to give a tertiary carbocation. This positive charge facilitates decarboxylation in the same way a β carbonyl does—by serving as an electron acceptor to facilitate decarboxylation.

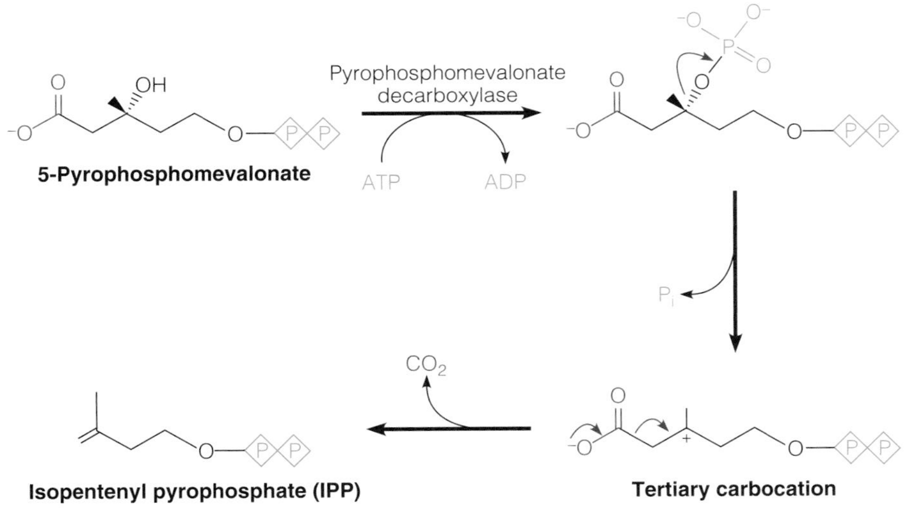

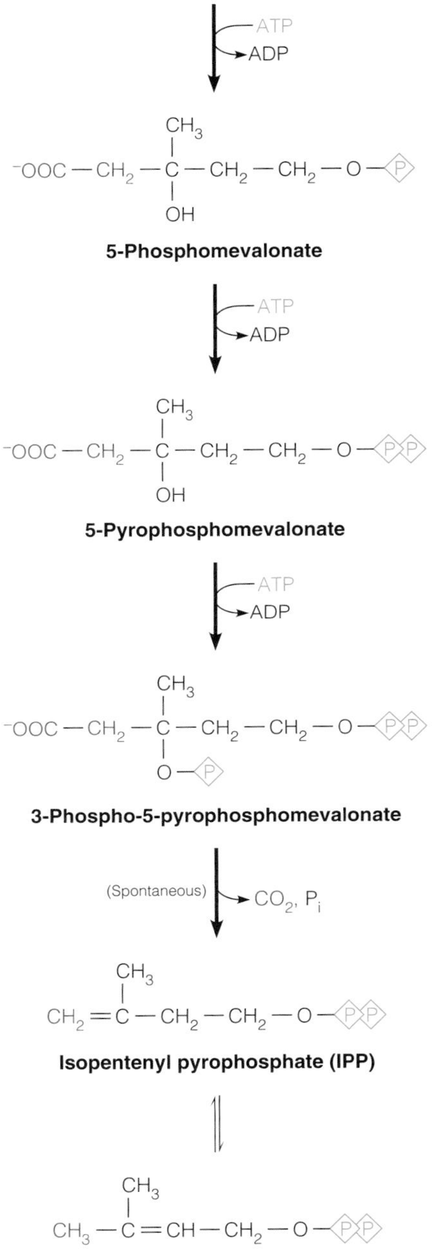

IPP isomerase catalyzes the isomerization of one molecule of the resulting isopentenyl pyrophosphate to the C_5 **dimethylallyl pyrophosphate**. The latter compound, as shown in Figure 19.20, reacts with a second molecule of isopentenyl pyrophosphate to give the C_{10} **geranyl pyrophosphate**, and still another molecule of isopentenyl pyrophosphate reacts with this product to give the C_{15} **farnesyl pyrophosphate**. Both of these reactions involve a tertiary carbocation intermediate in an S_N1 mechanism. Dissociation of PP_i from dimethylallyl pyrophosphate produces an allylic carbocation. The double bond of IPP behaves

FIGURE 19.19

Biosynthesis of mevalonate and conversion to isopentenyl pyrophosphate and dimethylallyl pyrophosphate. The two carbons of the third acetyl group are shown in magenta.

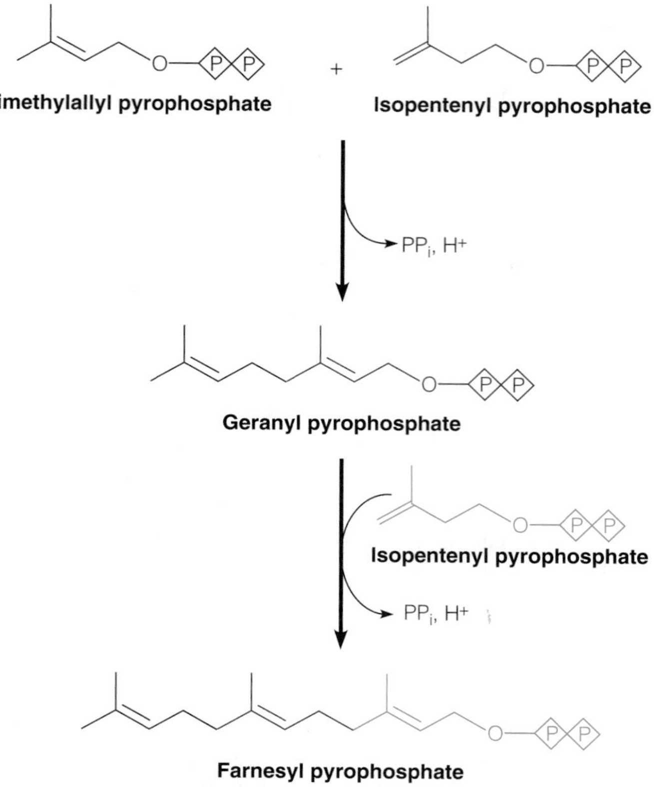

FIGURE 19.20

Conversion of isopentenyl pyrophosphate and dimethylallyl pyrophosphate to farnesyl pyrophosphate. Both of these head-to-tail condensations are catalyzed by the same prenyltransferase (farnesyl pyrophosphate synthase).

Dimethylallyl pyrophosphate   +   **Isopentenyl pyrophosphate**

PP$_i$, H$^+$

Geranyl pyrophosphate

Isopentenyl pyrophosphate

PP$_i$, H$^+$

Farnesyl pyrophosphate

Hydroxymethylglutaryl-CoA reductase, which catalyzes an early reaction in cholesterol biosynthesis, is the major control point for the overall process.

Cyclization of squalene, a C$_{30}$ hydrocarbon, creates the four-ring sterol nucleus.

as a nucleophile, condensing with the carbocation to give a second carbocation, which eliminates a proton to form the product.

Figure 19.21 shows the final reactions of squalene synthesis, catalyzed by **squalene synthase**, which is bound to membranes of the endoplasmic reticulum. This enzyme uses carbocation intermediates to join two molecules of farnesyl pyrophosphate in head-to-head fashion to give **presqualene pyrophosphate**. This activated cyclopropane intermediate then undergoes pyrophosphate elimination and rearrangement to a cyclopropylcarbinyl cation intermediate, which is reduced by NADPH to yield the C$_{30}$ squalene.

A remarkable feature of this part of the pathway is its stereochemistry. In the 1960s two British scientists, George Popják and John Cornforth, identified 14 "stereochemical ambiguities"—that is, 14 steps in the overall process that could go in either of two ways. For example, the triphosphorylated mevalonate derivative shown in Figure 19.19 could undergo decarboxylation by either a *cis* or *trans* elimination of the carboxyl and phosphoryl groups. Thus, 2^{14}, or 16,384, *different* stereochemical routes were possible for conversion of mevalonate to squalene. Remarkably, these scientists and their colleagues were able to identify the single stereochemical pathway that actually takes place among these 16,384 possibilities.

Stage 3: Cyclization of Squalene to Lanosterol and Its Conversion to Cholesterol

All subsequent reactions occur in the endoplasmic reticulum. The cyclization of squalene to lanosterol and the conversion of lanosterol to cholesterol are shown in Figure 19.22. The formation of lanosterol, which has the four-ring sterol nucleus, occurs in two steps. First, a mixed-function oxidase introduces an epoxide function at carbons 2 and 3. Protonation of this functional group initiates a series of *trans* 1,2 shifts of methyl groups and hydride ions, to produce lanosterol. A series of about 20 reactions follows, involving double-bond reductions and three demethylations; one methyl group is removed from C-14 and two from C-4. The penultimate product is 7-dehydrocholesterol, which undergoes a final reduction to yield cholesterol.

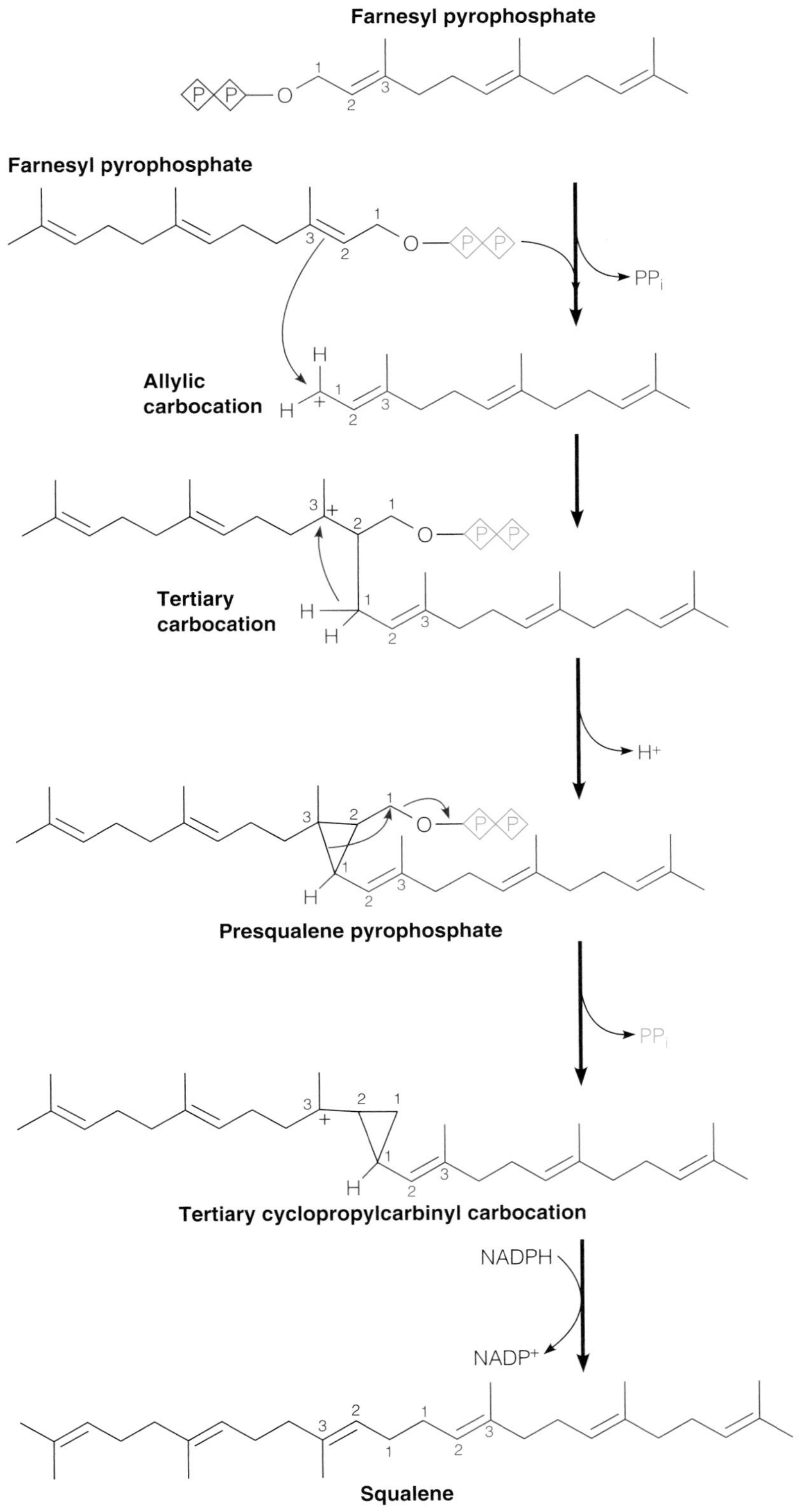

Farnesyl pyrophosphate

Farnesyl pyrophosphate

Allylic carbocation

Tertiary carbocation

Presqualene pyrophosphate

Tertiary cyclopropylcarbinyl carbocation

NADPH

NADP⁺

Squalene

FIGURE 19.21

Conversion of farnesyl pyrophosphate to squalene, catalyzed by squalene synthase. Dissociation of the pyrophosphate group on one of the molecules of farnesyl pyrophosphate (magenta) yields an allylic carbocation. The C2=C3 double bond nucleophilically attacks the carbocation, forming a tertiary cation at C3 of the first farnesyl pyrophosphate (black). Loss of a proton from C1 gives the activated cyclopropane intermediate, presqualene pyrophosphate. Dissociation of the second pyrophosphate leads to rearrangement and formation of another tertiary carbocation intermediate. Hydride transfer from NADPH completes the rearrangement, giving squalene.

Control of Cholesterol Biosynthesis

In Chapter 17, we learned that intracellular cholesterol levels are regulated by several mechanisms, including controlling de novo cholesterol biosynthesis, storing excess cholesterol in the form of cholesterol esters, and regulating the synthesis of the LDL receptor, which is responsible for endocytosis of cholesterol from the bloodstream. As noted earlier, HMG-CoA reductase, which catalyzes the committed reaction in cholesterol biosynthesis, represents a major target for regulation of the overall pathway (see Figure 17.9, page 718). It has long been known from feeding studies that dietary cholesterol efficiently suppresses the endogenous synthesis of cholesterol. But this regulation is not a simple case of feedback inhibition of the rate-limiting enzyme. The cholesterol pathway presents two complications. First, the end-product—cholesterol—is located entirely within the membrane. How does the cell monitor the levels of a membrane component? Second, the pathway from mevalonate produces several other important products besides cholesterol, including geranyl and farnesyl pyrophosphate for prenylation of proteins and synthesis of ubiquinone and dolichol. How does the cell coordinate the synthesis of all of these products? Control of HMG-CoA reductase occurs at both transcriptional and posttranscriptional levels, in a process that employs a set of protein–protein interactions that take place within the ER membrane. The central players in this elegant regulatory mechanism are **Insigs** (Insulin-induced growth response genes), **SREBPs** (sterol regulatory element

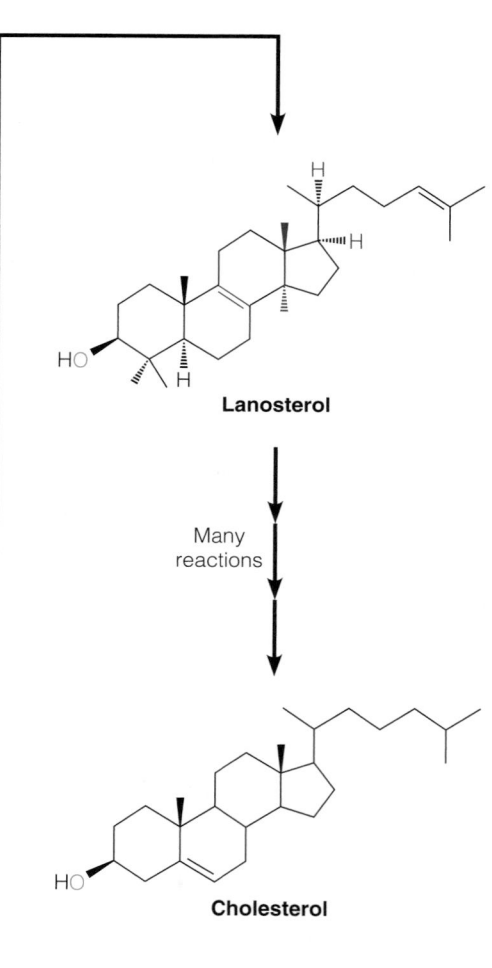

Squalene

Squalene epoxidase

$O_2 + NADPH + H^+$

$H_2O + NADP^+$

Enz—H

Squalene 2,3-epoxide

Cyclase

Cyclase

H^+

Lanosterol

Many reactions

Cholesterol

FIGURE 19.22

Conversion of squalene to cholesterol. Formation of squalene epoxide leads to a series of double-bond electron shifts that close the four rings, and a migration of a carbon atom from C-14 to C-13 yields the first steroid intermediate, lanosterol. Many subsequent reactions lead to 7-dehydrocholesterol, which undergoes reduction to cholesterol.

binding proteins), and **Scap** (SREBP cleavage-activating protein). All of these proteins are anchored in the ER membrane by multiple transmembrane segments. Insig was first identified as an mRNA whose abundance increased when cultured cells were treated with insulin, but the function of the protein was not known. We now understand that Insigs control cholesterol synthesis through sterol-induced protein–protein interactions within the ER membrane.

Mammalian HMG-CoA reductase is also anchored in the ER membrane, by a hydrophobic N-terminal domain composed of eight membrane-spanning segments. The catalytic C-terminal domain projects into the cytosol. In sterol-depleted cells, this protein is degraded quite slowly, with a half-life of more than 12 hours. However, when sterols accumulate in the ER membrane, HMG-CoA reductase is rapidly degraded ($t_{1/2} < 1\,hr$) by the ubiquitin–proteasome pathway (Chapters 20 and 28). Membrane sterols bind to a sterol-sensing domain within the N-terminal membrane domain of the enzyme (Figure 19.23). Sterol binding causes HMG-CoA reductase to bind to a population of Insigs that are associated with a ubiquitination complex comprising gp78, Ubc7, and VCP. Gp78 is an E3 ubiquitin ligase, which transfers ubiquitin from the E2 ubiquitin-conjugating enzyme Ubc7 to specific lysine residues of HMG-CoA reductase. This protein–protein interaction, stimulated by sterol binding, thus leads to ubiquitination of HMG-CoA reductase. VCP is an ATPase that extracts ubiquitinated reductase from membranes and delivers it to proteasomes for degradation. A nonsterol isoprenoid derived from mevalonate, such as geranylgeraniol, is also required for the extraction and degradation of ubiquitinated HMG-CoA reductase.

Several details of this regulatory process remain to be worked out, including the exact identity of the sterols that mediate it. In permeabilized cells, the most potent compounds are oxysterols (cholesterol derivatives that contain hydroxyl groups at various positions in the aliphatic side chain) and methylated sterols such as lanosterol (Figure 19.22) and 24,25-dihydrolanosterol. Cholesterol itself does not appear to regulate the degradation of HMG-CoA reductase.

Cholesterol metabolism is also controlled at the transcriptional level, and this is where SREBPs and Scap come in. As implied by their name (sterol regulatory element binding proteins), SREBPs are transcription factors that bind to the promoters of genes required to produce cholesterol, including the HMG-CoA reductase gene. Conversion of acetyl-CoA to cholesterol requires at least 20 enzymes, all of whose genes are activated by SREBP binding. SREBP also activates transcription of the gene for the LDL receptor, which mediates uptake of dietary cholesterol. However, as shown in Figure 19.24, SREBPs are synthesized as integral membrane proteins in the ER. When cells are depleted of cholesterol, the SREBPs translocate to the Golgi, where they are proteolytically processed to yield active transcription factors that enter the nucleus. This process requires Scap, which serves as an escort protein, and two specific proteases in the Golgi (S1P and S2P). When sterols accumulate in ER membranes (either from cellular uptake, or from de novo biosynthesis), exit of Scap–SREBP complexes from the ER is inhibited, blocking the proteolytic activation of SREBPs. This causes a decrease in the transcription of SREBP target genes, leading to lower rates of cholesterol synthesis and uptake. Like degradation of HMG-CoA reductase, the SREBP pathway is also controlled by membrane sterols and Insigs. In this case, cholesterol itself binds to a sterol-sensing domain within Scap, causing a conformational change in Scap that promotes Insig binding. This interaction blocks the binding of Scap to COPII proteins, which mediate the translocation of cargo molecules from the ER to the Golgi (Chapter 28). Thus, in sterol-replete cells, Insig sequesters Scap so that proteolytic processing of SREBP is prevented. In sterol-depleted cells, however, Scap–SREBP complexes no longer bind Insig, which becomes ubiquitinated and degraded. The Scap–SREBP complexes are free to translocate to the Golgi, where the SREBPs are processed to active transcription factors that stimulate gene expression, eventually leading to restoration of cholesterol levels.

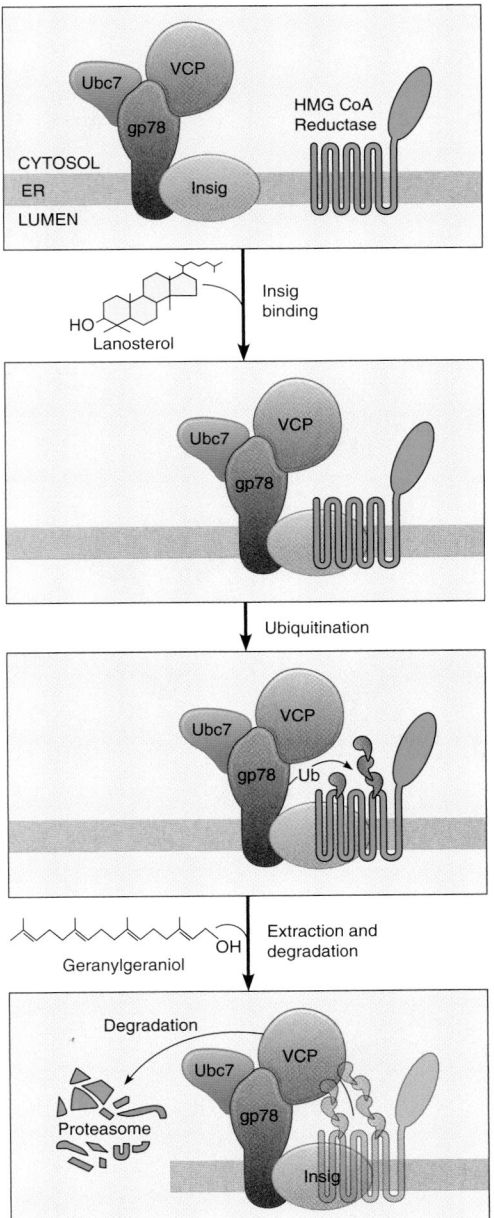

FIGURE 19.23

Regulation of HMG-CoA reductase by ubiquitin-mediated proteolysis. Sterol binding leads to rapid degradation of the enzyme. See text for details.

Reprinted from *Cell* 124:35–46, J. L. Goldstein, R. A. DeBose-Boyd, and M. S. Brown, Protein sensors for membrane sterols. © 2006, with permission from Elsevier.

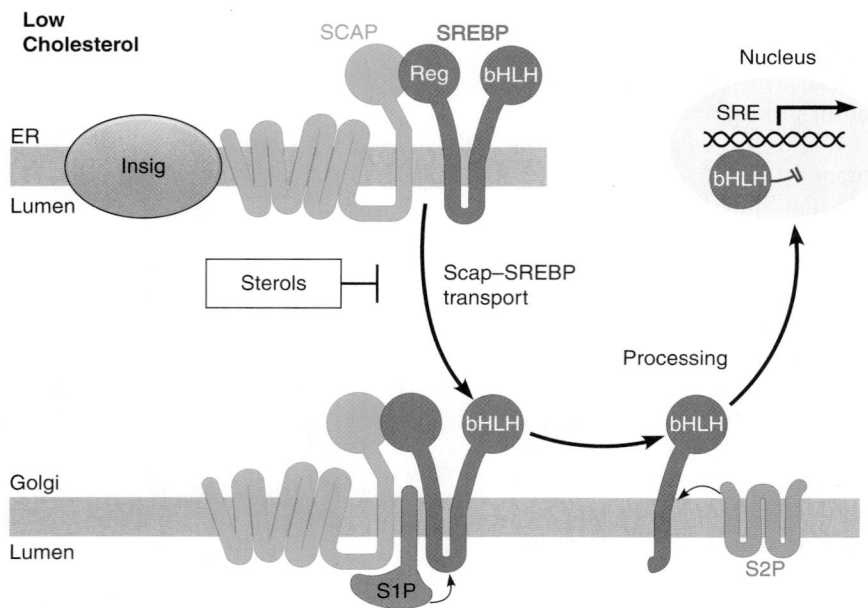

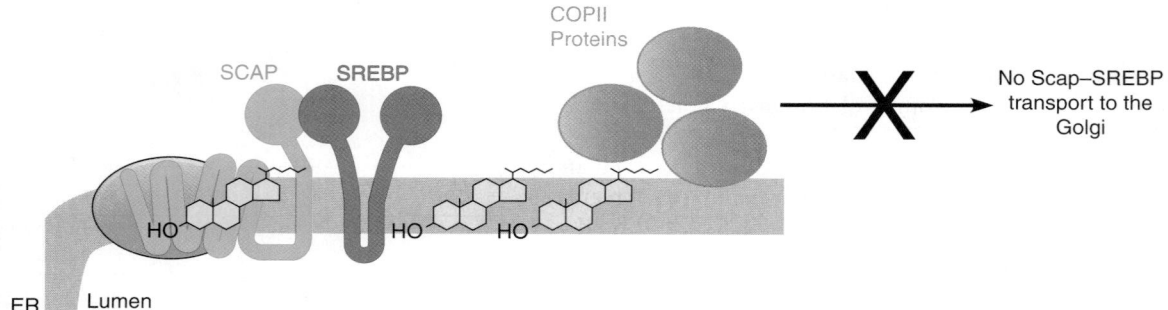

FIGURE 19.24

Insig-mediated regulation of SREBP activation. When cholesterol levels in the cell are low, Scap transports SREBP to the Golgi. Proteolytic processing by membrane-bound proteases S1P and S2P releases SREBP's transcription factor domain (bHLH), which enters the nucleus and binds to sterol regulatory elements (SRE) in the promoters of target genes, stimulating their transcription. High levels of cholesterol block this process by Insig-mediated retention of Scap–SREBP in the ER.

Modified from *Cell* (2006) 124:35–46, J. L. Goldstein, R. A. DeBose-Boyd, and M. S. Brown; and *Cell Research* (2008) 18:609–621, R. A. DeBose-Boyd.

Insig proteins integrate these transcriptional and posttranscriptional regulatory mechanisms. In both processes, sterol binding to a sterol-sensing domain (in HMG-CoA reductase or Scap) stimulates Insig binding. For HMG-CoA reductase, Insig binding means ubiquitination and destruction. For Scap, Insig binding turns off a transcriptional activation pathway. Together, these mechanisms ensure fine control over cellular cholesterol metabolism.

Finally, in some tissues, HMG-CoA reductase is subject to short-term regulation by reversible phosphorylation/dephosphorylation. Phosphorylation on a specific serine residue near the C-terminus of its catalytic domain inactivates the enzyme. This covalent modification is catalyzed by AMP-activated protein kinase (AMPK). Recall from Chapter 18 that the AMPK system acts as a sensor of cellular energy status—it is activated by increases in the cellular AMP:ATP ratio. Cholesterol biosynthesis is a particularly expensive pathway, requiring 36 moles of ATP and 16 moles of NADPH per mole of cholesterol. Thus, under conditions of low energy charge, activated AMPK switches off cholesterol synthesis by inhibiting HMG-CoA reductase. The enzyme is reactivated by dephosphorylation, catalyzed by a type 2A protein phosphatase.

In vertebrates cholesterol synthesis is controlled beautifully through the rate at which cholesterol enters cells from the bloodstream. As we discussed in Chapter 17, homeostasis is maintained by a mechanism that coordinates dietary intake of cholesterol, rate of endogenous cholesterol synthesis in the liver (and to a lesser extent in the intestine), and rate of cholesterol use by cells. That mechanism involves the LDL receptor, the agent most responsible for transporting cholesterol in the bloodstream. However, when these regulatory mechanisms break down in disease, or are overwhelmed by excessive dietary cholesterol, hypercholesterolemia, and eventually atherosclerosis, can result. Once the rate-limiting role of HMG-CoA reductase in cholesterol biosynthesis was understood, specific inhibitors were sought as a therapeutic approach to lowering blood cholesterol levels. Beginning in the 1970s, teams led by Akira Endo in Japan, and Alfred Alberts and Roy Vagelos in the U.S., screened several thousand fungal cultures for compounds that inhibited HMG-CoA reductase. Several compounds were discovered, collectively called **statins**, that act by competitively inhibiting HMG-CoA reductase. Shown in the margin are the structures of several widely used statins, including the fungal polyketides lovastatin and simvastatin and the synthetic atorvastatin (Lipitor). Each statin carries a mevalonate-like moiety (blue), explaining the competitive nature of their activity. Inhibition of HMG-CoA reductase depresses de novo cholesterol biosynthesis and, hence, intracellular cholesterol levels. This in turn leads to increased production of LDL receptors, allowing more rapid clearance of extracellular cholesterol from the blood, thus lowering blood cholesterol levels. Statins have proven to be spectacularly effective in the treatment of hypercholesterolemia, and are among the most widely prescribed drugs in the United States, Canada, and other developed nations.

Intermediates in Cholesterol Synthesis in Protein Prenylation

Prenylation involves transfer of C_{15} (farnesyl) or C_{20} (geranylgeranyl) groups from intermediates in cholesterol synthesis (see Figure 19.20) to cysteine residues four positions from the C-terminus of the target protein (Figure 19.25). Subsequent modifications involve the removal of the three C-terminal amino acid residues by an endoprotease and AdoMet-dependent methylation of the terminal carboxyl group. This results in a prenylated and carboxymethylated C-terminal cysteine residue, which is very hydrophobic, thus anchoring the protein in the bilayer. Most prenylated proteins are members of the small G protein superfamily, with the *ras* oncogene protein (see page 985) being the most prominent example. The enzyme **farnesyltransferase**, which transfers a C_{15} unit from **farnesyl pyrophosphate** to a substrate protein (Figure 19.25), has been recognized as an attractive target for inhibition in cancer chemotherapy. Inhibitors of this enzyme interfere with the growth of cancer cells in culture. Although the precise mechanisms of growth inhibition are not known, this is a promising area for molecular pharmacology.

Bile Acids

Now let us turn to the use of cholesterol for synthesis of other important metabolites—bile acids and steroid hormones. As mentioned in Chapter 17, bile acids are steroid derivatives with detergent properties, which emulsify dietary lipids in the intestine and thereby promote fat digestion and absorption. They are secreted from the liver, stored in the gallbladder, and passed through the bile duct and into the intestine. Biosynthesis of bile acids represents the major metabolic fate of cholesterol, accounting for approximately 90% of cholesterol catabolism in the normal human adult. By contrast, steroid hormone synthesis accounts for only about 50 mg of cholesterol metabolized per day.

Although about 400 to 500 mg of bile acids are synthesized daily, much more than this is secreted into the intestine. Most of the bile acids secreted into the upper small intestine are absorbed in the lower small intestine and returned to the liver for reuse, through the portal blood. This process, which handles 20 to 30 g of bile acids per day, is called the **enterohepatic circulation**. Daily elimination of bile acids in the feces amounts to just 0.5 g/day or less, which is compensated for by synthesis in the liver.

The most abundant bile acids in humans are **cholic acid** and **chenodeoxycholic acid** (shown in Figure 19.26 as the respective bile salts, cholate and chenodeoxycholate). These are usually conjugated in amide linkage with the amino acids

X = H   **Lovastatin (Mevacor)**
X = CH₃   **Simvastatin (Zocor)**

Atorvastatin (Lipitor)

HMG-CoA   **Mevalonate**

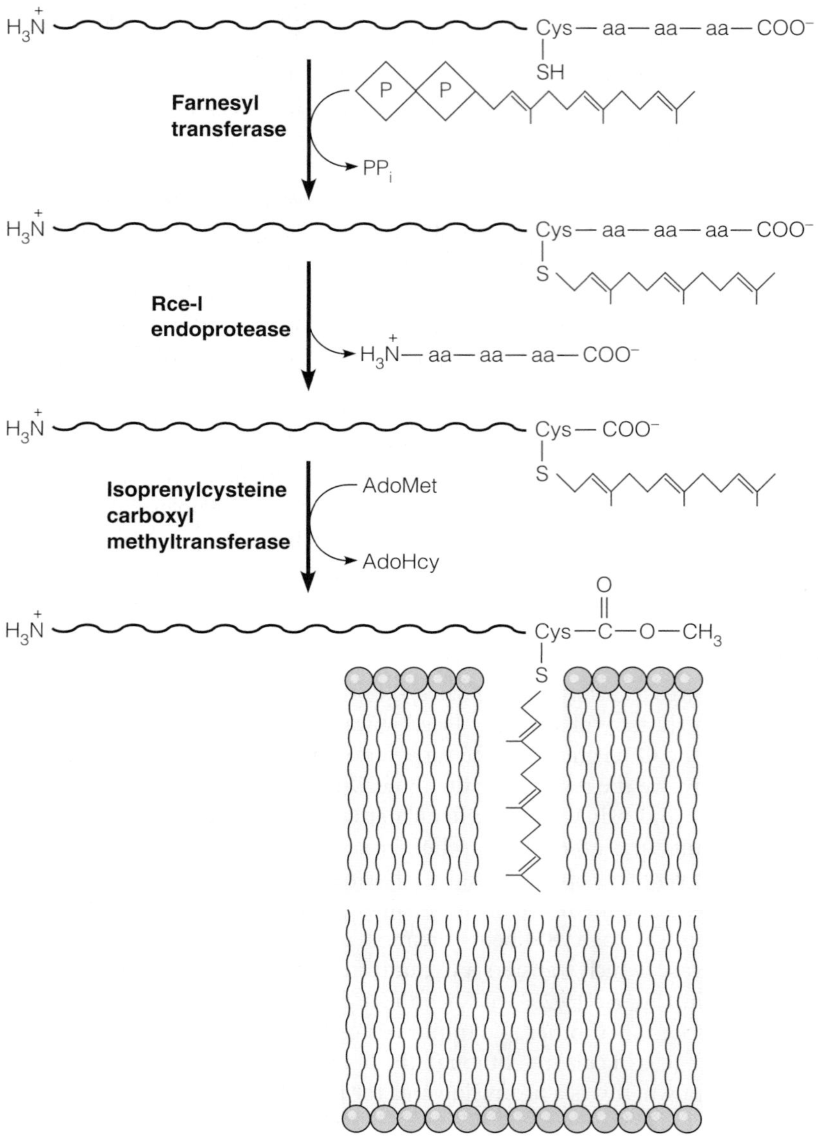

FIGURE 19.25

The protein prenylation pathway. aa = amino acid residue.

glycine or **taurine**, giving compounds called **bile salts**. The cholic acid conjugates with glycine and taurine are called **glycocholate** and **taurocholate**, respectively. Another bile acid, **deoxycholate**, is abundant in the bile of some other mammals. It is widely used as a laboratory reagent, to solubilize membrane proteins.

In the biosynthetic routes to cholate, glycocholate, and taurocholate outlined in Figure 19.26, a series of hydroxylations occurs, catalyzed by microsomal cytochrome P450 mixed-function oxidases. The first of these, catalyzed by cholesterol 7α-hydroxylase (CYP7A1), is rate-determining and hence plays the major role in controlling the overall pathway. The activity of this enzyme is suppressed by dietary bile acids. Most of the enzymes in bile acid synthesis also participate in other pathways, including oxysterol synthesis, steroid synthesis, or metabolism of very long-chain fatty acids.

Steroid Hormones

Cholesterol is the biosynthetic source of all steroid hormones, the extracellular messengers secreted by the gonads and the adrenal cortex, plus the placenta in pregnant females. In this chapter we introduce the biosynthetic pathways to steroid hormones; their actions are discussed in Chapter 23. In general, steroid

The principal categories of steroid hormones in vertebrates are progestins, glucocorticoids, mineralocorticoids, androgens, and estrogens.

hormones control metabolism at the gene level. They interact with intracellular protein receptors, and the hormone–receptor complexes bind to specific sites on the genome and affect transcription of neighboring genes.

Five major classes of hormone will concern us: (1) the **progestins** (progesterone), which regulate events during pregnancy; (2) the **glucocorticoids** (cortisol and corticosterone), which promote gluconeogenesis and, in pharmacological doses, suppress inflammation reactions; (3) the **mineralocorticoids** (aldosterone), which regulate ion balance by promoting reabsorption of K^+, Na^+, Cl^-, and HCO_3^- in the kidney; (4) the **androgens** (androstenedione and testosterone), which promote male sexual development and maintain male sex characteristics; and (5) the **estrogens** (estrone and estradiol), or female sex hormones, which support female characteristics. Most of these hormones are

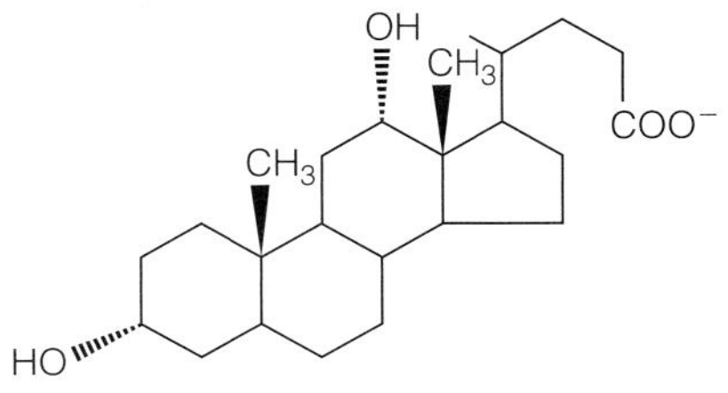

Deoxycholate

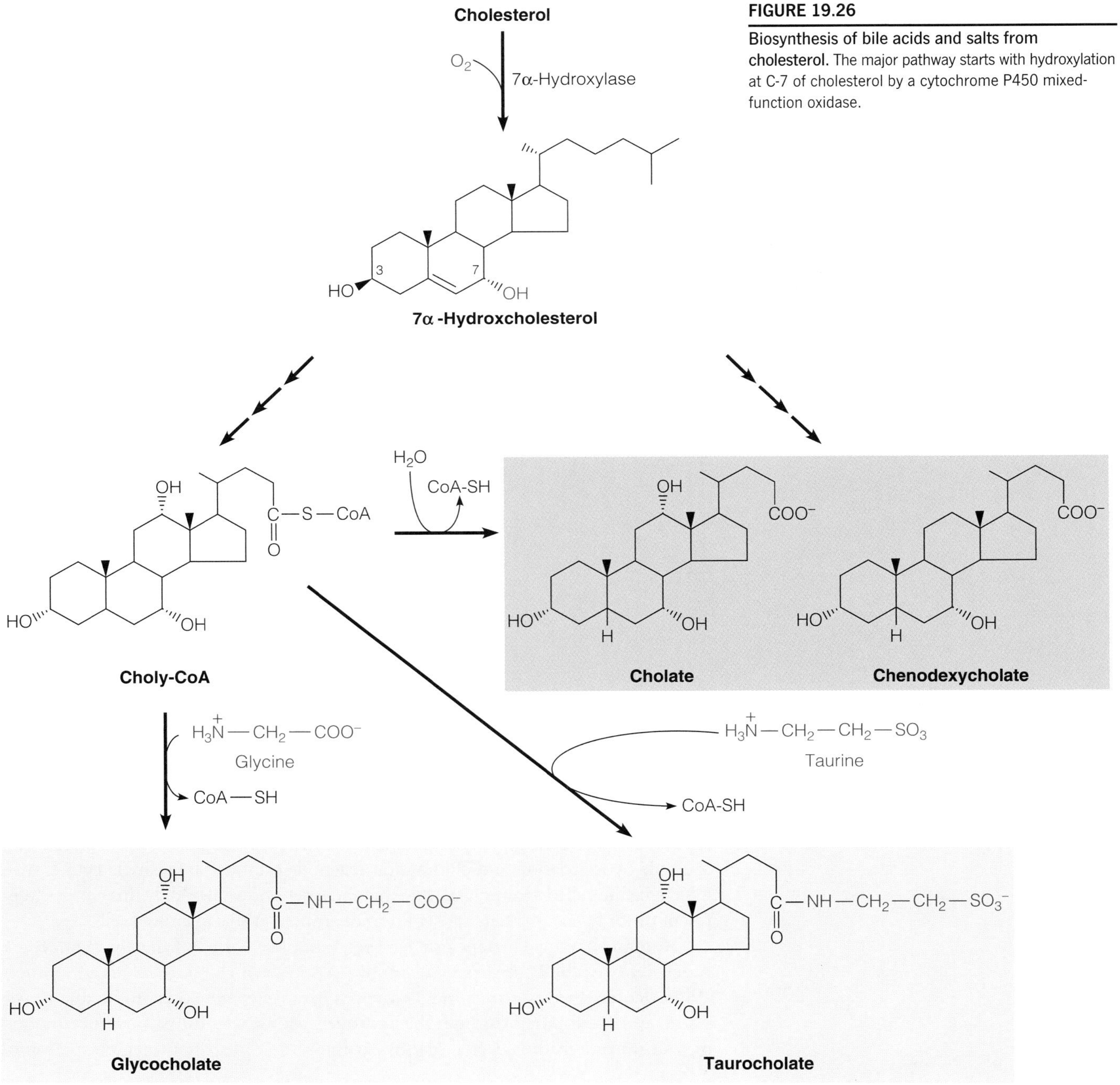

FIGURE 19.26

Biosynthesis of bile acids and salts from cholesterol. The major pathway starts with hydroxylation at C-7 of cholesterol by a cytochrome P450 mixed-function oxidase.

shown in Figure 19.27, which also summarizes their routes of synthesis (described below). In each case the side chain in cholesterol is either greatly shortened or nonexistent.

A general feature of steroid hormones is that they are not stored for release after synthesis. Therefore, the level of a circulating hormone is controlled primarily by its rate of synthesis, which is often controlled ultimately by signals from the brain. These signals usually act through intermediary hormones. For example, the peptide neurohormone **corticotropin releasing hormone (CRH)** is released from cells in the hypothalamus in response to central nervous system inputs (see Chapter 23). CRH stimulates release from the pituitary gland of **corticotropin**, or **adrenocorticotropic hormone (ACTH)**, which in turn stimulates the synthesis of glucocorticoids in adrenal cortex.

Activation of steroid hormone synthesis involves stimulation of both hydrolysis of cholesterol esters and uptake of cholesterol from cellular stores (ER and plasma membrane) into mitochondria of cells in the target organ. Translocation of cholesterol from the outer mitochondrial membrane to the inner mitochondrial membrane (IMM) is the rate-limiting step in the production of all steroids, and requires at least two proteins: steroidogenic acute regulatory (StAR) protein and translocator protein (TSPO). Once in the IMM, an integral membrane cytochrome P450 called **cholesterol side chain cleavage enzyme** hydroxylates the side chain of cholesterol at C-20 and C-22 and cleaves it, to yield **pregnenolone**, the precursor to all other steroid hormones.

> Pregnenolone is an intermediate en route from cholesterol to all other known steroid compounds.

Pregnenolone then moves to the ER, where its conversion to all of the other steroid hormones takes place, as shown in Figure 19.27. 3β-Hydroxysteroid dehydrogenase catalyzes oxidation of the C-3 hydroxyl to a ketone and isomerization of the double-bond. Hydroxylation at C-21 by an adrenal cortex enzyme, followed by two more hydroxylations and a dehydrogenation to form an aldehyde group, gives aldosterone, a mineralocorticoid. Hydroxylation of progesterone at C-17 gives 17α-hydroxyprogesterone, the precursor to all other steroids. Two hydroxylations of this intermediate give cortisol (a glucocorticoid), primarily in the adrenal gland. An enzyme in adrenal cortex and in gonads cleaves the side chain of 17-hydroxyprogesterone at C-17, giving **androstenedione**, a precursor to androgens and estrogens (and one of the now-banned "performance enhancing drugs" that have tainted professional baseball). All of these hydroxylations are catalyzed by cytochrome P450 enzymes, typically obtaining the necessary reducing equivalents from NADPH via P450 oxidoreductase (see Figure 15.32, page 664). Testosterone undergoes reduction at C-5, giving 5α-dihydrotestosterone, a somewhat more potent androgen. Androgens are converted to estrogens by aromatization of the A ring. This conversion involves multiple hydroxylation and elimination reactions catalyzed by a single cytochrome P450 enzyme called **aromatase**. Note that the reactions catalyzed by this enzyme represent the only known route for synthesis of aromatic rings in animal cells.

Human enzyme deficiencies have been described for all of the above enzymes. A deficiency of the 21-hydroxylase inhibits synthesis of glucocorticoids and mineralocorticoids. Progesterone and 17-hydroxyprogesterone accumulate due to the block, and are shunted through the androgen pathway, leading to overproduction of testosterone in the adrenal glands. At the same time, the underproduction of

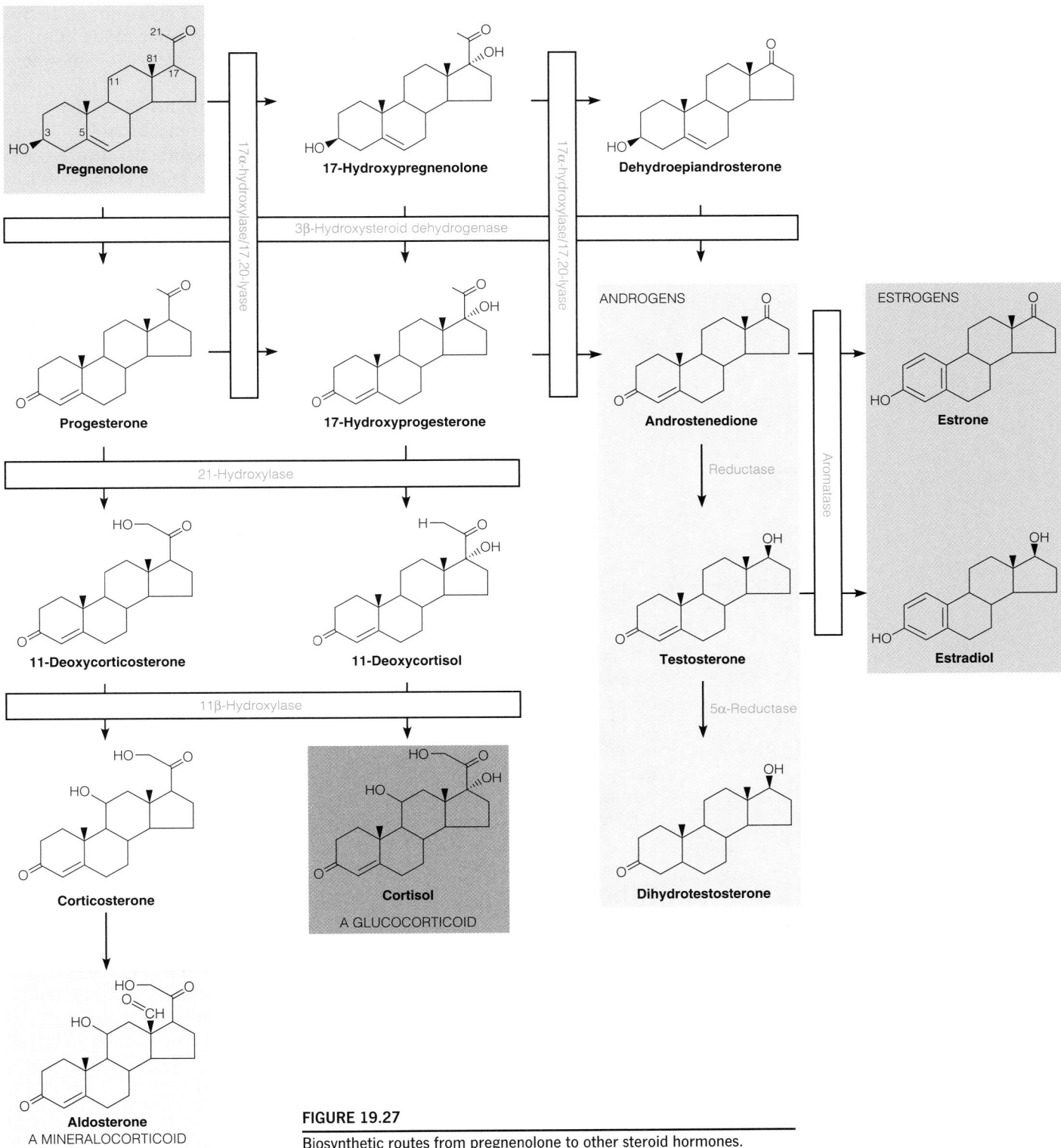

FIGURE 19.27

Biosynthetic routes from pregnenolone to other steroid hormones.

cortisol interferes with a feedback loop of hormonal control involving corticotropin releasing factor (Chapter 23) and ACTH; increased ACTH secretion stimulates the adrenals to grow (hyperplasia) and synthesize steroids, exacerbating the testosterone overproduction. This causes virilization (masculinization) of females. A deficiency of the 11β-hydroxylase also causes adrenal hyperplasia and is characterized by accumulation of the mineralocorticoid 11-deoxycorticosterone, causing hypertension, and the glucocorticoid 11-deoxycortisol. A deficiency of the 5α-reductase decreases effective androgen levels and leads to feminization of males. Fortunately, these and other steroid abnormalities can be treated with hormone replacement therapy, if detected early enough in life.

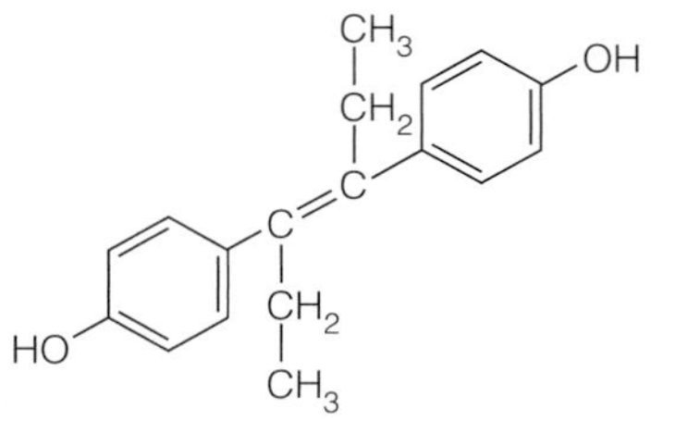

Hundreds of synthetic compounds with steroid hormone–like activity have been tested and used for various purposes. Widely used synthetic steroids are anti-inflammatory glucocorticoids. **Diethylstilbestrol**, a synthetic estrogen, was widely used to promote growth of beef cattle, until it was found to be potentially carcinogenic at the levels found in meat from treated cattle. Oral contraceptives are formulated with compounds containing progesterone and estrogen activities. Two widely used synthetic estrogens are **norethynodrel** and **mestranol**.

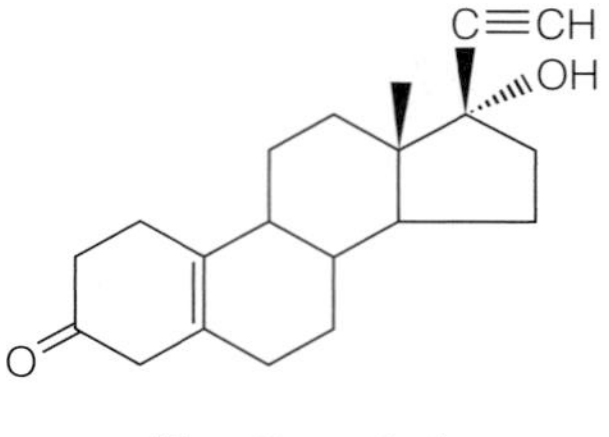

Diethylstilbestrol

Norethynodrel

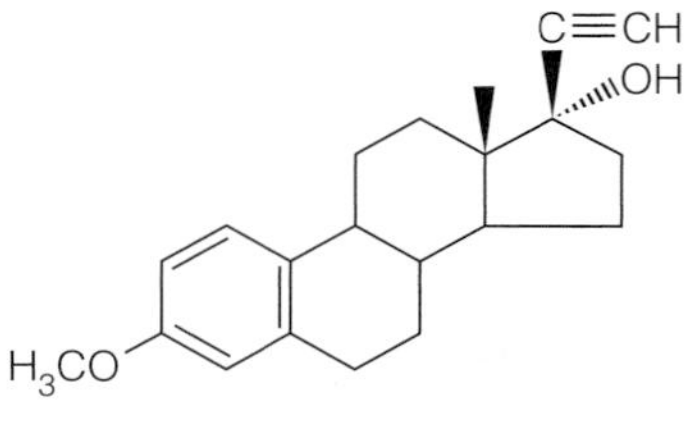

Mestranol

Mixtures of either drug with progesterone inhibit pituitary secretion of hormones that control the female reproductive cycle. This inhibition suppresses maturation of the ovarian follicle and ovulation.

Recent concern has focused upon environmental substances produced through human activity, such as certain pesticides, which happen to have estrogen-like activity. Some of these have been shown to interact with estrogen receptors and to stimulate similar biochemical responses. Solid evidence is accumulating that these "endocrine disrupters" may be responsible for declines in fertility of a number of animal species, possibly including humans, as well as increases in obesity and diabetes.

Mammalian cells lack the capacity for complete degradation of steroid compounds. Although a number of catabolic reactions do occur, most steroids and their metabolites become conjugated, through their hydroxyl groups, to glucuronate or sulfate. Either modification greatly increases solubility of the steroid and facilitates its elimination in the urine.

Other Isoprenoid Compounds

Lipid-Soluble Vitamins

The four lipid-soluble vitamins—A, D, E, and K—are all isoprenoid compounds. They are made up, like steroids, of activated five-carbon units. Thus, as a group they have a structural relatedness not seen among the water-soluble vitamins. On the other hand, the water-soluble vitamins have a *functional* uniformity in that all are designed to carry mobile metabolic groups, whereas the lipid-soluble vitamins are diverse in their functions.

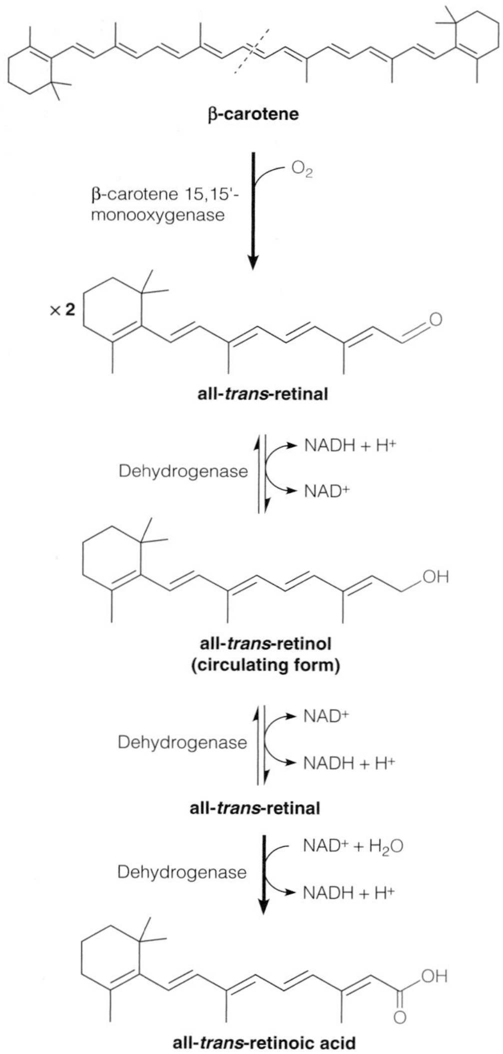

Vitamin A

There are three active forms of vitamin A: **all-*trans*-retinol, -retinal**, and **-retinoic acid**. Collectively, these are referred to as **retinoids**. The vitamin can be either consumed in the diet as esterified retinol, or biosynthesized from β-**carotene**, a plant isoprenoid especially abundant in carrots. β-carotene is cleaved in the intestine by a monooxygenase to form two molecules of all-trans-retinal (retinaldehyde), which are then reduced to retinol (Figure 19.28). All-*trans*-retinol is the form which circulates in the blood and which has the highest biological activity.

FIGURE 19.28

Metabolism of vitamin A.

Ingestion of this form will satisfy all the nutritional requirements for the vitamin. Oxidation of retinol (alcohol at C-15) to retinal (aldehyde at C-15) is reversible whereas its subsequent oxidation to retinoic acid is irreversible. These two reactions are catalyzed by specific dehydrogenases in peripheral tissues.

Vitamin A plays an important role in the visual process in rod cells of the retina, the cells primarily responsible for low-light vision, with relatively little color detection (Figure 19.29). The rod cell outer segments contain lamellar protein disks that are rich in the protein **opsin**. In the disks, all-*trans*-retinol undergoes isomerization followed by dehydrogenation to give **11-*cis*-retinal**. The chemical changes in photoreception are shown in Figure 19.30. 11-*cis*-Retinal forms a Schiff base with a lysine residue in opsin, giving **rhodopsin** (step 1). Rhodopsin has very strong light absorption in the 400- to 600-nm range (the visible wavelength of the spectrum). Absorption of a photon of light triggers a chain of events leading to neural excitation. The excited retinal in rhodopsin isomerizes to an all-*trans* form (step 2), followed by several conformational changes and release of all-*trans*-retinal (steps 3 and 4). After isomerization (step 5), the process can begin anew. Step 3 in Figure 19.30 is the key to the transduction of photon reception to a neural action potential. This process involves cyclic GMP and a G protein called **transducin**. Further details of this process are presented in Chapter 23.

Retinoids (mainly all-*trans*- and 9-*cis*-retinoic acid) also play important roles as developmental regulators, acting much like steroid hormones (see page 980). They interact with specific receptor proteins in the cell nucleus. The ligand–receptor complexes bind to specific DNA sequences, where they control the transcription of genes involved in embryonic development, reproduction, postnatal growth, differentiation of epithelia, and even immune responses. Indeed, the earliest effect of vitamin A deficiency is a keratinzation of epithelial tissues of the respiratory and urogenital tracts, in which the columnar epithelia become replaced by squamous epithelium. Lesions in the eyes (e.g., night blindness) occur much later in vitamin A deficiency.

Vitamin D

The most abundant form of vitamin D is vitamin D_3, or **cholecalciferol**. This is not truly a vitamin because it is not required in the diet. Instead, it arises by synthesis from 7-dehydrocholesterol, an intermediate in cholesterol biosynthesis (see margin, page 811). It is more accurate to think of vitamin D_3 as a prohormone because it is converted to a metabolite that acts analogously to a steroid hormone. Its action involves the regulation of calcium and phosphorus metabolism, particularly with respect to synthesis of the inorganic matrix of bone, which consists largely of calcium phosphate.

In skin cells, 7-dehydrocholesterol undergoes ultraviolet photolysis to give cholecalciferol. Because the UV rays come from sunlight, insufficient sunlight exposure can cause a deficiency of vitamin D_3 and result in the bone malformation known as **rickets**. Mild deficiency can lead to **osteoporosis**, loss of calcium from the bones. Vitamin D_3 is often added to dairy products as a dietary supplement because sunlight exposure is limited in many regions for much of the year.

Cholecalciferol undergoes two successive hydroxylations, each catalyzed by a mixed-function oxidase. The first, at carbon 25, involves a microsomal enzyme system in liver. 25-Hydroxycholecalciferol is then transported to the kidney, where a mitochondrial enzyme hydroxylates it at carbon 1. This reaction is activated by **parathyroid hormone**, which is secreted from the parathyroid gland when calcium levels are low. When calcium levels are adequate, the second hydroxylation occurs at C-24, instead of C-1, to give an inactive metabolite.

1,25-Dihydroxycholecalciferol, or 1,25(OH)D_3, is the hormonally active form of vitamin D. This compound migrates to target cells in the intestine and in osteoblasts (bone cells), where it binds to protein receptors that migrate to the cell nucleus. In intestine the hormone–receptor complex stimulates transcription,

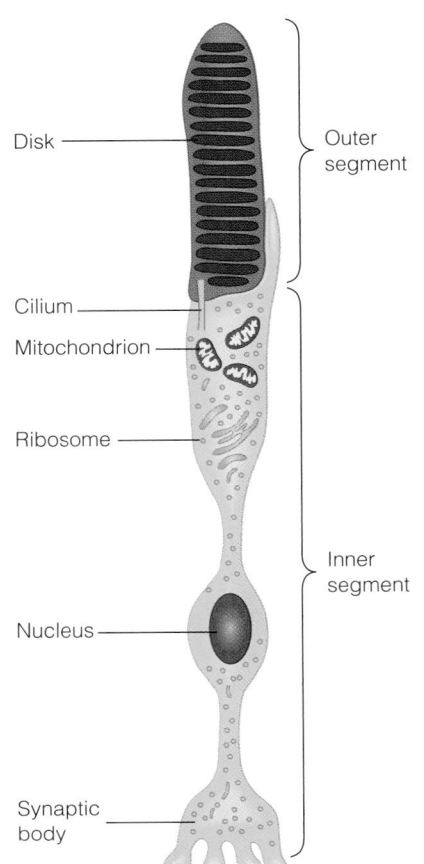

FIGURE 19.29

Schematic drawing of a rod cell. The outer segment is a stack of membranous disks, which contain the photoreceptive pigments. This segment is connected by a thin cilium to the inner segment, which contains the cell nucleus, cytosol, and synaptic body. The potential change produced in the outer segment travels to the synaptic body and is transmitted to one or more of the neurons of the retina.

Isomerization of a protein-bound form of vitamin A in the retina is the mechanism by which light energy is received in the eye.

1,25-Dihydroxycholecalciferol controls bone metabolism by regulating intestinal absorption of calcium.

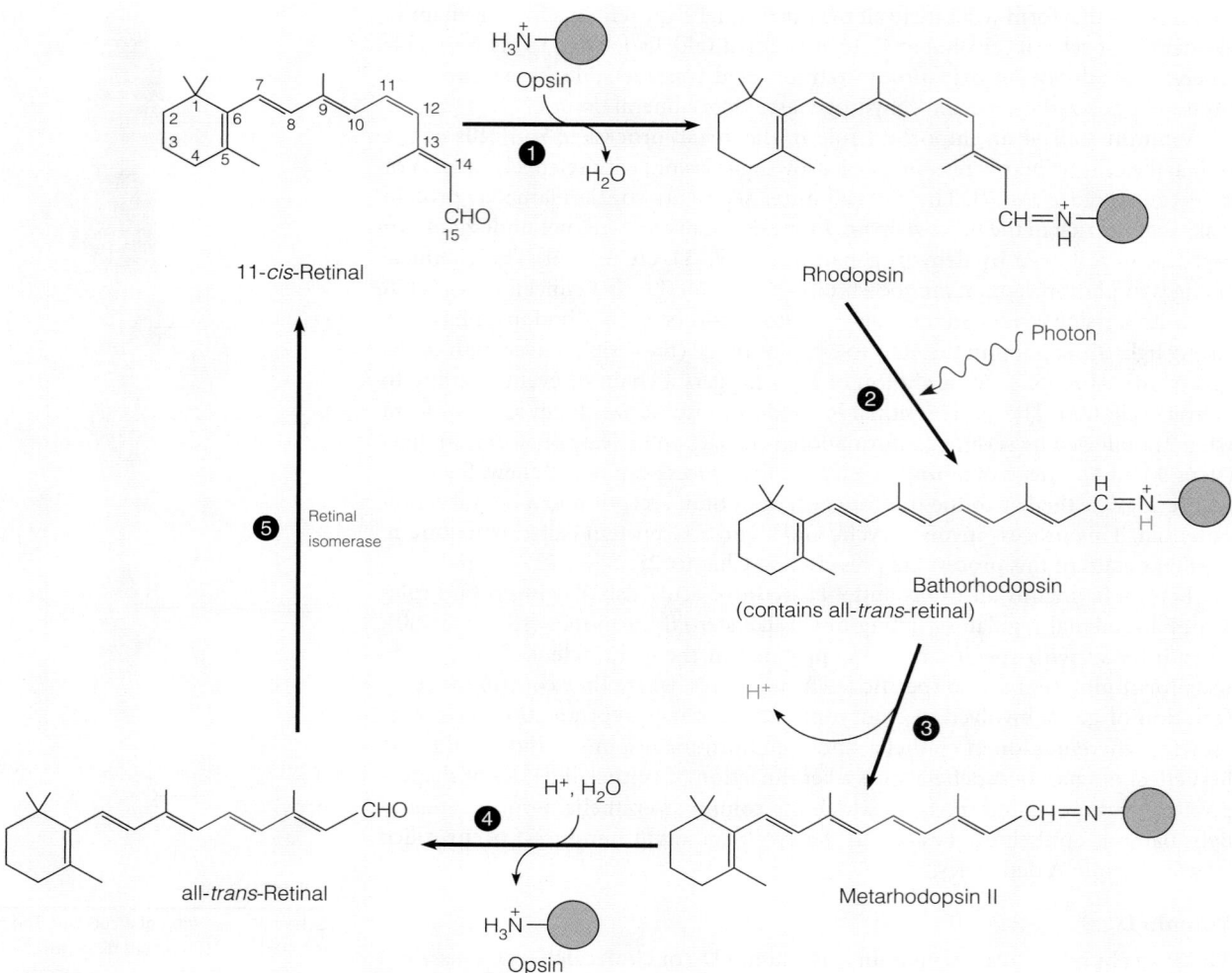

FIGURE 19.30

The chemical changes in photoreception.
11-*cis*-Retinal and opsin in a rod cell combine to form rhodopsin. Absorption of a photon of light leads to the chemical changes shown in steps 2 and 3. Step 3 involves at least 3 distinct conformational changes that occur in ~1 ms. Metarhodopsin II is the species that activates transducin (not shown) to initiate the visual cascade described in Chapter 23. After about 1 second, metarhodopsin II dissociates into all-*trans*-retinal, which isomerizes as the cycle begins again.

resulting in synthesis of a protein that stimulates calcium absorption into the bloodstream. In osteoblasts, 1,25(OH)D$_3$ stimulates calcium uptake for deposition as calcium phosphate.

Vitamin E

Vitamin E, also called α-**tocopherol**, was originally recognized in nutritional studies as an agent that prevented sterility in rats. The vitamin appears to play an antioxidant role, particularly in preventing attack of peroxides on unsaturated fatty acids in membrane lipids (see also Chapter 15, page 666). α-Tocopherol does prevent fatty acid peroxidation in vitro. However, vitamin E deficiency results in additional symptoms such as erythrocyte hemolysis and neuromuscular dysfunction that are not relieved by other antioxidants. Therefore, additional biological roles seem likely for this vitamin.

Vitamin K

Vitamin K was originally discovered as a lipid-soluble substance involved in blood coagulation. Vitamin K$_1$, or **phylloquinone**, is found in plants; the quinone portion of this molecule has a largely saturated side chain. Another form of the vitamin, vitamin K$_2$, or **menaquinone**, is found largely in animals and bacteria. Menaquinone has a partly unsaturated side chain. In animals, vitamin K$_2$ is essential for the carboxylation of glutamate residues in certain proteins, to give γ-**carboxyglutamate**. This modification allows the protein to bind calcium, an essential event in the blood clotting cascade (discussed in Chapter 11). Newborn children routinely receive vitamin K injections because most of our vitamin K comes from intestinal bacteria, which have not yet colonized the guts of newborns.

Carboxylation of glutamate residues occurs in other proteins that are active in the mobilization or transport of calcium. The carboxylating enzyme uses the reduced, hydroquinone, form of vitamin K. During the reaction, the hydroquinone becomes oxygenated to a quinone epoxide, and this reaction facilitates the attack of CO_2 on C-4 of a glutamate residue by deprotonating that carbon.

Other Terpenes

Terpene is a generic term for all compounds that are biosynthesized from isoprene precursors. Therefore, the compounds we have been discussing—cholesterol, bile acids, steroids, and lipid-soluble vitamins—are terpenes; they received special attention because of their importance in animal metabolism. Here we take a brief glimpse at the vast range of other terpene compounds. They include insect hormones and plant growth hormones, as well as the lipid-linked sugar carriers we encountered in Chapter 9.

Terpenes are biosynthesized ultimately from isopentenyl pyrophosphate (C_5) and dimethylallyl pyrophosphate (C_5). When these combine to yield geranyl pyrophosphate (C_{10}) (see Figure 19.20), any terpene formed thereby is called a **monoterpene**. When a compound is formed from 1 mole of farnesyl pyrophosphate (C_{15}), the product is called a **sesquiterpene. Triterpenes** (C_{30}) are formed from 2 moles of farnesyl pyrophosphate. Geranylgeranyl pyrophosphate (C_{20}) yields either **diterpenes** (C_{20}) or **tetraterpenes** (C_{40}). The dolichols and undecaprenol, introduced in Chapter 9, are examples of **polyprenols** (polyisoprenoid alcohols), which have more than 50 carbons. Structures of some common terpenes are given in Figure 19.31.

Eicosanoids: Prostaglandins, Thromboxanes, and Leukotrienes

We turn finally to a class of lipids that are distinguished by their potent physiological properties, low levels in tissues, rapid metabolic turnover, and common metabolic origin. The most important of these compounds are the **prostaglandins**; also included are the **thromboxanes** and **leukotrienes**. Collectively, these compounds are called **eicosanoids** because of their common origin from C_{20} polyunsaturated fatty acids, the eicosaenoic acids, particularly arachidonic acid, which is all-*cis*-5,8,11,14-eicosatetraenoic acid. Recall from Chapter 17 that arachidonic acid is synthesized from linolenic acid. Related C_{20} trienoic and pentaenoic acids serve as minor precursors to some prostaglandins and their relatives. In addition to the compounds discussed here, the eicosaenoic acids serve as precursors to another class of compounds, the hydroxyeicosaenoic acids and hydroperoxyeicosaenoic acids. The latter compounds are metabolic precursors to the leukotrienes. Related compounds are derived from polyunsaturated fatty acids shorter or longer than C_{20}, and the term **oxylipin** has been proposed as a generic term for this class of lipids, encompassing all chain lengths.

The prostaglandins and the closely related thromboxanes are derived from a common pathway; a different pathway leads from arachidonic acid to the leukotrienes. Like hormones, the eicosanoids exert specific physiological effects on target cells. However, they are distinct from most hormones in that they act locally, near their sites of synthesis, and they are catabolized extremely rapidly. Moreover, the actions of a given prostaglandin seem to vary in different tissues. The biological properties of the eicosanoids have led to great interest in their medical use and in uses of their analogs.

7-Dehydrocholesterol

↓ UV

*Sites of further hydroxylation

Cholecalciferol (vitamin D₃)

α-Tocopherol (vitamin E)

Phylloquinone (vitamin K₁)

Menaquinone (vitamin K₂)

A γ-carboxyglutamate residue complexed with calcium

The biologically active eicosanoids, derived from arachidonic acid, include prostaglandins, thromboxanes, and leukotrienes. They are short-lived, locally acting signaling molecules.

Class	Example	Function
Monoterpenes	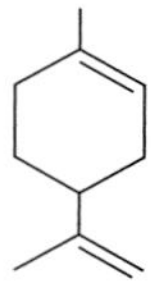 **Limonene**	Responsible for the characteristic odor of lemons
Sesquiterpenes	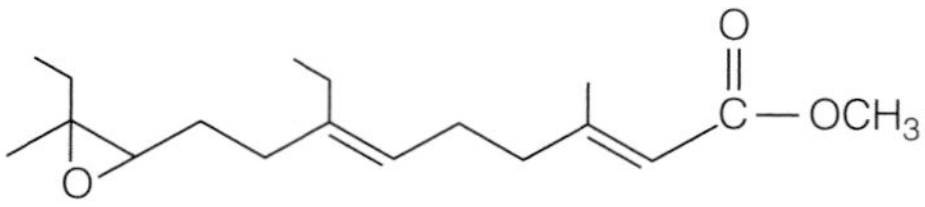 **Juvenile hormone I**	Controls metamorphosis in insects
Diterpenes	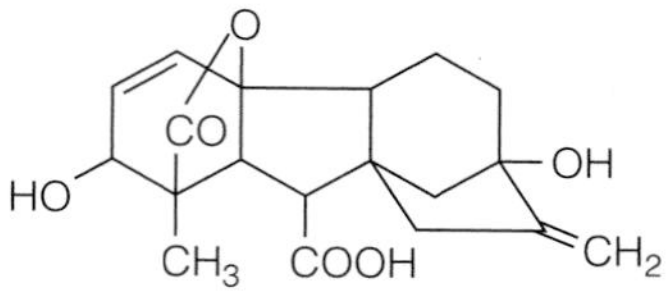 **Gibberellic acid**	Plant growth hormone
Triterpenes	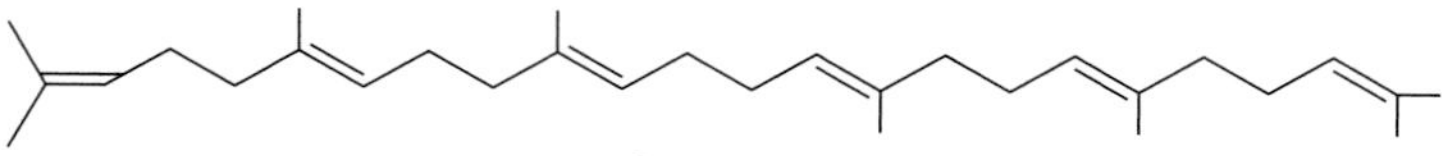 **Squalene**	Cholesterol precursor
Tetraterpenes	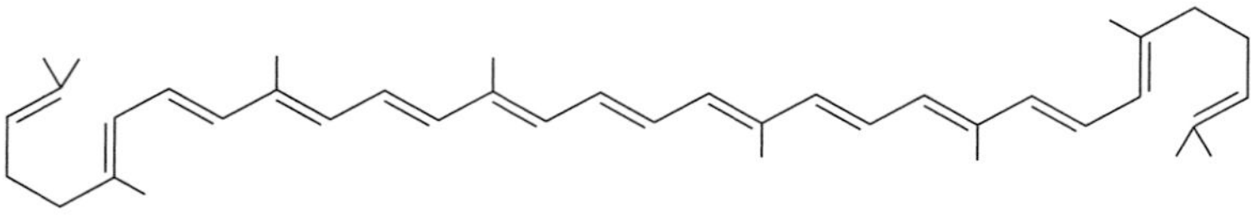 **Lycopene**	Tomato pigment
Polyprenols	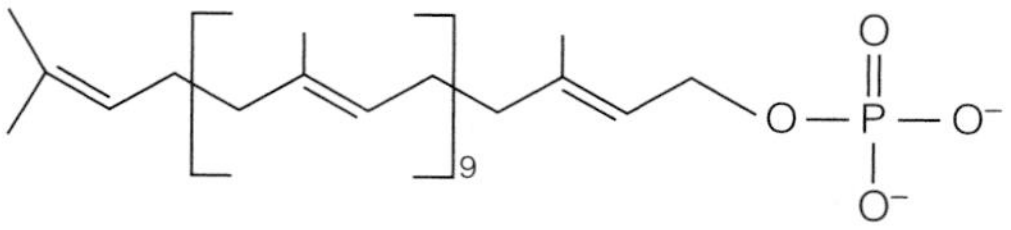 **Undecaprenol phosphate**	Sugar carrier for oligosaccharide synthesis
	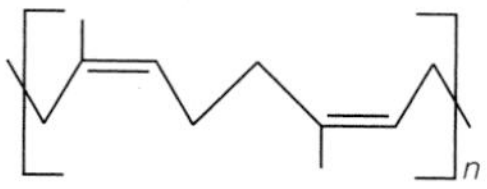 ***cis*-Polyisoprene**	Natural rubber

FIGURE 19.31

Some terpene compounds. These examples are representative of an enormous class of natural products.

Some Historical Aspects

The most important early chapters in prostaglandin research were written in Sweden. In the mid-1930s Ulf von Euler discovered that lipid extracts of human semen contained active compounds that, on injection into animals, stimulated smooth muscle contraction or relaxation and affected the blood pressure. Because of their presumed origin in the prostate gland, he named these compounds *prostaglandins.* Later it was realized that these compounds are widely distributed in animal tissues. The first structural elucidations were reported in the late 1950s, under the direction of Sune Bergström and Bengt Samuelsson, and biosynthetic pathways were described in the mid-1960s, in Sweden and in the Netherlands.

The biological properties of the prostaglandins attracted intense interest in the pharmaceutical industry, but initial progress was limited by the low availabilities of these compounds. Interest reached a peak in 1971 with the discovery that aspirin inhibits one of the enzymes in prostaglandin biosynthesis. This inhibition is now known as the major site of action of aspirin and other nonsteroidal anti-inflammatory drugs (NSAIDs). Later in the 1970s the thromboxanes and leukotrienes were discovered.

Structure

The first two prostaglandins to be isolated were called prostaglandins E and F, respectively, because of their preferential solubility in ether (E) or in phosphate buffer (F for *fosfat*, the Swedish word). We now denote these compounds by PGE and PGF, respectively; all other prostaglandins are denoted by a letter, for instance, PGA and PGH. Each prostaglandin has a cyclopentane ring and two side chains, with a carboxyl group in one side chain. A subscript numeral denotes the number of double bonds in the two chains. The most abundant prostaglandins, those synthesized from arachidonic acid, contain two double bonds; thus, PGE_2 would be the prostaglandin E derived from arachidonic acid. Finally, in the PGF series a subscript α indicates that the hydroxyl group at C-9 is *cis* to the 11-hydroxyl group, and β signifies a *trans* configuration. Structures of the most common prostaglandins are shown in Figure 19.32, along with that of thromboxane A_2 (TxA_2). TxA_2 was originally isolated from thrombocytes, or blood platelets, as a compound that stimulated platelet aggregation, an early step in blood clotting. Note its structural resemblance to PGE_2, except for the cyclic ether ring. Another thromboxane, TxB_2, is a hydrolysis product of TxA_2.

Biosynthesis and Catabolism

Here we discuss only the biosynthesis of the 2-series of prostaglandins. The 1- and 3-prostaglandins are synthesized identically, from related C_{20} fatty acids. The biosynthetic pathways, which occur in endoplasmic reticulum, are shown in

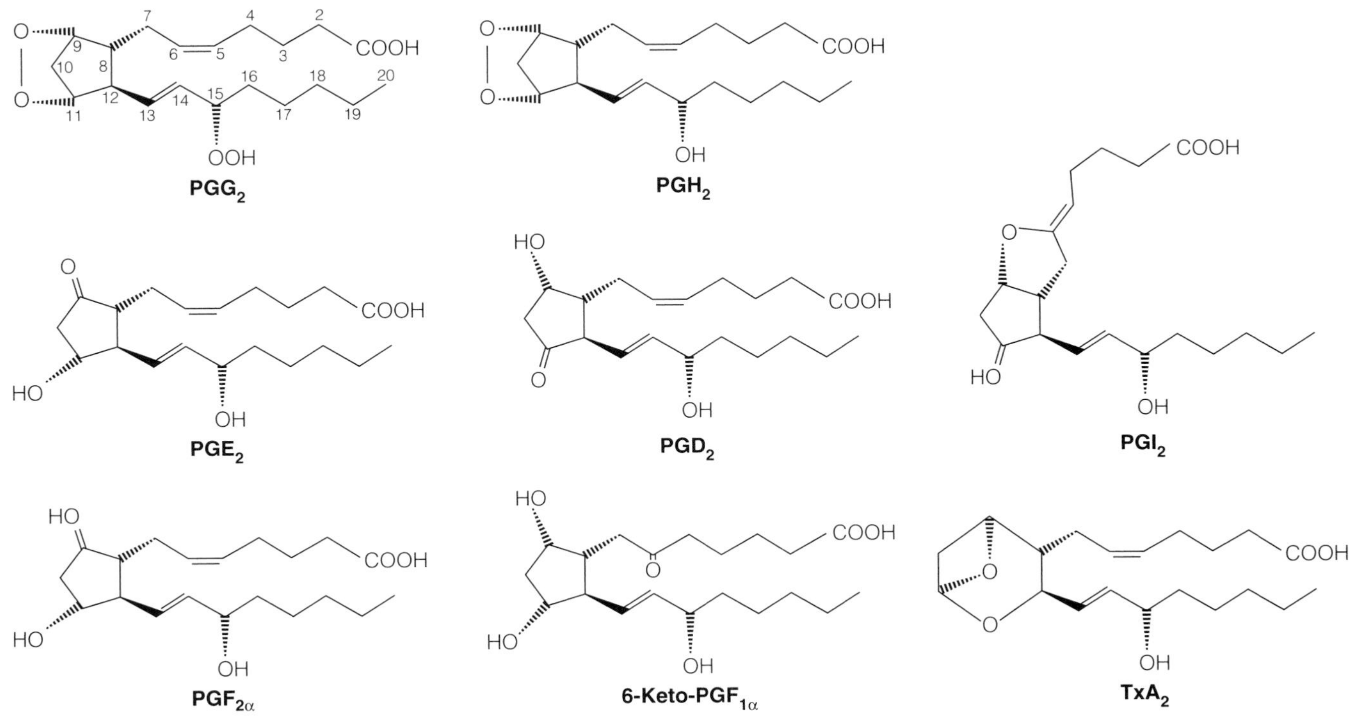

FIGURE 19.32

Structures of the major prostaglandins and thromboxane A_2. The figure shows the most abundant prostaglandins, those of the 2-series. They are derived from arachidonic acid, as is thromboxane A_2. Numbering of carbons begins with the carboxyl group as shown for the structure of PGG_2.

FIGURE 19.33

Summary of biosynthetic routes to the major prostaglandins and thromboxane A_2. PLA_2, phospholipase A_2; PLC, phospholipase C; DGL, diacylglycerol lipase; MGL, monoacylglycerol lipase.

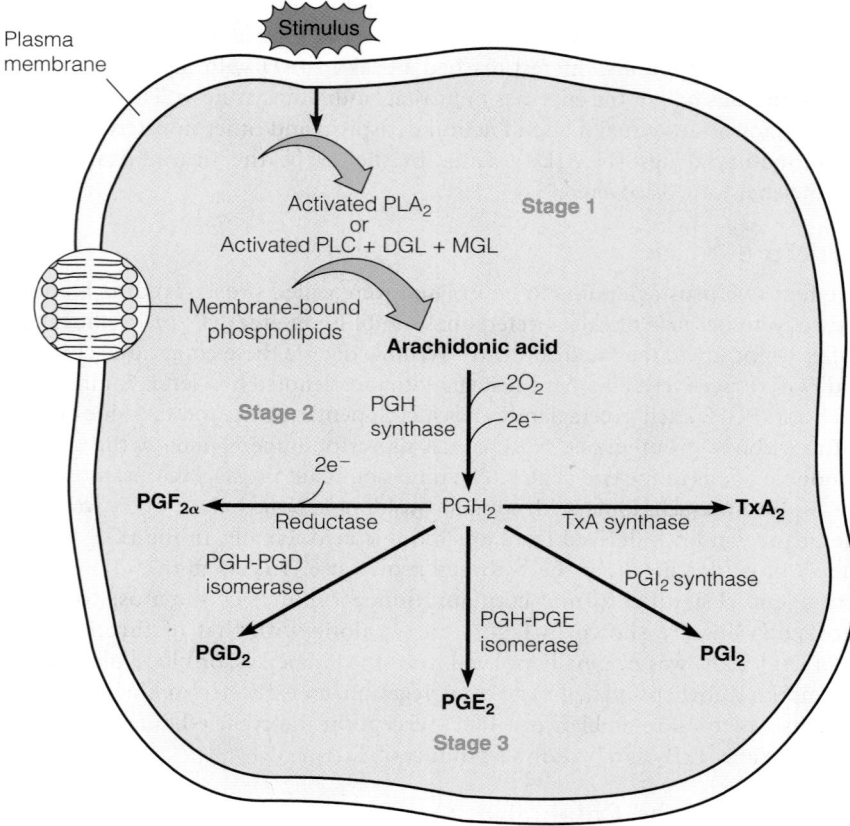

Figure 19.33. We can consider these pathways as occurring in three distinct stages: (1) release of arachidonic acid from membrane phospholipids; (2) oxygenation of arachidonate to yield PGH_2, a prostaglandin endoperoxide that serves as precursor to other prostaglandins; and (3) depending on the enzymes present in a cell, the conversion of PGH to other prostaglandins or to TxA_2.

The release of arachidonic acid from membrane phospholipids in stage 1 occurs as a result of tissue-specific stimuli by hormones such as **bradykinin** or epinephrine or by proteases such as thrombin. Pathological release can occur if membranes are perturbed. For example, the inflammation caused by bee stings is probably due to arachidonate release stimulated by the venom protein **melittin**. Arachidonic acid is commonly esterified at the glycerol C2 position in phosphatidylinositol and other membrane phospholipids and can be released by different enzymatic processes. One pathway involves the action of a cytosolic phospholipase A_2 on phosphatidylcholine or phosphatidylethanolamine, yielding arachidonate. A second route, prominent in brain, involves the action of phospholipase C on phosphatidylinositol, yielding diacylglycerol, which in turn undergoes cleavage by diacylglycerol lipase and then monoacylglycerol lipase (MGL; see page 722 in Chapter 17) to give free arachidonate.

Free arachidonate is acted on, in stage 2, by **PGH synthase**, a bifunctional endoplasmic reticulum membrane enzyme with two activities in a single heme-containing polypeptide chain. The first, a **cyclooxygenase**, introduces two molecules of O_2, one to form the ring and one to form a hydroperoxy group at C-15, giving PGG_2 (Figure 19.32). This complex reaction involves a tyrosyl radical, generated by the heme cofactor. The second activity, a **peroxidase**, involves a two-electron reduction of the peroxide to give PGH_2, with a hydroxyl group at C-15 (Figure 19.32). Mammalian cells contain two distinct forms of PGH synthase, called PGHS-1 and PGHS-2 (or COX-1 and COX-2, where COX stands for

The cyclooxygenase reaction, one of the first steps in eicosanoid synthesis, is the target site for aspirin action.

cyclooxygenase). COX-1 is constitutively expressed in most tissues, and is responsible for the physiological production of prostaglandins. COX-2 is induced by cytokines, mitogens, and endotoxins in inflammatory cells and is responsible for the elevated production of prostaglandins during inflammation. Both isoforms are covalently modified, and hence inactivated, by reaction with aspirin (acetylsalicylic acid). As shown, aspirin acetylates a specific serine residue, which in turn blocks access of the fatty acid substrate to the cyclooxygenase active site.

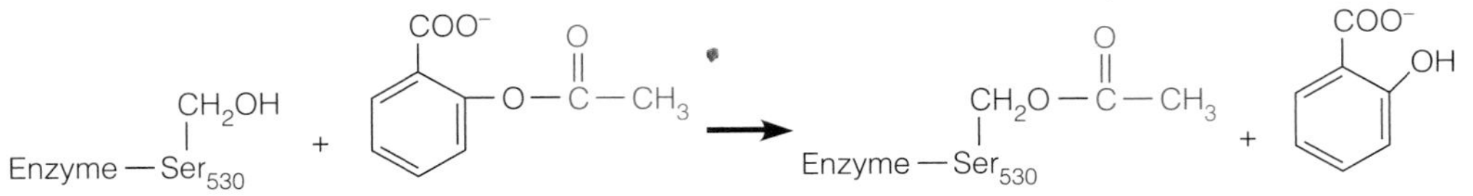

PGHS (COX)   **Acetylsalicylic acid**   →   **Acetylated PGHS**   **Salicylic acid**

The anti-inflammatory and analgesic properties of aspirin derive from the inhibition of COX-2. However, the inhibition of COX-1 has undesirable side effects on the gastrointestinal system, including ulceration. Another widely used NSAID, ibuprofen (Figure 19.34), acts somewhat more specifically upon COX-2 but is a less effective inhibitor than aspirin. Using structural data on the two COX isoforms, molecular pharmacologists designed aspirin analogs that selectively acylate COX-2, and these have the desirable properties of aspirin without the gastrointestinal side effects. Highly selective COX-2 inhibitors such as Vioxx and Celebrex (Figure 19.34), introduced in the late 1990s for relief of arthritis pain, bind much more tightly to COX-2 than to COX-1. Their selectivity for COX-2 is due the phenylsulfonamide moiety, which binds in a side pocket that is larger in COX-2 than in COX-1. As shown in Figure 19.35, COX-1 has the bulkier isoleucine in place of valine at position 523, essentially blocking the binding of drugs that would occupy this side pocket. Aspirin and other nonselective NSAIDs can still bind to COX-1 because they lack a phenylsulfonamide moiety that would occupy this side pocket. Despite the effectiveness of selective COX-2 inhibitors in the treatment of arthritis, more recent studies have revealed that long-term, high-dosage use of these drugs is associated with increased risk of heart attack and stroke (Vioxx was withdrawn from the market in 2004).

In stage 3 of Figure 19.33, a series of specific enzymes converts PGH_2 to other prostaglandins and to thromboxane A_2. Another pathway leads from arachidonate to the leukotrienes. Leukotriene C was originally discovered in the class of white blood cells called polymorphonuclear leukocytes and was named after the source (*leuko*cytes) and the *triene* structure (three double bonds). It is a potent muscle contractant that is involved in the pathogenesis of asthma, through constriction of the small airways in the lung. As shown in Figure 19.36, leukotrienes are formed from the initial attack on arachidonate of a **lipoxygenase**, which adds O_2 to C-5, giving 5-hydroperoxyeicosatetraenoic acid (5-HPETE). A dehydration to give the epoxide coupled with isomerization of double bonds gives leukotriene A_4. Hydrolysis of the epoxide ring yields leukotriene B_4. Transfer of the thiol group of glutathione yields leukotriene C_4. Subsequent modifications of the peptide chain (not shown) yield related compounds, leukotrienes D and E.

All of the eicosanoids are metabolized extremely rapidly, with most failing to survive a single pass through the circulatory system. The lung is a major site of prostaglandin catabolism. The many catabolic pathways all seem to start with conversion to 15-keto-13,14-dihydro derivatives.

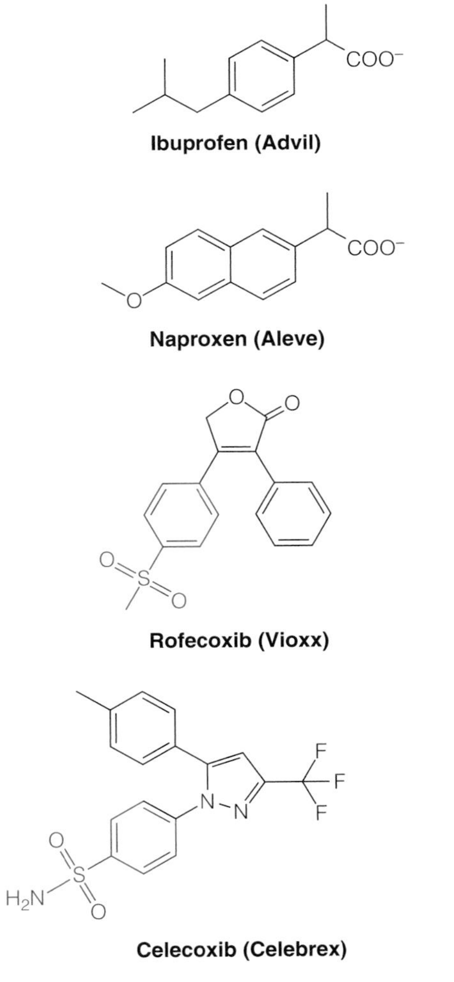

Biological Actions

As noted earlier, prostaglandins and their relatives can be considered as locally acting hormones. Evidently they act through binding to specific cellular receptors.

Ibuprofen (Advil)

Naproxen (Aleve)

Rofecoxib (Vioxx)

Celecoxib (Celebrex)

FIGURE 19.34

Nonsteroidal anti-inflammatory drugs (NSAIDs). Ibuprofen and naproxen are examples of nonselective COX Inhibitors. Rofecoxib (Vioxx) and celecoxib (Celebrex) are selective COX-2 inhibitors. The phenylsulphonamide moieties, which contribute to their selectivity, are highlighted in magenta.

Arachidonic acid

O_2 | 5-Lipoxygenase

5-HPETE

| 5-Lipoxygenase

Leukotriene A₄

H_2O | Hydrolase

Leukotriene B₄

| Glutathione

Leukotriene C₄

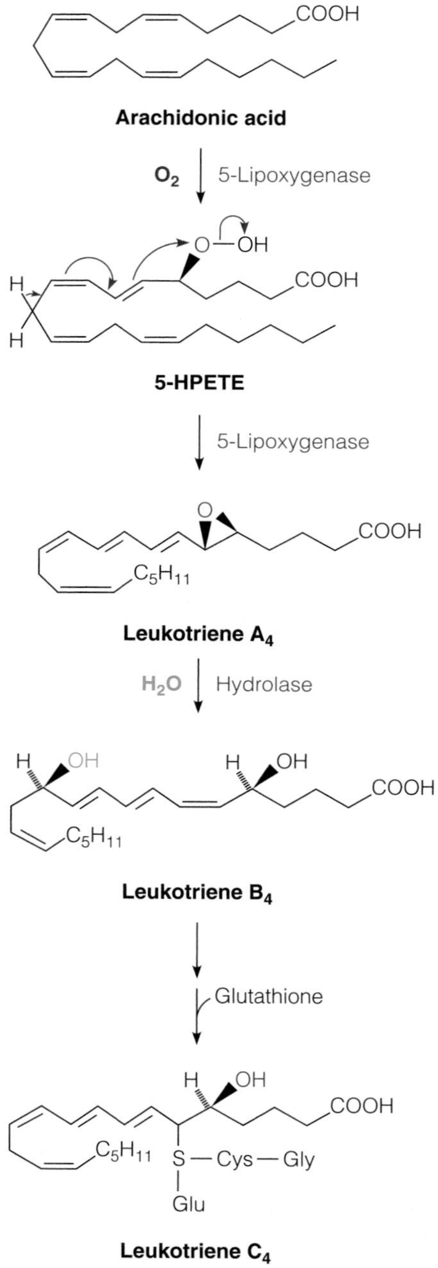

FIGURE 19.36

Biosynthesis of leukotrienes.

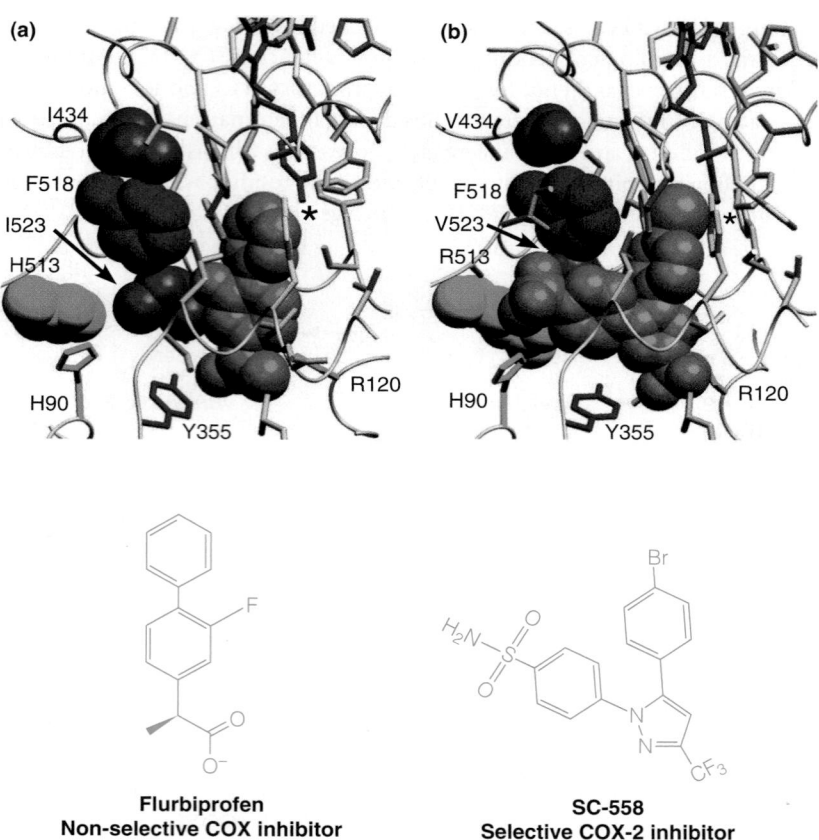

Flurbiprofen
Non-selective COX inhibitor

SC-558
Selective COX-2 inhibitor

FIGURE 19.35

Structural basis for selective inhibition of COX-2. **(a)** Ovine COX-1 with the nonselective NSAID flurbiprofen (blue) bound in the COX active site (PDB ID 1EQH). Residues Ile 434, His 513, Phe 518, Ile 523, and the inhibitor are shown in space-filling representations. Arg-120 (green), which is important in binding the normal substrate arachidonic acid, extends behind flurbiprofen to interact with its carboxylate. The heme cofactor is shown in red, and the tyrosine (pink) that forms the active site radical is indicated by an asterisk (*). **(b)** Mouse COX-2 with the selective inhibitor SC-588 (blue) bound in the COX active site (PDB ID 1CX2). Residues Val 434, Arg 513, Phe 518, Val 523, and the inhibitor are shown in space-filling representations. The phenylsulfonamide group of SC-588 extends into the side pocket made accessible by Val 523 (gray, barely visible behind the drug). The bulkier isoleucine at this position would prevent binding of SC-588 to COX-1. The chemical structures of the bound inhibitors are shown below.

Modified from *Annual Review of Biochemistry* 69:145–182, W. L. Smith, D. L. DeWitt, and R. M. Garavito, Cyclooxygenases: Structural, cellular, and molecular biology. © 2000 Annual Reviews.

We know relatively little about their subsequent effects at the molecular level, though there clearly are interactions with cyclic nucleotide metabolism. PGE stimulates adenylate cyclase in some cells, and PGF_2 has been reported to elevate levels of cyclic GMP in target cells. Although receptors for these compounds have not yet been characterized in detail, several compounds now under clinical study as anti-inflammatory agents apparently act by binding to receptors for leukotriene B_4. A number of the eicosanoids play roles in the inflammation process, as evidenced by the fact that the anti-inflammatory effects of aspirin derive, at least in part, from the inhibition of cyclooxygenase.

Other biological effects include inhibition of platelet aggregation and relaxation of coronary arteries by PGI_2, effects that are counteracted by TxA_2. It seems likely that the action of PGI_2 could prevent platelet binding to arterial walls. In damaged areas the inhibition of PGI_2 synthesis could allow TxA_2 to bind and cause platelet aggregation, thereby promoting the formation of a clot.

Although little is known so far about molecular actions of the eicosanoids, knowledge of their physiology is being applied in useful ways. Recent studies indicate that long-term aspirin administration reduces the risk of heart attacks. Presumably this effect is related to the reduced synthesis of eicosanoids (particularly TxA_2) that induce platelet aggregation, an early step in the clot formation that participates in myocardial infarction. Because prostaglandin release is involved in the uterine muscle contraction that occurs in labor, $PGF_{2\alpha}$ is used when it is necessary to induce labor in mothers at term. Related to this effect is that $PGF_{2\alpha}$ inhibits progesterone secretion and regression of the corpus luteum. $PGF_{2\alpha}$ and PGE_2 are used as well to induce abortion in the second trimester or to induce delivery in case of the death of a fetus. Prostaglandin derivatives are used in animal husbandry, to bring a group of female animals into heat simultaneously. PGI_2 is used to reduce the risk of blood clotting during cardiopulmonary bypass operations. PGE_1, a vasodilator, is being tested against various circulatory disorders, and forms of PGE, which also inhibit gastric secretion, are being used in the experimental treatment of stomach ulcers. Efforts in the pharmaceutical industry are devoted to developing longer-lived prostaglandin analogs.

SUMMARY

Glycerophospholipids, the predominant membrane lipids, are synthesized by routes that start from phosphatidic acid and intermediates activated by reaction with cytidine triphosphate. Retailoring of fatty acid side chains in phospholipids, exchange of polar head groups, and the actions of phospholipid exchange proteins, which insert phospholipids into membranes, all play roles in shaping the lipid composition of specific membranes. S-Adenosylmethionine is the methyl group donor for phosphatidylcholine synthesis.

Glycosphingolipids are assembled from ceramide and successive sugar additions involving glycosyltransferases and nucleotide-linked sugars. The pathway for turnover of these compounds in nervous tissue was elucidated from analysis of products that accumulated in cells of individuals with enzymatic defects in the pathway.

All steroid compounds—and indeed all isoprenoid compounds—are synthesized from acetate, by a pathway that proceeds through the six-carbon mevalonic acid and involves C_5, C_{10}, and C_{15} intermediates. Cyclization of the C_{30} hydrocarbon squalene leads to cholesterol, the precursor for all bile acids and steroid hormones. Synthesis of steroid hormones occurs in endocrine glands and involves cytochrome P450-dependent hydroxylations, oxidoreduction reactions, and side chain cleavage reactions. All steroid hormone synthesis proceeds from cholesterol through pregnenolone.

Arachidonic acid and other C_{20} unsaturated fatty acids are precursors to physiologically potent, locally acting hormones that include prostaglandins, thromboxanes, and leukotrienes. Although actions of these biological regulators are not yet understood at the molecular level, metabolism of these eicosanoids presents therapeutic targets for drugs used to control inflammation, blood clotting, and gastric secretion and to manipulate reproductive processes in various ways.

REFERENCES

Phospholipid Metabolism

Cornell, R. B., and I. C. Northwood (2000) Regulation of CTP:phosphocholine cytidylyltransferase by amphitropism and relocalization. *Trends Biochem. Sci.* 25:441–447. Describes the activation of this rate-limiting enzyme by membrane association.

Cronan, J. E. (2003) Bacterial membrane lipids: Where do we stand? *Annu. Rev. Microbiol.* 57:203–224. Summarizes recent data on the synthesis and function of phospholipids in bacteria.

Gelb, M. H., M. K. Jain, A. M. Hanel, and O. G. Berg (1995) Interfacial enzymology of glycerolipid hydrolases: Lessons from secreted phospholipase A$_2$. *Annu. Rev. Biochem.* 64:654–688. Enzymes that act upon lipid substrates in membranes operate at interfaces and follow rather different kinetic expressions.

Lessig, J., and B. Fuchs (2009) Plasmalogens in biological systems: Their role in oxidative processes in biological membranes, their contribution to pathological processes and aging and plasmalogen analysis. *Curr. Med. Chem.* 16:2021–2041.

Pan, Y. H., and B. J. Bahnson (2007) Structural basis for bile salt inhibition of pancreatic phospholipase A$_2$. *J. Mol. Biol.* 369:439–450.

Shindou, H., and T. Shimizu (2009) Acyl-CoA:lysophospholipid acyltransferases. *J. Biol. Chem* 284:1–5. A review of the phospholipid remodeling pathway (Lands' cycle) in mammals.

Vance, D. E., and J. E. Vance (2009) Physiological consequences of disruption of mammalian phospholipid biosynthetic genes. *J. Lipid Res.* 50 Suppl.:S132–S137. This review describes what we have learned about phospholipid biosynthesis from knockout mice.

Wanders, R. J., and H. R. Waterham (2006) Biochemistry of mammalian peroxisomes revisited. *Annu. Rev. Biochem.* 75:295–332. These organelles carry out several lipid metabolic pathways, notably ether phospholipid synthesis.

Zhang, Y. M., and C. O. Rock (2008) Membrane lipid homeostasis in bacteria. *Nat. Rev. Microbiol.* 6:222–233.

Sphingolipids

Bartke, N., and Y. A. Hannun (2009) Bioactive sphingolipids: Metabolism and function. *J. Lipid Res.* 50 Suppl.:S91–S96. This minireview discusses the biosynthetic pathways and the signaling functions of sphingolipids.

Gravel, R. A., M. M. Kaback, R. L. Proia, K. Sandhoff, K. Suzuki, and K. Suzuki (2001) The G$_{M2}$ Gangliosidoses. In *The Metabolic and Molecular Bases of Inherited Disease*, C. R. Scriver, A. L. Beaudet, W. S. Sly, D. Valle, B. Childs, K. W. Kinzler, and B. Vogelstein, eds., Vol. III, Ch. 153, pp. 3827–3876, McGraw-Hill, New York. This chapter from a four-volume series on inherited metabolic disorders covers many of the diseases illustrated in Figure 19.17.

Moser, H. W., T. Linke, A. H. Fensom, T. Levade, and K. Sandhoff (2001) Acid ceramidase deficiency: Farber lipogranulomatosis. In *The Metabolic and Molecular Bases of Inherited Disease*, C. R. Scriver, A. L. Beaudet, W. S. Sly, D. Valle, B. Childs, K. W. Kinzler, and B. Vogelstein, eds., Vol. III, Ch. 143, pp. 3573–3588, McGraw-Hill, New York. One of several chapters on lipid storage diseases in this four-volume series on inherited metabolic disorders.

Steroids and Isoprenoids

Clarke, P. R., and D. G. Hardie (1990) Regulation of HMG-CoA reductase: Identification of the site phosphorylated by the AMP-activated protein kinase in vitro and in intact rat liver. *EMBO J.* 9:2439–2446.

DeBose-Boyd, R. A. (2008) Feedback regulation of cholesterol synthesis: Sterol-accelerated ubiquitination and degradation of HMG CoA reductase. *Cell Res.* 18:609–621.

Gelb, M. H., L. Brunsveld, C. A. Hrycyna, S. Michaelis, F. Tamanoi, W. C. Van Voorhis, and H. Waldmann (2006) Therapeutic intervention based on protein prenylation and associated modifications. *Nat. Chem. Biol.* 2:518–528. A review that discusses recent developments in protein prenylation.

Ghayee, H. K., and R. J. Auchus (2007) Basic concepts and recent developments in human steroid hormone biosynthesis. *Rev. Endocr. Metab. Disord.* 8:289–300.

Goldstein, J. L., R. A. DeBose-Boyd, and M. S. Brown (2006) Protein sensors for membrane sterols. *Cell* 124:35–46. A review of the role of membrane-embedded proteins in the regulation of cholesterol biosynthesis.

Hotchkiss, A. K., C. V. Rider, C. R. Blystone, V. S. Wilson, P. C. Hartig, G. T. Ankley, P. M. Foster, C. L. Gray, and L. E. Gray (2008) Fifteen years after "Wingspread"—environmental endocrine disrupters and human and wildlife health: Where we are today and where we need to go. *Toxicol. Sci.* 105:235–259.

Miziorko, H. M. (2011) Enzymes of the mevalonate pathway of isoprenoid biosynthesis. *Arch. Biochem. Biophys.* 505:131–143.

Rone, M. B., J. Fan, and V. Papadopoulos (2009) Cholesterol transport in steroid biosynthesis: Role of protein-protein interactions and implications in disease states. *Biochim. Biophys. Acta* 1791:646–658. This review discusses the translocation of cholesterol into mitochondria for steroidogenesis.

Russell, D. W. (2009) Fifty years of advances in bile acid synthesis and metabolism. *J. Lipid Res.* 50 Suppl.:S120–S125.

Tabernero, L., D. A. Bochar, V. W. Rodwell, and C. V. Stauffacher (1999) Substrate-induced closure of the flap domain in the ternary complex structures provides insight into the mechanism of catalysis by 3-hydroxy-3-methylglutaryl-CoA reductase. *Proc. Natl. Acad. Sci. USA* 96:7167–7171. Mechanistic understanding of this important enzyme comes from crystallography.

Lipid-Soluble Vitamins

Booth, S. L. (2009) Roles for vitamin K beyond coagulation. *Annu. Rev. Nutr.* 29:89–110.

Clarke, M. W., J. R. Burnett, and K. D. Croft (2008) Vitamin E in human health and disease. *Crit. Rev. Clin. Lab. Sci.* 45:417–450. Reviews antioxidant and other properties of this vitamin.

DeLuca, H. F. (2008) Evolution of our understanding of vitamin D. *Nutr. Rev.* 66 (10 Suppl 2).:S73–S87. This minireview highlights the roles of vitamin D in skin, the immune system, and its protective role in some forms of cancer.

Fields, A. L., D. R. Soprano, and K. J. Soprano (2007) Retinoids in biological control and cancer. *J. Cell. Biochem.* 102:886–898. A review of the activities of vitamin A–related compounds in gene regulation.

von Lintig, J. (2010) Colors with functions: Elucidating the biochemical and molecular basis of carotenoid metabolism. *Annu. Rev. Nutr.* 30:35–56. This review summarizes the pathways of vitamin A synthesis from carotenoids in animals and some of the physiological functions of the vitamin.

Eicosanoids

Garavito, R. M., and A. M. Mulichak (2003) The structure of mammalian cyclooxygenases. *Annu. Rev. Biophys. Biomol. Struct.* 32:183–206. A recent review of the structure and function of the first enzymes in eicosanoid synthesis.

Shimizu, T. (2009) Lipid mediators in health and disease: Enzymes and receptors as therapeutic targets for the regulation of immunity and inflammation. *Annu. Rev. Pharmacol. Toxicol.* 49:123–150. This review covers prostaglandins, leukotrienes, platelet-activating factor, lysophosphatidic acid, sphingosine 1-phosphate, and other lipid mediators efficiently and completely.

PROBLEMS

1. Would you expect the reaction catalyzed by cardiolipin synthase to be strongly exergonic or strongly endergonic? Explain your reasoning.

*2. Phosphatidylserine (PS) is considered to be an intermediate in the biosynthesis of phosphatidylethanolamine (PE) in *E. coli*, yet PS is not found in appreciable amounts among *E. coli* membrane phospholipids. Because PS must be present in the membrane to serve as an intermediate, how might you explain its failure to accumulate to a significant extent? What kinds of experiments could test your proposed explanation?

3. Melittin is a protein in bee venom that activates phospholipase A_2. How might this effect contribute to the local inflammation that is caused by bee stings?

4. What would you expect to happen to levels of mevalonate in human plasma if an individual were to go from a meat-containing diet to a vegetarian diet?

5. Write a balanced equation for the synthesis of *sn*-1-stearoyl-2-oleylglycerophosphorylserine, starting with glycerol, the fatty acids involved, and serine.

6. If mevalonate labeled with ^{14}C in the carboxyl carbon were administered to rats, which carbons of cholesterol would become labeled?

7. Which step in lipid metabolism would you expect to be affected by 3,4-dihydroxybutyl-1-phosphonic acid (shown here)? Explain your answer.

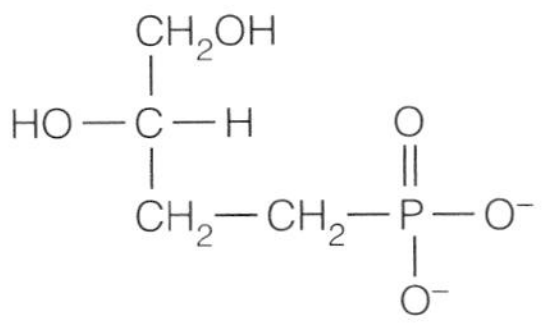

$$
\begin{array}{c}
CH_2OH \\
| \\
HO - C - H \quad\quad O \\
| \quad\quad\quad\quad || \\
CH_2 - CH_2 - P - O^- \\
| \\
O^-
\end{array}
$$

*8. Methyl-labeled [^{14}C]methionine at a specific activity of 2.0 millicuries per millimole was injected into rats. Six hours later the rats were killed. Phosphatidylcholine was isolated from the liver and found to have a specific activity of 1.5 millicuries per millimole. Calculate the proportions of phosphatidylcholine synthesized by the phosphatidylserine pathway and by the pathway starting from free choline. What further information would you need for your calculated values to reflect the true rates of these processes?

9. Identify a pathway for utilization of the four carbons of acetoacetate in cholesterol biosynthesis. Carry your pathway as far as the rate-determining reaction in cholesterol biosynthesis.

10. Explain why a deficiency of steroid 21-hydroxylase leads to excessive production of sex steroids (androgens and estrogens).

11. *cis*-Vaccenate is an 18-carbon unsaturated fatty acid abundant in *E. coli* membrane lipids. Propose a metabolic route for synthesis of this fatty acid, in light of the fact that stearic acid, the C_{18} saturated analogous fatty acid, is virtually absent from *E. coli* lipids.

12. The synthesis of phosphatidylcholine from phosphatidylethanolamine proceeds via three successive AdoMet-dependent methylation reactions. How might you determine experimentally whether these three reactions are catalyzed by the same enzyme or by three separate enzymes?

13. Briefly describe how cyclic AMP controls phospholipid synthesis.

*14. Assuming that all terpene compounds are synthesized by the HMG-CoA/mevalonate pathway, select a simple terpene (limonene, retinol, phylloquinone, or the like) and propose a route for its synthesis from isopentenyl pyrophosphate and dimethylallyl pyrophosphate.

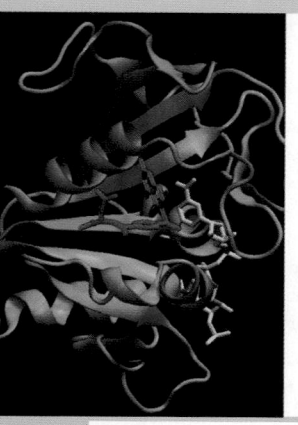

CHAPTER 20

Metabolism of Nitrogenous Compounds I: Principles of Biosynthesis, Utilization, and Turnover

Thus far our study of metabolism has concerned itself primarily with compounds that can be degraded completely to carbon dioxide and water—in other words, compounds containing only carbon, hydrogen, and oxygen. We turn now to nitrogen-containing compounds—amino acids and their derivatives, nucleotides, and the polymeric nucleic acids and proteins. This chapter and the next two describe the metabolism of low-molecular-weight nitrogenous compounds (Figure 20.1). The metabolism of specific amino acids is presented in Chapter 21, and nucleotide metabolism is covered in Chapter 22. This chapter focuses on how cells assimilate nitrogen and how they dispose of excess nitrogen. Because the biological availability of nitrogen is limited, and because the breakdown of nitrogenous compounds often yields toxic products, we will encounter some new metabolic principles.

The existence of 20 different amino acids in proteins implies the existence of 20 biosynthetic pathways and 20 degradative pathways. Although this might seem intimidating at first glance, the common features of pathways involving various amino acids should ease the task of learning. More important than many of the details of biosynthesis and degradation, however, are the numerous roles of amino acids other than as protein constituents, including their functions as precursors to hormones, vitamins, coenzymes, porphyrins, pigments, and neurotransmitters.

Chapters 20–22 also indicate how much we have learned from naturally occurring human mutations, as well as from mutations generated in the laboratory, in cultured cells, or in bacteria. Whereas a mutation that inactivates an enzyme in one of the central energy-generating or energy-storing pathways is likely to be lethal and, hence, not observed in living individuals, mutations that affect amino acid

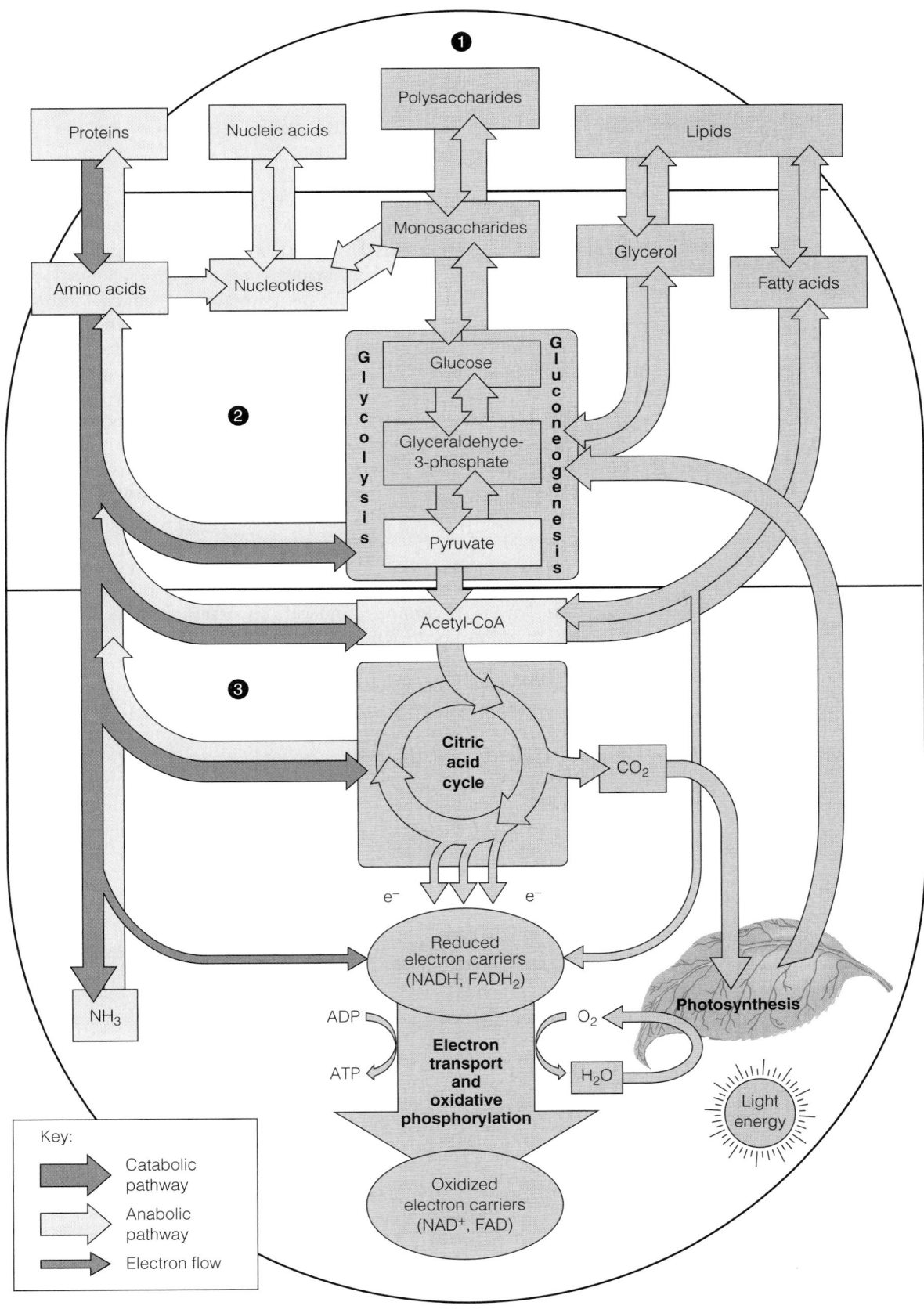

FIGURE 20.1

Pathways of nitrogen metabolism (purple) in the general pattern of intermediary metabolism.

metabolism are often not lethal and *are* found in living humans. The clinical consequences of these mutations are often tragic, but these inherited metabolic diseases have greatly enhanced our understanding of human biochemistry.

Utilization of Inorganic Nitrogen: The Nitrogen Cycle

Few organisms can use the N_2 in air and many soils are poor in nitrate. Thus, nitrogen bioavailability limits growth for most organisms.

For many organisms, growth and reproduction are limited by the availability of utilizable nitrogen, which in turn is limited by the abilities of organisms to utilize different inorganic forms of nitrogen. All organisms can convert ammonia (NH_3) to organic nitrogen compounds—that is, substances containing C—N bonds. However, not all organisms can synthesize ammonia from the far more abundant forms of inorganic nitrogen—dinitrogen gas (N_2), the most abundant component of the earth's atmosphere, and nitrate ion (NO_3^-), a soil constituent essential for the growth of most plants. The reduction of N_2 to NH_3, termed **biological nitrogen fixation**, is carried out only by certain microorganisms called **diazotrophs**. The reduction of NO_3^- to NH_3, by contrast, is widespread among both plants and microorganisms.

As in the consideration of any limited resource, it is useful to think about nitrogen metabolism in terms of an economy—a **nitrogen economy**—that focuses on questions of supply, demand, turnover, reuse, growth, and maintenance of a steady state. Within the biosphere, a balance is maintained between total inorganic and total organic forms of nitrogen. The conversion of inorganic to organic nitrogen, which starts with nitrogen fixation or nitrate reduction, is counterbalanced by catabolism, **denitrification**, and decay (Figure 20.2). Catabolism yields ammonia and various organic nitrogenous end products, which can in turn be metabolized by various bacteria: *Nitrosomonas* species oxidize ammonia to nitrite (NO_2^-), and *Nitrobacter* species oxidize nitrite to nitrate. These oxidations generate biological energy, just as other organisms derive energy from oxidation of carbohydrate or fat to CO_2. Other bacteria, the **denitrifying bacteria**, catabolize ammonia to N_2. Because of the toxicity of ammonia, there is much interest in using denitrifying bacteria and their enzymes in **bioremediation**, the use of living organisms to purify and detoxify environmental residues of human activity, such as manufacturing, or in waste disposal. Our concern here is with the use of nitrogen for amino acid and nucleotide biosynthesis, so our focus in the rest of this section will be on synthesis of ammonia from N_2 or from nitrate ion.

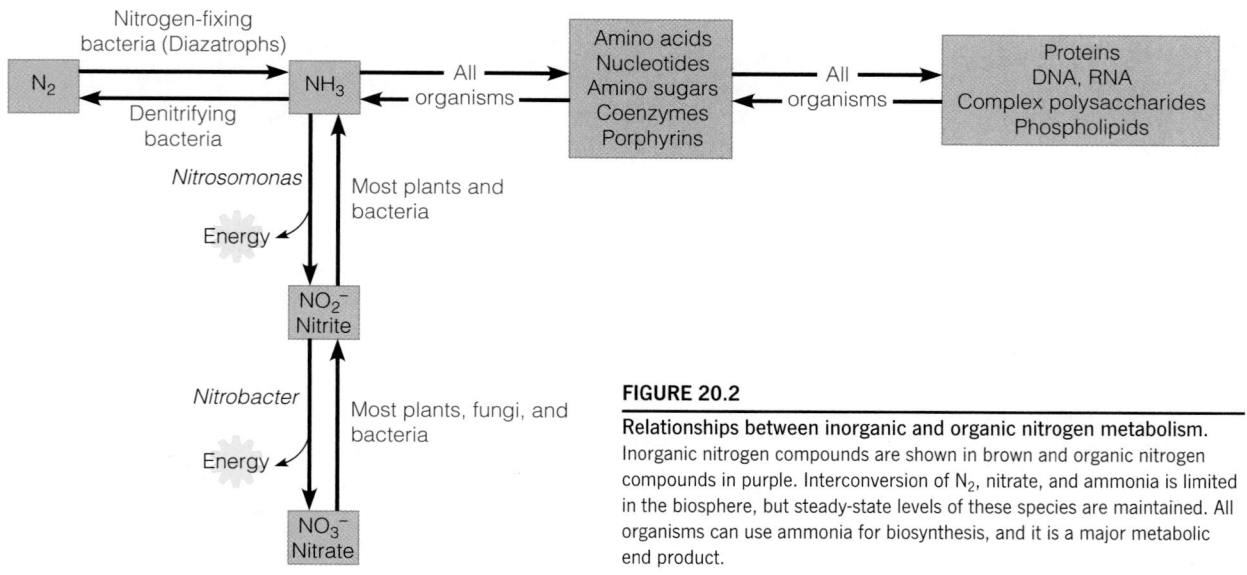

FIGURE 20.2

Relationships between inorganic and organic nitrogen metabolism. Inorganic nitrogen compounds are shown in brown and organic nitrogen compounds in purple. Interconversion of N_2, nitrate, and ammonia is limited in the biosphere, but steady-state levels of these species are maintained. All organisms can use ammonia for biosynthesis, and it is a major metabolic end product.

Biological Nitrogen Fixation

Although nitrogen gas makes up about 80% of the earth's atmosphere, its reduction to ammonia occurs in relatively few living systems—some free-living soil bacteria, such as *Klebsiella* and *Azotobacter*; photosynthetic cyanobacteria (blue-green algae); some archaea; and symbiotic nodules on the roots of leguminous plants, such as beans or alfalfa, that have been infected with certain bacteria, notably of the genus *Rhizobium* (Figure 20.3). The infecting bacterium assumes a modified form, called a **bacteroid**, inside the cells of infected plants. This symbiosis allows the host plant to grow without an exogenous nitrogen source. Some trees, such as alder, also form nitrogen-fixing nodules and thus have the capacity to fix nitrogen. Recent discoveries have revealed an impressive diversity of diazotrophs, including hyperthermophilic methane-producing archaea from hydrothermal vents and anaerobic methane-oxidizing archaea from deep-sea cold seeps.

Because nitrogen availability is the factor limiting the fertility of most soils, an understanding of biological nitrogen fixation is directly related to increasing the world's food supply. The triply bonded N_2 molecule, $N \equiv N$, with a bond energy of about 940 kJ/mol, is extraordinarily difficult to reduce. Industrially, the reduction is done by the Haber–Bosch process, a low-yield catalytic hydrogenation carried out at high temperature and pressure. This process is used in the manufacture of ammonia-based fertilizers. Interest in the molecular details of biological nitrogen fixation has derived partly from hopes of supplanting this energy-intensive process with a means of ammonia production that can take place under milder conditions.

Formally, nitrogen fixation can be compared with photosynthesis. Both N_2 and CO_2 are stable inorganic compounds whose reduction requires both energy and low-potential electrons—in the form of reduced electron carriers of very low E_0. As we saw in Chapter 16, photosynthesis uses light to generate both energy (through photophosphorylation) and low-potential electrons (as in reduced ferredoxin). Comparable mechanisms used in nitrogen fixation are still not completely understood, partly because the enzymes involved are extremely sensitive to oxygen and can be studied only under anaerobic conditions. The major reason nitrogen can be fixed in root nodules of plants infected with *Rhizobium* is that the nodules contain an abundant protein called **leghemoglobin**, which maintains an anaerobic environment by binding any O_2 that finds its way into the nodule and presenting it to respiratory enzymes—in a manner somewhat akin to the behavior of myoglobin in animals.

Biological N_2 reduction is catalyzed by the enzyme **nitrogenase**, of which four types are known. The most abundant and widely studied nitrogenase is the molybdenum (Mo)-dependent enzyme, such as that found in *Azotobacter vinelandii*. The stoichiometry of the overall reaction is as follows:

$$N_2 + 8H^+ + 16MgATP + 8e^- \longrightarrow 2NH_3 + H_2 + 16MgADP + 16P_i$$

Nitrogen fixation is a very expensive process, requiring hydrolysis of two ATPs per electron transferred. The ATP is generated through energy-yielding pathways of the organism, primarily carbohydrate catabolism. Although eight total electrons are required, reduction of N_2 to $2NH_3$ is a six-electron process. The other two electrons are "wasted" in the formation of H_2. Electrons for N_2 reduction are derived from low-potential carriers, either reduced ferredoxin or flavodoxin, a low-potential flavoprotein. H_2 is a by-product of nitrogen reduction. Some nitrogen-fixing species have the ability to "recycle" this hydrogen, to generate low-potential electron carriers for additional cycles of N_2 reduction.

The Mo-dependent nitrogenase consists of two separate metalloproteins. As outlined in Figure 20.4, one protein—called **molybdenum–iron (MoFe) protein**, **dinitrogenase**, or **component I**—catalyzes the reduction of N_2. The other—called **iron (Fe) protein**, **dinitrogenase reductase**, or **component II**—transfers electrons and protons, one at a time, to the MoFe protein, in a process coupled to the hydrolysis of two MgATPs. Both proteins contain iron–sulfur clusters, and MoFe protein also

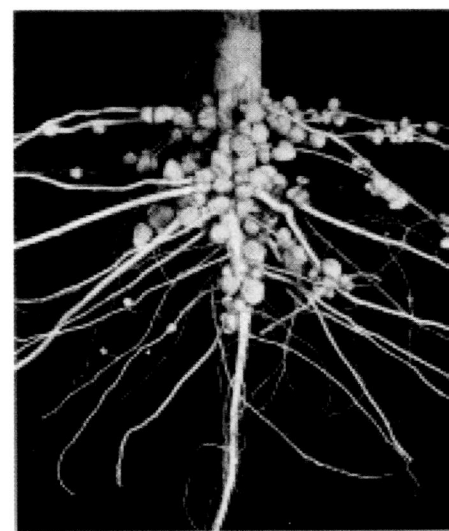

FIGURE 20.3

The site of nitrogen fixation in symbiotic root nodules. This root of a soybean plant is infected by nitrogen-fixing bacteria of the genus *Rhizobium*.

Courtesy of Novozymes BioAg.

$$N_2 + 3H_2 \xrightarrow[\text{450°C, 270 atm}]{\text{catalyst}} 2NH_3$$

The Haber–Bosch process

FIGURE 20.4

Mechanism of the two-component molybdenum-dependent nitrogenase reaction. MoFe protein (green) is molybdenum–iron protein, or dinitrogenase. Fe protein (pink) is iron protein, or dinitrogenase reductase. (These colors correspond to the protein structures in Figure 20.5.) MoFe protein catalyzes the reduction of N_2 and is reduced in turn by Fe protein. The Fe protein is believed to undergo a three-state cycle, and the MoFe protein a nine-state cycle. Binding of ATP to reduced Fe protein (FeP^{Red}—ATP) is thought to generate an altered conformation of Fe protein, with a very low reduction potential. Transfer of the electron from reduced Fe protein to MoFe protein is accompanied by hydrolysis of bound ATP to ADP + P_i. Oxidized Fe protein (FeP^{Ox}—ADP + P_i) then dissociates from the MoFe protein (this is the rate-limiting step for N_2 reduction) and is regenerated by exchange with ATP and reduction back to the +1 oxidation state. In the MoFe protein cycle, the transfer of an H^+ and an electron from Fe protein to MoFe protein occurs eight times during the reduction of one N_2 molecule. Sequential reduced states are represented by E_n, where n is the total number of electrons donated by the Fe protein. At least three e^-/H^+ must accumulate within the MoFe protein before N_2 binds, and binding of N_2 to MoFe protein occurs concomitantly with release of two bound hydrogens as H_2. Cleavage of the N—N bond occurs after $5\,e^-/H^+$ have been added.

Modified from *Annual Review of Biochemistry* 78:701–722, L. C. Seefeldt, B. M. Hoffman, and D. R. Dean, Mechanism of Mo-dependent nitrogenase. © 2009 Annual Reviews.

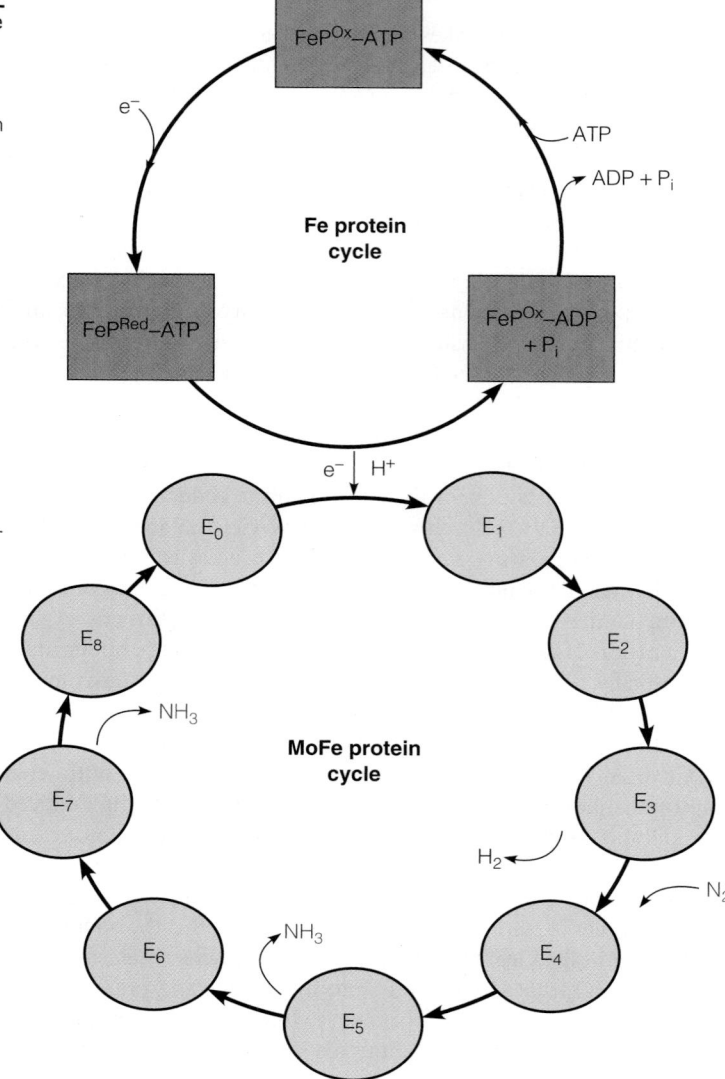

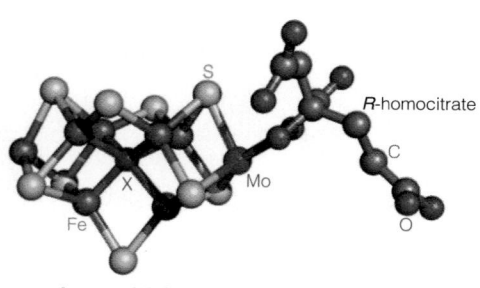

Iron–molybdenum cofactor (FeMo-co)

R-homocitrate

Modified from *Annual Review of Biochemistry* 78: 701–722, L. C. Seefeldt, B. M. Hoffman, and D. R. Dean, Mechanism of Mo-dependent nitrogenase. © 2009 Annual Reviews.

contains molybdenum, in the form of a tightly bound **iron–molybdenum cofactor** (FeMo-co). N_2 binds to this cofactor during its reduction, although the precise mode of binding is not yet known. The ultimate electron donor for the Fe protein depends on the organism, with reduced ferredoxin or flavodoxin being the most common donor.

Several crystal structures have been described for the individual Fe protein and MoFe protein components from *Azotobacter vinelandii* and, in 1997, a structure for the entire complex was reported. As shown in Figure 20.5, the Fe protein is a homodimer of ~32 kilodalton subunits with a single Fe_4S_4 cluster bridging the two subunits. Each subunit has one nucleotide binding site (bound by MgADP in Figure 20.5). Binding of ATP evidently forces a conformational change in the Fe protein, which drives its docking to the MoFe protein. The MoFe protein is an $\alpha_2\beta_2$-heterotetramer ($M_r \approx 250,000$) that contains two novel iron–sulfur complexes—the P cluster, which contains seven sulfurs and eight irons, and the iron–molybdenum cofactor (FeMo-co), which, as shown, contains nine sulfurs, seven irons, and one molybdenum atom coordinated to a molecule of **R-homocitrate** through its 2-hydroxy and 2-carboxyl groups (see margin). FeMo-co also contains an unidentified atom (X) (colored blue in the structures). As shown in Figure 20.5, electrons flow from ferredoxin or flavodoxin to the Fe_4–S_4 complex in

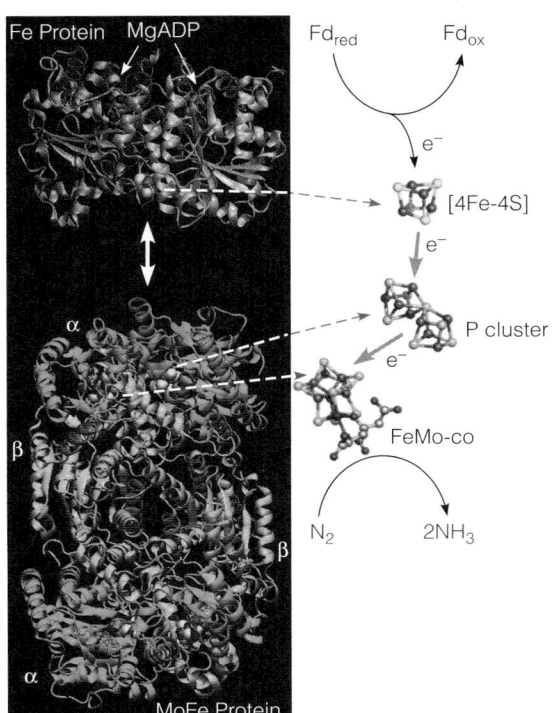

FIGURE 20.5

Structure of molybdenum-dependent nitrogenase from *Azotobacter vinelandii*.
Left: The two subunits of the homodimeric Fe protein (PDB ID 1FP6) are shown in shades of pink, each containing a bound MgADP and the bridging Fe_4–S_4 cluster. The α and β subunits of the $\alpha_2\beta_2$ tetramer of MoFe protein (PDB ID 1M1N) are shown in dark green and light green, respectively. (These colors correspond to those in Figure 20.4.) Each $\alpha\beta$ unit binds one P iron–sulfur cluster and one FeMo-co iron–sulfur cluster.
Right: The relative positions and structures of the Fe_4–S_4 cluster of the Fe protein, and the P cluster and the FeMo cofactor (FeMo-co) of the MoFe protein are shown. Sulfur atoms are yellow, iron atoms are orange, and the molybdenum atom is purple. The flow of electrons from reduced ferredoxin or flavodoxin (Fd) through the iron–sulfur clusters to N_2 is indicated. Hydrolysis of bound ATP is thought both to drive the reduction of P cluster by Fe protein and to trigger a conformational change in Fe protein that causes it to dissociate transiently from MoFe protein, assuring unidirectional electron flow.

Modified from *Annual Review of Biochemistry* 78:701–722, L. C. Seefeldt, B. M. Hoffman, and D. R. Dean, Mechanism of Mo-dependent nitrogenase. © 2009 Annual Reviews.

the Fe protein, and the hydrolysis of bound ATP somehow drives the electrons to the P cluster in the MoFe protein and then to FeMo-co. These three clusters are sufficiently close together in the complex to allow facile electron transfer.

Some bacteria have been found to contain more than one nitrogenase complex. *Azotobacter* contains three such systems, one of which uses vanadium instead of molybdenum, and one that has iron as the sole bound metal.

The genetics of nitrogen fixation is under intense study because of the desirability of transferring nitrogen-fixing capabilities to higher plants, thereby reducing the use of nitrogenous fertilizers. In *Klebsiella*, 13 genes known to be essential to the process are linked within one region of DNA some 24,000 base pairs in length, the *nif* gene cluster. Seven interspersed genes participate but are not essential for nitrogen fixation. Products of the essential genes include the two polypeptide subunits of MoFe protein, the one subunit of Fe protein, flavodoxin, and enzymes that synthesize FeMo-co.

A twist on the nitrogen-fixing symbiosis in leguminous plants was recently discovered. Although homocitrate is an essential component of the iron–molybdenum cofactor, the *nifV* gene that encodes homocitrate synthase is not present in most *Rhizobium* species that infect the host plant. Instead, the homocitrate synthase expressed in the root nodules is encoded by the host plant. Thus, functional nitrogenase requires homocitrate produced by the host plant cells, providing a molecular basis for the interdependent partnership between legumes and rhizobia in symbiotic nitrogen fixation.

Nitrate Utilization

The ability to reduce nitrate to ammonia is common to virtually all plants, fungi, and bacteria. The first step, reduction of nitrate (+5 oxidation state) to nitrite (+3 oxidation state) is catalyzed by **nitrate reductase**. The eukaryotic enzyme contains bound FAD, molybdenum, and a cytochrome b_5. The enzyme carries out the overall reaction:

$$NO_3^- + NAD(P)H + H^+ \longrightarrow NO_2^- + NAD(P)^+ + H_2O$$

Molybdopterin (Mo Cofactor)

Modified from *Annual Review of Plant Biology* 57: 623–647, G. Schwarz and R. R. Mendel, Molybdenum cofactor biosynthesis and molybdenum enzymes. © 2006 Annual Reviews.

Siroheme

NADH-specific enzymes are found most commonly in plants and algae; NADPH-specific enzymes are found only in fungi; and bispecific NAD(P)H forms are found in all three groups but most commonly in fungi. The electrons are transferred from NAD(P)H to enzyme-bound FAD, then to cytochrome b_5, then to molybdenum, and finally to the substrate. The molybdenum is bound to a cofactor containing a *pteridine* ring (see page 825), which is quite distinct from the structure of FeMo-co. In fact, all known molybdenum-requiring enzymes except nitrogenase contain a structure similar to that of this **molybdopterin**, in which the Mo center shows a pyramidal geometry.

Reduction of nitrite to ammonia is carried out in three steps

$$NO_2^- \longrightarrow NO^- \longrightarrow NH_2OH \longrightarrow NH_3$$

by one enzyme, **nitrite reductase**. Higher plants, algae, and cyanobacteria use ferredoxin as the electron donor in this six-electron reaction. This enzyme contains one Fe_4S_4 center and one molecule of **siroheme**, a partially reduced iron porphyrin.

Utilization of Ammonia: Biogenesis of Organic Nitrogen

Several ubiquitous enzymes use ammonia as a substrate for synthesis of glutamate, glutamine, asparagine, or carbamoyl phosphate.

Carbamoyl phosphate

Although plants, animals, and bacteria derive their nitrogen from different sources, virtually all organisms share a few common routes for utilization of inorganic nitrogen in the form of ammonia. Ammonia in high concentrations is quite toxic, but at lower levels it is a central metabolite, serving as substrate for four enzymes that convert it to various organic nitrogen compounds. At physiological pH the dominant ionic species is ammonium ion, NH_4^+ ($pK_a = 9.2$). However, the four reactions involve the unshared electron pair of NH_3, which is therefore the reactive species.

All organisms assimilate ammonia via reactions leading to glutamate, glutamine, asparagine, and **carbamoyl phosphate** (Figure 20.6). Because carbamoyl phosphate is used only in the biosynthesis of arginine, urea, and the pyrimidine nucleotides, most of the nitrogen that finds its way from ammonia to amino acids and other nitrogenous compounds does so via the two amino acids glutamate and glutamine. The α-amino nitrogen of glutamate and the side chain *amide* nitrogen of glutamine are both primary sources of N in biosynthetic pathways.

FIGURE 20.6

Reactions in assimilation of ammonia and major fates of the fixed nitrogen. ❶ Glutamate dehydrogenase; ❷ Glutamine synthetase; ❸ Asparagine synthetase; ❹ Carbamoyl phosphate synthetase; ❺ Glutamate synthase. In animals, glutamine, rather than NH_3, is the chief nitrogen source for pyrimidines.

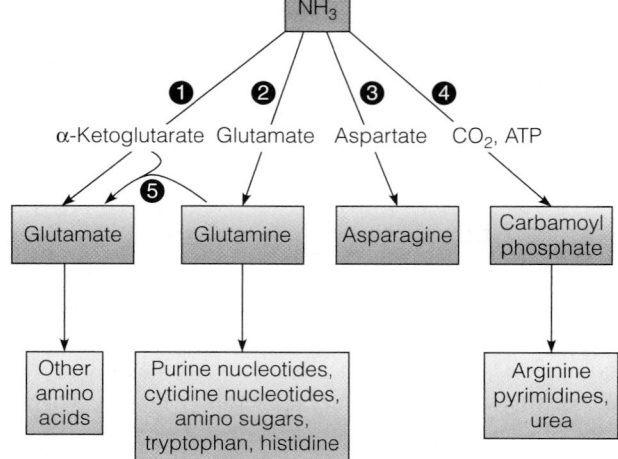

Glutamate Dehydrogenase: Reductive Amination of α-Ketoglutarate

Glutamate dehydrogenase catalyzes the reductive amination of α-ketoglutarate:

$$
\begin{array}{ccc}
\text{COO}^- & & \text{COO}^- \\
| & & | \\
\text{CH}_2 & & \text{CH}_2 \\
| & & | \\
\text{CH}_2 & & \text{CH}_2 \\
| & & | \\
\text{C}=\text{O} \quad + \text{NH}_3 + \text{NAD(P)H} + 2\text{H}^+ \rightleftharpoons \text{H}-\overset{+}{\underset{|}{\text{C}}}-\overset{+}{\text{NH}}_3 + \text{H}_2\text{O} + \text{NAD(P)}^+ \\
| & & | \\
\text{COO}^- & & \text{COO}^-
\end{array}
$$

α-Ketoglutarate **Glutamate**

The reaction is reversible, but the enzyme has a high K_M for ammonia (~1 mM). Most bacteria and many plants contain an NADPH-specific form of the enzyme, which acts primarily in the direction of glutamate formation. Consistent with these characteristics is that bacteria growing with ammonia as their sole nitrogen source use this reaction as the primary route for nitrogen assimilation. In animal cells, the reaction occurs in both directions, although the catabolic direction, supplying α-ketoglutarate for the citric acid cycle, probably predominates because intracellular ammonia concentrations are normally very low. The animal enzyme uses NAD^+ as its principal cofactor, but it can also use $NADP^+$. In animals, glutamate dehydrogenase, a hexamer of identical subunits, is located in the inner mitochondrial membrane, consistent with a primary role in energy generation. Moreover, the animal enzyme is allosterically controlled; α-ketoglutarate synthesis is inhibited by GTP and ATP and stimulated by ADP. Thus, the enzyme is activated under conditions of low energy charge. Yeasts and fungi contain both types of glutamate dehydrogenase—each appropriately regulated, with one tailored for nitrogen assimilation and one functioning primarily in catabolism.

A related enzyme that occurs only in microorganisms, plants, and lower eukaryotes, **glutamate synthase**, catalyzes a reaction comparable to that catalyzed by glutamate dehydrogenase but functions primarily in glutamate biosynthesis:

α-Ketoglutarate + glutamine + reductant $\longrightarrow$ 2 glutamate + oxidized reductant

In this reductive amination reaction, the amide N of glutamine replaces NH_3 as the N donor. Because glutamate dehydrogenase has a relatively high K_M for ammonia, glutamate synthase plays a larger role in glutamate synthesis in most cells where ammonia levels are lower (together with glutamine synthetase; see the next section). Glutamate synthase as isolated from several bacteria contains two types of subunits (α and β) in an 800-kilodalton holoenzyme; the $\alpha\beta$ protomer contains FAD, FMN, and several iron–sulfur centers. The enzyme from plants uses either NADPH, NADH, or ferredoxin, whereas glutamate synthase from other organisms uses NADH exclusively.

Glutamine Synthetase: Generation of Biologically Active Amide Nitrogen

Whether formed by action of glutamate dehydrogenase or glutamate synthase, or by transamination (see page 619, Chapter 14), glutamate can accept a second ammonia moiety to form glutamine in the reaction catalyzed by **glutamine synthetase**. Mn^{2+} is required.

Glutamate

ATP

ADP

γ-Glutamyl phosphate

NH₃

Pᵢ

Glutamine

Glutamate + NH₃ + ATP ⟶ **Glutamine** + ADP + Pᵢ

This enzyme is named a *synthetase*, rather than a *synthase* because the reaction couples bond formation with the energy released from ATP hydrolysis. Both enzymes are classified as ligases (see page 449), but a synthase does not require ATP.

The glutamine synthetase reaction occurs via an acyl phosphate intermediate. ATP phosphorylates the δ-carboxylate of glutamate to give a carboxylic-phosphoric acid anhydride (γ-glutamyl phosphate), which undergoes nucleophilic attack by the nitrogen of ammonia to give the amide product, glutamine.

As revealed originally by electron microscopy and more recently by X-ray crystallography, bacterial glutamine synthetase is a dodecamer, with 12 identical subunits forming two facing hexagonal arrays (Figure 20.7). The holoenzyme has a molecular weight of about 620,000. Each of the 12 catalytic sites is formed at an interface between subunits within a hexamer, and the reaction requires residues from two adjacent subunits.

Regulation of Glutamine Synthetase

Glutamine occupies a central role in nitrogen metabolism (see Figure 20.6). The amide nitrogen is used in biosynthesis of several amino acids (including glutamate, tryptophan, and histidine), purine and pyrimidine nucleotides, and amino sugars. In animals, glutamine synthetase is a key participant in detoxifying ammonia formed from amino acid catabolism, particularly in brain. In fact, glutamate and glutamine are two of the most abundant free amino acids in brain cells; accumulation of these amino acids can deplete α-ketoglutarate, their principal precursor, thereby interfering with the citric acid cycle and energy generation. In addition, glutamine participates in ammonia excretion in the kidney, and it is a major fuel for cells in the immune system. It is not surprising, therefore, that the glutamine synthetase reaction is tightly regulated.

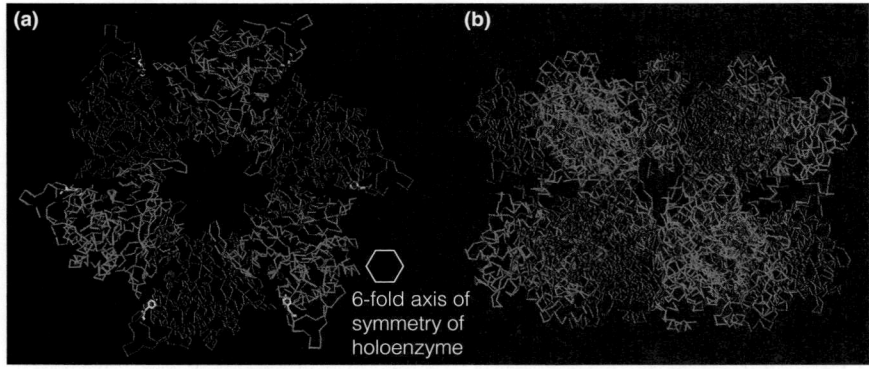

FIGURE 20.7

Structure of *Salmonella typhimurium* glutamine synthetase. **(a)** Top view of the enzyme, as revealed from the crystal structure (PDB ID 1FPY). Only six subunits can be seen from this viewpoint. The adenylylation site (Tyr 397) is shown in yellow and the two Mn²⁺ ions at each catalytic site are shown in red. **(b)** Side view of the dodecamer.

As revealed primarily in *E. coli*, several remarkable control mechanisms for this reaction interact with one another in complex ways. The activity of glutamine synthetase is controlled by two distinct but interlocking mechanisms: (1) allosteric regulation by **cumulative feedback inhibition** and (2) covalent modification of the enzyme brought about by a regulatory cascade.

Cumulative feedback inhibition involves the action of eight specific feedback inhibitors. Those eight inhibitors are either metabolic end products of glutamine (tryptophan, histidine, glucosamine-6-phosphate, carbamoyl phosphate, CTP, and AMP; see Figure 20.6) or indicators in some other way of the general status of amino acid metabolism (alanine, glycine). Remarkably, each ~50,000-dalton subunit of glutamine synthetase must contain binding sites for each of the eight inhibitors, as well as for substrates and products. Each of the eight compounds alone gives only partial inhibition, but in combination the degree of inhibition is increased until a mixture of all eight gives virtually complete blockage. This makes good metabolic sense because it ensures that an accumulation of the end product of one pathway does not shut off the supply of a substrate (glutamine) needed for other pathways.

Superimposed on the cumulative feedback inhibition is a mode of regulation involving covalent modification of the enzyme. Glutamine synthetase is regulated by **adenylylation**: A specific tyrosine residue in the enzyme (Tyr 397) reacts with ATP to form an ester between the phenolic hydroxyl group and the phosphate of the resultant AMP. That tyrosine residue lies very close to a catalytic site (Figure 20.7). Adenylylation inactivates the adjacent catalytic site. An enzyme molecule with all 12 sites adenylylated is completely inactive, whereas partial adenylylation yields partial inactivation.

Adenylylation and deadenylylation of glutamine synthetase involve a complex series of regulatory cascades (Figure 20.8). Both reactions are catalyzed by the same enzyme: a complex of **adenylyltransferase (AT)** and a small regulatory protein, P_{II}. The molecular form of P_{II}—uridylylated or deuridylylated—determines whether the complex catalyzes adenylylation or deadenylylation. Reversible uridylylation of P_{II} is catalyzed by yet another enzyme, a bifunctional **uridylyltransferase (UT)/uridylyl-removing enzyme (UR)**, in response to the intracellular glutamine concentration. The uridylyltransferase activity transfers a UMP residue to a specific tyrosine on the P_{II} molecule. The product, P_{II}-**UMP**, binds to adenylyltransferase to stimulate its *de*adenylylation of glutamine synthetase. P_{II}-UMP is deuridylylated by hydrolysis of the UMP residue, catalyzed by the uridylyl-removing activity of the bifunctional UT/UR. The non-UMP-containing form of P_{II} converts adenylyltransferase to an adenylylating enzyme. Activity of the uridylyltransferase is inhibited by glutamine; activity of the uridylyl-removing enzyme is stimulated by glutamine. α-Ketoglutarate concentration is sensed by the unmodified form of P_{II}—its ability to activate adenylyltransferase is antagonized by high concentrations of α-ketoglutarate.

These regulatory cascades provide a responsive mechanism ensuring that, when the supply of activated nitrogen (glutamine) is high, its further biosynthesis is shut down; the UMP-free form of P_{II} accumulates and activates the adenylylation activity of adenylyltransferase. This causes the AMP-containing, or less active, form of glutamine synthetase to accumulate. Conversely, when activated nitrogen supplies are low, α-ketoglutarate accumulates, stimulating the activity of glutamine synthetase by the opposite mechanism.

The modification state of P_{II} also controls the transcription of several nitrogen-regulated genes (Ntr), including the *glnA* gene encoding glutamine synthetase. This regulatory system involves interactions with a specific protein kinase and its transcription factor target. Thus, under conditions of nitrogen starvation (low glutamine), P_{II} is mostly in the P_{II}-UMP state, and *glnA* transcription is stimulated.

In contrast to the 12-subunit bacterial enzymes, mammalian glutamine synthetase is a homodecamer, composed of two pentameric rings stacked face-to-face.

Prokaryotic glutamine synthetase is controlled by cumulative feedback inhibition and covalent modification.

Adenylylated tyrosine residue

Uridylylated P_{II} tyrosine residue

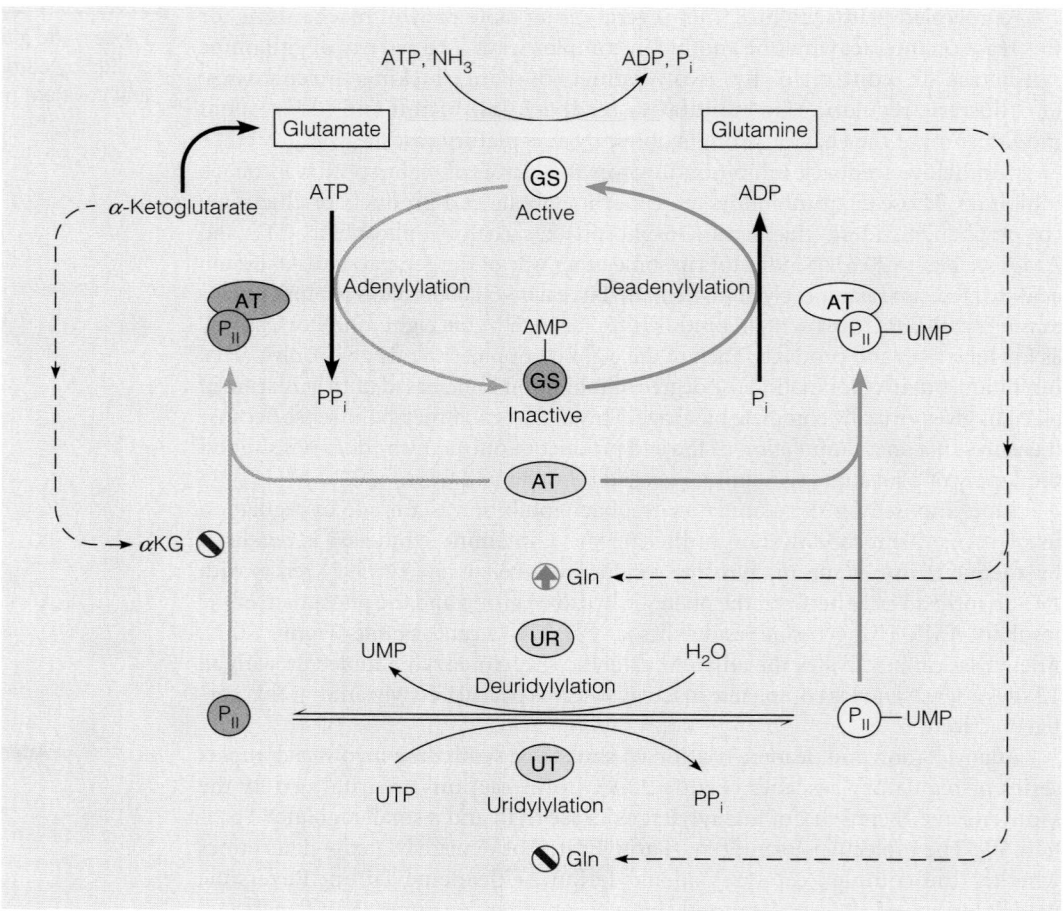

FIGURE 20.8

Regulation of the activity of bacterial glutamine synthetase. The complex of AT (adenylyltransferase) and P_{II} (a regulatory protein) catalyzes both the adenylylation and deadenylylation of glutamine synthetase (GS), depending on whether P_{II} is deuridylylated (left) or uridylylated (right). Uridylylation of P_{II} is catalyzed by uridylyltransferase (UT), and deuridylylation is catalyzed by uridylyl-removing enzyme (UR), both activities part of a bifunctional enzyme. Components shown in blue tend to promote activity of glutamine synthetase, whereas those shown in orange are associated with enzyme inactivation.

Despite sharing less than 20% amino acid sequence identity with the bacterial enzymes, the subunit folds and active site architectures are strikingly similar. Less is known about how glutamine synthetase is regulated in animal cells. In the liver, glutamine synthetase is feedback inhibited by carbamoyl phosphate, glutamine, and several other amino acids, whereas in the brain, the enzyme is inhibited only by carbamoyl phosphate. Mammalian glutamine synthetase does not appear to be regulated by covalent modification and, in fact, lacks the adenylylation loop of the bacterial enzymes.

To assimilate NH_4^+ in bacteria, glutamine synthetase functions together with glutamate dehydrogenase or glutamate synthase, depending on ammonia levels. If NH_4^+ levels are high, the glutamate dehydrogenase/glutamine synthetase pathway operates:

$$NH_3 + \alpha\text{-ketoglutarate} + NAD(P)H \xrightarrow{\text{Glutamate dehydrogenase}} \text{glutamate} + NAD(P)^+ + H_2O$$

$$NH_3 + \text{glutamate} + ATP \xrightarrow{\text{Glutamine synthetase}} \text{glutamine} + ADP + P_i$$

Net: $2NH_3 + \alpha\text{-ketoglutarate} + NAD(P)H + ATP \longrightarrow \text{glutamine} + ADP + P_i + NAD(P)^+ + H_2O$

When NH_4^+ levels are low, the glutamate synthase/glutamine synthetase pathway predominates:

$$2NH_3 + 2\text{ glutamate} + 2ATP \xrightarrow{\text{Glutamine synthetase}} 2\text{ glutamine} + 2ADP + 2P_i$$

$$\alpha\text{-ketoglutarate} + \text{glutamine} + \text{reductant} \xrightarrow{\text{Glutamate synthase}} 2\text{ glutamate} + \text{oxidized reductant}$$

Net: $2NH_3 + \alpha\text{-ketoglutarate} + \text{reductant} + 2ATP \longrightarrow \text{glutamine} + 2ADP + 2P_i + \text{oxidized reductant}$

In this pathway, only glutamine synthetase fixes ammonia—the role of glutamate synthase is to regenerate glutamate for the glutamine synthetase reaction.

Asparagine Synthetase: A Similar Amidation Reaction

Asparagine synthetase catalyzes a reaction comparable to that of glutamine synthetase. Although widespread, asparagine synthetase accounts for much less ammonia assimilation. This enzyme uses ammonia or glutamine in catalyzing the conversion of aspartate to asparagine.

Aspartate + ATP + NH_3 (Gln) $\longrightarrow$ **Asparagine** + AMP + (P)(P) + (Glu)

Note that asparagine synthetase cleaves ATP to yield AMP and PP_i, whereas glutamine synthetase yields ADP and P_i. Asparagine synthetase also differs in that glutamine is strongly preferred as a substrate over ammonia. The reactive species in bond formation is ammonia, which is generated at the active site by hydrolysis of the substrate glutamine.

Carbamoyl Phosphate Synthetase: Generation of an Intermediate for Arginine and Pyrimidine Synthesis

Recall from Figure 20.6 that the final route for assimilating ammonia first forms carbamoyl phosphate. The enzyme responsible is **carbamoyl phosphate synthetase**. Either ammonia or glutamine can serve as the nitrogen donor:

$$NH_3 + HCO_3^- + 2ATP \longrightarrow \text{carbamoyl phosphate} + 2ADP + P_i$$

$$\text{Glutamine} + H_2O + HCO_3^- + 2ATP \longrightarrow \text{carbamoyl phosphate} + 2ADP + P_i + \text{glutamate}$$

Carbamoyl phosphate

The bacterial enzyme can catalyze both reactions, although glutamine is the preferred substrate. Eukaryotic cells contain two forms of the enzyme. Form I, localized in mitochondria, has a preference for ammonia as substrate and is used in the arginine biosynthetic pathway and the urea cycle (see pages 841–842). Form II, present in the cytosol, has a strong preference for glutamine. It is inhibited by uridine triphosphate, consistent with its demonstrated involvement in pyrimidine nucleotide biosynthesis. As we will see in Chapter 22, the form II enzyme is part of a large multifunctional protein with three distinct catalytic sites, which catalyze the first three reactions of pyrimidine nucleotide synthesis.

The Nitrogen Economy: Aspects of Amino Acid Synthesis and Degradation

Metabolic Consequences of the Absence of Nitrogen Storage Compounds

Early biochemists and physiologists believed that the proteins of an adult animal were quite stable, while proteins in the diet were immediately metabolized to provide energy and the end products excreted. This dogma was challenged in the 1930s by Rudolf Schoenheimer. Schoenheimer had escaped Hitler's Germany and joined the Department of Biochemistry at Columbia University, where Harold Urey had discovered deuterium a few years earlier. This environment inspired Schoenheimer to explore the use of isotopic tracers to study metabolism in whole animals (see Tools of Biochemistry 12A). Urey succeeded in enriching nitrogen with ^{15}N, which Schoenheimer's group used to synthesize ^{15}N-labeled tyrosine. They discovered that following administration of ^{15}N-labeled tyrosine to a rat, only about 50% was recovered in the urine. Most of the remainder was incorporated into tissue proteins. Importantly, only a small amount of the ^{15}N found in the tissue proteins was attached to the original tyrosine carbon skeleton—most had been incorporated into other amino acids. Finally, they observed that an equivalent amount of unlabeled protein nitrogen was excreted, keeping the rat in **nitrogen equilibrium**.

Protein and nucleic acid metabolism differs significantly from the metabolism of carbohydrates and lipids. Whereas carbohydrates and lipids can be stored for mobilization as needed by an organism for energy generation or for biosynthesis, most organisms past the embryonic state have no polymeric nitrogen compounds whose function is to be stored and released on demand. Although plants store some nitrogenous compounds, such as asparagine in asparagus, and some insects have storage proteins in their blood, such compounds do not represent widely used nitrogen storage depots. The lack of such depots imposes special requirements on organisms, particularly because of the limited availability of utilizable nitrogen. Animals must continually replenish nitrogen supplies through the diet to replace nitrogen lost through catabolism. In much of the world, protein-rich foods cannot be produced in sufficient quantity to meet the nutritional needs of humans and domestic animals. When dietary protein is insufficient, proteins synthesized for other purposes, mostly muscle proteins, are broken down and not replaced. Such consequences occur even when the diet contains an adequate caloric content of protein, if that protein does not contain the needed amino acids (essential amino acids, see next section).

Just as we can think of a nitrogen economy for the biosphere, we can see it also in relation to individual organisms. Under optimal conditions, animals maintain nitrogen intake and excretion at equivalent rates. A well-nourished adult is said to be in nitrogen equilibrium or **normal nitrogen balance** if the daily intake of nitrogen through the diet is equal to that lost through urinary excretion and other processes, such as ammonia loss in sweat. Positive nitrogen balance, in which normal nitrogen intake exceeds nitrogen loss, is seen during pregnancy, growth of a juvenile, or recovery from starvation. In negative nitrogen balance, more nitrogen is lost than is taken in, a situation found in senescence, starvation, and certain disease states. Plants and microorganisms commonly excrete very little nitrogen. Microorganisms often grow so rapidly that nitrogen released by catabolism is reassimilated, and in plants nitrogen is often available in such severely limited amounts that this factor itself limits cellular growth rates.

Biosynthetic Capacities of Organisms

Organisms vary widely in their ability to synthesize amino acids. Many bacteria and most plants can synthesize all of their nitrogenous metabolites starting from a single nitrogen source, such as ammonia or nitrate. However, many microorganisms will

> Most organisms lack nitrogen storage depots.

use a preformed amino acid, when available, in preference to synthesizing that amino acid. Sometimes preformed amino acids are required. For example, over the course of evolution, the genus *Lactobacillus* has lost many biosynthetic capacities because they grow in milk, a very nutrient-rich environment. Some *Lactobacillus* species must therefore be provided with most of the 20 amino acids to be grown in the laboratory. Mammals are intermediate, being able to biosynthesize about half of the amino acids in quantities needed for growth and for maintenance of normal nitrogen balance.

The amino acids that must be provided in the diet to meet an animal's metabolic needs are called **essential amino acids** (Table 20.1). Those that need not be provided because they can be biosynthesized in adequate amounts are called **nonessential amino acids**. In general, the essential amino acids include those with complex structures, including aromatic rings and hydrocarbon side chains. The nonessential amino acids include those that are readily synthesized from abundant metabolites, such as intermediates in glycolysis or the citric acid cycle.

Although dieticians recommend a protein intake of 50 to 100 grams per day or more, a human can do quite well on a diet containing as little as 20 grams per day, if that protein is of high nutritional quality—that is, if it contains adequate proportions of essential amino acids. In general, the more closely the amino acid composition of ingested protein resembles the amino acid composition of the animal eating the protein, the higher the nutritional quality of that protein. For humans, mammalian protein is of the highest nutritional quality, followed by fish and poultry, and then by fruits and vegetables. (In this context, nutritional quality refers only to the single criterion of essential amino acid content.) Plant proteins in particular are often deficient in lysine, methionine, or tryptophan. However, a vegetarian diet provides adequate protein if it contains a variety of protein sources, with a deficiency in one source being compensated for by excess in another source.

Transamination

In Chapter 14 we introduced transamination, a process whereby amino acids can replenish citric acid cycle intermediates. Transamination plays a somewhat broader role in amino acid metabolism, in that it provides a route for redistribution of amino acid nitrogen. Because of the key role of glutamate in ammonia assimilation, it is a star player in transamination. In other words, glutamate is an abundant product of ammonia assimilation, and transamination uses glutamate nitrogen to synthesize other amino acids.

Transamination reactions are catalyzed by enzymes called **transaminases** or, more properly, **aminotransferases**. As shown below, transamination involves transfer of the α-amino group, usually of glutamate, to an α-keto acid, with formation of the corresponding amino acid plus the α-keto derivative of glutamate, which is α-ketoglutarate.

Essential amino acids cannot be biosynthesized in adequate amounts and must be provided in the diet.

| TABLE 20.1 | Nutritional requirements for amino acids in mammals |

Essential

Arginine*, histidine, isoleucine, leucine, lysine, methionine*, phenylalanine, threonine, tryptophan, valine

Nonessential

Alanine, asparagine, aspartate, cysteine, glutamate, glutamine, glycine, proline, serine, tyrosine

*Although mammals can synthesize arginine and methionine, the use of these amino acids for the production of urea and methyl groups, respectively, is greater than the capacity of their biosynthetic pathways.

Transamination is the reversible transfer of an amino group from an α-amino acid to an α-keto acid, with pyridoxal phosphate as a coenzyme.

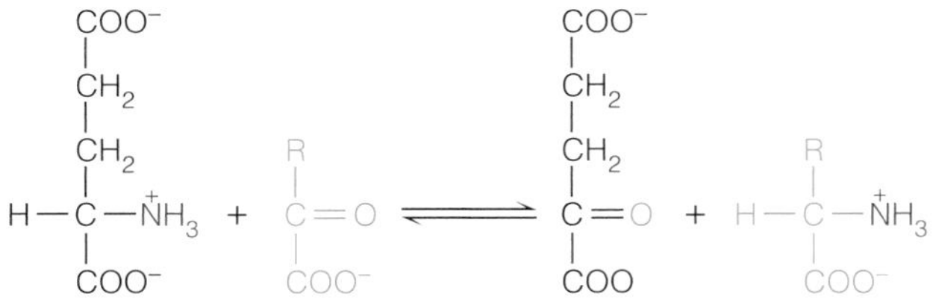

| Glutamate | α-Keto acid | α-Ketoglutarate | α-Amino acid |

Specific aminotransferases exist in animal cells for the synthesis of all of the amino acids found in proteins except threonine and lysine, so long as the corresponding α-keto acids are available. Thus, the inability of animal cells to synthesize most of the essential amino acids results from an inability to synthesize the carbon skeletons in the form of α-keto acids.

Pyridoxal phosphate

Schiff base between amino acid and pyridoxal phosphate

Aminotransferases utilize a coenzyme, **pyridoxal phosphate**, that is derived from vitamin B_6. The functional part of the cofactor is an aldehyde functional group, —CHO, attached to a pyridine ring. The catalytic cycle begins with condensation of this aldehyde with the α-amino group of an amino acid, to give a Schiff base, or aldimine, intermediate. Because pyridoxal phosphate participates in a diversity of reactions involving amino acids, we discuss catalytic mechanisms in a separate section, beginning on page 845.

Transamination reactions have equilibrium constants close to unity. Therefore, the direction in which a particular transamination proceeds is controlled in large part by the intracellular concentrations of substrates and products. This means that transamination can be used not only for amino acid synthesis but also for degradation of amino acids that accumulate in excess of need. In degradation the transaminase works in concert with glutamate dehydrogenase, as exemplified by the degradation of alanine:

$$\text{Alanine} + \alpha\text{-ketoglutarate} \xrightarrow{\text{Aminotransferase}} \text{pyruvate} + \text{glutamate}$$

$$\text{Glutamate} + NAD^+ + H_2O \xrightarrow{\text{Glutamate dehydrogenase}} \alpha\text{-ketoglutarate} + NADH + \overset{+}{N}H_4$$

$$\text{Net: Alanine} + NAD^+ + H_2O \longrightarrow \text{pyruvate} + NADH + \overset{+}{N}H_4$$

Thus, we see transamination as a mechanism for amino acid synthesis *or* degradation. Because the amino acids within a cell are rarely present in the proportions needed to synthesize the specific proteins of that cell, transamination plays an important role in bringing the amino acid composition into line with the organism's needs. It also participates in funneling excess amino acids toward catabolism and energy generation.

Most aminotransferases use glutamate/α-ketoglutarate as one of the two α-amino/α-keto acid pairs involved. Two such enzymes are important in the clinical diagnosis of human disease—serum glutamate-oxaloacetate transaminase (SGOT) and serum glutamate-pyruvate transaminase (SGPT):

$$\text{Glutamate} + \text{oxaloacetate} \xrightarrow{\text{SGOT}} \alpha\text{-ketoglutarate} + \text{aspartate}$$

$$\text{Glutamate} + \text{pyruvate} \xrightarrow{\text{SGPT}} \alpha\text{-ketoglutarate} + \text{alanine}$$

These enzymes, abundant in heart and in liver, are released from cells as part of the cell injury that occurs in myocardial infarction, infectious hepatitis, or other damage to either organ. Assays of these enzyme activities in blood serum can be used both in diagnosis and in monitoring the progress of a patient during treatment. Note the convention used in naming transaminases; the amino donor and the α-keto acid acceptor are named.

Protein Turnover

Proteins are subject to continuous biosynthesis and degradation, a process called **protein turnover**. For an intracellular protein whose total concentration does not change with time, the steady state level is maintained by synthesis of the protein at a rate just sufficient to replenish protein lost by degradation. Many of the amino acids released during protein turnover are reutilized in the synthesis of new proteins.

Quantitative Features of Protein Turnover

The macroscopic dimensions of protein turnover can be appreciated by considering a day in the life of a 70-kilogram person. That individual typically will consume 100 grams of protein during the day and, because he or she is in normal nitrogen balance, will excrete an equivalent amount of nitrogenous end products. Yet isotope labeling studies show that about 400 grams of protein are synthesized per day, and 400 grams broken down. About three-quarters of the released amino acids are reused in protein synthesis, with the remainder being degraded and the nitrogen excreted. Thus, the total amino acid pool consists of 500 g/day—100 ingested and 400 released via protein degradation. From this pool, 400 grams are used in protein synthesis and 100 grams are catabolized and excreted.

Individual proteins exhibit tremendous variability in their metabolic life-times, from a few minutes to many months. You should realize, however, that all of the proteins in the body are represented among the 400 grams broken down in a typical day. Extensive pulse-chase experiments in laboratory animals (see Tools of Biochemistry 12A) show that protein degradation follows first-order kinetics. For a particular protein, individual molecules are degraded at random, such that a semilogarithmic plot of isotope remaining in a protein versus time is linear. Thus, we can determine the metabolic half-life of a particular protein. In the rat the average protein has a half-life of 1 or 2 days. Table 20.2 gives information about the half-lives of some specific proteins.

As you might expect, proteins that are secreted into an extracellular environment, such as digestive enzymes, polypeptide hormones, and antibodies, turn over quite rapidly, whereas proteins that play a predominantly structural role, such as collagen of connective tissue, are much more stable metabolically. Enzymes catalyzing rate-determining steps in metabolic pathways are also short-lived. Indeed, for many enzymes the rate of breakdown is an important regulatory factor in controlling intracellular enzyme levels. By contrast, proteins that do not represent metabolic control points turn over relatively slowly. In the rat, cytochrome c has a half-life of nearly a week; in the human, hemoglobin is as long-lived as the erythrocyte in which it resides (about 120 days). But why should such proteins turn over at all, if their degradation does not represent a metabolic control mechanism? Isn't such turnover wasteful of energy? Let us consider that point.

Biological Importance of Protein Turnover

Like all other intracellular constituents, proteins are subjected to a barrage of environmental influences, primarily reactive oxygen species (see Chapter 15), which can affect their structure, conformation, and biological activity. The capacity of proteins to repair the resulting damage is limited. Protein turnover could be seen

TABLE 20.2 Half-lives and intracellular sites of degradation in protein turnover

Half-Life (hours)	Intracellular Location			
	Nucleus	Cytosol	Mitochondria	Endoplasmic Reticulum and Plasma Membrane
<2	Oncogene products	Ornithine decarboxylase, tyrosine aminotransferase, protein kinase C	δ-Aminolevulinic acid synthetase	HMG-CoA reductase
2–8	—	Tryptophan oxygenase, cAMP-dependent protein kinase	—	γ-Glutamyl transferase
9–40	Ubiquitin	Calmodulin, glucokinase	Acetyl-CoA carboxylase, alanine aminotransferase	LDL receptor, cytochrome P450
41–200	Histone H1	Lactate dehydrogenase, aldolase, dihydrofolate reductase, phytochrome P670	Cytochrome oxidase, pyruvate carboxylase, cytochrome c	Cytochrome b_5, cyt b_5 reductase
>200	Histones H2A, H2B, H3, H4	Hemoglobin, glycogen phosphorylase	—	Acetylcholine receptor

Reprinted from *Trends in Biochemical Sciences* 12:390–394, M. Rechsteiner, S. Rogers, and K. Rote. © 1987, with permission from Elsevier.

as an inefficient quality control system in which both normal and modified proteins are randomly degraded and replaced. However, it was discovered that proteolysis requires an input of energy, a surprising finding given that amide bond hydrolysis is exergonic. This energy requirement suggested that the process might be nonrandom and regulated. Indeed, protein molecules that have become chemically altered are preferentially degraded. A certain chemical change may mark a protein molecule, targeting it for degradation by a proteolytic enzyme that specifically recognizes the marker.

We know that in bacteria, mutant proteins are degraded much more rapidly than their wild-type counterparts. Evidently evolution has generated proteins whose conformation renders maximum stability in the intracellular environment, and most structural changes decrease this stability.

Protein turnover also represents a route for cellular adaptation to altered environmental conditions. For example, in many bacteria, extensive proteolysis is one of the metabolic events interlinked with **sporulation**, or spore formation. Spores represent a heat-stable form of the microorganism; they metabolize at negligible rates and can remain dormant for months or years. When metabolic conditions induce a growing cell to sporulate, extensive protein turnover occurs, with the amino acids released being used to synthesize proteins of the spore. This quasi-dormant state can be maintained indefinitely, with germination to vegetative cells occurring after improvement in the environmental conditions.

Except for specialized functions such as sporulation, it was thought until the early 1970s that protein turnover, meaning its complete degradation to amino acids, served just two major functions: (1) protein digestion, providing amino acids for synthesis of proteins and other metabolites derived from amino acids, and (2) ridding cells of defective proteins, including those that were mutationally or environmentally damaged. Accordingly, most intracellular proteases were thought to participate in protein turnover. Of course, it has long been recognized that specific endopeptidase protein cleavage reactions are involved in enzyme activation; an example is the blood clotting cascade, which we discussed in Chapter 11. Over the past three decades, it has become clear that this process of limited proteolysis—cleavage of a few specific peptide bonds in a protein—has a host of additional functions, including regulation of gene expression, response to environmental stress, and participation in cell signaling pathways. Of great current interest is the involvement of selective proteolytic reactions in signaling pathways leading to *apoptosis*, a process in normal development in which certain cells, having fulfilled their function in differentiation, undergo a programmed death. These functions are presented in detail in Chapter 29. Here we concern ourselves with those aspects of protein turnover that relate specifically to amino acid metabolism, to identifying major classes of intracellular proteases, and to describing some of the structural features that predispose certain proteins to degradation.

Intracellular Proteases and Sites of Turnover

Because most proteins are used intracellularly, most turn over within the cell. The earliest intracellular proteases to be characterized were in lysosomes. Although the lysosomal system is inside the cell, it functions primarily in the proteolysis of extracytoplasmic proteins, which enter the cell via endocytosis and are degraded within the vacuolar lumen. Lysosomes, which form by budding from the Golgi complex, are bags of digestive enzymes, containing proteases, nucleases, lipases, and carbohydrate-cleaving enzymes. As discussed further in Chapter 28, lysosomes play various cellular roles: secretion of digestive enzymes, digestion of organelles destined for destruction, digestion of food particles or bacteria engulfed by phagocytosis, and the nonselective engulfment and degradation of bulk cellular constituents (**autophagy**).

In eukaryotic cells two distinct classes of proteases have been found in the cytosol—a family of Ca^{2+}-activated neutral cysteine proteases called **calpains**,

All proteins are in a constant state of turnover and replacement, for repair of damage and for biological regulation.

and a large (2.5 megadalton) multisubunit ATP-dependent protease called the **proteasome** (Figure 20.9). Mammals have at least 14 different calpain proteases, which play important roles in necrotic cell death (**oncosis**). These enzymes are distinct from the lysosomal proteases, called **cathepsins**, which are designed to function in an acidic milieu. The proteasome, as we shall see, degrades proteins that have been modified by the attachment of the small protein ubiquitin.

In contrast to lysosomal enzymes, which are usually safely sequestered in their vesicles, any protease activity free in normal cytosol must be under strict control, so as to attack only those proteins whose destruction is needed—damaged, mutant, or otherwise dispensable proteins. The identification, or marking, of those proteins whose degradation suits the interests of the cell involves various tagging schemes described in the next section.

Chemical Signals for Turnover

The turnover rates for different proteins vary by as much as 1000-fold, whereas differences in protein stability, as measured by denaturation in vitro, may be much less. Four structural features are currently thought to be interrelated determinants of turnover rate: (1) **ubiquitination**, (2) metal-catalyzed oxidation of particular residues, (3) **PEST sequences**, and (4) particular N-terminal residues.

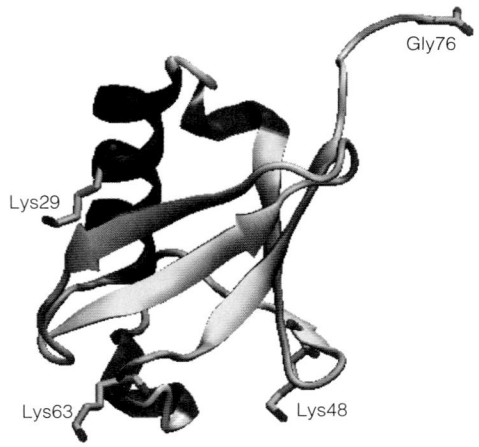

Human ubiquitin (PDB ID 1UBI)

Ubiquitination

Ubiquitin is a small (76-residue) protein expressed in all eukaryotic cells—it derives its name from its widespread (ubiquitous) distribution. Ubiquitin is covalently conjugated to specific cellular proteins in an ATP-dependent reaction, which condenses the terminal carboxyl group of ubiquitin with lysine amino groups on target proteins, as shown in Figure 20.10. In the first reaction, catalyzed by the ubiquitin-activating enzyme E1, a thioester is formed between the C-terminal Gly residue of ubiquitin and an internal Cys residue of E1 in a two-step process. Formation of this high-energy covalent linkage is driven by ATP hydrolysis, via a ubiquitin-adenylate intermediate. AMP is released upon nucleophilic attack by the Cys-SH of E1. The activated ubiquitin is next transferred to a Cys-SH of one of several E2 enzymes (ubiquitin-conjugating enzymes) in a trans-thiolation reaction. A ubiquitin-protein ligase, E3, is required for step 3—covalent attachment of ubiquitin to the $\varepsilon - NH_3^+$ group of one or more lysine residues on the target protein via an *isopeptide* bond. The use of high-energy thioester bonds to activate ubiquitin makes formation of the isopeptide bond thermodynamically favorable. Expenditure of the equivalent of two ATPs per ubiquitin added ensures that the reaction is both irreversible and specific. We will see an even more dramatic example of the cost of specificity when we discuss protein synthesis in Chapter 28 (see page 1211).

E3 can catalyze the successive addition of a ubiquitin moiety to a previously conjugated ubiquitin, forming a polyubiquitin chain in which a Lys of ubiquitin forms an isopeptide bond with the C-terminal Gly carboxyl group of the succeeding ubiquitin. Ubiquitin contains seven lysine residues, and polyubiquitin chains can involve one or more of these residues, although the functions of chains linked through Lys29, Lys48, and Lys63 are the best characterized. There appear to be just two E1 enzymes in mammals, which activate and transfer ubiquitin to all E2 enzymes. About 28 E2 enzymes are known in humans, each of which can transfer ubiquitin to several E3 proteins. Target protein selectivity resides with the E3 components, and based on commonly shared structural motifs, the human genome may contain as many as 1000 E3s. Two major families of E3 ubiquitin–protein ligases have been described: the HECT domain-containing E3s and the RING-finger domain-containing E3s. These two E3 families differ in structure and in mechanism (Figure 20.10). Because of the central role of E3 in

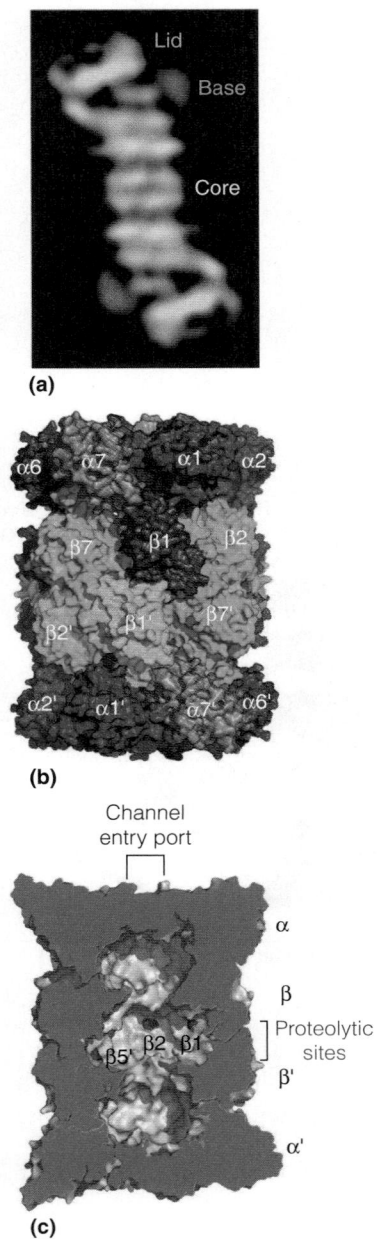

FIGURE 20.9

Structure of the proteasome. **(a)** This image was generated by averaging multiple electron micrographs of the 26S proteasome of the toad *Xenopus laevis*. The proteasome contains two major assemblies, a 28-subunit core particle (also known as the 20S particle) and a 19-subunit regulatory particle (also known as the 19S particle), which forms the base and lid assemblies. **(b)** and **(c)** The proteolytic active sites are located within the large internal space (~100×60 Å) of the 20S core particle. The lid and base assemblies of the 19S regulatory particle control substrate entry into the 20S core particle. Proteins tagged with the protein ubiquitin pass through the tube in an ATP-dependent fashion.

Reprinted from *Annual Review of Biochemistry* 78: 477–513, D. Finley, Recognition and processing of ubiquitin-protein conjugates by the proteasome. © 2009 Annual Reviews.

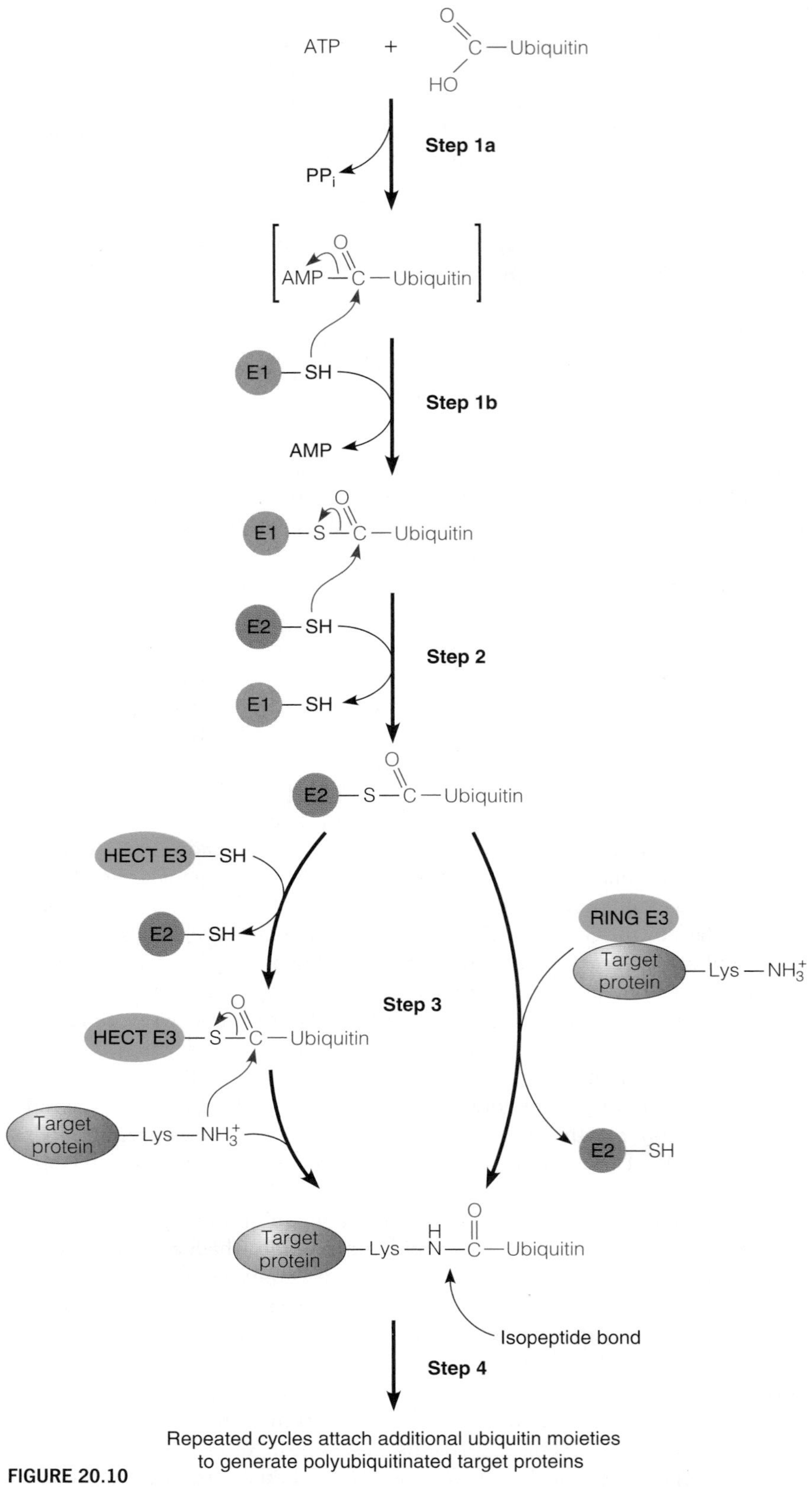

Repeated cycles attach additional ubiquitin moieties to generate polyubiquitinated target proteins

FIGURE 20.10

Enzymatic pathway for attachment of ubiquitin to target proteins. E1, E2, and E3 are proteins involved in the ubiquitin transfer to lysine residues on target proteins. **Steps 1a and 1b:** ATP-dependent formation of a thioester bond between the ubiquitin C-terminus and a cysteine thiol on E1. **Step 2:** Transfer of ubiquitin from E1 to E2. **Step 3:** Transfer of ubiquitin to lysine residue on the target protein. HECT domain-containing E3s (left pathway) catalyze formation of an E3-ubiquitin thioester intermediate before transferring ubiquitin to the substrate. RING-finger domain-containing E3s (right pathway) function as docking proteins, with E2 catalyzing nucleophilic attack of the substrate protein lysine amino group on the E2-ubiquitin thioester. **Step 4:** Formation of polyubiquitin chain on target protein.

determining the specificity and selectivity of the proteasome system, these proteins are implicated in a number of disease states, including neurodegenerative disorders, inflammatory diseases, muscle wasting disorders, and cancer.

Lys29- and Lys48-linked polyubiquitin chains serve as recognition markers for the 26S proteasome, and such tagged proteins dock at the proteasome via specific ubiquitin receptors. The target proteins are subsequently degraded to short peptides in an ATP-dependent process. The polyubiquitin chains are not degraded, but instead are released and disassembled by deubiquitinating enzymes in the lid of the proteasome. The free ubiquitin monomers can then be reused.

Until recently, the role of the ubiquitin system was thought to be limited to regulated proteolysis. It is now clear, however, that Lys63-linked polyubiquitin chains (and probably others as well) have nonproteasomal functions involved in the regulation of many critical processes, including cell cycle progression (Chapter 24), cholesterol synthesis (Chapter 19), inflammation, response to hypoxia (Chapter 18), and apoptosis (Chapter 28). As we have seen for other regulatory systems, ubiquitination is reversible. Ubiquitin removal is catalyzed by a family of **deubiquitinating enzymes** (Dubs)—the human genome encodes ~95 distinct Dubs. We also know of several other ubiquitin-like proteins, such as SUMO (small ubiquitin-like modifier) and ISG15 (interferon-stimulated gene 15kD), which use similar modification/deconjugation machinery and are involved in the regulation of numerous biological processes, including transcription, DNA repair, stress response, and innate immunity.

Although bacteria lack the 26S proteasome, they do possess functionally similar ATP-dependent proteolytic assemblies known as the **Lon** and **Clp proteases**. Both of these proteins are barrel-shaped, with the protease active sites located in a central cavity, like the 26S proteasome of eukaryotes. Homologs of Lon are found in archaea and in mitochondria of eukaryotes.

PEST Sequences

Examination of the amino acid sequences of short-lived proteins ($t_{1/2} < 2$ hours) shows that virtually all of these proteins contain one or more regions rich in proline, glutamate, serine, and threonine. From the one-letter designations for these amino acids (P, E, S, and T, respectively), these regions, some 12 to 60 residues long, have been called PEST sequences. Very few longer-lived proteins contain these regions. Moreover, creation of PEST sequences in long-lived proteins by site-directed mutagenesis increases their metabolic lability. PEST-containing proteins are degraded primarily by the ubiquitin–proteasome system. Current evidence indicates that phosphorylation of serine or threonine residues in the PEST sequence triggers ubiquitination and subsequent degradation of the protein. Thus, the PEST sequence represents another recognition scheme for targeting short-lived proteins to degradation by the 26S proteasome.

N-Terminal Amino Acid Residue

Experiments with bacteria carried out by Alexander Varshavsky revealed that the intracellular half-life of a particular protein varies depending on the identity of its N-terminal amino acid residue. An N-terminal residue of Phe, Leu, Tyr, Trp, Lys, or Arg is correlated with short metabolic lifetimes, whereas proteins with other amino termini are far longer-lived. These findings, originally made with natural proteins, are supported by experiments in which amino termini of proteins were altered by site-directed mutagenesis, with corresponding changes in metabolic half-lives of the mutant proteins. The same "N-end rule" has also been shown to apply to animal and plant cells, and represents yet another ubiquitin-dependent proteolytic system. The N-terminal residues of short-lived proteins function as degradation signals called **N-degrons**, which are recognized by specific E3 ligases called **N-recognins**. In prokaryotes, which lack the ubiquitin system, the N-degron is recognized and processed by the Clp protease system.

These and other observations indicate that specific structural features on proteins convey information about the metabolic stability of the proteins. The molecular nature of that information processing and the identities of the enzymes involved remain to be determined, and, as discussed further in Chapter 29, proteolysis is becoming recognized as comparable to phosphorylation in its importance as a metabolic regulatory mechanism.

Amino Acid Degradation and Metabolism of Nitrogenous End Products

Common Features of Amino Acid Degradation Pathways

In animals whose dietary protein intake exceeds the need for protein synthesis and other biosyntheses, the excess nitrogen is mostly degraded, with the carbon skeletons being metabolized in the citric acid cycle. Protein can thus be a significant contributor toward an animal's energetic requirements. In contrast, plants and bacteria generally can synthesize most of their own amino acids, and they regulate the anabolic pathways so that excesses rarely develop. Generally, microorganisms use preformed amino acids in preference to synthesizing their own, even though many bacteria can satisfy all of their requirements for nitrogen *and* carbon from a single amino acid. The degradative pathways active here are in general similar to those described in animals.

> Amino acid degradation usually begins with conversion to the corresponding α-keto acid by transamination or oxidative deamination.

With a few exceptions, the first step in amino acid degradation involves removal of the α-amino group to give the corresponding α-keto acid. This modification is usually effected by transamination, with the concomitant synthesis of glutamate from α-ketoglutarate. This is followed by the glutamate dehydrogenase reaction, as was shown on page 834. Thus, the net process is the deamination of the α-amino acid to the corresponding α-keto acid plus ammonia. The same net conversion can also be catalyzed by **L-amino acid oxidase**, a flavoprotein enzyme found in liver and kidney.

$$R-\underset{\underset{NH_3^+}{|}}{CH}-COO^- + FAD + H_2O \longrightarrow R-\underset{\underset{O}{\|}}{C}-COO^- + FADH_2 + NH_3$$

$$FADH_2 + O_2 \longrightarrow FAD + H_2O_2$$

$$2H_2O_2 \xrightarrow{\text{Catalase}} 2H_2O + O_2$$

The peroxide formed is decomposed by catalase. Kidney and liver are also rich in FAD-containing **D-amino acid oxidase**. The function of this enzyme in animals is unknown because the D isomers of amino acids are quite rare. However, bacterial cell walls contain D-amino acids, and a number of bacterial cells contain D-amino acid oxidases.

Once the nitrogen has been removed, the carbon skeleton can, depending on the physiological state of the organism, either proceed toward oxidation in the citric acid cycle or be used for biosynthesis of carbohydrate. Figure 20.11 shows the entry points into the citric acid cycle for the breakdown products of each of the amino acids. The individual pathways are presented in Chapter 21.

Amino acids whose skeletons generate pyruvate or oxaloacetate (examples are alanine and aspartate) are efficiently converted to carbohydrates via gluconeogenesis. Amino acids leading to acetyl-CoA or acetoacetyl-CoA, such as leucine, contribute heavily toward ketogenesis. The terms **glucogenic** and **ketogenic** have been used to classify amino acids as generators primarily of carbohydrate or ketone bodies, respectively.

Detoxification and Excretion of Ammonia

Although ammonia is a universal participant in amino acid synthesis and degradation, its accumulation in abnormal concentrations has toxic consequences.

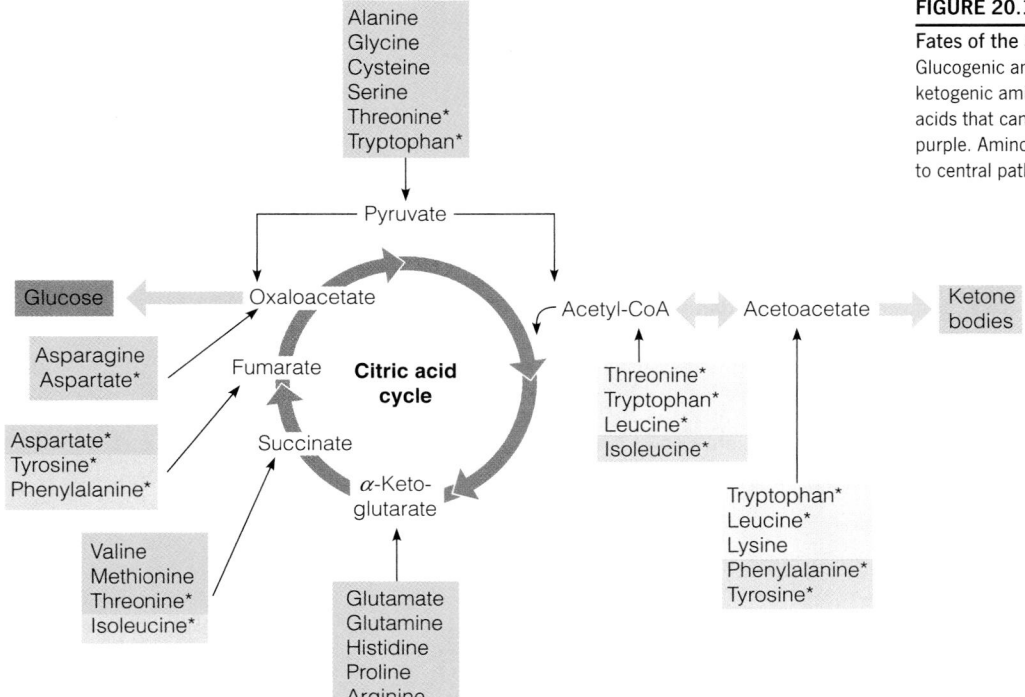

FIGURE 20.11

Fates of the amino acid carbon skeletons. Glucogenic amino acids are shown in orange, ketogenic amino acids in blue, and the few amino acids that can be both glucogenic and ketogenic in purple. Amino acids with more than one route of entry to central pathways are marked by an asterisk.

Therefore, cells undergoing active amino acid catabolism must be able to detoxify and/or excrete ammonia as fast as it is generated. For most aquatic animals, which can take in and pass out unlimited quantities of water, ammonia simply dissolves in the water and diffuses away. Because terrestrial animals must conserve water, they convert ammonia to a form that can be excreted without large water losses. Birds, terrestrial reptiles, and insects convert most of their excess ammonia to **uric acid**, an oxidized purine. Because uric acid is quite insoluble, it precipitates and can be excreted without a large water loss and without building up osmotic pressure. This is particularly important during the part of each animal's lifetime that is spent in the egg. The biosynthesis of uric acid occurs by the route used to synthesize purine nucleotides, a pathway presented in Chapter 22. Most mammals excrete the bulk of their nitrogen in the form of **urea** (an interesting exception is the Dalmatian dog, which excretes most nitrogen as uric acid). Urea is highly soluble and, lacking ionizable groups, does not affect the pH when it accumulates, as does ammonia.

The Krebs–Henseleit Urea Cycle

Urea is synthesized almost exclusively in the liver and then transported to the kidneys for excretion. The synthetic pathway, which is cyclic, was discovered by Hans Krebs and Kurt Henseleit in 1932, five years before the other cycle for which Krebs is famous. Krebs and Henseleit were investigating the pathway by adding possible precursors to liver slices and then measuring the amount of urea produced. When arginine was added, urea was produced in 30-fold molar excess over the amount of arginine administered. Similar results were seen if either of two structurally related amino acids, **ornithine** or **citrulline**, was substituted for arginine. Because these three amino acids seemed to function catalytically to promote urea synthesis, Krebs and Henseleit proposed the existence of a cyclic pathway (see margin).

That proposal was correct, as was subsequently confirmed by isolation of the enzymes involved and identification of the biosynthetic route to ornithine, which begins the pathway. These details are shown in Figure 20.12. Ornithine serves as a "carrier," upon which are assembled the carbon and nitrogen atoms that will

Animals have evolved pathways, adapted to their lifestyles, for excretion of ammonia, uric acid, or urea as the major nitrogenous end product.

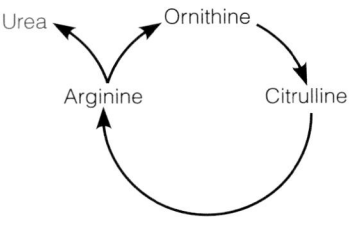

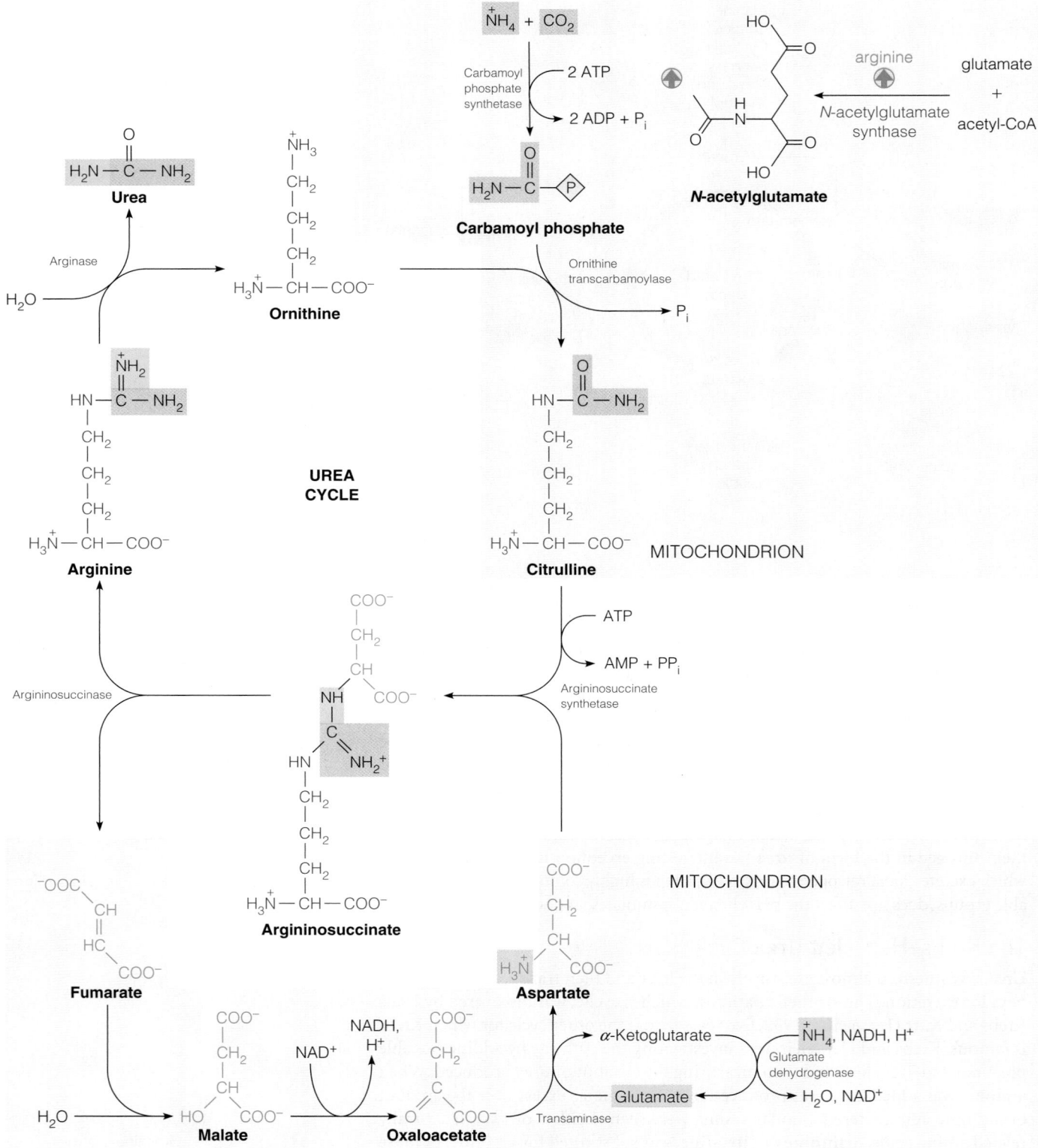

FIGURE 20.12

The Krebs–Henseleit urea cycle. Urea (upper left) contains a carbon and a nitrogen (orange) derived from carbamoyl phosphate and a nitrogen (purple) derived from aspartate. CO_2 and NH_4^+, the ultimate sources of these atoms, were incorporated in turn through the actions of carbamoyl phosphate synthetase (upper right) and glutamate dehydrogenase (lower right). Glutamate can serve directly as the source of some urea nitrogen. Yellow shading identifies reactions occurring in mitochondria. The rest of the pathway takes place in cytosol.

eventually constitute urea. Ornithine itself is synthesized from glutamate, by a pathway described in Chapter 21 (see Figure 21.31, page 904). The source of the carbon and one nitrogen atom in urea is carbamoyl phosphate, synthesized from NH_4^+ and CO_2 by carbamoyl phosphate synthetase I (CPS I; page 831). Carbamoyl phosphate reacts with ornithine, via the enzyme **ornithine transcarbamoylase**, to give citrulline. The second nitrogen comes from aspartate, which reacts with citrulline to form **argininosuccinate**, through the action of **argininosuccinate synthetase**. Next, **argininosuccinase** cleaves argininosuccinate in a nonhydrolytic, nonoxidative reaction to give arginine and fumarate. Arginine is cleaved hydrolytically by **arginase**, to regenerate ornithine and yield one molecule of urea.

The reactions of the urea cycle are compartmentalized in mitochondria and cytosol of liver cells. Glutamate dehydrogenase, the citric acid cycle enzymes, carbamoyl phosphate synthetase I, and ornithine carbamoyltransferase are localized in the mitochondrion, and the rest of the cycle occurs in the cytosol. This means that ornithine must be transported into mitochondria, and citrulline exported to the cytosol, in order for the cycle to proceed.

The enzyme arginase is responsible for the cyclic nature of the urea biosynthetic pathway. Virtually all organisms synthesize arginine from ornithine by the reactions shown in Figure 20.12. However, only **ureotelic** organisms (those excreting most of their nitrogen as urea) contain arginase and, hence, only those organisms carry out the cyclic pathway. Interestingly, the capacity to synthesize arginase develops in frogs at the same time that they undergo metamorphosis from the tadpole stage to the adult animal. Because the tadpole lives in water, it can excrete ammonia. The adult frog, being adapted to a terrestrial lifestyle, develops the ability to synthesize urea.

As noted already, one nitrogen atom in urea comes from aspartate. This atom is derived from ammonia, which is transferred to glutamate via the glutamate dehydrogenase reaction, then to aspartate by transamination. Note from the bottom of Figure 20.12 that a second cycle is used to maintain carbon balance by conversion of the fumarate produced from argininosuccinate cleavage to oxaloacetate in the citric acid cycle and then back to aspartate by transamination.

The use of aspartate as a nitrogen donor is a two-step process that deserves further comment. In the first step, catalyzed by argininosuccinate synthetase, the amide group of citrulline is condensed with the α-amino group of aspartate. The role of ATP in this reaction is to activate the amide for a nucleophilic acyl substitution by aspartate. The amide carbonyl of citrulline attacks the α-phosphate of ATP, displacing PP_i and producing an AMP-citrulline intermediate. This reaction is reversible, but is driven forward by hydrolysis of PP_i to $2P_i$. The nucleophilic amino group of aspartate then adds to the $C=N^+$ double bond to give a tetrahedral intermediate, which expels AMP to yield arginosuccinate. In the second step, catalyzed by argininosuccinase, the aspartate carbon skeleton is eliminated as fumarate. The net result of these two reactions is transfer of the amino group of aspartate to citrulline, producing arginine. We will see this same strategy used again in de novo purine biosynthesis (Chapter 22).

The net reaction for one turn of the urea cycle is as follows:

$CO_2 + NH_4^+ + 3ATP + aspartate + 2H_2O \rightarrow urea + 2ADP + 2P_i + AMP + PP_i + fumarate$

Two molecules of ATP are required to reconvert AMP to ATP, so really four (not three) high-energy phosphates are consumed in each turn of the cycle. Thus, the synthesis of this excretion product is energetically expensive. Excess ammonia is also produced when the animal is forced to catabolize amino acids in muscle as energy sources, for example during fasting. Animals have evolved both long-term and short-term mechanisms to regulate flux through the urea cycle. The levels of the four urea cycle enzymes and CPS I are increased in animals on high-protein diets and are decreased in animals fed protein-free diets. This long-term mechanism allows the animal to adjust its urea cycle capacity to changes in its diet.

Arginosuccinate synthetase mechanism

© 2005 Roberts and Company Publishers. Modified from *The Organic Chemistry of Biological Pathways*, John McMurry and Tadhg Begley.

Urea is synthesized by an energy-requiring cyclic pathway that begins and ends with ornithine.

Allosteric activation of CPS I by **N-acetylglutamate** provides short-term regulation of flux through the urea cycle. *N*-acetylglutamate is synthesized from acetyl-CoA and glutamate in a reaction catalyzed by **N-acetylglutamate synthase** (Figure 20.12, upper right). Mitochondrial levels of *N*-acetylglutamate are determined by the levels of glutamate, which rise with increased amino acid breakdown through transamination reactions. Thus, flux through the urea cycle is tied to the rate of amino acid degradation.

CPS I is also regulated by covalent modification—acetylation of specific lysine residues inactivates the enzyme. The enzyme responsible for acetylation is not known, but a specific mitochondrial sirtuin (SIRT5) catalyzes the deacetylation of CPS I. Recall from Chapter 18 (page 763) that sirtuins catalyze NAD^+-dependent deacetylation of proteins. Sirtuins are activated by intracellular NAD^+ levels, and mitochondrial NAD^+ levels rise during fasting. This rise activates SIRT5, which deacetylates CPS I to its active state, leading to increased carbamoyl phosphate synthesis and thus increased urea cycle flux. This sirtuin-mediated reversible acetylation of CPS I provides a mechanism to increase the capacity for ammonia disposal (via increased urea synthesis) under fasting conditions, when amino acids are being used as energy sources.

Several inherited disorders are known in humans, each caused by a defect in one of the five enzymes in the urea cycle. Although clinical symptoms vary, **hyperammonemia** (high blood NH_4^+ levels) is common to all of these diseases. Symptoms usually present in infants, and include lethargy, vomiting, irreversible brain damage, and even death. The ammonia toxicity is thought to be due to its effects on amino acid and energy metabolism—ammonia depletes citric acid cycle intermediates and NADH by overloading the glutamate dehydrogenase reaction, leading to an ATP deficiency. The brain is especially susceptible to ammonia toxicity.

Following its synthesis, urea is transported in the bloodstream to the kidneys, which filter it for excretion. Measurements of blood urea nitrogen (BUN) levels provide a sensitive clinical test of kidney function because filtration and removal of urea are impaired in cases of kidney malfunction. Analogously, blood ammonia measurements are a sensitive test of liver function. Liver damage, whether acute (hepatitis, poisoning) or chronic (alcoholic cirrhosis), reduces activity of the urea cycle. The accumulation of ammonia is toxic to the brain and thus is related to the comatose condition seen in advanced cases of chronic alcoholism.

Transport of Ammonia to the Liver

All animal organs degrade amino acids and produce ammonia. Two mechanisms are involved in transporting this ammonia from other tissues to liver for its eventual conversion to urea (Figure 20.13). Most tissues use glutamine synthetase to convert ammonia to the nontoxic, and electrically neutral, glutamine. The glutamine is transported in the blood to the liver, where, as noted also in Chapter 13, it is cleaved hydrolytically by glutaminase.

$$\text{Glutamine} + H_2O \longrightarrow \text{glutamate} + \text{ammonia}$$

Muscle, which derives most of its energy from glycolysis, uses a different route, the **glucose–alanine cycle**. Glycolysis generates pyruvate, which undergoes transamination with glutamate to give alanine and α-ketoglutarate. The glutamate in turn has acquired its nitrogen from ammonia, via glutamate dehydrogenase. The resultant alanine is transported to the liver, where it loses its nitrogen by a reversal of the previous processes. This reversal yields ammonia for urea synthesis, plus pyruvate. The pyruvate undergoes gluconeogenesis to give glucose, which is released to the blood for transport back to the muscle or for nourishment of the brain. This cyclic process helps muscle get rid of ammonia, with the carbon from the pyruvate being returned to the liver for gluconeogenesis.

At least five different diseases in humans result from inherited defects in urea cycle enzymes.

The glucose–alanine cycle removes toxic ammonia from muscle. Glutamine synthetase and glutaminase do the same for most other tissues.

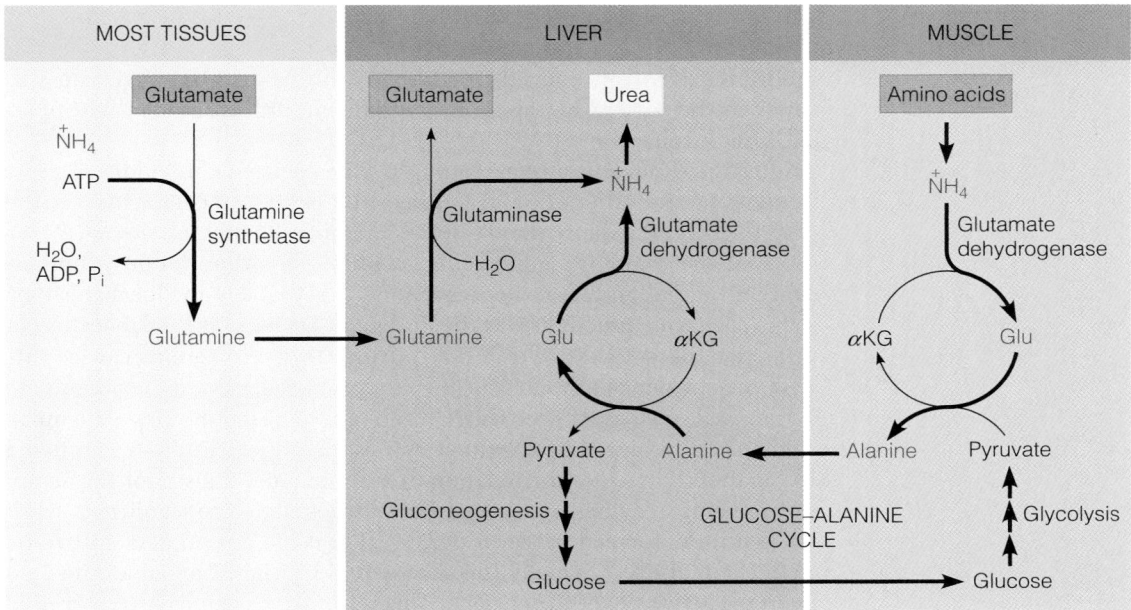

FIGURE 20.13

Transport of ammonia to the liver for urea synthesis. The carrier is glutamine in most tissues but is alanine in muscle.

Coenzymes Involved in Nitrogen Metabolism

Before presenting in detail the metabolism of amino acids and nucleotides, as we do in the next two chapters, we should consider three families of coenzymes that play major roles in amino acid and/or nucleotide metabolism. Although all have been mentioned before, we shall consider their actions in detail here. These cofactors include (1) pyridoxal phosphate, the cofactor for transamination and many other reactions of amino acid metabolism; (2) the folic acid coenzymes, which transfer single-carbon functional groups in synthesizing nucleotides and certain amino acids; and (3) the B_{12}, or cobalamin, coenzymes, which participate in the synthesis of methionine and, as noted in Chapter 17, the catabolism of methylmalonyl-CoA.

Pyridoxal Phosphate

Vitamin B_6 was discovered in the 1930s as the result of nutritional studies with rats fed vitamin-free diets. The vitamin as originally isolated is **pyridoxine**, named from its structural similarity to pyridine. Pyridoxine contains a hydroxymethyl group at position 4 of the pyridine ring. However, in the active coenzyme this group has been oxidized to an aldehyde and the hydroxymethyl group at position 5 is phosphorylated. Pyridoxal phosphate (which we shall abbreviate PLP) is the predominant coenzyme form, with pyridoxamine phosphate (PMP) being an intermediate in transamination reactions.

Marginal vitamin B_6 deficiency occurs frequently in humans and is associated with coronary artery disease, stroke, and an elevated risk of Alzheimer's disease. In addition, many drugs and poisons *induce* deficiency states, usually by reacting with the aldehyde group and thereby sequestering the coenzyme. A well-understood case is the induced B_6 deficiency arising during treatment of the mycobacterial infection known as tuberculosis. The antimycobacterial agent **isoniazid** (isonicotinic acid hydrazide) reacts covalently with pyridoxal

Pyridoxine

Pyridoxal

Pyridoxal phosphate (PLP)

Pyridoxamine phosphate (PMP)

All pyridoxal phosphate reactions involve initial Schiff base formation, followed by bond labilization caused by electron withdrawal to the coenzyme's pyridine ring.

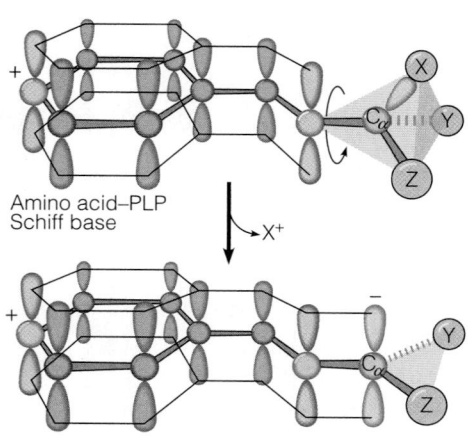

Enzyme-bound pyridoxal phosphate

Amino acid–PLP
Schiff base

Delocalized α carbanion

π Molecular orbital system of PLP-amino acid Schiff base

to make it unavailable for phosphorylation by pyridoxal kinase. Because the mycobacterium contains low levels of the kinase, its growth is effectively blocked by the agent. Prolonged treatment with this drug can generate a B_6 deficiency in the patient by the same mechanism, unless the diet is supplemented with the vitamin.

Pyridoxal phosphate is a remarkably versatile coenzyme. In addition to its involvement in transamination reactions, PLP serves as coenzyme for the majority of enzymes that catalyze some chemical change at the α-, β-, or γ-carbons of the common amino acids, including decarboxylations, eliminations, racemizations, and retro-aldol reactions. A key clue to the mechanism by which this coenzyme functions was the finding in Esmond Snell's laboratory in the 1940s that all of the known PLP-requiring enzyme reactions can be catalyzed, in the absence of any enzyme, by pyridoxal itself. Certain metal ions are required as well, such as Al^{3+} or Cu^{2+}. Although the reaction rates were much lower than those achieved by enzyme catalysis, the model studies permitted a detailed analysis that led to formulation of a unified mechanism for the action of PLP-requiring enzymes. The metal ion was postulated to stabilize a Schiff base or aldimine, formed between pyridoxal and the amino acid substrate, shown in the margin. Normally this role would be played by an amino acid residue in the active site of the enzyme. This is the same chemical strategy used in the fructose-1,6-bisphosphate aldolase reaction (Figure 13.5, page 528), in which a cationic imine (the Schiff base) lowers the energy barrier to the reaction. The ability of PLP to form a stable Schiff base is the key to its versatility in enzyme-catalyzed reactions.

Although pyridoxal phosphate is the coenzyme for all of these reactions, the reactive species is not the aldehyde group but rather an aldimine, formed between the coenzyme and an ε-amino group of a lysine residue in the active site. This bond can be reduced by sodium borohydride, to give irreversible bonding of the coenzyme to the active lysine residue. This finding allowed identification of the catalytic site on the enzyme and the specific lysine residue involved in coenzyme binding.

We now know that all pyridoxal phosphate–requiring enzymes act via the formation of a Schiff base between the amino acid and coenzyme (Figure 20.14). A cation, whether a metal (as in the nonenzymatic model system) or a proton (as in the enzymatic reaction), is essential to bridge the phenolate ion of the coenzyme and the imino nitrogen of the amino acid. This bridging maintains planarity of the structure allowing a large conjugated π molecular orbital system, which is essential for catalysis. The most important catalytic feature of the coenzyme is the electrophilic nitrogen of the pyridine ring, which acts as an *electron sink*, drawing electrons away from the amino acid and labilizing one of the three σ bonds on the α-carbon. The second important function is stabilizing the carbanion intermediate that results from bond cleavage. When an amino acid forms an imine with PLP, and the pyridinium N is protonated, all three σ bonds on the α-carbon become electron-deficient, and are susceptible to heterolytic cleavage. The σ bond that is aligned perpendicular to the plane of the π molecular orbital system is the one cleaved, and this is determined by the angle of rotation of the C_α—N bond, specified by interactions with the enzyme active site (see margin). *All of the known reactions of PLP enzymes can be described mechanistically in the same way*: formation of a planar Schiff base or aldimine intermediate, followed by bond cleavage and formation of a resonance-stabilized carbanion with a quinonoid structure, as shown in Figure 20.14. Depending on the bond labilized, formation of the aldimine can lead to transamination (as detailed in Figure 20.14), to decarboxylation, to racemization, or to retro-aldol cleavage (Figure 20.15).

For many years it was thought that PLP was the cofactor for all amino acid decarboxylases. However, a class of enzymes that uses *pyruvic acid* as a cofactor

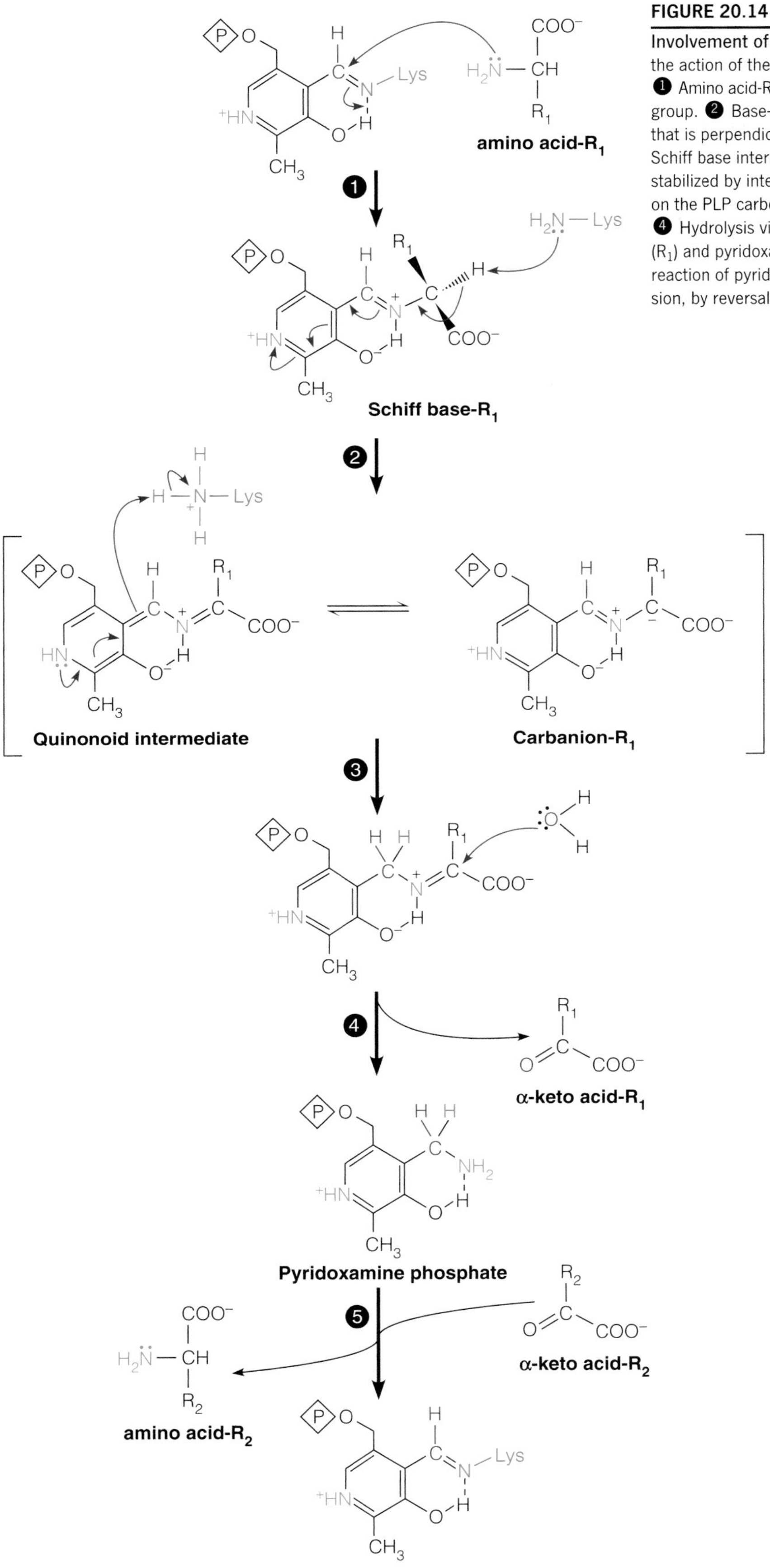

FIGURE 20.14

Involvement of pyridoxal phosphate in transamination. The figure shows the action of the positively charged pyridinium ion as an electron sink. ❶ Amino acid-R_1 reacts with the enzyme-bound PLP, displacing the lysine amino group. ❷ Base-catalyzed deprotonation (cleavage of the labile C—H σ bond that is perpendicular to the plane of the π molecular orbital system of the Schiff base intermediate) leads to formation of a carbanion, which is resonance stabilized by interconversion with a quinonoid intermediate. ❸ Reprotonation on the PLP carbon results in tautomerization of the imine C—N bond. ❹ Hydrolysis via a carbinolamine intermediate yields an α-keto acid product (R_1) and pyridoxamine phosphate. ❺ The transamination is completed by reaction of pyridoxamine phosphate with a second α-keto acid (R_2) and conversion, by reversal of steps 1–4, to enzyme-bound PLP and amino acid R_2.

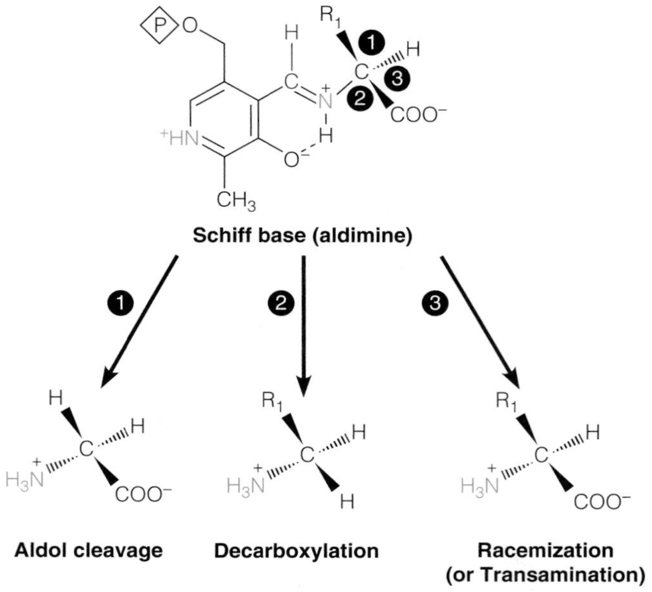

FIGURE 20.15

Versatility of the pyridoxal phosphate Schiff base (aldimine) in amino acid reactions. All three substituents at the α-carbon of the amino acid moiety can be labilized. Labilization of the R group ❶ leads to retro-aldol cleavage. Labilization of the carboxyl group ❷ leads to decarboxylation. Labilization of the hydrogen ❸ leads to racemization or transamination (as shown in Figure 20.14).

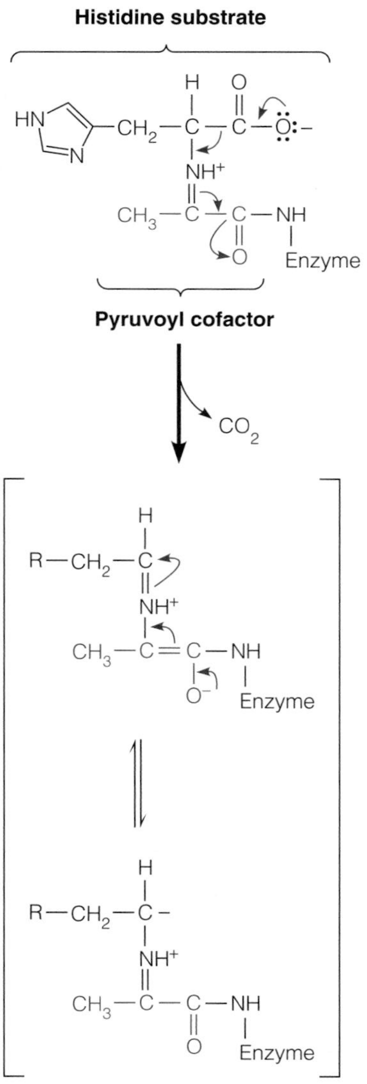

Schiff base (aldimine)

❶ **Aldol cleavage** ❷ **Decarboxylation** ❸ **Racemization (or Transamination)**

Histidine substrate

Pyruvoyl cofactor

is now known. Although structurally dissimilar to PLP, pyruvate serves a mechanistically similar role. The pyruvoyl group is formed from an internal serine residue in the proenzyme by a nonhydrolytic cleavage reaction and remains covalently bound to the new N-terminal residue via an amide linkage. During catalysis, the keto oxygen of pyruvate plays a role comparable to that of the PLP ring nitrogen, withdrawing electrons and resonance stabilizing the carbanion intermediate, as shown for the enzyme **histidine decarboxylase** of *Lactobacillus*.

Tetrahydrofolate Coenzymes and One-Carbon Metabolism

Discovery and Chemistry of Folic Acid

Coenzymes derived from the vitamin **folic acid** participate in the generation and utilization of single-carbon functional groups—methyl, methylene, and formyl. The vitamin was discovered by the British physician Lucy Wills in the 1930s, when she found that people with a certain type of **megaloblastic anemia** could be cured by treatment with yeast or liver extracts. The condition is characterized, like all anemias, by reduced levels of erythrocytes. The cells that remain are characteristically large and immature, suggesting a role for the vitamin in cell proliferation and/or maturation. The active component in the extracts was also shown to be essential for the growth of chicks and to be required in the growth media for certain bacteria, notably *Lactobacillus casei* and *Streptococcus faecium*. The latter findings allowed Esmond Snell to develop a rapid bioassay based on the growth of these bacteria, and isolation and structural identification soon followed. Snell and Herschel Mitchell, working with Roger Williams at the University of Texas, had to process *four tons* of spinach to obtain a few hundred micrograms of the active component. Because the vitamin was relatively abundant in leafy green vegetables such as spinach, it was named folic acid, from the Latin *folium* for leaf.

Chemically, folic acid is formed from three distinct moieties: (1) a bicyclic, heterocyclic **pteridine** ring, 6-methylpterin; (2) *p*-aminobenzoic acid (PABA), which is itself required for the growth of many bacteria; and (3) glutamic acid. These three moieties are shown in the overall structure:

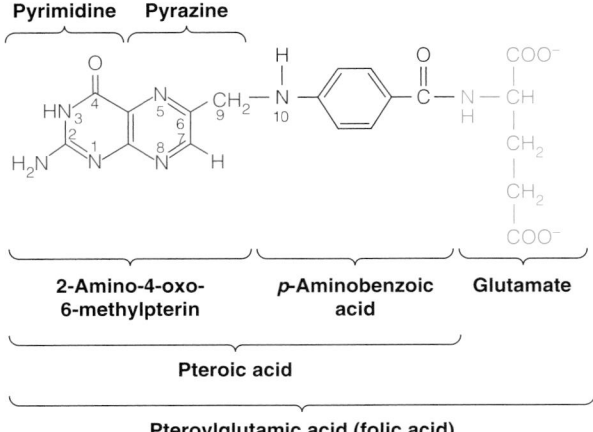

Pteroylglutamic acid (folic acid)

The pteridine ring was already known in nature, having been discovered in a large class of biological pigments. Insect wings and eyes contain pteridine pigments, as does the skin of amphibians and fish. Butterfly wings are particularly abundant in pteridines and were the first source from which any such compounds were identified structurally. These compounds are named after the Greek *pteron* ("wing"). Folic acid and its many derivatives are referred to as **folates**.

Conversion of Folic Acid to Tetrahydrofolate

Once inside a cell, folic acid is converted to active forms by two successive reductions of the pyrazine part of the pteridine ring. Both reactions are catalyzed by the NADPH-specific enzyme **dihydrofolate reductase**. The mechanism of this enzyme was described in Chapter 11 (page 430). The first reduction yields **7,8-dihydrofolate**, and the second reduction yields **5,6,7,8-tetrahydrofolate**.

Folate coenzymes contain multiple glutamate residues, which help them be retained within cells and bind more tightly to enzymes.

Folate (partial structure) → NADPH + H⁺ / NADP⁺ → **7,8-Dihydrofolate (DHF)** → NADPH + H⁺ / NADP⁺ → **Tetrahydrofolate (THF)**

For reasons that will become clear later, dihydrofolate is the preferred substrate, and hence its name is given to the enzyme. Note that reduction of the 5–6 double bond generates a new chiral center at C-6; the 6S-isomer of tetrahydrofolate is the naturally occurring form used by enzymes.

[6S]-Tetrahydrofolate

Pteridine | *p*-Aminobenzoic acid | Poly-γ-glutamate chain

Aminopterin (4-aminofolate)

Methotrexate (4-amino-10-methylfolate)

Trimethoprim

Pyrimethamine

Dihydrofolate reductase is the target for a number of useful anticancer, antibacterial, and antiparasitic drugs.

This structure illustrates two critical features of tetrahydrofolate. First, the N-5 and N-10 positions can carry single carbon units, designated as R_1 and R_2, for donation in biosynthetic processes (more about this below). Second, the naturally occurring intracellular folates contain a "tail" of glutamate residues, ranging from three to eight or more residues. These residues are linked to one another, not by the familiar peptide bond but rather by an amide bond between the γ-carboxyl group of the first glutamate and the α-amino group of the next. The glutamates are added, one at a time, in an ATP-dependent reaction catalyzed by **folylpoly-γ-glutamate synthetase**.

Most enzymes that use folate coenzymes bind more tightly to, and are more active with, polyglutamated forms than the monoglutamates. The additional glutamate residues are also important for intracellular retention of folates. Animal cells take up folates by active transport, but only the monoglutamated form is taken up efficiently. However, the monoglutamate forms can also be transported out of cells, so conjugation with additional glutamate residues converts the folate to a highly anionic form that cannot exit the cell.

Dihydrofolate reductase has been thoroughly studied because it is the target for action of a number of clinically useful **antimetabolites**. An antimetabolite is a synthetic compound, usually a structural analog of a normal metabolite, that interferes with the utilization of the metabolite to which it is related structurally. As early as 1948, two analogs of folate—**aminopterin** and **methotrexate**—had been synthesized and found to induce remissions in acute leukemias.

A decade later it was found that these compounds inhibit dihydrofolate reductase, binding to the enzyme at least 1000-fold more tightly than the normal substrates do. Thus, these analogs block the utilization of folate and dihydrofolate. We now know that their effectiveness derives from the involvement of dihydrofolate reductase in the biosynthesis of thymine nucleotides and, hence, of DNA. Inhibiting DNA synthesis blocks the proliferation of cancer cells, as discussed further in Chapter 22. Nearly two more decades passed before a detailed understanding of the mechanism of dihydrofolate reductase inhibition became available, through crystallization of enzyme–inhibitor complexes and determination of their three-dimensional structure (Figure 20.16).

Folate analogs such as methotrexate have been used in treating many different cancers in addition to leukemia. Other clinically useful dihydrofolate reductase inhibitors show selectivity among species-specific forms of the enzyme. **Trimethoprim** specifically inhibits bacterial dihydrofolate reductases and is widely used to treat bacterial infections, and **pyrimethamine** shows similar specificity against the enzyme of protozoal origin.

The entire concept of antimetabolites as drugs arose from early work on folate metabolism. Before World War II, one of the few effective antibacterial drugs available was **sulfanilamide**, one of the class of sulfonamide drugs ("sulfa drugs"). A British biochemist, D. D. Woods, noted a structural similarity between sulfanilamide and p-aminobenzoate, which was known to be essential for bacterial growth. Before anything was known about the relation between PABA and folic acid, Woods proposed that sulfanilamide acts by blocking the normal utilization of PABA, and he coined the term *antimetabolite*. PABA is not required for growth of animal cells, so the drug is not toxic to human cells. Years later, when the pathway of folate biosynthesis had been established, it was learned that Woods was right; the enzyme incorporating PABA is inhibited by sulfonamides. Because animal cells do

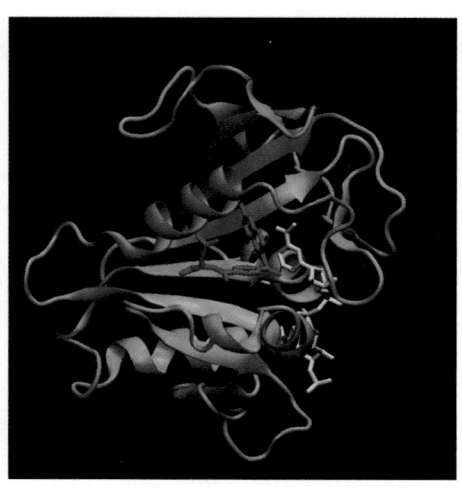

FIGURE 20.16

Human dihydrofolate reductase complexed with ligands. This figure shows the X-ray crystal structure of the human enzyme with bound NADPH (yellow) and methotrexate (red) (PDB ID 1U72). The additional amino group on methotrexate allows formation of an additional hydrogen bond to the enzyme, which increases its binding affinity to the folate binding site. These two ligands bind such that the pyridine ring of NADPH is in close proximity to the pteridine ring of the folate ligand, as required for the hydride transfer catalyzed by this enzyme. Notice that the folate binds in a bent, rather than linear, form.

not carry out the synthetic pathway but instead take up fully formed folate from the diet, they are not harmed by the drug. The concept elucidated by Woods, of seeking a metabolic difference between normal cells and pathological cells—infecting parasites, virus-infected cells, or cancer cells—and of exploiting that difference chemically, has had an enormous impact on the field of pharmacology.

Tetrahydrofolate in the Metabolism of One-Carbon Units

The coenzymatic function of tetrahydrofolate (**THF**) is the mobilization and utilization of single-carbon functional groups (one-carbon units). These reactions are involved in the metabolism of serine, glycine, methionine, and histidine, among the amino acids, and in the biosynthesis of purine nucleotides and the methyl group of thymine.

Tetrahydrofolate carries one-carbon units at the methyl, methylene, and formyl oxidation levels, equivalent in oxidation level to methanol, formaldehyde, and formic acid, respectively (Figure 20.17). One-carbon groups on THF can be carried on N-5 or N-10, or bridged between N-5 and N-10. The THF derivatives are named according to the oxidation state of the one-carbon unit and the nitrogen positions to which it is attached. Thus, **5,10-methylenetetrahydrofolate** (5,10-methylene-THF) carries a methylene group ($—CH_2—$) attached to N-5 and N-10:

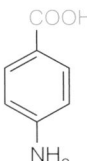

Sulfanilamide

***p*-Aminobenzoic acid (PABA)**

Tetrahydrofolate coenzymes transfer and interconvert one-carbon units at the methyl, methylene, and formyl oxidation levels.

5,10-methylenetetrahydrofolate

One-carbon units attached to tetrahydrofolate are activated for the formation of new carbon bonds ($C—S$, $C—N$, or $C—C$) in various biosynthetic reactions. The most reduced form, **5-methyltetrahydrofolate** (5-methyl-THF), donates its one-carbon unit to just one acceptor—homocysteine, forming the terminal $C—S$ bond of methionine (reaction 1 in Figure 20.17). The one-carbon unit carried by 5,10-methylenetetrahydrofolate is used to form new $C—C$ bonds, as in reactions 3, 4, and 5. The most oxidized form, **10-formyltetrahydrofolate** (10-formyl-THF), forms new $C—N$ bonds. Because the formyl group is being transferred from the N-10 of tetrahydrofolate to an N atom on the acceptor molecule, enzymes that use 10-formyltetrahydrofolate as a one-carbon donor are called **transformylases** or **formyltransferases**.

Tetrahydrofolate can acquire one-carbon units from diverse sources. For example, many cells carry out the ATP-dependent activation of formate to 10-formyltetrahydrofolate (reaction 8 in Figure 20.17). The degradation of histidine, in both bacterial and animal cells, provides one-carbon units at the level of methenyltetrahydrofolate via 5-formiminotetrahydrofolate (reactions 9 and 10). The bacterial fermentation of purines also produces 5-formiminotetrahydrofolate. However, most organisms derive the majority of their activated one-carbon units from the β-carbon of serine and the subsequent oxidation of glycine (reactions 3 and 4 in Figure 20.17). The first of these reactions is catalyzed by **serine hydroxymethyltransferase**:

$$serine + tetrahydrofolate \xrightleftharpoons{PLP} glycine + 5,10\text{-methylenetetrahydrofolate} + H_2O$$

This is a reversible reaction that, in the direction shown, yields glycine and 5,10-methylenetetrahydrofolate, but it can also be used for serine biosynthesis as needed. The enzyme also requires pyridoxal phosphate, so the actual substrate is the Schiff base formed from serine and PLP.

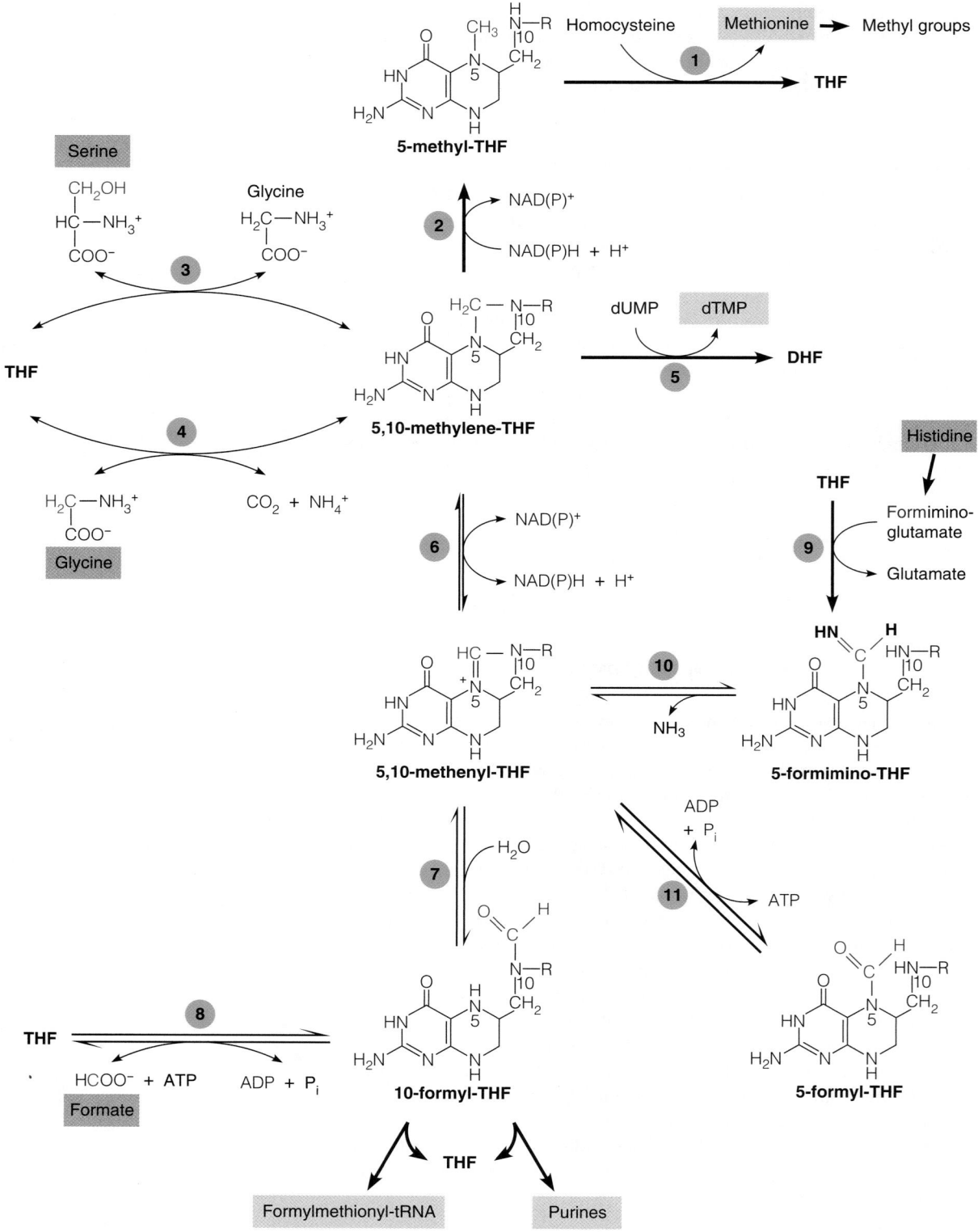

FIGURE 20.17

Metabolic reactions involving synthesis, interconversion, and utilization of one-carbon adducts of tetrahydrofolate. Major end products of one-carbon metabolism are highlighted in pink, and major sources of one-carbon units are highlighted in orange. The enzymes involved are ❶ homocysteine methyltransferase (also called methionine synthase), ❷ methylenetetrahydrofolate reductase, ❸ serine hydroxymethyltransferase, ❹ glycine cleavage system, ❺ thymidylate synthase, ❻ methylenetetrahydrofolate dehydrogenase, ❼ methenyltetrahydrofolate cyclohydrolase, ❽ 10-formyltetrahydrofolate synthetase, ❾ glutamate formiminotransferase, ❿ 5-formiminotetrahydrofolate cyclodeaminase, and ⓫ 5-formyltetrahydrofolate cycloligase (also called methenyltetrahydrofolate synthetase). THF = tetrahydrofolate, DHF = dihydrofolate, R = PABA-glutamate.

Glycine can yield an additional molecule of 5,10-methylenetetrahydrofolate through action of the **glycine cleavage system**, a multienzyme complex located in mitochondria:

$$\text{glycine} + \text{tetrahydrofolate} + NAD^+ \underset{}{\overset{\text{PLP, FAD, lipoamide}}{\rightleftharpoons}} \text{5,10-methylenetetrahydrofolate} + CO_2 + NH_3 + NADH + H^+$$

This reaction represents the chief catabolic route for glycine in most organisms. The overall reaction is as shown, and the pathway is mechanistically similar to that catalyzed by the pyruvate dehydrogenase complex (see Chapter 14). Four subunits are involved: P-protein, which catalyzes the PLP-dependent decarboxylation of glycine; T-protein, which catalyzes a THF-dependent aminomethyl transfer; L-protein, an FAD-dependent lipoamide dehydrogenase; and H-protein, a hydrogen carrier protein containing lipoic acid. The P-, T-, and H-protein subunits are unique to the glycine cleavage system, whereas the L-protein is shared with other mitochondrial α-ketoacid dehydrogenases, including the pyruvate dehydrogenase and α-ketoglutarate dehydrogenase complexes. The serine hydroxymethyltransferase and glycine cleavage reactions also compose that part of the photorespiration pathway that occurs in plant mitochondria (Chapter 16, page 701).

Once a one-carbon unit has been activated via its attachment to tetrahydrofolate, it can undergo interconversions, such as a change in oxidation state, or it can be used directly in a biosynthetic reaction. Figure 20.17 shows most of the known reactions involving tetrahydrofolate coenzymes. Note the reactions that involve a change in the oxidation level of the bound one-carbon unit: the reversible oxidation of 5,10-methylene-THF to 5,10-methenyl-THF, catalyzed by **5,10-methylenetetrahydrofolate dehydrogenase** (reaction 6), and the irreversible reduction of 5,10-methylene-THF to the 5-methyl derivative, brought about by a flavin-dependent **5,10-methylenetetrahydrofolate reductase** (reaction 2). Many organisms contain multifunctional enzymes or complexes that help channel these scarce and/or unstable intermediates. For example, reactions 6, 7, and 8 are combined into a single trifunctional protein called C_1-THF synthase in eukaryotes, and reactions 9 and 10 are found as a bifunctional enzyme in mammals. In addition, several of these reactions occur in both the cytoplasm and mitochondria.

As shown in Figure 20.17, one-carbon units derived from THF coenzymes are used in synthesis of purine nucleotides, thymine nucleotides (dTMP), and methionine, in addition to the reactions we have discussed. Furthermore, in prokaryotes and in eukaryotic mitochondria, 10-formyl-THF participates in synthesis of **N-formylmethionyl-tRNA**, which is involved in initiation of protein synthesis (discussed further in Chapter 28). In the synthesis of thymine nucleotides, catalyzed by **thymidylate synthase** (reaction 5), the THF coenzyme serves both as a one-carbon donor and as a source of reducing power. Because this enzyme generates the methyl group of thymine from 5,10-methylene-THF, it catalyzes both a one-carbon transfer and a reduction. The electrons come from the reduced pteridine ring, to give dihydrofolate as a product. Although dihydrofolate reductase can act on either folate or dihydrofolate, the reduction of dihydrofolate is more significant in vivo than that of folate because of the need for constant regeneration of tetrahydrofolate from dihydrofolate produced in the thymidylate synthase reaction. We will discuss the chemistry of the thymidylate synthase reaction in more detail in Chapter 22.

B$_{12}$ Coenzymes

Vitamin B$_{12}$ was discovered through studies of a formerly incurable disease, pernicious anemia. This condition begins with a megaloblastic anemia, which is virtually identical to that seen in folate deficiency but which leads to an irreversible degeneration of the nervous system if untreated. In 1926 two Harvard physicians, George Minot and William Murphy, found that symptoms of the disease could be alleviated by feeding patients large amounts of raw liver. The active material in the

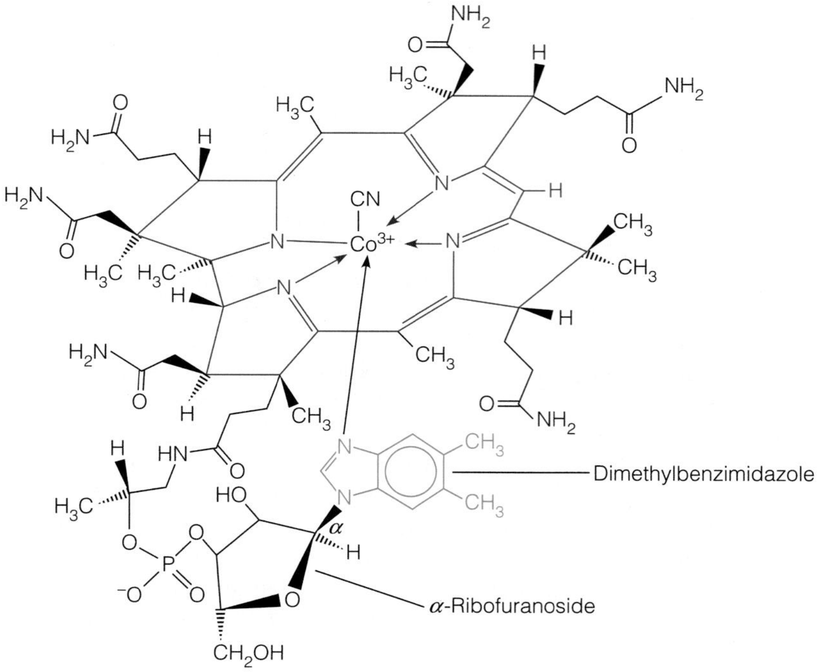

FIGURE 20.18

Structure of vitamin B_{12}. The molecule shown here is the cyanide-containing form originally isolated (cyanocobalamin). In cells, a water molecule or hydroxyl group takes the place of CN, forming the precursor to the coenzyme forms of B_{12}. The corrin ring is shown in magenta. 5,6-Dimethylbenzimidazole (DMB), which is linked to the cobalt, is shown in blue.

β-Methylaspartate

●

B_{12} coenzymes have either a methyl group or a 5′-deoxyadenosyl moiety linked to cobalt, making them the first known organometallics in metabolism.

liver, which was named vitamin B_{12}, was present in exceedingly small amounts, so many years passed until sufficient material had been isolated for characterization. In England in 1956, Dorothy Hodgkin and her colleagues used X-ray crystallography to complete the structure determination for this active substance. Hodgkin was awarded the Nobel Prize for this work.

The structure of vitamin B_{12} is shown in Figure 20.18. The metal cobalt is coordinated with a tetrapyrrole ring system, called a **corrin** ring, which is similar to the porphyrin ring of heme compounds. The cobalt is also linked to a heterocyclic base, 5,6-dimethylbenzimidazole (DMB). In the vitamin as isolated, the sixth coordination position of the cobalt is occupied by cyanide ion, but this ion is introduced during isolation. The vitamin as it exists in tissues contains either water or a hydroxyl moiety at this site. Because of the presence of cobalt and many amide nitrogens, B_{12} compounds are called **cobamides** or, more commonly but less accurately, **cobalamins**. B_{12} derivatives are also named in terms of the group occupying the sixth coordination position. Thus, the vitamin as isolated is **cyanocobalamin**, and the intracellular forms are **aquocobalamin** or **hydroxocobalamin**.

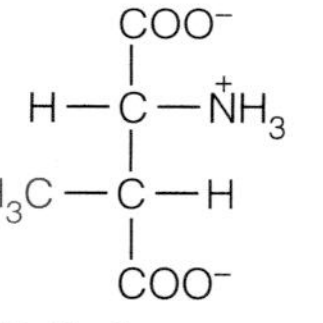

Coenzyme Forms of B_{12}

Two coenzymatically active forms of B_{12} are known to exist, in which the upper axial cobalt ligand is either a 5′-deoxyadenosyl or a methyl group (Figure 20.19). The first, **5′-deoxyadenosylcobalamin**, was discovered in 1958 by H. A. Barker. The bacterium *Clostridium cylindrosporum* ferments glutamate by first isomerizing it to β-methylaspartate. Barker and his colleagues showed that 5′-deoxyadenosylcobalamin is essential for this reaction to occur. Shortly afterward, studies of the enzymatic synthesis of methionine from **homocysteine** revealed the existence of a second coenzymatically active form, **methylcobalamin**, or methyl-B_{12}. Both 5′-deoxyadenosylcobalamin and methylcobalamin contain a covalent carbon–cobalt bond, making them true organometallics. In methionine synthesis the coenzyme transfers its methyl group and the cobalt cycles between the $+3$ (Co^{III}) and $+1$ (Co^{I}) oxidation states. Note that the methyl group is derived ultimately from 5-methyltetrahydrofolate (reaction 1, Figure 20.17).

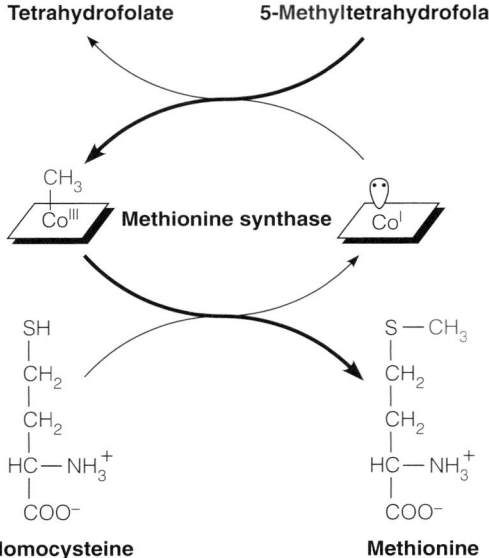

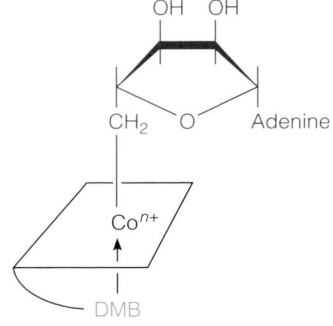

5'-Deoxyadenosyl-B$_{12}$
(5'-Deoxyadenosylcobalamin)

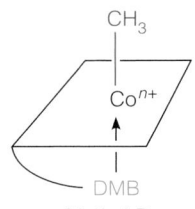

Methyl-B$_{12}$
(Methylcobalamin)

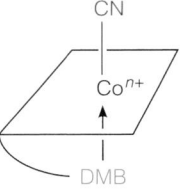

Vitamin B$_{12}$ as
originally isolated
(Cyanocobalamin)

FIGURE 20.19

Coenzymes derived from vitamin B$_{12}$. The corrin ring, identical in all known forms of B$_{12}$, is indicated here schematically. The Co bears a positive charge ($n = 1, 2,$ or 3), while each molecule is uncharged overall.

Three classes of B$_{12}$-dependent enzymes are now known—isomerases, methyltransferases, and reductive dehalogenases. Most of these occur in only a few bacterial species that carry out specialized fermentations. Also of interest is the involvement of methyl-B$_{12}$ in methane synthesis by methanogenic bacteria. Only two B$_{12}$-dependent reactions occur to a significant extent in mammalian metabolism: the synthesis of methionine from homocysteine, which we have shown here, and the isomerization of methylmalonyl-CoA to succinyl-CoA, which we introduced in Chapter 17 (page 732) as a key step in the oxidation of odd-chain fatty acids. As we saw with the methionine synthase reaction, B$_{12}$-dependent methyltransferases use methylcobalamin as the cofactor. The role of the B$_{12}$ coenzyme in the reductive dehalogenases is poorly understood. The isomerases catalyze 5'-deoxyadenosylcobalamin-dependent rearrangements involving exchange of a carbon-bound hydrogen with another carbon-bound functional group, as shown here; methylmalonyl-CoA mutase catalyzes a reaction of this type.

The migrating group (X) can be carbon, —NH$_2$, or —OH.

Action of Adenosylcobalamin

Carbon–carbon bonds are generally difficult to break and to form. B$_{12}$-dependent enzymes readily catalyze such reactions, and this activity, plus the novel cobalt–carbon bond, has focused attention on mechanisms of action of 5'-deoxyadenosylcobalamin. The carbon–cobalt bond is relatively weak, with a bond dissociation energy on the order of 140 kJ/mol (compared with 348 kJ/mol for a typical C—C bond). In fact, exposure to light easily breaks this Co—C bond, explaining the extreme light sensitivity of the B$_{12}$ coenzymes. 5'-deoxyadenosylcobalamin-dependent isomerases exploit the lability of the Co—C bond by generating free radical intermediates during catalysis. From studies on methylmalonyl-CoA mutase and other enzymes, the following facts have emerged: (1) Hydrogen transfer is stereospecific; some of the reactions proceed with inversion of configuration, and others do not. (2) Hydrogen that is transferred does not exchange with the protons of water; isotopic hydrogen present in a

FIGURE 20.20

The intramolecular isomerization catalyzed by methylmalonyl-CoA mutase. This proposed mechanism is consistent with experimental observations and is representative of the general mechanism for other B_{12}-dependent 1,2-isomerizations. **Step 1:** Homolytic cleavage of carbon–cobalt bond and formation of 5′-deoxyadenosyl radical. **Step 2:** Hydrogen abstraction and formation of substrate radical. **Steps 3 and 4:** 1,2 rearrangement involving a cyclopropyloxy radical. **Step 5:** Hydrogen abstraction from 5′deoxyadenosine to give succinyl-CoA and 5′-deoxyadenosyl radical. **Step 6:** Reformation of carbon–cobalt bond.

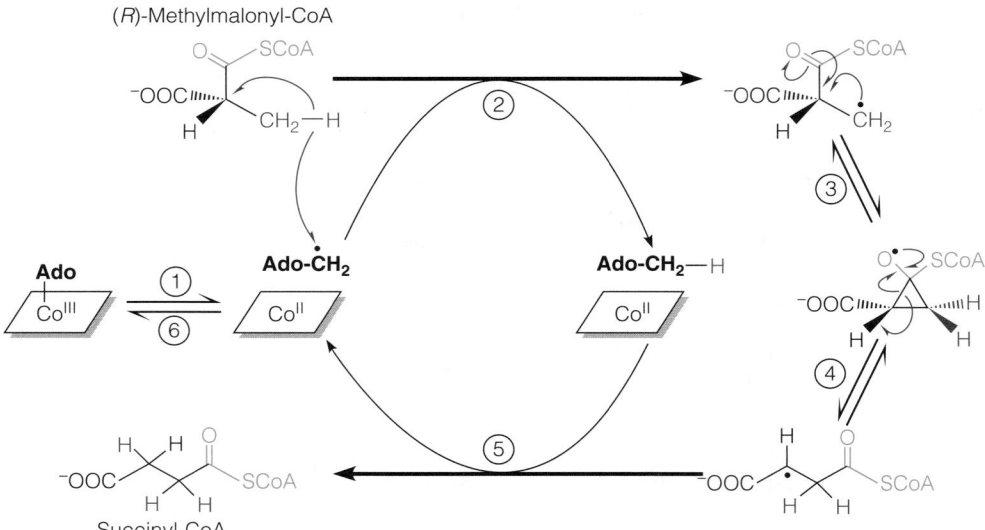

substrate is conserved in the product. (3) The transferred hydrogen is transiently carried on the 5′ carbon of the deoxyadenosyl moiety; labeling of this position with tritium leads to incorporation of that label into product. (4) Spectral studies indicate that the cobalt changes its oxidation state during catalysis.

All of these observations imply that the covalent carbon–cobalt bond on the 5′-deoxyadenosylcobalamin undergoes transient **homolytic** cleavage during catalysis. That is, the cobalt and the carbon each acquire one electron from the pair that formed the bond, creating a free radical at the deoxyadenosine C-5. Interaction with the substrate then creates a substrate radical, as shown for methylmalonyl-CoA mutase in Figure 20.20, leading to rearrangement of the substrate.

X-ray crystal structures of several B_{12}-dependent enzymes have revealed two different modes of coenzyme binding. In some enzymes, such as methylcobalamin-dependent methionine synthase and 5′-deoxyadenosylcobalamin-dependent methylmalonyl-CoA mutase, the coenzyme undergoes a major conformational change upon binding to the enzyme. As shown in Figure 20.21a, the DMB-containing "tail" is no longer coordinated to the cobalt. Instead, the tail has moved into a deep pocket, which helps bind the coenzyme tightly. Replacing the DMB in coordination with cobalt is a histidine residue on the enzyme, indicating a role for the enzyme in stabilizing reaction intermediates. In other B_{12}-dependent enzymes, such as diol dehydratase (Figure 20.21b) and ribonucleotide reductase, the DMB moiety remains coordinated to the cobalt in the lower axial position, as observed for the free coenzyme (Figure 20.18).

FIGURE 20.21

Two modes of coenzyme B_{12} binding. The cobalamin coenzyme is in red, with its corrin ring viewed from the edge. **(a)** The crystal structure of methylmalonyl-CoA mutase from *Propionibacterium shermanii* and bound coenzyme (PDB ID 4REQ), showing the DMB buried in a deep crevice. A histidine residue on the enzyme (yellow) contacts the upper face of the corrin ring, in place of DMB. **(b)** The crystal structure of diol dehydratase from *Klebsiella oxytoca* and bound coenzyme (PDB ID 1DIO), with DMB coordinated to the cobalt.

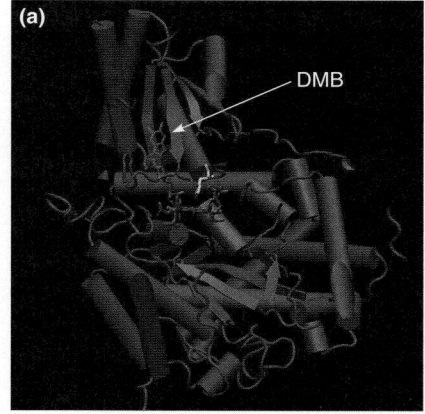

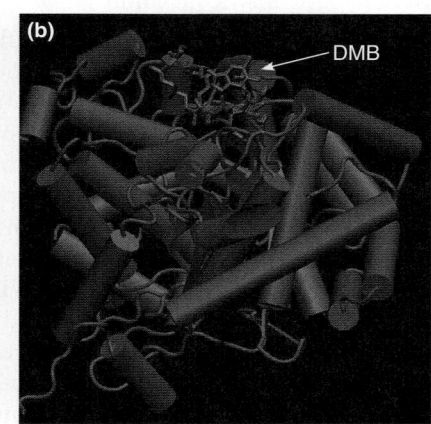

B$_{12}$ Coenzymes and Pernicious Anemia

Now let us return to the role of B$_{12}$ coenzymes in mammalian metabolism. Recall that vitamin B$_{12}$ was isolated as a factor that could cure pernicious anemia, which suggested that the disease is caused by B$_{12}$ deficiency. In fact, pernicious anemia is a disease of the stomach. Gastric tissue secretes a glycoprotein called **intrinsic factor**, which complexes with ingested B$_{12}$ in the digestive tract and promotes its efficient absorption through the terminal portion of the small intestine into the bloodstream. Pernicious anemia results from insufficient secretion of intrinsic factor. This is usually caused by an autoimmune process in which the body destroys the gastric lining cells that produce intrinsic factor. Indeed, patients who undergo surgical removal of the stomach for cancer or other problems can also develop the symptoms of pernicious anemia. The uncomplexed vitamin can be absorbed, but so poorly that massive doses must be administered to cure or prevent the disease.

But what is the relationship between failure to absorb B$_{12}$ and the defect in red blood cell formation that defines anemias? Given the hematologic similarity between anemias of folate deficiency and pernicious anemia, a relationship between folate and B$_{12}$ metabolism has long been suspected. In fact, the megaloblastic anemia seen in early stages of pernicious anemia is improved if folic acid is administered. However, this treatment only hastens the onset of the far more severe neurological symptoms of B$_{12}$ deficiency. Given that these neurological problems are never seen in simple folate deficiency, what is the metabolic relationship between these two vitamins?

Figure 20.22 outlines our current understanding: (1) When B$_{12}$ levels are low, flux through the methionine synthase reaction (reaction 1, Figure 20.17) decreases, but because adequate dietary methionine is usually available, there is no immediate disturbance of protein metabolism. (2) Reduction of 5,10-methylene-THF to 5-methyl-THF (reaction 2, Figure 20.17) continues because this reaction is virtually irreversible. (3) Because methionine synthase is the only mammalian enzyme known to use 5-methyl-THF, the decreased intracellular activity of this enzyme causes 5-methyl-THF to accumulate, at the expense of depleted pools of the other tetrahydrofolate coenzymes. In essence, the intracellular THF pool becomes "trapped" as 5-methyl-THF. Thus, even though total folate levels may seem ample, there is a *functional* folate deficiency, with insufficient levels of the formyl-THF and methylene-THF coenzymes needed for synthesis of nucleic acid precursors.

The methyl-trap hypothesis does not explain why untreated pernicious anemia progresses to a neurological disease because simple folate deficiency anemias show no such complications. Early observations suggested that abnormal fatty acid metabolism caused by inhibition of methylmalonyl-CoA mutase was responsible. The problem is extraordinarily difficult because dietary B$_{12}$ requirements are so low that generating animal models for B$_{12}$ deficiency is almost impossible. Some insight has been gained through the observation that the

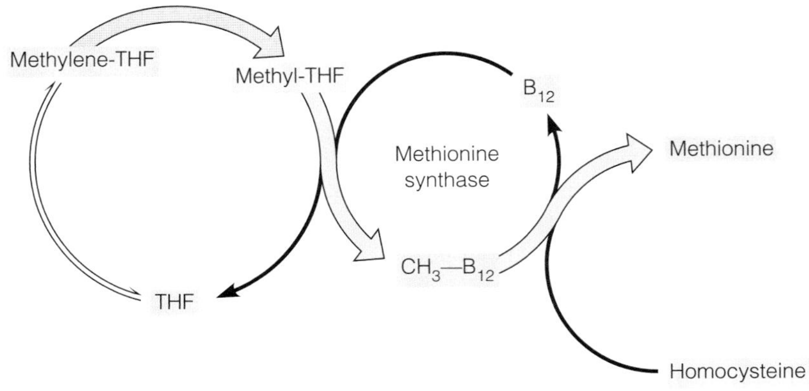

Pernicious anemia is caused by deficiency of a glycoprotein needed for intestinal absorption of vitamin B$_{12}$, leading to intracellular deficiencies of B$_{12}$ coenzymes.

B$_{12}$ deficiency causes 5-methyltetrahydrofolate to accumulate, with concomitant depletion of other folate coenzymes.

FIGURE 20.22

A relationship between folate and B$_{12}$ metabolism. This scheme is based on the apparent folate deficiency seen in early stages of B$_{12}$ deficiency as a result of decreased flux through the methionine synthase reaction. The diagram identifies intermediates that either accumulate (blue) or are depleted (gold).

commonly used anaesthetic nitrous oxide (N_2O), or laughing gas, induces a pernicious anemia–like state by inactivating methionine synthase. N_2O oxidizes the cobalt in B_{12} from the $+1$ (Co^I) to the $+2$ (Co^{II}) state and is converted to a potent oxidant that irreversibly damages the enzyme. Because methylmalonyl-CoA mutase does not use the Co^I oxidation state of cobalamin, it is resistant to N_2O treatment. These studies might argue that inhibition of methionine synthesis is responsible for the neurological dysfunction in pernicious anemia, but the biochemical mechanisms involved are not yet understood.

Folic Acid in the Prevention of Heart Disease and Birth Defects

In the mid-1990s, a series of clinical reports described correlations between folate deficiencies and increased risk of myocardial infarction. The same studies revealed that individuals at risk for heart attack also showed abnormally high levels of serum homocysteine. The simplest interpretation is that in folate-deficient individuals, decreased levels of tetrahydrofolate cofactors limit metabolic flux through the methionine synthase reaction (reaction 1, Figure 20.17), with consequent accumulation of homocysteine, the substrate for this enzyme. Elevated plasma homocysteine (homocysteinemia) is now considered to be a major independent risk factor for many types of cardiovascular disease, including coronary artery disease, stroke, and peripheral vascular occlusive disease. Homocysteine is presumed to be the toxic metabolite responsible for damage to the heart, although the mechanisms are not known, and, indeed, some studies have failed to show a correlation between folate status and heart disease. However, folate deficiencies have other recently recognized biological consequences, including abnormally high levels of uracil in DNA. As we discuss in Chapter 22, this phenomenon, which can lead to chromosome breakage, is a consequence of limitation of the biosynthesis of thymine nucleotides. Folate deficiency during embryogenesis causes a significant proportion of birth defects, including those affecting the craniofacies (e.g., cleft palate), the neural tube (e.g,. anencephaly and spina bifida), and the heart. The normal development of all three of these regions depends on the proper growth, differentiation, and migration of a set of multipotent neural crest cells. These cells apparently have a high demand for folate. Consequently, women are urged to take folic acid supplements throughout their pregnancies, but especially in the early stages, when the fetus's nervous system develops most rapidly. For example, neural tube development and closure is usually complete by 28 days post-conception. Neural tube defects (NTDs) occur with a prevalence of about 1 in 1,000 births, but periconceptional folic acid supplementation prevents 50%–75% of NTDs. Because the window during which folic acid supplementation is effective generally occurs before a woman discovers she is pregnant, the United States began fortifying enriched flour and other cereal grain products with folic acid to ensure that women of childbearing age have adequate folate levels. The prevalence of NTDs has indeed declined in the United States since 1998, when fortification became mandatory.

Folic acid deficiency increases the risk of cardiovascular disease and birth defects in humans.

SUMMARY

Although inorganic nitrogen is abundant, metabolism of most organisms is limited by nitrogen bioavailability. Reduction of N_2 in biological nitrogen fixation and reduction of nitrate in plant and bacterial metabolism generate ammonia, which all organisms can utilize. The capacity for amino acid synthesis varies greatly among organisms, with mammals requiring about half of the 20 common amino acids in the diet. Proteins are in a continual state of turnover and replacement, partly for replacement of damaged proteins and partly as the result of normal cellular regulatory mechanisms. Most amino acids released by protein turnover are reutilized for protein synthesis. When amino acids are degraded, either for catabolism of an oversupply or when needed for energy generation, the first step is usually removal of the α-amino group, either through transamination or oxidative deamination. The resultant ammonia is excreted directly (in fish), converted to uric acid (in most reptiles, insects, and birds), or converted to urea (in mammals). Urea synthesis is a cyclic pathway involving ornithine and arginine as intermediates. Transamination and numerous additional reactions undergone by amino acids use pyridoxal phosphate as a coenzyme. After condensation of the amino group of the amino acid with the aldehyde of the coenzyme to give a Schiff base, the pyridine ring of the coenzyme withdraws electrons transiently and destabilizes bonds that are broken in the reaction. Tetrahydrofolate binds one-carbon units at three different oxidation states, interconverts them, and transfers them in the synthesis of purine nucleotides, thymidine nucleotides, and several amino acids. B_{12} coenzymes include methylcobalamin, which participates in methionine biosynthesis, and 5'-deoxyadenosylcobalamin, the coenzyme for methylmalonyl-CoA mutase. Folate metabolism presents various chemotherapeutic targets, and folate and B_{12} deficiencies both have important clinical consequences.

REFERENCES

Inorganic Nitrogen Fixation

Hakoyama, T., K. Niimi, H. Watanabe, R. Tabata, J. Matsubara, S. Sato, Y. Nakamura, S. Tabata, L. Jichun, T. Matsumoto, K. Tatsumi, M. Nomura, S. Tajima, M. Ishizaka, K. Yano, H. Imaizumi-Anraku, M. Kawaguchi, H. Kouchi, and N. Suganuma (2009) Host plant genome overcomes the lack of a bacterial gene for symbiotic nitrogen fixation. *Nature* 462:514–517. Describes the molecular basis of the symbiotic partnership between legumes and rhizobia.

Masson-Boivin, C., E. Giraud, X. Perret, and J. Batut (2009) Establishing nitrogen-fixing symbiosis with legumes: How many rhizobium recipes? *Trends Microbiol.* 17:458–466. A contemporary view of the complexities of control of symbiotic nitrogen fixation.

Schwarz, G., and R. R. Mendel (2006) Molybdenum cofactor biosynthesis and molybdenum enzymes. *Annu. Rev. Plant Biol.* 57:623–647. Reviews the structure and chemistry of nitrate reductase.

Seefeldt, L. C., B. M. Hoffman, and D. R. Dean (2009) Mechanism of Mo-dependent nitrogenase. *Annu. Rev. Biochem.* 78:701–722. Reviews all of the known molybdenum-requiring reactions, with emphasis upon structures of the proteins involved.

General Aspects of Nitrogen Metabolism

Braissant, O. (2010) Current concepts in the pathogenesis of urea cycle disorders. *Mol. Genet. Metab.* 100 Suppl 1:S3–S12.

Brusilow, S. W., and A. L. Horwich (2001) Urea cycle enzymes. In *The Metabolic and Molecular Bases of Inherited Disease*, C. R. Scriver, A. L. Beaudet, W. S. Sly, D. Valle, B. Childs, K. W. Kinzler, and B. Vogelstein, eds., Vol. II, Ch. 85, pp. 1909–1963, McGraw-Hill, New York. Genetic disorders of urea cycle enzymes are reviewed in this chapter of the definitive work on inherited metabolic disorders.

Hallows, W. C., B. C. Smith, S. Lee, and J. M. Denu (2009) Ure(k)a! Sirtuins Regulate Mitochondria. *Cell* 137:404–406. This minireview describes recent discoveries of the role of reversible acetylation of CPS I in regulation of the urea cycle.

Ninfa, A. J., and P. Jiang (2005) PII signal transduction proteins: Sensors of alpha-ketoglutarate that regulate nitrogen metabolism. *Curr. Opin. Microbiol.* 8:168–173. This short review describes the role of covalent modification in the regulation of bacterial glutamine synthetase.

Smith, T. J., and C. A. Stanley (2008) Untangling the glutamate dehydrogenase allosteric nightmare. *Trends Biochem. Sci.* 33:557–564. This minireview discusses the evolution of the regulation of glutamate dehydrogenase in animals.

Walsh, C. T. (1979) *Enzymatic Reaction Mechanisms.* Freeman, San Francisco. A classic book, particularly valuable in the context of amino acid metabolism, one-carbon metabolism, cobalamin coenzymes, and oxygenases.

Protein Turnover

Ciechanover, A. (2009) Tracing the history of the ubiquitin proteolytic system: The pioneering article. *Biochem. Biophys. Res. Commun.* 387:1–10. This reminiscence by one of the discoverers of the ubiquitin system provides an interesting account of the prevailing views on protein turnover that had hindered progress, and the breakthrough that finally led to the discovery.

Finley, D. (2009) Recognition and processing of ubiquitin-protein conjugates by the proteasome. *Annu. Rev. Biochem.* 78:477–513. This article reviews the structure of the proteasome and how substrates are recognized and degraded by this intricate molecular machine.

Schoenheimer, R., S. Ratner, and D. Rittenberg (1939) Studies in protein metabolism. VII. The metabolism of tyrosine. *J. Biol. Chem.* 127:333–344. This classic paper describes one of the earliest uses of an isotopic tracer to study metabolism in whole animals (indeed, this experiment was performed on a single rat!).

Schwartz, A. L., and A. Ciechanover (2009) Targeting proteins for destruction by the ubiquitin system: Implications for human pathobiology. *Annu. Rev. Pharmacol. Toxicol.* 49:73–96. This review discusses the role of the ubiquitin proteolytic system in human disease.

Varshavsky, A. (2008) The N-end rule at atomic resolution. *Nat. Struct. Mol. Biol.* 15:1238–1240. This short article reviews the N-end rule and summarizes recent structural information on a bacterial N-end rule recognition component.

Welchman, R. L., C. Gordon, and R. J. Mayer (2005) Ubiquitin and ubiquitin-like proteins as multifunctional signals. *Nat. Rev. Mol. Cell. Biol.* 6:599–609. This review discusses the roles of some of the other ubiquitin-like proteins, including SUMO.

Folate and B$_{12}$ Coenzymes

Bailey, L. B., ed. (2010) *Folate in Health and Disease.* CRC Press, Taylor & Francis Group, Boca Raton, FL. A book-length review which covers the chemistry and mechanisms of action of folate coenzymes, one-carbon metabolism, and clinical aspects.

Banerjee, R., and S. W. Ragsdale (2003) The many faces of vitamin B$_{12}$: Catalysis by cobalamin-dependent enzymes. *Annu. Rev. Biochem.* 72:209–247. An excellent recent review of B$_{12}$ mechanisms, focused upon the chemistry and the structures of the enzymes involved.

Blom, H. J., G. M. Shaw, M. den Heijer, and R. H. Finnell (2006) Neural tube defects and folate: Case far from closed. *Nat. Rev. Neurosci.* 7:724–731. This article reviews the epidemiology, biochemistry, and genetics of the role of folic acid in preventing neural tube defects.

Fenton, W. A., R. A. Gravel, and D. S. Rosenblatt (2001) Disorders of propionate and methylmalonate metabolism. In *The Metabolic and Molecular Bases of Inherited Disease,* C. R. Scriver, A. L. Beaudet, W. S. Sly, D. Valle, B. Childs, K. W. Kinzler, and B. Vogelstein, eds., Vol. II, Ch. 94, pp. 2165–2193, McGraw-Hill, New York. Genetic disorders of B$_{12}$ metabolism are reviewed in this chapter of the definitive work on inherited metabolic disorders.

Gallagher, T., E. E. Snell, and M. L. Hackert (1989) Pyruvoyl-dependent histidine decarboxylase. Active site structure and mechanistic analysis. *J. Biol. Chem.* 264:12737–12743. This paper illustrates the similarity in function of the pyruvoyl and PLP cofactors.

Tibbetts, A. S., and D. R. Appling (2010) Compartmentation of mammalian folate-mediated one-carbon metabolism. *Ann. Rev. Nutr.* 30:57–81. Several of these reactions are catalyzed by multifunctional proteins or multienzyme complexes, and this has implications for optimal therapeutic use of folate antimetabolites.

PROBLEMS

1. Identify the most likely additional substrates, products, and coenzymes for each reaction in the following imaginary pathway.

2. The following diagram shows the biosynthesis of B$_{12}$ coenzymes, starting with the vitamin. DMB is dimethylbenzimidazole.
   (a) What one additional substrate or cofactor is required by enzyme B?
   (b) Genetic deficiency in animals of enzyme C would result in excessive urinary excretion of what compound?

   (c) Some forms of the condition described in (b) can be successfully treated by injection of rather massive doses of vitamin B$_{12}$. What kind of genetic alteration in the enzyme would be consistent with this result?
   (d) Genetic deficiency in animals of enzyme B will result in excessive urinary excretion of what amino acid?

*3. Using the principles described in the text regarding pyridoxal phosphate mechanisms, propose a mechanism for the reaction catalyzed by serine hydroxymethyltransferase.

4. Use numbers 1 to 5 to identify each carbon atom in the product of this reaction. What coenzyme is involved?

*5. Based on the mechanism for methylmalonyl-CoA mutase shown in Figure 20.20, propose a mechanism for the diol dehydrase reaction:

$$HO-CH_2-CH_2-OH \longrightarrow CH_3CHO$$

6. The precise mechanism of ammonia toxicity to the brain is not known. Speculate on a possible mechanism, based on possible effects of ammonia on levels of key intermediates in energy generation.

*7. Mutants of *Neurospora crassa* that lack carbamoyl phosphate synthetase I (CPS I) require arginine in the medium in order to grow, whereas mutants that lack carbamoyl-phosphate synthetase II (CPS II) require a pyrimidine, such as uracil. A priori, one would expect the active CPS II in the arginine mutants to provide sufficient carbamoyl phosphate for arginine synthesis, and the active CPS I in the pyrimidine mutants to "feed" the pyrimidine pathway. Explain these observations.

8. In some forms of leukemia the proliferating white blood cells contain very low levels of asparagine synthetase. Some years ago there was interest in treating these leukemias by purifying the enzyme asparaginase from *E. coli* and injecting it into the bloodstream of leukemic patients. Asparaginase catalyzes the hydrolysis of asparagine to aspartate plus ammonia. What is the rationale behind this mode of therapy, and why might you expect it not to work?

9. Indicate whether each of the following statements is true or false, and briefly explain your answer.
   (a) In general, the metabolic oxidation of protein in mammals is less efficient, in terms of energy conserved, than the metabolic oxidation of carbohydrate or fat.
   (b) Given that the nitrogen of glutamate can be redistributed by transamination, glutamate should be a good supplement for nutritionally poor proteins.
   (c) Arginine is a nonessential amino acid for mammals because the enzymes of arginine synthesis are abundant in liver.
   (d) Alanine is an essential amino acid because it is a constituent of every protein.

10. Write a series of balanced equations and a summary equation for the reactions of the glucose–alanine cycle.

11. Consider the following questions about glutamate dehydrogenase.
   (a) The reaction as shown on page 827 has NH_3 as a reactant, instead of NH_4^+, which is far more abundant at physiological pH. Why is NH_3 preferred?
   (b) Glutamate dehydrogenase has a K_M for ammonia (NH_3) of ~1 mM. However, at physiological pH the dominant ionic species is ammonium ion, NH_4^+ ($pK_a = 9.2$). Calculate the velocity (as a fraction of V_{max}) that would be achieved by glutamate dehydrogenase if the total intracellular ammonia concentration ($NH_3 + NH_4^+$) is 100 μM (approximate physiological concentration). Assume a mitochondrial matrix pH of 8.0.
   (c) The thermodynamic equilibrium for the reaction greatly favors α-ketoglutarate reduction, yet in mitochondria the enzyme acts primarily to oxidize glutamate to α-ketoglutarate. Explain.
   (d) Propose a reasonable mechanism for this reaction.

12. Explain the basis for the following statement: As a coenzyme, pyridoxal phosphate is covalently bound to enzymes with which it functions, yet during catalysis the coenzyme is not covalently bound.

13. Consider the regulation of *E. coli* glutamine synthetase, and explain the metabolic rationale for each of the following effects:
   (a) Inhibition of glutamine synthetase by carbamoyl phosphate
   (b) Activation of the deuridylylation of P_{II}-UMP by α-ketoglutarate
   (c) Activation of the uridylylation of P_{II} by ATP

14. Suppose that you wanted to determine the metabolic half-life of glutamine synthetase in HeLa cells (a line of human tumor cells) growing in tissue culture. Describe how this could be done experimentally.

*15. Folic acid is synthesized in bacteria as dihydrofolate, in a pathway starting from guanosine triphosphate. In this pathway, C-8 is lost as formate. From the structural similarities between guanine and pterin, predict which carbon and nitrogen atoms of GTP are the precursors to N-1, C-2, C-4, N-5, C-7, N-8, and C-9 of dihydrofolate.

Dihydrofolate

GTP

16. Which folate structure (from the list below)
   (a) Is the substrate for the enzyme that is inhibited by methotrexate and trimethoprim?
   (b) Has the most highly oxidized one-carbon substituent?
   (c) Is used in the conversion of serine to glycine?
   (d) Transfers its one-carbon substituent to a B_{12} coenzyme? What amino acid is synthesized as the end result of this reaction?
   (e) Is the coenzyme for the thymidylate synthase reaction?
   (f) Is not known to exist in nature?
   (g) Is used in purine nucleotide synthesis?

17. Glutamine affects the regulatory system for *E. coli* glutamine synthetase so as to promote the adenylylation of glutamine synthetase and inhibit the deadenylylation. Why do these effects make good metabolic sense?

18. Briefly discuss how a yeast cell might contain two glutamate dehydrogenases, one specialized for nitrogen assimilation and one for amino acid catabolism, and not dissipate energy in the futile cycle glutamate $\rightleftharpoons$ α-ketoglutarate.

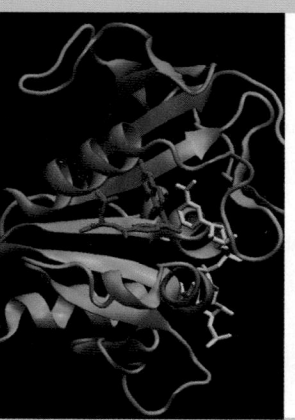

CHAPTER 21

Metabolism of Nitrogenous Compounds II: Amino Acids, Porphyrins, and Neurotransmitters

In Chapter 20 we presented unifying principles of amino acid metabolism—general features of synthetic and degradative pathways, common routes for utilizing and excreting ammonia, and coenzymes used in nitrogen metabolism. Here we consider the metabolism of the 20 standard amino acids, focusing on the fates and sources of their carbon skeletons. Our approach is to organize these amino acids into families that are metabolically related. We will begin with the degradation pathways and the major roles of each amino acid as an intermediate en route to metabolites other than protein. Some of the biological roles of amino acids—as neurotransmitters and neurotransmitter precursors and as porphyrin precursors—are so important that they are presented as separate major divisions within this chapter. Finally, we will discuss amino acid biosynthesis, focusing on the most common pathways leading to the nonessential amino acids. Biosynthesis of the essential amino acids, which is restricted to microorganisms and plants, will be touched on only briefly to illustrate important metabolic and mechanistic principles.

Pathways of Amino Acid Degradation

We learned in the preceding chapters that animals generate most of their metabolic energy from oxidation of carbohydrates and fats. Nevertheless, animals obtain 10%–15% of their energy from the oxidative degradation of amino acids. Chapter 20 described the processes by which amino acids are deaminated and the amino groups converted either to ammonia or to the amino group of aspartate for the production of urea. We now turn to the fates of the carbon skeletons. Because each of the 20 amino acids has a unique carbon skeleton, each amino acid requires its own degradation pathway. However, these 20 pathways all converge on just seven common metabolic intermediates: pyruvate, α-ketoglutarate, succinyl-CoA, fumarate, oxaloacetate, acetyl-CoA, and acetoacetate. We have already seen that several of these intermediates (pyruvate, α-ketoglutarate, succinyl-CoA, fumarate, oxaloacetate) are precursors for glucose synthesis (Chapters 13 and 14). Thus, amino acids whose carbon skeletons are degraded to one of these five intermediates are referred to as **glucogenic amino acids** (Figure 21.1). Acetyl-CoA and acetoacetate can be converted to ketone bodies (Chapter 17); amino acids whose carbon skeletons are degraded to either of these are referred to as **ketogenic amino acids**. Some amino acids are both glucogenic and ketogenic. Our discussion of these catabolic pathways will be organized around the seven common metabolic intermediates mentioned above.

Pyruvate Family of Glucogenic Amino Acids

Alanine, serine, and cysteine, each with three-carbon skeletons, are converted in one or two steps to the three-carbon metabolite pyruvate (Figure 21.2). Alanine is transaminated to pyruvate using the pyridoxal phosphate (PLP)-dependent chemistry we described in Chapter 20. Serine is dehydrated and deaminated to pyruvate by another PLP-dependent enzyme, **serine-threonine dehydratase**. Like all PLP-dependent enzymes, serine dehydratase catalyzes the formation of a planar Schiff base between PLP and the amino acid, but in this reaction, the substrate undergoes elimination of its C_α-hydrogen and C_β—OH (α,β elimination of H_2O) (see margin). The resulting enamine tautomerizes to an imine, which spontaneously hydrolyzes to form pyruvate and ammonia.

Cysteine is first transaminated by the same enzyme that transaminates aspartate. The resulting β-mercaptopyruvate can then undergo desulfuration by **mercaptopyruvate sulfurtransferase** to give pyruvate and H_2S, thiocyanate (SCN^-), sulfite (SO_3^{2-}), or thiosulfate ($S_2O_3^{2-}$). This reaction is one source of the gas H_2S in animals. Like another **gasotransmitter** molecule, nitric oxide (NO, see Chapter 23), H_2S is involved in the regulation of vascular blood flow and blood pressure. Indeed, the cardioprotective and antihypertensive effects of dietary garlic are mediated in large part by the production of H_2S from organic polysulfides that

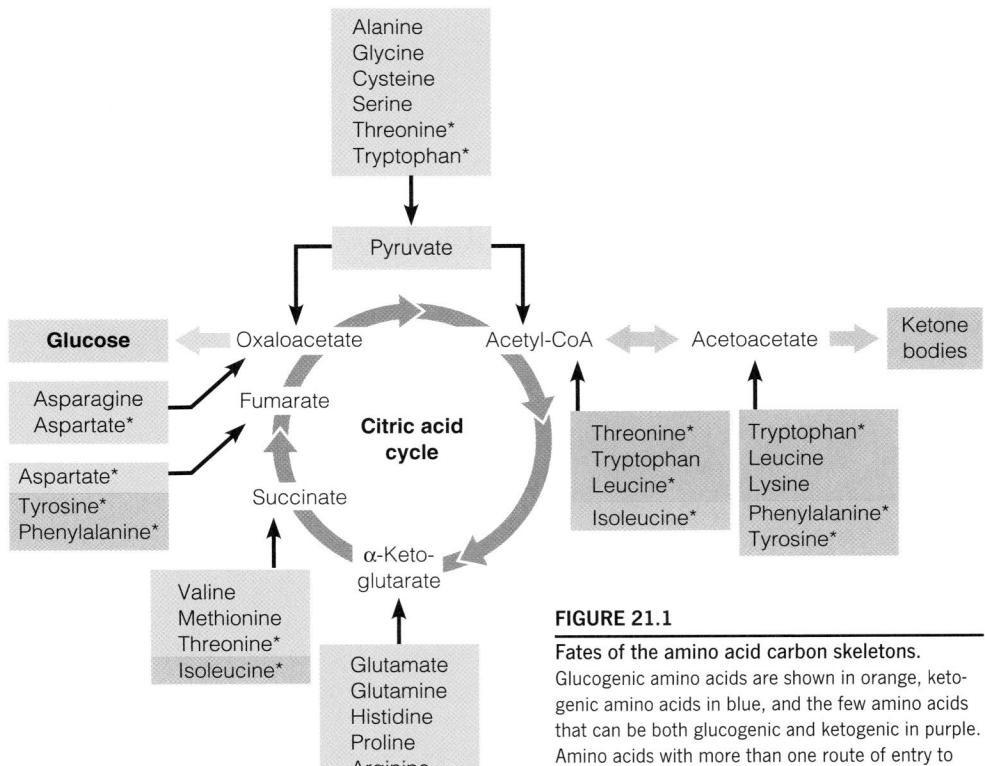

FIGURE 21.1

Fates of the amino acid carbon skeletons. Glucogenic amino acids are shown in orange, ketogenic amino acids in blue, and the few amino acids that can be both glucogenic and ketogenic in purple. Amino acids with more than one route of entry to central pathways are marked by an asterisk.

Mechanism of serine dehydratase reaction

FIGURE 21.2

Alanine, cysteine, glycine, serine, and threonine are degraded to pyruvate. The enzymes involved are ❶ threonine dehydrogenase, ❷ 2-amino-3-ketobutyrate ligase, ❸ glycine cleavage system, ❹ serine hydroxymethyltransferase, ❺ serine-threonine dehydratase, ❻ alanine aminotransferase, ❼ aspartate aminotransferase, and ❽ mercaptopyruvate sulfurtransferase. In humans, threonine is degraded by a different pathway, to succinyl-CoA (see Figure 21.4). All of these enzymes are located in mitochondria, except for the cytoplasmic serine dehydratase (reaction ❺).

Hydrogen sulfide (H_2S), derived from cysteine, is a powerful gaseous signaling molecule involved in the regulation of vascular blood flow and blood pressure.

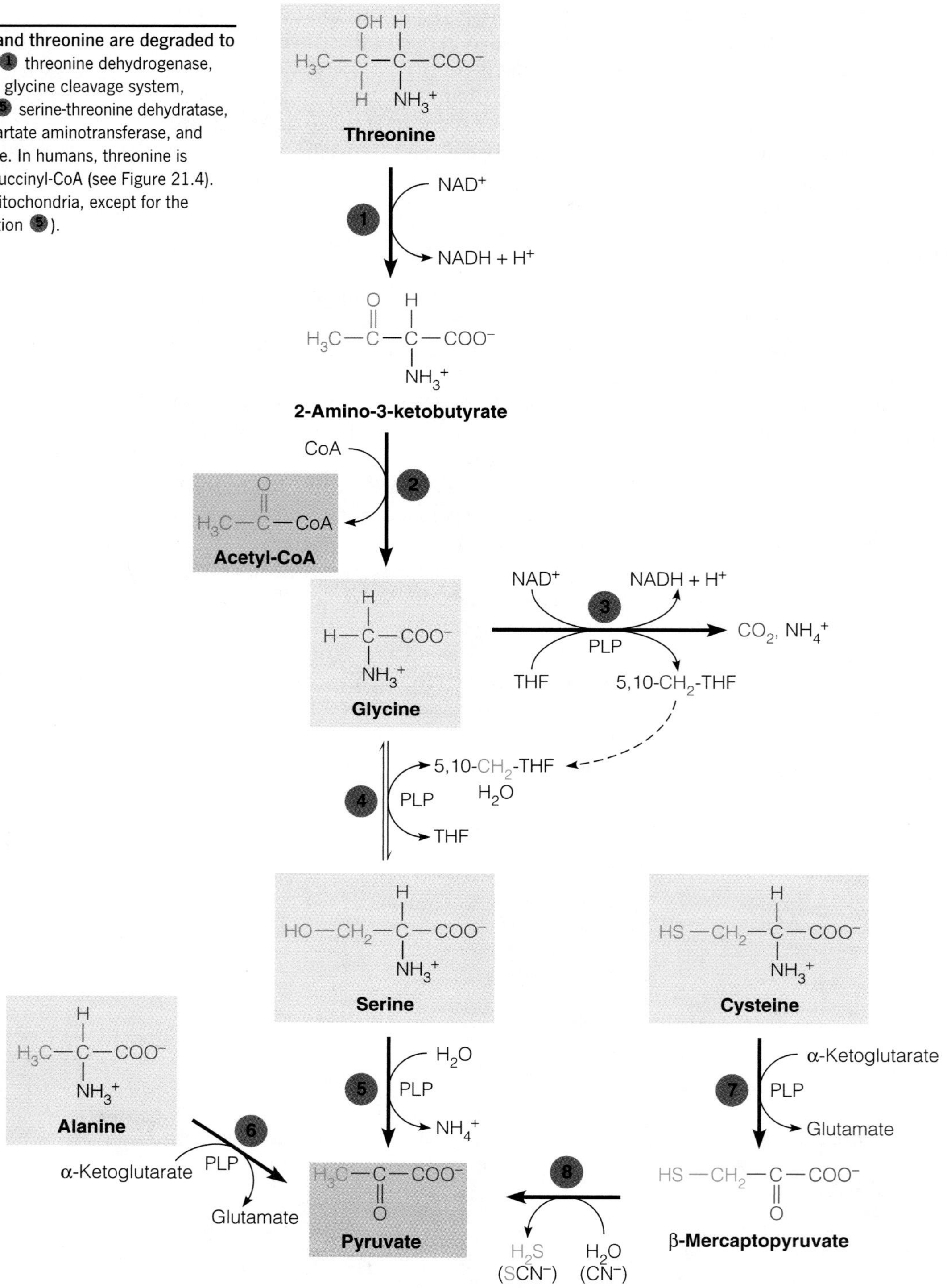

are abundant in garlic. Mercaptopyruvate sulfurtransferase can also transfer the sulfur to CN^- to give thiocyanate, an important mechanism for detoxifying cyanide.

In most organisms, threonine degradation begins with oxidation of its 2° alcohol, converting carbons 3 and 4 into an acetyl group. 2-Amino-3-ketobutyrate then undergoes a thiolytic cleavage analogous to that found in the β-oxidation of fatty acids (Chapter 17, page 728), releasing carbons 3 and 4 as acetyl-CoA, and carbons 1 and 2 as glycine. Glycine can be degraded to CO_2, NH_4^+, and 5,10-methylene-tetrahydrofolate by the PLP- and THF-dependent **glycine cleavage system** described in Chapter 20 (page 853). Alternatively, glycine can be converted to serine by the other PLP- and THF-dependent enzyme described in Chapter 20, **serine hydroxymethyltransferase**. The hydroxymethyl group of serine (C3) comes from 5,10-methylene-THF, which might derive from glycine cleavage (reaction 3) or some other one-carbon donor (see Figure 20.17, page 852). The importance of this pathway for glycine degradation is revealed by the inherited disease **nonketotic hyperglycinemia**. This is an autosomal recessive disease caused by defects in subunits of the glycine cleavage system, resulting in elevated glycine levels in cerebrospinal fluid, plasma, and urine. The disease typically occurs as a severe neonatal disorder with severe neurological symptoms, including mental retardation.

Threonine has a different fate in humans because we lack a functional threonine dehydrogenase (Figure 21.2, reaction 1). Instead, threonine is first dehydrated and deaminated by serine-threonine dehydratase to give a carbon skeleton that is eventually converted to succinyl-CoA (see Figure 21.4).

As was pointed out in Chapter 20, pyridoxal phosphate is a remarkably versatile coenzyme—the five PLP-dependent reactions in Figure 21.2 nicely illustrate this versatility.

Oxaloacetate Family of Glucogenic Amino Acids

The four-carbon skeletons of asparagine and aspartate are converted to oxaloacetate in a simple pathway (see margin). **Asparaginase** catalyzes hydrolytic cleavage of the asparagine amide to aspartate and ammonium. Aspartate is then transaminated directly to oxaloacetate by **aspartate transaminase**, the same enzyme that transaminates cysteine (Figure 21.2, reaction 7). Asparaginase is an interesting example of enzyme-based chemotherapy. The *E. coli* enzyme is widely used in the treatment of childhood acute lymphoblastic leukemia. Normal and malignant lymphocytes depend on the uptake of asparagine from the blood for growth—asparaginase depletes circulating asparagine, and this leads to complete remission in many cases.

α-Ketoglutarate Family of Glucogenic Amino Acids

The carbon skeletons of arginine, glutamine, histidine, and proline are all degraded to α-ketoglutarate via glutamate. Figure 21.3 presents the major catabolic routes for these five amino acids. Both proline and arginine are converted to glutamate through the intermediate **glutamate γ-semialdehyde**, which is then oxidized to glutamate. Histidine degradation begins with a nonoxidative deamination, followed by hydration and ring opening to yield **N-formiminoglutamic acid**, which is of interest because it serves as a donor of active one-carbon fragments (see Figure 20.17, page 852). The formimino group, derived from C2 and N3 of the imidazole ring, is transferred to tetrahydrofolate, yielding 5-formiminotetrahydrofolate and glutamate. Glutamine is hydrolyzed to glutamate by *glutaminase*, which in animals participates in transporting ammonia to the liver (see Figure 20.13, page 845). Finally, glutamate is oxidatively deaminated to α-ketoglutarate by glutamate dehydrogenase, which was discussed in depth in Chapter 20.

Histidine also undergoes decarboxylation to generate **histamine**, a substance with multiple biological actions. When secreted in the stomach, histamine promotes the secretion of hydrochloric acid and pepsin, both of which aid digestion. It is a potent vasodilator, released locally in sites of trauma, inflammation, or allergic reaction. The local enlargement of blood capillaries is the basis for the reddening that occurs in inflamed tissues. Release of histamine in trauma contributes to the dangerous lowering of blood pressure that can lead to shock. A large number of **antihistamines**, such as the two shown in the box, are in use to treat allergies and other inflammations. Typically, these drugs prevent the binding of histamine to its receptors.

Asparagine

Asparaginase H_2O → NH_4^+

Aspartate

Aspartate transaminase PLP α-Ketoglutarate → Glutamate

Oxaloacetate

Histidine

PLP → CO_2

Histamine

**Diphenhydramine
(Benadryl)**

**Desloratadine
(Clarinex)**

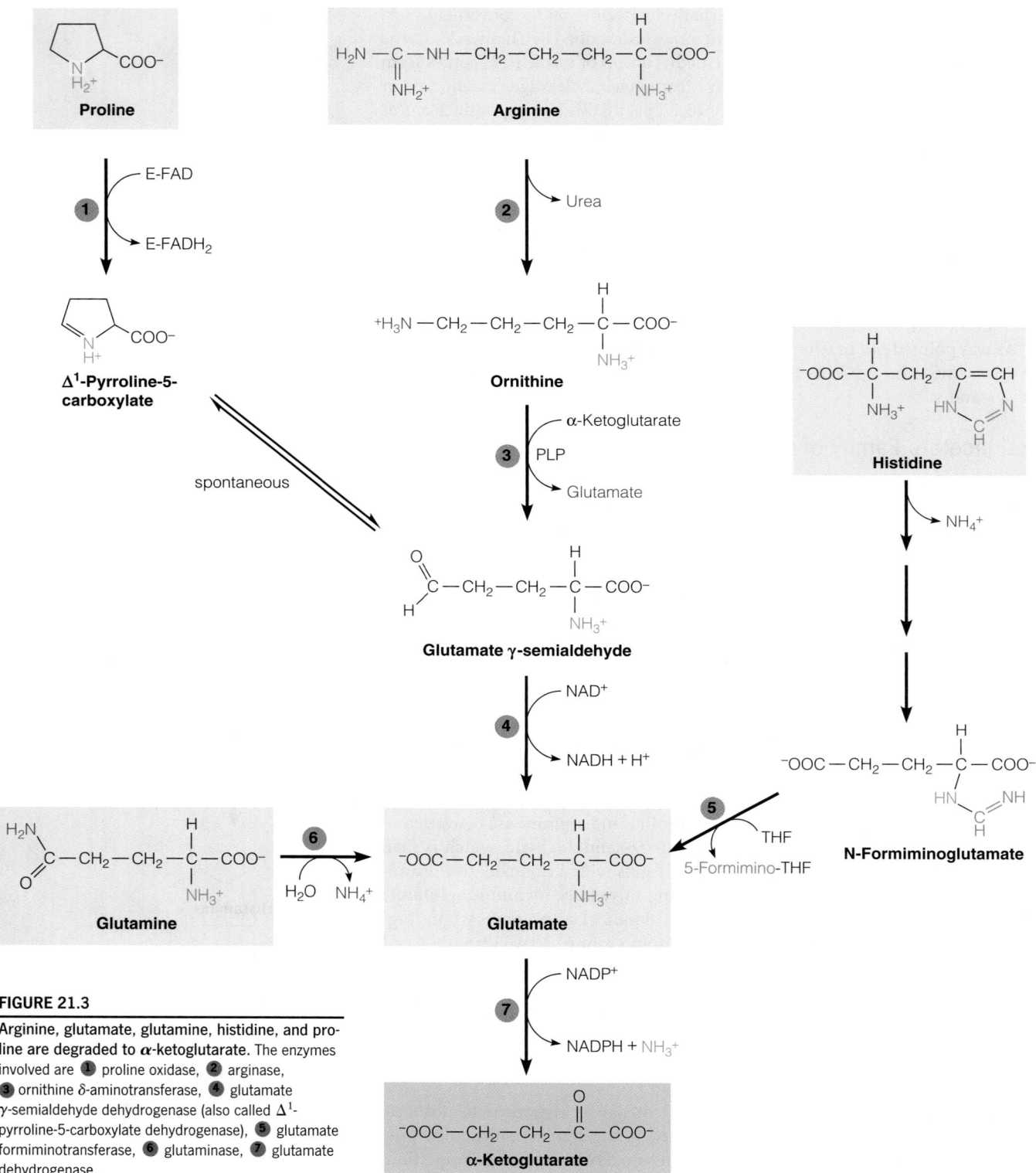

FIGURE 21.3

Arginine, glutamate, glutamine, histidine, and proline are degraded to α-ketoglutarate. The enzymes involved are ❶ proline oxidase, ❷ arginase, ❸ ornithine δ-aminotransferase, ❹ glutamate γ-semialdehyde dehydrogenase (also called Δ¹-pyrroline-5-carboxylate dehydrogenase), ❺ glutamate formiminotransferase, ❻ glutaminase, ❼ glutamate dehydrogenase.

Succinyl-CoA Family of Glucogenic Amino Acids

Isoleucine, valine, threonine, and methionine are degraded to the citric acid cycle intermediate, succinyl-CoA, by way of propionyl-CoA (Figure 21.4). The catabolism of methionine begins after it has donated its methyl group, via S-adenosylmethionine, in the methyl cycle (see Figure 21.9). The demethylated

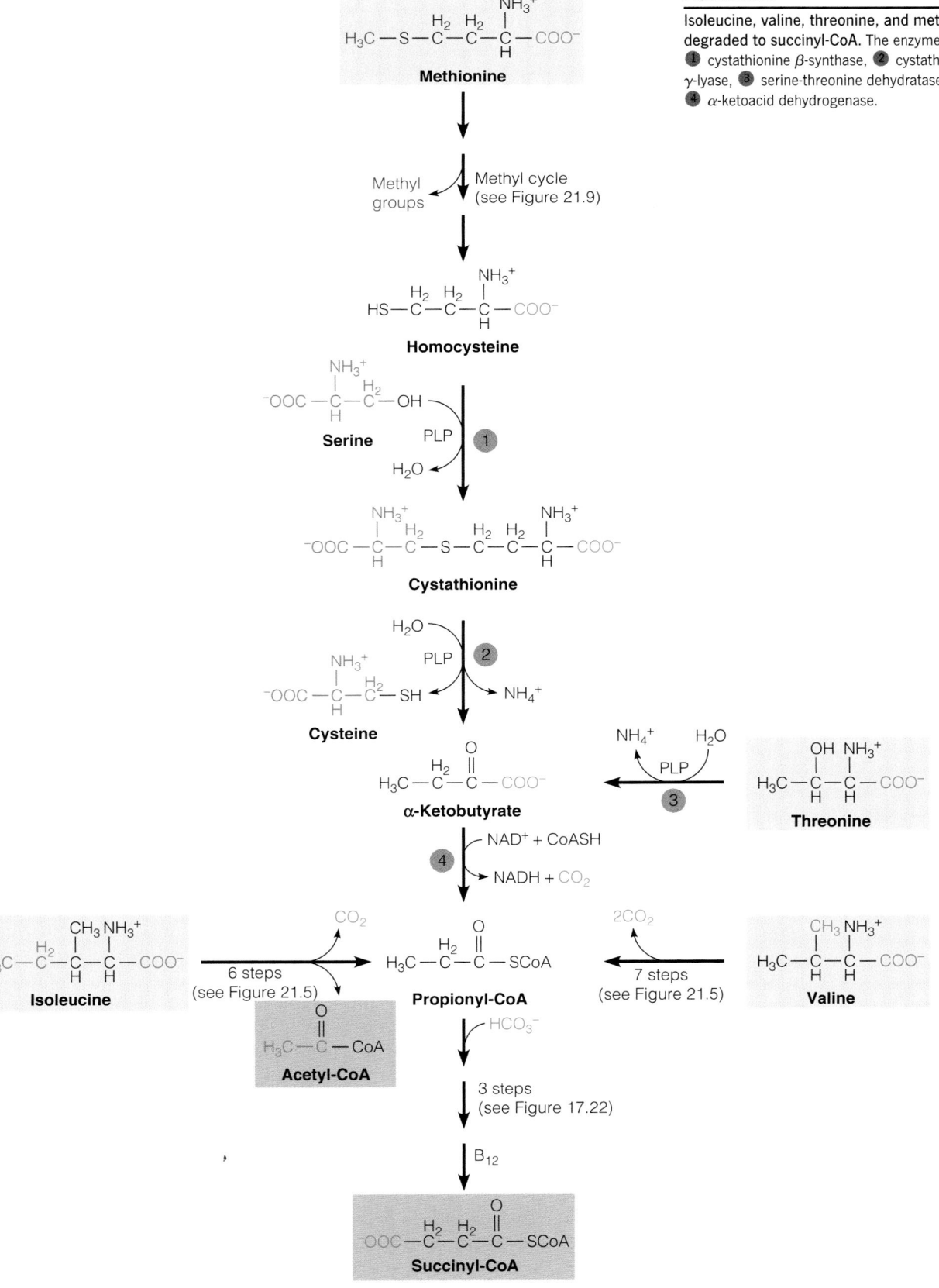

FIGURE 21.4

Isoleucine, valine, threonine, and methionine are degraded to succinyl-CoA. The enzymes involved are ❶ cystathionine β-synthase, ❷ cystathionine γ-lyase, ❸ serine-threonine dehydratase, and ❹ α-ketoacid dehydrogenase.

carbon skeleton, **homocysteine**, is converted to α-ketobutyrate and the —SH group replaces the —OH of serine to form cysteine. This is, in fact, the route by which animals synthesize cysteine (see Figure 21.29). In the first step of this two-step **transsulfuration** process, catalyzed by cystathionine β-synthase, homocysteine and serine condense to form the thioether **cystathionine** (Figure 21.4, reaction 1). In the second step, cystathionine γ-lyase catalyzes a γ-elimination to give cysteine, α-ketobutyrate, and ammonia (reaction 2). Thus, cleavage occurs on the other side of the S atom, resulting in transfer of the —SH group from the four-carbon backbone (homocysteine) to the three-carbon backbone (cysteine). Both transsulfuration enzymes are PLP-dependent; cystathionine β-synthase also requires a b-type heme cofactor. α-Ketobutyrate is oxidatively decarboxylated to propionyl-CoA (reaction 4), which is then converted to succinyl-CoA by the same B_{12}-requiring pathway we introduced in Chapter 17 as a key step in the oxidation of odd-chain fatty acids (page 732).

α-Ketobutyrate is also an entry point for carbon from threonine (Figure 21.4, reaction 3). Both threonine and serine are dehydrated and deaminated by the same dehydratase. This is the major pathway for threonine degradation in humans.

Isoleucine and valine are also degraded to the three-carbon propionyl-CoA. However, two of isoleucine's carbons are released as acetyl-CoA and one as CO_2; valine loses two carbons as CO_2. Oxidation of the branched-chain amino acids, including leucine, starts with a PLP-dependent transamination to the corresponding α-keto acid and then follows a common chemical strategy (Figure 21.5): (1) oxidative decarboxylation to give an acyl-CoA derivative; (2) FAD-dependent acyl-CoA dehydrogenation to introduce a double bond; (3) hydration of the double bond to introduce a hydroxyl group; and (4) NAD^+-dependent dehydrogenation to the corresponding keto derivative. All three branched-chain amino acids follow this same strategy, with only minor deviations. The final step in isoleucine degradation is a thiolytic cleavage that produces acetyl-CoA and propionyl-CoA. In the valine pathway, CoA is hydrolyzed off the carbon skeleton before the second dehydrogenation; a final decarboxylation gives propionyl-CoA. In the leucine pathway, the hydration step gives **3-hydroxy-3-methylglutaryl-CoA (HMG-CoA)**, an intermediate also involved in ketogenesis (Figure 17.24, page 735) and cholesterol biosynthesis (Figure 19.19, page 797). HMG-CoA is cleaved to acetoacetate plus acetyl-CoA by the same mitochondrial **HMG-CoA lyase** used in ketogenesis.

The chemical strategy used to oxidize the branched-chain amino acids should look familiar to you because you have seen it twice before. This same strategy forms the core of both the β-oxidation of fatty acids (Chapter 17) and the citric acid cycle (Chapter 14). These two pathways are shown alongside the branched-chain amino acid oxidation pathways in Figure 21.5. Once the amino acids are transaminated to the corresponding α-keto acids, they follow the same sequence we first learned in the citric acid cycle. β-oxidation adds a thiolytic cleavage at the end of this core pathway. This commonality illustrates a couple of important points about metabolic pathways. First, as we discussed in Chapter 12, metabolic pathways are the products of evolution, being built from enzymes and pathways that may have had other functions initially. It is thus likely that the branched-chain amino acid oxidation pathways arose from the evolutionarily ancient citric acid cycle pathway. Second, a cell can "economize" by using the same enzyme to metabolize several related substrates. Indeed, the first two steps of branched-chain amino acid degradation in Figure 21.5 are each catalyzed by a single enzyme that acts on all three carbon skeletons. **Branched-chain amino acid aminotransferase** uses α-ketoglutarate as the amino acceptor from all three amino acids and exists in two isozymes—cytoplasmic and mitochondrial—in mammals. This enzyme is at low levels in human liver but abundant in muscle, adipose, kidney, and brain, where branched-chain amino acids are important

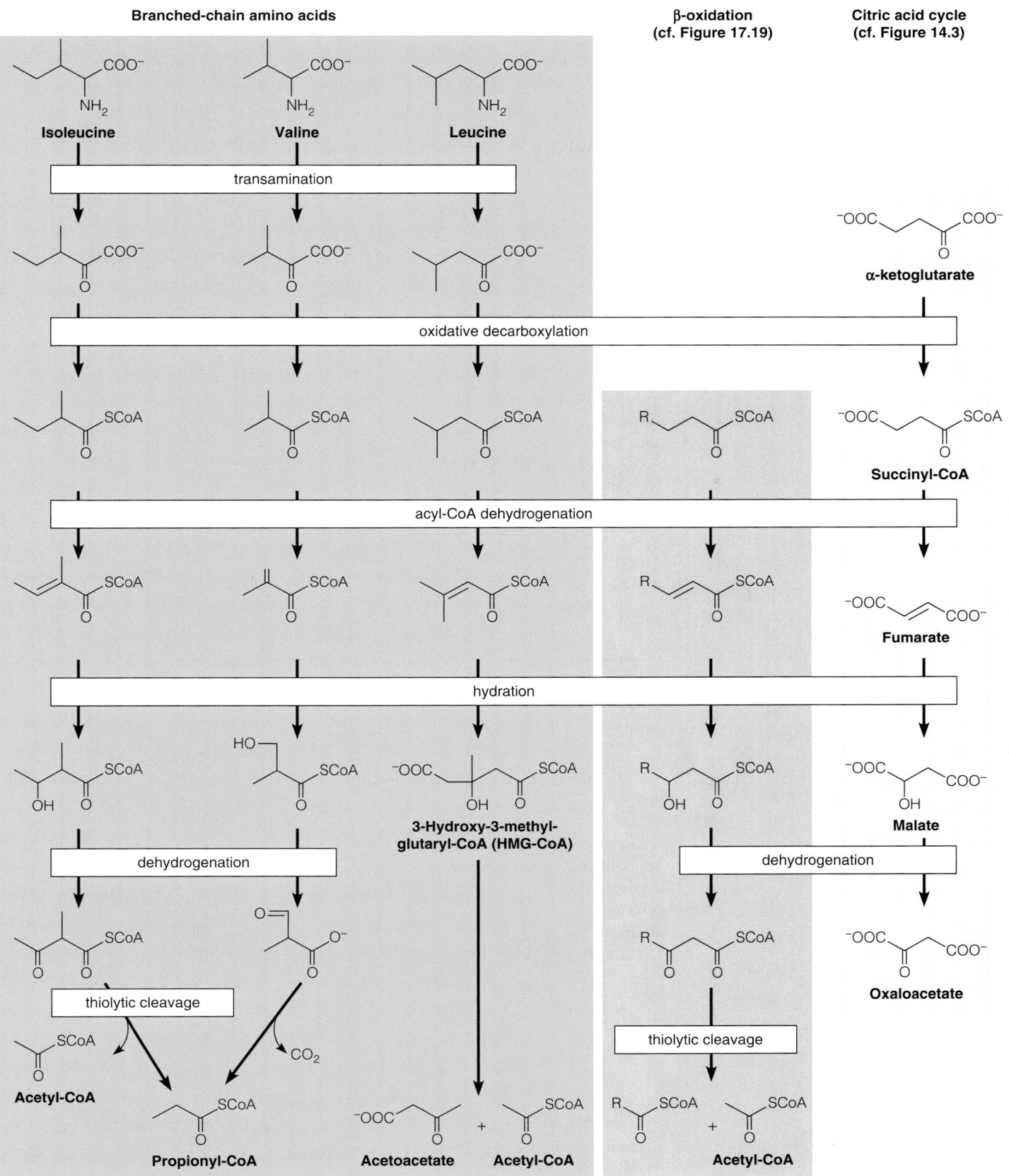

FIGURE 21.5

Branched-chain amino acid oxidation, fatty acid β-oxidation, and the citric acid cycle share a common chemical strategy.

A deficiency of branched-chain α-keto acid dehydrogenase complex, which metabolizes valine, leucine, and isoleucine in humans, leads to a severe mental developmental defect called maple syrup urine disease.

fuel sources during fasting. The subsequent oxidative decarboxylation of all three α-keto acid derivatives is carried out by the mitochondrial **branched-chain α-keto acid dehydrogenase complex**, which has the same E_1-E_2-E_3 multienzyme structure and mechanism of the pyruvate dehydrogenase and α-ketoglutarate dehydrogenase complexes (Chapter 14). Indeed, the E_3 subunits are identical in all three complexes from a given species. In a rare human disorder called **maple syrup urine disease**, or branched-chain ketoaciduria, this complex is defective. All three amino acids and their α-keto acid derivatives accumulate in the urine, and their characteristic odor gives the condition its name. The condition involves severe mental retardation.

Acetoacetate/Acetyl-CoA Family of Ketogenic Amino Acids

Lysine Degradation

We have already discussed the path by which leucine is degraded to acetoacetate and acetyl-CoA (Figure 21.5). Based on our experience with the branched-chain amino acids, we might predict that lysine degradation follows a similar strategy, once its two amino groups are removed. We know how to remove the α-amino group—by transamination—but what about lysine's ε-amino group? Although lysine is degraded by several pathways, the major route in mammals, the **saccharopine** pathway, does indeed follow this predicted strategy. Saccharopine is the intermediate in the two-step process that removes the ε-amino group from lysine:

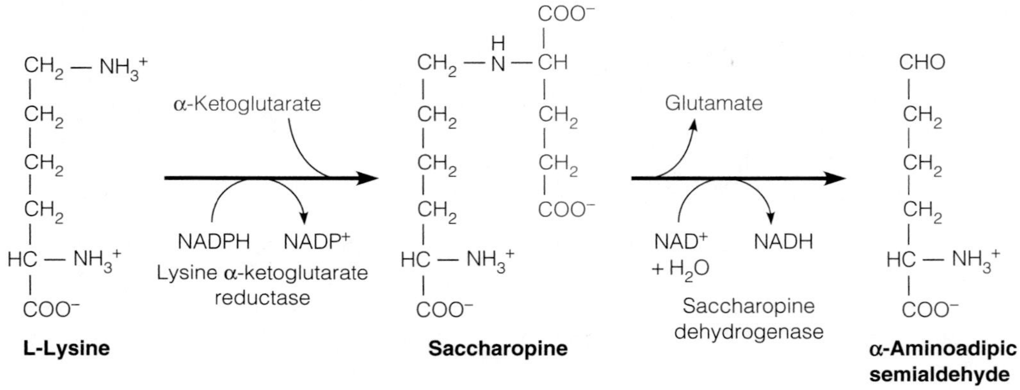

These two reactions are catalyzed by a bifunctional enzyme named **α-aminoadipic semialdehyde synthase**. The mammalian enzyme is a single protein composed of separate domains for the two catalytic activities. Although this process uses α-ketoglutarate as the amino acceptor, and produces glutamate, it is more similar to the argininosuccinate synthetase–argininosuccinase reactions of the urea cycle (Figure 20.12, page 842) than to a transamination:

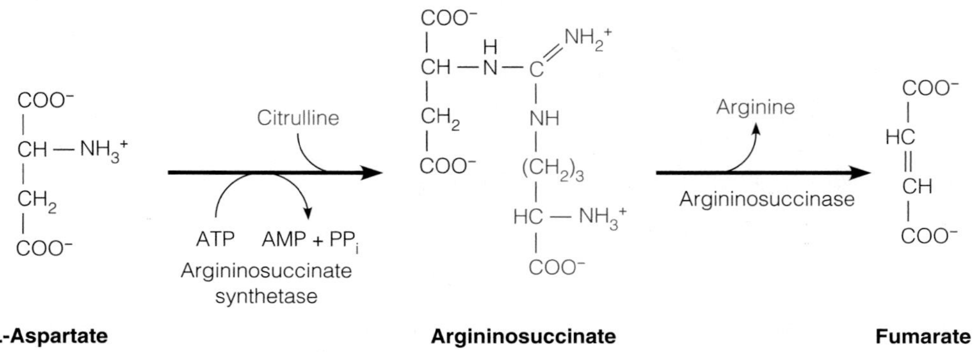

The rest of the pathway involves oxidation of the semialdehyde to the acid and transamination to remove the α-amino group, followed by the now familiar oxidative decarboxylation to give an acyl-CoA derivative; FAD-dependent acyl-CoA dehydrogenation to introduce a double bond; hydration of the double bond to introduce a hydroxyl group; and NAD^+-dependent dehydrogenation to the corresponding keto derivative. The only variation on this theme required by lysine is an extra decarboxylation after the acyl-CoA dehydrogenation step to remove the terminal carbon (the aldehyde of α-aminoadipic semialdehyde). The final degradation product is acetoacetate.

Tryptophan Degradation

Tryptophan is transformed by many pathways, but the major catabolic route proceeds via **kynurenine** to α-ketoadipate (Figure 21.6). The first reaction in the degradation of tryptophan is catalyzed by **tryptophan 2,3-dioxygenase**, a heme protein that incorporates both atoms of O_2 into the substrate. The fourth step, catalyzed by **kynureninase**, is a PLP-dependent cleavage of the C_β—C_γ bond, releasing the α and β carbons of the original tryptophan skeleton as alanine, which is readily transaminated to pyruvate. The other product of the cleavage reaction, 3-hydroxyanthranilic acid, is eventually converted to α-ketoadipate, an intermediate in the saccharopine pathway of lysine degradation. Indeed, the last seven steps of tryptophan and lysine degradation are identical, ending with acetoacetate.

FIGURE 21.6

Metabolic fates of tryptophan. The figure shows two major pathways—the major catabolic pathway that degrades most of the tryptophan molecule to acetoacetate and alanine, and the synthetic pathway to the nicotinamide nucleotides.

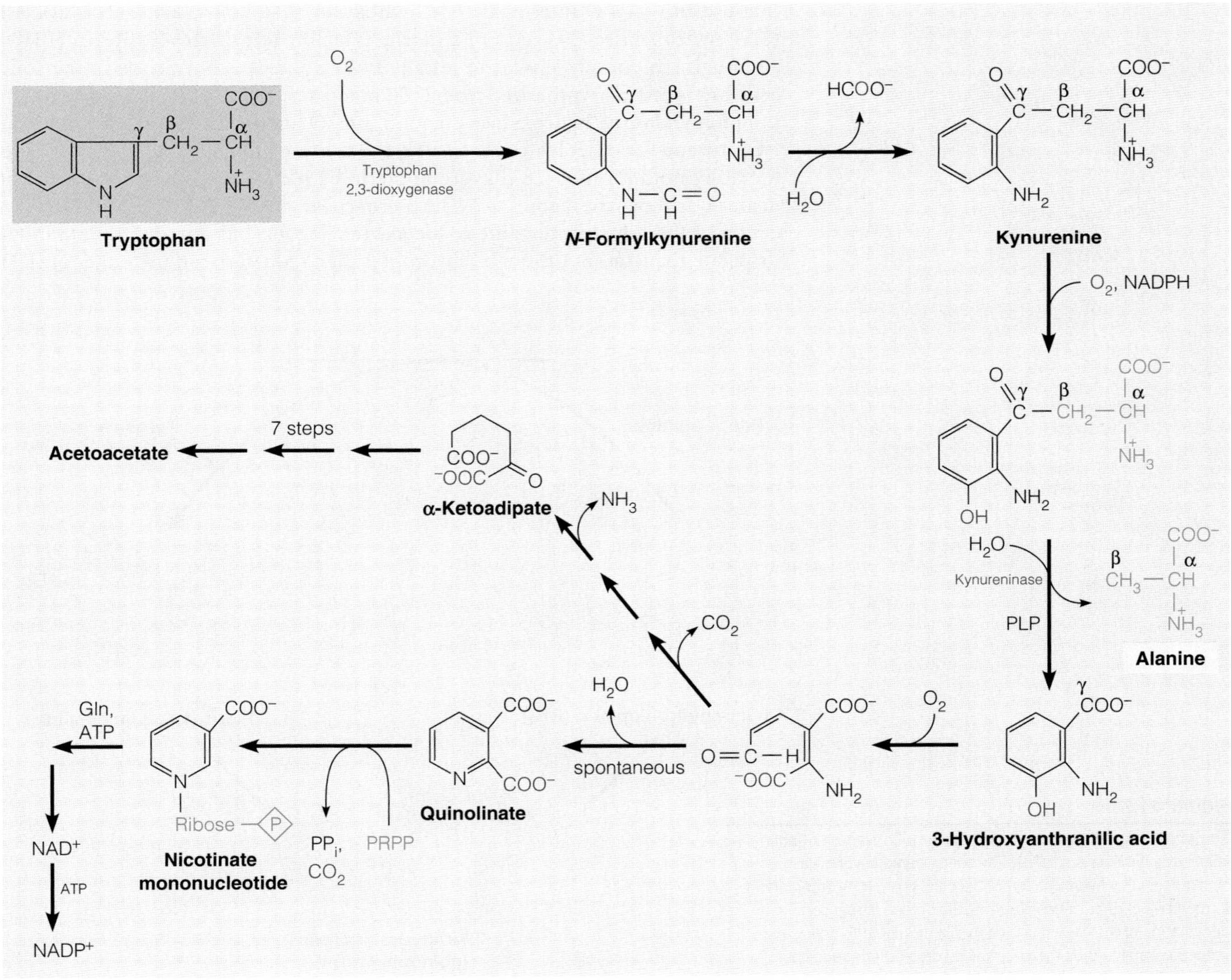

The amide nitrogen of glutamine is used in many amidotransferase reactions leading to purine and pyrimidine nucleotides, amino sugars, and nicotinamide nucleotides.

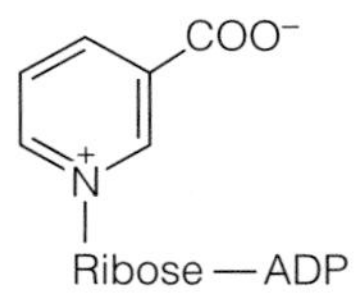

Nicotinate adenine dinucleotide

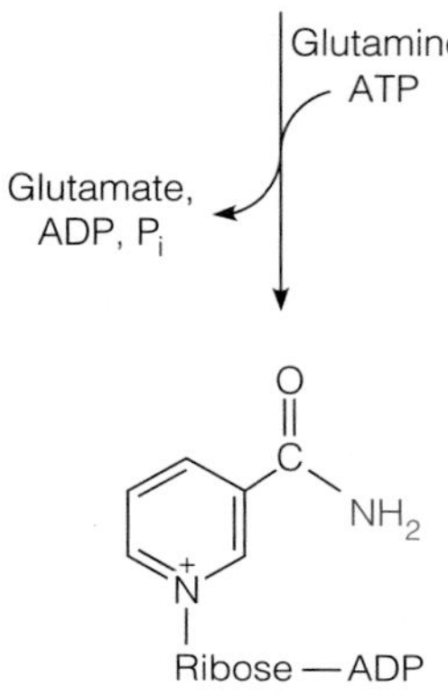

NAD⁺

Tryptophan is also an important precursor for the synthesis of nicotinamide nucleotides, accounting for as much as 50% of an animal's NAD$^+$. The rest comes from the vitamin niacin (see Table 11.8 on page 446). For both pathways the last reaction is catalyzed by a **glutamine-dependent amidotransferase** (see margin), in which glutamine donates its amide N and ATP hydrolysis provides the driving force for the reaction. Glutamine-dependent amidotransferases are widespread in metabolism, including reactions leading to other amino acids, to nucleotides, and to amino sugars and glycoproteins. We will look more closely at the mechanism of these enzymes in Chapter 22.

The role of tryptophan in NAD$^+$ biosynthesis can be inferred from studies of the nicotinamide deficiency disease **pellagra**. Pellagra was formerly endemic in regions where corn is a dietary staple such as the southern United States. Because corn proteins contain little tryptophan, deficiencies of this amino acid were common. The symptoms of tryptophan deficiency are identical to those of nicotinamide deficiency, as expected if a major role of tryptophan is as a nicotinamide substitute.

Phenylalanine and Tyrosine Degradation

The final two amino acids are degraded by a single pathway that begins with the hydroxylation of phenylalanine to tyrosine, catalyzed by **phenylalanine hydroxylase**. This interesting enzyme belongs to the **aromatic amino acid hydroxylase** family, along with tyrosine hydroxylase and tryptophan hydroxylase. In animals and microbes, all three enzymes are nonheme iron–containing monooxygenases that use an Fe(IV)O intermediate to hydroxylate their amino acid substrates. All three hydroxylases also require a pterin cofactor, **tetrahydrobiopterin** (BH$_4$). Although structurally similar to tetrahydrofolate, BH$_4$ is not derived from folic acid, but is instead synthesized from GTP in most mammalian cells and tissues. In the **phenylalanine hydroxylase** reaction, one O atom becomes the hydroxyl of tyrosine and the other hydroxylates BH$_4$ to a carbinolamine, which is converted to the quinonoid form of 7,8-dihydrobiopterin by **pterin-4a-carbinolamine dehydratase**, as shown in Figure 21.7. The pterin coenzyme is regenerated through the action of the **dihydropteridine reductase** (analogous, but not identical, to dihydrofolate reductase). Although dihydropteridine reductase can use either NADH

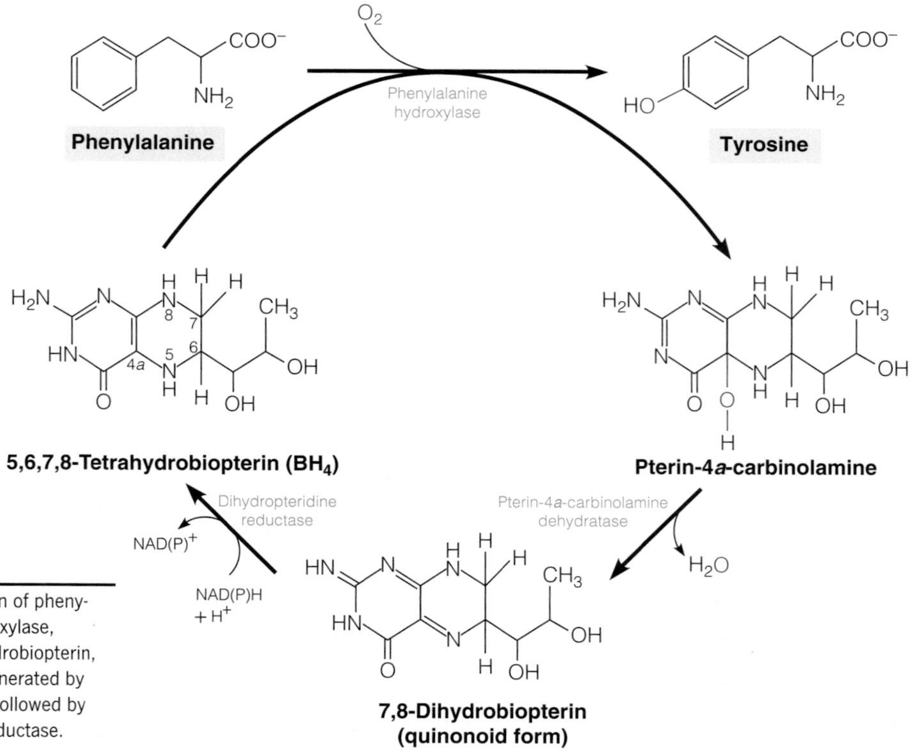

FIGURE 21.7

The phenylalanine hydroxylation system. Conversion of phenylalanine to tyrosine is catalyzed by phenylalanine hydroxylase, accompanied by hydroxylation of the cofactor, tetrahydrobiopterin, to pterin-4a-carbinolamine. Tetrahydrobiopterin is regenerated by dehydration of the carbinolamine to dihydrobiopterin, followed by reduction catalyzed by the enzyme dihydrobiopterin reductase.

or NADPH as an electron donor, it has a much lower K_M for NADH. This three-enzyme phenylalanine hydroxylation system occurs primarily in liver and, to a lesser extent, in kidney. The aromatic amino acid hydroxylases in plants use a tetrahydrofolate cofactor rather than tetrahydrobiopterin.

A hereditary deficiency of phenylalanine hydroxylase is responsible for **phenylketonuria (PKU)**, a condition that afflicts about 1 in 10,000 newborn infants in western Europe and the United States. PKU is an autosomal recessive trait, meaning that two parents heterozygous for the trait have 1 chance in 4 of having a phenylketonuric child. From the incidence of the disease, we can estimate that about 2% of the population are carriers. In phenylketonuria, phenylalanine accumulates to very high levels (**hyperphenylalaninemia**) because of the block in conversion to tyrosine, and much of this phenylalanine is metabolized via pathways that are normally little used—particularly transamination to phenylpyruvate (a phenyl*ketone*), and also subsequent conversion of phenylpyruvate to phenyllactate and phenylacetate. These compounds are excreted in urine in enormous quantities (1 to 2 grams per day).

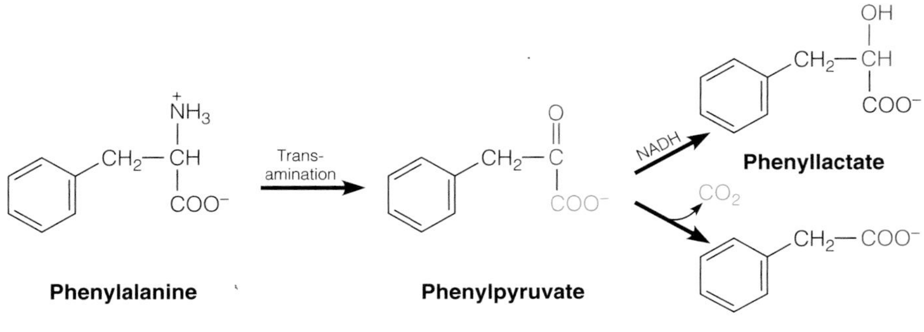

If undetected and untreated, PKU leads to profound mental retardation; the precise biochemical causation is not known, but evidence is accumulating that phenylalanine itself is the neurotoxic molecule. Fortunately, PKU can readily be detected at birth, and many hospitals carry out routine screening of newborns. If the condition is detected early, the onset of retardation can be prevented by feeding for several years a synthetic diet low in phenylalanine and rich in tyrosine, to allow normal development of the nervous system. Because the use of this synthetic diet is quite expensive, there has been much interest in prenatal diagnosis of PKU and in identification of heterozygous carriers. More than 500 distinct pathogenic mutations have been mapped on the human gene for phenylalanine hydroxylase. Most of these are missense mutations that lead to amino acid substitutions in the 452-amino acid protein. The structure of human phenylalanine hydroxylase has been determined by X-ray crystallography, and it is now possible to provide a molecular explanation for many of the PKU mutations.

PKU is by far the most common, but not the only, form of hyperphenylalaninemia. Hereditary deficiency of dihydropteridine reductase, pterin-4a-carbinolamine dehydratase, or any of the enzymes required for the biosynthesis of BH_4 lead to non-PKU hyperphenylalaninemia. Because tetrahydrobiopterin is involved in other hydroxylations, including those of tyrosine and tryptophan, defects in its synthesis or recycling cause more severe symptoms.

Degradation of tyrosine to fumarate and acetoacetate begins with its transamination by **tyrosine aminotransferase**, the rate-determining enzyme in this catabolic pathway (Figure 21.8). The product, *p*-hydroxyphenylpyruvate, is acted on by ***p*-hydroxyphenylpyruvate dioxygenase**, an unusual iron-containing enzyme, which catalyzes a ring hydroxylation, decarboxylation, and side chain migration, using ascorbate as a cofactor. Procollagen prolyl hydroxylase catalyzes the same chemistry (see page 906). This reaction involves a mechanism called the **NIH shift**, after scientists at the National Institutes of Health, who described a ring

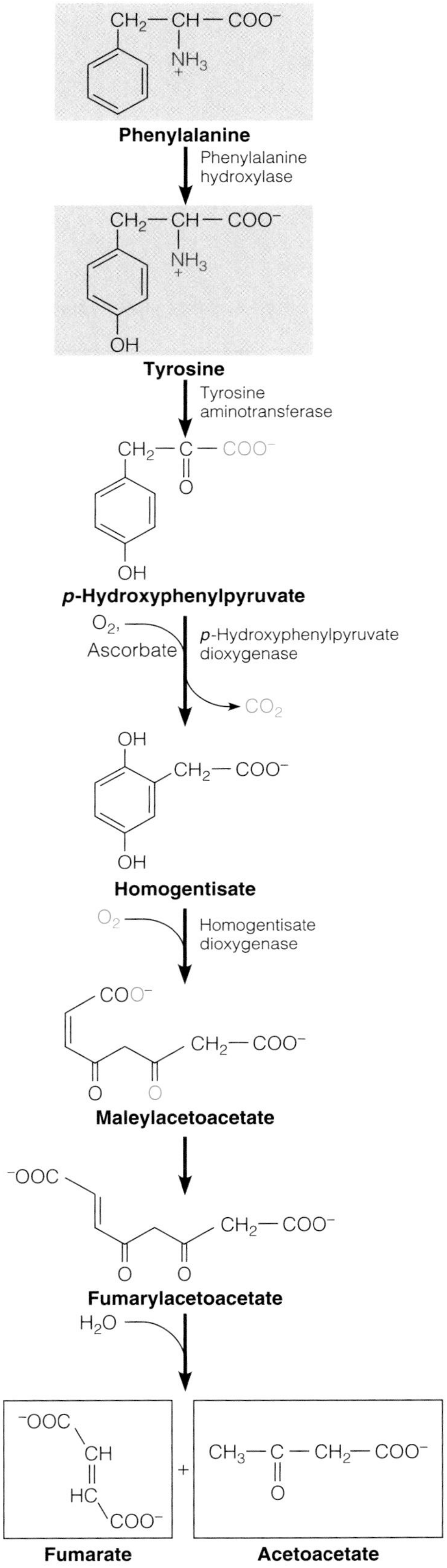

FIGURE 21.8

Catabolism of phenylalanine and tyrosine to fumarate and acetoacetate.

hydroxylation that proceeds via formation of an epoxide intermediate and migration of the alkyl side chain.

p-Hydroxyphenylpyruvate

Homogentisate

Phenylketonuria, if left untreated, leads to a severe mental deficiency arising from genetic insufficiency of phenylalanine hydroxylase.

The phenylalanine hydroxylase reaction also involves an NIH shift, with migration of the hydrogen at C-4 of phenylalanine to C-3 of tyrosine, without mixing with solvent.

The product of p-hydroxyphenylpyruvate oxidation, **homogentisic acid**, is oxidized by an iron-containing enzyme, **homogentisate dioxygenase**, that cleaves the ring to yield a straight-chain eight-carbon compound that isomerizes to **fumarylacetoacetate**. The latter ultimately cleaves to yield fumarate and acetoacetate, both of which are catabolized by standard energy-yielding pathways. In plants, homogentisate is the precursor to the plastoquinones used in photosynthesis (Chapter 16) and to the aromatic ring portion of vitamin E (see page 811).

A hereditary deficiency of the enzyme homogentisate dioxygenase in humans causes a condition that was known for centuries as the "dark urine disease" but is now called **alkaptonuria**. Homogentisate accumulates and is excreted in large amounts in the urine; its oxidation on standing causes the urine to become dark. Although the clinical symptoms of the disease are not severe, it is of considerable historical interest. Early in the twentieth century Sir Archibald Garrod examined pedigrees of the families of afflicted individuals, and in 1909 he wrote, "We may further conceive that the splitting of the benzene ring in normal metabolism is the work of a special enzyme, that in congenital alcaptonuria this enzyme is wanting, whilst in disease its working may be partially or even completely inhibited." In other words, Garrod proposed that one gene encodes one enzyme, long before the chemical nature of either genes or enzymes was known, and developed the concept of inheritable metabolic diseases.

Amino Acids as Biosynthetic Precursors

Beyond their incorporation into proteins, amino acids serve as precursors for a tremendous variety of other important metabolites, such as polyamines, glutathione, methyl groups, heme, neurotransmitters and other signaling molecules, and nucleotides (Chapter 22). In the following section, we highlight some of the most important pathways and products of amino acid metabolism.

S-Adenosylmethionine and Biological Methylation

In Chapter 19 we introduced S-adenosylmethionine (AdoMet) as a metabolically activated form of methionine, when we described the biosynthesis of phosphatidylcholine from phosphatidylethanolamine. Recall that the synthesis of AdoMet from methionine and ATP generates a sulfonium compound with a high group transfer potential. Most, though not all, of the group transfer reactions involving AdoMet are **transmethylations**, in which the methyl group is transferred to an acceptor, with the other product being S-adenosylhomocysteine (AdoHcy). The syntheses of creatine (see Figure 21.16) and of phosphatidylcholine (see page 784) are good examples of transmethylations.

Table 21.1 lists a number of biologically important AdoMet-dependent transmethylations. Note that the substrates can be polymeric proteins or nucleic acids. We

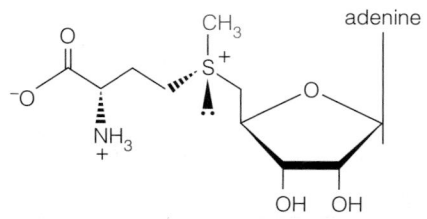

S-Adenosyl-L-methionine (AdoMet)

TABLE 21.1 Some AdoMet-dependent transmethylations

Methyl Group Acceptor	Methylated Product
Norepinephrine	Epinephrine (Figure 21.24)
Guanidinoacetic acid	Creatine (Figure 21.16)
Phosphatidylethanolamine	Phosphatidylcholine (page 784)
DNA-adenine or -cytosine	DNA-N-methyladenine or 5-methylcytosine
tRNA bases	Methylated tRNA bases
Nicotinamide	N^1-Methylnicotinamide
Protein amino acid residues	Methylated amino acid residues

discuss the functions of DNA and RNA methylation in later chapters. However, a brief discussion of protein methylation is in order here. Methylatable residues include—in different proteins—lysine, arginine, and residues containing free carboxyl groups. Histones are extensively methylated, with specific arginine and lysine residues being modified at particular times in the cell cycle. Lysine can be mono-, di-, or trimethylated; arginine can be dimethylated, in either a symmetric or asymmetric configuration. These post-translational modifications of histones play a central role in regulating transcription and maintaining genomic integrity, and they also contribute to the **epigenome** (inheritable, non-DNA chemical tags that influence gene expression; more about this in Chapter 29). We also know that ε-N-trimethyllysine, derived specifically from the enzymatic hydrolysis of methylated protein, serves as the precursor to carnitine, whose role in fatty acyl group transfer across membranes was presented in Chapter 17. Furthermore, in bacteria, protein methylation plays an important role in **chemotaxis**, the process whereby bacteria sense a concentration gradient of a chemical substance in the medium and move either toward or away from it. Chemotaxis is being studied as a rudimentary model for sensory transduction; it involves the cyclic methylation and demethylation of a group of proteins called **MCPs**, or methylatable chemotactic proteins. Finally, there are indications that protein methylation protects proteins, in at least two ways: (1) By blocking sites of ubiquitination (see Chapter 20), methylation evidently helps protect proteins from turnover. (2) Spontaneous damage of protein molecules during aging causes deamidation, isomerization, and racemization of their asparagine and aspartate residues. A methylation reaction can initiate the repair of these damaged residues.

The central metabolic role of AdoMet can be appreciated if you keep in mind that, except for a few reactions in bacterial metabolism, *the only known methyl group transfer that does not involve AdoMet is the synthesis of methionine itself.* As noted in Chapter 20, a methyl group is generated de novo through the reduction of 5,10-methylenetetrahydrofolate to 5-methyltetrahydrofolate (5-methyl-THF), catalyzed by a flavin-dependent methylenetetrahydrofolate reductase (MTHFR). The MTHFR reaction in mammalian liver uses NADPH as electron donor, and is consequently physiologically irreversible due to the high cytosolic NADPH/NADP$^+$ ratio and the large standard free energy change for the reduction of 5,10-methylene-THF. The methyl group is then transferred to homocysteine to yield methionine via methyl-B$_{12}$ and the action of methionine synthase (Figure 21.9). This methyl group is then activated for methyl transfers by the ATP-dependent conversion of methionine to AdoMet. *S*-adenosylhomocysteine, formed from AdoMet-dependent transmethylations, is hydrolyzed to yield adenosine and homocysteine by *S*-adenosylhomocysteine hydrolase. Homocysteine is remethylated to methionine, completing the *methyl cycle*. Homocysteine can be removed from the methyl cycle by conversion to cysteine, via the transsulfuration pathway we first described in Figure 21.4. Remethylation and transsulfuration each normally account for ~50% of homocysteine metabolism.

As shown in Figure 21.9, the methyl cycle is tightly connected to a *one-carbon cycle*—indeed, the ultimate source of all the methyl groups donated by AdoMet is the tetrahydrofolate one-carbon pool. Serine is the major donor of one-carbon

ε-*N*-Trimethyl-Lysine

Arginine

AdoMet
AdoHcy

Monomethyl-Arg

AdoMet
AdoHcy

Dimethyl-Arg (Symmetric)

Dimethyl-Arg (Asymmetric)

FIGURE 21.9

Methyl group metabolism and homocystinuria. Serine hydroxymethyltransferase ❶ catalyzes the entry of one-carbon units into the tetrahydrofolate (THF) pool. A deficiency of methylenetetrahydrofolate reductase (MTHFR) ❷ or methionine synthase ❸ blocks the conversion of homocysteine to methionine. X represents any methyl acceptor in an *S*-adenosylmethionine (AdoMet)-dependent methyltransferase reaction. A defect in cystathionine β-synthase ❹ causes homocysteine to accumulate because conversion to cystathionine is impaired. In any of these cases, excess homocysteine accumulates in the circulation (hyperhomocysteinemia) and urine (homocystinuria). Dashed lines indicate allosteric inhibition (magenta) or activation (green) of regulatory enzymes.

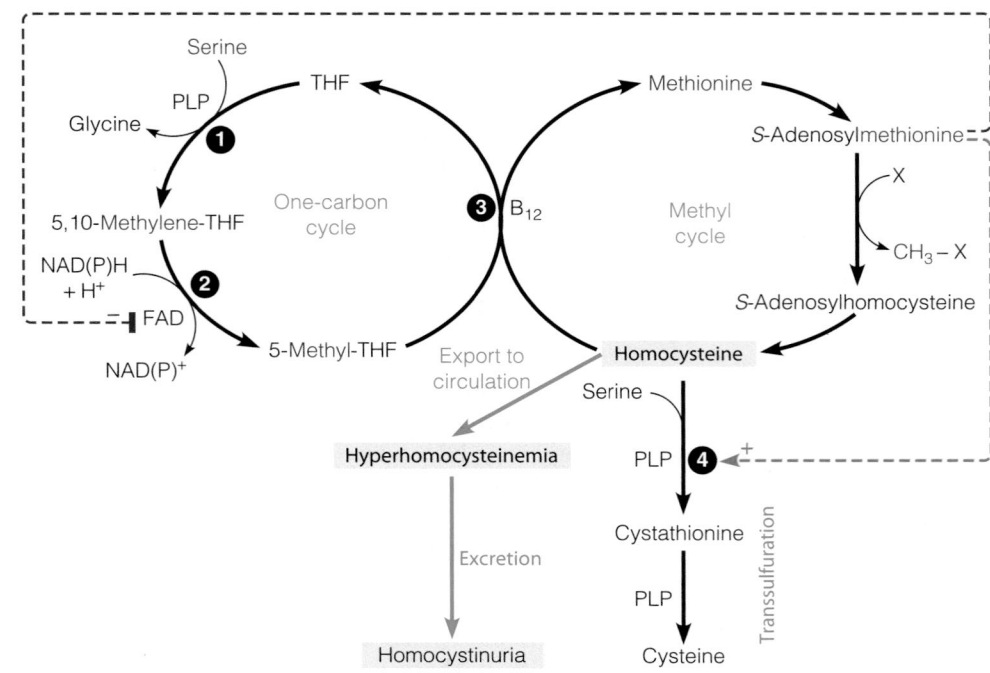

5-Methyltetrahydrofolate transfers a methyl group in methionine synthesis, but all other biological methyl transfers involve S-adenosylmethionine.

units, via the PLP-dependent serine hydroxymethyltransferase described in Chapter 20. Reduction of 5,10-methylene-THF to 5-methyl-THF commits this one-carbon unit to methyl group biogenesis because methionine synthase is the only enzyme known to utilize 5-methyl-THF in eukaryotes. Thus, if the methionine synthase reaction is blocked for some reason (e.g., a vitamin B_{12} deficiency), 5-methyl-THF accumulates, at the expense of depleted pools of the other THF coenzymes. This is the basis of the methyl trap we described in Chapter 20 (page 857).

As you might expect, regulatory mechanisms have evolved to maintain balance between the supply of one-carbon units and the production of methyl groups in the form of AdoMet. Because the MTHFR reaction catalyzes the committed step in methyl group biogenesis, the regulation of MTHFR is crucial for one-carbon metabolism in all organisms. Eukaryotic MTHFRs are feedback inhibited by AdoMet, a key regulatory feature that prevents CH_2-THF depletion when AdoMet levels are adequate (Figure 21.9). Most eukaryotic MTHFRs are composed of an N-terminal domain that contains the catalytic site and a C-terminal regulatory domain that contains an AdoMet allosteric binding site. Plant and bacterial MTHFRs are not inhibited by AdoMet, but those enzymes use NADH rather than NADPH. Due to the high cytosolic NAD^+/NADH ratio, the MTHFR reaction in these organisms is likely to be reversible, obviating a need for feedback regulation by AdoMet. The other important regulatory site is the first step of the transsulfuration pathway, where AdoMet is a positive allosteric effector of cystathionine β-synthase (Figure 21.9). Thus, under conditions of adequate methionine and methyl groups, AdoMet simultaneously turns off entry of one-carbon units into the methyl cycle, and diverts excess homocysteine into the transsulfuration pathway.

In humans, a genetic deficiency of cystathionine β-synthase leads to a condition called **homocystinuria**, in which homocysteine accumulates, as evidenced by excessive urinary excretion of homocystine (the oxidized disulfide derivative of homocysteine). Homocystinuria results in severe mental retardation, damage to blood vessels, and dislocation of the lens of the eye. In some patients, homocystinuria can be treated with vitamin B_6 (pyridoxine). Recall that cystathionine β-synthase is a PLP-dependent enzyme (Figure 21.4). The cystathionine β-synthase mutations in B_6-responsive patients typically produce enzymes with two- to five-fold lower affinity for PLP. These "K_M mutants" can often be rescued in vivo by administration of high doses of vitamin B_6.

After cystathionine β-synthase deficiency was described, it was found that similar symptoms resulted from deficiencies of either of two related enzymes—methionine synthase or 5,10-methylenetetrahydrofolate reductase. Deficiency of methionine synthase causes megaloblastic anemia and homocysteinemia, but is very rare in humans. MTHFR deficiency, however, is the most common inborn error of folate metabolism. Although a few cases of severe MTHFR deficiency are known, causing homocystinuria, much more common are patients that exhibit only moderate deficiency of MTHFR enzyme activity. In 1995, Rima Rozen and colleagues at McGill University in Montreal identified a C → T substitution in exon 4 of the human MTHFR gene, producing an Ala → Val substitution at position 222 in the catalytic domain of the enzyme (Figure 21.10). This missense, or nonsynonymous, codon change is quite common in North America, with up to 45% of the Caucasian population being heterozygous (carrying one T allele and one C allele), and 10%–15% being homozygous for the T allele. When a particular mutation occurs relatively frequently (>1% of the alleles in a population), it is called a *polymorphism*. Because this genetic change involves only a single nucleotide, it is called a single nucleotide polymorphism, or SNP. Patients who are homozygous for this C → T polymorphism in MTHFR exhibit mild homocysteinemia, but only if they also have low folate and riboflavin nutritional status. Heterozygotes generally have normal plasma homocysteine levels.

Biochemical analysis of the valine variant of the enzyme (product of the T allele) has provided a molecular explanation for the clinical findings. Compared to the normal alanine-containing enzyme (product of the C allele), the valine variant is *thermolabile*—it is completely inactivated at 46 °C, and is less stable than the Ala enzyme even at 37 °C. Lymphocytes taken from TT homozygotes have only ~ 50% of the MTHFR enzymatic activity of CC homozygotes in in vitro assays. How does this single amino acid substitution cause this thermolability? There is no crystal structure of a mammalian MTHFR yet, but structural studies of the *E. coli* homolog have shed much light on the problem. As mentioned above, bacterial MTHFRs lack the regulatory domain of the eukaryotic enzymes, but they are homologous to the catalytic domains. The *E. coli* enzyme is a homotetramer of 296 amino acid subunits. Each subunit is a classic $\beta_8\alpha_8$ barrel structure with the flavin cofactor (FAD) bound at the C-termini of the β-strands (Figure 21.11). Like the eukaryotic valine variant, the corresponding *E. coli* valine variant exhibits thermolability (Figure 21.12a). Kinetic studies of the Ala and Val versions of the *E. coli*

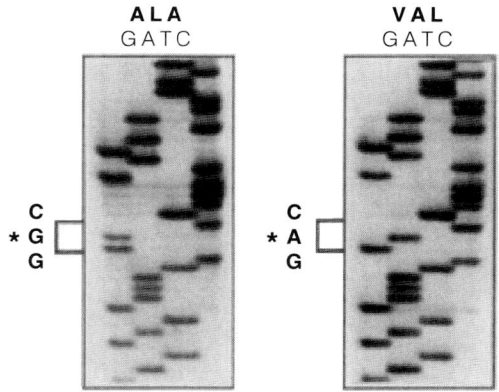

FIGURE 21.10

DNA sequencing reveals a common polymorphism in human methylenetetrahydrofolate reductase (MTHFR). The DNA sequence was obtained from the antisense strands of the MTHFR gene from two individuals, one carrying the C allele (left) and the other the T allele (right). The corresponding sense strands would encode a GCC codon (Ala) or GTC codon (Val), respectively.

Courtesy of Rowena Matthews, University of Michigan.

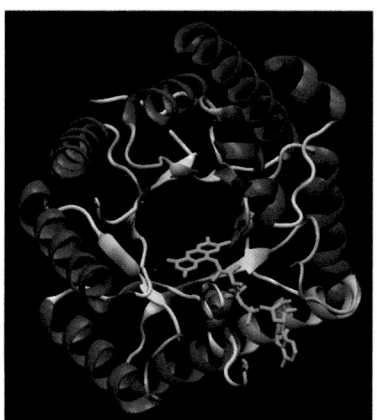 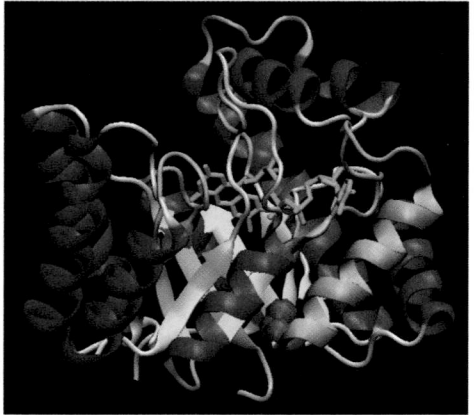

FIGURE 21.11

E. coli **MTHFR complexed with ligands.** This figure shows the X-ray crystal structure of the bacterial enzyme with bound FAD (green) (PDB ID 1B5T). The left panel shows a view down the axis of the $\beta_8\alpha_8$ barrel looking toward the C-terminal ends of the β strands (yellow). The right panel is a view perpendicular to the barrel axis. Alanine-177, shown in space-filling mode (red), is located at the N-terminal ends of the barrel. Substitution of this residue with a bulky valine perturbs packing of the adjacent α helix (cyan) that projects toward the FAD binding site.

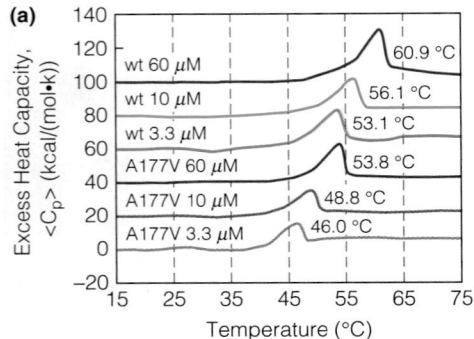

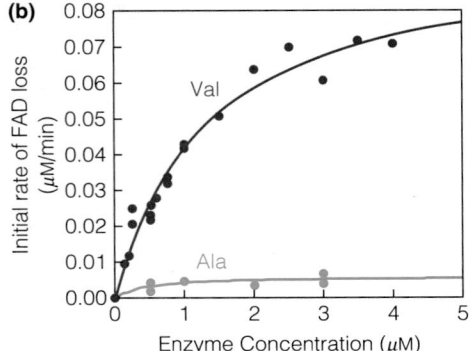

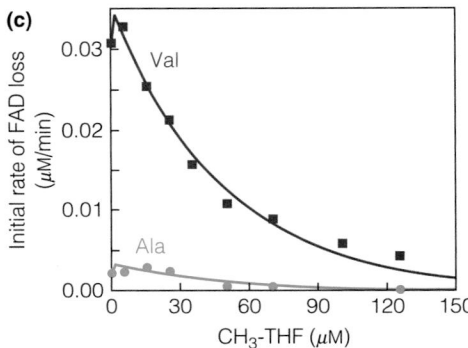

FIGURE 21.12

Biochemical analysis of Ala177Val mutant of E. coli MTHFR. (a) Differential-scanning calorimetry (see Tools of Biochemistry 6C) reveals the thermolability of the valine variant (A177V) as compared to the wild-type (wt) Ala177 enzyme. The melting temperatures (T_m) decrease for both enzymes as the protein concentrations decrease. **(b)** FAD dissociation was followed by observing the increase in flavin fluorescence (inset) over time after concentrated enzyme was diluted to the concentrations indicated on the x-axis. The Val177 enzyme loses FAD ~10 times faster than the Ala177 enzyme. **(c)** Folate cofactor (5-methyl-THF) protects both wild-type and mutant enzymes against FAD loss in a concentration-dependent fashion.

Courtesy of Rowena Matthews, University of Michigan.

Homocysteinemia can result from genetic defects in cystathionine β-synthase, methionine synthase, or 5,10-methylenetetrahydrofolate reductase, as well as from dietary deficiency of folic acid, vitamin B_6, or vitamin B_{12}.

enzyme revealed that the valine substitution did not affect K_M or k_{cat} values. Rather, the valine substitution greatly increases the propensity of the enzyme to lose its FAD cofactor (Figure 21.12b). Dissociation of the flavin cofactor decreases the thermal stability of the enzyme. Importantly, the folate substrate protects the enzyme against flavin loss (Figure 21.12c) and against thermal inactivation. In the *E. coli* enzyme, Ala177 (equivalent to Ala222 of human enzyme) is located at bottom of the β barrel, ~15 Å away from the catalytic site (Figure 21.11). Despite this distance, substitution of the larger valine residue disrupts the packing of an α helix that contributes to the FAD binding site. Propagation of this structural change through the protein decreases the binding affinity of the FAD cofactor compared to the normal alanine enzyme. Bound folate cofactor slows the loss of FAD. Thus, the biochemical properties of the enzyme predict the clinical phenotype: As long as cellular folate levels are adequate, TT homozygotes expressing the Val variant retain sufficient MTHFR activity. If their folate intake falls below a certain threshold level, however, MTHFR is no longer saturated with folate cofactor. FAD dissociation accelerates, leading to thermolability, decreased MTHFR activity, and decreased production of the 5-methyl-THF needed to remethylate homocysteine. This is a beautiful example of how structural and biochemical studies can lead to a better understanding of the molecular bases of inherited metabolic diseases.

From Figure 21.9, you can see how a deficiency in either cystathionine β-synthase, methionine synthase, or 5,10-methylenetetrahydrofolate reductase would cause homocysteine accumulation. In addition to these genetic causes, dietary deficiency of folic acid, vitamin B_6, or vitamin B_{12} can also lead to homocysteinemia.

Whereas the one-carbon and methyl cycles shown in Figure 21.9 are found in virtually all cells, liver and kidney possess an additional route for methionine synthesis that involves a different kind of transmethylation. Homocysteine can be remethylated using **betaine** as the methyl donor in a reaction catalyzed by the cytosolic betaine-homocysteine methyltransferase. Betaine, also known as trimethylglycine, is formed from the mitochondrial oxidation of choline.

CH3 CH3 Homocysteine S—CH3 H3C—N+—CH3 Oxidation H3C—N+—CH3 CH2 H2C—CH2OH H2C—COO− CH2 H3C N CH3 H—C—NH3+ H2C—COO COO−

Choline   **Betaine**   **Dimethylglycine**   **Methionine**

The betaine, which is a quaternary amine derivative of an amino acid, has a positive charge and a high group transfer potential, as does AdoMet itself. Because the methyl groups on betaine came originally from AdoMet during phosphatidyl choline synthesis (Chapter 19, page 784), the existence of this transmethylation process is consistent with the statement that all methyl groups come from methionine except those used in the synthesis of methionine.

Betaine synthesis is especially important in plants, where it accumulates to protect against the osmotic stresses of drought or high salinity. Betaine is one of several plant **osmoprotectants**, small, electrically neutral, nontoxic molecules that can stabilize proteins and membranes against denaturation by high salts and other solutes. The demand for methyl groups for betaine synthesis (via choline) often dwarfs that for all other methyl transfers in a stressed plant.

Methionine is also the source of dimethylsulfoniopropionate (DMSP), an osmoprotectant produced by many marine algae, coral-associated bacteria, and other phytoplankton. As shown in Figure 21.13, methionine is first transaminated and reduced. A second methyl group is donated by AdoMet, followed by an oxidative decarboxylation to give DMSP. This compound is catabolized by two

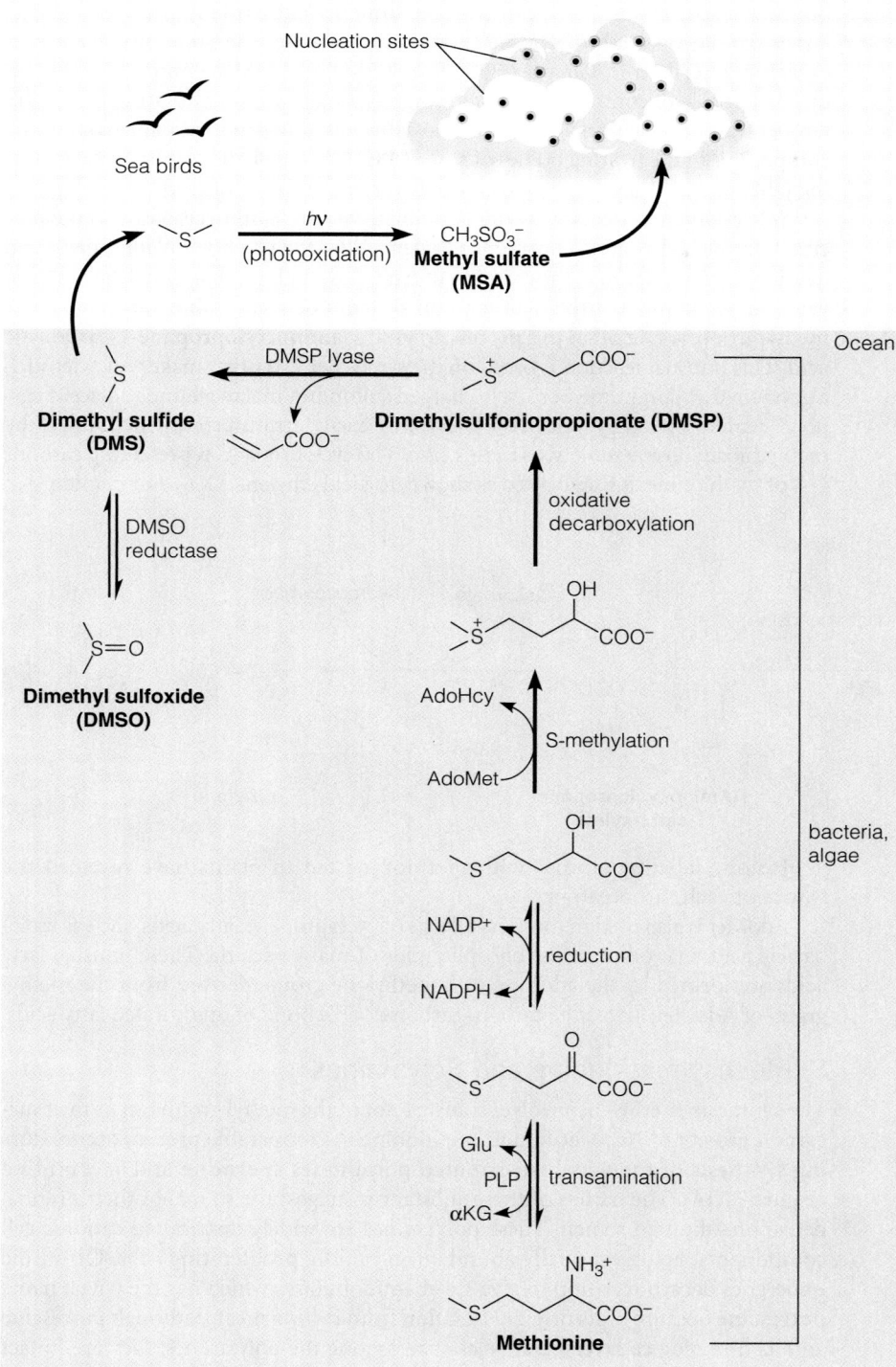

FIGURE 21.13

Environmental significance of dimethyl sulfide metabolism.

From *Science* 272:1599–1600, E. I. Stiefel, Molybdenum bolsters the bioinorganic brigade. © 1996. Reprinted with permission from Jeannette Stiefel.

competing pathways in marine bacteria. One pathway leads to methanethiol (CH_3SH) and acetic acid. In the pathway shown in Figure 21.13, DMSP is hydrolyzed to dimethyl sulfide (DMS) and acrylic acid. DMS is the principal volatile form of sulfur in the oceans, contributing approximately 1.5×10^{13} g of S to the atmosphere annually. The release of DMS to the atmosphere is probably responsible for the distinctive, and misnamed, "salt-air fragrance" that we associate with the sea. This smell attracts sea birds, who recognize it as an indicator of biological productivity in the ocean. Also, DMS in the atmosphere undergoes photooxidation to methyl sulfate (MSA), which serves as nucleation sites for the formation of clouds. DMS is also easily oxidized to dimethylsulfoxide (DMSO), but marine organisms posses an enzyme, **dimethylsulfoxide reductase**, that catalyzes the regeneration of DMS. DMSO reductase uses the same molybdopterin cofactor described in Chapter 20 (page 825). Thus, DMSO and DMS play an important role in controlling climate, contributing toward modulating the earth's temperature. A variation on this pathway is also used to produce sulfur-containing compounds in the fruiting bodies of some fungi. These volatile compounds, including DMS and H_2S, are important components of the aroma of truffles.

S-Adenosylmethionine plays an additional role in plant metabolism, as the precursor to the hydrocarbon hormone **ethylene**. Ethylene promotes plant growth and development and induces the ripening of fruit. Although mechanistic details are not yet clear, we know that the main carbon skeleton of methionine, rather than the methyl group, is split off in this process, to yield **1-aminocyclopropane-1-carboxylic acid**. This unusual reaction is based on the same chemistry that makes AdoMet such a good methyl donor: the positively charged sulfonium makes all three adjacent carbon centers strongly electrophilic and easily transferred, including the methythioadenosine moiety, as in this case. The cyclopropane, representing carbons 1–4 of methionine, is fragmented as shown, to yield ethylene, CO_2, and cyanide.

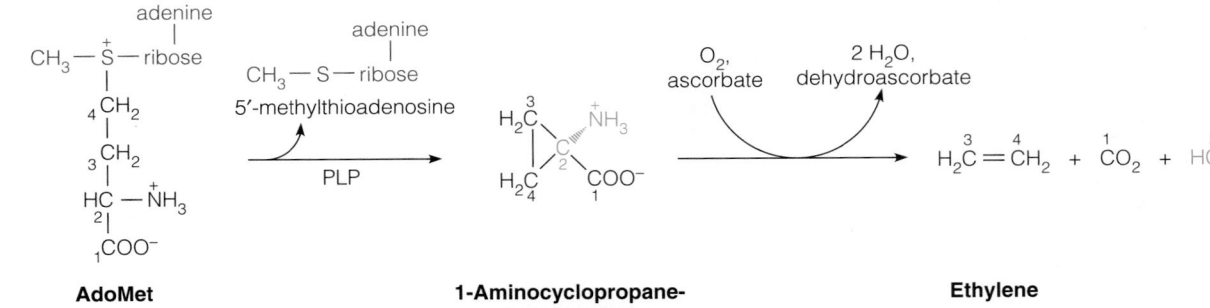

AdoMet   **1-Aminocyclopropane-**   **Ethylene**
                    1-carboxylate

Isotope labeling studies with methionine fed to plants have revealed the source of each carbon atom.

AdoMet is also used in the biosynthesis of cyclopropane fatty acids, such as lactobacillic acid, that occur in the phospholipids of many bacteria. These unusual fatty acids are formed by the addition of a methylene group, derived from the methyl group of AdoMet, across the carbon–carbon double bond of unsaturated fatty acids.

$$CH_3(CH_2)_5CH - CH(CH_2)_9COOH$$

Lactobacillic acid

S-Adenosylmethionine and Polyamines

The synthesis of ethylene involves transfer, not of the methyl group but of the four-carbon moiety of AdoMet-bound methionine. A comparable process occurs during synthesis of the widely distributed polyamines **spermine** and **spermidine** (Figure 21.14). The names of these substances suggest the source of their original detection—human semen. These polyamines are widely distributed cationic cell components, being especially abundant in rapidly proliferating cells. Ornithine undergoes decarboxylation to give 1,4-diaminobutane, which has the trivial name **putrescine** because of its original isolation from rotting meat. Although putrescine and its homolog **cadaverine** are classified among the polyamines, they are in fact diamines; they are synthesized by decarboxylation of ornithine and lysine, respectively. Ornithine, in turn, is synthesized from glutamate in the same pathway that leads eventually to arginine and proline (see Figure 21.31). **Ornithine decarboxylase** is a highly regulated enzyme whose activity responds to many hormonal stimuli. It has an extremely short metabolic half-life (approximately 10 minutes), and its intracellular activity is controlled largely at the level of protein degradation.

As polycations, the polyamines play multiple roles in stabilizing conformations of negatively charged nucleic acids. A polyamine molecule can bind to phosphates on both strands of a duplex, thereby stabilizing double-stranded DNA or a duplex region of RNA. In bacteriophage T4, for instance, about 40% of the negative charge on the viral DNA is neutralized by polyamines. Transfer RNA from some sources

Polyamines are required for cell proliferation because of their roles in stabilizing duplex DNA structures.

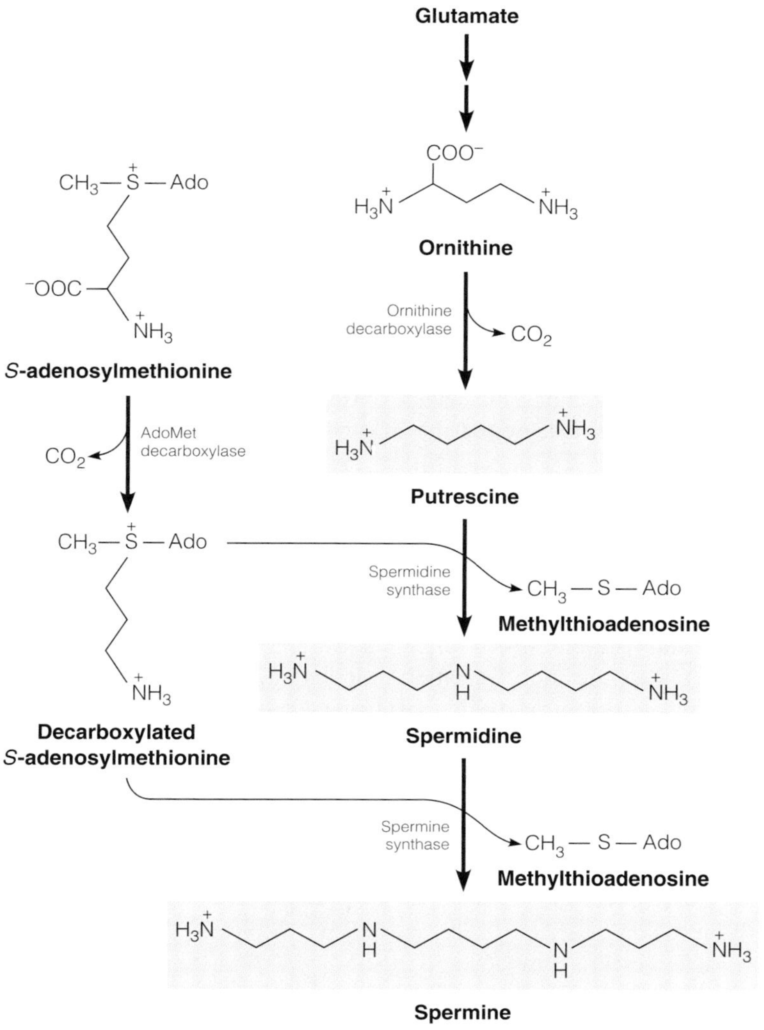

FIGURE 21.14

Biosynthesis of putrescine, spermidine, and spermine.

Difluoromethylornithine

contains two molecules of tightly bound spermine or spermidine per molecule of transfer RNA. Some proteins contain covalently bound polyamines, with the nitrogens linked to glutamate γ-carboxyl groups. In bacterial cells, polyamines also participate in regulating internal osmotic strength and they help to stabilize membranes. In animal cells, polyamines are involved in control of the electrical properties of excitable membranes. Also, polyamine biosynthesis is closely related to the proliferative state of the cell; when nucleic acid synthesis is activated, so is polyamine synthesis. Consequently, investigators are asking whether activation of polyamine synthesis early in tumorigenesis could be a marker for early cancer diagnosis or a target for cancer treatment. Consistent with the latter, the antiparasitic drug **difluoromethylornithine (DFMO)** arrests progression through the cell division cycle (see Chapter 28) by inhibiting ornithine decarboxylase, suggesting a role for polyamines in cell cycle regulation.

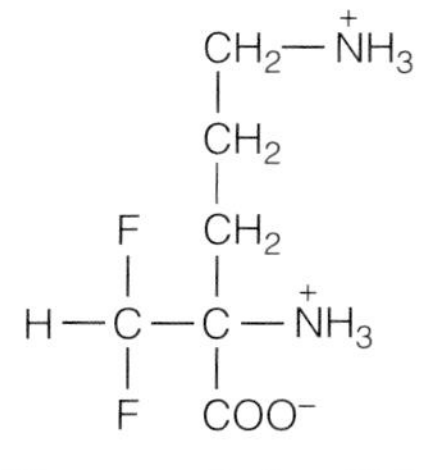

Putrescine serves as the precursor to spermidine, then to spermine, through the AdoMet-mediated transfer of active propylamino groups (Figure 21.14). First, AdoMet is decarboxylated by **AdoMet decarboxylase**. This enzyme is an example of a decarboxylase that contains as its cofactor not pyridoxal phosphate but covalently bound pyruvate (see page 848). The resulting propylamino group is then transferred to putrescine to give spermidine, and a second group is transferred to spermidine, catalyzed by spermine synthase, to give spermine. The other product of these reactions is 5′-methylthioadenosine, which undergoes phosphorolytic

Glutamate

CO_2

γ-Aminobutyric acid

cleavage to adenine and **5-methylthioribose-1-phosphate**. The latter compound is used for resynthesis of methionine by another pathway, not shown here.

Spermine is clearly required for normal development in mammals. In humans, the rare X-linked Snyder-Robinson syndrome is caused by mutation of the gene encoding spermine synthase. This syndrome is characterized by mental retardation, skeletal defects, osteoporosis, and facial asymmetry.

Other Precursor Functions of Glutamate

γ-Aminobutyric Acid

In addition to serving as a precursor for ornithine and the polyamines, glutamate is one of several amino acids serving as precursors to compounds that function in transmission of nerve impulses. Decarboxylation of glutamate yields **γ-aminobutyric acid**, or **GABA**. In addition, glutamate itself is a neurotransmitter. Functions of both glutamate and GABA in neurotransmission will be discussed later in this chapter.

Glutathione

Another major and ubiquitous metabolic fate of glutamate is the synthesis of **glutathione (GSH)**, or γ-glutamylcysteinylglycine. This tripeptide is synthesized by linking glutamate, cysteine, and glycine through amide bonds, but glutamate is linked via its γ-carboxyl rather than its α-carboxyl. GSH synthesis occurs as part of the **γ-glutamyl cycle**, which functions to transport amino acids into the cell (Figure 21.15). Amide bond formation is driven by ATP hydrolysis; each carboxyl group is activated for nucleophilic substitution by formation of an acyl phosphate intermediate (see margin, page 883). After its synthesis inside the cell, GSH is transported across the plasma membrane where it is hydrolyzed and its γ-glutamyl moiety is transferred to an extracellular amino acid (usually cysteine or methionine), forming a γ-glutamyl-amino acid conjugate. This reaction is catalyzed by **γ-glutamyl transpeptidase**, a plasma membrane protein whose active site faces the extracellular space. ATP is not required to drive this reaction because it involves the exchange of one amide bond for another. The γ-glutamyl-amino

FIGURE 21.15

Glutathione and the γ-glutamyl cycle. The enzymes involved are ❶ γ-glutamylcysteine synthase, ❷ glutathione synthase, ❸ γ-glutamyl transpeptidase, ❹ γ-glutamyl cyclotransferase, ❺ 5-oxoprolinase, ❻ dipeptidase.

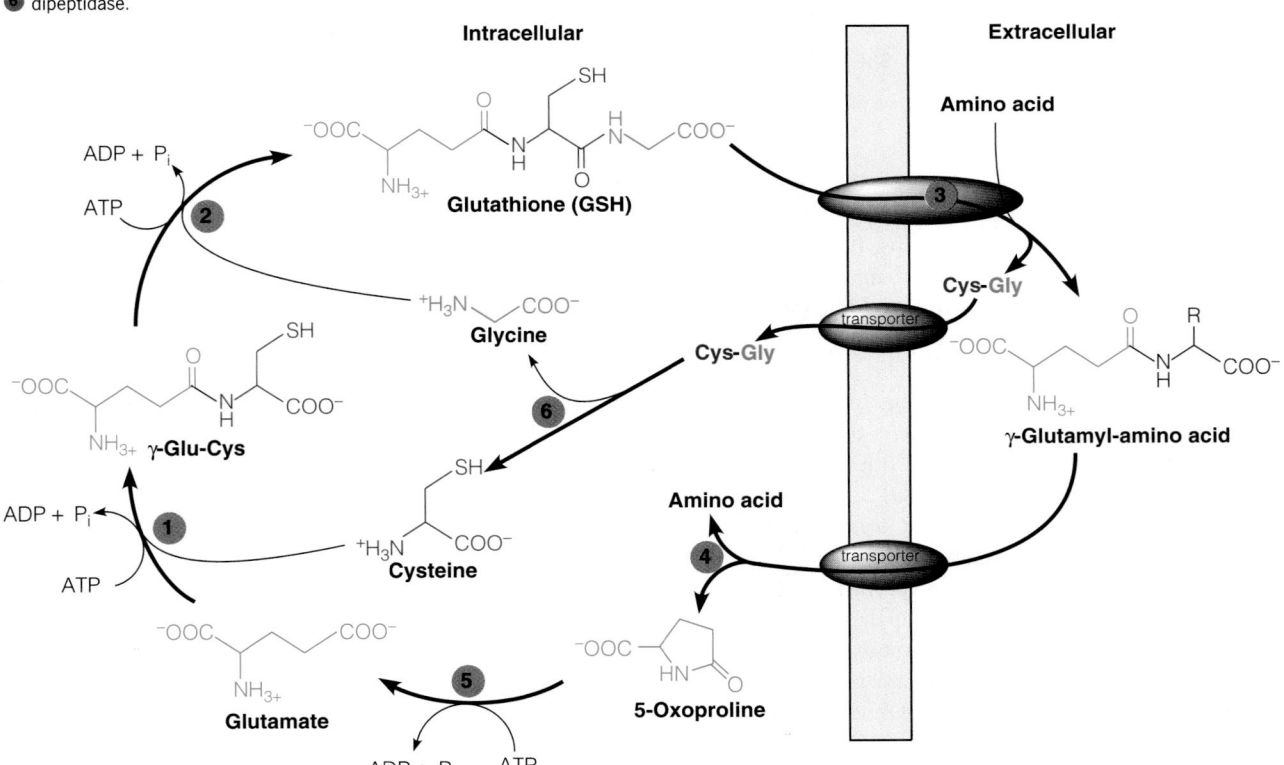

acid conjugate is transported into the cell, where it is hydrolyzed to liberate the free amino acid and 5-oxoproline from the glutamate moiety. 5-Oxoproline has a very stable internal amide bond, requiring ATP hydrolysis to cleave it to give glutamate. The other product of the transpeptidase reaction, the dipeptide cysteinylglycine, is hydrolyzed to the free amino acids, either extracellularly or intracellularly. These are then coupled to glutamate to regenerate GSH. The γ-glutamyl cycle is particularly important in cells with secretory or absorptive functions, such as kidney, pancreas, and intestine.

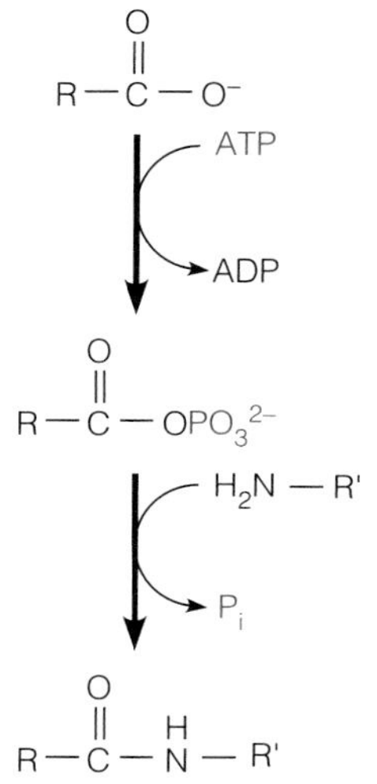

As first introduced in our discussion of the pentose phosphate pathway in Chapter 13, glutathione plays several important metabolic roles. This tripeptide, abundant in all cells, protects against two kinds of metabolic stress. First, it can nonenzymatically reduce a number of substances, such as peroxides or free radicals, which accumulate in cells under oxidizing conditions (Chapter 15, page 667). Glutathione maintains an intracellular reducing environment, which prevents intracellular protein thiols from oxidizing to disulfides. Second, through the action of widely distributed enzymes, the **glutathione S-transferases**, glutathione participates in detoxification of many substances, such as **xenobiotics** (foreign organic compounds not produced in metabolism) or electrophiles produced through the action of cytochrome P450–linked oxidases. Such compounds include organic halides, fatty acid peroxides derived from lipid oxidation, and products derived from radiation-damaged DNA. Glutathione reacts with such a compound (denoted RX) as shown here, followed by cleavage of the γ-glutamyl and glycyl residues and then acetylation by acetyl-CoA to give a **mercapturic acid**. This more soluble, less toxic derivative of the original compound can then be excreted in the urine.

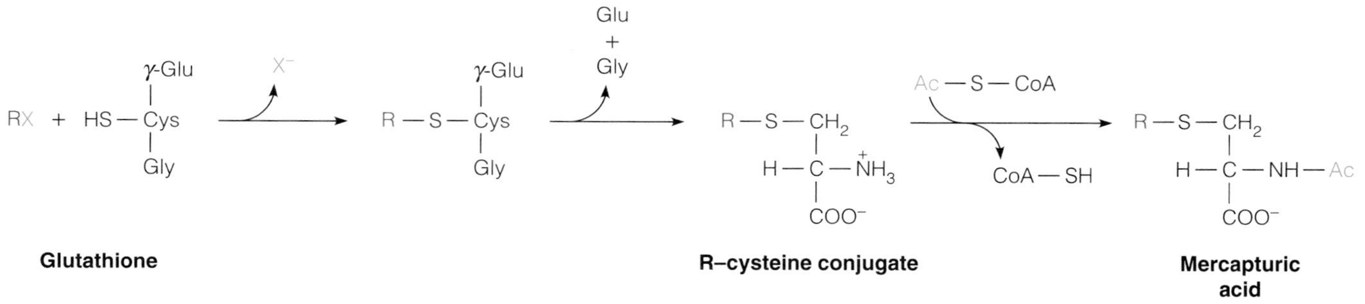

Alternatively, detoxification may involve catabolism of the cysteine conjugate to a methylthio compound or a glucuronide.

Another family of sulfur amino acids, the **ovothiols**, are found in fertilized eggs, where they play a role comparable to that of glutathione. These mercaptohistidine derivatives can posses one, two, or no methyl substituents on the α-amino group. Ovothiols, which are present at millimolar concentrations in eggs of marine echinoderms and mollusks, protect the egg against oxidative damage by peroxides produced at the egg surface early in fertilization. Ovothiol is in turn reduced by glutathione. Ovothiols also serve as antioxidants in trypanosomatid parasites such as *Leishmania* and *Trypanosoma*.

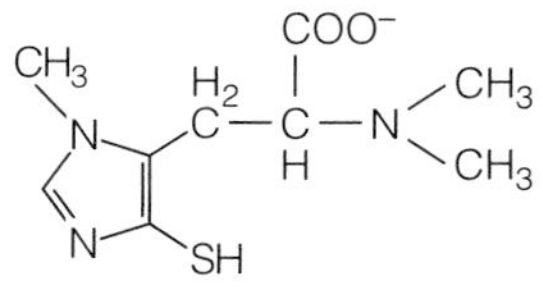

Finally, glutamate is involved in the synthesis, via an ATP-dependent conjugating system, of the polyglutamate tails of folic acid and its coenzymes (see Chapter 20, page 849). As in glutathione synthesis, the glutamate residues in the polyglutamate tails are linked through their γ-carboxyl groups.

Nitric Oxide and Creatine Phosphate

Beginning in the late 1980s, an unexpected role for arginine was described as precursor to a novel second messenger and neurotransmitter. This novel regulator has been identified as a gas, the free radical **nitric oxide** (NO), which is produced from arginine in an unusual reaction that also yields citrulline (Figure 21.16).

FIGURE 21.16

Biosynthesis of nitric oxide and creatine phosphate from arginine. In NO synthesis, the N^ω-hydroxy-L-arginine intermediate remains tightly bound to NO synthase (NOS).

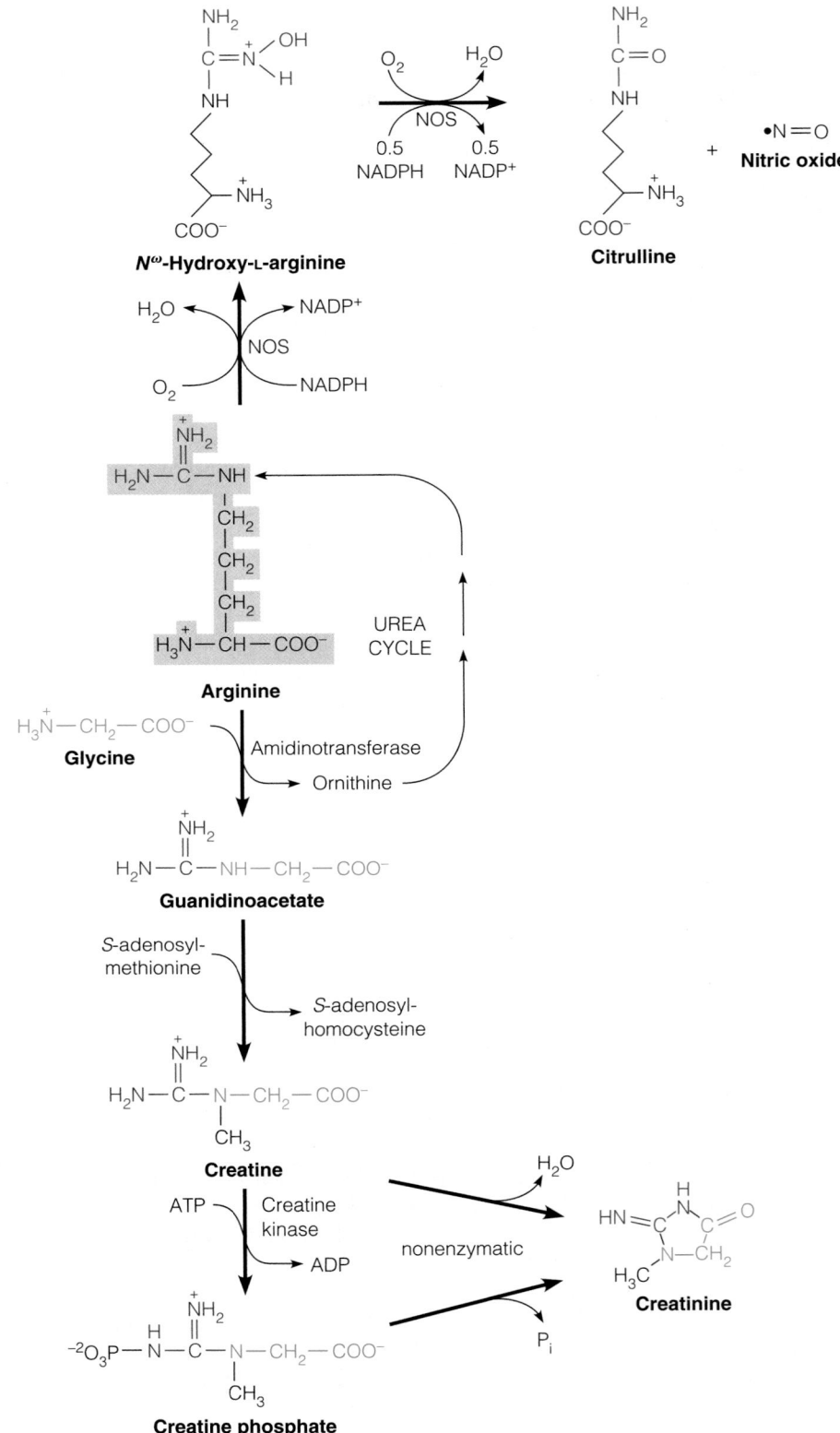

NO synthase (NOS), which catalyzes a two-step NADPH-dependent oxidation of arginine by O_2, contains bound FMN, FAD, heme iron, and tetrahydrobiopterin, the same cofactor involved in the synthesis of tyrosine from phenylalanine (Figure 21.7). The physiological role and mechanism of action of this gaseous signaling molecule is described in Chapters 7 and 23.

Arginine is also the precursor to the energy storage compound creatine phosphate (see Chapter 3, page 76). The guanidino group is transferred to glycine, releasing the rest of the arginine molecule as ornithine (Figure 21.16). Arginine can be regenerated from ornithine via the urea cycle (see Figure 20.12, page 842). The other product of the amidinotransferase reaction, **guanidinoacetic acid**, is methylated by S-adenosylmethionine to give creatine, followed by phosphorylation of the latter to creatine phosphate, catalyzed by creatine kinase. The amidinotransferase reaction is the rate-limiting step for creatine synthesis; its expression is repressed by creatine. Based on the tissue distribution of the enzymes involved, creatine biosynthesis appears to be an interorgan process in mammals. Guanidinoacetate is produced by the kidney and transported to the liver for methylation to creatine. Because creatine kinase is very low in liver, creatine is released into the bloodstream. Tissues with high energy demand, such as muscle and brain, take up creatine from blood and phosphorylate it to creatine phosphate.

Both creatine and creatine phosphate spontaneously cyclize to creatinine (Figure 21.16), which is excreted in the urine (1–2 gm per day in a normal adult). A typical western diet can supply about half of this lost creatine; the other half is from endogenous biosynthesis, and it has been estimated that adult humans synthesize ~4–8 mmol creatine per day. Indeed, the methylation of guanidinoacetic acid to creatine consumes more AdoMet than all other methylation reactions combined.

Tyrosine Utilization in Animals

Tyrosine plays several important roles in animal metabolism—as precursor to thyroid hormones, the biological pigments called **melanins**, and the *catecholamines*, which serve both as hormones and as neurotransmitters (discussed later in this chapter).

Thyroid Hormones

Thyroid hormones stimulate a number of metabolic processes, through activation of the transcription of particular genes (see Chapter 23). The synthesis of thyroid hormones, principally **thyroxine** (T_4) and **triiodothyronine** (T_3), occurs by an unusual pathway, at the level of tyrosine residues in a specific protein, **thyroglobulin**. This process occurs in the thyroid gland, which concentrates iodide ion from the blood serum for this purpose. Thyroglobulin is a large (2750 amino acids per monomer in humans) homodimeric glycoprotein synthesized by follicle cells of the thyroid gland and secreted into the follicular lumen. As many as 20% of the 140 tyrosine residues in thyroglobulin are iodinated in a reaction catalyzed by thyroperoxidase. In this H_2O_2-requiring reaction, iodide (I^-) is oxidized and covalently bound to tyrosyl side chains of thyroglobulin, forming mono- and di-iodotyrosyl residues. As shown in Figure 21.17, two iodinated tyrosine residues on the same polypeptide chain are oxidatively coupled (also catalyzed by thyroperoxidase) to give a residue of T_3 or T_4. Iodinated thyroglobulin is pinocytosed back into the follicle cell, where upon hormonal stimulation, it undergoes degradation to yield the free hormones, which are transported to their sites of action through the bloodstream. One result of iodine deficiency is **goiter**, a condition in which the thyroid gland grows abnormally large as it attempts to scavenge all available iodine. Before iodized salt came into widespread use, goiter was endemic in regions whose soil was deficient in iodine.

Melanins

The synthesis of melanins (Figure 21.18) occurs in pigment-producing cells, the **melanocytes**. Melanocytes in mammals and birds produce two chemically distinct types of melanin, **eumelanin** and **pheomelanin**. The biochemical machinery for melanin synthesis is contained within membrane-bound organelles

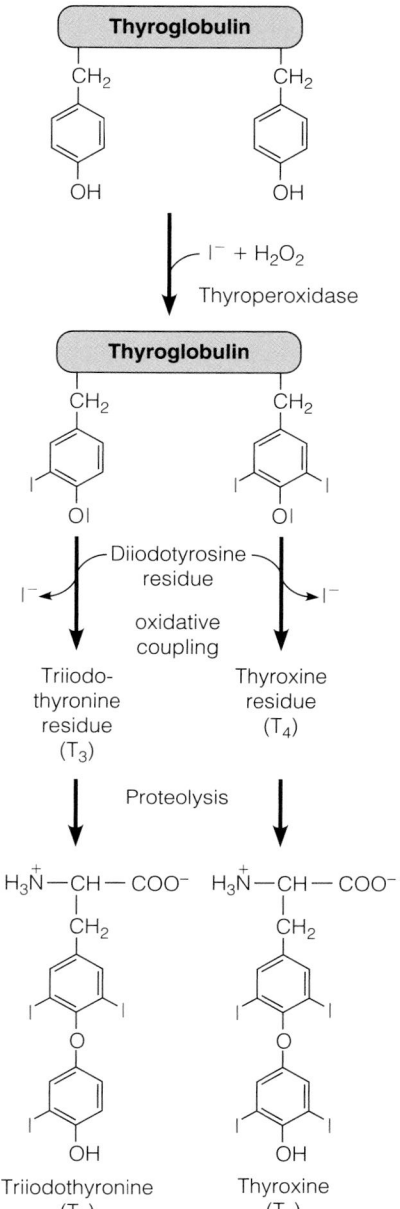

FIGURE 21.17

Biosynthesis of thyroid hormones as residues in the protein thyroglobulin. The iodinated forms of tyrosine—triiodothyronine (T_3) and thyroxine (T_4)—are released from these proteins by proteolytic degradation.

FIGURE 21.18

Biosynthetic pathways from tyrosine to melanins. Only a few steps are enzyme-catalyzed; most are spontaneous. Several steps involve a dopaquinone-dependent redox exchange; the dopa formed can be recycled to dopaquinone (DQ) via tyrosinase. The structures of eumelanin and pheomelanin are general ones; the arrows indicate sites for attachment to other units.

Photochemistry and Photobiology 84:582–592, S. Ito and K. Wakamatsu, Chemistry of mixed melanogenesis—pivotal roles of dopaquinone. © 2008 John Wiley & Sons, Inc. Modified with permission from John Wiley & Sons, Inc.

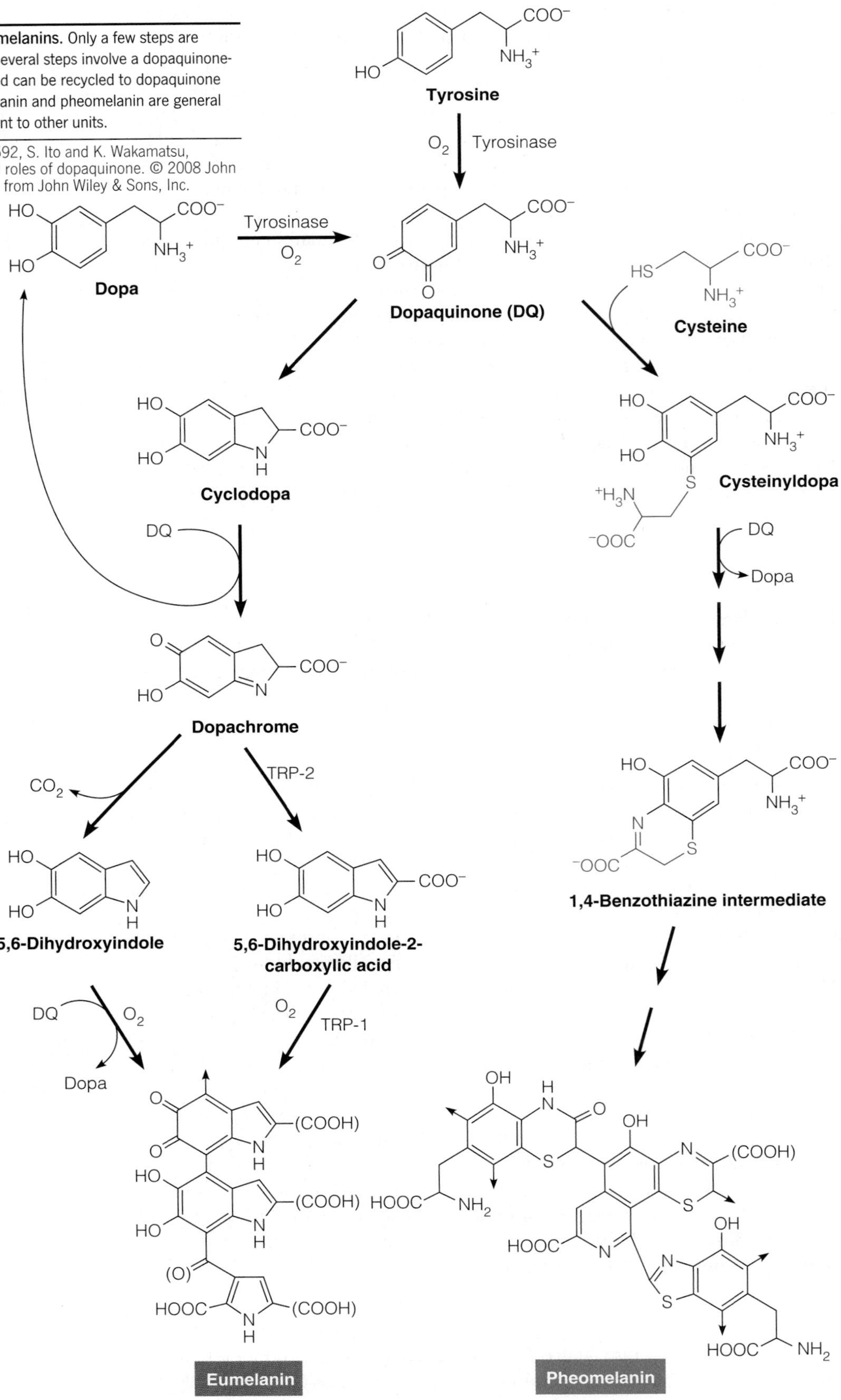

called melanosomes, to protect the cell against potentially toxic intermediates. Three major pigment enzymes are bound to the inner face of melanosome membranes: tyrosinase, tyrosinase-related protein-1 (TRP-1), and dopachrome tautomerase, also known as tyrosinase related protein-2 (TRP-2). The rate-limiting step is the first one, catalyzed by tyrosinase, a copper-containing oxygenase. It had been initially believed that **3,4-dihydroxyphenylalanine** (usually called **Dopa**) is formed first on the way to dopaquinone. Later work showed that dopaquinone is formed directly in the tyrosinase reaction. Dopaquinone (DQ) is highly reactive, and some of the subsequent reactions occur spontaneously. In the absence of sulfhydryl compounds, dopaquinone undergoes intramolecular addition of the amino group to produce cyclodopa. A nonenzymatic redox exchange between cyclodopa and dopaquinone then gives rise to dopachrome and dopa. Dopachrome then gradually rearranges to dihydroxyindoles, which are further oxidized and polymerized to produce the blackish-brown eumelanins.

In another branch of the pathway, dopaquinone reacts with cysteine en route to a related series of polymers, the yellowish- or reddish-brown pheomelanins. An individual's pigmentation is determined by the relative amounts of red and black melanins in the skin. These in turn result from the distribution and density of melanocytes in the basal layers of the skin, as well as the activities of the pathways leading to the different melanins. A genetic deficiency of tyrosinase causes an individual to lack pigmentation, a condition called **albinism**. The albino (white) mice and rats commonly used in research have an inherited defect in tyrosinase.

Humans exhibit a tremendous degree of variation in skin and hair color. Variations in skin color are adaptive, and are related to the amount of ultraviolet (UV) radiation that penetrates the skin. For the earliest *Homo sapiens* living in the equatorial tropics, dark skin provided protection against skin cancer. However, as early humans migrated out of Africa into more northern latitudes, skin pigmentation interfered with their ability to synthesize adequate levels of vitamin D, leading to rickets (Chapter 19). Lighter skinned variants would have had better reproductive success, and thus light skin became the predominant phenotype in these more northern populations. In addition to tyrosinase, polymorphisms in more than a dozen genes are now known to impact skin, hair, or eye color. One of the most prominent of these is the SLC24A5 gene, which encodes a potassium-dependent sodium–calcium ion exchanger. This protein is fully functional in African humans and most other vertebrates. However, virtually all lighter-skinned humans of European ancestry possess a nonsynonymous single nucleotide polymorphism (SNP) in the SLC24A5 gene that results in replacement of Ala with Thr at position 111 in the protein. Although the biochemical mechanism is not yet understood, this polymorphism impairs the sodium–calcium ion exchange activity of the protein, causing a significant reduction in melanin pigment production, and hence, lighter skin color.

Aromatic Amino Acid Utilization in Plants

Phenylalanine and tyrosine serve as precursors to an enormous number of plant substances, ranging from the polymeric lignin to tannins, pigments, and many of the flavor components of spices. In fact, the role of these amino acids as precursors to such substances in cinnamon oil, wintergreen oil, bitter almond, nutmeg, cayenne pepper, vanilla bean, clove, and ginger is related to their designation as *aromatic* amino acids. These are derived from coniferyl alcohol, which is also the central intermediate in lignin synthesis. L-Tyrosine is the starting point for the synthesis of opiates such as codeine and morphine in the poppy plant.

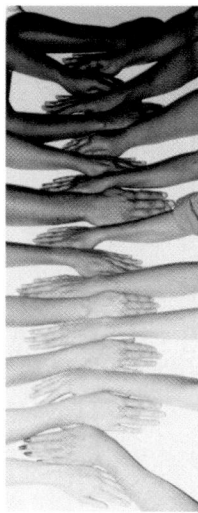

Human Molecular Genetics 18:R9–R17. R. A. Sturm, Molecular genetics of human pigmentation diversity. © 2009, by permission of Oxford University Press.

Coniferyl alcohol

Codeine

Morphine

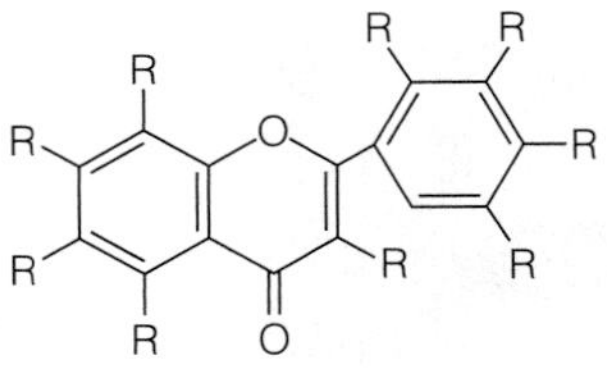

A flavonoid

Phenylalanine serves as the precursor to a large number of plant pigments and related polyphenolic compounds called **flavonoids**. These include many flower colorants, which serve in part as ultraviolet protectants, and also the respiratory inhibitor rotenone (see page 634). In the generic structure shown in the margin, the aromatic ring on the right comes from phenylalanine via the scheme shown below, and the ring to the left comes from malonyl-CoA, in a process akin to fatty acid and polyketide synthesis (Chapter 17). The substituent groups (R) are combinations of —H, —OH, and —OCH$_3$.

The same biosynthetic scheme also leads to a class of flavonoids called **anthocyanins**, which are common flower pigments. Substituents on the rings determine specific color, as shown. An offshoot from this pathway leads to synthesis of cocaine.

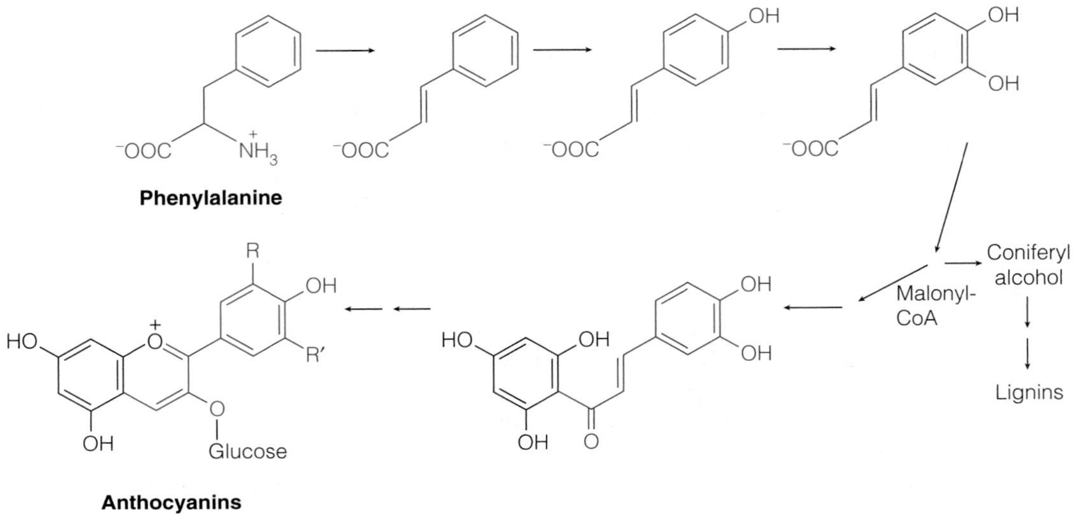

Phenylalanine

Anthocyanins

If R and R′ are both H, color is orange-red
If R is H and R′ is OH, color is crimson-purple
If R and R′ are both OH, color is blue

Coniferyl alcohol is the precursor to the complex and nearly inert **lignins**, a major constituent of woody tissue. Much of the effort in creating paper and textiles from wood involves degrading lignin to gain access to the cellulose fibers contained in woody tissues. An intriguing use of biotechnology involves the substitution of living organisms for the chemical pulping processes used to break down lignin, which release large quantities of sulfites to the environment. In these "biopulping" applications, fungi that produce lignin-degrading enzymes are being developed for production of cellulose without the accumulation of chemical pollutants.

Tryptophan is utilized for synthesis of a plant growth hormone. As shown here, the transamination product of tryptophan is decarboxylated to yield **indole-3-acetic acid**, or **auxin**.

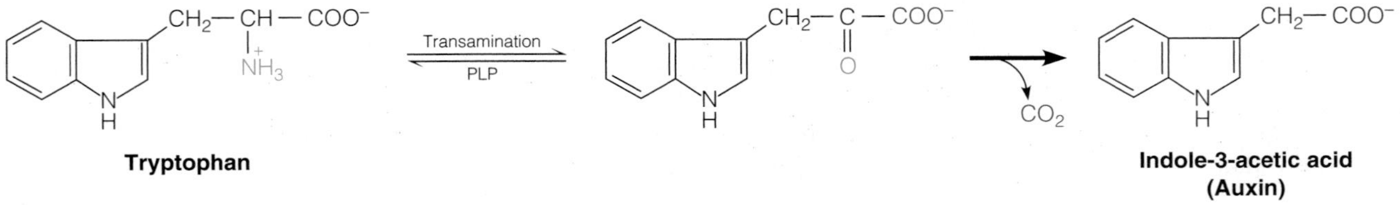

Tryptophan   **Indole-3-acetic acid (Auxin)**

Porphyrin and Heme Metabolism

Biosynthesis of Tetrapyrroles: The Succinate–Glycine Pathway

A major metabolic fate of glycine is its utilization for tetrapyrrole biosynthesis. *Tetrapyrrole* is a generic term for compounds containing four linked pyrrole rings. Four such classes of compounds are widespread in biology: the widely distributed iron *porphyrin* heme; the *chlorophylls* of plants and photosynthetic bacteria; the **phycobilins**, photosynthetic pigments of algae (Chapter 16); and the *cobalamins*, notably vitamin B_{12} and its derivatives (Chapter 20). Structures of most of these compounds have been shown previously. All tetrapyrroles are synthesized from a common precursor, **δ-aminolevulinic acid (ALA)** (also called 5-aminolevulinic acid). Figure 21.19 illustrates the relationships among the various synthetic pathways.

We shall concentrate here on the well-understood porphyrin synthetic pathway, which leads to heme. This pathway is widespread in animal tissues and, so far as is known, is similar in all organisms containing heme proteins, such as cytochromes. Eight reactions are involved, and they occur in two different cell compartments (Figure 21.20). The first reaction occurs in mitochondria, followed by four reactions in the cytosol, and finally, three more mitochondrial reactions. As we shall see, this compartmentation provides the opportunity for a novel control mechanism for the pathway.

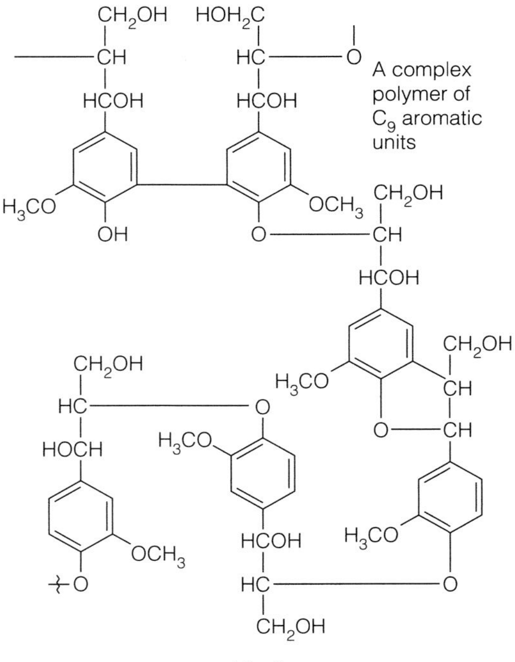

Early labeling studies in animals revealed that *all of the nitrogen of heme is derived from glycine, and all of the carbon is derived from succinate and glycine.* Hence, this synthesis is often called the **succinate–glycine pathway**. The first reaction is catalyzed by a pyridoxal phosphate–dependent enzyme, **δ-aminolevulinic acid synthase**, or ALA synthase. As shown in Figure 21.21, the binding of glycine to PLP (steps 1, 2) activates the α-carbon of glycine for an attack on the thioester carbon of succinyl-CoA (step 3). Elimination of CoASH (step 4) and decarboxylation (step 5) follow, to give the product ALA.

In most bacteria, archaea, and plants, ALA is formed by a completely different pathway, a three-step sequence beginning with glutamate (Figure 21.22). The first

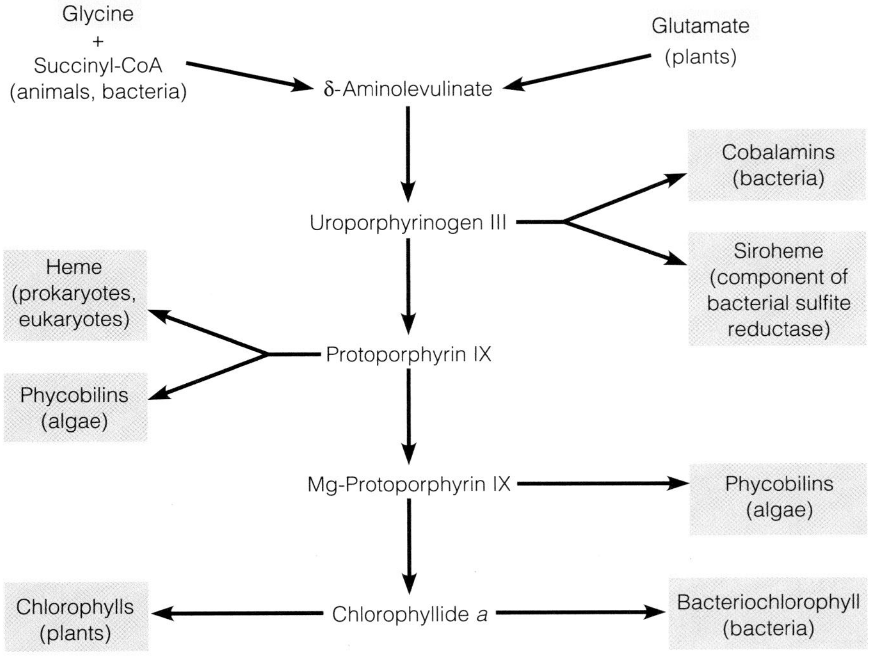

Lignin

δ-Aminolevulinic acid synthase brings together succinyl-CoA and glycine, which furnish all of the carbon and nitrogen in porphyrins, cobalamins, phycobilins, and chlorophylls.

FIGURE 21.19

Biosynthetic pathways to tetrapyrroles. Tetrapyrroles include heme, chlorophylls, phycobilins, and cobalamins. All are synthesized from δ-aminolevulinic acid, which is formed differently in plants from the way it is formed in bacterial and animal cells.

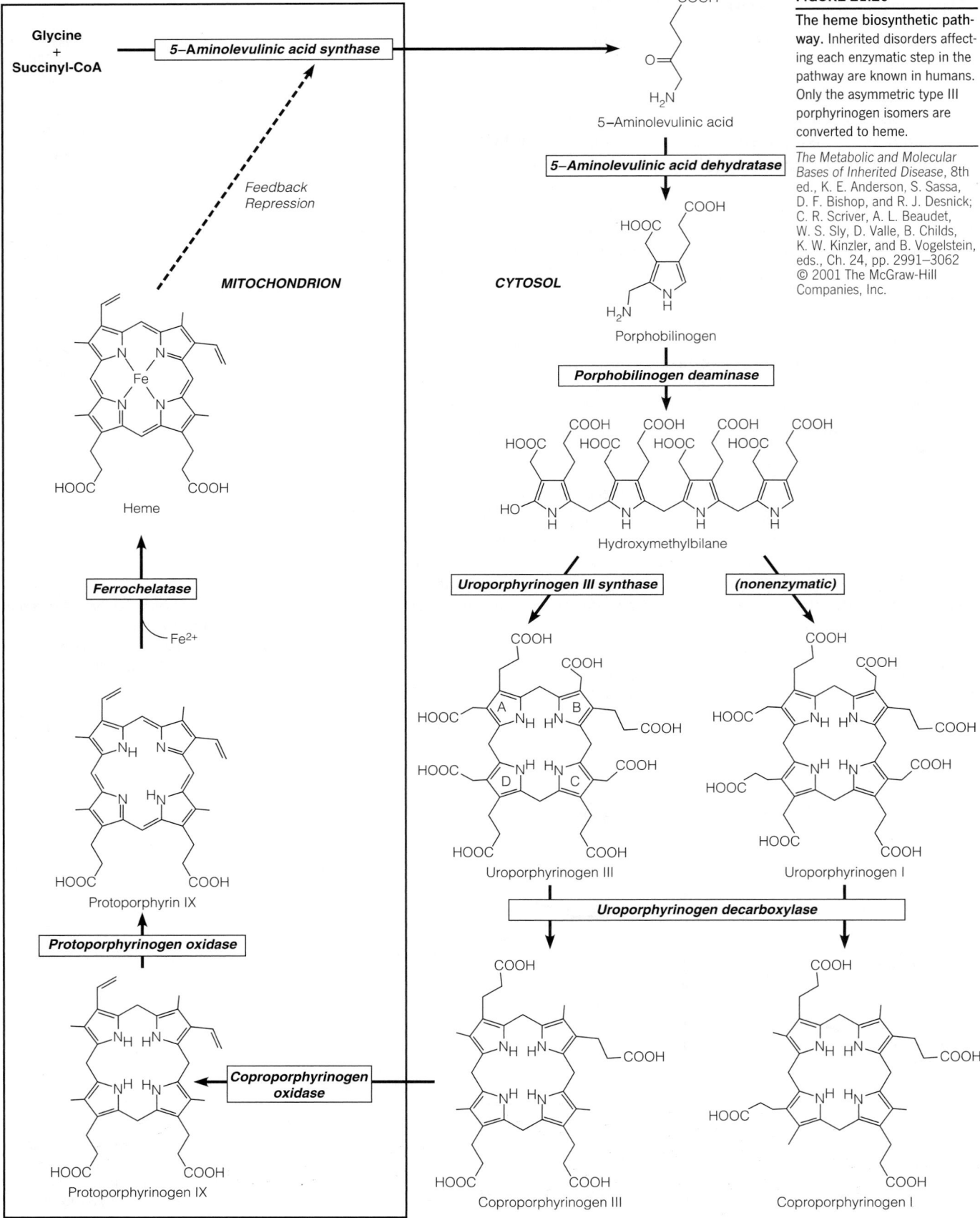

FIGURE 21.20

The heme biosynthetic pathway. Inherited disorders affecting each enzymatic step in the pathway are known in humans. Only the asymmetric type III porphyrinogen isomers are converted to heme.

The Metabolic and Molecular Bases of Inherited Disease, 8th ed., K. E. Anderson, S. Sassa, D. F. Bishop, and R. J. Desnick; C. R. Scriver, A. L. Beaudet, W. S. Sly, D. Valle, B. Childs, K. W. Kinzler, and B. Vogelstein, eds., Ch. 24, pp. 2991–3062 © 2001 The McGraw-Hill Companies, Inc.

FIGURE 21.21

The δ-aminolevulinic acid synthase reaction. The enzyme-bound pyridoxal phosphate (PLP) cofactor is shown in black. Like other PLP-requiring enzymes, the mechanism involves formation of a Schiff base between the amino acid and PLP, and a stabilized carbanion intermediate (compare with Figure 20.14, page 847).

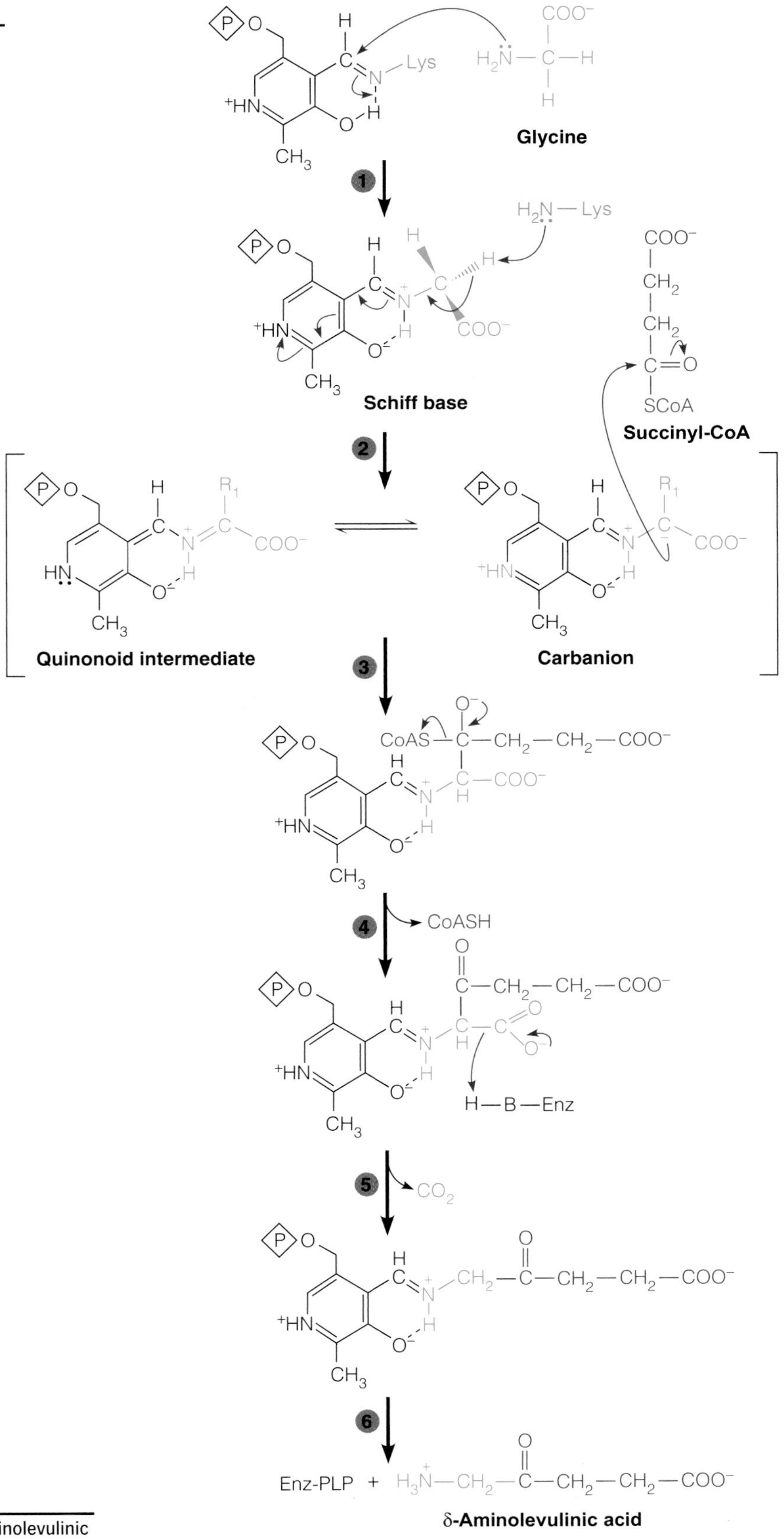

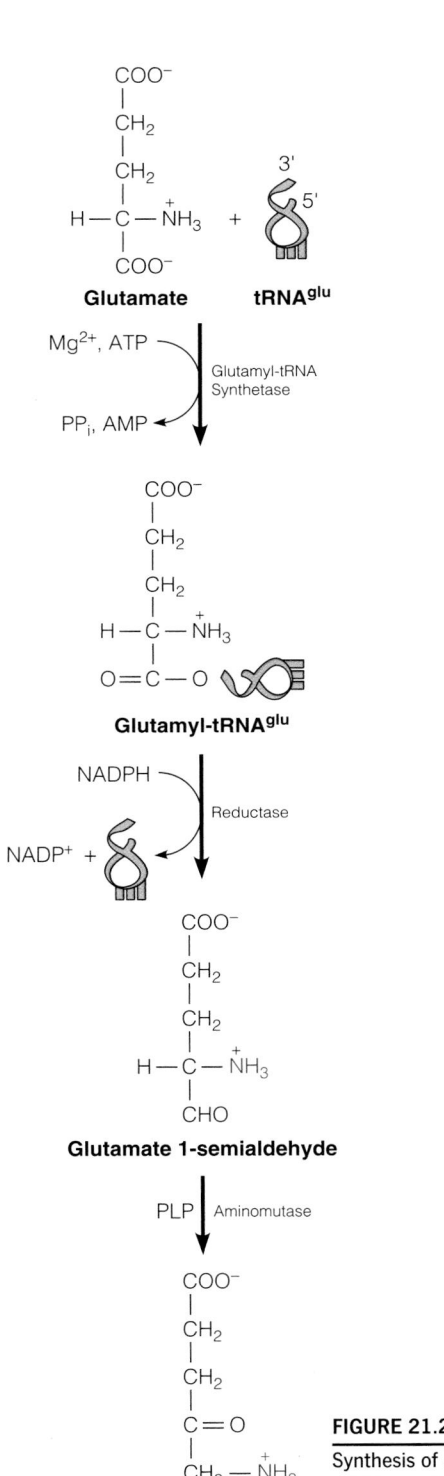

FIGURE 21.22

Synthesis of δ-aminolevulinic acid in most bacteria, archaea, and plants.

Pyridoxamine phosphate (PMP)

reaction of this novel pathway links glutamate, through its carboxyl group, to a specific transfer RNA, just as occurs in protein synthesis. The carboxyl group thus activated is then reduced by NADPH, giving glutamate 1-semialdehyde, which finally undergoes rearrangement to give ALA. The last enzyme, glutamate 1-semi-aldehyde aminomutase, catalyzes a PLP-dependent *internal* transamination. The proposed mechanism requires the PLP cofactor to be already aminated, as pyridoxamine 5′-phosphate (PMP). This amino group is transferred to the carbonyl carbon of glutamate 1-semialdehyde to give the intermediate 4,5-diaminovalerate and PLP. The C4 amino group is then transferred to PLP to yield ALA and regenerate PMP on the enzyme.

Glutamate 1-semialdehyde   **4,5-Diaminovalerate**   **ALA**

Despite using different substrates, glutamate 1-semialdehyde aminomutase is mechanistically and structurally related to the animal ALA synthase.

Because the major end product of the pathway in plants is chlorophyll, the synthesis of ALA is regulated by light. The identity of the specific light-regulated steps is now under active investigation.

Whether in plants, animals, or microorganisms, the remainder of the porphyrin synthetic pathway (Figure 21.20) involves three distinct processes: (1) synthesis of a substituted pyrrole compound, **porphobilinogen** from ALA; (2) condensation of four porphobilinogen molecules to yield a partly reduced precursor called a **porphyrinogen**; and (3) modification of the side chains, dehydrogenation of the ring system, and introduction of iron, to give the porphyrin product, heme. In the first stage, two molecules of ALA condense in the cytosol to form one molecule of porphobilinogen. The reaction is catalyzed by **ALA dehydratase**:

> Porphyrin biosynthesis involves (1) formation of a pyrrole ring; (2) condensation of four pyrrole moieties, giving a cyclic tetrapyrrole; and (3) side chain modifications and ring oxidations.

δ-ALA   **δ-ALA**   **Porphobilinogen**

Next, a PLP-requiring deaminase catalyzes the stepwise condensation of four molecules of porphobilinogen to give a linear tetrapyrrole, hydroxymethylbilane. **Uroporphyrinogen III synthase** then catalyzes an intramolecular rearrangement and ring closure to form the asymmetric **uroporphyrinogen III** (Figure 21.20; asymmetry refers to the arrangement of the acetate and propionate side chain substituents). In the absence of uroporphyrinogen III synthase, hydroxymethylbilane rapidly cyclizes nonenzymatically to the symmetrical **uroporphyrinogen I**. The symmetric compound and some metabolites derived from it are produced as nonfunctional side products, in low amounts.

Uroporphyrinogen III undergoes decarboxylation of its acetic acid side chains. The product then reenters the mitochondrion for further modifications: first, side chain modifications, then ring oxidation to yield a fully conjugated system, and finally the insertion of iron. The last reaction can proceed spontaneously, but it is catalyzed by **ferrochelatase**, an enzyme on the inner mitochondrial membrane that also requires a reducing agent. At this stage the completed heme is transported out of the mitochondrion for insertion into polypeptides to give completed heme proteins, including myoglobin and hemoglobin in vertebrates, and cytochromes and other heme proteins in all aerobic organisms.

In a rare hereditary condition called **congenital erythropoietic porphyria**, uroporphyrinogen III synthase is defective, and the symmetrical (and metabolically useless) type I porphyrins accumulate beyond the capacity of the body to excrete them. Their accumulation causes the urine to turn red, the skin to become acutely photosensitive, and the teeth to become fluorescent, all because of the deposition of the strongly light-absorbing porphyrins. In addition, erythrocytes are destroyed prematurely and insufficient heme is synthesized, making afflicted individuals quite anemic. It has been speculated that people labeled as vampires in medieval folktales suffered from this condition, which would explain their preference for the dark, their bizarre appearance, and their propensity for drinking blood. In fact, individuals with congenital erythropoietic porphyria can be treated by injections of heme.

Quite distinct from the above condition is **acute intermittent porphyria**, which results from deficiency in porphobilinogen deaminase. This deficiency causes ALA and porphobilinogen to accumulate in the liver. The condition is accompanied by episodes of acute abdominal pain and neurological disorders. It has been suggested that King George III of England suffered from this condition; evidence for this was put before the general public in the successful stage play and film *The Madness of King George*. Symptoms of porphyrias can also be acquired, most notably in lead poisoning. The crystal structure of ALA dehydratase shows that lead can displace zinc, the natural metal cofactor, and inhibit the enzyme, causing great accumulation of ALA.

Being the first committed step in heme synthesis, the ALA synthetase reaction (Figure 21.20) is the major control point. Heme and related compounds feedback-inhibit the enzyme. Heme also has two other important effects. At low concentrations, heme inhibits the *synthesis* of ALA synthetase at both transcriptional and translational levels. At higher levels, heme somehow blocks the *translocation* of ALA synthetase from the cytosol, where it is synthesized on ribosomes, into the mitochondrion, where it acts. Heme also inhibits the ferrochelatase reaction (Figure 21.20). A number of drugs and poisons cause excessive heme synthesis. In some cases, this effect results from stimulation of the synthesis of several cytochromes P450, which increases the demand for heme and hence activates ALA synthetase.

Tetrapyrrole biosynthesis is being exploited as a target for the action of weedkillers. The idea is to spray weeds in the dark with ALA. The pathway to chlorophyll begins, and when it becomes light, the pathway is completed, and chlorophyll is produced in such massive amounts that the plant weakens and dies.

Degradation of Heme in Animals

By far the most abundant porphyrin compound in vertebrates is the heme of hemoglobin. Therefore, the story of porphyrin degradation is largely the story of hemoglobin and heme degradation. Lacking nuclei, mammalian erythrocytes are incapable of renewal and self-destruct after characteristic intervals. In humans, the average erythrocyte life span is 120 days. Aged or damaged erythrocytes are destroyed upon passage through the spleen or liver (Figure 21.23).

Amino acids released from the globin portion of the hemoglobin molecule are catabolized or reused for protein synthesis. The heme portion undergoes degradation, starting with a mixed-function oxidase reaction that opens the ring and converts one of the methene bridge carbons to carbon monoxide. Iron is released from the resulting linear tetrapyrrole, called **biliverdin**, and is transported to storage pools in bone marrow for reuse in erythrocyte production. The tetrapyrrole is next reduced to **bilirubin**, which is excreted. Bilirubin is quite insoluble, and its

Porphyrias involve abnormal accumulations of heme precursors, either from overproduction of the unnatural type I porphyrins or from abnormally high flux through δ-ALA synthetase.

FIGURE 21.23

Catabolism of heme. Most of the heme comes from breakdown of aged erythrocytes, but some comes from cytochromes and other heme proteins. Letter abbreviations indicate side chains: P (propionic acid), —CH_2—CH_2—COO^-; A (acetic acid), —CH_2—COO^-; V (vinyl), —CH=CH_2; M (methyl), —CH_3.

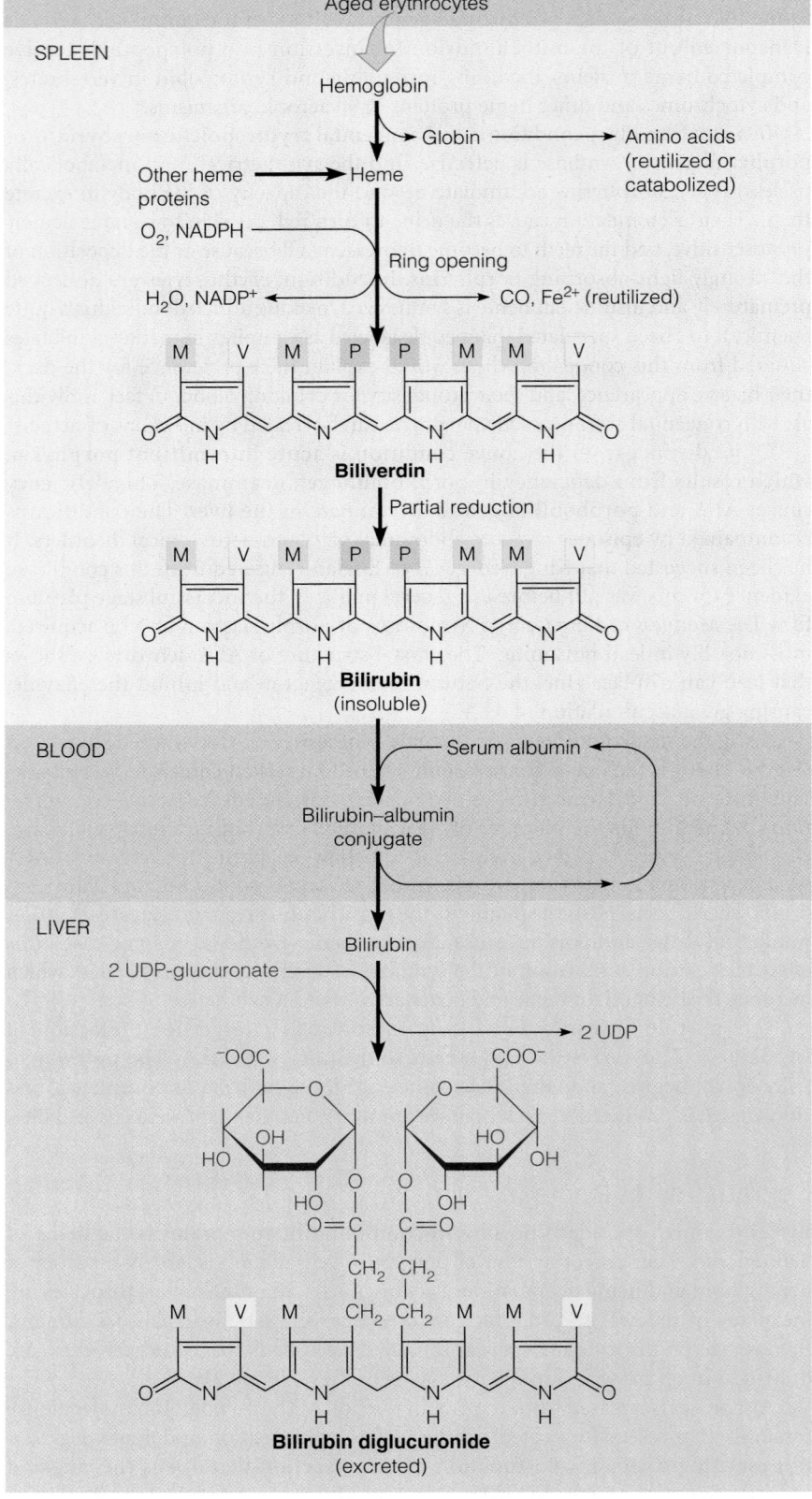

removal involves several organ systems. First, it complexes with serum albumin for transport to the liver. There, bilirubin is solubilized by conjugation with two molecules of **glucuronic acid**. The reaction is comparable to other glycosyltransferase reactions we have encountered (see Chapter 9), with the substrate being **UDP-glucuronate**. This solubilized compound, **bilirubin diglucuronide**, is secreted into the bile and ultimately excreted via the intestine.

Because several organ systems participate in the degradation of heme, there are numerous ways for things to go wrong. When heme catabolism is defective, bilirubin accumulates in the blood. This defect is first recognized because the distinctive color of bilirubin gives a yellow cast to the skin and the whites of the eyes. This condition, known as **jaundice**, is seen, for example, in acute or chronic liver disease, in which the glucuronate conjugating system is impaired and albumin synthesis might be defective; in bile duct obstruction (e.g., gallstone), when bilirubin diglucuronide cannot be secreted into the intestine; in Rh incompatibility reactions of infants, in which erythrocytes are destroyed by the immune system faster than the heme can be catabolized; or in premature infants, when the bilirubin conjugating system is not fully developed. Jaundiced infants are often placed under intense fluorescent light, which rearranges the structure of circulating bilirubin to more soluble products.

> Heme protein degradation in animals releases amino acids and iron, which are reused, and bilirubin, which must be solubilized for excretion.

Amino Acids and Their Metabolites as Neurotransmitters and Biological Regulators

Many amino acids and their metabolites participate in signal transduction processes—in hormonal control and in synaptic transmission of nervous impulses. As introduced in Chapter 10 and discussed further in Chapter 23, these two roles are comparable in that a low-molecular-weight substance released from one cell migrates to a target cell, where it interacts with specific receptors in the target cell membrane. The difference is that neurotransmission involves movement across a synapse, between two adjacent cells, whereas hormonal transmission occurs over a distance, with the hormonal messenger being transported through the bloodstream to the effector cell. The similarity of these two signal transduction processes is highlighted by the participation of compounds like epinephrine and histamine in both processes.

Among canonical amino acids that serve directly as neurotransmitters are glycine and glutamate. As noted earlier, GABA, the decarboxylation product of glutamate, is also a neurotransmitter. Several aromatic amino acid metabolites also function in neurotransmission. They include histamine, derived from histidine; **serotonin** (5-hydroxytryptamine), derived from tryptophan; and the **catecholamines**—epinephrine, **dopamine**, and **norepinephrine**—derived from tyrosine. We will describe the biosynthetic routes to these compounds here, and then we'll discuss their involvement in neurotransmission in Chapter 23.

> Glutamate, tyrosine, glycine, and tryptophan serve as neurotransmitters or precursors to neurotransmitters.

Biosynthesis of Serotonin and Catecholamines

The pathway to serotonin begins with hydroxylation of tryptophan by a tetrahydrobiopterin-dependent aromatic amino acid hydroxylase, similar to phenylalanine hydroxylase (see Figure 21.7). This reaction is followed by a PLP-dependent decarboxylation to yield serotonin.

Tryptophan hydroxylase and tyrosine hydroxylase are both tetrahydrobiopterin-dependent monooxygenases and are mechanistically and structurally related to phenylalanine hydroxylase. The mechanism of all three members of the aromatic amino acid hydroxylase family involves an NIH shift, with migration of a hydride accompanying the hydroxylation.

Tyrosine is hydroxylated to L-dopa by two distinct mechanisms in the synthesis of catecholamines and melanins.

Decarboxylation reactions of tryptophan, L-dopa, and histidine lead to a series of potent biological regulators.

Serotonin plays multiple regulatory roles in the nervous system, including neurotransmission. It is produced in the pineal gland, where it serves as precursor to **melatonin** (*O*-methyl-*N*-acetylserotonin). The pineal is known to regulate light–dark cycles in animals, and the levels of serotonin and melatonin undergo cyclic variations in phase with these cycles. Thus, although the cycle-related actions of these compounds are not yet known, they point to serotonin and melatonin as regulators of sleep and wakefulness. Many long-distance airline passengers take melatonin pills to escape jet lag by resetting their biological clocks. Serotonin is also secreted by cells in the small intestine, where it regulates intestinal peristalsis. Finally, serotonin is a potent vasoconstrictor, which helps regulate blood pressure. Several antiobesity treatments act by increasing serotonin levels, thereby creating a sense of satiety, or well-being with regard to food.

As shown in Figure 21.24, the pathway to catecholamines is similar, starting with another tetrahydrobiopterin-dependent hydroxylation (of tyrosine) followed by a decarboxylation. Tyrosine hydroxylase catalyzes the rate-limiting step of catecholamine synthesis, and it is feedback inhibited by the end products of the pathway—dopamine, norepinephrine, and epinephrine. The tyrosine hydroxylation product is L-dopa, which is formed by a quite different mechanism in melanin synthesis (see Figure 21.18, page 886). However, the latter pathway is localized to melanocytes, whereas most catecholamine synthesis occurs in the adrenal medulla and in the central nervous system.

Once formed, L-dopa undergoes a PLP-dependent decarboxylation (by the same enzyme that decarboxylates 5-hydroxytryptophan) to give dopamine. Dopamine serves in turn as substrate for a copper-containing monooxygenase, **dopamine β-hydroxylase**, giving norepinephrine (noradrenaline), which in turn is methylated by *S*-adenosylmethionine to give epinephrine (adrenaline). Although dopamine and norepinephrine are intermediates in epinephrine synthesis, each is a neurotransmitter in its own right, as discussed in Chapter 23.

FIGURE 21.24

Biosynthesis of the catecholamines—dopamine, norepinephrine, and epinephrine—from tyrosine.

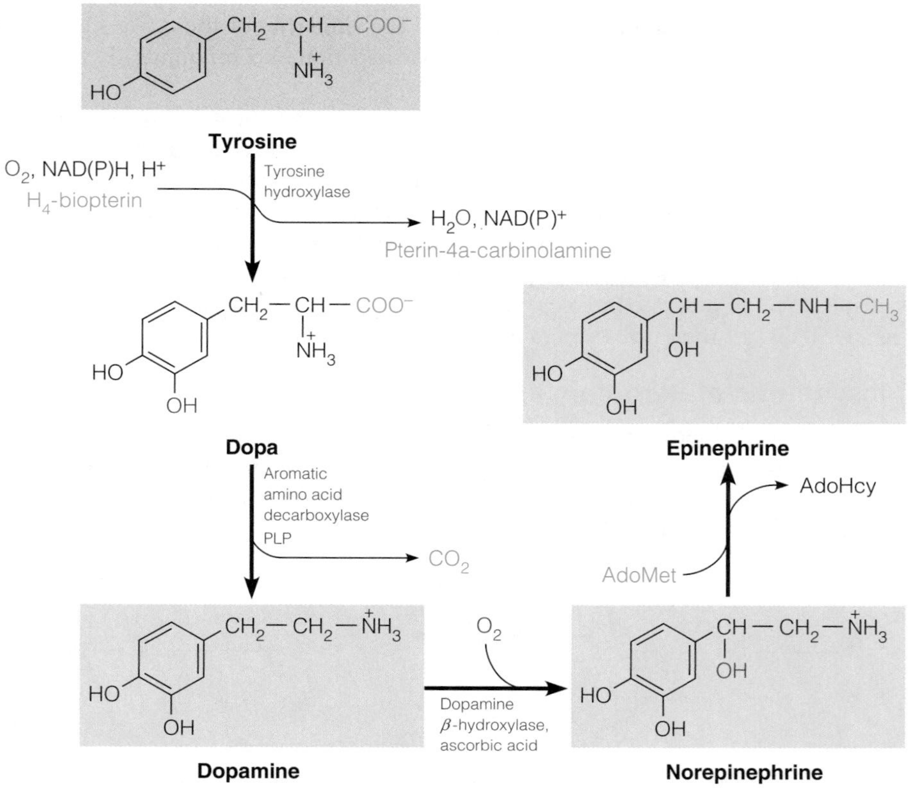

Amino Acid Biosynthesis

All amino acids can be synthesized from intermediates in glycolysis, the pentose phosphate pathway, or the citric acid cycle (Figure 21.25). About half are biosynthesized more or less directly from intermediates in the citric acid cycle or from pyruvate. We include in this family glutamate, aspartate, and alanine, which can be formed directly by transamination from α-ketoglutarate, oxaloacetate, and pyruvate, respectively. The family also includes glutamine and asparagine, which are formed directly from glutamate and aspartate, respectively; and proline and arginine, which are formed in short pathways from glutamate. Aspartate is also the starting point for threonine, methionine, and lysine. The carbon skeletons of serine, glycine, cysteine, and histidine, as well as the branched chain amino acids and the aromatic amino acids, are all derived from glycolytic intermediates.

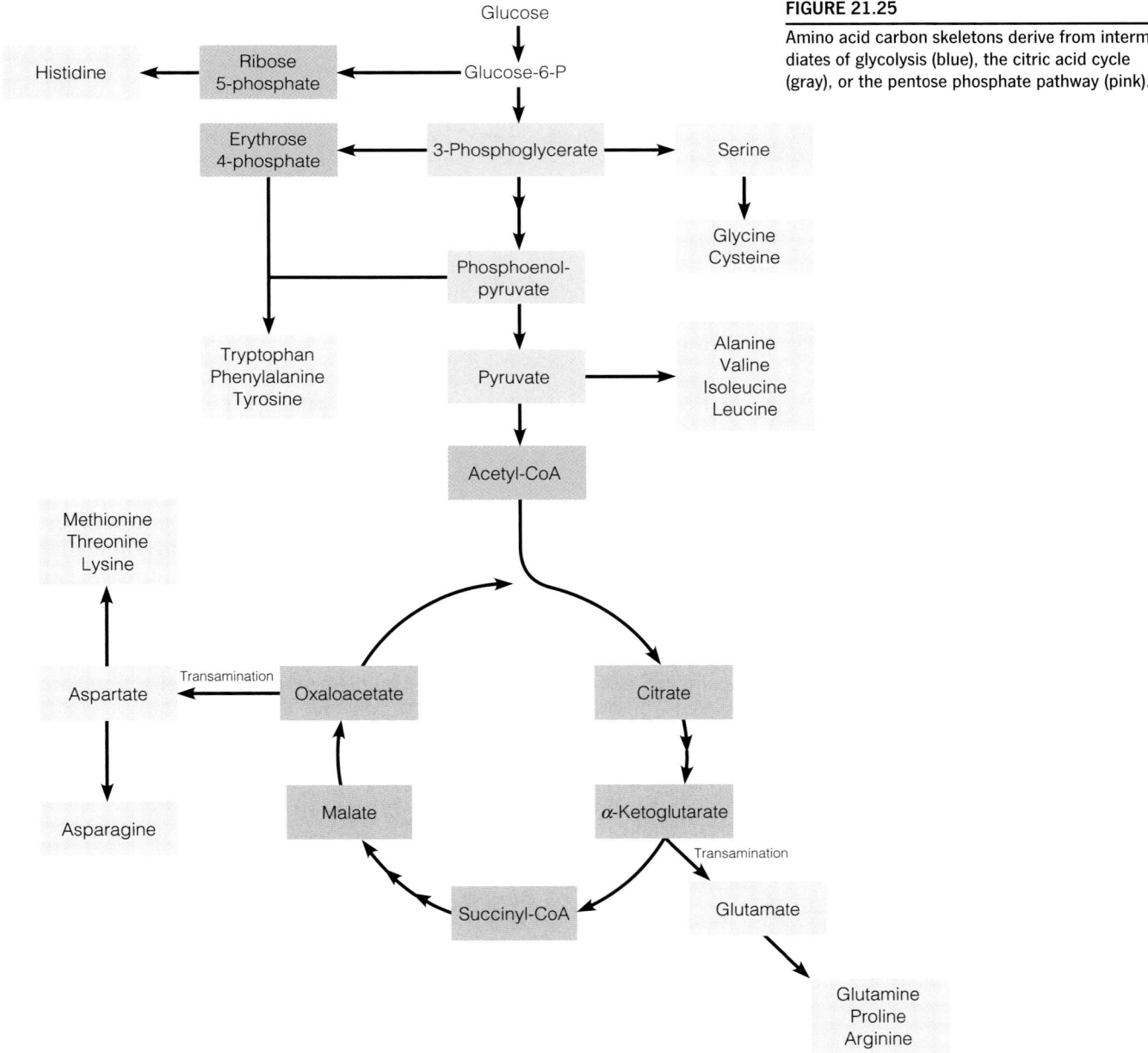

FIGURE 21.25

Amino acid carbon skeletons derive from intermediates of glycolysis (blue), the citric acid cycle (gray), or the pentose phosphate pathway (pink).

FIGURE 21.26

Synthesis of alanine, aspartate, glutamate, asparagine, and glutamine. ❶ asparagine synthetase, ❷ glutamine synthetase.

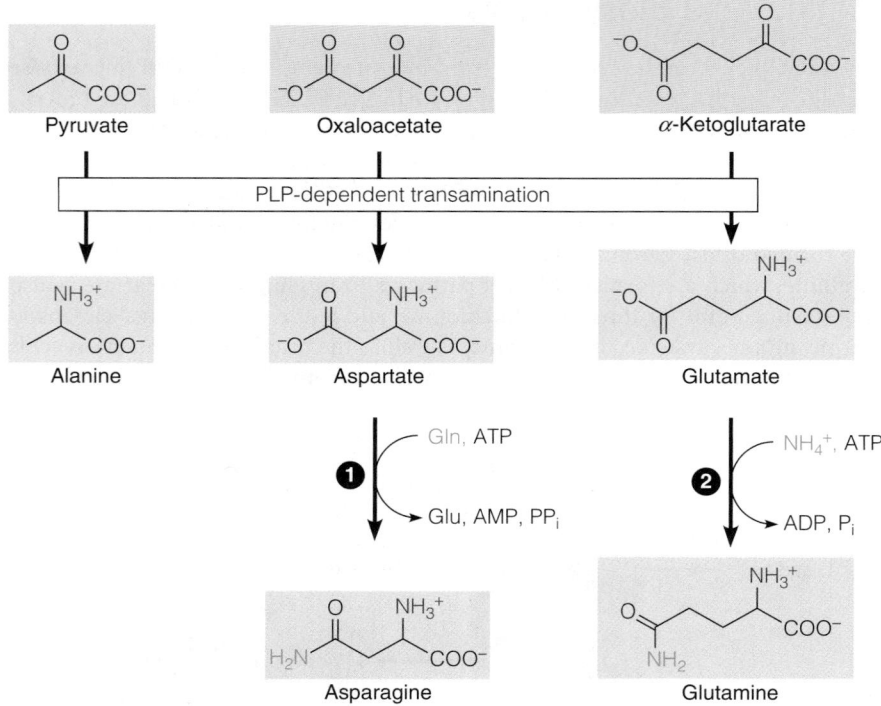

We will not cover all of the amino acid biosynthetic pathways in this final section of the chapter, but we will highlight the common pathways and those with particularly interesting chemistry.

Synthesis of Glutamate, Aspartate, Alanine, Glutamine, and Asparagine

PLP-dependent transamination provides the major route for the synthesis of glutamate, aspartate, and alanine, from α-ketoglutarate, oxaloacetate, and pyruvate, respectively (Figure 21.26). Reactions catalyzed by glutamate dehydrogenase and glutamate synthase, introduced in Chapter 20, present additional routes for glutamate synthesis from α-ketoglutarate. Asparagine is synthesized from aspartate in a reaction catalyzed by asparagine synthetase, a glutamine-dependent amidotransferase. Glutamine is synthesized from glutamate by a similar amidation reaction, except that ammonia provides the amide N. The γ-glutamyl phosphate intermediate in this glutamine synthetase reaction was described in Chapter 20 (page 828).

In animals, a major metabolic function of alanine is its role in the glucose–alanine cycle as a carrier of carbon for gluconeogenesis from muscle to liver (see Figure 20.13, page 845).

Synthesis of Methionine, Threonine, and Lysine from Aspartate

Aspartate leads via homoserine to threonine, lysine, and methionine.

The nitrogen of aspartate is used in the biosynthesis of arginine and urea, as noted in Chapter 20. Similar reactions are involved in purine nucleotide synthesis, and the entire aspartate molecule is used in pyrimidine nucleotide biosynthesis; both of these processes are discussed in Chapter 22. However, in plants and bacteria, aspartate is a precursor to three other amino acids via its conversion to **aspartate β-semialdehyde** and **homoserine**, as shown here. Separate pathways then lead from aspartate β-semialdehyde to lysine and from homoserine to methionine and threonine.

Aspartate   **Aspartyl-β-phosphate**   **Aspartate β-semialdehyde**   **Homoserine**

The first enzyme in this pathway, **aspartokinase**, is a major site for regulation of each biosynthetic pathway. Bacteria contain three distinct forms of aspartokinase, each with its own mode of allosteric regulation. Two of the isoenzymes are bifunctional enzymes, containing distinct domains that catalyze the first step (aspartokinase) and the third step (**homoserine dehydrogenase**). In one of the bifunctional isozymes, both catalytic activities are subject to allosteric inhibition by threonine, which can be thought of as an end product of the pathway. The third isoenzyme, a monofunctional aspartokinase, is subject to feedback inhibition by lysine. These observations suggest that the different isozymes are specialized to serve just one of the three end products. Bacteria contain a separate monofunctional enzyme, **aspartate-β-semialdehyde dehydrogenase**, that catalyzes the intervening reductive dephosphorylation of aspartyl-β-phosphate.

Lysine biosynthesis is distinguished by the fact that there are two distinct pathways. The most widespread pathway, found in bacteria, some lower fungi, algae, and higher plants, begins with aspartate β-semialdehyde. This pathway is named after its principal intermediate, **diaminopimelate**, which also serves an important function as a constituent of bacterial cell walls. The less common pathway, found in certain fungi and yeast, and in the protist *Euglena*, begins with α-ketoglutarate and involves **α-aminoadipic acid** and **saccharopine** (see page 870) as intermediates.

Homoserine has an interesting role in microbial metabolism unrelated to its function in amino acid synthesis. Bacterial cultures carry out certain processes only after reaching a particular cell density. This cell–cell communication phenomenon is called **quorum-sensing**, in which bacteria produce and detect extracellular chemicals to monitor when a certain population-density threshold has been crossed. Quorum sensing allows the bacteria to synchronously control gene expression in response to changes in cell density, triggering physiological responses as diverse as bioluminescence, biofilm production, antibiotic synthesis, virulence factor secretion, and conjugal gene transfer. For many bacteria the signaling molecule, called an autoinducer, is one of several long-chain *N*-acyl derivatives of homoserine (see margin). *N*-Acylhomoserine lactone is synthesized and secreted at a low, constant rate, and it diffuses back into cells. At a sufficiently high cell density, the extracellular and, hence, the intracellular concentrations of the lactone have risen high enough for it to bind to genetic regulatory proteins, which in turn stimulate transcription of genes required to activate the particular process.

***N*-Acylhomoserine lactone**

Methionine and Threonine Biosynthesis

In plants and bacteria, homoserine provides the carbon skeleton for methionine synthesis, with the sulfur coming from cysteine (Figure 21.27). Homoserine first reacts with succinyl-CoA to form **O-succinylhomoserine**. This reaction evidently is a control point because the enzyme catalyzing this reaction is feedback inhibited by methionine. *O*-Succinylhomoserine reacts with cysteine to give **cystathionine**,

FIGURE 21.27

Biosynthesis of methionine and threonine from homoserine, as it occurs in plants and bacteria.

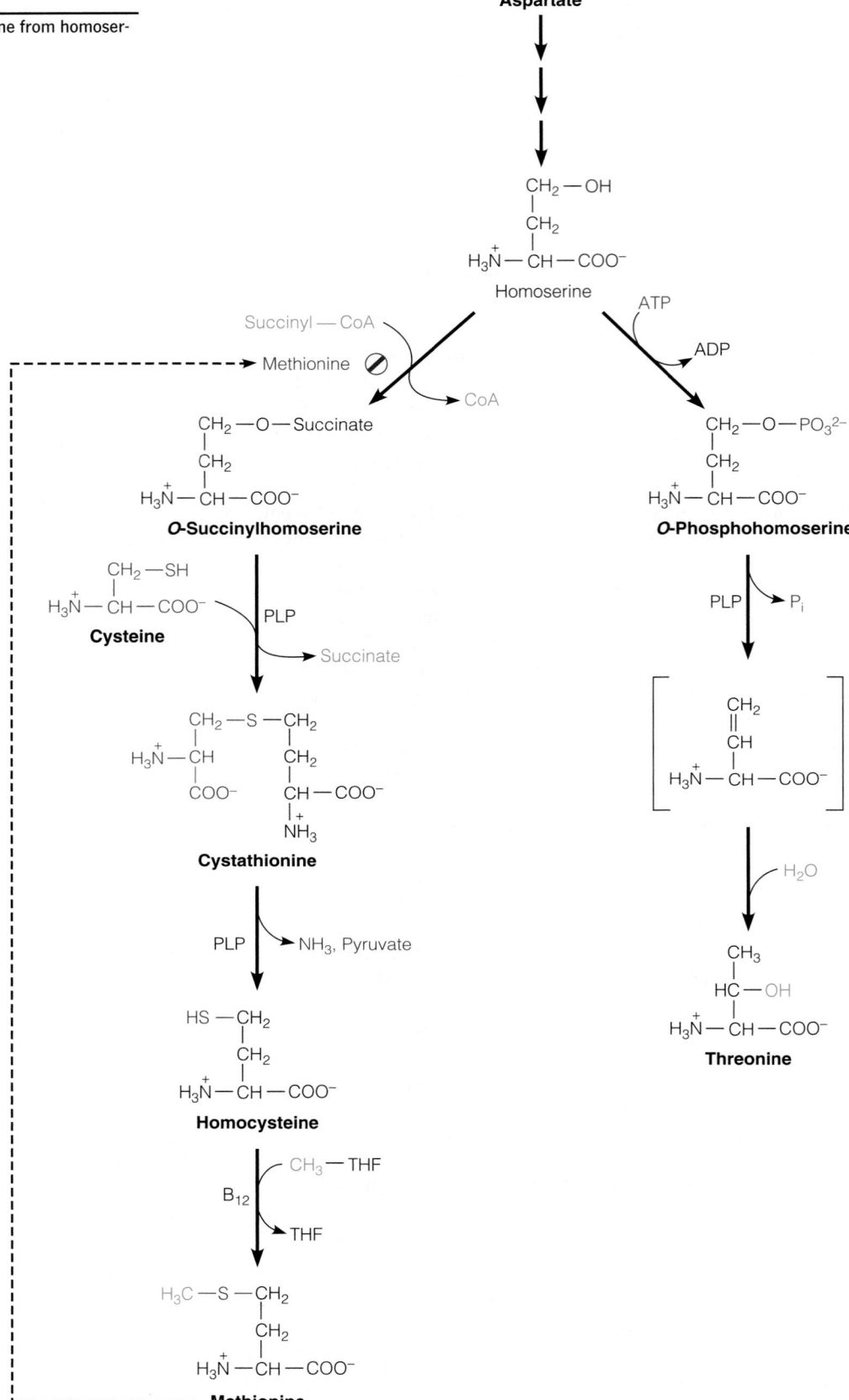

the same thioether compound we first saw in the methionine degradation pathway (Figure 21.4). Cleavage of cystathionine then occurs, with the sulfur becoming linked to the four-carbon side chain that started as homoserine. This is essentially the transsulfuration pathway running in the opposite direction from that in Figure 21.4. In this direction, the three-carbon pyruvate, rather than the four-carbon α-ketobutyrate, is the other product in this direction because a three-carbon donor (cysteine) was used to form cystathionine. The resulting sulfur-containing carbon skeleton, homocysteine, is the substrate for methionine synthase, which, as presented in Chapter 20, uses B_{12} to transfer a methyl group from 5-methyltetrahydrofolate, yielding methionine (see Figure 21.9).

Threonine is an essential amino acid for animals because its biosynthesis is limited to plants and prokaryotes. Threonine synthesis also begins with homoserine, which undergoes a phosphorylation, followed by a pyridoxal phosphate–dependent reaction that eliminates phosphate and rehydrates the resultant double bond with hydroxyl group migration to the β-carbon (Figure 21.27).

Metabolism of Sulfur-Containing Amino Acids

Reduction of Inorganic Sulfur

As shown in Figure 21.27, the S of methionine is derived from cysteine. How does cysteine obtain its sulfur atom? Like carbon and nitrogen, sulfur is made available to organisms largely in the form of inorganic compounds—principally sulfate, although some bacteria can synthesize organic compounds from elemental sulfur or from sulfite. Just as CO_2 and N_2 must undergo fixation to be utilized, the utilization of sulfate requires metabolic activation to a form that can readily undergo reduction. The process for sulfate is largely confined to plants and bacteria. The end product of this eight-electron reduction is S^{2-} (sulfide), and this is used for cysteine and methionine synthesis. The activated sulfate compound is **3-phosphoadenosine-5-phosphosulfate (PAPS)**. This nucleotide is formed in two steps from ATP and sulfate ion.

$$SO_4^{2-} + ATP \xrightarrow{\quad PP_i \quad} \text{Adenosine-5'-phosphosulfate} \xrightarrow{\quad ATP \quad} PAPS + ADP$$

PAPS is then reduced to sulfite and then sulfide by two NADPH-dependent reductases.

$$PAPS \xrightarrow{\quad NADPH \quad NADP^+ \quad} \text{3'-Phospho-AMP} + \underset{\text{Sulfite}}{SO_3^{2-}} \xrightarrow{\quad 3NADPH \quad 3NADP^+ \quad} \underset{\text{Sulfide}}{S^{2-}}$$

In all organisms, PAPS serves as an active agent for sulfate esterification, as in the synthesis of sulfated polysaccharides such as chondroitin sulfate (Chapter 9) and sulfated glycosphingolipids such as sulfogalactosylceramide (Chapter 19).

In bacteria, PAPS also serves as the substrate for sulfate reduction. Reduction of the sulfate in PAPS to sulfite (SO_3^{2-}) involves **thioredoxin**, a small protein ($M_r \cong 12,000$) that contains two reversibly oxidizable cysteine thiol groups. Thioredoxin is involved in several other intracellular redox reactions, as described in Chapters 16 and 22. The sulfite is subsequently reduced by **sulfite reductase**, a large and complex enzyme that catalyzes a six-electron transfer. The electrons are shuttled along a pathway involving NADPH, FAD, FMN, an iron–sulfur center, and the porphyrin siroheme (see page 826). No intermediates accumulate, just the product H_2S. In plants, adenosine 5-phosphosulfate, rather than PAPS, is the substrate for reduction.

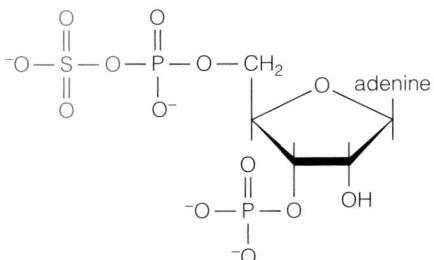

3'-Phosphoadenosine-5'-phosphosulfate (PAPS)

Phosphoadenosine phosphosulfate (PAPS) is an activated form of sulfate used both for sulfation reactions and as a substrate for sulfate reduction.

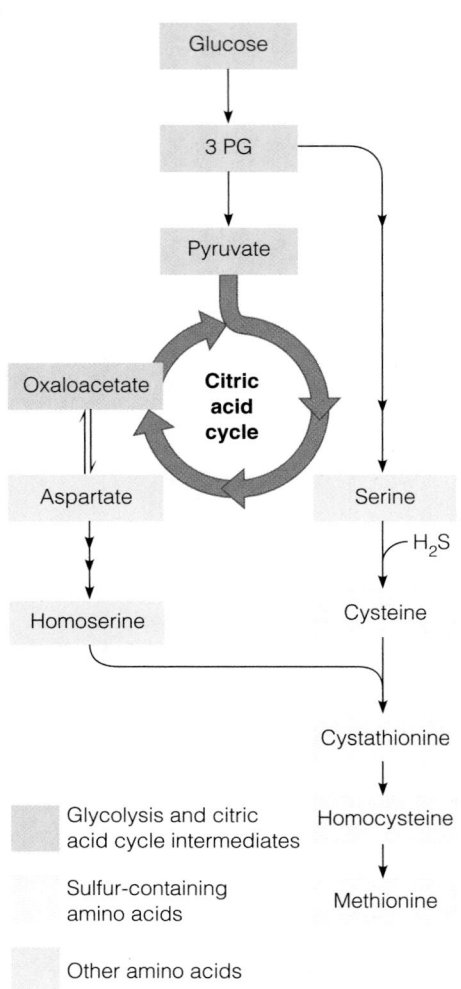

FIGURE 21.28

Outline of pathways for cysteine and methionine synthesis in plants and bacteria.

Plants and bacteria synthesize cysteine from inorganic sulfur and synthesize methionine from cysteine. Animals synthesize cysteine from dietary methionine.

Cystathionine

Synthesis of Cysteine in Plants and Bacteria

Bacteria and plants can synthesize cysteine by incorporation of H_2S, with serine providing the carbon skeleton (Figure 21.28). Animals, on the other hand, cannot directly incorporate sulfide, and so they derive cysteine either from the diet or from dietary methionine. Some bacteria can condense H_2S with serine directly, via a pyridoxal phosphate–dependent enzyme.

Serine + H_2S → **Cysteine** + H_2O

However, plants and most microorganisms use O-acetylserine as the substrate reacting with H_2S.

Serine → (Acetyl-CoA, CoASH) → **O-Acetylserine** → ($S^{2-} + H^+$, Acetate) → **Cysteine**

Methionine as the Source of Cysteine Sulfur in Animals

Methionine is classified as an essential amino acid for mammals, and cysteine is considered nonessential. Actually, the biosynthetic route in animals proceeds from methionine to cysteine, as summarized in Figure 21.29, so cysteine is nonessential only as long as the diet contains adequate methionine.

The synthesis of cysteine in animals resembles the reverse of the methionine biosynthetic pathway we just discussed, and which was described previously in the context of the pathway used to degrade methionine (refer to Figure 21.4). After methionine has donated its methyl group via S-adenosylmethionine (see Figure 21.9), the resulting homocysteine condenses with serine in the transsulfuration process to give cysteine. This two-step process results in the transfer of S from a four-carbon donor (homocysteine) to a three-carbon acceptor (serine), with the seven-carbon cystathionine as intermediate. The four-carbon skeleton of homocysteine is deaminated, forming α-ketobutyrate and ammonia. Plants and prokaryotes carry out the same pathway, with α-ketobutyrate used also for isoleucine biosynthesis.

Recall that homocysteine has two possible fates: remethylation or transsulfuration (Figure 21.9). As discussed previously (page 876), S-adenosylmethionine regulates these two fates by inhibiting remethylation at the methylenetetrahydrofolate reductase step and activating transsulfuration at the cystathionine β-synthase step. If the cell's methylation demands are being met, AdoMet levels will be high, and homocysteine will be directed down the transsulfuration pathway to cysteine (and glutathione) biosynthesis. Transsulfuration is also controlled by the cellular redox state: Oxidizing conditions increase flux from homocysteine to glutathione as a mechanism to protect the cell against oxidative stress. Cystathionine β-synthase is believed to sense the cellular redox state through its heme cofactor.

Note that cystathionine plays a central role both in the biosynthesis of methionine in plants and bacteria and in the biosynthesis of cysteine in animals. In methionine synthesis from cysteine (Figure 21.27), cystathionine cleavage accompanies the transfer of sulfur from a three-carbon to a four-carbon side chain. The converse occurs during the synthesis of cysteine from methionine (Figure 21.4).

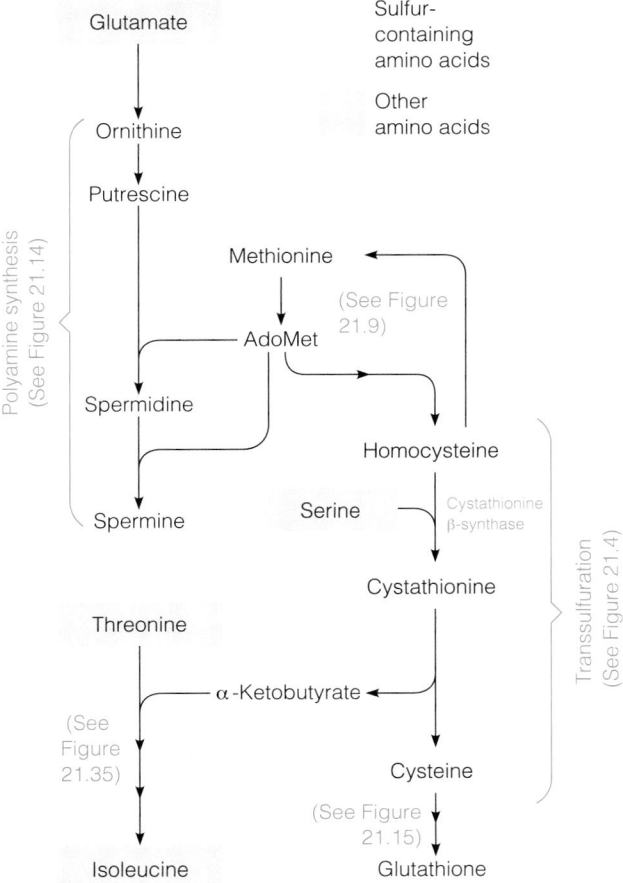

FIGURE 21.29

Outline of methionine metabolism. Except for the synthesis of isoleucine from threonine (see page 907), which is limited to plants and bacteria, these pathways occur in virtually all organisms.

Cysteine is also the precursor for a nonprotein amino acid, **taurine**, the most abundant free amino acid in animal tissues. The sulfhydryl of cysteine is oxidized to a sulfinate ($-SO_2^-$) and finally to a sulfonate ($-SO_3^-$) in taurine (Figure 21.30). This is an important pathway in liver, kidney, muscle, and brain, where taurine can reach an intracellular concentration of 25 mM. We do not know all the functions of taurine, but the main biological roles include the synthesis of the bile acid taurocholate (see Chapter 19), regulation of blood pressure, as an intracellular osmolyte, and as a potent antioxidant and anti-inflammatory agent. Interestingly, taurine is a major ingredient in Red Bull® energy drink—an 8 oz. can of the drink contains 1000 mg of taurine (compared with only 80 mg of caffeine). The health effects of such high intakes of taurine have not been carefully studied.

Synthesis of Proline, Ornithine, and Arginine from Glutamate

As we have seen, glutamate is perhaps the most active of all the amino acids in terms of its number of metabolic roles. Another important reaction of glutamate is the energy-requiring reduction of the γ-carboxyl group, to give **glutamate γ-semialdehyde** (Figure 21.31). Phosphorylation of the γ-carboxyl group to the unstable intermediate, γ-glutamyl phosphate, facilitates the endergonic reduction. This process, which leads toward both ornithine and proline, is comparable to the reduction of aspartate to aspartic semialdehyde (page 899). In plants and animals, this two-step reaction is catalyzed by a bifunctional enzyme composed of two domains—one catalyzes the ATP-dependent γ-carboxyl phosphorylation, the other catalyzes the NADPH-dependent reduction. The product, glutamate γ-semialdehyde, is in a nonenzymatic equilibrium with a cyclized tautomer, **Δ¹-pyrroline-5-carboxylic acid (P5C)**. This equilibrium is strongly pH-dependent,

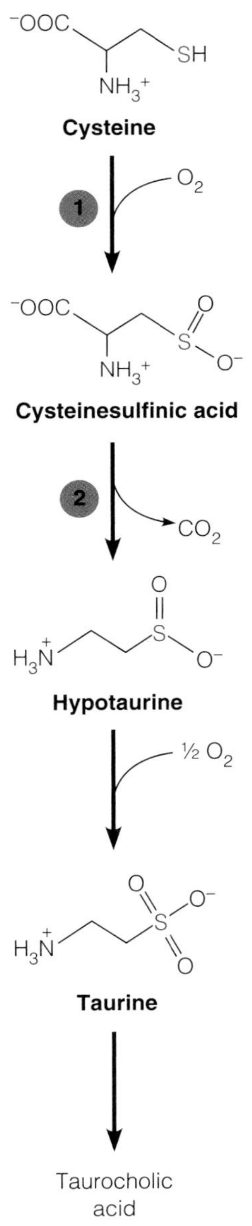

Cysteine

Cysteinesulfinic acid

Hypotaurine

Taurine

Taurocholic acid

FIGURE 21.30

Taurine biosynthesis. ❶ cysteine dioxygenase, ❷ cysteinesulfinate decarboxylase.

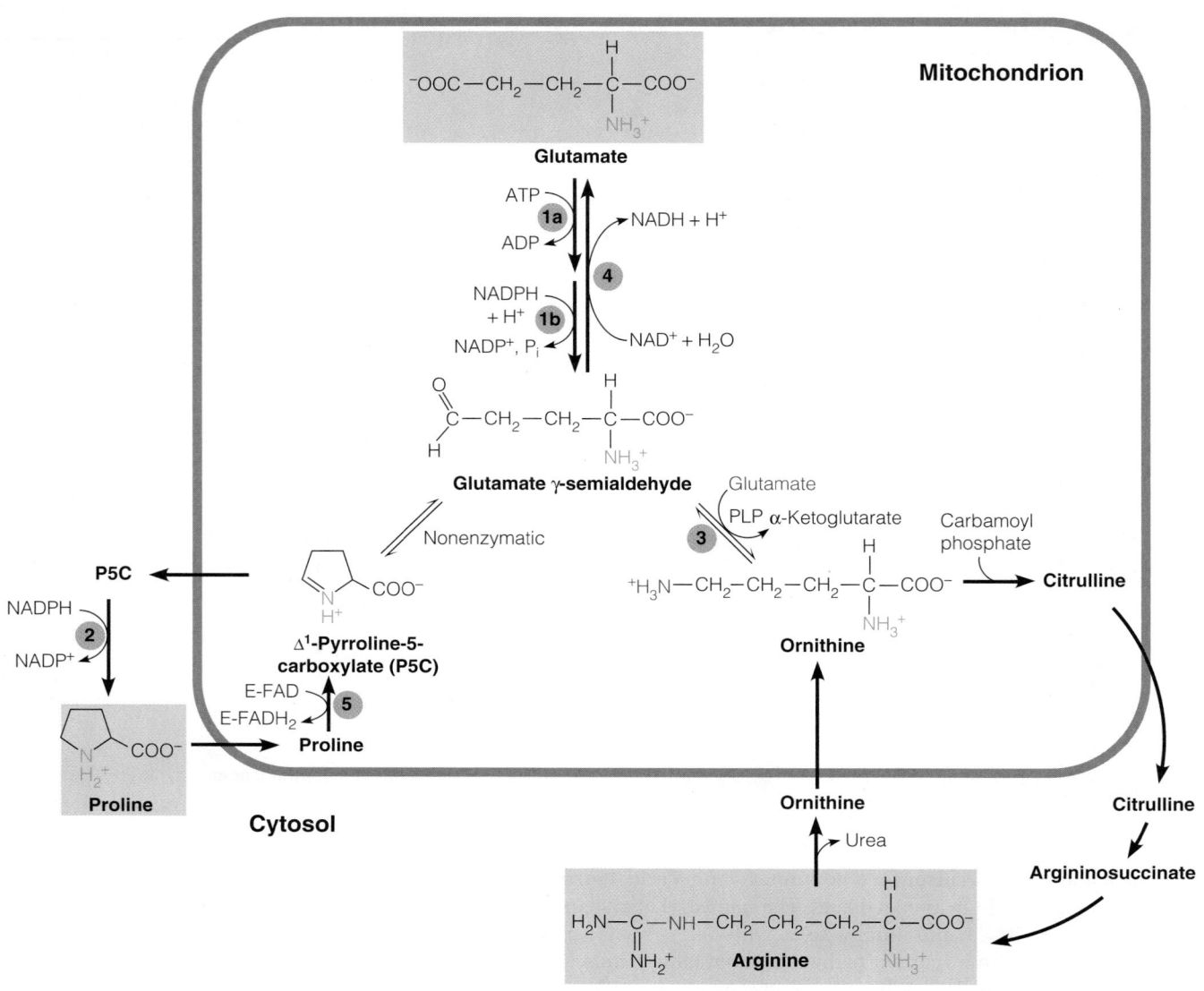

FIGURE 21.31

Proline and arginine are derived from glutamate. The enzymes involved are ❶ Δ^1-pyrroline-5-carboxylate (P5C) synthase, a bifunctional enzyme comprising glutamate kinase ❶ₐ and γ-glutamyl phosphate reductase ❶ᵦ activities; ❷ Δ^1-pyrroline-5-carboxylate reductase; ❸ ornithine δ-aminotransferase; ❹ glutamate γ-semi-aldehyde dehydrogenase (also called Δ^1-pyrroline-5-carboxylate dehydrogenase); ❺ proline oxidase.

γ-Glutamyl phosphate

Glutamate, one of the most metabolically active amino acids, is a precursor to glutamine, arginine, creatine phosphate, proline, hydroxyproline, polyamines, glutathione, and γ-aminobutyric acid.

Proline and arginine are synthesized and degraded by opposing pathways that utilize different enzymes and different cofactors, and occur in multiple compartments.

with P5C favored at pH above ~6.5. Thus, P5C is the physiological product, and the bifunctional enzyme is named Δ^1-pyrroline-5-carboxylic acid synthase, accordingly. Bacteria and lower eukaryotes use two separate enzymes encoded by two separate genes to carry out the glutamate kinase and γ-glutamyl phosphate reductase reactions. Sequence and structural analyses suggest that the bifunctional P5C synthase evolved from the two monofunctional enzymes, which are encoded by a single operon in bacteria. Finally, an NADPH-dependent reduction of Δ^1-pyrroline-5-carboxylic acid follows, to give proline.

Glutamate γ-semialdehyde leads not only to proline but also to ornithine and hence arginine. Ornithine is formed directly from glutamate γ-semialdehyde by transamination at the aldehyde group in a reaction catalyzed by **ornithine δ-aminotransferase**. Arginine is then synthesized from ornithine through the urea cycle, as we saw in Chapter 20.

The pathways of proline and arginine biosynthesis from glutamate are essentially the reverse of their degradation pathways (compare Figure 21.31 with

Figure 21.3), so how does the cell prevent these opposing pathways from operating at the same time? Several of the principles of bioenergetics and metabolic organization that we first outlined in Chapter 12 come into play here. As expected, different enzymes and coenzymes are used in the anabolic and catabolic directions. In particular, the biosynthetic directions have different ATP coupling coefficients from the catabolic directions (e.g., −1 vs. 0 for the proline pathways). Proline synthesis from glutamate, a four-electron reduction, uses NADPH, whereas proline oxidation to glutamate involves a flavin dehydrogenase and NAD^+. The only reversible enzymatic reaction is the transamination reaction catalyzed by ornithine δ-aminotransferase in the arginine pathway. Another factor that allows independent but coordinate control of the synthesis vs. degradation pathways is the intricate compartmentation of the enzymes in these pathways (Figure 21.31). Thus, P5C reductase, which catalyzes the reduction of P5C to proline, is cytoplasmic whereas proline oxidase, which catalyzes the oxidation of proline back to P5C, is mitochondrial. Both P5C synthase and proline oxidase are tightly bound to the matrix side of the inner membrane; ornithine δ-aminotransferase is a soluble matrix enzyme. Arginine metabolism, via the urea cycle, involves one mitochondrial reaction and two cytoplasmic reactions. Transport of P5C, proline, citrulline, and ornithine between the mitochondrial and cytosolic compartments is mediated by specific carriers in the mitochondrial inner membrane.

In bacteria, glutamate is acetylated to **N-acetylglutamate** before reduction, with the acetyl group being removed a couple of steps later (Figure 21.32). Acetylation prevents cyclization of the molecule after reduction, by condensation between the resulting aldehyde and the α-amino group as occurs nonenzymatically in proline synthesis. Ornithine is then converted to arginine by the same reactions found in the animal pathway (Figure 21.31).

In plants, glutamate γ-semialdehyde can also be synthesized from N-acetylglutamate γ-semialdehyde by transfer of the acetyl group on the latter to glutamate.

N-acetylglutamate γ-semialdehyde + glutamate ⇌ N-acetylglutamate + glutamate γ-semialdehyde

Hydroxyproline and Collagen

An important role of proline is its incorporation into polypeptide precursors to collagen and other connective tissue proteins, where it serves as a precursor to **hydroxyproline**. As mentioned in Chapter 5 (see page 144), hydroxyproline residues are generated by post-translational modification, following completion of the polypeptide chain. The nonhydroxylated collagen precursor is called **procollagen** (see Chapter 6). In this polypeptide, a proline residue two positions to the carboxyl side of a glycine residue is the preferred substrate for the action of **procollagen prolyl hydroxylase** (Figure 21.33). This enzyme belongs to the family of Fe(II)/α-ketoglutarate–dependent dioxygenases, which require L-ascorbic acid and molecular oxygen in addition to α-ketoglutarate and nonheme ferrous iron. α-Ketoglutarate is stoichiometrically decarboxylated during hydroxylation, with one atom of the O_2 molecule being incorporated into succinate and the other into the hydroxy group on the proline residue. Ascorbate is not needed for this reaction, and the enzyme can catalyze a number of reaction cycles in its absence. However, prolyl hydroxylases sometimes catalyze decarboxylation of α-ketoglutarate that is not coupled to hydroxylation of a proline residue. During this uncoupled turnover of α-ketoglutarate to succinate, Fe(II) becomes oxidized and the enzyme is inactivated. Although its precise role is not yet known, ascorbate is proposed to reduce the oxidized iron back to the Fe(II) state, reactivating the enzyme. In this role, ascorbate is required stoichiometrically and is oxidized to dehydroascorbate. This reaction is of particular interest because it represents one of the few well-defined roles for ascorbic acid, or vitamin C. As mentioned in Chapter 6, vitamin C deficiency, or scurvy, involves degeneration of connective tissue, and these problems derive from defective synthesis or maturation of collagen in connective tissue. The earliest recorded

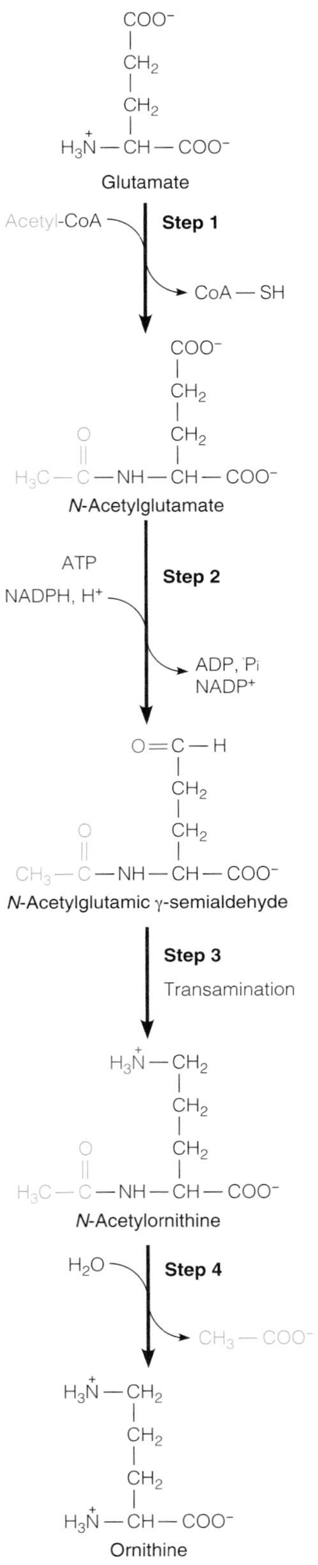

FIGURE 21.32

Biosynthesis of ornithine from glutamate in bacteria. The reduction of N-acetylglutamate (step 2) begins with phosphorylation of the carboxyl group by ATP, followed by NADPH-dependent reduction of the activated carboxyl group.

α-Ketoglutarate   Succinate

Procollagen peptide segment

Collagen hydroxyproline peptide segment

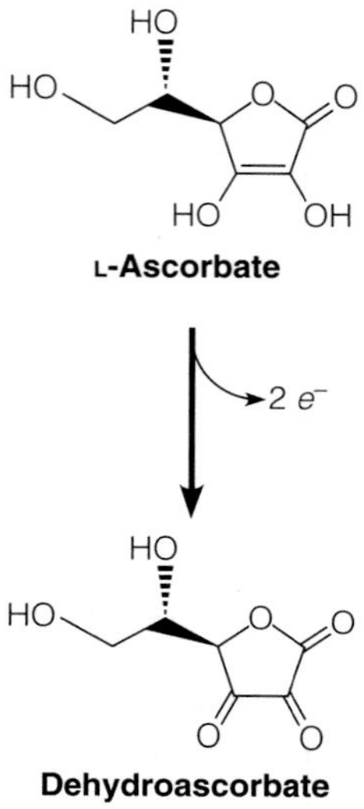

L-Ascorbate

Dehydroascorbate

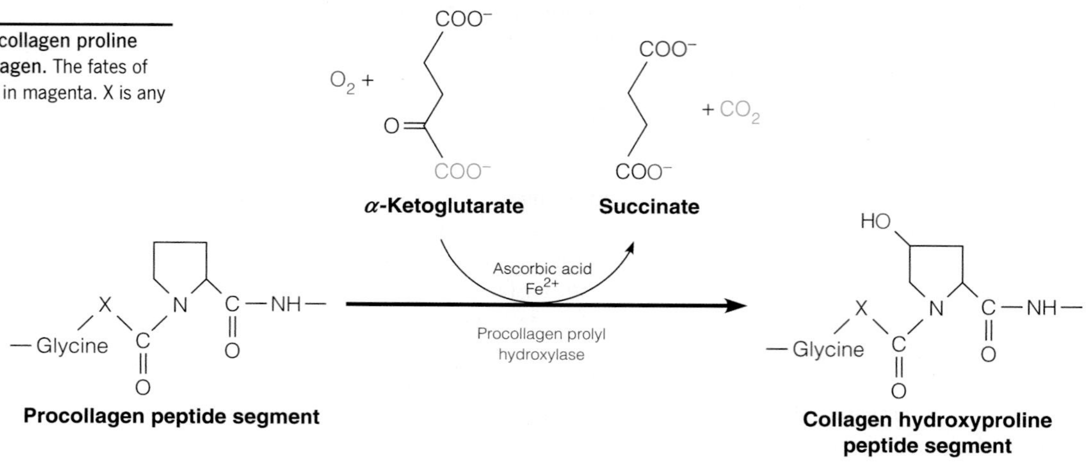

references to this deficiency disease go back at least 3500 years, but scurvy became a serious problem during the extended sailing voyages of exploration beginning in the 1500s. A breakthrough came when James Lind, a surgeon in the Royal Navy, was confronted with an outbreak of scurvy during a 1747 voyage. In what was the first controlled clinical trial in recorded history, Lind divided 12 scorbutic sailors into six pairs and treated each pair with a different "acidic" dietary supplement. The daily supplements he tested were a quart of cider; 25 drops of elixir of vitriol (sulfuric acid); 6 spoonfuls of vinegar (acetic acid); half a pint of seawater; two oranges plus one lemon; and a spicy paste plus a drink of barley water. The treatment of the citrus group stopped after six days when they ran out of fruit, but by that time both sailors showed significant improvement. The other groups continued to deteriorate. A longer trial of citrus fruit was conducted in 1794, and in 1795 the Royal Navy finally mandated lime or lemon juice for all British sailors on voyages of one month or longer (hence the nickname "limeys").

Synthesis of Serine and Glycine from 3-Phosphoglycerate

Serine and glycine are closely interconnected via the serine hydroxymethyltransferase reaction (see Figure 21.2). Although serine can be synthesized from glycine via this reaction, it proceeds more often in the reverse direction, as the principal biosynthetic route to glycine and to 5,10-methylenetetrahydrofolate. Most de novo serine biosynthesis occurs in a three-step sequence from the glycolytic intermediate 3-phosphoglycerate: oxidation of the alcohol to a ketone, transamination of the ketone to introduce the α-amino group, and finally dephosphorylation to give serine.

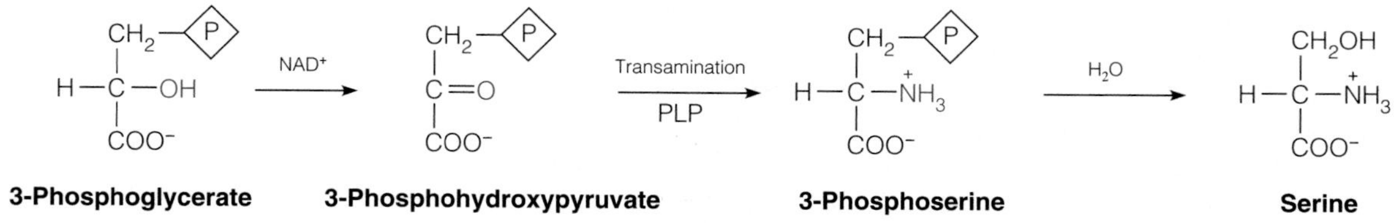

3-Phosphoglycerate   3-Phosphohydroxypyruvate   3-Phosphoserine   Serine

In bacteria and plants, the first committed step to serine synthesis, the NAD-dependent oxidation of 3-phosphoglycerate, is feedback inhibited by L-serine. Phosphoglycerate dehydrogenase is also an important control point in mammalian tissues with low serine concentrations. On the other hand, in tissues where serine levels are high, such as liver, it is the final enzyme, phosphoserine phosphatase, that is sensitive to feedback inhibition by L-serine. This is an interesting exception to the general concept that key regulatory enzymes are located at the beginning of a pathway.

Serine is quite active metabolically; we have already considered its roles in biosynthesis of phospholipids (Chapter 19) and cysteine (page 902), as well as

Serine is involved in glycine, phospholipid, and cysteine synthesis. Glycine is active in biosynthesis of purine nucleotides and porphyrins.

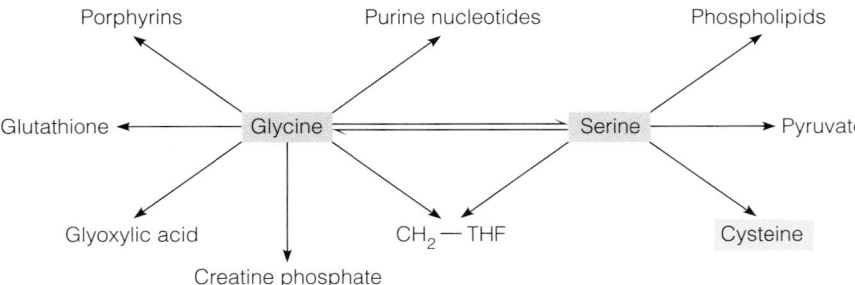

FIGURE 21.34

Metabolic interconversions and fates of serine and glycine.

its contribution of activated one-carbon units to the pool of tetrahydrofolate coenzymes (Chapter 20). Glycine also plays multiple roles, including contributions to the one-carbon pool and as a precursor to glutathione (Figure 21.15), to purine nucleotides (see Chapter 22), and to porphyrins (Figure 21.19). Figure 21.34 summarizes the metabolic fates of glycine and serine.

Serine and glycine are both major contributors to the pool of activated one-carbon groups, in the form of 5,10-methylenetetrahydrofolate.

Synthesis of Valine, Leucine, and Isoleucine from Pyruvate

The branched chain amino acids valine, leucine, and isoleucine are essential for mammals, and they are synthesized primarily in plant and bacterial cells. Furthermore, none of these amino acids is known to play significant metabolic roles other than as protein constituents and as substrates for their own degradation. The pathways involved are complex, and they are shown here only in outline.

Valine, leucine, and isoleucine are structurally related, and they share certain reactions and enzymes in their biosynthetic pathways (Figure 21.35). The last four reactions in valine and isoleucine biosynthesis are catalyzed by the same four enzymes. Valine biosynthesis begins with transfer of a two-carbon fragment from hydroxyethyl thiamine pyrophosphate to pyruvate. The two-carbon fragment derives from a second molecule of pyruvate in a TPP-dependent reaction similar to that catalyzed by pyruvate decarboxylase (Chapter 13). A similar TPP-dependent transfer of a two-carbon unit to α-ketobutyrate begins the pathway to isoleucine. The keto acid analog of valine is the input for a four-step pathway to leucine. In bacteria, each of these three amino acids controls its own synthesis by feedback inhibition of a different enzyme. In fact, the concept of allosteric control was developed largely in studies on the inhibition of threonine dehydratase by isoleucine.

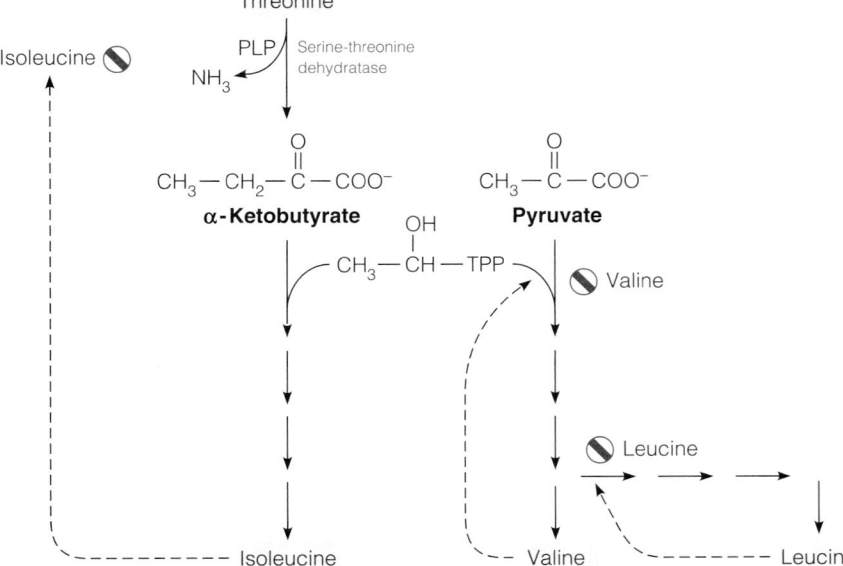

FIGURE 21.35

Biosynthesis of valine and isoleucine. After the serine–threonine dehydratase reaction (see Figure 21.4), one set of enzymes catalyzes the comparable reactions in valine and isoleucine synthesis. In bacteria, each end product regulates its own synthesis by inhibiting a specific enzyme.

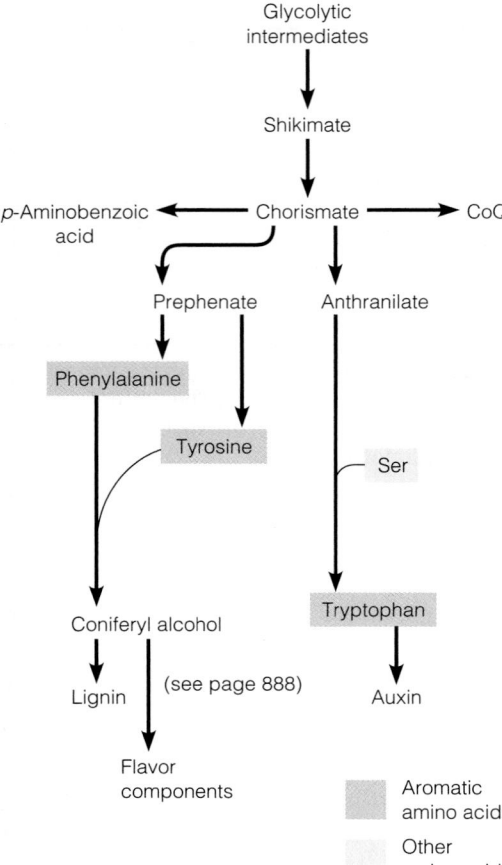

FIGURE 21.36

Overview of the biosynthesis of aromatic amino acids. The central pathways leading to the three amino acids are essentially the same in plants and bacteria. Details of the shikimic acid pathway are presented in Figures 21.37 and 21.38.

The shikimic acid pathway leads to synthesis of nearly all aromatic compounds, including lignin, and is thus one of the most productive pathways in biology.

$$HO-\overset{\overset{\displaystyle O}{\|}}{\underset{\underset{\displaystyle O^-}{|}}{P}}-CH_2-NH-CH_2-COO^-$$

Glyphosate

Synthesis of the Aromatic Amino Acids from Glycolytic Intermediates: The Shikimic Acid Pathway

Synthesis of these aromatic rings from noncyclic precursors involves complex chemistry. As with other lengthy biosynthetic pathways, such as the synthesis of vitamins, most of the aromatic biosynthetic capabilities have been lost during animal evolution. The synthetic pathways we shall discuss are limited to plants and bacteria—with one exception. That exception is the hydroxylation of phenylalanine to tyrosine, which we discussed in the context of phenylalanine degradation (Figure 21.7).

A single branched pathway in microorganisms and plants leads to synthesis of phenylalanine, tyrosine, tryptophan, and virtually all other aromatic compounds (Figure 21.36). Individual reactions of the pathway were established in bacteria. A large number of auxotrophic mutants could be isolated and characterized genetically (mapping), physiologically (identification of compounds that could satisfy growth requirements), and biochemically (identification of intermediates accumulating when a given step is blocked). So far as is known, the processes are quite similar in plants.

Several key findings were made. First, all of the carbon in phenylalanine and tyrosine was derived from erythrose-4-phosphate and the glycolytic intermediate phosphoenolpyruvate (PEP). Recall that erythrose-4-phosphate is also derived from glycolytic intermediates, via the pentose phosphate pathway (Chapter 13). Second, a class of mutants requiring phenylalanine, tyrosine, tryptophan, p-aminobenzoic acid, and p-hydroxybenzoic acid for growth could have all five requirements met by a single compound—**shikimic acid**. We now know that these mutants are blocked in the formation of shikimate (the fourth reaction in the part of the pathway shown in Figure 21.37) and that, when it is provided, all of the subsequent steps in the pathway can occur. Note that an unbranched pathway leads through shikimic acid to **chorismic acid**. From chorismic acid one pathway leads to **prephenic acid**, with subsequent branches leading to phenylalanine and tyrosine (Figure 21.38). Another pathway leads through **anthranilic acid** to tryptophan, still another pathway leads to p-aminobenzoic acid, and a final pathway leads, via p-hydroxybenzoic acid, to coenzyme Q. Thus, the shikimic acid pathway is responsible for biosynthesis of virtually all aromatic compounds because the products we have just mentioned serve in turn as precursors to other aromatic compounds.

Great attention has focused in recent years on the sixth reaction of the shikimic acid pathway, catalyzed by **5-enoylpyruvylshikimate-3-phosphate synthase (EPSP synthase)**, in higher plants. This enzyme is specifically inhibited by a compound called **glyphosate**, or glycine phosphonate (a *phosphonate* is a compound with a covalent bond linking carbon to phosphorus). The growth of most crop and weed plants is inhibited by glyphosate, which is an effective broad-spectrum herbicide sold as Roundup®. A recent achievement of biotechnology is the transfer of genes conferring resistance to glyphosate into crop plants. For example, "Roundup Ready" cotton seed was sold to farmers in the mid-1990s, and other crop plants, similarly modified, are now on the market. This modification allows simplified and effective weed control. Spraying a field should eliminate all plants except the genetically engineered species.

Interesting control mechanisms are involved in regulating these pathways. Studies on the genetic control of tryptophan biosynthesis, described in Chapter 29, have provided some of our most important insights into transcriptional regulation. More recent studies on the reactions leading to chorismic acid have revealed the existence of several multifunctional enzymes—single polypeptide chains containing two or more active sites for catalysis of sequential reactions—obviously, an efficient way to control several reactions jointly.

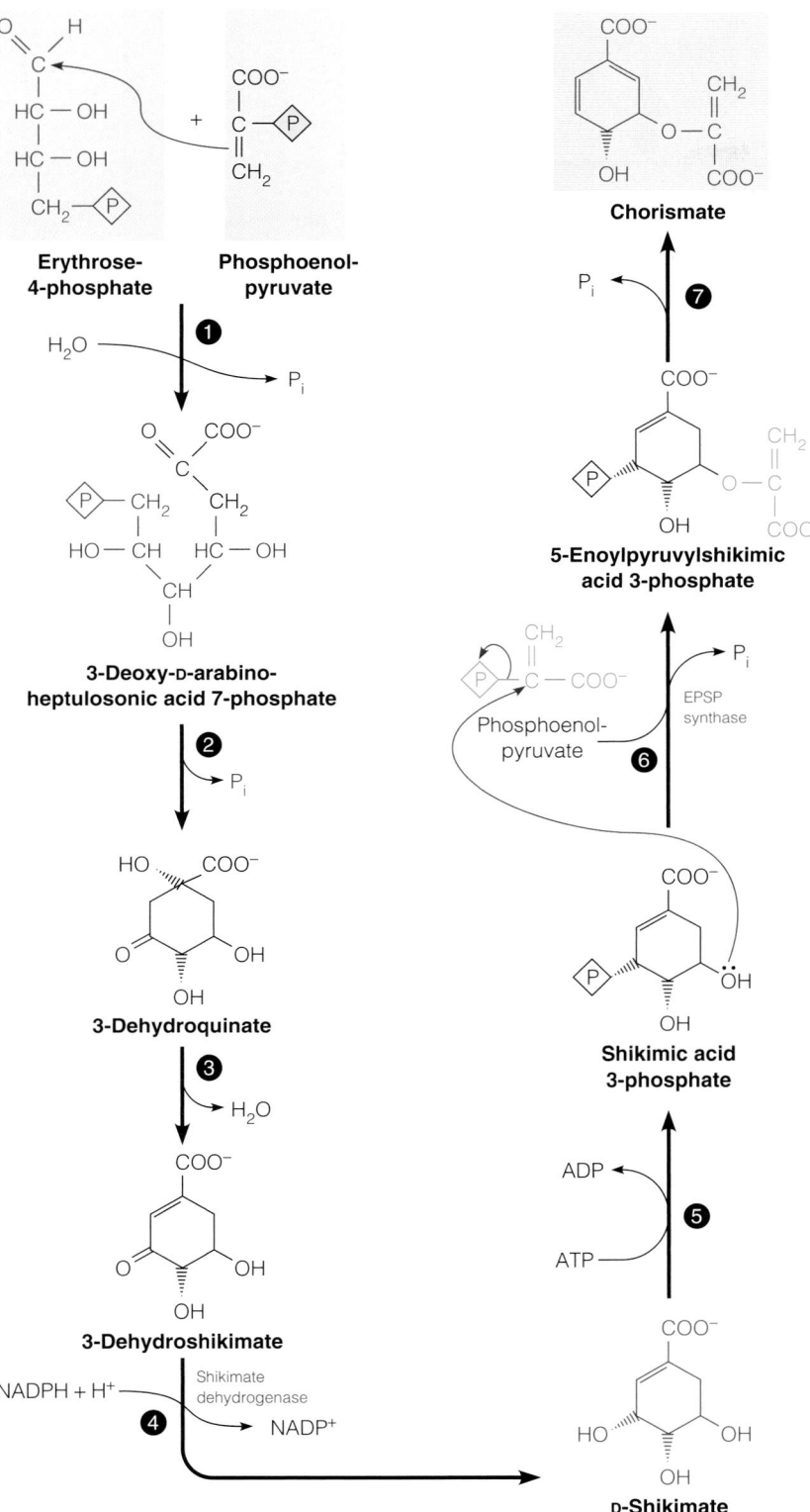

Chorismate

5-Enoylpyruvylshikimic acid 3-phosphate

EPSP synthase

Phosphoenol-pyruvate

Shikimic acid 3-phosphate

D-Shikimate

Shikimate dehydrogenase

3-Dehydroshikimate

3-Dehydroquinate

3-Deoxy-D-arabino-heptulosonic acid 7-phosphate

Erythrose-4-phosphate

Phosphoenol-pyruvate

FIGURE 21.37

Details of the shikimic acid pathway, I. This figure depicts the initial unbranched pathway from erythrose-4-phosphate and phosphoenolpyruvate to chorismic acid. The first reaction is driven by loss of phosphate from phosphoenolpyruvate. In the second reaction an unusual cobalt-requiring enzyme effects ring closure with dehydrogenation and loss of the second phosphate. Dehydration in the third step yields dehydroshikimate, which is then reduced by NADPH to shikimate. A three-carbon side chain is then attached via phosphorylation of shikimate and reaction with a second molecule of phosphoenolpyruvate. Dephosphorylation of this intermediate gives chorismate, the branch point of the pathway.

The first reaction in the pathway from anthranilic acid to tryptophan (Figure 21.38) involves an activated sugar derivative, **5-phosphoribosyl-1-pyrophosphate (PRPP)**, which plays its most widespread roles in nucleotide synthesis (see Chapter 22). In bacteria the genes encoding these enzymes are linked in a linear array, the **tryptophan operon**. As discussed in Chapter 27, an

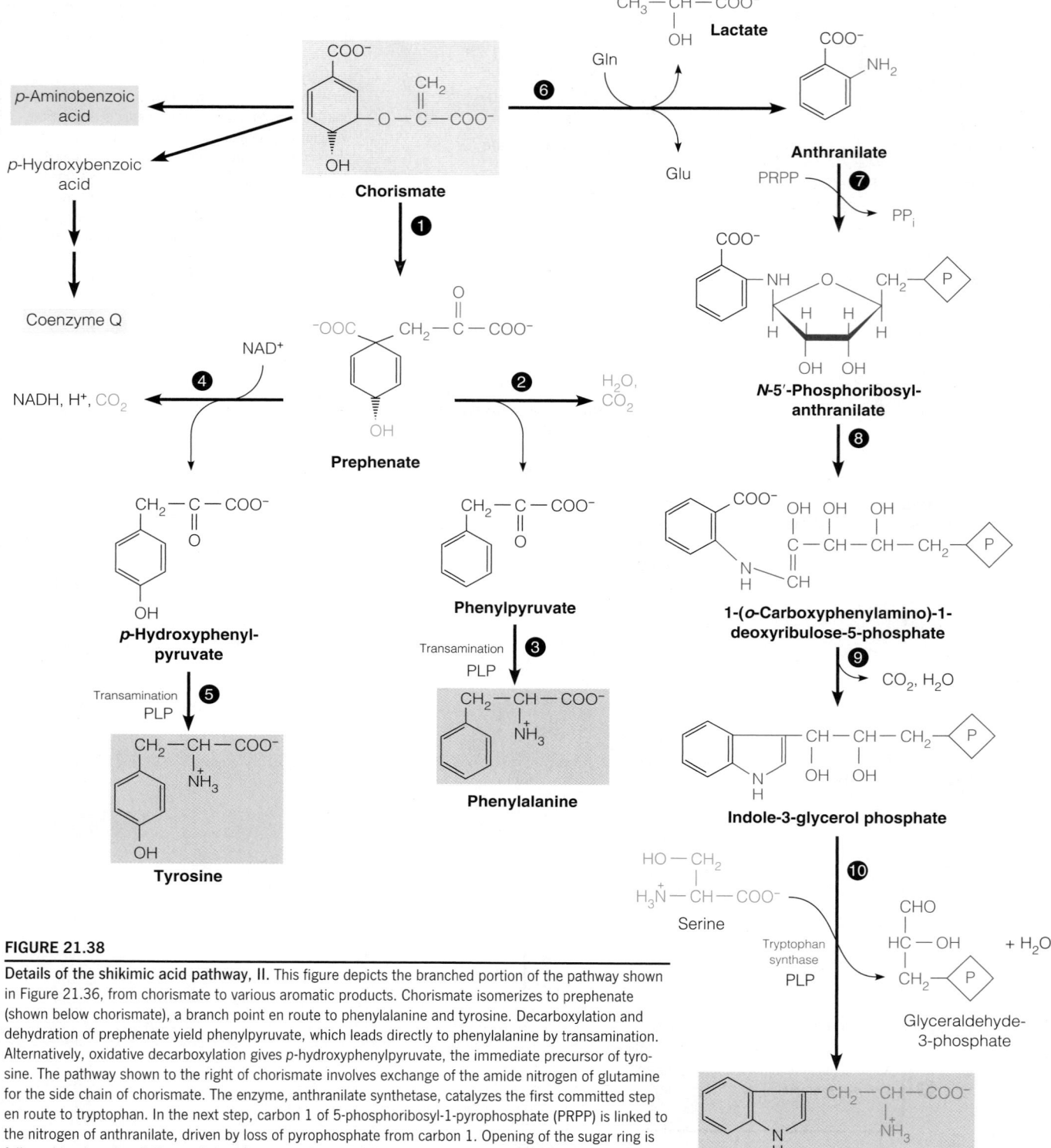

FIGURE 21.38

Details of the shikimic acid pathway, II. This figure depicts the branched portion of the pathway shown in Figure 21.36, from chorismate to various aromatic products. Chorismate isomerizes to prephenate (shown below chorismate), a branch point en route to phenylalanine and tyrosine. Decarboxylation and dehydration of prephenate yield phenylpyruvate, which leads directly to phenylalanine by transamination. Alternatively, oxidative decarboxylation gives p-hydroxyphenylpyruvate, the immediate precursor of tyrosine. The pathway shown to the right of chorismate involves exchange of the amide nitrogen of glutamine for the side chain of chorismate. The enzyme, anthranilate synthetase, catalyzes the first committed step en route to tryptophan. In the next step, carbon 1 of 5-phosphoribosyl-1-pyrophosphate (PRPP) is linked to the nitrogen of anthranilate, driven by loss of pyrophosphate from carbon 1. Opening of the sugar ring is followed by decarboxylation and ring closure to give indole-3-glycerol phosphate. In the final step, the three-carbon side chain of the indole compound is exchanged for that of serine, yielding tryptophan.

operon is a linked set of genes whose expression is regulated jointly at the level of transcription. In bacteria, yeasts, molds, and plants the final enzyme in the pathway, **tryptophan synthase**, is an $\alpha_2\beta_2$ dimer. Isolated α and β subunits catalyze the following partial reactions, with the holoenzyme catalyzing a concerted reaction in which indole does not dissociate from the enzyme surface but immediately reacts with serine to give tryptophan.

$$\text{Indole-3-glycerol phosphate} \xrightarrow{\alpha \text{ subunit}} \text{indole} + \text{3-phosphoglyceraldehyde}$$

$$\text{Indole} + \text{serine} \xrightarrow[\text{PLP}]{\beta \text{ subunit}} \text{tryptophan} + H_2O$$

Remarkably, X-ray crystallography shows that the intermediate, indole, is *channeled* from the α subunit active site to the β active site, a distance of 25 Å, through a tunnel in the interior of the protein molecule (Figure 21.39). Channeling refers to the direct transfer of a metabolite between sequential enzymes in a metabolic cycle. We will see another example of this in Chapter 22. Kinetic analysis, complementing more detailed structural analysis, suggests that each active site is covered intermittently by a "lid" (Figure 21.40), which acts to keep indole within the channel.

Synthesis of Histidine from Glycolytic Intermediates

The biosynthesis of histidine presents several parallels with the shikimic acid pathway, regarding methods used for its elucidation, complexity of the reactions involved, elegance of genetic pathway regulation, and practical applications of knowledge of the pathway. However, the histidine pathway is distinctive in being unbranched. As established largely in the laboratories of Bruce Ames and Philip

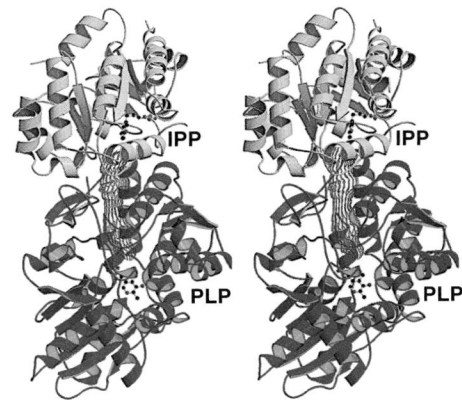

FIGURE 21.39

Structure of tryptophan synthase. This divergent (wall-eyed) stereo model shows the structure of one α—β functional unit from *Salmonella typhimurium* (PDB ID 1QOP). The α subunit is shown in yellow and the β subunit is shown in blue. A tunnel connecting the two active sites is indicated by the red chicken wire tube. Indole propanol phosphate (IPP, a competitive inhibitor of indole-3-glycerol phosphate) and pyridoxal phosphate (PLP) are shown as ball-and-stick models bound to the α and β subunits, respectively.

Reprinted with permission from *Accounts of Chemical Research* 36:539–548, F. M. Raushel, J. B. Thoden, and H. M. Holden, Enzymes with molecular tunnels. © 2003 American Chemical Society.

Histidine auxotrophic mutations have been useful both for defining the biosynthetic pathway and for analyzing environmental mutagenesis.

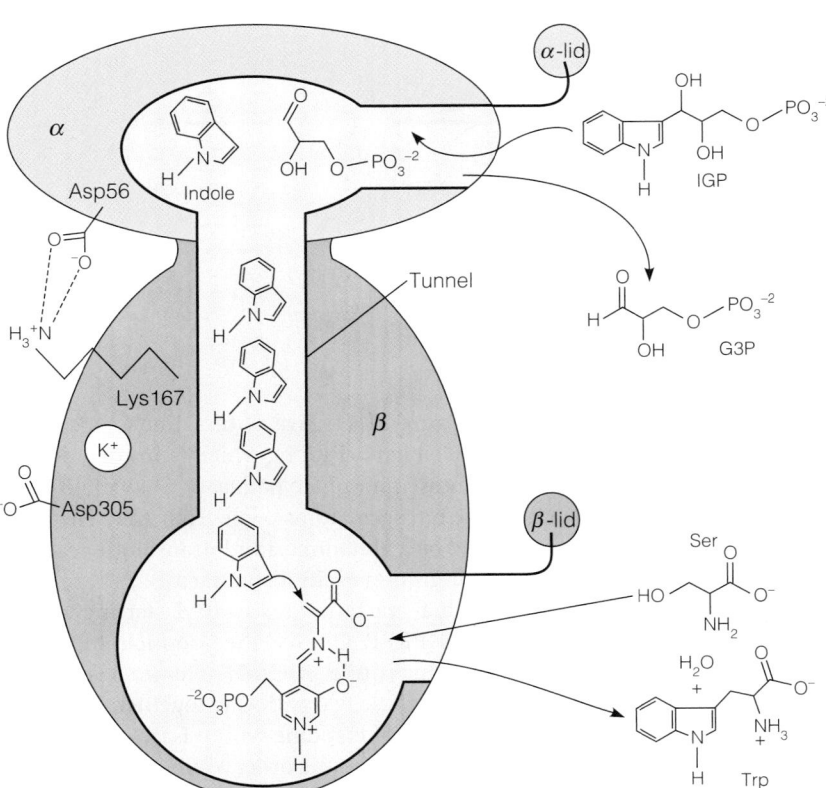

FIGURE 21.40

Schematic depiction of the action of tryptophan synthase in one α—β functional unit. Shown here are the active sites of the α (yellow) and β (blue) subunits, the interconnecting 25 Å tunnel filled with indole molecules, and the routes of substrate entry and product release for both subunits, with the covering "lids," whose alternate opening and closing restrains indole within the tunnel. Shown also are a binding site for the catalytically essential K^+ and a salt bridge known to provide allosteric linkage between the two subunits. IGP, indole-3-glycerol phosphate; G3P, glyceraldehyde-3-phosphate; ser, serine; trp, tryptophan.

Modified from *Trends in Biochemical Sciences* 22:22–27, P. Pan, E. Woehl, and M. F. Dunn, Protein architecture, dynamics and allostery in tryptophan synthase channeling. © 1997, with permission of Elsevier.

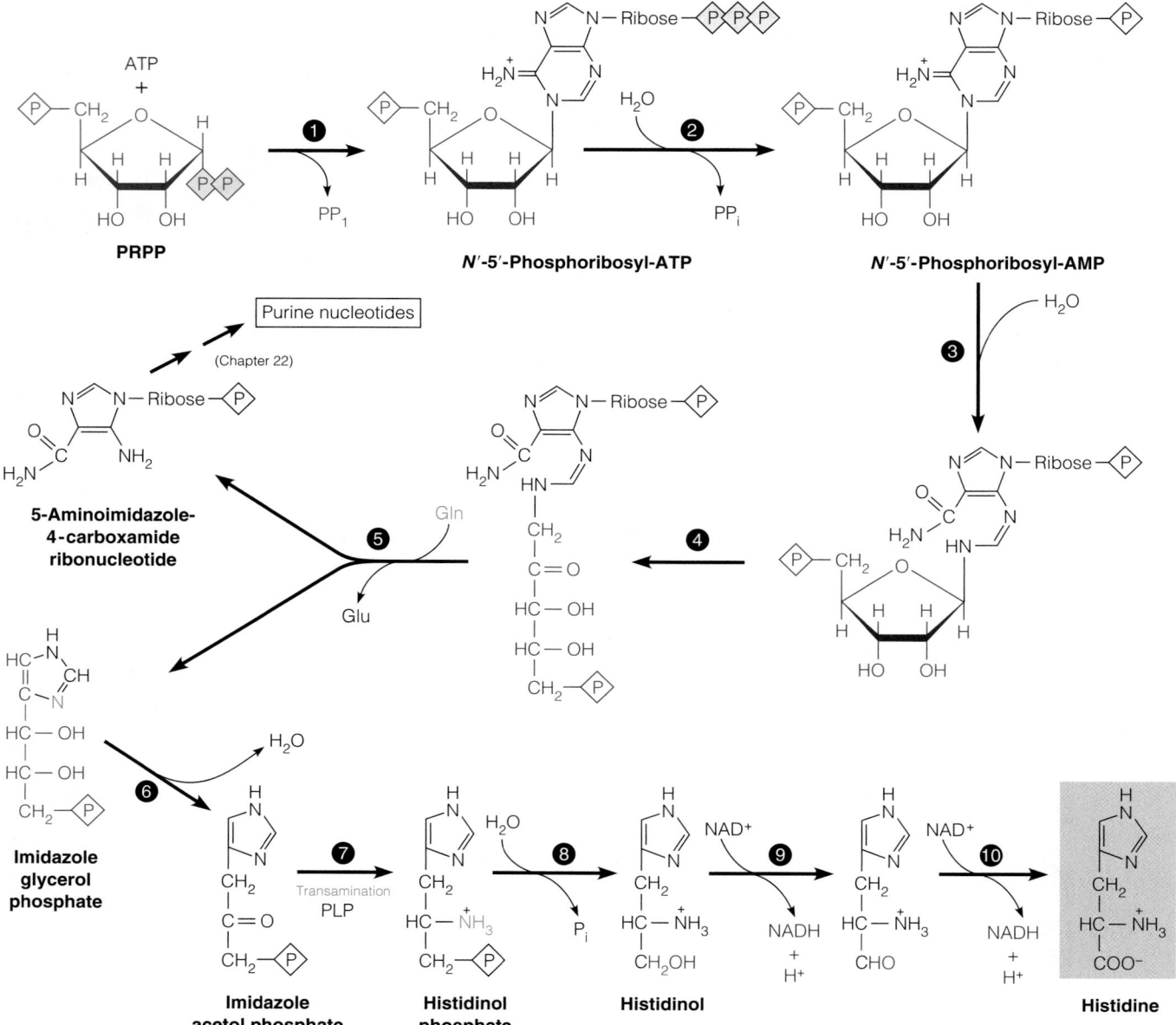

FIGURE 21.41

The pathway for histidine biosynthesis. After activation of a purine ring by reaction with 5-phosphoribosyl-1-pyrophosphate (PRPP, step ❶), the ring opens to give the third intermediate (steps ❷ and ❸). The ribose ring derived from PRPP then opens (step ❹). The transfer of amide nitrogen from glutamine (step ❺) is accompanied by cleavage and ring closure, giving the first imidazole compound, imidazole glycerol phosphate. The other product is an intermediate in the synthesis of purine nucleotides. Imidazole glycerol phosphate is transformed to histidine by a straightforward sequence involving dehydration (step ❻), transamination (step ❼), dephosphorylation (step ❽), and dehydrogenation (steps ❾ and ❿).

Hartman, the pathway is shown in Figure 21.41. Five of histidine's six carbons come from phosphoribosyl-1-pyrophosphate (PRPP), which is in turn derived from glucose-6-phosphate via the pentose phosphate pathway (Chapter 13). The sixth carbon and one of the imidazole nitrogen atoms come from ATP. The other imidazole nitrogen atom is introduced by a glutamine amidotransferase reaction, and the α-amino group comes from glutamate in a standard PLP-dependent aminotransferase reaction. Ten individual reactions are involved, starting with an unusual reaction that joins ATP and PRPP. One of the products of step 5, 5-aminoimidazole-4-carboxamide ribonucleotide (AICAR), is an intermediate in purine biosynthesis (Chapter 22), linking these two pathways together.

The 10 structural genes for the enzymes of histidine synthesis in enteric bacteria are linked to one another in the same order as the order of the reactions of the pathway. This set of genes, the **histidine operon**, is coordinately regulated at the transcriptional level, and all 10 genes are transcribed to give one large messenger

RNA, which is translated to give the 10 enzymes. This highly organized gene arrangement may facilitate regulation of the pathway.

Once the genes and gene products had been identified, Bruce Ames used the mutant bacteria that had been generated for these investigations in a novel way—to search for mutagens in the environment. Using the **Ames test**, researchers can count mutations simply by measuring the rate at which mutants that cannot synthesize histidine (histidine **auxotrophs**) mutate to a form that can synthesize the amino acid (**prototrophs**). The researcher treats a culture of auxotrophs with a suspected mutagen, plates the bacteria on medium containing no histidine, and counts the colonies that appear as the result of reversion mutations. Using this system, Ames and colleagues reported a very high correlation between compounds known to be carcinogenic in animals and those found to be mutagens in this test. Thus, the Ames test provides a quick and inexpensive way to search for suspected carcinogens in the environment. Moreover, these findings provided support for the idea that cancer arises as the result of a series of somatic cell mutations, an idea we discuss further in Chapter 23.

SUMMARY

Amino acids play numerous roles as intermediates in the biosynthesis of other metabolites, including purine nucleotides (glutamine, glycine, serine), pyrimidine nucleotides (aspartate, glutamine), polyamines and methyl groups (methionine), glutathione (glutamate, cysteine, glycine), creatine phosphate (arginine), neurotransmitters (tyrosine, tryptophan, glutamate, arginine), lignin, aromatic compounds and pigments (phenylalanine), hormones (tyrosine, histidine), porphyrins (glycine and glutamate in plants), and other amino acids. The roles of amino acids as neurotransmitters and neurotransmitter precursors are particularly important, as are their roles in porphyrin synthesis. In porphyrin synthesis, glycine condenses with succinate to give a heterocyclic ring compound, porphobilinogen, which is the precursor to all four pyrrole rings in heme and other porphyrins. Degradation of porphyrins in animals yields iron, which is reutilized, and bilirubin, an insoluble tetrapyrrole that is excreted. Amino acids are synthesized from intermediates in the citric acid cycle, glycolysis, and the pentose phosphate pathway.

REFERENCES

Glutamate, Aspartate, Alanine, Glutamine, Asparagine, Proline, Serine, Glycine, Threonine

Chaves, A. L. S., and P. C. de Mello-Farias. (2006) Ethylene and fruit ripening: From illumination gas to the control of gene expression, more than a century of discoveries. *Genet. Mol. Biol.* 29:508–515. A recent review of the history, synthesis, and actions of this important plant hormone.

Li, X., F. W. Bazer, H. Gao, W. Jobgen, G. A. Johnson, P. Li, J. R. McKnight, M. C. Satterfield, T. E. Spencer, and G. Wu (2009) Amino acids and gaseous signaling. *Amino Acids* 37:65–78. This review covers the biosynthesis and function of nitric oxide, as well as hydrogen sulfide.

Myllyharju, J. (2003) Prolyl 4-hydroxylases, the key enzymes of collagen biosynthesis. *Matrix Biol.* 22:15–24. A review of the mechanism of proline hydroxylation and its role in collagen synthesis.

Ng, W. L., and B. L. Bassler (2009) Bacterial quorum-sensing network architectures. *Annu. Rev. Genet.* 43:197–222. This comprehensive review describes the discovery, chemistry, and function of this cell–cell communication process.

Sommer-Knudsen, J., A. Bacic, and A. E. Clarke (1998) Hydroxyproline-rich plant glycoproteins. *Phytochemistry* 47:483–497. Hydroxyproline was long thought to be present only in animal connective tissue proteins. This article summarizes its occurrence and roles in plant structural proteins.

Tabatabaie, L., L. W. Klomp, R. Berger, and T. J. de Koning (2010) L-Serine synthesis in the central nervous system: A review on serine deficiency disorders. *Mol. Genet. Metab.* 99:256–262.

Sulfur-Containing Amino Acids

Becerra-Solano, L. E., J. Butler, G. Castaneda-Cisneros, D. E. McCloskey, X. Wang, A. E. Pegg, C. E. Schwartz, J. Sanchez-Corona, and J. E. Garcia-Ortiz (2009) A missense mutation, p.V132G, in the X-linked spermine synthase gene (SMS) causes Snyder-Robinson syndrome. *Am. J. Med. Genet. A* 149A:328–335. This paper describes how a defect in spermine synthesis leads to a serious human disease.

Brosnan, J. T., and M. E. Brosnan (2006) The sulfur-containing amino acids: An overview. *J. Nutr.* 136:1636S–1640S. This minireview summarizes the unique chemistry and biochemistry of the sulfur amino acids.

Cohen, S. S. (1998) *Biochemistry of the Polyamines.* Oxford University Press, New York. An all-encompassing review of the metabolism and functions of polyamines, written by a long-time leader in the field.

Gadalla, M. M., and S. H. Snyder (2010) Hydrogen sulfide as a gasotransmitter. *J. Neurochem.* 113:14–26; Lefer, D. J. (2007) A new gaseous signaling molecule emerges: Cardioprotective role of hydrogen sulfide. *Proc. Natl. Acad. Sci. USA* 104:17907–17908. These two articles describe the surprising discovery and function of yet another gaseous signaling molecule.

Giordano, M., A. Norici, and R. Hell (2005) Sulfur and phytoplankton: Acquisition, metabolism and impact on the environment. *New Phytol.* 166:371–382. A minireview of the chemistry and metabolism of sulfur-containing compounds in the marine ecosystem.

Wallace, H. M., A. V. Fraser, and A. Hughes (2003) A perspective of polyamine metabolism. *Biochem. J.* 376:1–14. This article covers recent advances in our understanding of the roles of polyamines in human disease, especially cancer.

S-Adenosylmethionine, Methylation, and Homocysteine

Brosnan, J. T., R. P. da Silva, and M. E. Brosnan (2011) The metabolic burden of creatine synthesis. *Amino Acids* 40:1325–1331. This review describes the quantitatively signficant demands that creatine synthesis places on methyl groups, arginine, and glycine in animals.

Clarke, S. (2003) Aging as war between chemical and biochemical processes: Protein methylation and the recognition of age-damaged proteins for repair. *Ageing Res. Rev.* 2:263–285. This article discusses spontaneous damage to proteins and the role of methylation in their repair.

da Silva, R. P., I. Nissim, M. E. Brosnan, and J. T. Brosnan (2009) Creatine synthesis: Hepatic metabolism of guanidinoacetate and creatine in the rat in vitro and in vivo. *Am. J. Physiol. Endocrinol. Metab.* 296:E256–E261. This paper presents evidence for the existence of an interorgan pathway for creatine biosynthesis in mammals.

How folate fights disease. *Nat. Struct. Biol.* 6:293–294 (1999). This editorial describes how biochemical and structural studies of a protein (methylenetetrahydrofolate reductase) lead to a better understanding of clinical results. It accompanied the research article that Figure 21.12 was taken from.

Kraus, J. P., and V. Kozich (2001) Cystathionine β-synthase and its deficiency. In *Homocysteine in Health and Disease*, R. Carmel and D. W. Jacobsen, eds., Ch. 20, pp. 223–243, Cambridge University Press, Cambridge. This review is part of a recent compendium on biochemical, genetic, and clinical aspects of homocysteine metabolism.

Roje, S., S. Y. Chan, F. Kaplan, R. K. Raymond, D. W. Horne, D. R. Appling, and A. D. Hanson (2002) Metabolic engineering in yeast demonstrates that *S*-adenosylmethionine controls flux through the methylenetetrahydrofolate reductase reaction in vivo. *J. Biol. Chem.* 277:4056–4061. A chimeric plant–yeast enzyme is constructed to study the role of AdoMet feedback regulation in vivo.

Rozen, R. (2001) Polymorphisms of folate and cobalamin metabolism. In *Homocysteine in Health and Disease*, R. Carmel and D. W. Jacobsen, eds., Ch. 22, pp. 259–269, Cambridge University Press, Cambridge. This chapter from a recent compendium on homocysteine metabolism describes biochemical and genetic aspects of methylenetetrahydrofolate reductase.

Aromatic Amino Acids

Brennan, M. M. (1998) New age paper and textiles. *Chem. Eng. News*, March 23 issue, pp. 39–47. A news-type article describing the use of biological reagents to degrade lignin and their applications in paper and textile production.

Dunn, M. F., D. Niks, H. Ngo, T. R. M. Barends, and I. Schlichting (2008) Tryptophan synthase: The workings of a channeling nanomachine. *Trends Biochem. Sci.* 33:254–264. The article reviews the structure and mechanism of one of the first enzymes for which channeling was recognized.

Fitzpatrick, P. F. (2003) Mechanism of aromatic amino acid hydroxylation. *Biochemistry* 42:14083–14091. This paper reviews the mechanism of this interesting enzyme family.

Garrod, A. E. (1909) *Inborn errors of metabolism. The Croonian lectures delivered before the Royal College of Physicians of London, in June, 1908.* Frowde, Hodder & Stoughton, London. In this classic text, Garrod develops his concept of inheritable metabolic diseases, based on his study of alkaptonuria.

Hayaishi, O. (2008) From oxygenase to sleep. *J. Biol. Chem.* 283:19165–19175. An autobiographical article by the discoverer of oxygenases and a pioneer in studies of tryptophan metabolism.

Ito, S., and K. Wakamatsu (2008) Chemistry of mixed melanogenesis—Pivotal roles of dopaquinone. *Photochem. Photobiol.* 84:582–592. This review summarizes the pathways of melanin synthesis.

Moens, A. L., and D. A. Kass (2006) Tetrahydrobiopterin and cardiovascular disease. *Arterioscler. Thromb. Vasc. Biol.* 26:2439–2444. This minireview focuses on the role of BH_4 as a cofactor for nitric oxide synthase.

Raushel, F. M., J. B. Thoden, and H. M. Holden (2003) Enzymes with molecular tunnels. *Acc. Chem. Res.* 36:539–548. This minireview discusses tryptophan synthase and other examples.

Scriver, C. R., and S. Kaufman (2001) Hyperphenylalaninemia: Phenylalanine hydroxylase deficiency. In *The Metabolic and Molecular Bases of Inherited Disease*, C. R. Scriver, A. L. Beaudet, W. S. Sly, D. Valle, B. Childs, K. W. Kinzler, and B. Vogelstein, eds., Vol. II, Ch. 77, pp. 1667–1734, McGraw-Hill, New York. This is the first of 14 chapters in this compendium that describe heritable metabolic disorders of amino acid metabolism.

Sturm, R. A. (2009) Molecular genetics of human pigmentation diversity. *Hum. Mol. Genet.* 18:R9–R17. This review summarizes new discoveries of genes involved in human melanin production.

Valine, Leucine, and Isoleucine

Chuang, D. T., and V. E. Shih (2001) Maple syrup urine disease (Branched-chain ketoaciduria). In *The Metabolic and Molecular Bases of Inherited Disease*, C. R. Scriver, A. L. Beaudet, W. S. Sly, D. Valle, B. Childs, K. W. Kinzler, and B. Vogelstein, eds., Vol. II, Ch. 87, pp. 1971–2005, McGraw-Hill, New York. This chapter describes the biochemistry and clinical consequences of deficiency of the branched-chain α-keto acid dehydrogenase.

Porphyrin Metabolism

Anderson, K. E., S. Sassa, D. F. Bishop, and R. J. Desnick (2001) Disorders of heme biosynthesis: X-linked sideroblastic anemia and the porphyrias. In *The Metabolic and Molecular Bases of Inherited Disease*, C. R. Scriver, A. L. Beaudet, W. S. Sly, D. Valle, B. Childs, K. W. Kinzler, and B. Vogelstein, eds., Vol. II, Ch. 124, pp. 2991–3062, McGraw-Hill, New York. A definitive review of the pathways and associated human disorders.

Jahn, D., E. Verkamp, and D. Söll (1992) Glutamyl-transfer RNA: A precursor of heme and chlorophyll biosynthesis. *Trends Biochem. Sci.* 17:215–218. This article discusses the unexpected role of tRNA in δ-ALA synthetase.

Schulze, J. O., W. D. Schubert, J. Moser, D. Jahn, and D. W. Heinz (2006) Evolutionary relationship between initial enzymes of tetrapyrrole biosynthesis. *J. Mol. Biol.* 358:1212–1220. This paper describes structural and mechanistic evidence for an evolutionary relationship between the two types of enzymes that produce δ-aminolevulinic acid.

Warren, M. J., J. B. Cooper, S. P. Wood, and P. M. Shoolingin-Jordan (1998) Lead poisoning, heme synthesis and 5-aminolaevulinic acid dehydratase. *Trends Biochem. Sci.* 23:217–221. The structural basis for the porphyria acquired in lead poisoning.

Warren, M. J., M. Jay, D. M. Hunt, G. H. Elder, and J. G. Rôhl (1996) The maddening business of King George III and porphyria. *Trends Biochem. Sci.* 21:229–234. A fascinating mixture of history and biochemistry, illustrated with scenes from the movie.

Neurotransmitters

Daubner, S. C., T. Le, and S. Wang (2011) Tyrosine hydroxylase and regulation of dopamine synthesis. *Arch. Biochem. Biophys.* 508:1–12.

Fernstrom, J. D., and M. H. Fernstrom (2007) Tyrosine, phenylalanine, and catecholamine synthesis and function in the brain. *J. Nutr.* 137: 1539S–1547S.

Windahl, M. S., C. R. Petersen, H. E. Christensen, and P. Harris (2008) Crystal structure of tryptophan hydroxylase with bound amino acid substrate. *Biochemistry* 47:12087–12094. This paper describes the structure of the rate-limiting enzyme in serotonin biosynthesis.

PROBLEMS

1. A clinical test sometimes used to diagnose folate deficiency or B_{12} deficiency is a histidine tolerance test, where one injects a large dose of histidine into the bloodstream and then carries out a series of biochemical determinations. What histidine metabolite would you expect to accumulate in a folate- or B_{12}-deficient patient, and why?

2. In bacteria much of the putrescine is synthesized, not from ornithine but from arginine, which decarboxylates to yield *agmatine*. Formulate a plausible pathway from arginine to putrescine, using this intermediate.

$$\overset{\overset{+}{N}H_2}{\underset{\parallel}{H_2N - C - NH - CH_2 - CH_2 - CH_2 - CH_2 - \overset{+}{N}H_3}}$$

Agmatine

3. The mitochondrial form of carbamoyl phosphate synthetase is allosterically activated by *N*-acetylglutamate. Briefly describe a rationale for this effect.

4. *Psilocybin* is a hallucinogenic compound found in some mushrooms. Present a straightforward pathway for its biosynthesis from one of the aromatic amino acids.

$$CH_2 - CH_2 - N(CH_3)_2$$

Psilocybin

*5. One can identify phenylketonurics and PKU carriers (heterozygotes) by means of a phenylalanine tolerance test. One injects a large dose of phenylalanine into the bloodstream and measures its clearance from the blood by measuring serum phenylalanine levels at regular intervals. Sketch curves showing relative blood phenylalanine concentration versus time that you would expect to be displayed by (a) a PKU patient, (b) a heterozygote, and (c) a normal individual. What kind of tolerance test could you devise to distinguish between PKU resulting from either phenylalanine hydroxylase deficiency or dihydropteridine reductase deficiency?

6. (a) Formaldehyde reacts nonenzymatically with tetrahydrofolate to generate 5,10-methylenetetrahydrofolate. [^{14}C]Formaldehyde can be used to prepare serine labeled in the β-carbon. What else would be needed?

(b) [^{14}C]Serine, prepared as described above, is useful for many things, but you would probably not want to use it for studies on protein synthesis because it would label nucleic acids, carbohydrates, and lipids, as well as proteins. Indicate how each of these classes of compounds could become labeled by this precursor.

7. If oxidation of acetyl-CoA yields 10 ATPs per mole through the citric acid cycle, how many ATPs will be derived from the complete metabolic oxidation of 1 mole of alanine in a mammal? Would the corresponding energy yield in a fish be higher or lower? Why? How much energy would be derived from the metabolic oxidation of 1 mole of isoleucine to CO_2, H_2O, and NH_3? Of tyrosine?

8. Some bacteria contain three different forms of aspartokinase, each with its own mode of regulation. Based on the roles of aspartokinase, as discussed in the text, propose a regulatory scheme applicable to each form of aspartokinase.

9. Proline betaine is a putative osmoprotectant in plants and bacteria, helping to prevent dehydration of cells.

$$\underset{CH_3 \quad CH_3}{\overset{+}{N}} - COO^-$$

Propose a plausible pathway for biosynthesis of this compound.

10. Most bacterial mutants that require isoleucine for growth also require valine. Why? Which enzyme or reaction would be defective in a mutant requiring only isoleucine (not valine) for growth?

11. Describe a series of allosteric interactions that could adequately control the biosynthesis of valine, leucine, and isoleucine.

12. The structure shown below is an intermediate in the synthesis of which biogenic amine? Use arrows to show how the next intermediate in this reaction is formed, and draw the structure of that intermediate.

$$\text{HN} \diagdown \text{N} - CH_2 - \overset{\overset{H}{|}}{C} - COO^-$$

13. Identify carbon atoms, by number, that are incorporated from this structure into the compounds listed in a–f.

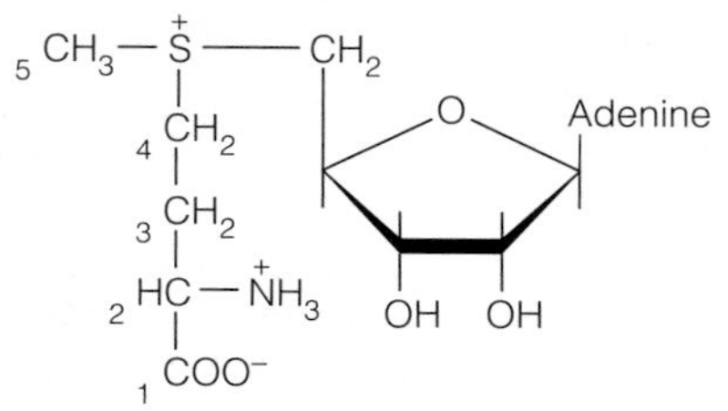

(a) Creatine phosphate _____
(b) Spermidine _____
(c) Ethylene _____
(d) Putrescine _____
(e) Glycine betaine _____
(f) Epinephrine _____

14. Why is phenylketonuria resulting from dihydropteridine reductase deficiency a more serious disorder than PKU resulting from phenylalanine hydroxylase deficiency?

*15. Propose a mechanism for the reaction catalyzed by dopamine β-hydroxylase. Assume that copper in the enzyme binds O_2 and that copper can change its oxidation state during the reaction.

16. The first two steps of ovothiol C biosynthesis are shown below. Propose the rest of the pathway, indicating the potential involvement of cofactors.

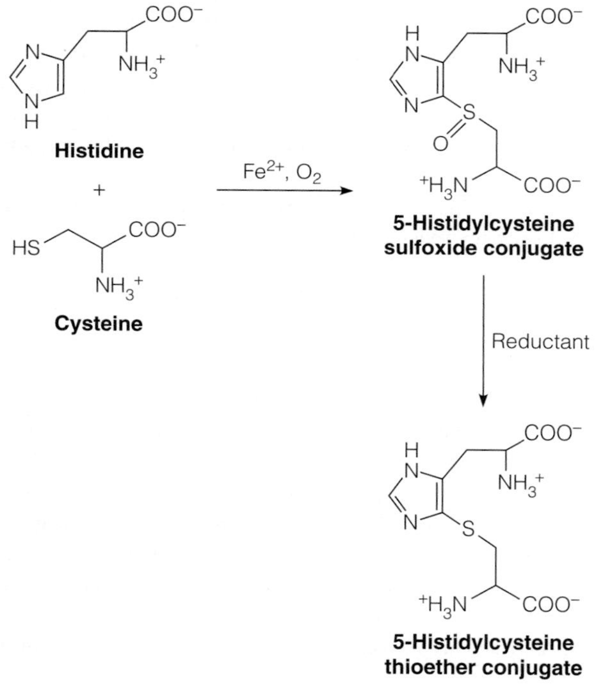

17. Glyphosate, which inhibits the EPSP synthase reaction of the shikimic acid pathway, is a phosphonic acid derivative of glycine; hence its name (glycine phosphonate). Based upon your knowledge of the EPSP synthase reaction (see pages 908–909), would you expect the inhibition to be competitive or noncompetitive with respect to each of the substrates? Briefly explain your answer.

18. In evaluating jaundiced patients, clinicians often determine whether the bilirubin that accumulates in the blood is primarily free bilirubin or the diglucuronide. Which form would accumulate primarily in chronic liver disease? In bile duct obstruction by a gallstone? In hemolytic anemia, such as that associated with glucose-6-phosphate dehydrogenase deficiency (see Chapter 13)? Briefly explain your answers.

19. Propose additional substrates and cofactors that might participate in each of the first three reactions in the flavonoid biosynthetic pathway shown on page 888.

20. Write a one-sentence answer to each question.
(a) Why does liver damage cause jaundice?
(b) Why do individuals with congenital erythropoietic porphyria become anemic?
(c) Why does a genetic deficiency of 5,10-methylenetetrahydrofolate reductase cause homocystinuria?

*21. Propose a chemical mechanism for the reaction catalyzed by the PLP-dependent glutamate 1-semialdehyde aminomutase.

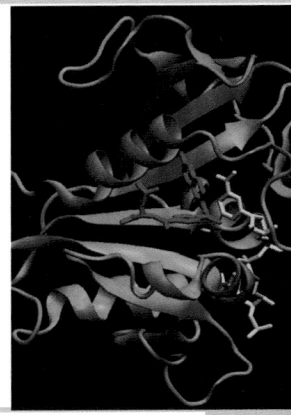

CHAPTER 22

Nucleotide Metabolism

We have encountered nucleotides repeatedly during our exploration of biochemistry. They serve as precursors to nucleic acids, as critical elements in energy metabolism, as carriers of activated metabolites for biosynthesis (such as nucleoside diphosphate sugars), as structural moieties of coenzymes, and, finally, as metabolic regulators and signal molecules (notably, cyclic AMP). In this chapter we discuss pathways of biosynthesis and degradation of purine and pyrimidine nucleotides, and we explore regulation of these processes—particularly critical in pathways leading to DNA replication. We discuss nucleotide biosynthetic enzymes as targets for the action of antimicrobial and anticancer drugs, and we describe the metabolic consequences of certain heritable alterations of nucleotide metabolism. The roles of nucleotides in metabolic or genetic regulation are discussed elsewhere in this book, in connection with the specific processes regulated.

Before starting this chapter, you may find it useful to review the information on nucleotide structure in Chapter 4. You should be aware also of the distinction between nucleosides and nucleotides. On complete hydrolysis, a mole of *nucleoside* yields at least 1 mole each of a sugar and a heterocyclic base, whereas a mole of *nucleotide* yields at least 1 mole each of a sugar, a base, and inorganic phosphate. A mole of *mononucleotide* contains only 1 mole each of base and sugar, but it may contain more than one phosphate. If it contains, for example, three phosphates, it is called a *nucleoside triphosphate.* The *deoxyribonucleotides,* which are used in DNA synthesis, are formed from *ribonucleotides* (RNA constituents) by pathways discussed later in this chapter.

Outlines of Pathways in Nucleotide Metabolism

Biosynthetic Routes: De Novo and Salvage Pathways

Unlike the other classes of metabolites we have encountered, neither nucleotides nor the bases and sugars from which they are formed are required to meet nutritional requirements, with the exception of some protozoan parasites.

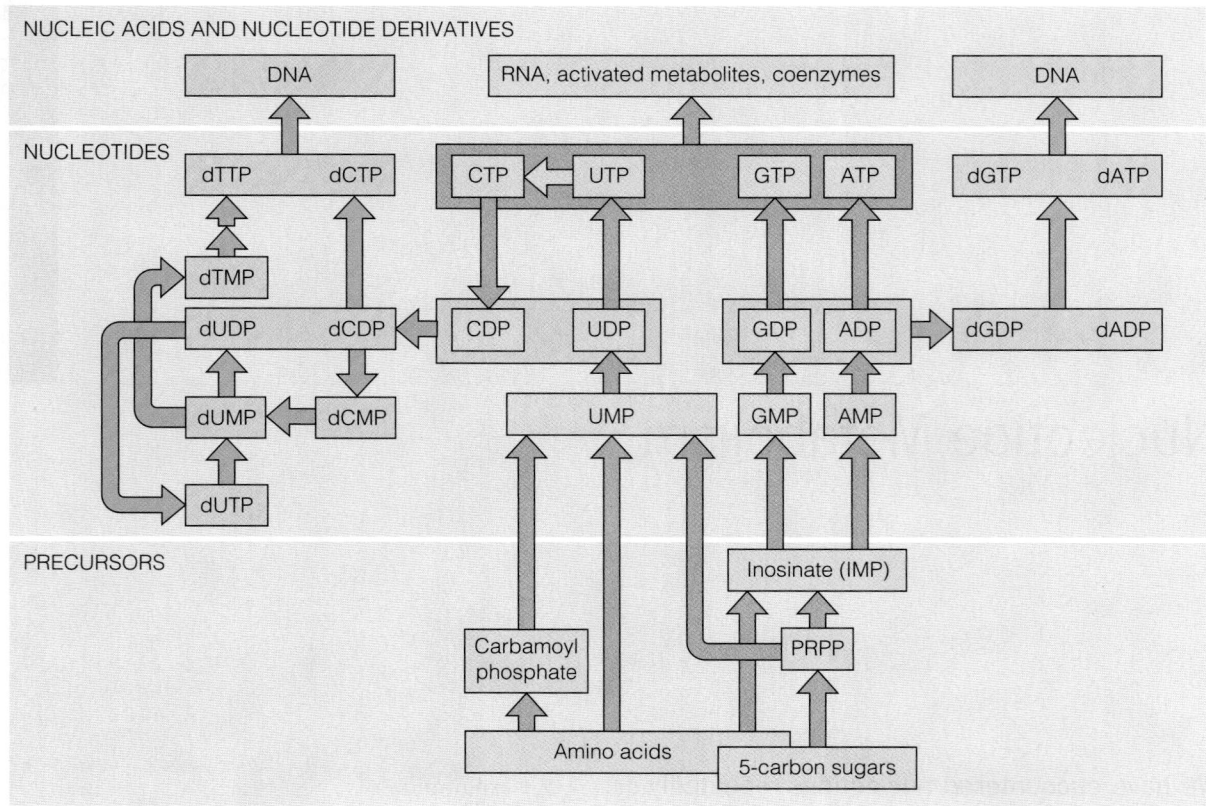

NUCLEIC ACIDS AND NUCLEOTIDE DERIVATIVES

FIGURE 22.1

Overview of nucleotide metabolism. De novo pathways for synthesis and utilization of ribonucleotides (orange) and deoxyribonucleotides (blue).

Nucleotides arise through de novo synthesis from low-molecular-weight precursors or through salvage of nucleosides or bases.

Most organisms can synthesize purine and pyrimidine nucleotides from low-molecular-weight precursors in amounts sufficient for their needs. These so-called **de novo pathways** are essentially identical throughout the biological world (Figure 22.1). Most organisms can also synthesize nucleotides from nucleosides or bases that become available either in the diet or through enzymatic break-down of nucleic acids. These processes are called **salvage pathways** because they involve the utilization of preformed purine and pyrimidine compounds that would otherwise be lost to biodegradation. As we shall see, salvage pathways represent important targets for treatment of microbial or parasitic diseases, sites for manipulation of biological systems (for example, in studies of mutagenesis or in preparation of monoclonal antibodies), and biological processes in which genetic alterations have severe and far-reaching consequences.

Nucleic Acid Degradation and the Importance of Nucleotide Salvage

Because salvage, or reuse, of purine and pyrimidine bases involves molecules released by nucleic acid degradation, let us begin by briefly considering these processes (Figure 22.2). Degradation can occur intracellularly (through the turnover of unstable messenger RNA species or through DNA repair pathways), as a result of cell death, or, in animals, through digestion of nucleic acids ingested in the diet.

In animals the extracellular hydrolysis of ingested nucleic acids represents the major route by which bases and nucleosides become available. The breakdown processes are comparable to those involved in protein digestion. Cleavage processes begin at internal linkages—in this case, phosphodiester bonds. Catalysis occurs via **endonucleases**, such as pancreatic ribonuclease or deoxyribonuclease, which function to digest nucleic acids in the small intestine. Endonucleolytic

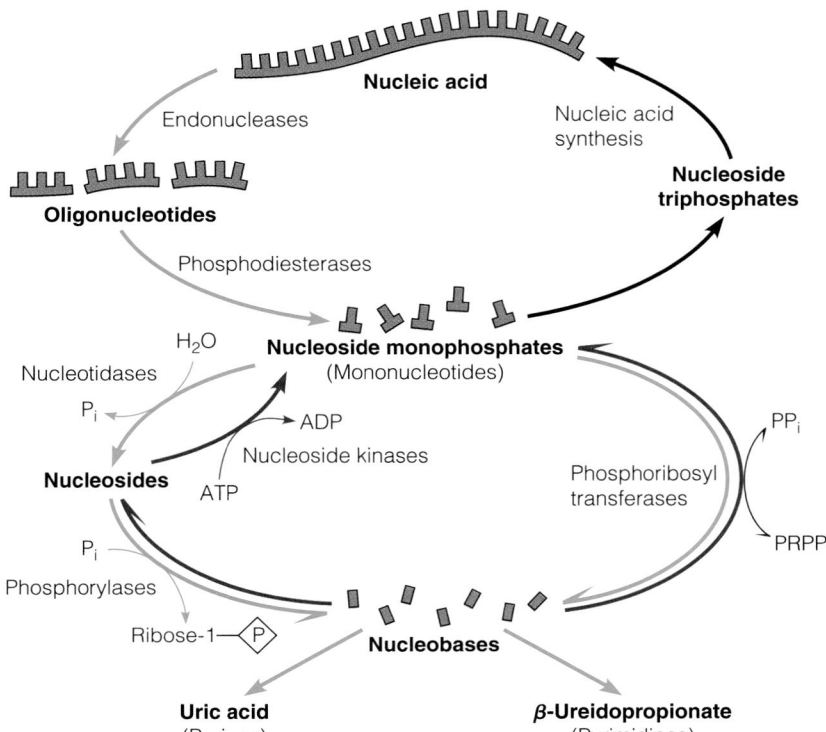

FIGURE 22.2

Reutilization of purine and pyrimidine bases. The figure shows relationships between nucleic acid catabolism (blue) and resynthesis of nucleotides by salvage pathways (magenta).

cleavages yield oligonucleotides, which are then cleaved **exonucleolytically** (at linkages near the ends of molecules) by nonspecific enzymes called **phosphodiesterases**. The products are mononucleotides—nucleoside 5'- or 3'- monophosphates, depending on the specificities of the enzymes involved. Nucleotides can then be cleaved hydrolytically, by a group of phosphomonoesterases called **nucleotidases**, to yield orthophosphate plus the corresponding nucleoside. Although hydrolytic cleavage of the resultant nucleoside does occur, the most common route for cleavage to the free base (nucleobase) involves the action of a **nucleoside phosphorylase**. Like glycogen phosphorylase, nucleoside phosphorylases cleave a glycosidic bond by adding across it the elements of inorganic phosphate, to yield the corresponding base plus ribose-1-phosphate (or deoxyribose-1-phosphate if the substrate is a deoxyribonucleoside):

Guanosine + phosphate $\rightleftharpoons$ guanine + **Ribose-1-phosphate**

(via Nucleoside phosphorylase)

These reactions are readily reversible, such that a nucleoside phosphorylase can also catalyze the first step in salvage synthesis of nucleotides from free nucleobases. When that occurs, the product nucleoside can be phosphorylated by ATP, through the action of a **nucleoside kinase** (Figure 22.2). Such enzymes are not universal. For example, animal cells contain neither a guanosine kinase nor a uridine phosphorylase, although these enzymes are found in other organisms.

If bases or nucleosides are not reused for nucleic acid synthesis via salvage pathways, the purine and pyrimidine bases are further degraded, to uric acid or β-ureidopropionate, respectively, as indicated in Figure 22.2. We consider these pathways later in this chapter.

PRPP: A Central Metabolite in De Novo and Salvage Pathways

An alternative salvage pathway, also shown in Figure 22.2, synthesizes nucleoside 5-phosphates directly from free bases. This route involves a class of enzymes called **phosphoribosyltransferases** and an activated sugar phosphate, **5-phospho-α-D-ribosyl-1-pyrophosphate (PRPP)**. Identified in Chapter 21 as an intermediate in histidine and tryptophan biosynthesis, PRPP is a key intermediate in the de novo synthesis of both purine and pyrimidine nucleotides. It is formed through the action of **PRPP synthetase**, which activates carbon 1 of ribose-5-phosphate by transferring to it the pyrophosphate moiety of ATP:

Ribose-5-phosphate   **5-Phospho-α-D-ribosyl-1-pyrophosphate (PRPP)**

A phosphoribosyltransferase reaction catalyzes the reversible transfer of a free base to the ribose of PRPP, displacing pyrophosphate and producing a nucleoside monophosphate. Because the deoxyribose analog of PRPP is absent from most cells, these enzymes are not involved directly in deoxyribonucleotide metabolism.

PRPP   + guanine   ⇌   Phospho-ribosyltransferase   **GMP**   + PP$_i$

In principle, such a reaction could participate in nucleotide breakdown. However, in vivo, pyrophosphate is rapidly cleaved by pyrophosphatase to give inorganic phosphate. This dictates that phosphoribosyltransferases most commonly act in the direction of nucleotide biosynthesis, as a way to salvage free nucleobases.

De Novo Biosynthesis of Purine Nucleotides

Early Studies on De Novo Purine Synthesis

The reactions of de novo purine nucleotide biosynthesis were identified in the 1950s, in the laboratories of John Buchanan and Robert Greenberg. Elucidation of the pathway began with the realization that birds excrete most of their excess nitrogen compounds in the form of *uric acid*, an oxidized purine (see Chapter 20). Thus, researchers were able to identify low-molecular-weight precursors to purines by administering isotopically labeled compounds to pigeons, crystallizing uric acid from the droppings, and, by selective chemical degradation, determining which positions were labeled by which precursors. This procedure yielded the pattern shown in Figure 22.3. At that time, 10-formyltetrahydrofolate was not known, but compounds such as formate, or serine labeled in the hydroxymethyl carbon, readily labeled C-2 and C-8 of uric acid.

Next, two related antibiotics, **azaserine** and **6-diazo-5-oxonorleucine (DON)**, were identified as potent inhibitors of purine nucleotide synthesis. Recognition that these compounds are structural analogs of glutamine led to the eventual realization

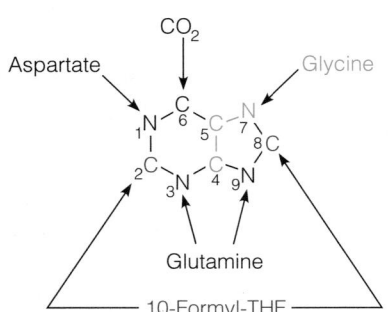

FIGURE 22.3

Low-molecular-weight precursors to the purine ring. The source of each atom in the ring, as established with isotopic tracer studies of uric acid synthesis.

that azaserine and 6-diazo-5-oxonorleucine are irreversible inhibitors of a class of enzymes called the **glutamine amidotransferases**, which catalyze the ATP-dependent transfer of the amido nitrogen of glutamine to an acceptor. Three such reactions occur in purine nucleotide synthesis (and two in pyrimidine nucleotide synthesis)—we will look more closely at the mechanism of these enzymes later in the chapter.

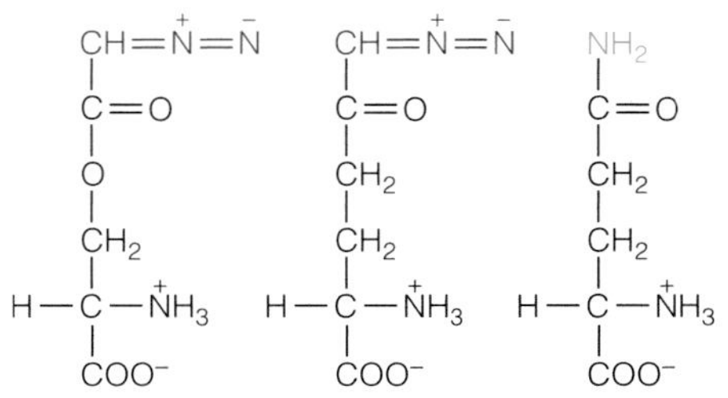

In later experiments, bacteria treated with sulfonamide drugs, such as sulfanilamide, excreted large quantities of a red compound. This compound was identified as an oxidation product of **5-aminoimidazole-4-carboxamide ribonucleotide (AICAR)**, which resembles an incomplete purine nucleotide. This finding suggested that AICAR is a biosynthetic intermediate whose use was somehow blocked by the drug. Because sulfonamides block the synthesis of folate coenzymes (see page 851), the accumulation of AICAR suggested that a folate coenzyme participates in the next reaction. Moreover, these observations suggested that the pathway proceeds at the nucleotide level—in other words, that the purine ring is assembled while already attached to the ribose-5-phosphate moiety.

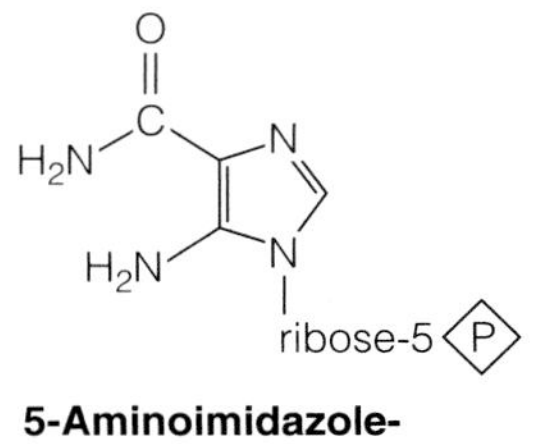

Purine Synthesis from PRPP to Inosinic Acid

Figure 22.4 summarizes the pathway leading from PRPP, the first intermediate, to the first fully formed purine nucleotide, **inosine 5'-monophosphate (IMP)**, also called **inosinic acid**. This compound is the 5'-ribonucleotide of the purine base **hypoxanthine**. Note that there are two glutamine amidotransferase reactions in this pathway, reactions 1 and 4. They differ mechanistically in that PRPP amidotransferase (reaction 1) does not require ATP because the substrate has been activated by ATP in the previous step. An inversion of configuration occurs in reaction 1, as the amido nitrogen displaces the pyrophosphate moiety. The latter is an excellent leaving group, giving a simple nucleotide (5-phosphoribosylamine), which carries the β-configuration at carbon 1 of the sugar, as do all the common nucleotides. This nitrogen will become N9 of the purine ring.

In reaction 2, the carboxyl group of glycine forms an amide with the amino group of phosphoribosylamine (PRA), giving glycinamide ribonucleotide (GAR). The role of ATP is to phosphorylate glycine's carboxyl group, activating it for nucleophilic attack by PRA's amino group. This is followed by a **transformylase** reaction, in which a formyl group is transferred from an N atom of 10-formyltetrahydrofolate (10-formyl-THF) to an N atom of the glycinamide—this carbon will become C8 of the purine ring. As we have noted, reaction 4 is catalyzed by an ATP-dependent amidotransferase, with glutamine providing N3 of the purine ring. Reaction 5 is an ATP-dependent ring closure, in which the formyl oxygen is activated by the γ-phosphate of ATP for nucleophilic attack by N1 of FGAR, giving the imidazole portion of the purine ring. Step 6, the carboxylation of aminoimidazole ribonucleotide (AIR) to form 4-carboxy-5-aminoimidazole ribonucleotide (CAIR), can occur by two distinct mechanisms. In higher organisms, AIR is reversibly carboxylated directly to CAIR by AIR carboxylase, as illustrated in Figure 22.4. In most bacteria, fungi, and plants, however, AIR is first converted to an unstable intermediate, N^5-CAIR. This ATP-dependent reaction, catalyzed by N^5-CAIR synthetase, is thought to occur via a carboxyphosphate intermediate, which is thereby activated for nucleophilic attack by the N^5-amino group of AIR. A mutase then catalyzes the reversible transfer of the CO_2 group from the carbamate of N^5-CAIR to C4, yielding CAIR.

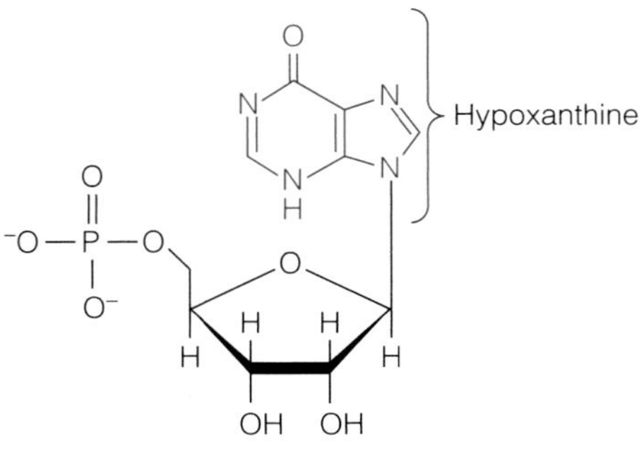

These carboxylation reactions are unusual in that neither AIR carboxylase nor the two-step process requires biotin. It has been proposed that the unstable

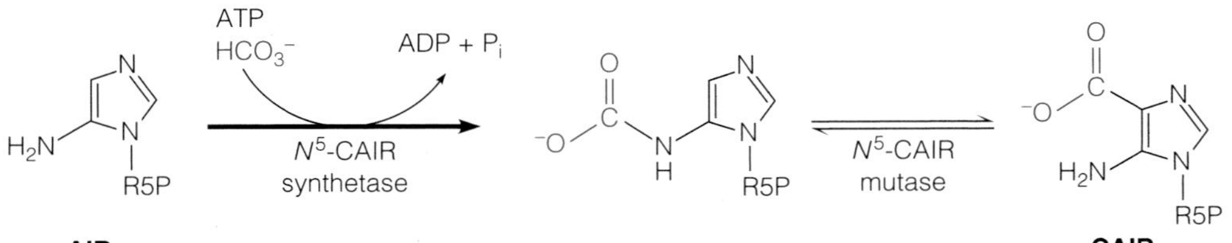

Azaserine   **6-Diazo-5-oxonorleucine (DON)**   **Glutamine**

5-Aminoimidazole-4-carboxamide ribonucleotide (AICAR)

Purines are synthesized at the nucleotide level, starting with PRPP conversion to phosphoribosylamine and purine ring assembly on the amino group.

Inosine 5'-monophosphate (Inosinic acid)

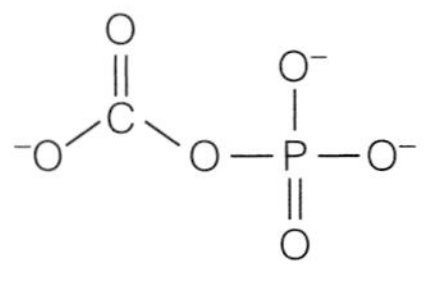

Carboxyphosphate

AIR N^5-**CAIR**   **CAIR**

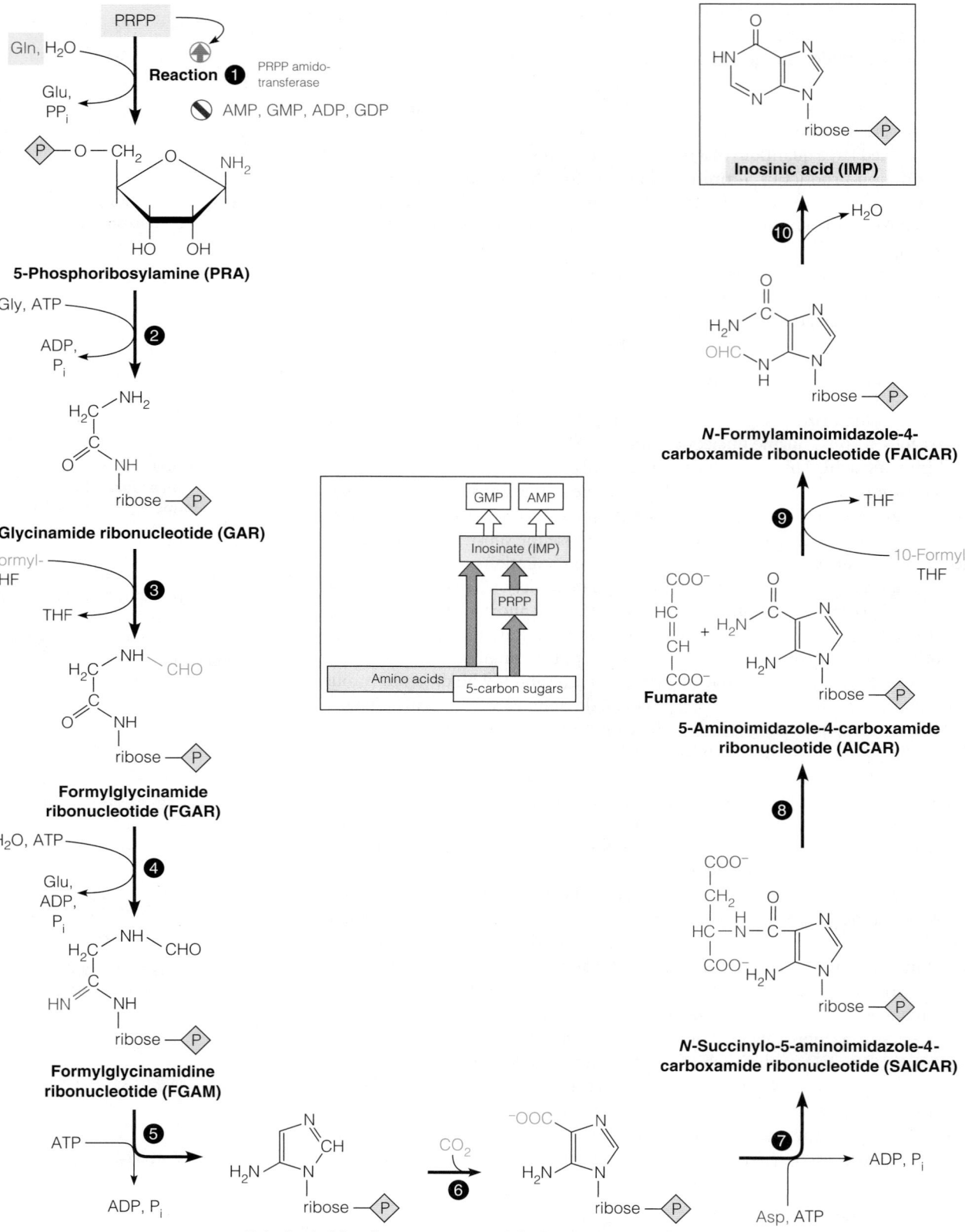

FIGURE 22.4

De novo biosynthesis of the purine ring. In this and subsequent figures ribose —(P) refers to a ribose-5-phosphate moiety in a nucleotide. The enzymes involved are ❶ PRPP amidotransferase, ❷ GAR synthetase, ❸ GAR transformylase, ❹ FGAR amidotransferase, ❺ AIR synthetase, ❻ AIR carboxylase, ❼ SAICAR synthetase, ❽ adenylosuccinate lyase, ❾ AICAR transformylase, and ❿ IMP cyclohydrolase.

N^5-CAIR intermediate functions as a CO_2 carrier like carboxybiotin (see Figure 14.20, page 619). The major function of the mutase then is to sequester the CO_2 released from N^5-CAIR decarboxylation for CAIR formation. In fact, N^5-CAIR mutase can catalyze the direct carboxylation of AIR, but its K_M for CO_2 is 110 mM. However, sequential action of N^5-CAIR synthetase and N^5-CAIR mutase produces CAIR with a K_M for CO_2 of less than 100 μM. In bacteria, these two enzymes are separate proteins, but in plants and yeast, the two activities are fused together in a single bifunctional enzyme, and it is likely that the unstable N^5-CAIR intermediate is channeled between the two domains. The mechanism of the AIR carboxylase reaction is not yet understood. CO_2, rather than bicarbonate, is the one-carbon donor, and ATP is not required to activate the CO_2. However the CO_2 is introduced, it becomes C6 of the purine ring.

Reactions 7 and 8 result in the transfer of a nitrogen from aspartate, by a mechanism identical to that used to convert citrulline to arginine in the urea cycle (Figure 20.12, page 842) and to that used to remove the ε-amino group of lysine (Chapter 21, page 870). First, the entire aspartate molecule is transferred to the carboxyl group of 4-carboxy-5-aminoimidazole ribonucleotide (reaction 7). An α,β-elimination reaction follows (reaction 8), yielding AICAR, the intermediate shown to accumulate in sulfonamide-treated bacteria. The enzyme catalyzing reaction 8, adenylosuccinate lyase, actually has dual substrate specificity—it also catalyzes the second step in the conversion of IMP to AMP (see Figure 22.6). Reaction 9 is another transformylase reaction, with a one-carbon group transferred from 10-formyl-THF, becoming C2 of the purine ring. As in the transformylation in reaction 3, the formyl group is transferred from N10 of 10-formyl-THF to an N atom of the acceptor (FAICAR). Finally, an internal condensation reaction (reaction 10) yields the first purine compound, inosinic acid (IMP).

Evolution has also produced divergent chemistry for the two formylation reactions (3 and 9). Some bacteria and archaea use formate rather than 10-formyl-THF as the formyl donor in ATP-dependent ligase reactions. In each case, the γ-phosphate of ATP is transferred to formate to give a formyl-phosphate intermediate, followed by nucleophilic attack at the activated formyl carbon by the GAR or AICAR amino group.

Vertebrate cells contain several of these activities in the form of multifunctional enzymes. This fact came to light when cloned genes for these enzymes were transferred into *E. coli*, and single cloned genes were found to complement (that is, to replace the function of) two or three different bacterial genes. For example, a single cloned cDNA (see Tools of Biochemistry 4B) allowed growth of bacterial purine auxotrophs defective in E2, E3, or E5. Subsequent analysis showed that the cloned vertebrate DNA encoded a single polypeptide that catalyzed these three reactions. Similar observations confirmed that reactions 6 and 7, and reactions 9 and 10, are catalyzed by bifunctional enzymes. Thus, the 10 activities of de novo purine biosynthesis are contained on only six proteins in vertebrates. Moreover, there is evidence that these six proteins associate together in the cytoplasm in a multiprotein complex called the **purinosome** (Figure 22.5). This evidence comes from cross-linking (see Tools of Biochemistry 13A), affinity chromatography (see Tools of Biochemistry 5A), colocalization (see Tools of Biochemistry 7A), and genetic studies. Additional enzymes may also be part of the purinosome, including serine hydroxymethyltransferase (SHMT) and the trifunctional enzyme, C_1-tetrahydrofolate (THF) synthase (see Figure 20.17, page 852). The combined activities of these two folate enzymes produce 10-formyl-THF, the formyl donor used by the two transformylases (reactions 3 and 9), from the one-carbon donor serine. The THF released from each transformylase reaction is recycled with another one-carbon unit from serine by SHMT (Figure 22.5).

Whether activities are carried on a multifunctional enzyme or a multienzyme complex such as the purinosome, several advantages of juxtaposing catalytic sites are readily apparent. Labile tetrahydrofolate coenzymes and purine intermediates can be protected. Intermediates whose concentrations are low can be "channeled" (that is, directly transferred from one catalytic site to the next). Finally, sequential enzyme activities can be jointly regulated under

Formyl phosphate

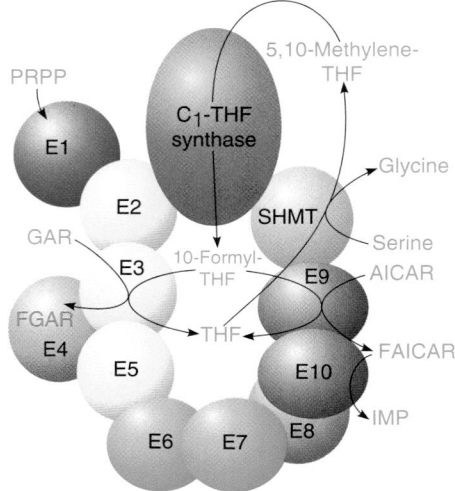

FIGURE 22.5

Model of the purinosome. The figure shows a hypothetical model for a multienzyme complex in animal cells that transforms PRPP to IMP. E1–E10 correspond to the enzymes catalyzing reactions 1–10 in Figure 22.4. Enzymes with the same coloring are contained on multifunctional enzymes. Serine hydroxymethyltransferase (SHMT) and the trifunctional enzyme C_1-tetrahydrofolate (THF) synthase, also proposed to be part of the complex, channel one-carbon units from serine into 10-formyl-THF for the transformylase reactions.

changing environmental and nutritional conditions. Recent evidence suggests that assembly of the purinosome is regulated by reversible phosphorylation-dephosphorylation of one or more of its component enzymes.

We noted at the beginning of this section that birds use the de novo purine synthesis pathway to excrete excess nitrogen. Some tropical legumes, such as soybean and cowpea, use this pathway to assimilate nitrogen. As described in Chapter 20, these plants possess symbiotic root nodules infected with *Rhizobium*. Inorganic nitrogen, fixed by bacterial nitrogenase activity, is secreted into the cytoplasm of the plant cell principally as NH_3 or NH_4^+, which is assimilated into the amide group of glutamine via glutamine synthetase and glutamate synthase (see page 831). The glutamine amide N is incorporated into the purine ring by enzymes located in cytoplasmic organelles, called plastids. The purines are oxidized to allantoin and allantoic acid (see Figure 22.9), and then exported from the nodule in xylem to provide the majority of the nitrogen for the plant's nutrition.

Synthesis of ATP and GTP from Inosinic Acid

Inosinic acid represents a branch point in purine nucleotide synthesis. IMP does not accumulate but is converted both to adenosine 5-monophosphate and to guanosine 5-monophosphate (Figure 22.6). The pathway to guanine nucleotides begins with an NAD^+-dependent oxidation of the hypoxanthine ring, yielding the nucleotide **xanthosine monophosphate (XMP)**; XMP contains the base xanthine. A glutamine-dependent amidotransferase reaction follows, yielding GMP. The route to AMP involves the transfer of nitrogen from aspartate to IMP, by a mechanism similar to that of reactions 7 and 8 in the de novo synthesis of the purine ring. First, a succinylonucleotide intermediate is formed, and then an α,β-elimination

IMP, the first fully formed purine nucleotide, is a branch point between adenine and guanine nucleotide biosynthesis.

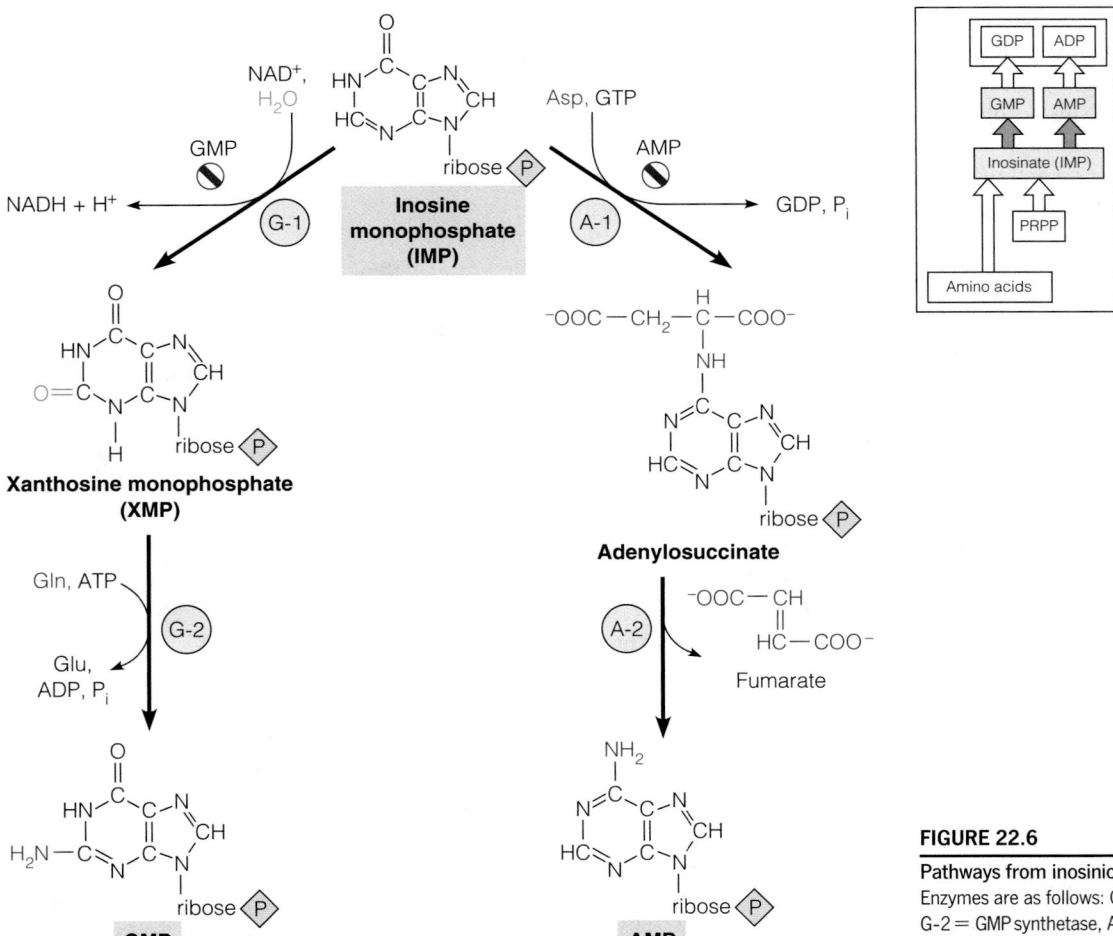

FIGURE 22.6

Pathways from inosinic acid (IMP) to GMP and AMP. Enzymes are as follows: G-1 = IMP dehydrogenase, G-2 = GMP synthetase, A-1 = adenylosuccinate synthetase, and A-2 = adenylosuccinate lyase.

reaction yields AMP plus fumarate. Indeed, the same enzyme (adenylosuccinate lyase) catalyzes both elimination reactions. Note from Figure 22.6 that the energy to drive the aspartate transfer reaction comes not from ATP but from GTP. This may represent a way to control the proportions of IMP that go to adenine and guanine nucleotide synthesis. GTP accumulation would tend to promote the pathway toward adenine nucleotides. Also, because the conversion of XMP to GMP is ATP-dependent, accumulation of ATP could promote guanine nucleotide synthesis.

Nucleotides are active in metabolism primarily as the nucleoside triphosphates. GMP and AMP are converted to their corresponding triphosphates through two successive phosphorylation reactions. Conversion to the diphosphates involves specific ATP-dependent kinases.

$$\text{GMP} + \text{ATP} \xrightarrow{\text{Guanylate kinase}} \text{GDP} + \text{ADP}$$

$$\text{AMP} + \text{ATP} \xrightarrow{\text{Adenylate kinase}} 2\text{ADP}$$

Phosphorylation of ADP to ATP occurs through energy metabolism—oxidative phosphorylation or substrate-level phosphorylations or (in plants) photophosphorylation. ATP can also be formed from ADP through the action of adenylate kinase, acting in the reverse of the direction shown here.

ATP is the phosphate donor for conversion of GDP (and other nucleoside diphosphates) to the triphosphate level through the action of **nucleoside diphosphate kinase**. This enzyme is highly active and has broad specificity, with regard to both phosphoryl group donor and acceptor.

$$\text{GDP} + \text{ATP} \xrightarrow{\text{NDP kinase}} \text{GTP} + \text{ADP} \qquad \Delta G^{\circ\prime} = 0$$

Because ATP is by far the most abundant nucleoside triphosphate in most cells, equilibrium and mass action considerations dictate that it is used most readily as the donor of the γ (outer) phosphate in synthesis of other nucleoside triphosphates.

In most organisms the biosynthesis of deoxyribonucleotides for DNA synthesis begins at the ribonucleoside diphosphate level, with reduction of the ribose moiety to $2'$ deoxyribose. This process is presented in detail later. In purine metabolism, subsequent phosphorylation by nucleoside diphosphate kinase yields the deoxyribonucleoside triphosphates, dATP and dGTP.

Nucleoside diphosphate kinase, an equilibrium-driven enzyme, transfers a phosphoryl group from ATP in the synthesis of all other nucleoside triphosphates.

Regulation of De Novo Purine Biosynthesis

Figure 22.7 summarizes the points of feedback regulation in de novo purine biosynthesis. Control over the biosynthesis of IMP is provided through feedback regulation of the early steps in purine nucleotide synthesis. PRPP synthetase (page 920) is inhibited by various purine nucleotides—particularly AMP, ADP, and GDP—and PRPP amidotransferase (reaction 1 of Figure 22.4) is inhibited allosterically by AMP, ADP, GMP, and GDP. This amidotransferase is also allosterically activated by PRPP, one of its substrates. Beyond the IMP branch point, GMP controls its own biosynthesis by inhibiting the conversion of IMP to XMP, and AMP controls its own formation by inhibiting the synthesis of adenylosuccinate (Figure 22.6). Additional control is exerted at the level of deoxyribonucleotide biosynthesis, as we discuss later. In bacteria, the expression of genes encoding these enzymes is controlled by a repressor protein (see Chapter 26), the product of the *purR* gene. This protein binds either hypoxanthine or guanine, and the resultant protein–purine base complex binds to DNA sites upstream of several genes of purine (and pyrimidine) synthesis, thereby inhibiting their transcription.

Utilization of Adenine Nucleotides in Coenzyme Biosynthesis

An important metabolic role of purine nucleotides is the synthesis of coenzymes, primarily those containing an adenylate moiety. These include the flavin

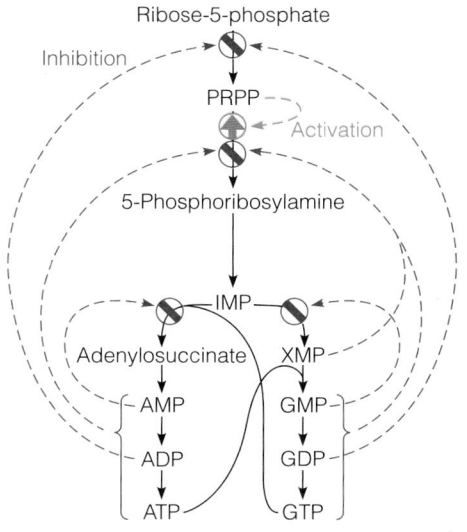

FIGURE 22.7

Regulation of de novo purine biosynthesis.

nucleotides, the nicotinamide nucleotides, and coenzyme A, as shown in the following summary:

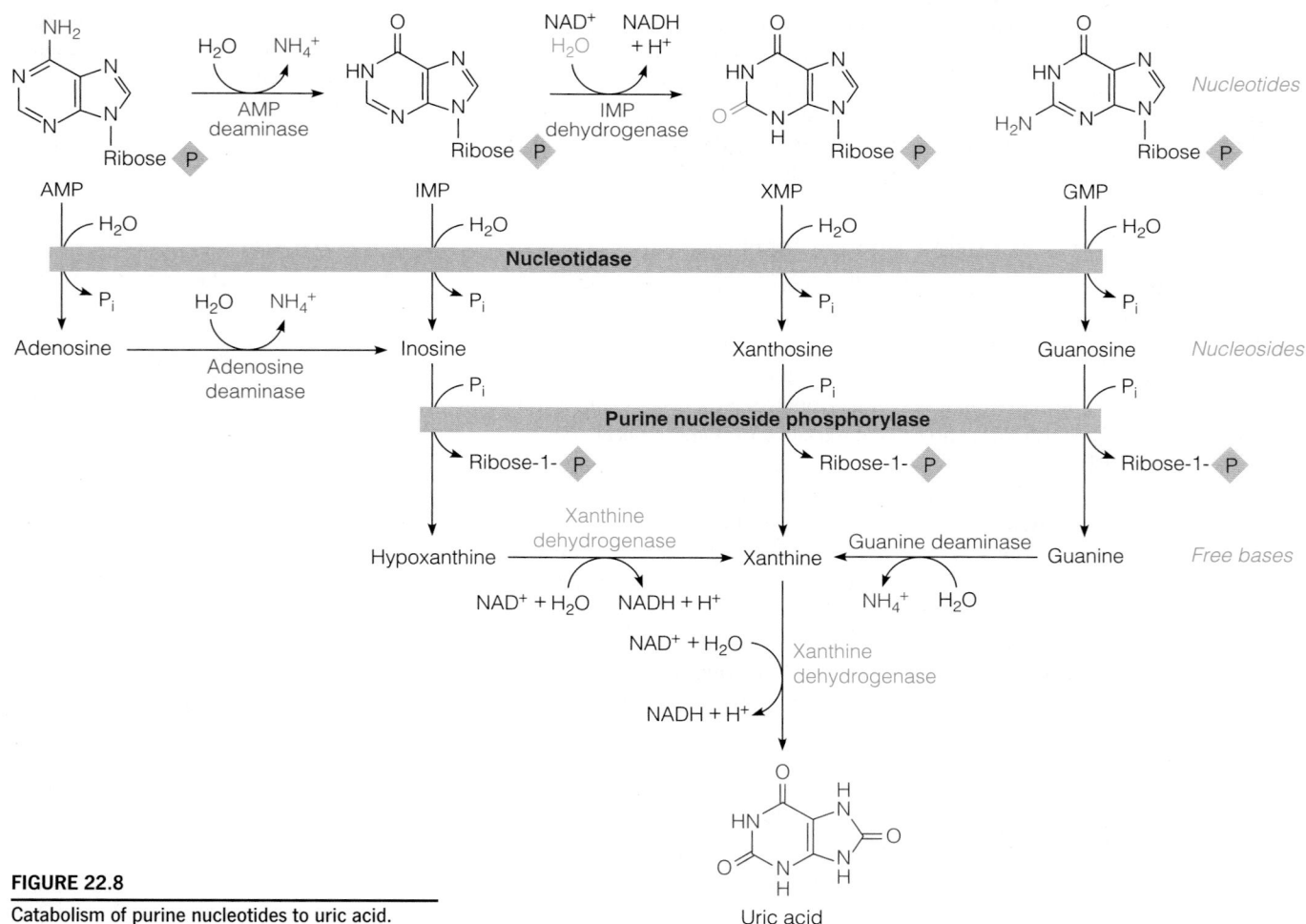

Purine Degradation and Clinical Disorders of Purine Metabolism

Formation of Uric Acid

> All purine degradation leads to uric acid. In some animals, additional degradation occurs.

Purine nucleotide catabolism yields uric acid, by routes shown in Figure 22.8. The specific pathways vary among organisms and among tissues of the same organisms, but the general strategy is conversion of the nucleotides to nucleosides, cat-

FIGURE 22.8

Catabolism of purine nucleotides to uric acid.

alyzed by **nucleotidases**, removal of the ribose-1-phosphate moiety, catalyzed by **purine nucleoside phosphorylase (PNP)**, and finally oxidation of the free bases to uric acid. Mammalian purine nucleoside phosphorylase cannot act on adenosine or deoxyadenosine, so adenosine must be first deaminated by **adenosine deaminase (ADA)** to give inosine. Alternatively, AMP is deaminated to inosinic acid (IMP). Inosine, xanthosine, and guanosine can be formed by hydrolysis of the respective nucleoside monophosphates, which are acted on by PNP to give hypoxanthine, xanthine, and guanine, respectively. Guanine is deaminated to xanthine by **guanine deaminase**, an enzyme abundant in mammalian brain and liver. Hypoxanthine is oxidized to xanthine, and xanthine to uric acid, by **xanthine oxidoreductase**. This enzyme, which oxidizes several other heterocyclic nitrogen compounds, contains bound FAD, two Fe_2S_2 clusters, and a molybdopterin complex. Two forms of the enzyme are found in mammals, referred to as xanthine dehydrogenase (XDH) and xanthine oxidase (XO), respectively. XDH predominates in healthy tissues, but under pathological conditions the protein is readily converted to XO through the reversible formation of a cysteine disulfide bond or by the irreversible proteolytic cleavage of XDH into three fragments. In both forms of the enzyme, oxidation of xanthine takes place at the molybdopterin center, and the electrons are rapidly transferred to FAD via the Fe_2S_2 clusters. The two forms differ, however, in the electron acceptor used to reoxidize the enzyme. XDH uses NAD^+ to reoxidize the reduced flavin on the enzyme, whereas XO uses molecular oxygen. Thus, the XO form of the enzyme ultimately reduces oxygen to H_2O_2, which is acted upon by catalase (see page 667).

$$\text{Xanthine} + O_2 + H_2O \rightarrow \text{Uric acid} + H_2O_2$$

In either mechanism, the keto oxygen atoms at C8 (and C2) derive from H_2O; the initial product of xanthine oxidation is the enol form of uric acid ($pK = 5.4$) that tautomerizes to the more stable keto form.

Purine catabolism in humans and the great apes, birds, reptiles, and insects ends with uric acid, which is excreted. However, most animals possess additional enzymes to further oxidize the purine ring to **allantoin** (excreted by other mammals) and then to **allantoic acid**, which is either excreted (in teleost fishes) or further catabolized to urea (in cartilaginous fishes, some molluscs, and amphibians) or ammonia (in most marine invertebrates). Figure 22.9 shows the pathway from uric acid to CO_2.

Excessive Accumulation of Uric Acid: Gout

Uric acid and its urate salts are quite insoluble. This property is advantageous in birds, reptiles, and insects because it provides a route for disposition of excess nitrogen that uses very little water: the waste material is excreted essentially as uric acid crystals. However, the insolubility of urates can present difficulties in mammalian metabolism, and most mammals possess an active urate oxidase to convert uric acid to allantoin (Figure 22.9). However, some 8–24 million years ago, several mutations accumulated in the urate oxidase gene during hominid evolution, leading to loss of function of this enzyme in humans and the great apes. Lack of urate oxidase activity results in >10-fold higher levels of uric acid in the blood of humans and most primates than in other mammals. Uric acid is a powerful scavenger of free radicals, providing protection against oxidative damage. Evolution might have selected for mutations in hominid urate oxidase that led to higher uric acid levels. On the other hand, these high concentrations predispose humans to crystalline uric acid deposition. Indeed, in North America and Europe, about 3 individuals in 1000 suffer from **hyperuricemia**—chronic elevation of blood uric acid levels well beyond the already high levels. Although the biochemical reasons for hyperuricemia vary, the condition goes by the single clinical name of **gout**. Prolonged or acute elevation of blood urate leads to its precipitation, as crystals of sodium urate, in the synovial fluid of joints. These precipitates cause inflammation, resulting in a painful arthritis, which, if untreated, leads ultimately to severe degeneration of the joints. Eating and drinking purine-rich foods are apt to

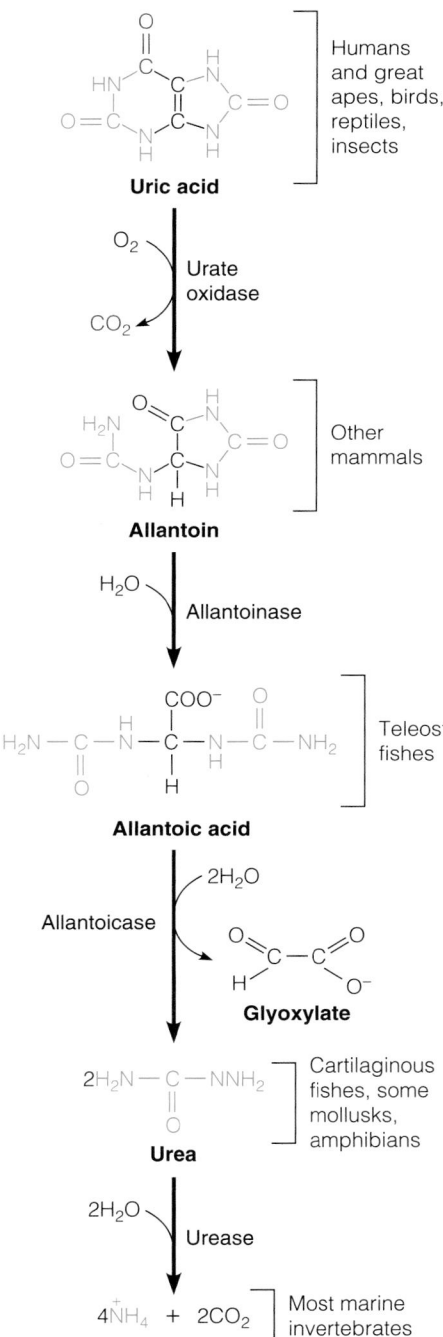

FIGURE 22.9

Catabolism of uric acid to ammonia and CO_2.

stimulate acute gouty attacks in susceptible individuals. Because such foods include "rich" items such as liver, sweetbreads, anchovies, and wine, gout is historically associated with an excess of high living.

Gout results either from overproduction of purine nucleotides, leading to excessive uric acid synthesis, or from impaired uric acid excretion through the kidneys. Several specific enzymatic defects can lead to excessive purine synthesis, as shown in Figure 22.10. One form of gout is characterized by elevated activity of PRPP synthetase (defect 1). Some inherited point mutations in the PRPP synthetase gene cause the enzyme to become resistant to feedback inhibition by purine nucleotides. Because the activity of PRPP amidotransferase (E1 in Figure 22.4) is controlled in part by the concentrations of substrates, an elevation of the steady-state pool of PRPP increases flux through the amidotransferase reaction, which represents a major control point in de novo purine biosynthesis (see Figure 22.7). Another form of gout results from deficiency of the salvage enzyme **hypoxanthine–guanine phosphoribosyltransferase** (**HGPRT**) (defect 2 in Figure 22.10).

$$\text{Hypoxanthine} + \text{PRPP} \underset{\text{HGPRT}}{\rightleftharpoons} \text{IMP} + \text{PP}_i$$
$$\text{Guanine} + \text{PRPP} \rightleftharpoons \text{GMP} + \text{PP}_i$$

HGPRT salvages both hypoxanthine and guanine. This is one of two phosphoribosyltransferases in animal purine metabolism; the other enzyme, **adenine phosphoribosyltransferase** (**APRT**), is specific for adenine.

$$\text{Adenine} + \text{PRPP} \underset{\text{APRT}}{\rightleftharpoons} \text{AMP} + \text{PP}_i$$

How does an HGPRT deficiency increase the rate of purine nucleotide synthesis? The HGPRT reaction, when active, consumes PRPP. Decreased flux through this reaction, when the enzyme is deficient, raises the steady-state level of PRPP. Elevated levels of PRPP increase flux through PRPP amidotransferase via mass action and allosteric activation, leading to increased purine biosynthesis.

Patients with glucose-6-phosphatase deficiency (glycogen storage disease type I; Chapter 13) are also gouty (defect 3 in Figure 22.10). Again, PRPP appears to be the key. Glucose-6-phosphate accumulation in the liver stimulates the

Several known genetic alterations in purine metabolism can lead to purine oversynthesis, uric acid overproduction, and gout.

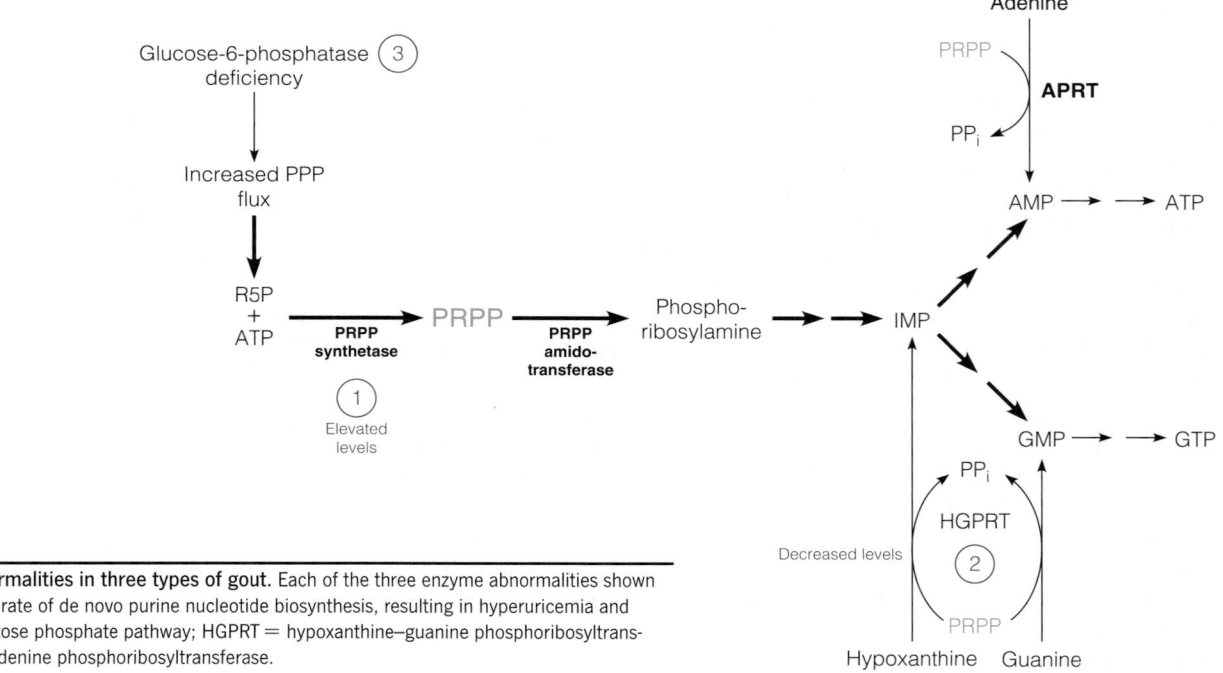

FIGURE 22.10

Enzymatic abnormalities in three types of gout. Each of the three enzyme abnormalities shown here elevates the rate of de novo purine nucleotide biosynthesis, resulting in hyperuricemia and gout. PPP = pentose phosphate pathway; HGPRT = hypoxanthine–guanine phosphoribosyltransferase; APRT = adenine phosphoribosyltransferase.

oxidative branch of the pentose phosphate pathway (Chapter 13), resulting in elevated ribose-5-phosphate, and thus PRPP, through the PRPP synthetase reaction. As noted earlier, gout also results from impaired uric acid excretion. In patients with glucose-6-phosphatase deficiency, prolonged hypoglycemia causes accumulation of organic acids (lactate and the like), and this accumulation interferes with tubular secretion of uric acid in the kidney. Gout is also a consequence of cancer chemotherapy, presumably resulting from an overload of purines caused by nucleic acid degradation after death of tumor cells.

Many cases of gout are successfully treated by the antimetabolite **allopurinol**, a structural analog of hypoxanthine in which the N7 and C8 positions are interchanged. Xanthine dehydrogenase hydroxylates allopurinol at C2 (as it does hypoxanthine), giving **alloxanthine**, which remains tightly bound to the reduced form of the enzyme. Allopurinol is thus a suicide substrate that strongly inhibits xanthine dehydrogenase. This inhibition causes accumulation of hypoxanthine and xanthine, both of which are more soluble and, hence, more readily excreted than is uric acid.

Salvage of Purines and Lesch–Nyhan Syndrome

Nucleic acids, particularly RNAs, are subject to degradation by nucleases which release nucleotides. These nucleotides are eventually enzymatically hydrolyzed to free purine and pyrimidine bases. Vertebrates can *salvage* free purine bases by converting them back into nucleotides for reuse in nucleic acid biosynthesis. The two major enzymes involved, adenine phosphoribosyltransferase (APRT) and hypoxanthine-guanine phosphoribosyltransferase (HGPRT), were described above. In humans, as much as 90% of free purines are salvaged and reused, rather than degraded or excreted. Careful analysis of patients with simple gout resulting from HGPRT deficiency reveals low but significant residual levels of the affected enzyme. Evidently the mutations involved alter the catalytic activity of the enzyme but do not abolish it completely. Far more serious consequences result from "null mutations" in the hypoxanthine-guanine phosphoribosyltransferase gene, which result in total absence of the HGPRT enzyme. Guanine and hypoxanthine are not salvaged, and are instead degraded by the pathways in Figure 22.8, leading to excess excretion of uric acid. This condition was first described in 1964 by medical student Michael Lesch and his faculty mentor, William Nyhan. Lesch–Nyhan syndrome is a sex-linked recessive trait because the structural gene for HGPRT is located on the X chromosome. Patients with this condition display a severe gouty arthritis, but they also have a dramatic malfunction of the nervous system, manifested as behavioural disorders, motor disability, learning disability, and hostile or aggressive behavior, often self-directed. In the most extreme cases, patients nibble at their fingertips or, if restrained, their lips, causing severe self-mutilation. Nyhan has likened this behavior to "nailbiting, with the volume turned up." The biochemical reason for this bizarre behavioural pattern is unknown, but the condition, even though rare, is of great interest because all of the aberrations derive ultimately from the single well-characterized enzyme deficiency affecting HGPRT levels. At present there is no successful treatment, and afflicted individuals have such severe gouty arthritis that they rarely live beyond 20 years. Overproduction of uric acid can be blocked by allopurinol, but this treatment has no efficacy against the neurological features of Lesch–Nyhan syndrome. However, the condition can be diagnosed prenatally through amniocentesis, by molecular genetic analysis of the HGPRT gene.

Unexpected Consequences of Defective Purine Catabolism: Immunodeficiency

A surprising feature of human purine metabolism came to light in 1972, through studies on a hereditary condition called **severe combined immunodeficiency disease (SCID)**. Patients with this condition are susceptible, often fatally, to infectious diseases because of a total inability to mount an immune response to antigenic challenge. This disease was made famous by the case of David Vetter, the "bubble boy"

Allopurinol
(enol form)

Hypoxanthine
(enol form)

Xanthine dehydrogenase

Alloxanthine
(enol form)

Xanthine
(enol form)

FIGURE 22.11

Metabolic consequences of adenosine deaminase (ADA) deficiency. AdoHcy, S-adenosylhomocysteine.

Modified with permission from Hershfield, M. S., and Mitchell, B. S. (2001) in *The Metabolic and Molecular Bases of Inherited Disease* (Scriver, C. R., Beaudet, A. L., Sly, W. S., Valle, D., Childs, B., Kinzler, K. W., and Vogelstein, B., eds.) Vol. II, Ch. 109, pp. 2585–2625. © The McGraw-Hill Companies, Inc.

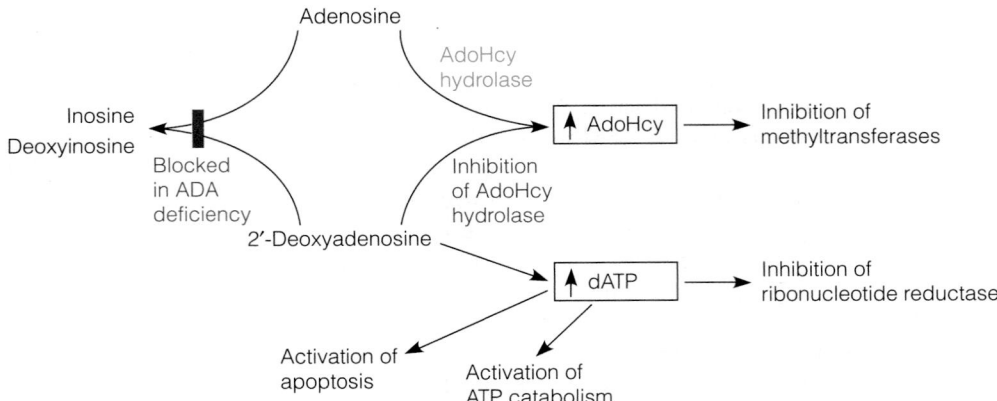

who lived his entire life in a sterile environment. In this condition, both B and T lymphocytes are affected; neither class of cells can proliferate as they must if antibodies are to be synthesized. In many such cases the immunodeficiency results from a heritable lack of the degradative enzyme adenosine deaminase (see Figure 22.8).

What is the basis for this unexpected relationship? First, adenosine deaminase (ADA) also acts on 2′-deoxyadenosine, which arises from the degradation of DNA (Figure 22.11). Second, white blood cells have abundant levels of salvage enzymes, including nucleoside kinases; thus, adenosine and 2′-deoxyadenosine that accumulate are readily converted in white cells to their respective nucleotides. These nucleotides include dATP, which is known to be a potent inhibitor of DNA replication because it inhibits the synthesis of deoxyribonucleotides from ribonucleotides (see page 940). White cells must proliferate for an immune response to occur. In turn, proliferation requires ample synthesis of DNA and its precursors. Additional mechanisms are involved because 2′-deoxyadenosine has been found to kill white cells even when they are not proliferating. One such mechanism came to light in 1997, when elevated dATP was reported to be a signaling agent that helped to trigger early metabolic events leading to apoptosis (Chapter 28). Finally, 2′-deoxyadenosine is an irreversible inhibitor of S-adenosylhomocysteine (AdoHcy) hydrolase, the enzyme in the methyl cycle (Figure 21.9, page 876) responsible for hydrolyzing AdoHcy to adenosine and homocysteine. This causes AdoHcy to accumulate, which inhibits S-adenosylmethionine-dependent methyltransferases, including those required for normal DNA and histone methylation reactions (Figure 22.11).

The standard treatment for ADA deficiency enzyme replacement requires frequent injections of purified bovine ADA covalently attached to the inert polymer polyethylene glycol (PEG-ADA). Although PEG-ADA often corrects the metabolic abnormalities, defects in the immune system usually persist. Partly because of the limitations of the standard therapy, adenosine deaminase deficiency was the first human disease to be treated by gene therapy. In 1990, a four-year-old girl with the condition was treated with a viral vector into which the gene for adenosine deaminase had been spliced by recombinant DNA technology, in hopes that the engineered virus would establish itself in enough cells to yield sufficient enzyme to degrade the accumulated deoxyadenosine compounds. This first gene therapy patient, now in her mid twenties, is still relatively healthy. However, she still receives periodic gene therapy and PEG-ADA to maintain the necessary levels of the enzyme in her blood. Thus, while the treatment proved safe, it is not clear how effective the gene therapy would have been alone.

A less severe immunodeficiency results from the lack of another purine degradative enzyme, purine nucleoside phosphorylase (PNP) (see Figure 22.8). Decreased activity of this enzyme leads to accumulation primarily of dGTP and 2′-deoxyguanosine. This accumulation also affects DNA replication, but less severely than does excessive dATP. Interestingly, the phosphorylase deficiency destroys only the T class of lymphocytes and not the B cells.

Pyrimidine Nucleotide Metabolism

De Novo Biosynthesis of the Pyrimidine Ring

Now we turn our attention to pyrimidine nucleotide biosynthesis. Pyrimidine biosynthesis is much simpler than formation of the structurally more complex purine nucleotides, with just two precursors, aspartate and carbamoyl phosphate, providing all the C and N atoms of the pyrimidine ring. As summarized in Figure 22.12, however, there are two major distinctions from the purine pathway. First, the pyrimidine ring is assembled as a free base, with conversion to a nucleotide occurring later in the pathway, when the base **orotic acid** is converted

Pyrimidine nucleotide synthesis occurs primarily at the free base level, with conversion to a nucleotide occurring late in the unbranched pathway.

FIGURE 22.12

De novo synthesis of pyrimidine nucleotides. Enzymes are as follows: ❶ carbamoyl phosphate synthetase, ❷ aspartate transcarbamoylase, ❸ dihydroorotase, ❹ dihydroorotate dehydrogenase, ❺ orotate phosphoribosyltransferase, ❻ OMP decarboxylase, ❼ UMP kinase, ❽ nucleoside diphosphate kinase, and ❾ CTP synthetase. Sites of allosteric control are indicated.

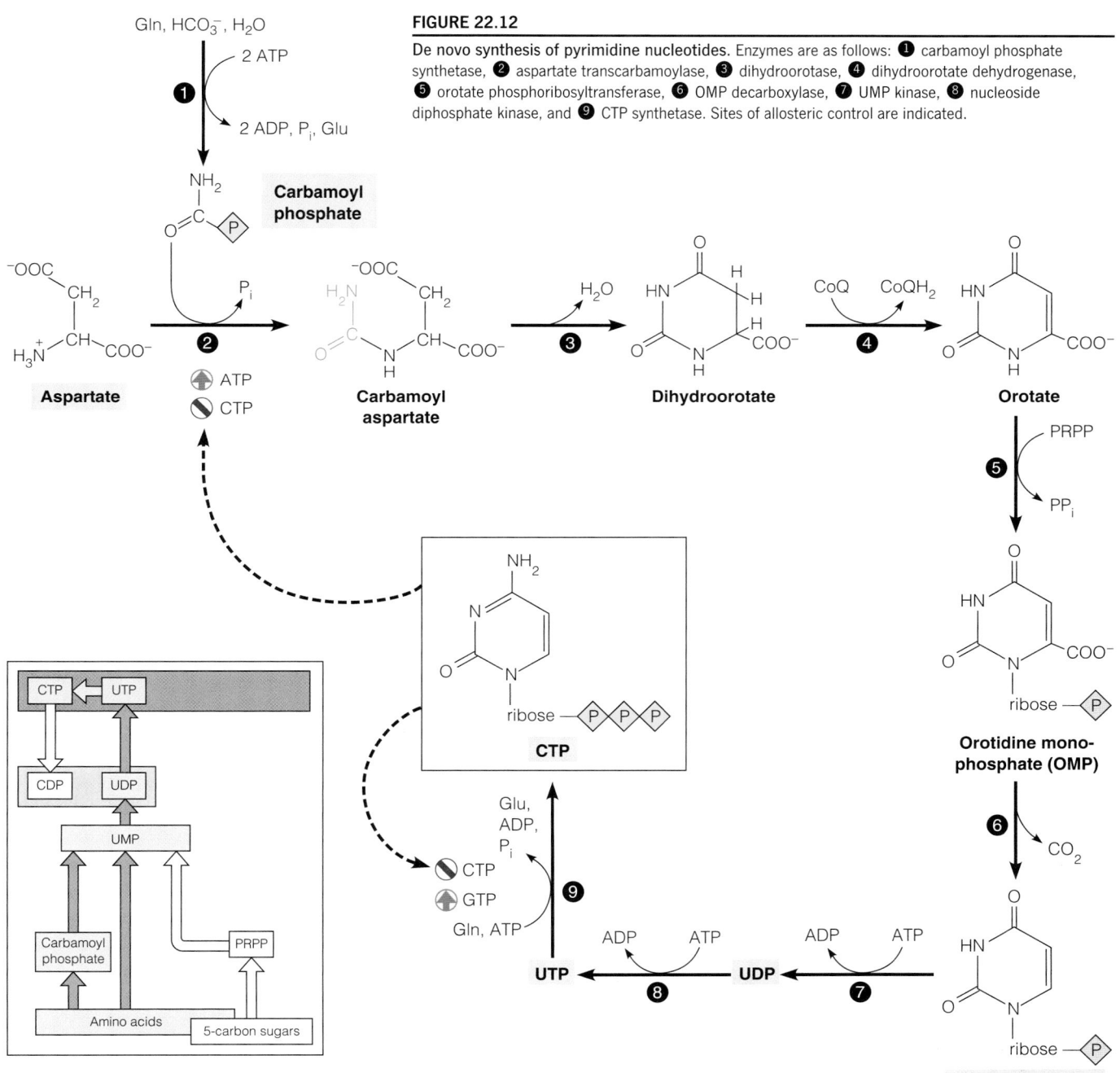

to orotidine monophosphate, or OMP. Second, the pyrimidine pathway is unbranched. Uridine triphosphate, one of the two common ribonucleoside triphosphates and hence an end product of the pathway, is also the substrate for formation of cytidine triphosphate, the other end product.

Pyrimidine synthesis begins with formation of carbamoyl phosphate, a reaction catalyzed by **carbamoyl phosphate synthetase** that was first presented in Chapter 20 (see page 831 and reaction 1 in Figure 22.12). Carbamoyl phosphate is formed from ATP, bicarbonate, and the amide nitrogen from glutamine. This amidotransferase reaction involves four separate chemical steps:

Bicarbonate ⇌(1) [**Carboxyphosphate**] ⇌(3) **Carbamate** ⇌(4) **Carbamoyl phosphate**

Glutamine ⇌(2) Glutamate
H_2O

Bacterial carbamoyl phosphate synthetase (CPS) is a heterodimer composed of a large subunit and a small subunit. The X-ray crystal structure of the *E. coli* CPS revealed the various substrate binding sites and the domains where the individual steps are catalyzed (Figure 22.13). The small subunit functions as a glutaminase, catalyzing the hydrolysis of glutamine to form ammonia, which is delivered to the large subunit. The large subunit contains two active sites: one binds bicarbonate, ATP, and ammonia and catalyzes carboxyphosphate and carbamate synthesis; the other active site binds a second ATP and catalyzes carbamoyl phosphate formation. The three unstable intermediates (ammonia, carboxyphosphate, and carbamate) do not diffuse from the enzyme until converted to the final product, carbamoyl phosphate. The crystal structure revealed a remarkable 96 Å tunnel in the interior of the enzyme through which the intermediates are channeled from one active site to the next (Figure 22.13). This direct transfer of reactive intermediates between active sites is a common feature in many enzymes, and is one mechanism by which enzymes increase reaction efficiency and avoid unwanted side reactions with unstable intermediates.

Recall from Chapter 20 that eukaryotic cells contain two forms of CPS. Form I, localized in mitochondria, has a preference for ammonia as substrate and is used in arginine biosynthesis (Chapter 21) and the urea cycle (Chapter 20). Form II, present in the cytosol, has a strong preference for glutamine as nitrogen donor, and is the enzyme that participates in pyrimidine biosynthesis.

Carbamoyl phosphate, preactivated by the two ATPs used in its synthesis, next condenses with aspartate to form carbamoylaspartate. This reaction, catalyzed by **aspartate transcarbamoylase (ATCase)**, involves nucleophilic attack by the α-amino

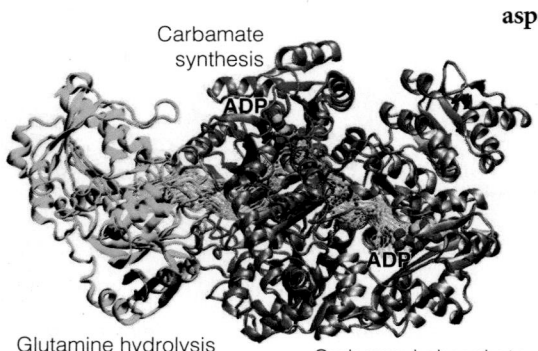

Carbamate synthesis
ADP
ADP
Glutamine hydrolysis
Carbamoyl phosphate synthesis

FIGURE 22.13

Channeling in carbamoyl phosphate synthetase. This figure shows the X-ray crystal structure of the heterodimeric *E. coli* enzyme (PDB ID 1C30). The small subunit (cyan) functions as a glutaminase. The two active sites in the large subunit are marked by bound ADP molecules. The N-terminal domain of the large subunit (pink) contains the active site for the synthesis of carboxyphosphate and carbamate. The C-terminal domain of the large subunit (blue) contains the active site for the synthesis of carbamoyl phosphate. The 96 Å tunnel connecting the three active sites is shown in green dotted lines.

Reprinted with permission from *Journal of the American Chemical Society* 132:3870–3878, L. Lund, Y. Fan, Q. Shao, Y. Q. Gao, and F. M. Raushel, Carbamate transport in carbamoyl phosphate synthetase: A theoretical and experimental investigation. © 2010 American Chemical Society.

group of aspartate on the electrophilic carbonyl C of carbamoyl phosphate, with release of P_i. The pyrimidine ring is closed in an intramolecular condensation catalyzed by **dihydroorotase**. Dihydroorotate is next oxidized to orotate (6-carboxyuracil) by **dihydroorotate dehydrogenase**. The bacterial enzymes are typically NAD^+-linked flavoproteins, containing both FAD and FMN. Eukaryotic dihydroorotate dehydrogenase is also a flavoprotein, but is located on the intermembrane face of the mitochondrial inner membrane, where it is reoxidized by coenzyme Q (ubiquinone), directly linking pyrimidine biosynthesis to electron transport. This localization means the substrate, dihydroorotate, must enter the mitochondrial intermembrane space and the product, orotate, must exit back into the cytoplasm. Inherited mutations in the dihydroorotate dehydrogenase gene have recently been identified as the cause of **Miller syndrome**. This rare disease is characterized by developmental anomalies including cleft lip or palate, absent digits, and ocular anomalies.

The ribose-5-phosphate moiety is attached next, using PRPP as the donor, to form orotidine monophosphate (OMP). This reaction is driven forward by hydrolysis of the eliminated PP_i. **Orotate phosphoribosyltransferase**, like all the other phosphoribosyltransferases we have discussed, is stereospecific for formation of a β glycosidic linkage. The final step of the pathway is a decarboxylation, yielding uridine monophosphate (UMP). UMP is phosphorylated to UTP by sequential action of nucleoside mono- and diphosphate kinases. CTP is synthesized from UTP by a glutamine-dependent amidotransferase reaction, catalyzed by **CTP synthetase**.

Control of Pyrimidine Biosynthesis in Bacteria

The first reaction committed solely to pyrimidine synthesis is the formation of carbamoyl aspartate from carbamoyl phosphate and aspartate, catalyzed by ATCase (reaction 2). In enteric bacteria, this enzyme represents a marvelous example of feedback control, as was discussed at length in Chapter 11. Recall that the enzyme is inhibited by the end product CTP and activated by ATP, the latter possibly representing a mechanism to keep purine and pyrimidine biosyntheses in balance. Recall also that the enzyme contains six each of two types of subunits, arranged as two catalytic trimers and three regulatory dimers.

Bacteria also regulate pyrimidine metabolism through control of the *synthesis* of ATCase and the other enzymes. The rate of transcription of an operon encoding both of the ATCase subunits can vary by as much as 150-fold, depending on the intracellular level of UTP. The higher the UTP concentration, the lower the rate of transcription of these genes.

Multifunctional Enzymes in Eukaryotic Pyrimidine Synthesis

Aspartate transcarbamoylase (reaction 2 in Figure 22.12) in eukaryotes is strikingly different from the *E. coli* enzyme. This came to light through analysis of ATCase inhibition by ***N*-phosphonoacetyl-L-aspartate (PALA)**.

PALA   **Putative transition state complex**

This compound, synthesized as an analog of the putative transition-state complex formed between the two substrates, inhibits pyrimidine synthesis in mammalian cells. However, cells eventually develop resistance to it, because levels of ATCase rise in these cells beyond the capacity of PALA to inhibit all of the activity. Surprisingly, these resistant cells contain similarly elevated levels of carbamoyl phosphate synthetase

In eukaryotes the first three reactions of pyrimidine synthesis are catalyzed by a trifunctional enzyme, the CAD protein. Similarly, the last two reactions are catalyzed by a bifunctional enzyme, UMP synthase.

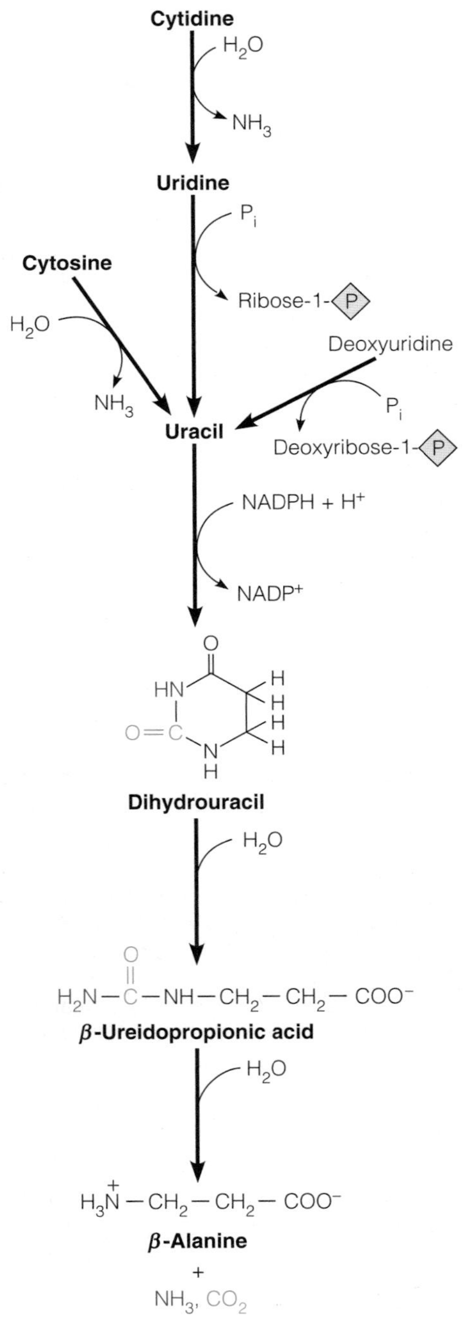

FIGURE 22.14

Catabolic pathways in pyrimidine nucleotide metabolism.

(reaction 1 in Figure 22.12) and dihydroorotase (reaction 3). The explanation for this observation came with the discovery of a single protein containing 3–6 identical polypeptide chains, each with M_r of about 220,000, that catalyzes all three reactions.

George Stark has given this trifunctional enzyme the acronym **CAD** (from the first letter in the name of each enzyme). He showed that the protein accumulates in PALA-resistant cells because the gene encoding the protein becomes amplified as a consequence of the selective pressure exerted by PALA; resistant cells contain many more copies of the gene than the normal complement of two copies per diploid cell. This phenomenon of gene **amplification** has now been observed many times in eukaryotic cells exposed to prolonged and specific stresses. We discuss the mechanisms involved in Chapter 26.

In mammalian cells, reactions 5 and 6 in Figure 22.12 are also catalyzed by a single protein, which has been named **UMP synthase**. A severe deficiency in this bifunctional enzyme is responsible for a rare genetic disease in humans, **hereditary orotic aciduria**. As the name implies, patients with this defect accumulate orotic acid, exactly what you would expect for a block at the orotate phosphoribosyltransferase reaction.

In contrast to bacteria, where the six reactions of pyrimidine biosynthesis are catalyzed by six distinct monofunctional enzymes, eukaryotes combine these activities on just three proteins (one trifunctional, one bifunctional, and one monofunctional). We don't know much about how this affects regulation of pyrimidine biosynthesis, but the juxtaposition of active sites may allow channeling, as mentioned for the multifunctional proteins in purine synthesis. Recall that dihydroorotate dehydrogenase (reaction 4 in Figure 22.12) is localized at the intermembrane space side of the inner mitochondrial membrane, whereas the bifunctional and trifunctional enzymes are both cytosolic. Given that intermediates must travel into and then out of the mitochondria, the kinetic advantage provided by channeling of the first and last steps would seem to be negated. However, both CAD and UMP synthase have been detected at the cytoplasmic face of the outer mitochondrial membrane, providing a mechanism to allow direct transfer of substrates and products across the outer membrane.

Another site for control of pyrimidine nucleotide synthesis is CTP synthetase (reaction 9). This enzyme is inhibited allosterically by its product CTP and is activated by GTP.

Salvage Synthesis and Pyrimidine Catabolism

Pyrimidine nucleotides are also synthesized by salvage pathways involving phosphorylases and kinases, comparable to those already discussed for purines. The catabolic pathways for pyrimidines, summarized in Figure 22.14, are simpler than those for purines. Because the intermediates are relatively soluble, there are few known derangements of pyrimidine breakdown. One of the breakdown products, β-alanine, is used in the biosynthesis of coenzyme A.

Glutamine-Dependent Amidotransferases

Glutamine-dependent aminations that utilize ATP hydrolysis as the driving force are a common theme in purine and pyrimidine biosynthesis. Five different reactions in these two pathways are catalyzed by glutamine-dependent amidotransferases: PRPP amidotransferase and FGAR amidotransferase (steps 1 and 4 in Figure 22.4); GMP synthetase (step G-2 in Figure 22.6); and carbamoyl phosphate synthetase and CTP synthetase (steps 1 and 9 in Figure 22.12). All these enzymes are believed to use a similar mechanism (summarized in Figure 22.15), in which the glutamine amido group (unreactive, non-nucleophilic) is activated by hydrolysis at the active site to deliver nucleophilic ammonia in high local concentrations sufficient to aminate the specific acceptor (usually the carbonyl group of an amide). ATP serves as an energy

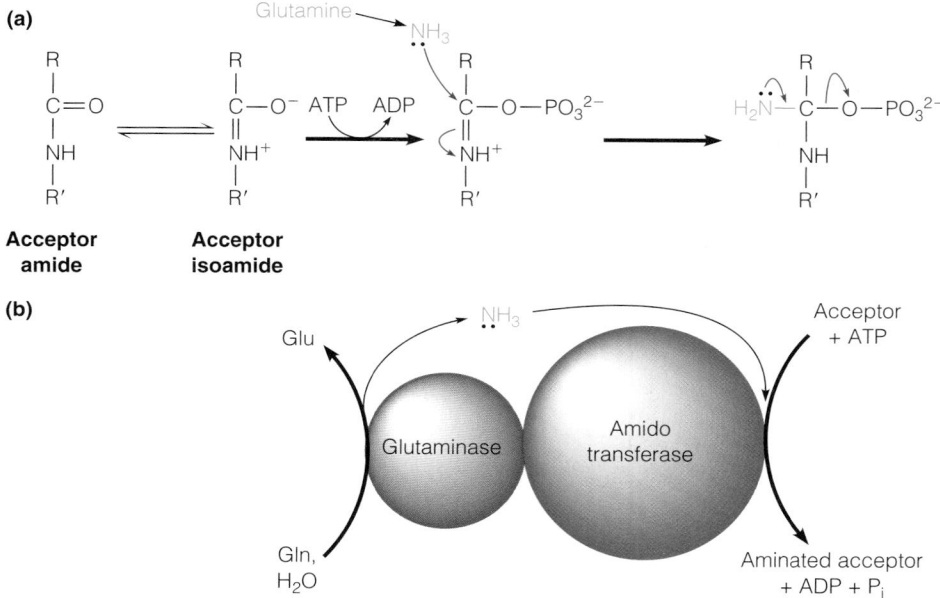

Acceptor amide

Acceptor isoamide

Product

FIGURE 22.15

Proposed mechanism of glutamine-dependent amidotransferases. **(a)** The acceptor, usually an amide, is in tautomeric equilibrium with the isoamide form and is activated for nucleophilic attack by phosphorylation. Attack by nucleophilic ammonia, generated from glutamine hydrolysis, forms a tetrahedral intermediate, which collapses to give P_i and the aminated product. **(b)** Schematic of the subunit (or domain) structure of a glutamine-dependent amidotransferase. The glutaminase subunit (or domain) delivers nucleophilic ammonia to the amidotransferase subunit (or domain), which catalyzes the ATP-dependent amination of the acceptor.

input to force the equilibrium in the amination direction, generally by forming a phosphorylated derivative of the substrate that can then act as a good leaving group upon attack by nucleophilic NH_3. Most of these enzymes can use NH_3 directly, but they exhibit a much higher K_m for ammonia than for glutamine. Note that PRPP amidotransferase, the first enzyme in purine biosynthesis, does not require ATP to aminate PRPP. However, the ribose C1 was previously activated by ATP in the PRPP synthetase reaction (page 920), as PP_i becomes an excellent leaving group in the subsequent nucleophilic substitution.

These enzymes are typically composed of two subunits (or domains) in which one catalyzes the glutaminase reaction (generation of nucleophilic NH_3) and the other catalyzes amination of the specific acceptor (recall CPS, Figure 22.13). In fact, the glutaminase subunits (or domains) are clearly evolutionarily related between various amidotransferases, within and across species.

Deoxyribonucleotide Biosynthesis and Metabolism

Most cells contain 5 to 10 times as much RNA as DNA. Moreover, as we have seen, ribonucleotides have multiple metabolic roles, while deoxyribonucleotides serve only as constituents of DNA. Therefore, most of the carbon that flows through nucleotide synthetic pathways goes into ribonucleoside triphosphate (rNTP) pools. However, the relatively small fraction that is diverted to the synthesis of deoxyribonucleoside triphosphates (dNTPs) is of paramount importance in the life of the cell. Consequently, there are especially close regulatory relationships between DNA synthesis and dNTP metabolism—more so than seen between other macromolecular biosyntheses and the pathways that provide their precursors. Overall pathways of dNTP biosynthesis are shown in Figure 22.16.

Recalling that DNA differs chemically from RNA in the nature of the sugar and in the identity of one of the pyrimidine bases, we can focus our discussion of deoxyribonucleotide biosynthesis on two specific processes—the conversion of ribose to deoxyribose, and the conversion of uracil to thymine. Both of these processes occur at the nucleotide level. Both processes are of great interest mechanistically, as target sites for chemotherapy for cancer or infectious diseases and from the standpoint of regulation. Accordingly, we shall present both processes in some detail.

FIGURE 22.16

Overview of deoxyribonucleoside triphosphate (dNTP) biosynthesis.

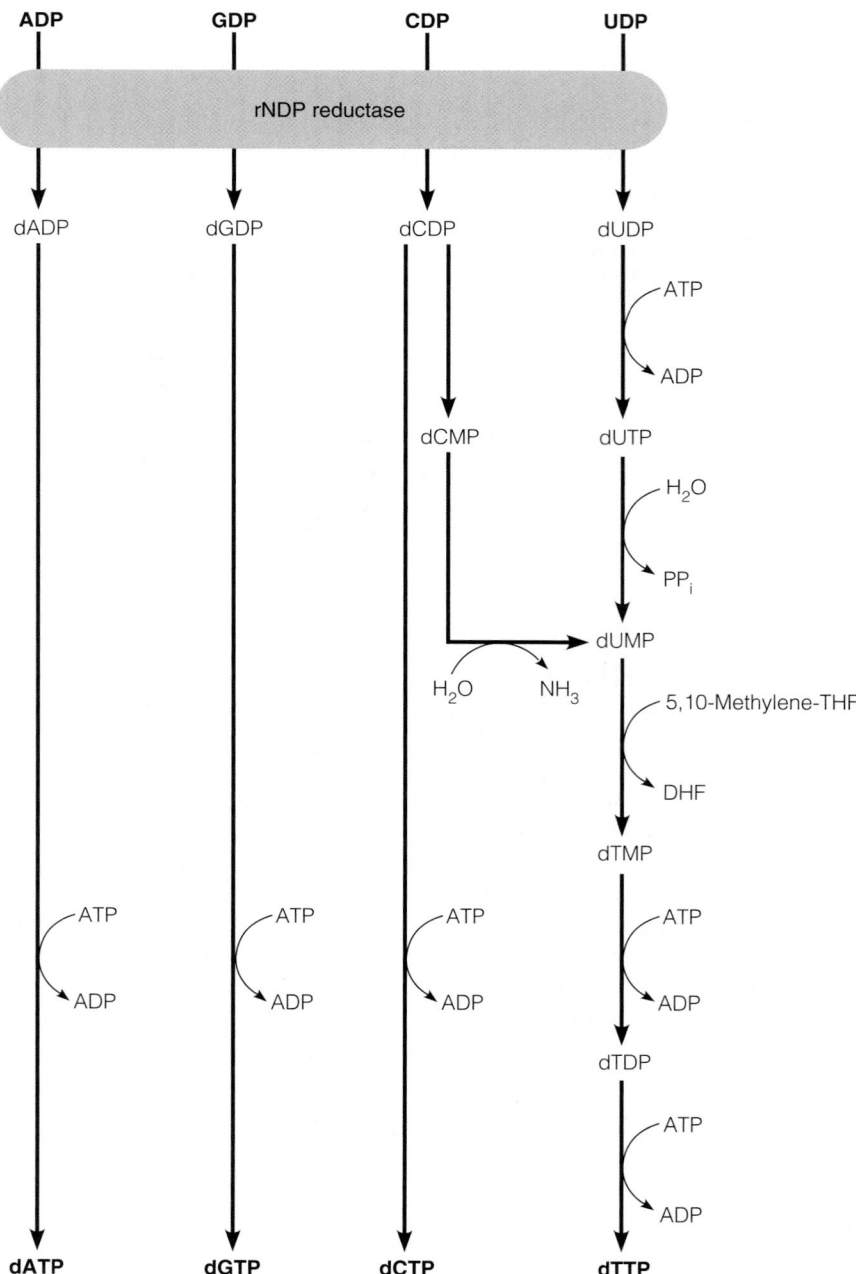

Reduction of Ribonucleotides to Deoxyribonucleotides

One enzyme, ribonucleotide reductase, reduces all four ribonucleotides to their deoxyribose derivatives.

Mechanistically, the reduction of ribose to deoxyribose involves replacement of the hydroxyl at C-2 by a hydrogen atom, with retention of configuration. Peter Reichard showed that this difficult reaction occurs at the nucleotide level, and this led to his discovery of the important enzyme **ribonucleotide reductase**. In all organisms studied thus far, a single enzyme reduces all four common ribonucleotide substrates to the corresponding 2-deoxyribonucleotides. A protein radical mechanism is involved in the reaction. Although evolution has created three widely diverse mechanisms for generating the protein radical, the three classes of ribonucleotide reductase evidently all use the same fundamental chemistry to reduce the substrates. The most widely distributed enzyme form, called class I ribonucleotide reductase, acts upon ribonucleoside diphosphate substrates, leading it to be also called **rNDP reductase**. This enzyme generates its radical on a specific tyrosine

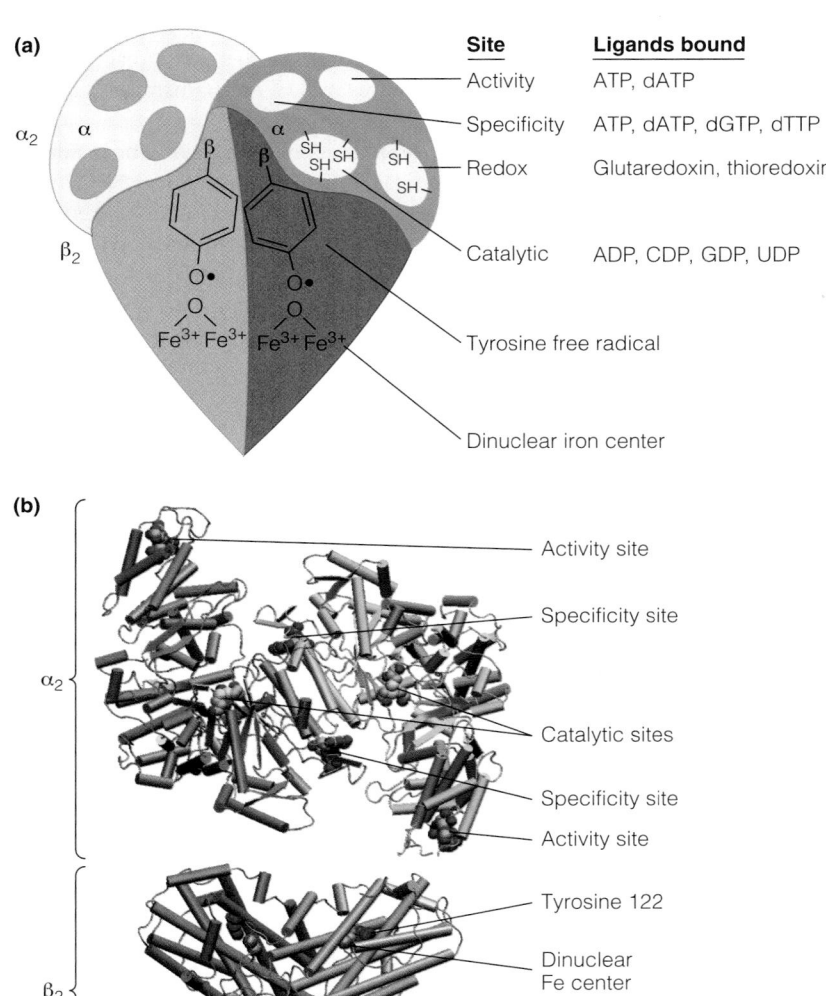

Site	Ligands bound
Activity	ATP, dATP
Specificity	ATP, dATP, dGTP, dTTP
Redox	Glutaredoxin, thioredoxin
Catalytic	ADP, CDP, GDP, UDP

Tyrosine free radical

Dinuclear iron center

Activity site

Specificity site

Catalytic sites

Specificity site

Activity site

Tyrosine 122

Dinuclear Fe center

FIGURE 22.17

Structure of *E. coli* ribonucleoside diphosphate reductase. **(a)** A schematic diagram identifying the functional center of β_2 (R2)—a tyrosine free radical and a dinuclear iron center with a bridged oxygen. Shown also are the ligand-binding sites in α_2 (R1): a redox site for interaction with an external electron donor, the catalytic site, the activity site, and the substrate specificity site. The activity and specificity sites are allosterically controlled. The redox site contains two functional cysteine residues, and the catalytic site contains three. **(b)** X-ray crystal structures of the homodimeric α_2 (blue and cyan) (PDB ID 3R1R, 4R1R) and β_2 (red and pink) (PDB ID 1PFR) proteins. The catalytic sites are marked by three cysteine residues in each α_2 subunit (Cys225, 462, and 439). The activity and specificity sites are marked by bound ATP and dTTP, respectively. In the β_2 dimer, Tyr 122 (blue), which carries the free radical, and the dinuclear iron center (orange) are shown. The precise orientation of the α_2 and β_2 proteins in the tetramer is not known.

residue, with the aid of a diferric oxygen bridge (see Figure 22.17). The class II enzymes, found in cyanobacteria, some bacteria, and *Euglena*, act upon ribonucleoside triphosphate substrates and use adenosylcobalamin, a B_{12} coenzyme (Chapter 20), to generate a free radical. The type III enzyme, found only in facultative or obligate anaerobes, also acts upon ribonucleoside triphosphate substrates. These enzymes use *S*-adenosylmethionine and an iron–sulfur center to generate the catalytically essential radical on a glycine residue. Our discussion will focus on class I, the most widespread form, whose action is shown at the top of Figure 22.16.

Structure of rNDP Reductase

As found in *E. coli* and mammalian cells, the class I rNDP reductases contain two nonidentical subunits, denoted α and β. In *E. coli*, the active form of the enzyme is composed of two 87,000-dalton α chains and two 43,000-dalton β chains. The active form of the mammalian enzyme is thought to be an $\alpha_6\beta_2$ oligomer at physiological ATP/dATP concentrations. However, the $\alpha_2\beta_2$ heterotetramer is fully active and is the form receiving the most attention regarding mechanism and regulation of the enzyme. As we shall see, regulation by allosteric modifiers involves changes in oligomerization state. The structure of the heterotetrameric form of the *E. coli* enzyme is illustrated in Figure 22.17. The catalytic site lies in the α_2 subunits. Within this site are three cysteine residues, which are conserved among

Ribonucleotide reductase contains catalytic residues on each of its subunits—redox-active thiols and a tyrosine free radical stabilized by an iron–oxygen complex.

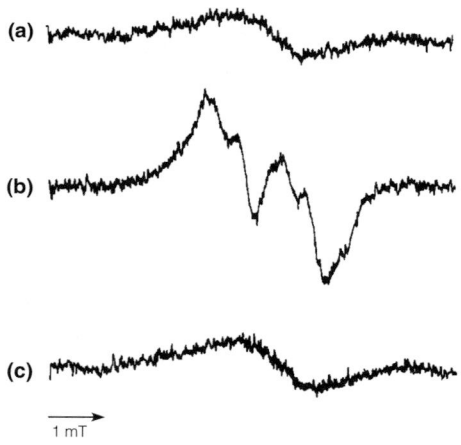

Hydroxyurea

FIGURE 22.18

Evidence that tyrosine 122 in *E. coli* β₂ protein carries the essential free radical. The figure shows electron paramagnetic resonance spectra of *E. coli* cells containing a cloned gene that overexpresses β₂. **(a)** Bacteria containing no clone (the normal level of β₂ protein is too low to create a significant spectral signal). **(b)** Bacteria containing the cloned wild-type gene. **(c)** Bacteria containing a mutant cloned gene in which tyrosine 122 has been changed to phenylalanine. Treatment of the purified enzyme with hydroxyurea yields a spectrum like that shown in (c).

different rNDP reductases. Two of the cysteine thiols are **redox-active**, so called because they undergo cyclic oxidation and reduction during the reaction. The third cysteine evidently functions as part of a free radical mechanism, as described in the next section.

The radical for ribonucleotide reduction is generated on a tyrosine residue in the β₂ subunits and transmitted via a radical transfer pathway over 35 Å to the subunits where the catalysis occurs. The β₂ subunits also contain an oxygen atom bridging two ferric ions near the tyrosine residues. This **dinuclear iron center** generates and stabilizes the tyrosyl radical. The α₂ subunits contain two classes of regulatory sites, which we shall discuss shortly. Finally, the α₂ subunits carry an additional pair of redox-active thiol groups, which interact with an external reductive cofactor. The mammalian enzyme is similar in structure.

Mechanism of Ribonucleotide Reduction

Although the precise mechanism of the rNDP reductase reaction is still under intense study, we can formulate a plausible mechanism based on the following observations. (1) Radiolabeling studies show that cleavage of the ribose C-3′ — H bond occurs during the reaction. (2) The reaction proceeds with retention of configuration at C-2′, which rules out displacement of the hydroxyl group by a hydride ion in an S_N2 reaction. (3) The thiol groups undergo oxidation during the reaction. (4) The tyrosine free radical participates in the reaction. This was shown first by the fact that **hydroxyurea**, an inhibitor of rNDP reductase, reversibly destroys the free radical. A more elegant demonstration is depicted in Figure 22.18. The free radical gives a characteristic electron paramagnetic resonance (EPR) spectrum. EPR spectroscopy detects the spin of an unpaired electron (such as that of an organic radical) as it interacts with the magnetic fields generated by the nuclei and other electrons of the molecule. Tyrosine 122, the residue that generates the radical in the *E. coli* enzyme, was changed to phenylalanine by site-directed mutagenesis of the cloned gene for β₂. The modified protein was inactive and showed no EPR spectrum—hence, no evidence for the existence of a free radical. However, because the tyrosyl radical is more than 35 Å from the catalytic site, as shown by crystallography (see Figure 22.17), one must postulate some sort of long-range process by which the unpaired electron in the tyrosine radical attracts an electron from an active site residue. Evidence indicates that (1) this residue is cysteine 439 in the *E. coli* enzyme and (2) a specific set of amino acid residues on α₂ and β₂ participates in the long-range electron transport process.

A plausible mechanism for rNDP reductase, based on the above observations, is shown in Figure 22.19. First, cysteine 439 in the active site is converted to a **thiyl radical**, by loss of an electron in a long range proton-coupled electron transport (PCET) process over 35 Å that results in reduction of tyrosine 122 (step 1). Next, the thiyl radical participates in abstraction of a hydrogen atom from C-3′ of the substrate (step 2). This is followed by loss of a hydroxide ion (water) from C-2′, and migration of the radical to C-2′ (step 3). Reduction at C-2′ by the redox-active cysteines by a proton-coupled electron transfer step generates a **disulfide radical anion**, which then reduces C-3′ (steps 4 and 5) in another PCET step. The PCET of steps 1 and 2 is then reversed (steps 6 and 7). The net result is transfer of an electron pair from the redox-active cysteine thiols (step 6), to give the nucleotide product, and regeneration of the tyrosyl radical on the β₂ subunit. The resultant cystine disulfide is now reduced by disulfide exchange with another pair of redox-active thiols in the C-terminus of the α₂ subunit; this is not shown in the figure. The resulting disulfide is reduced by an external cofactor (shown as Grx—see the next section), regenerating the active form of the enzyme (step 8). The rate-limiting step in class I rNDP reductases is a conformational change that controls the long-range proton-coupled electron transport process.

The amazing long-range proton-coupled electron transport process that results in radical propagation from tyrosine in β₂ to cysteine in α₂ is but the first example of what turns out to be a prevalent process in enzyme chemistry. When

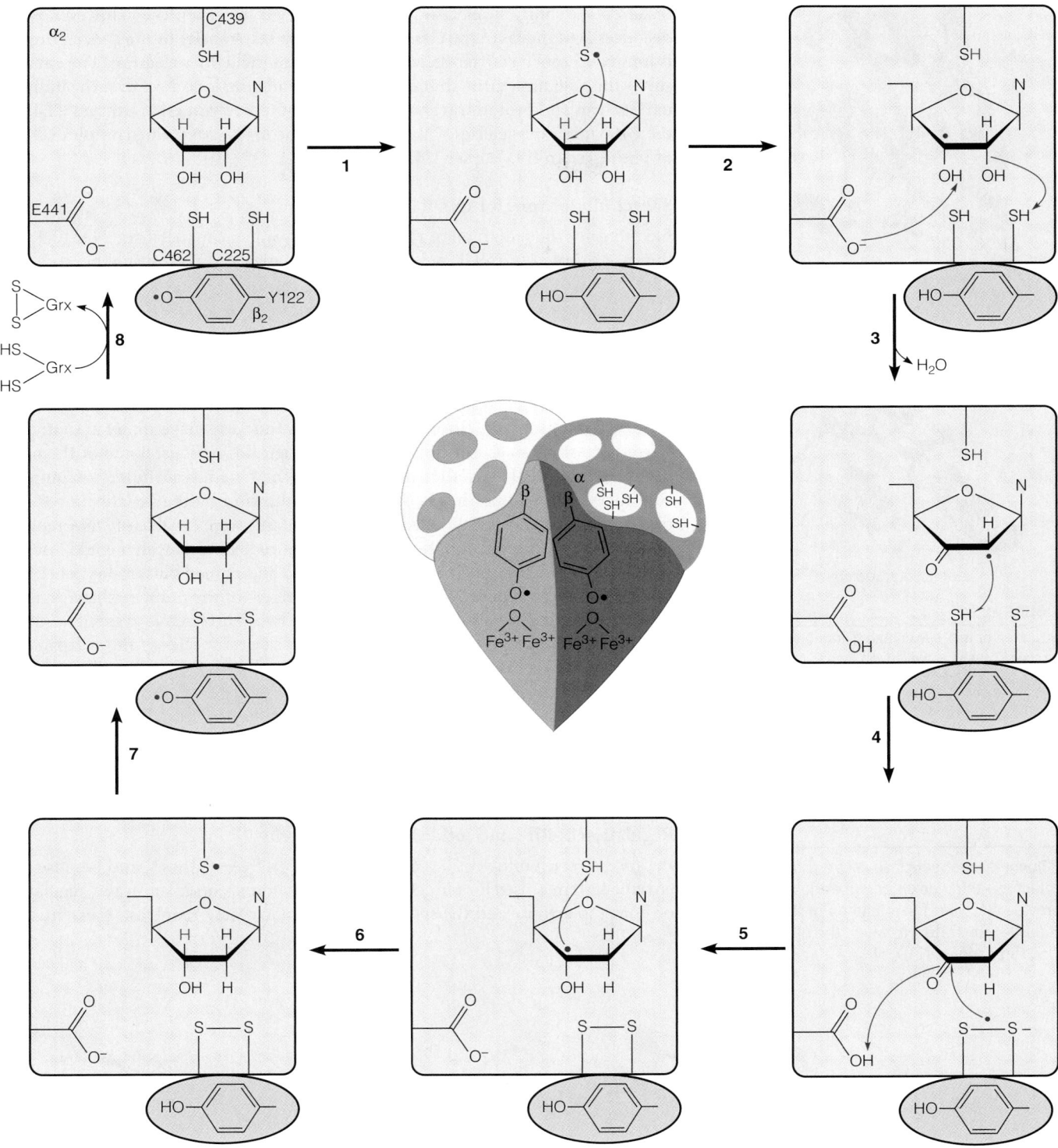

FIGURE 22.19

Reduction of a ribonucleoside diphosphate by rNDP reductase. This mechanism is consistent with available evidence. N is adenine, cytosine, guanine, or uracil. **Step 1:** The reaction is initiated by propagation of the tyrosyl radical (Y122) on the β_2 subunit to a cysteine in the α_2 catalytic site (Cys 439 in *E. coli*), generating a thiyl radical. In this process, called long-range proton-coupled electron transport, the tyrosyl radical is reduced to a tyrosine residue. **Step 2:** The substrate nucleotide interacts with the radical, creating a radical at C-3. **Step 3:** Protonation of the 2'-hydroxyl by one of the redox-active cysteines (225 and 462 in *E. coli*) and deprotonation of the 3'-hydroxyl by a glutamate (E441 in *E. coli*) facilitate dehydration at C-2' and migration of the radical to C-2'. **Steps 4 and 5:** Reduction at C-2' by the redox-active cysteines generates a disulfide radical anion, which then reduces C-3'. **Steps 6 and 7:** The long-range proton-coupled electron transport of steps 1 and 2 is reversed, producing the deoxyribonucleotide product and regenerating the tyrosyl radical on the β_2 subunit. **Step 8:** The Cys-Cys disulfide is reduced by a low-molecular-weight protein thiol, such as glutaredoxin (Grx); see text for further explanation. dNDP dissociation (not shown) allows rNDP substrate binding for another round of catalysis.

Modified with permission from *Journal of the American Chemical Society* 131:200–211, H. Zipse, E. Artin, S. Wnuk, G. J. S. Lohman, D. Martino, R. G. Griffin, S. Kacprzak, M. Kaupp, B. Hoffman, M. Bennati, J. Stubbe, and N. Lees, Structure of the nucleotide radical formed during reaction of CDP/TTP with the E441Q–$\alpha_2\,\beta_2$ of *E. coli* ribonucleotide reductase. © 2009 American Chemical Society.

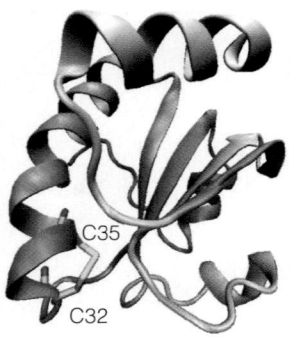

E. coli Thioredoxin

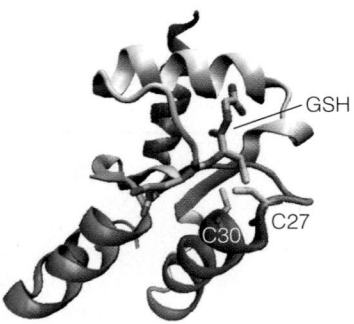

Yeast Glutaredoxin

FIGURE 22.20

Structure of thioredoxin and glutaredoxin. X-ray crystal structures of thioredoxin from *E. coli* (PDB ID 2TRX) and glutaredoxin from the yeast *Saccharomyces cerevisiae* (PDB ID 3C1S) show the locations of the redox-active cysteine residues near their surfaces. In the thioredoxin structure, Cys32 and Cys35 are in the oxidized disulfide state. In the glutaredoxin structure, Cys27 and Cys30 are reduced, and a molecule of glutathione (GSH) is bound with its cysteinyl-SH adjacent to the two cysteine residues.

Ribonucleotide reductase uses a protein cofactor—thioredoxin or glutaredoxin—to provide electrons for reduction of the ribonucleotide substrate. However, the ultimate electron donor is NADPH.

distances are short, the electron and proton may transfer together. However, when distances are long, as in the class I rNDP reductases, transfer of the heavier proton is limited to much shorter distances than that of the lighter electron. The enzyme solves this dilemma through thermodynamic and kinetic control over the individual electron and proton transport processes. Other enzymes that utilize PCET are the P450 monooxygenases (Chapter 15) and the oxygen-evolving complex (OEC) of photosystem II (Chapter 17).

Source of Electrons for rNDP Reduction

Electrons for the reduction of ribonucleotides come ultimately from NADPH, but they are shuttled to rNDP reductase by a coenzyme that is unusual because it is itself a protein (Figure 22.20). The first known member of this class of redox-active proteins is **thioredoxin**, a small protein ($M_r \cong 12,000$) with two thiol groups in the sequence Cys–Gly–Pro–Cys. These thiols undergo reversible oxidation to the disulfide, thereby reducing sulfurs in the active site of rNDP reductase (Figure 22.21). Oxidized thioredoxin is reduced by NADPH via the action of a flavoprotein enzyme, **thioredoxin reductase**.

Since its discovery, thioredoxin has been found to have many activities in vitro, suggesting an astonishing range of biological functions. Some of them are listed in Table 22.1. Whether thioredoxin is the true intracellular cofactor for ribonucleotide reduction was brought into question by the isolation of *E. coli* mutants lacking this protein. Because these mutants were capable of DNA replication, investigators looked in these cells for other redox proteins that could interact with rNDP reductase. Such a protein was found and named **glutaredoxin** because of its ability to be reduced by glutathione. Members of the thioredoxin superfamily, glutaredoxins are also small proteins with two redox-active cysteine residues (Figure 22.20). In fact, most organisms possess several species of thioredoxin and glutaredoxin. For example, *E. coli* has two thioredoxins, three glutaredoxins, and two glutaredoxin-like proteins. Eukaryotes possess distinct cytoplasmic and mitochondrial isoforms of these proteins. Recent evidence indicates that thioredoxin is the physiologically relevant electron donor for ribonucleotide reductase during DNA precursor synthesis in yeast. Whichever carrier is the principal cofactor for rNDP reductase, the ultimate electron source is NADPH.

Regulation of Ribonucleotide Reductase Activity

Because deoxyribonucleotides are used only for DNA synthesis, and because one enzyme system is used for reduction of all four ribonucleotide substrates, regulation of both the *activity* and the *specificity* of ribonucleotide reductase is essential to

FIGURE 22.21

Reductive electron transport sequences in the action of rNDP reductase. Either thioredoxin or glutaredoxin can reduce the oxidized form of the reductase. Oxidized thioredoxin and glutaredoxin are, in turn, reduced with electrons from NADPH in reactions catalyzed by thioredoxin reductase and glutathione reductase, respectively. Glutaredoxin reduction requires two molecules of glutathione (GSH), resulting in oxidized glutathione (GS-SG) (see Chapter 13, page 583).

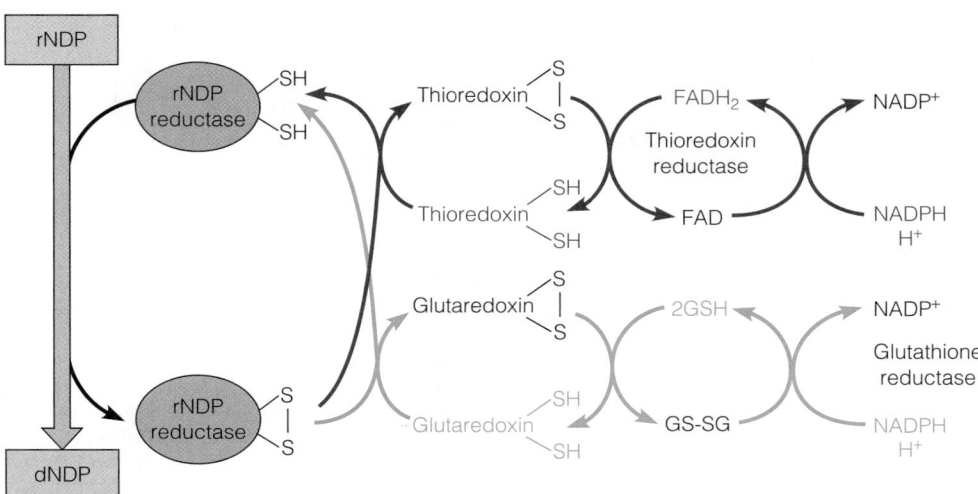

TABLE 22.1 Biological activities of thioredoxin

Activity	Organism
Cofactor for ribonucleotide reduction	All organisms
Protein folding (thioredoxin promotes correct disulfide bond formation)	All organisms
Possible control of insulin levels, through control of insulin reduction	Animals
Control of melanin formation (people with high levels of thioredoxin reductase tan easily)	Animals
Regulation of photosynthetic carbon fixation (see Chapter 16)	Plants
Sulfite reduction (see Chapter 21)	Plants, bacteria
Essential subunit of viral DNA polymerase	Bacteriophage T7
Maturation of filamentous phages by an unknown mechanism	Single-stranded DNA bacteriophages

maintain balanced pools of DNA precursors. This regulation is achieved through binding of nucleoside triphosphate effectors to two classes of regulatory sites on the α_2 subunits (two of each site per dimer in the E. coli enzyme—see Figure 22.17). The **activity sites** bind either ATP or dATP, with relatively low affinity, whereas the **specificity sites** bind ATP, dATP, dGTP, or dTTP, all with relatively high affinity. Binding of ATP at the activity sites tends to increase the catalytic efficiency of rNDP reductase for all substrates, whereas dATP acts as a general inhibitor of all four reactions. ATP and dATP binding affect the oligomerization status of the enzyme, shifting the equilibrium between active and inactive forms (Figure 22.22). Binding of nucleotides at the specificity sites modulates the activities of the enzyme toward different substrates, so as to maintain balanced rates of production of the four dNTPs. For example, binding of dTTP (with ATP bound in the activity site) activates the enzyme for reduction of GDP but decreases its ability to reduce either UDP or CDP. Table 22.2 summarizes the principal regulatory effects.

These effects are seen in vitro with purified enzymes. There is ample reason to conclude that similar regulatory effects also operate in intact cells. For example, either deoxyadenosine or thymidine will inhibit DNA synthesis when administered to intact cells. Measurements of intracellular pools of dNTPs show that in deoxyadenosine-treated cells the dATP pools expand (as expected from the effects of salvage pathways), while dTTP, dGTP, and dCTP pools shrink. This is probably why white blood cells cannot proliferate as needed in immunodeficiency states associated with a lack of adenosine deaminase (see page 930): dATP accumulation in these cells blocks deoxyribonucleotide synthesis and, hence, DNA replication.

Another example comes from cell biology. Researchers often **synchronize** cell cultures, that is, manipulate the cells so that all are brought to the same phase of the cell cycle. Synchrony can be accomplished by **thymidine block**, in which thymidine is added to the cells to inhibit DNA synthesis. This prevents the passage of cells from the G1 to the S phase of the cell cycle, and cells accumulate at this point, like car traffic stopped at a red light. Transferring the cells to a

Ribonucleotide reductase has two classes of allosteric sites. Activity sites influence catalytic efficiency, and specificity sites determine specificity for one or more of the four substrates.

Inhibition of DNA synthesis by thymidine or deoxyadenosine involves allosteric inhibition of ribonucleotide reductase by dTTP or dATP, respectively.

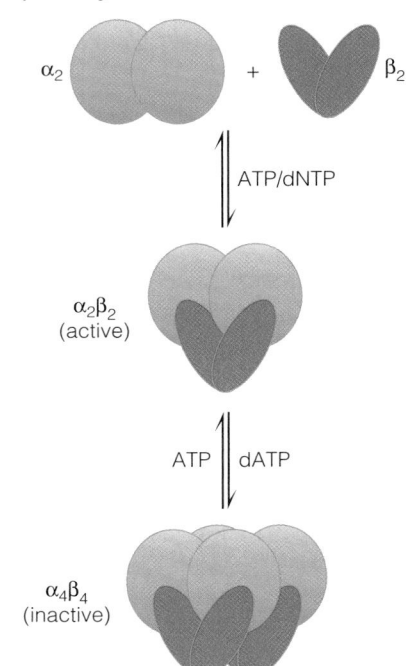

FIGURE 22.22

Regulation of E. coli ribonucleotide reductase by reversible oligomerization. All nucleotides stimulate formation of the active $\alpha_2\beta_2$ tetramer from α_2 and β_2 dimers. dATP binding at the activity sites shifts the equilibrium to formation of inactive $\alpha_4\beta_4$ octamer; ATP binding shifts the equilibrium back to active tetramer.

Journal of Biological Chemistry 283:35310–35318, R. Rofougaran, M. Crona, M. Vodnala, B.-M. Sjöberg, and A. Hofer, Oligomerization status directs overall activity regulation of the *Escherichia coli* class Ia ribonucleotide reductase. Modified with permission. © 2008 The American Society for Biochemistry and Molecular Biology. All rights reserved.

TABLE 22.2 Regulation of the activities of mammalian ribonucleotide reductase

Nucleotide Bound in		Activates Reduction of	Inhibits Reduction of
Activity Site	Specificity Site		
ATP	ATP or dATP	CDP, UDP	
ATP	dTTP	GDP	CDP, UDP
ATP	dGTP	ADP	CDP, UDP[a]
dATP	Any effector		ADP, GDP, CDP, UDP

[a]dGTP binding inhibits the reduction of pyrimidine nucleotides by the mammalian enzyme but not by the E. coli enzyme.

medium containing no thymidine is like a green light, reversing the inhibition and allowing the cells to initiate DNA replication synchronously. dNTP pool measurements in thymidine-blocked cells show that dTTP accumulates, as expected from salvage synthesis, while there is a specific depletion of dCTP, as expected from the effects of dTTP on ribonucleotide reductase activity. Indeed, addition of deoxycytidine restores normal dCTP pools (by salvage synthesis) and relieves the thymidine block.

Further support for the control of rNDP reductase activity in vivo comes from isolation of mammalian cell mutant lines whose growth is not inhibited by deoxyribonucleosides. rNDP reductase from these altered cells shows modifications in either activity or specificity sites, which render these enzymes less susceptible to inhibition by dNTP effectors. Some of these cell lines display both dNTP pool abnormalities and **mutator phenotypes**. That is, they show increased rates of spontaneous mutation at all genetic loci tested. Comparable observations have been made with mutant cells altered in either CTP synthetase or deoxycytidylate deaminase. These findings suggest that when dNTP concentrations are altered at DNA replication sites, the likelihood is increased for replication errors, which lead to mutations. This point is discussed further in Chapter 25.

The metabolic rationale for all of the effects shown in Table 22.2 is not immediately obvious. For example, why should dATP at the specificity site activate both CDP and UDP reduction? Part of the answer is that UDP reduction is a relatively minor pathway. Most dTTP comes from deoxycytidine nucleotides, via the dCMP deaminase reaction (see below).

Biosynthesis of Thymine Deoxyribonucleotides

The previous section explored the first metabolic reaction committed to DNA synthesis—the formation of deoxyribonucleoside diphosphates through the action of rNDP reductase. Once formed, three of the diphosphates—dADP, dGDP, and dCDP—are converted directly to the corresponding triphosphates by nucleoside diphosphate kinase. Biosynthesis of deoxythymidine triphosphate occurs partly from the dUDP produced via the reductase and partly from deoxycytidine nucleotides; the ratio varies in different cells and organisms. Note that we use the terms *thymidine* and *deoxythymidine* interchangeably. That is because thymine *ribonucleotides* are not normal metabolites, so the nucleoside containing thymine and deoxyribose need not be specifically identified as a deoxyribonucleoside.

The pathways are summarized in Figures 22.16 (page 936) and 22.23. Both of the de novo pathways shown in Figure 22.23 lead to deoxyuridine monophosphate (dUMP), the substrate for synthesis of thymine nucleotides: (1) dUDP is phosphorylated to dUTP, which is then cleaved by a highly active diphosphohydrolase, **dUTPase**. (2) dCDP is dephosphorylated to dCMP, which then undergoes deamination to dUMP by an aminohydrolase called **dCMP deaminase**. The enzyme requires dCTP as an allosteric activator and is inhibited by dTTP. *E. coli* and some other bacteria use a different route to dUMP. Deamination is carried out at the triphosphate level by **dCTP deaminase**, and the resultant dUTP is cleaved by dUTPase to dUMP and PP_i.

However it is formed, dUMP serves as substrate for formation of thymidine monophosphate (dTMP), catalyzed by **thymidylate synthase (TS)**. This enzyme transfers a one-carbon unit, at the methylene level of oxidation, and reduces it to the methyl level (see page 852 in Chapter 20). The one-carbon donor is 5,10-methylenetetrahydrofolate, which in this unusual reaction also serves as an electron donor, to give dihydrofolate (DHF) as the other reaction product (Figure 22.24). The folate cofactor must then be reduced, by dihydrofolate reductase, and it must acquire another one-carbon group, most commonly from serine via serine hydroxymethyltransferase. Interruption of any of these steps of the cycle interferes with thymine nucleotide formation. dTMP, once formed, is converted to dTTP by two successive phosphorylations (Figure 22.23).

---

dUMP, the substrate for thymidylate synthesis, can arise either from UDP reduction and dephosphorylation or from deamination of a deoxycytidine nucleotide.

In the reaction catalyzed by thymidylate synthase, 5,10-methylenetetrahydrofolate donates both a single-carbon methylene group and an electron pair to reduce that carbon to a methyl group.

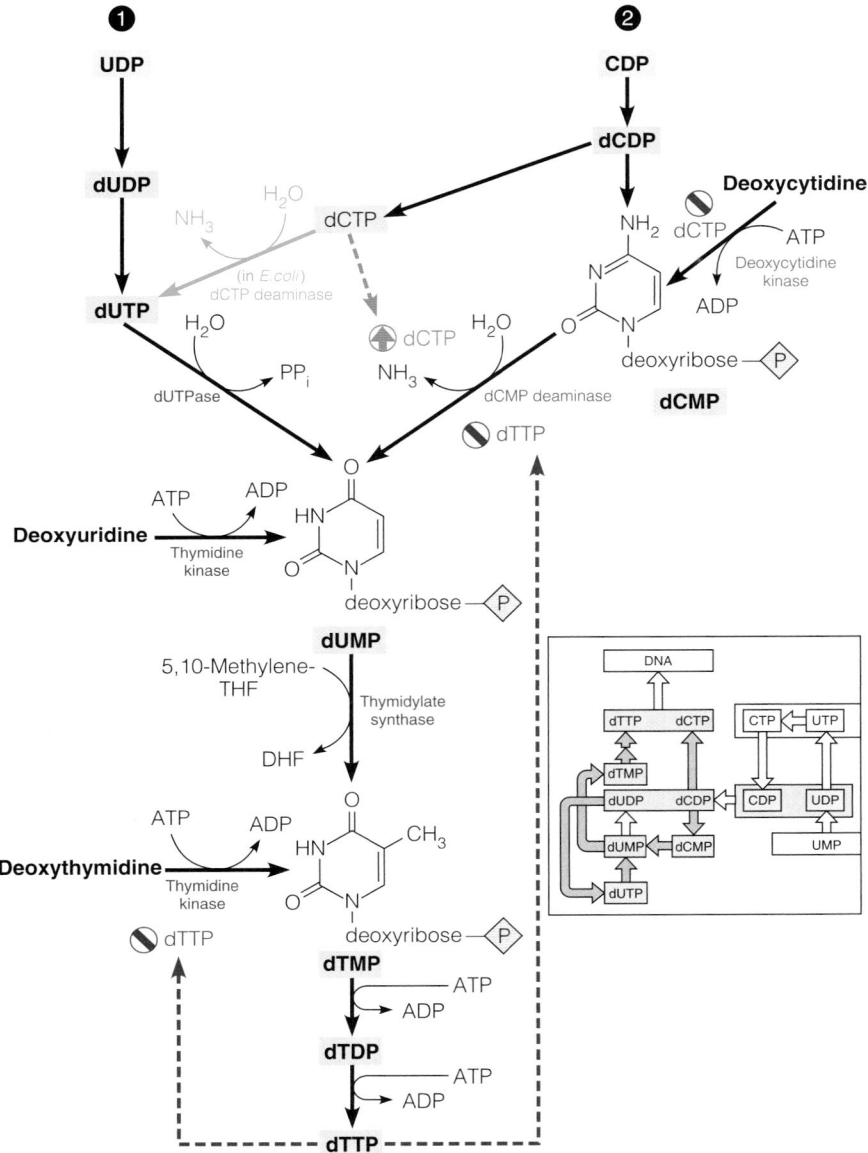

FIGURE 22.23

Salvage and de novo synthetic pathways to thymine nucleotides. The de novo pathways start with ❶ UDP or ❷ CDP, shown at the top. Dashed arrows indicate feedback regulatory loops.

Deoxyuridine Nucleotide Metabolism

In addition to the biosynthetic function of dUTPase in forming dUMP for thymine nucleotide formation, the enzyme plays an important role in excluding uracil from DNA. Were dUTP not rapidly degraded, it could serve as a satisfactory substrate for DNA polymerases. Indeed, dietary folate deficiency causes uracil misincorporation into DNA. Examination of Figure 22.24 reveals the mechanism of this effect: Folate deficiency leads to loss of 5,10-methylene-THF cofactor for the thymidylate synthase reaction, inhibiting the conversion of dUMP to dTMP. As the dUMP (and thus dUTP) levels rise at the expense of dTMP, DNA polymerase misincorporates dUTP into DNA. As discussed in Chapter 26, cells have a rather elaborate mechanism to ensure that any dUMP residues that do find their way into DNA are efficiently excised. Unfortunately, if the uracil misincorporation is severe, the repair process can lead to chromosome breaks. Because uracil is virtually identical to thymine with respect to its base-pairing properties, why is it so important for the cell to allow only thymine to be stably incorporated into DNA? The answer probably lies in the fact that dUMP residues can arise in DNA

FIGURE 22.24

Relationship between thymidylate synthase and enzymes of tetrahydrofolate metabolism. Sites of action of the chemotherapeutic drugs are indicated. See page 850 (Chapter 20) for structures of dihydrofolate reductase inhibitors. DHF, dihydrofolate; THF, tetrahydrofolate.

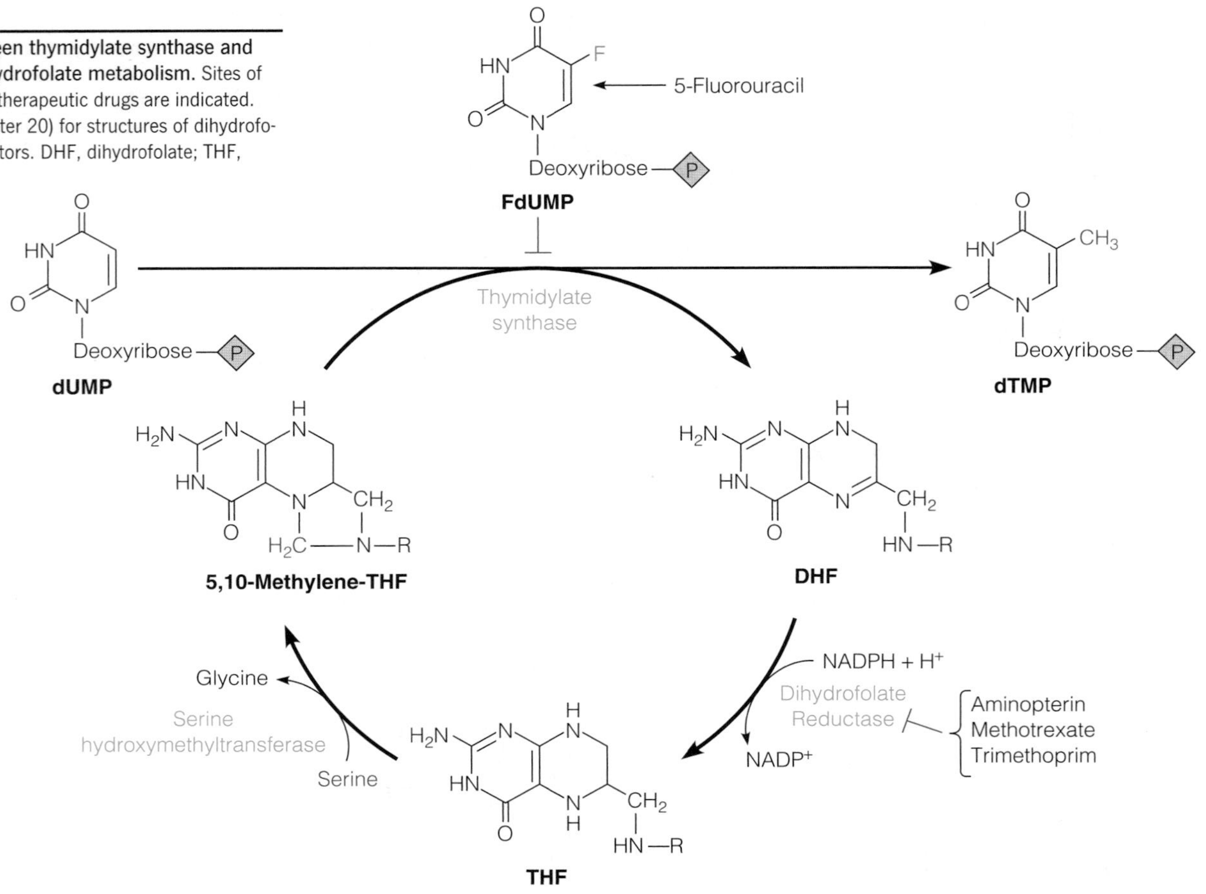

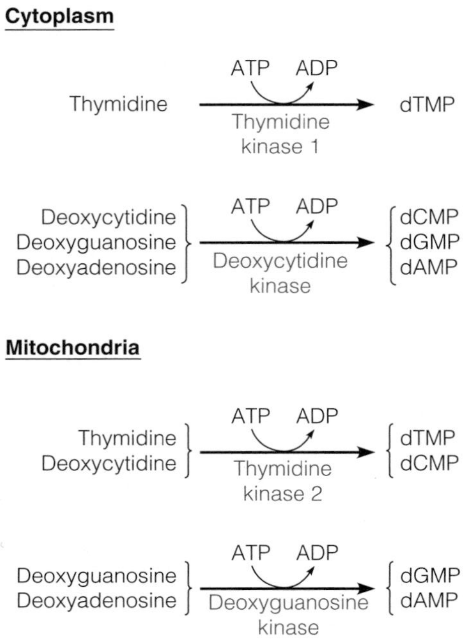

not only by dUTP incorporation but also by the spontaneous deamination of dCMP residues. The latter process, which occurs at an appreciable rate, would change the sense of the genetic message, so it seems advantageous for the cell to maintain genetic stability by having a surveillance system that excises dUMP residues no matter how they arise. Why thymidine was selected originally instead of uracil as a DNA base is a related question. Thermodynamic studies indicate that the methyl group of thymidine contributes significantly to hydrophobic interactions that stabilize double-helical DNA.

Salvage Routes to Deoxyribonucleotide Synthesis

As noted previously, purine salvage usually involves phosphoribosyltransferase reactions, which generate ribonucleoside monophosphates from the purine bases and PRPP. After phosphorylation to the diphosphate level, these compounds enter deoxyribonucleotide metabolism via ribonucleotide reductase. However, **deoxyribonucleoside kinases**, leading directly to deoxyribonucleoside monophosphates, are widely distributed, and they involve both purines and pyrimidines. Cells and organisms vary widely in their contents of ribo- and deoxyribonucleoside kinases. Human cells contain four different deoxyribonucleoside kinases—(1) thymidine kinase 1, which is located in the cytosol; (2) deoxycytidine kinase, also a cytosolic enzyme, which phosphorylates deoxyadenosine and deoxyguanosine as well as deoxycytidine, but only at higher concentrations; and two mitochondrial enzymes: (3) deoxyguanosine kinase, which phosphorylates deoxyadenosine and deoxycytidine as well as deoxyguanosine; and (4) thymidine kinase 2, which has a broader substrate specificity than the cytosolic enzyme, acting also upon deoxycytidine. The main supply of dNTPs for mitochondrial DNA (mtDNA) synthesis in resting cells comes from deoxynucleoside salvage, and the combined action of the two mitochondrial enzymes can supply all four dNTPs. Defects in either of these two enzymes causes mtDNA-depletion syndromes in humans, which are characterized by severe reduction in mtDNA copy number in affected tissues and progressive myopathy and encephalopathy.

As discussed later in this chapter, several nucleoside analogs are being used or tested in the treatment of cancer and several viral diseases. Invariably, these prodrugs must be converted to deoxyribonucleotides in order to be effective, and this has focused attention upon the deoxyribonucleoside kinases. For example, a side effect of **3-azido-2′,3′-dideoxythymidine** (AZT or zidovudine), the first drug to receive approval for treating human immunodeficiency virus (HIV) infections (see page 951), is cardiotoxicity—damage to the heart muscle. The major route for AZT utilization involves the mitochondrial isoform of thymidine kinase, also called TK2. The deoxyribonucleotides of AZT, and most of the other nucleoside analogs in clinical use, also interfere with mitochondrial function, and this is probably the basis for the cardiotoxicity. Most of these analogs act as alternative substrates for the mitochondrial DNA polymerase, thus inhibiting mitochondrial DNA replication. However, AZT is an even more potent inhibitor of TK2, and its effect on mitochondrial function may be related to alteration of dTTP pools. Current research is aimed at developing analogs whose metabolic activation does not occur in mitochondria.

Of the four deoxyribonucleoside kinases, three are synthesized constitutively, being produced at constant rates throughout the cell cycle. The exception is cytosolic thymidine kinase (TK1), the expression of which is highest during S phase, when DNA is being replicated; in this respect TK1 resembles enzymes of de novo deoxyribonucleotide synthesis, such as ribonucleotide reductase. For reasons not yet understood, TK1 salvages exogenous thymidine extremely efficiently. Experiments with radiolabeled precursors show that dTTP derived from salvage synthesis is usually incorporated into DNA in preference to thymidine nucleotides generated by de novo synthesis. This is the basis for the widespread technique of estimating rates of DNA replication by measuring incorporation of radiolabeled thymidine into DNA. As discussed in Tools of Biochemistry 12A, however, an accurate measurement of the rate of DNA synthesis from thymidine incorporation data requires measuring the specific radioactivity of the labeled dTTP pool. Nevertheless, because of the great efficiency with which thymidine is utilized in most cells, simple measurements of thymidine incorporation often yield fairly accurate estimates of DNA replication rates.

3′-Azido-2′,3′-dideoxythymidine (AZT or Zidovudine)

Thymidylate Synthase: A Target Enzyme for Chemotherapy

A goal of **chemotherapy**—the treatment of diseases with chemical agents—is to exploit a biochemical difference between the diseased tissue and the host tissue in order to interfere selectively with the disease process. Many chemotherapeutic agents were originally discovered by chance, through testing of analogs of normal metabolites. Most of these agents are limited in their effectiveness by unanticipated side effects, incomplete selectivity, and the development of resistance to the agent. One of the most exciting areas of modern biochemical pharmacology is rational drug design—the design of specific inhibitors based on knowledge of the molecular structure of the site to which the inhibitor will bind and the mechanism of action of the target molecule. For drugs whose target is an enzyme, it is necessary to know the three-dimensional structure of the enzyme and its mechanism of action. Obtaining this information requires a fusion of classical bioorganic chemistry, structural biology (X-ray crystallography and/or NMR), and site-directed mutagenesis. Thymidylate synthase presents an excellent example of the utility of these approaches.

As noted here and in Chapter 20, the goal of chemotherapy is to attack selectively a metabolic process that is specific to the pathological condition. Because thymidylate synthase (TS) participates in the synthesis of a deoxyribonucleotide, any disease that involves uncontrolled cell proliferation can in principle be treated with inhibitors of TS: Blocking the production of an essential DNA precursor should inhibit DNA replication with minimal effects on other processes. Cells that are not undergoing rapid proliferation should be relatively immune to such agents. Thus, cancer and a wide range of infectious diseases should be amenable to treatment by this approach.

Inhibition of thymidylate synthase is an approach to cancer chemotherapy, by causing specific inhibition of DNA synthesis.

None of this was recognized in the mid-1950s. In fact, TS had not yet been discovered. It was known, however, that certain tumor cells took up and metabolized uracil much more rapidly than normal cells. Without knowing the metabolic fates of uracil in detail, Charles Heidelberger hoped to kill tumor cells selectively by treatment with analogs that would block uracil metabolism in tumor cells. To that end, he undertook the chemical synthesis of **5-fluorouracil (FUra)** and its deoxyribonucleoside, **5-fluorodeoxyuridine (FdUrd)**. Both compounds were found to be potent inhibitors of DNA synthesis. FUra and FdUrd are prodrugs—their action as inhibitors involves their intracellular conversion to **5-fluorodeoxyuridine monophosphate (FdUMP)**, a dUMP analog that acts as an irreversible inhibitor of TS.

5-Fluorouracil (FUra) → Salvage pathways → 5-Fluorodeoxyuridine monophosphate (FdUMP) → Thymidine kinase (ADP, ATP) → 5-Fluorodeoxyuridine (FdUrd)

Both fluorouracil and fluorodeoxyuridine are used in cancer treatment. However, the fluorinated pyrimidines are not completely selective in their effects. For example, fluorouracil can be incorporated into RNA by salvage routes normally used for uracil, thereby interfering with the function of messenger RNA in both cancer and normal cells. Clearly, a detailed understanding of the active site of TS could lead to the design of completely specific enzyme inhibitors.

Analysis of the binding of 5-fluorodeoxyuridine monophosphate to TS has opened the door to understanding the enzyme's reaction mechanism and the structure of the active site. FdUMP is a true **mechanism-based inhibitor**, in that irreversible binding occurs only in the presence of the other substrate, 5,10-methylenetetrahydrofolate. Binding of the coenzyme induces a conformational change in the active site that duplicates early steps in the catalytic reaction and leads to irreversible FdUMP binding. Proteolytic digestion of the **ternary complex** containing FdUMP, methylene-THF, and enzyme led investigators in the laboratories of Heidelberger and Daniel Santi to isolate a peptide fragment of the enzyme containing both the inhibitor and the coenzyme. Eventually it was shown that FdUMP was linked to the methylene carbon of the THF coenzyme through C-5 of the pyrimidine ring, and to the enzyme through a cysteine sulfur covalently bonded to C-6 of the pyrimidine. The structure of this ternary complex suggested

Methylene-THF (partial structure)

Ternary complex between FdUMP, 5,10-methylene-THF, and thymidylate synthase

that the enzymatic reaction begins with nucleophilic attack by the cysteine thiol upon C-6 of the dUMP substrate.

The structure of the ternary complex between enzyme-bound FdUMP and 5,10-methylene-THF suggested the mechanism outlined in Figure 22.25. As noted previously, a cysteine thiolate ion on the enzyme initiates a nucleophilic attack on C-6 of dUMP (step 1). This step is facilitated by an active site general acid; C-5 now becomes a nucleophile, attacking the methylene carbon of the iminium cation that exists in equilibrium with 5,10-methylene-THF, forming the covalent enzyme-dUMP-THF ternary complex (step 2). Abstraction of a proton from C-5 by an active site general base (step 3) initiates an electron shift that leads to β-elimination of the cofactor (step 4), followed by hydrogen transfer from the C-6 of the cofactor to the pyrimidine (step 5), consistent with observations that this hydrogen is incorporated quantitatively into the thymidylate methyl group. This hydride transfer is accompanied by displacement of the cysteine thiolate, and also oxidizes the cofactor to dihydrofolate, which dissociates in step 5. Inhibition by FdUMP results from the electronegativity of fluorine, which generates a C—F bond at C-5 that cannot be broken. Thus, the reaction pathway cannot proceed to step 3.

Confirmation of this mechanism has come about with evidence that the substrate, dUMP, forms a covalent bond with the coenzyme during normal catalysis. More important evidence came from elucidation of a three-dimensional model of thymidylate synthase. In 1987 Robert Stroud, Daniel Santi, and their colleagues presented a model of the apoenzyme from *Lactobacillus casei*. TS is highly conserved evolutionarily, with about 20% of its residues being invariant among mammalian, bacterial, viral, fungal, and protozoal sequences. Shortly after the apoenzyme model was presented, two laboratories crystallized *E. coli* TS as a complex with dUMP and 5,10-methylene-THF (Figure 22.26). Analysis of these complexes confirmed the conformations of substrate and cofactor bound in the active site and the identification of active site residues in contact with these ligands. In these ternary complex structures, FdUMP is covalently bound to cysteine 146 (*E. coli* numbering) through C-6, and to the folate cofactor through C-5. Tyrosine 94 is thought to be the general base that abstracts the proton from C-5 in step 3, and glutamate 58 is the general acid that transfers a proton to and from the pyrimidine ring in steps 1–4. Several conserved lysine residues lie nearby and are thought to represent binding sites for the negatively charged polyglutamate tail on the folate coenzyme (see page 849). TS binds folate polyglutamates about 100-fold more tightly than folate monoglutamate.

In 1991 the crystallization of the human enzyme was reported, and this enzyme has been the focus of further drug development. Analysis of substrate binding interactions has led to design and synthesis of folate cofactor analogs (**antifolates**) that compete quite effectively with 5,10-methylene-THF for binding to TS (K_i values as low as 0.4 nM). Three such inhibitors are shown in the margin. Pemetrexed and raltitrexed are both used clinically to treat certain types of cancer. The same type of approach is underway in dozens of laboratories and companies, focused on drug targets that include membrane-bound or intracellular receptor proteins and nucleic acids, as well as enzymes. The development of specific HIV protease inhibitors as anti-AIDS drugs is an example of the potential of this approach.

Diseases other than cancer are also susceptible to attack by inhibition of thymidylate synthase. For example, parasitic protozoans, such as those that cause malaria, synthesize an unusual form of TS—a bifunctional enzyme, with both thymidylate synthase and dihydrofolate reductase activities. Knowing the structure of this enzyme's active site should allow development of inhibitors that would block this specific enzyme but not thymidylate synthase of the animal or human host.

Pemetrexed (Alimta)

Raltitrexed (Tomudex)

Nolatrexed (Thymitaq)

FIGURE 22.25

Mechanism for the reaction catalyzed by thymidylate synthase. A and B represent an active site general acid and base, respectively. DHF, dihydrofolate; THF, tetrahydrofolate.

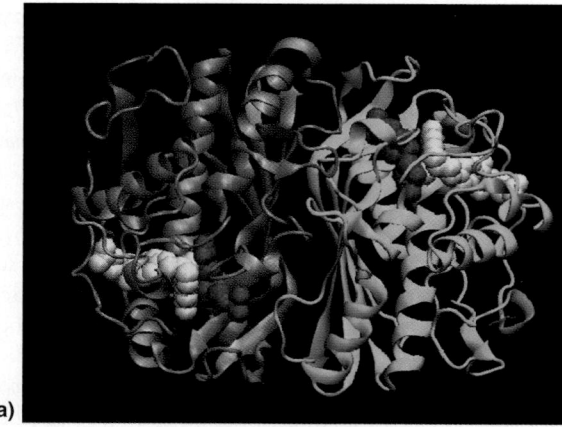

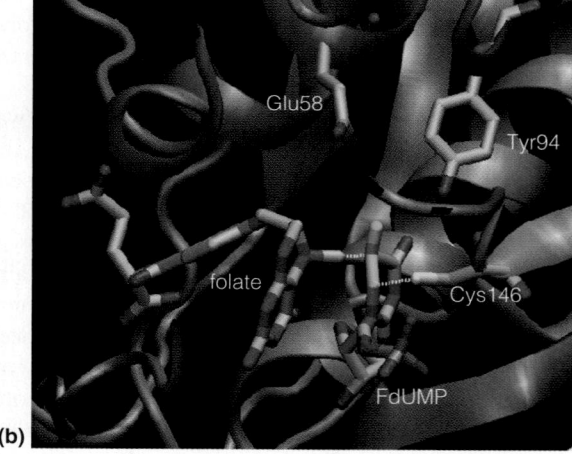

(a)

(b)

FIGURE 22.26

Structure of the homodimeric thymidylate synthase from *E. coli* (PDB IDs 1TLS, 1TSN). **(a)** Structure of the homodimer with bound folate cofactor (yellow) and FdUMP (red). **(b)** Active site region of the orange subunit in (a) showing the ternary complex formed between 5,10-methylene-THF, FdUMP, and Cys 146. The dashed lines indicate the covalent bonds of the ternary complex. The close proximity of Glu 58 (general acid) and Tyr 94 (general base) is also evident.

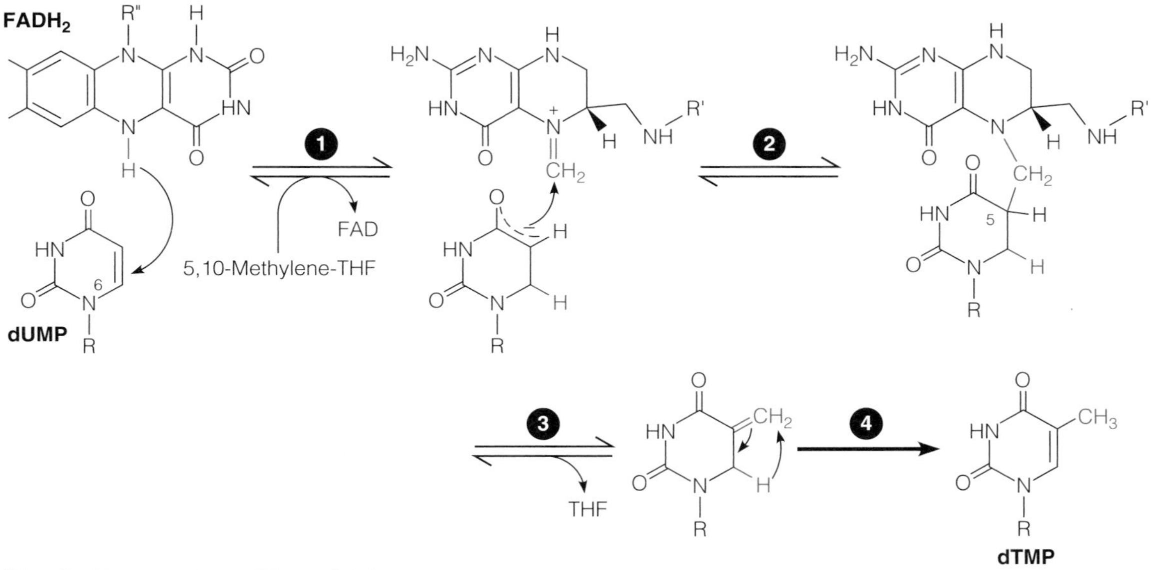

FIGURE 22.27

Proposed mechanism for the reaction catalyzed by flavin-dependent thymidylate synthase. THF, tetrahydrofolate.

Reprinted from *Archives of Biochemistry and Biophysics* 493:96–102, E. M. Koehn and A. Kohen, Flavin-dependent thymidylate synthase: A novel pathway towards thymine. © 2010, with permission from Elsevier.

Flavin-Dependent Thymidylate Synthase: A Novel Route to dTMP

It has long been thought that the only way cells can make thymidylate is the de novo pathway via thymidylate synthase, or salvage via thymidine kinase (Figure 22.23). However, in 2002, genomics studies led to the identification of several organisms that lack both the *thyA* and *tdk* genes (encoding thymidylate synthase and thymidine kinase, respectively). How do these organisms synthesize their thymine nucleotides for DNA synthesis? Biochemists soon discovered an alternate flavin-dependent thymidylate synthase (FDTS), encoded by the *thyX* gene. This gene is present in ~30% of all microorganisms, including several serious human pathogens. The flavin-dependent TSs share no structure or sequence homology with classical TS. Whereas classical TSs are homodimers with one active site per subunit (Figure 22.26a), FDTSs are homotetramers with four active sites, located at subunit interfaces. Like classical TS, FDTS uses 5,10-methylene-THF as the one-carbon donor, but FDTS was found to require NADPH as reductant. More importantly, structural studies with FDTSs revealed that they lack the active site cysteine nucleophile conserved in all the classical TSs, suggesting that they use different chemistry to catalyze conversion of dUMP to dTMP. Figure 22.27 shows the mechanism proposed for these flavin-dependent TSs. The reaction is initiated by a hydride transfer from the reduced flavin (FADH$_2$) directly to C-6 of the pyrimidine ring (step 1). The resulting enolate anion nucleophilically attacks the activated iminium form of 5,10-methylene-THF, forming a folate-pyrimidine adduct (step 2). THF is eliminated (step 3), and the resulting thymine isomer rearranges to give the product dTMP (step 4). No covalent enzyme intermediate forms in this mechanism, and THF is not oxidized to DHF. The role of NADPH is to reduce the enzyme-bound oxidized flavin (FAD) back to FADH$_2$, readying the enzyme for another reaction cycle.

This is a beautiful example of evolution taking different routes to solve the same problem. FDTS is in a sense a bifunctional enzyme, taking the place of both classical thymidylate synthase and dihydrofolate reductase:

$$\text{dUMP} + 5,10\text{-methylene-THF} \xrightleftharpoons{\text{classical TS}} \text{dTMP} + \text{DHF}$$

$$\text{DHF} + \text{NADPH} + \text{H}^+ \xrightleftharpoons{\text{DHFR}} \text{THF} + \text{NADP}^+$$

Net:

$$\text{dUMP} + 5,10\text{-methylene-THF} + \text{NADPH} + \text{H}^+ \rightleftharpoons \text{dTMP} + \text{THF} + \text{NADP}^+$$

$$\text{dUMP} + 5,10\text{-methylene-THF} + \text{NADPH} + \text{H}^+ \xrightleftharpoons{\text{FDTS}} \text{dTMP} + \text{THF} + \text{NADP}^+$$

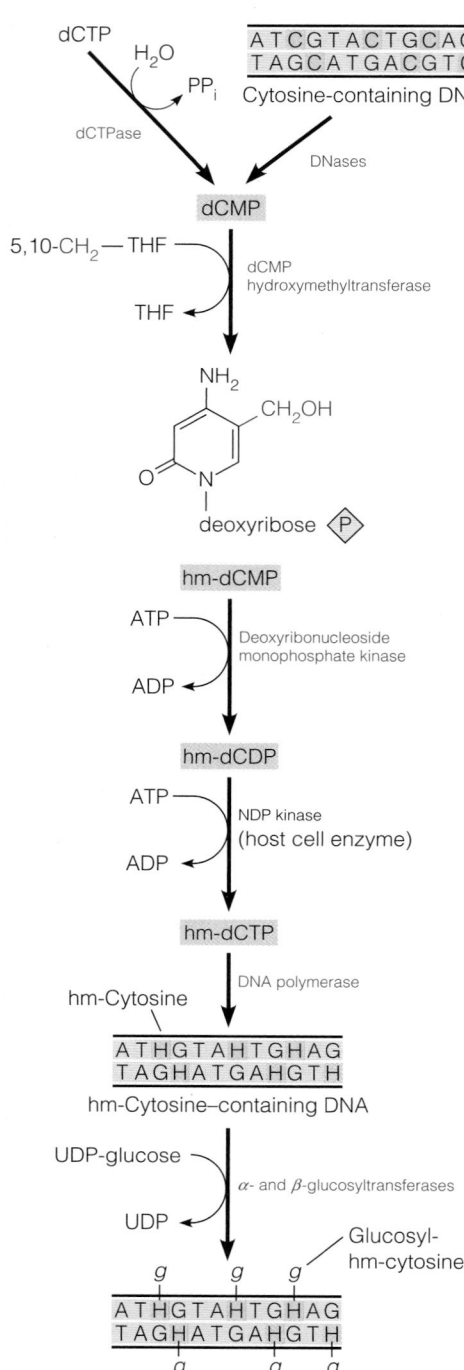

FIGURE 22.28

Metabolic pathways leading to nucleotide modifications in T-even phage–infected *E. coli*. Virus-coded enzymes are shown in red. hm = hydroxymethyl.

Novel metabolic processes induced by viruses are attractive chemotherapeutic targets.

The mechanistic differences between these two thymidylate synthases may provide an opportunity for the development of a new class of antibiotics. It should be possible to find inhibitors of FDTS that have little effect on the classical TS used by humans.

Virus-Directed Alterations of Nucleotide Metabolism

That viruses can redirect the metabolism of their host cells first came to light in 1957 through studies of nucleotide biosynthesis in *E. coli* bacteria infected by the T-even bacteriophages—T2, T4, and T6. G. R. Wyatt and Seymour Cohen had shown in 1952 that DNA of these viruses contains no cytosine but instead contains 5-hydroxymethylcytosine, with most of the hydroxymethyl groups further modified by being in glycosidic linkage with glucose moieties.

Cytosine   **5-Hydroxymethylcytosine**   **α-Glucosyl-5-hydroxymethylcytosine**

Continuing work showed that infection initiates the synthesis of virus-coded enzymes, which carry out these modifications (Figure 22.28). The principal enzymes in T-even phage–infected bacteria include a **dCTPase**, which cleaves dCTP to dCMP; a **dCMP hydroxymethyltransferase**, which transfers a one-carbon group from 5,10-methylene-THF to dCMP at the hydroxymethyl oxidation level; and a **deoxyribonucleoside monophosphate kinase**, which can phosphorylate the resultant 5-hydroxymethyl-dCMP. The dCMP hydroxymethylase uses the same mechanism as thymidylate synthase (Figure 22.25), involving a covalent enzyme-dCMP-THF ternary complex intermediate. Phosphorylation of 5-hydroxymethyl-dCDP to the triphosphate is catalyzed by nucleoside diphosphate kinase of the host cell. The glucosylation reactions occur after the modified nucleotide has been incorporated into DNA. In T4 phage–infected bacteria there are two **glucosyltransferase** reactions, one of which transfers glucose in the α configuration and one in the β. In addition, the viral genome specifies several deoxyribonucleases, which specifically cleave cytosine-containing DNA. This process helps the virus abolish expression of host cell genes, and it also provides a source of precursors for viral DNA synthesis.

Although base substitutions in DNA are rather unusual, about a dozen instances have now come to light in which one of the four common deoxyribonucleotides is replaced, wholly or in part, by a chemically altered derivative that retains the base-pairing specificity of the replaced nucleotide. Some *Bacillus subtilis* phages, for example, substitute uracil for thymine in their DNA. Other *B. subtilis* phages contain 5-hydroxymethyluracil in place of thymine. A phage of *Xanthomonas oryzae* substitutes 5-methylcytosine for every one of its DNA cytosine residues. In each case investigated, the virus directs the nucleotide modifications through synthesis of virus-encoded enzymes that create novel metabolic pathways in the cells they infect.

Plant and animal viruses do not contain extensive nucleic acid base modifications of the type found in bacteriophages. (Most organisms contain a significant proportion of methylated nucleic acid bases, but these modifications occur after polymerization; see Chapter 26.) However, virus-coded enzymes are often produced to help the infected cell augment its synthesis of nucleic acid precursors. In some cases the virus-specified enzyme differs from its host cell counterpart sufficiently that scientists can design selective enzyme inhibitors and thereby achieve a specific chemotherapy directed against the virus-infected cell. The best current example is the use of **acyclovir** and its relative, **ganciclovir**, in treating herpes virus infections. The herpes viruses are large DNA-containing viruses whose genomes encode several enzymes, including **thymidine kinase**. This virus-encoded enzyme (HSV-TK) has extremely broad substrate specificity, capable of phosphorylating both pyrimidine and purine nucleosides. Nucleoside analogs, such as acyclovir (also called **acycloguanosine**) and ganciclovir are converted ultimately to the 5′-triphosphate, which interferes with DNA replication. Uninfected cells do not efficiently phosphorylate acyclovir and ganciclovir, so DNA replication and, hence, virus growth are inhibited selectively in the infected cells.

Biological and Medical Importance of Other Nucleotide Analogs

The previous examples have made clear the utility of nucleoside and nucleotide analogs, primarily as drugs. In this section we discuss additional analogs of medical importance, as well as several that are useful as research reagents. Nucleotides are poorly transported into cells because of the negative phosphate charge, so most of the compounds we shall discuss are introduced into cells as nucleosides or nucleoside derivatives, where they are acted upon initially by nucleoside kinases (see page 919). After uptake and conversion to nucleotides, these compounds interfere with metabolism in various ways. In this section we shall use the terms *nucleotide analogs* and *nucleoside analogs* interchangeably.

Nucleotide Analogs as Chemotherapeutic Agents

The enzymes of nucleotide synthesis have been widely studied as target sites for the action of antiviral or antimicrobial drugs. As noted earlier, the aim is to identify a biochemical distinction between comparable processes of the uninfected host and the infected host.

Antiviral Nucleoside Analogs

One of the earliest antiviral drugs approved for use in humans is **arabinosyladenine (araA or Vidarabine)**. It is now being used to treat several viral diseases, including viral encephalitis, a neurological disease caused by another member of the herpes virus family. Unlike acycloguanosine, araA is phosphorylated to the triphosphate level by cellular kinases. The triphosphate, araATP, is a selective inhibitor of DNA polymerases encoded by herpes viruses. Thus, araA selectively interferes with viral DNA replication, even though all cells, infected and uninfected, form the triphosphate. Because araA is susceptible to degradation by adenosine deaminase, its effectiveness can be increased when administered with an inhibitor of the latter enzyme. The arabinose analog of deoxycytidine, **arabinosylcytosine (araC or Cytarabine)**, is being used in cancer chemotherapy. araCTP also interferes with DNA replication after conversion to the triphosphate.

Other analogs receiving considerable attention are those being used to combat acquired immune deficiency syndrome (AIDS) caused by HIV. One such analog, 3′-azido-2′,3′-dideoxythymidine (AZT; see page 945) is anabolized to the corresponding 5′ triphosphate, which is an inhibitor of viral reverse transcriptase (the enzyme that makes a DNA copy of the viral RNA—see Chapter 25). Other nucleoside analogs—**2′,3′-dideoxycytidine (ddC), 2′,3′-dideoxyinosine (ddI), 3′-thiacytidine (3TC)**, and **2′,3′-didehydro-3′-deoxythymidine (d4T)**—act by conversion to the corresponding triphosphate, which is incorporated into DNA but then blocks further replicative chain elongation because of the absence of a 3′ hydroxyl terminus. All four analogs have been approved for use in treating human HIV infections and both AZT and 3TC are components, along with HIV protease inhibitors, of the three-drug "cocktails" credited with long-term remission of HIV infections. In late 1994 two laboratories reported a synergistic effect of ddI and hydroxyurea against HIV infections, presumably starving the cell of dNTPs for reverse transcription (see Figure 25.46, page 1073).

Acycloguanosine (Acyclovir)

Ganciclovir

2′,3′-Dideoxycytidine (ddC)

2′,3′-Dideoxyinosine (ddI)

2′,3′-Didehydro-3′-deoxythymidine (d4T)

3′-Thiacytidine (3TC)

Formycin B

6-Mercaptopurine

6-Thioguanine

Cell proliferation rate is a determinant of the effectiveness of chemotherapy with dihydrofolate reductase inhibitors because thymidylate synthase reaction flux determines the rate at which tetrahydrofolate is oxidized to dihydrofolate.

Purine Salvage as a Target

An important biochemical anomaly is found in parasitic protozoans such as *Plasmodium*, which is responsible for malaria, and *Leishmania*, which causes a debilitating but usually nonfatal disease affecting the skin and visceral organs. Parasitic protozoans lack the capacity for de novo purine synthesis, and they depend entirely on salvage of nucleosides and bases provided by the host. Compounds such as allopurinol (see page 929) and **formycin B** inhibit the growth of these organisms in culture, partly through inhibition of salvage enzymes and partly through the ability of the salvage enzymes to anabolize the analog, an ability lacking in the corresponding host enzymes. For example, allopurinol is converted to an analog of inosinic acid and then to an AMP analog and is finally incorporated into RNA, where it interferes with messenger RNA coding in protein synthesis.

Purine salvage can also be taken advantage of, for example in the use of thiopurines as anti-cancer and immunosuppressive drugs. Like 5-fluorouracil, **6-mercaptopurine** and **6-thioguanine** are prodrugs that require metabolic activation in order to exert their cytotoxic effects. Both of these compounds are salvaged to nucleoside monophosphates by HGPRT (page 928) and then converted to triphosphates that can be incorporated into DNA or RNA. Incorporation of these thio analogs into DNA triggers cell cycle arrest and apoptosis, and these two drugs are mainstays of childhood leukemia treatment. 6-Mercaptopurine and 6-thioguanine were synthesized by Gertude Elion and George Hitchings, who also developed allopurinol and acyclovir—they received the 1988 Nobel Prize for Physiology or Medicine for their many contributions to chemotherapeutic drug development.

The thiopurines illustrate the importance of **pharmacogenetics**—understanding the genetically determined variations in responses to drugs in humans. 6-Mercaptopurine and 6-thioguanine can be metabolically inactivated by methylation of the thio group (*S*-methylation). This *S*-Adenosylmethionine-dependent methylation is catalyzed by thiopurine *S*-methyltransferase (TPMT), a cytoplasmic enzyme present in most human tissues. TPMT-deficient patients given the usual dose of thiopurines accumulate up to 10-fold higher cellular concentrations of the active thioguanine nucleotides and are thus at high risk for severe, and sometimes fatal, hematological toxicity. There are numerous variant TPMT alleles that are associated with low enzymatic activity in humans, and ~1 in 300 individuals will have low or undetectable levels of TPMT activity. These patients must be treated with much lower doses of the thiopurines to avoid these toxic side effects. There are now simple molecular genetic tests to determine a patient's TPMT genotype, and clinicians are beginning to use this information to optimize chemotherapy for each patient.

Folate Antagonists

Recall from Chapter 20 that the folic acid analog methotrexate was found long ago to induce remissions in certain acute leukemias. What is the basis for this selectivity? Given that folate cofactors play essential roles in synthesizing precursors to DNA, RNA, protein, *and* phospholipids, one would expect inhibition of tetrahydrofolate regeneration to be toxic to all cells. However, there is a rationale for the selective toxicity of folate antagonists against proliferating cells. Recall that the thymidylate synthase reaction oxidizes methylene-THF to dihydrofolate; this is the only known THF-requiring reaction that does not regenerate THF. From the reactions shown in Figure 22.24, one can predict that inhibition of dihydrofolate reductase blocks the recycling of DHF back to THF. Under these conditions the rate at which all of the intracellular reduced folates become oxidized is directly related to the intracellular activity of thymidylate synthase, which in turn is coordinated with the rate of DNA synthesis. Thus, proliferating cells, with rapid rates of DNA replication, will exhaust their tetrahydrofolate stores faster than nonproliferating cells.

Although antimetabolites such as fluorouracil or methotrexate do attack proliferating tissue selectively, they are still toxic to normal cells. Deleterious side effects are seen in tissues that proliferate as part of their normal function; such tissues include intestinal mucosa, hair cells, and components of the immune system. Equally serious is the development of drug-resistant cell variants. In such cases, levels of the target enzyme rise beyond the point that they can be controlled with inhibitors. Robert Schimke and his colleagues investigated the mechanism for the rise in dihydrofolate reductase activity, which can be several hundredfold in methotrexate-resistant cell lines. They found that prolonged incubation of cells in methotrexate often leads to selective amplification of the gene encoding dihydrofolate reductase; the number of DNA copies of this gene per cell increases many times, with a corresponding accumulation of the gene product, dihydrofolate reductase. Other resistance mutations involve more conventional mechanisms, such as altered transport of the drug into cells or alteration of the target enzyme, making it resistant to the antimetabolite.

Another class of dihydrofolate reductase inhibitors is exemplified by **trimethoprim**. This compound is a specific inhibitor of dihydrofolate reductases of prokaryotic origin. Trimethoprim and its relatives are widely used to treat both bacterial infections and certain forms of malaria. The success of these drugs derives from their being extremely weak inhibitors of vertebrate dihydrofolate reductases. Trimethoprim is often administered in conjunction with a sulfonamide drug to inhibit the synthesis of folate and hence to block sequential steps in the same pathway.

Folate antagonists have also been developed to target de novo purine biosynthesis. Lometrexol (5,10-dideazatetrahydrofolate) and LY309887 are both potent inhibitors (low nM K_i) of GAR transformylase (Figure 22.4). Pemetrexed, introduced previously as a thymidylate synthase inhibitor (page 947), is actually a **multi-targeted antifolate** because it also inhibits dihydrofolate reductase and GAR transformylase. It is used clinically to treat certain cancers.

Nucleotide Analogs and Mutagenesis

Some nucleotide analogs are excellent mutagens, useful both in the isolation of mutants and in studies on the mechanism of mutagenesis. Two such analogs are **2-aminopurine (2AP)** and **5-bromodeoxyuridine (BrdUrd)**. 2-Aminopurine is incorporated into DNA in place of adenine, but when a 2AP-containing template replicates, the analog occasionally base-pairs with cytosine rather than thymine. Thus, incorporation of 2AP changes an A-T pair in DNA to a G-C pair (Figure 22.29).

Bromodeoxyuridine functions similarly, but it has other uses as well. It is an excellent thymidine analog because the van der Waals radius of the bromine atom is close to that of the methyl group. Hence, it is efficiently incorporated into DNA. Mispairing of BrdUrd with a deoxyguanosine residue on replication of a BrdUrd-containing template can lead to mutagenesis by changing an A-T base pair to a G-C (see Figure 22.29). Alternatively, BrdUTP can compete with dCTP for incorporation opposite G in the template (not shown).

Because bromine is much heavier than a methyl group, it imparts an increased density to substituted DNA, thereby providing a physical basis for separating replicating from nonreplicating DNA (see Chapter 25). Finally, radiobiologists use BrdUrd as a radiosensitizing agent. Bromo-dUMP residues in DNA debrominate readily when the substituted DNA is irradiated with UV or near-UV light. This process generates free radicals, which cause various kinds of damage to the DNA structure.

Recent interest has focused upon a mutagenic nucleotide analog that is formed in normal metabolism. Recall from Chapter 15 that oxidative stress—generated, for example, by treating cells with hydrogen peroxide—damages DNA bases. The oxidation of guanine residues in DNA to 8-oxoguanine is strongly mutagenic because 8-oxoguanine base-pairs during replication with adenine almost as efficiently as it pairs with cytosine. This mispairing event could cause mutation by changing a G-C base pair to a T-A. Recently, an enzyme was discovered that specifically hydrolyzes the nucleotide 8-oxo-dGTP to 8-oxo-dGMP plus pyrophosphate (the *mutT* gene product; see Chapter 26). Mutations that inactivate this enzyme have a mutator phenotype, suggesting that a significant mutagenic pathway involves oxidation of guanine nucleotides to 8-oxoguanine nucleotides, followed by their incorporation into DNA, and that oxidative mutagenesis is minimized by the breakdown of 8-oxo-dGTP before this incorporation can occur.

Trimethoprim

LY309887

Lometrexol

2-Aminopurine   **5-Bromodeoxyuridine**

Nucleotide analogs with altered base-pairing properties are mutagenic because non-Watson–Crick base pairs form when the analog is either in the template or in an incoming nucleoside triphosphate.

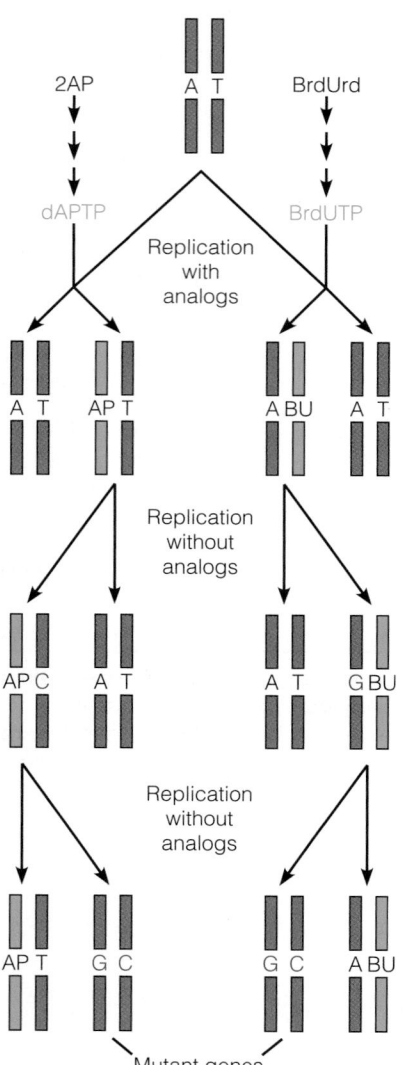

5-Fluoroorotic acid

dGTP

8-oxo-dGTP → 8-oxo-dGMP
mut T

8-oxo-dGTP

FIGURE 22.29

Mechanisms of mutagenesis by nucleotide analogs.
2-Aminopurine (2AP) is converted by salvage pathways
to deoxy aminopurine triphosphate (dAPTP), the dNTP
analog of dATP. BrdUrd is converted to BrdUTP, the
dNTP analog of dTTP. The first round of replication
occurs in the presence of the analog, and the second
and third rounds occur in its absence. Replication of
only the analog-containing duplex is shown in the second and third rounds. AP is a 2-aminopurine nucleotide
residue in DNA, and BU is a bromodeoxyuridine
nucleotide residue. In the pathways shown, both
analogs change an A-T base pair to a G-C base pair (red
letters). Other pathways can occur as well.

Nucleotide-Metabolizing Enzymes as Selectable Genetic Markers

Because most cells can synthesize nucleotides de novo, the enzymes of salvage synthesis are usually nonessential for cell viability. Moreover, as we have seen, many
inhibitors of these enzymes are available. Consequently, nucleotide-metabolizing
enzymes and the genes that encode them provide **selectable genetic markers**, which
have a variety of uses. As the term implies, one can devise selective growth conditions
such that only cells lacking a particular enzyme, or containing a particular enzyme,
will grow. For example, culture of cells in medium containing 6-thioguanine allows
growth of only the cells lacking active HGPRT (page 952). Similarly, one can isolate
cells lacking thymidine kinase (page 944) by selecting for a bromodeoxyuridine-
resistant phenotype because TK must be active in order to anabolize BrdUrd to a
toxic metabolite. Thus, one can measure forward mutation rates by observing the
appearance of these drug-resistant phenotypes.

Another useful selection strategy makes use of a pyrimidine analog, 5-fluo-
roorotic acid (5-FOA). In cells with an intact de novo pyrimidine biosynthesis pathway, this drug is a substrate of orotate phosphoribosyltransferase (Figure 22.12,
page 931), and it is metabolized to 5-fluoro-UTP and 5-fluoro-CTP. Incorporation
of these fluorinated nucleotides into DNA and RNA is lethal to the cell. However,
cells lacking active orotate phosphoribosyltransferase or OMP decarboxylase are
resistant to 5-FOA, providing a powerful selection for these cells.

By the same token, one can select for mutation in the reverse direction by adjusting culture conditions so that the capacity for salvage synthesis is essential to cell viability. A common technique, both in somatic cell genetic analysis and in preparation
of monoclonal antibodies (see Tools of Biochemistry 7A), is **cell fusion**. Two cell lines
of different origins are mixed under conditions in which some of them can physically
fuse, resulting in two different nuclei in one cytoplasm. One can select for these cell
hybrids by adjusting culture conditions so that only the hybrids will grow in **HAT
medium** (normal cell culture medium augmented with hypoxanthine, aminopterin,
and thymidine). Aminopterin inhibits dihydrofolate reductase (Figure 22.24)

and hence blocks de novo purine and thymidylate synthesis. Cells can survive only if they have active HGPRT, to utilize the hypoxanthine for purine synthesis, and thymidine kinase, to utilize the thymidine for thymidylate synthesis.

Genes with selectable phenotypes represent an important adjunct to the use of recombinant DNA technology for introducing novel genetic material into cells—animal, plant, or microbial—because one can introduce recombinant DNA molecules that contain a selectable marker in tandem with a gene whose transfer is desired. Application of selective conditions then forces the growth of only those cells that have acquired the new pair of genes. An intriguing variant on this approach is described in a report of a novel treatment of brain tumors. A recombinant DNA containing the herpes simplex virus deoxypyrimidine kinase gene was injected into the tumor. In experimental animals the proliferating (tumor) cells took up and replicated this DNA. After several days the animals were treated with ganciclovir, which led to selective killing of the tumor cells because of selective phosphorylation of the drug in those cells.

Genes for nucleotide-metabolizing enzymes are excellent selectable markers. Separate salvage and de novo pathways allow selection for survival or death of cells with particular metabolic traits.

SUMMARY

Nucleotides arise within cells from nucleic acid breakdown, from reuse (or salvage) of preformed nucleosides or nucleobases, or from de novo biosynthesis. Purine nucleotides are formed at the nucleotide level, in a 10-step pathway that leads from PRPP to inosinic acid. Beyond this branch point, separate pathways lead to adenine and guanine nucleotides. Purine catabolism yields uric acid, an insoluble compound that is formed in excess in a variety of disease states. Pyrimidines are synthesized at the base level, with conversion to a nucleotide occurring late in the pathway. An unbranched pathway leads to both UTP and CTP. In most organisms the ribonucleoside diphosphates are substrates for reduction of the ribose sugar in situ, yielding deoxyribonucleoside diphosphates, which in turn lead to the four dNTP DNA precursors. Ribonucleotide reductase is an important control site, inasmuch as it represents the first metabolic reaction committed to DNA synthesis. Biosynthesis of thymine nucleotides involves transfer of the methylene group of 5,10-methylenetetrahydrofolate to a deoxyuridine nucleotide, followed by reduction of the methylene group. Reactions of deoxyribonucleotide biosynthesis are target sites for enzyme inhibitors that have found use as anticancer, antimicrobial, antiviral, and antiparasitic drugs. Other nucleotide analogs have found use as research reagents—for example, in studies of mutagenesis or as DNA density labels.

REFERENCES

Enzymes of Nucleotide Metabolism

An, S., R. Kumar, E. D. Sheets, and S. J. Benkovic (2008) Reversible compartmentalization of de novo purine biosynthetic complexes in living cells. *Science* 320:103–106. This article describes the colocalization of GFP-tagged enzymes as evidence for a purinosome in mammalian cells.

An, S., M. Kyoung, J. J. Allen, K. M. Shokat, and S. J. Benkovic (2010) Dynamic regulation of a metabolic multi-enzyme complex by protein kinase CK2. *J. Biol. Chem.* 285:11093–11099. This article suggests that the purinosome is regulated by reversible phosphorylation-dephosphorylation.

Barlowe, C. K., and D. R. Appling (1990) Molecular genetic analysis of *Saccharomyces cerevisiae* C$_1$-tetrahydrofolate synthase mutants reveals a noncatalytic function of the ADE3 gene product and an additional folate-dependent enzyme. *Mol. Cell. Biol.* 10:5679–5687. Genetic evidence for the involvement of folate enzymes in a purinosome in yeast cells.

Blakley, R. L., and S. J. Benkovic (1984) *Folates and Pterins*, Vol. 1. Academic Press, New York. A multiauthored book containing reviews on dihydrofolate reductase, purine metabolism, and pyrimidine biosynthesis.

Elion, G. B. (1989) The purine path to chemotherapy. *Science* 244:41–47. Dr. Elion's Nobel Prize address, which described the development of allopurinol, acyclovir, 6-thioguanine, and other therapeutically valuable purine analogs.

Evans, D. R., and H. I. Guy (2004) Mammalian pyrimidine biosynthesis: Fresh insights into an ancient pathway. *J. Biol. Chem.* 279:33035–33038. A concise review on the organization and structures of the enzymes of de novo pyrimidine biosynthesis.

Hershfield, M. S., and B. S. Mitchell (2001) Immunodeficiency diseases caused by adenosine deaminase deficiency and purine nucleoside phosphorylase deficiency. In *The Metabolic and Molecular Bases of Inherited Disease*, C. R. Scriver, A. L. Beaudet, W. S. Sly, D. Valle, B. Childs, K. W. Kinzler, and B. Vogelstein, eds., Vol. II, Ch. 109, pp. 2585–2625, McGraw-Hill, New York.

Oda, M., Y. Satta, O. Takenaka, and N. Takahata (2002) Loss of urate oxidase activity in hominoids and its evolutionary implications. *Mol. Biol. Evol.* 19:640–653. Molecular genetic analysis of the urate oxidase gene during hominid evolution.

Raushel, F. M., J. B. Thoden, and H. M. Holden (1999) The amidotransferase family of enzymes: Molecular machines for the production and delivery of ammonia. *Biochemistry* 38:7891–7899. This paper reviews the chemistry and structure of this ubiquitous family of enzymes.

Smith, G. K., W. T. Mueller, G. F. Wasserman, W. D. Taylor, and S. J. Benkovic (1980) Characterization of the enzyme complex involving the folate-requiring enzymes of de novo purine biosynthesis. *Biochemistry* 19:4313–4321. This article describes the first evidence for a purinosome in mammalian cells, obtained from affinity purification and cross-linking.

Smith, P. M. C., and C. A. Atkins (2002) Purine biosynthesis. Big in cell division, even bigger in nitrogen assimilation. *Plant Physiol.* 128:793–802. This article reviews the role of the de novo purine pathway in leguminous plants.

Webster, D. R., D. M. O. Becroft, A. H. van Gennip, and A. B. P. Van Kuilenburg (2001) Hereditary orotic aciduria and other disorders of pyrimidine metabolism. In *The Metabolic and Molecular Bases of Inherited Disease*, C. R. Scriver, A. L. Beaudet, W. S. Sly, D. Valle, B. Childs, K. W. Kinzler, and B. Vogelstein, eds., Vol. II, Ch. 113, pp. 2663–2702, McGraw-Hill, New York. A thorough discussion of these rare diseases.

Yamamoto, S., K. Inoue, T. Murata, S. Kamigaso, T. Yasujima, J. Y. Maeda, Y. Yoshida, K. Y. Ohta, and H. Yuasa (2010) Identification and functional characterization of the first nucleobase transporter in mammals: Implication in the species difference in the intestinal absorption mechanism of nucleobases and their analogs between higher primates and other mammals. *J. Biol. Chem.* 285:6522–6531. For salvage pathways to occur, their substrates must get into cells. This paper describes the identification of the first mammalian nucleobase transporter.

Deoxyribonucleotide Biosynthesis

Finer-Moore, J. S., D. V. Santi, and R. M. Stroud (2003) Lessons and conclusions from dissecting the mechanism of a bisubstrate enzyme: Thymidylate synthase mutagenesis, function, and structure. *Biochemistry* 42:248–256.

Koc, A., C. K. Mathews, L. J. Wheeler, M. K. Gross, and G. F. Merrill (2006) Thioredoxin is required for deoxyribonucleotide pool maintenance during S phase. *J. Biol. Chem.* 281:15058–15063. This paper provides in vivo evidence for thioredoxin being a physiologically relevant electron donor for ribonucleotide reductase.

Logan, D. T. (2011) Closing the circle on ribonucleotide reductases. *Nat. Struct. Mol. Biol.* 18:251–253. This short commentary discusses new results on the regulation of mammalian ribonucleotide reductase.

Mathews, C. K. (1993) Enzyme organization in DNA precursor biosynthesis. *Prog. Nucleic Acid Res. Mol. Biol.* 44:167–203. This review summarizes evidence that dNTP biosynthetic enzymes are linked in multienzyme complexes, which may in turn be linked to DNA replication sites.

Mathews, C. K. (2006) DNA precursor metabolism and genomic stability. *FASEB J.* 20:1300–1314. This review describes the genetic consequences of deoxyribonucleotide pool imbalances.

Meyer, Y., B. B. Buchanan, F. Vignols, and J.-P. Reichheld (2009) Thioredoxins and glutaredoxins: Unifying elements in redox biology. *Annu. Rev. Genet.* 43:335–367.

Nordlund, P., and P. Reichard (2006) Ribonucleotide reductases. *Annu. Rev. Biochem.* 75:681–706. A comprehensive recent review, which focuses upon the structure and evolutionary significance of the existence of widely divergent classes of this important enzyme.

Nucleotide Analogs and Chemotherapy

Christopherson, R. I., S. D. Lyons, and P. K. Wilson (2002) Inhibitors of de Novo nucleotide biosynthesis as drugs. *Acc. Chem. Res.* 35:961–971.

Culver, K. W., Z. Ram, S. Wallbridge, H. Ishii, E. H. Oldfield, and R. M. Blaese (1992) In vivo gene transfer with retroviral vector—Producer cells for treatment of experimental brain tumors. *Science* 256:1550–1552. An exciting way to use herpes viral deoxypyrimidine kinase as a selective agent for tumor cell killing.

Gangjee, A., H. D. Jain, and S. Kurup (2007) Recent advances in classical and non-classical antifolates as antitumor and antiopportunistic infection agents. *Anticancer Agents Med. Chem.* 7:524–542; *Anticancer Agents Med. Chem.* 8:205–231, 2008. A recent two-part review.

Hardy, L. W., J. S. Finer-Moore, W. R. Montfort, M. O. Jones, D. V. Santi, and R. M. Stroud (1987) Atomic structure of thymidylate synthase: Target for rational drug design. *Science* 235:448–455. Describes determination of the crystal structure of this enzyme and its implications.

Krynetski, E., and W. E. Evans (2003) Drug methylation in cancer therapy: Lessons from the TPMT polymorphism. *Oncogene* 22:7403–7413. This paper discusses the important role of pharmacogenetics in optimizing chemotherapy.

Lee, H., J. Hanes, and K. A. Johnson (2003) Toxicity of nucleoside analogues used to treat AIDS and the selectivity of the mitochondrial DNA polymerase. *Biochemistry* 42:14711–14719.

Mitsuya, H. (ed.) (1997) *Anti-HIV Nucleosides: Past, Present, and Future.* R. G. Landes, Georgetown, Tex. This short book contains five articles by leading contributors to HIV-drug development.

PROBLEMS

1. Identify each reaction catalyzed by (a) a nucleotidase; (b) a phosphorylase; (c) a phosphoribosyltransferase.

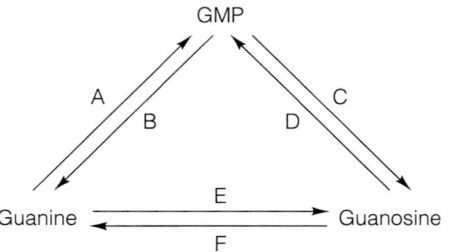

2. Describe the metabolic fate of 5-bromodeoxyuridine. Would you expect BrdUrd to *inhibit* DNA replication? Briefly explain your answer.

3. Predict the effects of the following compounds on intracellular nucleoside triphosphate levels. For each answer, plot percent initial nucleotide level as a function of time after administration of the agent, for each of the four nucleotides (semiquantitatively). Consider not only the primary effect of the compound but also any indirect effects on allosteric enzymes caused by nucleotide accumulation or depletion.
   (a) Effect of thymidine on dNTP levels
   (b) Effect of trimethoprim on bacterial rNTP levels
   (c) Effect of fluorodeoxyuridine on dNTP levels
   (d) Effect of hydroxyurea on dNTP levels
   (e) Effect of azaserine on rNTP levels

4. Radioactive uracil can be used to label all of the pyrimidine residues in DNA. Using either names or structures, present pathways for conversion of uracil to dTTP and to dCTP. For each reaction, show the involvement of cofactors, and identify sites of allosteric regulation.

5. Similarly, hypoxanthine (HX) can be used to label purine residues. As in Problem 4, write reactions showing the conversion of hypoxanthine to dATP and dGTP.

6. Leukemia is a neoplastic (cancerous) proliferation of white blood cells. Clinicians are currently testing deoxycoformycin, an adenosine deaminase inhibitor, as a possible antileukemic agent. Why might one expect this therapy to be effective?

*7. Under what conditions might one expect a deficiency of hypoxanthine–guanine phosphoribosyltransferase to affect the rate of *pyrimidine* nucleotide biosynthesis? How might one estimate the rate of pyrimidine nucleotide biosynthesis in living animals or people?

*8. A classic way to isolate thymidylate synthase–negative mutants of bacteria is to treat a growing culture with thymidine and trimethoprim. Most of the cells are killed, and the survivors are greatly enriched in thymidylate synthase–negative mutants.
(a) What phenotype would allow you to identify these mutants?
(b) What is the biochemical rationale for the selection? (That is, why are the mutants not killed under these conditions?)
(c) How would the procedure need to be modified to select mammalian cell mutants defective in thymidylate synthase?

9. As stated in the text, mammalian cells can become resistant to the lethal action of methotrexate by the selective survival of cells containing increases in dihydrofolate reductase gene copy number so that intracellular levels of the enzyme become very high. What other biochemical or genetic changes in cells could cause them to become resistant to methotrexate?

10. As stated in the text, bacteriophages have been discovered with the following base substitutions in their DNA:
(a) dUMP completely substituting for dTMP
(b) 5-hydroxymethyl-dUMP completely substituting for dTMP
(c) 5-methyl-dCMP completely substituting for dCMP

For any one of these cases, formulate a set of virus-coded enzyme activities that could lead to the observed substitution. Write a balanced equation for each reaction you propose.

*11. Radioisotope "suicide techniques" are often used in the selection of mutants. In one such technique, cells are grown in the presence of [³H]thymidine at very high specific activity. The cells are then stored frozen to allow for decay of some of the incorporated radioactivity. Decay of dTMP residues in DNA causes strand breakage and other potentially lethal events. Thus, cells that have incorporated thymidine into their DNA are very likely to be killed by this regimen.
(a) Identify an enzyme deficiency that would allow a mutant to survive such a regimen.
(b) Choose any enzyme of nucleotide metabolism, and devise a suicide procedure that could select for mutants deficient in that enzyme.

*12. For any of the multifunctional enzymes in purine or pyrimidine nucleotide biosynthesis, assume that the enzyme is available in purified form, and propose one or two experiments to determine whether the enzyme "channels" substrates through a multistep reaction sequence.

13. Write balanced equations for the three known reactions that transfer an amino group to a substrate by condensation with aspartate to give an intermediate that then undergoes α,β-elimination to give the product plus fumarate.

*14. CTP synthetase catalyzes the glutamine-dependent conversion of UTP to CTP. The enzyme is allosterically inhibited by the product, CTP. Mammalian cells defective in this allosteric inhibition are found to have a complex phenotype: They require thymidine in the growth medium, they have unbalanced nucleotide pools, and they have a mutator phenotype. Explain the basis for these observations.

15. If thymine nucleotides are degraded by the same enzymes as those catabolizing uridine nucleotides, give the structure of the thymine metabolite that corresponds to β-ureidopropionate.

16. (a) Explain the biochemical basis for the fact that one can synchronize cell populations by treating them with deoxythymidine.
(b) Explain the apparent paradox that dATP at low concentrations is an activator of ribonucleotide reductase, whereas at higher concentrations it becomes inhibitory.

17. The text states that ATP is synthesized primarily by energy metabolism, whereas other nucleoside triphosphates are formed from the action of nucleoside diphosphate kinase. What additional pathway exists for GTP synthesis?

18. A growing culture of yeast was fed with glycine labeled at the 2C position with ¹³C ([2-¹³C]glycine). After 24 hours, cells were harvested, and purine bases were extracted and analyzed by ¹³C-NMR. The following spectrum (top) was obtained (the bottom spectrum was obtained from cells grown with unlabeled glycine). Explain the metabolic pathways by which the 2-carbon of glycine is incorporated into the purine ring at the positions indicated by the resonances in the NMR spectrum.

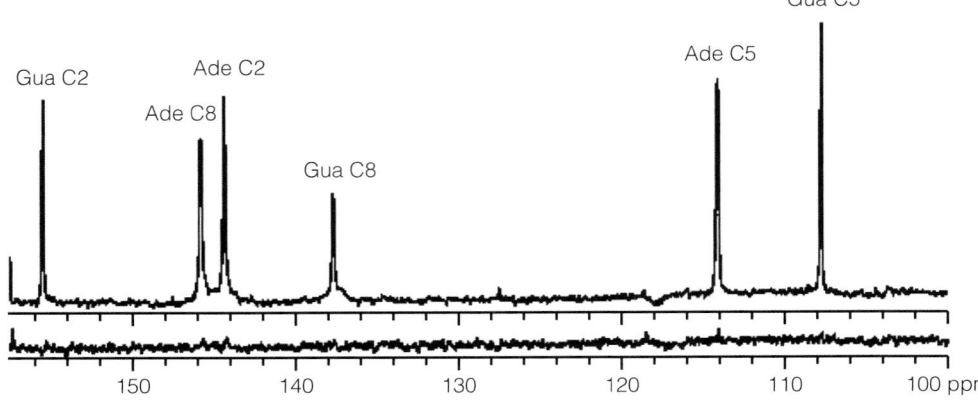

Reprinted with permission from *Biochemistry* 33:74–82, L. B. Pasternack, D. A. Laude Jr., and D. R. Appling, Carbon-13 NMR analysis of inter-compartmental flow of one-carbon units into choline and purines in *Saccharomyces cerevisiae.* © 1994 American Chemical Society.

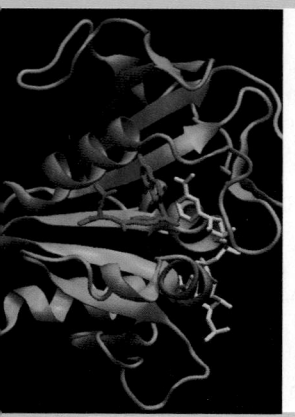

CHAPTER 23

Mechanisms of Signal Transduction

In previous chapters—particularly Chapters 12, 13, and 18—we have discussed actions of hormones in regulating metabolism. In this chapter we focus upon mechanisms of hormone action, and we see again that hormones represent but one class of extracellular chemical messengers. Hormones are molecules that are released from controller cells in the endocrine system and travel to target cells, usually some distance away, where they interact with specific receptors in the target cell, in the process known as signal transduction. In a broader sense, signal transduction refers to reception of an environmental stimulus by a cell, leading to metabolic change that adapts the cell to that stimulus. Interaction of a hormone with its receptor, by various mechanisms, stimulates or inhibits specific metabolic events within the target cell. The examples that we have treated in most detail so far involve the effects of epinephrine and glucagon upon metabolic cascades controlling the breakdown or synthesis of glycogen in animal tissues. In fact, these are the processes that gave our earliest understanding of molecular mechanisms in signal transduction.

Insulin, glucagon, and epinephrine are all *hormones* (from Greek, "to stir up or excite"). This term was coined in 1904 to describe **secretin**, a substance that is released in the upper small intestine and that acts in the stomach to stimulate the flow of gastric juice as an aid to digestion. Early research on hormones in animals revealed little about how they act, but it did show fundamental similarities among different hormones. First, they are secreted by specific tissues, the **endocrine glands**. Second, they are secreted directly into the bloodstream, rather than being excreted through ducts or stored in bladders. Thus, the response to a hormonal signal comes as a direct and rapid result of its secretion. Figure 23.1 shows locations of the major endocrine organs in the human body.

Hormones usually stimulate metabolic activities in tissues remote from the secretory organ. They are active at exceedingly low concentrations, in the micromolar to picomolar ranges. Furthermore, many hormones are metabolized

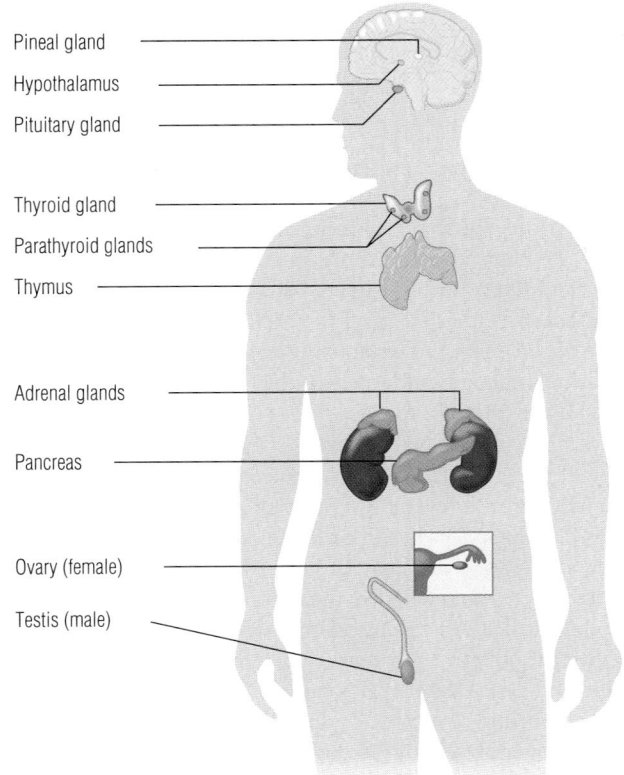

Pineal gland

Hypothalamus

Pituitary gland

Thyroid gland

Parathyroid glands

Thymus

Adrenal glands

Pancreas

Ovary (female)

Testis (male)

FIGURE 23.1

The major human endocrine glands and their central nervous system control centers. Some other tissues also produce hormones, such as the lining of parts of the gastrointestinal tract.

rapidly, so their effects are often short-lived, allowing rapid adaptations to metabolic changes. The low concentrations of hormones and their metabolic lability initially made it difficult to assay levels of any particular hormone. Until the 1960s it was usually necessary to use a bioassay. For example, **oxytocin**, which stimulates uterine contractions in labor, was assayed by adding hormone to strips of uterine muscle and measuring the length of the strips before and after hormone addition. The introduction of **radioimmunoassay** revolutionized the field of hormone analysis. In this technique an antibody to a hormone is mixed with a sample being studied, in the presence of radioactive hormone, and the amount of radioactivity precipitated by the antibody is related to the amount of hormone in the sample.

A special class of hormones includes the eicosanoids—prostaglandins, thromboxanes, and leukotrienes—which we discussed in Chapter 19. These mediators, also called paracrine hormones, act like hormones but are distinctive in their extreme metabolic lability, their synthesis in many cell types instead of just one endocrine gland, their lower target-organ specificity, and their actions primarily on cells close to those from which they were secreted.

Hormones differ in certain ways from other intercellular messengers—**pheromones**, which are transmitted between cells of different organisms; **neurotransmitters**, which act immediately across a synaptic junction from their sites of release; and **growth factors**, which differ from hormones in that their growth-stimulating activities are continuous, rather than being short-lived in response to a burst of secretion. Another class of signaling agents, **cytokines**, bind to specific receptors and stimulate cell growth and differentiation in the immune

Signal transduction involves cell-to-cell communication, via neurotransmitters, hormones, growth factors, and pheromones.

response. Distinctions among these classes of regulators are somewhat indefinite. For example, recall that catecholamines function both as neurotransmitters and as hormones, depending upon their sites of synthesis and release.

An Overview of Hormone Action

Until the 1950s we knew little about molecular mechanisms of hormone action. A popular theory was that a hormone stimulates a metabolic pathway by binding directly to the rate-determining enzyme for that pathway and activating it. Our current understanding emerged from research discussed in Chapters 12 and 13, namely, studies of the effect of epinephrine in stimulating glycogen mobilization. These investigations, carried out largely in the laboratories of Earl Sutherland and Edwin Krebs, showed that epinephrine does not enter cells, as it must if a rate-limiting enzyme is to be activated directly. Instead, as shown in Figure 13.31 (page 567), epinephrine binds to a macromolecular receptor at the cell surface and stimulates the formation of cyclic AMP, which acts as a second messenger and in turn stimulates the phosphorylation of target enzymes. The hormone itself is the first messenger. Today we know that all hormones so far investigated act through binding to specific receptors, whether those receptors are located inside the target cell or on the cell surface. An exception is gaseous signaling agents, such as nitric oxide, which interact directly with a target enzyme. The presence of specific receptors on specific cell types determines how hormones, secreted into the bloodstream, affect only certain tissues. For example, the preferential action of glucagon in stimulating glycogenolysis in liver derives from the density of glucagon receptors on the surface of liver cells. Second messengers are often used to transmit the message to the target metabolic pathway, though not all hormone actions involve a second messenger.

Hormone action can influence (1) enzyme activity (via second messengers), (2) synthesis of specific proteins, or (3) membrane permeability to ions or small metabolites.

Chemically, the hormones in vertebrate metabolism include (1) *peptides* or polypeptides, such as insulin or glucagon; (2) *steroids,* including glucocorticoids and the sex hormones; and (3) *amino acid derivatives,* including the catecholamines and thyroxine. Hormonal mechanisms include (1) enzyme activation or inhibition via second messengers, as noted for epinephrine and glucagon; (2) stimulation of the synthesis of particular proteins, through activation of specific genes; and (3) selective increases in the cellular uptake of certain metabolites. Among this last category are some receptors that serve directly as ion channels, with hormone binding causing a conformational change that opens the channel, and other receptors that stimulate uptake, such as the effects of insulin upon glucose uptake and subsequent utilization.

The receptor to which a hormone binds may be located either in the plasma membrane or inside the cell. Most hormones interacting with *intracellular receptors* (also called nuclear receptors) exert their effects at the gene level. The hormone–receptor complex migrates to the nucleus, where it interacts with specific DNA sites and affects rates of transcription of neighboring genes. These hormones include steroids, thyroid hormones, and the hormonal forms of vitamin D. In addition, **retinoids**, derived from retinoic acid (related to vitamin A), exert regulatory effects in embryonic development, through interactions with intracellular receptors (see Chapter 19).

Membrane receptors include (1) proteins that influence second-messenger synthesis, (2) ion channels, and (3) proteins with intrinsic enzyme activity.

We recognize three major classes of *membrane-bound receptors.* First are receptors, like those introduced in Chapter 13, that interact with G proteins and influence the synthesis of second messengers. Second are receptors that are themselves ion channels—comparable to the nicotinic acetylcholine receptor (page 991). Peptide hormones and epinephrine act primarily through these two classes of receptors. A third category, exemplified by the insulin receptor, is a transmembrane protein with a ligand-binding site on the extracellular side and a catalytic domain on the cytosolic side. In the insulin receptor, that catalyst is a protein kinase, which is stimulated by insulin binding to the extracellular domain to phosphorylate tyrosine residues on target proteins.

Mechanisms of hormones that act through membrane-bound receptors by the first and third mechanisms are summarized in Figure 23.2. Note that the end result of most interactions between a hormone and a membrane receptor is activation of one or more protein kinases, whether or not a second messenger is involved. When Edwin Krebs and Edmond Fischer described in the late 1950s the sequence of reversible protein phosphorylations in the epinephrine-induced glycogenolytic cascade, there was no indication of the extent to which protein phosphorylation would turn out to dominate cell signaling mechanisms. At present more than 500 different protein kinases have been shown to exist in human cells, all of them related, as determined by amino acid sequence homologies. The importance of protein phosphorylation was recognized with the 1992 Nobel Prize for Medicine or Physiology to Krebs and Fischer. More recent work is uncovering a host of specific protein *phosphatases*, also subject to control by cell-signaling mechanisms. As noted in Chapter 18, protein acetylation is becoming recognized as a reversible protein modification that might be comparable to protein phosphorylation in its wide usage.

The end result of many signal transduction events is the phosphorylation or dephosphorylation of target proteins.

FIGURE 23.2

Eukaryotic signal transduction systems involving membrane receptors (1–5) and/or second messengers (1–4). Nitric oxide is shown, even though it lacks a membrane receptor because it diffuses into the cell and interacts with second-messenger systems. The end result of each pathway is phosphorylation of one or more proteins, some but not all of which have been identified. Recent evidence shows that some pathways engage in cross-activation, as shown here by the dashed arrow indicating control of PK-G by cyclic AMP. PK-A, cyclic AMP–dependent protein kinase; PK-G, cyclic GMP–dependent protein kinase; PK-C, protein kinase C. Protein kinases are shown in dark orange, including the intrinsic protein tyrosine kinase domain in system 5.

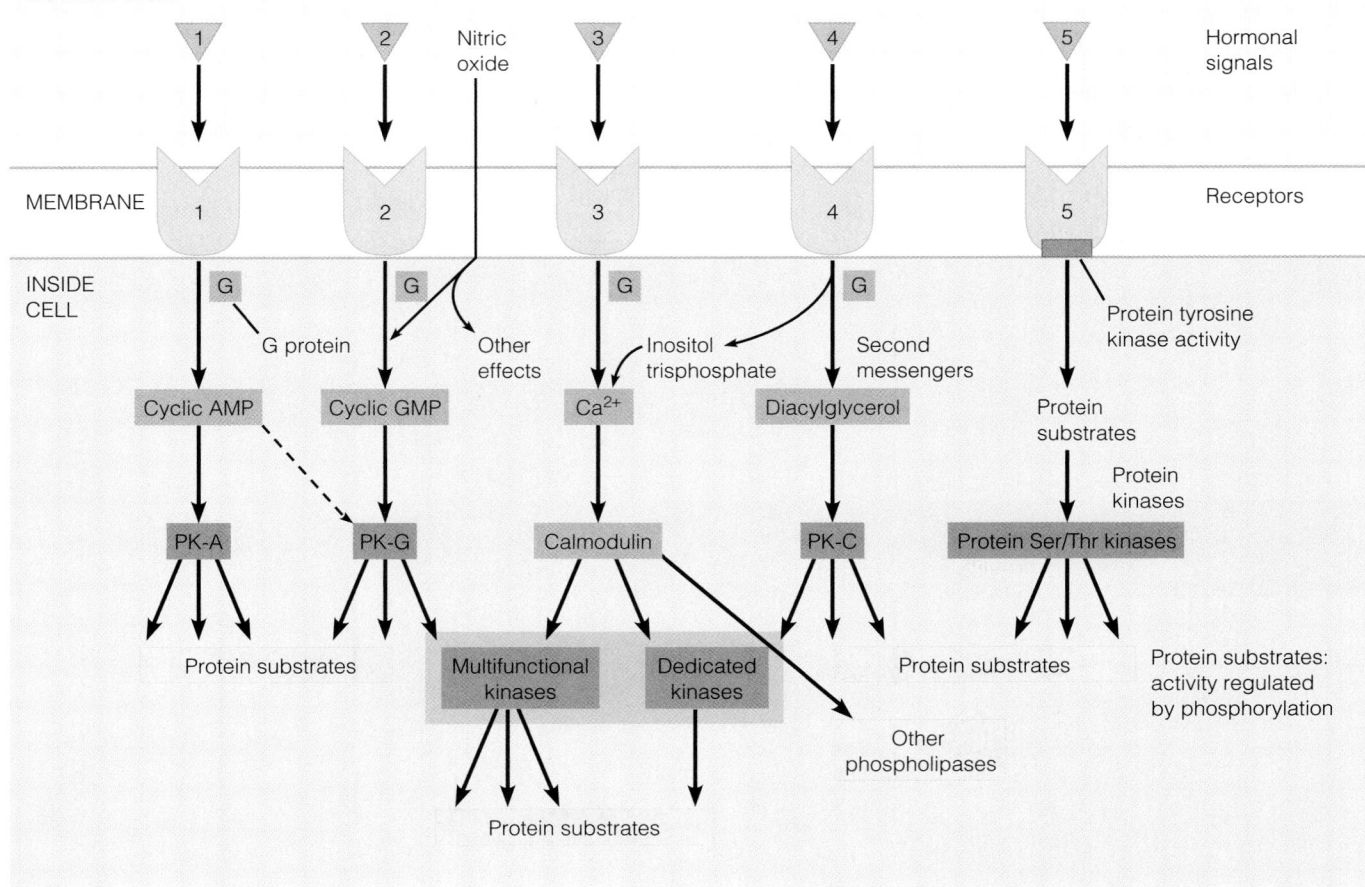

Hierarchical Nature of Hormonal Control

Specific hormone releasing factors from the hypothalamus control the release—and thus the action—of other hormones.

Hormonal regulation involves a hierarchy of cell types acting on each other either to stimulate or to modulate the release and action of a hormone. In vertebrates, the secretion of hormones from endocrine cells is stimulated by chemical signals from regulatory cells that occupy a higher position in this hierarchy (Figure 23.3). Hormonal action is controlled ultimately by the central nervous system. The master coordinator in mammals is the **hypothalamus**, a specialized center of the brain. The hypothalamus receives and processes sensory inputs from the environment via the central nervous system. In response it produces a number of hypothalamic hormones, some of them called **releasing factors**. These factors act on the pituitary, which is located just beneath the hypothalamus. Releasing factors stimulate the anterior portion of the pituitary to release specific hormones. Other hypothalamic hormones inhibit the secretion of particular pituitary hormones. Some pituitary hormones stimulate target tissue directly. For example, **prolactin** stimulates mammary glands to produce milk. However, most pituitary hormones act on endocrine glands that occupy an intermediate, or secondary, position in the

FIGURE 23.3

Hierarchical nature of hormone action in vertebrates. The pituitary represents the first target, being under hypothalamic control. Pituitary hormones then act on secondary targets, principally other endocrine glands, the hormone products of which collectively influence essentially all of the other organs and tissues. Neural stimulation of the adrenal medulla controls release of epinephrine.

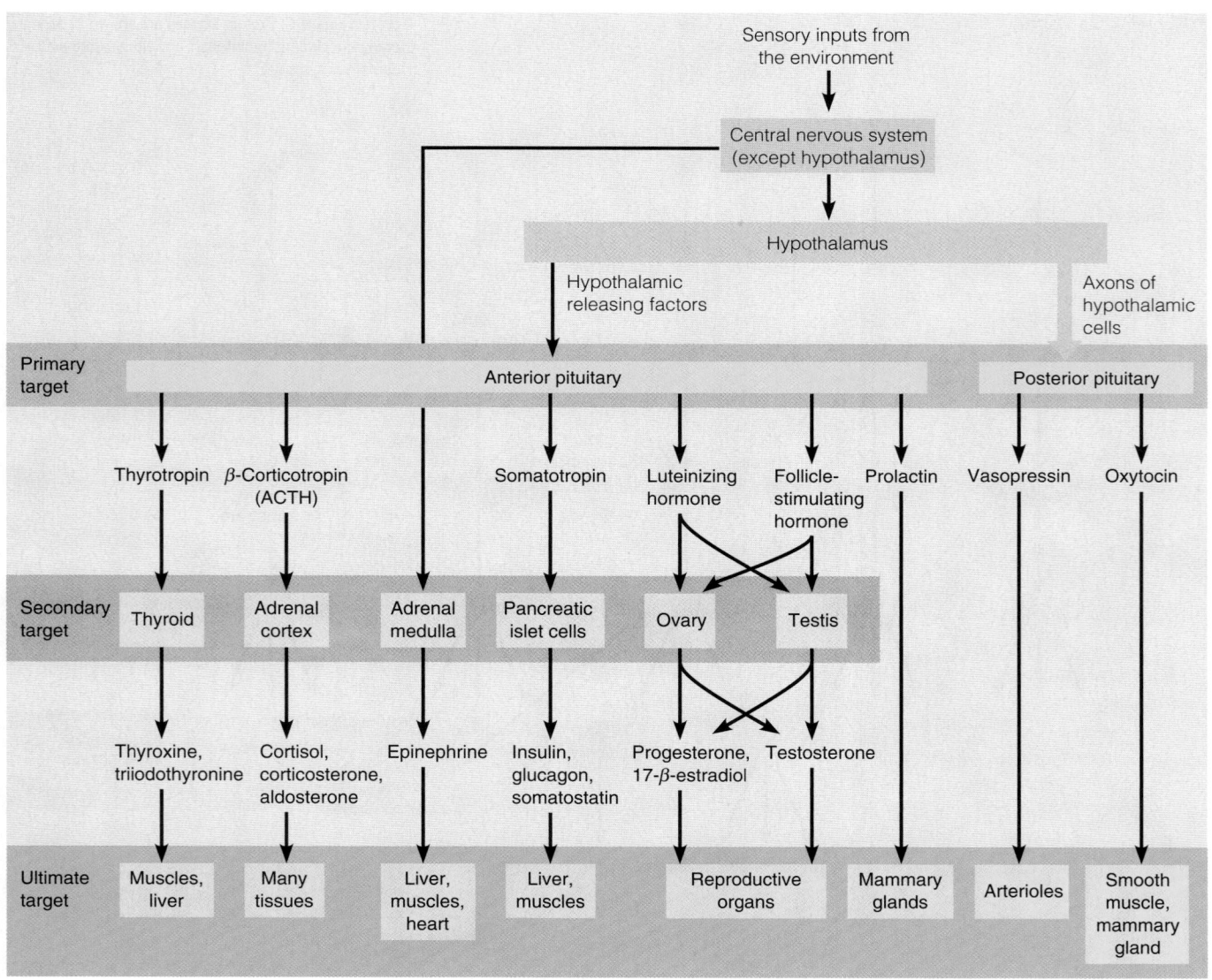

hierarchy, stimulating them to produce hormones that exert the ultimate actions on target tissues. Pituitary hormones that act on other endocrine glands are called **tropic hormones** or **tropins**. An example is **adrenal corticotropic hormone** (**ACTH**), also called **β-corticotropin**. This peptide is secreted from the anterior pituitary, and it stimulates the adrenal cortex to produce glucocorticoids and mineralocorticoids, which in turn act on a number of tissues, including kidney, muscle, and the immune system.

The action of a hormone is self-limiting because of the existence of feedback loops, in which secretion of a hormone sets in motion a series of events that leads to inhibition of that secretion. As shown in Figure 23.4, for example, the secretion of β-corticotropin from the pituitary is stimulated by **corticotropin releasing factor** (**CRF**), a hypothalamic hormone that is a 41-residue polypeptide. Hypothalamic cells contain glucocorticoid receptors, which sense the elevated levels of circulating glucocorticoids, such as cortisol, that result from stimulation of the adrenal cortex. Binding of glucocorticoids to these receptors inhibits the further release of CRF, thus completing a feedback loop.

Synthesis of Peptide Hormone Precursors

We have already presented the biosynthesis of steroid hormones (see Chapter 19) and of catecholamines and thyroid hormones (see Chapter 21), both of which occur via straightforward metabolic pathways. Nearly all peptide hormones are synthesized as inactive precursors and then converted to active hormones by proteolytic processing. Studies of the synthesis of insulin provided the first evidence for this phenomenon. Recall that this hormone contains two polypeptide chains, of 21 and 30 residues, with two interchain disulfide bridges and one intrachain bridge (see Figure 5.21, page 154). These chains are formed by cleavage from an 81-residue polypeptide, called **proinsulin**. The first product of translation of the insulin gene is the 105-residue **preproinsulin**. Cleavage from preproinsulin of a 24-residue N-terminal "signal sequence" gives proinsulin, which undergoes folding, disulfide bond formation, and cleavage to give the active hormone, insulin. The signal sequence is involved in transport of proteins through membranes (see Chapter 28).

All known polypeptide hormones are synthesized in "prepro" form, with a signal sequence and additional sequence(s) that are cleaved out during maturation of the hormone. In some cases a single precursor polypeptide sequence contains two or more distinct hormones. The most complex known example is a pituitary multihormone precursor that contains sequences for β- and γ-lipotropin, α-, β-, and γ-melanocyte-stimulating hormone (MSH), endorphin, and enkephalin, as well as ACTH. This precursor, called **pro-opiomelanocortin**, derives its name from its role as precursor to endogenous *opia*tes, *melano*cyte-stimulating hormone, and *corti*cotropin. A remarkable fact about pro-opiomelanocortin is that its cleavage at different sites in different cells means that different cell types produce distinct ensembles of hormones derived from this one precursor. Cleavage sites are shown in magenta in Figure 23.5. In the anterior pituitary, cleavage generates ACTH and β-lipotropin, and further processing in the central nervous system yields endorphin and enkephalin (page 994), among other products. Synthesis of peptide hormones almost always involves proteolytic cleavage of a protein precursor.

Signal Transduction: Receptors

As noted earlier, our first insights into molecular mechanisms of hormone action came from analysis of glycogenolysis, stimulated either by epinephrine or by glucagon. In Chapter 13 we described the modular nature of this system,

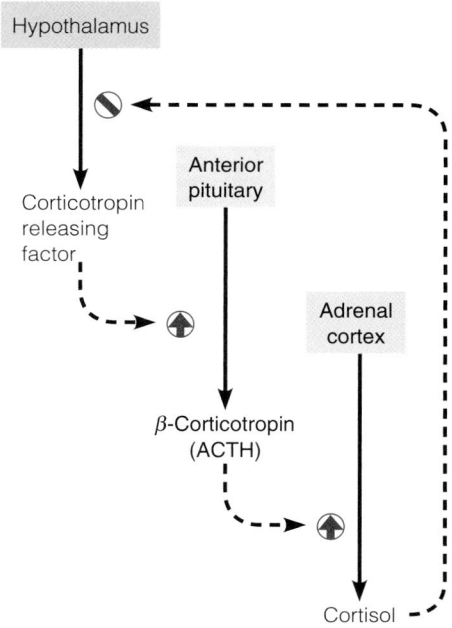

FIGURE 23.4

An example of feedback regulation of a hormone. Corticotropin releasing factor (CRF) stimulates the release of β-corticotropin (ACTH) from the anterior pituitary. ACTH stimulates the adrenal cortex to release cortisol, which feeds back on the hypothalamus to inhibit further release of CRF.

Hormone action is self-limiting because of the existence of feedback loops.

The central nervous system transmits signals to the hypothalamus, which produces releasing factors that act upon endocrine glands to control secretion of hormones with specific metabolic effects upon target tissues.

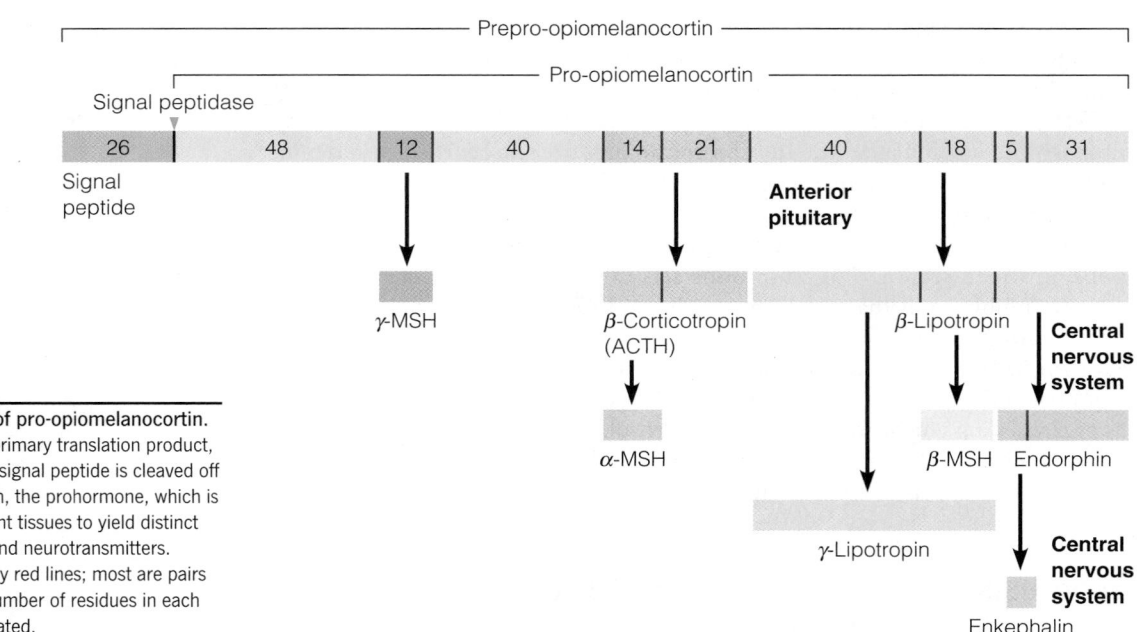

FIGURE 23.5

Structure and properties of pro-opiomelanocortin. The structure shown is the primary translation product, *prepro-opiomelanocortin*. A signal peptide is cleaved off to give pro-opiomelanocortin, the prohormone, which is cleaved differently in different tissues to yield distinct assemblages of hormones and neurotransmitters. Cleavage sites are marked by red lines; most are pairs of basic amino acids. The number of residues in each polypeptide product is indicated.

including three components—a receptor (the hormone receptor), a transducer (a G protein), and an effector (adenylate cyclase). In this section we discuss in more detail the nature and action of each of these components.

With the exception of gaseous signaling agents such as nitric oxide, hormones interact with target cells by initially binding to a macromolecular receptor, located either in the plasma membrane or in the interior of the cell. Because the receptor participates in transduction of the signal from the external messenger to some component of the metabolic machinery, it must have at least one additional functional site. The activity of this site is altered by hormone binding, just as the catalytic site of an allosteric enzyme is altered by the binding of effectors at remote sites.

Experimental Study of Receptors

Binding of hormones to receptors is saturable, comparable to the binding of substrates to enzymes.

Molecular interactions involving hormone receptors can be studied experimentally by methods comparable to those used in enzymology. Binding of a hormone to its receptor is saturable, and the kinetics resembles Michaelis–Menten kinetics. Most hormones bind tightly, with dissociation constants in the range of 0.1μM to 1.0 pM. The ability of a tissue to respond to hormonal stimulation is a function of the receptor density of cells in that tissue.

Binding of radioactive hormone or an analog can be used to identify and quantify receptors, either as an assay when purifying receptors or in determining receptor density in a given cell type. The tight binding between hormones and their receptors can also be exploited in designing purification protocols involving affinity chromatography (Chapter 5). In fact, one of the earliest applications of this technique was purification of the insulin receptor, using columns of immobilized insulin. This development eased a task that is normally difficult for two reasons. First, membrane-bound receptors must be solubilized prior to purification, without irreversible inactivation. Second, most hormone receptors are present in exceedingly small amounts. For example, adipose tissue contains only about 10^4 insulin receptor molecules per cell.

Another useful technique is photoaffinity labeling, which creates a covalent bond between a hormone analog and a receptor. In this technique, a hormone molecule is modified with a photoreactive group, such as an azido moiety

(—N_3). Mixing a radiolabeled hormone analog with a cell extract, followed by UV light irradiation, covalently binds the analog to its receptor, creating a radioactive tag that can be used to isolate the receptor.

Agonists and Antagonists

Often the ligand used in receptor-binding assays or in affinity chromatographic purification is not the hormone itself but an analog that happens to bind to the receptor, sometimes more tightly than the natural hormone. The analog may show little or no structural resemblance to the natural ligand, but it may show a stereochemical relationship that is evident in three-dimensional models. An example is the synthetic hormone analog **diethylstilbestrol**. This compound is 3 times more potent as an estrogen than 17-β-estradiol, which in turn is 10 times more potent than other natural estrogens. Diethylstilbestrol has been used in purifying and characterizing estrogen receptors. The synthetic hormone has had its uses outside the laboratory as well, including its addition to feed to stimulate growth of beef cattle, until it was found to be carcinogenic.

Diethylstilbestrol is an example of a hormone **agonist**—an analog that binds productively to a receptor and mimics the action of the endogenous hormone. An agonist is comparable to an alternative substrate for an enzyme: Its binding to a receptor is productive, in that it evokes a metabolic response comparable to that of binding the hormone. By contrast, a hormone **antagonist** binds to receptors but does not provoke the normal biological response. An antagonist is to a receptor as a competitive inhibitor is to an enzyme, in that both antagonists and competitive inhibitors compete with a normal ligand (hormone or substrate, respectively) for binding to a specific site on a protein and, by so binding, inhibit a normal biological process.

Agonists and antagonists have been useful in studies of the stereochemistry of binding sites on receptors. In turn, these investigations are useful in drug design, with a goal of activating or inactivating certain classes of receptors. For example, the agonist **isoproterenol** is used to treat asthma because it mimics the effects of catecholamines in relaxing bronchial muscles in the lung; it does so by interacting with one specific class of **adrenergic receptors** (so-called because they bind **adrenaline**, the old name for epinephrine). Another important drug, used to control blood pressure and pulse rate in cardiac patients, is **propranolol**, an antagonist of another class of adrenergic receptors, which control blood pressure and heartbeat rate.

Classes of Catecholamine Receptors

Studies of many agonists and antagonists of the catecholamines originally revealed the existence in vertebrates of four types of catecholamine receptors, each of which has a distinctive pattern of response to these analogs (now several more are recognized). The basic four are called the α_1-, α_2-, β_1-, and β_2-adrenergic receptors. Receptors of the β type are those we encountered before, in our discussions of epinephrine-induced lipolysis and glycogenolysis (see Chapters 13 and 17). Adrenergic receptors of different types, in different tissues, have various physiological effects, some of which are summarized in Table 23.1.

Receptors and Adenylate Cyclase as Distinct Components of Signal Transduction Systems

In the early stages of hormone research the epinephrine- or glucagon-stimulated mobilization of glycogen was the only defined biochemical response to hormone–receptor interaction. Adenylate cyclase, which synthesizes cyclic AMP in response to hormone secretion, was found to be a membrane-bound enzyme. Because the receptor is also membrane-bound, it was thought for some time that the receptor *was* adenylate cyclase and that epinephrine binding activated the enzyme. However, two observations argued against that interpretation. First, other hormones

Diethylstilbestrol

17-β-Estradiol

A hormone agonist mimics a hormone in binding productively to a receptor. An antagonist binds nonproductively, inhibiting the action of the natural hormone.

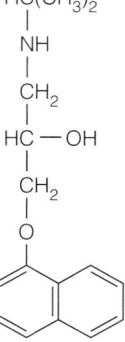

Isoproterenol

Epinephrine   **Propranolol**

TABLE 23.1   Some biological actions associated with adrenergic receptors

Receptor Class	Target Tissue	Hormone or Agonist Response
α_1	Iris of the eye	Contraction
	Intestine	Decreased motility
	Salivary glands	Potassium and water secretion
	Male sex organ	Ejaculation
	Bladder sphincter	Contraction
α_2	Pancreatic β cells	Decreased secretion
	Stomach	Decreased motility
	Adipocytes	Increased lipolysis
β_1	Heart	Increased contractility, rate, and depth
	Adipocytes	Decreased lipolysis
	Kidney	Increased renin secretion
β_2	Heart	Increased contractility, rate, and depth
	Lung	Relaxation
	Liver	Increased glycogenolysis and gluconeogenesis
	Pancreatic β cells	Increased pancreatic secretion
	Skeletal muscle	Increased contractility and glycogenolysis

Adapted from *Goodman and Gilman's Pharmacological Basis of Therapeutics*, 10th ed., J. G. Hardman, L. E. Limbird, and A. G. Goodman, eds., pp. 119–120. © 2001 The McGraw-Hill Companies, Inc.

Hormones that act through second messengers involve a three-protein module—receptor, transducer (G protein), and effector (adenylate cyclase or related enzyme).

were found to activate adenylate cyclase; more than a dozen are now known, including glucagon, ACTH, melanocyte-stimulating hormone, and luteinizing hormone. Adenylate cyclase seemed unlikely to have that many hormone-binding sites. Second, binding of catecholamines to the α_2 class of receptors was found to *inhibit* adenylate cyclase, suggesting that different kinds of proteins interact with adenylate cyclase to produce different metabolic effects.

Both of these observations indicated that the receptor and adenylate cyclase are distinct proteins. In fact, the resolution of β-adrenergic receptors from adenylate cyclase was observed experimentally in 1977, an important development because it showed that this hormonal response system has far more flexibility and versatility than previously thought. A wide variety of hormones could exert a multitude of biological effects through a common mechanism, namely, activation or inhibition of cyclic AMP synthesis. The diversity of signals and responses was built into both the diversity of receptors and the diversity of enzymes in target cells whose activities could be increased or inhibited by cyclic AMP–stimulated phosphorylation. It was soon learned that transduction of the hormonal signal to adenylate cyclase involved a third class of proteins—the G proteins, which we also introduced in Chapter 13. These developments were recognized by the award of the 1994 Nobel Prize to Martin Rodbell, who showed that receptors are distinct from adenylate cyclase, and Alfred Gilman, for the discovery of G proteins.

Because receptors are embedded in the membrane and are present in minute amounts, their isolation in quantities sufficient for structural analysis is a Herculean task. Cloning the receptor genes was crucial in gaining complete amino acid sequence information. Among the many receptor proteins whose amino acid sequences have now been determined—including both α_2- and β_2-adrenergic receptors—there are some remarkable structural similarities. The proteins are of comparable size, with 415 to 480 residues, including seven conserved regions that are rich in hydrophobic amino acids. It seems clear that these represent regions of α helix that are embedded in the membrane and linked by hydrophilic loops, projecting into both the extracellular environment and the cytosol, with the recognition site for the signaling agent on the extracellular side.

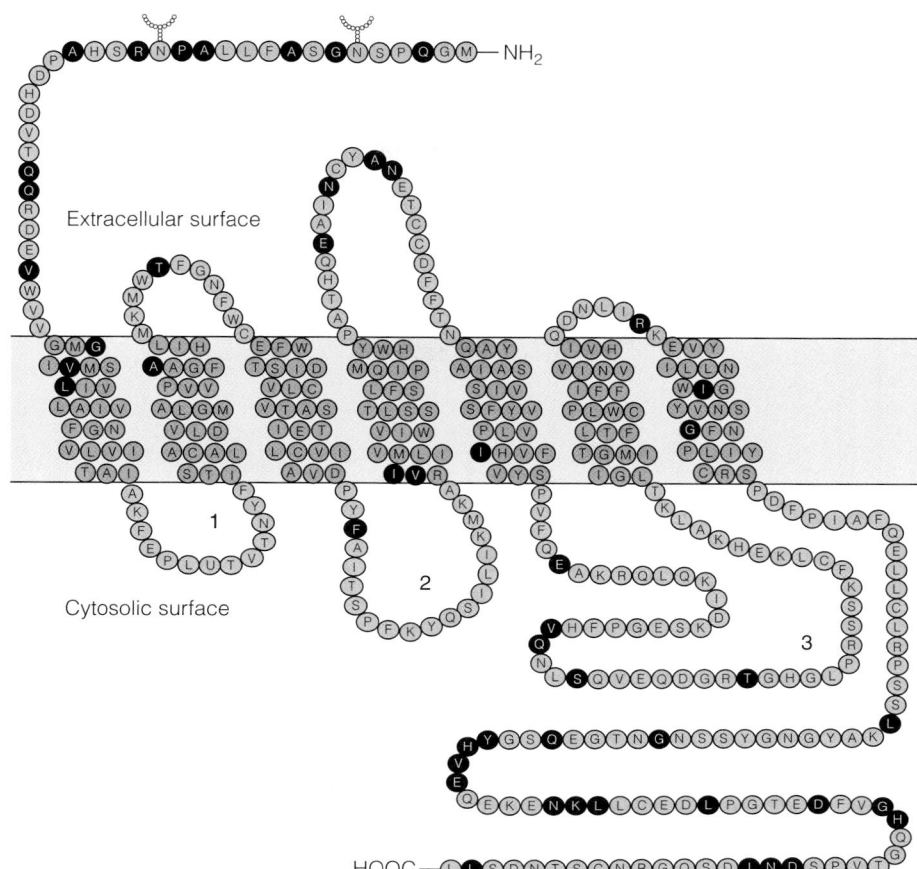

FIGURE 23.6

Amino acid sequence of the human **β₂**-adrenergic receptor. The seven conserved transmembrane domains are shown in orange. Note also the three extracellular and three cytoplasmic loops and the two N-linked oligosaccharide units on the extracellular side (bound to asparagine residues). Interaction of the receptor with G proteins is controlled in part by reversible phosphorylation of serine and threonine residues near the C-terminus. The amino acids colored black are the ones that are different in the hamster β₂-adrenergic receptor sequence.

From *Science* 238:615–616, J. L. Marx, Receptors highlighted at NIH symposium. © 1987. Reprinted with permission from AAAS.

The amino acid sequence of the β_2-adrenergic receptor (β_2AR) is depicted in Figure 23.6, along with the putative membrane-spanning domains. Because this class of receptor molecules snakes back and forth through the membrane, they are called serpentine receptors. Because nearly all of these receptors function in concert with G proteins they are also called G protein–coupled receptors, or GPCRs. It is estimated that half of all prescription, non-antibiotic drugs target members of this receptor class.

Structural analysis of GPCRs was challenging because of their membrane association. The first member of this class to be so characterized, in 2000, was rhodopsin, which is a G protein–coupled receptor, but which participates in light sensing in vision, rather than in hormone action (see page 809). The structure of the β_2-adrenergic receptor was described in 2007. Figure 23.7 shows this molecule in complex with a β-adrenergic antagonist, carazolol. The figure shows a remarkable similarity between the structures of rhodopsin and β_2AR, in placement of the seven transmembrane helices, and in the binding site for ligand (carazolol for β_2AR, 11-*cis*-retinal for rhodopsin). A significant difference is the involvement of a helical region of one of the extracellular domains of β_2AR in binding the ligand. Since 2007, a half dozen additional GPCR structures have been reported, and all follow the same general plan shown here for rhodopsin and β_2AR.

Δ^9-Tetrahydrocannabinol (THC), the active principle in marijuana, has been shown to act *via* binding to two G protein–coupled receptors. Recently, endogenous ligands for these receptors have been identified, and these molecules may play roles in adaptation to pain and inflammation. Activation of the receptors by endogenous ligands has various effects, including stimulation of appetite. Interestingly, these endogenous ligands are derivatives of arachidonic acid, with no evident structural relationship to THC.

Carazolol

Tetrahydrocannabinol

FIGURE 23.7

Structure of the human β_2-adrenergic receptor. **(a)** A model of two receptors embedded in the membrane and joined by cholesterol molecules (in yellow). The ligand carazolol is shown in green. **(b)** Comparison of the top (extracellular) views of rhodopsin (PDB ID 1F88) and the β_2-adrenergic receptor (PDB ID 2RH1), showing similarities in arrangement of the transmembrane helices. Each ligand (11-*cis*-retinal in rhodopsin, carazolol in β_2AR) is shown in red. A helical region of an extracellular domain, which helps to form the binding pocket for epinephrine, is shown in red.

(a) From *Science* 318 (5854), cover illustration. © 2007. Reprinted with permission from AAAS; (b) From *Science* 318:1253–1254, R. Ranganathan, Signaling across the cell membrane. © 2007. Reprinted with permission from AAAS.

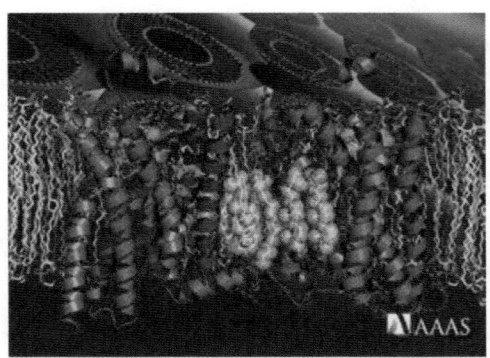
(a)

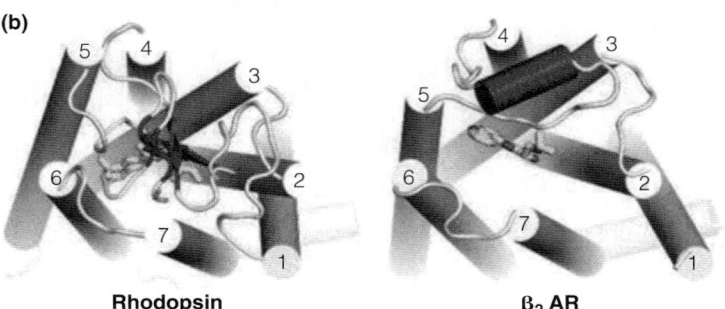
(b)

Rhodopsin β_2 AR

Transducers: G Proteins

We come now to the second component of the receptor–transducer–effector signaling system first described for the β-adrenergic response: the G proteins, so designated because of their ability to bind guanine nucleotides. In 1971, guanosine triphosphate was found to be required for activation of adenylate cyclase by β-adrenergic agonists, and late in the decade the basis for this requirement emerged: GTP-binding membrane proteins interact with receptor systems that activate or inhibit adenylate cyclase. Of the several known G proteins, the two best characterized are G_s, a family of G proteins involved in *stimulation* of adenylate cyclase, and G_i, a closely related family involved in responses that *inhibit* adenylate cyclase. Although both types of G proteins interact with other receptors as well (and with target proteins other than adenylate cyclase), it is useful first to describe their functions in terms of the adrenergic receptors.

Actions of G Proteins

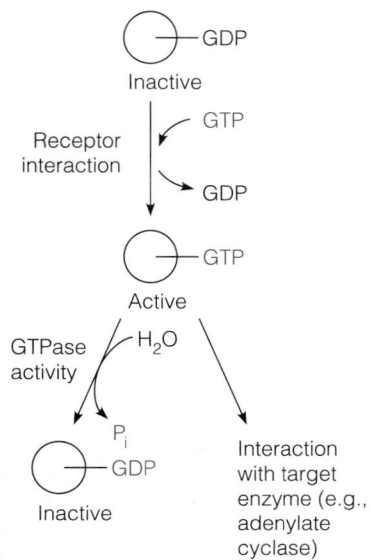

The G proteins are membrane proteins that in the *inactive* state bind guanosine diphosphate, GDP. As we discussed in Chapter 13, a hormone response leading to stimulation of adenylate cyclase—the binding of extracellular hormone or agonist to a receptor, typically a β-adrenergic receptor—causes a conformational change that stimulates the receptor to interact with a nearby molecule of G_s. This in turn stimulates an exchange of bound GDP for GTP—that is, the dissociation of GDP from G_s, to be replaced by GTP. G_s is thereby converted to a protein that activates adenylate cyclase, producing cyclic AMP from ATP. Cyclic AMP synthesis results in activation of cAMP-dependent protein kinase (protein kinase A), with consequent phosphorylation of target proteins, such as phosphorylase b kinase in cells that activate glycogen phosphorolysis.

To summarize, this signal transduction pathway involves (1) hormone binding to receptor; (2) receptor interaction with G_s, stimulating release of GDP and association of GTP with G_s; (3) stimulation of adenylate cyclase by the GTP-bound G_s; (4) stimulation by cAMP of protein phosphorylation; and (5) stimulation or inhibition of metabolic reactions.

The initial exchange reaction (top of Figure 23.8) is usually assisted by one of a class of proteins called **guanine nucleotide exchange factors** (GEFs). Continued activation of G_s depends on the presence of bound GTP. The hormonal response is limited, and hence is controlled, by the presence of a slow GTPase activity inherent to the G protein. Thus, bound GTP is slowly hydrolyzed to GDP, with concomitant loss of ability to stimulate adenylate cyclase. This process, like the initial activation, is protein-assisted, being helped by a **GTPase-activating protein** (GAP). The G_i protein functions similarly, but in response to extracellular signals whose function is the *inhibition* of adenylate cyclase, typically α_2 agonists. Here the binding of GTP provokes an inhibitory interaction of G_i with adenylate cyclase, which decreases the synthesis of cAMP.

Structure of G Proteins

G_s, G_i, and other G proteins have an $\alpha\beta\gamma$ trimeric structure (Figure 23.8): a 39- to 46-kilodalton α subunit, a 37-kilodalton β subunit, and an 8-kilodalton γ subunit. The human genome encodes at least 24 different α proteins, 5 β, and 6 γ, allowing for a great variety of different trimeric G proteins. In most of these the γ subunit is **prenylated**; that is, it contains a covalently bound C_{20} isoprenoid moiety at the C-terminal cysteine, which helps anchor the protein in the membrane and may facilitate protein–protein interactions (see Chapter 19). The α subunit is **myristylated** in two other G proteins, G_i and G_o, and palmitylated in G_s. That is, it contains a myristic acid or palmitic acid moiety in amide linkage with C-terminal glycine. The guanine nucleotide–binding site and its associated GTPase activity are both located on the α subunit. A hormonal stimulus leads to exchange of GDP for GTP and dissociation of the G protein, with the α–GTP complex moving along the membrane until it encounters a molecule of adenylate cyclase or effector molecule. The slow GTPase activity mentioned earlier eventually reconverts α–GTP to α–GDP, and the α–GDP complex dissociates from adenylate cyclase and rejoins the $\beta\gamma$ complex.

Consequences of Blocking GTPase

The importance of the GTPase activity in controlling the hormone response can be seen in the consequences of blocking it. Blocking can be achieved in vitro by substituting GTP with **GTPγS,** a GTP analog in which a sulfur atom substitutes for an oxygen on the γ phosphate of GTP, and which the GTPase activity cannot cleave. In a G_s-dependent system the result is irreversible activation of the target adenylate cyclase.

GTPγS

More dramatic are the effects of bacterial toxins that have G proteins as their biological targets. The toxin of *Vibrio cholerae* is an enzymatic protein with the ability to cleave NAD^+ and transfer its ADP-ribose moiety to a specific site in the α subunit of G_s. This modification of G_s inhibits its GTPase activity and converts α subunit to an irreversible activator of adenylate cyclase.

$$NAD^+ + \alpha_s \longrightarrow \text{nicotinamide} + \text{ADP-ribosyl-}\alpha_s$$

In the intestine the resultant cAMP accumulation promotes a physiological response controlled by cAMP—uncontrollable secretion of water and Na^+—and

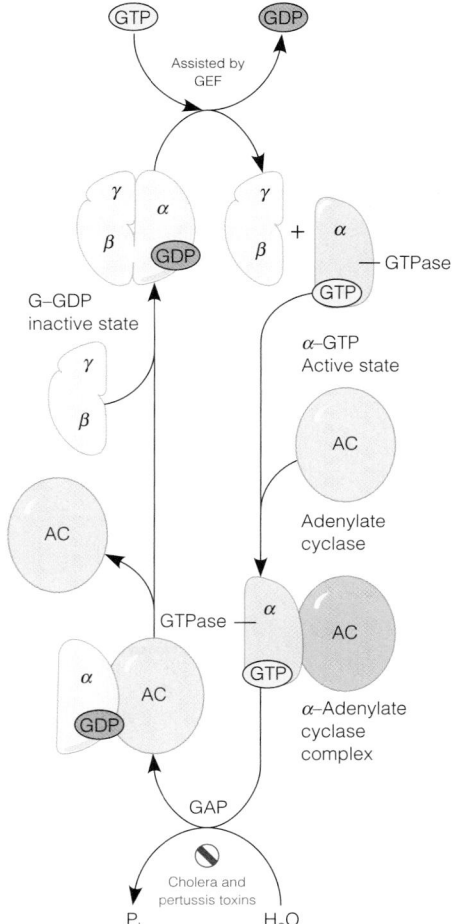

FIGURE 23.8

The cycle of G protein dissociation and reassociation. α, β, and γ are the three subunits of the G protein. The active form is the α–GTP complex (dark teal), while the inactive GDP complexes are shown in light teal. The sites of action of pertussis and cholera toxins are also shown. GEF, guanine nucleotide exchange factor; GAP, GTPase-activating protein.

Activation of G proteins involves GTP displacement of GDP bound to the α subunit and dissociation of the complex from $\beta\gamma$. Hormone action is limited by slow hydrolysis of the bound GTP.

Cyclic GMP

H_2O

5′ GMP

is responsible for the severe diarrhea and consequent dehydration and loss of salt that accompany cholera. A component of the toxin of *Bordetella pertussis,* which causes whooping cough, has a similar effect on the α subunit of the G_i protein, with different physiological effects—lowered blood glucose and hypersensitivity to histamine.

G Proteins in the Visual Process

There are remarkable similarities between the actions of G proteins in transmitting hormonal signals and their actions in the transmission of signals from light. Much of our understanding of G proteins in hormonal signal transduction came from studies of a G protein called **transducin** in the visual process. The extracellular stimulus and the biochemical end point are quite different in vision from those in hormone action, but the transmembrane signaling processes are almost identical. As mentioned in Chapter 19, the extracellular signal in vision is a photon of light, and the membrane receptor is **rhodopsin**, an abundant membrane protein in the outer segment of rod cells in the retina and, as noted earlier, structurally related to adrenergic receptors. A photochemical change in the structure of rhodopsin causes it to activate transducin so that it binds GTP. The transducin–GTP complex activates a specific **phosphodiesterase**, which hydrolyzes a cyclic nucleotide, **guanosine 3,5-monophosphate** (cyclic GMP or cGMP). Cleavage of cGMP, in turn, stimulates intracellular reactions that generate a visual signal to the brain. Thus, the stimulated *hydrolysis* of cGMP is the visual analog of the stimulated *synthesis* of cAMP in β-adrenergic responses.

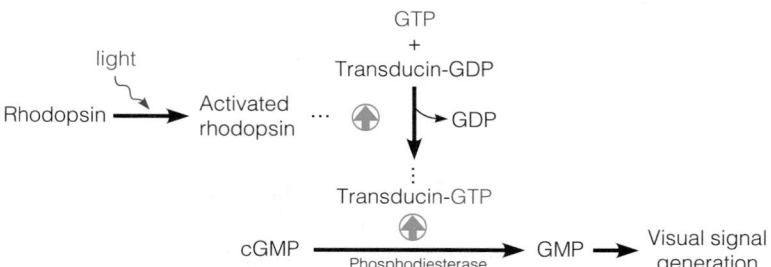

A Closer Look at G Protein Subunits

The G protein mechanism is used in many signal transduction pathways. Reassortment of the several α, β, and γ proteins means that a large number of different G proteins exists, giving great flexibility in response to this signal transduction element. Interaction with target enzymes is a function of the α subunits. Some interact with adenylate cyclase, some with ion channels, and some with phospholipases. One large subfamily of G proteins, called G_{olf}, is present in olfactory cells in the nose and functions with a large number of receptors involved in sensory reception of odors.

The α subunits of G proteins are part of a family of small GTP-binding proteins that are active when GTP is bound and inactive in the presence of GDP. This family includes the oncogene-specified Ras proteins (see page 985) and the GTP-binding elongation factors involved in protein synthesis (Chapter 28). In recent years much has been learned from crystallographic studies about how GTP binding activates this class of proteins and how the activated protein in turn interacts with its target. Figure 23.9a shows the structure of an $\alpha\beta\gamma$ heterotrimeric G protein–GDP complex, with the α subunit derived from a G_i protein. The "switch II" region, shown in red, is a domain that changes conformation when GTP is hydrolyzed. Panel b shows the structure of a $G_s\alpha$ protein superimposed on that of a $G_i\alpha$ subunit; both proteins are binding nonhydrolyzable GTPγS. Points of structural divergence are

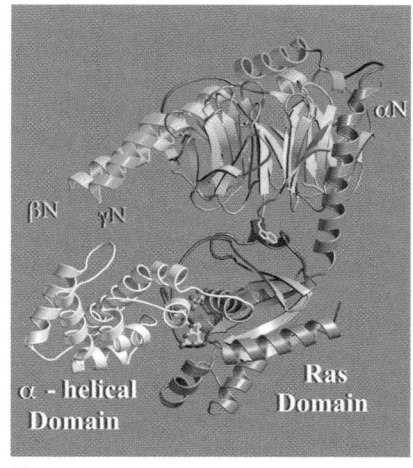

(a)

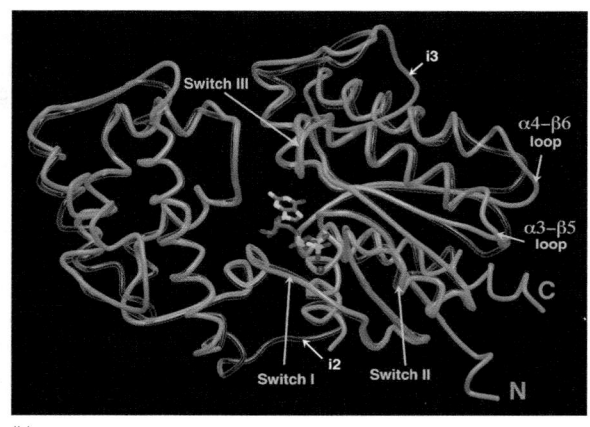

(b)

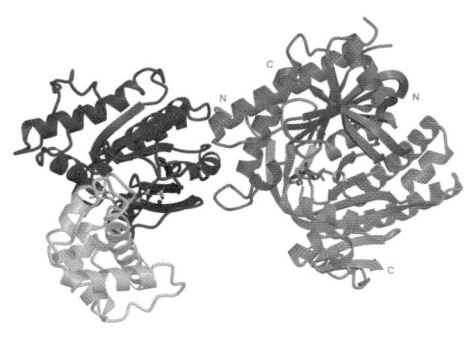

(c)

FIGURE 23.9

Structures of G proteins. **(a)** A heterotrimeric $\alpha\beta\gamma$ complex, with the α subunit derived from a G_i protein. The α subunit is in gray, β is in yellow, and γ is in green. The N-termini of the three proteins are shown; note that the lipid modifications at these sites are not present in these recombinant proteins. Shown in yellow are a GDP bound to the α subunit and the switch II region on α. **(b)** Superposition of an α protein from G_i (transparent rose) and a $G_s\alpha$ protein–GTPγS complex (solid gray). Two insertions in the $G_{\alpha s}$ protein relative to $G_{\alpha i}$ (i2 and i3) are shown with white arrows. The structures are identical in the GTP-binding regions. **(c)** A complex between $G_s\alpha$-GTPγS (to the left) and the catalytic core of adenylate cyclase (to the right). The upper side is believed to face the plasma membrane. The α protein consists of two major domains—a "helical" domain (ash gray) and a "ras-like" domain (charcoal; see panel a). Two domains of the adenylate cyclase (mauve and khaki) are also shown. GTPγS is shown as a red and green stick figure, and the switch II region of the α protein is shown in red.

(a) Reprinted from *Structure* 6:1169–1183, M. A. Wall, B. A. Posner, and S. R. Sprang, Structural basis of activity and subunit recognition in G protein heterotrimers. © 1998, with permission from Elsevier; (b) From *Science* 278:1943–1947, R. K. Sunahara, J. J. G. Tesmer, A. G. Gilman, and S. R. Sprang, Crystal structure of the adenylyl cyclase activator $G_s\alpha$ © 1997. Reprinted with permission from AAAS; (c) From *Science* 278:1907–1916, J. J. G. Tesmer, R. K. Sunahara, A. G. Gilman, and S. R. Sprang, Crystal structure of the catalytic domains of adenylyl cyclase in a complex with $G_s\alpha \cdot$ GTPγS. © 1997. Reprinted with permission from AAAS.

those expected to determine whether interaction of the α protein with a target enzyme, such as adenylate cyclase, activates or inhibits that enzyme. Panel c shows a complex between a $G_s\alpha$ bound to GTPγS and the catalytic core of adenylate cyclase. The switch II region of the α protein is in close contact with the target enzyme and participates in its activation.

Modulating the Hormonal Stimulus

Obviously, for hormonal signaling to be effective, the response must be modulated once conditions that created it have changed. Several processes are involved. First, as hormone secretion slows, hormone–receptor complexes dissociate by mass action. Second, as mentioned above, the intrinsic GTPase activity of G protein α subunit hydrolyzes bound GTP, leading to G protein inactivation. This process is stimulated by GAP protein. Third, the receptor itself undergoes inactivation by a process that begins with the action of **β-adrenergic receptor kinase** (β-ARK), which phosphorylates several C-terminal serines in the receptor molecule. This leads to binding of another protein, **β-arrestin**, which recognizes the phosphorylated sites and binds the receptor. This causes the receptor-β-ARK complex to become internalized in an endocytotic vesicle. Within the cell's interior the β-arrestin-β-ARK complex dissociates, and the receptor is dephosphorylated and returned to the cell surface, effectively resensitizing the receptor to subsequent hormone release. β-Arrestin also binds to other classes of receptors, leading to their regulation by similar mechanisms.

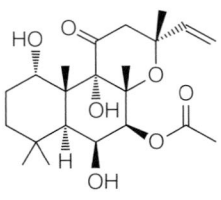

Forskolin

Effectors: Adenylate Cyclase

Although G proteins have several targets involved in signal transduction, we focus here upon adenylate cyclase (AC) because of its involvement in adrenergic signaling, which, as we have observed, represents a paradigm for signal transduction mechanisms. As noted earlier, AC catalyzes the conversion of ATP to cyclic AMP plus pyrophosphate. Mammalian cells contain 10 AC isoforms that are regulated by heterotrimeric G proteins, each consisting of two transmembrane domains, M1 and M2, and two homologous cytoplasmic domains, C1 and C2. Figure 23.10 shows the structure of the cytoplasmic domains crystallized in the presence of the α subunit of G_s and **forskolin**, a diterpene from an Indian plant that activates all but one of these adenylate cyclases. The action of forskolin, as shown partly by this structure, is to draw the two cytoplasmic domains together in a catalytically active conformation. Other studies indicate that the binding site for α_I is on the opposite side of the pseudo-symmetric catalytic domain, where it appears to prevent association of the catalytic domains (not shown). Both regulatory subunits bind to sites remote from the active site of the enzyme, indicating that complex allosteric processes must be involved in regulating AC activity.

The adenylate cyclase reaction involves activation of the 3' hydroxyl of ATP, for nucleophilic attack on the α (inner) phosphorus to create a phosphodiester bond, with pyrophosphate as a leaving group. The reaction is quite similar to that catalyzed by DNA polymerases, except that the AC reaction is intramolecular. Not shown in the figure is the fact that the constellation of amino acid residues in the active site, including strategically placed Asp residues, plus the requirement for two metal ions, supports a mechanism for adenylate cyclase that is quite similar to the two-metal mechanism now widely accepted for DNA polymerases (Chapter 25, page 1046).

Second-Messenger Systems

Cyclic AMP

As introduced in Chapter 13 and elaborated here, many signal transduction events involve the linked actions of receptor, G protein, and adenylate cyclase. These events either stimulate or inhibit the synthesis of a second messenger, cyclic AMP, inside the cell. Many intracellular processes are controlled in turn by the level of that second messenger. One such process, not yet mentioned, is the synthesis of receptor proteins themselves. Cyclic AMP binds to and activates cAMP-dependent protein kinase (PKA; see Chapter 13), which in turn phosphorylates a protein called CREB (**cAMP response element binding protein**), and the resultant phosphorylated protein controls transcription of genes, including those

FIGURE 23.10

Crystal structure of an adenylate cyclase catalytic domain. **(a)** The C1a and C2a catalytic domains (tan and green) were crystallized as a complex with forskolin (yellow) and α_s, the α subunit of G_s (blue-green lower left). The catalytic site where ATP is bound consists of residues from both domains. GTP bound to α_s is shown as well (gray). **(b)** Schematic diagram showing relationships of the catalytic domains to the transmembrane helical regions.

From *Science* 278:1907–1916, J. J. G. Tesmer, R. K. Sunahara, A. G. Gilman, and S. R. Sprang, Crystal structure of the catalytic domains of adenylyl cyclase in a complex with $G_s\alpha\cdot$GTPγS. © 1997. Reprinted with permission from AAAS. Adapted with permission from John J. G. Tesmer, University of Michigan.

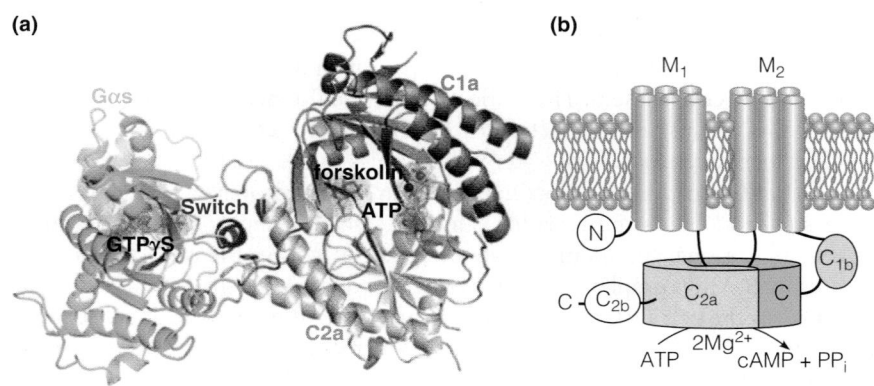

encoding particular receptors. Some of these actions represent adaptation of a cell to action of a hormone. In addition, cyclic AMP acts as a second messenger in the actions of many signaling agents other than epinephrine and glucagon, including dopamine, β-corticotropin (ACTH), histamine, serotonin, and prostaglandins.

With this diversity of signaling agents acting through cAMP, how does one explain the specificity of hormone action? Part of the answer lies in distribution of hormone receptors among tissues. Glucagon receptors, for example, are located in liver and adipose tissue, explaining the preferential effects of glucagon upon these tissues. Also, the distribution of stimulatory and inhibitory G proteins within different cells determines whether binding of a hormone to that cell will increase or decrease the intracellular cAMP concentration. More recently discovered is a class of proteins, called AKAPs (**A kinase anchoring proteins**), which are bound to specific sites within a cell and are controlled by localized pools of cAMP, thereby accounting for differential effects of cAMP within the same cell. Among the thirty or so human AKAPs are forms associated with microtubules, ion channels, or mitochondria and which bind protein kinase A at those specific sites, and hence, which localize the effects of cAMP within a single cell.

Cyclic GMP and Nitric Oxide

Cyclic AMP was the first second messenger known, but is far from the only one. We have briefly mentioned cyclic GMP (see page 970) and the phosphoinositide system (Chapter 19; see page 788). Much interest has focused on cyclic GMP, particularly with respect to its role in nitric oxide metabolism. Nitric oxide (NO·) is a gaseous signaling molecule, synthesized from arginine (Chapter 21), which plays important regulatory roles. Nitric oxide was originally identified as an agent in vasodilation of endothelial vascular cells and underlying smooth muscle. Signals that decrease blood pressure and inhibit platelet aggregation use nitric oxide as an intermediary. In inflammatory and immune responses an inducible form of nitric oxide synthase produces NO· at levels sufficient to be toxic to pathogenic organisms. NO· also regulates neurotransmission in the central nervous system.

The NO· synthase in endothelial vascular cells is acutely sensitive to calcium ion concentration; activation of the enzyme by Ca^{2+} causes NO· accumulation. Because NO· is a gas, it can diffuse rapidly into neighboring cells, where it exerts its control by binding to ferrous ion in a soluble form of guanylate cyclase and stimulating cyclic GMP formation. Guanylate cyclase is sensitive to inhibition by ATP at physiological concentrations, suggesting a regulatory link between nitric oxide signaling and the energetic state of the cell. Guanylate cyclase is not the only regulatory target for nitric oxide (it reacts readily with Fe^{2+}, a component of cytochrome oxidase), but cGMP elevation is probably the major cellular effect of NO· release.

Because NO· is unstable, its effects are short-lived. Because of its function in stimulating vasodilation, NO· plays a role in stimulating erection of the penis. The drug Viagra counteracts erectile dysfunction by inhibiting cyclic GMP phosphodiesterase, and hence, increasing the metabolic half-life of cyclic GMP.

Many cells contain a cGMP-stimulated protein kinase that, unlike the cAMP-activated enzyme, contains both catalytic and regulatory domains on one polypeptide chain of a homodimeric protein. Our understanding of the roles of cGMP in signal transduction has come more recently because its intracellular concentrations are 10- to 100-fold lower than those of cAMP.

Although the discovery of nitric oxide as a gaseous signaling molecule was unexpected, NO· is not the only gaseous signaling agent. It turns out that both hydrogen sulfide and carbon monoxide, released in small, subtoxic doses, also have anti-inflammatory and vasodilatory actions similar to those of nitric oxide, although different mechanisms are involved. Hydrogen sulfide, produced by desulfhydration of cysteine, activates an ATP-sensitive potassium channel in smooth muscle cells. Less is known about mechanisms of action of carbon monoxide.

Sildenafil (Viagra)

Calcium Ion

Calcium ion has also been considered a second messenger. Many cells respond to extracellular stimuli by altering their intracellular calcium concentration, which in turn exerts biochemical changes either by itself or through its interaction with calmodulin (see Chapter 13). Calcium levels themselves are controlled in large part by second messengers, including cAMP. In many nerve and muscle cells, the activation of adenylate cyclase results in an influx of extracellular calcium. cAMP activates a voltage-dependent calcium channel in the presynaptic nerve membrane, allowing calcium ions to flow into the cell and triggering synaptic transmission (see Chapter 10). That activation may involve phosphorylation, by cAMP-dependent protein kinase, of a protein component of the channel. In muscle cells, calcium influx triggers muscle contraction (see Chapter 8) and is responsible, for example, for the increased rate and force of heartbeats caused by β-adrenergic agonists. Because the second messenger cAMP regulates calcium influx, some have suggested calling calcium a third messenger rather than a second messenger.

Phosphoinositides

Cytosolic calcium levels can be increased also by release from *intracellular* calcium stores. Access to these intracellular stores is controlled by another set of messengers, the **phosphoinositide system**. Although similar in many respects to the adenylate cyclase system, the phosphoinositide system is distinctive in that the hormonal stimulus activates a reaction that generates *two* second messengers. The earliest experimental observations regarding this signal transduction system occurred in 1953, when Mabel and Lowell Hokin noted that administration of acetylcholine to pancreatic secretory cells led to rapid synthesis and turnover of the phosphatidylinositol fraction of membrane phospholipids. Similar observations were made in other systems stimulated by hormones, neurotransmitters, or growth factors. However, more than two decades elapsed before a unifying concept emerged to explain these observations.

We now know that a specific lipid in the phosphoinositide family, namely, **phosphatidylinositol 4,5-bisphosphate** (**PIP$_2$**), is a membrane-associated storage form for two second messengers. As shown in Figure 23.11, binding of an agonist to a receptor (step 1) stimulates a G protein to bind GTP (step 2), just as occurs during the adrenergic response. However, this G protein activates not adenylate cyclase but a different membrane-bound enzyme, **phospholipase C,** which in turn cleaves PIP$_2$ to yield two products (step 3)—**sn-1,2-diacylglycerol** (**DAG**) and **inositol 1,4,5-trisphosphate** (**InsP$_3$**). Both of these products act as second messengers. Therefore, the cleavage of PIP$_2$ by phospholipase C is the functional equivalent of the synthesis of cyclic AMP by adenylate cyclase.

The second-messenger role of inositol trisphosphate is to bind to and open calcium channels in the endoplasmic reticulum (ER), thereby releasing calcium from its intracellular stores in the ER (step 4 in Figure 23.11). This release has various effects on intracellular metabolism, as noted earlier, but it also contributes to the second-messenger role of diacylglycerol, which is the activation of membrane-bound protein kinase C (step 5). This enzyme requires for its activity *calcium* (hence the "C" designation) and a *phospholipid* (specifically, phosphatidylserine). The other second messenger, diacylglycerol, stimulates protein kinase C activity by greatly increasing the affinity of the enzyme for calcium ions. This requirement is specific for the *sn*-1,2-DAG; neither the 1,3- nor the 2,3-isomer is active. The enzyme phosphorylates specific serine and threonine residues in target proteins (step 6). As with cAMP-stimulated protein kinase, the specific cellular responses to protein kinase C activation, such as the phosphorylation of calmodulin shown in Figure 23.11, depend on the ensemble of target proteins that become phosphorylated in a given cell. Other known target proteins include the insulin receptor,

**Phosphatidylinositol
4,5-bisphosphate (PIP₂)**

**sn-1,2-Diacylglycerol
(DAG)**

**Inositol 1,4,5-trisphosphate
(InsP₃)**

β-adrenergic receptor, glucose transporter, HMG-CoA reductase, cytochrome P450, and tyrosine hydroxylase.

Now let us briefly consider the metabolism of inositol trisphosphate $(InsP_3)$ after its release from PIP_2. Three sequential hydrolytic steps yield inositol, which is then reincorporated into phosphatidylinositol, as discussed in Chapter 19, to

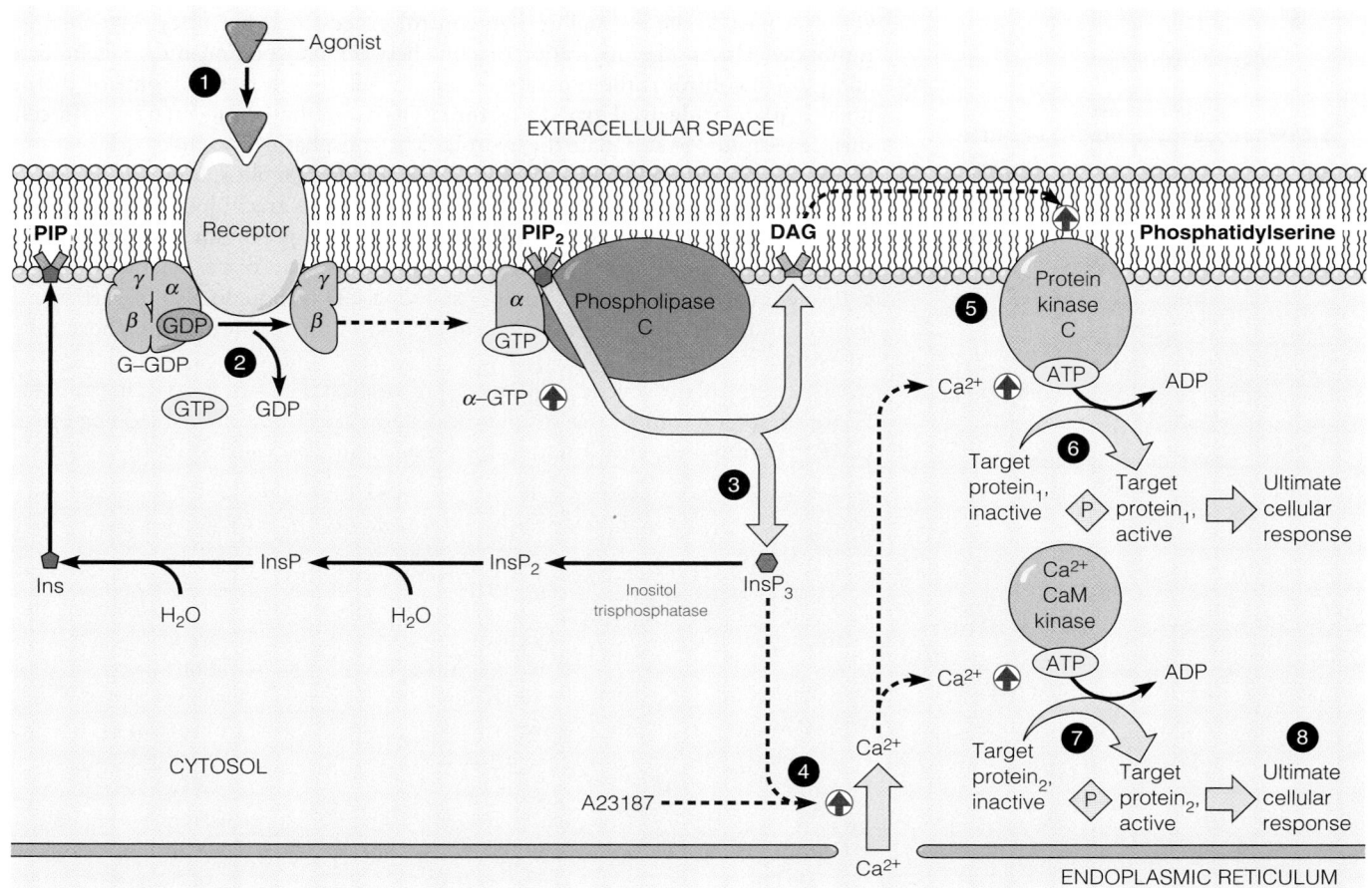

FIGURE 23.11

Signal transduction pathways involving phosphoinositide turnover. DAG, sn-1,2-diacylglycerol; Ins, inositol; InsP, inositol monophosphate; PIP, phosphatidylinositol-4-phosphate; PIP₂, phosphatidylinositol 4,5-bisphosphate; InsP₃, inositol 1,4,5-trisphosphate; InsP₂, inositol 1,4-bisphosphate. Most of the effects of calcium result from its binding to calmodulin (CaM). A23187 is a calcium ionophore, which can be used experimentally to release calcium from intracellular stores. The release of calcium ion stimulates both protein kinase C and calmodulin kinase.

regenerate PIP and PIP$_2$. The last hydrolytic step, the hydrolysis of inositol monophosphate to inositol, is specifically inhibited by **lithium ion**. This block inhibits the resynthesis of InsP$_3$ by depleting the cell of inositol.

$$\text{Inositol monophosphate} + H_2O \rightarrow \text{inositol} + P_i$$

Given that the phosphoinositide messenger system is widely used in nervous tissue, this action of lithium may be related to its efficacy in the treatment of bipolar syndrome, formerly called manic–depressive disorder.

Because many metabolic processes are controlled by calcium fluxes and by phosphorylation of specific proteins, the phosphoinositide system has great versatility as a control mechanism. The fact that a cell can use either DAG or InsP$_3$ or both mechanisms as the result of a single extracellular stimulus further increases this versatility. A partial listing of processes controlled by the phosphoinositide system is presented in Table 23.2.

Several observations imply a role for the phosphoinositide system not only in metabolic regulation but also in the control of cellular growth. First is the activity of a group of natural products called **phorbol esters**, part of whose structure resembles that of DAG (shown in magenta). These compounds are called **tumor promoters**. Not carcinogenic by themselves, they stimulate the formation of tumors when applied along with a carcinogen to experimental animals. Some phorbol esters have been found to activate protein kinase C independently of diacylglycerol. This finding is consistent with the hypothesis that protein kinase C activation is part of the normal growth control process that becomes perturbed in tumorigenesis. Another indication of a link between phosphoinositide metabolism and growth control is the function of certain cell growth factors. Some of them, notably **platelet-derived growth factor** (PDGF), are known to interact with cell surface receptors to stimulate the hydrolysis of phosphatidylinositol.

It is becoming evident that phospholipases other than phospholipase C are also stimulated by G proteins. Recall from Chapter 19 that arachidonic acid, released from phosphatidylcholine, is the major metabolic precursor to eicosanoids. Phospholipase A$_2$, which releases this fatty acid, is also part of a signal transduction pathway involving G proteins, and phospholipase D is thought also to participate in signal transduction via diacylglycerol formation. In addition, some phospholipases are controlled by Ca^{2+}.

To recapitulate: Cyclic AMP was the earliest known second messenger. However, several comparable second messengers are now known, including cyclic GMP, calcium ion, inositol trisphosphate, and diacylglycerol. In addition,

**A phorbol ester,
1-*O*-tetradecanoylphorbol-13-acetate**

***sn*-1,2-Diacylglycerol
(DAG)**

TABLE 23.2   Some cellular processes controlled by the phosphoinositide second-messenger system

Extracellular Signal	Target Tissue	Cellular Response
Acetylcholine	Pancreas	Amylase secretion
	Pancreas (islet cells)	Insulin release
	Smooth muscle	Contraction
Vasopressin	Liver	Glycogenolysis
Thrombin	Blood platelets	Platelet aggregation
Antigens	Lymphoblasts	DNA synthesis
	Mast cells	Histamine secretion
Growth factors	Fibroblasts	DNA synthesis
Spermatozoa	Eggs (sea urchin)	Fertilization
Light	Photoreceptors (*Limulus*)	Phototransduction
Thyrotropin-releasing hormone	Pituitary anterior lobe	Prolactin secretion

Courtesy of Slim Films. As published in *Scientific American* (1985) 253:142–152, M. J. Berridge, The molecular basis of communication within the cell.

phosphoinositides are far from the only phospholipids involved in cell signaling. In Chapter 19 we mentioned platelet-activating factor, which promotes platelet aggregation through its binding to a G protein-coupled receptor. Also, lysophosphatidic acid regulates the growth and development of cell types involved in wound healing, through its interaction with a G protein-coupled receptor. Finally, ceramide, although not a phospholipid, regulates processes involved in growth, differentiation, senescence, and apoptosis. It does this through activation of protein phosphatases, which activate growth and differentiation factors through cleavage of phosphate from specific phosphoproteins.

Receptor Tyrosine Kinases

We turn now to a family of related membrane receptors that are distinctive in having a single membrane-spanning domain, a ligand-binding site on the extracellular part of the molecule, and an intrinsic protein tyrosine kinase activity on the intracellular side. These include the **insulin receptor** (IR) and receptors for related peptide growth factors, including those for **epidermal growth factor** (EGF), **platelet-derived growth factor** (PDGF), **colony-stimulating factor 1** (CSF-1), **nerve growth factor** (NGF), and **fibroblast growth factor** (FGF), as well as a peptide **insulin-like growth factor 1** (IGF-1). These receptors represent a family of closely related proteins (see Figure 23.12), for the tyrosine kinase domains share amino acid sequence homology. With the exception of receptors for insulin and IGF-1, these are monomeric proteins that dimerize to activate the kinase activity in response to agonist binding. Some members of this class act through second messengers, but most act by initiating a protein phosphorylation cascade, similar to what we have seen in glycogenolysis. Receptors in this class include those for many peptide growth factors, as well as for insulin (which can also be considered a growth factor).

The first of this family to be characterized was the insulin receptor (Figure 23.12). This is a glycoprotein with an $\alpha_2\beta_2$ tetrameric structure, stabilized by interchain disulfide bonds. Both the α chain, of 735 residues, and the β chain, of 620 residues, are translated from a single mRNA, giving a polypeptide chain that then undergoes proteolytic processing. The α chain, which does not span the membrane, binds insulin near its C-terminus. The β chain has a single transmembrane domain, with its C-terminus in the cell interior. That C-terminal region is the site of a protein tyrosine kinase activity, which is stimulated by binding of insulin to the extracellular part of the receptor. The kinase activity is essential to the biological activity of the insulin receptor because some cases of non-insulin-dependent diabetes are associated with receptor mutations that abolish the kinase activity.

A number of more distantly related membrane receptors have other enzyme activities. Proteins of the **transforming growth factor β** (TGF-β) family bind to a receptor that has a protein serine/threonine kinase activity (similar to cAMP-dependent protein kinase). **Atrial natriuretic factor** (ANF), a peptide that controls blood volume, binds to a receptor that has both a guanylate cyclase activity and a predicted protein serine/threonine kinase activity. When excess blood volume stretches the atrium, ANF is released and travels to the kidney, where it activates cGMP synthesis, in turn increasing renal excretion of sodium and accompanying water, thereby reducing blood volume.

Signaling pathways involving receptor tyrosine kinases (RTKs) begin with binding of an agonist to the extracellular domain, and this triggers receptor dimerization (except for the insulin receptor and the IGF-1 receptor, each of which is already a dimer of $\alpha\beta$ subunits). Dimerization stimulates autophosphorylation of tyrosines in the cytoplasmic domain. Next, proteins are recruited to the intracellular domains of the RTK by virtue of containing either **SH2**

Second messengers include cyclic AMP, cyclic GMP, calcium ion, inositol trisphosphate, and diacylglycerol.

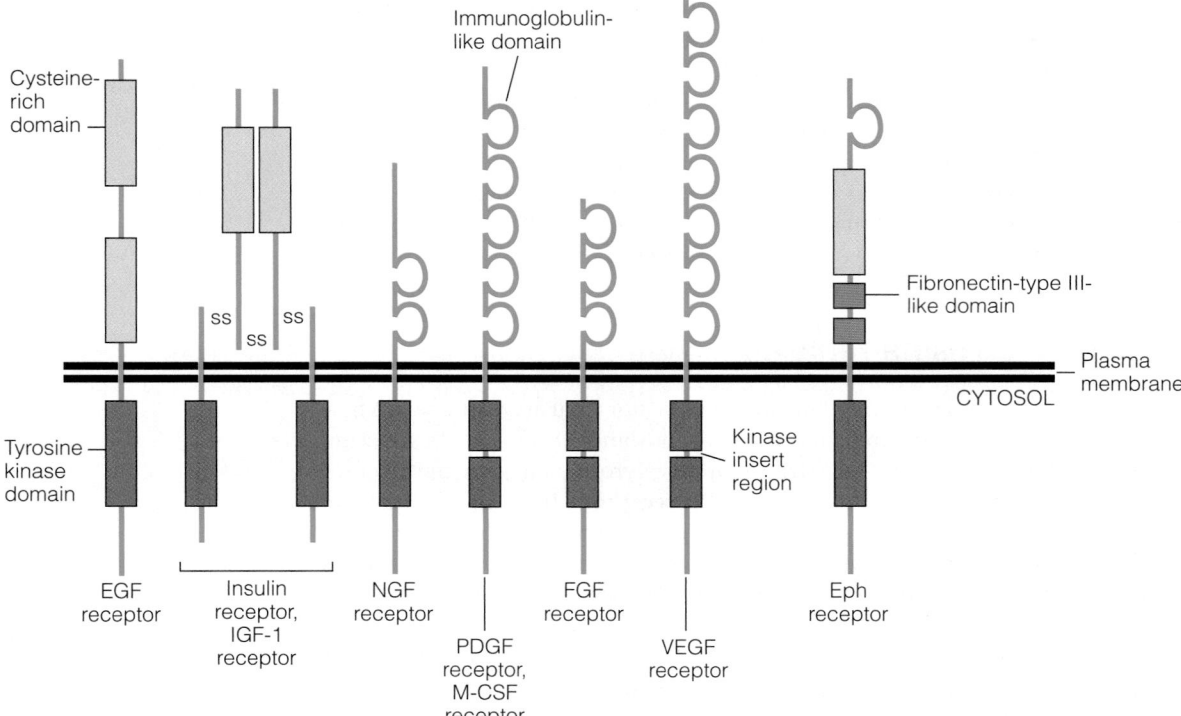

FIGURE 23.12

The insulin receptor and its structural relationship to other transmembrane receptors with protein tyrosine kinase activity. Representative proteins in each subfamily are shown. Some subfamilies have a "kinase insert domain" within the protein kinase module. VEGF, vascular endothelial growth factor; Eph, ephrins, regulators of cell movements during development. Other abbreviations presented in the text.

The insulin receptor and several related growth factor receptors contain one transmembrane domain per polypeptide chain and have an intrinsic protein tyrosine kinase activity.

(Src homology) domains or **PTB** (phosphotyrosine-binding) domains. Because of the numerous signaling pathways involved, receptor tyrosine kinases represent attractive drug targets, and the insulin receptor has been especially well characterized. Figure 23.13a shows the structure of the dimeric receptor kinase domain of the insulin receptor, complexed with one of its substrates, protein tyrosine phosphatase 1B (PTP1B), a negative regulator of insulin signaling.

Proteins associating with the tyrosine kinase domain may be either signaling agents or adaptor proteins, which create binding sites for downstream signaling proteins. As schematized in Figure 23.13b, a prominent substrate for the insulin receptor kinase is **insulin receptor substrate-1** (IRS-1), which binds to the kinase domain of the insulin receptor via its SH2 domains. One downstream effect of this phosphorylation is activation of **phosphoinositide 3-kinase**, which converts phosphatidylinositol 4,5-bisphosphate (PIP_2) to phosphatidylinositol 3,4,5-trisphosphate (PIP_3), a second messenger with a variety of intracellular targets. In the response to insulin, one effect of PIP_3 is activation of **protein kinase B** (PKB), also called Akt, by way of its attachment to the membrane. PKB then phosphorylates another protein kinase, glycogen synthase kinase (GSK3), which, in the phosphorylated form, promotes the inactivation of glycogen synthase. Therefore, as a result of insulin binding to its receptor, glycogen synthase cannot be inactivated, ensuring that glycogen synthesis continues. PKB also acts to relocalize the glucose transporter GLUT4 from internal membrane vesicles to the cell surface, where it promotes glucose uptake. Insulin also has long-term effects as a growth hormone, and these result from establishment of another phosphorylation cascade. As shown also in Figure 23.13, phosphorylated IRS-1 interacts with an adaptor protein called Grb2, which in turn activates Sos, which next binds to Ras (more about Ras on page 985). Ras is comparable to a G protein α subunit, although Ras is not a complete G protein. Sos acts like a guanine nucleotide exchange factor (GEF, page 969) in that it activates Ras by promoting binding of GTP, coupled with dissociation of bound GDP. GTP-bound Ras activates a protein kinase called Raf-1, which in turn activates another protein kinase, ERK.

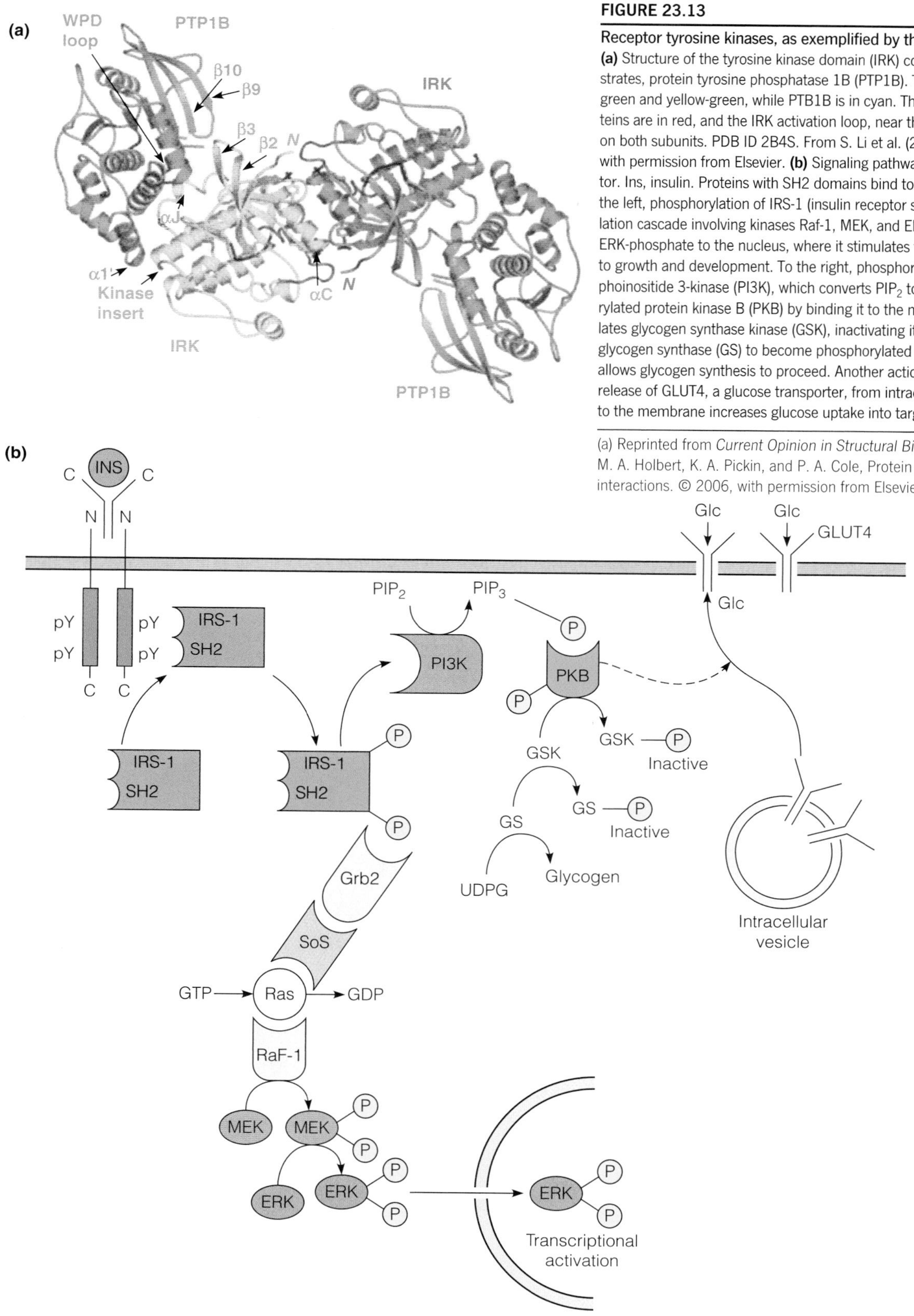

(a)

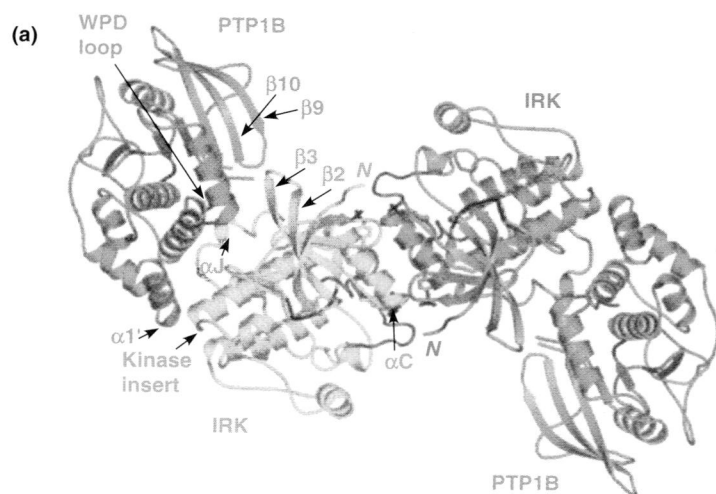

FIGURE 23.13

Receptor tyrosine kinases, as exemplified by the insulin receptor.
(a) Structure of the tyrosine kinase domain (IRK) complexed with one of its substrates, protein tyrosine phosphatase 1B (PTP1B). The two IRK subunits are in green and yellow-green, while PTB1B is in cyan. The catalytic loops of both proteins are in red, and the IRK activation loop, near the phosphotyrosine, is in gray on both subunits. PDB ID 2B4S. From S. Li et al. (2005) *Structure* 13:1643–1651, with permission from Elsevier. **(b)** Signaling pathways involving the insulin receptor. Ins, insulin. Proteins with SH2 domains bind to phosphotyrosine residues. To the left, phosphorylation of IRS-1 (insulin receptor substrate) initiates a phosphorylation cascade involving kinases Raf-1, MEK, and ERK, resulting in translocation of ERK-phosphate to the nucleus, where it stimulates transcription of genes related to growth and development. To the right, phosphorylated IRS-1 activates phosphoinositide 3-kinase (PI3K), which converts PIP$_2$ to PIP$_3$. PIP$_3$ activates phosphorylated protein kinase B (PKB) by binding it to the membrane, where it phosphorylates glycogen synthase kinase (GSK), inactivating it. This makes it impossible for glycogen synthase (GS) to become phosphorylated and, hence, inactivated; this allows glycogen synthesis to proceed. Another action of phosphorylated PKB is release of GLUT4, a glucose transporter, from intracellular stores. Its translocation to the membrane increases glucose uptake into target cells.

(a) Reprinted from *Current Opinion in Structural Biology* 16:668–675, R. Bose, M. A. Holbert, K. A. Pickin, and P. A. Cole, Protein tyrosine kinase–substrate interactions. © 2006, with permission from Elsevier.

(b)

Phosphorylation of ERK leads to its uptake, through nuclear pores, into the nucleus, where it phosphorylates additional proteins, which in turn bind to specific sites on the genome, promoting transcription of genes whose actions promote cell division and growth.

Steroid and Thyroid Hormones: Intracellular Receptors

Hormones acting through nuclear receptors generally have longer-lived effects than those interacting with membrane receptors.

The family of steroid receptors contains a conserved, zinc-containing DNA-binding sequence and a C-terminal hormone-binding domain.

Hormonal effects occurring via membrane receptors tend to be of short duration. Like the epinephrine-induced glycogenolytic cascade, they represent responses to rapid and urgent physiological demands, and they involve activation or inhibition of preexisting enzymes. By contrast, the effects of steroid hormones involve longer-term changes, such as the activation of a transport system or the conversion of a resting cell to a growing cell. Steroids and related hormones (thyroid, vitamin D, and retinoic acid hormones) act intracellularly. By virtue of their hydrophobic nature, they traverse the plasma membrane and exert their effects within the cell—actually, within the nucleus, where they control the activities of specific genes. In most cases, target genes are activated. Table 23.3 lists several proteins whose synthesis is regulated by these hormones.

These regulatory effects occur at the level of transcription of steroid-responsive genes. Steroid and related hormones act by binding in the cytosol to specific receptor proteins, which dimerize under the influence of the hormone. Binding in the cytosol is followed by movement of the hormone–receptor complex into the nucleus, where the complex interacts with specific DNA sites called **hormone-responsive elements (HREs)**. Binding of the complex to DNA affects transcription

TABLE 23.3 Target organs for steroid and thyroid hormones and major proteins whose synthesis is affected

Hormone Class	Target Organ	Protein[a]
Glucocorticoids	Liver	Tyrosine aminotransferase
		Tryptophan oxygenase
		α-Fetoprotein ($\downarrow$)
		Metallothionein
	Liver, retina	Glutamine synthase
	Kidney	phosphoenolpyruvate carboxykinase
	Oviduct	Ovalbumin
	Pituitary	Pro-opiomelanocortin
Estrogens	Oviduct	Ovalbumin
		Lysozyme
	Liver	Vitellogenin
		Apo-VLDL
Progesterone	Oviduct	Ovalbumin
		Avidin
	Uterus	Uteroglobin
Androgens	Prostate	Aldolase
	Kidney	β-Glucuronidase
	Oviduct	Albumin
1,25-Dihydroxyvitamin D$_3$	Intestine	Calcium-binding protein
Thyroid hormones	Liver	Carbamoyl phosphate synthetase
		Malic enzyme
	Pituitary	Growth hormone
		Prolactin ($\downarrow$)
Ecdysone (insects)	Epidermis	Dopa decarboxylase
	Fat body[b]	Vitellogenin

[a]Synthesis of each indicated protein is increased by the hormone, except for the two identified by ($\downarrow$).
[b]The fat body is an organ in insects that plays some of the same roles as liver and adipose tissue.

rates of nearby genes, by mechanisms now under intense study. Because of their site of action, members of this protein family are also called nuclear receptors.

Nuclear receptors exist at levels of only about 10^4 molecules per cell, which makes their purification difficult. However, because they bind to hormones quite tightly, it has been possible to purify these proteins by affinity chromatography. cDNA sequence analysis has revealed structural similarities among this class of receptors, and construction of hybrid receptors by recombinant DNA technology has allowed unambiguous identification of domains of function within the receptor molecule. Each receptor protein within this family contains a central conserved domain of about 80 residues, which is involved in DNA binding (Figure 23.14). On the N-terminal side of this domain is a region essential to transcriptional activation. Toward the C-terminus are domains involved in hormone binding, protein dimerization, and transcriptional activation.

All of the known receptors in this family contain bound zinc, which is essential for DNA binding, and the DNA-binding sequences show a completely conserved distribution of cysteine residues. These observations suggested that the zinc atoms could be complexed by the cysteine sulfurs in a pattern akin to the "zinc finger" structural motif associated with a number of other eukaryotic transcriptional regulatory proteins (see Chapter 27). That prediction has been supported by high-resolution NMR studies of receptor–DNA complexes (Figure 23.15).

The utility of a set of long-term-acting regulators is evident from a couple of examples. Estrogens and progesterone regulate the female reproductive cycle. In humans these hormones interact over a 4-week cycle to prepare the uterus for implantation of a fertilized ovum. Proliferation of the endometrium, the epithelial lining of the uterus, is the major event; clearly it requires new protein synthesis and increased blood flow to the uterus. These processes stop when a pituitary signal triggers decreased release of the hormones, causing sloughing off of cells in the uterine lining and the beginning of menstrual bleeding.

FIGURE 23.14

The conserved DNA-binding domain in steroid receptors. In the center are structural domains within steroid receptors, illustrated for the estrogen receptor. Above is the DNA-binding domain of the estrogen receptor, showing conserved cysteine residues that contact the bound zinc ions (a zinc finger DNA-binding motif; see Chapter 27). At the bottom are the DNA-binding domain sequences of related human receptors, with the conserved cysteine residues highlighted. ER, estrogen receptor; GR, glucocorticoid receptor; Trb, thyroid hormone receptor; PR, progesterone receptor; VitD, vitamin D receptor; RAR, retinoic acid receptor.

Reprinted from *Trends in Biochemical Sciences* 16: 291–296, J. W. R. Schwabe and D. Rhodes, Beyond zinc fingers: Steroid hormone receptors have a novel structural motif for DNA recognition. © 1991, with permission from Elsevier.

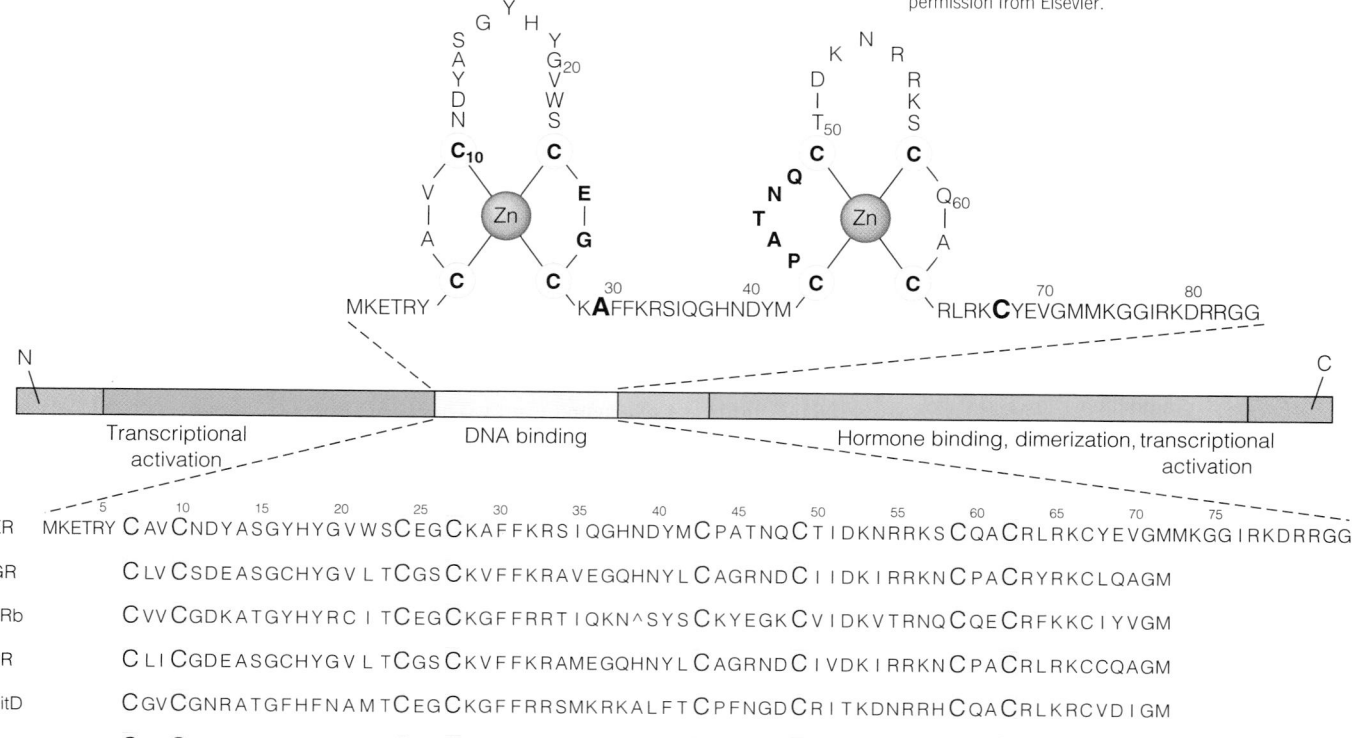

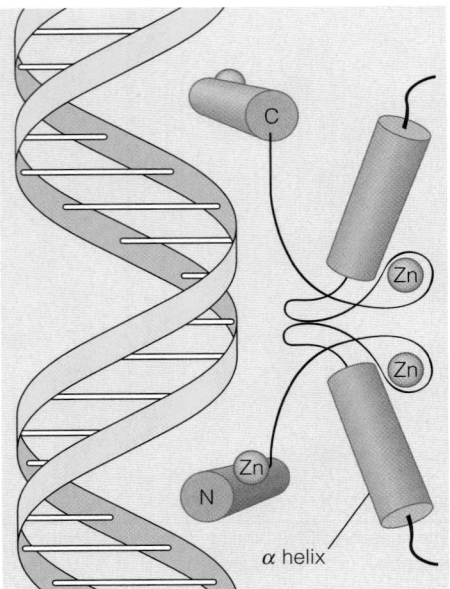

FIGURE 23.15

Binding of the estrogen receptor to DNA, as inferred from solution NMR spectroscopy. The dimeric receptor protein has two α-helical regions that bind to both ends of a symmetrical DNA sequence AGGTCAXXXTGACCT, within the major groove.

Reprinted from *Trends in Biochemical Sciences* 16: 291–296, J. W. R. Schwabe and D. Rhodes, Beyond zinc fingers: Steroid hormone receptors have a novel structural motif for DNA recognition. © 1991, with permission from Elsevier.

Tamoxifen

RU486

The actions of glucocorticoids are comparable, in that control of the synthesis of particular proteins allows for long-term metabolic adaptation. Whereas estrogens exercise control of reproductive metabolism over a several-week period, the secretion of glucocorticoids is a means of adaptation to longer-term stress. This adaptation involves stimulation of gluconeogenesis and synthesis of a variety of proteins, including some that counteract the effects of inflammation. Unlike estrogens, which act chiefly in reproductive tissues, the glucocorticoids influence cells in a wide variety of target tissues. Recent studies show that the rate of transcription of target genes following binding of the glucocorticoid-GR complex is acutely sensitive to the nucleotide sequence in the DNA binding region of the hormone-responsive element, making this sequence a crucial determinant of glucocorticoid sensitivity of target genes.

Investigations of the action of glucocorticoids as anti-inflammatory agents and as immunosuppressants have illuminated the action of another important signaling pathway, involving a transcriptional activator called NF-κB. This protein stimulates transcription of genes for a class of proteins called **cytokines**, which stimulate various reactions of the immune response, including proliferation of antibody-producing cells. Normally, NF-κB is bound to an inhibitory protein called IκBα, which prevents its translocation into the nucleus. Binding of an immune stimulator, such as **tumor necrosis factor** (TNF), to its plasma membrane receptor leads to the ubiquitination of IκBα and its subsequent degradation by the 26S proteasome (Chapter 20). This in turn allows NF-κB to translocate to the nucleus, where it activates transcription of about 300 genes, including those for cytokines (Figure 23.16a). Another gene activated is that for IκBα. This sets up a feedback loop, by escorting NF-κB out of the nucleus, followed by repeated cycles of IκBα degradation and translocation of NF-κB back into the nucleus. Recent fluorescent imaging studies show oscillations in NF-κB nuclear abundance, suggesting a timed response to what might also be a pulsatile stimulus (Figure 23.16b) and possibly explaining the diverse reactions, in different cells, to immune or inflammatory stimulation.

With respect to the anti-inflammatory properties of glucocorticoids, one of the target genes for activation by the glucocorticoid–receptor complex is the gene for IκBα. By stimulating synthesis of this protein, glucocorticoids prevent further translocation of NF-κB to the nucleus, thereby inhibiting expression of NF-κB-stimulated genes.

Steroid hormone receptors are target sites for several important drugs. **Tamoxifen** binds to estrogen receptors but does not activate estrogen-responsive genes. The growth of some breast tumor cells is activated by estrogen. Tamoxifen treatment of patients with such tumors after surgery or chemotherapy often antagonizes estrogen binding in residual tumor cells and retards their growth. However, patients taking tamoxifen after breast cancer surgery must be monitored carefully because there is also an increased risk of uterine cancer. **RU486**, which was developed in France, binds to progesterone receptors and blocks the events essential to implantation of a fertilized ovum in the uterus. Hence, RU486 is an effective contraceptive agent, even when taken after intercourse.

Signal Transduction, Oncogenes, and Cancer

One of the most exciting areas of current research involves investigations into genetic differences between cancer cells and the normal cells from which they are derived. These investigations have revealed, in a wide variety of tumor cells, mutationally altered forms or levels of proteins involved in signal transduction—including altered protein kinases, G proteins, nuclear receptors, growth factors, and growth factor receptors. Some tumor cells contain a normal signal transduction protein, but in excessive amounts. Genes responsible for such alterations are called **oncogenes**. Investigations of protein products of oncogenes, termed

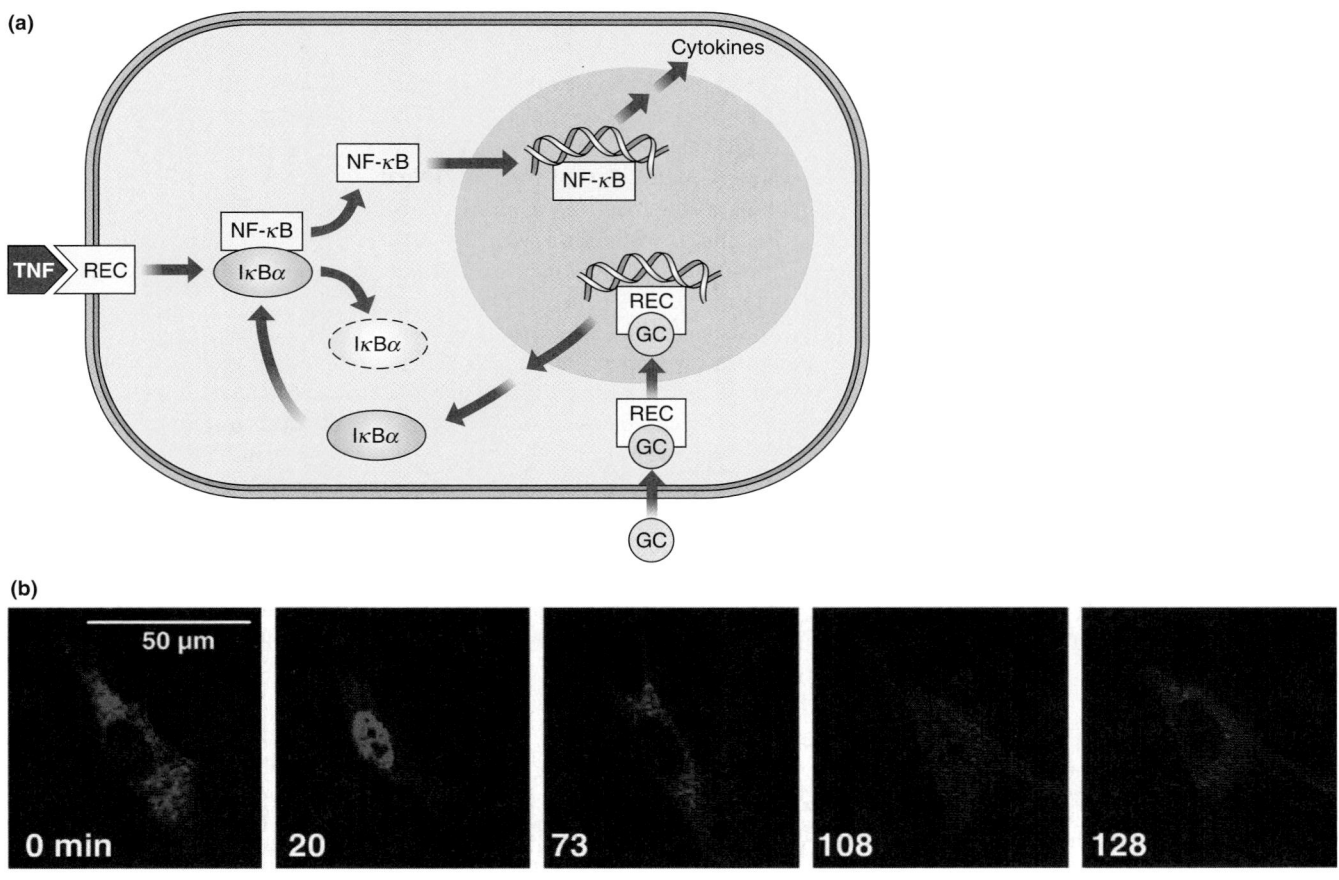

(a)

(b)

50 μm

0 min   20   73   108   128

FIGURE 23.16

Action of glucocorticoids (GCs) in suppressing immune and inflammatory reactions mediated by cytokines. **(a)** Action of glucocorticoids in counteracting NF-κB translocation to the nucleus; REC, receptor. **(b)** Oscillatory nature of the response to TNF signaling. Mouse embryo fibroblast cells transiently overexpressing a subunit of NF-κB treated with TNFα; cells were imaged by immunofluorescence microscopy at the indicated times in minutes.

From *Science* 342:242–246, L. Ashall, C. A. Horton, D. E. Nelson, P. Paszek, C. V. Harper, K. Sillitoe, S. Ryan, D. G. Spiller, J. F. Unitt, D. S. Broomhead, D. B. Kell, D. A. Rand, V. Sée, and M. R. H. White, Pulsatile stimulation determines timing and specificity of NF-κB-dependent transcription. © 2009. Reprinted with permission from AAAS.

oncoproteins, have illuminated roles of the normal forms of these proteins in regulating cell metabolism and growth and have spotlighted how normal control mechanisms go awry in a cancer cell.

Viral and Cellular Oncogenes

Two developments are particularly noteworthy in the history of the study of cancer, one involving tumor viruses and the other involving genetic analysis of human tumors. Regarding the first development, it has long been known that certain viruses cause cancer in infected animals. The first known tumor virus was **Rous sarcoma virus**, discovered in 1911 by Peyton Rous and shown by him to cause tumors in chickens.

Whether a virus contains RNA (like Rous sarcoma virus) or DNA, certain features are common in viral infections leading to cancer. First, cells become **transformed**. That is, they lose normal growth control mechanisms, and in cell culture they continue to proliferate under conditions that arrest the growth of normal cells. Second, the transformed cells are themselves tumorigenic; their injection into animals causes tumors. Third, part or all of the viral genome becomes linearly inserted into chromosomes of transformed cells. For RNA viruses like Rous sarcoma virus, the viral genome must be converted to double-stranded DNA before this insertion can occur. The viral enzyme that synthesizes DNA from a single-stranded RNA template is called **reverse transcriptase**, and viruses containing this enzyme are called **retroviruses** (see Chapter 25). The product of a reverse transcriptase reaction is called complementary DNA, or cDNA, because it is complementary in sequence to its RNA template.

Viral oncogenes are errant cellular proto-oncogenes, mostly encoding signal transduction elements that have been taken into viral genomes and have undergone subsequent mutations.

A number of nontumorigenic mutants of Rous sarcoma virus exist. Mapping mutations in these strains identified *src*, the viral oncogene responsible for transforming infected cells. Some of these mutants contain extensive deletions, which permitted Raymond Erikson to use nucleic acid hybridization techniques and, in 1978, to clone a cDNA corresponding to the viral *src* gene. Two surprising findings emerged. First, expression of the cloned gene yielded a protein with a protein tyrosine kinase activity. Thus, a specific enzyme activity, which might be associated with signal transduction, was also associated with the oncogene product. Second, further nucleic acid hybridization analysis showed that sequences corresponding to the viral *src* gene were present in normal cells. This finding suggested that viral oncogenes had their origins in normal cellular genes, or vice versa. One way to explain the transfer of an oncogene, or oncogene precursor, from cells to viruses is to postulate a rare genome excision event, as depicted in Figure 23.17. If an infection had caused insertion of the viral genome next to an oncogene precursor (or **proto-oncogene**), and if a subsequent excision event removed part or all of the proto-oncogene as well as the viral genome, then this faulty excision would have created a novel viral genome, containing a cellular gene. Subsequent evolution of the virus could change the cellular gene, creating an oncogene. Action of the oncogene would contribute toward transformation in a subsequent infection.

Sequence comparison of the *src* gene from viruses and cells revealed significant differences; the 19 C-terminal amino acids in the cellular gene are replaced by 11 different residues in the viral gene, causing the viral gene to be constantly turned on, promoting cell growth. Thus, we now speak of *v-src*, the viral form of the gene, and *c-src*, the cellular form. Analysis of many other tumor viruses yielded more than two dozen additional oncogenes. The corresponding proto-oncogenes encode a variety of proteins involved in cell signaling, some of which are identified in Table 23.4. Further analysis of infections leading to tumorigenesis showed that mutational alteration of the proto-oncogene is not always necessary. In some cases the viral genome is inserted adjacent to a proto-oncogene. Elements

FIGURE 23.17

Pathways by which proto-oncogenes can become oncogenes. A proto-oncogene is a normal cellular gene that can be converted to an oncogene and cause transformation to a cancer cell. This process can occur in two ways: (1) infection by a virus, which integrates into a chromosomal site next to a proto-oncogene and carries that gene along in its own genome when the virus replicates, or (2) mutation of the cellular proto-oncogene. In the first case, once cellular DNA becomes part of a viral genome, it can undergo mutation that converts the proto-oncogene to an oncogene. The oncogene can then cause transformation when this virus infects another cell.

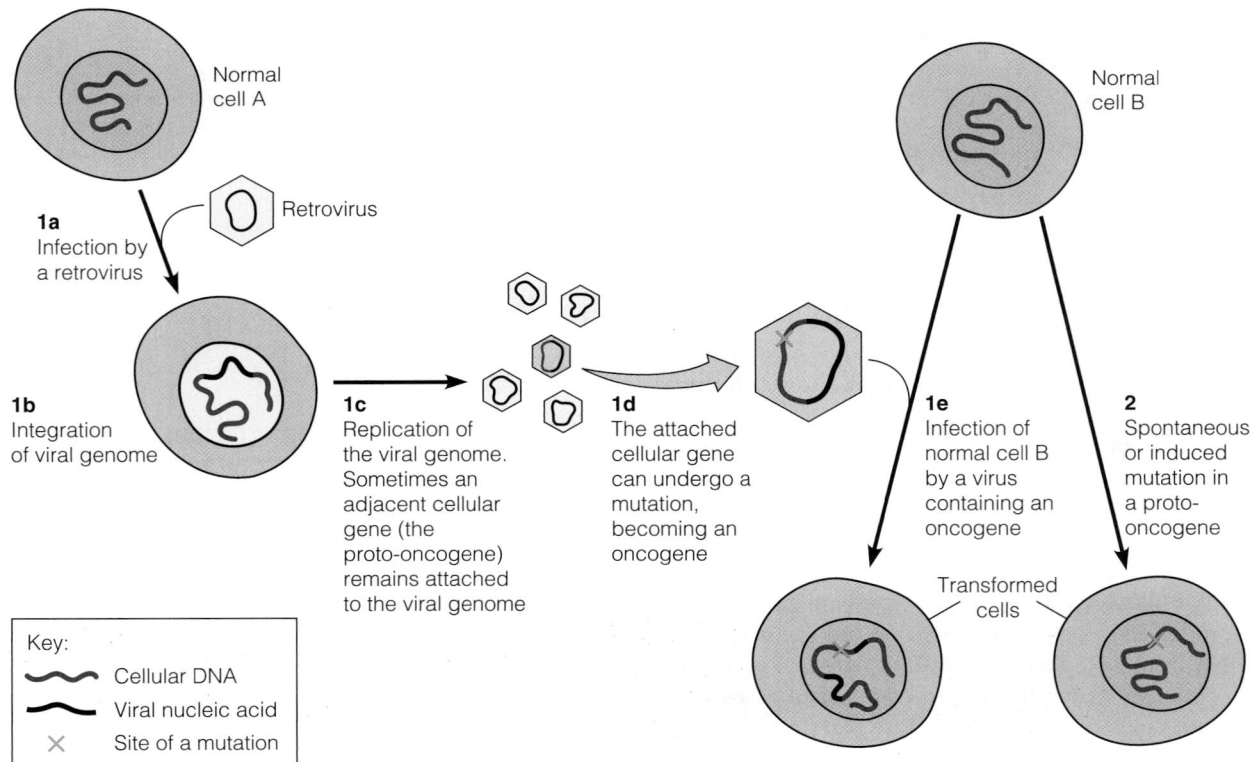

TABLE 23.4   Oncogene products as elements of signal transduction pathways

Signal Transduction Element	Oncogene	Isolated From	Gene Product
Growth factors	*sis*	Retrovirus	Platelet-derived growth factor
Growth factor receptors	*erbB, neu*	Retrovirus	Epidermal growth factor receptor
	fms	Retrovirus	Colony-stimulating factor 1 receptor
	trk	Tumor	Nerve growth factor receptor
	ros	Retrovirus	Insulin receptor
	kit	Retrovirus	PDGF receptor
	flg	Retrovirus	Fibroblast growth factor receptor
Intracellular transducers	*src*	Retrovirus	Protein tyrosine kinase
	abl	Retrovirus	Protein tyrosine kinase
	raf	Retrovirus	Protein serine kinase
	gsp	Tumor	G protein α subunit
	ras	Tumor, retrovirus	GTP/GDP-binding protein
Nuclear transcription factors	*jun*	Retrovirus	Transcription factor (AP-1)
	fos	Retrovirus	Transcription factor (AP-1)
	myc	Tumor, retrovirus	Transcription factor
	erbA	Retrovirus	Thyroid receptor

Data from J. D. Watson, M. Gilman, J. Witkowski, and M. Zoller, *Recombinant DNA*, 2nd ed., Scientific American Books, New York, p. 339. © 1992 James D. Watson.

of the viral genome stimulate transcription of the DNA sequences flanking the integration site. Thus, tumorigenesis can result from overexpression of normal genes encoding signal transduction machinery.

Although the Src protein is a protein tyrosine kinase, it is distinct from the receptor tyrosine kinases in that it is located in the cytoplasm, rather than at the membrane. Note that by convention, the name of the gene (*src*) is italicized, while the name of the corresponding protein (Src) is not.

Oncogenes in Human Tumors

Because human tumor viruses were formerly not known to exist, the relevance of viral oncogenes in animal viruses to an understanding of human cancer was not immediately apparent. That relevance came into sharp focus in the late 1980s, from work by Robert Weinberg and others on the isolation and analysis of transforming genes from human tumors. Weinberg isolated DNA from bladder cancer tissue and used it to **transfect** normal mouse fibroblasts (connective tissue precursor cells). That is, DNA was introduced into these cells, and transformed cells were isolated after outgrowth of the cells. DNA recovered from transformed cells was shown to contain human sequences. After additional rounds of transfection, the human DNA associated with the transformed mouse fibroblasts was sequenced. The transforming gene was shown to be nearly identical with a previously described oncogene from Harvey rat sarcoma virus, called the **H-*ras*** gene. Sequence analysis showed the *H-ras* gene sequence to be identical to *c-ras*, its counterpart in untransformed cells, with but a single difference—a mutation in the twelfth codon that changed a glycine codon in *c-ras* to a valine codon in the oncogene isolated from tumor tissue. Thus, human tumors were shown to contain an oncogene that is present in some tumor viruses, and in an altered form that presumably aided the tumorigenic process.

Ras genes are now known to encode a family of proteins—all of about 21 kilodaltons, with regions homologous to sequences in the α subunit of G proteins. Like the α subunit, the Ras proteins bind guanine nucleotides. Normal Ras proteins possess a GTPase activity, as do G_α proteins, whereas most *ras* oncogene proteins lack this activity. The GTPase activity suggested that normal Ras proteins function like G proteins in regulating metabolism. Lending support to this model

Activated oncogenes, closely related to viral oncogenes, have been isolated from human tumors.

FIGURE 23.18

Structural basis for Ras activation by GTP. **(a)** The complex of human c-H-*ras* protein with a nonhydrolyzable GTP analog. **(b)** The Ras-GDP complex. Large conformational changes in the protein result from displacement of GDP by GTP. Most prominent changes are in the regions called switch 1 (dark blue in the figure) and switch 2 (yellow-gold in the figure), both of which are in close contact with bound guanine nucleotide (in red). PDB ID 4Q21.

From *Science* 247:939–945, M. V. Milburn, L. Tong, A. M. deVos, A. Brunger, Z. Yamaizumi, S. Nishimura, and S. H. Kim, Molecular switch for signal transduction: Structural differences between active and inactive forms of protooncogenic ras proteins. © 1990. Reprinted with permission from AAAS and Sung-Hou Kim.

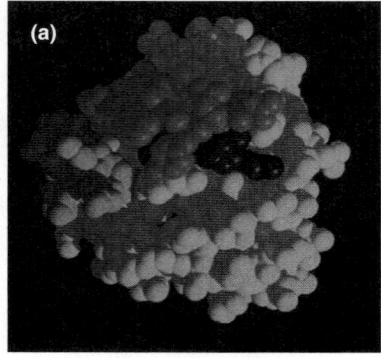
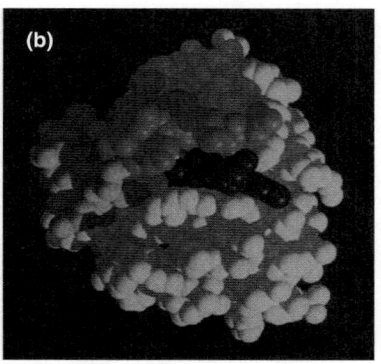

The Ras protein, which is mutationally altered in many human tumors, is a GTP-binding protein involved in signal transduction from growth factor receptors on the plasma membrane to specific gene activation events in the nucleus.

was the determination in 1988 of the three-dimensional structure of a Ras protein, crystallized as its complex with GDP (Figure 23.18). Amino acid residues known to be changed in mutations that generate *ras* oncogenes are positioned close to the bound guanine nucleotide. This positioning supports the idea that interactions between the proto-oncogene Ras protein and guanine nucleotides are important to metabolic control and that this control is lost when a normal cell is transformed to a cancer cell.

A major difference between Ras-type proteins and the related G_α proteins is the far higher GTPase activity of G_α proteins. As suggested earlier (page 969), a set of Ras-activating proteins is required to stimulate the GTPase activity of Ras. The basis for this difference was seen in 1994, with the first structural determination of a G_α protein (Figure 23.19). G_α, but not Ras, proteins contain a conserved arginine residue (R178), which interacts with the phosphates of bound GTP so as to stabilize the transition state for GTP hydrolysis.

Research on oncogenes has led to unifying theories of carcinogenesis. Work of Bruce Ames and others established that the great majority of chemical carcinogens are also mutagens. This finding suggested that chemical carcinogenesis involves mutagenesis of cellular proto-oncogenes, events that can occur in the absence of exogenous viruses (pathway 2 in Figure 23.17). Indeed, *ras* genes altered in codon 12, 13, or 61 have been detected in about 30% of spontaneous and chemically induced tumors, in both animals and humans.

A number of other genetic alterations have been detected in tumor tissue. Some are **tumor suppressor genes**. Unlike proto-oncogenes, these are genes that in the normal form suppress tumorigenesis. Loss of normal gene function, as in a deletion, leads to tumor formation because of deficient tumor suppression. One of these genes is called the **retinoblastoma gene, Rb**. Mutations in the two alleles of this gene cause a type of eye tumor, which has a familial association. The other most prominent tumor suppressor gene encodes a protein called **p53** (a protein of

FIGURE 23.19

GTP binding site on a G_α protein, $G_{i\alpha1}$. This figure shows bound GTPγS, a nonhydrolyzable GTP analog. Carbons on the bound ligand are in green, protein carbon in orange, nitrogen in blue, oxygen in red, sulfur and phosphorus in yellow, and magnesium in magenta. Stabilization of the phosphate negative charge by a conserved arginine (R178) facilitates hydrolysis of the bound nucleotide. PDB ID 1CIP.

From *Science* 265:1405–1412, D. E. Coleman, A. M. Berghuis, E. Lee, M. E. Linder, A. G. Gilman, and S. R. Sprang, Structures of active conformations of Gi alpha 1 and the mechanism of GTP hydrolysis. © 1994. Reprinted with permission from AAAS and Stephen R. Sprang.

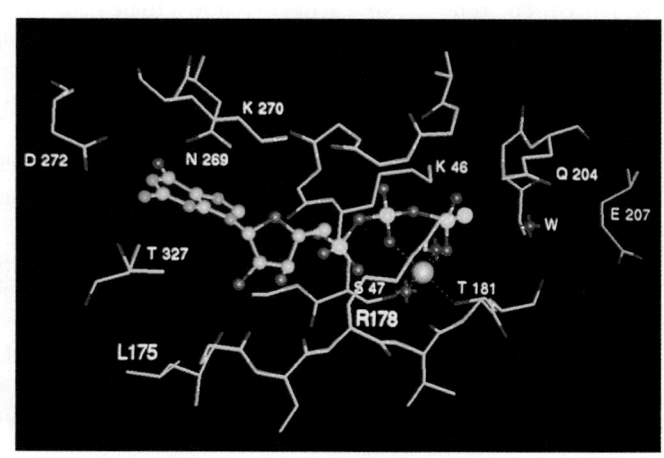

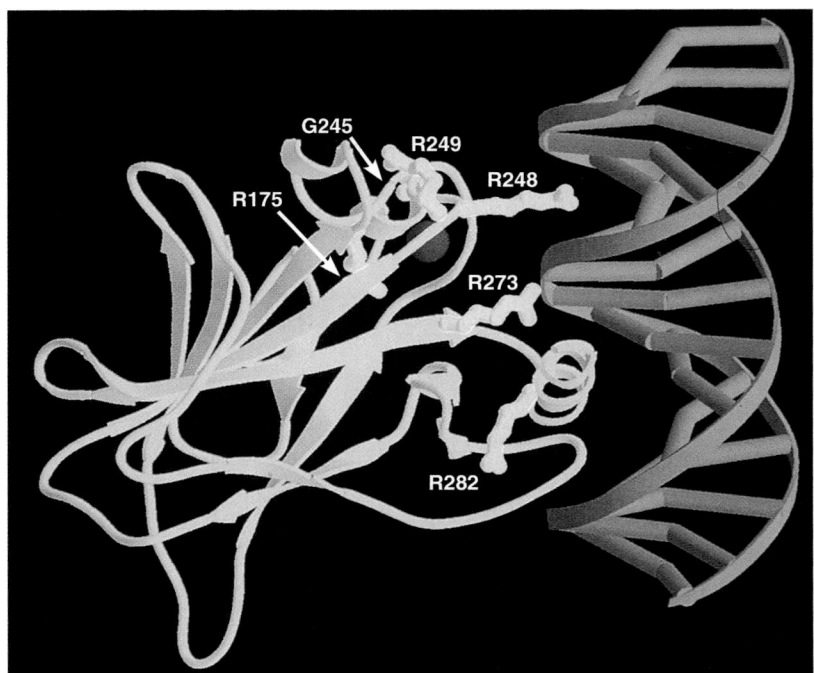

FIGURE 23.20

Structure of the p53–DNA complex. This ribbon drawing shows the DNA-binding domain of one subunit of the homotetrameric p53 (blue-green) complexed with an oligonucleotide pair containing the p53 binding site (blue). A bound zinc ion is shown in red. Shown in yellow are the six amino acid residues most often changed in mutant p53 proteins. PDB ID 1TUP.

From *Science* 265:346–355, Y. Cho, S. Gorina, P. D. Jeffrey, and N. P. Pavletich, Crystal structure of a p53 tumor suppressor-DNA complex: Understanding tumorigenic mutations. © 1994. Reprinted with permission from AAAS and Nikola Pavletich.

53 kilodaltons). Loss of p53 function leads to tumorigenesis, and at least half of all human tumors examined display p53 gene mutations. Although its biochemical actions are not completely defined, we know that p53 is a DNA-binding protein that regulates gene expression following DNA-damaging events. If damage is moderate, the effect is to retard progression through the cell cycle from G1 into S phase until the damage has been repaired. If damage is more extensive, p53 signals that the cell shall undergo **apoptosis**, or programmed cell death (Chapter 28, pages 1221–1222). The loss of such a checkpoint would involve inappropriate cell growth, contributing toward carcinogenesis. Binding to specific DNA sequences is critical to proper functioning of p53, as shown by the structure of a p53-DNA complex, and in particular by the fact that those amino acids in closest contact with DNA are the ones most often mutated in the *p53* gene analyzed in human tumors (Figure 23.20).

Human tumors contain a series of mutations, affecting both signal transduction components and tumor suppressor genes and gene products.

Oncogenes and the Central Growth Factor Activation Pathway

Beginning in the 1990s, several lines of work converged upon the realization that the Ras protein occupies a central role in an evolutionarily conserved pathway that conveys extracellular signals to the nucleus, where specific genes are activated for cell growth, division, and differentiation. As details of that pathway emerged, they fit into a unified framework that interrelates the biochemical properties of proto-oncogene products, underscores the dominance of protein phosphorylation as a control mechanism, and rationalizes the types of mutations that lead to cancer.

Ras-related proteins have been discovered in such diverse organisms as yeast, nematode worms, and *Drosophila*, in which they control aspects of mitotic and meiotic growth and embryonic development. Research on these organisms has illuminated a central control pathway in mammalian cells (Figure 23.21), in the process richly justifying the use of simple biological model systems for cancer research.

As noted earlier, growth factor receptors with tyrosine kinase activity undergo autophosphorylation (steps 1 and 2 in Figure 23.21). In the phosphorylated state, each receptor interacts with one or more protein exchange factors, which in turn activate Ras by stimulating the GDP–GTP exchange (step 3). Also interacting with

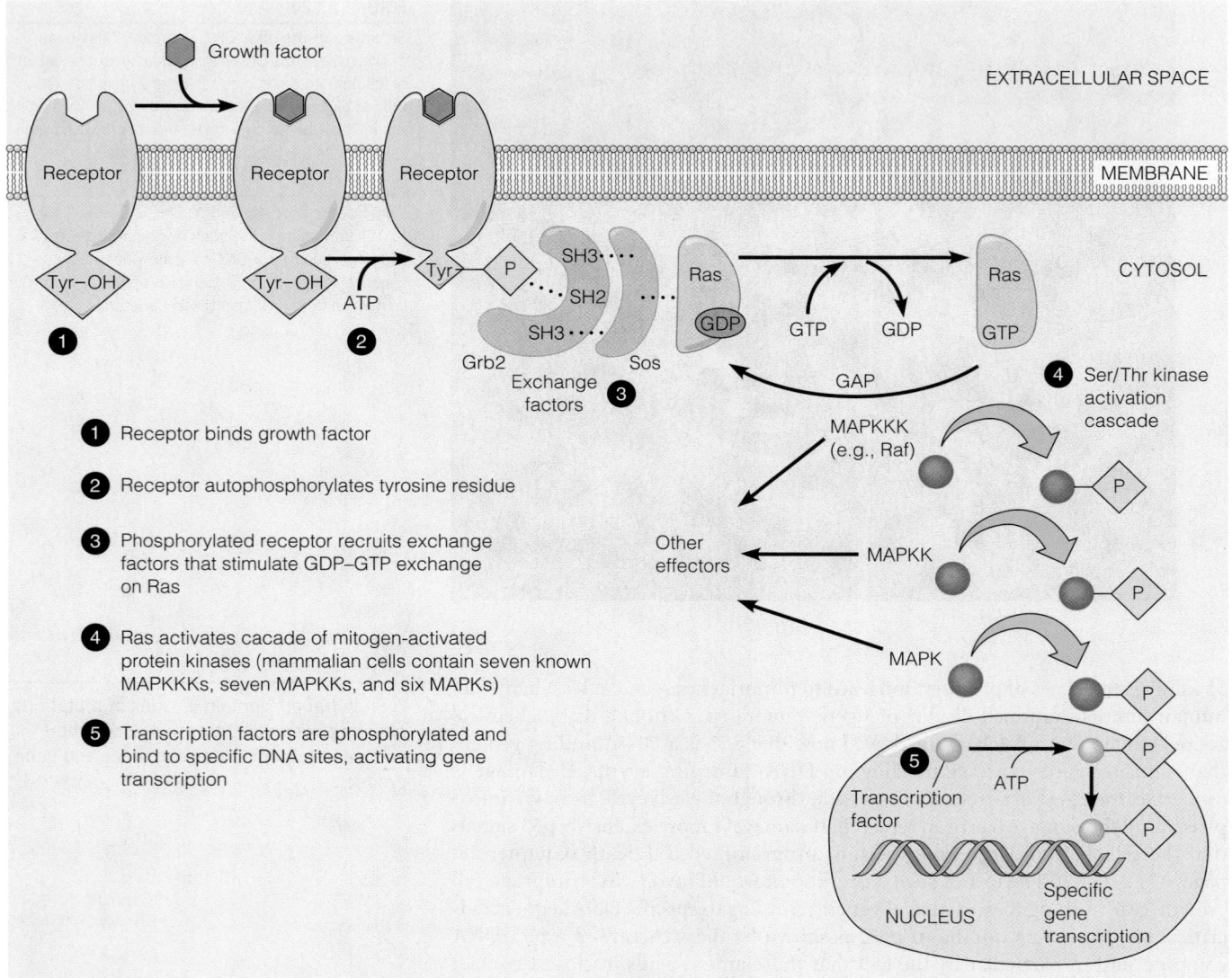

FIGURE 23.21

Role of Ras protein in a central growth factor activation pathway. Binding of a mitogen or growth factor to a receptor stimulates the receptor tyrosine kinase (RTK), usually with dimerization of the receptor (not shown). Autophosphorylation of RTK leads to recruitment of a series of proteins (exchange factors) that stimulate GDP–GTP exchange on Ras and thereby activate it. In the scheme shown, a protein called Grb2 interacts via SH2 domains with RTK and via SH3 domains with exchange factor Sos. Ras then activates proteins of a family called MAPKKK (mitogen-activated protein kinase kinase kinase), which in turn leads to activation of MAPKK proteins and then MAPK proteins. Eventually, specific transcription factors in the nucleus undergo phosphorylation, which activates them to bind specific DNA sites and activate transcription of particular genes. The activity of Ras is curtailed by a series of GTPase-activating proteins (GAPs) that promote the hydrolysis of bound GTP.

Based on S. E. Egan and R. A. Weinberg, *Nature* (1993) 365:782, unnumbered figure, p. 782.

Ras, and limiting its activity, are GTPase-activating proteins (GAPs, see page 969). Downstream of Ras is a cascade of further protein phosphorylations (step 4), which ultimately activates transcription factors (step 5), proteins that, like nuclear receptors, stimulate transcription of specific genes (see Chapter 27).

As seen in Figure 23.21, one class of downstream kinases is called **MAP kinase (MAPK)**. The acronym *MAP* stands for mitogen-activated protein. (A mitogen is a factor that stimulates mitosis.) Upstream of MAP kinases are another family of proteins, MAPKK (MAP kinase kinase), and still further upstream are the MAPKKK proteins (MAP kinase kinase kinase). One member of the MAPKKK class is the product of the *raf* proto-oncogene (see Table 23.4). In the growth factor pathway of insulin signaling (Figure 23.13b), Sos is the GTP–GDP exchange factor, Raf-1 is the MAPKKK, MEK is the MAPKK, and ARK is the MAP kinase.

Thus, we see the growth factor response as a cascade of protein phosphorylations, analogous to the well-known glycogen breakdown cascade. We see how blocking the Ras GTPase activity can lead to uncontrolled cell growth and cancer, by keeping the signaling pathway turned on and flooding the cell with growth-stimulatory signals. What isn't immediately obvious is how this pathway can explain the distinct responses of cells to different growth factors.

Analysis of the proteins that interact with autophosphorylated growth factor receptors has identified common sequences, the SH2 domains (pages 977–978), which are involved in these interactions. A protein that binds to a phosphorylated receptor molecule recognizes not only the phosphotyrosine residue but also a critical amino acid sequence on the carboxyl side of that tyrosine. Thus, receptor phosphorylation–dephosphorylation serves as a general on–off switch, with the more subtle protein interactions governed by the SH2 domain. Downstream participants are recruited via another set of sequences called SH3 domains. These domains identify the specific proteins that will be recruited to the signaling process in response to a particular growth stimulus.

The Cancer Genome Mutational Landscape

Given what we now know about cell growth control and the biochemical functions of oncoproteins and tumor suppressor proteins in cell signaling, what are we learning about the biology of cancer? Large-scale sequence analysis of DNA from human tumors yields what is called the "mutational landscape of the cancer genome." Figure 23.22 shows the results of combined sequence analysis of human colorectal tumors. In this plot the two horizontal axes denote chromosomal map position and the vertical axis depicts frequency with which a gene at that position is mutated in the colorectal cancers examined. The highest "mountains" on this landscape represent genes that are mutated in most cancers, which happen to include *H-ras*, *p53*, *APC*, and *PIK3* (overexpression of phosphoinositide 3-kinase, page 978). The smaller "hills" represent mutations seen in just a few of the tumors analyzed. The average colorectal cancer cell has about 80 mutations in protein-coding genes. What this landscape does not show is the considerable heterogeneity that also exists among individual cells in one tumor. Also not shown is that cancers from other tissues, such as breast, show completely different mutational landscapes. This type of information is discouraging in a sense because it suggests that there are numerous pathways by which a precursor cell population can undergo the mutations converting those cells to full-blown tumor cells, and, hence, complicating the search for single therapeutic targets that could lead to control of the process.

On the other hand, some notable successes have been achieved, based upon what has been learned. In 90% of chronic myelogenous leukemia cases, a chromosomal translocation has occurred, linking portions of chromosomes 9 and 22. This translocation fuses parts of two oncogenes, *c-abl* and *bcr*. The *c-abl* gene encodes a protein tyrosine kinase related to *src*. The gene fusion creates a novel protein kinase, which is overexpressed and stimulates growth-promoting pathways. The drug Gleevec is a specific inhibitor of this abnormal protein kinase and is a highly specific and effective treatment for this form of leukemia. Another drug, Herceptin, provides similarly specific effectiveness in treating certain forms of breast cancer. These examples suggest that the more we learn about genetic and biochemical changes in individual cancers, the better position science will be in to devise specific therapies.

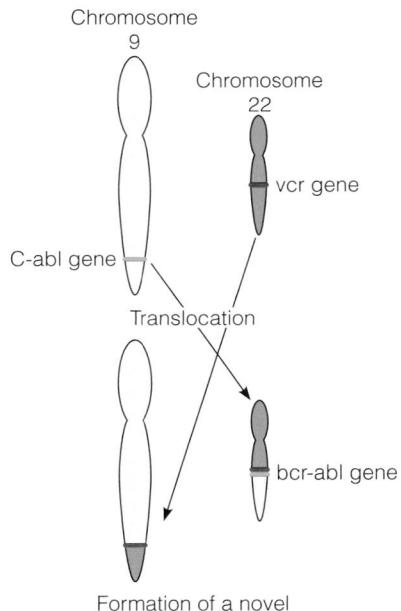

Formation of a novel
protein kinase gene
by chromosomal translocation

All colorectal cancers

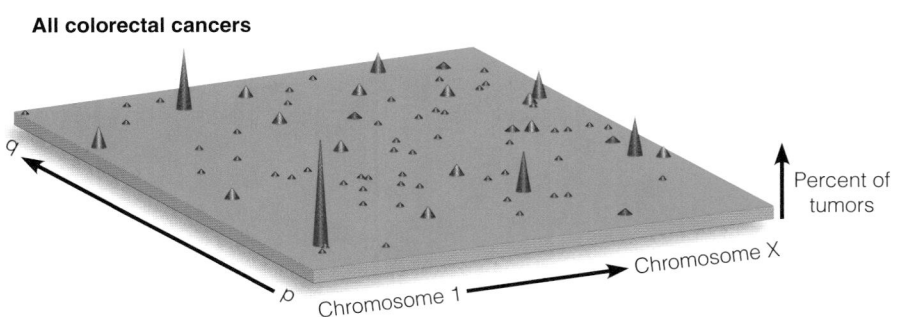

FIGURE 23.22

Mutational landscape of the human colorectal cancer genome. In this study, 11 tumors were subjected to analysis of expressed genes. Map position of each gene is plotted on the horizontal axes, and percentage of tumors with a mutation in a gene at each position is plotted on the vertical axis.

Adapted and reprinted by permission from the American Association for Cancer Research: E. J. Fox, J. J. Salk, and L. A. Loeb, Cancer genome sequencing—An interim analysis, *Cancer Research*, 2009, 69:4948–4950.

Neurotransmission

In Chapter 10 we described the propagation of an action potential along a neuron as a nervous impulse is transmitted. Recall that this involves a wave of depolarization as ion channels open to admit extracellular sodium and to allow efflux of intracellular potassium. It was long thought that transmission of the impulse across a synaptic junction, from neuron to neuron or from a neuron to a muscle cell, involved an electrical impulse triggered when the action potential reached the end of a neuron. We now know that fewer than 1% of all synaptic transmission events are mediated electrically, through gap junctions between cells. The vast majority involve release of a chemical substance, the neurotransmitter, from the upstream, or **presynaptic**, cell, followed by its diffusion through a synaptic cleft and its binding to receptors in the downstream, or **postsynaptic**, cell. The postsynaptic receptor is a ligand-gated ion channel; neurotransmitter binding triggers events that allow for propagation of the nervous impulse in the postsynaptic cell. About 100 different substances, including amino acids, biogenic amines, and peptides, have been identified as neurotransmitters in the brain. The most widely used, and best understood, neurotransmitters are acetylcholine, glutamate, glycine, GABA, dopamine, serotonin, norepinephrine, and epinephrine.

The Cholinergic Synapse

Our earliest biochemical understanding of neurotransmission involved the action of acetylcholine as the transmitter at the neuromuscular junction, where nervous stimulation of a muscle cell triggers its contraction. These studies generated the scheme for neurotransmission shown in Figure 23.23. When acetylcholine is the neurotransmitter, this event is called a **cholinergic synapse**. This terminology is generally used; for example, when the neurotransmitter is dopamine, we speak of a **dopaminergic** synapse. Acetylcholine is synthesized by the coenzyme A-dependent acetylation of choline, which in turn was formed by breakdown of phosphatidylcholine.

The synaptic cleft is about 20 nm wide. Neurotransmitter is stored in vesicles within a bulb at the end of the presynaptic cell, with about 5000 molecules per vesicle. Arrival of the action potential opens voltage-gated calcium channels, triggering an influx of calcium to the bulb. This in turn causes the vesicles to fuse with the plasma membrane at the end of the bulb, spilling acetylcholine into the extracellular space of the synaptic cleft. The neurotransmitter diffuses through the cleft

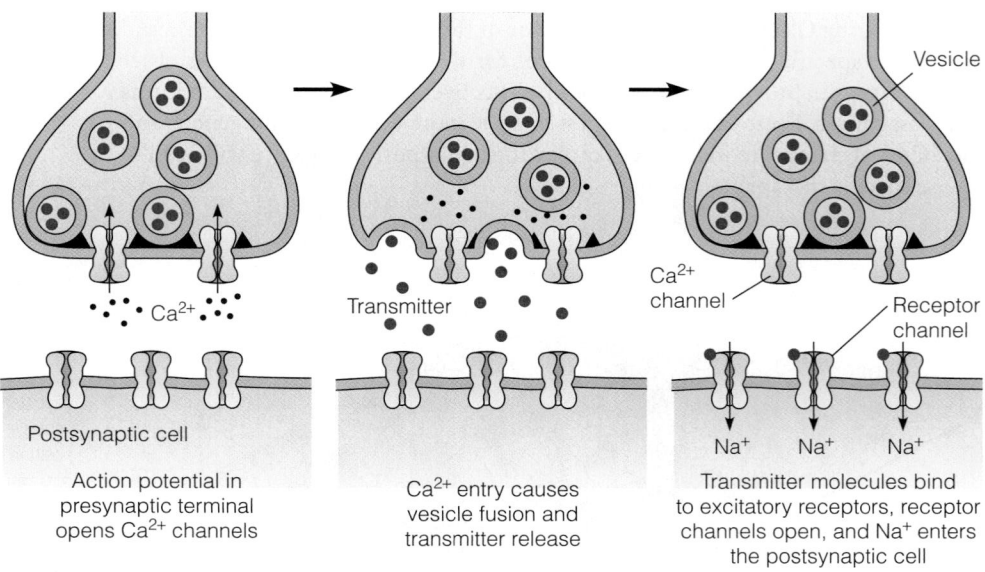

FIGURE 23.23

Transmission of a neural impulse across a synapse, such as a cholinergic synapse.

Action potential in presynaptic terminal opens Ca^{2+} channels

Ca^{2+} entry causes vesicle fusion and transmitter release

Transmitter molecules bind to excitatory receptors, receptor channels open, and Na^+ enters the postsynaptic cell

and binds to receptors in the postsynaptic cell membrane. This in turn triggers the opening of channels, allowing Na^+ influx and generating an action potential. If the postsynaptic cell is a muscle cell, the action potential opens Ca^{2+} channels, and the resultant rise in intracellular Ca^{2+} concentration triggers contraction. In the meantime, on the presynaptic side some of the released acetylcholine is taken back into the cell, and an acetylcholine transporter protein is refilling vesicles with neurotransmitter, exchanging protons for acetylcholine with the help of a **vacuolar ATPase**. The remainder of the acetylcholine in the cleft is inactivated by hydrolysis (see below), terminating the signaling event. The entire process occurs about 1000 times per second. Similar events occur during neuron-to-neuron transmission, except that here the increase in Na^+ in the postsynaptic cell triggers the establishment of an action potential and its transmission down that cell.

Structural studies of a receptor called the **nicotinic acetylcholine receptor**, because it binds the alkaloid nicotine, show it to be composed of five subunits, two of which are identical and each of which has five α-helical regions that span the membrane. As shown in Figure 23.24, the subunits interact to form a channel through the membrane. Reconstitution of this multisubunit protein into membrane vesicles showed that addition of acetylcholine causes ions to flow through it. Thus, this receptor is a gated channel, which undergoes a conformational change and opens a pore in response to binding of the neurotransmitter. Our understanding of the structure and function of this receptor originally involved isolation of the receptor from organs of the electric eel (*Electrophorus*) or electric ray (*Torpedo*), both of which have the receptor densely packed in an organ, called the electroplaque, which allows the animal to shock its prey by generating potentials of several hundred volts.

Within the intersynaptic cleft acetylcholine is rapidly hydrolyzed by acetyl-cholinesterase, thereby destroying excess neurotransmitter and restoring the resting

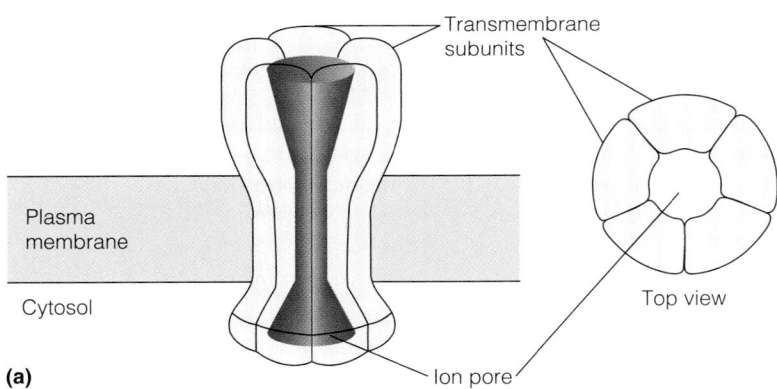

(a)

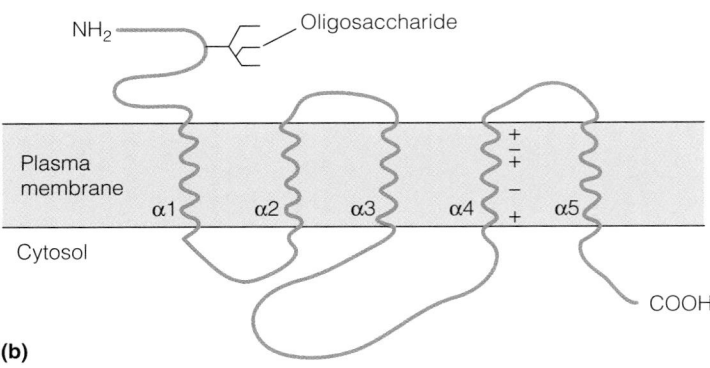

(b)

FIGURE 23.24

The nicotinic acetylcholine receptor. **(a)** Schematic model of the receptor. Five subunits combine to form a transmembrane structure with an ion pore in the center. **(b)** Structure of an individual subunit. There are four different kinds of subunits, but their sequences are all similar, and each individual subunit has the kind of structure depicted here. Five α helixes ($\alpha1$ to $\alpha5$) in each subunit traverse the membrane. The charged residues on helix $\alpha4$, which tend to be on one surface, probably line the wall of the pore.

Acetylcholine

H₂O

Acetylcholinesterase

CH₃COO⁻

Choline

Tubocurarine, an antagonist

Nicotine, an agonist

potential in the postsynaptic cell membrane. Acetylcholinesterase is the target for several organophosphate compounds such as the nerve gas **Sarin**, which was developed as a chemical warfare agent and which causes paralysis by reacting with an active site serine and irreversibly inactivating acetylcholinesterase (Table 11.7). Other agents act by binding to the receptor itself. **Tubocurarine**, one such agent, blocks the channel in the closed position and is considered a receptor antagonist, in contrast to **nicotine**, which activates the receptor and is an agonist. Tubocurarine is derived from curare, originally used by aboriginal people to tip poison arrows and paralyze human or animal prey. However, in the hands of qualified medical personnel, this is a useful muscle relaxant.

Fast and Slow Synaptic Transmission

For some time it was thought that the model of synaptic transmission obtained with acetylcholine receptors described all neuron-to-neuron transmission as well as transmission at the neuromuscular junction. An important difference is that some interneuronal transmission events are excitatory and some are inhibitory. In excitatory transmission, transmitter binding at the postsynaptic receptor causes sodium influx, which depolarizes the membrane and stimulates transmission of the action potential. Earlier we identified γ-aminobutyric acid, or GABA, as the principal inhibitory neurotransmitter. Binding of this agent triggers an influx of chloride ions, causing hyperpolarization and inhibiting the propagation of the action potential.

However, the situation is in fact considerably more complex. Paul Greengard has pointed out that each of the 100 billion cells in the human brain communicates directly with about 1000 other cells. The complexity is suggested in the electron micrograph in Figure 23.25, which shows multiple neuronal connections to the body of one nerve cell. Greengard shared the 2000 Nobel Prize in Physiology or Medicine with Arvid Carlsson and Eric Kandel, for their discoveries regarding slow synaptic transmission events, which occur with a timescale of hundreds of milliseconds to several minutes. In these cases neurotransmitter binding stimulates intracellular metabolic events comparable to those resulting from hormone action. Some of these events are mediated through second messengers, such as cyclic AMP, cyclic GMP, and phosphoinositides. For example, as schematized in Figure 23.26, some neurotransmitter–receptor interactions activate adenylate cyclase, causing activation of protein kinase A and triggering cyclic AMP–dependent metabolic responses in the post-synaptic cell, with time scales on the order of minutes. Persistent repetition of the transmitter–receptor interaction triggers genetic, as well as metabolic, modifications. In this longer-term response, with a time scale of days, second messenger-dependent protein kinases translocate to the nucleus,

FIGURE 23.25

Multiple synapses on the body of a single neuron. This scanning electron micrograph gives an idea of the complexity of the interconnection within the nervous system. Some of these synapses will be stimulatory and others inhibitory.

© Manfred Kage/Peter Arnold, Inc.

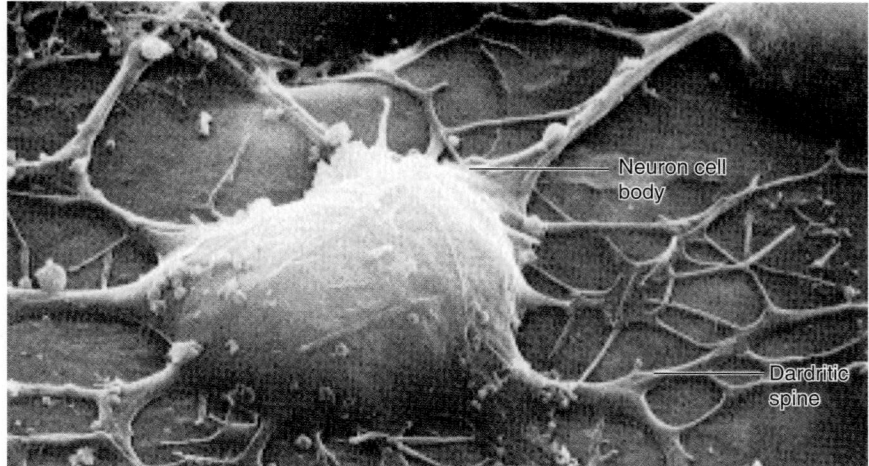

Neuron cell body

Dendritic spine

leading to stimulation of transcription of genes whose products help the cell to form new synaptic connections. Finally, prolonged neurotransmitter–receptor interaction can stimulate localized protein synthesis within the cell, to help stabilize that synaptic connection for long-term effects. Those effects that involve new protein synthesis are critically involved in learning and memory; the precise mechanisms involved are among the crucial issues of contemporary neuroscience.

Functions of Specific Neurotransmitters

Most of the fast excitatory synapses in brain use glutamate as the neurotransmitter, and most of the fast inhibitory synapses use GABA. Excessive firing of glutamatergic synapses, with consequent damage to the central nervous system, can result from overingestion of glutamate. For this reason monosodium glutamate, a major ingredient in soy sauce, was removed from baby food formulas some years ago, where it was being used as a flavor enhancer because the developing nervous system is particularly susceptible to this kind of damage.

Both glutamate and GABA participate in slow synapses as well, and it now seems clear that biogenic amines and peptide neurotransmitters participate only in slow synaptic transmission. Of particular interest are synapses for which dopamine is the neurotransmitter. Arvid Carlsson and others showed that several important psychiatric disorders involve abnormalities in dopamine signaling, including Parkinsonism, schizophrenia, drug addiction, and attention deficit–hyperactivity disorder (ADHD). Parkinsonism involves the death of dopamine-producing nerve cells in a portion of the brain called the **substantia nigra**. Success in treating some forms of Parkinsonism has resulted from treatment with massive doses of L-dopa, which can traverse a permeability blockade called the **blood-brain barrier** and undergo decarboxylation within the brain, to replenish depleted dopamine stores.

The relationship of dopamine with schizophrenia is seen by the fact that some effective anti-schizophrenic drugs have been shown to antagonize the effect of dopamine at one or more classes of receptors, by binding to the receptor and blocking the binding of dopamine. An example is **chlorpromazine**, which, although not closely related to dopamine structurally, does effectively antagonize dopamine-receptor binding. On the other hand, several drugs of abuse, including **mescaline** and **amphetamine**, are closely related to dopamine, and they act as

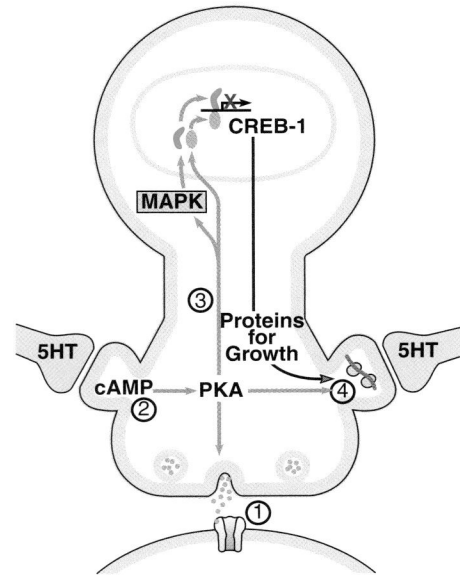

FIGURE 23.26

Four consequences of the action of neurotransmitters. (1) Rapid synapse, leading to opening of gated ion channels; **(2)** interaction of 5-hydroxytryptamine (5HT, or serotonin) with adenylate cyclase, causing activation of cAMP-activated protein kinase; **(3)** effect of longer-term stimulation, with protein kinases (e.g., MAPK) translocating to the nucleus and stimulating specific gene transcription by phosphorylating certain transcription factors such as CREB; **(4)** stimulation of localized protein synthesis by still undefined interactions with translation complexes.

From *Science* 294:1030–1038, E. R. Kandel, The molecular biology of memory storage: A dialogue between genes and synapses. © 2001. Reprinted with permission from AAAS.

Dopamine

Mescaline

Amphetamine

Chlorpromazine

Serotonin

Lysergic acid diethylamide

dopamine agonists, binding to dopamine receptors and mimicking its effects. Dopamine is considered a "pleasure agent," and many abused drugs stimulate the firing of dopaminergic synapses, either by raising dopamine levels or by mimicking the natural neurotransmitter. In a recent study, a targeted deletion in the mouse of D4, one of the four classes of dopamine receptors, caused the animals to become hypersensitive to ethanol, cocaine, and methamphetamine.

Synapses using serotonin as the neurotransmitter are also involved in the pathophysiology related to schizophrenia. **Lysergic acid diethylamide**, or LSD, is an indole derivative, like serotonin. LSD acts as a serotonin agonist, binding to a class of serotonin receptors and mimicking its effect. The popular (but illegal) recreational drug Ecstasy (3,4-methylenedioxy-N-methylamphetamine) acts by promoting the release of dopamine, serotonin, and norepinephrine from their storage sites.

Finally, **Ritalin**, used in treating ADHD (attention deficit–hyperactivity disorder), acts to promote dopamine release. This stimulant effect is paradoxical because the drug is used to calm hyperactive children. Recent research, however, suggests that its calming effect may result more from elevating levels of serotonin.

However, not all drugs of abuse involve slow synaptic transmission, and understanding dopamine and serotonin actions is only a small part of grasping the complexities of schizophrenia. Recent work has implicated glutamate receptors as well in control of addictive behavior. Inhibition of glutamate neurotransmission in rats has been shown to modulate the compulsive drug-seeking behavior after an initial drug experience, and neuroscientists are excited at the prospect of developing therapies involving glutamate antagonists, which might increase the likelihood that an addict would remain "clean" after treatment. Glutamate receptors are also strongly linked to the action of **phencyclidine** (PCP or "angel dust"). This compound blocks glutamate binding to the N-methyl-D-aspartate (NMDA) class of glutamate receptors, thereby inducing a schizophrenia-like state thought until recently to result from decreased glutamatergic neurotransmission. However, neuroscientists have found that lowering brain glutamate levels with another drug greatly diminishes the effectiveness of PCP, suggesting a possible new approach to the treatment of schizophrenia.

Drugs That Act in the Synaptic Cleft

So far, the psychopharmacological agents we have discussed act primarily as receptor agonists and antagonists. Other important classes of drugs affect neurotransmitter metabolism in the synaptic cleft. Catecholamines are catabolized in the cleft either by methylation (**catecholamine O-methyltransferase**, COMT) or by oxidation (**monoamine oxidase**, MAO). These enzymes limit the biological effects of catecholamine neurotransmitters, just as acetylcholinesterase limits the firing of cholinergic neurons. A number of drugs used to treat depression are inhibitors of either COMT or MAO, and their action increases the effective amounts of neurotransmitter by limiting its breakdown. A more recently developed drug, **fluoxetine**, marketed as Prozac, acts as a selective serotonin reuptake inhibitor (SSRI). Secreted neurotransmitter has three possible fates—binding to postsynaptic receptors, catabolism in the cleft, or reuptake into the presynaptic cell, for re-packaging into storage vesicles. Prozac selectively blocks the reuptake of serotonin, thereby increasing the amount that reaches the post-synaptic side and potentiating serotonergic synapses. Originally marketed as an antidepressant drug, Prozac has found use against a range of psychiatric disorders.

Peptide Neurotransmitters and Neurohormones

Finally, we mention a number of peptides, including somatostatin, neurotensin, and the **enkephalins**, which act as neurotransmitters (Table 23.5). The enkephalins, along with **β-endorphin**, function also as **neurohormones**, acting to

Ritalin
(methyl phenidylacetate)

Prozac
(fluoxetine)

TABLE 23.5 Some peptides that act as neurohormones (H) or neurotransmitters (T)

Name	H/T	Sequence[a]
β-Endorphin	H	YGGFMTSFKSQTPLVTLFKNAIIKNAYKKGE
Met-enkephalin	H, T	YGGFM
Leu-enkephalin	H, T	YGGFL
Neurotensin	T	pELYENKPRRPYIL
Somatostatin	T	AGCKNFFWKTFTSC

[a]The sub-sequence YGGF, common to β-endorphin and the enkephalins, appears to be essential for their narcotic effects. The p at the N-terminal end of neurotensin signifies that the glutamate has been cyclized to the "pyro" form.

modify the ways in which nerve cells respond to transmitters. The endorphins were discovered in the 1970s, as the result of Solomon Snyder's efforts to understand the effects of opiate drugs, such as **morphine**. After detecting morphine receptors in human brain, Snyder realized that the receptors must have a natural endogenous ligand because morphine is derived from a nonhuman source, the poppy. This work led to isolation of the endorphins, small peptides that function as natural analgesics. The modification of neural signals by these substances appears to be responsible for the insensitivity to pain that is experienced under conditions of great stress or shock. The effectiveness of opiate analgesics such as morphine is a consequence, perhaps accidental, of the recognition of these opiates by neurohormone receptors, despite their structural differences from neurohormones.

The endorphins and enkephalins are synthesized as part of the much longer hormone precursor *prepro-opiomelanocortin*. As described on page 964, this precursor is cleaved to release both the neurohormones and a number of other hormones with entirely different functions.

Signaling in Bacteria and Plants

Although the concept of signal transduction is somewhat different for single-celled than for multicellular organisms, the fact is that bacteria do respond to extracellular signals. In Chapter 21 we mentioned quorum sensing, in which bacteria respond to signals based upon cell density. In addition, bacteria respond to environmental factors, such as chemical attractants, oxygen concentration, or temperature gradients. The sensing mechanism is a **two-component system**, involving two protein kinases. The first component, a **receptor histidine kinase**, spans the plasma membrane. Interaction of this component with an environmental stimulus on the extracellular side stimulates autophosphorylation of a histidine residue on the intracellular domain. Next, the phosphate is transferred to an aspartate residue on the second protein, which is called a **response regulator**. The phosphorylated response regulator next interacts with the target; for example, if the signal is to direct response to a gradient of nutrient concentration, the flagellum undergoes a change in its rotational behavior, by mechanisms under active investigation, to move the bacterium toward a chemical attractant.

Our understanding of the molecular actions of plant hormones is less advanced than that of the vertebrate animal hormones. In part this is because some plant hormones are growth factors so that growth is the only readily measurable parameter. Furthermore, plant membranes are more difficult to isolate and study than are animal membranes.

We encountered some of the six major classes of plant hormones in earlier chapters about metabolism. To recapitulate, the six classes, illustrated in Figure 23.27, are (1) the diterpene **gibberellins**, derived from isopentenyl pyrophosphate; (2) the sesquiterpene **abscisic acid**, also derived from isopentenyl pyrophosphate; (3) the **cytokinins**, purine bases with a terpenoid side chain;

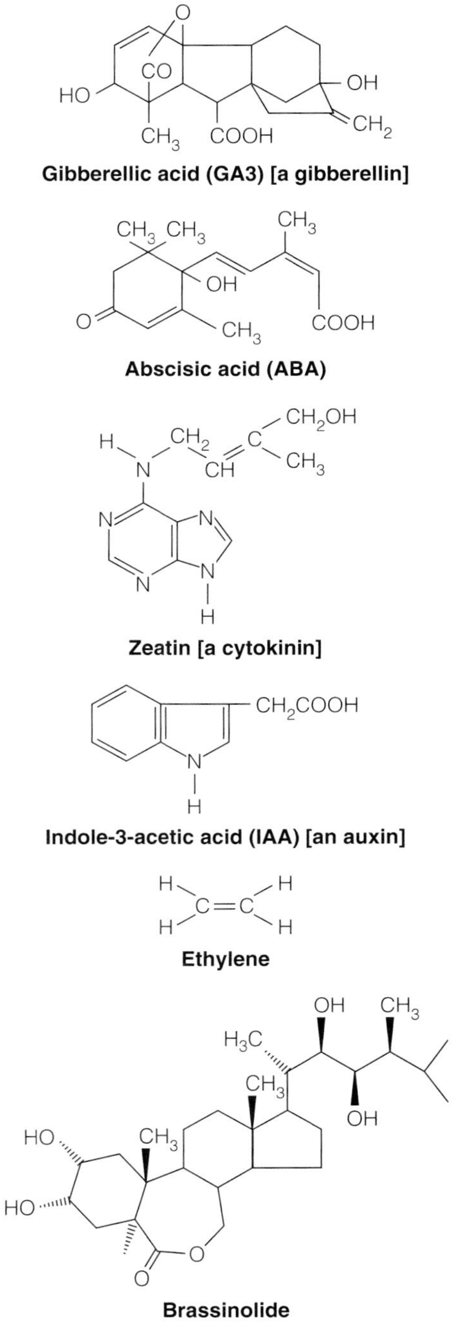

Gibberellic acid (GA3) [a gibberellin]

Abscisic acid (ABA)

Zeatin [a cytokinin]

Indole-3-acetic acid (IAA) [an auxin]

Ethylene

Brassinolide

FIGURE 23.27

Representatives of the six major classes of plant hormones.

(4) **auxins**, with the tryptophan metabolite *indole-3-acetic acid* being the most active; (5) **ethylene**, which comes from the methionyl moiety of *S*-adenosylmethionine; and (6) **brassinosteroids**, related to vertebrate steroid hormones. With the exception of brassinosteroids, these compounds are chemically quite distinct from animal hormones.

Additional differences from animal hormones lie in (1) the diversity of effects of a given plant hormone and (2) the hormonal activity of a large number of structurally related species. For example, at least a dozen different cytokinins have been characterized structurally. Though all contain adenine with a side chain on N-6, great variability occurs in the structure of that side chain.

Auxins are synthesized in apical buds (at the tip) of growing shoots. They stimulate growth of the main shoot and inhibit lateral shoot development. A class of auxin-binding membrane proteins may represent auxin receptors. Auxin action involves chemiosmotic proton gradients, which in turn establish auxin concentration gradients leading to different responses in different parts of the plant—partial cell wall degradation, necessary for growth to occur, and increased RNA and protein synthesis, necessary for differentiation. Recent evidence suggests that cyclic AMP is formed in plants and that it mediates the effects of auxin.

Cytokinins are produced in roots and promote growth and differentiation in many tissues. Cytokinins and auxins work together, and the ratio of cytokinin to auxin is often crucial in determining whether a plant will grow or differentiate. In many cases, single plant cells in tissue culture can regenerate whole plants. Here, the investigator must experiment with different proportions of cytokinins and auxins, searching empirically for the proportions that allow the best growth of normal plants.

About 100 forms of gibberellin are known. Some function as growth-promoting hormones and may stimulate expression of certain genes. Particular messenger RNAs are present in increased amounts as a result of gibberellin administration, supporting the idea that gibberellins act at the gene level.

Ethylene is considered a hormone of senescence. It stimulates fruit ripening and the aging of flowers, and it inhibits seedling growth. It also redirects auxin transport to promote transverse, rather than longitudinal, growth of plants. Abscisic acid (ABA) counteracts the effects of most other plant hormones. It inhibits germination, growth, budding, and leaf senescence and helps plants adjust to stresses such as cold or drought. Abscisic acid synthesis in leaves is stimulated by wilting. Physiological evidence suggests a role for ABA in ion and water balance. Brassinosteroids control several processes, including cell expansion, seed germination, and vascular development.

Recent work sheds light on signaling processes for these latter three hormones. Ethylene binds to a family of receptors in the ER membrane. The receptors then bind to a Raf-like protein kinase, CTR1 (Figure 23.28). This binding leads to inactivation of both receptors and CTR1, leading in turn to derepression of a positive regulatory molecule, EIN2. A MAP kinase cascade, shown in yellow, may be involved. EIN2 somehow stabilizes transcription factors EIN3 and EILs within the nucleus, leading to transcription of ethylene-responsive genes. Among these gene products are two proteins, EBF1 and EBF2, which regulate levels of EIN2, forming a feedback loop.

Brassinosteroids act similarly to steroid hormones in animals. An intracellular protein kinase called BIN2 phosphorylates two transcriptional regulators, BES1 and BZR1, targeting them for proteolytic destruction. Binding brassinosteroid to its receptor, BRI1, leads to inactivation of BIN2, which in turn leads to activation of BES1 and BZR1. BES1 joins with another protein, BIM, leading to transcriptional activation of some target genes, while BZR1 inhibits transcription of another set of target genes.

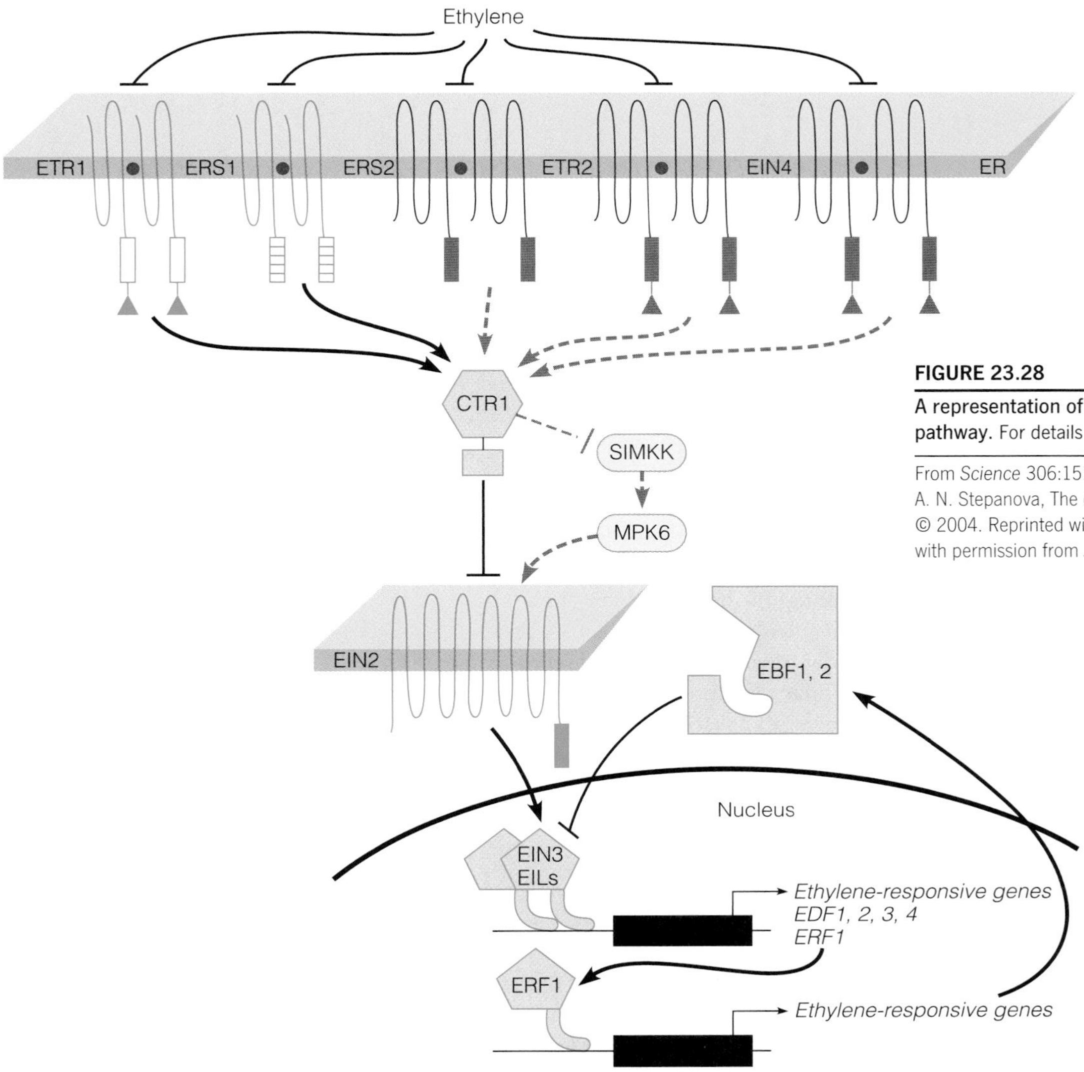

Ethylene

ETR1 • ERS1 • ERS2 • ETR2 • EIN4 • ER

CTR1

SIMKK

MPK6

EIN2

EBF1, 2

Nucleus

EIN3
EILs

Ethylene-responsive genes
EDF1, 2, 3, 4
ERF1

ERF1

Ethylene-responsive genes

FIGURE 23.28

A representation of the ethylene-signaling pathway. For details see text.

From *Science* 306:1513–1515, J. M. Alonso and A. N. Stepanova, The ethylene signaling pathway. © 2004. Reprinted with permission from AAAS. Redrawn with permission from Jose Alonso.

The ABA pathway begins with binding to an extracellular receptor, followed by release of calcium from vacuoles into the cytosol. Involved with this process is a signaling component called cyclic ADP-ribose (cADPR), already known to regulate calcium release in animal cells. As in the animal system, calcium release by cADPR opens ion channels in the plasma membrane. The release of intracellular ions reduces osmotic pressure and lowers cell turgor, also allowing for preservation of cell water when insufficient water is available in the soil to meet the needs of transpiration.

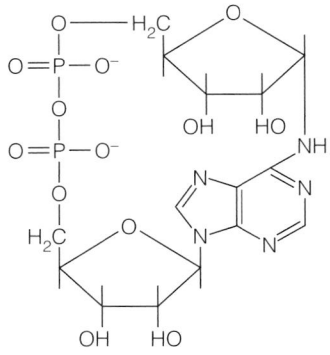

Cyclic ADP-ribose

SUMMARY

Hormone action is one element of signal transduction mechanisms, processes by which instructions are transmitted from cell to cell. Some hormones interact with intracellular receptors; these act at the gene level, with the hormone–receptor complex affecting transcription of specific genes in target tissue. Other hormones interact with plasma membrane receptors. Three types of such transmembrane receptors exist: (1) receptors that are ion channels, with hormone binding directly affecting membrane permeability to an ion; (2) receptors, such as the insulin receptor, that have a hormone-binding site on the exterior and an enzyme activity on the cytosolic side, with ligand binding stimulating that activity; and (3) receptors that act through G proteins to affect the levels of second messengers, including cyclic AMP, cyclic GMP, calcium ion, inositol trisphosphate, phosphatidylinositol bisphosphate, and diacylglycerol. Second messengers interact with a variety of intracellular metabolic processes, notably protein phosphorylation. G protein action is controlled by binding of guanine nucleotides. Stimulation by interaction with receptor causes GTP exchange for GDP, and hormone action is limited by the slow conversion of bound GTP to GDP.

Proto-oncogenes are cellular genes (most of which encode signal transduction proteins) that have been transferred to viral genomes and that subsequently undergo mutations converting them to oncogenes. Action of an oncogene, whether generated in a cell by viral infection or by mutation of a proto-oncogene, causes loss of metabolic and growth control that is associated with transformation of a normal cell to a cancer cell.

Neurotransmission involves release of neurotransmitter from storage vesicles in the presynaptic neuron, triggered when the wave of depolarization in the presynaptic neuron reaches the vesicles. Neurotransmitter is released into the synaptic cleft and a small portion of the released material is taken up in the postsynaptic neuron, where opening of ion channels leads to re-establishment of the action potential and its movement along the postsynaptic cell. At the neuromuscular junction, neurotransmitter uptake leads to muscle contraction.

Plant hormones include gibberellins, cytokinins, auxins, ethylene, brassinosteroids, and abscisic acid. Recent evidence indicates parallels between the actions of animal and plant hormones, although we know much less about plant hormones.

REFERENCES

General

Czech, M. P., and S. Corvera (1999) Signaling mechanisms that regulate glucose transport. *J. Biol. Chem.* 274:1865–1868. One of several JBC minireviews on the stimulation by insulin of glucose transport.

Protein Phosphorylation and Dephosphorylation

Dessauer, C. W. (2009) Adenylyl cyclase-A-kinase anchoring protein complexes: The next dimension in cAMP signaling. *Molec. Pharmacol.* 76:935–941. AKAPs and proteins that anchor adenylate cyclase.

Hafen, E. (1998) Kinases and phosphatases—A marriage is consummated. *Science* 280:1212–1213. Physical complexes between protein kinases and phosphatases facilitate the regulation of signal transduction pathways involving protein phosphorylation and dephosphorylation.

Hunter, T. (1995) Protein kinases and phosphatases: The yin and yang of protein phosphorylation and signaling. *Cell* 8D:225–238. One of nine still-timely reviews on signal transduction in this special issue.

Smith, F. D., L. K. Langenberg, and J. D. Scott (2006) The where's and when's of kinase anchoring. *Trends in Biochem. Sci.* 31:316–318. A brief historical review.

Synthesis of Peptide Hormones

Fisher, J. M., and R. H. Scheller (1988) Prohormone processing and the secretory pathway. *J. Biol. Chem.* 263:16515–16518. A review that describes how peptide hormones are formed by cleavage from high-molecular-weight precursors.

Receptors

Black, J. (1989) Drugs from emasculated hormones: The principle of syntopic antagonism. *Science* 245:486–493. Black's Nobel Prize address, which describes the development of drugs that are adrenergic receptor antagonists.

Boguth, C. A., P. Singh, C-c. Huang, and J. J. G. Tesmer (2010) Molecular basis for activation of G protein-coupled receptor kinases. *EMBO J* 29:3249–3259. Crystal structure of a G protein receptor kinase yields clues to allosteric interactions.

DiMarzo, V., and S. Petrosino (2007) Endocannabinoids and the regulation of their levels in health and disease. *Curr. Opin. Lipidol.* 18:129–140. A review of the endogenous ligand for the tetrahydrocannabinol receptors.

Lefkowitz, R. J., and S. K. Shency (2005) Transduction of receptor signals by β-arrestins. *Science* 308:512–517. A review describing the range of functions controlled by arrestins.

Mustafi, D., and K. Palczewski (2009) Topology of class A G protein-coupled receptors: Insights gained from crystal structures of rhodopsins, adrenergic, and adenosine receptors. *Molecular Pharm.* 75:1–12. A recent review of GPCR structures.

Pitcher, J. A., N. J. Freedman, and R. J. Lefkowitz (1998) G protein-coupled receptor kinases. *Annu. Rev. Biochem.* 67:653–692. A review describing this down-regulatory mechanism.

Rockman, H. A., W. J. Koch, and R. J. Lefkowitz (2002) Seven-transmembrane-spanning receptors and heart function. *Nature* 415:206–212. A comprehensive review, focusing upon receptor action in the heart.

Rosenbaum, D. M., S. G. F. Rasmussen, and B. K. Kobika (2009) The structure and function of G-protein-coupled receptors. *Nature* 459:356–363. Recent review showing remarkable structural similarity among four representatives of this receptor class.

Sprang, S. R. (2007) A receptor unlocked. *Nature* 450:355–356. Conveys the excitement emanating from the β-adrenergic receptor structure determination.

Xu, F., H. Wu, V. Katritch, G. W. Han, K. A., Jacobson, Z-G. Gao, V. Cherezov, and R. C. Stevens (2011) Structure of an agonist-bound human A_{2A} adenosine receptor. *Science* 332:322–327. One of the most recent GPCR structure determinations.

G Proteins

Huang, C-c, and J. J. G. Tesmer (2011) Recognition in the face of diversity: Interactions of heterotrimeric G proteins and G protein-coupled receptor (GPCR) kinases with activated GPCRs. *J. Biol. Chem.* 286:7715–7721. Structural evidence for common receptor binding modes for G proteins and GPCR kinases.

Scheffzek, K., M. R. Ahmadian, and A. Wittinghofer (1998) GTPase-activating proteins: Helping hands to complement an active site. *Trends Biochem. Sci.* 23:257–262. Crystal structures of several GAPs show how they interact with and activate the GTPase of Ras-related proteins.

Snyder, S. H., P. B. Sklar, and J. Pevsner (1988) Molecular mechanisms of olfaction. *J. Biol. Chem.* 263:13971–13975. This minireview describes evidence for G protein involvement in the sense of smell.

Sprang, S. R., and D. E. Coleman (1998) Invasion of the nucleotide snatchers: Structural insights into the mechanism of G protein GEFs. *Cell* 95:155–158. How proteins stimulate the GDP–GTP exchange in G protein activation.

Tesmer, J. J. G. (2010) The quest to understand heterotrimeric G protein signaling. *Nature Str. Biol.* 17:650–652. A recent short review with excellent figures.

Tesmer, J. J. G., R. K. Sunahara, A. G. Gilman, and S. R. Sprang (1997) Crystal structure of the catalytic domains of adenylyl cyclase in a complex with $G_{s\alpha}$-GTPγS. *Science* 278:1907–1916. Structural analysis of G proteins reveals the mechanism of adenylate cyclase activation.

Second-Messenger Systems

Berridge, M. (1993) Inositol trisphosphate and calcium signaling. *Nature* 361:315–325. The role of calcium as a second and/or third messenger.

Hatch, A. J., and J. D. York (2010) SnapShot: Inositol phosphates. *Cell* 143:1030–1030.e1. A one-page summary of the numerous roles of these compounds in signaling.

Hill, B. G., B. P. Dranka, S. M. Bailey, J. R. Lancaster, Jr., and V. M. Darley-Usmar (2010) What part of NO don't you understand? Some answers to the cardinal questions in nitric oxide biology. *J. Biol. Chem.* 285:19699–19704.

Hodgkin, M. N., T. R. Pettit, A. Martin, R. H. Mitchell, A. J. Pemberton, and M. J. O. Wakelam (1998) Diacylglycerols and phosphatidates: Which molecular species are intracellular messengers? *Trends Biochem. Sci.* 23:200–204. A short recent review of lipid-derived second messengers.

Hofmann, F. (2005) The biology of cyclic GMP-dependent protein kinases. *J. Biol. Chem.* 280:1–4. These enzymes are somewhat different from protein kinase A.

Hurley, J. H. (1999) Structure, mechanism, and regulation of adenylyl cyclase. *J. Biol. Chem.* 274:7599–7602. This review is particularly timely in view of recent structural insights into adenylate cyclase.

Majerus, P. W., M. V. Kisseleva, and F. A. Norris (1999) The role of phosphatases in inositol signaling reactions. *J. Biol. Chem.* 274:10669–10672. A brief review of the control of phosphoinositide synthesis and turnover.

Prentki, M., and S. R. M. Madiraju (2008) Glycerolipid signaling in health and disease. *Endocrine Reviews* 29:647–676.

Ruiz-Stewart, I., S. R. Tiyyagura, J. E. Lin, G. M. Pitari, S. Schulz, E. Martin, F. Murad, and S. A. Waldman (2004) Guanylyl cyclase is an ATP sensor coupling nitric oxide signaling to cell metabolism. *Proc. Natl. Acad. Sci. USA* 101:37–42. Evidence for a link between metabolism and a cell's energy status.

Singer, W. D., H. A. Brown, and P. C. Sternweis (1997) Regulation of eukaryotic phosphatidylinositol-specific phospholipase C and phospholipase D. *Annu. Rev. Biochem.* 66:475–509. A review of the synthesis of second messengers from phosphoinositides.

Wall, M. E., S. H. Francis, J. D. Corbin, K. Grimes, R. Richie-Jannetta, J. Kotera, B. A. Macdonald, R. R. Gibson, and J. Trewhella (2003) Mechanisms associated with cGMP binding and activation of cGMP-dependent protein kinase. *Proc. Natl. Acad. Sci. USA* 100:2380–2385. The mechanisms are not quite as well understood as the activation of protein kinase A is, but this paper describes a nice biophysical analysis.

Receptor Tyrosine Kinases

Claesson-Welsh, L. (1994) Platelet-derived growth factor receptor signals. *J. Biol. Chem.* 269:32023–32036. A discussion of all of the proteins known to interact with PDGF receptors.

Lemmon, M. A., and J. Schlessinger (2010) Cell signaling by receptor tyrosine kinases. *Cell* 141:1117–1134. A comprehensive recent review.

Schlessinger, J. (2004) Common and distinct elements in cellular signaling via EGF and FGF receptors. *Science* 306:1506–1507. A brief article showing the complete signaling systems initiated through these two receptors.

Xu, W., S. C. Harrison, and M. J. Eck (1997) Three-dimensional structure of the tyrosine kinase c-Src. *Nature* 385:595–602. Understanding the structure of this protein revealed how interaction with other proteins could activate the protein kinase activity of the Src protein.

Nuclear Receptors

Auwerx, J., and 39 coauthors (1999) A unified nomenclature system for the nuclear receptor superfamily. *Cell* 97:161–163. Forty leaders in this field propose a unifying classification scheme for this ever-growing family of receptors, with references to other review literature.

Meijsing, S. H., M. A. Pufall, A. Y. So, D. L. Bates, L. Chen, and K. Yamamoto (2009) DNA binding site sequence directs glucocorticoid receptor structure and activity. *Science* 324:407–410. Recent evidence for the extraordinary sensitivity of glucocorticoid signaling to DNA base sequence of the receptor binding site.

Tsai, M. J., and B. W. O'Malley (1994) Molecular mechanisms of action of steroid/thyroid receptor superfamily members. *Annu. Rev. Biochem.* 63:451–486. A comprehensive review of the older literature.

Oncogenes and Growth Factors

Ashall, L., and 13 coauthors (2009) Pulsatile stimulation determines timing and specificity of NF-κB-dependent transcription. *Science* 324:242–246. Unexpected complexity in this signaling system.

Birge, R. B., and H. Hanafusa (1993) Closing in on SH2 specificity. *Science* 262:1522–1524. A minireview describing how SH2 domains explain the specificity of cellular responses to growth factors.

Burley, S. K. (1994) p53: A cellular Achilles' heel revealed. *Structure* 2:789–792. A readable minireview describing the excitement created by the structural determination of the p53 DNA-binding domain.

Eilers, M., and R. N. Eisenman (2008) Myc's broad reach. *Genes & Development* 22:2755–2766. A review of an extraordinarily complex proto-oncogene.

Fox, E. J., J. J. Salk, and L. A. Loeb (2009) Cancer genome sequencing—an interim analysis. *Cancer Res.* 69:4948–4950. The complexity of cancer as revealed from DNA sequencing in individual tumors.

Green, D. R., and G. Kroemer (2009) Cytoplasmic functions of the tumour suppressor p53. *Nature* 458:1127–1130. A review of cytoplasmic functions, but also useful as a general recent review of p53.

Johnson, G. L., and Lapadet, R. (2002) Mitogen-activated protein kinase pathways mediated by ERK, JNK, and p38 protein kinases. *Science* 298:1911–1912. A brief article describing the parallels in a number of MAP kinase pathways.

Massagué, J. (1998) TGF-β signal transduction. *Annu. Rev. Biochem.* 67:753–791. A review describing the signaling pathways that involve receptor serine/threonine kinases.

Mishra, L., R. Derynck, and B. Mishra (2005) Transforming growth factor-β signaling in stem cells and cancer. *Science* 310:68–71. One of several readable articles in a special section of *Science* dealing with cell signaling.

Neurotransmission

Kandel, E., and L. Squire (2000) Neuroscience: Breaking down scientific barriers to the study of brain and mind. *Science* 290:1113–1120. Kandel's Nobel Prize lecture published alongside lectures by fellow Nobelists Arvid Carlsson and Paul Greengard.

Plant Hormones

Alonso, J. M., and A. N. Stepanova (2004) The ethylene signaling pathway. *Science* 306:1513–1515. One of several useful articles in a special section on cell signaling.

Heldt, H.-W. (1997) *Plant Biochemistry and Molecular Biology,* pp. 394–414. Oxford University Press, New York. Chapter 19 of this book contains a comprehensive treatment of plant growth hormones and their actions.

Yin, Y., D. Vafeados, Y. Tao, S. Yoshida, T. Asami, and J. Chory (2005) A new class of transcription factors mediates brassinosteroid-regulated gene expression in *Arabidopsis. Cell* 120:249–259. An article showing similarities of plant and animal steroid hormone signaling.

Yoo, S-H., Y. Cho, and J. Sheen (2009) Emerging connections in the ethylene signaling network. *Trends in Plant Sci.* 14:270–279. An up-to-date review of ethylene signaling.

PROBLEMS

1. G proteins show some amino acid sequence homology with bacterial elongation factors, proteins that are involved in protein synthesis. These proteins also bind and hydrolyze GTP. How might one determine whether plant tissues contain G proteins or similar entities?

2. List two or three factors that make it advantageous for peptide hormones to be synthesized as inactive prohormones that are activated by proteolytic cleavage.

3. Name a hormone whose concentration would probably not be amenable to analysis by radioimmunoassay, and explain your answer. (Hint: You must be able to generate an antibody to the hormone to develop such an assay.)

4. Describe a mechanism by which a steroid hormone might act to increase intracellular levels of cyclic AMP.

5. Signaling molecules interact with cells through specific macromolecular receptors. For each of the four receptors identified below, list all characteristics, by number, which accurately describe that receptor.
    (a) an adrenergic receptor _____
    (b) a steroid receptor _____
    (c) the LDL receptor _____
    (d) the insulin receptor _____
        1. Located at the cell surface
        2. Associated with the protein clathrin
        3. Ligand binding stimulates the activity of phospholipase C
        4. A transmembrane protein
        5. A DNA-binding protein
        6. Located in the cell interior
        7. Receptor–ligand complex moves to the lysosome
        8. Receptor–ligand complex becomes concentrated in the nucleus
        9. Receptor activation can inhibit the synthesis of glycogen
        10. The hormone–receptor complex activates specific gene transcription
        11. Internalization decreases the synthesis of cholesterol esters
        12. Action of this receptor diminishes the synthesis and activity of β-hydroxy-β-methylglutaryl-CoA reductase (HMG-CoA reductase)
        13. This receptor activates its own synthesis
        14. Biological activity of this receptor involves interaction with guanine nucleotide–binding proteins
        15. This receptor has a protein kinase activity
        16. Not known to act through a second messenger

6. Upon activation by a receptor, a G protein exchanges bound GDP for GTP, rather than phosphorylating GDP that is already bound. Similarly, the α subunit–GTP complex has a slow GTPase activity that hydrolyzes bound GTP, rather than exchanging it for GDP. Describe experimental evidence that would be consistent with these conclusions.

7. Answer each question in about one sentence.
    (a) What is the biochemical basis for the action of Viagra?
    (b) What is the biochemical basis for the action of Prozac?
    (c) Oral administration of *S*-adenosylmethionine has been reported to be effective in treating depression. Suggest a possible explanation.

8. Write a balanced equation for the hydrolysis of cyclic GMP, catalyzed by cGMP phosphodiesterase. Would you expect an inhibitor of this enzyme to potentiate or antagonize the action of Viagra? Explain.

9. Lithium ion inhibits the synthesis of inositol trisphosphate by inhibiting a reaction in the breakdown of inositol trisphosphate. Explain this apparent paradox.

10. Describe two general chemical features that distinguish plant hormones from animal hormones.

11. Suppose that you measured binding of EGF to the isolated EGF receptor at various concentrations. Would you expect the binding curve to be hyperbolic? Explain your answer.

12. Write balanced equations for reactions catalyzed by catechol O-methyltransferase and monoamine oxidase.

PART 5

Information

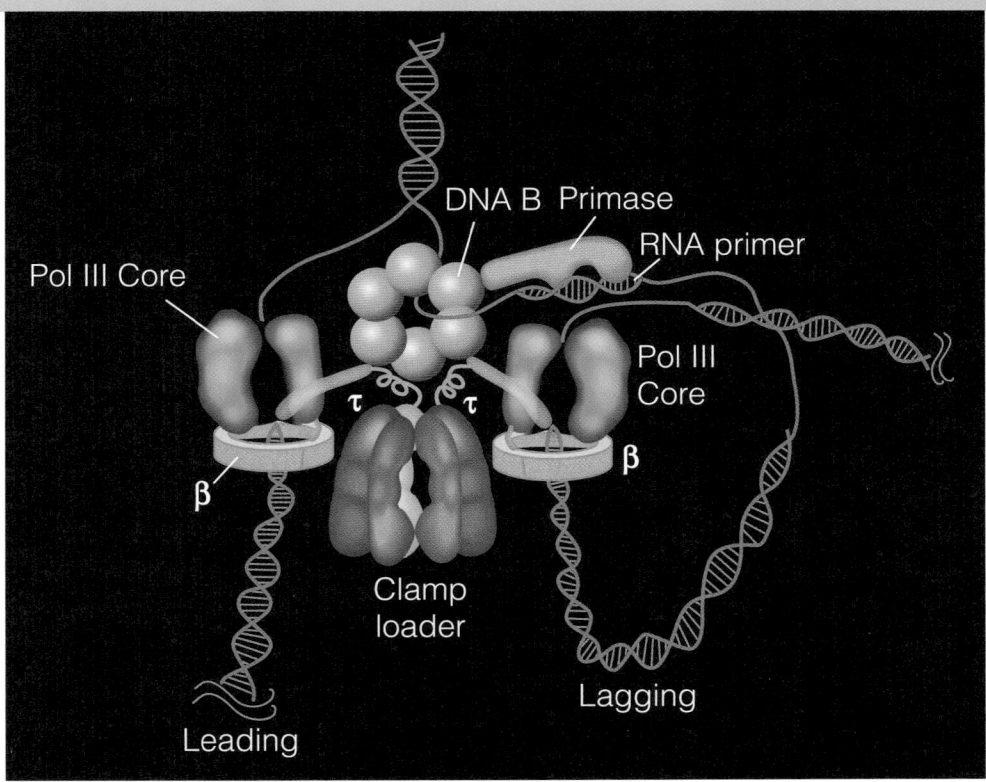

The replisome, a multicomponent protein machine that copies both strands of a DNA double helix. Source: Courtesy of Dr. Michael O'Donnell

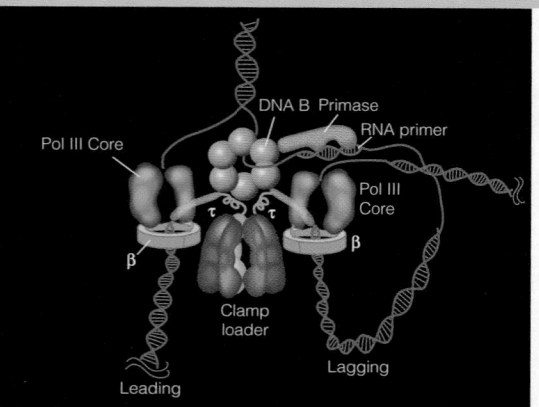

Pol III Core

DNA B Primase

RNA primer

Pol III Core

Clamp loader

β

β

Leading

Lagging

CHAPTER 24

Genes, Genomes, and Chromosomes

In this last major section of the book we are concerned with the storage, retrieval, processing, and transmission of biological information—processes we might call information metabolism, as distinguished from intermediary metabolism. In intermediary metabolism, the information specifying the nature of a chemical reaction lies within the three-dimensional structure of the enzyme involved. That structure determines which substrates are bound and which reactions are catalyzed. Of course, metabolic reactions are controlled ultimately by genetic information, which specifies the structures and properties of enzymes. However, the reactions we encounter from here on are distinguished by the direct involvement of genetic information—specifically, the requirement for a *template*, which functions along with an enzyme to specify the reaction catalyzed. The biological templates, nucleic acids, generally play a passive role, determining which substrates are bound and leaving catalysis to enzymes—although ribozymes (RNA enzymes) present an important exception to this generalization (Chapter 11). The basic processes—DNA replication, transcription (RNA synthesis), and translation (protein synthesis)—are described in detail in subsequent chapters. Here, however, we focus upon the organization and nature of the genes that carry and transmit that information, as well as the chromosomes and genomes within which individual genes reside. As discussed in Chapter 4, we define a gene as a chromosomal segment that encodes a single polypeptide chain or RNA molecule. Although in the past we might have called this section of our discipline genetic biochemistry, it may now be more appropriate to call it *genomic biochemistry*. That is because the powerful methods used to determine the complete nucleotide sequences of hundreds of organisms have made it possible to think of biological processes in a more global sense. Whereas the sequencing and cloning of individual genes, and the analysis of cloned genes and recombinant enzymes, allow great insight into metabolism at the level of individual or grouped reactions, we can now think of cellular and organismal function in terms of the coordinated expression of large blocks of genes—that is, as an integrated system.

Prokaryotic and Eukaryotic Genomes

Size of the Genome

With the exception of RNA viruses, the genomes of all organisms are made up of DNA sequences containing just four different nucleotides (plus small amounts of a few modified nucleotides; see pages 93 and 102). However, there is enormous variability in both genome size and in the physical state of the genome within its intracellular milieu. Single-celled organisms have relatively simple metabolic needs and so can thrive with as few as several hundred genes. Viruses, which co-opt the metabolic machinery of the cells they infect, are even simpler. The smallest DNA viruses have only 10 genes or fewer, while some RNA viruses have only three. As you can see in Figure 24.1, however, the largest viruses, such as variola, the smallpox virus, have 175,000 or more base pairs in their genomes, sufficient DNA to encode 150 proteins or more.

As you might expect, the genomes of the more complex multicellular organisms are far larger than those of bacteria or viruses. However, as you can see from Figure 24.1, there is no logical relationship between genome size and organismal complexity. The human genome, consisting of about three billion base pairs, is dwarfed by the genomes of some amphibians and some plants, which may be as much as 50-fold larger than those of humans. Because there is no basis for concluding that a bean plant is 50 times more complex than a human being, an inescapable conclusion is that eukaryotic genomes must contain considerable DNA that does not code for proteins or for the RNA machinery of protein synthesis. In the following sections, we describe several kinds of noncoding DNA sequences in eukaryotic genomes. In addition, as discussed in Chapter 29, much of the genomic DNA that does not encode proteins encodes regulatory RNA molecules.

Most eukaryotes require—and have—much larger genomes than prokaryotes.

Repetitive Sequences

The first indication that eukaryotic chromosomes contain noncoding DNA came as early as 1968, when Roy Britten and David Kohne developed a technique for analysis of DNA reassociation kinetics. In this method, the total DNA

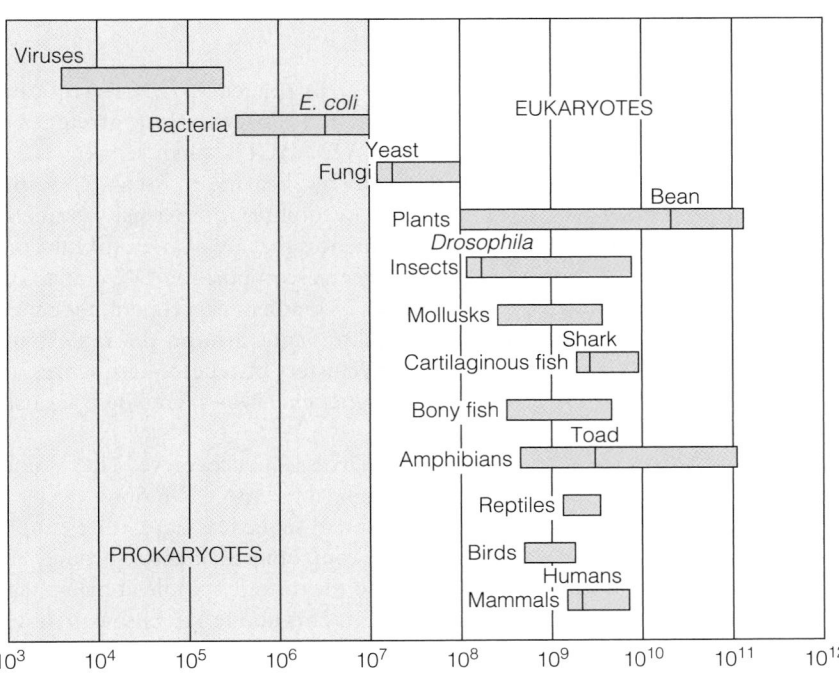

FIGURE 24.1

Genome size. The bars show the range of haploid genome sizes for different groups of organisms. A few specific organisms are marked with vertical lines, for example 3×10^9 for humans. Note that the genome-size scale is logarithmic and that many organisms have larger genomes than humans.

FIGURE 24.2

Reassociation kinetics of *E. coli* and bovine DNA. The abscissa corresponds to reassociation time, corrected for the difference in size between the *E. coli* and bovine genomes. The curve for *E. coli* corresponds to that expected for a collection of single-copy genes in a genome of the *E. coli* size—4.67×10^6 bp. The curve for bovine DNA exhibits two steps in reassociation. The slow step corresponds to single-copy DNA (nonrepeated sequences). The other corresponds to rapidly reassociating DNA made up of repeated sequences. Many classes of repeated DNA are represented in this phase of the reassociation.

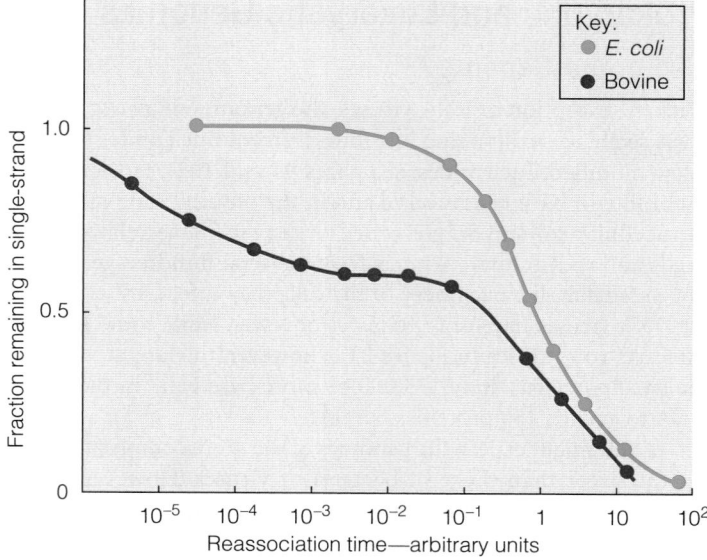

Repetitive DNA sequences in eukaryotic genomes include satellite DNAs and scattered duplicate sequences.

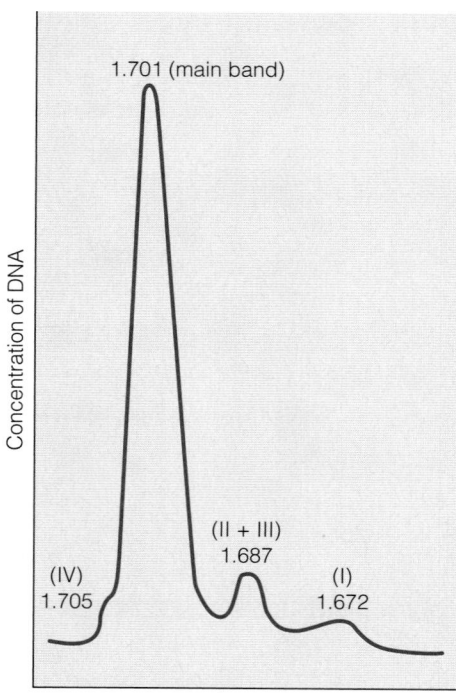

Buoyant density, g/cm³

FIGURE 24.3

Satellite DNA. Equilibrium density gradient centrifugation of total *Drosophila* DNA resolves satellite bands, surrounding the main band. These represent repetitive DNA fractions of differing base composition.

Adapted from *Journal of Molecular Biology* 96:665–674, S. A. Endow, M. L. Polan, and J. G. Gall, Satellite DNA sequences of *Drosophila melanogaster*. © 1975, with permission from Elsevier.

from an organism is cut by shear forces into pieces about 300 base pairs in length, then heated to cause strand separation, then slowly cooled, to allow strands of complementary sequence to reassociate. Sequences that are present in multiple copies will reassociate relatively quickly, whereas single sequences reassociate slowly because their abundance is low. When Britten and Kohne analyzed bovine DNA by this method, they were surprised to find that almost half the DNA reassociated much more quickly than expected for single-copy sequences in the large genome (Figure 24.2). To account for this rapid reassociation, they concluded that some DNA sequences are reiterated 10^5–10^6 times in one cell. Whereas practically all of the DNA of *E. coli* is single-copy, only about half of most mammalian DNAs and one third of plant DNAs fall into this category.

Satellite DNA

Further analysis has divided these reiterated DNA sequences into several categories. One type involves multiple tandem repetitions, over long stretches of DNA, of very short, simple sequences like $(ATAAACT)_n$. Such sequences can often be separated from the bulk of the DNA by shearing to break DNA into smaller fragments, followed by sedimentation to equilibrium in density gradients (see Chapter 4, page 102). Repetitive DNA sequences that are AT-rich (like the one just mentioned) have a lower density than average-composition DNA, and GC-rich sequences are more dense. Thus, in a density-gradient experiment, the reiterated, simple-sequence fragments form satellite bands around the main-band DNA (Figure 24.3). Because of this banding, clusters of repeated sequences are sometimes referred to as **satellite DNAs**. In higher eukaryotes, satellite DNA usually makes up 10%–20% of the total genome.

What function can such highly reiterated DNA sequences serve? They do not code for proteins, and most are not even transcribed into RNA. Some, at least, appear to play a structural role. Certain reiterated sequences have, for example, been found to be highly concentrated near the **centromeres** of chromosomes, the regions where sister chromatids are attached to the mitotic spindle at metaphase. In budding yeast, centromeres are AT-rich segments about 125 base pairs long, although most other organisms have shorter repeating sequences in centromeric DNA. Centromeres serve as binding sites for proteins that attach the spindle fibers in mitosis (page 1022).

Duplications of Functional Genes

There are many other classes of DNA sequences with varying degrees of repetition. Some of these sequences represent duplications of functional genes, and in some cases the repetitiveness seems to play a useful role, by allowing high levels of production of much-needed transcripts. Examples include the genes for ribosomal RNAs, of which up to several thousand copies may be present, and tRNA genes, with hundreds of copies of each type often found. The cell's continuous need for large quantities of ribosomes and tRNAs for translation is met by having multiple copies of these genes. The same is true for the genes for some much-used proteins, such as the histones that bind to eukaryotic DNA to form the chromatin structure (see page 1022). As pointed out in Chapter 26, even genes that are normally single-copy are sometimes amplified, either in response to environmental stress or in special tissues during embryonic development.

Alu Elements

Other kinds of repeated DNA sequences exist that do not code for proteins but whose true function remains mysterious. Such sequences are often scattered throughout the genome, rather than being clustered like the satellite DNAs. Some of these sequences may represent control elements of still unknown function. One of the most common such families in primates consists of the so-called *Alu* **sequences**. These sequences, of which there are more than one million copies in the human genome, are about 300 base pairs (bp) long. Their name reflects the common existence of a single site for the restriction endonuclease *Alu*I in most members of this class. The *Alu* sequences can be (inefficiently) transcribed into RNA, although they are not known to be translated. *Alu* sequences are examples of **SINES**, short interspersed elements. The human genome also contains up to one million **LINES**, long interspersed elements, sequences that can be as long as 10 kbp.

The function of the large number of *Alu* sequences remains uncertain, although some of them may contain origins for DNA replication. But it is also conceivable that many repetitive sequences such as these serve no useful function. They may simply exist in the genome as "molecular parasites." A way in which such sequences could spread through the genome has been proposed, on the basis of the observation that *Alu* sequences are flanked by short, repeated oligonucleotides resembling those of transposons (see Chapter 26). In this view, *Alu* sequences, like other mobile genetic elements, may be inserted at various places in the genome as reverse transcriptase copies of the RNA that is transcribed from them (page 1072). Recent studies suggest that the *Alu* sequences may have been derived from a small RNA (7SL RNA) involved in protein transport across membranes. We shall discuss this RNA in Chapter 28.

The function of many repeated sequences is unknown.

Introns

A second reason for the large size of eukaryotic genomes is that most eukaryotic genes are interrupted by introns. In Chapter 7 we pointed out that the coding regions of most eukaryotic genes—exons—are interrupted by noncoding regions—introns. We saw, as an example, that the β globin gene consists of three exons interrupted by two noncoding intron regions. This kind of structure is common in eukaryotes and is often more extreme than in the hemoglobin example. Consider the ovalbumin gene depicted in Figure 24.4a. This gene encodes a protein 386 amino acid residues in length, which could be accommodated by a message 1158 nucleotides long. Yet the total ovalbumin gene is about 7700 base pairs in length, containing eight exons interspersed by seven introns. The difference between the ovalbumin gene and its mRNA is dramatically displayed when hybrid DNA–RNA molecules are examined in the electron microscope (Figure 24.4b and c). The genomic DNA pairs to the mRNA along the exons, but the introns, which contain no counterpart in the mRNA molecule, are looped out, to form what are called **R loops**.

Most eukaryotic genes contain introns.

FIGURE 24.4

Exon–intron structure of the ovalbumin gene in chickens.
(a) Map of the 7700-bp gene, showing exons 1–7 plus an untranslated leader sequence (blue) and introns A–G (brown). **(b)** Electron micrograph of a hybrid formed by renaturing chicken genomic DNA with purified ovalbumin mRNA. **(c)** Diagram showing how the intron regions loop out in R loops in such a hybrid. The RNA is shown in red, the DNA exons in blue, and the DNA introns in brown.

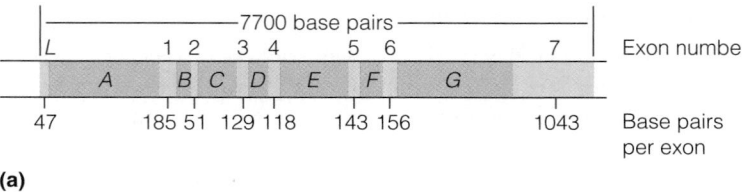

(a)

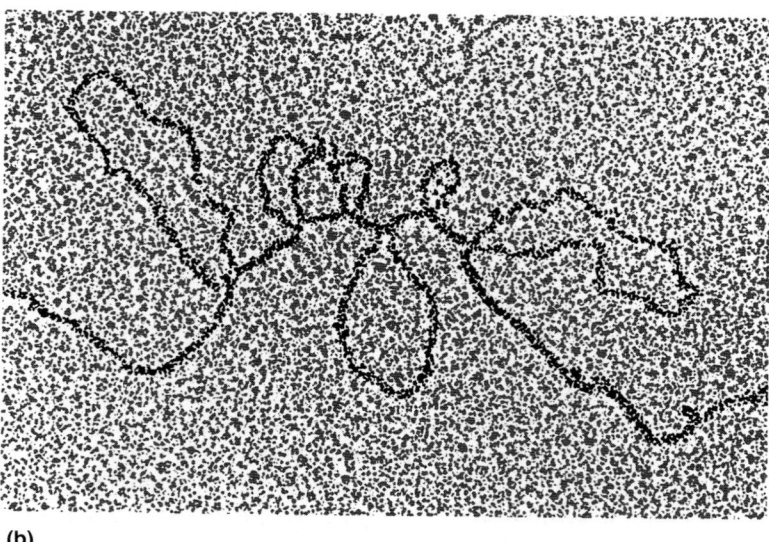

(b)

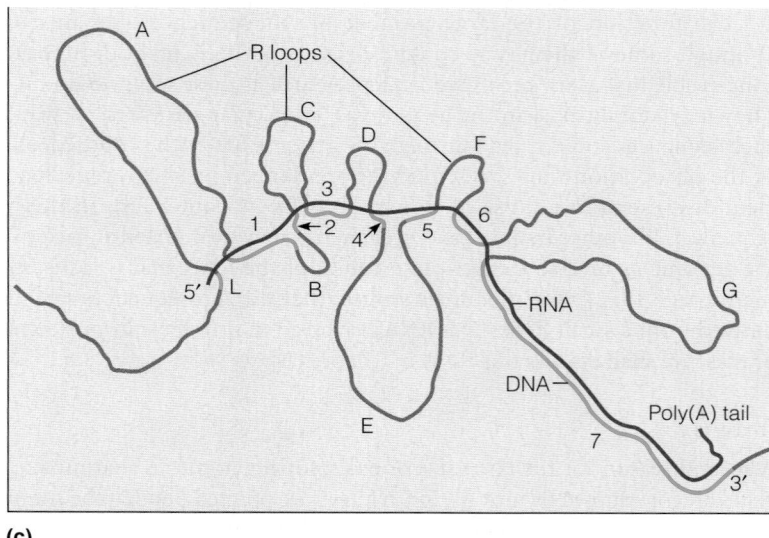

(c)

Sequence analysis shows that introns are present in most eukaryotic protein-coding genes and frequently exceed exons in total length. Lower eukaryotes like yeast usually have many fewer introns, and their genome size is correspondingly smaller. As in the case of prokaryotes, which also have few introns, the smaller yeast genome presumably reflects a need for more efficiency in cell replication.

As mentioned in Chapter 7, the function of introns is not yet wholly understood. It seems likely that they serve as loci for genetic recombination, allowing functional parts of proteins to be interchanged in evolution, a process called exon shuffling. Such loci also allow eukaryotes to make variants of a protein from a single gene, by splicing different exons together. This **alternative splicing**, discussed in Chapter 27, page 1161, is a more efficient way to store information than having a whole gene for each variant of the protein.

Gene Families

Multiple Variants of a Gene

Despite alternative splicing, in many cases complete variants of genes for the same type of protein are found to be expressed in different tissues or at different stages in development. We encountered an example in Chapter 7, where the embryonic (ζ and ε), fetal (α and γ), and adult (α, β, and δ) globins of mammals were described. For each of these proteins there exists a complete gene in every cell of the mammal.

Figure 7.30 (page 261) depicts the clusters of genes for the α and β classes of hemoglobins in humans. Each of these genes has the kind of exon–intron structure shown in Figure 7.25 (page 256). In addition, the genes themselves are separated by long stretches of nontranscribed DNA. Some portions of these intervening regions must contain control signals because the expression of the globin genes is under complex and subtle regulation. In the first place, although the globin gene clusters are found in all human cells, they are expressed *only* in *erythropoietic cells,* cells that give rise to red blood cells. Furthermore, as we saw in Chapter 7, expression of each variant is strictly constrained to certain developmental stages. For example, in the early embryo, only the ζ and ε genes are being transcribed; all other globin genes are turned off. As development proceeds, transcription switches first to the fetal α and γ genes, and at about the time of birth the adult β variant begins to dominate and transcription of γ ceases (see Figure 7.29, page 261). This kind of developmental regulation is peculiar to eukaryotes, although bacteria show simpler developmental changes in gene expression, for example, accompanying different growth phases or during the development of biofilms. The use of multiple variant genes is expensive to copy and store in DNA: The human genome devotes about 100,000 bp of DNA just to producing different variants of hemoglobin.

Many other gene families exist. Some, like the groups of genes encoding variants of the histones, seem to play developmental roles rather like those of the globin gene family. Others, like the immunoglobulin genes, appear to exist in multiple forms in order to satisfy a multiplicity of similar but distinct needs. In each case, it seems likely that the members of a particular gene family have evolved by successive duplications of an original, ancestral gene.

As our detailed knowledge of many different genomes has expanded, it has become obvious that the concept of "gene families" has a much broader sense than that used above. We deduce, from homologies in sequence and structure, that many genes we had first thought of as unrelated belong to extended families probably descended from one common ancestor. It may be that all existing genes descend from a relatively small number (several hundred?) sufficient to serve the needs of a very primitive ancestral organism.

Pseudogenes

Gene families often include as members one or more **pseudogenes**, or nonexpressed, genes. Pseudogenes can be recognized because they bear strong sequence similarity to expressed genes, from which they undoubtedly evolved, perhaps by a reverse transcriptase mechanism as has been proposed for *Alu* elements (see page 1005). They are no longer transcribed, however, because some element required for transcription (often a flanking control region, or **promoter**, a transcriptional initiation signal) is missing or defective. Because their sequences are not expressed, pseudogenes are no longer under strong selective control in evolution. In a sense, it does not matter what

TABLE 24.1   Properties of some bacterial and viral genomes

Organism or Virus	Genome Size, bp	Number of Genes	Physical Nature of the Genome
Escherichia coli	4,639,221	~4400	Circular duplex
Bacteriophage T4	168,889	~175	Linear duplex, circularly permuted[a]
Bacteriophage T7	39,936	~35	Linear duplex, small repetition each end
Bacteriophage λ	48,502	~50	Linear duplex, single-stranded ends
Influenza virus	~13,500	12	Single-stranded RNA
HIV	9,749	23	Single-stranded RNA
Bacteriophage φX174	5,387	11	Circular single-stranded DNA
Bacteriophage M13	6,407	11	Circular single-stranded DNA
Simian virus 40	5,226	6	Circular duplex DNA
Tobacco mosaic virus	~6,400	4	Single-stranded RNA
Bacteriophage MS2	3,689	4	Single-stranded RNA

[a]Circularly permuted means that all genomes have the same linear sequence of genes, but the end points differ (see Chapter 25).

happens to them, and as a consequence they can accumulate mutations that would be selected against in functional genes. Until recently it was thought that pseudogenes play no significant biological function—that they are evolutionary remnants. However, research on a tumor suppressor called PTEN showed that a pseudogene can bind a small inhibitory RNA molecule (siRNA, see Chapter 29) intended for the functional PTEN, and hence affect expression of the functional PTEN gene. Examples of pseudogenes can be seen in Figure 7.30 (page 261), which shows the arrangement of α and β globin gene variants in human DNA.

In summary, we find that a number of quite different explanations combine to account for the very large size of the genomes of eukaryotic organisms. At the same time, we still find it hard to rationalize the extreme variations in the amount of DNA that are sometimes observed even between closely related organisms. For example, among the amphibians alone there is more than a 100-fold range of genome sizes. The function, if any, of such enormous variation is still obscure, which suggests that we still do not understand some fundamental features of the eukaryotic genome.

By contrast, prokaryotic and viral genomes are much smaller, as befits the relative simplicity of the organisms involved, and more compact, as may befit organisms that are engaged in continued rapid growth during much of their lives. Introns exist in some viral and prokaryotic genomes, but they are quite uncommon. Indeed, in some cases adjacent genes overlap and are translated in different reading frames, to give different protein products. For many years the small size of these genomes and their accessibility for genetic manipulation, owing in part to their haploid nature, made bacteria (particularly *E. coli*) and bacteriophages the favored biological systems for elucidating details of genome replication and gene expression. Table 24.1 gives information about the genomes of some of the most useful systems and some other viruses of current interest.

Restriction and Modification

We turn next to the question of how the complete nucleotide sequence of a large genome is determined. Our ability to map and determine the nucleotide sequence of genomes depended crucially upon the existence of **restriction endonucleases**, which we identified in Chapter 4 as enzymes that catalyze double-stranded DNA cleavage in a sequence-specific fashion. Before discussing the sequence determination of large genomes, we digress here to describe these remarkable enzymes and the biological processes in which they are involved—**host-induced restriction**

and modification. Bacteria use site-specific DNA methylation to mark their own DNA and to cause DNA cleavage at the same sites to inactivate the DNA of invaders, such as viruses, which lack this marking; restriction–modification could be considered a bacterial immune system.

Biology of Restriction and Modification

Although restriction and modification were first described in 1952, it was not until Werner Arber's work in Switzerland in the mid-1960s that the biochemical basis for the phenomena became clear. The basic observation was as follows. Bacteriophage λ, when grown on the K12 strain of *E. coli* (which we shall call K), grows well in subsequent infections of the K strain, with every phage particle giving rise to a **plaque**. (A plaque is a cleared area on a Petri plate seeded with excess bacteria and a limited number of phages. It results from the multiplication of one virus particle and the localized lysis of bacteria. See Figure 24.5.) However, the phage grew poorly on *E. coli* strain B, with only 0.01% of the infecting phages giving rise to a plaque; in other words, the host strain *restricted* growth of the phage. Most of the infecting phage DNA was broken down in these nonproductive infections.

Phages purified from the few plaques that formed could infect strain B with high efficiency. The same phenomenon was seen in reverse: Phages grown on *E. coli* B infected strain K with low efficiency, but the few plaques that formed yielded phages that could infect strain K with high efficiency. These experiments suggested that although most of the phages exposed to a new strain of *E. coli* are destroyed, a few underwent an adaptation that allowed them to circumvent the bacterium's defense system and become infective to that strain. Arber coined the terms *restriction* and *modification* to explain these observations, and he showed that the biochemical basis for these phenomena is a pair of enzymes specific to each bacterial strain. When phages grown on strain K infect strain B, their DNA is almost always broken down by a B-specific restriction enzyme, and the bacterium survives. However, the B strain has an additional enzyme, which modifies its own DNA by methylating specific nucleotide residues. This pattern of methylation protects DNA against its own B-specific restriction enzyme. The methylation involves transfer of a methyl group from *S*-adenosylmethionine (AdoMet) to a specific base within the target sequence, as schematized in Figure 24.6.

Occasionally the modification enzyme of strain B methylates phage DNA as well, and in the process creates phage DNA resistant to breakdown by the B restriction system. The resistant phages are the small proportion of phages (0.01%) that plate on strain B. Because DNA of these phages is now fully protected from B-specific restriction, all of the phages that plate on B are able to infect the B strain in subsequent infections. Note that restriction and modification are **epigenetic** phenomena, in that the phenotype is preserved through the next round of phage growth, while phage DNA *sequences* are unaltered. The ability of a phage to overcome restriction in a particular host bacterium depends not upon any changes in the phage's genotype but upon the host strain in which that phage was previously grown. Note also that any DNA within a cell is subject to restriction and modification, including transforming and plasmid DNAs, as well as chromosomal or phage DNAs.

Restriction–modification systems are widespread among bacteria. Some are encoded by chromosomal genes, and some by plasmids. In 1970 Hamilton Smith observed that a restriction nuclease he was studying catalyzed double-stranded cleavage of DNA within a specific short nucleotide sequence. Shortly thereafter, it was found that modification has the same sequence specificity, as schematized in Figure 24.6. In this example, one nucleotide within a six-nucleotide sequence is the substrate for a specific DNA methylase. When that site is methylated, the DNA

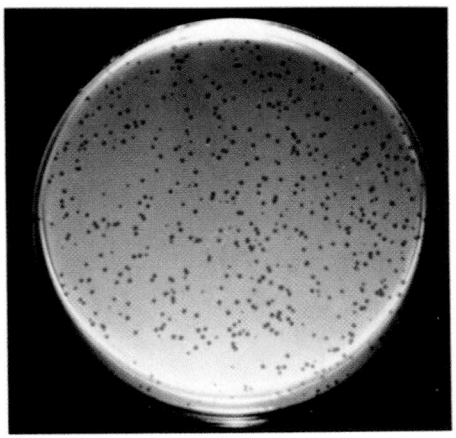

FIGURE 24.5

Bacteriophage plaques. A Petri plate was seeded with many bacteria (about 10[8]) and a small number of bacteriophage particles (about 200 T4 phages). Each phage particle generates a cleared area, or plaque, by infecting and lysing all of the bacteria growing near it.

Bacteria use restriction–modification, which involves nongenetic changes in DNA structure, to distinguish their own DNA from that of invaders.

is resistant to cleavage by a nuclease that recognizes the same hexanucleotide sequence; when that site is unmethylated, the DNA is susceptible to attack at that site. Hundreds of restriction endonucleases are now known to catalyze similar sequence-specific cleavages.

The importance of these developments was that, for the first time, scientists could isolate homogeneous DNA fragments of precisely defined length by treating DNA with a restriction nuclease in vitro and then resolving the fragments in the digest on an electrophoretic gel. These developments gave rise to gene cloning, as we discussed in Chapter 4. With respect to genome analysis, restriction fragments resolved by electrophoresis (Figure 24.7) can be arranged in order, so as to give physical maps of DNA molecules; the maps are called **restriction maps** because they show the physical locations of restriction sites.

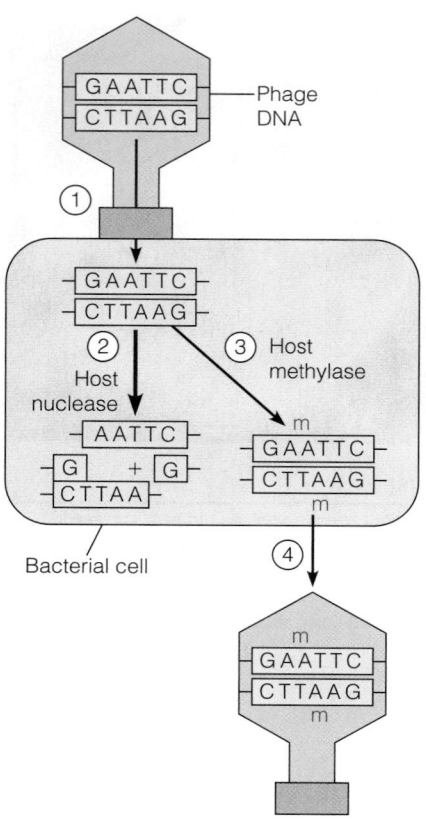

FIGURE 24.6

Host-induced restriction and modification.
A phage whose DNA is unmodified infects a bacterium with a restriction system that recognizes the DNA sequence 5′–GAATTC–3′ (step 1). Most phage DNA molecules are cleaved by the restriction nuclease (step 2), but the few that become methylated first on the innermost A are protected from attack (step 3). The phages that emerge contain modified (methylated) DNA (step 4). Because they are not vulnerable to restriction by the host nuclease, they are able to overcome the bacterium's defense system when they reinfect the same bacterial strain.

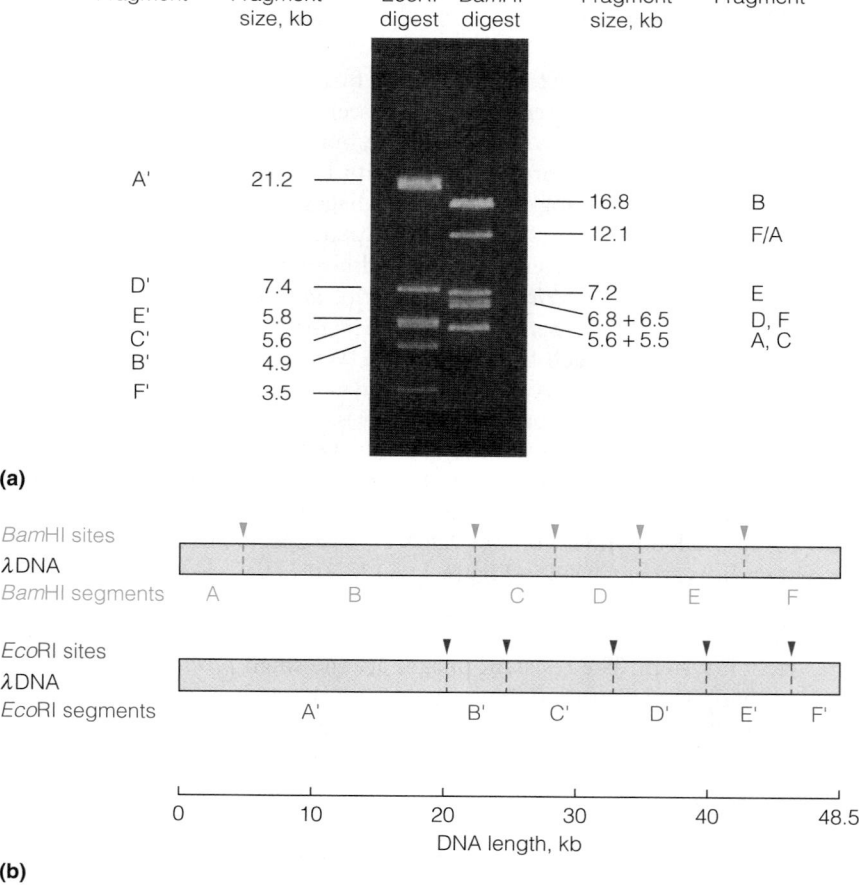

(a)

(b)

FIGURE 24.7

Fragmentation of bacteriophage λ DNA with restriction endonucleases EcoRI or BamHI.
(a) Experimental determination of fragmentation patterns, resulting from enzymatic digestion of the 48.5-kb linear DNA molecule from phage λ. Restriction digests are subjected to agarose gel electrophoresis, and the fragments are visualized under ultraviolet light after staining the gel with ethidium bromide, a fluorescent dye. Note that fragments with very similar sizes form but one band on a gel. **(b)** Maps of cleavage sites for each enzyme on the DNA molecule. By convention, the fragments are assigned letters as shown. In this experiment the 12.1 kb fragment in the BamHI digest resulted from linkage of the A and F (terminal) fragments by base pairing at their cohesive ends (page 1098). Mapping the restriction sites also requires data from partial digestion or with digestion of the DNA with EcoRI and BamHI together (not shown).

(a) Courtesy of Catherine Z. Mathews.

Properties of Restriction and Modification Enzymes

We now recognize three different types of restriction-modification systems—types I, II, and III. Each system consists of two distinct enzyme activities: a DNA methylase and an endonuclease that catalyzes a double-stranded DNA break. The sequence-specific endonucleases, those most widely used in molecular biology, are type II enzymes. Regardless of type, the enzymes are named with the first three letters denoting the bacterial species of origin and a fourth letter denoting an individual strain. For example, the restriction system from *E. coli* K is called **EcoK.** If more than one enzyme system is found in a given strain, the different enzymes are designated by Roman numerals. For instance, **EcoRI** is one of two known restriction systems in *E. coli* strain R, and **Hind**III is one of three enzymes from the d strain of *Haemophilus influenzae.*

Properties of each of the three types of restriction systems are discussed in the following sections and are summarized in Table 24.2.

Restriction enzymes of most use to biologists cleave both DNA strands site-specifically, depending on base methylation.

Type I

Type I enzymes have both methylase and nuclease activities in one protein molecule, which contains three subunits. One subunit contains the nuclease, one the methylase, and one a sequence-recognition determinant. The recognition site is not symmetrical, and cleavage occurs some distance (up to 10 kb) away from the recognition site, although methylation occurs within the recognition site. For cleavage, the enzyme remains bound to the recognition site, and DNA is looped out around it, with concomitant supercoiling. About 10^5 ATP molecules are hydrolyzed per cleavage event. Energy is probably needed for both translocation of the enzyme and supercoiling of the DNA. For reasons still not clear, both ATP and AdoMet are required for the cleavage activity. AdoMet may be an allosteric activator because it is not broken down during the reaction.

Type II

Type II restriction nucleases have been of greatest value to researchers because most of them cut within the recognition sequence, making the cleavage sequence-specific. Most of the type II enzymes are homodimers, with subunits of 30 to 40 kilodaltons. A divalent cation is required for cleavage, but ATP is not required. Each type II nuclease has a counterpart methylase, which binds to the same recognition sequence and methylates one nucleotide within that sequence. A **hemimethylated** DNA (with methyl group on one strand only) is a preferred substrate for the methylase but not for the nuclease, which generally cleaves only when the recognition site is unmethylated on both strands. Cleavage generates 3′ hydroxyl

TABLE 24.2 Properties of restriction-modification systems			
	Type I	**Type II**	**Type III**
Example	*Eco*B	*Eco*RI	*Eco*PI
Recognition site	TGAN$_8$TGCT	GAATTC	AGACC
Cleavage site	Up to 10 kbp away from recognition site	Between G and A (both strands)	24–26 base pairs 3′ to recognition site
Methylation site	$\overset{m}{\text{TGAN}_8\text{TGCT}}$ $\underset{m}{\text{ACTN}_8\text{ACGA}}$	$\overset{m}{\text{GAATTC}}$ $\underset{m}{\text{CTTAAG}}$	$\overset{m}{\text{AGACC}}$ (only one strand methylated)
Nuclease and methylase in one enzyme?	Yes	No	Yes
Requirements for cleavage	ATP, Mg^{2+}, AdoMet	Mg^{2+} or Mn^{2+}	Mg^{2+}, AdoMet
Requirements for methylation	ATP, Mg^{2+}, AdoMet	AdoMet	Mg^{2+}, AdoMet

Note: Each methylated base is identified with the letter m. All sequences read 5′ to 3′, left to right.

TABLE 24.3 Specificities of some type II restriction systems

Enzyme	Bacterial Source	Restriction and Modification Site[a]
*Bam*HI	*Bacillus amyloliquefaciens* H	G↓GATCC
*Bgl*II	*B. globiggi*	A↓GATCT
*Eco*RI	*Escherichia coli* RY13	G↓A̅A̅TTC (m over A)
*Eco*RII	*E. coli* R245	C̅C̅↓GG (m over C)
*Hae*III	*Haemophilus aegyptius*	GG↓C̅C̅ (m over C)
*Hga*I	*H. gallinarum*	GACGCNNNNN↓ CTGCGNNNNNNNNNN↓
*Hha*I	*H. haemolyticus*	GC̅G̅↓C (m over G)
*Hind*II	*H. influenzae* Rd	GTPy↓PuA̅C (m over A)
*Hind*III	*H. influenzae* Rd	A̅↓AGCTT (m over A)
*Hinf*I	*H. influenzae* Rf	G↓ANTC
*Hpa*I	*H. parainfluenzae*	GTT↓AAC
*Hpa*II	*H. parainfluenzae*	C↓C̅GG (m over C)
*Msp*I	*Moraxella* sp.	C↓CGG
*Not*I	*Nocardia rubra*	GC↓GGCCGC
*Ple*I	*Pseudomonas lemoignei*	GAGTCNNNN↓ CTCAGNNNNN↓
*Pst*I	*Providencia stuartii*	CTGCA↓G
*Sal*I	*Streptomyces albus* G	G↓TCGAC
*Sma*I	*Serratia marcescens* Sb	CC̅C̅↓GGG (m over C)
*Xba*I	*Xanthomonas badrii*	T↓CTAGA

[a]The methylated base in each site, where known, is identified with the letter m. All sequences read 5′ to 3′, left to right. The cleavage on the opposite strand in each case can be inferred from the symmetry of the site (except for *Hga*I and *Ple*I, each of which has an asymmetric site). Pu = purine, Py = pyrimidine, N = any base.

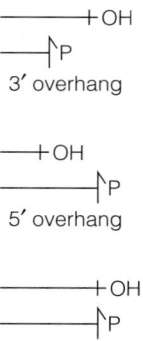

and 5′ phosphate termini. Cleavage sites on the two strands may be offset by as much as four nucleotides (as in *Eco*RI) or more, giving cuts with short, self-complementary, single-stranded termini. Some enzymes cleave to give a 5′-terminated single-stranded end ("overhang"), whereas others generate a 3′ overhang. Other type II nucleases, including *Sma*I and *Hind*II, generate blunt-ended fragments, in which the cutting sites are not offset. Most recognition sites are four, five, or six nucleotides in length, although a few type II enzymes recognize an eight-nucleotide sequence. Most show two-fold rotational sequence symmetry, suggesting that the two enzyme subunits are also arranged symmetrically. Table 24.3 shows the recognition sites for several widely used type II nucleases. Several hundred enzymes of this type have now been isolated. Not all type II nucleases are absolutely sequence specific. For example, *Hind*II recognizes four different hexanucleotide sequences, and some enzymes (such as *Hga*I) cleave at a site outside the recognition sequence.

The year 1986 saw the first crystallographic structural determination of a restriction nuclease (*Eco*RI) complexed with a double-stranded oligonucleotide containing its DNA recognition sequence. Figure 24.8 shows one polypeptide subunit of the dimeric enzyme in contact with its DNA recognition sequence. The DNA is bound in a cleft, and the protein has an N-terminal "arm" that wraps about the DNA. Sequence specificity is maintained by 12 hydrogen bonds, which link the purine residues in the site to a glutamate and two arginine residues (not shown in the figure). Binding of the DNA alters its structure to generate "kinks"; the sequences immediately flanking

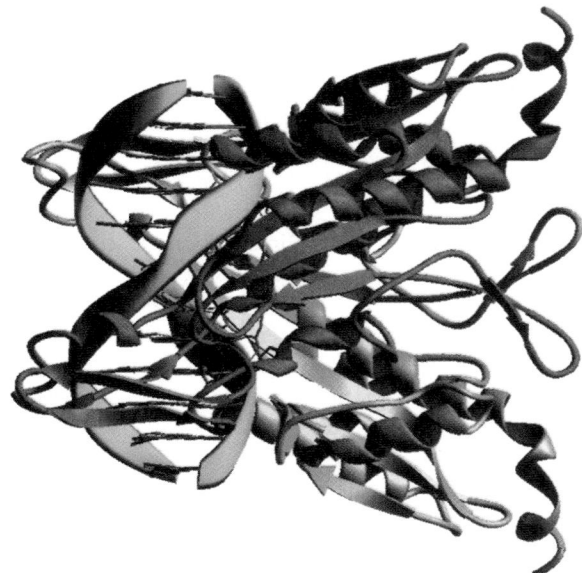

FIGURE 24.8

Structure of the *Eco*RI nuclease complexed with its DNA substrate. The DNA helix is shown in blue, while the two subunits of the protein are shown in orange and yellow, respectively. Note the "kink" in the DNA structure, resulting from the fact that the enzyme binds the central six-base-pair cutting site in the B conformation, while the flanking sequences are bound as A-form DNA. Note also the N-terminal "arm" on each protein subunit, which wraps around the DNA. PDB ID 1ERI.

Courtesy of John Rosenberg and colleagues, University of Pittsburgh.

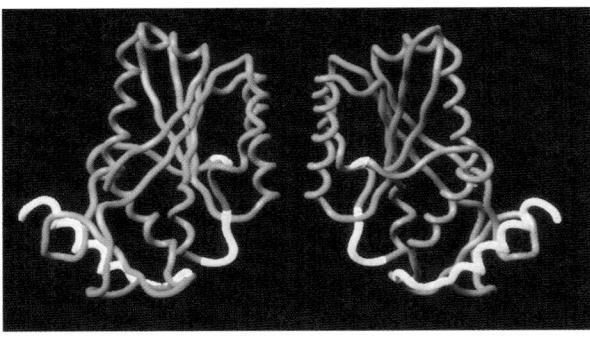

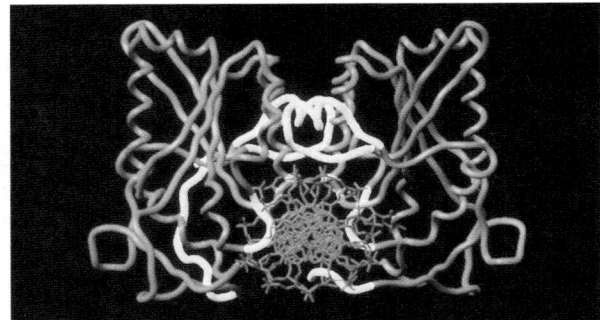

FIGURE 24.9

Structures of (a) free and (b) DNA-bound forms of *Bam*HI, with DNA shown end-on, in orange. Regions of the protein that undergo conformational change upon DNA binding, including the two C-terminal α helices, are shown in yellow. PDB ID 1HBM.

From *Science* 269:656–663, M. Newman, T. Strzelecka, L. F. Dorner, I. Schildkraut, and A. K. Aggarwal, Structure of *Bam*HI endonuclease bound to DNA: Partial folding and unfolding on DNA binding. © 1995. Reprinted with permission from AAAS.

the six-nucleotide cutting site (GAATTC) adopt the A duplex conformation, and the B structure is retained within the cutting site. The other subunit, not shown in the figure, contacts the substrate identically, accounting for the ability of the enzyme to catalyze symmetrical cleavages within the cutting site.

By contrast, *Bam*HI endonuclease does not kink its DNA substrate, which remains in the B form. However, as shown in Figure 24.9, the enzyme itself undergoes a major conformational change upon binding DNA. The C-terminal α helices from each subunit unwind and contact the DNA, one in the minor groove, and one along the sugar–phosphate backbone, thus introducing an unexpected element of asymmetry into the DNA–protein complex.

Structural studies on type II DNA methylases have been informative as well. Amazingly, structural studies on the *Hha* DNA methylase showed that the bases undergoing methylation rotate completely out of the DNA duplex and into a catalytic pocket within the enzyme, where methylation occurs (Figure 24.10). Since that demonstration in 1994, other enzymes that act upon specific DNA bases have been shown similarly to flip the target base, including both DNA methylases, as shown here, and glycosylases, enzymes involved in DNA repair (Chapter 26).

Modification methylases swing the target DNA base totally out of the helix in order to act upon it.

Type III

Type III enzymes resemble more closely the type I systems than the type II systems. Type III enzymes contain both nuclease and methylase activities in a two-subunit enzyme. They differ from type I enzymes in that they do not require ATP, they modify just one DNA strand, and the cleavage site is fairly close to the recognition site.

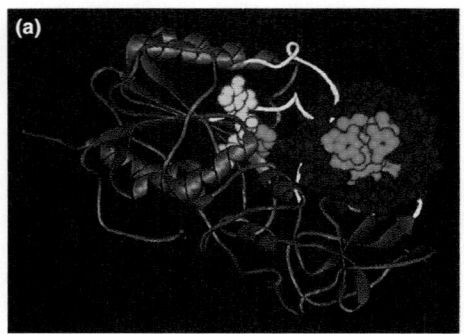

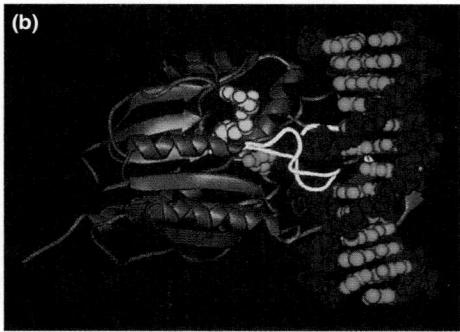

FIGURE 24.10

Structure of a complex of a type II DNA methylase with DNA. The structure is a ternary complex containing *Hha* methylase from *Haemophilus haemolyticus*, DNA, and *S*-adenosylhomocysteine. The loops containing the catalytic site are in white, and the rest of the enzyme is in orange. *S*-Adenosylhomocysteine is in yellow, the DNA backbone is magenta, and bases are green. In both views the flipped-out target cytosine base is clearly visible. **(a)** View looking down the helix. **(b)** Side view from the minor groove. PDB ID 1MHT.

Reprinted from *Cell* 76:357–369, S. Klimasauskas, S. Kumar, R. J. Roberts, and X. Cheng, *Hha*I methyltransferase flips its target base out of the DNA helix. © 1994, with permission from Elsevier.

Determining Genome Nucleotide Sequences

The ability to map restriction cleavage sites within a genome made it possible to determine the complete nucleotide sequence of small genomes. Once a restriction map of the genome is available, the genome is simply fragmented by restriction cleavage into relatively small pieces, each of which can be cloned into a suitable vector for subsequent sequence analysis (Tools of Biochemistry 4B). The collection of cloned genomic fragments is called a **library**, since it is a collection of separate information-containing units. Each "book" in the library is sequenced, with the final sequence being reassembled on the basis of the restriction map. Generally, digestion must be carried out with two or more restriction nucleases so that identification of overlap regions will allow alignment of adjacent fragments. Even simpler is "shotgun sequencing," a process that yields a random set of fragments, arising either from partial digestion of the genome by a restriction nuclease or by mechanical shear so that different clones in the library contain some identical sequences, representing regions of overlap between contiguous segments. Overlapping regions can be identified on the basis of restriction fragment length patterns, which allow all of the fragments to be aligned properly. Alternatively, the fragments can be sequenced individually and aligned by use of computer programs that look for regions of sequence identity. By these means it was possible for the 5386 bases in the ϕ X174 bacteriophage genome to be sequenced as early as 1977, with the 16,569 base pairs in human mitochondrial DNA being identified a few years later. A shotgun approach yielded the first sequence of a free-living organism, the bacterium *Haemophilus influenzae*, in 1995. This genome, of 1,830,137 base pairs, encodes about 1740 proteins. Many other bacterial genomes have since been sequenced by comparable approaches.

However, these simple approaches were not practical when applied to the much larger eukaryotic genomes. Even the unicellular yeast, *Saccharomyces cerevisiae*, contains a genome of 12 million base pairs, distributed among 16 chromosomes. Much more daunting is the human genome, with more than 3 billion base pairs in 24 chromosomes (22 autosomes plus the X and Y sex chromosomes). Obviously, different approaches were needed, both for alignment of genome fragments of contiguous sequence ("contigs") and for their assignment to specific chromosomes. In our discussion of sequence analysis of large genomes we refer primarily to the human genome, which was determined as a result of the Human Genome Project, begun in 1990 by a large international consortium under the direction of Francis Collins, and by a privately financed program directed by Craig Venter. These efforts yielded a preliminary "draft sequence" in 2000, with a "final" sequence following in 2003.

Mapping Large Genomes

What was needed first was a way to identify the chromosome from which each fragment in a total DNA digest was derived. One way to do this is **fluorescent in situ hybridization** (FISH). The investigator selects a restriction fragment containing a unique sequence, found nowhere else in the genome. This sequence, typically a few hundred nucleotides, is amplified by **polymerase chain reaction**, or PCR, a technique that allows selective amplification of any region of a genome, provided that one knows the base sequences surrounding the region of interest. PCR is described in Tools of Biochemistry 24A. The amplified DNA fragment is tagged with a fluorescent dye, then denatured and allowed to anneal in the presence of a preparation of human chromosomes arrested in metaphase, where paired, condensed chromosomes have not yet separated. Microscopic visualization under light of the appropriate wavelength for the fluorescent dye identifies the chromosome to which the probe has become annealed (Figure 24.11). FISH has other uses, including finding specific features in DNA for use in genetic counseling and medicine. FISH can also be used to detect and localize specific mRNAs within tissue samples. In this context, it can help define the spatial-temporal patterns of gene expression within cells and tissues.

By these means, sequence analysis of the human genome becomes 24 separate projects, each focused upon one chromosome. But each chromosome is still enormous, and additional landmarks are needed, landmarks that allow placement of fragments of sequence information as they became available. In classical genetics maps were constructed by recombination analysis of genes with recognizable phenotypes. For example, one could mate two strains of the fruitfly *Drosophila*, one a wild-type and the other with two mutations (both dominant)—one causing, for example, abnormal eye color and one causing abnormal wing development. If the two mutations lie in genes on different chromosomes, then they reassort randomly according to Mendel's laws, and each of the four possible phenotypes—wild-type, double mutant, and each of the two single mutants—is represented at 25 % among the progeny. On the other hand, if the two genes under study lie on the same chromosome, then analysis of hundreds of progeny would yield relatively few with just one of the two phenotypes—abnormal eye pigment or abnormal wing development. The two genes would be shown thereby to be *linked*—carried on the same chromosome. These outcomes are schematized in Figure 24.12.

Generating Physical Maps

Even though the two genes being analyzed may lie on the same chromosome, single mutants will arise among the progeny through genetic recombination, which involves breakage and joining of homologous chromosomes during meiosis. The percentage of recombinants arising under these circumstances is a measure of the distance along a chromosome between the genes; the greater the distance, the higher the frequency of recombination. By such analysis one can assemble a genetic map in which the distance between genes on the same chromosome is reported in terms of percent recombination between mutations in those genes.

For sequence analysis what is needed, of course, is not a map of genetic distances determined by biological measurements, which are essentially statistical, but a physical map, in which markers can be placed in terms of physical distance—kilobase pairs—from one another. Such markers are also indispensable for genetic analysis. Well before investigators had shown the audacity to propose sequencing the 3-billion-base-pair human genome, geneticists were trying to locate within the genome genes in which mutations give rise to inherited diseases. For example, cystic fibrosis results from deficiency or abnormality in a chloride channel protein. When the gene for this protein was in fact mapped in 1989, it was hoped that the information might lead to a cure for this dread disease. Unfortunately, that has not yet happened. However, identification of the gene and gene product related to a disease do help immeasurably in understanding the disease, seeking prevention, and guiding therapy.

The key to locating such disease genes was to have markers on a physical genome map, and then to use association studies in family pedigrees to find markers on the physical map that lie near the gene of interest. But what might those markers be? In 1980 David Botstein realized that random variations among human

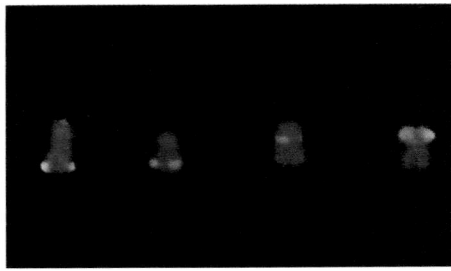

FIGURE 24.11

Mapping genes by fluorescent in situ hybridization. Four different DNA probes representing different genes were each tagged with a fluorescent dye and hybridized to human chromosome 21 in metaphase, showing hybridization signals (in yellow) at four distinct locations along the chromosome. Because metaphase chromosomes are made up of two nearly identical sister chromatids, each probe produced a pair of signals.

Reproduced from *The Human Genome Project: Deciphering the Blueprint of Heredity*, N. G. Cooper, ed., p. 112. © 1994 University Science Books, Mill Valley, CA.

Sequencing large genomes requires prior construction of a physical map of the genome.

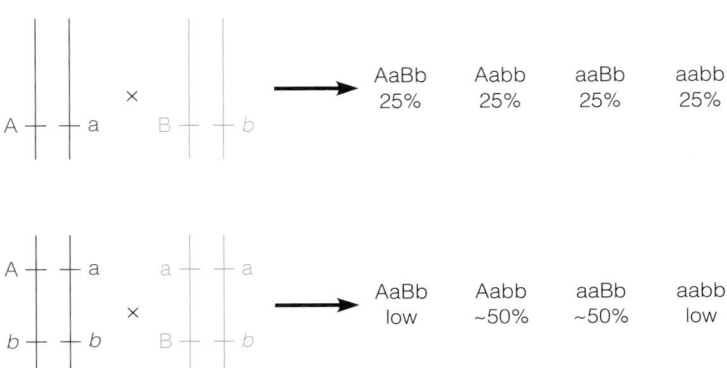

FIGURE 24.12

Segregation of two genetic markers, lying either on separate chromosomes (top) or linked on the same chromosome (bottom). A and B are dominant alleles of genes in which the genotype can be inferred from direct observation of a phenotype, such as eye color or wing shape. In each case, two heterozygous parents are mated, and the expected proportion of each genotype in the progeny is shown. When the genes being analyzed lie on different chromosomes, the markers assort randomly. When they lie on the same chromosome, wild-type (AaBb) or double-mutant (aabb) progeny arise only through relatively rare recombination events.

FIGURE 24.13

Analysis of a restriction fragment length polymorphism. For details see text. The thick line represents a hybridization probe—a labeled oligonucleotide complementary to part of the sequence toward the left end of each molecule as drawn.

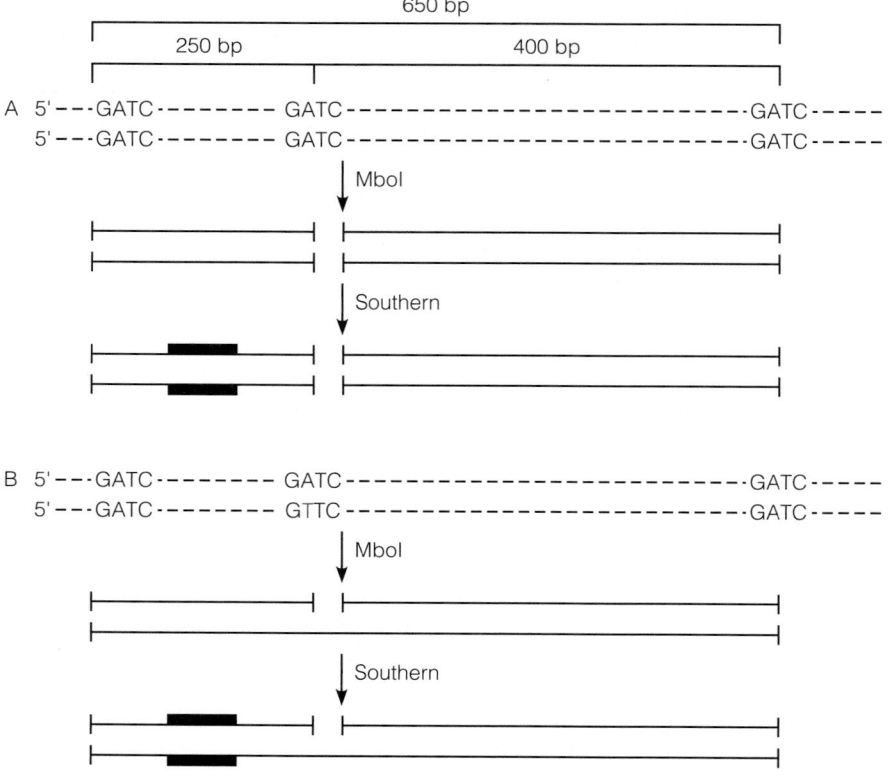

Southern Analysis

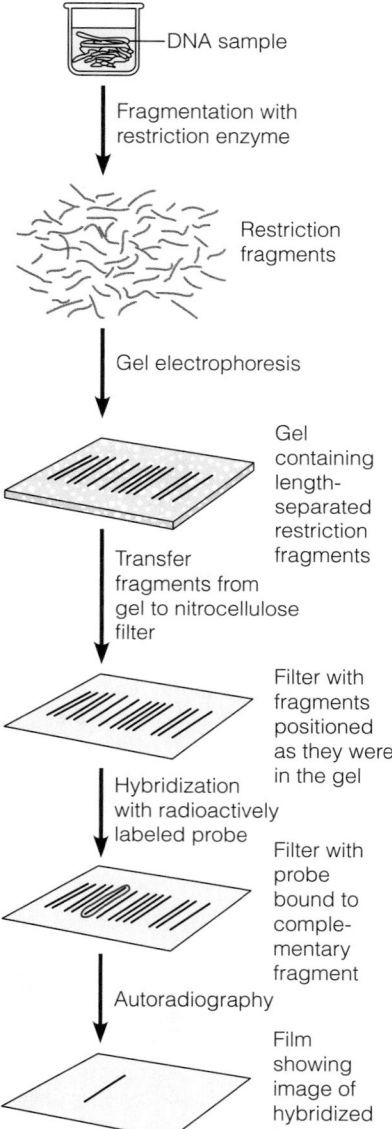

Gel containing length-separated restriction fragments

Filter with fragments positioned as they were in the gel

Filter with probe bound to complementary fragment

Film showing image of hybridized fragment

FIGURE 24.14

The principle of Southern transfer and hybridization.

Reproduced from *The Human Genome Project: Deciphering the Blueprint of Heredity*, N. G. Cooper, ed., p. 63. © 1994 University Science Books, Mill Valley, CA.

genome sequences could lead to the gain or loss of restriction cleavage sites, and that the resultant **polymorphisms**—variable sequences within the same region—could provide the markers needed to construct physical maps. Consider the four-base sequence 5'-GATC, the recognition site for the restriction endonuclease *Mbo*I. On average this site will occur at intervals of 256 base pairs ($1/4 \times 1/4 \times 1/4 \times 1/4$). Suppose that individual A has two homologous chromosomes identical within this region, containing *Mbo*I cleavage sites 650 base pairs apart. Both chromosomes contain an additional *Mbo*I site, spaced 250 nucleotides from the 5' end of this region as drawn (Figure 24.13). Now suppose that a second individual, B, has a **restriction fragment length polymorphism** (RFLP) within this region. One chromosome is identical to that just described, while the other has, 250 base pairs from the 5' end, not GATC, but GTTC. *Mbo*I will no longer cleave at this site. Now, if the genome of A is digested with *Mbo*I, this region will yield a 250-base-pair fragment and a 400-bp fragment, while the genome of B will yield the same 250- and 400-bp fragments from one chromosome and a 650-bp fragment from the chromosome not containing the internal *Mbo*I site. How might these fragments be detected against the enormous background resulting from *Mbo*I cleavage throughout the entire genome? A powerful technique called **Southern blotting**, or **Southern transfer**, after its inventor, Edwin Southern, permits this analysis.

The Principle of Southern Analysis

Consider the example shown in Figure 24.13. We would like to examine the *Mbo*I fragments in the region shown. But if we simply digest total human DNA with *Mbo*I and resolve the fragments by agarose gel electrophoresis, the pattern of fragments on the gel will be a smear, if the gel is analyzed by the standard method—ethidium bromide staining followed by examination of its fluorescence under ultraviolet light. There are far too many fragments, with each one present in minute amounts, for any one fragment to be visualized. In the Southern technique (Figure 24.14) the contents of the gel are transferred, or "blotted," to a sheet of nitrocellulose, under denaturing conditions. The initial Southern technique literally involved blotting; absorbent paper was placed between the electrophoretic gel and a sheet of nitrocellulose. Nitrocellulose binds irreversibly to single-stranded DNA, so what results is a replica of the agarose gel, with all DNA fragments tightly bound. Next one prepares a single-stranded DNA fragment identical to part of the region of interest, made highly radioactive by incorporation of ^{32}P. This is incubated with the nitrocellulose sheet under annealing conditions, allowing the radiolabeled DNA "probe" to find complementary sequences and yield renatured radioactive double-stranded DNA molecules. After this the nitrocellulose is washed to remove unbound probe and subjected to radioautography. Only those fragments in the DNA digest containing sequences homologous to those of the probe will be detected because of the bound radiophosphorus. More recently the use of radiolabeled hybridization probes has been supplanted by the use of fluorescent nucleotide analogs, which allows examination of gels after transfer under UV light, without the need for hazardous radioisotopes. Also, the transfer from agarose gel to nitrocellulose is now done through application of an electric current.

In the example shown in Figure 24.13, suppose that the labeled probe is complementary to a unique sequence in the 250-bp fragment. Then Southern analysis of an *Mbo*I digest will show a 250-bp fragment from the DNA of individual A, and fragments of 250 bp and 650 bp in the digest from individual B.

How are the radiolabeled probes produced? One method involves automated chemical synthesis of oligonucleotides in the presence of ^{32}P, or a fluorescent-tagged dNTP. More often, probes are prepared by polymerase chain reaction (see Tools of Biochemistry 24A, page 1034).

Polymorphisms within genomes include not only single-base changes, as shown in Figure 24.13, but also insertions, deletions, and repetitions of short sequences. Figure 24.15 shows typical variation in a 5-kbp region of the genome for 10 individuals, carrying 20 distinct copies of the human genome. Shown are 10 single-nucleotide polymorphisms (SNPs), an insertion–deletion polymorphism (indel), and a tetranucleotide-repeat polymorphism. The six common polymorphisms on the left side are strongly correlated. Although these six could in principle occur in 2^6 possible patterns, only three patterns are observed (shown

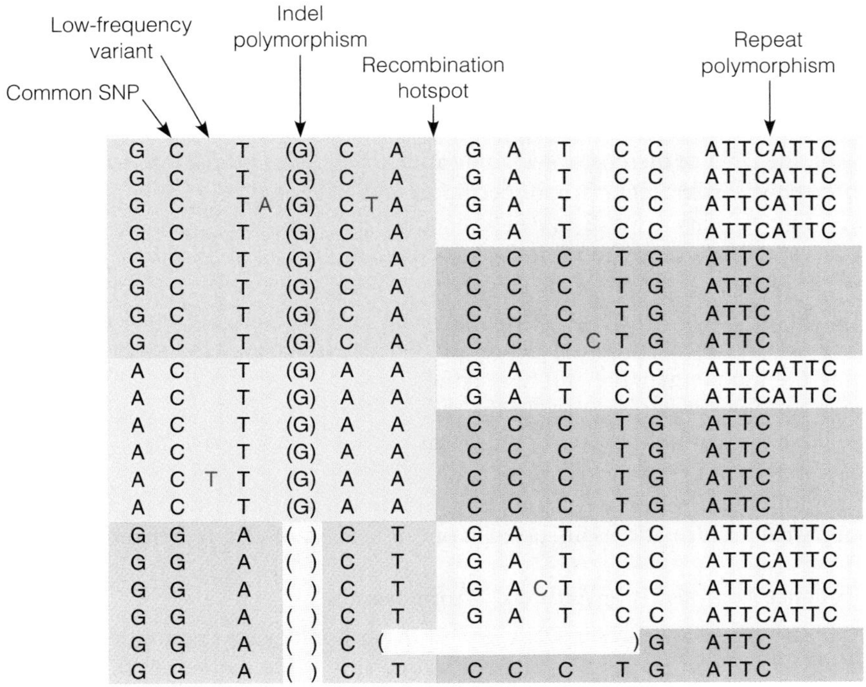

Southern blotting permits detection of minute amounts of DNA in the presence of a vast excess of nonspecific DNA.

FIGURE 24.15

DNA sequence variation in the human genome.
See text for description. The T insertion in the 13th line is found at low frequency.

From *Science* 322:881–888, D. Altshuler, M. J. Daly, and E. S. Lander, Genetic mapping in human disease. © 2008. Reprinted with permission from AAAS.

by pink, orange, and green). These patterns are called **haplotypes**. Similarly, the six common polymorphisms on the right side are strongly correlated and exist in only two haplotypes (shown by blue and purple). There is little or no correlation between the two groups of polymorphisms because there is a "hotspot," or region of high genetic recombination frequency, between them.

Southern Transfer and DNA Fingerprinting

Southern transfer and analysis has many applications, some of which are discussed elsewhere in this book. Best-known to the general public is DNA fingerprinting in forensic analysis. Restriction fragment–length polymorphisms (RFLPs, or "ruflups"), such as the variant sequences shown in Figures 24.13 and 24.15, exist throughout the genome. No two individuals, with the exception of identical twins, have the same pattern of RFLPs. Thus, a crime investigator who carries out Southern analysis using probes covering sufficient regions of the genome can generate a "DNA fingerprint" that establishes the identity of an individual with far more confidence than from the prints on our fingertips. Moreover, as we will see later, the power of PCR to amplify minute amounts of DNA means that DNA fingerprinting can be carried out with minuscule amounts of material left at a crime scene, such as the DNA in a single human hair. Most of the RFLPs used in forensic analysis are not single-base changes, but short tandem repeats (*repeat polymorphisms*), such as that shown in the blue portion of Figure 24.15.

Since DNA fingerprinting became routine in crime labs in the mid-1990s, hundreds of innocent criminal defendants have been acquitted and many previously convicted individuals have been exonerated. The large number of those freed, often after spending years in prison or having been sentenced to death, may yield satisfaction in the fact that molecular biology has contributed so successfully to the criminal justice system, but also sadness about years and lives unjustly sacrificed owing to deficiencies in that system.

Locating Genes on the Human Genome

The large number of known polymorphic regions in the human genome—those characterized by the presence of one or more RFLPs—provide many markers comparable to those of classical genes. Just as alleles of a gene can be recognized by observable phenotypes—eye color, for example—so also can alleles at polymorphic regions be recognized by the presence or absence of specific restriction sites. This allows disease genes to be mapped in terms of their proximity to a known polymorphic region. Consider the devastating inherited disease Huntington's disease, which took the life of folk singer Woody Guthrie. By carrying out RFLP analysis on the DNA of members of a family afflicted with this disease, one can identify those polymorphic regions, and specific sequence variants, that remain associated with susceptibility to the disease in several individuals. By these means, the physical location of the gene responsible for Huntington's disease can be located so that it can be sequenced, cloned, and investigated, with an eye toward identifying the mutant protein responsible for the disease and ways that this information could be used to seek treatment options. Thus, RFLPs have shown great value both in mapping the human genome and in seeking cures for genetic diseases based upon analysis of the causative genes. In the case of Huntington's disease genetic testing can help to inform decisions on child-bearing, but unfortunately, identification of the gene has not yet led to a cure for this devastating disease. Genetic testing is now being offered directly to the public by numerous companies. This has become somewhat problematic, as discussed in one of the end-of-chapter references.

Sequence Analysis Using Artificial Chromosomes

Although recent advances in DNA sequencing technology now permit many thousands of nucleotides to be sequenced in one run of the sequencer, the original Sanger method using fluorescent-tagged dideoxynucleotides was limited to several hundred nucleotides per sequencing run. Because a human chromosome might contain 100 million base pairs or more, there was a need for sequence analysis in DNA segments much smaller than individual chromosomes, but also much larger than typical restriction fragments. Yeast artificial chromosomes (YACs) were useful for this purpose. A YAC is a cloning vector that contains a centromere (page 1022), a DNA replication origin, telomeres (page 1022) at each end of the linear structure, a selectable marker such as a drug-resistance gene, and a cloning site. Inserts as large as 1000 kbp can be cloned into these constructs, which, after reintroduction into yeast cells, are maintained as the cell divides, just like natural chromosomes. A technique called **pulsed field gel electrophoresis**, which involves alternating voltage gradients during electrophoresis, permits resolution of the much larger DNA fragments resulting from analysis of these long DNA inserts. This approach subdivided each large sequencing project into sequence analysis of individual YACs, followed by assembly of sequences for a given chromosome. The approach is schematized in Figure 24.16, which also shows that each chromosome has a characteristic banding pattern when stained. It also illustrates that a very large number of individual restriction fragments must be cloned and sequenced to give the 130 million base pairs in the typical sequence shown here.

Size of the Human Genome

With all of these technologies in place, and with the aid of hundreds of workers, primarily from 20 institutions designated as genome centers, and from Craig Venter's privately financed effort, the entire human genome sequence, comprising some 3 billion base pairs, was determined. A surprise, given the large amount of DNA in the genome, was the relatively small number of genes that are expressed as proteins. From the size of the genome and the amount represented by noncoding structures such as satellite sequences, it was expected that the genome would contain about 100,000 genes. However, analysis of the draft sequence suggested that the true number of genes, identified as open reading frames with appropriate punctuation for initiation and termination of translation, was closer to 30,000. Further refinement of the sequence led the number of putative genes further downward, to a number between 20,000 and 25,000. However, phenomena such as alternative splicing and post-translational modification allow more than one protein product to be encoded by one gene. Nevertheless, the fact is that humans maintain organismic complexity and diversity using far fewer genes than had originally been expected.

From the first human sequence that was determined, and from analysis of RFLPs, it has become apparent that any two humans will show differences in their DNA sequences at about 1 base in 1000. In other words, any two humans, regardless of ethnic origin, are about 99.9% identical, genetically speaking.

It's interesting to consider the cost of large-scale genome sequence analysis, *vis à vis* technological innovations in sequencing methods. The government-sponsored Human Genome Project cost about $400 million. In 1999 sequence analysis of 1 million base pairs of DNA cost about $20,000; by 2010 that cost was 20 cents and dropping. By mid-2010, Illumina, the designer of one of the "second generation" sequence technologies (Tools of Biochemistry 4B), was offering complete sequence determination of one individual for $9500, and a race is under way among several companies to drop that price to $1000.

Physical Organization of Genes: The Nucleus, Chromosomes, and Chromatin

Chromosomes

As we see throughout this book, prokaryotic and eukaryotic cells differ from each other in fundamental ways. Among the most pronounced is the physical state of the genome. All organisms face the problem of compacting the genome, that is, reducing its physical dimensions to the point that it can fit into a cell of far smaller

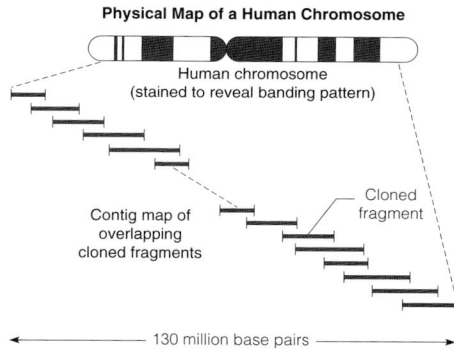

Physical Map of a Human Chromosome

Human chromosome (stained to reveal banding pattern)

Contig map of overlapping cloned fragments

Cloned fragment

130 million base pairs

The contig map spans the single DNA molecule contained in the chromosome.

FIGURE 24.16

Sequencing a large chromosome by individual sequence determination of cloned fragments in contiguous sequences.

Reproduced from *The Human Genome Project: Deciphering the Blueprint of Heredity*, N. G. Cooper, ed., p. 113. © 1994 University Science Books, Mill Valley, CA.

The human genome contains between 20,000 and 25,000 genes, far fewer than predicted from the amount of DNA in a human cell.

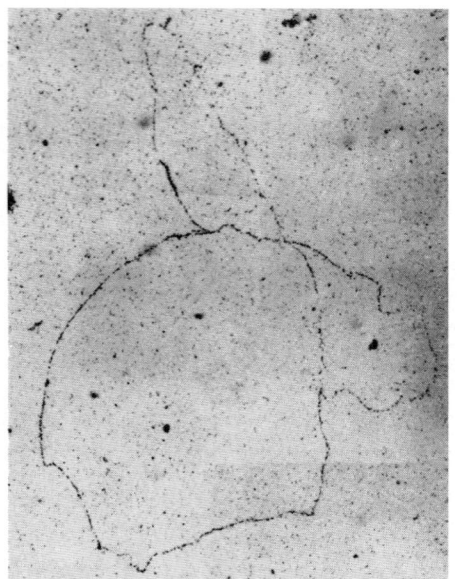

FIGURE 24.17

Radioautogram of a replicating *E. coli* chromosome after two generations of growth in [³H] thymidine.

Cold Spring Harbor Symposia on Quantitative Biology 28:44, J. Cairns. © 1963 Cold Spring Harbor Laboratory Press.

dimensions than the length of the DNA molecules in the genome—but they solve that problem in quite different ways. Bacteria are haploid organisms, with the entire genome represented by one DNA molecule, which is usually circular. With great care, this immensely thin and fragile molecule can be teased out of the cell without being broken, so that the circular structure can be visualized. Figure 24.17 shows an *E. coli* chromosome trapped in the midst of replication. The molecule was radiolabeled by growth of the bacteria in the presence of [³H]thymidine and visualized by radioautography. The figure shows a closed circular molecule with two Y-shaped junctions, each of which is a replication fork. Compaction of this molecule occurs largely by negative supercoiling and by organization into loops of supercoiled DNA about 50 kbp in length, with each loop bound to protein (Figure 24.18). This compacted structure, called a **nucleoid**, exists free within the cytosol of a bacterial cell, with a small number of attachment points to the membrane (see also Figure 1.8, page 15).

To the extent that the ultimate proof of a molecule is its validation by synthesis, the structure of bacterial chromosomes was confirmed in 2010, with the announcement by Craig Venter's group of the complete synthesis of a bacterial chromosome, whose capacity for reproduction and expression was confirmed by its ability to replicate after transfer into a bacterial cell from which the nucleoid had been previously removed.

The situation in eukaryotes is different. First, the eukaryotic cell is usually diploid, with two copies of each chromosome in each cell. An exception is the sex chromosome; female cells have two X chromosomes, while male cells contain one X and one Y chromosome. A eukaryotic cell's genome is divided into several or many chromosomes, each of which contains a single, very large, linear DNA molecule. Although their size varies greatly among organisms and even among different chromosomes in a given species, these DNA molecules are commonly of the order of 10^7 to 10^9 bp in length. Different eukaryotic species contain widely varying numbers of *distinguishable* chromosome pairs—from 1 (in an Australian ant) to 190 (in a species of butterfly). Whereas most prokaryotes are haploid, containing only one copy of their chromosome, most eukaryotic cells are diploid, carrying two

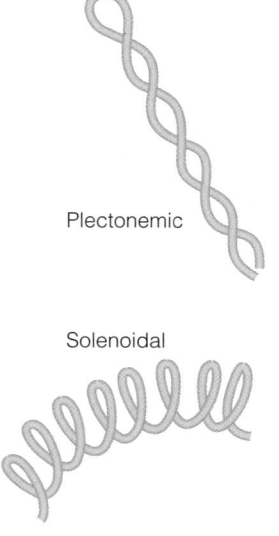

Plectonemic

Solenoidal

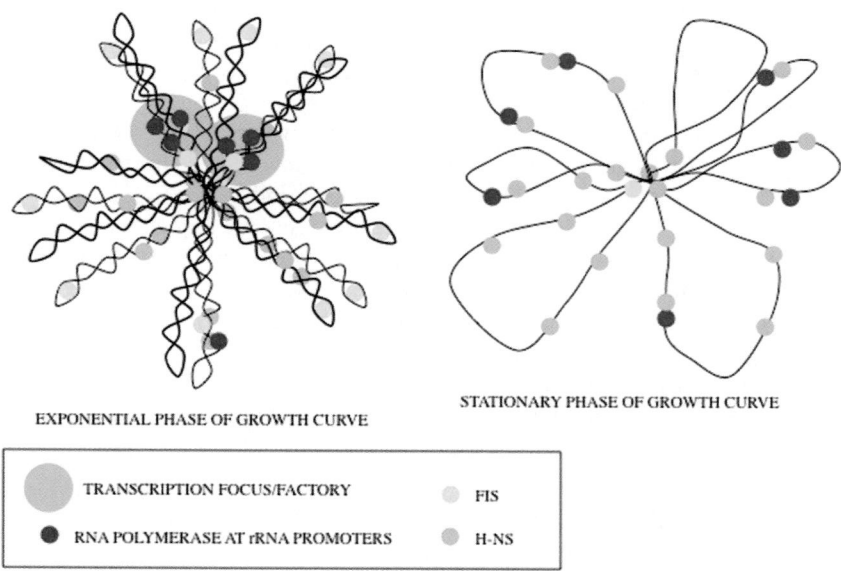

EXPONENTIAL PHASE OF GROWTH CURVE

STATIONARY PHASE OF GROWTH CURVE

TRANSCRIPTION FOCUS/FACTORY		FIS
RNA POLYMERASE AT rRNA PROMOTERS		H-NS

FIGURE 24.18

Structure of a bacterial nucleoid, showing independent domains of supercoiling, each stabilized by binding to protein. The term *plectonemic* refers to the type of supercoiling observed, with DNA strands intertwined in a regular way. The 1000-nm diameter of the structure allows it to fit within a bacterial cell that may be 2 to 5 μm in length. An alternative form of negative supercoiling, **solenoidal**, allows greater compaction and is seen in chromatin.

Reprinted by permission from Macmillan Publishers Ltd. *Nature Reviews Microbiology* 8:185-195, S. Dillon and C. J. Dorman, Bacterial nucleoid-associated proteins, nucleoid structure and gene expression. © 2010.

copies of each chromosome. As we have noted, the human genome is made up of 24 different chromosomes (22 autosomes plus the X and Y chromosomes) so that normal diploid human cells have a total chromosome number of 46.

An exception to the general statement that eukaryotic chromosomes are linear DNA molecules, folded into tightly packed chromatin structures, is organelle DNA—mitochondrial DNA in all eukaryotes, and chloroplast DNA in plants. Mitochondrial DNA is a circular duplex, as noted earlier, with the human mitochondrial genome containing 16,569 base pairs (see Figure 4.18, page 108). The genome is folded into a structure called the nucleoid, which is comparable to bacterial nucleoids. Human mitochondrial DNA encodes 13 proteins, all components of respiratory chain complexes. In addition, because these proteins are synthesized within the organelle, the genome also encodes a complete complement of ribosomal and transfer RNAs. Plant mitochondrial DNAs are much larger, up to two million base pairs or more. Chloroplast DNAs contain about 120,000 to 160,000 base pairs. It should not be surprising that organelle DNAs resemble prokaryotic DNAs in their structure and organization because there is now widespread agreement that these organelles evolved from primitive bacteria that infected primordial eukaryotic cells and evolved in tandem with their hosts.

Unlike the prokaryotic chromosome (nucleoid), the chromosomes of eukaryotes are not normally found free in the cytoplasm. In nondividing cells, the chromosomes are segregated within the nucleus (Figure 24.19) as an entangled mass of fibers of a DNA–protein complex called **chromatin**. The *nuclear envelope* is pierced by **nuclear pores**, which allow molecules as large as small protein and RNA molecules to pass in and out of the nucleus.

The nuclear pores are actually multisubunit complexes that provide openings of about 9 nm diameter. These allow the free diffusion of small molecules (including some small proteins) between nucleus and cytoplasm. However, the nuclear pore complex is also involved in selective transport of larger proteins and messenger RNA. In this process, two classes of assister proteins, called **exportins** and **importins**, are involved. Protein translation in eukaryotes occurs outside the nucleus, so transcription and translation cannot be directly coupled, as they are in prokaryotes (Chapter 27). Therefore, messenger RNA must be exported from the nucleus for translation. As we shall see later, considerable processing of the mRNA occurs before this export. During *mitosis*—the separation of daughter chromosomes following DNA replication—the nuclear envelope breaks down, and the diploid chromosomes condense into compact structures like the one shown in Figure 24.20. In this electron micrograph two newly replicated chromosomes

The multiple eukaryotic chromosomes are contained within the nucleus, except during mitosis when the nuclear envelope disintegrates.

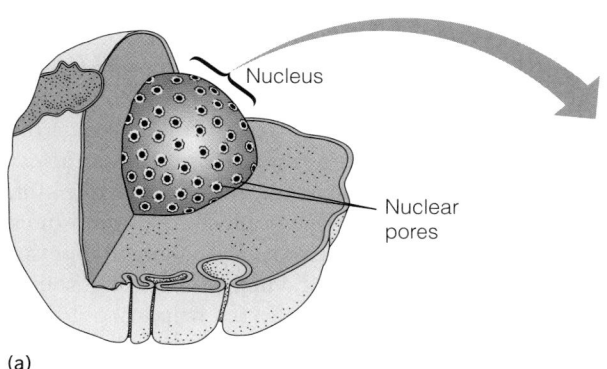

(a)

FIGURE 24.19

The nucleus. **(a)** Schematic, cutaway view of a typical animal cell, showing the position and relative size of the nucleus. **(b)** Electron micrograph of a section of a rat liver nucleus.

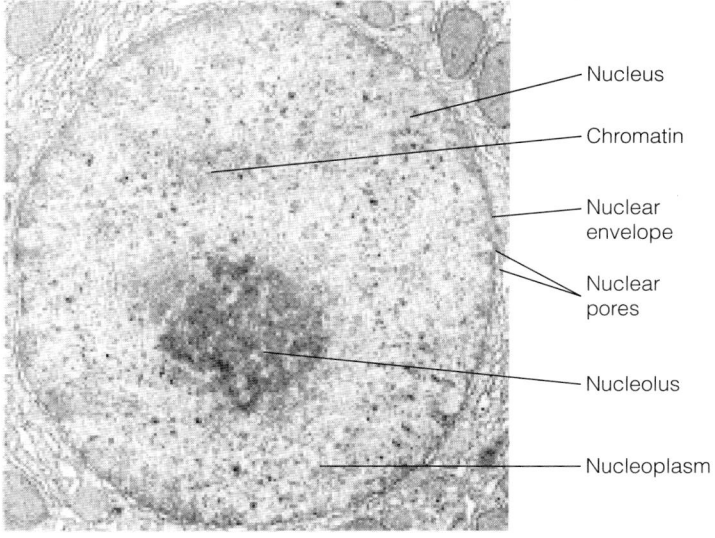

(b)

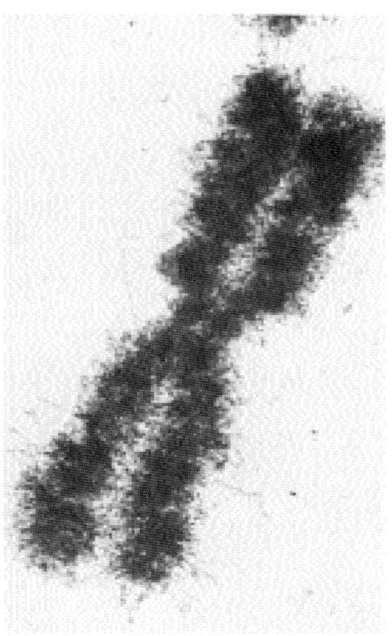

FIGURE 24.20

A mitotic chromosome. An electron microscope image of a human chromosome during the metaphase stage of mitosis. The constriction at the centromere and the lengthwise division into sister chromatids are clearly visible. The hairy-looking surface is made of loops of highly coiled chromatin.

Courtesy of G. F. Bahr, Armed Forces Institute of Pathology.

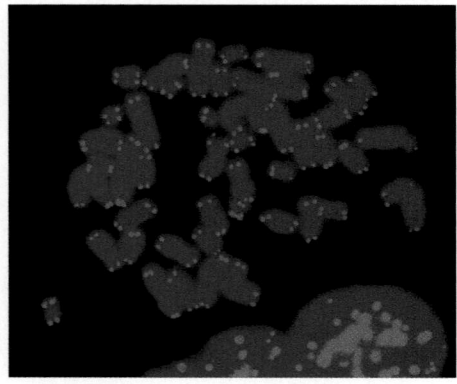

FIGURE 24.21

Mitotic human chromosomes in metaphase, stained separately by FISH for telomeres (pink).

Arturo Londono/ISM/Science Photo Library.

The chromatin of eukaryotes consists of DNA complexed with histones and nonhistone proteins.

(paired **chromatids**) are joined at the **centromere**, a structure that becomes linked to the spindle fibers as chromosome pairs are drawn apart in the latter stage of mitosis (page 1027).

Not seen in the electron micrograph of Figure 24.20 are the **telomeres**, structures at each end of eukaryotic chromosomes that protect the DNA from degradation and assure that each chromosome is completely copied during replication. A telomere is a simple tandemly repeated sequence that may be several kbp in length. Usually one strand is G-rich, and, as pointed out in Chapter 4, the dGMP residues often form a G-quartet structure. In humans the telomeric repeat sequence is 5′-TTAGGG. Figure 24.21 shows a spread of human metaphase chromosomes stained with fluorescent dyes showing telomeres (pink). We discuss telomeres and their biosynthesis in Chapter 25.

The enormous amount of DNA in a eukaryotic cell poses some serious problems. First, as noted earlier, there is the issue of *compaction*. The diploid DNA content of a human cell is more than 6×10^9 bp, corresponding to a total length of about 2 m. Somehow, all this DNA must be packed into a nucleus about $10 \mu m (10^{-5} m)$ in diameter. Second is the problem of *selective transcription*. In a typical differentiated eukaryotic cell, only a small fraction of the DNA is ever transcribed. Much, as we have seen, is nontranscribable. Many genes that undergo transcription do so only in certain cell lines in particular tissues, and then often only under special circumstances. To maintain and regulate such complex programs of selective transcription, the accessibility of the DNA to RNA polymerases must be under strict control. Both compaction and the control of gene expression in eukaryotes are achieved by having the DNA complexed with a set of special proteins to form the protein–DNA complex called chromatin.

Chromatin

The DNA-binding proteins of chromatin fall into two classes. The major class, **histones**, includes five types of protein, whose properties are outlined in Table 24.4. All histones are small, highly basic proteins. The histones are the basic building blocks of chromatin structure. Some have been remarkably well conserved in amino acid sequence throughout evolution. All known histone H4 molecules, for example, contain exactly 102 amino acid residues. Histone H4 shows only two substitutions between humans and peas and only eight substitutions between humans and yeast. The nucleoids of prokaryotic cells also have proteins associated with DNA, but these proteins are quite different from the histones and do not form a comparable chromatin structure. Thus, the histone-containing chromatin structure is a uniquely eukaryotic feature. In all kinds of eukaryotic nuclei, from yeast to human, the histones are present in an amount of about 1 gram per gram of DNA, and histones H2A, H2B, H3, and H4 are always found in equimolar quantities.

The histones are accompanied by a much more diverse group of DNA-binding proteins, rather unimaginatively named **nonhistone chromosomal proteins**. The total amount of the latter proteins varies greatly from one cell type to another, ranging from about 0.05 to 1 gram per gram of DNA. They include a bewildering variety of proteins, such as polymerases and other nuclear enzymes, nuclear receptor proteins, and regulatory proteins of many kinds. It is possible to count, on two-dimensional gels, approximately 1000 different nonhistone chromosomal proteins in a typical eukaryotic nucleus. Among the most abundant are topoisomerases and a class called **SMC proteins** (for structural maintenance of chromosomes). The major proteins in this class are **cohesins**, which help to hold sister chromatids together immediately after replication and continue to hold them together until chromosomes condense at metaphase, and **condensins**, which are essential to chromosome condensation as cells enter mitosis.

The association of proteins with eukaryotic DNA has long been recognized. As early as 1888 the German chemist Albrecht Kossel isolated histones from nuclei and recognized them as basic substances that would bind to the nucleic acid.

TABLE 24.4 Properties of the major histone types

Histone Type	Molecular Weight	Number of Amino Acid Residues	mol % Lys	mol % Arg	Role
H1	22,500	244	29.5	1.3	Associated with linker DNA; helps form higher-order structure
H2A	13,960	129	10.9	9.3	Two of each go to form the histone octamer core of the nucleosome.
H2B	13,774	125	16.0	6.4	
H3	15,273	135	9.6	13.3	
H4	11,236	102	10.8	13.7	

Note: All data are for calf thymus histones, except for H1, which is from rabbit.

Histones were, in fact, the first class of proteins to be recognized. However, the precise role of histones was not understood until about 1974. Then, research in a number of laboratories showed that these proteins combine in a specific way to form a repeating element of chromatin structure, the **nucleosome**.

The Nucleosome

If naked DNA (that is, DNA containing no bound proteins) is partially digested with a nonspecific endonuclease such as micrococcal nuclease, which cuts double strands almost randomly, a broad smear of polynucleotide fragments is produced. But in the early 1970s, researchers in a number of laboratories found that if the same experiment is conducted with chromatin, or even with whole nuclei (which some nucleases can penetrate through the nuclear pores), the DNA was cleaved in a quite specific, nonrandom way. On a polyacrylamide gel, the DNA from nuclease-digested chromatin gave a series of bands that were multiples of approximately 200 base pairs (Figure 24.22). This indicated that the nuclease could find easy access to the DNA only at regularly spaced points. At about the same time that these observations were made, other laboratories obtained electron micrographs of extended chromatin fibers, which revealed a regular "beaded" pattern in the chromatin structure, with one bead about every 200 bp (Figure 24.23). Still other researchers found that if nuclease digestion of chromatin was continued, it slowed down and nearly stopped when about 30% of the DNA had been consumed. The remaining protected DNA was found to be present in particles corresponding to the beads seen in the electron micrographs. These particles, called nucleosomes (or more precisely, *nucleosomal core particles*), were found to have a simple, definite composition that is practically invariant over the whole eukaryotic realm. They always contain 146 bp of DNA, wrapped about an octamer of histone molecules—two each of H2A, H2B, H3, and H4. This composition explains the equivalent amounts of these histones in chromatin. Both nucleosomes and nucleosome histone cores have been crystallized, and X-ray diffraction studies have revealed the structure shown in Figure 24.24. The DNA lies on the surface of the octamer and makes about 1.7 left-handed solenoidal superhelical turns about it. The structure of the octamer provides a left-handed helical "ramp" upon which the DNA is bound. The high-resolution data now available on the histone octamer reveal a commonality in histone structure—the *histone fold*—that was not evident from examination of sequences alone, and suggests an early common ancestor of these proteins.

Although the nucleosome itself is a nearly invariant structure in eukaryotes, the way in which nucleosomes are spaced along the DNA varies considerably among organisms and even among tissues in the same organism. The length of DNA between nucleosomes may vary from about 20 bp to over 100 bp. Exactly what determines the arrangement of nucleosomes along the DNA is still not wholly understood. However, it is now clear that at least some nucleosomes

The basic repeating structure in chromatin is the nucleosome, in which nearly two turns of DNA are wrapped about an octamer of histones.

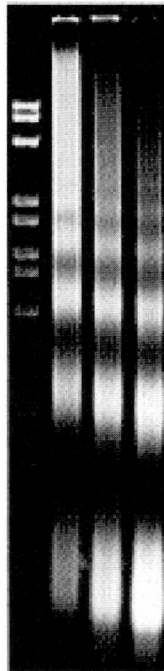

FIGURE 24.22

The kind of evidence that first suggested a repetitive structure in chromatin. In this gel, the three lanes to the right show DNA fragments obtained after three successively longer digestions of chicken erythrocyte chromatin by micrococcal nuclease. The column to the left contains DNA restriction fragments as size markers.

Courtesy of K. van Holde.

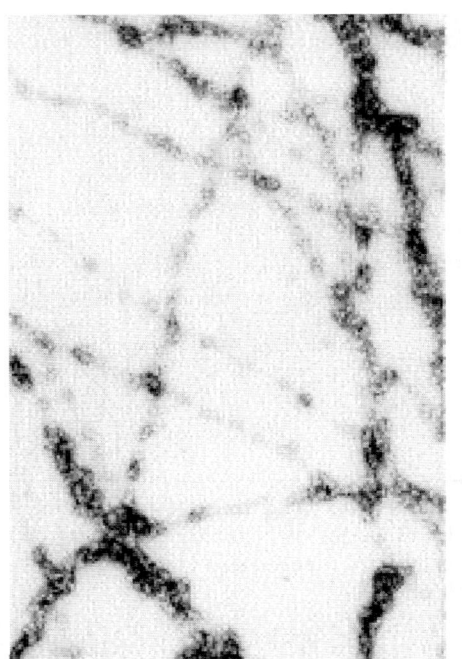

FIGURE 24.23

Beaded-fiber structure of chromatin. An electron micrograph of chromatin spread on a grid at low ionic strength and negatively stained. The spreading under these conditions unravels and stretches some of the condensed chromatin fibers to show the regularly spaced nucleosomes. This was one of the first photographs obtained showing this structure.

Courtesy of C. L. F. Woodcock, University of Massachusetts, Amherst.

occupy defined positions. The implications of this finding are discussed later. The internucleosomal, or *linker,* DNA is occupied by the H1-type (very lysine-rich) histones and nonhistone proteins. Figure 24.25 provides an overall schematic view of the fundamental elements of chromatin structure.

Histone molecules in chromatin are subject to numerous post-translational modifications, such as acetylation and methylation of lysine ε-amino groups. These modifications participate in regulating gene expression. In order to be transcribed, DNA must at least transiently undergo dissociation from the histone core, and histone amino acid modifications affect the strength of interactions between DNA and the histone core. Changes of this kind can bring about stable changes in the way genetic information is expressed. Inherited modifications that change genetic expression without changing DNA base sequence are termed **epigenetic.** As mentioned earlier in this chapter (page 1009), DNA methylation is an epigenetic modification because methylation patterns can be inherited. Although it has not been unequivocally established that histone modifications in chromatin are transmitted when cells divide, most workers consider histone modifications to be epigenetic. We return to these issues in Chapter 29.

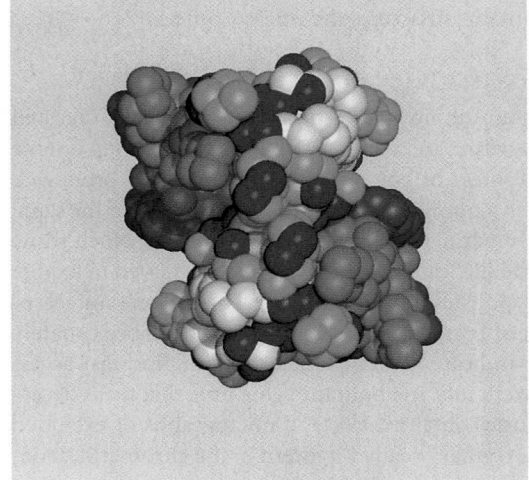

(a)

FIGURE 24.24

Structure of the nucleosome core particle as revealed by X-ray diffraction. (a) The histone octamer core. H3 is green, H4 is white, H2A is light blue, and H2B is dark blue. The lysine and arginine residues of the (H3/H4)$_2$ tetramer are shown in red. **(b)** A high-resolution (2.8 Å) model of the nucleosome core particle with DNA wrapped around it. Two views perpendicular to the twofold axis are shown. The histones are identified as follows: H3, blue; H4, green; H2A, yellow; H2B, red. The N-terminal tails of the histones are not completely resolved.

(a) Reprinted from *Proceedings of the National Academy of Sciences of the United States of America* 90:10489–10493, G. Arents and E. N. Moudrianakis, Topography of the histone octamer surface: Repeating structural motifs utilized in the docking of nucleosomal DNA. © 1993 National Academy of Sciences, U.S.A; (b) Redrawn from K. Luger et al., *Nature* (1997) 389:251–260. © 1997 Macmillan Magazines, Ltd. PDB ID 1AOI.

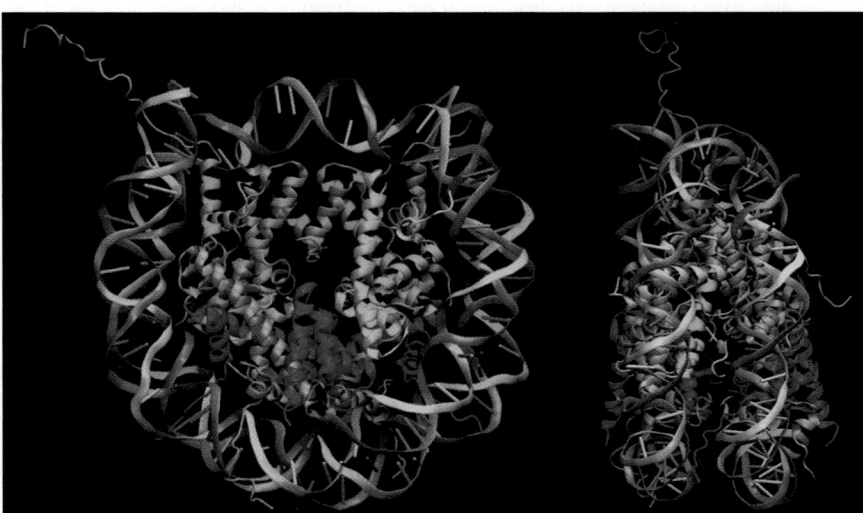

(b)

Higher-Order Chromatin Structure in the Nucleus

Wrapping DNA about histone cores to form nucleosomes accomplishes part of the compaction necessary to fit the eukaryotic DNA into the nucleus because the strand is thereby shortened several-fold. However, it is clear that much of the chromatin in the nucleus is even more highly compacted. The next stage in compaction involves folding the beaded fiber into a thicker fiber like that shown in Figure 24.26. These fibers are about 30 nm in diameter and must be further folded on themselves to make the thicker chromatin fibers visible in both metaphase chromosomes (see Figure 24.20) and the nuclei of nondividing (interphase) cells.

Evidence is emerging concerning the way in which the chromatin fiber is organized in both metaphase and interphase chromatin. Dye staining of the metaphase chromosomes from a particular organism gives a reproducible banding pattern (see Figure 24.16). In situ hybridization methods, in which specific sequences are located on chromosomes by hybridization with a complementary fluorescent nucleic acid probe and then visualized microscopically (Figure 24.11), show that particular DNA sequences are always located at the same places in specific chromosomes. Given that the DNA in a eukaryotic chromosome is one long continuous strand, this implies that some kind of regular folding must be present to preserve this order. Recent evidence indicates what this folding may be. If metaphase chromosomes are treated with polyanions like dextran sulfate, which strip off the histones and loosely bound nonhistone proteins, the DNA strands emerge as enormous loops from a *scaffold* of tightly bound protein. An electron micrograph of this structure is shown in Figure 1.4 (page 9). Individual loops vary in size but may range up to 100,000 bp in length—about the size of the β globin gene cluster, for example. Approximately 1000 such loops exist in the average chromosome.

Evidence also exists for a similar but more diffuse scaffold in the interphase nucleus. Removal of histones and weakly bound nonhistone proteins from intact nuclei by high salt concentrations or detergents, together with digestion of most of the DNA by nucleases, leaves a protein structure that has been called the **nuclear scaffold**, or **nuclear matrix** (Figure 24.26). This structure includes the laminar shell that lines the inside of the nuclear membrane, plus a network of fine fibers that seem to extend throughout the nucleus. When the chemical dissection is done gently, using the specific detergent lithium diiodosalicylate to remove the histones and most other proteins, the DNA connections to the nuclear matrix are undisturbed. Cleavage of the DNA with restriction endonucleases leaves specific fragments of DNA attached to the nuclear matrix. These fragments are spaced at rather long intervals along the genome and contain characteristic *matrix attachment regions (MARs)*. It appears that groups of coordinately expressed genes often lie between adjacent MARs, as Figure 24.27 illustrates for the repeated histone gene clusters in *Drosophila*.

The proteins that form the scaffold from which the loops extend include some interesting members, such as topoisomerases, enzymes that change DNA linking number (Chapter 4). It has been hypothesized that topoisomerase molecules at the base of a loop could bring about changes in the supercoiling on that particular loop. Such changes in supercoiling would be in addition to the coiling imposed by the nucleosomes. Changes in supercoiling may be involved in chromosome condensation and seem to be essential during replication and transcription. It seems likely that the structure of chromatin is dynamic, changing locally as the DNA is replicated or transcribed.

Although it has not yet been established with certainty, it is possible that at least portions of the loop-and-scaffold structures are identical in metaphase and interphase chromatin. If so, they would provide part of the mechanism necessary for the cell to "keep track" of chromatin structure through successive cell divisions. Furthermore, it now seems likely that this loop structure may play a role in the control of expression of groups of functionally linked genes.

The current view is that some loop domains—those involving the nontranscribed genes of a particular cell—may be permanently coiled into 30-nm fibers and perhaps supercompacted into even higher-order coiling. Such regions could correspond to the highly condensed regions of **heterochromatin** long recognized

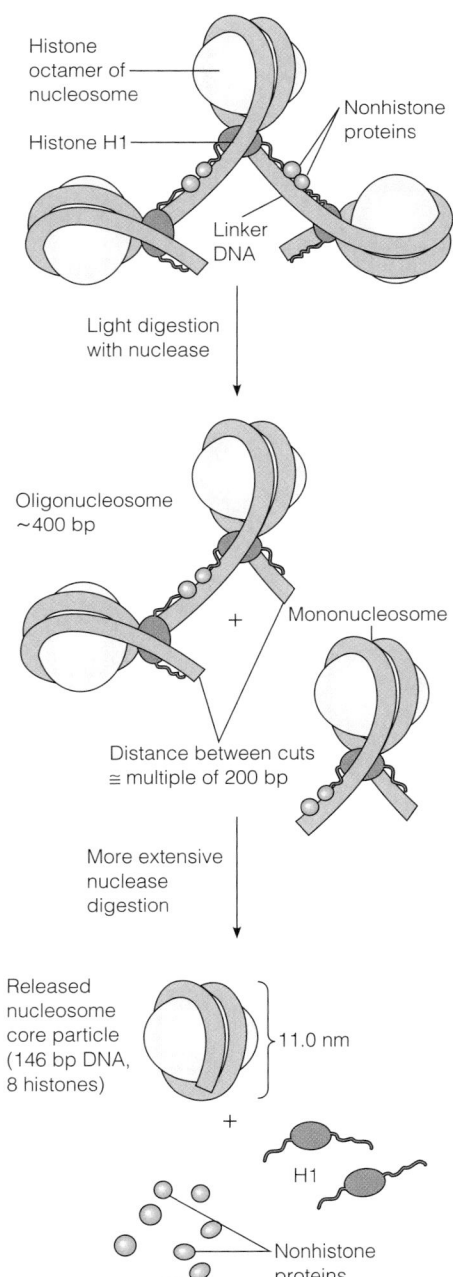

FIGURE 24.25

The elements of chromatin structure. At the top is our current understanding of the extended structure of a chromatin fiber. Light digestion with nuclease releases first mononucleosomes and oligonucleosomes. Then, as linker DNA is further digested, nonhistone proteins and H1 are released, to yield the core particle whose structure is shown in Figure 24.24.

The fiber formed by the nucleosomes is folded in vivo to form higher-order chromatin structure.

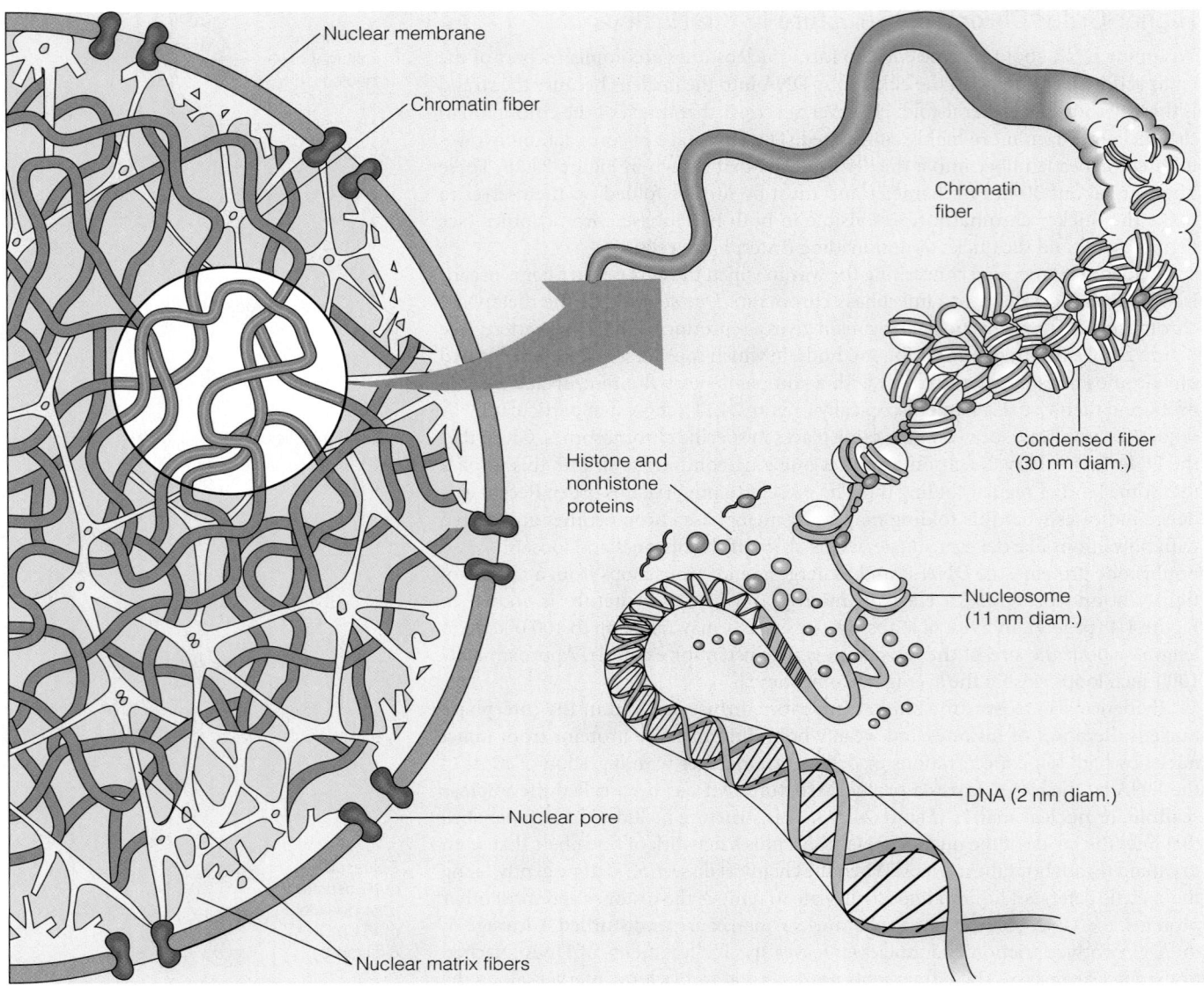

FIGURE 24.26

Levels of chromatin structure. To the left is a schematic view of a portion of the nucleus, with partially condensed chromatin fibers. A closer view (to the right) shows a chromatin fiber in which part is in the condensed (30 nm) form, and part is opened up, as for transcription.

Loops of chromatin, often containing individual gene clusters, are attached to the nuclear matrix.

by cytologists. The more-open chromatin regions, called **euchromatin**, may then correspond to relaxed domains within which transcription can occur. However, as we shall see in Chapters 27 and 29, the regulation of transcription is surely more subtle and complex than such a simple model would imply.

The Cell Cycle

The processes by which cells divide and DNA is replicated are somewhat more complicated in eukaryotes than in prokaryotes. As discussed further in Chapter 25, DNA replication in bacteria is an almost continuous process, at least during exponential growth, whereas in eukaryotic cells DNA replication occupies but a fraction of the time needed for a cell to divide into two identical daughter cells. In general the somatic cells of eukaryotes divide much less frequently, and some, in certain types of mature tissue, do not divide at all. Those that are dividing in growing tissues exhibit a well-defined *cell cycle,* which is almost always separated into several distinct phases, as shown in Figure 24.28. Let us follow a typical eukaryotic cell through one cycle. We can arbitrarily begin in what is called the *G1 phase* (or first gap phase), following cell division. At this point, the cell contains two copies of each chromosome, the normal diploid state of a eukaryotic cell. This state is indicated in Figure 24.28 by a DNA content of 2C, that is, twice the haploid amount. Sometime late in G1 phase, the commitment to divide is triggered. Because division will first require doubling of the DNA content, and the new DNA will need new histones to make chromatin, synthesis of histones is one of the first indications of incipient DNA replication (see Figure 24.28).

The cell then enters synthesis, or *S phase*. During this stage, the DNA is replicated and the histones and nonhistone proteins are deposited on the daughter DNA molecules to reproduce the chromatin structures. When replication is complete, the cell enters the second gap phase, or *G2 phase*. It now has a DNA content four times the haploid amount (4C). In most eukaryotic cells, the total time required for G1, S, and G2 phases is many hours. During this whole period, which is termed **interphase**, the chromatin is dispersed throughout the nucleus and is actively engaged in transcription.

Phases of Mitosis

At the end of G2, the cell is ready to enter M phase, or *mitosis*, during which it divides. Mitosis is a multistage process, and it has been subdivided for descriptive purposes into the phases depicted in Figure 24.29. In *prophase*, the replicated chromosomes condense into the typical *metaphase* chromosome structures so often pictured (see Figure 24.20). In *anaphase* the nuclear membrane disintegrates, and the *mitotic spindle* forms. The spindle consists of contractile microtubules that pull pairs of chromatids apart so that the daughter cells each receive identical sets of chromosomes. In *telophase* the nuclear membrane then re-forms about each daughter nucleus, and the cell itself divides. This cellular division is called **cytokinesis**. After division, the chromosomes of the daughter cells decondense, and a new G1 phase begins.

In many tissues of higher organisms, the G1 phase becomes very prolonged after growth and tissue differentiation are complete. The most extreme examples are fully differentiated nerve cells, most of which never divide again in mature organisms. Such nondividing cells are in a permanently arrested G1 phase, which is often called G_0. On the other hand, some specialized stem cells, such as those found in the bone marrow and intestinal epithelium, undergo continuous division throughout the life of the organism. These stem cells continually provide new differentiated cells to replace cells that have been lost or damaged.

The Centromere and Kinetochore

Earlier, we mentioned the centromere as the chromosomal site at which the chromosome becomes attached to the microtubules in spindle fibers, before sister chromatids become drawn apart. Figure 24.30 shows an electron micrograph of a chromosome attached to the microtubules of the mitotic spindle. The structure linking the centromere on the chromosome to the spindle fibers is the **kinetochore**.

What structural features describe the centromere? This is an important question because it is critical for correct chromosome segregation that each chromosome be drawn away at one, and only one, site. DNA sequences in centromeres differ from those in the rest of the chromosome, but in a nonconserved way. In budding yeast the centromere forms at a defined 125-bp sequence, while in higher eukaryotes centromeric sequences consist of much longer repetitions of a much shorter sequence. Chromatin structure is altered within the centromere, partly because of the replacement of histone H3 by a variant H3, called CenH3. Unlike the standard histones, CenH3 is quite divergent throughout evolution, evidently matching the variability seen among centromere DNA sequences. Although the structure of this modified chromatin has not yet been described in detail, there is evidence that DNA within the centromere is positively supercoiled, rather than negatively, as in the rest of the chromosome.

Further variability in centromeric chromatin is seen in the nucleosome structure. Often the replacement of H3 by CenH3 is the only change, yielding core particles with (CenH3/H4/H2A/H2B)$_2$ octamers. As suggested in Figure 24.31, modified

The eukaryotic cell passes through a cycle of phases: G1 (gap 1), S (DNA synthesis), G2 (gap 2), and M (mitosis).

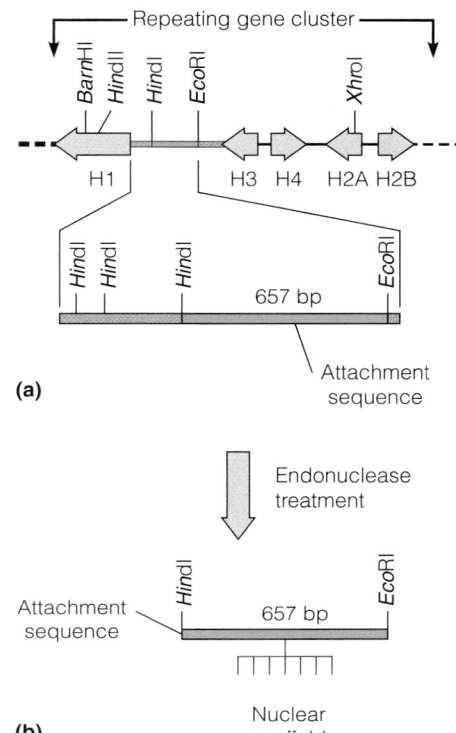

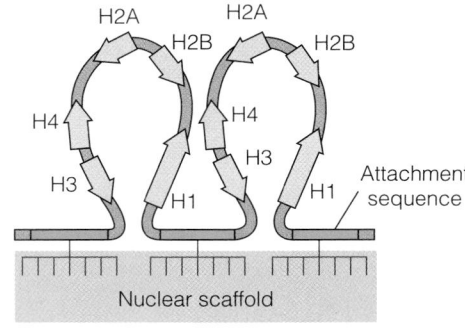

FIGURE 24.27

Attachment of gene clusters to the nuclear matrix. **(a)** A map of the repeating histone gene cluster in *Drosophila*, where each green arrow is a histone gene (arrowheads indicate direction of transcription). A number of restriction sites are also shown. **(b)** If *Drosophila* nuclei are extracted with lithium diiodosalicylate, to remove proteins gently, and are then digested with a collection of the restriction endonucleases shown, only the 657-bp HindI-EcoRI DNA fragments are left attached to the matrix. **(c)** The interpretation is that the gene clusters exist in individual loops, the bases of which are tied to the matrix.

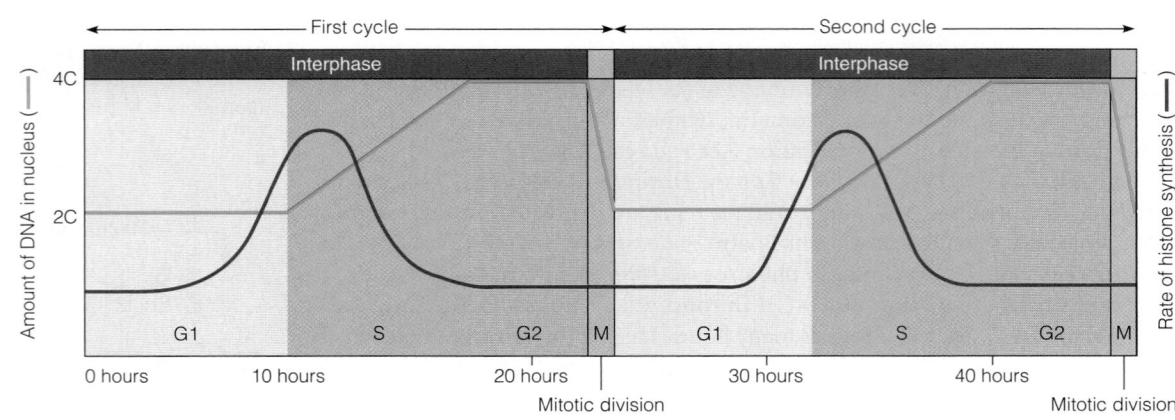

FIGURE 24.28

The eukaryotic cell cycle. Changes in the amount of DNA (blue line) and rate of histone synthesis (red line) with time during two cell cycles. The DNA content is measured in units of the haploid genome (C). The time scale is typical of many eukaryotic cells.

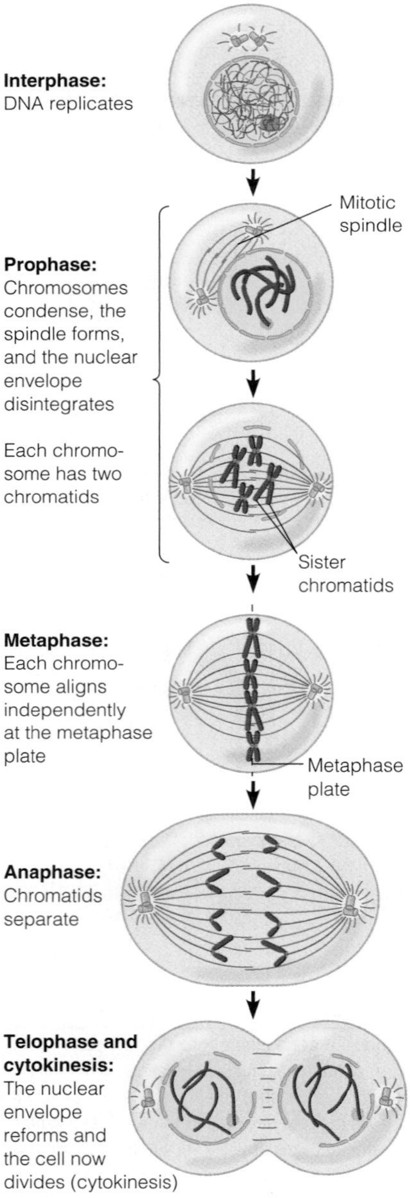

Interphase:
DNA replicates

Prophase:
Chromosomes condense, the spindle forms, and the nuclear envelope disintegrates

Each chromosome has two chromatids

Mitotic spindle

Sister chromatids

Metaphase:
Each chromosome aligns independently at the metaphase plate

Metaphase plate

Anaphase:
Chromatids separate

Telophase and cytokinesis:
The nuclear envelope reforms and the cell now divides (cytokinesis)

FIGURE 24.29

Mitosis. The cell entering the pathway was originally diploid. It has undergone DNA replication and is now in G2, with a DNA content of 4C. After the process is complete, each daughter cell will again be 2C.

chromatin is on the outside of the chromatin coil within the centromere, available to interact with proteins in the kinetochore, which will draw the two paired chromatids apart at anaphase. Unmodified nucleosomes are in the interface between sister chromatids, which are transiently held together by proteins called *cohesins*. The figure also shows predominant histone methylation patterns within and adjacent to the centromere.

The kinetochore is a complex structure consisting of two "plates," one in contact with the centromere and one with a microtubule. As shown in Figure 24.32, the inner and outer plates are joined by a protein complex called Ndc80. There is evidence that the protein complexes in the inner plate of each kinetochore contact just one CenH3 nucleosome. Because each kinetochore contacts just one microtubule, several kinetochores are involved in formation of each mitotic spindle.

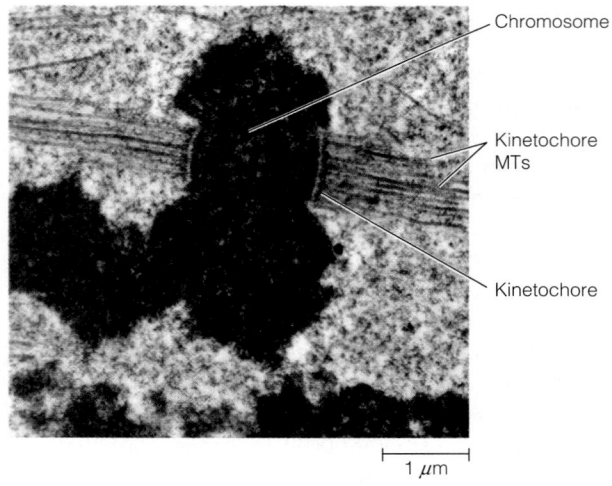

Chromosome

Kinetochore MTs

Kinetochore

1 μm

FIGURE 24.30

An electron micrograph of a condensed chromosome pair linked via their kinetochores to microtubules (MTs). The region of the chromosome in contact with the kinetochore is the centromere.

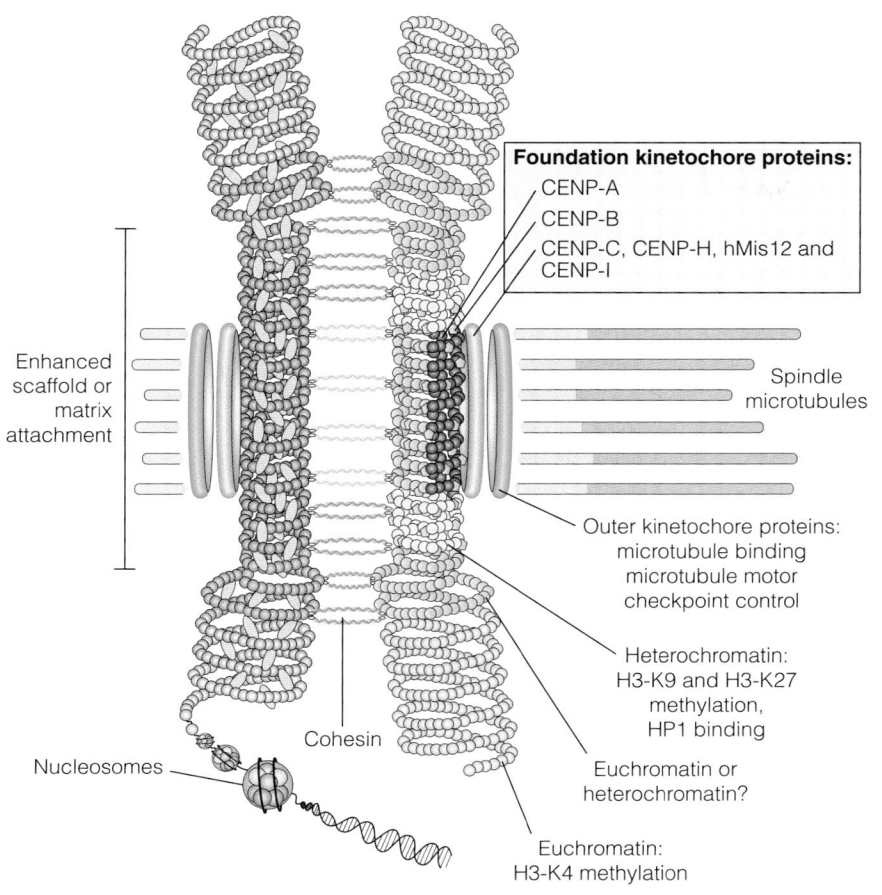

Foundation kinetochore proteins:
CENP-A
CENP-B
CENP-C, CENP-H, hMis12 and
CENP-I

Enhanced
scaffold or
matrix
attachment

Spindle
microtubules

Outer kinetochore proteins:
microtubule binding
microtubule motor
checkpoint control

Heterochromatin:
H3-K9 and H3-K27
methylation,
HP1 binding

Euchromatin or
heterochromatin?

Cohesin

Nucleosomes

Euchromatin:
H3-K4 methylation

FIGURE 24.31

Structural organization of chromatin in the human centromere. Interaction between the outer region of each of the paired chromatids with kinetochores and microtubules is schematized as well. H3-K9 and H3-K27 refer to histone modifications—methylation of lysine 9 and 27, respectively, on histone H3.

Reprinted from *Trends in Cell Biology* 14:359–368, D. J. Amor, P. Kalitsis, H. Sumer, and K. H. A. Choo, Building the centromere: From foundation proteins to 3D organization. © 2004, with permission from Elsevier.

Control of the Cell Cycle

Progression through the cell cycle is controlled by a series of protein phosphorylations. Major features of the regulatory program were elucidated by Lee Hartwell through studies on temperature-sensitive mutants of yeast blocked at high temperature in progression through the cycle. Key players in the program are small proteins called **cyclins** and the protein kinases that they activate, called **cyclin-dependent kinases** (Cdks). The system is designed so that the cell can assess at various stages whether conditions are right to move to the next phase. Each stage at which this assessment occurs is called a **checkpoint**. A checkpoint during G1 involves a G1-Cdk, and it acts to assure that the environment—availability of nutrients, and so on—is favorable for entry into S phase and commitment to DNA replication. A series of G1/S-Cdks and S-Cdks acts to detect DNA damage, exerting a hold on the cycle until that damage is repaired. The tumor suppressor protein p53 (Chapter 23) is involved in this process. A later checkpoint, involving M-Cdk, checks both for DNA damage and completion of replication, both of which are obviously essential before cells commit themselves to mitosis. Later, during mitosis, control is exercised by the **anaphase-promoting complex** (APC/C), which checks to ensure that all chromosomes are attached to spindles before the cell proceeds to anaphase. These processes are schematized in Figure 24.33.

The Cdks are protein serine/threonine kinases. Each one is activated by a specific cyclin, and then phosphorylates one or more target proteins. A major target of the G1-Cdk is the **Rb protein**, a molecule that controls the expression of proteins needed to pass the cell into S phase. Mutations in the gene for Rb give rise to **retinoblastoma**, a tumor affecting the eye. Once a checkpoint has been activated,

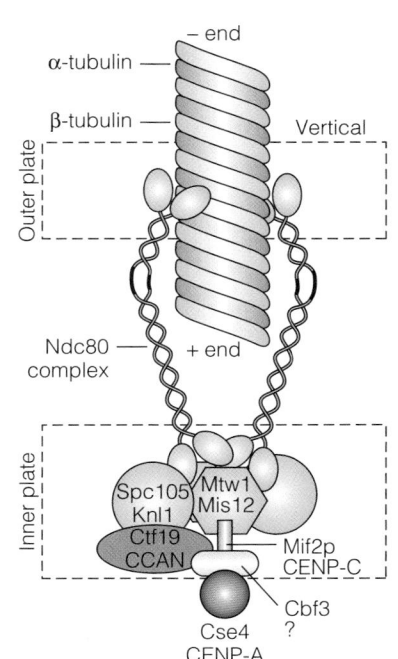

– end
α-tubulin
β-tubulin
Vertical
Outer plate
Ndc80
complex
+ end
Inner plate
Spc105
Knl1
Mtw1
Mis12
Ctf19
CCAN
Mif2p
CENP-C
Cbf3
?
Cse4
CENP-A

FIGURE 24.32

A model for kinetochore structure and organization. The model is based upon electron microscopic and proteomic analysis of kinetochores in yeast (*S. cerevisiae*). Cse4 and CENP-A are alternative names for the modified nucleosomes in the centromere.

Reprinted by permission from Macmillan Publishers Ltd. *The EMBO Journal* 28:2511–2531, S. Santaguida and A. Musacchio, The life and miracle of kinetochores. © 2009.

FIGURE 24.33

Control of the cell cycle by cyclin-dependent kinases, and the checkpoints at which they operate. **(a)** The major checkpoints, showing the conditions checked and the regulatory proteins involved at each checkpoint. **(b)** The pattern of synthesis and degradation of each cyclin throughout the cell cycle. Activities of the cyclin-dependent kinases remain constant.

Modified from *Molecular Biology of the Cell*, 5th ed., B. Alberts, A. Johnson, J. Lewis, M. Raff, K. Roberts, and P. Walter. © 2008 Garland Science/Taylor & Francis Group.

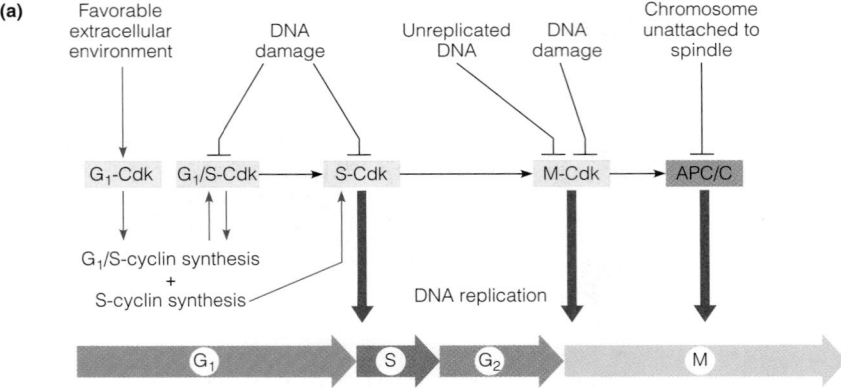

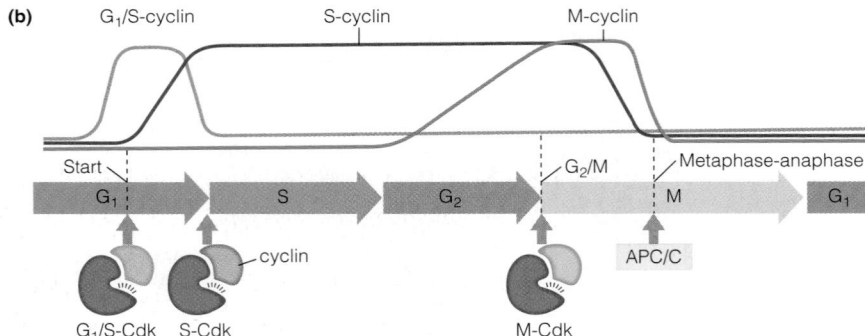

the cyclin–Cdk complex dissociates, and the cyclin is marked for degradation in the proteasome, a large protein complex, described in Chapter 20, which is involved in intracellular protein degradation. One of the functions of the APC complex is targeting proteins for degradation. Another Cdk target is p53, which becomes activated as a transcription factor by phosphorylation of key serine residues and blocks advance through the cycle until the DNA is repaired. If the damage is too great, then p53 promotes events leading to **apoptosis**, or programmed cell death (Chapter 28). Mutations in the gene for p53 have been implicated in Li-Fraumeni syndrome, which carries a predisposition to early breast cancer. And, as noted in Chapter 23, p53 gene mutations are found in a large proportion of all cancers.

SUMMARY

Most of the DNA in a prokaryotic genome encodes proteins or RNA or serves a regulatory function. However, 50% or more of the DNA in a eukaryotic genome consists of repeated sequences that do not have protein or RNA gene products. In eukaryotes there is no apparent relationship between the size of a genome and the complexity of the organism harboring that genome. Sequence determination of complex genomes, notably those of humans, depended crucially upon the development of new technologies, including use of type II restriction endonucleases, fluorescent in situ hybridization, Southern transfer, polymerase chain reaction, and use of yeast artificial chromosomes as vectors for cloning huge amounts of DNA. Viral genomes, either RNA or DNA, encode as few as 3 proteins or as many as 175. Bacterial genomes usually consist of one circular double-stranded DNA molecule encoding about 1500–2000 proteins, while the human genome encodes between 20,000 and 25,000 proteins, considerably fewer than predicted from the amount of DNA in the genome. Most eukaryotic cells are diploid, with two of each chromosome. An exception is the sex chromosomes, where a male typically has one X and one Y chromosome. Chromosomal DNA is compacted by folding about histone octamers, to form nucleosome core particles. These particles plus linker DNA make up chromatin, which refolds to form 30-nm fibers. Further compaction processes fold these fibers into the highly condensed form seen in chromatin. Eukaryotic cells divide *via* a well-defined cell cycle, involving G1, in which the chromosome number is 2; S, in which DNA replicates and histones are synthesized and chromatin assembled; G2, in which the chromosome number is 4; and M, a multistage process in which the cell undergoes mitosis. Structural analysis of centromeres and kinetochores is yielding insight into mechanisms of chromosome segregation during mitosis. The cell cycle is controlled by a complex series of protein phosphorylation events, in which protein kinase activities are constant throughout the cycle, but levels of kinase-activating proteins, the cyclins, rise and fall cyclically.

REFERENCES

Genes and Genomes

Altshuler, D., M. J. Daly, and E. S. Lander (2008) Genetic mapping in human disease. *Science* 322:881–888. A review of the methods used to map human disease-related genes.

Cooper, N. G., ed. (1994) *The Human Genome Project: Deciphering the Blueprint of Heredity.* University Science Books, Mill Valley, CA. Although this book was published several years before completion of the Human Genome Project, it describes the experimental approaches used quite clearly.

Gibson, D. G., and 23 coauthors (2010) Creation of a bacterial cell controlled by a chemically synthesized genome. *Science* 329:52–56. A big step in the creation of "synthetic life," from the J. Craig Venter Institute.

Green, R. E., and 56 coauthors (2010) A draft sequence of the Neandertal genome. *Science* 328:710–722. An article illustrating the power of PCR to amplify ancient DNAs.

Kayser, M., and P. de Kniff (2011) Improving human forensics through advances in genetics, genomics, and molecular biology. *Nature Rev. Genet.* 12:179–185. A contemporary review of the use of DNA technology in forensics.

Krimsky, S., and T. Simoncelli (2011) *Genetic Justice.* Columbia University Press, New York. A book-length treatment of DNA forensics, which offers cautions against misuses of the technology.

Lander, E. S. (2011) Initial impact of the sequencing of the human genome. *Nature* 470:187–197. A thoughtful retrospective article with 100 references.

Poliseno, L., L. Salmena, J. Zhang, B. Carver, W. J. Haveman, and P. P. Pandolfi (2010) A coding-independent function of gene and pseudogene mRNAs regulates tumor biology. *Nature* 465:1033–1040. The first evidence for function of a pseudogene.

Roache, J. C., and 14 coauthors (2010) Analysis of genetic inheritance in a family quartet by whole-genome sequencing. *Science* 328:636–639. A large amount of genetic and genomic information results from comparing genome sequences from closely related individuals.

Vashlishan Murray, A. B., M. J. Carson, C. A. Morris, and J. Beckwith (2010) Illusions of scientific legitimacy: Misrepresented science in the direct-to-consumer genetic-testing marketplace. *Trends in Genetics* 26:459–461. Skepticism about the proliferation of companies offering genetic and genomic testing to the public.

Zimmer, C. (2009) On the origin of eukaryotes. *Science* 325:666–668. A readable summary of evidence for the endosymbiont hypothesis, postulating that mitochondria and chloroplasts are highly evolved prokaryotes.

Restriction and Modification

Roberts, R. J., and X. Cheng (1998) Base flipping. *Annu. Rev. Biochem.* 67:181–198. DNA restriction methylases are not the only enzymes that flip bases in DNA substrates.

Roberts, R. J., and 46 coauthors (2003) A nomenclature for restriction enzyme, DNA methyltransferases, homing endonucleases, and their genes. *Nucl. Ac. Res.* 31:1805–1812. The large numbers of enzymes in these families and their diverse sources required that leaders in the field (47 of them!) establish some order.

Chromosomes and Chromatin

Black, B. E., and D. W. Cleveland (2011) Epigenetic centromere propagation and the nature of CENP-A nucleosomes. *Cell* 144:471–479. A model that connects assembly of centromeric chromatin with control of the cell cycle.

Bloom, K., and A. Joglekar (2010) Towards building a chromosome segregation machine. *Nature* 463:446–456. Contemporary review of centromeres and kinetochores.

Furuyama, T., and S. Henikoff (2009) Centromeric nucleosomes induce positive DNA supercoils. *Cell* 138:104–113. DNA–protein interactions are quite unusual in the centromere.

Hurtley, S. M., and E. Pennisi (2007) Journey to the center of the cell. *Science* 318:1399. Introduction to a special section of *Science* on structure and dynamics of the nucleus (pp. 1400–1416).

Luger, K., A. W. Mädes, R. K. Richmond, D. F. Sargent, and T. J. Richmond (1997) Crystal structure of the nucleosome core particle at 2.8 Å resolution. *Nature* 389:251–260.

Marx, J. (2002) Chromosome end game draws a crowd. *Science* 295:2348–2351. Early excitement about telomeres and telomerase.

Richmond, T. J., and C. A. Davey (2003) The structure of DNA in the nucleosome core. *Nature* 423:145–150. Associating with histone cores affects the shape of associated DNA.

Santos, S., and A. Musacchio (2009) The life and miracles of kinetochores. *EMBO J.* 28:2511–2531. A well-illustrated contemporary review.

Segal, E., and J. Widom (2009) What controls nucleosome positions? *Trends in Genetics* 25:335–343. Understanding the positioning of histone cores along DNA chains is crucial to understanding expression of chromatin-associated genes.

Sekulic, N., E. A. Bassett, D. J. Rogers, and B. E. Black (2010) The structure of (CENP-A-H4)₂ reveals physical features that mark centromeres. *Nature* 467:347–352. Closing in on centromere structure.

Torras-Llort, M., O. Moreno-Moreno, and F. Azorin (2009) Focus on the centre: The role of chromatin on the regulation of centromere identity and function. *EMBO J.* 28:2337–2348. This review deals mostly with the nature of CenH3.

Trun, N. J., and J. F. Marko (1998) Architecture of a bacterial chromosome. *ASM News* 64:276–283. A brief review of the bacterial nucleoid.

The Cell Cycle

Gasser, S. M. (2002) Visualizing chromatin dynamics in interphase nuclei. *Science* 296:1412–1416. Use of fluorescence microscopy for examining chromosome dynamics.

Nasmyth, K. (2001) A prize for proliferation. *Cell* 107:689–701. An appreciation of the work of Lee Hartwell, Paul Nurse, and Tim Hunt, who shared a Nobel Prize for discovering protein phosphorylation in control of the cell cycle.

Verdaasdonk, J. S., and K. Bloom (2011) Centromeres: Unique chromatin structures that drive chromosome segregation. *Nature Rev. Mol. Cell Bio.* 12:320–331. A comprehensive and timely review.

Weis, K. (2003) Regulating access to the genome: Nucleocytoplasmic transport throughout the cell cycle. *Cell* 112:441–451. A review of the nuclear pore complex, importins, exportins, and transport of macromolecules between nucleus and cytoplasm.

PROBLEMS

1. The exponential nature of PCR allows spectacular increases in the abundance of a DNA sequence being amplified. Consider a 10-kbp DNA sequence in a genome of 10^{10} base pairs. What fraction of the genome is represented by this sequence; i.e., what is the fractional abundance of this sequence in this genome? Calculate the fractional abundance of this target sequence after 10, 15, and 20 cycles of PCR, starting with DNA representing the whole genome and assuming that no other sequences in the genome undergo amplification in the process.

2. Of the restriction enzymes listed in Table 24.3, which enzymes generate flush, or blunt-ended, fragments? Of those that recognize offset sites and generate staggered cuts, which of these cuts cannot be converted to flush ends by the action of DNA polymerase? Why is this so?

3. The following diagram shows one-half of a restriction site.
   (a) Draw the other half.

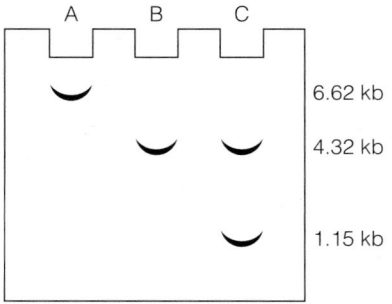

   (b) Use heavy arrows ($\uparrow\downarrow$) to identify type II cleavage sites that would yield blunt-ended duplex DNA products.
   (c) Use light arrows ($\uparrow\downarrow$) to identify type II cleavage sites yielding staggered cuts that could be converted directly to recombinant DNA molecules by DNA ligase, with no other enzymes involved.
   (d) If this were the recognition site for a type I restriction endonuclease, where would cutting of the duplex occur?
   (e) If DNA sequences were completely random, how large an interval (in kilobase pairs) would you expect between identical copies of this sequence in DNA?

4. pBR322 DNA (4.32 kb) was cleaved with *Hin*dIII nuclease and ligated to a *Hin*dIII digest of human mitochondrial DNA. One recombinant plasmid DNA was analyzed by gel electrophoresis of restriction cleavage fragments, with the following results: lane A = *Eco*RI-treated recombinant, lane B = *Hin*dIII-treated vector, and lane C = *Hin*dIII-treated recombinant.

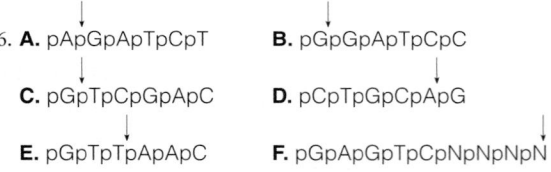

(a) Why was the recombinant plasmid treated with *Eco*RI for determination of its size?
(b) How might you explain the discrepancy between the size of the recombinant molecule and the sum of the sizes of the *Hin*dIII cleavage fragments?
(c) Draw a diagram of the recombinant showing the locations of the *Hin*dIII cleavage sites.

5. A small DNA molecule was cleaved with several different restriction nucleases, and the size of each fragment was determined by gel electrophoresis. The following data were obtained.

Enzyme	Fragment Size (kb)
*Eco*RI	1.3, 1.3
*Hpa*II	2.6
*Hin*dIII	2.6
*Eco*RI + *Hpa*II	1.3, 0.8, 0.5
*Eco*RI + *Hin*dIII	0.6, 0.7, 1.3

(a) Is the original molecule linear or circular?
(b) Draw a map of restriction sites, showing distances between sites, that is consistent with the data presented.
(c) How many additional maps are compatible with the data?
(d) What would have to be done to locate the cleavage sites unambiguously with respect to each other?

6. **A.** pApGpApTpCpT   **B.** pGpGpApTpCpC

   C. pGpTpCpGpApC   **D.** pCpTpGpCpApG

   E. pGpTpTpApApC   **F.** pGpApGpTpCpNpNpNpN

(a) Using the same shorthand as in Problem 3, show the complete structure of any one of the foregoing restriction cleavage sites (both strands and cutting sites).
(b) Which two of the cleavages shown will yield fragments that *cannot* be rejoined by *E. coli* DNA ligase? Why not? (Note: In vitro DNA ligase shows strong preference for sealing a staggered, rather than blunt-ended, cut.)
(c) Cleavage products from two of these reactions can readily be joined to one another by DNA ligase, even though the two enzymes recognize different sites. Which two?
(d) If you wished to linearize a newly isolated plasmid DNA, which one of the sites shown would be *least* likely to be represented only once in that DNA molecule? Assume that the DNA has equal proportions of the four nucleotides.

7. The average human chromosome contains about 1×10^8 bp of DNA.

(a) If each base pair has a mass of about 660 daltons, and there are about 2 grams of protein (histones plus nonhistones) per gram of DNA, how much does such a chromosome weigh, in grams?

(b) If the DNA were extended, how long would it be?

(c) An actual chromosome is about 5 mm in length. What is the approximate compaction ratio?

(d) You have about 10^{12} cells in your body. If you have 46 chromosomes in each cell, what is the approximate extended length of *all* of your DNA? For comparison, the distance from the earth to the sun is about 1.5×10^8 km.

8. Formation of nucleosomes and wrapping them into a 30-nm fiber provide part of the compaction of DNA in chromatin. If the fiber contains about six nucleosomes per 10 nm of length, what is the approximate compaction ratio achieved? Comment on the comparison of this answer with that of Problem 7.

*9. It is possible to "reconstitute" nucleosomes by mixing DNA and histone octamers in 2M NaCl and then dialyzing to low salt. When such experiments were carried out using a specific 208-bp fragment of sea urchin DNA, the following results were obtained: Digestion of the product with micrococcal nuclease gave quantitative production of 146-bp DNA. Upon cleavage of this DNA with a restriction nuclease having a single site in the fragment, several sharp bands were obtained, with sizes as follows: 29 bp, 39 bp, 107 bp, 117 bp. How would you interpret these data in terms of nucleosome positioning on this DNA?

10. A sample of chromatin was partially digested by the enzyme staphylococcal nuclease. The DNA fragments from this digestion were purified and resolved on a polyacrylamide gel. A set of DNA restriction fragments was used as markers. Distances of migration are given below. From these data, estimate the nucleosome repeat distance in the chromatin.

Marker DNA Fragment		Chromatin DNA Fragment
Size (bp)	d (cm)	d (cm)
94	40	30.5
145	34.2	19.2
263	25.2	14.4
498	16.7	11.5
794	11.5	—

*11. Pancreatic deoxyribonuclease I (DNase I) is a nuclease that makes single-strand nicks on double-stranded DNA. It has been observed that treatment of nucleosomal core particles with DNase I yields a peculiar result. When DNA from such a digestion is electrophoresed under denaturing conditions, the single-stranded fragments are observed to occur in a regular periodicity of about 10 bases. Suggest an explanation of this result in terms of the structure of the nucleosome.

12. Some viruses, like SV40, are closed circular DNAs carrying nucleosomes. If the SV40 virus is treated with topoisomerase, and the histone is then removed, it is found to still be supercoiled. However, if histones are removed *before* topoisomerase treatment, the DNA is relaxed. Explain.

13. Histone genes are unusual among eukaryotic genes in that they do not have introns, and histone mRNAs do not have poly(A) tails (see Chapter 27). Furthermore, in almost all eukaryotes, histone genes are arranged in multiple tandem domains, each domain carrying one copy of each of the five histone genes. Suggest an explanation for these features in terms of the special requirements for histone synthesis.

14. Consider the λ bacteriophage DNA molecule, as shown in Figure 24.7. Total digestion with the two restriction endonucleases used does not allow unambiguous placement of restriction sites, as shown in the map. Describe restriction patterns—either from partial digestion or digestion with both *Bam*H1 and *Eco*R1—that would help to establish the specific map of restriction sites shown in Figure 24.7(b).

15. RFLPs can include sequence changes created by single-base change, single-base insertion, single-base deletion, or tetranucleotide repeat, as shown in Figure 24.15. For each polymorphism, show a specific sequence change that would alter the restriction fragmentation pattern at that site and describe how the pattern would change—i.e., addition or loss of a site for a specific restriction endonuclease. Refer to Table 24.3 for the cleavage sites of several common restriction nucleases.

16.

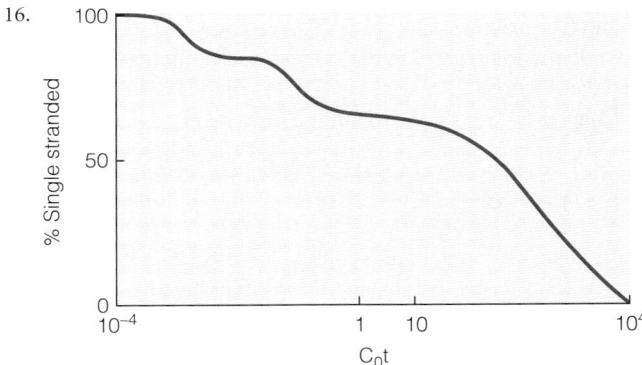

DNA renaturation curves occasionally show three distinct phases of renaturation. In this plot, DNA renaturation is plotted against "C_0t" (initial concentration times time of renaturation, essentially a measure of relative renaturation time).

(a) Identify each part of this plot that corresponds to reannealing of (1) unique sequences, (2) moderately repetitive sequences, and (3) highly repetitive sequences.

(b) Suppose that you cloned a single-copy gene, such as the gene for dihydrofolate reductase (DHFR), and subjected this to renaturation analysis. Sketch the curve you might expect.

(c) Suppose that you cloned the DHFR cDNA instead of genomic DNA. Would you expect this to reanneal (1) more slowly, (2) more rapidly, or (3) at the same rate as genomic DNA? Briefly explain your answer.

17. Refer to Figure 24.28, which shows the time spent in each of the four phases of the cell cycle for a typical eukaryotic cell. If a culture of these cells was inspected microscopically, what proportion of the cells would you expect to show condensed chromosomes?

TOOLS OF BIOCHEMISTRY 24A

POLYMERASE CHAIN REACTION

As we discuss in Chapter 4, gene cloning by recombinant DNA techniques revolutionized biology in the mid-1970s because it allows an investigator to isolate and amplify individual genes for analysis of their sequence, expression, and regulation. Cloning requires living cells, into which DNA molecules must be introduced for amplification. An equally revolutionary technique, **polymerase chain reaction (PCR)**, was invented by Kary Mullis in 1983. PCR allows the amplification of exceedingly small amounts of DNA in vitro, without prior transfer into living cells. This technique has facilitated the analysis of eukaryotic genes because it avoids some of the tedium involved in cloning DNA from very large genomes. In addition, the technique has dozens of practical applications.

To understand PCR, you must understand that DNA polymerase catalyzes the addition of deoxyribonucleoside triphosphate to a preexisting 3′-hydroxyl terminus of a growing daughter DNA strand (the primer). A template DNA strand is required, to instruct the polymerase regarding the correct nucleotide to insert at each step (see Figure 4.21, page 111). The reaction is presented in detail in Chapter 25, but shown in outline form here.

3′-pApTpTpCpApApGpApGpG..... + dTTP → 3′-pApTpTpCpApApGpApGpG.....
5′-pTpApApG-OH   5′-pTpApApGpT-OH

Most DNA polymerases also contain an associated 3′-exonuclease. This activity helps the enzyme to proofread errors, by removing nucleotides that occasionally are incorporated in non-Watson–Crick fashion.

PCR requires knowledge of sequences that flank the region to be amplified. Oligonucleotides complementary to these sequences are produced by automated chemical synthesis and are used as primers in a special series of DNA polymerase–catalyzed reactions (Figure 24A.1). First, DNA containing the sequences to be amplified is heat-denatured at 94–96 °K and then annealed to the primers, which are present in excess (steps 1 and 2). Annealing is carried out at 50–65 °K, depending upon the base composition of the DNAs involved. Next, polymerase chain extension is carried out from the primer termini (step 3). Then a second cycle of heat denaturation, annealing, and primer extension is carried out. Using a thermostable form of DNA polymerase, such as **Taq polymerase**, an enzyme from an organism that lives at high temperatures (in hot springs), avoids the need to add more polymerase at each cycle because the enzyme is not inactivated at DNA-denaturing temperature. Step 3 is typically carried out at 72 °K, the temperature optimum for *Taq* polymerase.

FIGURE 24A.1

Three cycles of the polymerase chain reaction. A segment within the region shown in blue is amplified, by use of primers (red) that are complementary to the ends of the blue segment. Note the exponential nature of the amplification process.

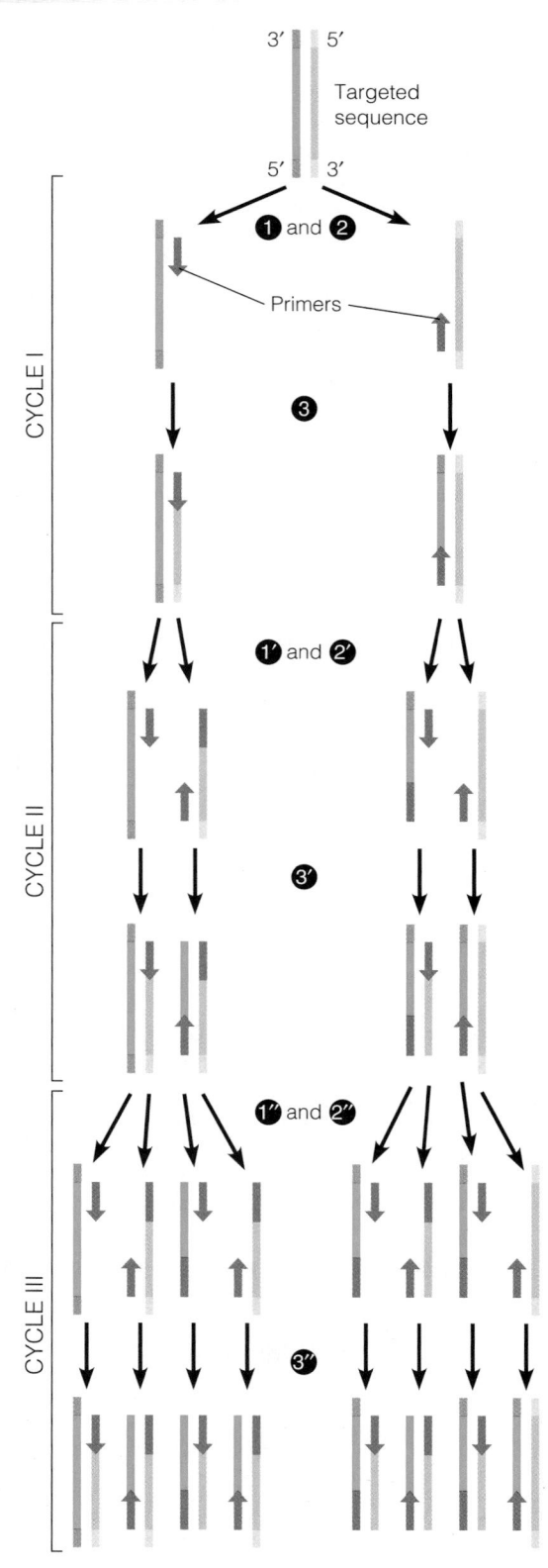

This cycle is repeated 30 or more times in an automated temperature-regulating device (a *thermal cycler*), with each cycle increasing the abundance of duplex DNA species bounded by the oligonucleotide primers. Eight such molecules have been formed by the end of cycle III, and the number doubles with each successive cycle. About 1 billion copies are present after 32 cycles (2^{n-1} to be exact, where n is the number of cycles).

Innumerable applications of this technique have been developed, including forensic analysis (as mentioned in the chapter), in which DNA can be amplified from exceedingly small samples of biological material (for instance, blood, semen, or hair) for identification of criminal suspects or of fathers in paternity cases. PCR-based forensic analysis focuses upon parts of the genome with variable numbers of tetranucleotide repeats (see Figure 24.15). Some parts of the human genome have a particular tetranucleotide sequence repeated variable numbers of times up to 30. Analyses of these regions by PCR can determine the number of repeats in one individual's genome. By analyzing several of these regions, an individual can be identified with great confidence, beginning with a DNA sample as small as one nanogram.

The field of "molecular anthropology" has developed from PCR and sequence analysis of human mitochondrial DNA, with the results used to formulate models of human evolution. Mitochondrial DNA is particularly useful because it undergoes mutation more rapidly than nuclear DNA, meaning that the amount of sequence variation is a more sensitive indicator of evolutionary time than is seen in nuclear DNA. A comparable field of "molecular paleontology" exists, with minute amounts of DNA being extracted from long-preserved biological samples, such as organisms frozen in ice or insects trapped in amber. The recent sequence analysis of DNA from specimens of Neandertals has generated new insights into human evolution. Of historical interest was the 1998 report describing PCR analysis of descendants of Thomas Jefferson, which showed that America's third president or a close relative fathered at least one child with Sally Hemings, one of his slaves. The DNA came from living people (descendants), but the results led to important historical conclusions. PCR is used in environmental microbiology, to detect microbial populations by searching for sequences unique to the organism being sought. In the same way, PCR can be used for diagnosis of microbial or viral infections. PCR is used also for prenatal diagnosis of genetic diseases, using primers specific for the gene sequence alteration responsible for a disorder. Similarly, oncogene mutations leading to cancer can be detected by PCR, and this has allowed extensive analysis of the sequence of genetic alterations leading from a precancerous lesion to a metastatic tumor.

PCR is not without its technical problems, however. For example, because some thermostable DNA polymerases, such as the commonly used *Taq* polymerase, lack a proofreading 3′ exonuclease, the DNA synthesis in PCR can be relatively inaccurate. This is not usually a problem if one wants to sequence the PCR product because errors are uniformly distributed over the length of DNA being amplified, with the abundance at each site being too low to affect sequencing operations. Also, proofreading polymerases from thermophilic organisms are now available. However, precautions must be taken (such as limiting the number of cycles) if one wants to clone PCR products, with the expectation that a natural sequence is being cloned. Another problem is the great sensitivity of the technique, which can lead to amplification of minute amounts of DNA contaminants in the sample. Again, a number of controls and modifications of the technique can be used to minimize this problem.

A number of variants of the original PCR technique have been devised. In quantitative, or "real-time," PCR, one of the nucleotides has a fluorescent label on it, which permits following the reaction in real time and relating the data to abundance of the target DNA in the original sample. Another variation of the technique, called RT-PCR, can be used to monitor gene expression, by analysis of the levels of a particular mRNA species. The sample is treated first with reverse transcriptase (RT), an RNA-dependent DNA polymerase (Chapter 25), which converts an mRNA molecule to a DNA of complementary sequence, which can then be amplified by PCR.

References

Erlich, H. A., and N. Arnheim (1992) Genetic analysis using the polymerase chain reaction. *Annu. Rev. Genet.* 26:479–506. A review by two developers of the technique.

Green, R. E., A. W. Briggs, A. Krause, K. Prüfer, H. A. Burbano, M. Siebauer, M. Lachmann, and S. Pääbo (2009) The Neandertal genome and ancient DNA authenticity. *EMBO J.* 28:2494–2502. A description of techniques used to rule out sources of error in sequence analysis of ancient DNAs.

Mullis, K., F. Ferre, and R. Gibbs, eds. (1994) *The Polymerase Chain Reaction.* Birkhäuser, Boston. The senior author of this book-length review collection is the inventor of PCR.

Mullis, K. B. (1997). *Nobel Lectures, Chemistry, 1991–1995,* B. G. Malmström, ed., World Scientific Publishing Co., Singapore, 1997. Mullis's Nobel Prize address makes fascinating reading. Also available online at: http://nobelprize.org/nobel_prizes/chemistry/laureates/1993/mullis-lecture.html

Nowak, R. (1994) Forensic DNA goes to court with O. J. *Science* 265:1352–1354. A contemporary news article, describing the use of PCR and restriction fragment–length polymorphisms as applied specifically to the O. J. Simpson murder trial.

Sambrook, J., and Russell, D. W. (2001) In vitro amplification of DNA by the polymerase chain reaction. Chapter in *Molecular Cloning: A Laboratory Manual,* Cold Spring Harbor Press, Cold Spring Harbor, NY. Specific instructions for using PCR.

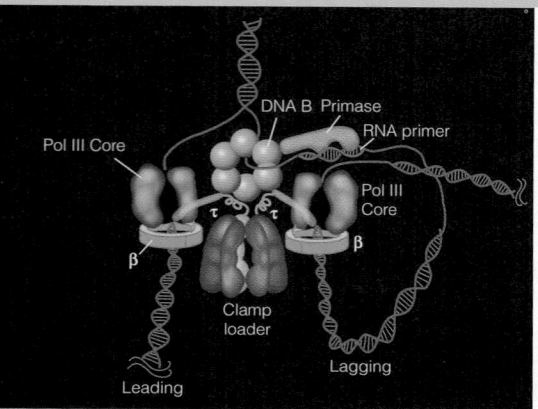

CHAPTER 25

DNA Replication

Biosynthesis of nucleic acids and proteins is carried out through the processes of replication, transcription, and translation. Most regulation occurs at the level of initiation.

Parental duplex

Intermediate in semiconservative replication

Two daughter duplexes

Our focus for the remainder of this book is upon processes in genome replication, genome maintenance, and gene expression. In this chapter we discuss DNA replication. Next, Chapter 26 focuses upon DNA repair and recombination and on various processes in genome rearrangement. In Chapter 27 we discuss the readout of information encoded in DNA—transcription and RNA processing. Chapter 28 focuses upon protein synthesis—how information encoded in the four-letter nucleic acid language is translated into the 20-letter amino acid language and how the polypeptides resulting from translation undergo processing and trafficking to their destinations as mature proteins, in the correct intracellular or extracellular location. Finally, Chapter 29 revisits these process-oriented chapters to consider how gene expression is regulated. Each process of gene expression involves a defined initiation step, followed by movement of an enzyme complex along a nucleic acid template as a product is elongated, followed in turn by a defined termination step. We will see that events controlling each process—replication, elongation, and translation—occur primarily at the level of initiation.

Early Insights into DNA Replication

Three central features of DNA replication were predicted in 1953 from Watson and Crick's model. First was their explicit prediction that DNA replication is semiconservative—that each of the two identical daughter DNA molecules contains one parental strand and one newly synthesized strand. This prediction was confirmed in 1957 by the elegant experiment of Meselson and Stahl, as we discussed in Chapter 4 (Figure 4.14, page 104). Not explicitly required by the model, but implied, was the idea that parental strand unwinding and synthesis of new DNA occur simultaneously, in the same microenvironment. In other words, replication occurs at a **fork**, in which parental strands are unwinding and both daughter strands are undergoing elongation, as suggested by the diagram. Also suggested by the Watson–Crick model was the premise that replication begins at one or more fixed sites—**replication origins**—on a chromosome.

The first images of replicating DNA were consistent with the idea that replication occurs within a fork structure. In Chapter 24 we showed John Cairns' radioautogram of a replicating *E. coli* DNA molecule (Figure 24.17, page 1020). That circular molecule contains two Y-shaped structures, either or both of which could be a replication fork; this is called a theta structure, from the Greek letter θ. The figure cannot indicate whether replication is unidirectional, with one Y junction representing the site of replication initiation, or bidirectional, with two forks, which are moving from a fixed origin halfway between the two Ys. These alternatives are shown at the top of Figure 25.1. An elegant variation of Cairns' experiment established for *E. coli* that replication is bidirectional from one fixed origin. In this experiment DNA was radiolabeled by growth of cells in the presence of [³H]thymidine. Just as cells were terminating one round of replication and initiating the next, the specific radioactivity of the thymidine was increased several-fold, so that areas of active DNA synthesis could be distinguished in a radioautogram by increased blackening of the film. If replication were unidirectional, termination and reinitiation would have to occur at the same site. Examination of individual DNA molecules, such as that shown in Figure 25.1, showed clearly that termination of one round of replication and initiation of the next occurred about 180° apart from each other, which could occur only if replication is bidirectional.

Another elegant radioautographic experiment gave further support to the idea of bidirectional replication from a fixed origin. Investigators wondered how *E. coli* cultures could double their cell numbers in 20 minutes, while the shortest time measured for replication of the entire circular chromosome was 40 minutes. The experiment, depicted in Figure 25.2, showed that *E. coli* responds to conditions favoring rapid growth by initiating a new round of chromosome replication while a previously initiated round is still in progress. Thus, when a cell divides, each daughter cell receives a chromosome that is far along in its next round of replication. Hence, the cell adapts to rapid growth by timing its replication events earlier in the cell cycle. This provides strong evidence that replication is controlled largely at the level of initiation.

Partially replicated DNA molecules can also be visualized in the electron microsocope. Bacteriophage λ has a linear genome, 48,502 base pairs long, as isolated from phage particles. However, at each 5′ end there is a 12-nucleotide single-stranded extension. During infection of *E. coli* these "cohesive ends," which are complementary in sequence, pair with each other to form a circular duplex, and the ends become covalently joined. Partially replicated molecules can be seen electron microscopically as theta structures, as depicted in Figure 25.3. Again, this picture by itself doesn't establish that replication is occurring in each fork, but a technique called denaturation mapping established that this molecule indeed was undergoing bidirectional replication at the time it was isolated.

What about eukaryotic cells, which have huge linear molecules in their chromosomes? Typically, a mammalian cell in culture will replicate all of its DNA in an eight-hour S phase (as compared to 40 minutes for doubling of the *E. coli* chromosome). Radioautographic and other evidence shows that individual

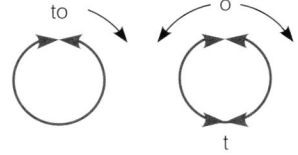

Unidirectional   Bidirectional

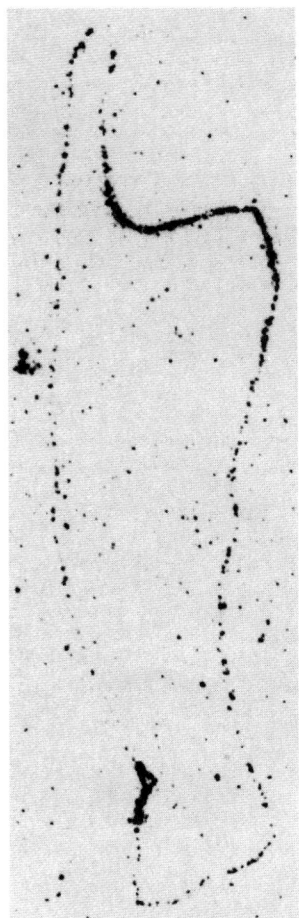

FIGURE 25.1

Demonstration of bidirectional replication by radioautography. This *E. coli* chromosome was labeled by long-term growth in [³H]thymidine, and the specific activity was increased four-fold just as a synchronized culture completed one round of replication and commenced another. The diagrams show the labeling patterns that would be expected for unidirectional replication, where the terminus (t) and origin (o) are adjacent, and for bidirectional replication, where they are on opposite sides of the chromosome.

Reprinted from *Journal of Molecular Biology* 74:599–604, R. L. Rodriguez, M. S. Dalbey, and C. I. Davern, Autoradiographic evidence for bidirectional DNA replication in *Escherichia coli*. © 1973, with permission from Elsevier.

FIGURE 25.2

Initiation of a new round of replication before completion of the preceding round. **(a)** Labeling with [³H]thymidine was started when the first round of replication began. The pattern of grain density shows that one of the reinitiated branches contains two labeled strands, while the other contains only one. **(b)** Solid lines denote radiolabeled DNA, and dashed lines denote unlabeled DNA. Black arrowheads depict first-generation replication forks, and blue arrowheads depict second-generation replication forks. Unreplicated DNA is nonradioactive, so it is not seen in the radioautogram.

Reprinted from *Journal of Molecular Biology* 73:55–58, E. B. Gyurasits and R. G. Wake, Bidirectional chromosome replication in *Bacillus subtilis.* © 1973, with permission from Elsevier.

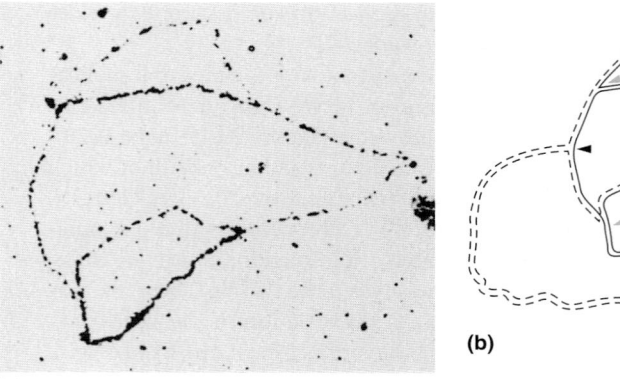
(a)

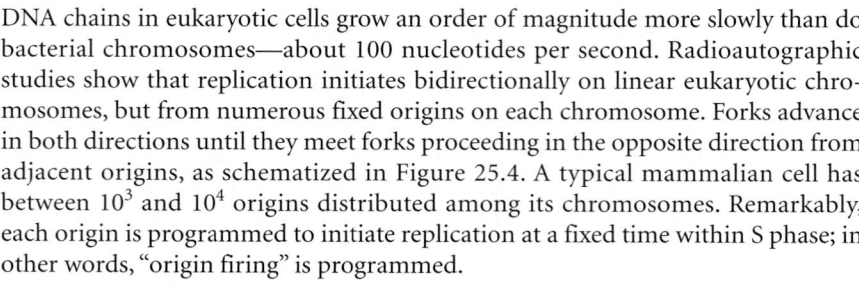
(b)

Origin

In bacteria DNA replication initiates from a fixed origin and proceeds in both directions from that origin.

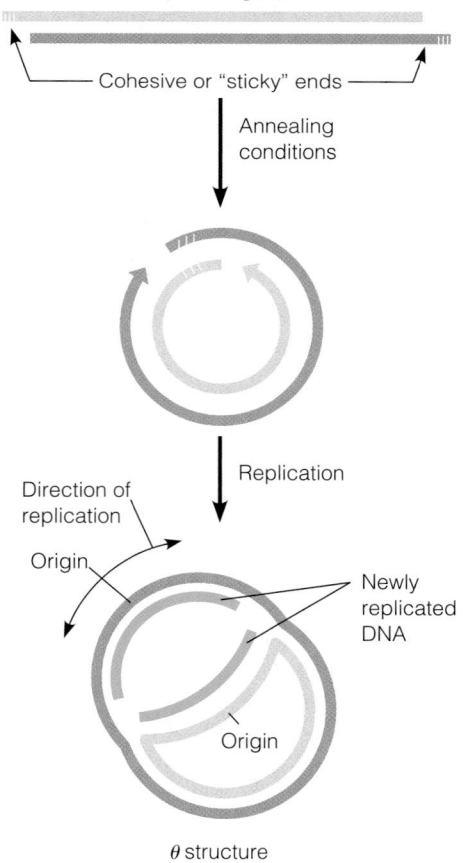

Replicating DNA

— Cohesive or "sticky" ends —

Annealing conditions

Replication

Direction of replication

Origin

Newly replicated DNA

Origin

θ structure

(a)

DNA chains in eukaryotic cells grow an order of magnitude more slowly than do bacterial chromosomes—about 100 nucleotides per second. Radioautographic studies show that replication initiates bidirectionally on linear eukaryotic chromosomes, but from numerous fixed origins on each chromosome. Forks advance in both directions until they meet forks proceeding in the opposite direction from adjacent origins, as schematized in Figure 25.4. A typical mammalian cell has between 10^3 and 10^4 origins distributed among its chromosomes. Remarkably, each origin is programmed to initiate replication at a fixed time within S phase; in other words, "origin firing" is programmed.

Because eukaryotic chromosomal DNA is linear, we must concern ourselves with events at the ends of these molecules; we deal with that later in this chapter. The main point, however, is that even before much had been learned about the enzymology of DNA replication, several important features of the process were evident—bidirectionality, fixed origins, simultaneous strand unwinding and chain elongation within a fork, and control at the level of chain initiation.

FIGURE 25.3

A partially replicated bacteriophage **λ** DNA molecule. **(a)** The original linear DNA circularizes by base pairing between the cohesive ends. After covalent closure replication begins. Partial replication yields a theta structure, resulting from bidirectional replication from an origin equidistant between the two forks. **(b)** An electron micrograph of partially replicated phage DNA. Examination of intermediates like these after partial denaturation showed that replication is bidirectional from a fixed origin.

Reprinted from *Journal of Molecular Biology* 51:61–73, M. Schnös and R. B. Inman, Position of branch points in replicating **λ** DNA. © 1970, with permission from Elsevier.

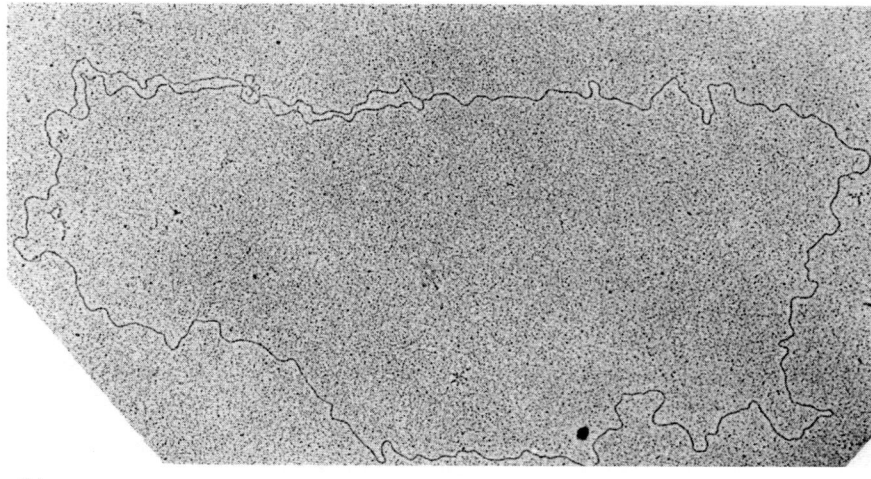

(b)

DNA Polymerases: Enzymes Catalyzing Polynucleotide Chain Elongation

The biochemical elucidation of DNA replication began in the mid-1950s, with Arthur Kornberg's discovery of DNA polymerase. Kornberg realized, from his experience in bioenergetics, that the substrates for DNA replication should be activated derivatives of the DNA nucleotide residues. Thus, the endergonic synthesis of phosphodiester bonds between nucleotide residues could be coupled to the exergonic breakdown of activated substrates, just as occurs in synthesis of other macromolecules, such as glycogen. Kornberg correctly predicted that these activated nucleotides would be the 2′-deoxyribonucleoside 5′-triphosphates. When he incubated radioactively labeled dNTPs with an extract of soluble proteins of *E. coli*, radioactivity was incorporated into high-molecular-weight material, which could be quantified by acid precipitation and radioactive counting; the radiolabeled product was shown to be DNA. The enzyme required added DNA, plus Mg^{2+}. As we have mentioned earlier (Chapter 4, page 111), two DNA molecules are required for the reaction—the template and the primer, the latter becoming covalently extended from its 3′ hydroxyl group. Also, the polarity of the product DNA strand (that is, the extended primer) is opposite to that of the template strand, as expected from the antiparallel orientation of DNA strands in the Watson–Crick model.

As shown in Figure 25.5, the DNA polymerase reaction involves nucleophilic attack by the 3′ hydroxyl group of the primer terminus on the α-phosphate of the deoxyribonucleotide substrate, which leads to covalent bond formation. The 3′ hydroxyl is a weak nucleophile, and the pyrophosphate provides a good leaving group. The reaction is readily reversible. In cells and in crude preparations, therefore, the reaction is drawn to the right by pyrophosphatase action on the other product of the reaction ($PP_i + H_2O \rightarrow 2 P_i$). Thus, two energy-rich phosphates are expended per nucleotide incorporated.

Structure and Activities of DNA Polymerase I

The DNA polymerase discovered by Kornberg in *E. coli* was later shown to be one of three different DNA polymerases in most bacterial cells (revised upward to five in 1999). The Kornberg enzyme is now called DNA polymerase I. We shall use this term henceforth and describe the other polymerases later in this chapter.

DNA Substrates for the Polymerase Reaction

As noted above, DNA polymerase I requires both template DNA and a primer, either DNA or RNA. In vitro, these roles can be played either by two distinct nucleic acids or by one molecule. As shown in Figure 25.6c, a single-strand DNA with self-complementary sequences can fold itself into a **hairpin** or **stem-loop** structure whose 3′ end can be extended by polymerase, using the 5′ end as the template. The figure shows other polymerase activities that can be demonstrated in vitro. The enzyme can copy around a circular single-strand template, such as the DNA extracted from small bacteriophages, like ϕ X174 or M13, so long as a primer is present, but it cannot join the ends (Figure 25.6a). When the template is linear, polymerase copies only to the 5′ end of the template and then it dissociates (Figure 25.6b and c). In similar fashion the enzyme can fill in a gap (Figure 25.6d), dissociating when the gap is reduced to a nick. Under some conditions the enzyme can also extend from a 3′ hydroxyl group at a nick. Typically, when this occurs, the 5′ end of the preexisting DNA is displaced in advance of the nick. This is called **strand displacement** synthesis (Figure 25.6e). Under conditions where the displaced strand is degraded, there is no net DNA accumulation, and the reaction is called **nick translation.**

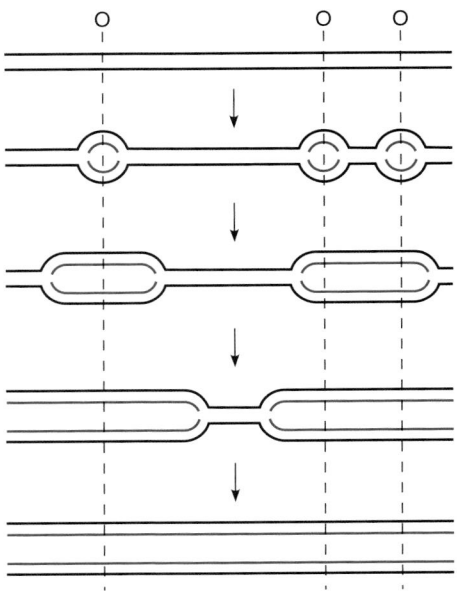

FIGURE 25.4

Bidirectional replication from several fixed origins (O) on a linear eukaryotic chromosome.

DNA polymerase catalyzes nucleophilic attack by the 3′ hydroxyl at the primer terminus upon the α-phosphate of an incoming dNTP, base-paired with its template.

FIGURE 25.5

The DNA polymerase reaction. Each incoming dNTP is positioned by base pairing with the appropriate template nucleotide, and a phosphodiester bond is created by nucleophilic attack of the primer-strand 3′ hydroxyl group on the α-phosphate of the dNTP.

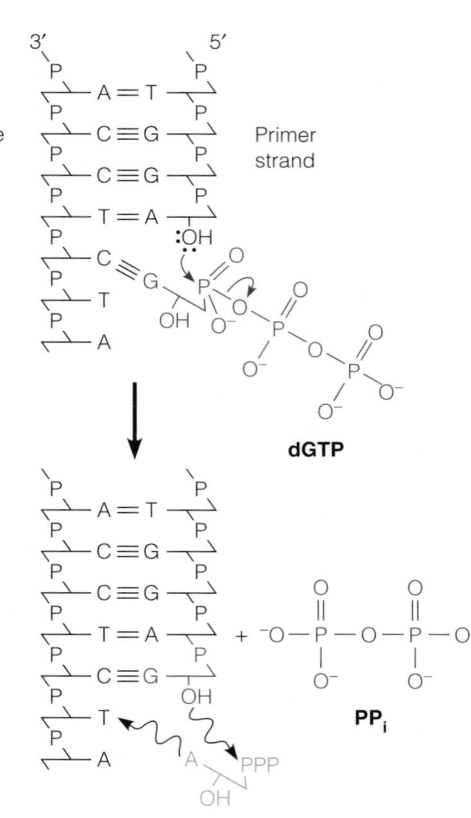

The DNA polymerase I molecule contains three active sites: a polymerase and two exonucleases.

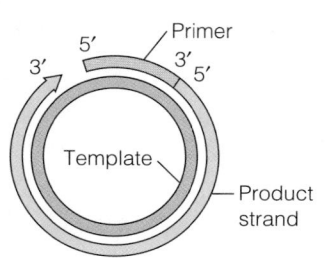

(a) Primed circular single strand

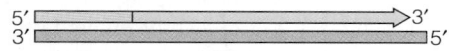

(b) Primed linear single strand

(c) Single-strand hairpin

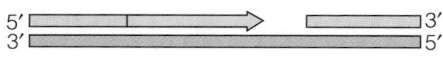

(d) Gapped duplex

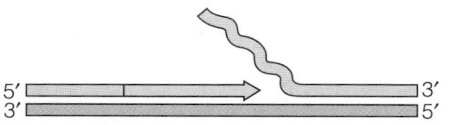

(e) Nicked duplex
  (strand displacement synthesis)

FIGURE 25.6

DNA substrates that can be acted upon by purified DNA polymerase. Each blue arrowhead marks a 3′ hydroxyl terminus at which chain extension is occurring.

Multiple Activities in a Single Polypeptide Chain

When DNA polymerase I was purified from *E. coli,* it was found to consist of a single polypeptide chain ($M_r = 103,000$). In addition to its polymerase activity, the purified enzyme has two nuclease activities: The **3′ exonuclease** degrades single-strand DNA from the 3′ end, and the **5′ exonuclease** degrades base-paired DNA from the 5′ terminus. The enzyme also cleaves RNA from a duplex containing one strand each of DNA and RNA. The 3′ exonuclease serves a "proofreading" function, to improve the accuracy with which a DNA template is copied. The activity will remove an improperly base-paired nucleotide from the growing 3′ end of a polydeoxynucleotide chain, giving the polymerase activity a second chance to insert the correct nucleotide specified by the template. We say more about the 3′ exonuclease later, when we discuss the fidelity of DNA replication. The 5′ exonuclease activity plays roles both in replication and in DNA repair as discussed later in this chapter and the next.

Structure of DNA Polymerase I

The three catalytic activities of DNA polymerase I have been localized to regions of the long polypeptide chain, partially through limited proteolysis of the enzyme by subtilisin or trypsin. As shown by Hans Klenow, this treatment splits the 103-kilodalton polypeptide into a small N-terminal fragment ($M_r = 35,000$) and a large C-terminal fragment ($M_r = 68,000$). The large fragment (also called the **Klenow fragment**) contains the polymerase and 3′ exonuclease domains; the small fragment contains the 5′ exonuclease domain. How the three catalytic sites are arranged in space is important to an understanding of how each of the two nuclease activities functions in concert with the polymerase.

Crystallographic study of the large fragment revealed a striking feature of the structure—a deep crevice, just large enough to accommodate B-form DNA, with a flexible subdomain that might allow bound DNA to be completely surrounded

(Figure 25.7). The protein molecule has been likened to a hand, with palm, thumb, and fingers. Cocrystallization of the Klenow fragment with a short duplex DNA shows that this is indeed the DNA-binding site and that bound DNA is almost completely surrounded by protein, which wraps around the DNA like a hand holding a cylinder. An interesting feature of the structure, revealed by study of mutant forms of the enzyme, is that the 3′ exonuclease active site is quite far—about 3 nm—from the polymerase active site. This suggests that about eight base pairs of DNA must unwind to move the 3′ terminal nucleotide from the polymerase to the 3′ exonuclease active site (see page 1071). Because the 5′ exonuclease activity is absent in the Klenow fragment, this unnatural enzyme is a useful laboratory reagent for synthesis of DNA in vitro when the researcher wishes specifically to avoid DNA degradation.

Functions of DNA Polymerase I

For his discovery of DNA polymerase, Arthur Kornberg was awarded the Nobel Prize in 1959. Although this discovery was monumental, several properties of the enzyme as isolated from *E. coli* were different from those expected for an enzyme catalyzing the major nucleotide incorporation reactions in biological DNA replication. First, the enzyme in vitro is too slow, with a V_{max} of about 20 nucleotides per second; in contrast, replicative chain growth occurs in vivo at around 800 nucleotides per second or more. Second, with about 400 molecules of enzyme per cell, the enzyme is present in great excess over the small number (fewer than 10) of replication forks per cell, suggesting that its function(s) is carried out elsewhere. Third, as noted earlier, DNA polymerase can extend chains only in a 5′ → 3′ direction. A complete understanding of replication requires knowing how both of the antiparallel chains of a DNA duplex are replicated within the same fork. Fourth, DNA polymerase cannot initiate the synthesis of new DNA chains but can only extend from preexisting 3′ hydroxyl termini. Finally, genetic evidence (page 1044) suggested the existence of other polymerases, as well as additional enzymes and proteins essential to DNA replication—proteins involved, for example, in DNA strand unwinding, chain initiation, and keeping polymerase bound to its template. Much of the genetics was done with *Escherichia coli* and bacteriophage T4, which infects *E. coli*. At this point we digress to say a bit about the genetics of these important biological systems.

A Brief Review of Microbial Genetics

To a greater extent than in intermediary metabolism, our present understanding of information metabolism has come from the field of genetics. Hence, it is useful to review several key genetic terms before more detailed discussions of mechanisms and control of replication. As discussed elsewhere in this book, the totality of genetic information in an organism, the **genome**, can be thought of most directly as the base sequences of all of the DNA molecules in a cell. These sequences consist of individual genes, each on a chromosome or on a small piece of extrachromosomal DNA. A **genetic map** identifies the relative positions of genes on chromosomes. As discussed in Chapter 24, one can generate a **linkage map** by measuring recombination frequencies or a **physical map**, which identifies the physical locations of genes on a chromosome.

Any characteristic of an organism that can be detected in terms of appearance, structure, or some measurable property is called the **phenotype**. The genetic composition of an individual, which specifies phenotypic properties, is the **genotype**. Individuals with different genotypes may have the same phenotype. In Chapter 17, for example, we learned that humans with different mutations affecting the LDL receptor can have a common phenotype—the elevated serum cholesterol levels that are associated with familial hypercholesterolemia.

An **allele** is a particular form of a gene. For example, phenylketonurics have two defective alleles of the gene for phenylalanine hydroxylase that encode inactive

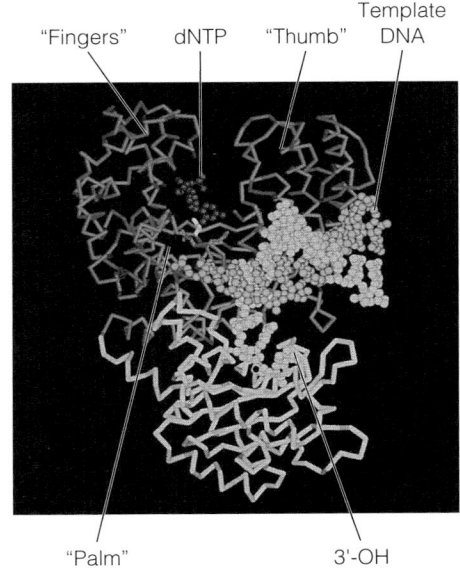

FIGURE 25.7

The Klenow fragment of *E. coli* DNA polymerase I. α-Carbon backbone representation of the Klenow fragment complexed with DNA, in an editing configuration. That is, the 3′ end of the growing strand (light blue) is bound at the 3′ exonuclease active site of the enzyme (yellow). The locations of the polymerase and 3′ exonuclease active sites have been identified by site-directed mutagenesis. The template strand is shown in green, and bound dCTP in red. PDB ID 1KFD.

From *Science* 260:352–355, L. Beese, V. Derbyshire, and T. A. Steitz, Structure of DNA polymerase I Klenow fragment bound to duplex DNA. © 1993. Reprinted with permission from AAAS and Lorena Beese.

forms of the enzyme—one on each chromosome. A **marker** is any allele whose frequency can be determined quantitatively. **Copy number** refers to the number of copies per cell of a gene or other DNA sequence. Most eukaryotic cells are **diploid**, meaning that the copy number of most chromosomal genes is two, while most prokaryotic cells have a **haploid** genotype, and thus a single allele for most genes. Extrachromosomal genes can have much higher copy numbers, whether they are carried on organelle DNA in eukaryotic cells or on an extrachromosomal DNA element such as a bacterial **plasmid**. A plasmid is usually a small circular DNA molecule whose copy number may be controlled independently of the regulatory mechanisms affecting chromosomal DNA replication.

A diploid individual in a natural environment usually has at least one functional copy, or allele, of each gene and normally displays a **wild-type** (normal) phenotype (unless the gene is recessive, in which case a homozygous individual has a mutant phenotype). A wild-type gene in an organism can undergo a mutation, thereby generating a mutant genotype, which may be detected as a mutant phenotype. For example, albinism represents a readily observable phenotype, resulting from mutations that inactivate tyrosinase (see Chapter 21). A mutant site within a gene can infrequently reverse the process that caused the mutation and restore the wild-type genotype. This type of mutation is called a **reversion**. More frequently, a wild-type phenotype can be restored by a mutation at a different site; this occurrence is called **second-site reversion**, or **suppression**. Suppression can occur either by mutation within the gene undergoing the original mutation (intragenic suppression) or within a different gene (intergenic suppression). Note that many mutations are **silent**; if a mutant protein retains some biological activity, there may not be an observable phenotype.

Because much of our discussion focuses upon replication in *E. coli* and its phages, you should understand some conventions in the naming of prokaryotic genes and gene products. Let us review these conventions in the context of the *E. coli* genetic map. Mapping genes in *E. coli* became possible after the discovery by Joshua Lederberg of sexual reproduction in bacteria. This process, termed **conjugation**, involves the transfer of DNA from a donor cell (the "male") to a recipient cell (the "female"), as schematized in Figure 25.8. About 100 minutes is required for complete chromosome transfer. Mapping is carried out by "interrupted mating" experiments. At intervals a conjugating culture is disrupted mechanically, to separate the donor and recipient cells. The researcher then analyzes the recipient cells, correlating the transfer of marker genes from the donor with the time at which mating was interrupted. Thus, map positions of genes were originally reported as "minutes," which identified the time at which a marker of interest was transferred by conjugation. Finer-scale mapping could be done by transduction, a process described in Chapter 26. With the complete genome sequence of *E. coli*, reported in 1997, we now know the physical locations of all *E. coli* genes.

Figure 25.9 shows part of the *E. coli* genetic map, identifying genes known to participate, either directly or indirectly, in DNA replication. The large number of these genes underscores the complexities of DNA replication, and we shall refer frequently to the genes shown on this map. Our purpose here is to describe the conventions. Each gene has an italicized lower-case three-letter designation that refers either to the gene product or to the phenotype of mutants defective in that gene, followed by a capital letter denoting the order of discovery of the gene or gene product. For example, *E. coli* was shown early to contain three DNA polymerases, named I, II, and III. The structural genes encoding these enzymes are *polA*, *polB*, and *polC*, respectively. Other genes were named from the defective phenotype. For example, all genes designated *dna* were originally identified from temperature-sensitive (*ts*) mutations in which DNA replication occurred normally when cells were grown at 30 °C but was blocked at 42 °C. Some of the gene products have been identified. For example, the *dnaG* gene product was found to be primase (page 1048). By convention, gene *products,* which are usually proteins, are given the corresponding nonitalicized and capitalized designations. For instance, DnaG is identical to primase; both terms refer to the product of the *dnaG* gene.

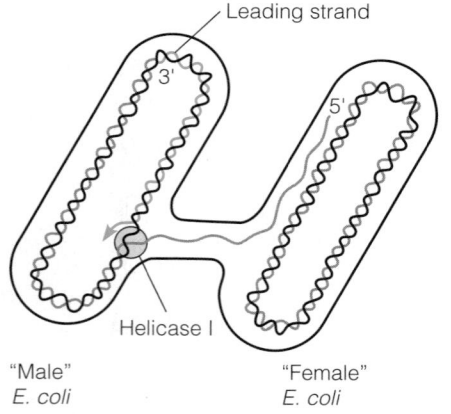

Leading strand

Helicase I

"Male"
E. coli

"Female"
E. coli

FIGURE 25.8

Bacterial conjugation. After a nick is made in the donor (male) chromosome, a helicase (page 1048) unwinds the two strands. The nicked strand (gray) is elongated (blue) at the 3' end, displacing the 5' end and transferring it into the recipient (female) cell. Later, the transferred DNA strand can undergo recombination with the recipient chromosome, thereby transferring genetic markers to the female.

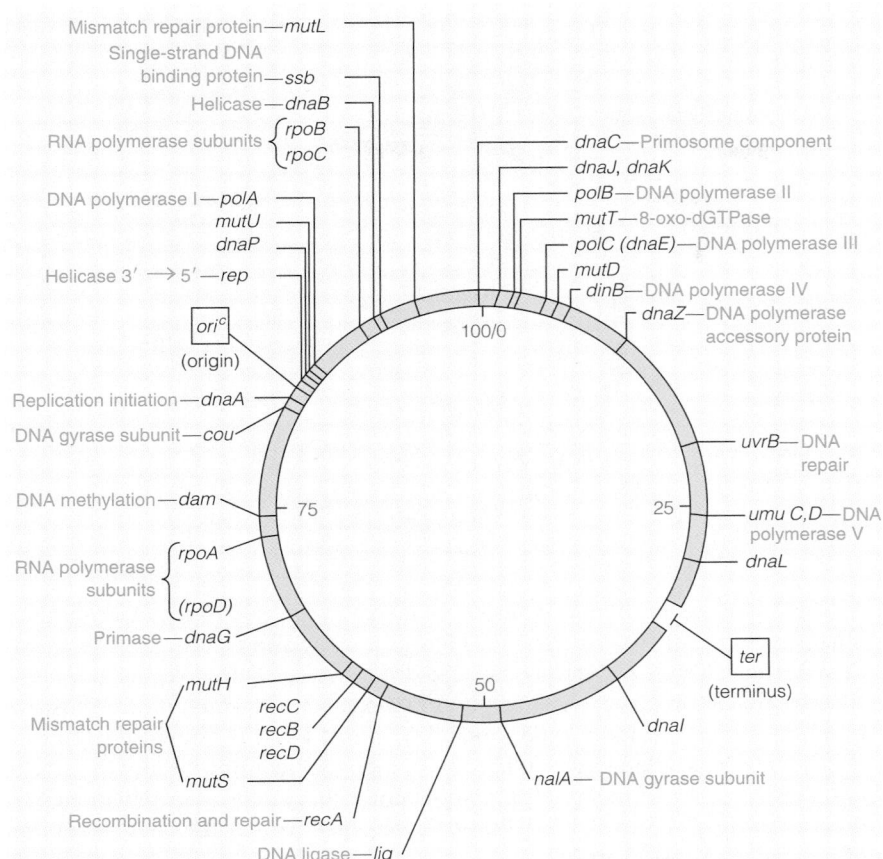

FIGURE 25.9

Partial genetic map of *E. coli*. Genes whose products are involved in DNA replication or repair are shown. Gene product names are given in blue. The *dna* genes play essential roles in DNA replication. The *mut* genes specify proteins that, when mutated, can cause elevated rates of spontaneous mutation. *ori* and *ter* are sites of initiation and termination of genome replication, respectively.

Much of what we know about the functions of other proteins in DNA replication has come from investigations with bacteriophage T4 (Figure 25.10). This virus contains a duplex DNA genome, about 169,000 base pairs in length. Although large for a virus, this genome is less than 5% of the size of the *E. coli* genome, making it simpler to work with. Phage T4 encodes nearly all of its own replicative proteins and enzymes. T4 mutants are readily isolated and mapped. In fact, T4 was the first biological system in which replication-defective mutants were described. This occurred in 1965, when gene 43 on the T4 map (Figure 25.11) was identified as the structural gene for a virus-coded DNA polymerase. Figure 25.11 shows locations on the T4 genome of 15 genes whose products play essential roles in DNA replication. Three of these gene products are enzymes that carry out synthesis of the modified DNA base 5-hydroxymethylcytosine (Chapter 22). However, the remaining proteins participate in replication of the T4 genome, by mechanisms similar to those in cellular DNA replication.

Multiple DNA Polymerases

In 1969 John Cairns isolated an *E. coli* mutant that was deficient in DNA polymerase I activity. With this very abundant enzyme absent from the cell, it became possible to detect two additional DNA polymerase activities in *E. coli* cells. These were named DNA polymerases II and III, and the original Kornberg polymerase was named DNA polymerase I. The properties of these three DNA polymerases are summarized in Table 25.1. Some time later two additional polymerases were described in *E. coli*, named IV and V. We say more about these enzymes in Chapter 26.

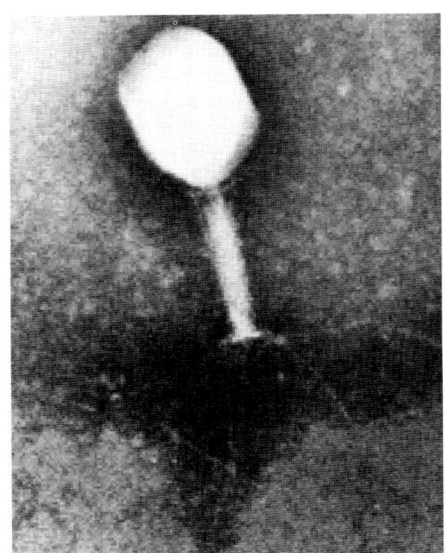

FIGURE 25.10

Bacteriophage T4. DNA is packed into the head of this virus particle. Infection begins with attachment of the particle to an *E. coli* cell, followed by contraction of the tail and injection of the DNA through the tail into the interior of the infected cell.

Reprinted from *Virology* 32:279–297, L. Simon and T. F. Anderson, The infection of *Escherichia coli* by T2 and T4 bacteriophages as seen in the electron microscope I. Attachment and penetration. © 1967, with permission from Elsevier.

FIGURE 25.11

Partial map of the T4 genome. Genes whose products participate in DNA metabolism are shown on the outside of the circle. The inner numbers represent distances in kbp. The reference point (0) represents the divide between two genes not shown—*rIIA* and *rIIB*.

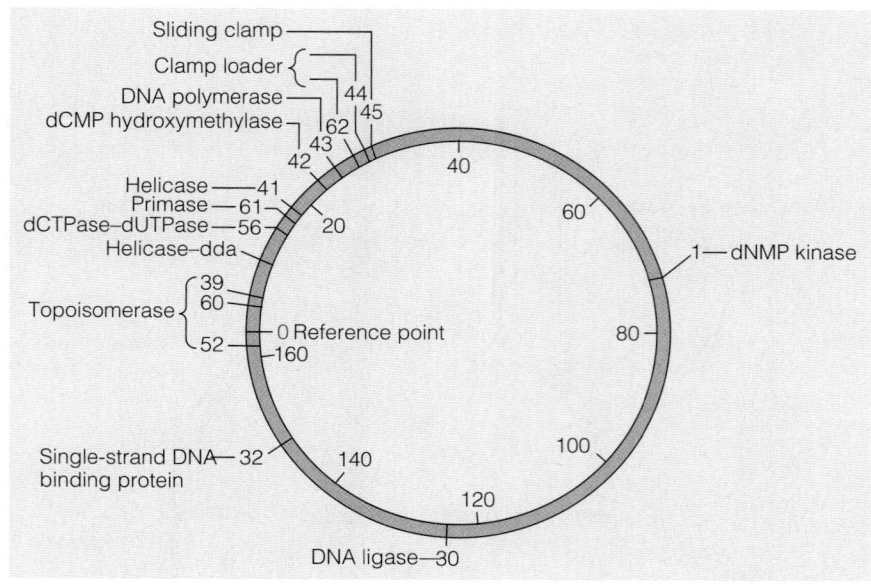

DNA polymerase III was assigned the major role in nucleotide incorporation during replication, on the basis of its high V_{max}. Note also that there are few molecules of the enzyme per cell, as expected if it functions only at replication forks. More definitive evidence of this role is the existence of temperature-sensitive mutant cells that contain a thermolabile form of DNA polymerase III and in which DNA replication in vivo is blocked at high temperature. Both of these phenotypes can be traced to a single mutation, powerful evidence that polymerase III plays an essential role in DNA replication. The gene encoding the catalytic subunit of polymerase III was named *polC*.

There are also *polB* mutants, lacking polymerase II. Although these mutants displayed no other obvious phenotype, recent evidence suggests that the enzyme participates in DNA repair (see Chapter 26). But what about polymerase I? In the original *polA* mutant described by Cairns, DNA replication occurred normally. However, the bacteria were abnormally sensitive to ultraviolet irradiation and to alkylating agents that react with DNA, suggesting a role for polymerase I in DNA repair. Further study of the Cairns mutant revealed a role in replication. Although lacking the *polymerase* activity of DNA polymerase I, the mutant

TABLE 25.1   The classical DNA polymerases of *E. coli*

Characteristic	Polymerase I	Polymerase II	Polymerase III
Structural gene	*polA*	*polB*	*polC*
Molecular weight	103,000	90,000	130,000
Number of molecules/cell	400	100	10
V_{max}, nucleotides/second	16–20	2–5	250–1000
3′ exonuclease	Yes	Yes	No[a]
5′ exonuclease	Yes	No	No
Processivity[b]	3–200	10,000	500,000
Mutant phenotype	[c]UV^{s}MMS^s	None	*dna*^{ts}
Biological function	DNA repair, RNA primer excision	SOS DNA repair?	Replicative chain extension

[a]The 3′ exonuclease is carried on a separate polypeptide chain, the DnaQ protein.
[b]The number of nucleotides incorporated per encounter between polymerase and DNA (see page 1050); *ts* means temperature-sensitive; *s* means sensitive.
[c]MMS (methylmethane sulfonate) is an agent that alkylates DNA.

bacteria did synthesize the small N-terminal fragment of the protein, which contains the 5′ exonuclease activity. Still later it was found that mutants lacking this 5′ exonuclease activity are also defective in DNA replication. Thus, it was established that two polymerases, I and III, play essential roles in DNA replication.

Structure and Mechanism of DNA Polymerases

Sequence and structural analysis of DNA polymerases has revealed that, although there is considerable diversity at the level of primary structure, all enzymes in this family show structural features in common. Like polymerase I, all of the polymerases described so far have a structure described for the Klenow fragment, with domains identified as "palm," "thumb," and "fingers," as shown in Figure 25.12. Analyses of DNA–enzyme complexes, conserved amino acid residues, and targeted mutations have indicated that the polymerase active site lies in the palm domain, and the 3′ exonuclease site lies at the base of the palm. High-resolution polymerase structures show two magnesium ions, bound to nucleotide phosphates and to conserved aspartate residues already known to be essential for catalysis. These features support a general polymerase mechanism proposed by Thomas Steitz, on the basis of the Klenow fragment structure (Figure 25.13). The mechanism is similar to that more recently proposed for adenylate cyclase, which we mentioned in Chapter 23. In the Steitz mechanism one metal ion polarizes the hydroxyl group at the 3′ primer terminus, facilitating nucleophilic attack of that moiety upon the α-phosphate of the

FIGURE 25.12

Comparison of DNA polymerase structures with primer and template bound. The four structures shown are oriented with respect to each other by superposition of the first two base pairs at the primer terminus. Palm, thumb, and fingers are shown for each structure. *Taq* polymerase is a thermostable enzyme used in PCR (PDB ID 1TAU). HIV-1 RT is the reverse transcriptase (RNA-directed DNA polymerase) from human immunodeficiency virus (PDB ID 3HVT; see page 1072). RB69 gp43 is the replicative DNA polymerase encoded by RB69, a bacteriophage closely related to T4 (PDB ID 1WAJ). pol β is a eukaryotic DNA polymerase involved in DNA repair (PDB ID 2BPG; see page 1062).

Journal of Biological Chemistry 274:17395–17398, T. A. Steitz, DNA polymerases: Structural diversity and common mechanisms. Reprinted with permission. © 1999 The American Society for Biochemistry and Molecular Biology. All rights reserved.

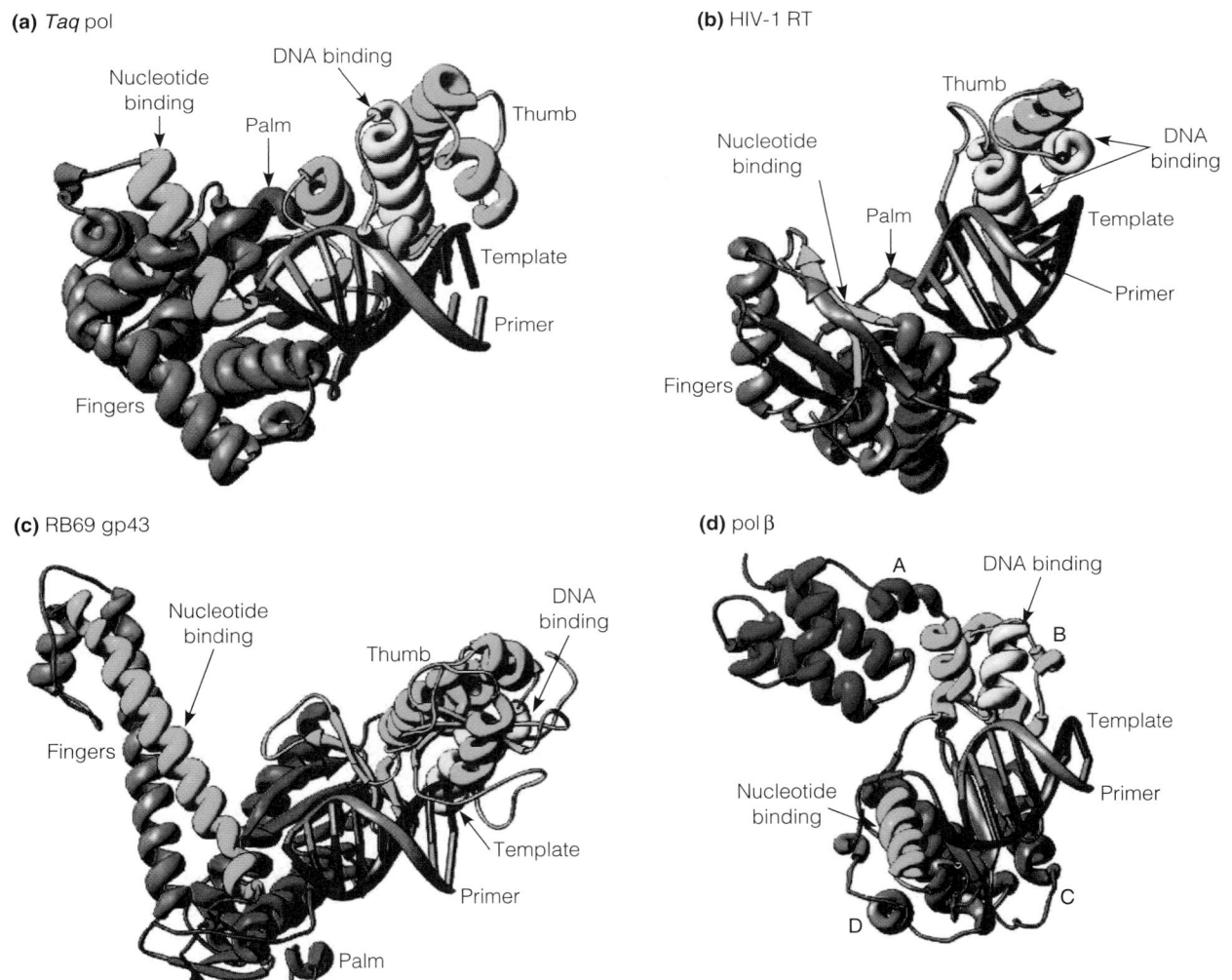

(a) *Taq* pol

(b) HIV-1 RT

(c) RB69 gp43

(d) pol β

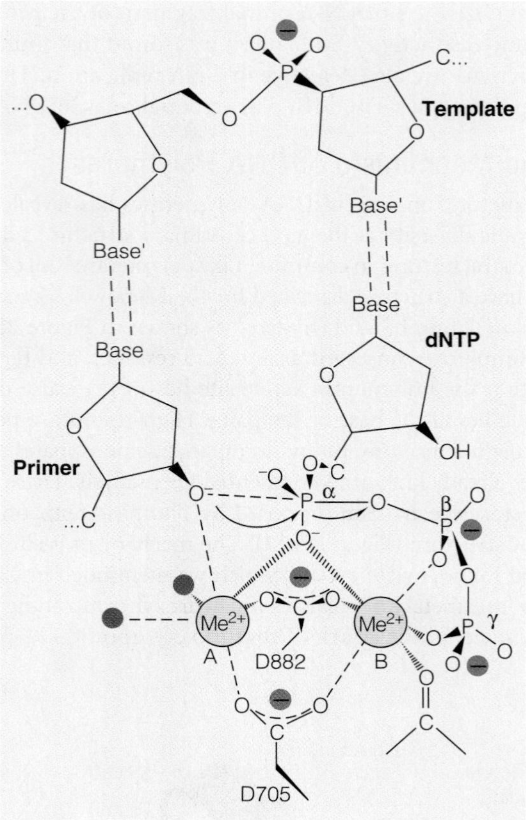

FIGURE 25.13

The two-metal mechanism of the DNA polymerase reaction. This mechanism was suggested originally from the structure of the T7 phage DNA polymerase–substrate complex. D705 and D882 are conserved aspartate residues, and the blue dots are water molecules bound to metal ion A.

Journal of Biological Chemistry 274:17395–17398, T. A. Steitz, DNA polymerases: Structural diversity and common mechanisms. Reprinted with permission. © 1999 The American Society for Biochemistry and Molecular Biology. All rights reserved.

dNTP substrate. Both metals stabilize a trigonal bipyramidal transition state, in which the α-phosphorus is linked to five oxygens, and the second metal facilitates leaving by the pyrophosphate. Extensive contacts occur between the enzyme and the DNA minor groove—contacts that could occur only with a properly base-paired duplex. Also, structures of polymerase–DNA complexes show that DNA near the primer terminus adopts a conformation more like that of A form than B form and that this conformational change facilitates the minor groove interactions. Moreover, the incoming dNTP was shown to fit snugly into a pocket that favors correct base-pairing with the template. Thus, these structures reveal both how the reaction is catalyzed and how the enzyme copies its template DNA with high accuracy.

We now know that most bacteria carry five DNA polymerases, and we see later in this chapter that human cells contain fifteen. Analysis of amino acid sequences and structural data from dozens of DNA polymerases revealed similarities that have led to their classification in seven families—A, B, C, D, X, Y, and reverse transcriptase (RT). Bacterial polymerase I is a family A enzyme, while polymerase III is in family C. RB69 gp43, shown in Figure 25.12, is in family B. Archaeal DNA polymerases are in family D, while family X and Y polymerases include enzymes involved in DNA repair, including pol β, shown also in Figure 25.12 (X), and bacterial pol IV and pol V, which we discuss in Chapter 26 (Y). Reverse transcriptase (page 1072) and telomerase (page 1068) are in the RT family.

Other Proteins at the Replication Fork

Although the discovery of DNA polymerase revealed the biochemical process by which DNA chains are elongated in replication, important questions remained. How are two antiparallel DNA chains extended in the same fork if polymerase can extend in only one direction? How is the synthesis of new chains initiated if polymerase is capable only of adding nucleotides to preexisting 3′ termini? How are parental DNA chains unwound?

Discontinuous DNA Synthesis

The first two questions were answered primarily through the work of Reiji Okazaki, who proposed that DNA replication could be discontinuous. In principle one parental strand (the **leading strand**) could be extended continuously, with polymerase moving 5′ to 3′ (from the 5′ terminus to the 3′ terminus), in the same direction as fork movement. Synthesis on the **lagging strand** would be discontinuous; chain extension along the leading strand would expose single-strand template on the lagging strand. This template could be copied in short fragments (later named Okazaki fragments), with polymerase moving opposite to the direction of fork movement. Thus, lagging strand synthesis would occur in short pieces. These could then be joined to high-molecular-weight DNA by the enzyme **DNA ligase**.

But how might these short DNA fragments be initiated, if DNA polymerase can only extend preexisting chains? Evidence suggested that Okazaki fragments could be initiated not with DNA, but with *RNA* oligonucleotides, which could serve as primers, for extension by DNA polymerase. The enzyme synthesizing these RNA primers was called **primase** and shown to be a special form of RNA polymerase, designed to function in a replication fork. These events are illustrated in Figure 25.14.

To support this model, Okazaki carried out a "pulse-chase" experiment with isotopic precursors (see Tools of Biochemistry 12A), which showed that DNA is synthesized as low-molecular-weight fragments (Okazaki fragments). Radiolabel was subsequently chased into high-molecular-weight DNA, presumably through the action of DNA ligase, an enzyme that closes nicks (single-strand breaks) in double-stranded DNA. In phage T4 DNA ligase is encoded by gene 30 (see Figure 25.11). When *E. coli* was infected at 42° with a gene 30 temperature-sensitive mutant, Okazaki fragments accumulated and were not incorporated into high-molecular-weight DNA until the temperature of the culture was dropped. This provided strong support for the postulated function of DNA ligase in processing Okazaki fragments.

DNA ligase requires that the nick to be sealed must contain 3′ hydroxyl and 5′ phosphoryl termini, and the nucleotides being linked must be adjacent in a duplex structure and properly base paired. DNA ligase is activated by adenylylation of a lysine residue in the active site (Figure 25.15). The enzyme in turn transfers the adenylyl moiety to the 5′ terminal phosphate of the DNA substrate, thereby activating it for nucleophilic attack by the 3′ hydroxyl group, with phosphodiester bond formation and displacement of AMP. The T4 phage enzyme uses ATP to adenylylate the enzyme, as do eukaryotic DNA ligases. However, the enzyme from *E. coli* and other bacteria uses NAD^+. Instead of being a redox cofactor, the dinucleotide in this enzyme is cleaved, to yield adenylylated enzyme plus nicotinamide mononucleotide (NMN).

RNA Primers

If short RNAs indeed serve as primers for lagging-strand replication, it should be possible to demonstrate covalent attachment of RNA to DNA during replication. Reiji and Tuneko Okazaki made *E. coli* cells permeable to exogenous nucleotides by brief treatment with a buffer containing toluene. DNA synthesis occurred when these treated bacteria were incubated with a mixture of α-[^{32}P]dNTPs and unlabeled ribonucleoside triphosphates (Figure 25.16). DNA was then isolated and treated with mild alkali, which hydrolyzes RNA to nucleoside 2′ and 3′ monophosphates, but does not degrade DNA (see Chapter 4, page 95). If DNA synthesis starts from 3′ RNA termini, then for each Okazaki fragment there should exist one RNA–DNA junction, or deoxyribonucleotide residue covalently linked via a radioactive phosphate group to a ribonucleotide residue. Treatment with mild alkali, which hydrolyzes RNA but not DNA (Chapter 4) would then *transfer* that phosphate from the 5′ position of the original deoxyribonucleotide substrate to the 3′ position of the ribonucleotide at the junction. Sure enough, for

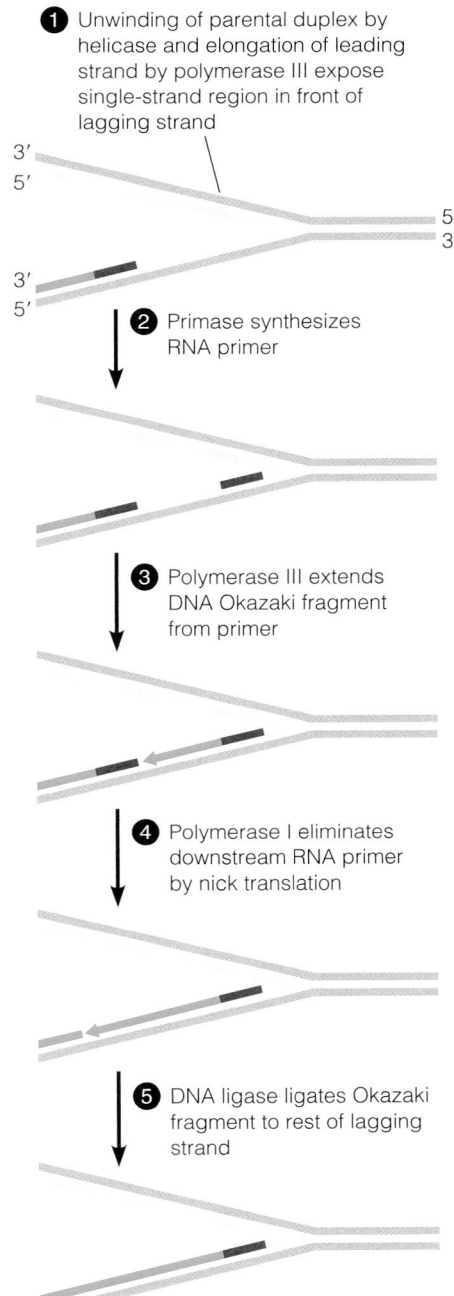

① Unwinding of parental duplex by helicase and elongation of leading strand by polymerase III expose single-strand region in front of lagging strand

② Primase synthesizes RNA primer

③ Polymerase III extends DNA Okazaki fragment from primer

④ Polymerase I eliminates downstream RNA primer by nick translation

⑤ DNA ligase ligates Okazaki fragment to rest of lagging strand

FIGURE 25.14

A model for DNA replication involving discontinuous synthesis on the lagging strand. A short RNA primer (red) is used to initiate synthesis of low-molecular-weight DNA fragments on the lagging strand (blue). RNA primers are replaced by deoxyribonucleotides (see text), and the fragments are sealed to high-molecular-weight DNA by DNA ligase. Each arrowhead on a nucleic acid strand identifies a 3′ hydroxyl group that can undergo polymerase-catalyzed chain extension. The replication fork is moving to the right.

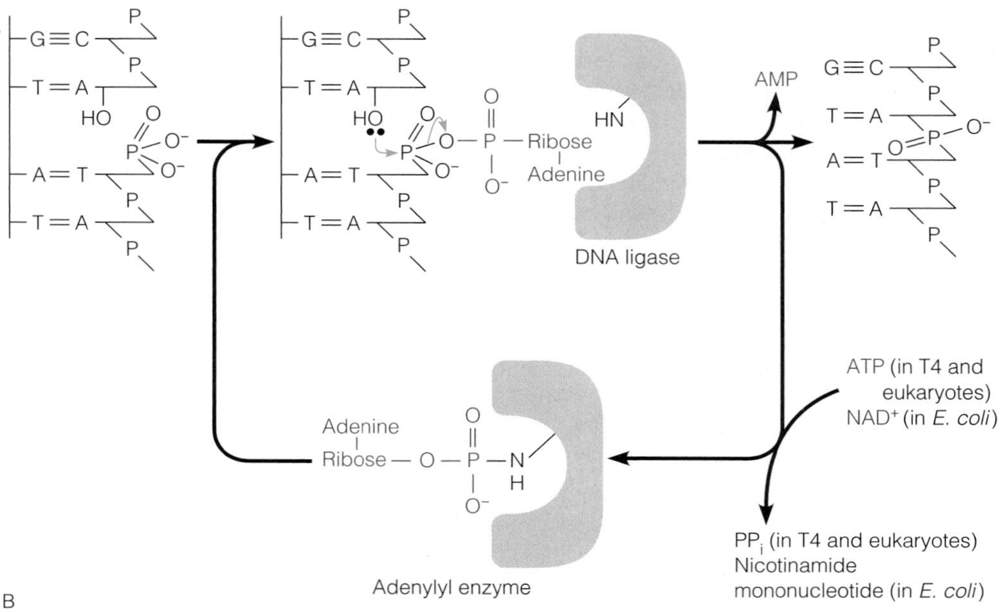

FIGURE 25.15

The reaction catalyzed by DNA ligase.

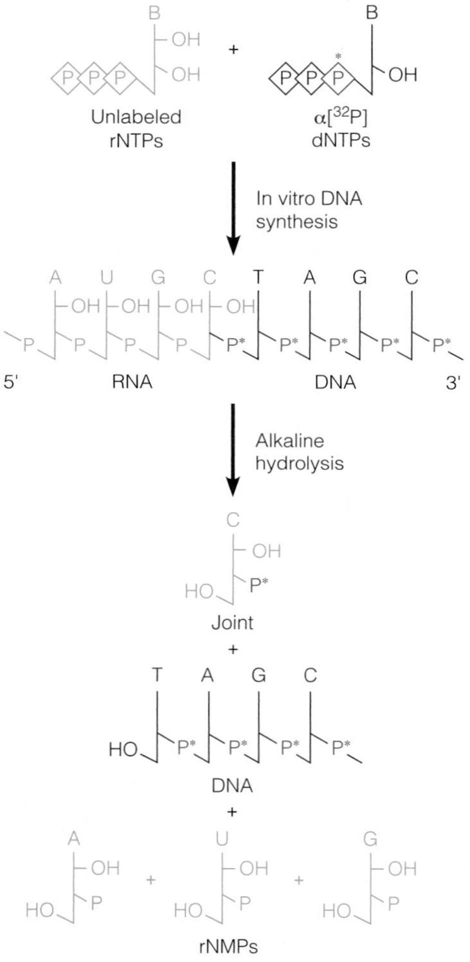

FIGURE 25.16

The transfer experiment that demonstrated the existence of RNA primers in DNA replication. Each Okazaki fragment generated one radiolabeled ribonucleotide from its RNA–DNA junction after alkaline hydrolysis.

each Okazaki fragment formed in this system, one radiolabeled phosphate was transferred to a ribonucleotide.

The enzyme that synthesizes RNA primers is called **primase**. In *E. coli* this enzyme is the product of the *dnaG* gene, and the comparable reaction in phage T4 is carried out by gp61, the product of gene 61. Priming involves pairing a ribonucleoside 5′-triphosphate opposite a deoxyribonucleotide residue in template DNA, followed by sequential ribonucleotide additions to the 3′ hydroxyl termini, just as occurs with DNA polymerases. At some point, RNA synthesis stops, and DNA polymerase continues to extend from the 3′ hydroxyl terminus of the RNA primer, but now with incorporation of deoxyribonucleotides. RNA primer structure has been described in some detail. For instance, in T4 and *E. coli* DNA replication the primer lengths are 5 and 11 nucleotides, respectively, and most primers have ATP as the 5′ terminal nucleotide.

For Okazaki fragments to be ligated to high-molecular-weight DNA, the RNA primers must be excised and replaced with corresponding deoxyribonucleotides. In *E. coli* DNA polymerase I is involved in this process, through its nick translation activity. The removal of ribonucleotides from the 5′ end of the primer, by the 5′ exonuclease of polymerase I, is coordinated with their replacement by deoxyribonucleotides (see Figure 25.17). An alternative means for digesting RNA primers is **ribonuclease H**, an enzyme that specifically hydrolyzes RNA base-paired with DNA of complementary sequence.

Now we are in a position to consider all of the proteins in a replication fork, as well as the functions of each. Figure 25.18, an idealized fork, includes proteins we have discussed, including DNA polymerases, primase, and ligase. Note that the replicative polymerase is dimeric, with one polymerase molecule each assigned to the leading and lagging strands. The primase is complexed with a **helicase**, one of several enzymes that can use the energy of ATP hydrolysis to drive parental strand unwinding. The helicase–primase complex is called a **primosome**. Attached to each polymerase molecule is a circular **sliding clamp**, which keeps polymerase bound to DNA over thousands of catalytic cycles. As DNA unwinding exposes single-strand template DNA, that DNA is coated with **single-strand DNA-binding protein** (SSB), which holds DNA in an extended conformation so that it can base-pair efficiently with incoming nucleotides. Finally, because unwinding the parental duplex creates torsional stress, a **topoisomerase** moves ahead of the fork, acting to relieve that stress. Ultimately, a topoisomerase plays a crucial role in introducing negative supercoiling into replicated DNA. Not shown in the figure is the **clamp-loading complex**, which both fastens and unfastens the sliding clamp to DNA. The complex of proteins supporting replication is called the **replisome**. However, it is unlike a static complex, such as pyruvate dehydrogenase because individual proteins within the replisome are in constant motion, dissociating from the complex and reassociating as parts of this dynamic machine.

The DNA Polymerase III Holoenzyme

The *E. coli polC* gene encodes a single polypeptide chain, with M_r of about 130,000. This protein has an intrinsic polymerase activity, but it is quite low. Within cells, the PolC protein functions as part of a multiprotein aggregate called the **DNA polymerase III holoenzyme**. As shown in Figure 25.19, the holoenzyme contains ten different polypeptide chains, each identified with a Greek letter. The α, ε, and θ subunits make up the "core polymerase," with α being the *polC* gene product, the protein with polymerase activity, and ε having a 3′ exonuclease activity comparable to the 3′ exonuclease domain of the DNA polymerase I polypeptide

Primase, a special class of RNA polymerase, synthesizes short RNA molecules as primers for lagging strand DNA replication.

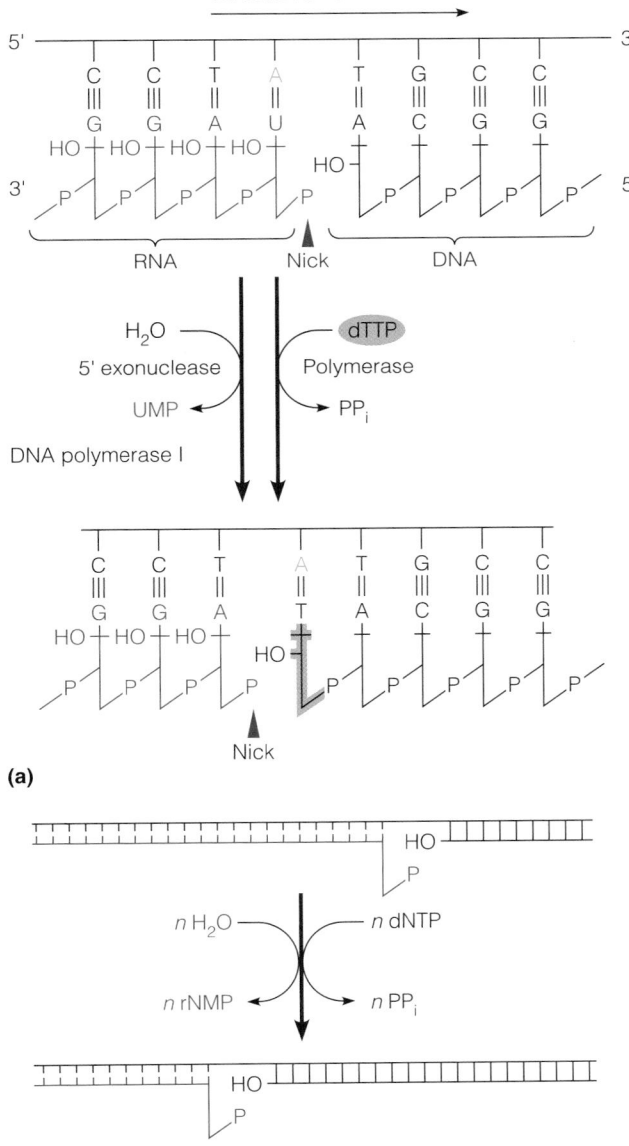

(a)

(b)

FIGURE 25.17

Nick translation in removal of RNA primers by coordinated action of the 5′ exonuclease and polymerase activities of DNA polymerase I. The figure shows replacement of base-paired UMP in the RNA primer by dTMP in the growing DNA chain. The template DNA is the lagging strand.

FIGURE 25.18

Schematic view of a replication fork. Note that polymerases catalyzing leading- and lagging-strand replication are linked together. See text for details.

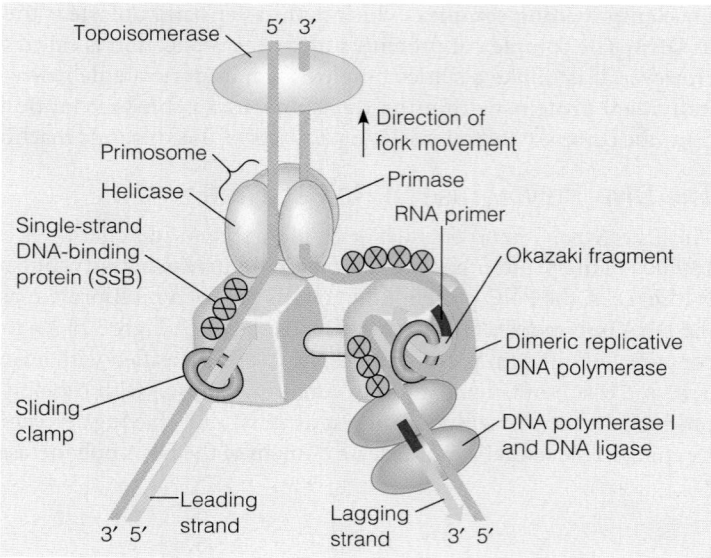

DNA polymerase III holoenzyme, a complex bacterial enzyme containing at least 10 subunits, plays the predominant role in replicative chain elongation.

chain. The function of θ is unknown, but it may act to improve the catalytic efficiency of α. The dimeric τ protein dimerizes the holoenzyme, holding leading and lagging strand polymerases together, ensuring that both DNA strands are elongated at the replication fork, even though the lagging strand polymerase moves in the direction opposite that of fork movement. χ mediates the switch from RNA primers to DNA.

Sliding Clamp

The β subunit was originally recognized as a protein essential for the *processivity* of DNA polymerase; in other words, its ability to remain bound to the template through many cycles of nucleotide addition. The core polymerase, once bound to template DNA, remains bound only long enough to extend a primer strand by 10–20 nucleotides. However, β tethers the enzyme to DNA, allowing it to incorporate several thousand nucleotides per binding event. We say that β converts DNA polymerase III from a highly *distributive* enzyme, which incorporates just a few nucleotides per binding event, to a highly *processive* enzyme, which remains bound through thousands of incorporation reactions. Crystallography showed β to be a circular molecule with a 3.5-nm opening, capable of completely surrounding double-stranded DNA. As shown in Figure 25.20, six α-helical domains (three per subunit) face the interior of the circle, with the hydrophobic residues in these helices having little attraction for DNA. Thus, the molecule acts as a sliding clamp, permitting polymerase to slide readily along DNA, but not to dissociate. The structure is remarkably well conserved evolutionarily, even though most other forms of this protein are trimers of dimers, not dimers of trimers.

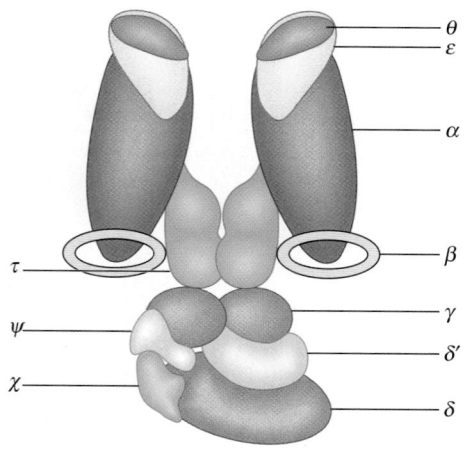

FIGURE 25.19

Subunit structure of the *E. coli* DNA polymerase III holoenzyme. Greek letter designations for the subunits are identified in the text.

Clamp-Loading Complex

How does a circular molecule wrap itself around DNA to begin processive synthesis? That is the function of the remaining five proteins, which form the γ complex, an assembly also called the **clamp loader**. This pentameric complex contains three copies of the γ protein and one copy each of the related δ and δ' subunits. χ and ψ are considered part of the clamp loader complex, but they do not participate directly. These proteins associate the γ complex with primase, and regulate the termination of RNA primer synthesis. The two N-terminal domains of each pentamer subunit form an *AAA + ATPase* module (ATPases associated with a variety of cellular activities).

Figure 25.21 shows the crystal structure of the γ complex bound to DNA and a schematic view of the clamp loading cycle. ATP is required, but not for opening the clamp. Instead, a conformational change driven by ATP binding leads the complex to bind the β clamp and to open it; this is followed by DNA binding. Once DNA has been encircled, the bound ATP is hydrolyzed, and the β ring closes. This event happens only once per round of replication on the leading strand. However, on the lagging strand, polymerase must rebind at the initiation of synthesis of each Okazaki fragment and must dissociate when the 5′ end of the preexisting daughter DNA strand is reached. In *E. coli* Okazaki fragments are 1 to 2 kb long. Because DNA chains are extended at about 800 nucleotides per second, this means that the clamp loading and unloading cycle must occur almost every second. And this remarkable process must all occur with the lagging strand core polymerase unit remaining bound to its leading strand partner, at the fork.

Single-Strand DNA-Binding Proteins: Maintaining Optimal Template Conformation

One of the earliest replication proteins to be identified, other than DNA polymerase itself, was a T4 protein called either *single-strand DNA-binding protein* (SSB) or *helix-destabilizing protein*. In an early application of affinity chromatography, Bruce Alberts immobilized DNA by binding it to cellulose and analyzed the T4 proteins that were retained by a column of this material. One protein was shown to be the product of gene 32 because the protein isolated from a *ts* gene 32 mutant was unable to bind to DNA at a restrictive temperature. Because gene 32 mutants were known to be defective in DNA repair and genetic recombination, as well as in DNA replication, it was clear that the protein played multiple roles in DNA metabolism.

Analysis of the purified gp32 showed that it binds specifically to single-strand DNA. Moreover, binding is strongly *cooperative*, meaning that the protein is far more likely to bind to DNA adjacent to a site already occupied than to an isolated site. In other words, binding of one gp32 molecule facilitates the binding of others, and the protein tends to bind in clusters. Thus, gp32 promotes the denaturation of DNA. Although it does not initiate denaturation, its presence lowers the melting temperature of DNA by as much as 40 °C.

The role of gp32 is to keep the template in an extended, single-strand conformation, with the purine and pyrimidine bases exposed so that they can base-pair readily with incoming nucleotides. This function is essential for DNA repair and genetic recombination, as well as for replication. Given that all three processes also

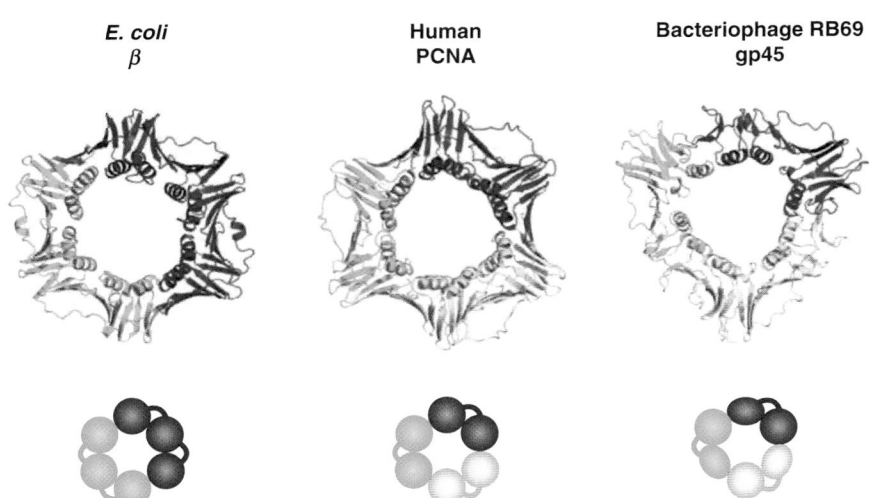

| *E. coli* β | Human PCNA | Bacteriophage RB69 gp45 |

FIGURE 25.20

Structure of the sliding clamp. Left, the *E. coli* β protein (PDB ID 2POL); center, human PCNA (proliferating cell nuclear antigen; PDB ID 1AXC); right, gp45, the sliding clamp of the T4-related phage, RB69 (PDB ID 1B77). Each protein forms a "doughnut" that can completely surround double-stranded DNA and thus keep polymerase associated with its DNA templates. The α helices on the inner surface of the subunit contact DNA but do not bind tightly enough to retard movement of the protein. The *E. coli* protein has two identical subunits with two DNA-associating domains, while the human and RB69 proteins have three subunits, each with two DNA-associating domains.

Reprinted from *DNA Repair* 8:570–578, L. B. Bloom, Loading clamps for DNA replication and repair. © 2009, with permission from Elsevier.

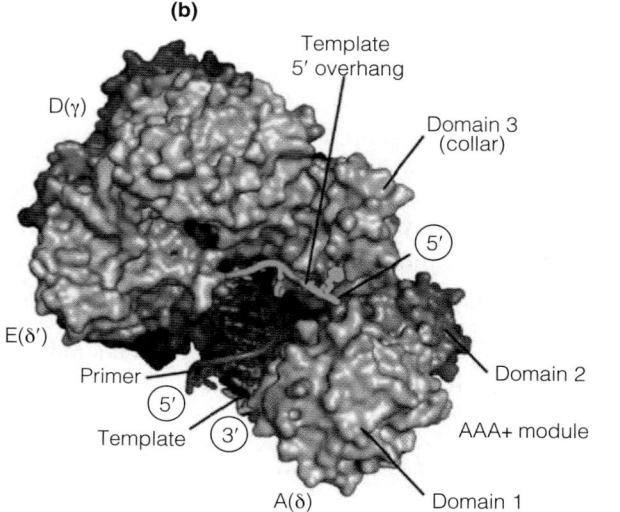

(a)

(b)

FIGURE 25.21

Scheme for action of the *E. coli* clamp loader **(a)** and structure of the γ complex bound to DNA **(b)**. The complex contains five protein subunits: three copies of γ (B, C, and D in the figure) and one each of δ and δ' (A and E, respectively). Only three of these subunits (A, D, and E) are visible in the structure; B and C are on the back side. Each of the five subunits contributes one domain toward a collar structure, and the remainder of each subunit contains an AAA + ATPase module. PDB ID 3glf.

Reprinted from *Cell* 137:659–671, K. R. Simonetta, S. L. Kazmirski, E. R. Goedken, A. J. Cantor, B. A. Kelch, R. McNally, S. N. Seyedin, D. L. Makino, M. O'Donnell, and J. Kuriyan, The mechanism of ATP-dependent primer-template recognition by a clamp loader complex. © 2009, with permission from Elsevier.

Single-strand DNA-binding proteins are essential in DNA replication, repair, and recombination by virtue of their ability to facilitate both DNA denaturation and renaturation.

involve the re-formation of duplex structures, with one strand each of parental and daughter DNA, it is of interest that gp32 facilitates the renaturation of single-strand DNA, as well as the denaturation of duplexes.

How can a single protein promote both duplex formation and duplex unwinding? The answer seems to lie in special design features of the gp32 molecule. Note that *gp32 does not itself unwind DNA strands*. Rather, it stabilizes single-strand DNA by binding *after* a region of DNA is unwound. Limited proteolysis of gp32 removes a C-terminal fragment. This modification renders the protein a stronger DNA denaturant in vitro, with a small increase in its equilibrium binding affinity for DNA. As suggested in Figure 25.22, the C-terminal domain provides a "flap" that partly covers the DNA-binding domain of the protein. When a short region of single-strand DNA is exposed, through partial and reversible unwinding of the duplex (sometimes called *breathing*), gp32 can bind with the flap down, but only to a small site (3 or 4 nucleotides). Occupancy of a complete site (7 to 10 nucleotides) is thereby inhibited, rendering binding both weak and noncooperative. If continued helix unwinding occurs, the flap can move up, permitting the complete site to be occupied and promoting further denaturation.

The partially proteolyzed gp32 lacks the ability to renature DNA in vitro, and it seems likely that the conformation of the native gp32 determines whether the protein functions primarily as a denaturant (flap-up position) or renaturant (flap-down position). The position of the flap is determined in part through interaction with other proteins at the replication fork. Thus, the protein can act both to stabilize a single-strand template as it pairs with incoming nucleotides and to facilitate re-formation of a duplex after DNA polymerase has passed by. Partial proteolysis can also remove an N-terminal domain, which has been shown to be essential for self-association of the protein and, hence, for cooperative DNA binding. The crystal structure of a gp32-ssDNA complex shows three domains of the protein to be involved in DNA binding, consistent with this model.

Single-strand DNA-binding proteins have now been found in many organisms. The protein of *E. coli* (specified by the *ssb* gene) also binds cooperatively to single-strand DNA. However, the mechanism of binding appears to be quite different from that in T4. In *E. coli,* DNA is wrapped about the outer surface of the tetrameric SSB protein. Moreover, under some conditions the SSB protein binding shows *negative* cooperativity, meaning that incoming SSB molecules avoid sites adjacent to bound molecules. It has been suggested that the protein binds in different modes, depending on whether it is participating in replication, DNA repair, or recombination.

In eukaryotic cells a heterotrimeric protein called replication factor A (RFA) serves the role of SSB in DNA replication. This protein undergoes phosphorylation during S phase or after DNA damage, suggesting a role in cellular coordination of DNA metabolism.

Helicases: Unwinding DNA Ahead of the Fork

Single-strand DNA-binding proteins do not themselves denature DNA. As noted, they stabilize single-strand DNA but cannot actively unwind duplex DNA strands. Such unwinding must occur if single-strand templates are to be exposed for polymerase action. The helicase proteins have this ability. They catalyze the ATP-dependent unwinding of double-strand DNA. *E. coli* cells contain at least a half dozen different helicases, some of which participate in DNA repair and some in bacterial conjugation. The principal helicase in DNA replication is DnaB (the protein product of the *dnaB* gene), which interacts with DnaG and other proteins to form the primosome (see Figure 25.18). The comparable roles in T4 DNA replication are played by gp41 (helicase) and gp61 (primase). In phage T7 a single protein (gp4) has both helicase and primase activities.

All known helicases are multimeric proteins. Most are homodimers, but a few, like DnaB, are homohexamers. In vitro, each helicase binds initially to single-stranded DNA, adjacent to a duplex region, and proceeds in a fixed direction ($5' \rightarrow 3'$ or $3' \rightarrow 5'$), displacing the unbound DNA strand as it moves, and with ATP hydrolysis coupled to movement. Although the helicases are homo-oligomers, they are structurally asymmetric when bound to DNA. For example, in the well-studied dimeric Rep helicase of *E. coli*, the ATP-binding and DNA-binding properties of the two subunits are quite different. This suggests a rolling or "hand-over-hand" mechanism, in which each subunit alternates between tight-binding and loose-binding conformations, depending upon whether ATP or ADP is bound at a given instant (Figure 25.23). Consistent with this model, the crystal structure of the Rep helicase complexed with single-strand DNA and ADP shows a dramatic structural difference between the subunits, resulting from a 130° rotation of one domain about a hinge region.

The hexameric helicases, exemplified by the gene 4 product of phage T7, may move by a quite different mechanism. This ring-shaped protein, which uses dTTP rather than ATP as its energy source, wraps about the single strand to which it binds and along which it moves. The strand that is not bound becomes displaced by helicase action and does not pass through the central hole. Three of the six subunits bind and hydrolyze dTTP, while the other three bind dTTP in noncatalytic fashion. Data on dTTP binding and hydrolysis suggest a rotary motion for the protein, comparable to that seen in the mitochondrial F_0F_1 ATP synthase (see Figure 25.24a,b), suggesting the intriguing possibility that the helicase moves along the DNA by rotation. The major difference is that the F_0F_1 ATP synthase uses the proton motive force to drive ATP *synthesis*, while the T7 helicase uses the energy of dTTP *hydrolysis* to drive a conformational change, giving

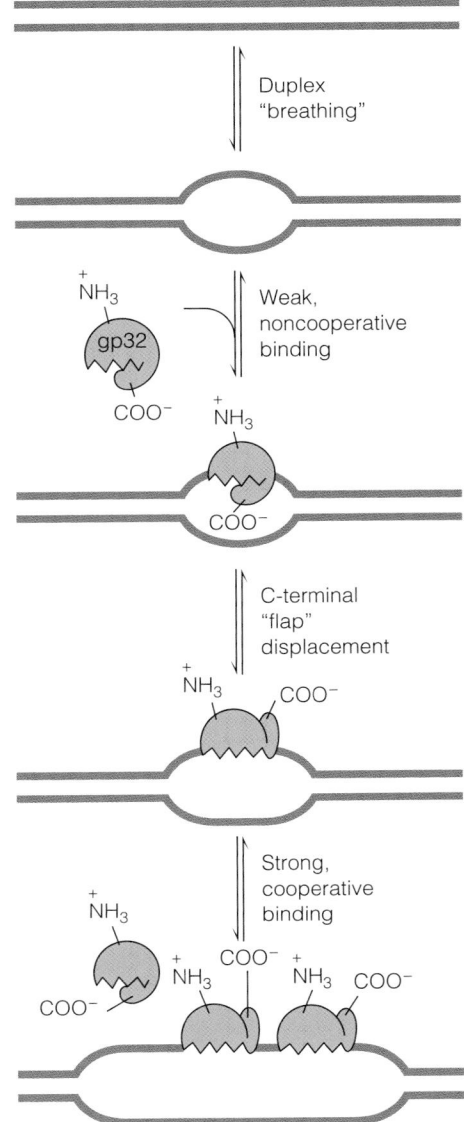

FIGURE 25.22

gp32 facilitation of both denaturation and renaturation of DNA. When only a short length of DNA is exposed, gp32 binds weakly. The folded-down configuration of the C-terminal domain prevents cooperative binding of other gp32 molecules, and hence, strand unwinding ("breathing"). If a longer region of ssDNA is exposed after the first gp32 molecule has bound, the C-terminal domain can shift to the "up" configuration, enlarging the DNA binding site and permitting cooperative binding of other gp32 molecules, thereby further extending the denatured region. For simplicity binding is shown on only one of the two denatured DNA strands.

Helicases are multimeric proteins that bind preferentially to one strand of a DNA duplex and use energy of ATP hydrolysis to actively unwind the duplex.

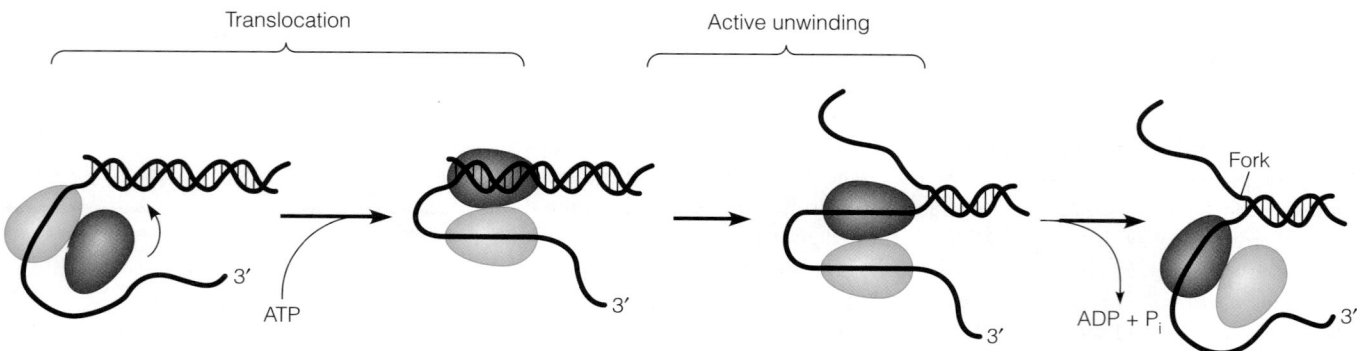

Translocation   Active unwinding

Fork

3′

ATP

3′

3′

ADP + P_i

3′

FIGURE 25.23

A model for helicase action. In this model a homodimeric enzyme, such as the *E. coli* Rep helicase, shows 3′ → 5′ polarity. In the first step, binding of ATP activates the subunit shown in red to bind to DNA at the double-strand/single-strand junction and, in the second step, to unwind a few base pairs. In the third step, the hydrolysis of bound ATP weakens the binding of the blue subunit to DNA, causing its dissociation, and thereby positioning it to begin the cycle anew, by invading the DNA duplex ahead of the fork that has been created, once another ATP has been bound. Thus, by alternating in their binding to DNA, the two subunits cause the enzyme to "roll" counterclockwise as shown in this model, unwinding the duplex as it moves.

Reprinted from *Cell* 90:635–647, S. Korolev, J. Hsieh, G. H. Gauss, T. M. Lohman, and G. Waksman, Major domain swiveling revealed by the crystal structures of complexes of *E. coli* rep helicase bound to single-stranded DNA and ADP. © 1997, with permission from Elsevier.

the homohexameric protein the structural asymmetry needed to drive motion along the DNA lattice. Figure 25.24c shows a model of the T7 gene product translocating along DNA.

As mentioned above, the DnaB helicase of *E. coli* is also a homohexamer, which wraps around single-stranded DNA. Each subunit has two DNA-binding domains, one strong and one weak, each of which occludes about twenty DNA nucleotide residues.

As mentioned earlier, the primosome interacts with the τ and χ subunits of the γ complex to coordinate the switch from primase to polymerase on the lagging strand. Based upon the protein–protein interactions demonstrated, Figure 25.25 shows how this might occur.

Considerable excitement was engendered by recent findings that two inherited human diseases, Werner's syndrome and Bloom's syndrome, result from helicase defects. Both conditions involve increased susceptibility to cancer, and Werner's syndrome patients also undergo premature aging, usually going gray in their twenties, developing cataracts, and dying of natural causes before age 50. Positional cloning (Chapter 24) of the defective genes responsible for these conditions revealed in both cases that the genes encode proteins related to the *E. coli* RecQ gene product. In *E. coli* this helicase participates in a homologous recombination pathway and may be involved in resumption of DNA replication after repair of radiation-induced DNA damage. Like the RecQ protein, the Werner's syndrome protein has been shown to have a 3′ → 5′ helicase activity. These findings provide intriguing clues to understanding relationships between genomic instability and both cancer and aging.

Topoisomerases: Relieving Torsional Stress

Bidirectional replication of the circular *E. coli* chromosome unwinds about 100,000 base pairs per minute. Were there not some mechanism for relieving this torsional stress, DNA ahead of the fork would become overwound as DNA at the fork became unwound, and replication could not be sustained. Topoisomerases, a group of enzymes that can interconvert different topological isomers of DNA (see Chapter 4), provide a "swivel" mechanism for relieving this stress. Topoisomerase action is demonstrated most simply in vitro by the relaxation of supercoiled DNA. One can incubate supercoiled DNA with a purified topoisomerase and observe, by gel electrophoresis, the intermediate stages in conversion of the supercoiled substrate to relaxed circular DNA containing no superhelical turns (Figure 25.26). This analysis reveals the existence of two general classes of topoisomerases—type I enzymes, which change the linking number in units of 1, and type II enzymes, which change the linking number in units of 2.

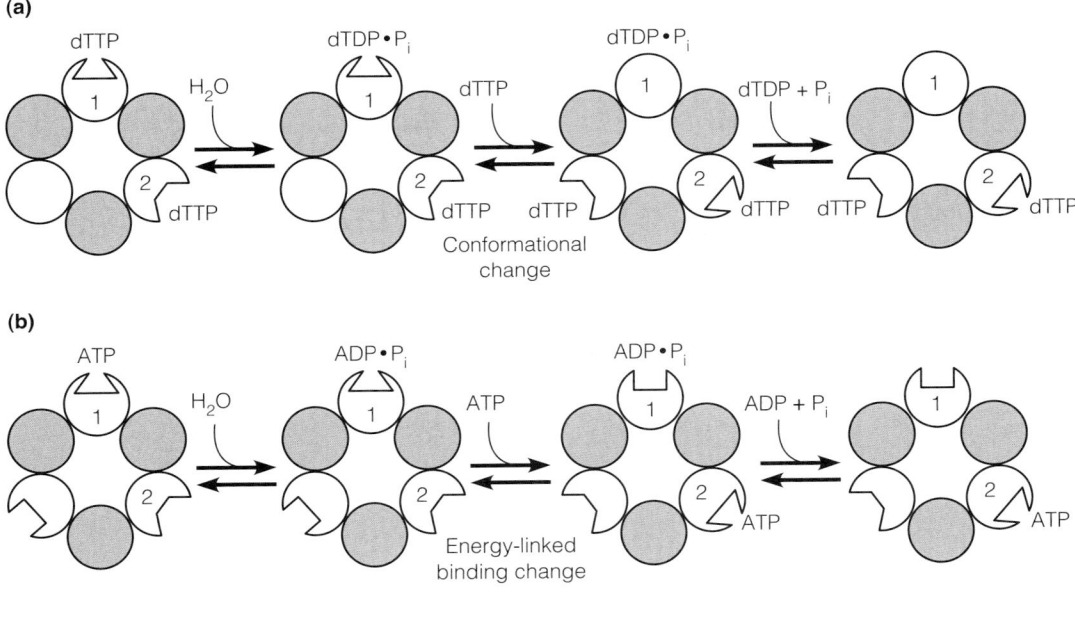

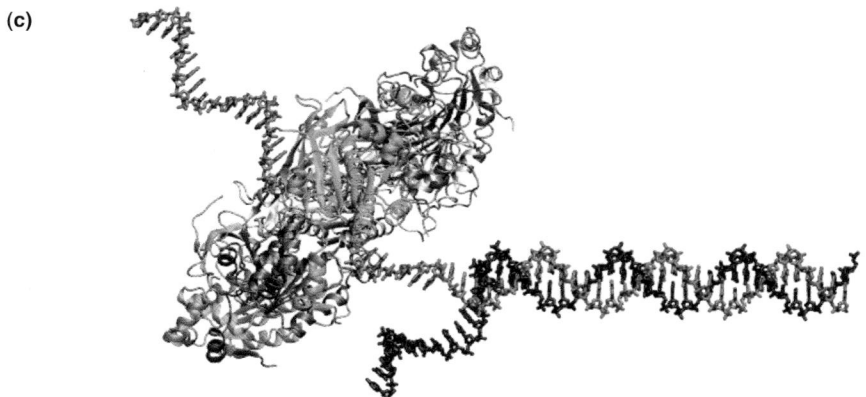

FIGURE 25.24

Structure and action of the T7 phage gene 4 helicase. **(a, b)** Comparison of T7 gp4 action with that of the ATP synthase rotary engine (see Figure 15.23, page 652). Shaded subunits in both enzymes depict noncatalytic sites. In the example shown the conformational change in T7 gp4 occurs directly after hydrolysis of dTTP at site 1. **(c)** A model of T7 gp4 helicase action. The protein is rotating along the blue DNA strand, excluding the red strand from the central channel (PDB ID 1CR1).

(a, b) Reprinted from *Proceedings of the National Academy of Sciences of the United States of America* 94:5012–5017, M. M. Hingorani, M. T. Washington, K. C. Moore, and S. S. Patel, The dTTPase mechanism of T7 DNA helicase resembles the binding change mechanism of the F₁-ATPase. © 1997 National Academy of Sciences, U.S.A.; (c) Reprinted from I. Donmez and S. S. Patel, Mechanisms of a ring shaped helicase, *Nucleic Acids Research* 34:4216–4224. © 2006, by permission of Oxford University Press.

Actions of Type I and Type II Topoisomerases

A type I topoisomerase breaks just one strand of the duplex (Figure 25.27). The enzyme remains covalently attached to the 5′ end of the broken strand, by forming a phosphodiester bond between the 5′ phosphate and a tyrosine hydroxyl. The 3′ end is then free to rotate (by one turn in the example shown). The hydroxyl group on the 3′ end then attacks the activated, covalently bound 5′ phosphate, closing the nick—in fact, *E. coli* type I topoisomerase was originally called nicking–closing enzyme. The result is that the linking number has been changed by 1. Eukaryotic topoisomerase I acts similarly, but the 3′ end, not the 5′ end, is immobilized during the reaction.

By contrast, a type II topoisomerase catalyzes a double-strand break, and the unbroken part of the duplex passes through the gap that is created (Figure 25.28). The most thoroughly studied type II topoisomerase is an *E. coli* enzyme also called **DNA gyrase** because it can not only relax a supercoiled molecule but also introduce negative superhelical turns into DNA. ATP hydrolysis is required for both activities of most type II enzymes. DNA gyrase is a tetramer, with two A and two B subunits. The A subunits bind and cleave DNA, while the B subunits carry out the energy transduction resulting from ATP hydrolysis.

As shown in Figure 25.28, gyrase action begins with DNA wrapping about the enzyme. The A subunit cleaves both DNA strands and immobilizes them, and

FIGURE 25.25

The primase–polymerase switch during lagging-strand synthesis. **(a)** DnaB helicase encircles the lagging strand, and primase has synthesized an RNA primer. Primase must contact SSB to remain bound. Core polymerase on the lagging strand is forcing that strand and the daughter strand to loop out. **(b)** The χ subunit of the γ complex interacts with SSB, leading to primase displacement. The γ complex is opening the clamp, and a newly completed Okazaki fragment is being released, along with its template. **(c)** Primase rebinds to single-strand template DNA to begin a new primer. In (b) the γ complex is also releasing a β clamp, and in (c) a new clamp is being attached. The periodic lengthening of the lagging strand loop has been likened to the playing of a trombone.

Courtesy of Dr. Michael O'Donnell.

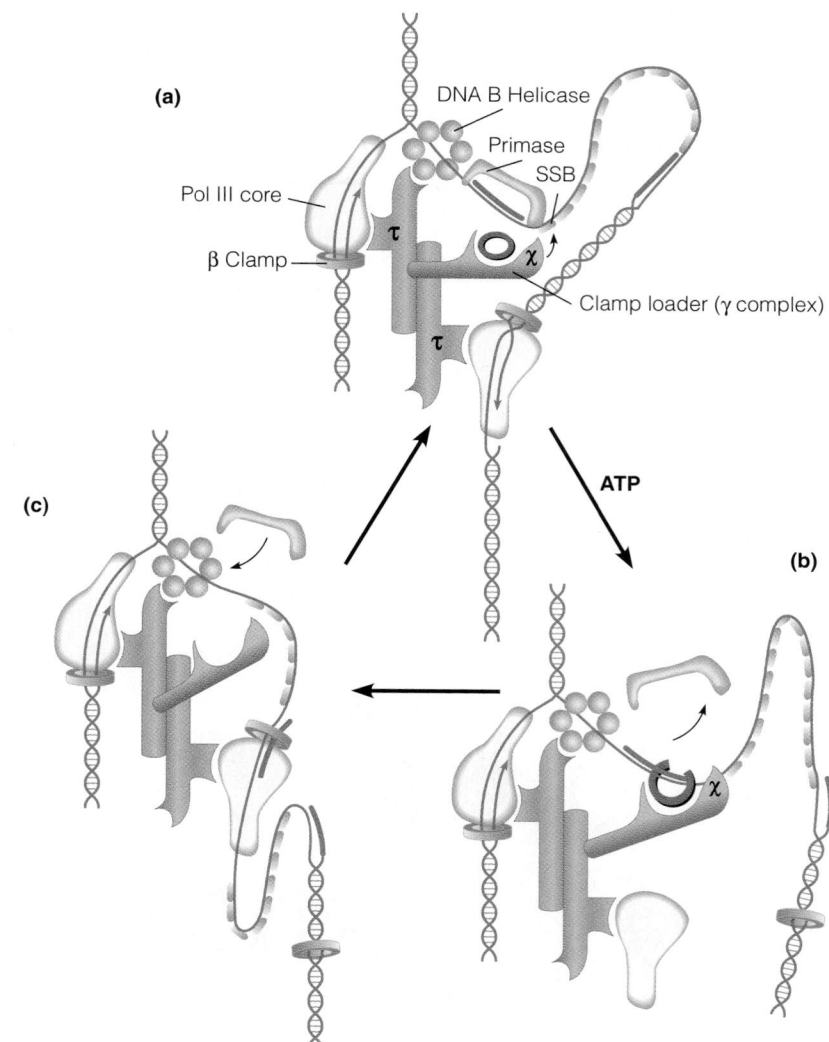

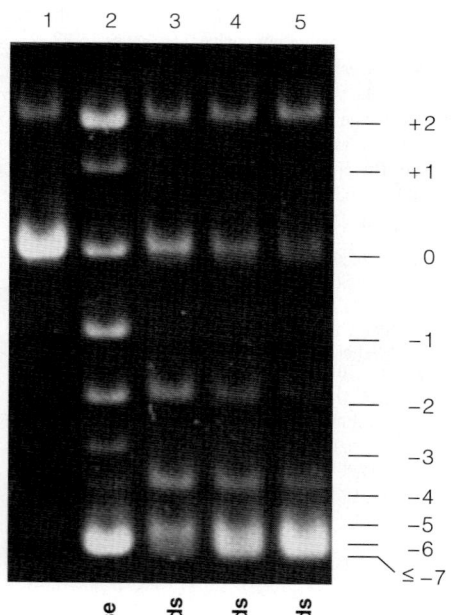

both strands of the duplex pass through the opening. This is followed by resealing of both strands of the duplex and dissociation of the enzyme. In the example shown, a circular DNA with one positive supercoil is converted to a product with one negative supercoil. Thus, the linking number is changed by 2, the distinction between a type I and a type II topoisomerase. In Figure 25.26, note that DNA treated with a type I topoisomerase (lane 2) shows twice as many intermediates as identical DNA treated with a type II enzyme (lanes 3–5) because the type I enzyme changes linking number in units of 1.

FIGURE 25.26

Action of type I and type II topoisomerases, as shown by gel electrophoresis. Lane 1 shows a relaxed circular DNA. Lane 2 shows the pattern from treatment of supercoiled DNA with type I topoisomerase. Lanes 3–5 show relaxed circles treated with DNA gyrase, a type II topoisomerase, for different lengths of time. Note that more different topoisomers can be seen in topoisomerase I reaction mixtures, as expected if changes in the linking number (L) occur in units of 1, whereas gyrase changes L in units of 2.

From *Science* 206:1081–1083, P. O. Brown and N. R. Cozzarelli, A sign inversion mechanism for enzymatic supercoiling of DNA. © 1979. Reprinted with permission from AAAS and Pat Brown.

The action of type I topoisomerases was clarified with publication of the crystal structure of human topoisomerase I. As shown in Figure 25.29, the enzyme completely wraps around its DNA substrate. Not evident from the figure is the fact that most of the DNA–protein contacts involve the DNA sugar–phosphate backbone rather than the bases, meaning that DNA is bound as an undistorted B-form helix. Also, the contacts on the "upstream" (5′) side of the scissile bond are far more numerous than those on the downstream (3′) side. Because the 5′ nucleotide in the nick is immobilized by its binding to a tyrosine residue, the structural data indicate that the free 3′ end is relatively free to rotate, as it must do if the supercoiled DNA is to become relaxed. A useful consequence of this structure determination was clarification of the mechanism of action of anticancer drugs that act by inhibiting topoisomerases because the structure reveals how such inhibitors can be bound, and it should lead to the design of more effective inhibitors. One inhibitor in current clinical use is **camptothecin**.

The Four Topoisomerases of *E. coli*

Since the discovery of topoisomerase I and DNA gyrase in the 1970s, *E. coli* has been shown to contain four different topoisomerases. The terminology is a bit confusing because the enzymes named topoisomerase I and topoisomerase III are both type I topoisomerases, and topoisomerase II (also called DNA gyrase) and topoisomerase IV are both type II topoisomerases. Of these four enzymes, DNA gyrase plays the dominant role during replicative chain elongation, both in relieving stress ahead of the fork and in introducing negative supercoils into newly synthesized DNA. We know of this role primarily through studies of the properties of gyrase inhibitors. The gyrase A subunit is the target for binding of **nalidixic acid**, a compound long known to inhibit DNA replication. Another replication inhibitor, **novobiocin**, binds to the B subunit and inhibits ATP cleavage. Inhibitors such as these are useful as antibacterial drugs. Mutant bacteria resistant to nalidixic acid or novobiocin show structural alterations in subunit A or B, respectively.

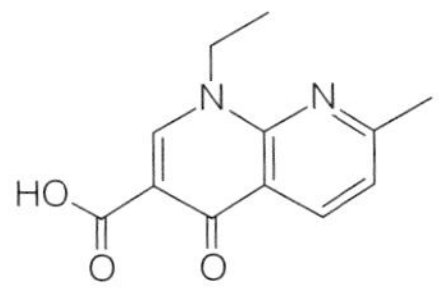

Topoisomerase IV plays a critical role in the completion of a round of replication. Type II topoisomerases catalyze a variety of topological interconversions,

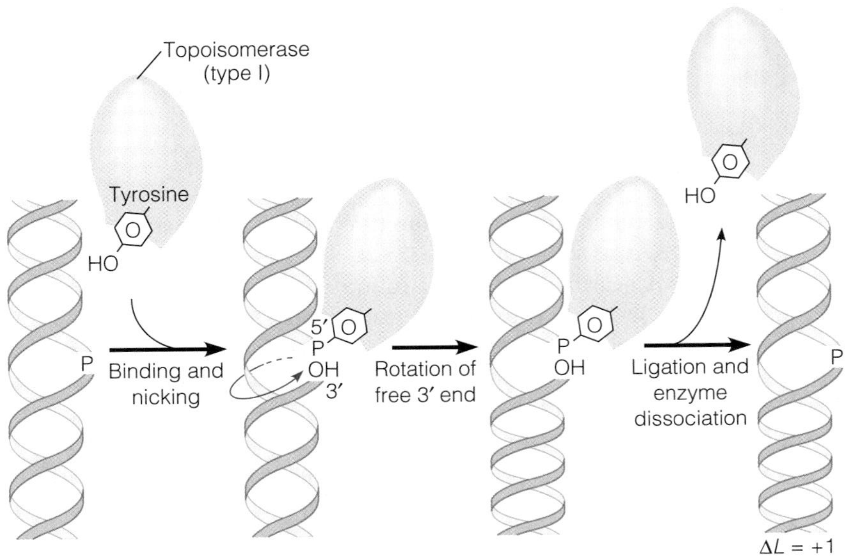

Type I topoisomerases break and reseal one DNA strand, and type II topoisomerases catalyze double-strand breakage and rejoining; hence, type I and type II enzymes change DNA linking number in units of 1 and 2, respectively.

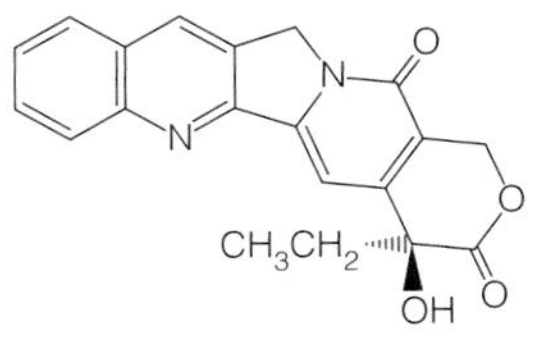

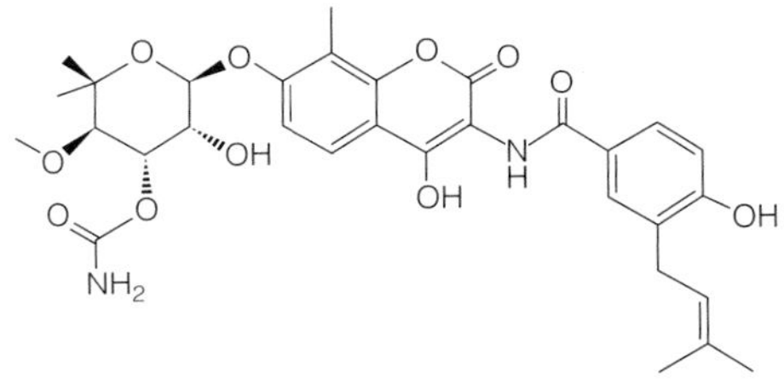

Camptothecin

Nalidixic Acid

Novobiocin

FIGURE 25.27

Action of a type I topoisomerase. The enzyme breaks one strand and immobilizes the 5′ end by a covalent bond between the DNA phosphate and a tyrosine residue (in *E. coli* topoisomerase I). Rotation of the 3′ end is followed by resealing. The linking number is increased by 1 in the example shown (an underwound DNA). Action of a type I topoisomerase on overwound DNA would decrease the linking number, by essentially the same mechanism.

Topoisomerase (type I)

Tyrosine

HO

P — Binding and nicking — 5′ / P / OH / 3′ — Rotation of free 3′ end — P / OH — Ligation and enzyme dissociation — P

HO

$\Delta L = +1$

FIGURE 25.28

Action of a type II topoisomerase. DNA gyrase of *E. coli*; the example shown is a tetrameric protein with two A and two B subunits. The enzyme is shown introducing two negative turns and changing the linking number from $+1$ to -1. The enzyme catalyzes a double-strand break, and the two DNA ends are bound by A subunits, which move the DNA ends apart so that the unbroken duplex can pass through the gap. Resealing converts the positive supertwist to a negative one, giving the overall molecule a ΔL of -2. Type II topoisomerases can relax underwound duplexes by the reverse of the above pathway.

Courtesy of Gary Carlton.

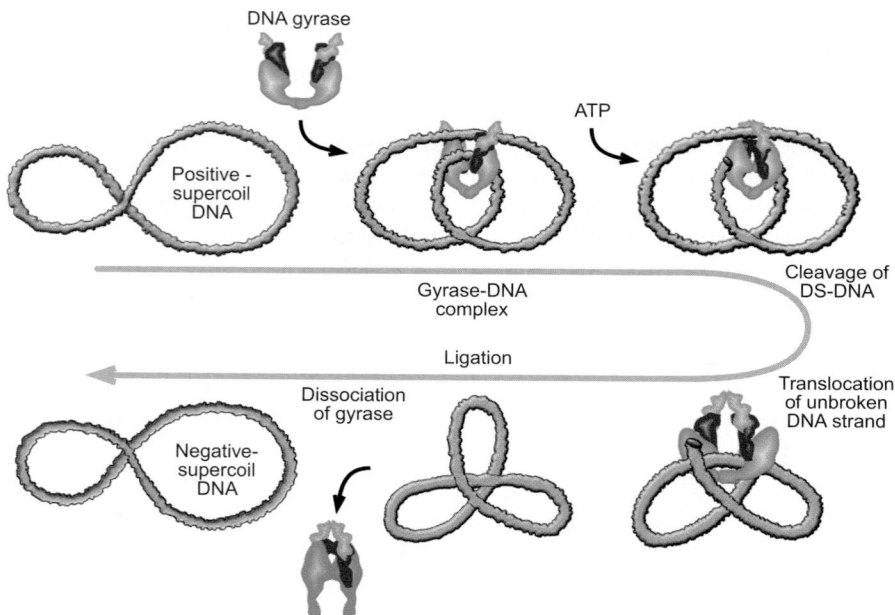

FIGURE 25.29

Crystal structure of human topoisomerase I in complex with a 22-bp DNA duplex. **(a)** The DNA strand nicked by the enzyme is shown in magenta upstream of the cleavage site and in pink downstream, and the intact strand is blue. For the protein, each individual domain appears in a different color. The protein wraps completely around the DNA, contacting 4 bp upstream and 6 bp downstream of the cleavage site. **(b)** End-on view of the topoisomerase–DNA complex. PDB ID 1A35. CPT, camptothecin.

From *Science* 279:1504–1513, M. R. Redinbo, L. Stewart, P, Kuhn, J. J. Champoux, and W. G. J. Hol, Crystal structures of human topoisomerase I in covalent and noncovalent complexes with DNA. © 1998. Reprinted with permission from AAAS.

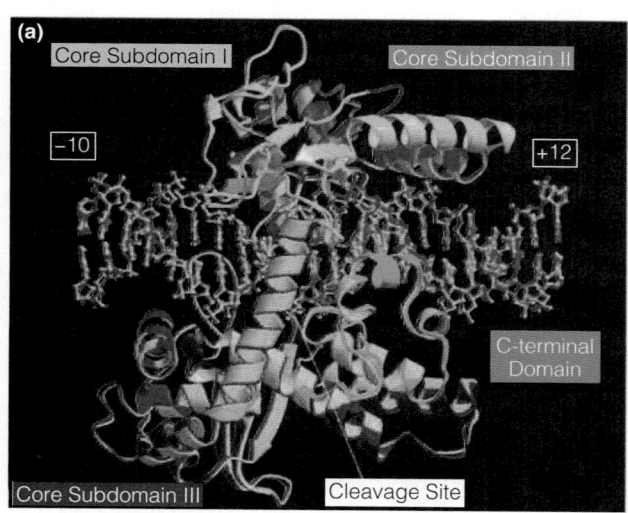

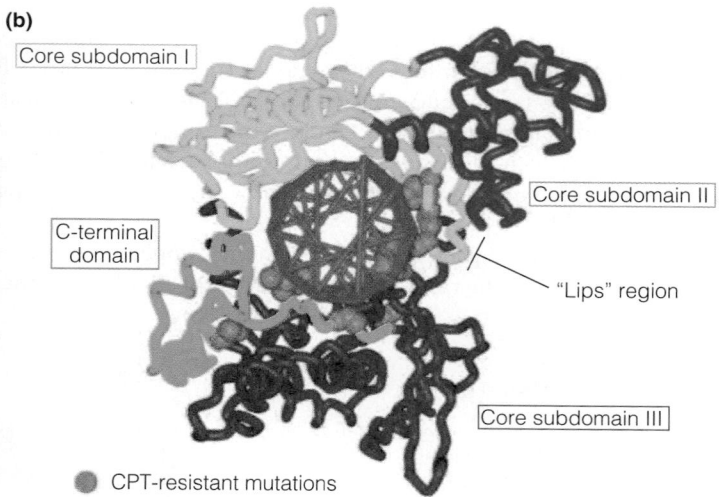

including knotting, unknotting, **catenation** (linking), and **decatenation** (unlinking) of circular DNAs, as shown in Figure 25.30.

A circular DNA nearing the end of a round of replication will generate two interlinked circles, so a type II topoisomerase action is necessary for separating the newly replicated molecules. As two forks approach each other at the replication terminus, steric barriers ultimately interfere with the unwinding activities of topoisomerases ahead of the two forks (Figure 25.31). At this stage the two incompletely replicated chromosomes are still interlinked. Topoisomerase IV plays a specific role in the decatenation process shown in Figure 25.31. This finding suggests a degree of specificity in topoisomerase action that is not immediately apparent from the idea that a strained DNA conformation is being relaxed. In accord with this suggestion, studies of type II topoisomerases show that they do not generate equilibrium mixtures of topoisomers. For example, a topoisomerase IV limit digest (DNA digested until no further reaction occurs) was shown to have far fewer knotted and catenated structures than expected for an equilibrium mixture; somehow, the enzyme selectively "disentangles" DNA molecules. Also, when a mixture of topoisomers generated by topoisomerase I action was subsequently treated with topoisomerase IV under the same conditions, the average linking number—the number of supercoils—remained the same, but the distribution of topoisomers about the mean became much tighter. ATP hydrolysis is required for this selectivity to be achieved, but remarkably, these observations indicate that a topoisomerase molecule can somehow scan a DNA molecule much larger than itself and direct the specific kinds of topological changes that occur.

Additional events are involved in termination of a round of replication in *E. coli*, evidently ensuring that both replisomes reach the same point before termination can occur. As schematized in Figure 25.32, the termination region contains ten homologous 23-base-pair *Ter* sites (*TerA–J*) arranged in oppositely oriented groups of five. Each *Ter* site binds a protein called Tus, a 36-kDa terminator protein. As each replisome approaches a *Ter*-Tus complex, it encounters either a "permissive" face, which allows the replisome to displace Tus and move on, or an oppositely oriented nonpermissive face, at which the replisome stalls because it cannot displace Tus. Thus, as shown in the figure, the replisome moving counterclockwise displaces Tus and passes the sites labeled *TerJ, TerG, TerF, TerB,* and *TerC* before stalling between *TerC* and the oppositely oriented *TerA*. A similar sequence of events leads the other replisome to the same point. This process evidently ensures that replication terminates at the same chromosomal site, even though the two replisomes may travel at different rates. The crystal structure of a Tus-*Ter* complex, shown in Figure 25.33, provides clues to the process of Tus displacement. The actual mechanism of termination is still under investigation.

A Model of the Replisome

Strictly speaking, topoisomerases are not components of the replisome because they act at a distance from the fork. However, now that we have discussed the major proteins involved in prokaryotic fork propagation, and as a prelude to discussion of

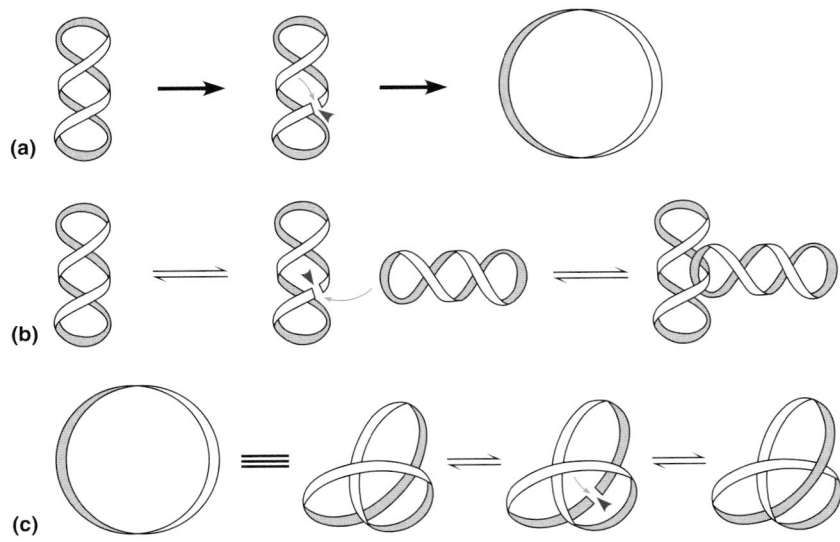

FIGURE 25.30

The types of topological interconversions catalyzed by type II topoisomerases.
(a) Relaxation. (b) Catenation and decatenation.
(c) Knotting and unknotting.

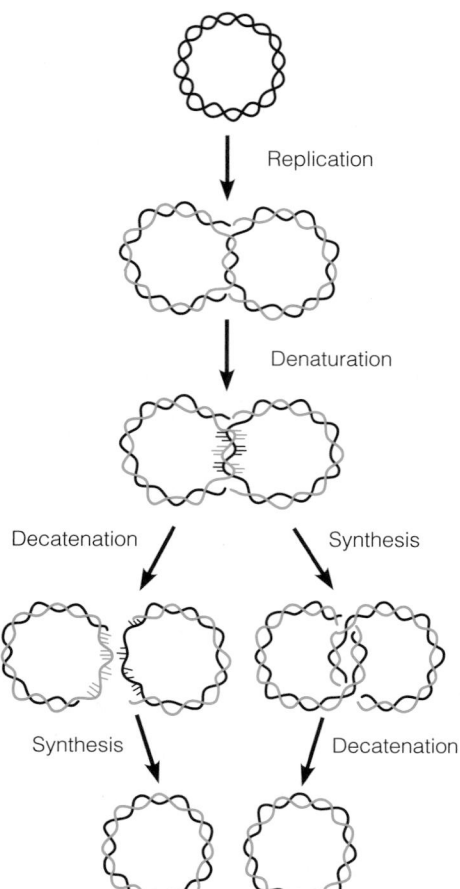

Replication

Denaturation

Decatenation   Synthesis

Synthesis   Decatenation

FIGURE 25.31

Topoisomerase action in termination of replication. In the absence of topoisomerases, steric forces would prevent gyrase from unwinding DNA as replication forks approached each other. Topoisomerase allows the circles to decatenate. It is not known whether decatenation occurs before or after the completion of replication. Both possibilities are shown here.

comparable eukaryotic proteins, we present here a model of the *E. coli* replisome, showing relationships among the constituent proteins (Figure 25.34).

Recently it has become possible to image the *E. coli* replisome in vivo, using fluorescent-tagged proteins and single-molecule techniques (Tools of Biochemistry 26A). These experiments indicate that the replisome contains three molecules of DNA polymerase, not the two shown in Figure 25.34. Only two of the three molecules are linked to sliding clamps, but three copies of τ are present, suggesting that the unlinked polymerase is waiting its turn to have a clamp loaded so that it can catalyze synthesis of the next Okazaki fragment.

Proteins in Eukaryotic DNA Replication

DNA Polymerases

The mechanism of replicative DNA chain elongation has been found to be remarkably constant through evolution. One striking difference between prokaryotic and eukaryotic replication is the requirement for three different DNA polymerases to propagate a eukaryotic replication fork. Early fractionation of DNA polymerases from yeast or mammalian cells yielded five different enzymes, three of which are involved in nuclear DNA replication, one in mitochondrial DNA replication, and one in repair. Later work revealed the existence of at least nine additional DNA polymerases in human cells, most of which participate in specialized repair processes. Properties of the five "classical" polymerases, which include the three

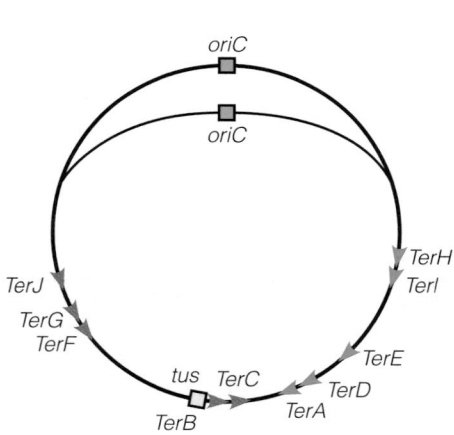

FIGURE 25.32

Polarized replication termination in *E. coli*. Replication initiates bidirectionally from *oriC*. The orientation of *Ter* sites, as discussed in the text, insures that both replisomes arrive at the same site (between *TerA* and *TerC*) before termination can begin.

Reprinted from *Cell* 125:1309–1319, M. D. Mulcair, P. M. Schaeffer, A. J. Oakley, H. F. Cross, C. Neylon, T. M. Hill, and N. E. Dixon, A molecular mousetrap determines polarity of termination of DNA replication in *E. Coli*. © 2006, with permission from Elsevier.

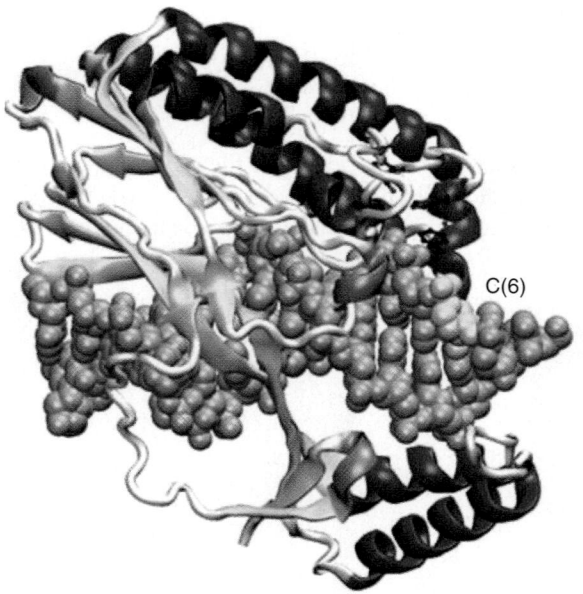

C(6)

FIGURE 25.33

Structure of the *E. coli* Tus protein, complexed with double-stranded DNA (in cyan). PDB ID 1ECR.

Reprinted from *Cell* 125:1309–1319, M. D. Mulcair, P. M. Schaeffer, A. J. Oakley, H. F. Cross, C. Neylon, T. M. Hill, and N. E. Dixon, A molecular mousetrap determines polarity of termination of DNA replication in *E. Coli*. © 2006, with permission from Elsevier.

replicative enzymes, are summarized in Table 25.2. The involvement of polymerases α, δ, and ε in replication was shown by the fact that all three are inhibited by **aphidicolin**, a fungal product with a steroid-like structure. Aphidicolin specifically inhibits eukaryotic DNA replication. Polymerases β and γ show low sensitivity to aphidicolin.

The large size and multi-subunit nature of eukaryotic DNA polymerases has made their structural elucidation more challenging than that for bacterial and phage polymerases. However, the structure of human pol γ was recently described and is of considerable interest. As shown in Table 25.2, the enzyme consists of a 137-kDa catalytic subunit (pol γA) and a 55-kDa accessory subunit (pol γB). Not shown is the fact that the holoenzyme contains two molecules of γB, each of which binds differently to pol γA. Each of the accessory subunits increases polymerase processivity, but by different mechanisms—one increasing affinity of γA for DNA and the other stimulating the rate of catalysis.

As shown in Figure 25.35, human pol γ fairly closely resembles T7 bacteriophage DNA polymerase. T7 pol also has a processivity-enhancing subunit, which is a protein of the host cell, thioredoxin, playing a function quite distinct from its role in redox reactions. Like other polymerases, both enzymes shown in Figure 25.35 contain a thumb domain (shown in green), palm (red), and fingers (blue). The pol γA structure shows the sites of several mutations known to be responsible for human genetic diseases. For example, the W748S mutation is associated with a condition called Alpers syndrome, which causes both cerebral cortical atrophy and liver failure in children.

Of particular interest is the fact that pol γ is responsible for the toxicity of several antiviral drugs used in HIV therapy. As pointed out in Chapter 22, nucleoside analogs such as dideoxycytidine act by conversion to their 5′ triphosphates, which interfere with HIV reverse transcriptase. Several of these analogs also interfere with mitochondrial DNA replication, by inhibiting pol γ. The structure of this enzyme provides information that should be useful in designing analogs with greater selectivity toward the viral polymerase.

Other Eukaryotic Replication Proteins

Much of what we know about the enzymology of *E. coli* DNA replication came from studies in which the replication of small bacteriophage genomes was followed in bacterial extracts or in purified protein systems. The rationale was to use as the replication template a small genome, which makes possible the isolation and biophysical characterization of replicative intermediates. In the same sense, much of what we know about proteins in human DNA replication comes from similar in vitro systems using the circular duplex DNA from the tumor virus SV40 as the replication template. These studies, plus earlier work on the phage T4 system and later work with yeast, reveal striking uniformity in the kinds of proteins at replication forks and the biochemical

FIGURE 25.34

The *E. coli* replisome. Note that the clamp loader, *via* the τ protein, serves as a link to dimerize the pol III holoenzyme.

Courtesy of Dr. Michael O'Donnell.

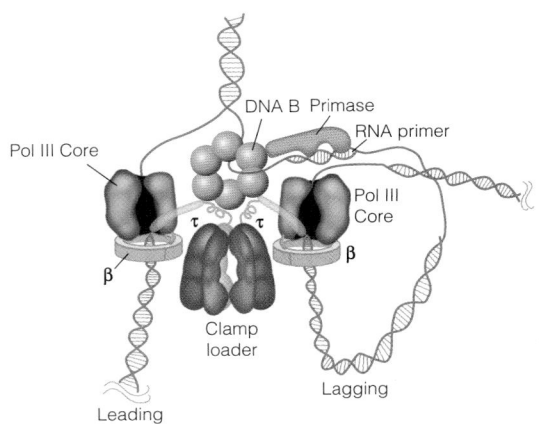

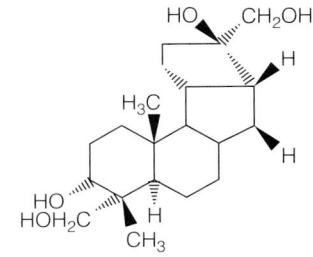

Aphidicolin

	Catalytic Subunit, kDa	Accessory Subunits, kDa	3′ Exonuclease	Fidelity	Processivity (with PCNA)	Biological Function
Pol α	160	49, 58, 70	No	$10^{-4} - 10^{-5}$	Moderate	Lagging strand primer synthesis
Pol β	37	None	No	5×10^{-4}	Low	DNA repair
Pol γ	137	55	Yes	10^{-5}	High	Mitochondrial DNA replication
Pol δ	122	12, 50, 68	Yes	$10^{-5} - 10^{-6}$	High	Lagging strand replication
Pol ε	251	12, 17, 59	Yes	$10^{-6} - 10^{-7}$	High	Leading strand replication

TABLE 25.2 Properties of the five "classical" eukaryotic DNA polymerases

functions of each, as shown in Table 25.3. One interesting difference is the requirement for two enzymes in eukaryotes for RNA primer removal— the 3′ exonuclease activity of Pol δ and the 5′ "flap" endonuclease activity of FEN 1 (flap endonuclease).

Replication of Chromatin

Eukaryotic cells face a DNA replication problem we have not encountered in our discussion so far focused upon bacterial and bacteriophage systems, namely, how the replisome deals with chromatin. Chromatin must be dismantled in advance of replication forks and reassembled on daughter DNA strands after a fork has passed through. And, as we discuss in later chapters, the distribution of core particles along DNA, and their modification patterns, are far from random, implying a higher-level control in the reassembly process. What we do know is summarized in

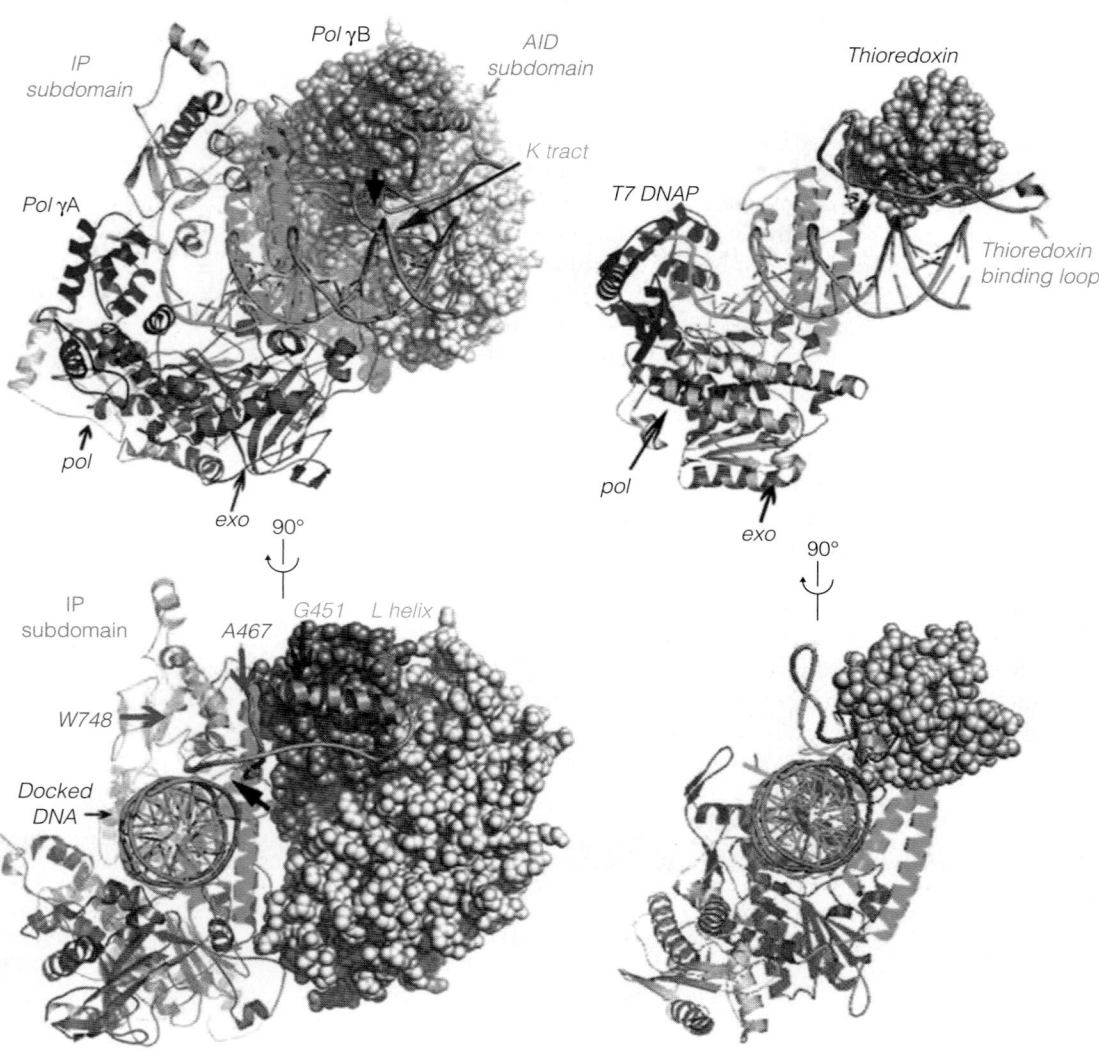

FIGURE 25.35

Structures of human DNA polymerase γ (left) and T7 phage DNA polymerase (right) holoenzymes. Both enzymes are shown complexed with DNA. The pol γ structure (PDB ID 3ILK and 3IKM) shows two unique domains (IP and AID, shown in gold), which are involved in DNA binding. Note the similarity in location of the polymerase and 3′ exonuclease catalytic domains. T7 DNA polymerase PDB ID 1T7P.

Reprinted from *Cell* 139:312–314, Y. S. Lee, W. D. Kennedy, and Y. W. Yin, Structural insight into processive human mitochondrial DNA synthesis and disease-related polymerase mutations. © 2009, with permission from Elsevier.

TABLE 25.3   Proteins that carry out analogous functions in DNA replication

Function	E. coli	Phage T4	SV40/Human	Yeast
DNA polymerase	Pol III core enzyme	gp43	Pol δ, Pol ε	Pol δ, Pol ε
Primase	DnaG	gp61	Pol α	Pol α
Helicase	DnaB	gp41	SV40 T antigen	MCM proteins
Proofreading	ε subunit of Pol III holoenzyme	gp43	Pol δ	Pol δ, Pol ε 3' exonuclease
Sliding clamp	β subunit of Pol III holoenzyme	gp45	PCNA	PCNA
Clamp loader	γ complex of Pol III holoenzyme	gp44/62	Replication factor C	Replication factor C
Single-strand DNA-binding protein	SSB	gp32	Replication protein A	Replication protein A
RNA primer removal	Pol I, RNase H	E. coli Pol I T4 RNase H	Pol δ, FEN 1	Pol δ, FEN 1

Figure 25.36. Nucleosomes are disassembled ahead of the replication fork and then reassembled on both daughter strands. Both preexisting and newly synthesized histones are used in the new nucleosomes, evidently with random deposition on the daughter strands. There is mixing of old and new histone molecules on the daughter strands, but the process is not entirely random. $(H3/H4)_2$ tetramers tend to remain intact, as do H2A/H2B dimers. This result conforms with results of in vitro studies; $(H3/H4)_2$ tetramers and H2A/H2B dimers are stable when released from nucleosomes, but the octamer is not.

The much more important question of how the precise arrangement of nucleosomes and nonhistone proteins is re-established after replication is still not understood. In fact, there may be changes in this arrangement in some instances. For example, in yeast newly synthesized histone H3 specifically undergoes acetylation of Lys-56, and this is important for nucleosome assembly and genome stability during replication. Such changes could account for the observation that differentiation of cells in embryonic development usually occurs at the time of cell division. However, such changes must be *programmed* ones, so their existence only makes the problem more complex.

Recent work indicates that the preservation of specific histone modifications after replication is perturbed under conditions of replication stress—for example, damage to the DNA template or depletion of DNA precursor pools. This can lead to epigenetic changes in gene expression, and it has been suggested that such changes could be partly responsible for oncogenic transformation.

Initiation of DNA Replication

When we ask how DNA replication is initiated from an origin, we are asking three interrelated questions: (1) What are the site-specific DNA–protein interactions that trigger initiation? (2) How do proteins act after binding to origin sequences? (3) How is the process controlled? Initiation seems to be the major target for control of replication. However, we know much less about the initiation of replication than we know about the initiation of transcription or translation. It is evident that intracellular contacts, of a still undefined type, link the replication apparatus to other cellular structures so that DNA replication is coordinated with the cell cycle. In prokaryotes it is likely that replication occurs at a site attached to the cell membrane, whereas eukaryotic DNA replication may occur at a DNA-and-protein-containing structure called the nuclear matrix (see Chapter 24, page 1025). The nature and significance of these physical linkages have remained elusive. Most of the events in

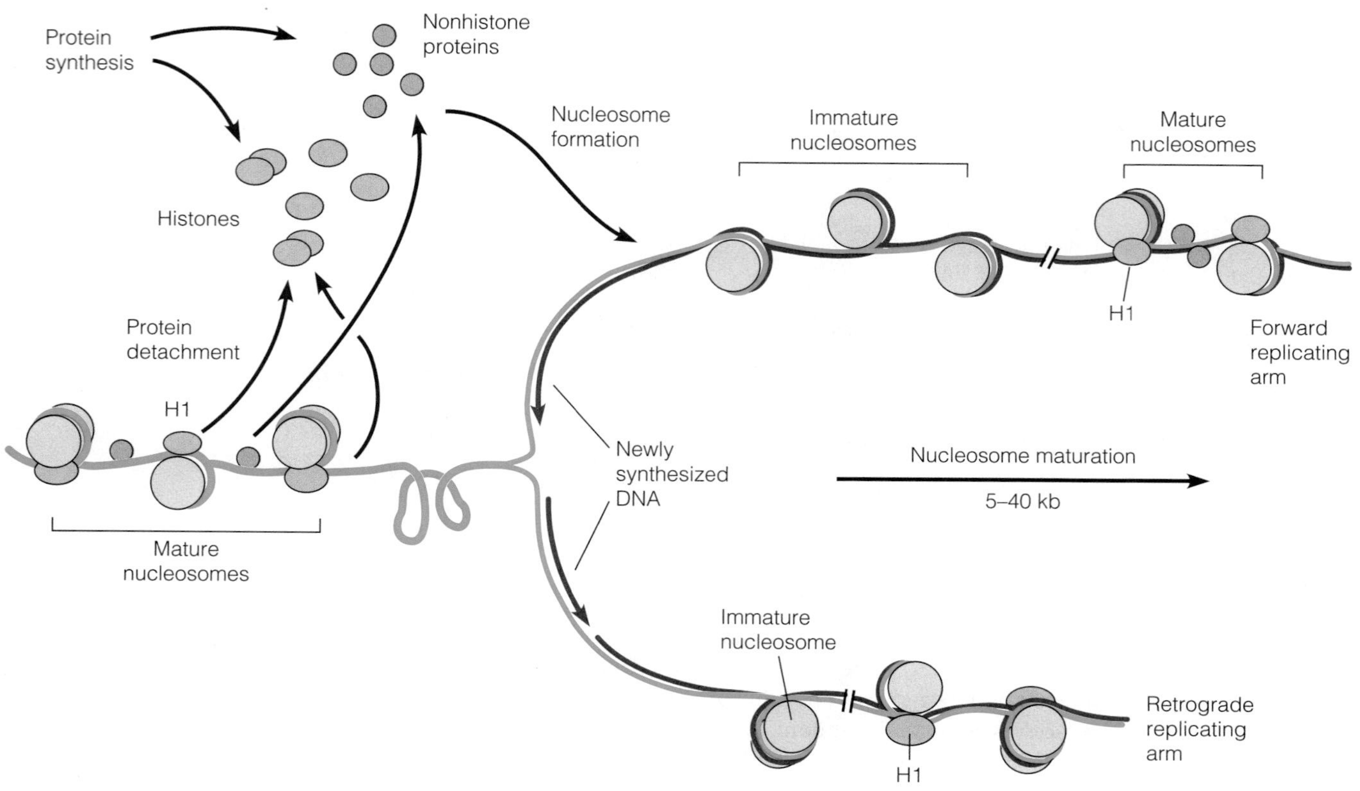

FIGURE 25.36

Model for chromatin replication. Nucleosomes on parental DNA are dissociated as the replication fork approaches and are reformed on newly synthesized daughter strands, with both old and newly synthesized histones being used. Maturation occurs slowly, with full organization not being re-established until many kilobases behind the moving fork. For simplicity this figure does not show proteins known to be active at the replication fork.

replication have been duplicated in soluble cell-free systems, without components of the membrane or nuclear matrix. Therefore, it has been difficult to define a biochemical role for the membrane or matrix in replication.

Requirements for Initiation of Replication

Because replication proceeds from fixed origins, there are two requirements for initiation: a nucleotide sequence that specifically binds initiation proteins, and a mechanism that generates a primer terminus to which nucleotides can be added by DNA polymerase. A number of phage, bacterial, plasmid, and organelle replication origins have been isolated by gene cloning, and their nucleotide sequences have been determined. In general, these origins include repeated sequences of either identical or opposite polarity (**direct repeats** or **inverted repeats**, respectively). This finding suggests that initiation proteins bind in multiple copies.

The two most straightforward ways to generate a primer terminus at the origin are first, nicking a strand of the parental duplex to expose a 3′ hydroxyl terminus and second, unwinding the parental duplex and synthesizing an RNA primer to expose a 3′ hydroxyl *ribonucleotide* terminus. Phages ϕX174 and G4, which replicate their single-stranded genomes *via* a circular duplex replicative intermediate, initiate the conversion of that intermediate DNA to single-strand progeny DNA by nicking with a site-specific endonuclease, the viral *cisA* gene product. By contrast, duplex DNA replication, where studied, occurs without nicking of the parental duplex and with the synthesis of RNA primers.

Initiation of *E. coli* DNA Replication at *ori^c*

Initiation of replication of the *E. coli* chromosome is reasonably well understood because the origin sequence has been cloned into plasmids whose replication from the origin can then be studied in vitro. This origin sequence, called *ori^c*, is 245 base pairs long. It contains four repeats of a 9-base-pair sequence that binds an initiation protein, the *dnaA* gene product. To the left of these sites, as shown in

Figure 25.37, are three direct repeats of a 13-base-pair sequence that is rich in A and T and thus is relatively easily denatured. The sequence also contains binding sites for several basic proteins (HU and IHF) that facilitate DNA bending, an important step in the sequence leading to initiation.

Step 1 is binding of 10 to 20 molecules of a complex of DnaA protein and ATP. The protein is activated for this step by reacting with the phospholipid cardiolipin, a process that may represent part of the coordination between DNA replication, membrane growth, and chromosome partitioning at cell division. Binding of DnaA, plus the basic proteins, bends the DNA rather sharply and creates negative superhelical tension. In turn this tension causes DNA unwinding in the 13-base-pair regions, opening up a short single-strand loop. Aided by DnaC, another initiation protein, the DnaB helicase in step 2 binds in both forks of this loop, and the helicase activity further unwinds this structure. In step 3, and subsequent steps (not shown), DnaG primase binds, and RNA primers are formed. Some of the first primers may be synthesized also by RNA polymerase, the transcription enzyme. However formed, the RNA primers are extended by DNA

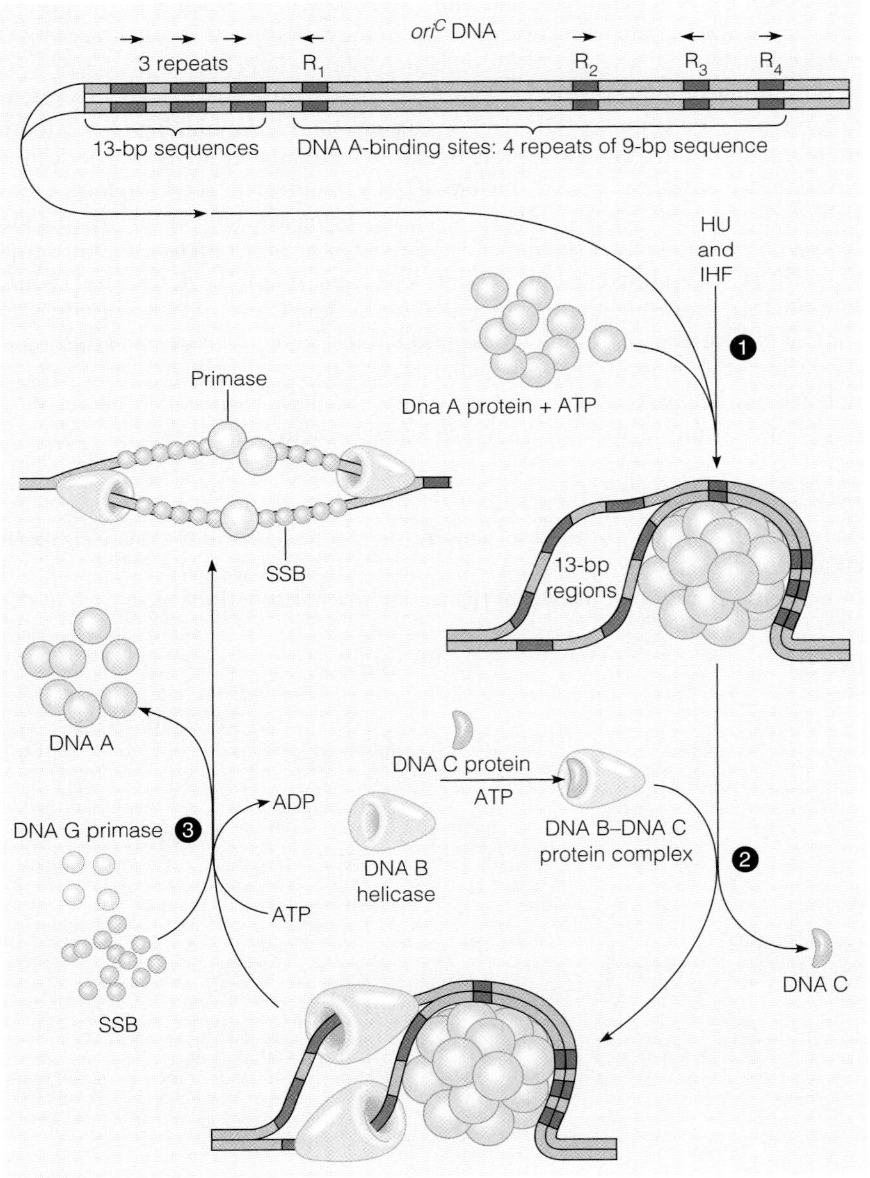

Protein binding, causing DNA to bend at the origin, initiates duplex DNA replication. Stress from bending causes nearby DNA to unwind, and primosomes assemble in the forks, forming RNA primers that are extended by DNA polymerases.

FIGURE 25.37

A model for initiation of *E. coli* DNA replication at *ori^c*. HU and IHF are double-strand DNA-binding proteins that facilitate DNA bending at the origin.

Redrawn from *Annual Review of Genetics*. 26:447–477, T. A. Baker and S. H. Wickner, Genetics and enzymology of DNA replication in *Escherichia coli*. © 1992 Annual Reviews.

polymerase III on both leading and lagging strands, and the two forks in the initiation complex mature, with a leading and a lagging strand in each fork. This may represent a general mechanism for initiation of chromosomal DNA replication.

Initiation of Eukaryotic Replication

With some exceptions prokaryotic organisms have a single origin of replication on a single chromosome. Far more complex is a typical eukaryotic cell, with thousands of replication origins distributed over multiple chromosomes, which "fire" at different times within S phase, and with not all origins firing within a single cell cycle. Moreover, there is the problem of "licensing" each origin so that it fires not more than once per cell cycle, to assure that all of the genome is replicated once and only once.

Early insight into yeast replication origins came from the identification of *autonomously replicating sequences* (ARS's), which are sequences essential for replication of plasmids after introduction into yeast cells. An ARS is typically several hundred base pairs in length, with subsequences carrying copies of the 11-base consensus sequence 5′ TTTTATATTTT 3′. The AT-rich composition suggests that, as with *E. coli ori^c*, initiation involves strand unwinding at the origin.

The complexity of eukaryotic initiation can be seen by considering the number of proteins involved in preparing origins to fire in the fission yeast, *Schizosaccharomyces pombe*, and the time needed for this process. As shown in Figure 25.38, this process begins in mitosis, hours before the onset of S phase, with binding of a six-protein origin replication complex (ORC). This prepares the site for binding of two more proteins, Cdc18 and Cdt1. These proteins in turn recruit the six Mcm proteins, which constitute a putative helicase for unwinding DNA strands at the origin, to form a pre-replicative complex (pre-RC). Licensing occurs during assembly of the pre-RC; the biochemistry of this step is still undefined. Several more initiation factors bind, some of which are phosphorylated by a cyclin-dependent kinase (CDK) or Hsk1-Dfp1. Phosphorylation is thought to be required for loading Csc45 onto the origin. This completes formation of the pre-initiation complex (Pre-IC). This prepares the origin for loading of primase and DNA polymerase, which occur in S phase, when replication actually begins. Once primase and polymerase have been loaded, all pre-RC's must be disassembled, to ensure that each licensed origin fires only once.

Mitochondrial DNA Replication

As noted earlier, the mitochondrial genome in mammalian cells is a relatively small circular duplex (16,569 bp in humans), which, unlike chromosomal DNA, should be amenable to rigorous analysis of replication mechanisms. Indeed, as early as the

FIGURE 25.38

Preparing a fission yeast replication origin for initiation. Steeper steps in this process represent sites of regulation. At each step, proteins that were bound in earlier steps are shown as fainter images. For further details see text.

Reprinted from *Cell* 136:812–814, E. Boye and B. Grallert, In DNA replication, the early bird catches the worm. © 2009, with permission from Elsevier.

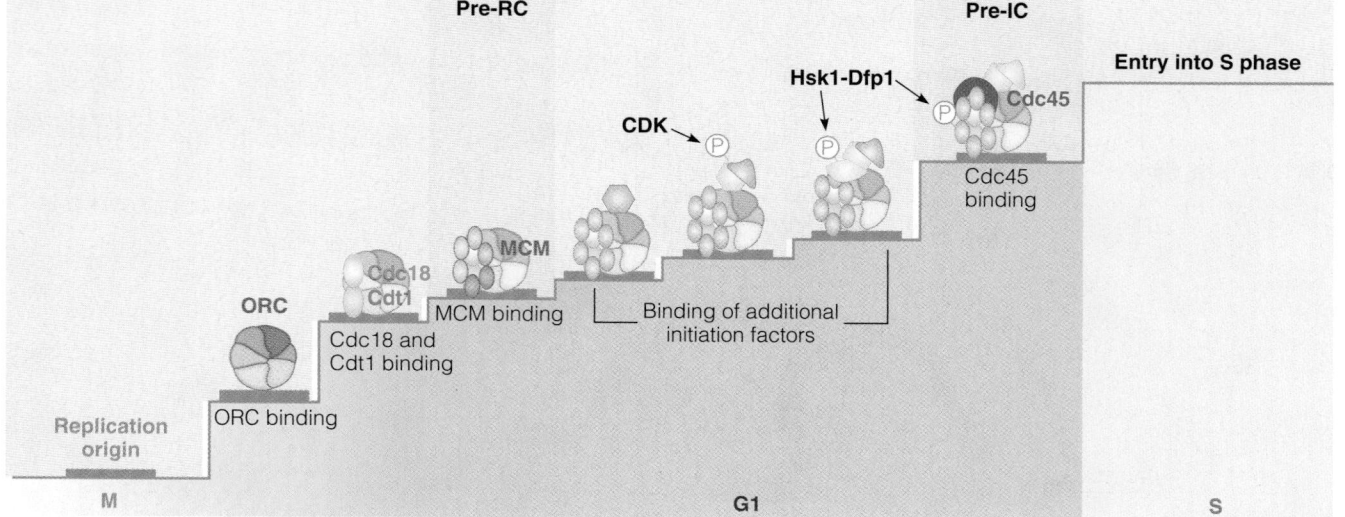

1970s purified intermediates in mtDNA replication were analyzed by electron microscopy. These studies supported an unusual asymmetric process, the strand displacement model, shown in Figure 25.39. In this model replication begins from a fixed origin on the L, or light, strand. However, only one strand is copied, with the other being displaced as single-strand DNA to form a D-loop. When the replication fork has moved about two thirds around the genome, an origin is exposed on the displaced H strand, and unidirectional L-strand synthesis begins backward from the direction of H strand synthesis, which continues. The entire process requires about one hour, comparable to the relatively slow rate of nuclear DNA replication in eukaryotic cells.

In 2002 the strand displacement model was challenged on the basis of analysis of mtDNA replication intermediates in two-dimensional agarose gel electrophoresis (Tools of Biochemistry 25A), introducing controversy into the field. More recently it was reported that *RNA* is extensively incorporated throughout the lagging strand during mtDNA replication. This RNA might well have been lost during purification of replicating mtDNA in the early electron microscopic analyses. Whether this represents primers for lagging strand replication remains to be seen, as does the possibility that mtDNA replicates by quite different mechanisms under different conditions.

Replication of Linear Genomes

Thus far we have discussed in detail only the replication of circular DNA genomes. Linear genomes, including those of several viruses as well as the chromosomes of eukaryotic cells, face a special problem—how to complete replication of the lagging strand (Figure 25.40). Excision of an RNA primer from the 5′ end of a linear molecule would leave a gap that cannot be filled by DNA polymerase action because of the absence of a primer terminus to extend. If this DNA could not be replicated, the chromosome would shorten a bit with each round of replication.

Linear Virus Genome Replication

Viruses reveal at least three strategies for dealing with this problem. Phages T4 and T7 exhibit **terminal redundancy**, the duplication of a small part of the genome at each end of the chromosome. Thus, recombination can occur at the incompletely replicated ends of two nascent DNA molecules without loss of genetic information (Figure 25.41a). This process is repeated in subsequent rounds of replication until the end-to-end linear aggregate (called a **concatemer**) is more than 20 times the length of a single phage chromosome. A virally specified nuclease then cuts this giant DNA into genome-length pieces when it is packaged into phage heads.

Bacteriophage ϕ29 and the adenoviruses have evolved a different strategy. The genomes of these viruses contain inverted repeat sequences at the ends. Replication begins at one end of the linear duplex, with a protein called **terminal**

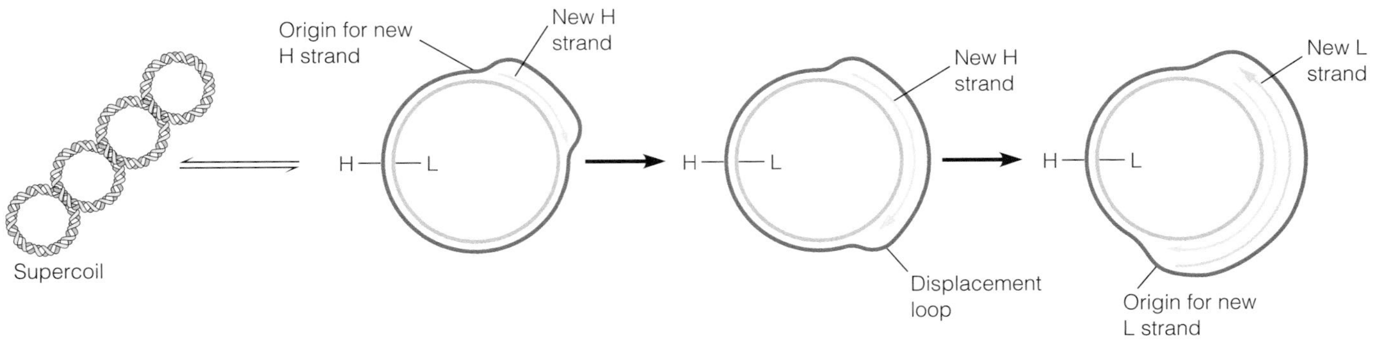

FIGURE 25.39

Strand displacement model for mitochondrial DNA replication. Parental heavy (H) and light (L) chains are shown in brown. New strands are in blue and pink.

Supercoil   Origin for new H strand   New H strand   New H strand   Displacement loop   New L strand   Origin for new L strand

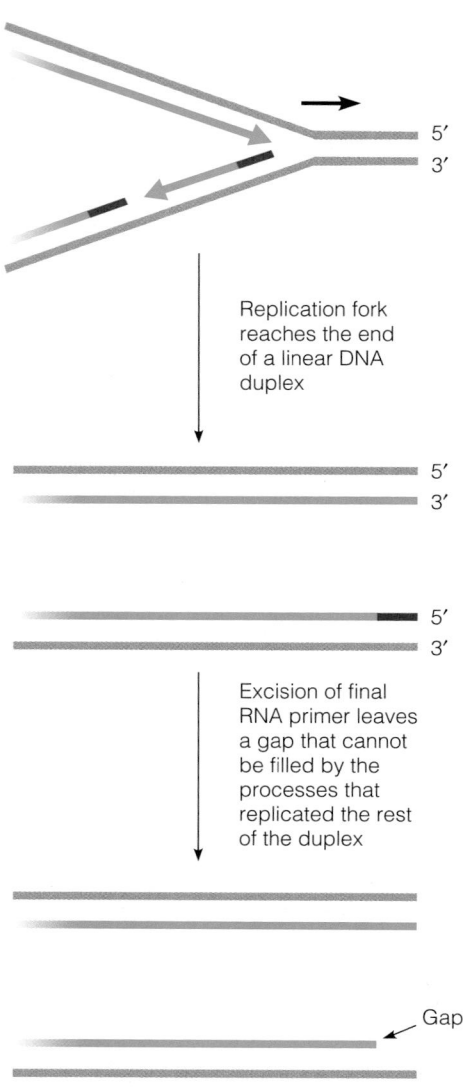

Replication fork reaches the end of a linear DNA duplex

Excision of final RNA primer leaves a gap that cannot be filled by the processes that replicated the rest of the duplex

Gap

FIGURE 25.40

The problem of completing the 5′ end in copying a linear DNA molecule.

Telomerase adds repeated short DNA segments to the ends of chromosomes.

protein serving as the primer. This protein in adenovirus reacts with dCTP to form a dCMP residue covalently linked through its phosphate to a serine residue (Figure 25.41b). The dCMP serves as primer for replication of the 3′-terminated strand, with the 5′ end being displaced as a single strand. Still another mechanism is involved in the replication of poxviruses, such as vaccinia. The two strands at each end of this linear genome are covalently linked together. Such a structure, when approached by a replication fork, can support movement of the leading strand *around* the link between the strands so that the final lagging strand primer can be replaced by DNA as in a circular duplex.

Telomerase

For the linear DNA molecules in eukaryotes the end replication problem has been solved by the addition of **telomeres** at the ends of each chromosome. As mentioned in Chapter 24, telomeric DNA consists of simple tandemly repeated sequences like those shown in Table 25.4. Typically, one strand is G-rich, the other C-rich. The G-rich strand forms a 3′-terminal overhang, typically about 15 residues in length. These sequences are repeatedly added to the 3′ termini of chromosomal DNAs by enzymes called *telomerases* (Figure 25.42). This elongation allows room for a primer to bind and initiate lagging-strand synthesis on the other strand, maintaining the approximate length of the chromosome and preventing the loss of coding sequences.

Note that the telomerase must add nucleotides without the use of a DNA primer. This is accomplished through the existence, in each telomerase molecule, of an essential RNA oligonucleotide that is complementary to the telomeric sequence being synthesized, and thus acts as a template. It has been speculated that telomerase is an evolutionary relic of a ribozyme that once served to catalyze DNA synthesis, a process long ago taken over by wholly protein polymerases. Because of its RNA-templated DNA synthesis, the protein portion of the enzyme, without its associated RNA, is called more descriptively TERT (Telomerase Reverse Transcriptase).

In 2008 the crystal structure of the catalytic subunit TERT was determined (Figure 25.43). Modeling of DNA and the RNA component into a large cleft in the protein shows a close fit, consistent with the mechanism outlined in Figure 25.42.

In recent years, it has become clear that telomeres and telomerase have wide-ranging significance in addition to their practical role in preventing chromosome shortening. First, it is possible that the G-rich strands typical of telomeres may aid in chromosome pairing by forming four-strand structures called G-quadruplexes (see Chapter 4, page 117). Secondary structure of this type has been observed many times in vitro with oligo-G, and it is known that a specific telomere-binding protein favors the formation of such structures. Second, there is a strong correlation between aging, cell senescence, and low levels of telomerase. Telomeres shorten with age, both in organisms and in cultured cells. Once telomeres disappear, chromosome ends are unprotected, and deleterious consequences may occur, such as end-to-end linking of two chromosomes.

Relevant to the relationship between telomerase and aging, cells in culture can be "immortalized" by introduction of active telomerase genes. These observations, together with the discovery that malignant tumor cells invariably have high levels of telomerase, have spurred intense interest in telomerase inhibition as a possible cancer therapy. The challenge, of course, is learning how to inhibit telomerase specifically in cancer cells.

Fidelity of DNA Replication

DNA replication is the most accurate of all known enzyme-catalyzed reactions, all the more remarkable given the speed of the reaction—approaching 1000 nucleotides incorporated per second in bacteria. From spontaneous mutation frequencies, which for all positions in a particular gene amount to about 10^{-6} per generation, we can estimate the chance that a particular nucleotide will be copied incorrectly at about 10^{-9} per base pair per round of replication. About two orders of magnitude of this specificity comes from **mismatch repair**, a process that recognizes and removes mismatched nucleotides and other aberrant structures, such as looped-out nucleotides (see Chapter 26). Still, this means that the DNA polymerase reaction *per se* can have a fidelity as high as 10^{-7} errors per nucleotide incorporated, and in fact, that is what is observed (see Table 25.2).

3' Exonucleolytic Proofreading

DNA polymerase accuracy is determined at both the nucleotide insertion step and the 3' exonucleolytic proofreading step. As mentioned earlier, the 3' exonuclease is positioned to recognize and remove mispaired nucleotides before the polymerase activity can add the next nucleotide. However, crystallographic analysis of DNA polymerases shows for most DNA polymerases that the exonuclease site is far from the polymerase site. In the Klenow fragment, for example (Figure 25.7), eight base pairs must be unwound to move the 3'-terminal nucleotide in the daughter strand from the polymerase site to the exonuclease site. How, then, is a mismatched nucleotide recognized? How is a high polymerization rate maintained if each newly incorporated nucleotide must be unwound from its template in order to be inspected?

The answer comes from steady-state kinetic analysis. Experiments of the kind shown in Figure 25.44 show that the K_M for chain extension from a mismatch is about one thousandfold higher than extension from a correctly matched terminus. At physiological dNTP concentrations this means that extension from a mismatch is very slow, and the delay in extension gives time for partial unwinding of the template-primer, a process that places the primer in the exonuclease site (Figure 25.45). Hence, the 3' exonuclease activity has no specificity for a mismatched nucleotide; rather, a mismatched nucleotide has a far higher probability of reaching that site than a matched nucleotide.

Polymerase Insertion Specificity

Implicit in the Watson–Crick model was the idea that the accuracy of DNA replication is governed by the hydrogen bonds that link A to T and G to C. However, the free energy of forming a Watson–Crick base pair differs from that for forming a mismatch (A-C or G-T) by only 4 to 13 kJ/mol. This corresponds to only a 100- to 1000-fold difference in binding energy between correct and incorrect base pairs. Therefore, if DNA polymerase were completely passive, only incorporating the nucleotide that spontaneously associates with a template base, errors would be made approximately 0.1% to 1% of the time, reflecting the relative abundance of correctly and incorrectly base-paired structures, error rates far higher than observed, even when proofreading is not occurring.

Pre-steady-state kinetic analysis of DNA polymerase reactions shows that dNTP binding to a polymerase–DNA complex precedes a major conformational change that must occur before "chemistry," that is, the formation and breakage of chemical bonds. A complete kinetic scheme for a single nucleotide addition is shown below.

Step 1
Step 2
dNTP
Step 3

$$E + DNA \underset{k_{-1}}{\overset{k_1}{\rightleftharpoons}} E{\cdot}DNA \underset{k_{-2}}{\overset{k_2}{\rightleftharpoons}} E{\cdot}DNA{\cdot}dNTP \underset{k_{-3}}{\overset{k_3}{\rightleftharpoons}} E^*{\cdot}DNA{\cdot}dNTP$$

$k_{-4} \big\Vert k_4$ **Step 4**

Step 5

PP_i

$$E{\cdot}DNA_{+1} \underset{k_{-6}}{\overset{k_6}{\rightleftharpoons}} E{\cdot}DNA_{+1}{\cdot}PP_i \underset{k_{-5}}{\overset{k_5}{\rightleftharpoons}} E^*{\cdot}DNA_{+1}{\cdot}PP_i$$

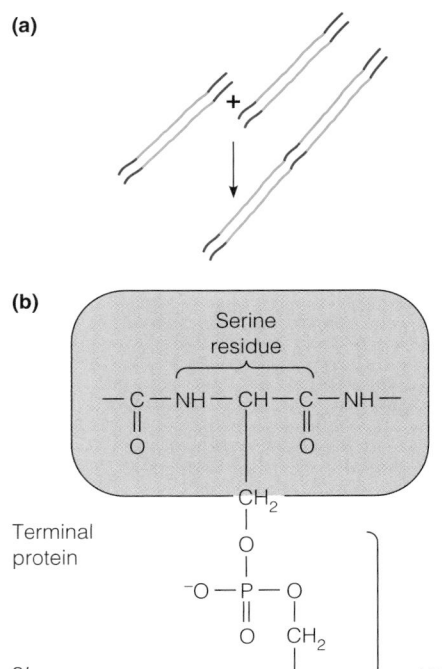

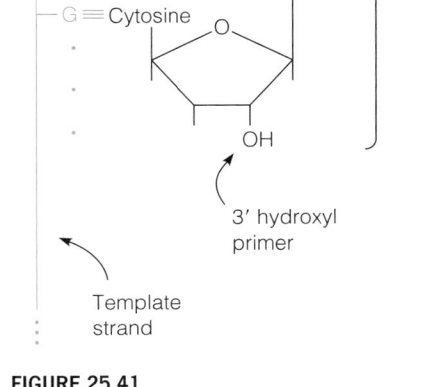

FIGURE 25.41

Strategies for replicating linear genomes. **(a)** Recombination between redundant chromosome ends. **(b)** Using a protein serine hydroxyl as a primer.

TABLE 25.4 Representative telomeric repeat sequences from several organisms	
Organism	**Repeat**[a]
Tetrahymena thermophila (protist)	TTGGGG
Saccharomyces cerevisiae (yeast)	$T(G)_{2-3}(TG)_{1-6}$[b]
Arabidopsis thaliana (plant)	TTTAGGG
Bombyx mori (silkworm)	TTAGG
Human	TTAGGG

[a]Written in $5' \rightarrow 3'$ direction.

[b]Yeasts are unusual in having somewhat variable telomeric repeats.

FIGURE 25.42

Extension of telomeric DNA by telomerase. **(a)** The overall reaction. Telomerase adds simple repeat sequences to the 3′ end of telomeric DNA, by the mechanism shown in part (b). Addition of an RNA primer allows lagging-strand synthesis, followed by ligation and RNA removal. **(b)** Proposed action of telomerase. The RNA carried by the telomerase matches the 3′ DNA end, and allows its extension. DNA loop formation then permits further extension. Dissociation of telomerase and its RNA follows after several rounds of extension.

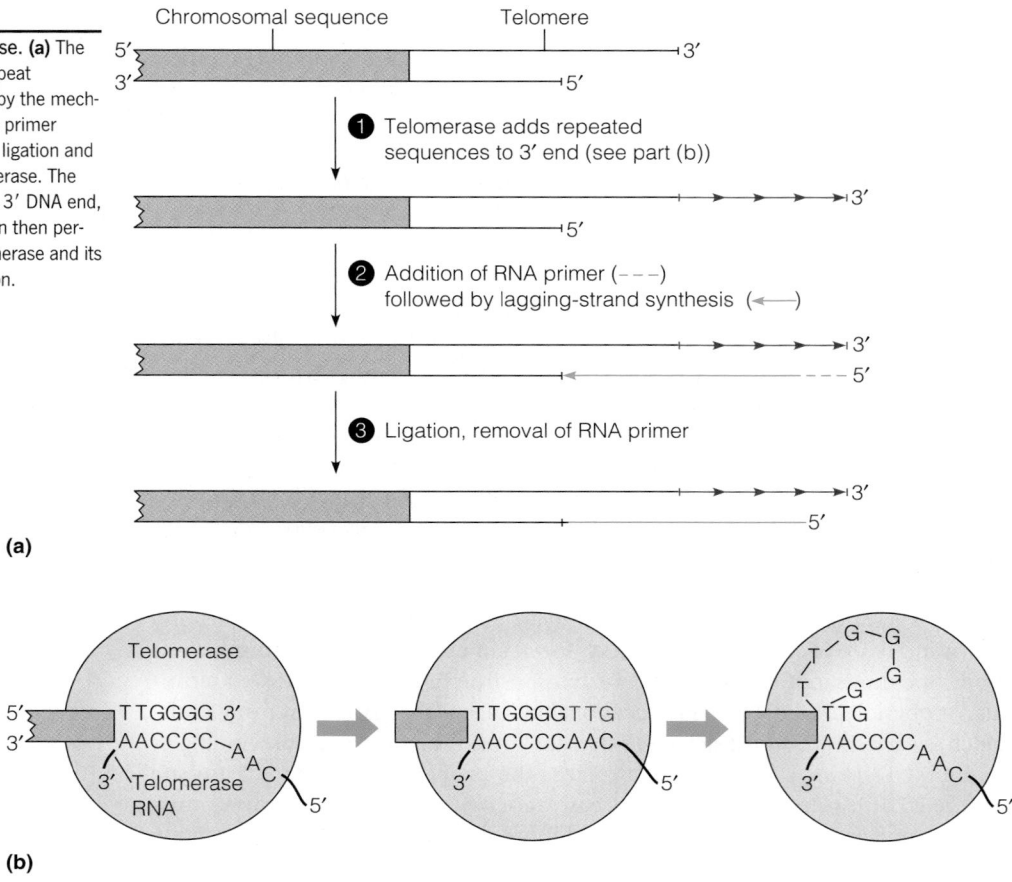

(a)

(b)

For a number of polymerases discrimination is shown to occur at both steps 2 (dNTP binding) and 3 (pre-chemistry conformational change). The chemistry step is not rate-limiting and does not contribute to discrimination. Evidence suggests that the conformational change occurring with a misaligned dNTP distorts catalytic residues on the enzyme so that dissociation of the misaligned dNTP becomes more likely.

Is hydrogen-bonding specificity sufficient to account for insertion specificity? This was tested by synthesis of dNTP analogs that are geometrically equivalent to natural dNTPs, but which lack hydrogen-bonding atoms. One example, a dTTP analog, is shown in the margin (next page). Although these analogs are generally poorer DNA polymerase substrates in terms of reaction rates, in many cases they show discrimination comparable to natural dNTPs; for example, the analog shown can compete with dTTP for incorporation opposite a dAMP residue in template DNA. Thus, specificity in DNA polymerase insertion reactions is seen to result from the shape of substrate molecules as well as their hydrogen bonding capacity. Base stacking interactions are probably involved as well.

Most attention in DNA replication fidelity has focused upon single-nucleotide substitution errors, in which one nucleotide is misincorporated. Less quantitative attention has been paid to other kinds of errors, including insertions and deletions. Of particular interest are errors resulting from oligonucleotide repeats along a template containing a short repeat sequence in tandem. A number of genetic diseases involve multiple tandem trinucleotide repeats within an affected gene. For example, Huntington's disease is an autosomal dominant neurological disease with a variable age of onset. In this condition a brain protein called **huntingtin** contains a stretch of consecutive glutamine residues, each encoded by 5′ CAG 3′. Normal individuals contain between 6 and 31 glutamines at this site, whereas affected individuals contain as many as 80 glutamines or

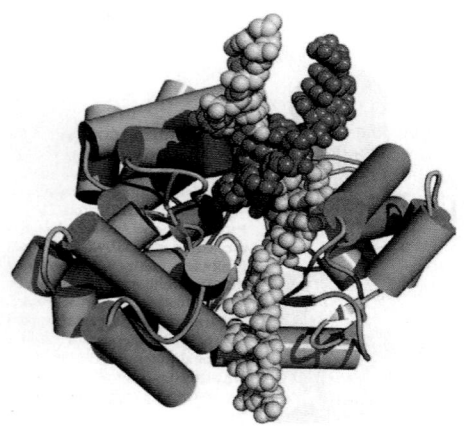

FIGURE 25.43

Crystal structure of telomerase catalytic subunit from the red flour beetle. DNA and telomerase RNA are modeled into the large central cleft. PDB ID 3du5.

Courtesy of Emmanuel Skordalakes, The Wistar Institute/UPENN.

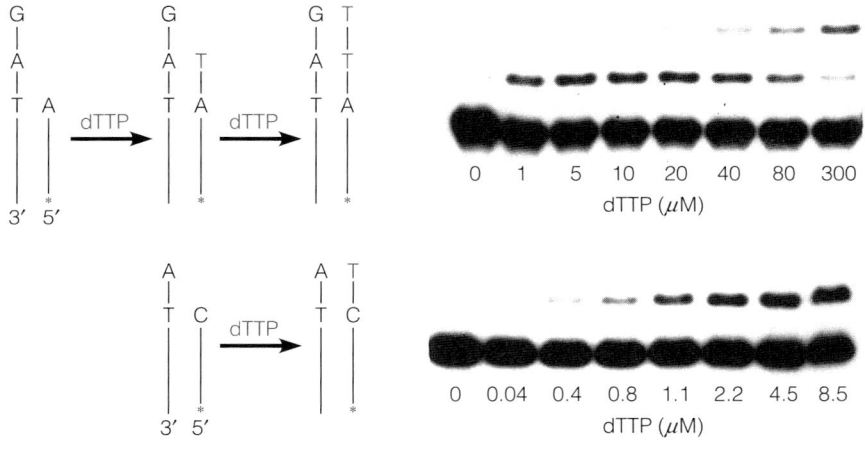

FIGURE 25.44

An in vitro system for studying the fidelity of DNA replication. The primer strand is radiolabeled at the 5′ end (marked with an asterisk) and is the most rapidly migrating radiolabeled species. The dTTP concentration dependence of extension from a correctly paired nucleotide (A in the A-T base pair) is shown at the top, and that of extension from a C-T mispair is shown at the bottom. dTTP was the only dNTP present during the incubations. In each case the products after incubation and denaturation were analyzed by SDS-polyacrylamide gel electrophoresis and autoradiography. The top panel shows that the rate of extension was maximal at $1\mu M$ dTTP and that further extension could be detected at higher dTTP levels. However, as shown in the bottom panel, dTTP concentrations in the millimolar range were required to see any chain extension from the mismatch.

Courtesy of Myron Goodman.

more. It seems likely that the number of repeats can increase if a product DNA strand undergoes slippage during an interval between successive nucleotide additions, forming a looped-out structure replication intermediate. In the example shown below, the product DNA strand (magenta) would have three more glutamine codons than the substrate.

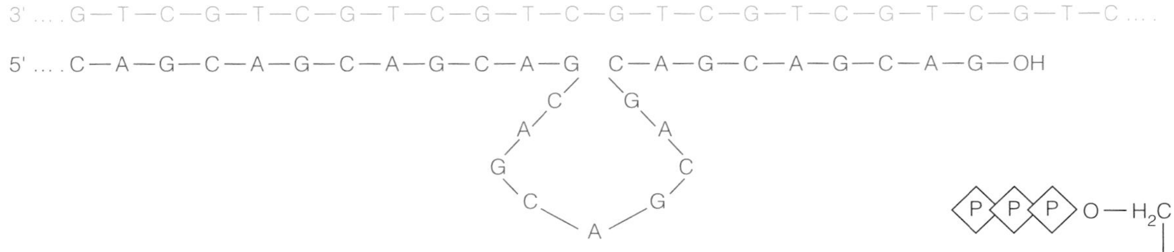

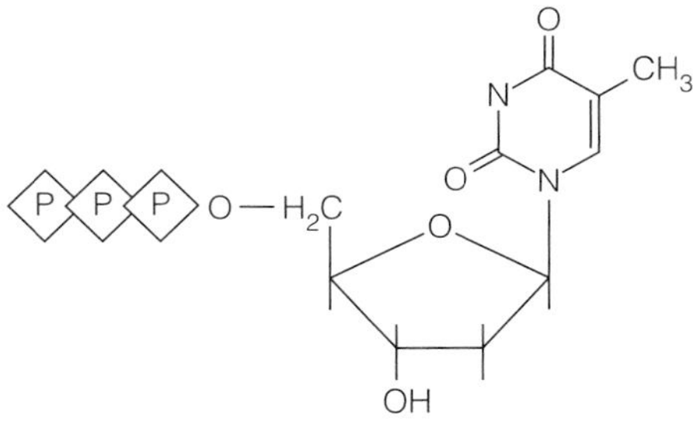

3′ .... G—T—C—G—T—C—G—T—C—G—T—C—G—T—C—G—T—C—G—T—C—G—T—C ....

5′ .... C—A—G—C—A—G—C—A—G—C—A—G   C—A—G—C—A—G—C—A—G—OH

In Huntington's disease the array of glutamine codons tends to become longer from one generation to the next, with a correspondingly earlier age of onset. The longer the array, the earlier the age of onset of this invariably fatal disease.

Until recently all investigations of DNA polymerase fidelity have focused upon incorporation of incorrect deoxyribonucleotides. However, in 2010 it was reported that yeast replicative DNA polymerases incorporate *ribo*nucleotides at significant rates. Given the discrimination ratios seen in vitro, plus the fact that intracellular ribonucleoside triphosphate concentrations are much higher than those of deoxyribonucleoside triphosphates, one can estimate that 10,000 ribonucleotide molecules might be incorporated into yeast DNA in each round of replication. The ribonucleotides are removed by an RNase H-dependent process, which is essential for maintenance of genome stability.

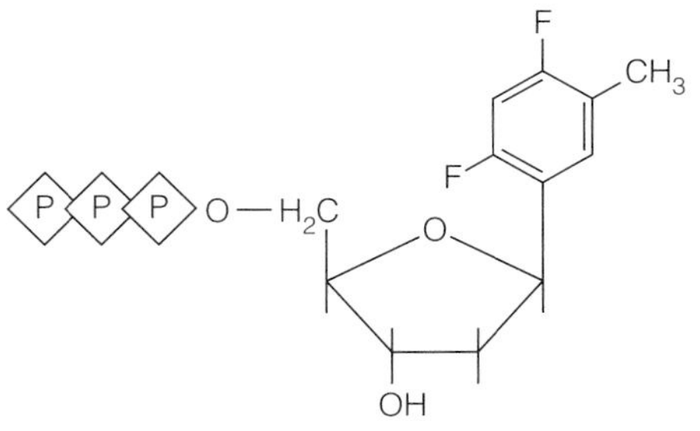

dTTP

Geometric analog of dTTP

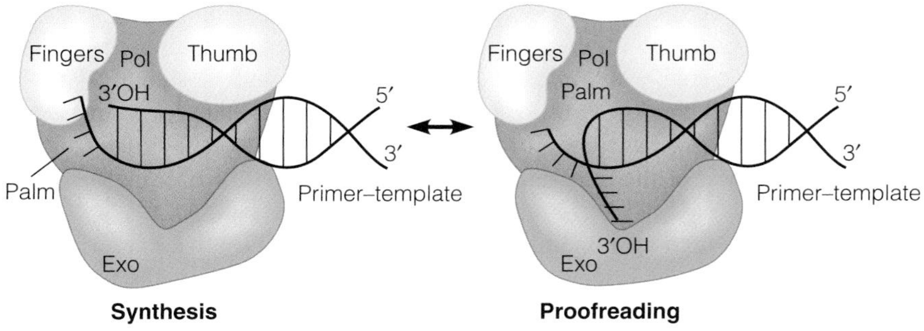

FIGURE 25.45

Kinetic basis for preferential excision of mismatched nucleotides by a 3′ exonuclease site distant from the polymerase site. The delay in extension from a mismatch allows for duplex unwinding at the primer terminus, placing the mismatched 3′ nucleotide in the exonuclease site.

Reprinted from *Cell* 92:295–305, T. A. Baker and S. P. Bell, Polymerases and the replisome: Machines within machines. © 1998, with permission from Elsevier.

Synthesis

Proofreading

RNA Viruses: The Replication of RNA Genomes

We conclude this chapter with a few words about the replication of viral genomes consisting of RNA. Virtually all known plant viruses contain RNA instead of DNA, as do several bacteriophages and many important animal viruses, including polio virus and influenza viruses. The **retroviruses**, which are responsible for many tumors and for acquired immune deficiency syndrome (AIDS), also contain RNA genomes.

RNA-Dependent RNA Replicases

Most RNA viruses contain a genome consisting of a single molecule of single-strand RNA. Usually that RNA is the "sense" strand for expression of genetic information at the level of translation. In other words, the RNA molecule that passes from the viral particle into the infected cell can serve directly as a messenger RNA, without requiring first the synthesis of complementary-strand RNA. One of the early products of translation of this input genome is the enzyme **replicase**, or RNA-dependent RNA polymerase. After complexing with required polypeptide subunits of host cell origin, this enzyme replicates the input RNA (plus strand), starting from the 3′ end. Thus, the new strand (minus strand) is laid down from its 5′ end to its 3′ end, the same direction in which DNA polymerases work. The newly replicated minus strands then serve as templates for synthesis of plus strands, which are packaged into progeny **virions**, or virus particles. More complex mechanisms come into play when the RNA genome is double-stranded or segmented (three or four separate RNA molecules) or when the virion RNA is itself the minus strand (viruses in this latter class are called **negative-strand viruses**).

The known RNA replicases lack proofreading activity, so it is no surprise that viral RNA replication is more error-prone than DNA replication, and RNA viruses undergo mutation and evolution far more rapidly than the organisms they infect. These characteristics are clearly related to viral pathogenesis, in part because a virus population infecting a plant or an animal can undergo change so rapidly that it can evade or counteract the host's defense mechanisms.

Replication of Retroviral Genomes

Different strategies for genome replication are involved in the action of retroviruses, so named because of the presence of a special enzyme, **reverse transcriptase**. In this class of viruses the single-strand RNA genome achieves latency—the ability to persist in a host cell for a long period without pathological effects—by making a DNA copy of itself and inserting that copy into the host cell genome. The DNA copy is made by reverse transcriptase, a multifunctional enzyme that is packaged in virions and enters the infected cell along with the viral genome. As shown in Figure 25.46, reverse transcriptase uses viral RNA as a template for synthesizing a complementary DNA strand, with a specific transfer RNA molecule serving as the primer (step 1). An RNase H activity of the enzyme then partially digests the RNA (step 2), and the structure circularizes by DNA–RNA base pairing (step 3). The nascent DNA chain is extended around the circle, and the tRNA primer is removed by RNase H activity (step 4). Strand displacement synthesis then occurs, with the RNA strand being displaced (step 5). The resultant double-strand DNA circle then recombines with a site on a chromosomal DNA and is linearly inserted into that DNA in the process. Under these conditions the integrated proviral genome can persist in a noninfectious state for many years, with most of its own genes turned off. Environmental stresses, still undetermined, can trigger excision of the integrated viral genome and return of the virus to an infectious state.

As discussed when we introduced azidothymidine (AZT) in Chapter 22, the reverse transcriptase of human immunodeficiency virus (HIV), the virus that causes AIDS, is an obvious target for antiviral therapy. Accordingly, intense

The rapid mutation rates of RNA viruses are largely due to RNA replication mechanisms that do not involve proofreading for mismatched nucleotides.

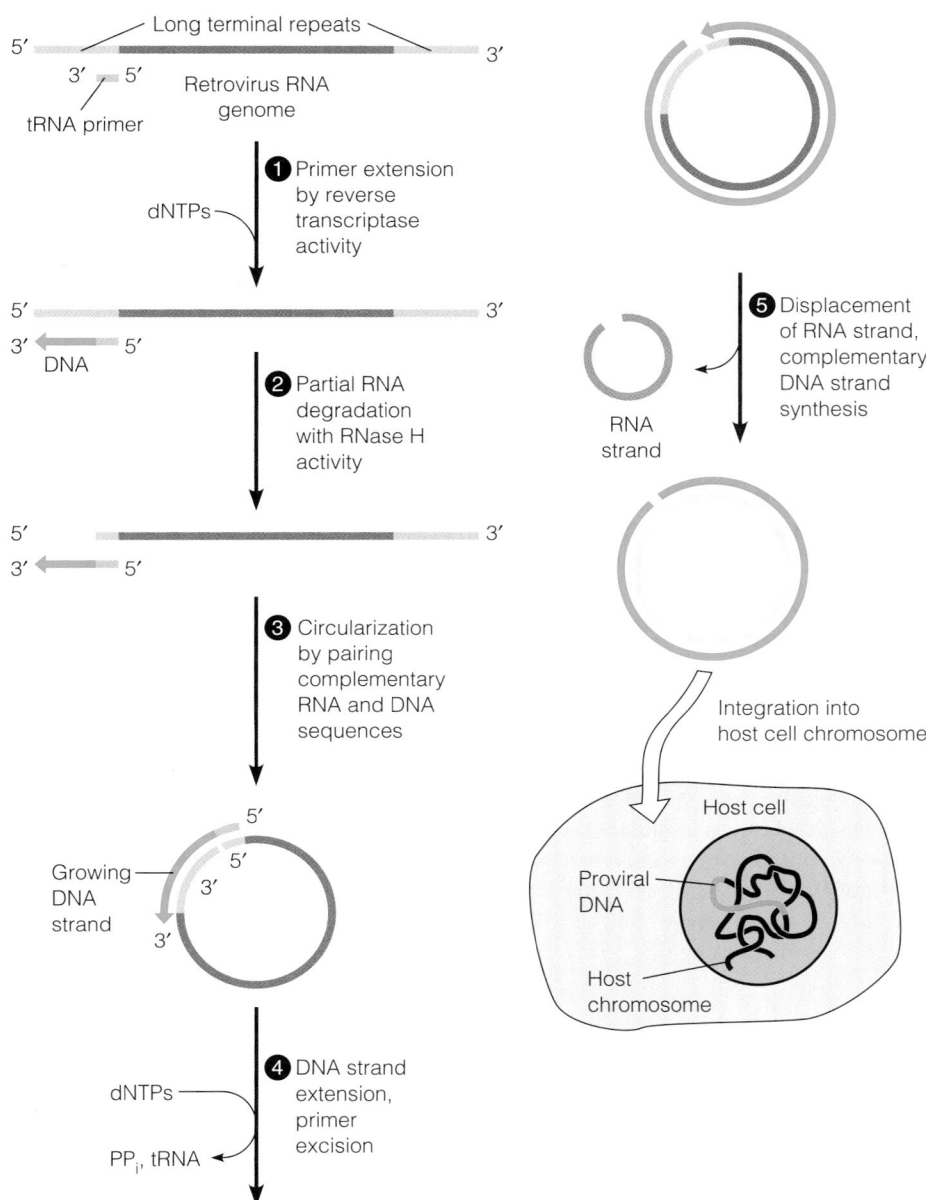

FIGURE 25.46

Simplified view of retrovirus life cycle. The RNA genome contains long terminal repeats, to one of which a tRNA molecule binds. Primer extension, partial RNA digestion, and circularization generate a substrate for extensive DNA synthesis. Ultimately, a circular duplex DNA molecule is formed, and it is the likely substrate for integration into a host chromosome. Evidence suggests that two viral RNA molecules must interact at step 2 or 3.

interest has focused on the structure of this enzyme, in order to design inhibitors based on the structure of the active site. The enzyme is a dimer, with one 51-kilodalton and one 66-kilodalton subunit. Both subunits are products of the same gene (see Figure 25.12). The smaller subunit, which results from proteolytic degradation of the larger subunit, collapses about the larger subunit, protecting it from degradation. As seen with DNA-dependent DNA polymerases, the large subunit forms a "hand," which grasps DNA. Analysis of the dimer structure shows the polymerase and RNase H sites to be about 20 nucleotides apart (Figure 25.47). This structural relationship coordinates the two activities, with insertion of a new nucleotide occurring concomitantly with excision of a ribonucleotide from the RNA strand of the hybrid. One of the great difficulties in devising vaccines against HIV derives from the absence of a proofreading exonuclease in HIV reverse transcriptase. This leads to frequent uncorrected replication errors and high rates of spontaneous mutagenesis, allowing the virus to generate variants that have resistance to antiviral antibodies produced via vaccination.

FIGURE 25.47

Structure of HIV reverse transcriptase. The enzyme was crystallized in the presence of a primer–template complex and a substrate analog (a dideoxyribonucleoside triphosphate) positioned at the 3′ primer terminus. The p66 subunit is shown in color, with the fingers in red, the palm in yellow, the thumb in orange, a connector domain in blue, and the RNase H domain in purple. The p51 subunit is in gray, and the nucleotide substrate (lower left) is in brown. PDB ID 1RTD.

From *Science* 282:1669–1675, H. Huang, R. Chopra, G. L. Verdine, and S. C. Harrison, Structure of a covalently trapped catalytic complex of HIV-1 reverse transcriptase: Implications for drug resistance. © 1998. Reprinted with permission from AAAS.

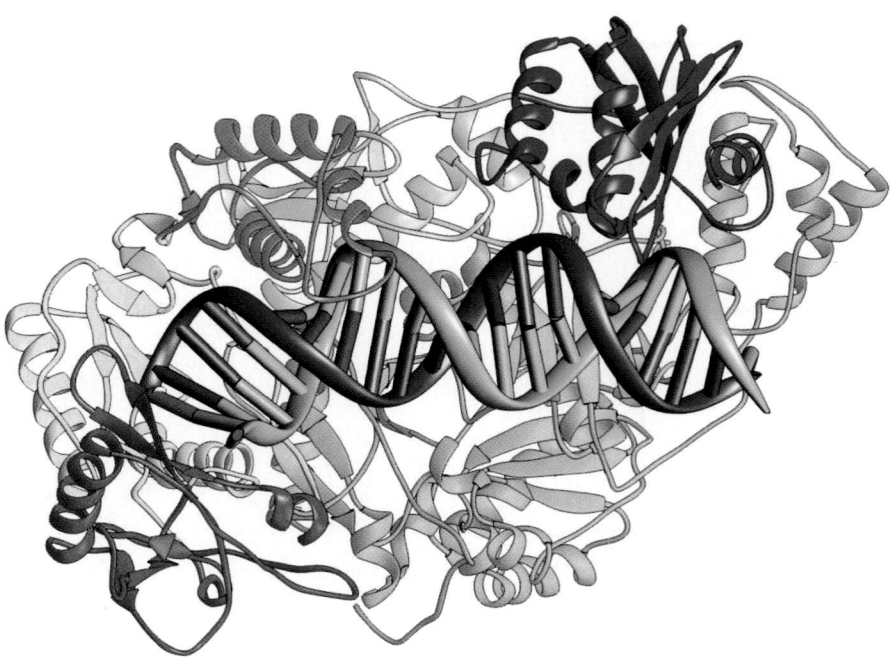

SUMMARY

DNA replication begins at fixed points on chromosomes. In most cases two replication forks are formed as a consequence of initiation. In each fork, parental DNA strands unwind, and each strand serves as template for DNA polymerase-catalyzed synthesis of a daughter strand. Both leading and lagging strands in a fork grow in a $5′ \rightarrow 3′$ direction, meaning that one strand must grow backward in short segments (Okazaki fragments) from the fork. That one strand, the lagging strand, elongates discontinuously, with each segment being initiated by a short RNA primer. Enzymes and proteins at the replisome, which propagates a replication fork, include DNA polymerase, which incorporates deoxyribonucleoside 5′-monophosphate moieties from dNTP substrates; sliding clamps, which enhance polymerase processivity; clamp loaders, which wrap the circular clamps about DNA and also carry out clamp unloading; helicases, which unwind parental DNA strands, using energy provided by ATP hydrolysis; primase, which synthesizes RNA primers; single-strand DNA-binding protein, which stabilizes DNA in the single-strand conformation to act as template and then facilitates renaturation with the nascent strands as deoxyribonucleotides are incorporated; and topoisomerases, which relieve torsional stress ahead of the fork and, in prokaryotes, introduce negative supercoiling into product DNA. Topoisomerases also decatenate (unlink) daughter DNA molecules at the end of a round of replication, thereby permitting chromosome segregation. Initiation of replication of duplex DNA involves binding of proteins to specific origin sequences, which leads to DNA bending and facilitates local DNA denaturation adjacent to the protein binding sites. This is followed by binding and activity of a helicase to further unwind parental DNA strands and then by primase to synthesize short RNA primers, which are extended by DNA polymerase. Replication of linear DNAs requires additional processes; in eukaryotic cells telomerase is responsible for copying chromosome ends, or telomeres. The great accuracy of DNA replication is maintained partly by substrate discrimination in the nucleotide insertion step and partly by a 3′ exonuclease activity that scans newly replicated DNA and excises mispaired 3′-terminal nucleotides. Replication of viral RNAs is far less accurate because of a lack of editing activity in RNA replicases.

REFERENCES

General

Bates, D. (2008) The bacterial replisome: Back on track? *Mol. Microbiol.* 69:1341–1348. A summary of evidence for spooling DNA through a stationary replisome.

Kornberg, A., and T. A. Baker (1992) *DNA Replication,* 2nd ed. W. H. Freeman and Co., San Francisco. Nearly two decades after its publication, still an excellent reference.

DNA Polymerases

Foti, J. J., and G. C. Walker (2010) SnapShot: DNA polymerases I prokaryotes, and SnapShot: DNA polymerases II mammals. *Cell* 141:192–193 and 370–371. Tabulated information about properties and functions of five prokaryotic and fourteen mammalian polymerases.

Johansson, E., and S. A. MacNeill (2010) The eukaryotic replicative DNA polymerases take shape. *Trends in Biochem. Sci.* 35:339–347. Current information about structures of eukaryotic polymerases.

Lee, Y.-S., W. D. Kennedy, and Y. W. Yin (2009) Structural insight into processive human mitochondrial DNA synthesis and disease-related polymerase mutations. *Cell* 139:312–324. Crystal structure of human DNA polymerase γ.

Loeb, L. A., and R. J. Monnat (2008) DNA polymerases and human disease. *Nature Rev. Genetics* 9:594–603. Excellent summary of the properties, fidelities, and functions of the multiple eukaryotic DNA polymerases.

Nick McElhinny, S. A., D. A. Gordenin, C. M. Stith, P. M. J. Burgers, and T. A. Kunkel (2008) Division of labor at the eukaryotic replication fork. *Molecular Cell* 30:137–144. Incisive experimental evidence establishing functions of Pol δ and Pol ε.

Other Replication Proteins

Bloom, L. B. (2009) Loading clamps for DNA replication and repair. *DNA Repair* 8:570–578. Mechanisms of clamp loading are discussed.

Donmez, I., and S. S. Patel (2006) Mechanisms of a ring shaped helicase. *Nucl. Ac. Res.* 34:4216–4224. The T7 gp4 helicase-primase may be the most thoroughly studied helicase from the standpoint of mechanism.

Froelich-Ammon, S. J., and N. Osheroff (1995) Topoisomerase poisons: Harnessing the dark side of enzyme mechanism. *J. Biol. Chem.* 270:21429–21432. A review of topoisomerases as targets for antimicrobial and anticancer drugs.

Indiani, C., and M. O'Donnell (2006) The replication clamp-loading machine at work in the three domains of life. *Nature Rev. Mol. Cell Biol.* 7:751–761. A complete discussion of clamp-loading complexes.

Karpel, R. L. (1990) T4 bacteriophage gene 32 protein. In *The Biology of Non-Specific DNA–Protein Interactions,* A. Rezvin, ed., pp. 103–130, CRC Press, Boca Raton, Fla. A detailed review of this important protein.

Langston, L. D., C. Indiani, and M. O'Donnell, (2009) Whither the replisome. *Cell Cycle* 8:2686–2691. A concise and readable minireview.

Lohman, T. M., K. Thorn, and R. D. Vale (1998) Staying on track: Common features of DNA helicases and molecular motors. *Cell* 93:9–12. Proteins that move along DNA are being likened, mechanically, to other proteins that cause physical movement within cells.

Nash, H. A. (1998) Topological nuts and bolts. *Science* 279:1490–1491. Commentary on two articles in the same issue of *Science* that describe the crystal structure and mechanism of human topoisomerase I.

Nelson, S. W., S. K. Perumal, and S. J. Benkovic (2009) Processive and unidirectional translocation of monomeric UvsW helicase on single-stranded DNA. *Biochemistry* 48:1036–1046. This paper provides entrée to the T4 phage system, which has also taught us much about the replisome.

Pulleyblank, D. E. (1997) Of topo and Maxwell's dream. *Science* 277: 648–649. Commentary on surprising findings reported in the same issue of *Science,* that type II topoisomerases have a mysterious ability to untangle DNA molecules, rather than achieving equilibrium topoisomerase distributions.

Simonetta, K. R., and eight coauthors (2009) The mechanism of ATP-dependent primer-template recognition by a clamp loader complex. *Cell* 137:659–671.

Wang, J. C. (2009) *Untangling the double helix: DNA entanglement and the action of the DNA topoisomerases.* Cold Spr. Hbr. Press, Cold Spring Harbor, NY. Up-to-date presentation of topoisomerases and DNA tertiary structure by the discoverer of topoisomerases.

Eukaryotic DNA Replication

Burgers, P. M. J. (2009) Polymerase dynamics at the eukaryotic replication fork. *J. Biol. Chem.* 284:4041–4045. An informative JBC minireview.

Cook, P. R. (1999) The organization of replication and transcription. *Science* 284:1790–1795. An engaging argument that DNA and RNA polymerases are organized within cells into multi-protein "factories" through which DNA templates are drawn.

De Lange, T. (2010) Telomere biology and DNA repair: Enemies with benefits. *FEBS Letters* 584:3673–3674. Introduction to a special issue on telomeres and telomerases.

Gilson, E., and V. Géli (2007) How telomeres are replicated. *Nature Rev. Mol. Cell Biol.* 8:825–838. A nice review of telomerase mechanisms.

Holt, I. J. (2009) Mitochondrial DNA replication and repair: All a flap. *Trends in Biochem. Sci.* 34:358–365. A summary of evidence for and against several models for mtDNA replication.

Jasencakova, Z., A. N. D. Scharf, K. Ask, A. Corpet, A. Imhof, G. Almouzni, and A. Groth (2010) Replication stress interferes with histone recycling and predisposition marking of new histones. *Mol. Cell* 37:736–743. A treatment of histone clearance and replacement as the replisome sweeps by.

Lowden, M. R., S. Flibotte, D. G. Moerman, and S. Ahmed (2011) DNA synthesis generates terminal duplications that seal end-to-end chromosome fusions. *Science* 332:468–471. Chromosomal aberrations that occur when telomerase activity is low or nonexistent.

Reyes-Lamothe, R., D. J. Sherratt, and M. C. Leake (2010) Stoichiometry and architecture of active DNA replication machinery in *Escherichia coli. Science* 328:498–501. Looking at the replisome in vivo with single-molecule approaches.

Sidorova, J. M. (2008) Roles of the Werner syndrome RecQ helicase in DNA replication. *DNA Repair* 7:1776–1786. Mechanistic study of the helicase that is defective in an important human disease.

Víglasky, V., L. Bauer, and K. Tlucková (2010) Structural features of intra- and intermolecular G-quadruplexes derived from telomeric repeats. *Biochemistry* 49:2110–2120. A biophysical analysis.

Initiation and Termination

Boye, E., and B. Grallert (2009) In DNA replication, the early bird catches the worm. *Cell* 136:812–814. A commentary, but in two pages a nice summary of the complexity of eukaryotic replication initiation.

Gilbert, D. M. (2010) Evaluating genome-scale approaches to eukaryotic DNA replication. *Nature Rev. Genet.* 11:673–684. Newer approaches to analyzing replication initiation in eukaryotic cells.

Mulcair, M. D., P. M. Schaefffer, A. J. Oakley, H. F. Cross, C. Neylon, T. M. Hill, and N. E. Dixon (2006) *Cell* 125:1309–1319. A molecular mousetrap determines polarity of termination of DNA replication in *E. coli.* An explanation for the polarity of termination.

Witz, G., and A. Stasiak (2010) DNA supercoiling and its role in DNA decatenation and unknotting. *Nucl. Aci. Res.* 38:2119–2133. Topological aspects of termination.

Fidelity of Nucleic Acid Synthesis

Goodman, M. F. (1997) Hydrogen bonding revisited: Geometric selection as a principal determinant of replication fidelity. *Proc. Natl. Acad. Sci. USA* 94:10493–10495. Commentary upon the research article, in the same issue of the journal, that described DNA synthesis carried out with a non-hydrogen-bonding dNTP analog.

Joyce, C. M., and S. J. Benkovic (2004) DNA polymerase fidelity: Kinetics, structure, and checkpoints. *Biochemistry* 43:14317–14324. A review pointing out that polymerases vary considerably in their use of the several specificity determinants.

Nick McElhinny, S. A., and eight coauthors (2010) Abundant ribonucleotide incorporation into DNA by yeast replicative polymerases. *Proc. Natl. Acad. Sci. USA* 107:4949–4954, and Genome instability due to ribonucleotide incorporation into DNA. *Nature Chem. Biol.* 6:774–781. Ribonucleotide incorporation, possibly the most abundant form of DNA damage.

Tsai, Y.-C., and K. A. Johnson (2006) A new paradigm for DNA polymerase specificity. *Biochemistry* 45:9675–9687. New mechanistic insights from the use of conformationally sensitive fluorophores attached to DNA polymerase.

Reverse Transcriptase

Hare, S., S. S. Gupta, E. Valkov, A. Engelman, and P. Cherepanov (2010) Retroviral intasome assembly and inhibition of DNA strand transfer. *Nature* 464:232–237. The structure of HIV integrase should aid in the search for new antiviral drugs.

Kohlstaedt, L. J., J. Wang, J. M. Friedman, P. A. Rice, and T. A. Steitz (1992) Crystal structure at 3.5 Å resolution of HIV-1 reverse transcriptase complexed with an inhibitor. *Science* 256:1783–1790. The first structural determination for a DNA polymerase other than *E. coli* polymerase I.

Peliska, J. A., and S. J. Benkovic (1992) Mechanism of DNA strand transfer reactions catalyzed by HIV-1 reverse transcriptase. *Science* 258:1112–1118. A mechanistic analysis of this important enzyme.

Temin, H. A. (1993) Retrovirus variation and reverse transcription: Abnormal strand transfers result in retrovirus genetic variation. *Science* 259:6900–6903. One of the co-discoverers of reverse transcriptase argues that genetic variability of HIV results from more than just a lack of a proofreading exonuclease.

PROBLEMS

1. Describe an experimental approach to determining the *processivity* of a DNA polymerase (that is, the number of nucleotides incorporated per chain per polymerase binding event).

*2. After Okazaki's proposal of discontinuous DNA chain growth, there was much controversy about whether both DNA strands are synthesized discontinuously, or only the lagging strand. Clearly, there is no need for the leading strand to be synthesized in fragments, but many workers found that pulse-labeled DNA fragments hybridized to both parental DNA strands. This finding indicated that both parental strands served as the template for synthesis of short DNA fragments. Propose an alternative mechanism by which leading strand replication could generate short fragments, and propose an experimental test of your suggestion. (Hint: Look at Chapter 26.)

3. The buoyant density of DNA can be increased by incorporation of either heavy stable isotopes, such as ^{15}N, or base analogs, such as 5-bromouracil. This problem asks you to calculate the density increment generated by each technique. In all calculations the density of "light" *E. coli* DNA is 1.710 g/mL. The G + C content is 51 mol % (that is, 51 moles of G + C per 100 moles of DNA nucleotides). Residue molecular weights of cesium salts of nucleotides are as follows: dAMP, 445; dTMP, 436; dGMP, 461; dCMP, 421; 5-bromo-dUMP, 501.
(a) Calculate the buoyant density of *E. coli* DNA when all ^{14}N atoms are replaced with ^{15}N atoms.
(b) Calculate the buoyant density of *E. coli* DNA when all thymine residues are replaced with 5-bromouracil.

4. A mixture of four α-[^{32}P]–labeled ribonucleoside triphosphates was added to permeabilized bacterial cells undergoing DNA replication,

and incorporation into high-molecular-weight material was followed over time, as shown in the accompanying graph. After 10 minutes of incubation a 1000-fold excess of unlabeled ribonucleoside triphosphates was added, with the results shown in the graph.

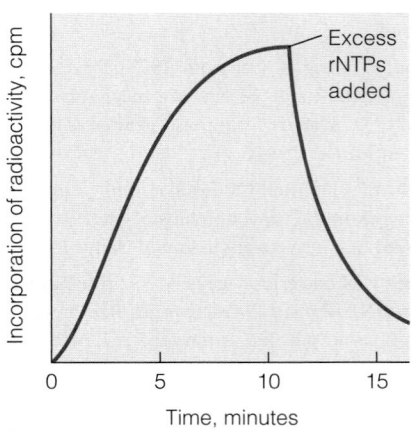

(a) Why was the excess of unlabeled rNTPs added?
(b) How could you tell that radioactivity is being incorporated as ribonucleotides, rather than an alternative such as reduction to deoxyribonucleotides, followed by incorporation?
(c) What does this experiment tell you about the process of DNA replication?

5. Deoxyadenylate residues in DNA undergo deamination fairly readily, as do deoxycytidylate residues.

   (a) What is the product of dAMP deamination?

   (b) The deamination product is known to base-pair with A, C, or T. What would be the genetic consequences if this deaminated site in DNA were not repaired and if it paired with C on the next round of replication?

6. The *E. coli* chromosome is 1.28 mm long. Under optimal conditions the chromosome is replicated in 40 minutes.

   (a) What is the distance traversed by one replication fork in 1 minute?

   (b) If replicating DNA is in the B form (10.4 base pairs per turn), how many nucleotides are incorporated in 1 minute in one replication fork?

   (c) If cultured human cells (such as HeLa cells) replicate 1.2 m of DNA during a 5-hour S phase and at a rate of fork movement one-tenth of that seen in *E. coli,* how many origins of replication must the cells contain?

   (d) What is the average distance, in kilobase pairs, between these origins?

7. DNA ligase has the ability to relax supercoiled circular DNA in the presence of AMP but not in its absence.

   (a) What is the mechanism of this reaction, and why is it dependent on AMP?

   (b) How might one determine that supercoiled DNA had in fact been relaxed?

*8. A recent paper reports that, in mammalian cells, genes that are expressed in a particular cell are replicated during the first half of S phase, and genes not expressed in that cell are replicated in the latter half of S phase. Briefly describe an experiment that could lead to this conclusion.

9. Although DNA polymerases require both a template and a primer, the following single-strand polynucleotide was found to serve as a substrate for DNA polymerase in the absence of any additional DNA.

   3′ HO-ATGGGCTCATAGCCGGAGCCCTAACC-
   GTAGACCACGAATAGCATTAGG-p 5′

   Give the structure of the product of this reaction.

10. The 3′ exonuclease activity of *E. coli* DNA polymerase I was found to show no discrimination between correctly and incorrectly base-paired nucleotides at the 3′ terminus; properly and improperly base-paired nucleotides are cleaved at equal rates there. How can this observation be reconciled with the fact that the 3′ exonuclease activity increases the accuracy with which template DNA is copied?

11. 2′,3′-Dideoxyinosine has been approved as an anti-HIV drug. Propose a mechanism by which it might block the growth of the HIV virus.

12. Aphidicolin-resistant mutants of mammalian cells often have alterations in the structure of DNA polymerase α that render the enzyme insensitive to inhibition. However, some mutants show alterations, not in DNA polymerase but in ribonucleotide reductase. How might a change in the latter enzyme cause an aphidicolin-resistant phenotype?

13. The *E. coli* chromosome requires 40 minutes for its complete replication, even in an optimally nourished cell. However, bacterial cells can divide as frequently as every 20 minutes. How can cells divide more rapidly, apparently, than their DNA can be copied?

*14. Supercoiled DNA is more compact than relaxed DNA of the same molecular weight, which gives supercoiled DNA greater electrophoretic mobility. Why, then, does positively supercoiled DNA migrate more slowly than relaxed in the experiment depicted in Figure 25.26? (Note: Read Tools of Biochemistry 25A before attempting to answer this question.)

15. An alternative to the model shown in Figure 25.23 for the action of homodimeric helicases is an "inchworming" model. In this model, the subunit bound to DNA moves forward by a few base pairs, unwinding as it goes, and with the 3′ end bound to that subunit as it moves toward the 5′ end of the bound strand. Then, ATP hydrolysis triggers a conformational change that leads to release of the 3′ end of the unwound DNA so that the process can repeat, as shown.

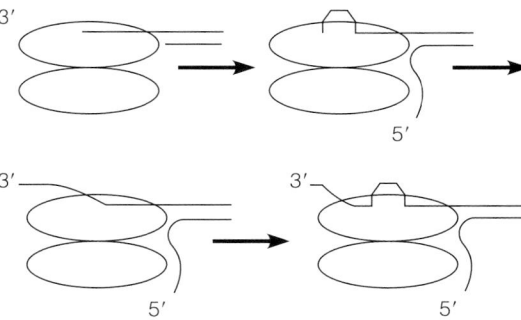

Identify a difference between this and the hand-over-hand model of Figure 25.23 that could be used experimentally to distinguish between the two models.

16. Propose a mechanism by which a type II topoisomerase could use the energy of ATP hydrolysis to scan a large DNA molecule and, thereby, to direct that the enzyme will catalyze largely "disentangling" reactions (decatenation and unknotting).

17. The exponential nature of PCR allows spectacular increases in the abundance of a DNA sequence being amplified. Consider a 10-kbp DNA sequence in a genome of 10^{10} base pairs. What fraction of the genome is represented by this sequence; i.e., what is the fractional abundance of this sequence in this genome? Calculate the fractional abundance of this target sequence after 10, 15, and 20 cycles of PCR, starting with DNA representing the whole genome and assuming that no other sequences in the genome undergo amplification in the process.

18. Isolated rat liver mitochondria are able to carry out incorporation of added deoxyribonucleoside triphosphates into mitochondrial DNA by a semiconservative process. With this system or using a cell culture system,

   (a) Outline an approach to determine whether RNA formed during mtDNA replication represents primers for DNA synthesis.

   (b) Outline an approach to determine whether mtDNA replication forks move unidirectionally or bidirectionally.

19. DNA precursor imbalances are mutagenic. For example, if dGTP accumulates, it can compete with dATP for incorporation opposite dTMP in the template, leading to a transition mutation. Recently it was reported that a modest balanced increase in all four dNTPs, three- or fourfold, stimulated mutagenesis out of proportion to the dNTP pool change. Describe a mechanism by which this could occur.

20. A plasmid DNA was isolated and shown by two-dimensional agarose gel electrophoresis to have a range of topoisomers, as shown in Tools of Biochemistry 25A.

   (a) Sketch the expected 2D gel pattern after treatment of this DNA with topoisomerase I.

   (b) Sketch the expected 2D gel pattern after treatment of this DNA with DNA gyrase and ATP.

TOOLS OF BIOCHEMISTRY 25A

TWO-DIMENSIONAL GEL ELECTROPHORETIC ANALYSIS OF DNA TOPOISOMERS

When one analyzes DNA molecules resolved by agarose gel electrophoresis (see Tools of Biochemistry 2A), DNA is visualized in the gel by soaking the gel with a fluorescent dye, such as **ethidium bromide**, or EtBr.

Ethidium bromide (EtBr)

This dye binds to duplex DNA by **intercalation**; the molecule is planar and about the size of one base pair, so it can fit between two adjacent DNA base pairs, forcing them apart. Intercalation greatly enhances the molecule's fluorescence, and DNA can then be visualized by observing an EtBr-treated gel under ultraviolet light.

But there is another way in which intercalators like EtBr are used in electrophoresis. If they are present *during* the electrophoresis, they can aid in distinguishing topoisomers. In forcing adjacent base pairs apart, EtBr tends to unwind a double helix. In B-form DNA, with about 10 base pairs per turn, two adjacent base pairs are rotated with respect to each other by 36° (360° per complete turn). One molecule of EtBr creates about 27° of unwinding, affecting the hydrodynamic properties of circular DNA in the following way. Recall from Chapter 4 that L, the linking number, is the sum of turns created by twist and writhe. EtBr binding decreases the number of turns needed to create an unconstrained duplex; it decreases T, the twist. Because L doesn't change unless the molecule is broken, the writhe component is affected: The number of turns devoted to writhe increases. In other words, EtBr increases the positive superhelicity of the molecule. Thus, an underwound molecule tends to

become relaxed because it loses negative superhelicity, whereas a relaxed closed duplex becomes overwound, and an overwound molecule becomes more so. Electrophoretic mobility through a gel is a function of compactness of the migrating molecule, with a relaxed duplex being least compact and, hence, most slowly migrating.

One can analyze the distribution of topoisomers in a circular DNA by two-dimensional gel electrophoresis, with the first dimension run in an ordinary agarose gel and the second dimension run in the presence of EtBr or a similar intercalating dye. For the plasmid DNA shown in Figure 25A.1a, these topoisomers include nicked duplexes, which are relaxed whether or not EtBr is present and which migrate slowly in both dimensions. For the slightly supercoiled molecules, both overwound and underwound topoisomers are present. For example, the species labeled 1 was originally overwound and has become more so, as is evident by its movement in the second dimension. Species 6, on the other hand, was underwound and moved more slowly in the second dimension, even though 1 and 6 originally had the same number of superhelical turns (writhe). The remaining species were present originally only as underwound topoisomers. At the bottom (species 18 and 22 and their neighbors) are topoisomers that could not be resolved in the first dimension but are resolved when the negative writhe is decreased, in the second dimension.

In Figure 25A.1b the plasmid analyzed was identical, except for the insertion of a 16-base-pair alternating G-C sequence, which under certain conditions forms Z-DNA (left-handed). The intercalating dye forces this DNA back to a right-hand helix, and this affects the number of superhelical turns. The topoisomers labeled 17 and 20 are species that contained the GC-rich insert in Z conformation before administration of the intercalating dye.

Reference

Peck, L. J., and J. C. Wang (1983) Energetics of B-to-Z transitions in DNA. *Proc. Natl. Acad. Sci. USA* 80:6206–6210.

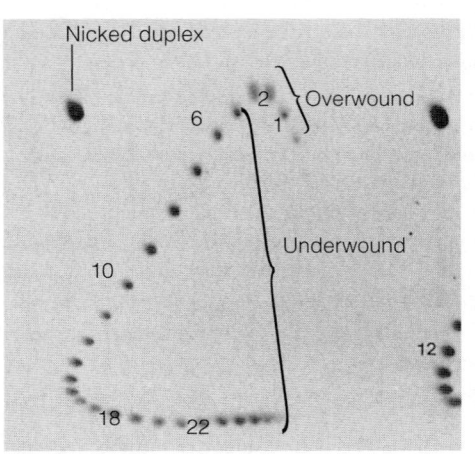
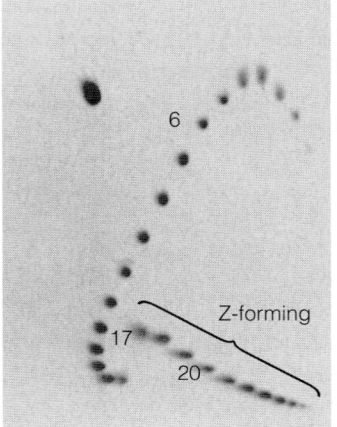

FIGURE 25A.1

Two-dimensional agarose gel electrophoretic analysis of circular DNA topoisomers. The same plasmid was employed for both analyses, except that the plasmid in **(b)** had a 16-base-pair insert that forms left-hand Z-DNA.

Courtesy of J. C. Wang.

(a)   (b)

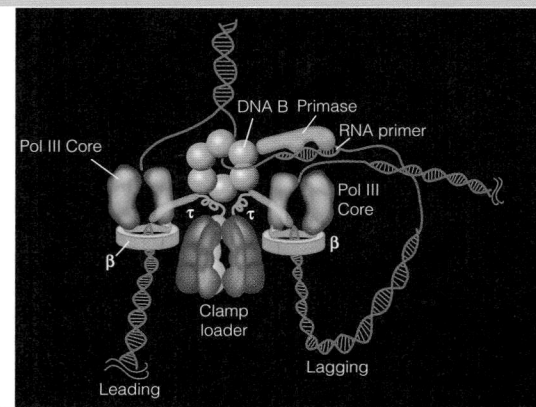

CHAPTER 26

DNA Restructuring:
Repair, Recombination,
Rearrangement, Amplification

Our focus now shifts from DNA as a *template* for its own replication to DNA as a *substrate*, in a number of processes that we categorize as **information restructuring**. DNA has long been thought to be particularly well suited to its role in storing genetic information because of its chemical stability. To be sure, the fact that DNA from prehistoric animals and humans can be isolated and sequenced attests to its stability. However, like all biomolecules, DNA is continuously exposed to damaging agents, including radiation, environmental chemicals, and endogenous agents such as reactive oxygen species. In addition, chemical change is introduced as errors during normal DNA replication. Meeting its function as a stable repository of genetic information requires that the cell be able to efficiently repair damage suffered by DNA. We will see that genetic recombination—the breakage and rejoining of DNA molecules, originally thought to be a process ensuring the diversification of the genome—is also an integral part of some repair mechanisms. Finally, we will consider the plasticity of the genome, in terms of the ability of DNA segments to move from point to point within a genome, or from genome to genome, or to undergo selective amplification.

From the biochemist's standpoint, information restructuring is less accessible than DNA replication. Replication is a central pathway that is a major metabolic event in the lifetime of all cells, and the activities of the enzymes and proteins involved are relatively high. As a result, their discovery and characterization, though elegant, were fairly straightforward. By contrast, information restructuring involves quantitatively minor pathways, with far less mass conversion per cell. Detection of the enzymes involved is correspondingly more difficult. For example, although DNA methylation has profound effects on gene integrity or expression (see Chapter 29), methylation may involve only one or a few sites per gene. Because of the low activities involved, analysis of DNA methylation requires experimental techniques much more sensitive than those used to analyze DNA replication.

The processes we consider in this chapter include (1) metabolic responses to DNA structural damage, principally mutagenesis and repair; (2) recombination,

FIGURE 26.1

A summary of the major processes in information restructuring. Restriction and modification were discussed in Chapter 24, and the other processes are presented in this chapter.

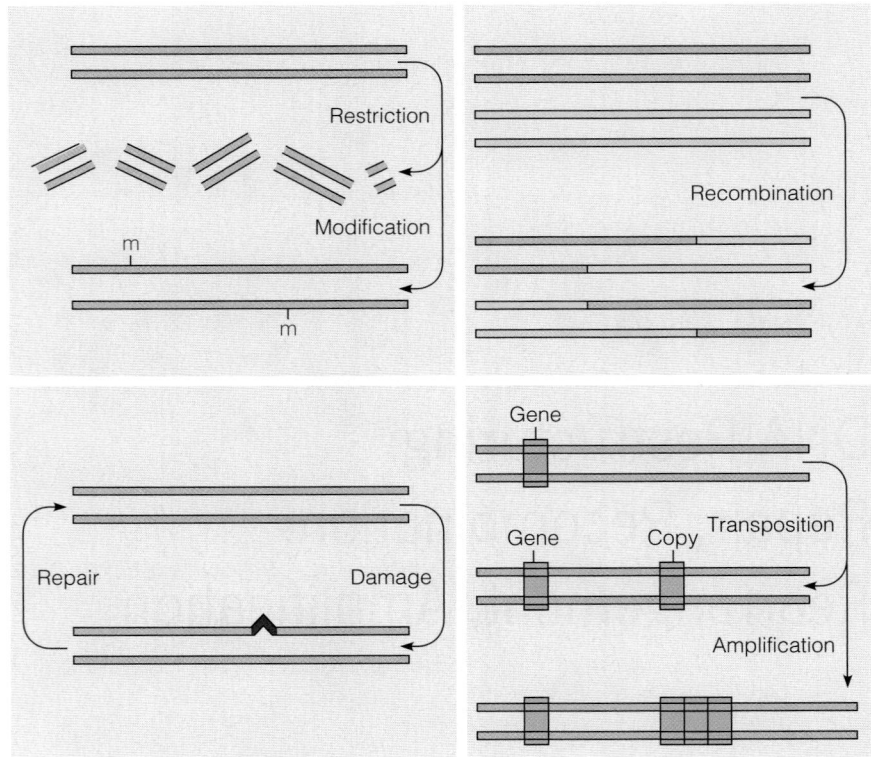

whereby the contents of a genome are redistributed, for example, during sexual reproduction; (3) gene rearrangements, including transpositions of DNA segments from one chromosomal integration site into another and the joining of DNA segments from distant parts of a genome; and (4) gene amplification, an increase in the copy number of individual segments of DNA, which occurs both as a normal developmental process and as a response to environmental stresses (Figure 26.1). Collectively, these processes are essential to the survival of cells. In broader terms, recombination and gene rearrangements are the source of most of the genetic variability in a population of cells or organisms and, along with mutation, form the basis for evolutionary changes. Other important information restructuring processes, which do not occur at the DNA level, are described elsewhere. They include alternative splicing and RNA editing (see Chapters 27 and 29).

DNA Repair

Types and Consequences of DNA Damage

Unprogrammed chemical changes occur in all biological macromolecules, either because of environmental damage or errors in synthesis. For most biomolecules, including RNA, protein, and membrane phospholipids, the effects of such changes are minimized by turnover and replacement of altered molecules. DNA is distinctive, however, in that its information content must be transmitted virtually intact from one cell to another during cell division or reproduction of an organism. Thus, DNA has a special need for metabolic stability. This stability is maintained in two ways—by a replication process of high accuracy, and by mechanisms for correcting genetic information when DNA suffers damage. In Chapter 25 we described mechanisms used to ensure high replication fidelity. Here we discuss the kinds of environmental and endogenous damage that occur and the several processes repairing that damage. These processes include (1) **direct repair**, in which a damaged DNA base undergoes a chemical or photochemical reaction to restore the original structure; (2) **nucleotide excision repair**, in which a section of DNA that contains a damaged site is excised and replaced with normal DNA; (3) **base excision repair**, which starts with cleavage of the glycosidic bond connecting a damaged base to the DNA sugar–phosphate backbone; (4) **mismatch repair**, a process that recognizes DNA mismatches created either by replication errors, nonhomologous recombination, or damage to one DNA base and corrects the error; (5) **daughter-strand gap repair**, in which newly replicated DNA duplexes undergo genetic recombination, with ultimate removal of the damaged DNA segment; (6) **translesion synthesis**, or error-prone repair, in which DNA damage that blocks replicative DNA polymerases can be copied by specialized, but often inaccurate, DNA polymerases; and (7) **double-strand break repair**, in which severed ends of a duplex can be rejoined.

Some of the most prominent endogenous DNA-damaging reactions include (1) depurination, the hydrolytic rupture of the glycosidic bond between deoxyribose and a purine base; (2) deamination, usually the hydrolytic conversion of a DNA-cytosine residue to uracil; (3) oxidation, notably, the oxidation of guanine to 8-oxoguanine or of thymine to thymine glycol; and (4) nonenzymatic methylation by *S*-adenosylmethionine. Some of these reactions are shown in Figure 26.2, with an indication of their frequency in mammalian cells. In addition, replication errors account for many thousands of mismatched DNA nucleotides per day.

Environmental DNA-damaging agents include ionizing radiation; ultraviolet radiation; DNA-methylating reagents, such as ***N*-methyl-*N*′-nitro-*N*-nitrosoguanidine** (MNNG); DNA-cross-linking reagents, such as the anticancer drug **cisplatin**; and

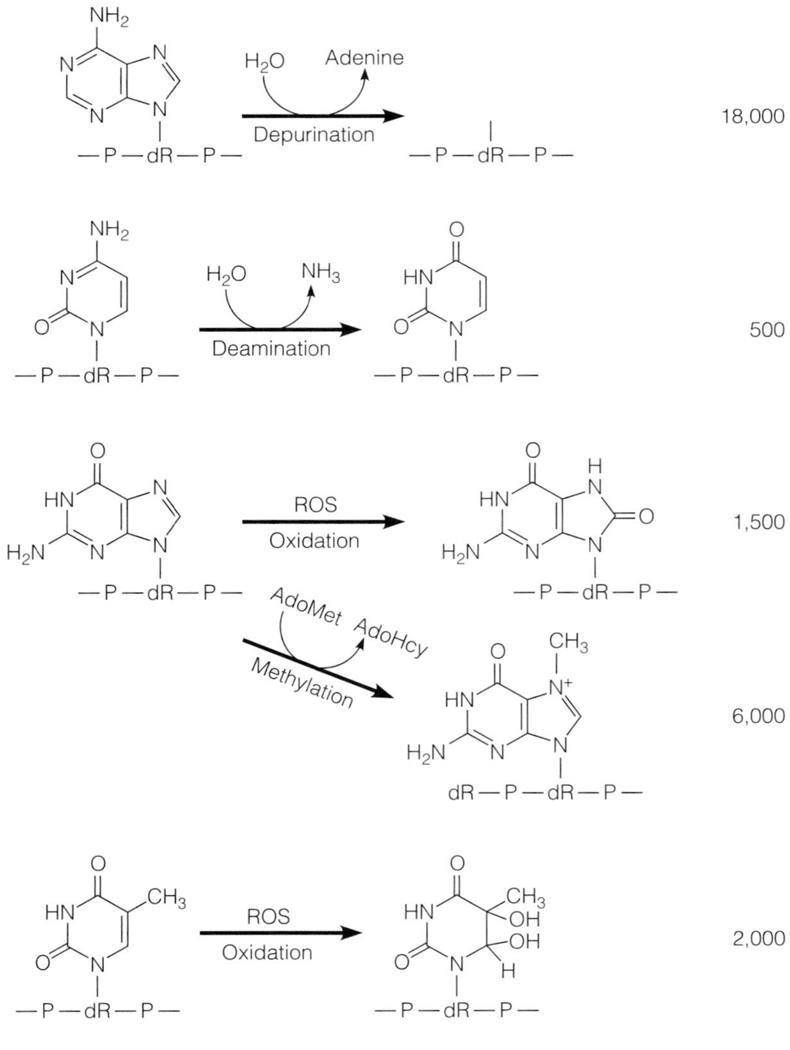

FIGURE 26.2

Endogenous DNA-damaging reactions. The approximate frequency of each reaction, in number of lesions per mammalian cell per day, is indicated. ROS, reactive oxygen species.

Data from *DNA Repair and Mutagenesis*, 2nd ed., E. C. Friedberg, G. C. Walker, W. Siede, R. D. Wood, R. A. Schultz, and T. Ellenberger. © 2006 ASM Press, Washington, DC.

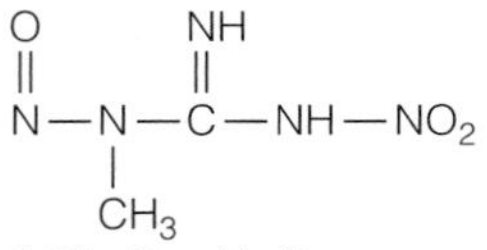

N-Methyl-N'-nitro-N-nitrosoguanidine (MNNG)

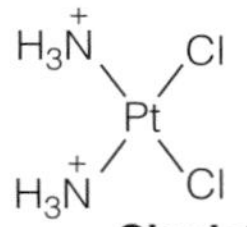

Cisplatin

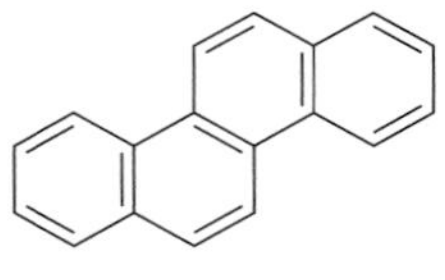

Benzo(a)pyrene

bulky electrophilic agents, such as **benzo[a]pyrene**, one of the carcinogenic hydrocarbons in tobacco smoke. We shall say more about the damage induced by such reagents and the repair processes that correct such damage. We will see that some repair processes are accurate, meaning that the original DNA sequence is restored, while some lead to mutations because the repair process is inaccurate. Because it is now clear that many cancers result from an accumulation of somatic cell mutations (see Chapter 23), the mechanisms of DNA repair are under intense study as determinants of an animal's susceptibility to cancer.

In studying mechanisms of DNA repair, one first needs to identify the chemical forms of DNA alteration that are subject to repair. Most of the early discoveries about DNA repair were made in studies with ultraviolet light-irradiated organisms. DNA was identified as the principal biological target for UV irradiation, partly through determination of **action spectra**—irradiation of bacteria or phages with UV light and determination of the wavelengths most effective in stimulating mutagenesis or death. Those wavelengths lie near 260 nm, where DNA light absorption is maximal.

When one examines either UV-irradiated DNA or the DNA extracted from a UV-irradiated organism, one detects small amounts of many different altered DNA constituents, called **photoproducts**. Prominent among them are intrastrand dimers consisting of two pyrimidine bases joined by a cyclobutane ring structure involving carbons 5 and 6 (Figure 26.3a). **Thymine dimers**, formed from two adjacent DNA thymine residues, were identified quite early as biologically significant photoproducts because the relative abundance of thymine dimers in irradiated DNA correlated most closely with death of irradiated phages or bacteria. Thus, the ability of an organism to survive ultraviolet irradiation was related directly to its ability to remove thymine dimers from its DNA. Dimerization draws the adjacent thymine residues together, distorting the helix such that replicative polymerization past this site is blocked.

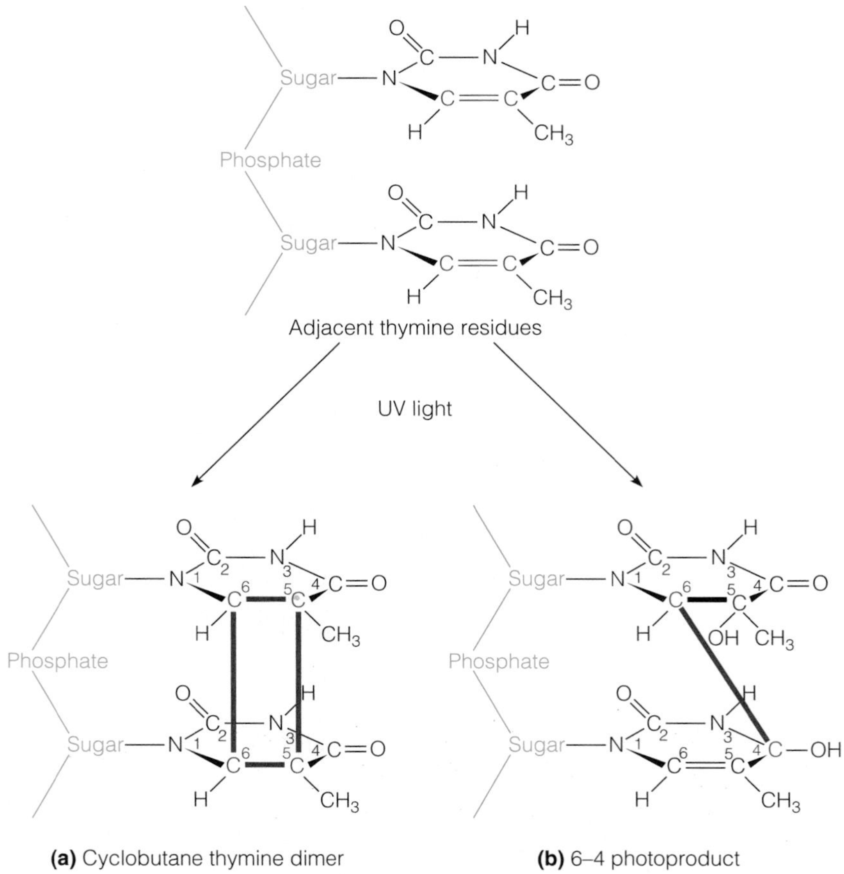

FIGURE 26.3

Structures of pyrimidine dimer photoproducts.

(a) Cyclobutane thymine dimer

(b) 6–4 photoproduct

For many years it was thought that ultraviolet light–induced *mutagenesis*, as well as cell death, was also caused primarily by cyclobutane thymine dimers. However, more recent data suggest that a different pyrimidine–pyrimidine dimer, called **6–4 photoproduct**, is probably the principal cause of UV-induced mutations. As shown in Figure 26.3b, these products are also dimers, linked via C-6 of the 5′ pyrimidine (either thymine or cytosine) and C-4 of the 3′ pyrimidine (usually cytosine, but occasionally thymine, as shown in Figure 26.3). The idea that 6–4 photoproducts are responsible for mutations in UV-irradiated DNA is supported by experiments in which cyclobutane thymine dimers were completely removed from UV-irradiated DNA by **photoreactivation** (see the next section). When this DNA was introduced into bacteria, the dimer removal had no effect on mutation frequency. However, the actual significance of 6–4 photoproducts in UV mutagenesis has not yet been established.

Direct Repair of Damaged DNA Bases: Photoreactivation and Alkyltransferases

Of the half dozen well-understood DNA repair processes, most involve removal of the damaged nucleotides, along with several adjacent residues, followed by replacement of the excised region using information encoded in the complementary (undamaged) strand. However, at least two processes involve reactions that *directly change* the damaged bases, rather than removing them.

Photoreactivation

A widely distributed enzyme, called **photoreactivating enzyme**, or **DNA photolyase**, repairs cyclobutane pyrimidine dimers in the presence of visible light. A wavelength of 370 nm is most effective. The enzyme binds to DNA in a light-independent process, specifically at the site of pyrimidine dimers. In the presence of visible-wavelength light, the bonds linking the pyrimidine rings are broken, after which the enzyme can dissociate in the dark. Clues to the mechanism have come with the finding that the enzyme contains two chromophores. (Recall from Chapter 16 that a chromophore is a structural moiety that absorbs light of characteristic wavelengths.) One chromophore is bound flavin adenine dinucleotide, deprotonated and in the reduced state (FADH$^-$; see Figure 26.4); the second in some photolyases is 5,10-methenyltetrahydrofolate (see Chapter 20) and in others is 8-hydroxy-5-deazaflavin. Mechanistic studies suggest a process akin to photosynthesis, with the second chromophore functioning as a light-harvesting factor, and transmitting light energy by fluorescence resonance energy transfer (FRET) to FADH$^-$, which functions like the photochemical reaction center, transferring an electron to the dimer and breaking the pyrimidine–pyrimidine bonds by a free radical mechanism. The crystal structure of the *E. coli* photolyase shows 5,10-methenyltetrahydrofolate bound at the surface, between N-terminal and C-terminal domains, with FADH$^-$ bound deeply within the C-terminal domain. Figure 26.4 shows the probable reaction pathway.

Although photolyase has been detected in numerous eukaryotic systems, recent evidence indicates that human cells do not contain an enzyme for photoreactivation. Thinning of the earth's ozone layer has been blamed for population declines in certain frog species, which have been shown to lack photolyase and, hence, to suffer damage from solar ultraviolet irradiation, particularly during embryonic development in clear lakes, which are transparent to UV rays. However, other factors have also been blamed.

O^6-Alkylguanine Alkyltransferase

Treatment of DNA with a methylating or ethylating reagent is comparable to ultraviolet irradiation in that various modified DNA bases are formed, some of which are lethal if not repaired and some of which are mutagenic. Some alkylating

Cyclobutane thymine dimers are the most lethal photoproduct in UV light–irradiated DNA. 6–4 photoproduct may be the strongest mutagenic product.

DNA can be repaired directly, by changing a damaged base to a normal one, or indirectly, by replacing a DNA segment containing the damaged nucleotide.

$$CH_3 - N - C - NH_2$$

Methylnitrosourea (MNU)

$$CH_3 - S - O - CH_2 - CH_3$$

Ethylmethanesulfonate (EMS)

$$CH_3 - N - C - NH - NO_2$$

***N*-Methyl-*N*′-nitro-*N*-nitrosoguanidine (MNNG)**

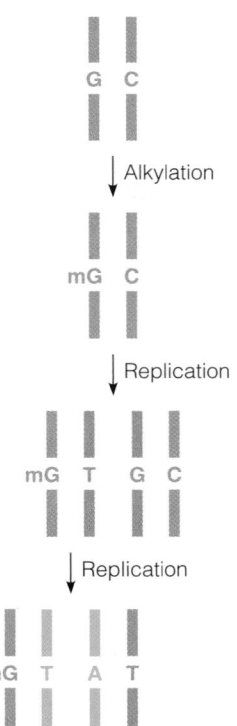

Guanine   **Cytosine**

Alkylation

O^6-**Methylguanine (mG)**   **Thymine**

FIGURE 26.5

Mispairing of O^6-methylguanine with thymine in a DNA duplex.

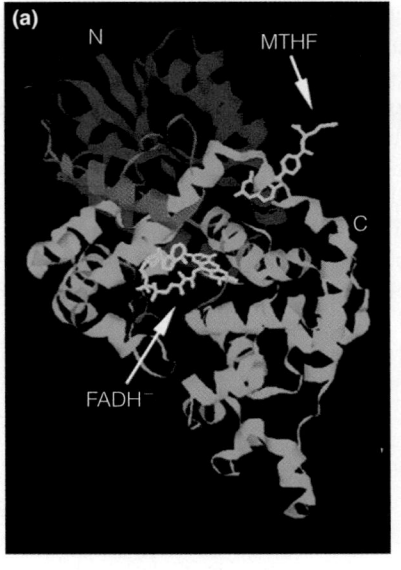

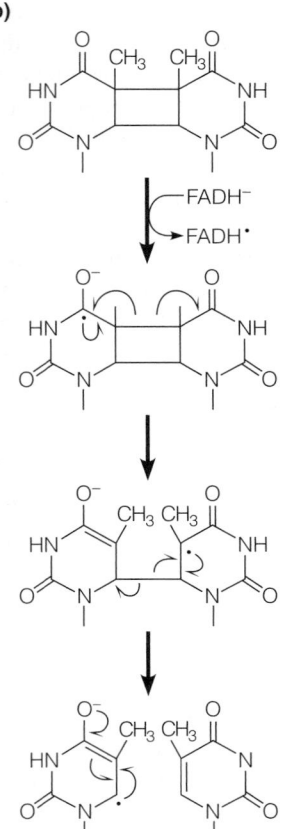

FIGURE 26.4

The thymine dimer photolyase. (a) structure of the *E. coli* enzyme, showing distinct N-terminal (red) and C-terminal (green) domains, with a linker in orange and showing bound folate and flavin cofactors. PDB ID 1DNP. **(b)** structures of FADH⁻ and FADH• and the likely reaction pathway.

agents are used in cancer chemotherapy because of their ability to block DNA replication and, hence, cell proliferation. Some are widely used as mutagens in the laboratory. Three methylating or ethylating reagents are shown in the margin on page 1083.

The bases altered by these reagents are primarily purines (phosphate oxygens are also targets), and the spectrum of products formed varies with the reagent used. The most highly mutagenic of these products, O^6-**alkylguanine**, is mutagenic because the modified base has a very high probability of pairing with thymine when the modified strand replicates (Figure 26.5). Thus, alkylation of a DNA-guanine stimulates a GC → AT transition (see margin, where mG is a methylguanine residue).

Repair of this type of damage involves an unusual enzyme, O^6-**alkylguanine alkyltransferase**, which transfers a methyl or ethyl group from an O^6-methylguanine or O^6-ethylguanine residue to a cysteine residue in the active site of the protein. Remarkably, this "catalyst," which is widely distributed in prokaryotes and eukaryotes, can function only once. Having become alkylated, it cannot remove the alkyl group, and the protein molecule turns over. Thus, the term *enzyme* is a misnomer. In bacteria the protein regulates both its own synthesis and that of another repair enzyme, a DNA-*N*-glycosylase that we shall discuss later. The regulation involves activation of the transcription of genes encoding these two proteins. There is evidence that the *alkylated* form of the alkyltransferase is the specific form of the transcriptional activator. This allows the cell to adapt to alkylation damage by using the alkylated protein as a specific signal to produce more of the proteins needed to repair the damage.

Nucleotide Excision Repair: Excinucleases

Nucleotide excision repair (NER) was originally discovered as an enzyme system capable of repairing thymine dimers created in DNA by UV irradiation. Unlike photoreactivation, this process can take place in the dark. The enzyme system involved, which in *E. coli* includes the products of genes *uvrA, uvrB,* and *uvrC,* is now known to act upon a number of DNA damaged sites containing lesions that may be quite bulky, such as those created by large alkyl groups, and that distort the DNA double helix. Very similar systems exist in mammalian cells and in yeast, so the process is probably universal.

As shown in Figure 26.6 for *E. coli,* the three-subunit UvrABC enzyme recognizes a lesion (a thymine dimer in the example shown) and, with the help of ATP hydrolysis, forces DNA to bend, leading to cleavage of the damaged strand at two sites—seven nucleotides to the 5′ side of the damaged site and four nucleotides to the 3′ side—leaving a potential gap eleven nucleotides in length, with a 3′ hydroxyl group and a 5′ phosphate at the ends. Polymerase and ligase action then replace the damaged 11-mer with undamaged DNA. Helicase II, the product of the *uvrD* gene, is also required, presumably to unwind and remove the excised oligonucleotide, which is ultimately broken down by other enzymes. The UvrABC enzyme is not a classical endonuclease because it cuts at two distinct sites, and the term **excinuclease** has been proposed for it, denoting its role in *exci*sion repair. This system is also involved in repairing another type of DNA damage—covalent cross-linking of the two strands to each other. Here, in order to preserve an intact template strand, the two strands are repaired sequentially, one after the other.

Excision repair also occurs in mammalian cells, as shown by the discovery of a human excinuclease that cleaves at positions -22 and $+6$ relative to a thymine dimer. A significant difference is the involvement of two different endonucleases in human excision repair—one for cutting on the 5′ side and one on the 3′ side. Excision repair in humans originally came to light through studies of a rare genetic disease called **xeroderma pigmentosum (XP)**. XP is actually a family of diseases, in which one or more enzymes of the excision pathway are deficient. In affected humans there is at present no known way to treat the condition. The biological consequences of XP include extreme sensitivity to sunlight and a high incidence of skin cancers. Although overexposure to the ultraviolet rays in sunlight increases the risk of skin cancer for all humans, the greatly increased skin cancer frequency in XP patients highlights the importance of UV repair pathways to mammals. Because there is no known treatment for XP, affected individuals must avoid sunlight. Two other diseases—**trichothiodystrophy** (TDD) and **Cockayne's syndrome** (CS)—result from other defects in NER proteins. Although these conditions do not involve abnormal carcinogenesis, other systems are profoundly affected, and mean lifespan is only 6 years for TTD and 12 for CS.

FIGURE 26.6

Excision repair of thymine dimers by the UvrABC excinuclease of *E. coli*. A complex of A and B proteins tracks along DNA until it reaches a thymine dimer or other damaged site, where it halts and forces the DNA to bend. UvrA (a "molecular matchmaker") then dissociates, allowing UvrC to bind to B. The BC complex cuts on both sides of the dimer. Helicase, polymerase, and ligase remove the damaged undecamer and replace it with new DNA. This system may use DNA polymerase II as well as pol I.

From *Science* 259:1415–1420, A. Sancar and J. E. Hearst, Molecular matchmakers. © 1993. Reprinted with permission from AAAS. Adapted with permission from Aziz Sancar.

Direct repair enzymes include photolyase, which uses light energy to repair pyrimidine dimers, and alkyltransferases, "enzymes" that are inactivated after just one catalytic cycle.

Excision repair involves endonuclease cleavage on both sides of a damaged site, followed by replacement synthesis.

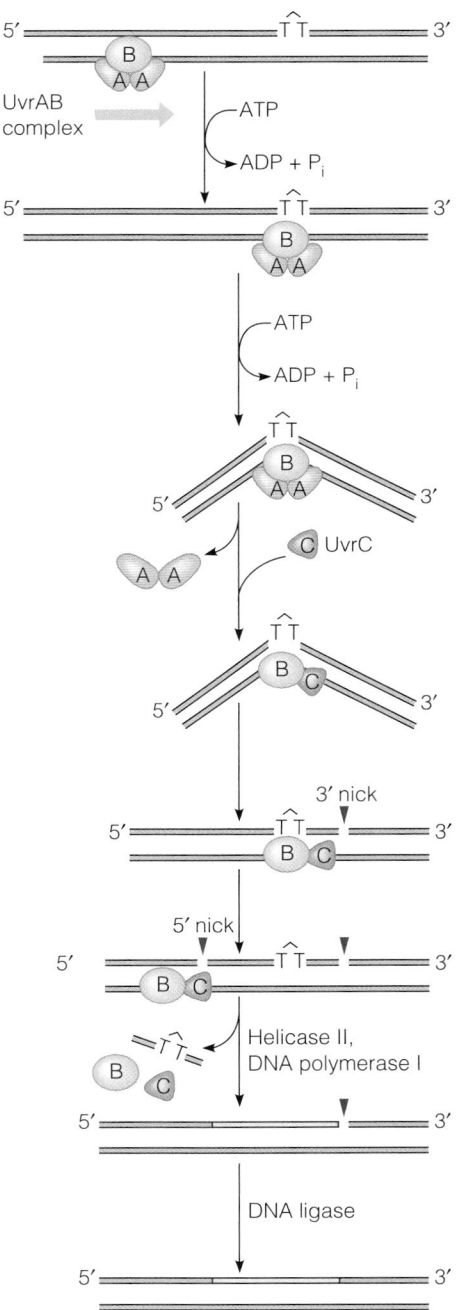

A significant difference in NER between prokaryotes and eukaryotes is that in the latter the substrate for repair is chromatin, not naked DNA. The amount of uncoated DNA required to assemble an NER complex is estimated to be about 100 base pairs. Because this is longer than the internucleosomal spacer region (20–80 base pairs), this must mean that core particles are either displaced or else they slide out of the way in order for damage to be repaired. These processes are under intense investigation.

Recent studies of nucleotide excision repair show that active genes (those undergoing transcription) are preferred substrates for excision repair, and within these genes the template DNA strand is preferentially repaired. This **transcription-coupled repair** may initiate when a transcribing RNA polymerase becomes stalled at the site of a DNA lesion. Coupling transcription with repair helps to ensure the integrity of genes that are actually used. In mammalian cells, transcription-coupled repair is a specialized mode of NER that requires additional proteins. Cockayne's syndrome results from genetic defects in one or more of the human enzymes of transcription-coupled repair. Another human gene, BRCA1, is also implicated in transcription-coupled excision repair. Mutations in BRCA1 are associated with increased risk of breast and ovarian cancer. One mode of mammalian NER requires both DNA polymerases δ and κ, while another mode requires only pol ε.

As noted above, NER also corrects DNA that has been damaged by formation of bulky DNA adducts. Many chemical carcinogens of environmental significance give rise to such adducts. In the absence of repair, or following error-prone repair, these adducts can lead to mutations. For example, polycyclic aromatic hydrocarbons (PAH) are prevalent organic pollutants formed by incomplete combustion, for example, by coal burning. As with many environmental carcinogens, PAH are not directly reactive with DNA. Indeed, their lack of chemical reactivity accounts for their persistence. However, PAH can undergo *metabolic activation,* that is, *biotransformation* leading to reactive electrophilic intermediates that can bind to DNA and other macromolecules. This metabolism can occur in the liver and in other organs. The best-characterized route for metabolic activation of PAHs is the pathway for benzo[*a*]pyrene metabolism, shown (in simplified form) in Figure 26.7. First, oxygenation catalyzed by cytochrome P450 generates an epoxide intermediate. This epoxide is hydrolyzed, in a reaction catalyzed by *epoxide hydrolase, to* give a dihydrodiol product. Finally, a second cytochrome P450-catalyzed oxygenation produces the highly reactive benzo[*a*]pyrene 7,8-dihydrodiol 9,10-epoxide (BPDE). BPDE is highly reactive with DNA, forming bulky covalent adducts primarily, but not exclusively, with guanine.

FIGURE 26.7

Metabolic activation of a carcinogenic polycyclic aromatic hydrocarbon, followed by reaction of the activated dihydrodiol epoxide with a DNA dGMP residue.

Courtesy of Dr. David Josephy.

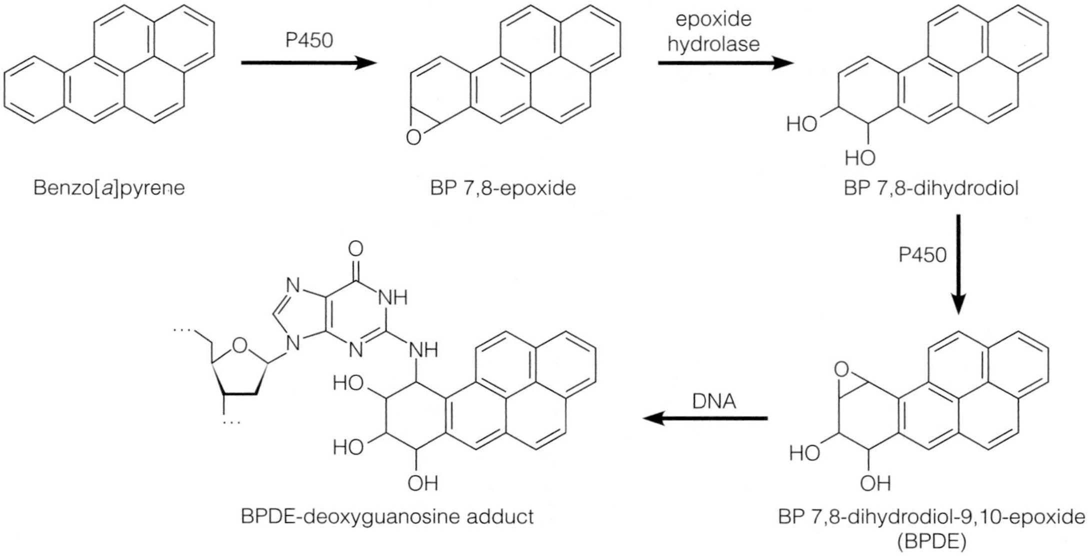

Benzo[*a*]pyrene → (P450) → BP 7,8-epoxide → (epoxide hydrolase) → BP 7,8-dihydrodiol → (P450) → BP 7,8-dihydrodiol-9,10-epoxide (BPDE) → (DNA) → BPDE-deoxyguanosine adduct

Base Excision Repair: DNA *N*-Glycosylases

The other form of excision repair, **base excision repair**, or BER, also removes one or more nucleotides from a site of base damage. However, this process initiates with enzymatic cleavage of the glycosidic bond between the damaged base and deoxyribose.

Replacement of Uracil in DNA by BER

One of the best-understood BER systems involves scanning of DNA, to remove uracil. Uracil can base-pair with adenine in a DNA duplex, and DNA polymerases readily accept deoxyuridine triphosphate as a substrate in place of thymidine triphosphate. Yet cells possess a rather elaborate two-stage mechanism that prevents deoxyuridylate residues from accumulating in DNA. The first stage of this process, described in Chapter 22, involves an active deoxyuridine triphosphatase that cleaves dUTP to dUMP and pyrophosphate, thereby minimizing the dUTP pool and its use as a replication substrate. The second stage involves **uracil-DNA *N*-glycosylase** (UNG), an enzyme that removes any dUMP residues that might have arisen through incorporation of a dUTP that escaped the action of dUTPase.

As shown in Figure 26.8, uracil-DNA *N*-glycosylase hydrolytically cleaves the glycosidic bond between N-1 of uracil and C-1 of deoxyribose. This yields free uracil and DNA with an **apyrimidinic site**, that is, a sugar residue lacking an attached pyrimidine. Another enzyme, **apyrimidinic endonuclease** (AP endonuclease) recognizes this site and cleaves the phosphodiester bond on the 5′ side of the deoxyribose moiety. This is followed in bacteria by the nick translation activity of DNA polymerase I, which inserts dTTP as a replacement for the dUMP that was removed, displacing the deoxyribose phosphate residue in the AP site. This is removed by a **deoxyribose-5′-phosphatase**. The resulting nick is sealed by DNA ligase.

Uracil *N*-glycosylase, like all DNA glycosylases examined to date, acts by flipping the target base out of the DNA duplex and binding it in a pocket where cleavage occurs. Structural analysis of human UNG shows that the pocket is small enough to exclude purine bases. More important, as shown in Figure 26.9, the pocket excludes thymine because of negative steric interaction between the thymine methyl group at C5 and Tyr-147. (Note: By convention eukaryotic genes and gene products are denoted with capital letters; thus, *UNG* is the human gene that encodes the protein UNG. In bacteria, gene *ung* encodes protein Ung.)

Why go to all this trouble just to replace a nucleotide that does not affect the information encoded in DNA? The likely answer is that uracil substituted for thymine (that is, base-paired with adenine) is not the true target of this DNA repair

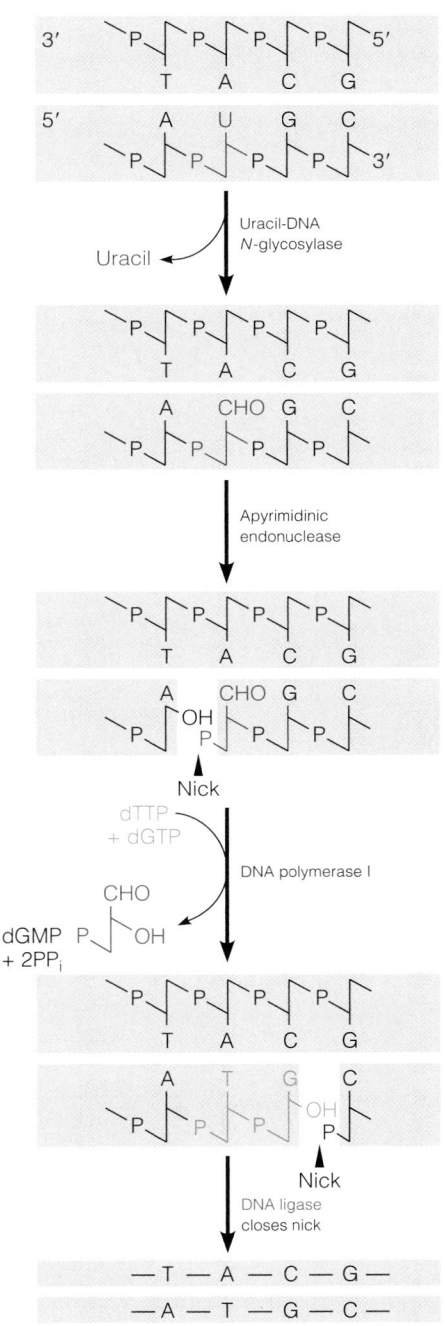

FIGURE 26.8

Action of the DNA uracil repair system. Uracil-DNA *N*-glycosylase (also called Ung) removes uracil, leaving an apyrimidinic site. A specific endonuclease recognizes this site and cleaves on the 5′ side. DNA polymerase I replaces the missing nucleotide, leaving deoxyribose-5-phosphate on the 5′ side of the nick. This is removed hydrolytically, and DNA ligase seals the nick.

FIGURE 26.9

The uracil-binding pocket of human UNG. The figure shows interactions with Asn-204, known to be involved in catalysis, and the negative steric interaction that would occur with Tyr-147, if the bound base were thymine instead of uracil.

Reprinted from *DNA Repair and Mutagenesis*, 2nd ed., E. C. Friedberg, G. C. Walker, W. Siede, R. D. Wood, R. A. Schultz, and T. Ellenberger. © 2006 ASM Press, Washington, DC.

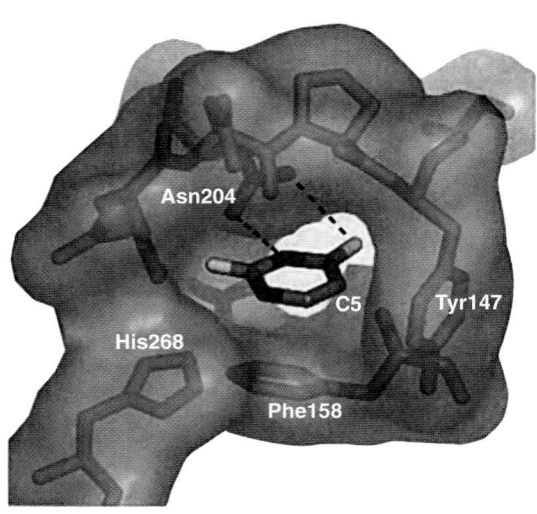

A base excision repair process removes uracil residues in DNA, whether they arose through deamination of cytosine residues or incorporation of deoxyuridine nucleotides instead of thymidine nucleotides.

system. Uracil residues in DNA can also arise through spontaneous deamination of cytosine residues. The latter alteration does change the genetic sense because it converts a G–C base pair to a G–U pair, and in a subsequent round of replication the U-containing strand would give rise to an A–T base pair. The uracil repair system prevents this mutation but does not discriminate between uracils paired with adenines or with guanines. Consistent with this model is the **hypermutable** phenotype displayed by mutants lacking an active uracil *N*-glycosylase. Such strains exhibit elevated rates of spontaneous mutagenesis, resulting from the accumulation of DNA dUMP residues paired with dGMP.

The replacement of uracil by thymine in DNA is one example of BER. Whereas the process always begins with cleavage of a glycosidic bond, BER proceeds via somewhat different pathways, as shown in Figure 26.10. In the center is the most common pathway, like that shown for DNA-uracil repair, with glycosylase action followed by action of an AP-endonuclease and deoxyribose-5-phosphatase. However, some DNA glycosylases have an associated **AP-lyase** activity (left pathway) in which the phosphodiester bond on the 3′ side of the repair site is cleaved hydrolytically. A phosphodiesterase then removes deoxyribose-5-phosphate, leaving a gap that is filled by DNA polymerase. To the right is shown a "long-patch" repair system seen in some eukaryotic BER pathways. Here, DNA polymerase catalyzes a strand-displacement reaction, with the displaced 5′-terminated "flap" eventually being cleaved by a "flap endonuclease" (FEN1) before final DNA ligase action.

Repair of Oxidative Damage to DNA

Most cells contain several DNA-*N*-glycosylases, including those specific for the alkylated bases *N*-methyladenine, 3-methyladenine, and 7-methylguanine. Of particular interest is the BER process used to repair oxidative DNA damage. As mentioned previously, one of the most abundant oxidation products resulting

FIGURE 26.10

Pathways of base excision repair. The square in the first reaction represents the excised base, and the oval represents a deoxyribose-5′-phosphate residue. For further details see text.

Adapted from *DNA Repair and Mutagenesis*, 2nd ed., E. C. Friedberg, G. C. Walker, W. Siede, R. D. Wood, R. A. Schultz, and T. Ellenberger. © 2006 ASM Press, Washington, DC.

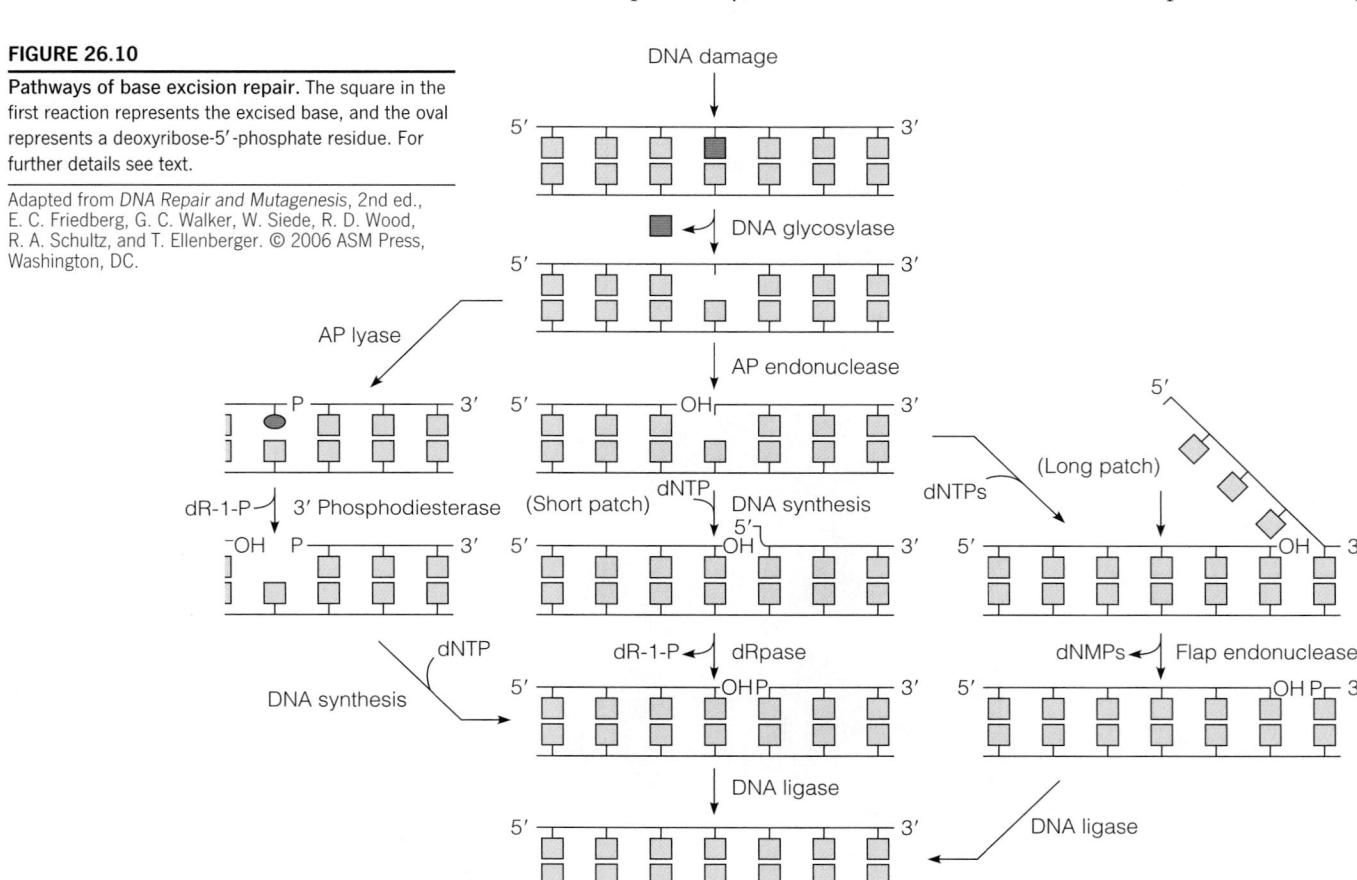

from DNA exposure to reactive oxygen species is 8-oxoguanine. This is a strongly mutagenic alteration because 8-oxoguanine pairs readily with adenine, as shown below.

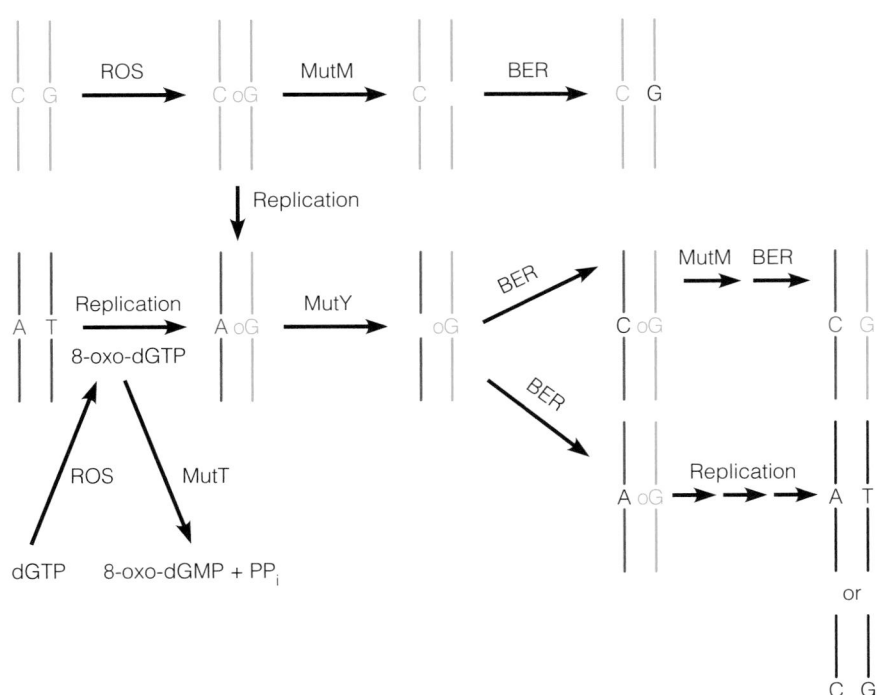

Adenine   8-Oxoguanine

The A-oxoG mispair can be an intermediate in a **transversion** mutation (AT → CG or CG → AT). In *E. coli* three genes encode proteins that minimize oxidative mutagenesis: *mutM*, *mutY*, and *mutT* (the "mut" designation indicates that mutations in any one of these genes have the effect of increasing the spontaneous mutation rate). All three gene products have mammalian homologs. MutM and MutY are both DNA-base glycosylases. MutM (or its mammalian counterpart, OGG1) cleaves 8-oxoguanine from DNA, to initiate BER. MutY cleaves adenine when it is matched with 8-oxoguanine in the other DNA strand. Interestingly, MutT has nothing to do with base excision repair. Instead, it is a nucleotidase, cleaving 8-oxo-dGTP, the dNTP of 8-oxoguanine, to the corresponding nucleoside monophosphate.

$$8\text{-oxo-dGTP} + H_2O \rightarrow 8\text{-oxo-dGMP} + \text{pyrophosphate}$$

This reaction is thought to "sanitize" cellular dNTP pools, by eliminating a nucleotide whose incorporation into DNA would be strongly mutagenic. Hence, as shown in Figure 26.11, two of the three *mut* gene products act to repair damage caused by the presence of 8-oxoguanine in DNA, while the third, MutT, acts to minimize incorporation of a damaged nucleotide into DNA.

FIGURE 26.11

Actions of *mutM*, *mutT*, and *mutY* gene products in countering the mutagenic effect of 8-oxoguanine (oG). Depending upon the route of introduction of oG to DNA, the pathways shown can cause either GC → AT or AT → GC transversions. MutT hydrolyzes 8-oxo-dGTP and prevents its incorporation during DNA replication. MutM excises oG from a C-G base pair to begin BER. MutY excises A from an A-oG base pair in another BER process, allowing an additional possibility for correcting an error in the next round of replication. Not all possible outcomes are shown in the figure. For example, 8-oxo-dGTP can be incorporated opposite template C, as well as opposite A. Also, MutY can excise A from an A-G base pair, thereby preventing spontaneous transversion mutagenesis. ROS, reactive oxygen species.

FIGURE 26.12

Structure of human OGG1 with either guanine or 8-oxoguanine bound flipped out and bound near the catalytic pocket. A combination of forces completely excludes guanine from the catalytic pocket. PDB IDm, 1YQK, 1YQR.

Based upon A. Banerjee, W. Yang, M. Karplus, and G. L. Verdine (2005) *Nature* 434:612–618, Structure of a repair enzyme interrogating undamaged DNA elucidates recognition of damaged DNA.

G complex   oxoG complex

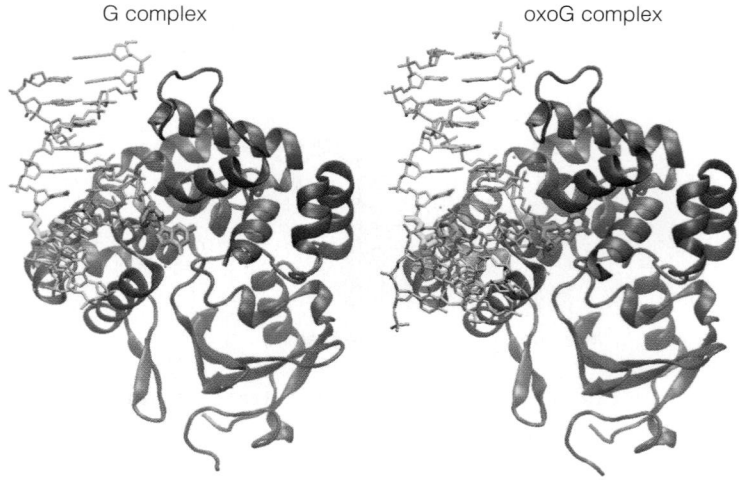

The question of how a DNA glycosylase "interrogates" its DNA substrate to distinguish target from nontarget bases is important because the enzyme must interact with thousands of dGMP residues for every 8-oxo-dGMP that it detects and removes. A cross-linking strategy was used with the human form of OGG1 to trap and determine the structures of complexes with either 8-oxoguanine or guanine flipped out from the double helix. As shown in Figure 26.12, guanine is completely excluded from the 8-oxoguanine-binding pocket. Both attractive and repulsive forces were found to contribute toward this exclusion. A remaining question pertains to the rate of the interrogation process. How much time is required to interrogate each nontarget dGMP compared to that needed for each target oxidized nucleotide?

Mismatch Repair

Mismatches, or non-Watson–Crick base pairs in a DNA duplex, can arise through replication errors, through deamination of 5-methylcytosine in DNA to yield thymine, or through recombination between DNA segments that are not completely homologous. In addition, mismatches result when DNA polymerase slides along its template, creating short loops or bulges in duplex DNA. We best understand the correction of replication errors so that is what we describe here.

If DNA polymerase introduces an incorrect nucleotide, creating a non-Watson–Crick base pair, the error is normally corrected by 3' exonucleolytic proofreading. If the error is not corrected immediately, the fully replicated DNA will contain a mismatch at that site. This error can be corrected by another process, called **mismatch repair.** In *E. coli* the proteins that participate include the products of genes *mutH, mutL,* and *mutS.* Another required gene product, originally called MutU, has now been identified as DNA helicase II (also identified as the product of the *uvrD* gene).

The mismatch correction system scans newly replicated DNA, looking for both mismatched bases and single-base insertions or deletions. MutS binds to DNA at the site of the mismatch, followed by the binding of MutL and then MutH. MutS is a "motor protein," which uses the energy of ATP hydrolysis to pull DNA from both directions until it reaches the site at which the repair process is to begin. When it finds an appropriate signal, part of one strand containing the mismatched region is cut out and replaced (Figure 26.13). How does the mismatch repair system recognize the right strand to repair? If it chose either strand randomly, it would choose incorrectly half the time and there would be no gain in replication accuracy. The answer is that the mismatch repair enzymes identify the newly replicated strand because for a short period that DNA is unmethylated. In *E. coli* the sequence —GATC— is crucial because that is the site methylated soon

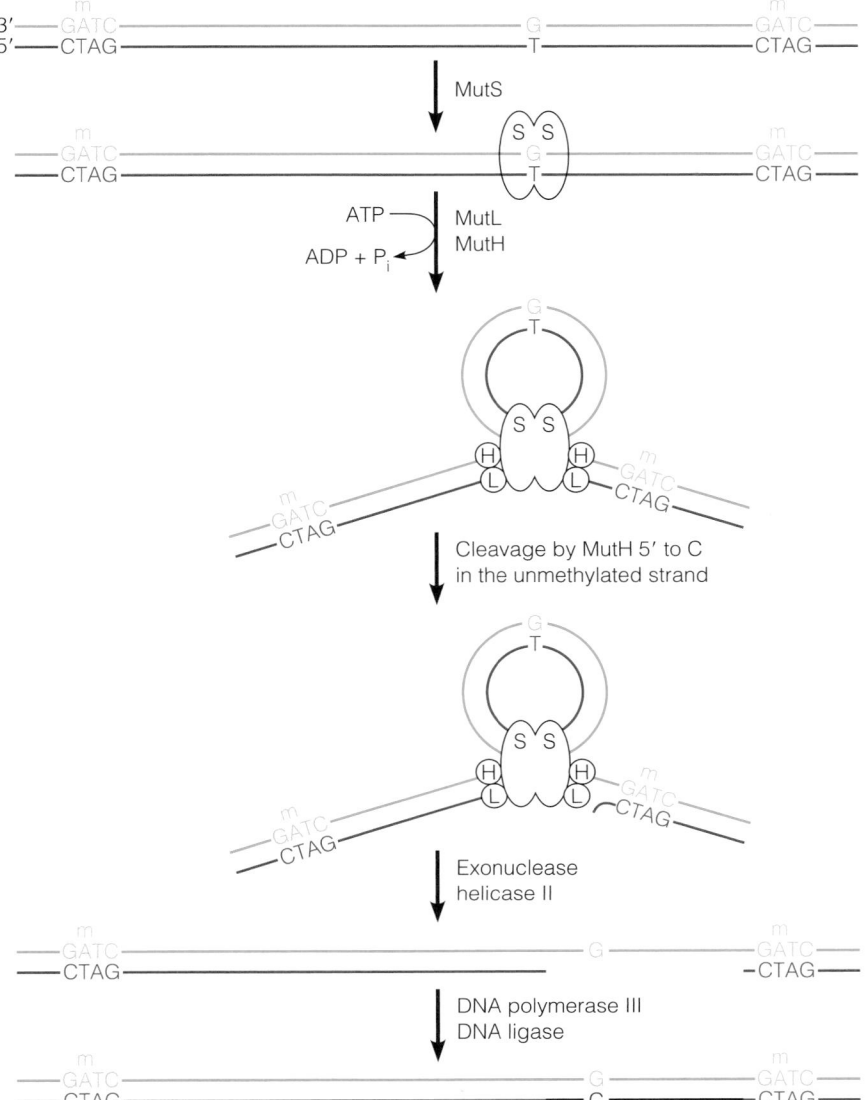

FIGURE 26.13

Methyl-directed mismatch repair in *E. coli*.
The newly replicated daughter strand (red) contains a T mismatched to G in the template strand (blue). The mismatch repair system identifies the daughter strand because it is not yet methylated. Thus, this system must function before the newly replicated daughter strand becomes methylated, through action of the Dam methylase on the A residue in the GATC sequence.

after replication, by action of the product of the *dam* gene (<u>D</u>NA <u>a</u>denine <u>m</u>ethylase). The mismatch repair enzymes look for —GATC— sequences that are not methylated. Recognition of an unmethylated GATC can target that strand for mismatch correction at a site as far as 1 kbp between the mismatch and the GATC site, in either direction. Once the methylation system has acted on all GATC sites in the daughter strand, it is too late for the mismatch repair system to recognize the more recently synthesized DNA strand, and any advantage in total DNA replication fidelity is lost. When the system functions properly, it has the effect of increasing overall replication fidelity by about 100-fold—from about 1 error in 10^7 base pairs replicated to about 1 in 10^9 or more.

As shown in Figure 26.13, the MutHLS complex moves along DNA in both directions until it encounters the nearest 5′-GATC sequence. An endonuclease activity of MutH then cleaves on the 5′ side of the G in the unmethylated strand. At that point helicase II unwinds the DNA, moving back past the mismatch, followed by an exonuclease that digests the displaced single strand—exonuclease VII or RecJ nuclease if the cut is 5′ to the mismatch, and exonuclease I or X if the cut is on the 3′ side (as shown in Figure 26.13). The resultant gap is filled by DNA polymerase III holoenzyme and DNA ligase, working in concert with SSB.

In bacteria the mismatch repair system uses DNA methylation to identify the strand that has a mispaired nucleotide.

A similar mismatch repair system exists in eukaryotic cells. However, the mismatch recognition step is somewhat more complex because three different MutS homologs (MSH proteins) are involved—MSH2, MSH3, and MSH6. These three proteins form heterodimers, with different mismatch specificities; that is, an MSH2–MSH6 complex recognizes single-base mismatches, insertions, and deletions, while an MSH2–MSH3 complex recognizes insertions and deletions of two to four nucleotides. Four heterodimeric MutL homologs exist also, but a MutH homolog has not been found. In fact, a major unanswered question is the mechanism by which the eukaryotic systems recognize and initiate repair on the newly replicated DNA strand; selective methylation is evidently not involved. Recent evidence suggests that the 5′ end of Okazaki fragments may involve the strand-identifying process.

Just as seen with bacteria, mutations in eukaryotic genes that control mismatch repair confer a mutator phenotype, raising spontaneous mutation rates at all loci. How do such mutations affect the biology of human cells? As mentioned in Chapter 23, Lawrence Loeb predicted in 1974 that the progression of a normal cell to a cancer cell involves the creation of a mutator phenotype because the natural mutation rate for somatic cells seemed too low to account for the number of heritable changes that occur during tumor cell progression. Two decades later, this prediction was confirmed, when mutations in mismatch repair proteins were found in tumor cells from individuals with an inherited cancer predisposition called HNPCC (heritable nonpolyposis colon cancer). To date, germ-line mutations in the genes for five different mismatch repair proteins have been found to be associated with HNPCC. The most common forms of the disease involve mutations affecting either hMLH1 (human Mut L homolog) or hMSH2 (human MutS homolog). The cancer predisposition is inherited in an autosomal dominant fashion, suggesting that most affected individuals are heterozygous, with one wild-type and one nonfunctional allele. Mismatch repair is normal until a somatic cell mutation inactivates the one functional allele, and mismatch repair capacity is essentially abolished, with a consequent increase in spontaneous mutagenesis. It is not clear why mutations affecting mismatch repair are associated specifically with tumors in the colon.

Tumor cells from those affected with HNPCC exhibit a phenomenon called **microsatellite instability**—a large number of mutations in regions of the genome containing repeats of single-, double-, and triple-nucleotide sequences, usually with large increases in the numbers of repeating units in such sequences. These data suggest that the product and template strands can normally slip at such sites so that DNA polymerase copies a short repeating sequence more than once, or else skips a segment. This creates a heteroduplex with a short loop, as shown for deletion mutagenesis in the margin. Normally a replication error of this type would be corrected by mismatch repair, but in a cell lacking normal mismatch repair, such errors would persist and accumulate. Studies of this type have given scientists enormous insight into the nature of cancer as a progressive genetic disease.

Daughter-Strand Gap Repair

Photoreactivation and excision repair are short-term metabolic responses. Both processes can occur within a few minutes after DNA suffers chemical or ultraviolet damage. However, if the photoreactivation and excision repair systems are defective, or become saturated by DNA damage too extensive for their capacities, longer-term systems can come into play in bacteria. One such system, **daughter-strand gap repair**, is comparable to genetic recombination, and indeed involves several of the same proteins. We shall say more about some of these proteins in the section on recombination, but we will briefly introduce the process here, as studied primarily in bacteria.

When a replicative polymerase encounters a thymine dimer or a bulky lesion, it cannot replicate past this site. If a thymine dimer is involved, polymerase III will

---

Genetic deficiencies in mammalian mismatch repair are often associated with colorectal cancer.

```
3′ —C—A—C—A—C—A—C—A—C—A—
5′ —G—T—G—T—G—T—OH
```
                2 template
                nucleotides
                loop out

```
            C—A
            |  |
 —C—A—C—A    C—A—C—A—
 —G—T—G—T—G—T
```
                DNA synthesis
                continues

```
            C—A
            |  |
 —C—A—C—A    C—A—C—A—
 —G—T—G—T—G—T—G—T
```
                Another round
                of replication

```
3′ —C—A—C—A—C—A—C—A—C—A—
5′ —G—T—G—T—G—T—G—T—G—T—

3′ —C—A—C—A—C—A—C—A—
5′ —G—T—G—T—G—T—G—T—
```
                2-base-pair deletion

incorporate dAMP opposite the first thymine base in the template. The double helix is distorted because of the thymine dimer, causing the structure to be recognized as a mismatch, and the 3′ exonucleolytic activity of DNA polymerase III cleaves out the newly incorporated dAMP. Thus, polymerase creates a futile cycle, "idling" at the damage site, converting dATP to dAMP by a continual process of insertion and exonucleolytic cleavage. Synthesis of an Okazaki fragment can commence on the 5′ side of the damaged site, leaving a gap starting on the 3′ side of the thymine dimer. This gap would be lethal if unrepaired, however, because it would generate a double-strand break in the next round of replication.

A process akin to genetic recombination can take place, not to repair the damage, but to allow replication to continue, saving the damaged site for subsequent repair by another process. Daughter-strand gap repair depends critically upon a multifunctional protein called **RecA**. Bacteria carrying mutations in *recA* were originally characterized as defective in general recombination and DNA repair. We now know that *recA⁻* bacteria, which specify a defective RecA protein, have a complex phenotype, including defective DNA repair. We shall describe the protein later in this chapter, but for our purposes here one property is important. RecA catalyzes **strand pairing**, or **strand assimilation**—the joining of two different DNAs by homologous base pairing with each other.

Because the gap that is generated at the site of a thymine dimer is created by faulty replication, the gap is close to the replication fork. Therefore, it is also close to the corresponding region on the other daughter duplex (Figure 26.14). If that region has itself not sustained damage, the RecA protein can initiate recombination between two homologous duplexes. The uninvolved parental strand, which is complementary to the damaged parental strand, recombines into the gap, opposite the damaged site. Additional proteins are involved, as shown in the figure and discussed in the section on recombination (see page 1097). A gap now exists in the previously undamaged arm, but because it lies opposite an undamaged template, it can be filled by action of DNA polymerase and DNA ligase. The thymine dimer itself is not repaired in this process, but the process allows time for an excision system to come in later and repair this damage. RecA protein is required for daughter-strand gap repair, particularly in the first reaction, where the undamaged parental strand undergoes pairing with the parental strand opposite the gap. RecA is a bacterial protein, but a related protein, Rad51, allows this process to occur in eukaryotic cells as well.

There is evidence that the response to a blocking lesion in the leading strand template is **replication fork regression** (see Figure 26.15). The fork moves backward instead of forward, to form a "chicken-foot" intermediate. In this structure the blocked leading strand can continue to replicate, using the other daughter strand as the template, and the lesion being bypassed, for repair to occur subsequently. When damage is on the lagging strand (not shown), it need not be corrected immediately, but a gap remains, which must be repaired—usually by daughter-strand gap repair. Note that both in this process and in daughter-strand gap repair the damaged site is not repaired, but replication can continue, and time is allowed for one of the other repair processes to take over and fix the damage.

Translesion Synthesis

As mentioned in the previous section, some types of DNA lesions are noninstructive, in the sense that when a replicative DNA polymerase encounters such a site, it stalls and replication ceases. Such lesions include abasic sites, thymine dimers, and bulky adducts. If damage is so extensive that true repair systems are overwhelmed, then damage-tolerant processes come into play, including the daughter-strand gap repair just described. Another set of processes has been elucidated more recently, with discovery of additional DNA polymerases that are able to replicate past a noninstructive lesion. Most of these enzymes have very low fidelity, partly because they lack 3′ exonucleolytic proofreading and partly because their catalytic sites are larger and more open than those of high-fidelity

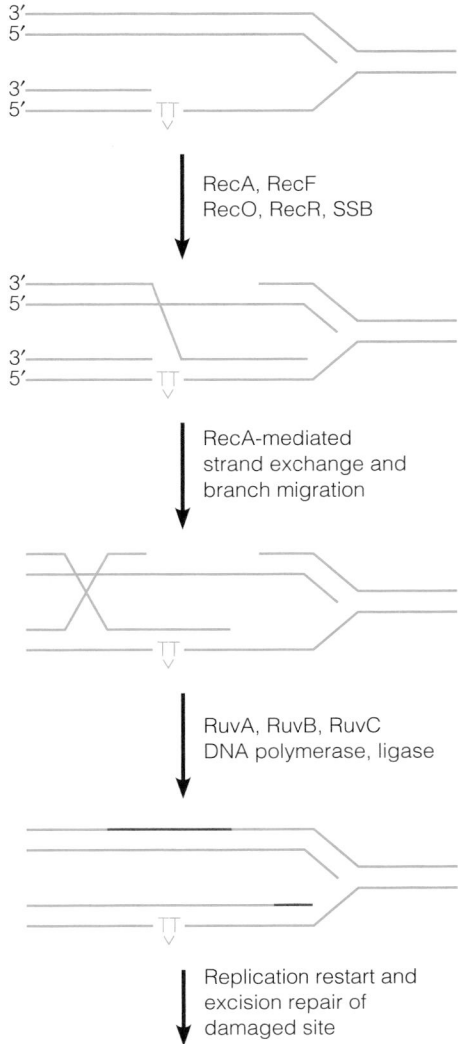

FIGURE 26.14

Daughter-strand gap repair. The undamaged parental DNA strand is transferred to the gap in the daughter strand, formed by the inability of DNA polymerase to replicate past it. The remaining steps occur by mechanisms similar or identical to those in homologous recombination, discussed later in this chapter.

FIGURE 26.15

Replication fork regression as a likely response to a blocking DNA lesion in the leading strand template. **(a), (b)** Regression pairs the damaged site (shown as a triangle) with its undamaged template (green). **(c), (e)** The daughter-strand complement of the damaged strand (brown) can then be copied from the other daughter strand (orange). The fork then reforms and replication can re-start, whether the damage has been repaired **(f)** or not **(d)**.

Adapted from *Nucleic Acids Research* 37:3475–3492, J. Atkinson and P. McGlynn, Replication fork reversal and the maintenance of genome stability. © 2007, by permission of Oxford University Press.

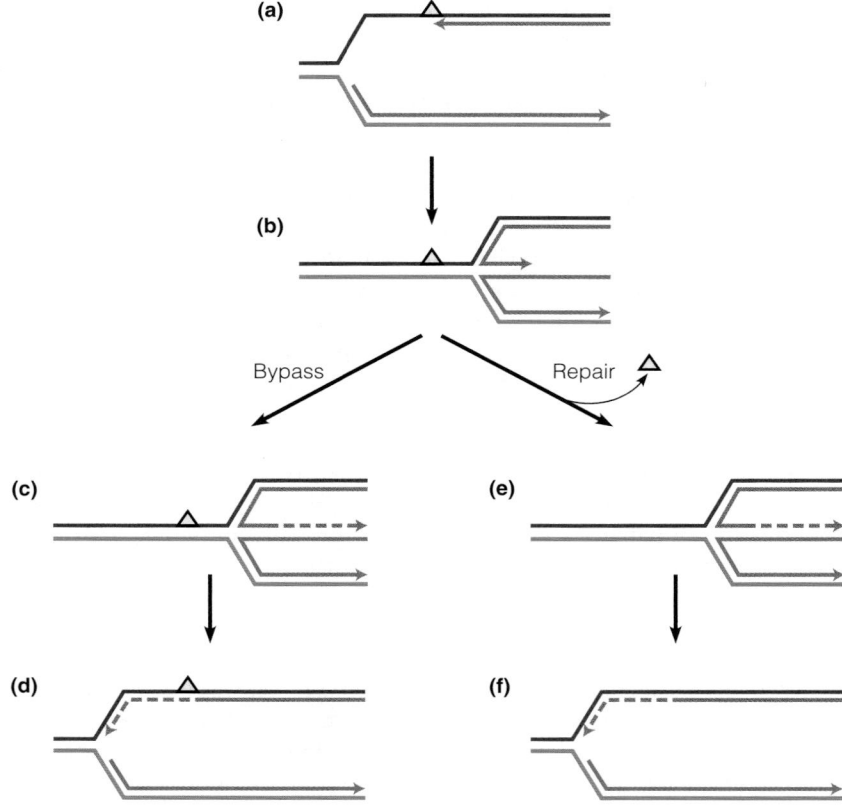

replicative polymerases. As mentioned in Chapter 25, DNA polymerases have been classified structurally into families—A, B, C, D, X, and Y. The more recently discovered damage-tolerant enzymes comprise the Y-family DNA polymerases. The enzymes are inducible as a result of DNA damage, but the control mechanisms involved differ markedly between prokaryotic and eukaryotic organisms. We discuss the bacterial system first.

The bacterial **SOS response** is a metabolic alarm system that helps the cell to save itself in the presence of potentially lethal stresses. Inducers of the SOS response include ultraviolet irradiation, thymine starvation, treatment with certain DNA-modifying reagents such as the cross-linker mitomycin C, and inactivation of genes essential to DNA replication. Responses include mutagenesis, filamentation (in which cells elongate by growth but don't divide), activated excision repair, and activation of latent bacteriophage genomes. Mutagenesis occurs because, under SOS conditions, the gaps that are formed opposite thymine dimers or other damaged sites can be filled by replication rather than by daughter-strand transfer, and, as noted above, this replication is extremely inaccurate. In fact, this process is the principal pathway by which ultraviolet light stimulates mutagenesis in bacteria.

Induction of the SOS response by UV light or a similar damaging condition activates transcription of about forty genes in *E. coli*. We discuss the mechanism of transcriptional activation in Chapter 29 (see Figure 29.16, page 1246). Among these SOS-inducible genes are *recA*, *dinB*, *umuC*, and *umuD*. Din stands for damage-inducible and Umu stands for ultraviolet mutagenesis. Bacteria bearing a *umu* mutation can be killed by UV light, but the survivors are not enriched for mutants; in other words, Umu function is required for mutagenesis by UV light. In recent years *dinB* has been shown to encode a new polymerase, pol IV, and the *umuC* and *umuD* genes to encode another new polymerase, pol V. Both of these enzymes are highly error-prone, and both have been categorized as Y-family polymerases. The structural gene for DNA polymerase II is also SOS-controlled, but pol II is an accurate enzyme and is not involved in translesion synthesis, so it is still an enzyme in search of a function.

After SOS activation, UmuD undergoes specific proteolysis to give UmuD′, and two molecules of this combine with one molecule of UmuC to give a pol V molecule. This associates with RecA bound to ATP, giving the active form of the enzyme. When the replisome containing pol III holoenzyme stalls at a damage site, pol V replaces it and binds to the sliding clamp. Once the damaged site has been copied (inaccurately), pol V steps aside, by a process under intense study, and pol III holoenzyme completes the replication process.

Considering that most mutations are probably deleterious, what is the advantage to the cell of accumulating mutations during DNA repair? Probably none, except that the alternative would be for the cell to die. In other words, extensive mutagenesis is seen as a worthwhile price for the cell to pay, if a massive UV dose has overwhelmed all other repair pathways and if error-prone replication past a dimer is necessary for the cell to stay alive.

As noted earlier, translesion synthesis is inducible in eukaryotic cells as well. Recall from Chapter 25 that eukaryotic cells were thought until recently to have only five DNA polymerases. Since the late 1990s a remarkable variety of new DNA polymerases has been discovered in eukaryotic cells, and, as mentioned in Chapter 25, the total number of human DNA polymerases now stands at 15 (including retroviral reverse transcriptases). Four of the new polymerases—Pol η, Pol ι, Pol κ, and REV1—are Y-family polymerases, and most are specialized to replicate past specific types of DNA damage. These enzymes all have generally low fidelity—high replicative error rates as shown *in vitro*. However, pol η is distinctive in being able to replicate past thymine dimers in an essentially error-free manner. Recent structural studies show how this can occur. As shown in Figure 26.16, Pol η has an active site spacious enough to accommodate a thymine dimer because the palm and fingers are rotated away from the "little finger" domain. In addition, van der Waals forces and hydrogen bonds hold the paired thymines so that both can partner with adenine on incoming dATP. All of the Y family polymerases have been shown to participate in translesion synthesis and to have additional functions as well; elucidation of these functions is an active area of current research.

Induction of some or all of the repair polymerases is part of the **DNA damage response**, a remarkably complex set of events in eukaryotic cells. Although we know little about specific regulation of each of the several Y-family polymerases, we briefly mention here some details of the DNA damage response. We will say more when we discuss events following double-strand DNA breakage.

Several different signaling pathways emanate from DNA-damaging events. The many proteins induced or activated by DNA damage have been classified as follows: (1) *sensors*, which are in contact with DNA and detect the damage, then recruit (2) *mediators*, which amplify the initial signal, and (3) *transducers*, which transmit the signal to (4) *effectors*, which stimulate cellular responses. Depending upon the severity of DNA damage, the cell type, and phase of the cell cycle, the cellular responses may lead to activation of DNA repair pathways, cell cycle arrest, or programmed cell death (**apoptosis**). We have previously mentioned p53 (Chapter 23), an effector that acts as a transcription factor whose effects include either cell cycle arrest, until damage has been repaired, or apoptosis if damage is too severe. Other effectors include cyclin-dedendent kinases, which regulate the cell cycle (Chapter 24).

The pathway that recruits translesion polymerases to damage sites involves first continued unwinding of DNA past a damaged site, with single-stranded DNA being coated with RPA, the eukaryotic single-strand DNA-binding protein, and next, recruitment by this DNA–protein complex of Rad17, the eukaryotic clamp loader. Binding of PCNA, the sliding clamp, follows, and this leads

DNA can be repaired after replication, either by recombination or by inducible error-prone repair. Both processes require RecA.

(a)
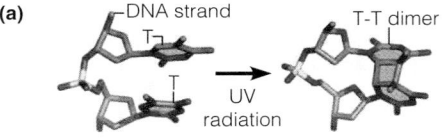

(b)
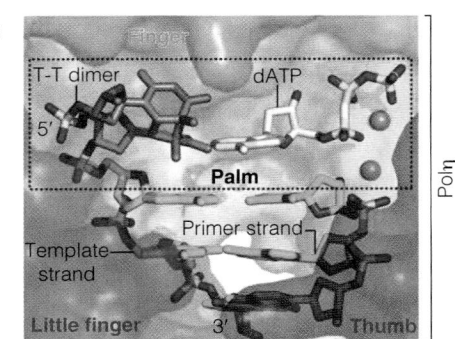

(c)
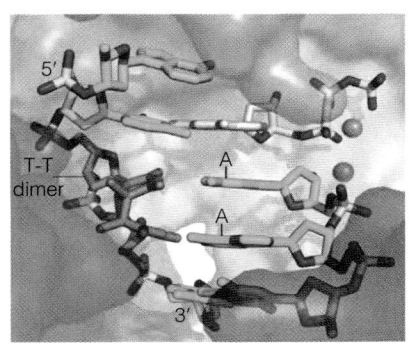

FIGURE 26.16

Accurate lesion bypass by Pol η. **(a)** Structural distortion created when two adjacent thymine bases undergo UV-induced dimerization. **(b)** Human Pol η in a ternary complex with dATP correctly positioned opposite the 3′ T in the dimer in the template strand. **(c)** Crystal structure of human Pol η in a ternary complex in which the DNA has been replicated by the addition of two nucleotides past the thymine dimer.

ultimately to recruitment of translesion polymerases to DNA damage sites. This complex network of reactions is under active investigation.

Double-Strand Break Repair

A double-strand break (DSB), a lesion that can be caused by ionizing radiation or as a consequence of replication stalling, is the most lethal form of DNA damage because it destroys the physical integrity of a chromosome. DSBs occur about 50 times per mammalian cell cycle. Moreover, as we see in the next section, double-strand DNA breaks occur naturally during meiotic recombination. DSBs can be repaired by either of two processes—**homologous recombination** (HR), using sequence information from an undamaged sister chromatid, or **nonhomologous end joining** (NHEJ). The latter process is more efficient, but if DNA ends are not rejoined at the precise sites where breakage occurred, genetic information will be lost or scrambled. By contrast, HR normally repairs the broken site precisely, but it can occur only during S or G2 cell cycle phases, when a homologous chromosome is available. Both processes begin with association of several signaling proteins at the severed ends, of which a principal player in vertebrates is a protein kinase called ATM (ataxia telangiectasia mutated). An early event in both HR and NHEJ is phosphorylation of a variant form of H2 histone, H2AX, in nucleosomes near the break. The target site is Ser-139, in the C-terminal tail of the histone. This may be part of a chromatin remodeling process that helps to move core particles out of the way, exposing DNA ends for processing.

In NHEJ a key participant is a heterodimeric protein called Ku70/80 (one 70-kDa and one 80-kDa protein). This associates with each of the severed ends. Additional end processing occurs, and a synapse is formed, which brings the two ends together. A specialized form of DNA ligase, DNA ligase IV, rejoins the ends. Ligase IV can join blunt ends or ends with compatible single-strand overhangs, and it has the ability to join DNA ends across short gaps. The actual reaction catalyzed in NHEJ has not yet been defined. The process is summarized in Figure 26.17.

Double-strand DNA breaks can be repaired either by homologous recombination or by nonhomologous end joining (NHEJ), a process that does not require DNA sequence homology at the ends being joined.

FIGURE 26.17

A pathway for nonhomologous end joining. The figure shows involvement of some proteins in addition to those identified in the text, including DNA-dependent protein kinase (DNA-PK), the processing protein Artemis, and the X4-L4 complex, which includes DNA ligase IV. For more information about ATM and MRN, see text.

Adapted with permission from *Biochemical Journal* (2009) 423:157–168, A. Hartlerode and R. Scully, Mechanisms of double-strand break repair in somatic mammalian cells. © The Biochemical Society.

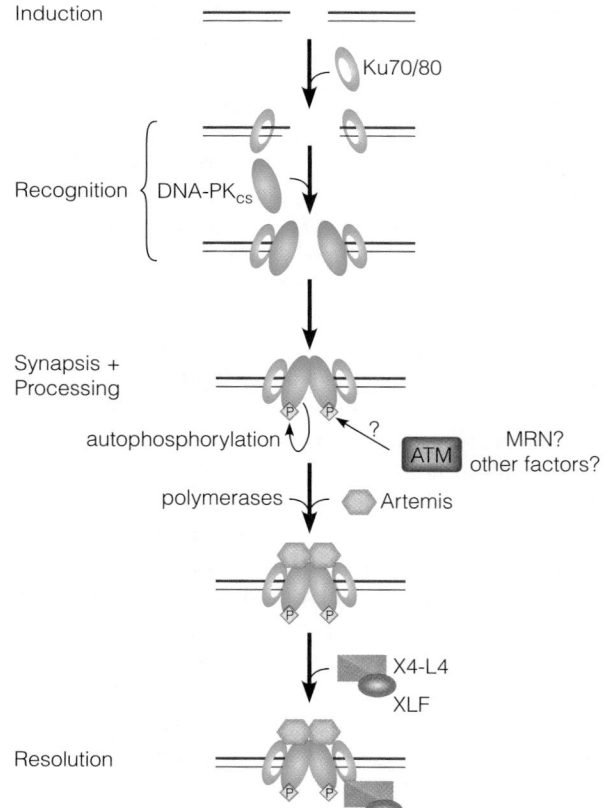

Homologous recombination is simple to visualize conceptually, as shown in Figure 26.18, but in fact is an extremely complex process, involving extensive signaling reactions as part of the DNA damage response. ATM activates a nuclease in a three-protein complex called MRN, as well as signaling downstream effectors via its protein kinase activity. The MRN-associated nuclease trims away 5′ ends, leaving 3′-terminated single-strand ends, which are coated with RPA (replication protein A; single-strand binding protein). The BRCA1 protein is involved in this process. Another protein kinase, ATR, recognizing the single-strand ends, binds and initiates more signaling. At this point RPA is replaced by Rad51, the eukaryotic counterpart to RecA. BRCA2 is involved in this process. The Rad51-ssDNA complex searches for homology, as mentioned earlier, and invades the homologous region of the undamaged sister chromatid. Resolution to give two undamaged chromatids can occur via several pathways, one of which is shown in Figure 26.18. As mentioned, some of these events, as well as regulation of cell cycle progression, are controlled by protein phosphorylation involving ATM and ATR, central regulators in the DNA damage response. These two enzymes have at least 25 known substrates, but the total number of proteins phosphorylated during the DNA damage response exceeds 700.

Human mutations are known in a number of the proteins involved in DSB repair. Mutations in either BRCA1 or BRCA2 were identified some time ago as risk factors for development of breast or ovarian cancer, and BRCA2 mutations are also associated with Fanconi's anemia. ATM received its name from the deficiency syndrome ataxia telangiectasia, which involves premature aging, cerebellar degeneration, and enhanced cancer susceptibility.

Recombination

Population genetics teaches us that the survival of a species depends upon its ability to maintain genetic diversity so that individuals can vary in their ability to respond to unforeseen environmental pressures. Diversity is maintained through both mutation, which alters single genes or small groups of genes in an individual, and recombination, which redistributes the contents of a genome among various individuals during reproduction. In classical biology, recombination is the outcome of crossing over between paired sister chromosomes during meiosis in eukaryotes, and in fact our earliest information about recombination came from cytological and cytogenetic observations in *Drosophila*. However, recombination encompasses more processes and biological functions than those involved in sexual reproduction. Strictly speaking, **recombination** is any process that involves the formation of new DNA from distinct DNA molecules, such that genetic information from each parental DNA molecule is present in the new molecules. The daughter-strand gap repair process described earlier is a form of recombination. So also is the integration of certain bacteriophage or plasmid genomes into the chromosomal DNA of a host bacterium; many viral genomes integrate into animal host cells, as seen for HIV in Chapter 25. Recombination is also involved in **transposition**, the movement of DNA from one chromosomal integration site to another. As we see later in this chapter, recombination of this type is involved in the vertebrate immune response, in generating antibody diversity.

Classification of Recombination Processes

Different recombination processes have quite distinct requirements, both for nucleotide sequence homology between the recombining partners and for proteins and enzymes to catalyze the process. Meiotic recombination in diploid organisms requires extensive sequence homology between the recombining partners and accordingly is called **homologous recombination**. This term also applies to certain recombinational events between bacterial chromosomes. New DNA can be introduced into a bacterial cell by various processes: (1) conjugation during bacterial

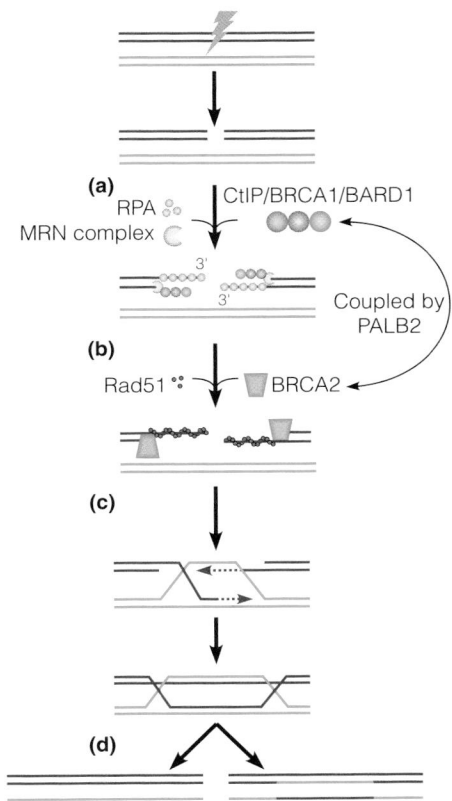

FIGURE 26.18

A pathway for homologous recombination to repair DNA double-strand breaks. **(a)** MRN, aided by other proteins, trims 5′ ends, and the single-strand 3′ ends are coated with replication protein A, single-strand binding protein. **(b)** RPA is replaced by Rad51 (RecA counterpart), aided by the BRCA2 protein. **(c)** Strand invasion of the undamaged homologous chromosome is followed by DNA synthesis and cutting, **(d)** which can yield either recombinant (right) or nonrecombinant (left) chromosomes.

Adapted with permission from *Biochemical Journal* (2009) 423:157–168, A. Hartlerode and R. Scully, Mechanisms of double-strand break repair in somatic mammalian cells. © The Biochemical Society.

Recombination is any process that creates end-to-end joining from two different DNA molecules.

mating; (2) transformation, when DNA is taken up by cells; or (3) **transduction**, when bacterial DNA that was packaged into a phage particle is introduced by infection. Transduction results when the assembly of phage particles in an infected cell goes awry, and bacterial DNA is assembled into a phage head. Reinfection of another bacterial cell introduces that packaged bacterial DNA into the new cell.

If the introduced DNA contains a replication origin, as in a plasmid, it can replicate autonomously once inside a new bacterial cell. More often the DNA does not contain an origin, and its information can be expressed and maintained only if the DNA is taken up into the resident chromosome by homologous recombination. Biochemical analysis of homologous recombination has focused on prokaryotic systems, which in turn have provided insight into meiotic recombination. Most bacterial homologous recombination processes share a common requirement for the RecA protein or its counterpart.

Site-specific recombination, by contrast, involves only limited sequence homology between recombining partners. Sites of breaking and joining are determined by specific DNA–protein interactions. The process was first described in bacteriophage λ. After infection the linear DNA undergoes circularization, guided by its cohesive ends (see Chapter 25, page 1037). The λ chromosome can either undergo multiple rounds of replication leading to virus production or undergo integration into a specific site on the host chromosome. In the latter condition, called **lysogeny**, most of the viral genes are inactivated, and the virus can maintain a long-term, nonlethal relationship with its host (Figure 26.19). Bacteriophages that can establish lysogeny are called **temperate phages**, in contrast to **virulent** phages, such as phage T4, which always lyse their host cells after infection. In most (not all) cases of lysogeny, integration occurs at a specific site. Integrative recombination in phage λ is being studied as a model for understanding integration of tumor virus genomes into DNA of infected cells. It was earlier studied as a model for homologous recombination between chromosomes, but the processes are different, as shown when it was learned that the phage and bacterial DNA sequences in the regions undergoing recombination have only 15 base pairs of homology. Moreover, the RecA protein is not required for this process. Rather, the virus encodes a site-specific enzyme, called **integrase**, and specific DNA–protein interactions between the enzyme and the recombining partners, rather than extensive DNA–DNA sequence homology, determine the site of recombination.

FIGURE 26.19

Site-specific recombination, establishing lysogeny in bacteriophage λ. The phage chromosome circularizes between genes *A* and *R*, and recombination takes place between the *attP* site and a corresponding region, *attB*, on the *E. coli* chromosome between the *gal* and *bio* markers. The enzyme integrase carries out the site-specific recombinational event with the help of a bacterial protein. *O* and *b* are additional genetic markers.

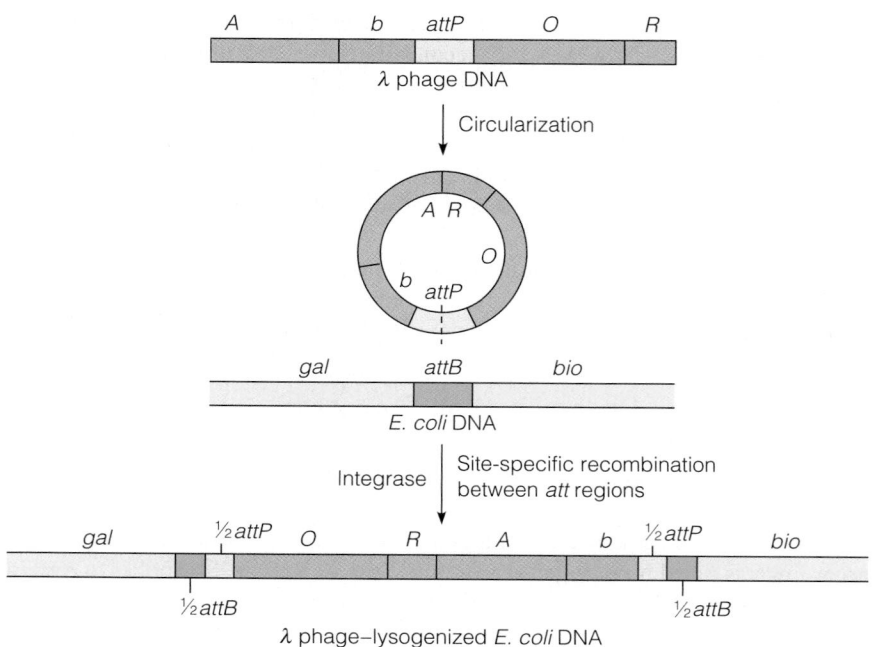

TABLE 26.1	Characteristics of different types of genetic recombination		
	Requirement		
Type	Sequence Homology	RecA Protein or Counterpart	Sequence-Specific Enzyme
Homologous	Yes	Yes	No[a]
Site-specific	Yes (about 15 bases)	No	Yes
Transposition	No	No	Yes
Illegitimate	No	No	Unknown

[a]The Chi site or its counterpart determines cutting sites (see page 1103), but initial recognition of pairing sites occurs by sequence homology.

We recognize two other forms of recombination. *Transposition* involves neither sequence homology nor the RecA protein but does require a special sequence on the donor DNA. This process is discussed in a later section. **Illegitimate recombination**, an extremely rare event that possibly occurs by chance, involves neither sequence homology nor the action of any known protein. Table 26.1 summarizes the main distinctions among the four major types of recombination.

Homologous Recombination

Breaking and Joining of Chromosomes

The most straightforward way to accomplish recombination is to break and rejoin DNA molecules. However, if recombination occurs this way, the sites of breakage must be precisely the same on both recombining chromosomes for intact genes to be regenerated. Some researchers favored alternative mechanisms, but in 1961 Matthew Meselson and Jean Weigle showed that recombination in fact occurs via breakage and rejoining of chromosomes. The demonstration, diagrammed in Figure 26.20, involved a Meselson–Stahl type of experiment (see Figure 4.14, page 104). *E. coli* was infected with two genetically marked λ phage populations, one of which had been density labeled by growth in $^{13}C-^{15}N$ medium. The phage particles resulting from this cross were centrifuged to equilibrium in a cesium chloride gradient. Phages with recombinant genotypes were recovered from all parts of the gradient, whereas nonrecombinant phages were uniformly light or heavy. This result could occur only if the recombinant phages contained DNA derived from both parents, by breaking and rejoining.

This experiment had one other important result. The phage output from the crosses was analyzed in standard fashion, by visual examination of plaques. Although each plaque arises from a single phage particle, many of the plaques in the Meselson–Weigle experiment contained phages of two different genotypes, even though all arose from a single infecting phage. This suggested that recombination involves the formation of a **heteroduplex DNA** region, in which one DNA strand comes from one parent and the other from the second parent. If the heteroduplex region contains a mismatch, subsequent replication of that DNA gives rise to two progeny DNA molecules of different genotypes, as shown in Figure 26.21.

Another early observation was that recombination is stimulated by processes that nick or break DNA strands, such as thymidine starvation or UV irradiation. This suggested a role for single-strand DNA, or free DNA ends, in initiating recombination.

Models for Recombination

Putting the preceding observations together along with data on recombination in fungi, Robin Holliday proposed in 1964 a model for homologous recombination between duplex DNA molecules. That model, detailed in Figure 26.22, continues to dominate thinking and experiments about this process. You will note a marked

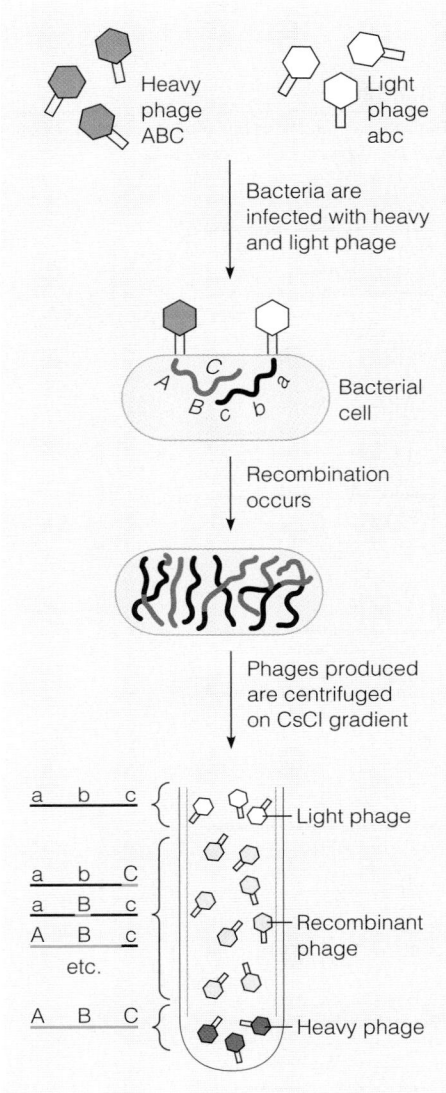

FIGURE 26.20

The Meselson–Weigle experiment. This experiment established that genetic recombination occurs by breakage and reunion of DNA strands.

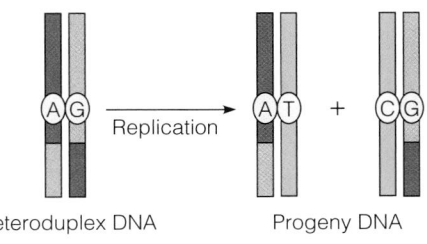

FIGURE 26.21

Generation of progeny phage containing two genotypes by replication of heteroduplex DNA.

FIGURE 26.22

The Holliday model for homologous recombination. A, a, Z, and z are genetic markers.

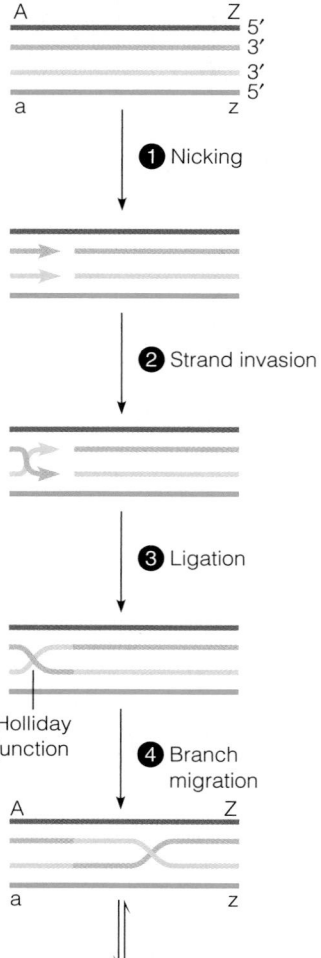

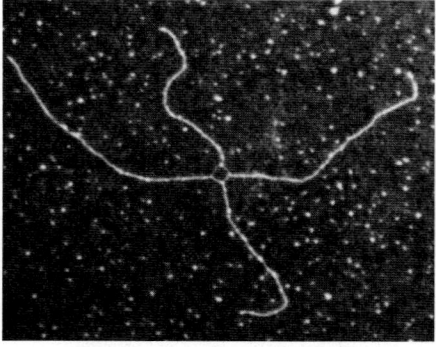

FIGURE 26.23

Electron microscopic visualization of a Holliday junction. This junction was created during recombination between two plasmid DNA molecules.

Science VU/H. Potter-D. Dressler/Visuals Unlimited, Inc.

resemblance of this model to the mechanism of homologous recombination in double-strand break repair (page 1097).

Holliday proposed that recombination begins with nicking at the same site on two paired chromosomes (step 1). Partial unwinding of the duplexes is followed by **strand invasion**, in which a free single-strand end from one duplex pairs with its unbroken complementary strand in the other duplex, and vice versa (step 2). Enzymatic ligation generates a crossed-strand intermediate, called a **Holliday junction** (step 3). The crossed-strand structure can move in either direction by duplex unwinding and rewinding (branch migration, step 4). The Holliday junction "resolves" itself into two unbroken duplexes, by a process of strand breaking and rejoining. The process leading to recombination begins with isomerization of the Holliday structure (step 5), followed by strand breakage so that the strands that break (in step 9) are those that were *not* broken in step 1. Resolution of the resulting structure (steps 10 and 11) generates two chromosomes recombinant for DNA flanking the region and each containing a heteroduplex region. However, if the original crossed strands (those that *were* broken in step 1) break and rejoin (steps 6–8), the products are nonrecombinant duplexes, each containing a heteroduplex region (that is, nonrecombinant with respect to the outside markers A and Z).

Considerable evidence now supports the central tenets of the Holliday model, particularly the electron microscopic visualization of Holliday junctions (Figure 26.23) and more recently crystallographic analysis of a synthetic Holliday junction (Figure 26.24). However, the model has been modified as new data have emerged. Matthew Meselson and Charles Radding proposed that recombination could start with a single nick (which eliminates the nagging question of how two

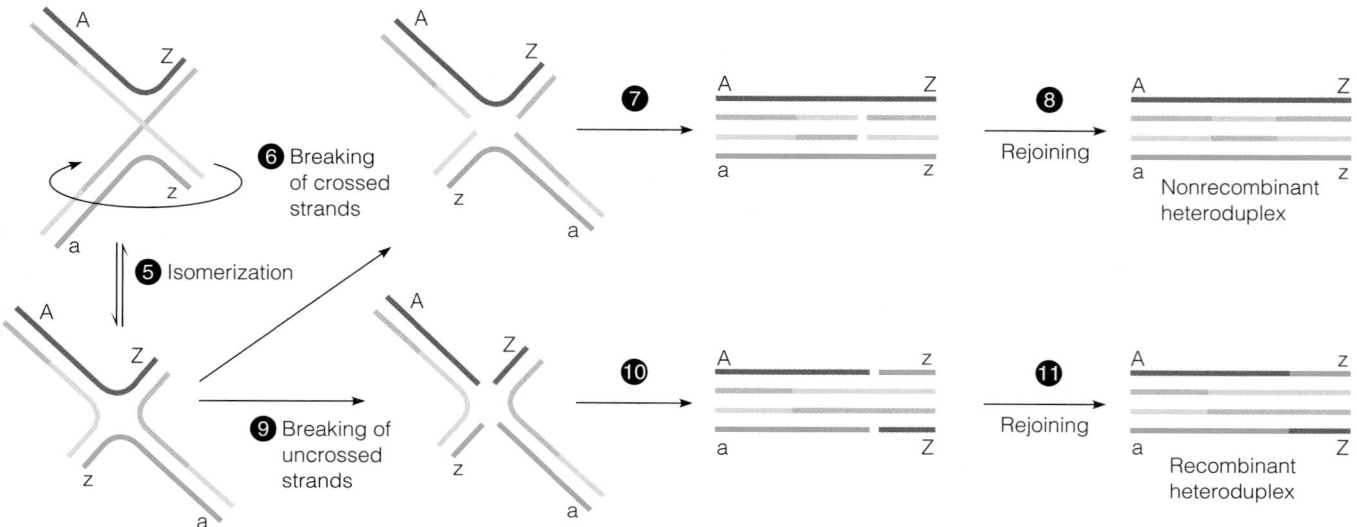

(a)

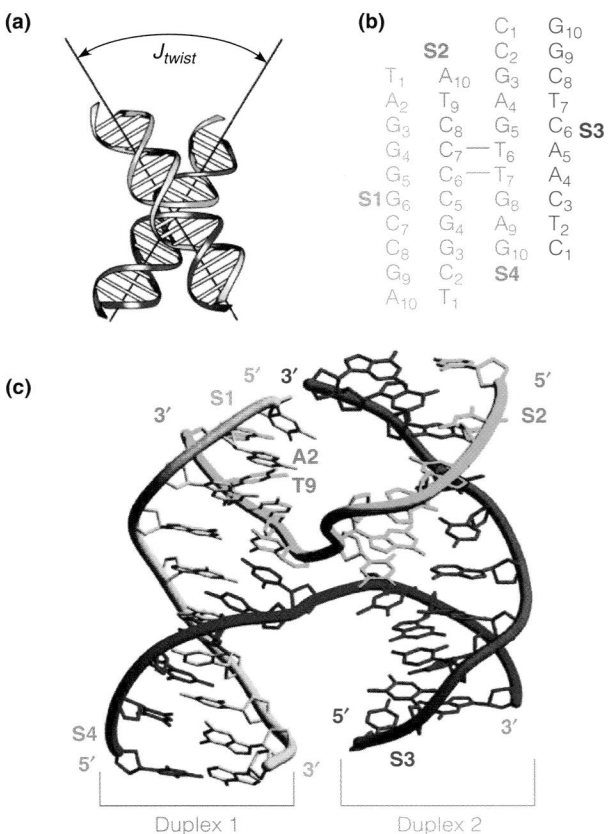

(b)

(c)

FIGURE 26.24

Crystal structure of a synthetic Holliday junction. This four-stranded structure was synthesized by annealing four chemically synthesized oligonucleotides [shown in **(b)**] designed to mimic a Holliday junction. **(a)** Shows the angle between the two duplexes forming the junction.

Adapted with permission from *Biochemistry* 48:7824–7832, P. Khuu, and P. S. Ho, A rare nucleotide base tautomer in the structure of an asymmetric DNA junction. © 2009 American Chemical Society.

duplexes could be nicked at precisely the same point). As shown in Figure 26.25, strand displacement synthesis occurs (magenta arrow, step 1), with a displaced single-strand end of the nicked duplex, A, invading the homologous region of the unbroken duplex, B (step 2). Eventually the displaced loop on duplex B is cleaved and partially degraded (step 3), and the displaced end from duplex A is ligated to B (step 4). Isomerization then occurs, as in the Holliday model, with the originally unbroken strands crossed (step 5). An additional feature of the Meselson–Radding model is the fact that branch migration can lead the final strand cutting to occur some distance from the site of either strand invasion or the original nick (steps 6 and 7). In principle, either a 5′ or a 3′ end could initiate the strand invasion process. As noted below, a 3′ end probably initiates recombination in *E. coli*.

Proteins Involved in Homologous Recombination

The Holliday and Meselson–Radding models explained most of the existing data on homologous recombination between paired chromosomes, particularly as studied in lower eukaryotes such as yeast. Moreover, the models could easily be adapted to explain daughter-strand gap repair or the recombination that occurs in bacteria after transformation or conjugation. Some of the proteins thought to participate, notably DNA polymerase, DNA ligase, and single-strand DNA-binding protein, had been characterized and shown to participate in recombination. What about other proteins that must function if the models are largely correct? To

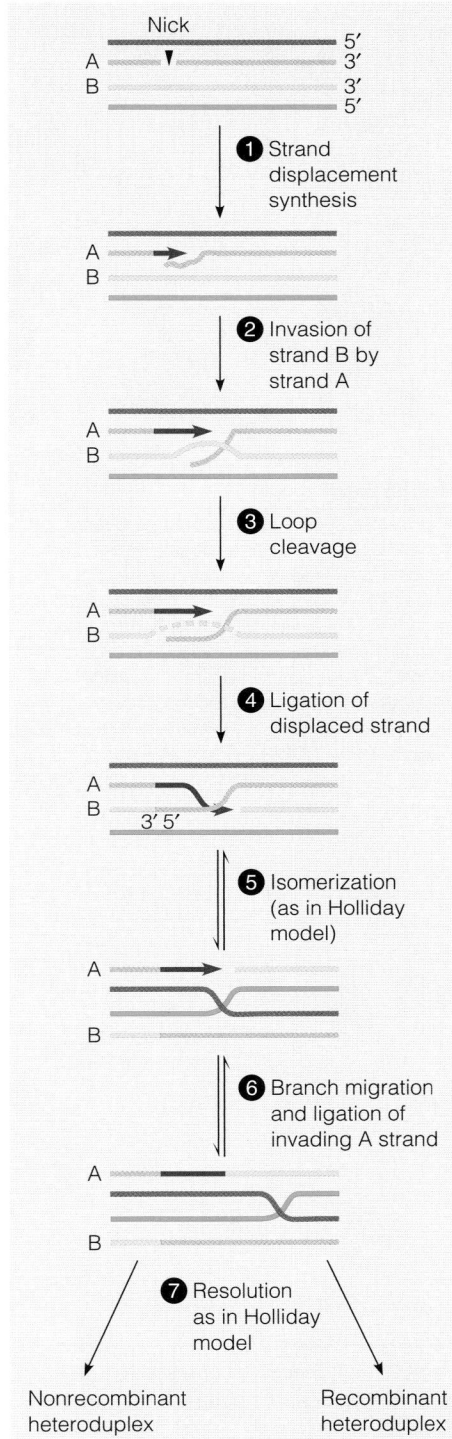

FIGURE 26.25

The Meselson–Radding model for homologous recombination.

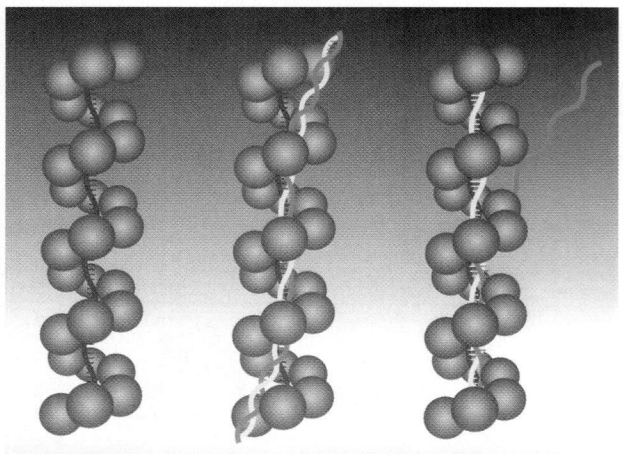

FIGURE 26.26

A model for RecA-mediated strand exchange. At the left, a RecA–ssDNA filament, with ssDNA shown in red. In the middle, a joint molecule, with triple-stranded DNA; the original ssDNA is wrapped in the minor groove of the duplex DNA (yellow and green strands). At the right, strand exchange is occurring. The red ssDNA is complementary in sequence to the yellow strand of the duplex, and RecA action is displacing the green strand, coincident with formation of a new red–yellow dsDNA.

Courtesy of M. Kubista et al., *Biological Structure and Dynamics*, R. H. Sarma and M. H. Sarma, eds. pp. 49–59. © 1996 Adenine Press, Inc.

RecA, a multifunctional bacterial enzyme, uses ATP to promote pairing of homologous DNA sequences.

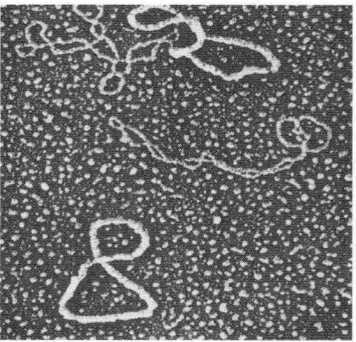

FIGURE 26.27

Electron microscopic visualization of joint molecules. For technical reasons, this image was obtained with the UvsX protein T4, which is similar in action to *E. coli* RecA. Single-strand DNA appears much thicker than duplex DNA because it is coated with UvsX protein. At the bottom is a circular single-strand DNA; in the middle, a circular duplex DNA; and at the top, a joint molecule involving one of each. Joint molecules can form between homologous sequences on circular molecules, but the strands are not interwound.

Journal of Biological Chemistry 262:9285-9292, L. D. Harris and J. Griffith. Reprinted with permission. © 1987 The American Society for Biochemistry and Molecular Biology. All rights reserved.

answer this question, we turn back to *E. coli* and its phages and the characteristics of bacterial mutants defective in recombination. Mutations conferring a recombination-defective (*rec⁻*) phenotype map in several loci, and two important gene products are responsible for most bacterial recombination events. One of these products, the RecA protein, was introduced earlier. The other protein is called exonuclease V or the RecBCD nuclease.

RecA is an amazing multifunctional protein with M_r of about 38,000. In recombination it promotes the ATP-dependent pairing of homologous strands, as described earlier in connection with daughter-strand gap repair. In vitro the protein catalyzes both single-strand exchanges, in which ssDNA invades a homologous duplex, and double-strand exchanges, in which both strands of an invading duplex essentially change partners. However, it has become evident that in vivo only three-strand reactions occur. As schematized in Figure 26.26 (left panel), the process begins with a reaction between RecA and single-stranded (ss) DNA, to give a characteristic nucleoprotein filament. As shown by scanning-tunneling microscopy of the filament and by crystallographic analysis of the protein (Figure 26.27), RecA wraps about ssDNA as a multisubunit right-handed helix, with six RecA monomers per turn. Once ssDNA is bound, the filament searches double-strand (ds) DNA, looking for sequences complementary to those in the single strand already bound. In this process, dsDNA is also taken up within the filament, giving a structure termed a joint molecule (Figure 26.26, center panel). Binding of dsDNA requires ATP, but that ATP need not be hydrolyzed in the process. By contrast, movement of the ssDNA-protein complex with respect to the dsDNA does require ATP hydrolysis. During this process, dsDNA is underwound and stretched to about 1.5 times its normal length. Movement of the complex is polarized, in a 5′-to-3′ direction along the initially bound ssDNA. During the movement, a triple-stranded structure transiently forms, with ssDNA (the red strand in Figure 26.26) wound in the minor groove of the dsDNA. The ssDNA continually tests the antiparallel strand in dsDNA (yellow) for sequence complementarity. It is not clear how this occurs, but evidently short oligonucleotide sequences swing out from the duplex structure and can pair with ssDNA if sequence complementarity is found. Once complementarity is established, branch migration occurs, with simultaneous strand exchange. A duplex is formed between the red strand (ssDNA originally) and the yellow strand (complementary to the red strand), while the displaced (green) strand is spooled out, away from the complex. ATP hydrolysis during this period may promote rotation of the DNA within the filament, facilitating the release of the displaced strand. A recent study at

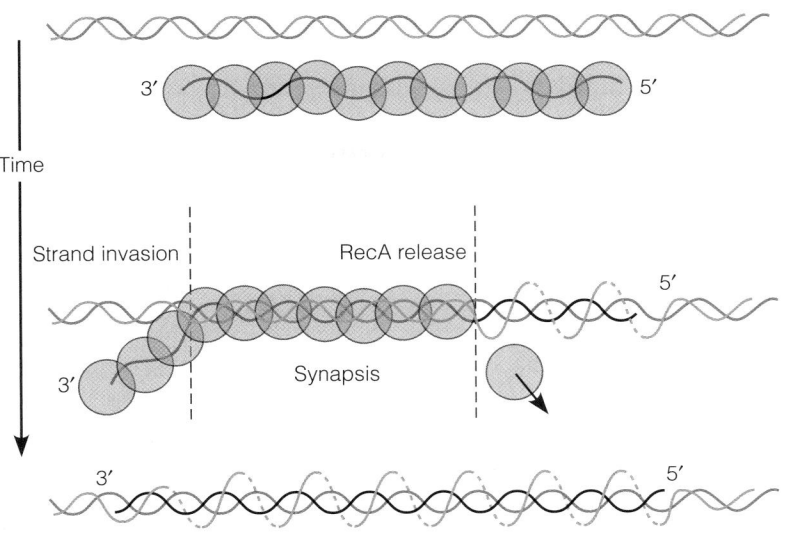

Time

Strand invasion | RecA release

3′ ... 5′

Synapsis

3′

5′

Final state

FIGURE 26.28

The RecA strand exchange reaction as shown from single-molecule studies. The black strand associated with RecA at the outset is complementary to the red strand in the duplex. Synapsis occurs 5′ to 3′ (rightward in the center diagram), with each RecA molecule displaced after its associated DNA has base-paired with its complement. In this model reaction the displaced blue strand is wrapped around the newly formed duplex.

Reprinted from *Molecular Cell* 30:530–538, T. van der Heijden, M. Modesti, S. Hage, R. Kanaar, C. Wyman, and C. Dekker, Homologous recombination in real time: DNA strand exchange by RecA. © 2008, with permission from Elsevier.

the single-molecule level confirms the general outline shown in Figure 26.26 and shows that about 80 base pairs are actively involved in the synapsis reaction at any time, even though the total region of homology may be much longer (Figure 26.28).

Recombination occurs preferentially at or near particular DNA sequences. In *E. coli* recombination is favored near a particular octanucleotide sequence, 5′-GCTGGTCC, called Chi (for crossover hotspot instigator). How does this site act to stimulate recombination? Another important enzyme, the RecBCD protein, a multifunctional heterotrimeric enzyme encoded by the *recB*, *recC*, and *recD* genes, displays sequence specificity for Chi. This enzyme binds at a double-strand break on duplex DNA and uses two helicase activities—RecB and RecD—to unwind and partially degrade the DNA. As shown from the crystal structure of a RecBCD-DNA complex, both helicases are in contact with DNA (Figure 26.29).

RecBCD, a multifunctional enzyme, unwinds and rewinds DNA, with one strand being unwound more rapidly and then converted to a single-strand 3′ end.

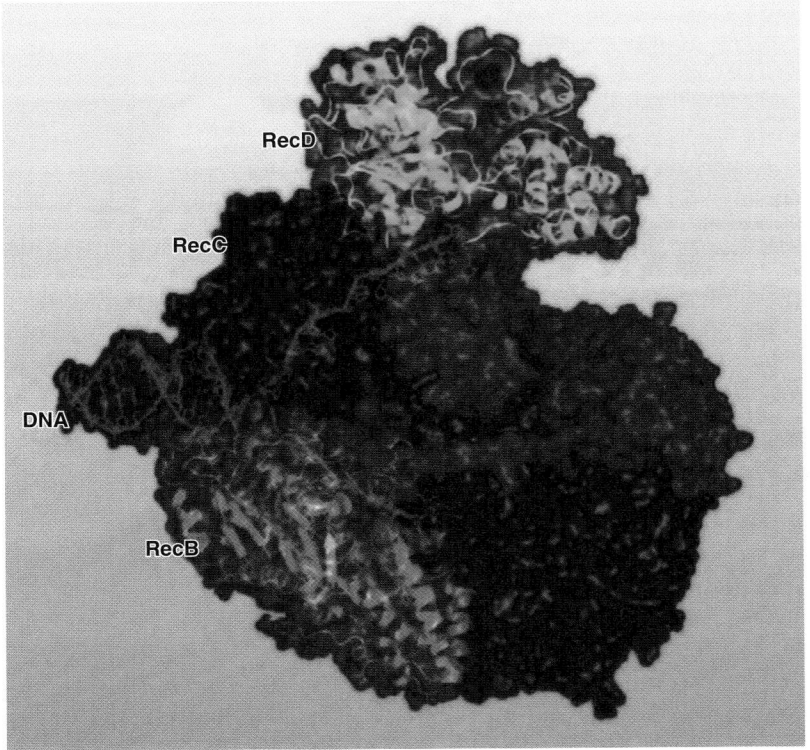

RecD

RecC

DNA

RecB

FIGURE 26.29

Crystal structure of *E. coli* RecBCD protein in complex with DNA. One strand of unwound DNA (magenta) is in contact with RecB and one with RecC. PDB ID 1W36.

Reprinted from *Cell* 131:651–653, D. B. Wigley, RecBCD: The supercar of DNA repair. © 2007, with permission from Elsevier.

The RecD helicase activity is higher than that of RecB, so as the protein moves, the 3' end is displaced as a single-strand loop ahead of RecB, and becomes coated with SSB protein. Both strands are degraded by associated nucleases, but because of the differential speeds of the protein motors, more of the 3' end is saved as the loop. When the enzyme reaches Chi, the protein pauses briefly, and a sequence-specific interaction causes RecBCD to switch speeds and switch its preferred polarity of DNA degradation. As B moves faster, the 3'-terminated loop is reeled in by RecB and coated with RecA. An associated nuclease releases the bound 3' end, as indicated in Figure 26.30, freeing it for strand invasion of a neighboring duplex.

In eukaryotic cells the RAD51 protein, as noted earlier, promotes homologous strand pairing in comparable fashion to RecA. The action of BRCA2 aids in RAD51-mediated recombination, and defective function of BRCA2 in this process may be related to the genomic instability that leads to breast or ovarian cancer in individuals with BRCA2 gene mutations.

Once a Holliday junction is formed, branch migration is essential for eventual formation of recombinant structures, as was shown in Figures 26.22 and 26.23. This is largely the responsibility of three other proteins. In *E. coli* these three proteins are products of the *ruvA*, *ruvB*, and *ruvC* genes. RuvA is a DNA-binding protein, whose specificity directs it toward the four-stranded Holliday structure. RuvB protein is an

FIGURE 26.30

A model for the action of RecBCD, Chi(χ) sites, and RecA in initiating homologous recombination. For details see text.

Reprinted from *Cell* 131:694–705, M.Spies, I. Amitani, R. J. Baskin, and S. C. Kowalczykowski, RecBCD enzyme switches lead motor subunits in response to χ recognition. © 2007, with permission from Elsevier.

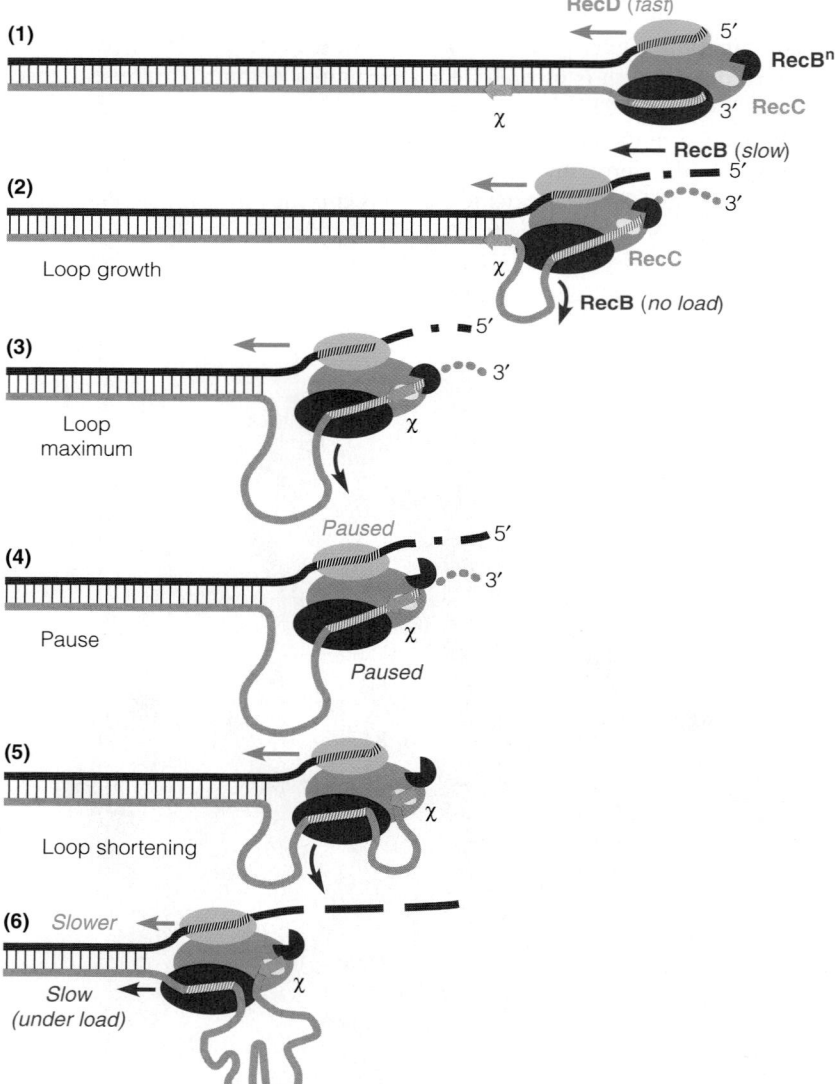

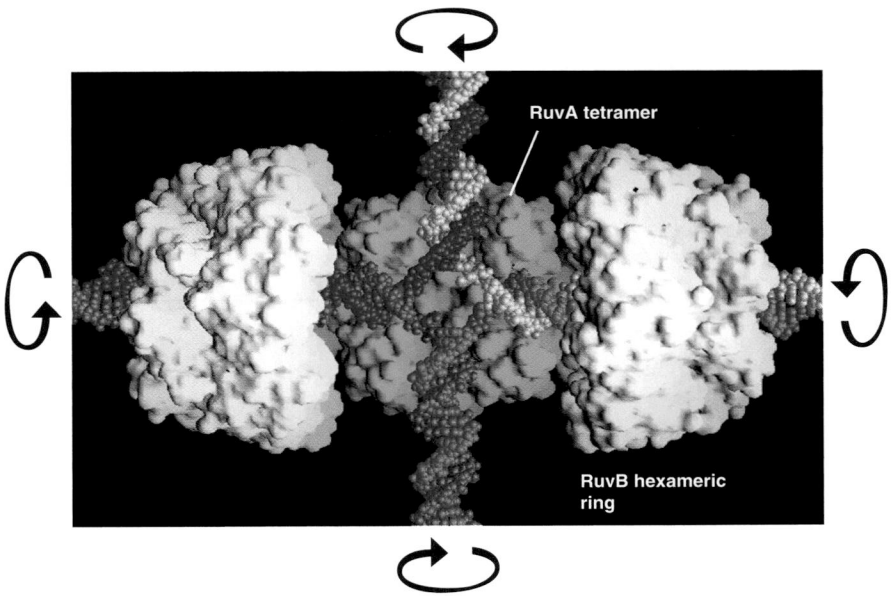

FIGURE 26.31

A model for the RuvA-RuvB-Holliday junction structure. This is based upon crystal structures of RuvA and RuvB. Branch migration is believed to involve spooling of DNA to left and right through the RuvB twin pumps, with the upper and lower arms being drawn into the center and eventually out through the pumps.

Courtesy of Peter Artymiuk, Krebs Institute, Sheffield.

ATP-requiring motor protein, which binds to two opposed arms of the junction. In the model of Figure 26.31, which is based upon crystal structures of the isolated RuvA and RuvB proteins, the two RuvB molecules act as twin pumps, rotating the two arms in opposite directions. This forces branch migration by driving the rotational movement of the other two strands toward the junction. Eventually RuvC binds and begins the resolution of the Holliday structure by nicking two strands.

Although Ruv protein homologs have not yet been detected in eukaryotic cells, much of the biochemistry of homologous recombination in eukaryotes is similar to what is described here. In particular, the RAD51 protein of both human cells and yeast has a strand-pairing activity similar (but not identical) to that of RecA, and the two proteins show extensive sequence homology. As noted earlier, in eukaryotic cells an essential function of homologous recombination is the repair of double-strand breaks, which can be created by ionizing radiation, oxidative stress, or other environmental damage, and would be lethal if not repaired. A broken chromosome can use the sequence information in its homolog to reconstruct the original DNA sequence at the site of the break, as was shown in Figure 26.18.

Site-Specific Recombination

As we have discussed, alignment of sites for homologous recombination is driven by DNA sequence homology. Another class of recombination reactions is directed by specific DNA–protein interactions, although a short stretch of DNA homology occurs at the actual site of cutting and resealing. Our biochemical understanding of this site-specific recombination is most advanced for the mechanism by which temperate phages such as λ become integrated at specific sites on the chromosome of the infected bacterium. This process provides an important model for studying counterpart reactions in higher organisms, such as integration of tumor virus genomes into the genomes of infected host cells.

The circularized chromosome integrates at a specific site on the *E. coli* chromosome, *attB*, which maps between genes involved in galactose utilization and biotin synthesis (the *gal* and *bio* markers), as was schematized in Figure 26.19. Integration occurs at a specific site on the phage chromosome, called *attP*. Two proteins are required for this site-specific recombination: (1) phage integrase (Int, the product of the *int* gene) and (2) an *E. coli* protein called IHF (integration host factor). IHF has been shown by X-ray crystallography to force a 90° turn when it binds to DNA (Figure 26.32a). Phage DNA must be supercoiled for the recombination to occur. This supercoiling, plus the distortion created by IHF binding to specific sites in

Recombination to integrate or excise temperate phage chromosomes involves specific DNA–protein interactions.

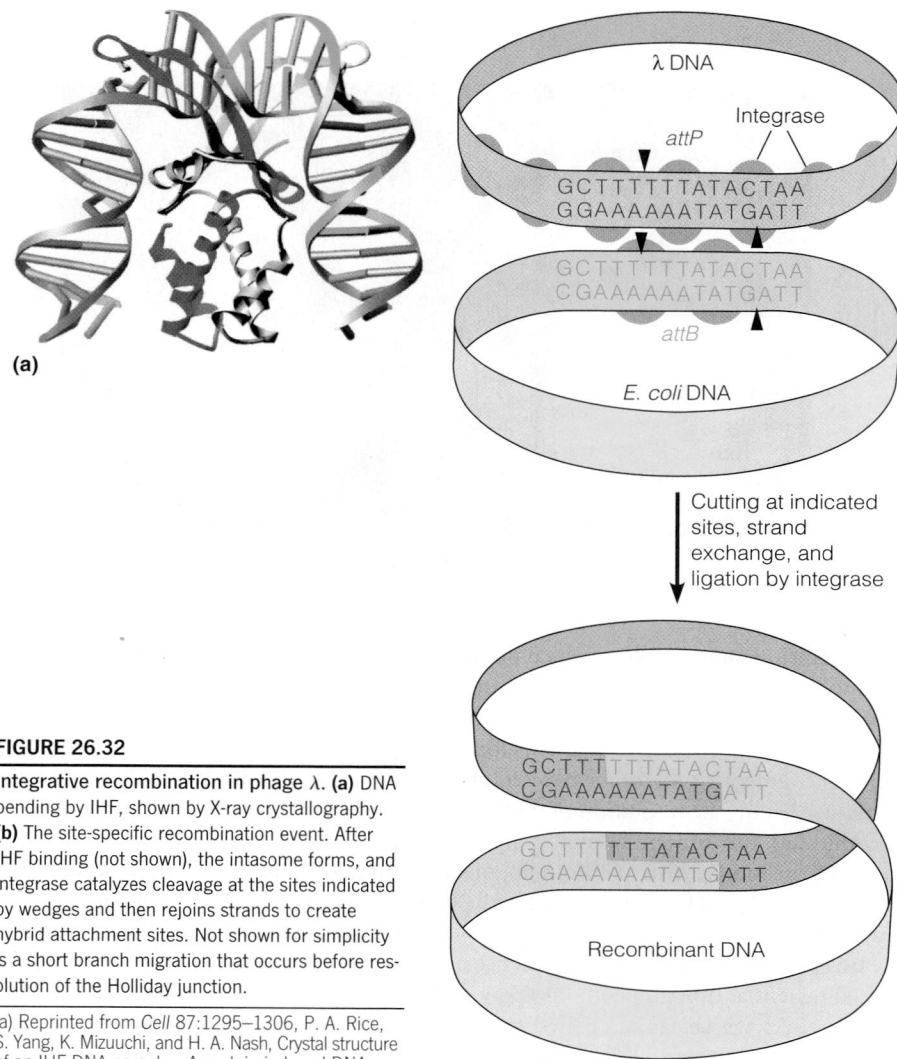

FIGURE 26.32

Integrative recombination in phage λ. (a) DNA bending by IHF, shown by X-ray crystallography. **(b)** The site-specific recombination event. After IHF binding (not shown), the intasome forms, and integrase catalyzes cleavage at the sites indicated by wedges and then rejoins strands to create hybrid attachment sites. Not shown for simplicity is a short branch migration that occurs before resolution of the Holliday junction.

(a) Reprinted from *Cell* 87:1295–1306, P. A. Rice, S. Yang, K. Mizuuchi, and H. A. Nash, Crystal structure of an IHF-DNA complex: A protein-induced DNA U-turn. © 1996, with permission from Elsevier.

attP, facilitates Int binding at adjacent sites. A specialized nucleoprotein structure called the intasome is formed, with the 230-base-pair *attP* region wrapped about seven Int molecules, each bound at a specific site. This structure becomes aligned with *attB*, which is only 23 base pairs long and which binds two molecules of Int. In the core of each site is a 15-base-pair region of complete homology (Figure 26.32b). In each of these sequences, Int creates a staggered cut, with a seven-nucleotide overlap. The ends then exchange to form a Holliday junction. The bacterial and phage core sequences yield two hybrid sequences, each of which contains both phage and bacterial DNA. The multifunctional Int protein, which has already catalyzed site-specific cutting and strand exchanges, now completes the process with a DNA ligase reaction to join the ends covalently.

When a phage chromosome becomes integrated, it is essentially dormant. Almost all of its genes are turned off, and it replicates only as a part of the bacterial chromosome in which it resides. However, changes in gene expression (see Chapter 29) can activate the integrated chromosome, or **prophage**. When this occurs, the above sequence of steps is reversed, with excision of a circular phage chromosome. An additional protein, called Xis, is required in addition to Int and IHF. Xis provides binding specificity for the hybrid phage-bacterial attachment sites that flank the prophage so that excision occurs only under conditions where both Int and Xis are present.

Gene Rearrangements

Until the mid-1970s the genetic information content of an organism or population was considered to be static. All cells of a differentiated organism were thought to have identical DNA contents, with variations among different cells arising at the level of gene expression. Supporting this idea was the fact that in some plants, such as carrots, a single differentiated cell can be manipulated in culture so as to give rise to a complete and normal plant. However, more recent developments have shown a plasticity to DNA that had not been expected. In normal eukaryotic development, segments of DNA can be deleted from the genome, can move from one site to another within a genome, or can duplicate themselves manyfold. In addition, mobile genetic elements have been described in both prokaryotes and eukaryotes. These segments of DNA can move from one chromosomal integration site to another, apparently unrelated to developmental processes. As noted earlier, these processes represent a specialized form of recombination.

Actually, the plasticity of DNA was predicted, but by very few scientists. Barbara McClintock's work on maize genetics, starting in the 1940s, led her to postulate genetic regulatory mechanisms effected through the action of mobile genetic elements. However, some three decades passed before the physical demonstration of such elements in bacteria focused attention on McClintock's pioneering work. For the remainder of this chapter we shall discuss three widely studied aspects of genome plasticity: the genetic basis for antibody variability in vertebrates, gene transposition, and gene amplification.

Immunoglobulin Synthesis: Generating Antibody Diversity

Recall from Chapter 7 that antibodies are proteins manufactured by vertebrate immune systems that aid in defense against infectious agents and other substances foreign to the animal. The immune response, resulting from introduction of an antigen, elicits the production of several highly specific antibodies. It is estimated that a human is capable of synthesizing more than 10 million distinct antibodies. Most of this great diversity is generated through the action of precisely controlled gene rearrangements, involving but a small fraction of the coding capacity of the genome. These rearrangements occur during differentiation of many individual clones of cells, each clone specialized for the synthesis of one and only one antibody. Other large protein families, notably T-cell receptors, are diversified by similar mechanisms. The immune response involves proliferation of clones of cells that produce antibodies specialized to bind the specific antigen, or immunogen, provoking that response. This clonal expansion allows large-scale production of specific antibodies, needed to combat infection or another challenge to the immune system.

To see how immunological diversity is generated, let us consider one type of antibody, the immunoglobulin G, or IgG, class. Recall from Chapter 7 that these proteins consist of two heavy chains and two light chains. Each chain comprises two distinct segments—a domain of variable polypeptide sequences and a constant domain, which is virtually invariant among different IgG light or heavy chains. We focus on the light chains and in particular the λ class of light chains. (Another class, κ, has somewhat different sequences in its constant region, but its development involves similar mechanisms.) Much of what we know about this process comes from the work of Susumu Tonegawa, who in 1987 was awarded the Nobel Prize in Physiology or Medicine for this work.

Figure 26.33 shows the organization of the precursor genes to κ chains in germ-line cells, undifferentiated for antibody formation, and the rearrangements leading to one such gene in a differentiated antibody-producing cell. Each light chain is encoded by DNA sequences that are noncontiguous in the genome of undifferentiated cells but are all in the same chromosome. These sequences are called V (variable), C (constant), and J (joining). The human genome contains

FIGURE 26.33

Gene rearrangements in antibody gene maturation. The rearrangements of C, V, and J sequences produce one mature κ light chain gene, and the transcription, processing, and translation of this gene produce an antibody κ light chain.

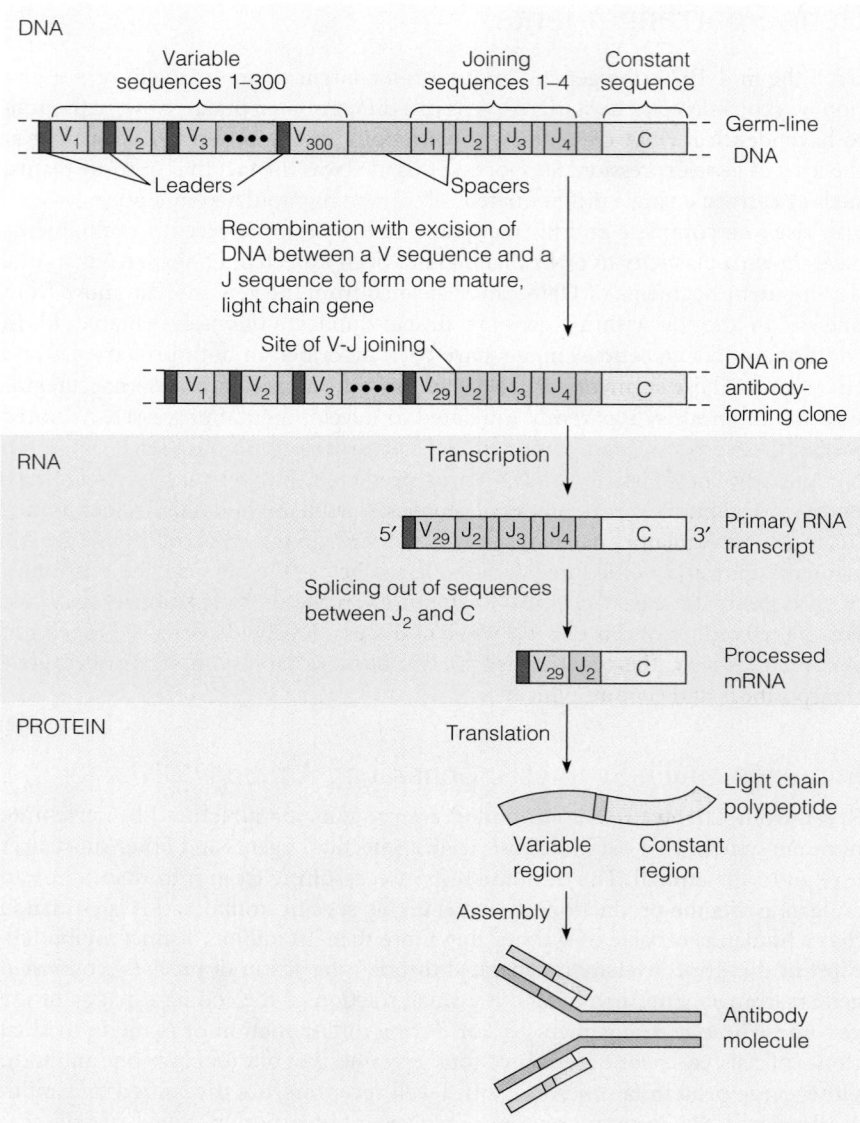

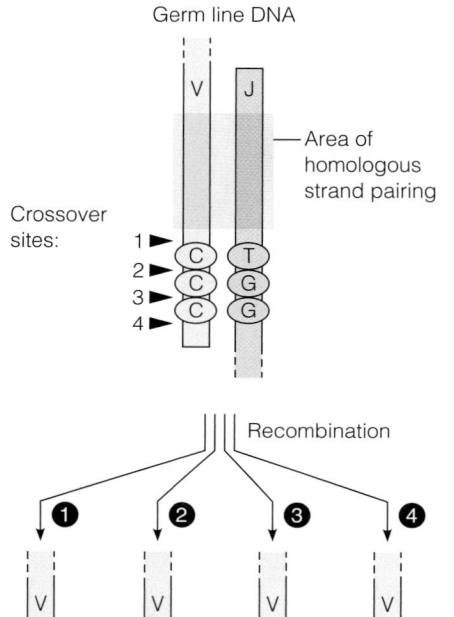

Possible recombinant forms in differentiated light chain genes

FIGURE 26.34

Generation of diversity by variable V-J joining mechanisms. Four crossover events are possible at a V-J junction, giving rise to codons for any three possible amino acids in this example. Only one DNA strand is shown.

about 300 different V sequences, each of which encodes the first 95 amino acids of the variable region; four different J sequences, each of which encodes the last 12 residues of the variable region and which join it to the constant region; and one C sequence, which encodes the constant region. In an embryonic cell the V sequences, each preceded by a leader sequence containing a transcriptional activator that is not expressed, form a tight cluster; the J sequences form another cluster some distance away; and the C sequence follows shortly after the J cluster. Each J sequence is flanked by nonexpressed spacer sequences.

In the differentiation of one antibody-forming clone of cells, a gene rearrangement links one of the approximately 300 V sequences with one of the 4 J sequences. All of the DNA that lies between these two spliced sequences is deleted in this rearrangement and disappears from all progeny of this cell line. Any upstream V sequences (on the 5′ side, to the left in Figure 26.33) and downstream J sequences (on the 3′ side, to the right) remain in these cells but are not used in antibody synthesis.

Additional diversity is provided by the way in which the V and J sequences recombine. The cutting and splicing can occur within the terminal trinucleotide sequences of V and J in any way that yields one trinucleotide sequence in the spliced product (Figure 26.34). This increases the total number of different light chain sequences by about 2.5 (the average number of different amino acids encoded by four random triplets). Thus, the total number of possible light chain sequences that can be formed from 300 V sequences and 4 J sequences is about 3000 (300 × 4 × 2.5).

Related DNA sequences are found to the 3′ side of each V sequence and to the 5′ side of each J sequence, and they represent recognition sites for the enzymes involved in the joining reaction. Those sequences, which are called recognition signal sequences, are as follows:

5′······V····CACAGTG····12 bases····ACAAAAAC····3′

3′······J····GTGTCAC····23 bases····TGTTTTTG····5′

Note that the homologous regions of these sequences are inverted repeats.

Recombination begins with two proteins called RAG1 and RAG2 (Figure 26.35). These proteins act similarly to the bacterial proteins involved in gene transposition (page 1110), beginning the process by catalyzing double-strand breaks between the two recognition signal sequences involved and the respective V and J coding sequences. Cellular DNA-repair proteins process the double-strand DNA breaks, creating a *coding joint*, in which the V and J sequences are fused in the appropriate reading frame, and a *signal joint*, which fuses the intervening DNA into a circular molecule. The nearly identical seven-base palindromic sequences and nearly complementary eight-base AT-rich regions in these segments allow alignment of distant regions of the chromosome, with a process akin to that in phage λ integration, recombining the sequences and excising the intervening DNA. In addition, sequence analysis indicates that the DNA joining reactions are imprecise, with nucleotides removed from one or both ends, creating additional diversity.

Still another mechanism acts to introduce sequence diversity into antibody chains—**somatic hypermutation**. The spontaneous mutation rate becomes very high, but specifically within sequences related to antibody formation. These sequences undergo a high rate of deamination of cytosine residues, leaving uracil residues paired incorrectly with guanine. Upon replication these mutant strands pair with adenine, completing the transition from G-C base pairs to A-T. The enzyme involved, **activation-induced deoxycytidine deaminase** (AID), evidently acts at paused transcription complexes, with the nontemplate DNA strand, which is single-stranded, serving as the deamination substrate.

The final step in producing a light chain polypeptide involves joining of the C and J segments (see Figure 26.33). This occurs not at the DNA level but at the level of messenger RNA. As discussed in Chapters 7 and 27, eukaryotic gene expression usually involves cutting and splicing of the messenger RNA, with excision of sequences that are not represented in the final gene product. In this case, transcription yields an RNA molecule extending from the 5′ side of the V gene that is spliced to J to the 3′ side of C. Depending on which J region has been spliced to V in this cell, the RNA excised during splicing may contain sequences corresponding to other J regions.

Heavy chains are formed similarly—from V sequences, J sequences, and a class of sequences called D. In addition, there are eight different C sequences, which are also involved in the synthesis of other antibody classes. The total number of possible IgG heavy chains is about 5000. Because any light chain can combine with any heavy chain to form a complete IgG, the total possible number of IgG molecules is 3000 × 5000, or $1.5 × 10^7$. In this way, enormous diversity can be generated from a very small fraction of the total DNA in germ-line cells. Even further diversity arises from the high rate of V sequence mutation by somatic hypermutation, during development of the antibody-producing cell. This allows the same V-J joining event to produce different IgGs.

It is not clear whether both of the homologous chromosomes in a diploid antibody-forming cell undergo identical rearrangements. However, given that each cell produces only one type of antibody, either that must occur, or else one chromosome is silenced after the other has completed its rearrangement.

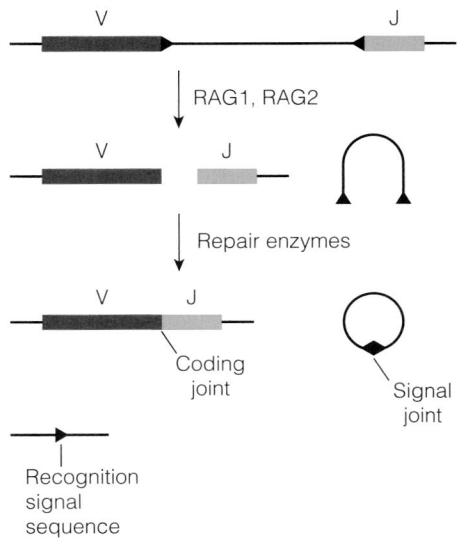

FIGURE 26.35

Site-specific recombination event catalyzed by RAG1 and RAG2.

The diversity of the immune response involves recombination among thousands of different DNA sequences, to yield a vast array of antibodies.

Transposable Genetic Elements

In this section we discuss transposable genetic elements—genes that do not have a fixed location in a genome but can move from place to place within the genome, albeit with low frequency. Transposition occurs without benefit of DNA sequence homology, but the enzymes catalyzing transposition recognize short sequences of about a half dozen nucleotides. Although the existence of gene transposition had been predicted by Barbara McClintock's work on maize genetics, the first physical characterization of transposable elements arose from studies of antibiotic-resistant strains of bacteria. By the early 1970s it was known that genes conferring resistance to drugs such as tetracycline or penicillin were usually carried on plasmids, whose DNA sequences bore no detectable homology with chromosomal DNA sequences of the host. Nevertheless, the genes for antibiotic resistance would appear, with low frequency, in the chromosome of the bacterium or in the DNA of a phage that had infected that cell. The presence of new DNA inserted into the host or phage chromosome could be confirmed by restriction cleavage analysis of the DNA or by heteroduplex analysis in the electron microscope. The existence of these "jumping genes," which move from one chromosome to another in seemingly random fashion, greatly altered our views on gene organization and evolution. The new concepts were of much more than academic interest because they relate to the use of antibiotics to treat bacterial infections—specifically, to the ease with which populations of antibiotic-resistant bacteria can arise.

Transposable elements have been demonstrated in many eukaryotes, including maize, *Drosophila*, and yeast. However, we shall concentrate on bacteria, whose physical structures and transposition mechanisms are best understood. Let us first point out several distinctions between bacterial transposition and other recombinational mechanisms we have discussed. First, transposition does not require extensive DNA sequence homology. Furthermore, transposition occurs normally in a *recA⁻* host, suggesting that homologous recombination events are not involved. Second, DNA synthesis is involved in bacterial transposition. Transposition always involves duplication of the target site, the short sequence (3–12 base pairs) at which the transposable element is inserted. In many instances the transposable element is itself replicated, with one copy being deposited in the new sequence and one remaining in the donor sequence. Finally, transposable elements can restructure a host chromosome. A transposable element can move from one site to another within the same chromosome, producing two homologous sequences resident in the same chromosome. Depending upon whether these sequences are oriented identically or in reverse, homologous recombination between them can yield a deletion or an inversion, as shown in Figure 26.36. Transposable elements also have other effects on the chromosomes they move to—either inactivation of any gene into which they move (where insertion interrupts the coding sequence) or activation of adjacent genes (where a promoter, or transcriptional activator, might be created next to the gene). Abortive transpositional events can cause deletions or inversions in the chromosome. Because such events are often lethal, evolution has selected for organisms with low rates of transposition. In the laboratory, insertional inactivation of genes is useful for isolating mutants defective in specific functions and for mapping genes.

We recognize three different classes of transposable elements in bacteria, with general structures as shown in Figure 26.37. In classifying these elements, we consider the involvement of two enzymes, **transposase** and **resolvase**, whose functions are discussed shortly. Class I elements, which encode a transposase but not a resolvase, are of two types. The simplest transposable element, called an **insertion sequence** (IS), consists simply of a gene for transposase, flanked by two short inverted repeat sequences of about 15 to 25 base pairs. A less simple structure called a **composite transposon** consists of a protein-encoding gene, such as a gene conferring antibiotic resistance, flanked by two insertion sequences, or IS-like elements. These elements may be in either identical or inverted orientations. Class II transposons contain only one set of short flanking direct repeat sequences. In addition to a protein-encoding gene (often conferring antibiotic resistance) and a transposase

Transposable genetic elements include insertion sequences, transposons, and certain bacteriophages that can insert at various locations.

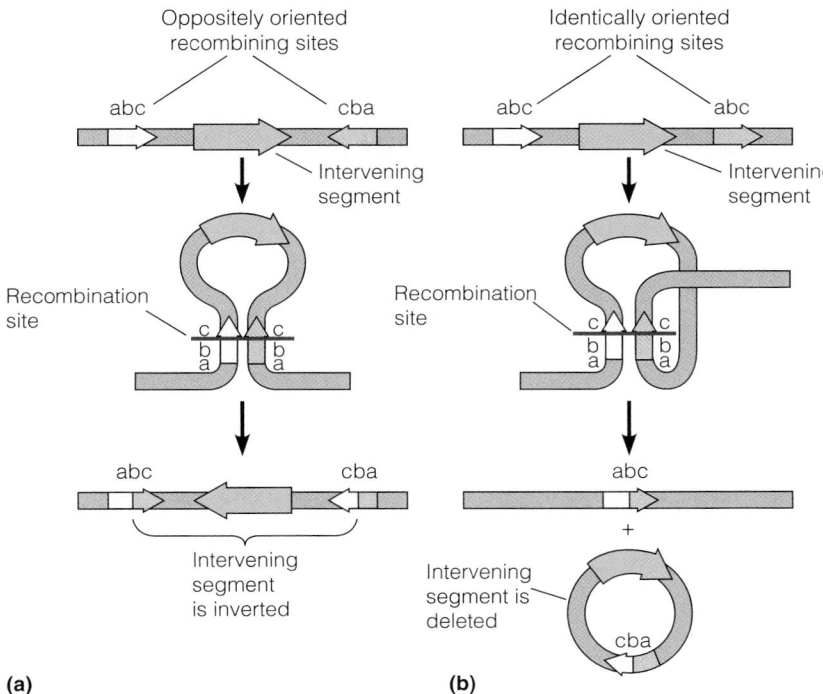

(a)   **(b)**

FIGURE 26.36

Genome rearrangements that can be promoted by homologous recombination between two copies of the same transposable element. Depending on the orientation of the two copies, either **(a)** inversion or **(b)** deletion can result.

CLASS I

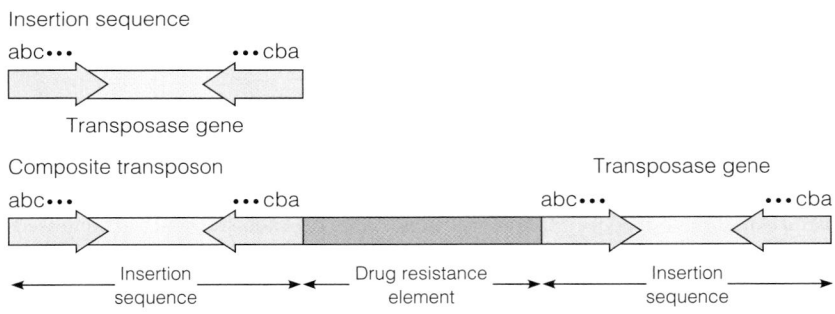

CLASS II

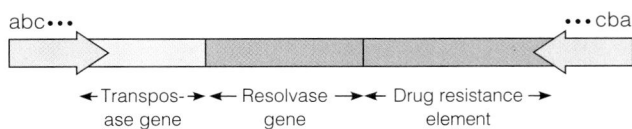

CLASS III

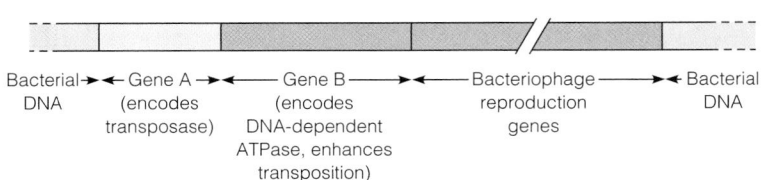

FIGURE 26.37

Structures of class I, class II, and class III mobile genetic elements. Inverted repeats are shown in purple. The "drug resistance element" may be one of a variety of protein-coding genes.

gene, these elements include a gene for resolvase. Finally, class III elements belong to a small group of bacteriophages, of which the best-known is phage Mu. This phage is known to insert its chromosome at random in the host chromosome by a transpositional mechanism and also to replicate its genome by a transpositional

TABLE 26.2   Structures of some transposable elements of *E. coli*

Element	Size (bp)	Target DNA (bp)	Resistance Conferred
Insertion sequences			
IS1	768	9	None
IS2	1327	5	None
IS10-R	1329	9	None
Composite transposons			
Tn5	5700	9	Kanamycin
Tn10	9300	9	Tetracycline
Tn2571	23,000	9	Chloramphenicol, fusidic acid, streptomycin, sulfonamides, and mercury
Class II transposons			
Tn3	4957	5	Ampicillin
Class III transposons			
Phage Mu	38,000	5	None

Excerpted from *Annual Review of Genetics* 15:341–404, N. Kleckner, Transposable elements in prokaryotes. © 1981 Annual Reviews.

mechanism, similar to that of type II elements. One gene of this phage, A, encodes a transposase. Another gene, B, encodes a protein with DNA-dependent ATPase activity. Whereas class I and class II transposable elements synthesize transposase at such low levels that transposition occurs at frequencies of only 10^{-7} to 10^{-5} per generation, phage Mu integrates about 100 times per lytic infection. The B gene product is partially responsible for this far greater efficiency of transposition. Other genes encode structural and other proteins of the virus.

Table 26.2 summarizes the properties of a number of transposons and insertion sequences. Note that each transposon (conventionally referred to with the abbreviation Tn) and IS inserts at a specific target sequence of five or nine base pairs in the examples shown. Insertion involves a duplication of that site, and it results in two copies of the target sequence, one on each side of the integrated element (Figure 26.38). It seems likely that this results from the action of transposase, which generates a staggered cut that brackets the target sequence.

FIGURE 26.38

Model of how direct repeats are generated during the insertion of a transposon or an insertion sequence.

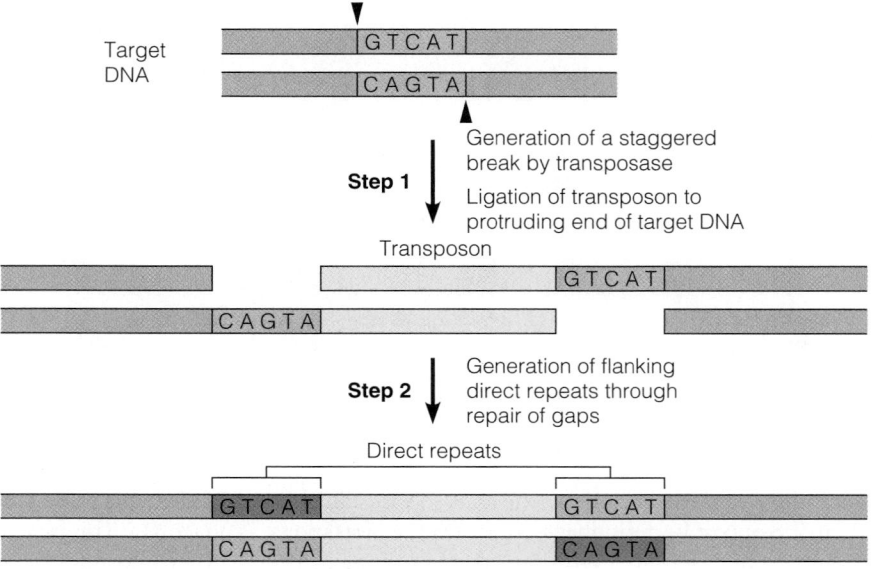

Attachment of the mobile element (Tn or Is) to each end results in gaps, which are then filled and ligated to generate the flanking direct repeats.

Because the transposable element never exists as free linear DNA, how are ends of the element generated, to join with the ends of the staggered cut? The currently favored model, shown in Figure 26.39, involves transposase introducing

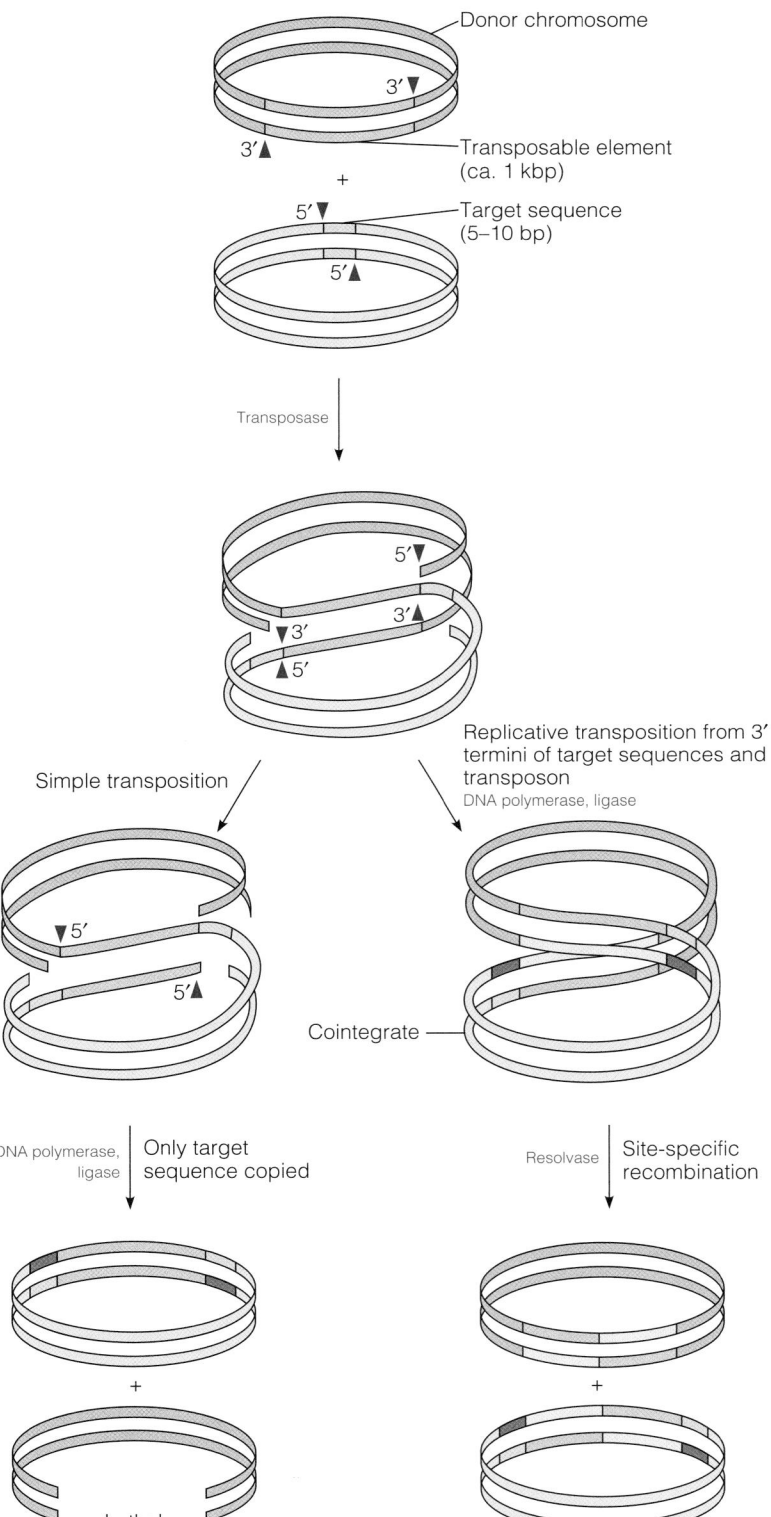

FIGURE 26.39

Models of simple transposition and replicative transposition.

both the staggered cuts in the target site and a nick at each of the 3′ ends of the element—precisely between the transposon sequence and the flanking direct repeat. Next, the free 5′ ends in the recipient DNA target sequence are joined to the 3′ ends of the element. Two outcomes are then possible. In simple transposition the joining is followed by cutting of the 5′ ends of the transposon, also immediately adjacent to the flanking sequences. This gives a gapped structure like that shown in Figure 26.38, which can be filled and closed by DNA polymerase and ligase. In this form of transposition, only the target sequence is copied; the donor chromosome suffers a lethal double-strand break. Tn10 (Table 26.2) transposes by a conservative mechanism, meaning that both original strands are somehow transferred to the new location.

The other process, **replicative transposition**, requires the enzyme resolvase, so it occurs only with class II and class III elements. The 3′ ends of the target chromosome, after the first cutting and splicing, serve as replicative primers for copying both the gaps, as shown in Figure 26.39, and the two strands of the transposable element itself. Ligase action generates a **cointegrate**, a large circular structure containing both donor and target chromosomes with two freshly replicated copies of the transposable element. The other enzyme, resolvase, now catalyzes site-specific recombination between the two elements, resulting in one copy of the transposable element inserted into each of the two chromosomes. Structural analyses of the *E. coli* enzyme, called $\gamma\delta$ resolvase, are yielding important clues to this complex process.

Retroviruses

Gene transposition in eukaryotic systems presents some strong similarities to and some distinct differences from transposition in bacteria. The first major distinction is that integration and excision are distinct processes in eukaryotes. Thus, the transposable element can be isolated in free form, often as a double-strand circular DNA. Second, replication of that DNA often involves the synthesis of an RNA intermediate. Both of these properties are seen in the retroviruses of vertebrates, perhaps the most widely studied class of eukaryotic transposable elements. As we noted in Chapter 25, these RNA viruses use reverse transcriptase to synthesize a circular duplex DNA, which can integrate into many sites of the host cell chromosome. The integrated retroviral genome bears remarkable resemblance to a bacterial composite transposon, as you can see by comparing Figure 26.40 with

FIGURE 26.40

Structure of retroviral genomes in the integrated state. **(a)** A nononcogenic virus. **(b)** An oncogenic virus such as Rous sarcoma virus, showing the viral oncogene downstream (rightward) from the viral replication genes. **(c)** A defective oncogenic virus, such as Moloney murine sarcoma virus, with the viral oncogene replacing part or all of a gene (*env*) essential to viral replication. In each case the LTRs are direct repeats, flanked by short inverted repeats.

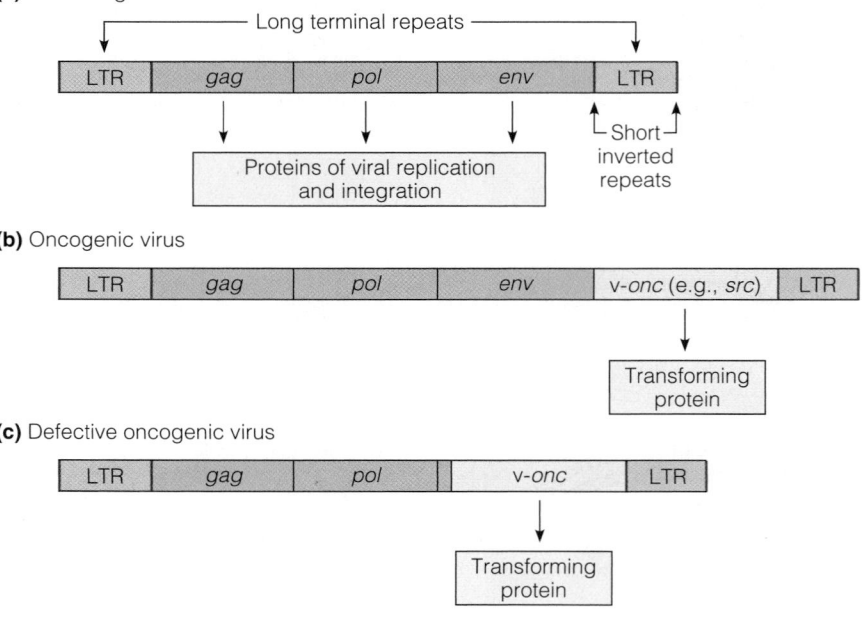

(a) Nononcogenic virus

Long terminal repeats

| LTR | gag | pol | env | LTR |

Proteins of viral replication and integration

Short inverted repeats

(b) Oncogenic virus

| LTR | gag | pol | env | v-onc (e.g., src) | LTR |

Transforming protein

(c) Defective oncogenic virus

| LTR | gag | pol | v-onc | LTR |

Transforming protein

Figure 26.37. The prototypical retroviral genome has three structural genes—*gag*, which encodes a polyprotein that undergoes cleavage to give virion core proteins; *pol*, which encodes the viral polymerase, or reverse transcriptase; and *env*, the major glycoprotein of the viral envelope. Flanking these structural genes are two direct repeats, the **long terminal repeats** (**LTRs**) of about 250 to 1400 base pairs each. Each LTR is flanked in turn by short inverted repeat sequences, 5 to 13 base pairs in length. Integration occurs by a mechanism that duplicates the target site so that the integrated viral gene, called a provirus, is flanked by direct repeats of 5 to 13 base pairs each of host cell DNA.

Just as bacterial transposons can carry passenger genes, so also can retroviruses. The earliest known retrovirus, Rous sarcoma virus, was also the first virus shown to contain an oncogene (see Chapter 23). Rous sarcoma virus was isolated in 1911 and shown to cause tumors in chickens. Not until 1978, however, was the src gene identified and shown to be responsible for tumorigenesis. The src gene product is a 60-kilodalton protein with a protein tyrosine kinase activity. A related but distinct sequence can be detected in the host genome. In Rous sarcoma virus the src oncogene lies to the 3′ side of the env gene. Other tumorigenic viruses contain the oncogene either inserted into or substituting for one of the genes gag, pol, and env. Because the loss of an essential gene makes it impossible for the virus to replicate, the latter class of virus can grow only in a cell coinfected with a *helper virus*, a related retrovirus that provides the missing function(s).

By various means, one can show that the action of an oncogene is essential for the virus to effect oncogenic transformation, the change of a normal cell to a cancerous cell (see pages 982–989 in Chapter 23). For example, mutational inactivation of the src gene of Rous sarcoma virus does not affect the ability of the virus to replicate, but it does render the virus unable to cause tumors in infected animals. Because each viral oncogene is related in sequence to a cellular counterpart, it is presumed that the viral oncogene originated in a cell many generations ago and underwent independent mutations after being picked up by a viral genome. Moreover, because many of the oncogene products, like the Src protein, have protein kinase activity, they may be aberrant forms of normal cellular regulatory elements, and their expression may be involved in the abnormal growth control that characterizes tumor cells. An additional mechanism may be related to activation of cellular genes by insertion of proviral DNA. The leftmost LTR in an integrated provirus (i.e., to the 5′ side as usually drawn) contains the transcriptional activator, or promoter, for the adjacent *gag* gene and the downstream *pol* and *env* genes. Because the LTRs are direct repeats, the rightmost LTR can activate transcription of cellular genes downstream from the integration site. If these genes include those involved in metabolic regulation, their overexpression may unbalance metabolism in some still undefined way and, hence, contribute to oncogenesis.

Transposable elements in eukaryotic cells show striking resemblances to retroviruses in sequence organization. Indeed, the term **retrotransposon** is used to denote this class of elements. These similarities are illustrated in Figure 26.41 for two retroviruses, plus Ty, a transposon of yeast; copia and 412, transposable elements in *Drosophila*; and IAP, a transposon found in the mouse genome.

Retroviral genomes and eukaryotic transposable elements have sequence similarities, to each other and to bacterial transposons.

Gene Amplification

The final process we discuss in information restructuring is the selective amplification of specific regions of the genome, principally in eukaryotic cells. This occurs in normal developmental processes and as a consequence of particular metabolic stress situations.

It has long been known that during oögenesis in certain amphibians the genes encoding ribosomal RNAs increase in copy number by some 2000-fold, in preparation for the large amount of protein synthesis that must occur in early development. The amplified DNA is in the form of extrachromosomal circles, each of which

FIGURE 26.41

Common sequence features in integrated retroviruses and other eukaryotic transposable elements. All these elements are bounded on the left and right sides by long terminal repeat sequences (LTR-L and LTR-R, respectively). Each LTR is flanked by short inverted repeat (IR) sequences (arrows). The approximate locations of various structural features are indicated, and the table gives some specific sequences associated with each feature in several transposable elements—IRs; promoter, a transcription start signal in each LTR; polyA, an adenine-rich sequence (absent from the yeast transposon Ty and the copia element in *Drosophila*); P, a sequence that anneals to an RNA that serves as a primer for replication; Pu, purine-rich sequence. 412, like copia, is a transposable element of *Drosophila*; IAP is a transposon found in the mouse genome; and RSV and MoMLV are retroviruses (Rous sarcoma virus and Moloney murine leukemia virus, respectively).

Reproduced from *Genes and Genomes*, M. Singer and P. Berg, p. 755 © 1991 University Science Books, Mill Valley, CA.

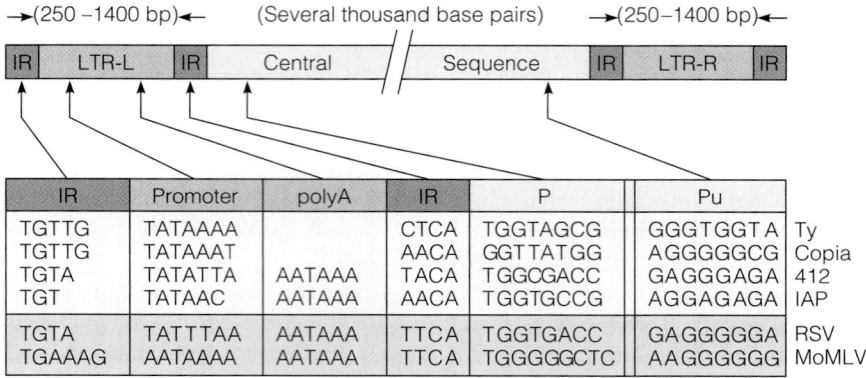

IR	Promoter	polyA	IR	P	Pu	
TGTTG	TATAAAA		CTCA	TGGTAGCG	GGGTGGTA	Ty
TGTTG	TATAAAT		AACA	GGTTATGG	AGGGGGCG	Copia
TGTA	TATATTA	AATAAA	TACA	TGGCGACC	GAGGGAGA	412
TGT	TATAAC	AATAAA	AACA	TGGTGCCG	AGGAGAGA	IAP
TGTA	TATTTAA	AATAAA	TTCA	TGGTGACC	GAGGGGGA	RSV
TGAAAG	AATAAAA	AATAAA	TTCA	TGGGGGCTC	AAGGGGGG	MoMLV

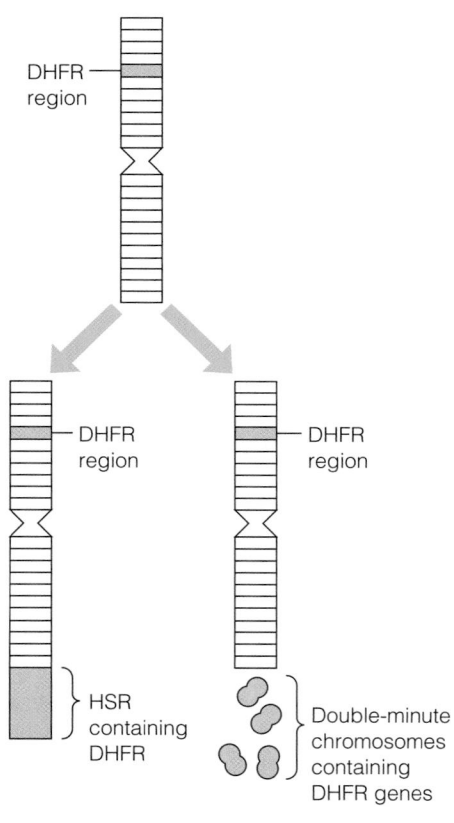

DHFR region

DHFR region

DHFR region

HSR containing DHFR

Double-minute chromosomes containing DHFR genes

FIGURE 26.42

Two modes of gene amplification leading to drug resistance.

Gene amplification generates multiple copies of DNA sequences at a separate site on the same chromosome. Recombination among homologous segments yields extrachromosomal amplified sequences.

contains several copies of the ribosomal DNA repeat and a replication origin. A similar situation has been analyzed in *Drosophila*, in which genes encoding egg proteins are amplified at a particular developmental stage. In the latter case, however, the amplification results from repeated rounds of replication initiation within the amplified region, and the amplified sequences remain within the chromosome of origin.

Both types of mechanisms apparently occur during development of certain drug-resistant mammalian cell lines in culture. This process has been studied most widely in cells that become resistant to methotrexate, a dihydrofolate reductase inhibitor. As discussed in Chapter 22, treatment of leukemia with methotrexate often leads to the emergence of drug-resistant leukemic cell populations, which contain vastly elevated levels of the target enzyme, dihydrofolate reductase (DHFR). As shown originally by Robert Schimke, overproduction of the enzyme usually results from specific amplification of a large DNA segment that includes the DHFR gene. In one process, tandem duplication of the DNA segment generates a giant chromosome with multiple gene copies, in what is called a **homogeneously staining region** (HSR), because it lacks the typical chromosome banding pattern. Alternatively, a DNA segment containing the DHFR gene can be excised, apparently by a recombinational process, to form minichromosomes called **double-minute chromosomes**, as shown schematically in Figure 26.42. Some resistant cells contain both types of amplified genes. Double-minute chromosomes are maintained within a cell only so long as selective pressure is maintained by growth of the cell in methotrexate. However, the chromosomally amplified phenotype is stable through many generations of cell growth. Figure 26.43 shows a fluorescence micrograph of metaphase chromosomes from a stably amplified Chinese hamster ovary cell line. DHFR sequences were visualized by in situ hybridization with a fluorescent-tagged DNA containing DHFR sequences. This technique is sufficiently sensitive to allow detection of single-copy sequences (white arrows). Note also the giant chromosome containing many gene-equivalents of DHFR gene sequences.

Amplification of genes under selective conditions has been widely observed—for example, in development of pesticide-resistant forms of insects. The mechanism of amplification is not yet clear. However, evidence such as that of Figure 26.43 shows that the amplified sequences are on the same chromosome as the original single-copy gene site, but at a significant distance away from that site. Such structures could arise either through recombination with unequal sister-chromatid exchange, schematized in Figure 26.44, or by a conservative transposition process. Later, homologous recombination within an amplified region can lead to excision of sequences containing one or more amplified sequences. In order to replicate autonomously, these excised sequences must have a centromere. Such elements probably represent the double-minute chromosomes.

The presence of selective pressure, such as the continuous presence of methotrexate, promotes specifically the survival of cells that can respond to that pressure, for instance, by overproducing DHFR. Once two or more copies of the gene are present

FIGURE 26.43

Chromosome structural changes that accompany dihydrofolate reductase gene amplification. The micrograph shows metaphase chromosomes from Chinese hamster ovary cells that are highly resistant to methotrexate. Chromosomal DNA was subjected to hybridization in situ with a fluorescence-labeled DHFR gene probe. White arrows point to single-copy genes. The amplified chromosomal sequences are on a giant form of the chromosome that also contains one of the original single-copy sequences.

Genes & Development 3:1913–1925, B. J. Trask and J. L. Hamlin, Early dihydrofolate reductase gene amplification events in CHO cells usually occur on the same chromosome arm as the original locus. © 1989 Cold Spring Harbor Laboratory Press. Reprinted by permission of Dr. Barbara Jo Trask and Dr. Joyce Hamlin.

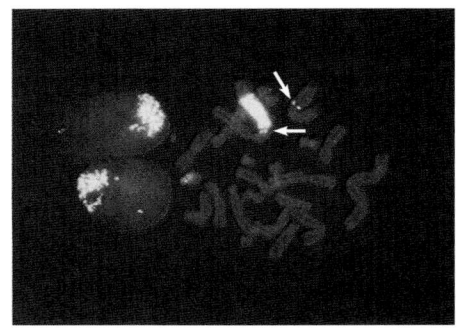

on a chromosome, additional copies can be generated by further recombinational events or by abnormalities of replication. Resistance is thus developed in stepwise fashion and occurs over many generations of growth. These findings have immense practical significance because cancer chemotherapy often involves long-term treatment with low doses of an antimetabolite—precisely the conditions most likely to nullify the effect of treatment by generating drug-resistant cells. Findings on gene amplification not only have changed the way in which anticancer drugs are administered but also have been extended to the tumorigenic process itself. Investigators have found that specific oncogenes become amplified during the clinical progression of certain human tumors. Thus, gene amplification is seen as a mechanism in normal development, in cellular adaptation to stress, and in abnormal developmental processes. Also, the gene duplications that have occurred during evolution (see Chapter 7, page 258) probably have taken place by similar pathways.

SUMMARY

A variety of information restructuring processes affect DNA, to protect it from environmental damage or foreign organisms, to diversify the individuals in a species, or to diversify the genetic constitution of somatic cells in differentiation. DNA repair encompasses several processes, including photoreactivation, in which light energy is captured to reverse pyrimidine dimer formation, and the removal of alkylguanines by "enzymes" that act only once. Nucleotide excision-repair systems recognize damaged nucleotides and cleave out and replace a patch containing 12 to 30 nucleotides that flanks the damaged segment. Base excision repair begins with an *N*-glycosylase reaction, followed by repair enzymes that excise one or more nucleotides at the abasic site. Damage can also be repaired by a recombinational process or by induced (error-prone) repair. Mismatch repair systems correct occasional errors arising during replication. Double-strand DNA breaks are repaired either by recombination or by nonhomologous end-joining. Genetic recombination between homologous DNA segments involves helicase-catalyzed duplex unwinding, site-specific DNA cutting, duplex-strand invasion by single-strand 3′ hydroxyl termini, chain extension, branch migration, and resolution of Holliday structures. Some recombinational events, such as integration of bacteriophage λ, are site-specific and governed by DNA–protein interactions. A wide range of gene rearrangements includes the joining reactions in differentiation of the immune response, gene transposition, retroviral genome integration, and gene amplification, which probably occurs by a recombinational mechanism. Gene amplification can either be a normal developmental process or occur in response to a specific environmental stress.

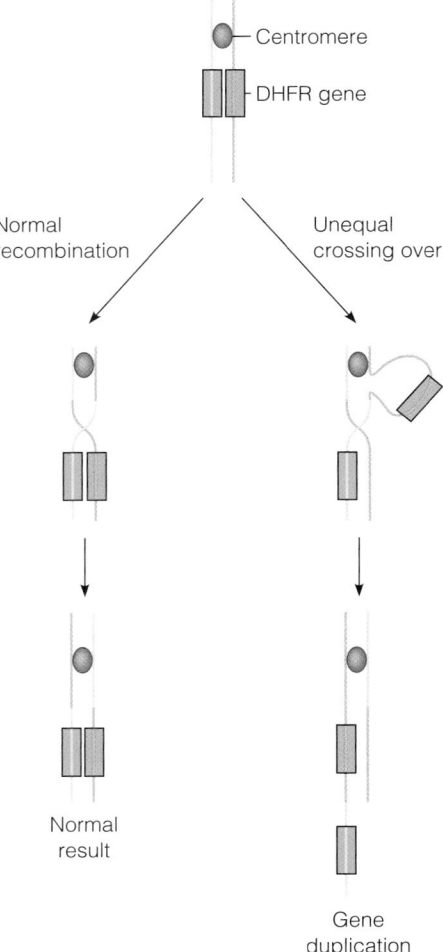

FIGURE 26.44

Unequal crossing over, as a mechanism to explain early steps in gene amplification.

REFERENCES

DNA Repair, General

Friedberg, E. C., G. C. Walker, W. Siede, R. D. Wood, R. A. Schultz, and T. Ellenberger (2006) *DNA Repair and Mutagenesis*, 2nd ed. ASM Press, Washington, D.C. The most authoritative contemporary book-length review of the subject.

DNA Damage Response

Derheimer, F. A., and M. B. Kastan (2010) Multiple roles of ATM in monitoring and maintaining DNA integrity. *FEBS Letters* 584:3675–3681. ATM is a central player in this process.

Jackson, S. P., and J. Bartek (2009) The DNA-damage response in human biology and disease. *Nature* 461:1071–1075. Nice correlation of human disease states with defects in the DNA damage response.

Petermann, E., and T. Helleday (2010) Pathways of mammalian replication fork restart. *Nat. Rev. Mol. Cell Biol.* 11:683–687. Restart is central to several modes of DNA repair.

Direct Repair

Sancar, A. (2009) Structure and function of DNA photolyase and *in vivo* enzymology: 50th anniversary. *J. Biol. Chem.* 283:32153–32157. A comprehensive recent review.

Zhang, Y., and 7 coauthors (2011) FTIR study of light-dependent activation and DNA repair processes of (6–4) photolyase. *Biochem.* 50:3591–3598. Biophysical analysis of photoreactivation.

Excision Repair

Cleaver, J. E., E. T. Lam, and I. Revet (2009) Disorders of nucleotide excision repair: The genetic and molecular basis of heterogeneity. *Nat. Rev. Genet.* 10:756–768. Xeroderma pigmentosum is not the only disease resulting from faulty NER.

Nag, R., and M. J. Smerdon (2009) Altering the chromatin landscape for nucleotide excision repair. *Mutat. Res.* 62:13–20. Nice emphasis on chromatin as the actual repair substrate.

Qi, Y., M. C. Spong, K. Nam, A. Banerjee, S. Jiralerspong, M. Karplus, and G. L. Verdine (2009) Encounter and extrusion of an intrahelical lesion by a DNA repair enzyme. *Nature* 462:762–768. How does MutM rapidly scan thousands of guanine nucleotides in DNA, targeting the few oxidized guanines for removal?

Sancar, A. (1996) DNA excision repair. *Annu. Rev. Biochem.* 65:43–82. This issue of *Annual Reviews* also contains articles on transcription-coupled repair, mismatch repair, and eukaryotic DNA repair.

Mismatch Repair

Jiricny, J. (1994) Colon cancer and DNA repair: Have mismatches met their match? *Trends Genet.* 10:164–168. Tremendous excitement resulted from the finding that an altered gene in some colon cancers is related to the MutS protein in *E. coli* mismatch repair.

Larrea, A. A., S. A. Lujan, and T. A. Kunkel (2010) SnapShot: DNA mismatch repair. *Cell* 141:730–730:e1. One of several capsule "microreviews" published by *Cell*.

McMurray, C. T. (2008) Hijacking of the mismatch repair system to cause CAG expansion and cell death in neurodegenerative disease. *DNA Repair* 7:1121–1134. Describes the involvement of mismatch repair systems in triplet expansion diseases.

Nick McElhinny, S. A., G. E. Kissling, and T. A. Kunkel (2010) Differential correction of lagging-strand replication errors made by DNA polymerases α and δ. *Proc. Natl. Acad. Sci. USA* 107:21070–21075. Insights into strand selection in eukaryotic mismatch repair.

Plucennik, A., L. Dzantiev, R. R. Iyer, N. Constantin, F. A. Kadyrov, and P. Modrich (2010) PCNA function in the activation and strand direction of MutLα endonuclease in mismatch repair. *Proc. Natl. Acad. Sci. USA* 107:16066-16071. Recent information about MutL function and PCNA involvement in eukaryotic mismatch repair.

Translesion Repair

Broyde, S., L. Wang, O. Rechkoblit, N. E. Geacintov, and D. J. Patel (2008) Lesion processing: High-fidelity versus lesion-bypass DNA poly-merases. *Trends Biochem. Sci.* 33:209–219. Reviews the more open substrate-binding sites in Y-family polymerases.

Furukohri, A., M. F. Goodman, and H. Maki (2008) A dynamic polymerase exchange with *Escherichia coli* DNA polymerase IV replacing DNA polymerase III on the sliding clamp. *J. Biol. Chem.* 283:11260–11269. The replacement of a replicative polymerase by a translesion enzyme presents interesting kinetic and mechanistic problems.

Loeb, L. A., and R. J. Monnat (2008) DNA polymerases and human disease. *Nature Rev. Genetics* 9:594–603. This review includes information about the multiple Y-family polymerases and their functions.

Patel, M., Q. Jiang, R. Woodgate, M. M. Cox, and M. F. Goodman (2010) A new model for SOS-induced mutagenesis: How RecA protein activates DNA polymerase V. *Crit. Revs. Biochem. & Mol. Biol.* 45:171–184. RecA is shown to be an integral part of pol V.

Silverstein, T. D., R. E. Johnson, R. Jain, L. Prakash, S. Prakash, and A. K. Aggarwal (2010) Structural basis for the suppression of skin cancers by DNA polymerase η. *Nature* 465:1039–1044. Structure reveals how Pol η replicates accurately past a thymine dimer.

Daughter-Strand Gap Repair

Atkinson, J., and P. McGlynn (2009) Replication fork reversal and the maintenance of genome stability. *Nucl. Ac. Res.* 37:3475–3492. Evidence that this process contributes substrates for recombinational repair.

Petermann, E., and T. Helleday (2010) Pathways of mammalian replication factor restart. *Nature Reviews Mol. Cell Biol.* 11:683–687. Use of single-molecule techniques to distinguish among the possible pathways.

Double-Strand Break Repair

Bzymek, M., N. H. Thayer, S. D. Oh, N. Kleckner, and N. Hunter (2010) Double Holliday junctions are intermediates of DNA break repair. *Nature* 464:937–942. A study that makes extensive use of two-dimensional DNA electrophoresis.

Carreira, A., and S. Kowalczykowski (2009) BRCA2. *Cell Cycle* 8:3445–3447. Description of the role of BRCA2 in targeting Rad51 in double-strand break repair by HR.

Flynn, R. L., and L. Zou (2011) ATR: A master conductor of cellular responses to DNA replication stress. *Trends Biochem. Sci.* 36:133–140.

Hartlerode, A. J., and R. Scully (2009) Mechanisms of double-strand break repair in somatic mammalian cells. *Biochem. J.* 423:157–168. A comprehensive recent review.

Mazón, G., E. P. Mimitou, and L. S. Symington (2010) SnapShot: Homologous recombination in DNA double-strand break repair. *Cell* 142:646–646e1. A two-page "microreview" that summarizes a lot of information.

Pandita, T. J., and C. Richardson (2009) Chromatin remodeling finds its place in the DNA double-strand break response. *Nucl. Ac. Res.* 37:1363–1377. Emphasizes that chromatin is the substrate for eukaryotic DNA repair.

Recombination

Bianco, P. R., R. B. Tracy, and S. C. Kowalczykowski (1998) DNA strand exchange proteins: A biochemical and physical comparison. *Frontiers Biosci.* 3:570–603. A detailed review of homologous recombination mechanisms and the proteins involved, in *E. coli*, phage T4, and yeast.

Chen, Z., H. Yang, and N. P. Pavletich (2008) Mechanism of homologous recombination from the RecA-ssDNA/dsDNA structures. *Nature* 453:489–496. RecA structure determination.

Haber, J. E. (1999) DNA recombination: The replication connection. *Trends Biochem. Sci.* 24:271–275. A review of double-strand break repair and its relationship to DNA replication and recombination.

Khuu, P., and P. S. Ho (2009) A rare nucleotide base tautomer in the structure of an asymmetric DNA junction. *Biochemistry* 48:7824–7832. One of several papers presenting crystal structures of synthetic Holliday junctions.

van der Heijden, T., M. Modesti, S. Hage, R. Kanaar, C. Wyman, and C. Dekker (2008) Homologous recombination in real time: DNA strand exchange by RecA. *Mol. Cell* 30:530–538. An analysis of RecA-catalyzed strand exchange by single-molecule technology.

Gene Rearrangements

Canugovi, C., M. Samaranayake, and A. S. Bhagwat (2009) Transcriptional pausing and stalling causes multiple clustered mutations by human activation-induced deaminase. *FASEB J.* 23:34–44. A model to explain the clustering of mutations in somatic hypermutagenesis.

Craig, N. L. (1997) Target site selection in transposition. *Annu. Rev. Biochem.* 66:437–474. Contains references to all aspects of gene transposition.

Lewin, B. (2008) *Genes IX.* Jones & Bartlett, Boston. Chapters 21 and 22 of this contemporary molecular genetics textbook present detailed discussions of transposons, retroviruses, and other transposable elements.

McClintock, B. (1984) The significance of responses of the genome to challenge. *Science* 226:792–801. McClintock's Nobel Prize address, giving the history of the first description of mobile genetic elements.

Milstein, C. (1986) From antibody structure to immunological diversification of the immune response. *Science* 231:1261–1268. Milstein was awarded the Nobel Prize for discovering monoclonal antibodies, but in this Nobel Prize lecture he discusses the generation of antibody diversity.

Murley, L. L., and N. D. F. Grindley (1998) Architecture of the resolvase synaptosome: Oriented heterodimers identify interactions for synapsis and recombination. *Cell* 95:553–562. Resolvase mechanisms explored by crystallography and site-directed mutagenesis.

Roth, D. B., and N. L. Craig (1998) VDJ recombination: A transposase goes to work. *Cell* 94:411–414. A relatively recent minireview describing the recombinational events in maturation of antibody-forming genes.

Schlissel, M. S., D. Schulz, and C. Vetterman (2009) A histone code for regulating V(D)J recombination. *Mol. Cell* 34:639-640. A recent minireview of antibody gene rearrangements.

Retroviruses

Varmus, H. (1988) Retroviruses. *Science* 240:1427–1435. Still one of the best reviews available.

Gene Amplification

Sharma, R. C., and R. T. Schimke (1994) The propensity for gene amplification: A comparison of protocols, cell lines, and selection agents. *Mutat. Res.* 304:243–260. Practical information from the laboratory that discovered dihydrofolate reductase gene amplification.

Smith, K. A., et al. (1995) Regulation and mechanisms of gene amplification. *Phil. Trans. Royal Soc. London Series B* 347:49–56. A readable and well-referenced review.

PROBLEMS

1. Predict whether a *dam* methylase deficiency would increase, decrease, or have no effect upon spontaneous mutation rates, and explain the basis for your prediction.

2. For each DNA repair process in column I, list *all* characteristics from column II that correctly describe that process.

I	II
(a) Nucleotide excision repair	1. RecA protein participates.
(b) Photoreactivation	2. Damaged nucleotides are removed by nick translation
(c) Base excision repair	3. A free radical mechanism is involved.
(d) Recombinational repair	4. The repair enzyme functions only once.
(e) SOS-driven error-prone repair	5. The key enzyme contains a bound folate cofactor.
(f) Alkyltransferase repair	6. No bases or nucleotides are removed from the DNA.
(g) Mismatch repair	7. Deficiency of this enzyme in humans greatly increases the risk of skin cancer.
(h) Double-strand break repair	8. This system is chiefly responsible for the mutagenic effect of ultraviolet light.
	9. The first enzyme in this pathway cleaves two phosphodiester bonds.
	10. This process begins up to 1 kbp away from the site to be repaired.
	11. DNA ligase catalyzes the final reaction.
	12. This process also occurs in meiotic recombination.
	13. Replication fork regression might occur during this process.

3.

A   **B**   **C**   **D**

For each of the following characteristics, list all of the bases (A, B, C, or D) to which they apply.

(a) A signal that identifies a parental DNA strand in the MutH,L,S mismatch correction system

(b) Most likely to be involved in cyclobutane dimer formation after ultraviolet irradiation of DNA

(c) A methylated base found immediately to the 5′ side of dGMP residues in eukaryotic DNA

(d) Created by treating DNA with alkylating agents that transfer methyl groups and repaired by an "enzyme" that functions only once in its lifetime

(e) Created by AdoMet-dependent methylation of a nucleotide residue in DNA

(f) A substrate for deamination at the DNA level, which would lead to a GC → AT transition

4. Homologous recombination in *E. coli* forms heteroduplex regions of DNA containing mismatched bases. Why are these mismatches not eliminated by the mismatch repair system?

5. Deficiencies in the activity of either dUTPase or DNA ligase stimulate recombination. Why?

6. Suppose that you wanted to study retroviral integration mechanisms by determining the nucleotide sequence at the integration site—several dozen nucleotides on each side of the viral–cellular DNA junction. Describe how to isolate DNA containing a junction site in amounts sufficient for sequence analysis.

7. Analysis of p53 gene mutations in human tumors shows that a large proportion of these mutations involve GC → AT transitions originating at sites of DNA methylation. Propose a model to explain preferential mutagenesis of this type at these sites.

8. A paper in the *Journal of Bacteriology* reported that a *mutT* mutant strain of *E. coli* displayed a mutator phenotype when the bacteria were cultured anaerobically. Is this a surprising result? Briefly explain your answer. Present one or two possible explanations for this observation.

9. It is thought that the signal for eukaryotic mismatch repair, for identifying the strand to be repaired, could be nicks in the DNA to be repaired. If so, would you expect the nicked strand or the intact strand to be the strand to be repaired? Briefly explain your answer.

10. Identify and briefly describe three of the processes by which deamination of DNA-cytosine residues by AID could lead to mutagenesis.

11. In what ways can insertion of a transposon affect the expression of genes in the neighborhood of the insertion site?

TOOLS OF BIOCHEMISTRY 26A

GENE TARGETING BY HOMOLOGOUS RECOMBINATION

From the earliest years of recombinant DNA research, in the early 1970s, it was apparent that potential applications for research and for practical purposes were limitless. In short order it became routine to express exogenous genes in bacterial cells at high levels and to introduce any desired mutation into those genes (see Tools of Biochemistry 4A). It soon became possible to inactivate any desired gene in a bacterial genome. These "gene knockout" techniques took advantage of high activities of recombination enzymes in bacteria. A vector could be designed with sequences homologous to the chromosomal site in which exogenous DNA was to be inserted. Depending upon the design of the vector, recombination between vector sequences and homologous chromosomal sequences could lead either to insertion of new sequences into the genome or excision of chromosomal sequences adjacent to the targeting sequence used. In either case, expression of the target gene would be knocked out.

Comparable applications in eukaryotic organisms were less straightforward, partly because of their diploid genomes and partly because of technical difficulties in introducing exogenous DNA into nuclei. However, in 1983 Ralph Brinster and Richard Palmiter and their colleagues caused a sensation when they engineered the rat gene for pituitary growth hormone into the germ line of mice, and expression of the gene caused those mice to double in size. A vector was prepared that contained the growth hormone gene downstream from the promoter for metallothionein, a protein involved in heavy metal detoxification. The vector was microinjected into fertilized mouse eggs. A small proportion of animals stably incorporated the vector into their genomes. Supplementation of these animals' diets with cadmium or zinc activated the metallothionein promoter, turning on the growth hormone gene.

Although this generation of the first transgenic animals was a signal achievement, it involved addition of an exogenous gene, inserted at random sites on the genome. Could similar technology be used to alter an animal's own genotype, by insertion of DNA sequences at the site of a targeted gene? Beginning in the late 1970s, Mario Capecchi and Oliver Smithies began experimenting with gene transfer to mouse embryonic stem cells, and their efforts were rewarded with the Nobel Prize in 2007. Keys to their success were both high rates of homologous recombination between exogenous vectors and endogenous genes, and powerful techniques for selection of the desired recombination events. What is done is to introduce a targeting vector into embryonic stem cells by microinjection or electroporation (application of an electric field). Those rare cells that undergo recombination of the vector with cognate genomic sequences are selected for in culture and then injected into an early mouse embryo, which is next surgically implanted in a foster mother. The offspring are mosaics because they contain cells derived both from the manipulated stem cells and the recipient embryo. Breeding these chimeras with wild-type mice leads to homozygous transgenic animals, showing that the transgene was transmitted through the germ line.

Early gene-targeting experiments in the Capecchi laboratory used a neomycin resistance gene and the gene for hypoxanthine phosphoribosyltransferase (*hprt*) as selectable markers. The *neo^r* marker confers resistance to the antibiotic G418 (an inhibitor of protein synthesis), while expression of *hprt* sensitizes cells to 6-thioguanine, an unnatural HPRT substrate (see Chapter 22). The approach is schematized in Figure 26A.1. The targeting vector contains the *neo^r* gene inserted within a fragment of the *hprt* gene that extends from exon 7 to exon 9. Homologous recombination

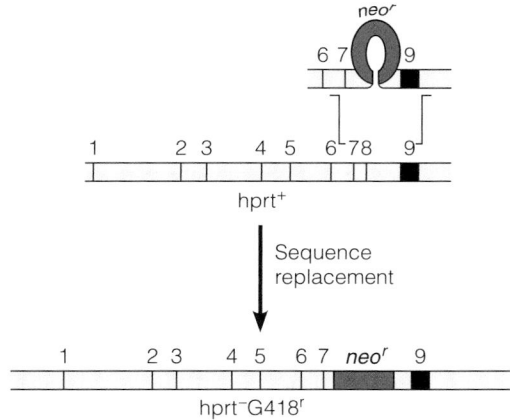

FIGURE 26.A.1

Targeted insertion of the *neo*^r gene into the chromosomal *hprt* gene.

Courtesy of M. R. Capecchi (2007) Nobel Lecture, © The Nobel Foundation.

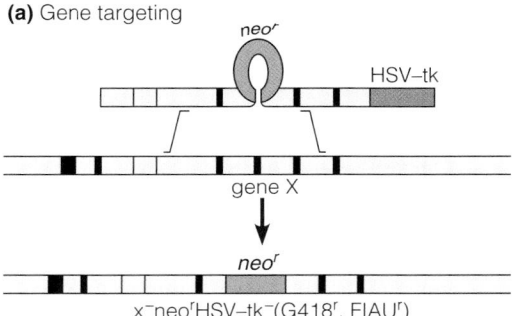

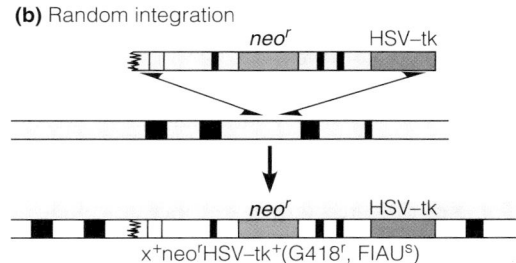

FIGURE 26.A.2

Double selection technique to insure insertion at the homologous chromosomal site.

Courtesy of M. R. Capecchi (2007) Nobel Lecture, © The Nobel Foundation.

between the vector and the chromosomal *hprt* gene leads to insertion of *neo*^r between exons 7 and 9 of the *hprt* gene. Thus, recombinants are resistant both to G418 (because they express *neo*^r) and to 6-thioguanine (because the *hprt* gene has been disrupted). The fact that the *hprt* gene is carried on the X chromosome makes this selection particularly effective because only one insertion event is required to generate resistance to 6-thioguanine.

However, analysis of this system showed that almost all of the insertion events occurred at random sites, not at the target site. Accordingly, a more rigorous selection was devised. This was based on the finding that most random insertion events occur at the ends of the targeting vector, while insertions via homologous recombination occurred at regions of sequence homology that were usually not at the ends. An approach to target an insertion in any gene "x" is shown in Figure 26A.2. Included at the end of the targeting vector is the herpes simplex virus gene for thymidine kinase, HSV-tk. Expression of this gene sensitizes a cell to the nucleoside analog 2′-fluoro-2′-deoxy-1-β-D-arabinosyl-5-iodouracil (FIAU); HSV TK phosphorylates this nucleoside, whereas the cellular thymidine kinase does not. As shown in part A of the figure, homologous insertion at the target site does not bring the HSV-tk gene into the recipient cell. Therefore, the recipient cell is resistant to G418 because of *neo*^r expression, and resistant to FIAU because HSV-tk is not being expressed. On the other hand, random integration, occurring at molecular ends, brings the HSV-tk gene into the recipient cell and converts the cell to FIAU sensitivity.

Powerful selection techniques such as that described here make it possible to introduce new genetic material into any gene in the mouse genome, and, by extension, into other vertebrate genomes. Normal gene expression can be knocked out, and in a mutant with defective gene expression, activity of the wild-type gene can be "knocked in" by similar approaches. There is excitement about using this approach to cure genetic diseases in humans. However, the principal applications to date involve targeted disruption of mouse genes. More than 11,000 mouse genes have been disrupted, in many different laboratories, to allow analysis of the function of

the respective missing gene. Many of these "knockout mouse" strains provide useful models of specific human diseases.

Gene disruption by homologous recombination has been used in other organisms. For example, in yeast a systematic study involved deletion of 6925 genes, one at a time, for analysis of the deleted gene function. In plants, by contrast, the principal means of genetic engineering do not involve homologous recombination. Most of the manipulations that generate, for example, "Roundup-Ready" seeds, use a bacterium, *Agrobacterium tumefaciens*, to provide a gene delivery vector (see Chapter 21). This bacterium causes tumors in infected plants because of expression of genes on a plasmid, the *Ti plasmid*. Part of the DNA of this plasmid, called T-DNA, becomes inserted into the infected plant genome as part of the tumorigenesis process. Transgenes are introduced into plant genomes by cloning them into vectors adjacent to T-DNA so that the transgene becomes inserted at the site of T-DNA insertion. For monocotyledinous plants, which cannot be infected by *Agrobacterium*, an alternate means of transformation involves the "gene gun," a device that physically fires DNA-coated pellets into cells. Not nearly as elegant as targeted gene transformation, but effective under certain circumstances.

References

Capecchi, M. R. (2007) Gene targeting, 1977–present. Nobel Prize lecture, available at http://nobelprize.org/nobel_prizes/medicine/laureates/2007/capecchi-lecture.html.

Winzeler, E. A., and 51 coauthors (1999) Functional characterization of the *S. cerevisiae* genome by gene deletion and parallel analysis. *Science* 285:901–906.

TOOLS OF BIOCHEMISTRY 26B

SINGLE-MOLECULE BIOCHEMISTRY

For many years chemists and biochemists alike have had to understand the behavior of molecules by observing large numbers of molecules and assuming that the behavior of individual molecules was similar to that of the average values observed. To be sure, it has long been possible to visualize single macromolecules, for example, by electron microscopy or atomic force microscopy (Tools of Biochemistry 1A). However, these are static measurements that do not allow analysis of molecular movements. Obviously, the ability to monitor single molecules might allow direct observation of fluctuations in behavior that are smoothed out by analysis of average behaviors. Within the past two decades a host of new techniques have become available that allow visualization of movements of single molecules. Some of these techniques use scanning electron microscopy, powerful fluorescence microscopes to detect fluorophores placed on protein molecules, or optical tweezers to immobilize single macromolecules. An optical tweezer is an instrument that uses a focused laser beam to provide a minute attractive or repulsive force, depending on the refractive index mismatch, to physically hold and move microscopic dielectric objects in solution.

An excellent example of single-molecule biochemistry is the experiment depicted in Chapter 15 (Figure 15.24, page 653), which provided direct evidence that the mitochondrial F_0F_1 complex is a rotary engine. In that experiment a molecule of the F_1 complex was immobilized in a microscopic field by means of a histidine tag. A fluorescent tag was attached to the γ subunit of the complex. Fluorescence microscopy allowed visualization of rotation of the γ subunit, its speed, direction, and dependence on ATP.

An early example of single-molecule visualization pertained to facilitated movement of a site-specific DNA-binding protein along a DNA lattice. It was observed in vitro that such proteins, including RNA polymerase, transcriptional repressors, and restriction enzymes, bind to a site more rapidly when the site is embedded in a long DNA molecule than a short one. Does this mean that the protein slides along the DNA, without dissociating, while it searches for its binding site? Or must other models be considered, such as intersegment transfer? As shown in Figure 26B.1, N. Shimamoto's laboratory devised an approach to the problem. DNA molecules containing a strong promoter, which binds RNA polymerase

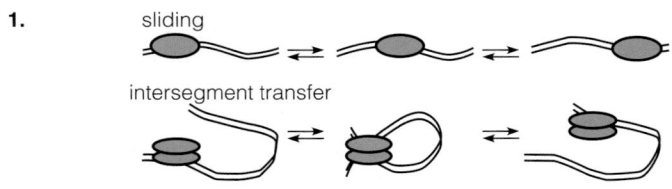

1.

sliding

intersegment transfer

2.

A. direct visualization

trace of a molecule

bulk flow

DNA

B. kinetic length effect

slow

fast

C. kinetic preference

fast

slow

D. kinetic non-preference

same velocity

E. rapid crosslinking

F. processive catalysis

FIGURE 26B.1

Analysis of RNA polymerase sliding along DNA containing one or more strong promoter sites. 1. In principle, facilitated movement along DNA can occur either by sliding or intersegment transfer. **2.** The experimental setup. Sliding is seen as horizontal deviation from the direction of bulk fluid flow. **B.–D.** Predicted kinetic behaviors. In **B** protein binds more rapidly to a site in a long than a short DNA molecule. In **C** protein binds more rapidly to a site adjacent to nonspecific DNA than to an identical site in the same molecule not adjacent to such DNA. In **D** protein binds equally rapidly to any site on the same molecule, so long as they are equally spaced. **3.** Experimental data showing either sliding (**right**) or simple drift through the DNA belt without binding. Interpretation of the data shown below.

(Top) *Journal of Biological Chemistry* 274:15293–15296, N. Shimamoto, One-dimensional diffusion of proteins along DNA: Its biological and chemical significance revealed by single-molecule measurements. Reprinted with permission. © 1999 The American Society for Biochemistry and Molecular Biology. All rights reserved; (Bottom) From *Science* 262:1561–1563, H. Kabata, O. Kurosawa, I. Arai, M. Washizu, S. A. Margarson, R. E. Glass, and N. Shimamoto, Visualization of single molecule dynamics of RNA polymerase sliding along DNA. © 1993. Reprinted with permission from AAAS.

3.

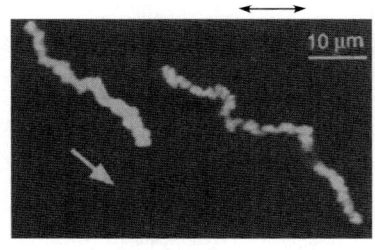

10 µm

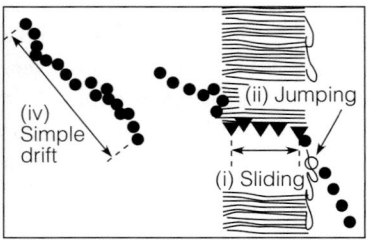

(iv) Simple drift

(ii) Jumping

(i) Sliding

tightly, were aligned in an electric field. Tangential to the DNA orientation, fluid flow carried across this "DNA belt" molecules of fluorescent-tagged RNA polymerase. Sliding would be seen as horizontal flow, as indicated in the diagram, and that in fact was seen.

Another excellent single-molecule study was mentioned in the body of this chapter (see Figure 26.28). In this study double-stranded DNA was immobilized at one end at the bottom of a flow cell. To the other end was attached a magnetic bead placed near a pair of magnets ("magnetic tweezers," see Figure 26B.2). Rotation

of the magnets can introduce supercoils, and movement parallel to the DNA length can stretch the molecule. The number of magnet rotations needed to keep the DNA fully stretched shows how supercoiling is changing, as shown in panel B of the figure. Binding RecA-coated single-strand DNA stretches the duplex, relieving some of the supercoils. Typical data are shown in Figure 26B.3. Binding a RecA-ssDNA filament in the presence of ATPγS, a nonhydrolyzable analog, stretched the DNA irreversibly (upper trace), while addition of ATP caused the stretched fiber to contract, con-

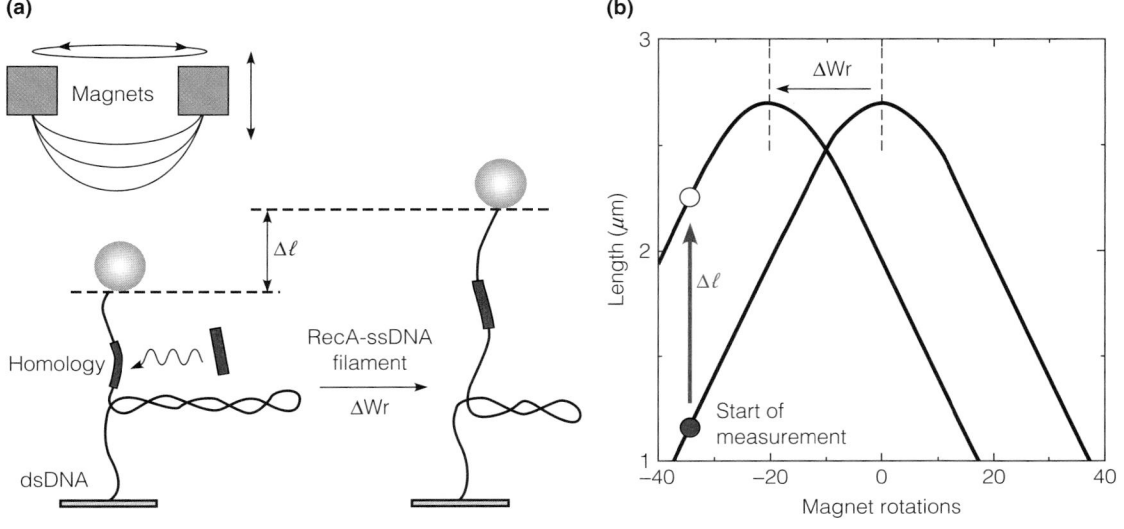

FIGURE 26B.2

Experimental setup for single-molecule analysis of RecA-mediated strand exchange. **(a)** The bead position, and hence, DNA length, is monitored by video microscopy and image analysis. The interaction of RecA-coated ssDNA (red bar) can be followed in real time because binding changes the end-to-end distance of the tethered DNA. **(b)** Binding RecA-ssDNA stretches the target DNA, thereby increasing its length; this can be determined from the number of rotations of the magnet needed to keep the target DNA fully stretched.

Reprinted from *Molecular Cell* 30:530–538, T. van der Heijden, M. Modesti, S. Hage, R. Kanaar, C. Wyman, and C. Dekker, Homologous recombination in real time: DNA strand exchange by RecA. © 2008, with permission from Elsevier.

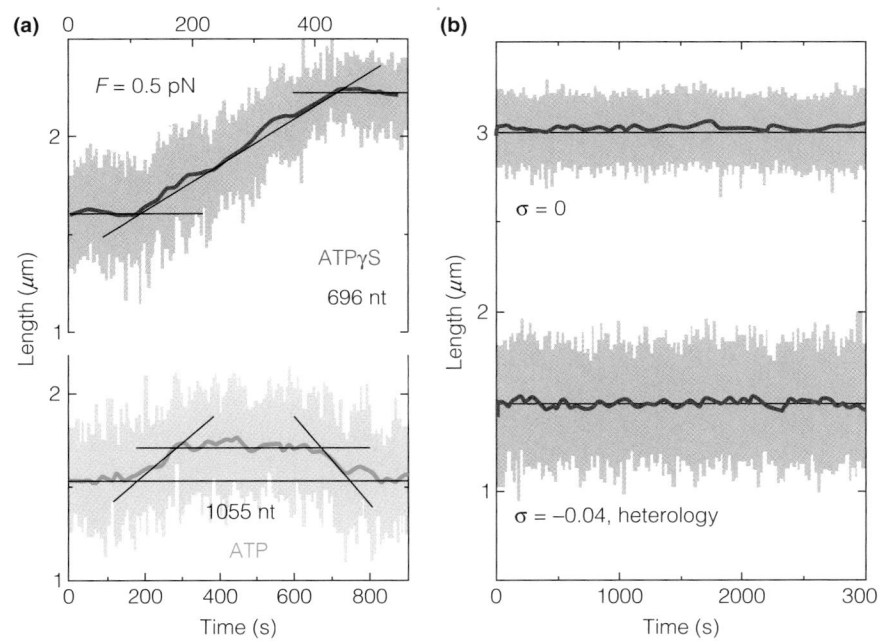

FIGURE 26B.3

Data from analysis of the RecA-DNA strand-exchange reaction. **(a)** Upper trace, ATP hydrolysis is prevented by use of a nonhydrolyzable analog. Lower trace, identical except that ATP is present; its hydrolysis allows completion of the strand-exchange reaction. **(b)** Control experiments showing no change in DNA length, as described in the text.

Reprinted from *Molecular Cell* 30:530–538, T. van der Heijden, M. Modesti, S. Hage, R. Kanaar, C. Wyman, and C. Dekker, Homologous recombination in real time: DNA strand exchange by RecA. © 2008, with permission from Elsevier.

sistent with the conclusion that ATP hydrolysis is not essential for strand invasion, but is needed for strand exchange. Controls (right panel) in which there was no supercoiling of the target DNA (upper trace) or no region of homology (lower trace) showed no changes in length. By varying the length of homologous regions and other parameters, it was possible to generate the schematic view of RecA-mediated strand exchange shown in Figure 26.26.

It should be apparent from the examples shown here that the use of single-molecule techniques to study DNA–protein interactions, motor proteins, and physical movements of proteins and enzymes is limited only by the experimenter's imagination.

References

Chen, C., and 10 coauthors (2011) Single-molecule fluorescence measurements of ribosomal translocation dynamics. *Mol. Cell* 42:367–377. Recent application of single-molecule technology to the elongation cycle in translation.

Shimamoto, N. (1999) One-dimensional diffusion of proteins along DNA. *J. Biol. Chem.* 274:15293–15296. Experiments that substantiated sliding of DNA-binding proteins along DNA.

Svoboda K., and S. M. Block (1994) Biological application of optical forces, *Ann. Rev. Biophys. and Biomol. Str.* 23:247–285. A description of the principles behind optical tweezers.

Van der Heijden, T., M. Modesti, S. Hage, R. Kanaar, C. Wyman, and C. Dekker (2008) Homologous recombination in real time: DNA strand exchange by RecA. *Mol. Cell* 30:530–538.

Van Holde, K. E. (1999) Biochemistry at the single-molecule level: Minireview series. *J. Biol. Chem.* 274:14515. Introduction to a series of four minireviews describing different single-molecule techniques.

Zlatanova, J., and K. van Holde (2006) Single molecule biology: What is it and how does it work? *Mol. Cell*, 24:317–329. A more contemporary review.

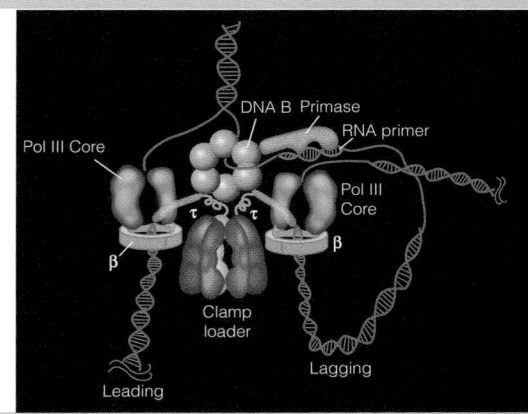

CHAPTER 27

Information Readout: Transcription and Post-transcriptional Processing

We turn now to *transcription*, in which information stored in the nucleotide sequence of DNA is read out, by the template-dependent synthesis of polyribonucleotides. Mechanistically, transcription is similar to DNA replication, particularly in the use of nucleoside triphosphate substrates and the template-directed growth of nucleic acid chains in a $5' \rightarrow 3'$ direction. There are, however, two major differences. First, with few known exceptions, only one DNA template strand is transcribed for a particular gene, and second, only part of the entire genetic potential of an organism is realized in one cell. In a differentiated eukaryotic cell, little of the total DNA is transcribed (although the proportion of DNA that is transcribed is larger than previously thought—see Chapter 29). Even in single-celled organisms, in which virtually all of the DNA sequences can be transcribed, only a small proportion of all genes may be transcribed at any time. Therefore, much of the interest in transcription involves regulation—the mechanisms used to select particular genes and template strands for transcription because this selection in large part governs the metabolic capabilities of a cell. Those mechanisms operate largely at the levels of initiation and termination of transcription through the actions of proteins that contact DNA in a highly site-specific manner (Figure 27.1). This figure shows a preview of what we have learned about transcriptional regulation through analysis of DNA-protein interactions. We discuss this important topic both here and in Chapter 29.

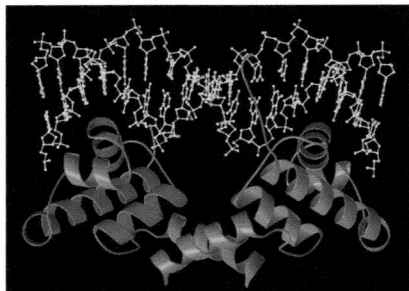

FIGURE 27.1

DNA–protein complexes that regulate transcription. Gene expression is controlled by proteins that recognize particular DNA sequences and bind at those sites. Shown here are two regulatory proteins of bacteriophage λ, which bind to the same *operator* sites in λ phage DNA and control the viral reproductive cycle by regulating transcription of genes adjacent to those sites. Above, the λ cI repressor; below, the Cro protein. In each graphic, DNA is in gold, and the protein is in green, except for the α-helical regions in contact with DNA in the major groove, which are in red. These proteins and their functions are described in Chapter 29.

Reprinted from *Proceedings of the National Academy of Sciences of the United States of America* 95:3431–3436, R. A. Albright and B. W. Matthews, How Cro and λ-repressor distinguish between operators: The structural basis underlying a genetic switch. © 1998 National Academy of Sciences, U.S.A.

The products of transcriptive RNA synthesis are rarely used directly. Almost always, transcription products are subject to further processing, such as trimming, cutting, splicing, and modifications at 3′ and 5′ ends. Therefore, this chapter deals also with post-transcriptional processing. We will be concerned here primarily with synthesis and processing of messenger RNA, ribosomal RNA, and transfer RNA, the species most directly involved in gene expression. However, we are learning that RNA is a far more versatile molecule than was thought until recently. We have already mentioned RNA enzymes—ribozymes—and their roles in RNA processing and splicing (Chapter 11), and we have described the involvement of RNA in the telomerase reaction (Chapter 25). In Chapter 29 we shall discuss the regulatory functions of RNA editing, riboswitches, small interfering RNAs, long noncoding RNAs, and microRNAs.

DNA as the Template for RNA Synthesis

The concept that RNAs are generated by template-directed copying of DNA base sequences is so firmly ingrained that we tend to forget about the critical experiments that led to our current understanding. This topic merits discussion because it involves a fascinating intellectual history, with many participants and some brilliant deductive reasoning. Even some of the false starts yielded important contributions. Also, the experimental systems used in the early work are some of those that continue to be useful, such as the lactose operon and bacteriophages T4 and λ.

The involvement of RNA in information transfer was suspected from the time DNA was first identified as the genetic storehouse—both from the chemical similarity of RNA to DNA and from the knowledge that proteins are synthesized on ribosomes. The latter fact meant that, in eukaryotic cells, information must somehow be transferred from the nucleus, where information is stored in DNA, to the cytoplasm, where most of the ribosomes reside. However, the nature of that information transfer was not clear until messenger RNA was shown to exist. This demonstration was far from straightforward because messenger RNA is quite unstable. Moreover, it constitutes such a small proportion of total cellular RNA (1% to 3% in bacteria) that in ordinary fractionations its presence is masked by the much more abundant ribosomal and transfer RNAs.

The Predicted Existence of Messenger RNA

Until about 1960 it was thought that ribosomal RNA (rRNA) represented the set of templates for protein synthesis. François Jacob and Jacques Monod, at France's Pasteur Institute, questioned this idea, in part because rRNAs are homogeneous in size (5S, 16S, and 23S in bacteria, as discussed in Chapter 28), whereas the molecular weights of proteins vary over at least two orders of magnitude. From analyzing *E. coli* mutants that are altered in the control of lactose metabolism, Jacob and Monod predicted the existence of messenger RNA (mRNA), an RNA species that is synthesized from a DNA template and is used in turn as the template for protein synthesis.

Lactose utilization in *E. coli* was known to be controlled by three enzymes, whose genes are adjacent on the chromosome. One of these is **β-galactosidase**, which hydrolyzes lactose and other β-galactosides. When bacteria are grown with glucose as the sole carbon source, the levels of the lactose-utilizing enzymes are very low, with less than one molecule of β-galactosidase per cell on average. However, substitution of lactose or a related β-galactoside for glucose in the medium leads to rapid enzyme **induction**, or synthesis of the three enzymes. β-Galactosidase ultimately represents as much as 6% of the total soluble protein of the cell. Removal of lactose from the culture slows the further synthesis of enzyme molecules. The rapid changes in β-galactosidase-forming capacity suggested that

Our original concepts of RNA synthesis came from genetic studies that predicted the existence of messenger RNA.

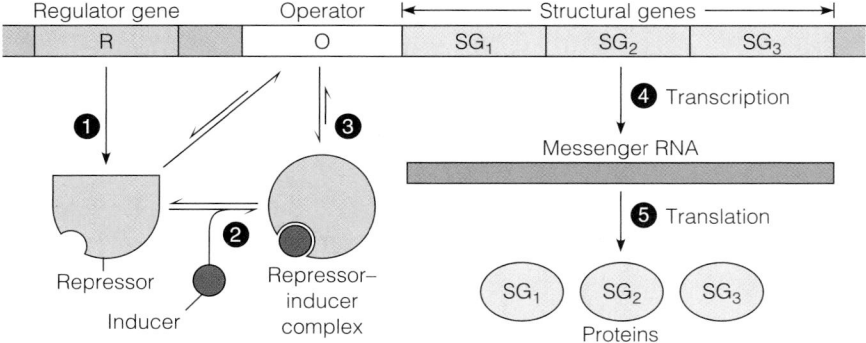

FIGURE 27.2

The operon model, as proposed in 1961 by Jacob and Monod. **Step 1:** The regulator gene R encodes a repressor molecule, which can bind to the operator (O) and thereby inhibit transcription of the adjacent structural genes $SG_{1,2,3}$. **Step 2:** A small-molecule inducer complexes with the repressor, thereby altering the equilibrium between the conformational states of the repressor. **Step 3:** The repressor–inducer complex binds less tightly to the operator. **Step 4:** This loosening facilitates transcription of the structural genes, resulting in production of messenger RNA—that is, an RNA copy of the structural genes. **Step 5:** The mRNA sequence is translated into proteins.

the template for synthesizing this enzyme is metabolically unstable—synthesized rapidly on demand and degraded when a continued stimulus to induction is absent (alternatively, control could be exercised at the translational level). Because rRNAs are quite stable, these species were unlikely to be intermediates in information transfer.

Jacob and Monod analyzed numerous *E. coli* mutants that displayed faulty control over induction of the lactose-utilizing enzymes. Some expressed all three genes at high levels even when lactose or a similar inducer was absent, and others could not induce any of the enzymes, even after addition of lactose. These experiments led to the notion of a macromolecular *repressor* that regulated the level of a messenger RNA by binding to a specific *operator* sequence on DNA and thereby turning off the synthesis of an RNA that encoded all three enzymes (see Chapter 29, pages 1233–1235).

Based on these studies, and on parallel work by Andre Lwoff with bacteriophage λ, Jacob and Monod in 1961 proposed a unifying hypothesis of gene regulation, in which transcription was regulated specifically at the level of initiation. Hypothetical regulatory elements called repressors and operators controlled the synthesis of other hypothetical entities called messenger RNAs (mRNAs). mRNA was postulated to be a complementary copy of the DNA that encompassed a set of structural genes, which encode proteins, as schematized in Figure 27.2. A set of contiguous genes plus adjacent regulatory elements that control their expression was termed an **operon**, and the Jacob–Monod hypothesis thus came to be known as the **operon model**.

Jacob and Monod correctly predicted several characteristics of the hypothetical messenger RNA. First, they predicted a high rate of mRNA synthesis followed by rapid degradation, which would explain the fast turn-on of the genes after induction and turn-off after removal of the inducer. Second, because of rapid synthesis and degradation, they expected mRNA to accumulate rapidly but not to high steady state levels. Third, because they thought that the messenger was a copy of two or more contiguous genes, they expected it to be fairly large and part of a heterogeneous size class of RNA. Finally, if the messenger RNA was a complementary copy of DNA, its nucleotide sequence should be identical to that of one of the template DNA strands.

Bacterial genetics predicted messenger RNA to be a collection of metabolically active RNAs that are present in low abundance, heterogeneous in size, and complementary in sequence to DNA.

T2 Bacteriophage and the Demonstration of Messenger RNA

The first physical demonstration of mRNA came from work with T2 and T4 bacteriophages. Infection by these large DNA viruses arrests all expression of host cell genes, and no accumulation of RNA can be detected after infection. However, in 1956 the use of radioisotopes led to detection of a distinctive RNA in T2-infected *E. coli*. When infected cultures were pulse-labeled with [^{32}P]orthophosphate for 3 or 4 minutes, about 2% of the total RNA became radioactive. This radiolabeled RNA had two properties that led to its eventual identification as viral messenger RNA.

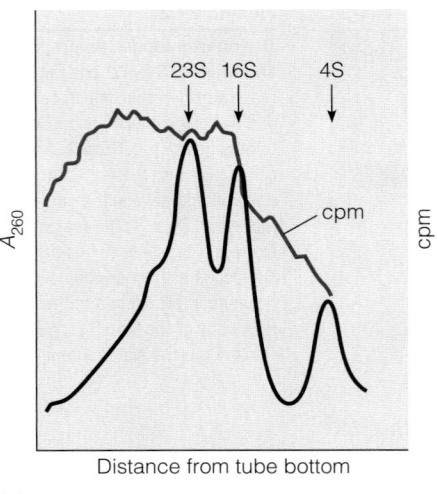

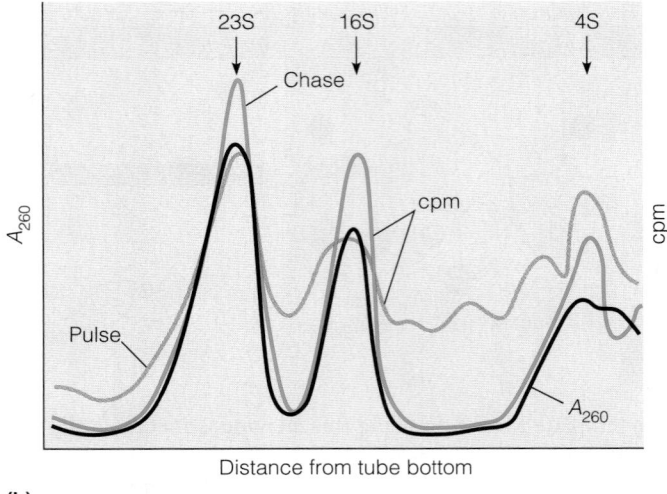

(a)   (b)

FIGURE 27.3

Demonstration of mRNA by pulse-labeling and sedimentation. **(a)** Sedimentation profile of total and pulse-labeled RNAs in T2 phage-infected *E. coli*. Total concentration of RNA in each fraction (black) was determined by ultraviolet absorbance (A_{260}). The peaks labeled 23S and 16S represent the major ribosomal RNA species. The radioactivity profile (red) shows the heterogeneous distribution of species synthesized during the pulse. **(b)** Pulse-labeled RNA species in uninfected bacteria and their fate in a chase. The orange line represents cells labeled for 3 minutes, the blue line depicts an identical culture with the label chased by 0.7 generation of growth in nonradioactive medium after the pulse, and the black line is the A_{260} profile. Again, the 23S and 16S peaks represent ribosomal RNA and the 4S peak is mostly transfer RNA.

(a) Data from *Journal of Molecular Biology* 14:71–84, K. Asano, Size heterogeneity of T2 messenger RNA. © 1965, with permission from Elsevier; (b) Data from *Proceedings of the National Academy of Sciences of the United States of America* (1961) 47:1564–1580 M. Hayashi and S. Spiegelman, The selective synthesis of informational RNA in bacteria.

The ability of T2 phage RNA to hybridize with T2 DNA and to associate with ribosomes made before infection was evidence for the existence of messenger RNA.

First, it was metabolically labile; in pulse-chase experiments (see Tools of Biochemistry 12A), radioactivity was rapidly lost from this RNA. Second, the RNA seemed to be a product of viral DNA metabolism because its nucleotide composition was close to that of T2 DNA; the RNA was rich in adenine and uracil and low in guanine and cytosine.

Important additional evidence came from sucrose gradient centrifugation of pulse-labeled RNA, which showed that the labeled material sediments heterogeneously and distinctly from any of the known rRNA or tRNA species (Figure 27.3a). Benjamin Hall and Sol Spiegelman established that this RNA is a viral gene product when they carried out the first DNA–RNA hybridization experiment, which showed that the labeled RNA was complementary in sequence to phage DNA. The earliest experiments were based on the fact that RNA is of higher density than DNA. Thus, a DNA–RNA hybrid could be detected in equilibrium gradient centrifugation as a species of intermediate density containing label derived from both DNA and RNA. Such a hybrid was formed when T2 RNA was heated and slowly cooled along with T2 DNA, but not when the DNA came from *E. coli*.

Additional support for the existence of phage messenger RNA came from a density shift experiment carried out by François Jacob, Matthew Meselson, and Sydney Brenner. *E. coli* was grown in dense medium ($^{13}C-^{15}N$) and infected with T2 in light medium. The rationale is similar to that of the Meselson–Stahl experiment on DNA replication (Chapter 4). After pulse labeling with radioactive amino acids, density analysis on CsCl gradients showed that phage proteins were synthesized on dense ribosomes, that is, ribosomes synthesized before infection. The fact that only phage proteins were being made ruled out any possibility that ribosomal RNA provided a template. Instead, this experiment pictured the ribosome as a nonspecific workbench, upon which any protein could be assembled, depending on which messenger template became associated with that workbench.

RNA Dynamics in Uninfected Cells

The experiments described above demonstrated the existence of mRNA in phage-infected *E. coli*. What about uninfected bacteria? Spiegelman and his colleagues showed that pulse-labeled RNA from uninfected *E. coli* hybridized to *E. coli* DNA. At very short labeling intervals the sedimentation pattern showed incorporation into both rRNA and tRNA species *and* a heterogeneously sedimenting species. After a chase (Figure 27.3b), the radioactivity profile followed the absorbance profile, showing that all RNA species had become labeled to equivalent specific activities. This finding is consistent with the postulated short lifetime of messenger RNA ($t_{1/2}$ of 2 or 3 minutes). mRNA would reach its maximal radioactivity within just a few minutes, but during the chase, mRNA turnover would release nucleotides that could flow into stable RNA species. Because those species do not turn over, label accumulates, and the fraction of total label in the stable RNA species continues to increase. Consistent with this idea, Spiegelman also showed that highly labeled ribosomal and transfer RNAs hybridize to *E. coli* DNA, demonstrating that all three major classes of RNA are synthesized from template DNA strands.

As noted previously, the earliest DNA–RNA hybridization experiments involved equilibrium gradient centrifugation, a time-consuming and expensive technique. Spiegelman and his colleagues made the important discovery that single-strand DNA binds irreversibly to membrane filters, made of material such as nitrocellulose. This technique allowed rapid hybridization analysis of a large number of samples because a radiolabeled RNA could hybridize to denatured DNA immobilized on a filter. After suitable treatment and washing of the filter, the extent of hybridization could be determined simply by placing the filter in a liquid scintillation counter and counting its radioactivity. The same principle—immobilization of nucleic acid on nitrocellulose followed by analysis of bound radioactivity—underlies Southern blotting and northern blotting (described in Chapter 24, page 1016), which are now more widely used to analyze gene organization and expression. These findings led also to microarray technology, in which thousands of DNA–RNA hybridization reactions are analyzed on a single DNA chip (see Tools of Biochemistry 27A).

The final chapter in the messenger RNA saga was the demonstration that isolated mRNAs have template activity in vitro. That is, they can program the synthesis of specific protein molecules in the presence of 20 amino acids and other factors. The translation of defined RNA templates, both synthetic and natural, was critical to deciphering the genetic code (see Chapter 28).

Enzymology of RNA Synthesis: RNA Polymerase

We now know that RNA synthesis involves the copying of a template DNA strand by RNA polymerase. However, the earliest description of an enzyme capable of synthesizing RNA in vitro occurred in the late 1950s, at about the same time as the discovery of DNA polymerase I. This RNA-synthesizing enzyme, called **polynucleotide phosphorylase**, was quite different from DNA polymerase. The enzyme required no template, and it used ribonucleoside *diphosphates* (rNDPs) as substrates to produce a random-sequence polynucleotide whose base composition matched the nucleotide composition of the reaction medium.

$$n \text{ rNDP} \rightleftharpoons (\text{rNMP})_n + n\text{P}_i$$

Initially it was thought that polynucleotide phosphorylase might be the major RNA-synthesizing enzyme, but the lack of a template requirement was troubling, as was the apparent absence of the enzyme in eukaryotic cells. Ultimately, polynucleotide phosphorylase turned out not to play a role in RNA synthesis in vivo but instead to participate in the *degradation* of bacterial messenger RNAs. However, the enzyme was of great value in the synthesis of polynucleotides used as templates for in vitro protein synthesis, when the genetic code was being elucidated (see Chapter 28).

Investigators continued to search for an enzyme that would copy a DNA template in vitro. In 1961 such an enzyme was discovered almost simultaneously in four different laboratories. The enzyme, **DNA-directed RNA polymerase**, resembled DNA polymerases in the nature of the reaction catalyzed.

$$n(\text{ATP} + \text{CTP} + \text{GTP} + \text{UTP}) \xrightarrow{\text{Mg}^{2+}, \text{DNA}} (\text{AMP–CMP–GMP–UMP})_n + n\text{PP}_i$$

The reaction product is a complementary RNA copy of the DNA template.

Polynucleotide phosphorylase catalyzes the reversible, template-independent synthesis of random-sequence polyribonucleotides.

Biological Role of RNA Polymerase

A single RNA polymerase catalyzes the synthesis of all three bacterial RNA classes—mRNA, rRNA, and tRNA. This was shown in experiments with **rifampicin** (Figure 27.4a), an antibiotic that inhibits RNA polymerase in vitro and blocks the synthesis of mRNA, rRNA, and tRNA in vivo. Rifampicin-resistant mutants of *E. coli* were found both to contain a rifampicin-resistant form of RNA polymerase and to be capable of synthesizing all three RNA classes in vivo in the presence of rifampicin. Because a single mutation affects both the RNA polymerase and the synthesis of all RNA types in vivo, RNA polymerase must be the one enzyme catalyzing all forms of transcription in bacteria.

In contrast, eukaryotes contain three distinct RNA polymerases, one each for synthesis of rRNA, mRNA, and small RNAs (tRNA plus the 5S species of rRNA)—RNA polymerases I, II, and III, respectively. The existence of separate enzymes was revealed partly because they differ in their sensitivity to inhibition by α-**amanitin** (Figure 27.4b), a toxin from the poisonous *Amanita* mushroom. RNA polymerase II is inhibited at low concentrations, RNA polymerase III is inhibited only at high concentrations, and RNA polymerase I is quite resistant.

Figure 27.4 shows the structures of two additional inhibitors. **Cordycepin**, or 3′-deoxyadenosine, is a transcription chain terminator because it lacks a 3′ hydroxyl group from which to extend. The nucleotide of cordycepin is incorporated into growing chains, confirming that transcriptional chain growth occurs in a 5′→ 3′ direction. Another important inhibitor is **actinomycin D**, which acts by binding to DNA. The tricyclic ring system (phenoxazone) intercalates between adjacent G-C base pairs, and the cyclic polypeptide arms fill the nearby narrow groove.

Because DNA polymerases and RNA polymerases catalyze similar reactions, it is interesting to compare some of their kinetic features. V_{max} for DNA polymerase III

FIGURE 27.4

Some inhibitors of transcription. Rifampicin is an inhibitor of bacterial transcription initiation, and α-amanitin is an inhibitor of eukaryotic RNA polymerases. Cordycepin is a transcription terminator because the 3′ position on the sugar is occupied by H (magenta), rather than a hydroxyl group. The tricyclic ring system of actinomycin (blue) intercalates between adjacent G-C base pairs in DNA; the R groups of the molecule (magenta) are cyclic polypeptides and fill in the narrow grooves of the helix.

(a) Rifampicin

(b) α-Amanitin

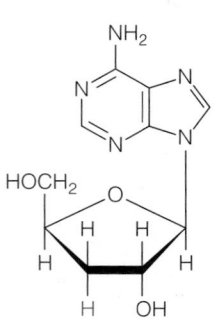

(c) Cordycepin (3'-deoxyadenosine)

(d) Actinomycin D

holoenzyme, at about 500 to 1000 nucleotides per second, is much higher than the chain growth rate for bacterial transcription—50 nucleotides per second, which is the same as V_{max} for purified RNA polymerase. Although there are only about 10 molecules of DNA polymerase III per *E. coli* cell, there are some 2000 molecules of RNA polymerase, of which half might be involved in transcription at any instant. This fits in with observations that replicative DNA chain growth is rapid but occurs at few sites, whereas transcription is much slower but occurs at many sites. The result is that far more RNA accumulates in the cell than DNA. Like the DNA polymerase III holoenzyme, the action of RNA polymerase is highly processive. Once past the initial stages of transcription, RNA polymerase rarely, if ever, dissociates from the template until the specific signal to terminate has been reached. Although this discussion is based primarily upon prokaryotic transcription, these characteristics of transcription are similar in eukaryotic cells.

Another important difference between DNA and RNA polymerases is the accuracy with which a template is copied. With an error rate of about 10^{-5}, RNA polymerase is less accurate than replicative DNA polymerase holoenzymes, although RNA polymerase is much more accurate than would be predicted from Watson–Crick base pairing alone. Given that RNA does not carry information from one cell generation to the next, an ultrahigh-fidelity template-copying mechanism is evidently not needed. However, recent observations suggest the existence of error-correction mechanisms. In *E. coli,* two proteins, called GreA and GreB, promote the hydrolytic cleavage of nucleotides at the $3'$ ends of nascent RNA molecules. These processes may be akin to $3'$ exonucleolytic proofreading by DNA polymerases, but there are important differences: First, cleavage of $3'$ ends of RNA molecules usually removes dinucleotides, rather than single nucleotides. Second, cleavage occurs within the polymerase catalytic site, whereas DNA polymerase proofreading occurs at a distinct exonuclease site. Finally, the rate of hydrolysis is much slower than the rate of RNA chain extension by RNA polymerase. As discussed below, $3'$ transcript cleavage probably plays a different role, in escaping blocks to continued elongation. However, there is evidence indicating that a misincorporated nucleotide is preferentially cleaved before continuation of chain elongation.

Structure of RNA Polymerase

When highly purified *E. coli* RNA polymerase is analyzed in denaturing electrophoretic gels, five distinct polypeptide subunits are observed. Their properties are summarized in Table 27.1. Two copies of the α subunit are present, along with one each of β, β', σ, and ω, giving M_r of about 450,000 for the holoenzyme. ω may be involved in regulation, but its precise role is not yet clear. However, much has been learned about the functions of the other subunits, partly from **mixed reconstitution** studies, in which the dissociated subunits are renatured and allowed to reassociate, with formation of active enzyme. For example, when β from a rifampicin-resistant

DNA replication involves rapid chain growth at few intracellular sites, and transcription involves slower growth at many sites. More RNA accumulates than DNA.

The functions of the various RNA polymerase subunits can be determined by reconstitution of active enzyme from isolated subunits.

TABLE 27.1 Subunit composition of *E. coli* RNA polymerase

Subunit	M_r	Number per Enzyme Molecule	Function
α	36,500	2	Chain initiation, interaction with regulatory proteins and upstream promoter elements
β	151,000	1	Chain initiation and elongation
β'	155,000	1	DNA binding
σ	70,000[a]	1	Promoter recognition
ω	11,000	1	Unknown

[a]The 70-kDa σ subunit is one of several alternative σ subunits.

form of RNA polymerase was recombined with α, β', and σ from wild-type cells, the reconstituted enzyme was seen to be rifampicin-resistant. This finding establishes β as the target for rifampicin inhibition. Furthermore, rifampicin is known to inhibit the initiation of transcription, so β must also play a role in initiation. The fact that β is also the target for inhibition by an inhibitor of elongation called **streptolydigin** points to β as a subunit directly involved in chain elongation.

The σ subunit is easily dissociated from RNA polymerase—for example, by passing the purified enzyme through a carboxymethylcellulose column. The σ-free enzyme, called **core polymerase**, is still catalytically active, but it binds to DNA at far more sites than does the RNA polymerase holoenzyme, and it shows no strand or sequence specificity. The σ subunit plays an essential role in directing RNA polymerase to bind to template at the proper site for initiation—the **promoter** site—and to select the correct strand for transcription. The addition of σ to core polymerase reduces the affinity of the enzyme for *nonpromoter* sites by about 10^4, thereby increasing the enzyme's specificity for binding to promoters.

These discoveries about σ, which were made in the early 1970s, suggested that gene expression might be regulated by having core polymerase interact with different forms of σ, which would in turn direct the holoenzyme to different promoters. In many instances, this does occur. For example, when *Bacillus subtilis* is induced to **sporulate** (that is, to generate metabolically inert cells capable of later outgrowth), new forms of σ are produced. These forms combine with core polymerase to redirect cellular metabolism for transcription of the genes involved in sporulation. Another example is apparent when an *E. coli* culture is stressed by a sudden temperature increase. In these heat-shocked cells a new form of σ appears and directs the modified RNA polymerase to a different set of promoters, thereby activating transcription of a block of genes called heat-shock genes. The most abundant σ in *E. coli*, and the one that will frame our discussions, is called σ^{70} because of its 70-kDa molecular weight. As shown in Table 27.2, seven different σ factors are known in *E. coli*, each designed to direct RNA polymerase to a functionally related set of genes.

Eukaryotic RNA polymerases show similarity with the prokaryotic enzymes, but they have a much more complex subunit structure. RNA polymerase II from yeast has twelve subunits. However, the common ancestry among RNA polymerases is evident, as can be seen in Figure 27.5. For example, the two proteins in the pol II catalytic core in all three eukaryotic multisubunit polymerases are related to β and β' of the bacterial enzyme. Shown also in the figure is the subunit composition of archaeal RNA polymerases, which shows them to be much more closely related to eukaryotic than to prokaryotic polymerases. Eukaryotic RNA polymerases have no direct counterpart to bacterial σ. Instead, a series of proteins called **transcription factors** functions in comparable fashion, helping to direct RNA polymerase to promoter sites and to form an initiation complex, as discussed later (page 1135). Bacteria have transcription factors also, as we describe in Chapter 29.

Although the multisubunit motif for RNA polymerases is the dominant structural theme, it is not universal. Exceptions include single-subunit polymerases

TABLE 27.2	σ Factors in *E. coli*	
Name	Structural Gene	Genes Whose Transcription Is Stimulated by Binding to Core Polymerase
σ^{70}	*rpoD*	"Housekeeping genes" (those involved in central pathways)
σ^{N}	*rpoN*	Genes involved in adaptation to nitrogen stress
σ^{S}	*rpoS*	Genes involved in adaptation to stationary phase
σ^{H}	*rpoH*	Genes involved in adaptation to heat shock
σ^{F}	*rpoF*	Genes related to flagella formation and chemotaxis
σ^{E}	*rpoE*	Adaptation to extreme heat shock; extracytoplasmic functions
σ^{FecI}	*fecI*	Ferric citrate transport; extracytoplasmic functions

Adapted from *Annual Review of Microbiology* 54:499–518, A. Ishihama, Functional modulation of *Escherichia coli* RNA polymerase. © 2000 Annual Reviews.

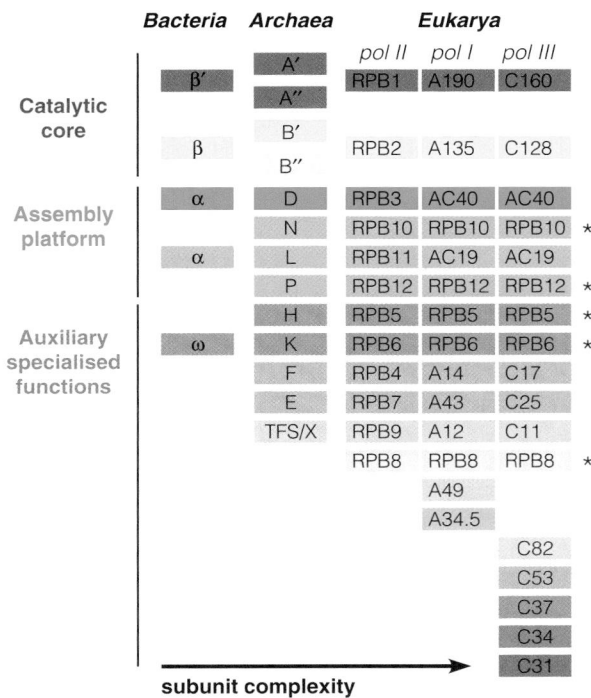

	Bacteria	Archaea	Eukarya		
			pol II	pol I	pol III
Catalytic core	β'	A'	RPB1	A190	C160
		A''			
	β	B'	RPB2	A135	C128
		B''			
Assembly platform	α	D	RPB3	AC40	AC40
		N	RPB10	RPB10	RPB10 ★
	α	L	RPB11	AC19	AC19
		P	RPB12	RPB12	RPB12 ★
Auxiliary specialised functions		H	RPB5	RPB5	RPB5 ★
	ω	K	RPB6	RPB6	RPB6 ★
		F	RPB4	A14	C17
		E	RPB7	A43	C25
		TFS/X	RPB9	A12	C11
			RPB8	RPB8	RPB8 ★
				A49	
				A34.5	
					C82
					C53
					C37
					C34
					C31

subunit complexity ⟶

FIGURE 27.5

RNA polymerase subunit structures in the three domains of life. The subunits are arranged by function rather than size. Homologous subunits are color coded. Subunits marked with an asterisk are conserved among the three eukaryotic RNA polymerases. These are core enzyme structures, not showing transiently bound proteins that aid in template recognition.

Molecular Microbiology 65:1395–1404, F. Werner, Structure and function of archaeal RNA polymerases. © 2007 John Wiley & Sons, Inc. Reproduced with permission from John Wiley & Sons, Inc.

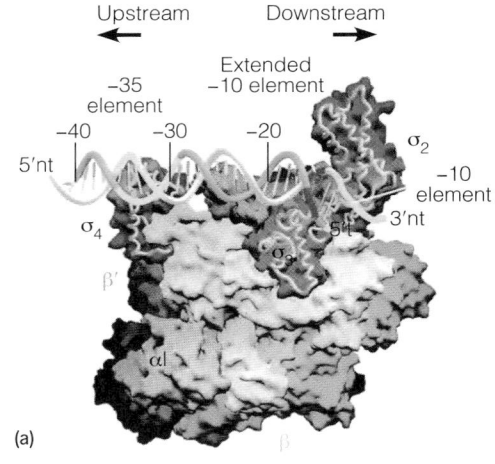

(a)

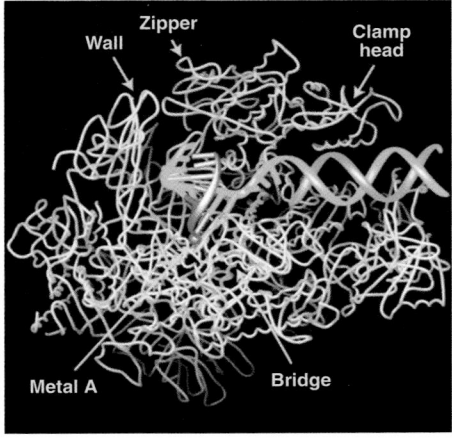

(b)

FIGURE 27.6

Crystal structures of RNA polymerases. **(a)** *Taq* RNA polymerase complexed with DNA. Both α subunits and ω are shown in gray; β, cyan; β', pink; σ, orange. The molecular surface of σ is partially transparent, showing the α-carbon backbone inside. The numbers next to the σ designations refer to domains on the protein molecule. Template DNA strand is green, except for promoter elements at −10 and −35 (see page 1139), with respect to position +1, where transcription will initiate. PDB ID, 1L9U. **(b)** Yeast RNA polymerase II. Parts of Rbp2 (counterpart to β) are removed to reveal bound nucleic acids. Bases of ordered nucleotides are shown as cylinders protruding from the backbone regions. Shown also are an active-site metal and several functional regions of the protein, which are described later. PDB ID, 116H.

From *Science* 296:1285–1290, K. S. Murakami, S. Masuda, E. A. Campblee, O. Muzzin, and S. A. Darst, Taq RNA polymerase; and *Science* 292:1876–1882, A. L. Gnatt, P. Cramer, F. Fu, D. A. Bushnell, and R. D. Kornberg, RNA polymerase II. © 2002 and 2001. Reprinted with permission from AAAS.

encoded by some bacteriophage genomes, including T7 and SP6, and a mitochondrial RNA polymerase found in vertebrates. Plant chloroplasts also contain an organelle-specific RNA polymerase. Figure 27.6 shows crystal structures of a prokaryotic multisubunit RNA polymerase, from *Thermus aquaticus*, and RNA polymerase II from yeast (pol II), as determined in the laboratories of Seth Darst and Roger Kornberg, respectively. We shall describe relations between these structures and the RNA polymerase reaction presently.

Mechanism of Transcription

Like DNA replication and protein synthesis, transcription occurs in three distinct phases—initiation, elongation, and termination. Initiation and termination signals in the DNA sequence punctuate the genetic message by directing RNA polymerase to particular genes and by specifying where transcription will start, where it will stop, and which strand will be transcribed. The signals involve both instructions encoded in DNA base sequences and interactions between DNA and proteins other than RNA polymerase. Most of our initial discussion focuses on prokaryotic RNA polymerases, exemplified by the widely studied *E. coli* enzyme, but the basic mechanics of transcription are similar in all organisms. Action of the eukaryotic enzymes is discussed later.

Initiation of Transcription: Interactions with Promoters

The overall processes of initiation and elongation are summarized in Figure 27.7. The first step in transcription is binding of RNA polymerase to DNA, followed by migration to an initiation DNA site, the promoter. In Tools of Biochemistry 27B, we describe *footprinting*, a technique used to locate DNA sites in contact with sequence-specific DNA-binding proteins, and we discuss promoter recognition on pages 1138–1140. For now, let us concentrate on the mechanistic details of synthesizing polyribonucleotide chains.

Bacterial RNA polymerase finds promoters by a search process (Figure 27.7, step 1), in which the holoenzyme binds nonspecifically to DNA, with low affinity, and then moves along the DNA until it reaches a promoter sequence, to which it

FIGURE 27.7

Initiation and elongation steps of transcription by bacterial RNA polymerase.

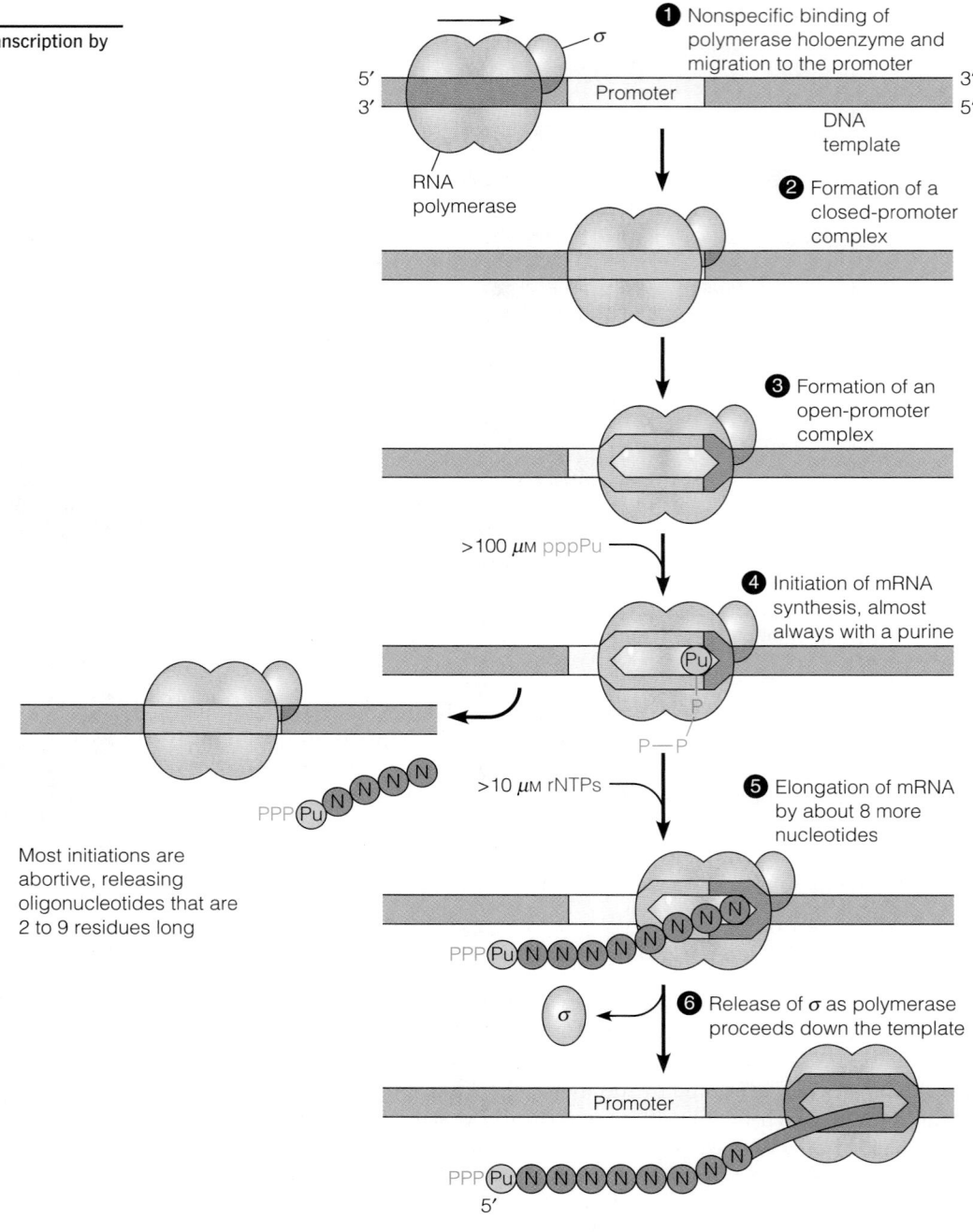

① Nonspecific binding of polymerase holoenzyme and migration to the promoter

② Formation of a closed-promoter complex

③ Formation of an open-promoter complex

>100 μM pppPu

④ Initiation of mRNA synthesis, almost always with a purine

Most initiations are abortive, releasing oligonucleotides that are 2 to 9 residues long

>10 μM rNTPs

⑤ Elongation of mRNA by about 8 more nucleotides

⑥ Release of σ as polymerase proceeds down the template

binds with much higher affinity. σ factor is essential for this search because, as noted earlier, the core enzyme does not bind to promoters more tightly than to nonpromoter sites. Biophysical evidence suggests that RNA polymerase slides along the DNA, without dissociating from it. Evidence for sliding of the enzyme along DNA comes partly from single-molecule studies, of the type described in Tools of Biochemistry 26B. Binding to DNA and then moving along it reduces the complexity of the search from three dimensions to one, just as finding a house becomes simpler once you find the street upon which that house is located.

The initial encounter between RNA polymerase holoenzyme and a promoter generates a **closed-promoter complex** (step 2). Whereas DNA strands unwind later in transcription, no unwinding is detectable in a closed-promoter complex. This complex forms with a K_a between 10^7 and 10^8 M^{-1} at 0.1 M NaCl. Binding is primarily electrostatic, for K_a is dependent on ionic strength. The complex is relatively labile, dissociating with a half-life of about 10 seconds. Footprinting studies show that polymerase is in contact with DNA from about nucleotide -55 to -5, where $+1$ represents the first DNA nucleotide to be transcribed.

Next, as shown by analysis with chemical reagents that react specifically with single-stranded DNA, RNA polymerase unwinds several base pairs of DNA, from about -10 to -1, giving an **open-promoter complex**, so-called because it binds DNA whose strands are open, or unwound (step 3). This highly temperature-dependent reaction occurs with half-times of about 15 seconds to 20 minutes, depending upon the structure of the promoter. The open-promoter complex is extremely stable; it is not easily disrupted by high ionic strength, and it forms with a K_a as high as $10^{12}\,M^{-1}$. An Mg^{2+}-dependent isomerization next occurs, giving a modified form of the open-promoter complex with the unwound DNA region now extending from -12 to $+2$. Analysis with scanning force microscopy (Tools of Biochemistry 1A), later confirmed by X-ray crystallography (Figure 27.8) indicates that DNA bending in the promoter region accompanies the transition from a closed-promoter to an open-promoter complex (Figure 27.7). This could be related to the fact, as noted later, that activation of the transcription of certain genes often involves interaction between the promoter and its associated proteins and with proteins bound at upstream regulatory sites, far to the 5′ side of the gene being transcribed.

Initiation and Elongation: Incorporation of Ribonucleotides

Having located a promoter and formed an open-promoter complex, the enzyme is ready to begin the synthesis of an RNA chain. Binding of the first ribonucleoside triphosphate (usually ATP or GTP) occurs with a half-saturating concentration of about $100\,\mu M$. Since binding of later rNTPs occurs with half-saturating concentrations of about $10\,\mu M$, this may represent a control process, favoring initiation only at relatively high rNTP concentrations. RNA polymerase contains two nucleotide-binding sites near the catalytic center. Current evidence suggests that rNTP binding at the low-affinity site is allosteric, and functions to drive the reaction in a forward direction.

Chain growth begins with binding of the template-specified rNTP (Figure 27.7, step 4), followed by binding of the next nucleotide. Next, nucleophilic attack by the 3′ hydroxyl of the first nucleotide on the α (innermost) phosphorus of the second nucleotide generates the first phosphodiester bond and leaves an intact triphosphate moiety at the 5′ position of the first nucleotide. Nucleotide incorporation occurs via the two-metal mechanism described for DNA polymerases in Chapter 25 (page 1046) and now thought to be universal for nucleic acid-synthesizing enzymes. The transcript is bound unstably during the first several phosphodiester bond-forming reactions, as shown by the fact that most initiations are abortive, with release of oligonucleotides two to nine residues long. The basis for this low efficiency of initiation is not fully understood, although structural analysis of RNA polymerases is yielding clues (page 1136).

During incorporation of the first 10 nucleotides, the σ subunit dissociates from the transcription complex, and the remainder of the transcription process is catalyzed by the core polymerase (steps 5 and 6). Once σ has dissociated, the elongation complex becomes quite stable. Transcription, as studied in vitro, can no longer be inhibited by adding rifampicin, and virtually all transcription events proceed to completion.

During *elongation* (steps 5 and 6), the core enzyme moves along the duplex DNA template. As it moves, it simultaneously unwinds the DNA, exposing a single-stranded template for base pairing with incoming nucleotides and with the nascent transcript (the most recently synthesized RNA), and it rewinds the template behind the 3′ end of the growing RNA chain, as suggested in Figure 27.9. In this model, about 18 base pairs of DNA are unwound to form a moving "transcription

Transcription begins with sequence-specific interaction between RNA polymerase and a promoter site, where duplex unwinding and template strand selection occur.

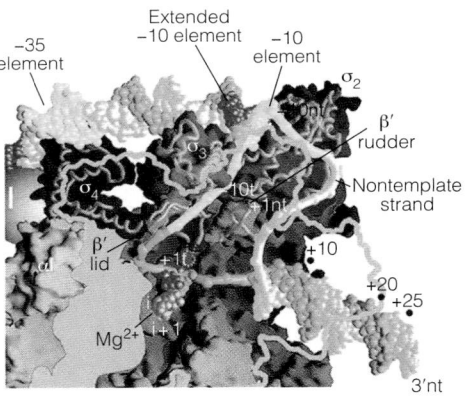

FIGURE 27.8

DNA bending in an open-promoter complex. This image is taken from the crystal structure of *Taq* RNA polymerase. Parts of β have been removed to show interior structure; outline of β is shown as a cyan line. The template DNA strand is shown in green, and the nontemplate strand is in yellow. Note the sharp bend in the template strand. The first two ribonucleotides to be incorporated (i and i + 1) are shown in orange and pink, respectively, close to a catalytically essential Mg^{2+}. Much of the α and β' subunits are rendered transparent, with the interior α-carbon chains shown in gold. The -10 and -35 regions of the DNA, which comprise the promoter, are shown in yellow. PDB ID, 1L9U.

From *Science* 296:1285–1290, K. S. Murakami, S. Masuda, E. A. Campblee, O. Muzzin, and S. A. Darst, Taq RNA polymerase. © 2002. Reprinted with permission from AAAS.

FIGURE 27.9

The transcription bubble. The lengths of unwound DNA and DNA–RNA hybrid were originally estimated from reactivities of transcription complexes with reagents such as KMnO$_4$, which oxidizes bases in single-strand nucleic acids. The length of DNA in contact with the enzyme is determined by footprinting (see Tools of Biochemistry 27A). Six or seven nucleotides of RNA behind the DNA hybrid are protected from ribonuclease attack by binding to the enzyme. Nt = nucleotide.

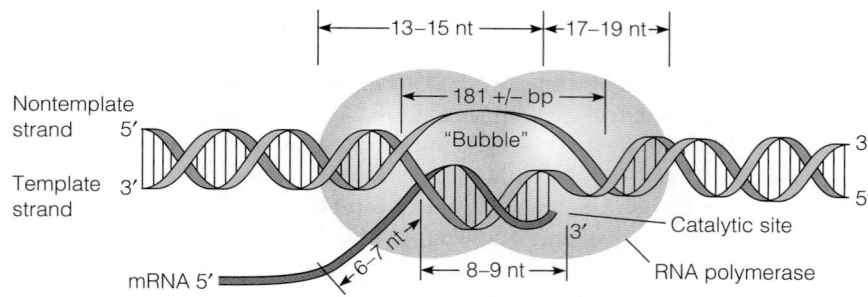

bubble." As one base pair becomes unwound in advance of the 3' end of the nascent RNA strand, one base pair becomes rewound near the trailing end of the RNA polymerase molecule. About 9 base pairs of the 3' end of the nascent transcript are hybridized to the template DNA strand. During elongation RNA polymerase functions as a true molecular motor. Techniques for analysis of single complexes show that RNA polymerase generates forces exceeding those of well-studied cytoskeletal motor proteins, such as myosin and kinesin (Chapter 8).

Figure 27.10 shows a stylized picture of an elongation complex. Although this model was derived from the structure of yeast RNA polymerase II, it shows generally applicable features of transcriptional elongation, such as the large wall, which forces DNA to bend, almost at right angles.

High-resolution structures of a bacterial RNA polymerase, from *Thermus thermophilus*, in the presence of a nonhydrolyzable rNTP analog, give a detailed picture of the nucleotide addition cycle. Critical to this process, as schematized in Figure 27.11, is a 20-residue sequence called the trigger, part of the bridge structure seen in Figure 27.10, which can exist either as a disordered loop or as two helices. After insertion of a nucleotide into the growing RNA chain followed by translocation, the loop is disordered and the rNTP entry channel is open. Entry of an rNTP is followed by folding of the trigger loop into two helices, constraining the rNTP and giving an opportunity for the enzyme to somehow test its shape and reject an incorrectly base-paired nucleotide. At this point the nucleotide, even though not yet incorporated, serves as a ratchet, to prevent backtracking by the enzyme. After catalysis the trigger helices unfold and the channel reopens.

FIGURE 27.10

Cutaway view of the RNA polymerase II elongation complex. Cut surfaces of the protein are lightly shaded, and the remainder, at the back, is darkly shaded. Direction of polymerase motion is left to right as shown. The template DNA strand is in blue. The nontemplate strand, in green, is not shown because it is disordered, except when paired with template DNA. RNA that is hybridized to the template DNA strand is in orange. DNA entering the enzyme is gripped by protein "jaws" (upper jaw not shown in this cutaway model). The 3' end of growing RNA is adjacent to one of the catalytically essential Mg^{2+} ions. The wall forces the DNA to turn. rNTPs probably enter the active site, as shown, through a funnel structure and pore. The 5' end of the growing RNA chain is diverted from the DNA template by a protein loop called the rudder, which limits the length of RNA hybridized to template DNA. The rudder and lid, which guide the exit of RNA, emanate from a large clamp that swings from back to front, as shown, over the catalytic site and contribute to the binding of nucleic acids, and hence, to the high processivity of transcription.

Courtesy of Dr. Roger Kornberg.

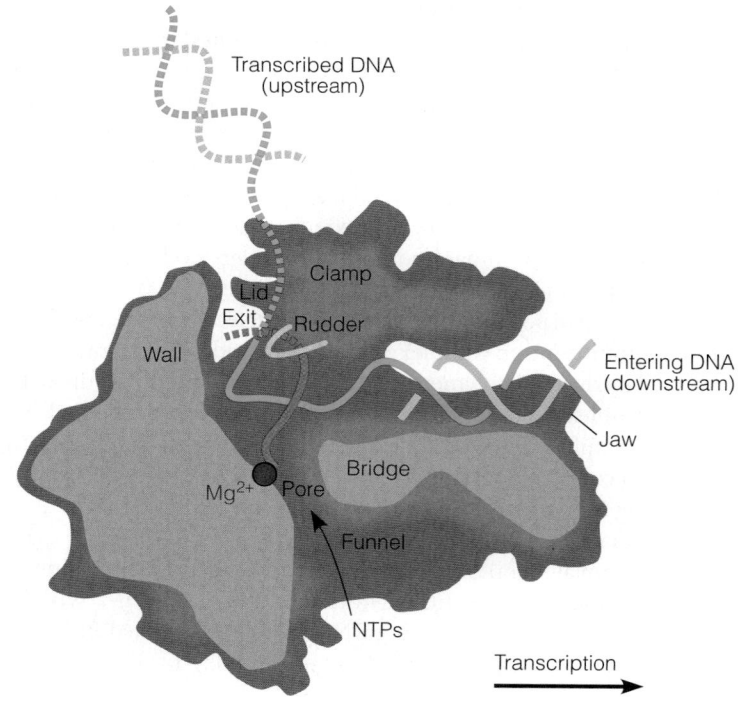

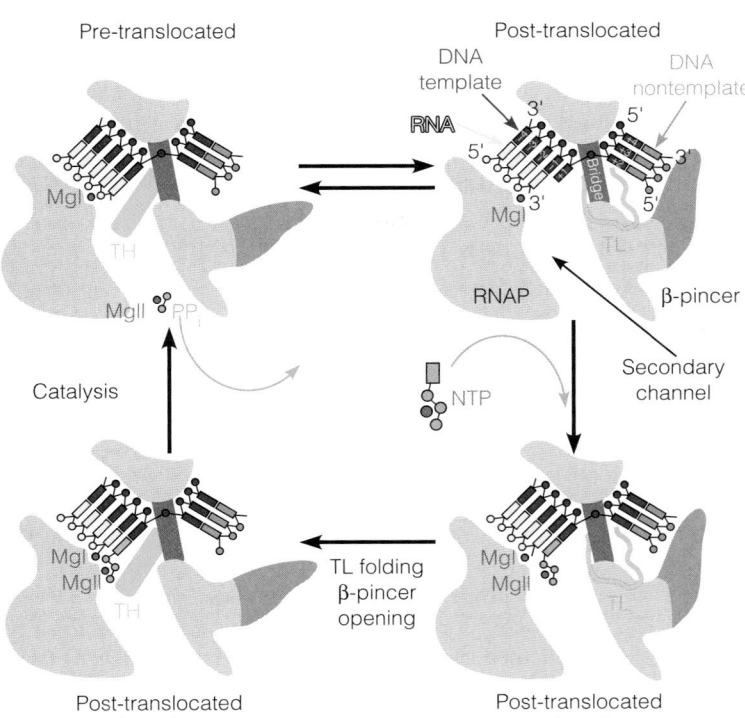

Pre-translocated

Post-translocated

DNA template

DNA nontemplate

RNA

MgI

TH

MgII PPi

Catalysis

RNAP

β-pincer

Secondary channel

NTP

Post-translocated insertion

TL folding β-pincer opening

Post-translocated preinsertion

FIGURE 27.11

The nucleotide addition cycle in *T. thermophilus* RNA polymerase. Opening of the pincer permits translocation of the DNA, with just one template nucleotide available for base-pairing. TH, trigger helix; TL, trigger loop. MgI and MgII are the two catalytically essential Mg^{2+} ions. For further details see text.

Reprinted by permission of Macmillan Publishers Ltd. *Nature* 448:163–168, D. G. Vassylyev, J. Zhang, M. Palangat, I. Artsimovitch, and R. Landick, Structural basis for substrate loading in bacterial RNA polymerase. © 2007.

Structural features of the single-subunit T7 phage RNA polymerase, as well as eukaryotic RNA polymerase II, suggest that this process is widespread. In this study the antibiotic streptolydigin, a well-known RNA polymerase inhibitor, was shown to freeze the enzyme in the preinsertion state.

The concept of a transcription bubble as a central intermediate in transcription, as depicted in Figure 27.9, suggests that the enzyme moves along the DNA template in register with the growing RNA transcript, with the footprint advancing by one base pair for each ribonucleotide incorporated into the transcript. In fact, footprinting of numerous initiation and elongation complexes has shown that the enzyme often advances discontinuously, holding its position for several cycles of nucleotide addition and then jumping forward by several base pairs along the template. Similar conclusions come from single-molecule studies and other contemporary approaches. These observations suggested that the means of RNA polymerase translocation are fundamentally different from the continuous movement implied by a picture of the transcription bubble. However, the "transcription bubble" paradigm can be at least partially reconciled with the footprinting data. It has long been known that some DNA sequences are difficult to transcribe, and RNA polymerase "pauses" when it reaches such a site in vitro, often sitting at the same site for several seconds before transcription is resumed. Recent experiments indicate that at such sites, RNA polymerase often translocates backward, and in the process the 3′ end of the nascent transcript is displaced from the catalytic site of the enzyme, giving a 3′ "tail," which may be several nucleotides long and is not base-paired to the template, protruding downstream of the enzyme (Figure 27.12). In order for transcription to resume, an RNA 3′ end must be positioned in the active site. This is evidently a primary function of the RNA 3′ cleavage reactions, which we mentioned on page 1131 as possible events in optimizing transcriptional fidelity. The "backtracking" shown in Figure 27.12 was observed initially only in transcription complexes obtained from a *greA greB* double mutant; otherwise, the displaced 3′ RNA end was cleaved and could not be detected. The GreA and GreB proteins have been shown to stimulate a transcript cleavage activity that is intrinsic to the polymerase itself. These observations suggest that RNA polymerase generally moves forward until one of these special sequences is reached or, perhaps, until a

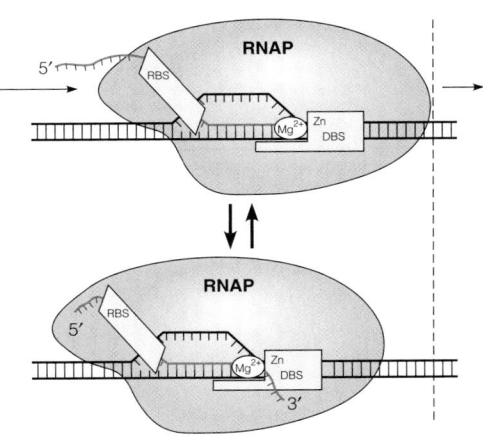

5′
RBS
RNAP
Mg^{2+}
Zn
DBS

5′
RBS
RNAP
Mg^{2+}
Zn
DBS
3′

FIGURE 27.12

Backtracking in an elongation complex. Above, the 3′ terminus of the transcript is in the active site (denoted by Mg^{2+}). Below, the enzyme has slipped backward, leaving the 3′ transcript terminus at the end of a non-base-paired RNA tail, some five nucleotides long. Transcription can resume either by forward sliding of polymerase back to the structure depicted above or, more likely, by cleavage of the non-base-paired part of the transcript, creating a new base-paired 3′ terminus. DBS and RBS, DNA-binding and RNA-binding site, respectively.

Reprinted from *Cell* 89:33–41, E. Nudler, A. Mustaev, E. Lukhtanov, and A. Goldfarb, The RNA–DNA hybrid maintains the register of transcription by preventing backtracking of RNA polymerase. © 1997, with permission from Elsevier.

transcription insertion error generates a DNA–RNA mispairing that weakens the hybrid and allows backtracking. Until recently the mechanism and significance of pausing were obscure. Single-molecule studies with defined DNA templates show general similarity between some pausing sites and other sites at which pausing plays a known regulatory function. And, as noted, pausing and 3′ cleavage evidently do function to improve transcription accuracy.

Punctuation of Transcription: Promoter Recognition

Promoter recognition is a crucial step in transcription, from the standpoint of regulation as well as mechanism. In *E. coli* it is estimated that the most frequently transcribed genes initiate transcription about once every 10 seconds, whereas some genes are transcribed as infrequently as once per generation (30 to 60 minutes). Promoter recognition is a rate-limiting step for transcription. Because all genes in bacteria are transcribed by the same core enzyme, variations in promoter structure must be responsible in large part for the great variation in the frequency of initiation. Information about promoter structure is of practical value in designing vectors for expression of cloned genes.

What structural features in DNA direct RNA polymerase to bind at a promoter site and to form an open-promoter complex? The first hint of an answer came in 1975, when David Pribnow and Heinz Schaller independently examined the limited DNA sequence data then available, which revealed that each gene transcribed in *E. coli* shared a short adenine- and thymine-rich sequence, centered about 10 nucleotides to the 5′ side of the transcriptional start site (Figure 27.13). (Identification of transcriptional start points is described in Tools of Biochemistry 27C.) There was some variation among the promoters analyzed, but a **consensus sequence** emerged within this conserved region. A consensus sequence comprises those bases that appear most frequently at each sequence position in a series of sequences thought to have a common function. Among the different initiation sequences analyzed in *E. coli*, that consensus sequence was TATAAT on the

> The frequency with which a gene is transcribed in *E. coli* genes is largely determined by the similarity between the gene's promoter sequence and the consensus sequence.

FIGURE 27.13

Conserved sequences in promoters recognized by *E. coli* RNA polymerase. Lengths of spacer sequences are also shown. Red arrows indicate transcription start site.

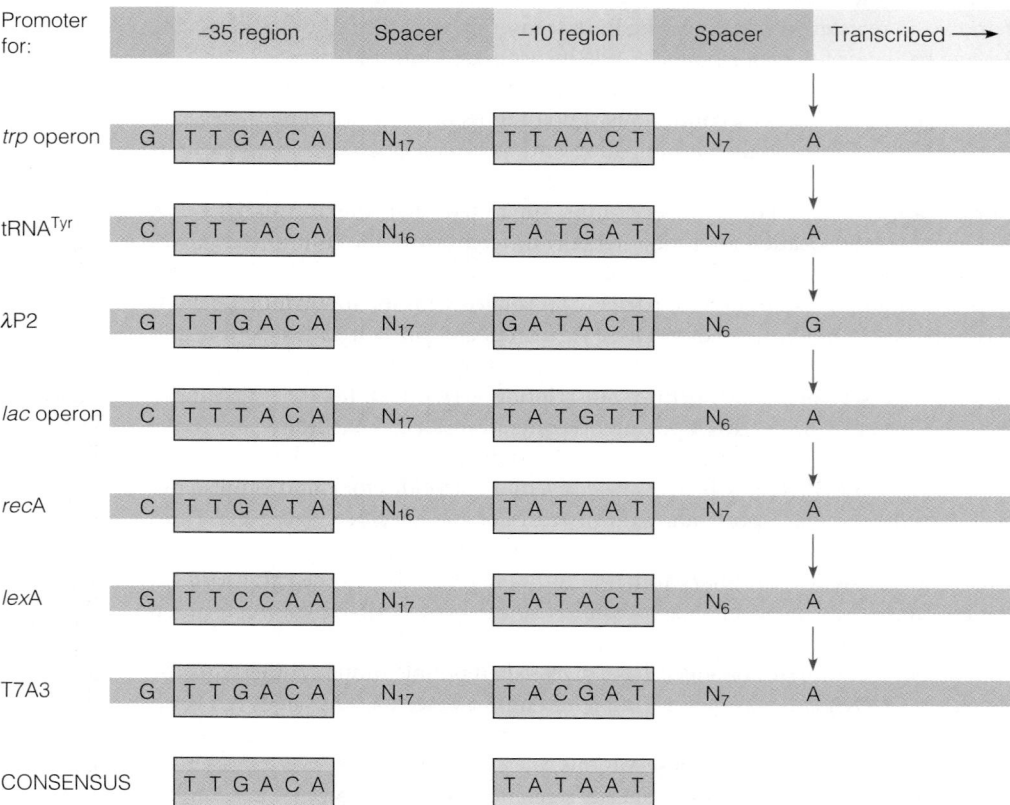

sense strand. The sense strand is the nontranscribed DNA strand. Because it is complementary in sequence to the template strand, the sense strand is identical in base sequence to the RNA product within the region transcribed, but with T instead of U. Later, another region of conserved nucleotide sequence was found centered at nucleotide −35, with a consensus sequence of TTGACA.

The two conserved sequences are called the −35 region and the −10 region (see Figure 27.13; the −10 region is also called the *Pribnow box*). No known natural promoter has −35 and −10 regions that are identical to the consensus sequences, but, in general, the more closely these regions in a promoter resemble the consensus sequences, the more efficient that promoter is in initiating transcription. Variations in promoter structure represent a simple way for the cell to vary rates of transcription from different genes. Figure 27.13 indicates the extent to which different nucleotides are conserved. Among 114 *E. coli* promoters examined in an early study, 6 or more of the 12 nucleotides in the two consensus sequences were found in more than 75% of the promoters.

What evidence points to a functional role for these conserved sequences in binding RNA polymerase and initiating transcription? First, a variety of mutations map in promoter regions and affect transcription efficiency in vivo. As shown in Figure 27.14, most of the promoter mutations that have been sequenced change the structure of either the −35 region or the −10 region, pointing directly to those sequences as having the greatest effect on transcriptional initiation efficiency. In general, the mutations that increase promoter strength (**up-promoter mutations**) change either the −35 or the −10 region to more closely resemble the consensus sequences. **Down-promoter mutations**, which decrease promoter strength, change the sequence away from the consensus. Similar conclusions have been drawn from site-directed mutagenesis of promoters and analysis of their efficiency in vitro. The latter studies have confirmed the importance of spacing between the two regions. Although most natural promoters have a 17-nucleotide spacer between the 35 and 10 regions, many have 16 or 18. In vitro studies show that a 17-nucleotide spacer yields the most efficient promoter structure.

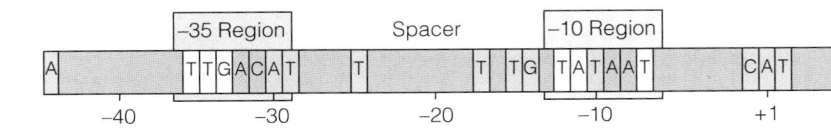

(a)

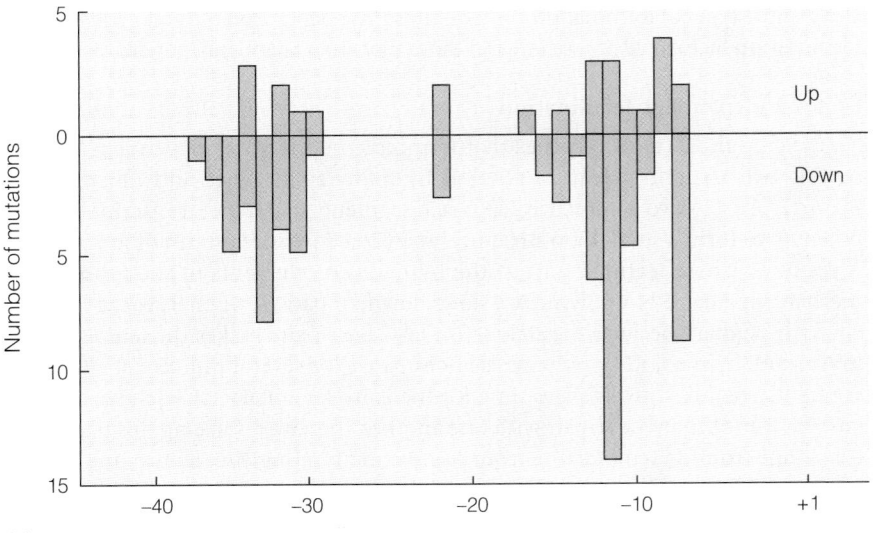

(b)

FIGURE 27.14

Survey of conserved nucleotides in *E. coli* promoters. Nucleotide sequences were compared among 114 known *E. coli* promoters. **(a)** Nucleotide positions that were invariant in at least 75% of the promoters are shown in yellow, moderately conserved nucleotides (50%–75%) in purple, and weakly conserved nucleotides (40%–50%) in blue. **(b)** The number of known promoter mutations affecting each site. Up-promoter mutations increase promoter strength; down-promoter mutations have the opposite effect.

From *Annual Review of Biochemistry* 54:171–204, D. Hawley and W. R. McClure, Mechanism and control of transcription initiation in prokaryotes. © 1985 Annual Reviews.

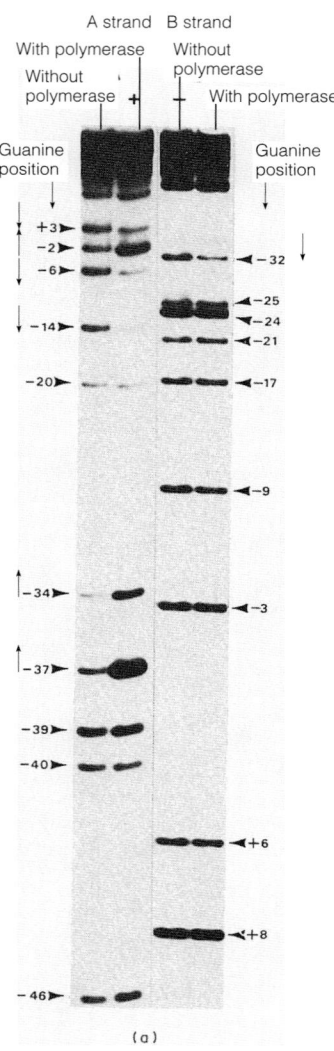

FIGURE 27.15

Identification of G residues that contact RNA polymerase in the *E. coli* tryptophan promoter. ↑ = enhancement of cleavage by binding of polymerase; ↓ = inhibition of cleavage.

Reprinted from *Journal of Molecular Biology* 144: 133–142, D. S. Oppenheim, G. N. Bennett, and C. Yanofsky, *Escherichia coli* RNA polymerase and *trp* repressor interaction with the promoter-operator region of the tryptophan operon of *Salmonella typhimurium*. © 1980, with permission from Elsevier.

DNA sequences that promote factor-independent termination include a run of four to eight A residues and a GC-rich region that forms a stem–loop.

The other evidence for the importance of the −35 and −10 regions is that most of the DNA nucleotides in close contact with RNA polymerase are those in or near these two conserved sequences. This was originally ascertained by comparing the susceptibilities of particular nucleotides to chemical modification, in the presence or absence of RNA polymerase. For example, dimethyl sulfate can be used to determine the susceptibility of guanine residues to methylation. First one treats a restriction fragment containing the sequence of interest with $\gamma[^{32}P]ATP$ and **polynucleotide kinase**, an enzyme that transfers phosphate to the free 5′-hydroxyl group on the fragment. This end-labeled fragment is then incubated with RNA polymerase and treated with dimethylsulfate, which reacts with guanine residues and can cleave the DNA at that site. The DNA fragment is then subjected to gel electrophoresis and radioautography. Cleavage products from the DNA–protein complex are compared with those from naked DNA. Figure 27.15 shows data from analysis of the *trp* promoter, which controls transcription of genes involved in tryptophan synthesis. Cleavage at some sites, −34 and −37, was enhanced by RNA polymerase binding, as seen by comparison of band intensities, while at other sites, −32, −14, and −6, reactivity was decreased. All of these susceptible guanines lie within the −10 and −35 boxes. Reagents causing chemical cleavage at other bases give similar results. Other techniques, such as DNA footprinting (Tools of Biochemistry 27A), lead to the same conclusion. Experiments of this type show that RNA polymerase also contacts bases upstream from the −35 box, in the −40 to −60 region. These contacts involve the polymerase α subunit, while the contacts with −35 and −10 boxes involve σ. The upstream contacts (−40 to −60) are particularly important in the promoters for ribosomal protein genes, which are very actively transcribed.

Another factor affecting transcriptional efficiency, in addition to the base sequence of the promoter, is the superhelical tension on the DNA template. The relation between DNA topology and transcriptional efficiency is now receiving considerable attention. The relationship is not clear because transcription of some genes is activated in vivo when the template is highly supercoiled—for example, by inactivating topoisomerase I. Transcription of other genes, by contrast, is inhibited under these conditions. Interestingly, the promoter for transcription of DNA gyrase subunits becomes activated when the gene is in a relaxed state. Given that gyrase introduces superhelical turns, this finding seems to represent a feedback mechanism in which the cell responds appropriately to a signal that intracellular DNA is becoming too relaxed.

Punctuation of Transcription: Termination

Because of the great stability of transcription complexes, termination of transcription, with release of the nascent transcript, is an involved process. In bacteria we recognize two distinct types of termination events—those that depend on the action of a protein **termination factor**, called ρ (rho), and those that are ρ factor-independent.

Factor-Independent Termination

Sequencing the 3′ ends of genes that terminate in a factor-independent manner reveals two structural features shared by many such genes and illustrated in Figure 27.16: (1) two symmetrical GC-rich segments that in the transcript have the potential to form a stem–loop structure, and (2) a downstream run of four to eight A residues. These features suggest the following as elements of the termination mechanism. First, RNA polymerase slows down, or pauses, when it reaches the first GC-rich segment because the stability of G-C base pairs makes the template hard to unwind. In vitro, RNA polymerase does pause for several minutes at a GC-rich segment. Second, pausing gives time for the complementary GC-rich parts of the nascent transcript to base-pair with one another, thereby displacing this part of the transcript from its template or from its enzyme binding site. Hence, the ternary complex of RNA polymerase, DNA template, and RNA is weakened. Further weakening, leading to dissociation, occurs when the A-rich segment is transcribed to give a series of A–U bonds (very weak), linking transcript to template.

The actual mechanism of termination is more complex than just described, in part because DNA sequences both upstream and downstream from the regions shown in Figure 27.16 also influence termination efficiency. Moreover, not all pause sites are termination sites. However, the scheme shown includes key elements of factor-independent termination.

Factor-Dependent Termination

Factor-dependent termination sites are less frequent, and the mechanism of this type of termination is still more complex. The ρ protein, a hexamer composed of identical subunits, was originally discovered in studies on termination of λ phage DNA transcription in vitro. This protein, which has been characterized as an RNA–DNA helicase, contains a nucleoside triphosphatase activity that is activated by binding to polynucleotides. Apparently ρ acts by binding to the nascent transcript at a C-rich site near the 3′ end, when RNA polymerase has paused (Figure 27.17). Then ρ moves along the transcript toward the 3′ end, with the helicase activity unwinding the 3′ end of the transcript from the template (and/or the RNA polymerase molecule) and causing its release.

In factor-dependent termination, ρ protein acts as an RNA–DNA helicase, unwinding the template–transcript duplex and facilitating release of the transcript.

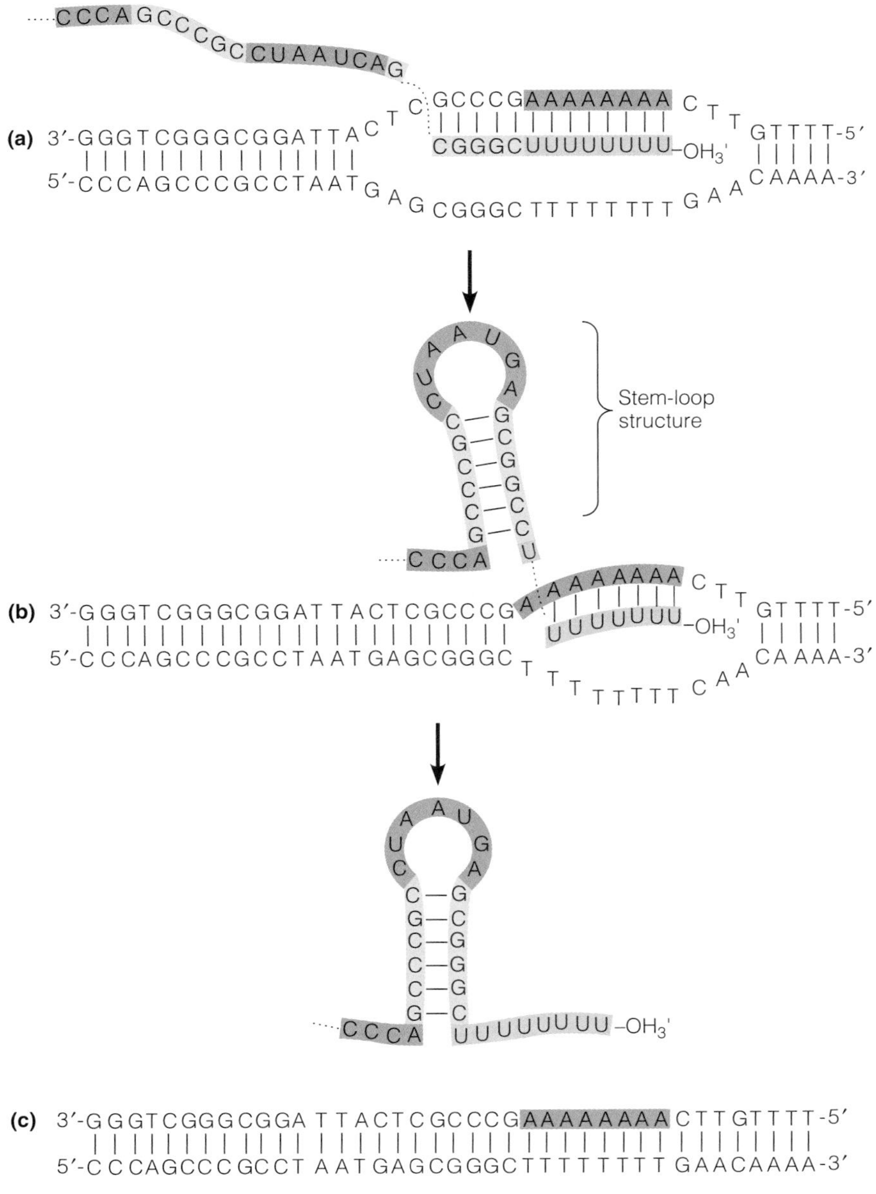

FIGURE 27.16

A model for factor-independent termination of transcription. **(a)** An A-rich segment of the template (orange segment on right) has just been transcribed into a U-rich mRNA segment. **(b)** RNA–RNA duplex, stabilized by G-C base pairs (yellow), eliminates some of the base pairing between template and transcript. **(c)** The unstable A–U bonds linking transcript to template hybrid dissociate, releasing the transcript.

ρ has been studied intensively as a model for understanding helicase action. Figure 27.18 shows the crystal structure of *E. coli* ρ complexed with RNA and an ATP analog (ADP + BeF$_3$). In the schematic (b) four of the six subunits bind ATP tightly. At the instant shown, the 5' RNA nucleotide (orange circle) is being released from the protein as the result of ATP hydrolysis at site D, and the nucleotide shown with a small circle and a dashed line is about to be bound, just as a catalytic site is being formed between the protein subunits labeled "free" and "locked," by insertion of an "arginine finger" motif into site E. The ρ structure is very similar to those of helicases of opposite polarity, and it appears that polarity

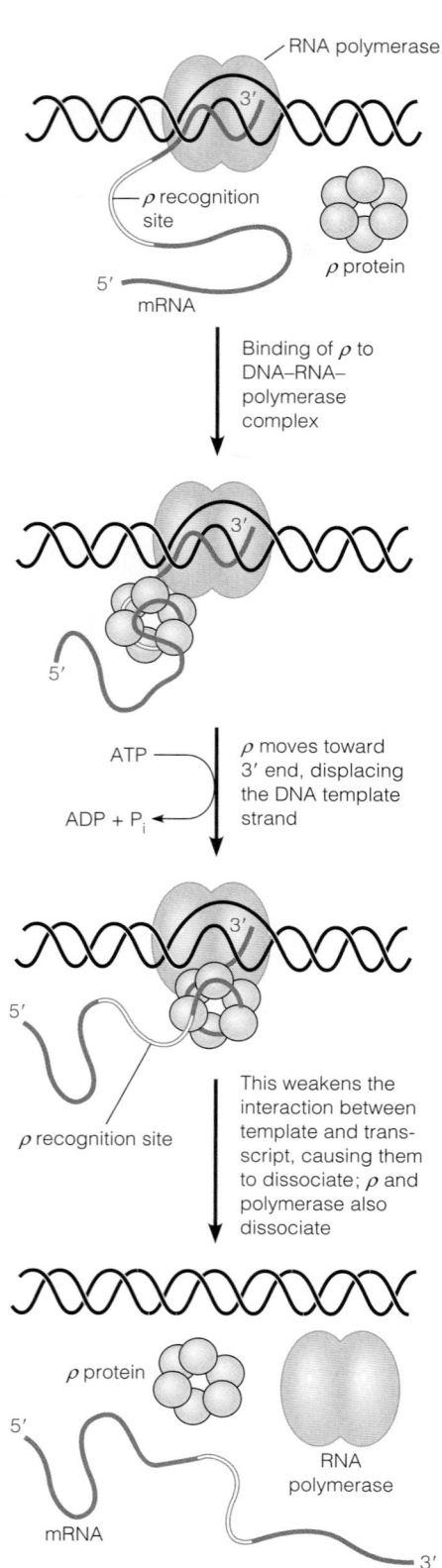

FIGURE 27.17

$\boldsymbol{\rho}$ **factor–dependent termination.** ρ binds to a site on the nascent transcript and unwinds the RNA–DNA duplex. Once ρ reaches RNA polymerase, interaction with bound NusA protein (not shown) leads to termination.

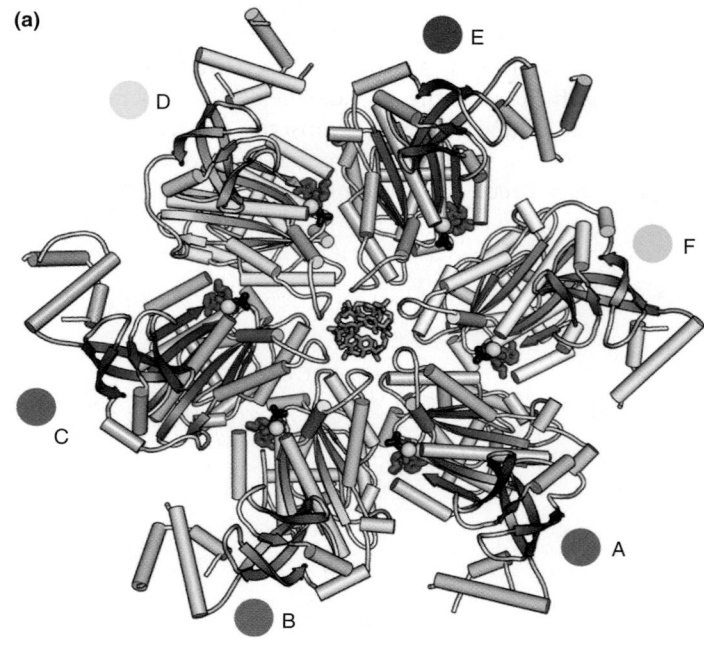

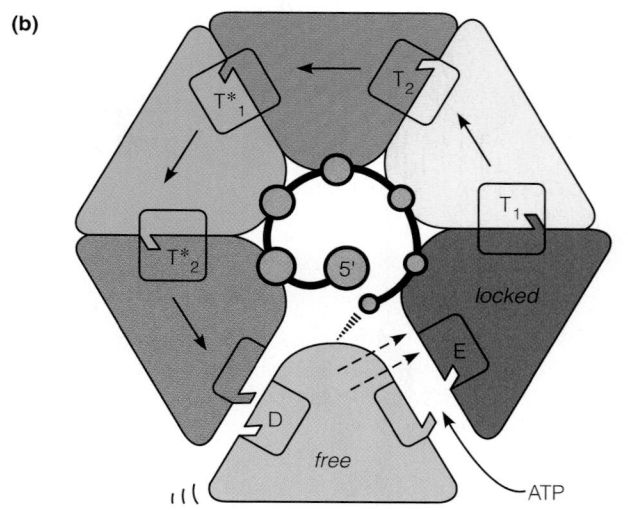

FIGURE 27.18

Crystal structure of *E. coli* $\boldsymbol{\rho}$. **(a)** RNA is shown in the central channel in orange. ADP is in magenta, BeF$_3$ is in black, and Mg^{2+} is in green. **(b)** Schematic diagram of the helicase reaction. For details see text. PDB ID, 3ICE.

is determined largely by the order of "ATP firing," that is, the order in which the individual protein subunits hydrolyze ATP. One ATP is hydrolyzed per RNA nucleotide unwound.

It is not clear what causes RNA polymerase to pause at ρ-dependent termination sites. The action of another protein, NusA, is somehow involved. NusA was discovered during studies of **antitermination** in phage λ. Antitermination occurs early in the phage transcription program, when two ρ-dependent termination sites are inactivated so that RNA polymerase can move past these sites and transcribe genes essential to phage development (see page 1239). For this inactivation to occur, a viral protein, the product of gene *N*, must interact with NusA (*N utilization substance*) by a mechanism as yet unknown. Mutations in either the *N* or the *E. coli nusA* gene interfere with antitermination and block phage development. The NusA protein is recruited to elongation complexes, where it binds to hairpin structures in nascent RNA, possibly facilitating termination this way.

Further insight into termination mechanisms has come from an extensively studied regulatory mechanism called **attenuation**. Attenuation controls the rate of transcription of certain operons by terminating the synthesis of a nascent transcript before RNA polymerase has reached the structural genes. Attenuation, which has been investigated as a model for end-of-transcript termination, is discussed in Chapter 29.

Transcription and Its Control in Eukaryotic Cells

Transcription in eukaryotes is a much more complex process than in prokaryotes. Not only is there much more discrimination in what is to be transcribed and what is not, but this transcription is precisely programmed during development and tissue differentiation. Furthermore, the transcription machinery must somehow deal with the complex levels of structure in eukaryotic chromatin. Reflecting this complexity is the fact that eukaryotic cells have several different RNA polymerases, each with a specialized function, as described earlier in this chapter. For each polymerase, several proteins must assemble at promoters and other upstream sites on the template DNA, along with RNA polymerase, in order to form a functional transcription complex. None of the three nuclear RNA polymerases (I, II, and III) has a counterpart to the σ factor of prokaryotic complexes. However, all three require a set of transcription factors that play roles comparable to that of σ, in addition to proteins that might be specialized for transcription of a particular gene. By convention, transcription factors are named TFI, TFII, or TFIII, depending upon whether they function with RNA polymerase I, II, or III, respectively. Within one class of transcription factors, each individual factor is identified with a letter; thus, TFIIA is one of several transcription factors functioning with RNA polymerase II.

Other differences pertain to the fact that bacterial genomes are organized into blocks of functionally related genes—operons, such as the lactose operon mentioned earlier in this chapter—that are cotranscribed to give multigenic mRNAs, while eukaryotic genes are almost always transcribed as monocistronic mRNAs. In addition, as we see later in this chapter, post-transcriptional processing is far more complex for eukaryotic than for prokaryotic genes.

> Eukaryotes have three kinds of nuclear RNA polymerases, each requiring additional protein factors to initiate transcription.

RNA Polymerase I: Transcription of the Major Ribosomal RNA Genes

The eukaryotic ribosome contains four rRNA molecules (see Chapter 28). The small subunit has an 18S rRNA, whereas the large subunit contains 28S, 5.8S, and 5S rRNA molecules. Of these, the 28S, 18S, and 5.8S subunits are all produced from an initial 45S pre-rRNA transcript, and it is the special function of RNA polymerase I (pol I) to carry out this transcription.

As was shown in Figure 27.5, pol I is a complex enzyme, containing 14 subunits totaling over 600,000 daltons. At least two transcription factors are known to be required (UBF1 and SL1). However, because only a single kind of gene is transcribed, there is no need for the elaborate apparatus, including multiple regulatory sites and multiple transcription factors, that we will find characteristic for pol II transcription.

The nucleolus is the site of ribosomal subunit assembly in eukaryotes. The gene for the 45S pre-rRNA is present in the nucleolus as multiple, tandemly arranged copies, as shown in Figure 27.19. After transcription, the 45S pre-rRNA is processed to yield 18S, 5.8S, and 28S rRNA molecules. About 6800 nucleotides are discarded in this process. The rRNAs are then combined with 5S rRNA from other regions of the nucleus and ribosomal proteins synthesized in the cytosol. The resulting ribosomal subunits are exported from the nucleolus back into the cytosol.

Transcription of the tandem copies of 45S pre-rRNA can be beautifully visualized in the electron microscope (Figure 27.19b). The structure of the nucleolar chromatin has been a subject of some controversy, but it appears most likely that nucleosomes are not present, at least in the transcribed regions. This absence of nucleosomes may be a specific chromatin modification to allow rapid and continuous transcription of these genes.

In some lower eukaryotes, like the protozoan *Tetrahymena*, the 28S rRNA contains an intron near its 3′ end. Excision of this intron and splicing of the RNA are carried out by a remarkable process in which the RNA *itself* acts as the catalyst, via the series of reactions described in Chapter 11 (see in particular Figure 11.41 on page 452). In higher eukaryotes we know less about the mechanism of rRNA processing, the assembly of ribosomes, or the coordination of synthesis of ribosomal proteins and ribosomal RNA.

RNA Polymerase III: Transcription of Small RNA Genes

RNA polymerase III (pol III) is the largest and most complex of the eukaryotic RNA polymerases. It contains 17 subunits, totaling 700,000 daltons. All of the genes it transcribes share certain features. The RNAs are small, they are not translated into proteins, and they are unique in that their transcription is regulated by

FIGURE 27.19

Transcription and processing of the major ribosomal RNAs in eukaryotes. The genes exist in tandem copies, separated by nontranscribed spacers. **(a)** The 45S transcripts first produced are processed by removal of portions shown in tan, to yield the 18S, 5.8S, and 28S products. These are then assembled into ribosomal subunits, through addition of proteins. **(b)** Electron micrograph of spread nucleolar rRNA genes undergoing transcription. Tandemly arranged genes are being transcribed from bottom to top.

(b) Courtesy of Oscar L. Miller, Jr. and Barbara Beaty, Oak Ridge National Laboratory, Oak Ridge, TN.

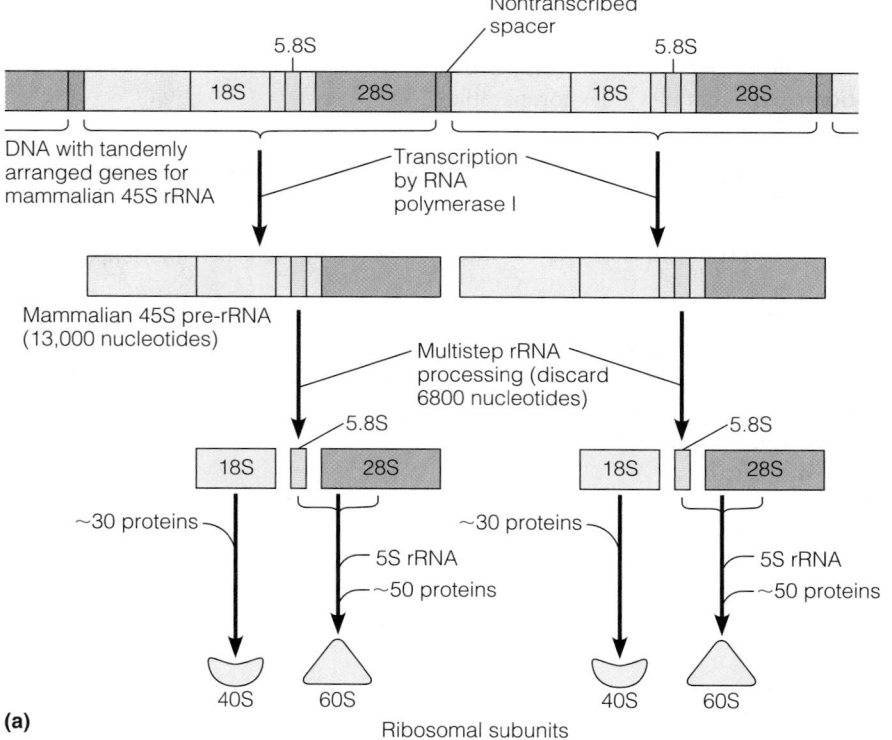
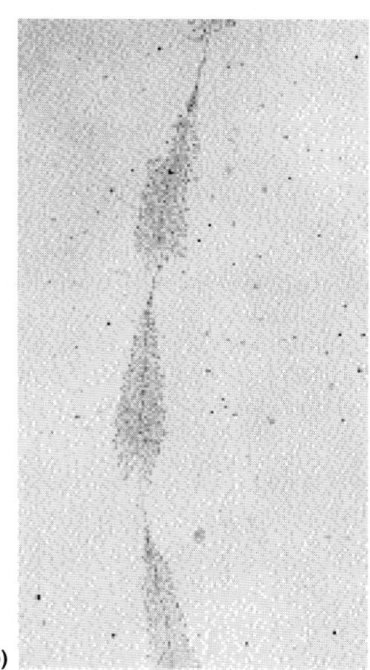

(a)

(b)

certain sequences that lie *within* the transcribed region. The major targets for pol III are the genes for all the tRNAs and for the 5S ribosomal RNA. Like the major ribosomal genes described in the previous section, these small genes are present in multiple copies, but they are usually not grouped together in tandem arrays, nor are they localized in one region of the nucleus. Rather, they are scattered over the genome and throughout the nucleus.

Of all the genes transcribed by pol III, the most thoroughly studied are those for 5S ribosomal RNA. In vitro experiments have revealed that at least three protein factors in addition to polymerase III are needed for expression of the 5S rRNA genes. Two of these transcription factors (TFIIIB and TFIIIC) appear to participate in the transcription of tRNA genes as well, but one, called TFIIIA, is specific for the 5S genes. The interaction of the three transcription factors, the polymerase, and the gene is shown in Figure 27.20. The molecule of TFIIIA

Pol I transcribes the major ribosomal RNA genes; pol III, small RNA genes; and pol II, protein-encoding genes and a few small RNA genes.

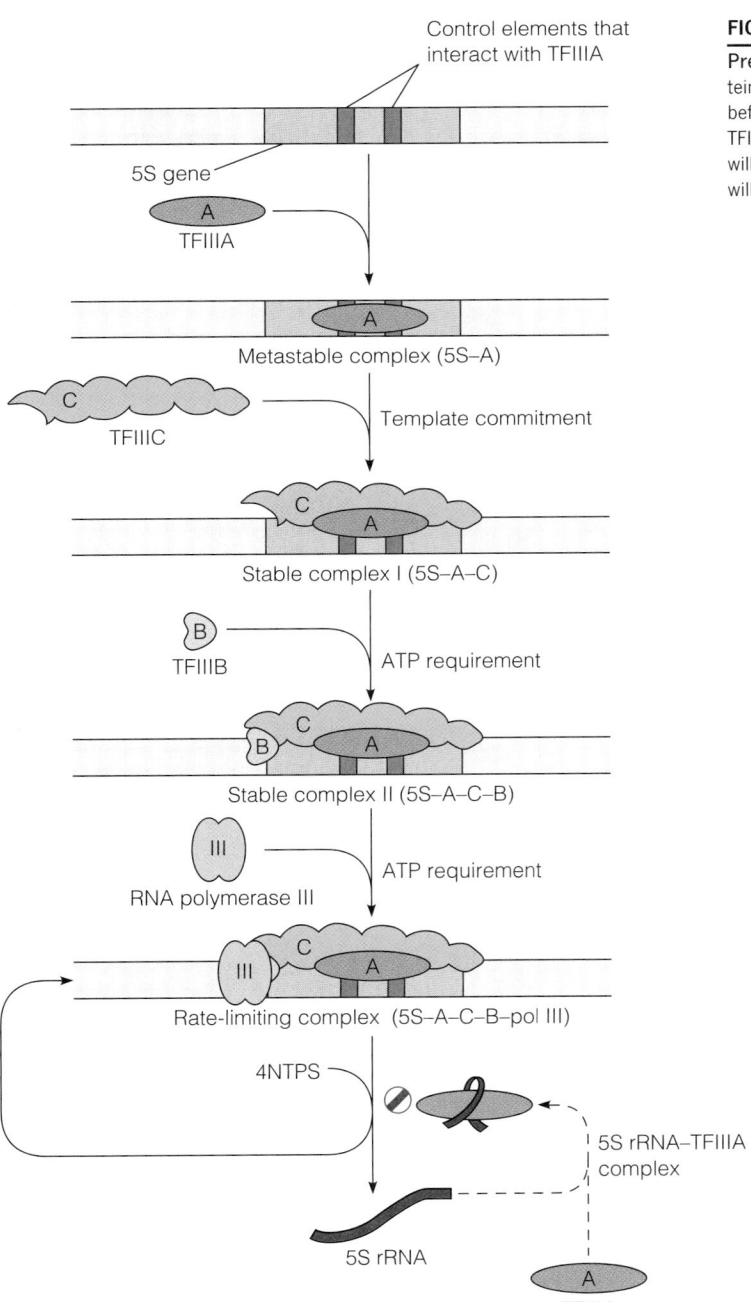

FIGURE 27.20

Preparation of a 5S rRNA gene for transcription. At least the three protein factors shown, plus RNA polymerase III, must assemble on the gene before transcription can occur. TFIIIA must bind to the gene before factors TFIIIC and TFIIIB can bind. Once the stable complex II has been formed, it will recycle with pol III to produce many RNA copies. An excess of 5S rRNA will form a complex with TFIIIA, inhibiting further transcription.

makes contact with DNA over a length of about 40 bp. Recognition of the sequence occurs in two blocks of about 12 bp each, at either end of the contact region. This somehow makes the gene accessible to TFIIIB, TFIIIC, and polymerase III. TFIIIA can also complex with 5S RNA. This propensity limits 5S RNA production when the RNA product is in excess, by removing TFIIIA from availability to bind to DNA.

TFIIIA is an example of an abundant class of sequence-specific DNA-binding proteins, in which metal-binding *zinc fingers* make contact with and identify DNA sequences (Figure 27.21). This class of proteins has conserved histidine and cysteine residues, which complex with zinc, as shown in Figure 27.21b. This DNA-binding protein motif was mentioned earlier, as a structural element in steroid hormone receptors (page 980).

Although TFIIIA is a monomer and TFIIIB is dimeric, TFIIIC is an enormous complex, involving six polypeptide chains and covering the whole 5S rRNA or tRNA gene. Just how the polymerase manages to repeatedly transcribe through such a protein complex is still unclear. Nonetheless, once pol III is attached, it can produce multiple transcripts before dissociating.

In later sections we shall encounter a variety of transcription factors, for the control of eukaryotic gene expression is almost entirely dependent on the site-specific interaction of these proteins with DNA. A typical transcription factor contains a DNA-binding domain and one or more regulatory domains, which can interact with other nuclear proteins to convey regulatory signals.

FIGURE 27.21

Zinc fingers. (a) The transcription factor TFIIIA binds to the 5S RNA gene via zinc fingers inserted into the major groove. The two major recognition regions, A block and C block, are contacted by fingers 7–9 and 1–3, respectively. **(b)** Structure of a zinc finger. The structure shown is for a synthetic polypeptide with sequences found in a zinc finger protein. The α helix and β-sheet motifs are shown in deep blue and green, respectively. The α helix binds within the major groove, as shown in panel (a). The two histidine residues and two cysteines that coordinate the zinc (red) are depicted in detail. Sulfur is in yellow.

From *Science* 245:635–637, M. S. Lee, G. P. Gippert, K. V. Soman, D. A. Case, and P. Wright, Three-dimensional solution structure of a single zinc finger DNA-binding domain. © 1989. Reprinted with permission from AAAS. Adapted with permission from Peter Wright.

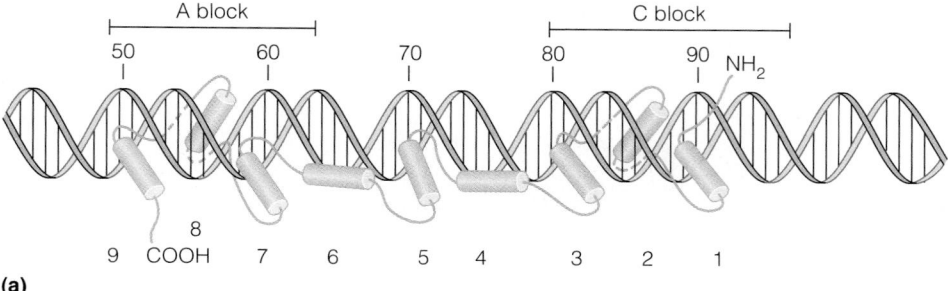

(a)

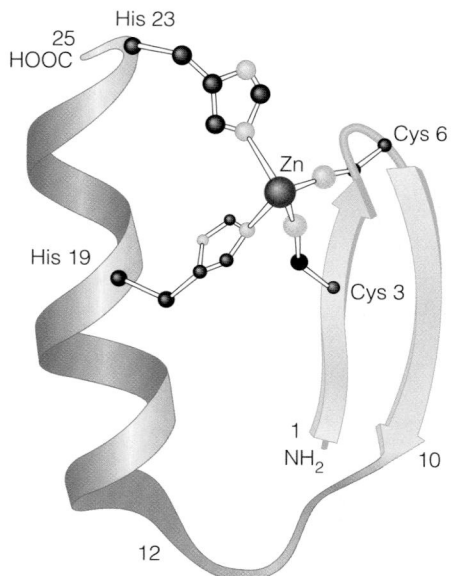

(b)

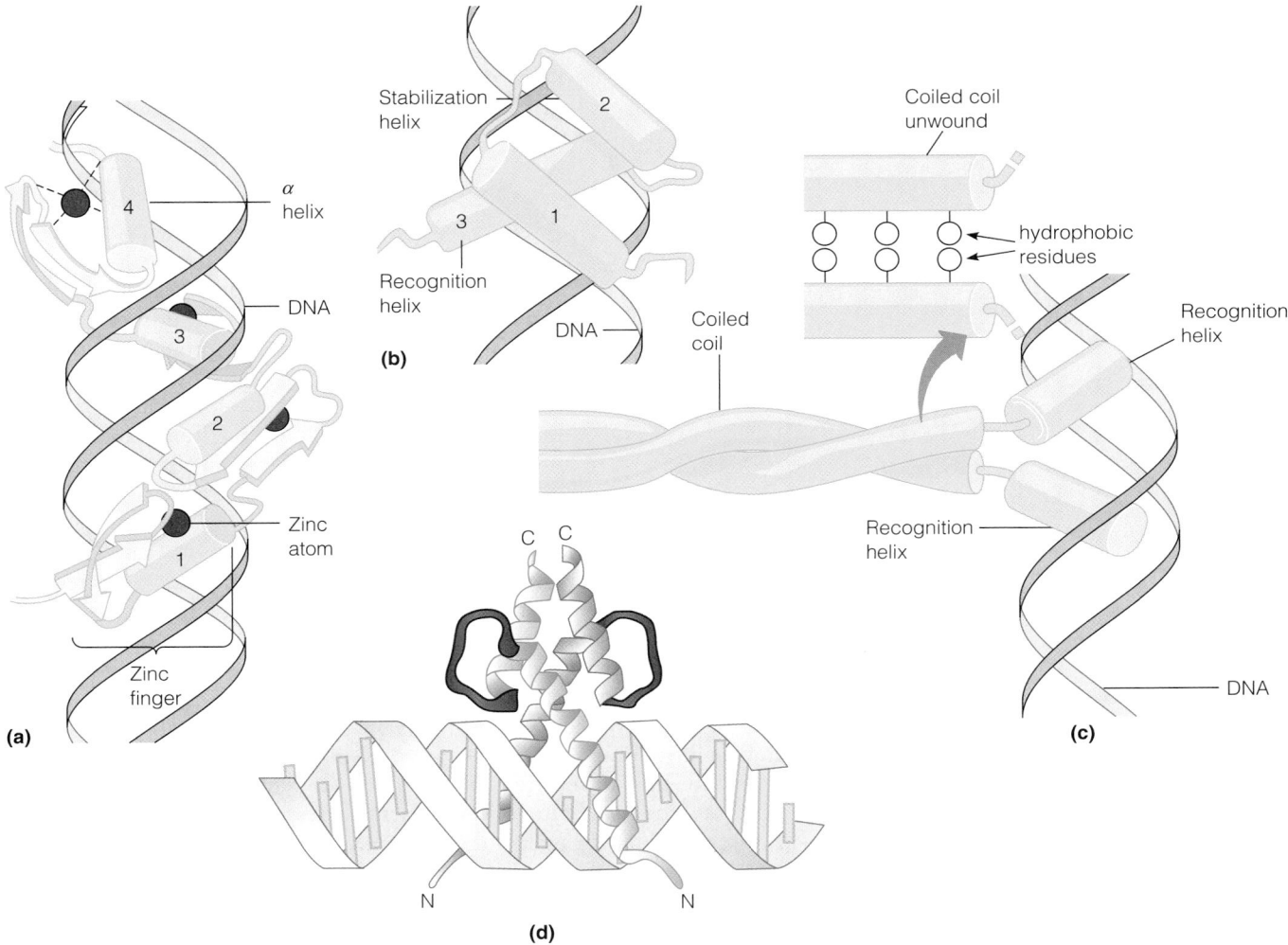

FIGURE 27.22

Structures of four common types of DNA-binding motifs from eukaryotic transcription factors.

(a) The zinc finger motif, showing the way the fingers fit along the major groove. **(b)** A helix–turn–helix motif. Note that the helix–turn–helix fits into the DNA major groove in a way similar to that of a zinc finger.
(c) A leucine zipper protein. The coiled coils are held together by hydrophobic residues, usually leucines. At their ends, they present a pair of recognition helices to the DNA major grooves. **(d)** A helix–loop–helix motif. The two monomers are held together in a four-helix bundle. Each monomer contributes parts of two α-helices held together by a flexible protein loop, shown in red. The N-termini of the two longer helices form specific contacts in the major groove of the DNA binding site.

In addition to the zinc finger proteins, three other major DNA-binding structural motifs are known, as depicted schematically in Figure 27.22 b, c, and d. In *helix–turn–helix* proteins, one α helix (called the recognition helix) lies in the major groove of the DNA, its side chains making specific contacts with the DNA bases. The helix–turn–helix motif was originally discovered in studies on sequence-specific prokaryotic transcriptional regulators, discussed in Chapter 29 and shown in Figure 27.1. A quite different class of DNA-binding proteins is called *leucine zipper* proteins. These are dimers, held together in a coiled-coil structure by hydrophobic interactions. They typically exhibit a regular pattern (seven-fold periodicity) of leucine or other hydrophobic residues in the helical tail regions, which favors side-by-side hydrophobic interaction. The N-terminal regions are recognition helices, lying in adjacent major grooves. The special feature of leucine zipper proteins is that they can form either homologous or heterologous dimers, thus allowing many combinatorial pairings between transcription factors. As seen with the leucine zipper proteins, the *helix–loop–helix* (HLH) motif is also one that allows either homologous or heterologous dimerization (Figure 27.22d). In this motif a short α helix is connected by a loop to a second, longer α helix. Because of the flexibility of the loop, one helix can fold back and pack against the other. As shown in the figure, the two-helix structure permits binding both to DNA and to the HLH motif of a second HLH protein, either the same (forming a homodimer) or different (forming a heterodimer).

RNA Polymerase II: Transcription of Structural Genes

All of the structural genes (those coding for protein products) in the eukaryotic cell are transcribed by polymerase II. This enzyme also transcribes some of the small nuclear RNAs involved in splicing (see page 1158). Like other RNA polymerases, pol II is a complex multisubunit enzyme. However, not even its 12 subunits are sufficient to allow pol II to initiate transcription on a eukaryotic promoter. Because the expression of many eukaryotic genes is either tissue-specific or developmental stage-specific or both, eukaryotic promoter structure is much more complex than that of prokaryotes. Protein factors in addition to RNA polymerase are required for promoter recognition, recruitment of RNA polymerase to a promoter, and generation of an active elongation complex. A typical eukaryotic promoter contains an initiator region (Inr) with the sequence YYAN^T_AYY, where N is any nucleotide, Y is a pyrimidine, and N represents the +1 initiation site. A counterpart to the bacterial Pribnow box, called the TATA box, positioned between −20 and −30, has the sequence TATAAAA. Upstream from that box are arrayed both general and gene-specific control elements—binding sites for transcription factors and other regulatory proteins, such as the hormone receptors that we introduced in Chapter 23. Additional regulatory sites may exist several hundred base pairs upstream from the initiation site; these are called **enhancer** regions. Although these far-upstream activator sites are involved in transcriptional regulation, they are not considered part of the promoter per se. Table 27.3 gives the sequence of several general and gene-specific promoter and enhancer elements, and Figure 27.23 shows the placement of these elements in several well-studied eukaryotic promoters.

Having introduced some of the control elements involved in eukaryotic promoters, let us now introduce or reintroduce the protein players—RNA polymerase II and the general transcription factors; these are presented in Table 27.4.

Before describing the roles of these proteins in transcription, we point out that TFIIH is an example of a "moonlighting protein," one with two or more different functions. TFIIH participates in transcription-coupled DNA repair (page 1086). When DNA synthesis is blocked at the site of a lesion, TFIIH can interact with DNA at that site and recruit other proteins of the nucleotide excision repair process. The significance of this function is seen in the fact that genetic deficiencies in one or more of the TFIIH subunits are responsible for some forms of xeroderma

---

TABLE 27.3 Some important pol II control elements and their corresponding transcription factors

Sequence Name	Consensus Sequence	Transcription Factor(s)	Comment
Some General Promoter and Enhancer Elements			
TATA box	TATAAAA	TBP, TFIID	The most common promoter element
CAAT box	GGCCAATCT	CP1	A common upstream element
GC box	GGGCGG	SP1	Often found in TATA-less promoters
Octamer	ATTTGCAT	Oct1, Oct2	Oct1 and Oct2 contain homeo domains[a]
Some Special Promoter and Enhancer Elements[b]			
HSE	GNNGAANNTCCNNG	Heat shock factor	Involved in heat-shock response
GRE	TGGTACAAATGTTCT	Glucocorticoid receptor	Mediates response to glucocorticoids
TRE	GAGGGACGTACCGCA	Thyroid receptor	Mediates response to thyroid hormones

[a]Homeo domains are discussed in Chapter 29 (page 1261).
[b]Abbreviations: HSE, heat-shock element; GRE, glucocorticoid response element; TRE, thyroid response element.

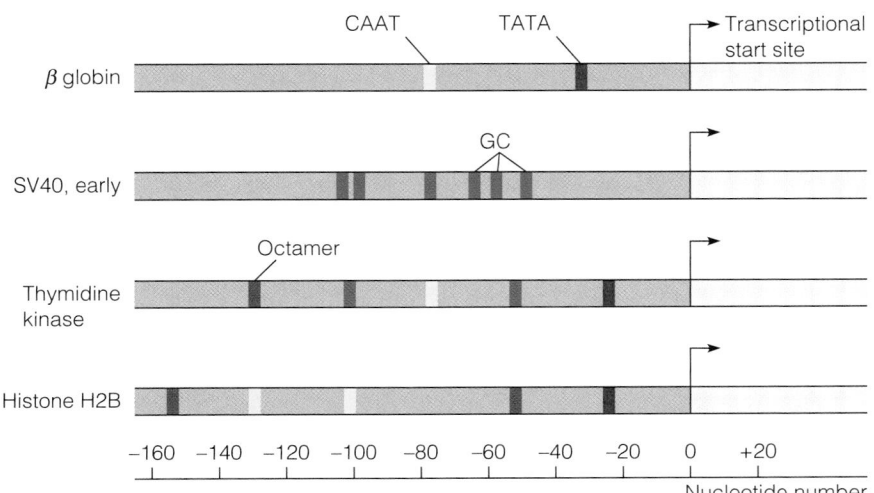

FIGURE 27.23

Structures of a few typical eukaryotic promoters. The colored boxes represent different regulatory elements: TATA = red, GC = blue, CAAT = yellow, and octamer = purple.

Based in part upon *Genes IV*, B. Lewin, Oxford: Oxford University Press, 1990.

pigmentosum (page 1085) and others for Cockayne's syndrome, which involves arrested development, neurological disorders, and photosensitivity.

To form a minimal complex capable of initiation, at least five additional protein factors are needed, as shown in Figure 27.24 and listed in Table 27.4. The order of addition is not known with certainty but starts, in the example shown, with binding to the TATA box, the most common initiation signal (and one that is used by all three nuclear RNA polymerases). As noted above, this usually lies about 20–30 bp upstream from the start site at Inr. Note that a *minimal* transcription unit involves the TATA-binding protein (TBP), whereas in vivo formation of an initiation complex probably always uses TFIID, a multisubunit structure incorporating both the TATA-binding protein and TATA-binding associated factors (TAFs). The TAFs interact with *activation factors* associated with upstream sites on specific genes,

Transcription can be modified by binding of trans-acting factors, either in the promoter or in distant enhancers.

TABLE 27.4   General transcription initiation factors from human cells[a,b]

Factor	Number of Subunits	Molecular Weight (kDa)	Function
TFIID — TBP[c]	1	38	Core promoter recognition (TATA); TFIIB recruitment
TFIID — TAFs	12	15–250	Core promoter recognition (non-TATA elements); positive and negative regulation
TFIIA	3	12, 19, 35	Stabilization of TBP binding; stabilization of TAF–DNA interactions; antirepression functions
TFIIB	1	35	RNA pol II–TFIIF recruitment; start-site selection by RNA pol II
TFIIF	2	30, 74	Promoter targeting of pol II; destabilization of nonspecific RNA pol II–DNA interactions
RNA pol II	12	10–220	Catalytic functions in RNA synthesis; recruitment of TFIIE
TFIIE	2	34, 57	TFIIH recruitment; modulation of TFIIH helicase, ATPase, and kinase activities; direct enhancement of promoter melting (?)
TFIIH	9	35–89	Promoter melting using helicase activity; promoter clearance (?) by CTD kinase activity

[a]The subunit compositions and polypeptide sizes are those described for the human factors, but homologs for virtually all have also been identified in rat, *Drosophila*, and yeast.

[b]Abbreviations used: CTD, carboxy-terminal domain of pol II; RNA pol II, RNA polymerase II; TAFs, TATA-binding protein–associated factors; TBP, TATA-binding protein.

[c]TBP, the TATA-binding protein, is also part of TFIIIB and of SL1, a pol I transcription factor.

Adapted from *Trends in Biochemical Sciences* 21:327–335, R. G. Roeder, The role of general initiation factors in transcription by RNA polymerase II. © 1996, with permission from Elsevier.

FIGURE 27.24

A model for formation of a minimal preinitiation complex (PIC) for pol II on a TATA promoter. In the simplest situation, binding of TATA-binding protein (TBP) initiates the sequence. Alternatively, in vivo TFIID, which includes both TBP and associated factors (TAFs), is used. This will also result in binding of TFIIA. The series of dots indicates phosphorylation of the C-terminal domain (CTD) of Rpb1, the largest subunit of pol II. Phosphorylation is necessary for release of the enzyme from the initiation site.

Reprinted from *Trends in Biochemical Sciences* 21: 327–335, R. G. Roeder, The role of general initiation factors in transcription by RNA polymerase II. © 1996, with permission from Elsevier.

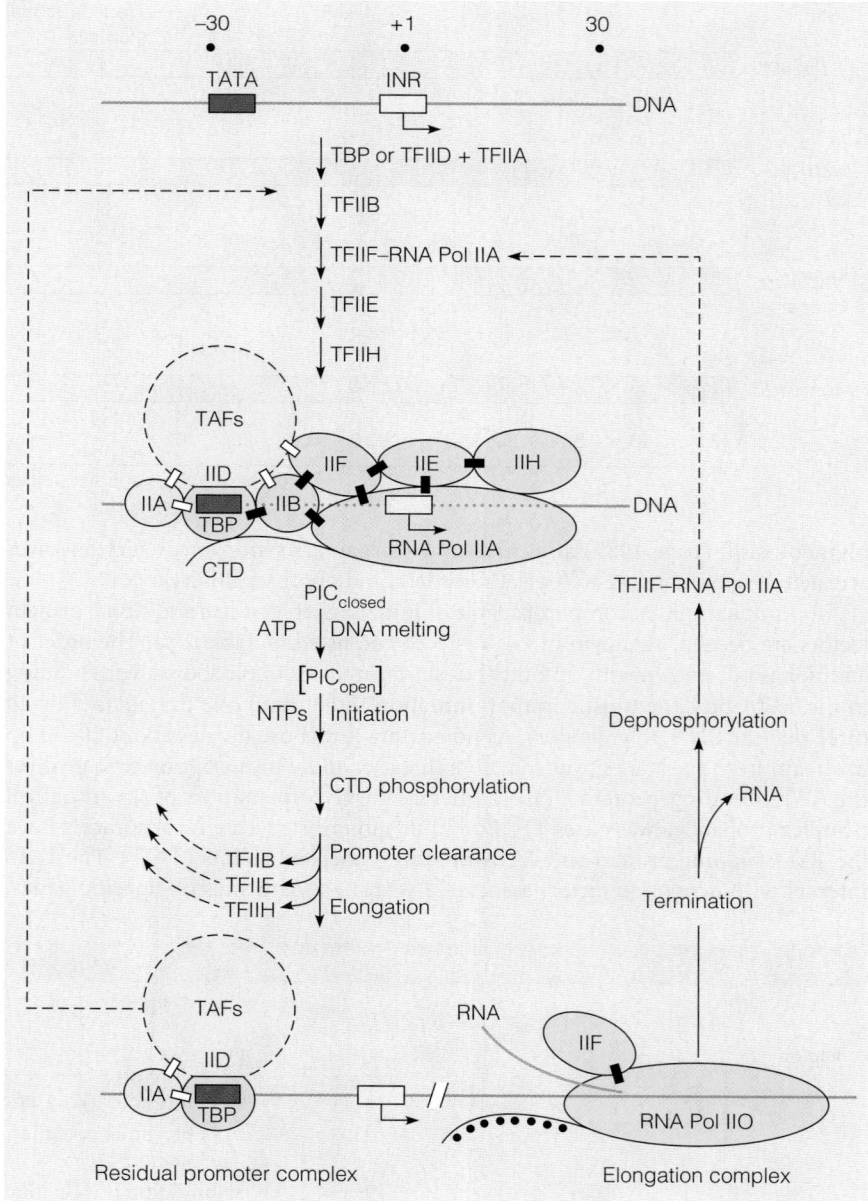

thereby providing communication for gene regulation. The structure of the complex containing the TATA site, TBP, TFIIB, and TFIIA has been deduced from X-ray diffraction and modeling studies. A remarkable feature that emerges is the pronounced kink induced in the DNA around the TATA site (Figure 27.25).

Essential to the formation of an elongation complex is phosphorylation of the carboxyl terminus of the largest subunit of RNA pol II. This protein contains several dozen repeats of the heptapeptide sequence –YSPTSPS–, as many as 52 repeats in the mammalian enzyme. Many or most of the serine residues in this sequence must undergo phosphorylation in order for RNA polymerase to escape the promoter. Recent studies show site-specific methylation of arginine residues in this domain as well.

TFIIB plays a particularly important role in converting the initial closed-promoter complex to an open-promoter complex, comparable to the complexes we have discussed for bacterial RNA polymerase. Recently Patrick Cramer and colleagues described a crystal structure for the complex between yeast RNA polymerase II and TFIIB, expanding our understanding of the biochemical steps

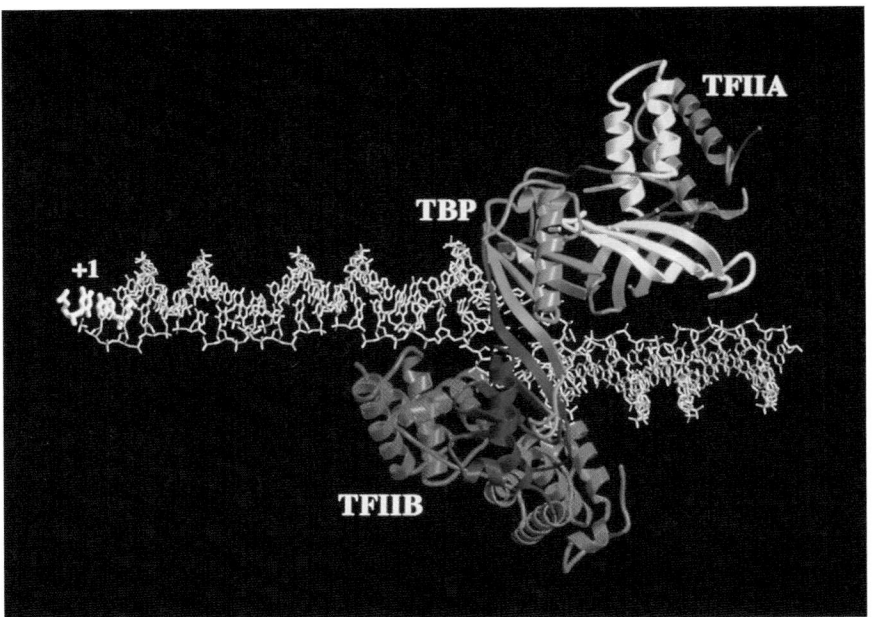

FIGURE 27.25

Computationally assembled model of the TFIIA–TBP–TFIIB–promoter complex based on crystal structures of TBP–TFIIB–TATA and TFIIA–TBP–ATA complexes. The amino- and carboxy-terminal direct-repeat domains of TBP are shown, respectively, in blue and purple; the amino- and carboxy-terminal direct-repeat domains of core TFIIB are shown, respectively, in red and magenta; and portions of the large and small subunits of yeast TFIIA are shown, respectively, in green and yellow. The transcription initiation start site (+1) is indicated in white. The view of the complex is from the top, showing TBP sitting astride the distorted TATA element and adjacent, but laterally displaced, upstream and downstream DNA segments (in standard B form) extending, respectively, leftward and below the plane of the figure and rightward and below the plane of the figure.

Reprinted from *Trends in Biochemical Sciences* 21: 327–335, R. G. Roeder, from the work of S.K. Burley, The role of general initiation factors in transcription by RNA polymerase II. © 1996, with permission from Elsevier.

involved. In Figure 27.26 the structure of the TATA box-binding protein (TBP) has been modeled in, to provide a picture of the initiation process. In forming the closed-promoter complex, TBP bends the DNA by 90 degrees. The C-terminal domain of TFIIB binds to TBP and flanking DNA regions. The N-terminal domain recruits RNA polymerase to promoter DNA near the transcription start site, forming the closed-promoter complex. Next a TFIIB structural element called the B-linker is involved in opening DNA before the the transcription start site, leading to the open-promoter complex. TFIIB is involved here in threading the template DNA strand into the active center. For this TFIIB uses another structural element, the B-reader, which consists of a helix followed by a mobile loop. Next, DNA is scanned for an initiator (Inr) motif. Following this, the first two ribonucleotide substrates are positioned opposite a conserved motif called Inr, and the first phosphodiester bond is formed. Just as seen with bacterial RNA polymerase, most of the early chain initiation events are abortive, possibly because of interference of the growing RNA chain with the B-reader loop. Finally, growth of RNA chains beyond seven nucleotides triggers release of TFIIB, and this completes the process of promoter escape. The process as described is actually similar to transcription initiation as studied with bacterial RNA polymerases, even though the proteins other than RNA polymerase are quite different.

FIGURE 27.26

A model of transcription initiation as deduced from the crystal structure of TFIIB complexed with RNA polymerase II RNA. Polymerase is in silver-gray, with other participants as shown. Once the closed- and open-promoter complexes have formed, as described in the text, a clash between the growing RNA chain and the B-reader helix, and between nontemplate DNA and the B-linker helix, are responsible for TFIIB dissociation as elongation begins.

Reprinted from *Biological Chemistry* 391:731–735, Patrick Cramer, Towards molecular systems biology of gene transcription and regulation. © 2010 Walter de Gruyter GmbH & Co. KG.

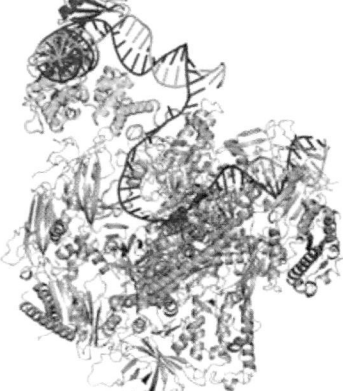

Closed promoter complex

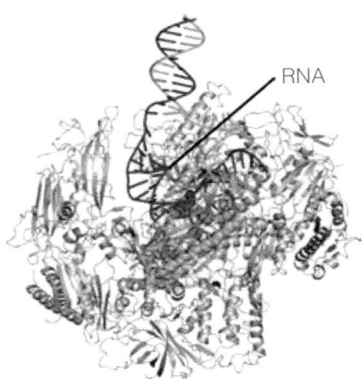

Open promoter complex

Elongation complex

FIGURE 27.27

A schematic representation of how DNA looping (perhaps mediated by nucleosomes) can bring enhancer-bound activator (or repressor) proteins into contact with TAFs associated with the core complex.

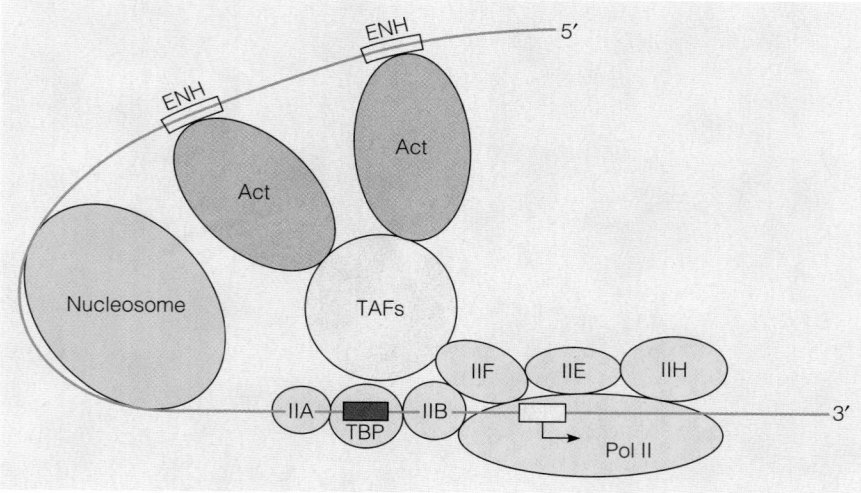

As mentioned earlier, trans-acting factors binding at sequences far removed from the promoter itself—by as much as several kilobase pairs—can influence transcription. Such sequences are called enhancers, and their mode of action appears to involve DNA looping, perhaps mediated by nucleosomes, which can bring enhancer-bound proteins into close physical contact with proteins bound to the promoter. Some of the transcription factors shown in Table 27.4 can bind in either promoter or enhancer regions. It is now evident that the TAF proteins shown in Figure 27.24 can act as intermediates between activator or repressor bound to enhancer and the core transcription complex (see Figure 27.27). Also involved in communication between upstream control elements and proteins bound at the promoter is a multiprotein complex called **mediator**. This will be discussed in more detail in Chapter 29.

The trans-acting factors tend to fall into a small number of classes, each defined by the kind of structural motif that interacts with the DNA. Schematic drawings of four of the better-known motifs were shown in Figure 27.22.

Chromatin Structure and Transcription

The complex interplay of transcription factors and polymerases we have described occurs not on naked DNA but on chromatin. The chromatin structure presents two major problems: First, how can the transcription factors and initiation complex bind to DNA in the presence of nucleosomes? Second, how can the actively transcribing polymerase pass through arrays of nucleosomes? This is an area of intense research interest. We present the current views of the first problem with a few examples and then briefly comment on the second problem.

The Problem of Initiation

As an example, consider the human β globin genes, which were described in Chapter 7. Although present in every human cell, these genes are expressed *only* in erythroid cells and in a fixed developmental sequence. In embryonic cells that have not yet begun synthesis of any globin, the chromatin of the β globin gene cluster appears much the same as in any other cell in the embryo and is quite densely covered with nucleosomes. But when differentiation of these cells commits them to globin synthesis, the whole β globin gene cluster undergoes changes in chromatin structure. One such change is the appearance of *hypersensitive sites*, regions particularly susceptible to digestion by nucleases. At early stages in developing human embryos, these sites appear in the 5' flanking regions of the embryonic genes, which are the first to be transcribed. Later, hypersensitive sites shift to

Nuclease hypersensitive sites disrupt chromatin to allow initiation.

the 5′ flanks of the adult genes. It is now clear that many of these sites represent regions a few tens or hundreds of base pairs in length in which nucleosomes have either been removed or "remodeled" so as to make the DNA contained therein more accessible. They provide points at which transcription factors and other trans-acting proteins can gain access to promoters and enhancers, thereby allowing the initiation and stimulation of transcription.

How are hypersensitive sites established in previously unresponsive genes? In some cases, for example the globin genes, it seems that the chromatin structure is rearranged at the time of replication. In other instances, protein factors seem to be able to interfere with chromatin structure at specific loci, opening hypersensitive sites.

Particularly interesting examples of this kind are found in hormonal regulation of transcription. A well-studied case involves the genes for the chicken egg-white proteins—ovalbumin, ovomucoid, and lysozyme. Transcription of these genes occurs only in the tubular cells of the hen oviduct. Even in immature chicks, the genomic domain containing the ovalbumin gene appears to have a somewhat different chromatin structure in these oviduct cells from that in other tissues. But only on stimulation by estrogen (either on sexual maturation of the chick or following hormone administration) does transcription of ovalbumin genes commence. Specific hypersensitive sites 5′ to some of the egg-white protein genes are opened by the presence of estrogen. Withdrawal of administered hormone from an immature chick leads to loss of hypersensitive sites and an immediate cessation of transcription of the genes.

There are many other examples of hormonal control of transcription. In each case the target cells contain specific proteins that are hormone receptors. When these proteins bind the hormone, they become capable of interacting with specific DNA sites or with nonhistone regulatory proteins bound to such sites. Thus, both positive and negative regulation is possible. In some cases, the hormone-binding receptor acts as a positive regulatory factor, for example, by binding to an enhancer element. In others, the hormone-binding receptor can interact with a repressor protein to augment or relieve the repression. Recent evidence suggests that the latter model may describe the response of chicken oviduct cells to estrogens.

Chromatin Remodeling

How are hypersensitive sites generated, and how is chromatin structure altered to make the DNA accessible? Only recently have pieces of evidence begun to fit together to answer these questions. First was the discovery, in yeast and then in higher eukaryotes, of **chromatin remodeling factors**. These are proteins that enable promoter regions to be able to accept the complex and bulky machinery depicted in Figure 27.24. The SWI/SNF and RSC complexes from yeast and the NURF complex from *Drosophila* are probably the best studied; significantly, all three require ATP hydrolysis to carry out their task. Exactly what such factors do is still unclear. They seem not to remove nucleosomes but rather to "open" them in some fashion. Recent in vitro work with the yeast RSC complex indicates that nucleosomes are, in fact, dissociated, at least transiently, to aid in the generation of transcription complexes. Other recent investigations using single-molecule techniques indicate that polymerase II does not separate template DNA from the histones in chromatin. Instead, polymerase takes advantage of conformational fluctuations in chromatin, pausing until partial opening allows it to move forward, acting like a ratchet in forcing its way around the nucleosome, a process that has been called "nudging through a nucleosome." Whether both processes—transient dissociation and nudging—occur regularly in eukaryotic transcription is an issue now being intensively pursued. We shall say more about chromatin remodeling complexes in Chapter 29.

Another and perhaps equally important role is played by histone acetyl-transferases and deacetylases. It has long been known that the histones of the

FIGURE 27.28

Acetylation of core histones. The general structure of each of the four core histones involves a helical "histone fold" domain plus an unstructured, highly basic N-terminal domain. Acetylation in nuclei occurs exclusively in the N-terminal domains, at the highly conserved sites indicated in magenta.

Adapted with permission from *Biochemistry* 37:17637–17641, J. C. Hansen, C. Tse, and A. P. Wolffe, Structure and function of the core histone N-termini: More than meets the eye. © 1997 American Chemical Society.

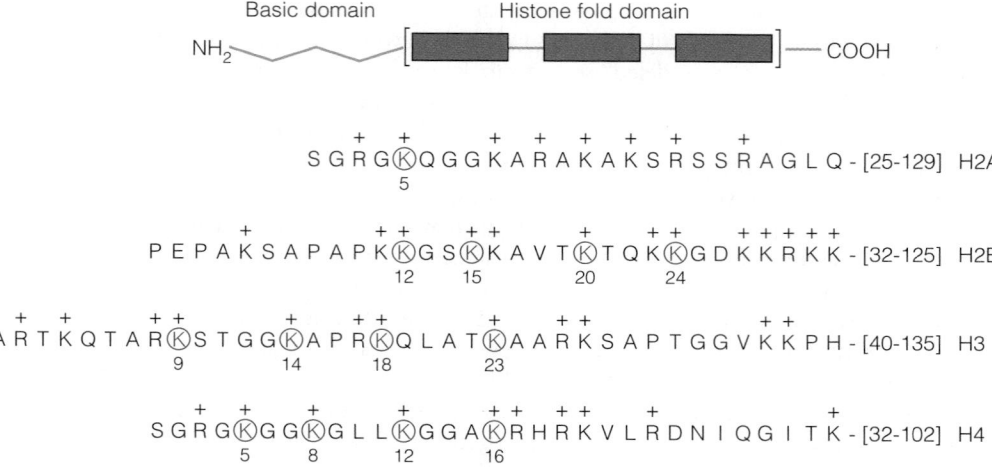

Pol II can transcribe through nucleosome arrays.

nucleosome core are subject to acetylation at specific lysine residues in the N-terminal tails (see Figure 27.28). Furthermore, high levels of acetylation are known to be correlated with high transcriptional activity, and vice versa. This stands to reason, as neutralization of histone basic residues by acetylation would loosen interactions between histones and DNA in chromatin. What is new is the discovery that a number of proteins recruited to the initiation complex by activators and TAFs (and, indeed, some of these proteins themselves) have histone acetylase activity. The fact that specific transcription factors are involved in this process may provide the long-sought explanation for how the chromatin of *specific* genes can be targeted for disruption.

Transcriptional Elongation

Formation of the preinitiation complex (Figure 27.24) is followed, in the presence of ribonucleoside triphosphates and ATP, by melting of a short region and initiation of transcription. As noted above, the C-terminal tail of pol II Rpb1 becomes strongly phosphorylated, leading to promoter release, and elongation begins, with a helicase activity clearing the way. A number of the core transcription factors are released, and pol II, together with TFIIF, moves along the DNA. A residual complex, containing TBP, TFIIA, TAFs, and probably activator proteins, remains at the start site, ready to initiate another round.

At this point the polymerase also acquires several special *elongation factors*. Some of these appear to assist the enzyme in traversing pause sites in the DNA. In in vitro experiments, using naked pol II, transcription is relatively slow and interrupted by frequent pauses, especially in T-rich regions. The phenomenon seems to be generally similar to that described earlier for prokaryotic transcription. The presence of elongation factors assists the enzyme in passing such sites. Nucleosomes form even more important obstacles to the progress of a polymerase II along the DNA. Although some prokaryotic polymerases can pass through nucleosomal arrays in in vitro studies, pol II is entirely blocked unless accessory proteins are present. These include nucleosome remodeling factors and a specific elongation factor called FACT.

Just how pol II transcribes through nucleosomes is still something of a mystery. Do the nucleosomes unfold and re-form as the polymerase passes? Are they temporarily displaced? Current evidence favors temporary displacement, but the issue is far from settled. One factor that may play a role in such displacement is the development of positive superhelical torsion ahead of a moving polymerase. A polymerase moving along a helical template must either continually rotate about the DNA or build up positive supercoils ahead (overwinding) to compensate for the unwinding it is doing. Such torsion would tend to destabilize nucleosomes because they contain negatively wrapped DNA.

Termination of Transcription

Even the termination of mRNA transcription is different in eukaryotes. Whereas the prokaryotic RNA polymerase recognizes terminator signals, which sometimes function with the aid of the ρ protein, the eukaryotic polymerase II usually continues to transcribe well past the end of the gene. In doing so, it passes through one or more AATAAA signals, which lie beyond the 3' end of the coding region (Figure 27.29). The pre-mRNA, carrying this signal as AAUAAA, is then cleaved by a special endonuclease that recognizes the signal and cuts at a site 11–30 residues 3' to it. At this point, a tail of polyriboadenylic acid, poly(A), as many as 300 bases long, is added by a special nontemplate-directed polymerase. The functions of the poly(A) tails of eukaryotic mRNAs include mRNA stabilization and facilitation of transport from nucleus to cytoplasm. We know that they cannot be essential for all messages because some mRNAs (for example, most histone mRNAs in higher eukaryotes) do not have them. However, as noted above, poly(A) tails relate to message stability, for the tail-less messages typically have much shorter lifetimes in the nucleus.

Post-transcriptional Processing

Bacterial mRNA Turnover

A major aspect of messenger RNA metabolism in eukaryotes is the events occurring *after* transcription, events that are necessary for messages to move from the nucleus to their sites of utilization in the cytosol. We discuss these events later in this chapter. In prokaryotes, by contrast, mRNAs are used in protein synthesis directly. In fact, a nascent mRNA can serve as a template for translation at its 5' end while still in the process of being synthesized toward the 3' end. In other words, transcription is coupled directly to translation.

The major post-transcriptional event in metabolism of prokaryotic mRNA is its own degradation, which in most cases is quite rapid. A few bacterial mRNAs, notably those encoding outer membrane proteins, are long-lived; however, most bacterial messages have half-lives of only 2 to 3 minutes. This short life span means that genes being expressed must be transcribed continuously and that most mRNA molecules are translated only a few times. Although this might seem energetically wasteful, it is consistent with prokaryotic lifestyles, which necessitate rapid adaptation to environmental changes. Earlier we noted the selective advantage to bacteria of expressing the genes for lactose utilization only when an inducer is present. By the same token, it would be wasteful for the cell to continue producing these proteins after lactose or a related sugar was exhausted from the milieu. Rapid degradation of *lac* mRNA ensures that the energetically wasteful synthesis of these proteins will cease soon after the need for these proteins is gone.

Although we have known about the instability of bacterial mRNA for more than five decades, we still understand surprisingly little about the pathway of degradation. There are probably overlapping mechanisms, involving hydrolysis by nucleases and phosphorolysis by polynucleotide phosphorylase. We do know that degradation starts from the 5' end, which is important because translation also starts from the 5' end. If degradation were to start from the 3' end, a ribosome starting from a 5' end might never reach an intact 3' end. There is reason to think that mRNA degradation sometimes starts with the action of ribonuclease III, an enzyme specific for duplex RNA, which could cleave in stem–loop structures and create sites for exonucleolytic attack. RNase III is actually involved in the maturation of certain phage mRNAs as they undergo post-transcriptional processing, but this involvement is not known to occur with bacterial mRNAs.

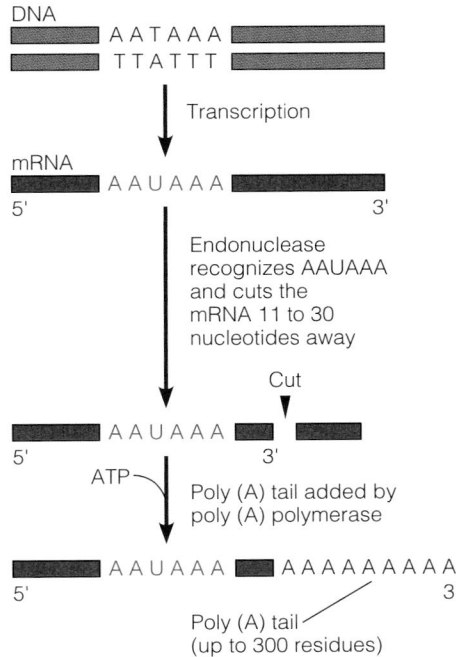

FIGURE 27.29

Termination of transcription in eukaryotes: addition of poly(A) tails. There is an AATAAA sequence near the 3' end of most eukaryotic genes. When this is transcribed to AAUAAA, it provides a signal for endonuclease cleavage and poly(A) tail addition.

Post-transcriptional Processing in the Synthesis of Bacterial rRNAs and tRNAs

Both ribosomal RNAs and transfer RNAs are synthesized in the form of larger transcripts (pre-rRNA and pre-tRNA, respectively), which undergo cleavage at both ends of the transcript, en route to becoming mature RNAs. This process is comparable to the processing of 45S pre-rRNA in eukaryotic cells (page 1144). However, as discussed in Chapter 28, the rRNA components in bacteria are somewhat smaller than in eukaryotes—23S, 16S, and 5S. The total amount of DNA encoding these rRNAs and tRNAs amounts to less than 1% of the *E. coli* genome, but because of the instability of mRNA (which is encoded by the remaining 99%), rRNA and tRNA constitute about 98% of the total RNA in a bacterial cell. It is important to realize also that transcription of rRNA genes is extremely efficient when cells are growing rapidly. The intracellular concentrations of ribonucleoside triphosphates are important control elements here; ATP, whose level is high in rapidly growing cells, activates rRNA gene transcription by stabilizing the relevant open-promoter complexes.

rRNA Processing

The *E. coli* genome contains seven different operons for rRNA species. Each one encodes, in a single 30S transcript, sequences for one copy each of 16S, 23S, and 5S rRNAs (Figure 27.30). Because the three species are used in equal amounts, the logic of this organization is apparent. Less easy to explain is that each transcript also includes sequences for one to four tRNA molecules, which vary among the seven different operons. Because rRNAs and tRNAs are all used in protein synthesis, the interspersion of rRNA and tRNA sequences may represent a means of coordinating the rates of synthesis of these RNAs, but specific mechanisms have not yet been revealed.

What enzymes are involved in processing the 30S pre-rRNA species? The abnormal accumulation of this species in bacterial strains defective in RNase III first suggested a role for this enzyme in rRNA processing. In fact, one double-strand cut in each of two giant stem–loop regions releases precursors to 16S and 23S rRNAs, and the same probably occurs for 5S rRNA. Further maturation steps require the presence of particular ribosomal proteins, which begin to assemble on the precursor RNAs while transcription is still in progress. The embedded tRNA sequences are processed to give mature tRNAs, along the same routes used for other tRNA species, as discussed next.

tRNA Processing

Aside from the tRNAs embedded in pre-rRNA transcripts, the other tRNAs are synthesized in transcripts that contain one to seven tRNAs each, all surrounded by lengthy flanking sequences. The maturation steps are summarized in Figure 27.31, using as an example the well-studied case of the *E. coli* tyrosine tRNA species (tRNA^{Tyr}). In this case, maturation starts (step 1) with an endonuclease that cleaves

Bacterial transcripts undergo post-transcriptional processing, involving both endonucleolytic and exonucleolytic cleavage.

FIGURE 27.30

Structure of *E. coli* 30S pre-rRNA. Sequences complementary to two promoter sites (P$_1$ and P$_2$), RNase III cleavage sites (RIII) that release 16S and 23S species, and the locations of tRNA sequences embedded within the transcript are shown.

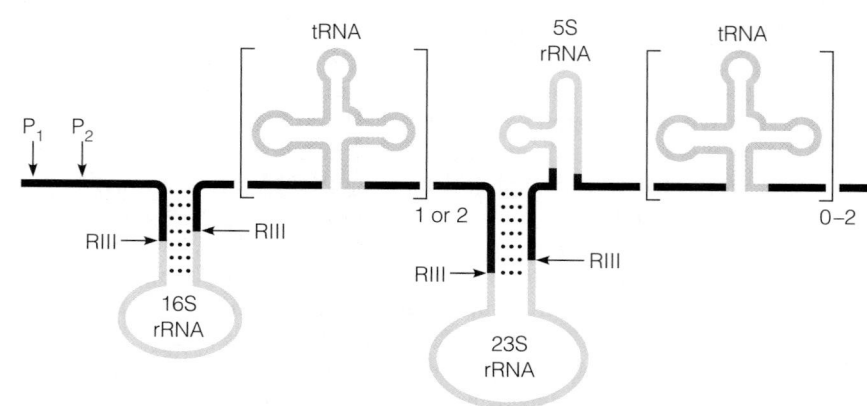

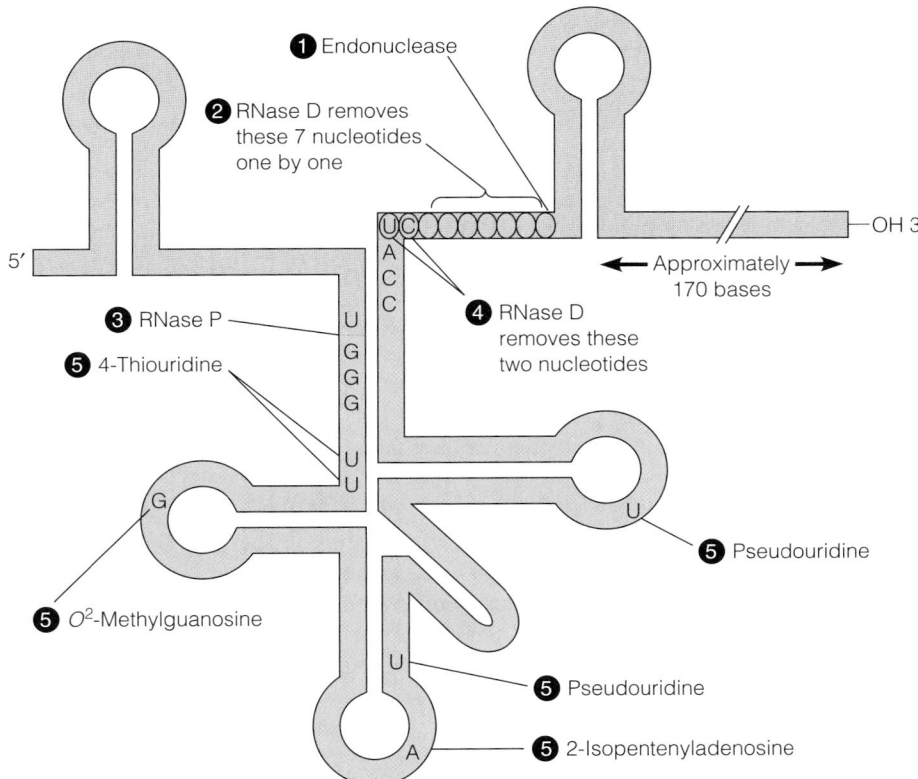

FIGURE 27.31

Modification steps (1–4) that occur in maturation of *E. coli* tRNA[Tyr] from its transcript and modified bases (5) seen in the mature tRNA. The sequence of the mature tRNA is shown in purple.

next to a stem–loop structure on the 3′ side of the tRNA sequence. This is followed by action of **ribonuclease D** (step 2), which carries out exonucleolytic cleavage to a point two nucleotides removed from the CCA sequence at the 3′ end. Next (step 3), the 5′ end is created by **ribonuclease P**, which cleaves to leave a phosphate on the 5′ terminal G. This enzyme creates the 5′ terminus of all tRNA molecules. It is not clear what structural features are recognized by RNase P, for different sequences are contained in the cleavage sites. As pointed out in Chapter 11, ribonuclease P was one of the first identified ribozymes. The enzyme consists of one RNA molecule of 377 nucleotides and one protein molecule with M_r of about 20,000. Both components are necessary for full catalytic activity, but under nonphysiological conditions the RNA molecule alone can catalyze accurate cleavage. Despite the involvement of RNA as the catalytic element in ribonuclease P, the enzyme in human mitochondria was recently found to consist entirely of three protein subunits.

Once the proper 5′ terminus has been created, ribonuclease D removes the remaining two nucleotides from the 3′ end (step 4). Should excessive "nibbling" occur through faulty control of RNase D activity, there is an enzyme (CCA nucleotidyltransferase) that will restore the CCA end to any tRNA in a nontranscriptive fashion. This enzyme specifically recognizes the 3′ terminus of tRNAs that lack the CCA end and catalyzes sequential reactions with a CTP, another CTP, and an ATP.

Creation of the modified bases (see Chapters 4 and 28) occurs at the final stage, including methylations, thiolations, reduction of uracil to dihydrouracil, and so forth. In the specific example shown, the modifications include formation of two pseudouridines, one 2-isopentenyladenosine, one O^2-methylguanosine, and one 4-thiouridine (step 5). These modifications serve to stabilize the tRNA molecules against intracellular degradation. The modifications are not essential for tRNA function in translation because many tRNAs lacking them are fully active in vitro. Pathways for eukaryotic tRNA synthesis are similar, including the involvement of ribonuclease P.

Despite the evident nonessentiality of these modifications for tRNA function in translation, cells expend a great deal of energy carrying them out. More than

90 different base modifications have been described. In the yeast *Saccharomyces cerevisiae,* more that 40 different tRNA-modifying enzymes, made up of nearly 100 protein subunits, carry out those modifications. The average tRNA molecule has more than 12 of its bases modified.

Gene Splicing in Prokaryotes

An additional post-transcriptional process, namely intron splicing, is almost exclusively confined to eukaryotes, although intron self-splicing has been well defined in a few bacteriophage genes. Interestingly, the phage genes subject to splicing are often those encoding enzymes of DNA or DNA precursor biosynthesis. For example, the three introns in the phage T4 genome are in genes encoding thymidylate synthase, the class I ribonucleotide reductase small subunit, and class III (anaerobic) ribonucleotide reductase. Also, prokaryotic introns sometimes encode a "homing endonuclease," an enzyme that in mixed infection can facilitate transfer of an intron to a cognate intronless gene.

Processing of Eukaryotic Messenger RNA

Prokaryotic and eukaryotic cells differ significantly in the ways that messenger RNAs for protein-coding genes are produced and processed. Recall that prokaryotic mRNAs are synthesized at the bacterial nucleoid in direct contact with the cytosol and are *immediately* available for translation. A specific nucleotide sequence at the 5′ end recognizes a site on the prokaryotic ribosomal RNA, allowing attachment of the ribosome and initiation of translation, often even before transcription of the message is completed. Hence, there is little or no post-transcriptional processing of bacterial mRNAs.

In eukaryotes, mRNA is produced in the nucleus and must be exported to the cytosol for translation. Furthermore, the initial product of transcription (*pre-mRNA*) includes all of the introns and substantial flanking regions; the introns must be removed before correct translation can occur. Finally, there is no ribosomal attachment sequence like the Shine–Dalgarno sequence in prokaryotes (see Chapter 28). For all these reasons, eukaryotic mRNA requires extensive processing before it can be used as a protein template. This processing takes place while mRNA is still in the nucleus.

Capping

The first modification occurs at the 5′ end of the pre-mRNA. First, one phosphate is removed hydrolytically from the triphosphate moiety at the 5′ terminal nucleotide. Next the resulting 5′ diphosphate end attacks the α (inner) phosphate of GTP; in essence the guanine nucleotide is added in *reverse* orientation (5′ → 5′). Together with the first two nucleotides of the chain, it forms what is known as a *cap* (Figure 27.32). The cap is further decorated by addition of methyl groups to the N-7 position of the guanine and to one or two sugar hydroxyl groups of the cap nucleotides. This cap structure serves to position the mRNA on the ribosome for translation, and it probably contributes also to stabilization of the message.

Splicing

After being capped, the pre-mRNA becomes complexed with a number of *small nuclear ribonucleoprotein particles (snRNPs,* often called "snurps"*),* which are themselves complexes of *small nuclear RNAs (snRNAs)* and special splicing proteins. The snRNAs are all less than 300 nucleotides in length. The snRNP–pre-mRNA complex is called a **spliceosome**, and it is here that the most elegant part of the processing takes place—the cutting and splicing that is necessary to excise introns from the pre-mRNA and join the ends of the two exons. In forming a spliceosome, snRNAs recognize and bind intron–exon splice sites by means of complementary sequences (Figure 27.33). Precise recognition of splice sequences is essential, for

---

Post-transcriptional processing in bacteria involves cleavage of the primary transcript, modification of bases (in tRNA synthesis), nontranscriptive nucleotide addition, and (in a *few cases*) intron splicing.

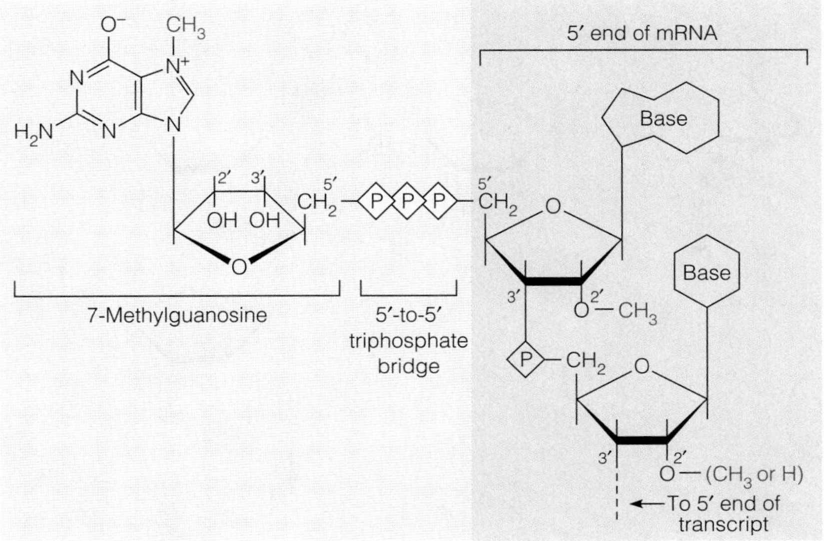

FIGURE 27.32

Structure of a processed mRNA 5′ end. Details of the 5′ cap region are shown. Methyl groups that are added are in magenta.

7-Methylguanosine

5′-to-5′ triphosphate bridge

5′ end of mRNA

Base

Base

O — CH₃

To 5′ end of transcript

O — (CH₃ or H)

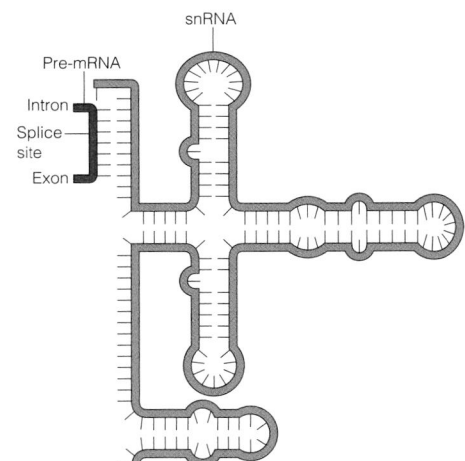

FIGURE 27.33

Structure of a small nuclear RNA (snRNA). Human U1 RNA is shown, together with the intron–exon boundary region to which it binds in forming the spliceosome.

even a single-base error would disrupt the sense of the genetic message. Table 27.5 shows some representative splice-site sequences and the consensus sequences common to most introns. A schematic view of the chemical aspects of the splicing mechanism is shown in Figure 27.34.

Excision of a single intron involves assembling and disassembling a spliceosome. Figure 27.35 depicts the overall process. The sequence begins with binding of the U1 snRNP to the G site at the 5′ end of the intron. The U2 snRNP then binds at the branch site. With continued assembly of the spliceosome, including addition of several more snRNPs, the lariat loop in the intron is formed and the two exons are joined. RNA in the U6 snRNP probably plays a catalytic role. Splicing has now been accomplished, and the products—a ligated mRNA and a looped intron—are released. As the spliceosome disintegrates, the looped intron is degraded, and the mRNA is exported from the nucleus. Exactly how this happens is still not known with certainty, but it may be in conjunction with some of the snRNP proteins.

TABLE 27.5 Representative sequences at splice junctions

Protein, Intron	5′ Splice Site Exon ↓	Branch Site Intron ↓	3′ Splice Site ↓ Exon
Ovalbumin, intron 3	···UCAG	GUACAG···A···UGUAUUCAG	UGUG
β globin, human, intron 1	···CGAG	GUUGGU···A···CACCCUUAG	GCUG
β globin, human, intron 2	···CAGG	GUGAGU···A···CCUCCACAG	CUCC
Immunoglobulin I, L-VI	···UCAG	GUCAGC···A···UGUUUCGAG	GGGC
Rat preproinsulin	···CAAG	GUAAGC···A···CCCUGGCAG	UGGC
Consensus sequences[a]	__AG	GURAGY···A···YYYYY____AG	———

[a]Here R stands for purine and Y for pyrimidine. Residues listed for the consensus sequence (in magenta) are those found in two-thirds or more of over 100 cases analyzed. The residues shown in magenta are invariant in all cases analyzed. Shown in blue is an AMP residue, the branch site, which is located 20 to 50 nucleotides upstream from the 3′ splice site. A pyrimidine-rich sequence lies just upstream from the 3′ splice site.

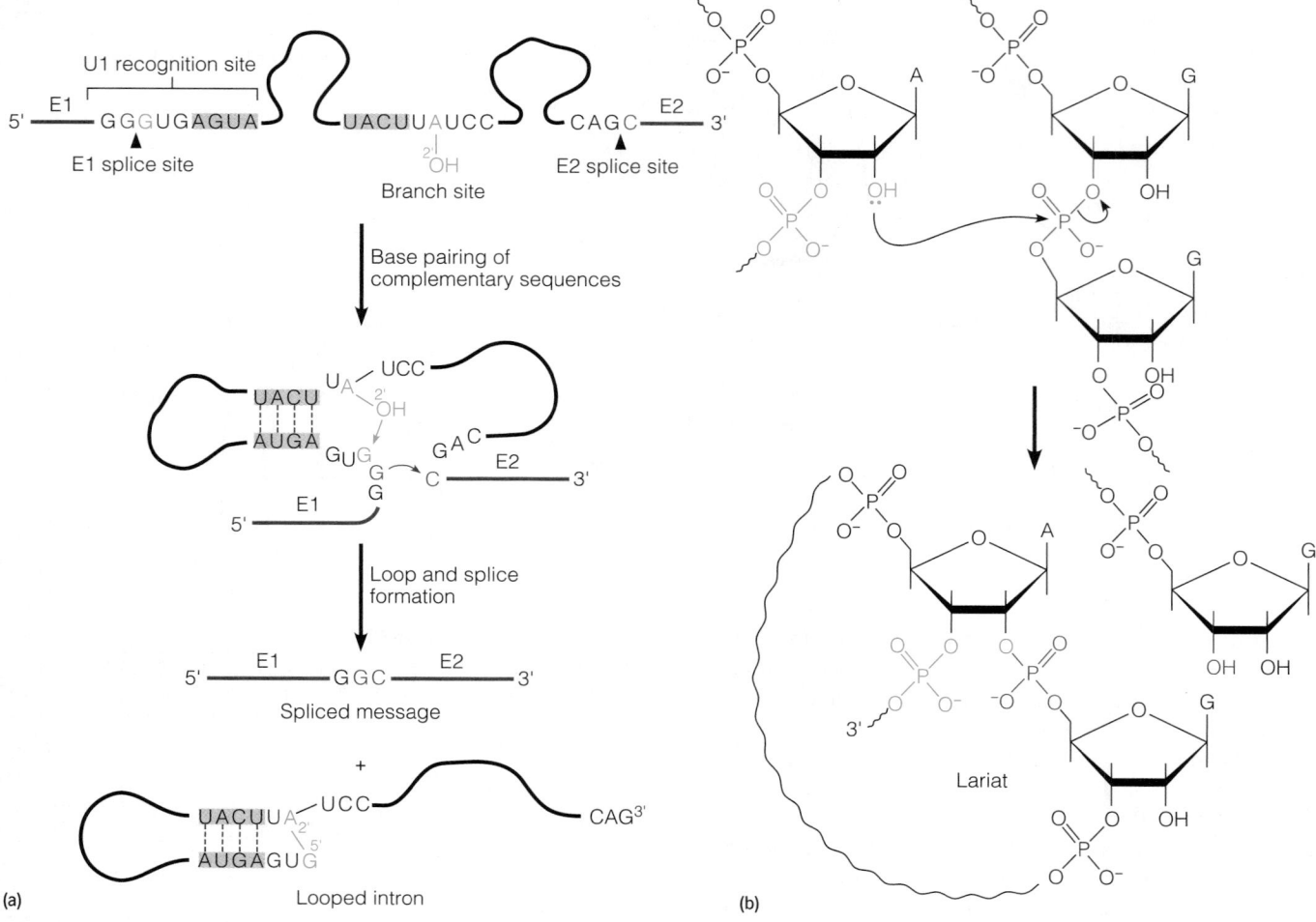

FIGURE 27.34

A schematic view of the mechanism of mRNA splicing. **(a)** The overall process. Exons (E1 and E2) are indicated by red lines, the intron by a black line or sequence. The E1 splice site, presumably with the aid of the small RNA U1, pairs with a sequence at the branch site to form a loop. The 2′ hydroxyl on the branch-site AMP carries out a transesterification reaction by attacking the phosphate of a GMP residue (blue) at the 3′ end of the exon 1 (the E1 splice site). This frees the adjacent G (red) to attack with its 3′ hydroxyl the phosphate 5′ to the C at the 5′ end of exon 2. The products are a spliced message and a looped intron "lariat" structure, which is then degraded. **(b)** The first transesterification reaction. The second reaction (not shown) involves nucleophilic attack of the GMP 3′-hydroxyl group (in red) upon the phosphate 5′ to a CMP residue, as seen schematically in part (a).

In 2011 Jeff Gelles and Melissa Moore and their colleagues described a novel technique for kinetic analysis of the individual steps in spliceosome formation. Key to their approach was development of intensely fluorescent tags for participants in spliceosome assembly, which enabled the investigators to follow individual steps at the single-molecule level and in real time. As shown in Figure 27.36, this approach described an ordered sequential pathway. Every step in spliceosome assembly was shown to be reversible, with no single step generating commitment to the entire pathway. Instead, commitment increases as assembly proceeds. The reversibility of each step may mean that alternative splicing can be regulated at any one of these steps. The approach described should help to illuminate poorly understood aspects of alternative splicing.

Because of the obvious importance of correct splicing for correct expression of genetic information, it should not be surprising that splicing errors are responsible for many genetic diseases; in fact, it is estimated that 15% of all genetic diseases arise from splicing errors. An example is some forms of thalassemia, a family of diseases arising from defective synthesis of hemoglobin chains (page 265). Mutations have been found in both the 5′ and 3′ splice sites of both genes for the β chain of human hemoglobin. Usually, an incorrect mRNA chain is formed, which leads to premature termination of translation of the message. In some cases a new 5′ splice site is created, and this may lead to a mixture of correctly and incorrectly spliced messages, a condition that may be less severe clinically.

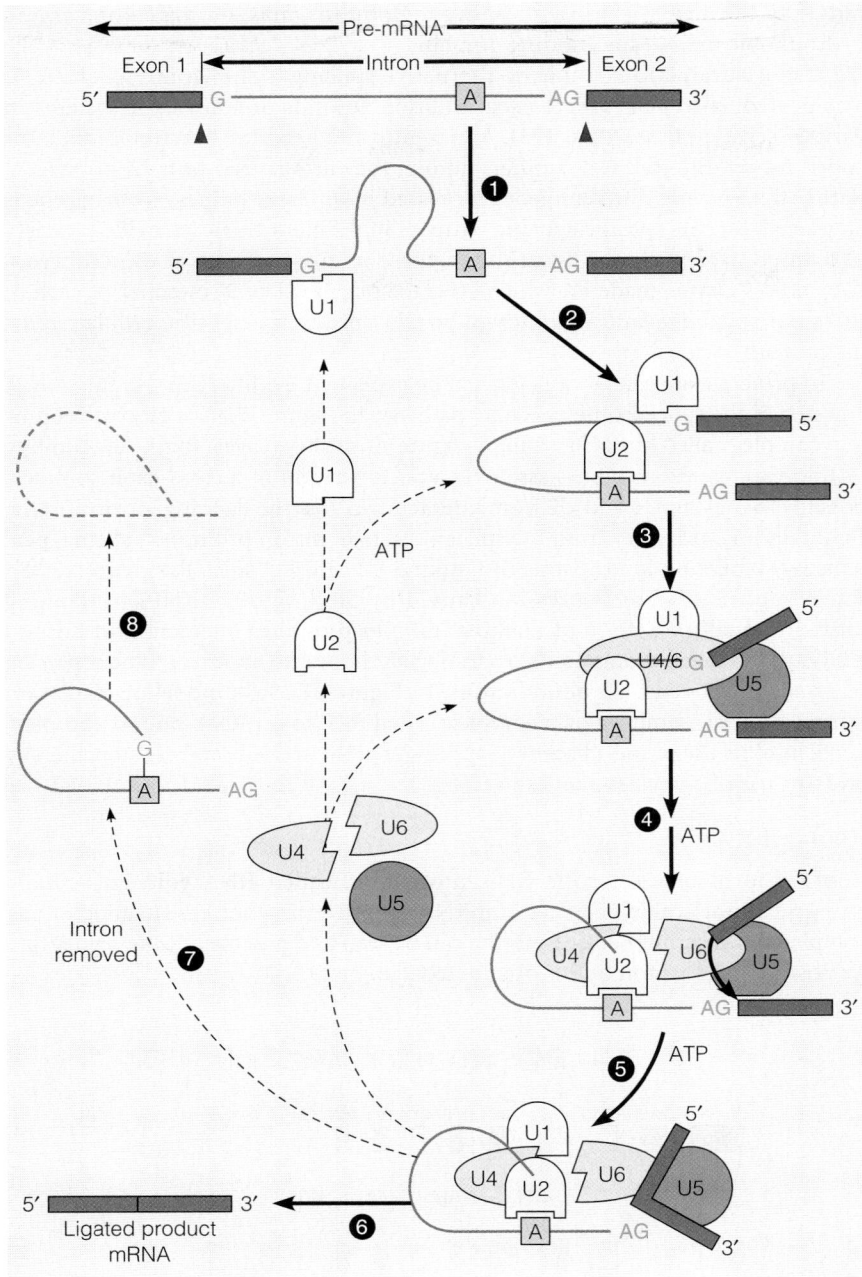

FIGURE 27.35

The overall process of splicing. The pre-mRNA plus assorted snRNPs assemble and disassemble a spliceosome, which carries out the splicing reaction. The snRNPs are designated U1, U2, and so on. In step 1 U1 is bound, which together with U2 binding (step 2) leads to a looped structure. Factors U4/6 and U5 then bind (step 3) and cleavage and transfer then occur (steps 4, 5). The spliceosome disassembles, releasing the ligated product (6) and the looped intron (7). This is degraded into small oligonucleotides (step 8).

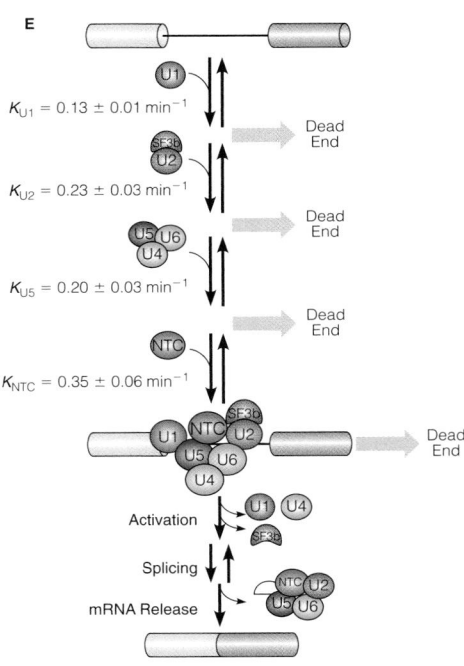

FIGURE 27.36

Kinetic analysis of spliceosome assembly. Each step in the assembly pathway was shown to be reversible, with rate constants in the forward direction as shown. Reversibility has not yet been detected in the activation step shown nor in mRNA release. SF3b is a protein splicing factor and NTC is a multiprotein complex called Prp19.

From *Science* 331:1289–1295, A. A. Hoskins, L. J. Friedman, S. S. Gallagher, D. J. Crawford, E. G. Anderson, R. Wombacher, N. Ramirez, V. W. Cornish, J. Gelles, and M. J. Moore, Ordered and dynamic assembly of single spliceosomes. © 2011. Reprinted with permission from AAAS.

Alternative Splicing

Once investigators had discovered and described mRNA splicing, they were surprised to learn that the same pre-mRNA can undergo splicing in several different ways. The existence of **alternative splicing** means that different combinations of exons from the same gene can be processed into different mature mRNAs and then undergo translation into quite different proteins in different tissues or at different developmental stages of the same organism. Alternative splicing made it easier to accept our surprise when the Human Genome Project revealed the existence of far fewer genes than had been expected given the size of the genome and the complexity of *Homo sapiens*. Alternative splicing greatly enlarges the repertoire of proteins that can be encoded by a genome. We mentioned one example of this phenomenon in Chapter 7, where we pointed

Alternative splicing allows one gene to specify several proteins.

out that the heavy chains of immunoglobulins may or may not carry a hydrophobic membrane-binding domain.

A more dramatic example of alternative splicing is shown in Figure 27.37. The protein α-*tropomyosin* is used in different kinds of contractile systems in various cell types (see page 293). Apparently, the need for functional domains coded for by different exons differs among the various uses of α-tropomyosin. Rather than having different genes expressed in different tissues, a single gene is employed, but the specific splicing patterns in different tissues provide a variety of α-tropomyosins. As the figure shows, there are two positions at which alternative choices can be made for which exon to splice in. The 3' member of each of these pairs is the *default exon;* it will be chosen unless a specific cellular signal dictates otherwise.

In principle alternative splicing can be explained by differential regulation of any step in the spliceosome assembly pathway. In Figure 27.38a a ribonucleoprotein complex called hnRNPL (L in the figure) is the key player. To the left, binding of L to variable exons blocks assembly after formation of a cross-exon complex, possibly to prevent U1 and U2 from interacting across the flanking introns. To the right, binding of L away from the splice sites in an intron promotes splicing, perhaps by stabilizing interactions of flanking U1 and U2 snRNPs. Figure 27.38b shows models of exon repression caused by binding of hnRNPs to flanking introns. Left, dimerization of flanking hnRNPs promotes interaction of U1 and U2 bound to distal exons. To the right hnRNP bound to an intron blocks pairing of proximal U1- and U2-bound exons. In Figure 27.38c, a model of regulation during catalysis, binding of a complex called SXL to an RNA-bound complex, SF45, inhibits use of the adjacent 3'-splice site in the second catalytic step, thereby favoring use of a downstream exon.

RNA Editing

Another form of eukaryotic RNA processing, called **RNA editing**, actually changes the nucleotide sequence of mRNA, by converting one base to another in a completed transcript. Because editing can be seen as a process in gene regulation, we postpone discussion of this process to Chapter 29.

FIGURE 27.37

α-Tropomyosin gene organization (rat) and seven alternative splicing pathways. Exons (red, constitutive; green, smooth muscle-specific; yellow, striated muscle-specific; white, variable) are indicated with their encoded amino acids (numbered). Experimentally documented splicing pathways (solid lines) and others (dotted lines) inferred from nuclease protection mapping are shown. The smooth (SM) and striated (STR) exons encoding amino acid residues 39–80 are mutually exclusive, and there are alternative 3'-terminal exons as well. UT signifies untranslated regions.

Adapted from *Annual Review of Biochemistry* 56:467–495, R. E. Breitbart, A. Andreadis, and B. Nadal-Ginard, Alternative splicing: A ubiquitous mechanism for the generation of multiple protein isoforms from single genes. © 1987 Annual Reviews.

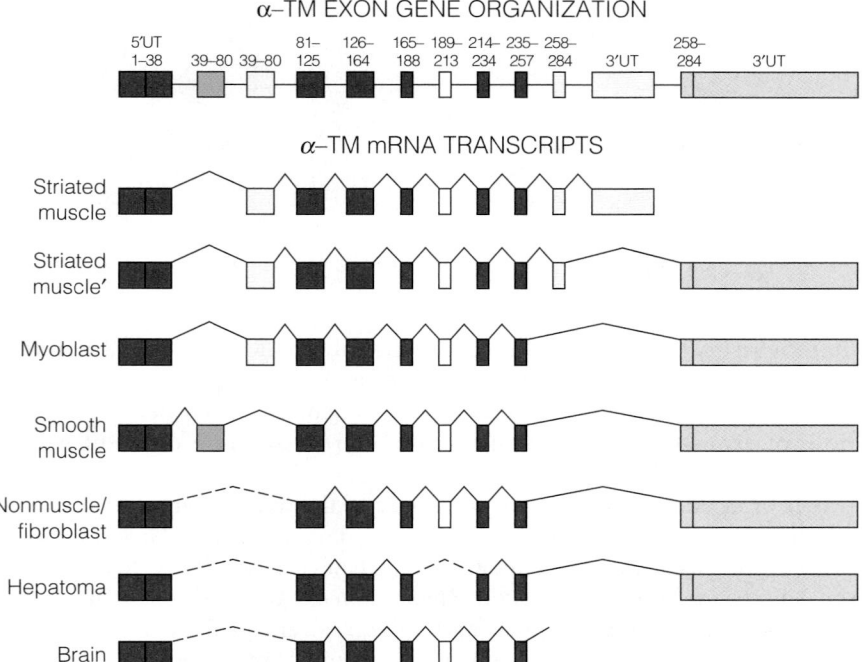

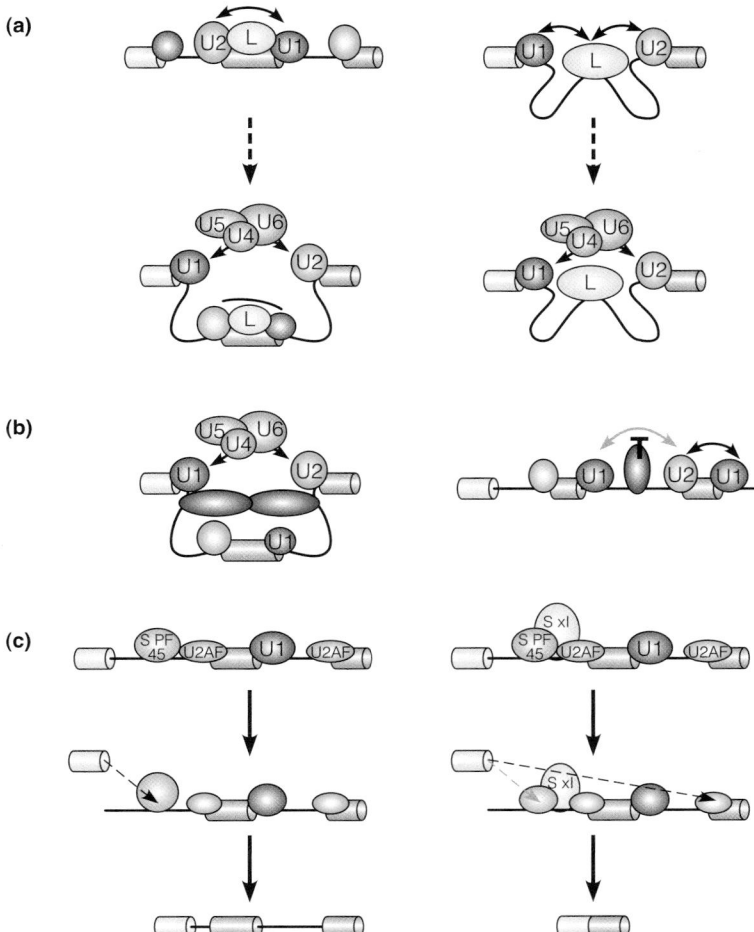

(a)

(b)

(c)

FIGURE 27.38

Possible mechanisms for alternative splicing by splice-site selection. For details see text.

Journal of Biological Chemistry 283:1217–1221, A. E. House and K. W. Lynch, Regulation of alternative splicing: More than just the ABCs. Reprinted with permission. © 2008 The American Society for Biochemistry and Molecular Biology. All rights reserved.

SUMMARY

All RNA is synthesized by the template-dependent copying of one DNA strand within a gene, catalyzed by RNA polymerase. RNA polymerases use 5′-ribonucleoside triphosphates as substrates, and they transcribe in a 5′ → 3′ direction. Prokaryotes synthesize all RNA classes with one polymerase, while eukaryotic cells have different polymerases—I, II, and III—for synthesis of ribosomal, messenger, and transfer RNA species, respectively.

Strand selection and duplex unwinding and rewinding are carried out by RNA polymerase. The enzyme binds at a promoter site, by formation of specific DNA–protein contacts, largely involving the enzyme's σ subunit in bacteria and by a host of transcription factors and regulatory proteins in eukaryotes. Most transcription initiations are abortive, but after a productive initiation, involving different factors in pro- and eukaryotes, elongation continues. In bacteria this is carried out by the core polymerase, $\alpha_2\beta\beta'\omega$. Transcription is highly processive and is terminated by specific DNA sequences, sometimes in bacteria with the participation of ρ protein. Structural analyses have identified common features in prokaryotic and eukaryotic RNA polymerases, which have helped to reveal common mechanistic features.

Post-transcriptional RNA processing includes cutting of pre-rRNA transcripts, which encode large and small ribosomal RNA components, and trimming of tRNA precursors in both prokaryotic and eukaryotic cells with the aid of the ribozyme ribonuclease P, followed by modification of several nucleotides in each tRNA molecule and nontranscriptive addition of the 3′ terminal CCA sequence. Bacterial mRNAs undergo little if any post-transcriptional processing, whereas eukaryotic messages are extensively processed, with polyadenylylation at the 3′ end, capping at the 5′ end with an inverted and modified guanine nucleotide residue, and splicing throughout the gene. Splicing is carried out by small nuclear ribonucleoprotein particles, guided by complementary base sequence interactions between splice sites and base sequences in the small nuclear RNA components. Alternative splicing is a process that expands the information content of a genome by directing quite different mRNA splicing patterns in different tissues and at different developmental stages.

REFERENCES

RNA Polymerase Structure and Function

Bustamante, C., M. Guthold, X. Zhu, and G. Yang (1999) Facilitated target location on DNA by individual *Escherichia coli* RNA polymerase molecules observed with the scanning force microscope operating in liquid. *J. Biol. Chem.* 274:16665–16668. Direct visualization of molecules indicates how RNA polymerase moves along DNA to promoters.

Cook, D. N., D. Ma, N. G. Pon, and J. E. Hearst (1992) Dynamics of DNA supercoiling by transcription in *Escherichia coli*. *Proc. Natl. Acad. Sci. USA* 89:10603–10607. The unsettled question of whether the act of transcription per se overwinds template DNA is explored.

Cramer, P., and E. Arnold (2009) Proteins: How RNA polymerases work. *Curr. Opinion in Struc. Biol.* 19:680–682. The introduction to a special issue of this journal, which has several excellent reviews on RNA polymerase structure and mechanism.

Cramer, P., D. A. Bushnell, and R. D. Kornberg (2001) Structural basis of transcription: RNA polymerase II at 2.8 Å resolution. *Science* 292:1863–1876. This paper and a companion paper from the same laboratory described the first high-resolution structure of a multi-subunit RNA polymerase.

Gelles, J., and R. Landick (1998) RNA polymerase as a molecular motor. *Cell* 93:13–16. A minireview summarizing ways to understand mechanochemical properties of RNA polymerase.

Murakami, K. S., S. Masuda, and S. A. Darst (2002) Structural basis of transcription initiation: RNA polymerase holoenzyme at 4Å resolution. *Science* 296:1280–1290. *Taq* RNA polymerase, the first eubacterial RNA polymerase structure determination.

Struhl, K. (1999) Fundamentally different logic of gene regulation in eukaryotes and prokaryotes. *Cell* 98:1–4. A concise description of the distinctions in transcription between higher and lower organisms.

Werner, F. (2007) Structure and function of archaeal RNA polymerases. *Molec. Microbiol.* 65:1395–1404. The archaeal enzymes resemble the eukaryotic polymerases more than they do the bacterial enzymes.

Promoter Recognition and Initiation

Durniak, K. J., S. Bailey, and T. A. Steitz (2008) The structure of a transcribing T7 RNA polymerase in transition from initiation to elongation. *Science* 322:553–557. The single-subunit phage T7 RNA polymerase was the first to have a high-resolution structure presented.

Huang, X., D. Wang, D. R. Weiss, D. A. Bushnell, R. D. Kornberg, and M. Levitt (2010) RNA polymerase II trigger loop residues stabilize and position the incoming nucleotide (*sic.*) triphosphate in transcription. *Proc. Natl. Acad. Sci. USA* 107:15745–15750. Recent progress in pol II structure and function.

Ishihama, A. (2010) Prokaryotic genome regulation: Multifactor promoters, multitarget regulators and hierarchic networks. *FEMS Microbiol. Revs.* 34:628–645. A systems biology approach to prokaryotic transcriptional regulation.

Ju, B-G., V. V. Lunyak, V. Perissi, I. Garcia-Bassets, D. W. Rose, C. K. Glass, and M. G. Rosenfeld (2006) A topoisomerase IIβ-mediated dsDNA break required for regulated transcription. *Science* 312:1798–1802. A topoisomerase requirement for transcription is shown to involve breaking the DNA template.

Kostrewa, D., M. E. Zeller, K-J. Armache, M. Seizl, K. Leike, M. Thomm, and P. Cramer (2009) RNA polymerase II-TFIIB structure and mechanism of transcription initiation. *Nature* 462:323–330. Structural insights into initiation.

McKenna, N. J., and B. W. O'Malley (2010) SnapShot: Nuclear receptors II. *Cell* 142:986–987. A microreview about nuclear receptors, an important family of transcription factors.

Revyakin, A., C. Liu, R. H. Ebright, and T. R. Strick (2006) Abortive initiation and productive initiation by RNA polymerase involve DNA scrunching. *Science* 314:1139–1147. Single-molecule approaches to the mechanism of abortive initiation.

Transcriptional Elongation

Buratkowski, S. (2008) Gene expression: Where to start? *Science* 322:1804–1805. A short introduction to four papers in this issue of *Science* reporting that RNA polymerase II catalyzes divergent transcription, giving normal transcripts plus short upstream antisense transcripts.

Buratkowski, S. (2009) Progression through the RNA polymerase II CTD cycle. *Mol. Cell* 36:541–546. A minireview dealing with phosphorylation of the C-terminus of pol II.

Larson, M. H., R. Landick, and S. M. Block (2011) Single-molecule studies of RNA polymerase: One singular sensation, every little step it takes. *Mol. Cell* 41:249–262. A recent review emphasizing the irregularity of many steps in transcription.

Pomerantz, R., and M. O'Donnell (2010) What happens when replication and transcription complexes collide? *Cell Cycle* 9:2537–2543. Replication and transcription apparati move in opposite directions along the same genome.

Thomsen, N. D., and J. M. Berger (2009) Running in reverse: The structural basis for translocation polarity in hexameric helicases. *Cell* 139:523–534. Structure and mechanism of *E. coli* ρ protein.

Vassylyev, D. G., M. N. Vassylyeva, A. Perederina, T. H. Tahirov, and I. Artsimovitch (2007) Structural basis for transcription elongation by bacterial RNA polymerase. *Nature* 448:157–162. A mechanistic analysis based upon a 2.5 Å structure of the *Thermus thermophilus* enzyme.

von Hippel, P. H. (1998) An integrated model of the transcription complex in elongation, termination, and editing. *Science* 281:660–665. Reviews events in transcription from a largely thermodynamic perspective.

Zenkin, N., Y. Yuzenkova, and K. Severinov (2006) Transcript-assisted transcriptional proofreading. *Science* 313:518–520. Insights into RNA polymerase fidelity mechanisms.

Pausing and Termination

Churchman, L. S., and J. S. Weissman (2011) Nascent transcript sequencing visualizes transcription at nucleotide resolution. *Nature* 469:368–375. Deep sequencing of transcript ends shows how extensive pausing is early in transcription.

Greenblatt, J., J. R. Nodwell, and S. W. Mason (1993) Transcriptional antitermination. *Nature* 364:401–406. A process discovered in phage λ, which has significance for eukaryotic and HIV gene expression.

Herbert, K. M., A. La Porta, B. J. Wong, R. A. Mooney, K. C. Neuman, R. Landick, and S. M. Block (2006) Sequence-resolved detection of pausing by single RNA polymerase molecules. *Cell* 125:1083–1094. Single-molecule approaches to understanding pausing.

Landick, R. (1999) Shifting RNA polymerase into overdrive. *Science* 284:598–599. A brief commentary, reviewing recent work on the mechanism of antitermination.

Park, J-S., and J. W. Roberts (2006) Role of DNA bubble rewinding in enzymatic transcription termination. *Proc. Natl. Acad. Sci. USA* 103:4870–4875. Evidence that rewinding the DNA helix upstream from the polymerase catalytic site contributes to termination.

Transcription, Chromatin, and the Cellular Milieu

Cook, P. R. (2010) A model for all genomes: The role of transcription factories. *J. Mol. Biol.* 395:1–10. A model of chromatin loops tethered to multipolymerase transcription factories.

Hodges, C., L. Bintu, L. Lubkowska, M. Kashlev, and C. Bustamante (2009) Nucleosomal fluctuations govern the transcription dynamics of RNA polymerase II. *Science* 325:626–628. Single-molecule studies support the concept of pol II nudging its way around the nucleosome.

Jørgensen, F. G., and M. H. Schierup (2009) Increased rate of human mutation where DNA and RNA molecules collide. *Trends Genet.* 25:523–527. One of several papers that explore the relationships between DNA and RNA polymerases acting on the same template.

Lorch, Y., B. Maier-Davis, and R. D. Kornberg (2006) Chromatin remodeling by nucleosome disassembly in vitro. *Proc. Natl. Acad. Sci. USA* 103:3090–3093. Evidence that a chromatin remodeling complex takes nucleosomes apart.

Venters, B. J., and B. F. Pugh (2009) *Crit. Rev. Biochem. Mol. Biol.* 44:117–141. How eukaryotic genes are transcribed. An excellent review, focusing upon the problem of chromatin transcription.

Post-transcriptional Processing

Altman, S., L. Kirsebom, and S. Talbot (1993) Recent studies of ribonuclease-P. *FASEB J.* 7:7–14. A discussion of one of the most interesting known ribozymes.

Chen, M., and J. L. Manley (2009) Mechanisms of alternative splicing regulation: Insights from molecular and genomics approaches. *Nature Rev. Mol. Cell Biol.* 10:741–753. A recent review of both splicing and alternative splicing.

Hoskins, A. A., and nine coauthors (2011) Ordered and dynamic assembly of single spliceosomes. *Science* 331:1289–1295. Single-molecule analysis of spliceosome assembly in living yeast cells carried out with a novel imaging technique.

Le Hir, H., A. Nott, and M. J. Moore (2003) How introns influence and enhance eukaryotic gene expression. *Trends Biochem. Sci.* 28:215–220. A summary of evidence indicating that intron processing enhances steps in gene expression.

Phizicky, E. M., and A. K. Hopper (2010) tRNA biology charges to the front. *Genes and Development* 24:1832–1860. Timely review on biochemistry and functions of tRNA post-transcriptional processing.

PROBLEMS

1. Outline an experimental approach to determining the average chain growth rate for transcription in vivo. Chain growth rate is the number of nucleotides polymerized per minute per RNA chain.

2. Outline an experimental approach to demonstrate the average RNA chain growth rate during transcription of a cloned gene in vitro.

3. Measurements of RNA chain growth rates are often led astray by the phenomenon of *pausing,* in which an RNA polymerase molecule stops transcription when it reaches certain sites, for intervals that may be as long as several minutes. How might pausing be detected?

4. Suppose you want to study the transcription in vitro of one particular gene in a DNA molecule that contains several genes and promoters. Without adding specific regulatory proteins, how might you stimulate transcription from the gene of interest relative to the transcription of the other genes on your DNA template? To make all of the complexes identical, you would like to arrest all transcriptional events at the same position on the DNA template before isolating the complex. How might you do this?

5. The *tac* promoter, an artificial promoter made from a chemically synthesized oligonucleotide, has been introduced into a plasmid. It is a hybrid of the *lac* and *trp* promoters, containing the −35 region of one and the −10 region of the other. This promoter directs transcription initiation more efficiently than either the *trp* or *lac* promoters. Why?

6. Would you expect actinomycin D to be a competitive inhibitor of RNA polymerase? What about cordycepin? Briefly explain your answers.

7. A restriction fragment was subjected to Maxam–Gilbert sequencing, with results as shown in the first four lanes of the radioautogram. S_1 nuclease mapping was carried out, for a gene whose transcription initiated within this sequence (see Tools of Biochemistry 27B). The transcript protected a fragment whose length was as shown in the fifth lane.

(a) Give the nucleotide sequence of both DNA strands and the first several RNA nucleotides. Identify all 3′ and 5′ ends.

(b) On your structure, show the approximate location of the Pribnow box (the −10 region).

(c) Assuming that the restriction fragment was created with a type II restriction endonuclease that recognizes a 6-base-pair site, show the structure of that site, and indicate where cleavage occurs.

8. Explain the basis for the following statement: Transcription of two genes on a plasmid can occur without the concomitant action of a topoisomerase, but only if those two genes are oriented in opposite directions.

9. Some years ago, it was suggested that the function of the poly(A) tail on a eukaryotic message may be to "ticket" the message. That is, each time the message is used, one or more residues is removed, and the message is degraded after the tail is shortened below a critical length. Suggest an experiment to test this hypothesis.

*10. It has been proposed that nucleosomes must be removed in order for transcription to proceed through chromatin. Suggest an experiment that might test this hypothesis.

11. Shown below is an R loop prepared for electron microscopy by annealing a purified eukaryotic messenger RNA with DNA from a genomic clone containing the full-length gene corresponding to the mRNA.

(a) How many exons does the gene contain? How many introns?

(b) Where in this structure would you expect to find a 5′,5′-internucleotide bond? Where would you expect to find a polyadenylic acid sequence?

12. Introns in eukaryotic protein-coding genes are rarely shorter than 65 nucleotides in length. What might be a rationale for this limitation?

13. Heparin is a polyanionic polysaccharide that blocks elongation by RNA polymerase. But heparin inhibits only when added before the onset of transcription and not if added after transcription begins. Explain this difference.

14. Estimate the time needed for the *E. coli* RNA polymerase at 37 °C to transcribe the entire gene for a 50-kilodalton protein. What assumption or assumptions must be made for this estimate to be accurate?

TOOLS OF BIOCHEMISTRY 27A

FOOTPRINTING: IDENTIFYING PROTEIN-BINDING SITES ON DNA

Transcription is controlled in large part through interactions of proteins with specific sites on DNA molecules, including operators and promoters. Such sites were initially identified through genetic analysis in the lactose system and phage λ. Biochemical analysis requires the identification and nucleotide sequence determination of a site, along with structural determination of the DNA–protein complexes that form. A technique called **footprinting**, usually involving protection against DNase I cleavage, is widely used to identify such sites. Footprinting can identify any DNA site that binds a protein specifically, as long as the protein binds sufficiently tightly.

The principle of the method, outlined in Figure 27A.1, is that binding of a protein to a specific DNA sequence should protect the protein-bound DNA from attack by DNase I (pancreatic deoxyribonuclease). The investigator first uses γ-[³²P]ATP and T4 polynucleotide kinase to prepare a 5′ end–labeled fragment of DNA containing a protein-binding site. One aliquot of the end-labeled DNA is mixed with the protein under study (step 1) and then incubated with DNase I under conditions in which most chains are cleaved only once (step 2). Another aliquot is incubated with DNase I under identical conditions, except that the protein is absent (step 2). Next, the two incubation mixtures are analyzed in adjacent lanes of a sequencing gel (step 3). The result is a ladder of fragments similar to that seen in a sequencing gel, except that there is little sequence specificity; the bands on the gel are spaced uniformly at one-nucleotide intervals, although band intensity varies considerably. Any sites protected from DNase attack because of interaction with

the DNA-binding protein under study yield either no band or a low-intensity band in the ladder from the DNA–protein complex, indicating that little or no cleavage occurred at that site. The blank region on the gel pattern from the DNA–protein complex (the "footprint") identifies the location and the size of the fragment in contact with the DNA-binding protein (such as RNA polymerase).

Recent improvements in footprinting technology involve cleavage with chemical agents such as **methidiumpropyl-EDTA-Fe²⁺(MPE — Fe²⁺).**

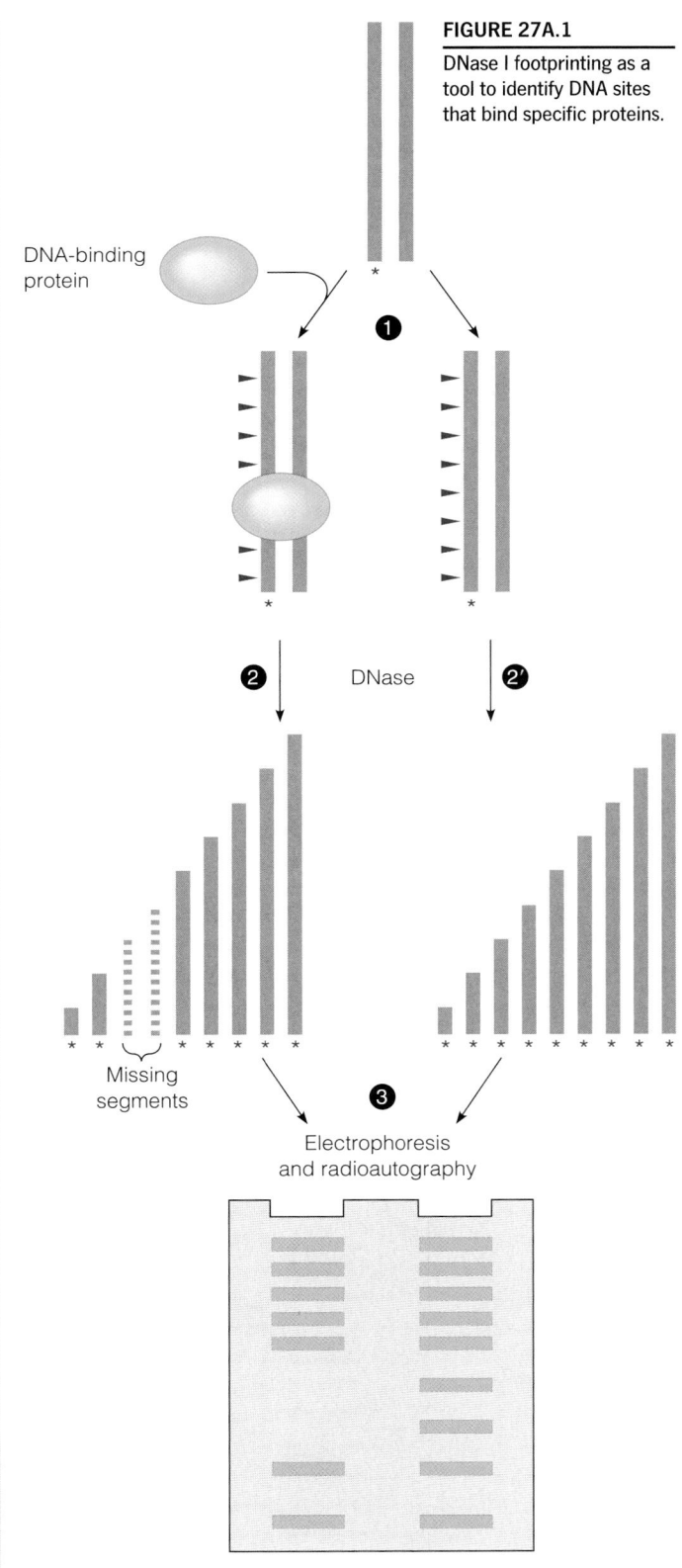

FIGURE 27A.1

DNase I footprinting as a tool to identify DNA sites that bind specific proteins.

DNA-binding protein

DNase

Missing segments

Electrophoresis and radioautography

This compound intercalates between DNA bases, as does ethidium bromide, and catalyzes oxidation leading to cleavage at a nearby site. Because there is some sequence selectivity in DNase I attack and virtually none with MPE — Fe^{2+}, the latter technique gives cleaner footprints. A related technique, described in the reference section following, generates hydroxyl radicals as a reactive but nonspecific DNA cleavage reagent.

Footprinting shows that *E. coli* RNA polymerase binds to a region about 60 base pairs long, extending from about 40 nucleotides upstream of (5′ to) the transcriptional start site to about 20 nucleotides past that site. In other words, the binding site extends from nucleotide −40 to +20, where the template for the first nucleotide in the transcript is +1. That first nucleotide, the 5′ end of the transcript, can be identified by various methods. A widely used technique is S$_1$ nuclease mapping, which is described in Tools of Biochemistry 27B.

Reference

Tullius, T. D., B. A. Dombroski, M. E. A. Churchill, and L. Kam (1989) Hydroxyl radical footprinting: A high-resolution method for mapping protein–DNA contacts. In *Recombinant DNA Methodology*, R. Wu, L. Grossman, and K. Moldave, eds., pp. 721–741. Academic Press, San Diego, Calif. A description of hydroxyl radical footprinting, with references to previously described methods.

TOOLS OF BIOCHEMISTRY 27B

MAPPING TRANSCRIPTIONAL START POINTS

Studies of transcriptional initiation and its control require methods for the accurate identification of transcriptional start points, specifically the DNA template nucleotide that encodes the 5′ nucleotide of the transcript. The low abundance of specific mRNAs and the high turnover of nearly all bacterial mRNAs make this a challenging task.

Prokaryotic transcripts all have a 5′ triphosphate terminus on the first nucleotide, as do eukaryotic transcripts before processing; in principle these could provide a handle for their identification. Given that this identification requires purification of the transcript, less laborious methods are preferable. One such method, **S₁ nuclease mapping**, uses the fungal enzyme S_1 nuclease, which specifically and quantitatively cleaves single-strand DNA and RNA (Figure 27B.1). The necessary materials are the cloned gene and a restriction fragment thought to contain the template for the 5′ end of the transcript. The fragment is 5′ end–labeled, as in the Maxam–Gilbert sequencing procedure, and cleaved asymmetrically with another restriction enzyme so that only the template DNA strand is labeled. In Maxam–Gilbert sequencing, 5′ end–labeled DNA is subjected to chemical reagents that cleave in a base-specific fashion. One reagent cleaves at sites occupied by either A or G, one cleaves in a strictly G-dependent fashion, one cleaves at T and C sites, and one cleaves specifically at C. Thus, by treating four aliquots of the same DNA with the four reagents and displaying the cleavage fragments on a sequencing gel, one can read the sequence of the DNA, as you can see in Figure 27B.1. Partly because Maxam–Gilbert sequencing is not completely base-specific, it has been largely supplanted by dideoxy sequence analysis, which uses replicative chain terminators to generate base-specific 3′ termini (see Tools of Biochemistry 4B). However, because S₁ mapping uses a 5′ end–labeled DNA fragment, it is useful to employ the Maxam–Gilbert approach, which also uses a 5′ end–labeled DNA fragment, to locate the 5′ end of the transcript precisely.

Next, the 5′ end–labeled DNA fragment is denatured and hybridized to mRNA, under conditions (high formamide) that favor the formation of DNA–RNA hybrids over DNA–DNA duplexes (Figure 27B.1, step 1). The only double-stranded nucleic acid, then, should be a DNA–RNA hybrid with 3′ single-stranded extensions from the 5′ end of the transcript and the labeled 5′ end of the restriction fragment. Treatment with S_1 nuclease (step 2) yields a fully duplex structure, with a labeled DNA strand whose length precisely measures the distance from the 5′ end of the transcript to the relevant restriction site. This distance can be identified by denaturing the DNA–RNA hybrid and running it on a sequencing gel, alongside a set of Maxam–Gilbert cleavage fragments (step 3).

The advent of PCR led to more reliable methods for isolating and identifying RNA 5′ ends. One such method is called rapid amplification of 5′ cDNA ends (5′-RACE). Partial sequence

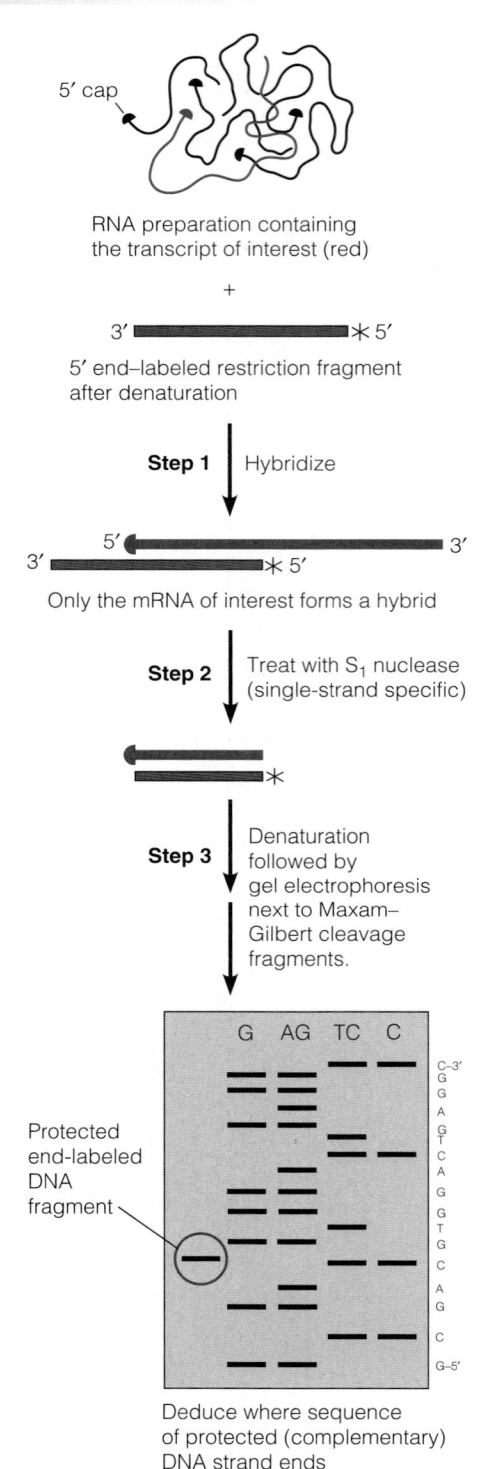

RNA preparation containing
the transcript of interest (red)

+

3′ ▬▬▬▬▬▬▬ ✳ 5′

5′ end–labeled restriction fragment
after denaturation

Step 1 | Hybridize

Step 2 | Treat with S₁ nuclease
(single-strand specific)

Step 3 | Denaturation
followed by
gel electrophoresis
next to Maxam–
Gilbert cleavage
fragments.

Protected
end-labeled
DNA
fragment

Deduce where sequence
of protected (complementary)
DNA strand ends

FIGURE 27B.1

S_1 nuclease mapping method for identifying the 5′ end of an RNA molecule.

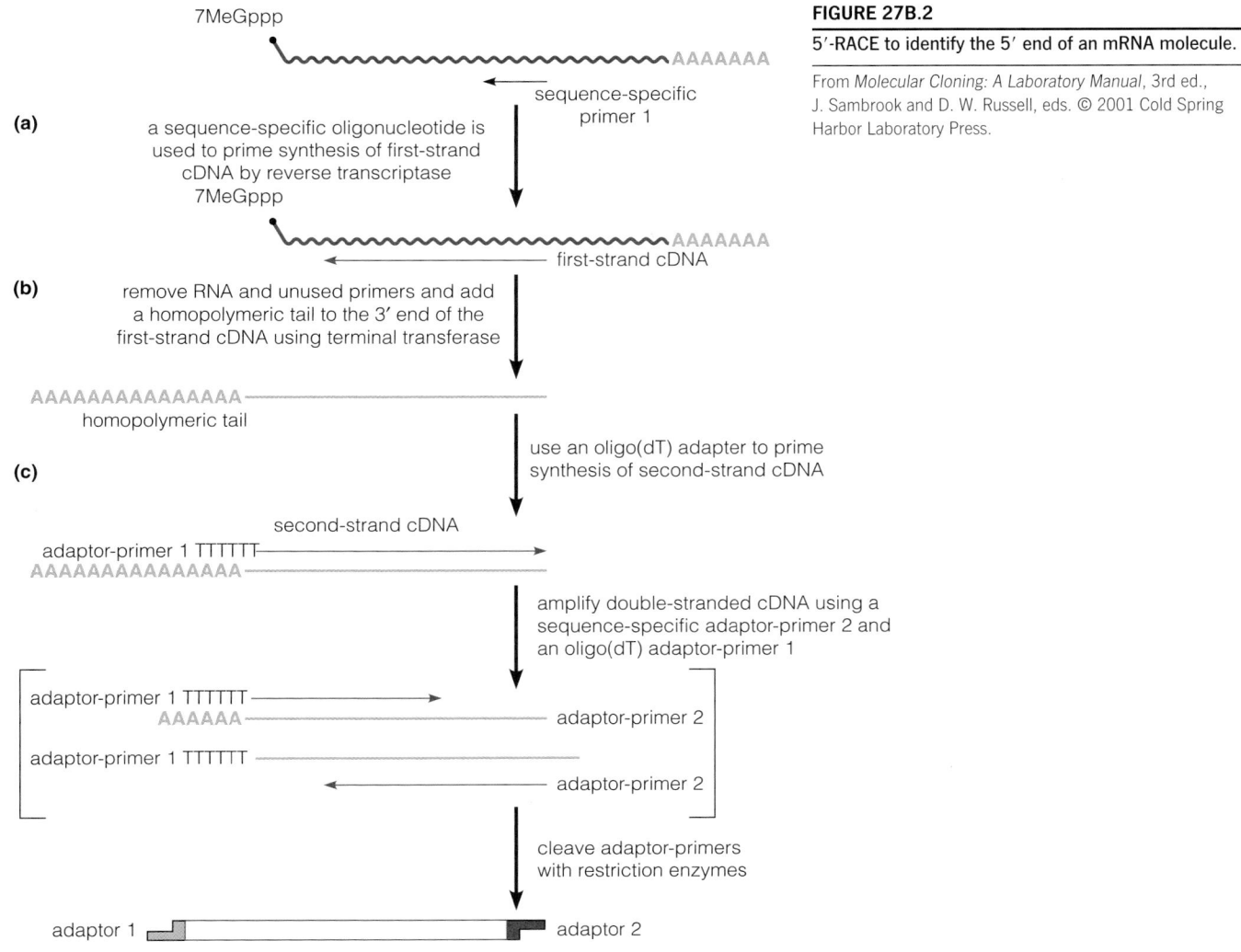

(a) a sequence-specific oligonucleotide is used to prime synthesis of first-strand cDNA by reverse transcriptase

sequence-specific primer 1

7MeGppp

first-strand cDNA

(b) remove RNA and unused primers and add a homopolymeric tail to the 3′ end of the first-strand cDNA using terminal transferase

homopolymeric tail

(c) use an oligo(dT) adapter to prime synthesis of second-strand cDNA

second-strand cDNA
adaptor-primer 1 TTTTTT

amplify double-stranded cDNA using a sequence-specific adaptor-primer 2 and an oligo(dT) adaptor-primer 1

adaptor-primer 1 TTTTTT
adaptor-primer 2
adaptor-primer 1 TTTTTT
adaptor-primer 2

cleave adaptor-primers with restriction enzymes

adaptor 1   adaptor 2

FIGURE 27B.2

5′-RACE to identify the 5′ end of an mRNA molecule.

From *Molecular Cloning: A Laboratory Manual*, 3rd ed., J. Sambrook and D. W. Russell, eds. © 2001 Cold Spring Harbor Laboratory Press.

information is needed to carry out this technique. As shown in Figure 27B.2, an oligonucleotide complementary to a region of the mRNA downstream from the 5′ end is annealed to the mRNA molecule. Reverse transcriptase is used to extend this primer to the 5′ end of the mRNA molecule. Excess primers are removed, and terminal deoxynucleotidyltransferase plus dATP adds a polyA tail to the 3′ end of this first-strand cDNA. Next, oligo(dT) is annealed to the polyA tail, and a heat-stable DNA polymerase extends this second-strand DNA to the 5′ end of the original

primer. Additional cycles of PCR follow until there is enough of the product to be cloned into a suitable vector for sequence analysis.

Reference

Sambrook, J., and D. W. Russell (2001) *Molecular Cloning: A Laboratory Manual*, Third Edition. Cold Spring Harbor Laboratory, Cold Spring Harbor, N.Y. Chapter 7 of this benchmark methods manual describes several techniques for RNA isolation and analysis.

TOOLS OF BIOCHEMISTRY 27C

DNA MICROARRAYS

The finding in the 1960s that single-stranded DNA binds irreversibly to membrane filters, and the development of recombinant DNA technology in the 1970s, led to a number of techniques for analysis of gene expression, that is, measuring levels of transcripts of particular genes in living cells. RNA could be radiolabeled in vivo and hybridized to gene-specific DNA—a cloned gene or a restriction fragment—and the bound radioactivity analyzed by radioautography or in a liquid scintillation counter.

Several techniques, such as northern analysis, were based upon these developments. However, such approaches allow the analysis of just one or a few genes in each experiment. With the availability of complete genome sequences, it became desirable to analyze levels of transcripts from many genes in a single experiment, that is, patterns of gene expression, that could be compared under different physiological conditions. Microarray technology allows such analysis.

FIGURE 27C.1

Flowchart: Performing a microarray experiment.

From *Molecular Cloning: A Laboratory Manual,* 3rd ed., J. Sambrook and D. W. Russell, eds. © 2001 Cold Spring Harbor Laboratory Press with permission from Vivek Mittal.

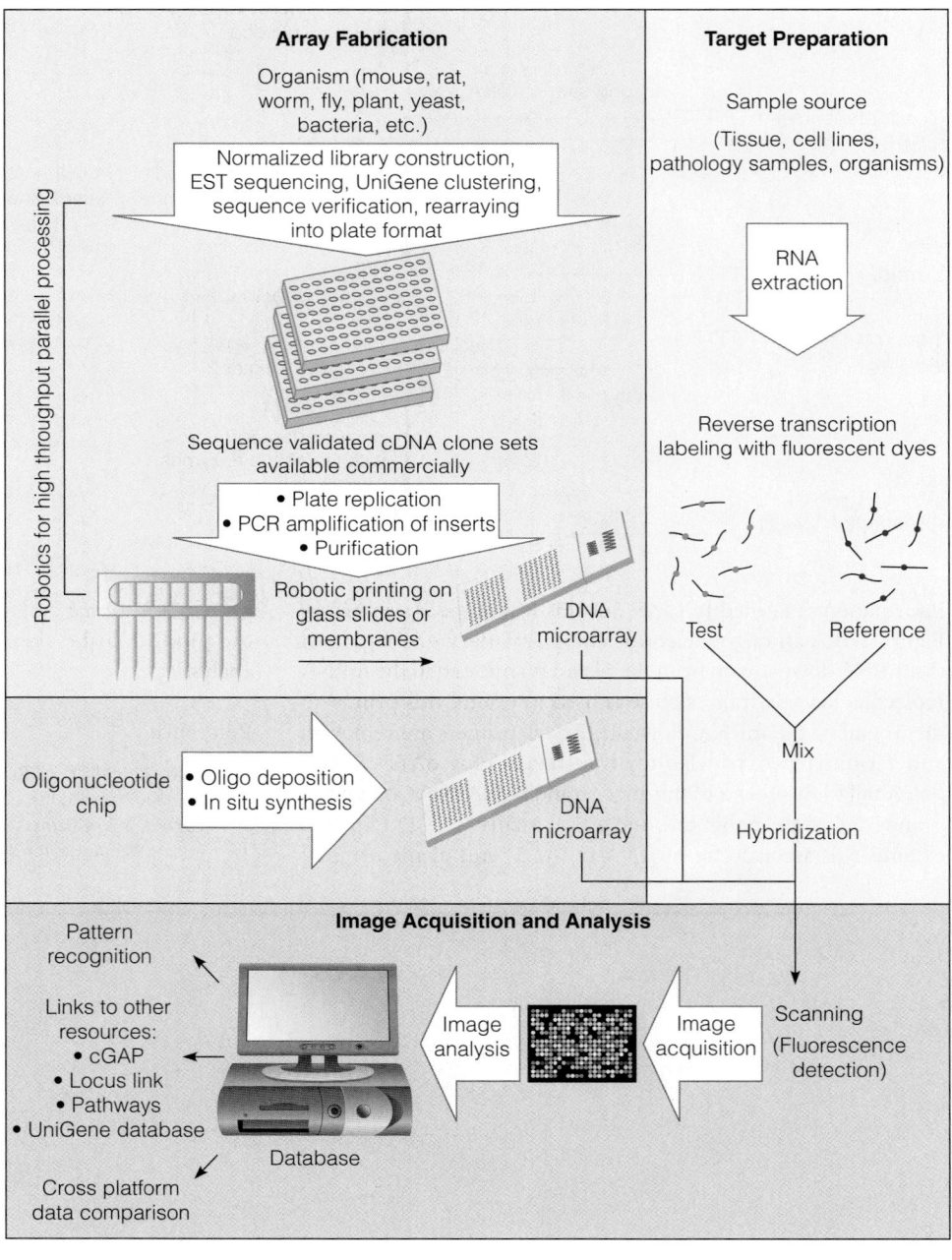

In a microarray experiment, minute amounts of gene-specific DNAs—usually several thousand—are immobilized on a substrate, such as glass or a membrane filter. The gene-specific DNAs are either cloned cDNAs or oligonucleotides. Using robotic technology, the investigator "prints" the DNAs onto the substrate, which may be a microscope slide, suitably coated to bind the applied DNAs. The DNAs are printed as a large array, which allows the investigator to identify each gene from its location on the array. The DNAs are fixed irreversibly on the substrate so that the "DNA chip" can be used repeatedly, by stripping the annealed RNA targets off the chip after each experiment.

Typically a microarray experiment involves comparison of gene expression profiles under different conditions—comparing a tumor with the tissue of origin, for example, or comparing a hormone-stimulated tissue with unstimulated tissue (see Figure 27C.1). The investigator wishes to learn which genes are activated under the conditions being analyzed, and which are repressed. Total mRNA is isolated from each tissue or cell culture and converted to a population of cDNAs by reverse transcriptase action. During the enzymatic synthesis of the cDNAs, one of the deoxyribonucleoside triphosphates is tagged with a fluorescent dye. Typically the reference sample is labeled with a red fluorophore and the test sample is labeled with a green fluorophore. After cDNA synthesis is complete, the two samples are mixed and subjected to annealing conditions in the presence of the microarray. Unhybridized cDNAs are washed off, and the array is then scanned. Scanning at wavelengths corresponding to emission maxima of the fluorophores reveals which transcripts are more abundant in the test than in the reference (more green fluorescence) and which are less abundant (more red fluorescence). Analysis of the image reveals which genes were stimulated and which repressed under the conditions being tested.

Microarray technology has numerous applications in addition to measuring patterns of gene expression. For example, by use of arrayed oligonucleotides representing different mutant forms of a gene of interest, one can carry out DNA–DNA hybridization on the gene chip and identify mutations or single-nucleotide polymorphisms in biological samples.

References

Ioannidis, J. P. A., and 15 coauthors (2009) Repeatability of published microarray gene expression analyses. *Nature Genetics* 41:149–155. Analyses of data in 18 articles emphasizes the importance of adequate controls, statistical analysis, and experimental details in generating and publishing meaningful microarray data.

Mittal, V. (2001) DNA array technology. In *Molecular Cloning: A Laboratory Manual,* 3rd ed., Vol. 3, J. Sambrook and D. W. Russell, eds., pp. A.10.1–A.10.19. Cold Spring Harbor Laboratory, Cold Spring Harbor, N.Y. Straightforward description of the technology.

TOOLS OF BIOCHEMISTRY 27D

CHROMATIN IMMUNOPRECIPITATION

So far the techniques we have discussed for identifying and characterizing binding sites for proteins such as RNA polymerase and repressors have involved single binding sites, whereas regulatory proteins such as nuclear hormone receptors act through a host of binding sites. Moreover, techniques such as footprinting are carried out in vitro, although our main interest is characterizing attachment sites for DNA-binding proteins in intact cells. Chromatin immunoprecipitation allows for identification of in vivo binding sites on a genome-wide basis.

The principle of the technique is that any DNA-binding protein can be covalently attached to its DNA-binding site(s) in vivo by use of a cross-linking reagent that can penetrate cell membranes and react covalently with both protein and DNA in a reversible manner. Formaldehyde is most commonly used, as shown in Figure 27D.1. After formaldehyde treatment of whole cells, chromatin is isolated and subjected to sonication under conditions reducing the length of each DNA molecule to fragments several hundred base pairs in length. The mixture is treated with antibody to the protein of interest, and the immunoprecipitated DNA–protein complexes collected. At this point the cross-links are broken, and the DNA that was precipitated along with the protein is subjected to sequence analysis. Originally this was done most often by PCR amplification of the DNA followed by conventional sequence analysis. But this approach allows analysis only of known and suspected DNA sequences. An alternate approach involves cloning all DNA fragments in the mixture and then using PCR primers corresponding to flanking sequences on the vector, followed by sequence analysis of each clone. With

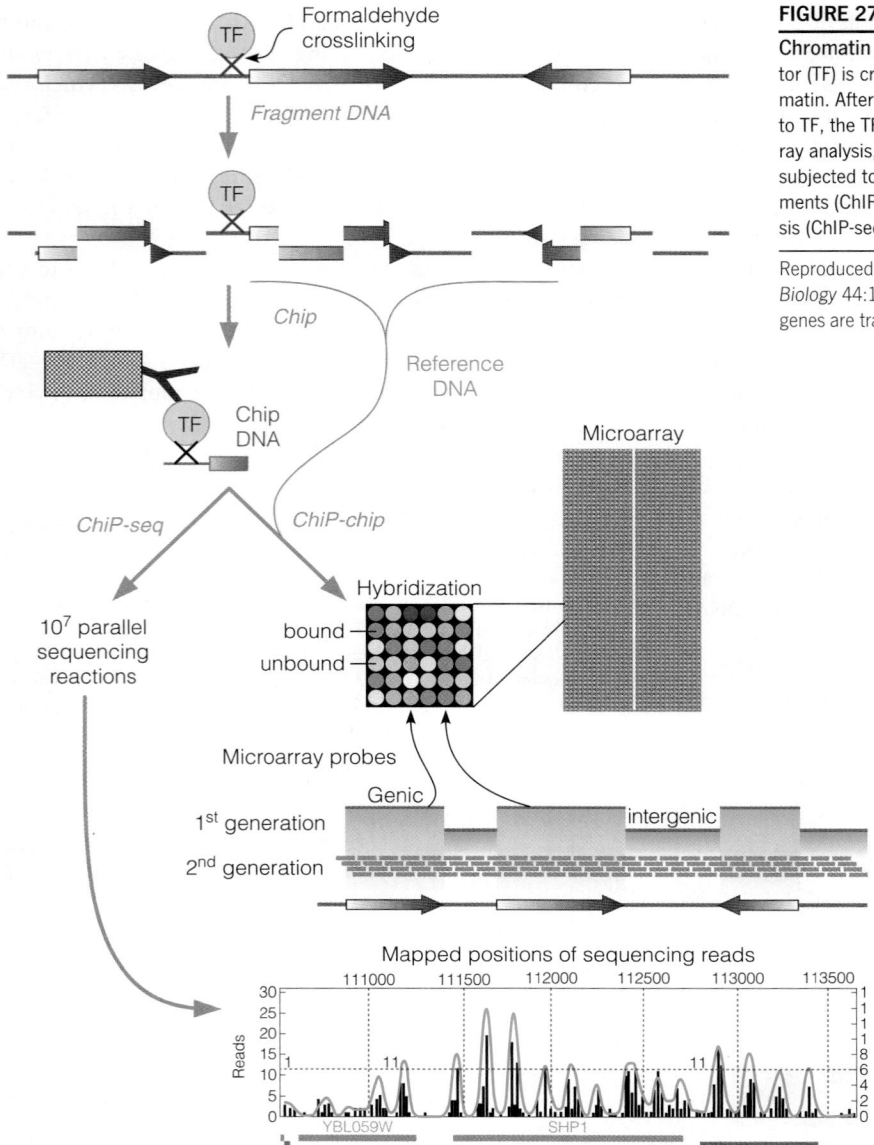

FIGURE 27D.1

Chromatin immunoprecipitation. In this figure a transcription factor (TF) is cross-linked to the DNA sites to which it binds in chromatin. After fragmentation and immunoprecipitation using antibody to TF, the TF-bound DNA fragments are identified either by microarray analysis, in which the DNA is bound to a red fluorophore and subjected to hybridization analysis with an array of genomic fragments (ChIP-chip), or subjected to massive parallel sequence analysis (ChIP-seq).

Reproduced from *Critical Reviews in Biochemistry and Molecular Biology* 44:117–141, B. J. Venters and B. F. Pugh, How eukaryotic genes are transcribed. © 2009 Informa Healthcare.

the advent of microarray technology it became possible to screen the DNA fragments against a DNA microarray containing hundreds or thousands of DNA sequences. This technique is called ChIP-chip because the immunoprecipitated DNA fragments are identified on a gene chip. A still more recent innovation, called ChIp-seq, involves parallel sequencing of all of the immunoprecipitated DNA using a next-generation sequencing technology that permits simultaneous sequence analysis of hundreds or thousands of DNA molecules.

References

Dedon, P. C., J. A. Soults, C. D. Allis, and M. A. Gorovsky, (1991) A simplified formaldehyde fixation and immunoprecipitation technique for studying protein-DNA interactions. *Anal. Biochem.*197:83–90. Some basic principles.

Orlando, V., H. Strutt, and R. Paro, (1997) Analysis of chromatin structure by in vivo formaldehyde cross-linking. *Methods* 11:205–214. A standard protocol.

Johnson D. S., A. Mortazavi, et al. (2007) Genome-wide mapping of in vivo protein-DNA interactions. *Science* 316:1497–1502. An early application of ChIP-seq. technology

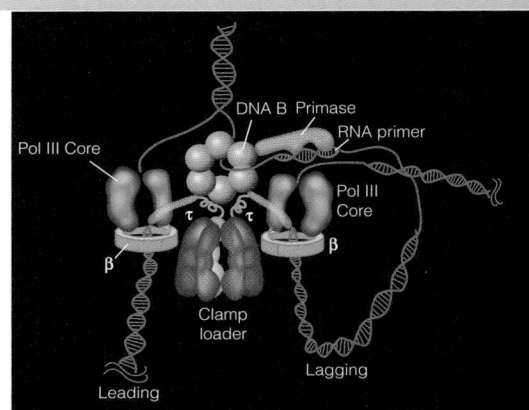

CHAPTER 28

Information Decoding: Translation and Post-translational Protein Processing

We turn now to the most complex process in biological information transfer—the decoding of genetic messages in the four-letter nucleic acid language to amino acid sequences in proteins, expressed in the 20-letter amino acid language. In DNA replication, transcription, and reverse transcription, information transfer is guided strictly by Watson–Crick base pairing between the template nucleic acid and the product, whether it be DNA or RNA. By contrast, when a messenger RNA sequence directs the synthesis of a specific protein, base sequence complementarity is still crucially involved, but a more complex overall process converts information encoded in a nucleotide sequence to information expressed as a specific amino acid sequence.

In terms of the number of components involved—ribosomal RNAs and proteins, transfer RNAs, amino acid activating enzymes, and soluble protein factors—and the number of different proteins in each cell—protein synthesis may well be the most complex of all metabolic processes, and it certainly involves the dominant fraction of a cell's metabolic effort. In a logarithmically growing bacterial cell, as much as 90% of the total metabolic effort may be devoted to protein biosynthesis, with the metabolic machinery for translation accounting for 35% of the cell's dry weight.

When we think of protein synthesis, we consider not only translation, which yields a specific amino acid sequence, but also post-translational processing and traffic, with each protein properly modified and transported to its ultimate intracellular or extracellular destination. We have already seen, for example, that protein processing involves cleavage, as in the conversion of preproinsulin to insulin, and modification of individual amino acids, as in the hydroxylation of proline residues in collagen synthesis or the phosphorylation of specific amino acid residues. We must also consider protein trafficking—how mature, or maturing, proteins are moved to their ultimate destinations. whether inside or outside the cell.

An Overview of Translation

The earliest evidence for the role of ribosomes in protein synthesis came from experiments in which radiolabeled amino acids were injected into rats, followed by isolation of the liver and fractionation of a liver homogenate. The label was shown to be incorporated earliest into ribosomes, either free or membrane-bound in the endoplasmic reticulum. This and related experiments established that the ribosome is the site of protein synthesis. Experiments with cell-free systems soon established the requirements for amino acid activation and for the involvement of small, stable RNAs (transfer RNAs).

In Chapters 4 and 5 we painted an introductory picture of translation. In Figure 4.23 (page 112) we showed how translation involves movement of a ribosome along an mRNA molecule, three nucleotides at a time, with each trinucleotide sequence in mRNA pairing with one specific transfer RNA charged with an amino acid, and with the polypeptide chain growing stepwise, one amino acid per step, from the N-terminus to the C-terminus. Figure 28.1 shows a somewhat

FIGURE 28.1

Translation of an RNA message into a protein. As the ribosome moves along the message, it accepts specific aminoacyl tRNAs in succession, selecting them by matching the trinucleotide anticodon on the tRNA to the trinucleotide codon on the RNA message **(step 1)**. The amino acid (in this example, the second one of the chain, Val) accepts the growing polypeptide chain (in this example the previously bound fMet) **(step 2)**, and the ribosome moves on to the next codon to repeat the process, while releasing the deacylated transfer RNA that held the growing peptide in the previous cycle (the tRNA for fMet, **step 3**). The preceding steps are repeated, adding more amino acids to the chain, until a stop signal is read **(step 4)**, whereupon a protein release factor causes both the polypeptide and the mRNA to be released. The polypeptide shown here is unrealistically short, to illustrate both initiation and termination.

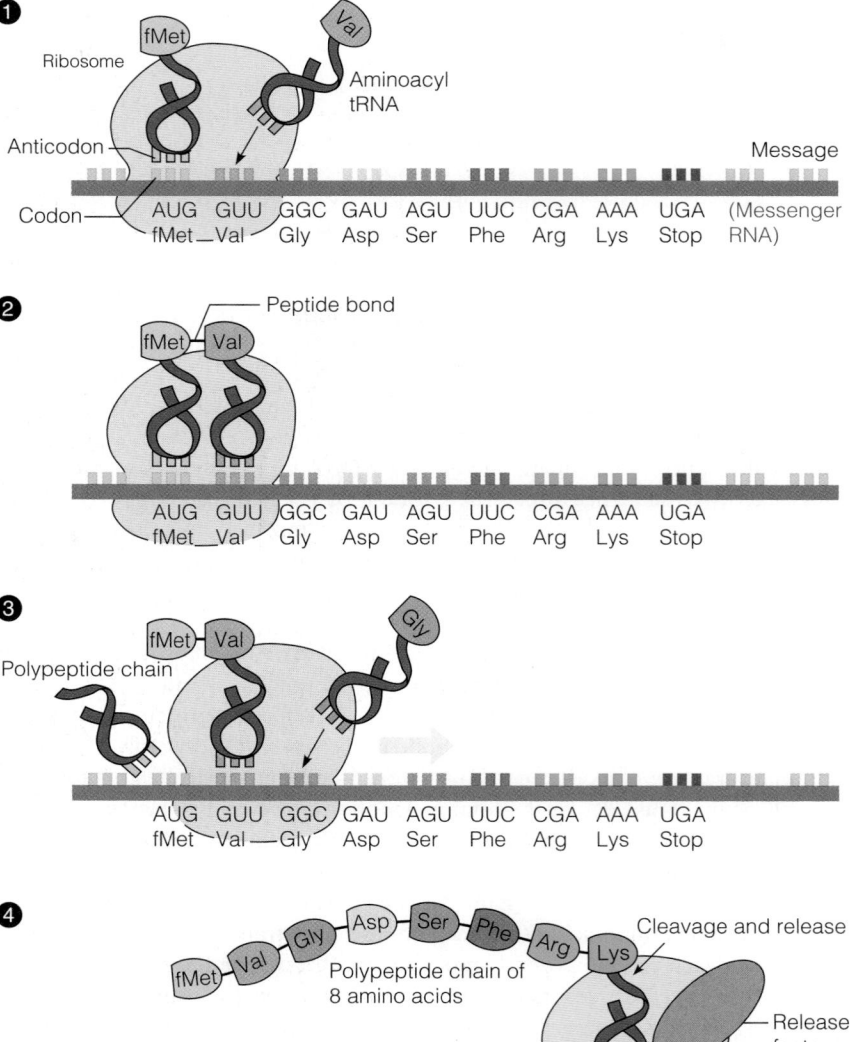

more detailed picture of this process and serves as an overview of translation. In Figure 5.18 (page 152) we showed the genetic code, that is, the correspondence between each of the 64 possible trinucleotide sequences and the 20 amino acids encoded by these sequences. In this chapter we expand upon both the process of protein synthesis and the elucidation and nature of the genetic code.

In 1958, several years before elucidation of the genetic code or the demonstration of messenger RNA, Francis Crick predicted the existence of *adaptor molecules*, each of which would function in translating the genetic message by binding to a specific amino acid and linking it to a molecular code word in the translation machinery. These adaptor molecules turned out to be transfer RNAs. As discussed in Chapter 4, each transfer RNA, or tRNA, molecule is 75–80 nucleotides in length (although some are as large as 93 nucleotides), folded by intramolecular hydrogen bonding into a three-loop structure. Each tRNA molecule is designed to bind one of the 20 amino acids, through the specificity of an amino acid–activating enzyme, more properly called an *aminoacyl-tRNA synthetase*. As shown in Figure 28.2, amino acid activation proceeds at the expense of ATP hydrolysis and results in esterification of the amino acid carboxyl group with the 3′ hydroxyl of the 3′ terminal tRNA nucleotide, yielding an aminoacyl-tRNA.

Each tRNA contains, in a region known as the **anticodon loop**, a trinucleotide sequence called the **anticodon** that is complementary to the appropriate trinucleotide codon in the message. Thus, the whole set of tRNAs contained in a cell composes a kind of molecular dictionary for the translation—it defines the correspondences between words in the four-letter nucleic acid language (gene sequence) and words in the 20-letter amino acid language (protein amino acid sequence).

The messenger RNA is bound to a ribosome, as shown in Figure 28.1. The aminoacyl tRNAs also bind here, one by one, matching their anticodons to the codons on the message, as shown in Figure 28.1, step 1. The growing peptide chain is transferred from the tRNA to which it is bound to the incoming aminoacyl-tRNA (step 2). The first tRNA is then released, and the ribosome moves one codon length along the message, allowing the next tRNA to come into place, carrying *its* amino acid (step 3). Again, expenditure of energy from high-energy phosphate hydrolysis is required at each step in the movement. As the ribosome moves along the messenger RNA, it eventually encounters a "stop" codon. At this point, the polypeptide chain is released. Step 4 shows a completed, although short, protein. In every cell, of every kind of organism, this remarkable machinery translates the information coded in thousands of different genes into thousands of different proteins. The cellular apparatus that binds all of these components and catalyzes the formation of peptide is the ribosome, a particle composed of both RNA and proteins. A ribosome can bind to mRNA and "read" it, as it moves along the RNA, accepting the charged tRNAs in the order dictated by the message and

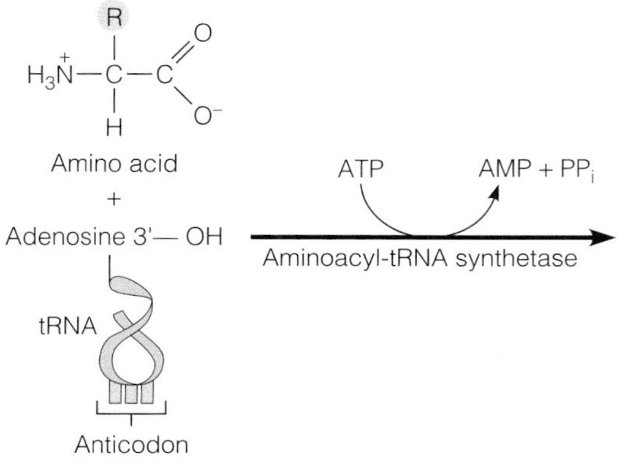

> Transfer RNAs are the adaptor molecules that match amino acid to codon.

FIGURE 28.2

Activation of amino acids for incorporation into proteins. A specific enzyme, aminoacyl-tRNA synthetase, recognizes both a particular amino acid and a tRNA carrying the corresponding anticodon. This synthetase catalyzes the formation of an aminoacyl tRNA, with accompanying hydrolysis of one ATP to AMP.

incorporating their amino acid residues one by one and in proper order within the growing polypeptide chain.

The mRNA message is always read in the $5' \rightarrow 3'$ direction, and the polypeptide chain is synthesized starting with its N-terminal residue. The direction of polypeptide synthesis was established in 1961 in a classic experiment. Howard Dintzis gave reticulocytes (hemoglobin-producing cells) a short pulse of [^{3}H] leucine and isolated the completed hemoglobin molecules at various times after the pulse. After cleaving these molecules into peptides with trypsin, he compared the radioactivity of peptides from various points in the chain. Immediately after the pulse label was added, radioactivity was seen only in chains that were undergoing synthesis before the pulse began and were just being completed during the pulse; the label at this time was found only in the C-terminal peptides. At longer times after the pulse, radioactivity was found to be incorporated into parts of the polypeptide closer and closer to the N-terminus, as synthesis of new protein molecules continued to be initiated. Dintzis therefore concluded that amino acids are added to a polypeptide chain starting at the N-terminus and working toward the C-terminus.

The simple picture of translation we have presented so far leaves a host of questions unanswered. How are tRNA and amino acid matched? How does the ribosome attach to the mRNA and move along it? How does it catalyze peptide bond formation? How does it start and stop translation correctly? How does it avoid making mistakes? Where does the energy for all of this activity come from? To answer such questions, we must dissect the whole process of translation, with careful examination of each of its parts. First, let us consider the genetic code in more detail.

The Genetic Code

We introduced the genetic code in Chapters 5 and 7. Here, we describe the key experiments that led to deciphering of the code, and we discuss some features of the code, including whether it is universal throughout biological systems.

By the late 1950s it was generally accepted that a protein's amino acid sequence is encoded by the sequence of bases in a nucleic acid template. A triplet code seemed most likely, with three nucleotides specifying one amino acid. Clearly, a doublet code wouldn't work because there are only 16 possible dinucleotide sequences (4×4), and we need at least 20 code words if each amino acid is to have its own code word. So a triplet code seemed the simplest; 64 possible trinucleotides ($4 \times 4 \times 4$) made it likely that some amino acids would have more than one code word.

Genetic experiments supported the idea of a triplet code and also a code that is *nonoverlapping* and *unpunctuated*. Figure 28.3 illustrates what we mean by these terms and suggests reasons why overlapping and punctuated codes were rejected.

How the Code Was Deciphered

Biochemical elucidation of the code began in 1961 by Marshall Nirenberg and Heinrich Matthaei, with their use of artificial RNA templates for in vitro protein synthesis. Recall from Chapter 27 (page 1129) that the enzyme polynucleotide phosphorylase will catalyze the nontemplate-dependent synthesis, from a mixture of ribonucleoside diphosphates, of a random-sequence RNA whose nucleotide composition matches that of the medium. Nirenberg and Matthaei polymerized UDP with the enzyme to synthesize polyU, a polyribonucleotide containing only UMP residues. When this artificial RNA was placed in a cell-free system containing a bacterial extract, ATP, GTP, and the 20 canonical amino acids (i.e., those commonly found in proteins), the product was a polypeptide containing only phenylalanine. Thus, the genetic code word for phenylalanine was shown to be a

Messenger RNA is read $5' \rightarrow 3'$. Polypeptide synthesis begins at the N-terminus.

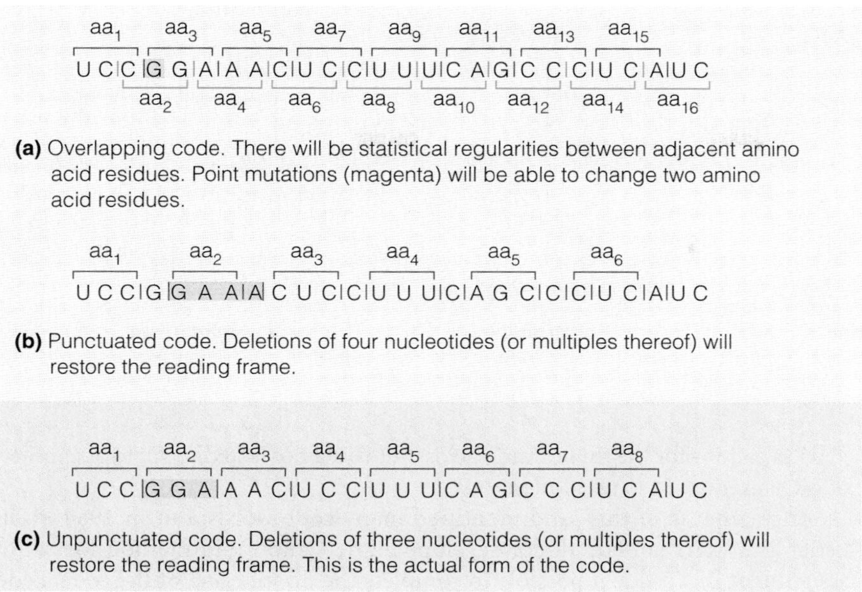

$aa_1 \quad aa_3 \quad aa_5 \quad aa_7 \quad aa_9 \quad aa_{11} \quad aa_{13} \quad aa_{15}$

U C|C |G| G|A|A A|C|U C|C|U U|U|C A|G|C C|C|U C|A|U C

$\quad aa_2 \quad aa_4 \quad aa_6 \quad aa_8 \quad aa_{10} \quad aa_{12} \quad aa_{14} \quad aa_{16}$

(a) Overlapping code. There will be statistical regularities between adjacent amino acid residues. Point mutations (magenta) will be able to change two amino acid residues.

$aa_1 \quad\quad aa_2 \quad\quad aa_3 \quad\quad aa_4 \quad\quad aa_5 \quad\quad aa_6$

U C C|G |G A A|A|C U C|C|U U U|C|A G C|C|C|U C|A|U C

(b) Punctuated code. Deletions of four nucleotides (or multiples thereof) will restore the reading frame.

$aa_1 \quad aa_2 \quad aa_3 \quad aa_4 \quad aa_5 \quad aa_6 \quad aa_7 \quad aa_8$

U C C|G G A|A A C|U C C|U U U|C A G|C C C|U C A|U C

(c) Unpunctuated code. Deletions of three nucleotides (or multiples thereof) will restore the reading frame. This is the actual form of the code.

FIGURE 28.3

Three conceivable kinds of genetic codes. Early research on the nature of the code quickly showed that a nonoverlapping, unpunctuated code **(c)** fit all experimental observations.

specific sequence of UMP residues—three if we are really dealing with a triplet code. In short order, polyC was shown to encode only proline and polyA only lysine.

Establishing code words for the other 17 amino acids was trickier. For example, consider the enzymatic synthesis of a polyribonucleotide from a nucleotide mixture containing CDP and ADP in a 5:1 molar ratio. Because the base composition of the polymer reflects the substrate ratio, the polymer contains eight trinucleotide codons, with CCC 125 times more abundant than AAA ($5 \times 5 \times 5$). Codons with two A and one C ($2A1C$ — AAC, ACA, CAA) are five-fold more abundant than AAA, and those with one A and two C ($2C1A$ — CCA, CAC, ACC) are 25-fold more abundant. When this polymer was used by Nirenberg and Matthaei, it stimulated the incorporation of proline, histidine, threonine, glutamine, asparagine, and lysine in molar ratios of 100, 23.4, 20, 3.3, 3.3, and 1, respectively. The data were best explained by assuming that the polymer contained two codons for Pro (CCC and $2C1A$) and two for Thr ($2A1C$ and $2C1A$). The codon for lysine was already known to be AAA. Codons assigned for Asp and Gln were both $2A1C$ and, for His, $1A2C$. This and other experiments with random-sequence polymers were able to establish the nucleotide *composition* of most codons, but not their *sequence*.

Two approaches led to the identification of codon sequences. First, H. Gobind Khorana synthesized polyribonucleotides of regular repeating sequence. For example, the polymer UCUCUCUC··· was shown to direct synthesis of an alternating copolymer, Ser-Leu-Ser-Leu-Ser-Leu ···. If the code is triplet and nonoverlapping, this means that UCU is the codon for either serine or leucine and CUC encodes the other amino acid. Because serine was known to have a $2U1C$ codon and leucine a $2C1U$ codon, this established UCU as a serine codon and CUC as a leucine codon. When a trinucleotide was used as the repeating unit, a different result was seen, as shown in Figure 28.4. The polymer AAGAAGAAG··· directed synthesis of three homopolypeptides—polylys, polyarg, and polyglu. This experiment didn't give codon sequences, but it did establish the triplet and nonoverlapping nature of the code. Here, the nature of the product was set by the initial **reading frame**—the trinucleotide sequence chosen for the first amino acid incorporation event. If GAA was selected, for example, every subsequent codon would also be GAA, making all amino acids in the product identical. The experiment did establish GAA, AGA, and

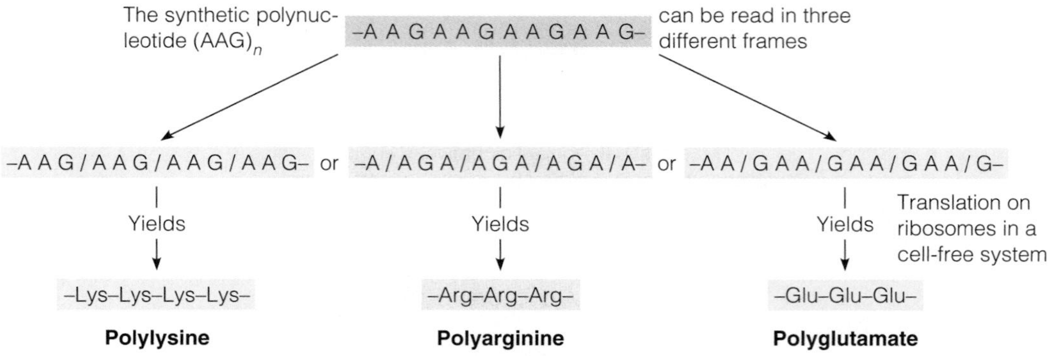

The synthetic polynucleotide (AAG)$_n$ —A A G A A G A A G A A G— can be read in three different frames

—A A G / A A G / A A G / A A G— or —A / A G A / A G A / A G A / A— or —A A / G A A / G A A / G A A / G—

Yields   Yields   Yields   Translation on ribosomes in a cell-free system

—Lys–Lys–Lys–Lys–   —Arg–Arg–Arg–   —Glu–Glu–Glu–

Polylysine   **Polyarginine**   **Polyglutamate**

FIGURE 28.4

Use of synthetic polynucleotides with repeating sequences to decipher the code. This example shows how polypeptides derived from the (AAG)$_n$ polymer were used to confirm the triplet code and help identify codons. The polymer (AAG)$_n$ can yield three different polypeptides, depending on which reading frame is employed.

The code is almost, but not quite, universal.

AAG as codons for the three amino acids, but could not lead directly to assignment of each amino acid to one codon.

Experiments of this kind identified many code words, but in 1964 Philip Leder and Marshall Nirenberg developed a new and rapid method for codon assignment that made it possible to complete the deciphering of the code. Leder and Nirenberg found that synthetic trinucleotides would bind to ribosomes and specify the binding of specific tRNAs. For example, UUU and UUC stimulated binding of phenylalanine tRNAs to ribosomes, and CCC and CCU stimulated binding of proline tRNA. Such experiments provided unequivocal evidence for the *redundancy* of the code because several different codons were found to correspond to a single amino acid. By the combined use of these techniques, the entire genetic code was established within a few years after demonstration of polyU-directed Phe incorporation.

Features of the Code

In the genetic code, as shown in Figure 28.5, 61 of the 64 trinucleotides are "sense" codons, that is, they code for one amino acid. The remaining three are normally "nonsense" codons in that they do not code for an amino acid (with some exceptions, see Table 28.1 and discussion on page 1180). When a ribosome encounters a nonsense codon (UAG, UAA, or UGA) in the correct reading frame, there is no aminoacyl-tRNA in the cell containing a matching anticodon, and translation ceases. As we see later, these codons are used as part of the normal machinery for terminating translation of a message. The code is *degenerate* (or redundant) in the sense that most amino acids have more than one codon and *unambiguous*, in the sense that a particular trinucleotide encodes one and only one amino acid. There are some exceptions to this generalization, as summarized in Table 28.1 and discussed later (see page 1180). In other words, the genetic code is almost, but not quite, universal.

Biological Validity of the Code

As described above, the assignment of genetic code words to amino acids was carried out strictly through the use of in vitro systems—amino acid incorporation directed by synthetic templates and assays of aminoacyl-tRNA binding to ribosomes. How could we be assured that these codon assignments are valid for translation of messages in living cells? Some of the validation came from amino acid sequence analysis of mutant human hemoglobins (Chapter 7). Most of the amino acid sequence changes could be accounted for by substitution mutations involving a single base, the most frequent spontaneous mutation. For example, the Glu → Val substitution seen with sickle-cell hemoglobin could be accounted for by changing a GAA Glu codon to a GUA Val codon, or GAG to GUG.

First position (5′ end)	Second position: U	C	A	G	Third position (3′ end)
U	UUU Phe	UCU Ser	UAU Tyr	UGU Cys	U
U	UUC Phe	UCC Ser	UAC Tyr	UGC Cys	C
U	UUA Leu	UCA Ser	UAA Stop	UGA Stop	A
U	UUG Leu	UCG Ser	UAG Stop	UGG Trp	G
C	CUU Leu	CCU Pro	CAU His	CGU Arg	U
C	CUC Leu	CCC Pro	CAC His	CGC Arg	C
C	CUA Leu	CCA Pro	CAA Gln	CGA Arg	A
C	CUG Leu	CCG Pro	CAG Gln	CGG Arg	G
A	AUU Ile	ACU Thr	AAU Asn	AGU Ser	U
A	AUC Ile	ACC Thr	AAC Asn	AGC Ser	C
A	AUA Ile	ACA Thr	AAA Lys	AGA Arg	A
A	AUG Met	ACG Thr	AAG Lys	AGG Arg	G
G	GUU Val	GCU Ala	GAU Asp	GGU Gly	U
G	GUC Val	GCC Ala	GAC Asp	GGC Gly	C
G	GUA Val	GCA Ala	GAA Glu	GGA Gly	A
G	GUG Val	GCG Ala	GAG Glu	GGG Gly	G

FIGURE 28.5

The genetic code (as written in RNA). We show here the genetic code as used in most organisms. Chain termination, or "stop," codons are shown in orange, and the usual start codon, AUG, is dark green. Other, rarely used, start codons are shown in light green. When AUG is used as a start codon, it codes for *N*-formylmethionine (fMet) in prokaryotes or methionine (Met) in eukaryotes; see page 1182. Otherwise, it codes for Met. Exceptions to these codon assignments are given in Table 28.1.

TABLE 28.1 Modifications of the genetic code

Codon	Usual Use	Alternate Use	Where Alternate Use Occurs
AGA AGG	Arg	Stop, Ser	Some animal mitochondria, some protozoans
AUA	Ile	Met	Mitochondria
CGG	Arg	Trp	Plant mitochondria
CUU CUC CUA CUG	Leu	Thr	Yeast mitochondria
AUU GUG UUG	Ile Val Leu	Start (*N*-fMet)	Some prokaryotes[a]
UAA	Stop	Glu	Some protozoans
UAG	Stop	Pyrrolysine Glu	Various archaea Some protozoans
UGA	Stop	Trp Selenocysteine Selenocysteine and Cys	Mitochondria, mycoplasmas Widespread[a] *Euplotes*

[a]Depends on context of message, other factors.

Proflavin

Selenocysteine

Pyrrolysine　　**Methionine sulfoxide**

The code is redundant. Several codons may correspond to a single amino acid, sometimes via wobble in the 5′ anticodon position.

Other important validation experiments were carried out by George Streisinger, using the T4 bacteriophage lysozyme system. Lysozyme is a phage-coded enzyme responsible for rupture of the host bacterium after a cycle of phage reproduction. Lysozyme mutants are easy to detect because they can produce phage but cannot lyse the host cell. Mutations were induced by treating phage-infected cells with **proflavin**, a large planar molecule that can fit, or *intercalate*, between successive base pairs in DNA and induce *frameshift* mutations, that is, additions or deletions of a single base, which alter the reading frame (Figure 7.26, page 257). In one such frameshift mutation, wild-type function was restored by a second mutation. Sequence analysis of the double-mutant lysozyme showed five changes from the wild-type sequence. The data were consistent only with the assumption that the double mutant was created by a single-base insertion and a single-base deletion, which restored the reading frame (Figure 28.6). All of the codons used experimentally to infer the nucleic acid sequence (long before DNA sequencing had been developed) were consistent with the codon assignments as determined in vitro.

Deviations from the Genetic Code

Why has the genetic code remained almost unchanged over so vast an evolutionary span? Perhaps it is simply because even small codon changes could be devastating. A single codon change could alter the sequence of nearly every protein made by the organism. Some of these changes would almost certainly have lethal effects. Therefore, codon changes have been opposed by intense selective pressure during evolution. They represent changes in the most basic rules of the game.

Yet significant deviations do occur, most notably, differences in the mitochondrial code and coding for the "21st and 22nd" amino acids, namely, selenocysteine and pyrrolysine (Chapter 5, page 144). A significant change in the mitochondrial code, as shown in Table 28.1, is the change in AUA from an isoleucine to a methionine codon. It has been argued that this represents an adaptation within mitochondria to oxidative stress. Methionine, whether free or as a residue in a protein, is readily oxidized, but just as readily reduced by methionine sulfoxide reductase. Hence, it is argued that it is advantageous for a mitochondrion to have increased methionine abundance in its proteins, to absorb reactive oxygen species that otherwise would be attacking less resilient targets. Indeed, mitochondrial proteins do have methionine in higher abundance than do proteins from other cell compartments.

Selenocysteine (21st amino acid) and pyrrolysine (22nd amino acid) are translated differently. Both use codons that are otherwise used in translation termination—UGA for selenocysteine (Sec) and UAG for pyrrolysine (Pyl). A special transfer RNA, tRNA^{Sec}, is a substrate for a seryl-tRNA^{Sec} synthetase, which charges serine directly, to give Ser-tRNA^{Sec}. (Note the convention; Ser refers to the amino acid bound and superscript Sec denotes the amino acid corresponding to the anticodon on that tRNA molecule.) The tRNA-linked Ser is then converted to Sec by a two-step process beginning with phosphorylation of the serine hydroxyl group. The resultant Sec-tRNA^{Sec} responds to a UGA codon. For a particular

FIGURE 28.6

Validation of the genetic code by amino acid sequence analysis of T4 phage lysozyme mutants. *e* is the gene for lysozyme, a portion of whose amino acid sequence is shown. One of two proflavin-induced mutations, either eJ42 or eJ44, disrupted the reading frame by deleting one base pair, and a second mutation restored the wild-type reading frame by inserting a base pair, but altering the amino acid sequence between the two mutant sites. The mRNA sequences encoding these five altered amino acids are inferred from the genetic code and the known action of the mutagen.

Adapted with permission from Eric Terzaghi from *Proceedings of the National Academy of Sciences of the United States of America* 56:500-507, E. Terzaghi, Y. Okada, G. Streisinger, J. Emrich, M. Inouye, and A. Tsugita, Change of a sequence of amino acids in phage T4 lysozyme by acridine-induced mutations, 1966.

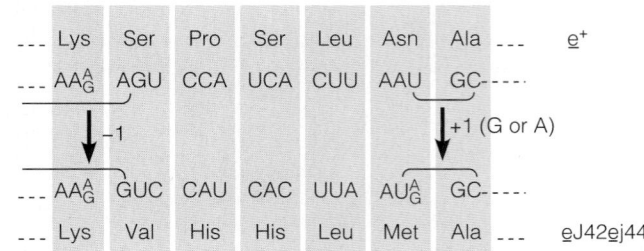

UGA to be translated as Sec rather than for termination, that UGA must have a special Sec Insertion Sequence (SECIS) available, usually in the 3′-untranslated region (3′UTR). Although selenocysteine is rather rare among proteins, the human proteome has been shown to contain 25 selenoproteins. As mentioned earlier (Chapter 15), some of these proteins participate in oxidant protection.

Pyrrolysine, by contrast, has a much narrower distribution, having been found so far in only about 1% of all sequenced genomes, mostly methanogenic archaea. Pyrrolysine is converted directly to a pyrrolysyl-tRNA by its own amino acyl-tRNA synthetase. The resultant Pyl-tRNA^{Pyl} has an anticodon that pairs with UAG, normally used in chain termination. So far it is not clear whether any of the UAGs in these genomes are read as stop codons or whether all encode Pyl.

Finally, although we say that the code is unambiguous, that concept may need revision in light of recent work with ciliated protozoa of the genus *Euplotes*. This organism uses UGA as one of three cysteine codons (the others are UGU and UGC). At least one gene in *E. crassus* has UGA triplets encoding both cysteine and selenocysteine. Sequence context is obviously key to ensuring correct insertional specificity. Because this organism also uses UGA as a tryptophan codon in its mitochondria, UGA is busy indeed.

The Wobble Hypothesis

If you examine the codon table shown in Figure 28.5, you will note that, in general, each amino acid is characterized by the first two codon letters. For example, all four Pro codons start with CC, and all four Val codons start with GU. Thus, redundancy is usually expressed in the third letter—ACU, ACC, ACA, and ACG all code for threonine. Soon after the code was deciphered, it was recognized that a single tRNA may recognize several different codons. The multiple recognition always involves the 3′ residue of the codon and therefore the 5′ residue of the anticodon.

In 1966, Francis Crick proposed that the 5′ base of the anticodon was capable of "wobble" in its position during translation, allowing it to make alternative (non-Watson–Crick) hydrogen-bonding arrangements with several different codon bases. An example is shown in Figure 28.7. G in the 5′ anticodon position can pair with either C or U in the codon, depending on the relative orientation of the pair. Considering both base-pairing possibilities and the observed selectivity of tRNAs, Crick proposed the set of "wobble rules" given in Table 28.2. This hypothesis nicely explains the frequently observed degeneracy in the 3′ site of the codon. The rather uncommon nucleoside, *inosine* (I, Chapter 22), is found in a number of anticodons, where it shows the ability to pair with A, U, or C.

Not all cases of multiple codon use involve translation of a single tRNA using wobble. As an example, consider the six leucine codons. Four of the six begin with CU and in principle could be translated by two different tRNAs, using wobble. However, the remaining two codons, UUA and UUG, will require a different anticodon, such as 3′-AAU-5′, which could translate both codons. In fact, *E. coli* contains five different leucine tRNAs and multiple **isoaccepting tRNAs**—tRNAs accepting and translating the same amino acid are common.

Codon Bias

Redundancy of the genetic code means that several nucleotide triplets can encode the same amino acid—leucine, for example, with six codons. In principle, a silent mutation, such as CUA → CUG, should have no biological consequences because both triplets encode leucine. Yet we find that use of degenerate codons by certain organisms is highly selective. In an extreme case, about half of all 64 codons are used either negligibly or not at all by the bacterium *Thermus thermophilus*. Although we don't know the evolutionary mechanisms leading to such

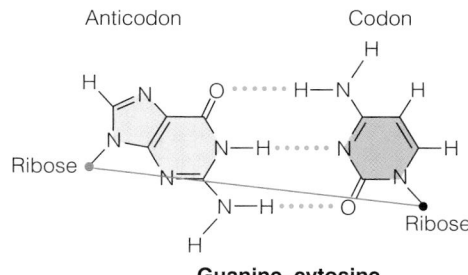

Guanine–cytosine

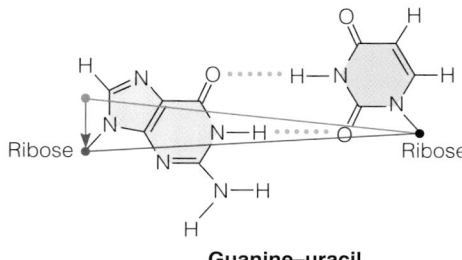

Guanine–uracil

FIGURE 28.7

The wobble hypothesis. As an example, we show how the anticodon base G can pair with either C or U in a codon. Movement ("wobble") of the base in the 5′ anticodon position is necessary for this capability (see arrow).

TABLE 28.2	Base-pairing capabilities in wobble pairs

Base at 5′ Position in Anticodon		Base at 3′ Position in Codon
G	pairs with	C or U
C	pairs with	G
A	pairs with	U
U	pairs with	A or G
I	pairs with	A, U, or C

asymmetry in codon selection, knowledge of codon bias has practical significance, for those wishing to express recombinant eukaryotic proteins in bacteria. Consider *E. coli*, a frequently used host for recombinant gene expression. Of the six arginine codons, two, AGA and AGG, are rarely used in *E. coli*, meaning that each of these triplets represents fewer than 1% of the arginine codons in the entire genome. Related to this is that the intracellular concentration of the tRNA with the anticodon, 3′-TCI-5′, which can translate these two rare codons, is quite low. This means that a recombinant gene with more abundant representation of these codons will be poorly expressed after transfer into *E. coli*. This situation can be remedied by site-directed mutagenesis of the recombinant genes to change these rare codons to arginine codons that are more abundant, and more efficiently translated, in the *E. coli* genome. An alternative approach is to engineer the *E. coli* host for overexpression of the rare tRNA so that the codons AGA and AGG can be efficiently translated.

An advantage of a code design that has synonymous codons with similar structures is that many mutations involving single-base changes are silent because a codon change such as the CUA → CUG mentioned above doesn't change the sense of the genetic message. Not only are many single-base changes silent, but many more are conservative, in the sense that a mutation may substitute a structurally similar amino acid that can be tolerated by the protein with no loss of function. For example, each of six leucine codons can be converted to a codon for the closely related valine by a single-base change. This suggests that the code has evolved to maximize genetic stability.

Stopping and Starting

Because the messenger RNA is invariably longer than the open reading frame that is to be translated, specific start and stop signals are required to begin and end translation. In almost all organisms, UAA, UAG, and UGA are used for stop signals and do not code for any amino acid (with the exceptions discussed above). A stop signal indicates that translation is to terminate and the polypeptide product is to be released by the ribosome. Clearly, three stop signals are more than is absolutely necessary, so it is not surprising to find that these codons are also used for designating amino acids in mitochondria and in other special cases (see Table 28.1).

Although nature has been generous in designating stop signals, it has been stingy in apportioning starts. The start signal commonly used in translation is AUG, which also serves as the single methionine codon. How does the ribosome know how to interpret this triplet properly so as to distinguish between internal Met sites and start sites? The answer is that the 5′ end of any message contains specific sequences to ensure that it is correctly attached to the ribosome (see page 1183). As the message begins to be read, the *first* AUG encountered is interpreted as a start signal, and translation begins. Although prokaryotic and eukaryotic cells handle this situation somewhat differently, the consequence is that N-formylmethionine (in prokaryotes) or methionine (in eukaryotes) is usually the first amino acid incorporated into a polypeptide chain. Therefore, all proteins start with N-fMet or Met, at least when they are first synthesized. However, in most cases this residue is either deformylated or removed as translation proceeds. Any AUG encountered after the start is treated as a signal to incorporate methionine within the sequence at that point. Very occasionally, GUG (normally valine), UUG (normally leucine), or AUU (normally isoleucine) serves as a prokaryotic start codon when located near the 5′ end of a message (see Table 28.1). When they do, however, they code for N-formylmethionine in the first position. In other positions these triplets are read as normal codons.

Prokaryotic messengers contain translational start and stop signals, as well as a sequence that aligns the mRNA on the ribosome.

N-formylmethionyl-tRNA

The Major Participants in Translation: mRNA, tRNA, and Ribosomes

mRNAs

As we indicated in Chapter 27, eukaryotic messenger RNAs are quite different from prokaryotic mRNAs. Prokaryotic mRNAs are more complex because many or most are *polycistronic*; they encode two or more polypeptide chains. This means that the mRNA sequence must be punctuated so that translation of the RNA corresponding to each gene is controlled by its own initiation and termination signals. Eukaryotic messages almost always encode just one protein, but the mRNA structure is the result of post-transcriptional processing far more extensive than that seen in prokaryotic systems.

As a good example of a prokaryotic mRNA, consider that produced by transcription of the *E. coli lac* operon, which was introduced in Chapter 27 and receives further attention in Chapter 29. This group of three linked genes—*lacZ*, *lacY*, and *lacA*—controls the utilization of lactose and related sugars by bacteria. As shown in Figure 28.8, these three genes are expressed as a single mRNA molecule some 5300 nucleotides in length. Within this mRNA are three **open reading frames**, corresponding to the *lacZ, Y*, and *A* genes. An open reading frame is a sequence within a messenger RNA, bounded by start and stop codons, that can be continuously translated. Each open reading frame has its own start and stop signals, and you can see that these signals vary considerably. There is extra, untranslated RNA between the reading frames and at the ends. The regions 5′ to each start signal contain sequences rich in A and G, which help to align the mRNA on the ribosome so that translation can begin at the proper points and in the correct reading frame. Such attachment sequences, found on all prokaryotic mRNAs, are called *Shine–Dalgarno sequences*, after J. Shine and L. Dalgarno, who first described them. A Shine–Dalgarno sequence can base-pair with a sequence contained in the ribosomal RNA, as shown in Table 28.3, to produce a proper alignment for starting translation. The different attachment sequences appear to have different affinities for ribosomes. For example, the three genes of the *lac* operon (Figure 28.8) are not translated to equal extents—*lacZ* is translated much more frequently than *lacY* or *lacA*.

Shine–Dalgarno sequences help align ribosomes on mRNAs to properly start translation.

FIGURE 28.8

The *lac* operon mRNA. The mRNA for the *E. coli lac* operon is about 5300 nucleotides long and contains the open reading frames for the *lacZ, lacY*, and *lacA* genes, each flanked appropriately by start, stop, and Shine–Dalgarno (SD) sequences.

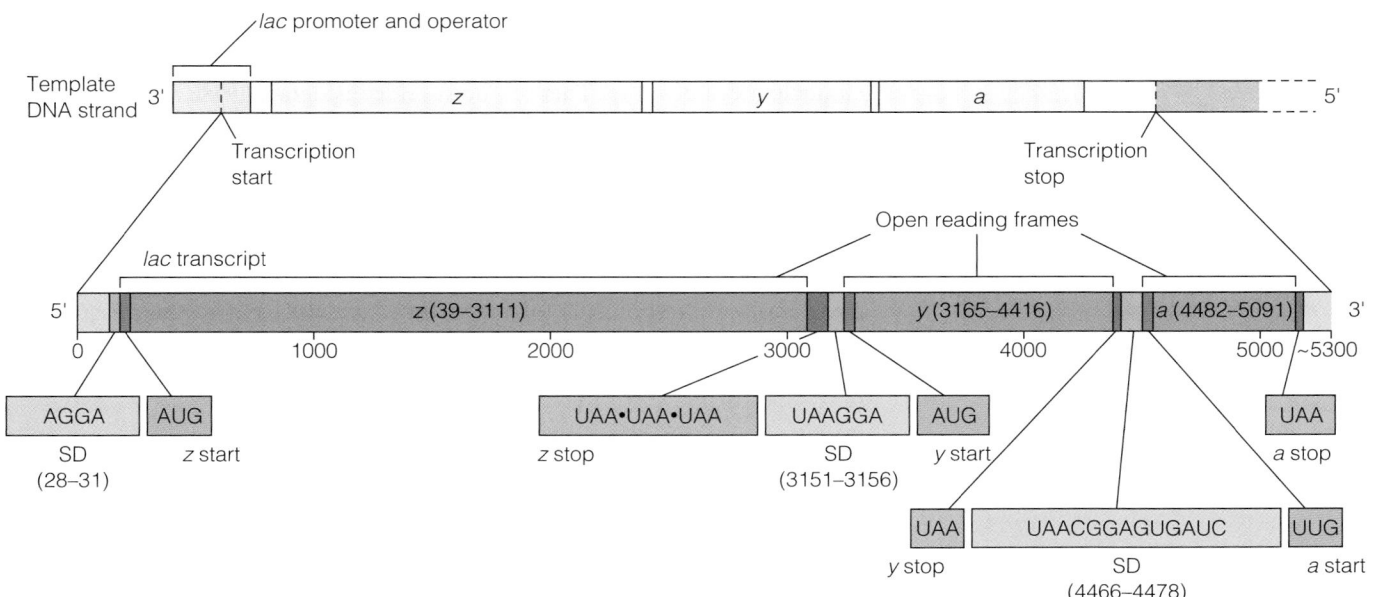

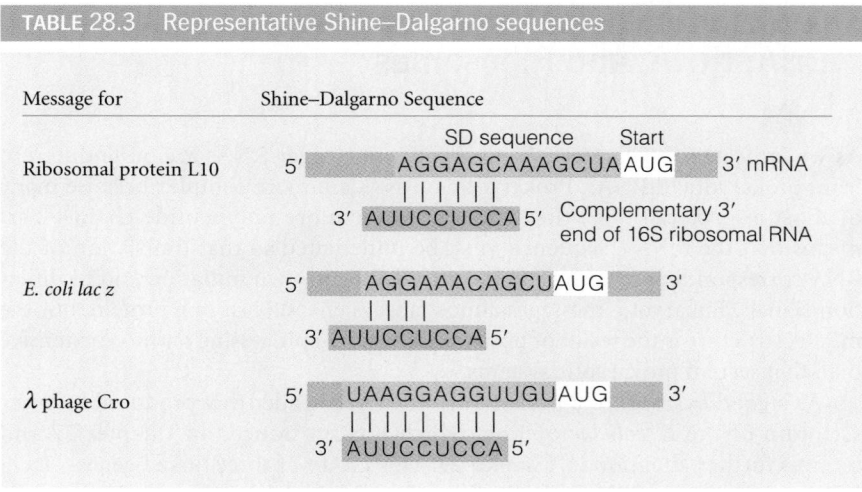

TABLE 28.3 Representative Shine–Dalgarno sequences

Message for	Shine–Dalgarno Sequence
Ribosomal protein L10	SD sequence Start 5′ AGGAGCAAAGCUA AUG 3′ mRNA | | | | | | 3′ AUUCCUCCA 5′ Complementary 3′ end of 16S ribosomal RNA
E. coli lac z	5′ AGGAAACAGCUAUG 3′ | | | | 3′ AUUCCUCCA 5′
λ phage Cro	5′ UAAGGAGGUUGUAUG 3′ | | | | | | | | | 3′ AUUCCUCCA 5′

The mRNA produced from the *lac* operon has all the basic elements necessary for its function: sequences to align it properly on the ribosome and sequences that start and stop translation at the proper points. Many mRNAs also have possibilities for forming three-dimensional secondary and tertiary structures, which can participate in regulating the relative production of the various protein products. We shall return to this point in Chapter 29.

Transfer RNA

Any cell, prokaryotic or eukaryotic, contains a battery of different types of tRNA molecules sufficient to incorporate all 20 amino acids into protein. This does not mean that there need be as many tRNA types as there are codons, for, as noted earlier, some tRNAs can recognize more than one codon, when the difference is in the third, or wobble, position. *E. coli*, for example, has about 40 different tRNAs—plenty to code for all amino acids, but not as many as the 61 amino acid codons. As noted on page 1180, the tRNA specific to a given amino acid is designated by writing the amino acid as a superscript, for example, tRNA^{Ala}.

Transfer RNA was the first natural polynucleotide sequence to be determined, in a pioneering study of yeast tRNA^{Ala} by Robert Holley in 1965. Since then, thousands of tRNAs have been sequenced. All have the general structure shown schematically in Figure 28.9a and have similar sequences of about 70 to 80 nucleotides or more. There is, however, considerable variation in detail, as shown in the examples in Figure 28.9b and c. Furthermore, the tRNAs are unique among RNA molecules in their high content of unusual and modified bases, three of which are shown in Figure 28.10. Biosynthesis of the modified bases always occurs *post-transcriptionally*, as mentioned in Chapter 27. For example, an isomerase converts a uridine residue (1-ribosyluracil) to the unusual C-glycoside pseudouridine (5-ribosyluracil), and *S*-adenosylmethionine-dependent methyltransferases are responsible for converting standard bases to their methylated derivatives.

Cloverleaf models of the kind shown in Figure 28.9 are useful for showing the general pattern of hydrogen bonding and denoting the functional parts of the tRNA. The *anticodon triplet* in the loop at the bottom is complementary to the mRNA codon and will make base pairs with it. Because the codon and anticodon, when paired, constitute a short stretch of double-stranded RNA, their directions must be antiparallel. In Figure 28.9 we have written the tRNA molecules with their 5′ ends to the left. Therefore, the messenger RNA, when shown in such figures, is written with its 5′ end to the right, opposite to the normal convention.

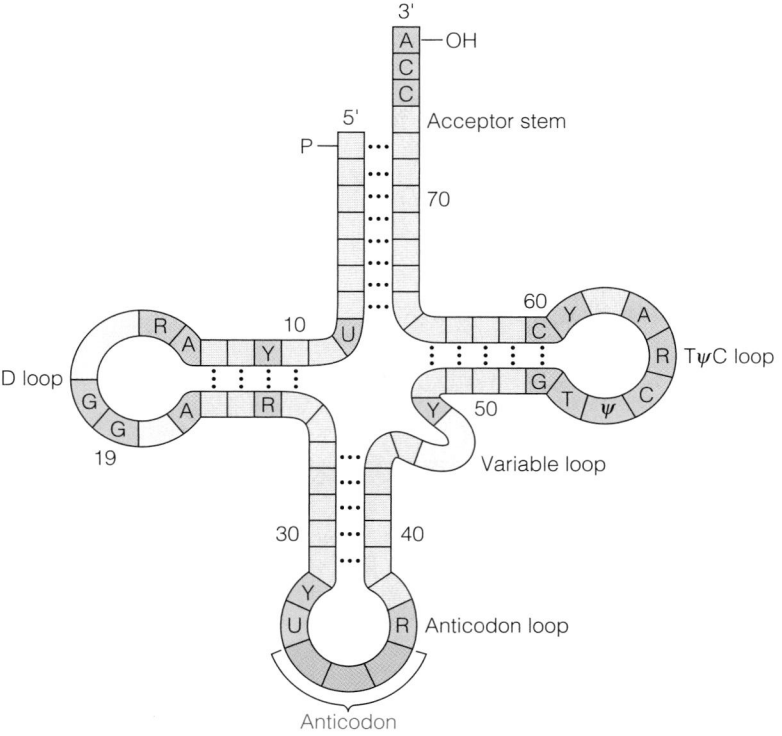

(a)

FIGURE 28.9

Structure of tRNAs. (a) Generalized tRNA structure. The positions of invariant and rarely varied bases are shown in purple. Regions in the D loop and the variable loop that can contain different numbers of nucleotides are shown in blue. The anticodon is shown in orange. **(b)** A leucine tRNA from *E. coli.* **(c)** A human mitochondrial tRNA for lysine. Code for bases: Y = pyrimidine, R = purine, ψ = pseudouridine, T = ribothymidine, and D = dihydrouridine (see Figure 28.10).

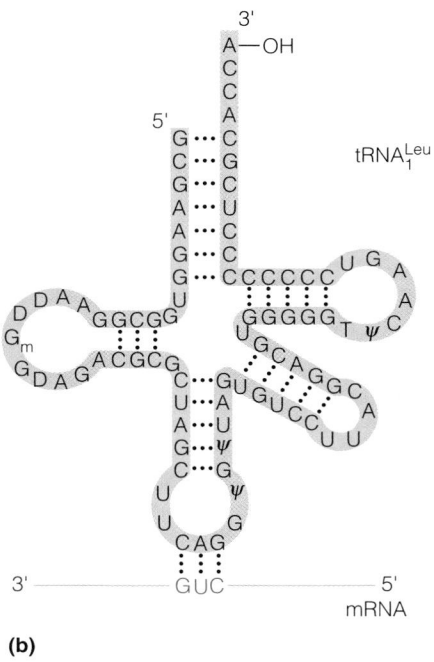

(b)

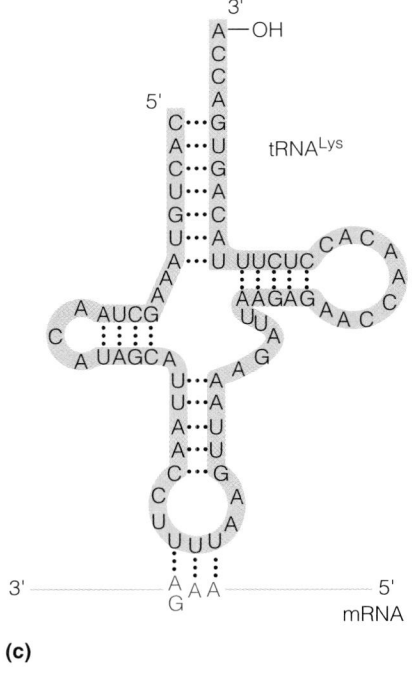

(c)

The *acceptor stem* at the top of the cloverleaf figure is where the amino acid will be attached, at the 3′ terminus of the tRNA. This stem always has the sequence 5′···CCA—OH 3′. Other common features of tRNA molecules are the *D loop* and the TψC *loop*, regions that contain a substantial fraction of invariant positions and frequently contain modified or unusual bases as well. The so-called *variable loop* is indeed variable, both in nucleotide composition and in length, as Figure 28.9 demonstrates.

Although cloverleaf models are convenient for depicting the primary structure and some elements of secondary structure, they are not good three-dimensional representations of tRNA molecules. X-ray diffraction studies of tRNA molecules

Pseudouridine (ψ)   **Ribothymidine (T)**   **Dihydrouridine (D)**

FIGURE 28.10

A sampling of the modified and unusual bases found in tRNAs.

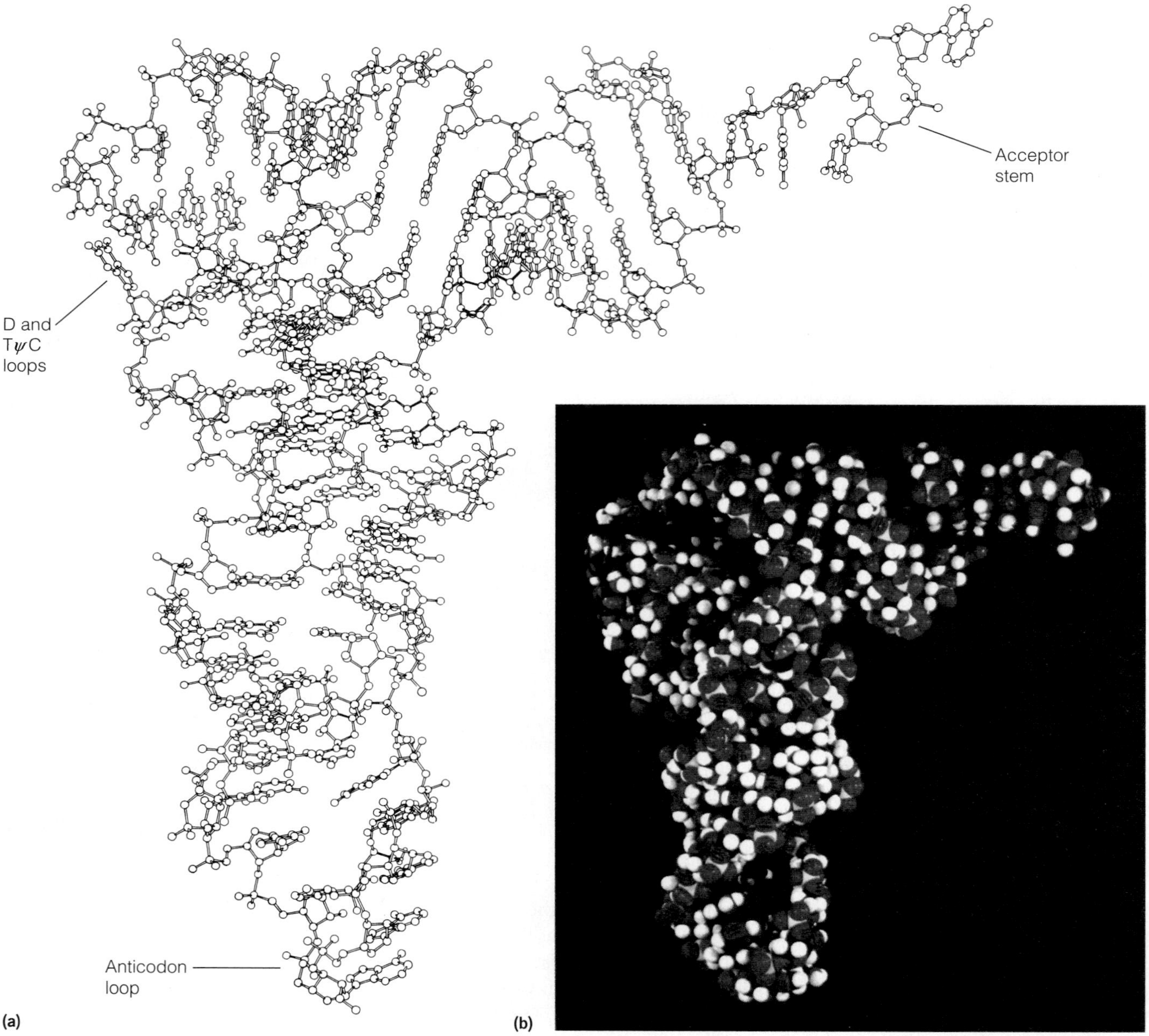

Acceptor
stem

D and
TψC
loops

Anticodon
loop

(a)

(b)

FIGURE 28.11

Model of yeast phenylalanine tRNA derived from X-ray diffraction studies. The anticodon is at the bottom, the 3′ acceptor stem at the upper right. **(a)** Drawing showing all atomic positions. **(b)** Space-filling model.

From *Science* 185:435–440, S. H. Kim, F. L. Suddath, G. J. Quigley, A. McPherson, J. L. Sussman, A. H. J. Wang, N. C. Seeman, and A. Rich, Three-dimensional tertiary structure of yeast phenylalanine transfer RNA. © 1974. Reprinted with permission from AAAS and Sung-Hou Kim.

All tRNAs share a general common structure that includes an anticodon loop, which pairs with codons, and an acceptor stem, to which the amino acid is attached.

have revealed that the real molecular shape is more complex, as you can see in Figures 28.11 and 4.20 (page 109). As these figures show, a tRNA molecule looks rather like a hand-held drill or soldering gun. The anticodon loop is at the bottom of the grip, and the acceptor stem is at the working tip. The D loop and the TψC loop are folded inward in a complex fashion near the top of the grip, to provide a maximum of hydrogen bonding and base stacking. Some of the hydrogen-bonding patterns required to produce this folding are rather unusual (Figure 28.12). The three-dimensional shapes of the tRNAs are highly conserved even though the primary structures vary. A likely explanation is that such conservation is necessary so that each tRNA can fit equally well onto the ribosome and carry out its function.

FIGURE 28.12

Unusual base pairings in tRNA. All are from the yeast tRNA^Phe shown in Figure 28.11. **(a, b)** Some unusual pair matches. **(c, d)** Some examples of triple interactions. R represents the ribosyl residue of the RNA chain. The bases prefixed by m are methylated at the carbon atom corresponding to the superscript. Numbers following the letters designating bases show the position in the sequence.

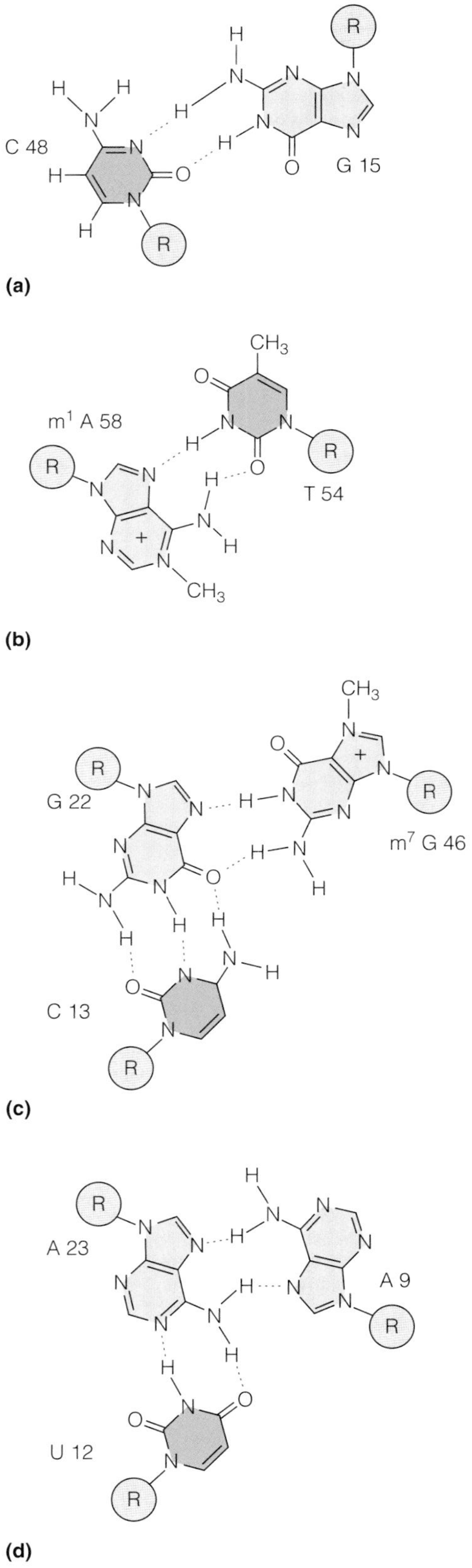

(a)

(b)

(c)

(d)

Coupling of tRNAs to Amino Acids and Formation of Aminoacylated tRNAs: The First Step in Protein Synthesis

Amino acids are attached to tRNAs by a covalent bond between the carboxylate of the amino acid and a ribose 3′ hydroxyl group of the invariant 3′ terminal adenosine residue on the tRNA. Pairing of the correct amino acid residues and the tRNAs is accomplished by a set of enzymes called **aminoacyl-tRNA synthetases** (abbreviated AARS). In *E. coli* there are 21 synthetases, each of which recognizes one amino acid and one or more tRNAs. Lysine is unique in having two synthetases. The reaction linking the two molecules, shown in Figure 28.13, proceeds in two steps. First, the amino acid, which is bound to the synthetase, is activated by ATP to form an **aminoacyl adenylate**. While still bound to the enzyme, this intermediate reacts with one of the correct tRNAs to form the covalent bond and release AMP.

Because all of the synthetases perform essentially the same function, one might expect them to represent minor variations on a common theme. This is, however, not the case; there are two general classes of aminoacyl-tRNA synthetases (I and II). Their active sites are completely different, and the two classes bind their cognate tRNAs from opposite sides. Furthermore, the class I enzymes tend to function as monomers, whereas the class II enzymes function as dimers or tetramers. Moreover, the enzymes differ mechanistically. Class II enzymes link the aminoacyl moiety in the aminoacyl adenylate intermediate directly to the 3′ hydroxyl in the tRNA acceptor, while class I enzymes synthesize first a 2′-aminoacyl-tRNA intermediate, which then undergoes intramolecular transesterification, giving the 3′-aminoacyl-tRNA product.

The reasons for these extreme differences are unknown, but they may reflect the utilization of some amino acids in proteins before others in the very early evolution of protein synthesis. A recent observation that may bear on this question is that some members of some classes of organisms (Gram-positive bacteria and archaea, for example), as well as some organelles, use an indirect transamidation route for charging some tRNAs. For example, tRNA^Gln is charged first with Glu, which is then replaced by Gln.

$$\text{Glu} + \text{tRNA}^{\text{Gln}} + \text{ATP} \xrightarrow{\overset{\text{glutamyl-tRNA}}{\text{synthetase}}} \text{Glu-tRNA}^{\text{Gln}} + \text{AMP} + \text{PP}_i$$

$$\text{Gln} + \text{Glu-tRNA}^{\text{Gln}} + \text{ATP} \xrightarrow{\overset{\text{Glu-tRNA}^{\text{Glu}}}{\text{amidotransferase}}} \text{Gln-tRNA}^{\text{Gln}} + \text{ADP} + \text{P}_i + \text{Glu}$$

Thus, these organisms do not require (although they may have) a Gln-tRNA^Gln synthetase, as do Gram-negative bacteria and eukaryotes. It has been suggested that glutamine was one of the last amino acids to be added to the protein repertoire and that it was initially incorporated by this route.

In higher organisms nine aminoacyl-tRNA synthetases are organized into a high-molecular-weight complex, along with three accessory proteins. The biological function of this complex is unknown, but it is assumed to contribute toward coordinating amino acid synthesis with protein synthesis.

You might expect that the synthetase would identify the correct tRNA on the basis of its anticodon, but many studies indicate that the identification process is more complex, and various nucleotides act as *identity elements*. In 1988, Ya-Ming Hou and Paul Schimmel showed that changing a single base pair (between residues 3 and 70 in the acceptor stem) of tRNA^Cys or tRNA^Phe to the G-U pair found in tRNA^Ala caused the alanine synthetase to accept the tRNA^Cys or tRNA^Phe and couple it to alanine. Other tRNAs appear to be recognized by their synthetases at many different locations (see Figure 28.14). No simple rule has emerged, although it is

Amino acids are coupled to their appropriate tRNAs by aminoacyl-tRNA synthetases.

FIGURE 28.13

Formation of aminoacyl tRNAs by aminoacyl tRNA synthetase. In step 1 the amino acid is accepted by the synthetase and is adenylylated, with the aminoacyl adenylate remaining bound to the enzyme. In step 2 the proper tRNA is accepted by the synthetase, and the amino acid residue is transferred to the 3′ OH of the 3′-terminal residue of the tRNA (class II enzymes) or to the 2′ hydroxyl, followed by isomerization to the 3′ aminoacyl-tRNA (class I enzymes). For class I enzymes the 2′ hydroxyl of the 3′-terminal AMP residue is the nucleophile for reaction 2.

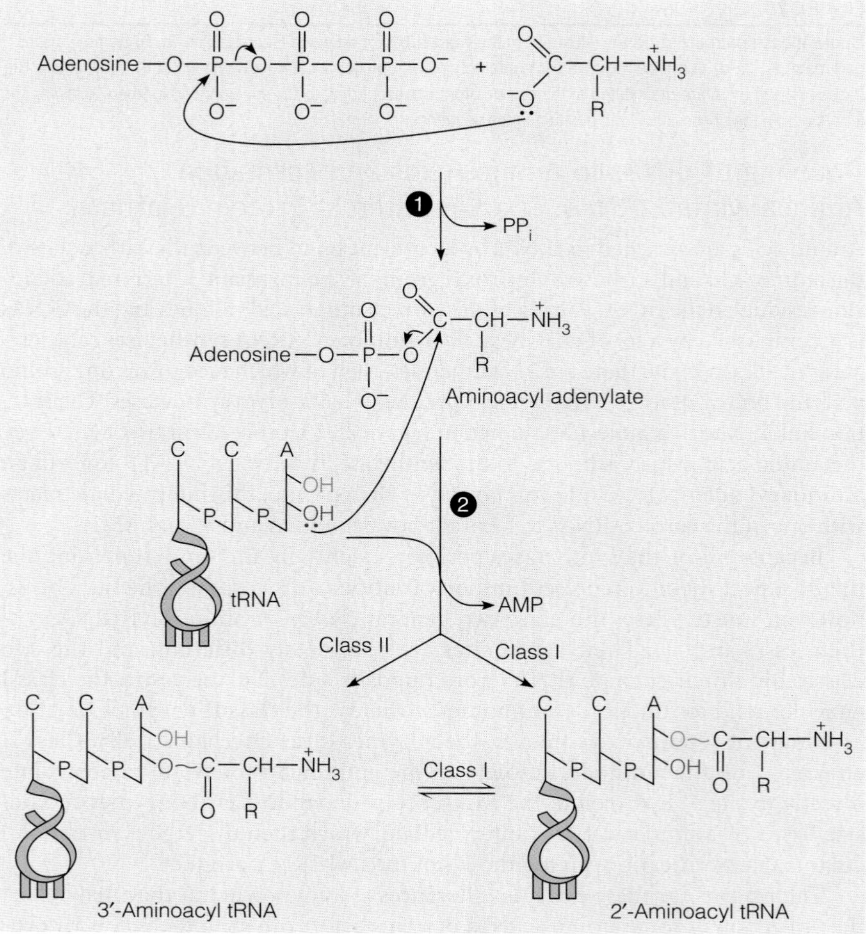

FIGURE 28.14

Major "identity elements" in some tRNAs. Red circles represent the positions that have been shown to identify the tRNA to its cognate synthetase. Shown also is a synthetic polynucleotide containing the G-U alanine identity element (in red), which is a good substrate for alanyl-tRNA synthetase.

From *Science* 240:1591–1592, L. Schulman and J. Abelson, Recent excitement in understanding transfer RNA identity. © 1988. Reprinted with permission from AAAS. Adapted with permission from John Abelson.

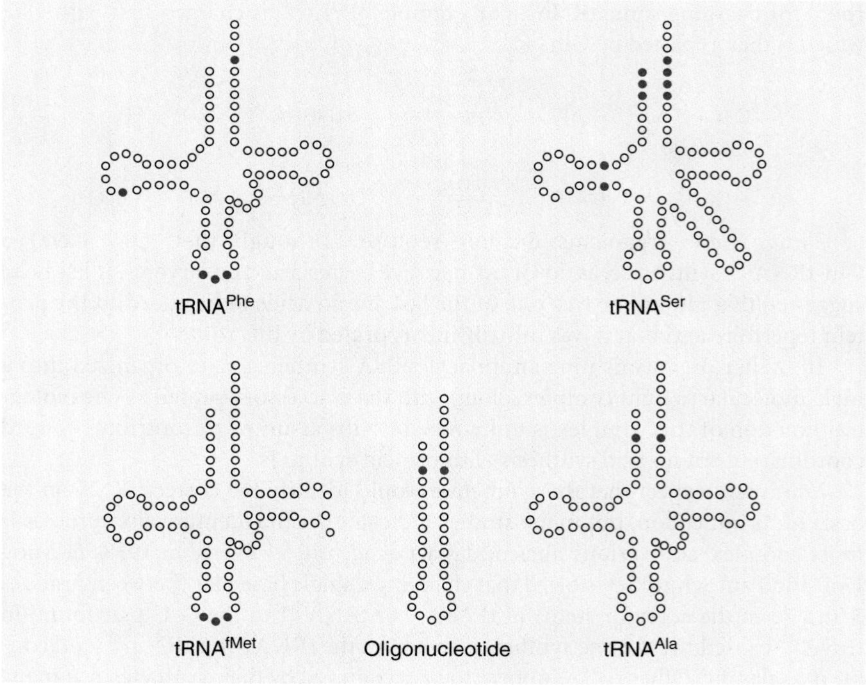

clear that identity elements are clustered in the anticodon loop and the acceptor stem. A dramatic illustration of the importance of identity elements is shown by the fact that the yeast tRNA^{Ala} shown in the figure can be trimmed to just a single-hairpin molecule, as shown, and, so long as a critical G-U base pair (shown in red) is present, the molecule can be efficiently and accurately aminoacylated.

Aminoacyl-tRNA synthetases contribute toward the fidelity of translation by a process akin to proofreading by DNA polymerases. In the instant between formation of an enzyme-bound aminoacyl adenylate and its conversion to aminoacyl-tRNA, the enzyme can sense the improper fit of the amino acid side chain and hydrolyze the intermediate before the amino acid can be linked to tRNA. Moreover, even if the wrong aminoacyl-tRNA is synthesized, the enzyme has a short time in which it can identify the mischarged amino acid as incorrect and hydrolyze it before it can be released to participate in translation. In these ways, aminoacyl-tRNAs contribute toward an overall error frequency for protein synthesis of about 10^{-4}, less accurate than DNA replication, to be sure, but with the consequences of error being much lower because the error is not propagated to the next generations. Most of the work on AARS proofreading has been done with isoleucyl-tRNA synthetase and its ability to mischarge valine, which differs from isoleucine by only one methylene group. Despite this small structural difference, the mischarging frequency is only 3×10^{-4}.

Insight into the recognition of tRNAs by their synthetases has been provided by crystallographic analysis of the complexes formed. Figure 28.15 shows the structure of a class I synthetase-tRNA complex, *E. coli* glutaminyl-tRNA. As shown in the figure, the tRNA lies across the protein, making a number of specific contacts, including crucial ones in the anticodon region and in the acceptor stem. Both of these regions are distorted in the complex, with the acceptor stem being elongated and inserted into the active site pocket. This pocket is formed by a common protein structural motif called the *dinucleotide fold,* which frequently acts as a nucleotide-binding region. In this case it also binds the ATP required for acylation. It provides a binding site for glutamine as well. Thus, all three participants in the reactions are grouped close together.

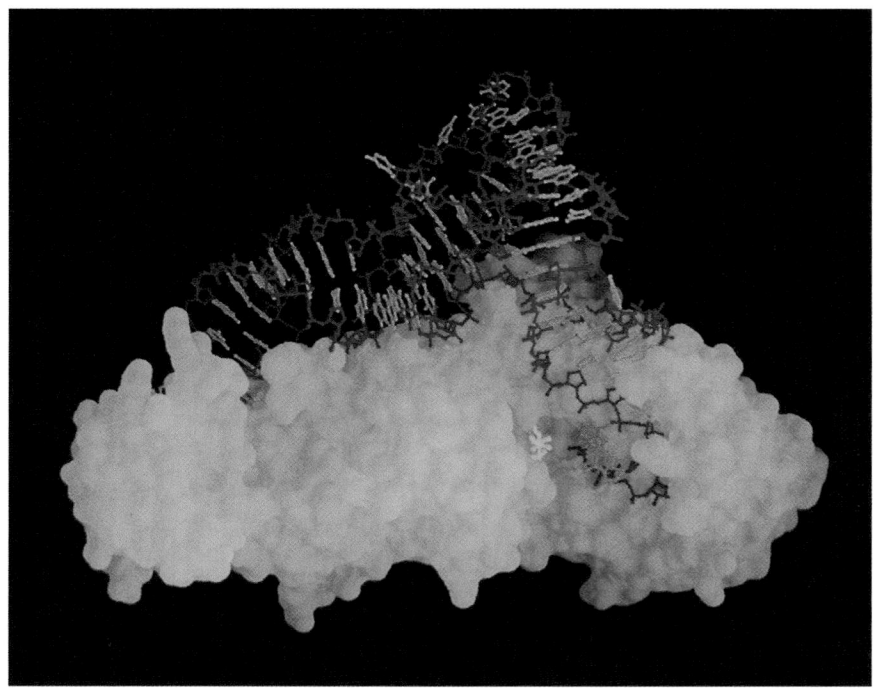

FIGURE 28.15

A model of the *E. coli* glutaminyl tRNA synthetase coupled with its tRNA and ATP. The tRNA is represented by a detailed atomic model, the protein by its solvent-accessible surface (blue). The ATP (green) and the 3′ acceptor stem of the tRNA fit into a deep cleft in the synthetase. This cleft will also accommodate the amino acid. This is a monomeric class I synthetase. PDB ID 1GSG.

From *Science* 246:1135–1142, M. A. Rould, J. J. Perona, D. Söll, and T. A. Steitz, Structure of *E. coli* glutaminyl-tRNA synthetase complexed with tRNA(Gln) and ATP at 2.8 Å resolution. © 1989. Reprinted with permission from AAAS and Thomas Steitz.

FIGURE 28.16

Yeast aspartyl tRNA synthetase complexed with two molecules of tRNA^Asp. This is a dimeric class II synthetase. Protein subunits are in white and pale green. tRNA molecules are in blue and gold. PDB ID 1ASY.

From *Science* 252:1682, M. Ruff, S. Krishnaswamy, M. Boeglin, A. Poterszman, A. Mitschler, A. Podjarny, B. Rees, J. C. Thierry, and D. Moras, Class II aminoacyl transfer RNA synthetases: Crystal structure of yeast aspartyl-tRNA synthetase complexed with tRNA(Asp). © 1991. Reprinted with permission from AAAS and Marc Ruff.

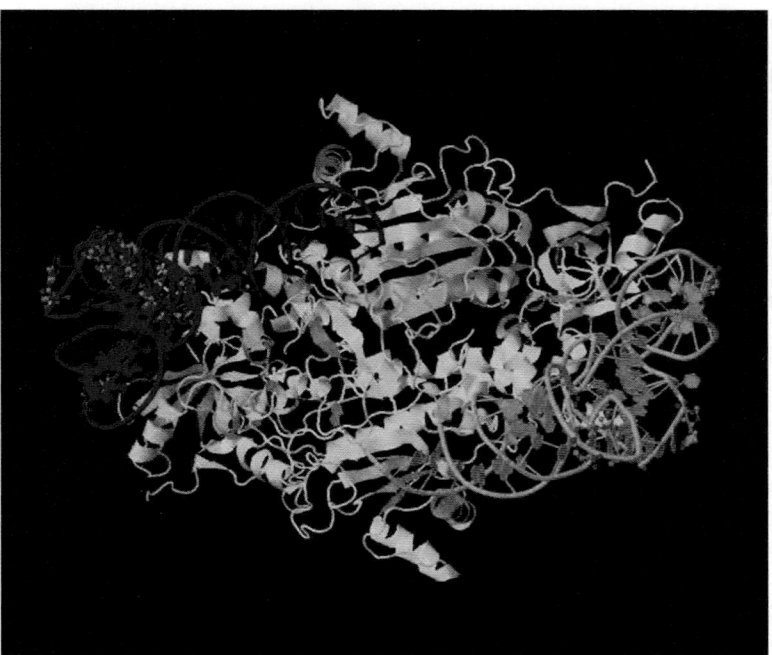

Similar interactions are seen with a class II synthetase. Figure 28.16 shows the dimeric yeast aspartyl tRNA synthetase complexed with two molecules of tRNA^asp. Note that the tRNA is bound to the enzyme in opposite orientation from that seen with class I synthetases. Only one of the two tRNA molecules is bound in a catalytically productive conformation.

One other feature of aminoacyl-tRNA synthetases deserves comment. In higher eukaryotes most of these enzymes are "moonlighting proteins"—proteins that evolved initially to play their well-known function in protein synthesis, but which in further evolution acquired additional functions. In humans, aminoacyl-tRNA synthetases are involved in functions as diverse as autoimmunity, control of apoptosis, regulation of ribosomal RNA synthesis, vascular development, and coordination of the DNA damage response. In all cases studied the catalytic machinery for aminoacyl-tRNA synthesis has remained undisturbed, and evolutionary modifications to convey additional functions occur elsewhere on the protein molecule.

The Ribosome and Its Associated Factors

We have now described two of the participants that must be brought together to carry out protein biosynthesis—the mRNA and the set of tRNAs charged with the appropriate amino acids. The actors are in the wings, and all that is needed is a proper director and a stage on which the events can unfold. Both are provided by the ribosome, and the typical cell requires many. An *E. coli* cell, for example, contains as many as 20,000 ribosomes, accounting for about 25% of the dried cell mass. Thus, a cell devotes a large part of its energy to producing ribosomes and to using them in protein synthesis.

Soluble Protein Factors in Translation

Before describing ribosomes in detail, however, we mention one more set of participants, whose functions will be described in detail later. These are the soluble proteins that participate in the three stages of translation—initiation factors, elongation factors, and release factors. Table 28.4 introduces these factors as initially studied in bacteria, as well as their eukaryotic counterparts. We shall refer back to the information in this table as we discuss mechanisms in translation.

TABLE 28.4 Soluble protein factors in translation

Function	Factor (Bacteria)	Factor (Eukaryotes)	Role in Translation
Initiation	IF1	eIF1, eIF1A	Promotes dissociation of preexisting 70S ribosome
	IF2	eIF2, eIF2B	Helps attach initiator tRNA
	IF3	eIF3, eIF4C	Similar to IF1; prepares mRNA for ribosome binding
		eIF4A, eIF4B, eIF4F	Same as eIF1, eIF1A
		eIF5	Helps dissociate eIF2, eIF3, eIF4C
		eIF6	Helps dissociate 60S subunit from inactive ribosomes
Elongation	EF-Tu	eEF1α	Helps deliver aminoacyl-tRNA to ribosomes
	EF-Ts	eEF1$\beta\gamma$	Helps recharge EF-Tu with GTP
	EF-G	eEF2	Facilitates translocation
Termination	RF1	eRF	Release factor (UAA,UAG)
	RF2		Release factor (UAA, UGA)
	RF3		A GTPase that promotes release

Components of Ribosomes

The ribosome is a large ribonucleoprotein particle containing 60–70% RNA and 30–40% protein. Ribosomes and their subunits are characterized in terms of their sedimentation coefficients in ultracentrifugation. Thus, the individual bacterial ribosome is called a 70S particle and has a molecular mass of about 2.5×10^6 Da. Eukaryotic ribosomes are somewhat larger, with a sedimentation coefficient of 80S and a molecular mass of 4.2×10^6 Da. When isolated ribosomes are placed in a buffer containing low Mg^{2+} ion, they dissociate into two smaller subunits. As shown in Figure 28.17, bacterial 70S ribosomes dissociate into 30S and 50S subunits. We shall see later that dissociation and reassociation of these subunits are crucially important during translation. Figure 28.17 also shows the number of RNA and protein components in each subunit. Note that the 50S bacterial subunit contains two rRNA molecules (5S and 23S) and 34 different proteins, while the

FIGURE 28.17

Components of bacterial and eukaryotic ribosomes. Bacterial (to the left) and eukaryotic (to the right) ribosomes are assembled along the same structural plan, with eukaryotic ribosomes being somewhat larger and more complex. The shapes of the ribosomal subunits were determined by electron microscopy.

Modified from *Molecular Biology of the Cell*, 4th ed., B. Alberts et al. Garland Science, New York, 2002.

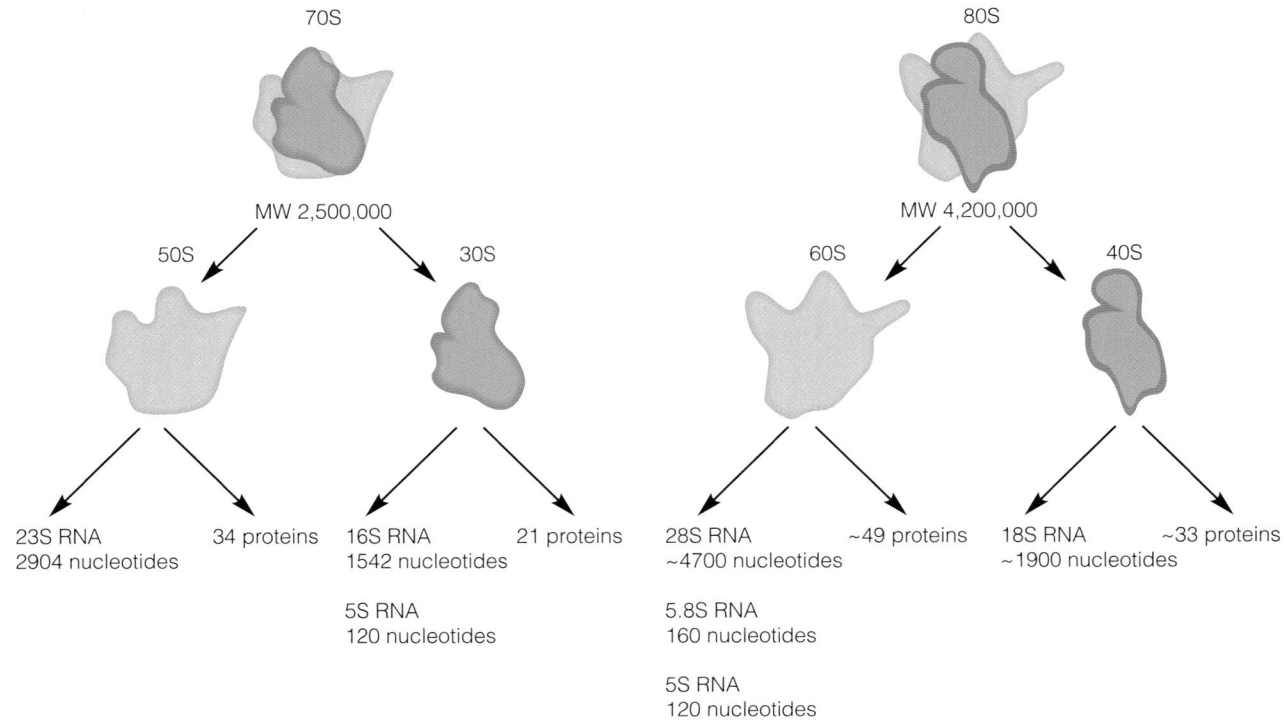

30S subunit contains just one rRNA (16S) and 21 proteins—all different from those in the 50S subunit. Proteins from the small subunit are called S1, S2, S3···S21, while those from the large subunit are called L1, L2, L3···L34. All proteins are present in one copy per ribosome, except for L12, which is present in four copies. Eukaryotic ribosomes are significantly larger, with larger rRNAs and more proteins. We shall be discussing primarily bacterial ribosomes, whose structures and functions are known in much greater detail.

Once the complexity of the ribosome was revealed, particularly the large numbers of proteins in each subunit, it seemed a daunting task to determine the structure of the particle and to understand the function of each protein. However, Peter Traub and Masayasu Nomura learned as early as 1968 that they could reassemble 30S ribosomal subunits from the separated RNA and protein components. The product, when combined with 50S subunits, was active for in vitro protein synthesis. An obligatory order of assembly was seen, with some proteins being incorporated only after binding of certain other proteins. As expected if transcription and translation are coupled, the proteins bound earliest in the pathway are those linked to the 5′ end. The ability to assemble ribosomes in vitro allowed analysis of the function of individual ribosomal proteins because ribosomal subunits could be assembled with specific proteins missing, followed by functional analysis of these deliberately altered particles.

More recent analysis shows that the pathway of ribosome assembly in vivo differs from the pathway in vitro pathway in significant ways. In this analysis, carried out by James Williamson and colleagues, assembly intermediates accumulated in *E. coli* treated with the antibiotic neomycin and were analyzed by ^{15}N pulse labeling and mass spectrometry. Additional information was obtained from electron micrographic analysis of individual assembly intermediates. Distinctions between the in vitro and in vivo pathways include the following. First, ribosome assembly in vivo occurs via parallel pathways, with one pathway adding proteins initially at the rRNA 5′ domain and another at the 3′ domain. Second, some proteins bound to central domains are incorporated before proteins binding to the ends of 16S rRNA. Third, the ribosomes are assembled both from newly synthesized proteins and from proteins that originated in previously synthesized intact subunits. Figure 28.18 summarizes principal features of the in vitro and in vivo pathways.

Sequence analyses reveal no significant homologies among the different proteins in a ribosome, but comparison of sequences between corresponding proteins in the ribosomes of different organisms reveals considerable evolutionary conservation. Thus, the ribosome is a complex object that evolved early in the history of life and has remained relatively unchanged. Although the ribosomes of eukaryotes differ significantly from those of prokaryotes, the evolutionary continuity is clear. The sequences of many ribosomal RNAs tell the same story. Indeed, because of their relatively slow evolutionary rates of change, rRNAs are useful as evolutionary yardsticks over vast phylogenetic distances. In fact, it was sequence analysis of 16S rRNAs that led Carl Woese to propose the existence of a third domain of life, the archaea.

Ribosomal RNA Structure

When the sequences of 16S rRNAs were originally determined, they were found to contain many regions of self-complementarity, which are capable of forming double-helical segments. A pattern like that shown in Figure 28.19 may seem so complex as to appear almost arbitrary, but comparison with other, even distantly related, 16S RNA sequences shows that the potentially double-stranded regions are highly conserved. Indeed, the secondary structure seems more highly conserved than is the primary structure, for it is often found that there are compensatory mutations in double-helical regions so as to maintain base pairing. A schematic illustration like that in Figure 28.19 is analogous to the cloverleaf

Despite their complexity, ribosomal subunits can be assembled in vitro.

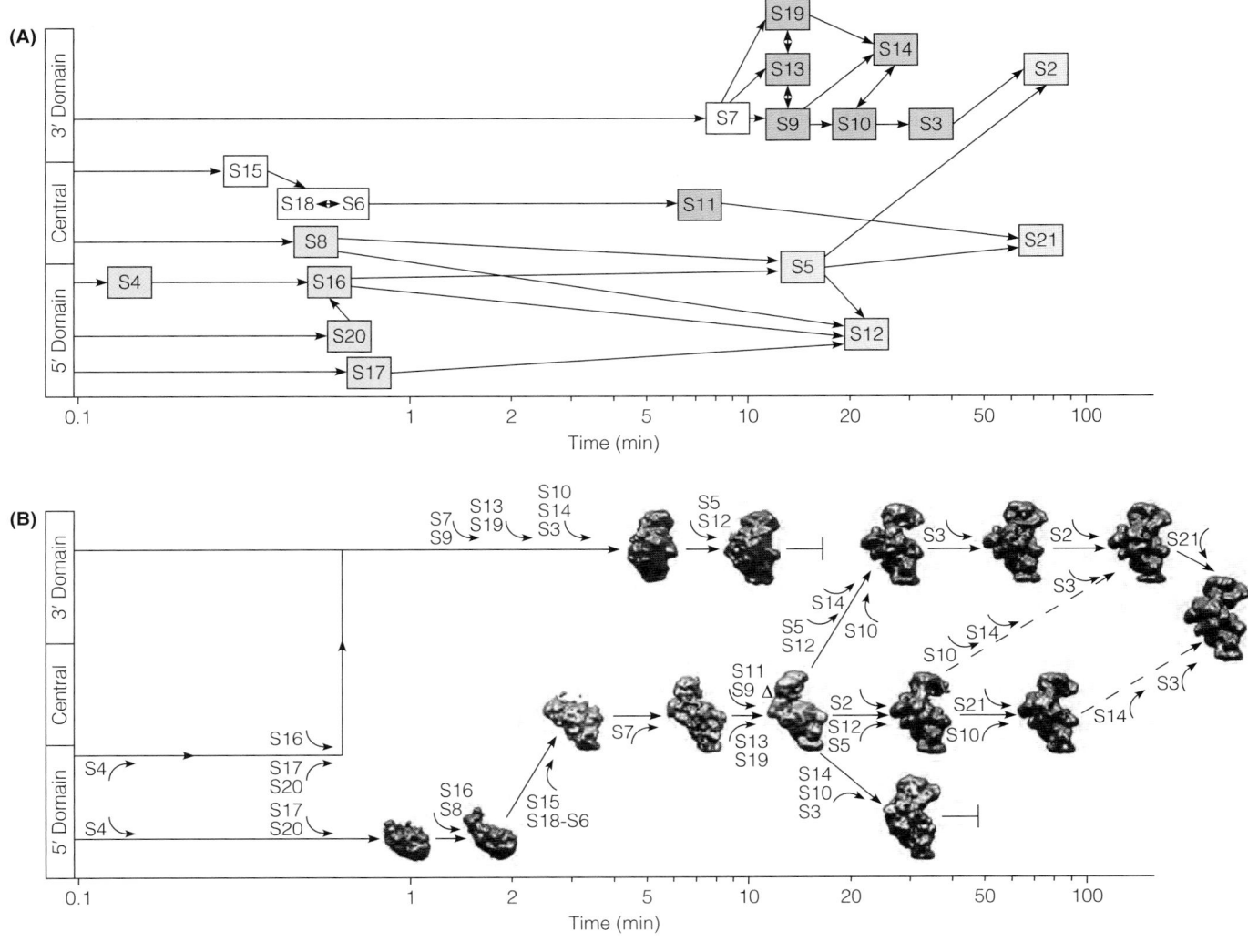

FIGURE 28.18

Assembly map for the 30S subunit. (a) Assembly pathway in vitro, as determined by Traub and Nomura. Arrows indicate the obligatory nature of some protein-binding events. For example, S7 must bind before S9, S13, or S19, but once S7 is bound, any of these three proteins can be added. The earliest protein-binding events occur near the 5′ end of the 16S rRNA, and an intermediate to which 5′- and central-domain proteins are bound must be formed before addition of 3′ domain proteins. **(b)** In vivo assembly map, as determined by Williamson and colleagues. Parallel pathways begin, with proteins added at either the 5′ domain or the 3′ domain of 16S rRNA.

From *Science* 330:673–677, A. M. Mulder, C. Yoshioka, A. H. Beck, A. E. Bunner, R. A. Milligan, C. S. Potter, B. Carragher, and J. R. Williamson, Visualizing ribosome biogenesis: Parallel assembly pathways for the 30S subunit. © 2010. Reprinted with permission from AAAS.

visualization of a tRNA (Figure 28.9). The actual rRNA is folded into a three-dimensional structure, just as is the tRNA. In the case of the ribosomal subunit, however, the structure is further complicated by the presence of ribosomal proteins bound to the RNA. However, it is now clear that the pattern shown in Figure 28.19 faithfully describes the secondary structure of 16S rRNA. 23S rRNA has a comparable secondary structure, actually more complex, reflecting its larger size.

Internal Structure of the Ribosome

Although EM images of intact ribosomes and their subunits were obtained some time ago, high resolution was difficult to achieve because of the necessity of staining or shadowing the particles. Nor could such techniques hope to tell us how the proteins and RNA were positioned inside the ribosome. Nevertheless, the overall

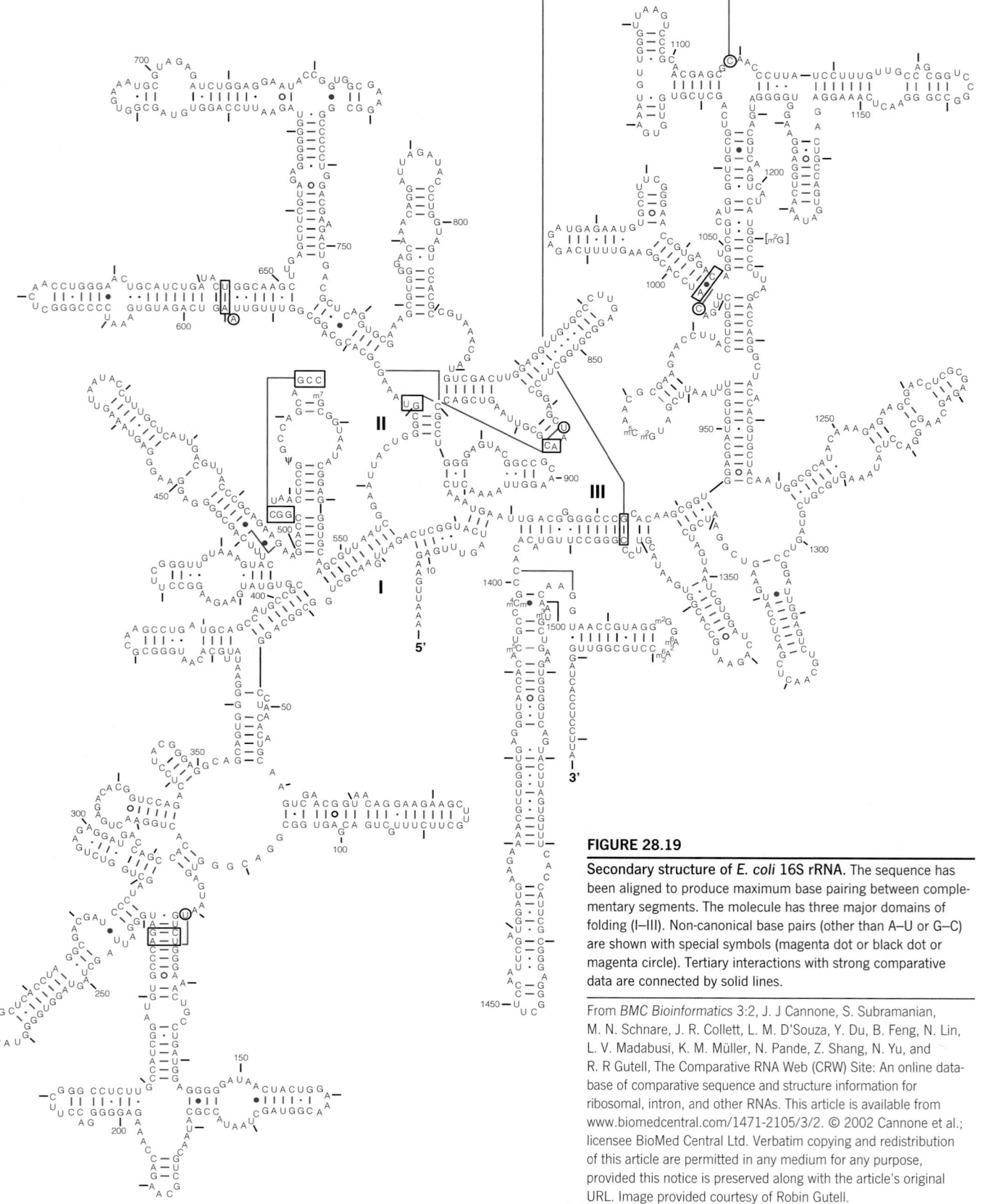

FIGURE 28.19

Secondary structure of *E. coli* 16S rRNA. The sequence has been aligned to produce maximum base pairing between complementary segments. The molecule has three major domains of folding (I–III). Non-canonical base pairs (other than A–U or G–C) are shown with special symbols (magenta dot or black dot or magenta circle). Tertiary interactions with strong comparative data are connected by solid lines.

shape of each subunit was seen as early as 1976 (Figure 28.20). Other techniques such as protein–protein or protein–RNA cross-linking, immunoelectron microscopy, cryoelectron microscopy, sequence analysis of the individual proteins, and neutron scattering gave much information about the placement of individual proteins within the overall structure.

Attempts to crystallize ribosomes began in the 1970s, but the size and complexity of these particles frustrated these early efforts. A key to the success of crystallization attempts, particularly in the laboratory of Ada Yonath, was the use of extremophilic bacteria as the source material. Although the *E. coli* ribosome was the most thoroughly studied to that point, the archaeal thermophile *Thermus thermophilus* and the halophile *Haloarcula marismortui* yielded the best crystals. Because the structure of the ribosome is well conserved evolutionarily, these bacteria made satisfactory models.

Figure 28.21 shows a model of the 70S ribosome based upon the first medium-resolution structures, in the late 1990s. Critical features known already or shown by this model are that the ribosome has three tRNA binding sites, that mRNA binding and decoding occur on the 30S subunit, that aminoacyl-tRNAs fill the gap between 30S and 50S subunits, that the newly synthesized polypeptide chain exits the ribosome through a tunnel in the 50S subunit, and that the peptidyltransferase reaction, which creates peptide bonds, occurs at a site on the 50S subunit.

Within a year after the medium-resolution structure was published, Thomas Steitz's laboratory published a high-resolution structure of the 50S subunit from *H. marismortui*, and shortly afterward Venki Ramakrishnan and colleagues described the 30S subunit from *T. thermophilus*, followed by the complete 70S ribosome from this organism. Figure 28.22 shows the Steitz 50S structure, and Figure 28.23 shows the Ramakrishnan 70S structure. Perhaps the most striking feature of both structures is that the peptidyltransferase site lies far from any protein. This structural work established conclusively that the ribosome is a *ribozyme*. We shall

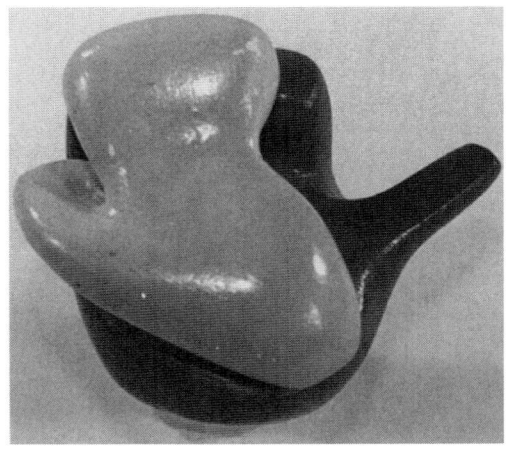

FIGURE 28.20

Images of ribosomal subunits as determined by electron microscopy. The 50S subunit is shown in black and the 30S subunit in light gray.

Reprinted from *Journal of Molecular Biology* 105: 131–159, J. A. Lake, Ribosome structure determined by electron microscopy of *Escherichia coli* small subunits, large subunits and monomeric ribosomes. © 1976, with permission from Elsevier.

FIGURE 28.21

A model of the 70S ribosome based upon early structural data. This model shows all three tRNA-binding sites occupied simultaneously, which does not normally occur. This view has the 30S subunit in front and the 50S subunit to the rear.

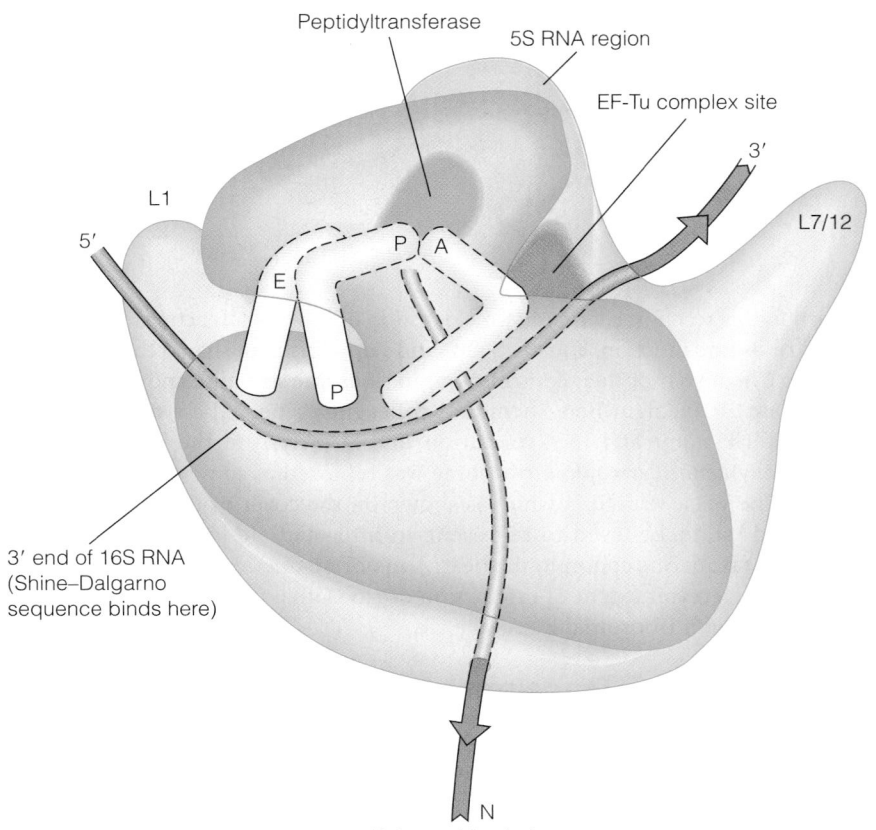

Peptidyltransferase

5S RNA region

EF-Tu complex site

L1

5′

3′

L7/12

E   P   A

P

3′ end of 16S RNA (Shine–Dalgarno sequence binds here)

N

Polypeptide chain

FIGURE 28.22

A high-resolution model of the 50S ribosomal sub-unit. This view shows the two stalks and central protuberance (CP), seen also in the early electron micrographs. In this image RNA is in gray, and proteins are in gold. The peptidyltransferase site, in green, is identified from the binding of an inhibitor. PDB ID 1FFK.

From *Science* 289:905–920, N. Ban, P. Nissen, J. Hansen, P. B. Moore, and T. A. Steitz, The complete atomic structure of the large ribosomal subunit at 2.4 Å resolution. © 2000. Reprinted with permission from AAAS.

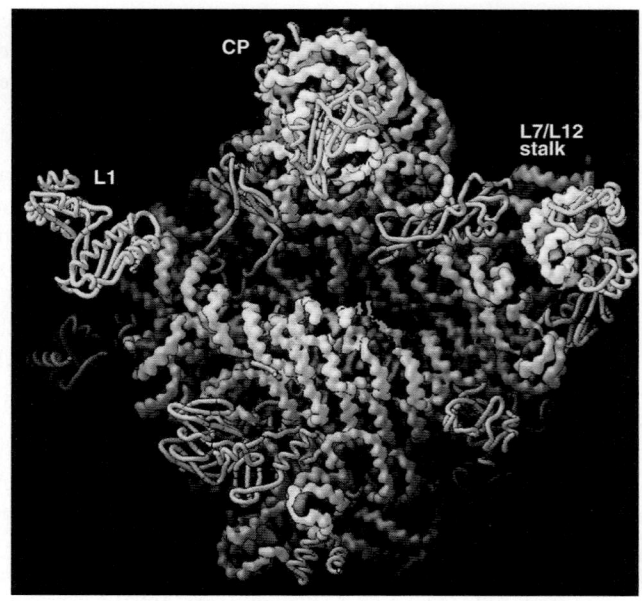

FIGURE 28.23

A model of the 70 ribosome, with mRNA and tRNA bound. The 30S subunit is in light blue-green (RNA) and blue (protein), and the 50S subunit is in orange (RNA) and brown (protein). Two bound tRNAs can be seen; peptidyl-tRNA in green and deacylated tRNA in yellow. mRNA is shown in gray. PDB ID 2j00 (30S-1), 2j01 (50S-1), 2j02 (30S-2), and 2j03 (50S-2).

Courtesy of V. Ramakrishnan (2009) Nobel Prize lecture. © The Nobel Foundation.

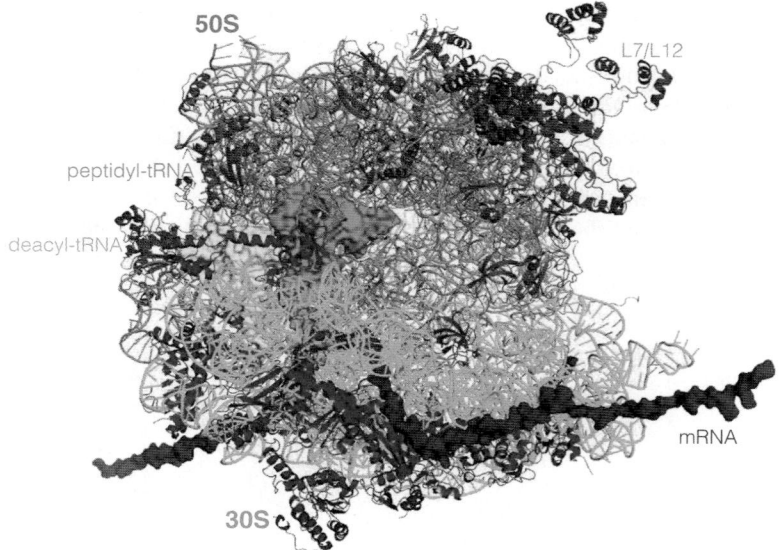

return to this reaction later and to the question of how this structural work illuminated ribosome function. Yonath, Steitz, and Ramakrishnan shared the 2009 Nobel Prize in Chemistry for their contributions to ribosome structure and function.

In late 2010 Adam Ben-Shem and colleagues reported a structure for the yeast 80S ribosome at 4.15 Å resolution, and in early 2011 the large subunit of the *Tetrahymena thermophila* ribosome was reported at somewhat higher resolution. The yeast structure, which is shown on the cover of this book, was in a "racheted" state, believed to represent an intermediate in translocation, the movement from one codon to the next. A major distinction from bacterial ribosomes is the greater extent of interaction of ribosomal proteins with each other, rather than with ribosomal RNAs.

Mechanism of Translation

We now have identified all of the major participants in the translation process: a messenger RNA, charged tRNAs, soluble protein factors, and the ribosome, where the actual translation events occur. As with transcription, we can divide translation into three stages: *initiation, elongation,* and *termination.* Here we describe these steps, primarily as they occur in prokaryotes, where our understanding is most detailed. Significant, though not fundamental, differences in eukaryotic protein synthesis are discussed next.

Each step in translation requires a number of specific proteins that interact with the major participants listed above. These proteins are referred to as *initiation factors* (IFs), *elongation factors* (EFs), and *release factors* (RFs). These factors, together with some of their properties and functions, are listed in Table 28.4.

Initiation

Initiation of translation is schematized in Figure 28.24. Initiation results in formation of a *70S initiation complex,* which consists of a ribosome bound to mRNA and to a charged *initiator tRNA.* In bacteria the initiator tRNA is charged with *N*-formylmethionine (fMet). First the mRNA and tRNA bind to a free 30S subunit, then the 50S subunit is added to form the entire complex. In most cases the initiation codon to which the initiator tRNA binds is AUG, also used as an internal methionine codon. As we indicated previously (page 1184), the initiator AUG is distinguished from internal methionine codons by the presence upstream of a Shine–Dalgarno sequence, which binds to a complementary sequence in 16S rRNA, thereby positioning the initiator AUG.

Binding of mRNA and initiator tRNA also requires binding of the three **initiation factors** (IF1, IF2, and IF3) to a free 30S subunit. The factors IF3 and IF1 promote dissociation of preexisting 70S ribosomes, thereby producing the free 30S subunits needed for initiation (Figure 28.24, step 1). The third factor, IF2, is bound carrying a molecule of GTP; it delivers the charged initiator tRNA in binding to the 30S subunit. IF2 is a G protein, similar to those involved in signal transduction. At about the same time that the IF2–fMet-tRNA^{fMet} complex is bound, the mRNA is bound (step 2). Although the order of these additions is still uncertain, it is clear that IF2–GTP is absolutely required for binding of the first (initiator) tRNA. With binding of the initiator tRNA and the mRNA, formation of the *30S initiation complex* is complete. The initiation complex has high affinity for a 50S subunit and binds one from the available pool (step 3), with concomitant release of IF3.

The initiator tRNA is special. It recognizes and binds to the AUG codon that would normally code for methionine, but it actually carries an *N*-formylmethionine. The formyl group is added *after* charging of the tRNA, by an enzyme *(transformylase)* that recognizes the particular tRNA^{fMet} and transfers a formyl group from 10-formyltetrahydrofolate (see Figure 20.17, page 852). *Only* tRNA^{fMet} is accepted to form the 30S initiation complex; all subsequent charged tRNAs require the fully assembled 70S ribosome. Therefore, most (not all) prokaryotic proteins are synthesized with the same N-terminal residue, *N*-formylmethionine. In almost all cases, the formyl group is removed during chain elongation. For many proteins the methionine itself is also cleaved off later.

10-Formyl-tetrahydrofolate   **Tetrahydrofolate**

transformylase

CHO
|
NH2   NH

S — O — tRNA   S — O — tRNA
    ‖   ‖
    O   O

Met-tRNA^{fMet} **fMet-tRNA**^{fMet}

The mRNA attaches to the 30S subunit near the 5′ end of the message, which is appropriate because all messages are translated in the 5′ → 3′ direction. As mentioned above, an AUG initiation codon is recognized by an upstream Shine–Dalgarno sequence, which is complementary to the sequence, 3′···UCCUCC···5′ in 16S rRNA. This will pair with any Shine–Dalgarno sequence (for example, those shown in Table 28.3). This pairing aligns the message correctly for the start of

Translation involves three steps—initiation, elongation, and termination—each aided by soluble protein factors.

In initiation, the correct attachment of mRNA to the ribosome is determined by binding of the Shine–Dalgarno sequence to a sequence on the 16S rRNA of the ribosome.

FIGURE 28.24

Initiation of protein biosynthesis in prokaryotes.
The ribosome contains three tRNA binding sites, shown here as E, P, and A; these are called the exit, peptidyl-tRNA, and aminoacyl-tRNA binding sites, respectively. The initiator AUG codon is positioned so that fMet-tRNA binds in the P site.

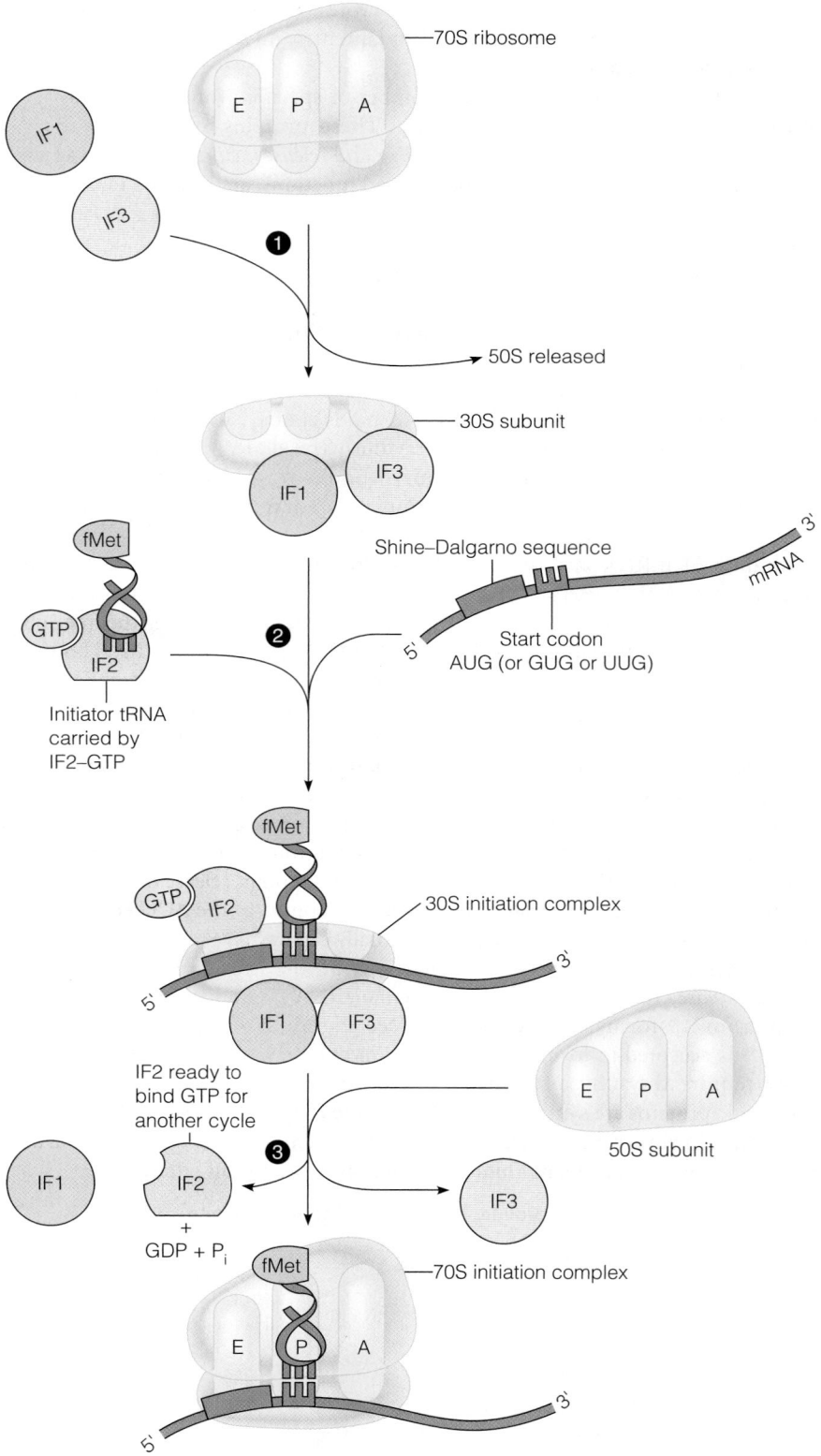

translation. In particular, it places the initiator codon next to the P site, one of three tRNA binding sites in the ribosome (see below).

Translation cannot start until the 50S subunit has bound to the 30S initiation complex. The ribosome has three sites for tRNA binding, called the P (peptidyl) site, the A (aminoacyl) site, and the E (exit) site. The AUG initiator codon, with its bound fMet-tRNA^{fMet}, aligns with the P site. At this point, the GTP molecule carried by IF2 is hydrolyzed, and IF2–GDP, P$_i$, and IF1 are all released. The 70S initiation complex so formed is ready to accept a second charged tRNA and begin elongation of the protein chain.

The locations of the P, A, and E binding sites for tRNAs were originally established by chemical cross-linking (Figure 28.25), but are now confirmed by X-ray crystallography (Figure 28.23). The anticodon ends of the tRNA molecules contact the 30S subunit, whereas the acceptor ends interact specifically with the 50S subunit. All of the ribosomal proteins contacted lie in the cavity between the 30S and 50S subunits. The tRNA molecules are oriented with their anticodons reaching the mRNA at the bottom of the cavity close to the 30S subunit and their acceptor ends contacting the peptidyltransferase region on the 50S subunit, near the top of the cavity.

Elongation

Growth of the polypeptide chain on the ribosome occurs by a cyclic process. Figure 28.26 illustrates a single round in this cycle. In this particular example, the fifth amino acid from the N-terminus is being linked to the sixth. However, all cycles are the same until a termination signal is reached.

At the beginning of each cycle, the nascent polypeptide chain is attached to a tRNA in the P (peptidyl) site, and the A (aminoacyl) and E (exit) sites are empty. Aligned with the A site is the mRNA codon corresponding to the *next* amino acid to be incorporated. The charged (aminoacylated) tRNA is escorted to the A site in a complex with a protein, the elongation factor EF-Tu, which also carries a molecule of GTP. (Note the parallel to IF2–GTP here.) EF-Tu plays an active role in ensuring that the correct aminoacyl-tRNA is fitted to its codon. As schematized in Figure 28.27, which results from work in the Ramakrishnan laboratory, the aminoacyl-tRNA is distorted in its complex with EF-Tu. Initial binding puts the tRNA anticodon loop into the decoding center on the 30S subunit, with the acceptor stem near the EF-Tu site. Nucleotides in the decoding site probe the major groove of the anticodon loop, specifically in positions 1 and 2. This structural work confirmed the wobble hypothesis, by showing that codon–anticodon fitting is more stringent in the first two positions. GTP hydrolysis by EF-Tu results in conformational changes that move the aminoacyl-tRNA entirely into the A site and cause dissociation of EF-Tu itself. The EF-Tu–GTP complex is then regenerated by the subsidiary cycle shown in Figure 28.28. After the charged tRNA is in place, it is checked both before and after the GTP hydrolysis and rejected if incorrect.

The next, and crucial, step is peptide bond formation (Figure 28.26, step 2). The polypeptide chain that was attached to the tRNA in the P site is now transferred to the amino group of the amino acid carried by the A-site tRNA. This step is catalyzed by *peptidyltransferase*, an integral part of the 50S subunit. As mentioned previously, the structure determination of the 50S subunit established conclusively that catalysis is carried out by the RNA portion of the subunit. This finding was crucial to acceptance of the "RNA world" model for the origin of life, for it indicates how progenitors to living cells could exist in the presence of RNA but the absence of protein.

Analysis of the 50S subunit in the Steitz laboratory showed that a conserved AMP residue (2486 in *H. marismortui* and 2451 in *E. coli*) exists in an environment that makes the purine ring unusually basic, probably resulting from hydrogen bonding to a nearby GMP. This suggests a process in which N3 abstracts a proton from the amino group of the aminoacyl-tRNA, converting the amino group to a better nucleophile that attacks the carboxyl carbon of the C-terminal amino acid, linked to peptidyl-tRNA, as shown in Figure 28.29. The protonated

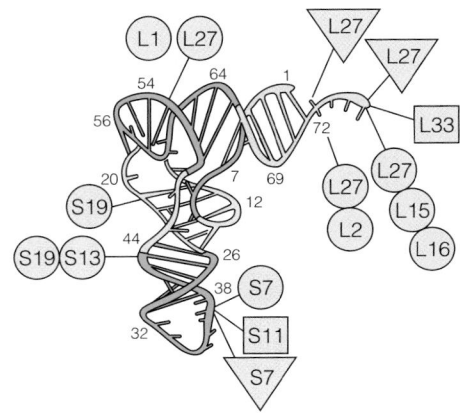

FIGURE 28.25

Environment of tRNAs at the ribosome as determined by cross-linking. Cross-links from defined nucleotide positions in the tRNA to ribosomal proteins are shown. Proteins were differentially cross-linked depending on the location of the tRNA (A site, triangles; P site, circles; E site, squares; S = small subunit, L = large subunit).

Biochimie 76:1235–1246, J. Wower, K. V. Rosen, S. S. Hixson, and R. A. Zimmermann, Recombinant photoreactive tRNA molecules as probes for cross-linking studies. Copyright © 1994 Société française de biochimie et biologie moléculaire/Elsevier Masson SAS. All rights reserved.

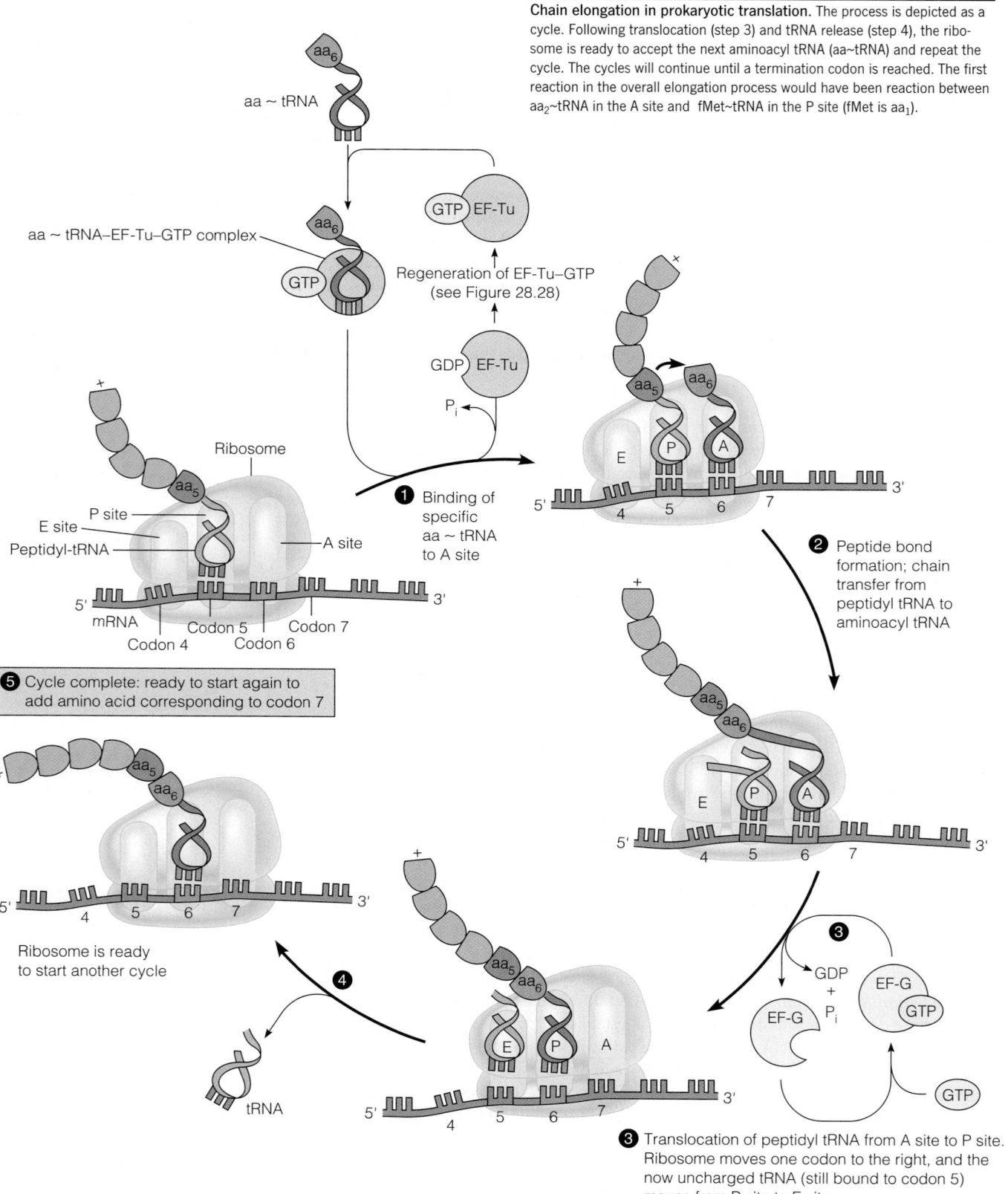

FIGURE 28.26

Chain elongation in prokaryotic translation. The process is depicted as a cycle. Following translocation (step 3) and tRNA release (step 4), the ribosome is ready to accept the next aminoacyl tRNA (aa~tRNA) and repeat the cycle. The cycles will continue until a termination codon is reached. The first reaction in the overall elongation process would have been reaction between aa_2~tRNA in the A site and fMet~tRNA in the P site (fMet is aa_1).

aa ~ tRNA

aa ~ tRNA–EF-Tu–GTP complex

Regeneration of EF-Tu–GTP (see Figure 28.28)

Ribosome

P site

E site

Peptidyl-tRNA

A site

mRNA

Codon 4   Codon 5   Codon 6   Codon 7

❶ Binding of specific aa ~ tRNA to A site

❷ Peptide bond formation; chain transfer from peptidyl tRNA to aminoacyl tRNA

❺ Cycle complete: ready to start again to add amino acid corresponding to codon 7

Ribosome is ready to start another cycle

tRNA

❸ Translocation of peptidyl tRNA from A site to P site. Ribosome moves one codon to the right, and the now uncharged tRNA (still bound to codon 5) moves from P site to E site

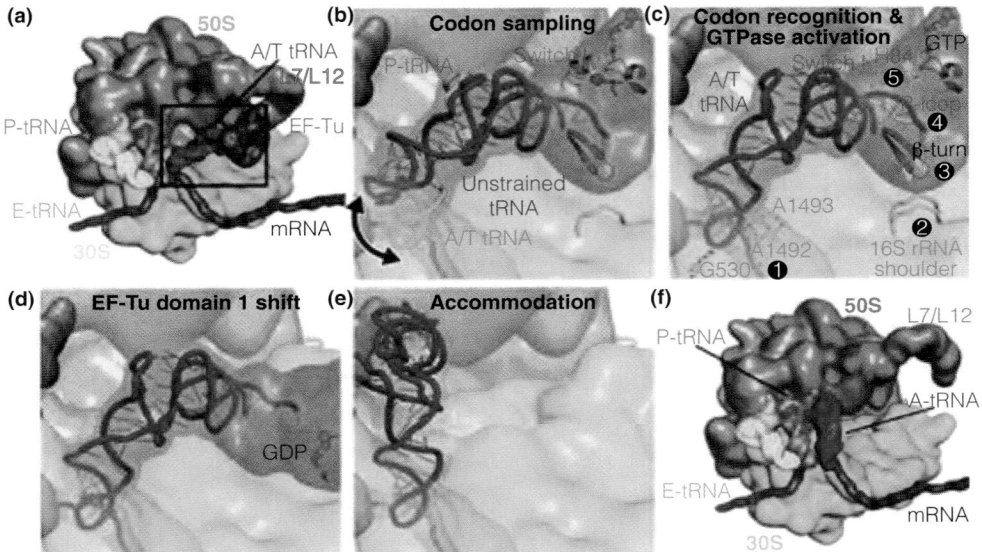

FIGURE 28.27

The ribosomal decoding pathway. (a) L7/L12 stalk on 50S subunit recruits ternary complex (aminoacyl-tRNA-EF-Tu-GTP) to a ribosome with deacylated tRNA in E site and peptidyl-tRNA in P site. The black frame is enlarged in subsequent panels. A/T tRNA is a tRNA molecule temporarily distorted, to interact simultaneously with the decoding center of the 30S subunit and EF-Tu, bound at a site in the intersubunit space. **(b)** tRNA samples codon–anticodon pairing. **(c)** The match is sensed by specific nucleotides in coding site (G530, A1492, A1493). Codon recognition triggers 30S subunit domain closure. A chain of conformational changes (shown as 2–5) opens a hydrophobic gate, allowing His84 on EF-Tu to initiate hydrolysis of GTP. **(d)** GTP hydrolysis and P$_i$ release cause conformational change in EF-Tu, leading to its release from the ribosome. **(e)** and **(f)** EF-Tu release leads to relaxation of aminoacyl-tRNA structure and its accommodation at both the coding site and the peptidyl-transferase site. PDB ID 2WRN, 2WRO,2WRQ, and 2WRR.

From *Science* 326:688–693, T. M. Schmeing, R. M. Voorhees, A. C. Kelley, Y.-G. Gao, F. V. Murphy IV, J. R. Weir, and V. Ramakrishnan, The crystal structure of the ribosome bound to EF-Tu and aminoacyl-tRNA to EF-Tu and aminoacyl-tRNA. © 2009. Reprinted with permission from AAAS.

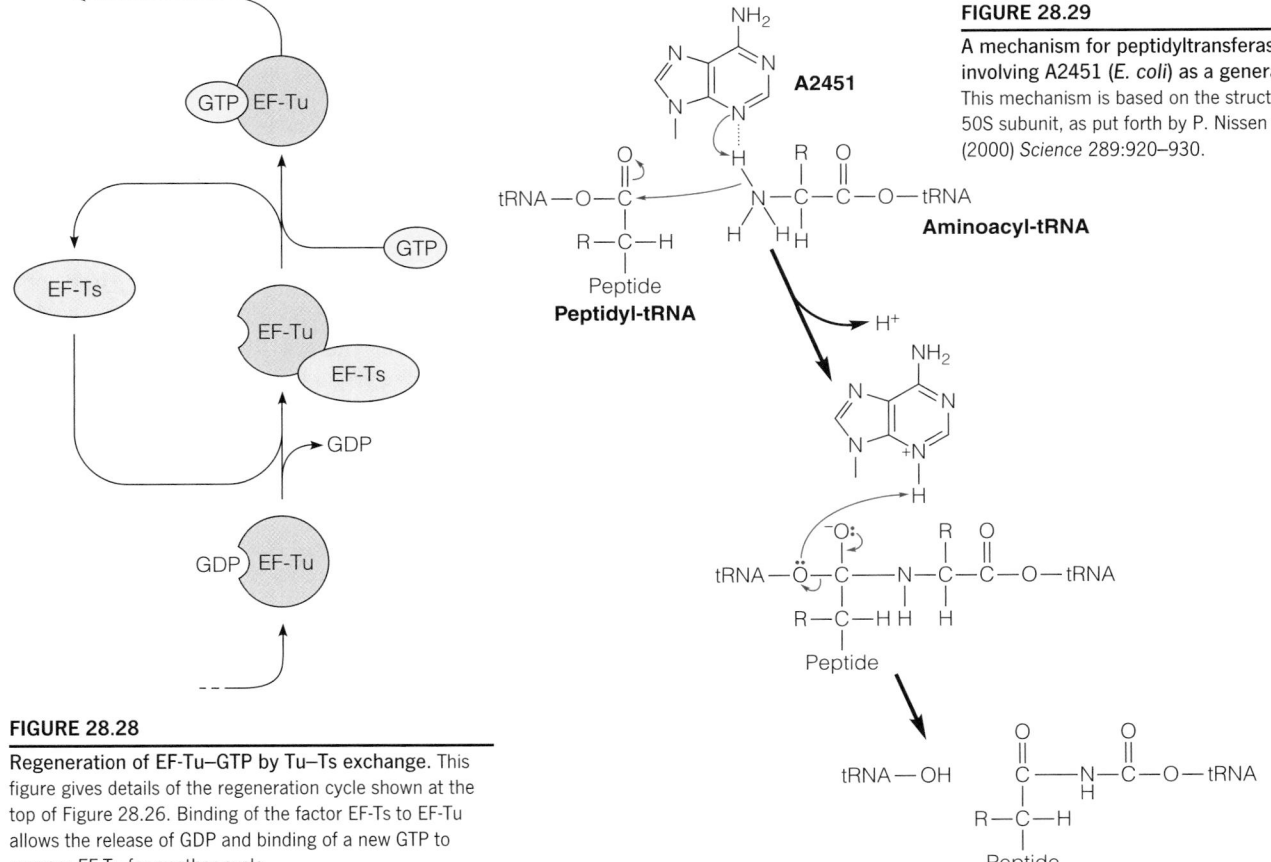

FIGURE 28.28

Regeneration of EF-Tu–GTP by Tu–Ts exchange. This figure gives details of the regeneration cycle shown at the top of Figure 28.26. Binding of the factor EF-Ts to EF-Tu allows the release of GDP and binding of a new GTP to prepare EF-Tu for another cycle.

FIGURE 28.29

A mechanism for peptidyltransferase involving A2451 (*E. coli*) as a general base. This mechanism is based on the structure of the 50S subunit, as put forth by P. Nissen et al. (2000) *Science* 289:920–930.

N3 then stabilizes a tetrahedral carbon intermediate by binding to the oxyanion. The proton is then transferred to the peptidyl-tRNA 3′ hydroxyl as the newly formed peptide deacylates. Concomitant with this transfer is a switch from the simple P and A states to hybrid states, in which the acceptor ends of the two tRNA molecules move into the leftward positions while the codon ends remain fixed as before. These hybrid sites are indicated as E/P and P/A. This can be considered the first half of the *translocation* step (step 3 in Figure 28.26).

In elongation, the growing peptide chain at the P site is transferred to the newly arrived aminoacyl tRNA in the A site. Translocation then moves this tRNA to the P site and the previous tRNA to the E site.

To complete the translocation step, the anticodon end of the now uncharged tRNA in the P site is transferred to the E site, and the tRNA in the A site (the tRNA that now has the nascent polypeptide chain attached to it) is moved completely to the P site. In the process, the ribosome moves a three-nucleotide step in the 3′ direction along the mRNA, placing a new codon adjacent to the now empty A site. Like peptidyl transfer, this step requires a protein factor (EF-G) bound to GTP and requires GTP hydrolysis. Crystallographic studies reveal a remarkable "molecular mimicry" between EF-G–GTP and the ternary complex aa-tRNA–EF-Tu–GTP. As Figure 28.30 shows, the protein and the RNA–protein complex have almost exactly the same shape, even though they differ entirely in composition and sequence. It is speculated that the reason for this similarity is to allow EF-G–GTP to move temporarily into the A site, facilitating the displacement of the peptidyl–tRNA complex. Structural studies support this model.

During translation the ribosome is "ratcheted" along the mRNA molecule by a process involving rotation of the two ribosomal subunits with respect to each other. Based on structural analysis of the ribosome in intermediate states of rotation, the ratcheting process has been schematized as shown in Figure 28.31.

At this point the E and P sites are occupied, but A is empty. As the deacylated tRNA is released from E (step 4, Figure 28.26), the A site gains high affinity and accepts the aminoacyl tRNA dictated by the next codon. A cycle of elongation is now complete. All is as it was at the start, except that now:

1. The polypeptide chain has grown by one residue.
2. The ribosome has moved along the mRNA by three nucleotide residues—one codon.
3. At least two molecules of GTP have been hydrolyzed.

The whole process is repeated again and again until a termination signal is reached, with the newly synthesized polypeptide chain being forced to exit the ribosome through the tunnel mentioned previously.

Termination

The completion of polypeptide synthesis is signaled by the translocation of one of the *stop codons* (UAA, UAG, or UGA) into the A site. Because there are no tRNAs that recognize these codons under normal circumstances, termination of the

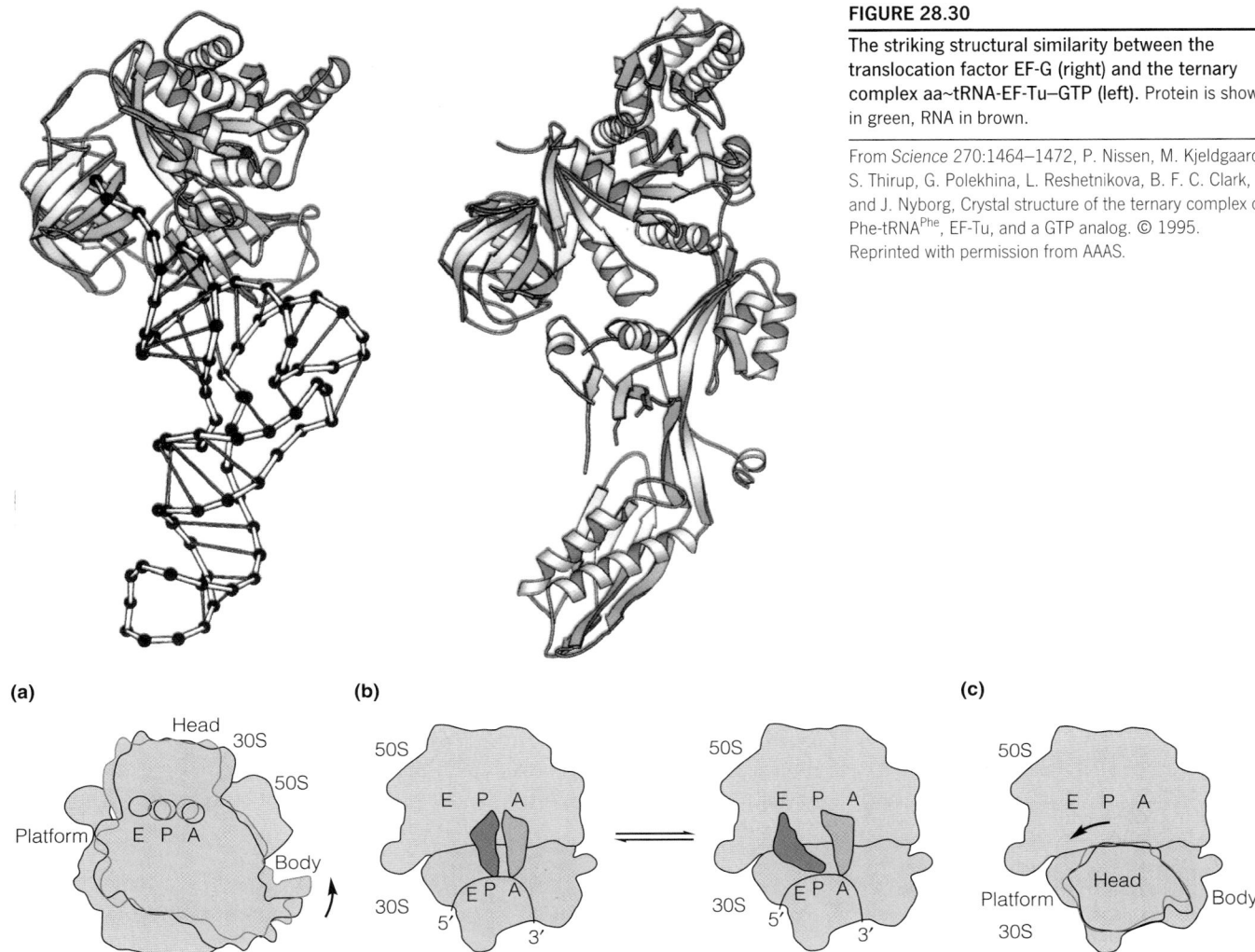

FIGURE 28.30

The striking structural similarity between the translocation factor EF-G (right) and the ternary complex aa~tRNA-EF-Tu–GTP (left). Protein is shown in green, RNA in brown.

From *Science* 270:1464–1472, P. Nissen, M. Kjeldgaard, S. Thirup, G. Polekhina, L. Reshetnikova, B. F. C. Clark, and J. Nyborg, Crystal structure of the ternary complex of Phe-tRNA[Phe], EF-Tu, and a GTP analog. © 1995. Reprinted with permission from AAAS.

(a) **(b)** **(c)**

FIGURE 28.31

A schematic view of ribosome subunit rotational motions, based on crystal structures of ribosomes in intermediate states. **(a)** View from the bottom. 30S subunit (blue) is shown in starting conformation after termination (outlined in red) to a fully rotated conformation seen during elongation (black outline). **(b)** Side view. During transition to the fully rotated state, tRNAs shift from binding in A/A and P/P sites (30S/50S) to occupying hybrid sites A/P and P/E. **(c)** Rotation in another plane can move the head domain of the 30S subunit as much as 14° toward the E site.

From *Science* 325:1014–1017, W. Zhang, J. A. Dunkle, and J. H. D. Cate, Structures of the ribosome in intermediate states of ratcheting. © 2009. Reprinted with permission from AAAS.

chain does not involve binding of a tRNA. Instead, protein *release factors* participate in the termination process. The three release factors found in prokaryotes are listed in Table 28.4. Two of these factors can bind to the ribosome when a stop codon occupies the A site: RF1 recognizes UAA and UAG, and RF2 recognizes UAA and UGA. The third factor, RF3, is a GTPase that appears to stimulate the release process, via GTP binding and hydrolysis. Structural analysis of the ribosome complexed with RF2 show that the release factor interacts directly with a UGA termination codon (Figure 28.32).

The sequence of termination events is as shown in Figure 28.33. After RF1 or RF2 has bound to the ribosome, the peptidyltransferase transfers the C-terminal residue of the polypeptide chain from the P-site tRNA to a water molecule, releasing the peptide chain from the ribosome. The chemistry of this reaction is similar to peptide bond formation (Figure 28.29), except that water replaces the α-amino group as the attacking nucleophile. The RF factors and GDP are then released, followed by the tRNA. The 70S ribosome is now unstable. Its instability is accentuated by the presence of a protein called *ribosome recycling factor,* and also by the initiation factors IF3 and IF1, and the ribosome readily dissociates to 50S and 30S subunits prepared for another round of translation.

When the ribosomal subunits separate, the 30S subunit may or may not dissociate from its mRNA. In some cases in which polycistronic messages are being translated, the 30S subunit may simply slide along the mRNA until the next

Termination requires protein release factors that somehow recognize stop codons.

50S
P-site tRNA
E-site tRNA
RF2
mRNA
30S
Domain 3
Domain 1
Domain 2
Domain 4

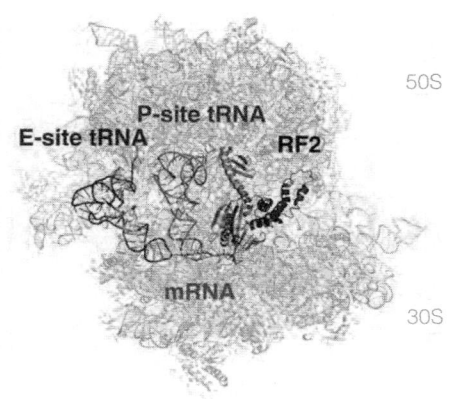

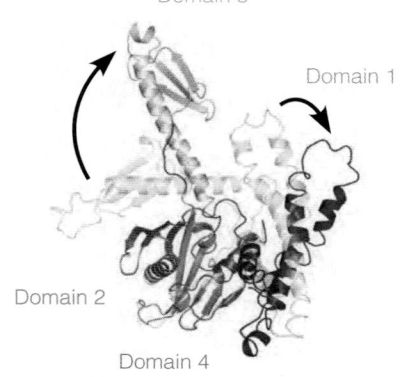

FIGURE 28.32

Interaction of RF2 with a UGA stop codon in the decoding center. Above, the ribosome in complex with RF2. UGA is in magenta, and RF2 is in green, except for those parts interacting directly with the codon, which are shown in red. RF2 helical domains that move significantly are shown in color, with the extent of the movements (domains 1 and 3), in the same color scheme, shown below. PDB ID 2jl5 to 2jl8.

From *Science* 322:953–956, A. Wexelbaumer, H. Jin, C. Neubauer, R. M. Voorhees, S. Petry, A. C. Kelley, and V. Ramakrishnan, Insights into translational termination from the structure of RF2 bound to the ribosome. © 2008. Reprinted with permission from AAAS.

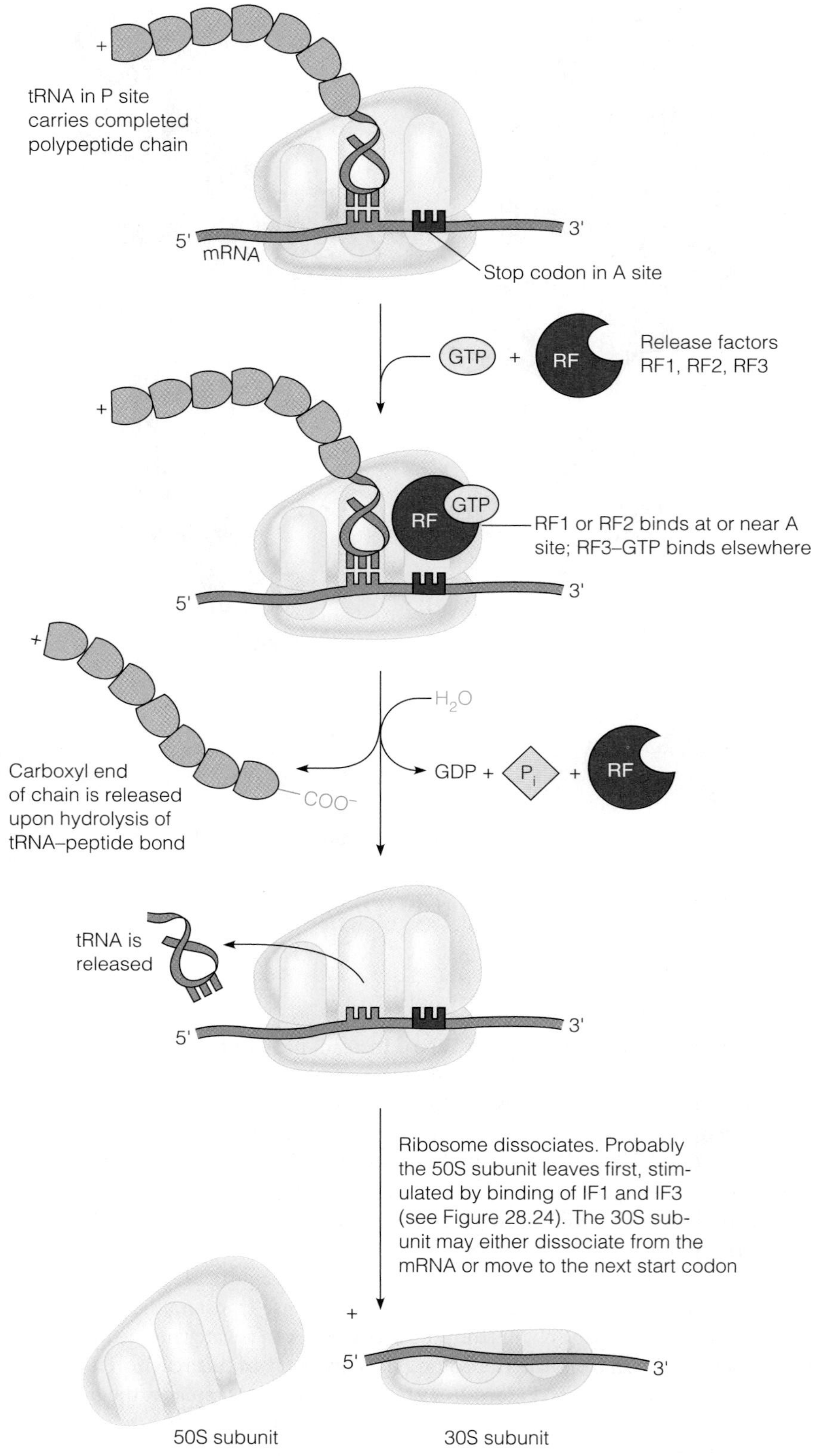

+
tRNA in P site
carries completed
polypeptide chain
5'
mRNA
3'
Stop codon in A site
GTP
+
RF
Release factors
RF1, RF2, RF3
+
RF
GTP
RF1 or RF2 binds at or near A
site; RF3–GTP binds elsewhere
5'
3'
+
H_2O
Carboxyl end
of chain is released
upon hydrolysis of
tRNA–peptide bond
COO^-
GDP + P_i + RF
tRNA is
released
5'
3'
Ribosome dissociates. Probably
the 50S subunit leaves first, stimulated by binding of IF1 and IF3
(see Figure 28.24). The 30S subunit may either dissociate from the
mRNA or move to the next start codon
+
5'
3'
50S subunit
30S subunit

FIGURE 28.33

Termination of translation in prokaryotes.

Shine–Dalgarno sequence and initiation codon are encountered and then begin a new round of translation. If the 30S subunit does dissociate from the message, it will soon reattach to another one.

Suppression of Nonsense Mutations

Understanding the process of termination helped clarify some peculiar observations concerning nonsense mutations. Recall from Chapter 7 that a *nonsense mutation* is one in which a codon for some amino acid has been mutated into a stop codon so that the polypeptide chain terminates prematurely. These mutations were originally discovered because their phenotypic expression could be *suppressed* by a class of mutations located in other genes. Recall from Chapter 25 that suppression is defined genetically as restoration of wild-type function by a second mutation at a different site. When this second mutation occurs in a different gene, the phenomenon is called **intergenic suppression**. Upon examination, the suppressors of nonsense mutations were found to lie in tRNA genes.

Consider the example shown in Figure 28.34. A nonsense mutation has changed a codon that normally specifies the amino acid tyrosine into a stop codon, causing premature termination of the polypeptide chain. If, however, one of the several tyrosine tRNAs mutates in its anticodon region so as to recognize

The effects of nonsense mutations can sometimes be suppressed by suppressor mutations, in which a tRNA mutates to recognize a stop codon and inserts an amino acid instead.

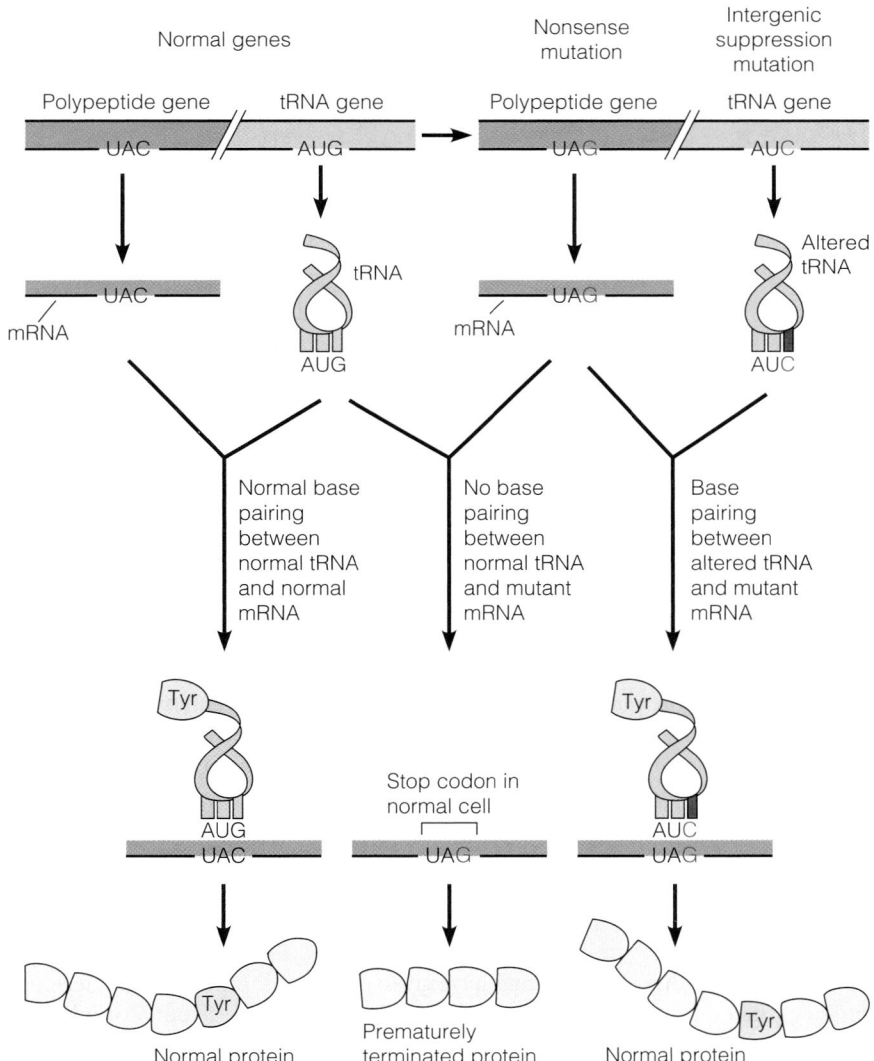

FIGURE 28.34

How an intergenic suppression mutation can overcome a nonsense mutation. A nonsense mutation in a protein-coding gene changes a codon for an amino acid into a stop codon, causing translation to terminate prematurely. Another mutation, in a tRNA gene, can circumvent the first mutation by altering the tRNA anticodon so that it will base-pair with the mutant mRNA. A functional protein is produced in this situation, even though suppression might not restore the original amino acid at that site.

the stop codon, translation can sometimes proceed in the normal fashion. Thus, a mutation that might otherwise be lethal can be suppressed by such a change, and the organism can survive. Clearly, it will still have problems, for the presence of such a mutated tRNA will interfere with the normal termination of other proteins. That they can survive at all depends on the fact that the suppressor mutation usually involves a minor tRNA species, little used in normal translation. Furthermore, such effects may be minimized by the frequent occurrence of two or more different stop signals in tandem in mRNAs. Even if the first stop codon is suppressed, the "emergency brake" still holds.

Suppressor mutations are by no means confined to correction of nonsense mutations. Some mutated tRNAs correct missense mutations, and some even contain two or four bases acting as the anticodon. These can therefore serve as **frameshift suppressors**.

Inhibition of Translation by Antibiotics

Many antibiotics have been shown to act by inhibiting specific steps in bacterial protein synthesis. Some of these antibiotics have proven to be useful reagents in analyzing mechanisms in translation, as well as in combating infections. We have already described the action of some kinds of antibiotics. In Chapter 9, we saw that the penicillins inhibit bacterial cell wall synthesis, and in Chapter 10 we discussed antibiotics such as gramicidin and valinomycin, which interfere with the ionic balance across membranes. Other antibiotics, such as rifampicin and streptolydigin (Chapter 27), block transcription in prokaryotes.

A host of naturally occurring substances interfere with various stages of protein synthesis. Some of these are shown in Figure 28.35. Each inhibits translation in a different way. Their importance to medicine stems largely from the fact that the translational machinery of eukaryotes is sufficiently different from that of prokaryotes that these antibiotics can be used safely in humans. In some cases (for example, the tetracyclines), antibiotics that would also inhibit eukaryotic translation are nevertheless harmless to eukaryotes because they cannot traverse the cell membranes of higher organisms.

A major problem with the therapeutic use of antibiotics is that microorganisms can develop resistance to many of them. An important example is *erythromycin* resistance. The erythromycin-binding site on the ribosome includes a specific region of the 23S RNA, and binding of the antibiotic can be inhibited by an enzyme that methylates a specific adenine residue in this region. Molecular biologists use erythromycin resistance in screening bacterial clones in recombinant DNA research. Resistance to erythromycin can be conferred to a bacterium by the insertion of a resistance gene coding for the methyltransferase on a bacterial plasmid. Bacteria containing the plasmid carrying the methylase gene will grow in an erythromycin-containing medium, whereas those lacking the plasmid will be killed. Thus, growth on such a medium automatically selects for only those clones that carry the plasmid. Because many such resistance genes are carried on plasmids, which are easily transferred from bacterium to bacterium, the frequency with which antibiotic-resistant strains arise is far higher than if the elements were carried on chromosomal genes. Another problem is the widespread use of antibiotics in animal husbandry, not to cure an infection, but rather to suppress any possible infections for improved animal weight maintenance and to prevent the spread of infection from animal to animal, under the crowded conditions in feedlots. That may be fine in the short run, but the resultant increased emergence of antibiotic-resistant strains makes many question the wisdom of this practice.

Structural studies on ribosomes have given enormous impetus to the development of new classes of antimicrobial agents to which resistance might not develop. In the same sense that knowledge of enzyme and receptor structure makes it possible to design entirely new inhibitors with therapeutic properties, the

A number of important antibiotics act by inhibiting translation in bacterial cells.

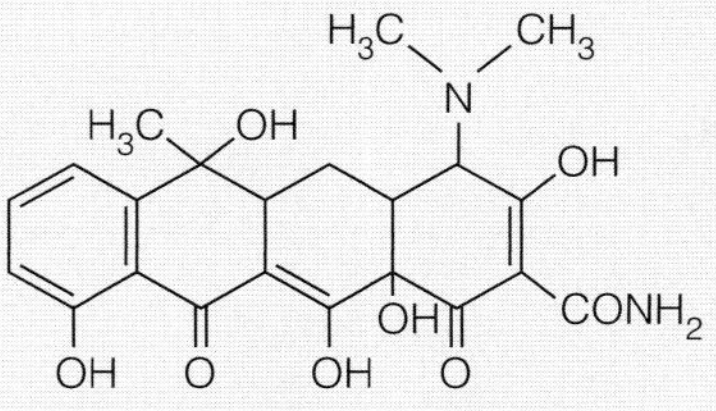

Tetracycline: Inhibits the binding of aminoacyl tRNAs to the ribosome and thereby blocks continued translation

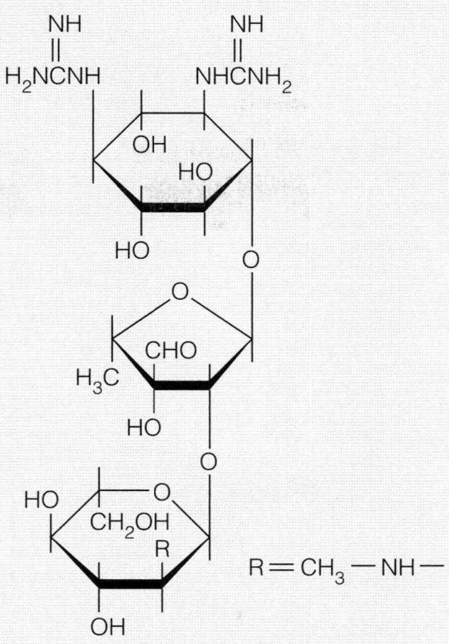

Streptomycin: Interferes with normal pairing between aminoacyl tRNAs and message codons, causing misreading, and thereby producing aberrant proteins

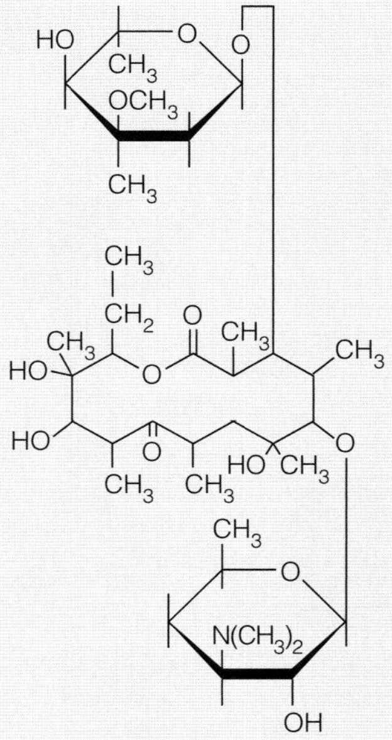

R = CH$_3$ — NH —

Erythromycin: Binds to a specific site on the 23S RNA and blocks elongation by interfering with the translocation step

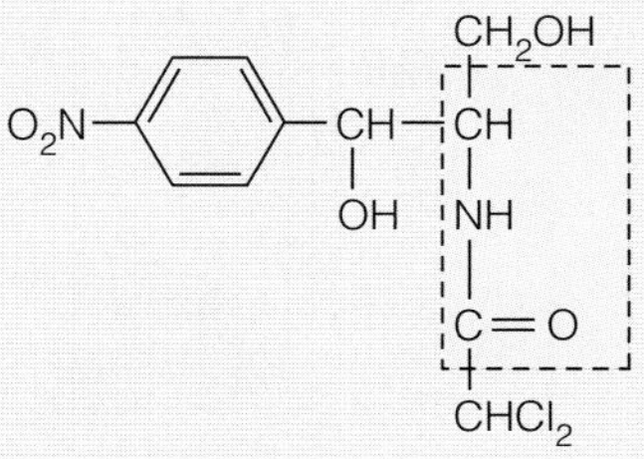

Chloramphenicol: Blocks elongation, apparently by acting as competitive inhibitor for the peptidyltransferase complex. The amide link (in blue) resembles a peptide bond

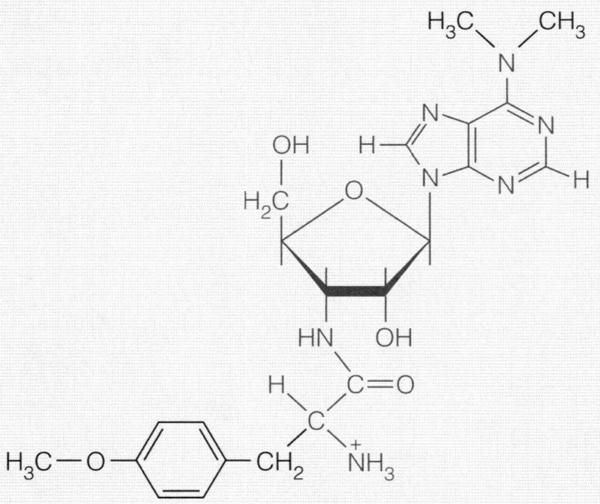

Puromycin: Causes premature chain termination. The red portion of the molecule resembles the 3′ end of the aminoacylated tRNA. It will enter the A site and transfer to the growing chain, causing premature chain release

FIGURE 28.35

Some antibiotics that act by interfering with protein biosynthesis. Erythromycin is one of the polyketide antibiotics whose biosynthesis was discussed in Chapter 17.

ribosome, because of its many activities and its structural conservation among bacteria, is an extremely attractive target for drug development.

Translation in Eukaryotes

The mechanism for translating messenger RNA into protein in eukaryotic cells is basically the same as in prokaryotes. In eukaryotes the ribosomes are larger and more complex, and virtually all mRNAs are monocistronic. There are more

In eukaryotes, translational initiation is more complex and requires more protein factors than in prokaryotes.

soluble protein factors, as we saw in Table 28.4, but the functions performed are comparable to those we have discussed for bacteria. The most significant differences are in initiation mechanisms; these are schematized for eukaryotic cells in Figure 28.36. Aside from the greater complexity of the ribosome and soluble

FIGURE 28.36

Initiation of translation in eukaryotes. Major differences from prokaryotic initiation are associated with cap binding and with the hunt for the first AUG.

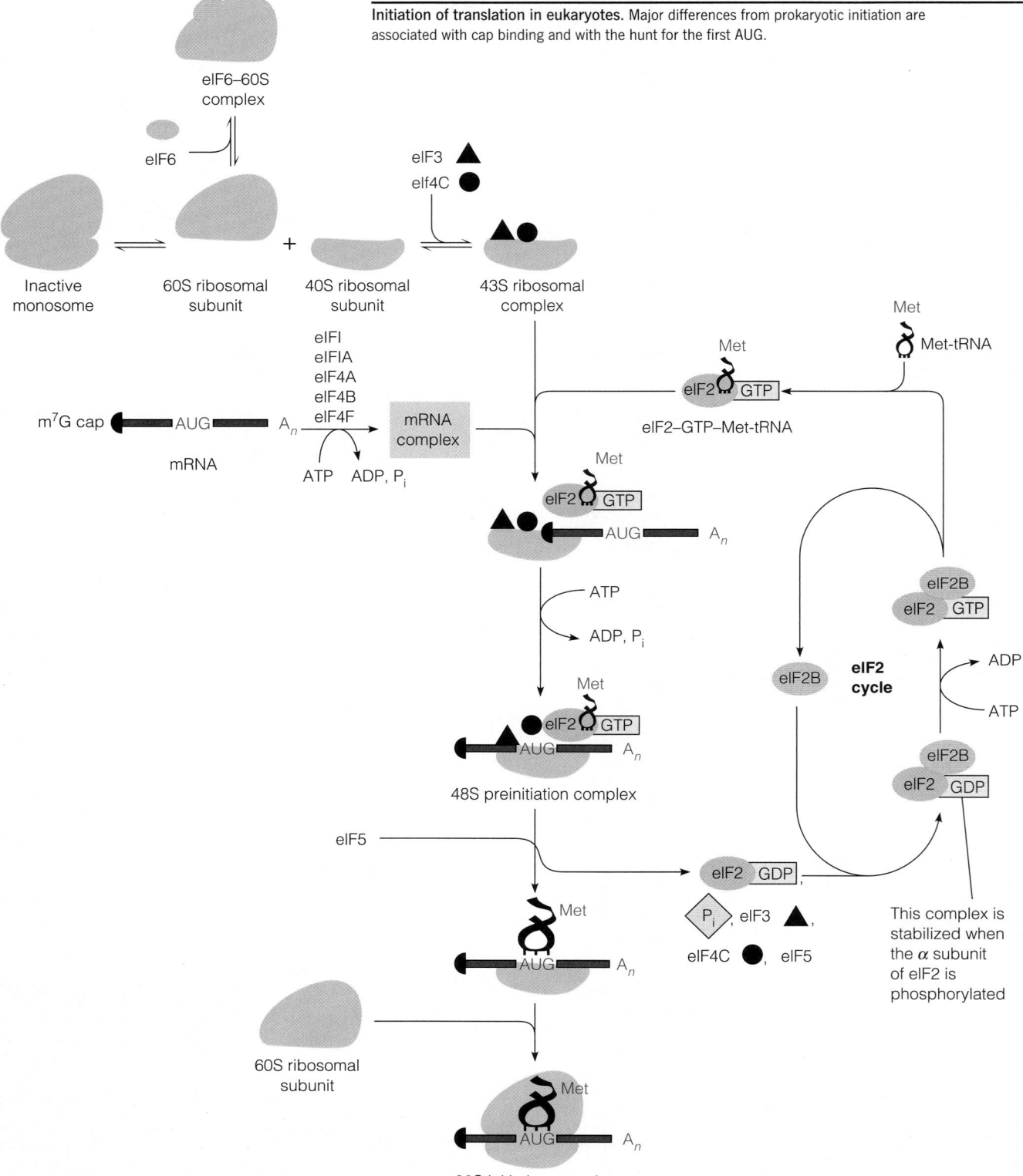

protein factors, the major differences are (1) that the 5′ end of a message is sensed not by a Shine-Dalgarno sequence, but by the 7-methylguanine cap, and (2) the N-terminal amino acid, inserted at the initiator AUG, is methionine, not *N*-formylmethionine. After detecting the 5′ cap, the ribosomal 40S subunit then scans along the mRNA (an ATP-dependent process) until the first AUG is found. At this point the initiation factors are released, and the 60S subunit is attached to begin translation.

A number of the common inhibitors of prokaryotic translation are also effective in eukaryotic cells. They include pactamycin, tetracycline, and puromycin. There are also inhibitors that are effective *only* in eukaryotes. Two important ones are *cycloheximide* and *diphtheria toxin.* Cycloheximide inhibits the translocation activity in the eukaryotic ribosome and is often used in biochemical studies when processes must be studied in the absence of protein synthesis. Diphtheria toxin is an enzyme, coded for by a bacteriophage that is lysogenic in the bacterium *Corynebacterium diphtheriae.* It catalyzes a reaction in which NAD^+ adds an *ADP-ribose* group to a specially modified histidine in the translocation factor eEF2, the eukaryotic equivalent of EF-G (Figure 28.37). Because the toxin is a catalyst, minute amounts can irreversibly block a cell's protein synthetic machinery; pure diphtheria toxin is one of the deadliest substances known.

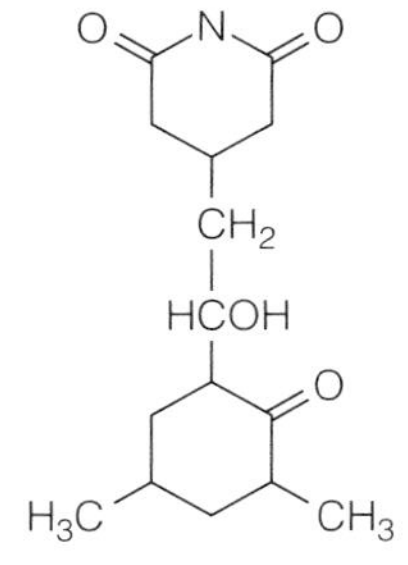

Protein Synthesis in Organelles

As described in Chapter 15, the mitochondrial genome (mtDNA) contains 37 genes that encode 13 proteins (in humans), all of which are subunits of the respiratory chain complexes. The remaining mtDNA genes encode 22 tRNAs and 2 rRNAs. These tRNAs and rRNAs are part of the mitochondrial protein synthesis machinery, required to translate the 13 proteins encoded in mitochondrial DNA. Reflecting its evolution from an ancient alpha-proteobacterium, the mitochondrial protein synthesis machinery is more closely related to the bacterial system than to the eukaryotic cytosolic system. Like the prokaryotic process, translation initiates with formylated Met-tRNA in mitochondria and requires only a handful of initiation and elongation factors. Also, mitochondrial protein synthesis is inhibited by some antibiotics that interfere with steps in bacterial protein synthesis. However, mitochondrial ribosomes have undergone a major remodeling during mitochondrial evolution. The rRNAs of mammalian mitochondrial ribosomes are smaller than their bacterial counterparts, and the large subunit of mitochondrial ribosomes completely lacks a 5S rRNA component. On the other hand, mitochondrial ribosomes have more protein subunits so that mitochondrial ribosomes have a protein:RNA ratio of 2:1, compared to a 1:2 ratio for the bacterial ribosome. Chloroplasts also possess their own protein synthesis machinery, but much less is known about the chloroplast system.

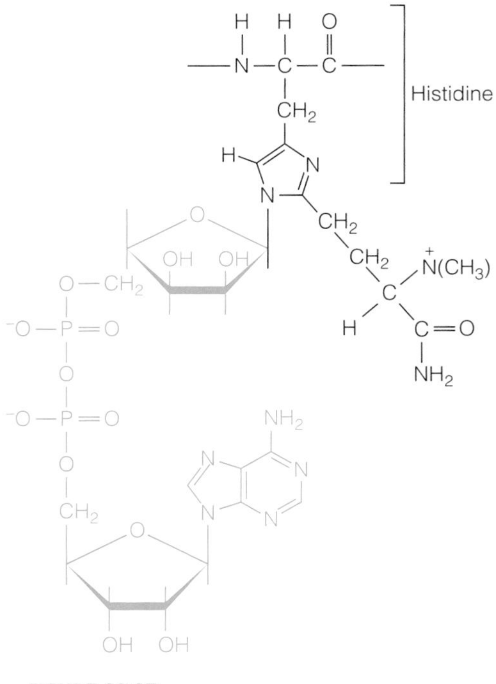

Rates and Energetics of Translation

Translation is a rapid process in prokaryotes. At 37 °C an *E. coli* ribosome can synthesize a 300-residue polypeptide chain in about 20 seconds. This means that a single ribosome passes through about 15 codons, or 45 nucleotides, in each second. This rate is almost exactly the same as our best estimates of the rate of prokaryotic *transcription,* which means that mRNA can be translated as fast as it is transcribed. That equality is not a coincidence. Recent studies with *E. coli* show that a ribosomal protein, NusE, interacts in the cell with an RNA polymerase component, NusG, and that through this interaction transcription and translation are physically coupled, as shown in Figure 28.38, with the rate of transcription being controlled by the rate of translation. Direct coupling of this type cannot occur in eukaryotic cells because the two processes take place in separate compartments.

Cycloheximide

FIGURE 28.37

ADP-ribosylated diphthamide derivative of histidine in eEF2. Synthesis of this derivative of a modified histidine in eEF2 using NAD^+ is catalyzed by diphtheria toxin. eEF2 is inactivated, and protein synthesis is therefore blocked. ADP ribose from NAD^+ is shown in blue. Diphthamide is in black.

But the rate mentioned above represents the growth of individual polypeptide chains and does not account for the total rate of protein synthesis in the cell because many ribosomes may be simultaneously translating a given message. In fact, if we were to carefully lyse *E. coli* cells, we would observe **polyribosomes** (also called polysomes) like those shown in Figure 28.39. Apparently, as soon as one

FIGURE 28.38

Coupling of transcription to translation in *E. coli* via the interaction between NusE and NusG.

From *Science* 328:436–437, J. W. Roberts, Syntheses that stay together. © 2010. Reprinted with permission from AAAS.

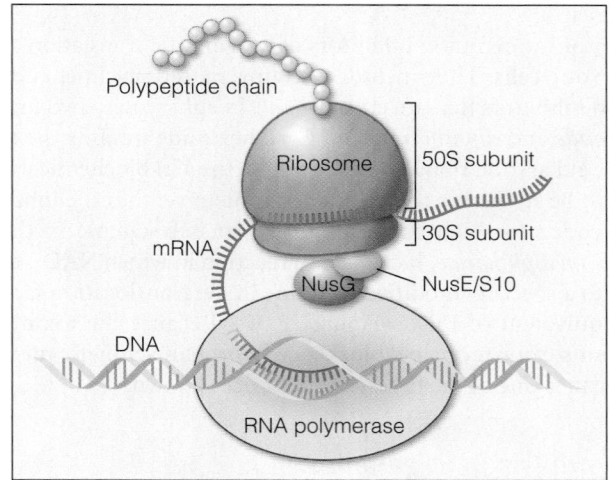

FIGURE 28.39

Polyribosomes. **(a)** Electron micrograph showing *E. coli* polyribosomes. The ribosomes are closely clustered on an mRNA molecule. **(b)** Schematic picture of a polyribosome like that shown in **(a)**. Each ribosome is to be imagined as moving from left to right.

(a) Courtesy of Barbara Hamkalo; (b) *Molecular Biology of the Gene*, 4th ed., James D. Watson, Nancy H. Hopkins, Jeffrey W. Roberts, Joan Argetsinger Steitz, and Alan M. Weiner. © 1987. Reprinted by permission of Pearson Education Inc., Upper Saddle River, NJ.

(a)

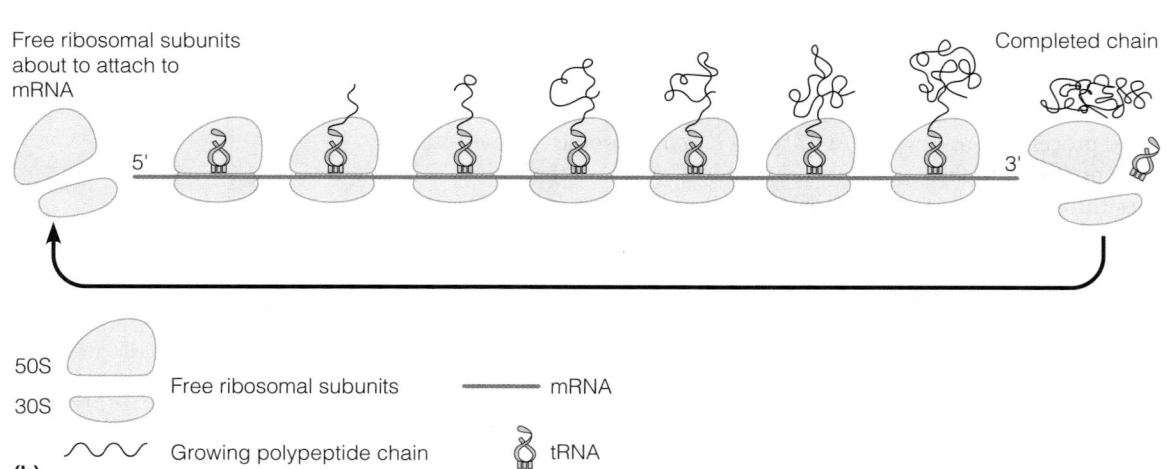

(b)

ribosome has moved clear of the 5′ region of the mRNA, another attaches. Under some conditions, as many as 50 ribosomes may be packed onto an mRNA, with one finishing translation every few seconds. Because each *E. coli* cell contains 15,000 ribosomes or more, all of them operating at full capacity can synthesize about 750 protein molecules of 300 residues each second.

The energy cost for this process is high. If we examine the individual steps in protein synthesis described earlier, we can make the following estimate of the total energy budget for synthesizing a protein of N residues:

The *equivalent* of

$2N$	ATPs are required to charge the tRNAs because the ATP is cleaved to AMP and PP_i, and PP_i is subsequently hydrolyzed.
1	GTP is needed for initiation.
N–1	GTPs are required to form the N—1 peptide bonds, in the EF-Tu–GTP hydrolysis step.
N–1	GTPs are necessary for the N—1 translocation steps.
1	GTP is required in termination.

Sum $= 4N$

Altogether, then, about $4N$ high-energy phosphate molecules must be hydrolyzed to complete a chain of N units. This is a minimal estimate, for it does not include the energy required to formylate methionine, nor any extra GTPs that may be expended in proofreading and replacing incorrectly bound tRNAs. Furthermore, there have been persistent, although debated, reports that *two* GTPs must be hydrolyzed for every aa-tRNA bound to the A site. But even at the conservative estimate, a typical protein of 300 residues costs the cell about 60,000 kJ of free energy per mole, if we assume ATP or GTP hydrolysis yields about 50 kJ/mol under cellular conditions. Proteins are expensive!

If we express the same data in terms of the energy requirement for synthesis per mole of *peptide bond,* we obtain a cost of about 200 kJ/mol. Given that the free energy change required to form a peptide bond in dilute aqueous solution is only about +20 kJ/mol, the price seems exorbitant. Why does the cell have no mechanism for making peptide bonds for a few dozen kilojoules each? Certainly, an input of even 40 kJ/mol would be enough to make the synthesis process very favorable—with an equilibrium constant of about 3000.

The key to this great energy expenditure is found in the fundamental nature of life. The cell is making polypeptides of *defined* sequence. If it were simply throwing together amino acids at random, the free energy price could be much cheaper. But a chain of 300 residues, made from 20 different amino acids, can be put together in 20^{300} different ways, whereas the cell needs *one* specific sequence. There is, in other words, a large entropy price to be paid in making specific sequences—and making them correctly. What this means at the mechanistic level is that every step in the assembly not only must be done with a free energy excess but also must involve a specific *choice*. Furthermore, the product must, at critical points, be checked by a proofreading mechanism, which in turn costs more energy. It is expensive to get a good translation of a book, for not only must the translators be expert and careful but their work must also be rechecked with great care.

Translation is fast but energy-expensive. About four ATP equivalents are needed for each amino acid added.

The Final Stages in Protein Synthesis: Folding and Covalent Modification

The polypeptide chain that emerges from the ribosome is not a completed, functional protein. It must fold into its tertiary structure, and it may have to associate with other subunits. In some cases, disulfide bonds must be formed, and other covalent modifications, such as hydroxylation of specific prolines and lysines must take place. Complexation with carbohydrate or lipid occurs after translation. In addition, many proteins are subjected to specific proteolytic cleavage to remove portions of the nascent chain.

Chain Folding

The cell need not wait until the entire chain is released from the ribosome to commence its finishing touches. The first portion of the nascent chain (about 30 residues) is protected as it passes through the tunnel in the ribosome. However, changes begin almost as soon as the N-terminal end emerges. There is good evidence that folding into the tertiary structure starts during translation and is nearly complete by the time the chain is released. For example, antibodies to *E. coli* β-galactosidase, which recognize the tertiary folding of the molecule, will attach to polyribosomes synthesizing this protein. This enzyme displays catalytic activity only as a tetramer. It has been demonstrated that nascent β-galactosidase chains, still attached to ribosomes, can associate with free subunits to form a functional tetramer. Thus, even quaternary structure can be partially established before synthesis is complete.

This behavior should not be surprising, if we recall (from Chapters 6 and 7) that formation of the secondary, tertiary, and quaternary levels of protein structure is thermodynamically favored. However, as we have seen in Chapter 6, in some cases this spontaneous folding must be aided by chaperone proteins.

Covalent Modification

Some of the covalent modifications of polypeptide chains also occur during translation. We mentioned earlier that the *N*-formyl group is removed from the initial *N*-fMet of most prokaryotic proteins. A specific *deformylase* catalyzes this reaction. In many cases, deformylation seems to happen almost as soon as the N-terminus emerges from the ribosome. Removal of the N-terminal methionine itself can also be an early event, but whether it happens or not apparently depends on the cotranslational folding of the chain. Presumably, in some cases this residue is "tucked away" and protected from proteolysis in the folded structure.

Some prokaryotic (and many eukaryotic) proteins experience much more severe proteolytic modifications. These proteins are usually the ones that are going to be exported from the cell or are destined for membrane or organelle locations. We discuss the more complicated eukaryotic protein processing in the next section and concentrate here on what happens in prokaryotes.

Bacterial proteins that are destined for secretion (**translocation** across the cell membrane) are characterized by highly hydrophobic **signal sequences** or **leader sequences** in the N-terminal regions. Representatives are listed in Table 28.5. After the protein has passed through the membrane, the leader sequence is cleaved off at the point indicated by the arrow in the table.

A current model for translocation in bacteria is shown in Figure 28.40. In many, but not all, cases the protein to be translocated (the pro-protein) is first complexed in the cytoplasm with a chaperone—the SecB protein in the example shown. This complexing keeps the protein from folding prematurely, which

Translation is immediately followed by various kinds of protein processing, including chain folding, covalent modification, and directed transport.

TABLE 28.5   N-terminal signal sequences of representative prokaryotic proteins

Protein	−20					−15					−10					−5					−1 ↓ +1		
Leucine-binding protein	M	K	A	N	A	K	T	I	I	A	G	M	I	A	L	A	I	S	H	T	A	M A	E E
Prealkaline phosphatase			M	K	Q	S	T	I	A	L	A	L	L	P	L	L	F	T	P	V	T	K A	R T
Prelipoprotein			M	K	A	T	K	L	V	L	G	A	V	I	L	G	S	T	L	L	A	G	C S

Note: Hydrophobic residues are in magenta. The cleavage site is designated by the arrow.

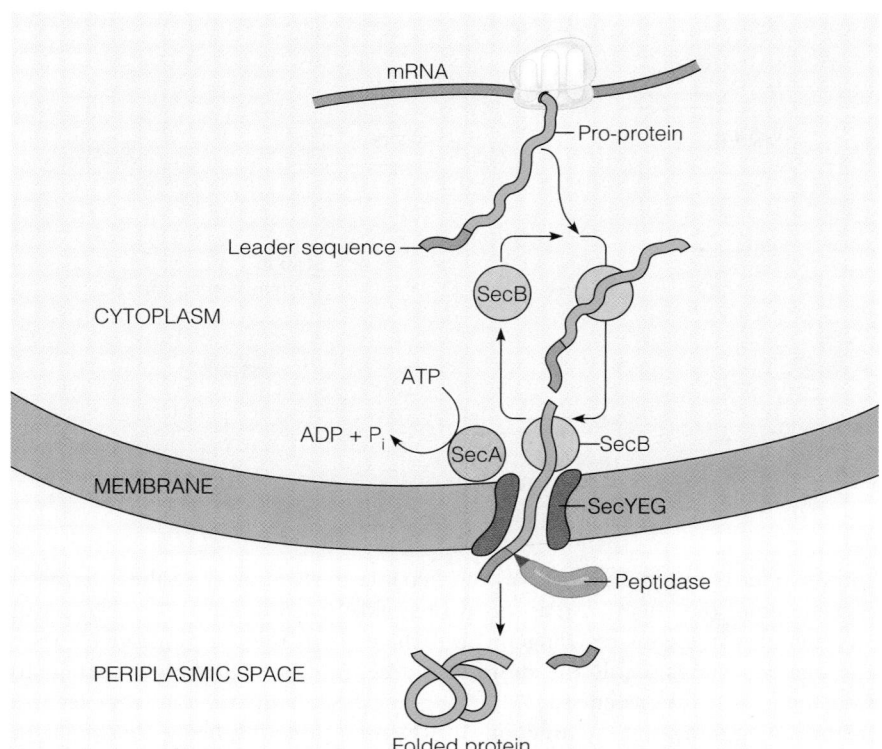

FIGURE 28.40

A current model for protein secretion by prokaryotes. The new polypeptide chain (the pro-protein) complexes with SecB, which prevents complete folding during transport to the membrane. At the membrane an ATPase, SecA, drives translocation through the membrane with the aid of SecYEG, which forms a membrane pore. The leader sequence is then cleaved off the secreted protein by a membrane peptidase.

would prevent it from being passed through a secretory pore in the membrane. This pore is composed of a heterotrimeric protein made up of SecE, SecY, and SecG—the "SecYEG translocon." The secretory pore is also a target for a fourth protein component, SecA. SecA is an ATPase, and both ATP hydrolysis and the electrochemical potential gradient across the membrane help drive translocation. Structural analysis of the SecA protein with and without bound adenine nucleotide suggests a mechanism comparable to that of DNA-dependent helicases in driving the protein through the membrane. After the pro-protein has been translocated, the leader peptide is cleaved off by a membrane-bound protease, and the protein can fold. The cleavage site, as shown in Table 28.5, usually lies between a small amino acid (often Gly or Ala) and an acidic or a basic one.

Protein Splicing

A small, but significant, number of proteins, mostly from single-celled organisms—bacteria, archaea, and eukaryotic microbes—undergo post-translational splicing, in a process comparable to the RNA splicing discussed in Chapter 27. In protein splicing an **intein** (internal protein segment) is cleaved from within the polypeptide sequence, yielding a mature protein, the **extein** (external protein). Embedding an intein into a normally nonspliced protein retains its splicing activity for the new protein host, indicating that amino acid residues for catalysis of splicing lie within the intein.

Although we know little about the biological functions of protein splicing, we do know that the mechanism is comparable to that of RNA splicing, with, of course, different functional groups involved. As shown in Figure 28.41, an N-terminal serine or threonine hydroxyl (or cysteine thiol; serine as shown) attacks the C-terminal peptide carbon of the upstream intein segment (N-extein), and this is followed by a transesterification involving a thiol or hydroxyl in the N-terminal residue of the downstream extein segment (C-extein) upon that same

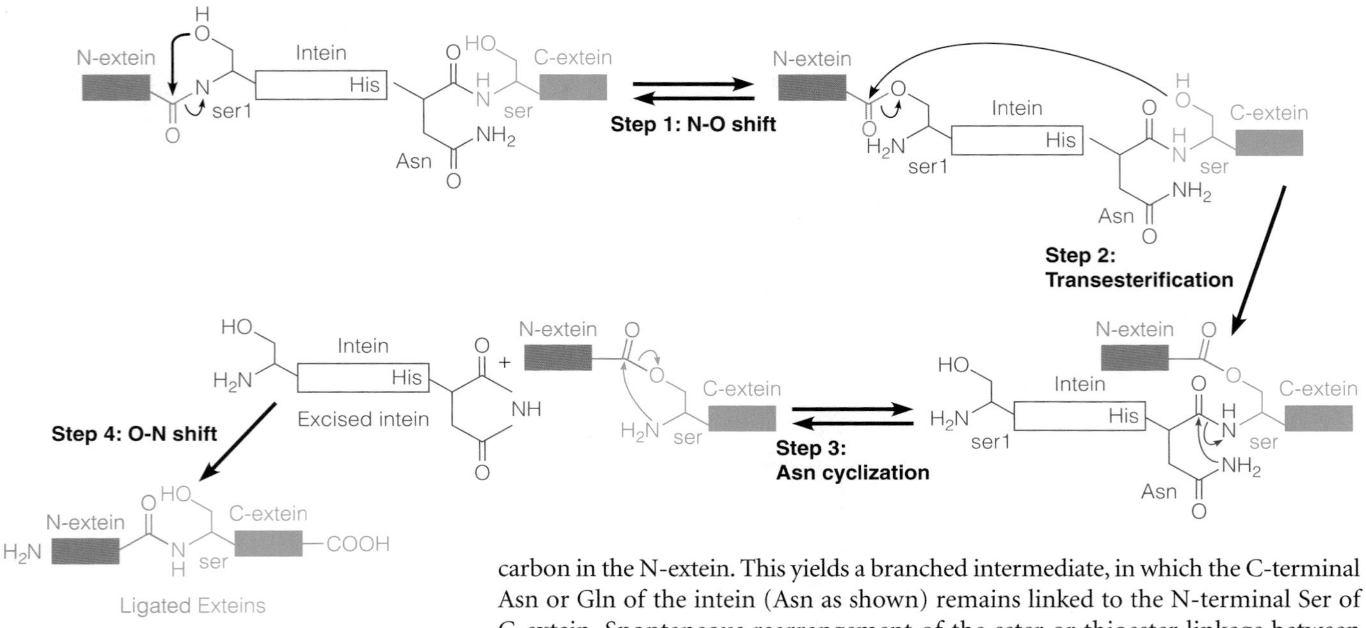

FIGURE 28.41

An outline of the mechanism of protein splicing.

carbon in the N-extein. This yields a branched intermediate, in which the C-terminal Asn or Gln of the intein (Asn as shown) remains linked to the N-terminal Ser of C-extein. Spontaneous rearrangement of the ester or thioester linkage between the ligated exteins yields the more stable peptide bond linking N-extein to C-extein. Most inteins encode a homing nuclease comparable to that seen in introns, suggesting that the functions of RNA and protein splicing are similar, in promoting their ability to move from gene to gene.

Protein Targeting in Eukaryotes

The eukaryotic cell is a multicompartmental structure. Its several organelles each require different proteins, only a few of which are synthesized within the organelles themselves. Most mitochondrial and chloroplast proteins, for example, are encoded by the nuclear genome and synthesized in the cytoplasm. They must be carefully distinguished from other newly synthesized proteins and selectively transported to their appropriate addresses. Other new proteins are destined for export out of the cell or into vesicles like lysosomes. The diversity of destinations for different proteins implies the existence of a complex system for labeling and sorting newly synthesized proteins and ensuring that they end up in their proper places. And, as seen with bacteria, there must be a process by which protein molecules, which may be hydrophilic, engage the hydrophobic membrane and find a way either to pass through or, as in the case of integral membrane proteins, to become embedded within the membrane.

Proteins Synthesized in the Cytoplasm

Proteins destined for the cytoplasm and those to be incorporated into mitochondria, chloroplasts, or nuclei are synthesized on polyribosomes free in the cytoplasm. The proteins targeted to organelles, as initially synthesized, contain specific signal sequences. These sequences probably aid in membrane insertion, but they also signal that these polypeptides will interact with a particular class of chaperones. These chaperones are members of the "heat-shock" Hsp70 family, and they act to ensure that the newly synthesized protein remains unfolded and is delivered to a receptor site on the organelle membrane. The unfolded protein then passes through membranes, through gates containing transport proteins that discriminate among proteins destined for the lumen, the membranes, or an organelle matrix. If it passes into an organelle matrix, the protein may be taken up by intraorganelle chaperones for final folding. The N-terminal targeting sequence is also cleaved off during this transport.

Transport of proteins into mitochondria is schematized in Figure 28.42. First, the Hsp70-bound protein attaches *via* a basic N-terminal signal sequence to a receptor protein as part of a structure called the TOM complex (translocation of outer membrane). An ATP-dependent reaction releases the protein from the receptor and inserts it into a pore, another part of the TOM complex. The signal sequence then interacts with another complex, the TIM complex (translocation of inner membrane), in the inner membrane. The electrochemical gradient across the inner membrane pulls the signal sequence through. A mitochondrial Hsp70 binds to the protein as it becomes exposed within the mitochondrial matrix and, in another energy-dependent reaction, pulls the rest of the protein through. The signal sequence is removed by a specific protease (MPP, matrix processing peptidase) within the matrix. Note that this process pulls the protein through both outer and inner mitochondrial membranes.

A quite different process occurs in nuclear transport. Originally it was thought that these proteins simply diffused into the nucleus through the nuclear pores and were then bound to chromatin. However, it is becoming clear that the nuclear pores are complex gates, rather than open channels. Proteins destined for the nucleus contain *nuclear localization sequences* (NLS) that help these proteins select the nucleus

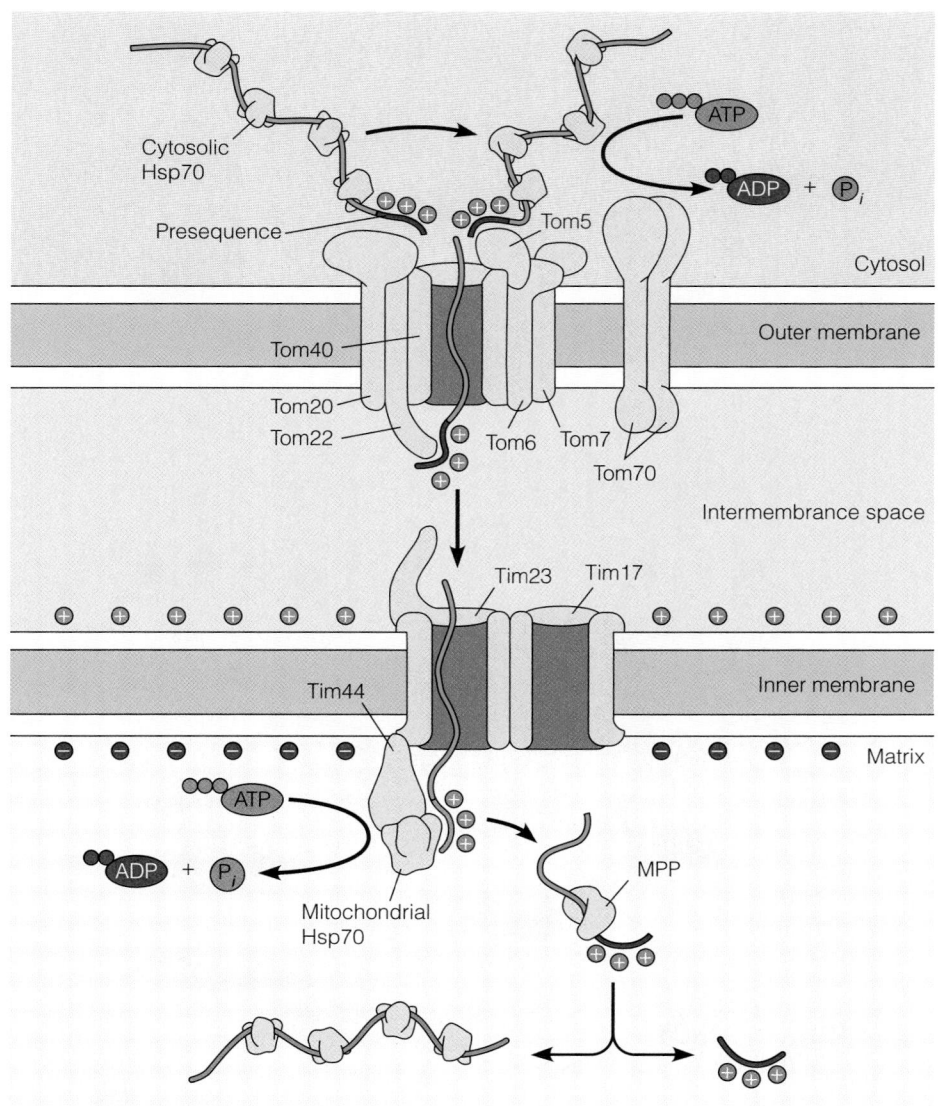

FIGURE 28.42

Transport of newly synthesized mitochondrial proteins into the matrix. Upper left, signal sequence of Hsp70-bound protein inserts into import receptor of the TOM complex (TOM20) in the outer membrane. Hsp70 dissociation is coupled to ATP hydrolysis. Insertion of the protein into the outer membrane (*via* TOM22) puts signal sequence in position to interact with TIM complex (TIM23) in the inner membrane. Potential across the inner membrane drives protein into intermembrane space. Signal sequence is cleaved off by MPP. Mitochondrial Hsp70 binds to protein in the matrix and uses the energy of ATP hydrolysis to pull the rest of the protein through.

Modified from *The Cell: A Molecular Approach*, 4th ed., G. M. Cooper and R. E. Hausmann (2007). American Society for Microbiology.

as their destination. Nuclear localization sequences can be found anywhere within the polypeptide sequence, not only the N-terminus. Moreover, the NLS is not removed as a consequence of transport. This is important because the nuclear membrane breaks down in each cell division cycle, and each nuclear protein must be re-transported into the nucleus after re-establishment of the nuclear envelope.

The nuclear localization signal on a prospective cargo protein interacts with a protein called **importin**, which carries the protein through the nuclear pore complex, as schematized in Figure 28.43. Energy for transport is provided by a monomeric G protein called **Ran** (<u>Ra</u>s-related <u>n</u>uclear protein). Ran is similar to other G proteins we have encountered, in that the protein becomes activated by exchange of Ran-bound GDP for GTP, by a guanine nucleotide exchange factor (GEF), and inactivated by GTPase-activating protein (GAP), which hydrolyzes bound GTP to GDP. Ran passes the nuclear pore freely. Because GEF is localized to the nucleus and GAP to the cytoplasm, Ran-GTP predominates in the nucleus and Ran-GDP in the cytoplasm.

Once the importin-cargo protein complex has passed into the nucleus, Ran-GTP binds to that complex, which displaces the cargo. The Ran-GTP complex is returned to the cytoplasm, where bound GTP is converted to GDP. Ran-GDP returns to the nucleus, where it exchanges GDP for GTP, and importin seeks a new NLS-containing cargo protein.

Proteins destined for the cytoplasm, nuclei, mitochondria, and chloroplasts are synthesized in the cytoplasm; those destined for organelles have specific targeting sequences.

FIGURE 28.43

A schematic view of delivery of a protein, synthesized in cytoplasm, into the nucleus. A protein with a nuclear localization signal (NLS) binds to importin. The complex binds to a nuclear pore and passes through. Within the nucleus Ran-GTP binds to the importin-cargo complex and displaces the cargo. The resultant Ran-importin complex is returned to the cytoplasm, where Ran-bound GTP is hydrolyzed to GDP. Ran-GDP returns to the nucleus, and its bound GDP is exchanged for GTP (not shown).

Modified from *The Cell: A Molecular Approach*, 4th ed., G. M. Cooper and R. E. Hausmann (2007). American Society for Microbiology.

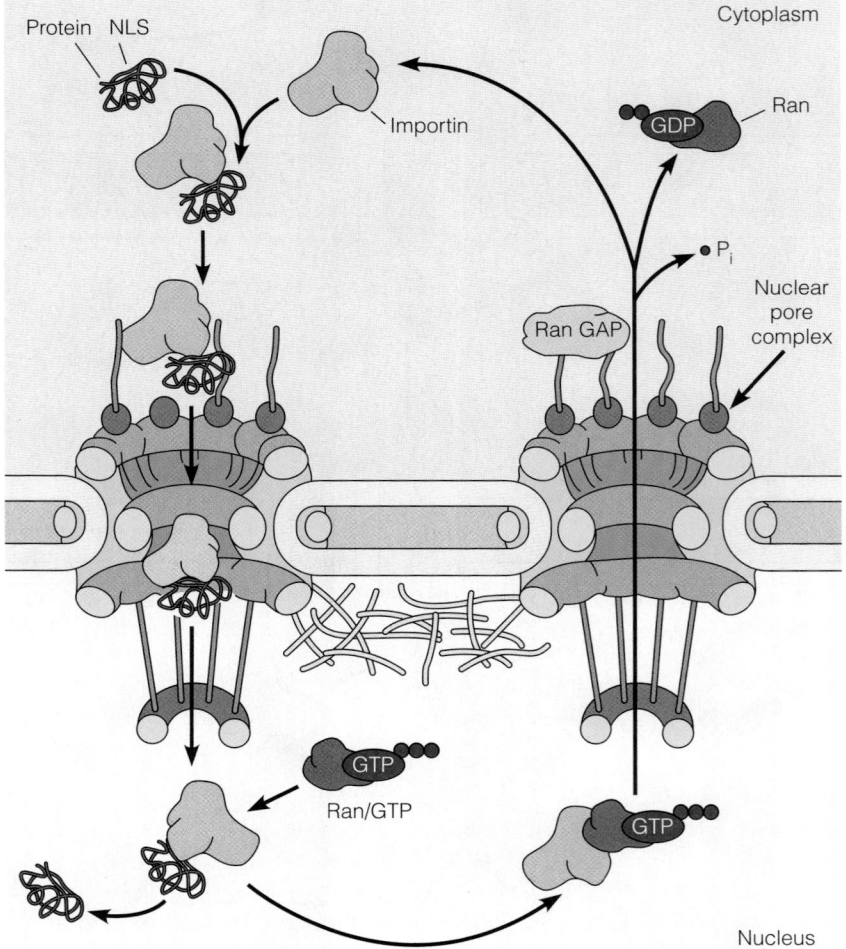

Proteins Synthesized on the Rough Endoplasmic Reticulum

Proteins destined for cellular membranes, lysosomes, or extracellular transport use a quite different distribution system. The key structures in this system are the **rough endoplasmic reticulum** (RER) and the Golgi complex (see also Chapter 9). The rough endoplasmic reticulum is a network of membrane-enclosed spaces within the cytoplasm. The RER membrane is heavily coated on the outer, cytosolic surface with polyribosomes; this coating is what gives the membrane its rough appearance. The Golgi complex resembles the RER in that it is a stack of thin, membrane-bound sacs. However, the Golgi sacs are not interconnected, nor do they carry polyribosomes on their surfaces. The role of the Golgi complex is to act as a "switching center" for proteins with various destinations.

Proteins that are to be directed to their destinations via the Golgi complex are synthesized by polyribosomes associated with the RER. Synthesis actually begins in the cytoplasm (Figure 28.44, step 1). The first sequence to be synthesized is an N-terminal *signal sequence,* part of a mechanism for attaching the ribosome and nascent protein to the RER. *Signal recognition particles (SRPs),* containing several proteins and a small (7S) RNA, recognize the signal sequences of the appropriate nascent proteins and bind to them as they are being extruded from the ribosomes (step 2).

The SRP has two functions. First, its binding temporarily halts translation so that no more than the N-terminal signal sequence extends from the ribosome. This pause prevents completion of the protein in the wrong place—that is, in the cytosol—and also inhibits premature folding of the polypeptide chain. Thus, the SRP is acting as a kind of chaperone. The second function of the SRP is to recognize a docking protein in the RER membrane. This is the trimeric Sec61 complex, homologous to the bacterial SecYEG. The docking protein binds the ribosome to the RER, and the signal sequence is inserted into the RER membrane (step 3). The SRP is then released (step 4), allowing translation to resume (step 5). The protein

Proteins destined for cell membranes, lysosomes, or export are synthesized on the rough endoplasmic reticulum, then modified and transported via the Golgi apparatus.

FIGURE 28.44

The sequence of events in synthesis of proteins on the rough endoplasmic reticulum. The time sequence of events is from left to right. Recent cryoelectron microscopic analysis of ribosome-Sec61 complexes shows that the Sec61 functions as a monomer. Model building allowed investigators to trace the path of newly synthesized protein from the tRNA through the ribosomal tunnel through the monomeric Sec61 embedded in the membrane, as shown in Figure 28.45.

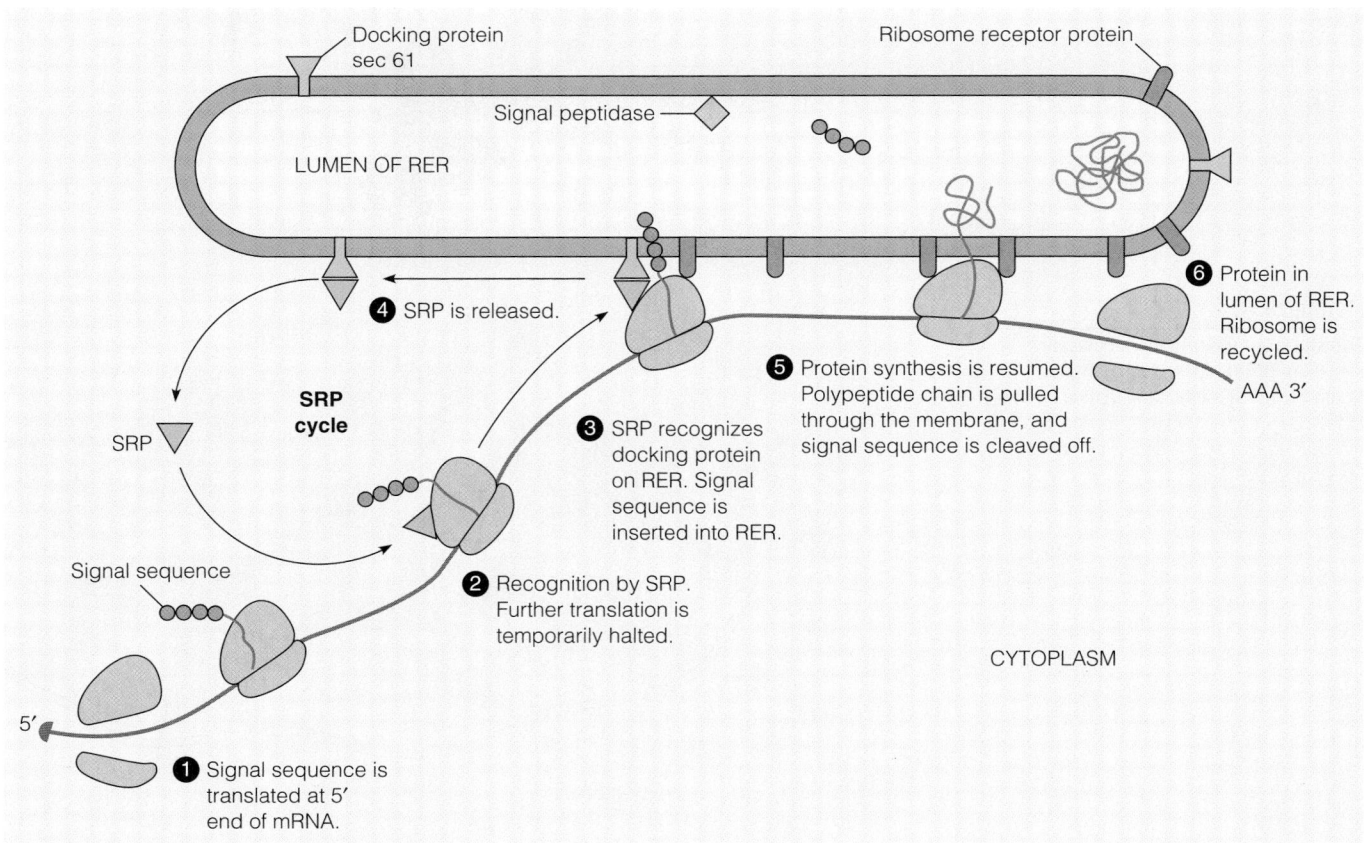

Docking protein sec 61

Ribosome receptor protein

Signal peptidase

LUMEN OF RER

❹ SRP is released.

SRP cycle

SRP

Signal sequence

❸ SRP recognizes docking protein on RER. Signal sequence is inserted into RER.

❺ Protein synthesis is resumed. Polypeptide chain is pulled through the membrane, and signal sequence is cleaved off.

❻ Protein in lumen of RER. Ribosome is recycled.

AAA 3′

❷ Recognition by SRP. Further translation is temporarily halted.

CYTOPLASM

5′

❶ Signal sequence is translated at 5′ end of mRNA.

being synthesized is actually *pulled* through the membrane by an ATP-dependent process. Before translation is complete, signal sequences are cleaved from some proteins by an RER-associated protease. These proteins are released into the lumen of the RER and further transported (step 6). Proteins that will remain in the endoplasmic reticulum have resistant signal peptides and thereby remain anchored to the RER membrane. A model of the protein translocation process, based upon the structure of Sec61, is shown in Figure 28.45.

Role of the Golgi Complex

The proteins that enter the lumen of the RER undergo the first stages of glycosylation at this point. Vesicles carrying these proteins then bud off the RER and move to the Golgi complex (Figure 28.46). Here the carbohydrate moieties of glycoproteins are completed (see pages 339–343 in Chapter 9 for details), and a final

FIGURE 28.45

Schematic representation of an actively translating and translocating eukaryotic ribosome-Sec61 complex. NC, nascent chain; PCC, protein-conducting channel (Sec61). P-tRNA, peptidyl-tRNA with its nascent chain. PDB ID 2ww9, 2wwa, and 2wwb.

From *Science* 326:1369–1372, T. Becker, S. Bhushan, A. Jarasch, J.-P. Armache, S. Funes, F. Jossinet, J. Gumbart, T. Mielke, O. Berninghausen, K. Schulten, E. Westhof, R. Gilmore, E. C. Mandon, and R. Beckmann, Structure of monomeric yeast and mammalian Sec61 complexes interacting with the translating ribosome. © 2009. Reprinted with permission from AAAS.

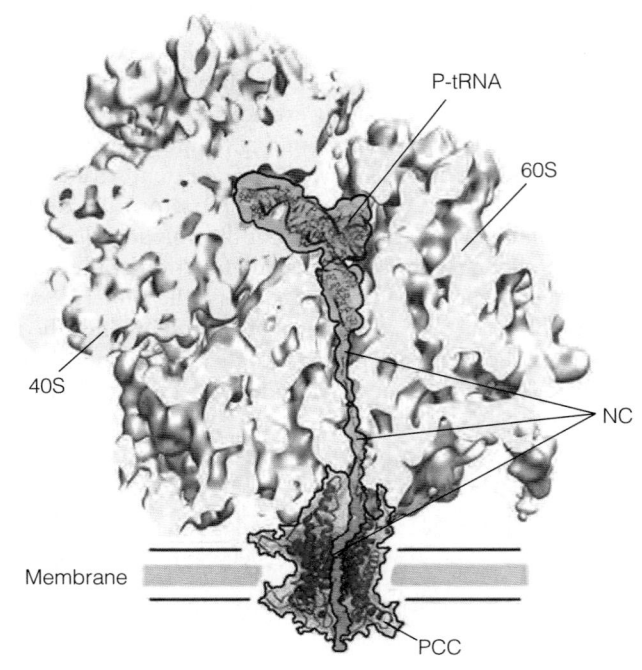

FIGURE 28.46

Transfer from the rough endoplasmic reticulum (RER) to the Golgi complex. Note that vesicles bud off the RER and move to the *cis* face of the Golgi. Primary lysosomal vesicles bud from the *trans* portion of the Golgi.

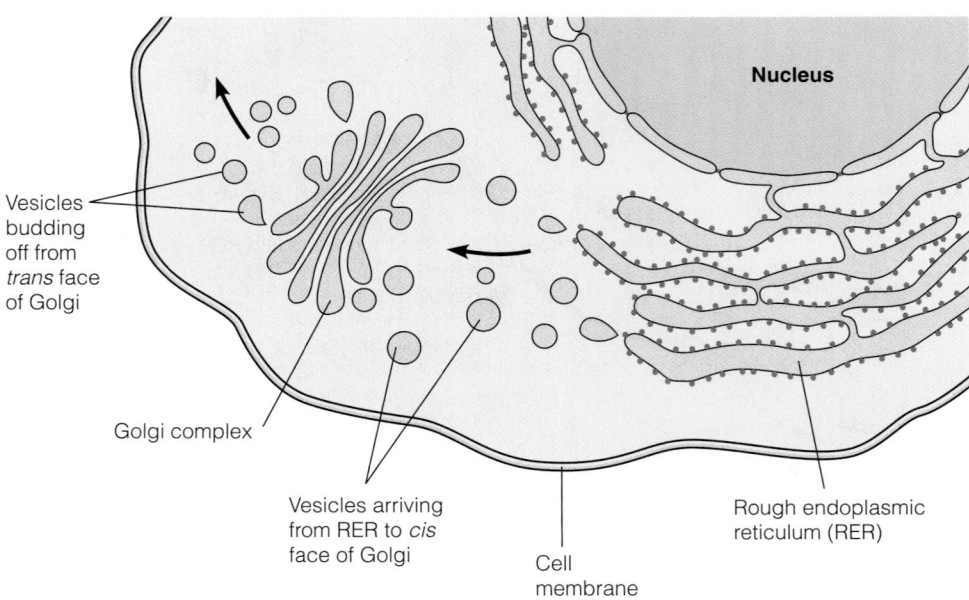

sorting occurs. The multiple membrane sacs that constitute the Golgi complex represent a multilayer arena for these processes. Vesicles from the RER enter at the *cis* face of the Golgi (that closest to the RER) and fuse with the Golgi membrane. Proteins are then passed, again via vesicles, to the intermediate layers. Finally, vesicles bud off from the *trans* face of the Golgi complex to form lysosomes, peroxisomes, or glyoxysomes or to travel to the plasma membrane. All of this transport of vesicles, from the RER to the *cis* face of the Golgi, to successive levels of the Golgi and on to their final destinations, requires high specificity in targeting. Transport of vesicles to the wrong destinations would cause cellular chaos. This sorting is accomplished by having each kind of protein cargo packed in a vesicle marked by specific vesicle membrane proteins. In some cases, the target membranes contain complementary proteins that interact with these and cause membrane fusion. These complementary pairs are called *SNARES* (soluble N-ethylmaleimide-sensitive factor attachment protein receptors)—v-SNARES on vesicles, t-SNARES on target membranes. The interaction of specific v- and t-SNARES, aided by cytosolic fusion proteins, leads to fusion of the vesicle and target membranes and delivery of the cargo (see Figure 28.47).

The Fate of Proteins: Programmed Destruction

In Chapter 11 we pointed out that one mechanism for control of enzymatic function is the selective degradation of certain enzymes. However, not only enzymes need to be destroyed in a programmed way. Regulatory proteins that are essential in certain parts of the cell cycle and deleterious in others must be eliminated at some point. Consider the cyclins (Chapter 24), for example, which must be broken down and resynthesized during each cell cycle. Proteins that have become damaged must also be removed. In some developmental processes, it is necessary to remove whole organelles or even entire cells and tissues.

Eukaryotic cells have two distinct methods for protein degradation. The lysosomes contain among their hydrolases proteolytic enzymes that will degrade any protein trapped within the organelle. Parallel to this process is a cytosolic degradation system, which is of necessity highly selective. The danger inherent in having nonspecific proteases loose in the cytosol should be evident. Both of these processes were described briefly in Chapter 20, and we supplement that information here.

The Lysosomal System

The lysosomal particles budded from the Golgi complex, known as **primary lysosomes**, are essentially bags of degradative enzymes. Over 50 different hydrolytic enzymes are contained in lysosomes, including proteases, nucleases, lipases, and carbohydrate-cleaving enzymes. The lysosomes play a number of important roles in cellular metabolism, as schematically depicted in Figure 28.48.

In some cell types, such as those in the pancreas that secrete degradative enzymes, primary lysosomes migrate to the cell surface and release their contents into the exterior medium (path A). Primary lysosomes may also fuse with *autophagic vesicles,* formed when smooth ER engulfs organelles destined for destruction (path B). The combined vesicle is called an *autophagic lysosome.* In some kinds of cells—mainly certain white blood cells—primary lysosomes may fuse with *phagocytic vacuoles* that have engulfed nutrient materials at the cell surface (path C). In these heterophagic lysosomes, the nutrients are digested and their amino acids, nucleotides, lipids, and other low-molecular-weight constituents released into the cytosol. Residual, undigested material is excreted when the heterophagic lysosomes and autophagic lysosomes find their way to the plasma membrane.

Cytosolic Protein Degradation

In contrast to the lysosomal enzymes, which are usually safely sequestered in their vesicles, any protease activity that is free in normal cytosol must be under rigid control. It must attack only the proteins whose destruction is needed. These may include

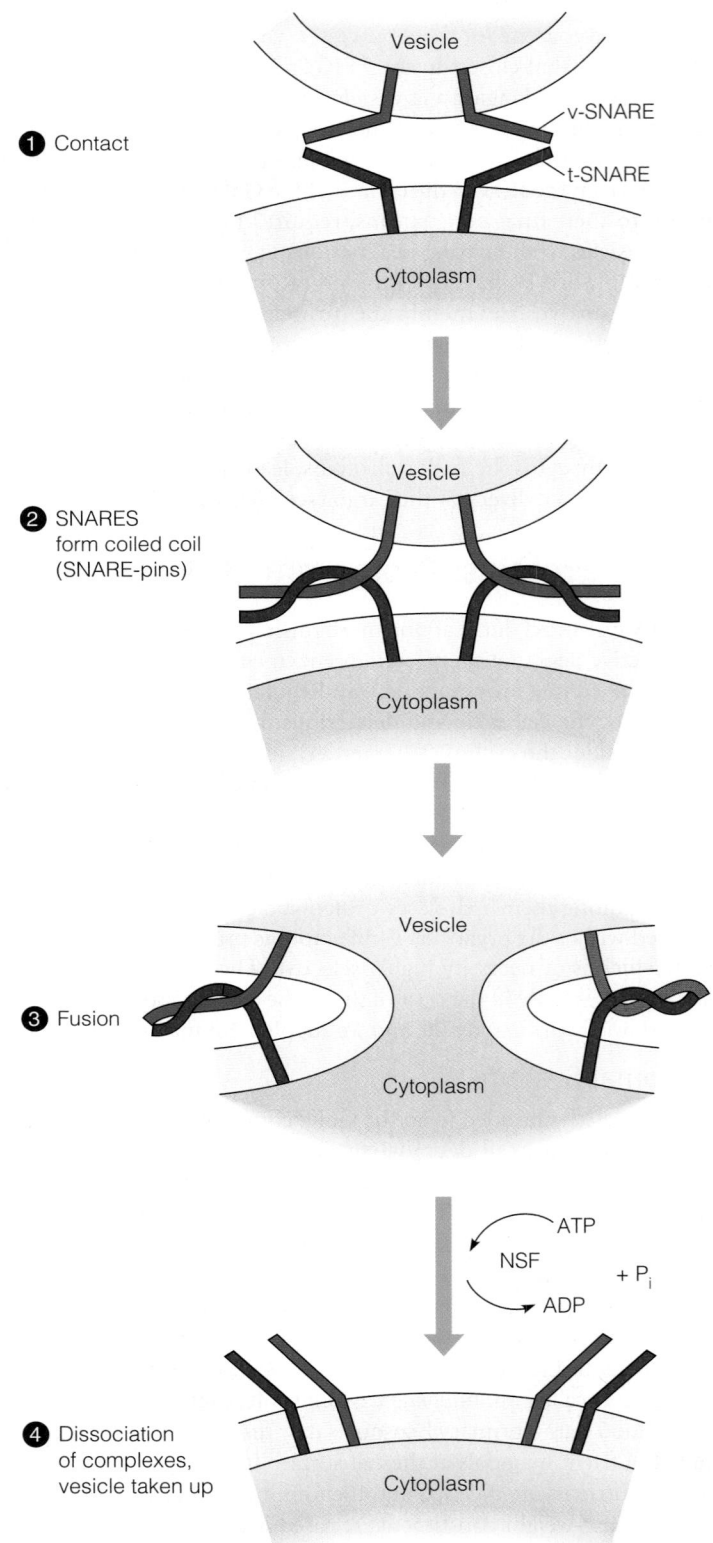

FIGURE 28.47

A schematic, and somewhat hypothetical, view of SNARE–pin fusion. Specific v-SNAREs and t-SNAREs dictate interaction, and form coiled-coil structures. After fusion, these are broken up by the factor NSF, using the energy of ATP hydrolysis.

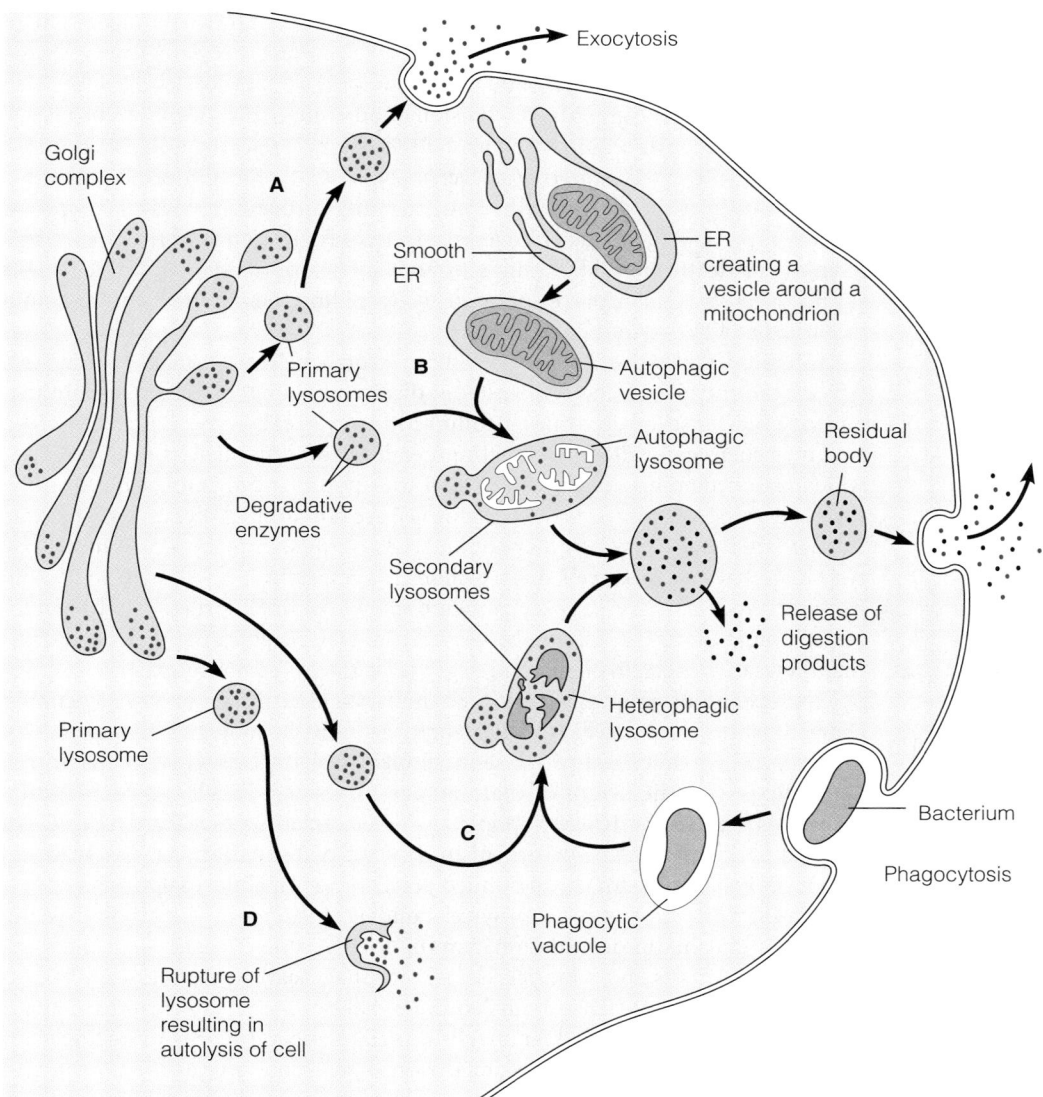

FIGURE 28.48

Formation of primary and secondary lysosomes and their role in cellular digestive processes. The primary lysosomes that bud from the Golgi can take several pathways.
Path A: Exocytosis—transport of enzymes to outside of cell.
Paths B and C: Phagocytosis—formation of phagic lysosomes for digesting organelles (autophagocytosis) or ingested matter (heterophagocytosis).
Path D: Autolysis—destruction of the cell itself.

damaged proteins, incorrectly synthesized proteins, or proteins no longer required at a particular stage in the cell cycle. If we recall that protein hydrolysis is a thermodynamically favored reaction, it becomes clear that the enzymes participating in such cytosolic degradation must be more than simple catalysts for the hydrolytic process—otherwise, destruction would be wholesale. Basically, there must be some means of distinguishing the proteins to be attacked from those to be left alone.

As described in Chapter 20, the major proteolytic system in cytosol uses ubiquitin to mark proteins destined for destruction. Recall that ubiquitin is a small, heat-stable protein that is transferred to lysine residues of proteins destined for breakdown or other processing. Although some ubiquitinated proteins are simply marked for translocation to specific cellular sites, and others are marked for reasons yet unknown, most are marked for ATP-dependent proteolytic digestion in the proteasome.

Apoptosis

Previously in this book we have mentioned *apoptosis*, which is a form of programmed cell death. It has long been known that cell death occurs on a substantial scale as part of normal embryonic development. For example, when a tadpole undergoes metamorphosis to a frog, its tail disintegrates as the result of the death

Protein degradation occurs via lysosomes—vesicles filled with hydrolytic enzymes—or intracellularly, often involving the marker ubiquitin and/or a multicatalytic complex.

of its constituent cells. Another example is development of toes and fingers, where the webbing that originally holds digits together also falls away as a result of programmed cell death. Both these and other developmental processes occur in close coordination with the growth and division of adjoining cells, such that the release of possibly toxic intracellular contents does not damage nearby cells. In addition, we now know that cells suffering irreversible damage, whose continued existence could threaten the organism, also undergo programmed cell death. Apoptosis (from Greek, "falling off," as of leaves from a tree) is distinguished from *necrosis*, a form of cell death caused by trauma or anoxia (lack of oxygen) or lack of blood supply. Necrotic cells rupture spontaneously, spilling their contents and causing inflammatory responses in neighboring cells.

By contrast, events in apoptosis occur on schedule. Receipt of appropriate signals causes collapse of the cytoskeleton, disassembly of the nuclear membrane, and condensation of chromatin, followed by its breakdown. The cell surface becomes chemically altered, attracting neighboring cells such as macrophages, which engulf membrane-enclosed bits of the apoptotic cell and further degrade them, ensuring that contents of the dying cell are not released directly to the extracellular environment, where they could, as in necrosis, provoke an inflammatory response. The principal cell surface change is migration of phosphatidylserine from the inner leaflet of the plasma membrane to the outer leaflet. This particular phospholipid on the cell surface serves as a signal to neighboring phagocytic cells that they can proceed to engulf and degrade fragments of the apoptotic cell.

Understanding apoptosis is critical to understanding cancer because a defining characteristic of tumor cells is their inability to undergo apoptosis. In Chapter 23, for example, we described p53, a sentinel protein that senses DNA damage and, depending upon the extent of that damage, either delays cell cycling until the damage is repaired, or triggers apoptosis, thereby insuring that the damaged cell will not survive to harm the organism. Loss of p53 function, as seen in many tumors, is responsible for the continued proliferation of abnormal tumor cells.

Among various ways to identify cells undergoing apoptosis is to observe the pattern of chromatin degradation. Among the degradative activities unleashed during apoptosis is an endonuclease activity that cleaves chromatin-associated DNA at linker regions between nucleosomes. Hence, analysis of chromatin by gel electrophoresis reveals a "ladder" pattern, with each band representing a particle containing an integral number of nucleosomes, as seen when purified chromatin is digested with an endonuclease (see Figure 24.22, page 1023). Another means of apoptotic detection involves the use of fluorescent dyes that register the loss of mitochondrial membrane potential.

Apoptosis can be triggered either extracellularly (*extrinsic* pathway) or intracellularly (*intrinsic* pathway). The extrinsic pathway predominates during normal development, whereas the intrinsic pathway is activated as a result of intracellular damage. Although the two pathways are triggered by different events, both pathways proceed via a series of conversions of inactive precursors to active proteolytic enzymes called **caspases**. The name refers to the fact that each of these enzymes has an active-site cysteine residue, and each attacks target proteins at specific aspartate residues; hence, c-asp-ases. These proteins are synthesized as inactive *procaspases*, which undergo proteolytic cleavage and dimerization, yielding enzymatically active $\alpha_2\beta_2$ heterotetramers.

The extrinsic pathway begins with recognition between a trimeric death receptor on the surface of the pro-apoptotic cell and a homotrimeric ligand on a cell, such as a killer lymphocyte. In the example schematized in Figure 28.49a, the ligand is a protein related to *tumor necrosis factor* (TNF), also mentioned in Chapter 23. An intracellular domain of the death receptor recruits an adaptor protein, which in turn recruits either procaspase-8 or -10, as shown, *via* a death effector domain, thereby forming DISC (<u>d</u>eath-<u>i</u>nducing <u>s</u>ignaling <u>c</u>omplex). This brings procaspase molecules into close proximity, activating them and allowing them to cleave one another, forming activated caspases. Next these cleave and activate other procaspases, setting in motion the chain of events that results in death of the cell.

Understanding apoptosis is crucial to understanding cancer, because tumor cells lose the ability to undergo apoptosis.

The intrinsic pathway, by contrast, is activated by signals that respond to intracellular events, such as DNA damage or lack of nutrients (Figure 28.49b). In mammalian cells the signaling agent is a protein, either *Bak* or *Bax*, which interacts with the mitochondrial outer membrane and triggers the release of cytochrome c from the intermembrane space, as we mentioned also in Chapter 15. Cytochrome c then binds to a protein called *Apaf1* (Apoptotic protease activating factor), triggering the hydrolysis of dATP bound to Apaf1. This in turn stimulates Apaf1 to oligomerize into the heptameric *apoptosome*, a pinwheel-like structure that recruits and activates procaspase-9. Cleavage and activation of procaspase-9 creates active caspase-9, which cleaves and activates downstream procaspases, ultimately activating the events of apoptosis. In both the intrinsic and extrinsic pathways, the downstream events resulting from the caspase cascade are areas of active current investigation.

Apoptosis, the major form of programmed cell death, is triggered by either an extrinsic or intrinsic pathway, both of which initiate a series of proteolytic activations of procaspases to active caspases.

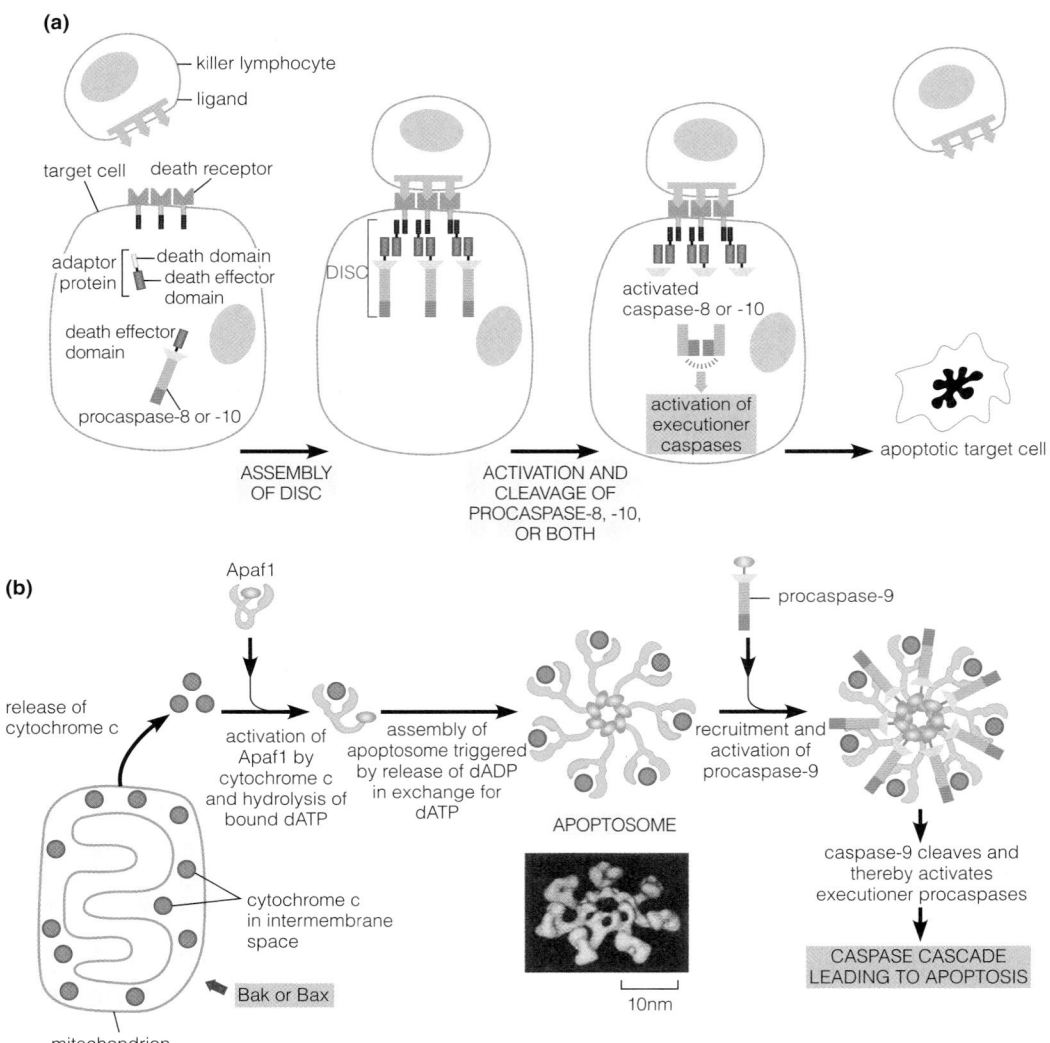

FIGURE 28.49

Extrinsic and intrinsic signaling pathways leading to apoptosis. (a) The extrinsic pathway. As shown, the interaction between a trimeric death cell receptor on the pro-apoptotic cell with a trimeric ligand in a killer cell stimulates assembly of DISC (the death-inducing signaling complex), which includes inactive procaspase-8 or -10 or both. Activation of the bound procaspase triggers a cascade of activation of downstream ("executioner") caspases. **(b)** The intrinsic pathway. A protein such as Bak or Bax interacts by unknown mechanisms with the mitochondrial outer membrane, releasing cytochrome c and other proteins in the intermembrane space. Cytochrome c binds to Apaf1, stimulating hydrolysis of bound dATP and triggering assembly of the heptameric apoptosome. Activation of procaspase-9 within the apoptosome stimulates the cleavage of downstream executioner caspases, as seen in the extrinsic pathway.

SUMMARY

The connection between gene (or mRNA) sequence and protein sequence is dictated by the genetic code. The code is almost, but not quite, uniform in all living creatures. The code is redundant, with multiple codons for most amino acids, and it includes start and stop signals.

Translation of mRNAs into polypeptide chains involves a number of steps. First, the appropriate tRNAs must be coupled to the corresponding amino acids, using a set of aminoacyl-tRNA synthetases. The site of translation is the ribosome—an RNA–protein complex composed of two subunits, each with its individual RNA and protein components. Translation of an mRNA on ribosomes involves three stages—initiation, elongation, and termination. Each stage requires certain protein factors in addition to the ribosome. In prokaryotic initiation the mRNA is fixed to the 30S subunit in proper register via its Shine–Dalgarno sequence, and the initiator N-formylmethionyl tRNA is matched to its AUG codon at the P site. The 50S subunit is then attached. The second tRNA enters the A site, and the incipient chain is transferred to it. Next, the mRNA is translocated so as to move the chain again into the P site, while the now unoccupied tRNA moves into the E site and is released. These steps involve intermediate hybrid states.

The chain continues to elongate in this fashion until a stop codon is reached, whereupon release factors bind to the ribosome and aid in releasing the polypeptide chain. The entire process of translation requires about 4 ATP equivalents for each amino acid added. Numerous antibiotics have their antibacterial effect by inhibiting various stages of the translation process.

Eukaryotic protein synthesis differs from that in bacteria in significant ways, starting with the fact that eukaryotic mRNAs are highly processed templates for single genes; ribosomes are larger and more complex, and more soluble protein factors are involved; different mechanisms are used for initiation because N-formylmethionine is not involved; and translation is coupled only indirectly with transcription because the two processes occur in different cell compartments.

As translation is being completed, folding and covalent modification of the polypeptide chain begin. The chain as synthesized may contain an N-terminal sequence that targets it for passage through membranes. Distinct transport processes involve uptake into organelles, uptake into the nucleus, or sorting for transport into lysosomes, membrane interiors, or the extracellular milieu.

REFERENCES

Of Historical Interest

Brenner, S., F. Jacob, and M. Meselson (1961) An unstable intermediate carrying information from genes to ribosomes for protein synthesis. *Nature* 190:576–581. Early evidence for the existence of mRNA.

Crick, F. H. C. (1958) On protein synthesis. *Symp. Soc. Exp. Biol.* 12:138–162. With great prescience, Crick foresees the essential nature of the translation mechanism.

Crick, F. H. C. (1966) Codon–anticodon pairing: The wobble hypothesis. *J. Mol. Biol.* 19:548–555.

Khorana, H. G. (1968) Nucleic acid synthesis in the study of the genetic code. In *Nobel Lectures, Physiology, and Medicine (1963–1970)*, pp. 341–343. American Elsevier, New York. A Nobel Prize winner's account of the deciphering of the code.

Traub, P., and M. Nomura (1968) Structure and function of *E. coli* ribosomes. V. Reconstitution of functionally active 30S ribosomal particles from RNA and proteins. *Proc. Natl. Acad. Sci. USA* 59:777–784.

The Code

Bender, A., P. Hajieva, and B. Moosmann (2008) Adaptive antioxidant methionine accumulation in respiratory chain complexes explains the use of a deviant genetic code in mitochondria. *Proc. Nat. Acad. Sci. USA* 105:16496–16501. Evidence for evolution of a different code in mitochondria.

Plotkin, J. B., and G. Kudla (2011) Synonymous but not the same: The causes and consequences of codon bias. *Nat. Rev. Genet.* 12:32–42. A recent review of codon bias.

Turanov, A. A., A. V. Lobanov, D. E. Fomenko, H. G. Morrison, M. L. Sogin, L. A. Klobutcher, D. L. Hatfield, and V. N. Gladyshev (2009) Genetic code supports targeted insertion of two amino acids by one codon. *Science* 323:259–261. Unexpected ambiguity within one gene.

Yuan, J., P. O'Donoghue, A. Ambrogelly, S. Gundllapalli, R. L. Sherrer, S. Paliora, M. Siminovic, and D. Söll (2010) Distinct genetic code expansion strategies for selenocysteine and pyrrolysine are reflected in different aminoacyl-tRNA formation systems. *FEBS Letters* 584:342–349. Coding for the 21st and 22nd amino acids.

Messenger RNA

Gesteland, R. F., R. B. Weiss, and J. F. Atkins (1992) Recoding: Reprogramming genetic decoding. *Science* 257:1640–1641. There are special signals in some mRNAs that alter code reading.

Shine, J., and L. Dalgarno (1974) The 3′-terminal sequence of *E. coli* 16S rRNA: Complementarity to nonsense triplets and ribosome binding sites. *Proc. Natl. Acad. Sci. USA* 71:1342–1346.

Transfer RNAs

Hatfield, D. L., B. J. Lee, and R. M. Pirtle (eds.) (1992) *Transfer RNA in Protein Synthesis*. CRC Press, Boca Raton, Fla. A collection of papers on diverse aspects of tRNA function.

Olejniczak, M., and O. C. Uhlenbeck (2006) tRNA residues that have coevolved with their anticodon to ensure uniform and accurate codon recognition. *Biochimie* 88:943–950. A phylogenetic analysis of tRNA base pairs that facilitate accurate codon recognition.

Söll, D., and V. RajBhandary (eds.) (1995) *tRNA: Structure, Biosynthesis, and Function*. ASM Press, Washington, D.C.

Aminoacyl-tRNA Synthetases and Aminoacyl-tRNA Coupling

Carter, C. W., Jr. (1993) Cognition, mechanism, and evolutionary relationships in aminoacyl-tRNA synthetases. *Annu. Rev. Biochem.* 62:715–748.

Curnow, A. W., K.-W. Hong, R. Yuan, S.-L. Kim, O. Martins, W. Winkler, T. M. Henkin, and D. Söll (1997) Glu-tRNA^{Gln}-amidotransferase: A novel heterotrimeric enzyme required for correct decoding of glutamine codons during translation. *Proc. Natl. Acad. Sci. USA* 94: 11819–11826. Charging tRNA^{Gln} with Gln in an indirect way.

Giegé, R., M. Sissler, and C. Florentz (1998) Universal rules and idiosyncratic features in tRNA identity. *Nucleic Acids Res.* 26:5017–5035.

Guo, M., P. Schimmel, and X-L. Yang (2010) Functional expansion of human tRNA synthetases achieved by structural inventions. *FEBS Lett.* 584:434–442. Aminoacyl-tRNA synthetases as moonlighting proteins.

Rould, M. A., J. J. Perona, D. Söll, and T. A. Steitz (1989) Structure of *E. coli* glutaminyl-tRNA synthetase complexed with tRNA^{Gln} and ATP at 2.8 Å resolution. *Science* 246:1135–1141. One of the first aminoacyl-tRNA synthetase structures.

Ribosomes

Ban, N., P. Nissen, J. Hansen, P. B. Moore, and T. A. Steitz (2000) The complete atomic structure of the large ribosomal subunit at 2.4 Å resolution. *Science* 289:905–919. The first high-resolution structure.

Kostelev, A., D. N. Ermolenko, and H. F. Noller (2008) Structural dynamics of the ribosome. *Curr. Opin. Chem. Biol.* 12:674–683. A review of movements made during ribosome function and the methods used to analyze them.

Mulder, A. M., C. Yoshioka, A. H. Beck, A. E. Bunner, R. A. Milligan, C. S. Potter, B. Carragher, and J. R. Williamson (2010) Visualizing ribosome biogenesis: Parallel assembly pathways for the 30S subunit. *Science* 330:673–677. Evidence for major distinctions between ribosome assembly pathways in vitro and in vivo.

Ramakrishnan, V. (2009) Decoding the genetic message: The 3D version. 2009 Nobel Lecture. http://nobelprize.org/nobel_prizes/chemistry/laureates/2009/ramakrishnan-lecture.html. This wide-ranging lecture includes a movie showing ribosomal movements during translation.

Ramakrishnan, V. (2011) The eukaryotic ribosome. *Science* 331:681–682. A commentary upon the first eukaryotic ribosome structure determinations.

Selmer, M., C. Dunham, F. V. Murphy IV, A. Wexelbaumer, S. Petry, A. C. Kelley, J. R. Weir, and V. Ramakrishnan (2006) Structure of the 70S ribosome complexed with mRNA and tRNA. *Science* 313:1935–1942. The title says it all.

Zimmerman, E., and A. Yonath (2009) Biological implications of the ribosome's stunning stereochemistry. *ChemBioChem* 10:63–72. A minireview coauthored by one of the 2009 Nobel laureates honored for ribosome structure determination.

The Translation Process

Gold, L., and G. Stormo (1987) Translational initiation. In Escherichia coli *and* Salmonella typhimurium: *Cellular and Molecular Biology*, Vol. 2, F. C. Neidhardt, J. L. Ingraham, B. Low, B. Magasanik, M. Schaechter, and H. E. Umbarger, eds., pp. 1302–1307. American Society for Microbiol., Washington, D.C.

Roberts, J. W. (2010) Syntheses that stay together. *Science* 328:436–437. A brief summary of two papers that described the transcription-translation coupling mechanism.

Schmeing, T. M., R. M. Voorhees, A. C. Kelley, Y-G. Gao, F. V. Murphy IV, J. R. Weir, and V. Ramakrishnan (2009). The crystal structure of the ribosome bound to EF-Tu and aminoacyl-tRNA. *Science* 326:688–694. This and a companion paper from the same laboratory present a structural analysis of events in translation.

Zaher, H. S., and R. Green (2009) Quality control by the ribosome following peptide bond formation. *Nature* 457:161–168. In addition to error correction by aminoacyl-tRNA synthetases, the ribosome has a process to minimize translational errors.

Zhong, W., J. A. Dunkle, and J. H. D. Cate (2009) Structures of the ribosome in intermediate states of ratcheting. *Science* 325:1014–1017. A structural analysis of ribosomal movements during translation.

Antibiotics

Cooperman, B. S., M. A. Buck, C. L. Fernandez, C. J. Weitzman, and B. F. D. Ghrist (1989) Antibiotic photoaffinity labeling probes of *E. coli* ribosomal structure and function. In *Photochemical Probes in Biochemistry*, P. E. Nielsen, ed., pp. 123–139. Kluwer, Dordrecht, Netherlands. A review of the use of antibiotics to probe ribosomes.

Steitz, T. A. (2009) From understanding ribosome structure and function to new antibiotics. 2009 Nobel Lecture in Chemistry. http://nobelprize.org/nobel_prizes/chemistry/laureates/2009/steitz-lecture.html. Steitz's Nobel lecture covers known antibiotic action from a structural standpoint and describes new antibiotics emerging from structural work.

Post-translational Modification

Ataide, S. F., N. Schmitz, K. Shen, A. Ke, S-o. Shan, J. A. Doudna, and N. Ban (2011) The crystal structure of the signal recognition particle in complex with its receptor. *Science* 331:881–886. Strucural insights into protein translocation.

Becker, T., and 13 coauthors (2009) Structure of monomeric yeast and mammalian Sec61 complexes interacting with the translating ribosome. *Science* 326:1369–1373. Structural analysis of the relationship between a ribosome and a protein-conducting channel.

Hunt, J. F., S. Weinkauf, L. Henry, J. J. Fak, P. McNicholas, D. B. Oliver, and J. Deisenhofer (2002) Nucleotide control of interdomain interactions in the conformational reaction cycle of SecA. *Science* 297:2018–2026. Structural analysis of energy coupling in bacterial protein secretion.

Portt, L., G. Norman, C. Clap, M. Greenwood, and M. T. Grenwood (2011) Anti-apoptosis and cell survival: A review. *Biochim. Biophys. Acta* 1813:238–259. How cells protect themselves against unprogrammed death.

Xue, M., and B. Zhang (2002) Do SNARE proteins confer specificity for vesicle fusion? *Proc. Natl. Acad. Sci. USA* 99:13359–13361. A minireview on SNARE proteins.

PROBLEMS

1. The following synthetic polynucleotide is synthesized and used as a template for peptide synthesis in a cell-free system from *E. coli*.

   ···AUAUAUAUAUAUAU···

   What polypeptide would you expect to be produced? Precisely what information would this give you about the code?

2. If the same polynucleotide described in Problem 1 is used with a *mitochondria-derived* cell-free protein-synthesizing system, the product is

   ···Met–Tyr–Met–Tyr–Met–Tyr···

   What does this say about differences between the mitochondrial and bacterial codes?

*3. When polynucleotides are synthesized with repeating triplets of nucleotide residues, from one to three kinds of polypeptide chains will be produced in cell-free synthesis.
   (a) Explain why these different results are possible.
   (b) Predict polypeptides produced when the following are used with an *E. coli* system: $(GUA)_n$; $(UUA)_n$.

*4. What kind of repeating polynucleotide would yield a single polypeptide with a tetrapeptide repeating unit?

5. Although the Shine–Dalgarno sequences vary considerably in different genes, they include examples like GAGGGG that could serve as code—in this case, for Glu–Gly. Does this imply that the sequence Glu–Gly cannot ever occur in a protein, lest it be read as a Shine–Dalgarno sequence? Speculate.

6. According to wobble rules, what codons should be recognized by the following anticodons? What amino acid residues do these correspond to?
   (a) 5′—ICC—3′
   (b) 5′—GCU—3′

*7. In the early days of ribosome research, before the exact role of ribosomes was clear, a researcher made the following observation. She could find, in sedimentation experiments on bacterial lysates, not only 30S, 50S, and 70S particles but also some particles that sedimented at about 100S and 130S. When she treated such a mixture with EDTA, everything dissociated to 30S and 50S particles. Upon adding divalent ions, she could regain 70S particles, but never 100S or 130S particles.

   (a) Suggest what the 100S and 130S particles might represent, in light of current knowledge of protein synthesis. What important discovery did the researcher miss?
   (b) Why do you think reassociation to 100S and 130S particles did not work?

8. The E site may not require codon recognition. Why?

9. What is the minumum number of tRNA molecules that a cell must contain in order to translate all 61 sense codons?

10. Suppose that the probability of making a mistake in translation at each translational step is a small number, δ. Show that the probability, p, that a given protein molecule, containing n residues, will be completely error-free is $(1-\delta)^n$.

11. Assume that the translational error frequency, δ, is 1×10^{-4}.
    (a) Calculate the probability of making a perfect protein of 100 residues.
    (b) Repeat for a 1000-residue protein.

*12. Devise an experiment, based on sucrose gradient sedimentation (see Figure 27.3, page 1128), that would demonstrate that proteins are synthesized on polyribosomes.

13. Assuming that glucose is burned to CO_2 as an energy source, how many amino acid residues can be incorporated into a protein molecule for each glucose consumed by a cell? Is this a maximum or minimum estimate?

14. Why might you expect a small dose of puromycin to be less effective in repressing bacterial growth than an equivalent number of molecules of erythromycin?

15. The antibacterial protein colicin E3 is an effective inhibitor of protein synthesis in prokaryotes. This protein is a nuclease, specifically attacking a phosphodiester bond near the 3′ end of the 16S RNA. Suggest a mechanism for the effect of colicin E3 on translation.

16. If a nonsense mutation can be suppressed by a tRNA mutation that changes an anticodon so that a nonsense codon can be translated, describe a mechanism for suppression of a frameshift mutation.

17. Compare and contrast the metabolic pathways leading to thymine in DNA and thymine as a modified base in tRNA.

*18. Refer to Figure 28.6, which showed restoration of a proper reading frame by a frameshift mutation. In another experiment with the T4 lysozyme system, the Streisinger group found that a second frameshift mutation in T4 gene *e* restored wild-type function. Sequence analysis of wild-type and double-mutant lysozymes were identical except for the amino acid sequences shown here. Describe the mutations that could have generated this result in terms of the base sequences in the *e* gene and known codon assignments.

   Wild-type: ···Lys-Ser-Pro-Ser-Leu-Asn-Ala···

   Mutant ···Lys-Ser-Val-His-His-Leu-Met-Ala···

19. The earliest work on the genetic code established UUU, CCC, and AAA as the codons for Phe, Pro, and Lys, respectively. Can you think of a reason why polyG was not used as a translation template in these experiments?

TOOLS OF BIOCHEMISTRY 28A

WAYS TO MAP COMPLEX MACROMOLECULAR STRUCTURES

As we continue to probe the structure of the cell, using ever gentler and more discriminating techniques, it becomes apparent that much of the cellular machinery is organized in complicated structures that are *assemblies* of macromolecules. The ribosome, with its several kinds of RNA and many kinds of proteins, is an excellent example. Although ribosomes, with their large size and complexity, have finally yielded to X-ray crystallography, some of the techniques used in earlier investigations are still important in learning about other complex biological stuctures. We describe several of these methods here.

Chemical Cross-linking

One way to learn about the arrangement of components in a particle is to investigate spatial relationships, through chemical **cross-linking** (see also Tools of Biochemistry 13A, page 588).

Consider an idealized particle (like that shown in Figure 28A.1) that contains three different protein molecules. If we have a bifunctional reagent (like one of those shown in Table 28A.1) that can react with side chain residues to form cross-links between adjacent protein molecules, we can allow them to react lightly, and then we can extract the protein as a mixture of cross-linked particles. Covalently linked partners that have formed can be identified in various ways. If we have antibodies to the different proteins, we can identify partners that have formed by using Western blotting (method 1 in Figure 28A.1; also see Tools of Biochemistry 7A). Alternatively, we might use one of the "cleavable" cross-linkers shown in Table 28A.1, together with two-dimensional gel electrophoresis (method 2 in Figure 28A.1). In any event, in the simple example shown, it is clear that protein A must lie between B and C because A can be linked to either of the other proteins but B and C do not form cross-linked dimers.

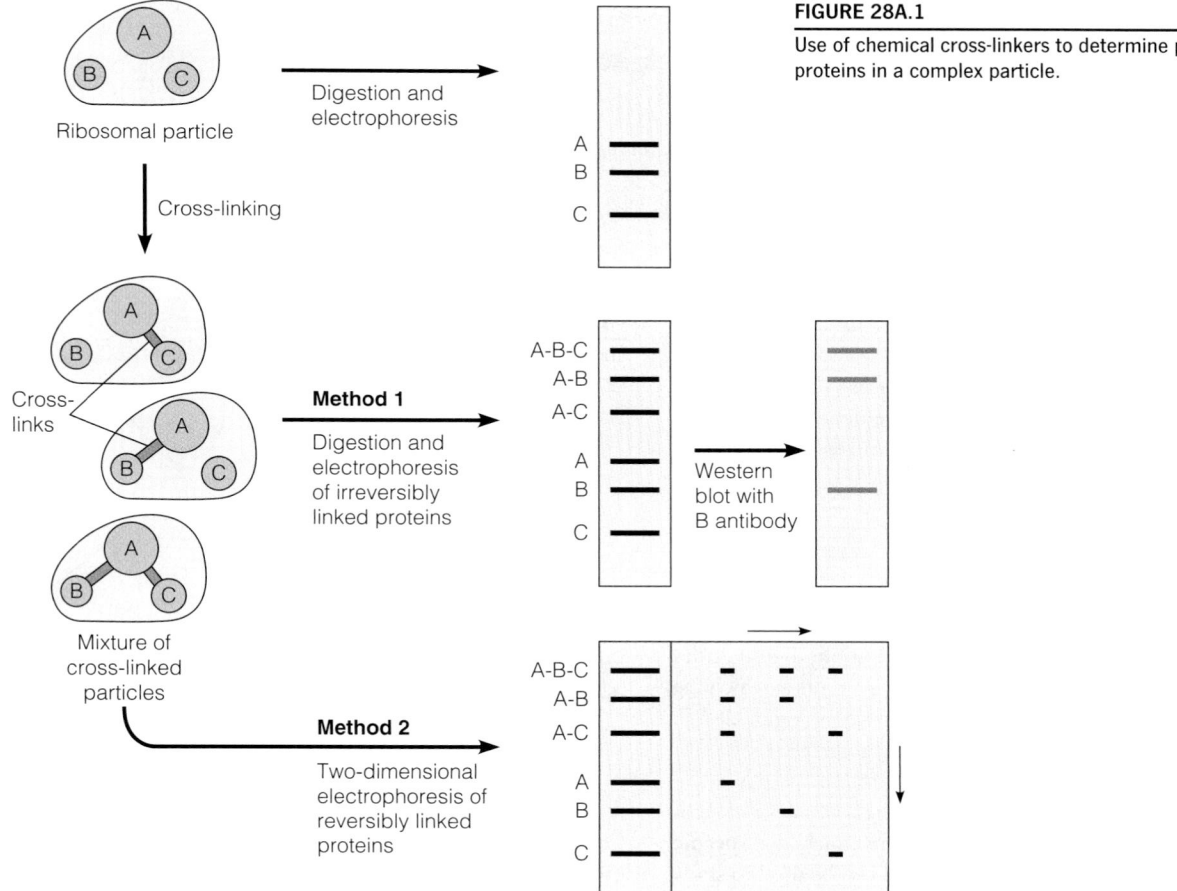

FIGURE 28A.1

Use of chemical cross-linkers to determine proximity of proteins in a complex particle.

TABLE 28A.1 Some protein cross-linking reagents

Reagent	Formula	Reacts Primarily with	Cleavable?
Bis(*N*-maleimidomethyl) ether		Sulfhydryl group	No
2,2′-Dicarboxy-4, 4′-azophenyldiisocyanate		Amino groups	Yes, by reduction of —N=N—
Dimethyl suberimidate	$CH_3O-\overset{NH}{\underset{\|}{C}}-(CH_2)_6-\overset{NH}{\underset{\|}{C}}-OCH_3$	Amino group	Yes, by ammonia
Tetranitromethane	$C(NO_2)_4$	Phenolic groups (tyrosine)	No
Methyl-4-azidobenzoimidate[a]		Amino groups and others	No
Methyl[3-(*p*-azidophenyl) dithio]propionimidate[a]		Amino groups and others	Yes, by reduction of —S—S—

[a]These reagents are photoactivatable. They may be first reacted, in the dark, through the imidate group to the right and then coupled to another group by activating the azide (N_3) with a flash of light.

In addition to the protein–protein cross-linking techniques described here, methods for RNA–protein and RNA–RNA cross-linking are now available (see References). These methods have played a large part in determining the detailed ribosomal models that led ultimately to crystal structures shown in this chapter.

Immunoelectron Microscopy

In **immunoelectron microscopy**, components that lie on the surface of a particle can be localized in a direct manner by using antibodies prepared against them. The Y-shaped antibody molecules form bridges between two particles, connecting points at which the particular component is accessible to the surface. In Figure 28A.2, we show the same idealized particle used in Figure 28A.1. The fact that protein B lies near the pointed end of the particle is made clear by the way in which anti-B antibodies tie two particles together. When appropriate hapten groups are attached at specific points on RNA molecules, the same method can be used to show

FIGURE 28A.2

Locating proteins on a particle surface by antibody binding.

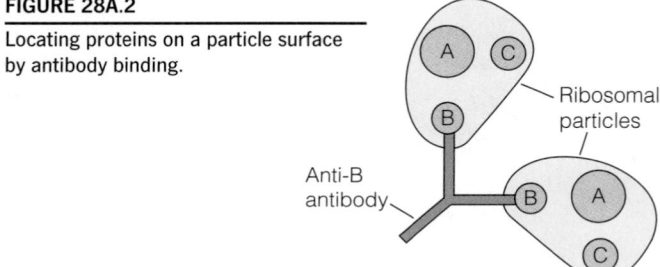

when these RNA sequences are near the ribosomal surface. The ends of the ribosomal RNAs have been located in this way, as has the position at which the nascent peptide emerges from the ribosome (see Figure 28A.3).

Cryoelectron Microscopy

Conventional techniques of transmission electron microscopy (Tools of Biochemistry 1A) suffer from a number of serious

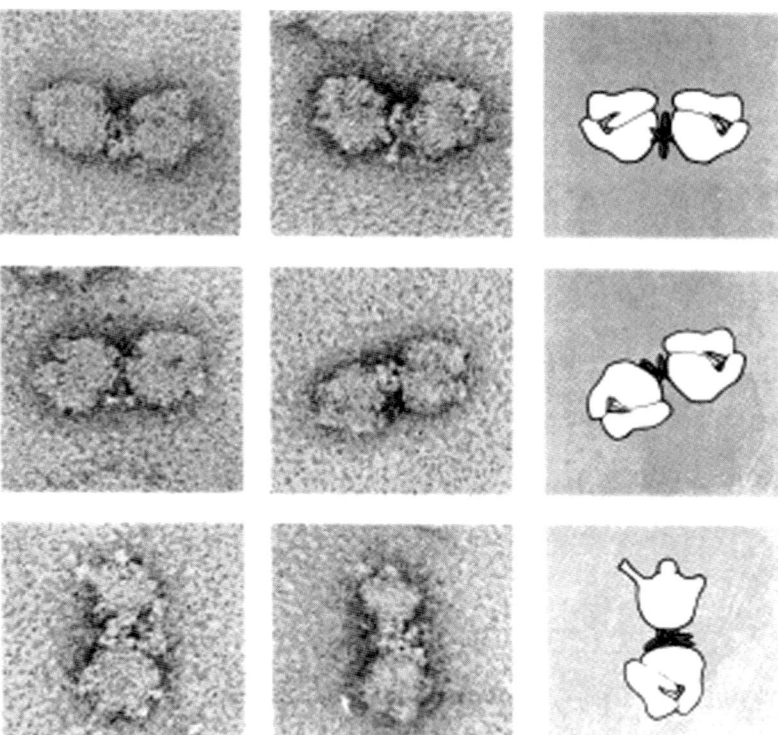

FIGURE 28A.3

Use of antibodies to detect where the polypeptide chain emerges from the ribosome. Antibodies to β-galactosidase were used to interact with β-galactosidase chains at the point where they emerge from the ribosomes. The 70S ribosomes were found to be tied together by such antibodies at sites near the back of the 50S subunit, at the location of a pore in this unit.

Courtesy of C. Bernabéu and J. A. Lake.

drawbacks in the study of delicate biological structures. First, the necessity of complete dehydration in the vacuum chamber of the microscope can greatly modify structures. Second, to produce enough contrast, samples have traditionally been stained with heavy metals or metal-shadowed, producing an inherent loss in resolution.

A relatively new technique that avoids these artifacts is *cryoelectron microscopy* (or *electron cryomicroscopy*). The idea is simple: A sample is very rapidly frozen in ice; the freezing is so rapid that the water does not crystallize but becomes *vitrified* (glassy). The sample can be vitrified in the interstices of an EM grid or vitrified in bulk and then thin-sectioned for examination. Because no stain is used, contrast is low, and methods of image enhancement are needed. For particles such as ribosomes, this is accomplished by combining the information from a number of faint images of particles in the ice to develop a picture. Because they will, in general, be randomly rotated, this random orientation must be taken into account in the computer analysis. The advantage, of course, is that the final generated image can be viewed "from all sides" and can be examined from any angle in computer display.

If particles (viruses, for example) have elements of symmetry, or if the structures of subunits are known, the analysis becomes easier. But even irregular particles like ribosomes can be analyzed with resolution approaching 1–2 nm. Last, but not least, the

sample has remained in an aqueous environment throughout the study, so damage to the structure should have been minimized.

Low-Angle Scattering of X-Rays and Neutrons

Although crystal diffraction studies of enormous particles like ribosomes are still challenging, we can learn much from studying the scattering of radiation from suspensions of such particles. When electromagnetic waves are scattered from a particle with dimensions much greater than the wavelength of the radiation, the scattering intensity depends on the angle of observation. As shown in Figure 28A.4, waves scattered from different regions within the particle at any angle other than 0° are out of phase, resulting in mutual interference and decreased intensity. At large angles the phase difference increases, causing partial cancellation of the scattered waves. This interference may be used to measure average dimensions of the particle. The intensity of scattering at angle θ (I_θ) compared with the scattering at angle 0 (I_0) is given (for small angles) by the following:

$$\frac{I_\theta}{I_0} = e^{-(16\pi^4 R_G^2/3\lambda^4)\sin^2(\theta/2)} \tag{28A.1}$$

Here λ is the wavelength of the radiation, and R_G is a quantity called the *radius of gyration*, a kind of average dimension of the

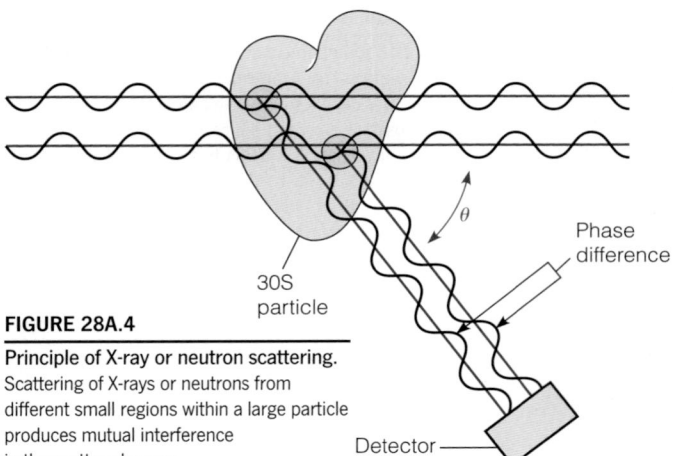

FIGURE 28A.4

Principle of X-ray or neutron scattering. Scattering of X-rays or neutrons from different small regions within a large particle produces mutual interference in the scattered waves.

particle. According to Equation 28A.1, a graph of $\ln(I_\theta/I_0)$ versus $\sin^2(\theta/2)$ should be a straight line at low angles, with initial slope of $(16\pi^4 R_G^2/3\lambda^4)$. Thus, measurement of the scattering at very low angles gives a measure of average particle size. At higher angles, the I_θ/I_0 curves have a more complex shape, with maxima and minima. These can be used to give additional information about the shape of the particle and its internal distribution of matter.

Although low-angle scattering of X-rays has been useful in studying particles in solution, a much more powerful technique is **low-angle neutron scattering**. It may seem strange at first to think of neutrons as radiation, but we must remember that, according to quantum mechanics, any elementary particle has wavelike properties as well. The wavelength of a particle with

mass m moving at velocity v is given by $\lambda = h/mv$, where h is Planck's constant. It turns out that "thermal neutrons" emerging from a nuclear reactor have a wavelength of a few tenths of a nanometer. Thus, they are of the right length for examination of details of macromolecular structure. Still more important, neutrons interact primarily with atomic nuclei and are therefore scattered differently by different atoms. Thus, nucleic acids and proteins scatter neutrons differently, and even hydrogen and deuterium have different scattering powers. Because H_2O and D_2O differ in scattering, it is possible to use as solvents H_2O/D_2O mixtures that match the neutron-scattering power of either the nucleic acid or protein portion of a nucleoprotein particle. Then, as shown in Figure 28A.5, we can make either nucleic acid or protein "disappear" into the background and measure the radius of gyration of either component. In the example shown, the greater R_G observed for nucleic acid than for protein tells us that the nucleic acid is concentrated on the outside of the particle.

An even more powerful variant of the same technique has been used to "map" the distances between particular pairs of proteins in complex particles. Suppose, as in Figure 28A.6, we have reconstituted particles containing just two proteins (among the many in the particle) that have been prepared from deuterium-fed bacteria. These two proteins will be heavily deuterated and will have a neutron-scattering power much different from that of the rest of the particle. If the H_2O/D_2O solvent is now mixed so as to match the average background in the nondeuterated portion of the particle, the two deuterated proteins will stand out in contrast. The neutron-scattering pattern obtained will be dominated by the interference in scattering between these two proteins and can be used to measure their separation. In addition, the

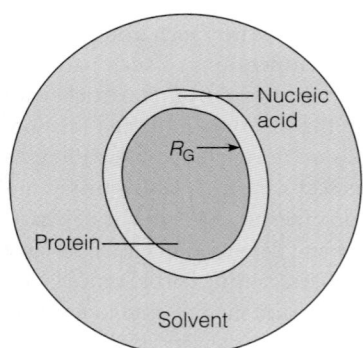

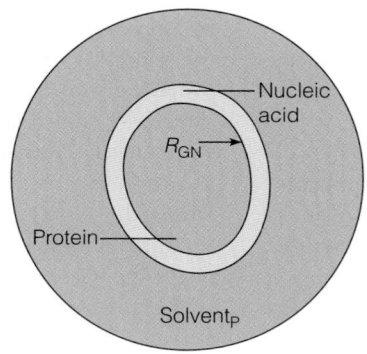

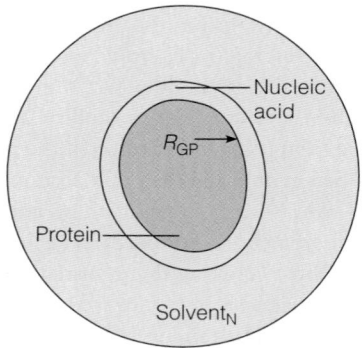

$Solvent_P$ = solvent that matches the scattering properties of the protein
$Solvent_N$ = solvent that matches the scattering properties of the nucleic acid
R_G = radius of gyration of the nucleic acid-protein complex
R_{GN} = radius of gyration of the nucleic acid
R_{GP} = radius of gyration of the protein

FIGURE 28A.5

Use of selective solvent matching to show, by neutron scattering, that the nucleic acid is on the outside of a nucleoprotein particle.

Principles of Biochemistry, 2nd ed., Kensal E. Van Holde, Curtis Johnson, and Pui Shing Ho, © 2006. Adapted by permission of Pearson Education Inc., Upper Saddle River, NJ.

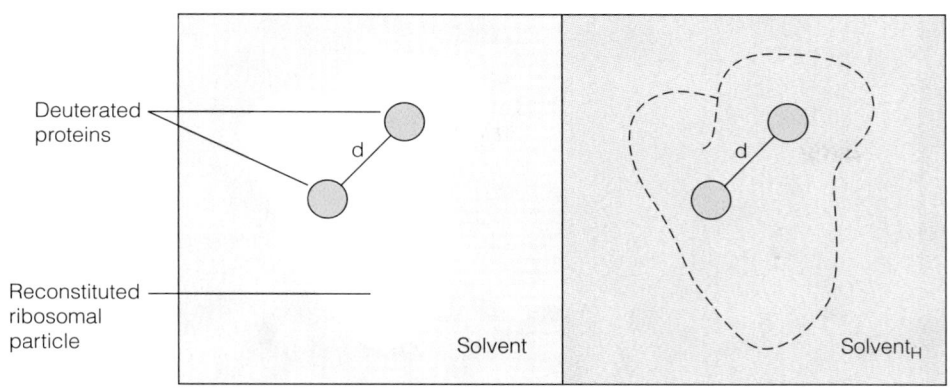

Solvent_H = solvent matching the nondeuterated part of particle

FIGURE 28A.6

Use of solvent matching to determine the distance between two selectively deuterated proteins in an undeuterated particle.

method allows measurement of the radius of gyration of a particular protein in situ.

Although neutron scattering can give much useful information, it is not a technique the average biochemist can use in the laboratory. Only a few locations in the world have large research reactors that are fitted to do neutron-scattering studies.

References

Boublik, M. (1990) Electron microscopy of ribosomes. In *Ribosomes and Protein Synthesis: A Practical Approach*, G. Spedding, ed., pp. 273–296. Oxford University Press, Oxford.

Brimacombe, R., B. Greuer, H. Gulle, M. Kasak, P. Mitchell, M. Oswald, K. Stade, and W. Stiege (1990) New techniques for the analysis of intra-RNA and RNA-protein cross-linking data from ribosomes. In *Ribosomes and Protein Synthesis: A Practical Approach*, G. Spedding, ed., pp. 131–159. Oxford University Press, Oxford.

Cornish, P. V., D. N. Ermolenko, D. W. Staple, L. Hoang, R. Hickerson, H. F. Noller, and T. Ha (2009) Following movement of the L1 stalk between three functional states in single ribosomes. *Proc. Natl. Acad. Sci. USA.* 106:2571–2576.

Serdyuk, I. N., M. Y. Pavlov, I. N. Rublevskaya, G. Zaccai, R. Leberman, and Y. M. Ostenavitch (1990) New possibilities for neutron scattering in the study of RNA–protein interactions. In *The Ribosome*, W. Hill et al., eds., pp. 194–202. American Society for Microbiology, Washington, D.C.

Stark, H., E. V. Orlova, J. Rinke-Appel, N. Jünke, F. Mueller, M. Rodnina, W. Wintermeyer, R. Brimacombe, and M. van Heel (1997) Arrangement of tRNAs in pre- and post-translocational ribosomes, revealed by electron cryomicroscopy. *Cell* 88:19–28. An elegant use of the method to follow conformational changes.

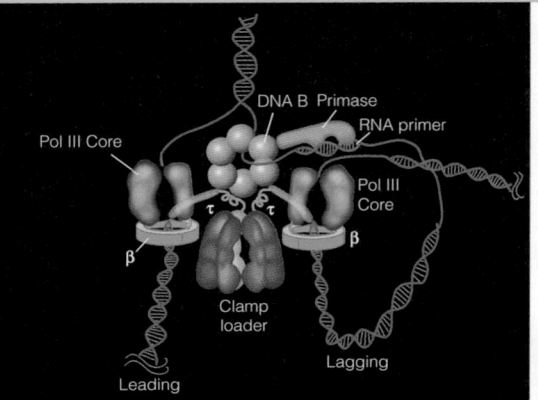

CHAPTER 29

Regulation of Gene Expression

A key to understanding life is understanding how the expression of genes is controlled. Why is hemoglobin present only in red blood cells? How does addition of lactose to a bacterial culture initiate a several thousand-fold increase in the rate of β-galactosidase synthesis? What causes enzymes of the urea cycle to appear during metamorphosis of a tadpole to an adult frog? What factors orchestrate the pattern of sequential gene expression during embryonic differentiation? These are just a few questions whose answers demand an understanding of gene regulation.

As we mentioned in Chapter 27, seminal insights into genetic control were presented in 1960, when François Jacob and Jacques Monod of France's Pasteur Institute proposed the operon model of gene regulation, based upon their genetic analysis of lactose utilization in *E. coli*. Contributing equally to the success of Jacob and Monod were parallel investigations by Andre Lwoff of the genetic regulation of reproduction of the temperate bacteriophage λ in *E. coli*. We shall return to this important biological system later. Based upon similarities in regulatory mechanisms between these two quite different systems, Jacob and Monod proposed that regulation of gene expression occurs primarily at the level of transcription, and specifically, at the level of transcriptional initiation. Their model was largely correct, but subsequent years have shown that regulation can occur at any stage in the expression of a gene. For example, we have seen in Chapter 26 that regulation can occur at the level of gene copy number, when environmental stresses cause amplification of genes whose products deal with the stress.

However, Jacob and Monod were correct; most regulation does occur at the transcriptional level, and that is what we focus upon in this chapter. We will begin with prokaryotic systems, which provide historical context, then move to the more complex eukaryotic regulatory processes. Along the way, we will present some examples of regulatory processes occurring at other levels of gene expression, particularly translation. We will also discuss recent discoveries that identified functions of small RNA molecules in gene regulation.

Regulation of Transcription in Bacteria

The Lactose Operon: Earliest Insights into Transcriptional Regulation

Recall from Chapter 27 that lactose utilization in *E. coli* is controlled by three contiguous genes—*lacZ* (β-galactosidase), *lacY* (β-galactoside permease, a transport protein), and *lacA* (thiogalactoside transacetylase, an enzyme of still unknown function). In the presence of an inducer, all three proteins accumulate simultaneously, but to different levels. Lactose itself leads to induction of the lactose operon, but the true intracellular inducer is **allolactose**, Galβ(1 → 6)Glc, a minor product of β-galactosidase action. In the laboratory one usually uses a synthetic inducer such as **isopropyl thiogalactoside (IPTG)**, which induces the lactose operon but is not cleaved by β-galactosidase. Hence, its concentration does not change during an experiment.

A mutation in a structural gene—*lacZ*, for example—can inactivate its product (β-galactosidase) without affecting control of the other two genes. However, mutations in the regulatory regions mapping *outside* genes *lacZ*, *lacY* and *lacA* can affect expression of all three structural genes. In their early work, Jacob and Monod recognized two distinct mutant phenotypes—**constitutive**, in which all three gene products are synthesized at high levels even when inducer is absent, and **noninducible**, in which all three enzyme activities remain low even after addition of an inducer. These mutations mapped in two sites, termed *o* and *i*. Importantly, Jacob and Monod were able to establish dominance relationships involving regulatory mutations, by using interrupted bacterial mating (Chapter 25) to create partial diploids. Problem 7 at the end of this chapter should help you to understand relationships among regulatory mutations, mutant phenotypes, and levels of *lac* operon enzymes.

The original Jacob–Monod model for gene regulation, based upon this system, was presented in Figure 27.2 (page 1127). A more contemporary map of the *lac* operon is shown in Figure 29.1. As Jacob and Monod correctly proposed, transcription of the three structural genes is initiated near an adjacent site, the **operator**. Transcription yields a single **polycistronic messenger RNA**; that is, it yields an RNA copy of all three genes. (The term **cistron** has genetic significance. For our purposes it is a region of a genome that encodes one polypeptide chain.) The *i* gene product is a macromolecular **repressor**, which in the active form binds to the operator, blocking transcription (Figure 29.2a).

The repressor protein also has a binding site for an inducer. Binding of IPTG, allolactose, or some other inducer at this site inactivates repressor by vastly decreasing its affinity for DNA (Figure 29.2b). This repressor inactivation stimulates transcription of *lacZ, Y,* and *A* because dissociation of the repressor–inducer complex from the operator removes a steric block to binding of RNA polymerase at the initiation site. Thus, the introduction of lactose activates synthesis of the gene products involved in its catabolism by removing a barrier to their transcription. This mode of regulation is essentially negative because the active regulatory element (the repressor) is an *inhibitor* of transcription. Positive control was later discovered, involving the CRP site shown in Figure 29.1. We shall discuss this shortly.

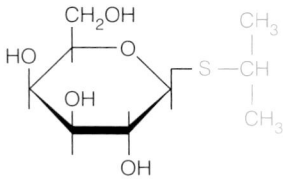

Lactose

Allolactose

Isopropyl β-thiogalactoside

FIGURE 29.1

A map of the lactose operon. The CRP site is the binding site for cAMP receptor protein, a regulatory factor (page 1237). The promoter also includes binding sites for RNA polymerase and *lac* repressor. Synthesis of repressor initiates at its own promoter (*i* promoter). Additional repressor-binding sites exist, 82 nucleotides upstream and 432 nucleotides downstream, respectively, from the transcriptional start point (not shown).

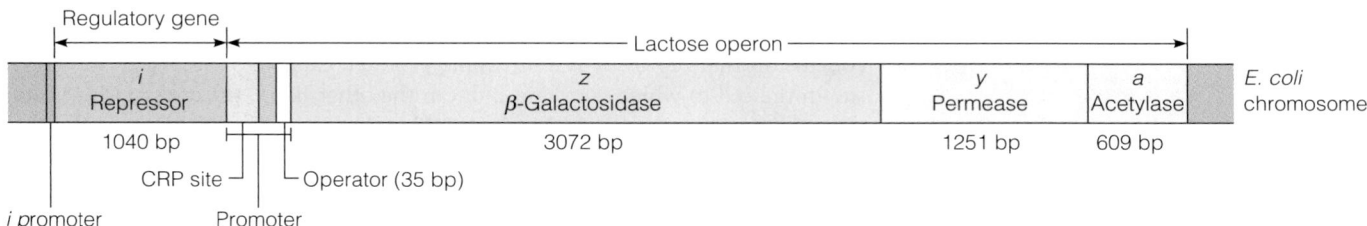

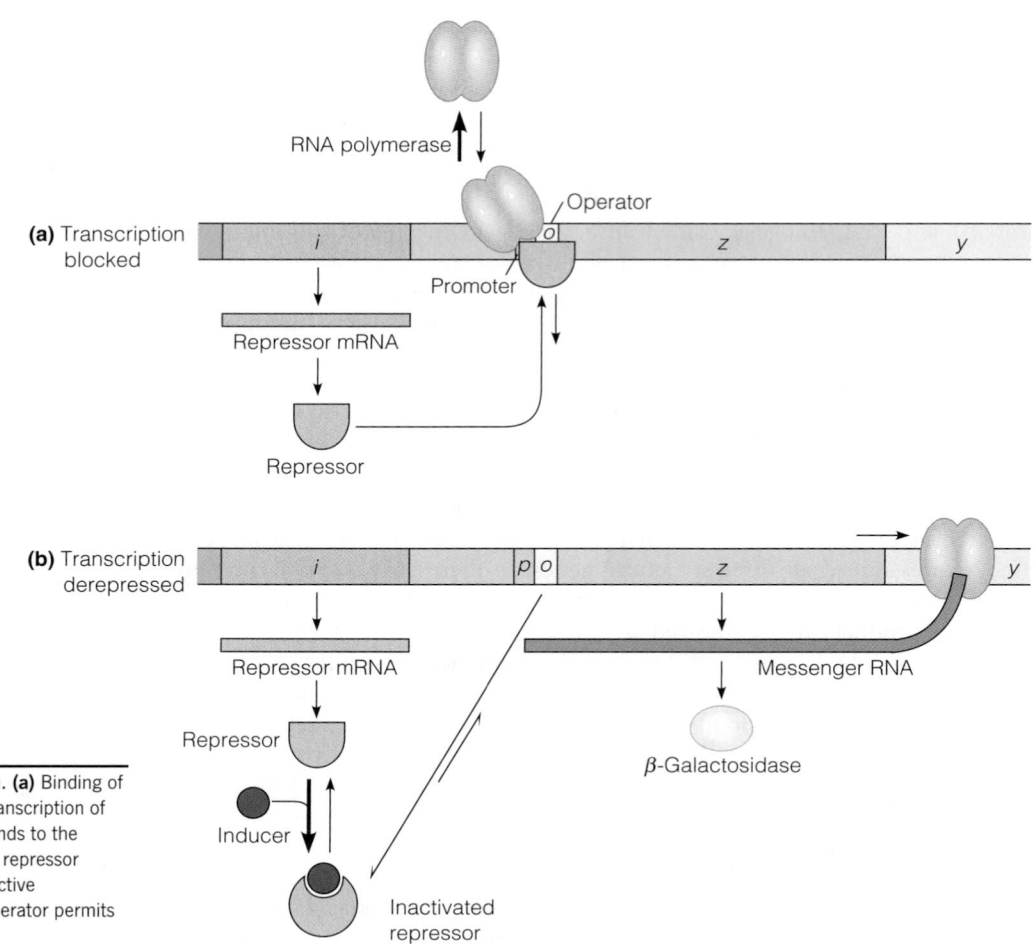

FIGURE 29.2

Configurations of the lactose operon. (a) Binding of the repressor to the operator inhibits transcription of the structural genes. **(b)** The inducer binds to the repressor, decreasing the affinity of the repressor for the operator. Dissociation of the inactive repressor–inducer complex from the operator permits transcription of the structural genes.

The phenotypes of partial diploids involving *lac* regulatory genes gave indispensable clues to mechanisms of transcriptional regulation.

As mentioned above, essential for developing the operon model was the ability to analyze regulation in partial diploids—bacteria containing one complete chromosome plus part of another, transferred in by conjugation. One copy of the operon resides in the chromosome, while another lies on an incomplete chromosome, introduced into a cell as part of the bacterial mating process. Noninducible mutations that mapped in *i* had a dominant phenotype, meaning that expression of the structural genes was low when both wild-type and mutant alleles of *i* were present. Jacob and Monod proposed that the mutant alleles give rise to mutant repressors that are unable to bind inducer. These mutant repressors would remain bound to DNA at operator sites on both the mutant and the normal chromosomes, even when inducer is present.

Constitutive mutations that mapped in *i* had a recessive phenotype. That is, they resulted in high gene expression, but only when two mutant alleles were present. These mutant alleles generated repressors that were defective in operator binding and thus could not turn off gene expression. Such mutations are recessive because a normal repressor in the same cytosol can bind to all operators and inhibit transcription. These observations showed that repressor mutations are **_trans_-dominant**, meaning that the *i* gene product encoded by one genome can affect gene expression from other genomes in the same cell. This finding led to the conclusion that repressor is a diffusible product, capable of acting on any DNA site in the cell to which it could bind. On the other hand, the constitutive mutations mapping in *o* had a **_cis_-dominant** effect. That is, in a cell with one wild-type operator and one mutant operator, only the genes on the same chromosome as the mutant operator were expressed constitutively. Given that a protein would be capable of diffusing through the cytosol and acting on other chromosomes, this finding suggested that the operator does not encode a gene product.

For a molecular mechanism based almost entirely on the indirect evidence of genetic analysis, the operon model as advanced by Jacob and Monod has stood the test of time remarkably well. Three major modifications to the model occurred as the system was subjected to further analysis. First, the promoter was discovered as an element distinct from the operator (although the two sites overlap). Second, although the repressor was first thought to be *i*-gene RNA, its isolation proved that it is protein. Third, Jacob and Monod proposed that all transcriptional regulation was negative; that is, binding a regulatory protein always inhibits transcription. However, the lactose operon, like many other regulated genes, also exhibits *positive* control of transcription (that is, *activation* of transcription by binding of a protein), as described on pages 1236–1238.

Isolation and Properties of Repressor

The *lac* repressor was isolated in 1966 by Walter Gilbert and Benno Müller-Hill. Because this repressor comprises only 0.001% of the total cell protein, Gilbert and Müller-Hill used mutants designed to overproduce it, so as to maximize its synthesis, to about 2% of total protein (this was several years before the gene could have been easily overexpressed by cloning). They then purified the protein on the basis of its ability to bind the synthetic inducer IPTG. The purified *lac* repressor is a tetramer, formed from four identical subunits, each with 360 amino acids ($M_r = 38,350$). The protein binds IPTG with a K_a of about $10^6 \, M^{-1}$, and it binds nonspecifically to duplex DNA with a K_a of about $3 \times 10^6 \, M^{-1}$. However, its specific binding at the *lac* operator is much tighter, with a K_a of $10^{13} \, M^{-1}$. Like RNA polymerase, repressor seeks its operator site by first binding to DNA at any site and then moving in one dimension along the DNA. It moves either by sliding or by transfer from one site to another, when the two sites are brought next to each other on adjacent loops of DNA.

Control by the *lac* repressor is exceedingly efficient, particularly in view of the minute amount of repressor present in an *E. coli* cell. The *i* gene is expressed at a low rate, giving about 10 molecules of repressor tetramer per cell. Although this corresponds to a concentration of only about $10^{-8} \, M$, this value is several orders of magnitude higher than the *dissociation* constant, meaning that in a *noninduced* cell the operator is bound by repressor more than 99.9% of the time—hence, the very low levels of *lac* operon proteins in uninduced cells (less than one molecule per cell). However, binding of inducer decreases the affinity of the repressor–inducer complex for operator by many orders of magnitude. Under these conditions, nonspecific binding of the repressor–inducer complex at other DNA sites becomes significant so that in *induced* cells the operator is occupied by repressor less than 5% of the time.

The Repressor Binding Site

The DNA site bound by *lac* repressor has been analyzed by footprinting and methylation-protection experiments of the type described earlier for RNA polymerase (see Figure 27.15, page 1140). As shown in the diagram below, the operator comprises 35 base pairs, including 28 base pairs of symmetrical sequence; that is, it includes a sequence that is identical in both directions (shaded in the diagram). Thus, the operator is an imperfect palindrome—imperfect because of the seven base pairs that do not show this symmetry.

A repressor–inducer system provides negative control of the *lac* operon. The repressor binds to operator, interfering with transcription initiation. An inducer binds to repressor, reducing its affinity for operator.

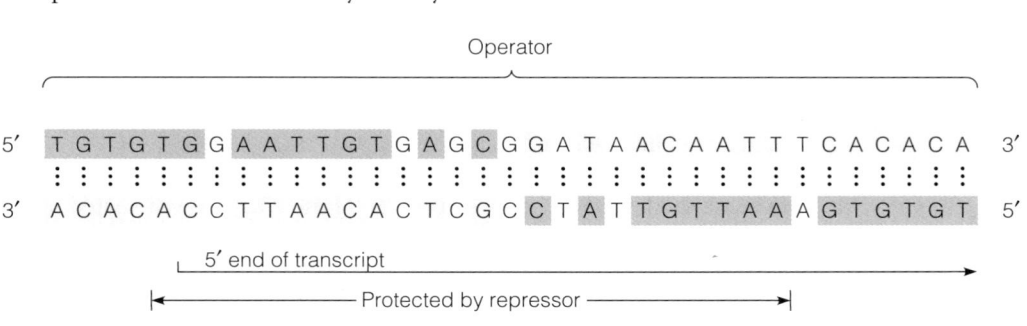

Operator

5′ T G T G T G G A A T T G T G A G C G G A T A A C A A T T T C A C A C A 3′

3′ A C A C A C C T T A A C A C T C G C C T A T T G T T A A A G T G T G T 5′

5′ end of transcript

⊢————— Protected by repressor —————⊣

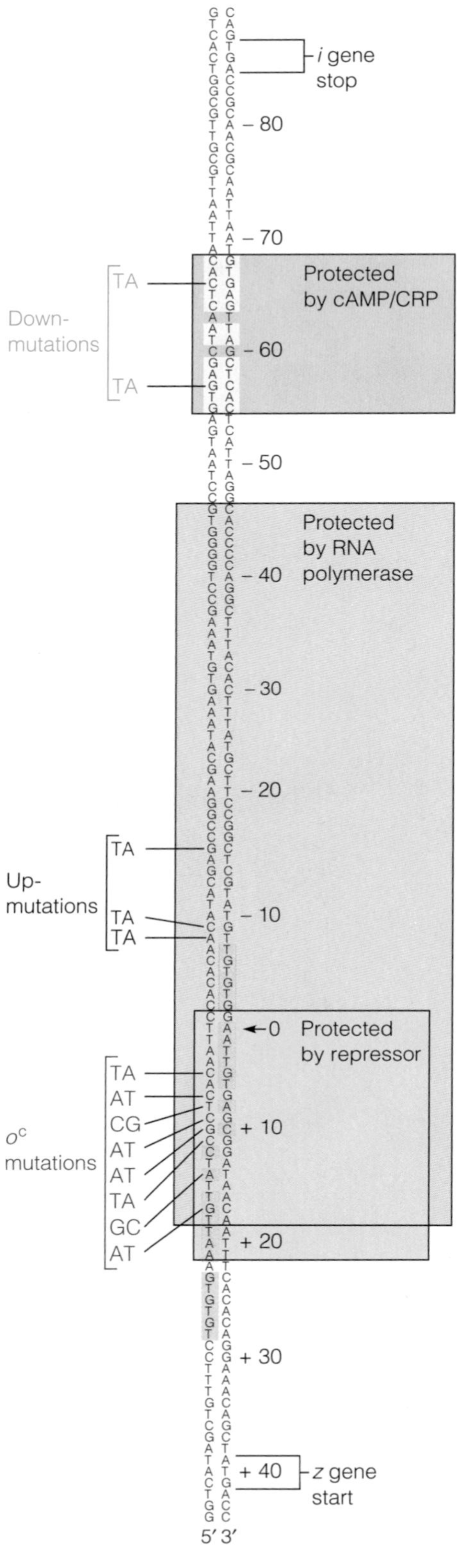

FIGURE 29.3

The 122-base *lac* regulatory region. The inverted repeat sequence in the CRP binding site is shaded in blue, and the palindromic areas of the operator are shaded in pink. The binding sites for all three proteins (CRP, repressor, and RNA polymerase), as determined by DNase I footprinting, are boxed. The sequence changes encountered in promoter and operator mutations are shown, as well as the start codon for the *lacZ* protein (Z gene) and the stop codon for repressor protein (*i* gene). Nucleotides are numbered from 0 at the *lac* transcriptional start point (the first mRNA nucleotide is +1).

The figure at left shows the *lac* regulatory region DNA sequence, numbered from −80 (near the *i* gene stop) to +40 (the *z* gene start), with the following boxed binding regions: "Protected by cAMP/CRP" (around −60), "Protected by RNA polymerase" (around −40 to −10), and "Protected by repressor" (around 0 to +20). Down-mutations (TA, TA) are indicated near the CRP region, Up-mutations (TA, TA, TA) near the RNA polymerase region, and o^c mutations (TA, AT, CG, AT, AT, TA, GC, AT) near the repressor region.

The transcriptional start point is included within the repressor-binding sequence, as shown in the diagram. Twenty-four of the 35 base pairs of the operator are protected from DNase attack by repressor binding. Operator-constitutive (o^c) mutations involve changes in the central portion of this sequence of nucleotides. Figure 29.3 shows how the operator and promoter overlap, as determined by the regions of DNA protected by binding either repressor or RNA polymerase, respectively.

Once the *lac* operon had been sequenced, it became apparent that two additional *lac* repressor-binding sites are located nearby, one centered upstream, at position −82, and one within the *lacZ* gene itself, at position +432; the original operator is centered at +11. Genetic analysis indicated that both additional sites participate in *lac* operon regulation. Although the significance of the downstream site is not clear, mutations affecting the upstream (−82) site led to incomplete repression of the operon. Evidence indicated that a looped DNA structure was essential for complete repression, with repressor contacting both the −82 and +11 sites. Confirmation came in 1996, with the crystal structure determination of the complete tetrameric *lac* repressor by itself, as a complex with a defined oligonucleotide fragment, and as a complex with IPTG. As shown in Figure 29.4a, the tetrameric protein consists of two dimeric units, joined by a hinge region. Each dimer binds DNA separately, suggesting that the tetrameric protein binds to both the +11 and −82 sites, creating a DNA loop of 93 base pairs between them. The DNA-binding domain of the protein is an α-helical region that contacts bases within the major groove of the operator DNA. This helical binding motif has been seen in other sequence-specific DNA-binding proteins, as we shall discuss shortly.

Figure 29.4b suggests the mechanism of induction. The repressor is an allosteric protein which, upon binding inducer, significantly increases the angle at which two monomeric units in a dimer relate to one another. This drives the DNA-binding helices apart by 3.5 Å so that they can no longer contact DNA binding sites, as they must in order to bind tightly.

Regulation of the *lac* Operon by Glucose: A Positive Control System

The *lac* repressor–operator system keeps the operon turned *off* in the absence of utilizable β-galactosides. An overlapping regulatory system, summarized in Figure 29.5, turns the operon *on* only when alternative energy sources are unavailable. *E. coli* has long been known to use glucose in preference to most other energy substrates. When grown in a medium containing both glucose and lactose, the cells metabolize glucose exclusively until the supply is exhausted. Then growth slows, and the lactose operon becomes activated in preparation for continued growth using lactose. This phenomenon, now known to involve a transcriptional *activation* mechanism, was originally called glucose repression or catabolite repression. Transcriptional activation occurs when glucose levels are low; control is exerted through intracellular levels of cyclic AMP.

Recall that in animal cells a rise in cAMP levels stimulates catabolic enzymes, which increase the levels of energy substrates. Those effects are mediated metabolically, through hormonal signals and triggering of metabolic cascades. In bacteria cAMP activation involves control of gene expression, but the end results are

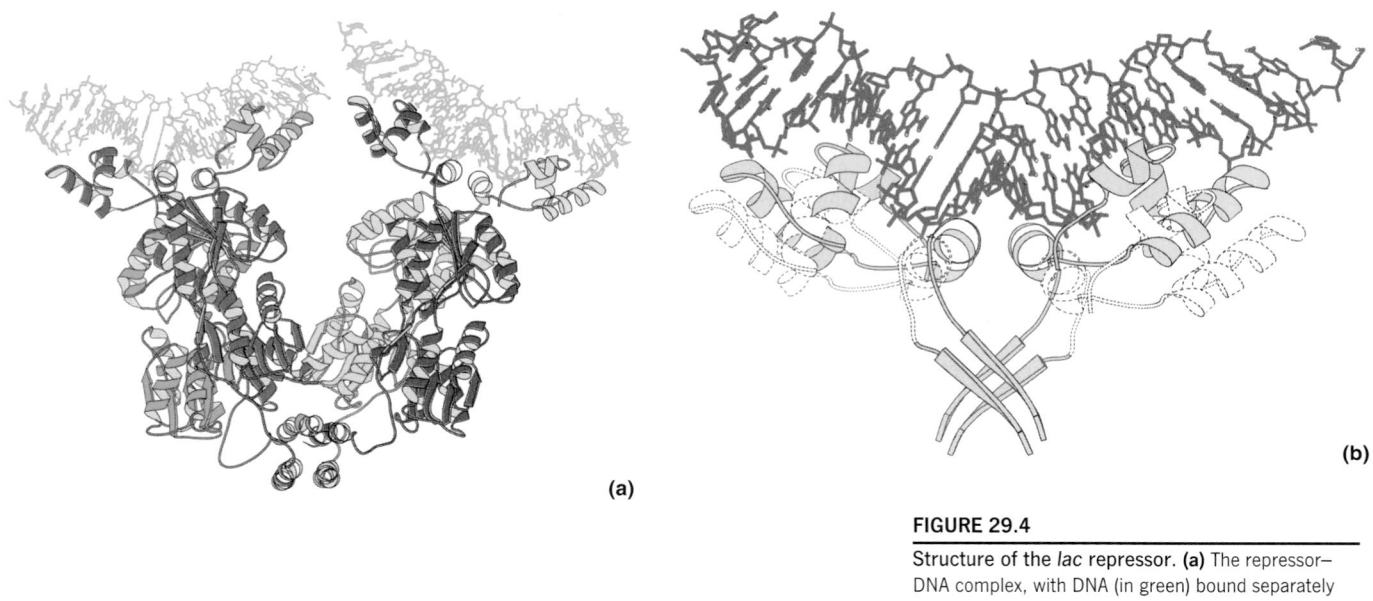

(a)

(b)

FIGURE 29.4

Structure of the *lac* repressor. (a) The repressor–DNA complex, with DNA (in green) bound separately to each of two dimeric units—one shown with the monomers in green and purple, and monomers of the other shown in yellow-green and red. **(b)** The effect of IPTG binding. Shown in yellow-green are the extreme N-termini of two monomeric units of the protein (residues 1–68), bound to DNA, in purple. The dashed structures represent the presumed positions of the DNA-binding helices after binding of inducer IPTG to the protein, which drives the DNA-binding helices apart. PDB ID 1LBG, 1LBH.

From *Science* 271:1247–1254, M. Lewis, G. Chang, N. C. Horton, M. A. Kercher, H. C. Pace, M. A. Schumacher, R. G. Brennan, and P. Lu, Crystal structure of the lactose operon repressor and its complexes with DNA and inducer. © 1996. Reprinted with permission from AAAS.

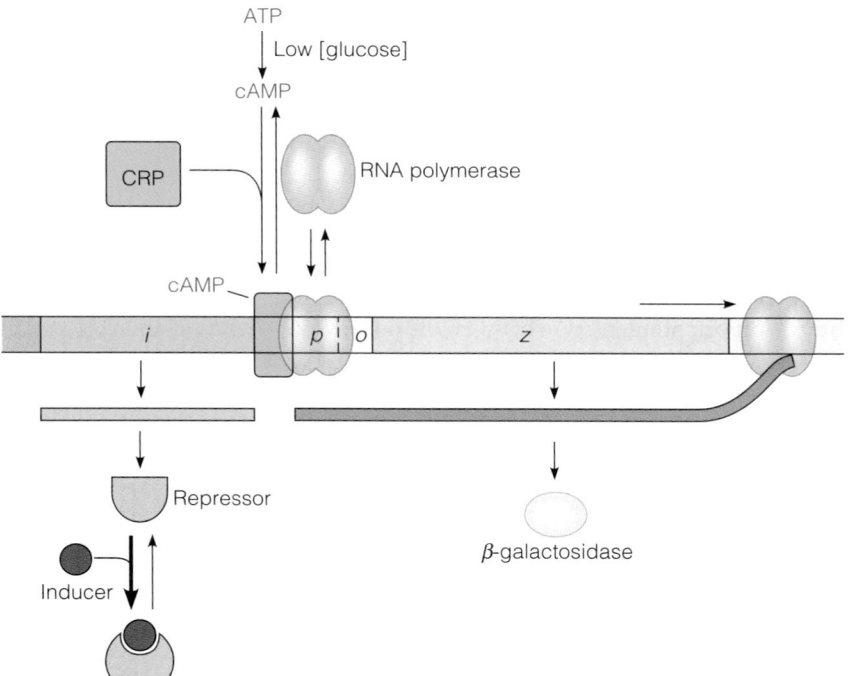

FIGURE 29.5

Activation of the *lac* operon. Repressor is inactivated by binding inducer, and cAMP receptor protein (CRP) is activated by binding cyclic AMP. Binding of the CRP–cAMP complex to DNA facilitates initiation of transcription by RNA polymerase.

similar. In *E. coli*, cAMP levels are low when intracellular glucose levels are high. The actual regulatory mechanism is not yet known. Adenylate cyclase apparently senses the intracellular level of an unidentified intermediate in glucose catabolism—hence, the current name for the regulatory process, **catabolite activation**. When glucose levels drop, as shown in Figure 29.5, cAMP levels rise, triggering activation of the lactose operon by its interaction with a protein called **cAMP receptor protein (CRP)**, formerly called **catabolite activator protein (CAP)**. This protein is a dimer, each of whose identical polypeptide chains contains 210 amino acid residues. When it binds cAMP, CRP undergoes a conformational change. The change greatly increases its affinity for certain DNA sites, including a site in the *lac* operon adjacent to the RNA polymerase binding site. The binding of cAMP–CRP at this site protects a DNA sequence from −68 to −55, as shown in Figure 29.3. This binding facilitates transcription of the *lac* operon by stimulating the binding of RNA polymerase to form a closed-promoter complex or by increasing the rate of open-promoter complex formation.

Our understanding of CRP action is still incomplete, partly because the cAMP–CRP complex activates several different gene systems in *E. coli,* all of them involved with energy generation. They include operons for utilization of other sugars, including galactose, maltose, arabinose, and sorbitol, and several amino acids. Among the operons that have been analyzed, the DNA-binding site of the cAMP-activated dimer varies considerably with respect to the transcriptional start point, suggesting that regulatory mechanisms involving this protein are complex.

Cyclic AMP receptor protein (CRP) provides positive control of *lac* and several other catabolite-repressible operons. The cAMP–CRP complex binds at the *lac* promoter when glucose levels are low and facilitates initiation of transcription.

The CRP–DNA Complex

The structure of the CRP–cAMP–DNA complex, as revealed by X-ray crystallography (Figure 29.6), shows how the protein binds to DNA. Each CRP subunit contains a characteristic pair of α helices, which are joined by a turn. One helix of each pair, shown as perpendicular to the plane of the figure, lies within the major groove of DNA. This **helix–turn–helix** structural motif, which was observed at about the same time as the structure of the λ phage Cro repressor (see Figure 27.1 and pages 1239–1244), is found in several DNA-binding regulatory proteins, suggesting common evolutionary origins for this family of proteins. Helix–turn–helix was the earliest-described DNA-binding structural motif, earlier than those we discussed in Chapter 27 (page 1147). We shall return to this motif and its regulatory significance when we discuss the repressors of phage λ.

Analysis of the DNA–protein complex shows also that CRP induces DNA to bend quite sharply when it binds. This bending may facilitate the initiation of transcription by bringing parts of RNA polymerase that are bound upstream into direct contact with the promoter or transcriptional start site. Evidence indicates that an important interaction between the polymerase α subunit and DNA occurs as a result of DNA bending.

FIGURE 29.6

Bending of DNA by binding to CRP–cAMP. This model was deduced from the crystal structure of the DNA–protein–cAMP complex. The DNA bases are in light blue, and the sugar–phosphate backbone is in yellow. The DNA-binding domains of the protein—two α helices that contact DNA bases in the major groove (see text)—are in purple and are perpendicular to the plane of the page. The cAMP-binding domain is in blue, and two bound cAMP molecules are in red. On the DNA molecules, those phosphates that are in closest contact with the protein and whose ethylation interferes with protein binding are in red. Those phosphates whose reactivity to ethylation is enhanced (on the outer edge of the bend) are in blue. PDB ID 1CGP.

From *Quarterly Reviews of Biophysics* 23:205–280, T. A. Steitz, Structural studies of protein–nucleic acid interaction: The sources of sequence-specific binding. © 1990 Cambridge University Press.

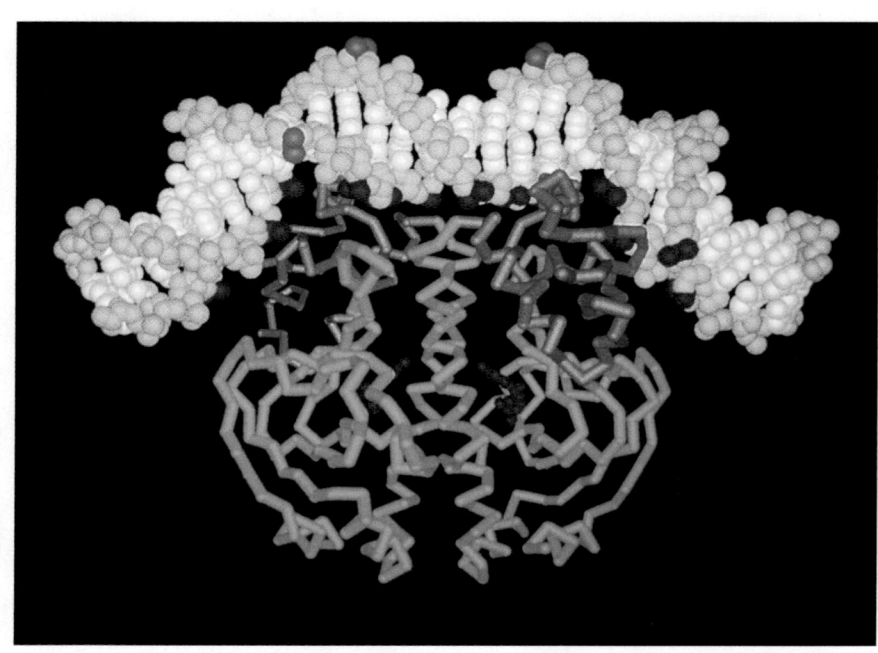

Bacteriophage λ: Multiple Operators, Dual Repressors, and Models for DNA Binding Specificity

We now return to phage λ, a larger and more complex genetic system than the lactose operon, but one that is regulated by similar factors—binding of proteins to specific DNA regulatory sites, just upstream from the genes that they control, with binding leading to either activation or inhibition of transcriptional initiation. However, because of the variety of relationships between the virus and its host bacterium, the specific controls used are more complex and more subtle than those we have been discussing. Recall from Chapter 26 that infection can have one of two possible outcomes—a lytic cycle of growth, comparable to that of phage T4, or lysogenization, in which the viral chromosome circularizes and undergoes site-specific integration into the host-cell chromosome (see Figure 26.19, page 1098), with consequent repression of nearly all viral genes. Once lysogeny has been established, the phage chromosome can be maintained as a transcriptionally repressed prophage for many generations. This state of repression can be broken, leading to excision of the viral chromosome as a circular DNA, followed by replication of viral DNA, followed by activation of genes needed to assemble virus particles. The virus must rely on four distinct patterns of gene expression needed for its four physiological states—(1) infection leading to lytic growth, (2) infection leading to establishment of lysogeny, (3) long-term maintenance of lysogeny, and (4) breaking of lysogeny with subsequent lytic growth.

The critical events in λ transcriptional regulation involve two different repressor proteins, called cI and Cro, each of which binds at two different operators. Each operator contains not one repressor-binding site (as in the lactose operon) but three, and each contains promoter sites interspersed with the repressor-binding sites. Transcription from the two promoter–operator sites takes place in opposite directions along the genome (see Figure 29.7). The repressors bind to each of the six operator sites with varying affinities, leading to varying occupancy of each binding site by each repressor under different physiological conditions. To add to the complexity, the cI repressor also serves, under certain conditions, as a transcriptional *activator*, promoting the expression of some genes while repressing that of others. To understand the biochemistry involved, we must first consider the genes and their positions on the λ genome (Figure 29.7), as well as the phenotypes of mutants altered in these regulatory genes.

Genes and Mutations in the λ System

Phage mutants defective in establishing or maintaining lysogeny have phenotypes comparable to those defective in *lac* regulation, and these similarities helped Jacob and Monod make the generalizations embodied in the operon model. The lysogenic

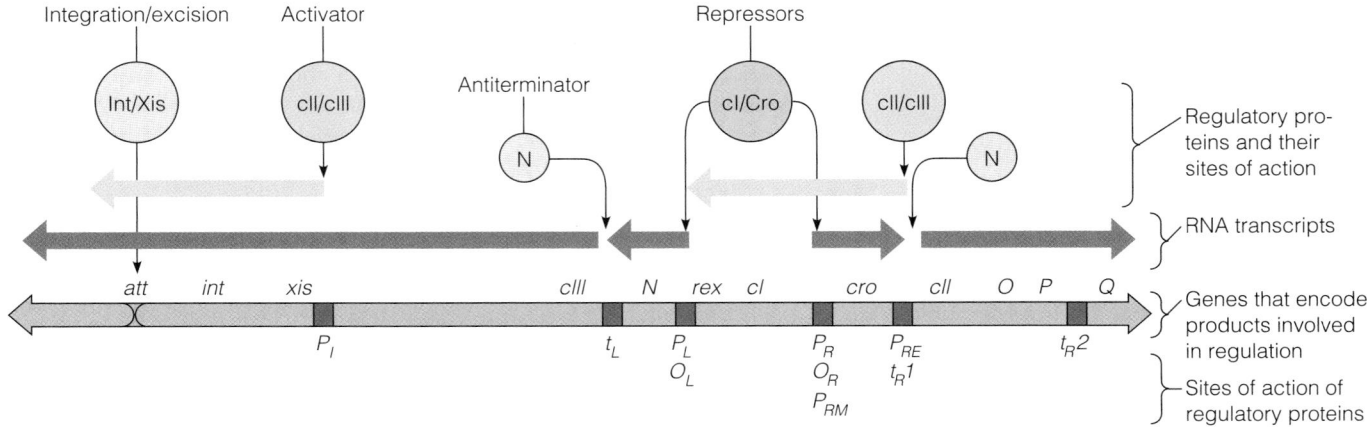

TABLE 29.1   Phenotypes of comparable *lac* and λ mutations

lac Phenotype	Corresponding λ Phenotype	Regulatory Abnormality
Inducer-constitutive, recessive	Clear-plaque; cannot establish lysogeny	Repressor defective in operator binding
Operator-constitutive, *cis*-dominant	Virulent; can replicate in a superinfected immune lysogen	Operator unable to bind repressor
Noninducible, *trans*-dominant	Noninducible (cannot be induced by UV or other treatments)	Repressor cannot bind inducer or be inactivated

response involves processes of viral gene inactivation, accomplished by the binding of a repressor to an operator, comparable to the one that keeps the *lac* operon turned off when inducer is absent. The major λ phenotypes and the counterpart *lac* mutations are summarized in Table 29.1.

When a defective repressor cannot bind to operators, the mutant phages give clear plaques when they are plated. Normally a λ plaque is turbid because it contains not only phage but also lysogenized bacteria, which continue to grow because they are *immune* to infection by additional phage. By contrast, mutants that cannot establish lysogeny give clear plaques because all of the cells are lysed. Clear-plaque mutants in λ map in three different genes—*cI, cII,* and *cIII. cI* is the structural gene for one of the repressors mentioned above, and the other two genes control the synthesis of the cI protein.

Virulence mutations, which map in operators, also exhibit a clear-plaque phenotype, but there is an important distinction. Bacteria lysogenic for λ are immune to infection by a second λ phage because repressor in the cell binds to the operators of any input phage. This is true for *cI* mutants because they contain normal operators, which can bind repressors. However, virulent mutants can produce progeny phage following infection of an immune lysogen because their operators cannot bind repressor after entry into the cell. These observations are similar in principle to those resulting from the experiments with partial diploids in the *lac* operon.

The λ *cI* Repressor and Its Operators

In 1967 Mark Ptashne isolated the λ repressor encoded by *cI*. The cI repressor is a dimeric protein, with a subunit M_r of 27,000; it binds through its N-terminal sequences to operator sites with a K_a of about $3 \times 10^{13}\,M^{-1}$ (Figure 29.8). Repressor–DNA interactions have been used to map and characterize what turned out to be two operator regions, one on each side of *cI*. The two operators control divergent transcriptional events from a central regulatory region—leftward (O_L) and rightward (O_R); see Figure 29.7. As shown by footprinting, each operator contains three separate repressor-binding sites, each about 17 base pairs long. The three repressor-binding sites are homologous, but not identical (Figure 29.9), and they are separated by spacer regions of three to seven base pairs. Mutations that confer virulence map within the repressor-binding regions. A fully virulent mutant has at least two mutations—one in O_L and one in O_R.

The λ operators are remarkable in several respects other than their multiple repressor-binding sites. (1) Mutations that affect promoter activity lie between the repressor-binding sites. Thus, operators and promoters are *interspersed,* so the regulatory regions are more properly called $O_L P_L$ and $O_R P_R$. (2) As seen in Figure 29.9, $O_R P_R$ controls transcription from *two distinct promoters*—one rightward (P_R) and one leftward (P_{RM}). (3) Transcription from $O_L P_L$ and $O_R P_R$ is controlled by *two different repressors*—cI and Cro (the *cro* gene is an acronym,

Regulation of lysogeny in λ phage is similar to regulation of the lactose operon but is more complex.

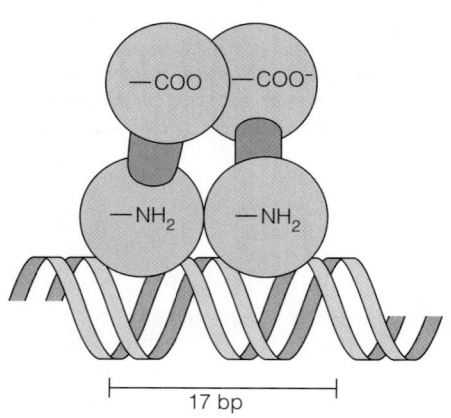

FIGURE 29.8

Model of λ cI repressor binding to DNA. The repressor is a dimer and binds to a 17-base-pair region in a λ operator.

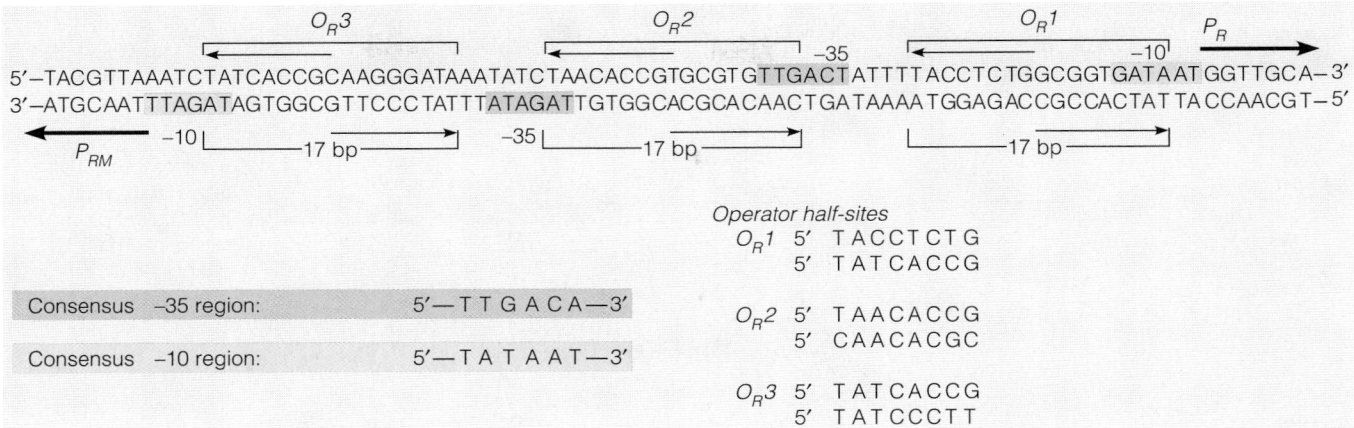

Operator half-sites

| O_R1 | 5′ | T A C C T C T G |
| | 5′ | T A T C A C C G |

| O_R2 | 5′ | T A A C A C C G |
| | 5′ | C A A C A C G C |

| O_R3 | 5′ | T A T C A C C G |
| | 5′ | T A T C C C T T |

Consensus −35 region: 5′—T T G A C A—3′

Consensus −10 region: 5′—T A T A A T—3′

FIGURE 29.9

The O_RP_R region. The upper diagram shows the nucleotide sequence of the O_RP_R region, including the three repressor-binding sites (O_R1, O_R2, and O_R3), the leftward promoter P_{RM}, and the −35 and −10 regions for the two promoters (shaded purple and blue, respectively). The thin horizontal arrows mark the operator half-sites. Below are the consensus sequences for the −35 and −10 regions and a chart that shows the partial homology among the operator half-sites (that is, the repressor-binding sites).

from cI repressor off). (4) Under certain conditions the cI repressor is a transcriptional *activator*, not an inhibitor. Another novel feature of this regulatory system is that *cI* transcription is initiated from different promoters under different physiological conditions. All of these complexities are related to the need for orderly and efficient phage gene control under quite different physiological conditions, as we shall see.

Early Genes in Phage λ

To understand the significance of *cI* gene regulation, we must identify several λ genes that are expressed early in infection (see Figure 29.7). *cI* and *cro* code for repressors, as just noted, and *cII* and *cIII* both stimulate cI synthesis. *rex* is a gene of still unknown function, the only gene aside from *cI* known to be expressed during lysogeny. We encountered *int, xis,* and *att* in Chapter 26, when we discussed site-specific recombination. *O* and *P* function in the initiation of λ DNA replication. The *N* gene product interacts with NusA (see page 1142) to prevent transcription termination. The *Q* product activates late gene transcription.

Interactions between the Two λ Repressors

The interspersed O_LP_L controls transcription of *N*, through interaction of its repressor-binding sites with the cI protein. However, most of the regulatory action occurs at O_RP_R, and it is here that the decision is made between lytic and lysogenic infection. This decision involves interactions of the two repressors, cI and Cro.

Quantitative footprinting experiments show that of the three repressor-binding sites in O_RP_R, cI binds most tightly to site O_R1, less tightly to O_R2, and still less tightly to O_R3. Moreover, cI binding is *cooperative*, so that when one repressor dimer is bound at O_R1, affinity for a second molecule is increased at O_R2. Cro protein is a dimer of identical 66-residue subunits. It binds considerably less tightly to any of the sites than does cI and in the reverse order of affinity. That is, site O_R3 is favored, followed by approximately equal binding at O_R2 and O_R1. Binding of Cro is noncooperative.

Although Cro is a repressor, it can also be considered an **antirepressor** because it antagonizes the action of cI in a very specific way. To understand how this works, we must first consider the transcriptional events occurring in the presence of varying levels of cI, illustrated in Figure 29.10. Because of the cooperative binding of cI to its operators, both sites O_R1 and O_R2 are usually occupied in the lysogenic state (Figure 29.10a), even though the intracellular concentration of cI is quite low (about 200 molecules per cell, or 10^{-7} M). This inhibits the rightward transcription of *cro* from its own promoter, but it *activates* the leftward transcription of *cI* from the promoter P_{RM} (the M stands for "maintenance," because this is the promoter from which *cI* is transcribed during maintenance of lysogeny).

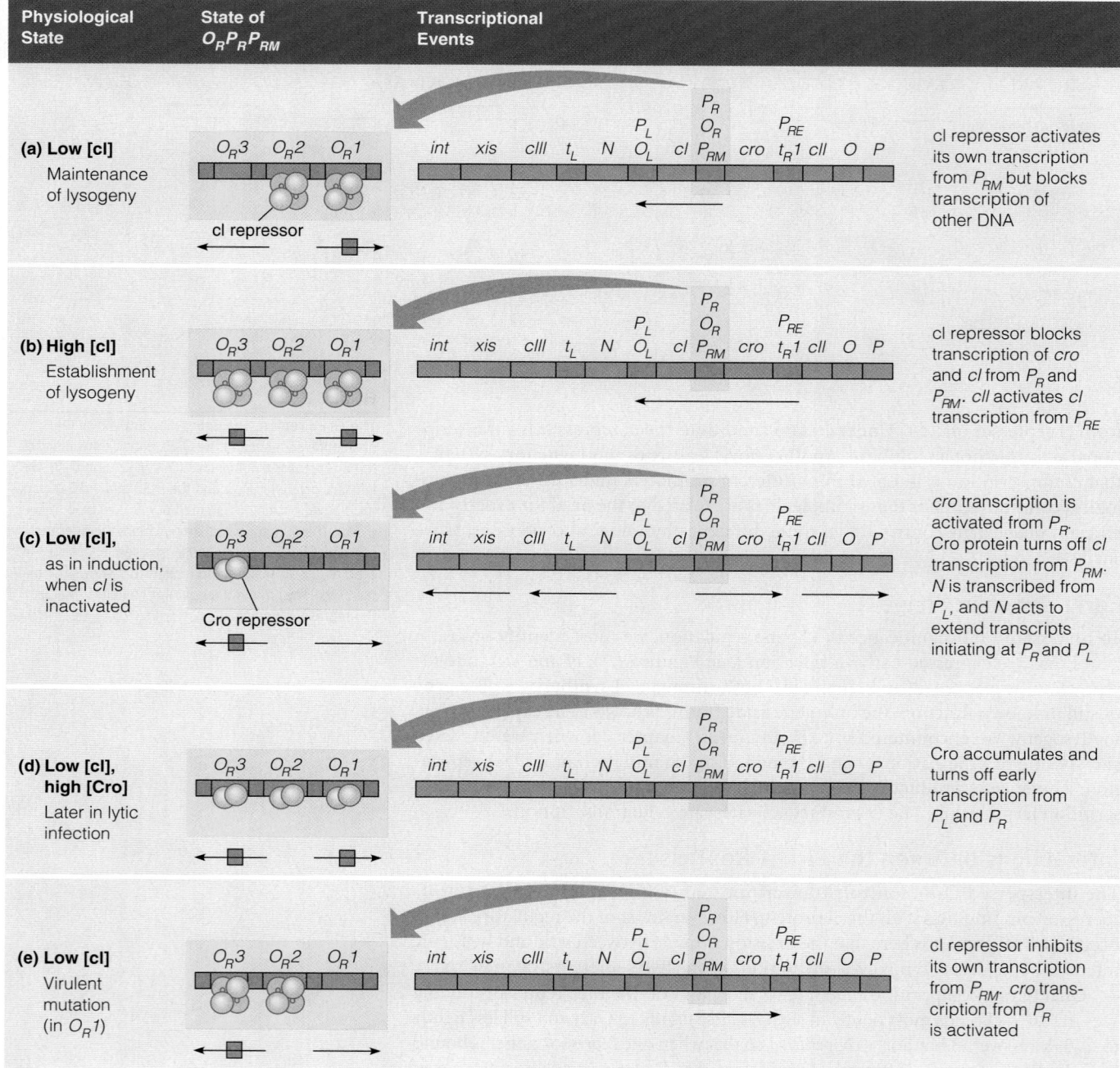

FIGURE 29.10

cl–Cro interactions in the $O_R P_{RM}$ region. Different physiological conditions lead to different interactions, with varying effects upon transcription of the rest of the λ genome. Gray boxes indicate blocked transcriptional events. Gene *rex*, which does not participate in regulation, is omitted for simplicity.

The −10 and −35 regions for the P_{RM} promoter lie within the operators. The evidence that cI really does activate its own transcription in the lysogenic state lies in the existence of a special class of *cI* mutants whose repressor can bind tightly at O_R1 and O_R2 but cannot stimulate *cI* transcription from P_{RM}.

During *establishment* of lysogeny (Figure 29.10b), when lytic and lysogenic genes are competing to determine the fate of the viral genome, there is a need for larger amounts of cI repressor than can be transcribed from P_{RM}. At this time a different *cI* promoter, called P_{RE}, is activated (E stands for "establishment"). In this activation the cII protein binds specifically at the −35 region of P_{RE} and stimulates RNA polymerase binding at that site. This transcriptional event yields a longer *cI* messenger RNA that is more efficiently translated than the message synthesized from P_{RM}. The result is sufficient cI repressor to bind all three sites in O_R and, hence, to block both transcriptional initiation events.

Now consider the events in prophage induction, when lysogeny is broken, leading to a lytic infection (Figure 29.10c). First, cI repressor is inactivated (we shall see shortly how this occurs), and the O_R sites become unoccupied. This permits transcription of *cro* from P_R, and the Cro protein blocks further transcription of *cI* from P_{RM}. At the same time, leftward transcription from P_L generates the N protein, blocking transcriptional termination at the sites indicated in Figure 29.10 as t_R1 and t_L. Thus, the two early transcripts for Cro and N are extended to activate new genes. Leftward transcription generates Int and Xis proteins, necessary for prophage excision. Rightward transcription generates O and P, necessary for DNA replication.

Subsequent regulatory events, including the action of the gene Q protein, activate transcription of late-acting genes, which encode structural proteins of the virus. At this time it is desirable to suppress early gene transcription, so that the late proteins can be made at maximal rates. This involves further action of the Cro protein, which by this time has accumulated to the point that it can bind to both O_R1 and O_L1, blocking transcription from P_R and P_L, respectively (Figure 29.10d). In infection by a virulent mutant (Figure 29.10e), *cI* transcription from P_{RM} is blocked, and this leads to activation of *cro* transcription from P_R.

Structure of Cro and cI Repressors and Related DNA-Binding Proteins

Studies of the three-dimensional structures of Cro and cI repressors have yielded remarkable insight into mechanisms by which proteins recognize specific DNA sequences. This in turn has greatly enhanced our understanding of how transcription is regulated through specific DNA–protein interactions.

In 1981 Brian Matthews determined the crystal structure of Cro protein. Cro is a homodimer of 66-residue subunits folded into three α-helical regions and three β strands (Figure 29.11). Model-building studies showed that two of the helices, numbered 2 and 3 in the figure, could fit within the major groove of the DNA double helix. These two helices are separated by a short β turn, forming a helix–turn–helix motif of the type we mentioned earlier for the CRP protein (see page 1238). In the Cro dimer, the two number 3 helices are 3.4 nm apart, the length of one turn of the DNA double helix. This distance suggested that the two subunits bind on the same side of the helix, in adjacent major groove sites, with the number 3 helices lying lengthwise in the grooves. This model was strongly supported by methylation- and ethylation-protection experiments, which identified the DNA functional groups in close contact with the protein. Note from Figure 29.9 that each operator site is an imperfect palindrome. Therefore, each of the two number 3 helices is in contact with a slightly different set of bases.

Lysogeny in λ phage is controlled by two repressors, cI and Cro, which bind with differing affinities to three operators in the O_RP_R region of interspersed operators and promoters.

The helix–turn–helix motif is widely used in prokaryotic transcriptional regulatory proteins. Specific contacts are made between DNA major groove bases and amino acids in a recognition helix.

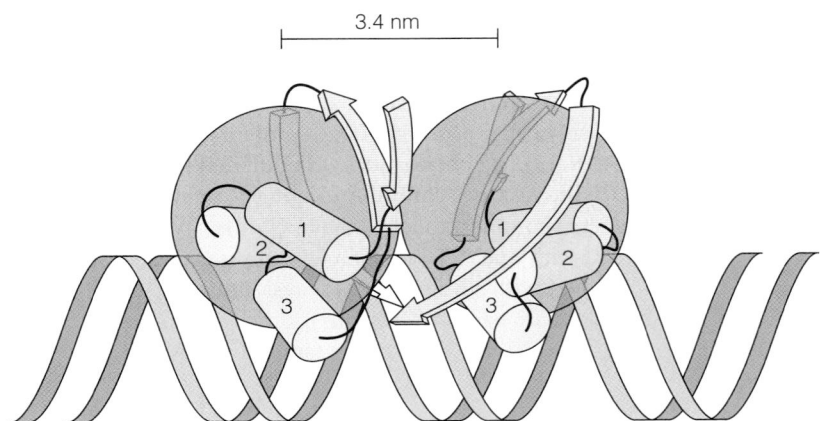

3.4 nm

FIGURE 29.11

Structural model of the Cro dimer–operator complex.

A Genetic Switch: Gene Control and Phage λ, M. Ptashne. © 1986 John Wiley & Sons Inc. Reproduced with permission of Blackwell Publishing Limited.

FIGURE 29.12

DNA binding faces of λ Cro, λ cI repressor, and CRP, showing the helix–turn–helix motif. The motif involves helices 2 and 3 in Cro and cI and helices E and F in CRP. The black ellipses mark centers of symmetry.

Courtesy of T. A. Steitz and I. T. Weber.

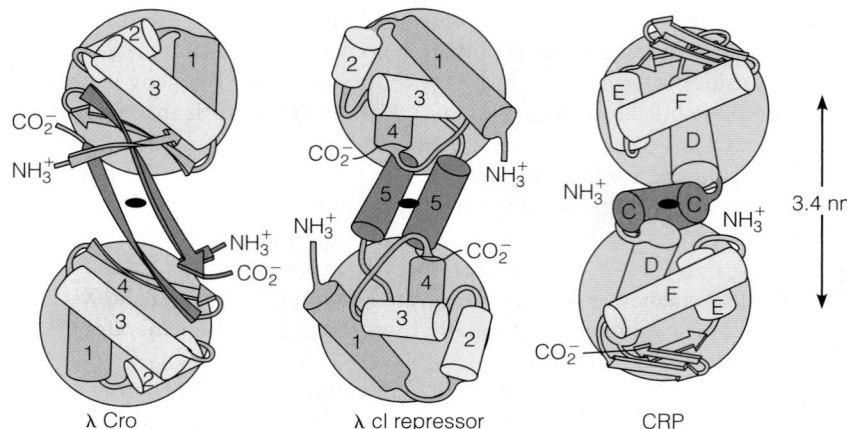

λ Cro   λ cI repressor   CRP

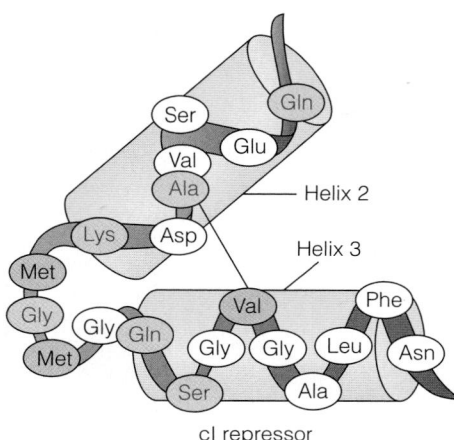

cI repressor

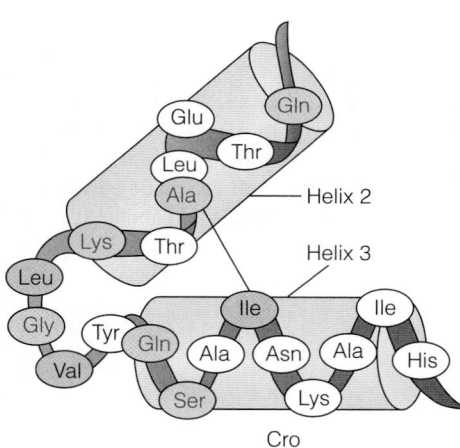

Cro

FIGURE 29.13

Conserved residues in the DNA-binding helices of λ cI repressor and Cro. Conservative substitutions are shown in purple and identities in pink. In both proteins the alanine in helix 2 contacts a residue in helix 3, which helps to position the helices with respect to each other.

The amino acid sequence within and between helices 2 and 3 shows remarkable homology with corresponding sequences in a large family of sequence-specific DNA-binding proteins but *not* with DNA-binding proteins that showed no sequence preference. This observation suggested that the helix–turn–helix motif is a commonly evolved structural element in transcriptional regulatory proteins, at least in prokaryotes. Note that the *lac* repressor also contacts DNA in a helix–turn–helix motif (Figure 29.4). As indicated earlier, this was the first structural motif elaborated for a sequence-specific DNA-binding protein. In Chapter 27 we identified several other such motifs, more recently discovered, such as the zinc finger, leucine zipper, and the helix–loop–helix motif.

Helices E and F of the cAMP receptor protein adopt the helix–turn–helix motif, as mentioned on page 1147. Once the three-dimensional structure of λ cI repressor was shown also to have the helix–turn–helix motif, it seemed likely that the helix–turn–helix structure is involved in DNA binding in all of these proteins. The relevant structural similarities for Cro, CRP, and cI repressor are shown in Figure 29.12. Amino acid sequence homologies among these regions are shown for Cro and cI in Figure 29.13. Note that the sequences, though similar, are not identical. If they were identical, we would not be able to explain how Cro and cI repressors differ in their relative affinities for different operators. cI repressor contains an additional binding determinant—a pair of "arms," or short polypeptide segments, that extend from helix 1 and can be seen in Figure 29.14 extending around the helix and establishing contacts on the other side of the DNA duplex. These arms probably explain why cI binds more tightly to its operators than Cro does.

The α-3 helix is called the **recognition helix** because its position deep within the major groove allows it to contact specific DNA bases and hence to determine sequence specificity of binding. The α-2 helix is in contact primarily with DNA phosphates. These electrostatic contacts strengthen binding but do not contribute to specificity. Supporting the concept of α-3 as a recognition helix is that most *cI* mutations that reduce specific binding of repressor to operator DNA alter the amino acid sequence in this region of the protein.

Crystallographic analysis of the respective DNA–protein complexes explains how Cro and cI can bind to the same operator sites with different binding affinities. As shown in Figure 29.15, the residues common to both proteins are in contact with DNA sequence elements common to all of the operators. In both proteins a glutamine residue interacts with one A-T base pair. cI repressor establishes specificity through a contact in O_R1 with a unique alanine residue, whereas Cro can be in contact with three specific base pairs in O_R3 with unique asparagine and lysine residues. Also, because the two α-3 helices lie closer together in Cro (2.9 nm) than in cI (3.4 nm), the orientations of these helices with respect to the major grooves of operator DNA are quite different.

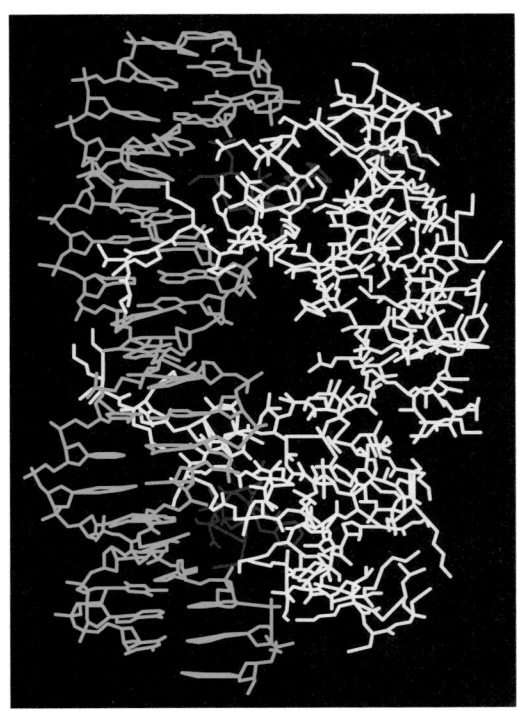

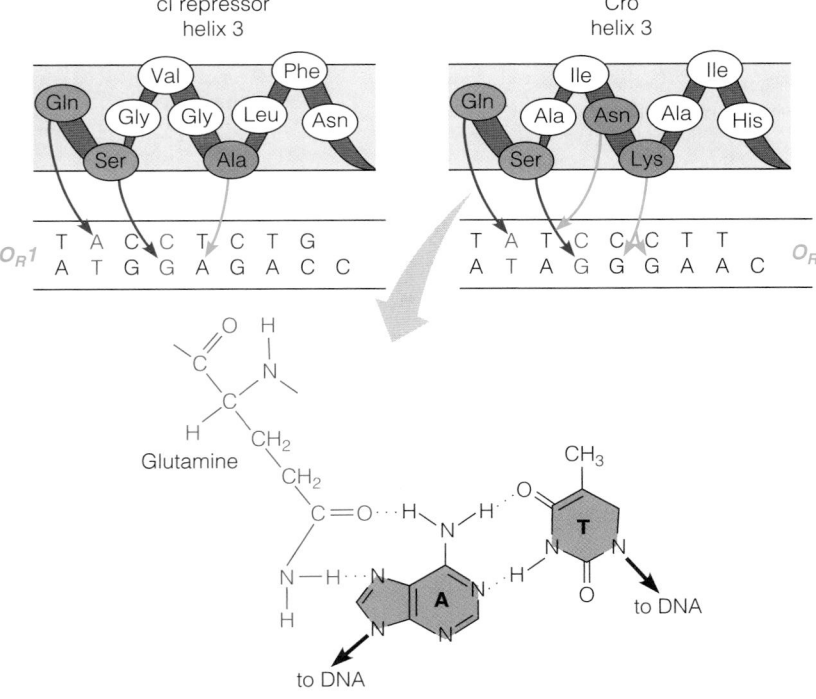

FIGURE 29.14

Structure of the λ cI repressor-DNA complex showing the recognition helices (in red) and the arms wrapped around the back side of the helix.

From *Science* 242:893–907, S. R. Jordan and C. O. Pabo, Structure of the lambda complex at 2.5 Å resolution: Details of the repressor-operator interactions. © 1988. Reprinted with permission from AAAS and Carl O. Pabo.

FIGURE 29.15

Specific amino acid–nucleotide contacts for cI and Cro repressors. The conserved residues (orange) bind to nucleotides common to all of the operators, and unique residues (purple) bind to non-conserved nucleotides in the operators. Also shown is the structure of a glutamine residue in contact with an A-T base pair. Note that the two repressors are shown binding to different operators.

A Genetic Switch: Gene Control and Phage λ, M. Ptashne. © 1986 John Wiley & Sons Inc. Reproduced with permission of Blackwell Publishing Limited.

The SOS Regulon: Activation of Multiple Operons by a Common Set of Environmental Signals

How is the λ cI repressor inactivated when the prophage is excised and begins a cycle of lytic growth? Various DNA-damaging treatments are known to induce λ prophages, including ultraviolet irradiation, inhibition of DNA replication, and chemical damage to DNA. Evidently the virus finds it advantageous to leave a damaged cell, like rats leaving a sinking ship. Because of the similarity in genetic control between the λ and *lac* systems, investigators sought a small molecule, perhaps a nucleotide, that would accumulate after these treatments and that might be the ligand that binds to cI and inactivates it. Surprisingly, the λ repressor was found to be inactivated by a quite different mechanism—proteolytic cleavage. Analysis of this cleavage reaction revealed the SOS system described in Chapter 26 as one of the elements in error-prone DNA repair, in which the genes are controlled by a single repressor–operator system. Such a set of unlinked genes, regulated by a common mechanism, is called a **regulon**. The heat-shock genes, all activated by a transient temperature rise, make up another regulon.

The control elements in the *E. coli* SOS regulon are the products of genes *lexA* and *recA*. We have encountered RecA protein before, in its role of stimulating DNA strand pairing during recombination. Strikingly, this small protein has an enzymatic activity in addition to the activities involved in recombination. When bound to single-stranded DNA, it can stimulate proteolytic cleavage of the proteins encoded by *cI*, *lexA*, and *umuD*. LexA is a repressor that binds to nearly three dozen operators scattered about the *E. coli* genome (Figure 29.16). Each operator controls the transcription of one or more proteins that help the cell respond after

The SOS regulon. The figure shows locations on the *E. coli* chromosome of some of the genes controlled by the LexA repressor. *dinA* is the structural gene for DNA polymerase II, while *dinB* encodes DNA polymerase IV and *dinF* is a damage-inducible gene of unknown function. *umuC,D* (so-called because mutants in this gene could not undergo *u*ltraviolet *mu*tagenesis) encodes the highly error-prone DNA polymerase V. LexA repressor (pink) is inactivated by proteolysis, which is somehow enhanced by a complex of RecA protein (blue) and single-strand DNA.

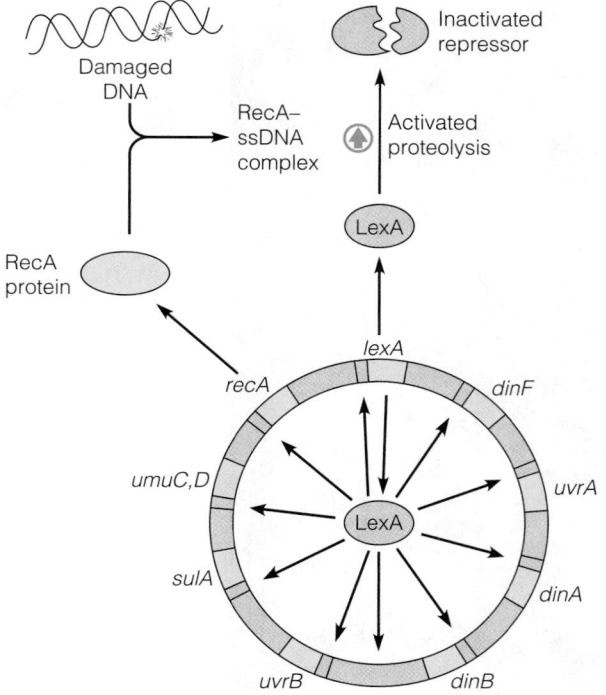

The SOS regulon is activated by DNA damage, which stimulates RecA to cause proteolytic cleavage of the LexA and λ cI repressors.

environmental damage that might harm the genetic apparatus. About 40 known proteins are induced as a result of LexA inactivation. These proteins include the gene products of *uvrA* and *uvrB,* involved in excision repair; *umuC,D,* the genes for the error-prone DNA polymerase V; *sulA,* involved in cell division control; *dinA,* the structural gene for DNA polymerase II; *recA* itself; *lexA* itself; *dinB,* the structural gene for error-prone DNA polymerase IV; and *dinF* (function unknown).

In a healthy cell, *lexA* and *recA* are expressed at low levels, with sufficient LexA protein to turn off the synthesis of the other SOS genes completely. LexA protein does not completely abolish either *lexA* transcription or that of *recA.* The trigger that activates the SOS system after damage is single-stranded DNA. As we have seen, UV irradiation generates gapped DNA structures, and so do other conditions that induce the SOS system. RecA binding within a gap activates LexA proteolysis by a mechanism not yet clear. Intracellular levels of LexA decrease, removing the LexA barrier to *recA* transcription. RecA protein accumulates in large amounts. Simultaneously, cleavage of the LexA protein activates transcription of all genes under *lexA* control. In a λ lysogen, cleavage of λ cI repressor is stimulated as well, activating prophage excision and replication, as discussed earlier.

Sequencing of operators that respond to LexA has yielded a consensus sequence, with 7 highly conserved bases in a 20-base-pair region. However, different LexA-sensitive genes have this sequence located quite differently with respect to the transcriptional start site. The location of each LexA operator, and its similarity to the consensus sequence, control the time after DNA damage at which a particular gene is expressed. Thus, the SOS response appears to be a coordinated series of events, controlled in part by the nature and severity of the initial DNA-damaging event.

Biosynthetic Operons: Ligand-Activated Repressors and Attenuation

The lactose operon is involved with catabolism of a substrate. Therefore, the gene products are not needed unless the substrate is also present to be consumed. A different situation is encountered with genes whose products catalyze biosynthesis—of

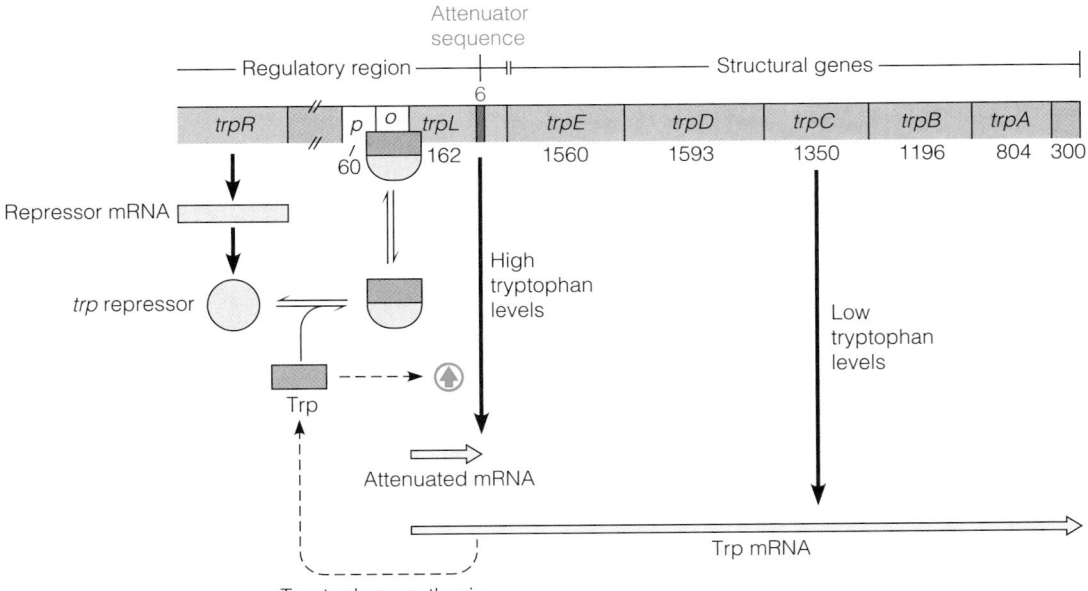

FIGURE 29.17

The *trp* operon. The figure shows regulation by *trp* repressor and by attenuation. *trpa*, the attenuator site, is shown in red.

an amino acid, for example. Because biosynthesis consumes energy, it is to the cell's advantage to use the preformed amino acid, if it is available. Therefore, the regulatory goal is to repress gene activity, by turning *off* the synthesis of enzymes in the pathway when the end product is available. Regulation of the *E. coli trp* operon, which controls the five reactions transforming chorismic acid into tryptophan (see Figure 21.38, page 910), demonstrates two ways of accomplishing this shutdown: A repressor design in which binding of a small-molecule ligand *activates* the repressor, rather than inactivating it, and premature termination of transcription.

The *trp* operon consists of five adjacent structural genes whose transcription is controlled from a common promoter–operator regulatory region (Figure 29.17). The *trp* repressor, a 58-kilodalton protein encoded by the nonadjacent *trpR* gene, binds a low-molecular-weight ligand, namely tryptophan. However, in this case the protein–ligand complex is the *active* form of the protein, which binds to the operator and blocks transcription. When intracellular tryptophan levels decrease, the ligand–protein complex dissociates and the free protein ("aporepressor") leaves the operator so that transcription is activated. If we call lactose an inducer in a catabolic system, it is appropriate to call tryptophan a **corepressor** in this anabolic system.

The crystal structure of the *trp* repressor–DNA complex shows a helix–turn–helix motif, comparable to that seen with the λ cI, Cro, and *lac* repressors; binding tryptophan to this protein reorients the helices to activate binding to DNA. Remarkably, this model shows no direct contacts between residues in the recognition helices and specific DNA bases. It has been proposed that bound water molecules make sequence-specific contacts between amino acids in the recognition helix and nucleotides in the operator.

The *trp* operon has an additional regulatory feature, now known to be involved in controlling numerous biosynthetic operons. Charles Yanofsky found that activities of the *trp* enzymes varied over a 600-fold range under different physiological conditions, more than could be accounted for by a repressor–operator mechanism alone. Analysis revealed a second mechanism, called *attenuation*, that involves early termination of *trp* operon transcription under conditions of tryptophan abundance. Note from Figure 29.17 a 162-nucleotide sequence called *trpL*, the *trp* leader region. A site called *a*, the attenuator, is 133 nucleotides from the 5′ end of the *trpL* sequence. When tryptophan levels are high, transcription

The *trp* repressor inhibits tryptophan synthesis by binding as the repressor–tryptophan complex to the *trp* operator, blocking transcription.

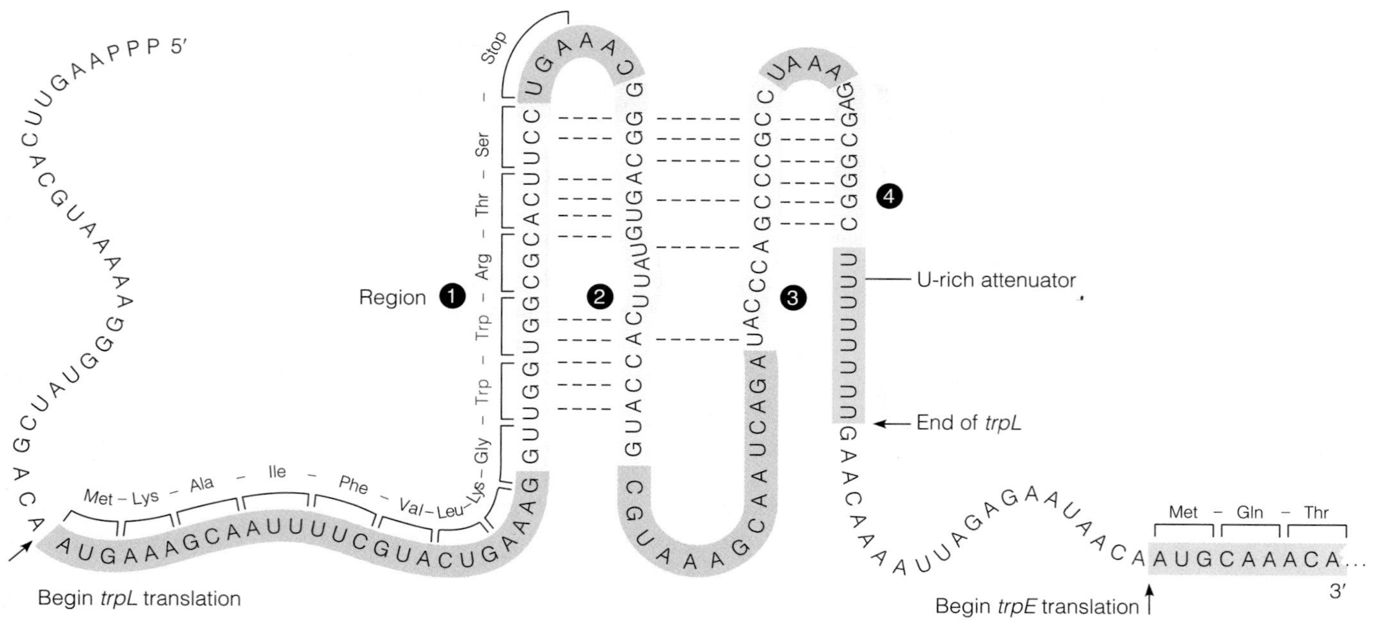

FIGURE 29.18

RNA base sequence of the *trp* leader region. The four internally complementary sequences that participate in attenuation (yellow) are shown, as well as the two *trp* codons (magenta) in region 1 that act as a pause site for RNA polymerase. The translational stop codon after region 1 (see Chapter 28) may serve to prevent needless translation of those few full-length messages that are produced despite attenuation.

Attenuation is a regulatory mechanism in which ribosome positioning on an mRNA determines whether transcription of an operon will terminate before transcription of the structural genes begins.

terminates at *a*, to give a truncated 133-nucleotide transcript rather than the complete 7000-nucleotide *trp* mRNA. The structural genes are not transcribed, so tryptophan is not synthesized.

Critical to understanding the mechanism of attenuation is the presence of four oligonucleotide sequences in the *trp* leader region that are capable of base-pairing to form stem–loop structures in the RNA transcript (Figure 29.18). In the most stable conformation (Figure 29.19a), region 1 pairs with 2, and region 3 pairs with 4, to give two stem–loops. The 3–4 structure, being followed by eight Us, is an efficient transcription terminator because it resembles the factor-independent terminator structure shown in Figure 27.16 (page 1141).

When tryptophan levels are low (Figure 29.19b), formation of the 3–4 stem–loop is inhibited, and termination does not occur at the attenuation site. Note that region 1 contains two tryptophan codons (see Figure 29.18). In prokaryotes, translation is coupled to transcription, so a ribosome can begin translating a message from its 5′ end while the message is still being synthesized at its 3′ end. In this case, the ribosome stalls when it reaches the two tryptophan codons because there is insufficient tryptophanyl-tRNA to translate them. The presence of the bulky ribosome prevents region 1 from base-pairing with 2, leaving region 2 free to base-pair with 3. Once region 3 is unavailable to base-pair with 4, the 3–4 stem–loop transcriptional terminator cannot form, and the entire message is synthesized. Conversely, when tryptophan is abundant (Figure 29.19c), the ribosome does not stall, thereby occluding region 2 and allowing the 3–4 stem–loop structure to form, which leads to transcription termination on the 3 side of 3–4.

Neither the *trpR* system nor the attenuator is simply an on–off system. Both respond in graded fashion to the intracellular tryptophan level. Even though both systems are controlled by the same signal, the action of two distinct control systems greatly extends the possible range of transcription rates of the *trp* operon, giving maximum efficiency to regulation of these genes. At low tryptophan concentration the repressor–operator interaction is the principal regulatory mechanism, whereas the effects of attenuation are more significant at moderate to high tryptophan levels.

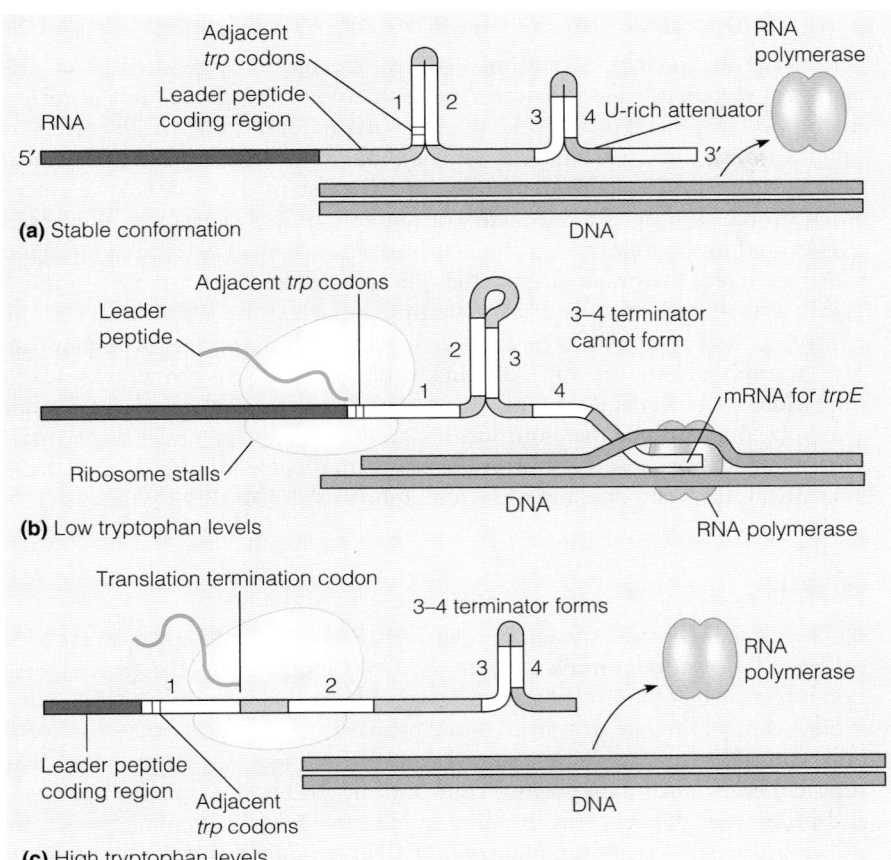

FIGURE 29.19

Mechanism of attenuation in the *trp* operon. **(a)** Most stable conformation for leader mRNA. **(b)** Conformation for leader mRNA at low tryptophan levels. **(c)** Conformation for leader mRNA at high tryptophan levels.

Although the foregoing model was originally proposed simply by inspection of the *trp* leader sequence, it is now supported by several lines of evidence. Significant confirmation comes from the existence in other attenuation-controlled operons of "stalling sequences"—sequences at which movement of a ribosome is inhibited at a low concentration of the product of the operon. These include bacterial operons for synthesis of leucine, with four adjacent leucine codons in the leader sequence, and of histidine, with seven.

Transcriptional control via a quite different termination–antitermination mechanism in the leader region has been described in *Bacillus subtilis*. Synthesis of aminoacyl-tRNA synthetases is controlled by the level of amino-acylation of the cognate tRNA. For example, in cells starved for tyrosine, synthesis of tyrosyl-tRNA synthetase is activated by an antitermination mechanism that allows transcription to proceed past a potential termination site in a leader region. By contrast, termination proceeds efficiently when most of the appropriate tRNA is charged with tyrosine, and there is no need to synthesize more of the aminoacyl-tRNA synthetase. The mechanism involves a complex secondary structure of the upstream leader sequence for *tyrS*, the structural gene for tyrosyl-tRNA synthetase. As shown in Figure 29.20, uncharged tRNA^{Tyr} can stabilize an antiterminator structure in leader RNA, by means of base-pairing interactions between the anticodon and mRNA, as well as the 3' end of the uncharged tRNA. These latter interactions are hindered when the 3' end of tRNA is aminoacylated. Thus, readthrough occurs, and the gene is transcribed. By contrast, the aminoacylated tRNA cannot form these base-pairing interactions. Thus, the leader forms a terminator, which prevents transcription of the gene.

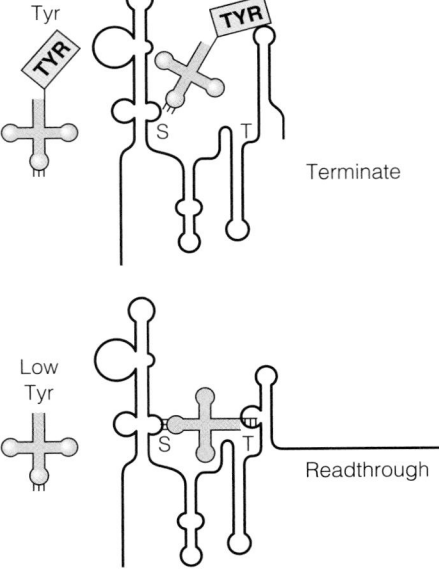

FIGURE 29.20

Model for induction of *B. subtilis* *tyrS* gene by uncharged tRNA^{Tyr}. Readthrough leading to gene expression is favored when tRNA^{Tyr} is in the uncharged state, as described in the text.

Molecular Microbiology 13:381–387, T. M. Henkin, Micro review tRNA-dircted transcription antitermination. © 1994 John Wiley & Sons, Inc. Reproduced with permission from John Wiley & Sons, Inc.

Control of rRNA Synthesis: The Stringent Response

The last prokaryotic transcriptional control mechanism presented here—the **stringent response**—was discovered more than four decades ago but is not yet understood in detail. The synthesis of ribosomal and transfer RNAs in bacteria is inhibited when protein synthesis is decreased by amino acid starvation. This inhibition evidently prevents the unnecessary production of translational machinery when protein synthesis is blocked. During amino acid starvation uncharged tRNAs bind to the ribosome A site. This leads to ribosomal binding of a protein called **stringent factor**, an enzyme that phosphorylates GTP at the 3' position, giving a nucleotide that we abbreviate pppGpp. Next the γ phosphate of the 5'-triphosphate moiety is cleaved off, giving a regulatory nucleotide, **guanosine 3',5'-tetraphosphate** (ppGpp). Binding of ppGpp to RNA polymerase leads to inhibition of ribosomal RNA synthesis. A class of mutants, called *relaxed*, cannot synthesize pppGpp, and therefore does not accumulate ppGpp under these conditions and does not show the stringent response. In other words, for these mutants amino acid starvation does not cause inhibition of rRNA or tRNA synthesis.

Guanosine 3',5'-tetraphosphate

Guanosine 3',5'-tetraphosphate senses amino acid starvation and responds by decreasing ribosomal RNA synthesis.

Applicability of the Operon Model: Variations on a Theme

Biochemical analyses of the *lac*, λ phage, *trp*, and SOS regulatory systems have confirmed the central features proposed by Jacob and Monod—that gene expression is regulated at the level of transcription and that specific protein–DNA interactions control the rate of transcription, primarily by regulating transcriptional initiation. These analyses have also revealed several important variations on that simple theme—including positive control of initiation, interspersed operators and promoters, dual proteins binding to the same site, multiple operons controlled by the same repressor, induction of DNA bending by regulatory proteins, and early termination as a regulatory mechanism. References at the end of this chapter will direct you to information about other well-studied operons, particularly the galactose and arabinose operons.

However, although the operon concept is well established, more recent analysis by genomic techniques, including microarray analysis and chromatin immunoprecipitation, has shown transcriptional regulation in bacteria to be more complex than earlier postulated. Many operons are controlled by more than one transcription factor—σ factors, repressors, and transcriptional activators. For example, the transcription of *nrdA* and *nrdB*, the structural genes for ribonucleotide reductase subunits, is controlled by at least five different proteins, including CRP. Eight proteins control transcription of *sodA*, the structural gene for superoxide dismutase. In addition, more than two dozen multitarget transcription factors are known, including the CRP and LexA discussed in this chapter. Findings such as these suggest hierarchical networks of transcriptional regulation.

In addition, although the operon model provides the basis for our understanding of prokaryotic transcription, developments in recent years have highlighted important regulatory processes occurring at other levels of gene expression—particularly translation. We deal with some of these processes later in this chapter.

Transcriptional Regulation in Eukaryotes

From Chapter 27 you are already aware that transcription and its regulation are far more complex in eukaryotic cells than in bacteria and viruses. Much of the added complexity stems from the fact that the transcription template in eukaryotes is chromatin, not naked DNA. Also, the complexity associated with multicellularity and highly differentiated states demands higher orders of regulation. Among the eukaryotes complexity of regulation increases with complexity of the organism. The genome of the nematode worm, *Caenorhabditis elegans*, contains about 20,000 genes, nearly as many as humans, while the fruitfly *Drosophila* has about 14,000 genes. As we mentioned earlier, alternative splicing helps to account for the increased information content of the human genome, despite surprisingly small differences in numbers of genes. Differences in transcriptional regulation contribute as well. Both worms and flies are estimated to encode about 1000 transcription factors, while the human genome encodes about 3000. But the difference in complexity is far more than three-fold because regulation by transcription factors is *combinatorial*. Because many different factors are used in assembling a transcription initiation complex, and because each transcription factor can participate in controlling multiple genes, a nearly infinite number of transcription factor combinations is possible. By contrast, bacteria have about a half dozen RNA polymerase σ subunits, which can be considered as transcription factors because they bind to RNA polymerase and direct it to certain groups of promoters. However, a single bacterial RNA polymerase molecule contains just one σ factor. So, even though bacterial genes are regulated by multiple transcription factors, as mentioned in the previous section, the multiplicity of factors controlling transcription of a single gene is far greater in eukaryotic cells.

The importance of transcription factors in generating biological diversity is illustrated by recent studies showing interspecies variations in distribution of transcription factor binding sites. In one such study, five vertebrate species were compared by ChIP-seq with respect to distribution in one gene, *PCK1* (which encodes an isoform of protein kinase C), of binding sites for one transcription factor, CEBPA (CCAAT/enhancer-binding protein α). As shown in Figure 29.21, the binding-site distribution varied dramatically among these five species. Less dramatic, but comparable, variation has been seen in transcription factor binding among different humans, showing this as a feature both in speciation and in interindividual variation within a species.

Additional regulatory complexity is brought about by the more recently discovered existence of small regulatory RNA molecules, called microRNA, or miRNA (page 1267). As we shall see, the number of microRNA molecules encoded by a genome is closely related to the complexity of the organism.

A large number of transcription factors helps to explain the complexity of higher eukaryotes, even though the number of genes in a genome may be smaller than originally expected.

Chromatin and Transcription

Recall from Chapter 24 that chromatin consists of individual units—nucleosomes—with each nucleosome containing 200 or more base pairs of DNA, 147 of which are wrapped about a histone core. Within these 147 base pairs, the central 80 or so are organized by a heterotetramer of histones H3 and H4. About 40 base pairs on each side are more loosely associated with H2A/H2B dimers. Spacer DNA, which is not associated with the core histones, associates with histone H1 and other proteins. The core histones (H2A, H2B, H3, and H4) have a common helical core structure, called the *histone fold*, joined to a relatively unstructured N-terminal tail. Specific amino acid residues within the tails are subject to modification—acetylation, methylation

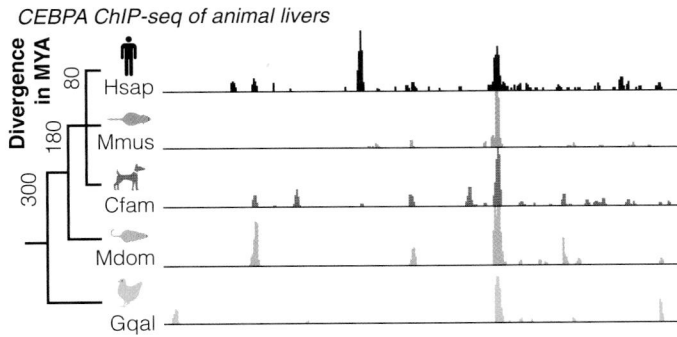

CEBPA ChIP-seq of animal livers

FIGURE 29.21

Binding site variation for transcription factor CEBPA along gene *PCK1* as determined by ChIP-seq. There is one highly conserved binding site, but essentially all of the other sites are variable. To the left is an evolutionary tree showing divergence in millions of years ago (MYA) for the five species—Hsap (human), Mmus (mouse), Cfam (dog), Mdom (opossum), and Cqal (chicken).

From *Science* 328:1036–1040, D. Schmidt, M. D. Wilson, B. Ballester, P. C. Schwalie, G. D. Brown, A. Marshall, C. Kutter, S. Watt, C. P. Martinez-Jimenez, S. Mackay, I. Talianidis, P. Flicek, and D. T. Odom, Five-vertebrate ChIP-seq reveals the evolutionary dynamics of transcription factor binding. © 2010. Reprinted with permission from AAAS.

(mono-, di-, and tri-), ubiquitylation, sumoylation, ADP-ribosylation, and phosphorylation. Acetylation usually involves lysine amino groups. By neutralizing the charge, acetylation loosens the ionic bonds that link histones to DNA, and generally activates the target nucleosome as a transcription template, whereas deacetylation, conversely, tends to inhibit transcription.

Chromatin exists within cells either in a highly compacted form, called **heterochromatin**, or a more loosely structured form, called **euchromatin**. Heterochromatin is transcriptionally inactive, and genes in particular tissues, which are permanently silenced in those tissues as a result of differentiation, are often found in heterochromatin. The repressed chromatin structure is often associated with methylation of histone H3 at lysine 9 (H3K9). Methylation at this site leads to binding of HP1 (Heterochromatin Protein 1) at that methyl mark. Repression is also induced by other histone modifications, such as H3K27 or ubiquitylation of H3K119. Whether these modifications are causes or effects of the repressive structure is not yet clear. By contrast, the H3K4 mark is a transcriptional activating signal. Many biochemists believe in a "histone code" that would relate specific chromatin modifications to specific regulatory effects in a predictable and rational way, as seen with the genetic code. So far, however, the histone code has proved to be elusive, and recent evidence indicates that the effect of a particular histone modification upon gene expression is dependent upon the context, that is, the modification status of adjacent or nearby histones. This has led some investigators to suggest that the relationship between histone modification and gene expression is more akin to a "language" than a more restrictive "code."

In order for a gene to be transcribed, the promoter must be relatively clear of nucleosomes; stated otherwise, it should lie in a nucleosome-free region (NFR). The nuclease-hypersensitive sites that we discussed in Chapter 27 probably represent NFRs. To some extent, clearing a promoter of histones is a function of chromatin remodeling complexes (Chapter 27; also see page 1254). In addition, chromatin remodeling and the binding of certain transcription factors are evidently responsible for the replacement of the canonical histones H2 and H3 by variants called H3.3 and H2A.Z. It is not clear how these replacements lead to transcriptional activation, but circumstantial evidence for an activating role is strong. Use of ChIP-seq (Tools of Biochemistry 27D) has shown that H2A.Z is specifically associated with nucleosomes near the transcription start site (Figure 29.22).

Transcriptional Control Sites and Genes

As in prokaryotic gene expression, transcription in eukaryotes is regulated by the interplay of *trans*-acting proteins that interact with *cis*-acting sites on the DNA template. *Cis*-acting sites within the promoter include NFR (nucleosome-free region), TSS (transcription start site), Inr (initiation region), TATA box, and BRE (TFIIB recognition element). This list is not exhaustive. Sites farther upstream are generally called enhancers in metazoans, and UAS (upstream activator sites) or

---

Several factors combine, including chromatin remodeling and histone modifications, to render eukaryotic transcription start sites relatively free of nucleosomes.

FIGURE 29.22

Distribution of H2A.Z-containing nucleosomes near a transcription start site (TSS). ChIP-seq was used to demonstrate that in humans, *Drosophila*, and yeast, the modified histone H2A.Z is most abundant in nucleosomes at positions +1 and −1 relative to the transcription start site (TSS).

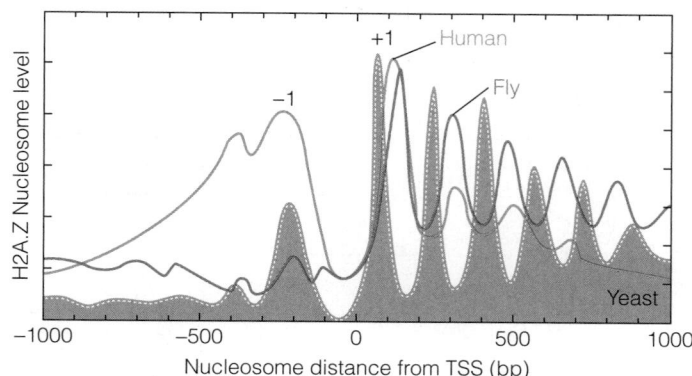

URS (upstream repressor sites) in yeast. Although the TATA box, as we indicated in Chapter 27, is the most widely distributed core promoter element, only about 20% of yeast genes actually contain a TATA box. Because of the large number of proteins bound in each initiation complex, there is tremendous variability among different genes in promoter and enhancer sequences.

Trans-acting factors include not only general transcription factors but also chromatin remodeling factors, chromatin modifying factors (such as histone acetyltransferases, deacetylases, methylases, or demethylases), and a multiprotein complex called *Mediator*, which we briefly mentioned in Chapter 27. Mediator is a large multisubunit complex found in all eukaryotes. Mediator links upstream activating (or repressing) elements with RNA polymerase II—specifically, with the C-terminal domain of the largest subunit, which undergoes phosphorylation as a prerequisite to activation. Mediator in yeast contains 21 subunits plus a 4-subunit subassembly that, when linked to Mediator, represses transcription, as indicated in Figure 29.23b. The large size of Mediator can be seen in Figure 29.24, which shows image-processed electron micrographs of the complex in association with RNA polymerase II.

> Mediator is a multisubunit complex that links upstream regulatory sequences, such as enhancers, with RNA polymerase II and general transcription factors at the promoter site.

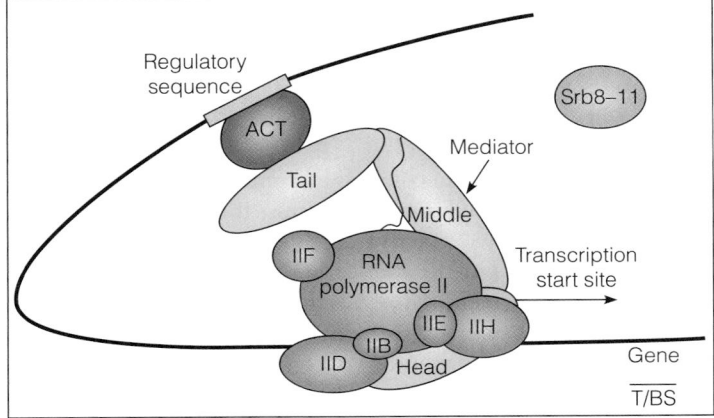

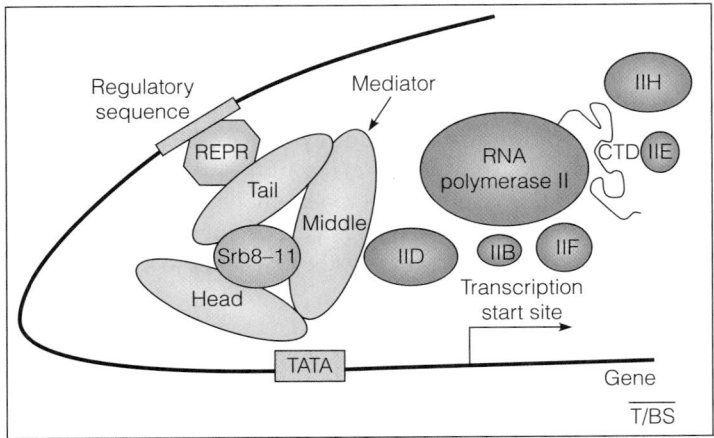

FIGURE 29.23

Mediator as a bridge between gene-specific regulatory factors and the general transcription machinery at the pol II promoter. (a) Activation. Mediator has three distinct structural domains—head, tail, and middle. The tail domain interacts with transcriptional activators (ACT, in red) and links Mediator with RNA polymerase. **(b)** Repression. The four-subunit complex containing Srb8, Srb9, Srb10, and Srb11, when bound to Mediator, prevents its interaction with RNA polymerase II and the basal transcription machinery.

Reprinted from *Trends in Biochemical Sciences* 30:240–244, S. Björklund and C. M. Gustafson, The yeast mediator complex and its regulation. © 2005, with permission from Elsevier.

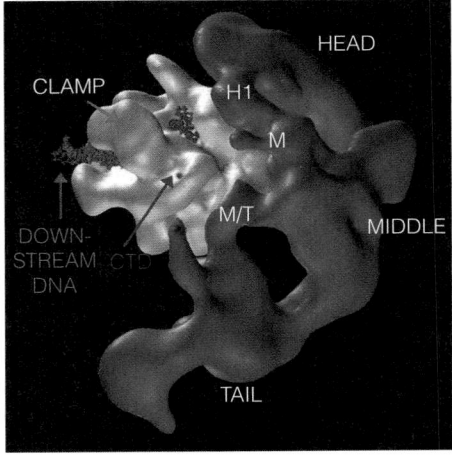

FIGURE 29.24

The structure of yeast Mediator in complex with RNA polymerase II. The structure of Mediator (dark blue) was determined by image processing of cryoelectron micrographs, and that of RNA polymerase II (light blue) from its crystal structure. The red dot marks the location where the C-terminal domain of the largest pol II subunit (the one that undergoes phosphorylation) emerges from the surface of the enzyme.

Reprinted from *Molecular Cell* 10:409–415, J. A. Davis, Y. Takagi, R. D. Kornberg, and F. J. Asturias, Structure of the yeast RNA polymerase II holoenzyme. © 2002, with permission from Elsevier.

FIGURE 29.25

Chromatin remodeler families and conserved domains of the ATPase-containing subunit.

Reproduced from *Critical Reviews in Biochemistry and Molecular Biology* 44:117–141, B. J. Venters and B. F. Pugh, How eukaryotic genes are transcribed. © 2009 Informa Healthcare.

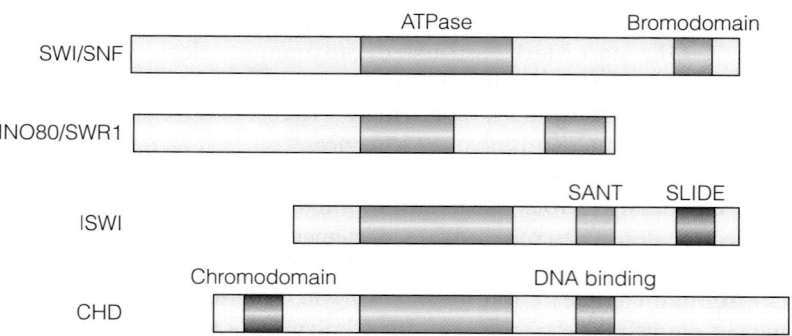

Nucleosome Remodeling Complexes

These complexes couple the energy of ATP hydrolysis to changes in chromatin structure, usually resulting in transcriptional activation. In most cases the energy of ATP hydrolysis is coupled to translocation of the histone core particle along DNA, with the core histones either sliding along DNA or being displaced. Four families of chromatin remodelers are known, classified in terms of the structure of the ATPase-containing subunit. Additional remodelers are known, but have not yet been well characterized in terms of structure or biological function. Figure 29.25 shows the domain organization of the ATPase-containing subunit in each of the four described families.

The SWI/SNF ATPase subunit contains a **bromodomain**, a domain that interacts specifically with acetylated lysines. This draws the complex to chromatin that has already become somewhat activated by histone acetylation. Chromatin remodeler RSC, which was mentioned earlier (page 1153), is a member of the SWI/SNF family. Recent evidence indicates that this complex completely dissociates histones from DNA, beginning with disruption of DNA-histone bonds by the remodeler, followed by ATP-dependent translocation. Presumably, once transcription has been initiated and RNA polymerase has moved downstream, chromatin is reassembled. In yeast the RSC complex also functions at promoters for RNA polymerases I and III.

The INO80 catalytic subunit is distinctive in that the ATPase domain is split. This complex plays a broader role in cellular metabolism than the other remodelers in that it is found in DNA repair complexes and at sites of resolution of stalled replication forks. Recent evidence points to additional roles in telomere regulation, chromosome segregation, and cell checkpoint control. A principal action of this remodeler is replacement in chromatin of canonical histones with histone variants, specifically, exchange of H2A for H2A.Z. As mentioned earlier, H2A.Z is particularly abundant in those nucleosomes nearest to a transcription start site.

Unlike the complexes just described, the ISWI family of remodelers is associated with transcriptional repression. The domains identified in Figure 29.25 as SANT and SLIDE are thought to bind histone tails and linker DNA, respectively. Although little is known about the function of this family, recent evidence in *Drosophila* identifies a role in maintaining the higher-order structure of the male X chromosome.

Chd1, the ATPase subunit of the CHD remodeler, contains a **chromodomain**, which interacts specifically with methylated histones. In vitro studies show strong interaction with methylated H3K4, a transcription-activating methyl mark. Although CHD has been found to be associated with transcriptionally active chromatin, its specific function has not been identified. However, intense interest has focused upon the relationship between Chd1 and the ability of embryonic stem cells to maintain the pluripotent state, that is, the ability to develop into any kind of differentiated cell. So the current picture of this family of remodelers is that, by helping to maintain all chromatin in an open conformation, CHD helps to retain in that cell the ability to express any combination of genes needed for a specific developmental pathway.

Transcription Initiation

As noted earlier, most sequence-specific regulatory sequences, such as nuclear receptor–binding sites, are found some distance upstream from transcription-start sites, and Mediator plays a key role in connecting these sites to the downstream promoters. It is estimated that each eukaryotic gene contains about five specific regulatory sites. Because of the large number of sequence-specific regulatory proteins, and their uses in controlling multiple genes, this complexity is probably necessary to prevent the accidental expression of a gene that might result if only one or two specific regulatory events needed to occur.

The action of chromatin-remodeling complexes, the modifications of histones near promoter sites, the binding of upstream sequence-specific transcriptional activators, and their relationship to Mediator set the stage for final assembly of a transcription complex—binding of RNA polymerase II and the several general transcription factors that we identified in Chapter 27, including TATA box-binding protein and TFII proteins A, B, D, E, F, and H. TFIIH contains at least two enzymatic activities—an ATP-dependent helicase that unwinds template DNA strands to expose the transcription template, and a protein kinase that converts the initiation complex to an elongation complex, by phosphorylating serine residues in the C-terminal domain of the largest pol II subunit.

Recent studies with ChIP techniques indicate that RNA polymerase II is bound at most genes, whether expressed or not, although not all genes are fully occupied by pol II. However, this finding suggests that binding of these general transcription factors, not binding of pol II itself, is rate-limiting for initiation. Although these events are largely independent of the gene being transcribed, there is some gene specificity. For example, Rap1 is a protein that binds directly to TFIID, and this interaction helps to drive the transcription of the highly expressed genes for ribosomal protein synthesis. In any event, the sequence of binding general transcription factors is as we indicated in Chapter 27 (Figure 27.24), and the final preinitiation complex can be schematized as in Figure 29.26.

As is seen also with prokaryotic transcription, many eukaryotic initiation complexes pause soon after initiation. This may involve a site for specific regulation because several protein factors are known both to promote and to overcome pausing. One such positive regulator is called P-TEFb (positive transcriptional elongation factor b). This protein phosphorylates the second serine in each of the heptapeptide carboxy terminal repeats in RNA polymerase II—perhaps augmenting the activity of the TFIIH protein kinase, which acts at the fifth serine residue in each repeat.

Chromatin-remodeling complexes use the energy of ATP hydrolysis to move nucleosomes out of the way for transcription initiation, but they participate in other functions as well.

Most, but not all, events involving general transcription-factor binding at promoters are nonspecific with regard to the gene being activated.

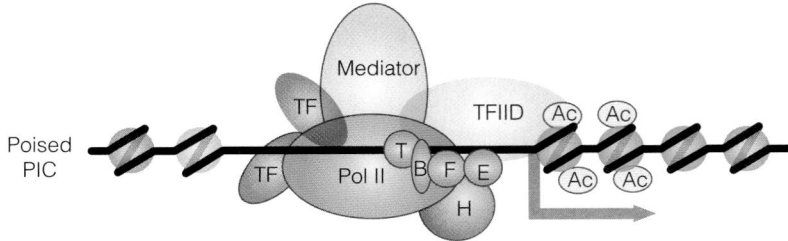

FIGURE 29.26

Schematic view of a transcription preinitiation complex.

Reproduced from *Critical Reviews in Biochemistry and Molecular Biology* 44:117–141, B. J. Venters and B. F. Pugh, How eukaryotic genes are transcribed. © 2009 Informa Healthcare.

Ph
Ph Ph
Ph
YSPTSPS
1234567
YSPTSPS
YSPTSPS
YSPTSPS
CTD status
Initiation
Elongation
Termination
Kin28
hCdk7
Srb10/Cdk8
Ctk1
hCdk9
Srb10/Cdk8
CTD writers
Mediator
Capping factors
Set1
Paf1
Nrd1
Ess1/Pin1
Set2
Pcf11
Spt6
Rtt103
CTD readers
Ssu72
Scp1
Fcp1
CTD erasers

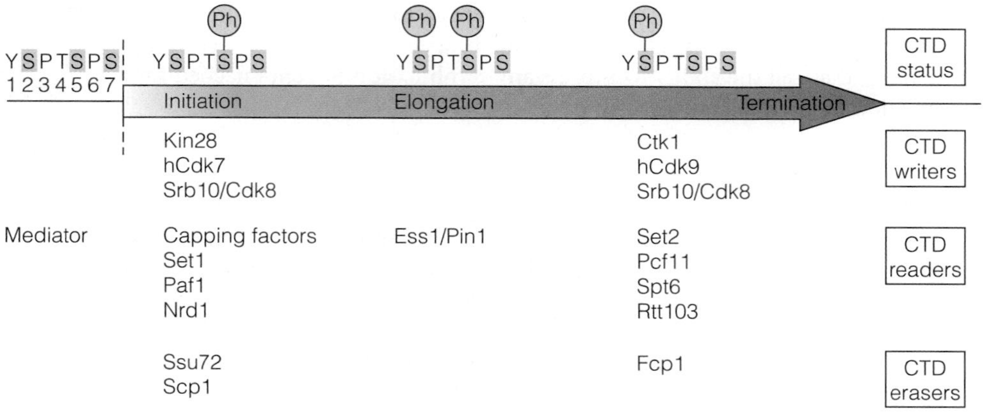

FIGURE 29.27

Writers, readers, and erasers of the CTD phosphorylation code. Writers (serine kinases), readers, and erasers (phosphatases) are identified under the specific serine phosphorylation pattern that most strongly stimulates binding of the respective protein to the CTD. The pattern shown here is oversimplified because evidence suggests that S7, Y1, and T4 also undergo phosphorylation and dephosphorylation during the transcription cycle.

Regulation of the Elongation Cycle by RNA Polymerase Phosphorylation

The heptad repeats in CTD, the C-terminal domain of the large RNA polymerase subunit, are unphosphorylated when the enzyme interacts with Mediator for placement at the promoter. The phosphorylation status of these repeats changes with time after initiation, in a functionally significant way that suggests a "CTD phosphorylation code." As indicated in Figure 29.27, proteins that interact with the CTD can be classified as code "writers" (serine kinases), "readers" (proteins that interact with specific phosphorylation patterns), or "erasers" (serine phosphatases). Among the proteins binding early are those involved in modifying the RNA 5′ end by synthesizing the 7-methyl guanylyl cap. Other CTD readers are enzymes that change the histone modification status, and the patterns change as the predominant CTD phosphorylation changes from Ser-5 to Ser-2 as termination approaches. Figure 29.28 shows the predominant histone and CTD modification patterns as transcription proceeds. The late modifications help to recruit protein factors involved in transcription termination. Strikingly, the CTD itself appears to participate in these changes through *cis-trans* isomerization of the proline residues in the CTD, catalyzed by the proline isomerase Ess1, which binds preferentially to the doubly phosphorylated CTD.

The shifting pattern of CTD phosphorylation as RNA polymerase moves through a gene controls events such as mRNA capping, histone modifications, and the recruitment of transcription termination factors.

Time-dependent changes in CTD phosphorylation are of intense interest to investigators trying to understand the "histone code," or "language," mentioned earlier as a presumed relationship between specific histone modification patterns and functional consequences, such as activation or repression of specific genes, or epigenetic phenomena. Epigenetics refers to heritable changes in gene expression that do not involve changes in DNA base sequence. We discuss epigenetics in more detail in the next section, in connection with DNA methylation.

H3K4me3
H3K14ac
Ser5P CTD
H3K4me2
H3K79me3
Ser2P CTD
H3R2me2a
H3K4me
H3K36me
Density
TSS
Average gene

FIGURE 29.28

Histone and pol II CTD modifications as a function of gene position. The nucleosome distribution relative to the TSS is shown in gray. The green, black, and red traces represent genome-wide histone and CTD modification patterns for the respective modifications shown.

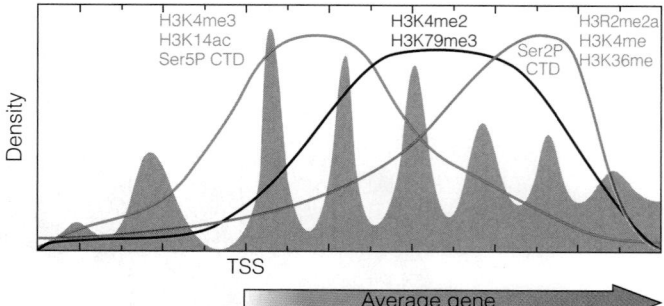

DNA Methylation, Gene Silencing, and Epigenetics

DNA Methylation in Eukaryotes

As we discussed in Chapter 24, DNA methylation in prokaryotes is well understood. The target base for methylation is always adenine, and the processes involving methylation include restriction/modification and methyl-directed mismatch repair. The situation in eukaryotes is quite different. The only eukaryotic DNA base to undergo methylation is cytosine. In eukaryotes DNA mismatch repair does not involve methylation, and there appears to be no direct eukaryotic counterpart to restriction and modification. Instead, DNA methylation participates in still incompletely understood processes involving gene silencing.

Those cytosine residues undergoing methylation in eukaryotes are usually C's that are immediately 5' to G's, i.e., C's in a CpG dinucleotide. CpG is relatively underrepresented in eukaryotic genomes, possibly because of the high frequency with which C can undergo deamination to give U. Base excision repair will normally replace the U with T, which becomes part of a G-C to A-T transition mutational pathway. Deamination of 5-methyl-C gives T directly, making the G-mC deamination an even stronger event in effecting a transition mutation.

Although CpG dinucleotides are underrepresented in DNA, there are regions of eukaryotic genomes in which CpG dinucleotides are found at the statistically expected frequency. These regions are called **CpG islands**. CpG islands are generally longer than about 500 base pairs, and they have a total GC content of 55% or greater. Generally, C's in CpG islands are undermethylated, while most CpG methylation occurs in CpG-poor regions of the genome. For unknown reasons, this distribution of methylated sites is often reversed in cancer cells, and interest is focused upon the question of whether this is related to altered patterns of gene expression seen in cancer, such as decreased expression of tumor suppressor genes.

DNA methylation patterns are heritable, making DNA methylation the best understood example of an epigenetic process. As mentioned earlier, epigenetics refers to the heritable transmission of a gene expression pattern that does not involve a change in DNA base sequence. The pattern of methylation in a particular genome is established early in embryonic development. Mammalian cells contain three different DNA-cytosine methyltransferase enzymes—Dnmt1, Dnmt3a, and Dnmt3b. *De novo* methylation, early in development, is carried out by Dnmt3a and 3b. How these enzymes generate specific methylation patterns is not yet clear. However, once an embryonic methylation pattern is established, that pattern can be faithfully reproduced by the maintenance methylase, Dnmt1, which tracks with the replication apparatus and methylates every C in the daughter strand that was methylated in the parental strand. This process is illustrated in Figure 29.29. A key finding leading to understanding this phenomenon, also shown in Figure 29.29, is the effect of **5-azacytidine** upon maintenance methylation. Azacytidine is a cytidine analog that can be metabolized to the 5-aza analog of dCTP and be incorporated into DNA. However, because the pyrimidine ring of 5-azacytidine contains N instead of C at position 5, it cannot undergo stable methylation. Hence, in the presence of 5-azacytidine, a 5-mCpG dinucleotide is converted to CpG in three rounds of replication. This treatment was found to activate previously repressed genes, a finding that constituted important early evidence that correlated DNA methylation with gene silencing. For example, adult bone marrow cells were found to reactivate the synthesis of fetal hemoglobin, which is normally repressed during development.

Recent structural analysis of Dnmt1 reveals an autoinhibitory mechanism, in which unmethylated CpG dinucleotides are occluded from the active site. Hence, only hemimethylated CpG nucleotides can be methylated, explaining how this enzyme can function only in maintenance methylation.

Abnormal DNA methylation patterns are a hallmark of cancer cells.

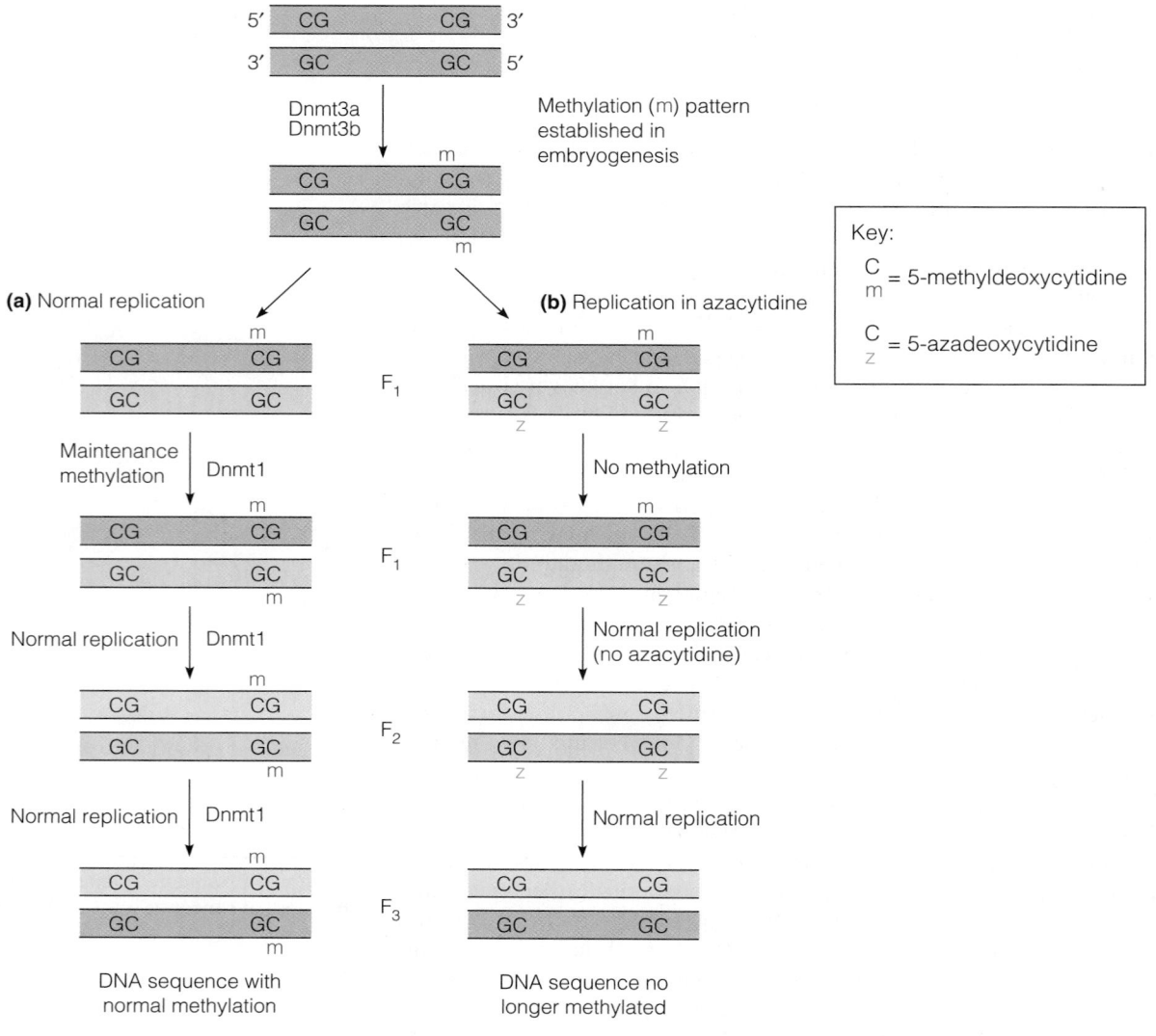

FIGURE 29.29

De novo and maintenance methylation of DNA and the effect of 5-azacytidine upon DNA methylation.

5-Azacytidine

5-Azadeoxycytidine triphosphate

More recent evidence points to a natural role of DNA demethylation in activation of repressed genes. In early 2010 two laboratories reported that deamination of DNA-methylcytosine bases by AID, a DNA-cytosine deaminase involved in antibody maturation (page 1109), also plays a role in the induced dedifferentiation of somatic cells to induced pluripotent cells, capable of differentiation in culture to any state. This is an important development in attempts to use stem cell biology for therapeutic purposes.

There seems not to be a single mechanism accounting for gene repression as a consequence of methylation. In some cases binding of transcription factors to methylated DNA is inhibited. In other cases there are effects upon histone modifications that promote transcriptional inactivation. For example, as we have seen, H3K4 methylation is associated with transcriptionally active chromatin. Some H3K4 methyltransferases (these are *protein* methyltransferases) are targeted to sites containing high levels of CpG dinucleotides, which in turn interfere with methylation of these histones and, hence, with transcriptional activation.

DNA Methylation and Gene Silencing

Whatever the mechanism in terms of effects upon chromatin structure, it is clear that DNA-cytosine methylation is responsible for permanent gene silencing. Two such phenomena are well established—*X chromosome inactivation* and establishment and maintenance of *gene imprinting*. During embryonic development in mammals, one of the two X chromosomes in female cells is permanently inactivated by DNA-cytosine methylation. As described above, this involves *de novo* methylation, followed by maintenance methylation throughout life. The significance of this modification is that the level of expression of genes carried on the X chromosome is then approximately equal for both male and female cells. In most mammals the choice of a chromosome to inactivate is random. Gene imprinting is similar, in that it involves permanent gene inactivation during embryonic development. For some genes only a single parental allele is expressed— some from the father, some from the mother. Again, DNA methylation is responsible for shutting off expression of the imprinted gene. Structural analysis of the Dnmt3a DNA methylase suggests an explanation for the specificity of this latter effect. Dnmt3a functions in concert with a regulatory protein, Dnmt3L. These proteins were found to co-crystallize as a tetramer with the structure Dnmt3L-Dnmt3a-Dnmt3a-Dnmt3L. Binding this protein to DNA showed the two methylase active sites to be about one DNA helical turn apart (Figure 29.30). Examination of several maternally imprinted genes showed them to have a periodicity in spacing of CpG dinucleotides, some 8 to 10 nucleotides apart, suggesting a basis for the specificity in choice of the genes to be imprinted.

Imprinting is affected in a rare condition called *Prader-Willi syndrome*. In this condition a region of paternal chromosome 15 is deleted or unexpressed. This region includes a gene that controls imprinting; normally paternal genes from this region are expressed, while maternal genes are silenced. The defect in imprinting means that, while most people have one working copy of each gene in this region, affected individuals have none. The condition is characterized by short stature, obesity, and delayed puberty. In a comparable condition, *Angelman syndrome*, the same region of chromosome 15 is affected in the maternally derived genetic material.

Genomic Distribution of Methylated Cytosines

Recently, the availability of second- and third-generation DNA sequencing procedures, which allow rapid generation of vast amounts of sequence data, has made it possible to analyze the entire human "DNA methylome," that is, the distribution of 5-methylcytosine bases throughout the genome. The approach is to treat DNA with sodium bisulfite, which quantitatively deaminates cytosine to uracil, but does not affect methylcytosine. This is followed by high-throughput sequence analysis. The most intriguing finding, when this was carried out recently, was that embryonic stem cells have about one-quarter of their methylcytosines in a non-CpG

> DNA methylation can lead to permanent gene inactivation by processes involving chromatin structure and histone modification.

> DNA CpG methylation is responsible for at least two developmental gene inactivation processes—X chromosome inactivation and gene imprinting.

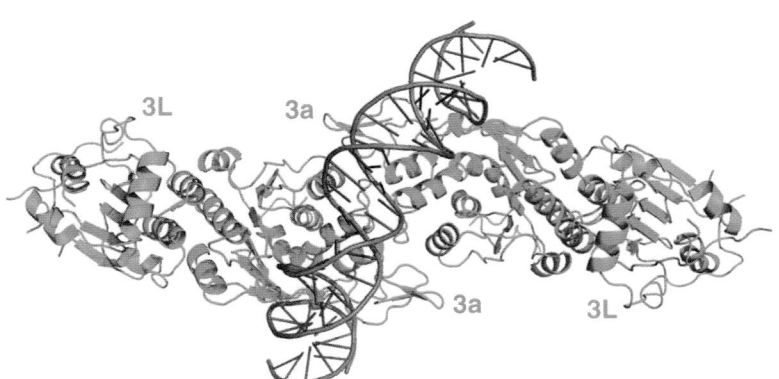

FIGURE 29.30

Structure of the Dnmt3a heterotetramer complexed with a DNA 25-mer. Although the two Dnmt3a active sites are not specifically shown in this model, they were reported to be about ten base pairs apart. Binding of multiple methylase molecules along the DNA duplex could account for preferential methylation of a gene that contained CpG base pairs at 8- to 10-base pair spacings. The two identical a subunits are shown in green and the L subunits in light blue. PDB ID 2QRV.

Reprinted from *Structure* 16:341–350, X. Cheng and R. M. Blumenthal, Mammalian DNA methyltransferases: A structural perspective. © 2008, with permission from Elsevier.

context. When embryonic stem cells underwent differentiation, non-CpG methylation disappeared. Moreover, it is known that certain cell lines can be induced to return to a pluripotent state (able to differentiate into any state), and these induced pluripotent cell lines were found once again to contain a large proportion of non-CpG methylated sites. These findings provide important clues to the potential use of embryonic stem cells in treating diseases such as type 1 diabetes or Parkinsonism. Incidentally, techniques are now available for determining DNA base sequence and cytosine methylation status in a single operation, without a need for bisulfite treatment.

Other Proposed Epigenetic Phenomena

5-Hydroxymethylcytosine

In 2009 it was reported that DNAs from some cells, mostly in nervous tissue, contain significant proportions not only of 5-methylcytosine but also of *5-hydroxymethylcytosine*. As we noted in Chapter 22, bacteriophage T4 and its relatives have 5-hydroxymethylcytosine completely substituted for cytosine (page 950). In phages the modification occurs at the nucleotide level. The situation in mammalian cells is quite different; hydroxymethylcytosine arises through oxidation of methyl groups in 5-methylcytosine. Recent evidence suggests that DNA-cytosine oxidation is a process in DNA demethylation, leading to reactivation of silenced genes.

Chromatin Histone Modifications

Intense interest is focusing upon other possible mechanisms of epigenetic inheritance, including processes involving chromatin histone side chain modification and noncoding RNA molecules. In particular, the premise that modification of histone amino acid residues is an epigenetic phenomenon is widely accepted. There is no doubt that histone side chain modifications affect gene expression, and that no changes in DNA base sequence are involved. The issue is whether patterns of histone modification are inherited from one cell generation to the next. There is evidence that daughter chromosomes after mitosis both have histone modifications characteristic of the parent cell, a concept that supports histone modification as an epigenetic phenomenon. However, as stated in a recent review article, "To date ... only DNA methylation has been shown to be stably inherited between cell divisions" (R. Margueron and D. Reinberg [2010] *Nature Reviews Genetics* 11:285–296). One possible way to account for histone-mark transmission would be for histone dimers to mix during mitosis such that, for example, the two H2A/H2B dimers in a newly replicated nucleosome would contain one parental dimer and one newly synthesized, such that the parental dimer could somehow instruct the new dimer regarding its modification pattern. However, although density labeling experiments are consistent with this concept of histone mixing, we know nothing about how a histone molecule in parental chromatin might serve as a template for the corresponding histone after replication.

Control of Higher-Order Developmental Patterns: Homeotic Genes

Throughout this chapter we have emphasized how carefully the life of an organism, whether single-celled or multicellular, is programmed in its DNA. We now realize that for eukaryotes *much* more information is stored in DNA than simply the recipes for a collection of proteins and special nucleic acids. A variety of signals are hidden within the genes themselves—signals to determine how transcripts will be cut and spliced, where gene products will go, and even how long they will last. In addition, a vast amount of information, usually coded in the sequences surrounding certain genes, specifies when, either in the course of

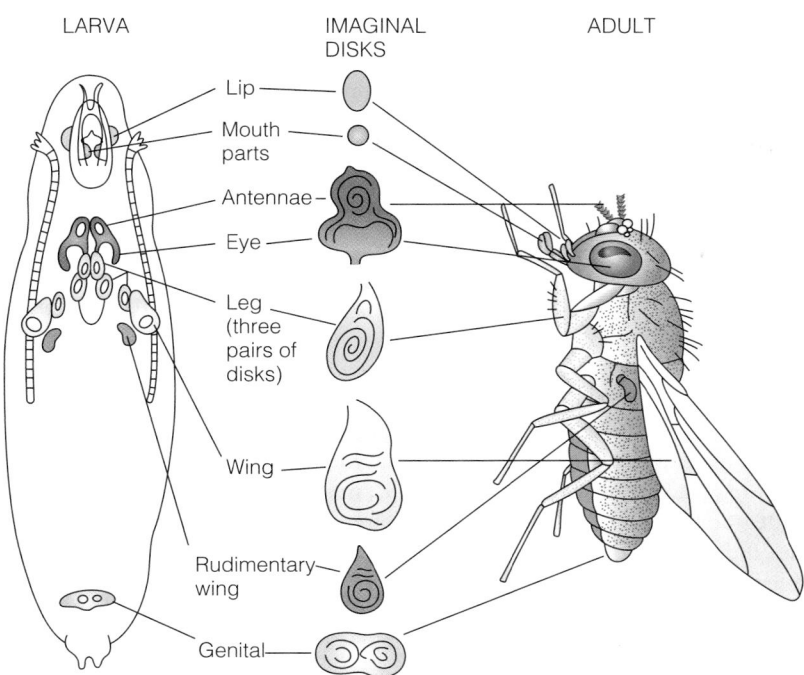

LARVA   IMAGINAL DISKS   ADULT

Lip

Mouth parts

Antennae

Eye

Leg (three pairs of disks)

Wing

Rudimentary wing

Genital

FIGURE 29.31

Imaginal disks in the development of *Drosophila.* Each of the imaginal disks shown in the larva develops into a specific segment of the adult.

From J. W. Fristrom, R. Raikow, W. Petri, and D. Stewert (1970). In *Problems in Biology: RNA in Development,* E. W. Hanly, ed. University of Utah Press. Reprinted by permission.

development or in response to environmental stresses, certain genes are to be transcribed in certain cells.

At a higher level, further genetic instructions *coordinating* these transcriptional instructions must exist because the development of an organism requires the programmed differentiation of some cells, the proliferation of certain tissues, and the death of selected cells. We have only glimpses of how such information may be encoded in the genome. Much of what we know comes from studies of development of the fruitfly, *Drosophila melanogaster.* Developmental biologists have long known that as the larva of this insect develops, groups of cells are set apart as disklike structures called **imaginal disks** (Figure 29.31). These groups of cells will form specific parts of the adult fly. As the larva metamorphoses, larval cells are destroyed by apoptosis, and each imaginal disk develops into a different portion of the adult.

Geneticists working with *Drosophila* have long recognized classes of **homeotic mutations**—mutations that scramble the whole developmental pattern in defined ways. One, called the *Antennapedia* mutation, causes perfectly formed legs to grow in the places near the eye where antennae are normally formed. Another group, termed *bithorax* mutations, leads to abnormal development of thoracic segments, producing, for example, extra pairs of wings (Figure 29.32). Recent molecular biological studies have revealed large clusters of **homeotic genes** that control these developmental processes and are the sites of homeotic mutations. Remarkably, a common sequence element of about 180 bp is repeated many times in these gene clusters. This sequence, now called the *homeo box,* codes for a 60-residue polypeptide sequence called the *homeo domain.* Proteins containing this domain are nuclear, DNA-binding proteins of the helix–loop–helix class. It seems likely that each acts as a regulator of transcription for a coordinated group of proteins.

Most remarkable is that the homeo box is not confined to insects but is found in many other organisms, including amphibians and mammals. The sequences are highly conserved over this phylogenetic range. This discovery hints at a quite unexpected uniformity in developmental mechanisms between very distantly related organisms.

FIGURE 29.32

A bithorax mutation of *Drosophila.* **Top**: Normal fly. **Bottom**: A mutant in which the thoracic segment that normally produces a pair of halteres, or rudimentary wings, has been transformed into one that produces a fully developed set of wings.

Pascal Goetgheluck/Science Photo Library.

Homeotic mutations that modify developmental pathways involve special protein factors carrying the homeo domain.

Regulation of Translation

As noted earlier, the Jacob–Monod paradigm dominated our thinking for many years about gene expression and its control. We have seen in the previous section how much of the operon model applies also to the more complex processes of eukaryotic gene expression, yet how many striking differences we have encountered—mRNA splicing, mRNA capping and tailing, multiple transcription factors in initiation complexes, chromatin as the transcription template, chromatin remodeling, the involvement of Mediator, control by phosphorylation of RNA polymerase, and the absence of multicistronic mRNAs, to cite a few examples. Equally striking are recent discoveries of important regulatory processes acting at the translational level. One of the most active research arenas, aided by second-generation DNA sequencing (also called deep sequencing; see Tools of Biochemistry 4A), is the discovery of many classes of RNA molecules that are themselves genetic regulators. We present examples of these regulatory processes first for prokaryotes, then for eukaryotes. Some of these RNA regulatory phenomena, such as RNA interference, operate at the level of mRNA degradation, and are treated in a separate section (page 1267). Other processes, including riboswitches (page 1269) and RNA editing (page 1270), are also treated separately.

Regulation of Prokaryotic Translation

Prokaryotic translation is regulated by at least three mechanisms—ribosome occlusion due to mRNA tertiary structure, translational repression caused by protein binding to mRNA, and actions of regulatory RNA molecules, which base-pair with mRNA molecules. Particularly with respect to regulatory RNA molecules, this is a currently active research frontier, so new regulatory processes and mechanisms are likely to emerge.

Ribosome Binding Site Occlusion

This kind of control appears to explain the regulation of translation of a number of polycistronic messages. An example is found in the messenger RNA of bacteriophage MS2, shown in Figure 29.33. This complex and economical mRNA encodes four essential proteins—one of them coded from an alternate, overlapping reading frame. Consider the different requirements that translation of this message must satisfy: To make new viruses, *many* copies of the coat protein and a significant number of replicase subunits are needed. But only *one* copy of the A protein (which is used in virus assembly) and only a small amount of the lysis protein (used in viral release) are needed per virus. To simply translate the whole polycistronic message equally in all its parts would be woefully inefficient. Instead,

FIGURE 29.33

The RNA of bacteriophage MS2. This RNA molecule serves as a message for all four of the proteins required by this virus. The coding sequences for the A protein, coat protein, and replicase subunit are shown in gray. The message for the lysis protein (the L protein, shown in blue) actually overlaps the coat protein and replicase messages. It is translated in a different reading frame.

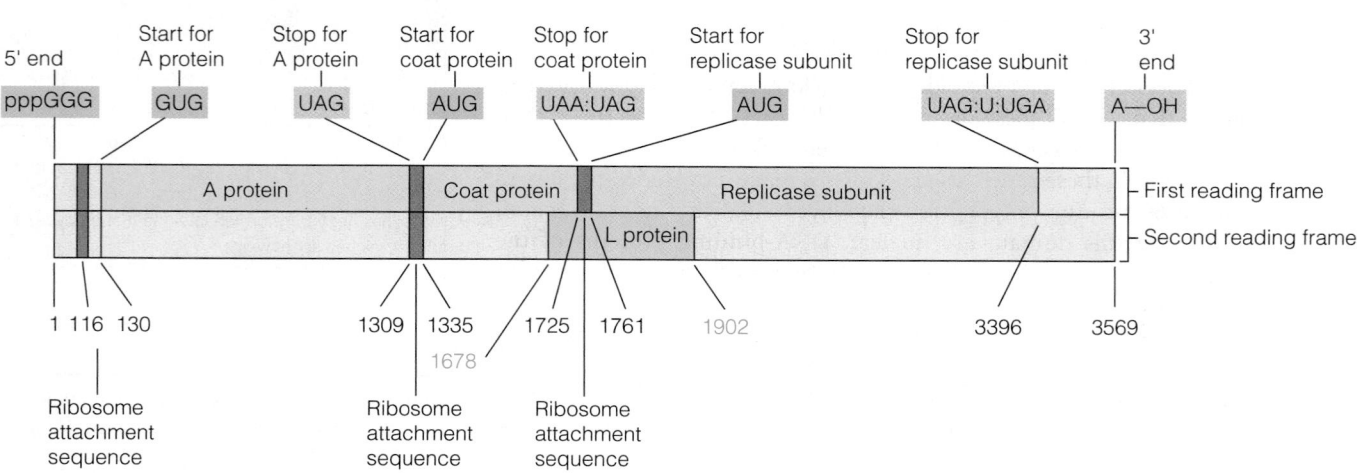

the ribosome-binding sequence at the 5′ end of the mRNA (which would start translation of the A protein message) is normally blocked by tertiary folding of the mRNA molecule. Therefore, the ribosome normally begins translation by binding to an attachment sequence near the start site of the coat-protein message, translating this message efficiently and proceeding sometimes, but not always, to the replicase message. As replicase is made, it catalyzes replication of the viral mRNA itself. Copies of new plus strands that are still being transcribed have not yet folded into their final conformation. Therefore their 5′ ends are still open for initiation and can attach to ribosomes to translate A protein. But this happens only once in the lifetime of each mRNA—at the time it is newly synthesized. The message for the lysis protein, which overlaps the coat protein message but in a different reading frame, is apparently translated only occasionally, as a consequence of a frameshift slip during translation of the coat protein.

Translational Repression, by Protein Binding to mRNA

An elegant example of this kind of control is found in the synthesis of the bacterial ribosomal proteins themselves. As Figure 29.34 shows, *E. coli* ribosomal proteins are encoded by polycistronic messages. In each group encoded by such a message, there is *one* protein that is capable of binding at or near the 5′ end of the message and blocking its translation. This binding seems to occur because the mRNA has a tertiary structure similar to the normal binding site for that protein on rRNA.

The beauty of the system can be seen by comparing Figure 29.34 with Figure 28.18. The proteins that control ribosomal protein synthesis—S4, S7, and S8—are also among the first to bind onto the 16S rRNA in ribosome assembly. These proteins are the keys to ribosome construction, and they have a very high affinity for the target rRNAs. If ribosomal RNA is abundant in the cell, which would signal need for ribosomal proteins, the "control" proteins are incorporated into ribosomes, and synthesis of all ribosomal proteins can proceed. But if rRNA is in short supply, the unused "control" proteins will bind to the target mRNAs, shutting down the synthesis of other, currently unneeded ribosomal proteins. In other words, these proteins can be considered to act as "translational repressors." This control of ribosome production acts in concert with stringent control of rRNA synthesis, exercised by ppGpp (page 1250) to give complementary processes by which ribosome assembly is inhibited under nutrient-limiting conditions.

Actions of Regulatory RNA Molecules

The idea that RNA molecules themselves could serve as genetic regulators has been known since the early 1980s, but the broad significance of this mechanism has only recently become evident. An early finding was the discovery of antisense

Ribosomal protein binding to mRNA is a site for regulation of translation.

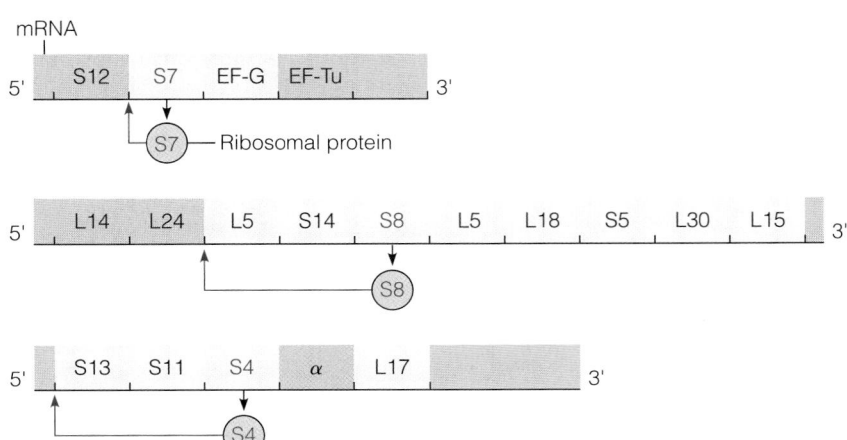

FIGURE 29.34

Regulation of synthesis of ribosomal proteins. Shown are three polycistronic messages encoding ribosomal proteins in *E. coli*. Note that these mRNAs also code for EF-G, EF-Tu, and the α subunit (in purple) of RNA polymerase. Translation of each polycistronic message is controlled by one of the ribosomal proteins it controls, shown in magenta. Arrows show the binding sites for each protein. Control extends *only* over the genes shown in teal.

FIGURE 29.35

Blocking of translational initiation by antisense RNA. The transposase gene contains an antisense sequence at its 3′ end. When transcribed, this antisense segment can fold back and base-pair to block initiation of translation.

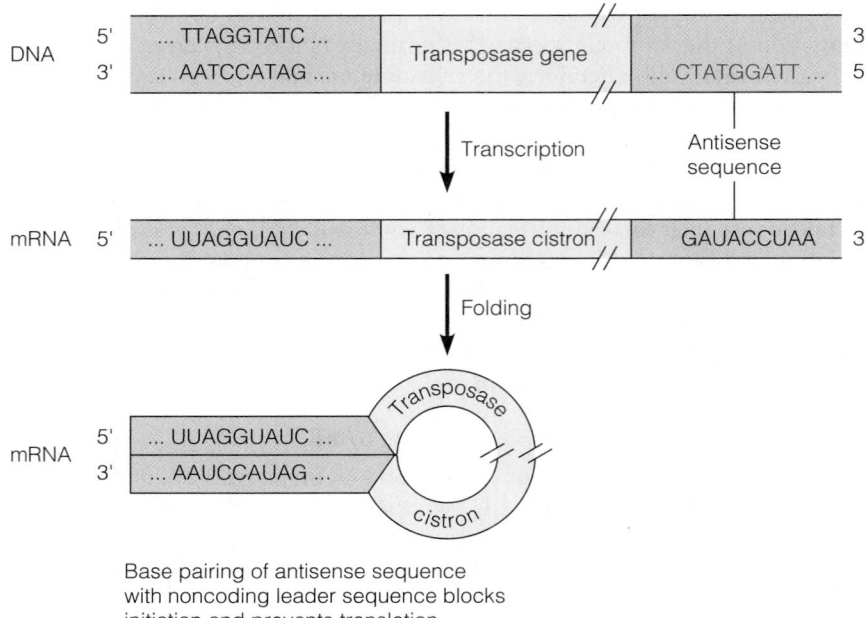

Base pairing of antisense sequence with noncoding leader sequence blocks initiation and prevents translation

regulation, in which the mRNA for the transposase of the Tn10 transposon can base-pair with an upstream sequence, thereby blocking initiation of transposase mRNA translation (Figure 29.35).

Another example of antisense regulation involves *ompC* and *ompF*, two genes that encode outer membrane proteins in *E. coli*. These genes are osmoregulated: Cells respond to growth in a medium of high osmolarity by shutting down the synthesis of OmpF protein and activating the synthesis of OmpC so that the total amount of protein is constant and the cell's internal environment is maintained. The postulated mechanism of the *ompF* shutoff is shown in Figure 29.36. High osmolarity in some way triggers synthesis of an antisense RNA, the product of the *micF* gene. This RNA is partly complementary to sequences in the 5′ end of *ompF* mRNA. The *micF* RNA inactivates the *ompF* message, by annealing to it, and thereby forming a duplex RNA in vivo. The translational initiation sequences of OmpF mRNA, which must be single-stranded to direct translation, are included in this duplex. This is responsible for blocking translation of the message. Another gene regulated by antisense RNA is *crp*, the structural gene for cAMP receptor protein.

More recently, analysis of whole genome sequences followed by confirming experimental evidence has revealed that regulation by small RNAs (now called sRNAs) is much more widespread than originally realized. In *E. coli* alone, some 80 sRNAs have been discovered. Some, like the MicF RNA mentioned above, bind

FIGURE 29.36

Inactivation of *ompF* mRNA by pairing with antisense RNA from the *micF* gene. A change in osmolarity stimulates transcription of the *micF* gene. The transcript is largely complementary to a region in *ompF* RNA that includes the translational start site. Hairpin loops within the sequences allow base pairing between complementary regions on the two mRNAs, and in this way both transcripts are prevented from serving as templates for protein synthesis.

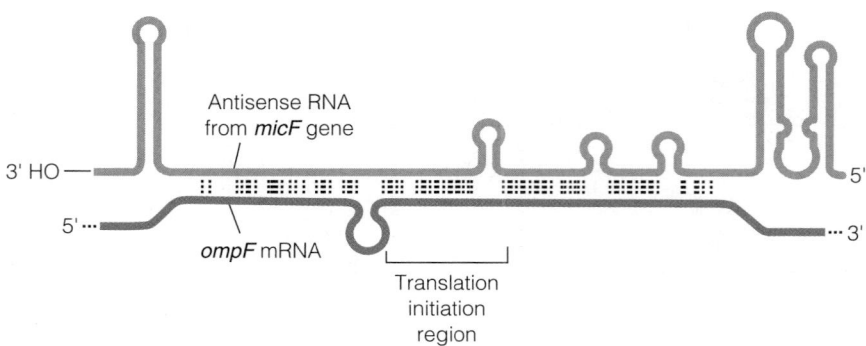

to just one target sequence. Others bind several different sequences. Some, like the antisense examples mentioned here, act by sequence-specific binding to RNA. Others act by binding specific target proteins. Most of the regulatory effects mentioned here are exerted at the level of initiation. Again, this makes the most sense energetically. Regulation by RNAs instead of proteins also makes good metabolic sense because the energy-using reactions of translation can be avoided.

Synthesis of designed antisense RNAs is being widely used as an approach to **gene knockdown**, when an investigator wishes to block expression of a specific gene without the extensive efforts involved in creating knockout organisms by targeted gene interruption. A synthetic oligonucleotide can be introduced into cells, usually following treatment to transiently increase membrane permeability, with a sequence targeted to the gene whose expression is to be inhibited. Often the knockdown reagent is not an RNA molecule but an RNA analog, one that has been modified so as to avoid enzymatic degradation within the cell. A popular family of antisense reagents is the *morpholinos*, oligomers or polymers that use the same bases as found in natural RNA, but which have morpholine instead of ribose and a phosphorodiamidate bond to link adjacent "nucleotide" units; see the margin.

Antisense analogs such as morpholinos are being developed as therapeutic reagents, where a target nucleic acid can be identified, such as a viral genome, to become bound and be thereby inactivated. The stumbling block to developing this approach is the difficulty of making these reagents in a form permeable to cell membranes, something that is far easier to do with cells in culture than with cells in living organisms.

Regulation of Eukaryotic Translation

Comparable translational regulatory processes occur in prokaryotic and eukaryotic cells, although there are some significant differences, such as control by phosphorylation of initiation factors (see below). Most striking, however, are parallel developments in prokaryotic and eukaryotic cells regarding regulatory roles of noncoding RNAs. In prokaryotes these RNA molecules are generally short, as implied by the name sRNA. By contrast, this family of RNA molecules in eukaryotes is of higher molecular weight, normally over 200 nucleotide residues in length; these are usually called long noncoding RNAs, or ncRNAs. We discuss these after presenting a couple of better-understood examples of translational regulation.

Phosphorylation of Eukaryotic Initiation Factors

A number of the soluble protein-translation factors in eukaryotes are subject to control by phosphorylation. We give a couple of examples here. eIF2 is a G protein, involved in binding Met-tRNA to the P site as an essential step in initiation (Table 28.4). Four different protein kinases are known to phosphorylate the α subunit of eIF2 at Ser-51. Phosphorylation of eIF2 increases its affinity for its guanine nucleotide exchange factor, eIF2B, leading to formation of an inactive eIF2B-eIF2-GDP complex, thereby shutting down translation initiation. The four protein kinases (not all of which are universally distributed) include (1) the protein 2(GCN2), a kinase that is activated by uncharged tRNA and thereby is a sensor for amino acid starvation; (2) a double-stranded RNA-activated protein kinase, which responds to viral infection; (3) an endoplasmic reticulum protein kinase, which is activated by unfolded proteins in the ER; and (4) a heme-regulated inhibitor kinase (HRI), which is activated under conditions of heme deprivation, thereby shutting off synthesis of hemoglobin when the heme cofactor is scarce. This effect is important in reticulocytes (immature erythrocytes), which are enucleated, but which have ample mRNA for globin synthesis. As shown in Figure 29.37, the biosynthesis of hemoglobin is regulated by this process and, hence, the translation of globin mRNA is shut off unless adequate heme is available to complex with the protein. Another example, mentioned in Chapter 18, is the phosphorylation of 4EBP1 by

Morpholine

Adenine

Cytosine

Uracil

Morpholino Oligo

FIGURE 29.37

Regulation of translation in erythropoietic cells by heme levels. If heme levels fall, the heme-controlled kinase becomes active and phosphorylates eIF2 (magenta arrow). This blocks further translation by tying this factor into a stable complex with eIF2B. When heme levels are adequate, the kinase is inhibited, and eIF2 is available for translation initiation.

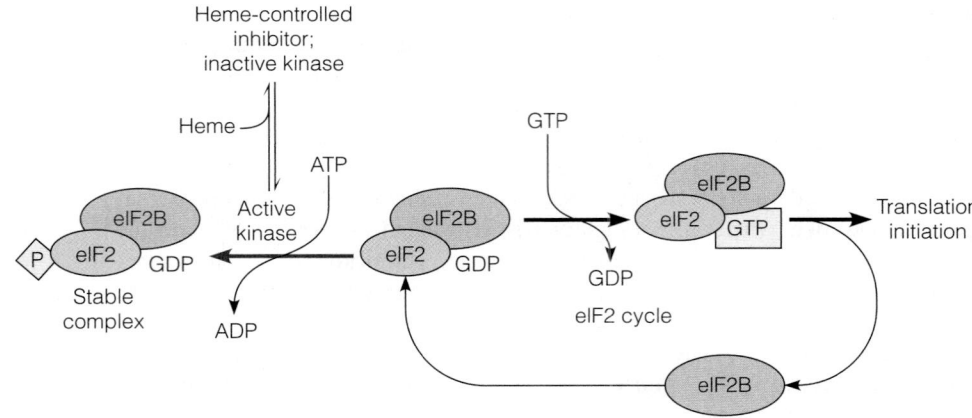

Phosphorylation of initiation factors is commonly used for translation-level control in eukaryotes.

mTOR. Phosphorylation relieves the binding of 4EBP1 to eIF4E, thereby freeing the initiation factor to participate in protein synthesis.

Another example is eIF4E, which is involved in recognition of the 5′ mRNA cap, essential to identification of the initiating AUG codon. In eukaryotic cells one or more MAP kinase signaling pathways (Chapter 23) leads to phosphorylation (of Ser-209 in the human protein), with resultant loss of ability of eIF4E to recognize the cap. eIF4E is subject to additional regulation, involving translational repression (see above) by a family of proteins called 4E-BPs (IF4E-binding proteins). Binding of these proteins is controlled by phosphorylation, with the phosphorylation state, again, being controlled by amino acid availability, cellular energy status, or effects of growth factors or other stimuli. Overexpression of eIF4E causes oncogenic transformation of some cell lines in culture, and levels of the protein are abnormally high in some human cancers.

Long Noncoding RNAs

Until recently it was thought that only a small proportion of the human genome, and hence, of eukaryotic genomes in general, undergoes transcription. This conclusion was compatible with results from the Human Genome Project, when it was revealed that protein-coding open-reading frames accounted for less than 2% of all genomic sequences. However, several approaches, including microarray analysis of the human transcriptome, revealed that more than 90% of the human genome is likely to be transcribed. This means that the mammalian transcriptome contains very large numbers of noncoding RNA molecules. Although some of these molecules may represent transcriptional "noise," it is likely that many of the newly discovered transcripts are, in fact, functional. Indeed, for some of these ncRNAs that have been characterized, there is evidence for antisense as a regulatory mechanism, as we have seen for sRNAs in bacteria. However, to the extent analyzed to date, eukaryotic ncRNAs appear to act at the transcriptional level. Transcription of a noncoding region in the vicinity of a protein-coding gene can inhibit transcription of that gene, by interfering with the binding of transcription factors. An interesting example is transcription of human *DHFR*, the structural gene for dihydrofolate reductase. In this case a noncoding RNA forms a triplex structure with the *DHFR* promoter, which results in disruption of the preinitiation complex. Of course, this pushes the regulatory issue back one step; what controls expression of this regulatory ncRNA?

To date there are few indications that ncRNAs act at the translational level, for example, by occluding ribosome-binding sites, as we have seen for bacterial sRNAs. However, the phenomenon of RNA interference, which we consider next, establishes that noncoding RNAs regulate translation by controlling *degradation* of mRNA molecules.

RNA Interference

A series of striking discoveries, beginning in the late 1990s, revealed unexpected processes involving RNA-based genetic regulatory and cellular defense mechanisms. The original observations stemmed from attempts to use genetic engineering to increase purple pigmentation in petunias, by introducing into purple flowers additional genes encoding the pigment synthesis pathway. Surprisingly, the resultant transgenic plants were not purple; flowers either had highly variegated pigmentation or were white. Somehow, it appeared, the pigment-forming genes had switched each other off. At first it was thought that this could be resulting from formation of *antisense RNA*, that is, RNA transcribed from the nontemplate DNA strand, which would pair with normal sense RNA, creating nontranslatable double-stranded RNAs. Further experiments, primarily by Craig Mello and Andrew Fire, led to a different conclusion and to the discovery of **RNA interference**, or RNAi. The term covers two distinct processes, both of which involve small RNA molecules, some 21 to 24 nucleotides in length. One class of these molecules, **microRNA**, or miRNA, is involved in gene regulation, while **small interfering RNAs**, or siRNAs, are formed primarily as a cellular defense mechanism.

MicroRNAs

MicroRNAs are specific gene regulatory products; it is estimated that nearly one third of all human genes are regulated by miRNA molecules. The number of specific miRNA molecules in cells of a particular organism is related to the evolutionary complexity of the organism. In a recent study the total numbers of known miRNAs were reported as 677 for humans and 491 for the mouse. By contrast, *Drosophila* had 147, while the sponge had only 8. miRNAs are derived from partially palindromic RNA molecules. As shown in Figure 29.38, these precursor molecules result from transcription by RNA polymerase II, followed by 5′ capping and 3′ polyadenylylation, just as in mRNA synthesis, yielding a primary transcript (pri-miRNA). Ends of the molecule meet to give a stem–loop hairpin, and complementary sequences pair up. Processing begins in the nucleus, with cleavage by an enzyme complex called *Drosha,* some 22 nucleotides from the stem–loop junction, to give a partial hairpin with a short 3′ overhang. This is recognized by an exportin complex (like importin in reverse), which carries the miRNA precursor to the cytoplasm, where it is bound by another complex called *Dicer.* In conjunction with another protein called *Argonaute* (AGO), Dicer degrades one strand of the partial duplex miRNA precursor and transfers the remaining single strand, which is now a completed miRNA molecule, along with the Argonaute protein, to still another complex, called RISC (RNA-induced silencing complex). RISC then binds to the 3′ untranslated sequences of target mRNA molecules, using sequence complementarity with the miRNA as a guide. The region along which sequence complementarity is sought is about seven nucleotides. If the miRNA sequence is completely complementary to that of the target mRNA within those seven nucleotides, that mRNA is completely degraded (not shown in the figure). This process is catalytic, in that once an mRNA has been degraded, RISC can seek further targets.

If the miRNA-mRNA sequences match only partially, then activity of the mRNA is slowed, by various mechanisms. There may be inhibition of binding of translational initiation factors, or ribosome stalling, or activation of enzymes (CCR4-NOT in the figure) that remove the 3′ polyA tail. There may not always be immediate mRNA degradation, but its translational activity is slowed.

Eventually, the inhibited mRNA is transferred to a cytosolic site called the **P-body** (P for processing), where it is sequestered from ribosomes, and hence, translationally inactive. The P-body is the site for ultimate degradation of all mRNA molecules, whether or not they arrived there accompanied by RISC. By this process a single miRNA can participate in regulating as many as several hundred

Small RNAs produced from processing of double-stranded RNAs are widely used in gene regulation (miRNAs) and as a defense mechanism (siRNAs).

FIGURE 29.38

Biogenesis of miRNA. The mRNA sequence selected for processing can be either in an exon or an intron. For further details see text.

Reprinted by permission from Macmillan Publishers Ltd. *Nature Reviews Molecular Cell Biology* 11:252–263, M. Inui, G. Martello, and S. Piccolo, MircoRNA control of signal transduction. © 2010.

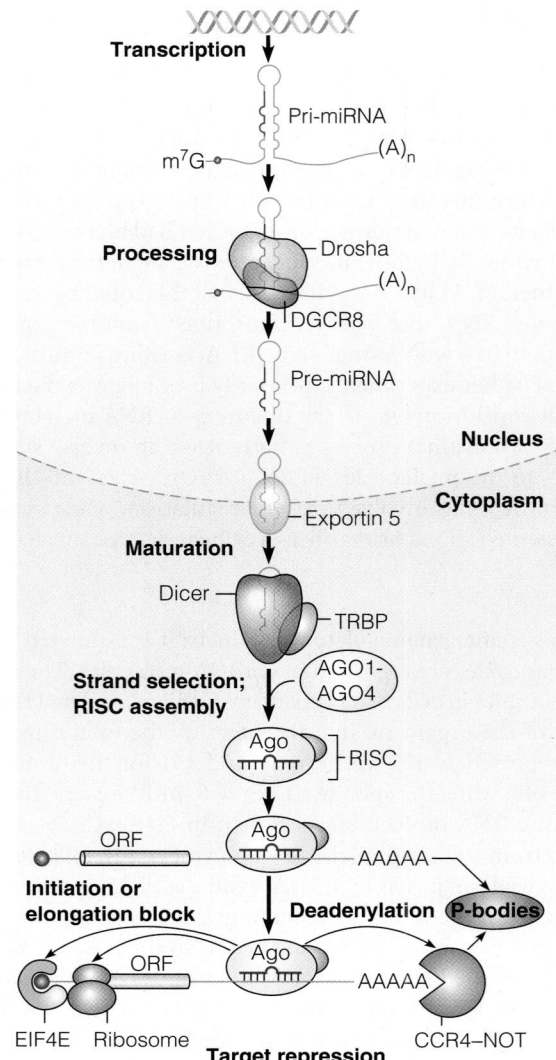

different mRNAs. Also, regulation by miRNAs can be combinatorial, in that the 3′ UTR of an mRNA can be bound by two or more miRNAs, with each binding of a second miRNA yielding further repression of the translational activity of that mRNA. Although we have learned much about these processes, our understanding of what they mean in terms of specific metabolic control is still at an early stage. One fairly recent finding is that some miRNAs can *activate* translation, with the effect of the same miRNA oscillating between inhibitory and activating during the cell division cycle.

Small Interfering RNAs

RNAi as a defense mechanism arises normally as a result of virus infection of a cell. If the virus has an RNA genome but is not a retrovirus, perfectly matched double-stranded RNA molecules arise in the infected cell as intermediates in viral genome replication. These dsRNAs are cleaved in the cytosol by Dicer, to give a series of perfectly matched double-stranded RNA molecules, each about 23 base pairs long, called siRNAs (small interfering RNAs). These are targeted to viral RNA molecules, and because the nucleotide match is perfect, the RNA molecules are cleaved as indicated in our discussion of miRNA synthesis. This process is particularly effective in plants, whose cells are connected by fine channels. Hence, the RNA interference activity can be spread from cell to cell, leading an entire plant to become virus-resistant, even though only a few cells might have been infected initially.

Although we have much to learn about particular aspects of miRNA regulation, we do know that the synthesis of specific miRNAs is tissue- and developmental-stage-specific. Initial transcription of the pre-miRNA molecule is subject to the same regulatory processes as those we have discussed for mRNA biosynthesis, and each following step is controlled, although many details remain to be elucidated.

With respect to miRNA function, we know that these molecules are involved in regulating processes as diverse as cellular proliferation, control of development, apoptosis, homeostasis, and tumorigenesis. With respect to apoptosis, an interesting connection was recently described. In *C. elegans,* Dicer is directly involved; action of a particular caspase upon Dicer converts it from a ribonuclease to a deoxyribonuclease, responsible for the chromosome fragmentation that is a hallmark of apoptosis. Another recent study described a specific role of an miRNA molecule in controlling activity of a tumor-suppressor gene called PTEN. In this case a PTEN pseudogene (Chapter 24) is transcribed, and the miRNA that would otherwise downregulate the tumor suppressor is drawn to, and inactivates, the pseudogene mRNA instead. This action of the pseudogene, in "titrating away" the PTEN miRNA, suggests an interplay between pseudogene and functional gene activities and helps to explain why "nonfunctional" pseudogenes might have been retained through evolution—to act as "decoys," protecting the functional mRNA.

It should come as no surprise, therefore, to learn that there is tremendous interest in harnessing miRNA biology to treat or prevent diseases such as cancer. Although this field is in its infancy, RNAi has found widespread use in the laboratory for specific gene knockdown. In Tools of Biochemistry 26A, we described how genes can be knocked out in living cells or organisms by targeted gene replacement. As we mentioned also on page 1120, the most complete and specific and permanent way to study gene function involves ablation of that gene, for example, by targeted deletion, followed by investigation of the deletion mutant. However, use of RNAi is a much faster, albeit less complete and less specific, way to achieve the same goal. In this approach a short hairpin RNA (shRNA) can be prepared by chemical synthesis and introduced into target cells. These hairpin molecules are metabolized essentially identically to siRNAs, which leads to degradation of the target mRNA. In cultured mammalian cells there are several reagents that can be used to transiently increase cell permeability, allowing direct uptake into cells, followed by specific gene inactivation. With other biological systems, such as the nematode, *Caenorhabditis elegans*, the shRNA can simply be injected into the intestine, where it is more or less efficiently taken up into cells. Gene knockdown by RNAi is rarely as complete or as specific as is the creation of knockout organisms, but, as noted above, it is much simpler, and, hence, can be used to investigate the functions of several different genes as part of the same study.

Riboswitches

The discovery of gene regulatory mechanisms involving RNA molecules, along with previously mentioned findings that RNA can serve as an enzyme, strengthened belief in an RNA world—a primordial biosphere existing before proteins, in which biological functions later assumed by proteins could be carried out by primitive RNA molecules. Supporting this concept is the fairly recent discovery of **riboswitches**—mRNA molecules whose translation is controlled by specific binding of a metabolic end product of the pathway in which that mRNA is involved. These molecules were originally discovered in bacteria and have recently been found in some plants and fungi as well.

A riboswitch usually has an **aptamer** at or near its 5′ end. An aptamer is a specific binding site created from an oligonucleotide or polynucleotide sequence. Aptamers were originally recognized on the basis of a laboratory technique called

SELEX, which creates such binding sites in vitro through repeated cycles of a selection process. With the discovery of riboswitches, in about 2002, it was found that aptamers of much greater binding specificity had been created by evolution. As noted above, a riboswitch RNA has a binding site for a particular metabolite at the 5′ end of an mRNA that encodes an enzyme in the metabolic pathway leading to that metabolite. Many riboswitches contain sites for binding nucleotides or coenzymes—thiamine pyrophosphate, flavin adenine dinucleotide, *S*-adenosylmethionine, and so on. The earliest known riboswitch has a binding site for adenosylcobalamin (B_{12} cofactor). Specificity of the binding sites is high, as shown by the finding that a riboswitch for *S*-adenosylmethionine was shown to bind that particular target molecule at least 100-fold more tightly than it binds the closely related *S*-adenosylhomocysteine. Figure 29.39 shows the crystal structure of a riboswitch binding site for thiamine pyrophosphate. The structure shown is part of the 3′ end of the mRNA encoding one of the enzymes in TPP biosynthesis. Binding of TPP to the riboswitch evidently abolishes the template activity of this mRNA, thereby shutting off TPP biosynthesis. It is evident that a complex folding pattern has evolved to completely enfold the target molecule. The dissociation constant for the complex shown is in the range of 50 nM. It is essential that riboswitches bind their target molecules tightly because they are controlling the biosynthesis of nucleotides and coenzymes that exist within cells at very low concentrations.

In each case, binding of the target molecule has the effect of shutting off the expression of genes involved in synthesis of the target molecule. This can occur at the level of either transcription or translation, as shown in Figure 29.40. A specific example is shown in Figure 29.41. Here the riboswitch is for a bacterial second messenger, cyclic di-guanosine monophosphate (c-di-GMP). This nucleotide controls synthesis of proteins in the flagellae of *Clostridium difficile*. The figure shows how binding the messenger molecule to the riboswitch alters RNA conformation to create a transcriptional terminator, causing premature termination of transcription of an operon that encodes 13 proteins, in the same sense that creation of a premature terminator controls transcription of the tryptophan operon (page 1247).

RNA Editing

RNA editing was another unanticipated process when it was first described in 1986. In this process the nucleotide sequence of an RNA molecule is actually changed post-transcriptionally. It was found that in trypanosomes the mRNAs for some mitochondrial proteins were modified by insertion and occasional removal

> A riboswitch is an mRNA molecule that has a specific binding site for a metabolic end product, which, when bound, blocks either transcription or translation of downstream genes.

FIGURE 29.39

Structure of the thiamine pyrophosphate riboswitch from the plant *Arabidopsis thaliana*. **(a)** Secondary structure diagram for the TPP-binding domain of the riboswitch. Residues involved in pyrophosphate binding are shown with green asterisks and those involved in thiamine binding with red asterisks. Conserved nucleotides are shown in red. **(b)** Crystal structure of the TPP riboswitch, with bound TPP shown in yellow and the "sensor helices," which bind portions of the TPP molecule as shown. PDB ID 2CKY.

From *Science* 312:1208–1211, S. Thore, M. Leibundgut, and N. Ban, Structure of the eukaryotic thiamine pyrophosphate riboswitch with its regulatory ligand. © 2006. Reprinted with permission from AAAS.

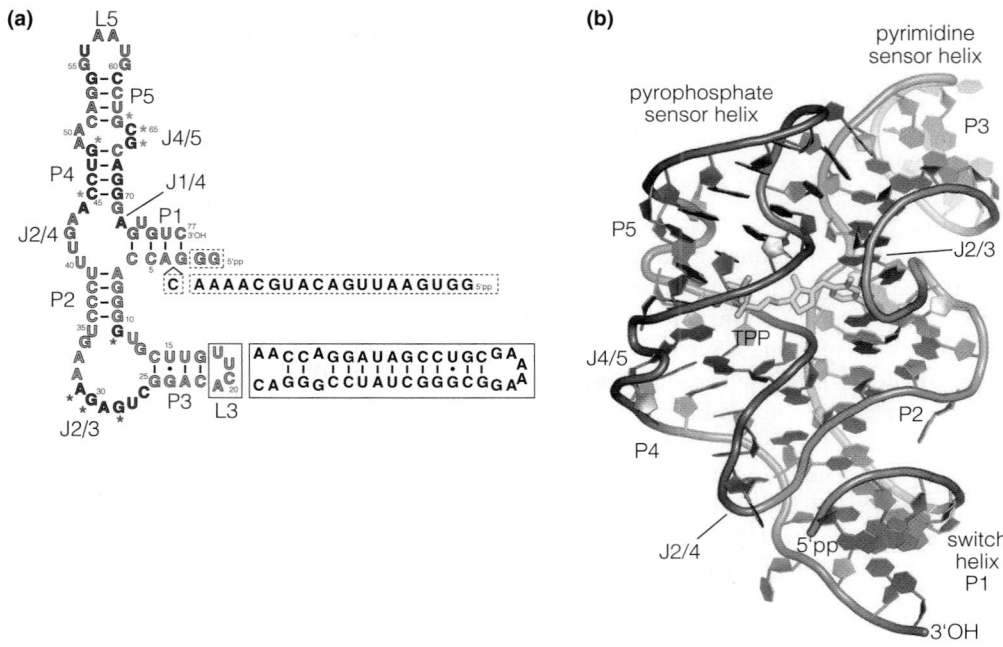

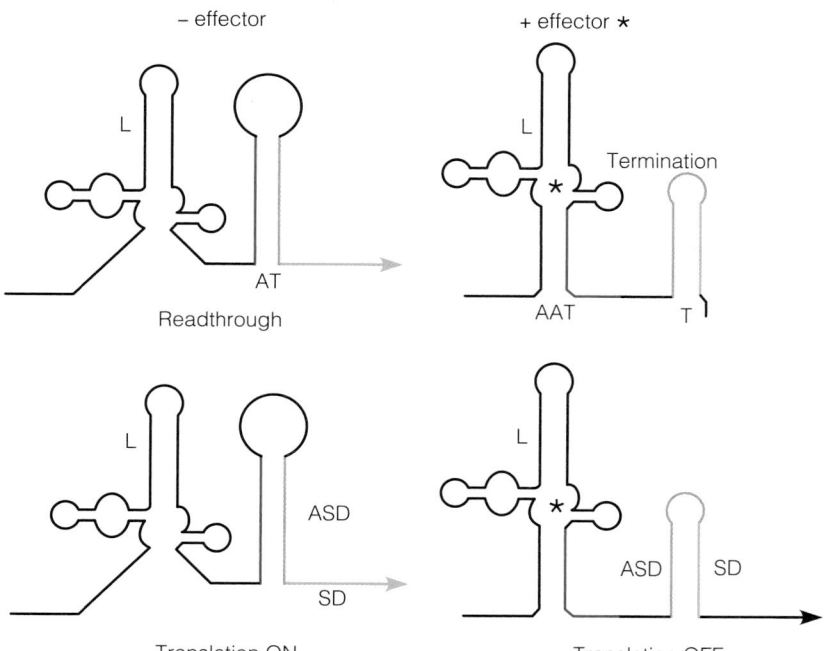

- effector

+ effector ★

Readthrough

Translation ON

Termination

AAT   T

Translation OFF

FIGURE 29.40

Mechanisms of action of a riboswitch. In the absence of effector (upper left) the ligand-binding site (L) is unoccupied, and a transcriptional antiterminator (AT) can form (see Figure 29.19, mechanism of attenuation). Alternatively, if the riboswitch functions at the level of translation (lower left), a region containing the Shine–Dalgarno sequence (SD) is not paired with its complement (anti-SD, or ASD), and translation can occur. In the presence of an effector, the mRNA tertiary structure can change to pair the antiterminator sequence with its complement (AAT), allowing a terminator loop (T) to form, leading to transcriptional termination (upper right). Alternatively (lower right), binding the ligand permits the Shine–Dalgarno sequence to pair with its complement (ASD) so that translation cannot initiate.

Genes & Development 22:3383–3390, T. M Henkin, © 2008 Cold Spring Harbor Laboratory Press. Modified by permission of Tina M. Henkin.

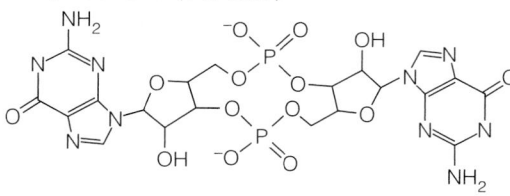

Cyclic di-GMP (c-di-GMP)

FIGURE 29.41

Control of transcription of a flagellar protein operon by a riboswitch for the regulatory nucleotide c-di-GMP. Binding c-di-GMP to the riboswitch changes long-range interactions in the 5′ end of the riboswitch, involving sequences P2, P3, and the antiterminator, creating a transcription terminator and blocking transcription of an operon that encodes proteins involved in producing the flagellum.

Modified with permission from *Microbe* 5:13–20, R. R. Breaker, RNA second messengers and riboswitches: Relics from the RNA world? © 2010 American Society for Microbiology.

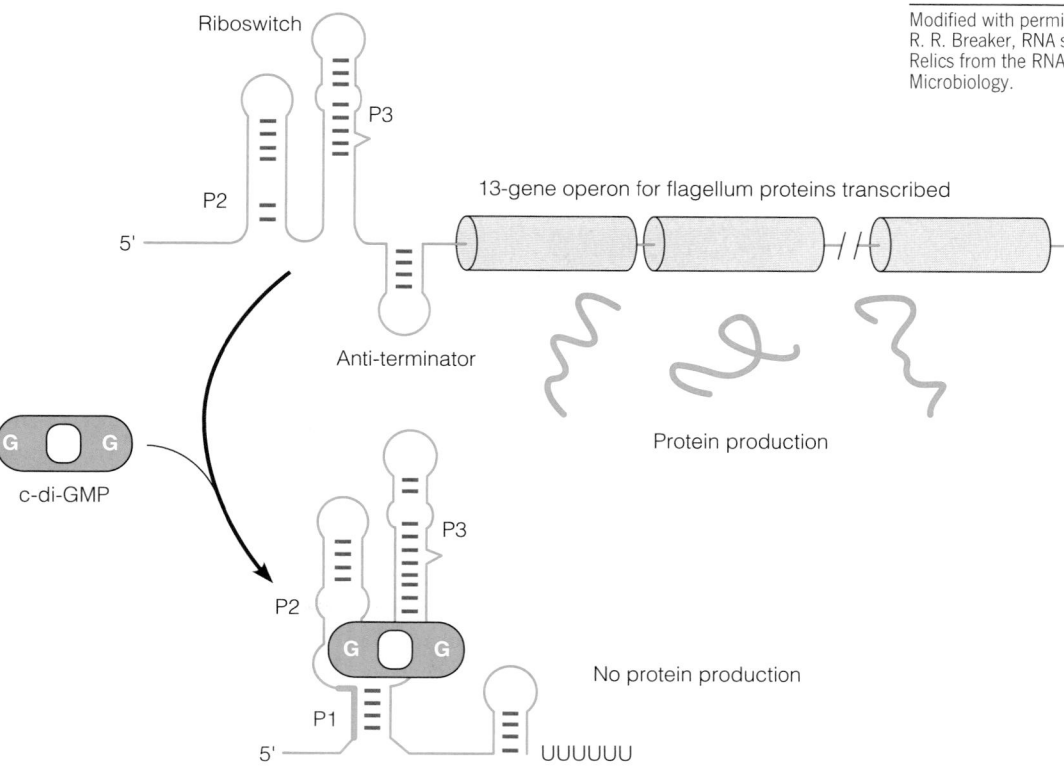

Riboswitch

P3

P2

5′

13-gene operon for flagellum proteins transcribed

Anti-terminator

Protein production

c-di-GMP

No protein production

P3

P2

P1

5′

Terminator

UUUUUU

of UMP residues. In some cases as many as half of the nucleotides in the mature mRNA molecule were U's that had been inserted. Obviously the complete sense of the genetic message is changed. However, the process is not random. The process is directed by *guide RNAs*, which have 5′ ends complementary to one end of the molecule to be edited, followed by a sequence of nucleotides identical to that which is to be inserted. Each insertion step of a single nucleotide involves cleavage, insertion, and re-ligation of the RNA chain. The biological significance of this seemingly wasteful process is still not known.

RNA editing in mammalian cells is much different, simply involving enzymatic deamination of selected AMP or CMP RNA nucleotides. As many as 1000 mammalian genes may be subject to RNA editing. The deaminase enzymes involved recognize a double-stranded RNA that is formed between the site that is to be edited and a complementary sequence elsewhere in the molecule, typically in a downstream intron. As we described in Chapter 17, the mRNA for apolipoprotein B undergoes a CMP to UMP deamination at a site that yields a premature stop codon. Hence, the apolipoprotein product is shorter in the gut (where editing occurs) than in the liver (where it does not). As a result, the lipoproteins produced in these two tissues have somewhat different properties (see Figure 17.5, page 713).

An interesting example of RNA-AMP deamination occurs in the mRNA for a neurotransmitter-gated ion channel in the brain. The edit converts a glutamine to an arginine codon; the affected amino acid lies on the inner wall of the channel and affects the calcium permeability of the channel. Although the reason for the edit is not clear, its significance was shown in knockout mice lacking the deaminase enzyme responsible for the edit. These mice were subject to repeated epileptic seizures and early death. When the gene for the ion channel was modified to contain the arginine without need for deamination of a Glu codon, the mice developed normally. So, even though the purpose of the editing step is unknown, its indispensability is established.

Editing is used by lymphocytes as a defense against HIV infection. These cells carry out large-scale deamination of CMP residues to UMP, using an enzyme similar to activation-induced deoxycytidine deaminase (AID), which is responsible for somatic cell hypermutation (page 1109). Because the RNA in this case is the genome of the virus, the extensive deamination creates multiple mutations, which cripple or kill the virus. However, the virus has evolved a defense mechanism of its own—a protein inhibitor of the deaminase, which is carried into the infected cell along with the viral genome.

SUMMARY

Gene expression is controlled primarily at the level of transcription. In bacteria, control is exercised by binding of repressors and transcriptional activators at promoter sites, thereby influencing the binding of RNA polymerase. Many or most genes are expressed as operons, in which as many as a dozen proteins or more can be translated from the same multicistronic mRNA. A common structural motif in sequence-specific prokaryotic proteins is the helix–turn–helix, in which amino acids in a recognition helix form specific contacts with nucleotides in the major groove of target DNA. The helix–turn–helix is also involved in eukaryotic gene regulation, along with additional motifs, including helix–loop–helix, leucine zipper, and zinc finger. Gene regulation in eukaryotes also occurs at the level of transcription initiation, but regulation is much more complex. As many as 50 proteins may have to assemble at a promoter and upstream enhancer in order for RNA polymerase II to bind and transcribe a protein-coding gene. These proteins include sequence-specific activators, bound at enhancers; Mediator, which connects enhancers to the transcription machinery; chromatin-remodeling complexes, which clear the transcription start region for binding RNA polymerase; and general transcription factors. Time-dependent changes in the phosphorylation pattern of the RNA polymerase II carboxy terminal domain control timing of events in transcription. Important RNA-based regulatory processes have come to light in recent years, most notably RNA interference. Processing of small RNAs to short duplex RNAs and then to single-stranded miRNAs is becoming widely recognized as a translational regulatory mechanism. Other short RNAs, siRNAs, are formed by similar processes, but are used to degrade invading RNAs, such as those introduced by viral infection. Riboswitches are RNAs that control biosynthetic pathways by specific and tight binding of metabolic end products of those pathways.

REFERENCES

Prokaryotic Transcription

Albright, R. A., and B. W. Matthews (1998) How Cro and λ-repressor distinguish between operators: The structural basis underlying a genetic switch. *Proc. Natl. Acad. Sci. USA* 95:3431–3436. Structural analysis of DNA–protein complexes containing the same operator but different repressors.

Henkin, T. (1994) tRNA-directed transcription termination. *Mol. Microbiol.* 13:381–387. A description of a novel form of transcriptional regulation.

Ishihama, A. (2010) Prokaryotic gene regulation: Multifactor promoters, multitarget regulators and hierarchic networks. *FEMS Microbiol. Rev.* 34:628–645. Complexities of prokaryotic transcription revealed by genomic analysis.

Landick, R. (1999) Shifting RNA polymerase into overdrive. *Science* 284:598–599. A brief commentary, reviewing recent work on the mechanism of antitermination.

Lewis, M., G. Chang, N. C. Horton, M. A. Kercher, H. C. Pace, M. A. Schumacher, R. G. Brennan, and P. Lu (1996) Crystal structure of the lactose operon repressor and its complexes with DNA and inducer. *Science* 271:1247–1254. A significant achievement, given the size and importance of the protein.

Ptashne, M. (2004) *A Genetic Switch, Third Edition. Phage Lambda Revisited.* Cold Spring Harbor Laboratory Press, Cold Spring Harbor, NY. A clear account of the complexities of phage λ regulation and the importance of understanding it.

Sorek, R., and P. Cossart (2010) Prokaryotic transcriptomics: A new view on regulation, physiology, and pathogenicity. *Nature Rev. Gen.* 11:9–16. Applications of next-generation DNA sequencing to bacterial transcriptional regulation.

Yaniv, M. (2011) The 50th anniversary of the publication of the operon theory in the *Journal of Molecular Biology*: Past, present, and future. *J. Mol. Biol.* 409:1–6. A retrospective on a famous 1961 paper by Jacob and Monod, which was the first detailed presentation of the operon model.

Yanofsky, C. (2003) Using studies on tryptophan to answer basic biological questions. *J. Biol. Chem.* 278:10859–10878. A brief autobiography in which the author describes the many ways that the *trp* operon has illuminated genetic regulation.

Eukaryotic Transcriptional Regulation

Cook, P. R. (2010) A model for all genomes: The role of transcription factories. *J. Mol. Biol.* 395:110. Not a study of regulation *per se*, but a source of important insights into the intracellular organization of the transcription machinery.

D'Alessio, J. A., K. J. Wright, and R. Tjian (2009) Shifting players and paradigms in cell-specific transcription. *Mol. Cell* 36:924–931. New information about transcription factors.

Kornberg, R. D. (2005) Mediator and the mechanism of transcriptional activation. *Trends in Biochem. Sci.* 30:235–239. One of several articles in a special issue devoted to Mediator.

Pomerantz, R. T., and M. O'Donnell (2010) Direct restart of a replication fork stalled by a head-on RNA polymerase. *Science* 327:590–592. A study of encounters between DNA and RNA polymerases.

Ptashne, M., and A. Gann (2002) *Genes and Signals*. Cold Spring Harbor Laboratory Press, Cold Spring Harbor, NY. A clear account of the complexities of eukaryotic transcription, with emphasis on yeast.

Schmid, D., and 12 coauthors (2010) Five-vertebrate ChIP-seq reveals the evolutionary dynamics of transcription factor binding. *Science* 328:1036–1040. Transcription factor binding shown to be a major factor in interspecies variation.

Venters, B. J., and B. F. Pugh (2009) How eukaryotic genes are transcribed. *Crit. Rev. Biochem. Mol. Biol.* 44:117–141. A comprehensive and timely review.

Weake, V. M., and J. L. Workman (2010) Inducible gene expression: Diverse regulatory mechanisms. *Nature Rev. Genet.* 11:426–437. Another comprehensive review.

Regulatory RNA Molecules

Nagano, T., and P. Fraser (2011) No-nonsense functions for long noncoding RNAs. *Cell* 145:178–181. A recent review.

Sharp, P. A. (2009) The centrality of RNA. *Cell* 136:577–580. A brief review summarizing all of the recently discovered classes of regulatory RNAs.

Waters, L. S., and G. Storz (2009) Regulatory RNAs in bacteria. *Cell* 136:615–628. Comprehensive focus on sRNAs and other prokaryotic regulatory RNAs.

Wilusz, J. E., H. Sunwoo, and D. L. Spector (2009) Long noncoding RNAs: Functional surprises from the RNA world. *Genes. Dev.* 23:1494–1504. Surprising findings, beginning with the realization that most of the eukaryotic genome is transcribed.

Translation-Level Regulation

Hernández, G., M. Altmann, and P. Lasko (2009) Origins and evolution of the mechanisms regulating translation initiation in eukaryotes. *Trends in Biochem. Sci.* 35:63–73. A timely review.

Chromatin and Gene Regulation

Clapier, C. R., and B. R. Cairns (2009) The biology of chromatin remodeling complexes. *Ann. Rev. Biochem.* 78:273–304. Contemporary discussion of the functions in which these complexes participate.

Gaspar-Maia, A., and 10 coauthors (2009) Chd1 regulates open chromatin and pluripotency of embryonic stem cells. *Nature* 460:863–870. Newly discovered relationship between a chromatin-remodeling complex and the pluripotency of embryonic stem cells.

Ho, L., and G. R. Crabtree (2010) Chromatin remodeling during development. *Nature* 463:474–484. A complete and contemporary review.

Lorch, Y., B. Maier-Davis, and R. D. Kornberg (2010) Mechanism of chromatin remodeling. *Proc. Natl. Acad. Sci. USA* 107:3458–3462. Evidence supporting a model of disruption of DNA–histone links in chromatin remodeling.

Oliver, S. S., and J. M. Denu (2011) Dynamic interplay between histone H3 modifications and protein interpreters: Emerging evidence for a "histone language." *Chem. Bio. Chem.* 12:299–307. Thoughtful discussion of the histone code.

Smith, E., and A. Shilatifard (2010) The chromatin signaling pathway: Diverse mechanisms of recruitment of histone-modifying enzymes and varied biological outcomes. *Mol. Cell* 40:689–701. Critical discussion of the concept of a histone code.

Talbert, P. B., and S. Henikoff (2010) Histone variants—ancient wrap artists of the epigenome. *Nature Rev. Mol. Cell Biol.* 11:264–275. Excellent review of histone variants.

DNA Methylation and Epigenetics

Bhutani, N., J. J. Brady, M. Damian, A. Sacco, S. Y. Corbel, and H. M. Blau (2010) Reprogramming toward pluripotency requires AID-dependent DNA demethylation. *Nature* 463:1042–1048. Recent evidence for DNA demethylation as a crucial event in gene reactivation.

Bonasio, R., S. Tu, and D. Reinberg (2010) Molecular signals of epigenetic states. *Science* 330:612–616. Critical discussion of the meaning of "epigenetics," in a special issue of *Science* devoted to epigenetics.

Iqbal, K., S-G. Jin, G. P. Peifer, and P. E. Szabo (2011) Reprogramming of the paternal genome upon fertilization involves genome-wide oxidation of 5-methylcytosine. *Proc. Natl. Acad. Sci. USA* 108:3542–3647. Title is self-explanatory.

Jia, D., R. Z. Jurkowska, X. Zhang, A. Jelsch, and X. Cheng (2007) Structure of Dnmt3a bound to Dnmt3L suggests a model for *de novo* DNA methylation. *Nature* 449:248–253. Structural insights into DNA methylation specificity.

Karberg, S. (2009) Switching on epigenetic therapy. *Cell* 139:1029–1031. One of several short reviews indicating how understanding epigenetic mechanisms might help in treating and preventing disease.

Law, J. A., and S. E. Jacobsen (2010) Establishing, maintaining and modifying DNA methylation patterns in plants and animals. *Nature Rev. Genet.* 11:204–220. A recent and comprehensive review of the functions of DNA methylation.

Lister, R., and 17 coauthors (2009) Human DNA methylomes at base resolution show widespread epigenomic differences. *Nature* 462:315–322. Methylation at the sequence level reveals important information about pluripotent stem cells.

Popp, C., and seven coauthors (2010) Genome-wide erasure of DNA methylation in mouse primordial germ cells is affected by AID deficiency. *Nature* 463:1101–1106. DNA demethylation is shown to be crucial for induction of a dedifferentiated state.

Song, J., O. Rechkoblit, T. H. Bestor, and D. J. Patel (2011) Structure of DNMT1-DNA complex reveals a role for autoinhibition in maintenance DNA methylation. *Science* 331:1036–1039. Title is self-explanatory.

Homeotic Genes and the Homeo Box

Gehring, W. J., M. Affolter, and T. Buerglin (1994) Homeodomain proteins. *Annu. Rev. Biochem.* 63:487–526. A still-timely review.

Kessel, M., and P. Gruss (1990) Murine developmental control genes. *Science* 249:373–379. Homeo boxes in mammalian development.

RNA Interference

Bonetta, L. (2009) RNA-based therapeutics: Ready for delivery? *Cell* 136:581–584. Excitement about therapeutic uses of RNAi.

Djuranovic, S., A. Nahvi, and R. Green (2011) A parsimonious model for gene regulation by miRNAs. *Science* 331:550–553. A short contemporary review.

Inui, M., G. Martello, and S. Piccolo (2010) MicroRNA control of signal transduction. *Nature Rev. Mol. Cell Biol.* 11:252–263. Specific roles for some miRNAs.

Krol, J., I. Loedige, and W. Filipowicz (2010) The widespread regulation of microRNA biogenesis, function and decay. *Nature Rev. Genet.* 11:597–610. Comprehensive recent review of RNAi.

Poliseno, L., L. Salmena, J. Zhang, B. Carver, W. J. Haveman, and P. P. Pandolfi (2010) A coding-independent function of gene and pseudogene mRNAs regulates tumour biology. *Nature* 465:1033–1040. An evolutionarily retained function for pseudogenes.

Voorhoeve, P. M. (2010) MicroRNAs: Oncogenes, tumor suppressors, or master regulators of cancer heterogeneity? *BBA Reviews of Cancer* 1805:72–86. Relationships between RNAi and cancer.

Williams, A. H., and 8 coauthors (2009) MicroRNA-206 delays ALS progression and promotes regeneration of neuromuscular synapses in mice. *Science* 326:1549–1553. Identification of a specific microRNA that regulates nervous system function.

Wum L., H. Zhou, Q. Zhang, J. Zhang, F. Ni, C. Liu, and Y. Qi (2010) DNA methylation mediated by a microRNA pathway. *Mol. Cell* 38:465–475. A specific role for miRNA in plants.

Riboswitches

Breaker, R. R. (2010) RNA second messengers and riboswitches: Relics from the RNA world? *Microbe* 5:13–20. The author argues that riboswitches constitute evidence for a primordial RNA world.

Henkin, T. M. (2008) Riboswitch RNAs: Using RNA to sense cellular metabolism. *Genes and Development* 22:3383–3390. A comprehensive review.

Roth, A., and R. R. Breaker (2009) The structural and functional diversity of metabolite-binding riboswitches. *Ann. Rev. Biochem.* 78:305-334. A recent review of riboswitches from their discoverer.

Zhang, J., M. W. Lau, and A. R. Ferré-D'Amare (2010) Ribozymes and riboswitches: Modulation of RNA functions by small molecules. *Biochemistry* 49:9123–9131. Focus on structural features of small molecule binding to RNA.

RNA Editing

Brennicke, A., A. Marchfelder, and S. Binder (1999) RNA editing. *FEMS Microbiol. Revs.* 23:297–316. An excellent review of the older literature.

Maas, S., S. Patt, M. Schrey, and A. Rich (2001) Underediting of glutamate receptor GluR-B mRNA in malignant glioma. *Proc. Natl. Acad. Sci. USA* 98:14687–14692. Severe medical consequences of defective RNA editing in humans.

Wulff, B-E., M. Sakurai, and K. Nishikura (2011) Elucidating the inosinome: Global approaches to adenosine-to-inosine signaling. *Nature Rev. Genet.* 12:81–85. Review of an important process in RNA editing.

PROBLEMS

1. The active form of lactose repressor binds to the operator with a dissociation constant of 10^{-13} M for the reaction R+O $\rightleftharpoons$ RO. About 10 molecules per *E. coli* cell suffice to keep the operon turned off in the absence of inducer.
   (a) If the average *E. coli* cell has an intracellular volume of 0.3×10^{-12} mL, calculate the approximate intracellular concentration of repressor.
   (b) If the average cell contains two copies of the *lac* operon, calculate the approximate intracellular concentration of operators.
   (c) Calculate the average intracellular concentration of *free* operators under these conditions.
   (d) Explain how a cell with a haploid chromosome could contain an average of two copies of the *lac* operon.

2. Is attenuation likely to be involved in eukaryotic gene regulation? Briefly explain your answer.

*3. Suppose you want to study the transcription in vitro of one particular gene in a DNA molecule that contains several genes and promoters. Without adding specific regulatory proteins, how might you stimulate transcription from the gene of interest relative to the transcription of the other genes on your DNA template? To make all of the complexes identical, you would like to arrest all transcriptional events at the same position on the DNA template before isolating the complex. How might you do this?

4. For some time it was not clear whether *lac* repressor inhibits *lac* operon transcription by inhibiting the binding of RNA polymerase to its promoter or by allowing transcription initiation but blocking elongation past the site of bound repressor. How might you distinguish between these possibilities?

5. A *lac* operon containing one mutation was cloned into a plasmid, which was introduced by transformation into a bacterium containing a wild-type *lac* operon. The three genes of the chromosomal operon were rendered noninducible in the presence of the plasmid.
   (a) What kind of mutation in the plasmid operon could have this effect?
   (b) Suppose the result of transformation was to cause the three plasmid *lac* genes to be expressed constitutively, at a high level. What type of plasmid gene mutation could have this result?

*6. Several new genes in the SOS regulon were identified by an ingenious use of "Mud" phages. These are derivatives of phage Mu that have a promoterless β-galactosidase gene inserted at a particular point in this phage genome. How might these phages be used to identify genes whose expression is turned on after ultraviolet irradiation of bacteria?

7. Partial diploid forms of *E. coli* were created, each of which contained a complete lactose operon at its normal chromosomal site and the regulatory sequences only *(i, p, o)* on a plasmid. Predict the effect of each mutation upon the activity of β-galactosidase before and after the addition of inducer. Use $-$, $+$ or $++$ to indicate approximate activity levels.

Mutation	Before	After
(a) No mutations in either chromosomal or plasmid genes	_____	_____
(b) A mutation in the plasmid operator, which abolishes its binding to repressor	_____	_____
(c) A mutation in the chromosomal promoter, which reduces affinity of promoter for RNA polymerase by 10-fold	_____	_____
(d) An *i* gene mutation in the chromosome, which abolishes binding of the *i* gene product to inducer	_____	_____
(e) An *i* gene mutation in the plasmid, which abolishes binding of the repressor to the inducer	_____	_____
(f) A chromosomal *o* mutation, which abolishes its binding to repressor	_____	_____
(g) A mutation in the gene for CRP, which abolishes its binding to cyclic AMP	_____	_____

*8. What type of mutation of the *lac* repressor might be both constitutive and *trans*-dominant?

9. It has been proposed that thiogalactoside transacetylase (LacA in the lactose operon) plays a role in detoxification—ridding the cell of potentially toxic β-galactosides by acetylating them to facilitate their diffusion out of the cell. How might you test this proposal?

10. Consider Figure 29.34. Note that accumulation of EF-G and the α subunit of RNA polymerase are subject to translational repression by ribosomal proteins S7 and S4, respectively. This would seem to be disadvantageous to the organism because S4, S7, and S8 binding to ribosomal mRNAs are mechanisms to control ribosome assembly. Why might this phenomenon *not* be disadvantageous?

11. Riboswitches are generally considered to have been discovered in about 2002. But a comparable regulatory process was described much earlier, when Nomura et al. ([1980] *Proc. Natl. Acad. Sci. USA* 77:7084) described the regulation of ribosomal protein synthesis carried out by binding of ribosomal proteins to their own mRNAs (Figure 29.35). Was the riboswitch actually discovered years earlier? Discuss similarities and differences in control of ribosomal protein synthesis and riboswitch regulation as discussed in this chapter.

12. Describe experimental evidence that would indicate that most or nearly all of the DNA sequences in a mammalian genome are transcribed.

ANSWERS TO PROBLEMS

CHAPTER 2

1. (a) 3.54 kJ/mol.
   (b) 151 kJ/mol.

2. $CCl_4 < H_2S < H_3\overset{+}{N}CH_2COO^- < H_3\overset{+}{N}CH_2CH_2CH_2COO^-$. CCl_4 is symmetrical, $\mu = 0$; H_2S will be comparable to H_2O; the latter two involve separation of whole charges, and in the last, separation is greatest.

3. The graph will be a reflection of the graph given, across the $E = 0$ line.

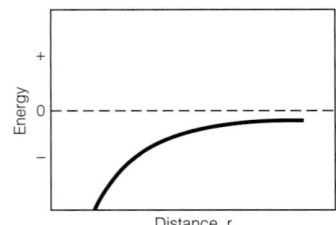

4. (a) $K_a = \dfrac{[H^+][A^-]}{[HA]} = \dfrac{x^2}{A_0 - x}$

   $$K_aA_0 - K_ax = x^2$$

   or $x^2 + K_ax - K_aA_0 = 0$

   $$x = \dfrac{-K_a \pm \sqrt{K_a^2 + 4K_aA_0}}{2}$$

   (b) First, assume $x << A_0$ and obtain

   $$x^2 \cong K_aA_0, \qquad x \cong \sqrt{K_aA_0}$$

   Use this approximate x to obtain a better approximation to the denominator $(A_0 - x)$. Repeat calculation until change in each step is less than required precision.

   If K_a is very small (acid weak), either the above or direct approximation yields

   $$[H^+] \cong \sqrt{K_aA_0}$$

5. (a) pH = 0.456.
   (b) pH = 2.608 by approximation, 2.609 exact.
   (c) pH = 3.108 by approximation, 3.113 exact.

6. (a) $K_a = 8.16 \times 10^{-5}$.
   (b) pH = 2.40.

7. Choose a series of about 10 points at which volumes of KOH have been added to neutralize increasing fractions of the acid. Use the Henderson–Hasselbalch equation to calculate the pH at each point.

8. (a) pH = 4.46.
   (b) pH = 2.57.

9. (a) 0.138 M.
   (b) $[KH_2PO_4] = 0.126$ M; $[Na_2HPO_4] = 0.174$ M.

10. pH = 3.93.

11. The best choice would be a mixture of $H_2PO_4^-$ and HPO_4^{2-}, which has $pK_a = 6.86$.

12. One will dissolve 2×7.507 g of glycine in the 2 L to make a 0.1 M solution. To obtain a pH of 9.0, the Henderson–Hasselbalch equation states that the ratio of $Gly^-/Gly^{\pm}$ must be 0.251. This can be achieved by adding NaOH to a concentration of 0.02 M. For 2 L, this would require 0.04 mol, or 40 mL of 1 M NaOH.

13. 8.5% as H_2CO_3. Very little CO_3^{2-}.

14. (a) $B = (A_0/2.303)f(1 - f)$.
    (b) When $f = 0.5$.
    (c) $A_0/9.2/2$.
    (d) Increasing A_0 increases B.

15. Protein molecules in aqueous solution become increasingly protonated as the pH decreases. Thus, proteins become more positively charged because carboxylic acids become *less negatively charged* as pH drops, whereas amines become *more positively charged*. Proteins become more negatively charged as pH increases, because acidic groups become more negatively charged while the basic groups become less positively charged. These effects are summarized in Table 2.9.

16. (a) pI = 11.1. Form I will be present at insignificant concentration at pH values above 7.
    (b) +0.38
    (c) Yes. When pH < pI the molecule is predicted to carry a positive charge.

17. Because of the proton-attracting power of the dianion produced when two of the citric acid protons have dissociated, the third is held anomalously strongly. In fact, it has a pK_a of 6.86.

18. The ionic strength of 1 M $(NH_4)_2SO_4$ is 3.0.

19. Lysine has a pI of about 9.5. Thus, it would not migrate at pH 9.5, whereas cysteine with a side chain pK_a of 8.3 would carry a partial negative charge at this pH, and arginine, with its very high side chain pK_a (12.5) would be strongly positively charged.

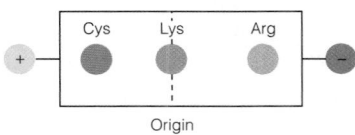

20. (a) Isoelectric focusing. The single charge difference would make little difference in mobility, but a detectable difference in pI.
    (b) At a pH below 6.4 (e.g., pH 5) such that the His sidechain is highly (+) charged, whereas the Val sidechain would carry no charge.

CHAPTER 3

1. $\Delta H_f - \Delta U_f = -0.165$ J/mol, thus, $\Delta U_f = 6,010.165$ J/mol,
$\left(\dfrac{\Delta H_f - \Delta U_f}{\Delta U_f}\right) \times 100 = 0.003\%$.

2. (a) $+926$ kJ/mol.
(b) 463 kJ/mol.

3. (a) $\Delta S° = +470$ J/K·mol.
(b) A solid material is being transformed entirely into gases. The greater freedom of motion of gas molecules results in an entropy increase.
(c) $\Delta U° = 103.4$ kJ/mol.
(d) Because the system does work on the surroundings, in expanding from a solid to a gas.

4. (a) $\Delta G° = -2872$ kJ/mol.
(b) $\Delta G° = -1896$ kJ/mol.
(c) 34%.

5. (a) 1.18×10^{-4} mM.
(b) $\Delta G°' = -16.7$ kJ/mol.
(c) 24.6 M. It is never reached because glucose-6-phosphate is continually consumed in other reactions. The system never reaches true thermodynamic equilibrium.

6. (a) $K = 5.44 \times 10^{-2}$; $(f_{G3P})_{eq} = 0.052$.
(b) $\Delta G = -4.37$ kJ/mol.

7. (a) ΔS must be positive because the increase in available states corresponds to an increase in entropy.
(b) Since $\Delta G = \Delta H - T\Delta S$, a positive ΔS yields a negative contribution to ΔG (T is always a positive number). Thus, for proteins to be stable, which requires ΔG for the above to be positive, denaturation must involve a large positive ΔH and/or an additional negative contribution to ΔS. As we shall see in Chapter 6, both occur.

8. This process should correspond to an entropy decrease ($\Delta S < 0$), since order is being established in the H_2O structure.

9. (a) We expect ΔS for denaturation to be positive order is lost. If ΔH is also positive (energy required to break internal bonds), then we have the $+$, $+$ situation described in the first entry in Table 3.3 (p. 68).
(b) If many hydrophobic residues are present in a protein, ΔS could be negative for unfolding (see Problem 8). If ΔH were also negative, we would have the $-$, $-$ situation in Table 3.3.

10. Given $\Delta G° = \Delta H° - T\Delta S°$ and $\Delta G° = -RT \ln K$, combination and rearrangement gives the desired equation, which is called the van't Hoff equation. According to the van't Hoff equation, if $\Delta H°$ and $\Delta S°$ are independent of temperature, a graph of $\ln K$ versus $1/T$ should be a straight line with a slope of $-\Delta H°/R$.

11. (a) $\Delta H° = +59.0$ kJ/mol.
(b) $\Delta S° = -103$ J/K·mol. To obtain the true K, so as to get $\Delta G°$, one must divide K_W by 55.5, the molar concentration of pure water.

12. (a) $K = 17.6$.
(b) Toward glucose-1-phosphate.

13. (a) 2.18×10^3 kJ.
(b) ~71.5 moles.

14. 217 g.

15. For a mole of protein molecules, $\Delta S = R \ln W - R \ln 1$, where W is the number of conformations available to each, and R is the gas constant, 8.314 J/K·mol. Because there are 99 bonds between 100 residues, $W = 3^{99}$.
(a) $\Delta S = 9.04 \times 10^2$ J/K·mol.
(b) $\Delta H = 292$ kJ/mol.
(c) Increase, because ΔS and ΔH are both positive.

16. (a) 5.93×10^{-6} kJ.
(b) -5.93×10^{-6} kJ.
(c) In (a) 1.94×10^{-7} moles ATP, in (b) none.

17. (a) 0.527 M.
(b) -62.4 kJ/mol.

CHAPTER 4

1. (a) It must be a single-strand DNA, since Chargaff's rules are not obeyed.
(b) It should be mostly random coil, with perhaps some self-bonding into hairpins, etc.

2. (a) 3′TGGCATTCCGAAATC5′.
(b) 5′pApCpCpGpTpApApGpGpCpTpTpTpApGp3′.
(c) 3′UGGCAUUCCGAAAUC5′.

3. The center of symmetry is indicated on the figure at right by a large black dot.

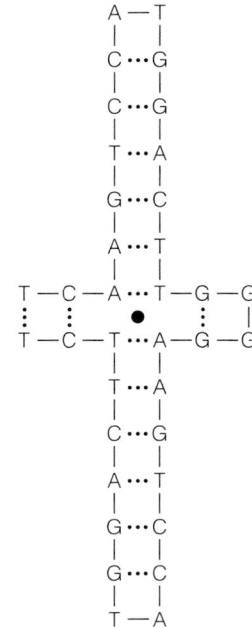

4. $\sigma = \Delta L/L_0 = -0.06$ $\Delta L = -0.06\, L_0$

In *E. coli* $L_0 = 4.639 \times 10^6$ bp (1 turn $= 10.5$ bp) $= 441.8 \times 10^3$ turns

$\Delta L = \Delta T + \Delta W, \qquad \Delta T = 0$

$\Delta W = \Delta L = 0.06\, L_0 = -0.06(441.8 \times 10^3 \text{ turns}) = -2.65 \times 10^4$

5. ^{5}G—U—C—C—A—G—C—C—A—U—U—G—C—G—U

     ^{3}C—G—G—U—A—A—C—G—C—U

6. 5150 bp $\times$ 0.255 nm/bp $= 1313$ nm (A form)

7. (a) $L_0 = 2100$ bp $\times$ 1 turn/10.5 bp $= 200$ turns.

(b) $\sigma = -\Delta L/L_0 = -12/200 = -0.06$.

8. 20 base pairs corresponds to 2 turns of B-DNA at $+1$ twist/turn, and 1.7 turns of Z-DNA at -1 twist/turn. Therefore, (a) $\Delta T = -3.7$, (b) $\Delta L = 0$, (c) $\Delta W = +3.7$.

9. The Epstein–Barr virus should be highest; the bacteriophage T4 should be the lowest.

10. (a) The DNA must be single-stranded.

(b) There must be small regions of self-complementary base pairing.

11. Approximately 60% GC.

12. (a) It must require 3 turns in one repeat, to give an integral number. The number $= 28$ bp/repeat.

(b) More tightly.

(c) Disfavor.

13. The distribution indicates the presence of superhelical molecules with $\Delta L = \pm 1, \pm 2, \pm 3$, etc. These could form as a consequence of fluctuations in structure at the moment when the circle is sealed. They pro-vide a direct demonstration of the Boltzmann distribution of molecular energies.

14. (a) Positive, because L is fixed ($= 0$), and ΔT is made negative by the ethidium.

(b) None.

(c) Positive writing caused by ethidium decreases the negative writing until $W = 0$; this structure will have a minimum mobility because it is more extended than the writhed forms. As more ethidium is added, the DNA writhes positively, again increasing mobility.

15. Initial attack of OH^- on a P atom in a phosphodiester bond gives rise to formation of a cyclic $2', 3'$ phosphodiester intermediate. This unstable intermediate undergoes hydrolysis, giving either a $2'$ or $3'$ phosphomonoester. Thus, alkaline hydrolysis of RNA gives a mixture of ribonucleoside $2'$ and $3'$ phosphates. Because DNA lacks a $2'$ hydroxyl group, the cyclic phosphodiester intermediate cannot form, and the phosphodiester bond is not hydrolyzed.

16. Consider pH 3.8, the pK_a for the AMP amino group. Use the Henderson–Hasselbalch equation to calculate approximate proportions of charged and uncharged amino groups for each nucleotide at that pH.

Nucleotide	Charge on amino group (approx.)	Charge on phosphate group	Net charge
AMP	+0.5	−1.0	−0.5
CMP	+0.7	−1.0	−0.3
GMP	+0.1	−1.0	−0.9
UMP	—	−1.0	−1.0

All migrate toward the anode UMP>GMP>AMP>CMP

CHAPTER 5

1. 109.5.

2. (a) SYSMEHFRWGKPV.

(b) 1623.93 g/mol.

3. (a) The curve will exhibit inflections corresponding to two groups titrating near 4 (carboxyl terminus and glutamic side chain), one group near 7 (histidine), two near 9 (N-terminus and lysine), one near 10 (tyrosine), and one near 12 (arginine).

(b) At pH $= 1$ the overall charge will be $+3.998$; at pH $= 5$ the overall charge will be $+2.104$; at pH $= 11$ the overall charge will be -1.827.

(c) pI $= 8.813$.

4. (a) SYSMEHFR, WGKPV.

(b) SYSM*, EHFRWGKPV; M* $=$ homoserine lactone.

(c) SYS, MEHFRWGKP, V.

5. DSGPYKMEHFRWGSPPKD.

6. (a) ~ -1 at pH 7; ~ -4 at pH 12.

(b) (1) 2; (2) no cleavage; (3) 3.

(c) Electrophoresis, or chromatography on a cationic column, at pH 7.

7. Asp–Arg–Val–Tyr–Ile–Met–Pro–Phe.

8. AC.

9. (a) 2.

(b) Between the first and second, and third and fourth cysteines.

10. (a) One possibility, of many:

$\cdots$ UGUAAUUGUAAAGCGCCCGAGACCGCGCUUU

         GUGCUCGACGAUGUCAACAACAU$\cdots$

(b) Because it does not have an N-terminal methionine, at least some proteolytic cleavage must be involved in its synthesis.

11. Met–Phe–Pro–Ser–Tyr–Pro–Lys–Asp–Lys–Lys–Glu–

12. (a) Either the N-terminus is blocked (as by acetylation, for example) or the peptide is cyclic.

(b) If the peptide is not cyclic, there should be a free C-terminus. In most cases (except if it is Pro) this can be attacked by carboxypeptidase A. Alternatively, if the peptide is cyclic, some protease can probably be found that will cleave it only once, giving a free N-terminus to allow sequencing.

13. Because the A chain contains no basic residues and the B chain has several, either isoelectric focusing, electrophoresis, or ion-exchange chromatography should work well. Note that separation should be best at about pH $= 5$, where the histidines on the B chain are positively charged and glutamates on the A chain are negative.

14. (a) The Glu codons GAG and GAA can be converted to Val codons by substituting a T for the second base to give GTG and GTA, which are Val codons.

(b) The pI of sickle hemoglobin (HbS) is higher than the pI of normal hemoglobin (HbA). The Glu to Val mutation reduces the overall negative charge on the protein; thus, there will be an excess of (+) charge on HbS when the pH of the solution $=$ pI for HbA. If a protein is (+) charged the pH must be BELOW the pI of the protein; thus, the pI of HbS must be above the pI of HbA.

15. Since the mutations replace non-ionizable sidechains with ionizable sidechains, we do not need to know anything about the charge state of the Gln and Phe that are replaced (since they carry no charge in the wt protein). We can then assume that any difference in charge between the wt and the mutant at pH $= 5.5$ is due to the presence of partial charges on the mutant sidechains. Using the Henderson–Hasselbalch equation with the pK_as given in the problem, the charges on the His and Glu sidechains can be calculated at pH $= 5.5$ (the pI of the wild-type

protein). This is done because the charge on the wt protein is zero when pH = 5.5. This analysis yields an overall charge on the mutant of

+0.559 at pH = pI. If a protein is positively charged the pH must be BELOW the pI; thus, the pI of the mutant is greater than 5.5.

CHAPTER 6

1. (a) Left-handed.
   (b) 3.0.

2. If the helix were distorted so as to give 3.5 residues/turn, a leucine would project on the same face every 2 turns. The leucines on the two chains would provide hydrophobic faces for interaction. This structure is called a *leucine zipper*.

3. (a) The four helices could be arranged so that the hydrophobic side chains would all point toward the center of the bundle and would pack together there (see Problem 2). This would give a stabilizing hydrophobic core.
   (b) A proline at this point would break the helix near the Fe_2 binding sites. This would probably mean that Fe_2 could not be bound, and the mutant protein would be nonfunctional.

4. They are spaced about three to four residues apart. Therefore, they will all lie on the same side of the α helix. This suggests that this side of the helix may face the interior of the protein.

5. In the unfolded form of the protein, most of its hydrogen bond donors and acceptors can make H bonds to water. Therefore, it is only the *difference* in hydrogen bonding energy between the folded and unfolded states that contributes to protein stability. The energy required to break the H-bonds in the unfolded state must be subtracted from the energy released when new H-bonds form in the folded state.

6. (a) $3^{200} = 2.7 \times 10^{95}$.
   (b) Not all of these conformations will be sterically possible. But even if only 0.1% of these are allowed, there are still 2.7×10^{92}, a very large number.

7. (a) $\Delta S_{folding} = S_{folded} - S_{unfolded}$
   $$= R \ln W_{folded} - R \ln W_{unfolded}$$
   $$= 8.314 \text{ J/K} \cdot \text{mol} \times [\ln 1 - \ln(2.7 \times 10^{92})]$$
   $$= -1769 \text{ J/K} \cdot \text{mol} = -1.77 \text{ kJ/K} \cdot \text{mol}$$
   (b) $\Delta H_{folding} = 96 \times (-5 \text{ kJ/mol}) = -480 \text{ kJ/mol}$
   (c) $\Delta G_{folding} = \Delta H_{folding} - T \Delta S_{folding}$
   $$= -480 \text{ kJ/mol} - 298 \text{ K}(-1.77 \text{ kJ/K} \cdot \text{mol})$$
   $$= +47 \text{ kJ/mol}$$

   Since $\Delta G_{folding} > 0$ at 25 °C, the protein would not be stable. It would be stable below 0 °C. This points out the importance of sources of stabilization other than backbone H bonds.

8. An α helix from residues 4–11, a β sheet between 14–19 and 24–30. There is very probably a β turn involving residues 20–23.

9. (a) C_4, held together by heterologous interaction, or D_2, held together by isologous interaction.
   (b) C_2, since each $\alpha\beta$ dimer forms an asymmetric unit.

10. (a) C_8, D_4.
    (b) D_4, because it involves more subunit–subunit interactions.
    (c) Both. There must be heterologous interactions about the four-fold axis and isologous interactions about the two-fold axes.

11. The measurement by circular dichroism will detect only the unfolding of tertiary structure in the subunits. Calorimetry will also pick up the energy required to break down the quaternary structure.

12. (a) Using data in Table 5.1, we calculate the molecular weight to be 1072. A 1 mg/cm^3 solution will be 9.33×10^{-4} M. From Figure 5.6, we see that the molar extinction coefficient of Tyr is ~1000 M^{-1}cm^{-1}. Phe does not contribute appreciably. Therefore, ε is 0.93 cm^2/mg.

(b) 2.8 mg/cm^3.
(c) 5%.

13. In the absence of BME, a single band of $M \cong 70,000$ is obtained. This suggests, but does not prove, that there are two identical, noncovalently linked subunits. However, the addition of BME removes this band and gives two bands of $M \cong 30,000$ and $\cong 40,000$, respectively. The sum of these is 70,000, strongly suggesting that the native molecule contains four subunits, two of $M \cong 30,000$, two of $M \cong 40,000$. The 30,000 and 40,000 units are paired by at least one disulfide bond.

14. The fact that disulfide reduction has little effect means that the protein is a single chain. It must have an extended structure, which cleavage at a critical Arg residue can relax, giving faster migration; the fragments are still held together by disulfide bonds. Cleavage of these, after thrombin cleavage, yields two fragments.

15. Assuming that X-ray diffraction is not practical, we have:
    (a) Analysis by CD or by NMR.
    (b) More than one secondary/tertiary folding can be observed for the same sequence. Therefore, sequence alone cannot dictate folding in all cases, and sequence-based predictions must sometimes fail badly.

16. (a) No. DNA is charged and therefore polar, so most of the DNA-binding helix is likely to be composed of polar residues that interact with either the DNA or the solvent (this is supported by sequence analysis of the bHLH family of DNA binding proteins).
    (b) The N-terminus is interacting with the DNA. Two reasons might be given for this: (1) the α-amino group of the N-terminus is positively charged and will interact favorably with the negative charge on the phosphodiester backbone of the DNA; (2) this orientation also situates the "partial positive" end of the helical macrodipole for favorable electrostatic interactions with the negatively charged phosphodiester backbone of the DNA.

17. Consider how the mutation would affect the relative free energies in the folded and unfolded states of the protein. Because the mutation is on the surface, loss of hydrophobic contacts in the protein core is not an issue and the amino acid is likely to be interacting with solvent to similar extents in both folded and unfolded states; thus, effects on ΔH are predicted to be minimal. For the same reason the $\Delta S_{solvent}$ is likely to be small because side chain solvation is predicted to be similar in both the folded and unfolded states (see Equation 6.2b). $\Delta S_{protein}$ changes the most due to the conformational flexibility of Gly compared to Pro. Gly will stabilize both the folded and the unfolded states; however, it stabilizes the unfolded state more due to the dramatic increase in conformational entropy of the unfolded state as a result of this mutation. The stabilization of the unfolded state for the mutant means that $\Delta G_{folding}$ (wt) $< \Delta G_{folding}$(mutant); thus, the mutation is destabilizing.

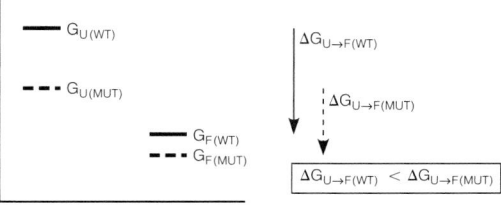

18. The orientation of four contiguous residues is required to initiate one turn of a helical structure (Figure 6.3a). Likewise a minimum of three contiguous residues must be ordered to initiate a turn (Figure 6.21), which could then initiate H-bonding between antiparallel strands.

Nucleation of an antiparallel sheet can therefore be faster than nucleation of a helix. The initiation of a parallel sheet requires a noncontiguous sequence to form H-bonding interactions. Because the effective concentration of contiguous residues is higher than that of noncontiguous residues, the nucleation of a parallel sheet will be significantly slower.

CHAPTER 7

1. (a) ~2.6 mm Hg.
   (b) About 94%.

2. (a) Decrease P_{50}.
   (b) Increase P_{50}.
   (c) Decrease P_{50}.

3. (a) (1) 98%; (2) 56%.
   (b) About 42%.
   (c) About 63%.

4. In the absence of O_2 from the lungs, the residual O_2 bound to hemoglobin is used for continued energy production in respiring cells (such as cardiac and skeletal muscle). Underwater, the crocodile continues to generate CO_2 from metabolic activity. The CO_2 is rapidly converted to HCO_3^- by carbonic anhydrase. As $[HCO_3^-]$ increases, more will bind to hemoglobin, shifting it to greater T-state conformation. This favors O_2 release; thus, as the crocodile stays underwater its hemoglobin delivers most of the bound O_2 as a result of increased binding of HCO_3^- to the T conformation.

5. (a) $P_{50} = 4.467$ mm. (Note: This very low value is a consequence of stripping, plus the fact that the temperature is lower than physiological [37 °C].)
   (b) $h \cong 3.5$.
   (c) $P_{50T} \sim 20.3$ mm Hg; $P_{50R} \sim 1.16$ mm Hg. These are obtained by extrapolating the limiting lines tangential to the extremes of the Hill plot.

6. Hemoglobin concentration can influence affinity only if protein dissociation or association is occurring. Most likely is dissociation to $\alpha\beta$ dimers, which will have lost some of the interactions that stabilize the T state. Thus, ligand-binding affinity should rise and P_{50} should decrease.

7. (a) $P_{50} = 126$ mm Hg.
   (b) $h = \sim 2.6$ (the slope of the Hill plot near $Y_{O2} = 0$ is 2.588).
   (c) Since $h \sim 2.6$, there must be at least 3 sites because $h \leq n$. (Actually, this is an immense molecule, with 24 binding sites.)

8. (a) Because H146 lies in the α/β interface (see Figure 7.15), mutation should be expected to interfere with the $T \rightleftarrows R$ transition. The effect is to increase affinity. This mutation is known; it is hemoglobin Hiroshima.
   (b) Because F8 is involved in heme binding (see Figure 7.17), the heme should be unstable. Leucine will not ligate to the heme iron.
   (c) β_2 His is involved in BPG binding (see Figure 7.23). Changing to Asp would weaken this.
   In each case, a single base change could suffice.

9. (a) Faster.
   (b) Probably same.
   (c) Faster.

10. (a) It would begin with a slope of about 1 at low values of log P_{O_2}, would then show a region with slope less than 1, and approach a line with a slope of 1.
    (b) MWC theory involves switching an equilibrium of all sites between two states—weak and strong. But in KNF theory, one site can modify an adjacent site in either direction—to stronger or weaker binding.

11. A likely explanation: In native Hb, the binding of oxygen is actually hindered by the fact that pulling on helix F must move it against constraints within the molecule. In the imidazole replacement, there is no need to do the extra work of moving helix F. This difference shows up as a more favorable free energy for binding.

12. (a) $\Delta Y_{O_2} = Y_{O_2(\text{lungs})} - Y_{O_2(\text{capillaries})} = \left(\dfrac{(85)^{3.2}}{(31)^{3.2} + (85)^{3.2}} \right) -$
    $\left(\dfrac{(25)^{3.2}}{(31)^{3.2} + (25)^{3.2}} \right) = 0.961 - 0.334 = 0.627$
    (b) Living at 1460 m (~4600 ft), the Dalai Lama has adapted to a lower atmospheric P_{O_2}. These adaptations included elevated [2,3-BPG], which will shift P_{50} to higher values.

13. (a) Yes, because the linkage between subunits is such that forcing one pair of helices apart favors moving the other pair apart, making O_2 binding easier in the second pair.
    (b) Probably so. Deprotonation of His 13 destroys the salt bridge, allowing easier opening of the O_2 binding site.
    (c) The molecule would exhibit higher O_2 affinity, and probably lesser cooperativity because the O_2 sites would be opened further. Possibly, the whole structure would become unstable.

14. (a) If one multiplies both sides of the first equation by $1 + Kc$, divides through by c, and rearranges, the second equation results.
    (b) According to the second equation, the slope of a graph of r/c versus r should be $-K$, and its intercept at $r/c = 0$ should give n. Using the data in this way, we get $K = 2.2 \times 10^4 \, M^{-1}$ and $n = 2.1$. Because the true number of sites must be integral, we would choose $n = 2$.

15. The inter-chain disulfide bond would not form, thus the H and L chains would not be stabilized by a *covalent* bond between them. Depending on the strength of the non-covalent interactions between the chains, the loss of the disulfide could result in dissociation of the chains and loss of the antigen binding.

CHAPTER 8

1. (a) In the relaxed state, there is about 0.6 μm overlap on each side; in the contracted state, about 0.75 μm.
   (b) About 10 steps.

2. (a) 0.0033 min $\cong$ 0.2 s.
   (b) 0.021 min = 1.2 s.
   (c) That creatine phosphate must be continually produced in active muscle.

3. That the growth of actin filaments (which occur at the ends) is somehow essential for cell-shape modification by nonmuscle actin.

4. If the GTPase activity is slow, microtubules that happen to have GTP ends will be able to pick up GTP-tubulin and grow for a considerable time, until the GTP at the end is hydrolyzed. These will then tend to shrink as long as release of GDP-tubulin continues to reveal new GDP ends. Thus, depending on its "end state" (GTP or GDP), a

given microtubule will either grow or shrink in competition with its neighbors.

5. (a) Release of the myosin headpiece from the thin filament requires ATP binding. Until ATP binds, the myosin-actin cross-bridge will remain intact, thereby preventing extension of the sarcomeres.
(b) Without the action of the Ca^{2+} transporter, Ca^{2+} will leak across the membrane from the side of high Ca^{2+} concentration (in the transverse tubule) to the side of lower Ca^{2+} concentration (inside the sarcomere). As $[Ca^{2+}]$ increases, it will bind TnC, thereby stimulating myosin binding to actin. The lack of ATP will

result in a persistent cross-bridge (see part [a]), characteristic of the *rigor* state.
(c) Decomposition includes cleavage of actin and myosin by intracellular proteases (e.g., enzymes such as trypsin and chymotrypsin described in Chapter 5).

6. The observation would suggest that several power strokes may occur in succession. These may, however, be on different myosin heads—the actin filament can slide a considerable distance past a single head during the portion of the ATP cycle when it is disengaged.

CHAPTER 9

1. (a)

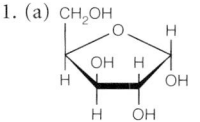

α-D-**Xylofuranose**

(b)

(c)

(d)

2. Galactitol has the following structure:

Because it has a plane of symmetry between C-3 and C-4, it is optically inactive. Such compounds, which contain asymmetric carbons but have no net optical activity, are called *meso* forms.

3. In the chair form, there is more steric clash between the 2-OH and the 1-OH in the α form of glucose and in the β form of mannose. (Compare glucose and mannose as glucose is pictured in Figure 9.13.) Furthermore, dipole–dipole interactions will be more favorable in β-D-glucose and α-D-mannose.

4. For the reaction

α-D-Glucopyranose $\rightleftharpoons$ β-D-glucopyranose

we have $K = 64/36 = 1.78$. Therefore

$$\Delta G° = -RT \ln K = -8.314 \, J/K \cdot mol \times 313 \, K \times 0.577$$
$$= -1.50 \, kJ/mol$$

The explanation relates to Problem 3. As Figure 9.12 shows, the 2-OH and 1-OH come quite close in α-D-glucose. Furthermore, the dipole moments of the two OH groups are more nearly parallel in the α-anomer, and antiparallel in the β-anomer. All of these are essentially energetic considerations, so ΔG arises primarily from an enthalpy contribution.

5.

6. Reducing: maltose, cellobiose, lactose, gentiobiose. Nonreducing: trehalose, sucrose.

7.

Note: For simplicity, H atoms are represented by vertical bars.

8. Hyaluronic acid.

9. (a) *S*.
(b) *R*.

10. Cleavage of the pyranose form will occur twice, producing formic acid. Cleavage of the furanose form will also occur twice (between C2 and C3 and between C5 and C6), but formaldehyde will be produced from C6.

11.

Xylan

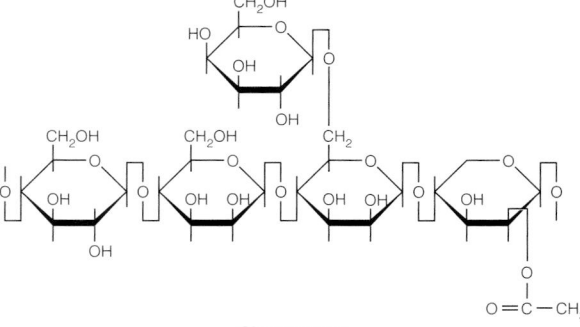

Glucomannan

12. Perhaps the Gal–GlcNAC linkage is not β, or not 1–4.

13. (a) a; (b) f; (c) a; (d) c; (e) d, f.

14. The influenza hemagglutinin binds to the surface of any cell that contains sialic acid. When it binds to the surface of erythrocytes, this causes the cells to aggregate, or clump.

15. Six possible tripeptide sequences involving A, B, and C.

Some possible trisaccharides: $Glc\alpha(1 \to 4)GlcUA\alpha(1 \to 4)GlcNAc$
$Glc\beta(1 \to 4)GlcUA\beta(1 \to 4)GlcNAc$
$Glc\alpha(1 \to 3)GlcUA\alpha(1 \to 4)GlcNAc$
$Glc\alpha(1 \to 4)GlcUA\beta(1 \to 4)GlcNAc$
$Glc\alpha(1 \to 4)GlcUA\beta(1 \to 6)GlcNAc$
$Glc\beta(1 \to 4)GlcUA\beta(1 \to 6)GlcNAc$
$Glc\alpha(1 \to 2)GlcUA\beta(1 \to 4)GlcNAc$
$Glc\beta(1 \to 3)GlcUA\beta(1 \to 4)GlcNAc$
$Glc\alpha(1 \to 4)GlcUA\beta(1 \to 3)GlcNAc$
$Glc\alpha(1 \to 4)GlcUA\beta(1 \to 2)GlcNAc$

This is a partial list of the trisaccharides with the sequence Glc—GlcUA—GlcNAc. There will be comparable numbers of trisaccharides with each of the other possible trisaccharide sequences—a very large number altogether.

16. Mannose and galactose: no, they differ in configuration at C2 and C4; Allose and altrose: yes, they differ only at C2; Gulose and talose: no, they differ at C2 and C3; Ribose and arabinose: yes, they differ only at C2.

17.

18.

UDP-GlcNAc   **UDP-GalNAc**

CHAPTER 10

1. (a) $CH_3(CH_2)_5CH=CH(CH_2)_9COOH$.
   (b) $CH_3(CH_2)_5CH=CH(CH_2)_9COOH$. (This is called *cis*-vaccenic acid.)
   (c) According to Table 10.1, any with fewer than 10 carbons will melt below 30 °C. An example would be $CH_3(CH_2)_6COOH$, *n*-octanoic acid.

2. (a) Fatty acid, long-chain alcohol.
   (b) Glycerol, fatty acid.
   (c) Carbohydrate, long-chain alcohol.

3. From the data, 4.74×10^9 cells would have a total surface area of $4.74 \times 10^{11} \, (\mu m)^2$, or $0.474 \, m^2$. The ratio of monolayer area to cell surface is $0.89 \, m^2/0.474 \, m^2$, or 1.89, very close to 2.00.

4. (a) 20.
   (b) The sequence MVGALLLLVVALGIGILFM is 19 residues long, is very hydrophobic, and has a number of helix-forming residues.

5. The concentration ratio is 10^5. Therefore
   (a) $\Delta G = 29.7$ kJ/mol.
   (b) $\Delta G = 36.4$ kJ/mol.
   In either case, hydrolysis of 1 mole of ATP (at cellular concentration) would suffice to transport 1 mole of ion.
   (c) Yes, because ΔG for ATP hydrolysis is -53.4 kJ/mol under these conditions.

6. $J = -P(C_2 - C_1)$; from Table 10.6, we have $P = 2.4 \times 10^{-10}$ cm/s. If we express $C_2 - C_1$ in mol/cm^3, J will have dimensions of mol/cm$^2 \cdot$s. Calculation yields $J = 2.04 \times 10^{-14}$ mol/cm$^2 \cdot$s.

Therefore, the amount transferred in 1 min across $100 \, (\mu m)^2$ ($=100 \times 10^{-8}$ cm^2) will be as follows:

$$M = -1.224 \times 10^{-18} \, mol$$

7. The initial concentration inside $=100$ mM $= 0.1 \times 10^{-3}$ mol/cm^3. The cell volume is $100 \, (\mu m)^3$, or 100×10^{-12} cm^3. The amount initially present is as follows:

$$M = 0.1 \times 10^{-3} \, mol/cm^3 \times 1 \times 10^{-10} \, cm^3$$
$$= 1 \times 10^{-14} \, mol$$

Therefore, the percentage escaping in 1 min will be

$$\frac{1.224 \times 10^{-18} \, mol}{1 \times 10^{-14} \, mol} \times 100 = 1.224 \times 10^{-2}\%$$

Thus, unless facilitated transport is available, "leakage" of K$^+$ from cells is very slow.

8. Consider mass flow from one direction only, with ligand A at concentration [A] outside the membrane. It reacts with carrier X at the membrane surface to give XA:

$$X + A \longrightarrow XA \qquad K_{bind} = [XA]/([X][A])$$

If we say that J is proportional to the fraction of carriers occupied, then

$$J = -P(C_2 - C_1)\frac{[XA]}{[X] + [XA]}$$

and we get

$$J = -P(C_2 - C_1)\frac{K_{bind}[A]}{1 + K_{bind}[A]}$$

This approaches a limiting value $(J = -P(C_2 - C_1)$ at high [A].

9. If the calcium ion is *maintained* at this concentration difference, we must have, from equation (10.2), $V_{in-out} = -92$ mV. Because V_{in-out} is defined as the potential inside minus the potential outside, the inside of the organelle is negative with respect to the outside.

10. (a) -61.5 mV (right $-$).
    (b) $+61.5$ mV (right $+$).
    (c) 0.

11. No in (a) and (b). Yes in (c). Concentrations will equalize.

12. (a) 0.27.
    (b) -72.8 mV.

13. The subunits traversing the membrane usually need to present different domains to the two sides. Thus, no symmetry involving twofold axes parallel to the membrane surfaces (like D_n) should be expected.

14. 2.71 kg of K^+, 2.39 kg of Na^+.

15. The maximum value of $[G6P]_{in}$ will be achieved when ΔG for the glucose phosphorylation reaction inside the cell is zero. If we assume the concentrations of free glucose outside and inside the cell are equal (i.e., 5 mM), the maximum value of $[G6P]_{in} = 162$ M. This absurdly high value is never achieved, as G6P is rapidly consumed in other metabolic reactions (see Chapter 13).

16. (a) -21.0 kJ/mol H^+
    (b) $ADP + P_i + 3H^+_{out} \rightarrow ATP + H_2O + 2H^+_{in}$
    (c) 2.7 mM

CHAPTER 11

1. Given $\ln([A]/[A]_0) = -kt$. When $t = t_{1/2}$, we have

$$\ln\left(\frac{1}{2}\right) = -0.693 = -kt_{1/2}$$

Therefore, $t_{1/2} = 0.693/k$. The constant of proportionality is 0.693.

2. Graphing $\ln([A]/[A]_0)$ versus t gives a curved line. Therefore, kinetics are not first-order.

3. About 1.1×10^{14}.

4. The Asp must be protonated to act as a general acid catalyst; thus, activity will be higher when $pH < pK_a$ and lower when $pH > pK_a$. At $pH = pK_a$, expect 50% of maximal activity because the Asp will be 50% protonated.

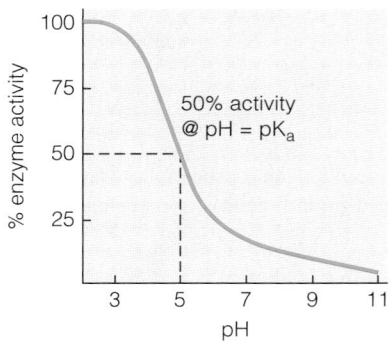

5. If the protein can bind to a random location on the DNA and slide along until it finds the specific site, a more efficient one-dimensional search replaces an entropically less favorable three-dimensional search.

6. Given $d[A]/dt = -k_1[A] + k_{-1}[B]$. Noting that $[B] = [A]_0 - [A]$, we get

$$d[A]/dt = -(k_1 + k_{-1})[A] + k_{-1}[A]_0 \tag{1}$$

The second term on the right can be evaluated from the equilibrium condition

$$-k_1[A]_e + k_{-1}[B]_e = 0 \tag{2}$$

which gives, after further substitutions,

$$k_{-1}[A]_0 = (k_1 + k_{-1})[A]_e \tag{3}$$

Inserting into (1), and noting that $d[A]/dt = d([A] - [A]_e)/dt$, we get

$$\frac{d([A] - [A]_e)}{([A] - [A]_e)} = -(k_1 + k_{-1})dt \tag{4}$$

After integrating, this yields the result

$$[A] - [A]_e = ([A]_0 - [A]_e)e^{-(k_1+k_{-1})t} \quad \text{Q.E.D.} \tag{5}$$

7. (a) From equation (11.7) we may write

$$-d[A]/dt = k[A]^2$$

Rearranging to give:

$$d[A]/[A]^2 = -kdt$$

and integrating from $[A]_0$ to $[A]_t$ and $t = 0$ to t, we obtain

$$1/[A] - 1/[A]_0 = kt$$

(b) One way to do it is to plot $1/[A]$ versus t.
(c) The data yield a linear graph, so the reaction is second-order.

8. (a) It is quite difficult to estimate V_{max} in this way. A value of about 150 $(\mu mol/L)min^{-1}$ might be guessed at. Then, by taking K_M as the substrate concentration at $V_{max}/2$, we would estimate $K_M \sim 25\ \mu mol/L$.
   (b) $V_{max} = 164\ (\mu mol/L)\ min^{-1}$; $K_M = 32\ \mu mol/L$. These values are considerably more reliable than those found in (a).
   (c) Best fit gives $V_{max} = 158\ (\mu mol/L)\ min^{-1}$; $K_M = 30\ \mu mol/L$.

9. (a) $V_{max} = 164\ (\mu mol/L)\ min^{-1}$, $k_{cat} = 1.64 \times 10^5 min^{-1} = 2733\ s^{-1}$.
   (b) Using $K_M = 32\ \mu mol/L$, $k_{cat}/K_M = 8.5 \times 10^7\ (mol/L)^{-1}s^{-1}$. This compares favorably with the largest values of k_{cat}/K_M.

10. (a) It is necessary only to rearrange the equation to

$$\left(\frac{K_M}{[S]} + 1\right)d[S] = -V_{max}dt$$

and then integrate from $t = 0$ ([S]) = [S]$_0$) to a finite time t. We obtain

$$[S]_0 - [S] - K_M \ln \frac{[S]}{[S]_0} = V_{max}t$$

(b) When $[S]_0 \gg K_M$, the first two terms on the left will be much larger than the third, so

$$[S_0] - [S] \cong V_{max}t$$

or

$$[S] = [S]_0 - V_{max}t$$

This corresponds to a situation in which the enzyme is saturated with substrate, so substrate molecules are simply being consumed at the maximum rate.

11. (a) E270 acts as a GBC in step one and as a GAC in step two.
(b) R145 provides specific ion–ion interactions with the C-terminal carboxylate of the substrate. This confers specificity for cleavage of the C-terminal residue from the peptide substrate.

12. The change in k_{cat} between pH 6 and pH 7 must involve loss of a proton in the active site. The best candidate is His 57. The increase in K_M at higher pH must involve a change in the binding site. The group involved is probably the N-terminus on Ile 16, created by the cleavage that activates chymotrypsin.

13. (a) If we take k_{cat}/K_M ratios, we find that PAPAF would be digested most rapidly and PAPAG most slowly.
(b) A hydrophobic residue C-terminal to the bond cleaved seems to be favored. Elastase always requires a small residue (like Ala) to the N-terminal side.
(c) Serine and histidine.

14. (a) The enzyme must be stable both to the presence of detergents and to moderately high temperatures.
(b) Replace the methionine, by site-directed mutagenesis, with another residue. Because methionine is quite hydrophobic, a hydrophobic replacement would seem appropriate. A single base change in the codon could yield Phe, Leu, Ile, or Val.
(c) The oxyanion is formed after S binds; thus, for a mutation of a residue that only interacts with the oxyanion intermediate one would not expect K_M to change significantly (it does not, according to Wells et al.); however, k_{cat} should be reduced due to the loss of enthalpic stabilization of the transition state (it is—by factor of 2500).
(d) Using Equation 11.11, and assuming that the value of γ is the same for both mutants:

$$\text{rate enhancement} = \frac{k_{catN155}}{k_{catT155}}$$

$$= \frac{\gamma_{N155}\left(\frac{k_B T}{h}\right)e^{\left(\frac{-\Delta G^{o\ddagger}_{N155}}{RT}\right)}}{\gamma_{T155}\left(\frac{k_B T}{h}\right)e^{\left(\frac{-\Delta G^{o\ddagger}_{T155}}{RT}\right)}} = \left(\frac{\gamma_{N155}}{\gamma_{T155}}\right)e^{\left(\frac{\Delta\Delta G^{o\ddagger}}{RT}\right)}$$

$$\frac{k_{catN155}}{k_{catT155}} = 2500 = (1)e^{\left[\frac{\Delta\Delta G^{o\ddagger}}{(0.008314 \text{ kJ/mol K})(310 \text{ K})}\right]}$$

$$\Delta G^{\ddagger} = 20 \text{ kJ/mol}$$

(e) The dielectric constant, ε, is lower in the enzyme active site than it is in water; thus, Coulomb's law (Equation 2.2) predicts a stronger interaction between the bond donor and acceptor.

15. (a) Mixed.
(b) No inhibitor: $K_M = 2.5$ mmol/L, $V_{max} = 5.1$ (mmol/L)min^{-1}. With inhibitor: $K_M = 7.6$ mmol/L, $V_{max} = 3.0$ (mmol/L)min^{-1}.

16. (a) Competitive.
(b) $V_{max} \sim 5.1$ (mmol/L)min^{-1} at all inhibitor concentrations.

[I] (mmol/L)	K_M^{app} (mmol/L)
0	2.5
3	3.8
5	5.4

(c) Using the data when [I] = 3 mM, the apparent $K_I = 5.5$ mM; using the [I] = 5 mM data, $K_I = 4.2$ mM. $K_I = 4.9$ mM is a reasonable estimate, given the variance in the experimental data.

17. According to equation (11.33), we can write the Michaelis–Menten equation as $V = V_{max} - K_M V/[S]$, so we plot V versus $V/[S]$. The intercept at $V/[S] = 0$ will give V_{max}; the slope will be K_M^{app}. Therefore, the graphs will look like the following sketches:

(a) Competitive inhibition $[I]_1 < [I]_2$

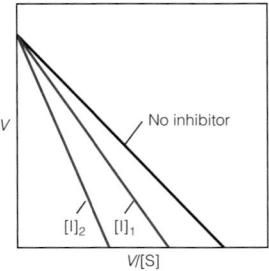

(b) Noncompetitive inhibition $[I]_1 < [I]_2$

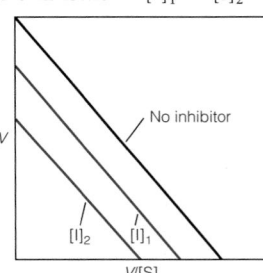

18. One should replace the hydrophobic group in TPCK with a positively charged group. A good candidate would be

$$\text{H}_3\overset{+}{\text{N}}-(\text{CH}_2)_4-\overset{\overset{\displaystyle H}{|}}{\underset{\underset{\displaystyle O=\overset{\displaystyle |}{\underset{\displaystyle |}{S}}=O}{\underset{\displaystyle NH}{|}}}{C}}-\overset{\overset{\displaystyle O}{||}}{C}-\text{CH}_2\text{Cl}$$

19. Find K_{eq} for $F \rightleftharpoons U$ for the mutant from the ratio of rate constants; then, calculate $\Delta G^{o\prime}_{F \rightarrow U}$:

$$K_{eq} = \frac{[U]}{[F]} = \frac{k_{F \rightarrow U}}{k_{U \rightarrow F}} = \frac{3.62 \times 10^{-5}\text{s}^{-1}}{255\text{s}^{-1}} = 1.42 \times 10^{-7}$$

$$\Delta G^{o\prime}_{F \rightarrow U} = -RT\ln(1.42 \times 10^{-7}) = 39.1 \text{ kJ/mol}$$

The mutant is more stable than the wild-type myoglobin.

20. (a) If we make a graph of v versus [S], the data show a slightly sigmoidal curve. A more convincing demonstration is to use a Lineweaver–Burk plot or an Eadie–Hofstee plot. Neither is linear.
(b) From the v versus [S] plot, we can estimate V_{max} to be approximately 5 (mmol/L) sec^{-1}.

21. The equation is

$$\log\left(\frac{v/V_{max}}{1-v/V_{max}}\right) = \log[S] - \log K_M$$

The Hill plot of $\log\left(\dfrac{v/V_{max}}{1-v/V_{max}}\right)$ versus $\log[S]$ will approach straight lines with slope $= 1$ at low and high values of $\log[S]$. The former will yield K_M^T, the latter K_M^R.

22.

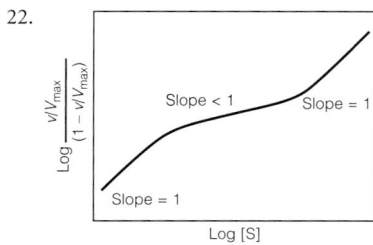

23. In the MWC theory, the more favorable binding to the R state makes the equilibrium shift as [S] is increased; the shift can only be toward R. In the KNF theory, binding to one site can have any kind of effect on another, including inhibition.

24. (a) The key to the derivation is to note that

$$Y = \frac{\text{number of sites occupied}}{\text{total number of sites}}$$

$$= \frac{[RS] + 2[RS_2]}{2([R] + [R] + [RS] + [RS_2])}$$

Expressions given for [T], [RS], and [RS$_2$] are inserted and the equation simplified.
(b) This must be done graphically. A large L and a small K ensure a sigmoidal curve.

25. (a) Activation of the molecule by trypsin will be greatly reduced due to removal of the R that trypsin binds in its active site.
(b) If activation occurs, the N-terminal peptide will no longer be constrained by an S–S bond and may be released.
(c) Automodification of π-chymotrypsin to α-chymotrypsin could be reduced, although this mutation is relatively conservative (loss of a methyl group); thus, the effect on activation might be slight.

26. The simplest technique would be to incubate the ribozyme with a radiolabeled pentanucleotide and collect samples from time to time for electrophoresis. Autoradiography should initially show only the pentanucleotide, but with time both longer and shorter labeled oligomers should appear on the gel.

27. Stable transition state mimics of amide bond hydrolysis are challenging to synthesize. A number of haptens with tetrahedral oxyanions have been used: phosphonates, sulfonamides, boronic esters. These molecules have been used to raise antibodies that show rate enhancements for amide bond hydrolysis of ~10^3 (see R. Aggarwal et al. (2003) *Chem. Eur. J.* 9:3132).

28. (a) $[E]_t = [E] + [ES] + [ESI]$. Using the expressions for α', K_M, and K_I':

$$[E]_t = [ES]\left(\frac{K_M}{[S]} + 1 + \frac{[I]}{K_I'}\right) = [ES]\left(\frac{K_M}{[S]} + \alpha'\right)$$

$$V = k_{cat}[ES] = \frac{k_{cat}[E]_t}{\dfrac{K_M}{[S]} + \alpha'} = \frac{V_{max}[S]}{K_M + [S]\alpha'}$$

(b) $[E]_t = [E] + [EI] + [ES] + [ESI]$. Using the expressions for α, α', K_M, K_I and K_I'

$$[E]_t = [ES]\left(1 + \frac{[I]}{K_I'}\right) + [E]\left(1 + \frac{[I]}{K_I}\right) = [ES](\alpha') + [E](\alpha)$$

$$[E]_t = [ES]\left(\alpha' + \frac{\alpha[K_M]}{[S]}\right)$$

$$V = k_{cat}[ES] = \frac{k_{cat}[E]_t}{\alpha' + \dfrac{\alpha[K_M]}{[S]}} = \frac{V_{max}[S]}{\alpha K_M + [S]\alpha'}$$

29. (a) Enalaprilat is too polar to cross membranes; whereas enalapril is less polar and can cross membranes to get from the gut to circulation.
(b):

$$10 = \frac{\alpha K_M + [S]}{K_M + [S]}$$

$$v_{o(inhibited)} = \frac{0.1 V_{max}[S]}{K_M + [S]} = \frac{V_{max}[S]}{\alpha K_M + [S]}$$

$$\alpha = \frac{10(K_M + [S]) - [S]}{K_M} = \frac{10(12\,\mu M + 75\,\mu M) - 75\,\mu M}{12\,\mu M}$$

$$= 66.25 = 1 + \frac{[I]}{K_I}$$

$$K_I = \frac{2.4 \times 10^{-9}\,M}{66.25} = 3.7 \times 10^{-11}\,M$$

CHAPTER 12

1.

Equation	RQ
(a) $C_2H_5OH + 3O_2 \rightarrow 2CO_2 + 3H_2O$	0.67*
(b) $CH_3COOH + 2O_2 \rightarrow 2CO_2 + 2H_2O$	1.0
(c) $CH_3(CH_2)_{16}COOH + 26O_2 \rightarrow 18CO_2 + 18H_2O$	0.69
(d) $CH_3(CH_2)_{14}(CH)_2COOH + 25\frac{1}{2}O_2 \rightarrow 18CO_2 + 17H_2O$	0.71
(e) $CH_3(CH_2)_{12}(CH)_4COOH + 25O_2 \rightarrow 18CO_2 + 16H_2O$	0.71

*Example: RQ = 2CO$_2$/3O$_2$.

2. (a) >1.
(b) <1.

Because NAD$^+$-dependent enzymes usually act to dehydrogenate (oxidize) substrates, an [NAD$^+$]/[NADH] ratio greater than unity tends to drive reactions in that direction. Similarly, [NADP$^+$]/[NADPH] ratios less than unity provide concentrations that tend to drive these reactions in the direction of substrate reduction.

3. (a) 0.
(b) 1.
(c) 4.

4. for glucose ($C_6H_{12}O_6$): MW $= 6(12 \text{ g mol}^{-1})$
$+12(1 \text{ g mol}^{-1}) + 6(16 \text{ g mol}^{-1}) = 180 \text{ g mol}^{-1}$
Energy yield/mole glucose
$= (15.64 \text{ kJ g}^{-1})(180 \text{ g mol}^{-1}) = 2815.2 \text{ kJ mol}^{-1}$ glucose
Energy yield/mole carbon
$= (2815.2 \text{ kJ mol}^{-1}$ glucose$)/(6 \text{ mol carbon mol}^{-1}$ glucose$)$
$= 469.2 \text{ kJ mol}^{-1}$ carbon
for palmitic acid ($C_{16}H_{32}O_2$): MW $= 16(12 \text{ g mol}^{-1})$
$+ 32(1 \text{ g mol}^{-1}) + 2(16 \text{ g mol}^{-1}) = 256 \text{ g mol}^{-1}$
Energy yield/mole palmitic acid
$= (38.90 \text{ kJ g}^{-1})(256 \text{ g mol}^{-1}) = 9958.4 \text{ kJ mol}^{-1}$ PA
energy yield/mole carbon
$= (9958.4 \text{ kJ mol}^{-1}$ PA$)/(16 \text{ mol carbon mol}^{-1}$ PA$)$
$= 622.4 \text{ kJ mol}^{-1}$ carbon

5. $\Delta G = \Delta G^{\circ\prime} + RT \ln([\text{ADP}][\text{P}_i]/[\text{ATP}])$
$= -30500 \text{ J/mol} + (8.315 \text{ J/mol} \cdot \text{K})(310 \text{ K})$
$\times \ln(1 \times 10^{-3} \text{ M})(1 \times 10^{-3} \text{ M})/(3 \times 10^{-3} \text{ M})$
$= -51134 \text{ J/mol} = -51.14 \text{ kJ/mol}$

6. (a)

Treatment	% Change in flux
Inhibitor that reduces activity of X by 10%	10% decrease
Activator that increases activity of X by 10%	10% increase

(b) Metabolic flux rate $=$ rate of X $-$ rate of Y $= 100 - 80$
$= 20 \text{ pmol}/10^6 \text{ cells/sec}$ in the direction of $B \rightarrow C$

(c)

Change in flux	Treatment	Direction of flux	% Change in flux
		($B \rightarrow C$ or $C \rightarrow B$)	
$90 - 80 = 10$	Inhibitor that reduces activity of X by 10%	$B \rightarrow C$	50% decrease
$110 - 80 = 30$	Activator that increases activity of X by 10%	$B \rightarrow C$	50% increase
$100 - 160 = -60$	Doubling the activity of enzyme Y	$C \rightarrow B$	400% change (decrease)

(d) A substrate cycle allows amplification of a small change in an enzyme's activity; e.g., a 10% change in activity gives a 50% change in flux.

7. (a) $K = e^{\left(\frac{-\Delta G^{\circ\prime}}{RT}\right)} = e^{\left(\frac{16700 \text{ J/mol}}{(8315 \text{ J/mol} \cdot \text{K})(298 \text{ K})}\right)}$

$= 845 = \dfrac{[\text{ADP}][\text{glucose-6-phosphate}]}{[\text{ATP}][\text{glucose}]}$

Since [ATP] $=$ [ADP]: [glucose-6-P]/[glucose] $= 845$ under physiological conditions.

(b) K $= [\text{glucose}][\text{P}_i]/[\text{glucose-6-P}] = 262$
$[\text{glucose}]/[\text{glucose-6-P}] = 262/(1 \times 10^{-3} \text{ M}) = 262{,}000$
(c) Any condition where the [glucose]/[glucose-6-P] ratio is between 262,000 and 0.0012 (1/845).
(d) The relative concentrations of allosteric regulators that exert kinetic control over the enzymes in the opposing pathways.

*8. The most direct way is to purify the enzyme to homogeneity and determine both its molecular weight and turnover number. Then, from the activity observed in a crude extract, one can calculate the number of active enzyme molecules needed to achieve that activity (assuming that the extract does not contain inhibitors of the activity). Another approach is to treat the extract with a specific antibody to the enzyme of interest and quantitate the protein immunoprecipitated. One still needs to know the molecular weight to convert this value to number of molecules of enzyme.

*9. (a) 20% (600 cpm/pmol/3000 cpm/pmol).
(b) 1500 cpm dTTP incorporated per minute per 10^6 cells, corrected for the 20% fraction calculated in part (a) $(1500/0.2) = 7500$ cpm/10^6 cells/min.

$$\frac{7500 \text{ cpm}}{10^6 \text{cells} \cdot \text{min}} \times \frac{1 \text{ pmol}}{3000 \text{ cpm}} \times \frac{6.02 \times 10^{11} \text{ molecules}}{1 \text{ pmol}}$$
$$= 1.5 \times 10^6 \text{ molecules/min/cell}$$

(c) Prepare an acid extract of the cells (e.g., 5% trichloroacetic acid), separate the nucleotides by ion-exchange HPLC, and determine in the dTTP fraction its radioactivity and its mass, the latter from UV absorbance.

10. (a) N $= 1 \mu\text{mol}$ orthophosphate $= 6.02 \times 10^{17}$ molecules/μmol

$\lambda = \dfrac{0.693}{t_{1/2}} = \dfrac{0.693}{14.2 \text{ days}} = \dfrac{0.693}{2.04 \times 10^4 \text{ min}} = 3.39 \times 10^{-5} \text{ min}^{-1}$

dN/dt $= (6.02 \times 10^{17} \text{ molecules}/\mu\text{mol})(3.39 \times 10^{-5} \text{min}^{-1})$
$= 2.04 \times 10^{13} \text{ dpm}/\mu\text{mol}$

$(2.04 \times 10^{13} \text{ dpm}/\mu\text{mol})(1 \text{ mCi}/2.2 \times 10^9 \text{ dpm})$
$= 9274 \text{ mCi}/\mu\text{mol}$

(b) $(10 \text{ mCi}/\mu\text{mol})(3.67 \times 10^7 \text{ dps/mCi})$
$= 3.67 \times 10^8 \text{ dps}/\mu\text{mol} = \text{dN/dt} = \lambda\text{N}$

$\lambda = \dfrac{0.693}{t_{1/2}} = \dfrac{0.693}{12.1 \text{ years}} = \dfrac{0.693}{3.8 \times 10^8 \text{ sec}} = 1.8 \times 10^{-9} \text{s}^{-1}$

N $= (3.67 \times 10^8 \text{ dps}/\mu\text{mol})/\lambda$
$= (3.67 \times 10^8 \text{ dps}/\mu\text{mol})/(1.8 \times 10^{-9} \text{ s}^{-1})$
$= 2.04 \times 10^{17}$ atoms

$(1 \mu\text{mol leucine})(6.02 \times 10^{17} \text{ molecules}/\mu\text{mol})$
$(13 \text{ H/molecule}) = 7.83 \times 10^{18}$ H atoms
(at physiological pH)
% radioactive $= (100\%) \times$
$(2.04 \times 10^{17} \text{ atoms})/(7.83 \times 10^{18} \text{H atoms}) = 2.6\%$

CHAPTER 13

1. $\Delta G = \Delta G^{\circ\prime} + RT\ln([\text{FBP}][\text{ADP}]/[\text{F6P}][\text{ATP}])$. With the concentrations given, $\Delta G = \Delta G^{\circ\prime}$, since the natural log term is zero. Therefore, the reaction is neither more nor less exergonic in muscle than under standard conditions.

2. Ethanol will generate NADH, through the action of alcohol dehydrogenase, and this will reduce formaldehyde back to methanol, also through alcohol dehydrogenase. The acetaldehyde formed can be metabolized further to acetate. Ethanol may also compete with methanol for binding to alcohol dehydrogenase.

3. Probably F1,6BP. As triose phosphate began to accumulate, the unfavorable equilibrium for the forward reaction might drive both DHAP and GAP back to F1,6BP. Of course, GAP would accumulate first, but probably not to significantly increased levels.

4. (a) 4 (2 each from glucose and fructose).
   (b) 5 (3 from G6P + 2 from fructose).

5. C-3 or C-4. Both become C-1 of pyruvate, which is lost as CO_2 in the pyruvate decarboxylase reaction.

6. (a) Sucrose + $5P_i$ + 5ADP → 4 lactate + 5ATP + $4H_2O$ + $4H^+$
   (b) Maltose + $4P_i$ + 4ADP + $4NAD^+$ → 4 pyruvate +

   $$4ATP + 3H_2O + 4NADH + 4H^+$$

   (c) (Glucose)residue + $2P_i$ + 2ATP → 2 ethanol +

   $$2ATP + 2CO_2 + H_2O + 2H^+$$

7. Glyceraldehyde-3-phosphate dehydrogenase. The acyl arsenate analog of 1,3-bisphosphoglycerate spontaneously hydrolyzes, preventing the production of 1,3-bisphosphoglycerate.

8. MW of ethanol (C_2H_6O) = 2(12 g/mol) + 6(1 g/mol) + 16 g/mol = 46 g/mol. 10% w/v EtOH = 100 g/L. Convert to molar: (100 g/L)/ (46 g/mol) = 2.18 M

   Because fermentation of 1 mole of glucose generates 2 moles of ethanol, the glucose source must have been present at 0.5 (2.18 M) = 1.09 M. No, because glucose is not an abundant free component of most fermentable plant sources. The majority of their carbon sources exist as starch, sucrose, or some other oligosaccharide or polysaccharide.

9.

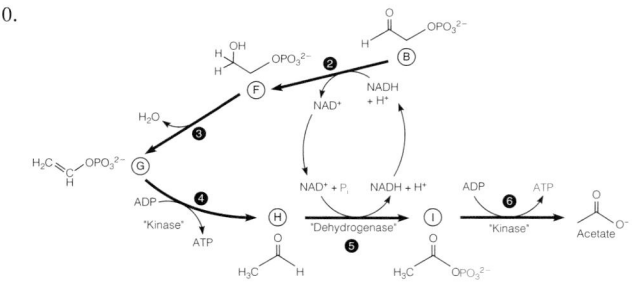

10.

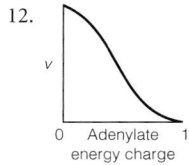

11. Glc → G6P → G1P → UDP-Glc → UDP-Gal → lactose
                                        Glc
    2Glucose + ATP + UTP → lactose + ADP + UDP + PP_i

12.

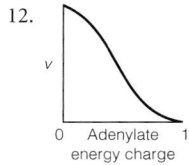

13. UDP-glucose is used catalytically in the sense that for each molecule of Gal-1-P converted to G1P, one molecule of UDP-Glc is converted

to UDP-Gal, which is immediately reconverted to UDP-Glc, to react with another molecule of Gal-1-P.

14.

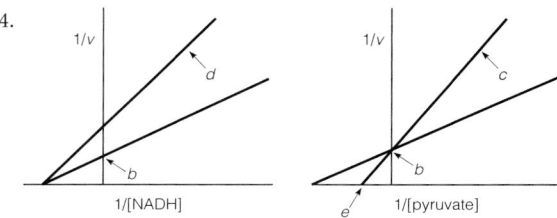

   (f) 10^{-3} M.

15. (a) 1: 2DHAP → F1, 6BP → F6P → G6P → G1P $\xrightarrow{1ATP}$ UDPG
                                                    → glucosyl residue.

   (b) 1.
   (c) 7 (see Figure 13.11).
   (d) 1.

16. (a) 2 generated (glycolysis).
   (b) 6 consumed (gluconeogenesis requires ATP).

17. (a) because lactate is the only substrate that must go through the pyruvate carboxylase reaction.

18. None. The CO_2 that is fixed comes off in the PEP carboxykinase reaction.

19. (a) Fructose-6-phosphate + ATP → fructose-2,6-bisphosphate + ADP

   (b) 2 Oxaloacetate + 2ATP + 2GTP + 2NADH + $2H^+$

   + $4H_2O$ → glucose + $2CO_2$ + $2NAD^+$ + 2ADP + 2GDP + $4P_i$

   (c) Glucose + ATP + UTP → UDP-Glc + PP_i + ADP

   (d) 2 Glycerol + 2ATP + $2NAD^+$ + $2H_2O$ → glucose + 2ADP + 2NADH + $2H^+$ + $2P_i$

   (e) 2 Malate + 2ATP + 2GTP + $3H_2O$ → glucose-6-phosphate + $2CO_2$ + 2ADP + 2GDP + $3P_i$

20.

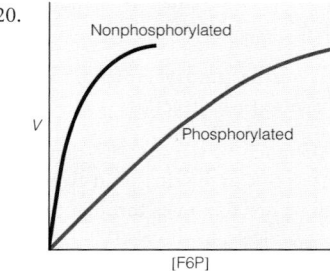

21.

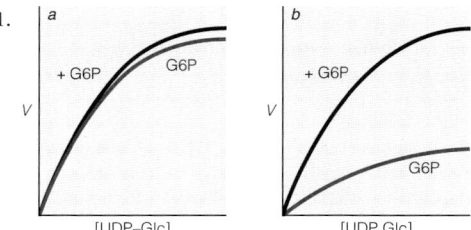

22. AMP activation of glycogen phosphorylase *b*; glucose-6-P activation of glycogen synthase *b*.

23. Because pyruvate carboxylase has two metabolic roles—replenishment of citric acid cycle intermediates (Chapter 14) and initiation of gluconeogenesis. PEPCK catalyzes the first reaction committed to gluconeogenesis.

24. (1) The liver is the most active gluconeogenic tissue. Because alanine is an important gluconeogenic precursor, its accumulation in liver is a signal that gluconeogenesis should be activated, with simultaneous inhibition of glycolysis. Also, alanine accumulates in muscle under conditions of high metabolic demand, where glycolysis must still function to provide ATP.

    (2) The function of glucose-6-phosphatase is production of glucose, for exit from the cell and transport to other tissues. Therefore, the enzyme plays a role only in those tissues (primarily liver) that manufacture glucose for export.

25. PP1 inactivates phosphorylase b kinase, by catalyzing its dephosphorylation. This blocks the conversion of glycogen phosphorylase b to a, thereby inhibiting glycogenolysis. Inhibitor 1, by inhibiting PP1, has the opposite effect.

26. (a) The function of glucagon is to increase blood glucose concentration. Stimulation of liver glycogen breakdown and gluconeogenesis and inhibition of liver glycolysis are consistent with this function.

    (b) The glucose-6-phosphatase deficiency would interfere with release of glucose from the liver for export to other tissues.

    (c) Phosphorylase b kinase converts phosphorylase b to a, which activates glycogen breakdown, and it converts glycogen synthase a to b, the less active form of this enzyme.

    (d) Glucose-6-phosphate must be hydrolyzed in order for glucose to exit the liver cell and be exported to other tissues.

27. (a) $G6P + 2NADP^+ + H_2O \rightarrow R5P + CO_2 + 2NADPH + 2H^+$

    (b) $G6P + 12NADP^+ + 7H_2O \rightarrow 6CO_2 + 12NADPH + 12H^+$

28. C-1 and C-3 of fructose-6-phosphate should be labeled. Erythrose-4-phosphate should be unlabeled.

29.

(see Fig. 14.20, p. 619)

CHAPTER 14

1. Administer separately 1-[^{14}C]glucose and 6-[^{14}C]glucose, and measure initial rates of $^{14}CO_2$ formation. The ratio gives relative flux rates through the two pathways. For example, if the flux rates are equal, the ratio from C^1-labeled glucose to C^6-labeled glucose is 2:1.

2. C-1: all released as CO_2. C-2 and C-3: all retained in oxaloacetate.

3. First turn: one-quarter. Second turn: three-eighths.

4. The action of pyruvate carboxylase on the labeled pyruvate would yield oxaloacetate labeled such that, when these carbons proceed through the citric acid cycle, C-5 of isocitrate would be labeled.

5. Addition to isolated glyoxysomes of citrate, isocitrate, glyoxylate, malate, or oxaloacetate would stimulate succinate formation out of proportion to the amount added.

6. C-3 and C-4, because these become the carboxyl group of pyruvate, which is lost in the pyruvate dehydrogenase reaction.

7. First, the cytosolic location of much of the NADP$^+$-dependent enzyme raises suspicions about its role in the citric acid cycle. More important, cells that have high or low flux rates through the cycle have high or low activities, respectively, of the NAD$^+$-dependent enzyme. Activity of the NADP$^+$-dependent enzyme does not vary in coordination with activities of other cycle enzymes.

8. E_1: pyruvate + lipoyllysine $\rightarrow$ acetyl–lipoyllysine + CO_2

    E_2: acetyl-lipoyllysine + CoA-SH $\rightarrow$ acetyl-CoA + dihydrolipoyllysine

    E_3: dihydrolipoyllysine + NAD$^+$ $\rightarrow$ NADH + H$^+$ + lipoyllysine

9. (a) This is a signal that levels of substrates are adequate for citric acid cycle oxidation, so that pyruvate can be shunted into gluconeogenesis instead of being oxidized. In addition, if acetyl-CoA is accumulating due to unbalanced fat vs. carbohydrate metabolism, it will divert pyruvate to OAA until balance is achieved.

    (b) This tends to inactivate pyruvate dehydrogenase (by phosphorylation) when levels of reduced electron carriers (NADH) are sufficient for ATP production via the respiratory chain and, hence, to make pyruvate available for other purposes.

    (c) This is a signal to reduce flux through the citric acid cycle when levels of reduced electron carriers are adequate for energy generation.

    (d) When the energy charge is low, the accumulation of ADP provides a signal to activate the citric acid cycle and thereby increase the oxidation of nutrients for ATP production.

    (e) This is an example of product inhibition, but it also serves as a general indicator that when an energy-rich substrate (succinyl-CoA) is abundant, flux through the citric acid cycle can be reduced.

    (f) Ca^{2+} mediates stimulation of PDH activity during muscle contraction. Ca^{2+} is a critical signaling molecule for contraction in vertebrate muscle, which places a huge demand on ATP production.

10. (a) An oxidation is required to convert the aldehyde to an acid. An oxidation requires an electron acceptor (e.g., NAD$^+$ or E-FAD):

    (b) The modified pathway bypasses the succinyl-CoA synthetase reaction of the standard citric acid cycle, which produces one ATP (GTP) by a substrate-level phosphorylation. The modified pathway therefore generates one less ATP than the standard pathway. There is no difference in the number of reduced cofactors generated (four oxidations per cycle).

11. Some possible mechanisms: substrate-level control of citrate synthase, activation of citrate lyase by acetyl-CoA or fatty acids, inhibition of isocitrate lyase by succinate (to ensure adequate flux through the citric acid cycle).

12. $2 \text{ acetyl-CoA} + 2\text{NAD}^+ + \text{E}-\text{FAD} + 3\text{H}_2\text{O} \rightarrow \text{oxaloacetate} + 2\text{NADH} + \text{E-FADH}_2 + 2\text{CoA-SH} + 4\text{H}^+$

13. The redox potential of the flavin, which does not dissociate from its enzyme, depends on its protein environment. In lipoamide dehydrogenase, its redox potential is held more negative than in other flavin dehydrogenases, so that electrons can be passed onto NAD^+ under physiological conditions.

14. $[\text{NAD}^+]/[\text{NADH}]$ should be high, so that it can promote the oxidation of substrates, e.g., $\text{malate} + \text{NAD}^+ \rightleftharpoons \text{oxaloacetate} + \text{NADH} + \text{H}^+$. Conversely, because NADPH and NADP^+ usually promote reduction of substrates, we expect $[\text{NADP}^+]/[\text{NADPH}]$ to be low.

CHAPTER 15

1. $+28.95 \text{ kJ/mol}$ (from the equation $\Delta G^{\circ\prime} = -nF\Delta E^{\circ\prime}$).
   $\Delta E^{\circ\prime} = -0.32 - (-0.17)$.

2. (a) $\text{Cyt } c \rightarrow \text{Cu}_A \rightarrow \text{heme } a \rightarrow \text{heme } a_3 - \text{Cu}_B \rightarrow \text{O}_2$
   (i.e., complex IV)
   (b) To block oxidation of endogenous substrates and to inhibit reverse electron flow (i.e., to prevent electrons flowing back up the respiratory chain).
   (c) Because ATP is formed, with a P/O ratio of ~1.0, complex IV must be a coupling site (i.e., a site of proton pumping).
   (d) From Table 15.1:

   $\text{Cyt } c_{\text{red}} \rightarrow \text{Cyt } c_{\text{ox}} + \text{e}^- \ (X\,2)$
   $1/2\,\text{O}_2 + 2\text{H}^+ + 2\text{e}^- \rightarrow \text{H}_2\text{O}$
   ───────────────────────────────
   $2\text{Cyt } c_{\text{red}} + 1/2\,\text{O}_2 + 2\text{H}^+ \rightarrow 2\text{Cyt } c_{\text{ox}} + \text{H}_2\text{O}$

   Coupled to:
   $\text{ADP} + \text{P}_i \rightarrow \text{ATP} + \text{H}_2\text{O}$
   Net: $2\text{Cyt } c_{\text{red}} + \dfrac{1}{2}\text{O}_2 + \text{ADP} + \text{P}_i + 2\text{H}^+ \rightarrow 2\text{Cyt } c_{\text{ox}} + \text{ATP} + 2\text{H}_2\text{O}$
   $[\text{Cyt c (Fe}^{2+})]$ $[\text{Cyt c (Fe}^{3+})]$

   (e) -78.1 kJ/mol for overall reaction (calculated as in Problem 1).

3. (a) β-hydroxybutyrate $\rightarrow$ NADH $\rightarrow$ complex I $\rightarrow$ CoQ $\rightarrow$ complex III $\rightarrow$ cytochrome c.
   (b) 2, because cytochrome c oxidase is bypassed.
   (c) Because NADH cannot freely enter the mitochondrion.
   (d) To block cytochrome oxidase, so that electrons exit the chain at cytochrome c.
   (e) β-hydroxybutyrate $+ 2 \text{ cyt } c_{\text{ox}} + 2\,\text{ADP} + 2\text{P}_i + 4\text{H}^+ \rightarrow$ acetoacetate $+ 2 \text{ cyt } c_{\text{red}} + 2\text{ATP} + 2\text{H}_2\text{O}$
   (f) -53.8 kJ/mol β-hydroxybutyrate (calculated as in Problem 2).

4. To block succinate dehydrogenase and measure phosphorylation resulting only from the α-ketoglutarate dehydrogenase reaction. P/O ratio $=$ ~3.5 (2.5 ATP from NADH and 1 from the succinyl-CoA synthetase reaction).

5. $+67.6 \text{ kJ/mol}$ (solved as in Problem 1). Minimum $[\text{NAD}^+]/[\text{NADH}]$ ratio $= 2.5 \times 10^{10}$, calculated from Equation 3.23 (Chapter 3):
   $$\Delta G = \Delta G^{\circ\prime} + RT\ln\left(\frac{[\text{fumarate}][\text{NADH}]}{[\text{succinate}][\text{NAD}^+]}\right)$$
   where $\Delta G < 0$.

6. $[\text{ADP}] = 1.44 \text{ mM}$ and $[\text{AMP}] = 0.29 \text{ mM}$, calculated from two simultaneous equations: the defining equation for energy charge, and $[\text{ADP}] = 5 \times [\text{AMP}]$.
   $\Delta G = -27.8 \text{ kJ/mol}$
   from $\Delta G = \Delta G^{\circ\prime} + RT\ln ([\text{ADP}][\text{P}_i])/[\text{ATP}]$.

7. 1.8×10^{17}. Calculate $\Delta G^{\circ\prime}$ as in Problem 1, and then apply $\Delta G^{\circ\prime} = -RT\ln K_{\text{eq}}$. Use $E^{\circ\prime} = +0.30 \text{ V}$ for H_2O_2 (see Table 15.1).

8. Add ADP in limiting amount and measure O_2 uptake. The ratio of μmol ADP consumed to μatom oxygen taken up is identical to the P/O ratio (see Figure 15.27).

9. Because the energy not used for ATP was dissipated as heat, and the subjects developed uncontrollable fevers.

10. 1.

11. (1) NADH is oxidized by FMN, not FAD.
    (2) Reduced flavin (FADH_2 or FMNH_2) is oxidized by CoQ, not CoQH_2.
    (3) cyt c_1 accepts e^- from cyt b, and is then oxidized by cyt c.
    (4) O_2 is reduced to H_2O, not H_2O_2.

12. (a) $\Delta G^{\circ\prime} = -nF\Delta E^{\circ\prime} = -2\,(96485)$
    $\times[-0.23 - (-0.32)] = -17.37 \text{ kJ/mol}$
    (b) The same, because $E^{\circ\prime}$ for NAD^+/NADH is the same as for $\text{NADP}^+/\text{NADPH}$.
    (c) ΔG would probably be positive in vivo for an NADH-linked enzyme because the high NAD^+/NADH concentration ratio would promote the oxidation of GSH to GSSG.

13. (a) Succinate $\rightarrow$ FAD $\rightarrow$ Q $\rightarrow$ cyt b $\rightarrow$ cyt c_1 $\rightarrow$ cyt c
    (b) succinate $+ 2 \text{ cyt } c_{\text{ox}} + \text{ADP} + \text{P}_i \rightarrow$ fumarate $+ 2 \text{ cyt } c_{\text{red}} + \text{ATP} + \text{H}_2\text{O}$.
    (c) For the redox reaction, $\Delta G^{\circ\prime} = -nF\Delta E^{\circ\prime} = -2\,(96485)(0.2 - 0.03) = -42.5 \text{ kJ/mol}; \Delta G^{\circ\prime}$ for synthesis of ATP is $+30.5 \text{ kJ/mol}$; net $\Delta G^{\circ\prime} = -42.5 + 30.5 = -12 \text{ kJ/mol}$.
    (d) To block cytochrome oxidase and force electrons to exit the respiratory chain at cytochrome c.
    (e) 0. No ATP would be produced because 2,4-dinitrophenol would dissipate the proton gradient.

14. DNP acts catalytically in the sense that one DNP molecule can mediate the transport of multiple protons into the matrix: after deprotonating in the matrix (due to the higher pH), DNP cycles back across the membrane to bind and bring another proton into the matrix. One important implication is that DNP is toxic in very small amounts.

15. (a) $\Delta G^{\circ\prime} = -nF\Delta E^{\circ\prime} = -2(96485)(0.82 - 0.03)$
$= -151 \text{ kJ/mol}.$
(b) For protons (eqn 15.5, p. 647): $\Delta G^{\circ\prime} = 2.3\,RT\,\Delta\text{pH} + F\Delta\psi = 24.9\,\text{kJ/mol protons}.$
$151/24.9 = 6$ protons maximum
(c) Complexes III and IV.

16. All 3 inhibitors leave d oxidized, so d must be last; all three inhibitors leave c reduced, so c must be first:

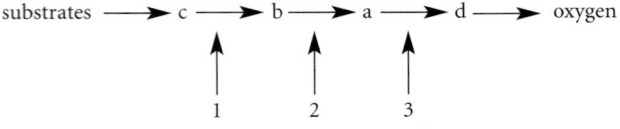

17. (a) A, 1; B, 2; C, 3; D, 4; E, 5.
(b) 3 divided by 4 (ratio of uptakes in presence and absence of ADP, with substrate present for both measurements).
(c) 5; O_2 depleted.
(d) 4; ADP level low because ATP level is high.
(e) 3; rapid ATP production and turnover demand rapid and continuous O_2 uptake.

CHAPTER 16

1. $\Delta G^{\circ} = 164 \text{ kJ/mol}$. An einstein of 700 nm light would yield 171 kJ.

2. 14.7%.

3. 1050 nm.

4. (a) After being attached to C-2 of ribulose-1,5-bisphosphate, the ^{14}C will become the carboxylate (C-1) of *one* of the two molecules of 3-phosphoglycerate. This becomes the carbonyl carbon (C-1) in G3P, or C-1 in DHAP. Upon condensation to form fructose-1,6-bisphosphate, the ^{14}C will show up in carbons 3 and/or 4.
(b) No. They may carry 0, 1, or 2 labeled carbons, depending on what combination of labeled and unlabeled trioses has been used in their formation.

5. Assuming 48 photons per mole of hexose, the leaf could theoretically produce 0.0263 mol, or 4.73 g, of hexose in 1 hour. It will, in fact, produce only a small fraction of this, for not all photons are absorbed, nor do all absorbed photons serve to pass electrons through the photosynthetic pathway.

6. (a) No, because plastoquinones are not involved in this process.
(b) Addition of ferricyanide as an electron donor allows a Hill reaction.

7. By using illumination at both 700 and 680 nm, either singly or together. Observe the oxidation state of plastoquinones and plastocyanin. If only p700 is involved, these will be oxidized. If only p680, they will be reduced. If both are involved, simultaneous illumination at the two wavelengths will give enhanced oxidation of the donor.

8. Carbon 3.

9. Competitive.

10. The data suggest the strong binding of an inhibitor. It is apparently heat stable, so it is probably not a protein, but it is released on ammonium sulfate precipitation.

11. (a) Decrease.
(b) Increase.
(c) Increase.

12. (a) Because both O_2 and CO_2 are being consumed and produced by the opposing processes of photorespiration and photosynthesis, a steady-state ratio will be attained.
(b) The relative affinity of rubisco for CO_2 and O_2.

13. Most of the 3PG that is initially labeled is reused in replacing RuBP, which then reenters the cycle.

CHAPTER 17

1. Palmitic acid, 106 ATP; stearic acid, 120 ATP; linoleic acid, 116 ATP; oleic acid, 118.5 ATP. The lower values for linoleic and oleic acids vs. stearic acid are due to the presence of double bonds in the former.

2. 34% $[(108\,\text{ATP} \times 30.5)/(9788) \times 100\%]$.

3. 336.5 (106 from each palmitate, 18.5 from glycerol). 336.5 ATP/51 carbons $= 6.6$ ATPs per carbon atom (5.3 for glucose).

4. Example for palmitic acid (c) follows.
$\text{CoA-SH} + \text{palmitate} + \text{ATP} \rightarrow \text{palmitoyl-CoA} + \text{AMP} + PP_i$
$PP_i + H_2O \rightarrow 2P_i$
$\text{Palmitoyl-CoA} + 7\,\text{CoA-SH} + 7\text{FAD} + 7\text{NAD}^+ + 7H_2O \rightarrow$
$8\,\text{acetyl-CoA} + 7\text{FADH}_2 + 7\text{NADH} + 7H^+$
$8\,\text{acetyl-CoA} + 16H_2O + 24\text{NAD}^+ + 8\text{FAD} + 8\text{ADP}$

$+ 8P_i \rightarrow 16CO_2 + 24\text{NADH} + 24H^+ + 8\text{ATP} + 8\text{FADH2}$
$+ 8\,\text{CoA-SH}$

Sum: Palmitate $+ 24H_2O + 31\text{NAD}^+ + 15\text{FAD} + 8\text{ADP}$
$+ 6P_i \rightarrow 16CO_2 + 31\text{NADH} + 31H^+ + 15\text{FADH}_2$
$+ 7\text{ATP} + \text{AMP}$

Now add the equations for the metabolic oxidation of NADH and FADH_2.

$31\text{NADH} + 31H^+ + 15.5O_2 + 77.5\text{ADP} + 77.5P_i \rightarrow 31\text{NAD}^+$
$+ 77.5\text{ATP} + 108.5H_2O$

$15\text{FADH}_2 + 7.5O_2 + 22.5\text{ADP} + 22.5P_i \rightarrow 15\text{FAD} + 22.5\text{ATP}$
$+37.5H_2O$

Sum: Palmitate $+ 23O_2 + 108\text{ADP} + 106P_i \rightarrow 16CO_2$
$+ 107\text{ATP} + \text{AMP} + 122H_2O$

5.

Acetoacetyl-CoA $\longrightarrow$ 2 acetyl-CoA		0 ATP
2 Acetyl-CoA $\longrightarrow$ 4CO$_2$		20 ATP (two turns of citric acid cycle)

Sum: 20 ATP

Propionoacetyl-CoA $\longrightarrow$ propionyl-CoA + acetyl-CoA		4 ATP (one cycle of β-oxidation)
Propionyl-CoA $\longrightarrow$ succinyl-CoA		−1 ATP
Succinyl-CoA $\longrightarrow$ 4CO2		20 ATP (two turns of citric acid cycle)
Acetyl-CoA $\longrightarrow$ 2CO2		10 ATP (one turn of citric acid cycle)

Sum: 34 ATP

6. Carnitine acyltransferase I. If inhibition occurred at a later step, then palmitoylcarnitine oxidation would be inhibited, as well as that of palmitoyl-CoA.

7. The acyl-ACP produced by one subunit undergoes the next round of reductive two-carbon addition on the other subunit.

8. Acetyl-CoA $\xrightarrow{\text{Citric acid cycle}}$ oxaloacetate $\xrightarrow{\text{PEP carboxykinase}}$

PEP $\xrightarrow{\text{Gluconeogenesis}}$ glucose

The main point is that the carbons lost in one turn of the citric acid cycle are not the ones that entered as acetyl-CoA in that cycle.

9. Propionyl-CoA $\longrightarrow$ methylmalonyl-CoA $\longrightarrow$ succinyl-CoA $\longrightarrow$ oxaloacetate $\longrightarrow$ phosphoenolpyruvate $\longrightarrow$ glucose

10. 7 (during each cycle, one tritium atom is lost at the 3-hydroxyacyl-ACP dehydratase step, and the resulting double bond is reduced by unlabeled NADPH).

11. Increase in citrate levels would increase generation of acetyl-CoA in cytosol, hence stimulating fatty acid synthesis.

12. 1. Glucagon activates phosphorylation of pyruvate dehydrogenase, which inhibits the formation of acetyl-CoA, the substrate for the enzyme.
    2. Glucagon promotes triacylglycerol breakdown, yielding increased levels of fatty acids, which, as acyl-CoAs, could prevent polymerization and activation of acetyl-CoA carboxylase.

13. Malonyl-CoA, a key intermediate in fatty acid synthesis, inhibits carnitine acyltransferase I, thereby blocking the entry of fatty acyl units into the mitochondrion for oxidation. Fatty acyl-CoAs, the substrates for fatty acid oxidation, inhibit fatty acid synthesis by interfering with the polymerization of acetyl-CoA carboxylase. Hormonal effects on adipocytes are opposed; insulin promotes fatty acid synthesis by several mechanisms, while glucagon promotes fat breakdown and fatty acid oxidation.

14. This could be a way for a cell to inhibit fatty acid synthesis under conditions where substrates are needed for oxidation, to provide ATP.

15. Increased glucose levels in the cytosol stimulate glycolysis, which provides pyruvate for oxidation in the mitochondrion. The resultant acetyl-CoA can return to the cytosol (as citrate) and generate precursors for increased fatty acid synthesis. The increased intracellular glucose also stimulates adipocytes do synthesize glycogen.

CHAPTER 18

1. It is presumed that starch increases blood glucose levels less than simple sugars do. Thus, there is less stimulation of insulin secretion. Insulin would tend to retard energy mobilization from intracellular stores—something not desirable during a marathon.

2. About 2.6 hours. Normal glycogen reserves are about 6800 kJ for a 70 kg person (see p. 710 in Chapter 17). Thus,

   (6800 kJ) × (1 kcal/4.183 kJ) = 1625.6 kcal

   For running,

   (15000 kcal/day) × (1 day/24 hr) = 625 kcal/hr

   Time = (1625.6 kcal)/(625 kcal/hr) = 2.6 hr

3. About 28% at rest, about one-tenth that value during a marathon.

4. The measured proteolysis represents the sum of rates of protein synthesis and breakdown. The actual rate of protein breakdown does not rise, but in the early stage of a fast, the utilization of amino acids in catabolic pathways reduces the concentrations needed to support protein synthesis at rates that counterbalance breakdown. Later, as fatty acids and ketones are used more for energy, amino acids are spared for this purpose and are more readily available to be used for protein synthesis.

5. Glucosylated hemoglobin accumulates in the blood of diabetics and can easily be measured. Because the glucosylation reaction is covalent, the level of glucosylated hemoglobin reflects the level of blood glucose over a period of about four months (the life of an erythrocyte), whereas a single determination of glucose reflects only the value at the time of blood sampling. Thus, longer-term diabetes control can be monitored with greater accuracy if glucosylated hemoglobin is measured.

6. Liver contains low levels of the enzyme that synthesizes acetoacetyl-CoA from acetoacetate, ATP, and CoA-SH. Therefore, when liver synthesizes ketone bodies, they cannot readily be activated for catabolism within the hepatocyte. Instead, they are released and ultimately utilized by other tissues.

7. Adipose tissue lacks glycerol kinase. Glycolysis generates dihydroxyacetone phosphate, which is reduced to glycerol 3-phosphate (see Chapter 17).

8. (a) Malonyl-CoA at high levels inhibits carnitine acyltransferase I, and this inhibits ketogenesis by blocking the transport of fatty acids into mitochondria, both for β-oxidation and for ketogenesis (see Chapter 17).

(b) The high $K_{0.5}$ of hexokinase IV, a liver-specific enzyme, allows the liver to control the rate of glucose phosphorylation over a wide range of glucose concentrations. Accumulation of glucose-6-phosphate activates the b form of glycogen synthase and promotes glycogen deposition. By several mechanisms the liver also senses when blood glucose levels are low and mobilizes its glycogen reserves accordingly (see Chapter 13).

CHAPTER 19

1. ΔG is probably ~0 (K_{eq} close to unity) because the bond broken is identical to the bond created (a transesterification).

2. If phosphatidylserine synthase and phosphatidylserine decarboxylase (E_1 and E_2 in Figure 19.4) were kinetically and physically coupled, PS could never accumulate because once formed by E_1, it would immediately be converted to PE by E_2. If the two enzymes are tightly coupled in the membrane, then addition of radiolabeled PS to an enzyme system would not label PE because E_2 would act only on PS generated by E_1.

3. By stimulating release of arachidonic acid from membrane phospholipids. These in turn would be converted to prostaglandins, which contribute to inflammation.

4. Substitution of vegetable fats for animal fats could decrease cholesterol levels. This would ultimately decrease inhibition of HMG-CoA reductase levels by cholesterol, which could result in increased mevalonate levels.

5. glycerol + ATP → ~~glycerol-3-P~~ + ADP
stearic acid + ~~CoASH~~ + ATP → ~~stearoyl-CoA~~ + AMP + PP$_i$
oleic acid + ~~CoASH~~ + ATP → ~~oleyl-CoA~~ + AMP + PP$_i$
~~glycerol-3-P~~ + ~~stearoyl-CoA~~ + ~~oleyl-CoA~~ →
~~phosphatidic acid~~ + 2 CoASH
~~phosphatidic acid~~ + CTP → ~~CDP-diacylglycerol~~ + PP$_i$
~~CDP-diacylglycerol~~ + serine → sn-1-stearoyl-2-
phosphatidic acid + 2 CoASH

Net: glycerol + stearic acid + oleic acid + 3ATP + CTP + serine → sn-1-stearoyl-2-oleylglycerophosphorylserine + ADP + 2AMP + 3PP$_i$ + CMP

6. None, because the carboxyl C is lost in conversion to the isopentenyl pyrophosphate intermediate.

7. Because this is a nonhydrolyzable analog of glycerol-3-phosphate, you might expect it to be acylated without difficulty to give the phosphonate analog of diacylglycerol. By acting as an analog of diacylglycerol, this could competitively inhibit the synthesis of CDP-diacylglycerol from phosphatidic acid.

8. 25% from the choline utilization pathway, 75% from the phosphatidylserine pathway. If all the PC came from the PS pathway, you would expect the specific activity of PC to be 3× that of the methionine (3 × 2 = 6 mCi/mmol). Instead, it is only 1.5/6 = 25%, suggesting that 75% of PC was synthesized from free choline. These calculations assume that the endogenous methionine pool is small and becomes virtually completely labeled. To know true intracellular rates of synthesis, you must also know the specific radioactivity of the final intermediate in each pathway, whether pools of intermediates are compartmentalized, and the rate of degradation of the product (see Tools of Biochemistry 12A).

9. Phosphorylation of pyruvate kinase by cyclic AMP–dependent protein kinase. The phosphorylated form of the enzyme is far less active than the dephosphorylated form (see Chapter 14).

9. Acetoacetate + succinyl-CoA → acetoacetyl-CoA + succinate
acetoacetyl-CoA + acetyl-CoA → HMGCoA + CoASH
HMGCoA + 2NADPH + 2H$^+$ → mevalonate + 2NADP$^+$ + CoASH

10. By shutting down the pathway leading to aldosterone, this deficiency increases the supply of progesterone available for conversion to sex steroids.

11. The most straightforward route is desaturation of C16 palmitoyl-CoA to palmitoleoyl-CoA, followed by a C_2 elongation at the carboxyl group, both described on page 744, to give cis-vaccenyl-CoA. Alternatively, stearoyl-CoA could be an enzyme-bound intermediate in a process starting with elongation and followed by desaturation.

12. The most straightforward way is to purify one of the enzymes, such as PE methylase, to homogeneity and then to ask whether that enzyme can catalyze three methylations to give PC, or whether it can act upon the monomethylated and dimethylated intermediates. Alternatively, one could follow all three activities through a fractionation procedure.

13. Cyclic AMP promotes triacylglycerol breakdown, through activation of hormone-sensitive lipase. This probably increases intracellular levels of diacylglycerol, which could in turn increase flux through the last reactions in the salvage pathways to PE and PC (see Figure 19.7).

14. Example: 2 isopentenyl pyrophosphate to limonene:

CHAPTER 20

1. 5′-Deoxyadenosyl-B_{12}; tetrahydrofolate; ATP + glutamine; α-ketoglutarate + pyridoxal phosphate; S-adenosylmethionine.

2. (a) 5-Methyltetrahydrofolate.
   (b) Methylmalonate.
   (c) Decreased affinity of enzyme C for B_{12} coenzyme (K_M mutant).
   (d) Homocysteine (or homocystine) (see Chapter 21).

3.

4.

adenosylcobalamin

5.

6. One plausible theory proposes that ammonia depletes pools of α-ketoglutarate through the glutamate dehydrogenase and glutamine synthetase reactions, which convert αKG to glutamate and glutamine, and that this diminishes ATP production by reducing flux through the citric acid cycle.

7. CPS I is in mitochondria, and CPS II is in cytosol. Apparently, carbamoyl phosphate cannot cross the mitochondrial membrane, so that which is formed in mitochondria can be used only for arginine synthesis, and that formed in cytosol is used only for pyrimidine synthesis.

8. The injected enzyme should convert all circulating asparagine to aspartate. Normal cells would take up aspartate and resynthesize asparagine. However, leukemic cells would be unable to synthesize asparagine and would starve. On the other hand, an enzyme injected into the bloodstream would probably have too short a half-life to be effective. Moreover, being a foreign protein, it might cause an immunological reaction.

9. (a) True. The complete catabolism of amino acids yields CO_2, H_2O, and ammonia. However, the ammonia must be converted to urea for detoxification and excretion, and this requires ATP, which decreases the net ATP yield.
(b) False. Glutamate can be used to synthesize essential amino acids only if the carbon skeletons are available as keto acids.
(c) False. Arginine is used *catalytically* in the urea cycle in liver, and thus most of the arginine that is formed is cleaved to urea and ornithine. Little arginine is left over to meet the needs of other tissues.
(d) False. Although it is present in all proteins, alanine can be synthesized by mammalian cells and is not required in the diet.

10. Refer to Figure 20.13:

$\overset{+}{N}H_4$ (muscle) + α-ketoglutarate + NADH + H^+ $\longrightarrow$
$\qquad\qquad\qquad$ glutamate + H_2O + NAD^+

Glutamate + pyruvate $\longrightarrow$ α-ketoglutarate + alanine

Alanine + α-ketoglutarate $\longrightarrow$ glutamate + pyruvate

Glutamate + H_2O + $\overset{+}{N}AD^+$ $\longrightarrow$
$\qquad\qquad$ $\overset{+}{N}H_4$ (liver) + α-ketoglutarate + NADH + H^+

Sum: $\overset{+}{N}H_4$ (muscle) $\longrightarrow$ $\overset{+}{N}H_4$ (liver)

11. (a) NH_3 has an unshared electron pair that can initiate nucleophilic attack on the electron-poor carbonyl carbon atom.
(b) $[NH_3]$ + $[NH_4^+]$ = 100 μM. From the Henderson–Hasselbalch equation, at pH 8.0 $[NH_3]/[NH_4^+]$ = $10^{-1.2}$ = 0.063. Thus, $[NH_3]$ = 6.7 μM (and $[NH_4^+]$ = 93.3 μM). 6.7 μM $[NH_3]$ is ($6.7 \times 10^{-3})K_M$. The Michaelis–Menten equation can be rearranged to

$$\frac{V}{V_{max}} = \frac{[S]}{K_M + [S]}$$

Substituting $[S]$ = ($6.7 \times 10^{-3})K_M$ into the equation gives

$$\frac{V}{V_{max}} = \frac{6.7 \times 10^{-3} K_M}{K_M + (6.7 \times 10^{-3})K_M} = 0.0067$$

(c) The direction of a reaction depends both upon the equilibrium constant and the concentrations of reactants and products. The intramitochondrial $[NAD^+]/[NADH]$ ratio is high, and this drives the reaction toward α-ketoglutarate.

(d)

12. PLP forms a covalent Schiff base between the aldehyde carbon of the coenzyme and an ε-amino group of a lysine residue. Obviously, this bond must be broken for the coenzyme to form a Schiff base with an amino acid substrate.

13. (a) This will control the synthesis of carbamoyl phosphate by regulating the synthesis of glutamine, the preferred substrate for carbamoyl phosphate synthetase.
(b) This will have the effect of activating the deadenylylation of glutamine synthetase, which in turn activates glutamine synthetase. This would promote the utilization of α-ketoglutarate for glutamate and glutamine synthesis.
(c) This also has the effect of activating glutamine synthetase, under conditions where the ATP needed for the reaction is abundant.

14. First, purify the enzyme from HeLa cells or another human tissue and prepare antiserum against the enzyme. Then label HeLa cell cultures by growth in a radioactive amino acid. At intervals following removal of the labeled amino acid (a pulse-chase experiment), prepare a cell-free protein extract, treat an aliquot with antiserum, and count the radioactivity in the immunoprecipitate.

15.

Dihydrofolate:	GTP:
N-1	N-3
C-2	C-2
C-4	C-6
N-5	N-7
C-7	C-1'
N-8	N-9
C-9	C-3'

16. (a) B. $\quad$ (b) C. $\quad$ (c) D. $\quad$ (d) E, methionine.
(e) A. $\quad$ (f) F. $\quad$ (g) C.

17. Adenylylation tends to inactivate glutamine synthetase. Therefore, the effect of glutamine is to inhibit its own synthesis when it is present in abundance.

18. The most straightforward answer is to place the two enzymes in different cell compartments—the catabolic enzyme in mitochondria and the assimilative enzyme in cytosol.

CHAPTER 21

1. Formiminoglutamate, because the next reaction in its catabolism requires tetrahydrofolate.

2.

3. *N*-Acetylglutamate is an intermediate in ornithine biosynthesis. Activity of the urea cycle requires both ornithine and carbamoyl phosphate. If insufficient carbamoyl phosphate is available, ornithine will accumulate, and this could cause accumulation of the precursor, *N*-acetylglutamate. This accumulation acts as a signal to stimulate carbamoyl phosphate synthesis to increase urea cycle flux.

4.

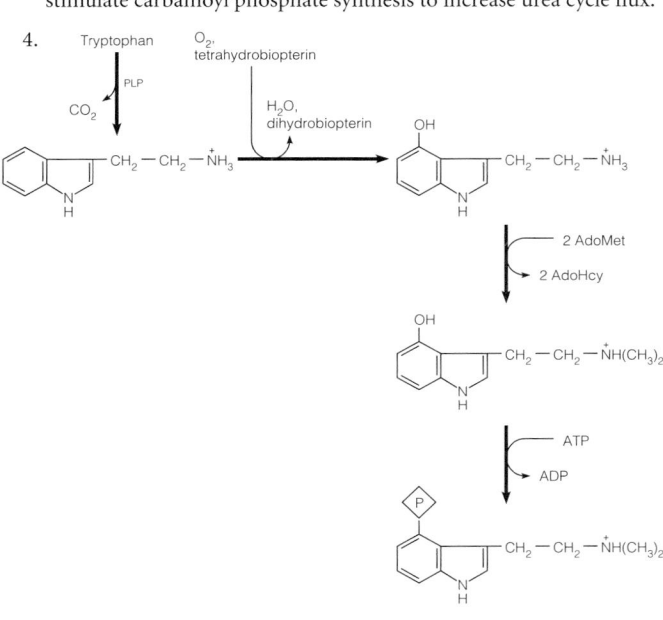

5.

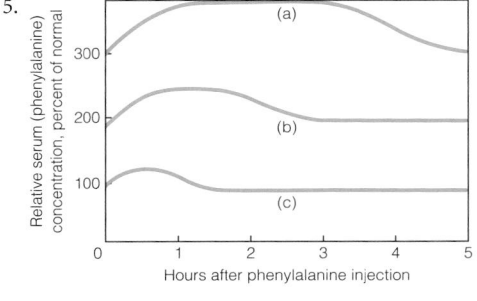

A tryptophan tolerance test

6. (a) Tetrahydrofolate, glycine, and serine hydroxymethyltransferase.
   (b)

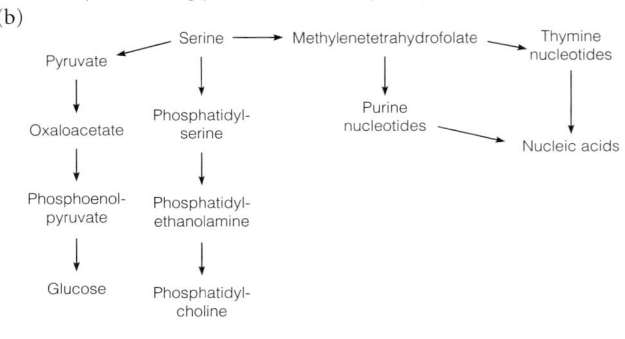

7. Alanine = 13 ATPs

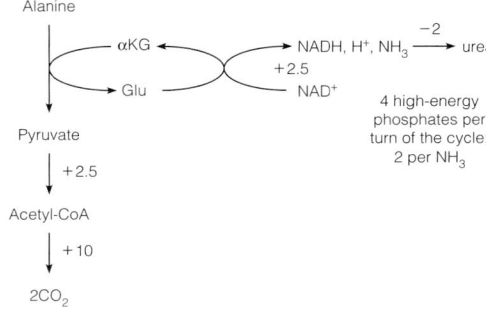

4 high-energy phosphates per turn of the cycle; 2 per NH₃

You might expect all of these pathways to generate a little more energy in a fish because it is not necessary to consume ATP in converting ammonia to urea for excretion.

Isoleucine = 36 ATPs (see Figure 21.5 and Problem 5 in Chapter 17)

Tyrosine = 40.5 ATPs (see Figure 21.8)

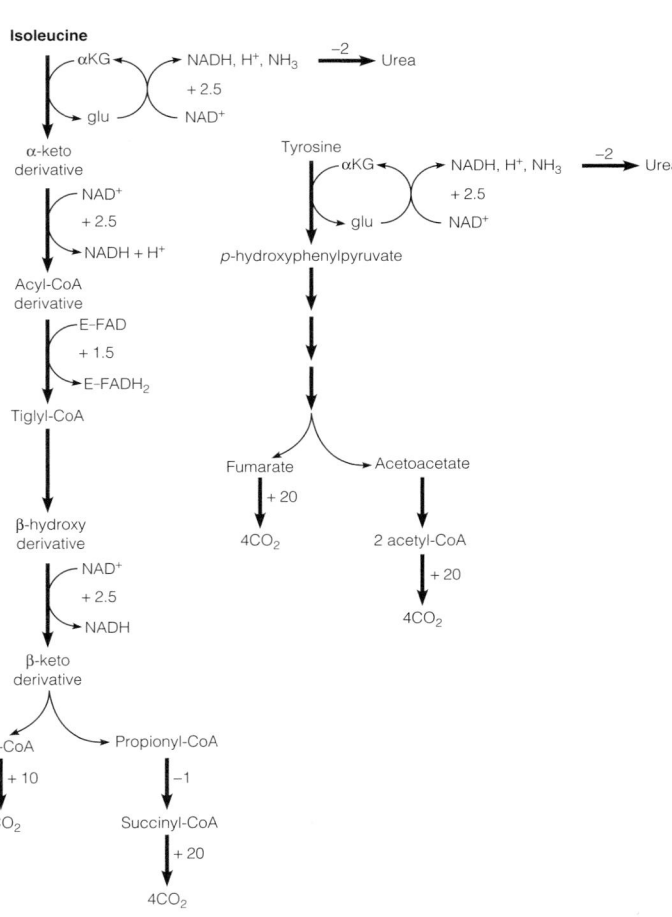

8. One should be feedback inhibited or have its synthesis repressed by threonine, a second by methionine, and a third by lysine because the aspartokinase reaction is involved in separate biosynthetic pathways leading to each of these three amino acids.

9.

10. Because the same enzymes are involved in comparable steps of both isoleucine and valine biosynthesis. Threonine dehydratase.

11. Threonine $\rightarrow \alpha$-ketobutyrate: inhibited by isoleucine

α-ketoisovalerate + acetyl-CoA $\rightarrow \beta$-isopropyl malate: inhibited by leucine

Control of valine synthesis is more complicated because three of the enzymes are involved in synthesis of all three amino acids. One could look for cumulative feedback inhibition—by valine, isoleucine, and leucine—of the first committed reaction:

Pyruvate + hydroxyethyl-TPP $\rightarrow \alpha$-acetolactate

12. Histamine.

R = imidazole

13. (a) 5.   (d) None.
    (b) 2, 3, 4.   (e) 5 (X3).
    (c) 3, 4.   (f) 5.

14. Because a pteridine reductase deficiency would impair all tetrahydrobiopterin-dependent reactions, which include the synthesis of catecholamines, serotonin, and nitric oxide, as well as tyrosine.

15.

16.

17. The structure of glyphosate is similar to that of phosphoenolpyruvate, suggesting that it acts as a competitive inhibitor with respect to PEP and a noncompetitive inhibitor with respect to shikimic acid 3-phosphate.

18. Because the synthesis of bilirubin diglucuronide occurs in the liver, a chronically diseased liver would be unable to effect this conversion, and unconjugated bilirubin would accumulate. In either a bile duct obstruction or an elevated heme destruction resulting from hemolysis, there would not initially be an impairment of the conjugating system, so most of the accumulated bilirubin would be in the conjugated form.

19. Reaction 1: pyridoxal phosphate
Reaction 2 and reaction 3: O_2 and NADPH (or another two-electron donor)

20. (a) Conjugation of glucuronate with bilirubin is carried out by a liver enzyme, and liver damage reduces capacity for this reaction, which is essential for solubilization and excretion of bilirubin.

(b) By diverting intermediates in heme synthesis to the nonutilizable type I porphyrins, these individuals are unable to synthesize sufficient heme for hemoglobin in red blood cells.

(c) Deficiency of this enzyme reduces flux through methionine synthase, and this leads to accumulation of more homocysteine than can be metabolized via cystathionine.

21.

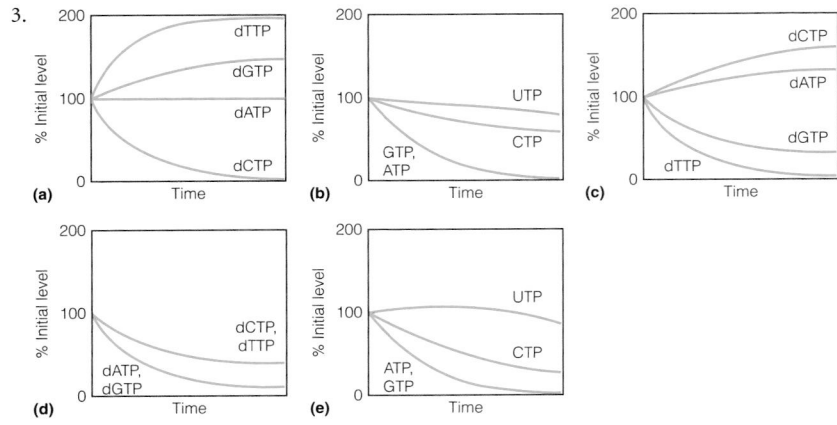

Modified from *Proceedings of the National Academy of Sciences of the United States of America* 103:13688–13693, J. Stetefeld, M. Jenny, and P. Burkhard, Intersubunit signaling in glutamate-1-semialdehyde-aminomutase. © 2006 National Academy of Sciences, U.S.A.

CHAPTER 22

1. (a) C.
   (b) E, F.
   (c) A, B.

2. $\text{BrdUrd} \xrightarrow[\text{ATP}]{\text{TK}} \text{Br-dUMP} \xrightarrow{\text{ATP}} \text{Br-dUDP} \xrightarrow[\text{ATP}]{\text{Nucleoside diphosphate kinase}} \text{Br-dUTP}$

   Br-dUTP can be incorporated into DNA in place of dTTP. However, it might also inhibit DNA synthesis by acting as a false feedback inhibitor of CDP reduction. Because dTTP is an allosteric inhibitor of CDP reduction by ribonucleotide reductase, Br-dUTP might have a similar effect. This could inhibit DNA replication by causing a dCTP deficiency.

3.

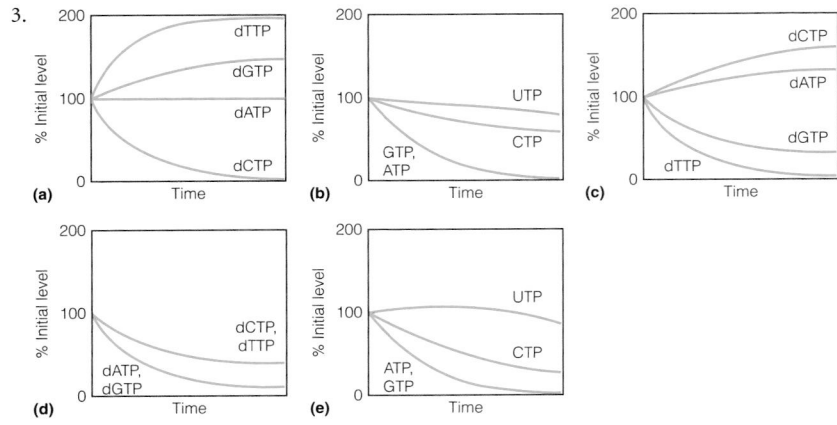

4. Outline answer (* = sites of allosteric regulation).

Uracil → uridine → UMP → UDP → UTP $\overset{*}{\to}$ CTP → CDP $\overset{*}{\to}$ dCDP → dCTP
   ↓*
dUDP → dUTP → dUMP → dTMP → dTDP → dTTP

5. Outline answer (* = sites of allosteric regulation).

HX → IMP $\overset{*}{\to}$ XMP → GMP → GDP $\overset{*}{\to}$ dGDP → dGTP
   ↓*
adenylosuccinate → AMP → ADP → dADP → dATP

6. Deoxycoformycin treatment might lead to adenosine and deoxyadenosine accumulation, as in adenosine deaminase deficiency. This would cause dATP to accumulate and shut off all activities of ribonucleoside diphosphate reductase.

7. Decreased flux through HGPRT could cause the substrates to accumulate, including PRPP. This could increase pyrimidine synthesis at the orotate phosphoribosyltransferase reaction, if PRPP levels are normally subsaturating for that enzyme. Orotate incorporation into nucleotide pools or nucleic acids would give a reasonable estimate of the rate of de novo pyrimidine nucleotide synthesis.

8. (a) Either a deficiency of thymidylate synthase activity, measured in a cell-free extract, or a growth requirement of cells for thymidine would confirm that the mutants are deficient in this enzyme.
(b) Because the mutants have negligible flux through the thymidylate synthase reaction, they do not deplete their intracellular tetrahydrofolate. Thymidine satisfies the need for synthesis of thymine nucleotides.
(c) Use methotrexate instead of trimethoprim because trimethoprim is not an effective inhibitor of mammalian dihydrofolate reductase.

9. Mutation in DHFR gene that makes the enzyme resistant; mutation in transcriptional or translational control mechanism that causes overproduction of normal DHFR; transport defect that causes failure of methotrexate to be taken up by cell.

10. (a) Inhibitors of thymidylate synthase and dUTPase specified by the viral genome.
(b) Virus-specified dUMP hydroxymethylase and hm-dUMP kinase:

5,10–CH$_2$ — THF   THF   ATP   ADP
dUMP ⟶ hm-dUMP ⟶

           ATP   ADP
hm-dUDP ⟶ hm-dUTP

(c) Virus-specified dCMP methylase, e.g.,

5,10–CH$_2$ — THF   DHF   ATP   ADP
dCMP ⟶ CH$_3$-dCMP ⟶

           ATP   ADP
CH$_3$-dCDP ⟶ CH$_3$-dCTP

11. (a) Mutants with a defect in DNA synthesis (no thymidine incorporated into DNA).
(b) Mutants defective in thymidine kinase, thymidylate synthase, dihydrofolate reductase, or serine hydroxymethyltransferase can be selected for by their resistance to decay of incorporated [^{3}H]deoxyuridine.

12. If intermediates in a pathway are restricted in their ability to diffuse away from enzyme surfaces, then a multistep pathway can be catalyzed efficiently with very low steady state levels of intermediates. Thus, you could set up an assay system for such a sequence and analyze levels of intermediates at various times. Another test for channeling involves running a multistep sequence beginning with a radiolabeled substrate and asking whether addition of a large excess of an unlabeled intermediate will dilute the radioactivity of the product. A pathway channeled by an enzyme complex preferentially uses intermediates that are produced by that complex.

13. 4-carboxy-5-aminoimidazole ribonucleotide + aspartate + ATP → N-succinylo-5-aminoimidazole-4-carboxamide ribonucleotide + ADP + P$_i$

IMP + aspartate + GTP → adenylosuccinate + GDP + P$_i$

Citrulline + aspartate + ATP → argininosuccinate + AMP + PP$_i$

14. Uncontrolled conversion of UTP to CTP elevates pools of cytidine and deoxycytidine nucleotides, while pools of uridine and thymidine nucleotides are diminished. The depletion of endogenous thymidine nucleotide explains the growth requirement for exogenous thymidine, and the perturbed dNTP pool imbalance (dCTP/dTTP ratio is elevated) causes replication errors, principally C incorporated opposite A, that lead to mutations.

15.
$$\underset{}{H_2N-\overset{\overset{O}{\|}}{C}-NH-CH_2-\overset{\overset{CH_3}{|}}{CH}-COO^-}$$

16. (a) Feeding thymidine stimulates salvage pathway synthesis and accumulation of ATP, which inhibits CDP reduction by ribonucleotide reductase and inhibits DNA replication by depleting the cell of dCTP.
(b) dATP binds to both activity and specificity sites of ribonucleotide reductase, with low and high affinities, respectively. When bound at the specificity site, dATP activates the reduction of CDP and UDP. At higher levels, dATP binds at activity sites and inhibits all four activities of ribonucleotide reductase.

17. Succinyl-CoA synthetase.

18. Glycine is incorporated into purines as an intact molecule at the GAR synthetase reaction (step 2), accounting for labeling of the C5 position. Label at C2 and C5 comes from 10-formyl-THF (steps 3 and 9). Glycine donates its C2 to the tetrahydrofolate one-carbon pool in the mitochondrial glycine cleavage reaction (see Chapter 21), yielding 5,10-methylene-THF, which is readily converted to 10-formyl-THF for incorporation into purines. This experiment is from Pasternack, L. B., D. A. Laude, Jr., and D. R. Appling, (1994) ^{13}C NMR analysis of intercompartmental flow of one-carbon units into choline and purines in Saccharomyces cerevisiae. *Biochemistry* 33:74–82.

CHAPTER 23

1. Using recombinant DNA techniques (see Chapter 4), one could prepare nucleic acid sequences that encode either an entire subunit of a G protein or a GTP-binding domain and ask whether that "probe" can hybridize with DNA from a plant species of interest. Alternatively, one could ask whether antibodies against mammalian G proteins cross-react with any proteins in a plant extract. If so, the protein could be isolated and its properties analyzed (including its ability to bind guanine nucleotides). Another approach is to look for proteins that can be ADP-ribosylated by cholera toxin or pertussis toxin.

2. Some possible factors: storage, control of activity, proper folding of the polypeptide chain, signal sequences for direction to the proper part of the cell.

3. Ethylene. It would probably be difficult to prepare an antibody against it, as it is a gas.

4. Because steroid hormone–receptor complexes activate the transcription of specific genes, a steroid hormone could conceivably activate the transcription of the gene for adenylate cyclase and, hence, increase the steady state level of this enzyme.

5. (a) 1, 4, 9, 14.
   (b) 5, 6, 8, 10, 16.
   (c) 1, 2, 7, 12, 16.
   (d) 1, 4, 15, 16.

6. One could treat a G protein–GDP complex with γ-[^{32}P]ATP and ask whether radiolabeled GTP is synthesized. The fact that it is not indicates that the G protein cannot phosphorylate bound GDP. One can also show that radiolabeled GTP can displace bound, unlabeled GDP under various conditions. One can also show that activation of the G protein requires the presence of GTP. Finally, one can demonstrate that the isolated α subunit has GTPase activity, by showing its ability to convert GTP to GDP in the absence of GDP.

7. (a) Viagra inhibits cyclic GMP phosphodiesterase, so cGMP levels remain high, thereby prolonging the effects of nitric oxide upon regional blood flow.
   (b) Prozac selectively inhibits reuptake of serotonin by presynaptic neurons, thereby increasing the amount of the neurotransmitter that reaches the postsynaptic cell, increasing synaptic transmission in areas related to regulation of mood.
   (c) If AdoMet could cross the blood–brain barrier, it might conceivably increase methylation of norepinephrine to give epinephrine. Because that seems unlikely, maybe one should view the initial report with skepticism.

8. cGMP + $H_2O \rightarrow$ 5′-GMP. Because Viagra acts by inhibiting this enzyme, any other compound acting similarly should potentiate the effect of Viagra.

9. Because inositol trisphosphate is derived from phosphatidylinositol, which in turn is derived from diacylglycerol and inositol, inositol phosphates must be completely dephosphorylated in order to be reincorporated into phosphatidylinositol for another round of synthesis.

10. 1. Most plant hormones are terpenoid compounds.
    2. There are no known peptide hormones in higher plants. Steroid-like compounds have been described in plants, but it is not clear that they act like steroid hormones.

11. The reaction of EGF with its receptor should be trimolecular, because EGF binding should promote dimerization of the monomeric receptor. Therefore, a simple hyperbolic curve would not be expected.

12. Catechol O-methyltransferase
    Norepinephrine + AdoMet $\rightarrow$ Epinephrine + AdoHcy

    Monoamine oxidase
    Dopamine + $O_2 \rightarrow$ 3,4-Dihydroxyphenylacetic acid + NH_3

CHAPTER 24

1. 10×10^3 bp/10^{10} bp in genome $= 10^{-6}$
   Amplification $= 2^{n-1}$ where n is the number of cycles
   After 10 cycles: $2^9 \times 10^{-6} = 1.02 \times 10^{-3}$
   After 15 cycles: $2^{14} \times 10^{-6} = 3.3 \times 10^{-2}$
   After 20 cycles: $2^{19} \times 10^{-6} =$ essentially 1.0

2. Flush ends: *Eco*RII, *Hae*III, *Hind*II, *Hpa*I, *Sma*I. Not made flush by DNA polymerase (because the recessed end lacks a 3′ hydroxyl): *Hga*I, *Hha*I, *Ple*I, *Pst*I.

3. (a)
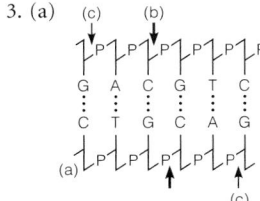
   (d) Up to 1 kb away from this site.
   (e) A particular hexanucleotide sequence would arise at intervals of $1/4^6$, or every 4056 base pairs.

4. (a) To linearize circular DNA, so that its electrophoretic mobility can be compared with those of other linear DNA fragments.
   (b) The recombinant DNA contains two tandem 1.15-kb inserts.
   (c)
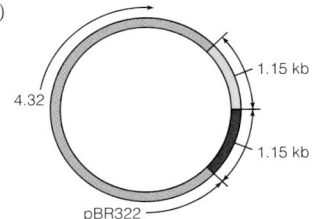

5. (a) Circular.
   (b)
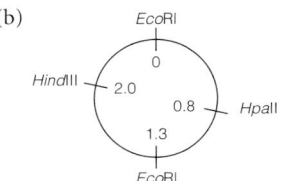
   (c) 7.
   (d) Cleave with *Hpa*II plus *Hind*III.

6. (a)

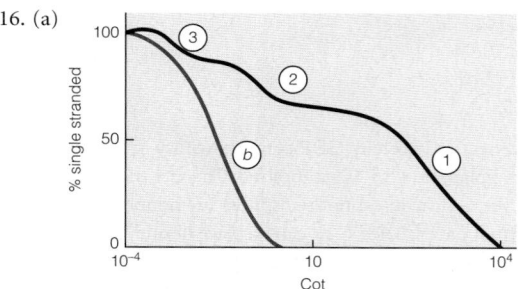

(b) A and B.
(c) E and F.
(d) F.

7. (a) 6.6×10^2 g/mol $\times 2 \times 10^8$ nuc. molecules/chromosome/6.02×10^{23} molecules/mol $= 2.2 \times 10^{-13}$ g DNA/ chromosome $+ 4.4$ g protein/chromosome $= 0.66$ pg
(b) 10^8 bp $\times 3.4$Å /bp $\times 10^{-8}$ cm/Å $= 3.4$ cm
(c) 3.4×10^{-2}m/5×10^{-6} m $= 6.8 \times 10^3$
(d) 46 chrom./cell $\times 10^{12}$ cells $\times 3.4 \times 10^{-2}$ m/chromosome $= 1.56 \times 10^{12}$ m

8. If we take the average nucleosome (including linker) to contain about 200 bp, we get a compaction ratio of 40 for the 30-nm chromatin fiber. Obviously, this fiber must be greatly folded on itself to yield the overall ratio of over 7000 found in the metaphase chromosome.

9. If the nucleosome were exactly positioned (all occupying the same site on the 208-bp fragment), only a single pair of bands (summing to 146 bp) would have been obtained. The fact that two such pairs $(29 + 117, 39 + 107)$ were observed means that there are two alternative locations, 10 bp apart.

10. Using a graph of log (bp) vs. d for the markers, one can interpolate the chromatin fragments to obtain sizes of 185, 380, 578, and 772 bp. A repeat of about 195 bp is indicated, with the monosome being slightly degraded.

11. The simplest explanation comes from the hypothesis (since proved) that the DNA lies on the surface of the particle. Thus, each strand is maximally exposed approximately every 10 residues and is most susceptible to nicking at these periodically spaced points.

12. The DNA makes 1.75 left-hand superhelical turns about each nucleosome. These turns are "constrained" and cannot be removed while the histones are still present. However, if the histones are first removed, the DNA so produced will have unconstrained supercoils, which can be relaxed by topoisomerase.

13. Histones are required in very large amounts during only one brief period (the beginning of S phase) in the cell cycle. Furthermore, they are always required in equivalent amounts, to make nucleosomes. The presence of multiple copies aids in rapid transcription, and an equal number in each domain should help maintain balance in production. The absence of introns and poly(A) tails means that two of the major steps involved in processing most eukaryotic genes are avoided, so that histone mRNAs can be delivered rapidly to the cytoplasm.

14. For example, cutting with both *EcoRI* and *BamHI* would yield a digest in which fragment B in the *BamHI* digest had disappeared. That places an *EcoRI* site within the B fragment. Sizes of all of the fragments in the digest would suggest that only one *EcoRI* site lies within fragment B. Now if a partial digest with *BamHI* yielded a fragment corresponding to A + B and digestion of this fragment with *EcoRI* yielded a fragment nearly as large as A + B, this would locate the *BamHI* site toward the 5′ end of B.

15. An addition or deletion or substitution of one or more bases in any symmetrical restriction site would inactivate that site.
As an example, for a tetranucleotide repeat –AATT- within the *EcoRI* site: –GAATTC– → –GAATTAATTC–
The *EcoRI* site is lost.

16. (a)

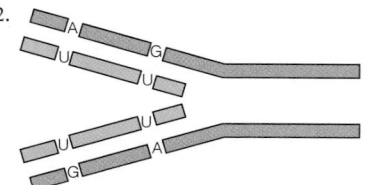

(b) The red line shown in the graph.
(c) The genomic DNA would renature more slowly than cDNA because it has greater sequence complexity among the 300-bp fragments used for the analysis.

17. Measurement of the approximate length of each phase of the cell cycle indicates that M, when cells are undergoing mitosis, accounts for about 3% of the total time in one cell cycle. That should represent the proportion of cells undergoing mitosis in a random (non-synchronized) culture, and, hence, the percentage of total cells showing condensed chromosomes.

CHAPTER 25

1.

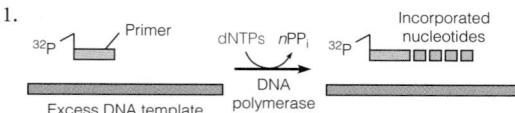

Use a 5′ end-labeled primer, and run a polymerase reaction in DNA excess (so that any polymerase that dissociates will rebind to a new chain). Stop reactions after a brief incubation and determine the molecular weight of radioactive material by gel electrophoresis. M_r tells how many nucleotides were incorporated per chain, which gives the processivity (number of nucleotides incorporated per unit time per chain).

2.

Incorporation of dUMP generates sites for chain breakage, as the uracil replacement process begins. If the single-strand breaks had not been fully repaired when DNA was isolated, short fragments would be seen. Fewer of these would be seen if the experiment were done with a mutant defective in uracil-DNA-*N*-glycosylase—and more if the experiment were done with a dUTPase-negative mutant.

3. (a) 1.725 g/cm^3.
   (b) 1.772 g/cm^3.

4. (a) To eliminate virtually all further incorporation of radioisotope.
   (b) If incorporated as ribonucleotides, the labeled product will be hydrolyzed by mild alkali. If incorporated as deoxyribonucleotides, the product is alkali stable.
   (c) RNA primers are metabolically unstable. They are rapidly degraded, as they are replaced by deoxyribonucleotides in DNA.

5. (a) Deoxyinosinic acid (dIMP).
   (b) Conversion of an A–T to a G–C base pair.

6. (a) 0.016 mm.
   (b) 9.79 × 10^4 nucleotides per fork per minute.
   (c) 1250 origins (2500 replication forks).
   (d) 2937 kilobase pairs.

7. (a) It's the reversal of the DNA ligase reaction. Single-strand interruption relaxes supercoiled DNA.

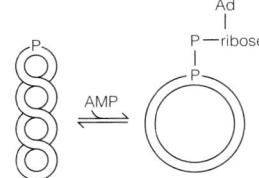

   (b) Analyze the DNA by gel electrophoresis.

8. Synchronize cells and label with 5-bromodeoxyuridine either early or late in S phase. Separate replicated DNA by CsCl equilibrium centrifugation. This can be analyzed by hybridization with cloned DNAs from genes known to be expressed or not expressed in those cells.

9.

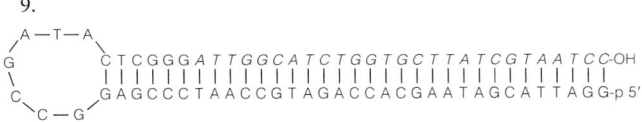

   Note: The 3′ exonuclease removes the mismatched A and T before polymerase action begins.

10. The polymerase activity was found to extend more slowly from an improperly paired terminus. Thus, a mismatched base has a longer residence time at the 3′ terminus than a matched base, increasing the likelihood of cleavage by the exonuclease activity.

11. Dideoxyinosine is converted to the corresponding dideoxyribonucleoside triphosphate, ddITP. Incorporation of this nucleotide in place of dGTP would block further chain elongation because of the absence of a 3′-OH terminus. If the HIV reverse transcriptase incorporates ddITP more readily than the cellular DNA polymerases do, then replication of the viral genome would be selectively inhibited.

12. If the mutation affected one of the feedback control sites of ribonucleotide reductase, the cell could overproduce dNTPs to the extent that competitive inhibition in vivo of replicative DNA polymerases by aphidicolin could be overcome.

13. They initiate rounds of replication at intervals shorter than 40 min. Thus, a cell may have two identical chromosomes, each at the same stage of replication.

14. Because ethidium bromide in the gel relaxes the positive supercoils, converting an overwound duplex to a relaxed circle, with a consequent loss of mobility.

15. The inchworming model suggests that each subunit of helicase is continuously in contact with just one DNA strand, while in the hand-over-hand model, continuous strand exchange occurs.

16. One mechanism would have the enzyme bind at two distant sites on DNA and then slide along DNA so that the two DNA contact points come closer and closer together. If any knots or links are encountered, they can be removed by topoisomerase cutting and resealing.

17. Fractional abundance: 10^{-6}.
    Abundance after 10 cycles, 1.02 × 10^{-3}; 15 cycles, 3.3 × 10^{-2}; 20 cycles, 1.0 (essentially 100%).

18. What is needed here is an in vitro system that will replicate mtDNA with purified proteins (polymerase, helicase-primase, single-strand DNA-binding protein, etc.) so that replication can be initiated synchronously in a population of purified mtDNA molecules. Such a system may be available soon.
    (a) Incubate mtDNA in this system with unlabeled dNTPs and radiolabeled rNTPs. Determine the ability of labeled RNA to anneal to denatured mtDNA restriction fragments. If the RNA represents replication primers it should hybridize only to fragments representing replication origins. Another approach is to carry out a transfer experiment comparable to the Okazaki experiment outlined in Figure 25.16, and determine what proportion of incorporated ribonucleotides become linked to deoxyribonucleotides.
    (b) If mtDNA replicates by the strand displacement model described in Figure 25.36, then all DNA synthesis is unidirectional. Label replicating DNA with radiolabeled dNTPs in the above in vitro system. Because complete mtDNA replication requires an hour, all labeled DNA formed in the first ten minutes or so should hybridize only with L strand DNA. Later in the process, DNA that becomes labeled will hybridize with both strands. Doing this experiment requires that one separate the strands of mtDNA, so that annealing to each strand can be measured separately.

19. It is likely that increasing dNTP concentrations, even modestly, will increase rates of DNA chain extension from a mismatch at the 3′ terminus. This will have the effect of inhibiting proofreading because a mismatch that occurred because of incorporation of an incorrect deoxyribonucleotide will have a greater probability of undergoing chain extension than nucleolytic removal, and this seals the mismatch in place, where a mutation will occur in the next round of replication.

20. (a) Extensive treatment with topoisomerase I will relax all DNA molecules, so they should all migrate like species 6 in Figure 25A.1, a circular molecule that is neither underwound nor overwound.
    (b) Treatment of underwound DNA with gyrase and ATP would make an already underwound population of molecules even more underwound. So the pattern of spots shown in Figure 25A.1 might be expected to shift downward as the figure is oriented. The most strongly underwound molecules might not change their migration because there is a thermodynamic limit to the number of superhelical turns that gyrase can introduce.

CHAPTER 26

1. Increase, because mismatch repair would repair the "correct" mismatched nucleotide half the time.

2. (a) 2, 7, 9, 11.
   (b) 3, 5, 6.
   (c) 2, 11.
   (d) 1, 11.
   (e) 1, 8, 11.
   (f) 4, 6.
   (g) 10, 11.
   (h) 6, 11, 12.

3. (a) A.
   (b) B.
   (c) C.
   (d) D.
   (e) A, C.
   (f) C.

4. Because both strands are methylated.

5. dUTPase deficiency increases dUMP incorporation into DNA and increases subsequent excision repair. Ligase deficiency increases the mean lifetime of Okazaki fragments. Both conditions increase the number of single-strand interruptions, the structures that initiate recombination.

6. First, clone a large number of DNA fragments from cells containing the integrated virus. Next, carry out colony hybridization on these clones, using labeled viral DNA as the hybridization probe. From Southern blot analysis of clones containing viral DNA, identify viral DNA fragments that are larger than corresponding fragments from the nonintegrated double-strand viral DNA. These should represent viral sequences fused to adjacent cellular sequences. These can now be subcloned and sequenced.

7. Nonenzymatic deamination of the cytosine in a G·C base pair yields a G·U base pair, which is readily repaired by the uracil-N-glycosylase base excision repair system. However, if the C in that base pair has been methylated before deamination, then deamination of G·mC yields a G·T base pair, which is not as easily corrected. Also, it is possible that methylation of C in a G·C base pair facilitates its deamination.

8. This is a surprising result because the function of the MutT protein is thought to be hydrolysis of 8-oxo-dGTP, an oxidation product of dGTP. It is not clear how this oxidation could occur in the absence of O_2. Perhaps MutT acts also on another mutagenic nucleotide not yet identified, which can be formed during anaerobic growth.

9. It's likely that repair would occur on the nicked strand, for that would be the strand containing 3′ hydroxyl termini from which repair polymerases could initiate DNA synthesis.

10. The most likely pathway involves deamination of G-C to G-U, with the U specifying A on the next round of DNA synthesis. Another possibility is for UNG to create an abasic site after removing U created by deamination, with any base then being inserted opposite that site at the next round of replication, before the repair process can be completed. Or, during repair DNA synthesis after deamination of C in a G-C base pair, dUTP could be incorporated at an A-T base pair, leading to A-U, which then undergoes erroneous UNG repair, as described above.

11. By inserting within a gene the transposon would interrupt expression of the gene. By inserting into an adjacent regulatory region, the result could be either enhanced or reduced transcription of the gene, depending upon the nature of the regulatory region.

CHAPTER 27

1. Pulse-label cells with a radioactive RNA precursor, isolate RNA, and hydrolyze with mild alkali. This will yield one nucleoside per 3′ terminus. Total picomoles of labeled nucleotide in the hydrolysate divided by the number of picomoles of nucleosides gives the number of nucleotides incorporated per chain in the labeling interval.

2. First, devise conditions under which only the cloned gene is synthesized. You could initiate transcription by adding three ribonucleotides. At time zero, add the fourth nucleotide and stop incubation after a fixed time. Use a small, measured amount of template so that the number of transcribing genes is roughly equal to the number of template molecules. Determine the average length of transcripts synthesized by gel electrophoresis.

3. By gel electrophoresis of radiolabeled RNA products. Any paused species will accumulate and can be detected as a heavier-than-expected band on a radioautograph of the gel.

4. A dinucleotide, complementary to the first two template nucleotides, can bypass the first nucleotide incorporation step and permit efficient transcription at low nucleotide concentrations. Then one can arrest all transcription events at one nucleotide by

knowing the nucleotide sequence of the gene being transcribed and adding only two or three rNTPs, instead of all four, with the missing nucleotide being the one that would have been inserted at the site at which you desire to arrest transcription.

5. Because it uses the −35 region of *trp* and the −10 region of *lac*, the hybrid *tac* promoter more closely matches the consensus −35 and −10 sequences than does either *lac* or *trp*.

	Number of identities with consensus sequence
lac	9/12
trp	8/12
tac	10/12

6. Actinomycin is a noncompetitive inhibitor because it binds to DNA, not to a substrate-binding site on the enzyme. Cordycepin acts as the nucleotide, 3′-deoxyATP. This should compete with ATP for utilization by RNA polymerase.

7. (a) RNA $3'$ CUAGGAUUCAGUAAGG $5'$

    DNA $5'$ GATCCTAAGTCATTCCCATTAC<u>CAGTAT</u>AG $3'$
    $3'$CTAGGATTCAGTAAGGGTAATGGTCATATC $5'$

(b) Pribnow box is underlined in (a)

(c)   ↓

TAGCTA
ATCGAT

↑

8. If two genes on a plasmid are oriented in the same direction and both are being transcribed, then the overwinding created ahead of gene 1 can be compensated by the underwinding created behind gene 2, and vice versa.

9. One way would be to measure the distribution of poly(A) tail lengths in bulk mRNA. If a distribution were found that cut off at a distinct minimal value, the hypothesis would be supported; all mRNA with shorter tail lengths would have been degraded.

10. There are many possible ways. One is to simply cross-link DNA to histones in an in vitro transcription system. If this blocks transcription, which proceeds in the presence of un-cross-linked nucleosomes, this provides evidence of the necessity of removal, or at least displacement. A more critical test would be to carry out transcription in vitro in the presence of an excess of radiolabeled, nongenomic DNA. If nucleosomes are found on this after transcription, removal from the genomic DNA is indicated.

11. (a) 7 exons, 6 introns.

(b) These two structures will be at opposite ends of the RNA (shorter strand). However, one cannot unequivocally identify the $5'$ and $3'$ ends from such a figure, unless something is known of its sequence.

12. It may be difficult to form loops shorter than 65 nucleotides.

13. Once the transcription complex is formed, it is extremely stable, and in accessible to heparin.

14. A 50-kDa protein would contain about 450 amino acid residues, and therefore would be encoded by a gene containing 3×450, or 1350 nucleotides. If V_{max} for RNA polymerase is 50 nucleotides per second, this gene could be transcribed in 27 seconds, assuming that the gene contains no introns. Also the time needed for initiation and formation of a stable transcription complex would lengthen the time needed.

CHAPTER 28

1. · · · Ile–Tyr–Ile–Tyr–Ile–Tyr– · · · . Either AUA codes for Ile and UAU for Tyr or vice versa.

2. The mitochondrial code uses AUA to code for *Met*, not *Ile*. The experiment also tells us that UAU is a Tyr codon.

3. (a) First, codons such as UUU or AAA can yield only one polypeptide. Second, some reading frames in some polynucleotides will be read as repeats of stop codons.

(b) $(GUA)_n$ Val_n or Ser_n (third reading frame is UAG = stop).
$(UUA)_n$ Leu_n or Tyr_n or Ile_n

4. Any polynucleotide built from a tetranucleotide repeat, unless a stop codon is involved.

5. The Shine–Dalgarno sequence, like the AUG that is read uniquely as fMet, occurs near the $5'$ end of the message. The implication is that, in forming the initiation complex, only a region near the $5'$ end of mRNA can bind. That is, a Shine–Dalgarno sequence occurring by chance within a message could not bind at this site to form an initiation complex. Other explanations may be possible.

6. (a) $5'GGU3'$, $5'GGC3'$, or $5'GGA3'$. All are Gly codons.

(b) $5'AGU3'$ or $5'AGC3'$. Both are Ser codons.

7. (a) The simplest explanation for particles of 100S and 130S is that they represent some kind of dimers and trimers of 70S particles (note the nonlinearity of S with mass). From what we know now, these represented small oligoribosomes on mRNA. The research missed discovering the basic mechanism of protein synthesis.

(b) Dissociation would have produced ribosomal subunits and free RNA. Because these were only fragments of polyribosomes, the chance that they would reassociate onto the RNA is small. Furthermore, mRNA is extremely unstable, unless precautions are taken to eliminate traces of ribonucleases (a fact that this experimenter did not appreciate at that time). Thus, the mRNA was probably quickly degraded.

8. Because the tRNA is going to be released from the E-site, codon recognition and pairing with the anticodon are in no way advantageous and may, in fact, be detrimental to smooth release.

9. 32 different tRNAs with different anticodons, using wobble rules, could translate all 61 sense codons.

10. If δ is the probability that an error is committed at each step, then $1 - \delta$ is the probability that an error has not been made in any one step. The probability that no error has been made in *any* of the n steps is then $(1 - \delta)^n$. This is the probability that the protein is entirely error-free.

11. (a) 0.990.

(b) 0.904. In this case, nearly 10% of the proteins would contain one or more errors.

12. Bacteria are given a pulse of radioactive amino acids (e.g., [^{14}C] leucine) to label new polypeptide chains. The cells are broken; nucleoids, cell walls, and membranes are removed by centrifugation; and the lysate is placed on a sucrose gradient. The appearance of ^{14}C in fractions sedimenting more rapidly than 70S indicates that newly synthesized protein is associated with such structures. Samples taken some time *after* the pulse will have little label in these fractions, showing that it was protein in the process of translation that was associated with the polysome fraction. Ribonuclease digestion will shift the radiolabel to 70S or lower.

13. 36 (ATP/glucose)/4 (ATP/residue added) = 9 residues added/glucose.

This is a maximum because any inefficiency in ATP production, or use of extra ATP in proofreading and so forth, will give less efficiency.

14. Each puromycin molecule becomes bound to the polypeptide chain it has aborted and is hence used only once. An erythromycin molecule can block an entire polyribosome and can be used repeatedly.

15. Cutting a phosphodiester bond near the 3′ end of 16S RNA might (and in fact does) remove the sequence that binds the Shine–Dalgarno sequence on the messenger RNA.

16. If a frameshift mutagen creates a one-base insertion adjacent to a tRNA anticodon, then that tRNA might translate an mRNA that also has a one-base insertion at the corresponding codon site.

17. Thymine in DNA occurs as the result of thymidylate synthase creating dTMP, which is phosphorylated to dTTP and incorporated into DNA by DNA polymerase.
    Thymine in tRNA arises post-transcriptionally, by an S-adenosyl-methionine-dependent methylation of a UMP residue.

18. The assumption is that the frameshift mutagen created a two-base insertion and that the second mutation created an additional one-base insertion. This restored the reading frame, but with a net increase of one amino acid residue—an acceptable modification of the protein. Or, the first mutation could have been a one-base insertion, with the reading frame restored by a two-base insertion. The inferred codon assignments were as follows.

Wild-type: AA^A/$_G$ AGU CCA UCA CUU AAU GC.....

$\downarrow$ +2 (GU or UG) $\searrow$ +1 (G or A)

Double mutant: AA^A/$_G$AGU GUC CAU CAC UUA AU^A/$_G$ GC......

19. PolyG is insoluble, and, hence, cannot be used as a translation template in vitro.

CHAPTER 29

1. (a) $\dfrac{(10 \text{ molecules/cell})/6.02 \times 10^{23} \text{ molecules/mol}}{0.3 \times 10^{-15} \text{ L/cell}}$

   $= 5.5 \times 10^{-8}$ M

   (b) $\dfrac{10 \text{ molecules/cell}}{5.5 \times 10^{-8} \text{ M}} = \dfrac{2 \text{ molecules/cell}}{1.1 \times 10^{-8}\text{M}}$

   (c) Approximately 2.5×10^{-14} M.
   (d) A rapidly growing cell could have two copies, on average, because each single chromosome is partly replicated.

2. It's unlikely because in eukaryotes transcription occurs in the nucleus and translation in the cytosol. Thus, direct coupling between transcription and translation, essential for attenuation, is absent.

3. A dinucleotide, complementary to the first two template nucleotides, can bypass the first nucleotide incorporation step and permit efficient transcription at low nucleotide concentrations. Then one can arrest all transcription events at one nucleotide by knowing the nucleotide sequence of the gene being transcribed and adding only two or three rNTPs, instead of all four, with the missing nucleotide being the one that would have been inserted at the site at which you desire to arrest transcription.

4. Using footprinting, carry out competitive binding experiments to ask whether repressor inhibits RNA polymerase binding and vice versa. Also, examine transcription in vitro to determine whether transcription extends as far as the repressor binding site.

5. (a) The plasmid *i* gene encodes a repressor that cannot bind inducer.
   (b) The plasmid operator cannot bind repressor.

6. Mu inserts randomly in the genome (see Chapter 26). If it is inserted downstream from a damage-inducible promoter, UV irradiation of that cell will activate β-galactosidase synthesis from the integrated phage genome.

7.

	Before	After
(a)	−	+ +
(b)	−	+ +
(c)	−	+
(d)	−	−
(e)	−	−
(f)	+ +	+ +
(g)	−	+

8. A mutation that abolishes binding to repressor. Subunits of the mutant repressor could interact with those of the wild-type repressor synthesized in the same cell, to form mixed tetramers with reduced affinity for operator DNA.

9. The simplest test would be to treat a *lacA* mutant with a number of potentially toxic β-galactosides and determine their ability to kill the bacteria or to slow their rate of growth. If transacetylase activity is required to eliminate these potentially toxic compounds from the cell, then the inability to carry out this reaction should result in demonstrable toxicity.

10. The repression of EF-G could be seen as part of a coordinated shutdown of protein synthesis as a result of ribosomal protein accumulation. Thus, both ribosome assembly and a key step in translation would be inhibited. With respect to an RNA polymerase α subunit, this is a catalytic function, whereas inhibition of ribosome assembly is a stoichiometric function. Presumably one subunit of the catalytic RNA polymerase could be decreased a bit, with only modest attenuation of intracellular RNA polymerase flux rates, while ribosome assembly would be inhibited in proportion to the decreased availability of ribosomal subunits.

11. The main similarity is that binding to the 5' end of an mRNA inhibits expression of that RNA, through evolution of a potential binding site on the RNA for a regulatory compound. Main differences: The regulatory compound is a macromolecule (ribosomal gene expression) or a small metabolite (riboswitch). Binding to mRNA affects translation of that RNA, thereby controlling synthesis of a gene product (ribosomal gene expression) or expression of an early gene in a biosynthetic pathway (riboswitch). Finally, riboswitch control can be expressed either at the level of translation (like ribosomal protein synthesis) or transcription.

12. One approach: label all RNA species by incubation in the presence of a radiolabeled RNA precursor. Use the labeled RNA library to probe a microarray representing all genes of the cell. Another approach: Reverse transcribe the RNA species in a total RNA extract, then carry out deep sequencing to detect all DNA sequences present.

absorbance (A_l) A dimensionless number that indicates how well a solution of a substance absorbs light of a given wavelength. It is defined as the negative logarithm of the fraction of light of wavelength l that passes through a sample of the solution; its value depends on the length of the light path, the concentration of the solution, and the extinction coefficient of the substance at that wavelength.

acetylcholinesterase An enzyme found in cholinergic synapses that breaks down acetylcholine and thus terminates its action on the postsynaptic cell.

action potential A wave of transient depolarization that travels along the membrane of a nerve cell (or any other kind of excitable cell, such as a muscle cell) as a result of fluxes of ions across the membrane. A nerve impulse.

activated state With respect to a chemical reaction, a transient high-energy state of a reactant molecule (such as an unfavorable electron configuration or strained conformation) that enables the molecule to undergo the reaction.

active site The site on an enzyme molecule where the substrate binds and where the reaction is facilitated. It is often a cleft or pocket in the surface of the enzyme.

active transport The transport of a substance across a biological membrane by a mechanism that can work against a concentration (or electrochemical) gradient. It always requires the expenditure of cellular energy. Compare *facilitated transport, passive transport.*

adenylate energy charge A measure of the energy status of a cell, calculated as the ratio of the ATP concentration plus one half the ADP concentration divided by the sum of the AMP, ADP, and ATP concentrations. See *energy charge.*

adenylylation In cells, the transfer of an adenylyl moiety from ATP to another molecule. Some enzymes are regulated by reversible adenylylation; often incorrectly referred to as adenylation.

adipocytes Fat cells; cells that are specialized for storing triacylglycerols and for releasing them to the blood in the form of fatty acids and glycerol as required.

adrenergic receptors Cell-surface receptors that bind epinephrine or norepinephrine. There are several different types with somewhat different ligand specificities and effects. (The term comes from *adrenaline,* the old name for epinephrine.)

affinity chromatography A protein purification technique in which a specific ligand for the protein of interest is immobilized upon a chromatographic support, so that enrichment of that protein is based upon its specific affinity for the immobilized ligand.

affinity constant See *association constant.*

agonist In molecular biology, a substance that mimics the cellular effects of a natural compound (such as a hormone or neurotransmitter) by binding to and activating the same cellular receptor. Compare *antagonist.*

A helix Referring to nucleic acids, a right-handed helix structure of nucleic acid duplexes that has a smaller pitch and a larger diameter than the B-DNA helix. It is the structure adopted by RNA duplexes and RNA–DNA hybrid molecules.

AIDS (acquired immune deficiency syndrome) A disease caused by prolonged infection with human immunodeficiency virus (HIV) and characterized by crippling of the immune system. Victims die of infections or cancers that their immune system cannot control.

alditols Compounds that are produced by reducing the carbonyl group on a monosaccharide.

aldose A monosaccharide in which the carbonyl group comes at the end of the chain and thus represents an aldehyde group. Compare *ketose.*

alkaloids A large group of nitrogenous basic substances found in plants. Most of them taste bitter, and many are pharmacologically active. The term can also be used for synthetic compounds of the same type. Morphine, caffeine, and nicotine are familiar alkaloids.

alignment Referring to sequence alignment, the comparison of nucleic acid or protein sequences by lining up identical or closely related portions of sequence in the two sequences being compared.

allele A specific version of a gene that occupies a particular location in the genome. It is distinguished from other alleles of the same gene by differences in nucleotide sequence.

allosteric With respect to enzymes, an effect that is produced on the activity of one part of an enzyme (such as an active site) by the binding of an effector to a different part of the enzyme.

alternative splicing The splicing of a eukaryotic RNA transcript in different ways, to include or exclude certain exons from the final mRNA.

Alu Elements DNA sequences about 300 base pairs long that occur in many copies scattered throughout the genome of primates; the human genome has hundreds of thousands of them. They may serve an unknown function, or they may be purely "parasitic," spreading as mobile elements through the genome.

Ames test A test for the mutagenicity of a substance. A strain of the bacterium *Salmonella typhimurium* having a mutation that disables an enzyme necessary for histidine utilization is exposed to the substance in question and plated on a medium lacking histidine. A reversion mutation that activates the mutant enzyme causes the cells to grow on this medium.

amino terminus See *N-terminus.*

amphipathic For a molecule, the property of having both hydrophobic and hydrophilic portions. Usually one end or side of the molecule is hydrophilic and the other end or side is hydrophobic.

ampholyte A substance whose molecules have both acidic and basic groups.

anabolism The sum of all the metabolic processes by which complex biomolecules are built up from simpler ones. In general, these processes consume rather than produce cellular energy. Compare *catabolism.*

anaerobic Refers to the absence of oxygen or the absence of a need for it; processes that must or can occur without oxygen are called anaerobic processes.

anaplerotic Referring to pathways that replenish pools of citric acid cycle intermediates.

androgens The male sex hormones; specifically, the steroid hormones testosterone, dihydrotestosterone, and androstenedione, which act mainly to promote male sexual development and maintain male sex characteristics.

anneal Referring to a process (usually slow cooling after heat denaturation) that allows complementary nucleic acid single strands to find each other and pair up, reforming a duplex molecule.

anomers Stereoisomers of cyclized monosaccharide molecules differing only in the configuration of the substituents on the carbonyl carbon. (This carbon is a center of chirality in the cyclized but not in the open-chain form of the molecule.)

antagonist In biochemistry, a substance that counteracts the cellular effects of a natural compound (such as a hormone or neurotransmitter) by binding to the cellular receptor for the compound and blocking its action. Compare *agonist.*

antibodies (also called immunoglobulins) A set of related proteins that are produced by B lymphocytes and can bind with specificity to antigens. Some types are released into body fluids and mediate humoral immunity; other types are retained on the surface of the B cell or are taken up and displayed by some other cell types.

anticodon The nucleotide triplet on a tRNA that binds to a complementary codon on mRNA during protein synthesis and thereby mediates the translation of the codon into a specific amino acid.

antigen A substance that can elicit a specific immune response.

antigenic determinant See *epitope*.

antimetabolite A substance that is a structural analog of a normal metabolite or otherwise resembles it and that interferes with the utilization of the metabolite by the cell.

antioxidant A strongly reducing compound, such as ascorbic acid, which counteracts the tendency of a metabolite to undergo oxidation to a potentially toxic or harmful species.

antiport A membrane transport process that couples the transport of a substance in one direction across a membrane to the transport of a different substance in the other direction. Compare *symport*.

antisense RNA An RNA molecule that is complementary to an mRNA; it can block translation of the mRNA by forming a duplex with it. Gene expression can be regulated by the production of antisense RNAs.

antiserum Serum that contains antibodies against a particular antigen.

apolipoproteins The specific proteins that constitute the protein fraction of lipoproteins; they mediate the interactions of lipoproteins with tissues.

apoprotein For a protein with a prosthetic group (which see) the polypeptide portion of the molecule without the prosthetic group.

apoptosis Programmed cell death (as distinguished from necrosis; see *autolysis*).

archaea A group of prokaryotes that are biochemically distinct from the true bacteria (Eubacteria) and that separated from them early in the history of life. Modern archaea live mostly in extreme environments, such as acid hot springs.

association constant (*K*) (also called affinity constant) An equilibrium constant that indicates the tendency of two chemical species to associate with each other; it is equal to the concentration of the associated form divided by the product of the concentrations of the free species at equilibrium.

asymmetric carbon A carbon molecule that carries four different substituents and therefore acts as a center of chirality, meaning that the substance can occur in two different enantiomers (stereoisomers that are nonsuperimposable mirror images of each other).

atherosclerotic plaques The protruding masses that form on the inner walls of arteries in atherosclerotic disease. A mature plaque consists partly of lipid, mainly cholesterol esters, which may be free or contained in lipid-engorged macrophages called foam cells, and partly of an abnormal proliferation of smooth-muscle and connective-tissue cells.

attenuation A mechanism for regulating prokaryotic gene expression in which the synthesis of a nascent RNA transcript is terminated before RNA polymerase has reached the structural genes.

autocatalytic Refers to a reaction that an enzyme catalyzes on part of its own structure, such as a cleavage performed by a protease on its own polypeptide precursor.

autoimmunity A condition in which the body mounts an immune response against one of its own normal components.

autolysis Programmed cell death; the orderly self-destruction of a cell in a multicellular organism. It is the process by which unwanted cells are eliminated in the body. Also called *apoptosis* (which see).

autonomously replicating sequences (ARSs) Sequences in yeast chromosomes that, when incorporated into an artificial plasmid, enable the plasmid to replicate efficiently in yeast cells.

autotrophs Organisms that can synthesize their organic compounds entirely from inorganic precursors, in particular needing only CO_2 as a carbon source. Compare *heterotrophs*.

auxotrophs Microorganism strains that require as a nutrient a particular substance that is not required by the prototype (wild-type) strain.

Usually the requirement results from a mutation that disables an enzyme necessary for the endogenous synthesis of the substance.

axis of symmetry An imaginary axis through a structure, such that rotating the structure around the axis through an appropriate angle leaves the appearance of the structure unchanged.

axon A threadlike process extending from a nerve cell by which impulses are transmitted to other nerve cells or to effector cells such as muscle or gland cells. Most nerve cells have one axon; shorter processes that function in receiving impulses from other neurons are called dendrites.

base excision repair (BER) A DNA repair process that begins with cleavage of a glycosidic bond between a damaged base and the deoxyribose to which it is linked.

B-DNA A DNA duplex with a specific right-handed helix structure. It is the usual form of DNA duplexes in vivo.

Beer's law The equation that relates the absorbance of a solution sample at a given wavelength to the length of the light path, the concentration of the dissolved substance, and the extinction coefficient of the substance at that wavelength. See *extinction coefficient*.

bile acids A family of amphipathic cholesterol derivatives that are produced in the liver and excreted in the bile; salts of the bile acids emulsify fat in the intestine.

biogenic amines A set of low-molecular-weight amino acid derivatives that contain a basic amino group and function in the body as intercellular mediators. Examples are serotonin, histamine, and epinephrine.

bioinformatics The quantitative and computational analysis of biological data.

blood–brain barrier A selective permeability barrier that is found in the walls of blood vessels in the brain and that prevents most large or polar molecules from readily entering brain tissue. Physically the barrier consists of tight junctions between endothelial cells; these cells have transporters for polar substances such as glucose that need to enter the brain.

blood group antigens A group of oligosaccharides that are carried in the form of glycoproteins and glycolipids on the surface of cells, including blood cells; they are encoded by a large number of polymorphic gene loci and can provoke an immune response in an individual with different blood group antigens.

Bohr effect The effect of pH on oxygen binding by hemoglobin, by which a decrease in pH causes a decrease in oxygen affinity. The effect promotes both the release of oxygen from hemoglobin in the tissues and the release of CO_2 from the blood to the air in the lungs.

branch migration During homologous recombination, the migration of a crossover point (Holliday junction) by simultaneous unwinding and rewinding in both duplexes.

buffering The ability of a mixture of an acid and its conjugate base at a pH near their pK_a to minimize pH changes caused by an influx of acid or base.

calorie A unit of energy defined as that amount of heat energy that will raise the temperature of 1 gram of water by 1 °C. 1 calorie = 4.182 joules (which see).

Calvin cycle The cycle of photosynthetic dark reactions by which CO_2 is fixed, reduced, and converted to glyceraldehyde-3-phosphate (the precursor of hexose monophosphates).

carbohydrates In general, substances that have the stoichiometric formula $(CH_2O)_n$, where n is at least 3, or that are derived from such a substance by the addition of functional groups.

carboxyl terminus See *C-terminus*.

carnitine A low-molecular-weight lysine derivative that shuttles fatty acids through the inner mitochondrial membrane to the matrix. The fatty acyl moiety is transferred from CoA to carnitine for transit through the membrane and is then transferred back to CoA; the carnitine released on the matrix side of the membrane is shuttled back for reuse.

caspases A family of proteases involved in apoptosis (which see); (<u>c</u>ysteine <u>asp</u>artate prote<u>ases</u>) have an active site cysteine and attack bonds containing an asp residue.

catabolism The sum of all the metabolic processes by which complex molecules are broken down to simpler ones, including the processes by which molecules are broken down to yield cellular energy. Compare *anabolism.*

catabolite activation In bacteria, a transcriptional control system that induces the synthesis of enzymes for the catabolism of energy substrates other than glucose when glucose levels are low. It involves an activator protein, CRP, that binds cyclic AMP under conditions of low glucose; this complex then binds to DNA sites and promotes transcription of the appropriate genes.

cathepsins Lysosomal proteases that function in degrading proteins in lysosomes and are also released into the cell at large during cell autolysis (programmed cell death).

cell wall A rigid, protective wall around a plant, bacterial, or fungal cell that is secreted by the cell and is located outside the plasma membrane.

center of chirality With respect to organic compounds, a carbon atom that has four different substituents attached to it; such a group cannot be superimposed on its own mirror image and therefore can occur in two enantiomers.

centromere The region of a chromosome where the two sister chromatids are attached together. It is also the site of attachment for spindle fibers during mitosis and meiosis.

C4 cycle A cycle in some plants that minimizes the wasteful effects of photorespiration by using an enzyme other than rubisco to perform the initial fixation of CO_2. This enzyme is found in mesophyll cells, where it fixes CO_2 into a four-carbon compound (hence *C4*). This fixed carbon is shuttled into sheltered bundle-sheath cells, where it is released as CO_2 and enters the Calvin cycle.

chaotropic The property of being able to disrupt the hydrogen bonding structure of water. Substances that are good hydrogen bonders, such as urea or guanidine hydrochloride, are chaotropic. Concentrated solutions of these substances tend to denature proteins because they reduce the hydrophobic effect.

chaperonins Proteins that are involved in managing the folding of other proteins. Some of them help proteins to fold correctly; some prevent premature folding; and some prevent polypeptides from associating with other polypeptides until they have folded properly.

checkpoint A point during the cell cycle at which the cell monitors its readiness to proceed into the next phase.

chemical cross-linking A technique for investigating the mutual arrangement of components in a complex. The complex is exposed to a reagent that can form chemical cross-links between adjacent components and is then disaggregated and analyzed. Components that are linked together can be assumed to be neighbors in the complex.

chemical potential (also called partial molar free energy) In a system, the free energy that resides in a chemical component per mole of the component present. For example, in a system consisting of *a* moles of component A and *b* moles of component B, the total free energy G would be the sum of the free energy in the two components: $G = aGw_A + bGw_B$.

chemiosmotic coupling The coupling of an enzyme-catalyzed chemical reaction to the transport of a substance across a membrane either with or against its concentration gradient. The outstanding example is the coupling of ATP synthesis to the movement of protons across a membrane in response to a proton gradient.

chemotaxis The process by which bacteria sense a concentration gradient of a particular substance in the medium and move either up or down the gradient.

chemotherapy The treatment of disease with chemical agents.

chiral With respect to a molecule or other object, the property of being non-superimposable on its mirror image. An atom that makes a molecule chiral, such as a carbon with four different substituents, is called a chiral atom or center of chirality.

chloroplasts The organelles in plant and algal cells that carry out photosynthesis.

chromatin The filamentous material of eukaryotic chromosomes, consisting of DNA with associated histones and other proteins. During interphase it is dispersed and fills most of the nucleus; during nuclear division it condenses into compact chromosomes.

chromophore A chemical group that absorbs light at characteristic wavelengths.

chylomicron A type of lipoprotein that is produced in the intestinal villi and serves to transport dietary lipids in the circulation.

circular dichroism The property of absorbing right circularly polarized light and left circularly polarized light to different extents. Stereoisomers exhibit circular dichroism. Also, some types of secondary structure, such as α helices and β sheets in proteins, exhibit a predictable circular dichroism at specific wavelengths.

circular dichroism spectrum (CD spectrum) An optical spectrum obtained using circularly polarized light; it gives the circular dichroism of the substance over a range of wavelengths.

cis-**dominant** Refers to a mutation in a genetic regulatory element that affects the expression of appropriate genes *only* on the same chromosome, not on another homologous chromosome present in the same cell. *Cis*-dominance demonstrates that a regulatory element does not code for a diffusible factor.

cistron The smallest unit of DNA that must be intact to code for the amino acid sequence of a polypeptide; thus, the coding part of a gene, minus $5'$ and $3'$ untranslated sequences and regulatory elements.

citric acid cycle (also called tricarboxylic acid cycle and Krebs cycle) A cycle of reactions that takes place in the mitochondrial matrix and results in the oxidation of acetyl units to CO_2 with the production of reducing equivalents and ATP. It is a central pathway in oxidative respiration. Other substrates besides acetyl-CoA can enter the cycle at intermediate points.

clamp loader A multiprotein complex that opens the sliding clamp in DNA replication and allows binding and release of the sliding clamp.

clathrate structure The cagelike structure of organized water molecules that forms around a hydrophobic molecule in solution. The structure has lower entropy than liquid water, which helps explain why hydrophobic substances dissolve poorly in water.

clonal selection theory A model (proved correct) describing how the body is able to produce specific immune responses against a vast array of antigens. The B and T cells produced by the body have randomly generated antigen specificities. When a particular antigen enters the body, it induces proliferation only in B and T cells that happen to be specific for it. Thus, the antigen selects the cells that will mount an immune response against it and stimulates them to undergo clonal proliferation.

clone A group of cells, organisms, or DNA sequences that are genetically identical because they are all derived from a single ancestor.

closed promoter complex A pre-transcriptive complex of RNA polymerase and a promoter in which the promoter DNA strands have not yet unwound.

coated pit A cell membrane pit that is lined on its cytosolic side by a meshwork of the protein clathrin. Coated pits participate in the mechanism of receptor-mediated endocytosis, in which surface receptors that have bound specific extracellular substances are gathered into coated pits, which pinch off to become cytoplasmic vesicles.

codon A trinucleotide sequence corresponding to one specific amino acid in the genetic code; a "code word."

coenzyme An organic small molecule that binds to an enzyme and is essential for its activity but is not permanently altered by the reaction. Most coenzymes are derived metabolically from vitamins.

colony hybridization A technique that is used to screen bacteria for the presence of a specific recombinant DNA sequence. Colonies of the bacteria are transferred to a filter, treated to lyse the cells and denature the DNA, and then exposed to a labeled DNA probe that is complementary to part of the sequence in question. Colonies that bind the probe possess the sequence.

competitive inhibitor A substance that inhibits an enzyme-catalyzed reaction by competing with the substrate for the active site; the inhibitor can reversibly occupy the active site but does not undergo the reaction.

concatemer A DNA molecule that consists of a tandem series of complete genomes. Some phage genomes form concatemers during replication as part of a strategy for replicating the full length of a linear DNA duplex.

confocal microscopy A light-microscopy technique that allows high resolution in thick samples.

conjugation With respect to bacteria, a process in which two bacterial cells pair up and one of them passes a copy (usually partial) of its chromosome to the other. Also called bacterial mating.

consensus sequence For a group of nucleotide or amino acid sequences that show similarity but are not identical (for example, the sequences for a family of related regulatory gene sequences), an artificial sequence that is compiled by choosing at each position the residue that is found there most often in the sequences under study.

constitutive With respect to gene expression, refers to proteins that are synthesized at a fairly steady rate at all times instead of being induced and repressed in response to changing conditions.

cooperative transition A transition in a multipart structure such that the occurrence of the transition in one part of the structure makes the transition likelier to happen in other parts.

copy number The number of copies per cell of a particular gene or other DNA sequence.

Cori cycle The metabolic cycle by which lactate produced by tissues engaging in anaerobic glycolysis, such as exercising muscle, is regenerated to glucose in the liver and returned to the tissues via the bloodstream.

counterion atmosphere A cloud of oppositely charged small ions (*counterions*) that collects around a macroion dissolved in a salt solution. Counterion atmospheres partly shield macroions from each other's charges and thus affect their interactions.

cristae Folds in the inner mitochondrial membrane that project into the mitochondrial matrix. The enzymes of the electron transport chain and oxidative phosphorylation are located mainly on the cristae.

crossover point A specific site of inhibition at which an overall pathway is blocked.

cruciform In a DNA duplex, a structure that can be adopted by a palindromic sequence, in which each strand base-pairs with itself to form an arm that projects from the main duplex and terminates in a hairpin loop. The two arms form a "cross" with the main duplex.

cryoelectron microscopy A variation of electron microscopy (which see) in which samples are frozen in a glassy ice matrix.

C-terminus (also called carboxyl terminus) The end of a polypeptide chain that carries an unreacted carboxyl group. See also *N-terminus*.

curie The basic unit of radioactive decay; an amount of radioactivity equivalent to that produced by 1 g of radium, namely 2.22×10^{12} disintegrations per minute.

cyclic photophosphorylation In photosynthesis, photophosphorylation (light-dependent ATP synthesis) that is linked to a cyclic flow of electrons from photosystem II down an electron transport chain and back to photosystem II; it is not coupled to the oxidation of H_2O or to the reduction of $NADP^+$. Compare *noncyclic photophosphorylation*.

cyclins Proteins that regulate the cell cycle by binding to and activating specific nuclear protein kinases. Cyclin-dependent kinase activations occur at three points during the cell cycle, thus providing three decision points as to whether the cycle will proceed.

cytokines Compounds that regulate cell processes during the immune response.

cytokinesis The division of a eukaryotic cell to form two cells. It usually accompanies nuclear division, although nuclear division can occur without cytokinesis.

cytoskeleton An organized network of rodlike and fiberlike proteins that pervades a cell and helps give it its shape and motility. The cytoskeleton includes actin filaments, microtubules, and a diverse group of filamentous proteins collectively called intermediate filaments.

cytosol The fluid medium that is located inside a cell but outside the nucleus and organelles (for eukaryotes) or the nucleoid (for prokaryotes). It is a semiliquid concentrated solution or suspension.

dark reactions The photosynthetic subprocesses that do not depend *directly* on light energy; specifically, the synthesis of carbohydrate from CO_2 and H_2O. Compare *light reactions*.

denaturation For a nucleic acid or protein, the loss of tertiary and secondary structure so that the polymer becomes a random coil. For DNA, this change involves separation of the two strands. Denaturation can be induced by heating and by certain changes in chemical environment.

depurination Cleavage of the glycosidic bond between C-1 of deoxyribose and a purine base in DNA. A common form of DNA damage.

diabetes mellitus A disease caused by a deficiency in the action of insulin in the body, resulting either from low insulin levels or from inadequate insulin levels combined with unresponsiveness of the target cells to insulin. The disease is manifested primarily by disturbances in fuel homeostasis, including hyperglycemia (abnormally high blood glucose levels).

dialysis The process by which low-molecular-weight solutes are added to or removed from a solution by means of diffusion across a semipermeable membrane.

diastereomers Molecules that are stereoisomers but not enantiomers of each other. Isomers that differ in configuration about two or more asymmetric carbon atoms and are not complete mirror images.

dielectric constant A dimensionless constant that expresses the screening effect of an intervening medium on the interaction between two charged particles. Every medium (such as a water solution or an intervening portion of an organic molecule) has a characteristic dielectric constant.

difference spectrum With respect to absorption spectra, a spectrum obtained by loading the sample cuvette with the substances under study and a reference cuvette with an equimolar sample of the same substances in a known state (for example, fully oxidized) and recording the difference between the two spectra.

diffraction pattern The pattern that is produced when electromagnetic radiation passes through a regularly repeating structure; it results because the waves scattered by the structure interact destructively in most directions (creating dark zones) but constructively in a few directions (creating bright spots). For the pattern to be sharp, the radiation wavelength must be somewhat shorter than the repeat distance in the structure. See also *X-ray diffraction*.

diffusion coefficient (*D*) A coefficient that indicates how quickly a particular substance will diffuse in a particular medium under the influence of a given concentration gradient.

diploid For a cell or an organism, the possession of two homologous sets of chromosomes per nucleus (with the possible exception of sex chromosomes, which may be present in only one copy). Compare *haploid*.

dismutation A reaction in which two identical substrate molecules have different fates; particularly, a reaction in which one of the substrate molecules is oxidized and the other reduced.

dispersion forces Weak intermolecular attractive forces that arise between molecules that are close together because the fluctuating electron distributions of the molecules become synchronized so as to produce a slight electrostatic attraction. These forces play a role in the internal packing of many biomolecules.

dissociation constant For an acid, the equilibrium constant K_a for the dissociation of the acid into its conjugate base and a proton. For a complex of two biomolecules, the equilibrium constant K_d for dissociation into the component molecules.

DNA damage response A coordinated series of events following DNA-damaging events in eukaryotic cells.

DNA gyrase An enzyme that is able to introduce negative superhelical turns into a circular DNA helix.

DNAzyme A catalytic DNA molecule, so far available only as a synthetic construct.

domain A portion of a polypeptide chain that folds on itself to form a compact unit that remains recognizably distinct within the tertiary structure of the whole protein. Large globular proteins often consist of several domains, which are connected to each other by stretches of relatively extended polypeptide.

double-strand break repair Processes in eukaryotic cells that allow repair of double-stranded DNA breakages.

drug resistance factors Bacterial plasmids that carry genes coding for resistance to antibiotics.

dyad axis A two-fold axis of symmetry.

editing See *RNA editing*.

eicosanoids A family of compounds, including prostaglandins and thromboxanes, that are derived from arachidonic acid, an eicosaenoic fatty acid.

einstein One mole of photons.

electron microscopy A form of microscopy in which electrons are used as radiation. Capable of very high resolution.

electron spin resonance (also called electron paramagnetic resonance, or EPR) A form of spectroscopy that is sensitive to the environment of unpaired electrons in a sample.

electron transport chain A sequence of electron carriers of progressively higher reduction potential in a cell that is linked so that electrons can pass from one carrier to the next. The chain captures some of the energy released by the flow of electrons and uses it to drive the synthesis of ATP.

electrophoresis A method for separating electrically charged substances in a mixture. A sample of the mixture is placed on a supporting medium (a piece of filter paper or a gel), to which an electrical field is applied. Each charged substance migrates toward the cathode or the anode at a speed that depends on its net charge and its frictional interaction with the medium. See also *gel electrophoresis*.

elongation factors Nonribosomal protein factors that are necessary participants in the chain-elongation cycle of polypeptide synthesis; they interact with the ribosome–mRNA complex or with other major cycle participants.

enantiomers (also called optical isomers) Stereoisomers that are nonsuperimposable mirror images of each other. The term *optical isomers* comes from the fact that the enantiomers of a compound rotate polarized light in opposite directions.

endergonic In a nonisolated system, a process that is accompanied by a positive change in free energy (positive *G*) and therefore is thermodynamically not favored. Compare *exergonic*.

endocrine glands Glands that synthesize hormones and release them into the circulation. The hormone-producing gland cells are called endocrine cells.

endocytosis Uptake into a cell.

endonuclease An enzyme that hydrolytically cleaves a nucleic acid chain at an internal phosphodiester bond.

endoplasmic reticulum (ER) A highly folded membranous compartment within the cytoplasm that is responsible for a great variety of cellular tasks, including the glycosylation and trafficking of proteins destined for secretion or for the cell membrane or some organelles. It also functions in lipid synthesis, and the enzymes of many pathways of intermediary metabolism are located on its surface.

endorphins A class of endogenous brain peptides that exert analgesic effects in the central nervous system by binding to opiate receptors. They are produced by cleavage of the large polypeptide pro-opiomelanocortin.

energy The ability to do work. See *internal energy*.

energy charge A quantity that indicates the state of a cell's energy reserves. It is equal to the cell's reserves of the free energy sources ATP and ADP (taking into account that ADP stores less free energy than ATP) divided by the total supply of ATP and its breakdown products ADP and AMP: ([ATP] + 1/2[ADP])/([ATP] + [ADP] + [AMP]).

enhancer sequence A DNA sequence that is distant from a gene but to which a protein factor that affects the gene's transcription can bind to exert its action. Enhancers are linked to RNA polymerase at the transcriptional start site by Mediator.

enthalpy (*H*) A thermodynamic quantity (function of state) that is equal to the internal energy of a system plus the product of the pressure and volume: *H* = *E* + *PV*. It is equal to the heat change in constant-pressure reactions, such as most reactions in biological systems.

entropy (*S*) A thermodynamic quantity (function of state) that expresses the degree of disorder or randomness in a system. According to the second law of thermodynamics, the entropy of an open system tends to increase unless energy is expended to keep the system orderly.

enzyme A catalyst for a biological reaction. Usually a protein molecule, but some enzymes are RNA molecules or ribonucleoproteins.

epigenetics The study of hereditary changes in gene expression that do not involve changes in DNA nucleotide sequence.

epimers Sugar isomers differing in configuration about one asymmetric carbon.

episomes Plasmids that can undergo integration into the bacterial chromosome.

epitope (also called antigenic determinant) The specific portion of an antigen particle that is recognized by a given antibody or T-cell receptor.

equilibrium A condition in which there are no net currents of matter or energy.

essential amino acids Amino acids that must be obtained in the diet because they cannot be synthesized in the body (at least not in adequate amounts).

essential fatty acids Fatty acids that must be obtained in the diet because they cannot be synthesized in the body in adequate amounts. Examples are linoleic acid and linolenic acid.

eukaryotes Organisms whose cells are compartmentalized by internal cellular membranes to produce a nucleus and organelles. Compare *prokaryotes*.

exergonic In a nonisolated system, a process that is accompanied by a negative change in free energy (negative *G*) and therefore is thermodynamically favored. Compare *endergonic*.

exocrine cell A cell that secretes a substance that is excreted through a duct either into the alimentary tract or to the outside of the organism. Exocrine cells are grouped together in exocrine glands.

exocytosis Movement out of a cell.

exon A region in the coding sequence of a gene that is translated into protein (as opposed to introns, which are not). The name comes from the fact that exons are the only parts of an RNA transcript that are seen outside the nucleus. Compare *intron*.

exportins A class of proteins involved in transporting materials out of nuclei. See *importins*.

extinction coefficient A coefficient that indicates the ability of a particular substance in solution to absorb light of wavelength *l*. The molar extinction coefficient, e_M, is the absorbance that would be displayed by a 1.0 M solution in a 1-cm light path.

facilitated transport (also called facilitated diffusion) The movement of a substance across a biological membrane in response to a concentration or electrochemical gradient where the movement is facilitated by membrane pores or by specific transport proteins. Compare *active transport, passive transport*.

fatty acid A carboxylic acid with a long hydrocarbon chain.

fermentations Processes in which cellular energy is generated from the breakdown of nutrient molecules where there is no net change in the oxidation state of the products as compared with that of the reactants; fermentation can occur in the absence of oxygen.

fibrous proteins Proteins of elongated shape, often used as structural materials in cells and tissues. Compare *globular proteins.*

first law of thermodynamics The law stating that energy cannot be created or destroyed and that it is therefore possible to account for any change in the internal energy of a system E by an exchange of heat (q) and/or work (w) with the surroundings. $E = q + w$.

first-order reaction A reaction whose rate depends on the first power of the concentration of a reactant. Compare *second-order reaction.*

Fischer projection A convention for representing stereoisomers in a plane. The tetrahedron of bonds on a carbon is represented as a plane cross, where the bonds to the right and left are assumed to be pointing toward the viewer and the bonds to the top and bottom are assumed to be pointing away from the viewer. Fischer projections of monosaccharides are oriented with the carbonyl group at the top; the chiral carbon farthest from the carbonyl group (which is the one that determines whether the sugar is the D or the L form) is then drawn with its hydroxyl to the right for the D form and to the left for the L form.

flavin adenine dinucleotide (FAD), flavin mononucleotide (FMN) Coenzymes derived from vitamin B_2 (riboflavin) that function as electron acceptors in enzymes that catalyze electron transfer reactions.

flavonoids A family of plant compounds, of diverse biological functions, derived from phenylalanine.

fluid mosaic model A model describing cellular membrane structure, according to which the proteins are embedded in a phospholipid bilayer and are free to move in the plane of the membrane.

fluorescence The phenomenon by which a substance that absorbs light at a given wavelength reradiates a portion of the energy as light of a longer wavelength.

flux With reference to a chemical pathway, the rate (in moles per unit time) at which reactant "flows through" the pathway to emerge as product. The term can be used for the rate at which particles undergo any process in which they either flow or can be thought of metaphorically as flowing.

flux control coefficient For a reaction in a pathway the ratio of change in flux rate through that reaction to change in flux rate through the entire pathway.

F_0F_1 complex (also called F_0F_1 ATP synthase or complex V) The enzyme complex in the inner mitochondrial membrane that uses energy from the transmembrane proton gradient to catalyze ATP synthesis. The F_0 portion of the complex spans the membrane, and the F_1 portion, which performs the ATP synthase activity, projects into the mitochondrial matrix.

footprinting With respect to molecular genetics, a technique used to identify the DNA segment in contact with a given DNA-binding protein. The DNA–protein complex is subjected to digestion with a nonspecific nuclease, which cleaves at the residues that are not protected by the protein, and the protected region is identified by gel electrophoresis.

frameshift mutation A mutation that changes the reading frame for a gene by adding or deleting one or two nucleotides, thereby reducing the remainder of the message $3'$ to the mutation to gibberish.

frameshift suppressor A mutant tRNA that contains either two or four bases in the anticodon loop and can suppress the effects of a particular frameshift mutation in a gene.

free energy (G) (also called Gibbs free energy) A thermodynamic quantity (function of state) that takes into account both enthalpy and entropy: $G = H–TS$, where H is enthalpy, S is entropy, and T is absolute temperature. The *change in free energy* (G) for a process, such as a chemical reaction, takes into account the changes in enthalpy and entropy and indicates whether the process will be thermodynamically favored at a given temperature.

frictional coefficient A coefficient that determines the frictional force on a particular particle (such as a molecule) in a particular medium at a given velocity. In the context of electrophoresis or centrifugation, it determines how fast a chemical species will move in a particular medium in response to a given electrical field or centrifugal force.

furanose A sugar with a five-membered ring, akin to furan.

fusion proteins Genetically engineered proteins that are made by splicing together coding sequences from two or more genes. The resulting protein thus combines portions from two different parent proteins.

futile cycle A pair of reactions that interconvert the same substrates and products in opposite directions, one of which consumes ATP and the other of which generates ATP. Uncontrolled operation of both reactions results in breakdown of ATP with no useful work performed.

G proteins A family of membrane-associated proteins that transduce signals received by various cell-surface receptors. They are called G proteins because binding of GTP and GDP is essential to their action.

gated channel A membrane ion channel that can open or close in response to signals from outside or within the cell.

gel electrophoresis A type of electrophoresis in which the supporting medium is a thin slab of gel held between glass plates. The technique is widely used for separating proteins and nucleic acids. See also *electrophoresis, isoelectric focusing.*

genetic code The code by which the nucleotide sequence of a DNA or RNA molecule specifies the amino acid sequence of a polypeptide. It consists of three-nucleotide codons that either specify a particular amino acid or tell the ribosome to stop translating and release the polypeptide. With a few minor exceptions, all living things use the same code.

genetic recombination Any process that results in the transfer of genetic material from one DNA molecule to another. In eukaryotes, it can refer specifically to the exchange of matching segments between homologous chromosomes by the process of crossing over during meiosis.

genome The total genetic information contained in a cell, an organism, or a virus.

genomics The investigation of hereditary features of biological processes and mechanisms focused upon the whole genome, as compared with focusing on the individual genes.

genotype The genetic constitution of an individual organism. Compare *phenotype.*

gibberellins A family of diterpene plant growth hormones.

Gibbs free energy See *free energy.*

globular proteins Proteins whose three-dimensional folded shape is relatively compact. Compare *fibrous proteins.*

glucan A glucose polymer.

glucocorticoids The steroid hormones cortisol and corticosterone, which are secreted by the adrenal cortex. In addition to other functions, they promote gluconeogenesis in response to low blood sugar levels.

glucogenic In fuel metabolism, refers to substances (such as some amino acids) that can be used as substrates for glucose synthesis.

gluconeogenesis The processes by which glucose is synthesized from noncarbohydrate precursors, such as glycerol, lactate, some amino acids, and (in plants) acetyl-CoA.

glucose transporter A membrane protein that is responsible for transporting glucose across a cell membrane. Different tissues may have glucose transporters with different properties.

glycan Another name for oligo- or polysaccharide.

glycocalyx The polysaccharide coat found on many eukaryotic cells.

glycolipids Lipids that have saccharides attached to their head groups.

glycolysis The initial pathway in the catabolism of carbohydrates, by which a molecule of glucose is broken down to two molecules of pyruvate, with a net production of ATP molecules and the reduction of two NAD^+ molecules to NADH. Under aerobic conditions, these NADH molecules are reoxidized by the electron transport chain; under anaerobic conditions, a different electron acceptor is used.

glycoprotein A protein with a covalently linked oligosaccharide.

glycosaminoglycans (also called mucopolysaccharides) Polysaccharides composed of repeating disaccharide units in which one sugar is either *N*-acetylgalactosamine or *N*-acetylglucosamine. Typically the disaccharide unit carries a carboxyl group and often one or more sulfates, so that most glycosaminoglycans have a high density of negative charges. Glycosaminoglycans are often combined with protein to form proteoglycans and are an important component of the extracellular matrix of vertebrates.

glycosidase An enzyme that hydrolytically cleaves a glyosidic bond.

glycosidic bond The covalent link between the carbonyl carbon of a sugar molecule and another molecule, for example, a purine or pyrimidine in a nucleoside or another sugar in an oligosaccharide.

glyoxysome A specialized type of peroxisome found in plant cells. It performs some of the reactions of photorespiration, and it also breaks down fatty acids to acetyl-CoA by β-oxidation and converts the acetyl-CoA to succinate via the glyoxylate cycle, thus enabling plants to convert fatty acids to carbohydrates.

Golgi complex A stack of flattened membranous vesicles in the cytoplasm. It serves as a routing center for proteins destined for secretion or for lysosomes or the cell membrane; it performs similar functions for membrane lipids, and it also modifies and finishes the oligosaccharide moieties of glycoproteins.

G-quadruplex A four-stranded DNA structure stabilized by groupings of four deoxyguanylate residues, called G-quartets.

growth factors Peptide mediators that influence the growth and/or differentiation of cells; they differ from growth hormones in being produced by many tissues and in acting locally.

half-life (also called half-time) For a chemical reaction, the time at which half the substrate has been consumed and turned into product. The term can also refer to the analogous point in other processes, such as the radioactive decay of an isotope.

haploid For a cell or an organism, the possession of only one copy of each chromosome per nucleus. Compare *diploid*.

haplotype Referring to a specific grouping of restriction fragment length polymorphisms (which see) in a particular genome.

hapten A molecule that is too small to stimulate an immune response by itself but can do so when coupled to a larger, immunogenic carrier molecule (usually a protein).

Haworth projection A conventional planar representation of a cyclized monosaccharide molecule. The hydroxyls that are represented to the right of the chain in a Fischer projection are shown below the plane in a Haworth projection.

heat-shock proteins A group of chaperonins that accumulate in a cell after it has been subjected to a sudden temperature jump or other stress. They are thought to help deal with the accumulation of improperly folded or assembled proteins in stressed cells.

helicases Enzymes that catalyze the energy-dependent unwinding of duplex nucleic acids.

helix–loop–helix motif A binding motif that is found in calmodulin and some other calcium-binding proteins as well as in some DNA-binding proteins. It consists of two α helix segments connected by a loop.

helix–turn–helix motif A DNA-binding motif that is responsible for sequence-specific DNA binding in many transcription factors. It consists of two α helix segments connected by a β turn; one of the helices occupies the DNA major groove and makes specific base contacts.

helper T cells T lymphocytes whose role is to recognize antigens and help other defensive cells to mount an immune response. They help activate antigen-stimulated B cells (resulting in production of specific antibodies) and/or antigen-stimulated cytotoxic T cells (resulting in attack on antigenic cells), and they also produce immune mediators that stimulate nonspecific defense responses.

heme A molecule consisting of a porphyrin ring (either protoporphyrin IX or a derivative) with a central complexed iron; it serves as a prosthetic group in proteins such as myoglobin, hemoglobin, and cytochromes.

hemimethylated With respect to DNA, refers to the condition in which one strand of the duplex is methylated and the other is not. Newly replicated DNA is hemimethylated; normally a methylase enzyme then methylates appropriate bases in the new strand.

Henderson–Hasselbalch equation An equation relating pH to the concentrations of two species contributing to buffering and to the pK_a of the relevant protonic equilibrium.

heteropolymer A polymer made up of differing monomeric units; native nucleic acids and proteins are heteropolymers, as compared with synthetic homopolymers (which see), which are assembled from identical monomeric units.

heterotrophs Organisms that cannot synthesize their organic compounds entirely from inorganic precursors but must consume at least some organic compounds made by other organisms. In particular, these organisms cannot use CO_2 as a carbon source. Compare *autotrophs*.

heterotropic For proteins with multiple subunits, referring to interactions among subunits that are different in structure.

heterozygous In a diploid organism, the possession of two different alleles for a given gene (as opposed to two copies of the same allele). Compare *homozygous*.

hexose A six-carbon sugar.

high-density lipoprotein (HDL) A type of lipoprotein particle that functions mainly to scavenge excess cholesterol from tissue cells and transport it to the liver, where it can be excreted in the form of bile acids.

Hill coefficient A coefficient that indicates the degree of cooperativity of a cooperative transition; it is the maximum slope of a Hill plot of the transition.

histones The proteins that participate in forming the nucleosomal structure of chromatin. Four of the five kinds of histones make up the core particle of the nucleosome; the fifth is associated with the linker DNA between nucleosomes. All histones are small, very basic proteins.

Holliday junction An intermediate during homologous recombination; a four-armed structure in which each of the participating DNA duplexes has exchanged one strand with the other duplex.

homeo box A common sequence element of about 180 base pairs that is found in homeotic genes. It codes for a sequence-specific DNA-binding element of the helix–loop–helix class. See also *homeotic genes*.

homeostasis A steady state in living systems, usually maintained at conditions far from equilibrium by performing work upon the surrounding environment.

homeotic genes Genes that contain homeo box elements and typically are involved in controlling the pattern of organismal development. Homeotic mutations, which scramble portions of this pattern, affect homeotic genes. The nuclear DNA-binding proteins encoded by these genes presumably serve as transcriptional regulators for the coordinated expression of groups of genes. See also *homeo box*.

homologous recombination Genetic recombination that requires extensive sequence homology between the recombining DNA molecules. Meiotic recombination by crossing over in eukaryotes is an example.

homology Relatedness of two nucleic acid or protein sequences.

homolytic reaction A bond-cleavage reaction in which each of the two shared electrons in the cleaved bond ends up in a different product.

homopolymer A polymer that is made of only one kind of monomer. Starch, made only of glucosyl units, is an example. Polymers that include more than one kind of monomer, like polypeptides and nucleic acids, are called *heteropolymers* (which see).

homotropic For a multisubunit protein, refers to interactions among identical subunits.

homozygous In a diploid organism, the possession of two identical alleles for a given gene. Compare *heterozygous*.

hormone A substance that is synthesized and secreted by specialized cells and carried via the circulation to target cells, where it elicits specific changes in the metabolic behavior of the cell by interacting with a hormone-specific receptor.

hormone-responsive element A DNA site that binds an intracellular hormone–receptor complex; binding of the complex to a hormone-responsive element affects the transcription of specific genes.

host-induced restriction and modification A genetic system found in bacteria whereby a genetic element (often a plasmid) encodes both an enzyme for the methylation of DNA at a specific base sequence and an endonuclease that cleaves unmethylated DNA at that sequence. The system thus *restricts* the DNA that can survive in the cell to DNA that is *modified* by methylation at the correct sequences.

hybridomas Cultured cell lines that are made by fusing antibody-producing B lymphocytes with cells derived from a mouse myeloma (a type of lymphocyte cancer). Like B cells, they produce specific antibodies, and like myeloma cells, they can proliferate indefinitely in culture.

hydrogen bond An attractive interaction between the hydrogen atom of a donor group, such as —OH or ═NH, and a pair of nonbonding electrons on an acceptor group, such as O═C. The donor group atom that carries the hydrogen must be fairly electronegative for the attraction to be significant.

hydrophilic Refers to the ability of an atom or a molecule to engage in attractive interactions with water molecules. Substances that are ionic or can engage in hydrogen bonding are hydrophilic. Hydrophilic substances are either soluble in water or, at least, wettable. Compare *hydrophobic*.

hydrophobic The molecular property of being unable to engage in attractive interactions with water molecules. Hydrophobic substances are nonionic and nonpolar; they are nonwettable and do not readily dissolve in water. Compare *hydrophilic*.

hydrophobic effect With respect to globular proteins, the stabilization of tertiary structure that results from the packing of hydrophobic side chains in the interior of the protein.

hypochromism With respect to DNA, a reduction in the absorbance of ultraviolet light of wavelength of about 260 nm that accompanies the transition from random-coil denatured strands to a double-strand helix. It can be used to track the process of denaturation or renaturation.

immunoglobulins See *antibodies*.

importins A class of proteins involved in importing molecules into the nucleus. See *exportins*.

inborn errors of metabolism Human mutations that result in specific derangements of intermediary metabolism. Usually the problem is an enzyme that is inactive, overactive, too scarce, or too abundant; symptoms may result from the insufficient production of a necessary metabolite and/or from the accumulation of another metabolite to toxic levels.

induced dipole A molecule has an induced dipole if an external electric field induces an asymmetric distribution of charge within it.

induced fit model A model for how enzymes interact with substrates to achieve catalysis. According to this model, the empty active site of the enzyme only roughly fits the substrate(s), and the entry of substrate causes the enzyme to change its shape so as to both tighten the fit and cause the substrate to adopt an intermediate state that resembles the transition state of the uncatalyzed reaction. This is currently the dominant model for enzymatic catalysis.

induction In cellular metabolism, the synthesis of a particular protein in response to a signal; for example, the synthesis of an enzyme in response to the appearance of its substrate.

in situ hybridization A technique for finding the chromosomal location of a particular DNA sequence by probing the chromosomes with a radiolabeled or fluorescent-tagged sequence that will hybridize with the sequence in question. The location of the probe is then visualized with radioautography or by fluorescence microscopy.

intercalation With respect to DNA, refers to the fitting (intercalation) of a small molecule between adjacent bases in a DNA helix.

interferon An antiviral glycoprotein produced in response to viral infection.

intermediary metabolism All the reactions in an organism that are concerned with storing and generating metabolic energy and with the biosynthesis of low-molecular-weight compounds and energy-storage compounds. It does not include nucleic acid and protein synthesis.

internal energy (E) The energy contained in a system. For the purposes of biochemistry, the term encompasses all the types of energy that might be changed by chemical or nonnuclear physical processes, including the kinetic energy of motion and vibration of atoms and molecules and the energy stored in bonds and noncovalent interactions.

intron A region in the coding sequence of a gene that is not translated into protein. Introns are common in eukaryotic genes but are rarely found in prokaryotes. They are excised from the RNA transcript before translation. Compare *exon*.

in vitro (Latin, *in glass*) Referring to the analysis of biological processes carried out outside of a living cell or organism, as in a glass test tube.

in vivo (Latin, *in life*) Referring to a biological process as it occurs within a living cell or organism.

ion-exchange resins Polycationic or polyanionic polymers that are used in ion-exchange column chromatography to separate substances on the basis of electrical charge.

ionic strength A quantity that reflects the total concentration of ions in a solution and the stoichiometric charge (charge per atom or molecule) of each ion. It is defined as $I = \frac{1}{2} \sum_i M_i Z_i^2$, where M_i and Z_i are respectively the molarity and stoichiometric charge of ion i. It is used, for example, in calculating the effective radius of a counterion atmosphere.

ionophore A substance that selectively permeabilizes membranes to particular ions or molecules.

ion pore A pore in a cellular membrane through which ions can diffuse. It is formed by a transmembrane protein and can discriminate among ions to some degree on the basis of size and charge. Many ion pores are gated, meaning that they can open and close in response to signals.

isoelectric focusing A version of gel electrophoresis that allows ampholytes to be separated almost purely on the basis of their isoelectric points. The ampholytes are added to a gel that contains a pH gradient and are subjected to an electric field. Each ampholyte migrates until it reaches the pH that represents its isoelectric point, at which point it ceases to have a net electric charge and therefore comes to a halt and accumulates. See also *gel electrophoresis, isoelectric point*.

isoelectric point (pI) The pH at which the net charge on an ampholyte is, on average, zero.

isoenzymes (also called isoforms or isozymes) Different but related forms of an enzyme that catalyze the same reaction. Often differ in only a few amino acid substitutions.

isomorphous replacement The replacement of one atom in a macromolecule with a heavy metal atom in such a way that the structure of the macromolecule does not change. It is used in the determination of molecular structure by X-ray crystal diffraction.

isozymes See *isoenzymes*.

joule (J) A unit for energy or work, defined as the work done by a force of 1 newton when its point of application moves 1 meter in the direction of the force. It is the unit of energy used in the Système Internationale (SI).

α-keratins A class of keratins that are the major proteins of hair. They consist of long α-helical polypeptides, which are wound around each other to form triplet helices.

ketogenic Referring to fuel molecules whose catabolism generates acetyl-CoA, and, hence, which tend to increase levels of ketone bodies.

ketone bodies The substances acetoacetate, β-hydroxybutyrate, and acetone, which are produced from excess acetyl-CoA in the liver when the rate of fatty acid β-oxidation in liver mitochondria exceeds the rate at which acetyl-CoA is used for energy generation or fatty acid synthesis.

ketose A monosaccharide in which the carbonyl group occurs within the chain and hence represents a ketone group. Compare *aldose*.

kinetic isotope effect A change in a rate-determining step of enzyme catalysis brought about by substitution of an isotope, usually a heavy atom, at a specific position in a substrate.

kinetochore The structure that connects the centromere to the microtubules that will carry out chromosome separation during mitosis and meiosis.

Krebs cycle See *citric acid cycle*.

lagging strand During DNA replication, the strand that is synthesized in the opposite direction to the direction of movement of the replication fork; it is synthesized as a series of fragments that are subsequently joined. Compare *leading strand*.

leader sequence (also called signal sequence) For an mRNA, the non-translated sequence at the 5′ end of the molecule that precedes the initiation codon. For a protein, a short N-terminal hydrophobic sequence that causes the protein to be translocated into or through a cellular membrane.

leading strand During DNA replication, the strand that is synthesized in the same direction as the direction of movement of the replication fork; it is synthesized continuously rather than in fragments. Compare *lagging strand*.

leukotrienes A family of molecules that are synthesized from arachidonic acid by the lipoxygenase pathway and function as local hormones, primarily to promote inflammatory and allergic reactions (such as the bronchial constriction of asthma).

library With respect to molecular genetics, a large collection of random cloned DNA fragments from a given organism, sometimes representing the entire nuclear genome.

ligand In general, a small molecule that binds specifically to a larger one—for example, a hormone that binds to a receptor. The term can also be used to mean a chemical species that forms a coordination complex with a central atom, which is usually a metal atom.

light reactions The photosynthetic subprocesses that depend *directly* on light energy; specifically, the synthesis of ATP by photophosphorylation and the reduction of $NADP^+$ to NADPH via the oxidation of water. Compare *dark reactions*.

lignins Complex polymers, derived from phenylalanine, that compose the major constituents of woody tissue.

Lineweaver–Burk plot A plot that allows one to derive the rate constant k_{cat} and the Michaelis constant K_M for an enzyme-catalyzed reaction. It is constructed by measuring the initial reaction rate V at various substrate concentrations [S] and plotting the values on a graph of $1/V$ versus $1/[S]$.

linkage map A map showing the arrangement of genes on a chromosome; it is constructed by measuring the frequency of recombination between pairs of genes.

linking number (L) The total number of times the two strands of a closed, circular DNA helix cross each other by means of either twist or writhe; this equals the number of times the two strands are interlinked. It reflects both the winding of the native DNA helix and the presence of any supercoiling. See also *twist, writhe*.

lipid bilayer A membrane structure that can be formed by amphipathic molecules in an aqueous environment; it consists of two back-to-back layers of molecules, in each of which the polar head groups face the water and the nonpolar tails face the center of the membrane. The fabric of cellular membranes is a lipid bilayer.

lipid raft A membrane microdomain enriched in sphingolipids and cholesterol.

lipids A chemically diverse group of biological compounds that are classified together on the basis of their generally apolar structure and resulting poor solubility in water.

lipoproteins Any lipid–protein conjugate. Specifically refers to lipid–protein associations that transport lipids in the circulation. Each consists of a core of hydrophobic lipids surrounded by a skin of amphipathic lipids with embedded apolipoproteins. Different kinds of lipoproteins play different roles in lipid transport.

long terminal repeats (LTRs) A pair of direct repeats several hundred base pairs long that are found at either end of a retroviral genome. They are involved in integration into the host genome and in viral gene expression.

low-angle neutron scattering A set of techniques that can be used to find the size of a particle in solution or to find the size or spacing of internal regions that can be distinguished by different neutron scattering power, such as the protein and nucleic acid components of a nucleoprotein particle or labeled proteins within a multisubunit complex.

low-density lipoprotein (LDL) A type of lipoprotein particle that functions mainly to distribute cholesterol from the liver to other tissues. Its protein component consists of a single molecule of apoprotein B-100.

lysogeny A latent state that can be achieved by some bacteriophages, in which the phage genome is integrated into the host bacterial chromosome and few if any viral genes are expressed.

mass spectrometry A method for determining the molecular mass from the velocity of ions in a vacuum.

mechanism-based inhibitor An enzyme inhibitor whose action depends on the enzyme's catalytic mechanism. Typically it is a substrate analog that irreversibly modifies the enzyme at a particular step in the catalytic cycle.

mediator A large multiprotein complex that connects upstream transcriptional activators and enhancers with transcription machinery bound near a transcriptional start site.

membrane electrical potential With respect to biological membranes, a voltage difference that exists across a membrane owing to differences in the concentrations of ions on either side of the membrane.

messenger RNA (mRNA) RNA molecules that act as templates for the synthesis of polypeptides by ribosomes.

metabolism The totality of the chemical reactions that occur in an organism. Compare *anabolism, catabolism, intermediary metabolism*.

metabolome The entire complement of low-molecular-weight compounds (metabolites) within a cell or organism.

metabolomics The investigation of biological processes conducted through analysis of the metabolome (which see).

metabolon A complex of enzymes, usually linked noncovalently, which carry out functionally related reactions and, hence, can facilitate catalysis of multistep reaction pathways.

metastability For a system, the condition of being in a state that does not represent thermodynamic equilibrium but is nearly stable at the time scale of interest because progress toward equilibrium is slow.

micelles Tiny droplets that form when an amphipathic substance that has a polar head group and a nonpolar tail region (such as a fatty acid) is added to an aqueous medium and shaken. Each droplet consists of a spherical cluster of amphipathic molecules arranged with their polar head groups facing out toward the water and their nonpolar tails facing in toward the center.

Michaelis–Menten equation An equation that gives the rate of an enzyme-catalyzed reaction in terms of the concentrations of substrate and enzyme as well as two constants that are specific for a particular combination of enzyme and substrate: a rate constant, k_{cat}, for the catalytic production of product when the enzyme is saturated, and the Michaelis constant, K_M.

microarray Also called a gene chip, a collection of several thousand gene fragments, either oligonucleotides or cDNAs, immobilized on a small slide as a precise array, so that a pattern of total RNAs from a cell can be subjected to annealing conditions, with the extent of hybridization to each gene fragment in the chip giving a global picture of transcription patterns.

microRNA An RNA molecule, 21 to 23 nucleotides long, also called miRNA, which regulates the activity of a specific mRNA that contains partially or completely homologous nucleotide sequences.

microtubules Fiberlike cytoplasmic structures that consist of units of the protein tubulin arranged helically to form a hollow tube. They are involved in various kinds of cellular motility, including the beating of cilia and flagella and the movement of organelles from one part of the cell to another.

microtubule-associated proteins (MAPs) A class of proteins associated with microtubules that assist in dynamic processes.

minus strand In viral genomes, a nucleic acid strand that is complementary to the RNA strand that serves as mRNA. Compare *plus strand*.

mismatch repair A system for the correction of mismatched nucleotides or single-base insertions or deletions produced during DNA replication; it scans the newly replicated DNA, and when it finds an error, it removes and replaces a stretch of the strand containing the error.

missense mutation A mutation that alters a DNA codon so as to cause one amino acid in a protein to be replaced by a different one.

mitochondria The organelles whose chief task it is to supply the cell with ATP via oxidative phosphorylation. They contain the enzymes for pyruvate oxidation, the citric acid cycle, the β-oxidation of fatty acids, and oxidative phosphorylation, as well as the electron transport chain.

mixed-function oxidase An oxygenase enzyme that catalyzes a reaction in which two different substrates are oxidized, one by the addition of an oxygen atom from O_2 and the other by supplying two hydrogen atoms to reduce the remaining oxygen atom to H_2O.

molten globule A hypothetical intermediate state in the folding of a globular protein, in which the overall tertiary framework has been established but internal side chains (especially hydrophobic ones) are still free to move about.

mucopolysaccharides See *glycosaminoglycans*.

multicatalytic proteinase complex (MPC) A massive complex of proteolytic enzymes that is found in the cytosol of many eukaryotic cells and seems to function in the programmed destruction of cellular proteins.

muscarinic acetylcholine receptors A class of receptors for the neurotransmitter acetylcholine that are characterized by an ability to bind the toadstool toxin muscarine. Synapses that have these receptors may be either excitatory or inhibitory. Compare *nicotinic acetylcholine receptors*.

mutagen A substance that is capable of causing mutations.

mutarotation The spontaneous racemization about the carbonyl carbon in a sugar.

mutation Any inheritable change in the nucleotide sequence of genomic DNA (or genomic RNA, in the case of an RNA virus).

myofibrils A unit of thick and thin filaments of muscle fibers.

Nernst equation An equation that relates the electrical potential across a membrane to the concentrations of ions on either side of the membrane.

neurohormones Substances that are released from neurons and modulate the behavior of target cells, which are often other neurons. Unlike neurotransmitters, they do not act strictly across a synapse. Most neurohormones are peptides.

neurotoxin A toxin that acts by disrupting nerve cell function. Fast-acting neurotoxins often act by blocking the action of an ion gate necessary for the development of an action potential.

neurotransmitter A low-molecular-weight substance that is released from an axon terminal in response to the arrival of an action potential and then diffuses across the synapse to influence the postsynaptic cell, which may be either another neuron or a muscle or gland cell.

nick translation A process beginning at a nick in a DNA duplex, in which a DNA polymerase adds nucleotides one at a time to the 3′ terminal nucleotide in the nick, in synchrony with cleavage of the nucleotide on the 5′ side of the nick. The process causes the location of the nick to migrate (hence the origin of the term).

nicotinic acetylcholine receptors A class of receptors for the neurotransmitter acetylcholine that are characterized by their ability to bind nicotine. Synapses with this kind of receptor are excitatory. Compare *muscarinic acetylcholine receptors*.

nitrogen equilibrium Also called normal nitrogen balance, a condition in which daily nitrogen excretion equals total dietary nitrogen intake.

noncompetitive inhibitor An inhibitor of an enzyme-catalyzed reaction that acts by binding to a site on the enzyme different from the active site and reducing the enzyme's catalytic efficiency. Irreversible binding of a substrate analog at the active site is kinetically similar to noncompetitive inhibition.

noncovalent interactions All the kinds of interactions between atoms and molecules that do not involve the actual sharing of electrons in a covalent bond; they include electrostatic interactions, permanent and induced dipole interactions, and hydrogen bonding.

noncyclic photophosphorylation In photosynthesis, photophosphorylation (light-dependent ATP synthesis) that is linked to a one-way flow of electrons from water through photosystems II and I and finally to NADPH; it is thus coupled to the oxidation of H_2O and the reduction of $NADP^+$. Compare *cyclic photophosphorylation*.

nonhomologous end joining A process in double-strand break repair in which severed ends are linked together without a requirement for sequence homology at the severed ends.

nonsense mutation A mutation that creates an abnormal stop codon and thus causes translation to terminate prematurely; the resulting truncated protein is usually nonfunctional.

Northern blotting A technique for detecting the presence of a specific RNA sequence in a cell and determining its size. The total RNA of the cell is extracted, resolved by gel electrophoresis, and blotted onto a filter. There it is incubated under annealing conditions with a radiolabeled probe for the sequence in question, and heteroduplexes of the probe with RNA are detected by radioautography.

N-terminus (also called amino terminus) The end of a polypeptide chain that carries an unreacted amino group. A ribosome synthesizes a polypeptide in the direction from the N-terminus to the C-terminus. See also *C-terminus*.

nuclear envelope The double membrane that encloses the nucleus. It is pierced by nuclear pores that allow even quite large molecules, such as mRNAs and nuclear proteins, to enter or leave the nucleus.

nuclear magnetic resonance (NMR) spectroscopy A type of spectroscopy that depends on the fact that isotope nuclei having the property of spin will resonate with specific frequencies of microwave radiation when placed in a magnetic field of given strength. The resonance energy is sensitive to the local molecular environment, so NMR spectroscopy can be used to explore molecular structure. Also, different living tissues have characteristic overall NMR spectra, which are sensitive to changes in the tissue environment. NMR can thus be used in the study of tissue metabolism and the diagnosis of disease.

nuclear matrix (also called nuclear scaffold) A protein web that is left in the nucleus when histones and other weakly bound proteins are removed and most of the DNA is digested away. It is presumed to act as an organizing scaffold for the chromatin.

nuclease An enzyme that cleaves nucleic acids.

nucleoid The large, circular DNA molecule of a prokaryotic cell, along with its associated proteins; also sometimes called the bacterial chromosome.

It is supercoiled and forms a dense mass within the cell, and the term *nucleoid* is often used for the cell region occupied by this mass.

nucleolus A region in the nucleus of a cell where ribosomal RNAs are transcribed and processed and ribosomes are assembled.

nucleoside A molecule that, upon complete hydrolysis, yields 1 mole per mole of a purine or pyrimidine base and a sugar.

nucleosome The first-order structural unit for the packing of DNA in chromatin, consisting of 146 bp of DNA wrapped 1.75 times around a core octamer of histone proteins. Successive nucleosomes are connected by stretches of "linker" DNA.

nucleotide A molecule that, upon complete hydrolysis, yields at least 1 mole per mole of a purine or pyrimidine base, a sugar, and inorganic phosphate.

nucleotide excision repair A DNA repair process in which a stretch of DNA that includes a damaged site is cut out endonucleolytically and replaced.

nucleus The membrane-bound structure in a eukaryotic cell that contains the chromosomal genetic material and associated components. It is also the place where RNA molecules are processed and ribosomes are assembled.

Okazaki fragments The discontinuous stretches in which the lagging strand is initially synthesized during DNA replication; these fragments are later joined to form a continuous strand.

oligonucleotide A molecule composed of nucleotide residues linked 3′ to 5′ but smaller than a nucleic acid, which is called a polynucleotide; often refers to a product of chemical synthesis.

oligopeptide A molecule composed of amino acid residues linked by peptide bonds that is smaller than a protein, which is called a polypeptide.

oncogene A gene that, in a mutated version, can help transform a normal cell to a cancer cell. Many oncogenes code for mutant proteins that are involved in the reception and transduction of growth factor signals.

oncoprotein The protein product of an oncogene.

open-promoter complex A complex between RNA polymerase holoenzyme and a promoter that has undergone initial unwinding (has "opened") preparatory to the start of transcription. It is preceded by a much less stable *closed-promoter complex,* in which the promoter has not unwound, that may either fall apart or proceed to an open-promoter complex.

open reading frame A sequence within a messenger RNA that is bounded by start and stop codons and can be continuously translated. It represents the coding sequence for a polypeptide.

operator A DNA site where a repressor protein binds to block the initiation of transcription from an adjacent promoter.

operon A set of contiguous prokaryotic structural genes that are transcribed as a unit, along with the adjacent regulatory elements that control their expression.

optical isomers See *enantiomers.*

organelles Membrane-bound compartments in the cytoplasm of eukaryotic cells. Each kind of organelle carries out a specific set of functions. Examples are mitochondria, chloroplasts, and nuclei.

orthologs Homologous genes in different species that originated by vertical descent from a common ancestral gene by speciation. Compare paralogs.

orthophosphate The ionic form of phosphoric acid as it exists at physiological pH.

osteoporosis Loss of calcium from the bones resulting in porous, brittle bones.

oxidase An enzyme that catalyzes the oxidation of a substrate with oxygen as the electron acceptor.

oxidative phosphorylation The phosphorylation of ADP to ATP that occurs in conjunction with the transit of electrons down the electron transport chain in the inner mitochondrial membrane.

oxidative stress Actual or potential damage to cells and tissues caused by prolonged high levels of reactive oxygen species (which see).

oxygenase An enzyme that catalyzes the incorporation of oxygen into a substrate.

palindrome With respect to DNA, a segment in which the sequence is the same on one strand read right to left as on the other strand read left to right; thus, a back-to-back pair of inverted repeats.

paralogs Homologous genes in a single organism that originated by gene duplication. Paralogous genes often have the same or similar function, but not always. As long as one copy continues to perform the original function, the other copy can escape the original selective pressure and is free to mutate and acquire new functions. Compare orthologs.

partial molar free energy See *chemical potential.*

partition coefficient (K) A coefficient that indicates how a particular substance will distribute itself between two media if allowed to diffuse to equilibrium between them; it is equal to the ratio of the solubilities of the substance in the two media.

passive transport (also called passive diffusion) With respect to membrane transport, the movement of a substance across a biological membrane by molecular diffusion through the lipid bilayer. Compare *active transport, facilitated transport.*

Pasteur effect The inhibition of glycolysis by oxygen; discovered by Pasteur when he found that aerobic yeast cultures metabolize glucose relatively slowly.

pentose A five-carbon sugar.

peptide bond The bond that links successive amino acids in a peptide; it consists of an amide bond between the α-carboxyl group of one amino acid and the α-amino group of the next.

peptidoglycan A polysaccharide linked covalently to peptide or polypeptide chains.

peptidyltransferase During ribosomal polypeptide synthesis, the enzyme complex that transfers the polypeptide chain from the tRNA in the P site to the amino acid carried by the tRNA in the A site, thereby adding another amino acid to the chain. The active site is carried within the RNA portion of the large ribosomal subunit.

permanent dipole In chemistry, a molecule that has a permanent, asymmetric distribution of charge such that one end is negative and the other end positive. The water molecule is an example: The oxygen end has a partial negative charge, and the hydrogen end has a partial positive charge.

peroxisome A small, vesicular organelle that specializes in carrying out cellular reactions involving the transfer of hydrogen from a substrate to O_2. These reactions produce the by-product H_2O_2, which is split to H_2O and O_2 by the peroxisomal enzyme catalase.

PEST sequences A family of amino acid sequences that have been found on cellular proteins that undergo rapid turnover; they may target proteins for rapid proteolysis. They consist of a region about 12 to 60 residues long that is rich in proline, glutamate, serine, and threonine (P, E, S, and T in the one-letter abbreviation system).

phenotype The appearance and other measurable characteristics of an organism; it results from the interaction of the organism's genetic makeup with the environment. Compare *genotype.*

phenylketonuria One of the earliest-understood inborn errors of metabolism, characterized by accumulation of phenylalanine and its catabolic products.

pheromones Intercellular mediator compounds that are released from one organism and influence the metabolism or behavior of another organism, usually of the same species. Sex attractants, which elicit reproductive behavior in suitable recipients, are an example.

phorbol esters A group of natural substances that resemble *sn*-1,2-diacylglycerol (DAG) in part of their structure and can act as tumor promoters. This effect suggests that the DAG second-messenger system may be involved in growth factor action.

phosphate transfer potential Refers to the thermodynamic tendency of an organic phosphate compound to transfer its phosphate to an acceptor; literally, $-\Delta G^{\circ\prime}$ of hydrolysis for the phosphate bond.

phosphodiester link The linkage that connects the nucleotide monomers in a nucleic acid. It consists of a phosphate residue that links the sugar moieties of successive monomers by forming an ester bond with the $5'$ carbon of one sugar and the $3'$ carbon of the next.

photophosphorylation Phosphorylation of ADP to ATP that depends directly on energy from sunlight. The light energy is captured by a pigment such as chlorophyll and is passed in the form of excited electrons to an electron transport chain; the electron transport chain uses energy from the electrons to create a proton gradient across a membrane, which drives the synthesis of ATP.

photoproducts The products that result when light energy causes a chemical reaction to occur in a substance. With respect to DNA, the term refers to the types of damaged DNA that can be caused by UV irradiation.

photoreactivation A DNA repair process in which an enzyme uses light energy to break cyclobutane pyrimidine dimers created by UV irradiation and to restore the correct bonding.

photorespiration The cycle of reactions that occurs in place of the Calvin cycle when the photosynthetic enzyme rubisco adds O_2 rather than CO_2 to ribulose bisphosphate carboxylase. It takes place partly in chloroplasts, partly in peroxisomes, and partly in mitochondria; it expends ATP energy and loses a previously fixed CO_2 molecule in the process of regenerating the Calvin cycle intermediate 3-phosphoglycerate.

photosynthesis The process by which energy from light is captured and used to drive the synthesis of carbohydrates from CO_2 and H_2O.

photosystem A structural unit in a cellular membrane that captures light energy and converts a portion of it to chemical energy. The photosynthesis practiced by plants, algae, and cyanobacteria involves two types of photosystem, both of which capture energy in the form of high-energy electrons and transduce it via an electron transport chain.

plaque A clear area that is formed by a local phage infection in a lawn of cultured bacteria in a Petri dish; for purposes of experimentation, it is the phage equivalent of a bacterial colony. The term *plaque* refers also to atherosclerotic plaques (which see)—fatty deposits that line the inner surfaces of coronary arteries.

plasma membrane The lipid bilayer membrane that encloses the cytoplasm; it is surrounded by the cell wall if one is present.

plasmids Small, extrachromosomal circular DNA molecules found in many bacteria. They replicate independently of the main chromosome and may occur in multiple copies per cell.

plus strand In viral genomes, a nucleic acid strand that can serve as mRNA or (for a DNA strand) that is homologous to one that can; as distinct from the complementary (minus) strand. Most viruses with single-strand genomes package only the plus or minus strand in virions; the other strand is made transiently during replication. Compare *minus strand*.

polyamines A family of compounds, most derived from ornithine, that have several functions in growth control, including stabilization of nucleic acids and membranes.

polymer A large molecule that is made by linking together prefabricated molecular units (monomers) that are similar or identical to each other. The number of monomers in a polymer may range up to millions.

polymerase chain reaction (PCR) A technique that is used to amplify the number of copies of a specific DNA sequence through repeated cycles of heat denaturation and DNA polymerase–catalyzed replication.

polyribosome Several ribosomes bound to and translating the same mRNA molecule; also called a polysome.

polytene chromosome An extra-thick chromosome that includes many parallel copies of the original DNA molecule; it is produced by repeated rounds of DNA replication without separation of the resulting copies.

Polytene chromosomes are found in various cell types, notably *Drosophila* salivary gland cells; they are useful in chromosome mapping because they are large and because the genes on the strands are arranged in strict register.

P/O ratio In oxidative phosphorylation, the ratio of P atoms consumed (in ATP synthesis) to atoms of oxygen consumed (in respiration).

porphyrins A family of tetrapyrrole compounds, derived from succinate and glycine, that include heme and chlorophyll.

primary structure For a nucleic acid or a protein, the sequence of the bases or amino acids in the polynucleotide or polypeptide. Compare *quaternary structure, secondary structure, tertiary structure*.

primer A short piece of DNA or RNA that is base-paired with a DNA template strand and provides a free $3'$–OH end from which a DNA polymerase can extend a DNA strand. Also refers to DNA oligomers used in the polymerase chain reaction.

primosome An enzyme complex that is located in the replication fork during DNA replication; it synthesizes the RNA primers on the lagging strand and also participates in unwinding the parental DNA helix.

prion An infectious agent that contains protein but no nucleic acid.

processivity For a DNA or an RNA polymerase, the average number of nucleotides incorporated per event of binding between the polymerase and a $3'$ primer terminus. It describes the tendency of a polymerase to remain bound to a template.

prochiral Referring to a symmetrical molecule that becomes asymmetric as a result of binding to an enzyme.

prokaryotes Primitive single-celled organisms that are not compartmentalized by internal cellular membranes; the eubacteria and archaea. Compare *eukaryotes*.

promoter A DNA sequence that can bind RNA polymerase, resulting in the initiation of transcription.

prophage An inactive phage genome that is present in a bacterial cell and its progeny. It is integrated into the host chromosome.

prostaglandins A family of compounds that are derived from certain long-chain unsaturated fatty acids (particularly arachidonic acid) by a cyclooxygenase pathway and that function as local hormones.

prosthetic group A metal ion or small molecule (other than an amino acid) that forms part of a protein in the protein's native state and is essential to the protein's functioning; its attachment to the protein may be either covalent or noncovalent.

proteases Enzymes that cleave peptide bonds in a polypeptide. Many show specificity for a particular amino acid sequence.

proteasome A large, ATP-dependent protease complex that is found in the cytosol of cells and is involved in the selective degradation of short-lived cytoplasmic proteins.

proteoglycans Glycoproteins in which carbohydrate is the dominant element. The carbohydrate is in the form of glycosaminoglycan polysaccharides, which are connected to extended core polypeptides to form huge, feathery molecules. Proteoglycans are important components of the intercellular matrix.

proteome The entire complement of proteins within a cell or organism.

proteomics Investigation of biological processes carried out through analysis of the proteome (which see).

protofilaments The 13 linear columns of tubulin units that can be visualized in the structure of a microtubule; they result because each turn of the microtubule helix contains exactly 13 tubulin units. Each protofilament consists of alternating α and β tubulin subunits.

proton motive force (pmf) An electrochemical H^+ gradient that is set up across a cellular membrane by membrane-bound proton pumps, such as the ones in the inner mitochondrial membrane or thylakoid membrane.

As the protons flow back down their gradient across the membrane, they can drive processes such as ATP synthesis.

proton pumping The active pumping of protons across a cellular membrane to form a proton gradient. For example, the electron transport chains of the inner mitochondrial and thylakoid membranes incorporate proton pumps, which create the proton gradient that powers the ATP synthases of these membranes.

provirus An animal virus genome that is integrated into a host chromosome.

pseudogenes Generally nontranscribed stretches of DNA that bear a strong sequence similarity to functioning genes and obviously arose from them during evolution. Many gene families contain pseudogene members. Recent evidence indicates that some pseudogenes are, in fact, transcribed.

pulsed field gel electrophoresis A type of gel electrophoresis in which the orientation of the electric field is changed periodically. This technique makes it possible to separate very large DNA molecules, up to the size of whole chromosomes.

pyranose A sugar that forms a six-carbon ring.

quantum efficiency (Q) With respect to photosynthesis, the ratio of oxygen molecules released to photons absorbed.

quaternary structure For a protein, the level of structure that results when separate, folded polypeptide chains (subunits) associate in a specific way to produce a complete protein. Compare *primary structure, secondary structure, tertiary structure.*

quorum sensing A bacterial phenomenon, such as biofilm formation, that is dependent upon cell density.

radioautography A technique in which an item containing radioactively labeled elements (for example, a tissue slice or a chromatography gel) is laid against a photographic film; the radioactivity exposes the film to form an image of the labeled elements. Also called autoradiography.

Ramachandran plot A plot that constitutes a map of all possible backbone conformations for an amino acid in a polypeptide. The axes of the plot consist of the rotation angles of the two backbone bonds that are free to rotate (ϕ and ψ, respectively); each point ϕ, ψ on the plot thus represents a conceivable amino acid backbone configuration.

random coil Refers to a linear polymer that has no secondary or tertiary structure but instead is wholly flexible with a randomly varying geometry. This is the state of a denatured protein or nucleic acid.

rate constant With respect to chemical reactions, a constant that relates the reaction rate for a particular reaction to substrate concentrations.

rate enhancement The factor by which the rate of a catalyzed reaction is increased relative to the same reaction carried out without a catalyst or with a nonbiological catalyst.

rate equation An equation, such as the Michaelis–Menten equation (which see), that relates velocity of an enzyme-catalyzed reaction to measurable parameters.

reaction center In photosynthesis, a specific pair of chlorophyll molecules in a photosystem that collect light energy absorbed by other chlorophyll molecules and pass it to an electron acceptor, normally the first compound of an electron transport chain.

reaction coordinate A measure of the progress of a reaction toward completion.

reactive oxygen species (ROS) Oxygen species intermediate in oxidation level between O_2 and H_2O, which are more reactive than O_2; ROS include superoxide, peroxide, peroxynitrite, and hydroxyl radical.

receptor A protein that binds selectively to a specific molecule (such as an intercellular mediator or antigen) and initiates a biological response.

recognition helix In a helix–turn–helix DNA binding motif, the α-3 helix, which fits deep in the major groove and is responsible for the sequence specificity of binding.

recombinant DNA molecule A DNA molecule that includes segments from two or more precursor DNA molecules.

recombination A process in an organism in which two parent DNA molecules give rise to daughter DNA that combines segments from both parent molecules. It may involve the integration of one DNA molecule into another, the substitution of a DNA segment for a homologous segment on another DNA molecule, or the exchange of homologous segments between two DNA molecules.

reducing equivalent An amount of a reducing compound that donates the equivalent of 1 mole of electrons in an oxidation–reduction reaction. The electrons may be expressed in the form of hydrogen atoms.

regulon A group of unlinked (nonadjacent) genes that are all regulated by a common mechanism.

release factors Independent protein factors that are necessary participants in the release of a finished polypeptide chain from a ribosome.

replicon A unit in the genome that consists of an origin of replication and all the DNA that is replicated from that origin.

replisome A complex of DNA polymerase and other enzymes and proteins that catalyzes the multiple processes involved in replicative DNA chain elongation.

repressor In molecular genetics, a protein that inhibits the transcription of a gene by binding to an operator.

residue An amino acid or nucleotide as it exists within a polypeptide or polynucleotide chain, respectively. The structure is the same as that of the respective free compound, minus the elements of one water molecule.

respiration With respect to energy metabolism, the process in which cellular energy is generated through the oxidation of nutrient molecules with O_2 as the ultimate electron acceptor. This type of respiration is also called *cellular respiration* to distinguish it from respiration in the sense of breathing.

respiratory chain The electron transport chain that is employed during cellular respiration and has O_2 as the ultimate electron acceptor.

respiratory quotient For a metabolite undergoing complete oxidation to CO_2 and water, the ratio of CO_2 molecules generated to O_2 molecules consumed.

resting potential The voltage difference that exists across the membrane of an excitable cell, such as a nerve cell, except in places where an action potential is in progress. It is a consequence of the ion gradients that are maintained across the membrane.

restriction endonucleases Enzymes that catalyze the double-strand cleavage of DNA often at specific base sequences. Many restriction endonucleases with different sequence specificities have been found in bacteria; they are used extensively in molecular genetics.

restriction fragment length polymorphisms (RFLPs) A type of genetic polymorphism that is readily detected by Southern blotting and can be used to screen for genetic diseases. It is based on the fact that alleles often have different restriction endonuclease cleavage sites and therefore produce different arrays of fragments upon cleavage with appropriate endonucleases.

retinoids Substances that are derived from retinoic acid (a form of vitamin A) and act as intercellular mediators; they are particularly important in regulating development.

retroviruses A family of RNA viruses that possess reverse transcriptase. After the virus infects a cell, this enzyme transcribes the RNA genome into a double-strand DNA version, which integrates into a host chromosome. Human immunodeficiency virus (HIV) is a retrovirus.

reverse transcriptase An enzyme found in retroviruses that synthesizes a double-strand DNA molecule from a single-strand RNA template. It is an important tool in molecular genetics and is the target for several antiviral drugs.

ribosomes Large protein–RNA complexes that are responsible for synthesizing polypeptides under the direction of mRNA templates.

riboswitch The 5′ end of certain mRNAs, which can fold around specific metabolites, usually end-products of a pathway, thereby inhibiting translation of that mRNA and shutting down the pathway.

ribozyme An enzyme containing RNA as its catalytic entity; can be either an RNA molecule or a ribonucleoprotein.

RNA editing A type of RNA processing that has been found in the mitochondrial mRNAs of certain eukaryotes, in which the RNA sequence is altered by the insertion of uridine residues at specific sites; more generally, any process that involves alteration of an mRNA sequence.

RNAi Short for RNA interference, a collection of processes in which small RNA molecules (21 to 23 nucleotides) attach to particular mRNAs by homologous base pairing and limit translation of the target RNA by various means, including stimulated degradation.

RNA primer During DNA replication, the short stretch of RNA nucleotides that is laid down at the beginning of each Okazaki fragment; it provides a 3′ –OH end from which DNA polymerase can extend the fragment. It is later replaced with DNA.

rubisco (ribulose bisphosphate carboxylase-oxygenase) The enzyme that accomplishes carbon fixation in photosynthesis by adding CO_2 to ribulose-1,5-bisphosphate. It can also add O_2 in place of CO_2, initiating photorespiration.

saponification Base-catalyzed hydrolysis of a triacylglycerol.

sarcoplasmic reticulum A network of membranous tubules that surrounds each myofibril in a skeletal muscle cell. It is a specialized region of endoplasmic reticulum; its main function is to sequester and then release the Ca^{2+} that triggers myofibril contraction.

satellite DNA DNA consisting of multiple tandem repeats of very short, simple nucleotide sequences. It typically makes up 10% to 20% of the genome of higher eukaryotes; at least some of it may play a role in chromosome structure.

scanning electron microscopy (SEM) A type of electron microscopy in which a beam of electrons is scanned across an object, and the pattern of reflected electrons is analyzed to create an image of the object's surface. Compare *transmission electron microscopy*.

secondary structure Local folding of the backbone of a linear polymer to form a regular, repeating structure. The B- and Z-forms of the DNA helix and the α-helix and β-sheet structures of polypeptides are examples. Compare *primary structure, quaternary structure, tertiary structure*.

second law of thermodynamics The law that states that the entropy in a closed system never decreases. An alternative statement is that processes that are thermodynamically favored at constant temperature and pressure involve a decrease in free energy.

second messenger An intracellular substance that relays an extracellular signal (such as a hormonal signal) from the cell membrane to intracellular effector proteins.

second-order reaction A reaction in which two reactant molecules must come together for the reaction to occur. The reaction is called second-order because the reaction rate depends on the square of reactant concentration (for two molecules of the same reactant) or on the product of two reactant concentrations (for two different reactants). Compare *first-order reaction*.

sedimentation coefficient (S) A coefficient that determines the velocity at which a particular particle will sediment during centrifugation; it depends on the density of the medium, the specific density of the particle, and the size, shape, and mass of the particle.

sedimentation equilibrium A technique for using centrifugation to measure the mass of a large molecule such as a protein. A solution of the substance is centrifuged at low speed until the tendency of the substance to sediment is balanced by its tendency to diffuse to uniform concentration; the resulting concentration gradient is used to measure the molecular mass.

semiconservative replication A mode of DNA replication in which each daughter duplex contains one strand from the parent duplex and one newly synthesized strand. This is the way DNA replication actually occurs.

sense strand For a gene, the DNA strand that is homologous to an RNA transcript of the gene—that is, it carries the same sequence as the transcript, except with T in place of U. It is thus complementary to the strand that served as a template for the RNA.

sex factors Plasmids that specify gene products that enable bacteria to engage in conjugation (bacterial mating).

sickle-cell disease A genetic disease resulting from a hemoglobin mutation. It produces fragile erythrocytes, leading to anemia.

signal recognition particles (SRPs) Cytoplasmic particles that dock ribosomes on the surface of the endoplasmic reticulum (ER) if the nascent polypeptide is destined to be processed by the ER. The SRP recognizes and binds to a specific N-terminal signal sequence on the nascent polypeptide.

signal sequence See *leader sequence*.

siRNA Also known as small interfering RNA, a short (21 to 23 nucleotides) double-stranded RNA that targets other RNAs with homologous sequences for degradation; involved in resistance by plants to virus infection.

sirtuins (For silent information regulators) A family of proteins with NAD^+-dependent protein deacetylase activity that are involved in regulating energy metabolism.

site-directed mutagenesis A technique by which a specific mutation is introduced at a specific site in a cloned gene. The gene can then be introduced into an organism and expressed.

6–4 photoproduct A type of DNA damage caused by UV irradiation in which a bond forms between carbon-6 of one pyrimidine base and carbon-4 of an adjacent pyrimidine base. This type of photoproduct appears to be the chief cause of UV-induced mutations.

somatic hypermutation A large increase in the spontaneous mutation rate of genes in the immune system, participating in the terminal stages of antibody diversification.

somatic mutation A mutation that occurs in a cell of an organism other than a germ-line cell; it may affect the organism in which it occurs, but it cannot be passed on to progeny.

SOS response A bacterial response to various potentially lethal stresses, including severe UV irradiation. It involves the coordinated expression of a set of proteins that carry out survival maneuvers, including an error-prone type of repair for thymine dimers in DNA.

Southern blotting A technique for detecting the presence of a specific DNA sequence in a genome. The DNA is extracted, cleaved into fragments, separated by gel electrophoresis, denatured, and blotted onto a nitrocellulose filter. There it is incubated under annealing conditions with a radiolabeled probe for the sequence in question, and heteroduplexes of the probe with genomic DNA are detected by radioautography.

spectrophotometer An instrument that exposes a sample to light of defined wavelengths and measures the absorbance. Different types of spectrophotometers operate in different wavelength ranges, such as ultraviolet, visible, and infrared.

spin label A substance that has an unpaired electron detectable by electron spin resonance and that is used as a chemical label.

spliceosome A protein–RNA complex in the nucleus that is responsible for splicing introns out of RNA transcripts.

standard reduction potential (E_0) For a given pair consisting of an electron donor and its conjugate acceptor, the reduction potential under standard conditions (25 °C; donor and acceptor both at 1 M concentration). Sometimes called redox potential.

standard state A reference state, with respect to which thermodynamic quantities (such as chemical potentials) are defined. For substances in solution, standard state indicates 1 M concentration at 1 atm pressure and 25 °C.

statins A family of drugs that lower cholesterol levels by inhibiting HMG-CoA reductase.

stop codons RNA codons that signal a ribosome to stop translating an mRNA and to release the polypeptide. In the normal genetic code, they are UAG, UGA, and UAA.

stringent response A mechanism that inhibits the expression of all structural genes in bacteria under conditions of amino acid starvation. It involves inhibition of the synthesis of ribosomal and transfer RNAs.

substrate A reactant in an enzyme-catalyzed reaction.

substrate-level phosphorylation Synthesis of a nucleoside triphosphate (usually ATP) driven by the breakdown of a compound with higher phosphate transfer potential.

suicide inhibitor An enzyme inhibitor on which the enzyme can act catalytically but which irreversibly alters the active site of the enzyme in the process. (It is called a suicide inhibitor because the enzyme "commits suicide" by acting on it.)

supercoiling For a DNA double helix, turns of the two strands around each other that either exceed or are fewer than the number of turns in the most stable helical conformation. Only a helix that is circular or else fixed at both ends can support supercoiling. See *twist, writhe*.

superhelix density (s) A measure of the superhelicity of a DNA molecule. It is equal to the change in linking number caused by the introduction of supercoiling divided by the linking number the DNA molecule would have in its relaxed state.

suppression With respect to mutations, a mutation that occurs at a different site from that of an existing mutation in a gene but restores the wild-type phenotype.

Svedberg unit (S) In ultracentrifugation, a unit used for the sedimentation coefficient; it is equal to 10^{-13} second.

swinging cross-bridge model The periodic attachment and release of crossbridges, with a cross-bridge conformational change, slide the thin and thick filaments past one another.

symport A membrane transport process that couples the transport of a substrate in one direction across a membrane to the transport of a different substrate in the same direction. Compare *antiport*.

systems biology A form of biological investigation that focuses upon the operation and interactions of entire systems, such as all of the metabolic reactions occurring in a cell, as compared with the analysis of individual steps and processes.

tautomers Structural isomers that differ only in the location of their hydrogens and double bonds.

Tay–Sachs disease A genetic disease caused by a deficiency of the lysosomal enzyme *N*-acetylhexosaminidase A, which is involved in sphingolipid degradation. The deficiency results in accumulation of the ganglioside sphingolipid GM_2, particularly in the brain.

telomerase A DNA polymerase that adds a short repeating sequence to the 3′ end at either end of a chromosomal DNA molecule, thus creating a single-strand overhang. This overhang gives room for priming the origin of a final Okazaki fragment during DNA replication so that the full length of the chromosome can be copied.

telomeres Special DNA sequences at the ends of eukaryotic chromosomes.

temperate phages Bacteriophages that can establish a condition of lysogeny. See also *lysogeny*.

template strand A DNA or an RNA strand that directs the synthesis of a complementary nucleic acid strand.

terpenes A family of compounds, including the steroids, that are assembled from five-carbon units related to isoprene.

tertiary structure Large-scale folding structure in a linear polymer that is at a higher order than secondary structure. For proteins and RNA molecules, the tertiary structure is the specific three-dimensional shape into which the entire chain is folded. Compare *primary structure, quaternary structure, secondary structure*.

thioredoxin A low-molecular-weight protein with a redox-active dithiol grouping that plays a variety of roles, mostly as a biological reductant.

thylakoids The membrane-bound sacs within a chloroplast (which see) that contain the photosystems (which see).

topoisomerases Enzymes that change the supercoiling of DNA helices by either allowing the superhelical torsion to relax (thus reducing the supercoiling) or adding more twists (thus increasing the supercoiling).

topoisomers With respect to DNA, closed circular DNA molecules that are identical except in their sense or degree of supercoiling. DNA topoisomers can be interchanged only by cutting one or both strands using topoisomerases.

transamination In the cell, the enzymatic transfer of an amino group from an amino acid to a keto acid. The keto acid becomes an amino acid and vice versa.

transcription The synthesis of an RNA molecule complementary to a DNA strand; the information encoded in the base sequence of the DNA is thus "transcribed" into the RNA version of the same code. Compare *translation*.

transcription factors Proteins that influence the transcription of particular genes, usually by binding to specific promoter sites.

transduction A gene-exchange process that involves packaging of bacterial DNA during virus infection followed by its transfer to another cell in a subsequent round of infection.

transfer RNA (tRNA) A class of small RNA molecules that transfer amino acids to ribosomes to be incorporated into proteins.

transgenic Refers to an organism whose genome contains one or more DNA sequences from a different species (transgenes). Genetic engineering can be used to create transgenic animals.

transition In genetics, a substitution mutation in which a purine–pyrimidine base pair gives rise to another purine–pyrimidine base pair.

transition state In any chemical reaction, the high-energy or unlikely state that must be achieved by the reacting molecule(s) for the reaction to occur.

translation The synthesis of a polypeptide under the direction of an mRNA, so that the nucleotide sequence of the mRNA is "translated" into the amino acid sequence of the protein. Compare *transcription*.

translesion synthesis A form of DNA repair in which specialized DNA polymerases participate by replicating past a damaged site in the undamaged template strand.

transmission electron microscopy (TEM) A type of electron microscopy in which a beam of electrons passes through the object to be viewed and creates an image on a photographic plate or screen. Compare *scanning electron microscopy*.

transposable genetic elements Genetic elements that are able to move from place to place within a genome. A transposon is one type of transposable element.

transposase An enzyme that is involved in the transposition, the movement of a bacterial transposon into a target site.

transversion In genetics, a substitution mutation in which a purine–pyrimidine base pair gives rise to a pyrimidine–purine base pair.

triacylglycerol Often called neutral fat or triglyceride, a lipid that upon hydrolysis yields one glycerol and three fatty acids.

tricarboxylic acid cycle See *citric acid cycle*.

tumor promoter A compound that functions beyond the initial stage of carcinogesis, often acting as a growth stimulator.

turnover number With respect to an enzyme-catalyzed reaction, the number of substrate molecules one enzyme molecule can process (turn over) per second when saturated with substrate. It is equivalent to the catalytic rate constant, k_{cat}.

twist (*T*) With respect to a DNA double helix, the total number of times the two strands of the helix cross over each other, excluding writhing. It is a measure of how tightly the helix is wound. See also *linking number, writhe*.

ubiquitin A widely distributed low-molecular-weight protein which, when linked to the C-terminal residue of another protein, marks that protein for metabolic degradation.

ultrafiltration The technique of filtering a solution under pressure through a semipermeable membrane, which allows water and small solutes to pass through but retains macromolecules.

uncompetitive inhibition A form of enzyme inhibition in which the inhibitor binds reversibly to the enzyme–substrate complex but not to the free enzyme.

uricotelic Referring to organisms, such as birds or reptiles, that excrete most of their nitrogen in the form of uric acid.

van der Waals radius (*r*) The effective radius of an atom or a molecule that defines how close other atoms or molecules can approach; it is thus the effective radius for closest molecular packing.

vector In genetic engineering, a DNA molecule that can be used to introduce a DNA sequence into a cell where it will be replicated and maintained. Usually a plasmid or a viral genome.

very low-density lipoprotein (VLDL) A type of lipoprotein particle that is manufactured in the liver and functions mainly to carry triacylglycerols from the liver to adipose and other tissues.

vesicle A spherical particle formed from a phospholipid bilayer.

virion A single virus particle.

viruses Infectious entities that contain the nucleic acid to code for their own structure but that lack the enzymatic machinery of a cell; they replicate by invading a cell and using its machinery to express the viral genome. Most viruses consist of little but nucleic acid enclosed in a protein coat; some viruses also have an outer lipid-bilayer envelope.

wax An ester formed from a long-chain fatty acid and a long-chain fatty alcohol.

Western blotting A technique for identifying proteins or protein fragments in a mixture that react with a particular antibody. The mixture is first resolved into bands by one-dimensional denaturing gel electrophoresis. The protein bands are then "blotted" onto a nitrocellulose sheet, the sheet is treated with the antibody, and any bands that bind the antibody are identified. More accurately called immunoblotting.

wild-type Refers to the normal genotype found in free-living, natural members of a group of organisms.

writhe (*W*) With respect to a supercoiled DNA helix, the number of times the helix as a whole crosses over itself—that is, the number of superhelical turns that are present. See also *linking number, twist*.

xenobiotic An organic compound that is not produced by the organism in which it is found.

X-ray diffraction A technique that is used to determine the three- dimensional structure of molecules, including macromolecules. A crystal or fiber of the substance is illuminated with a beam of X-rays, and the repeating elements of the structure scatter the X-rays to form a diffraction pattern that gives information on the molecule's structure. See also *diffraction pattern*.

yeast artificial chromosomes (YACs) Artificial chromosomes used for cloning and maintaining large fragments of genomic DNA for investigational purposes. A YAC is constructed by recombinant DNA techniques from a yeast centromere, two telomeres (chromosome ends), selectable markers, and cloned DNA in the megabase range.

Z-DNA A DNA duplex with a specific left-hand helical structure. In vitro, it tends to be the most stable form for DNA duplexes that have alternating purines and pyrimidines, especially under conditions of cytosine methylation or negative supercoiling.

Ab	antibody		F	phenylalanine
Ac-CoA	acetyl-coenzyme A		F	Faraday constant
ACP	acyl carrier protein		F_{ab}	antibody molecule fragment that binds antigen
ADH	alcohol dehydrogenase		FAD	flavin adenine dinucleotide
AdoMet	S-adenosylmethionine		$FADH_2$	reduced flavin adenine dinucleotide
ADP	adenosine diphosphate		FBP	fructose-1,6-bisphosphate
Ag	antigen		FBPase	fructose bisphosphatase
AIDS	acquired immune deficiency syndrome		Fd	ferredoxin
Ala	alanine		fMet	N-formylmethionine
AMP	adenosine monophosphate		FMN	flavin mononucleotide
Arg	arginine		F1P	fructose-1-phosphate
ARS	autonomously replicating sequence		F6P	fructose-6-phosphate
Asn	asparagine		G	Gibbs free energy
Asp	aspartic acid		GABA	γ-aminobutyric acid
atm	atmosphere		Gal	galactose
ATP	adenosine triphosphate		GAP	glyceraldehyde-3-phosphate
bp	base pair		GC-MS	gas chromatography-mass spectrometry
BPG	bisphosphoglycerate		GDP	guanosine diphosphate
cal	calorie		Glc	glucose
cAMP	cyclic 3′,5′-adenosine monophosphate		Gln	glutamine
CD	circular dichroism		Glu	glutamic acid
cDNA	complementary DNA		Gly	glycine
CDP	cytidine diphosphate		GMP	guanosine monophosphate
Chl	chlorophyll		G1P	glucose-1-phosphate
CMP	cytidine monophosphate		GS	glutamine synthetase
CoA or CoA-SH	coenzyme A		GSH	glutathione (reduced glutatione)
CoQ	coenzyme Q		G6P	glucose-6-phosphate
cpm	counts per minute		GSSG	glutathione disulfide (oxidized glutathione)
CRP	cAMP receptor protein (catabolite activator protein)		GTP	guanosine triphosphate
			h	hour
CTP	cytidine triphosphate		h	Planck's constant
Cys	cysteine		Hb	hemoglobin
d	deoxy		HDL	high-density lipoprotein
Da	dalton		HIV	human immunodeficiency virus
dd	dideoxy		hnRNA	heterogeneous nuclear RNA
DEAE	diethylaminoethyl		HPLC	high-pressure (or high-performance) liquid chromatography
DHAP	dihydroxyacetone phosphate			
DHF	dihydrofolate		HX	hypoxanthine
DHFR	dihydrofolate reductase		Hyl	hydroxylysine
DNA	deoxyribonucleic acid		Hyp	hydroxyproline
DNP	dinitrophenol		IDL	intermediate-density lipoprotein
dopa	dihydroxyphenylalanine		IF	initiation factor
dTDP	thymidine diphosphate		IgG	immunoglobulin G
dTMP	thymidine monophosphate or thymidylate		Ile	isoleucine
dTTP	thymidine triphosphate		IMP	inosine monophosphate
E	reduction potential		$InsP_3$	inositol 1,4,5-trisphosphate
EF	elongation factor		IPTG	isopropylthiogalactoside
EGF	epidermal growth factor		IR	infrared
EPR	electron paramagnetic resonance		ITP	inosine triphosphate
ER	endoplasmic reticulum		J	joule

MAKEUP IS ART

Professional techniques for creating original looks

"Dedicated to my loving Grandmother Mavis Kenneally. Without you this would have not been possible."

JANA

This edition published in 2013
by Carlton Books Limited
20 Mortimer Street
London W1T 3JW

10 9 8 7 6 5 4 3 2 1

A CIP catalogue record for this book is available from the British Library.
ISBN 978 1 78097 295 4

Printed and bound in China

Author: Academy of Freelance Makeup

Creative Directors: Jana Ririnui and Lan Nguyen

Senior Executive Editor: Lisa Dyer

Copy Editor: Sarah-Jane Corfield-Smith

Makeup artists: Jana Ririnui, Sandra Cooke, Rachel Wood, Jose Bass, Jo Sugar, Lina Dahlbeck, Philippe Milleto, Christina Iravedra, Carolyn Roper, Jade Hunka, Nat Van Zee, Maria Papaodpoulou, Elsbeth Tan, Sharka, Karla M Barchieri.

Contributing writers: Jana Ririnui, Sandra Cooke, Rachel Wood, Jo Sugar, Barbra Villasenor, Lina Dahlbeck, Jose Bass, Sarah-Jane Corfield-Smith, Yvonne Chung, Carolyn Roper.

Photographers: Camille Sanson, Catherine Harbour, Fabrice Lachant, Keith Clouston, Jason Ell, Desmond Murray, Conrad Atton, Zoe Barling, Mo Saito, Han Lee de Boer, Allan Chiu, Roberto Aguliar, Daniel Nadel, Lou Denim, Dez Mighty.

Hair Stylists: Jana Ririnui, Marc Eastlake, Desmond Murray, Zoe Irwin, Andrea Cassolari, Nathalie Malbert, Natasha Mygdal, Cyndia Harvey, Fabio Vivan, Tim Furssedonn, Zach Jardin.

Stylists: Rebekah Roy, Karl Willett, Shyla Hassan, Nasrin Jean-Batiste, Svetlana Prodanic, David Hawkins, Sampson Soboye, Zed-Eye, Claudia Behnke.

Front cover Photography by Camille Sanson, makeup by Lan Nguyen, head piece by Marc Eastlake Design using Swarovski. **Overleaf** Photography by Lou Denim, makeup by Jade Hunka using MAC, model Jodie @ Nevs. **Pages 4–5** Photography by Lou Denim.

2009 book design by Disciple Productions; 2010, 2013 design by Carlton Books Ltd.

MAKEUP IS ART

Professional techniques for creating original looks

THE ACADEMY OF FREELANCE MAKEUP

CARLTON
BOOKS

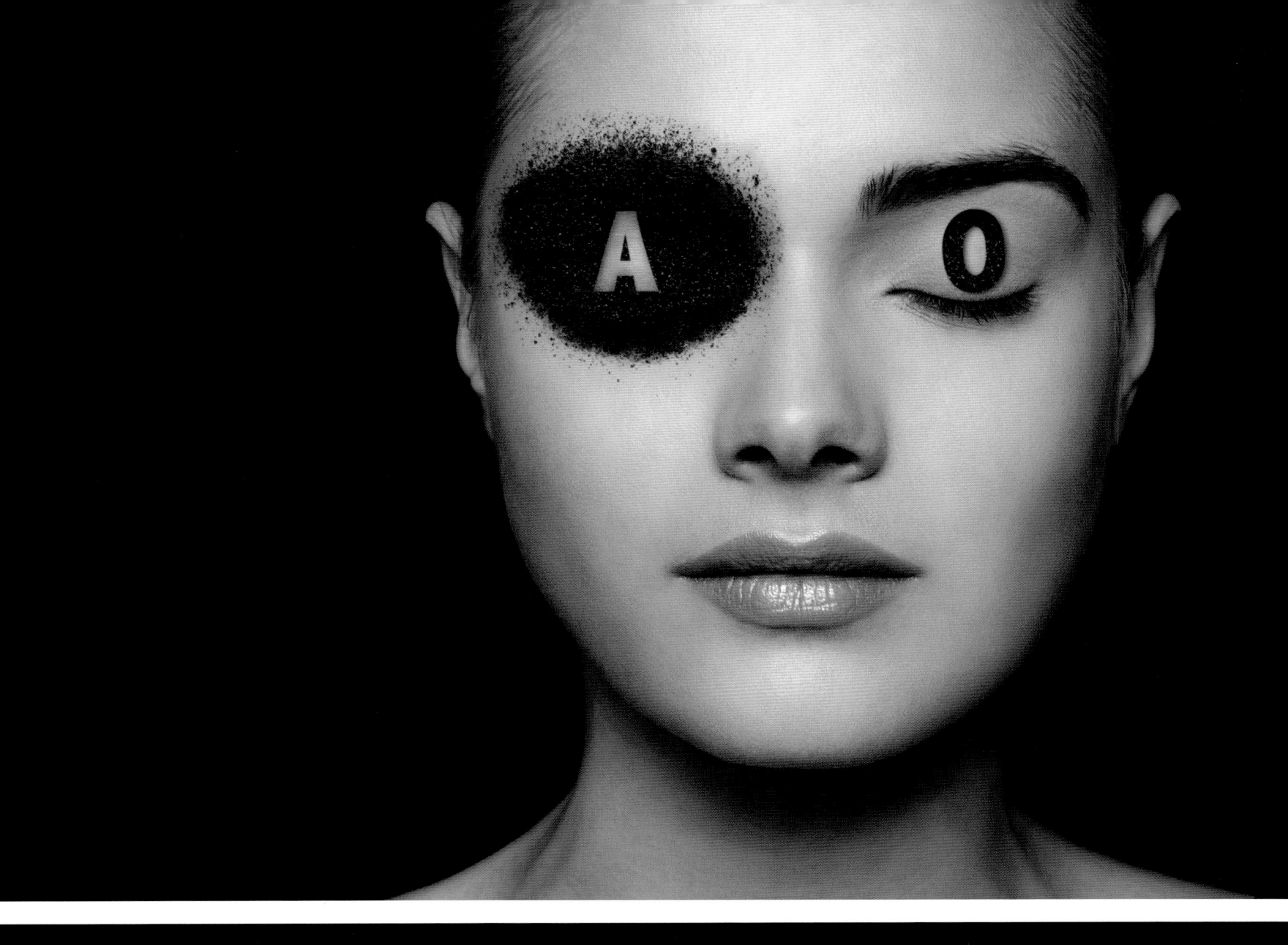

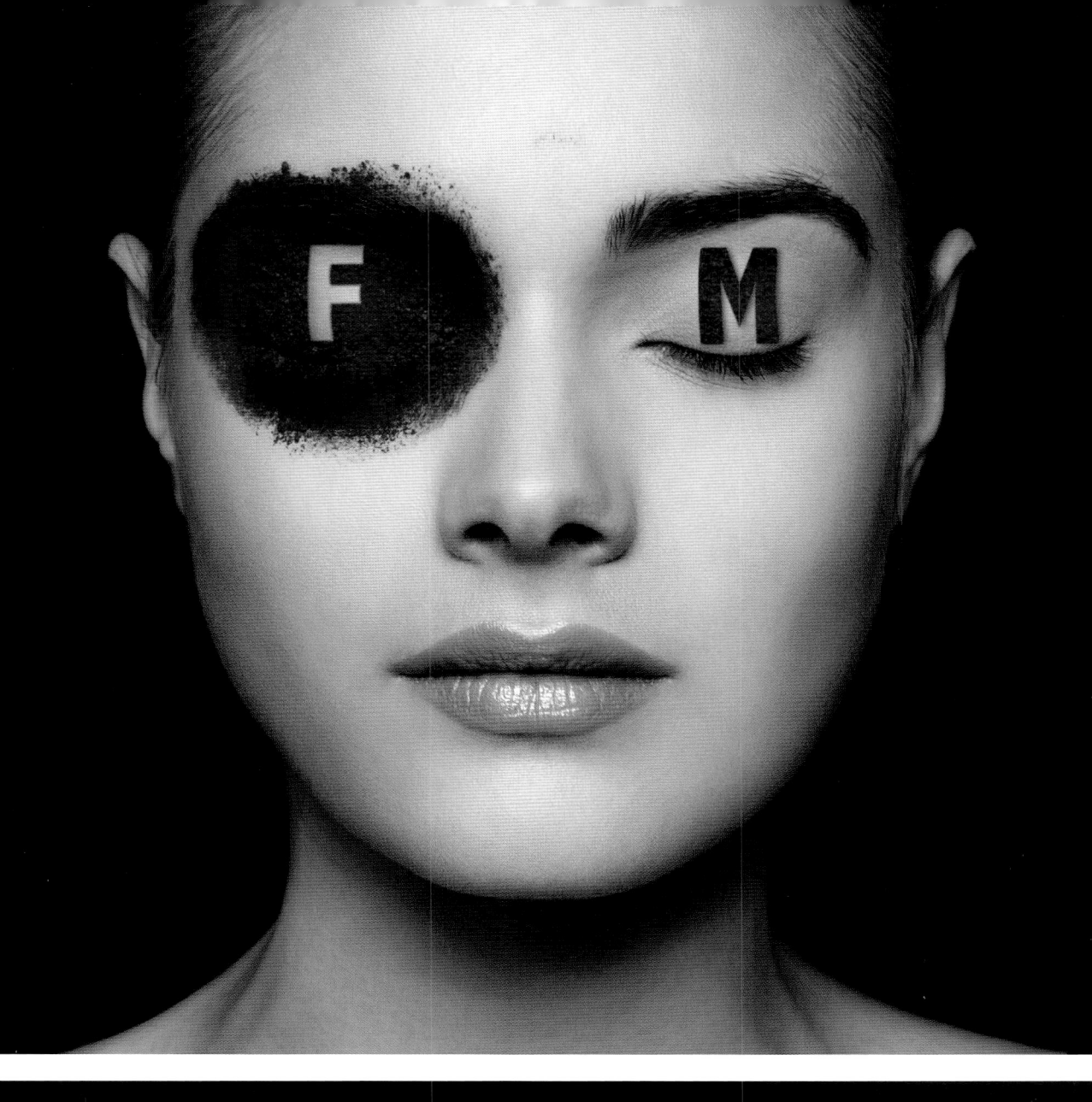

CONTENTS

JANA RIRINUI

Jana Ririnui, Makeup Artist and Founder of AOFM

"I was born in New Zealand in a small town called Invercargill, far away from the fast pace of a fashion capital like London. I started my career at 14, working for free at a hair salon after school. My dream was to become a hairdresser and I was prepared to work hard to make this a reality. I moved to Australia where there were more opportunities, allowing me to build up my hairdressing skills. It was here that I was spotted and approached by a modelling agency and started to model. Although it took me away from my hairdressing, modelling gave me a wider perspective of the fashion world and experience on the other side of the camera. It was here that I realized that makeup artistry is where my heart belonged. I decided to move to London in 1999, knowing it would be the perfect place to combine my skills as a hairdresser with my potential as a makeup artist.

I studied at one of London's well-known schools, but upon completion of my courses as a makeup student, I felt a little lost as no-one had really prepared me for the hard road ahead or how competitive the industry was going to be. Despite the lack of guidance I worked really hard on my career and after painstaking work and testing with photographers I eventually found my feet as a makeup artist. I took every opportunity to work – so long as I was applying makeup, I was happy. I worked on cosmetic counters, shooting on my days off with photographers and assisting for free just to gain more experience. I do not think I really began to learn about the industry until I started assisting other makeup artists. This all helped lead me to become a successful artist. I have worked on a number of famous magazines along with some big names, but my passion has always been the desire to teach and provide help to aspiring makeup artists who were in the same situation as me when I finished my training.

While talking to friends and fellow makeup artists about setting up a school, it was felt there was a need to offer training conducted by professional, working makeup artists. As the industry is so small and competitive it seemed clear to us that the best people to teach makeup skills and industry know-how are those who are already successful artists themselves. I didn't get to learn about the industry until I was undertaking jobs, so I wanted to ensure my students would have the opportunity to understand the workings of the profession while developing their skills.

So the Academy was born – the first in the industry to be owned and operated by freelance makeup artists with a love of the industry. My desire is to encourage the next generation of makeup artists, giving them the tools, knowledge and understanding that is essential to succeed. My philosophy has seen AOFM's reputation grow as one of the leading makeup schools and our students have gone on to become some of the finest and most successful international freelance makeup artists, as well as top fashion and beauty editors for major magazines and newspapers.

I formed the AOFM Pro team, which has become a well-known creative force within the industry, driven by the team's flair and conceptual ideas. We have come together with other top industry professionals to create a book dedicated to inspire other makeup artists, people wishing to enter this exciting career and anyone who appreciates fine art and beauty.

This book is packed full of creative ideas, inside knowledge from our artists' experiences and advice on how this forever-changing industry operates. It will keep you one step ahead of the competition and give you a true insight into the creative world of makeup artistry."

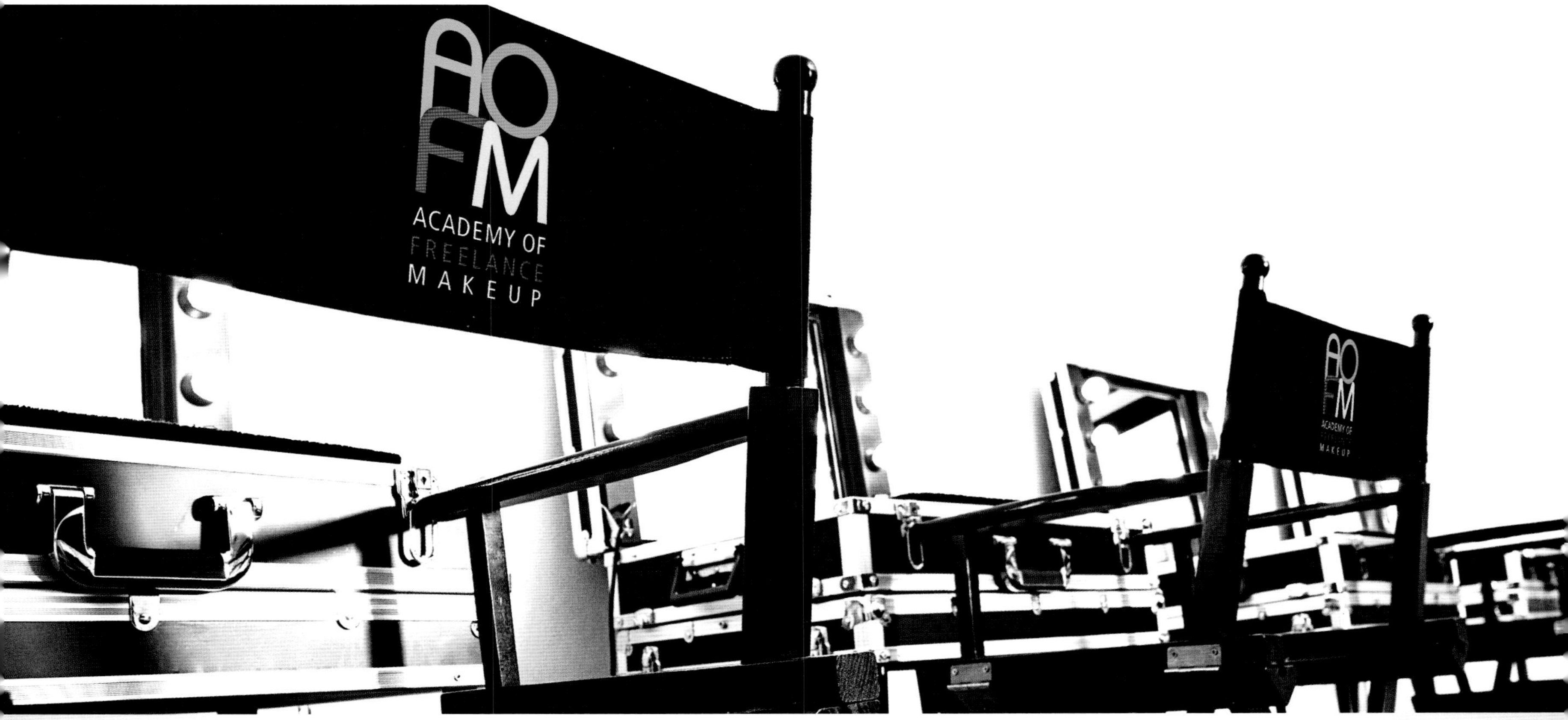
AO
FM
ACADEMY OF
FREELANCE
MAKEUP

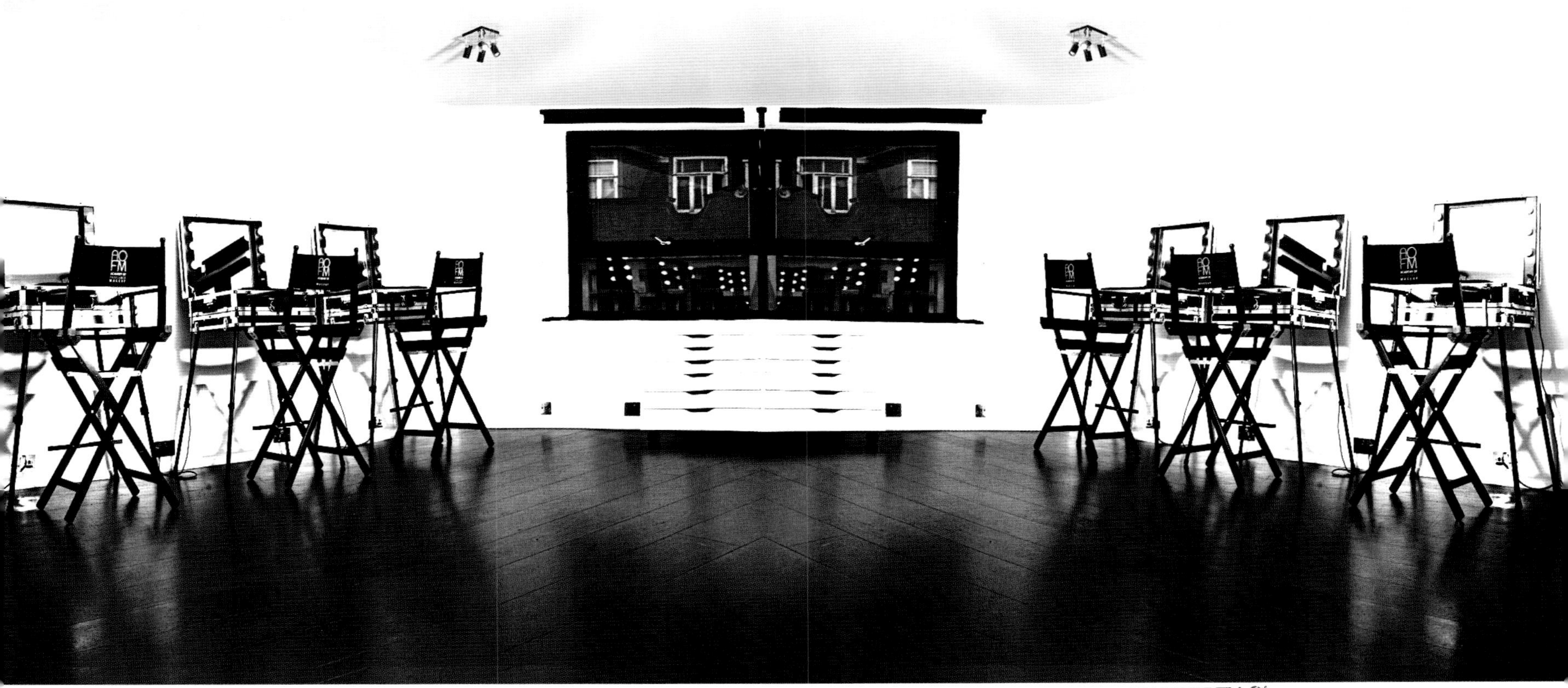

SKINCARE

As a makeup artist, it helps to understand skin, as it is the canvas that you work on. If you prepare and look after the skin well, then your makeup will sit better and last longer. It is fully recommended to study the skin to the level of a beautician, as this will help you understand makeup application more fully.

It is not necessary to carry a skincare range for each skin type in your makeup kit. However, it is important to be able to manage and control each skin type to your advantage. Many tutors at AOFM and makeup artists are big fans of Dermalogica. It is a brand that acknowledges the importance of skin and creates products that make the makeup artist's job easier, as well as offering ongoing training for makeup artists lucky enough to be included on their artist list. The following pages will guide you through the essential skincare products for an artist's kit.

When choosing a skincare range, opt for one that is not highly perfumed and is suitable for use on sensitive skin. Models have makeup applied and removed so frequently that their skin becomes easily sensitized. It is also important to carry a good-quality range if you plan to work with celebrities – which is why, for so many makeup artists, Dermalogica ticks the boxes.

When doing pre-recorded television, videos, bridal makeup, beauty and fashion shoots, and celebrity makeup, always use a full range of products, including cleanser, toner and moisturizer, unless the celebrity requests their own preferred brand and products. Live television or shooting on location often requires wipes and moisturizer to save time. On fashion shows, if time allows, use a full range, but when pushed for time it is essential to carry wipes.

Photographer:
Catherine Harbour
Makeup: Lancôme
Model: Kasia Z @
First Management

It is important to pay attention to dry and oily skin. Dry skin needs oil and moisture put back into it, so skin preparation and products containing oil and moisture will help the makeup last longer. If you do not treat dry skin you may experience flaking, which will ruin your foundation. Oily skin needs products that help reduce shine and oil. Skincare and makeup products containing oil will cause your makeup to slide, and shine will be visible very quickly, making the makeup difficult to manage.

Cleansers

An ultra-calming cream cleanser is an effective and gentle cleanser that is mild enough to be used to remove eye makeup. Is also convenient on location as it can be removed without water. Face wipes, such as MAC and Simple, can be used instead of cleanser and toner for speed and convenience. These are commonly used backstage and on photo shoots to remove makeup quickly and efficiently.

Toners

Use a convenient spray toner that is gentle, moisturizing and refreshing. It helps prepare the skin for moisturizer, and does not need to be removed.

Treatments for Eyes and Lips

A firming vitamin-packed treatment that contains silicone, which acts like a primer, will smooth wrinkles and moisturize the sensitive skin around the eyes and lips. Lucas Pawpaw Ointment is great to use on dry, chapped lips.

Treating the Problem (optional)

Some skins may require additional treatment for problem areas. For dry skin or to add moisture use a hydrating booster before moisturizing. This can also be used on the lips. For sensitive skin use a gentle treatment for reducing redness and calming the skin. For mature skin or premature ageing, a firming booster or serum can help reduce the appearance of fine lines and wrinkles. For skin prone to breakouts and spots, a clearing spot treatment is effective to help reduce breakouts. Dermalogica has pre-moisturizer boosters for all these conditions.

Moisturizers

Choose your moisturizer according to your skin type. For chronically dry skin, choose a rich, concentrated moisturizer. For normal to dry skin, a medium-weight moisturizer such as Dermalogica's Skin Smoothing Cream is ideal to hydrate and smooth the skin. For oily skin, you will need a light-hydration formula that reduces the appearance of oil and helps close the pores.

Tinted Moisturizers

A tinted moisturizer is used to moisturize the skin as well as provide sheer coverage to help even out skin tone. It is ideal for people that do not want to wear a foundation, but still want a little coverage.

Beauty Serums

Serums are used as part of a facial treatment. Similar to a moisturizer, a serum is a liquid for treating skin conditions such as dehydration, redness, discoloration and fine lines. They contain highly concentrated ingredients that are absorbed into the skin quickly and more deeply for an intensive effect. The high concentration of skin-benefiting ingredients penetrates into the skin, producing dramatic and long-lasting results.

To use, apply a drop of serum to your fingertips and massage it gently into skin in the morning and at night, then follow with your usual moisturizer. You do not need to use more than a drop or two to cover the whole face and neck. You will probably feel an immediate change in your skin the first time you use a serum – it may feel softer, smoother and perhaps even a little tighter.

Exfoliants

The skin is always generating new skin cells at the lower layer and sending them to the surface. As the cells rise to the surface they gradually die and become filled with keratin. These keratinized skin cells are important because they protect the skin during the creation of new skin cells. As we age, however, the process of cell turnover slows down. Cells start to pile up unevenly on the skin's surface, giving it a dry, rough, dull appearance. Exfoliation is beneficial because it removes the older cells and reveals the fresher, younger skin cells below. It also makes expensive facial products like serums easier to penetrate into the skin.

Only use products designed for the face, as some of those designed for the body can be too abrasive and irritating for the face. A fine powder exfoliant that becomes a paste when mixed with water will smooth out dry, flaky skin and remove dead skin cells. The best method is to use exfoliating gloves or a synthetic scrubbing sponge designed for the face. To do this, gently rub your face with the sponge using circular motions. Avoid scrubbing under the eyes, where the skin is thin and can easily be damaged. The residue can be removed with warm water or damp cotton wool pads.

Exfoliating can cause the skin to dry out, and dry skin is an invitation for skin to wrinkle. By over-exfoliating you run the risk of bursting the delicate blood vessels under the skin as well, and if the vessels burst, your skin may appear permanently flushed.

Photographer:
Catherine Harbour
Makeup: Lancôme
Model: Rosanne F @
Premier

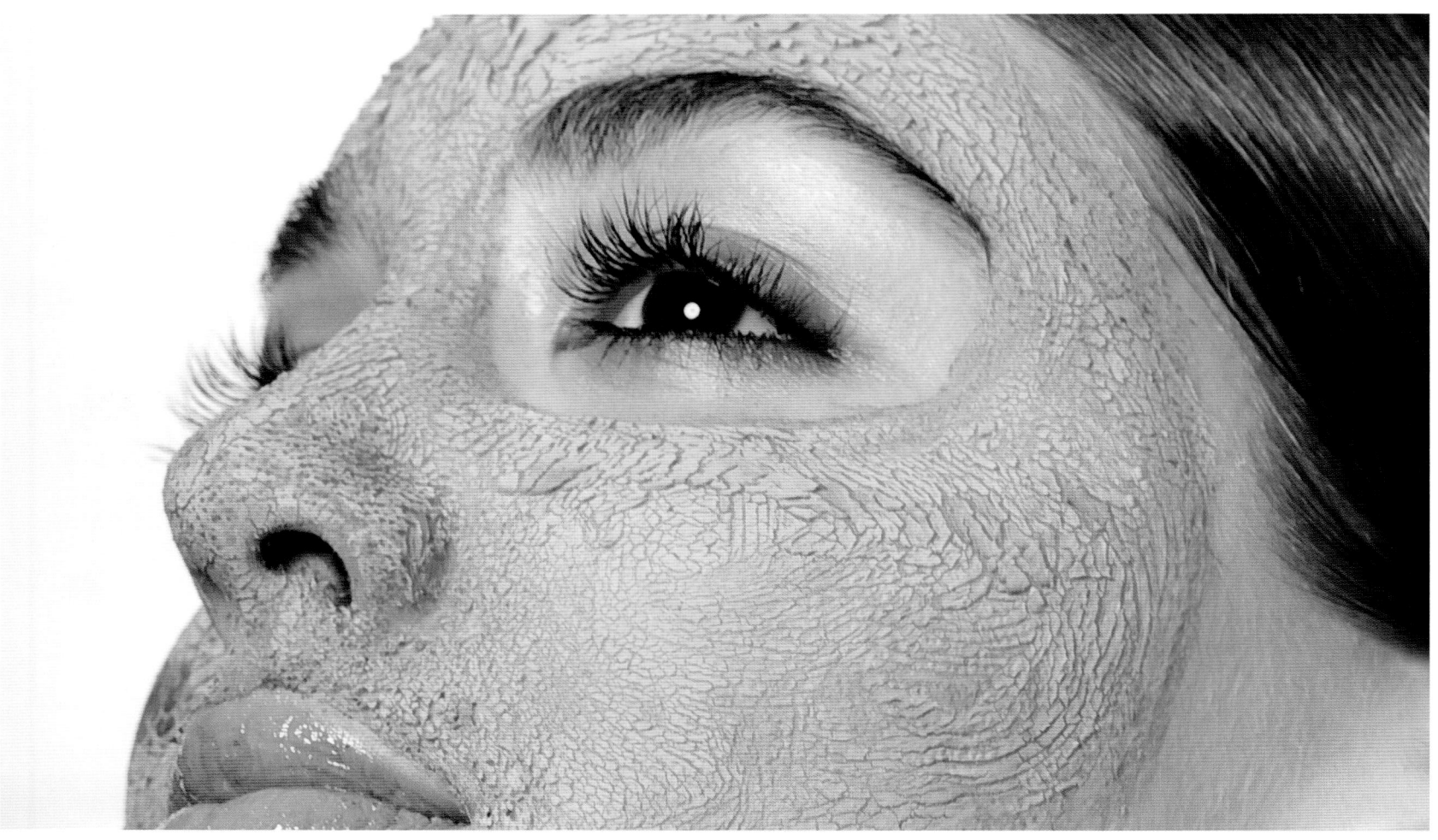

Face Masks

Masks can make the difference between good skin and great skin. There is a different one appropriate for every skin type. Facial masks will exfoliate your skin and remove any buildup of sluggish cells. A face mask, along with daily skin washes, will keep your pores unclogged and help clear up blemishes by stimulating blood circulation. When the seasons change your skin also changes, so you may need to switch your mask with the change in weather. Most are best applied once or twice weekly.

Clay Masks

Best for those with oily skin, clay masks draw out toxins and impurities and are essential for keeping the skin clear.

Peel-off Masks

Intended for the gentle removal of dead skin cells, peel-off masks leave your complexion replenished and radiant. They work with all skin types.

Paraffin Masks

For hydrating and softening the skin, paraffin wax masks are appropriate for all skin types. They aid blood circulation, which will brighten the complexion.

Pore Strips

For the instant removal of blackheads, oil and dirt, and to minimize pores, use pore strips. They have twice the effectiveness of pore-cleansing facial washes and are handy to have in your kit to prevent clogged pores and breakouts. You achieve instant results and the strips can be used up to three times a week.

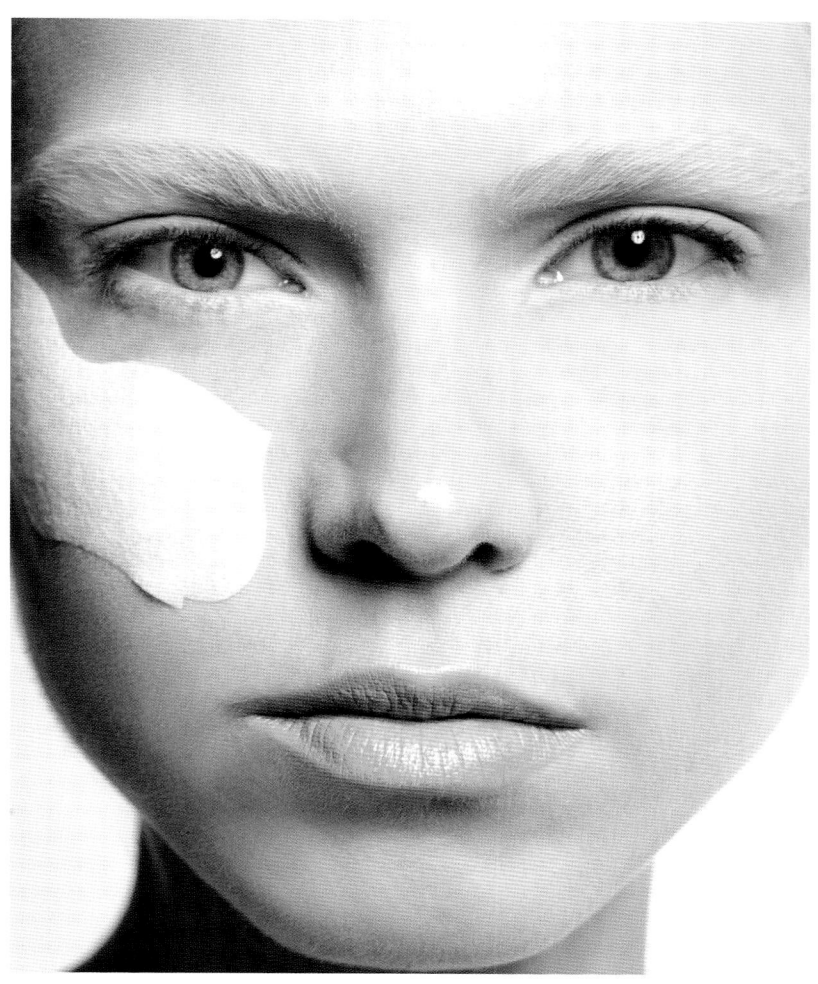

Applying a Mask

Try to use a facial mask at least three times a month. Applying a mask requires a specific technique, but most of us do not know that – we just plaster it all over the place. The first step is to use a flat, wide brush or spatula. Avoid using your fingers!

1 Start applying the mask with a brush or spatula from the centre of the face outward, using even strokes. Be careful to avoid the lips and the sensitive area around the eyes.

2 Leave the mask on for at least 20 minutes, unless a particular time frame is mentioned on the packaging. Never leave your mask on the skin for more time than recommended by the manufacturer. Some ingredients could harm the skin if they stay on for longer than necessary.

3 To remove the mask, moisten with a little water and gently rub it off the skin, working one area at a time. In the case of peel-off masks, you will have to peel! Finish by rinsing the face with plenty of water, using circular strokes to remove any residue.

Photographer:
Catherine Harbour
Makeup: Jade Hunka
using Lancôme
Models: Kathleen B
(left) and Gabriele D
(right) @ Premier

THE FACE

In today's society, how you look matters. When creating the perfect look for your client, skin, makeup, hair and nails are all areas that need attention. It is important for an artist to be able to cover skills in these areas confidently, even if there are specialists available. It also helps keep the job diverse and interesting.

Cosmetic products are constantly changing. New products become available and existing ones are improved. It is important to know the ingredients and to be guided by the specialist suppliers. It is also beneficial to test the products first in case a model has become sensitive or allergic. Ultimately it will save you time and avoid costly mistakes.

The freedom of style, that is so prevalent now, allows us to experiment by mixing the classic with the new, and by adding your own signature twist to gives it an edge. Trends always get repeated but as products change they can produce a completely different look.

There are so many timeless beauty icons to use as inspiration but the important rule is that you enhance the facial features to make the model look beautiful. For example, by having a flawless base and then just a touch of colour you can create something that is fun and strong but still gives a beautiful effect.

Photographer: Jason Ell **Makeup:** Jose Bass using Shu Uemura **Hair:** Natasha Mygdal using Bumble & Bumble **Model:** Hannah @ Next

FACE SHAPES

The oval face shape is no longer considered the ideal "perfect" face shape. With today's cultural diversity, we embrace the beauty within everyone. As much as it is useful to know your face shape, it no longer determines what makeup suits you. Instead, use your face shape as a guide to maximizing your natural beauty. Along with cosmetics, your hair style will also help accentuate your face shape.

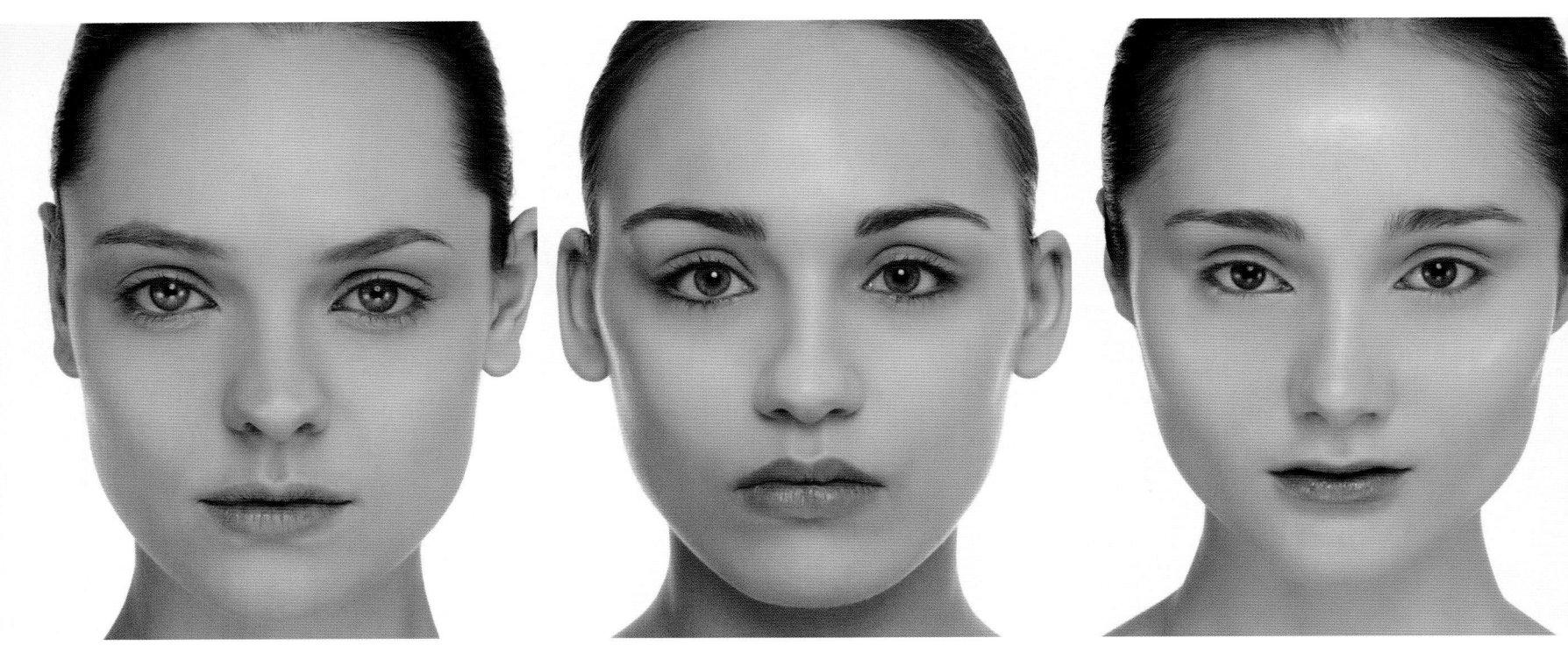

Oval
Once considered the ideal face shape because of its perfect proportions and symmetry, the oval face is generally slightly broader at the cheeks and narrower at the forehead and chin. Little (if any) contouring is needed as the face is well balanced. A good makeup tip is to accentuate your best features. A celebrity example is Kylie Minogue.

Round
This shape face is usually wider than an oval face, with fuller cheeks and often a rounded or non-prominent chin. Round faces tend to have that "baby-face" look. Contour under the cheekbones, on the temples and under the jawline to give the illusion of a more defined bone structure. A celebrity example is Cameron Diaz.

Heart/Inverted Triangle
Wider at the forehead , the heart shape curves to a narrow, pointed chin. For those with an exaggerated heart, the hairline mimics the rounded two bumps of the heart. As it is difficult to soften the prominent chin, accentuate your best feature, such as the eyes or cheeks, to draw attention away from the chin. A celebrity example is Reese Witherspoon.

Photographers:
Keith Clouston and
Catherine Harbour
Makeup: Jade Hunka
and Jo Sugar using
Bobbi Brown **Models:**
Melinte, Natalie, Elena
@ Oxygen, Rosanne F,
Larissa H, Gabriele D
@ Premier

Square
Square faces tend to be a similar width between the forehead, the cheeks and the jawline. Soften square features by contouring under the cheekbones and temples. A celebrity example is Demi Moore.

Triangle
This face type is characterized by a dominant jawline that narrows at the cheekbones and forehead. Contouring under the cheekbones will help add some width to the face. A celebrity example is Victoria Beckham.

Diamond
The forehead and chin are narrower, with wide and inevitably high cheekbones. Little or no contouring is required, simply play up the best features. A celebrity example is Scarlett Johansson.

PRIMERS

A primer is a product that is used to prepare the skin before foundation. For the most part they are all silicon based. The silicon in the product is used to fill in pores and fine lines, to create a smooth base for the foundation or other makeup to be applied. The main benefit of using a primer is that it helps the makeup last longer on the skin.

There are many types of primers available, in the same way that there are numerous moisturizers for different skin types. Primers work best on cleansed, prepared skin, and, if the skin is exceptionally dry, then a primer can be used on the skin after moisturizer.

Types of primers include hydrating ones used for moisturizing dry skin, mattifying for oily skin to help control shine, and primers to plump the skin for a more youthful appearance. Many also contain SPF. Primers can come in a range of textures, such as lotions, mousse, silicon gels, as well as mineral powder primers, specially suited for sensitive skin.

Professional makeup artists, particularly those in the fashion industry, tend to have a range of primers in their kits for use on different jobs. Although in the fashion industry we generally look after models with young, plump skin, and are able to be on set to do touch-ups in between shots, there are still occasions when primers are needed, such as shooting an outdoor editorial where a primer with SPF is essential. Models with tired or dehydrated skin (we see this a lot during fashion week) may require a hydrating primer to help sooth and smooth the skin.

Primers are popular in the television and film industry, where a range of different aged actors and presenters are used. In addition, being on set or on location and under warm lights for long periods of time requires makeup to last longer in between touch-ups.

With the growing industry of high-definition television and film, the amount of makeup worn by actors and presenters needs to be more natural, and yet still effective. As powders can look heavy on television and film, mattifying primers are ideal to help prevent shine and reduce the need for excess powder and continual touch-ups. In addition, on more mature actors and presenters, a silicon-based primer, such as Prep +Prime Skin produced by MAC, will help smooth over fine lines to form a smooth base for foundation to go more flawlessly on to skin. The Becca Line and Pore Corrector is a skin-tone, oil-free primer that can help minimize the pores and lines around the eyes and nose for a smoother base application.

Photographer:
Keith Clouston
Hair and Makeup:
Jana Ririnui using
Becca **Model:**
Yu @ Oxygen

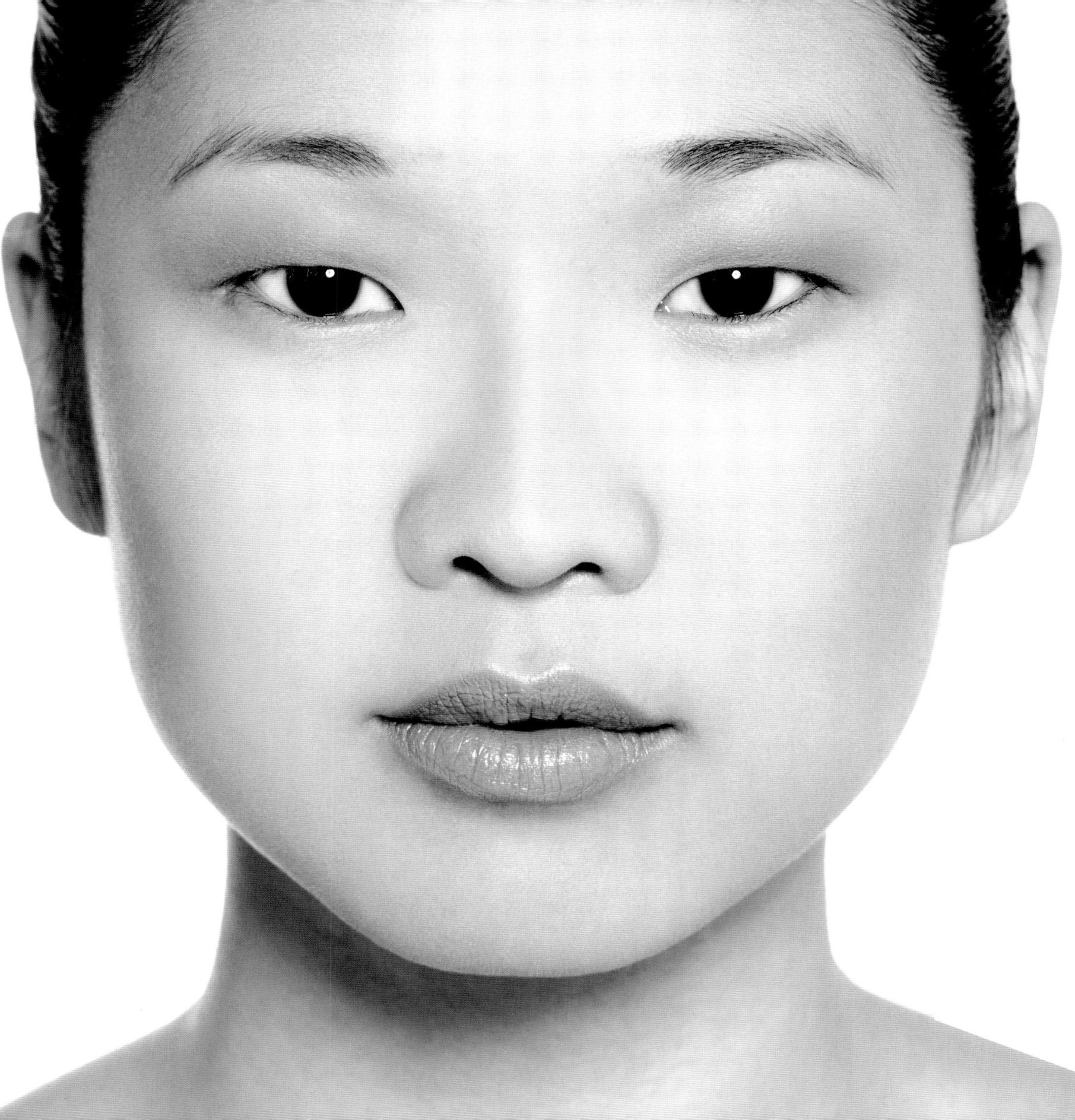

FOUNDATIONS & CONCEALERS

Foundations

The main point of using a foundation or base is to even out the tone and colour of the skin. How a makeup artist wants the skin to appear varies from job to job – all creative briefs differ. On one project a makeup artist may want the skin to have a glow, while on other jobs, an old-fashioned matte, full-coverage base may be needed or the makeup artist may be required to create a natural-looking finish, as if there is no foundation applied at all.

Choosing what type of product to use is one of the biggest challenges a new makeup artist can face. Here is a guide to help sort out the confusion.

Full Coverage

This type of foundation gives skin the appearance of flawlessness. A good full coverage can effectively cover scars, discoloration and blemishes – almost like a concealer – over the entire face. These foundations also work well on Asian and black skin tones.

• To get a flawless base on any skin type, dilute the full coverage or stick foundation by mixing it with a sheer fluid tinted moisturizer.

Sheer Coverage

Sheer foundation lets the skin look natural and unmade. It will even out skin tone, but allows freckles, beauty marks, and so on, to be visible, giving the effect of a makeup-free face.

Oil Free/Oil Control

Great for those with oily skin, this formula can help keep shine at bay by creating a matte-textured finish. Usually "oil free" means it contains no lanolin or mineral oils (these are the oils that tend to clog pores and cause sensitivity).

Illuminating

This is a foundation that leaves a glow and luminous finish on the skin. Some illuminating foundations can contain tiny pearl and glitter particles. When working on photo shoots, it can make the skin look greasy.

Moisturizing

The best base for dry to normal skin types, it gives extra hydration and works well on mature skin.

Cream to Powder

This formula has a more traditional texture and usually comes in a compact. It goes on creamy and dries to a powder finish. Cream-to-powder formulas give a medium finish, which is ideal for skin that needs more coverage than a liquid foundation, but not as much as full coverage.

Powder Based

Best used on oily skin, as it has no liquid or moisture in it, powder-based foundation gives a matte, nonreflective finish to the skin. Another advantage is that it can be used with a brush to set makeup.

Mousse

A whipped product that gives light coverage and a natural glow or matte finish, mousse foundations feel light on the skin.

Tinted Moisturizer

A combination of a moisturizer and some colour, the tinted moisturizer is perfect for evening out the skin tone. It gives a very light coverage and sinks easily into the skin.

• Try using tinted moisturizers if your complexion is dewy, and you are really looking for a light, sheer wash to even out any blotchiness.

Concealers

Used to cover blemishes and imperfections once a foundation has been applied, concealers offer heavier coverage than foundations and also help to cover dark circles under the eyes. When choosing a concealer, you should consider both consistency and colour. It should be lightweight and creamy. The product needs to glide on smoothly and not require a lot of rubbing. Find the right shade and blend it in so it looks flawless, not obvious – it should match your skin exactly. Concealer looks most natural when you work in sheer layers and build it up gradually.

If you are covering major undereye circles, you are going to want a formula that provides heavier coverage. If your skin is sensitive then you should choose a formula that contains only pure mineral pigment colour and no dyes, oils, talc or starch.

Photographer: Keith
Clouston **Hair and
Makeup:** Jana Ririnui
using Bobbi Brown
Model: Leanne @
Oxygen

Applying Foundation

1 Before applying foundation, make sure the skin is cleansed and moisturized. Wait a few minutes for the moisturizer to settle into the skin before application. If the face is damp and sticky, it can cause the foundation to clog up and look uneven. Try to use a non-sticky lotion that is not greasy or use a facial oil, such as Bio Oil.

2 Dab a small amount of foundation on the face in little dots on the forehead, down the nose and on the chin and cheeks.

3 Always apply in downward strokes to minimize pores. Sweep over ears and under jaw and do not cover up natural freckles or rosy cheeks.

4 Blend the foundation with a brush, sponge or your fingertips. Make sure it is blended completely to create a natural, flawless look. Always apply foundation quickly and lightly. Smooth, even strokes will prevent the colour looking clogged.

5 Check application in natural light to make sure the foundation is properly blended. If you need more coverage, apply concealer in specific areas rather than adding more foundation.

Concealing Dark Circles

Use a yellow-tinted concealer to hide dark purple circles and mauve or tan concealers to hide brownish circles; mauve or tan are also best for blending into dark or black skin. There are even green and light blue products to conceal red under-the-eye circles.

1 Apply several dots of concealer under the eyes, then tap and press in using your finger, never rub. Apply concealer on other uneven spots on the face – including the chin, and around the nose and mouth if need be – and tap in.

2 Pay close attention to the area where your eye meets the bridge of your nose. This area tends to have the darkest circles and consequently may need more concealer.

3 Apply foundation you normally wear to even out your complexion. Blend it as you usually would, without paying any special attention to under the eye. Once your foundation is in place, you should be able to clearly see any dark circles still evident. Then tap in a heavier coverage of concealer as in step 1.

4 Powder the concealed areas with a translucent powder, using just enough so that the concealer no longer looks sticky or shiny. Using a soft brush helps application.

• For more staying power and better coverage, let the concealer set for five to ten seconds after application and before blending.

• Mix a highlighter with the concealer to add brightness under the eye or use a brightening concealer. It will camouflage like a concealer, while adding flattering highlights to the face.

Photographer:
Catherine Harbour
Makeup: Jana Ririnui
using MAC **Model:**
Larissa H @ Premier

FOUNDATIONS & SKIN TONES

Photographer:
Catherine Harbour
Makeup: Giorgio
Armani
Model: Naiyana

Photographer: Keith
Clouston **Makeup:**
Giorgio Armani
Model: Yu @ Oxygen

Matching makeup to skin tone is a tricky business. The perfect base foundation should even up the skin and highlight its natural radiance. Olive skin and dark skin naturally glows and rarely requires much concealer. It is important not to get stuck using the same makeup all year round as skin changes with the season and can be either dry or oily. It also gets slightly darker in the summer, then paler in the winter. Also in summer, the face and neck can be different colours where the sun catches the face rather than the neck. So it is a good idea to match the foundation to the neck or add a bronzer to even up the skin after you have finished your makeup application.

Ethnic women commonly have oily skin. Oil-control products work well, but skin tends to dry out if they are used on a regular basis. This is because these products typically contain alcohol. By matching the product to your skin type you can limit this problem.

Dark Skin Tones
Black skin has a tendency to become dull, grey and ashy-looking with the wrong foundation. It is also very rare for black or dual heritage skin to be uniform all over – darker foreheads, lighter cheekbones, and so on. Concealer should only be used if there are imperfections like scars, acne

and large pores. It is extremely important to apply powder over foundation to set it and stop shine, rather than apply a heavy foundation. Other ways to limit shine are to use an oil-free primer before adding foundation and applying foundation with a brush or sponge instead of your fingers.

For dark skins, foundation should be somewhere between the natural skin colour and half a shade above or below. Use a tone-on-tone foundation or tinted moisturizer and add colour by using a blush or bronzer. It is best to choose a blush that highlights the skin tone, like peach, apricot, tan or even bronze, instead of the rosy pinks that are more suitable for pale skin. Layer the blush with more than one colour for a more dynamic look, using cream or powder blush. Shimmer on under-eye circles and wrinkles can be used lightly to brighten and contour the cheeks.

Latin, Indian and Asian Skin Tones
These skin tones are more olive-toned, so makeup with yellow undertones works best. Be forewarned that foundations sometimes turn a shade darker on oily skin because it mixes with the skin's natural oils; a lighter shade can be used to avoid foundation looking too dark. If cheeks have a natural, pink warmth, just dust a little powder (with pink undertones) on top of the foundation.

HIGHLIGHTING

Highlighting products are essential to help sculpt and contour the face. Whereas contour colours are used to create shadow and depth (see pages 32–3), highlighting products are used to reflect light, which adds an additional dimension of definition to the face. Highlighting is particularly popular in the fashion industry, in both editorial and catwalk shoots, where makeup is used to tell stories, create illusions and complement the fashion designer's vision.

Highlighting products come in a range of different textures, from shimmer powders and cream sticks to liquids and gels. There are different names for these products, including illuminizers, shimmers, iridescent and pearlized powders, but they all do the same thing.

An illuminizer (such as the Becca's Shimmering Skin Perfector) is great when mixed with a foundation to create a dewy, radiant glow and to even out skin tone. A slightly bronzed illuminzer mixed with foundation can add a subtle, bronzed, healthy glow.

Shimmer powder or liquids placed above the cheekbone reflect light and further enhance cheekbones. The Yves Saint Laurent Touche Éclat was originally designed for use as an illuminizer, but has proved popular as a concealer for under the eye. The highlighting effect reflects light and brightens dark circles.

Applying shimmer in a straight line down the centre of the nose creates the illusion of a straighter, smaller nose. Highlighters can also be applied to the centre of the lips to create the illusion of plumped, fuller lips.

On the catwalks, cream shimmers are often applied to the face first, then a powder shimmer is applied on top to create a strong, reflective effect. Highlighting is not just restricted to the face, but can also be used on the body to create the illusion of longer, more slender limbs. Highlighters can be brushed on to the collarbones and down the arms and the centre of the shins to give the illusion of length. Alternatively, oils can be used on the body to create a more reflective "wet" shine. MAC Strobe Cream is a popular choice for highlighting the body and can be mixed in with foundations.

Photographer: Zoe Barling **Makeup:** Sandra Cooke using Giorgio Armani **Hair:** Natasha Mygdal using Bumble & Bumble **Model:** Emma C @ IMG

Blusher is the one product that can instantly make you look younger, healthier and prettier. A good blusher replaces the natural colours in your cheeks that drain away with age.

There are two types: cream and powder. Recently cosmetic brands have developed sculpting blushers that can be combined to create a strong and contoured cheekbone. Some powder blushes can be highly pigmented so you end up with excess product on the brush and it can look overdone, especially if placed too near to the nose.

Blushers can also be used for contouring, as can a wide range of products, such as eyeshadows, bronzers, and liquid and cream foundations, depending on the result you want to achieve. MAC cosmetics produce a range of sculpt and shading colours in the form of blushers that are specifically designed for contouring. The colours are all various shades of matte brown, cream and nude and are very much staples in a makeup artist kit.

Black and white photo shoots require clever use of contouring and highlighting, as the contrast of light and dark shadows affect the tone and shade of the makeup. Cool colours will seem lighter than they are and warm colours will appear grey and darker. Black pencil liner or eyeshadow will look dark grey, so a generous application is needed.

Photographer:
Fabrice Lachant
Makeup: Sandra
Cooke using Nars
Hair: Nathalie Malbert
Model: Olena @ Nevs

Blushers

For healthy, natural colour, nothing beats blusher. However, it tends to be overused. Powder blush is typically best for oily and combination skin. Cream is great for dry skin and liquid and gel are best for oily skin. For even better results, combine cream and powder together – it helps blusher to stay on longer and looks more luminous. Cream blush is great for mature skin as it blends easily for a very natural look. Blush stains are ideal for well-moisturized skin but best avoided on dry skin as they tend to dry very fast. Blending cream blushers with a powder blush can also look very effective and beautiful on the skin.

Choosing a Colour

Start by choosing a colour, using nature as your guide. Find a colour that matches your cheeks when they are flushed after exercise or from being out in the cold. Fair skin looks great in rose, olive in peach and dark skin in apricot or even red. A dusky pink blusher will warm up any tired-looking skin. Another trick is to find one that matches your lip colour.

Applying Blusher

For best results, the skin should be prepped and foundation applied carefully. Apply the blush onto the high apple of the cheekbones and softly blend backward, following the cheekbone and ending at the hairline. Use a professional full brush for applying powder blush, tapping off any excess. For extra contouring and a more chiselled look, a bronzed or darker colour can be used underneath.

Blush should never go below the bottom of your nose or any closer in to the centre of your face than the iris of your eye. Apply a small amount of blush to your forehead – on the spot where the sun normally hits the face – if you look very pale. Applying blush near the eye can help give your eyes a sparkle, too.

- For over-applied powder blush, use translucent powder on top to calm down the colour. For over-applied cream blush, blot colour off with a tissue.

Applying Gels and Creams

For a natural, healthy flush, keep the colour light and blended. Dab a dot of gel or cream on the apple of the cheeks only, using the middle finger, then blend with ring and middle fingers. The clean finger will pick up any excess.

- A little foundation can be used to lighten over-applied gel or liquid blush, but as these formulas "stain" the cheeks, the only real way to correct the colour is to wash the face, moisturize and reapply.

Sun-Kissed Look

For a sun-kissed sheen dab bronzer powder on your forehead, chin and nose before applying a blush using a large brush and light strokes. Those with darker skin tones should try a caramel-coloured bronzer and avoid orange. After applying the blusher, use a sheer highlighting powder in a C-shape from the centre of your browbone to the centre of your cheeks. You can also dab a little shimmery blush on the highest point of the cheekbone, nearest the eye.

Working with Face Shapes

Accentuate the cheekbones by using a slightly darker shade in that area, while blending it down the cheek for a more natural look. If you have a full face, focus your blush near your hairline. If you have high cheekbones, apply the blush to in the centre of your face and do not apply to the underside of your cheeks.

Round faces: Give an illusion of slenderness by applying blush in a sideways V on the cheekbones. Blend it well and add a little to the chin as well.

Square faces: Soften the square angles of the face with a dab of blush on the forehead and chin. Apply blush from the centre of the eyes toward the cheekbones and blend.

Rectangular faces: Applying blush below the outer corners of the eyes will reduce the elongated square shape.

Oval faces: Apply blush to the prominent part of the cheekbone and blend it carefully toward the temples.

Photographer:
Catherine Harbour
Makeup: Becca
Model: Larissa H @
Premier

Photographer:
Roberto Aguilar
Makeup: Sandra
Cooke using MAC **Hair:**
Heath Grout using Tigi
Model: Darya @ FM
Models

Contouring

Contouring creates illusions on the face. It is the art of using shadow and highlight to sculpt, emphasize and accentuate features. A prime example of this is with cheekbones: highlighting above the bone and shadowing just below dramatically draws out the cheekbone.

To truly understand contouring, you need to understand the structure of the face. Symmetry is actually what makes a face "beautiful" and more photogenic, and so contouring can be used as a tool to create the illusion of balance and symmetry in the face. Even the most subtle form of contouring has the power to transform, creating a natural look without the need to add colour and "makeup". Contouring can also be done on a far more obvious level, particularly in high fashion where it is a deliberate feature of the makeup. Contouring can do the following:

* Give more defined cheekbones.
* Define the jawline/chin and minimize a "double" chin.
* Minimize a large forehead.
* Create fuller lips.
* Straighten and/or narrow a wider/larger nose.
* Lift sagging eyes.

• The key point to remember is that the dark colours draw back and make things appear smaller and lighter colours make features appear larger and closer.

• Always use a matte shade when creating shadows. For achieving a natural contour look, this should be between two and three shades darker than the skin tone.

• You can use a matte or shimmer product to highlight, although for more mature skins it is recommended that a matte shade is used, as shimmer products will accentuate fine lines and wrinkles.

• The level of blending you do will determine how natural the contouring looks. Always check all angles of your face in a natural light before completion, to ensure there are no obvious streaks.

• Highlighting is the opposite of contouring. With darker shades, you create a shadow, such as under the cheekbone, to help create the "shape", whereas the highlight colour, applied where the light would hit the bone – above the cheekbone, not on the cheekbone, helps create the "depth" of the shape.

POWDERS

Sheer and Translucent Powders

There are many different types of powder, ranging from pressed to loose powders and sheer to full coverage. The suitability of each depends on whether you are using it for everyday wear, bridal, fashion/editorial or television and film. Sheer powders or blot powders are great for setting foundation and reducing shine on the skin. The nice thing about them is that they tend not build up or get "cakey" after a few applications. This makes them great for all-day shoots or everyday wear. You may also use loose translucent powder, which gives a sheer finish. A colourless powder can be used on a wide range of skin tones from dark to light. A good one is the matte Shu Uemura loose powder, as the particles are very fine and absorb a maximum amount of oil on the skin.

- You can use a dark-coloured sheer powder as a bronzer on lighter tones of skin.

Coverage Powders

You can also purchase powders with a bit more coverage to them. These are useful to take out any redness in the skin or level out skin tone. MAC Mineralize Skin Finish powders are great for this, as they melt right into the skin and even out any discoloration. If you have a slightly incorrectly coloured foundation, these powders can deepen the skin tone or lighten it up a shade, depending on what colour is used.

There are full-coverage powders on the market as well. These can be used like a foundation. Some people prefer to use these (as opposed to liquid foundation) because they are faster to apply and easier to carry around for touching up. They are good for the everyday person, but would be far too heavy to use constantly on a bride or during a photo shoot.

You can get away with coverage powders as a "quick fix" backstage if a model needs a bit more coverage before she heads out on the catwalk. A popular full coverage powder is Studio Fix by MAC.

Loose and Pressed Powders

For the everyday person, the choice between loose and pressed powder is usually down to personal preference. Things to consider are whether or not you will need to be carrying it around for touch-ups (in which case a loose powder can be more bulky and messy) and what kind of coverage you would like to get out of a powder, as most loose powders tend to be more sheer in style.

Backstage at a fashion show, pressed powders are usually preferred. This is because they are far less messy to run around with and it is less likely the powder will spill or blow onto the clothes. During a fashion or editorial shoot, the choice of powders can be left to the artist's preference. Here, they will have more space and hopefully a station, so there will be less worries about being bumped into and spilling powder on the clothes.

- Wait to powder the face of your model on the shoot until the photographer has taken the first shot. This is because you may not need to powder the face at all, depending on what type of look you are going for, or you may only need minimal powder in certain areas.

- At a wedding, a pressed powder may be more beneficial as you can carry it around easily all day and evening for touch-ups. If you are not staying around for the full day, you can give the powder to the bride so she can use it herself without worrying about making a mess.

Photographer:
Catherine Harbour
Makeup: Shu Uemura
Model: Larissa H @ Premier

Being tanned or bronzed gives people a healthy glow that is generally associated with a relaxing holiday in the sun. It also gives the appearance of being slimmer and helps hide skin imperfections. However, people are now aware, more than ever, that real tans can prematurely age skin and cause skin cancer – hence the beauty industry's fascination with fake tanning products.

There are so many products available that give a good tan effect that there is no excuse for anybody to go out and soak up the sun and burn their skin. The colour choices are vast and suit everyone from the whitest of white to olive skin tones. They do require a little effort, but in the long term they will save your skin and allow you to walk round with a year-long healthy glow.

Temporary Bronzing

Considering spring/summer fashion collections are shot during the winter, makeup artists need to use a great deal of fake tans and bronzers. On a photo shoot, artists have a number of options to tan the skin. An airbrush machine can be used to apply temporary tans to models. Alcohol-based airbrush paints do not rub off on clothes or in water. Instead, they are quick to dry, and can easily be applied with a compressor and a specialized body air gun. This process, however, can be messy and requires specialist equipment, product and remover.

To create that sun-kissed glow without an air gun, the makeup artist has a variety of different products to consider. At fashion week it is common to see makeup artists using body foundation, often mixed with moisturizer, to cover blemishes and darken the legs, arms, back and any other skin showing on the models. There are also numerous bronzing products available, such as bronzed oils, gels, lotions and creams, that all provide an instant tan, but which can easily be washed off at the end of a shoot. These kinds of product are not ideal to use with certain clothes as they can stain. Powdered bronzers are also great, but are generally better used on smaller areas, such as the face and chest, rather than the whole body. Shimmer bronzers can also be used to accentuate and highlight points, such as the front of the shins, thighs, shoulders and collar bone.

For those wishing to be bronzed for a special one-off event, there are numerous body and face bronzers that will wash off with water at the end of the night. The only downside with this option is that if it is extremely hot or raining, your beautiful healthy glow may run off or, even worse, cause streaks! Bronzers can also tend to get on clothes, although as most are water-based, they are easily washed off. For makeup artists, bronzers are ideal for changing the skin tone on photo shoots, as it is quick, easy and can be applied exactly where it is needed.

Again, the bronzers come in a variety of different mediums, such as creams, lotions, sprays and powders. Facial bronzers can be very simply applied all over the face or, to give a realistic, sun-kissed glow, applied only on the places where the sun would fall, as it would if you were tanning naturally, along the cheek bones, temples, forehead and lightly along the length of the nose and the chin. This soft wash of colour will lift the face, giving an extra-healthy glow.

Photographer:
Han Lee De Boer
Makeup: Sandra
Cooke using St Tropez
Hair: Cyndia Harvey
Model: Cat @ FM

Lasting Fake Tans

For a lasting tan, the quickest and most efficient way to get that all-over bronzed glow is to get a spray tan, applied by a trained therapist using a spray gun. You can choose from a light golden glow to very dark tan and the colour is dependent on the number of coats you wish to have applied. Most tans will develop over a couple of hours. Prior to the tanning, it is recommend that a full-body exfoliation is done so that all rough skin is removed – this helps the colour go on evenly. Spray tans usually last about a week and can be maintained with proper body exfoliation and moisturizing, as well as weekly top-ups.

There are other at-home fake tan options, such as aerosol sprays, creams, gels and lotions. These can be messy and take time to dry, but they do provide a good alternative to the spray tan. The newest trend is the gradual tan – body moisturizers with a hint of fake tan that gradually build up to create and maintain a healthy glow.

Getting the Perfect Tan

Fake tans usually take between 2 to 4 hours to develop. Make sure you read the instructions for the particular product you are using, and always wait the recommended time before applying a second coat if you want to go darker.

1 To ensure an even, lasting tan, ensure any hair removal is done a day prior to application.

2 Before starting, exfoliate to ensure the skin is silky smooth.

3 Make sure drier areas, such as elbows, knees and ankles, are well moisturized to help avoid unsightly product buildup on these areas.

4 Apply the tan with a tanning mitt so as not to stain your hands and to create a more even coverage. You may need to get a friend to apply the product in hard-to-reach places.

5 Before the product dries, use a soft cloth to gently wipe the knees, ankles, elbows, and in between the fingers and toes to remove extra product, which tends to cling to these areas.

6 Wear old, loose clothing while you are waiting for the product to dry.

• To maintain your tan, top up the application about once a week. Participating in intense sports or water activities will reduce the life of your tan, and you may need to apply more coats more regularly.

• Try not to shower for 24 hours after application of the tan.

• Regular light exfoliation and moisturizing will help prolong and even the tan.

Photographer:
Roberto Aguilar
Makeup: Sandra
Cooke using St Tropez
Hair: Heath Grout
using Tigi **Model:** Darya
@ FM Models

THE *EYES*

Photographer: Roberto
Aguliar **Makeup:** Jose
Bass using Shu Uemura

EYEBROWS

Eyebrows frame the face. Therefore, it is good to have them clean, visible and shaped to match the face. For a natural, symmetrical look, the length of the eyebrows must be slightly longer than the eye. This means that the brow needs to start just above the tear duct and end approximately 5mm (³⁄₁₆ in) after the outer corner of the eye. This will create a young and fresh look.

Eyebrow Shaping

When it comes to the arch of the eyebrow be very careful to find the correct position of the arch. To do this, look straight into a mirror, place the handle of a thin makeup brush on the outside of your nostril and point it so that it is in line with the outside of your iris. Where the brush crosses the brow is where the highest point of the arch should be.

To define the brow, remove all the hairs on top of the eyebrow. If you are worried, imagine a straight line from the start of your eyebrow up to the highest point of the arch. All the hairs above this line are unnecessary so pluck them away. This will leave you with a sharp and defined brow shape.

Repeat the same process underneath the brow. Imagine a straight line from beneath to the highest point of the arch, everything under this line can be taken away. Be very careful as you pluck, just three hairs plucked away or left can make a huge difference.

Another way of shaping the eyebrows is to trim them. Where there is excess hair, or any hairs that are particularly unruly, brush them upward and trim off any length that goes over the imaginary lines mentioned previously. If your brows still need more definition, then you can apply a little brow shadow to outline the new defined shape. Use an ashy brown on lighter brows, and for really dark brows try a plum tone.

- If you want a cat-like look, make your brows longer and straighter – this is a great look for older women who want to look a little younger.

- If you want to disguise your own brow and draw on a completely new one, then a great tip is to wet some soap, apply it like a paste over the eyebrow and then carefully add a good concealer on top. You can then draw on any style of eyebrow that you want.

- The best way to get really good at styling eyebrows is to practice on paper – try drawing different styles and shapes.

- Avoid having the arch in the middle of the brow. It will make the eyes look narrow. By having the arch a little further along, the eyes will look open and full.

- Do not pluck the start of the brow too much, if the brows look too far apart from each other it will make the eyes look as if they are on the side of the head – like a fish!

- The best tweezers have pointy ends and only take one hair out at a time. Tweezers with a slanted edge will take lots of hairs out in one go, which can lead to over-plucking, resulting in the wrong shape brow.

Photographer: Allan Chiu **Makeup:** Jose Bass using Shu Uemura **Model:** Anna @ FM

3 To make shaping easier, draw in the eyebrows before plucking them. It is a good idea to first draw different brow shapes on a piece of paper. You can make the arch angle pointy, or more curvaceous and natural. Repeat this as much as you need to because it will help you visualize the eyebrow before drawing it onto the face. It will also ensure that you do not get distracted by the natural (or unnatural) shape of the eyebrows.

4 Once you have defined the line, draw in your eyebrows with an eyebrow pencil or eyebrow powder. A powder can be easily erased if you make a mistake.

5 Now pluck every hair above and below your drawn line.

6 If the hair is too long at the start of the brow, brush the hair directly upward and trim it to the top drawn line. Leave a tiny extra so you do not over-trim. Sometimes the hair at the end of the eyebrows is long and stubborn, in which case brush it down and trim it according to the line you have drawn.

7 To make the brows more defined and full, fill them in with powder and brush them softly to make them look natural. The result should be well-defined brows which draws more attention to a clean eye.

Enhancing Brows

To give shape and definition to the eyebrows it is as necessary to clear strays from above as it is from below the natural eyebrow. The widest part of the eyebrow should be the starting point in the centre, and it should get progressively thinner until the thinnest point at the end.

1 The eyebrow should be slightly longer than the outer corner of the eye. The starting point of the eyebrow should be aligned with your tear duct. Hold a thin makeup brush vertically alongside the nose to see a straight line from the tear duct to the start of the eyebrow.

2 The arch should normally be where the browbone stands out – the part of the eyebrow starting at the start and meeting the highest point should be longer than the part that stretches from the arch outward to the end. To find the ideal arch, angle a thin makeup brush from the side of your nose across the outer part of your iris. This will give you an idea where the arch should be.

Photographer: Keith Clouston **Makeup:** Jade Hunka using Shu Uemura **Model:** Elena @ Oxygen

Photographer: Keith Clouston **Makeup:** Sharka using Shu Uemura **Model:** Melinte @ Oxygen

EYESHADOW TEXTURES

How the colour of the eyeshadow appears on the eyelid depends on the cosmetic brand used, its texture and its application. Using an eye primer or eyeshadow base, such as Benefit's Lemon Aid, can boost colour, eliminate creasing and allow the makeup to last longer.

Matte

Matte eyeshadow is a flat colour without shine. It is a great choice as a contour or crease colour on a photo shoot, as it gives a better appearance of depth than shimmery shades. A contour eyeshadow brush is a useful tool to get this desired effect.

Cream

Cream shadows have a moist or wet texture. Some creams that stay moist and appear wet are prone to creasing, while others go on wet, but set dry. Bobbi Brown's Longwear cream shadow has a great crease-proof formula. Cream shadows also work well as a base on the eyelid to intensify the colour of a powder shadow.

Frost

Frost eyeshadow has a crystal-like shine to it. When applying it, use a stiff brush, which will give a stronger application of the frosted pigment.

Glitter

Containing small particles of sparkle to reflect light, glitters are great for music videos, pop promos and night shoots, as they add a real "wow" factor. Loose glitter can be applied on the eye area like an eyeshadow, but it needs a sticky surface to grip onto, such as a cream shadow or eyelash glue.

Loose Pigment

A loose-pigmented eyeshadow is made up of tiny particles of colour and usually comes in a small pot with a lid, rather than a flat casing. You can get varying degrees of effects with loose pigment shadow. If applied to the eye area with a fluffy brush, it can create a sheer wash of colour. If loose pigment is applied to the eyelid with a denser tool, such as a basic eyeshadow pad, then you get a more opaque and stronger tone. Loose pigment can also be mixed into a liquid, such as MAC's Mixing Medium to create a liquid makeup that can be painted onto the skin. Barry M's range of loose pigments come in a terrific colour palette.

Photographer: Fabrice
Lachant **Makeup:**
Sandra Cooke using
MAC **Hair:** Nathalie
Malbert **Model:**
Olena @ Nevs

Photographer:
Keith Clouston
Makeup: Jana Ririnui
using Bobbi Brown
Model: Leanne @
Oxygen

Photographer: Daniel Nadel **Makeup:** Jo Sugar using MAC **Model:** Sophie Willing

Photographer: Lou
Denim **Makeup:** Karla
M Barchieri using Inglot
Model: Jodie @ Nevs

There are many ways to create a smoky eye, but the most popular is by using black powder eyeshadow all the way around the eye and then blending upward and out above the crease. You can also use cream shadows, greasepaint or gel eyeliner. Once you have mastered the art of blending this bold look, you can start using strong colours or pigments and playing around with different shapes such as the "cat eye" (great for nights out), a "panda eye" (which is very trendy and grungy, see also page 85) and smokier eyes using cream shadow and a kohl pencil.

To create a good smoky eye, you must invest in quality brushes, but you will only need two. The first is an application brush to sweep on the eyeshadow, such as a Kolinsky sable-hair brush, and the second is a blending or crease brush to blend out hard lines.

- When blending from the crease of the eye, use very small circular motions and only use the tip of the brush, about 2–3 mm. Avoid using too much pressure.

- If the brush is separating when you are blending shadow, you are working too agressively and will create a patchy effect. Use the blending brush like a feather tickling the face.

Powder Shadow Smoky Look

This fashion look, opposite, can be worked in various colour combinations, using light, mid- and dark tones of powder shadow on the eyes. Here the light tone is pale pink, the mid-tone is soft purple and the dark tones are grey and black.

1 Using a peach or nude eyeshadow, start with a neutral wash of colour applied from the eyelash line all the way up to the brow.

2 Now take a matte pale pink, the lightest colour, and apply it to the lid from the lash line to the fold of the eye using a stiff brush. Do not go any higher or you won't have space to blend out.

3 Take a soft purple, the mid-tone, and apply it from the outer corner of the eye and up over top of the crease toward the bridge of the nose, blending softly upward as you go.

4 Once you have created the correct shape, take the smoky grey and deepen the crease, again using a stiff brush.

5 Using the same brush with the darkest tone, here black, bring some of the colour underneath the eye, below the bottom lashes. This colour can be placed from the outer corner of the eye right to the inner corner.

6 Next, take a softer brush with a bit of mid-tone colour on it and blend the shadow along the lower lashes downward, softening any hard edges.

7 Using a light, shimmery shadow with a satin finish (in the same tones you have been using) and a soft brush, add a bit of highlight to the inner corner of the eye and right underneath the brow. This will brighten, lift and open up the eye.

8 Now, add black gel liner to the top lid using an angle brush and line the lower waterline with a black kohl pencil to deepen the look.

9 To finish, apply a generous amount of black mascara to the top and lower lashes.

(above)
Photographer: Lou Denim **Makeup:** Jana Ririnui and Jade Hunka using MAC **Models:** Natalia Gal @ Models 1 (left) and Viktorija Skyte @ Storm (right)

(left) **Photographer:** Catherine Harbour **Makeup:** Jana Ririnui using Bobbi Brown **Model:** Jodie @ Nevs

Greasepaint Smoky Eyes

The deep, glossy look, opposite, was created using black greasepaint with black powder shadow applied over the top to set the cream. Suitable for fashion work, such as editorials and fashion shows, this is not ideal if you need to get long use out of the look, as it will crease over time.

- Using pigment rather than powder shadow on top of the greasepaint or wet gel liner will give a different effect.

- Add clear lip gloss over the top to finish a smoky eye, but only apply at the last minute or the makeup will crease quickly. This effect looks great on men for editorial shots and the more you blend out the gloss, the better.

Gel Smoky Eyes

Longwear gel eyeliner, used in the images above, is a good, quick way to do a smoky eye but it will require a bit more skill. With this method you need to work fast as the gel dries quickly, and once it has, can be tricky to blend or correct. Make sure you apply it to one eye at a time and only to one section at a time as you work.

First apply the gel eyeliner to the upper lid from the lashline to the fold of the socket. Wait until the glossy texture has dried, usually in about 2–3 minutes. Don't blend any shadow over the top until the gel is dry or you will have an uneven result.

Use the tip of a crease brush to blend and don't apply too much pressure. You can add black kohl on the inner rim of the eye and soften with black shadow under the lower lash line.

Electric Nude

Contour the upper eyelids with warm bronze tones and use an electric blue khol pencil to extend the bottom eyeliner. Striking and simply simple!

Romantic Goth

Using red and vibrant pink under the eye and on the eyelid can make the eyes look tired and severe, but adding a matte black eyeshadow into the crease of the eye – combined with velvety red lips – creates a sophisticated, romantic goth look.

Photographer: Lou
Denim **Makeup:** Jose
Bass **Model:** Joanna
Stubbs @ Models 1

Radical Vamp

Brighten black lips by pairing them with pinks and purples. Blend the colours upward and outward, starting with lavender in the inner corners brushed up into the brows and bleeding into red and pink. If you are daring, add lime green as a brow highlight. No mascara is necessary with the black lips.

Pussycat Doll

First apply green shadow on top and blue below. For feline definition, extend black cream eyeliner on the outer edge of the eye and blend it into the crease with plenty of mascara and black khol on the upper and lower rims.

The pencil is one of the most popular forms of eyeliner. There is a huge choice, as they are available in every colour of the rainbow and in metallic, matte and satin finishes. Pencils are generally the easiest type of eyeliner to apply and, when sharpened, produce a nice clean line that is easy to correct. A common trend with makeup companies today is to produce kohl pencils with softer leads that are gentle on the eye and can be smudged to create a more smoky effect.

Using a black pencil in the inner rim of the eye will make it appear smaller and more sultry. A white pencil in the inner rim opens up the eyes, reduces the appearance of redness and makes eyes appear more bright and alert. Waterproof eyeliners should not be used on the inner rim of the eyes.

Liquid Eyeliners

A wet, almost ink-like liner, these can be applied using a brush or often come packaged with a wand or soft, pen-like nib. They range in size, from very thin brushes that create fine lines to thicker, wider brushes that create a more striking, eyeliner effect. Liquid eyeliners can take a minute or so to dry and any movement while wet can cause them to smudge. They can be tricky to use and mistakes are often messy to correct but, once dry, your liner is set all day (or night). Liquid eyeliners tend to have a shiny finish once dry.

Gel Eyeliners

As the name implies, gels have a soft, gel-like texture and need a brush to apply. They dry reasonably quickly, but can also be moved or smudged. When almost dry they give an eyeshadow effect, but with more staying power. They are less messy than their liquid counterparts and by using a fine brush you can create really precise lines.

Cake Eyeliners

A flat, pressed powder, cake eyeliner is used with a wet brush to create a bold line. This can intensify the lashes or "smoky eye" look. If the pigment is not strong enough, the eyeliner will not be as striking. A thin, flat brush or angled brush is essential. Makeup artists sometimes expand on this idea to create an eyeliner using a highly pigmented powder or eyeshadow and a wet brush dipped in a mixing medium or water. This is mainly used to create more unusual colours.

MASCARAS

Ask any makeup artist and they will tell you that the quality of a mascara is largely based on its wand. The trend in recent years is to go back to the old-school 1950s-era wands. Mascaras come in a variety of formulas and colours – from the usual blacks, browns and clear to blue, green, purple, gold, silver and white.

There are mascaras that claim to lengthen, thicken, curl, or a combination of all the above. Some come with a primer, to coat the lashes before the colour is applied. This is normally to help thicken the lashes. Then there are the new formulations of mascara, which form tubes over individual lashes, which can be removed by wiping off with water. Mascaras are also available in a waterproof formula, although these can be harder to remove at the end of the day. Although there are many great mascaras on the market, a good makeup artist can make most mascaras work for them using the following tips.

* Always curl the lashes.
* Make sure the mascara has not dried out – mascaras have a 2–3 month shelf life after opening.
* Apply a few coats of mascara, concentrating it at the base of the lashes and making sure that every lash has been coated.
* Use an eyelash comb to make sure there are no clumps.

In addition to mascara for eyelashes, there are also mascaras for brows. They are either used to slightly lighten or darken the eyebrows and have a more natural effect than filling in eyebrows with powders or pencils. They usually come in colours that match hair colour, such as blonde, brown, dark brown and red. Clear mascara can be used to comb and hold eyebrows in place.

Photographer:
Catherine Harbour
Makeup: Sandra
Cooke using MAC
Hair: Nathalie Malbert
Model: Robyn @ FM

Photographer:
Camille Sanson **Hair
and Makeup:** Cristina
Iravedra using Bobbi
Brown **Model:** Claudia
@ Nevs

65

EYELASHES

Photographer:
Keith Clouston
Makeup: Sandra
Cooke using Swarovski
Model: Michelle Easter
@ Red Models

Photographer:
Catherine Harbour
Makeup: Shu Uemura
Hair: Andrea Cassolari
Model: Adrienn Densi
@ Nevs

Eyelashes have a significant presence in the industry and can be customized by adding feathers, diamanté beads and glitter. There are so many choices available to help you be creative. Doubling up on lashes can give a different dimension to the eyes, as they can make them look bigger or more sultry. Individual eyelashes can be added to the corners, or you can cut them from a strip to give a cat-like look.

In a more creative way, placing fake lashes under the eyes and in the sockets can give an interesting look on a photo shoot. Also, by mixing pigments or glitter with clear mascara, a lash-mixing medium or clear glue, you can alter the texture, colour and shape of the lashes. In fact, there is no limit to what you can achieve.

Photographer: Camille Sanson **Makeup:** Shu Uemura **Hair:** Andrea Cassolari **Stylist:** Nasrin Jean-Baptiste **Model:** Erin @ Nevs

Photographer: Han Lee de Boer **Makeup:** Nat van Zee using Shu Uemura **Model:** Sophie @ Select

Green Feather Lashes

1 Using a stiff fan-shaped brush, create random brushmarks with white and black cream eyeshadow on the cheek and forehead.

2 Using black cream eyeshadow, paint an outline of a wing shape from the inner corner of the eye outward along the top lash line, and back toward the inner corner under the lower lash line. Leave a similar, minimized area on the moving part of the eyelid clear, creating a wing shape inside the larger wing shape.

3 Fill in the outer wing shape with the black cream eyeshadow and the inner wing shape with a dark green liquid makeup, such as Aquarelle Makeup Forever. Allow to dry.

4 Cover the outline of the black wing shape with various-sized feather lashes, such as those from Shu Uemura, sticking them on with eyelash adhesive. Alternatively, customize ordinary false lashes with real feathers or paper cut-outs – the more you experiment, the more individual the look can be.

Sticker Effect

The effect seen above is created by first using liquid foundation and beige contouring for the base, as in step one of the Rainbow Lash, opposite. Eyebrows are groomed with a brow gel, and neon stickers are cut into random sizes of triangles and arranged on the forehead.

Photographer: Han
Lee de Boer **Makeup:**
Nat van Zee using Shu
Uemura **Model:** Kim
Glaser @ Next

Rainbow Lashes

1 For the base, brush a liquid foundation all over the face, and use a beige contouring powder such as MAC Sculpt & Shape in bone beige. Contour along the cheekbones to the ear and around the sides of the neck. Highlight the nose and contour along the sides of the nose.

2 Using a bright orange cream eyeshadow, line the top lashes of the eyes to create an angle at the inner corner, sloping upward at the outside corner. Build up the colour to a thick line.

3 Now glue rainbow lashes close to the top lash line. Spread the glue evenly along the edge of the lashes and let them part-dry. Then press in place gently, especially at the inner and outer edges, to get the closest fit possible.

4 For the lips, cover the outer corners with concealer and use a turquoise gel liner to create rosebud lips with an angled cupid's bow. For a perfect lip line, use a small, straight, angled brush and draw using a dot-to-dot motion with your brush as the guide line. Concealer can hide any mistakes. You can also stretch the skin taut, which will allow you to draw a clean line without gaps or bumps.

- If you find it difficult to get false lashes close enough to the natural lash line, fill in the gap with an eyeliner or eyeshadow and blend to hide the join.

Photographer:
Catherine Harbour
Makeup: Jana Ririnui
using Shu Uemura
Model: Kasia @ First

Photographer:
Fabrice Lachant
Makeup: Shu Uemura
Hair and skullcaps:
Marc Eastlake
Model: Monika R
@ Next

THE *LIPS*

Photographer: Conrad Atton **Makeup:** MAC

LIPSTICK TEXTURES

When choosing a lipstick (and this rule applies to most makeup), remember that what a product looks like in it is packaging is not always how it appears on the skin. This is because of the product's texture. In fashion and styling, makeup textures are as important as the colour.

To get a true sense of a lipstick colour, neutralize the lip with a concealer or lip primer. Lip Plump by Benefit or Lip Erase by MAC are both makeup artist favourites. Beware of cold sores on the lip area – using a lip brush on an infected area and then touching the lipstick with the brush can infect your product. Disposable lip wands, like those available from the Pro Makeup Shop (www. thepromakeupshop.com), are essential.

Preparing the Lips

It is important to prep the lip before applying lipstick, just as you would prep the skin before applying makeup to the face. Dry, flaky or cracked lips can ruin the finished result. You can gently exfoliate the lips using a treatment like Hollywood Lips Sweet Sugar Scrub, then apply their Soothing Day Relief to heal and protect. Glam Balm by Rodial also buffs and plumps the lips.

Sheer Lipstick

This is a transparent veil of colour that allows the natural lip tone to come through by creating a see-through effect. No matter how much product you apply to the lips when using a sheer tone, the effect will be very limited in colour. Although sheers are not the best choice if you want to build a strong colour, they are a great choice when you want to create a natural makeup look with only a hint of colour. A popular example of a natural-looking sheer lipstick is Laura Mercier's Bare Lips, which is a pinky brown shade. A lip stain like Benefit's Benetint goes one step further as it colours the lips without having a coating.

Matte Lipstick

Matte lipstick gives a flat colour without shine or gloss, and most are dense in colour. As they are less likely to bleed into the fine lines around the lip area, they are great for more mature models. Matte lipsticks are also good for creating period looks from the 1920s to the 1950s (see pages 96–103). For example, if you wanted to create a 1940s lip, a matte red like MAC's Russian Red would be perfect.

Cream or Satin Lipstick

Lipsticks that are cream- or satin-textured have a smooth appearance with a touch of sheen that appears almost moist. Cream and satin lipsticks can have different levels of transparency depending on the brand. They are a good choice for bridal makeup, as they have a hint of moisture and more colour than a sheer lipstick. Ladies Choice lipstick by Benefit is a pert-toned pink with a satin/cream texture that is perfect for brides.

Frost Lipstick

Frosted lipsticks have a touch of glimmer or metallic added to the colour to create a shimmer effect. Frosted lipsticks can also have different variations of transparency, like cream or satin tones. They are a good choice if you want to create lips with reflective shine without using lip gloss. Revlon's Silver City Pink is a terrific choice for a 1960s retro lip.

Opaque Lipstick

Opaque lipstick has a dense colour and is used to fully cover the natural lips. An opaque lipstick can have a matte, satin, frost or gloss finish, but the colour is non-transparent. An example of an opaque lipstick with frost and a touch of sheen is CB96 by MAC – it is an orange lipstick that also has gold frost running through it, but the colour is full-on. If you want to create lips with high colour it is best to choose one that is opaque.

Applying Lip Colour

Always block out the lips with concealer or foundation before applying lipstick, to create a clean base – the colours will also appear truer. Lipliners are useful to recreate a different lip shape or to enhance the existing fullness. By colouring the whole lip you can build a solid base, and then add a lipstick to intensify the colour. Other products, such as eyeshadow, cream eyeshadow, loose powder pigment, greasepaints and blush creams, can also be used on the lips.

• Pressing the product onto the lips firmly with your finger is sometimes easier to control. Adding gloss on top will give volume to the lips, and by mixing colours you can create interesting tones. Using shimmer or glitter on top gives high shine for photo shoots.

Photographer:
Catherine Harbour
Makeup: Lancôme
Model: Ilize Bajane @
Next

VAMPire Lips

The inspiration for vamp lips comes from Bram Stoker's 1897 novel *Dracula* and films such as *Lost Boys* (1987), *Nosferatu* (1922) and the cult classic *Vamp* (1986) starring Grace Jones.

1 Prepare the lips with a lip primer, then block out the colour with a foundation appropriate for your skin tone.

2 Choose a shade of red lipstick that suits your complexion: a blue-based red if your skin is pink-beige or a warm red if your skin is golden-beige. Apply with a lip brush for precision.

3 Line your lips top and bottom with a black pencil or black cream eyeliner. Using a lip brush, blend the black lipliner into the lipstick.

4 For more dimension dab a bit of clear lipgloss on the top and bottom middle of your lips.

Blood-Red Stained Lips

Stained lips look sophisticated, effortless and can even appear to be something of an accident. Creating the look, however, is far from accidental!

1 Prepare by applying a lip serum for smoothness, and then a lip balm for moisturizing.

2 Choose a shade of light or dark red (preferably satin or matte, nothing glossy) that suits your skin tone. If your skin is pink-beige go for a cool red; if it is golden-beige reach for a warm red.

3 Squeeze a bit of lip balm on a mixing palette and add a few scrapes of the lipstick. Mix them together with a spatula.

4 With a lip brush or, even better, with your ring finger, take a bit of the moisturizer/lipstick mix and dab it on the pout of the lips, pushing the colour in. Concentrate the colour application on the middle of the lips and work your way to the outer corners. Keep dabbing your colour mix until your lips are fully stained.

5 With a tissue or a cotton swab go around the mouth to clean any excess product and give more of that unfinished "work in progress" look. To finish, top up with a bit more lip balm.

Bubble-Gum Lips

A bubble-gum pout is a luscious, juicy pout, inspired by the mouthwatering sensation of chewing gum and popping bubbles. Think pink and lots of gloss.

1 Prepare the lips by drenching them in lip balm. Apply a bit of concealer that is a shade or two lighter than your foundation around your mouth so the gloss, when applied, will not "bleed".

2 For staying power, line your lips with a pink lipliner. Remember to choose a cool undertone pink if your skin is pink-beige or a warm undertone pink if your skin is golden-beige.

3 Brush on a pink lipstick that is suitable for your skin, depending on whether you have a cool undertone or a warm undertone.

4 Pick three lip glosses – light, medium and dark – so you can add dimension, again choosing between cool and warm tones.

5 Apply the lightest lip gloss in the middle of your pout, the medium lip gloss from the middle to the outer corners and the dark focusing on the corners of the lips.

6 Smack your lips, and pop you go!

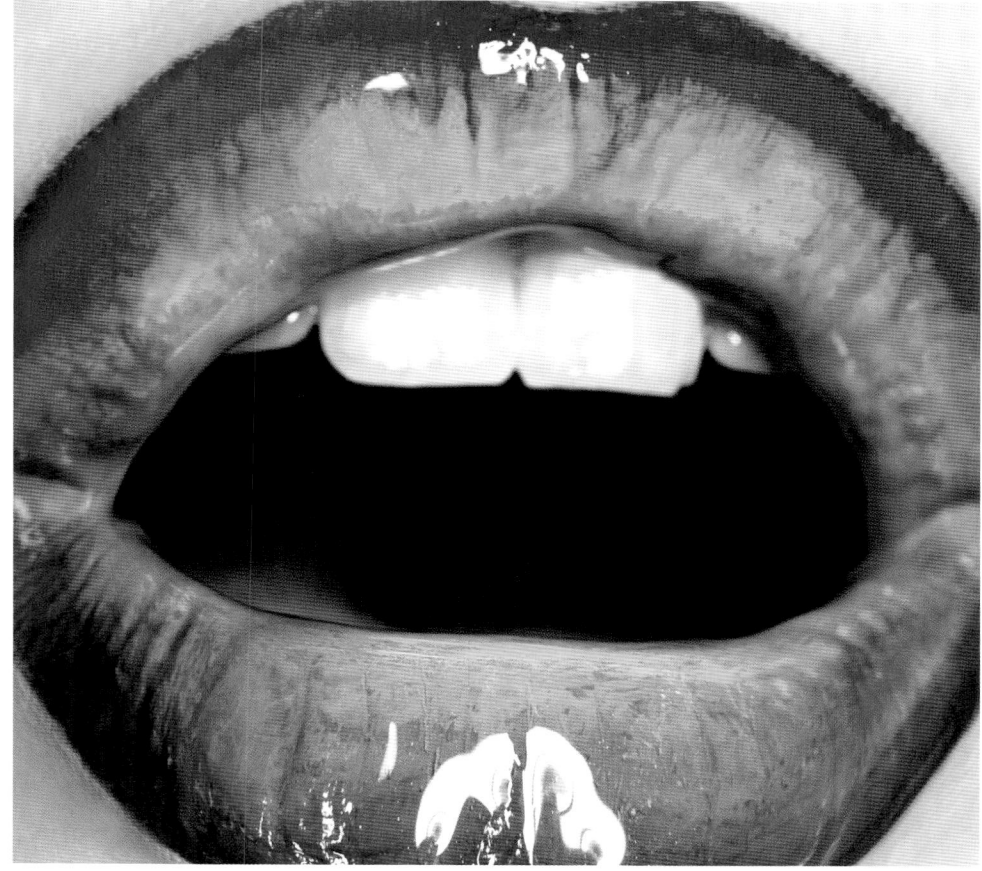

(red and purple lips, left) **Photographer:** Catherine Harbour **Makeup:** Dior. (orange and yellow lips, left) **Photographer:** Keith Clouston **Make-up:** Sandra Cooke using MAC **Model:** Lina O'Connor @ Nevs

(bubble-gum lips, top right) **Photographer:** Catherine Harbour **Makeup:** Rachel Wood using MAC. (rainbow lips, right) **Photographer:** Catherine Harbour **Makeup:** Maria Papadopoulou using MAC **Model:** Heather West @ Nevs

Photographer: Conrad
Atton **Makeup:** MAC
Model: Hannah @ Next

Strong Eyes and Lips

1 Apply a medium-coverage foundation to the face using a duo-fibre foundation brush and buff it into the skin for more natural, even coverage.

2 Next, apply concealer only to areas needed, using a small, synthetic brush and blending in with the fingers using a stipple/patting motion. The heat from your fingers will help push the product into the skin, while the stipple motion allows you to build coverage without moving the product outward from the area needed.

3 Powder more oily areas on the face, such as the forehead, nose and chin. Also powder underneath the eye to set the concealer and prevent it from creasing, and the lid, too, to set the foundation and create a smooth base for eyeshadow.

4 For the eyes, start off with a flat application brush and pick up some dark green shadow. Apply from the lash line up to the crease.

5 Take a soft blending brush and some mid-tone green shadow and buff the hard edge out and upward from the crease. Using two greens will give a soft gradation from dark to light, therefore creating that "smoky" look.

6 Next, take a kohl pencil in the deepest green you can find and apply generously to the top lash line. Take a stiff, round-tip brush and blend the line out to deepen the look and thicken the line.

7 Take an angle brush and a bright yellow shadow and apply colour densely and evenly directly underneath the lower lashes. You can blend the shadow out by pulling it downward to create a softer look if preferred.

8 Take a light gold/champagne shadow or other soft highlight colour and apply it to the inner corner of the eye and underneath the eyebrow. This will help lift and open up the eye.

9 Finally, apply a volumizing mascara to the top and bottom lashes and build up the intensity by increasing the amount of coats you put on.

10 Now move on to the brows. Using a clean angle brush and a shadow that matches the brows, softly fill in the hairs in a short, sweeping motion. If you apply too much product, comb through the brow using a disposable mascara wand to lift the excess colour out.

11 For the cheeks, take an angled contour brush and a cool-toned contour colour. Apply this right underneath the cheekbone, pulling from the hairline toward the nose.

12 Next, take a soft blush brush and apply a deep pinkish/red colour on the apples of the cheek, blending out on the rest of the cheekbone.

13 Highlight the top of the cheek bone by using a light-coloured shimmer and a smaller brush.

14 On the lips, use a dark purple/burgundy lip pencil to follow the natural lip line. Fill in the rest of the lips with the liner to prolong the wear of the lipstick and intensify the colour.

15 Using a synthetic lip brush and a lipstick that closely matches the lipliner colour, apply colour right up to the lip line but do not over-apply.

16 To finish the look, mix gold glitter with clear gloss and apply all over the lips. Alternatively, for a more subtle look, use a shiny, gold gloss.

Photographer: Lou Denim **Makeup:** Sharka **Model:** Kat @ mandpmodels

Photographer:
Fabrice Lachant
Makeup: MAC

Photographer:
Catherine Harbour
Makeup: Sandra Cooke
using Illamasqua
Model: Robyn @ FM

MEN

The cosmetic industry is no longer the sole domain of women. There has been a major shift in attitudes toward men's grooming in the last decade, which has been reflected in the growing number of cosmetic ranges designed and targeted toward men. It initially started with products for shaving, and has since expanded into skincare and makeup.

"Grooming" is the term commonly used to describe hair and makeup for men at a photo shoot. The art of grooming is to make the male model look fresh, with clear skin and groomed hair, without appearing to be made up. In the days of film and soft lighting, grooming was less important, but in today's high-definition environment it is important for male models look their "natural" best.

In addition, the prominence of male models in the fashion industry is also increasing. There are more dedicated male fashion magazines (such as *GQ* and *Maxim*), men's fashion editorial stories (featuring male models and menswear designers) and also men's fashion weeks that are held in Milan and Paris each season.

Grooming generally involves cleansing and moisturizing the skin, correcting skin tone, concealing blemishes, reducing shine, combing the brows and shaving or plucking facial hair if required. If you are at a photo shoot with no stylist, then grooming also involves cutting and trimming as well as styling the hair. Grooming is not just limited to above the neck, but also includes ensuring the hands and nails are in good shape, that the body is well moisturized and the skin tone is correct.

Increasingly though, for catwalk and editorial pieces, grooming can also include applying eye makeup, such as eyeshadows and eyeliners, contouring and highlighting the face, and even applying crazy paints and colours. At men's fashion weeks, there are dedicated teams of makeup artists, hair stylists and nail technicians to groom and apply makeup for the male models to mirror the visions of the designers. So, it seems that the future of grooming will no longer be grooming only, but rather the art of makeup.

Photographer:
Catherine Harbour
Grooming: Kiehl's
Model: Nicholas
Robinson

Photographer: Fabrice
Lachant **Grooming:**
Lancôme Homme
Stylist: Karl Willett
Model: Max @
Dynamite Hosts.
Crown by Fred
Butler, underwear
by Calvin Klein

Photographer: Fabrice Lanchant **Makeup:** MAC **Styling:** Karl Willett, head piece by Fred Butler

Photographer: Fabrice Lanchant **Makeup:** MAC (left), Lina Dahlbeck using MAC (right)
Styling: Karl Willett
Models: Alex Beer and Anthony Lowther. Skull by Butler & Wilson, head piece stylist's own

ICONIC ERAS

Prior to 1920, wearing makeup could be a dangerous proposition. It contained chemicals such as lead, sulphur and mercury. Beyond the health risks, it was not proper for nice girls to wear makeup. Those who did wore pale, muted colours and their supplies were hidden away from disapproving fathers and husbands. Women who feared the consequences of wearing makeup often vigorously pinched their cheeks and lips to give them colour. The prevailing attitude was that a lady had no reason to be in the sun and therefore should be pale. Some historians even suspect that the look became popular because of the prevalence of tuberculosis, which made its sufferers look very pallid.

The 1920s flapper, the 1940s brow, the 1950s Marilyn Monroe vamp and the 1960s Twiggy: all these, and other iconic makeup styles, are used over and over in film and fashion. Every year there seems to be another revival. Fashion designers take inspiration from the past, foreign cultures and street fashion and bring old trends to life with a new twist. The same goes for the makeup companies. New colours, products and techniques are brought back and re-introduced. We see brightly coloured lips all over the catwalks one season and Gothic black eyes the next.

The 1920s and 1930s were all about the dark and defined, with smoky eyes and bee-stung lips. In recent years, we have seen this look on numerous catwalk shows, like those created by Pat McGrath at Galliano for Dior. In the 1920s, eyebrows were kept natural and unplucked, while eyes were dark and smoky. The "smoky eye" is still one of the most popular makeup looks today. The incredibly beautiful actress Louise Brooks, called LuLu, was one of the fashion and beauty icons of the decade with her trademark Dutch bob and dramatic makeup, while Greta Garbo set the trend in the 1930s with plucked and drawn-in brows.

The 1940s, the swing era but more remembered for World War II, saw women join the workforce when their men were off at war. Looks were simplified; the red lipstick was a necessity. Lips were true red with the top lip overdrawn and rounded, heavy eyeliner was flicked at the outer corners and eyebrows were kept natural. A good example of the look of the time was the movie star Veronica Lake.

The glamorous Hollywood look of the 1950s is still current and trendy, with the pin-up, Marilyn Monroe look being adapted by the likes of Christina Aguilera, Scarlett Johansson and Dita von Teese, to name just a few. During this decade there was a transformation from the ponytailed, innocent teenage look to heavily-styled, beehive hair and defined eyes. Starlets wore red lipstick, eyeliner, blusher and lots of mascara; eyebrows were kept natural and grown out. Teenagers started wearing pink- and peach-coloured lipstick, as parents didn't approve of the heavily made-up looks.

The 1960s saw a complete change in makeup trends. Twiggy was discovered in 1966 and changed the world of fashion and beauty with her short, androgynous hair-do and Mod fashions created by such designers as Mary Quant. Twiggy wore light pastel lipstick and eyeshadow, false eyelashes, thick eyeliner drawn into her socket lines, and lots and lots of mascara. Mary Quant is just one of many designers who take credit for inventing the miniskirt and hot pants, but by the middle of the decade, she had brought out a new, trendy makeup line that offered multi-use products and palettes. The end of the 1960s was the beginning of the flower-power era and women's liberation, and makeup was minimal and natural. Hair was left long and unstyled and, not surprisingly, cosmetics companies struggled.

The 1970s saw a mix of glam rock, disco and punk music that soon spiced up the world of fashion and beauty. Clothes were fitted, trousers flared and unisex looks were introduced. Makeup was bold and garish and lots of colour and glitter was used. French and Italian *Vogue* showed

Photographer: Fabrice Lachant **Makeup artist:** Lina Dahlbeck using MAC **Model:** Emily S @ Nevs

The 1920s Look

The look, as re-created right, was dark, mysterious and vamp, with heavy kohl-rimmed eyes and a cupid bow-mouth. Round shapes were created in the brow and eyes, with pink blush on the apple of the cheeks and rosebud dark lips. The brow shape starts straight, then rounds.

models wearing blood-red lipstick, black eye makeup and pencil-thin eyebrows. David Bowie and Roxy Music were the faces of glam rock. The makeup was correspondingly bold, colourful and above all, theatrical. Punk music and the anarchy movement dominated the end of the decade's social scene: safety pins became nose and ear jewellery and hair was harshly dyed, styled and sculpted into mohawks or spikes. Vivienne Westwood and her partner Malcolm McLaren created the iconic Sex Pistols look.

The 1980s saw an explosion of colour. Big and eccentric hairstyles and makeup looks were popularized by television, film and music stars. Shows such as *Dynasty*, *Miami Vice* and *Fame*, coupled with major artists like Michael Jackson and Madonna, were just a few of the influences. Duran Duran and Adam and the Ants were the faces of the New Romantic scene and experimented with androgynous looks.

Hair was worn big and frizzy by both sexes throughout the 1980s and lots of hair products were used, while makeup was colourful and bold. The fluoro colours of fashion were replicated in the makeup of the time, with women wearing brightly coloured mascaras and bold eyeshadows, blended up to the brow. Blushers were also applied in excess and blended to the hairline.

Throughout this chapter you will see a range of different retro makeup looks from the 1920s through to the current trends. You can imitate them if you like, or simply be inspired by the amazing way the past can inject energy into a look.

Photographer: Lou
Denim **Makeup artist:**
Lina Dahlbeck
using MAC

The 1930s Look

The look for this decade, see left, was elegant and sophisticated, with softer eyeshadow, a blended socket with heavy lashes and mascara. Blush is sculpted on the cheek. Fuller lips slope from the bow. The brow shape is rounded from start to finish.

1 Apply a tinted moisturizer or light-coverage foundation all over face. Touch up any imperfections or dark circles with a concealer. Apply concealer or eyeshadow primer on the eyelids as a base.

2 Using a soft black eyeliner, draw a thick line along the upper lash line. You can also use a black cream eyeshadow for this. Using your fingers or a synthetic brush, smudge the line up to the socket then apply a black eyeshadow on top to set. Using a soft brush, blend the colour toward the socket.

3 Apply a light grey or dusky brown in the sockets to help the black blend into the browbone area.

4 Leave the browbone as it is and set with a translucent powder, or apply a highlighter up to the eyebrow.

5 Curl the lashes. Apply plenty of black mascara.

6 Brush the eyebrows into place and define them using an eyebrow shadow or pencil. Set the shape using a clear brow fix.

7 The look is all about the dramatic eyes and lips, so keep the cheeks as natural as possible with just a touch of peach blusher on the outer corners of the apples of the cheeks. Blend it up toward the temples and follow with a pressed powder contouring shadow in natural in the hollows of the cheeks to create depth and shape.

8 Line the lips with a dark burgundy lipliner, then fill in lips with the same pencil. Blend with a clean lip brush before applying a matching matte lipstick on top. Alternatively, for a more modern twist, use gloss or a lipstick with a slight sheen instead of the matte lipstick.

Photographer:
Catherine Harbour
Makeup: Rachel
Wood using Chanel
Hair: Marc Eastlake
using Bumble & Bumble
Model: Sabrina @ Next

The 1940s Look

Period makeup was minimal, classic and elegant. To re-create the look, taupe socket eyeshadow was used with liner smudged through the fine lashes. Blush was sculpted on the cheek and red lips were fuller and wider at the bow. The brows were long and arched.

The 1950s Look

Glamorous, clean and fresh, the eyes were pastel – blue, pink and violet. Eyeliner was used in an Egyptian shape for a classic 1950s flick. In this decade false eyelashes were used, and lips were a lighter red, pink or orange. Blush in pinks and peaches were sculpted on the apples of the cheeks and the brow was strong and arched.

The look here was inspired by the original Barbie-doll look, mixed with *I Love Lucy's* Lucille Ball. Wide eyes were painted with a powder-blue shadow, white eyeliner was applied on the inside of the eye and black liquid liner – thicker in the middle of the eye – was applied across the lid. A touch of white loose shimmer eyeshadow on the middle of the lid created a touch of high glamour. False eyelashes were applied to the top eyelash.

A classic red pout was created with the bottom lip drawn on slightly smaller than the top lip, producing an innocent, Barbie-doll pout. The warm tone foundation was kept matte and heavy so the look has a B-movie, colourama effect.

Photographer: Keith Clouston **Makeup:** Rachel Wood using Benefit **Makeup assistant:** Lauren Amps **Hair:** Fabio Vivian **Model:** Invlid @ mandpmodels

The 1960s Dollybird Look

The pale face "dollybird" look reigned supreme. Eyes were pale with strong liner applied top and bottom and charcoal upper lids, with heavy false lashes. Heavy highlighter and sculpted blush were used, and lips were in pale brown, soft pinks and peaches.

The look here was inspired by Andy Warhol, Twiggy and a touch of over-the-top Austin Powers. A combination of black wet/dry powder liner and matte white eyeshadow were used to create a graphic eye. Four layers of black mascara were applied to the eyelashes, then a very thick pair of fluffy false eyelashes were added on top to finish the look.

Photographer:
Catherine Harbour
Makeup: Rachel
Wood using Benefit
Hair: Marc Eastlake
using Bumble & Bumble
Model: Viktoria @
Oxygen

Photographer: Lou Denim **Makeup:** Cristina Iravedra using Bobbi Brown **Model:** Eline @ manadp models

The 1960s Pop-Art Look

Based on the iconic image of the 1960s model Twiggy by Richard Avedon, this great style, seen above, can be adapted for a party look.

1 Apply a light foundation, such as MAC Face and Body, all over the face, making sure to cover any dark circles with a medium-to-heavy cream concealer. Blend foundation on the ears and into the neck and chest area.

2 Next, apply loose translucent powder to set the foundation and avoid shine later on.

3 Apply a light/pale grey matte eyeshadow all over eyelids, then add a dark grey/black matte eyeshadow over the eye crease to create a "banana" effect. Blend the dark colour well, but make sure that it is still well defined. Only use the dark colour on the right eye, as the left one will be painted.

4 Curl the eyelashes.

5 Using a thin, short eyeliner brush, apply black gel eyeliner on the right eye, contouring the eye and making sure the tear duct area is free of product. Flick the line upwards at the outside top corner. Along the bottom make a few spikes or flicks downward to create a false eyelash effect.

6 Apply eyeliner to the left eye, but do not add flicks along the bottom line.

7 Apply a white eye pencil all over the inner line of both eyes. This will make the eyes look dramatically bigger and will intensify the liner and the eye shape.

8 Add spiked false eyelashes on top of the natural ones, sticking them on with surgical eyelash glue. If the eyelashes are too big, cut them to size before adding the glue.

9 Apply black mascara with a zigzag motion to blend both real and false lashes.

10 Apply a peach blusher on the cheekbones with a circular motion.

11 Line the lips with a peach/orange lipliner, following the natural contour of the lips. Using a lip brush, fill in with peach lip gloss.

12 To draw the flower, use acrylic brushes (available from art shops) and blue, orange and pale green body paints. Mix the dry paint with water until you get a creamy consistency.

13 First paint the blue inner colour, creating the main flower shape (be really careful around the eye). Contour the shape with orange colour, using a thinner brush. Draw on the green leaf with a fine brush.

The 1960s Op-Art Look

Inspired by a Paco Rabanne earring created in the late 1960s, this period look, see right, is made contemporary with the addition of a graphic design. On the face peach powder blush is used with beige sculpting powder and a creamy neutral coral beige lipstick is used on the lips. Gel/cream liners in pure white and black are used with matte white and black eyeshadow to create the eye design.

Photographer: Alexandros Papanikolopoulos **Beauty Editor:** Maria Papadopoulou for Vimadonna magazine using MAC **Model:** Elodie @ Action

The 1970s Look

Colourful, natural face painting created a tanned look and highlighting and shading were used to sculpt the look, as seen left. Eyes were colourful, with pencils used for definition, and in iridescent shadow textures. Lips were very glossy and brows were thin with a slight arch.

1 Apply a tinted moisturizer or light-coverage foundation all over face, making sure to touch up any imperfections or dark circles with a concealer. Use a concealer or eyeshadow primer on the eyelids.

2 Next add a light green eyeshadow to the lids and, using a soft blending brush, blend toward the socket.

3 Apply a tan/brown eyeshadow to the sockets of the eye and blend up toward the middle of the browbone area. Highlight the eyes using a shimmery gold shadow on the browbone under the brows.

4 Give the eyes, and the whole look, more depth and drama by applying black eyeliner along the upper and lower lash lines Blend with a small brush and set with dark green eyeshadow.

5 Curl the lashes and apply black mascara.

6 Brush the eyebrows in place and define them using an eyebrow shadow or pencil, then set the shape using a clear brow fix.

7 Define the cheekbones by applying bronzer and blending up toward the temples. Add a pressed powder contouring shadow in a natural tone to the hollows of the cheeks to create depth and shape.

8 Line the lips with a barely-there beige lipliner. Blend the edges using your fingers or a lip brush and apply a clear lip gloss on top.

9 Finish with a dusting of translucent powder.

Photographer: Lou Denim
Makeup: Lina Dahlbeck
using MAC **Model:** Emma
Cantaloup @ Union

Photographer: Keith Clouston Makeup: Rachel Wood using Chanel (right), Lancôme (left) Hair: Jana Ririnui using Babyliss Pro and Redken Stylist: Karl Willett Models: Pippa and Kelly @ Oxygen. Catsuit by Dior (vintage), bangle by Freedom

The 1980s Look

Matte, strong, serious and glamorous, 1980s makeup uses definition on eyes with a deep blended socket and liner top and bottom. Power red or shocking pink lipstick with natural heavy brows complete the look.

Photographer: Keith Clouston **Makeup and Hair:** MAC and L'Oréal Professional **Stylist:** Karl Willett **Model:** Pippa @ Oxygen. Dress by Lucy Wrightwick, vintage gloves by Gucci

MODERN MILLENNIUM

Today in the twenty-first century, there are no particular trends to follow, but mainly creative takes on the past. From season to season, it can be a focus on one area of the face, such as the eyes, eyebrows or lips. Kept to a minimal look, it is easy to see the importance of great skin, whether you are creating a matte or dewy finish. This is the key element in fashion shoots as a base to which colours and accessories are added.

Looks are widely varied and can be cold or warm, period or futuristic, strong and full-on or soft and natural, finished or unfinished. It is the way in which makeup is applied and played with by the artist that defines what the look will be. It is importance is to achieve the right balance between being artistic and developing a commercial look that is beautiful, wearable and successful.

As new products come on the market, makeup artists must keep up to date with trends and develop an understanding of the new formulas. This ensures that the look you create is on trend. You can also be inspired to create interesting and new looks by mixing the colours you already have and using them differently on the face. For example, a bright-coloured cheek cream base, such as Shock Pink blush by MAC, could also be used on the lips and eyes to create a modern feel.

Photographer:
Catherine Harbour
Makeup: MAC
Model: Ilize Bajane
@ Next

The Millennium Look

Today's looks are eclectic, varied and a combination of all the eras mixed into one. Metallic cream eyeshadows or greasepaints, glitters and strong colours give impact and foundation approaches, whether dewy or matte, can vary. Nowadays looks can be dictated by a designer, stylist, artist, magazine editor, art director or even a PR agent, as well as celebrities and the music industry.

1 Using a foundation brush, apply full-coverage foundation all over for a clean, flawless base.

2 With an eyeliner brush apply black gel liner on the entire eyelid up to the socket. Soften the edges and set with black shadow. Blend well.

3 Now line the inner eye rims with a white pencil. Add white gel liner, such as MAC Pure White, on the lower lash line and under the lash line from the inner corner to the outer corner to achieve a thick white line.

4 Draw black gel liner under the white liner and soften the edges.

5 Attach false lashes on upper lash only, sticking them on with eyelash glue.

6 Neaten eyebrows and fill in with an eyebrow pencil or shadow if necessary.

7 Softly contour under the cheekbones with a natural blush colour.

8 Apply a matte red lipstick to the lips and add a slick of clear gloss over the top.

Photographer: Lou Denim **Makeup:** Elsbeth Tan using MAC **Styling:** Claudia Behnke **Model:** Lucy Jacques @ Nevs

DESIGNER FACES IN THE CITY

Photographer: Lou
Denim **Makeup:** AOFM
New York graduates
using Bobbi Brown
Styling: Jules Wood

STREETS OF MANHATTAN

The Designer Faces in the City narrative was about devising creative looks around the busy lifestyle of city girls with New York as a background through seasonal changes. For the Streets of Manhattan story, here, the look is all about creating sexy, wearable makeup for a day into evening transition, which will last until the morning. Different scenes were created episodically from evening until early morning: the girls meet up, have a run-in with the police later on and finally end up grabbing breakfast on the street in the early morning.

Photographer: Lou Denim **Makeup:** AOFM New York graduates using Bobbi Brown **Styling:** Jules Wood

Photographer: Lou
Denim **Makeup:**
Breianne Zellinsky using
Bobbi Brown **Styling:**
Jules Wood

This scene, Taxi, was set up to be a 'bright lights, big city' look in the deep heat of the summer. Gold shimmer and glitter with black kohl were used to create a smoky eye with a twist. The focus of the glitter here is just on just the eyelid, so that there are flashes of sparkle as you move your eyes. It is important to use an eyeshadow formula with very fine glitter particles and apply an eyeshadow in the same colour beforehand. Shadow under the glitter will help to soften the effect, and prevent it from looking uneven. You could also apply a soft, shimmery highlighter to the browbone or upper cheekbone, to link with the glitter.

TAXI

Photographer: Lou
Denim **Makeup:**
Breianne Zellinsky
using Bobbi Brown
Styling: Jules Wood

Springtime in the city was the theme for this scene and the versions on the following pages, with a pretty makeup look to complement the colour of the floral setpieces. The narrative was a walk through the cobblestone streets of New York City, the spring colours emerging and painting the landscape in vibrant hues. Bursting blooms of purple lilacs, lavender and wisteria elegantly sprawl through the streets. Tulips and daffodils explode with bright colours that brighten a busy day ahead. Through Central Park the trees transform, the spring air is fresh and beauty, where every day something new appears, is delighted in.

Springtime is the perfect season to explore colour when applying makeup. Using Bobbi Brown cosmetics we have used a series of vibrant colours within their range to create these exquisite makeup looks. It is all about the eyes; exploding through the pages as a flower would fresh in bloom. The colours are swept across the lid and extend out, mimicking the petals of a blooming flower. The skin is glowing and radiant as if the sun had been shining upon the model all day. Using Bobbi Brown cream blushers we were able to keep the dewy effect on the cheeks and lips using those perfect petal pinks.

NEW YORK IN BLOOM

Photographer: Lou Denim **Makeup:** Breianne Zellinsky using Bobbi Brown **Styling:** Jules Wood

Photographer: Lou
Denim **Makeup:**
Breianne Zellinsky
using Bobbi Brown
Styling: Jules Wood

I LOVE NYC

Contouring and highlighting was used to create beautiful, soft, wearable makeup that is great for every day. With the cold winter as a theme, matt browns and shimmer golds were used on the eye with a shimmer highlighter on the cheekbones and the centre of the lips. A touch of rose lipgloss was applied with the fingers and bronze contoured on the cheekbones. Eyeliner was avoided – the look is all about the structure of the face.

Photographer: Lou Denim **Makeup:** Jill Freeman using Bobbi Brown **Styling:** Jules Wood

RED CARPET

When asked to do "red carpet" and celebrity makeup, there are many factors to take into consideration as the makeup is often dictated by the celebrity's personal preference or the image that they wish to project. It is essential to research your celebrity, or client, to see if they maintain a constant image or prefer to change their look regularly. Look at the makeup they use and see where you can improve their overall look and style.

Find out what event they will be attending, and, if possible, what they are wearing. Speak with the hairdresser and stylist to see what their ideas are and meet the client before the event to discuss what they want.

Be armed with tearsheets and printouts of your client and have a range of examples of what you will do so you can show the visuals and ensure you are using the same terminology. You may be called to the client's house, a hotel or a salon to do the makeup. Either way, it is a good idea to do a mini-facial before starting the makeup process. It also helps to wait about 10 minutes for the creams to be fully absorbed. This is an ideal time for the hairdresser to start working on the hair.

You should be able to complete your makeup within 30 minutes to an hour. The makeup is nearly always glamorous and you will often be asked to recreate the classic Hollywood look, so know your periods and what looks are popular with other celebrities. Be aware of current trends and be prepared to work at the same time as the hairdresser – celebrities are often late or working to a tight schedule. Check your work with your own camera. The makeup should be done to last the entire event, as you will not be on hand to do touch-ups later on.

Carry empty pots with you so you can give samples of the makeup you have used to the client, in case they want to touch-up themselves later. When you are working with a celebrity over a period of time, or if you are asked, be ready to advise them on which products they should have in their personal makeup bag. Companies will often give you free products for celebrities they want to be associated with.

Treat this type of makeup the same as you would photography work, as the paparazzi will be taking pictures at the event.

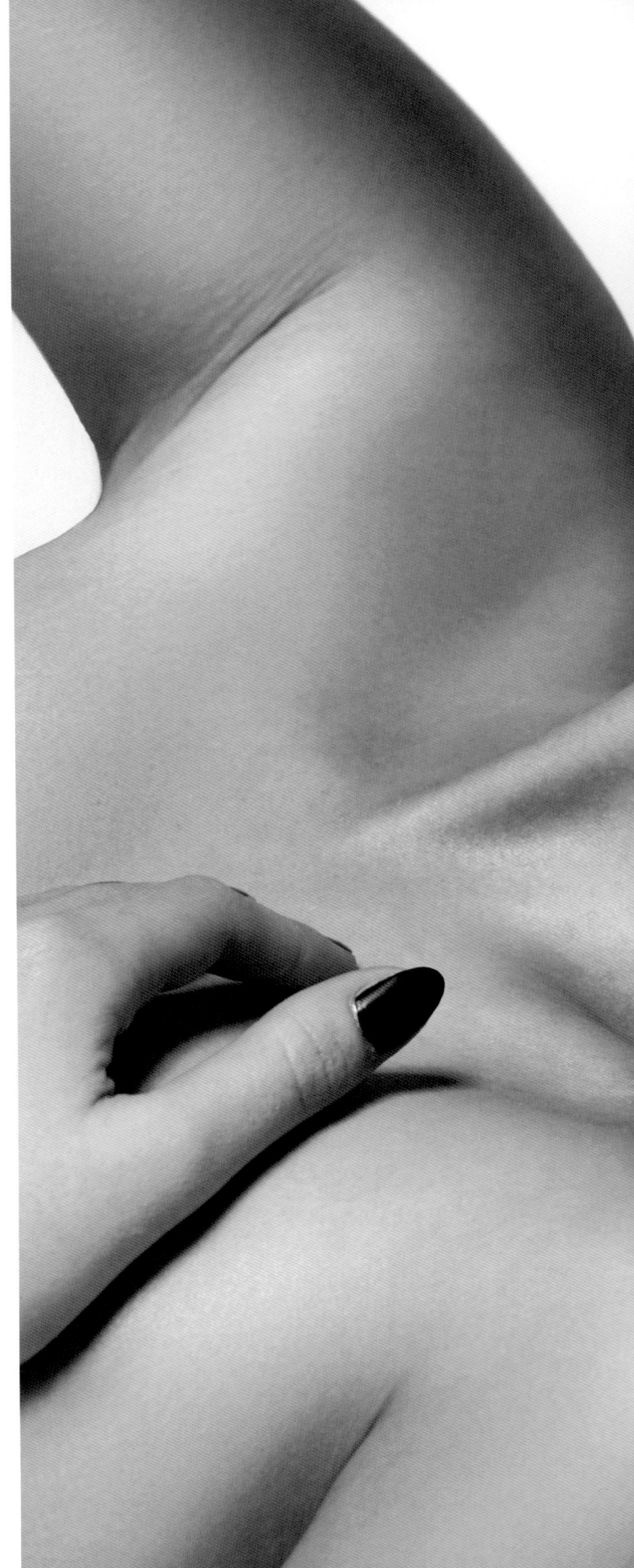

Photographer:
Keith Clouston
Makeup: Sandra
Cooke using Giorgio
Armani **Hair:** Natasha
Mygdal using Kérastase
Model: Hollie @ IMG

WORKING WITH AN ACTRESS

Tamsin Egerton

This shot is from an editorial shoot for *Urban Life* magazine, featuring actress and model Tamsin Egerton, best known for her role as Chelsea Parker in the *St Trinian's* films (2007 and 2009). The magazine wanted to show her as a sophisticated woman and they had very clear, strong views on how to get this across. The team met before the shoot to discuss the ideas and how they would work together; each member had creative input to finalize a storyboard. The looks for the day, and the order to shoot them in, were talked through.

Photographer: Dez Mighty **Stylist:** Zed-Eye **Makeup:** Jo Sugar using Dermalogica and MAC **Hair:** Zach Jardin @ Neville

EDITORIAL

Photographer: Mo
Saito **Makeup:** Sandra
Cooke using Becca
Hair: Zoe Irwin @ Frank
Stylist: David Hawkins
@ Frank **Model:**
Hannah @ Next

LIGHTING & MAKEUP

"There are so many elements that need to come together to produce a high-standard fashion editorial. Lots of planning and art direction, a good makeup artist, a stylist with the right clothing and accessories and the right models that are able to take direction and move well.

Some of the best editorial models are successful because they have a face like a blank canvas that can be moulded easily into any look. Lighting needs to be well thought-out, executed and tested on the model to make sure it complements her bone structure and shape.

Makeup is an integral part of the shoot, as it brings to life the concept and creates the look and feel of the model. Photographers will use the way the light bounces off the makeup to create a polished and beautiful result. A talented makeup artist will also know how to create perfect skin and reduce the amount of retouching needed later, thus creating a better-quality image. Editorial makeup often calls for a little more creativity and is a vehicle for pushing boundaries, new looks and techniques."

Camille Sanson, Photographer

"I am inspired by really strong makeup and one of the most important components of a fashion156.com shoot is to work with hugely creative makeup artists. Someone who really looks at the model's face and the clothing and then decides if their ideas are actually going to work. Sometimes initial ideas need to be scrapped and an equally strong idea needs to be conceptualized and produced within minutes. So many artists I meet want to play it safe. For me, eyes, lips and cheeks can all be strong on some (many!) occasions. Preconceived ideas and rules sometimes need to be broken!"

Guy Hipwell, Editor and Creative Director fashion156.com

Photographer: Camille Sanson **Makeup:** MAC **Hair:** Andrea Cassolari **Model:** Kate Willing. Blue dress by Inbarspectar, jacket by Manish Arora

STYLISTS & MAKEUP ARTISTS

"Let's just assume the makeup artist can do the basics – beautiful skin. And they are really creative. Then, the other qualities I think a makeup artist has to have is to be well organized, have a team they can trust, and they need to know about trends. Not just makeup trends, but also fashion and colour trends. A makeup artist, like the stylist, needs to be able to look at a selection of clothes and offer a variety of makeup looks. Often a junior makeup artist will ask 'What do you want?' I like a makeup artist who can offer suggestions and is not just an order taker. I'll make a mood board to give the team a vision and direction, but often the makeup artist will make one too. A beauty story can start with the vision of the makeup artist and they'll present a mood board. From there I can suggest the jewellery, accessories, etc.

I like to work with the same team because I can trust them. My work can be quite varied, from commercial to catwalk, so I need a makeup artist who is strong on many levels. He or she also needs to be personable. Once you know someone who can do the job it becomes all about personality – do you get along with them, are they the right fit? There are so many good makeup artists, so it can come down to having a similar vision – do they contribute to the team, are they responsible and fun to work with?"

Rebekah Roy, Fashion Stylist

Photographer: Camille Sanson **Makeup:** MAC **Hair:** Andrea Cassolari **Stylist:** Nasrin Jean-Baptiste **Model:** Elena @ Oxygen. Dress by Bernard Chandran, bracelet by J.W. Anderson, shoes by Modernist, all headwear customized by the stylist from a selection at Angels

Lace bodycon
jumpsuit, stylist's own,
skirt by Topshop,
gloves by Gloved Up

Jacket by Louise
Amstrup, dress by
Ashish, bracelets
by Disaya

Photographer:
Camille Sanson
Makeup: Philippe
Miletto @ Terri Tanaka
Hair: Andrea Cossolari
Model: Erin @ Nevs

CHATEAU DE CHANTILLY

Opulence and decadence were the ideas behind this editorial here, shot at the Château de Chantilly in France and published in the French magazine *Blanc*.

The editorial was titled The Escape, and it is a story of two girls stuck in a manor house during an apocalypse. They appear glamorous and wonderful but, if you take a closer look, they desire to break free from the confines of luxury, almost as if they are trapped. The styling and location for this shoot was high-end and grand, so the makeup had to match – everything had to be super glossy. Using the Chanel Spring/Summer 2013 collection we went with deep red and cherry lips, a hint of blush on the cheeks, super-highlighted glowing skin and to finish it off a smoky 1920s-style eye. As the shoot went on we added a little more drama to the makeup, paled down the skin, added more colour and gloss to the cheeks and eyes and even added a little beauty mark to finish the look. The styling, makeup and hair all complemented the location and lighting perfectly.

Producer: Jana Ririnui
Makeup: Michelle Webb using Chanel Les Beiges S/S 2013
Production assistants: Breianne Zellinsky and Amy Macdonald
Photographer: Thomas Nights **Styling:** Oliver Vaughn **Fashion Assistants:** Frances Knee and Chantelle James **Hair:** Jan Przemyk **Models:** Anastasia and Laura

Anastasia (left) wears dress by Marina Quereshi, hat by Atsuko Kudo, shoes by Ursula Mascaro. Lauren (right) wears dress by Roberto Cavalli, jewellery by Fope Gioielli, gloves by Corlette, shoes by Miu Miu

(left) Lauren wears dress by Christian Dior, headpiece by John Rocha, gloves by Hasan Hejazi

(right) Lauren (left) wears dress by Yiquing Yin, bag and shoes by Christian Louboutin, bracelet by Fope Gioielli. Anastasia (right) wears dress by Felder Felder, shoes by Christian Louboutin, bracelet by Lola Rose, ring by Joubi

(left) Dresses by Bibi Bachtadze. Anastasia (left) wears shoes by Manolo Blahnik, headpiece by Omiru, earrings by Lanvin, bracelet by Bottega Veneta. Lauren (right) wears eye mask by Soft Paris, shoes by Ursula Mascaro, Delphine jewellery by Charlotte Parmentier

(above) Headpiece by Atsuko Kudo. Lauren (left) wears dress by Nina Naustdal Couture, Delphine necklace by Charlotte Parmentier, shoes by Manolo Blahnik. Anastasia (right) wears dress by Luisa Beccaria, gloves by Aspinal, necklace by Isabel Marant

Anastasia (left)
wears dress by Katya
Shehurina, necklace
by Isabel Marant,
headpiece by Lara
Jensen, belt by Felder
Felder. Lauren (right)
wears dress by
Carven, headpiece
by Atsuko Kudo

BEAUTY *EDITORIAL*

POST-PRODUCTION

A photographer's job extends beyond standing behind the lens and capturing the shots. When everyone's working day is over, the photographer must start work on the post-production process. Post-production is the general term used to describe the process of turning the raw shots from the camera on the day of the photo shoot into the polished, glossy images we see in magazines.

The first step of post-production is the selection of images. Often hundreds of images are taken in any one day of shooting, so narrowing down to the best is a big job, often with only miniscule differences between the shots. The team, including hair, makeup and styling, may help with this process and narrow them down to a smaller number of potential final images. Once the final images are selected, retouching is carried out to perfect and tweak the images.

Retouching Images

A widely standardized practice in magazines and advertising campaigns, in order to smooth out wrinkles, fine lines, lumps, bumps and blemishes, retouching can also be used to stretch out models, slim models down and remove cellulite, which can result in an unrealistic picture of beauty.

Retouching allows small errors to be rectified, for example, if a model's lipstick has faded in one corner, a few stray hairs need to be removed or there are clumps of mascara. Knowing photographs can be retouched can sometimes lead to sloppiness, as some makeup artists may feel they do not have to work hard to perfect their makeup application.

Although the responsibility of retouching belongs to the photographer, the makeup artist should review the images, make sure they are happy with them and that they are projecting their work, and ensure that the makeup has not been completely altered.

Photographer: Camille Sanson **Makeup:** Dior
Hair: Kuni Kohzaki
Model: Danielle Foster

ADVERTISING

Makeup artists are almost always used in advertising. An example of this may be the appearance of sweat on the brow of a sportsman in a Nike commercial, or the pigtails on a little girl in a toy advert.

From the commercial advertisements you see on television to billboard posters; a makeup artist was needed on the shoot to ensure the people being photographed or filmed were looking their best. The look or type of makeup a makeup artist will do on the "talent" (how the model or actor is referred to on set) is very job-specific. The makeup look not only has to fit in with the story of the advertisement, but also has to be appropriate for the product it is advertising.

Photographer:

Camille Sanson

Makeup: Lancôme

Hair: Marc Eastlake

Stylist: Shyla Hassan

Model: Sabrina @ Next

Dress by Bora Aksu

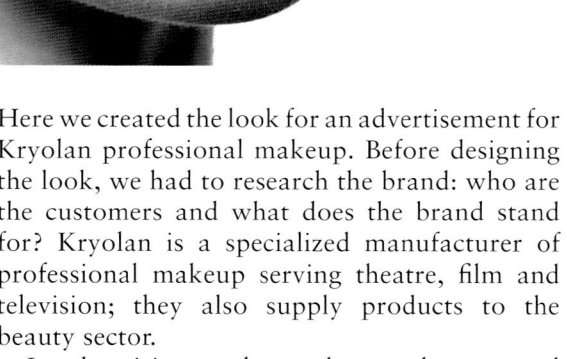

COLOUR EFFECTS

Here we created the look for an advertisement for Kryolan professional makeup. Before designing the look, we had to research the brand: who are the customers and what does the brand stand for? Kryolan is a specialized manufacturer of professional makeup serving theatre, film and television; they also supply products to the beauty sector.

In advertising work, you have to be prepared with a wide variety of looks in case the original idea does not work. Ask yourself whether the look should be natural or strong, and what products should be used to promote the brand. Having the right balance of looks to attract a customer is important, as is knowing what is going to make the image stand out. Sometimes campaigns are modified to make the image stand out more, but it is important that the look shown is its truest form, according to the aims of the brand.

Once the skin had been prepped, a full-coverage foundation was used to create a flawless airbrushed effect. Cheeks were kept soft, contoured with a brightly coloured blush, applied lightly. For the eyes, strongly pigmented greasepaints were used, placing colours in sections on the eyelid – the colour was kept in solid blocks to show the true colour pigments. Blending at the sockets with another bright colour opened up the eye shape. The eye was all about strong, graphic colours, rather than smudging, which would have made it look softer.

Eyebrows were filled in to create a modern twist and to help give a more graphic feel. Lips were also blocked in with a solid colour to complement the rest of the look. Adding highly coloured gloss to the lips helped to keep it looking modern. Various colours were used in different areas of the face, and the model was shot at different angles to provide variation.

Photographer: Catherine Harbour
Makeup: Kryolan
Models: Gabriele D and Larissa H @ Premier

CONCEPTUAL

Avant-garde makeup is as infinite as one's imagination. It is all about pushing the boundaries and encourages you to think outside the box. Prime examples of this type of makeup look can often be seen on the catwalk at Paris couture fashion week. Designers John Galliano and Jean-Paul Gaultier are two of the leading trendsetters of this look.

Many designers showcase their work to demonstrate their individuality and cutting-edge values. Often, their clothes are considered to be more like wearable art. Therefore, the makeup designs that complement the clothes can also be seen as quirky and beautiful art works.

The concept of the makeup designs, the colours used and how and where the makeup is applied, can influence the commercial market, too, and inspire the creation of new makeup products. Brands including Charles Fox, MAC and Shu Uemura are constantly developing interesting makeup ideas for professional artists. This constant stream of development means there is no limit to what you can create.

The catwalk shows also inspire the future trends and avant-garde editorials in magazines. High fashion magazines such as *Numero*, *10*, *Dazed & Confused* and *Pop* are just some of the many that take inspiration from the catwalk and create their own images for the public domain. They ultimately feed the makeup and styling world with even more creative imagery and new looks.

Photographer: Zoe Barling
Make-up: Sandra Cooke using MAC
Hair: Natasha Mygdal using Bumble & Bumble
Model: Alex C @ IMG

Photography and Hair:
Desmond Murray
Make-up: Jo Sugar
using Illamasqua
& Kryolan
Stylist:
Sampson Sobage
Models: Ani Grigorian
& Lucy Flower

Photographer:
Catherine Harbour
Make-up: Philippe
Miletto using MAC
Hair: Andrea Cassolari
Model: Iliza @ Next

THROUGH THE WORMHOLE

"For this shoot I wanted to take makeup into the future – it seems technology is moving forward so why shouldn't makeup? I have always had a love for strange and crazy makeup, working with texture, glitter paint and turning someone into more of a walking art piece, so when I was brainstorming for this shoot I knew I wanted it to be something amazing and impactful. We decided to paint with light and lasers on top of the makeup and in and around the background, bringing technology into the application process.

The story was titled Through the Wormhole because it was as if the models were being transported to the future, where everything is surreal; I used a lot of paints and pigments and bright colours to change their features and make them look almost cartoon-like. The looks are also a little deconstructed, imperfect. Sometimes things are more beautiful that way.

Once the basic makeup was completed, colour torches and lasers were shone on top of the face to create more depth to the makeup and add a surreal feel."

Michelle Webb

Photographer: Tanausu Herrera **Makeup:** Michelle Webb

Photographer: Tanausu
Herrera **Makeup:**
Michelle Webb

Photographer: Tanausu
Herrera **Makeup:**
Michelle Webb

Photographer: Camille Sanson **Makeup:** MAC **Hair:** Marc Eastlake using Bumble & Bumble **Styling:** Shyla Hassan **Model:** Sara Amos @ INC

(far left) Blue dress by Bernard Chandran, jewellery by Erickson Beamon. (left) Mesh top with gold spots by Topshop, peach blouse, stylist's own, skirt by William Tempest. (right top) Black dress by William Tempest, jewellery by Erickson Beamon. (right bottom) Navy waistcoat by Modernist, cream shorts by Bernard Chandran, jewellery by Erickson Beamon. All head pieces by Marc Eastlake

Salvador Dali
Photography and Styling: Camille Sanson
Makeup: MAC **Hair:** Andrea Cassolari
Models: Hannah @ Next and Claudia @ Nevs

**Photography and
Styling:** Camille Sanson
Makeup: MAC and
L'Oréal Professional
Model: Claudia @ Nevs

Photography and Styling: Fabrice Lachant **Makeup:** MAC **Hair:** Marc Eastlake using Bumble & Bumble **Model:** Renee Mansbridge

BODY ART

Body decoration is one of the oldest forms of art. Historically it has been a tradition of many ancient tribes across the globe, using materials such as coloured clay and charcoal. It has now entered the mainstream and is used in mediums such as ad campaigns, album covers and fashion shows.

Body painting is an increasingly competitive industry. It is important to have the skills needed before taking on any professional jobs, as good-quality work is not something you can easily achieve without plenty of practise. Know the paint's limitations as well as your own. Work a design through with a client from their initial concept to the final approved design. This way, you will be sure that what you are producing is according to plan, you will feel confident about the job and, ultimately, you will produce an amazing piece of art. To get repeat business, remember that you are only ever as good as your last job.

There are two main techniques: brush and sponge painting and airbrushing. It is easiest to start with a few simple brushes, sponges and face paints. A basic kit should consist of primary and neutral colours, such as black, white, red, yellow and blue. You do not want to spend all your time mixing, so green, purple, pink, orange, grey and brown are also handy staples.

Brushes
When choosing brushes look for ones with dense bristles, as they will cause less streaking. Below is a selection of basic brushes to have in your tool kit.

- A brush with a curved edge for smooth lines.
- A flat, straight-edged brush to create well-defined lines.
- A large flat brush that is at least 5cm (2in) wide to cover large areas of the body quickly.
- A medium-sized flat-edged brush.
- A small, flat-edged brush, square to rectangular in shape, to create straight lines and crisp edges.
- A fine-pointed brush for detailed work, and for working from a thick line to a thin one.
- Face painting sponges to blend colours on the body, cover underwear in paint and get into awkward areas, especially around the eye area.

Using Paints
Body paints are made according to stringent guidelines – they are non-toxic and non-allergic. Because the paints are activated with water, you will find that when you try to layer your colours they may start to blend into one another. For example, if you paint white detail over a black background, you will quickly find that the bright white turns grey. It is much easier to cover a lighter colour with a darker colour than the other way around. When necessary however, fix your base coat with a fixing spray or strong-hold hairspray. This will help eliminate blending problems.

Airbrushing
It is expensive buying all the necessary equipment, but there are other tools that can be used to create dramatic effects when a paintbrush is not enough. An airbrushing kit consists of a compressor, an airbrush gun and special paints. Airbrushing is an extremely useful skill to have when you work on jobs that require a lot of stencilling and when you want very softly blended highlights and shadows.

To use an airbrush gun you push down on the trigger to start the airflow and pull back on the trigger to release the paint. The lighter you press it, the softer the airflow, and the more you pull back, the more paint is released.

Once you get used to the way the gun works you will be able to create anything from a line as thin as a hair to a smooth, perfectly blended background. Practise on cheap canvases and with acrylic paint to build your skills. When you can, practise on a model so you can get used to working on a living, breathing and sometimes moving canvas.

Working with Models
When planning your work, think about the order in which you want to cover the body. When body paint is dry it is relatively touch-proof, but it will smudge easily if rubbed. It is often best, if working with a female model, to cover their breasts first so she feels less exposed and more comfortable.

Leave certain areas of the body such as the mouth, hands and bottom, clean or without detail initially, as most detailed body paints will take between three to six hours to be completed and during this time, your model will probably need to sit down, eat, drink and use the bathroom.

Some models will struggle having to stand up for such long periods of time without moving, so make sure you have sugary snacks and drinks on hand to keep energy levels up. Some might not want to drink a lot of water because they are worried about needing to visit the toilet and smudging the paint, so be sure to reassure them that it is a lot easier to clean up smudged paint than having to clean off an entire section because your model has fainted!

Photographer: George Kuchler
Makeup: Carolyn Roper using MAC
Models: Carolyn Roper and Craig Tracy

Photographer:
Catherine Harbour
Makeup: MAC
Stencilling: Carolyn
Roper **Hair:** Andrea
Callosari **Model:** Kate
Willing. Headpiece by
Dominic Elvin

Stencilled Snowflakes

Here stencils were used to create an image of a girl trapped inside a snow globe, but instead of snow falling all around her the snowflakes are on her. White-coloured hairspray was painted through snowflake stencils to create an airbrushed effect. White glitter was then applied on top using a touch of Duo eyelash glue to keep it in place. A mixture of pink, pearl and lavender eyeshadows were then blended on the face to create a cold look to the model's skin. MAC Strobe Cream was also blended into a pale tone of MAC Face and Body Foundation.

Photographer:
Keith Clouston
**Makeup and
Stencilling:** Rachel
Wood using MAC
Makeup Assistant:
Lauren Amps **Hair:** Fabio
Vivian **Model:** Invlid @
mandpmodels

CONCEPTUAL BODY PAINTING

This is a modern interpretation of an Art Deco-style painting, inspired by the work of Polish artist Tamara de Lempicka. Taking the colours and main characters from some of the original paintings, the design was worked around the shape of the body.

The main design was created using Mehron airbrush paints, and is complemented by simple beauty makeup using pigments in the same shades as the paint. The fine lines that fan out around the two main figures were created using very fine glitter glue, also used to frame the face. Crystals around the eyes and cheekbones, attached with hydro-mastic body glue, completed the design. The steps below can be adapted for any body-art idea.

1 When planning the design, remember that your main canvas areas are the front and back of the torso, so this is where you want to position your main details. The arms and legs are a good place to get creative with colour blending.

2 Always start by painting the basic outline of the design in the colour that relates to the finished image. This makes it easier to work out where different colours start and finish and makes blocking in the rest of the colours less confusing.

3 Once the base lines are drawn, start putting on the base coat. Always mix the paint to a really creamy consistency, as this will ensure the base coat doesn't streak. When the base coat is finished you can start to add the details.

4 Body paints sink into each other over time and will not look as vibrant, so it is important to start with the shadows and leave the highlights until the end. Adding a layer of sealant in between the base coat and your detail is the best way of minimizing colour corruption.

5 Airbrushing creates beautiful, soft highlights and shadows and is used at the end of the painting to create a real sense of depth.

• To avoid tell-tale join marks, cover the largest areas, such as the chest and back, first and finish at the sides where joins will be far less noticeable to clients and photographers.

Photographer: Lou Denim **Makeup:** Carolyn Roper using Mehron **Model:** Femke @ Nevs

186

Photographer: Fabrice
Lachant **Makeup:**
Carolyn Roper
using Mehron

COLLABORATING

"Collaborating with makeup artists is always different, depending on what the brief is. For instance, when I am shooting for the British Hairdressing Awards, it is predominately about the hair, so the relationship is with a team of people and the mood board. The mood board will dictate the feeling of the whole shoot, ensuring the makeup artist, photographer, fashion stylist as well as the hairdresser are all singing from the same hymn sheet. Sometimes the hairdresser will know what they want from the makeup, other times they will not, so the makeup artist will have more input in the design of the makeup.

Sometimes, on a hair shoot, it may not be clear what is wanted with the makeup, so there will either be a prep day in advance or the team will arrive early and play with the makeup to see what works. This way there is plenty of scope for individual creativity and freedom.

Fashion shoots are totally different, especially if it is a test shoot. Everyone experiments and is pushing the boundaries in their own field, but the team comes together for the overall look – it is not just about one person, it is a collective effort.

When doing fashion shows, the focus is on the clothes. Before the show there will be a fitting where the head hair stylist and makeup artist create their looks on a model and present them to the designer to see if it works. If the designer is happy, then the heads go back to their teams and brief them with the look. This needs to be done very quickly, so there is very little time for prepping and often the hair and makeup is being done at the same time. The makeup artist needs to work in tandem with the hairdresser and vice versa.

When working with celebrities, they dictate their image so you have to liaise with makeup and styling. It is all about what the celebrity wants and creating an overall image, so you may be recreating a style they are known for rather than having creative input."

Desmond Murray, Celebrity Photographer and Hair Stylist

Photography and Hair: Desmond Murray **Makeup:** Jo Sugar using MAC and Kryolan **Model:** Sophie Willing

Creating the Look

For this shoot the models were to be dark and appear a bit like a silhouette, with light catching parts of the body. After considering black models, the makeup artist decided to achieve the look purely with cosmetics, first experimenting with different mediums.

1 For the face, Studio Fix Fluid by MAC, a matte medium-coverage foundation, was applied, with contouring done using MAC's Studio Sculpt; highlighting was achieved with Touche Éclat.

2 Black eyeliner was applied on the inside rim of the eyes, and a smoky eye created over the mobile lid into the socket using black gel eyeliner and matte black eyeshadow.

3 False lashes were applied on the upper and lower lashes and the eyebrows were drawn in black gel eyeliner.

4 Lips were blocked out, then highlighted using loose iridescent powder.

5 Next the model changed into a black G-string that could be thrown away afterwards and nipple covers to make her feel more comfortable. A disposable gown was on hand to keep the model warm. The body was prepped with Barrier Cream from Charles Fox.

6 Kryolan greasepaint in black was applied to the whole body using a sponge and smoothed with a large synthetic brush; the face was also outlined in greasepaint. The hands and feet were done last.

7 After the shoot the greasepaint was removed with a cream remover and baby oil; soap and water will only make the makeup more difficult to remove.

• Body hair on the model is undesirable, as the hair will show through the makeup. Ensure your model is hair-free in advance of a bodypainting shoot.

Photography and Hair: Desmond Murray
Makeup: Jo Sugar using MAC and Kryolan **Model:** Sophie Willing

UNDERWATER

There are many different scenarios where waterproof makeup may be required. You could be shooting on a beach, in a swimming pool or filming in a water tank on a film set. In most cases, the model will need to be retouched after just one shot in the water.

Working with water can be unpredictable and it is not always easy to guess how the skin is going to look or react when makeup and water meet, so be prepared to make changes. The key to successfully using and working with waterproof makeup is how you apply it and set it afterwards. It is important that all the skin on show is covered with foundation or a waterproof fake tan. Finish the skin with a specialist fixer spray and powder, which will help the makeup stay on longer in water. Sometimes a few blasts of hairspray can also do the trick.

Water reflects light and, depending on the type of shoot you are working on, the water can alter the colour of the skin, making it looked washed-out and colourless. This is especially true when working in deep water tanks. For best results and to combat this, use cream-based products and strong, vibrant colours to bring out the features. By mixing mediums with eye pigments it will give you a longer-lasting look, but these will tend to crack when worn all day.

Photographer: Fabrice Lachant **Makeup:** MAC **Model:** Kate Willing **Makeup Assistants:** Kelly Mendiola and Karla B @ AOFM Agency

ADORNMENT

Makeup is evolving all the time and continues to be very experimental. By adorning the skin and using the technique of mixing paint and pigments you can create mask-like effects, which are visually inspiring. This can usually be seen in creative magazines and in jewellery advertisements where the face and piece of jewellery is the focus of the image.

There are alternative materials that can look interesting when applied to areas of the face, such as fabrics, beads, crystals, feathers and most art and craft materials. Crystals add a touch of glamour to a look, and when applied sparsely around the eye area can be wearable for a special occasion. For the best results, only use a small amount of glue, as too much will cause the stone to move and become messy. Duo adhesive glue can be used to stick these materials to the skin, as it is safe and easy to remove afterwards.

Photographer: Camille Sanson **Makeup:** MAC and Swarovski **Model:** Ida @ Oxygen. Jewellery piece by Louis Mariette

Working with glitter can often be a bit messy and tricky – loose glitter and pigments stick for a while, but will eventually fall off onto the face, depending on how much you use.

When you are using glitter around the eyes, use a cream-based eyeshadow, greasepaint or any product that has a wet consistency and place the glitter directly on top using your finger or a flat eyeshadow brush. You can use a mixing medium, such as a gel eyeliner, to stick glitter on small areas such as under the eyes. For larger areas around the eye, an option is to use a face-and-body mixing medium, which is easily spread and mixed in with the glitter to create a metallic effect. It stays wet a little longer so you can build the depth of glitter, layer by layer. To remove all traces of glitter use a water-based cleanser, such as Lancôme Bi-Facil instant cleanser or baby oil.

Natural Look

To achieve a natural shimmer and sparkle around the eye, dab the glitter onto the base lightly and put more on the inner eye for a brightening effect. Some fine shimmer glitters can be used on the cheekbones to create highlights.

Disco Look

For a party look like the one top left, make sure the base eyeshadow is strong and blended on top and underneath the eye. dab a little mixing medium on all over and quickly place the multi-tone glitter to set. You can control the intensity of the glitter by slowly adding more, bit by bit.

- Use masking tape to remove stray glitter from the face.

- Completing the eyes before applying the foundation will save time and avoid smearing.

(far left) **Photographer:** Camille Sanson **Makeup:** MAC **Model:** Olga @ Profile. Hat by Louis Mariette

(left top and bottom) **Photographer:** Catherine Harbour **Makeup:** MAC **Model:** Gabriele D @ Premier

Photographer:
Camille Sanson
Makeup: MAC
Stylist: Nasrin Jean-
Baptiste **Models:** Elena
@ Oxygen (left) and
Kate Willing (right)

WORKING WITH DIAMANTÉS

Using diamantés to adorn the face calls for plenty of patience. Picking each individual diamanté and placing it can be a long process – the trick is to use good tweezers. Choose flat-backed diamantés, otherwise they will not stick to the skin.

To save time, make sure you have a good idea of your design before you begin. The diamantés should be the last product to be applied once you have finished creating the rest of the look. Sketch out where you plan to place the stones; this is particularly important if you are using different colours and sizes of diamanté. For example, a big gem on the eyelid may not be a good idea as it can restrict eye movement and make the eye look sleepy. When working with coloured diamantés it is best to use a base colour makeup similar to the gems to give the look a solid finish, as the gaps between the stones can be visible and sometimes look untidy. If you want to achieve contrast though, then you need not worry about this.

Applying Diamantés

Use a transparent glue that dries to a rubbery consistency, such as Duo, which makes it easier to rectify mistakes and pull off badly placed diamantés. Alcohol-based glue takes longer to dry and can irritate the skin. Place a dab of glue on a tray or the back of your hand and use the tweezers to pick up a diamanté; dip the flat side lightly into the glue. You only need a dot, as too much glue can be messy and the diamanté will slide around, making it hard to position accurately. Once you have placed the first diamanté, continue positioning them side-by-side, ensuring that they are in line and evenly dispersed. Continue in this way until you have finished.

(overleaf left) Dress by Richard Sorger, leather floral head piece from Erickson Beamon, Swarovski leather choker from Renush

(overleaf right) Head piece by Louis Mariette, jacket by Louis de Gama, necklace by Raris, necklace by Swarovski Crystallized, necklace (worn on hands) from Erickson Beamon, selection of rings from Erickson Beamon and Swarovski Crystallized

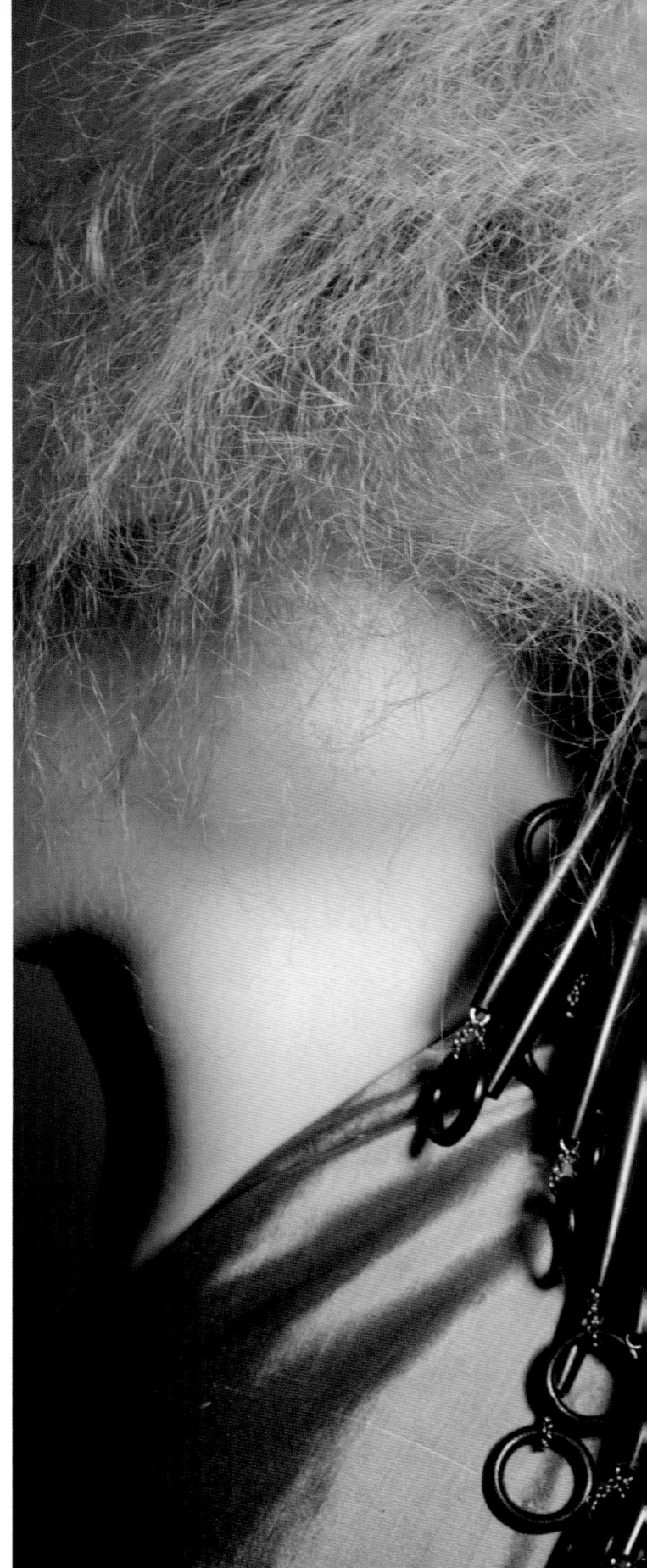

Photographer:
Catherine Harbour
Makeup: MAC Pro
Makeup Assistants:
Elsbeth Tan and Sharka
Hair: Tim Furssedonn @
Toni & Guy using Label
M **Stylist:** Rebekah Roy
Models: Rea and
Larissa H @ Boss.
Location: Inc Space.
Dress by Yan To,
necklace by Raris,
bracelet by Bex Rox

Crystal eyewear by
Louis Mariette, leather
top by Louis de Gama.
(right) Dress by Bryce
D'Anice Aime, hat
from House of Flora,
rings from Bex Rox and
Swarovski, necklace
from Erickson Beamon

TOOLS

A makeup artist's kit is not just made up of makeup, but also includes a large selection of tools. These are essential to make the job easier and allow for the correct application of makeup.

Eyelash Curlers

An essential item in a makeup artist's kit, an eyelash curler needs to be used before mascara and, for maximum impact, used as close to the eyelid as possible (without pinching the skin). Heated versions are also available and can be used to curl stubborn, straight lashes. It is possible to heat metal eyelash curlers using a hairdryer, but these need to be tested for temperature on the inner forearm prior to applying to the eye. The most popular brand of eyelash curler is by Shu Uemura, which also manufactures mini eyelash curlers, great for curling the odd, stubborn, straight lash.

Tweezers

Besides tidying up eyebrows, tweezers are needed for applying false lashes and working with latex glue, such as Duo eyelash adhesive. Makeup artists will also use tweezers and Duo for applying intricate crystal-, bead- and feather work to the face.

Masking Tape

Handy to remove glitter fallout that may be stuck to the face or clothes, masking tape can also be used to create precise straight lines or graphic shapes on the face.

Bags

The most efficient way to transport all the materials a professional makeup artist needs for a job is in a suitcase on wheels. Not only does it hold all the kit you need (and spares),

but it is a far easier and safer way to carry the heavy load. Most makeup artists will organize their kit into clear makeup bags, with similar products grouped together, to be able to see what products are available and to know where to find the products easily and quickly.

Other Essentials

For a full kit, you will also need disposable items, such a tissues, cotton buds, cotton pads, baby wipes (handy for cleaning hands and kit), and basic face wipes for when makeup needs to be removed quickly.

Photographer:
Camille Sanson
Makeup: Shu Uemura
Model: Erin @ Nevs

MAKEUP BRUSHES

Brushes are probably the most important tools in a makeup artist's kit. Not only do they help apply makeup with accuracy and precision, but they also allow the artist to blend and move the product to create the desired effect. Brushes come in a variety of different hair types, as well as sizes and shapes for various application use. The product formula determines which brush is suitable for use. Most makeup artists will have a large selection, and sometimes several of the same of their favourite brushes.

Cleaning Brushes

Cleaning brushes between applications is essential, both to remove product and to sanitize the brushes. A good, simple, alcohol-based brush cleaner, such as isopropyl alcohol, will kill bacteria instantly and dries in seconds, making the brushes ready to reuse.

All brushes AOFM
Pro Tool.

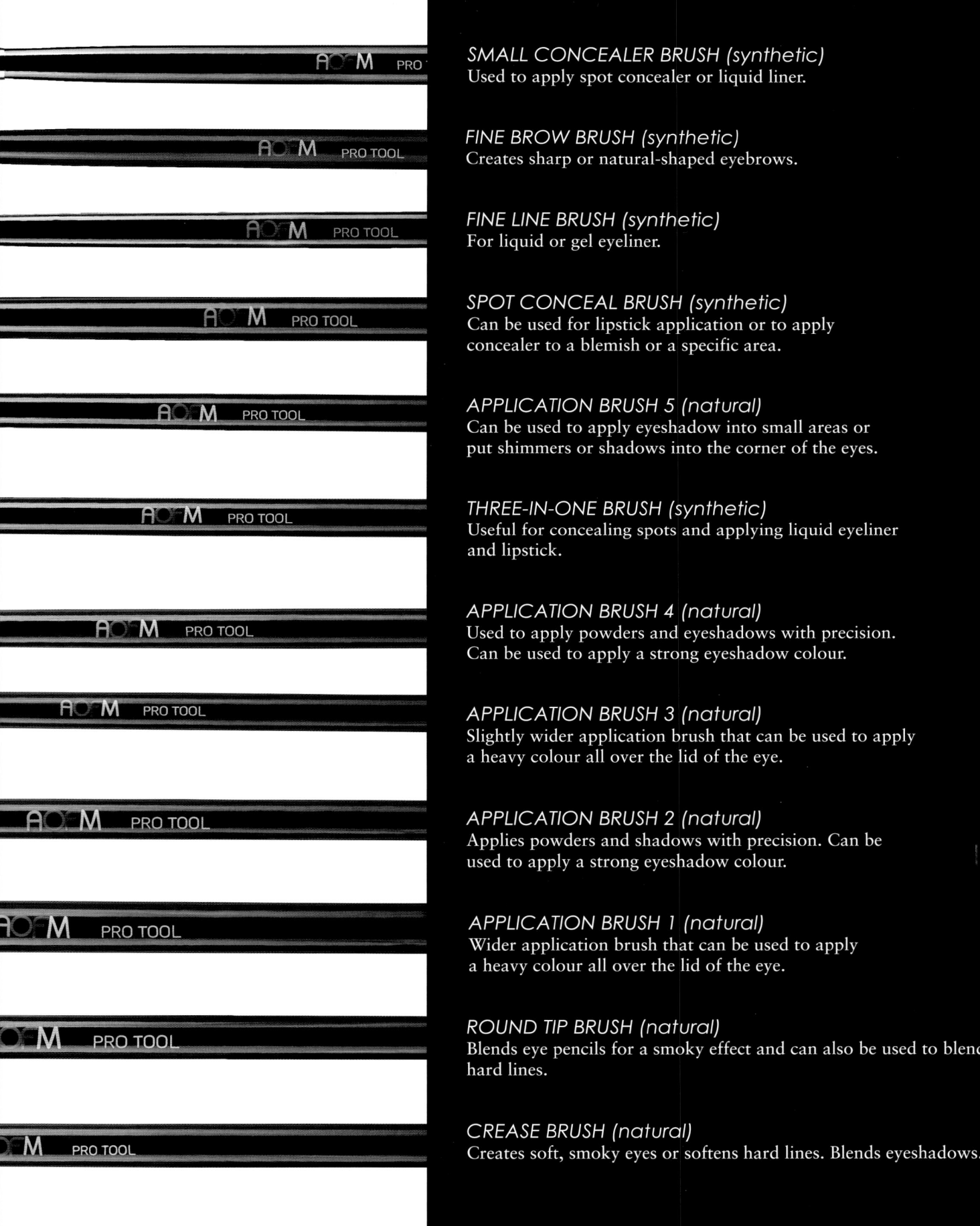

SMALL CONCEALER BRUSH (synthetic)
Used to apply spot concealer or liquid liner.

FINE BROW BRUSH (synthetic)
Creates sharp or natural-shaped eyebrows.

FINE LINE BRUSH (synthetic)
For liquid or gel eyeliner.

SPOT CONCEAL BRUSH (synthetic)
Can be used for lipstick application or to apply
concealer to a blemish or a specific area.

APPLICATION BRUSH 5 (natural)
Can be used to apply eyeshadow into small areas or
put shimmers or shadows into the corner of the eyes.

THREE-IN-ONE BRUSH (synthetic)
Useful for concealing spots and applying liquid eyeliner
and lipstick.

APPLICATION BRUSH 4 (natural)
Used to apply powders and eyeshadows with precision.
Can be used to apply a strong eyeshadow colour.

APPLICATION BRUSH 3 (natural)
Slightly wider application brush that can be used to apply
a heavy colour all over the lid of the eye.

APPLICATION BRUSH 2 (natural)
Applies powders and shadows with precision. Can be
used to apply a strong eyeshadow colour.

APPLICATION BRUSH 1 (natural)
Wider application brush that can be used to apply
a heavy colour all over the lid of the eye.

ROUND TIP BRUSH (natural)
Blends eye pencils for a smoky effect and can also be used to blend
hard lines.

CREASE BRUSH (natural)
Creates soft, smoky eyes or softens hard lines. Blends eyeshadows.

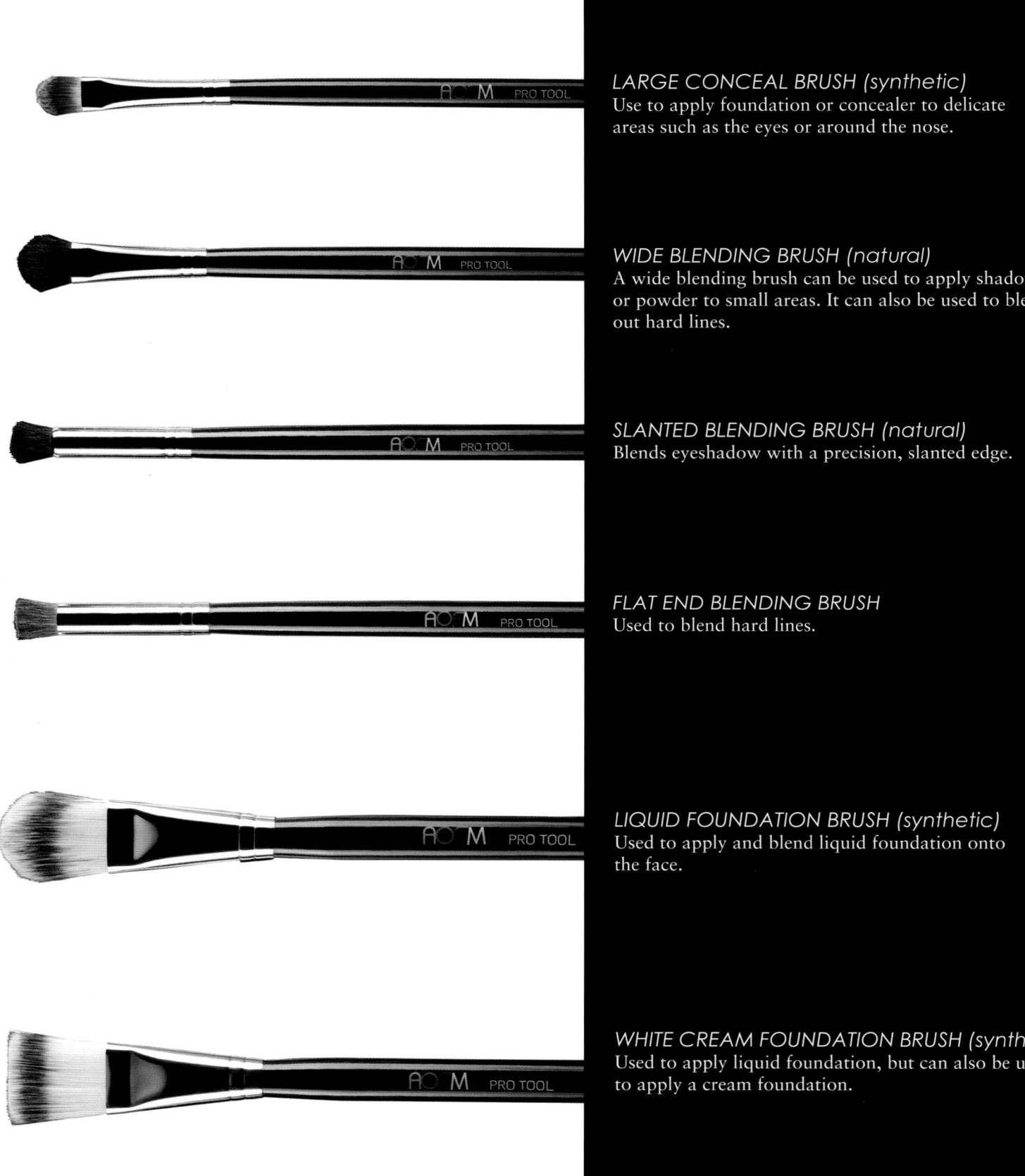

LARGE CONCEAL BRUSH *(synthetic)*
Use to apply foundation or concealer to delicate areas such as the eyes or around the nose.

WIDE BLENDING BRUSH *(natural)*
A wide blending brush can be used to apply shadows or powder to small areas. It can also be used to blend out hard lines.

SLANTED BLENDING BRUSH *(natural)*
Blends eyeshadow with a precision, slanted edge.

FLAT END BLENDING BRUSH
Used to blend hard lines.

LIQUID FOUNDATION BRUSH *(synthetic)*
Used to apply and blend liquid foundation onto the face.

WHITE CREAM FOUNDATION BRUSH *(synthetic)*
Used to apply liquid foundation, but can also be used to apply a cream foundation.

What really happens before you see a runway show from either the front row, on the television or in the pages of a glossy magazine, can sometimes be taken for granted if you are not actually involved in the process. It takes dedication and months of hard work to prepare for such an event.

A long time before a model even sets foot on the catwalk there are meetings to discuss the trend forecasts, concepts and designs. Models, locations and crews have to be picked and tested and then the PR teams go into overdrive to make sure that key press teams come and cover the event. It is a cycle that has been going on for many seasons and one that will never change. As a makeup artist you cannot help but want to be a part of it all.

A makeup artist has an important role in the catwalk show, as it is their responsibility to make sure that the look complements the designer's collection without overpowering it and taking the attention away from the garments. The look needs to be beautiful and fresh, but it also should be as new and exciting as the clothes because these looks will be the guidelines for next season's trends. Beauty editors from across the world will be using snapshots of the catwalk shows as references in their editorial, and let's not forget the celebrities that are in the front row – if they see a look they like, then you never know what may happen.

The head makeup artist for the show will have spent weeks designing mood boards and collaborating with the chief designer, fashion stylist and hair stylist to come up with the various looks. Mood boards are a kind of collage of images that designers put together to create a visual guide and inspiration for their collection. Very often the designer and stylists will email them to the key makeup and hair stylists to give an idea of the feel of the new collection of garments and what kind of atmosphere they want to create for the show. This in turn gives the makeup a guideline and basis on what they are going to design in a way of a makeup look for the show.

A "test" is where they will have a practise run on a model. The hair and makeup will try out their designs and tweak it on the model until everyone is happy with the final look. The designer will try an outfit or two on the model from the new collection to check that it all meshes well with the hair and the makeup and the stylist is there to ensure it's all put together well – they are the ones who often source the shoes, tights, jewellery, and so on. They also help in deciding which outfit goes best on each model for the show.

Also before the show, the designer and stylist hold a "casting", a kind of interview/audition where models are seen by the designer, asked to try on an outfit and walk for them to see how they would potentially move on a runway.

Products and Brands

Product sponsorship can play a major role in how a makeup artist designs the look for a show. A cosmetic brand could be the one who pays the makeup artist for using their product at the show and also supplies the makeup for the team backstage, so the artist has to ensure that the look is created with products from that particular cosmetic brands line. For example, if designer David Koma's show is sponsored by Benefit and he asked for red-coloured eyes, then it's up to the makeup designer to create this look using a Benefit product such as Benetint (which is a red-cheek stain). Often a cosmetic brand will arrange press interviews backstage during the show to ask the artist what brand products they used for the show.

In preparation for the show, the makeup artist will be in contact with their product sponsor and order the amount of products he or she will need for the show. This is usually estimated by how many models there are and how many makeup assistants the key artist will have. Every makeup artist works differently but the average is one makeup assistant for every two to three models.

Makeup assistants can be pulled in from the head makeup artist's agents or a brand such as MAC can offer a team to work backstage. Sometimes if a brand such as AOFM is sponsoring the show, they can provide makeup assistants for the artist. Most makeup artists will have a sort of "first" assistant. This is a makeup artist who works with them often. This artists knows the head artist's makeup style and what products they like to use best, as well as how their kit is set up. They can help run things for the head artist when it gets very busy backstage.

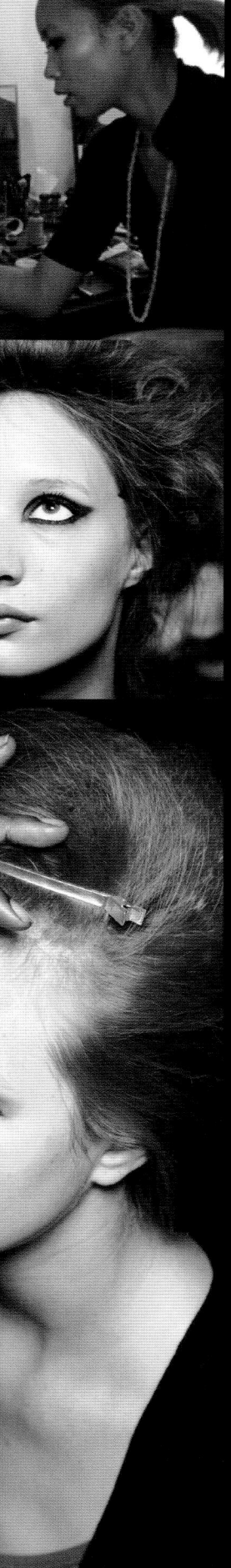

Both the head makeup artist and the makeup assistants are required to bring their makeup kits to ensure that they have everything they need to create the looks.

Four hours is the average time to prep before a fashion show and this is used for both the makeup and hair and last-minute clothing fittings on the models. At the start of the working day, the key makeup artist will use one of the models and do a "demo" on a model to demonstrate the look for the show. The head artist will show the key products to be used on all the models. It's imperative that the makeup assistants show up on time to watch this demo so they know what the look is for the show. They in turn will re-create this look on the models. Each model is then checked over by the head artist and any corrections or finishing touches of the look are done. It is important that the looks are finished on time and are on spec, as there is nothing worse than seeing mistakes on stage. Finally there is a makeup test and full dress rehearsal. with the fashion show producer for the models so they know exactly what to do for the show. The makeup head and a few assistants will check how the makeup is looking on the models on the runway, this helps see if there are any imperfections, blending or changes needed.

Just before the show, there are final checks. Often each assistant is given a small job such as powdering or lip touch-ups. This is often during the line up just before a show starts backstage – where all the models are queued in order and wearing their first outfits just as they are to step out onto the catwalk.

Finally, it is important to be focused when working on a catwalk show, as backstage can be very chaotic with television crews interviewing stylists and the designers, beauty editors wanting to get a preview, photographers getting close-ups of the makeup and models running in and out, getting their hair done and having clothes fittings. There will always be elements that you have not planned for, but you will have to deal with them in best way possible, such as a model arriving late or having bad skin. However, when the music starts, the lights dim and the first model walks down the runway, everyone will be looking at something that you have helped to create and you realize it has all been worth it.

One of the best things a new makeup artist can do to advance their career is to assist an established makeup artist. They can learn new creative ideas, innovative ways to use their brushes and see what products other makeup artists favour in their kits. Most importantly, they learn the proper etiquette for being backstage at a fashion show or what is really required from the makeup artist on a photo shoot.

Being a Good Makeup Assistant

Show up with clean brushes and a basic makeup kit. Most of the time, especially on fashion shows, the head makeup artist will provide or have the key products required to complete the look.

Do not be late and try to be early if possible; a makeup artist will usually need to be the first person on a job. It is important to be set up and ready to go when the models arrive. If the shoot is for a magazine cover with a celebrity, you do not want to keep the star waiting for you! If a makeup assistant is working on a fashion show, they must be there on time to learn the brief or the look for the job.

Listen to instructions from the head makeup artist, the most frustrating thing for the head makeup artist is for their assistant to go off and do their own thing. If a client has asked for a particular look on a job and the makeup artist does beautiful makeup, but it doesn't fulfil the brief, they can lose the job or never be hired again.

If you are standing around doing nothing, always ask the makeup artist what else can be done. Cleaning brushes, making a cup of tea and holding hair pins are all tedious jobs, but need to be done and are always much appreciated on set. Doing those extra little things will make you stand out from other less-helpful assistants.

If you are assisting, you are working on the head makeup artist's job and with their clients. Behaving with a mature and professional attitude is essential. You should not be giving the client your makeup card for future work; that is seen as trying to steal a client from the head makeup artist and is definitely frowned upon. A makeup assistant's behaviour is a reflection on the makeup artist themselves.

If you are booked to assist on a job and something comes up so you are unable to do it, always call the makeup artist as soon as possible. Do not text or send an email. Picking up the phone is the professional thing to do, better yet suggest a replacement if possible. Keeping a good network of other makeup artists is a great way to boost your own career – often makeup artists get double booked and having a good circle of contacts generates more work for everyone.

Most established makeup artists receive around three emails a week from new makeup artists or students enquiring about assisting. So never forget that assistants are easily replaced, but a great one is invaluable.

QUESTIONS & ANSWERS

How do I become a successful makeup artist?

Training, assisting, testing and practising are all key to becoming successful, however being dedicated, having great contacts and continually networking are also essential elements to ensure you have a good chance of doing well as a makeup artist.

Do I need a qualification to work in the makeup industry?

In the UK, there is no standard form of accepted qualification in the freelance makeup industry. Many schools have their own form of certificate or offer NVQ, BABTAC and BTEC certificates. As there is no standard governing body, qualifications are not required by the industry to work as a professional makeup artist and they do not guarantee work. It really depends on the type of work you are pursuing.

If you are looking to join a makeup agency, work on a photo shoot with a new photographer or on a catwalk show, then you will be asked for your portfolio as proof of work and skills. Sometimes you will be asked to demonstrate your ideas or test the look before the actual shoot day. Your portfolio is really the only form of qualification accepted, especially within the fashion industry.

If you are seeking work with a makeup brand on a counter, companies prefer their sales staff to have some makeup training, but again it is not essential, as they would get the brand's makeup artist tutor to go through the techniques and concept of the looks that they want you to sell to the customer. Often there will be a few training days to show you the new products and trends for the season.

If you are working in the beauty, film and television industry in licensed premises, you may be asked to show both your qualifications and a CV/résumé. In film and television, a show reel will often be requested.

ance is the same as being self-
equires you to set up your own
register your chosen name as a
pany. You pay your own salary
ectly with the client; you are in
g your taxes and insurance. Being
have the freedom to do whatever
d work when you want to work.
have to be aware that you do
rks of being signed to a contract,
uld usually get holiday leave and

t work, you do not get paid, it is
at. You always need to plan ahead
ob so your diary is filled and you
It is easy to get de-motivated and
you have to do all the paperwork
d work and also to be creative at

d to a contract with a makeup
ou the security of having someone
you grow and build your profile.
the administrative side of your
invoices and payments. They will
ur diary and keep you in the heart
, which is not easy since there are
trying to do the same thing.

a new makeup artist
ent?

nt can be very competitive. There
aces per agency, but usually once
st lands one, it is far easier to get
A good way for a new makeup
o an agency is to call them up and
they have just finished a makeup
e available for assisting work. It
for you to have worked on a few

tests first before approaching an agent, as they
will probably ask to see your work. Very often,
once a makeup artist is on an assistant list and
they do well, there is room to move up into the
agency in a shorter space of time.

How do I know where to get the best training possible?

If you are looking for a makeup school, there are
several points to consider. The makeup industry
is hard and very competitive; this cannot be
stressed or highlighted enough. There are many
schools claiming to promise you the world,
however, it is up to you to create a successful
career for yourself as a working makeup artist.
No school can promise every student success.

Many academies claim to offer working
makeup artists as tutors. Always ask to see the
school tutor's current portfolios. If they cannot
supply one, then it is doubtful they are current
working fashion makeup artists. If they do
supply one, then look at it closely – if they are
not credited in a magazine or their work is not
of magazine-quality, then they are not really top
working makeup artists. It is no guarantee that
schools with great pictures on their website use
good makeup tutors who are active in the industry.
Ask who will be teaching you and then do your
own research on the tutor on the Internet.

Find out who supports the school – this can
include big name cosmetic brands and hair
companies. Any good school will be affiliated
and will work with big name companies, as
this shows they are highly regarded within the
industry. If you are looking at schools claiming
to offer international certificates that will ensure
you will get work abroad afterward, be wary
as there is no such thing as a fully qualified
makeup artist, whether this is on a national or
international basis.

A certificate alone will not guarantee work. There is no recognized international makeup association or qualification. The only association recognized by the leading UK schools is the National Association of Screen Makeup and Hair, (NASMAH, www.nasmah.co.uk.)

Schools are set up to act as a means of access to fast-track knowledge and an understanding of the industry – they do not qualify you to work. There are a number of top makeup artists in the industry, working on magazines like *Vogue* and television and feature films, with no previous makeup training, but who have become very successful.

Find out what cosmetic brands the school offers you to work with during your training. If a school only offers one make, you will be stuck in the industry, as no makeup artist will ever use just one brand. Your training should give you an understanding of all the products a professional would use. Be cautious of schools that offer kits, as in many cases they are not professional products and you can end up in a situation where you are not confident or comfortable using these products. Once you complete your training, you should have a good working knowledge of all beauty products.

It is important to see what past students have achieved from the school. Ask to meet current students, as this will give you a good insight into whether the school offers the right education for you. Student testimonials mean nothing and can be manipulated to sound better, so you must get all the facts right before you decide to part with your money.

Often on a course, you will be offered the chance to shoot your own portfolio to get you started, which will mean working with a model and photographer. It is important that you see the quality of images from past students, because this shoot will be the first chance you have to showcase your work. Can you visualize it in a magazine? Does the model look professional?

As a student, you will be paying a lot of money for your training, so you deserve to get the best teachers, products to work with and aftercare that money can buy. Many students assume that a career within the makeup industry comes straight after training. This is often not the case, as you still need to build up your practical experience.

Now I have trained, what is the best way to build a professional makeup kit?

Putting together a professional makeup kit when starting out can seem daunting and costly, but build it up slowly and it will not be as scary as you may think. When starting out, it is important not to buy the most expensive materials until you need them. It is advisable that students buy good-quality foundations, concealers and eyeshadows with good, strong pigment as a base first. As for lip glosses, lipsticks and other bits and pieces, it is better to go for a cheaper version to start with and invest in them as you get more experienced. When you start working with big-name designers, magazines and high-profile events, many press agents for cosmetic companies will give you products in return for a credit, which is also a form of advertising for the brand.

How does a new makeup artist build up their portfolio?

After training, new makeup artists work on tests. These are unpaid photo shoots organized between the photographer, stylist, makeup artist and sometimes a hairdresser. The aim of a test is to produce new photographs that showcase the work of all of those involved. On a test, model agencies often supply one of their new faces for the shoot to help build up their book. From there, it is up to the individual makeup artist to contact the modeling agencies, letting them know that they have just finished a makeup course and that they are available to show their work from the shoot. In turn, the agencies will often call new makeup artists to work on tests with their models.

There are also lots of great Internet sites with postings from photographers looking for makeup artists to collaborate with them on tests. New makeup artists can also post their work credits and portfolio pictures here. Remember, your portfolio is your tool in guaranteeing you future work, but it is just as important to have business cards to hand out while networking, as you never know who you might meet. A large volume of the work freelancers get are usually by chance meetings, word of mouth and through friends.

It is also good practice to have a website to showcase your work. The more people that know you exist, the more work you will get and a better profile you will build up.

How can I get to assist a makeup artist?

Usually, when you have finished your training, you will have been given experience to assist your tutors on jobs. Some schools have an after care service where they will direct you to the right places for contacts and makeup agencies. It is up to you to motivate yourself and offer your time and help to a current working makeup artist.

It is important to try and get work with as many different artists as possible, to gain tips from them. Every makeup artist has their own style and steps to create a look, even if the outcome is the same. There are so many different techniques and clever shortcuts that you can only learn through experience.

Does a makeup assistant get paid?

Many makeup assisting jobs are unpaid, but the experience you can gain from them is priceless. A makeup artist can learn and develop new skills and build up their speed and confidence. After assisting makeup artists for free, small paid jobs and clients are often passed on to the assistant.

How long will it take until a new makeup artist starts making money?

It is hard to say exactly, as each makeup artist's career varies. It really is a mixture of luck, determination and talent. Some makeup artists may already have industry contacts, while others will have to knock on a lot of doors to be given chances, but if you are willing to do the legwork it usually pays off.

The majority of makeup artists will say it took them two to three years to start getting some better-paid makeup jobs. Remember that during this time, a new makeup artist is testing for free and spending money building up their kit, so a part-time job in the industry would be a good idea to help fund this.

How can I ensure I get work in such a competitive industry?

It is best to focus on a certain area in the industry and excel in that. As a makeup artist you will get jobs in other areas, but when you are starting out, you need to realize where your skills are best suited within the industry. The retail industry, your personality and ability to understand the needs of a customer, who does not know much about makeup, plays a vital role. It is important to understand that you are working with all different types of skin and ages. Product knowledge of the brand is essential and you need to know how to sell the brand by recommending the right products for your customer. You will need to have a CV/résumé and an interview with the store manager, then if you are successful you will receive formal training. There are plenty of brands to work with, so choose one that you can see yourself working with and contact the manager directly.

In television, film and theatre, the work is regular and you get perks, such as a regular salary and benefits. These jobs are less widely available as they tend to be long-term contracts of at least six months. To get into this field, you need to send your CV/résumé to the head makeup designer and then offer to assist. In theatre, you would also be expected to groom wigs, cut hair and work with the wardrobe mistress. Offering your services and working in other fields, such as production, where you just help out and do what you can, will help establish you as a member of the team. You may just get a lucky break, too.

The fashion and catwalk industry is renowned as being very competitive. Your success depends on your own creativity, how much exposure your profile has had and your previous experience in working with big names. Do not be overwhelmed. It can be frightening, as your skills and portfolio are often judged before you have even had the chance to meet the client. The best way is to research makeup agencies and find a few artists that you aspire to be like, then offer your assistance. It really is who you know that can get you the big jobs because they are rarely advertised. A lot of artists get their work from being recommended by a photographer, stylist or a client they have worked with before. Networking at exclusive launches and parties are all vital parts of your work, as people like to put a face to a name.

Bridal makeup is done primarily for the camera and only secondly for the naked eye (the guests). The bride will continuously look back at the pictures for the rest of her life, so it is important that she looks perfect in them. A traditional church wedding will require a look with soft colours, often a mix of pastel pinks and peaches, while ethnic weddings often require more vibrant colours.

It is important to have a wedding trial approximately one month before the wedding to determine what the bride wants. Work with tear sheets as the general public often do not understand makeup. These tear sheets will help the bride picture what is suitable for her. As always, you should know what the current trends are, be able to make celebrity references, know your period makeup and understand different cultures. Keep a written and visual record of what products you have used in the past, so that everything runs smoothly on the day.

You will either work with a hairdresser or be required to do the hair yourself, and you may also be asked to do nails. You may be asked to bring assistants with you, depending on how many people you are required to makeup, so do not forget to factor this into your fee. It is normal to expect a deposit as weddings are often booked up to a year in advance. Make sure you are paid for the wedding trial separately, in case your bride changes her mind and cancels the wedding (it does happen). Ensure you have allowed enough time for makeup application and that you have talked through your schedule with the rest of the team.

On the day of the wedding, arrive early, be calm and be organized in order to put your bride at ease. Remember this is one of the most important days of her life, so make it a special experience for her. Prepare a touch-up kit, either for yourself or for the bride to use, and be prepared for last-minute changes. You may be asked to do additional makeup on the day, or you may be required to change the makeup into an evening look for the reception.

When you are starting out, contacts and networking are really important. Ask your friends if you can do their wedding makeup and always carry business cards. If you do a good job, you will get most of your clients through referrals. You will often get referrals from the bride or meet people at the wedding who love the makeup and are getting married themselves, or have a friend who is. Advertising in local or bridal press can help and having a website can be a real advantage. Get yourself into an environment where you can meet clients easily, a make over studio or makeup counter work can put you in direct contact with potential clients.

SOURCES & SUPPLIERS

MAKEUP

Barry M
1 Bittacy Business Centre
Mill Hill East
London NW7 1BA
UK
Tel: 020 8346 7773
www.barrym.com

BECCA
Becca (London) Ltd
91A Pelham Street
London SW7 2NJ
UK
Tel: 020 7225 2501

Becca Inc.
132 Ninth Street
2nd Floor
San Francisco, CA 94103
USA
Tel: 415 553 8972

Becca Cosmetics Australia
Unit 4/36 O'Riordan Street
Alexandria
Sydney NSW 2015
Australia
Tel: (61) 2 8399 1274
www.beccacosmetics.com

Benefit
685 Market Street
7th Floor
San Francisco
CA 94105
USA
Tel: 1 800 781 2336 (phone orders)
www.benefitcosmetics.com

Greenwood House
91–99 New London Road
Chelmsford CM2 0PP
UK
Tel: 01245 347 138,
0800 496 1084 (phone orders)
www.benefitcosmetics.co.uk

Bobbi Brown
Estée Lauder
73 Grosvenor Street
London W1K 3BQ
UK
Tel: 0800 054 2988
(customer service)
www.bobbibrown.co.uk

The Estée Lauder Companies Inc.
767 Fifth Avenue
New York, NY 10153
USA
Tel: 1 877 310 9222
www.bobbibrowncosmetics.com

Chanel
Chanel Beauty Products
Rotherwick House
19–21 Old Bond Street
London W1S 4PX
UK
Tel: 020 7493 3836
www.chanel.com/en_GB

Chanel Inc.
15 East 57th Street
New York, NY 10022
USA
Tel: 1 800 550 0005
www.chanel.com

Charles Fox
22 Tavistock Street
Covent Garden
London WC2E 7PY
UK
Tel: 020 7240 3111
www.charlesfox.co.uk
see also Kryolan entry

Christian Dior
Marble Arch House
66–68 Seymour Street
London W1H 5AF
UK
Tel: 020 7563 6300
www.dior.com

Giorgio Armani
16, Place Vendôme
75001 Paris
France
www.giorgioarmanibeauty.com

UK
www.giorgioarmanibeauty.co.uk

USA
Tel: 1 877 276 2643 (customer service)
www.giorgioarmanibeauty-usa.com

JESSICA
Gerrard International Limited
NNC House
47 Theobald Street
Borehamwood
Hertfordshire WD6 4RT

UK
Tel: 0845 2171360, 0845 2171360 (customer service)
www.jessica-nails.co.uk
www.jessicacosmetics.co.uk

Jessica Cosmetics International
12747 Saticoy Street
North Hollywood, CA 91605
USA
Tel: 818 759 1050, 1 800 582 4000 (customer service)
www.jessicacosmetics.com

Lancôme
14 rue Royale
75008 Paris, France
www2.lancome.com/index.aspx

UK
www.lancome.co.uk

USA
www.lancome-usa.com

Laura Mercier
Gurwitch Products, LLC
13259 North Promenade Blvd
Stafford, TX 77477
USA
Tel: 1 888 637 2437
www.lauramercier.com

L'Oréal Ltd
255 Hammersmith Road
London W6 8AZ
UK
Tel: 020 8762 4000
www.loreal.co.uk

L'Oréal USA
575 Fifth Avenue
New York, NY 10017
Tel: 212 818 1500
www.lorealusa.com

L'Oréal International
41 Rue Martre
92217 Clichy Cedex
France
Tel: (33) 14756 7000
www.loreal.fr

MAC
73 Grosvenor Street
London W1K 3BQ
UK
Tel: 0870 034 6700,
0870 034 2676 (phone orders)
www.maccosmetics.co.uk

767 Fifth Avenue
New York, NY 10153
USA
Tel: 1 800 588 0070
www.maccosmetics.com

Make Up Forever
6 Goldhawk Mews
London W12 8PA
UK
Tel: 020 8740 6788
www.makeupforever.com

USA
www.makeupforeverusa.com
Available at Sephora, see www. sephora.com

Nars
900 Third Avenue
New York, NY 10022
USA
Tel: 1 888 788 NARS
www.narscosmetics.com

UK
www.narscosmetics.co.uk

The Pro Makeup Shop
20 Mortlake High Street
London SW14 8JN
UK
Tel: 020 3178 2960
www.thepromakeupshop.com

Revlon
Greater London House
Hampstead Road
London NW1 7QX
Tel: 020 7391 7400
www.revlon.co.uk

Revlon Inc.
237 Park Avenue
New York, NY 10017
USA
Tel: 212 527 4000
www.revlon.com

Screen Face
20 Powis Terrace
off Westbourne Park Road
London W11 1JH
UK
Tel: 020 7221 8289
www.screenface.com

Shu Uemura
55 Neal Street
Charing Cross
London WC2H 9PJ
UK
020 7836 5588
www.shuuemura.com

Yves Saint Laurent Beauty
34–36 Perrymount Road
Haywards Heath
West Sussex RH16 3DN
UK
Tel: 01444 255700
www.ysl.com

USA
www.yslbeautyus.com

MAKEUP ARTISTS

Carolyn Roper
www.getmadeup.com

Jo Sugar
www.josugar.com

Sandra Cooke
www.sandracooke.net

Rachel Wood
www.rachelmakeup.co.uk

HAIR PRODUCTS

Bumble and Bumble
415 13th Street
New York, NY 10014
USA
Tel: 866 513 0498
www.bumbleandbumble.com

Babyliss Pro
The Conair Group Ltd
PO BOX 356
Fleet GU51 3ZQ
UK
Tel: 08705 133 191
www.babyliss.co.uk

USA
www.babylissus.com

KMS California
KPSS Inc.
981 Corporate Boulevard
Linthicum Heights MD 21090
USA
Tel: 410 850 7555,
1 800 342 5567 (phone orders)
www.kmscalifornia.com

KPSS UK Ltd
6 Park View
Alder Close
Eastbourne
East Sussex BN23 6QE
UK
Tel: 01323 413 200

KPSS Australia Pty Ltd
1A The Crescent
Kingsgrove NSW 2208
Australia
Tel: 612 9554 1900

Label M
Mascolo Group Ltd
Innovia House, Marish Wharf
St Mary's Road, Langley, Berkshire
SL3 6DA
UK
Tel: 0870 770 8080
www.labelm.co.uk

Redken
565 Fifth Avenue
New York, NY 10017
USA
www.redken.com

UK
www.redken.co.uk

TIGI
Tigi Australia Ltd
21/39 Herbert Street
St Leonards
NSW 2065
Australia
Tel: (61) 2 9439 9666
www.tigihaircare.com

Tigi Haircare Ltd
Tigi House, Bentinck Road
West Drayton UB7 7RQ
UK
Tel: 01895 458550

Tigi Linea
1655 Waters Ridge Drive
Lewisville, TX 75057-6013
USA
Tel: 469 528 4300

HAIR STYLISTS

Desmond Murray
www.desmondmurray.com

Fabio Vivan
www.fabiovivan.com

Marc Eastlake
www.myspace.com/thegentrysalon

Natasha Mygdal
www.natashamygdal.com

Zoe Irwin
http://head1st.net/zoe/

SKINCARE

Alpha-H
18 Millennium Circuit
Helensvale Q 4212
Gold Coast, Australia
www.alpha-h.com
Tel: (61) 7 5529 4866
www.alpha-h.com

Bio Oil
Union Swiss
66 Long Street
Cape Town 8001
South Africa
Tel: (27) 21 424 4230
www.bio-oil.com

Dermalogica
1535 Beachey Place
Carson, CA 90746
USA
Tel: 310 900 4000
www.dermalogica.com

Caxton House Randalls Way
Leatherhead
Surrey KT22 7TW
UK
Tel: 01372 363600
www.dermalogica.com/uk

111 Chandos Street
Crows Nest NSW 2065
Australia
Tel: 1 800 659 118 or
(61) 2 8437 9600

Kiehl's
54 Columbus Avenue
New York, NY 10023
USA
Tel: 1 800 543 4572 or
732 951 4545
www.kiehls.com

186A King's Road
London SW3 5XP
UK
Tel: 020 7751 5950
www.kiehls.co.uk

Rodial
The Plaza
535 Kings Road
London SW10 0SZ
UK
Tel: 020 7351 1720
www.rodial.co.uk

St Tropez
Beauty Source Limited Unit
4C Tissington Close
Chilwell NG9 6QG
UK
Tel: 0115 983 6363
www.st-tropez.com

PO Box 800876
Santa Clarita, CA 91380
USA
Tel: 1 800 366 6383
www.st-tropez.com

PO Box 334
Terrey Hills NSW 2084
Australia
Tel: 1 800 358 999
www.sttropeztan.com.au

Simple
Simple Health and Beauty Ltd
4th Floor Chadwick House
Blenheim Court
Solihull B91 2AA
UK
Tel: 0121 712 6523
www.simple.co.uk

FASHION & ACCESSORIES

Erickson Beamon
www.ericksonbeamon.com

Louis Mariette
www.louismariette.com

Swarovski
www.swarovski.com

PHOTOGRAPHERS

Atton Conrad
www.attonconrad.com

Camille Sanson
www.camillesanson.com

Catherine Harbour
www.catherineharbour.com

Dez Mighty
www.dezmighty.com

Fabrice Lachant
www.photobyfabrice.com

Han Lee De Boer
www.hanleedeboer.com

Jason Ell
www.jasonell.com

Keith Clouston
www.keithclouston.com

Lou Denim
www.loudenim.com

Roberto Aguilar
www.photoaguilar.com

Zoe Barling
www.zoebarling.com

STYLISTS

Karl Willett
www.ilovemystylist.com

Narin Jean-Baptiste
www.nasrinjeanbaptiste.com

Rebekah Roy
www.fashion-stylist.net

Samson Soboye
www.samson-soboye.com

Shyla Hassan
www.shylahassan.com

Svetlana Prodanic
www.svetlanastyling.com

SPACES and SERVICES

Balcony Jump Management
www.balconyjump.co.uk

Blow PR
www.blow.co.uk

Disciple Productions
www.disciple-productions.com

Escala Music
www.escalamusic.com/gb

First Model Management
www.firstmodelmanagement.
co.uk

FM Agency
www.fmmodelagency.com

IMG Models
www.imgmodels.com

I.N.C. Space
www.inc-space.com

International Collective
www.internationalcollective.co.uk

M&P Models
www.mandpmodels.com

Modus Dowal Walker PR
www.moduspublicity.com

Nevs Model Agency
www.nevsmodels.co.uk

Next Models
www.nextmodels.com

No. 5 Cavendish Square
www.no5ltd.com

Oceanfall Talent Agency
www.oceanfall.co.uk

Octagon
www.octagon-uk.com

Purple PR
www.purplepr.com

Salon Sensations
www.salonsensations.co.uk

PRODUCTS USED FOR THE LOOKS

SKINCARE

Instant Results
Alpha-H White Gold Skin Brightening
Solution
Alpha-H Liquid Gold
Alpha-H Phase Three Clearing Gel
Elizabeth Eight-Hour Cream
Shu Uemura Deapsea Hydrability
Moisturizing Lip Balm

Exfoliators
Neutrogena Deep Clean Foaming
Scrub
Clarins Gentle Exfoliator Brightening
Toner
Dermalogica Daily Microfoliant
Alpha-H Micro Cleanse Exfoliant
Origins Modern Friction Face Scrub

Cleansers
Shu Uemura Balancing Cleansing Oil
Lancôme Gel Eclat
Chanel Gel Pureté Anti Pollution
Rinse Off Foaming Cleanser

Face Masks
Dermalogica Skin Hydrating Masque
Kiss My Face Organics Pore Shrink
Deep Pore Cleansing Mask
Clean & Clear Morning Burst Shine
Control Facial Scrub with Bursting
Beads
Chanel Precision Masque Destressant
Purete Purifying Cream Mask
Lancôme Hydra-Intense Masque
Alpha-H 15% Glycolic Hydrating
Mask with Lavender

Pore Strips
Biore Pore Cleansing Strips.
Tea Tree and Witch Hazel Nose
Pore Strips

Serums
Boots No.7 Protect & Perfect
Estée Lauder Advanced Night Repair
Christian Dior Capture Totale Multi-
Perfection Concentrated Treatment
Serum
L'Oréal Age Perfect Intensive
Reinforcing Serum
Shiseido Benefiance NutriPerfect Eye
Serum

MAKEUP

Foundations
Full Coverage
Bobbi Brown Foundation Stick
Becca Foundation Stick
MAC Full Coverage Foundation

Medium to Full Coverage
MAC Studio Sculpt Foundation

Sheer Coverage
MAC Face and Body Foundation

Oil Free/Oil Control
Nars Oil Free Foundation

Illuminating
Chanel Vitalumière Foundation
Giorgio Armani Foundation

Moisturizing
Bobbi Brown Luminous
Moisturizing Foundation

Cream to Powder
MAC Studio Tech Foundation
Shu Uemura Nobara Cream
    Foundation

Powder-Based
MAC Studio Fix

Liquid
RMK Liquid Foundation

Mousse
Lancôme Magie Matte Mousse
Max Factor Miracle Touch
    Foundation
Maybelline Dream Matte Mousse
    Foundation

Tinted Moisturizer
Becca Luminous Skin Colour
Laura Mercier Tinted Moisturizer

For Dark Skins
Armani Fluid Sheer in Golden
    Bronze or Sienna
Bobbi Brown Foundation Stick
Becca Luminous Skin Colour and
    Stick Foundation
Becca Shimmering Skin Perfector
    in Bronze or Topaz
Giorgio Armani Face Fabric
    Second Skin Nude Makeup
MAC Studio Tech Foundation

For Asian/Latin/Indian Skins
Bobbi Brown Foundation Stick
Becca Tinted Moisturizer and
    Stick Foundation
Giorgio Armani Designer Shaping
    Cream Foundation SPF 20
Lancôme Photogenic Lumessence
    Foundation
MAC Mineralize Foundation

Concealer
Bobbi Brown Creamy Concealer
Estée Lauder Smoothing Skin
    Concealer
Laura Mercier Secret Camouflage
Revlon New Complexion Concealer

Blusher
Becca Lip and Cheek Cream
Benefit Posie Tint
MAC Cheek Cream
Maybelline Dream Mousse Blush

Highlighter
MAC Strobe Cream
Yves Saint Laurent Touche Éclat

Contouring
MAC Sculpt & Shape Powder

Powder
Full Coverage
MAC Mineralize Skin Finish Powder
MAC Studio Fix

Sheer Powder
Ben Nye Loose Powder
Shu Uemura Loose Powder

Loose Pigment Powders
Barry M Dazzle Dust
Kryolan Living Color

Eye Primer
Benefit Lemon Aid

Eyeshadow
Bobbi Brown Longwear cream
    eyeshadow
Shu Uemura

Eyeliner
MAC Fluidline gel eyeliner

Lips
Lip Primers/Concealers
Benefit Lip Plump
MAC Lip Erase
MAC Prep + Prime Lip

Exfoliators
Hollywood Lip Sweet Sugar Scrub
    and Soothing Day Relief
Rodail Glam Balm

Sheer Colour
Laura Mercier Bare Lips
Benefit Benetint

Matte Colour
MAC Russian Red

Cream Texture
Benefit Lady's Choice

Frost
Revlon Silver City Pink

Opaque
MAC CB96

Gloss
MAC Clear Lipglass

ABOUT AOFM

The Academy of Freelance Makeup is set in the buzzing and exciting surroundings of London's Soho. AOFM has become London's number one makeup school in fashion, editorial, catwalk and bridal makeup – it is the place to learn the skills and gain the experience you need to start your career as a budding makeup artist.

AOFM firmly believes the only way to truly understand how the industry operates is to learn from the people who are working in the business right now. All of their tutors are current working artists, many of them from renowned agencies. This enables them to demonstrate the latest styles and techniques from fashion and catwalk shows and advertising campaigns from all over the world.

With Versace, Armani, Dior and Italian *Vogue* among the names on their client lists, AOFM tutors are some of the most talented in the game. Each tutor has exceptional creative flair in their respective fields, providing students with exposure to a variety of different styles and techniques.

The Academy is one of the only institutes sponsored and supported by a number of famous cosmetic and hair companies. This gives students the opportunity to work with these companies and their products, giving them exposure to a variety of different brands.

AOFM are also proud to offer their students invaluable work experience and jobs within the industry after completing their courses due to their strong relationships and contacts within the industry. Students leave the academy well prepared and with a thorough understanding of the business, giving them an invaluable headstart to their career.

Unique assisting opportunities include prestigious fashion events such as London Fashion Week, where AOFM is a sponsor of many shows (heading up on average 60 shows a year), New York Fashion Week, the *Clothes Show Live* and *Britain's Next Top Model*. Advertorial and editorial shoots within a number of high-profile media channels include British and Italian *Vogue*, MTV and X Factor. The AOFM Pro Team are well-known experts within the industry and are often asked to provide seasonal trends, expert quotes and hints and tips for fashion and beauty brands such as Chloé, Versace and L'Oréal. Celebrity clients include Girls Aloud, the Pussycat Dolls and Giselle.

AOFM is in a league of its own, offering students and makeup wannabes all the advantages of learning from top working makeup artists. They have built a unrivalled reputation in training professional artists and their world-renowned creative Pro Team travel all over the globe demonstrating their skills.

Academy of Freelance Makeup
63 Dean Street
Soho, London W1D 4QG
Tel: 020 7434 4488
www.aofmakeup.com
email info@aofmakeup.com

ACKNOWLEDGMENTS

We would like to say a special thank you to the following model agencies and PR for all their support:

Robert Hannan at Next Models, Teoh Gold at Nevs, Kai at FM Agency, Maxine Henshilwood at Oxygen, Trix Stephenson at First Model Management, Khalid El Awad and James Clark at IMG Models, Russell and Charlie at M&P Model Management.

Aude at Christian Dior, Christina Aristodemou and Jo Scicluna at MAC Artist Relations, Claire Nash and Helen at Shu Uemura, Jenna at Becca, Claire and Dafna at Bobbi Brown, Samantha at KMS, Caroline Young at St Tropez, Raj Kaur and the team at Lancôme, Jess at Purple PR, Clair at Benefit Cosmetics, Alison and the marketing team at Dermalogica, Nicky at Redken, the team at Dowal Walker PR for all their support, Jane at Illamasqua, Mary at Jessica Nails, Zoe at Chanel, Candice at Balcony Jump management, the team at Blow PR, Sophie Stanbury and the team at Swarovski, Chris Manoe and the team at International Collective, Mark Whittel at No. 5 Cavendish for the use of location and hospitality. And finally a special thanks to Jaon Mallett and Ioannis Pagonis.

AFTERWORD

"After years of writing about beauty, directing beauty and fashion shoots and working with so many talented makeup artists, I decided I wanted to put my pen down, pick up my blusher brush and get a little more hands on with makeup. This is what led me to the Academy of Freelance Makeup to train as a makeup artist. That was a few years ago, and today I have my pen back in my hand to scribble what I hope have been some helpful notes and corrections to this fabulous book, and to write these final words.

Makeup artists are a thing of wonder to me. Over the years, I have watched them transform models, real women and celebrities into works of art. It always has – and always will – amaze me how beauty products can hold such power. They have the ability to instantly boost confidence, hide all manner of sins, highlight beautiful features and be the tools to create something so special that it can take your breath away.

This book shows how, with creative thinking, ability and sheer determination, you can become a makeup artist and when the imagination is set free, beauty can be created from beauty.

So, although I have my pen back in one hand, I will always have my blusher brush in the other."

Sarah-Jane Corfield-Smith,
Lifestyle Journalist

INDEX

Figures in italics indicate captions.

(overleaf)
Photographer:
Camille Sanson
Makeup: MAC and
Swarovski. Jewellery
by Louis Mariette